FEATURES AND BENEFITS
Glencoe Advanced Mathematical Concepts
Precalculus with Applications

1. **NCTM Standards**   As the correlation on pages T18–T19 shows, strict a[...] *NCTM Standards 2000* in developing the program. The first interleaf pag[...] the Teacher's Wraparound Edition contains a list of *NCTM Standards* co[...] in the chapter (see page 4a).

2. **Applications and Problem Solving**   Because the ability to grasp concepts and skills is greatly enhanced when they are tied to applications, every lesson opens with a real-world application (page 653). Real-world application examples are also included in every lesson (page 655, Example 2). Exercises in the *Applications and Problem Solving* section of every exercise set provide problem-based learning opportunities, along with *Critical Thinking* exercises (page 660, Exercises 33–38). The *Career Choices* feature adds another dimension (page 662) by informing students of careers that use mathematics as an essential professional skill.

3. **Graphing Calculator**   The *Graphic Calculator Explorations* (pages 26, 106, 169) provide an opportunity for students to investigate mathematical concepts by using a graphing calculator as a tool for discovery. *Graphing Calculator* exercises are also included in the exercise sets (page 166, Exercises 26–31). Keystroke instructions for the basic computation and graphing features of a graphing calculator are provided in the *Graphing Calculator Appendix* (pages A2–A24). Margin features reference students to the appropriate help in the appendix (page 147). Additional hints on using the graphing calculator can be found in the *Graphing Calculator Tip* margin features (page 164).

4. **Review**   *Mixed Review* is included in every exercise set, which also includes an *SAT/ACT Practice* question (page 146, note Exercise 46). Each exercise in the *Mixed Review* is referenced to the lesson in which the concept was first introduced. The *Mid-Chapter Quiz* in each chapter provides a review of the concepts taught thus far in the chapter (page 235). The *Extra Practice* (pages A26–A55) in the back of the textbook provides additional exercises for each lesson. Practice *Chapter Tests* are also included to help students prepare for assessment of each chapter (pages A56–A70). *Self-Check Quizzes* at **www.amc.glencoe.com** give students the opportunity to check their progress in each lesson.

5. **SAT/ACT Preparation**   The exercise set for each lesson concludes with an *SAT/ACT Practice* question (page 228, Exercise 53). At the end of each chapter, there is instruction in mastering one of the strands on these tests (page 272) and 10 practice problems (page 273), which include multiple-choice and grid-in questions. Additional practice is offered through the Glencoe Web site (page 273).

6. **interNET CONNECTION**   There is a wide variety of Internet connections that are provided by the Glencoe Web site: **www.amc.glencoe.com**

   Self-Check Quizzes (page 156)   *History of Mathematics* (page 319)
   Data Updates (page 283)   *Graphing Calculator Programs* (page 331)
   Research (page 357)   *Review Activities* (page 335)
   Career Choices (page 312)   *SAT/ACT Practice* (page 341)
   Internet Projects (page 339)

7. **Calculus Concepts and Skills**   Students will be better prepared for a course in calculus because this program integrates concepts and skills related to calculus throughout the program, such as the concept of critical points and limits (pages 171–179). Chapter 15 presents a formal introduction to calculus (pages 940–983).

8. **Discrete Mathematics**   Unit 4 contains three chapters (Chapters 12–14) that give students the opportunity to explore sequences and series, combinatorics, probability, and statistical analysis.

9. **Teacher's Classroom Resources**   This package contains an extensive set of supplementary materials that are designed to meet the complete spectrum of student needs and interests (page ii).

STUDENT EDITION
Contents in Brief

Glencoe ADVANCED
Mathematical Concepts

Precalculus with Applications

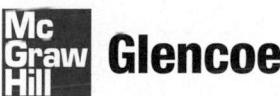

McGraw Hill **Glencoe**

New York, New York   Columbus, Ohio   Chicago, Illinois   Peoria, Illinois   Woodland Hills, California

Glencoe

ADVANCED
Mathematical Concepts
Precalculus with Applications

Student Edition
Interactive Student Edition
Teacher Wraparound Edition

Applications
Math and Science Activities Masters
Problem Solving and Applications Masters

Meeting Individual Needs
Enrichment Masters
Prerequisite Skills Masters
Study Guide and Practice Masters

Technology/Multimedia
Graphing Calculator Keystroke Masters
Spreadsheet Applications Masters
TI-92 and CBL Interdisciplinary Activities Masters
Advanced MathPASS CD-ROM
Multimedia Applications CD-ROM

Assessment and Evaluation
Assessment and Evaluation Masters
TestCheck and Worksheet Builder
SAT Practice and Sample Test Workbook
ACT Practice and Sample Test Workbook
Standardized Test Preparation CD-ROM

Teaching Aids
Answer Key Maker
Answer Key Masters
Answer Key Transparencies
Block Schedule Planning Guide
5-Minute Check Transparencies
Lesson Planning Guide
Solutions Manual
TeacherWorks CD-ROM
Teaching Transparencies

interNET CONNECTION **Glencoe Internet Support:** www.amc.glencoe.com

- Self-Check Quizzes
- Research Helps
- Data Updates
- Career Data
- Graphing Calculator Programs
- Chapter Review Activities
- SAT/ACT Practice
- History of Mathematics
- Internet Projects

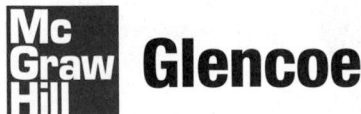

Glencoe

The McGraw-Hill Companies

Send all inquiries to:
Glencoe/McGraw-Hill
8787 Orion Place
Columbus, OH 43240

ISBN: 0-07-860861-9 (Student Edition)
0-07-860862-7 (Teacher Wraparound Edition)

1 2 3 4 5 6 7 8 9 10 071/055 12 11 10 09 08 07 06 05 04 03

Authors

Berchie Holliday
Mathematics Teacher
Northwest Local School District
Cincinnati, Ohio

Berchie Holliday has taught mathematics at every level from middle school to college and has served as coordinator in the areas of mathematics, science, testing, computer science. Dr. Holliday received her B.S. in mathematics from Emory University in Atlanta, Georgia, her M.A.T. in education from Northwestern University in Evanston, Illinois, and her Ed.D. in curriculum and instruction at the University of Cincinnati in Ohio. Emphasizing active student engagement, she has developed and conducted numerous local and national inservice workshops in mathematics, graphing calculator, and computer applications.

Gilbert J. Cuevas
Professor of Mathematics
Education
University of Miami
Miami, Florida

Gibert Cuevas received his B.A. in mathematics and M.Ed. and Ph.D., both in educational research, from the University of Miami. He also holds a M.A.T. in mathematics from Tulane University in New Orleans, Louisiana. Dr. Cuevas is a member of many mathematics, science, and research associations on the local, state, and national levels and has been an author and editor of several NCTM publications. He is also a frequent speaker at NCTM conferences, particularly on the topics of equity and mathematics for all students.

Melissa S. McClure
Mathematics Writer and
Consultant
Teaching for Tomorrow
Fort Worth, Texas

Melissa McClure has taught in middle school, high school, and college. She received her bachelor's degree in mathematics from Angelo State University in San Angelo, Texas. Dr. McClure received her master's degree in secondary education and Ph.D. in curriculum and instruction both from the University of North Texas in Denton. Her research interests lie in the development of effective instructional strategies for technological advancement within secondary and postsecondary mathematics.

John A. Carter
Mathematics Department
Chairperson
Community High School
West Chicago, Illinois

John Carter obtained his bachelor's degree in mathematics, his M.Ed. in education, and his C.A.S. in educational administration all from the University of Illinois Urbana-Champaign. He is currently working on his doctorate degree in mathematics education at Illinois State University in Normal. Mr. Carter has received the Presidential Award for Excellence in Mathematics Teaching. He served as president for the Mathematics Department Heads of Western Chicago Suburbs and is serving on the Board of Directors for the Illinois Council of Teachers of Mathematics.

Daniel Marks
Associate Professor of
Mathematics
Auburn University
Montgomery
Montgomery, Alabama

Daniel Marks received his bachelor's degree and his master's degree, both in mathematics, from California Institute of Technology in Pasadena. He received his Ed.D. in mathematics education from Auburn University. Dr. Marks was elected to the honor society of Phi Kappa Phi. He is a member of Mathematical Association of America and Phi Delta Kappa.

ACADEMIC CONSULTANTS AND REVIEWERS

Each of the Academic Consultants read all 15 chapters, while each Teacher Reviewer read at least two chapters. The Consultants and Reviewers gave suggestions for improving the Student Edition and the Teacher's Wraparound Edition.

ACADEMIC CONSULTANTS

Glenn Gould
Mathematics
  Teacher/Educational
  Consultant
Liberty High School
Bealeton, Virginia

Dr. Roland Minton
Co-author of *McGraw-Hill
  Calculus*
Professor of Mathematics
Roanoke College
Salem, Virginia

Barbara Nunn
Secondary Mathematics
  Curriculum Specialist
Broward County Schools
Fort Lauderdale, Florida

Margene Ryberg
Mathematics
  Teacher/Department
  Chairperson
Northwest High School
Cincinnati, Ohio

Dr. Robert Smith
Co-author of *McGraw-Hill
  Calculus*
Professor of Mathematics and
  Department Chair
Millersville University of
  Pennsylvania
Millersville, Pennsylvania

Dr. Paul Zitzewitz
Professor of Physics
University of Michigan,
  Dearborn
Dearborn, Michigan

TEACHER REVIEWERS

Debra K. Anderson
Mathematics Teacher
Renton High School
Renton, Washington

Sue Ellen Baker
Mathematics Teacher
Hoover High School
Hoover, Alabama

Melody Boring
Mathematics Coordinator,
  Grades 7–12
St. Joseph School District
St. Joseph, Missouri

Mary S. Burkholder
Mathematics Department
  Chairperson
Chambersburg Area Senior
  High School
Chambersburg, Pennsylvania

Rod Camden
Mathematics Department
  Chairperson
E. C. Glass High School
Lynchburg, Virginia

Robyn W. Carlin
La SIP Site Coordinator
East Baton Rouge Parish
  Schools
Baton Rouge, Louisiana

Mark Daniels
Mathematics Department
  Chairperson
McNeil High School
Austin, Texas

Dr. Nestor Diaz
Mathematics and Computer
  Science Department
  Chairperson
Coral Gables Senior High School
Coral Gables, Florida

James R. Ebbert
Mathematics Teacher
Deland High School
Deland, Florida

Everett J. Gaston
Mathematics Department
  Chairperson
Bullard High School
Fresno, California

Nancy Donlon Grigassy
Mathematics Department
  Chairperson
Stephen F. Austin High School
Sugar Land, Texas

1
RELATIONS, FUNCTIONS, AND GRAPHS

Unit 1 *inter*NET Projects

TELECOMMUNICATION
Introduction, 3
Projects, 61, 123, 201, 271

Matrices, 97

GRAPHING CALCULATOR EXPLORATIONS

TABLE OF CONTENTS

Chapter 11   EXPONENTIAL AND LOGARITHMIC FUNCTIONS   694

Lesson 10-7, page 675

Lesson 12-5, page 798

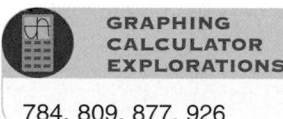

GRAPHING CALCULATOR EXPLORATIONS

UNIT

5 CALCULUS

HISTORY of **MATHEMATICS**

_inter_NET CONNECTION

GRAPHING CALCULATOR EXPLORATIONS

Student Handbook

Teacher's Wraparound Edition
Table of Contents

Teacher's Handbook

Chapter Overviews

Student Handbook

Customizing Your Precalculus Course

Glencoe Advanced Mathematical Concepts includes a wide variety of topics, more than can normally be studied in one year. The multitude of topics allows you the flexibility to easily customize your course of study. The versatility of *Glencoe Advanced Mathematical Concepts* and its many components allows you to customize your mathematics course to meet the needs of your local district's curriculum. Several possible courses of study are described on pages T2-T4 and summarized below.

Year-Long Courses
Precalculus Emphasis
Trigonometric Emphasis

Semester Courses
Trigonometry
Analytic Geometry
Algebra and Elementary Functions
Algebra Review and Discrete Mathematics

For the one-year courses, there are two scheduling possibilities—standard 45/50 minute periods and 90-minute blocks. Year-long courses are based on 9-week grading periods. Block schedules often meet every day for a semester. However, some schedules may meet on alternate days throughout the year.

Precalculus Emphasis

This course of study is for students who are well-prepared for advanced mathematics and who will probably go on to take another mathematics course. Previous experience with trigonometry is helpful, but not a prerequisite. The Suggested Pacing assumes that no review of second-year algebra is needed. However, you may wish to use the Study Guide and Assessment materials at the end of Chapters 1-3 to assess your students' familiarity with these topics.

YEAR-LONG SCHEDULE
45/50-minute periods

Grading Period	Chapter	Days
1	Review Chapters 1-3	Optional
	4	15
	5	13
	6 (through Lesson 6-4)	7
2	6 (beginning with Lesson 6-5)	7
	7	13
	8	16
3	9	15
	10	15
	11 (through Lesson 11-4)	6
4	11 (beginning with Lesson 11-5)	7
	13	11
	14	10
	15	8
	TOTAL	**160**

BLOCK SCHEDULE
90-minute periods

Chapter	Days
4	7.5
5	7
6	7
7	7
8	8
9	7
10	8
11	7
12	8
13	5
14	5
15	5.5
TOTAL	**82**

The total number of days in the Suggested Pacing is less than the typical 180-day school year and 90-day semester to allow for flexibility in planning due to testing, school cancellation, or shortened class periods.

Trigonometric Emphasis

This course of study is for students who may need more review of concepts from previous mathematics courses. This may be the last mathematics course these students will take in high school. Students taking this course will concentrate on those chapters that review algebraic concepts and contain trigonometric applications.

YEAR-LONG SCHEDULE 45/50-minute periods		
Grading Period	**Chapter**	**Days**
1	1	14
	2	14
	3	15
2	4	15
	5	14
	6	14
3	7	13
	8	16
	9 (through Lesson 9-4)	6
4	9 (beginning with Lesson 9-5)	9
	11	13
	12	16
	TOTAL	**159**

BLOCK SCHEDULE 90-minute periods	
Chapter	**Days**
1	8.5
2	7
3	7
4	7
5	7
6	7
7	7
8	7
9	8
10	7
11	7
12	8.5
TOTAL	**81**

The total number of days in the Suggested Pacing is less than the typical 180-day school year and 90-day semester to allow for flexibility in planning due to testing, school cancellation, or shortened class periods.

Daily Planning

- A more detailed Suggested Pacing chart appears in the interleaf preceding each chapter in the *Teacher's Wraparound Edition.*

- The *Lesson Planning Guide* and *Block Scheduling Guide* offer further suggestions for the materials to be covered each day.

- The *TeacherWorks: All-in-One Lesson Planner and Resource Center* CD-ROM enables you to customize an entire course of study to meet your specific needs.

Semester Courses

Many schools offer semester courses that earn one-half credit. You can use the chapters of *Glencoe Advanced Mathematical Concepts* to create individual semester courses for your specialized curriculum. The following tables offer suggestions of four possible courses that utilize 45-50 minute periods for the entire semester.

TRIGONOMETRY

Grading Period	Chapter	Days
1	3	15
	5	14
	6	13
2	7	14
	8	13
	9	15
TOTAL		84

ANALYTIC GEOMETRY

Grading Period	Chapter	Days
1	1	13
	2	13
	3	16
2	5	14
	9	14
	10	14
TOTAL		84

ALGEBRA and ELEMENTARY FUNCTIONS

Grading Period	Chapter	Days
1	1	8
	2	8
	3	12
	4	13
2	10	15
	11	13
	12	14
TOTAL		83

ALGEBRA REVIEW and DISCRETE MATHEMATICS

Grading Period	Chapter	Days
1	1	9
	2	10
	3	10
	4	12
2	12	17
	13	13
	14	12
TOTAL		83

The total number of days in the Suggested Pacing is less than the typical 90-day semester to allow for flexibility in planning due to testing, school cancellation, or shortened class periods.

Bringing Mathematics to Life!

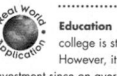

Real World Application

Glencoe Advanced Mathematical Concepts brings mathematics to life through **Real-World Applications** central to students' own lives like music, sports, and nutrition. Application problems also extend students' own experiences through mathematical situations involving consumer topics and citizenship.

Each lesson begins with a real-world application that introduces students to problem-based learning. As students study the lesson, they have the opportunity to revisit and solve the problem that was introduced. The homework exercises contain many additional applications that illustrate the concepts taught in the lesson. In this way, students can see that math is a powerful tool that can help them succeed in school and in the future.

Modeling real-world data using mathematics is interwoven throughout *Glencoe Advanced Mathematical Concepts*. Students examine data such as population growth. They plot points, look for patterns and trends in graphs, and write linear, quadratic, polynomial, exponential, logarithmic, and trigonometric equations to model the data. These graphs and equations are then used to predict the outcomes of the particular situation. Graphing calculators provide powerful support in the modeling of the data.

CAREER CHOICES

Career Choices in each chapter provide students with a brief glance at various careers, both technical and non-technical. Students learn some basic information about each career, what kind of degree is required, academic courses needed, and the outlook for the career. In addition, a graphic is included that highlights an aspect of the career.

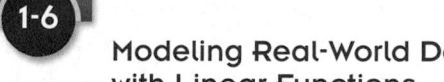

1-6

Modeling Real-World Data with Linear Functions

OBJECTIVES
- Draw and analyze scatter plots.
- Write a prediction equation and draw best-fit lines.
- Use a graphing calculator to compute correlation coefficients to determine goodness of fit.
- Solve problems using prediction equation models.

Real World Application

Education The cost of attending college is steadily increasing. However, it can be a good investment since on average, the higher your level of education, the greater your earning potential. The chart shows the average tuition and fees for a full-time resident student at a public four-year college. Estimate the average college cost in the academic year beginning in 2006 if tuition and fees continue at this rate. *This problem will be solved in Example 1.*

Academic Year	Tuition and Fees
1990–1991	2159
1991–1992	2410
1992–1993	2349
1993–1994	2537
1994–1995	2681
1995–1996	2811
1996–1997	2975
1997–1998	3111
1998–1999	3243

Source: The College Board and National Center for Educational Statistics

As you look at the college tuition costs, it is difficult to visualize how quickly the costs are increasing. When real-life data is collected, the data graphed usually does not form a perfectly straight line. However, the graph may approximate a linear relationship. When this is the case, a **best-fit line** can be drawn, and a **prediction equation** that models the data can be determined. Study the **scatter plots** below.

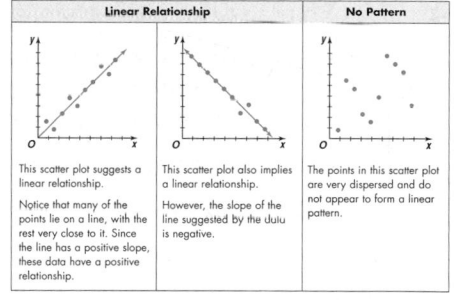

Linear Relationship		No Pattern
This scatter plot suggests a linear relationship. Notice that many of the points lie on a line, with the rest very close to it. Since the line has a positive slope, these data have a positive relationship.	This scatter plot also implies a linear relationship. However, the slope of the line suggested by the data is negative.	The points in this scatter plot are very dispersed and do not appear to form a linear pattern.

38 Chapter 1 Linear Relations and Functions

page 38

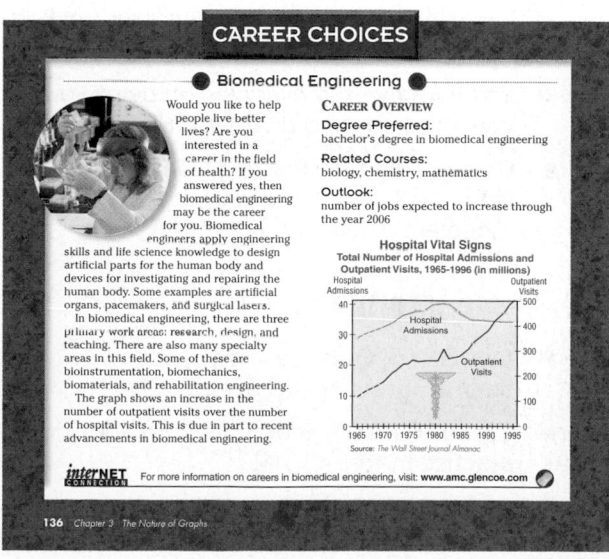

CAREER CHOICES

● Biomedical Engineering ●

Would you like to help people live better lives? Are you interested in a career in the field of health? If you answered yes, then biomedical engineering may be the career for you. Biomedical engineers apply engineering skills and life science knowledge to design artificial parts for the human body and devices for investigating and repairing the human body. Some examples are artificial organs, pacemakers, and surgical lasers.

In biomedical engineering, there are three primary work areas: research, design, and teaching. There are also many specialty areas in this field. Some of these are bioinstrumentation, biomechanics, biomaterials, and rehabilitation engineering.

The graph shows an increase in the number of outpatient visits over the number of hospital visits. This is due in part to recent advancements in biomedical engineering.

CAREER OVERVIEW

Degree Preferred:
bachelor's degree in biomedical engineering

Related Courses:
biology, chemistry, mathematics

Outlook:
number of jobs expected to increase through the year 2006

Hospital Vital Signs
Total Number of Hospital Admissions and Outpatient Visits, 1965-1996 (in millions)

Source: The Wall Street Journal Almanac

internet CONNECTION For more information on careers in biomedical engineering, visit: **www.amc.glencoe.com**

136 Chapter 3 The Nature of Graphs

page 136

GLENCOE'S ASSESSMENT ADVANTAGE

Assessing Your Students

Glencoe Advanced Mathematical Concepts helps you assess students' ability to organize information, apply previously learned information, and make conjectures based on gathered data. The following features and components will help you accurately assess each student's achievement.

In the Student Edition...

■ **Look Back** features throughout the text help students connect previously learned concepts to new information.

■ **Guided Practice** exercises present practice problems similar to those in the exercises to assess students' understanding of the concepts presented in the lesson.

■ Every lesson has a **Mixed Review** that includes **SAT/ACT Practice**.

■ Every chapter has a **Mid-Chapter Quiz.**

■ The end matter for every chapter includes:

   ■ a **Study Guide and Assessment** that includes vocabulary review, review exercises for each lesson, applications and problem-solving exercises, and open-ended questions.

   ■ **SAT and ACT Preparation**, a unique two-page testing lesson.

■ The **Student Handbook** in the back of the text includes:

   ■ **Extra Practice** exercises for each lesson.

   ■ a practice **Chapter Test.**

In the Teacher's Wraparound Edition...

■ Every lesson includes a **5-Minute Check** that covers the previous lesson or chapter.

■ **Error Analysis** in every lesson helps you help your students avoid common errors.

■ **Open-Ended Assessment** in every lesson helps student solidify daily learning by modeling, speaking, or writing.

On the Internet...

■ **Self-Check Quizzes** allow students to check their progress in each lesson.

■ **Chapter Test Practice** provides additional practice in test taking.

■ **SAT/ACT Practice** simulates questions that appear on standardized tests.

In the supplementary materials...

The **Assessment and Evaluation Masters** include the following for each chapter.

- 3 multiple-choice tests (Honors, Average, Basic)
- 3 free-response tests (Honors, Average, Basic)
- 1 open-ended assessment
- 1 mid-chapter test
- 4 quizzes
- 1 SAT and ACT practice
- 1 cumulative review

Also included are 2 semester tests and 1 final test for each of the two curricula prescribed in the Teacher's Wraparound Edition, Precalculus Emphasis and Trigonometric Emphasis.

 The **TestCheck and Worksheet Builder** (Windows/Macintosh) is a state-of-the-art *networkable* CD-ROM that has three integrated modules.

- The **Worksheet Builder** creates customized workseets, tests, and quizzes with any combination of free-response, multiple-choice, short-answer, and open-ended items.
- The **Student Module** gives you the option of having students take tests on-screen and get immediate feedback.
- Use the optional **Management System** to keep detailed student records.

The TestCheck and Worksheet Builder also contains special banks of questions for SAT, ACT, and TIMSS.

 5-Minute Check Transparencies provide a quick review of the previous lesson or chapter. There is one full-color transparency for every lesson. The 5-Minute Check is also printed in the Teacher's Wraparound Edition to make your lesson plans easier to prepare.

 The **Advanced MathPASS** CD-ROM reviews and reinforces important concepts through a unique Pretest-Tutorial-Guided Practice-Posttest format. Self-paced and easy-to-use, it is also an excellent tool for standardized test preparation.

 SAT Practice and Sample Test Workbook
This workbook contains a complete review of all of the types of problems found on the SAT. Sample tests are also included.

 ACT Practice and Sample Test Workbook
This workbook contains a complete review of all of the types of problems found on the ACT. Sample tests are also included.

 Glencoe Internet Support
SAT/ACT Practice is available at **www.amc.glencoe.com**.

Preparing for the SAT and ACT

Teachers asked for a precalculus program that helps to prepare their students for success on the SAT and ACT. Glencoe provides an innovative solution.

A Complete SAT and ACT Prep Course

The two-page *SAT and ACT Preparation* lesson at the end of each chapter (two lessons in Chapter 1) is a unique tool for helping students prepare for SAT and ACT success.

- The 16 two-page lessons can be used as a stand-alone SAT and ACT prep course. All SAT and ACT problem types are covered.

- Each lesson has at least two completely worked-out examples.

- Each lesson contains 10 test items – 9 multiple choice and 1 grid in.

- Each lesson covers a specific type of SAT and ACT question.

Listed below are the sixteen SAT and ACT Preparation lessons.

pages 340-341

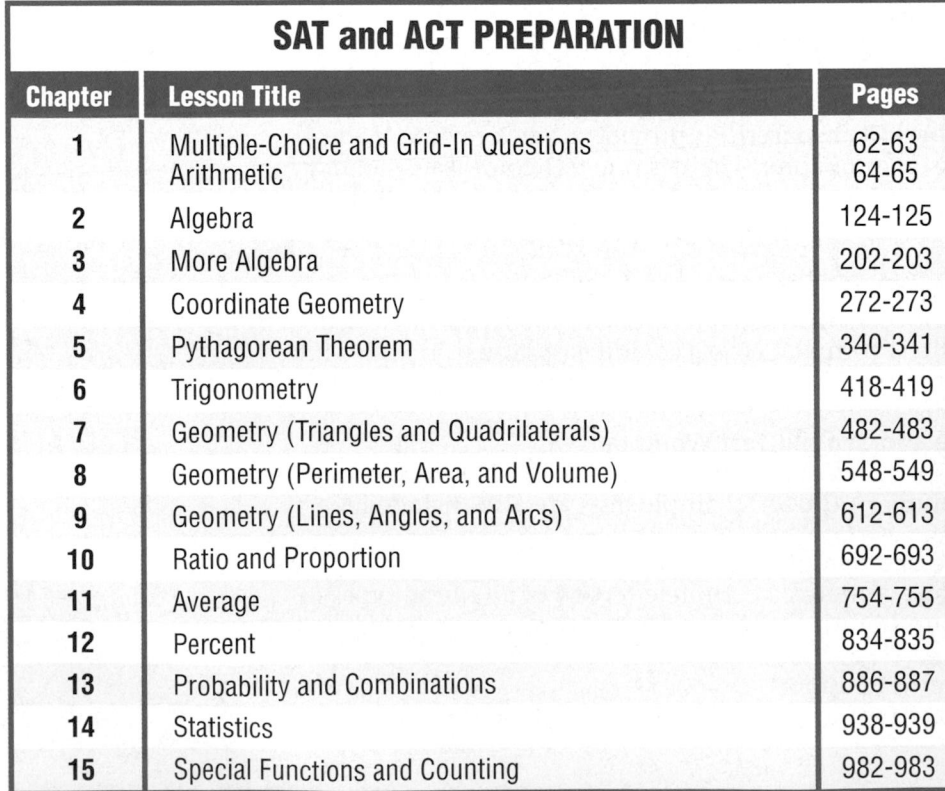

SAT and ACT PREPARATION		
Chapter	**Lesson Title**	**Pages**
1	Multiple-Choice and Grid-In Questions	62-63
	Arithmetic	64-65
2	Algebra	124-125
3	More Algebra	202-203
4	Coordinate Geometry	272-273
5	Pythagorean Theorem	340-341
6	Trigonometry	418-419
7	Geometry (Triangles and Quadrilaterals)	482-483
8	Geometry (Perimeter, Area, and Volume)	548-549
9	Geometry (Lines, Angles, and Arcs)	612-613
10	Ratio and Proportion	692-693
11	Average	754-755
12	Percent	834-835
13	Probability and Combinations	886-887
14	Statistics	938-939
15	Special Functions and Counting	982-983

The Princeton Review

Test-taking tips and hints in the *SAT and ACT Preparation* lessons are provided by The Princeton Review, the #1 test prep company in the United States.

TEST-TAKING TIP
- Draw diagrams.
- Locate points on the grid or number line.
- Eliminate any choices that are clearly incorrect.

page 272

Teacher Support

- The Solutions Manual has complete solutions for all 150 test items.

- The *Teacher's Wraparound Edition* correlates each test question to the *SAT and ACT Preparation* lesson where that particular type of question is taught. In this way, intervention can be provided.

- The *Assessment and Evaluation Masters* include additional SAT/ACT Practice tests and a bubble-in answer sheet typical of standardized tests.

Additional Standardized Test Practice Resources

- **Student Edition** An SAT/ACT Practice question is included at the end of the Mixed Review of each lesson. In addition to multiple choice, grid-in items are included.
- *inter*NET CONNECTION Online SAT and ACT test practice is located at **www.amc.glencoe.com**.
- **Workbooks** In partnership with The Princeton Review, Glencoe publishes workbooks that prepare students for success on the SAT, ACT, and state and local proficiency tests.

- **Standardized Test Preparation CD-ROM** (Windows/Macintosh) This CD ROM contains blocks of test items from state proficiency tests, the SAT, the ACT, TIMSS, and NAEP.

TIMSS Test Items

The Third International Mathematics and Science Study (TIMSS) included an advanced mathematics test that was given to a sample of students in 17 countries. The test measured achievement in numbers, equations, functions, calculus, geometry, statistics, and probability.

In each chapter of the *Teacher's Wraparound Edition*, there is one test item, similar to an actual TIMSS item, whose solution requires knowledge of the content of that chapter. These test items are located on pages 65, 125, 203, 273, 341, 419, 483, 549, 613, 693, 755, 835, 887, 939, and 983.

> *"I want an advanced mathematics program that gives my students the opportunity to use the Internet."*

Using the Internet

The use of the Internet is integrated as a problem-solving tool, a research tool, and as a review and test tool, throughout *Glencoe Advanced Mathematical Concepts.*

*inter*NET **Projects** connected to the mathematics of each unit in *Glencoe Advanced Mathematical Concepts* provide students with the opportunity to use the Internet to conduct research to complete a long-term project. Each project is an excellent opportunity for students to apply what they are learning as they complete their project individually or as part of a cooperative group.

*inter*NET Projects

Chapter	Unit 1: Telecommunication
1	***Is Anybody Listening?*** *Use the Internet and a graph to determine the best telephone service.*
2	***You've Got Mail!*** *Determine the best Internet Service Provider by using a system of equations.*
3	***Sorry, You Are Out of Range for Your Telephone Service…*** *Analyze mobile telephone offers using graphs of equations.*
4	***The Pen Is Mightier Than the Sword!*** *Make a presentation using Internet tools, a word-processor, graphics, spreadsheets, or presentation software.*
	Unit 2: The Cyberclassroom
5	***Does Anybody Out There Know Anything About Trigonometry?*** *Use Internet sites to expand your study of trigonometry.*
6	***What Is Your Sine?*** *Find applications of the sine and cosine functions.*
7	***That's as Clear as Mud!*** *Design your own web page that includes instruction on trigonometry.*
8	***Vivid Vectors*** *Investigate applications of vectors.*
	Unit 3: Space - The Final Frontier
9	***From Point A to Point B*** *Research alternatives to the rectangular coordinate system.*
10	***Out in Orbit!*** *Investigate the conic sections that model the orbits of objects in space.*
11	***Kepler is Still King!*** *Study Kepler's Laws. Use information gathered on the Internet to verify Kepler's Third Law.*
	Unit 4: The United States Census Bureau
12	***That's a Lot of People!*** *Develop arithmetic and geometric models to represent population data.*
13	***Radically Random!*** *Find probabilities involving different groups of U.S. citizens.*
14	***More and More Models!*** *Examine population growth using various statistical models.*
	Unit 5: Diseases
15	*Investigate medical data and predict future trends.*

Internet Connections throughout the Student Edition refer students to the *Glencoe Advanced Mathematical Concepts* Web site **www.amc.glencoe.com**.

- **Data Update** Students can find up-to-date information on subjects discussed in *Glencoe Advanced Mathematical Concepts* problems.

  Data Update For the latest information about the minimum wage, visit **www.amc. glencoe.com.**
  page 167

- **Graphing Calculator Programs** help students use TI-83 Plus calculators to investigate the mathematics they are learning.

  Graphing Calculator Programs To download a graphing calculator program that performs complex iteration, visit: **www.amc. glencoe.com**
  page 582

- **Research** connections guide students to the information they need to complete problems.

  Research For data about amount of daylight, average temperatures, or tides, visit **www.amc. glencoe.com**
  page 388

- **Self-Check Quizzes** Students can take a randomized quiz for every lesson that is graded online.

   www.amc.glencoe.com/self_check_quiz

- Students can find up-to-date information on subjects discussed in *Advanced Mathematical Concepts* problems.

  3. Find out more about the personalities referenced in this article and others who contributed to the history of matrices. Visit **www.amc.glencoe.com**
  page 97

- **CAREER CHOICES** Career connections allow students to investigate careers that use the mathematics they are learning.

   For more information on careers in aerospace engineering, visit: **www.amc.glencoe.com**
  page 499

- **SAT/ACT Practice and Chapter Review Activities** Prepare for Chapter Tests and SAT, ACT, and TIMSS standardized tests using Glencoe's online Test Practice.

   SAT/ACT Practice For additional test practice questions, visit: **www.amc.glencoe.com**
  page 65

   For additional review and practice for each lesson, visit: **www.amc.glencoe.com**
  page 57

Reaching All Learners

People learn in many different ways. There are several different learning styles that help us approach and solve problems. Everyone possesses varying degrees of each of these learning styles, but the ways in which they combine and blend are as varied as the personalities of the individuals. The Teacher's Wraparound Edition of *Glencoe Advanced Mathematical Concepts* provides you with ways to accommodate students with these diverse learning styles.

Learning Style	Characteristics of Students	Where Can I Find This?
verbal/ linguistic	read regularly, write clearly, and easily understand the written word	**Communicating Mathematics** exercises ask students to describe, write, and explain mathematical concepts. Students also express what they have learned in their **Math Journals**.
logical	use numbers, logic, and critical thinking skills	Clearly-written **Examples** present important concepts, and **Critical Thinking** exercises extend those concepts. **You Decide** exercises help students formulate convincing arguments.
visual/ spatial	think in terms of pictures and images	Mathematical concepts are presented in multiple representations both in the **Examples** and **Exercises**, stressing the connection between the visual and the abstract. In the **Career Choices** feature, students are asked to interpret graphs and drawings.
auditory/ musical	have "good ears" and can produce rhythms and melodies	**Real-World Applications** in the lessons, examples, and exercises, show students how the mathematics they are learning relates to music and sound.
kinesthetic	learn from touch, movement, and manipulating objects	**Internet Projects** allows students to explore the virtual world through Internet connections to data, activities, and graphics not easily accessible in the classroom.
interpersonal	understand and workwell with other people	**Graphing Calculator Explorations** offer opportunities for students to collaborate with others.
intrapersonal	have a realistic understanding of their strengths and weaknesses	**Math Journals** and **Portfolio** activities help students personalize mathematics.
naturalist	can distinguish among, classify, and use features of the environment	**Real-World Applications** in the lessons, examples, and exercises show students how the mathematics they are learning relates to the world around them.

As a mathematics teacher, you may want to assign activities to students that accommodate their strongest learning styles, but frequently ask them to use their weakest learning styles. Additional activities entitled **Addressing Individual Needs** are provided in the bottom margins of the lesson notes in the Teacher's Wraparound Edition. These resources guarantee that your classroom will be a multisensory environment, providing multiple paths for student learning.

Researching the Program

Glencoe Advanced Mathematical Concepts, as well as Glencoe's entire mathematics series, is the product of ongoing classroom-oriented research that involves students, teachers, curriculum supervisors, administrators, parents, and college-level mathematics educators.

The programs that make up the Glencoe mathematics series are currently being used by millions of students and tens of thousands of teachers. The key reason for the success of these programs in the classroom is the fact that each Glencoe author team is a mix of practicing classroom teachers, curriculum supervisors, and college-level mathematics educators. Glencoe's balanced author teams help ensure that Glencoe mathematics programs are both practical and progressive.

Glencoe Advanced Mathematical Concepts was first published in 1963. Now in its tenth revision, this book has been used to instruct literally millions of mathematics students. To this solid foundation, Glencoe has added up-to-date educational research to build a valid program you can use to educate students in the twenty first century.

Prior to the publication of **Glencoe Advanced Mathematical Concepts,** research activities included:

- a review of educational research and recommendations by groups such as NCTM,

- in-depth analyses of manuscript by a wide range of reviewers and consultants,

- discussion groups involving mathematics teachers, department heads, and supervisors,

- focus groups involving mathematics edectors,

- face-to-face interviews with mathematics educators,

- telephone surveys of mathematics educators, and

- mail surveys of mathematics educators.

Feedback from teachers, curriculum supervisors, and even students who currently use **Glencoe Advanced Mathematical Concepts** or other Glencoe mathematics programs was also incorporated as Glencoe planned and published this revision. For example, *From the Classroom of...* features, which are printed in the Teacher's Wraparound Edition, are one result of this feedback.

This research makes it possible for Glencoe's authors and editors to publish outstanding instructional resources.

Implementing the NCTM Standards 2000

In 1989, the National Council of Teachers of Mathematics (NCTM) published their *Curriculum and Evaluation Standards for School Mathematics*, which gave mathematics teachers their first set of goals toward a national mathematics curriculum. Teachers and supervisors have embraced these Standards and developed state standards based on this framework. In 2000, the National Council of Teachers of Mathematics released a revision of these guidelines entitled *NCTM Principles and Standards for School Mathematics* on which this correlation has been based.

Standard 1: Number and Operation

- Understand numbers, ways of representing numbers, relationships among numbers, and number systems
- Understand the meaning of operations and how they relate to each other
- Use computational tools and strategies fluently and estimate appropriately

    Pages 67-118, 265-266, 757-826, 837-880

Standard 2: Patterns, Functions, and Algebra

- Understand various types of patterns and functional relationships
- Use symbolic forms to represent and analyze mathematical situations and structures
- Use mathematical models and analyze change in both real and abstract contexts

    Pages 5-37, 45-56, 67-86, 88-118, 127-196, 205-266, 299-312, 485-519, 574-579, 615-686, 704-748, 757-826, 941-976

Standard 3: Geometry & Spatial Sense

- Analyze characteristics and properties of two- and three-dimensional geometric objects
- Select and use different representational systems, including coordinate geometry and graph theory
- Recognize the usefulness of transformations and symmetry in analyzing mathematical situations
- Use visualization and spatial reasoning to solve problems both within and outside of mathematics

    Pages 20-37, 45-56, 67-77, 87-96, 127-188, 229-242, 258-264, 277-334, 343-412, 485-542, 553-606, 615-686, 738-739, 961-976

Standard 4: Measurement

- Understand attributes, units, and systems of measurements
- Apply a variety of techniques, tools, and formulas for detecting measurements

    Pages 258-264, 343-412, 493-504, 738-739, 961-976

Standard 5: Data Analysis, Statistics, and Probability

- Pose questions and collect, organize, and represent data to answer those questions
- Interpret data using methods of exploratory data analysis
- Develop and evaluate inferences, predictions, and arguments that are based on data
- Understand and apply basic notions of chance and probability

Pages 852-880, 889-932

Standard 6: Problem Solving

- Build new mathematical knowledge through their work with problems
- Develop a disposition to formulate, represent, abstract, and generalize in situations within and outside mathematics
- Apply a wide variety of strategies to solve problems and adapt the strategies to new situations
- Monitor and reflect on their mathematical thinking in solving problems

Pages 67-86, 98-118, 127-196, 205-266, 277-334, 343-412, 421-476, 500-542, 553-606, 662-669, 695-748, 757-781, 784-826, 837-880, 889-932, 941-976

Standard 7: Reasoning and Proof

- Recognize reasoning and proof as essential and powerful parts of mathematics
- Make and investigate mathematical conjectures
- Develop and evaluate mathematical arguments and proofs
- Select and use various types of reasoning and methods of proof as appropriate

Pages 38-44, 67-77, 87, 106-118, 159-168, 171-179, 320-329, 421-476, 738-739, 820-826

Standard 8: Communication

- Organize and consolidate their mathematical thinking to communicate with others
- Express mathematical ideas coherently and clearly to peers, teachers, and others
- Extend their mathematical knowledge by considering the thinking and strategies of others
- Use the language of mathematics as a precise means of mathematical expression

Pages 67-86, 88-118, 333-334, 738-739

Standard 9: Connections

- Recognize and use connections among different mathematical ideas
- Understand how mathematical ideas build on one another to produce a coherent whole
- Recognize, use, and learn about mathematics in context outside of mathematics

Pages 67-86, 88-118, 305-332, 404, 733-739, 927-932

Standard 10: Representation

- Create and use representations to organize, record, and communicate mathematical ideas
- Develop a repertoire of mathematical representations that can be used purposefully, flexibly, and appropriately
- Use representations to model and interpret physical, social, and mathematical phenomena

Pages 13-56, 67-118, 127-188, 229-242, 258-266, 313-318, 359-412, 520-533, 553-606, 615-686, 704-748, 792-798, 804-826, 846-851, 859-874, 889-932, 961-976

Using a Graphing Calculator in Your Classroom

Since its invention in 1990, the graphing calculator has become a powerful tool in the mathematics classroom. The features of a graphing calculator enable students not only to graph functions, but explore the nature of those functions. A graphing calculator can be used to study the relationship among algebraic expressions as well as performing complex computations. It also provides a platform for statistical analysis and graphing.

Throughout *Glencoe Advanced Mathematical Concepts,* **Graphing Calculator Explorations** give students guidance in exploring specific mathematical topics. These activities pose an idea for students to explore using their graphing calculators. Then they are asked to make conjectures and write about their findings during the activity. These activities may be embedded in the lesson text or they may be a complete stand-along lesson.

GRAPHING CALCULATOR EXPLORATION

Use a graphing calculator to find each ratio for the 22.6° angle in each triangle. Record each ratio as a decimal. Make sure your calculator is in degree mode.

$R_1 = \dfrac{\text{side opposite}}{\text{hypotenuse}}$ $R_2 = \dfrac{\text{side adjacent}}{\text{hypotenuse}}$

$R_3 = \dfrac{\text{side opposite}}{\text{side adjacent}}$

Find the same ratios for the 67.4° angle in each triangle.

TRY THESE 1–4, 6. See margin.

1. Draw two other triangles that are similar to the given triangles.

2. Find each ratio for the 22.6° angle in each triangle.

3. Find each ratio for the 67.4° angle in each triangle.

WHAT DO YOU THINK?

4. Make a conjecture about R_1, R_2, and R_3 for any right triangle with a 22.6° angle.

5. Is your conjecture true for any 67.4° angle in a right triangle? **yes**

6. Do you think your conjecture is true for any acute angle of a right triangle? Why?

page 284

Graphing Calculator Exercises are included in many of the lessons and are signified by the graphing calculator logo.

A **Graphing Calculator Appendix**, beginning on page A2 at the back of the Student Edition, gives keystroke-specific instructions on how to perform some of the basic graphing calculator procedures students will use in this text.

The commands shown in this text are those for the TI-83 Plus calculator. The TI-83 Plus operates almost identically to the TI-83. However, the TI-83 Plus has the capability of downloading additional applications not available on the TI-83.

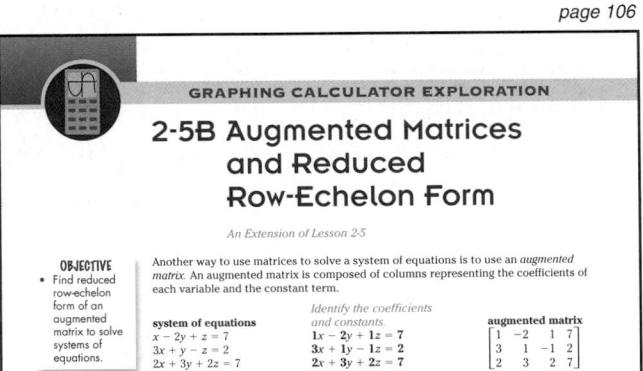

page 106

GRAPHING CALCULATOR EXPLORATION

2-5B Augmented Matrices and Reduced Row-Echelon Form

An Extension of Lesson 2-5

OBJECTIVE
• Find reduced row-echelon form of an augmented matrix to solve systems of equations.

Another way to use matrices to solve a system of equations is to use an *augmented matrix.* An augmented matrix is composed of columns representing the coefficients of each variable and the constant term.

system of equations	*Identify the coefficients and constants.*	augmented matrix
$x - 2y + z = 7$	$1x - 2y + 1z = 7$	$\begin{bmatrix} 1 & -2 & 1 & 7 \\ 3 & 1 & -1 & 2 \\ 2 & 3 & 2 & 7 \end{bmatrix}$
$3x + y - z = 2$	$3x + 1y - 1z = 2$	
$2x + 3y + 2z = 7$	$2x + 3y + 2z = 7$	

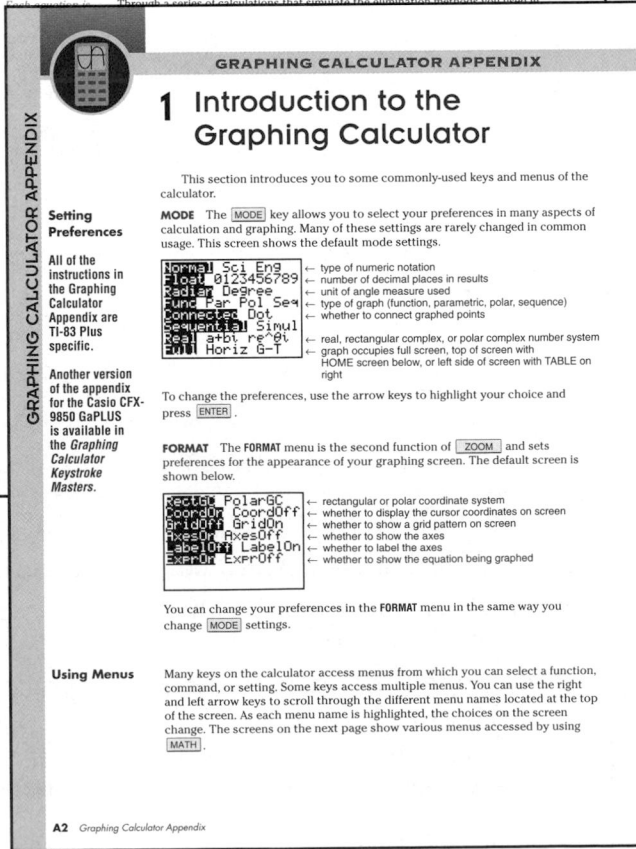

GRAPHING CALCULATOR APPENDIX

1 Introduction to the Graphing Calculator

This section introduces you to some commonly-used keys and menus of the calculator.

Setting Preferences

All of the instructions in the Graphing Calculator Appendix are TI-83 Plus specific.

Another version of the appendix for the Casio CFX-9850 GaPLUS is available in the *Graphing Calculator Keystroke Masters.*

MODE The MODE key allows you to select your preferences in many aspects of calculation and graphing. Many of these settings are rarely changed in common usage. This screen shows the default mode settings.

```
Normal Sci Eng     ← type of numeric notation
Float 0123456789   ← number of decimal places in results
Radian Degree      ← unit of angle measure used
Func Par Pol Seq   ← type of graph (function, parametric, polar, sequence)
Connected Dot      ← whether to connect graphed points
Sequential Simul
Real a+bi re^θi    ← real, rectangular complex, or polar complex number system
Full Horiz G-T     ← graph occupies full screen, top of screen with
                     HOME screen below, or left side of screen with TABLE on
                     right
```

To change the preferences, use the arrow keys to highlight your choice and press ENTER.

FORMAT The FORMAT menu is the second function of ZOOM and sets preferences for the appearance of your graphing screen. The default screen is shown below.

```
RectGC PolarGC   ← rectangular or polar coordinate system
CoordOn CoordOff ← whether to display the cursor coordinates on screen
GridOff GridOn   ← whether to show a grid pattern on screen
AxesOn AxesOff   ← whether to show the axes
LabelOff LabelOn ← whether to label the axes
ExprOn ExprOff   ← whether to show the equation being graphed
```

You can change your preferences in the FORMAT menu in the same way you change MODE settings.

Using Menus

Many keys on the calculator access menus from which you can select a function, command, or setting. Some keys access multiple menus. You can use the right and left arrow keys to scroll through the different menu names located at the top of the screen. As each menu name is highlighted, the choices on the screen change. The screens on the next page show various menus accessed by using MATH.

A2 *Graphing Calculator Appendix*

page A2

interNET
CONNECTION

To find
keystroke
instructions for
each Graphing
Calculator
Exploration, visit
**www.amc.
glencoe.com**

**These sections do
not correspond
with the order of
topics in this
textbook.**

In *Glencoe Advanced Mathematical Concepts,* the graphing calculator is an important tool in exploring precalculus topics. A graphing calculator is actually a specialized hand-held computer. **Graphing Calculator Explorations** throughout this text offer opportunities for you to use a graphing calculator to explore the mathematical topics you are studying. You will also see graphing calculator screens actually generated by a graphing calculator. If you do not recognize the keys, commands, or screens shown in this test, you may have a different model of calculator. Refer to your owner's manual to verify the proper commands and keystrokes to use.

Graphing Calculator Appendix

At the back of this text, you will find a **Student Handbook.** The **Graphing Calculator Appendix** begins on page A2. This appendix introduces you to some of the common functions of the graphing calculator most widely used in high school classrooms. It is divided into nine sections:

1 Introduction to the Graphing Calculator

2 Graphing Functions

3 Analyzing Functions

4 Graphing Inequalities

5 Matrices

6 Graphing Trigonometric Functions

7 Graphing Special Functions

8 Parametric and Polar Equations

9 Statistics and Statistical Graphs

In each section, you will find keystroke instructions for specific examples that demonstrate some of the basic operations and most common menus available on the calculator. Whenever you have specific questions about commands or procedures, you should refer to your owner's manual for instruction.

SAT and ACT Preparation

It is important to note that graphing calculators are permitted when taking the ACT or SAT and are required for the AP Calculus test. Through your experiences with the graphing calculator in studying *Glencoe Advanced Mathematical Concepts,* you should become comfortable with its operation and appreciate the power of this tiny computer.

Unit Overview

Many of the topics in Unit 1 may be a review of topics students have learned in previous courses. If you feel that your students have a good background in algebra, you may wish to use the Unit 1 Review in the *Assessment and Evaluation Masters* (pp. 393–394) as a placement test.

GLENCOE'S ASSESSMENT ADVANTAGE **Assessment**

A review and a test on the concepts in Unit 1 are provided in the *Assessment and Evaluation Masters* (pp. 393–398). The Unit 1 Review is shown below. The Unit 1 Test is shown on the next page.

Assessment and Evaluation Masters, pp. 393–394

NAME _____ DATE _____ PERIOD _____

Unit **1**

Unit 1 Review, Chapters 1–4

Given that x is an integer, state the relation representing each equation as a set of ordered pairs. Then, state whether the relation is a function. Write yes or no.
1. $y = 3x + 1$ and $-1 \le x \le 3$
2. $y = |2 - x|$ and $-2 \le x \le 3$

Find $[f \circ g](x)$ and $[g \circ f](x)$ for each $f(x)$ and $g(x)$.
3. $f(x) = 3x + 1$
   $g(x) = x + 3$
4. $f(x) = 4x^2$
   $g(x) = -x^3$
5. $f(x) = x^2 - 25$
   $g(x) = 2x - 4$

Find the zero of each function.
6. $f(x) = 4x - 10$
7. $f(x) = 15x$
8. $f(x) = 0.75x + 3$

Write the slope-intercept form of the equation of the line through the points with the given coordinates.
9. $(4, -4)$, $(6, -10)$
10. $(1, 2)$, $(5, 4)$

Write the standard form of the equation of each line described below.
11. parallel to $y = 3x - 1$
    passes through $(-1, 4)$
12. perpendicular to $2x - 3y = 6$
    x-intercept 2

The table below shows the number of T-shirts sold per day during the first week of a senior-class fundraiser.

Day	Number of Shirts Sold
1	12
2	21
3	32
4	43
5	56

13. Use the ordered pairs $(2, 21)$ and $(4, 43)$ to write the equation of a best-fit line.
14. Predict the number of shirts sold on the eighth day of the fundraiser. Explain whether you think the prediction is reliable.

Graph each function.
15. $f(x) = [[x + 2]]$
16. $f(x) = [2x] - 1$
17. $f(x) = \begin{cases} x - 2 \text{ if } x \le -1 \\ 2x \text{ if } -1 < x < 1 \\ -x \text{ if } x \ge 2 \end{cases}$

Graph each inequality.
18. $x + 3y < 12$
19. $y \ge -\frac{2}{3}x + 5$

Solve each system of equations.
20. $y = -4x$
    $x - y = 5$
21. $x + y = 12$
    $2x - y = -4$
22. $7x - z = 13$
    $y + 3z = 18$
    $11x + y = 27$

Use matrices A, B, C, and D to find each sum, difference, or product.

$A = \begin{bmatrix} 6 & 2 \\ 3 & -3 \end{bmatrix}$ $B = \begin{bmatrix} -4 & 6 \\ 5 & 7 \end{bmatrix}$ $C = \begin{bmatrix} 3 & 2 & -1 \\ -5 & -8 & 1 \end{bmatrix}$

$D = \begin{bmatrix} 2 & 0 \\ 6 & -3 \\ -5 & -1 \end{bmatrix}$ $E = \begin{bmatrix} -3 & 1 & 5 \\ -1 & 4 & -2 \\ 3 & 2 & -1 \end{bmatrix}$

23. $A + B$
24. $2A - B$
25. CD
26. $AB + CD$

Use matrices A, B, and E above to find the following.
27. Evaluate the determinant of matrix A.
28. Evaluate the determinant of matrix E.
29. Find the inverse of matrix B.

Solve each system of inequalities by graphing. Name the coordinates of the vertices of each polygonal convex set. Then, find the maximum and minimum values for the function $f(x, y) = 2y - 2x - 3$.
30. $x \ge 0$
    $y \ge 0$
    $2y + x \le 1$
31. $x \ge 2$
    $y \ge -3$
    $y \le 5 - x$
    $y + 2x \le 8$

© Glencoe/McGraw-Hill   393   Advanced Mathematical Concepts

Relations, Functions, and Graphs

Throughout this text, you will see that many real-world phenomena can be modeled by special relations called functions that can be written as equations or graphed. As you work through Unit 1, you will study some of the tools used for mathematical modeling.

2 Unit 1   *Relations, Functions, and Graphs*

Resource Manager

📁 **Reproducible Masters**
• *Math and Science Activities,* pp.1–5
• Unit 1 Review, *Assessment and Evaluation,* pp. 393–394
• Unit 1 Test, *Assessment and Evaluation,* pp. 395–398

Unit 1 *inter*NET Projects

TELECOMMUNICATION

In today's world, there are various forms of communication, some that boggle the mind with their speed and capabilities. In this project, you will use the Internet to help you gather information for investigating various aspects of modern communication. At the end of each chapter, you will work on completing the Unit 1 Internet Project. Here are the topics for each chapter.

CHAPTER 1
(page 61)

Is Anybody Listening? Everyday that you watch television, you are bombarded by various telephone service commercials offering you the best deal for your dollar.
Math Connection: How could you use the Internet and graph data to help determine the best deal for you?

CHAPTER 2
(page 123)

You've Got Mail! The number of homes connected to the Internet and e-mail is on the rise. Use the Internet to find out more information about the types of e-mail and Internet service providers available and their costs.
Math Connection: Use your data and a system of equations to determine if any one product is better for you.

CHAPTER 3
(page 201)

Sorry, You Are Out of Range for Your Telephone Service ... Does your family have a cell phone? Is its use limited to a small geographical area? How expensive is it? Use the Internet to analyze various offers for cellular phone service.
Math Connection: Use graphs to describe the cost of each type of service. Include initial start-up fees or equipment cost, beginning service offers, and actual service fees.

CHAPTER 4
(page 271)

The Pen is Mightier Than the Sword! Does anyone write letters by hand anymore? Maybe fewer people are writing by pen, but most people use computers to write letters, reports, and books. Use the Internet to discover various types of word processing, graphics, spreadsheet, and presentation software that would help you prepare your Unit 1 presentation.
Math Connection: Create graphs using computer software to include in your presentation.

*inter*NET
CONNECTION

For more information on the Unit Project, visit:
www.amc.glencoe.com

Using the *inter*NET Projects

When you or your students visit **www.amc.glencoe.com** you will find links to outside Web sites relating to these Internet projects. These links are checked frequently to ensure that they are still functioning and that the information is current.

Glencoe makes every effort to ensure that the Web site links for *Advanced Mathematical Concepts* are current and contain appropriate content. However, these Web sites are not under Glencoe's control.

In the Alternative Assessment section of each chapter's Study Guide and Assessment, you can find follow-up activities for the Internet Projects.

Assessment and Evaluation Masters, pp. 395–398

Linear Relations and Functions

Instructional Objectives

Lesson (pages)	Lesson Objectives	NCTM Standards 2000	State/Local Objectives
1-1 (5–12)	• Determine whether a given relation is a function. • Identify the domain and range of a relation or function. • Evaluate functions.	2	
1-2 (13–19)	• Perform operations with functions. • Find composite functions. • Iterate functions using real numbers.	2, 10	
1-3 (20–25)	• Graph linear equations. • Find the x- and y-intercepts of a line. • Find the slope of a line through two points. • Find zeros of linear functions.	2, 3, 10	
1-3B (26)	• Investigate the effect of changing the value of m or b in $y = mx + b$.	2, 3, 10	
1-4 (27–31)	• Write linear equations.	2, 3, 10	
1-5 (32–37)	• Write equations of parallel and perpendicular lines.	2, 3, 10	
1-6 (38–44)	• Draw and analyze scatter plots. • Write a prediction equation and draw best-fit lines. • Use a graphing calculator to compute correlation coefficients to determine goodness of fit. • Solve problems using prediction equation models.	3, 7, 10	
1-7 (45–51)	• Identify and graph piecewise functions including greatest integer, step, and absolute value functions.	2, 3, 10	
1-8 (52–56)	• Graph linear inequalities.	2, 3, 10	

Suggested Pacing *See pages T2–T4 for additional options in course-planning.*

COURSE	Lesson 1-1	Lesson 1-2	Lesson 1-3	Lesson 1-4	Lesson 1-5
Precalculus Emphasis Chapters 4–15 (45–50 minutes)	OPTIONAL—Chapter 1 is a review of topics students have learned in previous courses. You may wish to use the Study Guide and Assessment to assess students' familiarity with these topics.				
Precalculus Emphasis (block)	OPTIONAL				
Trigonometric Emphasis Chapters 1–9, 11–12 (45–50 minutes)	1.5 days	1.5 days	(with Lesson 1-3B) 1.5 days	1.5 days	1.5 days
Trigonometric Emphasis (block)	1.5 days		(with Lesson 1-3B) 1.5 days	1.5 days	

Instructional Resources

Lesson	Graphing Calculator Applications	Study Guide and Practice	Enrichment	Assessment and Evaluation	Graphing Calculator Keystrokes	Problem Solving & Applications	Spreadsheet Applications	TI-92 and CBL Interdisciplinary Activities	Transparencies A and B
	Student Edition	Blackline Masters (page numbers)							
1-1	11	1–2	1				1		1-1
1-2	13	3–4	2	17	1				1-2
1-3	22	5–6	3						1-3
1-3B	26				2–3				
1-4		7–8	4	16, 17					1-4
1-5		9–10	5						1-5
1-6	40–43	11–12	6	18		1			1-6
1-7	46–51	13–14	7			2			1-7
1-8		15–16	8	18					1-8
Study Guide/ Assessment				3–15, 21					
SAT/ACT Practice				19–20					

Lesson 1-6	Lesson 1-7	Lesson 1-8	Study Guide & Assessment	Chapter Test	TOTAL
1.5 days	1.5 days	1.5 days	1 day	1 day	14 days
1.5 days		2 days		0.5 day	8.5 days

The worksheets shown on this page are a small sample of the materials available on *TeacherWorks: All-in-One Lesson Planner and Resource Center.*

This CD-ROM contains all of the blackline masters and transparencies available for this program.

It also includes a lesson planner and interactive Teacher Edition, so you can customize lesson plans and reproduce classroom resources quickly and easily, from just about anywhere.

Applications

Problem Solving and Applications Masters, pp. 1, 2

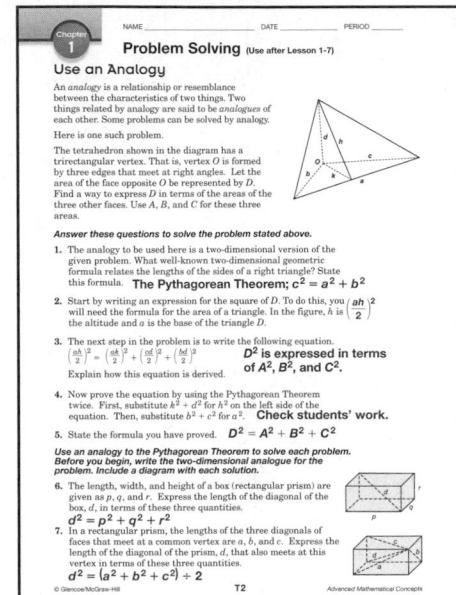

Technology/Multimedia

Spreadsheet Applications Masters, p. 1

Assessment Resources

Type	Student Edition	Teacher's Wraparound Edition	Assessment and Evaluation Masters
Ongoing Assessment	Mid-Chapter Quiz, p. 31	5-Minute Check, pp. 5, 13, 20, 27, 32, 38, 45, 52	Mid-Chapter Test, p. 16 Quizzes A–D, pp.17–18
Mixed Review	Mixed Review, pp. 19, 25, 31, 37, 44, 51, 56		Cumulative Review, p. 21
Error Analysis	You Decide, pp. 9, 48	Error Analysis, pp. 8, 17, 23, 29, 35, 41, 48, 54	
Standardized Test Practice	SAT/ACT Practice, pp. 12, 25, 31, 37, 44, 51, 56 SAT & ACT Preparation, pp. 62–65	TIMSS, p. 65	SAT/ACT Practice, pp. 19–20
Open-Ended Assessment	Math Journal, pp. 17, 29 Open-Ended Assessment, p. 61 Portfolio, p. 61	Modeling: pp. 25, 26, 37, 56 Speaking: pp. 19, 31 Writing: pp. 12, 44, 51	Open-Ended Assessment, p. 15
Chapter Assessment	Study Guide and Assessment, pp. 57–61 Chapter Test, p. A56		Multiple-Choice Tests (Forms 1A, 1B, 1C), pp. 3–8 Free-Response Tests (Forms 2A, 2B, 2C), pp. 9–14

TestCheck and Worksheet Builder is a networkable CD-ROM with three integrated modules. The **Worksheet Builder** creates customized worksheets, tests, and quizzes of free-response, multiple-choice, short-answer, and open-ended items. The **Student Module** gives you the option of having students take tests on-screen and get immediate feedback on their performance. Use the optional **Management System** to keep detailed student records.

Additional Chapter Resources

Student Edition
Career Choices, p. 12
Internet Project, p. 61
Extra Practice, pp. A26–A27

Teacher's Classroom Resources

Technology
Graphing Calculator
Keystrokes Masters

Teaching Aids
Answer Key Transparencies
Block Schedule Planning Guide
Lesson Planning Guide
Solutions Manual

Glencoe Instructional Technology

Instructional
Advanced MathPASS CD-ROM
Multimedia Applications CD-ROM

Assessment
TestCheck and Worksheet Builder

Visit **www.amc.glencoe.com** for data updates, career information, graphing calculator programs, and other interactive activities.

Chapter Overview

This chapter reviews functions and relations and operations with functions. Students evaluate, find the zeros of, write, and graph linear functions and inequalities. Students also write equations of parallel and perpendicular lines. In addition, they use scatter plots and piecewise functions to model real-world data.

 Using a Graphing Calculator

There are numerous opportunities, all optional, to use a graphing calculator in this chapter.
- Using **TABLE** to find the domain of a function (p. 11)
- Exploring the sum of two functions (p. 13)
- Finding the zeros of a linear function (p. 22)
- Investigating families of linear graphs (p. 26)
- Finding equations of regression lines and Pearson product-moment correlation values (pp. 40–43, 51)
- Graphing greatest integer functions (p. 46)

The **Graphing Calculator Appendix**, which begins on page A2, contains keystroke instructions on many of these topics.

LINEAR RELATIONS AND FUNCTIONS

CHAPTER OBJECTIVES ....●
- **Determine whether a given relation is a function and perform operations with functions.** *(Lessons 1-1, 1-2)*
- **Evaluate and find zeros of linear functions using functional notation.** *(Lesson 1-1, 1-3)*
- **Graph and write functions and inequalities.** *(Lessons 1-3, 1-4, 1-7, 1-8)*
- **Write equations of parallel and perpendicular lines.** *(Lesson 1-5)*
- **Model data using scatter plots and write prediction equations.** *(Lesson 1-6)*

4 *Chapter 1 Linear Relations and Functions*

 SAT & ACT Preparation

This textbook contains a complete SAT/ACT preparation course described on pages T8–T9 of the *Teacher's Handbook*. This chapter contains:
- an SAT or ACT practice question at the end of each lesson
- examples of SAT and ACT problems involving arithmetic problems on page 64
- ten SAT and ACT practice problems on page 65
- additional SAT and ACT practice in the *Assessment and Evaluation Masters,* pp. 19–20
- SAT and ACT test practice banks in the *TestCheck and WorksheetBuilder*

1-1

Relations and Functions

OBJECTIVES
- Determine whether a given relation is a function.
- Identify the domain and range of a relation or function.
- Evaluate functions.

METEOROLOGY Have you ever wished that you could change the weather? One of the technologies used in weather management is cloud seeding. In cloud seeding, microscopic particles are released in a cloud to bring about rainfall. The data in the table show the number of acre-feet of rain from pairs of similar unseeded and seeded clouds.

An acre-foot is a unit of volume equivalent to one foot of water covering an area of one acre. An acre-foot contains 43,560 cubic feet or about 27,154 gallons.

Acre-Feet of Rain	
Unseeded Clouds	**Seeded Clouds**
1.0	4.1
4.9	17.5
4.9	7.7
11.5	31.4
17.3	32.7
21.7	40.6
24.4	92.4
26.1	115.3
26.3	118.3
28.6	119.0

Source: Wadsworth International Group

We can write the values in the table as a set of ordered pairs. A pairing of elements of one set with elements of a second set is called a **relation.** The first element of an ordered pair is the *abscissa.* The set of abscissas is called the **domain** of the relation. The second element of an ordered pair is the *ordinate.* The set of ordinates is called the **range** of the relation. *Sets D and R are often used to represent domain and range.*

Relation, Domain, and Range	A relation is a set of ordered pairs. The domain is the set of all abscissas of the ordered pairs. The range is the set of all ordinates of the ordered pairs.

Example

1 **METEOROLOGY** State the relation of the rain data above as a set of ordered pairs. Also state the domain and range of the relation.

Relation: {(28.6, 119.0), (26.3, 118.3), (26.1, 115.3), (24.4, 92.4), (21.7, 40.6), (17.3, 32.7), (11.5, 31.4), (4.9, 17.5), (4.9, 7.7), (1.0, 4.1)}

Domain: {1.0, 4.9, 11.5, 17.3, 21.7, 24.4, 26.1, 26.3, 28.6}

Range: {4.1, 7.7, 31.4, 17.5, 32.7, 40.6, 92.4, 115.3, 118.3, 119.0}

There are multiple representations for each relation. You have seen that a relation can be expressed as a set of ordered pairs. Those ordered pairs can also be expressed as a table of values. The ordered pairs can be graphed for a pictorial representation of the relation. Some relations can also be described by a rule or equation relating the first and second coordinates of each ordered pair.

Lesson 1-1 Relations and Functions **5**

Resource Manager

Reproducible Masters
- *Study Guide and Practice,* pp. 1–2
- *Enrichment,* p. 1
- *Spreadsheet Applications,* p. 1

Transparencies
- 5-Minute Check, 1-1
- Teaching, 1-1

5-Minute Check

1. Find the value of the expression $x^2 + 4x + 4$ if $x = -2$. **0**
2. Solve $5n + 6 = -3n - 10$. **−2**
3. Evaluate $|x - 2y| - |2x - y| - xy$ if $x = -2$ and $y = 7$. **19**
4. Factor $8xy^2 - 4xy$. **$4xy(2y - 1)$**
5. Simplify $\frac{12x^5y^3}{21xy^4}$. $\frac{4x^4}{7y}$

Motivating the Lesson
Discussion Ask students to explain what the word *function* implies in: *The distance that Annette traveled was a function of her speed and the amount of time she drove.* **How far she drives depends on her speed and the time driven.** Explain that the value of a mathematical function *depends on* the value of one or more variables.

In-Class Example
Example 1
Meteorology State the relation of the windchill data as a set of ordered pairs. Also state the domain and range of the relation.

Windchill Factors at 20° F	
Wind Speed (mph)	**Windchill Temperature (°F)**
5	19
10	3
15	−5
20	−10
25	−15
30	−18

Source: *Time Almanac*

Relation: {(5, 19), (10, 3), (15, −5), (20, −10), (25, −15), (30, −18)}
D = {5, 10, 15, 20, 25, 30}
R = {19, 3, −5, −10, −15, −18}

In-Class Examples

Example 2

The domain of a relation is all consecutive integers between −2 and 2. The range *y* of the relation is 2 less than twice *x*, where *x* is a member of the domain. Write the relation as a table of values and as an equation. Then graph the relation.

x	y
−1	−4
0	−2
1	0

Equation:
$y = 2x - 2$

Example 3

State the domain and range of each relation.

a.

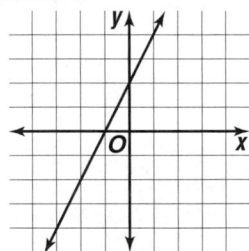

D = all real numbers;
R = all real numbers

b.

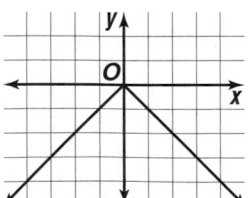

D = all real numbers;
R = all negative numbers and zero

Example 4

State the domain and range of each relation. Then state whether the relation is a function.

a. {(−2, 0), (3, 2), (4,5)}
   D = {−2, 3, 4}; R = {0, 2, 5}; function

b. {(−3, −2), (−2, −2), (−1, −2), and (0, −2)} **D = {−3,−2,−1,0}; R = {−2}; function**

Example ❷ The domain of a relation is all positive integers less than 6. The range *y* of the relation is 3 less *x*, where *x* is a member of the domain. Write the relation as a table of values and as an equation. Then graph the relation.

Table:

x	y
1	2
2	1
3	0
4	−1
5	−2

Graph:

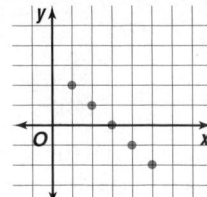

Equation: $y = 3 - x$

You can use the graph of a relation to determine its domain and range.

Example ❸ State the domain and range of each relation.

a.

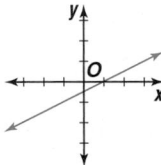

It appears from the graph that all real numbers are included in the domain and range of the relation.

b.

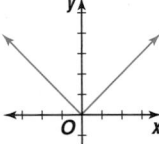

It appears from the graph that all real numbers are included in the domain. The range includes the non-negative real numbers.

The relations in Example 3 are a special type of relation called a **function**.

Function	A function is a relation in which each element of the domain is paired with exactly one element in the range.

Example ❹ State the domain and range of each relation. Then state whether the relation is a function.

a. {(−3, 0), (4, −2), (2, −6)}

   The domain is {−3, 2, 4}, and the range is {−6, −2, 0}. Each element of the domain is paired with exactly one element of the range, so this relation is a function.

b. {(4, −2), (4, 2), (9, −3), (−9, 3)}

   For this relation, the domain is {−9, 4, 9}, and the range is {−3, −2, 2, 3}. In the domain, 4 is paired with two elements of the range, −2 and 2. Therefore, this relation is *not* a function.

📁 Prerequisite Skills

A review of the prerequisite skills for this lesson is provided in the *Prerequisite Skills Masters.*

Skill Number and Content	Pages
1 Properties of Real Numbers	3–4
3 Order of Operations	7–8

An alternate definition of a function is a set of ordered pairs in which no two pairs have the same first element. This definition can be applied when a relation is represented by a graph. If every vertical line drawn on the graph of a relation passes through no more than one point of the graph, then the relation is a function. This is called the **vertical line test.**

a relation that is a function	a relation that is not a function
	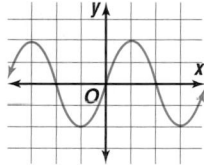

Example 5 Determine if the graph of each relation represents a function. Explain.

a.

b.

No, the graph does not represent a function. A vertical line at $x = 1$ would pass through infinitely many points.

Every element of the domain is paired with exactly one element of the range. Thus, the graph represents a function.

x is called the independent variable, and y is called the dependent variable.

Any letter may be used to denote a function. In **function notation,** the symbol $f(x)$ is read "f of x" and should be interpreted as the value of the function f at x. Similarly, $h(t)$ is the value of function h at t. The expression $y = f(x)$ indicates that for each element in the domain that replaces x, the function assigns one and only one replacement for y. The ordered pairs of a function can be written in the form (x, y) or $(x, f(x))$.

Every function can be evaluated for each value in its domain. For example, to find $f(-4)$ if $f(x) = 3x^3 - 7x^2 - 2x$, evaluate the expression $3x^3 - 7x^2 - 2x$ for $x = -4$.

Example 6 Evaluate each function for the given value.

a. $f(-4)$ if $f(x) = 3x^3 - 7x^2 - 2x$

$$f(-4) = 3(-4)^3 - 7(-4)^2 - 2(-4)$$
$$= -192 - 112 - (-8) \text{ or } -296$$

b. $g(9)$ if $g(x) = |6x - 77|$

$$g(9) = |6(9) - 77|$$
$$= |-23| \text{ or } 23$$

In-Class Examples

Example 5
Determine whether the graph of each relation represents a function. Explain.

a.
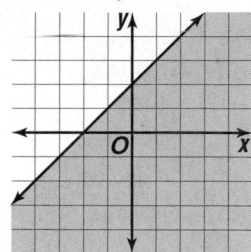

No; does not pass vertical line test.

b.
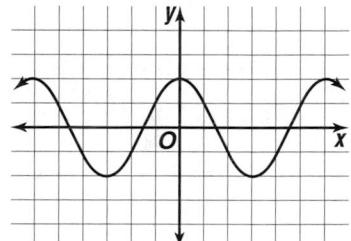

Yes; each element of the domain is paired with exactly one element of the range.

Example 6
Evaluate each function for the given value.
a. $f(-1)$ if $f(x) = -x^3 - 1$.   **0**
b. $f(3)$ if $f(x) = |3x^2|$.   **27**

Addressing Individual Needs

Kinesthetic Have students search their homes for an appliance that models a function machine. **Samples could include a pasta machine, a coffee maker, or a sausage grinder.** Explain that a function acts on a variable as a machine acts on raw materials. They each take input and apply a process to it, producing an output. Have students bring models or diagrams of their machines to class and explain how they work like a function.

Functions can also be evaluated for another variable or an expression.

Example 7 Evaluate each function for the given value.

a. $h(a)$ if $h(x) = 3x^7 - 10x^4 + 3x - 11$
$$h(a) = 3(a)^7 - 10(a)^4 + 3(a) - 11 \quad x = a$$
$$= 3a^7 - 10a^4 + 3a - 11$$

b. $j(c - 5)$ if $j(x) = x^2 - 7x + 4$
$$j(c - 5) = (c - 5)^2 - 7(c - 5) + 4 \quad x = c - 5$$
$$= c^2 - 10c + 25 - 7c + 35 + 4$$
$$= c^2 - 17c + 64$$

When you are given the equation of a function but the domain is not specified, the domain is all real numbers for which the corresponding values in the range are also real numbers.

Example 8 State the domain of each function.

a. $f(x) = \dfrac{x^3 + 5x}{x^2 - 4x}$

Any value that makes the denominator equal to zero must be excluded from the domain of f since division by zero is undefined. To determine the excluded values, let $x^2 - 4x = 0$ and solve.

$$x^2 - 4x = 0$$
$$x(x - 4) = 0$$
$$x = 0 \text{ or } x = 4$$

Therefore, the domain includes all real numbers except 0 and 4.

b. $g(x) = \dfrac{1}{\sqrt{x - 4}}$

Any value that makes the radicand negative must be excluded from the domain of g since the square root of a negative number is not a real number. Also, the denominator cannot be zero. Let $x - 4 \le 0$ and solve for the excluded values.

$$x - 4 \le 0$$
$$x \le 4$$

The domain excludes numbers less than or equal to 4. The domain is written as $\{x \mid x > 4\}$, which is read "the set of all x such that x is greater than 4."

CHECK FOR UNDERSTANDING

Communicating Mathematics

Read and study the lesson to answer each question.   1–3. See margin.

1. **Represent** the relation $\{(-4, 2), (6, 1), (0, 5), (8, -4), (2, 2), (-4, 0)\}$ in two other ways.

2. **Draw** the graph of a relation that is not a function.

3. **Describe** how to use the vertical line test to determine whether the graph at the right represents a function.

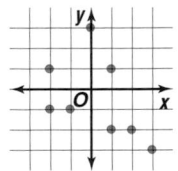

Additional Answers

1.

x	y
−4	2
6	1
0	5
8	−4
2	2
−4	0

2. Sample answer:

4. You Decide Keisha says that all functions are relations but not all relations are functions. Kevin says that all relations are functions but not all functions are relations. Who is correct and why? **See margin.**

Guided Practice

5. The domain of a relation is all positive integers less than 8. The range y of the relation is x less 4, where x is a member of the domain. Write the relation as a table of values and as an equation. Then graph the relation. **See margin.**

State each relation as a set of ordered pairs. Then state the domain and range.

6.

x	y
−3	4
0	0
3	−4
6	−8

$\{(-3, 4),$ $(0, 0), (3, -4),$ $(6, -8)\};$ $D =$ $\{-3, 0, 3, 6\};$ $R = \{-8, -4,$ $0, 4\}$

7.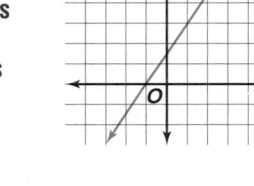

$\{(-6, 1),$ $(-4, 0),$ $(-2, -4),$ $(1, 3), (4, 3)\};$ $D = \{-6, -4, 2,$ $1, 4\}; R = \{-4,$ $0, 1, 3\}$

Given that x is an integer, state the relation representing each equation by making a table of values. Then graph the ordered pairs of the relation.

8–9. See margin.

8. $y = 3x + 5$ and $-4 \leq x \leq 4$ **9.** $y = -5$ and $1 \leq x \leq 8$

State the domain and range of each relation. Then state whether the relation is a function. Write *yes* or *no*. Explain.

10–11. See margin for domains, ranges, and explanations.

10. $\{(1, 2), (2, 4), (-3, -6), (0, 0)\}$ **yes** **11.** $\{(6, -2), (3, 4), (6, -6), (-3, 0)\}$ **no**

12. Study the graph at the right.
   a. State the domain and range of the relation.   **domain: all reals; range: all reals**
   b. State whether the graph represents a function. Explain.   **Yes; the graph passes vertical line test.**

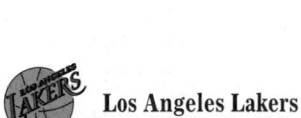

Evaluate each function for the given value.

13. $f(-3)$ if $f(x) = 4x^3 + x^2 - 5x$ **−84**

14. $g(m + 1)$ if $g(x) = 2x^2 - 4x + 2$ **$2m^2$**

15. State the domain of $f(x) = \sqrt{x + 1}$.   **$x \geq -1$**

16. Sports The table shows the heights and weights of members of the Los Angeles Lakers basketball team during a certain year.
   a. State the relation of the data as a set of ordered pairs. Also state the domain and range of the relation.
   b. Graph the relation. **See Answer Appendix.**
   c. Determine whether the relation is a function.   **no**

16a. $\{(83, 240),$ $(81, 220),$ $(82, 245),$ $(78, 200),$ $(83, 255),$ $(73, 200),$ $(80, 215),$ $(77, 210),$ $(78, 190),$ $(73, 180),$ $(86, 300),$ $(77, 220),$ $(82, 260)\};$ $\{78, 73, 77, 80,$ $81, 82, 83, 86\};$ $\{180, 190, 200,$ $210, 215, 220,$ $240, 245, 255,$ $260, 300\}$

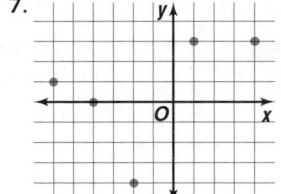

Los Angeles Lakers

Height (in.)	Weight (lb)
83	240
81	220
82	245
78	200
83	255
73	200
80	215
77	210
78	190
73	180
86	300
77	220
82	260

Source: *Preview Sports*

 www.amc.glencoe.com/self_check_quiz

Lesson 1-1 Relations and Functions **9**

Lesson 1-1 **9**

Additional Answers

17.

x	y
1	3
2	6
3	9
4	12
5	15
6	18
7	21
8	24
9	27

18.

x	y
−6	−11
−5	−10
−4	−9
−3	−8
−2	−7
−1	−6

19.

x	y
−4	4
−3	5
−2	6
−1	7
0	8
1	9
2	10
3	11
4	12

20. $\{(-5, -5), (-3, -3), (-1, -1), (1, 1)\}$; $D = \{-5, -3, -1, 1\}$; $R = \{-5, -3, -1, 1\}$

21. $\{(-10, 0), (-5, 0), (0, 0), (5, 0)\}$; $D = \{-10, -5, 0, 5\}$; $R = \{0\}$

22. $\{(4, 0), (5, 1), (8, 0), (13, 1)\}$; $D = \{4, 5, 8, 13\}$; $R = \{0, 1\}$

23. $\{(-3, -2), (-1, 1), (0, 0), (1, 1)\}$; $D = \{-3, -1, 0, 1\}$; $R = \{-2, 0, 1\}$

24. $\{(-5, 5), (-3, 3), (-1, 1), (2, -2), (4, -4)\}$; $D = \{-5, -3, -1, 2, 4\}$; $R = \{-4, -2, 1, 3, 5\}$

25. $\{(3, -4), (3, -2), (3, 0), (3, 1), (3, 3)\}$; $D = \{3\}$; $R = \{-4, -2, 0, 1, 3\}$

EXERCISES

Practice

17–19. See margin for tables and graphs.

Write each relation as a table of values and as an equation. Graph the relation.

17. the domain is all positive integers less than 10, the range is 3 times x, where x is a member of the domain $y = 3x$

18. the domain is all negative integers greater than −7, the range is x less 5, where x is a member of the domain $y = x - 5$

19. the domain is all integers greater than −5 and less than or equal to 4, the range is 8 more than x, where x is a member of the domain $y = 8 + x$

State each relation as a set of ordered pairs. Then state the domain and range.

20–31. See margin.

20.

x	y
−5	−5
−3	−3
−1	−1
1	1

21.

x	y
−10	0
−5	0
0	0
5	0

22.

x	y
4	0
5	1
8	0
13	1

interNET CONNECTION

Graphing Calculator Programs
For a graphing calculator program that plots points in a relation, visit www.amc.glencoe.com

B **23.**   **24.**   **25.**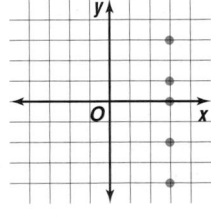

Given that x is an integer, state the relation representing each equation by making a table of values. Then graph the ordered pairs of the relation.

26. $y = x - 5$ and $-4 \le x \le 1$ **26–31.** See Answer Appendix.   **27.** $y = -x$ and $1 \le x < 7$

28. $y = |x|$ and $-5 \le x \le 1$ **29.** $y = 3x - 3$ and $0 < x < 6$

30. $y^2 = x - 2$ and $x = 11$ **31.** $|2y| = x$ and $x = 4$

State the domain and range of each relation. Then state whether the relation is a function. Write *yes* or *no*. Explain.   **35.** no

32. $\{(4, 4), (5, 4), (6, 4)\}$ **yes**   **33.** $\{(1, -2), (1, 4), (1, -6), (1, 0)\}$ **no**

34. $\{(4, -2), (4, 2), (1, -1), (1, 1), (0, 0)\}$ **no**   **35.** $\{(0, 0), (2, 2), (2, -2), (5, 8), (5, -8)\}$

36. $\{(-1.1, -2), (-0.4, -1), (-0.1, -1)\}$ **yes**   **37.** $\{(2, -3), (9, 0), (8, -3), (-9, 8)\}$ **yes**

32–37. See margin for domains, ranges, and explanations.

For each graph, state the domain and range of the relation. Then explain whether the graph represents a function.   **38–40.** See margin.

38.   **39.**   **40.**

Evaluate each function for the given value.

41. $f(3)$ if $f(x) = 2x + 3$ **9**

42. $g(-2)$ if $g(x) = 5x^2 + 3x - 2$ **12**

43. $h(0.5)$ if $h(x) = \frac{1}{x}$ **2**

44. $j(2a)$ if $j(x) = 1 - 4x^3$ **$1 - 32a^3$**

45. $f(n - 1)$ if $f(x) = 2x^2 - x + 9$

46. $g(b^2 + 1)$ if $g(x) = \frac{3 - x}{5 + x}$ **$\frac{2 - b^2}{6 + b^2}$**

47. Find $f(5m)$ if $f(x) = |x^2 - 13|$. **$|25m^2 - 13|$**

45. $2n^2 - 5n + 12$

49. $x \le -3$ or $x \ge 3$

50. $x < -\sqrt{7}$ or $x > \sqrt{7}$

State the domain of each function.

48. $f(x) = \frac{3x}{x^2 - 5}$ **$x \neq \pm \sqrt{5}$**

49. $g(x) = \sqrt{x^2 - 9}$

50. $h(x) = \frac{x + 2}{\sqrt{x^2 - 7}}$

Graphing Calculator

51. You can use the table feature of a graphing calculator to find the domain of a function. Enter the function into the Y= list. Then observe the y-values in the table. An error indicates that an x-value is excluded from the domain. Determine the domain of each function.

a. $f(x) = \frac{3}{x - 1}$ **$x \neq 1$**

b. $g(x) = \frac{3 - x}{5 + x}$ **$x \neq -5$**

c. $h(x) = \frac{x^2 - 12}{x^2 - 4}$ **$x \neq -2, 2$**

Applications and Problem Solving

Real World Application

52. **Education**   The table shows the number of students who applied and the number of students attending selected universities.

a. State the relation of the data as a set of ordered pairs. Also state the domain and range of the relation.   **See margin.**

b. Graph the relation.   **See margin.**

c. Determine whether the relation is a function. Explain.

c. Yes; no member of the domain is paired with more than one member of the range.

University	Number Applied	Number Attending
Auburn University	9244	3166
University of California, Davis	18,584	3697
University of Illinois-Champaign-Urbana	18,140	5805
University of Maryland	16,182	3999
State University of New York – Stony Brook	13,589	2136
The Ohio State University	18,912	5950
Texas A&M University	13,877	6233

Source: *Newsweek, "How to get into college, 1998"*

53. **Critical Thinking**   If $f(2m + 1) = 24m^3 + 36m^2 + 26m$, what is $f(x)$? (*Hint:* Begin by solving $x = 2m + 1$ for m.)   **$3x^3 + 4x - 7$**

54. **Aviation**   The temperature of the atmosphere decreases about 5°F for every 1000 feet that an airplane ascends. Thus, if the ground-level temperature is 95°F, the temperature can be found using the function $t(d) = 95 - 0.005d$, where $t(d)$ is the temperature at a height of d feet. Find the temperature outside of an airplane at each height.

a. 500 ft   b. 750 ft   c. 1000 ft   d. 5000 ft   e. 30,000 ft

54a. 92.5°F
54b. 91.25°F
54c. 90°F
54d. 70°F
54e. −55°F

55. **Geography**   A global positioning system, GPS, uses satellites to allow a user to determine his or her position on Earth. The system depends on satellite signals that are reflected to and from a hand-held transmitter. The time that the signal takes to reflect is used to determine the transmitter's position. Radio waves travel through air at a speed of 299,792,458 meters per second. Thus, the function $d(t) = 299,792,458t$ relates the time t in seconds to the distance traveled $d(t)$ in meters.

a. Find the distance a sound wave will travel in 0.05, 0.2, 1.4, and 5.9 seconds.

b. If a signal from a GPS satellite is received at a transmitter in 0.08 seconds, how far from the transmitter is the satellite?   **23983396.64 m**

55a.
14,989,622.9 m;
59,958,491.6 m;
419,709,441.2 m;
1,768,775,502 m

Extra Practice   See p. A26.

Lesson 1-1   Relations and Functions   **11**

Teaching Tip   Remind students that they can edit their functions after entering them into the **Y =** list. Move the cursor to the function that needs editing. The functions can be defined so that when an expression evaluates to a non-real number, the value is not plotted, and no error message appears.

Additional Answers *(pp. 10–11)*

32. {4, 5, 6}; {4}; each x-value is paired with exactly one y-value.

33. {1}; {−6, −2, 0, 4}; the x-value 1 is paired with more than one y-value.

34. {0, 1, 4}; {−2, −1, 0, 1, 2}; the x-values 1 and 4 are paired with more than one y-value.

35. {0, 2, 5}; {−8, −2, 0, 2, 8}; the x-values 2 and 5 are paired with more than one y-value.

36. {−1.1, −0.4, −0.1}; {−2, −1}; each x-value is paired with exactly one y-value.

37. {−9, 2, 8, 9}; {−3, 0, 8}; each x-value is paired with exactly one y-value.

38. Domain: all reals; range: all reals; not a function because it fails the vertical line test.

39. Domain: {−3, −2, −1, 1, 2, 3}; range: {−1, 1, 2, 3}; a function because each x-value is paired with exactly one y-value.

40. Domain: $\{x \mid -8 \le x \le 8\}$; range: $\{y \mid -8 \le y \le 8\}$; not a function because it fails the vertical line test.

52a. {(9244, 3166), (18,584, 3697), (18,140, 5805), (16,182, 3999), (13,589, 2136), (18,912, 5950), (13,877, 6233)}; {9244, 18,584, 18,140, 16,182, 13,589, 18,912, 13,877}; {3166, 3697, 5805, 3999, 2136, 5950, 6233}

52b.

Number Attending (thousands)

Number Applied (thousands)

Extra Credit

If $f(3k + 2) = 54k^3 + 18k^2 + 12k$, what is $f(x)$?   **$2x^3 - 10x^2 + 20x - 16$**

Lesson 1-1   **11**

Open-Ended Assessment
Writing Have students write an example of a relation that is a function and an example of one that is not. Then, have them write a sentence or two explaining the difference between the two examples.

56. **Critical Thinking** $P(x)$ is a function for which $P(1) = 1$, $P(2) = 2$, $P(3) = 3$, and $P(x + 1) = \dfrac{P(x - 2) P(x - 1) + 1}{P(x)}$ for $x \geq 3$. Find the value of $P(6)$. **$\frac{4}{7}$**

57. **SAT Practice** What is the value of $7^2 - (3^2 + 4^2)$? **B**
   - **A** 56
   - **B** 24
   - **C** 0
   - **D** -24
   - **E** -56

Enrichment Masters, p. 1

CAREER CHOICES

Veterinary Medicine

If you like working with animals and have a strong interest in science, you may want to consider a career in veterinary medicine. Many veterinarians work with small animals, such as pets, maintaining their good health and treating illnesses and injuries. Some veterinarians work with large animals, such as farm animals, to ensure the health of animals that we depend upon for food. Still other veterinarians work to control diseases in wildlife.

Duties of veterinarians can include administering medications to the animals, performing surgeries, instructing people in the care of animals, and researching genetics, prevention of disease, and better animal nutrition.

Many veterinarians work in private practice, but jobs are also available in industry and governmental agencies.

CAREER OVERVIEW

Degree Preferred:
D.V.M. (doctor of veterinary medicine) consisting of six years of college

Related Courses:
biology, chemistry, mathematics

Outlook:
number of jobs expected to increase through 2006

Spending on Pet Health Care

Dollars (billions)
$7.83
$4.15
1991   1997

Source: American Veterinary Medical Association

interNET **CONNECTION**   For more information on careers in veterinary medicine, visit: **www.amc.glencoe.com**

CAREER CHOICES   **VETERINARY MEDICINE**

Math Connection
- If the graph depicts part of the information that can be expressed as a relation, what type of information is contained in the domain and the range? **D = {years}, R = {amount spent}**
- Based on the graph, would you expect spending on pet health care to increase or decrease in the year 2000? **increase**

1-2

Composition of Functions

OBJECTIVES
* Perform operations with functions.
* Find composite functions.
* Iterate functions using real numbers.

BUSINESS Each year, thousands of people visit Yellowstone National Park in Wyoming. Audiotapes for visitors include interviews with early settlers and information about the geology, wildlife, and activities of the park. The revenue $r(x)$ from the sale of x tapes is $r(x) = 9.5x$. Suppose that the function for the cost of manufacturing x tapes is $c(x) = 0.8x + 1940$. What function could be used to find the profit on x tapes? *This problem will be solved in Example 2.*

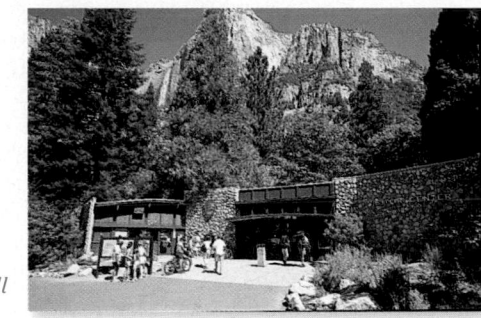

To solve the profit problem, you can subtract the cost function $c(x)$ from the revenue function $r(x)$. If you have two functions, you can form new functions by adding, subtracting, multiplying, or dividing the functions.

GRAPHING CALCULATOR EXPLORATION

Use a graphing calculator to explore the sum of two functions.

♦ Enter the functions $f(x) = 2x - 1$ and $f(x) = 3x + 2$ as **Y1** and **Y2**, respectively.

♦ Enter **Y1** + **Y2** as the function for **Y3**. To enter **Y1** and **Y2**, press VARS, then select **Y-VARS**. Then choose the equation name from the menu.

♦ Use **TABLE** to compare the function values for **Y1**, **Y2**, and **Y3**.

TRY THESE
Use the functions $f(x) = 2x - 1$ and $f(x) = 3x + 2$ as **Y1** and **Y2**. Use **TABLE** to observe the results for each definition of **Y3**.

1. **Y3** = **Y1** − **Y2**   2. **Y3** = **Y1** · **Y2**

3. **Y3** = **Y1** ÷ **Y2**   **1–3. See Answer Appendix.**

WHAT DO YOU THINK?

4. Repeat the activity using functions $f(x) = x^2 - 1$ and $f(x) = 5 - x$ as **Y1** and **Y2**, respectively. What do you observe?

5. Make conjectures about the functions that are the sum, difference, product, and quotient of two functions.   **4–5. See margin.**

The Graphing Calculator Exploration leads us to the following definitions of operations with functions.

Operations with Functions

Sum:	$(f + g)(x) = f(x) + g(x)$
Difference:	$(f - g)(x) = f(x) - g(x)$
Product:	$(f \cdot g)(x) = f(x) \cdot g(x)$
Quotient:	$\left(\dfrac{f}{g}\right)(x) = \dfrac{f(x)}{g(x)}, \ g(x) \neq 0$

Resource Manager

 Reproducible Masters
* *Study Guide and Practice,* pp. 3–4
* *Enrichment,* p. 2
* *Assessment and Evaluation,* p. 17
* *Graphing Calculator Keystrokes,* p. 1

 Transparencies
* 5-Minute Check, 1-2
* Teaching, 1-2

---

1 FOCUS

5-Minute Check
Lesson 1-1

1. State the domain and range of the relation {(−2, 2), (1, 2), (2, 3)}. **D = {−2, 1, 2}; R = {2, 3}**

2. Is the relation in Exercise 1 a function? Explain. **Yes; each point in the domain is paired with exactly one point in the range.**

Evaluate each function for the given value $f(x)$.

3. $f(-2)$ if $f(x) = 6 - x^2$ **2**

4. $f(m)$ if $f(x) = \dfrac{5}{2 - x}$ $\dfrac{5}{2 - m}$

5. $f(3a)$ if $f(x) = \sqrt{x^2 - 4}$ $\sqrt{9a^2 - 4}$

Motivating the Lesson
Real-World Application Have students read the application. Ask them to describe what they think the two equations represent. For example, in $r(x) = 9.5x$, what does 9.5 represent? **Each audiotape costs $9.50.** In $c(x) = 0.8x + 1940$, what does 0.8 represent? **Each tape costs $0.80 to manufacture.**

 Graphing Calculator Exploration

Remind students that **Y-VARS** is a menu used to choose the name for the type of function needed. Students can choose **Function, Parametric, Polar,** or **Sequence.**

Additional Answers
Graphing Calculator Exploration

4. Sample answer: the (sum/difference/product/quotient) of the function values is the (sum/difference/product/quotient) of the functions.

5. Sample answer: for functions $f(x)$ and $g(x)$, $(f + g)(x) = f(x) + g(x)$; $(f - g)(x) = f(x) - g(x)$; $(f \cdot g)(x) = f(x) \cdot g(x)$; and $\left(\dfrac{f}{g}\right)(x) = \dfrac{f(x)}{g(x)}, \ g(x) \neq 0.$

In-Class Examples

Example 1

Given $f(x) = 2x - 1$ and $g(x) = x^2$, find each function.

a. $(f + g)(x)$ $x^2 + 2x - 1$

b. $(f - g)(x)$ $-x^2 + 2x - 1$

c. $(f \cdot g)(x)$ $2x^3 - x^2$

d. $\left(\dfrac{f}{g}\right)(x)$ $\dfrac{2x - 1}{x^2},\ x \neq 0$

Example 2

Business For the Lotsa Coffee Shop, the revenue $r(x)$ in dollars from selling x cups of coffee is $r(x) = 1.5x$. The cost $c(x)$ for making and selling the coffee is $c(x) = 0.2x + 110$.

a. Write the profit function.
$P(x) = 1.3x - 110$

b. Find the profit on 100, 200, and 500 cups of coffee sold.
$20; $150; $540

Teaching Tip When discussing composition of functions, point out that $[f \circ g](x)$ is found by substituting $g(x)$ for x in $f(x)$.

For each new function, the domain consists of those values of x common to the domains of f and g. The domain of the quotient function is further restricted by excluding any values that make the denominator, $g(x)$, zero.

Example **1** Given $f(x) = 3x^2 - 4$ and $g(x) = 4x + 5$, find each function.

a. $(f + g)(x)$

$$
\begin{aligned}
(f + g)(x) &= f(x) + g(x) \\
&= 3x^2 - 4 + 4x + 5 \\
&= 3x^2 + 4x + 1
\end{aligned}
$$

b. $(f - g)(x)$

$$
\begin{aligned}
(f - g)(x) &= f(x) - g(x) \\
&= 3x^2 - 4 - (4x + 5) \\
&= 3x^2 - 4x - 9
\end{aligned}
$$

c. $(f \cdot g)(x)$

$$
\begin{aligned}
(f \cdot g)(x) &= f(x) \cdot g(x) \\
&= (3x^2 - 4)(4x + 5) \\
&= 12x^3 + 15x^2 - 16x - 20
\end{aligned}
$$

d. $\left(\dfrac{f}{g}\right)(x)$

$$
\begin{aligned}
\left(\dfrac{f}{g}\right)(x) &= \dfrac{f(x)}{g(x)} \\
&= \dfrac{3x^2 - 4}{4x + 5},\ x \neq -\dfrac{5}{4}
\end{aligned}
$$

You can use the difference of two functions to solve the application problem presented at the beginning of the lesson.

Example **2** **BUSINESS** Refer to the application at the beginning of the lesson.

a. Write the profit function.

b. Find the profit on 500, 1000, and 5000 tapes.

a. Profit is revenue minus cost. Thus, the profit function $p(x)$ is $p(x) = r(x) - c(x)$.

The revenue function is $r(x) = 9.5x$. The cost function is $c(x) = 0.8x + 1940$.

$$
\begin{aligned}
p(x) &= r(x) - c(x) \\
&= 9.5x - (0.8x + 1940) \\
&= 8.7x - 1940
\end{aligned}
$$

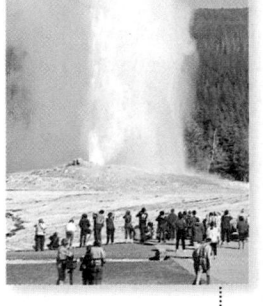

b. To find the profit on 500, 1000, and 5000 tapes, evaluate $p(500)$, $p(1000)$, and $p(5000)$.

$p(500) = 8.7(500) - 1940$ or 2410

$p(1000) = 8.7(1000) - 1940$ or 6760

$p(5000) = 8.7(5000) - 1940$ or 41,560

The profit on 500, 1000, and 5000 tapes is $2410, $6760, and $41,560, respectively. *Check by finding the revenue and the cost for each number of tapes and subtracting to find profit.*

Functions can also be combined by using **composition.** In a composition, a function is performed, and then a second function is performed on the result of the first function. You can think of composition in terms of manufacturing a product. For example, fiber is first made into cloth. Then the cloth is made into a garment.

14 *Chapter 1 Linear Relations and Functions*

Prerequisite Skills

A review of the prerequisite skills for this lesson is provided in the *Prerequisite Skills Masters.*

Skill Number and Content	Pages
2 Fractions	5–6
4 Algebraic Expressions	9–10

In composition, a function g maps the elements in set R to those in set S. Another function f maps the elements in set S to those in set T. Thus, the range of function g is the same as the domain of function f. A diagram is shown below.

R	S
x	$g(x) = \frac{1}{4}x$
4	1
8	2
12	3

S	T
x	$f(x) = 6 - 2x$
1	4
2	2
3	0

domain of g(x)   *The range of g(x) is the domain of f(x).*   *range of f(x)*

The function formed by composing two functions f and g is called the **composite** of f and g. It is denoted by $f \circ g$, which is read as "f composition g" or "f of g."

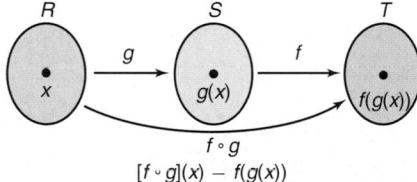

$f \circ g$
$[f \circ g](x) - f(g(x))$

Composition of Functions

Given functions f and g, the composite function $f \circ g$ can be described by the following equation.

$$[f \circ g](x) = f[g(x)]$$

The domain of $f \circ g$ includes all of the elements x in the domain of g for which $g(x)$ is in the domain of f.

Example ③ Find $[f \circ g](x)$ and $[g \circ f](x)$ for $f(x) = 2x^2 - 3x + 8$ and $g(x) = 5x - 6$.

$[f \circ g](x) = f(g(x))$

        $= f(5x - 6)$ *Substitute $5x - 6$ for $g(x)$.*

        $= 2(5x - 6)^2 - 3(5x - 6) + 8$ *Substitute $5x - 6$ for x in $f(x)$.*

        $= 2(25x^2 - 60x + 36) - 15x + 18 + 8$

        $= 50x^2 \quad 135x \ | \ 98$

$[g \circ f](x) = g(f(x))$

        $= g(2x^2 - 3x + 8)$ *Substitute $2x^2 - 3x + 8$ for $f(x)$.*

        $= 5(2x^2 - 3x + 8) - 6$ *Substitute $2x^2 - 3x + 8$ for x in $g(x)$.*

        $= 10x^2 - 15x + 34$

The domain of a composed function $[f \circ g](x)$ is determined by the domains of both $f(x)$ and $g(x)$.

Teaching Tip Students will use
iterates again when they study
iterative sequences in Lesson 12-8.

Example **4** State the domain of $[f \circ g](x)$ for $f(x) = \sqrt{x - 4}$ and $g(x) = \dfrac{1}{x^2}$.

$f(x) = \sqrt{x - 4}$ *Domain: $x \ge 4$*

$g(x) = \dfrac{1}{x^2}$ *Domain: $x \ne 0$*

If $g(x)$ is undefined for a given value of x, then that value is excluded from the domain of $[f \circ g](x)$. Thus, 0 is excluded from the domain of $[f \circ g](x)$.

The domain of $f(x)$ is $x \ge 4$. So for x to be in the domain of $[f \circ g](x)$, it must be true that $g(x) \ge 4$.

$g(x) \ge 4$

$\dfrac{1}{x^2} \ge 4$ $g(x) = \dfrac{1}{x^2}$

$1 \ge 4x^2$ *Multiply each side by x^2.*

$\dfrac{1}{4} \ge x^2$ *Divide each side by 4.*

$\dfrac{1}{2} \ge |x|$ *Take the square root of each side.*

$-\dfrac{1}{2} \le x \le \dfrac{1}{2}$ *Rewrite the inequality.*

Therefore, the domain of $[f \circ g](x)$ is $-\dfrac{1}{2} \le x \le \dfrac{1}{2}, x \ne 0$.

The composition of a function and itself is called iteration. Each output of an iterated function is called an iterate. To iterate a function $f(x)$, find the function value $f(x_0)$, of the initial value x_0. The value $f(x_0)$ is the first iterate, x_1. The second iterate is the value of the function performed on the output; that is, $f(f(x_0))$ or $f(x_1)$. Each iterate is represented by x_n, where n is the iterate number. For example, the third iterate is x_3.

Example **5** Find the first three iterates, x_1, x_2, and x_3, of the function $f(x) = 2x - 3$ for an initial value of $x_0 = 1$.

To obtain the first iterate, find the value of the function for $x_0 = 1$.

$x_1 = f(x_0) = f(1)$

         $= 2(1) - 3$ or -1

To obtain the second iterate, x_2, substitute the function value for the first iterate, x_1, for x.

$x_2 = f(x_1) = f(-1)$

         $= 2(-1) - 3$ or -5

Now find the third iterate, x_3, by substituting x_2 for x.

$x_3 = f(x_2) = f(-5)$

         $= 2(-5) - 3$ or -13

Thus, the first three iterates of the function $f(x) = 2x - 3$ for an initial value of $x_0 = 1$ are -1, -5, and -13.

Addressing Individual Needs

Interpersonal Have students work in pairs. Each student thinks of a function. Then both work together to find the sum, difference, product, and quotient of their functions and the composite of the two functions.

CHECK FOR UNDERSTANDING

Communicating Mathematics

Read and study the lesson to answer each question.   1–4. See margin.

1. **Write** two functions $f(x)$ and $g(x)$ for which $(f \cdot g)(x) = 2x^2 + 11x - 6$. Tell how you determined $f(x)$ and $g(x)$.

2. **Explain** how iteration is related to composition of functions.

3. **Determine** whether $[f \circ g](x)$ is always equal to $[g \circ f](x)$ for two functions $f(x)$ and $g(x)$. Explain your answer and include examples or counterexamples.

4. *Math Journal*   Write an explanation of function composition. Include an everyday example of two composed functions and an example of a realworld problem that you would solve using composed functions.

Guided Practice

5. Given $f(x) = 3x^2 + 4x - 5$ and $g(x) = 2x + 9$, find $f(x) + g(x)$, $f(x) - g(x)$, $f(x) \cdot g(x)$, and $\left(\frac{f}{g}\right)(x)$. $3x^2 + 6x + 4; 3x^2 + 2x - 14;$
$6x^3 + 35x^2 + 26x - 45; \frac{3x^2 + 4x - 5}{2x + 9}, x \neq -\frac{9}{2}$

Find $[f \circ g](x)$ and $[g \circ f](x)$ for each $f(x)$ and $g(x)$.

6. $f(x) = 2x + 5$ $2x + 11;$
   $g(x) = 3 + x$ $2x + 8$

7. $f(x) = 2x - 3$ $2x^2 - 4x - 3;$
   $g(x) = x^2 - 2x$ $4x^2 - 16x + 15$

8. State the domain of $[f \circ g](x)$ for $f(x) = \frac{1}{(x-1)^2}$ and $g(x) = x + 3$. $x \neq -2$

9. Find the first three iterates of the function $f(x) = 2x + 1$ using the initial value $x_0 = 2$.   **5, 11, 23**

10. **Measurement** In 1954, the Tenth General Conference on Weights and Measures adopted the kelvin K as the basic unit for measuring temperature for all international weights and measures. While the kelvin is the standard unit, degrees Fahrenheit and degrees Celsius are still in common use in the United States. The function $C(F) = \frac{5}{9}(F - 32)$ relates Celsius temperatures and Fahrenheit temperatures. The function $K(C) = C + 273.15$ relates Celsius temperatures and Kelvin temperatures.

10a. $K(F) = \frac{5}{9}(F - 32) + 273.15$

10b. 233.15, 248.71, 255.37, 273.15, 373.15

   a. Use composition of functions to write a function to relate degrees Fahrenheit and kelvins.

   b. Write the temperatures $-40°F$, $-12°F$, $0°F$, $32°F$, and $212°F$ in kelvins.

EXERCISES

Practice

Find $f(x) + g(x)$, $f(x) - g(x)$, $f(x) \cdot g(x)$, and $\left(\frac{f}{g}\right)(x)$ for each $f(x)$ and $g(x)$.

11–13.
See margin.

11. $f(x) = x^2 - 2x$
    $g(x) = x + 9$

12. $f(x) = \frac{x}{x + 1}$
    $g(x) = x^2 - 1$

13. $f(x) = \frac{3}{x - 7}$
    $g(x) = x^2 + 5x$

14. If $f(x) = x + 3$ and $g(x) = \frac{2x}{x - 5}$, find $f(x) + g(x)$, $f(x) - g(x)$, $f(x) \cdot g(x)$, and $\left(\frac{f}{g}\right)(x)$. $\frac{x^2 - 15}{x - 5}, x \neq 5; \frac{x^2 - 4x - 15}{x - 5}, x \neq 5; \frac{2x^2 + 6x}{x - 5}, x \neq 5;$
$\frac{x^2 - 2x - 15}{2x}, x \neq 0 \text{ or } 5$

 www.amc.glencoe.com/self_check_quiz

Lesson 1-2 Composition of Functions **17**

3 PRACTICE/APPLY

Error Analysis

Watch for students who evaluate composite functions incorrectly by making the wrong substitutions.

Prevent by re-emphasizing that in $(g \circ h)(x)$, the second function is the one substituted into the first.

Assignment Guide

Core: 11–29 odd, 31–38
Enriched: 12–26 even, 28–38

Additional Answers

11. $x^2 - x + 9; x^2 - 3x - 9;$
    $x^3 + 7x^2 - 18x; \frac{x^2 - 2x}{x + 9},$
    $x \neq -9$

12. $\frac{x^3 + x^2 - 1}{x + 1}, x \neq -1;$
    $\frac{-x^3 - x^2 + 2x + 1}{x + 1}, x \neq -1;$
    $x^2 - x, x \neq -1; \frac{x}{x^3 + x^2 - x - 1},$
    $x \neq -1 \text{ or } 1$

13. $\frac{x^3 - 2x^2 - 35x + 3}{x - 7}, x \neq 7;$
    $-\frac{x^3 - 2x^2 - 35x - 3}{x - 7}, x \neq 7;$
    $\frac{3x^2 + 15x}{x - 7}, x \neq 7;$
    $\frac{3}{x^3 - 2x^2 - 35x}, x \neq -5, 0, \text{ or } 7$

Study Guide and Practice Masters, pp. 3–4

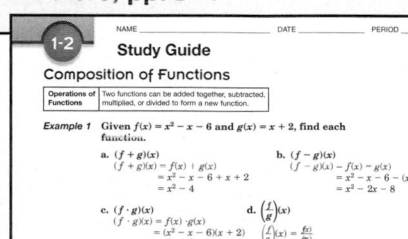

Additional Answers

1. **Sample answer:** $f(x) = 2x - 1$ and $g(x) = x + 6$; sample explanation: factor $2x^2 + 11x - 6$.

2. Iteration is composing a function on itself by evaluating the function for a value and then evaluating the function on that function value.

3. No; $[f \circ g](x)$ is the function $f(x)$ performed on $g(x)$ and $[g \circ f](x)$ is the function $g(x)$ performed on $f(x)$. See students' counterexamples.

4. **Sample answer:** composition of functions is performing one function after another. An everyday example is putting on socks and then putting shoes on top of the socks. Buying an item on sale is an example of when a composition of functions is used in a real-world situation.

Lesson 1-2 **17**

Find $[f \circ g](x)$ and $[g \circ f](x)$ for each $f(x)$ and $g(x)$.

 B

15. $f(x) = x^2 - 9$ $x^2 + 8x + 7; x^2 - 5$ **16.** $f(x) = \frac{1}{2}x - 7$ $\frac{1}{2}x - 4; \frac{1}{2}x - 1$
$g(x) = x + 4$ $g(x) = x + 6$

17. $f(x) = x - 4$ $3x^2 - 4; 3x^2 -$ **18.** $f(x) = x^2 - 1$ $25x^4 - 1;$
$g(x) = 3x^2$ $24x + 48$ $g(x) = 5x^2$ $5x^4 - 10x^2 + 5$

19. $f(x) = 2x$ $2x^3 + 2x^2 + 2;$ **20.** $f(x) = 1 + x$ $x^2 + 5x + 7;$
$g(x) = x^3 + x^2 + 1$ $8x^3 + 4x^2 + 1$ $g(x) = x^2 + 5x + 6$ $x^2 + 7x + 12$

21. What are $[f \circ g](x)$ and $[g \circ f](x)$ for $f(x) = x + 1$ and $g(x) = \frac{1}{x-1}$? $\dfrac{x}{x-1}, x \neq 1;$
$\dfrac{1}{x}, x \neq 0$

State the domain of $[f \circ g](x)$ for each $f(x)$ and $g(x)$.

 C

22. $f(x) = 5x$ **all reals**   **23.** $f(x) = \frac{1}{x}$ $x \neq 7$ **24.** $f(x) = \sqrt{x - 2}$
$g(x) = x^3$ $g(x) = 7 - x$ $g(x) = \dfrac{1}{4x}$
                                                      $x \leq \frac{1}{8}, x \neq 0$

Find the first three iterates of each function using the given initial value.

25. $f(x) = 9 - x; x_0 = 2$ **26.** $f(x) = x^2 + 1; x_0 = 1$ **27.** $f(x) = x(3 - x); x_0 = 1$
    7, 2, 7   **2, 5, 26**   **2, 2, 2**

Applications and Problem Solving

28. Retail   Sara Sung is shopping and finds several items that are on sale at 25% off the original price. The items that she wishes to buy are a sweater originally at $43.98, a pair of jeans for $38.59, and a blouse for $31.99. She has $100 that her grandmother gave her for her birthday. If the sales tax in San Mateo, California, where she lives is 8.25%, does Sara have enough money for all three items? Explain.   **Yes; the total with the discount and tax is $93.01.**

29. Yes; see margin for justification.

29. Critical Thinking   Suppose the graphs of functions $f(x)$ and $g(x)$ are lines. Must it be true that the graph of $[f \circ g](x)$ is a line? Justify your answer.

30. Physics   When a heavy box is being pushed on the floor, there are two different forces acting on the movement of the box. There is the force of the person pushing the box and the force of friction. If W is work in joules, F is force in newtons, and d is displacement of the box in meters, $W_p = F_p d$ describes the work of the person, and $W_f = F_f d$ describes the work created by friction. The increase in kinetic energy necessary to move the box is the difference between the work done by the person W_p and the work done by friction W_f.

   a. Write a function in simplest form for net work. $W_n = d(F_p - F_f)$

   b. Determine the net work expended when a person pushes a box 50 meters with a force of 95 newtons and friction exerts a force of 55 newtons.   **2000 J**

31. Finance   A sales representative for a cosmetics supplier is paid an annual salary plus a bonus of 3% of her sales *over* $275,000. Let $f(x) = x - 275,000$ and $h(x) = 0.03x$.

31a. $h[f(x)]$, because you must subtract before figuring the bonus.

   a. If x is greater than $275,000, is her bonus represented by $f[h(x)]$ or by $h[f(x)]$? Explain.

   b. Find her bonus if her sales for the year are $400,000.   **$3750**

32. Critical Thinking   Find $f\left(\frac{1}{2}\right)$ if $[f \circ g](x) = \dfrac{x^4 + x^2}{1 + x^2}$ and $g(x) = 1 - x^2$. $\frac{1}{2}$

Additional Answer

29. If $f(x)$ and $g(x)$ are both lines, they can be represented as $f(x) = m_1 x + b_1$ and $g(x) = m_2 x + b_2$. Then, $[f \circ g](x) = m_1(m_2 x + b_2) + b_1$
$$= m_1 m_2 x + m_1 b_2 + b_1.$$

Since m_1 and m_2 are constants, $m_1 m_2$ is a constant. Similarly, m_1, b_2, and b_1 are constants, so $m_1 b_2 + b_1$ is a constant. Thus, $[f \circ g](x)$ is a linear function if $f(x)$ and $g(x)$ are both linear.

33. **International Business** Value-added tax, VAT, is a tax charged on goods and services in European countries. Many European countries offer refunds of some VAT to non-resident based businesses. VAT is included in a price that is quoted. That is, if an item is marked as costing $10, that price includes the VAT.

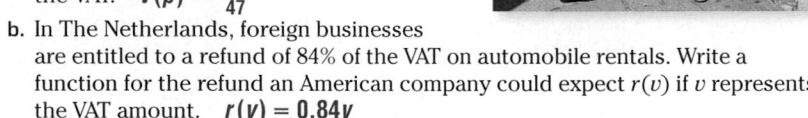

a. Suppose an American company has operations in The Netherlands, where the VAT is 17.5%. Write a function for the VAT amount paid $v(p)$ if p represents the price including the VAT. $v(p) = \dfrac{7p}{47}$

b. In The Netherlands, foreign businesses are entitled to a refund of 84% of the VAT on automobile rentals. Write a function for the refund an American company could expect $r(v)$ if v represents the VAT amount. $r(v) = 0.84v$

c. Write a function for the refund expected on an automobile rental $r(p)$ if the price including VAT is p. $r(p) = \dfrac{147p}{1175}$

d. Find the refunds due on automobile rental prices of $423.18, $225.64, and $797.05.   **$52.94, $28.23, $99.72**

Mixed Review

34. **Finance** The formula for the simple interest earned on an investment is $I = prt$, where I is the interest earned, p is the principal, r is the interest rate, and t is the time in years. Assume that $5000 is invested at an annual interest rate of 8% and that interest is added to the principal at the end of each year. *(Lesson 1-1)*

a. Find the amount of interest that will be earned each year for five years.

b. State the domain and range of the relation.   **a–c. See margin.**

c. Is this relation a function? Why or why not?

35. **{(−1, 8), (0, 4), (2, −6), (5, −9)}; $D =$ {−1, 0, 2, 5}; $R =$ {−9, −6, 4, 8}**

35. State the relation in the table as a set of ordered pairs. Then state the domain and range of the relation. *(Lesson 1-1)*

x	y
−1	8
0	4
2	−6
5	−9

36. **$D =$ {1, 2, 3, 4}; $R =$ {5, 6, 7, 8}; yes, every element in the domain is paired with exactly one element of the range.**

36. What are the domain and the range of the relation {(1, 5), (2, 6), (3, 7), (4, 8)}? Is the relation a function? Explain. *(Lesson 1-1)*

37. Find $g(-4)$ if $g(x) = \dfrac{x^3 + 5}{4x}$. *(Lesson 1-1)* $3\dfrac{11}{16}$

38. Given that x is an integer, state the relation representing $y = |-3x|$ and $-2 \le x \le 3$ by making a table of values. Then graph the ordered pairs of the relation. *(Lesson 1-1)*   **See margin.**

39. **SAT/ACT Practice** Find $f(n - 1)$ if $f(x) = 2x^2 - x + 9$.   **C**

A $2n^2 - n + 9$

B $2n^2 - n + 8$

C $2n^2 - 5n + 12$

D 9

E $2n^2 + 4n + 8$

Extra Practice See p. A26.

Extra Credit

Find $f\left(\dfrac{1}{8}\right)$ if $f(x) - g(x) = x^3 - x^2 + x - 2$ and $g(x) = \dfrac{x}{x - 1}$. $-\dfrac{7281}{3584}$

5-Minute Check
Lesson 1-2

Given $f(x) = 3x$ and $g(x) = x^2 - 1$, find each function.
1. $(f + g)(x)$ $x^2 + 3x - 1$
2. $(f - g)(x)$ $-x^2 + 3x + 1$
3. $[f \circ g](x)$ $3x^2 - 3$
4. $[g \circ f](x)$ $9x^2 - 1$

5. Find the first three iterates, x_1, x_2, and x_3, of the function $f(x) = 2x - 3$ for an initial value of $x_0 = 1$.
$-1, -5, -13$

Motivating the Lesson

Discussion On the chalkboard or the overhead projector, graph a nonvertical line crossing the x-axis at 2. Say that the line is the graph of the function $f(x)$. Ask students to give the value of x for which $f(x) = 0$. **2**

Graphing Linear Equations

AGRICULTURE American farmers produce enough food and fiber to meet the needs of our nation and to export huge quantities to countries around the world. In addition to raising grain, cotton and other fibers, fruits, or vegetables, farmers also work on dairy farms, poultry farms, horticultural specialty farms that grow ornamental plants and nursery products, and aquaculture farms that raise fish and shellfish. In 1900, the percent of American workers who were farmers was 37.5%. In 1994, that percent had dropped to just 2.5%. What was the average rate of decline? *This problem will be solved in Example 2.*

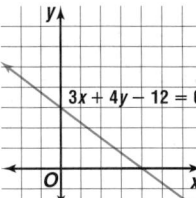

The problem above can be solved by using a linear equation. A **linear equation** has the form $Ax + By + C = 0$, where A and B are not both zero. Its graph is a straight line. The graph of the equation $3x + 4y - 12 = 0$ is shown.

The solutions of a linear equation are the ordered pairs for the points on its graph. An ordered pair corresponds to a point in the coordinate plane. Since two points determine a line, only two points are needed to graph a linear equation. Often the two points that are easiest to find are the **x-intercept** and the **y-intercept.** The x-intercept is the point where the line crosses the x-axis, and the y-intercept is the point where the graph crosses the y-axis. In the graph above, the x-intercept is at $(4, 0)$, and the y-intercept is at $(0, 3)$. *Usually, the individual coordinates 4 and 3 called the x- and y-intercepts.*

Example **Graph $3x - y - 2 = 0$ using the x-and y-intercepts.**

Substitute 0 for y to find the x-intercept. Then substitute 0 for x to find the y-intercept.

x-intercept	**y-intercept**
$3x - y - 2 = 0$	$3x - y - 2 = 0$
$3x - (0) - 2 = 0$	$3(0) - y - 2 = 0$
$3x - 2 = 0$	$-y - 2 = 0$
$3x = 2$	$-y = 2$
$x = \dfrac{2}{3}$	$y = -2$

The line crosses the x-axis at $\left(\dfrac{2}{3}, 0\right)$ and the y-axis at $(0, -2)$. Graph the intercepts and draw the line.

Resource Manager

 Reproducible Masters
- *Study Guide and Practice,* pp. 5–6
- *Enrichment,* p. 3

 Transparencies
- 5-Minute Check, 1-3
- Teaching, 1-3

 Technology/Multimedia
- Advanced MathPASS, Lesson 1

The **slope** of a nonvertical line is the ratio of the change in the ordinates of the points to the corresponding change in the abscissas. The slope of a line is a constant.

| Slope | The slope, m, of the line through (x_1, y_1) and (x_2, y_2) is given by the following equation, if $x_1 \neq x_2$.

$$m = \frac{y_2 - y_1}{x_2 - x_1}$$ | |

The slope of a line can be interpreted as the rate of change in the y-coordinates for each 1-unit increase in the x-coordinates.

Example **AGRICULTURE** Refer to the application at the beginning of the lesson. What was the average rate of decline in the percent of American workers who were farmers?

The average rate of change is the slope of the line containing the points at $(1900, 37.5)$ and $(1994, 2.5)$. Find the slope of this line.

$$m = \frac{y_2 - y_1}{x_2 - x_1}$$

$$= \frac{2.5 - 37.5}{1994 - 1900} \quad \begin{array}{l} \text{Let } x_1 = 1900, \ y_1 = 37.5, \\ x_2 = 1994, \text{ and } y_2 = 2.5. \end{array}$$

$$= \frac{-35}{94} \text{ or about } -0.37$$

On average, the number of American workers who were farmers decreased about 0.37% each year from 1900 to 1994.

A linear equation in the form $Ax + By = C$ where A is positive is written in **standard form.** You can also write a linear equation in **slope-intercept form.** Slope-intercept form is $y = mx + b$, where m is the slope and b is the y-intercept of the line. You can graph an equation in slope-intercept form by graphing the y-intercept and then finding a second point on the line using the slope.

| Slope-Intercept Form | If a line has slope m and y-intercept b, the slope-intercept form of the equation of the line can be written as follows.

$$y = mx + b$$ |

Example ③ Graph each equation using the y-intercept and the slope.

a. $y = \dfrac{3}{4}x - 2$

The y-intercept is -2. Graph $(0, -2)$.
Use the slope to graph a second point.
Connect the points to graph the line.

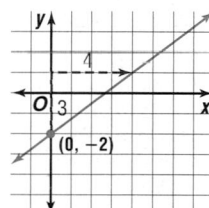

Lesson 1-3 Graphing Linear Equations **21**

📂 **Prerequisite Skills**

A review of the prerequisite skills for this lesson is provided in the *Prerequisite Skills Masters.*

Skill Number and Content	Pages
5 Solving Equations and Inequalities	11–12

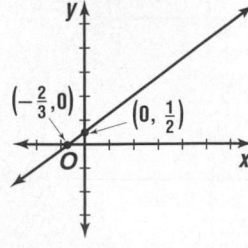
Graphing Calculator Appendix

For keystroke instruction on how to graph linear equations, see page A5.

b. $2x + y = 5$

Rewrite the equation in slope-intercept form.

$$2x + y = 5 \rightarrow y = -2x + 5$$

The *y*-intercept is 5. Graph (0, 5). Then use the slope to graph a second point. Connect the points to graph the line.

There are four different types of slope for a line. The table below shows a graph with each type of slope.

A line with undefined slope is sometimes described as having "no slope."

Types of Slope			
positive slope	**negative slope**	**0 slope**	**undefined slope**

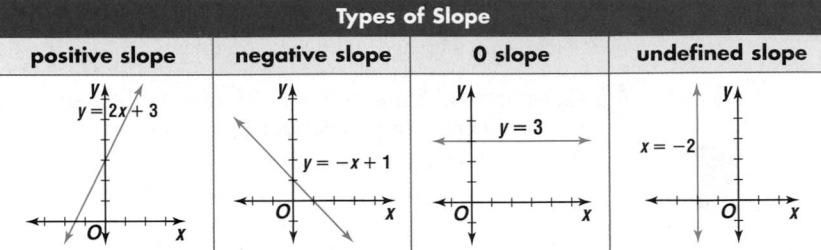

Notice from the graphs that not all linear equations represent functions. A linear function is defined as follows. *When is a linear equation not a function?*

Linear Functions	A linear function is defined by $f(x) = mx + b$, where m and b are real numbers.

Values of *x* for which $f(x) = 0$ are called **zeros of the function** *f*. For a linear function, the zeros can be found by solving the equation $mx + b = 0$. If $m \neq 0$, then $-\dfrac{b}{m}$ is the only zero of the function. The zeros of a function are the *x*-intercepts. Thus, for a linear function, the *x*-intercept has coordinates $\left(-\dfrac{b}{m}, 0\right)$.

In the case where $m = 0$, we have $f(x) = b$. This function is called a **constant function** and its graph is a horizontal line. The constant function $f(x) = b$ has no zeros when $b \neq 0$ or every value of *x* is a zero if $b = 0$.

Example **4** **Find the zero of each function. Then graph the function.**

Graphing Calculator Appendix

For keystroke instruction on how to find the zeros of a linear function using the **CALC** menu, see page A11.

a. $f(x) = 5x + 4$

To find the zeros of $f(x)$, set $f(x)$ equal to 0 and solve for *x*.

$$5x + 4 = 0 \Rightarrow x = -\frac{4}{5}$$

$-\dfrac{4}{5}$ is a zero of the function. So the coordinates of one point on the graph are $\left(-\dfrac{4}{5}, 0\right)$. Find the coordinates of a second point. When $x = 0$, $f(x) = 5(0) + 4$, or 4. Thus, the coordinates of a second point are (0, 4).

b. $f(x) = -2$

Since $m = 0$ and $b = -2$, this function has no x-intercept, and therefore no zeros. The graph of the function is a horizontal line 2 units below the x-axis.

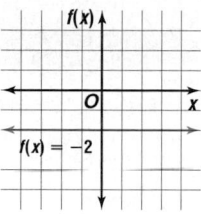

CHECK FOR UNDERSTANDING

Communicating Mathematics

Read and study the lesson to answer each question. 3–4. See margin.

1. **Explain** the significance of m and b in $y = mx + b$.

2. **Name** the zero of the function whose graph is shown at the right. Explain how you found the zero.

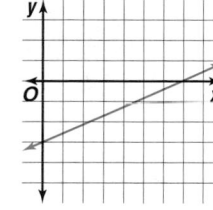

3. **Describe** the process you would use to graph a line with a y-intercept of 2 and a slope of -4.

4. **Compare and contrast** the graphs of $y = 5x + 8$ and $y = -5x + 8$.

1. m represents the slope of the graph and b represents the y-intercept

2. 7; the line intercepts the x-axis at (7, 0)

Guided Practice

Graph each equation using the x- and y-intercepts. 5–6. See margin.

5. $3x - 4y + 2 = 0$

6. $x + 2y - 5 = 0$

Graph each equation using the y-intercept and the slope. 7–8. See margin.

7. $y = x + 7$

8. $y = 5$

Find the zero of each function. If no zero exists, write *none*. Then graph the function. 9–10. See margin for graphs.

9. $f(x) = \frac{1}{2}x + 6$ **−12**

10. $f(x) = 19$ **none**

11. Archaeology Archaeologists use bones and other artifacts found at historical sites to study a culture. One analysis they perform is to use a function to determine the height of the person from a tibia bone. Typically a man whose tibia is 38.500 centimeters long is 173 centimeters tall. A man with a 44.125-centimeter tibia is 188 centimeters tall. **a. (38.500, 173), (44.125, 188)**

a. Write two ordered pairs that represent the function.

b. Determine the slope of the line through the two points. **2.667**

c. Explain the meaning of the slope in this context. **See margin.**

www.amc.glencoe.com/self_check_quiz

Lesson 1-3 Graphing Linear Equations **23**

Addressing Individual Needs

Logical Students learn that $y = 0$ at the point where the graph of a linear equation crosses the x-axis. Point out that, for similar reasons, $x = 0$ where the graph crosses the y-axis. For a given equation, students can easily find these points and use them to draw the graph of the equation. Encourage students to use their knowledge of opposites when graphing with slope. For example, 2 units down and 3 units right is equivalent to 2 units up and 3 left.

Additional Answers

24.

$f(x) = 9x + 5$

$\left(-\frac{5}{9}, 0\right)$

25.

$f(x) = 4x - 12$

$(3, 0)$

26.

$f(x) = 3x + 1$

$\left(-\frac{1}{3}, 0\right)$

27.

$f(x) = 14x$

$(0, 0)$

28.

$f(x) = 12$

29.

$f(x) = 5x - 8$

$\left(\frac{8}{5}, 0\right)$

31.

$y = -\frac{3}{2}x + 3$

$(2, 0)$

35b. For each 1-degree increase in the temperature, there is a $\frac{1}{4}$ - pascal increase in the pressure.

EXERCISES

Practice

Graph each equation.   **12–23. See Answer Appendix.**

A
12. $y = 4x - 9$ **13.** $y = 3$ **14.** $2x - 3y + 15 = 0$

15. $x - 4 = 0$ **16.** $y = 6x - 1$ **17.** $y = 5 - 2x$

B
18. $y + 8 = 0$ **19.** $2x + y = 0$ **20.** $y = \frac{2}{3}x - 4$

21. $y = 25x + 150$ **22.** $2x + 5y = 8$ **23.** $3x - y = 7$

Find the zero of each function. If no zero exists, write *none*. Then graph the function.

24. $f(x) = 9x + 5$ $-\frac{5}{9}$ **25.** $f(x) = 4x - 12$ **3**   **26.** $f(x) = 3x + 1$ $-\frac{1}{3}$

27. $f(x) = 14x$ **0**   **28.** $f(x) = 12$ **none**   **29.** $f(x) = 5x - 8$ $\frac{8}{5}$

30. Find the zero for the function $f(x) = 5x - 2$. $\frac{2}{5}$

31. 2; see margin for graph.
31. Graph $y = -\frac{3}{2}x + 3$. What is the zero of the function $f(x) = -\frac{3}{2}x + 3$?

C
32. Write a linear function that has no zero. Then write a linear function that has infinitely many zeros.   **Sample answer: $f(x) = 5$; $f(x) = 0$**

Applications and Problem Solving

33. Electronics   The voltage V in volts produced by a battery is a linear function of the current i in amperes drawn from it. The opposite of the slope of the line represents the battery's effective resistance R in ohms. For a certain battery, $V = 12.0$ when $i = 1.0$ and $V = 8.4$ when $i = 10.0$.
  a. What is the effective resistance of the battery?   **0.4 ohms**
  b. Find the voltage that the battery would produce when the current is 25.0 amperes.   **2.4 volts**

34. Critical Thinking   A line passes through $A(3, 7)$ and $B(-4, 9)$. Find the value of a if $C(a, 1)$ is on the line.   **24**

35. Chemistry   According to Charles' Law, the pressure P in pascals of a fixed volume of a gas is linearly related to the temperature T in degrees Celsius. In an experiment, it was found that when $T = 40$, $P = 90$ and when $T = 80$, $P = 100$.
  a. What is the slope of the line containing these points? $\frac{1}{4}$
  b. Explain the meaning of the slope in this context.   **See margin.**
  c. Graph the function.   **See margin.**

36. Critical Thinking   The product of the slopes of two non-vertical perpendicular lines is always -1. Is it possible for two perpendicular lines to both have positive slope? Explain.   **See margin.**

37. Accounting   A business's capital costs are expenses for things that last more than one year and lose value or wear out over time. Examples include equipment, buildings, and patents. The value of these items declines, or depreciates over time. One way to calculate depreciation is the straight-line method, using the value and the estimated life of the asset. Suppose $v(t) = 10{,}440 - 290t$ describes the value $v(t)$ of a piece of software after t months.

37a–c. See margin.
  a. Find the zero of the function. What does the zero represent?
  b. Find the slope of the function. What does the slope represent?
  c. Graph the function.

Additional Answers

35c.

36. No; the product of two positives is positive, so for the product of the slopes to be –1, one of the slopes must be positive and the other must be negative.

37a. 36; the software has no monetary value after 36 months.

38. Critical Thinking How is the slope of a linear function related to the number of zeros for the function? **See margin.**

39. Economics Economists call the relationship between a nation's disposable income and personal consumption expenditures the marginal propensity to consume or MPC. An MPC of 0.7 means that for each $1 increase in disposable income, consumption increases $0.70. That is, 70% of each additional dollar earned is spent and 30% is saved.

a. Suppose a nation's disposable income, x, and personal consumption expenditures, y, are shown in the table at the right. Find the MPC. **0.86**

39b. $1552.30
b. If disposable income were to increase $1805 in a year, how many additional dollars would the average family spend?

x (billions of dollars)	y (billions of dollars)
56	50
76	67.2

c. The marginal propensity to save, MPS, is $1 - $ MPC. Find the MPS. **0.14**

d. If disposable income were to increase $1805 in a year, how many additional dollars would the average family save? **$252.70**

Mixed Review

40. $x^2 + 2x - 4$, **40.** Given $f(x) = 2x$ and $g(x) = x^2 - 4$, find $(f + g)(x)$ and $(f - g)(x)$. *(Lesson 1-2)*
$-x^2 + 2x + 4$

41. Business Computer Depot offers a 12% discount on computers sold Labor Day weekend. There is also a $100 rebate available. *(Lesson 1-2)*

a. Write a function for the price after the discount $d(p)$ if p represents the original price of a computer. $d(p) = 0.88p$

b. Write a function for the price after the rebate $r(d)$ if d represents the discounted price. $r(d) = d - 100$

c. Use composition of functions to write a function to relate the selling price to the original price of a computer. $r(p) = 0.88p - 100$

d. Find the selling prices of computers with original prices of $799.99, $999.99, and $1499.99. **$603.99, $779.99, $1219.99**

42. 50, 24
42. Find $[f \circ g](-3)$ and $[g \circ f](-3)$ if $f(x) = x^2 - 4x + 5$ and $g(x) = x - 2$. *(Lesson 1-2)*

43. Given $f(x) = 4 + 6x - x^3$, find $f(9)$. *(Lesson 1-1)* **-671**

44. No; the graph **44.** Determine whether the graph at the right
fails the vertical represents a function. Explain. *(Lesson 1-1)*
line test.

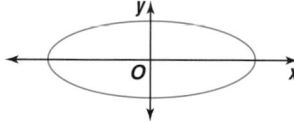

45. {(−3, 14), **45.** Given that x is an integer, state the relation representing $y = 11 - x$ and
(−2, 13), $3 \le x \le 0$ by listing a set of ordered pairs. Then state whether the relation
(−1, 12), (0, 11)}, is a function. *(Lesson 1-1)*
yes

46. SAT/ACT Practice What is the sum of four integers whose average is 15? **D**

A 3.75
B 15
C 30
D 60
E cannot be determined

Extra Practice See p. A26.

Extra Credit

A graph passes through the point at $(-3, 2)$ and has a slope of $\frac{2}{3}$. Find the x- and y-intercepts. **−6, 4**

Open-Ended Assessment
Modeling Have students give a real-world example of a linear function. Then, have them graph the function and draw where on the graph they would find the zeros of the function. **Graph by connecting two points that satisfy the equation. The zero is the x-intercept.**

Additional Answers
(pp. 24–25)
37b. **−290; For every 1-month change in the number of months, there is a $290 decrease in the value of the software.**

37c.

38. A function with a slope of 0 has no zeros if its y-intercept is not 0; a function with a slope of 0 has an infinite number of zeros if its y-intercept is 0; a function with any slope other than 0 has exactly 1 zero.

Enrichment Masters, p. 3

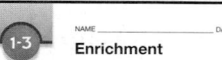

1 FOCUS

Motivating the Lesson
Hands-On Activity Have students draw several nonvertical lines that appear to be parallel on a coordinate plane. Then, have them write the equations of the lines in slope-intercept form. What do they notice about the group of graphs?

2 TEACH

A reproduction of this lesson with graphing calculator keystrokes provided can be found in the *Graphing Calculator Keystrokes Masters*, pp. 2–3.

Teaching Tip If parallel lines appear too close together on the screen, suggest that students change the viewing window to a shorter range of values.

3 PRACTICE/APPLY

Assignment Guide
All: 1–3

4 ASSESS

Open-Ended Assessment
Modeling Have students create other families of graphs using graphing calculators and describe the relationship that defines each family. Examples might include lines perpendicular to a given line, vertical lines, horizontal lines, and so on.

GRAPHING CALCULATOR EXPLORATION

1-3B Analyzing Families of Linear Graphs

An Extension of Lesson 1-3

OBJECTIVE
- Investigate the effect of changing the value of m or b in $y = mx + b$.

A **family of graphs** is a group of graphs that displays one or more similar characteristics. For linear functions, there are two types of families of graphs. Using the slope-intercept form of the equation, one family is characterized by having the same slope m in $y = mx + b$. The other type of family has the same y-intercept b in $y = mx + b$.

You can investigate families of linear graphs by graphing several equations on the same graphing calculator screen.

Example

Graph $y = 3x - 5$, $y = 3x - 1$, $y = 3x$, and $y = 3x + 6$. Describe the similarities and differences among the graphs.

Graph all of the equations on the same screen. Use the viewing window, $[-9.4, 9.4]$ by $[-6.2, 6.2]$.

Notice that the graphs appear to be parallel lines with the same positive slope. They are in the family of lines that have the slope 3.

The slope of each line is the same, but the lines have different y-intercepts. Each of the other three lines are the graph of $y = 3x$ shifted either up or down.

$[-9.4, 9.4]$ scl:1 by $[-6.2, 6.2]$ scl:1

equation	slope	y-intercept	relationship to graph of $y = 3x$
$y = 3x - 5$	3	−5	shifted 5 units down
$y = 3x - 1$	3	−1	shifted 1 unit down
$y = 3x$	3	0	same
$y = 3x + 6$	3	6	shifted 6 units up

TRY THESE

1–3. See margin.

1. Graph $y = 4x - 2$, $y = 2x - 2$, $y = -2$, $y = -x - 2$, and $y = -6x - 2$ on the same graphing calculator screen. Describe how the graphs are similar and different.

WHAT DO YOU THINK?

2. Use the results of the Example and Exercise 1 to predict what the graph of $y = 3x - 2$ will look like.

3. Write a paragraph explaining the effect of different values of m and b on the graph of $y = mx + b$. Include sketches to illustrate your explanation.

26 Chapter 1 Linear Relations and Functions

Additional Answers
1. See students' graphs. All of the graphs are lines with y-intercept at $(0, -2)$. Each line has a different slope.
2. A line parallel to the ones graphed in the Example and passing through $(0, -2)$.
3. Sample answer: the graphs of lines with the same value for m are parallel. The graphs of lines with the same value for b have the same y-intercept.

OBJECTIVE
- Write linear equations.

Writing Linear Equations

ECONOMICS Each year, the U.S. Department of Commerce publishes its *Survey of Current Business*. Included in the report is the average personal income of U.S. workers.

Personal income is one indicator of the health of the U.S. economy. How could you use the data on average personal income for 1980 to 1997 to predict the average personal income in 2010? *This problem will be solved in Example 3.*

Years since 1980	Average Personal Income ($)
0	9916
5	13,895
10	18,477
11	19,100
12	19,802
13	20,810
14	21,846
15	23,233
16	24,457
17	25,660

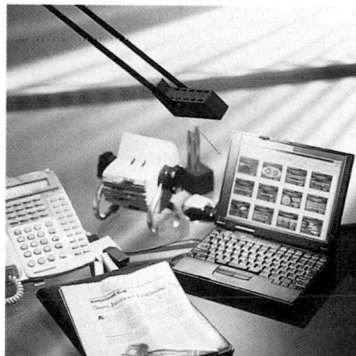

A mathematical **model** may be an equation used to approximate a real-world set of data. Often when you work with real-world data, you know information about a line without knowing its equation. You can use characteristics of the graph of the data to write an equation for a line. This equation is a model of the data. Writing an equation of a line may be done in a variety of ways depending upon the information you are given. If one point and the slope of a line are known, the slope-intercept form can be used to write the equation.

Example **1** Write an equation in slope-intercept form for each line described.

a. a slope of $-\frac{3}{4}$ and a *y*-intercept of 7

Substitute $\frac{3}{4}$ for *m* and 7 for *b* in the general slope-intercept form.

$$y = mx + b \rightarrow y = -\frac{3}{4}x + 7$$

The slope-intercept form of the equation of the line is $y = -\frac{3}{4}x + 7$.

b. a slope of -6 and passes through the point at $(1, -3)$

Substitute the slope and coordinates of the point in the general slope-intercept form of a linear equation. Then solve for *b*.

$$y = mx + b$$
$$-3 = -6(1) + b \quad \textit{Substitute } -3 \textit{ for } y, 1 \textit{ for } x, \textit{ and } -6 \textit{ for } m.$$
$$3 = b \quad \textit{Add 6 to each side of the equation.}$$

The *y*-intercept is 3. Thus, the equation for the line is $y = -6x + 3$.

1 FOCUS

 5-Minute Check
Lesson 1-3

Graph each equation using the *x*- and *y*-intercepts.
1. $2x + y - 4 = 0$

2. $4x + 8y - 16 = 0$

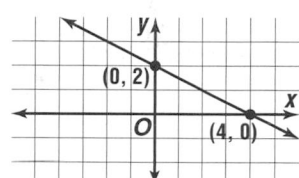

3. Find the zero of the function $f(x) - 9x - 3$. Then graph the function. $\frac{1}{3}$

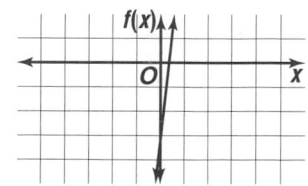

Motivating the Lesson

Real-World Application Have students read the application. Ask them to estimate the average personal income of U.S. workers in 2010. Then have students explain their reasoning. **Sample answer: for each 1 year increase, the average personal income increases by about $1000. At that rate, a good estimate would be about $39,000.**

Teaching Tip Make sure that students understand that personal income represents income received by workers less their contributions to social insurance such as Social Security.

Teaching Tip Before working Example 1, explain why b is the y-intercept in the equation $y = mx + b$. **The y-intercept is the point where $x = 0$. Substituting $x = 0$ in $y = mx + b$ gives $y = m(0) + b$ or b.**

In-Class Examples

Example 1
Write an equation in slope-intercept form for each line described.

a. a slope of $\frac{1}{2}$ and a y-intercept of 4 $y = \frac{1}{2}x + 4$

b. a slope of 2 and passes through the point at (−3, 2) $y = 2x + 8$

Example 2
Teaching Mrs. Alvarez is preparing her art classroom for a new year of school. She placed a purchase order for $800 to buy paper, chalk, and clay. In addition, she expects the expenses each month to be $120. Write an equation that models the total expense y after x months. $y = 120x + 800$

Example 3
Voting Over time, the percent of registered voters turning out to vote in the United States has been declining. The table shows the percent of registered voters who voted in presidential elections from 1964 to 1996.

Elections since 1964	Total Voter Turnout (% of registered voters)
0	95.8
1	89.7
2	79.9
3	77.6
4	76.5
5	74.6
6	72.5
7	78.2
8	63.4

Source: International IDEA

(continued on next page)

Example **2 BUSINESS**   Alvin Hawkins is opening a home-based business. He determined that he will need $6000 to buy a computer and supplies to start. He expects expenses for each following month to be $700. Write an equation that models the total expense y after x months.

The initial cost is the y-intercept of the graph. Because the total expense rises $700 each month, the slope is 700.

$y = mx + b$

$y = 700x + 6000$ *Substitute 700 for m and 6000 for b.*

The total expense can be modeled by $y = 700x + 6000$.

When you know the slope and a point on a line, you can also write an equation for the line in **point-slope form.** Using the definition of slope for points (x, y) and (x_1, y_1), if $\frac{y - y_1}{x - x_1} = m$, then $y - y_1 = m(x - x_1)$.

Point-Slope Form	If the point with coordinates (x_1, y_1) lies on a line having slope m, the point-slope form of the equation of the line can be written as follows. $$y - y_1 = m(x - x_1)$$

If you know the coordinates of two points on a line, you can find the slope of the line. Then the equation of the line can be written using either the slope-intercept or the point-slope form.

Example **3 ECONOMICS**   Refer to the application at the beginning of the lesson.

a. Find a linear equation that can be used as a model to predict the average personal income for any year.

b. Assume that the rate of growth of personal income remains constant over time and use the equation to predict the average personal income for individuals in the year 2010.

c. Evaluate the prediction.

a. Graph the data. Then select two points to represent the data set and draw a line that might approximate the data. Suppose we chose (0, 9916) and (17, 25,660). Use the coordinates of those points to find the slope of the line you drew.

$m = \dfrac{y_2 - y_1}{x_2 - x_1}$

$= \dfrac{25{,}660 - 9916}{17 - 0}$ $x_1 = 0, y_1 = 9916, x_2 = 17, y_2 = 25{,}660$

≈ 926 *Thus for each 1-year increase, average personal income increases $926.*

Addressing Individual Needs

 Auditory/Musical   Give students measures of music and ask which most closely resemble a linear relationship. Ask them to describe why the music has a linear pattern.   **Sample answers: a scale or an arpeggio moves from one note to another note through other points in a linear manner. Likewise a passage moving from *piano* to *forte* increases in volume in a linear fashion.**

Use point-slope form.

$$y - y_1 = m(x - x_1)$$
$$y - 9916 = 926(x - 0)$$ *Substitute 0 for x_1, 9916 for y_1, and 926 for m.*
$$y = 926x + 9916$$

The slope-intercept form of the model equation is $y = 926x + 9916$.

b. Evaluate the equation for $x = 2010$ to predict the average personal income for that year. The years since 1980 will be $2010 - 1980$ or 30. So $x = 30$.

$$y = 926x + 9916$$
$$y = 926(30) + 9916$$ *Substitute 30 for x.*
$$y = 37{,}696$$

The predicted average personal income is about \$37,696 for the year 2010.

c. Most of the actual data points are close to the graph of the model equation. Thus, the equation and the prediction are probably reliable.

CHECK FOR UNDERSTANDING

Communicating Mathematics

Read and study the lesson to answer each question.   **3, 5. See margin.**

1. **List** all the different sets of information that are sufficient to write the equation of a line.   **slope and y-intercept; slope and any point; two points**

2. **Demonstrate** two different ways to find the equation of the line with a slope of $\frac{1}{4}$ passing through the point at $(3, -4)$.   **See Answer Appendix.**

3. **Explain** what 55 and 49 represent in the equation $c = 55h + 49$, which represents the cost c of a plumber's service call lasting h hours.

4. $y = \frac{1}{2}x - 3$

4. **Write** an equation for the line whose graph is shown at the right.

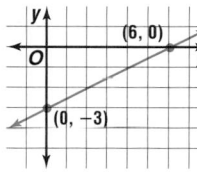

5. *Math Journal*   Write a sentence or two to describe when it is easier to use the point-slope form to write the equation of a line and when it is easier to use the slope-intercept form.

Guided Practice

Write an equation in slope-intercept form for each line described.

6. $y = -\frac{1}{4}x - 10$

7. $y = 4x - 10$

8. $y = \frac{7}{2}x - \frac{31}{2}$

9. $y = 2$

6. slope $= -\frac{1}{4}$, y-intercept $= -10$ 7. slope $= 4$, passes through $(3, 2)$

8. passes through $(5, 2)$ and $(7, 9)$ 9. horizontal and passes through $(-9, 2)$

10. **Botany**   Do you feel like every time you cut the grass it needs to be cut again right away? Be grateful you aren't cutting the Bermuda grass that grows in Africa and Asia. It can grow at a rate of 5.9 inches per day! Suppose you cut a Bermuda grass plant to a length of 2 inches.   **a. $y = 5.9x + 2$**
   a. Write an equation that models the length of the plant y after x days.
   b. If you didn't cut it again, how long would the plant be in one week?   **43.3 in.**
   c. Can this rate of growth be maintained indefinitely? Explain.   **See margin.**

 www.amc.glencoe.com/self_check_quiz   *Lesson 1-4 Writing Linear Equations* **29**

Additional Answers

3. **55 represents the hourly rate and 49 represents the fee for coming to the house.**

5. **Sample answer: when given the slope and the y-intercept, use slope-intercept form. When given the slope and a point, use point-slope form. When given two points, find the slope, then use point-slope form.**

10c. **Sample answer: no; the grass could not support its own weight if it grew that tall.**

In-Class Example
Example 3 (continued)

a. Find a linear equation that can be used as a model to predict the average percent of registered-voter turnout for any U.S. presidential election.   **Sample answer: using points (2, 79.9) and (5, 74.6) gives $y = -1.77x + 83.4$.**

b. Assume that the rate of decrease of voter turnout remains constant over time and use the equation to predict voter turnout in the year 2008.   **about 5.6%**

c. Evaluate the prediction.   **Most of the points are close to the graph of the model equation. Thus, the prediction is probably reliable.**

3 PRACTICE/APPLY

Error Analysis

Watch for students who confuse horizontal and vertical line definitions.

Prevent by reminding students that a horizontal line has an equation $y = c$ and a vertical line has an equation $x = c$, where c is a real number.

Study Guide and Practice Masters, pp. 7–8

Assignment Guide

Core: 11–23 odd, 25–27, 29–35
Enriched: 12–24 even, 25–35
All: Mid-Chapter Quiz, 1–10

Additional Answers

(pp. 30–31)

11. $y = 5x - 2$

12. $y = 8x + 61$

13. $y = -\frac{3}{4}x$

14. $y = -12x + \frac{1}{2}$

15. $y = 6x - 19$

16. $x = 12$

17. $y = -\frac{4}{9}x + \frac{49}{9}$

18. $y = \frac{5}{8}x + 5$

19. $y = 1$

20. $x = -4$

21. $x = 0$

22. $y = 0.25x - 6$

27a. Sample answer: using (20, 28) and (27, 37),

$y = \frac{9}{7}x + \frac{16}{7}$.

27b. using sample answer from part a, 26.7 mpg

27c. Sample answer: the estimate is close but not exact since only two points were used to write the equation.

28b. Sample answer: only two points were used to make the prediction equation, so many points lie off of the line.

29. Yes; the slope of the line through (5, 9) and (–3, 3) is

$\frac{3 - 9}{-3 - 5}$ or $\frac{3}{4}$. The slope of the line through (–3, 3) and (1, 6)

is $\frac{6 - 3}{1 - (-3)}$ or $\frac{3}{4}$. Since these

two lines would have the same slope and would share a point, their equations would be the same. Thus, they are the same line and all three points are collinear.

30.

$3x - 2y - 5 = 0$

Practice

11–22. See margin.

Write an equation in slope-intercept form for each line described.

11. slope = 5, y-intercept = -2

12. slope = 8, passes through $(-7, 5)$

13. slope = $-\frac{3}{4}$, y-intercept = 0

14. slope = -12, y-intercept = $\frac{1}{2}$

15. passes through $A(4, 5)$, slope = 6

16. no slope and passes through $(12, -9)$

17. passes through $A(1, 5)$ and $B(-8, 9)$

18. x-intercept = -8, y-intercept = 5

19. passes through $A(8, 1)$ and $B(-3, 1)$

20. vertical and passes through $(-4, -2)$

21. the y-axis

22. slope = 0.25, x-intercept = 24

23. Line ℓ passes through $A(-2, -4)$ and has a slope of $-\frac{1}{2}$. What is the standard form of the equation for line ℓ? $x + 2y + 10 = 0$

24. Line m passes through $C(-2, 0)$ and $D(1, -3)$. Write the equation of line m in standard form. $x + y + 2 = 0$

Applications and Problem Solving

25. Sports   Skiers, hikers, and climbers often experience altitude sickness as they reach elevations of 8000 feet and more. A good rule of thumb for the amount of time that it takes to become acclimated to high elevations is 2 weeks for the first 7000 feet. After that, it will take 1 week more for each additional 2000 feet of altitude.

   a. Write an equation for the time t to acclimate to an altitude of f feet.

   b. Mt. Whitney in California is the highest peak in the contiguous 48 states. It is located in Eastern Sierra Nevada, on the border between Sequoia National Park and Inyo National Forest. About how many weeks would it take a person to acclimate to Mt. Whitney's elevation of 14,494 feet?   **about 5.7 weeks**

25a.
$t = 2 + \frac{x - 7000}{2000}$

26. Critical Thinking   Write an expression for the slope of a line whose equation is $Ax + By + C = 0$. $m = -\frac{A}{B}$

27. Transportation   The mileage in miles per gallon (mpg) for city and highway driving of several 1999 models are given in the chart.
a–c. See margin.

Model	City (mpg)	Highway (mpg)
A	24	32
B	20	29
C	20	29
D	20	28
E	23	30
F	24	30
G	27	37
H	22	28

   a. Find a linear equation that can be used to find a car's highway mileage based on its city mileage.

   b. Model J's city mileage is 19 mpg. Use your equation to predict its highway mileage.

   c. Highway mileage for Model J is 26 mpg. How well did your equation predict the mileage? Explain.

Extra Credit

Write the ordered pair for the y-intercept of a line which can be modeled by the equation $\frac{A}{B}x + \frac{1}{C}y - Q = 0$.   **(0, CQ)**

28. **Economics** Research the average personal income for the current year.
    a. Find the value that the equation in Example 2 predicts.
    b. Is the average personal income equal to the prediction? Explain any difference. **See margin.**

29. **Critical Thinking** Determine whether the points at $(5, 9)$, $(-3, 3)$, and $(1, 6)$ are collinear. Justify your answer. **See margin.**

Mixed Review

31a. **$6111 billion**

31b. **The rate is the slope.**

33. $x^5 - 3x^4 + 7x^3$, $\dfrac{x^3}{x^2 - 3x + 7}$, where $g(x) \neq 0$

30. Graph $3x - 2y - 5 = 0$. *(Lesson 1-3)* **See margin.**

31. **Business** In 1995, retail sales of apparel in the United States were $70,583 billion. Apparel sales were $82,805 billion in 1997. *(Lesson 1-3)*
    a. Assuming a linear relationship, find the average annual rate of increase.
    b. Explain how the rate is related to the graph of the line.

32. If $f(x) = x^3$ and $g(x) = 3x$, find $g[f(-2)]$. *(Lesson 1-2)* **−24**

33. Find $(f \cdot g)(x)$ and $\left(\dfrac{f}{g}\right)(x)$ for $f(x) = x^3$ and $g(x) = x^2 - 3x + 7$. *(Lesson 1-2)*

34. Given that x is an integer, state the relation representing $y = x^2$ and $-4 \leq x \leq -2$ by listing a set of ordered pairs. Then state whether this relation is a function. *(Lesson 1-1)* **{(−4, 16), (−3, 9), (−2, 4)}, yes**

35. **SAT/ACT Practice** If $xy = 1$, then x is the reciprocal of y. Which of the following is the arithmetic mean of x and y? **A**

A $\dfrac{y^2 + 1}{2y}$ B $\dfrac{y + 1}{2y}$ C $\dfrac{y^2 + 2}{2y}$

D $\dfrac{y^2 + 1}{y}$ E $\dfrac{x^2 + 1}{y}$

MID-CHAPTER QUIZ

1. What are the domain and the range of the relation $\{(-2, -3), (-2, 3), (4, 7), (2, -8), (4, 3)\}$? Is the relation a function? Explain. (Lesson 1-1) **See margin.**

2. Find $f(4)$ for $f(x) = 7 - x^2$. (Lesson 1-1) **−9**

3. If $g(x) = \dfrac{3}{x - 1}$, what is $g(n + 2)$? (Lesson 1-1) **See margin.**

4. **Retail** Amparo bought a jacket with a gift certificate she received as a birthday present. The jacket was marked 33% off, and the sales tax in her area is 5.5%. If she paid $45.95 for the jacket, use composition of functions to determine the original price of the jacket. (Lesson 1-2) **$65**

5. If $f(x) = \dfrac{1}{x - 1}$ and $g(x) = x + 1$, find $[f \circ g](x)$ and $[g \circ f](x)$. (Lesson 1-2) **$\dfrac{1}{x}$, $\dfrac{x}{x - 1}$, $x \neq 1$**

Graph each equation. (Lesson 1-3)

6. $2x - 4y = 8$ 7. $3x = 2y$
6–7. See margin.

8. Find the zero of $f(x) = 5x - 3$. (Lesson 1-3) **$\dfrac{3}{5}$**

9. Points $A(2, 5)$ and $B(7, 8)$ lie on line ℓ. What is the standard form of the equation of line ℓ? (Lesson 1-4) **$3x - 5y + 19 = 0$**

10. **Demographics** In July 1990, the population of Georgia was 6,506,416. By July 1997, the population had grown to 7,486,242. (Lesson 1-4)
    a. If x represents the year and y represents the population, find the average annual rate of increase of the population. **about 139,975**
    b. Write an equation to model the population change. **$y = 139,975x - 272,043,834$**

Open-Ended Assessment
Speaking Have students compare and contrast the slope-intercept and point-slope forms of linear equations. Ask students to name situations for which one form is preferable to the other.

Mid-Chapter Quiz
The Mid-Chapter Quiz provides students with a brief review of the concepts and skills in Lessons 1-1 through 1-4. Lesson numbers are given to the right of exercises or instruction lines so students can review concepts not yet mastered.

 Chapter 1, Quiz B (Lessons 1-3 and 1-4) is available in the *Assessment and Evaluation Masters*, p. 17.

 Chapter 1 Mid-Chapter Test (Lessons 1-1 through 1-4) is available in the *Assessment and Evaluation Masters*, p. 16.

Enrichment Masters, p. 4

Additional Answers
Mid-Chapter Quiz
1. {−2, 2, 4}, {−8, −3, 3, 7}; no, −2 in the domain is paired with more than one element of the range.

3. $\dfrac{3}{n + 1}$

6.

7.

Writing Equations of Parallel and Perpendicular Lines

E-COMMERCE Have you ever made a purchase over the Internet? Electronic commerce, or e-commerce, has changed the way Americans do business. In recent years, hundreds of companies that have no stores outside of the Internet have opened.

Suppose you own shares in two Internet stocks, Bookseller.com and WebFinder. One day these stocks open at $94.50 and $133.60 per share, respectively. The closing prices that day were $103.95 and $146.96, respectively. If your shares in these companies were valued at $5347.30 at the beginning of the day, is it possible that the shares were worth $5882.03 at closing? *This problem will be solved in Example 2.*

1 FOCUS

5-Minute Check
Lesson 1-4

Write an equation in slope-intercept form for each line described.

1. $6x - 2y + 7 = 0$
   $y = 3x + \frac{7}{2}$

2. passes through $(-6, -9)$,
   slope $= \frac{2}{3}$ $y = \frac{2}{3}x - 5$

3. passes through $A\left(-\frac{1}{2}, 4\right)$
   and $B\left(\frac{1}{2}, 6\right)$ $y = 2x + 5$

4. the *x*-axis $y = 0$

5. vertical and passes through $(-2, 1)$ $x = -2$

This problem can be solved by determining whether the graphs of the equations that describe the situation are parallel or coincide. Two lines that are in the same plane and have no points in common are **parallel lines.** The slopes of two nonvertical parallel lines are equal. The graphs of two equations that represent the same line are said to **coincide.**

Parallel Lines	Two nonvertical lines in a plane are parallel if and only if their slopes are equal and they have no points in common. Two vertical lines are always parallel.

Motivating the Lesson
Discussion Have students work in small groups. Each group should graph the following equations on the same coordinate plane.
$y = 2x - 5$ $y = 2x - 1$
$y = 2x + 2$ $y = 2x + 4$
Ask students what conjecture they might make. **It appears that lines with the same slope are parallel.**

We can use slopes and *y*-intercepts to determine whether lines are parallel.

Example Determine whether the graphs of each pair of equations are *parallel, coinciding,* or *neither.*

a. $3x - 4y = 12$
   $9x - 12y = 72$

Write each equation in slope-intercept form.

$3x - 4y = 12$ $9x - 12y = 72$
$\quad y = \frac{3}{4}x - 3$ $\quad y = \frac{3}{4}x - 6$

The lines have the same slope and different *y*-intercepts, so they are parallel. The graphs confirm the solution.

2 TEACH

In-Class Example
Example 1
Determine whether the graphs of each pair of equations are *parallel, coinciding,* or *neither.*
a. $2x + 4y = 4$
   $x + 2y = 5$ **parallel**
b. $5x - 3y = 12$
   $10x - 6y = 24$ **coinciding**

Resource Manager

Reproducible Masters
• *Study Guide and Practice,* pp. 9–10
• *Enrichment,* p. 5

Transparencies
• 5-Minute Check, 1-5
• Teaching, 1-5

b. $15x + 12y = 36$
$5x + 4y = 12$

Write each equation in slope-intercept form.

$15x + 12y = 36$ $5x + 4y = 12$

$y = -\dfrac{5}{4}x + 3$ $y = -\dfrac{5}{4}x + 3$

The slopes are the same, and the y-intercepts are the same. Therefore, the lines have all points in common. The lines coincide. *Check the solution by graphing.*

You can use linear equations to determine whether real-world situations are possible.

Example

2 **FINANCE** Refer to the application at the beginning of the lesson. Is it possible that your shares were worth $5882.03 at closing? Explain.

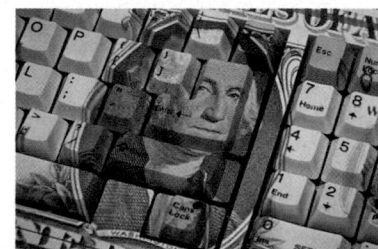

Let x represent the number of shares of Bookseller.com and y represent the number of shares of WebFinder. Then the value of the shares at opening is $94.50x + 133.60y = 5347.30$. The value of the shares at closing is modeled by $103.95x + 146.96y = 5882.03$.

Write each equation in slope-intercept form.

$94.50x + 133.60y = 5347.30$

$y = \dfrac{945}{1336}x + \dfrac{53{,}473}{1336}$

$103.95x + 146.96y = 5882.03$

$y = -\dfrac{945}{1336}x + \dfrac{53{,}473}{1336}$

Since these equations are the same, their graphs coincide. As a result, any ordered pair that is a solution for the equation for the opening value is also a solution for the equation for the closing value. Therefore, the value of the shares could have been $5882.03 at closing.

In Lesson 1-3, you learned that any linear equation can be written in standard form. The slope of a line can be obtained directly from the standard form of the equation if B is not 0. Solve the equation for y.

$$Ax + By + C = 0$$
$$By = -Ax - C$$
$$y = -\frac{A}{B}x - \frac{C}{B}. \quad B \neq 0$$
$$\qquad\quad \uparrow \qquad\quad \uparrow$$
$$\qquad slope \qquad y\text{-}intercept$$

So the slope m is $-\dfrac{A}{B}$, and the y-intercept b is $-\dfrac{C}{B}$.

In-Class Examples

Example 3
Write the standard form of the equation of the line that passes through the point at (−2, 10) and is parallel to the graph of $2x + 5y + 4 = 0$. **$2x + 5y − 46 = 0$**

Example 4
Write the standard form of the equation of the line that passes through the point at (2, −3) and is perpendicular to the graph of $6x − 8y − 5 = 0$.
$4x + 3y + 1 = 0$

Teaching Tip For Example 4, remind students to solve the standard form of an equation for the $y = mx + b$ form. Then the slope can easily be determined and used to find the equation of the line parallel or perpendicular to the given line.

Example ❸ **Write the standard form of the equation of the line that passes through the point at (4, −7) and is parallel to the graph of $2x − 5y + 8 = 0$.**

Any line parallel to the graph of $2x − 5y + 8 = 0$ will have the same slope. So, find the slope of the graph of $2x − 5y + 8 = 0$.

$$m = -\frac{A}{B}$$
$$= -\frac{2}{(-5)} \text{ or } \frac{2}{5}$$

Use point-slope form to write the equation of the line.

$$y - y_1 = m(x - x_1)$$
$$y - (-7) = \frac{2}{5}(x - 4) \quad \textit{Substitute 4 for } x_1, -7 \textit{ for } y_1, \textit{ and } \frac{2}{5} \textit{ for m.}$$
$$y + 7 = \frac{2}{5}x - \frac{8}{5}$$
$$5y + 35 = 2x - 8 \quad \textit{Multiply each side by 5.}$$
$$2x - 5y - 43 = 0 \quad \textit{Write in standard form.}$$

There is also a special relationship between the slopes of perpendicular lines.

Perpendicular Lines

Two nonvertical lines in a plane are perpendicular if and only if their slopes are opposite reciprocals.

A horizontal and a vertical line are always perpendicular.

You can also use the point-slope form to write the equation of a line that passes through a given point and is perpendicular to a given line.

Example ❹ **Write the standard form of the equation of the line that passes through the point at (−6, −1) and is perpendicular to the graph of $4x + 3y − 7 = 0$.**

The line with equation $4x + 3y − 7 = 0$ has a slope of $-\frac{A}{B} = -\frac{4}{3}$. Therefore, the slope of a line perpendicular must be $\frac{3}{4}$.

$$y - y_1 = m(x - x_1)$$
$$y - (-1) = \frac{3}{4}[x - (-6)] \quad \textit{Substitute −6 for } x_1, -1 \textit{ for } y_1, \textit{ and } \frac{3}{4} \textit{ for m.}$$
$$y + 1 = \frac{3}{4}x + \frac{9}{2}$$
$$4y + 4 = 3x + 18 \quad \textit{Multiply each side by 4.}$$
$$3x - 4y + 14 = 0 \quad \textit{Write in standard form.}$$

You can use the properties of parallel and perpendicular lines to write linear equations to solve geometric problems.

34 *Chapter 1 Linear Relations and Functions*

Addressing Individual Needs

 Verbal/Linguistic Discuss the informal meanings of the terms *parallel* and *perpendicular.* Then review the geometric meanings of the terms. Ask students to give everyday examples of parallel and perpendicular lines.

Example **5** **GEOMETRY** Determine the equation of the perpendicular bisector of the line segment with endpoints $S(3, 4)$ and $T(11, 18)$.

Recall that the coordinates of the midpoint of a line segment are the averages of the coordinates of the two endpoints. Let S be (x_1, y_1) and T be (x_2, y_2). Calculate the coordinates of the midpoint.

$$\left(\frac{x_1 + x_2}{2}, \frac{y_1 + y_2}{2}\right) = \left(\frac{3 + 11}{2}, \frac{4 + 18}{2}\right)$$
$$= (7, 11)$$

The slope of $\overline{ST}$ is $\frac{18 - 4}{11 - 3}$ or $\frac{7}{4}$.

The slope of the perpendicular bisector of $\overline{ST}$ is $-\frac{4}{7}$. The perpendicular bisector of $\overline{ST}$ passes through the midpoint of $\overline{ST}$, $(7, 11)$.

$y - y_1 = m(x - x_1)$	*Point-slope form*
$y - 11 = -\frac{4}{7}(x - 7)$	*Substitute 7 for x_1, 11 for y_1, and $\frac{-4}{7}$ for m.*
$7y - 77 = -4x + 28$	*Multiply each side by 7.*
$4x + 7y - 105 = 0$	*Write in standard form.*

CHECK FOR UNDERSTANDING

Communicating Mathematics

Read and study the lesson to answer each question.

1. **Describe** how you would tell that two lines are parallel or coincide by looking at the equations of the lines in standard form. **See margin.**

2. **Explain** why vertical lines are a special case in the definition of parallel lines.

2. They have no slope.

3. $-\frac{4}{3}, \frac{3}{4}$

3. **Determine** the slope of a line that is parallel to the graph of $4x + 3y + 19 = 0$ and the slope of a line that is perpendicular to it.

4. **Write** the slope of a line that is perpendicular to a line that has undefined slope. Explain. **See margin.**

Guided Practice

Determine whether the graphs of each pair of equations are *parallel, coinciding, perpendicular,* or *none of these.*

5. $y = 5x - 5$
   $y = -5x + 2$ **none of these**

6. $y = -6x - 2$
   $y = \frac{1}{6}x - 8$ **perpendicular**

7. $y = x - 6$
   $x - y + 8 = 0$ **parallel**

8. $y = 2x - 8$
   $4x - 2y - 16 = 0$ **coinciding**

9. Write the standard form of the equation of the line that passes through $A(5, 9)$ and is parallel to the graph of $y = 5x - 9$. **$5x - y - 16 = 0$**

10. Write the standard form of the equation of the line that passes through $B(-10, -5)$ and is perpendicular to the graph of $6x - 5y = 24$. **$5x + 6y + 80 = 0$**

Lesson 1-5 Writing Equations of Parallel and Perpendicular Lines **35**

Additional Answers *(p. 37)*

32. We are given $y = m_1x + b_1$ and $y = m_2x + b_2$ with $m_1 = m_2$ and $b_1 \neq b_2$. Assume that the lines intersect at point (x_1, y_1). Then, $y_1 = m_1x_1 + b_1$ and $y_1 = m_2x_1 + b_2$. Substitute $m_1x_1 + b_1$ for y_1 in $y_1 = m_2x_1 + b_2$. Then, $m_1x_1 + b_1 = m_2x_1 + b_2$. Since $m_1 = m_2$, substitute m_1 for m_2. The result is $m_1x_1 + b_1 = m_1x_1 + b_2$. Subtract m_1x_1 from each side to find $b_1 = b_2$. However, this contradicts the given information that $b_1 \neq b_2$. Thus, the assumption is incorrect and the lines do not share any points.

33a. No; the lines that represent the situation do not coincide.

33b. Yes; the lines that represent the situation coincide.

---

11. Geometry A quadrilateral is a parallelogram if both pairs of its opposite sides are parallel. A parallelogram is a rectangle if its adjacent sides are perpendicular. Use these definitions to determine if the *EFGH* is a *parallelogram*, a *rectangle*, or *neither*. **parallelogram**

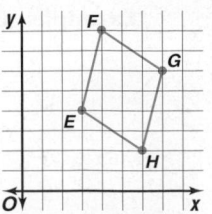

EXERCISES

Practice

Determine whether the graphs of each pair of equations are *parallel*, *coinciding*, *perpendicular*, or *none of these*.

 A

12. perpendicular

14. none of these

17. perpendicular

12. $y = 5x - 18$
$2x + 10y + 10 = 0$

13. $y - 7x + 5 = 0$ **parallel**
$y - 7x - 9 = 0$

14. $y = \frac{1}{3}x + 11$
$y = 3x - 9$

B

15. $y = -3$
$x = 6$ **perpendicular**

16. $y = 4x - 3$ **coinciding**
$4.8x - 1.2y = 3.6$

17. $4x - 6y = 11$
$3x + 2y = 9$

18. $y = 3x - 2$
$3x + y = 2$ **none of these**

19. $5x + 9y = 14$
$y = -\frac{5}{9}x + \frac{14}{9}$ **coinciding**

20. $y + 4x - 2 = 0$
$y + 4x + 1 = 0$
parallel

21. Are the graphs of $y = 3x - 2$ and $y = -3x + 2$ *parallel, coinciding, perpendicular,* or *none of these*? Explain. **None of these; the slopes are not the same nor opposite reciprocals.**

Write the standard form of the equation of the line that is parallel to the graph of the given equation and passes through the point with the given coordinates.

22. $2x - y - 8 = 0$

23. $4x - 9y - 183 = 0$

24. $y + 11 = 0$

25. $x + 5y + 15 = 0$

26. $x + 6y + 5 = 0$

27. $y + 13 = 0$

22. $y = 2x + 10; (0, -8)$

23. $4x - 9y = -23; (12, -15)$

24. $y = -9; (4, -11)$

Write the standard form of the equation of the line that is perpendicular to the graph of the given equation and passes through the point with the given coordinates.

C

25. $y = 5x + 12; (0, -3)$

26. $6x - y = 3; (7, -2)$

27. $x = 12; (6, -13)$

28. The equation of line ℓ is $5y - 4x = 10$. Write the standard form of the equation of the line that fits each description.
   a. parallel to ℓ and passes through the point at $(-15, 8)$ **$4x - 5y + 100 = 0$**
   b. perpendicular to ℓ and passes through the point at $(-15, 8)$ **$5x + 4y + 43 = 0$**

29. The equation of line m is $8x - 14y + 3 = 0$.
   a. For what value of k is the graph of $kx - 7y + 10 = 0$ parallel to line m? **4**
   b. What is k if the graphs of m and $kx - 7y + 10 = 0$ are perpendicular? **$-\frac{49}{4}$**

Applications and Problem Solving

30. Critical Thinking Write equations of two lines that satisfy each description.
   a. perpendicular and one is vertical **Sample answer: $y - 1 = 0, x - 1 = 0$**
   b. parallel and neither has a *y*-intercept **Sample answer: $x - 7 = 0, x - 9 = 0$**

31. Geometry An altitude of a triangle is a segment that passes through one vertex and is perpendicular to the opposite side. Find the standard form of the equation of the line containing each altitude of $\triangle ABC$.
$x - 5y - 29 = 0; x = 7; x + 5y + 15 = 0$

www.amc.glencoe.com/self_check_quiz

32. Critical Thinking The equations $y = m_1x + b_1$ and $y = m_2x + b_2$ represent parallel lines if $m_1 = m_2$ and $b_1 \neq b_2$. Show that they have no point in common. (*Hint:* Assume that there is a point in common and show that the assumption leads to a contradiction.) **See margin.**

33. Business The Seattle Mariners played their first game at their new baseball stadium on July 15, 1999. The stadium features Internet kiosks, a four-story scoreboard, a retractable roof, and dozens of espresso vendors. Suppose a vendor sells 216 regular espressos and 162 large espressos for a total of $783 at a Monday night game. **a–b. See margin.**

a. On Thursday, 248 regular espressos and 186 large espressos were sold. Is it possible that the vendor made $914 that day? Explain.

b. On Saturday, 344 regular espressos and 258 large espressos were sold. Is it possible that the vendor made $1247 that day? Explain.

34. Economics The table shows the closing value of a stock index for one week in February, 1999.

34a. See margin.

a. Using the day as the x-value and the closing value as the y-value, write equations in slope-intercept form for the lines that represent each value change.

34b. parallel lines; no

b. What would indicate that the rate of change for two pair of days was the same? Was the rate of change the same for any of the days shown?

c. Use each equation to predict the closing value for the next business day. The actual closing value was 1241.87. Did any equation correctly predict this value? Explain. **See margin.**

Stock Index February, 1999	
Day	**Closing value**
8	1243.77
9	1216.14
10	1223.55
11	1254.04
12	1230.13

Mixed Review

35. Write the slope-intercept form of the equation of the line through the point at $(1, 5)$ that has a slope of -2. (*Lesson 1-4*) $y = -2x + 7$

36. Business Knights Screen Printers makes special-order T-shirts. Recently, Knights received two orders for a shirt designed for a symposium. The first order was for 40 T-shirts at a cost of $295, and the second order was for 80 T-shirts at a cost of $565. Each order included a standard shipping and handling charge. (*Lesson 1-4*)

a. Write a linear equation that models the situation. $y = 6.75x + 25$

b. What is the cost per T-shirt? **$6.75**

c. What is the standard shipping and handling charge? **$25**

37. Graph $3x - 2y - 6 = 0$. (*Lesson 1-3*) **See margin.**

38. Find $[g \circ h](x)$ if $g(x) = x - 1$ and $h(x) = x^2$. (*Lesson 1-2*) $x^2 - 1$

39. Write an example of a relation that is not a function. Tell why it is not a function. (*Lesson 1-1*) **See margin.**

40. SAT Practice Grid-In If $2x + y = 12$ and $x + 2y = -6$, what is the value of $2x + 2y$? **4**

Extra Practice See p. A27.

Lesson 1-5 Writing Equations of Parallel and Perpendicular Lines **37**

Extra Credit

Write the standard form of the equations of the lines that pass through the given point $(a, -c)$ and are parallel and perpendicular, respectively, to the graph of $4x + y - 3 = 0$.
$4x + y - 4a + c = 0$; $-x + 4y + 4c + a = 0$

Lesson 1-5 **37**

1-6

1 FOCUS

5-Minute Check
Lesson 1-5

Determine whether the graphs of each pair of equations are *parallel, coinciding, perpendicular,* or *none of these.*

1. $x - 2y = 8$
   $3x - 6y = 24$ **coinciding**

2. $2x - y = 9$
   $-4x - 2y = 18$ **none of these**

3. Write the standard form of the equation of the line that passes through the point at $(2, -5)$ and is parallel to the graph of $3x - 2y + 12 = 0$.
   $3x - 2y - 16 = 0$

4. Write the standard form of the equation of the line that passes through the point at $(-4, 2)$ and is perpendicular to the graph of $4x - 2y + 5 = 0$. $x + 2y = 0$

Motivating the Lesson

Discussion Discuss how correlations are important to decision-making. For example, income is usually related positively to the level of education; performance on a test is usually related to how much effort is put into preparing for the test; sales performance is related to marketing strategy in business; and so on.

Modeling Real-World Data with Linear Functions

OBJECTIVES
- Draw and analyze scatter plots.
- Write a prediction equation and draw best-fit lines.
- Use a graphing calculator to compute correlation coefficients to determine goodness of fit.
- Solve problems using prediction equation models.

Education The cost of attending college is steadily increasing. However, it can be a good investment since on average, the higher your level of education, the greater your earning potential. The chart shows the average tuition and fees for a full-time resident student at a public four-year college. Estimate the average college cost in the academic year beginning in 2006 if tuition and fees continue at this rate. *This problem will be solved in Example 1.*

Academic Year	Tuition and Fees
1990–1991	2159
1991–1992	2410
1992–1993	2349
1993–1994	2537
1994–1995	2681
1995–1996	2811
1996–1997	2975
1997–1998	3111
1998–1999	3243

Source: The College Board and National Center for Educational Statistics

As you look at the college tuition costs, it is difficult to visualize how quickly the costs are increasing. When real-life data is collected, the data graphed usually does not form a perfectly straight line. However, the graph may approximate a linear relationship. When this is the case, a **best-fit line** can be drawn, and a **prediction equation** that models the data can be determined. Study the **scatter plots** below.

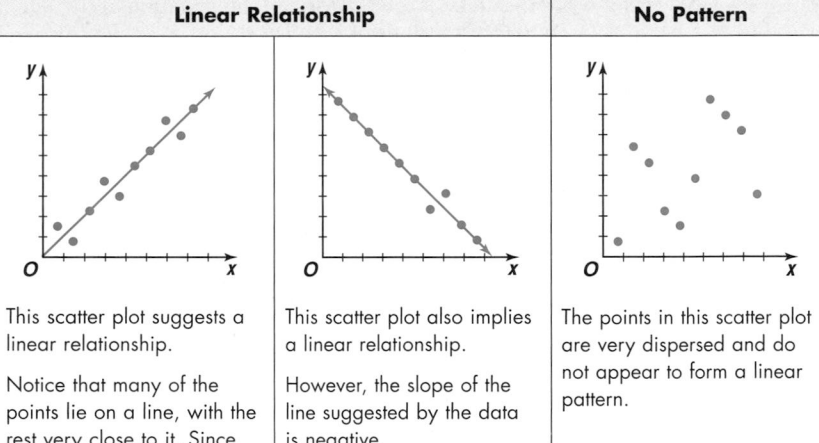

Linear Relationship		No Pattern
This scatter plot suggests a linear relationship. Notice that many of the points lie on a line, with the rest very close to it. Since the line has a positive slope, these data have a positive relationship.	This scatter plot also implies a linear relationship. However, the slope of the line suggested by the data is negative.	The points in this scatter plot are very dispersed and do not appear to form a linear pattern.

38 *Chapter 1   Linear Relations and Functions*

Resource Manager

Reproducible Masters
- *Study Guide and Practice,* pp. 11–12
- *Enrichment,* p. 6
- *Assessment and Evaluation,* p.18
- *Problem Solving and Applications,* p. 1

Transparencies
- 5-Minute Check, 1-6
- Teaching, 1-6

A prediction equation can be determined using a process similar to determining the equation of a line using two points. The process is dependent upon your judgment. You decide which two points on the line are used to find the slope and intercept. Your prediction equation may be different from someone else's. A prediction equation is used when a rough estimate is sufficient.

Example **1** **EDUCATION** Refer to the application at the beginning of the lesson. **Predict the average college cost in the academic year beginning in 2006.**

Graph the data. Use the starting year as the independent variable and the tuition and fees as the dependent variable.

Average Tuition and Fees

Select two points that appear to represent the data. We chose (1992, 2349) and (1997, 3111). Determine the slope of the line.

$$m = \frac{y_2 - y_1}{x_2 - x_1} \qquad \textit{Definition of slope}$$

$$= \frac{3111 - 2349}{1997 - 1992} \qquad (x_1, y_1) = (1992, 2349), (x_2, y_2) = (1997, 3111)$$

$$= \frac{762}{5} \text{ or } 152.4$$

Now use one of the ordered pairs, such as (1992, 2349), and the slope in the point-slope form of the equation.

$$y - y_1 = m(x - x_1) \qquad \textit{Point-slope form of an equation}$$

$$y - 2349 = 152.4(x - 1992) \qquad (x_1, y_1) = (1992, 2349), \text{ and } m = 152.4$$

$$y = 152.4x - 301,231.8$$

Thus, a prediction equation is $y = 152.4x - 301,231.8$. Substitute 2006 for x to estimate the average tuition and fees for the year 2006.

$$y = 152.4x - 301,231.8$$

$$y = 152.4(2006) - 301,231.8$$

$$y = 4482.6$$

According to this prediction equation, the average tuition and fees will be $4482.60 in the academic year beginning in 2006. *Use a different pair of points to find another prediction equation. How does it compare with this one?*

2 TEACH

Teaching Tip Remind students to graph the data in Example 1 before choosing which points to use in determining the slope.

In-Class Example

Example 1
Income The table summarizes the total U.S. personal income from the years 1986 to 1997. Predict personal income in the year 2001. **Sample answer: $7920.60 billion**

Personal Income	
Year	Income ($ billion)
1986	3658.4
1987	3888.7
1988	4184.6
1989	4501.0
1990	4804.2
1991	4981.6
1992	5277.2
1993	5519.2
1994	5757.9
1995	6072.1
1996	6425.2
1997	6784.0

Source: U.S. Commerce Department, Bureau of Economic Analysis

Teaching Tip Point out that the costs given in Example 1 are *only* tuition and fees. They do not include the cost of housing and food.

From the Classroom of...

Sheri Gwinn Miller
Gaston Christian Senior High School
Belmont, North Carolina

"I have my math students poll other students in the school about their height and shoe size. We then use this information to make a prediction equation. Students use this equation to predict a parent's shoe size given the parent's height. At home, students compare this prediction with the parent's actual shoe size. The more data collected for this activity the better!"

Data that are linear in nature will have varying degrees of **goodness of fit** to the lines of fit. Various formulas are often used to find a **correlation coefficient** that describes the nature of the data. The more closely the data fit a line, the closer the correlation coefficient r approaches 1 or -1. Positive correlation coefficients are associated with linear data having positive slopes, and negative correlation coefficients are associated with negative slopes. Thus, the more linear the data, the more closely the correlation coefficient approaches 1 or -1.

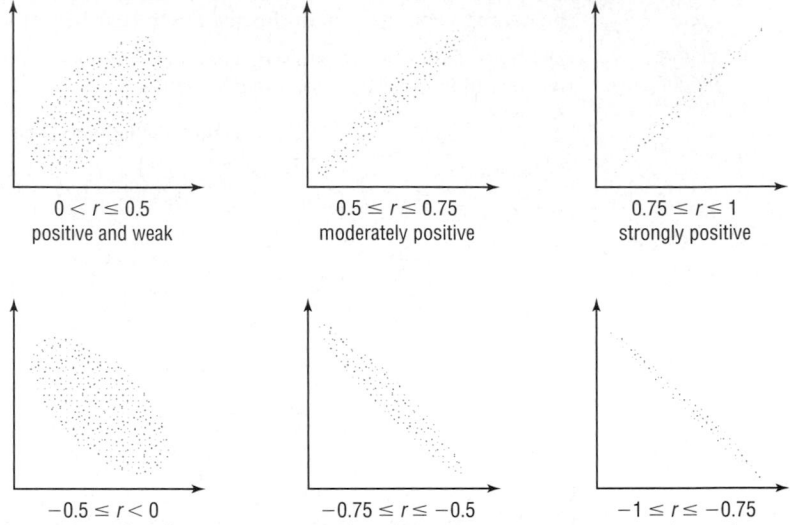

| $0 < r \leq 0.5$ positive and weak | $0.5 \leq r \leq 0.75$ moderately positive | $0.75 \leq r \leq 1$ strongly positive |

| $-0.5 \leq r < 0$ | $-0.75 \leq r \leq -0.5$ | $-1 \leq r \leq -0.75$ |

Statisticians normally use precise procedures, often relying on computers to determine correlation coefficients. The graphing calculator uses the **Pearson product-moment correlation,** which is represented by r. When using these methods, the best fit-line is often called a **regression line.**

Example ② **NUTRITION** The table contains the fat grams and Calories in various fast-food chicken sandwiches.

a. Use a graphing calculator to find the equation of the regression line and the Pearson product-moment correlation.

b. Use the equation to predict the number of Calories in a chicken sandwich that has 20 grams of fat.

Chicken Sandwich (cooking method)	Fat (grams)	Calories
A (breaded)	28	536
B (grilled)	20	430
C (chicken salad)	33	680
D (broiled)	29	550
E (breaded)	43	710
F (grilled)	12	390
G (breaded)	9	300
H (chicken salad)	5	320
I (breaded)	26	530
J (breaded)	18	440
K (grilled)	8	310

Graphing Calculator Appendix

For keystroke instruction on how to enter data, draw a scatter plot, and find a regression equation, see pages A22–A25.

a. Enter the data for fat grams in list L1 and the data for Calories in list L2. Draw a scatter plot relating the fat grams, *x*, and the Calories, *y*.

Then use the linear regression statistics to find the equation of the regression line and the correlation coefficient.

The Pearson product-moment correlation is about 0.98. The correlation between grams of fat and Calories is strongly positive. Because of the strong relationship, the equation of the regression line can be used to make predictions.

b. When rounding to the nearest tenth, the equation of the regression line is $y = 11.6x + 228.3$. Thus, there are about $y = 11.6(20) + 228.3$ or 460.3 Calories in a chicken sandwich with 20 grams of fat.

[0, 45] scl: 1 by [250, 750] scl: 50

```
LinReg
y=ax+b
a=11.62131837
b=228.3159505
r²=.9539063457
r=.9766812918
```

It should be noted that even when there is a large correlation coefficient, you cannot assume that there is a "cause and effect" relationship between the two related variables.

CHECK FOR UNDERSTANDING

Communicating Mathematics

Read and study the lesson to answer each question.   **2. See margin.**

1. **Explain** what the slope in a best-fit line represents.   **the rate of change**

2. **Describe** three different methods for finding a best-fit line for a set of data.

3. **Write** about a set of real-world data that you think would show a negative correlation.   **Sample answer: age of a car and its value**

Guided Practice

4a. See Answer Appendix.

4b. Sample answer: Using (1991, 1800) and (1994, 2266), $y = 155.3x - 307,402.3$

4c. $y = 98.3x - 193,819.4$; $r \approx 0.95$

4d. See margin.

Complete parts a–d for each set of data given in Exercises 4 and 5.

a. Graph the data on a scatter plot.

b. Use two ordered pairs to write the equation of a best-fit line.

c. Use a graphing calculator to find an equation of the regression line for the data. What is the correlation coefficient?

d. If the equation of the regression line shows a moderate or strong relationship, predict the missing value. Explain whether you think the prediction is reliable.

4. **Economics**   The table shows the average amount that an American spent on durable goods in several years.

Personal Consumption Expenditures for Durable Goods									
Year	1990	1991	1992	1993	1994	1995	1996	1997	2010
Personal Consumption ($)	1910	1800	1881	2083	2266	2305	2389	2461	?

Source: U.S. Dept. of Commerce

Lesson 1-6 Modeling Real-World Data with Linear Functions **41**

Additional Answer (p. 42)

5a.   **Computers in Schools**

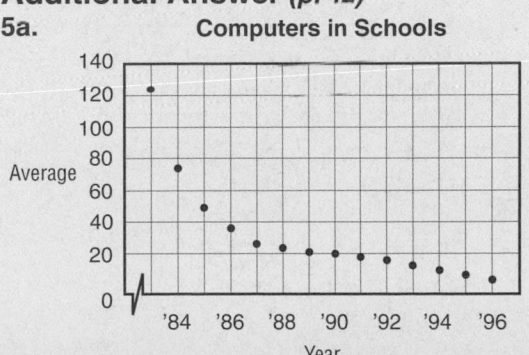

5. Education Do you share a computer at school? The table shows the average number of students per computer in public schools in the United States.

	Students per Computer							
Academic Year	1983–1984	1984–1985	1985–1986	1986–1987	1987–1988	1988–1989	1989–1990	1990–1991
Average	125	75	50	37	32	25	22	20

Academic Year	1991–1992	1992–1993	1993–1994	1994–1995	1995–1996	1996–1997	?
Average	18	16	14	10.5	10	7.8	1

Source: QED's Technology in Public Schools

EXERCISES

Applications and Problem Solving

Complete parts a–d for each set of data given in Exercises 6–11.   6–11. See margin.

a. Graph the data on a scatter plot.

b. Use two ordered pairs to write the equation of a best-fit line.

c. Use a graphing calculator to find an equation of the regression line for the data. What is the correlation coefficient?

d. If the equation of the regression line shows a moderate or strong relationship, predict the missing value. Explain whether you think the prediction is reliable.

  6. Sports The table shows the number of years coaching and the number of wins as of the end of the 1999 season for selected professional football coaches.

NFL Coach	Years	Wins
Don Shula	33	347
George Halas	40	324
Tom Landry	29	270
Curly Lambeau	33	229
Chuck Noll	23	209
Chuck Knox	22	193
Dan Reeves	19	177
Paul Brown	21	170
Bud Grant	18	168
Steve Owen	23	153
Marv Levy	17	?

Source: World Almanac

  7. Economics Per capita personal income is the average personal income for a nation. The table shows the per capita personal income for the United States for several years.

Year	1990	1991	1992	1993	1994	1995	1996	1997	2005
Personal Income ($)	18,477	19,100	19,802	20,810	21,846	23,233	24,457	25,660	?

Source: U.S. Dept. of Commerce

 www.amc.glencoe.com/self_check_quiz

6a.
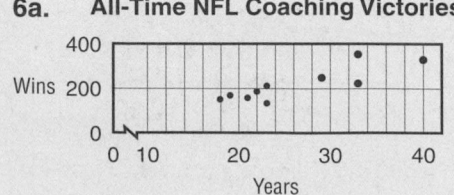
All-Time NFL Coaching Victories

7a.

Personal Income

8a, 9a, 10a, 11a. See Answer Appendix.

8. Transportation   Do you think the weight of a car is related to its fuel economy? The table shows the weight in hundreds of pounds and the average miles per gallon for selected 1999 cars.

Weight (100 pounds)	17.5	20.0	22.5	22.5	22.5	25.0	27.5	35.0	45.0
Fuel Economy (mpg)	65.4	49.0	59.2	41.1	38.9	40.7	46.9	27.7	?

Source: U.S. Environmental Protection Agency

9. Botany   Acorns were one of the most important foods of the Native Americans. They pulverized the acorns, extracted the bitter taste, and then cooked them in various ways. The table shows the size of acorns and the geographic area covered by different species of oak.

Acorn size (cm³)	0.3	0.9	1.1	2.0	3.4	4.8	8.1	10.5	17.1
Range (100 km²)	233	7985	10,161	17,042	7900	3978	28,389	7646	?

Source: Journal of Biogeography

10. Employment   Women have changed their role in American society in recent decades. The table shows the percent of working women who hold managerial or professional jobs.

Percent of Working Women in Managerial or Professional Occupations									
Year	1986	1988	1990	1992	1993	1994	1995	1996	1997	2008
Percent	23.7	25.2	26.2	27.4	28.3	28.7	29.4	30.3	30.8	?

Source: U.S. Dept. of Labor

11. Demographics   The world's population is growing at a rapid rate. The table shows the number of millions of people on Earth at different years.

World Population							
Year	1	1650	1850	1930	1975	1998	2010
Population (millions)	200	500	1000	2000	4000	5900	?

Source: World Almanac

12. Critical Thinking   Different correlation coefficients are acceptable for different situations. For each situation, give a specific example and explain your reasoning.   **a–c. See margin.**

a. When would a correlation coefficient of less than 0.99 be considered unsatisfactory?

b. When would a correlation coefficient of 0.6 be considered good?

c. When would a strong negative correlation coefficient be desirable?

Lesson 1-6   Modeling Real-World Data with Linear Functions   **43**

Open-Ended Assessment

Writing Write a short paragraph describing the correlation graphs on page 40. Give examples of what type of real-world data might correspond to each graph type.

 Chapter 1, Quiz C (Lessons 1-5 and 1-6) is available in the *Assessment and Evaluation Masters,* p. 18.

Additional Answers

13. The rate of growth, which is the slope of the graphs of the regression equations, for the women is less than that of the men's rate of growth. If that trend continues, the men's median salary will always be more than the women's.

16a.

$y = 0.82x + 24$

16c. If the nation had no disposable income, personal consumption expenditures would be $24 billion. For each $1 billion increase in disposable income, there is a 0.82 billion dollar increase in personal consumption expenditures.

Enrichment Masters, p. 6

1-6 **Enrichment**

Significant Digits

All measurements are approximations. The **significant digits** of an approximate number are those which indicate the results of a measurement.

For example, the mass of an object, measured to the nearest gram, is 210 grams. The measurement 21̲0 g has 3 significant digits. The mass of the same object, measured to the nearest 100 g, is 200 g. The measurement 200 g has one significant digit.

Several identifying characteristics of significant digits are listed below, with examples.

1. Non-zero digits and zeros between significant digits are significant. For example, the measurement 9.071 m has 4 significant digits, 9, 0, 7, and 1.
2. Zeros at the end of a decimal fraction are significant. The measurement 0.050 mm has 2 significant digits, 5 and 0.
3. Underlined zeros in whole numbers are significant. The measurement 104,0̲0̲0 km has 5 significant digits, 1, 0, 4, 0, and 0.

In general, a computation involving multiplication or division of measurements *cannot* be more accurate than the least accurate measurement of the computation. Thus, the result of computation involving multiplication or division of measurements should be rounded to the number of significant digits in the least accurate measurement.

Example The mass of 37 quarters is 21̲0 g. Find the mass of one quarter.

mass of 1 quarter = 21̲0 g ÷ 37   21̲0 has 3 significant digits.
= 5.68 g   37 does not represent a measurement. Round the result to 3 significant digits. Why?

Write the number of significant digits for each measurement.

1. 8314.20 m — 6
2. 30.70 cm — 4
3. 0.01 mm — 1
4. 0.0605 mg — 3
5. 37̲0,000 km — 3
6. 370,0̲0̲0 km — 5
7. 9.7 × 10⁴ g — 2
8. 3.20 × 10⁻² g — 3

Solve each problem. Round each result to the correct number of significant digits.

9. 23 m × 1.54 m — 35 m²
10. 12,0̲00 ft ÷ 520 ft — 23.1
11. 2.5 cm × 25 — 63 cm
12. 11.01 mm × 11 — 121.1 mm
13. 908 yd ÷ 0.5 — 1,820 yd
14. 38.6 m × 4.0 m — 150 m²

© Glencoe/McGraw-Hill   T6   *Advanced Mathematical Concepts*

---

13. Critical Thinking The table shows the median salaries of American men and women for several years. According to the data, will the women's median salary ever be equal to the men's? If so, predict the year. Explain. **See margin.**

	Median Salary ($)				
Year	Men's	Women's	Year	Men's	Women's
1985	16,311	7217	1991	20,469	10,476
1986	17,114	7610	1992	20,455	10,714
1987	17,786	8295	1993	21,102	11,046
1988	18,908	8884	1994	21,720	11,466
1989	19,893	9624	1995	22,562	12,130
1990	20,293	10,070	1996	23,834	12,815

Source: U.S. Bureau of the Census

Mixed Review

14. Business During the month of January, Fransworth Computer Center sold 24 computers of a certain model and 40 companion printers. The total sales on these two items for the month of January was $38,736. In February, they sold 30 of the computers and 50 printers. (*Lesson 1-5*)

a. Assuming the prices stayed constant during the months of January and February, is it possible that their February sales could have totaled $51,470 on these two items? Explain. **No; the lines do not coincide.**

b. Assuming the prices stayed constant during the months of January and February, is it possible that their February sales could have totaled $48,420 on these two items? Explain. **Yes; the lines coincide.**

15. Line ℓ passes through $A(-3, -4)$ and has a slope of -6. What is the standard form of the equation for line ℓ? (*Lesson 1-4*) **$6x + y + 22 = 0$**

16. Economics The equation $y = 0.82x + 24$, where $x \geq 0$, models a relationship between a nation's disposable income, x in billions of dollars, and personal consumption expenditures, y in billions of dollars. Economists call this type of equation a *consumption function*. (*Lesson 1-3*)

a. Graph the consumption function. **See margin.**

b. Name the y-intercept. **$24 billion**

c. Explain the significance of the y-intercept and the slope. **See margin.**

17. $x^3 + 3x^2 + 3x + 1; x^3 + 1$ **17.** Find $[f \circ g](x)$ and $[g \circ f](x)$ if $f(x) = x^3$ and $g(x) = x + 1$. (*Lesson 1-2*)

18. Determine if the relation $\{(2, 4), (4, 2), (-2, 4), (-4, 2)\}$ is a function. Explain. (*Lesson 1-1*) **Yes; each domain value is paired with exactly one range value.**

19. SAT/ACT Practice Choose the equation that is represented by the graph. **C**

A $y = 3x - 1$

B $y = \frac{1}{3}x - 1$

C $y = 1 - 3x$

D $y = 1 - \frac{1}{3}x$

E none of these

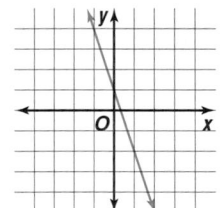

Extra Practice See p. A27.

Extra Credit

Consider a company's data about income level and number of years of education at the secondary or college level, in the table at right. Find a best-fit line for the data. Then, make a statement about how representative the line is of the data. **The line is strongly, positively correlated to the data.**

Years	Income($)
2	10,000
4	15,500
6	26,000
8	38,000
10	49,500
12	62,000

1-7

Piecewise Functions

ACCOUNTING The Internal Revenue Service estimates that taxpayers who itemize deductions and report interest and capital gains will need an average of almost 24 hours to prepare their returns. The amount that a single taxpayer owes depends upon his or her income. The table shows the tax brackets for different levels of income for a certain year.

Single Individual Income Tax	
Limits of Taxable Income	**Tax Bracket**
$0 to $25,350	15%
$25,351 to $61,400	28%
$61,401 to $128,100	31%
$128,101 to $278,450	36%
over $278,450	39.6%

Source: *World Almanac*

A problem related to this will be solved in Example 3.

The tax table defines a special function called a **piecewise function.** For piecewise functions, different equations are used for different intervals of the domain. The graph below shows a piecewise function that models the number of miles from home as a function of time in minutes. Notice that the graph consists of several line segments, each of which is a part of a linear function.

Brittany traveled at a rate of 30 mph for 8 minutes. She stopped at a stoplight for 2 minutes. Then for 4 minutes she traveled 15 mph through the school zone. She sat at the school for 3 minutes while her brother got out of the car. Then she traveled home at 25 mph.

A Trip to and from School

When graphing piecewise functions, the partial graphs over various intervals do not necessarily connect. The definition of the function on the intervals determines if the graph parts connect.

Lesson 1-7 Piecewise Functions **45**

5-Minute Check
Lesson 1-6

Refer to the data in the table.

Total World Population by Decade	
Year	**Total World Population (in millions)**
1950	2556
1960	3039
1970	3707
1980	4454
1990	5279

Source: U.S. Bureau of the Census, International Data Base

1. Predict the world population in the year 2000. **Sample answer: 5959 million people**
2. Graph the data and draw a best-fit line. Make a statement about how representative the line is of the data.

Total Population of the World by Decade

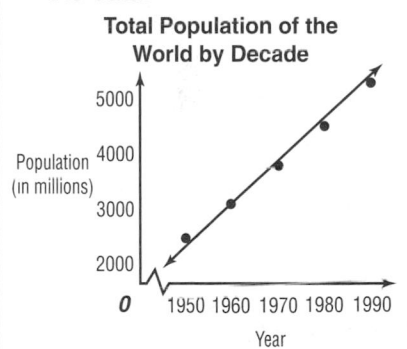

These points cluster very closely to the best-fit line, so it is a good model of the data.

Motivating the Lesson
Hands-On Activity Have students read the tax table showing the tax rates for a single individual in the Application. Then, have students go to the local or school library to research the tax tables for married couples. How do these tables compare? **Sample answer: a joint return allows for a larger income range per tax bracket.**

Teaching Tip When graphing a greatest integer function, it is helpful to make a table of the domain of *x*-values and the range, *f*(*x*).

In-Class Examples

Example 1
Graph
$$f(x) = \begin{cases} 2 \text{ if } x \le -5 \\ x + 4 \text{ if } -5 < x \le 4. \\ -\frac{1}{2}x \text{ if } x > 4 \end{cases}$$

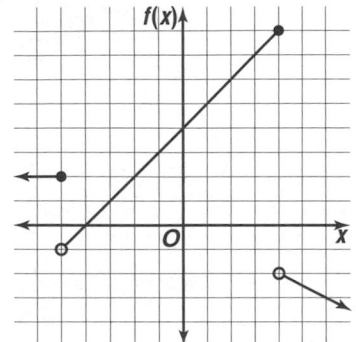

Example 2
Graph $f(x) = 3[\![x]\!]$.

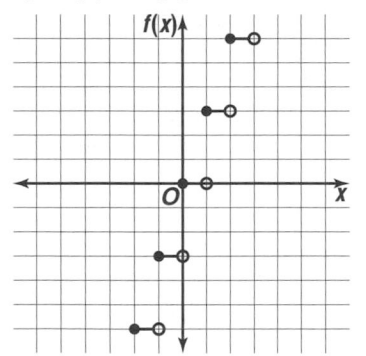

Teaching Tip You may want to tell students that the greatest integer function is often called the *floor function*, because it rounds any number down to the nearest integer. A similar step function called the *ceiling function* rounds any number up to the nearest integer.

Example ❶ Graph $f(x) = \begin{cases} 1 \text{ if } x \le -2 \\ 2 + x \text{ if } -2 < x \le 3. \\ 2x \text{ if } x > 3 \end{cases}$

First, graph the constant function $f(x) = 1$ for $x \le -2$. This graph is part of a horizontal line. Because the point at $(-2, 1)$ is included in the graph, draw a closed circle at that point.

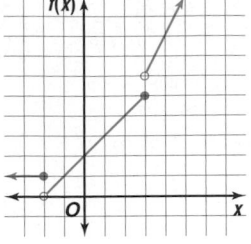

Second, graph the function $f(x) = 2 + x$ for $-2 < x \le 3$. Because $x = -2$ is not included in this part of the domain, draw an open circle at $(-2, 0)$. $x = 3$ is included in the domain, so draw a closed circle at $(3, 5)$ since for $f(x) = 2 + x, f(3) = 5$.

Third, graph the line $y = 2x$ for $x > 3$. Draw an open circle at $(3, 6)$ since for $f(x) = 2x, f(3) = 6$.

Graphing Calculator Tip

On a graphing calculator, **int (X)** indicates the greatest integer function.

A piecewise function where the graph looks like a set of stairs is called a **step function.** In a step function, there are breaks in the graph of the function. You cannot trace the graph of a step function without lifting your pencil. One type of step function is the **greatest integer function.** The symbol $[\![x]\!]$ means *the greatest integer not greater than x.* This does not mean to round or truncate the number. For example, $[\![8.9]\!] = 8$ because 8 is the greatest integer not greater than 8.9. Similarly, $[\![-3.9]\!] = -4$ because -3 is greater than -3.9. The greatest integer function is given by $f(x) = [\![x]\!]$.

Example ❷ Graph $f(x) = [\![x]\!]$.

Make a table of values. The domain values will be intervals for which the greatest integer function will be evaluated.

x	f(x)
$-3 \le x < -2$	-3
$-2 \le x < -1$	-2
$-1 \le x < 0$	-1
$0 \le x < 1$	0
$1 \le x < 2$	1
$2 \le x < 3$	2
$3 \le x < 4$	3
$4 \le x < 5$	4

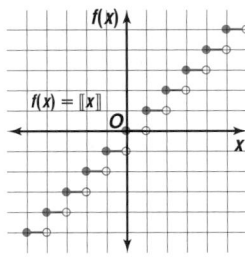

Notice that the domain for this greatest integer function is all real numbers and the range is integers.

The graphs of step functions are often used to model real-world problems such as fees for cellular telephones and the cost of shipping an item of a given weight.

From the Classroom of...

Doug Glackin
Glidden-Ralston School
Glidden, Iowa

"I have students create jack-o-lantern designs on graph paper using functions. They then use *Sketchpad* to put the designs on the computer. Once the design is finished, they make a printout for their template, and carve the design using the pattern they create."

3 Refer to the application at the beginning of the lesson.

a. Graph the tax brackets for the different incomes.

b. What is the tax bracket for a person who makes **$70,000**?

a.

Income ($100,000)

b. $70,000 falls in the interval $61,401 to $128,100. Thus, the tax bracket for $70,000 is 31%.

The **absolute value function** is another piecewise function. Consider $f(x) = |x|$. The absolute value of a number is always nonnegative. The table lists a specific domain and resulting range values for the absolute value function. Using these points, a graph of the absolute value function can be constructed. Notice that the domain of the graph includes all real numbers. However, the range includes only nonnegative real numbers.

table		graph	piecewise function
f(x) = |x|		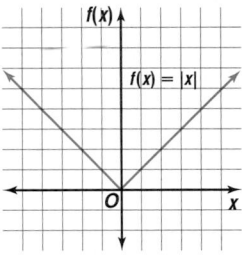	$f(x) = \begin{cases} -x \text{ if } x < 0 \\ x \text{ if } x \geq 0 \end{cases}$

x	**f(x)**
−3	3
−2.4	2.4
0	0
0.7	0.7
2	2
3.4	3.4

Example **4** Graph $f(x) = 2|x| - 6$.

Use a table of values to determine points on the graph.

x	**2|x| − 6**	**(x, f(x))**
−6	2|−6| − 6 = 6	(−6, 6)
−3	2|−3| − 6 = 0	(−3, 0)
−1.5	2|−1.5| − 6 = −3	(−1.5, −3)
0	2|0| − 6 = −6	(0, −6)
1	2|1| − 6 = −4	(1, −4)
2	2|2| − 6 = −2	(2, −2)

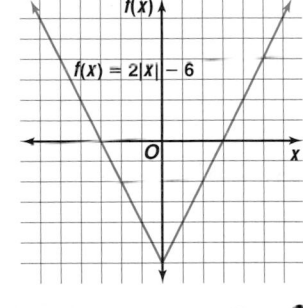

Lesson 1-7 Piecewise Functions **47**

3 PRACTICE/APPLY

Teaching Tip In Example 3, point out that the "steps" do not have to be the same width.

In-Class Examples

Example 3

Taxes The amount of tax that a corporation pays depends on its taxable income. The tax table shows the federal tax rate brackets for various levels of corporate income.

Federal Tax Brackets	
Taxable Income($)	Federal Tax Rate(%)
0–50,000	15
50,001–75,000	25
75,001–100,000	34
100,001–335,000	39

Source: *Time Almanac,* 1999

a. Graph the tax brackets for the different incomes.

Federal Corporation Taxes

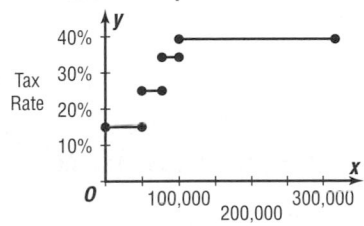

b. What is the tax bracket for a corporation with a taxable income of $90,000? **34%**

Example 4
Graph $f(x) = -2|x| + 3$.

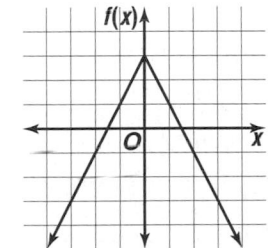

Teaching Tip Remind students that all absolute value functions have a general "v" shape. The coefficient in front of the absolute value symbol determines whether the graph opens upward or downward and whether it is narrow or wide.

In-Class Example

Example 5
Identify the type of function that models each situation. Then write a function for the situation.

a. **Manufacturing** The stated weight of a box of cereal is 15.5 ounces. If the discrepancy is more than 0.4 ounce, the production line is stopped for adjustments. **Absolute value function; let *w* represent the weight and *d(w)* represent the discrepancy. Then, *d(w)* = |15.5 − w|.**

b. **Business** InfoSystem charges $250 a month to organize a company's payroll for up to 25 employees and an additional $100 a month for each 25 employees over 25. **Greatest integer function; let *n* represent the number of employees and *f(n)* = the cost in dollars:**

$$f(n) = \begin{cases} 250,\ 1 \le n \le 25 \\ 250 + 100\left[\!\!\left[\dfrac{n-25}{25}\right]\!\!\right], \\ n > 25 \end{cases}$$

---

3 PRACTICE/APPLY

Error Analysis

Watch for students who confuse types of piecewise functions.

Prevent by stressing that step functions look like stairsteps and absolute value functions look like a normal or inverted capital "A".

Additional Answer (p. 49)

9. Greatest integer function; *h* is hours, *c(h)* is the cost,
$$c(h) = \begin{cases} 50h \text{ if } [\![h]\!] = h \\ 50[\![h+1]\!] \text{ if } [\![h]\!] < h \end{cases}.$$

Many real-world situations can be modeled by a piecewise function.

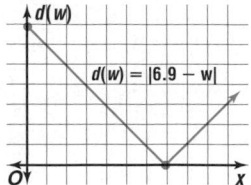

Example **⑤** Identify the type of function that models each situation. Then write a function for the situation.

a. **Manufacturing** The stated weight of a box of rice is 6.9 ounces. The company randomly chooses boxes to test to see whether their equipment is dispensing the right amount of product. If the discrepancy is more than 0.2 ounce, the production line is stopped for adjustments.

The situation can be represented with an absolute value function. Let *w* represent the weight and *d(w)* represent the discrepancy. Then $d(w) = |6.9 - w|$.

b. **Business** On a certain telephone rate plan, the price of a cellular telephone call is 35¢ per minute or fraction thereof.

This can be described by a greatest integer function.

Let *m* represent the number of minutes of the call and *c(m)* represent the cost in cents.

$$c(m) = \begin{cases} 35m \text{ if } [\![m]\!] = m \\ 35[\![m+1]\!] \text{ if } [\![m]\!] < m \end{cases}$$

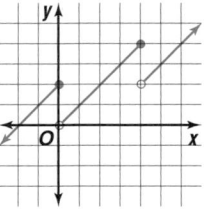

CHECK FOR UNDERSTANDING

Communicating Mathematics

Read and study the lesson to answer each question.

1. **Write** $f(x) = |x|$ as a piecewise function. $f(x) = \begin{cases} x \text{ if } x \ge 0 \\ -x \text{ if } x < 0 \end{cases}$

2. **State** the domain and range of the function $f(x) = 2[\![x]\!]$. **reals, even integers**

3. **Write** the function that is represented by the graph.

3. $f(x) = \begin{cases} x + 2 \text{ if } x \le 0 \\ x \text{ if } 0 < x \le 4 \\ x - 2 \text{ if } x > 4 \end{cases}$

4. **You Decide** Misae says that a step graph does not represent a function because the graph is not connected. Alex says that it does represent a function because there is only one *y* for every *x*. Who is correct and why? **Alex is correct because he is applying the definition of a function.**

Guided Practice Graph each function. **5–8. See Answer Appendix.**

5. $f(x) = \begin{cases} 2x \text{ if } 0 \le x \le 4 \\ 8 \text{ if } 4 < x \le 7 \end{cases}$

6. $f(x) = \begin{cases} 6 \text{ if } x \le -6 \\ |x| \text{ if } -6 < x < 6 \\ 6 \text{ if } x > 6 \end{cases}$

7. $f(x) = -[\![x]\!]$

8. $f(x) = |x - 3|$

48 Chapter 1 Linear Relations and Functions

9. **Business** Identify the type of function that models the labor cost for repairing a computer if the charge is $50 per hour or fraction thereof. Then write and graph a function for the situation.   **See margin.**

10. **Consumerism** Guillermo Lujan is flying from Denver to Dallas for a convention. He can park his car in the Denver airport long-term parking lot at the terminal or in the shuttle parking facility closeby. In the long-term lot, it costs $1.00 per hour or any part of an hour with a maximum charge of $6.00 per day. In shuttle facility, he has to pay $4.00 for each day or part of a day. Which parking lot is less expensive if Mr. Lujan returns after 2 days and 3 hours?   **shuttle facility**

EXERCISES

Practice

Graph each function.   **11–22. See Answer Appendix.**

 A

11. $f(x) = \begin{cases} 2x + 1 \text{ if } x < 0 \\ 2x - 1 \text{ if } x \geq 0 \end{cases}$

12. $g(x) = |x - 5|$

 B

13. $h(x) = [\![x]\!] + 2$

14. $g(x) = |2x + 3|$

15. $f(x) = [\![x - 1]\!]$

16. $h(x) = \begin{cases} 3 \text{ if } -1 \leq x \leq 1 \\ 4 \text{ if } 1 < x \leq 4 \\ x \text{ if } x > 4 \end{cases}$

17. $g(x) = 2|x - 3|$

18. $f(x) = [\![-3x]\!]$

19. $h(x) = \begin{cases} x + 3 \text{ if } x \leq 0 \\ 3 - x \text{ if } 1 < x \leq 3 \\ 3x \text{ if } x > 3 \end{cases}$

20. $f(x) = \begin{cases} -2x \text{ if } x < 1 \\ 3 \text{ if } x = 1 \\ 4x \text{ if } x > 1 \end{cases}$

C

21. $j(x) = \dfrac{2}{[\![x]\!]}$

22. $g(x) = |9 - 3|x||$

Identify the type of function that models each situation. Then write and graph a function for the situation.   **23–25. See margin.**

23. **Tourism** The table shows the charge for renting a bicycle from a rental shop on Cumberland Island, Georgia, for different amounts of time.

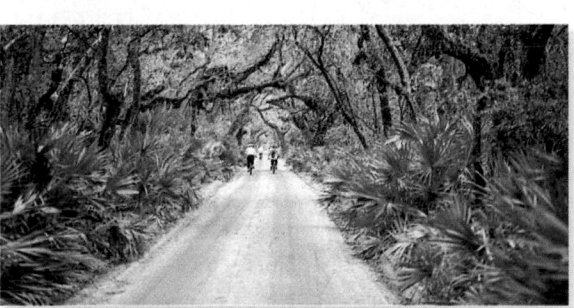

Island Rentals	
Time	**Price**
$\frac{1}{2}$ hour	$6
1 hour	$10
2 hours	$16
Daily	$24

24. **Postage** The cost of mailing a letter is $0.33 for the first ounce and $0.22 for each additional ounce or portion thereof.

25. **Manufacturing** A can of coffee is supposed to contain one pound of coffee. How does the actual weight of the coffee in the can compare to 1 pound?

 www.amc.glencoe.com/self_check_quiz

24. greatest integer; w is the weight in ounces, $c(w)$ is the cost in dollars,
$$c(w) = \begin{cases} 0.33 + 0.22(w - 1) \text{ if } [\![w]\!] = w \\ 0.33 + 0.22[\![w]\!] \text{ if } [\![w]\!] < w \end{cases}.$$

25. absolute value; w is the weight in pounds, $d(w)$ is the discrepancy, $d(w) = |1 - w|$.

***Study Guide and Practice Masters*, pp. 13–14**

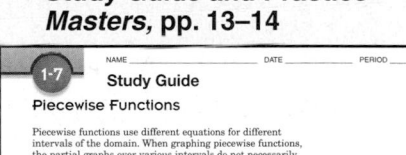

Additional Answer

23. Step; t is the time in hours, $c(t)$ is the cost in dollars,

$$c(t) = \begin{cases} 6 \text{ if } t \leq \frac{1}{2} \\ 10 \text{ if } \frac{1}{2} < t \leq 1 \\ 16 \text{ if } 1 < t \leq 2 \\ 24 \text{ if } 2 < t \leq 24. \end{cases}$$

26b. v is the value of the order, s(v) is the shipping,

$$s(v) = \begin{cases} 3.50 \text{ if } 0.00 \le v \le 25.00 \\ 5.95 \text{ if } 25.01 \le v \le 75.00 \\ 7.95 \text{ if } 75.01 \le v \le 125.00 \\ 9.95 \text{ if } 125.01 \le v \end{cases}$$

26c.

Shipping (dollars)

Value of Order (dollars)

28c.

$d(t) = |65 - t|$

29b.

$$t(x) = \begin{cases} 6\% \text{ if } x \le \$10{,}000 \\ 8\% \text{ if } \$10{,}000 < x \le \$20{,}000 \\ 9.5\% \text{ if } x > \$20{,}000 \end{cases}$$

29c.

Tax Rate (percent)

Income (thousands of dollars)

30. No; the functions are the same if x is positive. However, if x is negative, the functions yield different values. For example,

$[g \circ f](1.5) = 1$ and
$[f \circ g](1.5) = 1$;
$[g \circ f](-1.5) = 2$ and
$[f \circ g](-1.5) = 1$.

Real World Application

26. Retail Sales   The table shows the shipping charges that apply to orders placed in a catalog.

a. What type of function is described?   **step**

b. Write the shipping charges as a function of the value of the order.

c. Graph the function.
   b–c. See margin.

	Shipping to or within the United States
Value of Order	**Shipping, Packing, and Handling Charge**
$0.00–25.00	$3.50
$25.01–75.00	$5.95
$75.01–125.00	$7.95
$125.01 and up	$9.95

27. If n is any integer, then all ordered pairs (x, y), where x and y are both in the interval [n, n + 1) are solutions.

27. Critical Thinking   Describe the values of x and y which are solutions to $[\![x]\!] = [\![y]\!]$.

28. Engineering   The degree day is used to measure the demand for heating or cooling. In the United States, 65°F is considered the desirable temperature for the inside of a building. The number of degree days recorded on a given date is equal to the difference between 65 and the mean temperature for that date. If the mean temperature is above 65°F, cooling degree days are recorded. Heating degree days are recorded if the mean temperature is below 65°F.

a. What type of function can be used to model degree days?   **absolute value**

b. Write a function to model the number of degree days d(t) for a mean temperature of t°F. $d(t) = |65 - t|$

c. Graph the function.   **See margin.**

d. The mean temperature is the mean of the high and low temperatures for a day. How many degree days are recorded for a day with a high of temperature of 63°F and a low temperature of 28°F? Are they heating degree days or cooling degree days?   **19.5 heating degree days**

29. Accounting   The income tax brackets for the District of Columbia are listed in the tax table.

Income	Tax Bracket
up to $10,000	6%
more than $10,000, but no more than $20,000	8%
more than $20,000	9.5%

a. What type of function is described by the tax rates?   **step**

b. Write the function if x is income and t(x) is the tax rate.   **See margin.**

c. Graph the tax brackets for different taxable incomes.   **See margin.**

d. Alicia Davis lives in the District of Columbia. In which tax bracket is Ms. Davis if she made $36,000 last year?   **9.5%**

30. Critical Thinking   For $f(x) = [\![x]\!]$ and $g(x) = |x|$, are $[f \circ g](x)$ and $[g \circ f](x)$ equivalent? Justify your answer.   **See margin.**

31. Transportation The table shows the percent of workers in different cities who use public transportation to get to work. *(Lesson 1-6)*

a. Graph the data on a scatter plot.

b. Use two ordered pairs to write the equation of a best-fit line.

c. Use a graphing calculator to find an equation for the regression line for the data. What is the correlation value?

d. If the equation of the regression line shows a moderate or strong relationship, predict the percent of workers using public transportation in Baltimore, Maryland. Is the prediction reliable? Explain.

City	Workers 16 years and older	Percent who use Public Transportation
New York, NY	3,183,088	53.4
Los Angeles, CA	1,629,096	10.5
Chicago, IL	1,181,677	29.7
Houston, TX	772,957	6.5
Philadelphia, PA	640,577	28.7
San Diego, CA	560,913	4.2
Dallas, TX	500,566	6.7
Phoenix, AZ	473,966	3.3
San Jose, CA	400,932	3.5
San Antonio, TX	395,551	4.9
San Francisco, CA	382,309	33.5
Indianapolis, IN	362,777	3.3
Detroit, MI	325,054	10.7
Jacksonville, FL	312,958	2.7
Baltimore, MD	307,679	22.0

Source: U.S. Bureau of the Census

32. Write the standard form of the equation of the line that passes through the point at (4, 2) and is parallel to the line whose equation is $y = 2x - 4$. *(Lesson 1-5)* **$2x - y - 6 = 0$**

33. Sports During a basketball game, the two highest-scoring players scored 29 and 15 points and played 39 and 32 minutes, respectively. *(Lesson 1-3)*

a. Write an ordered pair of the form (minutes played, points scored) to represent each player. **(39, 29), (32, 15)**

b. Find the slope of the line containing both points. **2**

c. What does the slope of the line represent?

34. Business For a company, the revenue $r(x)$ in dollars, from selling x items is $r(x) = 400x - 0.2x^2$. The cost for making and selling x items is $c(x) = 0.1x + 200$. Write the profit function $p(x) = (r - c)(x)$. *(Lesson 1-2)* **$399.9x - 0.2x^2 - 200$**

35. Retail Winston bought a sweater that was on sale 25% off. The original price of the sweater was \$59.99. If sales tax in Winston's area is 6.5%, how much did the sweater cost including sale tax? *(Lesson 1-2)* **\$47.92**

36. State the domain and range of the relation {(0, 2), (4, −2), (9, 3), (−7, 11), (−2, 0)}. Is the relation a function? Explain. *(Lesson 1-1)* **See margin.**

37. SAT Practice Which of the following expressions is *not* larger than 5×6^{12}? **A**

A $5 + 6^{12}$

B 7×6^{12}

C 5×8^{12}

D 5×6^{14}

E 10^{13}

Extra Practice See p. A27.

Extra Credit

Graph $x = |y + 1|$.

Graphing Linear Inequalities

1 FOCUS

5-Minute Check
Lesson 1-7

1. Graph
$$f(x) = \begin{cases} 2 \text{ if } x \leq 0 \\ x + 1 \text{ if } 0 < x < 2. \\ -2x \text{ if } x \geq 2 \end{cases}$$

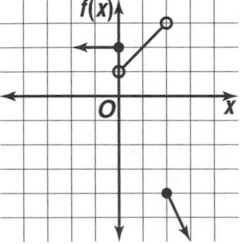

2. Graph $f(x) = \dfrac{|x|}{2}$.

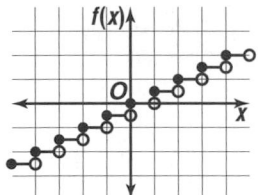

3. Graph $f(x) = -3|x| + 2$.

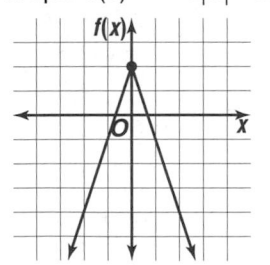

Motivating the Lesson

Hands-On Activity Have students roll two different-colored number cubes and record each result as an ordered pair, with one color representing the abscissa and another the ordinate. Have the class graph all of the ordered pairs and the line $x + y = 3$ on the chalkboard. Then, have students determine which ordered pairs are represented by points above the line and what characteristic they have in common. $x + y > 3$

OBJECTIVE
• Graph linear inequalities.

NUTRITION Arctic explorers need endurance and strength. They move sleds weighing more than 1100 pounds each for as much as 12 hours a day. For that reason, Will Steger and members of his exploration team each burn 4000 to 6000 Calories daily!

An *endurance diet* can provide the energy and nutrients necessary for peak performance in the Arctic. An endurance diet has a balance of fat and carbohydrates and protein. Fat is a concentrated energy source that supplies nine calories per gram. Carbohydrates and protein provide four calories per gram and are a quick source of energy. What are some of the combinations of carbohydrates and protein and fat that supply the needed energy for the Arctic explorers? *This problem will be solved in Example 2.*

This situation can be described using a **linear inequality.** A linear inequality is not a function. However, you can use the graphs of linear functions to help you graph linear inequalities.

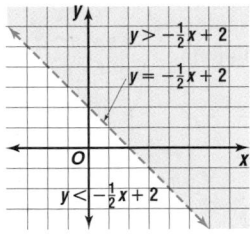

The graph of $y = -\frac{1}{2}x + 2$ separates the coordinate plane into two regions, called **half planes.** The line described by $y = -\frac{1}{2}x + 2$ is called the **boundary** of each region. If the boundary is part of a graph, it is drawn as a solid line. A boundary that is not part of the graph is drawn as a dashed line. The graph of $y > -\frac{1}{2}x + 2$ is the region above the line. The graph of $y < -\frac{1}{2}x + 2$ is the region below the line.

When graphing an inequality, you can determine which half plane to shade by testing a point on either side of the boundary in the original inequality. If it is not on the boundary, the origin $(0, 0)$ is often an easy point to test. If the inequality statement is true for your test point, then shade the half plane that contains the test point. If the inequality statement is false for your test point, then shade the half plane that does not contain the test point.

52 *Chapter 1 Linear Relations and Functions*

Resource Manager

 Reproducible Masters
• *Study Guide and Practice,* pp. 15–16
• *Enrichment,* p. 8
• *Assessment and Evaluation,* p. 18

 Transparencies
• 5-Minute Check, 1-8
• Teaching, 1-8

 Technology/Multimedia
• Advanced MathPASS, Lesson 2

Example **1** **Graph each inequality.**

a. $x > 3$

The boundary is not included in the graph. So the vertical line $x = 3$ should be a dashed line.

Testing $(0, 0)$ in the inequality yields a false inequality, $0 > 3$. So shade the half plane that does not include $(0, 0)$.

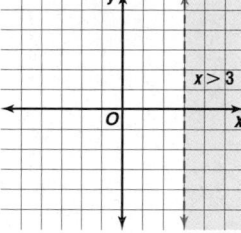

b. $x - 2y - 5 \leq 0$

$$x - 2y - 5 \leq 0$$
$$-2y \leq -x + 5$$
$$y \geq \frac{1}{2}x - \frac{5}{2} \quad \text{\textit{Reverse the inequality when you divide or}}$$
$$\text{\textit{multiply by a negative.}}$$

The graph does include the boundary. So the line is solid.

Testing $(0, 0)$ in the inequality yields a true inequality, so shade the half plane that includes $(0, 0)$.

c. $y > |x - 2|$

Graph the equation $y = |x - 2|$ with a dashed boundary.

Testing $(0, 0)$ yields the false inequality $0 > 2$, so shade the region that does not include $(0, 0)$.

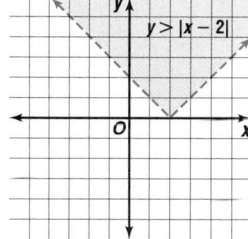

You can also graph relations such as $-1 < x + y \leq 3$. The graph of this relation is the intersection of the graph of $-1 < x + y$ and the graph of $x + y \leq 3$. Notice that the boundaries $x + y = 3$ and $x + y = -1$ are parallel lines. The boundary $x + y = 3$ is part of the graph, but $x + y = -1$ is not.

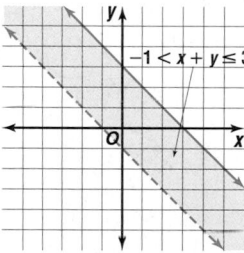

Lesson 1-8 Graphing Linear Inequalities **53**

2 TEACH

Teaching Tip When working Example 1, remind students that dashed lines on their graphs mean $<$ or $>$. Solid lines represent $\leq$ or $\geq$.

In-Class Example

Example 1
Graph each inequality.
a. $x \leq 2$

b. $x - 2y < 8$

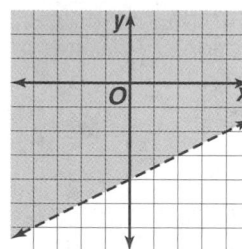

c. $y \leq |x + 1|$

Addressing Individual Needs

Visual/Spatial Have students graph the linear inequality $y > -\frac{3}{4}x + 5$. Ask them to describe what each area of the graph represents. Then, have them write a paragraph describing why a linear inequality is not a function. **Half plane; sample answer: in a linear inequality, each point in the domain may have more than one corresponding point in the range, violating the rules for functions.**

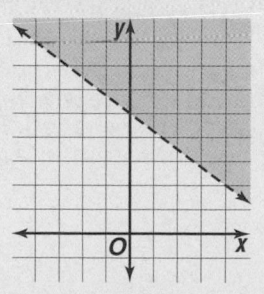

Teaching Tip In Example 2, discuss with students the possibility that other restrictions besides Calorie requirements may be enforced. For example, there is a minimum amount of protein or fats needed to keep the body running properly.

In-Class Example

Example 2

Manufacturing A small craft business makes red-trimmed baskets and blue-trimmed baskets. Each blue-trimmed basket requires 2 feet of ribbon and each red-trimmed basket requires 1 foot of ribbon. The company wants to use between 54 and 72 feet of ribbon each day. Let x represent the number blue-trimmed baskets and y represent the number of red-trimmed baskets.

a. Draw a graph that models the combinations of baskets that can be made.

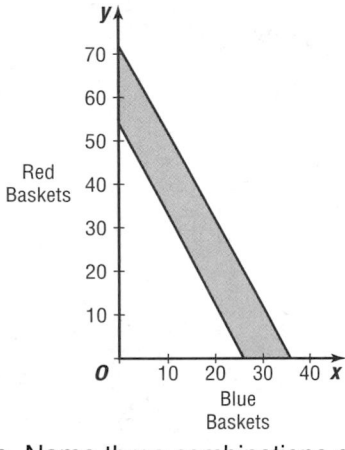

b. Name three combinations of red-trimmed and blue-trimmed baskets that meet the requirements.
36 red-trimmed, 18 blue-trimmed; 19 red-trimmed, 20 blue-trimmed; 20 red-trimmed, 25 blue-trimmed

---

3 PRACTICE/APPLY

Error Analysis

Watch for students who have difficulty determining which region to shade in the graph of an inequality.

Prevent by showing students how to test points in each region.

Example **NUTRITION** Refer to the application at the beginning of the lesson.

a. **Draw a graph that models the combinations of grams of fat and carbohydrates and protein that the arctic team diet may include to satisfy their daily caloric needs.**

Let x represent the number of grams of fat and y represent the number of grams of carbohydrates and protein. The team needs at least 4000, but no more than 6000, Calories each day. Write an inequality.

$$4000 \leq 9x + 4y \leq 6000$$

You can write this compound inequality as two inequalities, $4000 \leq 9x + 4y$ and $9x + 4y \leq 6000$. Solve each part for y.

$$4000 \leq 9x + 4y \quad \text{and} \quad 9x + 4y \leq 6000$$

$$4000 - 9x \leq 4y \qquad\qquad 4y \leq 6000 - 9x$$

$$1000 - \frac{9}{4}x \leq y \qquad\qquad y \leq 1500 - \frac{9}{4}x$$

Graph each boundary line and shade the appropriate region. The graph of the compound inequality is the area in which the shading overlaps.

Grams of Carbohydrate and Protein

Grams of Fat

b. **Name three combinations of fat or carbohydrates and protein that meet the Calorie requirements.**

Any point in the shaded region or on the boundary lines meets the requirements. Three possible combinations are (100, 775), (200, 800), and (300, 825). These ordered pairs represent 100 grams of fat and 775 grams of carbohydrate and protein, 200 grams of fat and 800 grams of carbohydrate and protein, and 300 grams of fat and 825 grams of carbohydrate and protein.

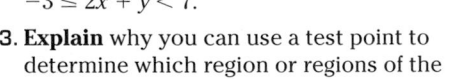

CHECK FOR UNDERSTANDING

Communicating Mathematics

Read and study the lesson to answer each question.

1. **Write** the inequality whose graph is shown.

2. **Describe** the process you would use to graph $-3 \leq 2x + y < 7$.

3. **Explain** why you can use a test point to determine which region or regions of the graph of an inequality should be shaded.

1. $y \geq 2x - 6$

2–3. See margin.

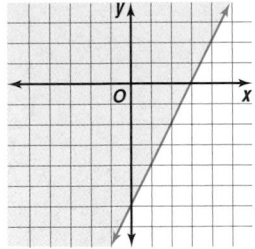

Additional Answers

2. Graph the lines $-3 = 2x + y$ and $2x + y = 7$. The graph of $-3 = 2x + y$ is solid and the graph of $2x + y = 7$ is dashed. Test points to determine which region(s) should be shaded. Then shade the correct region(s).

3. Sample answer: the boundaries separate the plane into regions. Each point in a region either does or does not satisfy the inequality. Using a test point allows you to determine whether all of the points in a region satisfy the inequality.

Guided Practice

Graph each inequality.   **4–7. See Answer Appendix.**

4. $x + y < 4$

5. $3x - y \leq 6$

6. $7 < x + y \leq 9$

7. $y < |x + 3|$

8. **Business**   Nancy Stone has a small company and has negotiated a special rate for rental cars when she and other employees take business trips. The maximum charge is $45.00 per day plus $0.40 per mile. Discounts apply when renting for longer periods of time or during off-peak seasons.

   a. Write a linear inequality that models the total cost of the daily rental $c(m)$ as a function of the total miles driven, m.   **$c(m) \leq 45 + 0.4m$**

   b. Graph the inequality.   **See margin.**

   c. Name three combinations of miles and total cost that satisfy the inequality.   **Sample answer: (0, 45), (10, 49), (20, 50)**

EXERCISES

Practice

Graph each inequality.   **9–22. See Answer Appendix.**

 A

9. $y < 3$

10. $x - y > -5$

11. $2x + 4y \geq 7$

12. $-y < 2x + 1$

B

13. $2x - 5y + 19 \leq 0$

14. $-4 \leq x - y \leq 5$

15. $y > |x|$

16. $-2 \leq x + 2y \leq 4$

C

17. $y > |x| + 4$

18. $y < |2x + 3|$

19. $-8 \leq 2x + y < 6$

20. $y - 1 > |x + 3|$

21. Graph the region that satisfies $x \geq 0$ and $y \geq 0$.

22. Graph $2 < |x| \leq 8$.

Applications and Problem Solving

23. **Manufacturing**   Many manufacturers use inequalities to solve production problems such as determining how much of each product should be assigned to each machine. Suppose one bakery oven at a cookie manufacturer is being used to bake chocolate cookies and vanilla cookies. A batch of chocolate cookies bakes in 8 minutes, and a batch of vanilla cookies bakes in 10 minutes.

   a. Let x represent the number of batches of chocolate cookies and y represent the number of batches of vanilla cookies. Write a linear inequality for the number of batches of each type of cookie that could be baked in one oven in an 8-hour shift.   **$8x + 10y \leq 480$**

   b. Graph the inequality.   **See margin.**

   c. Name three combinations of batches of chocolate cookies and vanilla cookies that satisfy the inequality.   **Sample answer: (0, 48), (60, 0), (45, 6)**

   d. Often manufacturers' problems involve as many as 150 products, 218 facilities, 10 plants, and 127 customer zones. Research how problems like this are solved.   **Sample answer: Using complex computer programs and systems of inequalities.**

24. **Critical Thinking**   Graph $|y| \geq x$.   **See margin.**

www.amc.glencoe.com/self_check_quiz   *Lesson 1-8   Graphing Linear Inequalities*   **55**

Additional Answers

23b.

24.

Assignment Guide

Core: 9–21 odd, 23–25, 27–32
Enriched: 10–22 even, 23–32

Additional Answer

8b.

Study Guide and Practice Masters, pp. 15–16

25a. points in the first and third quadrants

25. Critical Thinking Suppose $xy > 0$.
 a. Describe the points whose coordinates are solutions to the inequality.
 b. Demonstrate that for points whose coordinates are solutions to the inequality, the equation $|x + y| = |x| + |y|$ holds true. **See margin.**

26. Engineering Mechanics The production cost of a job depends in part on the accuracy required. On most sheet metal jobs, an accuracy of 1, 2, or 0.1 mils is required. A mil is $\frac{1}{1000}$ inch. This means that a dimension must be less than $\frac{1}{1000}, \frac{2}{1000},$ or $\frac{1}{10,000}$ inch larger or smaller than the blueprint states. Industrial jobs often require a higher degree of accuracy.

26a.
$$|8 - x| \leq \frac{1}{500};$$
$$\left|4\frac{1}{4} - y\right| \leq \frac{1}{500}$$

 a. Write inequalities that models the possible dimensions of a part that is supposed to be 8 inches by $4\frac{1}{4}$ inches if the accuracy required is 2 mils.
 b. Graph the region that shows the satisfactory dimensions for the part. **See Answer Appendix.**

27. Exercise The American College of Sports Medicine recommends that healthy adults exercise at a target level of 60% to 90% of their maximum heart rate. You can estimate your maximum heart rate by subtracting your age from 220.
 a. Write a compound inequality that models age, *a*, and target heart rate, *r*.
 b. Graph the inequality. **See Answer Appendix.**
 a. $0.6(220 - a) \leq r \leq 0.9(220 - a)$

Mixed Review

28. Business Gatsby's Automotive Shop charges $55 per hour or any fraction of an hour for labor. *(Lesson 1-7)*
 a. What type of function is described? **step**
 b. Write the labor charge as a function of the time. **See margin.**
 c. Graph the function. **See Answer Appendix.**

29. The equation of line ℓ is $3x - y = 10$. *(Lesson 1-5)*
 a. What is the standard form of the equation of the line that is parallel to ℓ and passes through the point at $(0, -2)$? $3x - y - 2 = 0$
 b. Write the standard form of the equation of the line that is perpendicular to ℓ and passes through the point at $(0, -2)$. $x + 3y + 6 = 0$

30. Write the slope-intercept form of the equation of the line through $(1, 4)$ and $(5, 7)$. *(Lesson 1-4)* $y = \frac{3}{4}x + 3\frac{1}{4}$

31. Temperature The temperature in Indianapolis on January 30 was 23°F at 12:00 A.M. and 48°F at 4:00 P.M. *(Lesson 1-3)* **a. (0, 23), (16, 48); 1.5625**
 a. Write two ordered pairs of the form (hours since midnight, temperature) for this date. What is the slope of the line containing these points?

31b. the average change in the temperature per hour

 b. What does the slope of the line represent?

32. SAT/ACT Practice Which expression is equivalent to $\frac{9^5 - 9^4}{8}$? **E**
 A $\frac{1}{8}$ B $\frac{9}{8}$ C $\frac{9^3}{8}$
 D $\frac{9^9}{8}$ E 9^4

Extra Credit

Graph the given inequalities on the same grid and describe the solution set of points satisfying both inequalities.
 $y \leq 2x + 3$ $y \geq x^2 - 4$
The solution set is the region bounded by the line $y = 2x + 3$ on the top and the curve $y = x^2 - 4$ on the bottom, including the boundaries.

VOCABULARY

abscissa (p. 5)
absolute value function (p.47)
boundary (p. 52)
coinciding lines (p. 32)
composite (p. 15)
composition of functions (pp. 14-15)
constant function (p. 22)
domain (p. 5)
family of graphs (p. 26)
function (p. 6)
function notation (p. 7)
greatest integer function (p. 46)
half plane (p. 52)

iterate (p. 16)
iteration (p. 16)
linear equation (p. 20)
linear function (p. 22)
linear inequality (p. 52)
ordinate (p. 5)
parallel lines (p. 32)
perpendicular lines (p. 34)
piecewise function (p. 45)
point-slope form (p. 28)
range (p. 5)
relation (p. 5)
slope (pp. 20-21)
slope-intercept form (p. 21)
standard form (p. 21)
step function (p. 46)

vertical line test (p. 7)
x-intercept (p. 20)
y-intercept (p. 20)
zero of a function (p. 22)

Modeling

best-fit line (p. 38)
correlation coefficient (p. 40)
goodness of fit (p. 40)
model (p. 27)
Pearson-product moment correlation (p. 40)
prediction equation (p. 38)
regression line (p. 40)
scatter plot (p. 38)

UNDERSTANDING AND USING THE VOCABULARY

Choose the letter of the term that best matches each statement or phrase.

1. for the function f, a value of x for which $f(x) = 0$ **c**

2. a pairing of elements of one set with elements of a second set   **f**

3. has the form $Ax + By + C = 0$, where A is positive and A and B are not both zero   **d**

4. $y - y_1 = m(x - x_1)$, where (x_1, y_1) lies on a line having slope m **g**

5. $y = mx + b$, where m is the slope of the line and b is the y-intercept   **i**

6. a relation in which each element of the domain is paired with exactly one element of the range   **a**

7. the set of all abscissas of the ordered pairs of a relation   **h**

8. the set of all ordinates of the ordered pairs of a relation   **j**

9. a group of graphs that displays one or more similar characteristics   **e**

10. lie in the same plane and have no points in common   **b**

a. function
b. parallel lines
c. zero of a function
d. linear equation
e. family of graphs
f. relation
g. point-slope form
h. domain
i. slope-intercept form
j. range

interNET **CONNECTION**   For additional review and practice for each lesson, visit: **www.amc.glencoe.com**

TEST OPTIONS

Six forms of Chapter 1 Test are available in the *Assessment and Evaluation Masters* as shown in the chart.

Chapter 1 Test, Form 1B, is shown at the right. Chapter 1 Test, Form 2B, is shown on the next page.

Form	Pages	Type
1A	3–4	Honors
1B	5–6	Average
1C	7–8	Basic
2A	9–10	Honors
2B	11–12	Average
2C	13–14	Basic

Form 1 tests are multiple choice.
Form 2 tests are free response.

STUDY GUIDE AND ASSESSMENT

VOCABULARY

This section provides a listing of the new terms, properties, and phrases that were introduced in this chapter. The exercises check students' understanding of the terms by using a variety of verbal formats including matching, completion, and true/false.

Glossary A complete glossary of terms begins on page A71.

Assessment and Evaluation Masters, pp. 5–6

SKILLS AND CONCEPTS

Objectives and Examples This section reviews the skills and concepts of the chapter and shows completely worked examples.

Review Exercises These exercises provide practice for the corresponding objectives.

Assessment and Evaluation Masters, pp. 11–12

SKILLS AND CONCEPTS

OBJECTIVES AND EXAMPLES	REVIEW EXERCISES

Lesson 1-1 Evaluate a function.

• Find $f(-2)$ if $f(x) = 3x^2 - 2x + 4$.
Evaluate the expression $3x^2 - 2x + 4$ for $x = -2$.

$$f(-2) = 3(-2)^2 - 2(-2) + 4$$
$$= 12 + 4 + 4$$
$$= 20$$

16. $16c^2 + 8c - 4$

17. $|m^2 + 5m + 4|$

Evaluate each function for the given value.

11. $f(4)$ if $f(x) = 5x - 10$ **10**

12. $g(2)$ if $g(x) = 7 - x^2$ **3**

13. $f(-3)$ if $g(x) = 4x^2 - 4x + 9$ **57**

14. $h(0.2)$ if $h(x) = 6 - 2x^3$ **5.984**

15. $g\left(\dfrac{1}{3}\right)$ if $g(x) = \dfrac{2}{5x}$ $\dfrac{6}{5}$

16. $k(4c)$ if $k(x) = x^2 + 2x - 4$

17. Find $f(m + 1)$ if $f(x) = |x^2 + 3x|$.

Lesson 1-2 Perform operations with functions.

• Given $f(x) = 4x + 2$ and $g(x) = x^2 - 2x$, find $(f + g)(x)$ and $(f \cdot g)(x)$.

$$(f + g)(x) = f(x) + g(x)$$
$$= 4x + 2 + x^2 - 2x$$
$$= x^2 + 2x + 2$$

$$(f \cdot g)(x) = f(x) \cdot g(x)$$
$$= (4x + 2)(x^2 - 2x)$$
$$= 4x^3 - 6x^2 - 4x$$

18–23. See margin.

Find $(f + g)(x)$, $(f - g)(x)$, $(f \cdot g)(x)$, and $\left(\dfrac{f}{g}\right)(x)$ for each $f(x)$ and $g(x)$.

18. $f(x) = 6x - 4$
    $g(x) = 2$

19. $f(x) = x^2 + 4x$
    $g(x) = x - 2$

20. $f(x) = 4 - x^2$
    $g(x) = 3x$

21. $f(x) = x^2 + 7x + 12$
    $g(x) = x + 4$

22. $f(x) = x^2 - 1$
    $g(x) = x + 1$

23. $f(x) = x^2 - 4x$
    $g(x) = \dfrac{4}{x - 4}$

Lesson 1-2 Find composite functions.

• Given $f(x) = 2x^2 + 4x$ and $g(x) = 2x - 1$, find $[f \circ g](x)$ and $[g \circ f](x)$.

$$[f \circ g](x) = f(g(x))$$
$$= f(2x - 1)$$
$$= 2(2x - 1)^2 + 4(2x + 1)$$
$$= 2(4x^2 - 4x + 1) + 8x + 4$$
$$= 8x^2 + 6$$

$$[g \circ f](x) = g(f(x))$$
$$= g(2x^2 + 4x)$$
$$= 2(2x^2 + 4x) - 1$$
$$= 4x^2 + 8x - 1$$

24–29. See margin.

Find $[f \circ g](x)$ and $[g \circ f](x)$ for each $f(x)$ and $g(x)$.

24. $f(x) = x^2 - 4$
    $g(x) = 2x$

25. $f(x) = 0.5x + 5$
    $g(x) = 3x^2$

26. $f(x) = 2x^2 + 6$
    $g(x) = 3x$

27. $f(x) = 6 + x$
    $g(x) = x^2 - x + 1$

28. $f(x) = x^2 - 5$
    $g(x) = x + 1$

29. $f(x) = 3 - x$
    $g(x) = 2x^2 + 10$

30. State the domain of $[f \circ g](x)$ for $f(x) = \sqrt{x - 16}$ and $g(x) = 5 - x$. $x \le -11$

TestCheck and Worksheet Builder

This **networkable** software has three modules.
• **Worksheet Builder** to make worksheets and tests
• **Student Module** to take tests on-screen
• **Management System** to keep student records
Three special banks are included for SAT, ACT, and TIMSS.

OBJECTIVES AND EXAMPLES

Lesson 1-3 Graph linear equations.

• Graph $f(x) = 4x - 3$.

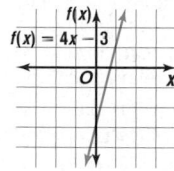

Lesson 1-4 Write linear equations using the slope-intercept, point-slope, and standard forms of the equation.

• Write the slope-intercept form of the equation of the line that has a slope of 24 and passes through the point at $(1, 2)$.

$$y = mx + b \qquad \textit{Slope-intercept form}$$
$$2 = -4(1) + b \quad y = 2, x = 1, m = -4$$
$$6 = b \qquad \textit{Solve for b.}$$

The equation for the line is $y = -4x + 6$.

Lesson 1-5 Write equations of parallel and perpendicular lines.

• Write the standard form of the equation of the line that is parallel to the graph of $y = 2x - 3$ and passes through the point at $(1, -1)$.

$$y - y_1 = m(x - x_1) \quad \textit{Point-slope form}$$
$$y - (-1) = 2(x - 1) \qquad y_1 = -1, m = 2, x = 1$$
$$2x - y - 3 = 0$$

Write the standard form of the equation of the line that is perpendicular to the graph of $y = 2x - 3$ and passes through the point at $(6, -1)$.

$$y - y_1 = m(x - x_1) \quad y_1 = -1,$$
$$y - (-1) = -\frac{1}{2}(x - 6) \quad m = -\frac{1}{2}, x = 6$$
$$x + 2y - 2 = 0$$

REVIEW EXERCISES

Graph each equation. **31–38. See Answer Appendix**

31. $y = 3x + 6$
32. $y = 8 - 5x$
33. $y - 15 = 0$
34. $0 = 2x - y - 7$
35. $y = 2x$
36. $y = -8x - 2$
37. $7x + 2y = -5$
38. $y = \frac{1}{4}x - 6$

41. $y = \frac{1}{2}x + \frac{9}{2}$
42. $y = \frac{1}{2}x + 4$

Write an equation in slope-intercept form for each line described.

39. slope = 2, y-intercept = -3 $y = 2x - 3$
40. slope = -1, y-intercept = 1 $y = -x + 1$
41. slope = $\frac{1}{2}$, passes through the point at $(-5, 2)$
42. passes through $A(-4, 2)$ and $B(2, 5)$
43. x-intercept = 1, y-intercept = -4 $y = 4x - 4$
44. horizontal and passes through the point at $(3, -1)$ $y = -1$
45. the x-axis $y = 0$
46. slope = 0.1, x-intercept = 1 $y = 0.1x - 0.1$

Write the standard form of the equation of the line that is parallel to the graph of the given equation and passes through the point with the given coordinates.

47. $y = x + 1$; $(1, 1)$ $x - y = 0$
48. $y = \frac{1}{3}x - 2$; $(-1, 6)$ $x - 3y + 19 = 0$
49. $2x + y = 1$; $(-3, 2)$ $2x + y + 4 = 0$

Write the standard form of the equation of the line that is perpendicular to the graph of the given equation and passes through the point with the given coordinates.

50. $y = -2x + \frac{1}{4}$; $(4, -8)$ $x - 2y - 20 = 0$
51. $4x - 2y + 2 = 0$; $(1, 4)$ $x + 2y - 9 = 0$
52. $x = -8$; $(4, -6)$ $y + 6 = 0$

Additional Answers (p. 58)

18. $6x - 2$; $6x - 6$; $12x - 8$; $3x - 2$
19. $x^2 + 5x - 2$; $x^2 + 3x + 2$; $x^3 + 2x^2 - 8x$; $\frac{x^2 + 4x}{x - 2}$, $x \neq 2$
20. $4 + 3x - x^2$; $4 - 3x - x^2$; $12x - 3x^3$; $\frac{4 - x^2}{3x}$, $x \neq 0$
21. $x^2 + 8x + 16$; $x^2 + 6x + 8$; $x^3 + 11x^2 + 40x + 48$; $x + 3$, $x \neq -4$
22. $x^2 + x$; $x^2 - x - 2$; $x^3 + x^2 - x - 1$; $x - 1$, $x \neq -1$
23. $\frac{x^3 - 8x^2 + 16x + 4}{x - 4}$, $x \neq 4$; $\frac{x^3 - 8x^2 + 16x - 4}{x - 4}$, $x \neq 4$; $4x$, $x \neq 4$; $\frac{x^3 - 8x^2 + 16x}{4}$, $x \neq 4$
24. $4x^2 - 4$; $2x^2 - 8$
25. $1.5x^2 + 5$; $0.75x^2 + 15x + 75$
26. $18x^2 + 6$; $6x^2 + 18$
27. $x^2 - x + 7$; $x^2 + 11x + 31$
28. $x^2 + 2x - 4$; $x^2 - 4$
29. $-2x^2 - 7$; $2x^2 - 12x + 28$

Additional Answers

53b. Sample answer: using (1987, 10,434) and (1996, 12,909), $y = 275x - 535,991$.

53c. $y = 147.8x - 282,157.4$; $r \approx 0.61$

53d. 14,181,600 visitors; sample answer: this is not a good prediction, because the r value does not indicate a strong relationship.

54.

55.

56.

57.

58.

OBJECTIVES AND EXAMPLES

Lesson 1-6 Draw and analyze scatter plots.

This scatter plot implies a linear relationship. Since data closely fits a line with a positive slope, the scatter plot shows a strong, positive correlation.

This scatter plot implies a linear relationship with a negative slope.

The points in this scatter plot are dispersed and do not form a linear pattern.

REVIEW EXERCISES

53. a. Graph the data below on a scatter plot.

  b. Use two ordered pairs to write the equation of a best-fit line.   **b–d. See margin.**

  c. Use a graphing calculator to find an equation of the regression line for the data. What is the correlation value?

  d. If the equation of the regression line shows a moderate or strong relationship, predict the number of visitors in 2005. Explain whether you think the prediction is reliable.   **53a. See Answer Appendix.**

Overseas Visitors to the United States (thousands)					
Year	1987	1988	1989	1990	1991
Visitors	10,434	12,763	12,184	12,252	12,003
Year	1992	1993	1994	1995	1996
Visitors	11,819	12,024	12,542	12,933	12,909

Source: U.S Dept. of Commerce

Lesson 1-7 Identify and graph piecewise functions including greatest integer, step, and absolute value functions.

Graph $f(x) = |3x - 2|$.

This is an absolute value function. Use a table of values to find points to graph.

x	(x, f(x))
0	(0, 2)
1	(1, 1)
2	(2, 4)
3	(3, 7)
4	(4, 10)

$f(x) = |3x - 2|$

Graph each function.   54–58. See margin.

54. $f(x) = \begin{cases} x \text{ if } 0 \le x \le 5 \\ 2 \text{ if } 5 < x \le 8 \end{cases}$

55. $h(x) = \begin{cases} -1 \text{ if } -2 \le x \le 0 \\ -3x \text{ if } 0 < x \le 2 \\ 2x \text{ if } 2 < x \le 4 \end{cases}$

56. $f(x) = [\![x]\!] + 1$

57. $g(x) = |4x|$

58. $k(x) = 2|x| + 2$

Lesson 1-8 Graph linear inequalities.

Graph the inequality $2x - y < 4$.

$2x - y < 4$
$\quad y > 2x - 4$

The boundary is dashed. Testing (0, 0) yields a true inequality, so shade the region that includes (0, 0).

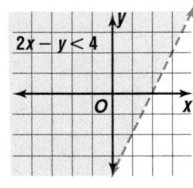

$2x - y < 4$

Graph each inequality.   59–66. See Answer Appendix.

59. $y > 4$ **60.** $x \le 5$

61. $x + y \le 1$ **62.** $2y - x < 4$

63. $y \le |x|$ **64.** $y - 3x > 2$

65. $y > |x| - 2$ **66.** $y < |x - 2|$

60   *Chapter 1   Linear Relations and Functions*

APPLICATIONS AND PROBLEM SOLVING

67. Aviation   A jet plane start from rest on a runway. It accelerates uniformly at a rate of 20 m/s². The equation for computing the distance traveled is $d = \frac{1}{2}at^2$. *(Lesson 1-1)*

   a. Find the distance traveled at the end of each second for 5 seconds.

   b. Is this relation a function? Explain.

   67a–b. See margin.

68. Finance   In 1994, outstanding consumer credit held by commercial banks was about $463 billion. By 1996, this amount had grown to about $529 billion. *(Lesson 1-4)*

   a. If *x* represents the year and *y* represents the amount of credit, find the average annual increase in the amount of outstanding consumer credit.

   b. Write an equation to model the annual change in credit.   **$y = 33x - 65{,}339$**

   68a. about $33 billion

69. Recreation   Juan wants to know the relationship between the number of hours students spend watching TV each week and the number of hours students spend reading each week. A sample of 10 students reveals the following data.

Watching TV	Reading
20	8.5
32	3.0
42	1.0
12	4.0
5	14.0
28	4.5
33	7.0
18	12.0
30	3.0
25	3.0

Find the equation of a regression line for the data. Then make a statement about how representative the line is of the data. *(Lesson 1-6)*   **$y = -0.284x + 12.964$; The correlation is moderately negative, so the regression line is somewhat representative of the data.**

ALTERNATIVE ASSESSMENT

OPEN-ENDED ASSESSMENT

1. If $[f \circ g](x) = 4x^2 - 4$, find $f(x)$ and $g(x)$. Explain why your answer is correct.

2. Suppose two distinct lines have the same *x*-intercept.

   a. Can the lines be parallel? Explain your answer.

   b. Can the lines be perpendicular? Explain your answer.

3. Write a piecewise function whose graph is the same as each function. The function should not involve absolute value.

   a. $y = x + |4 - x|$

   b. $y - 2x + |x + 1|$

   1–3. See margin.

Additional Assessment   See p. A56 for Chapter 1 Practice Test.

Unit 1 *inter*NET Project

TELECOMMUNICATION

Is Anybody Listening?

- Research several telephone long-distance services. Write and graph equations to compare the monthly fee and the rate per minute for each service.

- Which service would best meet your needs? Write a paragraph to explain your choice. Use the graphs to support your choice.

📁 PORTFOLIO

Select one of the functions you graphed in this chapter. Write about a real-world situation this type of function can be used to model. Explain what the function shows about the situation that is difficult to show by other means.   **See students' work.**

Internet Address Book

Record useful Internet addresses in the space at right for quick reference.

www.amc.glencoe.com

Multiple-Choice and Grid-In Questions

Overview Pages 62–64 are part of a complete SAT/ACT preparation course included in the Student Edition. Each of the 15 chapters ends with two pages that expose students to the types of problems that occur on the SAT and ACT tests.

The following chart outlines the topics covered in each SAT & ACT Preparation feature.

Chapter/Topic	
1	Arithmetic
2	Algebra
3	More Algebra
4	Coordinate Geometry
5	Pythagorean Theorem
6	Trigonometry
7	Geometry: Triangles and Quadrilaterals
8	Geometry: Perimeter, Area, and Volume
9	Geometry: Lines, Angles, and Arcs
10	Ratio and Proportion
11	Average
12	Percent
13	Probability and Combinations
14	Statistics
15	Special Functions and Counting

About the Tests The ACT contains 60 multiple-choice questions. The SAT contains two types of items—multiple choice (about 45 questions), and grid in (10 questions).

Pages 62 and 63 have been included to give students an overview of the two types of items on the SAT—multiple choice and grid in.

At the end of each chapter in this textbook, you will find practice for the SAT and ACT tests. Each group of 10 questions contains nine multiple-choice questions, and one grid-in question.

MULTIPLE CHOICE

The majority of questions on the SAT are multiple-choice questions. As the name implies, these questions offer five choices from which to choose the correct answer.

The multiple choice sections are arranged in order of difficulty, with the easier questions at the beginning, average difficulty questions in the middle, and more difficult questions at the end.

Every correct answer earns one raw point, while an incorrect answer results in a loss of one fourth of a raw point. Leaving an answer blank results in no penalty.

The test covers topics from numbers and operations (arithmetic), algebra 1, algebra 2, functions, geometry, statistics, probability, and data analysis. Each end-of-chapter practice section in this textbook will cover one of these areas.

Arithmetic
Six percent of 4800 is equal to 12 percent of what number?

A 600
B 800
C 1200
D 2400
E 3000

Write and solve an equation.
$0.06(4800) = 0.12x$
$288 = 0.12x$
$\dfrac{288}{0.12} = x$
$2400 = x$

Choice **D** is correct.

Algebra
If $(p + 2)(p^2 - 4) = (p + 2)^q(p - 2)$ for all values of p, what is the value of q?

A 1
B 2
C 3
D 4
E It cannot be determined from the given information.

Factor the left side.

$$(p + 2)(p^2 - 4) = (p + 2)^q(p - 2)$$
$$(p + 2)(p + 2)(p - 2) = (p + 2)^q(p - 2)$$
$$(p + 2)^2(p - 2) = (p + 2)^q(p - 2)$$
$$(p + 2)^2 = (p + 2)^q$$

If $a^m = a^n$, then $m = n$.
$$2 = q$$
Answer choice **B** is correct.

Geometry
In the figure, what is the value of x?

A 25
B 30
C 45
D 90
E 135

This is a multi-step problem. Use vertical angle relationships to determine that the two angles in the triangle with x are 80° and 55°. Then use the fact that the sum of the measures of the angles of a triangle is 180 to determine that x equals 45. The correct answer is choice **C**.

GRID IN

Another section on the SAT includes questions in which you must mark your answer on a grid printed on the answer sheet. These are called *Student Produced Response* questions (or Grid-Ins), because you must create the answer yourself, not just choose from five possible answers.

Every correct answer earns one raw point, but there is no penalty for a wrong answer; it is scored the same as no answer.

These questions are *not* more difficult than the multiple-choice questions, but you'll want to be extra careful when you fill in your answers on the grid, so that you don't make careless errors. Grid-in questions are arranged in order of difficulty.

The instructions for using the grid are printed in the SAT test booklet. *Memorize* these instructions before you take the test.

The grid contains a row of four boxes at the top, two rows of ovals with decimal and fraction symbols, and four columns of numbered ovals.

After you solve the problem, always write your answer in the boxes at the top of the grid.

Start with the left column. Write one numeral, decimal point, or fraction line in each box. Shade the oval in each column that corresponds to the numeral or symbol written in the box. Only the shaded ovals will be scored, so work carefully. Don't make any extra marks on the grid.

Suppose the answer is $\frac{2}{3}$ or 0.666 … . You can record the answer as a fraction or a decimal. For the fraction, write $\frac{2}{3}$. For a decimal answer, you must enter the most accurate value that will fit the grid. That is, you must enter as many decimal place digits as space allows. An entry of .66 would not be acceptable.

There is no 0 in bubble column 1. This means that you do *not* enter a zero to the left of the decimal point. For example, enter .25 and not 0.25.

Here are some other helpful hints for successfully completing grid-in questions.

• You don't have to write fractions in simplest form. Any equivalent fraction that fits the grid is counted as correct. If your fraction does not fit (like 15/25), then either write it in simplest form or change it to a decimal before you grid it.

• There is no negative symbol. Grid-in answers are never negative, so if you get a negative answer, you've made an error.

• If a problem has more than one correct answer, enter just one of the answers.

• Do not grid mixed numbers. Change the mixed number to an equivalent fraction or decimal. If you enter 11/2 for $1\frac{1}{2}$, it will be read as $\frac{11}{2}$. Enter it as 3/2 or 1.5.

THE PRINCETON REVIEW *SAT & ACT Preparation* **63**

Scoring of Multiple-Choice Questions One raw point is given for every correct answer. No points are given for a blank answer. There is a penalty of one fourth of a raw point for an incorrect answer.

Scoring of Grid-In Questions
The scoring of grid-in questions is different from multiple-choice questions. There is no penalty for a wrong answer; it is scored the same as no answer. This means, of course that leaving a question blank is just as costly as writing a wrong answer. For this reason, students should answer as many grid-ins as they can.

Chapter 1 Skill: Arithmetic

Pages 64–65 are part of a complete SAT/ACT preparation course that is described in detail on pages T8–T9 of the *Teacher's Handbook*. The test items were written in the same style as those in the SAT and ACT.

 These questions were aligned and verified by The Princeton Review, the nation's leader in test preparation.

Teaching Tip Remind students that prime factorization is a common tool used in simplifying fractions and radicals. Knowledge about prime factorization would help students solve Exercises 1, 2, 3, and 8 of the SAT and ACT Practice for Chapter 1 on page 65.

Diagnosis & Prescription

Each of the 10 SAT and ACT practice problems on page 65 is cross-referenced to the chapter where that SAT or ACT skill is covered. If students miss a particular type of problem, you can have them study that skill.

Problem Number and Skill	Page
1–10 Arithmetic	64

Arithmetic Problems

All SAT and ACT tests contain arithmetic problems. Some are easy and some are difficult. You'll need to understand and apply the following concepts.

odd and even	factors	divisibility
positive, negative	integers	fractions
scientific notation	exponents	roots
prime numbers	decimals	inequalities

Several concepts are often combined in a single problem.

THE PRINCETON REVIEW

TEST-TAKING TIP
Know the properties of zero and one. For example, 0 is even, neither positive nor negative, and not prime. 1 is the only integer with only one divisor. 1 is not prime.

SAT EXAMPLE

1. What is the sum of the positive even factors of 12?

HINT Look for words like *positive, even,* and *factor.*

Solution First find all the factors of 12.

$$1 \quad 2 \quad 3 \quad 4 \quad 6 \quad 12$$

Re-read the question. It asks for the sum of *even* factors. Circle the factors that are even numbers.

$$1 \quad ② \quad 3 \quad ④ \quad ⑥ \quad ⑫$$

Now add these even factors to find the sum.

$$2 + 4 + 6 + 12 = 24 \quad \text{The answer is 24.}$$

This is a grid-in problem. Record your answer on the grid.

ACT EXAMPLE

2. $(-2)^3 + (3)^{-2} + \frac{8}{9}$

A -7 B $-1\frac{7}{9}$ C $\frac{8}{9}$

D $1\frac{7}{9}$ E 12

HINT Analyze what the $-$ (negative) symbol represents each time it is used.

Solution Use the properties of exponents to simplify each term.

$(-2)^3 = (-2)(-2)(-2)$ or -8

$(3)^{-2} = \frac{1}{3^2}$ or $\frac{1}{9}$

Add the terms.

$(-2)^3 + (3)^{-2} + \frac{8}{9} = -8 + \frac{1}{9} + \frac{8}{9}$
$$= -8 + 1 \text{ or } -7$$

The answer is choice **A.**

Always look at the answer choices before you start to calculate. In this problem, three (incorrect) answer choices include fractions with denominators of 9. This may be a clue that your calculations may involve ninths.

Never assume that because three answer choices involve ninths and two are integers, that the correct answer is more likely to involve ninths. Also don't conclude that because the expression contains a fraction that the answer will necessarily have a fraction in it.

64 Chapter 1 *Linear Relations and Functions*

💿 TestCheck and Worksheet Builder

Special banks of SAT, ACT, and TIMSS items are included in this CD-ROM.

After you work each problem, record your answer on the answer sheet provided or on a piece of paper.

Multiple Choice

1. Which of the following expresses the prime factorization of 54? **D**

   A 9×6

   B $3 \times 3 \times 6$

   C $3 \times 3 \times 2$

   D $3 \times 3 \times 3 \times 2$

   E 5.4×10

2. If 8 and 12 each divide K without a remainder, what is the value of K? **E**

   A 16

   B 24

   C 48

   D 96

   E It cannot be determined from the information given.

3. After $\dfrac{4\frac{1}{3}}{2\frac{3}{5}}$ has been simplified to a single fraction in lowest terms, what is the denominator? **B**

   A 2   B 3   C 5

   D 9   E 13

4. For a class play, student tickets cost $2 and adult tickets cost $5. A total of 30 tickets were sold. If the total sales must exceed $90, what is the minimum number of adult tickets that must be sold? **E**

   A 7   B 8   C 9

   D 10   E 11

5. $-|-7| - |-5| - 3|-4| =?$ **A**

   A -24 B -11 C 0

   D 13   E 24

6. $(-4)^2 + (2)^{-4} + \dfrac{3}{4}$ **A**

   A $16\frac{13}{16}$ B $16\frac{3}{4}$ C $-15\frac{7}{32}$

   D $15\frac{7}{32}$ E 16

7. Kerri subscribed to four publications that cost $12.90, $16.00, $18.00, and $21.90 per year. If she made an initial down payment of one half of the total amount and paid the rest in 4 equal monthly payments, how much was each of the 4 monthly payments? **A**

   A $8.60

   B $9.20

   C $9.45

   D $17.20

   E $34.40

8. $\sqrt{64 + 36} =?$ **A**

   A 10   B 14   C 28

   D 48   E 100

9. What is the number of distinct prime factors of 60? **C**

   A 12

   B 4

   C 3

   D 2

   E 1

10. **Grid-In** There are 24 fish in an aquarium. If $\dfrac{1}{8}$ of them are tetras and $\dfrac{2}{3}$ of the remaining fish are guppies, how many guppies are in the aquarium? **14**

 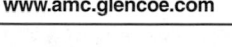 **SAT/ACT Practice** For additional test practice questions, visit: **www.amc.glencoe.com**

THE PRINCETON REVIEW *SAT & ACT Preparation* **65**

A bubble-in answer sheet for these practice problems is available on p. 1 of the *Assessment and Evaluation Masters*.

Additional Practice

Additional SAT and ACT practice questions and a bubble-in answer sheet are available in the *Assessment and Evaluation Masters*, pp. 2, 19, and 20.

Assessment and Evaluation Masters, pp. 19–20

Extra Credit

TIMSS The vertices of the triangle PQR are the points $P(1, 2)$, $Q(4, 6)$, and $R(-4, 12)$. Which one of the following statements about triangle PQR is true? **B**

   A PQR is a right triangle with right angle P.

   B PQR is a right triangle with right angle Q.

   C PQR is a right triangle with right angle R.

   D PQR is not a right triangle.

Systems of Linear Equations and Inequalities

Instructional Objectives

Lesson (pages)	Lesson Objectives	NCTM Standards 2000	State/Local Objectives
2-1 (67–72)	• Solve systems of equations graphically. • Solve systems of equations algebraically.	1, 2, 3, 6, 7, 8, 9, 10	
2-2 (73–77)	• Solve systems of equations involving three variables algebraically.	1, 2, 3, 6, 7, 8, 9, 10	
2-3 (78–86)	• Model data using matrices. • Add, subtract, and multiply matrices.	1, 2, 6, 8, 9, 10	
2-4A (87)	• Determine the effect of matrix multiplication on a vertex matrix.	1, 3, 7, 10	
2-4 (88–96)	• Use matrices to determine the coordinates of polygons under a given transformation.	1, 2, 3, 8, 9, 10	
2-5 (98–105)	• Evaluate determinants. • Find inverses of matrices. • Solve systems of equations by using inverses of matrices.	1, 2, 6, 8, 9, 10	
2-5B (106)	• Find reduced row-echelon form of an augmented matrix to solve systems of equations.	1, 2, 6, 7, 8, 9, 10	
2-6 (107–111)	• Graph systems of inequalities. • Find the maximum or minimum value of a function defined for a polygonal convex set.	1, 2, 6, 7, 8, 9, 10	
2-7 (112–118)	• Use linear programming procedures to solve applications. • Recognize situations where exactly one solution to a linear programming application may not exist.	1, 2, 6, 7, 8, 9, 10	

Suggested Pacing   See pages T2–T4 for additional options in course-planning.

COURSE	Lesson 2-1	Lesson 2-2	Lesson 2-3	Lesson 2-4	Lesson 2-5
Precalculus Emphasis Chapters 4–15 (45–50 minutes)	OPTIONAL—Chapter 2 is a review of topics students have learned in previous courses. You may wish to use the Study Guide and Assessment to assess students' familiarity with these topics.				
Precalculus Emphasis (block)	OPTIONAL				
Trigonometric Emphasis Chapters 1–9, 11, 12 (45–50 minutes)	1.5 days	1.5 days	(with Lesson 2-4A) 2.5 days	1.5 days	(with Lesson 2-5B) 2 days
Trigonometric Emphasis (block)	0.5 day	(with Lesson 2-4A) 1.5 days		1.5 days	

Instructional Resources

	Student Edition	Blackline Masters (page numbers)							
Lesson	Graphing Calculator Applications	Study Guide and Practice	Enrichment	Assessment and Evaluation	Graphing Calculator Keystrokes	Problem Solving & Applications	Spreadsheet Applications	TI-92 and CBL Interdisciplinary Activities	Transparencies A and B
2-1	67, 69	21–22	13		4				2-1
2-2		23–24	14	43		4			2-2
2-3	80, 86	25–26	15			3			2-3
2-4A	87				5–6				
2-4		27–28	16	42, 43					2-4
2-5	99, 100, 103	29–30	17				2	1–4	2-5
2-5B	106				7–8				
2-6		31–32	18	44					2-6
2-7		33–34	19	44					2-7
Study Guide/ Assessment				29–41, 47					
SAT/ACT Practice				45–46					

Lesson 2-6	Lesson 2-7	Study Guide & Assessment	Chapter Test	TOTAL
1.5 days	1.5 days	1 day	1 day	14 days
(with Lesson 2-5B) 1.5 days	1.5 days		0.5 day	7 days

The worksheets shown on this page are a small sample of the materials available on *TeacherWorks: All-in-One Lesson Planner and Resource Center.*

This CD-ROM contains all of the blackline masters and transparencies available for this program.

It also includes a lesson planner and interactive Teacher Edition, so you can customize lesson plans and reproduce classroom resources quickly and easily, from just about anywhere.

Applications

Problem Solving and Applications Masters, pp. 3, 4

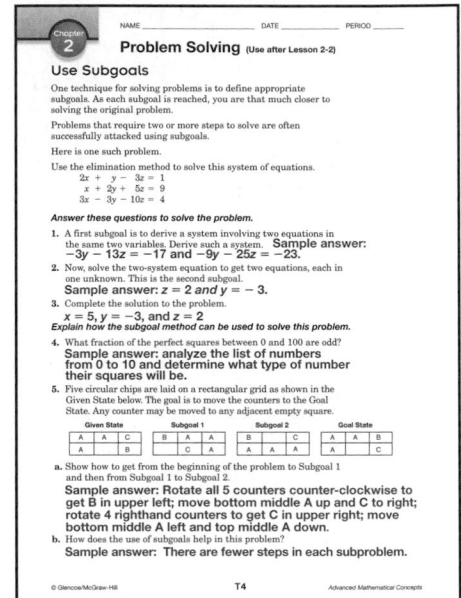

Technology/Multimedia

Spreadsheet Applications Masters, p. 2

TI-92 & CBL Interdisciplinary Activities Masters, pp. 1–4

Type	Student Edition	Teacher's Wraparound Edition	Assessment and Evaluation Masters
Ongoing Assessment	Mid-Chapter Quiz, p. 96	5-Minute Check, pp. 67, 73, 78, 88, 98, 107, 112	Mid-Chapter Test, p. 42 Quizzes A–D, pp. 43–44
Mixed Review	Mixed Review, pp. 72, 77, 86, 96, 104, 111, 118		Cumulative Review, p. 47
Error Analysis	You Decide, pp. 82, 109	Error Analysis, pp. 70, 76, 82, 92, 102, 109, 115	
Standardized Test Practice	SAT/ACT Practice, pp. 72, 77, 86, 96, 105, 111, 118 SAT/ACT Preparation, pp. 124–125	TIMSS, p.125	SAT/ACT Practice, pp. 45–46
Open-Ended Assessment	Math Journal, pp. 70, 92 Open-Ended Assessment, p. 123 Portfolio, p. 123	Modeling: pp. 87, 96, 105, 111 Speaking: pp. 77, 118 Writing: pp. 72, 86, 106	Open-Ended Assessment, p. 41
Chapter Assessment	Study Guide and Assessment, pp. 119–123 Chapter Test, p. A57		Multiple-Choice Tests (Forms 1A, 1B, 1C), pp. 29–34 Free-Response Tests (Forms 2A, 2B, 2C), pp. 35–40

TestCheck and Worksheet Builder is a networkable CD-ROM with three integrated modules. The **Worksheet Builder** creates customized worksheets, tests, and quizzes of free-response, multiple-choice, short-answer, and open-ended items. The **Student Module** gives you the option of having students take tests on-screen and get immediate feedback on their performance. Use the optional **Management System** to keep detailed student records.

Additional Chapter Resources

Student Edition
Career Choices, p. 105
History of Mathematics, p. 97
Internet Project, p. 123
Extra Practice, pp. A28–A29

Teacher's Classroom Resources

Technology
Graphing Calculator Keystrokes Masters

Teaching Aids
Answer Key Transparencies
Block Schedule Planning Guide
Lesson Planning Guide
Solutions Manual

Glencoe Instructional Technology

Instructional

Advanced MathPASS CD-ROM

Multimedia Applications CD-ROM

Assessment

TestCheck and Worksheet Builder

Visit **www.amc.glencoe.com** for data updates, career information, graphing calculator programs, and other interactive activities.

Chapter Overview

This chapter reviews the concepts of solving systems of linear equations and inequalities and operations with matrices.

 Using a Graphing Calculator

There are numerous opportunities, all optional, to use a graphing calculator in this chapter.

- Using the **TRACE** and **CALC** functions to estimate the solution for a system of equations (pp. 67, 69)
- Investigating the operational properties of matrices (p. 86)
- Graphing transformations of geometric figures (p. 87)
- Finding the determinant and inverse of a matrix (pp. 99, 100, 103)
- Solving systems of equations by using inverse matrices or reduced row-echelon form (pp. 100, 103, 106)

The **Graphing Calculator Appendix**, which begins on page A2, contains keystroke instructions on many of these topics.

SYSTEMS OF LINEAR EQUATIONS AND INEQUALITIES

CHAPTER OBJECTIVES ....●

- **Solve systems of equations and inequalities.**
  (Lessons 2-1, 2-2, 2-6)
- **Define matrices.** *(Lesson 2-3)*
- **Add, subtract, and multiply matrices.** *(Lesson 2-3)*
- **Use matrices to model transformations.** *(Lesson 2-4)*
- **Find determinants and inverses of matrices.** *(Lesson 2-5)*
- **Use linear programming to solve problems.** *(Lesson 2-7)*

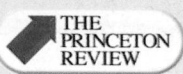 **SAT & ACT Preparation**

This textbook contains a complete SAT/ACT preparation course described on pages T8–T9 of the *Teacher's Handbook.* This chapter contains:

- an SAT or ACT practice question at the end of each lesson
- examples of SAT and ACT problems involving algebra problems on page 124
- ten SAT and ACT practice problems on page 125
- additional SAT and ACT practice in the *Assessment and Evaluation Masters,* pp. 45–46
- SAT and ACT test practice banks in the *TestCheck and Worksheet Builder*

2-1

Solving Systems of Equations in Two Variables

OBJECTIVES
- Solve systems of equations graphically.
- Solve systems of equations algebraically.

CONSUMER CHOICES Madison is thinking about leasing a car for two years. The dealership says that they will lease her the car she has chosen for $326 per month with only $200 down. However, if she pays $1600 down, the lease payment drops to $226 per month. What is the break-even point when comparing these lease options? Which 2-year lease should she choose if the down payment is not a problem? *This problem will be solved in Example 4.*

The *break-even point* is the point in time at which Madison has paid the same total amount on each lease. After finding that point, you can more easily determine which of these arrangements would be a better deal. The break-even point can be determined by solving a system of equations.

A **system of equations** is a set of two or more equations. To "solve" a system of equations means to find values for the variables in the equations, which make all the equations true at the same time. One way to solve a system of equations is by graphing. The intersection of the graphs represents the point at which the equations have the same x-value and the same y-value. Thus, this ordered pair represents the solution common to both equations. This ordered pair is called the **solution** to the system of equations.

Example **Solve the system of equations by graphing.**

$$3x - 2y = -6$$
$$x + y = -2$$

Graphing Calculator Tip

You can estimate the solution to a system of equations by using the **TRACE** function on your graphing calculator.

First rewrite each equation of the system in slope-intercept form by solving for y.

$$\begin{array}{c} 3x - 2y = -6 \\ x + y = -2 \end{array} \quad \text{becomes} \quad \begin{array}{c} y = \dfrac{3}{2}x + 3 \\ y = -x - 2 \end{array}$$

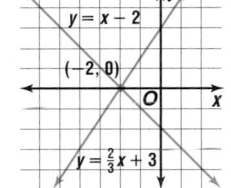

Since the two lines have different slopes, the graphs of the equations are intersecting lines. The solution to the system is $(-2, 0)$.

As you saw in Example 1, when the graphs of two equations intersect there is a solution to the system of equations. However, you may recall that the graphs of two equations may be parallel lines or, in fact, the same line. Each of these situations has a different type of system of linear equations.

A **consistent** system of equations has at least one solution. If there is exactly one solution, the system is **independent.** If there are infinitely many solutions, the system is **dependent.** If there is no solution, the system is **inconsistent.** By rewriting each equation of a system in slope-intercept form, you can more easily determine the type of system you have and what type of solution to expect.

Lesson 2-1 Solving Systems of Equations in Two Variables **67**

1 FOCUS

 5-Minute Check
Chapter 1

1. State the domain and range of the relation {(−2, 4), (1, 3), (−2, −2)}. Then state whether the relation is a function.
   {−2, 1}; {4, 3, −2}; no

2. Find $[f \circ g](x)$ and $[g \circ f](x)$ for $f(x) = 2x − 4$ and $g(x) = x^2$.
   $2x^2 − 4$; $4x^2 − 16x + 16$

3. Write the equation in slope-intercept form for the line with slope 6 and passing through the point at (−3, 4).
   $y = 6x + 22$

4. Graph $f(x) = |4 − x|$.

5. Graph $2x + y \le 7$.

Motivating the Lesson

Real-World Application Discuss *break-even point* as it relates to putting on a school dance. How many tickets must be sold to break even, or cover expenses? The expenses may include the cost of the food, the cost of hiring a disc jockey to play music, the cost of decoration, and so on.

In-Class Examples

Example 1
Solve the system of equations by graphing.
$y = 4x - 18$
$y = \dfrac{3}{4}x - 5$ **(4, –2)**

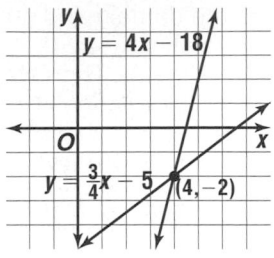

Example 2
Use the elimination method to solve the system of equations.
$5x + 2y = 340$
$3x - 4y = 360$ **(80, –30)**

Teaching Tip In Example 2, point out that either variable can be eliminated first when using the elimination method. If possible, choosing to eliminate a variable whose coefficient in one equation is a multiple or a negative multiple of its coefficient in the other equation makes elimination simpler.

Stress the importance of substituting the first value obtained from elimination into the original equations. The final solution should also be checked with the original equations.

The chart below summarizes the characteristics of these types of systems.

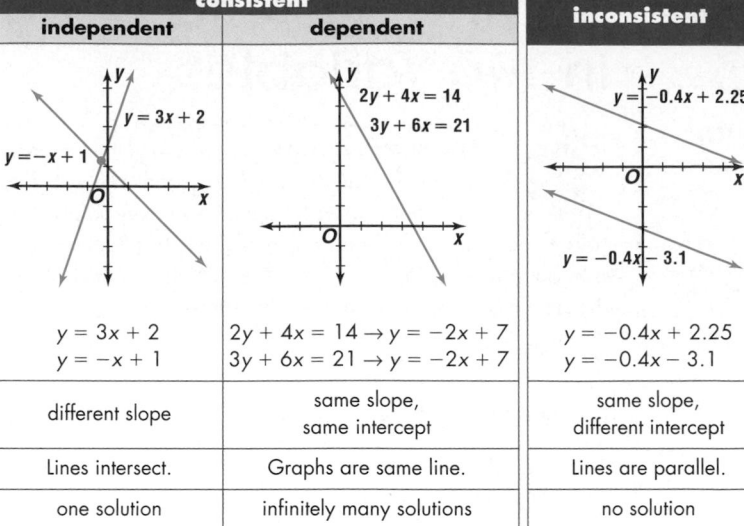

consistent		inconsistent
independent	**dependent**	
$y = 3x + 2$ $y = -x + 1$	$2y + 4x = 14 \rightarrow y = -2x + 7$ $3y + 6x = 21 \rightarrow y = -2x + 7$	$y = -0.4x + 2.25$ $y = -0.4x - 3.1$
different slope	same slope, same intercept	same slope, different intercept
Lines intersect.	Graphs are same line.	Lines are parallel.
one solution	infinitely many solutions	no solution

When graphs result in lines that are the same line, we say the lines coincide.

Often, graphing a system of equations is not the best method of finding its solution. This is especially true when the solution to the system contains non-integer values. Systems of linear equations can also be solved algebraically. Two common ways of solving systems algebraically are the **elimination method** and the **substitution method**. In some cases, one method may be easier to use than the other.

Example **Use the elimination method to solve the system of equations.**

$1.5x + 2y = 20$
$2.5x - 5y = -25$

One way to solve this system is to multiply both sides of the first equation by 5, multiply both sides of the second equation by 2, and add the two equations to eliminate *y*. Then solve the resulting equation.

$5(1.5x + 2y) = 5(20)$
$2(2.5x - 5y) = 2(-25)$

$\Rightarrow$

$\begin{array}{r} 7.5x + 10y = 100 \\ \underline{5x - 10y = -50} \\ 12.5x \quad\quad = 50 \\ x = 4 \end{array}$

Now substitute 4 for *x* in either of the *original* equations.

$1.5x + 2y = 20$
$1.5(4) + 2y = 20 \quad x = 4$
$2y = 14$
$y = 7$

The solution is (4, 7). *Check it by substituting into $2.5x - 5y = -25$. If the coordinates make both equations true, then the solution is correct*

If one of the equations contains a variable with a coefficient of 1, the system can often be solved more easily by using the substitution method.

Prerequisite Skills

A review of the prerequisite skills for this lesson is provided in the *Prerequisite Skills Masters*.

Skill Number and Content	Pages
1 Properties of Real Numbers	3–4
4 Algebraic Expressions	9–10
5 Solving Equations & Inequalities	11–12

Example **3** Use the substitution method to solve the system of equations.

$$2x + 3y = 8$$
$$x - y = 2$$

You can solve the second equation for either y or x. If you solve for x, the result is $x = y + 2$. Then substitute $y + 2$ for x in the first equation.

$$2x + 3y = 8$$
$$2(y + 2) + 3y = 8 \quad x = y + 2$$
$$5y = 4$$
$$y = \frac{4}{5}$$

Now substitute $\frac{4}{5}$ for y in either of the original equations, and solve for x.

$$x - y = 2$$
$$x - \frac{4}{5} = 2 \quad y = \frac{4}{5}$$
$$x = \frac{14}{5}$$

The solution is $\left(\frac{14}{5}, \frac{4}{5}\right)$.

GRAPHING CALCULATOR EXPLORATION

3. accurate to a maximum of 8 digits   4–5. See margin.

You can use a graphing calculator to find the solution to an independent system of equations.

- Graph the equations on the same screen.

- Use the **CALC** menu and select **5:intersect** to determine the coordinates of the point of intersection of the two graphs.

WHAT DO YOU THINK?

3. How accurate are solutions found on the calculator?

4. What type of system do the equations $5x - 7y = 70$ and $10x - 14y = 120$ form? What happens when you try to find the intersection point on the calculator?

5. Graph a system of dependent equations. Find the intersection point. Use the **TRACE** function to move the cursor and find the intersection point again. What pattern do you observe?

TRY THESE

Find the solution to each system.

1. $y = 500x - 20$
   $y = -20x + 500$
   (1, 480)

2. $3x - 4y = 320$
   $5x + 2y = 340$
   (76.923077, −22.30769)

You can use a system of equations to solve real-world problems. Choose the best method for solving the system of equations that models the situation.

Example **4** **CONSUMER CHOICES**   Refer to the application at the beginning of the lesson.

a. What is the break-even point in the two lease plans that Madison is considering?

b. If Madison keeps the lease for 24 months, which lease should she choose?

a. First, write an equation to represent the amount she will pay with each plan. Let C represent the total cost and m the number of months she has had the lease.

Lease 1 ($200 down with monthly payment of $326): $C = 326m + 200$

Lease 2 ($1600 down with monthly payment of $226): $C = 226m + 1600$

Now, solve the system of equations. Since both equations contain C, we can substitute the value of C from one equation into the other.

(continued on the next page)

Addressing Individual Needs

Interpersonal Have pairs of students compare tables of values that they create for a system of equations. Lead them to develop a set of rules for determining the solution set.
- **If an ordered pair is on both lists, it is the solution.**
- **If the slopes are equal, there is no solution.**
- **If for x_1 and x_2, where the y-value for x_1 of one equation is greater than the y-value for x_1 of the other equation and the y-value for x_2 of the first equation is less than the y-value for x_2 of the second equation, a solution exists between x_1 and x_2.**

Error Analysis

Watch for students who confuse the terms inconsistent and dependent.

Prevent by reminding students that the system is inconsistent if there are no ordered pairs that satisfy both equations. In that case, no points are common to the graphs of the lines.

Additional Answers

1. **Sample answer:** $4x + 7y = 21$, $y = 2x - 1$. The substitution method is usually easier to use whenever one or both of the equations are already solved for one variable in terms of the other.

2. **Sample answer:** Madison might consider whether the large down-payment would strap her financially; if she wants to buy the car at the end of the lease, then she might also consider which lease would offer the best buyout.

Study Guide and Practice Masters, pp. 21–22

$$C = 326m + 200$$
$$226m + 1600 = 326m + 200 \quad C = 226m + 1600$$
$$1400 = 100m$$
$$14 = m$$

With the fourteenth monthly payment, she reaches the break-even point.

b. The graph of the equations shows that after that point, Lease 1 is more expensive for the 2-year lease. So, Madison should probably choose Lease 2.

Communicating Mathematics

Read and study the lesson to answer each question.   1–3. See margin.

1. **Write** a system of equations in which it is easier to use the substitution method to solve the system rather than the elimination method. Explain your choice.

2. Refer to the application at the beginning of the lesson. **Explain** what factors Madison might consider before making a decision on which lease to select.

3. *Math Journal* **Write** a description of the three different possibilities that may occur when graphing a system of two linear equations. Include examples and solutions that occur with each possibility.

Guided Practice

4. State whether the system $2y + 3x = 6$ and $4y = 16 - 6x$ is *consistent and independent, consistent and dependent,* or *inconsistent.* Explain your reasoning. **Inconsistent; see margin for explanation.**

Solve each system of equations by graphing.   **See margin for graphs.**

5. $y = 5x - 2$
   $y = -2x + 5$ **(1, 3)**

6. $x - y = 2$
   $2x = 2y + 10$ **no solution**

Solve each system of equations algebraically.

7. $7x + y = 9$
   $5x - y = 15$ **(2, −5)**

8. $3x + 4y = -1$
   $6x - 2y = 3$ $\left(\dfrac{1}{3}, -\dfrac{1}{2}\right)$

9. $\dfrac{1}{3}x - \dfrac{3}{2}y = -4$
   $5x - 4y = 14$ **(6, 4)**

10. **Sales**   HomePride manufactures solid oak racks for displaying baseball equipment and karate belts. They usually sell six times as many baseball racks as karate-belt racks. The net profit is $3 from each baseball rack and $5 from each karate-belt rack. If the company wants a total profit of $46,000, how many of each type of rack should they sell?   **12,000 baseball, 2000 karate**

 www.amc.glencoe.com/self_check_quiz

Additional Answers

3. **Sample answer:** consistent systems of equations have at least one solution. A consistent, independent system has exactly one solution; a consistent, dependent system has an infinite number of solutions. An inconsistent system has no solution. See students' work for examples and solutions.

4. **Sample answer:** The graphs of the equations are lines with slope $-\dfrac{3}{2}$, but each equation has a different y-intercept. Therefore, the graphs of the two equations do not intersect and the system has no solution.

Practice

State whether each system is *consistent and independent*, *consistent and dependent*, or *inconsistent*.

A 11. $x + 3y = 18$
$-x + 2y = 7$
consistent and independent

12. $y = 0.5x$
$2y = x + 4$
inconsistent

13. $-35x + 40y = 55$
$7x = 8y - 11$
consistent and dependent

Solve each system of equations by graphing. **See margin for graphs.**

14. $x = 5$
$4x + 5y = 20$ **(5, 0)**

15. $y = -3$
$2x = 8$ **(4, −3)**

16. $x + y = -2$
$3x - y = 10$ **(2, −4)**

B 17. $x + 3y = 0$
$2x + 6y = 5$ **no solution**

18. $y = x - 2$
$x - 2y = 4$ **(0, −2)**

19. $3x - 2y = -6$
$x = 12 - 4y$ **(0, 3)**

20. Determine what type of solution you would expect from the system of equations $3x - 8y = 10$ and $16x - 32y = 75$ without graphing the system. Explain how you determined your answer. **See margin.**

Solve each system of equations algebraically.

21. $5x - y = 16$
$2x + 3y = 3$ **(3, −1)**

22. $3x - 5y = -8$
$x + 2y = 1$ **(−1, 1)**

23. $y = 6 - x$
$x = 4.5 + y$ **(5.25, 0.75)**

24. $2x + 3y = 3$
$12x - 15y = -4$ $\left(\dfrac{1}{2}, \dfrac{2}{3}\right)$

25. $-3x + 10y = 5$
$2x + 7y = 24$ **(5, 2)**

26. $x = 2y - 8$
$2x - y = -7$ **(−2, 3)**

C 27. $2x + 5y = 4$
$3x + 6y = 5$ $\left(\dfrac{1}{3}, \dfrac{2}{3}\right)$

28. $\dfrac{3}{5}x - \dfrac{1}{6}y = 1$
$\dfrac{1}{5}x + \dfrac{5}{6}y = 11$ **(5, 12)**

29. $4x + 5y = -8$
$3x - 7y = 10$ $\left(-\dfrac{6}{43}, -\dfrac{64}{43}\right)$

30. (−5, −6)

30. Find the solution to the system of equations $3x - y = -9$ and $4x - 2y = -8$.

31. Explain which method seems most efficient to solve the system of equations $a - b = 0$ and $3a + 2b = -15$. Then solve the system. **See margin.**

Applications and Problem Solving

32. **Sports** Spartan Stadium at San Jose State University in California has a seating capacity of about 30,000. A newspaper article states that the Spartans get four times as many tickets as the visiting team. Suppose S represents the number of tickets for the Spartans and V represents the number of tickets for the visiting team's fans.

a. Which system could be used by a newspaper reader to determine how many tickets each team gets? **B**

A $4S + 4V = 30,000$
$S = 4V$

B $S - 4V = 0$
$S + V = 30,000$

C $S + V = 30,000$
$V - 4S = 0$

32b. Spartans: 24,000; visitors: 6000

b. Solve the system to find how many tickets each team gets.

33. **Geometry** Two triangles have the same perimeter of 20 units. One triangle is an isosceles triangle. The other triangle has a side 6 units long. Its other two sides are the same lengths as the base and leg of the isosceles triangle.

a. What are the dimensions of each triangle? **6, 6, 8; 6, 6, 8**
b. What type of triangle is the second triangle? **isosceles**

Additional Answers (p. 70)

5.

6.

Assignment Guide

Core: 11–31 odd, 33–37, 39–45
Enriched: 12–30 even, 32–45

Additional Answers

14.

15.

16.

17.

18.

19.

20. **Consistent and independent; if each equation is written in slope-intercept form, they have different slopes, which means that they will intersect at some point.**

(continued on page 72)

Writing Have students write a brief paragraph discussing the different possibilities for the graphs of two linear equations in two variables.

Additional Answers
(pp. 71–72)

31. Sample answer: elimination could be considered easiest since the first equation multiplied by 2 added to the second equation eliminates *b*; substitution could also be considered easiest since the first equation can be written as *a* = *b*, making substitution very easy; (–3, –3).

35b. If you drink 7 servings of soft drink, the price for each option is the same. If you drink fewer than 7 servings of soft drink during that week, the disposable cup price is best. If you drink more than 7 servings of soft drink, the refillable mug price is best. See students' choices.

35c. Over a year's time, the refillable mug would be more economical.

Enrichment Masters, p. 13

NAME _____ DATE _____ PERIOD _____

2-1 **Enrichment**

Set Theory and Venn Diagrams

Set theory, which was developed by the nineteenth century German logician and mathematician, Georg Cantor, became a fundamental unifying principle in the study of mathematics in the middle of the twentieth century. The use of sets permits the precise description of mathematical concepts.

The intersection of two sets determines the elements common to the two sets. Thus, the intersection of two lines in a system of equations refers to the point or points that are common to each of the lines belonging to each of the lines. We can use Venn diagrams, which are named for the British logician, John Venn, to visually represent the intersection of two sets.

Example Let *U* = the set of all points in the Cartesian coordinate plane.
Let *A* = the set of all points that satisfy the equation *x* = 4.
Let *B* = the set of all points that satisfy the equation *y* = −2.
Draw a Venn diagram to represent *U*, *A*, and *B*.

The shaded region represents the intersection of sets *A* and *B*, written *A* ∩ *B*. In this example, *A* ∩ *B* = {(4, −2)}.

Since the solution sets of two parallel lines have no points in common, the sets are called **disjoint sets**. In a Venn diagram, such sets are drawn as circles that do not overlap.

A ∩ *B* = ∅.

Use the diagram at the right to answer the following questions.
Let *A* = the set of all points that satisfy the equation for line *p*.
Let *B* = the set of all points that satisfy the equation for line *q*.

1. In which numbered region do the points that satisfy only the equation for line *p* lie? **1**
2. In which numbered region do the points that satisfy only the equation for line *q* lie? **3**
3. In which numbered region do the points that satisfy the equations for neither line *p* nor line *q* lie? **4**
4. If the equation of line *p* is 2*x* − *y* = 4 and the equation of line *q* is 3*x* + *y* = 6, which point lies in region 2? **(2,0)**

© Glencoe/McGraw-Hill   T13   *Advanced Mathematical Concepts*

34.
$y = -\frac{1}{4}x - 2,$
$y = 4x - 19$

34. Critical Thinking The solution to a system of two linear equations is (4, −3). One equation has a slope of 4. The slope of the other line is the negative reciprocal of the slope of the first. Find the system of equations.

35. Business The first Earth Day was observed on April 22, 1970. Since then, the week of April 22 has been Earth Week, a time for showing support for environmental causes. Fans Café is offering a reduced refill rate for soft drinks during Earth Week for anyone purchasing a Fans mug. The mug costs $2.95 filled with 16 ounces of soft drink. The refill price is 50¢. A 16-ounce drink in a disposable cup costs $0.85.
  a. What is the approximate break-even point for buying the mug and refills in comparison to buying soft drinks in disposable cups? **(7, 5.95)**
  b. What does this mean? Which offer do you think is best? **See margin.**
  c. How would your decision change if the refillable mug offer was extended for a year? **See margin.**

36. Critical Thinking Determine what must be true of *a, b, c, d, e,* and *f* for the system $ax + by = c$ and $dx + ey = f$ to fit each description.
  a. consistent and independent $\frac{a}{b} \neq \frac{d}{e}$
  b. consistent and dependent $\frac{a}{b} = \frac{d}{e}, c = f$
  c. inconsistent $\frac{a}{b} = \frac{d}{e}, c \neq f$

37. Incentive Plans As an incentive plan, a company stated that employees who worked for four years with the company would receive $516 and a laptop computer. Mr. Rodriquez worked for the company for 3.5 years. The company pro-rated the incentive plan, and he still received the laptop computer, but only $264. What was the value of the laptop computer? **$1500**

38. Ticket Sales In November 1994, the first live concert on the Internet by a major rock'n'roll band was broadcast. Most fans stand in lines for hours to get tickets for concerts. Suppose you are in line for tickets. There are 200 more people ahead of you than behind you in line. The whole line is three times the number of people behind you. How many people are in line for concert tickets?
603 people

Mixed Review

39. Graph $-2x + 7 \geq y$. *(Lesson 1-8)*   **See Answer Appendix.**

40. Graph $f(x) = 2|x| - 3$. *(Lesson 1-7)*   **See Answer Appendix.**

41. Write an equation of the line parallel to the graph of $y = 2x + 5$ that passes through the point at (0, 6). *(Lesson 1-5)* $y = 2x + 6$

42. $12,500
42. Manufacturing The graph shows the operational expenses for a bicycle shop during its first four years of business. How much was the startup cost of the business? *(Lesson 1-3)*

Tru–Ride Bicycle Shop

43. Find $[f \circ g](x)$ if $f(x) = 3x - 5$ and $g(x) = x + 2$. *(Lesson 1-2)*   **3x + 1**

44. {18}, {−3, 3}; no, because there are two range values paired with a single domain value.
44. State the domain and range of the relation {(18, −3), (18, 3)}. Is this relation a function? Explain. *(Lesson 1-1)*

45. SAT/ACT Practice $\sqrt{\frac{\sqrt{25}}{5}} = $ **A**
  A 1   B $\sqrt{2}$ C 2   D 5   E $5\sqrt{2}$

Extra Practice   See p. A28.

Extra Credit

Solve the system of linear equations.
$x + y + z = 5$
$-x + y - 2z = -5$
$3x - 2y + z = -3$
(−1, 2, 4)

Solving Systems of Equations in Three Variables

OBJECTIVE
- Solve systems of equations involving three variables algebraically.

 Real World Application

SPORTS In 1998, Cynthia Cooper of the WNBA Houston Comets basketball team was named Team Sportswoman of the Year by the Women's Sports Foundation. Cooper scored 680 points in the 1998 season by hitting 413 of her 1-point, 2-point, and 3-point attempts. She made 40% of her 160 3-point field goal attempts. How many 1-, 2-, and 3-point baskets did Ms. Cooper complete? *This problem will be solved in Example 3.*

You will learn more about graphing in three-dimensional space in Chapter 8.

This situation can be described by a system of three equations in three variables. You used graphing to solve a system of equations in two variables. For a system of equations in three variables, the graph of each equation is a plane in space rather than a line. The three planes can appear in various configurations. This makes solving a system of equations in three variables by graphing rather difficult. However, the pictorial representations provide information about the types of solutions that are possible. Some of them are shown below.

Systems of Equations in Three Variables		
Unique Solution	**Infinite Solutions**	**No Solution**
(x, y, z)		
The three planes intersect at one point.	The three planes intersect in a line.	The three planes have no points in common.

You can solve systems of three equations more efficiently than graphing by using the same algebraic techniques that you used to solve systems of two equations.

Example ❶ Solve the system of equations by elimination.

$x - 2y + z = 15$
$2x + 3y - 3z = 1$
$4x + 10y - 5z = -3$

One way to solve a system of three equations is to choose pairs of equations and then eliminate one of the variables. Because the coefficient of x is 1 in the first equation, it is a good choice for eliminating x from the second and third equations. *(continued on the next page)*

(continued on the next page)

---

1 FOCUS

 5-Minute Check
Lesson 2-1

1. Solve the system of equations by graphing.
$2x + 3y + 3 = 0$
$3x + y = -8$ **(–3, 1)**

Solve each system of equations algebraically.

2. $y = -2x + 5$
$y = -\frac{5}{4}x + 2$ **(4, –3)**

3. $y - 2x + 4$
$3x + 4y = 10$ $\left(-\frac{6}{11}, \frac{32}{11}\right)$

4. $3x - 2y = -10.5$
$x + 4y = 38.5$
(2.5, 9)

5. Sweats Unlimited sells sweatshirts for $28 and sweatpants for $18. The daily revenue of the company is $21,200, and the daily fixed cost is $9800. The cost of producing the shirts is $18 each, and the cost for the pants is $6 each. Find the break-even point for the daily sales. **500 sweatshirts and 400 sweatpants**

Motivating the Lesson
Real-World Application Discuss situations in which there are three unknowns in an equation instead of two. For example, a company may sell three products, or a manager may have three employees to schedule. How would you solve a system of equations to represent these situations?

 Resource Manager

 Reproducible Masters
- *Study Guide and Practice,* pp. 23–24
- *Enrichment,* p. 14
- *Assessment and Evaluation,* p. 43
- *Problem Solving and Applications,* p. 4

 Transparencies
- 5-Minute Check, 2-2
- Teaching, 2-2

Teaching Tip In Example 1, point out that an alternate method of solution could be to eliminate z using the first and second and the first and third equations. Ask students whether the solution is the same using either method.

In-Class Examples

Example 1
Solve the system of equations by elimination.
$5x - 2y + z = 11$
$2x + y + 3z = 0$
$6x - 2y - 2z = 16$ **(2, –1, –1)**

Example 2
Solve the system of equations by substitution.
$3y = -9z$
$4x + 2y - 2z = 0$
$-3x - y + 4z = -2$ **(–4, 6, –2)**

Teaching Tip In Example 2, point out that the equation $4x = -8z$ could also be solved for z as $z = -\frac{1}{2}x$. Then, $-\frac{1}{2}x$ would be substituted for z in each of the other two equations.

To eliminate x using the first and second equations, multiply each side of the first equation by -2.

$$-2(x - 2y + z) = -2(15)$$
$$-2x + 4y - 2z = -30$$

Then add that result to the second equation.

$$\begin{array}{r} -2x + 4y - 2z = -30 \\ 2x + 3y - 3z = 1 \\ \hline 7y - 5z = -29 \end{array}$$

To eliminate x using the first and third equations, multiply each side of the first equation by -4.

$$-4(x - 2y + z) = -4(15)$$
$$-4x + 8y - 4z = -60$$

Then add that result to the third equation.

$$\begin{array}{r} -4x + 8y - 4z = -60 \\ 4x + 10y - 5z = -3 \\ \hline 18y - 9z = -63 \end{array}$$

Now you have two linear equations in two variables. Solve this system. Eliminate z by multiplying each side of the first equation by -9 and each side of the second equation by 5. Then add the two equations.

$$\begin{array}{l} -9(7y - 5z) = -9(-29) \\ 5(18y - 9z) = 5(-63) \end{array} \Rightarrow \begin{array}{r} -63y + 45z = 261 \\ 90y - 45z = -315 \\ \hline 27y = -54 \\ y = -2 \quad \text{The value of } y \text{ is } -2. \end{array}$$

By substituting the value of y into one of the equations in two variables, we can solve for the value of z.

$$\begin{array}{l} 7y - 5z = -29 \\ 7(-2) - 5z = -29 \quad y = -2 \\ z = 3 \quad \text{The value of } z \text{ is 3.} \end{array}$$

Finally, use one of the original equations to find the value of x.

$$\begin{array}{l} x - 2y + z = 15 \\ x - 2(-2) + 3 = 15 \quad y = -2, z = 3 \\ x = 8 \quad \text{The value of } x \text{ is 8.} \end{array}$$

The solution is $x = 8$, $y = -2$, and $z = 3$. This can be written as the **ordered triple** $(8, -2, 3)$. *Check by substituting the values into each of the original equations.*

The substitution method of solving systems of equations also works with systems of three equations.

Example **2** **Solve the system of equations by substitution.**

$4x = -8z$
$3x - 2y + z = 0$
$-2x + y - z = -1$

You can easily solve the first equation for x.

$4x = -8z$
$x = -2z$ *Divide each side by 4.*

Then substitute $-2z$ for x in each of the other two equations. Simplify each equation.

$$3x - 2y + z = 0$$
$$3(-2z) - 2y + z = 0 \quad x = -2z$$
$$-2y - 5z = 0$$

$$-2x + y - z = -1$$
$$-2(-2z) + y - z = -1 \quad x = -2z$$
$$y + 3z = -1$$

Solve $y + 3z = -1$ for y. $\quad y + 3z = -1$
$$y = -1 - 3z \quad \textit{Subtract 3z from each side.}$$

Substitute $-1 - 3z$ for y in $-2y - 5z = 0$. Simplify.
$$-2y - 5z = 0$$
$$-2(-1 - 3z) - 5z = 0 \quad y = -1 - 3z$$
$$z = -2$$

Now, find the values of y and x. Use $y = -1 - 3z$ and $x = -2z$. Replace z with -2.

$$y = -1 - 3z$$
$$y = -1 - 3(-2) \quad z = -2$$
$$y = 5$$

$$x = -2z$$
$$x = -2(-2) \quad z = -2$$
$$x = 4$$

The solution is $x = 4$, $y = 5$, *and* $z = -2$. *Check each value in the original system.*

Many real-world situations can be represented by systems of three equations.

Example ③ **SPORTS** Refer to the application at the beginning of the lesson. Find the number of 1point free throws, 2-point field goals, and 3-point field goals Cynthia Cooper scored in the 1998 season.

Write a system of equations. Define the variables as follows.

x = the number of 1-point free throws
y = the number of 2-point field goals
z = the number of 3-point field goals

The system is:

$x + 2y + 3z = 680$ *total number of points*

$x + y + z = 413$ *total number of baskets*

$\dfrac{z}{160} = 0.40$ *percent completion*

The third equation is a simple linear equation. Solve for z.
$\dfrac{z}{160} = 0.40$, so $z = 160(0.40)$ or 64.

Now substitute 64 for z to make a system of two equations.

$$x + 2y + 3z = 680$$
$$x + 2y + 3(64) = 680 \quad z = 64$$
$$x + 2y = 488$$

$$x + y + z = 413$$
$$x + y + 64 = 413 \quad z = 64$$
$$x + y = 349$$

Solve $x + y = 349$ for x. Then substitute that value for x in $x + 2y = 488$ and solve for y.

$$x + y = 349$$
$$x = 349 - y$$

$$x + 2y = 488$$
$$(349 - y) + 2y = 488 \quad x = 349 - y$$
$$y = 139$$

(continued on the next page)

Teaching Tip While working Example 3, encourage students to check their solutions in all *three* equations to verify accuracy. It is possible that a solution may work for one or two equations, but not all three.

In-Class Example
Example 3
Sports In the 1998 season, Cynthia Cooper's teammate Sheryl Swoopes made 83% of her 86 attempted free throws. She made 244 of her 1-point, 2-point, and 3-point attempts, resulting in 453 points. Find the number of 1-point free throws, 2-point field goals, and 3-point field goals Swoopes made in the 1998 season.
71 1-point free throws, 137 2-point field goals, 36 3-point field goals

---

Addressing Individual Needs

 Intrapersonal Have students identify and evaluate different methods for solving a system of equations in three variables. For example, the first step to solving the system below could be to eliminate x, to eliminate y, or to eliminate z. You could also substitute an expression for x, y, or z. What are the advantages of solving each way? Which method would you choose and why?

$$x + 3y - 7z = 14$$
$$5x - 3y + 2z = -3$$
$$-2x + 4y - z = 7 \quad \textbf{(1, 2, -1)}$$

Error Analysis

Watch for students who eliminate different variables from equations.

Prevent by stressing that the goal is to reduce the system to a system of two equations in two variables. Students need to combine two pairs of equations to eliminate the same variable.

Assignment Guide

Core: 9–17 odd, 19–28
Enriched: 8–18 even, 19–28

Additional Answers

1. Solving a system of three equations involves eliminating one variable to form two systems of two equations. Then solving is the same.

2. The solution would be an equation in two variables. Sample example: the system
$2x + 4y + 6z = 12$,
$x + 2y + 3z = 6$, and
$3x - 5y - 6z = 27$ has a solution of all values of x and y that satisfy $5x - y = 39$.

Study Guide and Practice Masters, pp. 23–24

NAME _____ DATE _____ PERIOD _____

Study Guide

Solving Systems of Equations in Three Variables

You can solve systems of three equations in three variables using the same algebraic techniques you used to solve systems of two equations.

Example Solve the system of equations by elimination.

$4x + y + 2z = -8$
$2x - 3y - 4z = 4$
$-7x + 2y + 3z = 2$

Choose a pair of equations and then eliminate one of the variables. Because the coefficient of y is 1 in the first equation, eliminate y from the second and third equations.

To eliminate y using the first and second equations, multiply both sides of the first equation by 3.

$3(4x + y + 2z) = 3(-8)$
$12x + 3y + 6z = -24$

Then add the second equation to that result.

$\begin{array}{r} 12x + 3y + 6z = -24 \\ 2x - 3y - 4z = 4 \\ \hline 14x + 2z = -20 \end{array}$

To eliminate y using the first and third equations, multiply both sides of the first equation by -2.

$-2(4x + y + 2z) = -2(-8)$
$-8x - 2y - 4z = 16$

Then add the third equation to that result.

$\begin{array}{r} -8x - 2y - 4z = 16 \\ -7x + 2y + 3z = 2 \\ \hline -15x - z = 18 \end{array}$

Now you have two linear equations in two variables. Solve this system. Eliminate z by multiplying both sides of the second equation by 2. Then add the two equations.

$2(-15x - z) = 2(18)$ $14x + 2z = -20$
$-30x - 2z = 36$ $-30x - 2z = 36$
                        $\overline{-16x = 16}$
                        $x = -1$ The value of x is -1.

By substituting the value of x into one of the equations in two variables, we can solve for the value of z.

$14x + 2z = -20$
$14(-1) + 2z = -20$ $x = -1$
$z = -3$

The value of z is -3.

Finally, use one of the three original equations to find the value of y.

$4x + y + 2z = -8$
$4(-1) + y + 2(-3) = -8$ $x = -1, z = -3$
$y = 2$

The value of y is 2.

The solution is $x = -1, y = 2, z = -3$. This can be written as the **ordered triple** $(-1, 2, -3)$.

© Glencoe/McGraw-Hill   23   *Advanced Mathematical Concepts*

Solve for x.
$x = 349 - y$
$x = 349 - 139$ $y = 139$
$x = 210$

In 1998, Ms. Cooper made 210 1-point free throws, 139 2-point field goals, and 64 3-point field goals. *Check your answer in the original problem.*

CHECK FOR UNDERSTANDING

Communicating Mathematics

Read and study the lesson to answer each question. **1–3. See margin.**

1. **Compare and contrast** solving a system of three equations to solving a system of two equations.

2. **Describe** what you think would happen if two of the three equations in a system were consistent and dependent. Give an example.

3. **Write** an explanation of how to solve a system of three equations using the elimination method.

Guided Practice

Solve each system of equations.

4. $4x + 2y + z = 7$
   $2x + 2y - 4z = -4$
   $x + 3y - 2z = -8$
   $(3, -3, 1)$

5. $x - y - z = 7$
   $-x + 2y - 3z = -12$
   $3x - 2y + 7z = 30$
   $(7, -1, 1)$

6. $2x - 2y + 3z = 6$
   $2x - 3y + 7z = -1$
   $4x - 3y + 2z = 0$
   no solution

7. **Physics** The height of an object that is thrown upward with a constant acceleration of a feet per second per second is given by the equation $s = \frac{1}{2}at^2 + v_0t + s_0$. The height is s feet, t represents the time in seconds, v_0 is the initial velocity in feet per second, and s_0 is the initial height in feet. Find the acceleration, the initial velocity, and the initial height if the height at 1 second is 75 feet, the height at 2.5 seconds is 75 feet, and the height at 4 seconds is 3 feet.
acceleration: -32 ft/s², initial velocity: 56 ft/s, initial height: 35 ft

---

EXERCISES

Practice

Solve each system of equations.

8. (1, −4, 4) **A**

8. $x + 2y + 3z = 5$
   $3x + 2y - 2z = -13$
   $5x + 3y - z = -11$

9. (−2, 2, 4)

9. $7x + 5y + z = 0$
   $-x + 3y + 2z = 16$
   $x - 6y - z = -18$

10. (10, −7, 1)

10. $x - 3z = 7$
    $2x + y - 2z = 11$
    $-x - 2y + 9z = 13$

B

11. (−6, −4, 7)

11. $3x - 5y + z = 9$
    $x - 3y - 2z = -8$
    $5x - 6y + 3z = 15$

12. (2, −7, 12)

12. $8x - z = 4$
    $y + z = 5$
    $11x + y = 15$

13. no solution

13. $4x - 3y + 2z = 12$
    $x + y - z = 3$
    $-2x - 2y + 2z = 5$

17. (−4, 10, 7)

14. $36x - 15y + 50z = -10$
    $2x + 25y = 40$
    $54x - 5y + 30z = -160$
    $(-5, 2, 4)$

15. $-x - 3y + z = 54$
    $4x + 2y - 3z = -32$
    $2y + 8z = 78$
    $(11, -17, 14)$

16. $1.8x - z = 0.7$
    $1.2y + z = -0.7$
    $1.5x - 3y = 3$
    $(0.5, -0.75, 0.2)$

C 17. If possible, find the solution of $y = x + 2z$, $z = -1 - 2x$, and $x = y - 14$.

18. What is the solution of $\frac{1}{8}x - \frac{2}{3}y + \frac{5}{6}z = -8$, $\frac{3}{4}x + \frac{1}{6}y - \frac{1}{3}z = -12$, and $\frac{3}{16}x - \frac{5}{8}y - \frac{7}{12}z = -25$? If there is no solution, write *impossible*. **(−16, 24, 12)**

76 *Chapter 2 Systems of Linear Equations and Inequalities*   www.amc.glencoe.com/self_check_quiz

---

Extra Credit

A parabola with equation $y = ax^2 + bx + c$ passes through points at $(-2, -2)$, $(1, 1)$, and $(-3, -1)$. Use a system of three linear equations in three variables to find the values of a, b, and c. **$a = 1, b = 2, c = -2$**

Additional Answer

3. Sample answer: use one equation to eliminate one of the variables from the other two equations. Then eliminate one of the remaining variables from the resulting equations. Solve for a variable and substitute to find the values of the other variables.

Applications and Problem Solving

19. Finance Ana Colón asks her broker to divide her 401K investment of $2000 among the International Fund, the Fixed Assets Fund, and company stock. She decides that her investment in the International Fund should be twice her investment in company stock. During the first quarter, the International Fund earns 4.5%, the Fixed Assets Fund earns 2.6%, and the company stock falls 0.2%. At the end of the first quarter, Ms. Colón receives a statement indicating a return of $58 on her investment. How did the broker divide Ms. Colón's initial investment?

19. International Fund = $1200; Fixed Assets Fund = $200; company stock = $600

20. Critical Thinking Write a system of equations that fits each description.
a. The system has a solution of $x = -5$, $y = 9$, $z = 11$. **a–c. See margin.**
b. There is no solution to the system.
c. The system has an infinite number of solutions.

21. Physics Each year the Punkin' Chunkin' contest is held in Lewes, Delaware. The object of the contest is to propel an 8- to 10-pound pumpkin as far as possible. Steve Young of Hopewell, Illinois, set the 1998 record of 4026.32 feet. Suppose you build a machine that fires the pumpkin so that it is at a height of 124 feet after 1 second, the height at 3 seconds is 272 feet, and the height at 8 seconds is 82 feet. Refer to the formula in Exercise 7 to find the acceleration, the initial velocity, and the initial height of the pumpkin. **(−32, 138, 2)**

22. Critical Thinking Suppose you are using elimination to solve a system of equations. **a–b. See margin.**
a. How do you know that a system has no solution?
b. How do you know when it has an infinite number of solutions?

23. Number Theory Find all of the ordered triples (x, y, z) such that when any one of the numbers is added to the product of the other two, the result is 2. **(1, 1, 1), (−2, −2, −2)**

24. (45, 60)

Mixed Review

24. Solve the system of equations, $3x + 4y = 375$ and $5x + 2y = 345$. *(Lesson 2-1)*

25. Graph $y \leq -\frac{1}{3}x + 2$. *(Lesson 1-8)* **See margin.**

26. Show that points with coordinates $(-1, 3)$, $(3, 6)$, $(6, 2)$, and $(2, -1)$ are the vertices of a square. *(Lesson 1-5)* **See margin.**

27b. $2000, $50
27c. See Answer Appendix.

27. Manufacturing It costs ABC Corporation $3000 to produce 20 of a particular model of color television and $5000 to produce 60 of that model. *(Lesson 1-4)*
a. Write an equation to represent the cost function. $C(x) = 50x + 2000$
b. Determine the fixed cost and variable cost per unit.
c. Sketch the graph of the cost function.

28. SAT/ACT Practice In the figure, the area of square $OXYZ$ is 2. What is the area of the circle? **C**

A $\frac{\pi}{4}$ B $\pi\sqrt{2}$ C 2π

D 4π E 8π

Extra Practice See p. A28.

Open-Ended Assessment
Speaking Have students list the steps they would follow in order to solve a system of three linear equations in three variables.

Chapter 2, Quiz A (Lessons 2-1 and 2-2) is available in the *Assessment and Evaluation Masters,* p. 43.

Additional Answers
20a. Sample answer:
$x + y + z = 15$; $2x + z = 1$; $2y - z = 7$
20b. Sample answer:
$4x + y + z = 12$;
$-4x - y - z = -10$;
$5y - z = 9$
20c. Sample answer:
$x + y + z = 6$;
$2x - y - 2z = 8$;
$x - 2y - 3z = 2$
22a. Sample answer: a system has no solution when you reach a contradiction, such as $1 = 0$, as you solve.
22b. Sample answer: a system has an infinite number of solutions when you reduce the system to two equivalent equations such as $x + y = 1$ and $2x + 2y = 2$.

Enrichment Masters, p. 14

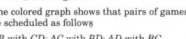

2-2 Enrichment

Graph Coloring

The student council is scheduling a volleyball tournament for teams from all four classes. They want each class team to play every other class team exactly once. How should they schedule the tournament?

If we call the teams A, B, C, and D, all of the possible games among the four teams can be represented as AB, AC, AD, BC, BD, and CD.

Draw a graph to represent the problem. Then color the graph. To **color** a graph means to color the vertices of that graph so that no two vertices connected by an edge have the same color.

Let the vertices represent the possible games. Let the edges represent games that cannot be scheduled at the same time. For example, if team B is playing team C, then team C cannot play team D.

Choose a vertex at which to begin.
- Color AB red. Since CD is not connected to AB, color it red as well. All of the other vertices are connected to AB, so do not color them red.
- Color AC blue. Since BD is not connected to AC, color it blue.
- Color AD green. Since BC is not connected to AD, color it green.

The colored graph shows that pairs of games can be scheduled as follows
AB with CD; AC with BD; AD with BC

The **chromatic number** of a graph is the least number of colors necessary to color the graph. The chromatic number of the graph above is 3.

1. What does the chromatic number of the graph in the example above represent? **the number of game times scheduled**

2. Draw and color a graph to represent the same type of tournament, but with 6 teams playing. **See students' graphs.**

3. What is the chromatic number for your graph? **5**

© Glencoe/McGraw-Hill   T14   *Advanced Mathematical Concepts*

Lesson 2-2 **77**

Additional Answers

25.

$y \leq -\frac{1}{3}x + 2$

26. $AB = BC = CD = AD = 5$ units; $ABCD$ is a rhombus.
Slope of $\overline{AB} = -\frac{4}{3}$ and slope of $\overline{BC} = \frac{3}{4}$, so $\overline{AB} \perp \overline{BC}$.
A rhombus with a right angle is a square.

Modeling Real-World Data with Matrices

1 FOCUS

5-Minute Check
Lesson 2-2

Solve each system of equations algebraically.

1. $x + y = -1$
   $2x - y = -2$
   $4x + y - z = 8$ **(–1, 0, –12)**

2. $3x - 5y + 9z = 20$
   $4x - y + 2z = 18$
   $2x + 2y + 3z = 18$ **(4, 2, 2)**

3. $6x - 2y - 4z = -2$
   $7x + 3y + 2z = 8$ $\left(\frac{1}{2}, 1, \frac{3}{4}\right)$
   $3x + y + 6z = 7$

4. The North High Music Boosters are selling T-shirts, hats, and sweatshirts as a fundraiser for a band trip. Monday they sold 34 T-shirts, 44 hats, and 17 sweatshirts for a total of $1134. The Boosters' sales on Tuesday included 27 T-shirts, 23 hats, and 24 sweatshirts for a total of $1036. Sales on Wednesday were $1418 for 31 T-shirts, 51 hats, and 29 sweatshirts. What are the prices for the Boosters' T-shirts, hats, and sweatshirts? **T-shirts, $12; hats, $8; sweatshirts, $22**

Motivating the Lesson

Real-World Application Ask students to imagine that they are running their own business. Sales results are recorded by item number and by month. How could the information be organized so it could be read easily?

OBJECTIVES
- Model data using matrices.
- Add, subtract, and multiply matrices.

TRAVEL Did you ever go on a vacation and realize that you forgot to pack something that you needed? Sometimes purchasing those items while traveling can be expensive. The average cost of some items bought in various cities is given below.

$4.03 Atlanta
$4.21 Los Angeles
$7.08 Tokyo
$3.97 Mexico City

$6.78 Atlanta
$7.41 Los Angeles
$7.43 Mexico City
$36.57 Tokyo

$18.98 Atlanta
$20.49 Los Angeles
$32.25 Mexico City
$63.71 Tokyo

Source: Runzheimer International

Data like these can be displayed or modeled using a matrix. *A problem related to this will be solved in Example 1.*

The plural of matrix is matrices.

A **matrix** is a rectangular array of terms called **elements.** The elements of a matrix are arranged in rows and columns and are usually enclosed by brackets. A matrix with m rows and n columns is an $m \times n$ **matrix** (read "m by n"). The **dimensions** of the matrix are m and n. Matrices can have various dimensions and can contain any type of numbers or other information.

2 × 2 matrix

$$\begin{bmatrix} -\dfrac{3}{5} & \dfrac{1}{2} \\ 3 & -\dfrac{3}{4} \end{bmatrix}$$

2 × 5 matrix

$$\begin{bmatrix} 0.2 & 3.4 & -1.1 & 2.5 & 6.7 \\ 3.4 & -3.4 & -22 & 0.5 & 7.2 \end{bmatrix}$$

3 × 1 matrix

$$\begin{bmatrix} -2 \\ 3 \\ 11 \end{bmatrix}$$
The element 3 is in row 2, column 1.

Special names are given to certain matrices. A matrix that has only one row is called a **row matrix,** and a matrix that has only one column is called a **column matrix.** A **square matrix** has the same number of rows as columns. Sometimes square matrices are called matrices of **nth order,** where n is the number of rows and columns. The elements of an $m \times n$ matrix can be represented using double subscript notation; that is, a_{24} would name the element in the second row and fourth column.

$$\begin{bmatrix} a_{11} & a_{12} & a_{13} & \cdots & a_{1n} \\ a_{21} & a_{22} & a_{23} & \cdots & a_{2n} \\ a_{31} & a_{32} & a_{33} & \cdots & a_{3n} \\ \vdots & \vdots & \vdots & \ddots & \vdots \\ a_{m1} & a_{m2} & a_{m3} & \cdots & a_{mn} \end{bmatrix}$$
a_{ij} is the element in the ith row and the jth column.

Resource Manager

 Reproducible Masters
- *Study Guide and Practice,* pp. 25–26
- *Enrichment,* p. 15
- *Problem Solving and Applications,* p. 3

 Transparencies
- 5-Minute Check, 2-3
- Teaching, 2-3

 Technology/Multimedia
- Advanced MathPASS, Lessons 4 and 5

Example 1

TRAVEL Refer to the application at the beginning of the lesson.

a. Use a matrix to represent the data.

b. Use a symbol to represent the price of pain reliever in Mexico City.

a. To represent data using a matrix, choose which category will be represented by the columns and which will be represented by the rows. Let's use the columns to represent the prices in each city and the rows to represent the prices of each item. Then write each data piece as you would if you were placing the data in a table.

	Atlanta	Los Angeles	Mexico City	Tokyo
film (24 exp.)	$4.03	$4.21	$3.97	$7.08
pain reliever (100 ct)	$6.78	$7.41	$7.43	$36.57
blow dryer	$18.98	$20.49	$32.25	$63.71

Notice that the category names appear outside of the matrix.

b. The price of pain reliever in Mexico City is found in the row 2, column 3 of the matrix. This element is represented by the symbol a_{23}.

Just as with numbers or algebraic expressions, matrices are equal under certain conditions.

Equal Matrices | Two matrices are equal if and only if they have the same dimensions and are identical, element by element.

Example 2 Find the values of x and y for which the matrix equation $\begin{bmatrix} y \\ x \end{bmatrix} = \begin{bmatrix} 2x - 6 \\ 2y \end{bmatrix}$ is true.

Since the corresponding elements are equal, we can express the equality of the matrices as two equations.
$y = 2x - 6$
$x = 2y$

Solve the system of equations by using substitution.

$y = 2x - 6$
$y = 2(2y) - 6$ *Substitute 2y for x.*
$y = 2$ *Solve for y.*

$x = 2(2)$ *Substitute 2 for y in*
$x = 4$ *the second equation to find x.*

The matrices are equal if $x = 4$ and $y = 2$. *Check by substituting into the matrices.*

Matrices are usually named using capital letters. The sum of two matrices, $A + B$, exists only if the two matrices have the same dimensions. The ijth element of $A + B$ is $a_{ij} + b_{ij}$.

Addition of Matrices | The sum of two $m \times n$ matrices is an $m \times n$ matrix in which the elements are the sum of the corresponding elements of the given matrices.

2 TEACH

Teaching Tip When discussing the dimensions of matrices, emphasize the difference between a row and a column in a matrix. Stress that a 3 × 2 matrix is not the same as a 2 × 3 matrix.

Also, help students understand double-subscript notation by putting a 4 × 4 or a 5 × 5 matrix on the chalkboard and asking them to identify various elements by their positions in the matrix.

In-Class Examples

Example 1
Agriculture During the summer of 2000, Ms. Robbins received several types of grains on her farm to feed her livestock.

a. Use a matrix to represent the data. June—15,000 bushels corn, 2000 bushels soybeans, 500 bushels oats; July—13,500 bushels corn, 6500 bushels soybeans, 1000 bushels oats; August—14,000 bushels corn, 5500 bushels soybeans, 1500 bushels oats

Sample answer:

	Corn	Soybeans	Oats
Jun.	15,000	2000	500
Jul.	13,500	6500	1000
Aug.	14,000	5500	1500

b. Use a symbol to represent the number of bushels of soybeans in August. a_{32}

Example 2
Find the values of x and y for which the matrix equation $\begin{bmatrix} y \\ y - 3 \end{bmatrix} = \begin{bmatrix} 4x \\ 2x + 1 \end{bmatrix}$ is true.

$x = 2, y = 8$

Example ③ Find $A + B$ if $A = \begin{bmatrix} -2 & 0 & 1 \\ 0 & 5 & -8 \end{bmatrix}$ and $B = \begin{bmatrix} -6 & 7 & -1 \\ 4 & -3 & 10 \end{bmatrix}$.

Graphing Calculator Appendix

For keystroke instruction on entering matrices and performing operations on them, see pages A16–A17.

$A + B = \begin{bmatrix} -2 + (-6) & 0 + 7 & 1 + (-1) \\ 0 + 4 & 5 + (-3) & -8 + 10 \end{bmatrix}$

$= \begin{bmatrix} -8 & 7 & 0 \\ 4 & 2 & 2 \end{bmatrix}$

You know that 0 is the additive identity for real numbers because $a + 0 = a$. Matrices also have additive identities. For every matrix A, another matrix can be found so that their sum is A. For example,

if $A = \begin{bmatrix} a_{11} & a_{12} \\ a_{21} & a_{22} \end{bmatrix}$, then $\begin{bmatrix} a_{11} & a_{12} \\ a_{21} & a_{22} \end{bmatrix} + \begin{bmatrix} 0 & 0 \\ 0 & 0 \end{bmatrix} = \begin{bmatrix} a_{11} & a_{12} \\ a_{21} & a_{22} \end{bmatrix}$.

The matrix $\begin{bmatrix} 0 & 0 \\ 0 & 0 \end{bmatrix}$ is called a **zero matrix**. The $m \times n$ zero matrix is the **additive identity matrix** for any $m \times n$ matrix.

You also know that for any number a, there is a number $-a$, called the additive inverse of a, such that $a + (-a) = 0$. Matrices also have additive inverses. If $A = \begin{bmatrix} a_{11} & a_{12} \\ a_{21} & a_{22} \end{bmatrix}$, then the matrix that must be added to A in order to have a sum of a zero matrix is $\begin{bmatrix} -a_{11} & -a_{12} \\ -a_{21} & -a_{22} \end{bmatrix}$ or $-A$. Therefore, $-A$ is the additive inverse of A. The additive inverse is used when you subtract matrices.

Subtraction of Matrices	The difference $A - B$ of two $m \times n$ matrices is equal to the sum $A + (-B)$, where $-B$ represents the additive inverse of B.

Example ④ Find $C - D$ if $C = \begin{bmatrix} 9 & 4 \\ -1 & 3 \\ 0 & -4 \end{bmatrix}$ and $D = \begin{bmatrix} 8 & -2 \\ -6 & 1 \\ 5 & -5 \end{bmatrix}$.

$C - D = C + (-D)$

$= \begin{bmatrix} 9 & 4 \\ -1 & 3 \\ 0 & -4 \end{bmatrix} + \begin{bmatrix} -8 & 2 \\ 6 & -1 \\ -5 & 5 \end{bmatrix}$

$= \begin{bmatrix} 9 + (-8) & 4 + 2 \\ -1 + 6 & 3 + (-1) \\ 0 + (-5) & -4 + 5 \end{bmatrix}$ or $\begin{bmatrix} 1 & 6 \\ 5 & 2 \\ -5 & 1 \end{bmatrix}$

You can multiply a matrix by a number; when you do, the number is called a **scalar.** The product of a scalar and a matrix A is defined as follows.

Scalar Product	The product of a scalar k and an $m \times n$ matrix A is an $m \times n$ matrix denoted by kA. Each element of kA equals k times the corresponding element of A.

Example **5** If $A = \begin{bmatrix} -4 & 1 & -1 \\ 3 & 7 & 0 \\ -3 & -1 & 8 \end{bmatrix}$, find $3A$.

$$3\begin{bmatrix} -4 & 1 & -1 \\ 3 & 7 & 0 \\ -3 & -1 & 8 \end{bmatrix} = \begin{bmatrix} 3(-4) & 3(1) & 3(-1) \\ 3(3) & 3(7) & 3(0) \\ 3(-3) & 3(-1) & 3(8) \end{bmatrix}$$ *Multiply each element by 3.*

$$= \begin{bmatrix} -12 & 3 & -3 \\ 9 & 21 & 0 \\ -9 & -3 & 24 \end{bmatrix}$$

You can also multiply a matrix by a matrix. For matrices A and B, you can find AB if the number of columns in A is the same as the number of rows in B.

$$\begin{bmatrix} 3 & -8 & 1 \\ 1 & 0 & 2 \end{bmatrix} \cdot \begin{bmatrix} 0 & 2 & 0 & 1 \\ -4 & 0 & -2 & 1 \\ 1 & -3 & -1 & 6 \end{bmatrix} \qquad \begin{bmatrix} 5 & 3 & 1 & 0 \\ 6 & 0 & 2 & -3 \\ -5 & 3 & 2 & 2 \end{bmatrix} \cdot \begin{bmatrix} 1 & 0 & 12 & 9 \\ 0 & 0 & -4 & -8 \\ -2 & 3 & 4 & 3 \end{bmatrix}$$

    2 × 3   *3 × 4*   *3 × 4*   *3 × 4*

Since 3 = 3, multiplication is possible.   *Since 4 ≠ 3, multiplication is **not** possible.*

The product of two matrices is found by multiplying columns and rows.
Suppose $A = \begin{bmatrix} a_1 & b_1 \\ a_2 & b_2 \end{bmatrix}$ and $X = \begin{bmatrix} x_1 & y_1 \\ x_2 & y_2 \end{bmatrix}$. Each element of matrix AX is the product of one row of matrix A and one column of matrix X.

$$AsX = \begin{bmatrix} a_1 & b_1 \\ a_2 & b_2 \end{bmatrix} \cdot \begin{bmatrix} x_1 & y_1 \\ x_2 & y_2 \end{bmatrix} = \begin{bmatrix} a_1x_1 + b_1x_2 & a_1y_1 + b_1y_2 \\ a_2x_1 + b_2x_2 & a_2y_2 + b_2y_2 \end{bmatrix}$$

In general, the product of two matrices is defined as follows.

Product of Two Matrices	The product of an $m \times n$ matrix A and an $n \times r$ matrix B is an $m \times r$ matrix AB. The ijth element of AB is the sum of the products of the corresponding elements in the ith row of A and the jth column of B.

Example **6** Use matrices $A = \begin{bmatrix} 7 & 0 \\ 5 & 3 \end{bmatrix}$, $B = \begin{bmatrix} 3 & -3 & 6 \\ 5 & 4 & -2 \end{bmatrix}$, and $C = \begin{bmatrix} 6 & -1 & 4 \\ 2 & -2 & -1 \end{bmatrix}$ to find each product.

a. AB

$$AB = \begin{bmatrix} 7 & 0 \\ 5 & 3 \end{bmatrix} \cdot \begin{bmatrix} 3 & -3 & 6 \\ 5 & 4 & -2 \end{bmatrix}$$

$$AB = \begin{bmatrix} 7(3) + 0(5) & 7(-3) + 0(4) & 7(6) + 0(-2) \\ 5(3) + 3(5) & 5(-3) + 3(4) & 5(6) + 3(-2) \end{bmatrix} \text{ or } \begin{bmatrix} 21 & -21 & 42 \\ 30 & -3 & 24 \end{bmatrix}$$

b. BC

B is a 2×3 matrix and C is a 2×3 matrix. Since B does not have the same number of columns as C has rows, the product BC does not exist. BC is undefined.

In-Class Example

Example 7

Tuition At The Ohio State University, professional students pay different tuition rates based on the programs they have chosen. For the 1997–98 school year, in-state students in the school of medicine paid $3665 per quarter, dental students paid $3064 per quarter, and veterinary medicine students paid $2906 per quarter. The chart lists the total student enrollment in those programs for each quarter of the 1997–98 school year. Use matrix multiplication to find the amount of tuition paid for each of these four quarters.

Quarter	Enrollment		
	Med.	Dent.	Vet.
Autumn	855	380	535
Winter	842	379	530
Spring	826	372	525
Summer	217	180	133

Source: The Ohio State University Registrar

Dollars

Autumn	5,852,605
Winter	5,282,166
Spring	5,197,148
Summer	1,733,323

3 PRACTICE/APPLY

Error Analysis

Watch for students who multiply matrices by multiplying the corresponding elements.

Prevent by stressing the definition of matrix multiplication.

Example **7** **SPORTS** In football, a player scores 6 points for a touchdown (TD), 3 points for kicking a field goal (FG), and 1 point for kicking the extra point after a touchdown (PAT). The chart lists the records of the top five all-time professional football scorers (as of the end of the 1997 season). Use matrix multiplication to find the number of points each player scored.

Scorer	TD	FG	PAT
George Blanda	9	335	943
Nick Lowery	0	383	562
Jan Stenerud	0	373	580
Gary Anderson	0	385	526
Morten Andersen	0	378	507

Source: *The World Almanac and Book of Facts,* 1999

Write the scorer information as a 5×3 matrix and the points per play as a 3×1 matrix. Then multiply the matrices.

$$
\begin{array}{c}
\text{Blanda} \\ \text{Lowery} \\ \text{Stenerud} \\ \text{Anderson} \\ \text{Andersen}
\end{array}
\begin{array}{ccc}
\text{TD} & \text{FG} & \text{PAT} \\
\left[\begin{array}{ccc}
9 & 335 & 943 \\
0 & 383 & 562 \\
0 & 373 & 580 \\
0 & 385 & 526 \\
0 & 378 & 507
\end{array}\right]
\end{array}
\cdot
\begin{array}{c}
 \\ \text{TD} \\ \text{FG} \\ \text{PAT}
\end{array}
\begin{array}{c}
\text{pts} \\
\left[\begin{array}{c}
6 \\ 3 \\ 1
\end{array}\right]
\end{array}
=
$$

$$
\begin{array}{c}
\text{Blanda} \\ \text{Lowery} \\ \text{Stenerud} \\ \text{Anderson} \\ \text{Andersen}
\end{array}
\begin{array}{c}
\text{pts} \\
\left[\begin{array}{c}
9(6) + 335(3) + 943(1) \\
0(6) + 383(3) + 562(1) \\
0(6) + 373(3) + 580(1) \\
0(6) + 385(3) + 526(1) \\
0(6) + 378(3) + 507(1)
\end{array}\right]
\end{array}
=
\begin{array}{c}
\text{Blanda} \\ \text{Lowery} \\ \text{Stenerud} \\ \text{Anderson} \\ \text{Andersen}
\end{array}
\begin{array}{c}
\text{pts} \\
\left[\begin{array}{c}
2002 \\ 1711 \\ 1699 \\ 1681 \\ 1641
\end{array}\right]
\end{array}
$$

CHECK FOR UNDERSTANDING

Communicating Mathematics

Read and study the lesson to answer each question. **1, 3, 4. See margin.**

1. **Write** a matrix other than the one given in Example 1 to represent the data on travel prices.

2. **Tell** the dimensions of the matrix $\begin{bmatrix} 4 & 0 & -2 & 4 \\ -1 & 3 & -1 & 5 \end{bmatrix}$. **2 × 4**

3. **Explain** how you determine whether the sum of two matrices exists.

4. **You Decide** Sarah says that $\begin{bmatrix} 3 & 2 & 3 \\ -4 & 2 & 0 \\ 0 & -1 & 2 \\ 2 & 2 & 5 \end{bmatrix}$ is a third-order matrix. Anthony disagrees. Who is correct and why?

Guided Practice

Find the values of *x* and *y* for which each matrix equation is true.

5. (7, 2)

5. $\begin{bmatrix} 2y \\ x \end{bmatrix} = \begin{bmatrix} x - 3 \\ y + 5 \end{bmatrix}$ 6. $[18 \quad 24] = [4x - y \quad 12y]$ 7. $[16 \quad 0 \quad 2x] = [4x \quad y \quad 8 - y]$
   (5, 2)   **(4, 0)**

82 Chapter 2 *Systems of Linear Equations and Inequalities*

Additional Answers

1. Sample answer:

	Film	Pain Reliever	Blow Dryer
Atlanta	$4.03	$6.78	$18.98
Los Angeles	$4.21	$7.41	$20.49
Mexico City	$3.97	$7.43	$32.25
Tokyo	$7.08	$36.57	$63.71

3. The sum of two matrices exists if the matrices have the same dimensions.

4. Anthony is correct. A third order matrix has 3 rows and 3 columns. This matrix has 4 rows and 3 columns.

8. $\begin{bmatrix} 3 & 4 \\ -2 & 4 \end{bmatrix}$

9. impossible

10. $\begin{bmatrix} -5 & 2 \\ 2 & -8 \end{bmatrix}$

11. $\begin{bmatrix} 16 & 4 \\ -8 & 24 \end{bmatrix}$

12. impossible

13. $[6 \quad -18]$

Use matrices *X*, *Y*, and *Z* to find each of the following. If the matrix does not exist, write *impossible*.

$$X = \begin{bmatrix} 4 & 1 \\ -2 & 6 \end{bmatrix} \qquad Y = [0 \quad -3] \qquad Z = \begin{bmatrix} -1 & 3 \\ 0 & -2 \end{bmatrix}$$

8. $X + Z$ 9. $Z - Y$ 10. $Z - X$ 11. $4X$ 12. ZY 13. YX

14. **Advertising** A newspaper surveyed companies on the annual amount of money spent on television commercials and the estimated number of people who remember seeing those commercials each week. A soft-drink manufacturer spends \$40.1 million a year and estimates 78.6 million people remember the commercials. For a package-delivery service, the budget is \$22.9 million for 21.9 million people. A telecommunications company reaches 88.9 million people by spending a whopping \$154.9 million. Use a matrix to represent this data. **See margin.**

EXERCISES

Practice

Find the values of *x* and *y* for which each matrix equation is true.

 15. $\begin{bmatrix} y \\ x \end{bmatrix} = \begin{bmatrix} 2x - 1 \\ y - 5 \end{bmatrix}$ **(6, 11)**

16. $[9 \quad 13] = [x + 2y \quad 4x + 1]$ **(3, 3)**

17. $\begin{bmatrix} 4x \\ 5 \end{bmatrix} = \begin{bmatrix} 15 + x \\ 2y \end{bmatrix}$ **(5, 2.5)**

18. $[x \quad y] = [2y \quad 2x - 6]$ **(4, 2)**

19. $\begin{bmatrix} 27 \\ 8 \end{bmatrix} = \begin{bmatrix} 3y \\ 5x - 3y \end{bmatrix}$ **(7, 9)**

20. $\begin{bmatrix} 4x - 3y \\ x + y \end{bmatrix} = \begin{bmatrix} 11 \\ 1 \end{bmatrix}$ **(2, −1)**

21. $[2x \quad y \quad -y] = [-10 \quad 3x \quad 15]$ **(−5, −15)**

22. $\begin{bmatrix} -12 \\ 2 \\ 12y \end{bmatrix} = \begin{bmatrix} 6x \\ y + 1 \\ 10 - x \end{bmatrix}$ **(−2, 1)**

 23. $\begin{bmatrix} x + y & 3 \\ y & 6 \end{bmatrix} = \begin{bmatrix} 0 & 2y - x \\ y^2 & 4 - 2x \end{bmatrix}$ **(−1, 1)**

24. $\begin{bmatrix} x^2 + 1 & 5 - y \\ x + y & y - 4 \end{bmatrix} = \begin{bmatrix} 2 & x \\ 5 & 2 \end{bmatrix}$ **(−1, 6)**

25. Find the values of *x*, *y*, and *z* for $3\begin{bmatrix} x & y - 1 \\ 4 & 3z \end{bmatrix} = \begin{bmatrix} 15 & 6 \\ 6z & 3x + y \end{bmatrix}$. **(5, 3, 2)**

26. Solve $-2\begin{bmatrix} w + 5 & x - z \\ 3y & 8 \end{bmatrix} = \begin{bmatrix} -16 & -4 \\ 6 & 2x + 8z \end{bmatrix}$ for *w*, *x*, *y*, and *z*. **(3, 0, −1, −2)**

Use matrices *A*, *B*, *C*, *D*, *E*, and *F* to find each of the following. If the matrix does not exist, write *impossible*. **27–46. See margin.**

$$A = \begin{bmatrix} 5 & 7 \\ -6 & 1 \end{bmatrix} \quad B = \begin{bmatrix} 3 & 5 \\ -1 & 8 \end{bmatrix} \quad C = \begin{bmatrix} 4 & -2 & 3 \\ 5 & 0 & -1 \\ 9 & 0 & 1 \end{bmatrix} \quad D = \begin{bmatrix} 0 & 1 & 2 \\ -2 & 3 & 0 \\ 4 & 4 & -2 \end{bmatrix}$$

$$E = \begin{bmatrix} 8 & -4 & 2 \\ 3 & 1 & -5 \end{bmatrix} \quad F = \begin{bmatrix} -6 & -1 & 0 \\ 1 & 4 & 0 \end{bmatrix}$$

27. $A + B$ 28. $A + C$ 29. $D + B$ 30. $D + C$ 31. $B - A$

32. $C - D$ 33. $4D$ 34. $-2F$ 35. $F - E$ 36. $E - F$

37. $5A$ 38. BA 39. CF 40. FC 41. ED

42. AA 43. $E + FD$ 44. $-3AB$ 45. $(BA)E$ 46. $F - 2EC$

Additional Answer

14.

	Budget ($ million)	Viewers (million)
Soft-drink	40.1	78.6
Package delivery	22.9	21.9
Telecommunications	154.9	88.9

Assignment Guide

Core: 15–53 odd, 55–64
Enriched: 16–48 even, 49–64

Additional Answers

27. $\begin{bmatrix} 8 & 12 \\ -7 & 9 \end{bmatrix}$

28. impossible

29. impossible

30. $\begin{bmatrix} 4 & -1 & 5 \\ 3 & 3 & -1 \\ 13 & 4 & -1 \end{bmatrix}$

31. $\begin{bmatrix} -2 & -2 \\ 5 & 7 \end{bmatrix}$

32. $\begin{bmatrix} 4 & -3 & 1 \\ 7 & -3 & -1 \\ 5 & -4 & 3 \end{bmatrix}$

33. $\begin{bmatrix} 0 & 4 & 8 \\ -8 & 12 & 0 \\ 16 & 16 & -8 \end{bmatrix}$

34. $\begin{bmatrix} 12 & 2 & 0 \\ -2 & -8 & 0 \end{bmatrix}$

35. $\begin{bmatrix} -14 & 3 & -2 \\ -2 & 3 & 5 \end{bmatrix}$

36. $\begin{bmatrix} 14 & -3 & 2 \\ 2 & -3 & -5 \end{bmatrix}$

37. $\begin{bmatrix} 25 & 35 \\ -30 & 5 \end{bmatrix}$

38. $\begin{bmatrix} -15 & 26 \\ -53 & 1 \end{bmatrix}$

39. impossible

40. $\begin{bmatrix} -29 & 12 & -17 \\ 24 & -2 & -1 \end{bmatrix}$

41. $\begin{bmatrix} 16 & 4 & 12 \\ -22 & -14 & 16 \end{bmatrix}$

42. $\begin{bmatrix} -17 & 42 \\ -36 & -41 \end{bmatrix}$

43. $\begin{bmatrix} 10 & -13 & -10 \\ -5 & 14 & -3 \end{bmatrix}$

44. $\begin{bmatrix} -24 & -243 \\ 57 & 66 \end{bmatrix}$

45. $\begin{bmatrix} -42 & 86 & -160 \\ -421 & 213 & -111 \end{bmatrix}$

46. $\begin{bmatrix} -66 & 31 & -60 \\ 57 & 16 & -6 \end{bmatrix}$

47. Find $3XY$ if $X = \begin{bmatrix} 2 & 4 \\ 8 & -4 \\ -2 & 6 \end{bmatrix}$ and $Y = \begin{bmatrix} 3 & -3 & 6 \\ 5 & 4 & -2 \end{bmatrix}$. $\begin{bmatrix} 78 & 30 & 12 \\ 12 & -120 & 168 \\ 72 & 90 & -72 \end{bmatrix}$

48. If $K = \begin{bmatrix} 1 & -7 \\ 3 & 2 \end{bmatrix}$ and $J = \begin{bmatrix} -4 & 5 \\ 1 & -1 \end{bmatrix}$, find $2K - 3J$. $\begin{bmatrix} 14 & -29 \\ 3 & 7 \end{bmatrix}$

Applications and Problem Solving

Real World Application

49. Entertainment How often do you go to the movies? The graph below shows the projected number of adults of different ages who attend movies at least once a month. Organize the information in a matrix. **See margin.**

Projected Movie Attendance

Age group	Year		
	1996	2000	2006
18 to 24	8485	8526	8695
25 to 34	10,102	9316	9078
35 to 44	8766	9039	8433
45 to 54	6045	6921	7900
55 to 64	2444	2741	3521
65 and older	2381	2440	2572

Source: American Demographics

interNET CONNECTION

Data Update For the latest National Endowment for the Arts survey, visit www.amc.glencoe.com

50. Music The National Endowment for the Arts exists to broaden public access to the arts. In 1992, it performed a study to find what types of arts were most popular and how they could attract more people. The matrices below represent their findings.

Percent of People Listening or Watching Performances

1982

	TV	Radio	Recording
Classical	25	20	22
Jazz	18	18	20
Opera	12	7	8
Musicals	21	4	8

1992

	TV	Radio	Recording
Classical	25	31	24
Jazz	21	28	21
Opera	12	9	7
Musicals	15	4	6

Source: National Endowment for the Arts

a. Find the difference in arts patronage from 1982 to 1992. Express your answer as a matrix. **See margin.**

b. Which areas saw the greatest increase and decrease in this time? **Classical radio and TV musicals**

51. Critical Thinking Consider the matrix equation $\begin{bmatrix} -2 & 3 \\ 4 & -5 \end{bmatrix} \cdot \begin{bmatrix} a & b \\ c & d \end{bmatrix} = \begin{bmatrix} -2 & 3 \\ 4 & -5 \end{bmatrix}$.

51a. $a = 1$, $b = 0$, $c = 0$, $d = 1$

a. Find the values of a, b, c, and d to make the statement true.

b. If any matrix containing two columns were multiplied by the matrix containing a, b, c, and d, what would you expect the result to be? Explain. **a matrix equal to the original matrix**

84 Chapter 2 Systems of Linear Equations and Inequalities

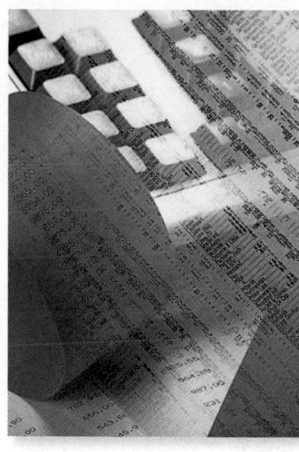

52. Finance Investors choose different stocks to comprise a balanced portfolio. The matrix below shows the prices of one share of each of several stocks on the first business day of July, August, and September of 1998.

	July	August	September
Stock A	$33 $\frac{13}{16}$	$30 $\frac{15}{16}$	$27 $\frac{1}{4}$
Stock B	$15 $\frac{1}{16}$	$13 $\frac{1}{4}$	$8 $\frac{3}{4}$
Stock C	$54	$54	$46 $\frac{7}{16}$
Stock D	$52 $\frac{1}{16}$	$44 $\frac{11}{16}$	$34 $\frac{3}{8}$

a. Mrs. Numkena owns 42 shares of stock A, 59 shares of stock B, 21 shares of stock C, and 18 shares of stock D. Write a row matrix to represent Mrs. Numkena's portfolio. **[42 59 21 18]**

b. Use matrix multiplication to find the total value of Mrs. Numkena's portfolio for each month to the nearest cent. **July, $4379.94; Aug, $4019.50; Sep, $3254.69**

53. Critical Thinking Study the matrix at the right. In which row and column will 2001 occur? Explain your reasoning. **See margin.**

$$\begin{bmatrix} 1 & 3 & 6 & 10 & 15 & \dots \\ 2 & 5 & 9 & 14 & 20 & \dots \\ 4 & 8 & 13 & 19 & 26 & \dots \\ 7 & 12 & 18 & 25 & 33 & \dots \\ 11 & 17 & 24 & 32 & 41 & \dots \\ 16 & 23 & 31 & 40 & 50 & \dots \\ \vdots & \vdots & \vdots & \vdots & \vdots & \vdots \end{bmatrix}$$

54. Discrete Math Airlines and other businesses often use *finite graphs* to represent their routes. A finite graph contains points called *nodes* and segments called *edges*. In a graph for an airline, each node represents a city, and each edge represents a route between the cities. Graphs can be represented by square matrices. The elements of the matrix are the numbers of edges between each pair of nodes. Study the graph and its matrix at the right.

$$\begin{array}{c} \\ R \\ S \\ T \\ U \end{array} \begin{array}{cccc} R & S & T & U \\ \begin{bmatrix} 0 & 2 & 0 & 1 \\ 2 & 0 & 0 & 0 \\ 0 & 0 & 0 & 1 \\ 1 & 0 & 1 & 0 \end{bmatrix} \end{array}$$

a. Represent the graph with nodes *A, B, C,* and *D* at the right using a matrix. **a–b. See margin.**

b. Equivalent graphs have the same number of nodes and edges between corresponding nodes. Can different graphs be represented by the same matrix? Explain your answer.

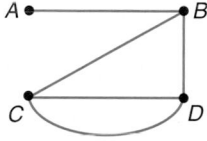

53. The numbers in the first row are the triangular numbers. If you look at the diagonals in the matrix, the triangular numbers are the end numbers. To find the diagonal that contains 2001, find the smallest triangular number that is greater than or equal to 2001. The formula for the *n*th triangular number is $\frac{n(n+1)}{2}$. Solve $\frac{n(n+1)}{2} \geq 2001$. The solution is 63. So the 63rd entry in the first row is $\frac{63(63+1)}{2} = 2016$. Since 2016 – 2001 = 15, we must count 15 places backward along the diagonal to locate 2001 in the matrix. This movement takes us from the position (row, column) = (1, 63) to (1 + 15, 63 – 15) = (16, 48).

54a.

$$\begin{array}{c} \\ A \\ B \\ C \\ D \end{array} \begin{array}{cccc} A & B & C & D \\ \begin{bmatrix} 0 & 1 & 0 & 0 \\ 1 & 0 & 1 & 1 \\ 0 & 1 & 0 & 2 \\ 0 & 1 & 2 & 0 \end{bmatrix} \end{array}$$

54b. No; since the matrix shows the number of nodes and the numbers of edges between each pair of nodes, only equivalent graphs will have the same matrix.

Extra Credit

Find the values of *x* and *y* for which the matrix equation $\begin{bmatrix} 2x & y \\ -x & 3y \end{bmatrix} \cdot \begin{bmatrix} 3 \\ -1 \end{bmatrix} = \begin{bmatrix} 25 \\ 12 \end{bmatrix}$ is true. **(3, –7)**

Open-Ended Assessment

Writing Have students give an example of two matrices whose product is commutative. Then have them provide a counter-example to prove that the multiplication of matrices is not commutative in general. Ask them to write a short paragraph restating this proof in sentence form.

 Graphing Calculator Exploration

Students may assume that since matrices contain real numbers, the properties of real numbers must hold true for matrices. Discuss why the properties are not necessarily true for matrices.

Additional Answers
Graphing Calculator Exploration

1. All of the properties except for the Commutative Property of Multiplication hold true. When multiplying matrices, the order of the multiplication produces different results. However, in addition of matrices, order is not important.

3. All properties except the commutative property of multiplication will hold for square matrices. A proof similar to the ones in Exercises 2a–2d can be used to verify this conjecture.

Enrichment Masters, p. 15

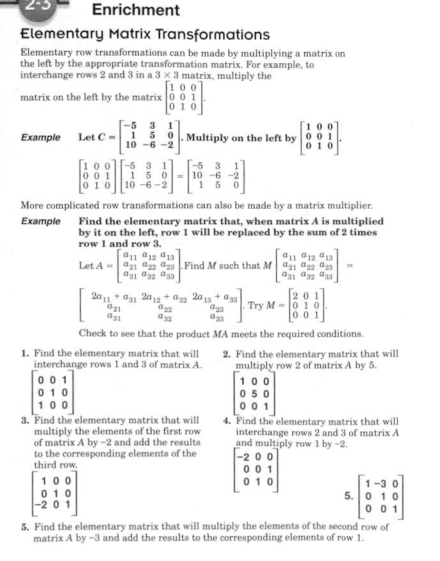

Mixed Review

55. $\left(\frac{1}{2}, \frac{1}{3}, \frac{1}{4}\right)$

56. consistent and dependent

59. Sample answer: using (60, 83) and (10, 65), $y = 0.36x + 61.4$

55. Solve the system $2x + 6y + 8z = 5$, $-2x + 9y + 12z = 5$, and $4x + 6y - 4z = 3$. *(Lesson 2-2)*

56. State whether the system $4x - 2y = 7$ and $-12x + 6y = -21$ is *consistent and independent, consistent and dependent,* or *inconsistent.* *(Lesson 2-1)*

57. Graph $-6 \leq 3x - y \leq 12$. *(Lesson 1-8)* **See margin.**

58. Graph $f(x) = |3x| + 2$. *(Lesson 1-7)* **See margin.**

59. **Education** Many educators believe that taking practice tests will help students succeed on actual tests. The table below shows data gathered about students studying for an algebra test. Use the data to write a prediction equation. *(Lesson 1-6)*

Practice Test Time (minutes)	15	75	60	45	90	60	30	120	10	120
Test scores (percents)	68	87	92	73	95	83	77	98	65	94

60. Write the slope-intercept form of the equation of the line through points at $(1, 4)$ and $(5, 7)$. *(Lesson 1-4)* $y = \frac{3}{4}x + 3\frac{1}{4}$

61. Find the zero of $f(x) = 5x - 3$ *(Lesson 1-3)* $\frac{3}{5}$

62. Find $[f \cdot g](x)$ if $f(x) = \frac{2}{5}x$ and $g(x) = 40x - 10$. *(Lesson 1-2)* $16x^2 - 4x$

63. Given $f(x) = 4 + 6x - x^3$, find $f(14)$. *(Lesson 1-1)* -2656

64. **SAT/ACT Practice** If $\frac{2x - 3}{x} = \frac{3 - x}{2}$, which of the following could be a value of x? **A**

    A -3 **B** -1 **C** 37 **D** 5 **E** 15

GRAPHING CALCULATOR EXPLORATION

Remember the properties of real numbers:

Properties of Addition
Commutative $a + b = b + a$
Associative $(a + b) + c = a + (b + c)$

Properties of Multiplication
Commutative $ab = ba$
Associative $(ab)c = a(bc)$

Distributive Property
$a(b + c) = ab + ac$

Do these properties hold for operations with matrices?

TRY THESE
Use matrices A, B, and C to investigate each of the properties shown above.

$$A = \begin{bmatrix} -1 & 2 \\ 3 & 0 \end{bmatrix} \quad B = \begin{bmatrix} 4 & 0 \\ -2 & -3 \end{bmatrix} \quad C = \begin{bmatrix} -2 & -3 \\ 4 & 2 \end{bmatrix}$$

1, 3. See margin.
WHAT DO YOU THINK?

1. Do these three matrices satisfy the properties of real numbers listed at the left? Explain.

2. Would these properties hold for any 2×2 matrices? Prove or disprove each statement below using variables as elements of each 2×2 matrix.
    a. Addition of matrices is commutative.
    b. Addition of matrices is associative.
    c. Multiplication of matrices is commutative.
    d. Multiplication of matrices is associative.
    See Answer Appendix.

3. Which properties do you think hold for $n \times n$ matrices? Explain.

Extra Practice   See p. A28.

Additional Answers

57.

58.

2-4A Transformation Matrices

An Introduction to Lesson 2-4

OBJECTIVE
- Determine the effect of matrix multiplication on a vertex matrix.

The coordinates of a figure can be represented by a matrix with the *x*-coordinates in the first row and the *y*-coordinates in the second. When this matrix is multiplied by certain other matrices, the result is a new matrix that contains the coordinates of the vertices of a different figure. You can use **List** and **Matrix** operations on your calculator to visualize some of these multiplications.

TRY THESE

Step 1 Enter $A = \begin{bmatrix} 0 & -1 \\ 1 & 0 \end{bmatrix}$, $B = \begin{bmatrix} -1 & 0 \\ 0 & -1 \end{bmatrix}$, and $C = \begin{bmatrix} 0 & 1 \\ -1 & 0 \end{bmatrix}$.

Step 2 Graph the triangle *LMN* with $L(1, -1)$, $M(2, -2)$, and $N(3, -1)$ by using **STAT PLOT**. Enter the *x*-coordinates in **L1** and the *y*-coordinates in **L2**, repeating the first point to close the figure. In **STAT PLOT**, turn **Plot 1** on and select the connected graph. After graphing the figure, press $\boxed{\text{ZOOM}}$ 5 to reflect a less distorted viewing window.

[−4.548 ..., 4.548 ...] scl: 1 by
[−3, 3] scl: 1

The List▶matr and Matr▶list commands transfer the data column for column. That is, the data in List 1 goes to Column 1 of the matrix and vice versa.

Step 3 To transfer the coordinates from the lists to a matrix, use the **9:List▶matr** command from the **MATH** submenu in the **MATRX** menu. Store this as matrix *D*. Matrix *D* has the *x*-coordinates in Column 1 and *y*-coordinates in Column 2. But a vertex matrix has the *x*-coordinates in *Row* 1 and the *y*-coordinates in *Row* 2. This switch can be easily done by using the **2:^{T}** (transpose) command found in the **MATH** submenu as shown in the screen.

Step 4 Multiply matrix *D* by matrix *A*. To graph the result we need to put the ordered pairs back into the **LIST** menu.
- This means we need to transpose *AD* first. Store as new matrix E.
- Use the **8:Matr▶list** command from the math menu to store values into **L3** and **L4**.

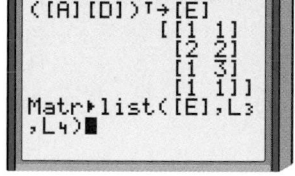

Step 5 Assign **Plot 2** as a connected graph of the **L3** and **L4** data and view the graph.

WHAT DO YOU THINK?

1–3. See Answer Appendix.

1. What is the relationship between the two plots?
2. Repeat Steps 4 and 5 replacing matrix *A* with matrix *B*. Compare the graphs.
3. Repeat Steps 4 and 5 replacing matrix *A* with matrix *C*. Compare the graphs.
4. Select a new figure and repeat this activity using each of the 2 × 2 matrices. Make a conjecture about these 2 × 2 matrices. **See margin.**

Lesson 2-4A Transformation Matrices **87**

---

1 FOCUS

Motivating the Lesson
Discussion The term matrix is used in other fields of study as well. Encourage students to discuss the use of matrices in business. Have them talk about how a graphing calculator or computer program can efficiently calculate data.

2 TEACH

  A reproduction of this lesson with graphing calculator keystrokes provided can be found in the *Graphing Calculator Keystrokes Masters*, pp. 5–6.

Teaching Tip Remind students that transformations are one-to-one mappings with pre-images and images.

3 PRACTICE/APPLY

Assignment Guide
All: 1–4

4 ASSESS

Open-Ended Assessment
Modeling Provide students with a large wallclock with a removable plastic face. Have students attach a triangle to the end of the minute hand. As each five-minute interval passes, have students draw how the rotation of the triangle changes as it moves clockwise.

Additional Answer

4. See students' work for graphs. Multiplying a vertex matrix by $\begin{bmatrix} 0 & -1 \\ 1 & 0 \end{bmatrix}$ results in a vertex matrix for a figure that is a 90° counterclockwise rotation of the original figure. Multiplying a vertex matrix by $\begin{bmatrix} -1 & 0 \\ 0 & -1 \end{bmatrix}$ results in a vertex matrix for a figure that is a 180° counterclockwise rotation of the original figure. Multiplying a vertex matrix by $\begin{bmatrix} 0 & 1 \\ -1 & 0 \end{bmatrix}$ results in a vertex matrix for a figure that is a 270° counterclockwise rotation of the original figure.

1 FOCUS

5-Minute Check
Lesson 2-3

1. One day, a theater had the following ticket sales. *Autumn Sky*, 423 adult, 135 student; *Joe, the Teenage Ape*, 413 adult, 389 student; and *Young Moses*, 472 adult, 402 student. Use a matrix to represent the data.
Sample answer:

	Adult	Student
Autumn	423	135
Joe	413	389
Moses	472	402

2. Find the values of x and y for which the equation
$$\begin{bmatrix} y + 4 \\ y \end{bmatrix} = \begin{bmatrix} x - 1 \\ 3x \end{bmatrix}$$ is true.
$(-2.5, -7.5)$

Use the matrices *A*, *B*, and *C* to find each of the following.
$$A = \begin{bmatrix} 0 & 5 \\ 3 & -1 \end{bmatrix}, B = \begin{bmatrix} -3 & 5 \\ 1 & 4 \end{bmatrix},$$

$$C = \begin{bmatrix} 0 & 9 & -3 \\ 4 & 11 & 4 \end{bmatrix}$$

3. $A + B$ $\begin{bmatrix} -3 & 10 \\ 4 & 3 \end{bmatrix}$

4. $-2C$ $\begin{bmatrix} 0 & -18 & 6 \\ -8 & -22 & -8 \end{bmatrix}$

5. BC $\begin{bmatrix} 20 & 28 & 29 \\ 16 & 53 & 13 \end{bmatrix}$

Motivating the Lesson
Hands-On Activity Have students draw a triangle on grid paper. Then cut out the triangle and place it on a coordinate plane made out of the same sized grid. Outline the triangle. Then slide it to a new position on the grid and outline again. Write matrices to represent the coordinates of the vertices of each triangle. Ask students how the matrices compare.

Modeling Motion with Matrices

OBJECTIVE
• Use matrices to determine the coordinates of polygons under a given transformation.

COMPUTER ANIMATION In 1995, animation took a giant step forward with the release of the first major motion picture to be created entirely on computers. Animators use computer software to create three-dimensional computer models of characters, props, and sets. These computer models describe the shape of the object as well as the motion controls that the animators use to create movement and expressions. The animation models are actually very large matrices.

Even though large matrices are used for computer animation, you can use a simple matrix to describe many of the motions called **transformations** that you learned about in geometry. Some of the transformations we will examine in this lesson are **translations** (slides), **reflections** (flips), **rotations** (turns), and **dilations** (enlargements or reductions).

An n-gon is a polygon with n sides.

A $2 \times n$ matrix can be used to express the vertices of an *n*-gon with the first row of elements representing the *x*-coordinates and the second row the *y*-coordinates of the vertices.

Triangle *ABC* can be represented by the following **vertex matrix.**

$$\begin{array}{cc} & A \quad B \quad C \\ \text{\textit{x}-coordinate} & \begin{bmatrix} -2 & 0 & 1 \\ 4 & 6 & -1 \end{bmatrix} \\ \text{\textit{y}-coordinate} & \end{array}$$

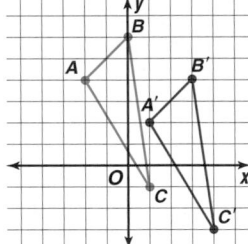

Triangle *A′B′C′* is congruent to and has the same orientation as △*ABC,* but is moved 3 units right and 2 units down from △*ABC*'s location. The coordinates of △*A′B′C′* can be expressed as the following vertex matrix.

$$\begin{array}{cc} & A' \quad B' \quad C' \\ \text{\textit{x}-coordinate} & \begin{bmatrix} 1 & 3 & 4 \\ 2 & 4 & -3 \end{bmatrix} \\ \text{\textit{y}-coordinate} & \end{array}$$

Compare the two matrices. If you add $\begin{bmatrix} 3 & 3 & 3 \\ -2 & -2 & -2 \end{bmatrix}$ to the first matrix, you get the second matrix. Each 3 represents moving 3 units right for each *x*-coordinate. Likewise, each -2 represents moving 2 units down for each *y*-coordinate. This type of matrix is called a **translation matrix.** In this transformation, △*ABC* is the **pre-image,** and △*A′B′C′* is the **image** after the translation.

Resource Manager

 Reproducible Masters
• *Study Guide and Practice,* pp. 27–28
• *Enrichment,* p. 16
• *Assessment and Evaluation,* pp. 42, 43

 Transparencies
• 5-Minute Check, 2-4
• Teaching, 2-4

Example **1** Suppose quadrilateral *ABCD* with vertices *A*(−1, 1), *B*(4, 0), *C*(4, −5), and *D*(−1, −3) is translated 2 units left and 4 units up.

a. Represent the vertices of the quadrilateral as a matrix.

b. Write the translation matrix.

c. Use the translation matrix to find the vertices of *A′B′C′D′*, the translated image of the quadrilateral.

d. Graph quadrilateral *ABCD* and its image.

Note that the image under a translation is the same shape and size as the pre-image. The figures are congruent.

a. The matrix representing the coordinates of the vertices of quadrilateral *ABCD* will be a 2 × 4 matrix.

$$\begin{array}{cc} & \begin{array}{cccc} A & B & C & D \end{array} \\ \begin{array}{c} x\text{-coordinate} \\ y\text{-coordinate} \end{array} & \begin{bmatrix} -1 & 4 & 4 & -1 \\ 1 & 0 & -5 & -3 \end{bmatrix} \end{array}$$

b. The translation matrix is $\begin{bmatrix} -2 & -2 & -2 & -2 \\ 4 & 4 & 4 & 4 \end{bmatrix}$.

c. Add the two matrices.

$$\begin{bmatrix} -1 & 4 & 4 & -1 \\ 1 & 0 & -5 & -3 \end{bmatrix} + \begin{bmatrix} -2 & -2 & -2 & -2 \\ 4 & 4 & 4 & 4 \end{bmatrix} = \begin{array}{c} \begin{array}{cccc} A' & B' & C' & D' \end{array} \\ \begin{bmatrix} -3 & 2 & 2 & -3 \\ 5 & 4 & -1 & 1 \end{bmatrix} \end{array}$$

d. Graph the points represented by the resulting matrix.

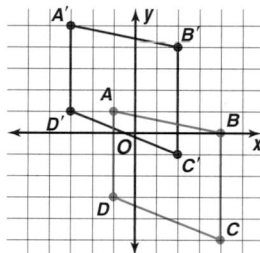

There are three lines over which figures are commonly reflected.
- the *x*-axis
- the *y*-axis, and
- the line *y* = *x*

The preimage and the image under a reflection are congruent.

In the figure at the right, △*P′Q′R′* is a reflection of △*PQR* over the *x*-axis. There is a 2 × 2 **reflection matrix** that, when multiplied by the vertex matrix of △*PQR*, will yield the vertex matrix of △*P′Q′R′*.

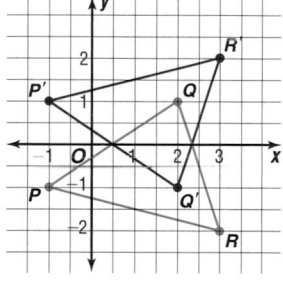

Let $\begin{bmatrix} a & b \\ c & d \end{bmatrix}$ represent the unknown square matrix.

Thus, $\begin{bmatrix} a & b \\ c & d \end{bmatrix} \cdot \begin{bmatrix} -1 & 2 & 3 \\ -1 & 1 & -2 \end{bmatrix} = \begin{bmatrix} -1 & 2 & 3 \\ 1 & -1 & 2 \end{bmatrix}$, or

$\begin{bmatrix} -a-b & 2a+b & 3a-2b \\ -c-d & 2c+d & 3c-2d \end{bmatrix} = \begin{bmatrix} -1 & 2 & 3 \\ 1 & -1 & 2 \end{bmatrix}$.

Lesson 2-4 Modeling Motion with Matrices **89**

Teaching Tip In Example 1, it may help students to picture sliding the quadrilateral across the coordinate plane.

Teaching Tip When translating matrices, stress that a movement *right* or *up* is a positive change in the coordinates and a movement *left* or *down* is a negative change.

In-Class Example

Example 1
Suppose quadrilateral *RSTU* with vertices *R*(3, 2), *S*(7, 4), *T*(9, 8), and *U*(5, 6) is translated 2 units right and 3 units down.

a. Represent the vertices of the quadrilateral as a matrix.

$$\begin{bmatrix} 3 & 7 & 9 & 5 \\ 2 & 4 & 8 & 6 \end{bmatrix}$$

b. Write the translation matrix.

$$\begin{bmatrix} 2 & 2 & 2 & 2 \\ -3 & -3 & -3 & -3 \end{bmatrix}$$

c. Use the translation matrix to find the vertices of *R′S′T′U′*, the translated image of the quadrilateral.

$$\begin{bmatrix} 5 & 9 & 11 & 7 \\ -1 & 1 & 5 & 3 \end{bmatrix}$$

d. Graph the quadrilateral *RSTU* and its image.

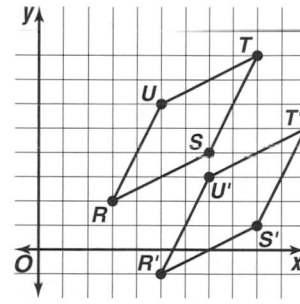

Since corresponding elements of equal matrices are equal, we can write equations to find the values of the variables. These equations form two systems.

$$-a - b = -1 \qquad\qquad -c - d = 1$$
$$2a + b = 2 \qquad\qquad\quad 2c + d = -1$$
$$3a - 2b = 3 \qquad\qquad 3c - 2d = 2$$

When you solve each system of equations, you will find that $a = 1$, $b = 0$, $c = 0$, and $d = -1$. Thus, the matrix that results in a reflection over the *x*-axis is $\begin{bmatrix} 1 & 0 \\ 0 & -1 \end{bmatrix}$. This matrix will work for any reflection over the *x*-axis.

The matrices for a reflection over the *y*-axis or the line $y = x$ can be found in a similar manner. These are summarized below.

Reflection Matrices		
For a reflection over the:	**Symbolized by:**	**Multiply the vertex matrix by:**
x-axis	$R_{x\text{-axis}}$	$\begin{bmatrix} 1 & 0 \\ 0 & -1 \end{bmatrix}$
y-axis	$R_{y\text{-axis}}$	$\begin{bmatrix} -1 & 0 \\ 0 & 1 \end{bmatrix}$
line $y = x$	$R_{y = x}$	$\begin{bmatrix} 0 & 1 \\ 1 & 0 \end{bmatrix}$

Example

Real World Application

2 **ANIMATION** To create an image that appears to be reflected in a mirror, an animator will use a matrix to reflect an image over the *y*-axis. Use a reflection matrix to find the coordinates of the vertices of a star reflected in a mirror (the *y*-axis) if the coordinates of the points connected to create the star are (−2, 4), (−3.5, 4), (−4, 5), (−4.5, 4), (−6, 4), (−5, 3), (−5, 1), (−4, 2), (−3, 1), and (−3, 3).

First write the vertex matrix for the points used to define the star.

$$\begin{bmatrix} -2 & -3.5 & -4 & -4.5 & -6 & -5 & -5 & -4 & -3 & -3 \\ 4 & 4 & 5 & 4 & 4 & 3 & 1 & 2 & 1 & 3 \end{bmatrix}$$

Multiply by the *y*-axis reflection matrix.

$$\begin{bmatrix} -1 & 0 \\ 0 & 1 \end{bmatrix} \cdot \begin{bmatrix} -2 & -3.5 & -4 & -4.5 & -6 & -5 & -5 & -4 & -3 & -3 \\ 4 & 4 & 5 & 4 & 4 & 3 & 1 & 2 & 1 & 3 \end{bmatrix} =$$
$$\begin{bmatrix} 2 & 3.5 & 4 & 4.5 & 6 & 5 & 5 & 4 & 3 & 3 \\ 4 & 4 & 5 & 4 & 4 & 3 & 1 & 2 & 1 & 3 \end{bmatrix}$$

The vertices used to define the reflection are (2, 4), (3.5, 4), (4, 5), (4.5, 4), (6, 4), (5, 3), (5, 1), (4, 2), (3, 1), and (3, 3).

The preimage and the image under a rotation are congruent.

You may remember from geometry that a rotation of a figure on a coordinate plane can be achieved by a combination of reflections. For example, a 90° counterclockwise rotation can be found by first reflecting the image over the *x*-axis and then reflecting the reflected image over the line $y = x$. The **rotation matrix**, Rot_{90}, can be found by a composition of reflections. Since reflection matrices are applied using multiplication, the composition of two reflection matrices is a product. Remember that $[f \circ g](x)$ means that you find $g(x)$ first and then evaluate the result for $f(x)$. So, to define Rot_{90}, we use

$$Rot_{90} = R_{y\,=\,x} \circ R_{x\text{-axis}} \text{ or } Rot_{90} = \begin{bmatrix} 0 & 1 \\ 1 & 0 \end{bmatrix} \cdot \begin{bmatrix} 1 & 0 \\ 0 & -1 \end{bmatrix} \text{ or } \begin{bmatrix} 0 & -1 \\ 1 & 0 \end{bmatrix}.$$

Similarly, a rotation of 180° would be rotations of 90° twice or $Rot_{90} \circ Rot_{90}$. A rotation of 270° is a composite of Rot_{180} and Rot_{90}. The results of these composites are shown below.

Rotation Matrices		
For a counterclockwise rotation about the origin of:	**Symbolized by:**	**Multiply the vertex matrix by:**
90°	Rot_{90}	$\begin{bmatrix} 0 & -1 \\ 1 & 0 \end{bmatrix}$
180°	Rot_{180}	$\begin{bmatrix} -1 & 0 \\ 0 & -1 \end{bmatrix}$
270°	Rot_{270}	$\begin{bmatrix} 0 & 1 \\ -1 & 0 \end{bmatrix}$

Example

3 ANIMATION Suppose a figure is animated to spin around a certain point. Numerous rotation images would be necessary to make a smooth movement image. If the image has key points at (1, 1), (−1, 4), (−2, 4), and, (−2, 3) and the rotation is about the origin, find the location of these points at the 90°, 180°, and 270° counterclockwise rotations.

First write the vertex matrix. Then multiply it by each rotation matrix.

The vertex matrix is $\begin{bmatrix} 1 & -1 & -2 & -2 \\ 1 & 4 & 4 & 3 \end{bmatrix}$.

$$Rot_{90} \quad \begin{bmatrix} 0 & -1 \\ 1 & 0 \end{bmatrix} \cdot \begin{bmatrix} 1 & -1 & -2 & -2 \\ 1 & 4 & 4 & 3 \end{bmatrix} = \begin{bmatrix} -1 & -4 & -4 & -3 \\ 1 & -1 & -2 & -2 \end{bmatrix}$$

$$Rot_{180} \quad \begin{bmatrix} -1 & 0 \\ 0 & -1 \end{bmatrix} \cdot \begin{bmatrix} 1 & -1 & -2 & -2 \\ 1 & 4 & 4 & 3 \end{bmatrix} = \begin{bmatrix} -1 & 1 & 2 & 2 \\ -1 & -4 & -4 & -3 \end{bmatrix}$$

$$Rot_{180} \quad \begin{bmatrix} 0 & 1 \\ -1 & 0 \end{bmatrix} \cdot \begin{bmatrix} 1 & -1 & -2 & -2 \\ 1 & 4 & 4 & 3 \end{bmatrix} = \begin{bmatrix} 1 & 4 & 4 & 3 \\ -1 & 1 & 2 & 2 \end{bmatrix}$$

Addressing Individual Needs

Auditory/Musical You can relate translations to musical scales. Play a tune on a keyboard. Then translate the notes up or down an octave and play the tune again. How does the change in the music relate to a translation?

Teaching Tip In Example 4 in the Student Edition, point out that the colored type corresponds to the colored portions of the graph.

3 PRACTICE/APPLY

Error Analysis

Watch for students who confuse matrices of reflections and rotations.

Prevent by having students graph the results of matrix multiplication to check that the correct transformation is represented.

Study Guide and Practice Masters, pp. 27–28

All of the transformations we have discussed have maintained the shape and size of the figure. However, a dilation changes the size of the figure. The dilated figure is similar to the original figure. Dilations using the origin as a center of projection can be achieved by multiplying the vertex matrix by the scale factor needed for the dilation. *All dilations in this lesson are with respect to the origin.*

Example **4** A trapezoid has vertices at $L(-4, 1)$, $M(1, 4)$, $N(7, 0)$, and $P(-3, -6)$. Find the coordinates of the dilated trapezoid $L'M'N'P'$ for a scale factor of 0.5. Describe the dilation.

First write the coordinates of the vertices as a matrix. Then do a scalar multiplication using the scale factor.

$$0.5 \begin{bmatrix} -4 & 1 & 7 & -3 \\ 1 & 4 & 0 & -6 \end{bmatrix} = \begin{bmatrix} -2 & 0.5 & 3.5 & -1.5 \\ 0.5 & 2 & 0 & -3 \end{bmatrix}$$

The vertices of the image are $L'(-2, 0.5)$, $M'(0.5, 2)$, $N'(3.5, 0)$, and $P'(-1.5, -3)$.

The image has sides that are half the length of the original figure.

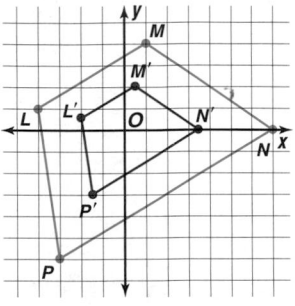

CHECK FOR UNDERSTANDING

Communicating Mathematics

Read and study the lesson to answer each question. **1–3. See margin.**

1. **Name** all the transformations described in this lesson. Tell how the pre-image and image are related in each type of transformation.

2. **Explain** how 90°, 180°, and 270° counterclockwise rotations correspond to clockwise rotations.

3. *Math Journal* **Describe** a way that you can remember the elements of the reflection matrices if you forget where the 1s, −1s, and 0s belong.

4. **Match** each matrix with the phrase that best describes its type.

a. $\begin{bmatrix} -1 & -1 \\ 1 & 1 \end{bmatrix}$ **6**

b. $\begin{bmatrix} -1 & 0 \\ 0 & 1 \end{bmatrix}$ **2**

c. $\begin{bmatrix} 0 & 1 \\ 1 & 0 \end{bmatrix}$ **3**

d. $\begin{bmatrix} 0 & -1 \\ 1 & 0 \end{bmatrix}$ **4**

(1) dilation of scale factor 2

(2) reflection over the y-axis

(3) reflection over the line $y = x$

(4) rotation of 90° counterclockwise about the origin

(5) rotation of 180° about the origin

(6) translation 1 unit left and 1 unit up

Additional Answers

1. Translation, reflection, rotation, dilation; translations do not affect the shape, size, or orientation of figures; reflections and rotations do not change the shape or size of figures; dilations do not change the shape, but do change the size of figures.

2. 90° counterclockwise = (360 − 90)° or 270° clockwise; 180° counterclockwise = (360 − 180)° or 180° clockwise; 270° counterclockwise = (360 − 270)° or 90° clockwise.

Guided Practice

Use matrices to perform each transformation. Then graph the pre-image and the image on the same coordinate grid. **5–9. See Answer Appendix for graphs.**

5. Triangle *JKL* has vertices *J*(−2, 5), *K*(1, 3), and *L*(0, −2). Use scalar multiplication to find the coordinates of the triangle after a dilation of scale factor 1.5. **J'(−3, 7.5), K'(1.5, 4.5), L'(0, −3)**

6. A'(−2, 1), B'(2, 1), C'(2, −3), D'(−2, −3)

6. Square *ABCD* has vertices *A*(−1, 3), *B*(3, 3), *C*(3, −1), and *D*(−1, −1). Find the coordinates of the square after a translation of 1 unit left and 2 units down.

7. Square *ABCD* has vertices at (−1, 2), (−4, 1), (−3, −2), and (0, −1). Find the image of the square after a reflection over the *y*-axis.

7. A'(1, 2), B'(4, 1), C'(3, −2), D'(0, −1)

8. Triangle *PQR* is represented by the matrix $\begin{bmatrix} 3 & -1 & 1 \\ 2 & 4 & -2 \end{bmatrix}$. Find the image of the triangle after a rotation of 270° counterclockwise about the origin.

8. P'(2, −3), Q'(4, 1), R'(−2, −1)

9. Find the image of △*LMN* after $Rot_{180} \circ R_{y\text{-axis}}$ if the vertices are *L*(−6, 4), *M*(−3, 2), and *N*(−1, −2). **L'(−6, −4), M'(−3, −2), N'(−1, 2)**

10. **Physics** The wind was blowing quite strongly when Jenny was baby-sitting. She was outside with the children, and they were throwing their large plastic ball up into the air. The wind blew the ball so that it landed approximately 3 feet east and 4 feet north of where it was thrown into the air.

a. Make a drawing to demonstrate the original location of the ball and the translation of the ball to its landing spot. **See margin.**

b. If $\begin{bmatrix} x \\ y \end{bmatrix}$ represents the original location of the ball, write a matrix that represents the location of the translated ball. $\begin{bmatrix} x + 3 \\ y + 4 \end{bmatrix}$

EXERCISES

Practice

11. A'(3, 3), B'(3, 12), C'(15, 3) **A**

Use scalar multiplication to determine the coordinates of the vertices of each dilated figure. Then graph the pre-image and the image on the same coordinate grid. **11–13. See margin for graphs.**

11. triangle with vertices *A*(1, 1), *B*(1, 4), and *C*(5, 1); scale factor 3

12. triangle with vertices *X*(0, 8), *Y*(−5, 9), and *Z*(−3, 2); scale factor $\frac{3}{4}$

12. X'(0, 6), Y'(−3$\frac{3}{4}$, 6$\frac{3}{4}$), Z'(−2$\frac{1}{4}$, 1$\frac{1}{2}$)

13. quadrilateral *PQRS* with vertex matrix $\begin{bmatrix} -3 & -2 & 1 & 4 \\ 0 & 2 & 3 & 2 \end{bmatrix}$; scale factor 2

14. Graph a square with vertices *A*(−1, 0), *B*(0, 1), *C*(1, 0), and *D*(0, −1) on two separate coordinate planes.

13. P'(−6,0), Q'(−4, 4), R'(2, 6), S'(8, 4)

a. On one of the coordinate planes, graph the dilation of square *ABCD* after a dilation of scale factor 2. Label it *A'B'C'D'*. Then graph a dilation of *A'B'C'D'* after a scale factor of 3. **See margin.**

b. On the second coordinate plane, graph the dilation of square *ABCD* after a dilation of scale factor 3. Label it *A'B'C'D'*. Then graph a dilation of *A'B'C'D'* after a scale factor of 2. **See margin.**

14c. The final results are the same image.

c. Compare the results of parts **a** and **b**. Describe what you observe.

Assignment Guide

Core: 11–27 odd, 29–42
Enriched: 12–28 even, 29–42
All: Mid-Chapter Quiz, 1–10

Additional Answers

10a.

11.

12.

13.

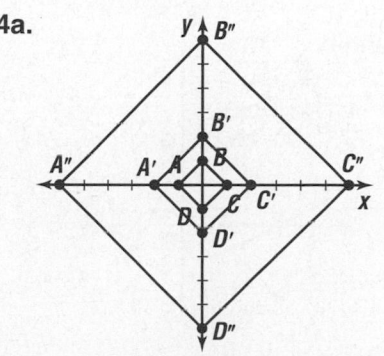

14a.

(continued on page 94)

Additional Answer *(p. 92)*

3. Sample answer: the first row of the reflection matrix affects the *x*-coordinates and the second row affects the *y*-coordinates. A reflection over the *x*-axis changes (*x*, *y*) to (*x*, −*y*), so the first row needs to be [1 0] so that the *x* is unchanged, and the second row needs to be [0 −1] so that the *y*-coordinates are the opposite. Similar reasoning can be used for a reflection over the *y*-axis, which changes (*x*, *y*) to (−*x*, *y*), and a reflection over the line *y* = *x*, which interchanges the values for *x* and *y*.

14b.

15.

16.

17.

18a–b.

19.

15–17. See margin for graphs.

Use matrices to determine the coordinates of the vertices of each translated figure. Then graph the pre-image and the image on the same coordinate grid.

15. triangle WXY with vertex matrix $\begin{bmatrix} -2 & 1 & 3 \\ 0 & 5 & -1 \end{bmatrix}$ translated 3 units right and 2 units down **$W'(1, -2)$, $X'(4, 3)$, $Y'(6, -3)$**

16. quadrilateral with vertices $O(0, 0)$, $P(1, 5)$, $Q(4, 7)$, and $R(3, 2)$ translated 2 units left and 1 unit down **$O'(-2, -1)$, $P'(-1, 4)$, $Q'(2, 6)$, $R'(1, 1)$**

17. square $CDEF$ translated 3 units right and 4 units up if the vertices are $C(-3, 1)$, $D(1, 5)$, $E(5, 1)$, and $F(1, -3)$ **$C'(0, 5)$, $D'(4, 9)$, $E'(8, 5)$, $F'(4, 1)$**

18. Graph $\triangle FGH$ with vertices $F(4, 1)$, $G(0, 3)$, and $H(2, -1)$.

a. Graph the image of $\triangle FGH$ after a translation of 6 units left and 2 units down. Label the image $\triangle F'G'H'$. **a–b. See margin.**

b. Then translate $\triangle F'G'H'$ 1 unit right and 5 units up. Label this image $\triangle F''G''H''$.

c. What translation would move $\triangle FGH$ to $\triangle F''G''H''$ directly? **translation of 5 units left and 3 units up**

19–21. See margin for graphs.

Use matrices to determine the coordinates of the vertices of each reflected figure. Then graph the pre-image and the image on the same coordinate grid.

▶ B 19. $\triangle ABC$ with vertices $A(-1, -2)$, $B(0, -4)$, and $C(2, -3)$ reflected over the x-axis **$A'(-1, 2)$, $B'(0, 4)$, $C'(2, 3)$**

20. $D'(-2, 4)$, $E'(-6, 2)$, $F'(-3, -4)$, $G'(1, -2)$

20. $R_{y\text{-axis}}$ for a rectangle with vertices $D(2, 4)$, $E(6, 2)$, $F(3, -4)$, and $G(-1, -2)$

21. a trapezoid with vertices $H(-1, -2)$, $I(-3, 1)$, $J(-1, 5)$, and $K(2, 4)$ for a reflection over the line $y = x$ **$H'(-2, -1)$, $I'(1, -3)$, $J'(5, -1)$, $K'(4, 2)$**

22–24. See margin for graphs.

Use matrices to determine the coordinates of the vertices of each rotated figure. Then graph the pre-image and the image on the same coordinate grid.

22. Rot_{90} for $\triangle LMN$ with vertices $L(1, -1)$, $M(2, -2)$, and $N(3, -1)$

22. $L'(1, 1)$, $M'(2, 2)$, $N'(1, 3)$

23. square with vertices $O(0, 0)$, $P(4, 0)$, $Q(4, 4)$, $R(0, 4)$ rotated 180°

23. $O'(0, 0)$, $P'(-4, 0)$, $Q'(-4, -4)$, $R'(0, -4)$

24. pentagon $STUVW$ with vertices $S(-1, -2)$, $T(-3, -1)$, $U(-5, -2)$, $V(-4, -4)$, and $W(-2, -4)$ rotated 270° counterclockwise **$S'(-2, 1)$, $T'(-1, 3)$, $U'(-2, 5)$, $V'(-4, 4)$, $W'(-4, 2)$**

▶ C 25. **Proof** Suppose $\triangle ABC$ has vertices $A(1, 3)$, $B(-2, -1)$, and $C(-1, -3)$. Use each result of the given transformation of $\triangle ABC$ to show how the matrix for that reflection or rotation is derived. **See Answer Appendix.**

a. $\begin{bmatrix} 1 & -2 & -1 \\ -3 & 1 & 3 \end{bmatrix}$ under $R_{x\text{-axis}}$ b. $\begin{bmatrix} -1 & 2 & 1 \\ 3 & -1 & -3 \end{bmatrix}$ under $R_{y\text{-axis}}$

c. $\begin{bmatrix} 3 & -1 & -3 \\ 1 & -2 & -1 \end{bmatrix}$ under $R_{y=x}$ d. $\begin{bmatrix} -3 & 1 & 3 \\ 1 & -2 & -1 \end{bmatrix}$ under Rot_{90}

e. $\begin{bmatrix} -1 & 2 & 1 \\ -3 & 1 & 3 \end{bmatrix}$ under Rot_{180} f. $\begin{bmatrix} 3 & -1 & -3 \\ -1 & 2 & 1 \end{bmatrix}$ under Rot_{270}

94 Chapter 2 *Systems of Linear Equations and Inequalities*

20.

21.

22.

Given △JKL with vertices J(−6, 4), K(−3, 2), and L(−1, −2). Find the coordinates of each composite transformation. Then graph the pre-image and the image on the same coordinate grid.   26–28. See margin for graphs.

26. J″(4, 1), K″(1, 3), L″(−1, 7)

26. rotation of 180° followed by a translation 2 units left 5 units up

27. $R_{y\text{-axis}} \circ R_{x\text{-axis}}$ J″(6, −4), K″(3, −2), L″(1, 2)

28. $Rot_{90} \circ R_{y\text{-axis}}$ J″(−4, 6), K″(−2, 3), L″(2, 1)

Applications and Problem Solving

29. Games   Each of the pieces on the chess board has a specific number of spaces and direction it can move. Research the game of chess and describe the possible movements for each piece as a translation matrix.

    a. bishop   **b.** knight   **c.** king **29a–c. See Answer Appendix.**

30. Critical Thinking   Show that a dilation with scale factor of −1 is the same result as Rot_{180}.   **See margin.**

31. Entertainment   The Ferris Wheel first appeared at the 1893 Chicago Exposition. Its axle was 45 feet long. Spokes radiated from it that supported 36 wooden cars, which could hold 60 people each. The diameter of the wheel itself was 250 feet. Suppose the axle was located at the origin. Find the coordinates of the car located at the loading platform. Then find the location of the car at the 90° counterclockwise, 180°, and 270° counterclockwise rotation positions.   (0, −125); (125, 0), (0, 125), (−125, 0)

32. Critical Thinking $R_{y\text{-axis}}$ gives a matrix for reflecting a figure over the y-axis. Do you think a matrix that would represent a reflection over the line y = 4 exists? If so, make a conjecture and verify it.   **See margin.**

33. Animation   Divide two sheets of grid paper into fourths by halving the length and width of the paper. Draw a simple figure on one of the pieces. On another piece, draw the figure dilated with a scale factor of 1.25. On a third piece, draw the original figure dilated with a scale factor of 1.5. On the fourth piece, draw the original figure dilated with a scale factor of 1.75. Continue dilating the original figure on each of the remaining pieces by an increase of 0.25 in scale factor each time. Put the pieces of paper in order and flip through them. What type of motion does the result of these repeated dilations animate?   **See margin.**

34. Critical Thinking   Write the vertex matrix for the figure graphed below.

    a. Make a conjecture about the resulting figure if you multiply the vertex matrix by $\begin{bmatrix} 3 & 0 \\ 0 & 2 \end{bmatrix}$.   **a–c. See margin.**

    b. Copy the figure on grid paper and graph the resulting vertex matrix after the multiplication described.

    c. How does the result compare with your conjecture? This is often called a *shear*. Why do you think it has this name?

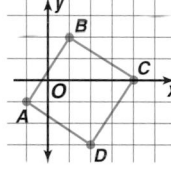

Lesson 2-4   Modeling Motion with Matrices   **95**

Additional Answers *(pp. 94–95)*

23.

24.

26.

Additional Answers

27.

28.

30. Consider $\begin{bmatrix} a & b \\ c & d \end{bmatrix}$. **Dilation with scale factor −1**

$$-1\begin{bmatrix} a & b \\ c & d \end{bmatrix} = \begin{bmatrix} -a & -b \\ -c & -d \end{bmatrix}$$

Rotation of 180°

$$\begin{bmatrix} -1 & 0 \\ 0 & -1 \end{bmatrix}\begin{bmatrix} a & b \\ c & d \end{bmatrix} = \begin{bmatrix} -a & -b \\ -c & -d \end{bmatrix}$$

The vertex matrices for the images of a dilation with scale factor −1 and a rotation of 180° are the same, so the images are the same.

32. Sample answer: there is no single matrix to achieve this. You could reflect over the x-axis and then translate 2(4) or 8 units upward.

33. See students' work; the repeated dilations animate the growth of something from small to larger, similar to a lens zooming into the origin.

34a. $\begin{bmatrix} -1 & 1 & 4 & 2 \\ -1 & 1 & 2 & 0 & -3 \end{bmatrix}$; Sample answer: the figure would be enlarged disproportionately.

34b.

34c. See students' work; the figure appears as if blown out of proportion.

Lesson 2-4   **95**

Open-Ended Assessment

Modeling Have students work in pairs. They should graph a figure together. Then one student describes a motion and the other models the motion using a matrix. Finally, they should exchange roles and model a different motion.

Mid-Chapter Quiz

The Mid-Chapter Quiz provides students with a brief review of the concepts and skills in Lessons 2-1 through 2-4. Lesson numbers are given to the right of exercises or instruction lines so students can review concepts not yet mastered.

 Chapter 2, Quiz B (Lessons 2-3 and 2-4) is available in the *Assessment and Evaluation Masters*, p. 43.

 Chapter 2 Mid-Chapter Test (Lessons 2-1 through 2-4) is available in the *Assessment and Evaluation Masters*, p. 42.

Additional Answers
Mid-Chapter Quiz

7. $\begin{bmatrix} 1 & 13 & -1 \\ 4 & -9 & 14 \end{bmatrix}$

8. impossible

9. $\begin{bmatrix} -11 & -7 & 27 \\ 8 & -9 & -2 \end{bmatrix}$

Enrichment Masters, p. 16

 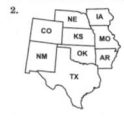

Mixed Review

35. Find $A + B$ if $A = \begin{bmatrix} 3 & 8 \\ -2 & 4 \end{bmatrix}$ and $B = \begin{bmatrix} 1 & 5 \\ -2 & 8 \end{bmatrix}$. *(Lesson 2-3)* $\begin{bmatrix} 4 & 13 \\ -4 & 12 \end{bmatrix}$

36. Solve the system of equations. *(Lesson 2-2)*
$x - 2y = 4.6$
$y - z = -5.6$ **(0.4, −2.1, 3.5)**
$x + y + z = 1.8$

37. **Sales** The Grandview Library holds an annual book sale to raise funds and dispose of excess books. One customer bought 4 hardback books and 7 paperbacks for $5.75. The next customer paid $4.25 for 3 hardbacks and 5 paperbacks. What are the prices for hardbacks and for paperbacks? *(Lesson 2-1)* **hardbacks $1, paperbacks $0.25**

38. Of (0, 0), (3, 2), (−4, 2), or (−2, 4), which satisfy $x + y \geq 3$? *(Lesson 1-8)* **(3, 2)**

39. Write the standard form of the equation of the line that is parallel to the graph of $y = 4x - 8$ and passes through (−2, 1). *(Lesson 1-5)* **4x − y + 9 = 0**

40. Write the slope-intercept form of the equation of the line that passes through the point at (1, 6) and has a slope of 2. *(Lesson 1-3)* **y = 2x + 4**

41. $x^5 - 3x^4 + 7x^3$, $\dfrac{x^3}{x^2 - 3x + 7}$ 41. If $f(x) = x^3$ and $g(x) = x^2 - 3x + 7$, find $(f \cdot g)(x)$ and $\left(\dfrac{f}{g}\right)(x)$. *(Lesson 1-2)*

42. **SAT/ACT Practice** If $2x + y = 12$ and $x + 2y = -6$, find the value of $2x + 2y$. **B**
A 0   B 4   C 8   D 12   E 14

MID-CHAPTER QUIZ

1. Use graphing to solve the system of equations $\frac{1}{2}x + 5y = 17$ and $3x + 2y = 18$. (Lesson 2-1) **(4, 3)** 2. $\left(\frac{1}{2}, 6\right)$

2. Solve the system of equations $4x + y = 8$ and $6x - 2y = -9$ algebraically. (Lesson 2-1)

3. **Sales** HomeMade Toys manufactures solid pine trucks and cars and usually sells four times as many trucks as cars. The net profit from each truck is $6 and from each car is $5. If the company wants a total profit of $29,000, how many trucks and cars should they sell? (Lesson 2-1) **4000 trucks, 1000 cars**

Solve each system of equations. (Lesson 2-2)

4. $2x + y + 4z = 13$
$3x - y - 2z = -1$
$4x + 2y + z = 19$

5. $x + y = 1$
$2x - y = -2$
$4x + y + z = 8$

6. Find the values of x and y for which the matrix equation $\begin{bmatrix} y - 3 \\ y \end{bmatrix} = \begin{bmatrix} x \\ 2x \end{bmatrix}$ is true. (Lesson 2-3) **(3, 6)**

Use matrices A and B to find each of the following. If the matrix does not exist, write *impossible*. (Lesson 2-3) **7–9. See margin.**

$A = \begin{bmatrix} 3 & 5 & -7 \\ -1 & 0 & 4 \end{bmatrix}$ $B = \begin{bmatrix} -2 & 8 & 6 \\ 5 & -9 & 10 \end{bmatrix}$

7. $A + B$ 8. BA 9. $B - 3A$

10. What is the result of reflecting a triangle with vertices at $A(a, d)$, $B(b, e)$, and $C(c, f)$ over the x-axis and then reflecting the image back over the x-axis? Use matrices to justify your answer. (Lesson 2-4) **See Answer Appendix.**

4. **(2, 5, 1)** 5. $\left(-\frac{1}{3}, 1\frac{1}{3}, 8\right)$

Extra Credit

In a *glide reflection*, a figure is reflected and then translated in the direction of the line of reflection. Suppose that a glide reflection of triangle ABC with vertices $A(3, 2)$, $B(4, 8)$, and $C(-1, 5)$ is over the x-axis and 5 units in the positive direction. What are the coordinates of the vertices of $\triangle A'B'C'$? **A′(8, −2), B′(9, −8), and C′(4, −5)**

MATRICES

Computers use matrices to solve many types of mathematical problems, but matrices have been around for a long time.

Early Evidence Around 300 B.C., as evidenced by clay tablets found by archaeologists, the Babylonians solved problems that now can be solved by using a system of linear equations. However, the exact method of solution used by the Babylonians has not been determined.

Margaret H. Wright

About 100 B.C.–50 B.C., in ancient China, Chapter 8 of the work *Jiuzhang suanshu (Nine Chapters of the Mathematical Art)* presented a similar problem and showed a solution on a counting board that resembles an augmented coefficient matrix.

There are three types of corn, of which three bundles of the first type, two of the second, and one of the third make 39 measures. Two of the first, three of the second, and one of the third make 34 measures. And one of the first, two of the second, and three of the third make 26 measures. How many measures of grain are contained in one bundle of each type?

Author's table		
1	2	3
2	3	2
3	1	1
26	34	39

The Chinese author goes on to detail how each column can be operated on to determine the solution. This method was later credited to **Carl Friedrich Gauss.**

The Renaissance The concept of the determinant of a matrix, which you will learn about in the next lesson, appeared in Europe and Japan at almost identical times. However, **Seki** of Japan wrote about it first in 1683 with his *Method of Solving the Dissimulated Problems.* Seki's work contained matrices

written in table form like those found in the ancient Chinese writings. Seki developed the pattern for determinants for 2×2, 3×3, 4×4, and 5×5 matrices and used them to solve equations, but not systems of linear equations.

In the same year in Hanover (now Germany), **Gottfried Leibniz** wrote to **Guillaume De l'Hôpital** who lived in Paris, France, about a method he had for solving a system of equations in the form $C + Ax + By = 0$. His method later became known as **Cramer's Rule.**

Modern Era In 1850, the word *matrix* was first used by **James Joscph Sylvester** to describe the tabular array of numbers. Sylvester actually was a lawyer who studied mathematics as a hobby. He shared his interests with **Arthur Cayley,** who is credited with the first published reference to the inverse of a matrix.

Today, computer experts like **Margaret H. Wright** use matrices to solve problems that involve thousands of variables. In her job as Distinguished Member of Technical Staff at a telecommunications company, she applies linear algebra for the solution of real-world problems.

1. See margin.

ACTIVITIES

1. Solve the problem from the *Jiuzhang suanshu* by using a system of equations.

2. Research the types of problems solved by the Babylonians using a system of equations.   **2–3. See students' work.**

3. **interNET CONNECTION** Find out more about the personalities referenced in this article and others who contributed to the history of matrices. Visit **www.amc.glencoe.com**

History of Mathematics **97**

Additional Answer
1. 9.25 of the first type, 4.25 of the second type, and 2.75 of the third type

This feature provides students with an opportunity to see how the mathematics that they are studying was developed. It also gives students a glimpse of how mathematical discoveries are made.

More About the Mathematicians

Seki was born in Fujioka, Kozuke, Japan in 1642. He was self-educated in mathematics, having been introduced to it by a servant of his adopted family. Seki was the first to study determinants.

Gottfried Leibniz was born in 1646 in Leipzig, Saxony (now Germany). Dissatisfied with Aristotle's logic, he developed his own. Later he said that he then learned the logic that was the basis of rigorous mathematical proofs.

Guillaume De l'Hôpital was born in 1661 in Paris, France. l'Hôpital served as a cavalry officer but resigned due to nearsightedness. From then on he directed his attention to mathematics.

Carl Friedrich Gauss was born in Brunswick, Germany, in 1777. Known as "the prince of mathematics," Gauss made contributions to astronomy, geodesy, and electricity as well as mathematics.

James Joseph Sylvester was born in 1814 in London, England. Sylvester taught physics, worked as a lawyer and actuary, and gave private math instruction. His pupils included Florence Nightingale.

Arthur Cayley was born in 1821 in Surrey, England. He worked at the courts with James Joseph Sylvester. The two discussed mathematics as they walked.

Margaret H. Wright holds a B.S. degree, an M.S. degree, and a Ph.D. from Stanford University. Her work involves using matrices to find practical and optimal solutions to real-world problems.

1 FOCUS

5-Minute Check
Lesson 2-4

Use matrices to perform each transformation. Then graph the pre-image and the image on the same coordinate grid.

1. triangle *RST* with vertices $R(-1, 1)$, $S(3, 3)$, and $T(5, 0)$, dilated by a scale factor of 2

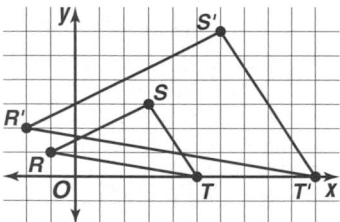

2. square *SQAR* with vertices $S(4, 1)$, $Q(7, 3)$, $A(9, 0)$, and $R(6, -2)$, translated 4 units right and 2 units down

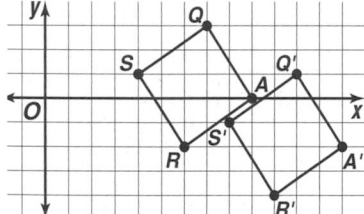

3. parallelogram *ABCD* with vertices $A(-4, 2)$, $B(1, 5)$, $C(6, 6)$, and $D(1, 3)$, reflected across the *y*-axis

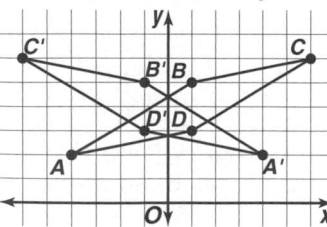

4. triangle *MNP* with vertices $M(5, -3)$, $N(0, 2)$, and $P(2, 2)$, rotated 180°

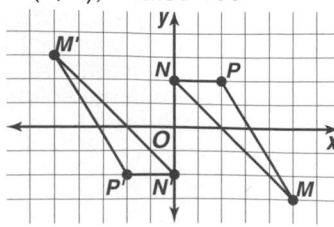

2-5

Determinants and Multiplicative Inverses of Matrices

OBJECTIVES
- Evaluate determinants.
- Find inverses of matrices.
- Solve systems of equations by using inverses of matrices.

Real World Application

INVESTMENTS Marshall plans to invest $10,500 into two different bonds in order to spread out his risk. The first bond has an annual return of 10%, and the second bond has an annual return of 6%. If Marshall expects an 8.5% return from the two bonds, how much should he invest into each bond? *This problem will be solved in Example 5.*

This situation can be described by a system of equations represented by a matrix. You can solve the system by writing and solving a matrix equation.

The term determinant is often used to mean the value of the determinant.

Each square matrix has a **determinant**. The determinant of $\begin{bmatrix} 8 & 4 \\ 7 & 6 \end{bmatrix}$ is a number denoted by $\begin{vmatrix} 8 & 4 \\ 7 & 6 \end{vmatrix}$ or $\det \begin{bmatrix} 8 & 4 \\ 7 & 6 \end{bmatrix}$. The value of a second-order determinant is defined as follows. *A matrix that has a nonzero determinant is called nonsingular.*

Second-Order Determinant	The value of $\det \begin{bmatrix} a_1 & b_1 \\ a_2 & b_2 \end{bmatrix}$, or $\begin{vmatrix} a_1 & b_1 \\ a_2 & b_2 \end{vmatrix}$, is $a_1 b_2 - a_2 b_1$.

Example **1** Find the value of $\begin{vmatrix} 8 & 4 \\ 7 & 6 \end{vmatrix}$.

$$\begin{vmatrix} 8 & 4 \\ 7 & 6 \end{vmatrix} = 8(6) - 7(4) \text{ or } 20$$

The **minor** of an element of any nth-order determinant is a determinant of order $(n - 1)$. This minor can be found by deleting the row and column containing the element.

$$\begin{vmatrix} a_1 & b_1 & c_1 \\ a_2 & b_2 & c_2 \\ a_3 & b_3 & c_3 \end{vmatrix} \quad \textit{The minor of } a_1 \textit{ is } \begin{vmatrix} b_2 & c_2 \\ b_3 & c_3 \end{vmatrix}.$$

One method of evaluating an *n*th-order determinant is expanding the determinant by minors. The first step is choosing a row, any row, in the matrix. At each position in the row, multiply the *element* times its *minor* times its *position sign,* and then add the results together for the whole row. The position signs in a matrix are alternating positives and negatives, beginning with a positive in the first row, first column.

$$\begin{bmatrix} + & - & + & \cdots \\ - & + & - & \cdots \\ + & - & + & \cdots \\ \vdots & \vdots & \vdots & \vdots \end{bmatrix}$$

Resource Manager

 Reproducible Masters
- *Study Guide and Practice,* pp. 29–30
- *Enrichment,* p. 17
- *TI-92 & CBL Interdisciplinary Activities,* pp. 1–4
- *Spreadsheet Applications,* p. 2

 Transparencies
- 5-Minute Check, 2-5
- Teaching, 2-5

| Third-Order Determinant | $\begin{vmatrix} a_1 & b_1 & c_1 \\ a_2 & b_2 & c_2 \\ a_3 & b_3 & c_3 \end{vmatrix} = a_1 \begin{vmatrix} b_2 & c_2 \\ b_3 & c_3 \end{vmatrix} - b_1 \begin{vmatrix} a_2 & c_2 \\ a_3 & c_3 \end{vmatrix} + c_1 \begin{vmatrix} a_2 & b_2 \\ a_3 & b_3 \end{vmatrix}$ |

Example ② Find the value of $\begin{vmatrix} -4 & -6 & 2 \\ 5 & -1 & 3 \\ -2 & 4 & -3 \end{vmatrix}$.

Graphing Calculator Tip

You can use the **det(** option in the **MATH** listings of the **MATRX** menu to find a determinant.

$$\begin{vmatrix} -4 & -6 & 2 \\ 5 & -1 & 3 \\ -2 & 4 & -3 \end{vmatrix} = -4 \begin{vmatrix} -1 & 3 \\ 4 & -3 \end{vmatrix} - (-6) \begin{vmatrix} 5 & 3 \\ -2 & -3 \end{vmatrix} + 2 \begin{vmatrix} 5 & -1 \\ -2 & 4 \end{vmatrix}$$
$$= -4(-9) + 6(-9) + 2(18)$$
$$= 18$$

For any $m \times m$ matrix, the identity matrix, 1, must also be an $m \times m$ matrix.

The **identity matrix for multiplication** for any square matrix A is the matrix I, such that $IA = A$ and $AI = A$. A second-order matrix can be represented by $\begin{bmatrix} a_1 & b_1 \\ a_2 & b_2 \end{bmatrix}$. Since $\begin{bmatrix} a_1 & b_1 \\ a_2 & b_2 \end{bmatrix} \cdot \begin{bmatrix} 1 & 0 \\ 0 & 1 \end{bmatrix} = \begin{bmatrix} 1 & 0 \\ 0 & 1 \end{bmatrix} \cdot \begin{bmatrix} a_1 & b_1 \\ a_2 & b_2 \end{bmatrix} = \begin{bmatrix} a_1 & b_1 \\ a_2 & b_2 \end{bmatrix}$, the matrix $\begin{bmatrix} 1 & 0 \\ 0 & 1 \end{bmatrix}$ is the identity matrix for multiplication for any second-order matrix.

| Identity Matrix for Multiplication | The identity matrix of nth order, I_n, is the square matrix whose elements in the main diagonal, from upper left to lower right, are 1s, while all other elements are 0s. |

Multiplicative inverses exist for some matrices. Suppose A is equal to $\begin{bmatrix} a_1 & b_1 \\ a_2 & b_2 \end{bmatrix}$, a nonzero matrix of second order.

The term inverse matrix generally implies the multiplicative inverse of a matrix.

The **inverse matrix** A^{-1} can be designated as $\begin{bmatrix} x_1 & y_1 \\ x_2 & y_2 \end{bmatrix}$. The product of a matrix A and its inverse A^{-1} must equal the identity matrix, I, for multiplication.

$$\begin{bmatrix} a_1 & b_1 \\ a_2 & b_2 \end{bmatrix}\begin{bmatrix} x_1 & y_1 \\ x_2 & y_2 \end{bmatrix} = \begin{bmatrix} 1 & 0 \\ 0 & 1 \end{bmatrix}$$

$$\begin{bmatrix} a_1x_1 + b_1x_2 & a_1y_1 + b_1y_2 \\ a_2x_1 + b_2x_2 & a_2y_1 + b_2y_2 \end{bmatrix} = \begin{bmatrix} 1 & 0 \\ 0 & 1 \end{bmatrix}$$

From the previous matrix equation, two systems of linear equations can be written as follows.

$$a_1x_1 + b_1x_2 = 1 \qquad\qquad a_1y_1 + b_1y_2 = 0$$
$$a_2x_1 + b_2x_2 = 0 \qquad\qquad a_2y_1 + b_2y_2 = 1$$

Lesson 2-5 Determinants and Multiplicative Inverses of Matrices **99**

By solving each system of equations, values for $x_1, x_2, y_1,$ and y_2 can be obtained.

$$x_1 = \frac{b_2}{a_1b_2 - a_2b_1} \qquad\qquad y_1 = \frac{-b_1}{a_1b_2 - a_2b_1}$$

$$x_2 = \frac{-a_2}{a_1b_2 - a_2b_1} \qquad\qquad y_2 = \frac{a_1}{a_1b_2 - a_2b_1}$$

If a matrix A has a determinant of 0 then A^{-1} does not exist.

The denominator $a_1b_2 - a_2b_1$ is equal to the determinant of *A*. If the determinant of *A* is not equal to 0, the inverse exists and can be defined as follows.

Inverse of a Second-Order Matrix	If $A = \begin{bmatrix} a_1 & b_1 \\ a_2 & b_2 \end{bmatrix}$ and $\begin{vmatrix} a_1 & b_1 \\ a_2 & b_2 \end{vmatrix} \neq 0$, then $A^{-1} = \dfrac{1}{\begin{vmatrix} a_1 & b_1 \\ a_2 & b_2 \end{vmatrix}} \begin{bmatrix} b_2 & -b_1 \\ -a_2 & a_1 \end{bmatrix}$.

$A \cdot A^{-1} = A^{-1} \cdot A = I$, where I is the identity matrix.

Example **3** Find the inverse of the matrix $\begin{bmatrix} 2 & -3 \\ 4 & 4 \end{bmatrix}$.

Graphing Calculator Appendix

For keystroke instruction on how to find the inverse of a matrix, see pages A16-A17.

First, find the determinant of $\begin{bmatrix} 2 & -3 \\ 4 & 4 \end{bmatrix}$.

$\begin{vmatrix} 2 & -3 \\ 4 & 4 \end{vmatrix} = 2(4) - 4(-3)$ or 20

The inverse is $\dfrac{1}{20}\begin{bmatrix} 4 & 3 \\ -4 & 2 \end{bmatrix}$ or $\begin{bmatrix} \frac{1}{5} & \frac{3}{20} \\ -\frac{1}{5} & \frac{1}{10} \end{bmatrix}$ *Check to see if $A \cdot A^{-1} = A^{-1} \cdot A = 1$.*

Just as you can use the multiplicative inverse of 3 to solve $3x = -27$, you can use a matrix inverse to solve a matrix equation in the form $AX = B$. To solve this equation for *X*, multiply each side of the equation by the inverse of *A*. *When you multiply each side of a matrix equation by the same number or matrix, be sure to place the number or matrix on the left or on the right on each side of the equation to maintain equality.*

$$AX = B$$
$$A^{-1}AX = A^{-1}B \quad \text{Multiply each side of the equation by } A^{-1}.$$
$$IX = A^{-1}B \quad A^{-1} \cdot A = I$$
$$X = A^{-1}B \quad IX = X$$

Example **4** Solve the system of equations by using matrix equations.

$$2x + 3y = -17$$
$$x - y = 4$$

Write the system as a matrix equation.

$$\begin{bmatrix} 2 & 3 \\ 1 & -1 \end{bmatrix} \cdot \begin{bmatrix} x \\ y \end{bmatrix} = \begin{bmatrix} -17 \\ 4 \end{bmatrix}$$

To solve the matrix equation, first find the inverse of the coefficient matrix.

$$\dfrac{1}{\begin{vmatrix} 2 & 3 \\ 1 & -1 \end{vmatrix}}\begin{bmatrix} -1 & -3 \\ -1 & 2 \end{bmatrix} = -\dfrac{1}{5}\begin{bmatrix} -1 & -3 \\ -1 & 2 \end{bmatrix} \quad \begin{vmatrix} 2 & 3 \\ 1 & -1 \end{vmatrix} = 2(-1) - (1)(3) \text{ or } -5$$

Now multiply each side of the matrix equation by the inverse and solve.

$$-\frac{1}{5}\begin{bmatrix} -1 & -3 \\ -1 & 2 \end{bmatrix} \cdot \begin{bmatrix} 2 & 3 \\ 1 & -1 \end{bmatrix} \cdot \begin{bmatrix} x \\ y \end{bmatrix} = -\frac{1}{5}\begin{bmatrix} -1 & -3 \\ -1 & 2 \end{bmatrix} \cdot \begin{bmatrix} -17 \\ 4 \end{bmatrix}$$

$$\begin{bmatrix} x \\ y \end{bmatrix} = \begin{bmatrix} -1 \\ -5 \end{bmatrix}$$

The solution is $(-1, -5)$.

Example **5** **INVESTMENTS** Refer to the application at the beginning of the lesson. How should Marshall divide his $10,500 investment between the bond with a 10% annual return and a bond with a 6% annual return so that he has a combined annual return on his investments of 8.5%?

First, let x represent the amount to invest in the bond with an annual return of 10%, and let y represent the amount to invest in the bond with a 6% annual return. So, $x + y = 10,500$ since Marshall is investing $10,500.

Write an equation in standard form that represents the amounts invested in both bonds and the combined annual return of 8.5%. That is, the amount of interest earned from the two bonds is the same as if the total were invested in a bond that earns 8.5%.

$$\begin{array}{ll} 10\%x + 6\%y = 8.5\%(x + y) & \textit{Interest on 10\% bond} = 10\%x \\ 0.10x + 0.06y = 0.085(x + y) & \textit{Interest on 6\% bond} = 6\%y \\ 0.10x + 0.06y = 0.085x + 0.085y & \textit{Distributive Property} \\ 0.015x - 0.025y = 0 & \\ \quad\quad 3x - 5y = 0 & \textit{Multiply by 200 to simplify the coefficients.} \end{array}$$

Now solve the system of equations $x + y = 10,500$ and $3x - 5y = 0$. Write the system as a matrix equation and solve.

$$\begin{array}{l} x + y = 10,500 \\ 3x - 5y = 0 \end{array} \longrightarrow \begin{bmatrix} 1 & 1 \\ 3 & -5 \end{bmatrix} \cdot \begin{bmatrix} x \\ y \end{bmatrix} = \begin{bmatrix} 10,500 \\ 0 \end{bmatrix}$$

Multiply each side of the equation by the inverse of the coefficient matrix.

$$-\frac{1}{8}\begin{bmatrix} -5 & -1 \\ -3 & 1 \end{bmatrix} \cdot \begin{bmatrix} 1 & 1 \\ 3 & -5 \end{bmatrix} \cdot \begin{bmatrix} x \\ y \end{bmatrix} = -\frac{1}{8}\begin{bmatrix} -5 & -1 \\ -3 & 1 \end{bmatrix} \cdot \begin{bmatrix} 10,500 \\ 0 \end{bmatrix}$$

$$\begin{bmatrix} x \\ y \end{bmatrix} = \begin{bmatrix} 6562.5 \\ 3937.5 \end{bmatrix}$$

The solution is (6562.5, 3937.5). So, Marshall should invest $6562.50 in the bond with a 10% annual return and $3937.50 in the bond with a 6% annual return.

In-Class Example

Example 5
Metallurgy A metallurgist wants to make 32 kilograms of an alloy with 60% iron. She has quantities of two metals, one with an iron content of 44% and another with an iron content of 92%. How much of each metal should she use? **about 21.3 kg of the 44% metal and about 10.7 kg of the 92% metal**

Addressing Individual Needs

 Visual/Spatial Students who are visual learners may find the lattice method of finding a determinant easier to master. The lattice method uses diagonals to find a determinant. The diagram below shows how to use this method to find the determinant in Example 2 in the Student Edition.
Add the products of the "positive" diagonals and subtract the products of the "negative" diagonals to get
$-12 + 36 + 40 - 4 + 48 - 90 = 18$.

Error Analysis

Watch for students who try to find the multiplicative inverse of the matrix by finding the reciprocal of each element of the matrix.

Prevent by having students check an inverse matrix by multiplying it by the original matrix. The result of the multiplication should be the identity matrix.

Assignment Guide
Core: 15–45 odd, 46, 47, 49, 51, 53–60
Enriched: 14–44 even, 46–60

Additional Answers

2. Sample answer: $\begin{bmatrix} 3 & 2 & 0 \\ 4 & -3 & 5 \end{bmatrix}$ is

not a square matrix; $\begin{bmatrix} 2 & -5 \\ 1 & -1 \\ 0 & 9 \end{bmatrix}$

also has no determinant.

3. Sample answer: $\begin{bmatrix} 1 & 0 & 0 & 0 \\ 0 & 1 & 0 & 0 \\ 0 & 0 & 1 & 0 \\ 0 & 0 & 0 & 1 \end{bmatrix}$.

Study Guide and Practice Masters, pp. 29–30

CHECK FOR UNDERSTANDING

1. Sample answer: a matrix with a nonzero determinant.

Read and study the lesson to answer each question. **2–4. See margin.**

1. **Describe** the types of matrices that are considered to be nonsingular.

2. **Explain** why the matrix $\begin{bmatrix} 3 & 2 & 0 \\ 4 & -3 & 5 \end{bmatrix}$ does not have a determinant. Give another example of a matrix that does not have a determinant.

3. **Describe** the identity matrix under multiplication for a fourth-order matrix.

4. **Write** an explanation as to how you can decide whether the system of equations, $ax + cy = e$ and $bx + dy = f$, has a solution.

Guided Practice   Find the value of each determinant.

5. $\begin{vmatrix} 4 & -1 \\ -2 & 3 \end{vmatrix}$ **10**

6. $\begin{vmatrix} 12 & -26 \\ -15 & 32 \end{vmatrix}$ **−6**

7. $\begin{vmatrix} 4 & 1 & 0 \\ 5 & -15 & -1 \\ -2 & 10 & 7 \end{vmatrix}$ **−413**

8. $\begin{vmatrix} 6 & 4 & -1 \\ 0 & 3 & 3 \\ -9 & 0 & 0 \end{vmatrix}$ **−135**

Find the inverse of each matrix, if it exists.

9. $\begin{bmatrix} -2 & 3 \\ 5 & 7 \end{bmatrix}$ $-\dfrac{1}{29}\begin{bmatrix} 7 & -3 \\ -5 & -2 \end{bmatrix}$

10. $\begin{bmatrix} 4 & 6 \\ 6 & 9 \end{bmatrix}$ **does not exist**

Solve each system of equations by using a matrix equation.

11. $5x + 4y = -3$
$-3x - 5y = -24$ $\left(-\dfrac{111}{13}, \dfrac{129}{13}\right)$

12. $6x - 3y = 63$
$5x - 9y = 85$ **(8, −5)**

13. **Metallurgy**   Aluminum alloy is used in airplane construction because it is strong and lightweight. A metallurgist wants to make 20 kilograms of aluminum alloy with 70% aluminum by using two metals with 55% and 80% aluminum content. How much of each metal should she use?   **8 kg of the metal with 55% aluminum and 12 kg of the metal with 80% aluminum**

EXERCISES

Practice   Find the value of each determinant.

A

14. $\begin{vmatrix} 3 & 4 \\ 2 & 5 \end{vmatrix}$ **7**

15. $\begin{vmatrix} -4 & -1 \\ 0 & -1 \end{vmatrix}$ **4**

16. $\begin{vmatrix} 9 & 12 \\ 12 & 16 \end{vmatrix}$ **0**

17. $\begin{vmatrix} -2 & 3 \\ -2 & 1 \end{vmatrix}$ **4**

18. $\begin{vmatrix} 13 & 7 \\ -5 & -8 \end{vmatrix}$ **−69**

19. $\begin{vmatrix} -6 & 5 \\ 0 & -8 \end{vmatrix}$ **48**

20. $\begin{vmatrix} 4 & -1 & -2 \\ 0 & 2 & 1 \\ 2 & 1 & 3 \end{vmatrix}$ **26**

21. $\begin{vmatrix} 2 & -1 & 3 \\ 3 & 0 & -2 \\ 1 & -3 & 0 \end{vmatrix}$ **−37**

22. $\begin{vmatrix} 8 & 9 & 3 \\ 3 & 5 & 7 \\ -1 & 2 & 4 \end{vmatrix}$ **−90**

23. $\begin{vmatrix} 4 & 6 & 7 \\ 3 & -2 & -4 \\ 1 & 1 & 1 \end{vmatrix}$ **1**

24. $\begin{vmatrix} 25 & 36 & 15 \\ 31 & -12 & -2 \\ 17 & 15 & 9 \end{vmatrix}$ **−3183**

25. $\begin{vmatrix} 1.5 & -3.6 & 2.3 \\ 4.3 & 0.5 & 2.2 \\ -1.6 & 8.2 & 6.6 \end{vmatrix}$ **175.668**

26. Find det A if $A = \begin{bmatrix} 0 & 1 & -4 \\ 3 & 2 & 3 \\ 8 & -3 & 4 \end{bmatrix}$.   **112**

Additional Answer

4. **Sample answer:** the system has a solution if $ad - bc \ne 0$, since you can use the inverse of the matrix $\begin{bmatrix} a & c \\ b & d \end{bmatrix}$ to find the solution.

Find the inverse of each matrix, if it exists. **27–32. See margin.**

27. $\begin{bmatrix} 2 & -3 \\ -2 & -2 \end{bmatrix}$

28. $\begin{bmatrix} 2 & 0 \\ 1 & 0 \end{bmatrix}$

29. $\begin{bmatrix} 4 & 2 \\ 1 & 2 \end{bmatrix}$

30. $\begin{bmatrix} 6 & 7 \\ -6 & 7 \end{bmatrix}$

31. $\begin{bmatrix} -4 & 6 \\ 8 & -12 \end{bmatrix}$

32. $\begin{bmatrix} 9 & 13 \\ 27 & 36 \end{bmatrix}$

33. What is the inverse of $\begin{bmatrix} \frac{3}{4} & -\frac{1}{8} \\ \frac{1}{5} & \frac{1}{2} \end{bmatrix}$? $\begin{bmatrix} \frac{1}{2} & \frac{1}{8} \\ -5 & \frac{3}{4} \end{bmatrix}$

Solve each system by using a matrix equation. **36. (−9, 7)**

34. $4x - y = 1$
    $x + 2y = 7$ **(1, 3)**

35. $9x - 6y = 12$
    $4x + 6y = -12$ **(0, −2)**

36. $x + 5y = 26$
    $3x - 2y = -41$

37. $4x + 8y = 7$ $\left(\frac{7}{12}, \frac{7}{12} \right)$
    $3x - 3y = 0$

38. $3x - 5y = -24$
    $5x + 4y = -3$ **(−3, 3)**

39. $9x + 3y = 1$ $\left(\frac{1}{3}, -\frac{2}{3} \right)$
    $5x + y = 1$

Solve each matrix equation. The inverse of the coefficient matrix is given.

40. $\left(\frac{2}{3}, 1, -\frac{4}{3} \right)$

41. $\left(\frac{2}{9}, -\frac{4}{3}, -\frac{1}{3} \right)$

40. $\begin{bmatrix} 3 & -2 & 3 \\ 1 & 2 & 2 \\ -2 & 1 & -1 \end{bmatrix} \cdot \begin{bmatrix} x \\ y \\ z \end{bmatrix} = \begin{bmatrix} -4 \\ 0 \\ 1 \end{bmatrix}$, if the inverse is $\frac{1}{9} \begin{bmatrix} -4 & 1 & -10 \\ -3 & 3 & -3 \\ 5 & 1 & 8 \end{bmatrix}$.

41. $\begin{bmatrix} -6 & 5 & 3 \\ 9 & -2 & -1 \\ 3 & 1 & 1 \end{bmatrix} \cdot \begin{bmatrix} x \\ y \\ z \end{bmatrix} = \begin{bmatrix} -9 \\ 5 \\ -1 \end{bmatrix}$, if the inverse is $-\frac{1}{9} \begin{bmatrix} -1 & -2 & 1 \\ -12 & -15 & 21 \\ 15 & 21 & -33 \end{bmatrix}$.

Graphing Calculator

Use a graphing calculator to find the value of each determinant.

42. $\begin{vmatrix} -2 & -4 & 2 & -3 \\ 2 & 3 & 6 & 0 \\ 0 & 9 & 4 & -5 \\ 4 & -7 & 1 & 8 \end{vmatrix}$ **−216**

43. $\begin{vmatrix} 2 & -9 & 1 & 8 & 4 \\ -10 & -1 & 2 & 7 & 0 \\ 0 & 4 & -6 & 1 & -8 \\ 6 & -14 & 11 & 0 & 3 \\ 5 & 1 & -3 & 2 & -1 \end{vmatrix}$ **30,143**

Use the algebraic methods you learned in this lesson and a graphing calculator to solve each system of equations.

44. $0.3x + 0.5y = 4.74$
    $12x - 6.5y = -1.2$
    (3.8, 7.2)

45. $x - 2y + z = 7$
    $6x + 2y - 2z = 4$
    $4x + 6y + 4z = 14$ **(2, −1, 3)**

46. 150,000 in 1995 and 240,000 in 1994

Applications and Problem Solving

46. **Industry** The Flat Rock auto assembly plant in Detroit, Michigan, produces three different makes of automobiles. In 1994 and 1995, the plant constructed a total of 390,000 cars. If 90,000 more cars were made in 1994 than in 1995, how many cars were made in each year?

47. **Critical Thinking** Demonstrate that the expression for A^{-1} is the multiplicative inverse of A for any nonsingular second-order matrix. **See margin.**

48. **Chemistry** How many gallons of 10% alcohol solution and 25% alcohol solution should be combined to make 12 gallons of a 15% alcohol solution? **8 gal of 10% and 4 gal of 25%**

Teaching Tip For Exercises 42 and 43, have students recheck the entry of each element in a matrix before solving for the determinant.

Additional Answers

27. $-\frac{1}{10} \begin{bmatrix} -2 & 3 \\ 2 & 2 \end{bmatrix}$

28. does not exist

29. $\frac{1}{6} \begin{bmatrix} 2 & -2 \\ -1 & 4 \end{bmatrix}$

30. $\frac{1}{84} \begin{bmatrix} 7 & -7 \\ 6 & 6 \end{bmatrix}$

31. does not exist

32. $-\frac{1}{27} \begin{bmatrix} 36 & -13 \\ -27 & 9 \end{bmatrix}$

47. Let $A = \begin{bmatrix} a_1 & b_1 \\ a_2 & b_2 \end{bmatrix}$ and $I = \begin{bmatrix} 1 & 0 \\ 0 & 1 \end{bmatrix}$.

$A^{-1} = \begin{bmatrix} \dfrac{b_2}{a_1 b_2 - a_2 b_1} & \dfrac{-b_1}{a_1 b_2 - a_2 b_1} \\ \dfrac{-a_2}{a_1 b_2 - a_2 b_1} & \dfrac{a_1}{a_1 b_2 - a_2 b_1} \end{bmatrix}$

$AA^{-1} = \begin{bmatrix} \dfrac{a_1 b_2 - a_2 b_1}{a_1 b_2 - a_2 b_1} & \dfrac{-a_1 b_1 + b_1 a_1}{a_1 b_2 - a_2 b_1} \\ \dfrac{a_2 b_2 - a_2 b_2}{a_1 b_2 - a_2 b_1} & \dfrac{a_1 b_2 - a_2 b_1}{a_1 b_2 - a_2 b_1} \end{bmatrix}$

$= \begin{bmatrix} 1 & 0 \\ 0 & 1 \end{bmatrix} = I$

Thus, $AA^{-1} = I$.

49. Critical Thinking If $A = \begin{bmatrix} a & b \\ c & d \end{bmatrix}$, does $(A^2)^{-1} = (A^{-1})^2$? Explain. **See Answer Appendix.**

50. Geometry The area of a triangle with vertices at (a, b), (c, d), and (e, f) can

50. $8\frac{1}{2}$ or 8.5 square units

be determined using the equation $A = \frac{1}{2} \begin{vmatrix} a & b & 1 \\ c & d & 1 \\ e & f & 1 \end{vmatrix}$. What is the area of a

triangle with vertices at $(1, -3)$, $(0, 4)$, and $(3, 0)$? (*Hint:* You may need to use the absolute value of the determinant to avoid a negative area.)

51. Retail Suppose that on the first day of a sale, a store sold 38 complete computer systems and 53 printers. During the second day, 22 complete systems and 44 printers were sold. On day three of the sale, the store sold 21 systems and 26 printers. Total sales for these items for the three days were $49,109, $31,614, and $26,353 respectively. What was the unit cost of each of these two selected items? **computer system: $959, printer: $239**

52. Education The following type of problem often appears on placement tests or college entrance exams.

> *Jessi has a total of 179 points on her last two history tests. The second test score is a 7-point improvement from the first score. What are her scores for the two tests?* **first test: 86, second test: 93**

Mixed Review

53. Geometry The vertices of a square are $H(8, 5)$, $I(4, 1)$, $J(0, 5)$, and $K(4, 9)$. Use matrices to determine the coordinates of the square translated 3 units left and 4 units up. (*Lesson 2-4*) **$H'(5, 9)$, $I'(1, 5)$, $J'(-3, 9)$, $K'(1, 13)$**

54. $\begin{bmatrix} 6 & -\frac{21}{4} \\ -3 & 0 \end{bmatrix}$

54. Multiply $\begin{bmatrix} 8 & -7 \\ -4 & 0 \end{bmatrix}$ by $\frac{3}{4}$. (*Lesson 2-3*)

55. Solve the system $x - 3y + 2z = 6$, $4x + y - z = 8$, and $-7x - 5y + 4z = -10$. (*Lesson 2-2*) **infinitely many solutions**

56. Graph $g(x) = -2[\![x + 5]\!]$. (*Lesson 1-7*) **See margin.**

57. Write the standard form of the equation of the line that is perpendicular to $y = -2x + 5$ and passes through the point at $(2, 5)$. (*Lesson 1-5*) **$x - 2y + 8 = 0$**

58. Write the point-slope form of the equation of the line that passes through the points at $(1, 5)$ and $(2, 3)$. Then write the equation in slope-intercept form. (*Lesson 1-4*) **$(y - 5) = -2(x - 1)$ or $(y - 3) = -2(x - 2)$; $y = -2x + 7$**

59. Safety In 1990, the Americans with Disabilities Act (ADA) went into effect. This act made provisions for public places to be accessible to all individuals, regardless of their physical challenges. One of the provisions of the ADA is that ramps should not be steeper than a rise of 1 foot for every 12 feet of horizontal distance. (*Lesson 1-3*)

a. What is the slope of such a ramp? **$\frac{1}{12}$ or approximately 0.0833**

b. What would be the maximum height of a ramp 18 feet long? **1.5 ft**

60. Find $[f \circ g](x)$ and $[g \circ f](x)$ if $f(x) = x^2 + 3x + 2$ and $g(x) = x - 1$. (*Lesson 1-2*) **$x^2 + x$, $x^2 + 3x + 1$**

Extra Practice See p. A29.

Extra Credit

What would be the order of a minor for a sixth-order determinant? What would be the minor of a_1 for the following determinant?

$\begin{vmatrix} a_1 & b_1 & c_1 & d_1 & e_1 & f_1 \\ a_2 & b_2 & c_2 & d_2 & e_2 & f_2 \\ a_3 & b_3 & c_3 & d_3 & e_3 & f_3 \\ a_4 & b_4 & c_4 & d_4 & e_4 & f_4 \\ a_5 & b_5 & c_5 & d_5 & e_5 & f_5 \\ a_6 & b_6 & c_6 & d_6 & e_6 & f_6 \end{vmatrix}$

5;

$\begin{vmatrix} b_2 & c_2 & d_2 & e_2 & f_2 \\ b_3 & c_3 & d_3 & e_3 & f_3 \\ b_4 & c_4 & d_4 & e_4 & f_4 \\ b_5 & c_5 & d_5 & e_5 & f_5 \\ b_6 & c_6 & d_6 & e_6 & f_6 \end{vmatrix}$

61. No, more than one element of the range is paired with the same element of the domain.

61. Determine if the set of points whose coordinates are $(2, 3)$, $(-3, 4)$, $(6, 3)$, $(2, 4)$, and $(-3, 3)$ represent a function. Explain. *(Lesson 1-1)*

62. SAT Practice The radius of circle E is 3. Square $ABCD$ is inscribed in circle E. What is the best approximation for the difference between the circumference of circle E and the perimeter of square $ABCD$? **B**

A 3
B 2
C 1
D 0.5
E 0

4 ASSESS

Open-Ended Assessment
Modeling Have each student create a problem that can be solved by using a 2 × 2 system of equations. Then ask for a demonstration of how to use matrices to solve the problem.

CAREER CHOICES

● Agricultural Manager ●

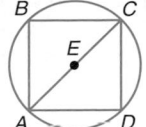

When you hear the word *agriculture,* you may think of a quaint little farmhouse with chickens and cows running around like in the storybooks of your childhood, but today Old McDonald's farm is big business. Agricultural managers guide and assist farmers and ranchers in maximizing their profits by overseeing the day-to-day activities. Their duties are as varied as there are types of farms and ranches.

An agricultural manager may oversee one aspect of the farm, as in feeding livestock on a large dairy farm, or tackle all of the activities on a smaller farm. They also may hire and supervise workers and oversee the purchase and maintenance of farm equipment essential to the farm's operation.

CAREER OVERVIEW

Degree Preferred:
Bachelor's degree in agriculture

Related Courses:
mathematics, science, finance

Outlook:
number of jobs expected to decline through 2006

Number of Farms and Average Farm Size 1975–1998

Source: NASS, Livestock & Economics Branch

interNET CONNECTION For more information on careers in agriculture, visit: **www.amc.glencoe.com**

Lesson 2-5 Determinants and Multiplicative Inverses of Matrices **105**

CAREER CHOICES — AGRICULTURAL MANAGER

Math Connection
- In what year did the two graphs intersect? **Actually, the graphs do not intersect because there are different vertical scales.**
- What trends do the graphs represent? How are these trends related?
  The graphs show the number of farms decreasing each year. If the number of acres of farmland stays approximately the same and the number of farms decreases, the size of the farms will increase.

Enrichment Masters, p. 17

Motivating the Lesson

Discussion Ask students to estimate how long it would take to write out the reduced row-echelon form for the system of equations in Exercise 1. Then discuss the advantages of using a graphing calculator. Point out that a system of equations with fractions would take even less time on a graphing calculator compared with the handwritten method.

A reproduction of this lesson with graphing calculator keystrokes provided can be found in the *Graphing Calculator Keystrokes Masters*, pp. 7–8.

Teaching Tip Remind students that zero coefficients in the system of equations need to be represented in the graphing calculator matrix.

Assignment Guide

All: 1–5

Open-Ended Assessment

Writing Have students solve the system of equations in Exercise 1 by the reduced row-echelon form using paper and pencil. Ask students to write which operation was used for each step of the reduction.

Then have students determine how many different combinations of steps were used by their class. Students should then write a brief summary of the most popular method used by their classmates.

2-5B Augmented Matrices and Reduced Row-Echelon Form

An Extension of Lesson 2-5

OBJECTIVE
• Find reduced row-echelon form of an augmented matrix to solve systems of equations.

Another way to use matrices to solve a system of equations is to use an *augmented matrix*. An augmented matrix is composed of columns representing the coefficients of each variable and the constant term.

system of equations	Identify the coefficients and constants.	augmented matrix
$x - 2y + z = 7$	$1x - 2y + 1z = 7$	$\begin{bmatrix} 1 & -2 & 1 & 7 \\ 3 & 1 & -1 & 2 \\ 2 & 3 & 2 & 7 \end{bmatrix}$
$3x + y - z = 2$	$3x + 1y - 1z = 2$	
$2x + 3y + 2z = 7$	$2x + 3y + 2z = 7$	

Each equation is always written with the constant term on the right.

Through a series of calculations that simulate the elimination methods you used in algebraically solving a system in multiple unknowns, you can find the *reduced row-echelon form* of the matrix, which is $\begin{bmatrix} 1 & 0 & 0 & c_1 \\ 0 & 1 & 0 & c_2 \\ 0 & 0 & 1 & c_3 \end{bmatrix}$, where c_1, c_2, and c_3

A line is often drawn to separate the constants column.

represent constants. The graphing calculator has a function **rref(** that will calculate this form once you have entered the augmented matrix. It is located in the **MATH** submenu when **MATRX** menu is accessed.

For example, if the augmented matrix above is stored as matrix A, you would enter the matrix name after the parenthesis and then insert a closing parenthesis before pressing $\boxed{\text{ENTER}}$. The result is shown at the right.

Use the following exercises to discover how this matrix is related to the solution of the system.

TRY THESE

Write an augmented matrix for each system of equations. Then find the reduced row-echelon form. 1–3. See margin for matrices.

1. $2x + y - 2z = 7$
$x - 2y - 5z = -1$
$4x + y + z = -1$

2. $x + y + z - 6 = 0$
$2x - 3y + 4z - 3 = 0$
$4x - 8y + 4z - 12 = 0$

3. $w + x + y + z = 0$
$2w + x - y - z = 1$
$-w - x + y + z = 0$
$2x + y = 0$

4. (equations) $x = -1, y = 5, z = -2;$ $x = 7, y = 1, z = -2; w = 1, x = -1, y = 2, z = -2$

WHAT DO YOU THINK?

4. Write the equations represented by each reduced row-echelon form of the matrix in Exercises 1-3. How do these equations related to the original system? **See above for equations; they are the solutions for the system.**

5. What would you expect to see on the graphing calculator screen if the constants were irrational or repeating decimals? **The calculator would show the first part of the number and follow it by ... .**

Additional Answers

1. $\begin{bmatrix} 2 & 1 & -2 & 7 \\ 1 & -2 & -5 & -1 \\ 4 & 1 & 1 & -1 \end{bmatrix}$,

$\begin{bmatrix} 1 & 0 & 0 & -1 \\ 0 & 1 & 0 & 5 \\ 0 & 0 & 1 & -2 \end{bmatrix}$,

$(-1, 5, -2)$

2. $\begin{bmatrix} 1 & 1 & 1 & 6 \\ 2 & -3 & 4 & 3 \\ 4 & -8 & 4 & 12 \end{bmatrix}$,

$\begin{bmatrix} 1 & 0 & 0 & 7 \\ 0 & 1 & 0 & 1 \\ 0 & 0 & 1 & -2 \end{bmatrix}$,

$(7, 1, -2)$

3. $\begin{bmatrix} 1 & 1 & 1 & 1 & 0 \\ 2 & 1 & -1 & -1 & 1 \\ -1 & -1 & 1 & 1 & 0 \\ 0 & 2 & 1 & 0 & 0 \end{bmatrix}$,

$\begin{bmatrix} 1 & 0 & 0 & 0 & 1 \\ 0 & 1 & 0 & 0 & -1 \\ 0 & 0 & 1 & 0 & 2 \\ 0 & 0 & 0 & 1 & -2 \end{bmatrix}$,

$(1, -1, 2, -2)$

Solving Systems of Linear Inequalities

OBJECTIVES
- Graph systems of inequalities.
- Find the maximum or minimum value of a function defined for a polygonal convex set.

SHIPPING Package delivery services add extra charges for oversized parcels or those requiring special handling. An oversize package is one in which the sum of the length and the *girth* exceeds 84 inches. The *girth* of a package is the distance around the package. For a rectangular package, its girth is the sum of twice the width and twice the height. A package requiring special handling is one in which the length is greater than 60 inches. What size packages qualify for both oversize and special handling charges?

The situation described in the problem above can be modeled by a **system of linear inequalities.** To solve a system of linear inequalities, you must find the ordered pairs that satisfy both inequalities. One way to do this is to graph both inequalities on the same coordinate plane. The intersection of the two graphs contains points with ordered pairs in the solution set. If the graphs of the inequalities do not intersect, then the system has no solution.

Example ❶ **SHIPPING** What size packages qualify for both oversize and special handling charges when shipping?

First write two inequalities that represent each type of charge. Let ℓ represent the length of a package and g represent its girth.
Oversize: $\ell + g > 84$
Special handling: $\ell > 60$

Neither of these inequalities includes the boundary line, so the lines are dashed. The graph of $\ell + g > 84$ is composed of all points above the line $\ell + g = 84$. The graph of $\ell > 60$ includes all points to the right of the line $\ell = 60$. The green area is the solution to the system of inequalities. That is, the ordered pair for any point in the green area satisfies both inequalities. For example, (90, 20) is a length greater than 90 inches and a girth of 20 inches which represents an oversize package that requires special handling.

Not every system of inequalities has a solution. For example, $y > x + 3$ and $y < x - 1$ are graphed at the right. Since the graphs have no points in common, there is no solution.

Resource Manager

 Reproducible Masters
- *Study Guide and Practice,* pp. 31–32
- *Enrichment,* p. 18
- *Assessment and Evaluation,* p. 44

 Transparencies
- 5-Minute Check, 2–6
- *Teaching, 2–6*

5-Minute Check
Lesson 2-5

Find the value of each determinant.

1. $\begin{vmatrix} 8 & -4 \\ -3 & 2 \end{vmatrix}$ 4

2. $\begin{vmatrix} -2 & -3 & 1 \\ 9 & -7 & -4 \\ 2 & 3 & 0 \end{vmatrix}$ 41

3. Find the inverse of the matrix $\begin{bmatrix} 4 & 2 \\ -1 & -9 \end{bmatrix}$, if it exists.

$-\dfrac{1}{34}\begin{bmatrix} -9 & -2 \\ 1 & 4 \end{bmatrix}$

4. Solve the system of equations by using a matrix equation.
$x + 4y = -6$
$3x + 5y = 8$ $\left(\dfrac{62}{7}, -\dfrac{26}{7}\right)$

5. Solve the matrix equation
$\begin{bmatrix} 0 & -1 & 1 \\ 1 & 2 & 1 \\ 2 & 1 & -1 \end{bmatrix} \cdot \begin{bmatrix} x \\ y \\ z \end{bmatrix} = \begin{bmatrix} 0 \\ -1 \\ 3 \end{bmatrix}$

if the inverse is

$-\dfrac{1}{6}\begin{bmatrix} -3 & 0 & -3 \\ 3 & -2 & 1 \\ -3 & -2 & 1 \end{bmatrix}$.

$\left(1\dfrac{1}{2}, -\dfrac{5}{6}, -\dfrac{5}{6}\right)$

Motivating the Lesson

Hands-On Activity Use spaghetti or toothpicks to create various polygons on grid paper. Have students try to determine the coordinates of the points of intersection of the sides of the polygons. If possible, also have them determine the equations of the sides.

Teaching Tip Point out that it is important to consider which scale to use when graphing inequalities. The application in Example 1 on page 107 uses a large scale because the x- and y- intercepts for one of the equations are both 84.

In-Class Examples

Example 1

Sales Belan Chu is a graphic artist who makes greeting cards. Her startup cost will be $1500 plus $ 0.40 per card. In order for her to remain competitive with large companies, she must sell her cards for no more than $1.70 each. How many cards must Ms. Chu sell in order to make a profit?

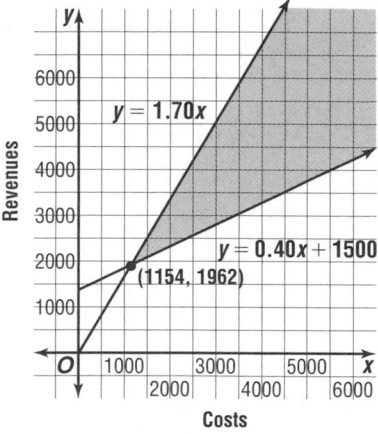

The shaded area of the graph represents the solutions. She must sell more than 1200 cards in order to make a profit.

Example 2

a. Solve the system of inequalities by graphing.
$x \geq 0$
$y \geq 0$
$x + y \leq 5$

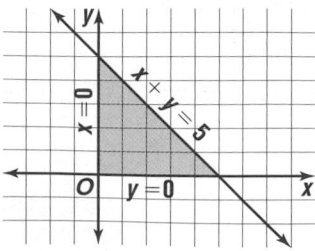

b. Name the coordinates of the polygonal convex set.

A system of more than two linear inequalities can have a solution that is a bounded set of points. A bounded set of all points on or inside a convex polygon graphed on a coordinate plane is called a **polygonal convex set.**

Example **a.** Solve the system of inequalities by graphing.
$x \geq 0$
$y \geq 0$
$2x + y \leq 4$

b. Name the coordinates of the vertices of the polygonal convex set.

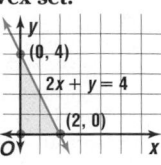

a. Since each inequality includes an equality, the boundary lines will be solid. The shaded region shows points that satisfy all three inequalities.

b. The region is a triangle whose vertices are the points at $(0, 0)$, $(0, 4)$ and $(2, 0)$.

An expression whose value depends on two variables is a function of two variables. For example, the value of $6x + 7y - 9$ is a function of x and y and can be written $f(x, y) = 6x + 7y - 9$. The expression $f(3, 5)$ would then stand for the value of the function f when x is 3 and y is 5.

$$f(3, 5) = 6(3) + 7(5) - 9 \text{ or } 44.$$

Sometimes it is necessary to find the maximum or minimum value that a function has for the points in a polygonal convex set. Consider the function $f(x, y) = 5x - 3y$, with the following inequalities forming a polygonal convex set.

$$y \geq 0 \qquad -x + y \leq 2 \qquad 0 \leq x \leq 5 \qquad x + y \leq 6$$

You may need to use algebraic methods to determine the coordinates of the vertices of the convex set.

By graphing the inequalities and finding the intersection of the graphs, you can determine a polygonal convex set of points for which the function can be evaluated. The region shown at the right is the polygonal convex set determined by the inequalities listed above. Since the polygonal convex set has infinitely many points, it would be impossible to evaluate the function for all of them. However, according to the **Vertex Theorem,** a function such as $f(x, y) = 5x - 3y$ need only be evaluated for the coordinates of the vertices of the polygonal convex boundary in order to find the maximum and minimum values.

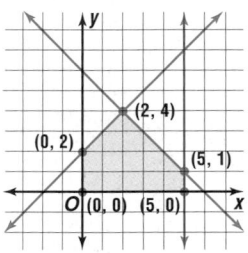

Vertex Theorem	The maximum or minimum value of $f(x, y) = ax + by + c$ on a polygonal convex set occurs at a vertex of the polygonal boundary.

The value of $f(x, y) = 5x - 3y$ at each vertex can be found as follows.

$f(x, y) = 5x - 3y$
$f(0, 0) = 5(0) - 3(0) = 0$
$f(0, 2) = 5(0) - 3(2) = -6$

$f(2, 4) = 5(2) - 3(4) = -2$
$f(5, 1) = 5(5) - 3(1) = 22$
$f(5, 0) = 5(5) - 3(0) = 25$

Therefore, the maximum value of $f(x, y)$ in the polygon is 25, and the minimum is -6. The maximum occurs at $(5, 0)$, and the minimum occurs at $(0, 2)$.

From the Classroom of...

Todd Parker
West Jefferson High School
West Jefferson, Ohio

"We use waffles, syrup, and whipped cream to graph inequalities. Students draw an inequality out of a hat and then graph the inequality on a waffle using the whipped cream to draw the boundary equation and the syrup for shading. Students are not allowed to eat their creations until the inequalities are graphed correctly."

Example ③ Find the maximum and minimum values of $f(x, y) = x - y + 2$ for the polygonal convex set determined by the system of inequalities.

$$x + 4y \leq 12 \qquad 3x - 2y \geq -6 \qquad x + y \geq -2 \qquad 3x - y \leq 10$$

First write each inequality in slope-intercept form for ease in graphing the boundaries.

Boundary **a**	Boundary **b**	Boundary **c**	Boundary **d**
$x + 4y \leq 2$	$3x - 2y \geq -6$	$x + y \geq -2$	$3x - y \leq 10$
$4y \leq -x + 12$	$-2y \geq -3x - 6$	$y \geq -x - 2$	$-y \leq -3x + 10$
$y \leq -\frac{1}{4}x + 3$	$y \leq \frac{3}{2}x + 3$		$y \geq 3x - 10$

You can use the matrix approach from Lesson 2-5 to find the coordinates of the vertices.

Graph the inequalities and find the coordinates of the vertices of the resulting polygon.

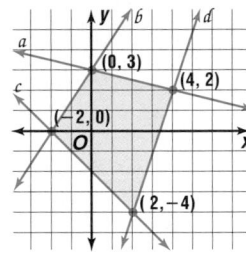

The coordinates of the vertices are $(-2, 0), (2, -4), (4, 2), (0, 3)$.

Now evaluate the function
$f(x, y) = x - y + 2$ at each vertex.
$f(-2, 0) = -2 - 0 + 2$ or 0
$f(2, -4) = 2 - (-4) + 2$ or 8
$f(4, 2) = 4 - 2 + 2$ or 4
$f(0, 3) = 0 - 3 + 2$ or -1

The maximum value of the function is 8, and the minimum value is -1.

CHECK FOR UNDERSTANDING

Communicating Mathematics

Read and study the lesson to answer each question.

1. Refer to the application at the beginning of the lesson. **a–b. See margin.**

   a. **Define** the *girth* of a rectangular package.

   b. **Name** some objects that might be shipped by a package delivery service and classified as oversized *and* requiring special handling.

2. Tomas is correct. There are functions in which the coordinates of more than one vertex will yield the same value for the function.

2. **You Decide** Marcel says there is only one vertex that will yield a maximum for any given function. Tomas says that if the numbers are correct, there could be two vertices that yield the same maximum. Who is correct? Explain your answer.

3. **Determine** how many vertices of a polygonal convex set you might expect if the system defining the set contained five inequalities, no two of which are parallel.
   3–4. See margin.

Guided Practice

4. Solve the system of inequalities by graphing. $x + 2y \geq 4 \qquad x - y \leq 3$

5. See margin for graph; $(-1, 0)$, $(-1, 3)$, $(0, 4)$, $(7, 0.5)$, $(7, 0)$

5. Solve the system of inequalities by graphing. Name the coordinates of the vertices of the polygonal convex set.
   $$y \geq 0 \qquad -1 \leq x \leq 7 \qquad -x + y \leq 4 \qquad x + 2y \leq 8$$

Find the maximum and minimum values of each function for the polygonal convex set determined by the given system of inequalities.

6. $f(x, y) = 4x + 3y$
   $4y \leq x + 8$
   $x + y \geq 2$
   $y \geq 2x - 5$ **25, 6**

7. $f(x, y) = 3x - 4y$
   $x - 2y \geq -7$
   $x + y \geq 8$
   $2x - y \leq 7$ **3, −11**

Addressing Individual Needs

Interpersonal Have students work in cooperative groups to make up examples of inequalities. Have them graph the inequalities and apply the Vertex Theorem with the function $f(x, y) = x + 2y + 1$.

Teaching Tip In Examples 2 and 3, have students verify that the vertices of the polygon satisfy each inequality of the system.

In-Class Example

Example 3
Find the maximum and minimum values of $f(x) = y - 2x + 5$ for the polygonal convex set determined by the system of inequalities.
$x \geq 1 \qquad y \geq 2$
$y \leq 8 \qquad x + y \geq 5$
$2x + y \leq 14$

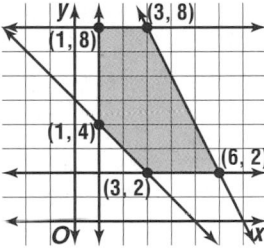

Maximum is 11 at (1, 8); minimum is −5 at (6, 2).

3 PRACTICE/APPLY

Error Analysis
Watch for students who graph inequalities incorrectly.

Prevent by having students use a check point to verify that the correct region is shaded.

Study Guide and Practice Masters, pp. 31–32

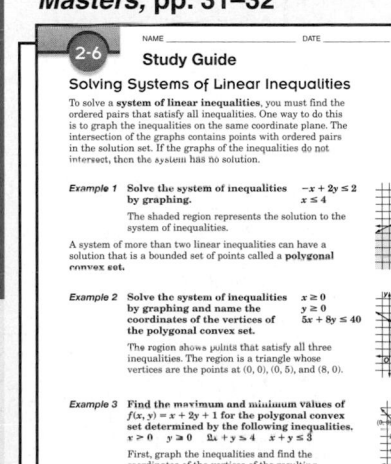

Additional Answers *(p. 109)*

1a. the sum of twice the width and twice the height

1b. Sample answer: skis, fishing rods

3. You might expect five vertices; however, if the equations were dependent or if they did not intersect to form the sides of a convex polygon, there would be fewer vertices.

4.

5.

9.

10.

---

8. Business Gina Chuez has considered starting her own custom greeting card business. With an initial start-up cost of $1500, she figures it will cost $0.45 to produce each card. In order to remain competitive with the larger greeting card companies, Gina must sell her cards for no more than $1.70 each. To make a profit, her income must exceed her costs. How many cards must she sell before making a profit?   **at least 1200 cards**

EXERCISES

Practice

 A

Solve each system of inequalities by graphing.   **9–11. See margin.**

9. $y + x \geq 1$
$y - x \leq -1$

10. $y > 1$
$y < -3x + 3$
$y > -3x + 1$

11. $2x + 5y < 25$
$y < 3x - 2$
$5x - 7y < 14$

12. Determine if $(3, -2)$ belongs to the solution set of the system of inequalities $y < \frac{1}{3}x + 5$ and $y < 2x + 1$. Verify your answer.   **Yes; see margin for verification.**

Solve each system of inequalities by graphing. Name the coordinates of the vertices of the polygonal convex set.   **13–15. See Answer Appendix for graphs.**

13. $\left(\frac{3}{5}, 3\frac{1}{5}\right)$,
$\left(-\frac{2}{5}, 1\frac{1}{5}\right)$,
$\left(1\frac{3}{5}, \frac{1}{5}\right)$

 B

13. $y \geq -0.5x + 1$
$y \leq -3x + 5$
$y \leq 2x + 2$

14. $x \leq 0$
$y + 3 \geq 0$
$x \geq y$
(0, 0), (0, −3), (−3, −3)

15. $y \geq 0$
$y - 5 \leq 0$
$y + x \leq 7$
$5x + 3y \geq 20$

15. **(2, 5), (7, 0), (4, 0), (1, 5)**

16. Find the maximum and minimum values of $f(x, y) = 8x + y$ for the polygonal convex set having vertices at $(0, 0)$, $(4, 0)$, $(3, 5)$, and $(0, 5)$.   **32, 0**

Find the maximum and minimum values of each function for the polygonal convex set determined by the given system of inequalities.

17. $f(x, y) = 3x + y$
$x \leq 5$
$y \geq 2$
$2x - 5y \geq -10$ **19, 2**

18. $f(x, y) = y - x$
$y \leq 4 - 2x$
$x + 2 \geq 2$
$y \geq 0$ **4, −2**

19. $f(x, y) = x + y$
$y \leq 6$
$4x - 5y \geq -10$
$2x - 5y \leq -10$ **16, 2**

17–22. See Answer Appendix for graphs.

C

20. $f(x, y) = 4x + 2y + 7$
$x \geq 0$
$y \geq 1$
$x + y \leq 4$ **21, 9**

21. $f(x, y) = 2x - y$
$y \leq 4x + 6$
$x + 4y \leq 7$
$2x + y \leq 7$
$x - 6y \leq 10$ **9, −4**

22. $f(x, y) = -2x + y + 5$
$2 \leq y \leq 8$
$x \geq 1$
$2x + y + 2 \leq 16$
$y \geq 5 - x$ **11, −5**

Applications and Problem Solving

23. Geometry Find the system of inequalities that will define a polygonal convex set that includes all points in the interior of a square whose vertices are $A(4, 4)$, $B(4, -4)$, $C(-4, -4)$, and $D(-4, 4)$.   **$x \leq 4$, $x \geq -4$, $y \leq 4$, $y \geq -4$**

24. Critical Thinking Write a system of more than two linear inequalities whose set of solutions is not bounded.   **Sample answer: $y \geq 3$, $x \geq 4$, $4x + 3y \geq 12$**

25. Critical Thinking A polygonal convex set is defined by the following system of inequalities.   **a–b. See margin.**

$y \leq 16 - x$
$0 \leq 2y \leq 17$

$3y \geq -2x + 11$
$y \leq 3x + 1$

$y \geq 2x - 13$
$y \geq 7 - 2x$

a. Determine which lines intersect and solve pairs of equations to determine the coordinates of each vertex.

b. Find the maximum and minimum values for $f(x, y) = 5x + 6y$ in the set.

11.

12. It is true for both inequalities:

$y < \frac{1}{3}x + 5$ $y < 2x + 1$

$-2 < \frac{1}{3}(3) + 5$ $-2 < 2(3) + 1$

$-2 < 6$ **true** $-2 < 7$ **true**

26. Business Christine's Butter Cookies sells large tins of butter cookies and small tins of butter cookies. The factory can prepare at most 200 tins of cookies a day. Each large tin of cookies requires 2 pounds of butter, and each small tin requires 1 pound of butter, with a maximum of 300 pounds of butter available each day. The profit from each day's cookie production can be estimated by the function $f(x, y) = \$6.00x + \$4.80y$, where x represents the number of large tins sold and y the number of small tins sold. Find the maximum profit that can be expected in a day. **$1080**

27d. The maximum number of customers can be reached by renting 120 ft² at Main St.

27. Fund-raising The Band Boosters want to open a craft bazaar to raise money for new uniforms. Two sites are available. A Main Street site costs $10 per square foot per month. The other site on High Street costs $20 per square foot per month. Both sites require a minimum rental of 20 square feet. The Main Street site has a potential of 30 customers per square foot, while the High Street site could see 40 customers per square foot. The budget for rental space is $1200 per month. The Band Boosters are studying their options for renting space at both sites. **a. See Answer Appendix.**

a. Graph the polygonal convex region represented by the cost of renting space.

b. Determine what function would represent the possible number of customers per square foot at both locations. $f(x, y) = 30x + 40y$

c. If space is rented at both sites, how many square feet of space should the Band Boosters rent at each site to maximize the number of potential customers? **80 ft² at the Main St. site and 20 ft² at the High St. site**

d. Suppose you were president of the Band Boosters. Would you rent space at both sites or select one of the sites? Explain your answer.

28a. 3 is $3 profit on each batch of garlic dressing and 2 is $2 profit on each batch of raspberry dressing

28b. 3 batches garlic dressing, 4 batches raspberry dressing

28. Culinary Arts A gourmet restaurant sells two types of salad dressing, garlic and raspberry, in their gift shop. Each batch of garlic dressing requires 2 quarts of oil and 2 quarts of vinegar. Each batch of raspberry dressing requires 3 quarts of oil and 1 quart of vinegar. The chef has 18 quarts of oil and 10 quarts of vinegar on hand for making the dressings that will be sold

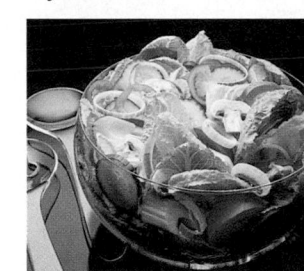

in the gift shop that week. If x represents the number of batches of garlic dressing sold and y represents the batches of raspberry dressing sold, the total profits from dressing sold can be expressed by the function $f(x, y) = 3x + 2y$.

a. What do you think the 3 and 2 in the function $f(x, y) = 3x + 2y$ represent?

b. How many batches of each types of dressing should the chef make to maximize the profit on sales of the dressing?

Mixed Review

29. Find the inverse of $\begin{bmatrix} 2 & 1 \\ -3 & 2 \end{bmatrix}$. *(Lesson 2-5)* $\frac{1}{7}\begin{bmatrix} 2 & -1 \\ 3 & 2 \end{bmatrix}$

30. Graph $y < -2x + 8$. *(Lesson 1-8)* **30–31. See Answer Appendix.**

31. **Scuba Diving** Graph the equation $d + 33 = 33p$, which relates atmospheres of pressure p to ocean depth d in feet. *(Lesson 1-3)*

32. State the domain and range of the relation $\{(16, -4), (16, 4)\}$. Is this relation a function? Explain. *(Lesson 1-1)* **{16}, {−4, 4}; no, two *y*-values for one *x*-value**

33. **SAT Practice Grid-In** What is the sum of four integers whose mean is 15? **60**

Extra Practice See p. A29.

Extra Credit

Could the graph of a polygonal region ever contain more than one maximum value of a function? Explain. **Yes; if the graph of the function also contains a side of the polygonal region, the function will have more than one maximum or minimum value.**

---

4 ASSESS

Open-Ended Assessment
Modeling Use a grid transparency and spaghetti or toothpicks to construct a polygon on the coordinate plane. Explain how the polygon relates to the lesson. What do the interior points of the polygon represent?

Chapter 2, Quiz C (Lessons 2-5 and 2-6) is available in the *Assessment and Evaluation Masters,* p. 44.

Additional Answers (p. 110)

25a. vertices: $\left(5\frac{1}{2}, 0\right)$, $\left(6\frac{1}{2}, 0\right)$, $\left(9\frac{2}{3}, 6\frac{1}{3}\right)$, $\left(7\frac{1}{2}, 8\frac{1}{2}\right)$, $\left(2\frac{1}{2}, 8\frac{1}{2}\right)$, $\left(1\frac{1}{5}, 4\frac{3}{5}\right)$, $\left(2\frac{1}{2}, 2\right)$

25b. max at $\left(7\frac{1}{2}, 8\frac{1}{2}\right) = 88\frac{1}{2}$; min at $\left(2\frac{1}{2}, 2\right) = 24\frac{1}{2}$

Enrichment Masters, p. 18

NAME _____ DATE _____ PERIOD _____

2-6 Enrichment

Reading Mathematics: Reading the Graph of a System of Inequalities

Like all other kinds of graphs, the graph of a system of linear inequalities is useful to us only as long as we are able to "read" the information presented on it.

The line of each equation separates the xy-plane into two **half-planes.** Shading is used to tell which half-plane satisfies the inequality bounded by the line. Thus, the shading below the boundary line $y = 4$ indicates the solution set of $y \le 4$. By drawing horizontal or vertical shading lines or lines with slopes of 1 or −1, and by using different colors, you will make it easier to "read" your graph.

The shading in between the boundary lines $x \ge 0$ and $x \le 5$, indicate that the inequality $0 \le x \le 5$ is equivalent to the *intersection* $x \ge 0 \cap x \le 5$.

Similarly, the solution set of the system, indicated by the cross-hatching, is the intersection of the solutions of all the inequalities in the system. We can use set notation to write a description of this solution set:

$\{(x, y) | 0 \le x \le 5 \cap y \le 4 \cap x + 2y \ge 7\}$.

We read this " The set of all ordered pairs, (x, y), such that x is greater than or equal to 0 and x is less than or equal to 5 and y is less than or equal to 4 and x + 2y is greater than or equal to 7."

The graph also shows where any maximum and minimum values of the function f occur. Notice, however, that these values occur only at points belonging to the set containing the *union* of the vertices of the convex polygonal region.

Use the graph at the right to show which numbered region or regions belong to the graph of the solution set of each system.

1. $x \ge 0$
   $y \le x$
   5, 6, 7

2. $0 \le x \le 4$
   $y \le x$
   $x + y \ge 4$
   5

3. The domain of $f(x, y) = 2x + y$ is the xy-plane. If we graphed this function, where would we represent its range? **On a 3rd axis, perpendicular to the xy-plane.**

© Glencoe/McGraw-Hill   T18   *Advanced Mathematical Concepts*

1 FOCUS

5-Minute Check
Lesson 2-6

1. Solve the system of inequalities by graphing.
$x - y < -2$
$x - y > 4$ **no solution**

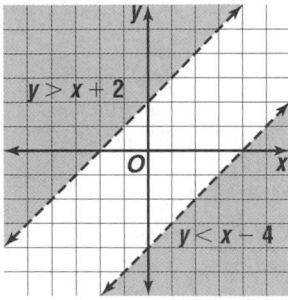

2. Solve the system of inequalities by graphing. Name the coordinates of the vertices of the polygonal convex set.
$x \geq 0$
$y \geq 0$
$x + y \leq 6$ **(0, 0), (6, 0), (0, 6)**

3. Find the maximum and minimum values of $f(x, y) = 2x + y + 1$ for the polygonal convex set determined by the given system of inequalities.
$x \geq 0$ $x + y \leq 5$
$y \geq 0$ $2x + 3y \leq 12$
max 11 at (5, 0); min 1 at (0, 0)

Motivating the Lesson
Discussion Ask students how they think business owners determine how much of each of their products to make. Also, ask how work schedules might be created.

OBJECTIVES
- Use linear programming procedures to solve applications.
- Recognize situations where exactly one solution to a linear programming application may not exist.

Linear Programming

MILITARY SCIENCE
When the U.S. Army needs to determine how many soldiers or officers to put in the field, they turn to mathematics. A system called the *Manpower Long-Range Planning System* (MLRPS) enables the army to meet the personnel needs for 7- to 20-year planning periods. Analysts are able to effectively use the MLRPS to simulate gains, losses, promotions, and reclassifications. This type of planning requires solving up to 9,060 inequalities with 28,730 variables! However, with a computer, a problem like this can be solved in less than five minutes.

The Army's MLRPS uses a procedure called **linear programming.** Many practical applications can be solved by using this method. The nature of these problems is that certain **constraints** exist or are placed upon the variables, and some function of these variables must be maximized or minimized. The constraints are often written as a system of linear inequalities.

The following procedure can be used to solve linear programming applications.

Linear Programming Procedure

1. Define variables.
2. Write the constraints as a system of inequalities.
3. Graph the system and find the coordinates of the vertices of the polygon formed.
4. Write an expression whose value is to be maximized or minimized.
5. Substitute values from the coordinates of the vertices into the expression.
6. Select the greatest or least result.

In Lesson 2-6, you found the maximum and minimum values for a given function in a defined polygonal convex region. In linear programming, you must use your reasoning abilities to determine the function to be maximized or minimized and the constraints that form the region.

Example **1** **MANUFACTURING** Suppose a lumber mill can turn out 600 units of product each week. To meet the needs of its regular customers, the mill must produce 150 units of lumber and 225 units of plywood. If the profit for each unit of lumber is $30 and the profit for each unit of plywood is $45, how many units of each type of wood product should the mill produce to maximize profit?

Define variables.   Let x = the units of lumber produced.
Let y = the units of plywood produced.

112   *Chapter 2   Systems of Linear Equations and Inequalities*

Resource Manager

 Reproducible Masters
- *Study Guide and Practice,* pp. 33–34
- *Enrichment,* p. 19
- *Assessment and Evaluation,* p. 44

 Transparencies
- 5-Minute Check, 2-7
- Teaching, 2-7

Write inequalities.	$x \geq 150$	*There cannot be less than 150 units of lumber produced.*
	$y \geq 225$	*There cannot be less than 225 units of plywood produced.*
	$x + y \leq 600$	*The maximum number of units produced is 600.*

Graph the system.

The vertices are at (150, 225), (375, 225), and (150, 450).

Write an expression. Since profit is $30 per unit of lumber and $45 per unit of plywood, the profit function is $P(x, y) = 30x + 45y$.

Substitute values.
$P(150, 225) = 30(150) + 45(225)$ or 14,625
$P(375, 225) = 30(375) + 45(225)$ or 21,375
$P(150, 450) = 30(150) + 45(450)$ or 24,750

Answer the problem. The maximum profit occurs when 150 units of lumber are produced and 450 units of plywood are produced.

In certain circumstances, the use of linear programming is not helpful because a polygonal convex set is not defined. Consider the graph at the right, based on the following constraints.

$x \geq 0$
$y \geq 0$
$x \leq 3$
$2x - 3y \geq 12$

The constraints do not define a region with any points in common in Quadrant I. When the constraints of a linear programming application cannot be satisfied simultaneously, then the problem is said to be **infeasible.**

Sometimes the region formed by the inequalities in a linear programming application is **unbounded.** In that case, an optimal solution for the problem may not exist. Consider the graph at the right. A function like $f(x, y) = x + 2y$ has a *minimum* value at (5, 3), but it is not possible to find a maximum value.

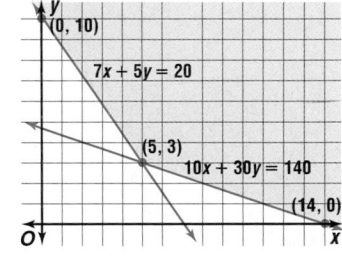

2 TEACH

Teaching Tip In Example 1, point out that the constraints come from the conditions in the problem. Note that every set of constraints is sufficient to solve a problem.

In-Class Example

Example 1
Manufacturing The profit on each set of cassettes that is manufactured by MusicMan, Inc., is $8. The profit on a single cassette is $2. Machines A and B are used to produce both types of cassettes. Each set takes nine minutes on Machine A and three minutes on Machine B. Each single takes one minute on Machine A and one minute on Machine B. If Machine A is run for 54 minutes and Machine B is run for 42 minutes, determine the combination of cassettes that can be manufactured during the time period that most effectively generates profit within the given constraints.   **2 sets and 36 singles**

It is also possible for a linear programming application to have two or more optimal solutions. When this occurs, the problem is said to have **alternate optimal solutions.** This usually occurs when the graph of the function to be maximized or minimized is parallel to one side of the polygonal convex set.

Example **SMALL BUSINESS** The Woodell Carpentry Shop makes bookcases and cabinets. Each bookcase requires 15 hours of woodworking and 9 hours of finishing. The cabinets require 10 hours of woodworking and 4.5 hours of finishing. The profit is $60 on each bookcase and $40 on each cabinet. There are 70 hours available each week for woodworking and 36 hours available for finishing. How many of each item should be produced in order to maximize profit?

Define variables.
Let b = the number of bookcases produced.
Let c = the number of cabinets produced.

Write inequalities.

$b \geq 0, c \geq 0$ *There cannot be less than 0 bookcases or cabinets.*
$15b + 10c \leq 70$ *No more than 70 hours of woodworking are available.*
$9b + 4.5c \leq 36$ *No more than 36 hours of finishing are available.*

Graph the system.

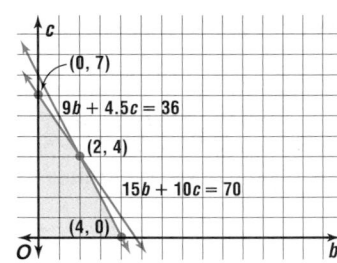

The vertices are at $(0, 7)$, $(2, 4)$, $(4, 0)$, and $(0, 0)$.

Write an expression.
Since profit on each bookcase is $60 and the profit on each cabinet is $40, the profit function is $P(b, c) = 60b + 40c$.

Substitute values.
$P(0, 0) = 60(0) + 40(0)$ or 0
$P(0, 7) = 60(0) + 40(7)$ or 280
$P(2, 4) = 60(2) + 40(4)$ or 280
$P(4, 0) = 60(4) + 40(0)$ or 240

Answer the problem.
The problem has alternate optimal solutions. The shop will make the same profit if they produce 2 bookcases and 4 cabinets as it will from producing 7 cabinets and no bookcases.

Addressing Individual Needs

 Logical Students who are logical learners may benefit from using charts to organize the information given in application problems. Have students make a chart of the vertices of the polygonal convex set. Write the ordered pairs in one column and the value of the function in another.

Communicating Mathematics

Read and study the lesson to answer each question.   1–3. See margin.

1. **Explain** why the inequalities $x \geq 0$ and $y \geq 0$ are usually included as constraints in linear programming applications.

2. **Discuss** the difference between the graph of the constraints when a problem is infeasible and a graph whose constraints yield an unbounded region.

3. **Write,** in your own words, the steps of the linear programming procedure.

Guided Practice

4. Graph the system of inequalities. In a problem asking you to find the maximum value of $f(x, y)$, state whether this situation is *infeasible,* has *alternate optimal solutions,* or is *unbounded.* Assume that $x \geq 0$ and $y \geq 0$.
$0.5x + 1.5y \geq 7$
$3x + 9y \leq 2$
$f(x, y) = 30x + 20y$ **See margin for graph; infeasible.**

5. **Transportation**   A package delivery service has a truck that can hold 4200 pounds of cargo and has a capacity of 480 cubic feet. The service handles two types of packages: small, which weigh up to 25 pounds each and are no more than 3 cubic feet each; and large, which are 25 to 50 pounds each and are 3 to 5 cubic feet each. The delivery service charges $5 for each small package and $8 for each large package. Let x be the number of small packages and y be the number of large packages in the truck.

5g. No; if revenue is maximized, the company will not deliver any large packages, and customers with large packages to ship will probably choose another carrier for all of their business.

   a. Write an inequality to represent the weight of the packages in pounds the truck can carry. $25x + 50y \leq 4200$

   b. Write an inequality to represent the volume, in cubic feet, of packages the truck can carry. $3x + 5y \leq 480$

   c. Graph the system of inequalities.   **See margin.**

   d. Write a function that represents the amount of money the delivery service will make on each truckload. $P(x, y) = 5x + 8y$

   e. Find the number of each type of package that should be placed on a truck to maximize revenue.   **160 small packages, 0 large packages**

   f. What is the maximum revenue per truck?   **$800**

   g. In this situation, is maximizing the revenue necessarily the best thing for the company to do? Explain.

Solve each problem, if possible. If not possible, state whether the problem is *infeasible,* has *alternate optimal solutions,* or is *unbounded.*

6. 50 brochures, 150 fliers

6. **Business**   The manager of a gift store is printing brochures and fliers to advertise sale items. Each brochure costs 8¢ to print, and each flier costs 4¢ to print. A brochure requires 3 pages, and a flier requires 2 pages. The manager does not want to use more than 600 pages, and she needs at least 50 brochures and 150 fliers. How many of each should she print to minimize the cost?

7. **Manufacturing**   Woodland Bicycles makes two models of off-road bicycles: the Explorer, which sells for $250, and the Grande Expedition, which sells for $350. Both models use the same frame, but the painting and assembly time required for the Explorer is 2 hours, while the time is 3 hours for the Grande Expedition. There are 375 frames and 450 hours of labor available for production. How many of each model should be produced to maximize revenue?   **225 Explorers, 0 Grande Expeditions**

Lesson 2-7   Linear Programming   **115**

Error Analysis
Watch for students who assume that the point farthest from the origin maximizes the function.

Prevent by stressing that each vertex must be tested in order to find the maximum and minimum values of the function for the polygonal convex set.

Additional Answers
4.

5c.

Study Guide and Practice Masters, pp. 33–34

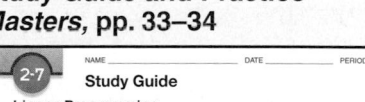

2-7   Study Guide

Linear Programming

The following example outlines the procedure used to solve linear programming problems.

Example   The B & W Leather Company wants to add handmade belts and wallets to its product line. Each belt nets the company $18 in profit, and each wallet nets $12. Both belts and wallets require cutting and sewing. Belts require 2 hours of cutting time and 6 hours of sewing time. Wallets require 3 hours of cutting time and 3 hours of sewing time. If the cutting machine is available 12 hours a week and the sewing machine is available 18 hours per week, what mix of belts and wallets will produce the most profit within the constraints?

Define variables.   Let b = the number of belts.
Let w = the number of wallets.

Write inequalities. $b \geq 0$
$w \geq 0$
$2b + 3w \leq 12$ cutting
$6b + 3w \leq 18$ sewing

Graph the system.

Write an equation.   Since the profit on belts is $18 and the profit on wallets is $12, the profit function is $B(b, w) = 18b + 12w$.

Substitute values. $B(0, 0) = 18(0) + 12(0) = 0$
$B(0, 4) = 18(0) + 12(4) = 48$
$B(1.5, 3) = 18(1.5) + 12(3) = 63$
$B(3, 0) = 18(3) + 12(0) = 54$

Answer the problem.   The B & W Company will maximize profit if it makes and sells 1.5 belts for every 3 wallets.

When constraints of a linear programming problem cannot be satisfied simultaneously, then **infeasibility** is said to occur.

The solution of a linear programming problem is **unbounded** if the region defined by the constraints is infinitely large.

© Glencoe/McGraw-Hill   33   Advanced Mathematical Concepts

Additional Answers

9.

10.

11.

12c.

13b.
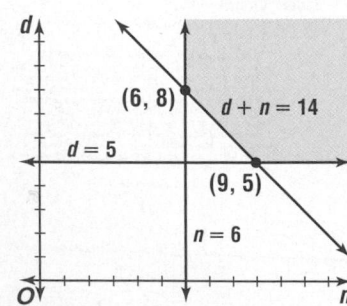

8. **Business** The Grainery Bread Company makes two types of wheat bread, light whole wheat and regular whole wheat. A loaf of light whole wheat bread requires 2 cups of flour and 1 egg. A loaf of regular whole wheat uses 3 cups of flour and 2 eggs. The bakery has 90 cups of flour and 80 eggs on hand. The profit on the light bread is $1 per loaf and on the regular bread is $1.50 per loaf. In order to maximize profits, how many of each loaf should the bakery make? **alternate optimal solutions**

EXERCISES

Practice

Graph each system of inequalities. In a problem asking you to find the maximum value of $f(x, y)$, state whether the situation is *infeasible*, has *alternate optimal solutions*, or is *unbounded*. In each system, assume that $x \geq 0$ and $y \geq 0$ unless stated otherwise. **See margin for graphs.**

9. $y \geq 6$
   $5x + 3y \leq 15$
   $f(x, y) = 12x + 3y$
   infeasible

10. $2x + y \geq 48$
    $x + 2y \geq 42$
    $f(x, y) = 2x + y$
    unbounded

11. $4x + 3y \geq 12$
    $y \leq 3$
    $x \leq 4$
    $f(x, y) = 3 + 3y$
    alternate optimal solutions

Applications and Problem Solving

12. **Veterinary Medicine** Dr. Chen told Miranda that her new puppy needs a diet that includes at least 1.54 ounces of protein and 0.56 ounce of fat each day to grow into a healthy dog. Each cup of Good Start puppy food contains 0.84 ounce of protein and 0.21 ounce of fat. Each cup of Sirius puppy food contains 0.56 ounce of protein and 0.49 ounce of fat. If Good Start puppy food costs 36¢ per cup and Sirius costs 22¢ per cup, how much of each food should Miranda use in order to satisfy the dietary requirements at the minimum cost?

   a. Write an inequality to represent the ounces of protein required.

   b. Write an inequality to represent the ounces of fat required.   **$0.21g + 0.49s \geq 0.56$**

   c. Graph the system of inequalities.

   d. Write a function to represent the daily cost of puppy food.   **$C(g, s) = 36g + 22s$**

   e. How many cups of each type of puppy food should be used in order to minimize the cost?

   f. What is the minimum cost?   **60.5¢**

12a. Let $g =$ the number of cups of Good Start food and $s =$ the number of cups of Sirius food.
$0.84g + 0.56s \geq 1.54$

12c. See margin.

12e. 0 cups of Good Start and 2.75 cups of Sirius

13. **Management** Angela's Pizza is open from noon to midnight each day. Employees work 8-hour shifts from noon to 8 P.M. or 4 P.M. to midnight. The store manager estimates that she needs at least 5 employees from noon to 4 P.M., at least 14 employees from 4 P.M. to 8 P.M., and 6 employees from 8 P.M. to midnight. Employees are paid $5.50 per hour for hours worked between noon and 4 P.M. The hourly pay between 4 P.M. and midnight is $7.50.

   a. Write inequalities to represent the number of day-shift workers, the number of night-shift workers, and the total number of workers needed.

   b. Graph the system of inequalities.   **See margin.**

   c. Write a function to represent the daily cost of payroll.   **$C(n, d) = 52d + 60n$**

   d. Find the number of day-shift workers and night-shift workers that should be scheduled to minimize the cost.   **8 day-shift and 6 night-shift workers**

   e. What is the minimal cost?   **$776**

13a. Let $d =$ the number of day-shift workers and $n =$ the number of night-shift workers.
$d \geq 5$
$n \geq 6$
$d + n \geq 14$

 www.amc.glencoe.com/self_check_quiz

Solve each problem, if possible. If not possible, state whether the problem is infeasible, has alternate optimal solutions, or is unbounded.

B

14. Agriculture The county officials in Chang Qing County, China used linear programming to aid the farmers in their choices of crops and other forms of agricultural production. This led to a 12% increase in crop profits, a 54% increase in animal husbandry profits, while improving the region's ecology. Suppose an American farmer has 180 acres on which to grow corn and soybeans. He is planting at least 40 acres of corn and 20 acres of soybeans. Based on his calculations, he can earn $150 per acre of corn and $250 per acre of soybeans. **a. 120 acres of corn, 60 acres of soybeans**
   a. If the farmer plants at least 2 acres of corn for every acre of soybeans, how many acres of each should he plant to earn the greatest profit?
   b. What is the farmer's maximum profit? **$33,000**

15. 10 section I questions, 2 section II questions

15. Education Ms. Carlyle has written a final exam for her class that contains two different sections. Questions in section I are worth 10 points each, and questions in section II are worth 15 points each. Her students will have 90 minutes to complete the exam. From past experience, she knows that on average questions from section I take 6 minutes to complete and questions from section II take 15 minutes. Ms. Carlyle requires her students to answer at least 2 questions from section II. Assuming they answer correctly, how many questions from each section will her students need to answer to get the highest possible score?

16. Manufacturing Newline Recyclers processes used aluminum into food or drink containers. The recycling plant processes up to 1200 tons of aluminum per week. At least 300 tons must be processed for food containers, while at least 450 tons must be processed for drink containers. The profit is $17.50 per ton for processing food containers and $20 per ton for processing drink containers. What is the profit if the plant maximizes processing? **$23,250**

17. $4000 in First Bank, $7000 in City Bank

17. Investment Diego wants to invest up to $11,000 in certificates of deposit at First Bank and City Bank. He does not want to deposit more than $7,500 at First Bank. He will deposit at least $1,000 but not more than $7,000 at City Bank. First Bank offers 6% simple interest on deposits, while City Bank offers $6\frac{1}{2}$% simple interest. How much should Diego deposit into each account so he can earn the most interest possible in one year?

18. Human Resources Memorial Hospital wants to hire nurses and nurse's aides to meet patient needs at minimum cost. The average annual salary is $35,000 for a nurse and $18,000 for a nurse's aide. The hospital can hire up to 50 people, but needs to hire at least 20 to function properly. The head nurse wants at least 12 aides, but the number of nurses must be at least twice the number of aides to meet state regulations. How many nurses and nurse's aides should be hired to minimize salary costs? **24 nurses, 12 nurse's aides**

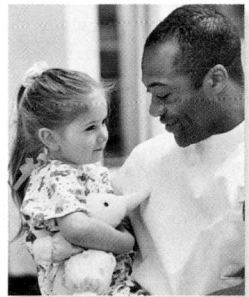

19. Manufacturing A potato chip company makes chips to fill snack-size bags and family-size bags. In one week, production cannot exceed 2,400 units, of which at least 600 units must be for snack-size bags and at least 900 units must be for family size. The profit on a unit of snack-size bags is $12, and the profit on a unit of family-size bags is $18. How much of each type of bag must be processed to maximize profits? **600 units for snack-size, 1800 units for family-size**

From the Classroom of...

Carmella Ettaro
Westlake High School
Westlake Village, California

"I incorporate a unit on the simplex method that I designed. Students work in pairs using the computer lab and matrix software."

Open-Ended Assessment
Speaking Have students explain the differences among an infeasible region, an unbounded region, and a region that has alternate optimal solutions.

 Chapter 2, Quiz D (Lesson 2-7)
is available in the *Assessment and Evaluation Masters*, p. 44.

Additional Answer
26.

$y = 3|x - 2|$

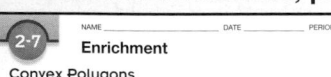

2-7 NAME _____ DATE _____ PERIOD _____
Enrichment

Convex Polygons

You have already learned that over a closed convex polygonal region, the maximum and minimum values of any linear function occur at the vertices of the polygon. To see why the values of the function at any point on the boundary of the region must be between the values at the vertices, consider the convex polygon with vertices P and Q.

Let W be a point on $\overline{PQ}$.
If W lies between P and Q, let $\frac{PW}{PQ} = w$.

Then $0 < w < 1$ and the coordinates of W are $((1-w)x_1 + wx_2, (1-w)y_1 + wy_2)$. Now consider the function $f(x, y) = 3x - 5y$.

$f(W) = 3[(1-w)x_1 + wx_2] - 5[(1-w)y_1 + wy_2]$
$= (1-w)3x_1 + 3wx_2 + (1-w)(-5y_1) - 5wy_2$
$= (1-w)(3x_1 - 5y_1) + w(3x_2 - 5y_2)$
$= (1-w)f(P) + wf(Q)$

This means that $f(W)$ is between $f(P)$ and $f(Q)$, or that the greatest and least values of $f(x, y)$ must occur at P or Q.

Example If $f(x, y) = 3x + 2y$, find the maximum value of the function over the shaded region at the right.
The maximum value occurs at the vertex (6, 3). The minimum value occurs at (0, 0). The values of $f(x, y)$ at W_1 and W_2 are between the maximum and minimum values.
$f(Q) = f(6, 3) = 24$
$f(W_1) = f(2, 1) = 8$
$f(W_2) = f(5, 2.5) = 20$
$f(P) = f(0, 0) = 0$

Let P and Q be vertices of a closed convex polygon, and let W lie on $\overline{PQ}$. Let $f(x, y) = ax + by$.
1. If $f(Q) = f(P)$, what is true of f? of $f(W)$? **It produces alternate optimal solutions; $f(W) = f(P)$.**
2. If $f(Q) = f(P)$, find an equation of the line containing P and Q. **Sample answer: $ax + by = f(P)$**

© Glencoe/McGraw-Hill   T19   *Advanced Mathematical Concepts*

118 *Chapter 2*

---

20. 3 batches of soap and 2 batches of shampoo

24. minimum: $-2\frac{1}{2}$, maximum: 10

25. (0, 6)

27. Sample answer: $C = \$13.65 + \$0.15(n - 30)$; $\$15.45$

---

20. Crafts Shelly is making soap and shampoo for gifts. She has 48 ounces of lye and 76 ounces of coconut oil and an ample supply of the other needed ingredients. She plans to make as many batches of soap and shampoo as possible. A batch of soap requires 12 ounces of lye and 20 ounces of coconut oil. Each batch of shampoo needs 6 ounces of lye and 8 ounces of coconut oil. What is the maximum number of batches of *both* soap and shampoo possible?

21. Manufacturing An electronics plant makes standard and large computer monitors on three different machines. The profit is $40 on each monitor. Use the table below to determine how many of each monitor the plant should make to maximize profits.   **alternate optimal solutions**

Manufacturing Analysis

Machine	Hours Needed to Make		Total Hours Available
	Small Monitor	Large Monitor	
A	1	2	16
B	1	1	9
C	1	4	24

22. Critical Thinking Find the area enclosed by the polygonal convex set defined by the system of inequalities $y \geq 0$, $x \leq 12$, $2x + 6y \leq 84$, $2x - 3y \leq -3$, and $8x + 3y \geq 33$.   **69 square units**

23. Critical Thinking Mr. Perez has an auto repair shop. He offers two bargain maintenance services, an oil change and a tune-up. His profit is $12 on an oil change and $20 on a tune-up. It takes Mr. Perez 30 minutes to do an oil change and 1 hour to do a tune-up. He wants to do at least 25 oil changes per week and no more than 10 tune-ups per week. He can spend up to 30 hours each week on these two services.

  a. What is the most Mr. Perez can earn performing these services?   **$720**

  b. After seeing the results of his linear program, Mr. Perez decides that he must make a larger profit to keep his business growing. How could the constraints be modified to produce a larger profit?   **Sample answer: Spend more than 30 hours per week on these services.**

Mixed Review

24. A polygonal convex set is defined by the inequalities $y \leq -x + 5$, $y \leq x + 5$, and $y + 5x \geq -5$. Find the minimum and maximum values for $f(x, y) = \frac{1}{3}x - \frac{1}{2}y$ within the set. *(Lesson 2-6)*

25. Find the values of x and y for which $\begin{bmatrix} 4x + y \\ x \end{bmatrix} = \begin{bmatrix} 6 \\ 2y - 12 \end{bmatrix}$ is true. *(Lesson 2-3)*

26. Graph $y = 3|x - 2|$. *(Lesson 1-7)*   **See margin.**

27. Budgeting Martina knows that her monthly telephone charge for local calls is $13.65 (excluding tax) for service allowing 30 local calls plus $0.15 for each call after 30. Write an equation to calculate Martina's monthly telephone charges and find the cost if she made 42 calls. *(Lesson 1-4)*

28. SAT/ACT Practice If $\frac{2x - 3}{x} = \frac{3 - x}{2}$, which could be a value for x?   **A**
   A −3   B −1   C 37   D 5   E 15

Extra Credit

Let x and y be positive integers so that $x + y \leq 9$ and $3x + 2y \leq 22$. What is the maximum value of $3x + 4y$? of $x + 5y$?
36; 45

VOCABULARY

additive identity matrix (p. 80)
column matrix (p. 78)
consistent (p. 67)
dependent (p. 67)
determinant (p. 98)
dilation (p. 88)
dimensions (p. 78)
element (p. 78)
elimination method (p. 68)
equal matrices (p. 79)
identity matrix for multiplication (p. 99)
image (p. 88)
inconsistent (p. 67)
independent (p. 67)

infeasible (p. 113)
inverse matrix (p. 99)
$m \times n$ matrix (p. 78)
matrix (p. 78)
minor (p. 98)
nth order (p. 78)
ordered triple (p. 74)
polygonal convex set (p. 108)
pre-image (p. 88)
reflection (p. 88)
reflection matrix (p. 89)
rotation (p. 88)
rotation matrix (p. 91)
row matrix (p. 78)
scalar (p. 80)
solution (p. 67)
square matrix (p. 78)

substitution method (p. 68)
system of equations (p. 67)
system of linear inequalities (p. 107)
transformation (p. 88)
translation (p. 88)
translation matrix (p. 88)
vertex matrix (p. 88)
Vertex Theorem (p. 108)
zero matrix (p. 80)

Modeling

alternate optimal solutions (p. 114)
constraints (p. 112)
linear programming (p. 112)
unbounded (p. 113)

UNDERSTANDING AND USING THE VOCABULARY

Choose the correct term from the list to complete each sentence.

1. Sliding a polygon from one location to another without changing its size, shape, or orientation is called a(n) __?__ of the figure. **translation**

2. Two matrices can be __?__ if they have the same dimensions. **added**

3. The __?__ of $\begin{bmatrix} -1 & 4 \\ 2 & -3 \end{bmatrix}$ is -5. **determinant**

4. A(n) __?__ system of equations has no solution. **inconsistent**

5. The process of multiplying a matrix by a constant is called __?__. **scalar multiplication**

6. The matrices $\begin{bmatrix} 2x \\ 0 \\ 16 \end{bmatrix}$ and $\begin{bmatrix} 8 - y \\ y \\ 4x \end{bmatrix}$ are __?__ if $x = 4$ and $y = 0$. **equal matrices**

7. A region bounded on all sides by intersecting linear inequalities is called a(n) __?__. **polygonal convex set**

8. A rotation of a figure can be achieved by consecutive __?__ of the figure over given lines. **reflections**

9. The symbol a_{ij} represents a(n) __?__ of a matrix. **element**

10. Two matrices can be __?__ if the number of columns in the first matrix is the same as the number of rows in the second matrix. **multiplied**

added
consistent
determinant
dilation
divided
element
equal matrices
identity matrices
inconsistent
inverse
multiplied
polygonal convex set
reflections
scalar multiplication
translation

 interNET CONNECTION   For additional review and practice for each lesson, visit: **www.amc.glencoe.com**

TEST OPTIONS

Six forms of Chapter 2 Test are available in the *Assessment and Evaluation Masters* as shown in the chart.

Chapter 2 Test, Form 1B, is shown at the right. Chapter 2 Test, Form 2B, is shown on the next page.

 GLENCOE'S ASSESSMENT ADVANTAGE

Form	Pages	Type	
1A	29–30	Honors	Form 1 tests are multiple choice. Form 2 tests are free response.
1B	31–32	Average	
1C	33–34	Basic	
2A	35–36	Honors	
2B	37–38	Average	
2C	39–40	Basic	

STUDY GUIDE AND ASSESSMENT

VOCABULARY

This section provides a listing of the new terms, properties, and phrases that were introduced in this chapter. The exercises check students' understanding of the terms by using a variety of verbal formats including matching, completion, and true/false.

Glossary A complete glossary of terms begins on page A71.

Assessment and Evaluation Masters, pp. 31–32

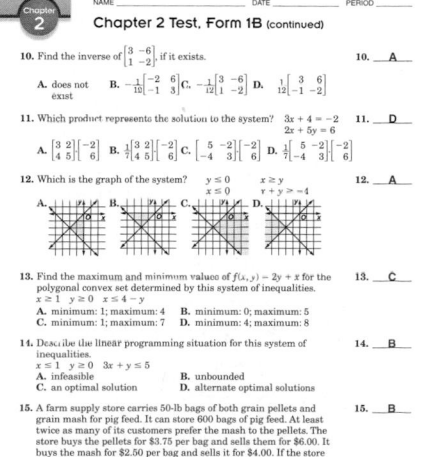

SKILLS AND CONCEPTS

Objectives and Examples This section reviews the skills and concepts of the chapter and shows completely worked examples.

Review Exercises These exercises provide practice for the corresponding objectives.

Assessment and Evaluation Masters, pp. 37–38

SKILLS AND CONCEPTS

OBJECTIVES AND EXAMPLES

Lesson 2-1 Solve systems of two linear equations.

- Solve the system of equations.
  $y = -x + 2$
  $3x + 4y = 2$
  Substitute $-x + 2$ for y in the second equation.

  $3x + 4y = 2$ $y = -x + 2$
  $3x + 4(-x + 2) = 2$ $y = -6 + 2$
  $x = 6$ $y = -4$

  The solution is $(6, -4)$.

Lesson 2-2 Solve systems of equations involving three variables algebraically.

- Solve the system of equations.
  $x - y + z = 5$
  $-3x + 2y - z = -8$
  $2x + y + z = 4$
  Combine pairs of equations to eliminate z.

$x - y + z = 5$	$-3x + 2y - z = -8$
$-3x + 2y - z = -8$	$2x + y + z = 4$
$-2x + y = -3$	$-x + 3y = -4$

  $-5y = 5$
  $y = -1$
  $-2x + (-1) = -3 \rightarrow x = 1$
  $(1) - (-1) + z = 5 \rightarrow z = 3$ *Solve for z.*
  The solution is $(1, -1, 3)$.

Lesson 2-3 Add, subtract, and multiply matrices.

- Find the difference.
  $\begin{bmatrix} -4 & 7 \\ 6 & -3 \end{bmatrix} - \begin{bmatrix} 8 & -3 \\ 2 & 0 \end{bmatrix} = \begin{bmatrix} -4 - 8 & 7 - (-3) \\ 6 - 2 & -3 - 0 \end{bmatrix}$

  $= \begin{bmatrix} -12 & 10 \\ 4 & -3 \end{bmatrix}$

  Find the product.
  $-2\begin{bmatrix} 1 & -4 \\ 0 & 3 \end{bmatrix} = \begin{bmatrix} -2(1) & -2(-4) \\ -2(0) & -2(3) \end{bmatrix}$

  $= \begin{bmatrix} -2 & 8 \\ 0 & -6 \end{bmatrix}$

REVIEW EXERCISES

Solve each system of equations algebraically.

11. $2y = -4x$
    $y = -x - 2$
    (2, -4)

12. $y = x - 5$
    $6y - x = 0$
    (6, 1)

13. $2x = 5y$
    $3y + x = -1$

14. $y = 6x + 1$
    $2y - 15x = -4$
    (2, 13)

15. $3x - 2y = -1$
    $2x + 5y = 12$
    (1, 2)

16. $x + 5y = 20.5$
    $3y - x = 13.5$
    (-0.75, 4.25)

13. $\left(-\dfrac{5}{11}, -\dfrac{2}{11}\right)$

Solve each system of equations algebraically.

17. $x - 2y - 3z = 2$
    $-3x + 5y + 4z = 0$
    $x - 4y + 3z = 14$ **(-10, -6, 0)**

18. $-x + 2y - 6z = 4$
    $x + y + 2z = 3$
    $2x + 3y - 4z = 5$ **(-1, 3, 0.5)**

19. $x - 2y + z = 7$
    $3x + y - z = 2$
    $2x + 3y + 2z = 7$ **(2, -1, 3)**

Use matrices *A, B,* and *C* to find each of the following. If the matrix does not exist, write *impossible.*   **20–27. See margin.**

$A = \begin{bmatrix} 7 & 8 \\ 0 & -4 \end{bmatrix}$ $B = \begin{bmatrix} -3 & -5 \\ 2 & -2 \end{bmatrix}$ $C = \begin{bmatrix} 2 \\ -5 \end{bmatrix}$

20. $A + B$

21. $B - A$

22. $3B$

23. $-4C$

24. AB

25. CB

26. $4A - 4B$

27. $AB - 2C$

120   *Chapter 2 Systems of Equations and Inequalities*

TestCheck and Worksheet Builder

This **networkable** software has three modules.
- **Worksheet Builder** to make worksheets and tests
- **Student Module** to take tests on-screen
- **Management System** to keep student records

Three special banks are included for SAT, ACT, and TIMSS.

OBJECTIVES AND EXAMPLES

Lesson 2-4 Use matrices to determine the coordinates of polygons under a given transformation.

Reflections	**Rotations** (counterclockwise about the origin)
$R_{x\text{-axis}} = \begin{bmatrix} 1 & 0 \\ 0 & -1 \end{bmatrix}$	$Rot_{90} = \begin{bmatrix} 0 & -1 \\ 1 & 0 \end{bmatrix}$
$R_{y\text{-axis}} = \begin{bmatrix} -1 & 0 \\ 0 & 1 \end{bmatrix}$	$Rot_{180} = \begin{bmatrix} -1 & 0 \\ 0 & -1 \end{bmatrix}$
$R_{y=x} = \begin{bmatrix} 0 & 1 \\ 1 & 0 \end{bmatrix}$	$Rot_{270} = \begin{bmatrix} 0 & -1 \\ -1 & 0 \end{bmatrix}$

REVIEW EXERCISES

Use matrices to perform each transformation. Then graph the pre-image and image on the same coordinate grid. **28–33. See margin.**

28. $A(-4, 3)$, $B(2, 1)$, $C(5, -3)$ translated 4 units down and 3 unit left

29. $W(-2, -3)$, $X(-1, 2)$, $Y(0, 4)$, $Z(1, -2)$ reflected over the x-axis

30. $D(2, 3)$, $E(2, -5)$, $F(-1, -5)$, $G(-1, 3)$ rotated $180°$ about the origin.

31. $P(3, -4)$, $Q(1, 2)$, $R(-1, 1)$ dilated by a scale factor of 0.5

32. triangle ABC with vertices at $A(-4, 3)$, $B(2, 1)$, $C(5, -3)$ after $Rot_{90} \circ R_{x\text{-axis}}$

33. What translation matrix would yield the same result on a triangle as a translation 6 units up and 4 units left followed by a translation 3 units down and 5 units right?

Lesson 2-5 Evaluate determinants.

Find the value of $\begin{vmatrix} -7 & 6 \\ 5 & -2 \end{vmatrix}$.

$\begin{vmatrix} -7 & 6 \\ 5 & -2 \end{vmatrix} = (-7)(-2) - 5(6)$

$= -16$

Find the value of each determinant.

34. $\begin{vmatrix} -3 & 5 \\ -4 & 7 \end{vmatrix}$ **−1**   35. $\begin{vmatrix} 8 & -4 \\ -6 & 3 \end{vmatrix}$ **0**

36. $\begin{vmatrix} 3 & -1 & 4 \\ 5 & -2 & 6 \\ 7 & 3 & -4 \end{vmatrix}$ **24**   37. $\begin{vmatrix} 5 & 0 & -4 \\ 7 & 3 & -1 \\ 2 & -2 & 6 \end{vmatrix}$ **160**

38. Determine whether $\begin{bmatrix} 2 & -4 & 1 \\ 3 & 8 & -2 \end{bmatrix}$ has a determinant. If so, find the value of the determinant. If not, explain. **no, not a square mat**

Lesson 2-5 Find the inverse of a 2×2 matrix.

Find X^{-1}, if $X = \begin{bmatrix} 2 & -1 \\ 1 & -3 \end{bmatrix}$.

$\begin{vmatrix} 2 & -1 \\ 1 & -3 \end{vmatrix} = -6 - (-1)$ or -5 and $-5 \neq 0$

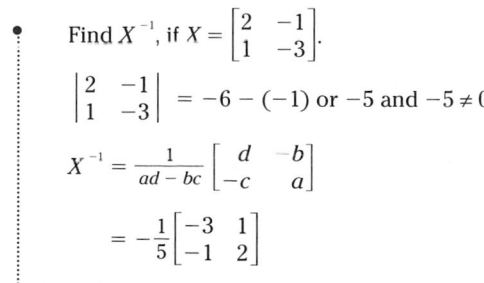

$X^{-1} = \frac{1}{ad - bc} \begin{bmatrix} d & -b \\ -c & a \end{bmatrix}$

$= -\frac{1}{5} \begin{bmatrix} -3 & 1 \\ -1 & 2 \end{bmatrix}$

Find the inverse of each matrix, if it exists. **39–44. See Answer Appendix.**

39. $\begin{bmatrix} 3 & 8 \\ -1 & 5 \end{bmatrix}$ 40. $\begin{bmatrix} 5 & 2 \\ 10 & 4 \end{bmatrix}$

41. $\begin{bmatrix} -3 & 5 \\ 1 & -4 \end{bmatrix}$ 42. $\begin{bmatrix} 3 & 2 \\ 5 & 7 \end{bmatrix}$

43. $\begin{bmatrix} 2 & -5 \\ 6 & 1 \end{bmatrix}$ 44. $\begin{bmatrix} 2 & -4 \\ -1 & 2 \end{bmatrix}$

Chapter 2   Study Guide and Assessment   **121**

Additional Answers

28. $A'(-7, -1)$, $B'(-1, -3)$, $C'(2, -7)$

29. $W'(-2, 3)$, $X'(-1, -2)$, $Y'(0, -4)$, $Z'(1, 2)$

30. $D'(-2, -3)$, $E'(-2, 5)$, $F'(1, 5)$, $G'(1, -3)$

31. $P'(6, -8)$, $Q'(2, 4)$, $R'(-2, 2)$

(continued on page 122)

Additional Answers *(p. 120)*

20. $\begin{bmatrix} 4 & 3 \\ 2 & -6 \end{bmatrix}$ 21. $\begin{bmatrix} -10 & -13 \\ 2 & 2 \end{bmatrix}$ 24. $\begin{bmatrix} -5 & -51 \\ -8 & 8 \end{bmatrix}$ 25. **impossible**

22. $\begin{bmatrix} -9 & -15 \\ 6 & -6 \end{bmatrix}$ 23. $\begin{bmatrix} -8 \\ 20 \end{bmatrix}$ 26. $\begin{bmatrix} 40 & 52 \\ -8 & -8 \end{bmatrix}$ 27. **impossible**

Additional Answers (p. 121)

32. $A'(3, -4)$, $B'(1, 2)$, $C'(-3, 5)$

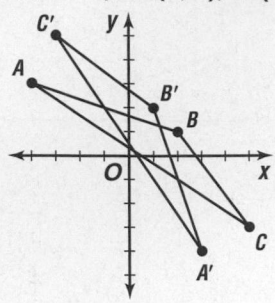

33. $\begin{bmatrix} 1 & 1 & 1 \\ 3 & 3 & 3 \end{bmatrix}$

OBJECTIVES AND EXAMPLES	**REVIEW EXERCISES**

Lesson 2-5 Solve systems of equations by using inverses of matrices.

- Solve the system of equations $3x - 5y = 1$ and $-2x + 2y = -2$ by using a matrix equation.

$$\begin{bmatrix} 3 & -5 \\ -2 & 2 \end{bmatrix} \cdot \begin{bmatrix} x \\ y \end{bmatrix} = \begin{bmatrix} 1 \\ -2 \end{bmatrix}$$

$$-\frac{1}{4}\begin{bmatrix} 2 & 5 \\ 2 & 3 \end{bmatrix} \cdot \begin{bmatrix} 3 & -5 \\ -2 & 2 \end{bmatrix} \cdot \begin{bmatrix} x \\ y \end{bmatrix} = -\frac{1}{4}\begin{bmatrix} 2 & 5 \\ 2 & 3 \end{bmatrix}\begin{bmatrix} 1 \\ -2 \end{bmatrix}$$

$$\begin{bmatrix} x \\ y \end{bmatrix} = \begin{bmatrix} 2 \\ 1 \end{bmatrix}$$

Solve each system by using a matrix equation.

45. $2x + 5y = 1$
$\quad -x - 3y = 2$ **(13, −5)**

46. $3x + 2y = -3$
$\quad -6x + 4y = 6$ **(−1, 0)**

47. $-3x + 5y = 1$
$\quad -2x + 4y = -2$ **(−7, −4)**

48. $4.6x - 2.7y = 8.4$
$\quad 2.9x + 8.8y = 74.61$ **(5.7, 6.6)**

Lesson 2-6 Find the maximum or minimum value of a function defined for a polygonal convex set.

- Find the maximum and minimum values of $f(x, y) = 4y + x - 3$ for the polygonal convex set graphed at the right.

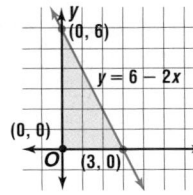

$(0, 6)$
$y = 6 - 2x$
$(0, 0)$
$O \quad (3, 0)$

$f(0, 0) = 4(0) + 0 - 3 = -3$ *minimum*
$f(3, 0) = 4(0) + 3 - 3 = 0$
$f(0, 6) = 4(6) + 0 - 3 = 21$ *maximum*

Find the maximum and minimum values of each function for the polygonal convex set determined by the given system of inequalities.

49. $f(x, y) = 2x + 3y$
$\quad x \geq 1$
$\quad y \geq -2$
$\quad y + x \leq 6$
$\quad y \leq 10 - 2x$
17, −4

50. $f(x, y) = 3x + 2y + 1$
$\quad x \geq 0$
$\quad y \geq 4$
$\quad y + x \leq 11$
$\quad 2y + x \leq 18$
$\quad x \leq 6$
29, 9

Lesson 2-7 Use linear programming procedures to solve applications.

- **Linear Programming Procedure**
  1. Define variables.
  2. Write the constraints as a system of inequalities.
  3. Graph the system and find the coordinates of the vertices of the polygon formed.
  4. Write an expression to be maximized or minimized.
  5. Substitute values from the coordinates of the vertices into the expression.
  6. Select the greatest or least result.

Use linear programming to solve.

51. Transportation Justin owns a truck and a motorcycle. He can buy up to 28 gallons of gasoline for both vehicles. His truck gets 22 miles per gallon and holds up to 25 gallons of gasoline. His motorcycle gets 42 miles per gallon and holds up to 6 gallons of gasoline. How many gallons of gasoline should Justin put in each vehicle if he wants to travel the most miles possible? **22 gallons in the truck and 6 gallons in the motorcycle**

APPLICATIONS AND PROBLEM SOLVING

52. Sports In a three-team track meet, the following numbers of first-, second-, and third-place finishes were recorded.

School	First Place	Second Place	Third Place
Broadman	2	5	5
Girard	8	2	3
Niles	6	4	1

Use matrix multiplication to find the final scores for each school if 5 points are awarded for a first place, 3 for second place, and 1 for third place. (Lesson 2-4)
Broadman 30; Girard 49; Niles 43

53. Geometry The perimeter of a triangle is 83 inches. The longest side is three times the length of the shortest side and 17 inches more than one-half the sum of the other two sides. Use a system of equations to find the length of each side. (Lesson 2-5) **39 in., 31 in., 13 in.**

54. Manufacturing A toy manufacturer produces two types of model spaceships, the *Voyager* and the *Explorer.* Each toy requires the same three operations. Each *Voyager* requires 5 minutes for molding, 3 minutes for machining, and 5 minutes for assembly. Each *Explorer* requires 6 minutes for molding, 2 minutes for machining, and 18 minutes for assembly. The manufacturer can afford a daily schedule of not more than 4 hours for molding, 2 hours for machining, and 9 hours for assembly. (Lesson 2-7)

a. If the profit is $2.40 on each *Voyager* and $5.00 on each *Explorer,* how many of each toy should be produced for maximum profit? **18 Voyagers, 25 Explorers**

b. What is the maximum daily profit? **$168.20**

ALTERNATIVE ASSESSMENT

OPEN-ENDED ASSESSMENT

1. Suppose that a quadrilateral *ABCD* has been rotated 90° clockwise about the origin twice. The resulting vertices are $A'(2, 2)$, $B'(-1, 2)$, $C'(-2, -1)$, and $D'(3, 0)$. **a–b. See margin.**

   a. State the original coordinates of the vertices of quadrilateral *ABCD* and how you determined them.

   b. Make a conjecture about the effect of a double rotation of 90° on any given figure.

2. If the determinant of a coefficient matrix is 0, can you use inverse matrices to solve the system of equations? Explain your answer and illustrate it with such a system of equations. **See margin.**

Additional Assessment See p. A57 for Chapter 2 Practice Test.

Unit 1 *interNET* Project

TELECOMMUNICATION

You've Got Mail!

• Research several Internet servers and make graphs that reflect the cost of using the server over a year's time period.
• Research various e-mail servers and their costs. Write and graph equations to compare the costs.
• Determine which Internet and e-mail servers best meet your needs. Write a paragraph to explain your choice. Use your graphs to support your choice.

PORTFOLIO

Devise a real-world problem that can be solved by linear programming. Define whether you are seeking a maximum or minimum. Write the inequalities that define the polygonal convex set used to determine the solution. Explain what the solution means.

Chapter 2 Study Guide and Assessment **123**

Internet Address Book

Record useful Internet addresses in the space at right for quick reference.

www.amc.glencoe.com

Chapter 2 Skill: Algebra

Pages 124–125 are part of a complete SAT/ACT preparation course that is described in detail on pages T8–T9 of the *Teacher's Handbook*. The test items were written in the same style as those in the SAT and ACT.

 These questions were aligned and verified by The Princeton Review, the nation's leader in test preparation.

Teaching Tip Students may find word problems easier to understand if they practice expressing other problems with words. Familiarity with the verbal forms of terms and operations will reduce stress and speed up calculations.

Diagnosis & Prescription

Each of the 10 SAT and ACT practice problems on page 125 is cross-referenced to the chapter where that SAT or ACT skill is covered. If students miss a particular type of problem, you can have them study that skill.

Problem Number and Skill	Page
1 Algebra	124
2 Arithmetic	64
3 Algebra	124
4 Arithmetic	64
5 Arithmetic	64
6 Algebra	124
7 Arithmetic	64
8 Algebra	124
9 Algebra	124
10 Arithmetic	64

Algebra Problems

About one third of SAT math problems and many ACT math problems involve algebra. You'll need to simplify algebraic expressions, solve equations, and solve word problems.

The word problems often deal with specific problems.
- consecutive integers
- age
- motion (*distance* = *rate* × *time*)
- investments (*principal* × *rate* = *interest income*)
- work
- coins
- mixtures

THE PRINCETON REVIEW

TEST-TAKING TIP
Review the rules for simplifying algebraic fractions and expressions. Try to simplify expressions whenever possible.

ACT EXAMPLE

1. $\dfrac{(xy)^3 z^0}{x^3 y^4} =$

   A $\dfrac{1}{y}$ **B** $\dfrac{z}{y}$ **C** z **D** xy **E** xyz

HINT Review the properties of exponents.

Solution Simplify the expression. Apply the properties of exponents.

$(xy)^3 = x^3 y^3$.

Any number raised to the zero power is equal to 1. Therefore, $z^0 = 1$.

Write y^4 as $y^3 y^1$.

Simplify.

$$\dfrac{(xy)^3 z^0}{x^3 y^4} = \dfrac{x^3 y^3 \cdot 1}{x^3 y^3 y^1}$$

$$= \dfrac{1}{y}$$

The answer is choice **A**.

SAT EXAMPLE

2. The sum of two positive consecutive integers is x. In terms of x, what is the value of the smaller of these two integers?

   A $\dfrac{x}{2} - 1$ **B** $\dfrac{x - 1}{2}$ **C** $\dfrac{x}{2}$

   D $\dfrac{x + 1}{2}$ **E** $\dfrac{x}{2} + 1$

HINT On multiple-choice questions with variables in the answer choices, you can sometimes use a strategy called "plug-in."

Solution The "plug-in" strategy uses substitution to test the choices. Suppose the two numbers were 2 and 3. Then $x = 2 + 3$ or 5. Substitute 5 for x and see which choice yields 2, the smaller of the two numbers.

Choice A: $\dfrac{5}{2} - 1$ This is not an integer.

Choice B: $\dfrac{5 - 1}{2} = 2$ This is the answer, but check the rest just to be sure.

Alternate Solution You can solve this problem by writing an algebraic expression for each number. Let a be the first (smallest) positive integer. Then $(a + 1)$ is the next positive integer. Write an equation for "The sum of the two numbers is x" and solve for a.

$$a + (a + 1) = x$$
$$2a + 1 = x$$
$$2a = x - 1$$
$$a = \dfrac{x - 1}{2}$$ The answer is choice **B**.

124 *Chapter 2 Systems of Equations and Inequalities*

 ## TestCheck and Worksheet Builder

Special banks of SAT, ACT, and TIMSS items are included in this CD-ROM.

After you work each problem, record your answer on the answer sheet provided or on a piece of paper.

Multiple Choice

1. If the product of $(1 + 2)$, $(2 + 3)$, and $(3 + 4)$ is equal to one half the sum of 20 and x, then $x =$ **D**

   A 10 **B** 85 **C** 105 **D** 190 **E** 1,210

2. $5\frac{1}{3} - 6\frac{1}{4} = ?$ **A**

   A $-\frac{11}{12}$

   B $-\frac{1}{2}$

   C $-\frac{2}{7}$

   D $\frac{1}{2}$

   E $\frac{9}{12}$

3. Mia has a pitcher containing x ounces of root beer. If she pours y ounces of root beer into each of z glasses, how much root beer will remain in the pitcher? **D**

   A $\frac{x}{y} + z$

   B $xy - z$

   C $\frac{x}{yz}$

   D $x - yz$

   E $\frac{x}{y} - z$

4. Which of the following is equal to 0.064? **D**

   A $\left(\frac{1}{80}\right)^2$ **B** $\left(\frac{8}{100}\right)^2$ **C** $\left(\frac{1}{8}\right)^2$

   D $\left(\frac{2}{5}\right)^3$ **E** $\left(\frac{8}{10}\right)^3$

5. A plumber charges $75 for the first thirty minutes of each house call plus $2 for each additional minute that she works. The plumber charged Mr. Adams $113 for her time. For what amount of time, in minutes, did the plumber work? **C**

   A 38 **B** 44 **C** 49 **D** 59 **E** 64

6. If $\frac{2 + x}{5 + x} = \frac{2}{5} + \frac{2}{5}$, then $x =$ **E**

   A $\frac{2}{5}$ **B** 1 **C** 2 **D** 5 **E** 10

7. Which of the following must be true?

   I. The sum of two consecutive integers is odd.

   II. The sum of three consecutive integers is even.

   III. The sum of three consecutive integers is a multiple of 3. **D**

   A I only

   B II only

   C I and II only

   D I and III only

   E I, II, and III

8. Jose has at least one quarter, one dime, one nickel, and one penny in his pocket. If he has twice as many pennies as nickels, twice as many nickels as dimes, and twice as many dimes as quarters, then what is the least amount of money he could have in his pocket? **D**

   A $0.41 **B** $0.64 **C** $0.71

   D $0.73 **E** $2.51

9. Simplify $\dfrac{\frac{3}{2}}{\left(\frac{3}{2}\right)^2}$. **C**

   A $\frac{27}{8}$

   B $\frac{3}{2}$

   C $\frac{2}{3}$

   D $\frac{1}{2}$

   E $\frac{1}{3}$

10. **Grid-In** At a music store, the price of a CD is three times the price of a cassette tape. If 40 CDs were sold for a total of $480 and the combined sales of CDs and cassette tapes totaled $600, how many cassette tapes were sold? **30**

inter NET
CONNECTION **SAT/ACT Practice** For additional test practice questions, visit: **www.amc.glencoe.com**

THE PRINCETON REVIEW *SAT & ACT Preparation* **125**

A bubble-in answer sheet for these practice problems is available on p. 1 of the *Assessment and Evaluation Masters.*

Additional Practice

Additional SAT and ACT practice questions and a bubble-in answer sheet are available in the *Assessment and Evaluation Masters,* pp. 2, 45, and 46.

***Assessment and Evaluation Masters,* pp. 45–46**

Extra Credit

TIMSS A translation maps $A(2, -3)$ onto $A'(-3, -5)$. Under the same translation, find the coordinates of B', the image of $B(1, 4)$. **$(-4, 2)$**

The Nature of Graphs

Instructional Objectives

Lesson (pages)	Lesson Objectives	NCTM Standards 2000	State/Local Objectives
3-1 (127–136)	• Use algebraic tests to determine whether the graph of a relation is symmetrical. • Classify functions as even or odd.	2, 3, 6, 10	
3-2 (137–145)	• Identify transformations of simple graphs. • Sketch graphs of related functions.	2, 3, 6, 10	
3-3 (146–151)	• Graph polynomial, absolute value, and radical inequalities in two variables. • Solve absolute value inequalities.	2, 3, 6, 10	
3-4 (152–158)	• Determine inverses of relations and functions. • Graph functions and their inverses.	2, 3, 6, 10	
3-5 (159–168)	• Determine whether a function is continuous or discontinuous. • Identify the end behavior of functions. • Determine whether a function is increasing or decreasing on an interval.	2, 3, 6, 7, 10	
3-5B (169–170)	• Construct and graph functions with gap discontinuities.	2, 3, 6, 10	
3-6 (171–179)	• Find the extrema of a function.	2, 3, 6, 7, 10	
3-7 (180–188)	• Graph rational functions. • Determine vertical, horizontal, and slant asymptotes.	2, 3, 6, 10	
3-8 (189–196)	• Solve problems involving direct, inverse, and joint variation.	2, 6	

Suggested Pacing   *See pages T2–T4 for additional options in course-planning.*

COURSE	Lesson 3-1	Lesson 3-2	Lesson 3-3	Lesson 3-4	Lesson 3-5
Precalculus Emphasis Chapters 4–15 (45–50 minutes)	OPTIONAL—Chapter 3 is a review of topics students have learned in previous courses. You may wish to use the Study Guide and Assessment to assess students' familiarity with these topics.				
Precalculus Emphasis (block)	OPTIONAL				
Trigonometric Emphasis Chapters 1–9, 11–12 (45–50 minutes)	1.5 days	1.5 days	1.5 days	1.5 days	(with Lesson 3-5B) 2.5 days
Trigonometric Emphasis (block)	0.5 day	1.5 days		1.5 days	

Instructional Resources

Lesson	Student Edition — Graphing Calculator Applications	Blackline Masters (page numbers) — Study Guide and Practice	Enrichment	Assessment and Evaluation	Graphing Calculator Keystrokes	Problem Solving & Applications	Spreadsheet Applications	TI-92 and CBL Interdisciplinary Activities	Transparencies A and B
3-1	133	39–40	25		9	6			3-1
3-2	138, 143	41–42	26	69					3-2
3-3	147–149	43–44	27						3-3
3-4	153	45–46	28	68, 69					3-4
3-5	164, 166	47–48	29						3-5
3-5B	169–170				10–12				
3-6	172–174, 176–177	49–50	30	70		5		5–8	3-6
3-7	181	51–52	31						3-7
3-8		53–54	32	70			3		3-8
Study Guide/ Assessment				55–67, 73					
SAT/ACT Practice				71–72					

Lesson 3-6	Lesson 3-7	Lesson 3-8	Study Guide & Assessment	Chapter Test	TOTAL
1.5 days	1.5 days	1.5 days	1 day	1 day	15 days
(with Lesson 3-5B) 1.5 days		1.5 days		0.5 day	7 days

Resource Manager

TeacherWorks™

The worksheets shown on this page are a small sample of the materials available on *TeacherWorks: All-in-One Lesson Planner and Resource Center.*

This CD-ROM contains all of the blackline masters and transparencies available for this program.

It also includes a lesson planner and interactive Teacher Edition, so you can customize lesson plans and reproduce classroom resources quickly and easily, from just about anywhere.

Applications

Problem Solving and Applications Masters, pp. 5, 6

Technology/Multimedia

Spreadsheet Applications Masters, p. 3

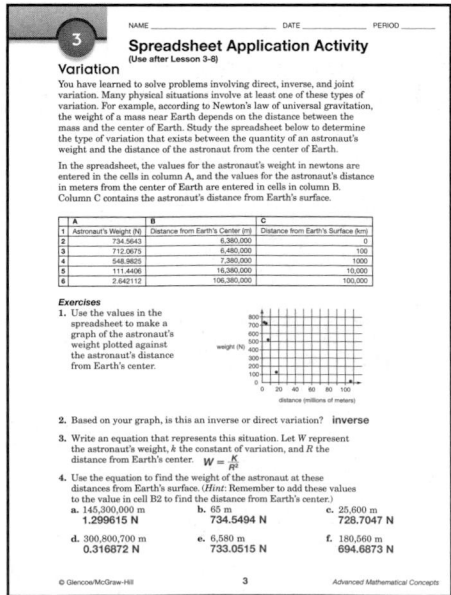

TI-92 & CBL Interdisciplinary Activities Masters, pp. 5–8

Assessment Resources

Type	Student Edition	Teacher's Wraparound Edition	Assessment and Evaluation Masters
Ongoing Assessment	Mid-Chapter Quiz, p. 151	5-Minute Check, pp. 127, 137, 146, 152, 159, 171, 180, 189	Mid-Chapter Test, p. 68 Quizzes A–D, pp. 69–70
Mixed Review	Mixed Review, pp. 135–136, 145, 151, 158, 168, 179, 188, 196		Cumulative Review, p. 73
Error Analysis	You Decide, pp. 134, 156	Error Analysis, pp. 133, 142, 149, 155, 165, 176, 185, 193	
Standardized Test Practice	SAT/ACT Practice, pp. 136, 145, 151, 158, 168, 179, 188, 196 SAT & ACT Preparation, pp. 202–203	TIMSS, p. 203	SAT/ACT Practice, pp. 71–72
Open-Ended Assessment	Math Journal, pp. 149, 165 Open-Ended Assessment, p. 201 Portfolio, p. 201	Modeling: pp. 145, 158, 179 Speaking: pp. 151, 188 Writing: pp. 136, 168, 170, 196	Open-Ended Assessment, p. 67
Chapter Assessment	Study Guide and Assessment, pp. 197–201 Chapter Test, p. A58		Multiple-Choice Tests (Forms 1A, 1B, 1C), pp. 55–60 Free-Response Tests (Forms 2A, 2B, 2C), pp. 61–66

TestCheck and Worksheet Builder is a networkable CD-ROM with three integrated modules: The **Worksheet Builder** creates customized worksheets, tests, and quizzes of free-response, multiple-choice, short-answers, and open-ended items. The **Student Module** gives you the option of having students take tests on-screen and get immediate feedback on their performance. Use the optional **Management System** to keep detailed student records.

Additional Chapter Resources

Student Edition
Career Choices, p. 136
Internet Project, p. 201
Extra Practice, pp. A30–A31

Teacher's Classroom Resources

Technology
Graphing Calculator
Keystrokes Masters

Teaching Aids
Answer Key Transparencies
Block Schedule Planning Guide
Lesson Planning Guide
Solutions Manual

Glencoe Instructional Technology

Instructional
Advanced MathPASS CD-ROM
Multimedia Applications CD-ROM

Assessment
TestCheck and Worksheet Builder

Visit **www.amc.glencoe.com** for data updates, career information, graphing calculator programs, and other interactive activities.

Chapter Overview

This chapter reviews the concepts of symmetry of graphs. Students identify and sketch families of graphs that include polynomials, absolute value functions, greatest integer functions, rational functions, and square root functions.

Students also find and graph inverses of functions and identify asymptotes associated with rational functions. They identify the critical points and end behavior of the graphs of polynomial functions and solve problems involving direct, inverse, and joint variation.

 Using a Graphing Calculator

There are numerous opportunities, all optional, to use a graphing calculator in this chapter.

- Using **TRACE** and **TABLE** to analyze a function (pp. 133, 172–174, 181)
- Finding *x*-intercepts of a set of functions (p. 143)
- Determining intervals for which a function is increasing or decreasing (p. 166)
- Constructing and graphing functions with gap discontinuities (pp. 169–170)
- Determining and classifying a function's extrema (pp. 172–174, 176, 177)

The **Graphing Calculator Appendix**, which begins on page A2, contains keystroke instructions on many of these topics.

THE NATURE OF GRAPHS

CHAPTER OBJECTIVES

- **Graph functions, relations, inverses, and inequalities.** *(Lessons 3-1, 3-3, 3-4, 3-7)*
- **Analyze families of graphs.** *(Lesson 3-2)*
- **Investigate symmetry, continuity, end behavior, and transformations of graphs.** *(Lessons 3-1, 3-2, 3-5)*
- **Find asymptotes and extrema of functions.** *(Lessons 3-6, 3-7)*
- **Solve problems involving direct, inverse, and joint variation.** *(Lesson 3-8)*

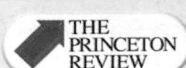 **SAT & ACT Preparation**

This textbook contains a complete SAT/ACT preparation course described on pages T8–T9 of the *Teacher's Handbook*. This chapter contains:

- an SAT or ACT practice question at the end of each lesson
- examples of SAT and ACT problems involving quadratic expressions and equations on page 202
- ten SAT and ACT practice problems on page 203
- additional SAT and ACT practice in the *Assessment and Evaluation Masters,* pp. 71–72
- SAT and ACT test practice banks in the *TestCheck and Worksheet Builder*

Symmetry and Coordinate Graphs

OBJECTIVES
- Use algebraic tests to determine if the graph of a relation is symmetrical.
- Classify functions as even or odd.

PHARMACOLOGY Designing a drug to treat a disease requires an understanding of the molecular structures of the substances involved in the disease process. The substances are isolated in crystalline form, and X rays are passed through the symmetrically-arranged atoms of the crystals. The existence of symmetry in crystals causes the X rays to be diffracted in regular patterns. These symmetrical patterns are used to determine and visualize the molecular structure of the substance. *A related problem is solved in Example 4.*

Like crystals, graphs of certain functions display special types of symmetry. For some functions with symmetrical graphs, knowledge of symmetry can often help you sketch and analyze the graphs. One type of symmetry a graph may have is **point symmetry**.

Point Symmetry	Two distinct points P and P' are symmetric with respect to point M if and only if M is the midpoint of $\overline{PP'}$. Point M is symmetric with respect to itself.

When the definition of point symmetry is extended to a set of points, such as the graph of a function, then each point P in the set must have an **image point** P' that is also in the set. A figure that is symmetric with respect to a given point can be rotated $180°$ about that point and appear unchanged. Each of the figures below has point symmetry with respect to the labeled point.

Symmetry with respect to a given point M can be expressed as symmetry about point M.

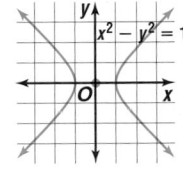

The origin is a common point of symmetry. Observe that the graphs of $f(x) = x^3$ and $g(x) = \frac{1}{x}$ exhibit symmetry with respect to the origin. Look for patterns in the table of function values beside each graph.

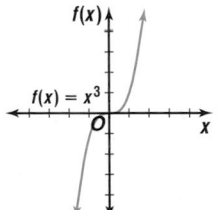

$f(x) = x^3$			
x	$f(x)$	$f(-x)$	$-f(x)$
1	1	-1	-1
2	8	-8	-8
3	27	-27	-27
4	64	-64	-64

Note that $f(-x) = -f(x)$.

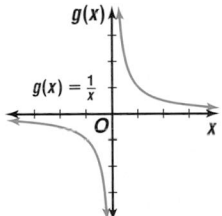

$g(x) = \frac{1}{x}$			
x	$g(x)$	$g(-x)$	$-g(x)$
1	1	-1	-1
2	$\frac{1}{2}$	$-\frac{1}{2}$	$-\frac{1}{2}$
3	$\frac{1}{3}$	$-\frac{1}{3}$	$-\frac{1}{3}$
4	$\frac{1}{4}$	$-\frac{1}{4}$	$-\frac{1}{4}$

Note that $g(-x) = -g(x)$.

Lesson 3-1 *Symmetry and Coordinate Graphs* **127**

Resource Manager

📁 **Reproducible Masters**
- *Study Guide and Practice,* pp. 39–40
- *Enrichment, p. 25*
- *Graphing Calculator Keystrokes, p. 9*
- *Problem Solving and Applications,* p. 6

💻 **Transparencies**
- 5-Minute Check, 3-1
- Teaching, 3-1

🕐 **5-Minute Check**
Chapter 2

1. Solve the system of equations algebraically.
   $x + y = 2$
   $3x = y - 10$ $(-2, 4)$

2. If $A = \begin{bmatrix} -3 & 4 \\ 2 & 7 \end{bmatrix}$ and
   $B = \begin{bmatrix} -1 & 5 \\ -3 & 2 \end{bmatrix}$, find $A + B$.
   $\begin{bmatrix} -4 & 9 \\ -1 & 9 \end{bmatrix}$

3. Find the value of $\begin{vmatrix} 5 & -4 \\ -3 & 9 \end{vmatrix}$.
   33

4. Solve the system of inequalities by graphing. Then, name the coordinates of the vertices of the polygonal convex set.
   $x \geq 0, y \geq 0$
   $y \leq x + 2$
   $y \leq -x + 8$
   (0, 0), (0, 2), (3, 5), (8, 0)

5. A triangle has vertices $A(6, 2)$, $B(-4, -4)$, and $C(-4, 8)$. Find the coordinates of the triangle after a dilation of scale factor of 0.25. $A'\left(1\frac{1}{2}, \frac{1}{2}\right)$, $B'(-1, -1)$, $C'(-1, 2)$

Motivating the Lesson
Hands-On Activity *Symmetry* is a familiar notion to many students. Open the lesson by bringing magazines to class and having students pick out logos from the advertisements that display various types of symmetry.

Teaching Tip After discussing the figures with symmetry on page 127, encourage students to sketch other examples of symmetry about a given point M. Include a discussion of any symmetries in arbitrary curves that students may draw.

Teaching Tip After defining *symmetry with respect to the origin,* students may need to see several numerical examples to help them understand the difference between $f(-x)$ and $-f(x)$. For example, if $f(x) = x^2 - 2x - 1$, then $f(-x) = x^2 + 2x - 1$, while $-f(x) = -x^2 + 2x + 1$.

In-Class Example

Example 1
Determine whether each graph is symmetric with respect to the origin.

a.

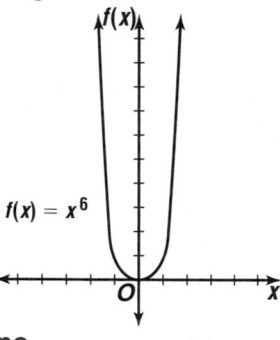

$f(x) = x^6$

no

b.

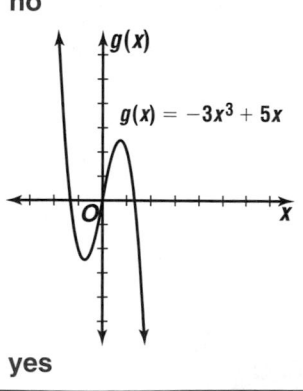

$g(x) = -3x^3 + 5x$

yes

The values in the tables suggest that $f(-x) = -f(x)$ whenever the graph of a function is symmetric with respect to the origin.

Symmetry with Respect to the Origin	The graph of a relation S is symmetric with respect to the origin if and only if $(a, b) \in S$ implies that $(-a, -b) \in S$. A function has a graph that is symmetric with respect to the origin if and only if $f(-x) = -f(x)$ for all x in the domain of f.

$(a, b) \in S$ means the ordered pair (a, b) <u>belongs to</u> the solution set S.

Example 1 demonstrates how to algebraically test for symmetry about the origin.

Example ❶ Determine whether each graph is symmetric with respect to the origin.

a. $f(x) = x^5$

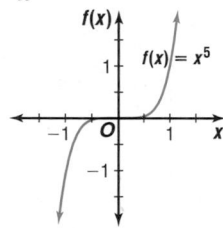

The graph of $f(x) = x^5$ appears to be symmetric with respect to the origin.

b. $g(x) = \dfrac{x}{1 - x}$

The graph of $g(x) = \dfrac{x}{1 - x}$ does not appear to be symmetric with respect to the origin.

We can verify these conjectures algebraically by following these two steps.

1. Find $f(-x)$ and $-f(x)$.
2. If $f(-x) = -f(x)$, the graph has point symmetry.

a. $f(x) = x^5$

Find $f(-x)$.

$f(-x) = (-x)^5$ *Replace x with $-x$.*

$f(-x) = -x^5$ $(-x)^5 = (-1)^5 x^5$
$\qquad\qquad\qquad = -1x^5 \text{ or } -x^5$

Find $-f(x)$.

$-f(x) = -x^5$ *Determine the opposite of the function.*

The graph of $f(x) = x^5$ is symmetric with respect to the origin because $f(-x) = -f(x)$.

b. $g(x) = \dfrac{x}{1 - x}$

Find $g(-x)$.

$g(-x) = \dfrac{-x}{1 - (-x)}$ *Replace x with $-x$.*

$\qquad = \dfrac{-x}{1 + x}$

Find $-g(x)$.

$-g(x) = -\dfrac{x}{1 - x}$ *Determine the opposite of the function.*

$\qquad = \dfrac{-x}{1 - x}$

The graph of $g(x) = \dfrac{x}{1 - x}$ is not symmetric with respect to the origin because $g(-x) \neq -g(x)$.

128 *Chapter 3 The Nature of Graphs*

📁 Prerequisite Skills

A review of the prerequisite skills for this lesson is provided in the *Prerequisite Skills Masters.*

Skill Number and Content	Pages
4 Algebraic Expressions	9–10

Another type of symmetry is **line symmetry.**

Line Symmetry	Two distinct points P and P' are symmetric with respect to a line ℓ if and only if ℓ is the perpendicular bisector of $\overline{PP'}$. A point P is symmetric to itself with respect to line ℓ if and only if P is on ℓ.

Teaching Tip Ask students how many lines of symmetry a circle would have. **infinite number**

Each graph below has line symmetry. The equation of each line of symmetry is given. Graphs that have line symmetry can be folded along the line of symmetry so that the two halves match exactly. Some graphs, such as the graph of an ellipse, have more than one line of symmetry.

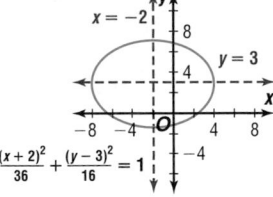

Some common lines of symmetry are the x-axis, the y-axis, the line $y = x$, and the line $y = -x$. The following table shows how the coordinates of symmetric points are related for each of these lines of symmetry. Set notation is often used to define the conditions for symmetry.

Symmetry with Respect to the:	Definition and Test	Example
x-axis	$(a, -b) \in S$ if and only if $(a, b) \in S$. *Example:* $(2, \sqrt{6})$ and $(2, -\sqrt{6})$ are on the graph. *Test:* Substituting (a, b) and $(a, -b)$ into the equation produces equivalent equations.	$x = y^2 - 4$ $(2, \sqrt{6})$ $y = 0$ $(2, -\sqrt{6})$
y-axis	$(-a, b) \in S$ if and only if $(a, b) \in S$. *Example:* $(2, 8)$ and $(-2, 8)$ are on the graph. *Test:* Substituting (a, b) and $(-a, b)$ into the equation produces equivalent equations.	$y = -x^2 + 12$ $(-2, 8)$ $(2, 8)$ $x = 0$

(continued on the next page)

From the Classroom of...

Ann Conway-Konzelman
Villa Victoria Academy
Ewing, New Jersey

"To help students appreciate the use of line symmetry in art, I have students cut out pictures from magazines. Students cut a picture in half and then attempt to reproduce the other side using the idea of line symmetry. I've had some students turn in projects so good that from a moderate distance other students viewing their work were not sure which side of the picture was the magazine photo!"

130 Chapter 3

In-Class Example

Example 2
Determine whether the graph of $x^2 + y = 3$ is symmetric with respect to the x-axis, the y-axis, the line $y = x$, the line $y = -x$, or none of these. **y-axis**

Symmetry with Respect to the Line:	Definition and Test	Example
$y = x$	$(b, a) \in S$ if and only if $(a, b) \in S$. *Example:* $(2, 3)$ and $(3, 2)$ are on the graph. *Test:* Substituting (a, b) and (b, a) into the equation produces equivalent equations.	graph with points $(2, 3)$ and $(3, 2)$, curve $xy = 6$, line $y = x$
$y = -x$	$(-b, -a) \in S$ if and only if $(a, b) \in S$. *Example:* $(4, -1)$ and $(1, -4)$ are on the graph. *Test:* Substituting (a, b) and $(-b, -a)$ into the equation produces equivalent equations.	graph with ellipse $17x^2 + 16xy + 17y^2 = 225$, points $(4, -1)$ and $(1, -4)$, line $y = -x$

You can determine whether the graph of an equation has line symmetry without actually graphing the equation.

Example **2** **Determine whether the graph of $xy = -2$ is symmetric with respect to the x-axis, y-axis, the line $y = x$, the line $y = -x$, or none of these.**

Substituting (a, b) into the equation yields $ab = -2$. Check to see if each test produces an equation equivalent to $ab = -2$.

x-axis	$a(-b) = -2$	*Substitute $(a, -b)$ into the equation.*
	$-ab = -2$	*Simplify.*
	$ab = 2$	*Not equivalent to $ab = -2$*
y-axis	$(-a)b = -2$	*Substitute $(-a, b)$ into the equation.*
	$-ab = -2$	*Simplify.*
	$ab = 2$	*Not equivalent to $ab = -2$*
$y = x$	$(b)(a) = -2$	*Substitute (b, a) into the equation.*
	$ab = -2$	*Equivalent to $ab = -2$*
$y = -x$	$(-b)(-a) = -2$	*Substitute $(-b, -a)$ into the equation.*
	$ab = -2$	*Equivalent to $ab = -2$*

Therefore, the graph of $xy = -2$ is symmetric with respect to the line $y = x$ and the line $y = -x$. A sketch of the graph verifies the algebraic tests.

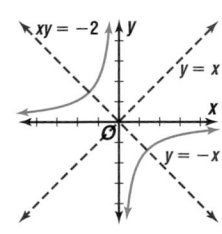

You can use information about symmetry to draw the graph of a relation.

Example **③** **Determine whether the graph of $|y| = 2 - |2x|$ is symmetric with respect to the *x*-axis, the *y*-axis, both, or neither. Use the information about the equation's symmetry to graph the relation.**

Substituting (a, b) into the equation yields $|b| = 2 - |2a|$. Check to see if each test produces an equation equivalent to $|b| = 2 - |2a|$.

x-axis $|-b| = 2 - |2a|$ *Substitute $(a, -b)$ into the equation.*
      $|b| = 2 - |2a|$ *Equivalent to $|b| = 2 - |2a|$ since $|-b| = |b|$.*

y-axis $|b| = 2 - |-2a|$ *Substitute $(-a, b)$ into the equation.*
      $|b| = 2 - |2a|$ *Equivalent to $|b| = 2 - |2a|$, since $|-2a| = |2a|$.*

Therefore, the graph of $|y| = 2 - |2x|$ is symmetric with respect to both the *x*-axis and the *y*-axis.

To graph the relation, let us first consider ordered pairs where $x \geq 0$ and $y \geq 0$. The relation $|y| = 2 - |2x|$ contains the same points as $y = 2 - 2x$ in the first quadrant.

Therefore, in the first quadrant, the graph of $|y| = 2 - |2x|$ is the same as the graph of $y = 2 - 2x$.

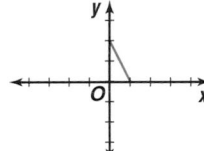

Since the graph is symmetric with respect to the *x*-axis, every point in the first quadrant has a corresponding point in the fourth quadrant.

Since the graph is symmetric with respect to the *y*-axis, every point in the first and fourth quadrants has a corresponding point on the other side of the *y*-axis.

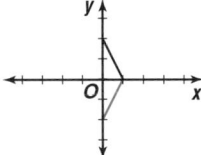

In-Class Example
Example 3
Determine whether the graph of $|y| = |x| + 1$ is symmetric with respect to the *x*-axis, the *y*-axis, both, or neither. Use the information about the equation's symmetry to graph the relation.
both

Example **④ CRYSTALLOGRAPHY** A crystallographer can model a cross-section of a crystal with mathematical equations. After sketching the outline on a graph, she notes that the crystal has both x-axis and y-axis symmetry. She uses the piecewise function $y = \begin{cases} 2 & \text{if } 0 \le x \le 1 \\ 3 - x & \text{if } 1 \le x \le 3 \end{cases}$ to model the first quadrant portion of the cross-section. Write piecewise equations for the remaining sides.

Look Back
Refer to Lesson 1-7 for more about piecewise functions.

Since the graph has x-axis symmetry, substitute $(x, -y)$ into the original equation to produce the equation for the fourth quadrant portion.

$$y = \begin{cases} 2 & \text{if } 0 \le x \le 1 \\ 3 - x & \text{if } 1 \le x \le 3 \end{cases}$$

$$-y = \begin{cases} 2 & \text{if } 0 \le x \le 1 \\ 3 - x & \text{if } 1 \le x \le 3 \end{cases} \quad \text{\textit{Substitute } (x, -y)\text{\textit{ into the equation.}}}$$

$$y = \begin{cases} -2 & \text{if } 0 \le x \le 1 \\ 3 + x & \text{if } 1 \le x \le 3 \end{cases} \quad \text{\textit{Solve for y.}}$$

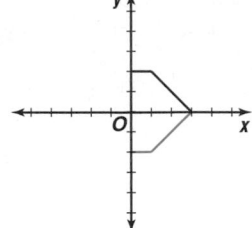

The equation $y = \begin{cases} -2 & \text{if } 0 \le x \le 1 \\ x - 3 & \text{if } 1 \le x \le 3 \end{cases}$ models the fourth quadrant portion of the cross section.

Since the graph has y-axis symmetry, substitute $(-x, y)$ into the first and fourth quadrant equations to produce the equations for the second and third quadrants.

$$y = \begin{cases} 2 & \text{if } 0 \le x \le 1 \\ 3 - x & \text{if } 1 \le x \le 3 \end{cases} \quad \text{\textit{Start with the first quadrant equation.}}$$

$$y = \begin{cases} 2 & \text{if } 0 \le -x \le 1 \\ 3 - (-x) & \text{if } 1 \le -x \le 3 \end{cases} \quad \text{\textit{Substitute } (-x, y)\text{\textit{ into the equation.}}}$$

$$y = \begin{cases} 2 & \text{if } 0 \ge x \ge -1 \\ 3 + x & \text{if } -1 \ge x \ge -3 \end{cases} \quad \text{\textit{This is the second quadrant equation.}}$$

$$y = \begin{cases} -2 & \text{if } 0 \le x \le 1 \\ x - 3 & \text{if } 1 \le x \le 3 \end{cases} \quad \text{\textit{Start with the fourth quadrant equation.}}$$

$$y = \begin{cases} -2 & \text{if } 0 \le -x \le 1 \\ -(-x) - 3 & \text{if } 1 \le -x \le 3 \end{cases} \quad \text{\textit{Substitute } (-x, y)\text{\textit{ into the equation.}}}$$

$$y = \begin{cases} -2 & \text{if } 0 \ge x \ge -1 \\ x - 3 & \text{if } -1 \ge x \ge -3 \end{cases} \quad \text{\textit{This is the third quadrant equation.}}$$

The equations $y = \begin{cases} 2 & \text{if } -1 \le x \le 0 \\ 3 + x & \text{if } -3 \le x \le -1 \end{cases}$ and $y = \begin{cases} -2 & \text{if } -1 \le x \le 0 \\ x - 3 & \text{if } -3 \le x \le -1 \end{cases}$ model the second and third quadrant portions of the cross section respectively.

Addressing Individual Needs

Kinesthetic Have students create shapes out of waxed paper that have either line or point symmetry. They should use the paper-folding method or turn the shapes about a fixed point to verify the type of symmetry. Students may use straight pins to hold shapes at a fixed point while they rotate them.

Functions whose graphs are symmetric with respect to the *y*-axis are **even functions.** Functions whose graphs are symmetric with respect to the origin are **odd functions.** Some functions are neither even nor odd. From Example 1, $f(x) = x^5$ is an odd function, and $g(x) = \dfrac{x}{1-x}$ is neither even nor odd.

even functions		odd functions	
$f(-x) = f(x)$		$f(-x) = -f(x)$	
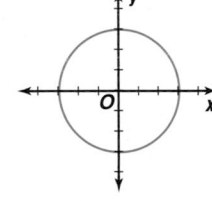 symmetric with respect to the *y*-axis		symmetric with respect to the origin	

GRAPHING CALCULATOR EXPLORATION

You can use the **TRACE** function to investigate the symmetry of a function.

▶ Graph the function.

▶ Use ⌐TRACE⌐ to observe the relationship between points of the graph having opposite *x*-coordinates.

▶ Use this information to determine the relationship between $f(x)$ and $f(-x)$.

TRY THESE Graph each function to determine how $f(x)$ and $f(-x)$ are related.

1. $f(x) = x^8 - 3x^4 + 2x^2 + 2$ $f(-x) = f(x)$
2. $f(x) = x^7 + 4x^5 - x^3$ $f(-x) = -f(x)$

WHAT DO YOU THINK?

3. Identify the functions in Exercises 1 and 2 as *odd, even,* or *neither* based on your observations of their graphs.   **even; odd**

4. Verify your conjectures algebraically.

5. How could you use symmetry to help you graph an even or odd function? Give an example.

6. Explain how you could use the **ASK** option in **TBLSET** to determine the relationship between $f(x)$ and $f(-x)$ for a given function.

4–6. See margin.

CHECK FOR UNDERSTANDING

Communicating Mathematics

Read and study the lesson to answer each question.

1. Refer to the tables on pages 129–130. **Identify** each graph as an even function, an odd function, or neither. Explain.   **1–2. See margin.**

2. **Explain** how rotating a graph of an odd function 180° will affect its appearance. Draw an example.

3. Consider the graph at the right.
   a. **Determine** four lines of symmetry for the graph.
   b. How many other lines of symmetry does this graph possess?   **infinitely many**
   c. What other type of symmetry does this graph possess?

3a. Sample answer: $y = 0$, $x = 0$, $y = x$, $y = -x$;

3c. point symmetry about the origin

4. **Write** an explanation of how to test for symmetry with respect to the line $y = -x$.   **See margin.**

Lesson 3-1   Symmetry and Coordinate Graphs   **133**

Additional Answers

1. **The graph of $y = -x^2 + 12$ is an even function. The graph of $xy = 6$ is an odd function. The graphs of $x = y^2 - 4$ and $17x^2 + 16xy + 17y^2 = 225$ are neither.**

2. **The graph of an odd function is symmetric with respect to the origin. Therefore, rotating the graph 180° will have no effect on its appearance. See students' work for example.**

4. **Substitute (a, b) into the equation. Substitute $(-b, -a)$ into the equation. Check to see whether both substitutions result in equivalent equations.**

Additional Answers

5. Graphically: if a graph has origin symmetry, then any portion of the graph in Quadrant I has an image in Quadrant III. If the graph is then symmetric with respect to the y-axis, the portion in Quadrants I and II have reflections in Quadrants II and IV, respectively. Therefore, any piece in Quadrant I has a reflection in Quadrant IV and the same is true for Quadrants II and III. Therefore, the graph is symmetric with respect to the x-axis.

Algebraically: substituting $(-x, -y)$ into the equation followed by substituting $(-x, y)$ is the same as substituting $(x, -y)$.

11.

12.

20. $g(x) = \frac{x^2 - 1}{x}$

Replace x with −x.

$g(-x) = \frac{(-x)^2 - 1}{(-x)}$

$g(-x) = -\frac{x^2 - 1}{x}$

Determine the opposite of the function.

$-g(x) = -\left(\frac{x^2 - 1}{x}\right)$

$g(-x) = -g(x)$

---

5. You Decide Alicia says that any graph that is symmetric to the origin and to the y-axis must also be symmetric to the x-axis. Chet disagrees. Who is correct? Support your answer graphically and algebraically. **Alicia; see margin for explanation.**

Guided Practice

Determine whether the graph of each function is symmetric with respect to the origin.

6. $f(x) = x^6 + 9x$ **no**

7. $f(x) = \frac{1}{5x} - x^{19}$ **yes**

Determine whether the graph of each equation is symmetric with respect to the x-axis, y-axis, the line $y = x$, the line $y = -x$, or none of these.

8. $6x^2 = y - 1$ **y-axis**

9. $x^3 + y^3 = 4$ **y = x**

10. Copy and complete the graph at the right so that it is the graph of an even function.

Determine whether the graph of each equation is symmetric with respect to the x-axis, the y-axis, both, or neither. Use the information about symmetry to graph the relation.

11. $y = \sqrt{2 - x^2}$ **y-axis**

12. $|y| = x^3$ **x-axis**

11–12. See margin for graphs.

13. Physics Suppose the light pattern from a fog light can be modeled by the equation $\frac{x^2}{25} - \frac{y^2}{9} = 1$. One of the points on the graph of this equation is at $\left(6, \frac{3\sqrt{11}}{5}\right)$, and one of the x-intercepts is -5. Find the coordinates of three additional points on the graph and the other x-intercept.

other points: $\left(-6, \frac{3\sqrt{11}}{5}\right), \left(6, -\frac{3\sqrt{11}}{5}\right), \left(-6, -\frac{3\sqrt{11}}{5}\right)$; **x-intercept: (5, 0)**

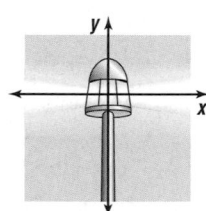

EXERCISES

Practice

Determine whether the graph of each function is symmetric with respect to the origin.

A
14. $f(x) = 3x$ **yes**

15. $f(x) = x^3 - 1$ **no**

16. $f(x) = 5x^2 + 6x + 9$ **no**

17. $f(x) = \frac{1}{4x^7}$ **yes**

18. $f(x) = -7x^5 + 8x$ **yes**

19. $f(x) = \frac{1}{x} - x^{100}$ **no**

B
20. Is the graph of $g(x) = \frac{x^2 - 1}{x}$ symmetric with respect to the origin? Explain how you determined your answer. **Yes; see margin for explanation.**

Determine whether the graph of each equation is symmetric with respect to the x-axis, y-axis, the line $y = x$, the line $y = -x$, or none of these.

21. $y = x$ and $y = -x$

26. x-axis and y-axis

21. $xy = -5$

22. $x + y^2 = 1$ **x-axis**

23. $y = -8x$ **none of these**

24. $y = \frac{1}{x^2}$ **y-axis**

25. $x^2 + y^2 = 4$ **all**

26. $y^2 = \frac{4x^2}{9} - 4$

27. Which line(s) are lines of symmetry for the graph of $x^2 = \frac{1}{y^2}$? **x-axis and y-axis, y = x and y = −x**

www.amc.glencoe.com/self_check_quiz

Extra Credit

Suppose $f(x)$ is an even function and $g(x)$ is an odd function. Is $h(x) = f(x) \cdot g(x)$ an even function, an odd function, or neither? Justify your answer. **It is odd because** $h(-x) = f(-x) \cdot g(-x) = f(x) \cdot (-g(x)) = -f(x) \cdot g(x)$. **This is** $-h(x)$.

For Exercises 28–30, refer to the graph.

28. Complete the graph so that it is the graph of an odd function.

29. Complete the graph so that it is the graph of an even function.

30. Complete the graph so that it is the graph of a function that is neither even nor odd. **See margin.**

Determine whether the graph of each equation is symmetric with respect to the *x*-axis, the *y*-axis, both, or neither. Use the information about symmetry to graph the relation. **31–36. See Answer Appendix for graphs.**

31. $y^2 = x^2$ **both**   32. $|x| = -3y$ **y-axis**   33. $y^2 + 3x = 0$ **x-axis**

 34. $|y| = 2x^2$ **both**   35. $x = \pm\sqrt{12 - 8y^2}$ **both** 36. $|y| = xy$ **neither**

37. Graph the equation $|y| = x^3 - x$ using information about the symmetry of the graph. **See margin.**

Applications and Problem Solving

38a. origin, *x*- and *y*-axis symmetry

38c. $(2, -\sqrt{5})$, $(-2, \sqrt{5})$, $(-2, -\sqrt{5})$

41. $(4\sqrt{2}, 6)$ or $(-4\sqrt{2}, 6)$

Mixed Review

38. **Physics** The path of a comet around the Sun can be modeled by a transformation of the equation $\dfrac{x^2}{8} + \dfrac{y^2}{10} = 1$.

a. Determine the symmetry in the graph of the comet's path.

b. Use symmetry to graph the equation $\dfrac{x^2}{8} + \dfrac{y^2}{10} = 1$. **See margin.**

c. If it is known that the comet passes through the point at $(2, \sqrt{5})$, name the coordinates of three other points through which it must pass.

39. **Critical Thinking** Write the equation of a graph that is symmetric with respect to the *x*-axis. **Sample answer: $y = 0$**

40. **Geometry** Draw a diagram composed of line segments that exhibits both *x*- and *y*-axis symmetry. Write equations for the boundaries. **See margin.**

41. **Communication** Radio waves emitted from two different radio towers interfere with each other's signal. The path of interference can be modeled by the equation $\dfrac{y^2}{12} - \dfrac{x^2}{16} = 1$, where the origin is the midpoint of the line segment between the two towers and the positive *y*-axis represents north. Juana lives on an east-west road 6 miles north of the *x*-axis and cannot receive the radio station at her house. At what coordinates might Juana live relative to the midpoint between the two towers?

42. **Critical Thinking** Must the graph of an odd function contain the origin? Explain your reasoning and illustrate your point with the graph of a specific function. **No; see margin for explanation.**

43. **Manufacturing** A manufacturer makes a profit of $6 on a bicycle and $4 on a tricycle. Department A requires 3 hours to manufacture the parts for a bicycle and 4 hours to manufacture parts for a tricycle. Department B takes 5 hours to assemble a bicycle and 2 hours to assemble a tricycle. How many bicycles and tricycles should be produced to maximize the profit if the total time available in department A is 450 hours and in department B is 400 hours? *(Lesson 2-7)* **50 bicycles, 75 tricycles**

Extra Practice See p. A30.

Lesson 3-1 Symmetry and Coordinate Graphs **135**

Additional Answers
38b.

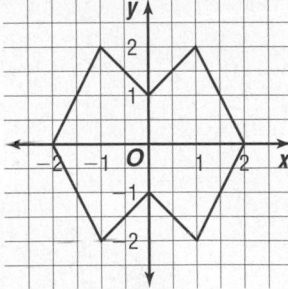

40. Sample answer:
$y = x + 1$ $y = -x - 1$
$y = x - 1$ $y = 2x + 4$
$y = -2x + 4$ $y = 2x - 4$
$y = -2x - 4$ $y = -x + 1$

42. If an odd function has a *y*-intercept, then it must be the origin. If it were not, say it were (0, 1), then the graph would have to contain (–1, 0). This would cause the relationship to fail the vertical line test and would therefore not be a function. But, not all odd functions have a *y*-intercept. Consider the graph of $y = \dfrac{1}{x}$.

Additional Answers

30. Sample answer:

37. The equation $|y| = x^3 - x$ is symmetric about the *x*-axis.

Lesson 3-1 **135**

Open-Ended Assessment

Writing Write a paragraph discussing when a relation is symmetric with respect to the x-axis, the y-axis, the line $y = x$, the line $y = -x$, or none of these.

Additional Answer

47.

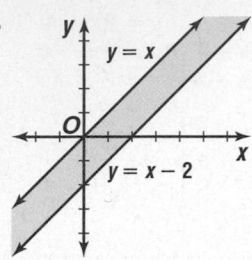

44. $\begin{bmatrix} 59 & 38 \\ 74 & 47 \end{bmatrix}$

44. Find AB if $A = \begin{bmatrix} 4 & 3 \\ 7 & 2 \end{bmatrix}$ and $B = \begin{bmatrix} 8 & 5 \\ 9 & 6 \end{bmatrix}$. *(Lesson 2-3)*

45. Solve the system of equations, $2x + y + z = 0$, $3x - 2y - 3z = -21$, and $4x + 5y + 3z = -2$. *(Lesson 2-2)* **$(-2, -3, 7)$**

46. State whether the system, $4x - 2y = 7$ and $-12x + 6y = -21$, is *consistent and independent, consistent and dependent,* or *inconsistent.* *(Lesson 2-1)* **consistent and dependent**

47. Graph $0 \le x - y \le 2$. *(Lesson 1-8)* **See margin.**

48. Write an equation in slope-intercept form for the line that passes through $A(0, 2)$ and $B(-2, 16)$. *(Lesson 1-4)* **$y = -7x + 2$**

49. $-2x + 23$, $-2x + 5$

49. If $f(x) = -2x + 11$ and $g(x) = x - 6$, find $[f \circ g](x)$ and $[g \circ f](x)$. *(Lesson 1-2)*

50. **SAT/ACT Practice** What is the product of 75^3 and 75^7? **B**

A 75^5 **B** 75^{10} **C** 150^{10} **D** 5625^{10} **E** 75^{21}

CAREER CHOICES

● Biomedical Engineering ●

Would you like to help people live better lives? Are you interested in a career in the field of health? If you answered yes, then biomedical engineering may be the career for you. Biomedical engineers apply engineering skills and life science knowledge to design artificial parts for the human body and devices for investigating and repairing the human body. Some examples are artificial organs, pacemakers, and surgical lasers.

In biomedical engineering, there are three primary work areas: research, design, and teaching. There are also many specialty areas in this field. Some of these are bioinstrumentation, biomechanics, biomaterials, and rehabilitation engineering.

The graph shows an increase in the number of outpatient visits over the number of hospital visits. This is due in part to recent advancements in biomedical engineering.

CAREER OVERVIEW

Degree Preferred:
bachelor's degree in biomedical engineering

Related Courses:
biology, chemistry, mathematics

Outlook:
number of jobs expected to increase through the year 2006

Hospital Vital Signs
Total Number of Hospital Admissions and Outpatient Visits, 1965-1996 (in millions)

Source: *The Wall Street Journal Almanac*

inter NET CONNECTION For more information on careers in biomedical engineering, visit: **www.amc.glencoe.com**

CAREER CHOICES   BIOMEDICAL ENGINEERING

Math Connection
- Explain what the graph is showing. **The graph shows the number of hospital admissions and outpatient visits for the years from 1965 through 1995.**
- What trends are shown in hospital admissions and outpatient visits from 1965 through 1995? **The number of hospital admissions increased during the 1970s, leveled off during the early 1980s, declined in the late 1980s, and then remained about constant until 1995. The number of outpatient visits has steadily increased since 1965, with a particularly sharp increase and then a sudden decrease around 1983.**

3-2

Families of Graphs

OBJECTIVES
- Identify transformations of simple graphs.
- Sketch graphs of related functions.

ENTERTAINMENT At some circuses, a human cannonball is shot out of a special cannon. In order to perform this death-defying feat safely, the maximum height and distance of the performer must be calculated accurately. Quadratic functions can be used to model the height of a projectile like a human cannonball at any time during its flight. The quadratic equation used to model height versus time is closely related to the equation of $y = x^2$. *A problem related to this is solved in Example 5.*

All parabolas are related to the graph of $y = x^2$. This makes $y = x^2$ the **parent graph** of the family of parabolas. Recall that a *family of graphs* is a group of graphs that displays one or more similar characteristics.

A parent graph is a basic graph that is transformed to create other members in a family of graphs. Some different types are shown below. Notice that with the exception of the constant function, the coefficient of x in each equation is 1.

constant function

identity function

polynomial functions

square root function

absolute value function

greatest integer function

rational function

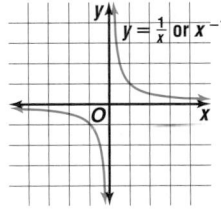

Look Back
Refer to Lesson 2-4 for more about reflections and translations.

Reflections and translations of the parent function can affect the appearance of the graph. The transformed graph may appear in a different location, but it will resemble the parent graph. A reflection flips a figure over a line called the *axis of symmetry. The axis of symmetry is also called the line of symmetry.*

Lesson 3-2 Families of Graphs **137**

3-2

1 FOCUS

5-Minute Check
Lesson 3-1

Determine whether the graph of each equation is symmetric with respect to the *x*-axis, the *y*-axis, the line $y = x$, the line $y = -x$, or none of these.

1. $4x^2 + 9y^2 = 36$

   x-axis, y-axis, origin

2. $2x + \frac{y}{3} = 4$ **none of these**

Determine whether each graph is symmetric with respect to the origin.

3. **no**

4. **yes**

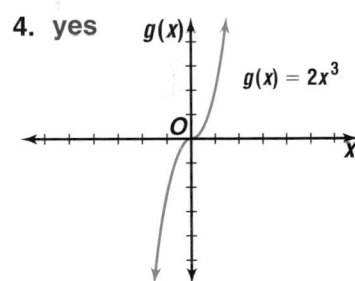

5. Complete the graph so that it is the graph of an odd function.

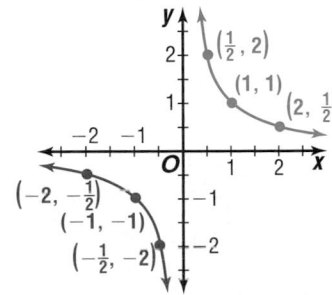

Lesson 3-2 **137**

Motivating the Lesson

Discussion Display the graphs of $y = x^2$, $y = (x-2)^2$, and $y = (x+3)^2$ on separate transparencies using equivalent axes and scales. Ask students to make conjectures about the three graphs and then use overlapping transparencies to verify their conjectures.

2 TEACH

In-Class Examples

Example 1

Graph $f(x) = x^3$ and $g(x) = -x^3$. Describe how the graphs of $g(x)$ and $f(x)$ are related.

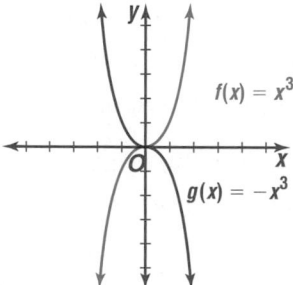

They are reflections of each other over the x-axis.

Example 2

Use the parent graph $y = x^3$ to sketch the graph of each function.

a. $y = x^3 - 1$

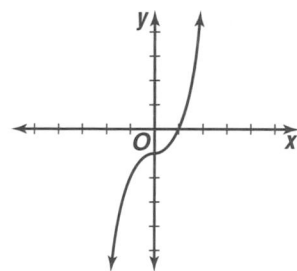

b. $y = (x - 1)^3$

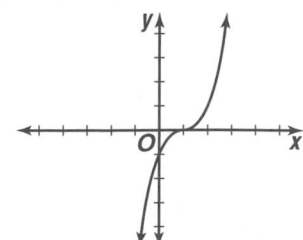

c. $y = (x - 1)^3 + 3$

Example  **Graph $f(x) = |x|$ and $g(x) = -|x|$. Describe how the graphs of $f(x)$ and $g(x)$ are related.**

Graphing Calculator Tip

You can use a graphing calculator to check your sketch of any function in this lesson.

x	$f(x) = |x|$	$g(x) = -|x|$
-2	2	-2
-1	1	-1
0	0	0
1	1	-1
2	2	-2

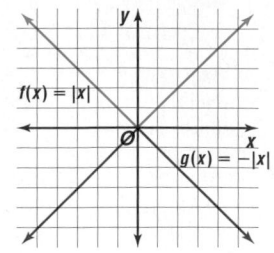

To graph both equations on the same axis, let $y = f(x)$ and $y = g(x)$.

The graph of $g(x)$ is a reflection of the graph of $f(x)$ over the x-axis. The symmetric relationship can be stated algebraically by $g(x) = -f(x)$, or $f(x) = -g(x)$. *Notice that the effect of multiplying a function by -1 is a reflection over the x-axis.*

When a constant c is added to or subtracted from a parent function, the result, $f(x) \pm c$, is a translation of the graph up or down. When a constant c is added or subtracted from x before evaluating a parent function, the result, $f(x \pm c)$, is a translation left or right.

Example ❷ **Use the parent graph $y = \sqrt{x}$ to sketch the graph of each function.**

a. $y = \sqrt{x} + 2$

This function is of the form $y = f(x) + 2$. Since 2 is added to the parent function $y = \sqrt{x}$, the graph of the parent function moves up 2 units.

b. $y = \sqrt{x - 4}$

This function is of the form $y = f(x - 4)$. Since 4 is being subtracted from x before being evaluated by the parent function, the graph of the parent function $y = \sqrt{x}$ slides 4 units right.

c. $y = \sqrt{x + 3} - 1$

This function is of the form $y = f(x + 3) - 1$. The addition of 3 indicates a slide of 3 units left, and the subtraction of 1 moves the parent function $y = \sqrt{x}$ down 1 unit.

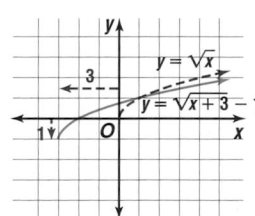

Remember that a dilation has the effect of shrinking or enlarging a figure. Likewise, when the leading coefficient of x is not 1, the function is expanded or compressed.

138 Chapter 3 *The Nature of Graphs*

📁 Prerequisite Skills

A review of the prerequisite skills for this lesson is provided in the *Prerequisite Skills Masters*.

Skill Number and Content	Pages
6 Polynomials	13–14

Example ③ **Graph each function. Then describe how it is related to its parent graph.**

a. $g(x) = 2[\![x]\!]$

The parent graph is the greatest integer function, $f(x) = [\![x]\!]$. $g(x) = 2[\![x]\!]$ is a vertical expansion by a factor of 2. The vertical distance between the steps is 2 units.

b. $h(x) = -0.5[\![x]\!] - 4$

$h(x) = -0.5[\![x]\!] - 4$ reflects the parent graph over the x-axis, compresses it vertically by a factor of 0.5, and shifts the graph down 4 units. *Notice that multiplying by a positive number less than 1 compresses the graph vertically.*

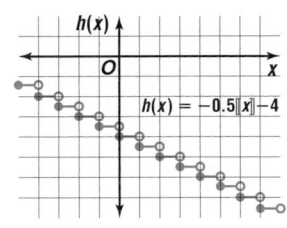

The following chart summarizes the relationships in families of graphs. The parent graph may differ, but the transformations of the graphs have the same effect. Remember that more than one transformation may affect a parent graph.

Change to the Parent Function $y = f(x)$, $c > 0$	Change to Parent Graph	Examples
Reflections $y = -f(x)$ $y = f(-x)$	Is reflected over the x-axis. Is reflected over the y-axis.	$y = f(x)$, $y = f(-x)$, $y = -f(x)$
Translations $y = f(x) + c$ $y = f(x) - c$	Translates the graph c units up. Translates the graph c units down.	$y = f(x) + c$, $y = f(x)$, $y = f(x) - c$
$y = f(x + c)$ $y = f(x - c)$	Translates the graph c units left. Translates the graph c units right.	$y = f(x + c)$, $y = f(x)$, $y = f(x - c)$

(continued on the next page)

From the Classroom of...

Susan Singleton
Fitzgerald High School
Fitzgerald, Georgia

"I have students model with their bodies the shapes of the families of graphs (arms up in a U-shape for quadratics, V-shape for absolute value, etc). Then they move up, down, right, left, narrower, or wider to show the various transformations."

Teaching Tip Graph the function in Example 3a in the Student Edition as a class, making sure that students see how the step function develops. Make sure that they understand the difference between the open and closed circles.

In-Class Example

Example 3
Graph each function. Then describe how it is related to the parent graph.
a. $g(x) = 4x^{-1}$

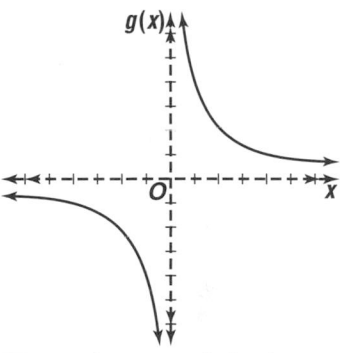

The parent graph is the rational function, $f(x) = \frac{1}{x}$. $g(x) = 4x^{-1}$ is a vertical expansion by a factor of 4.

b. $h(x) = -0.25x^{-1} + 3$

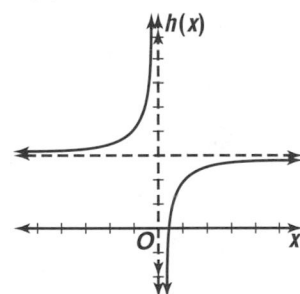

$h(x) = -0.25x^{-1} + 3$ reflects the parent graph over the x-axis, compresses it vertically by a factor of 0.25, and shifts the graph up 3 units.

Teaching Tip After discussing the parent graphs, encourage students to memorize them. It is much easier to do transformations of a function if the parent graph is known. Students should consider the eight functions given on page 137 as a *basic toolkit*.

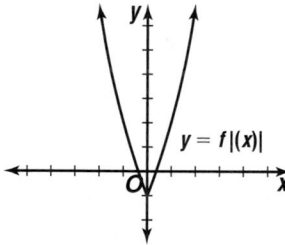
Change to the Parent Function $y = f(x)$, $c > 0$	Change to Parent Graph	Examples
Dilations		
$y = c \cdot f(x)$, $c > 1$ $y = c \cdot f(x)$, $0 < c < 1$	Expands the graph vertically. Compresses the graph vertically.	
$y = f(cx)$, $c > 1$ $y = f(cx)$, $0 < c < 1$	Compresses the graph horizontally. Expands the graph horizontally.	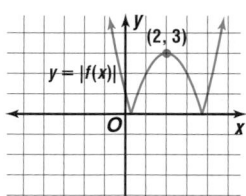

Other transformations may affect the appearance of the graph. We will look at two more transformations that change the shape of the graph.

Example **4** **Observe the graph of each function. Describe how the graphs in parts b and c relate to the graph in part a.**

a. $f(x) = (x - 2)^2 - 3$

The graph of $y = f(x)$ is a translation of $y = x^2$. The parent graph has been translated right 2 units and down 3 units.

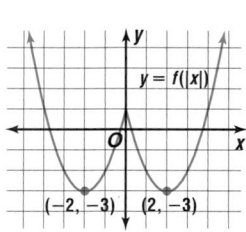

b. $y = |f(x)|$

$$|f(x)| = |(x - 2)^2 - 3|$$

This transformation reflects any portion of the parent graph that is below the *x*-axis so that it is above the *x*-axis.

c. $y = f(|x|)$

$$f(|x|) = (|x| - 2)^2 - 3$$

This transformation results in the portion of the parent graph on the left of the *y*-axis being replaced by a reflection of the portion on the right of the *y*-axis.

⑤ ENTERTAINMENT A traveling circus invites local schools to send math and science teams to its Science Challenge Day. One challenge is to write an equation that most accurately predicts the height of the flight of a human cannonball performer at any given time. Students collect data by witnessing a performance and examining time-lapse photographs of the flight. Using the performer's initial height of 15 feet and the photographs, one team records the data at the right. Write the equation of the related parabola that models the data.

Time (seconds)	Height (feet)
0	15
1	39
2	47
3	39
4	15

A graph of the data reveals that a parabola is the best model for the data. The parent graph of a parabola is the graph of the equation $y = x^2$. To write the equation of the related parabola that models the data, we need to compare points located near the vertex of each graph. An analysis of the transformation these points have undergone will help us determine the equation of the transformed parabola.

From the graph, we can see that parent graph has been turned upside-down, indicating that the equation for this parabola has been multiplied by some negative constant c. Through further inspection of the graph and its data points, we can see that the vertex of the parent graph has been translated to the point (2, 47). Therefore, an equation that models the data is $y = c(x - 2)^2 + 47$.

Height (ft) — Time (s)

Points: (0, 15), (1, 39), (2, 47), (3, 39), (4, 15)

To find c, compare points near the vertex of the graphs of the parent function $f(x)$ and the graph of the data points. Look at the relationship between the differences in the y-coordinates for each set of points.

x	$f(x) = x^2$
−2	4
−1	1
0	0
1	1
2	4

differences: ⟩ −3, ⟩ −1, ⟩ 1, ⟩ −3

x Time (seconds)	h(x) Height (feet)
0	15
1	39
2	47
3	39
4	15

differences: ⟩ 24, ⟩ 8, ⟩ 8, ⟩ 24

These differences are in a ratio of 1 to −8. This means that the graph of the parent graph has been expanded by a factor of −8. Thus, an equation that models the data is $y = -8(x - 2)^2 + 47$.

Lesson 3-2 Families of Graphs **141**

Addressing Individual Needs

Interpersonal Have students work in groups to determine whether families of functions have the same symmetries as the parent function. Encourage students to use computers or graphing calculators to help them discuss and test their conjectures.

Error Analysis

Watch for students who have difficulty evaluating an absolute value function.

Prevent by having students make tables of values that are both positive and negative and evaluate the function in order to form ordered pairs for graphing.

Additional Answers

2. The graph of $y = (x + 3)^2$ is a translation of $y = x^2$ three units to the left. The graph of $y = x^2 + 3$ is a translation of $y = x^2$ three units up.

4. When $c > 1$, the graph of $y = f(x)$ is compressed horizontally by a factor of c.

   When $c = 1$, the graph of $y = f(x)$ is unchanged.

   When $0 < c < 1$, the graph is expanded horizontally by a factor of $\frac{1}{c}$.

8a. expanded horizontally by a factor of 5

8b. translated right 5 units and down 2 units

8c. expanded vertically by a factor of 3, translated up 6 units

Study Guide and Practice Masters, pp. 41–42

CHECK FOR UNDERSTANDING

Communicating Mathematics

Read and study the lesson to answer each question.

1. **Write** the equation of the graph obtained when the parent graph $y = x^3$ is translated 4 units left and 7 units down. $\quad y = (x + 4)^3 - 7$

2. **Explain** the difference between the graphs of $y = (x + 3)^2$ and $y = x^2 + 3$. **See margin.**

3. **Name** two types of transformations for which the pre-image and the image are congruent figures.   **reflections and translations**

4. **Describe** the differences between the graphs of $y = f(x)$ and $y = f(cx)$ for $c > 0$. **See margin.**

5. **Write** equations for the graphs of $g(x)$, $h(x)$, and $k(x)$ if the graph of $f(x) = \sqrt[3]{x}$ is the parent graph.

5a. $g(x) = \sqrt[3]{x} + 1$

5b. $h(x) = -\sqrt[3]{x} - 1$

5c. $k(x) = \sqrt[3]{x + 2} + 1$

   a.   b.   c.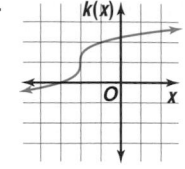

Guided Practice

Describe how the graphs of $f(x)$ and $g(x)$ are related.

6. The graph of $g(x)$ is the graph of $f(x)$ translated left 4 units.

7. The graph of $g(x)$ is the graph of $f(x)$ compressed horizontally by a factor of $\frac{1}{3}$ and then reflected over the x-axis.

6. $f(x) = |x|$ and $g(x) = |x + 4|$

7. $f(x) = x^3$ and $g(x) = -(3x)^3$

Use the graph of the given parent function to describe the graph of each related function.   8–9. See margin.

8. $f(x) = x^2$
   a. $y = (0.2x)^2$
   b. $y = (x - 5)^2 - 2$
   c. $y = 3x^2 + 6$

9. $f(x) = x^3$
   a. $y = |x^3 + 3|$
   b. $y = -(2x)^3$
   c. $y = 0.75(x + 1)^3$

Sketch the graph of each function.   10–11. See margin.

10. $f(x) = 2(x - 3)^3$

11. $g(x) = (0.5x)^2 - 1$

12. **Consumer Costs**   The cost of labor for servicing cars at B & B Automotive is $50 for each whole hour or for any fraction of an hour.   **a–b. See Answer Appendix.**

   a. Graph the function that describes the cost for x hours of labor.

   b. Graph the function that would show a $25 additional charge if you decide to also get the oil changed and fluids checked.

   c. What would be the cost of servicing a car that required 3.45 hours of labor if the owner requested that the oil be changed and the fluids be checked?   **$225**

 www.amc.glencoe.com/self_check_quiz

142   *Chapter 3   The Nature of Graphs*

Additional Answers

9a. translated up 3 units, portion of graph below x-axis reflected over the x-axis

9b. reflected over the x-axis, compressed horizontally by a factor of $\frac{1}{2}$

9c. translated left 1 unit, compressed vertically by a factor of 0.75

10.

11.

EXERCISES

Practice

Describe how the graphs of $f(x)$ and $g(x)$ are related.   **13–18. See margin.**

13. $f(x) = x$ and $g(x) = x + 6$

14. $f(x) = x^2$ and $g(x) = \frac{3}{4}x^2$

15. $f(x) = |x|$ and $g(x) = |5x|$

16. $f(x) = x^3$ and $g(x) = (x - 5)^3$

17. $f(x) = \frac{1}{x}$ and $g(x) = \frac{3}{x}$

18. $f(x) = [\![x]\!] + 1$ and $g(x) = -[\![x]\!] - 1$

19. Describe the relationship between the graphs of $f(x) = \sqrt{x}$ and $g(x) = -\sqrt{0.4x} + 3$.   **The graph of $g(x)$ is the graph of $f(x)$ reflected over the x-axis, expanded horizontally by a factor of 2.5, and translated up 3 units.**

Use the graph of the given parent function to describe the graph of each related function.   **20–25. See margin.**

20. $f(x) = x^2$
   a. $y = -(1.5x)^2$
   b. $y = 4(x - 3)^2$
   c. $y = \frac{1}{2}x^2 - 5$

21. $f(x) = |x|$
   a. $y = |0.2x|$
   b. $y = 7|x| - 0.4$
   c. $y = -9|x + 1|$

22. $f(x) = x^3$
   a. $y = (x + 2)^3 - 5$
   b. $y = -(0.8x)^3$
   c. $y = \left(\frac{5}{3}x\right)^3 + 2$

23. $f(x) = \sqrt{x}$
   a. $y = \frac{1}{3}\sqrt{x + 2}$
   b. $y = \sqrt{-x} - 7$
   c. $y = 4 + 2\sqrt{x - 3}$

24. $f(x) = \frac{1}{x}$
   a. $y = \frac{1}{0.5x}$
   b. $y = \frac{1}{6x} + 8$
   c. $y = \frac{1}{|x|}$

25. $f(x) = [\![x]\!]$
   a. $y = \left[\!\left[\frac{5}{2}x\right]\!\right] - 3$
   b. $y = -0.75[\![x]\!]$
   c. $y = [\![\,|x| - 4\,]\!]$

26. Name the parent graph of $m(x) = |-9 + (0.75x)^2|$. Then sketch the graph of $m(x)$.   **$y = x^2$; see Answer Appendix.**

27. Write the equation of the graph obtained when the graph of $y = \frac{1}{x}$ is compressed vertically by a factor of 0.25, translated 4 units right, and then translated 3 units up. $y = \frac{0.25}{x - 4} + 3$

Sketch the graph of each function.   **28–34. See Answer Appendix.**

28. $f(x) = -(x + 4) + 5$

29. $g(x) = |x^2 - 4|$

30. $h(x) = (0.5x - 1)^3$

31. $n(x) = -2.5[\![x]\!] + 3$

32. $q(x) = -4|x - 2| - 1$ **33.** $k(x) = -\frac{1}{2}(x - 3)^2 - 4$

34. Graph $y = f(x)$ and $y - f(|x|)$ on the same set of axes if $f(x) = (x + 3)^2 - 8$.

Graphing Calculator

Use a graphing calculator to graph each set of functions on the same screen. Name the x-intercept(s) of each function.

35. a. $y = x^2$ **0**
   b. $y = (4x - 2)^2$ **0.5**
   c. $y = (2x + 3)^2$ **−1.5**

36. a. $y = x^3$ **0**
   b. $y = (3x - 2)^3$ **0.66̄**
   c. $y = (4x + 1)^3$ **−0.25**

37. a. $y = \sqrt{x}$ **0**
   b. $y = \sqrt{2x + 5}$ **−2.5**
   c. $y = \sqrt{5x - 3}$ **0.6**

35–37. See Answer Appendix for graphs.

Lesson 3-2 Families of Graphs **143**

Additional Answers

21a. expanded horizontally by a factor of 5

21b. expanded vertically by a factor of 7, translated down 0.4 units

21c. reflected across the x-axis, translated left 1 unit, expanded vertically by a factor of 9

22a. translated left 2 units and down 5 units

22b. expanded horizontally by a factor of 1.25, reflected over the x-axis

22c. compressed horizontally by a factor of $\frac{3}{5}$, translated up 2 units

23a. translated left 2 units, compressed vertically by a factor of $\frac{1}{3}$

23b. reflected over the y-axis, translated down 7 units

23c. expanded vertically by a factor of 2, translated right 3 units and up 4 units

24a. expanded horizontally by a factor of 2

24b. compressed horizontally by a factor of $\frac{1}{6}$, translated 8 units up

24c. The portion of parent graph on the left of the y-axis is replaced by a reflection of the portion on the right of the y-axis.

25a. compressed horizontally by a factor of $\frac{2}{5}$, translated down 3 units

25b. reflected over the y-axis, compressed vertically by a factor of 0.75

25c. The portion of the parent graph on the left of the y-axis is replaced by a reflection of the portion on the right of the y-axis. The new image is then translated 4 units right.

Additional Answers

13. The graph of $g(x)$ is a translation of the graph of $f(x)$ up 6 units.

14. The graph of $g(x)$ is the graph of $f(x)$ compressed vertically by a factor of $\frac{3}{4}$.

15. The graph of $g(x)$ is the graph of $f(x)$ compressed horizontally by a factor of $\frac{1}{5}$.

16. The graph of $g(x)$ is a translation of $f(x)$ right 5 units.

17. The graph of $g(x)$ is the graph of $f(x)$ expanded vertically by a factor of 3.

18. The graph of $g(x)$ is the graph of $f(x)$ reflected over the x-axis.

20a. reflected over the x-axis, compressed horizontally by a factor of $0.\overline{6}$

20b. translated right 3 units, expanded vertically by a factor of 4

20c. compressed vertically by a factor of $\frac{1}{2}$, translated down 5 units

38. Technology Transformations can be used to create a series of graphs that appear to move when shown sequentially on a video screen. Suppose you start with the graph of $y = f(x)$. Describe the effect of graphing the following functions in succession if n has integer values from 1 to 100.

a. $y_n = f(x + 2n) - 3n$. **The graph would continually move left 2 units and down 3 units.**

b. $y_n = (-1)^n f(x - n)$. **See margin.**

39. Critical Thinking Study the coordinates of the x-intercepts you found in the related graphs in Exercises 35–37. Make a conjecture about the x-intercept of $y = (ax + b)^n$ if $y = x^n$ is the parent function. **See margin.**

40. Business The standard cost of a taxi fare is $1.50 for the first unit and 25 cents for each additional unit. A unit is composed of distance (one unit equals 0.2 mile) and/or wait time (one unit equals 75 seconds). As the cab moves at more than 9.6 miles per hour, the taxi's meter clocks distance. When the cab is stopped or moving at less than 9.6 miles per hour, the meter clocks time. Thus, traveling 0.1 mile and then waiting at a stop light for 37.5 seconds generates one unit and a 25-cent charge.

a. Assuming that the cab meter rounds up to the nearest unit, write a function that would determine the cost for x units of cab fare, where $x > 0$.

40a. $y = \begin{cases} 0.25[\![x - 1]\!] + 1.50 \text{ if } [\![x]\!] = x \\ 0.25[\![x]\!] + 1.50 \text{ if } [\![x]\!] < x \end{cases}$

b. Graph the function found in part a. **See margin.**

41a–c. See margin for graphs and comparisons.

41. Geometry Suppose $f(x) = 5 - |x - 6|$.

a. Sketch the graph of $f(x)$ and calculate the area of the triangle formed by $f(x)$ and the positive x-axis. **25 units²**

b. Sketch the graph of $y = 2f(x)$ and calculate the area of the new triangle formed by $2f(x)$ and the positive x-axis. How do the areas of part a and part b compare? Make a conjecture about the area of the triangle formed by $y = c \cdot f(x)$ in the first quadrant if $c \geq 0$. **50 units²; 25c units²**

c. Sketch the graph of $y = f(x - 3)$ and recalculate the area of the triangle formed by $f(x - 3)$ and the positive x-axis. How do the areas of part a and part c compare? Make a conjecture about the area of the triangle formed by $y = f(x - c)$ in the first quadrant if $c \geq 0$. **25 units²; 25 units²**

42a. (1) $y = x^2$
(2) $y = x^3$
(3) $y = -x^2$
(4) $y = -x^3$

42. Critical Thinking Study the parent graphs at the beginning of this lesson.

a. Select a parent graph or a modification of a parent graph that meets each of the following specifications.
   (1) positive at its leftmost points and positive at its rightmost points
   (2) negative at its leftmost points and positive at its rightmost points
   (3) negative at its leftmost points and negative at its rightmost points
   (4) positive at its leftmost points and negative at its rightmost points

b. Sketch the related graph for each parent graph that is translated 3 units right and 5 units down. **See margin.**

c. Write an equation for each related graph. **(1)** $y = (x - 3)^2 - 5$; **(2)** $y = (x - 3)^3 - 5$; **(3)** $y = -(x - 3)^2 - 5$; **(4)** $y = -(x - 3)^3 - 5$

144 *Chapter 3 The Nature of Graphs*

43. **Critical Thinking** Suppose a reflection, a translation, or a dilation were applied to an even function. **b. horizontal translation**

   a. Which transformations would result in another even function?

   b. Which transformations would result in a function that is no longer even?
   a. reflection over the *x*-axis, reflection over the *y*-axis, vertical translation, horizontal compression or expansion, and vertical expansion or compression

Mixed Review

44. Is the graph of $f(x) = x^{17} - x^{15}$ symmetric with respect to the origin? Explain. *(Lesson 3-1)* **yes; $f(-x) = -f(x)$**

45. **Child Care** Elisa Dimas is the manager for the Learning Loft Day Care Center. The center offers all day service for preschool children for $18 per day and after school only service for $6 per day. Fire codes permit only 50 children in the building at one time. State law dictates that a child care worker can be responsible for a maximum of 3 preschool children and 5 school-age children at one time. Ms. Dimas has ten child care workers available to work at the center during the week. How many children of each age group should Ms. Dimas accept to maximize the daily income of the center? *(Lesson 2-7)*
   30 preschoolers and 20 school-age

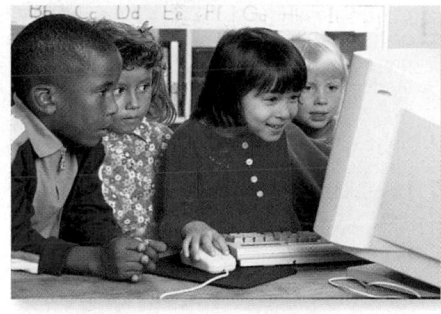

46. **Geometry** Triangle ABC is represented by the matrix $\begin{bmatrix} 5 & 1 & -2 \\ -4 & 3 & -1 \end{bmatrix}$. Find the image of the triangle after a $90°$ counterclockwise rotation about the origin. *(Lesson 2-4)* **$A'(4, 5)$, $B'(-3, 1)$, $C'(1, -2)$**

47. Find the values of x, y, and z for which $\begin{bmatrix} x^2 & 7 & 9 \\ 5 & 12 & 6 \end{bmatrix} = \begin{bmatrix} 25 & 7 & y \\ 5 & 2z & 6 \end{bmatrix}$ is true. *(Lesson 2-2)* **$x = \pm5$, $y = 9$, $z = 6$**

48. Solve the system of equations algebraically. *(Lesson 2-1)*
   $6x + 5y = -14$
   $5x + 2y = -3$ **$(1, -4)$**

49. The graph implies a negative linear relationship.

49. Describe the linear relationship implied in the scatter plot at the right. *(Lesson 1-6)*

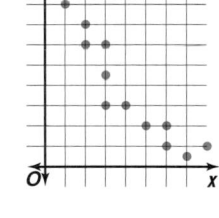

50. Find the slope of a line perpendicular to a line whose equation is $3x - 4y = 0$. *(Lesson 1-5)* $-\dfrac{4}{3}$

51. **Fund-Raising** The Band Boosters at Palermo High School are having their annual doughnut sale to raise money for new equipment. The equation $5d - 2p = 500$ represents the amount of profit p in dollars the band will make selling d boxes of doughnuts. What is the p-intercept of the line represented by this equation? *(Lesson 1-3)* **-250**

52. $\dfrac{2}{3}x^2 - 4x + 4$, $\dfrac{4}{9}x^2 - \dfrac{20}{3}x + 25$

52. Find $[f \circ g](x)$ and $[g \circ f](x)$ if $f(x) = \dfrac{2}{3}x - 2$ and $g(x) = x^2 - 6x + 9$. *(Lesson 1-2)*

53. **SAT/ACT Practice** If $d = m - \dfrac{50}{m}$ and m is a positive number that increases in value, then d **A**
   A increases in value.
   B increases, then decreases.
   C remains unchanged.
   D decreases in value.
   E decreases, then increases.

Extra Practice See p. A30.

Extra Credit

Write a rule to predict whether a function has *y*-axis symmetry. **If $f(x) = f(-x)$, then $f(x)$ has *y*-axis symmetry. Furthermore, any time a function has *x*-terms with only even powers, $f(x) = f(-x)$.**

Open-Ended Assessment

Modeling Have students work in pairs to select a parent function and graph it. Then, have students take turns graphing a related function and explaining how the two functions are related. Students should include absolute value, square root, and greatest integer functions.

📁 **Chapter 3, Quiz A (Lessons 3-1 and 3-2)** is available in the *Assessment and Evaluation Masters*, p. 69.

Enrichment Masters, p. 26

3-3

Graphs of Nonlinear Inequalities

OBJECTIVES
- Graph polynomial, absolute value, and radical inequalities in two variables.
- Solve absolute value inequalities.

PHARMACOLOGY Pharmacists label medication as to how much and how often it should be taken. Because oral medication requires time to take effect, the amount of medication in your body varies with time. Suppose the equation $m(x) = 0.5x^4 + 3.45x^3 - 96.65x^2 + 347.7x$ for $0 < x \leq 6$ models the number of milligrams of a certain pain reliever in the bloodstream x hours after taking 400 milligrams of it. The medicine is to be taken every 4 hours. At what times during the first 4-hour period is the level of pain reliever in the bloodstream above 300 milligrams? *This problem will be solved in Example 5.*

Problems like the one above can be solved by graphing inequalities. Graphing inequalities in two variables identifies all ordered pairs that will satisfy the inequality.

Example ① Determine whether $(3, -4)$, $(4, 7)$, $(1, 1)$, and $(-1, 6)$ are solutions for the inequality $y \leq (x - 2)^2 - 3$.

Substitute the x-value and y-value from each ordered pair into the inequality.

$$y \leq (x - 2)^2 - 3$$
$$-4 \stackrel{?}{\leq} (3 - 2)^2 - 3 \quad (x, y) = (3, -4)$$
$$-4 \leq -2 \checkmark \qquad \text{true}$$

$$y \leq (x - 2)^2 - 3$$
$$1 \stackrel{?}{\leq} (1 - 2)^2 - 3 \quad (x, y) = (1, 1)$$
$$1 \leq -2 \qquad \text{false}$$

$$y \leq (x - 2)^2 - 3$$
$$7 \stackrel{?}{\leq} (4 - 2)^2 - 3 \quad (x, y) = (4, 7)$$
$$7 \leq 1 \qquad \text{false}$$

$$y \leq (x - 2)^2 - 3$$
$$6 \stackrel{?}{\leq} (-1 - 2)^2 - 3 \quad (x, y) = (-1, 6)$$
$$6 \leq 6 \checkmark \qquad \text{true}$$

Of these ordered pairs, $(3, -4)$ and $(-1, 6)$ are solutions for $y \leq (x - 2)^2 - 3$.

Look Back
Refer to Lesson 1-8 for more about graphing linear inequalities.

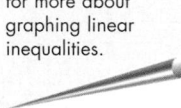

Similar to graphing linear inequalities, the first step in graphing nonlinear inequalities is graphing the boundary. You can use concepts from Lesson 3-2 to graph the boundary.

Example ② Graph $y \geq (x - 4)^3 - 2$.

The boundary of the inequality is the graph of $y = (x - 4)^3 - 2$. To graph the boundary curve, start with the parent graph $y = x^3$. Analyze the boundary equation to determine how the boundary relates to the parent graph.

$$y = (x - 4)^3 - 2$$

move 4 units right   *move 2 units down*

---

1 FOCUS

5-Minute Check
Lesson 3-2

1. Describe how $f(x) = |x|$ and $g(x) = -3|x + 2|$ are related.
   $g(x)$ is the graph of $f(x)$ translated left 2 units, reflected over the x-axis, and expanded vertically by a factor of 3.

Use the graph of $y = x^2$ to describe the graph of each related function.

2. $y = 0.5x^2 + 2$ **The parent graph is compressed vertically by a factor of 0.5 and translated up 2 units.**

3. $y = (x - 4)^2 - 3$ **The parent graph is translated right 4 units and down 3 units.**

Sketch the graph of each function.

4. $y = 2x^2$

5. $y = \dfrac{x^2 - 1}{-2}$

Motivating the Lesson
Real-World Application Give students antacid tablets and ask them to determine their initial weight. Allow students to lower their antacid tablets resting on slotted spoons into cups of water. Have students remove the spoons from the water at 10 second intervals and weigh the tablets. Repeat until the tablets are dissolved. Have students guess what kind of curve would be modeled by the weight of a tablet (y-axis) versus time (x-axis).

Resource Manager

📁 **Reproducible Masters**
- *Study Guide and Practice,* pp. 43–44
- *Enrichment,* p. 27

💿 **Transparencies**
- 5-Minute Check, 3-3
- Teaching, 3-3

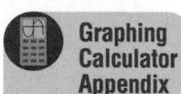
Since the boundary is included in the inequality, the graph is drawn as a solid curve.

The inequality states that the y-values of the solution are greater than the y-values on the graph of $y = (x - 4)^3 - 2$. For a particular value of x, all of the points in the plane that lie above the curve have y-values greater than $y = (x - 4)^3 - 2$. So this portion of the graph should be shaded.

To verify numerically, you can test a point not on the boundary. *It is common to test (0, 0) whenever it is not on the boundary.*

$y \geq (x - 4)^3 - 2$

$0 \overset{?}{\geq} (0 - 4)^3 - 2$ *Replace (x, y) with (0, 0).*

$0 \geq -66$ ✓   *True*

Since $(0, 0)$ satisfies the inequality, the correct region is shaded.

The same process used in Example 2 can be used to graph inequalities involving absolute value.

Example **3** Graph $y > 3 - |x + 2|$.

Begin with the parent graph $y = |x|$.

It is easier to sketch the graph of the given inequality if you rewrite it so that the absolute value expression comes first.

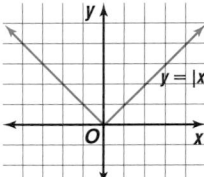

$y = 3 - |x + 2|$ → $y = -|x + 2| + 3$

This more familiar form tells us the parent graph is reflected over the x-axis and moved 2 units left and three units up. The boundary is not included, so draw it as a dashed line.

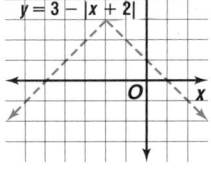

The y-values of the solution are greater than the y-values on the graph of $y = 3 - |x + 2|$, so shade above the graph of $y = 3 - |x + 2|$.

Verify by substituting $(0, 0)$ in the inequality to obtain $0 > 1$. Since this statement is false, the part of the graph containing $(0, 0)$ should not be shaded. Thus, the graph is correct.

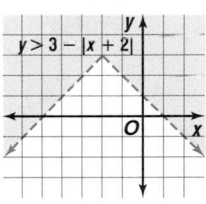

2 TEACH

In-Class Examples

Example 1
Determine whether $(-2, 5)$, $(3, -1)$, $(-4, 2)$, and $(-1, -1)$ are solutions for the inequality $y \geq 2x^3 + 7$. **$(-2, 5)$ and $(-4, 2)$ are solutions.**

Example 2
Graph $y \leq \sqrt[3]{x - 1} + 2$.

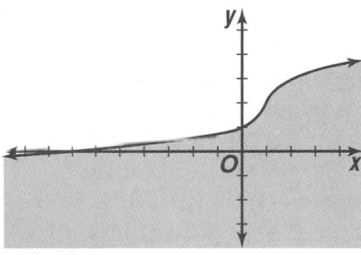

Example 3
Graph $y > |x - 5| + 4$.

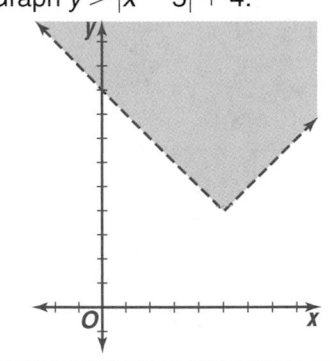

Teaching Tip While presenting Examples 2 and 3, you may wish to show each step in the sequence of transformations leading up to the final graph.

 Prerequisite Skills

A review of the prerequisite skills for this lesson is provided in the *Prerequisite Skills Masters.*

Skill Number and Content	Pages
3 Order of Operations	7–8

To solve absolute value inequalities algebraically, use the definition of absolute value to determine the solution set. That is, if $a < 0$, then $|a| = -a$, and if $a \geq 0$, then $|a| = a$.

Example 4 Solve $|x - 2| - 5 < 4$.

There are two cases that must be solved. In one case, $x - 2$ is negative, and in the other, $x - 2$ is positive.

Case 1
$$(x - 2) < 0$$
$$|x - 2| - 5 < 4$$
$$-(x - 2) - 5 < 4 \quad |x - 2| = -(x - 2)$$
$$-x + 2 - 5 < 4$$
$$-x < 7$$
$$x > -7$$

Case 2
$$(x - 2) > 0$$
$$|x - 2| - 5 < 4$$
$$x - 2 - 5 < 4 \quad |x - 2| = (x - 2)$$
$$x - 7 < 4$$
$$x < 11$$

The solution set is $\{x \mid -7 < x < 11\}$. *$\{x \mid -7 < x < 11\}$ is read "the set of all numbers x such that x is between −7 and 11."*

Verify this solution by graphing.

First, graph $y = |x - 2| - 5$. Since we are solving $|x - 2| - 5 < 4$ and $|x - 2| - 5 = y$, we are looking for a region in which $y = 4$. Therefore, graph $y = 4$ and graph it on the same set of axes.

Identify the points of intersection of the two graphs. By inspecting the graph, we can see that they intersect at $(-7, 4)$ and $(11, 4)$.

Now shade the region where the graphs of the inequalities $y > |x - 2| - 5$ and $y < 4$ intersect. This occurs in the region of the graph where $-7 < x < 11$. Thus, the solution to $|x - 2| - 5 < 4$ is the set of x-values such that $-7 < x < 11$.

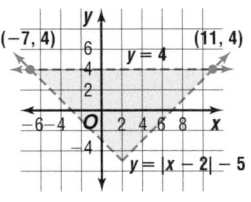

Nonlinear inequalities have applications to many real-world situations, including business, education, and medicine.

Example 5 **PHARMACOLOGY** Refer to the application at the beginning of the lesson. At what times during the first 4-hour period is the amount of pain reliever in the bloodstream above 300 milligrams?

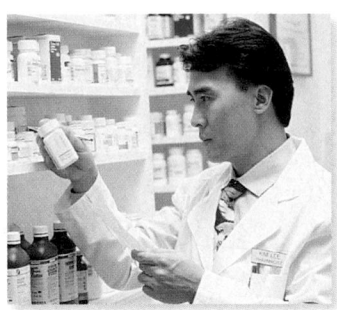

Since we need to know when the level of pain reliever in the bloodstream is above 300 milligrams, we can write an inequality.
$$0.5x^4 + 3.45x^3 - 96.65x^2 + 347.7x > 300$$
for $0 < x \leq 6$

Addressing Individual Needs

 Logical Have students draw a graph of $y < x^2 - 2x + 1$. Ask them:
- What is the first step in drawing the graph?
- Which side of the boundary should be shaded? Are the boundary points solutions?
- What is a convenient point to test?

Let $y = 0.5x^4 + 3.45x^3 - 96.65x^2 + 347.7x$ and $y = 300$. Graph both equations on the same set of axes using a graphing calculator.

Calculating the points of intersection, we find that the two equations intersect at about $(1.3, 300)$ and $(3.1, 300)$. Therefore, when $1.3 < x < 3.1$, the amount of pain reliever in the bloodstream is above 300 milligrams. That is, the amount exceeds 300 milligrams between about 1 hour 18 minutes and 3 hours 6 minutes after taking the medication.

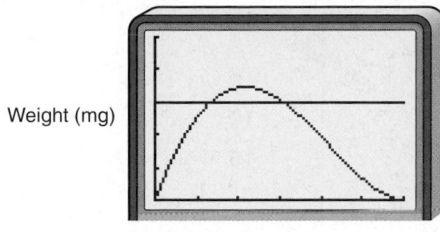

Weight (mg)

Time (h)
[0, 6] scl: 1 by [0, 500] scl: 100

CHECK FOR UNDERSTANDING

Communicating Mathematics

Read and study the lesson to answer each question. **2–4. See margin.**

1. **Describe** how knowledge of transformations can help you graph the inequality $y \le 5 + \sqrt{x - 2}$.

1. A knowledge of transformations can help determine the graph of the boundary of the shaded region, $y = 5 + \sqrt{x - 2}$.

2. **State** the two cases considered when solving a one-variable absolute value inequality algebraically.

3. **Write** a procedure for determining which region of the graph of an inequality should be shaded.

4. *Math Journal* **Sketch** the graphs of $y = |x - 3| + 2$ and $y = 1$ on the same set of axes. Use your sketch to solve the inequality $|x - 3| + 2 < 1$. If no solution exists, write *no solution*. Write a paragraph to explain your answer.

Guided Practice

Determine whether the ordered pair is a solution for the given inequality. Write *yes* or *no*.

5. $y \ge -5x^4 + 7x^3 + 8, (-1, -3)$ **yes**   6. $y < |3x - 4| - 1, (0, 3)$ **no**

Graph each inequality. **7–9. See margin.**

7. $y \le (x + 1)^3$ 8. $y \le 2(x - 3)^2$ 9. $y > -|x - 4| + 2$

Solve each inequality.

10. $|x + 6| > 4$
$\{x \mid x < -10 \text{ or } x > -2\}$

11. $|3x - 4| \le x$
$\{x \mid 1 \le x \le 2\}$

12. **Manufacturing** The AccuData Company makes compact disks measuring 12 centimeters in diameter. The diameters of the disks can vary no more than 50 micrometers or 5×10^{-3} centimeter.

a. Write an absolute value inequality to represent the range of diameters of compact disks. $|x - 12| \le 0.005$

b. What are the largest and smallest diameters that are allowable?
12.005 cm, 11.995 cm

 www.amc.glencoe.com/self_check_quiz   *Lesson 3-3 Graphs of Nonlinear Inequalities* **149**

3 PRACTICE/APPLY

Error Analysis
Watch for students who inappropriately change the direction of the inequality symbol when solving the inequality.

Prevent by reminding students that the sign only changes when multiplying or dividing by a negative number.

Additional Answers
4. **This inequality has no solution since the two graphs do not intersect.**

7.

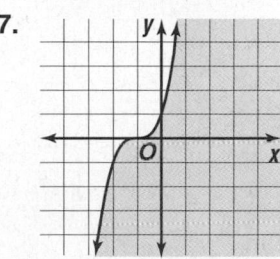

(continued on page 150)

Study Guide and Practice Masters, pp. 43–44

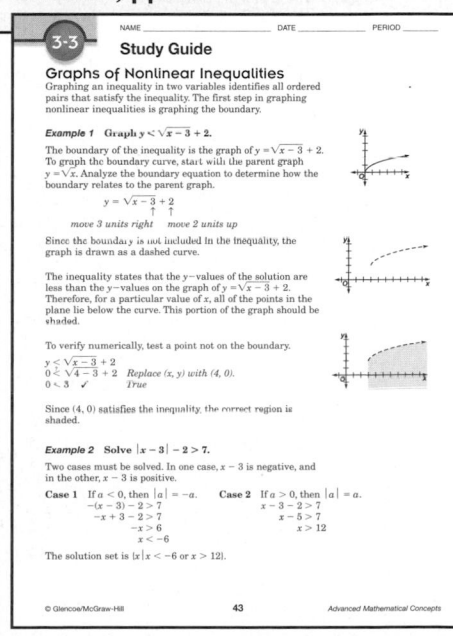

Additional Answers
(pp. 149–151)

8.

9.

32.

50.

51.
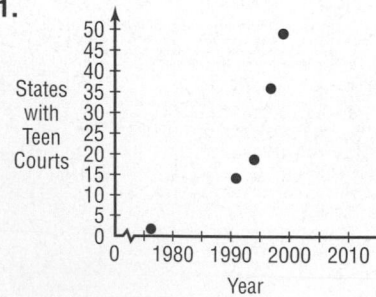

EXERCISES

Practice

Determine whether the ordered pair is a solution for the given inequality. Write *yes* or *no*.

A

13. $y < x^3 - 4x^2 + 2$, $(1, 0)$ **no**

14. $y < |x - 2| + 7$, $(3, 8)$ **no**

15. $y > -\sqrt{x + 11} + 1$, $(-2, -1)$ **yes**

16. $y < -0.2x^2 + 9x - 7$, $(10, 63)$ **no**

17. $y \le \dfrac{x^2 - 6}{x}$, $(-6, -9)$ **yes**

18. $y \ge 2|x|^3 - 7$, $(0, 0)$ **yes**

19. Which of the ordered pairs, $(0, 0)$, $(1, 4)$, $(1, 1)$, $(-1, 0)$, and $(1, -1)$, is a solution for $y \le \sqrt{x} + 2$? How can you use these results to determine if the graph at the right is correct?   **(0, 0), (1, 1), and (1, −1); if these points are in the shaded region and the other points are not, then the graph is correct.**

Graph each inequality.   **20–31. See Answer Appendix.**

B

20. $y \le x^2 - 4$

21. $y > \sqrt{0.5x}$

22. $y < |x - 9|$

23. $y > |2x| + 3$

24. $y < (x - 5)^2$

25. $y \ge -x^3$

26. $y > -(0.4x)^2$

27. $y \le |3(x - 4)|$

28. $y < \sqrt{x + 3} + 5$

29. $y \ge (x - 1)^2 - 3$

30. $y \ge (2x + 1)^3 + 2$

31. $y \le -3|x - 2| + 4$

32. Sketch the graph of the inequality $y \ge x^3 - 6x^2 + 12x - 8$.   **See margin.**

Solve each inequality.   **34.** $\{x \mid x \le -18 \text{ or } x \ge 10\}$ **35.** $\{x \mid -2 < x < 9\}$

C

33. $|x + 4| > 5$

34. $|3x + 12| \ge 42$

35. $|7 - 2x| - 8 < 3$

33. $\{x \mid x < -9$ or $x > 1\}$

36. $|5 - x| \le x$
   $\{x \mid x \ge 2.5\}$

37. $|5x - 8| < 0$
   no solution

38. $|2x + 9| - 2x \ge 0$
   all real numbers

39. Find all values of x that satisfy $-\dfrac{2}{3}|x + 5| \ge -8$. $\{x \mid -17 \le x \le 7\}$

Applications and Problem Solving

40. Chemistry   Katie and Wes worked together on a chemistry lab. They determined the quantity of the unknown in their sample to be 37.5 ± 1.2 grams. If the actual quantity of unknown is x, write their results as an absolute value inequality. Solve for x to find the range of possible values of x.

41. Critical Thinking   Solve $3|x - 7| < |x - 1|$. $\{x \mid 5.5 < x < 10\}$

40. $|x - 37.5| \le 1.2$; $36.3 \le x \le 38.7$

42. Critical Thinking   Find the area of the region described by $y \ge 2|x - 3| + 4$ and $x - 2y \ge -20$.   **30 units²**

43. Education   Amanda's teacher calculates grades using a weighted average. She counts homework as 10%, quizzes as 15%, projects as 20%, tests as 40%, and the final exam as 15% of the final grade. Going into the final, Amanda has scores of 90 for homework, 75 for quizzes, 76 for projects, and 80 for tests. What grade does Amanda need on the final exam if she wants to get an overall grade of at least an 80? $x \ge 83\frac{2}{3}$

44. Critical Thinking   Consider the equation $|(x - 3)^2 - 4| = b$. Determine the value(s) of b so that the equation has

a. no solution. $b < 0$

b. one solution.   **none**

c. two solutions. $b = 0$ or $b > 4$

d. three solutions. $b = 4$

e. four solutions. $0 < b < 4$

Additional Answers (p. 151)

45a.

45b. The shaded region shows all points (x, y) where x represents the number of cookies sold and y represents the possible profit made for a given week.

45. Business After opening a cookie store in the mall, Paul and Carol Mason hired an consultant to provide them with information on increasing their profit. The consultant told them that their profit P depended on the number of cookies x that they sold according to the relation $P(x) \le -0.005(x - 1200)^2 + 400$. They typically sell between 950 and 1000 cookies in a given day.

    a. Sketch a graph to model this situation.   **a–b. See margin.**

    b. Explain the significance of the shaded region.

46. the graph of $g(x)$ is the graph of $f(x)$ reflected over the x-axis and expanded vertically by a factor of 2

Mixed Review

46. How are the graphs of $f(x) = x^3$ and $g(x) = -2x^3$ related? *(Lesson 3-2)*

47. Determine whether the graph of $y = -\dfrac{1}{x^4}$ is symmetric with respect to the x-axis, y-axis, the line $y = x$, the line $y = -x$, or none of these. *(Lesson 3-1)*
y-axis

48. Find the inverse of $\begin{bmatrix} 8 & -3 \\ 4 & -5 \end{bmatrix}$. *(Lesson 2-5)* $-\dfrac{1}{28} \begin{bmatrix} -5 & 3 \\ -4 & 8 \end{bmatrix}$

49. $\begin{bmatrix} 6 & \dfrac{-21}{4} \\ -3 & 0 \end{bmatrix}$

49. Multiply $\begin{bmatrix} 8 & -7 \\ -4 & 0 \end{bmatrix}$ by $\dfrac{3}{4}$. *(Lesson 2-3)*

50. Graph $y = 3|x| + 5$. *(Lesson 1-7)*.   **See margin.**

51. See margin.

51. Criminal Justice The table shows the number of states with teen courts over a period of several years. Make a scatter plot of the data. *(Lesson 1-6)*

52. Find $[f \circ g](4)$ and $[g \circ f](4)$ for $f(x) = 5x + 9$ and $g(x) = 0.5x - 1$. *(Lesson 1-2)*   **14, 13.5**

53. SAT Practice Grid-In Student A is 15 years old. Student B is one-third older. How many years ago was student B twice as old as student A?   **10**

Year	States with Teen Courts
1976	2
1991	14
1994	17
1997	36
1999	47*

Source: American Probation and Parole Association.

*Includes District of Columbia

MID-CHAPTER QUIZ

Determine whether each graph is symmetric with respect to the x-axis, the y-axis, the line $y = x$, the line $y = -x$, the origin, or none of these. *(Lesson 3-1)* **1–2. See margin.**

1. $x^2 + y^2 - 9 = 0$ **2.** $5x^2 + 6x - 9 = y$

3. $x = \dfrac{7}{y}$ **4.** $y = |x| + 1$

$y = x, y = -x,$ origin y-axis

Use the graph of the given parent function to describe the graph of each related function. *(Lesson 3-2)* **5–6. See margin.**

5. $f(x) = [\![x]\!]$ **6.** $f(x) = x^3$

  a. $y = [\![x]\!] - 2$ a. $y = 3x^3$

  b. $y = -[\![x - 3]\!]$ b. $y = (0.5x)^3 - 1$

  c. $y = \dfrac{1}{4}[\![x]\!] + 1$ c. $y = (x + 1)^3 + 4$

7. Sketch the graph of $g(x) = -0.5(x - 2)^2 + 3$. *(Lesson 3-2)* **See Answer Appendix.**

8. Graph the inequality $y \ge \left(\dfrac{1}{3}x\right)^2 + 2$. *(Lesson 3-3)* **See Answer Appendix.**

9. Find all values of x that satisfy $|2x - 7| < 15$. *(Lesson 3-3)* **$-4 < x < 11$**

10. Technology In September of 1999, a polling organization reported that 64% of Americans were "not very" or "not at all" concerned about the Year-2000 computer bug, with a margin of error of 3%. Write and solve an absolute value inequality to describe the range of the possible percent of Americans who were relatively unconcerned about the "Y2K bug." *(Lesson 3-3)* $|x - 64| < 3;$
Source: The Gallup Organization   **$61 < x < 67$**

Extra Credit

Graph the solution to the following system of inequalities.

$y > |x - 2|$
$y < |x + 1|$

$y < |x + 1|$ $y > |x - 2|$

---

4 ASSESS

Open-Ended Assessment

Speaking Ask students to explain how to graph an inequality. Have them describe how to determine where the graph should be shaded.

Mid-Chapter Quiz

The Mid-Chapter Quiz provides students with a brief review of the concepts and skills in Lessons 3-1 through 3-3. Lesson numbers are given to the right of exercises or instruction lines so students can review concepts not yet mastered.

Additional Answers
Mid-Chapter Quiz

1. x-axis, y-axis, $y = x$, $y = -x$, origin

2. none of these

5a. translated down 2 units

5b. reflected over the x-axis, translated right 3 units

5c. compressed vertically by a factor of $\dfrac{1}{4}$, translated up 1 unit

6a. expanded vertically by a factor of 3

6b. expanded horizontally by a factor of 2 and translated down 1 unit

6c. translated left 1 unit and up 4 units

Enrichment Masters, p. 27

Inverse Functions and Relations

1 FOCUS

5-Minute Check
Lesson 3-3

1. Determine whether (2, 44) is a solution for $y \geq 2x^4 + x^3 + 2x - 1$. **no**

Graph each inequality.

2. $y > 2x^3 - 3$

3. $y < |x + 4|$

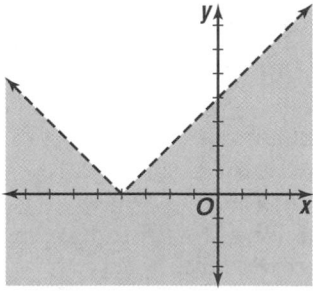

4. $y \geq \sqrt{x + 2}$

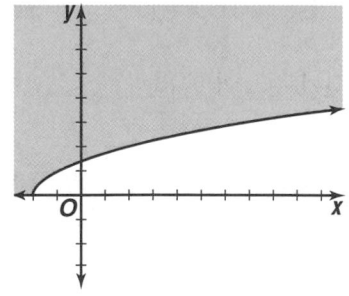

5. Solve $|x - 1| > 4$.
$\{x | x > 5 \text{ or } x < -3\}$

Motivating the Lesson

Hands-On Activity Have students practice reflecting various graphs of known functions over the line $y = x$. Then, have them try to state the equations of the reflected graphs. Tell them that the objective of the lesson is to learn how to determine these equations.

OBJECTIVES
- Determine inverses of relations and functions.
- Graph functions and their inverses.

Real World Application

METEOROLOGY The hottest temperature ever recorded in Montana was 117° F on July 5, 1937. To convert this temperature to degrees Celsius C, subtract 32° from the Fahrenheit temperature F and then multiply the result by $\frac{5}{9}$. The formula for this conversion is $C = \frac{5}{9}(F - 32)$. The coldest temperature ever recorded in Montana was $-57°$ C on January 20, 1954. To convert this temperature to Fahrenheit, multiply the Celsius temperature by $\frac{9}{5}$ and then add 32°. The formula for this conversion is $F = \frac{9}{5}C + 32$.

The temperature conversion formulas are examples of **inverse functions.** Relations also have inverses, and these inverses are themselves relations.

Inverse Relations	Two relations are inverse relations if and only if one relation contains the element (b, a) whenever the other relation contains the element (a, b).

If $f(x)$ denotes a function, then $f^{-1}(x)$ denotes the inverse of $f(x)$. However, $f^{-1}(x)$ may not necessarily be a function. To graph a function or relation and its inverse, you switch the x- and y-coordinates of the ordered pairs of the function. This results in graphs that are symmetric to each other with respect to the line $y = x$.

Example ① Graph $f(x) = -\frac{1}{2}|x| + 3$ and its inverse.

To graph the function, let $y = f(x)$. To graph $f^{-1}(x)$, interchange the x- and y-coordinates of the ordered pairs of the function.

Note that the domain of one relation or function is the range of the inverse and vice versa.

$f(x) = -\frac{1}{2}|x| + 3$		$f^{-1}(x)$	
x	**f(x)**	**x**	**$f^{-1}(x)$**
−3	1.5	1.5	−3
−2	2	2	−2
−1	2.5	2.5	−1
0	3	3	0
1	2.5	2.5	1
2	2	2	2
3	1.5	1.5	3

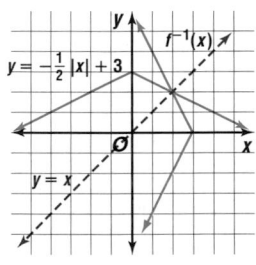

The graph of $f^{-1}(x)$ is the reflection of $f(x)$ over the line $y = x$.

Note that the inverse in Example 1 is not a function because it fails the vertical line test.

Resource Manager

Reproducible Masters
- *Study Guide and Practice,* pp. 45–46
- *Enrichment,* p. 28
- *Assessment and Evaluation,* pp. 68, 69

Transparencies
- 5-Minute Check, 3-4
- Teaching, 3-4

You can use the **horizontal line test** to determine if the inverse of a relation will be a function. If every horizontal line intersects the graph of the relation in at most one point, then the inverse of the relation is a function.

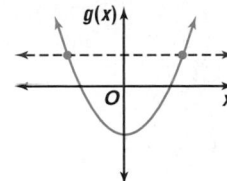

The inverse of f(x) is a function.   *The inverse of g(x) is not a function.*

You can find the inverse of a relation algebraically. First, let $y = f(x)$. Then interchange x and y. Finally, solve the resulting equation for y.

Example **2** Consider $f(x) = (x + 3)^2 - 5$.

    a. Is the inverse of $f(x)$ a function?

    b. Find $f^{-1}(x)$.

    c. Graph $f(x)$ and $f^{-1}(x)$ using a graphing calculator.

Graphing Calculator Tip

You can differentiate the appearance of your graphs by highlighting the symbol in front of each equation in the **Y=** list and pressing [ENTER] to select line (\\), thick (\\), or dot (·.).

    a. Since the line $y = -2$ intersects the graph of $f(x)$ at more than one point, the function fails the horizontal line test. Thus, the inverse of $f(x)$ is not a function.

    b. To find $f^{-1}(x)$, let $y = f(x)$ and interchange x and y. Then, solve for y.

$$y = (x + 3)^2 - 5 \quad \textit{Let } y = f(x).$$
$$x = (y + 3)^2 - 5 \quad \textit{Interchange } x \textit{ and } y.$$
$$x + 5 = (y + 3)^2 \quad \textit{Isolate the expression containing } y.$$
$$\pm\sqrt{x + 5} = y + 3 \quad \textit{Take the square root of each side.}$$
$$y = -3 \pm \sqrt{x + 5} \quad \textit{Solve for } y.$$
$$f^{-1}(x) = -3 \pm \sqrt{x + 5} \quad \textit{Replace } y \textit{ with } f^{-1}(x).$$

    c. To graph $f(x)$ and its inverse, enter the equations
$y = (x + 3)^2 - 5$,
$y = -3 + \sqrt{x + 5}$, and
$y = -3 - \sqrt{x + 5}$ in the same viewing window.

$y = (x + 3)^2 - 5$

$y = -3 \pm \sqrt{x + 5}$

[−15.16, 15.16] scl: 2 by [−10, 10] scl: 2

Teaching Tip For Example 1, remind students that the inverse function is symmetric about the line $y = x$.

In-Class Examples
Example 1
Graph $f(x) = x^2$ and its inverse.

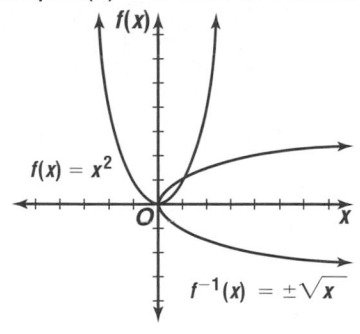

$f(x) = x^2$

$f^{-1}(x) = \pm\sqrt{x}$

Example 2
Consider $f(x) = x^2 - 4$.

a. Is the inverse of $f(x)$ a function? **no**

b. Find $f^{-1}(x)$. $f^{-1}(x) = \pm\sqrt{x + 4}$

c. Graph $f(x)$ and $f^{-1}(x)$ using a graphing calculator.

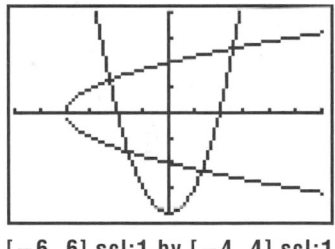

[−6, 6] scl:1 by [−4, 4] scl:1

Teaching Tip Before Example 2, point out to students that the graph of an inverse relation will be a reflection of the original graph over the line $y = x$.

 Prerequisite Skills

A review of the prerequisite skills for this lesson is provided in the *Prerequisite Skills Masters.*

Skill Number and Content	Pages
5 Solving Equations & Inequalities	11–12

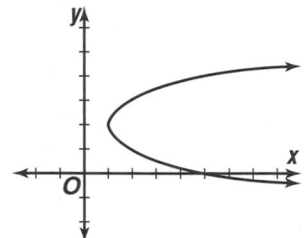
You can graph a function by using the parent graph of an inverse function.

Example ③ Graph $y = 2 + \sqrt[3]{x - 7}$.

The parent function is $y = \sqrt[3]{x}$ which is the inverse of $y = x^3$.

To graph $y = \sqrt[3]{x}$, start with the graph of $y = x^3$. Reflect the graph over the line $y = x$.

To graph $y = 2 + \sqrt[3]{x - 7}$, translate the reflected graph 7 units to the right and 2 units up.

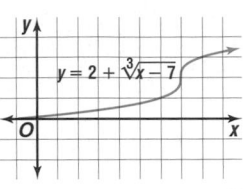

To find the inverse of a function, you use an **inverse process** to solve for y after switching variables. This inverse process can be used to solve many real-world problems.

Example ④ **FINANCE** When the Garcias decided to begin investing, their financial advisor instructed them to set a goal. Their net pay is about 65% of their gross pay. They decided to subtract their monthly food allowance from their monthly net pay and then invest 10% of the remainder.

a. Write an equation that gives the amount invested *I* as a function of their monthly gross pay *G* given that they allow $450 per month for food.

b. Determine the equation for the inverse process and describe the real-world situation it models.

c. Determine the gross pay needed in order to invest $100 per month.

a. One model for the amount they will invest is as follows.

$$I = 0.10 \cdot (0.65G - 450)$$

b. Solve for *G*.

$$I = 0.10(0.65G - 450)$$
$$10I = 0.65G - 450 \qquad \textit{Multiply each side by 10.}$$
$$10I + 450 = 0.65G \qquad \textit{Add 450 to each side.}$$
$$\frac{10I + 450}{0.65} = G \qquad \textit{Divide each side by 0.65.}$$

This equation models the gross pay needed to meet a monthly investment goal *I* with the given conditions.

c. Substituting 100 for I gives $G = \dfrac{10I + 450}{0.65}$, or about \$2231. So the Garcias need to earn a monthly gross pay of about \$2231 in order to invest \$100 per month.

Look Back
Refer to Lesson 1-2 to review the composition of two functions.

If the inverse of a function is also a function, then a composition of the function and its inverse produces a unique result.

Consider $f(x) = 3x - 2$ and $f^{-1}(x) = \dfrac{x + 2}{3}$. You can find functions $[f \circ f^{-1}](x)$ and $[f^{-1} \circ f](x)$ as follows.

$$\begin{aligned}[f \circ f^{-1}](x) &= f\left(\dfrac{x + 2}{3}\right) & f^{-1}(x) &= \dfrac{x + 2}{3} \\ &= 3\left(\dfrac{x + 2}{3}\right) - 2 & f(x) &= 3x - 2 \\ &= x \end{aligned}$$

$$\begin{aligned}[f^{-1} \circ f](x) &= f^{-1}(3x - 2) & f(x) &= 3x - 2 \\ &= \dfrac{(3x - 2) + 2}{3} & f^{-1}(x) &= \dfrac{x + 2}{3} \\ &= x \end{aligned}$$

This leads to the formal definition of inverse functions.

Inverse Functions	Two functions, f and f^{-1}, are inverse functions if and only if $[f \circ f^{-1}](x) = [f^{-1} \circ f](x) = x$.

Example **5** Given $f(x) = 4x - 9$, find $f^{-1}(x)$, and verify that f and f^{-1} are inverse functions.

$$\begin{aligned}y &= 4x - 9 & &f(x) = y \\ x &= 4y - 9 & &\textit{Interchange } x \textit{ and } y. \\ x + 9 &= 4y & &\textit{Solve for } y. \\ \dfrac{x + 9}{4} &= y \\ \dfrac{x + 9}{4} &= f^{-1}(x) & &\textit{Replace } y \textit{ with } f^{-1}(x).\end{aligned}$$

Now show that $[f \circ f^{-1}](x) = [f^{-1} \circ f](x) = x$.

$$\begin{aligned}[f \circ f^{-1}](x) &= f\left(\dfrac{x + 9}{4}\right) \\ &= 4\left(\dfrac{x + 9}{4}\right) - 9 \\ &= x \end{aligned}$$

$$\begin{aligned}[f^{-1} \circ f](x) &= f^{-1}(4x - 9) \\ &= \dfrac{(4x - 9) + 9}{4} \\ &= x \end{aligned}$$

Since $[f \circ f^{-1}](x) = [f^{-1} \circ f](x) = x$, f and f^{-1} are inverse functions.

CHECK FOR UNDERSTANDING

Communicating Mathematics

Read and study the lesson to answer each question.   **1. See margin.**

1. **Write** an explanation of how to determine the equation for the inverse of the relation $y = \pm\sqrt{x - 3}$.

2. **Determine** the values of n for which $f(x) = x^n$ has an inverse that is a function. Assume that n is a whole number.   **n is odd.**

Lesson 3-4   Inverse Functions and Relations   **155**

From the Classroom of...

Gail M. Leach
St. Cloud High School
St. Cloud, Florida

"To help visualize inverse relations, I have students use 'patty paper' to make their own transfer paper. They can trace the original graph, flip and do a quarter turn, and trace over the image. This transfers the inverse onto the xy-plane and also illustrates symmetry with respect to the line $y = x$."

---

3 PRACTICE/APPLY

Additional Answers

4. Sample answer: if you draw a horizontal line through the graph of the function and it intersects the graph more than once, then the inverse is not a function.

9. $f^{-1}(x) = -\frac{1}{3}x + \frac{2}{3}$; a function

10. $f^{-1}(x) = \frac{1}{\sqrt[3]{x}}$; a function

11. $f^{-1}(x) = -2 \pm \sqrt{x-6}$; not a function

12.

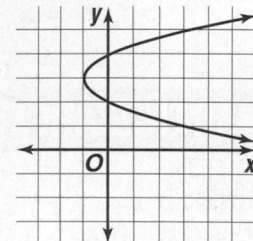

13. $[f \circ f^{-1}](x) = f(2x + 10)$

$= \frac{1}{2}(2x + 10) - 5$

$= x$

$[f^{-1} \circ f](x) = f^{-1}\left(\frac{1}{2}x - 5\right)$

$= 2\left(\frac{1}{2}x - 5\right) + 10$

$= x$

Since $[f \circ f^{-1}](x) = [f^{-1} \circ f](x) = x$, f and f^{-1} are inverse functions.

24.

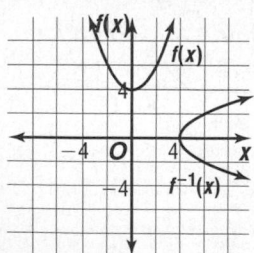

25. $f^{-1}(x) = \frac{x-7}{2}$; a function

26. $f^{-1}(x) = -x - 2$; a function

27. $f^{-1}(x) = \frac{1}{x}$; a function

28. $f^{-1}(x) = \pm\sqrt{-\frac{1}{x}}$; not a function

29. $f^{-1}(x) = 3 \pm \sqrt{x-7}$; not a function

30. $f^{-1}(x) = 2 \pm \sqrt{x+1}$; not a function

31. $f^{-1}(x) = \frac{1}{x} - 2$; a function

32. $f^{-1}(x) = 1 \pm \frac{1}{\sqrt{x}}$; not a function

---

3. Find a counterexample to this statement: The inverse of a function is also a function. **Sample answer:** $f(x) = x^2$

4. Show how you know whether the inverse of a function is also a function without graphing the inverse. **See margin.**

5. You Decide Nitayah says that the inverse of $y = 3 \pm \sqrt{x+2}$ cannot be a function because $y = 3 \pm \sqrt{x+2}$ is not a function. Is she right? Explain. **She is wrong. The inverse is $f^{-1}(x) = (x-3)^2 - 2$, which is a function.**

Guided Practice Graph each function and its inverse. **6–8. See Answer Appendix.**

6. $f(x) = |x| + 1$ **7.** $f(x) = x^3 + 1$ **8.** $f(x) = -(x-3)^2 + 1$

Find $f^{-1}(x)$. Then state whether $f^{-1}(x)$ is a function. **9–11. See margin.**

9. $f(x) = -3x + 2$ **10.** $f(x) = \frac{1}{x^3}$ **11.** $f(x) = (x+2)^2 + 6$

12. See margin.   **12.** Graph the equation $y = 3 \pm \sqrt{x+1}$ using the parent graph $p(x) = x^2$.

13. $f^{-1}(x) = 2x + 10$; see margin for verification.   **13.** Given $f(x) = \frac{1}{2}x - 5$, find $f^{-1}(x)$. Then verify that f and f^{-1} are inverse functions.

14. Finance If you deposit \$1000 in a savings account with an interest rate of r compounded annually, then the balance in the account after 3 years is given by the function $B(r) = 1000(1 + r)^3$, where r is written as a decimal.

14a. $r = -1 + \frac{\sqrt[3]{B}}{10}$

a. Find a formula for the interest rate, r, required to achieve a balance of B in the account after 3 years.

b. What interest rate will yield a balance of \$1100 after 3 years? **3.23%**

EXERCISES

Practice

Graph each function and its inverse. **15–23. See Answer Appendix.**

 A

15. $f(x) = |x| + 2$ **16.** $f(x) = |2x|$ **17.** $f(x) = x^3 - 2$

18. $f(x) = x^5 - 10$ **19.** $f(x) = [\![x]\!]$ **20.** $f(x) = 3$

21. $f(x) = x^2 + 2x + 4$ **22.** $f(x) = -(x+2)^2 - 5$ **23.** $f(x) = (x+1)^2 - 4$

 B

24. For $f(x) = x^2 + 4$, find $f^{-1}(x)$. Then graph $f(x)$ and $f^{-1}(x)$. **$f^{-1}(x) = \pm\sqrt{x-4}$; see margin for graph.**

Find $f^{-1}(x)$. Then state whether $f^{-1}(x)$ is a function. **25–33. See margin.**

25. $f(x) = 2x + 7$ **26.** $f(x) = -x - 2$ **27.** $f(x) = \frac{1}{x}$

28. $f(x) = -\frac{1}{x^2}$ **29.** $f(x) = (x-3)^2 + 7$ **30.** $f(x) = x^2 - 4x + 3$

31. $f(x) = \frac{1}{x+2}$ **32.** $f(x) = \frac{1}{(x-1)^2}$ **33.** $f(x) = -\frac{2}{(x-2)^3}$

34. If $g(x) = \frac{3}{x^2 + 2x}$, find $g^{-1}(x)$. **$g^{-1}(x) = -1 \pm \sqrt{\frac{3}{x} + 1}$**

35–38. See Answer Appendix. Graph each equation using the graph of the given parent function.

 C

35. $f(x) = \sqrt{x+5}$, $p(x) = x^2$ **36.** $y = 1 \pm \sqrt{x-2}$, $p(x) = x^2$

37. $f(x) = -2 - \sqrt[3]{x+3}$, $p(x) = x^3$ **38.** $y = 2\sqrt[5]{x} - 4$, $p(x) = x^5$

39–40. See margin for verification.

Given $f(x)$, find $f^{-1}(x)$. Then verify that f and f^{-1} are inverse functions.

39. $f(x) = -\frac{2}{3}x + \frac{1}{6}$ $f^{-1}(x) = -\frac{3}{2}x + \frac{1}{4}$ **40.** $f(x) = (x-3)^3 + 4$ $f^{-1}(x) = 3 + \sqrt[3]{x-4}$

www.amc.glencoe.com/self_check_quiz

Additional Answers

33. $f^{-1}(x) = 2 - \sqrt[3]{\frac{2}{x}}$; a function

39. $[f \circ f^{-1}](x) = f\left(-\frac{3}{2}x + \frac{1}{4}\right)$

$= -\frac{2}{3}\left(-\frac{3}{2}x + \frac{1}{4}\right) + \frac{1}{6}$

$= x - \frac{1}{6} + \frac{1}{6}$

$= x$

$[f^{-1} \circ f](x) = f^{-1}\left(-\frac{2}{3}x + \frac{1}{6}\right)$

$= -\frac{3}{2}\left(-\frac{2}{3}x + \frac{1}{6}\right) + \frac{1}{4}$

$= x - \frac{1}{4} + \frac{1}{4}$

$= x$

Since $[f \circ f^{-1}](x) = [f^{-1} \circ f](x)$, f and f^{-1} are inverse functions.

41. Analytic Geometry The function $d(x) = |x - 4|$ gives the distance between x and 4 on the number line. **a–c. See margin.**

a. Graph $d^{-1}(x)$.

b. Is $d^{-1}(x)$ a function? Why or why not?

c. Describe what $d^{-1}(x)$ represents. Then explain how you could have predicted whether $d^{-1}(x)$ is a function without looking at a graph.

42. Fire Fighting The velocity v and maximum height h of water being pumped into the air are related by the equation $v = \sqrt{2gh}$ where g is the acceleration due to gravity (32 feet/second²).

a. Determine an equation that will give the maximum height of the water as a function of its velocity. $h = \dfrac{v^2}{64}$

b. The Mayfield Fire Department must purchase a pump that is powerful enough to propel water 80 feet into the air. Will a pump that is advertised to project water with a velocity of 75 feet/second meet the fire department's needs? Explain.   **Yes; the pump can propel water to a height of about 88 ft.**

43. Critical Thinking

a. Give an example of a function that is its own inverse.   **Sample answer: $y = -x$.**

b. What type of symmetry must the graph of the function exhibit?

43b. The graph of the function must be symmetric about the line $y = x$.

c. Would all functions with this type of symmetry be their own inverses? Justify your response.   **Yes, because the line $y = x$ is the axis of symmetry *and* the reflection line.**

44. Consumer Costs A certain long distance phone company charges callers 10 cents for every minute or part of a minute that they talk. Suppose that you talk for x minutes, where x is any real number greater than 0.   **a–e. See margin.**

a. Sketch the graph of the function $C(x)$ that gives the cost of an x-minute call.

b. What are the domain and range of $C(x)$?

c. Sketch the graph of $C^{-1}(x)$.

d. What are the domain and range of $C^{-1}(x)$?

e. What real-world situation is modeled by $C^{-1}(x)$?

45. Critical Thinking Consider the parent function $y = x^2$ and its inverse $y = \pm\sqrt{x}$. If the graph of $y = x^2$ is translated 6 units right and 5 units down, what must be done to the graph of $y = \pm\sqrt{x}$ to get a graph of the inverse of the translated function? Write an equation for each of the translated graphs. **See margin.**

46. Physics The formula for the kinetic energy of a particle as a function of its mass m and velocity v is $KE = \frac{1}{2}mv^2$.

46a. $v = \pm\sqrt{\dfrac{2KE}{m}}$

a. Find the equation for the velocity of a particle based on its mass and kinetic energy.

b. Use your equation from part a to find the velocity in meters per second of a particle of mass 1 kilogram and kinetic energy 15 joules.   **±5.5 m/s**

c. Explain why the velocity of the particle is not a function of its mass and kinetic energy.   **There are always two velocities.**

Lesson 3-4 Inverse Functions and Relations **157**

Open-Ended Assessment

Modeling Have students work in pairs. Have each student write an equation and graph it. Then, ask the students to trade equations and write the corresponding inverse functions. Students should then graph the inverse functions.

 Chapter 3, Quiz B (Lessons 3-3 and 3-4) is available in the *Assessment and Evaluation Masters,* p. 69.

 Chapter 3 Mid-Chapter Test (Lessons 3-1 through 3-4) is available in the *Assessment and Evaluation Masters,* p. 68.

Additional Answer
53.

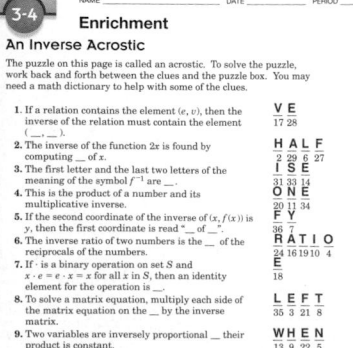
---

47. Cryptography One way to encode a message is to assign a numerical value to each letter of the alphabet and encode the message by assigning each number to a new value using a mathematical relation.

a. Does the encoding relation have to be a function? Explain.

b. Why should the graph of the encoding function pass the horizontal line test?

c. Suppose a value was assigned to each letter of the alphabet so that $1 = A, 2 = B, 3 = C, …, 26 = Z$, and a message was encoded using the relation $c(x) = -2 + \sqrt{x + 3}$. What function would decode the message?

d. Try this decoding function on the following message:

| 1 | 2.899 | 2.123 | 0.449 | 2.796 | 1.464 | 2.243 | 2.123 | 2.690 |
| 0 | 2.583 | 0.828 | 1 | | 2.899 | 2.123 | | |

FUNCTIONS ARE FUN

47a. Yes. If the encoded message is not unique, it may not decode properly.

47b. So the encoded message may be decoded.

47c. $c^{-1}(x) = (x + 2)^2 - 3$

Mixed Review

48. Solve the inequality $|2x + 4| \le 6$. *(Lesson 3-3)* $\{x \mid -5 \le x \le 1\}$

49. State whether the figure at the right has point symmetry, line symmetry, neither, or both. *(Lesson 3-1)* **both**

50. **Retail** Arturo Alvaré, a sales associate at a paint store, plans to mix as many gallons as possible of colors A and B. He has 32 units of blue dye and 54 units of red dye. Each gallon of color A requires 4 units of blue dye and 1 unit of red dye. Each gallon of color B requires 1 unit of blue dye and 6 units of red dye. Use linear programming to answer the following questions. *(Lesson 2-7)*

a. Let a be the number of gallons of color A and let b be the number of gallons of color B. Write the inequalities that describe this situation.

b. Find the maximum number of gallons possible. **14 gallons**

50a. $a \ge 0, b \ge 0, 4a + b \le 32, a + 6b \le 54$

51. Solve the system of equations $4x + 2y = 10, y = 6 - x$ by using a matrix equation. *(Lesson 2-5)* **(−1, 7)**

52. Find the product $\frac{1}{2}\begin{bmatrix} 9 & -3 \\ -6 & 6 \end{bmatrix}$. *(Lesson 2-3)* $\begin{bmatrix} \frac{9}{2} & -\frac{3}{2} \\ -3 & 3 \end{bmatrix}$

53. Graph $y < -2x + 8$. *(Lesson 1-8)* **See margin.**

54. Line ℓ_1 has a slope of $\frac{1}{4}$, and line ℓ_2 has a slope of 4. Are the lines *parallel, perpendicular,* or *neither*? *(Lesson 1-5)* **neither**

55. Write the slope-intercept form of the equation of the line that passes through points at (0, 7) and (5, 2). *(Lesson 1-4)* $y = -x + 7$

56. **SAT/ACT Practice** In the figure at the right, if $\overline{PQ}$ is perpendicular to $\overline{QR}$, then $a + b + c + d = ?$ **C**

A 180   B 225   C 270

D 300   E 360

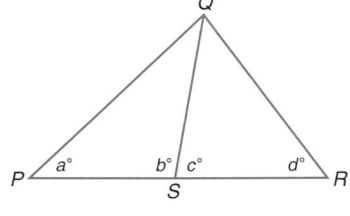

Extra Practice See p. A30.

Extra Credit

State whether the function $f(x) = [\![x]\!]$ has an inverse which is a function. Explain. **No; the graph does not pass the horizontal line test. Therefore, for the inverse, an element in the domain may be paired with more than one element in the range.**

3-5

Continuity and End Behavior

OBJECTIVES
- Determine whether a function is continuous or discontinuous.
- Identify the end behavior of functions.
- Determine whether a function is increasing or decreasing on an interval.

POSTAGE On January 10, 1999, the United States Postal Service raised the cost of a first-class stamp. After the change, mailing a letter cost $0.33 for the first ounce and $0.22 for each additional ounce or part of an ounce. The graph summarizes the cost of mailing a first-class letter. *A problem related to this is solved in Example 2.*

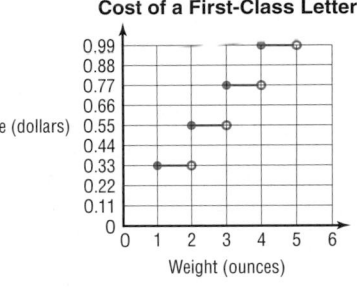

Cost of a First-Class Letter

Price (dollars)

Weight (ounces)

Most graphs we have studied have been smooth, continuous curves. However, a function like the one graphed above is a **discontinuous** function. That is, you cannot trace the graph of the function without lifting your pencil.

There are many types of discontinuity. Each of the functions graphed below illustrates a different type of discontinuity. That is, each function is discontinuous at some point in its domain.

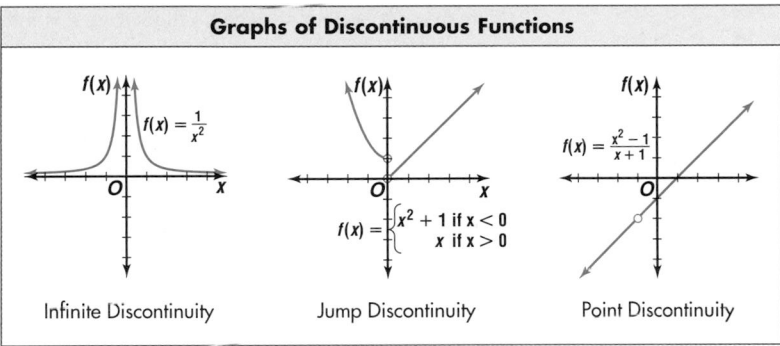

Graphs of Discontinuous Functions

$f(x) = \frac{1}{x^2}$ $f(x) = \begin{cases} x^2 + 1 \text{ if } x < 0 \\ x \text{ if } x > 0 \end{cases}$ $f(x) = \frac{x^2 - 1}{x + 1}$

Infinite Discontinuity   Jump Discontinuity   Point Discontinuity

The postage graph exhibits jump discontinuity.

- **Infinite discontinuity** means that $|f(x)|$ becomes greater and greater as the graph approaches a given x-value.
- **Jump discontinuity** indicates that the graph stops at a given value of the domain and then begins again at a different range value for the same value of the domain.
- When there is a value in the domain for which the function is undefined, but the pieces of the graph match up, we say the function has **point discontinuity.**

There are functions that are impossible to graph in the real number system. Some of these functions are said to be **everywhere discontinuous.** An example of such a function is $f(x) = \begin{cases} 1 \text{ if } x \text{ is rational} \\ -1 \text{ if } x \text{ is irrational} \end{cases}$.

Lesson 3-5 Continuity and End Behavior **159**

Resource Manager

 Reproducible Masters
- *Study Guide and Practice,* pp. 47–48
- *Enrichment,* p. 29

 Transparencies
- 5-Minute Check, 3-5
- Teaching, 3-5

---

1 FOCUS

5-Minute Check
Lesson 3-4

Graph each function and its inverse.

1. $y = -3x + 4$.

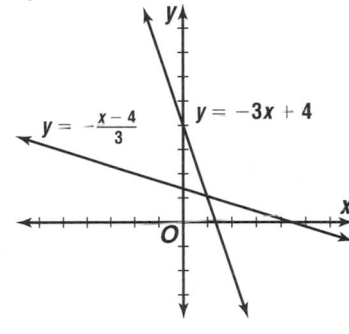

$y = -\frac{x-4}{3}$ $y = -3x + 4$

2. $y = x^2 + 1$.

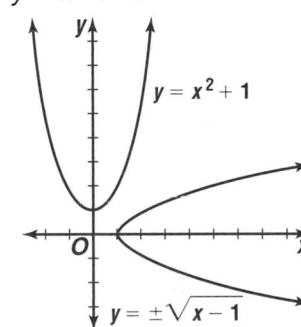

$y = x^2 + 1$

$y = \pm\sqrt{x - 1}$

3. Find $f^{-1}(x)$ for the relation $f(x) = 2x^2 + x - 3$. Then state whether $f^{-1}(x)$ is a function.

$f^{-1}(x) = \frac{-1 \pm \sqrt{8x + 25}}{4}$;

$f^{-1}(x)$ **is not a function.**

4. Graph the equation using the graph of the given parent function.
$f(x) = 2 + \sqrt{x - 1},\ p(x) = x^2$.

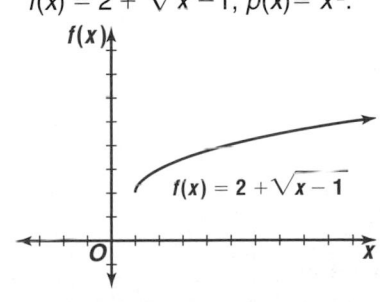

$f(x) = 2 + \sqrt{x - 1}$

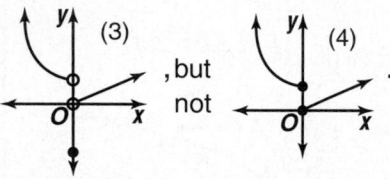
If a function is not discontinuous, it is said to be **continuous**. That is, a function is continuous at a number c if there is a point on the graph with x-coordinate c and the graph passes through that point without a break.

Linear and quadratic functions are continuous at all points. If we only consider x-values less than c as x approaches c, then we say x is approaching c from the left. Similarly, if we only consider x-values greater than c as x approaches c, then we say x is approaching c from the right.

Continuity Test

A function is continuous at $x = c$ if it satisfies the following conditions:
(1) the function is defined at c; in other words, $f(c)$ exists;
(2) the function approaches the same y-value on the left and right sides of $x = c$; and
(3) the y-value that the function approaches from each side is $f(c)$.

Example **1** **Determine whether each function is continuous at the given x-value.**

a. $f(x) = 3x^2 + 7; x = 1$

Check the three conditions in the continuity test.

(1) The function is defined at $x = 1$. In particular, $f(1) = 10$.

(2) The first table below suggests that when x is less than 1 and x approaches 1, the y-values approach 10. The second table suggests that when x is greater than 1 and x approaches 1, the y-values approach 10.

x	$y = f(x)$
0.9	9.43
0.99	9.9403
0.999	9.994003

x	$y = f(x)$
1.1	10.63
1.01	10.0603
1.001	10.006003

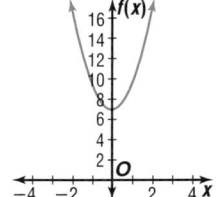

(3) Since the y-values approach 10 as x approaches 1 from both sides and $f(1) = 10$, the function is continuous at $x = 1$. This can be confirmed by examining the graph.

b. $f(x) = \dfrac{x - 2}{x^2 - 4}; x = 2$

Start with the first condition in the continuity test. The function is not defined at $x = 2$ because substituting 2 for x results in a denominator of zero. So the function is discontinuous at $x = 2$. *This function has point discontinuity at $x = 2$.*

160 Chapter 3 *The Nature of Graphs*

c. $f(x) = \begin{cases} \dfrac{1}{x} & \text{if } x > 1 \\ x & \text{if } x \le 1 \end{cases}; x = 1$

The function is defined at $x = 1$.
Using the second formula we find $f(1) = 1$.

The first table suggests that $f(x)$ approaches 1 as x approaches 1 from the left. We can see from the second table that $f(x)$ seems to approach 1 as x approaches 1 from the right.

x	f(x)	x	f(x)
0.9	0.9	1.1	0.9091
0.99	0.99	1.01	0.9901
0.999	0.999	1.001	0.9990

Since the $f(x)$-values approach 1 as x approaches 1 from both sides and $f(1) = 1$, the function is continuous at $x = 1$.

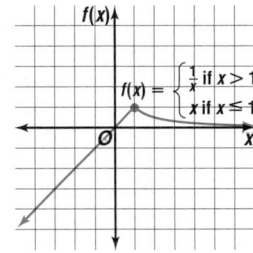

A function may have a discontinuity at one or more x-values but be continuous on an interval of other x-values. For example, the function $f(x) = \dfrac{1}{x^2}$ is continuous for $x > 0$ and $x < 0$, but discontinuous at $x = 0$.

Continuity on an Interval	A function $f(x)$ is continuous on an interval if and only if it is continuous at each number x in the interval.

In Chapter 1, you learned that a piecewise function is made from several functions over various intervals. The piecewise function

$f(x) = \begin{cases} 3x - 2 & \text{if } x > 2 \\ 2 - x & \text{if } x \le 2 \end{cases}$ is continuous for $x > 2$

and $x < 2$ but is discontinuous at $x = 2$. The graph has a jump discontinuity. This function fails the second part of the continuity test because the values of $f(x)$ approach 0 as x approaches 2 from the left, but the $f(x)$-values approach 4 as x approaches 2 from the right.

Example ❷ **POSTAGE** Refer to the application at the beginning of the lesson.

a. **Use the continuity test to show that the step function is discontinuous.**

b. **Explain why a continuous function would not be appropriate to model postage costs.**

a. The graph of the postage function is discontinuous at each integral value of *w* in its domain because the function does not approach the same value from the left and the right. For example, as *w* approaches 1 from the left, *C*(*w*) approaches 0.33 but as *w* approaches 1 from the right, *C*(*w*) approaches 0.55.

b. A continuous function would have to achieve all real *y*-values (greater than or equal to 0.33.) This would be an inappropriate model for this situation since the weight of a letter is rounded to the nearest ounce and postage costs are rounded to the nearest cent.

$x \to \infty$ is read as "x approaches infinity."

Another tool for analyzing functions is **end behavior.** The end behavior of a function describes what the *y*-values do as $|x|$ becomes greater and greater. When *x* becomes greater and greater, we say that *x* approaches infinity, and we write $x \to \infty$. Similarly, when *x* becomes more and more negative, we say that *x* approaches negative infinity, and we write $x \to -\infty$. The same notation can also be used with *y* or *f*(*x*) and with real numbers instead of infinity.

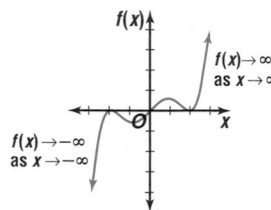

Example ❸ Describe the end behavior of $f(x) = -2x^3$ and $g(x) = -x^3 + x^2 - x + 5$.

Use your calculator to create a table of function values so you can investigate the behavior of the *y*-values.

$f(x) = -2x^3$		$g(x) = x^3 + x^2 - x + 5$	
x	**f(x)**	**x**	**g(x)**
−10,000	2×10^{12}	−10,000	1.0001×10^{12}
−1000	2×10^9	−1000	1,001,001,005
−100	2,000,000	−100	1,010,105
−10	2000	−10	1115
0	0	0	5
10	−2000	10	−905
100	−2,000,000	100	−990,095
1000	-2×10^9	1000	−999,000,995
10,000	-2×10^{12}	10,000	-9.999×10^{11}

Notice that both polynomial functions have *y*-values that become very large in absolute value as *x* gets very large in absolute value. The end behavior of *f*(*x*) can be summarized by stating that as $x \to \infty$, $f(x) \to -\infty$ and as $x \to -\infty$, $f(x) \to \infty$. The end behavior of *g*(*x*) is the same. *You may wish to graph these functions on a graphing calculator to verify this summary.*

In general, the end behavior of any polynomial function can be modeled by the function comprised solely of the term with the highest power of x and its coefficient. Suppose for $n \geq 0$

$$p(x) = a_n x^n + a_{n-1} x^{n-1} + a_{n-2} x^{n-2} + \cdots + a_2 x^2 + a_1 x + a_0.$$

Then $f(x) = a_n x^n$ has the same end behavior as $p(x)$. The following table organizes the information for such functions and provides an example of a function displaying each type of end behavior.

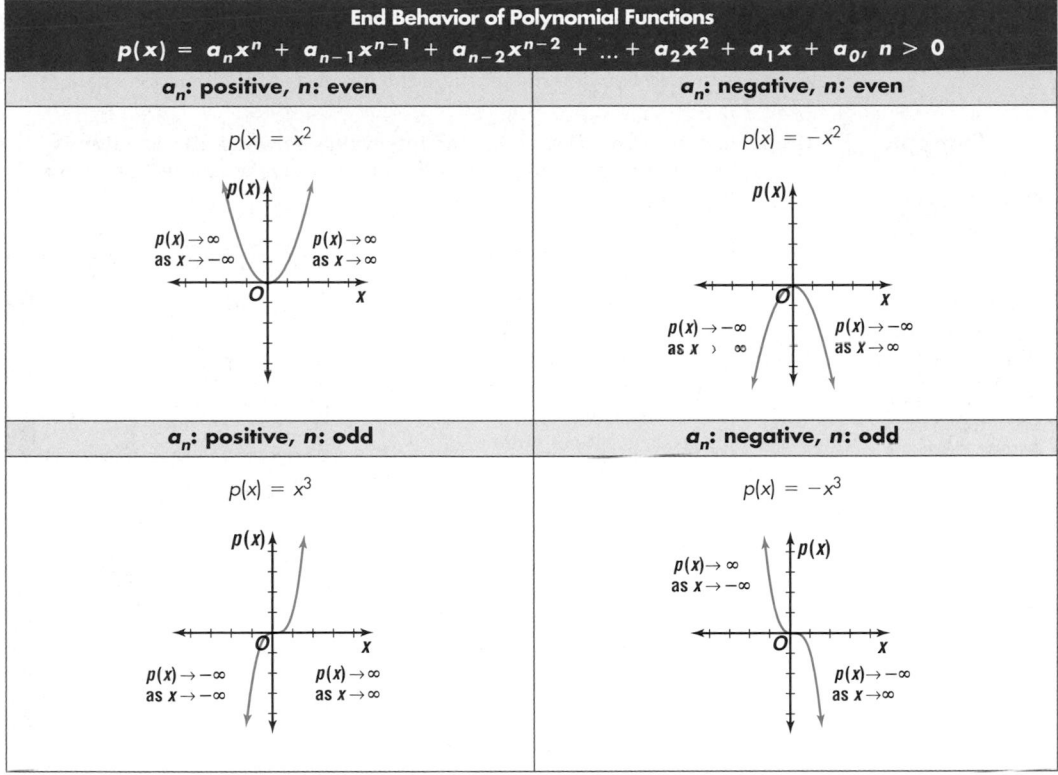

End Behavior of Polynomial Functions

$p(x) = a_n x^n + a_{n-1} x^{n-1} + a_{n-2} x^{n-2} + \ldots + a_2 x^2 + a_1 x + a_0,\ n > 0$

a_n: positive, n: even	a_n: negative, n: even
$p(x) = x^2$ $p(x) \to \infty$ as $x \to -\infty$ $p(x) \to \infty$ as $x \to \infty$	$p(x) = -x^2$ $p(x) \to -\infty$ as $x \to -\infty$ $p(x) \to -\infty$ as $x \to \infty$
a_n: positive, n: odd	**a_n: negative, n: odd**
$p(x) = x^3$ $p(x) \to -\infty$ as $x \to -\infty$ $p(x) \to \infty$ as $x \to \infty$	$p(x) = -x^3$ $p(x) \to \infty$ as $x \to -\infty$ $p(x) \to -\infty$ as $x \to \infty$

Another characteristic of functions that can help in their analysis is the **monotonicity** of the function. A function is said to be monotonic on an interval I if and only if the function is increasing on I or decreasing on I.

Whether a graph is increasing or decreasing is always judged by viewing a graph from left to right.

The graph of $f(x) = x^2$ shows that the function is decreasing for $x < 0$ and increasing for $x > 0$.

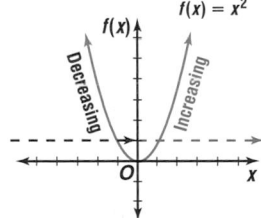

$f(x) = x^2$

Lesson 3-5 Continuity and End Behavior **163**

Teaching Tip After discussing Example 3, stress that the end behavior of the polynomials depends on the end behavior of the leading term when terms are listed in order from the highest power of x.

Teaching Tip A graphing calculator can be used to determine the end behavior of a function. Students should reset the range values to successively larger values and view the function. The **ZOOM** feature of some calculators may also be helpful.

In-Class Example

Example 4
Graph each function. Determine the interval(s) on which the function is increasing and the interval(s) on which the function is decreasing.

a. $f(x) = x^2 - 7$

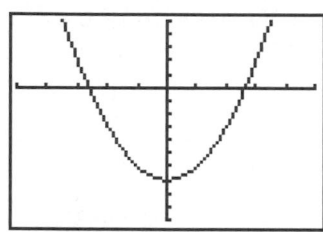

$[-5, 5]$ scl:1 by $[-10, 5]$ scl:1

For $x < 0$, the function is decreasing. For $x > 0$, the function is increasing.

b. $f(x) = -\dfrac{1}{x}$

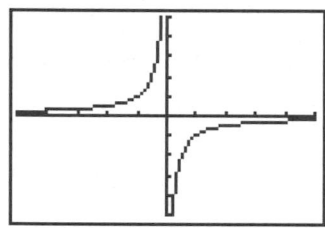

$[-5, 5]$ scl:1 by $[-5, 5]$ scl:1

For $x < 0$, the function is increasing. For $x > 0$, the function is increasing.

c. $h(x) = 5x^3 + x^2 - x + 4$

$[-5, 5]$ scl:1 by $[-2, 10]$ scl:1

For $x < -1$, the graph is increasing. For $-1 < x < \dfrac{1}{5}$, the function is decreasing. For $x > \dfrac{1}{5}$, the function is increasing.

Increasing, Decreasing, and Constant Functions	A function f is increasing on an interval I if and only if for every a and b contained in I, $f(a) < f(b)$ whenever $a < b$. A function f is decreasing on an interval I if and only if for every a and b contained in I, $f(a) > f(b)$ whenever $a < b$. A function f remains constant on an interval I if and only if for every a and b contained in I, $f(a) = f(b)$ whenever $a < b$.

Points in the domain of a function where the function changes from increasing to decreasing or vice versa are special points called *critical points*. You will learn more about these special points in Lesson 3-6. Using a graphing calculator can help you determine where the direction of the function changes.

Example **4** Graph each function. Determine the interval(s) on which the function is increasing and the interval(s) on which the function is decreasing.

Graphing Calculator Tip

By watching the *x*- and *y*-values while using the **TRACE** function, you can determine approximately where a function changes from increasing to decreasing and vice versa.

a. $f(x) = 3 - (x - 5)^2$

The graph of this function is obtained by transforming the parent graph $p(x) = x^2$. The parent graph has been reflected over the *x*-axis, translated 5 units to the right, and translated up 3 units. The function is increasing for $x < 5$ and decreasing for $x > 5$. At $x = 5$, there is a critical point.

$[-10, 10]$ scl:1 by $[-10, 10]$ scl:1

b. $f(x) = \dfrac{1}{2}|x + 3| - 5$

The graph of this function is obtained by transforming the parent graph $p(x) = |x|$. The parent graph has been vertically compressed by a factor of $\dfrac{1}{2}$, translated 3 units to the left, and translated down 5 units. This function is decreasing for $x < -3$ and increasing for $x > -3$. There is a critical point when $x = -3$.

$[-10, 10]$ scl:1 by $[-10, 10]$ scl:1

c. $f(x) = 2x^3 + 3x^2 - 12x + 3$

This function has more than one critical point. It changes direction at $x = -2$ and $x = 1$. The function is increasing for $x < -2$. The function is also increasing for $x > 1$. When $-2 < x < 1$, the function is decreasing.

$[-5, 5]$ scl:1 by $[-30, 30]$ scl:5

164 Chapter 3 The Nature of Graphs

Addressing Individual Needs

 Naturalist Have students use the edge of a leaf to create a graph. Have them determine where the graph is increasing or decreasing. Then tell them that when the value of *y* increases, the distance from the central vein to the edge increases. As the value of *y* decreases, the distance also decreases.

Communicating Mathematics

Read and study the lesson to answer each question.   1–4. See margin.

1. **Explain** why the function whose graph is shown at the right is discontinuous at $x = 2$.

2. **Summarize** the end behavior of polynomial functions.

3. **State** whether the graph at the right has infinite discontinuity, jump discontinuity, or point discontinuity, or is continuous. Then describe the end behavior of the function.

4. *Math Journal*   **Write** a paragraph that compares the monotonicity of $f(x) = x^2$ with that of $g(x) = -x^2$. In your paragraph, make a conjecture about the monotonicity of the reflection over the x-axis of any function as compared to that of the original function.

Guided Practice

Determine whether each function is continuous at the given x-value. Justify your answer using the continuity test.   5–6. See margin.

5. $y = \dfrac{x - 5}{x + 3}; x = -3$

6. $f(x) = \begin{cases} x^2 + 2 \text{ if } x < -2 \\ 3x \text{ if } x \geq -2 \end{cases}; x = -2$

Describe the end behavior of each function.

8. $y \to -\infty$ as $x \to \infty$,
$y \to -\infty$ as $x \to -\infty$

7. $y = 4x^5 + 2x^4 - 3x - 1$
$y \to \infty$ as $x \to \infty$, $y \to -\infty$ as $x \to -\infty$

8. $y = -x^6 + x^4 - 5x^2 + 4$

Graph each function. Determine the interval(s) for which the function is increasing and the interval(s) for which the function is decreasing.

9. $f(x) = (x + 3)^2 - 4$

10. $y = \dfrac{x}{x^2 + 1}$ 9–10. See Answer Appendix.

11. **Electricity**   A simple electric circuit contains only a power supply and a resistor. When the power supply is off, there is no current in the circuit. When the power supply is turned on, the current almost instantly becomes a constant value. This situation can be modeled by a graph like the one shown at the right. I represents current in amps, and t represents time in seconds.

Current (amps) / Time(s)

a. At what t-value is this function discontinuous? $t = 4$

b. When was the power supply turned on?   when $t = 4$

c. If the person who turned on the power supply left and came back hours later, what would he or she measure the current in the circuit to be?   **10 amps**

 www.amc.glencoe.com/self_check_quiz   *Lesson 3-5   Continuity and End Behavior*   **165**

Error Analysis

Watch for students who test only one condition in the continuity test.

Prevent by showing students examples of functions that fail only one condition but pass the others, such as

$$f(x) = \begin{cases} 2x - 3, \text{ if } x \neq 1 \\ 0, \text{ if } x = 1 \end{cases}.$$

Additional Answers

1. Sample answer: the function approaches 1 as x approaches 2 from the left, but the function approaches -4 as x approaches 2 from the right. This means that the function fails the second condition in the continuity test.

3. infinite discontinuity; $f(x) \to \infty$ as $x \to -\infty$, $f(x) \to -\infty$ as $x \to \infty$.

4. $f(x) = x^2$ is decreasing for $x < 0$ and increasing for $x > 0$. $g(x) = -x^2$ is increasing for $x < 0$ and decreasing for $x > 0$. Reflecting a graph switches the monotonicity. In other words, if $f(x)$ is increasing, the reflection will be decreasing and vice versa.

5. No; y is undefined when $x = -3$.

6. No; $f(x)$ approaches 6 as x approaches -2 from the left but $f(x)$ approaches -6 as x approaches -2 from the right.

Study Guide and Practice Masters, pp. 47–48

Additional Answer

2.

a_n	n	$x\to$	$p(x)\to$	a_n	n	$x\to$	$p(x)\to$
positive	even	∞	∞	negative	even	∞	$-\infty$
positive	even	$-\infty$	∞	negative	even	$-\infty$	$-\infty$
positive	odd	∞	∞	negative	odd	∞	$-\infty$
positive	odd	$-\infty$	$-\infty$	negative	odd	$-\infty$	∞

Teaching Tip For Exercises 26–31, review basic shapes of cubic, quadratic, absolute value, and $\frac{1}{x}$ functions so that students will set viewing windows on their graphing calculators appropriately.

Additional Answers

12. The function is defined when $x = 1$; the function approaches –3 as x approaches 1 from both sides; and $y = -3$ when $x = 1$.

13. The function is undefined when $x = 2$.

14. The function is defined when $x = -3$; the function approaches 0 as x approaches –3 from both sides; and $f(-3) = 0$.

15. The function is defined when $x = 3$; the function approaches 1 (in fact is equal to 1) as x approaches 3 from both sides; and $y = 1$ when $x = 3$.

16. $f(x)$ approaches –7 as x approaches –4 from the left, but $f(x)$ approaches 6 as x approaches –4 from the right.

17. The function is defined when $x = 1$; $f(x)$ approaches 3 as x approaches 1 from both sides; and $f(1) = 3$.

20. $y \to \infty$ as $x \to \infty$, $y \to -\infty$ as $x \to -\infty$.

21. $y \to -\infty$ as $x \to \infty$, $y \to -\infty$ as $x \to -\infty$.

22. $y \to \infty$ as $x \to \infty$, $y \to \infty$ as $x \to -\infty$.

23. $y \to \infty$ as $x \to \infty$, $y \to \infty$ as $x \to -\infty$.

24. $y \to 0$ as $x \to \infty$, $y \to 0$ as $x \to -\infty$.

25. $f(x) \to 2$ as $x \to \infty$, $f(x) \to 2$ as $x \to -\infty$.

26.

$[-6, 6]$ scl:1 by $[-30, 30]$ scl:5

increasing for $x < -3$ and $x > 1$; decreasing for $-3 < x < 1$

Practice

Determine whether each function is continuous at the given x-value. Justify your answer using the continuity test. **12–17. See margin for justifications.**

 12. $y = x^3 - 4$; $x = 1$ **yes**

13. $y = \dfrac{x+1}{x-2}$; $x = 2$ **no**

 14. $f(x) = \dfrac{x+3}{(x-3)^2}$; $x = -3$ **yes**

15. $y = \left[\!\left[\frac{1}{2}x\right]\!\right]$; $x = 3$ **yes**

16. $f(x) = \begin{cases} 3x + 5 \text{ if } x \leq -4 \\ -x + 2 \text{ if } x > -4 \end{cases}$; $x = -4$ **no**

17. $f(x) = \begin{cases} 2x + 1 \text{ if } x \geq 1 \\ 4 - x^2 \text{ if } x < 1 \end{cases}$; $x = 1$ **yes**

18. Determine whether the graph at the right has infinite discontinuity, jump discontinuity, or point discontinuity, or is continuous. **jump discontinuity**

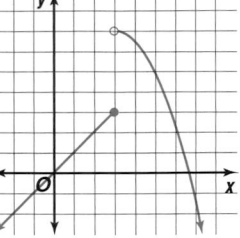

19. Find a value of x at which the function $g(x) = \dfrac{x-4}{x^2 - 3x}$ is discontinuous. Use the continuity test to justify your answer. **Sample answer: $x = 0$. $g(x)$ is undefined when $x = 0$.**

Describe the end behavior of each function. **20–25. See margin.**

 20. $y = x^3 + 2x^2 + x - 1$

21. $y = 8 - x^3 - 2x^4$

22. $f(x) = x^{10} - x^9 + 5x^8$

23. $g(x) = |(x-3)^2 - 1|$

24. $y = \dfrac{1}{x^2}$

25. $f(x) = -\dfrac{1}{x^3} + 2$

Graphing Calculator

26–31. See margin.

Graph each function. Determine the interval(s) for which the function is increasing and the interval(s) for which the function is decreasing.

26. $y = x^3 + 3x^2 - 9x$

27. $y = -x^3 - 2x + 1$

28. $f(x) = \dfrac{1}{x+1} - 4$

29. $g(x) = \dfrac{x^2 + 5}{x - 2}$

30. $y = |x^2 - 4|$

31. $y = (2|x| - 3)^2 + 1$

Applications and Problem Solving

32. Physics The gravitational potential energy of an object is given by $U(r) = -\dfrac{GmM_e}{r}$, where G is Newton's gravitational constant, m is the mass of the object, M_e is the mass of Earth, and r is the distance from the object to the center of Earth. What happens to the gravitational potential energy of the object as it is moved farther and farther away from Earth? **$U(r)$ approaches 0.**

33. Critical Thinking A function $f(x)$ is increasing when $0 < x < 2$ and decreasing when $x > 2$. The function has a jump discontinuity when $x = 3$ and $f(x) \to -\infty$ as $x \to \infty$. **a–b. See margin.**

a. If $f(x)$ is an even function, then describe the behavior of $f(x)$ for $x < 0$. Sketch a graph of such a function.

b. If $f(x)$ is an odd function, then describe the behavior of $f(x)$ for $x < 0$. Sketch a graph of such a function.

27.

$[-7.6, 7.6]$ scl:1 by $[-5, 5]$ scl:1

decreasing for all x

28.

$[-7.6, 7.6]$ scl:1 by $[-8, 2]$ scl:1

decreasing for $x < -1$ and $x > -1$

34. **Biology** One model for the population P of bacteria in a sample after t days is given by $P(t) = 1000 - 19.75t + 20t^2 - \frac{1}{3}t^3$.
   a. What type of function is $P(t)$? **polynomial**
   b. When is the bacteria population increasing? **$0.5 < t < 39.5$**
   c. When is it decreasing? **$0 < t < 0.5$ and $t > 39.5$**

35. **Employment** The graph shows the minimum wage over a 43-year period in 1996 dollars adjusted for inflation. **a–b. See margin.**

Value of The Federal Minimum Wage, 1954-1996
In 1996 Dollars Adjusted for Inflation

Source: Department of Labor

interNET
C O N N E C T I O N

Data Update
For the latest information about the minimum wage, visit **www.amc. glencoe.com.**

   a. During what time intervals was the adjusted minimum wage increasing?
   b. During what time intervals was the adjusted minimum wage decreasing?

36. **Analytic Geometry** A line is *secant* to the graph of a function if it intersects the graph in at least two distinct points. Consider the function $f(x) = -(x - 4)^2 - 3$.
   a. On what interval(s) is $f(x)$ increasing? **$x < 4$**
   b. Choose two points in the interval from part a. Determine the slope of the secant line that passes through those two points. **Answers will vary.**
   c. Make a conjecture about the slope of any secant line that passes through two points contained in an interval where a function is increasing. Explain your reasoning. **The slope is positive; see margin for explanation.**
   d. On what interval(s) is $f(x)$ decreasing? **$x > 4$**
   e. Extend your hypothesis from part c to describe the slope of any secant line that passes through two points contained in an interval where the function is decreasing. Test your hypothesis by choosing two points in the interval from part d. **The slope is negative; see students' work.**

37. **Critical Thinking** Suppose a function is defined for all x-values and its graph passes the horizontal line test. **a. The function must be monotonic.**
   a. What can be said about the monotonicity of the function?
   b. What can be said about the monotonicity of the inverse of the function?
   The inverse must be monotonic.

Lesson 3-5 Continuity and End Behavior **167**

(continued on page 168)

Lesson 3-5 **167**

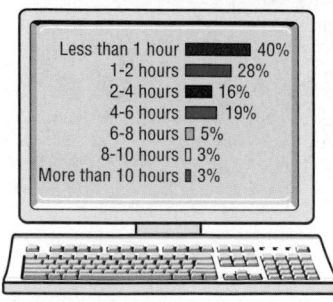

4 ASSESS

Open-Ended Assessment

Writing Have students write a paragraph that explains how to determine the end behavior of a function.

Additional Answers
(pp.167–168)

36c. In an interval where a function is increasing, for any two points on the graph, the x- and y-coordinates of one point will be greater than those of the other point, ensuring that the slope of the line through the two points will be positive.

38a.

38b. $0 < x < 1$, $1 < x < 2$, $2 < x < 4$, $4 < x < 6$, $6 < x < 8$, $x > 8$

41. The graph of $g(x)$ is the graph of $f(x)$ translated left 2 units and down 4 units.

168 *Chapter 3*

---

38. Computers The graph at the right shows the amount of school computer usage per week for students between the ages of 12 and 18. **a–b. See margin.**

a. Use this set of data to make a graph of a step function. On each line segment in your graph, put the open circle at the right endpoint.

b. On what interval(s) is the function continuous?

Student Computer Usage

Less than 1 hour	40%
1-2 hours	28%
2-4 hours	16%
4-6 hours	19%
6-8 hours	5%
8-10 hours	3%
More than 10 hours	3%

Source: Consumer Electronics Manufacturers Association

39. Critical Thinking Determine the values of a and b so that f is continuous.

$$f(x) = \begin{cases} x^2 + a \text{ if } x \ge 2 \\ bx + a \text{ if } -2 < x < 2 \\ \sqrt{-b - x} \text{ if } x \le -2 \end{cases} \quad a = 4, b = 2$$

Mixed Review

40. Find the inverse of the function $f(x) = (x + 5)^2$. *(Lesson 3-4)* $f^{-1}(x) = -5 \pm \sqrt{x}$

41. Describe how the graphs of $f(x) = |x|$ and $g(x) = |x + 2| - 4$ are related. *(Lesson 3-2)* **See margin.**

42. Find the maximum and minimum values of $f(x, y) = x + 2y$ if it is defined for the polygonal convex set having vertices at $(0, 0)$, $(4, 0)$, $(3, 5)$, and $(0, 5)$. *(Lesson 2-6)* **13, 0**

43. Find the determinant of $\begin{bmatrix} 5 & -4 \\ 8 & 2 \end{bmatrix}$. *(Lesson 2-5)* **42**

44. Consumer Costs Mario's Plumbing Service charges a fee of $35 for every service call they make. In addition, they charge $47.50 for every hour they work on each job. *(Lesson 1-4)*

a. Write an equation to represent the cost c of a service call that takes h hours to complete. $c = 47.5h + 35$

b. Find the cost of a $2\frac{1}{4}$-hour service call. **$141.88**

45. Find $f(-2)$ if $f(x) = 2x^2 - 2x + 8$. *(Lesson 1-1)* **20**

46. SAT Practice One box is a cube with side of length x. Another box is a rectangular solid with sides of lengths $x + 1$, $x - 1$, and x. If $x > 1$, how much greater is the volume of the cube than that of the other box? **A**
A x
B $x^2 - 1$
C $x - 1$
D 1
E 0

Extra Practice See p. A31.

Extra Credit

Find values of m and b so that
$$f(x) \begin{cases} -x^2 \text{ if } x \le 0 \\ mx + b \text{ if } 0 < x < 1 \\ x^2 \text{ if } x \ge 1 \end{cases} \text{ is continuous.}$$

Sample answer: $m = 1$, $b = 0$

3-5B Gap Discontinuities

An Extension of Lesson 3-5

OBJECTIVE

• Construct and graph functions with gap discontinuities.

A function has a *gap discontinuity* if there is some interval of real numbers for which it is not defined. The graphs below show two types of gap discontinuities. The first function is undefined for $2 < x < 4$, and the second is undefined for $-4 \leq x \leq -2$ and for $2 \leq x \leq 3$.

These are also called Boolean operators.

The relational and logical operations on the **TEST** menu are primarily used for programming. Recall that the calculator delivers a value of 1 for a true equation or inequality and a value of 0 for a false equation or inequality. Expressions that use the logical connectives ("and", "or", "not", and so on) are evaluated according to the usual truth-table rules. Enter each of the following expressions and press $\boxed{\text{ENTER}}$ to confirm that the calculator displays the value shown.

Expression	Value
$3 \leq 7$	1
$-2 > -5$	1
$-4 > 6$	0

Expression	Value
$(-1 < 8)$ and $(8 < 9)$	1
$(2 > 4)$ or $(-7 \geq -3)$	0
$(4 < 3)$ and $(1 < 12)$	0

Relational and logical operations are also useful in defining functions that have point and gap discontinuities.

Example ● **Graph $y = x^2$ for $x \leq -1$ or $x \geq 2$.**

Enter the following expression as **Y1** on the $\boxed{Y=}$ list.

$$X^2/((X \leq -1) \text{ or } (X \geq 2))$$

The function is defined as a quotient. The denominator is a Boolean statement, which has a value of 1 or 0. If the x-value for which the numerator is being evaluated is a member of the interval defined in the denominator, the denominator has a value of 1. Therefore, $\frac{f(x)}{1} = f(x)$ and that part of the function appears on the screen.

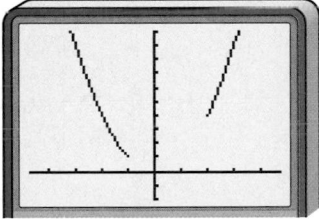

$[-4.7, 4.7]$ scl:1 by $[-2, 10]$ scl:1

(continued on the next page)

---

1 FOCUS

Motivating the Lesson
Discussion Ask students to use what they have learned about functions to determine why a function might be undefined.
Sample answer: Dividing by zero is undefined.

2 TEACH

📁 A reproduction of this lesson with graphing calculator keystrokes provided can be found in the *Graphing Calculator Keystrokes Masters,* pp. 10–12.

Teaching Tip Remind students to count parentheses when entering equations into graphing calculators in order to avoid an error message.

In-Class Example

Graph $y = -x^3$ for $x \leq -1$ or $x \geq 1$.

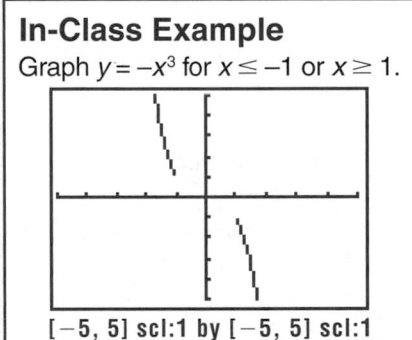

$[-5, 5]$ scl:1 by $[-5, 5]$ scl:1

---

3 PRACTICE/APPLY

Assignment Guide

All: 1–14

4 ASSESS

Open-Ended Assessment

Writing Ask students to write a paragraph explaining why the square root functions in Exercises 9 and 10 are undefined under certain intervals. **Sample answer: a negative sign under the radical yields an imaginary number that cannot be graphed in the *xy*-plane.**

Additional Answers

9.

[−3, 6.4] scl:1 by [−2, 8] scl:1

10.

[−9.4, 9.4] scl:1 by [−6.2, 6.2] scl:1

11. Sample answer: $y =$

$$\frac{x^2}{((x \le 2) \text{ or } ((x \ge 5) \text{ and } (x \le 7)) \text{ or } (x \ge 8))}$$

12. Yes; sample justification: if $f(x)$ is a polynomial function, then the graph of $y =$

$$\frac{f(x)}{(|x - [\![x]\!]| < 0.25)} \text{ is like the}$$

graph of $f(x)$, but with an infinite number of "interval bites" removed.

13. yes; sample justification: The equation $y =$

$$\frac{x^2(x \le -2) + (2x - 4)(x \ge 4)}{((x \le -2) \text{ or } (x \ge 4))}$$

is a possible equation for the function described.

If the *x*-value is not part of the interval, the value of the denominator is 0. At these points, $\frac{f(x)}{0}$ would be undefined. Thus no graph appears on the screen for this interval.

When you use relational and logical operations to define functions, be careful how you use parentheses. Omitting parentheses can easily lead to an expression that the calculator may interpret in a way you did not intend.

1–8. See Answer Appendix for graphs.

TRY THESE

3. {all real numbers $x \mid x < -3$ or $x \ge 1$}

4. {all real numbers $x \mid x \le -3$ or $x > -2$}

6. {all real numbers $x \mid x < -6$ or $x > -2$}

Graph each function and state its domain. You may need to adjust the window settings.

1. $y = \dfrac{x^2 - 2}{(x > 3)}$ {all real numbers $x \mid x > 3$}

2. $y = \dfrac{0.5x + 1}{((x \ge -2) \text{ and } (x \le 4))}$

2. {all real numbers $x \mid -2 \le x \le 4$}

3. $y = \dfrac{-2x}{((x < -3) \text{ or } (x \ge 1))}$

4. $y = \dfrac{-0.2x^3 + 0.3x^2 - x}{((x \le -3) \text{ or } (x > -2))}$

5. $y = \dfrac{|x|}{(|x| > 1)}$ {all real numbers $x \mid x < -1$ or $x > 1$}

6. $y = \dfrac{|x - 1| - |x - 3|}{(|x + 4| > 2)}$

7. $y = \dfrac{1.5x}{([\![x]\!] \ne 3)}$ {all real numbers $x \mid x < 3$ or $x \ge 4$}

8. $y = \dfrac{0.5x^2}{(([\![x]\!] \ne -2) \text{ and } ([\![x]\!] \ne 1))}$ {all real numbers $x \mid x < -2$ or $-1 \le x < 1$ or $x \ge 2$}

Relational and logical operations are not the only tools available for defining and graphing functions with gap discontinuities. The square root function can easily be used for such functions. Graph each function and state its domain.

9. $y = \sqrt{(x - 1)(x - 2)(x - 3)(x - 4)}$ {all real numbers $x \mid x \le 1$ or $2 \le x \le 3$ or $x \ge 4$}

10. $y = \dfrac{x}{\dfrac{\sqrt{(x - 1)(x - 2)}}{\sqrt{(x - 1)(x - 2)}}}$ {all real numbers $x \mid x < 1$ or $x > 2$}

9–10. See margin for graphs.

WHAT DO YOU THINK?

11–13. See margin.

11. Suppose you want to construct a function whose graph is like that of $y = x^2$ except for "bites" removed for the values between 2 and 5 and the values between 7 and 8. What equation could you use for the function?

12. Is it possible to use the functions on the **MATH NUM** menu to take an infinite number of "interval bites" from the graph of a function? Justify your answer.

13. Is it possible to write an equation for a function whose graph looks just like the graph of $y = x^2$ for $x \le -2$ and just like the graph of $y = 2x - 4$ for $x \ge 4$, with no points on the graph for values of x between -2 and 4? Justify your answer.

14. Use what you have learned about gap discontinuities to graph the following piecewise functions. **See margin for graphs.**

a. $f(x) = \begin{cases} -2x & \text{if } x < 0 \\ -x^4 + 2x^3 + 3x^2 + 3x & \text{if } x \ge 0 \end{cases}$

b. $g(x) = \begin{cases} (x + 4)^3 - 2 & \text{if } x < -2 \\ x^2 + 2 & \text{if } -2 \le x \le 2 \\ -(x - 4)^3 - 2 & \text{if } x > 2 \end{cases}$

Additional Answers

14a.

[−15.2, 15.2] scl:1 by [−10, 10] scl:1

14b.

[−9.1, 9.1] scl:1 by [−6, 6] scl:1

3-6

Critical Points and Extrema

OBJECTIVE
• Find the extrema of a function.

 BUSINESS America's 23 million small businesses employ more than 50% of the private workforce. Owning a business requires good management skills. Business owners should always look for ways to compete and improve their businesses. Some business owners hire an analyst to help them identify strengths and weaknesses in their operation. Analysts can collect data and develop mathematical models that help the owner increase productivity, maximize profit, and minimize waste. *A problem related to this will be solved in Example 4.*

Optimization is an application of mathematics where one searches for a maximum or a minimum quantity given a set of constraints. When maximizing or minimizing quantities, it can be helpful to have an equation or a graph of a mathematical model for the quantity to be optimized.

Critical points are those points on a graph at which a line drawn tangent to the curve is horizontal or vertical. A polynomial may possess three types of critical points. A critical point may be a **maximum**, a **minimum**, or a **point of inflection**. When the graph of a function is increasing to the left of $x = c$ and decreasing to the right of $x = c$, then there is a maximum at $x = c$. When the graph of a function is decreasing to the left of $x = c$ and increasing to the right of $x = c$, then there is a minimum at $x = c$. A point of inflection is a point where the graph changes its curvature as illustrated below.

Recall from geometry that a line is tangent to a curve if it intersects a curve in exactly one point.

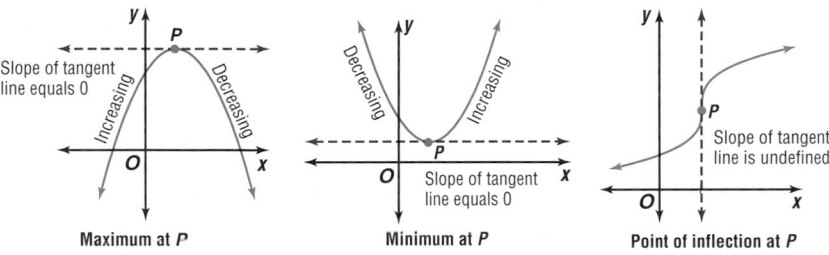

Maximum at P Minimum at P Point of inflection at P

Maxima is the plural of maximum and minima is the plural of minimum. Extrema is the plural of extremum.

The graph of a function can provide a visual clue as to when a function has a maximum or a minimum value. The greatest value that a function assumes over its domain is called the **absolute maximum**. Likewise the least value of a function is the **absolute minimum**. The general term for maximum or minimum is **extremum**. The functions graphed below have absolute extrema.

Absolute minimum

Absolute maximum

Lesson 3-6 Critical Points and Extrema **171**

 5-Minute Check
Lesson 3-5

Determine whether each function is continuous at the given x-value.

1. $y = x^2 + x - 5$; $x = 7$ **yes**

2. $y = \frac{2x - 3}{x - 4}$; $x = 4$ **no**

3. $f(x) = \begin{cases} x + 6 & \text{if } x \geq 2 \\ x^2 + 4 & \text{if } x < 2 \end{cases}$; $x = 2$ **yes**

4. Describe the end behavior of $f(x) = -6x^4 + 3x^3 - 17x^2 - 5x + 12$. **As $x \to \infty$, $f(x) \to -\infty$ and as $x \to -\infty$, $f(x) \to -\infty$.**

5. Graph the function $f(x) = 1 + (x + 3)^2$. Determine the interval(s) on which the function is increasing and the interval(s) on which the function is decreasing. **The function is increasing for $x > -3$ and decreasing for $x < -3$.**

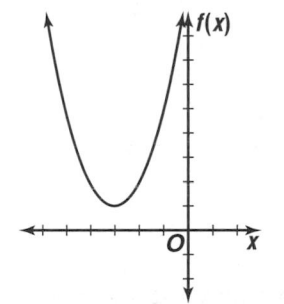

Motivating the Lesson
Hands-On Activity Have cooperative groups of students toss tennis balls to each other outside of school. One student in each group should estimate the maximum height of the ball. Another student should draw the path of the ball. Point out that the maximum height is an extreme value of the "height function" they have modeled.

Resource Manager

▼ **Reproducible Masters**
• *Study Guide and Practice,* pp. 49–50
• *Enrichment,* p. 30
• *Assessment and Evaluation,* p. 70
• *Problem Solving and Applications,* p. 5
• *TI-92 & CBL Interdisciplinary Activities,* pp. 5–8

📖 **Transparencies**
• 5-Minute Check, 3-6
• Teaching, 3-6

💿 **Technology/Multimedia**
• Advanced MathPASS, Lesson 7

Functions can also have **relative extrema**. A **relative maximum** value of a function may not be the greatest value of *f* on the domain, but it is the greatest *y*-value on some interval of the domain. Similarly, a **relative minimum** is the least *y*-value on some interval of the domain. The function graphed at the right has both a relative maximum and a relative minimum.

Note that extrema are *values of the function;* that is, they are the *y*-coordinates of each maximum and minimum point.

Example ❶ Locate the extrema for the graph of $y = f(x)$. Name and classify the extrema of the function.

The function has a relative minimum at $(−3, −1)$.

The function has a relative maximum at $(1, 5)$.

The function has a relative minimum at $(4, −4)$.

Since the point $(4, −4)$ is the lowest point on the graph, the function appears to have an absolute minimum of $−4$ when $x = 4$. This function appears to have no absolute maximum values, since the graph indicates that the function increases without bound as $x \to \infty$ and as $x \to −\infty$.

A branch of mathematics called calculus can be used to locate the critical points of a function. You will learn more about this in Chapter 15. A graphing calculator can also help you locate the critical points of a polynomial function.

Example ❷ Use a graphing calculator to graph $f(x) = 5x^3 − 10x^2 − 20x + 7$ and to determine and classify its extrema.

Use a graphing calculator to graph the function in the standard viewing window. Notice that the *x*-intercepts of the graph are between −2 and −1, 0 and 1, and 3 and 4. Relative maxima and minima will occur somewhere between pairs of *x*-intercepts.

$[−10, 10]$ scl:1 by $[−10, 10]$ scl:1

For a better view of the graph of the function, we need to change the window to encompass the observed *x*-intercepts more closely.

One way to do this is to change the *x*-axis view to $-2 \leq x \leq 4$. Since the top and bottom of the graph are not visible, you will probably want to change the *y*-axis view as well. The graph at the right shows $-40 \leq y \leq 20$. From the graph, we can see there is a relative maximum in the interval $-1 < x < 0$ and a relative minimum in the interval $1 < x < 3$.

[−2,4] scl:1 by [−40, 20] scl:10

There are several methods you can use to locate these extrema more accurately.

Method 1: Use a table of values to locate the approximate greatest and least value of the function. (*Hint:* Revise **TBLSET** to begin at -2 in intervals of 0.1.)

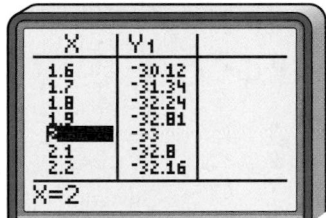

There seems to be a relative maximum of approximately 14.385 at $x = -0.7$ and a relative minimum of -33 at $x = 2$.

You can adjust the **TBLSET** increments to hundredths to more closely estimate the *x*-value for the relative maximum. A relative maximum of about 14.407 appears to occur somewhere between the *x*-values -0.67 and -0.66. A fractional estimation of the *x*-value might be $x = -\frac{2}{3}$.

Method 2: Use the **TRACE** function to approximate the relative maximum and minimum.

[−2, 4] scl:1 by [−40, 20] scl:10

[−2, 4] scl:1 by [−40, 20] scl:10

There seems to be a maximum at $x \approx -0.66$ and a minimum at $x \approx 2.02$.

Method 3: Use **3:minimum** and **4:maximum** options on the **CALC** menu to locate the approximate relative maximum and minimum.

[−2,4] scl:1 by [−40, 20] scl:10

[−2,4] scl:1 by [−40, 20] scl:10

The calculator indicates a relative maximum of about 14.4 at $x \approx -0.67$ and a relative minimum of -33 at $x \approx 2.0$.

All of these approaches give approximations; some more accurate than others. From these three methods, we could estimate that a relative maximum occurs near the point at $(-0.67, 14.4)$ or $(-\frac{2}{3}, 14.407)$ and a relative minimum near the point at $(2, -33)$.

If you know a critical point of a function, you can determine if it is the location of a relative minimum, a relative maximum, or a point of inflection by testing points on both sides of the critical point. The table below shows how to identify each type of critical point.

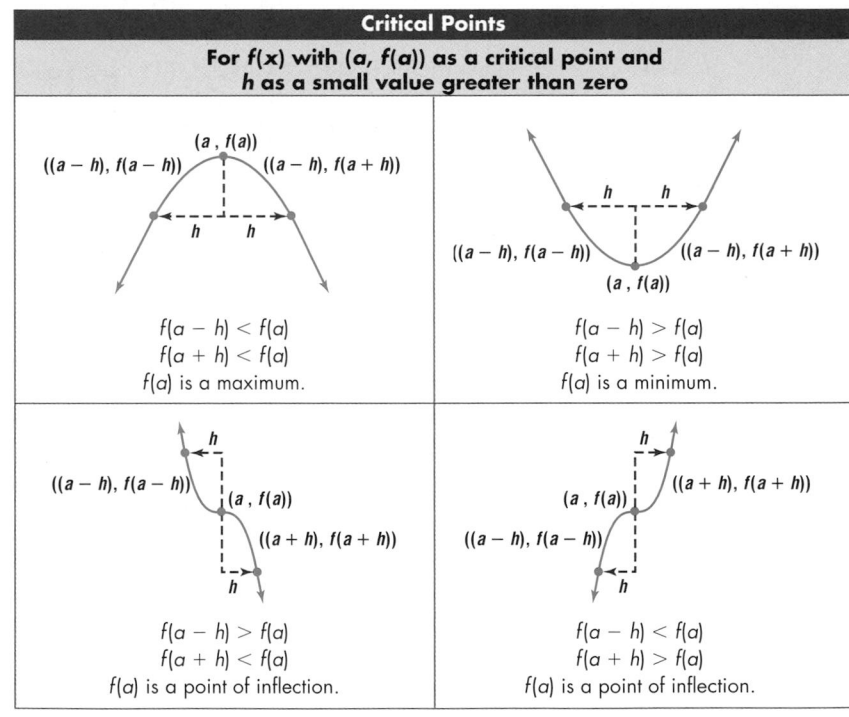

Critical Points
For $f(x)$ with $(a, f(a))$ as a critical point and h as a small value greater than zero

$(a , f(a))$
$((a − h), f(a − h))$ $((a − h), f(a + h))$

$f(a − h) < f(a)$
$f(a + h) < f(a)$
$f(a)$ is a maximum.

$((a − h), f(a − h))$ $((a − h), f(a + h))$
$(a , f(a))$

$f(a − h) > f(a)$
$f(a + h) > f(a)$
$f(a)$ is a minimum.

$((a − h), f(a − h))$
$(a , f(a))$
$((a + h), f(a + h))$

$f(a − h) > f(a)$
$f(a + h) < f(a)$
$f(a)$ is a point of inflection.

$(a , f(a))$
$((a + h), f(a + h))$
$((a − h), f(a − h))$

$f(a − h) < f(a)$
$f(a + h) > f(a)$
$f(a)$ is a point of inflection.

You can also determine whether a critical point is a maximum, minimum, or inflection point by examining the values of a function using a table.

Example **③** The function $f(x) = 2x^5 - 5x^4 - 10x^3$ has critical points at $x = -1$, $x = 0$, and $x = 3$. Determine whether each of these critical points is the location of a maximum, a minimum, or a point of inflection.

Evaluate the function at each point. Then check the values of the function around each point. Let $h = 0.1$.

x	x − 0.1	x + 0.1	f(x − 0.1)	f(x)	f(x + 0.1)	Type of Critical Point
−1	−1.1	−0.9	2.769	3	2.829	maximum
0	−0.1	0.1	0.009	0	−0.010	inflection point
3	2.9	3.1	−187.308	−189	−187.087	minimum

You can verify this solution by graphing f(x) on a graphing calculator.

You can use critical points from the graph of a function to solve real-world problems involving maximization and minimization of values.

Example **④** **BUSINESS** A small business owner employing 15 people hires an analyst to help the business maximize profits. The analyst gathers data and develops the mathematical model $P(x) = \frac{1}{3}x^3 - 34x^2 + 1012x$. In this model, P is the owner's monthly profits, in dollars, and x is the number of employees. The model has critical points at $x = 22$ and $x = 46$.

a. Determine which, if any, of these critical points is a maximum.

b. What does this critical point suggest to the owner about business operations?

c. What are the risks of following the analyst's recommendation?

a. Test values around the points. Let $h = 0.1$.

x	x − 0.1	x + 0.1	P(x − 0.1)	P(x)	P(x + 0.1)	Type of Critical Point
22	21.9	22.1	9357.21	9357.33	9357.21	maximum
46	45.9	46.1	7053.45	7053.33	7053.45	minimum

The profit will be at a maximum when the owner employs 22 people.

b. The owner should consider expanding the business by increasing the number of employees from 15 to 22.

c. It is important that the owner hire qualified employees. Hiring unqualified employees will likely cause profits to decline.

Addressing Individual Needs

Kinesthetic To help students understand that the slope of the tangent line at a maximum or minimum point of a graph is 0, have them draw a graph on grid paper and then fold the paper so that the crease is a line tangent to the maximum or minimum point of the graph. Determine two points on the line and calculate its slope.

Error Analysis

Watch for students who cannot find all maxima and minima for a graph.

Prevent by reminding students to enlarge the viewing window or plot a wide range of points as a check in order to make sure that they have accounted for all extrema.

Additional Answers

1. Check value of the function at *x*-values very close to the critical point. Be sure to check values on both sides. If the function values change from increasing to decreasing, the critical point is a maximum. If the function values change from decreasing to increasing, the critical point is a minimum. If the function value continues to increase or to decrease, the critical point is a point of inflection.

2. rel. min.; $f(0.99) \approx -3.9997$; $f(1) \approx -4$; $f(1.01) \approx -3.9997$.

   By testing points on either side of the critical point, it is evident that the graph of the function is decreasing as *x* approaches 1 from the left and increasing as *x* moves away from 1 to the right. Therefore, on the interval $0.99 < x < 1.01$, $(1, 4)$ is a relative minimum.

Study Guide and Practice Masters, pp. 49–50

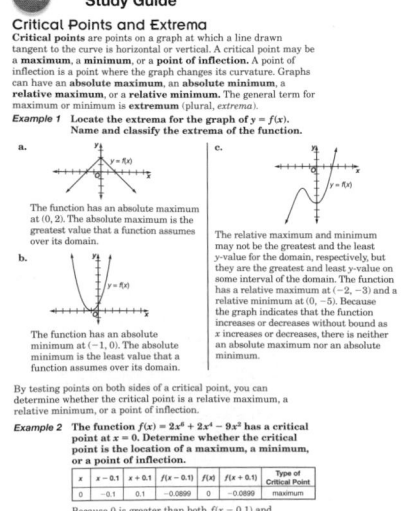

Communicating Mathematics

Read and study the lesson to answer each question.   1–3. See margin.

1. **Write** an explanation of how to determine if a critical point is a maximum, minimum, or neither.

2. **Determine** whether the point at $(1, -4)$, a critical point of the graph of $f(x) = x^3 - 3x - 2$ shown at the right, represents a relative maximum, a relative minimum, or a point of inflection. Explain your reasoning.

[−10, 10] scl:1 by [−10, 10] scl:1

3. **Sketch** the graph of a function that has a relative minimum at $(0, -4)$, a relative maximum at $(-3, 1)$, and an absolute maximum at $(4, 6)$.

Guided Practice

Locate the extrema for the graph of $y = f(x)$. Name and classify the extrema of the function.

4.

5.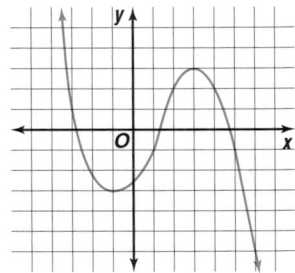

rel. min.:$(-3, -2)$; rel. max.: $(1, 6)$ rel. min.: $(-1, -3)$; rel. max.: $(3, 3)$

Use a graphing calculator to graph each function and to determine and classify its extrema.   6–7. See Answer Appendix for graphs.

6. $f(x) = 2x^5 - 5x^4$
   rel. max.: $(0, 0)$; rel. min.: $(2, -16)$

7. $g(x) = x^4 + 3x^3 - 2$
   rel. min.: $(-2.25, -10.54)$

Determine whether the given critical point is the location of a *maximum*, a *minimum*, or a *point of inflection*.

8. $y = 3x^3 - 9x - 5, x = -1$ max.

9. $y = x^2 + 5x - 6, x = -2.5$ min.

10. $y = 2x^3 - x^5, x = 0$ pt. of inflection

11. $y = x^6 - 3x^4 + 3x^2 - 1, x = 0$ min.

12. **Agriculture**   Malik Davis is a soybean farmer. If he harvests his crop now, the yield will average 120 bushels of soybeans per acre and will sell for $0.48 per bushel. However, he knows that if he waits, his yield will increase by about 10 bushels per week, but the price will decrease by $0.03 per bushel per week.

12a. $P(x) = (120 + 10x)(0.48 - 0.03x)$; see margin for graph.

a. If *x* represents the number of weeks Mr. Davis waits to harvest his crop, write and graph a function $P(x)$ to represent his profit.   **2 weeks**

b. How many weeks should Mr. Davis wait in order to maximize his profit?

c. What is the maximum profit?   **$58.80 per acre**

d. What are the risks of waiting?   **Rain or other bad weather could delay harvest and/or destroy part of the crop.**

www.amc.glencoe.com/self_check_quiz

Additional Answers

3. Sample answer:

12a.

Practice

Locate the extrema for the graph of $y = f(x)$. Name and classify the extrema of the function.

14. abs. max.: $(-1, 3)$; rel. min.: $(0.5, 0.5)$; rel. max.: $(1.5, 2)$

A

13.

14.

15.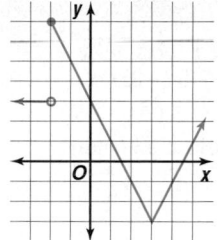

15. rel. max.: $(-2, 7)$, abs. min.: $(3, -3)$

abs. max.: $(-4, 1)$

16.

17.

18.

17. abs. min.: $(3, -8)$; rel. max.: $(5, -2)$; rel. min.: $(8, -5)$

rel. max.: $(-6, 4)$, rel. min.: $(-2, -3)$

no extrema

Use a graphing calculator to graph each function and to determine and classify its extrema.

19–24. See margin for graphs.

B

19. $f(x) = -4 + 3x - x^2$ abs. max.: $(1.5, -1.75)$

20. $V(w) = w^3 - 7w - 6$

20. rel. max.: $(-1.53, 1.13)$; rel. min.: $(1.53, -13.13)$

21. $g(x) = 6x^3 + x^2 - 5x - 2$

22. $h(x) = x^4 - 4x^2 - 2$

21. rel. max.: $(-0.59, 0.07)$, rel. min.: $(0.47, -3.51)$

23. $f(x) = 2x^5 + 4x^2 - 2x - 3$ rel. max.: $(-1, 1)$; rel. min.: $(0.25, -3.25)$

24. $D(t) = t^3 + t$ **no extrema**

25. Determine and classify the extrema of $f(x) = x^4 + 5x^3 + 3x^2 - 4x$.
abs. min.: $(-3.18, -15.47)$; rel. min.: $(0.34, -0.80)$; rel. max.: $(-0.91, 3.04)$

22. abs. min.: $(-1.41, -6)$, $(1.41, -6)$; rel. max.: $(0, -2)$

Determine whether the given critical point is the location of a *maximum*, a *minimum*, or a *point of inflection*.

26. $y = x^3, x = 0$ **pt. of inflection**

27. $y = -x^2 + 8x - 10, x = 4$ **max.**

28. $y = 2x^2 + 10x - 7, x = -2.5$ **min.**

29. $y = x^4 - 2x^2 + 7, x = 0$ **max.**

C

30. $y = \frac{1}{4}x^4 - 2x^2, x = 2$ **min.**

31. $y = x^3 - 9x^2 + 27x - 27, x = 3$ **pt. of inflection**

32. $y = \frac{1}{3}x^3 + \frac{1}{2}x^2 - 2x + 1, x = -2$ **max.**

33. $y = x^3 - x^2 + 3, x = \frac{2}{3}$ **min.**

34. A function f has a relative maximum at $x = 2$ and a point of inflection at $x = -1$. Find the critical points of $y = -2f(x + 5) - 1$. Describe what happens at each new critical point. **The point of inflection is now at $x = -6$ and there is now a relative minimum at $x = -3$.**

Assignment Guide

Core: 13–37 odd, 38–51
Enriched: 14–34 even, 35–51

 Teaching Tip For Exercises 6, 7, and 19–24, remind students that it is possible for a graph to have no extrema.

Additional Answers

21.

$[-5, 5]$ scl:1 by $[-5, 5]$ scl:1

22.

$[-5, 5]$ scl:1 by $[-8, 2]$ scl:1

23.

$[-5, 5]$ scl:1 by $[-5, 5]$ scl:1

24.

$[-5, 5]$ scl:1 by $[-5, 5]$ scl:1

Additional Answers

19.

$[-5, 5]$ scl:1 by $[-8, 2]$ scl:1

20.

$[-5, 5]$ scl:1 by $[-16, 4]$ scl:2

Additional Answer

39. The particle is at rest when $t \approx 0.14$ and when $t \approx 3.52$. Its positions at these times are $s(0.14) \approx -8.79$ and $s(3.52) \approx -47.51$.

Applications and Problem Solving

35a. $V(x) = 2x$ $(12.5 - 2x)$ $(17 - 2x)$

35b. 2.37 cm by 2.37 cm

35. Manufacturing A 12.5 centimeter by 34 centimeter piece of cardboard will have eight congruent squares removed as in the diagram. The box will be folded to create a take-out hamburger box.

a. Find the model for the volume $V(x)$ of the box as a function of the length x of the sides of the eight squares removed.

b. What are the dimensions of each of the eight squares that should be removed to produce a box with maximum volume?

c. Construct a physical model of the box and measure its volume. Compare this result to the result from the mathematical model. **See student's work.**

36. Business The Carlisle Innovation Company has created a new product that costs $25 per item to produce. The company has hired a marketing analyst to help it determine a selling price for the product. After collecting and analyzing data relating selling price s to yearly consumer demand d, the analyst estimates demand for the product using the equation $d = -200s + 15,000$.

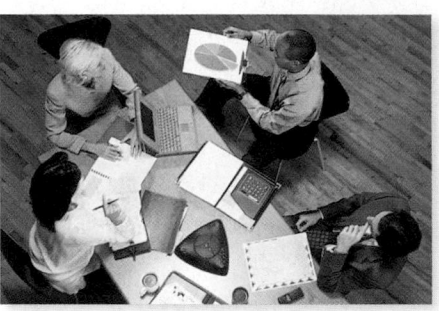

a. If yearly profit is the difference between total revenue and production costs, determine a selling price s, $s \geq 25$, that will maximize the company's yearly profit, P. (Hint: $P = sd - 25d$) **$50**

b. What are the risks of determining a selling price using this method? **Sample answer: The company's competition might offer a similar product at a lower cost.**

37. Telecommunications A cable company wants to provide service for residents of an island. The distance from the closest point on the island's beach, point A, directly to the mainland at point B is 2 kilometers. The nearest cable station, point C, is 10 kilometers downshore from point B. It costs $3500 per kilometer to lay the cable lines underground and $5000 per kilometer to lay the cable lines under water. The line comes to the mainland at point M. Let x be the distance in kilometers from point B to point M.

a. Write a function to calculate the cost of laying the cable. **a. $f(x) = 5000(\sqrt{x^2 + 4}) + 3500(10 - x)$**

b. At what distance x should the cable come to shore to minimize cost? **1.96 km from point B**

38. Critical Thinking Which families of graphs have points of inflection but no maximum or minimum points? **equations of the form $y = x^n$ or $y = \sqrt[n]{x}$, where n is odd**

39. Physics When the position of a particle as a function of time t is modeled by a polynomial function, then the particle is at rest at each critical point. If a particle has a position given by $s(t) = 2t^3 - 11t^2 + 3t - 9$, find the position of the particle each time it is at rest. **See margin.**

Extra Credit

For the graph of $y = \sin \theta$, describe the extrema of the function. **The maxima of the function occur at every positive and negative even multiple of π. The minima of the function occur at every positive and negative odd multiple of π.**

40. Critical Thinking A cubic polynomial can have 1 or 3 critical points. Describe the possible combinations of relative maxima and minima for a cubic polynomial. **See margin.**

Mixed Review

41. Is $y = \dfrac{5x}{x^2 - 3x - 10}$ continuous at $x = 5$? Justify your answer by using the continuity test. *(Lesson 3-5)* **No; the function is undefined when $x = 5$.**

42. Graph the inequality $y \le \dfrac{1}{5}(x - 2)^3$. *(Lesson 3-3)* **See margin.**

43. 120 units of notebook and 80 units of newsprint

43. Manufacturing The Eastern Minnesota Paper Company can convert wood pulp to either newsprint or notebook paper. The mill can produce up to 200 units of paper a day, and regular customers require 10 units of notebook paper and 80 units of newsprint per day. If the profit on a unit of notebook paper is $400 and the profit on a unit of newsprint is $350, how much of each should the plant produce? *(Lesson 2-7)*

44.
$-3 \le x \le 2$,
$-1 \le y \le 4$

44. Geometry Find the system of inequalities that will define a polygonal convex set that includes all points in the interior of a square whose vertices are $A(-3, 4)$, $B(2, 4)$, $C(2, -1)$, and $D(-3, -1)$. *(Lesson 2-6)*

45. Find the determinant for $\begin{bmatrix} 1 & 3 \\ 2 & 5 \end{bmatrix}$. Does an inverse exist for this matrix? *(Lesson 2-5)* **-1, yes**

46. Find $3A + 2B$ if $A = \begin{bmatrix} 4 & -2 \\ 5 & 7 \end{bmatrix}$ and $B = \begin{bmatrix} -3 & 5 \\ -4 & 3 \end{bmatrix}$. *(Lesson 2-3)* $\begin{bmatrix} 6 & 4 \\ 7 & 27 \end{bmatrix}$

47. Sports Jon played in two varsity basketball games. He scored 32 points by hitting 17 of his 1-point, 2-point, and 3-point attempts. He made 50% of his 18 2-point field goal attempts. Find the number of 1-point free throws, 2-point field goals, and 3-point field goals Jon scored in these two games. *(Lesson 2-2)* **5 free throws, 9 2-point field goals, 3 3-point field goals**

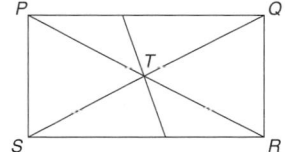

48. Graph $y + 6 \ge 4$. *(Lesson 1-8)* **See margin.**

49. Determine whether the graphs of $2x + 3y = 15$ and $6x = 4y + 16$ are *parallel, coinciding, perpendicular,* or *none of these*. *(Lesson 1-5)* **perpendicular**

50. Describe the difference between a relation and a function. How do you test a graph to determine if it is the graph of a function? *(Lesson 1-1)* **See margin.**

51. SAT/ACT Practice Refer to the figure at the right. What percent of the area of rectangle $PQRS$ is shaded? **D**

A 20%

B $33\dfrac{1}{3}\%$

C 30%

D 25%

E 40%

Extra Practice   See p. A31.

Lesson 3-6   Critical Points and Extrema **179**

Additional Answers

40. If a cubic has one critical point, then it must be a point of inflection. If it were a relative maximum or minimum, then the end behavior for a cubic would not be satisfied. If a cubic has three critical points, then one must be a maximum, another a minimum, and the third a point of inflection.

42.

Open-Ended Assessment

Modeling Have each student make an open box from a sheet of 8.5 x 11 inch paper. Discuss how to find the dimensions that result in the maximum volume of the box.

Chapter 3, Quiz C (Lessons 3-5 and 3-6) is available in the *Assessment and Evaluation Masters*, p. 70.

Additional Answers

48.

50. A relation relates members of a set called the domain to members of a set called the range. In a function, the relation must be such that each member of the domain is related to one and only one member of the range. You can use the vertical line test to determine whether a graph is the graph of a function.

Enrichment Masters, p. 30

1 FOCUS

Locate the extrema for each graph of $y = f(x)$. Name and classify the extrema of the function.

1.

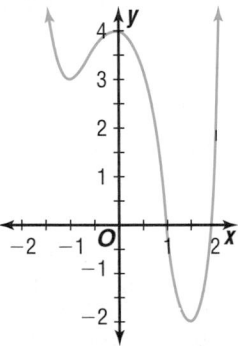

abs. min. : (1.5, −1.9);
rel. max. : (0, 4);
rel. min. : (−1, 3)

2.

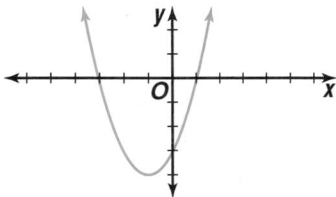

abs min. : (−1, −4)

3. Use a graphing calculator to graph $f(x) = x^3 - 3x + 4$ and to determine and classify its extrema.
rel. max. : (−1, 6);
rel. min. : (1, 2)

[−5, 5] scl:1 by [−10, 15] scl:5

4. Determine whether the given critical point is the location of a *maximum*, a *minimum*, or a *point of inflection*.

$y = x^2 + 2x - 3, x = -1$
abs. min.

3-7

Graphs of Rational Functions

OBJECTIVES
- Graph rational functions.
- Determine vertical, horizontal, and slant asymptotes.

CHEMISTRY The creation of standard chemical solutions requires precise measurements and appropriate mixing. For example, a chemist may have 50 liters of a 10-molar sodium chloride (NaCl) solution (10 moles of NaCl per liter of water.) This solution must be diluted so it can be used in an experiment. Adding 4-molar NaCl solution (4 moles of NaCl per liter of water) to the 10-molar solution will decrease the concentration. The concentration, C, of the mixture can be modeled by $C(x) = \frac{500 + 4x}{50 + x}$, where x is the number of liters of 4-molar solution added. The graph of this function has characteristics that are helpful in understanding the situation. *A problem related to this will be solved in Example 3.*

The concentration function given above is an example of a **rational function.** A rational function is a quotient of two polynomial functions. It has the form $f(x) = \frac{g(x)}{h(x)}$, where $h(x) \neq 0$. The parent rational function is $f(x) = \frac{1}{x}$.

The graph of $f(x) = \frac{1}{x}$ consists of two branches, one in Quadrant I and the other in Quadrant III. The graph has no x- or y-intercepts. The graph of $f(x) = \frac{1}{x}$, like that of many rational functions, has branches that approach lines called **asymptotes.** In the case of $f(x) = \frac{1}{x}$, the line $x = 0$ is a **vertical asymptote.** Notice that as x approaches 0 from the right, the value of $f(x)$ increases without bound toward positive infinity (∞). As x approaches 0 from the left, the value of $f(x)$ decreases without bound toward negative infinity ($-\infty$).

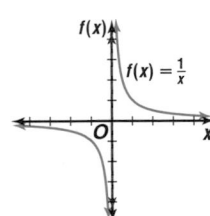

$$f(x) = \frac{1}{x}$$

A vertical asymptote in the graph of a rational function is also called a pole of the function.

Vertical Asymptote	The line $x = a$ is a vertical asymptote for a function $f(x)$ if $f(x) \to \infty$ or $f(x) \to -\infty$ as $x \to a$ from either the left or the right.

Also notice for $f(x) = \frac{1}{x}$ that as the value of x increases and approaches positive infinity, the value of $f(x)$ approaches 0. The same type of behavior can also be observed in the third quadrant: as x decreases and approaches negative infinity, the value of $f(x)$ approaches 0. When the values of a function approach a constant value b as $x \to \infty$ or $x \to -\infty$, then the function has a **horizontal asymptote.** Thus, for $f(x) = \frac{1}{x}$, the line $f(x) = 0$ is a horizontal asymptote.

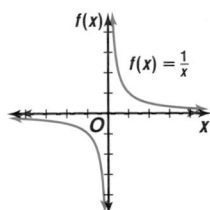

$$f(x) = \frac{1}{x}$$

Horizontal Asymptote	The line $y = b$ is a horizontal asymptote for a function $f(x)$ if $f(x) \to b$ as $x \to \infty$ or as $x \to -\infty$.

Resource Manager

There are several methods that may be used to determine if a rational function has a horizontal asymptote. Two such methods are used in the example below.

Example **1** **Determine the asymptotes for the graph of** $f(x) = \dfrac{3x-1}{x-2}$.

Since $f(2)$ is undefined, there may be a vertical asymptote at $x = 2$. To verify that $x = 2$ is indeed a vertical asymptote, you have to make sure that $f(x) \to \infty$ or $f(x) \to -\infty$ as $x \to 2$ from either the left or the right.

x	$f(x)$
1.9	-47
1.99	-497
1.999	-4997
1.9999	-49997

Graphing Calculator Tip

You can use the TRACE or TABLE function on your graphing calculator to approximate the vertical or horizontal asymptote.

The values in the table suggest that $f(x) \to -\infty$ as $x \to 2$ from the left, so there is a vertical asymptote at $x = 2$.

Two different methods may be used to find the horizontal asymptote.

Method 1

Let $f(x) = y$ and solve for x in terms of y. Then find where the function is undefined for values of y.

$$y = \frac{3x-1}{x-2}$$

$y(x-2) = 3x - 1$ *Multiply each side by $(x-2)$.*

$xy - 2y = 3x - 1$ *Distribute.*
$xy - 3x = 2y - 1$
$x(y-3) = 2y - 1$ *Factor.*
$x = \dfrac{2y-1}{y-3}$ *Divide each side by $y - 3$.*

The rational expression $\dfrac{2y-1}{y-3}$ is undefined for $y = 3$. Thus, the horizontal asymptote is the line $y = 3$.

Method 2

First divide the numerator and denominator by the highest power of x.

$$y = \frac{3x-1}{x-2}$$

$$y = \frac{\dfrac{3x}{x} - \dfrac{1}{x}}{\dfrac{x}{x} - \dfrac{2}{x}}$$

$$y = \frac{3 - \dfrac{1}{x}}{1 - \dfrac{2}{x}}$$

As the value of x increases positively or negatively, the values of $\dfrac{1}{x}$ and $\dfrac{2}{x}$ approach zero. Therefore, the value of the entire expression approaches $\dfrac{3}{1}$ or 3. So, the line $y = 3$ is the horizontal asymptote.

Method 2 is preferable when the degree of the numerator is greater than that of the denominator.

The graph of this function verifies that the lines $y = 3$ and $x = 2$ are asymptotes.

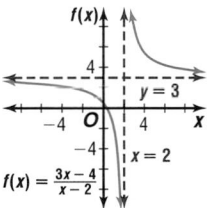

$f(x) = \dfrac{3x-4}{x-2}$

Prerequisite Skills

A review of the prerequisite skills for this lesson is provided in the *Prerequisite Skills Masters.*

Skill Number and Content	Pages
2 Fractions	5–6
9 Factoring by Grouping and Greatest Common Factor	19–20
10 Factoring Special Products of Polynomials	21–22
11 Factoring by Completing the Square	23–24

Motivating the Lesson
Hands-On Activity Give each group of students a piece of string. Have them cut the string in half, measure the pieces, and record the results. Have them repeat these actions until they can no longer measure the pieces. Then, have them observe their data. As the number of pieces grows, what lengths do the pieces approach? 0

2 TEACH

Teaching Tip For Example 1, you may need to review division of polynomials with students.

In-Class Example
Example 1
Determine the asymptotes for the graph of $f(x) = \dfrac{x}{x-1}$.
$x = 1, y = 1$

Teaching Tip In Example 1, show students how to examine the quotient of the leading terms of the numerator and denominator in order to find the horizontal asymptote. As x approaches ∞, these are the terms that will grow most quickly.

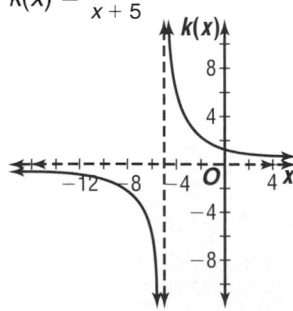
Example **2** Use the parent graph $f(x) = \frac{1}{x}$ to graph each function. Describe the transformation(s) that take place. Identify the new location of each asymptote.

a. $g(x) = \frac{1}{x + 5}$

To graph $g(x) = \frac{1}{x + 5}$, translate the parent graph 5 units to the left. The new vertical asymptote is $x = -5$. The horizontal asymptote, $y = 0$, remains unchanged.

b. $h(x) = -\frac{1}{2x}$

To graph $h(x) = -\frac{1}{2x}$, reflect the parent graph over the x-axis, and compress the result horizontally by a factor of 2. This does not affect the vertical asymptote at $x = 0$. The horizontal asymptote, $y = 0$, is also unchanged.

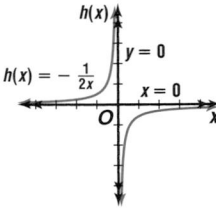

c. $k(x) = \frac{4}{x - 3}$

To graph $k(x) = \frac{4}{x - 3}$, stretch the parent graph vertically by a factor of 4 and translate the parent graph 3 units to the right. The new vertical asymptote is $x = 3$. The horizontal asymptote, $y = 0$, is unchanged.

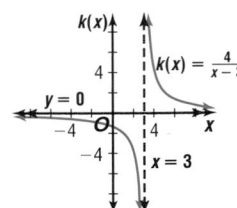

d. $m(x) = -\frac{6}{x + 2} - 4$

To graph $m(x) = -\frac{6}{x + 2} - 4$, reflect the parent graph over the x-axis, stretch the result vertically by a factor of 6, and translate the result 2 units to the left and 4 units down. The new vertical asymptote is $x = -2$. The horizontal asymptote changes from $y = 0$ to $y = -4$.

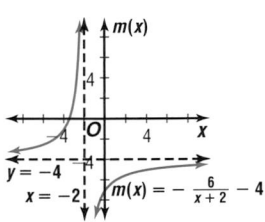

Many real-world situations can be modeled by rational functions.

Teaching Tip In Example 2, explain that $f(x) = \frac{1}{x}$ is the parent graph for *all* rational functions. Therefore, students can determine whether a function is a rational function by examining it to see whether it is a transformation of $f(x) = \frac{1}{x}$.

Example

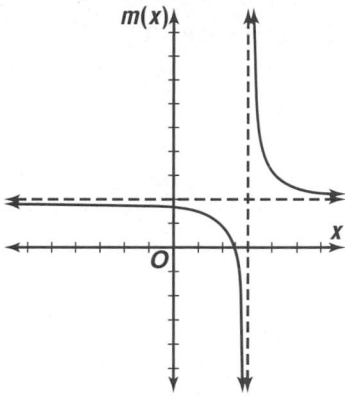

3 CHEMISTRY Refer to the application at the beginning of the lesson.

a. Write the function as a transformation of the parent $f(x) = \frac{1}{x}$.

b. Graph the function and identify the asymptotes. Interpret their meaning in terms of the problem.

a. In order to write $C(x) = \frac{500 + 4x}{50 + x}$ as a transformation of $f(x) = \frac{1}{x}$, you must perform the indicated division of the linear functions in the numerator and denominator.

$$x + 50 \overline{\smash{\big)}\, 4x + 500} \quad \Rightarrow \quad 4 + \frac{300}{x + 50} \text{ or } 300\left(\frac{1}{x + 50}\right) + 4$$
$$\underline{4x + 200}$$
$$300$$

Therefore, $C(x) = 300\left(\frac{1}{x + 50}\right) + 4$.

b. The graph of $C(x)$ is a vertical stretch of the parent function by a factor of 300 and a translation 50 units to the left and 4 units up. The asymptotes are $x = -50$ and $y = 4$.

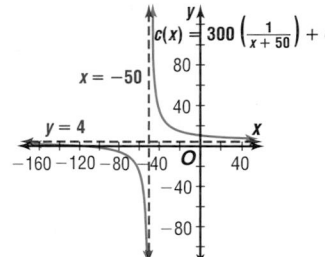

For the purposes of our model, the domain of $C(x)$ must be restricted to $x \geq 0$ since the number of liters of the 4-molar solution added cannot be negative. The vertical asymptote $x = -50$ has no meaning in this application as a result of this restriction. The horizontal asymptote indicates that as the amount of 4-molar solution that has been added grows, the overall concentration will approach a molarity of 4.

A third type of asymptote is a **slant asymptote.** Slant asymptotes occur when the degree of the numerator of a rational function is *exactly* one greater than that of the denominator. For example, in $f(x) = \frac{x^3 + 1}{x^2}$, the degree of the numerator is 3 and the degree

of the denominator is 2. Therefore, this function has a slant asymptote, as shown in the graph. Note that there is also a vertical asymptote at $x = 0$. *When the degrees are the same or the denominator has the greater degree, the function has a horizontal asymptote.*

Lesson 3-7 Graphs of Rational Functions **183**

Example 2 (continued)

d. $m(x) = \frac{1}{x - 3} + 2$

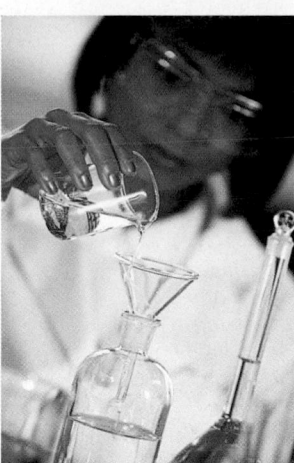

The parent graph is translated 3 units to the right and 2 units up with new vertical asymptote at $x = 3$ and new horizontal asymptote at $y = 2$.

Example 3

Physics In an electronics laboratory, an engineer determined that the current I in an electric circuit was $I(x) = \frac{-2x^2 + 10x + 1}{5 - x}$, where x represents the time that it takes for the current to pass through a circuit.

a. Write the function as a transformation of the parent $f(x) = \frac{1}{x}$.

$$I(x) = 2x + \frac{1}{5 - x}$$

b. Graph the function and identify the asymptotes. Interpret their meaning in terms of the problem.

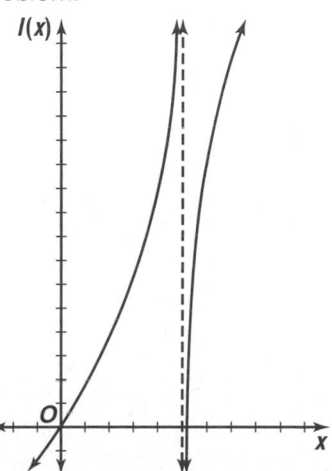

Vertical asymptote at $x = 5$; as x approaches 5, the current approaches infinity, which will most likely burn out the circuit.

In-Class Example

Example 4
Determine the slant asymptote
for $f(x) = \frac{4x^2 + 6x - 37}{x + 4}$.

the line $y = 4x - 10$

Slant Asymptote	The oblique line ℓ is a slant asymptote for a function $f(x)$ if the graph of $y = f(x)$ approaches ℓ as $x \to \infty$ or as $x \to -\infty$.

Example ④ Determine the slant asymptote for $f(x) = \frac{2x^2 - 3x + 1}{x - 2}$.

This method can also be used to determine the horizontal asymptote when the degree of the numerator is equal to or greater than that of the denominator.

First, use division to rewrite the function.

$$\begin{array}{r} 2x + 1 \\ x - 2 \overline{)\ 2x^2 - 3x + 1} \\ \underline{2x^2 - 4x} \\ x + 1 \\ \underline{x - 2} \\ 3 \end{array} \quad \Rightarrow \quad f(x) = 2x + 1 + \frac{3}{x - 2}$$

As $x \to \infty$, $\frac{3}{x - 2} \to 0$. So, the graph of $f(x)$ will approach that of $y = 2x + 1$. This means that the line $y = 2x + 1$ is a slant asymptote for the graph of $f(x)$. *Note that $x = 2$ is a vertical asymptote.*

The graph of this function verifies that the line $y = 2x + 1$ is a slant asymptote.

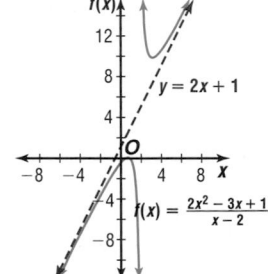

There are times when the numerator and denominator of a rational function share a common factor. Consider $f(x) = \frac{(x + 2)(x - 3)}{x - 3}$. Since an x-value of 3 results in a denominator of 0, you might expect there to be a vertical asymptote at $x = 3$. However, $x - 3$ is a common factor of the numerator and denominator.

Numerically, we can see that $f(x)$ has the same values as $g(x) = x + 2$ except at $x = 3$.

x	2.9	2.99	3.0	3.01	3.1
f(x)	4.9	4.99	—	5.01	5.1
g(x)	4.9	4.99	5.0	5.01	5.1

The y-values on the graph of f approach 5 from both sides but never get to 5, so the graph has point discontinuity at $(3, 5)$.

Whenever the numerator and denominator of a rational function contain a common linear factor, a point discontinuity may appear in the graph of the function. If, after dividing the common linear factors, the same factor remains in the denominator, a vertical asymptote exists. Otherwise, the graph will have point discontinuity.

Addressing Individual Needs

Logical Fill in the boxes in the polynomial division problem in order to predict the asymptotes for the graph of

$$f(x) = \frac{\frac{1}{2}x^2 - \boxed{\frac{1}{2}}x + 1\frac{27}{32}}{x - \boxed{\frac{5}{4}}}.$$

$$y = \frac{1}{2}x + \frac{1}{8}; \ x = \frac{5}{4}$$

Example **5** Graph $y = \dfrac{(x+3)(x+1)}{x(x+3)(x-2)}$.

$$y = \dfrac{(x+3)(x+1)}{x(x+3)(x-2)}$$

$$= \dfrac{x+1}{x(x-2)}, x \neq -3$$

Since $x + 3$ is a common factor that does not remain in the denominator after the division, there is point discontinuity at $x = -3$. Because y increases or decreases without bound close to $x = 0$ and $x = 2$, there are vertical asymptotes at $x = 0$ and $x = 2$. There is also a horizontal asymptote at $y = 0$.

The graph is the graph of $y = \dfrac{x+1}{x(x-2)}$, except

for point discontinuity at $\left(-3, -\dfrac{2}{15}\right)$.

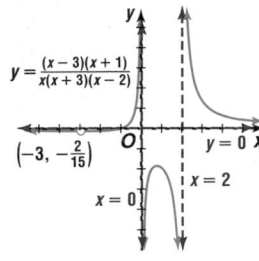

$y = \dfrac{(x-3)(x+1)}{x(x+3)(x-2)}$

$\left(-3, -\dfrac{2}{15}\right)$ $y = 0$ $x = 2$ $x = 0$

CHECK FOR UNDERSTANDING

Communicating Mathematics

Read and study the lesson to answer each question. **1, 3, 4. See margin.**

1. **Graph** $f(x) = \dfrac{1}{x}$ after it has been translated 2 units to the right and down 6 units.

   a. What are its asymptotes?

   b. Write an equation of the translated graph.

2. **Draw** a graph to illustrate each type of asymptote discussed in this lesson: vertical, horizontal, and slant. **See Answer Appendix.**

3. **Write** an equation of a rational function that has point discontinuity.

4. **True or False:** If a value of x causes a zero in the denominator of a rational function, then there is a vertical asymptote at that x-value. Explain.

Guided Practice

Determine the equations of the vertical and horizontal asymptotes, if any, of each function.

5. $f(x) = \dfrac{x}{x-5}$ $x = 5, y = 1$

6. $g(x) = \dfrac{x^3}{(x-2)(x+1)}$ $x = 2, x = -1$

7. The graph at the right shows a transformation of $f(x) = \dfrac{1}{x}$. Write an equation of the function.

$f(x) = \dfrac{1}{x+1} - 2$

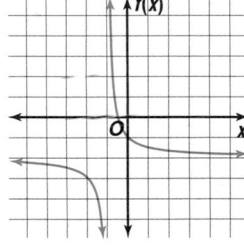

Lesson 3-7 Graphs of Rational Functions **185**

Additional Answers

1.

1a. $x = 2, y = -6$

1b. $y = \dfrac{1}{x-2} - 6$

3. Sample answer: $f(x) = \dfrac{x(x+1)}{x+1}$

4. False; sample explanation: if that x-value also causes the numerator to be 0, there could be a hole instead of a vertical asymptote.

In-Class Example
Example 5

Graph $y = \dfrac{(x+2)(x-3)}{x(x-4)^2(x+2)}$.

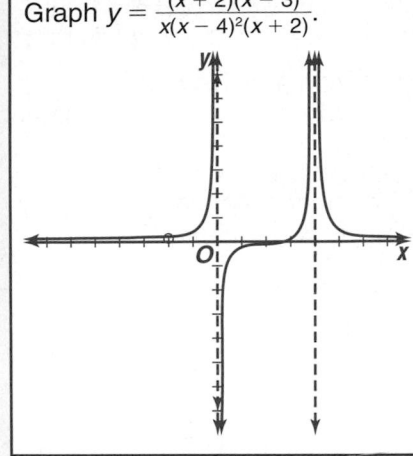

3 PRACTICE/APPLY

Error Analysis
Watch for students who experience difficulty with division of polynomials.

Prevent by having students express remainders as rational expressions instead of using R with a number to represent the remainder.

Study Guide and Practice Masters, pp. 51–52

Lesson 3-7 **185**

Additional Answers

8. The parent graph is translated 4 units right. The vertical asymptote is now at $x = 4$. The horizontal asymptote, $y = 0$, is unchanged.

9. The parent graph is translated 2 units left and down 1 unit. The vertical asymptote is now at $x = -2$ and the horizontal asymptote is now $y = -1$.

11.

12.

13a.

Use the parent graph $f(x) = \dfrac{1}{x}$ to graph each equation. Describe the transformation(s) that have taken place. Identify the new locations of the asymptotes.   **8–9. See margin.**

8. $y = \dfrac{1}{x - 4}$

9. $y = \dfrac{1}{x + 2} - 1$

10. Determine the slant asymptote for the function $f(x) = \dfrac{3x^2 - 4x + 5}{x - 3}$. $y = 3x + 5$

Graph each function.   **11–12. See margin.**

11. $y = \dfrac{x + 2}{(x + 1)(x - 1)}$

12. $y = \dfrac{x^2 + 4x + 4}{x + 2}$

13. Chemistry The Ideal Gas Law states that the pressure P, volume V, and temperature T, of an ideal gas are related by the equation $PV = nRT$, where n is the number of moles of gas and R is a constant.

a. Sketch a graph of P versus V, assuming that T is fixed.   **See margin.**

b. What are the asymptotes of the graph?   **b. $P = 0$, $V = 0$**

c. What happens to the pressure of the gas if the temperature is held fixed and the gas is allowed to occupy a larger and larger volume?   **The pressure approaches 0.**

EXERCISES

Practice   Determine the equations of the vertical and horizontal asymptotes, if any, of each function.

14. $f(x) = \dfrac{2x}{x + 4}$ $x = -4$, $y = 2$

15. $f(x) = \dfrac{x^2}{x + 6}$ $x = -6$

16. $g(x) = \dfrac{x - 1}{(2x + 1)(x - 5)}$

16. $x = -\dfrac{1}{2}$, $x = 5$, $y = 0$

17. $g(x) = \dfrac{x - 2}{x^2 + 4x + 3}$ $x = -1$, $x = -3$, $y = 0$

18. $h(x) = \dfrac{x^2}{x^2 + 1}$ $y = 1$

19. $h(x) = \dfrac{(x + 1)^2}{x^2 - 1}$ $x = 1$, $y = 1$

B   **20.** What are the vertical and horizontal asymptotes of the function $y = \dfrac{x^3}{(x - 2)^4}$? $x = 2$, $y = 0$

Each graph below shows a transformation of $f(x) = \dfrac{1}{x}$. Write an equation of each function.

21. $f(x) = \dfrac{1}{x + 3} + 1$

22. $f(x) = \dfrac{1}{x - 2} - 3$

23. $f(x) = -\dfrac{1}{x} + 1$

 www.amc.glencoe.com/self_check_quiz

Additional Answers (p. 187)

24. The parent graph is translated 3 units up. The vertical asymptote, $x = 0$, is unchanged. The horizontal asymptote is now $y = 3$.

25. The parent graph is translated 4 units right and expanded vertically by a factor of 2. The vertical asymptote is now $x = 4$. The horizontal asymptote, $y = 0$, is unchanged.

Use the parent graph $f(x) = \dfrac{1}{x}$ to graph each equation. Describe the transformation(s) that have taken place. Identify the new locations of the asymptotes. **24–29. See margin.**

24. $y = \dfrac{1}{x} + 3$

25. $y = \dfrac{2}{x-4}$

26. $y = \dfrac{2}{x+3} - 1$

27. $y = -\dfrac{3}{x} + 2$

28. $y = \dfrac{3x+1}{x-3}$

29. $y = \dfrac{-4x+2}{x+5}$

Determine the slant asymptote of each function.

30. $f(x) = \dfrac{x^2 + 3x - 3}{x+4}$ $y = x - 1$

31. $f(x) = \dfrac{x^2 + 3x - 4}{x}$ $y = x + 3$

32. $f(x) = \dfrac{x^3 - 2x^2 + x - 4}{x^2 + 1}$ $y = x - 2$

33. $f(x) = \dfrac{x^2 - 4x + 1}{2x - 3}$ $y = \dfrac{1}{2}x - \dfrac{5}{4}$

34. Does the function $f(x) = \dfrac{x^3 - 4x^2 + 2x - 6}{x+3}$ have a slant asymptote? If so, find an equation of the slant asymptote. If not, explain.   **No; the degree of the numerator is 2 more than that of the denominator.**

Graph each function.   **35–40. See Answer Appendix.**

35. $y = \dfrac{(x-2)(x+1)}{x}$

36. $y = \dfrac{x^2 - 4}{x - 2}$

37. $y = \dfrac{x+2}{x^2 - 4}$

38. $y = \dfrac{(x-2)^2(x+1)^2}{(x-2)(x-1)}$

39. $y = \dfrac{x^2 - 6x + 9}{x^2 - x - 6}$

40. $y = \dfrac{x^2 - 1}{x^2 - 2x + 1}$

Applications and Problem Solving

41. Chemistry Suppose the chemist in the application at the beginning of the lesson had to dilute 40 liters of a 12-molar solution by adding 3-molar solution.

   a. Write the function that models the concentration of the mixture as a function of the number of liters t of 3-molar solution added. $C(t) = \dfrac{480 + 3t}{40 + t}$

   b. How many liters of 3-molar solution must be added to create a 10-molar solution?   **11.43L**

42. Electronics Suppose the current I in an electric circuit is given by the formula $I = t + \dfrac{1}{10 - t}$, where t is time. What happens to the circuit as t approaches 10?

42. Sample answer: The circuit melts or one of the components burns up.

43. Critical Thinking Write an equation of a rational function whose graph has all of the following characteristics.   **See margin.**
- x-intercepts at $x = 2$ and $x = -3$;
- a vertical asymptote at $x = 4$; and
- point discontinuity at $(-5, 0)$.

44. Geometry The volume of a rectangular prism with a square base is fixed at 120 cubic feet.

   a. Write the surface area of the prism as a function $A(x)$ of the length of the side of the square x. $A(x) = \dfrac{480}{x} + 2x^2$

   b. Graph the surface area function.   **See Answer Appendix.**

   c. What happens to the surface area of the prism as the length of the side of the square approaches 0?   **The surface area approaches infinity.**

45. Critical Thinking The graph of a rational function cannot intersect a vertical asymptote, but it is possible for the graph to intersect its horizontal asymptote for small values of x. Give an example of such a rational function. (*Hint:* Let the x-axis be the horizontal asymptote of the function.)

   Sample answer: $f(x) = \dfrac{x}{x^2 + 1}$

---

26. The parent graph is translated 3 units left. The translated graph is then expanded vertically by a factor of 2 and translated 1 unit down. The vertical asymptote is now $x = -3$ and the horizontal asymptote is now $y = -1$.

---

27. The parent graph is expanded vertically by a factor of 3, reflected about the x-axis, and translated 2 units up. The vertical asymptote, $x = 0$, is unchanged. The horizontal asymptote is now $y = 2$.

28. The parent graph is translated 3 units right. The translated graph is expanded vertically by a factor of 10 and then translated 3 units up. The vertical asymptote is $x = 3$ and the horizontal asymptote is $y = 3$.

29. The parent graph is translated 5 units left. The translated graph is expanded vertically by a factor of 22 and then translated 4 units down. The vertical asymptote is $x = -5$ and the horizontal asymptote is $y = -4$.

43. Sample answer:
$$f(x) = \dfrac{(x-2)(x+3)(x+5)^2}{(x-4)(x+5)}$$

Open-Ended Assessment
Speaking Present students with an example of a rational function. Ask them to explain how to predict the equations of its asymptotes.

Additional Answers
46a. a vertical asymptote at $r = 0$ and a horizontal asymptote at $F = 0$

46b. The force of repulsion increases without bound as the charges are moved closer and closer together. The force of repulsion approaches 0 as the charges are moved farther and farther apart.

46. **Physics** Like charges repel and unlike charges attract. Coulomb's Law states that the force F of attraction or repulsion between two charges, q_1 and q_2, is given by $F = \frac{kq_1q_2}{r^2}$, where k is a constant and r is the distance between the charges. Suppose you were to graph F as a function of r for two positive charges.   **a–b. See margin.**

   a. What asymptotes would the graph have?

   b. Interpret the meaning of the asymptotes in terms of the problem.

47. **Analytic Geometry** Recall from Exercise 36 of Lesson 3-5 that a *secant* is a line that intersects the graph of a function in two or more points. Consider the function $f(x) = x^2$.

   47a. $\dfrac{a^2 - 9}{a - 3}$

   a. Find an expression for the slope of the secant through the points at $(3, 9)$ and (a, a^2).

   b. What happens to the slope of the secant as a approaches 3?
   The slope approaches 6.

Mixed Review

48. Find and classify the extrema of the function $f(x) = -x^2 + 4x - 3$. *(Lesson 3-6)* **abs. max.: (2, 1)**

49. Find the inverse of $x^2 - 9 = y$. *(Lesson 3-4)* $y = \pm\sqrt{x + 9}$

50. Find the maximum and minimum values of $f(x, y) = y - x$ defined for the polygonal convex set having vertices at $(0, 0)$, $(4, 0)$, $(3, 5)$, and $(0, 5)$. *(Lesson 2-6)* **5, −4**

51. Find $-4\begin{bmatrix} -6 & 5 \\ 8 & -4 \end{bmatrix}$. *(Lesson 2-3)* $\begin{bmatrix} 24 & -20 \\ -32 & 16 \end{bmatrix}$

52. **Consumer Awareness** Bill and Liz are going on a vacation in Jamaica. Bill bought 8 rolls of film and 2 bottles of sunscreen for $35.10. The next day, Liz paid $14.30 for three rolls of film and one bottle of sunscreen. If the price of each bottle of sunscreen is the same and the price of each roll of film is the same, what is the price of a roll of film and a bottle of sunscreen? *(Lesson 2-1)* **$3.25, $4.55**

53. Of $(0, 0)$, $(3, 2)$, $(-4, 2)$, or $(-2, 4)$, which satisfy $x + y \geq 3$? *(Lesson 1-8)* **(3, 2)**

54. Write the equation $15y - x = 1$ in slope-intercept form. *(Lesson 1-4)* $y = \dfrac{1}{15}x + \dfrac{1}{15}$

55. Find $[f \circ g](x)$ and $[g \circ f](x)$ for $f(x) = 8x$ and $g(x) = 2 - x^2$. *(Lesson 1-2)* $16 - 8x^2, 2 - 64x^2$

56. **SAT/ACT Practice** Nine playing cards from the same deck are placed to form a large rectangle whose area is 180 square inches. There is no space between the cards and no overlap. What is the perimeter of this rectangle? **B**

   A 29 in.   **B** 58 in.   **C** 64 in.

   D 116 in.   **E** 210 in.

188   *Chapter 3   The Nature of Graphs*

Extra Practice See p. A31.

Extra Credit

Determine the horizontal asymptotes of $y = \dfrac{x}{x - 4}$, $y = \dfrac{x^2}{x^2 - 4}$, and $y = \dfrac{x^3}{x^3 - 4}$.
Then make a conjecture about the horizontal asymptotes of $y = \dfrac{x^n}{x^n - 4}$, for any positive integer n.   **Each will have a horizontal asymptote of $y = 1$.**

Direct, Inverse, and Joint Variation

OBJECTIVE
• Solve problems involving direct, inverse, and joint variation.

AUTO SAFETY The faster you drive, the more time and distance you need to stop safely, and the less time you have to react. The stopping distance for a vehicle is the sum of the reaction distance d_1, how far the car travels before you hit the brakes, and the braking distance d_2, how far the car travels once the brakes are applied. According to the National Highway Traffic Safety Administration (NHTSA), under the best conditions the reaction time t of most drivers is about 1.5 seconds. Once brakes are applied, the braking distance varies directly as the square of the car's speed s. *A problem related to this will be solved in Example 2.*

The relationship between braking distance and car speed is an example of a **direct variation.** As the speed of the car increases, the braking distance also increases at a constant rate.

A direct variation can be described by the equation $y = kx^n$. The k in this equation is called the **constant of variation.** To express a direct variation, we say that y varies directly as x^n.

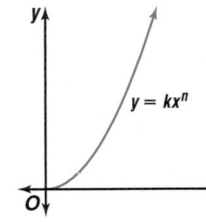

$y = kx^n$

Direct Variation	y varies directly as x^n if there is some nonzero constant k such that $y = kx^n$, $n > 0$. k is called the *constant of variation*.

Notice when $n = 1$, the equation of direct variation simplifies to $y = kx$. This is the equation of a line through the origin written in slope-intercept form. In this special case, the constant of variation is the slope of the line.

To find the constant of variation, substitute known values of x^n and y in the equation $y = kx^n$ and solve for k.

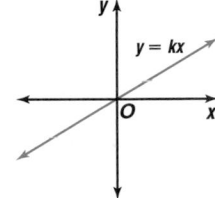

$y = kx$

Example **1** Suppose y varies directly as x and $y = 27$ when $x = 6$.

a. Find the constant of variation and write an equation of the form $y = kx^n$.

b. Use the equation to find the value of y when $x = 10$.

a. In this case, the power of x is 1, so the direct variation equation is $y = kx$.

$y = kx$
$27 = k(6)$ $y = 27, x = 6$
$\frac{27}{6} = k$ *Divide each side by 6.*
$4.5 = k$

The constant of variation is 4.5. The equation relating x and y is $y = 4.5x$.

Lesson 3-8 Direct, Inverse, and Joint Variation **189**

 Resource Manager

📁 **Reproducible Masters**
• *Study Guide and Practice*, pp. 53–54
• *Enrichment*, p. 32
• *Assessment and Evaluation*, p. 70
• *Spreadsheet Applications*, p. 3

📺 **Transparencies**
• 5-Minute Check, 3-8
• Teaching, 3-8

⏱ **5-Minute Check**
Lesson 3-7

1. Determine the equations of the horizontal, vertical, or slant asymptotes in the graph of $y = \frac{(x-2)(x+3)}{x(x+2)}$.
$x = -2$, $x = 0$, $y = 1$

2. Use the parent graph $f(x) = \frac{1}{x}$ to graph $h(x) = \frac{5}{x-3} - 2$. Describe the transformations that have taken place. Identify the new locations of the asymptotes.
The parent graph is translated 3 units right, expanded vertically by a factor of 5, and translated 2 units down. The vertical asymptote is now $x = 3$ and the horizontal asymptote is now $y = -2$.

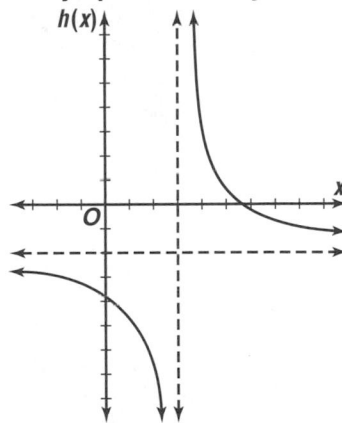

$h(x)$

3. Graph $y = \frac{(x-3)(x+1)}{x(x+1)}$.

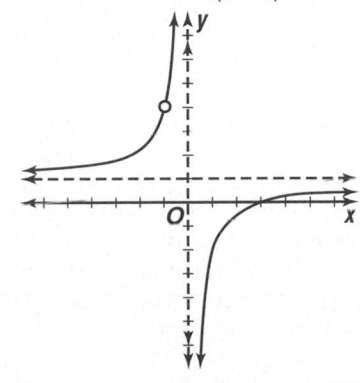

Real-World Application Inflate a balloon to demonstrate Boyle's Law: *If the temperature of a gas remains constant, the pressure P of the gas varies inversely as the volume V of the gas, or PV = k.* If $k = 130$, use a graphing calculator to demonstrate the graph of $P = \frac{k}{V}$.

2 TEACH

In-Class Examples

Example 1
Suppose y varies directly as x and $y = 45$ when $x = 2.5$.
a. Find the constant of variation and write an equation of the form $y = kx^n$.   **18; $y = 18x$**
b. Use the equation to find the value of y when $x = 4$.   **72**

Example 2
Safety When an object such as a car is accelerating, twice the distance d it travels varies directly with the square of the time t elapsed. One car accelerating for 4 minutes travels 1440 feet.
a. Write an equation of direct variation relating travel distance to time elapsed. Then graph the equation.
   $2d = 180t^2$

b. Use the equation to find the distance traveled by the car in 8 minutes.   **5760 ft**

Teaching Tip While working Example 2, explain that although speed and distance are expressed in different units, they do not need to be converted to the same unit, because the constant of variation summarizes the calculations needed for the conversion relation. In contrast to multiplication or addition problems, the answer in this type of problem uses both units in relation to one another.

---

b. $y = 4.5x$
   $y = 4.5(10)$ *$x = 10$*
   $y = 45$

When $x = 10$, the value of y is 45.

Direct variation equations are used frequently to solve real-world problems.

Example **2** **AUTO SAFETY**   Refer to the application at the beginning of this lesson. The NHTSA reports an average braking distance of 227 feet for a car traveling 60 miles per hour, with a total stopping distance of 359 feet.

a. Write an equation of direct variation relating braking distance to car speed. Then graph the equation.

b. Use the equation to find the braking distance required for a car traveling 70 miles per hour.

c. Calculate the total stopping distance required for a car traveling 70 miles per hour.

a. First, translate the statement of variation into an equation of the form $y = kx^n$.

The braking distance varies directly as the square of the car's speed.

$$d_2 \qquad = \qquad k \qquad \cdot \qquad s^2$$

Then, substitute corresponding values for braking distance and speed in the equation and solve for k.

$d_2 = ks^2$
$227 = k(60)^2$ *$d_2 = 227, s = 60$*
$\dfrac{227}{(60)^2} = k$ *Divide each side by $(60)^2$.*
$0.063 \approx k$

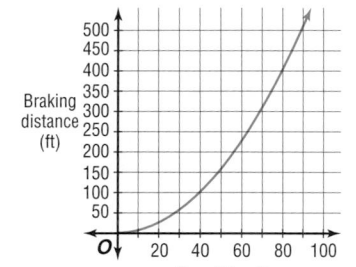

The constant of variation is approximately 0.063. Thus, the equation relating braking distance to car speed is $d_2 = 0.063s^2$. Notice the graph of the variation equation is a parabola centered at the origin and opening up.

b. Evaluate the equation of variation for $s = 70$.

$d_2 = 0.063s^2$
$d_2 = 0.063(70)^2$ *$s = 70$*
$d_2 = 308.7$ or about 309

The braking distance for a car traveling 70 mph is about 309 feet.

c. To find the total stopping distance, first calculate the reaction distance given a reaction time of 1.5 seconds and a speed of 70 miles per hour. Since time is in seconds, we need to change miles per hour to feet per second.

$$\frac{70 \text{ mi}}{1 \text{ h}} \cdot \frac{1 \text{ h}}{60 \text{ min}} \cdot \frac{1 \text{ min}}{60 \text{ s}} \cdot \frac{5280 \text{ ft}}{1 \text{ mi}} \approx 103 \text{ ft/s}$$

$d_1 = st$ *reaction distance = rate of speed · time*

$d_1 = (103)(1.5)$ or about 155 feet   *$s \approx 103$ ft/s; $t = 1.5$ s*

The total stopping distance is then the sum of the reaction and braking distances, $155 + 309$ or 464 feet.

If you know that y varies directly as x^n and one set of values, you can use a proportion to find the other set of corresponding values.

Suppose $y_1 = kx_1{}^n$ and $y_2 = kx_2{}^n$. Solve each equation for k.

$$\frac{y_1}{x_1} = k \text{ and } \frac{y_2}{x_2} = k \quad \Rightarrow \quad \frac{y_1}{x_1} = \frac{y_2}{x_2} \quad \begin{array}{l}\textit{Since both ratios equal } k, \\ \textit{the ratios are equal.}\end{array}$$

Using the properties of proportions, you can find many other proportions that relate these same x^n- and y-values. *You will derive another of these proportions in Exercise 2.*

Example ③ **If y varies directly as the cube of x and $y = -67.5$ when $x = 3$, find x when $y = -540$.**

Use a proportion that relates the values.

$$\frac{y_1}{x_1} = \frac{y_2}{x_2}$$

$$\frac{-67.5}{(3)^3} = \frac{-540}{(x_2)^3} \qquad \textit{Substitute the known values.}$$

$$-67.5(x_2)^3 = -540(3)^3 \qquad \textit{Cross multiply.}$$

$$(x_2)^3 = 216$$

$$x_2 = \sqrt[3]{216} \text{ or } 6 \qquad \textit{Take the cube root of each side.}$$

When $y = -540$, the value of x is 6. *This is a reasonable answer for x, since as y decreased, the value of x increased.*

Many quantities are **inversely proportional** or are said to *vary inversely* with each other. This means that as one value increases the other decreases and vice versa. For example, elevation and air temperature vary inversely with each other. When you travel to a higher elevation above Earth's surface, the air temperature decreases.

Inverse Variation	y varies inversely as x^n if there is some nonzero constant k such that $x^n y = k$ or $y = \dfrac{k}{x^n}$, $n > 0$.

Teaching Tip In Example 3 in the Student Edition, students may need to write out $-67.5 = k(3)^3$ and $-540 = kx^3$ before setting up the proportion by solving both equations for k and setting them as equal.

In-Class Example

Example 3
If y varies directly as the square of x and $y = 30$ when $x = 4$, find x when $y = 270$. **12**

Teaching Tip For problems such as Example 3, students may find it helpful to write out each of the $y = a$ when $x = b$ statements when grouping the numbers.

Suppose y varies inversely as x such that $xy = 4$ or $y = \frac{4}{x}$. The graph of this equation is shown at the right. Notice that in this case, k is a positive value, 4, so as the value of x increases, the value of y decreases.

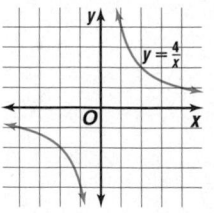

Just as with direct variation, a proportion can be used with inverse variation to solve problems where some quantities are known. The proportion shown below is only one of several that can be formed.

$$x_1{}^n y_1 = k \text{ and } x_2{}^n y_2 = k$$
$$x_1{}^n y_1 = x_2{}^n y_2 \qquad \textit{Substitution property of equality}$$
$$\frac{x_1{}^n}{y_2} = \frac{x_2{}^n}{y_1} \qquad \textit{Divide each side by } y_1 y_2.$$

Other possible proportions include $\frac{y_1}{y_2} = \frac{x_2{}^n}{x_1{}^n}$, $\frac{y_2}{x_1{}^n} = \frac{y_1}{x_2{}^n}$, and $\frac{x_2{}^n}{x_1{}^n} = \frac{y_1}{y_2}$.

Example **4** If y varies inversely as x and $y = 21$ when $x = 15$, find x when $y = 12$.

Use a proportion that relates the values.

$$\frac{x_1}{y_2} = \frac{x_2}{y_1} \qquad n = 1$$
$$\frac{15}{12} = \frac{x_2}{21} \qquad \textit{Substitute the known values.}$$
$$12x_2 = 315 \qquad \textit{Cross multiply.}$$
$$x_2 = \frac{315}{12} \text{ or } 26.25 \quad \textit{Divide each side by 15.}$$

When $y = 12$, the value of x is 26.25.

Another type of variation is **joint variation.** This type of variation occurs when one quantity varies directly as the product of two or more other quantities.

Joint Variation y varies jointly as x^n and z^n if there is some nonzero constant k such that $y = kx^n z^n$, where $x \neq 0$, $z \neq 0$, and $n > 0$.

Example **5** **GEOMETRY** The volume V of a cone varies jointly as the height h and the square of the radius r of the base. Find the equation for the volume of a cone with height 6 centimeters and base diameter 10 centimeters that has a volume of 50π cubic centimeters.

Read the problem and use the known values of V, h, and r to find the equation of joint variation.

Addressing Individual Needs

Verbal/Linguistic   Have students choose one of the Practice exercises to work. As they solve the exercise, have them provide a written explanation for each step. Ask them to make it detailed and clear enough that an absent student could follow the process. Then, have students exchange explanations and check the details.

The volume varies jointly as the height and the square of the radius.

$$V = k \cdot h \cdot r^2$$

$V = khr^2$

$50\pi = k(6)(5)^2$ *V = 50π, h = 6, r = 5*

$50\pi = 150k$

$\dfrac{\pi}{3} = k$ *Solve for k.*

The equation for the volume of a cone is $V = \dfrac{\pi}{3}hr^2$.

CHECK FOR UNDERSTANDING

Communicating Mathematics

Read and study the lesson to answer each question.

1. **Describe** each graph below as illustrating a direct variation, inverse variation, or neither.

a.

b.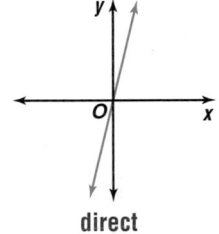

c.   direct

inverse   neither

2. **Derive** a different proportion from that used in Example 3 relating x^n and y if y varies directly as x.   **See margin.**

3. The line does not go through the origin, therefore its equation is not of the form $y = kx^n$.

3. **Explain** why the graph at the right does not represent a direct variation.

4. **Write** statements relating two quantities in real life that exemplify each type of variation.

     a. direct   b. inverse   c. joint
     a–b. See margin for sample answers.

Guided Practice

Find the constant of variation for each relation and use it to write an equation for each statement. Then solve the equation.

5. If y varies inversely as x and $y = 3$ when $x = 4$, find y when $x = 15$.   **12, $xy = 12$; $\dfrac{4}{5}$**

6. If y varies directly as the square of x and $y = -54$ when $x = 9$, find y when $x = 6$. $-\dfrac{2}{3}$; $y = -\dfrac{2}{3}x^2$; -24

7. If y varies jointly as x and the cube of z and $y = 16$ when $x = 4$ and $z = 2$, find y when $x = -8$ and $z = -3$.   **0.5; $y = 0.5xz^3$; 108**

8. 0.4,
$y = 0.4\dfrac{xz}{w^2}$; 2

8. If y varies jointly as x and z and inversely as the square of w, and $y = 3$ when $x = 3$, $z = 10$, and $w = 2$, find y when $x = 4$, $z = 20$, and $w = 4$.

Write a statement of variation relating the variables of each equation. Then name the constant of variation.   **9–11. See margin.**

9. $\dfrac{x^4}{y} = 7$

10. $A = \ell w$

11. $x = \dfrac{-3}{y}$

Lesson 3-8   Direct, Inverse, and Joint Variation   **193**

Additional Answers

25. C varies directly as d; π.

26. y varies directly as x; $\frac{1}{4}$.

27. y varies jointly as x and the square of z; $\frac{4}{3}$.

28. V varies directly as the cube of r; $\frac{4}{3}\pi$.

29. y varies inversely as the square of x; $\frac{5}{4}$.

30. y varies inversely as the square root of x; 2.

31. A varies jointly as h and the quantity $b_1 + b_2$; 0.5.

32. y varies directly as x and inversely as the square of z; $\frac{1}{3}$.

33. y varies directly as x^2 and inversely as the cube of z; 7.

---

12. **Forestry** A lumber company needs to estimate the volume of wood a load of timber will produce. The supervisor knows that the volume of wood in a tree varies jointly as the height h and the square of the tree's girth g. The supervisor observes that a tree 40 meters tall with a girth of 1.5 meters produces 288 cubic meters of wood.

    a. Write an equation that represents this situation. **$V = 3.2hg^2$**

    b. What volume of wood can the supervisor expect to obtain from 50 trees averaging 75 meters in height and 2 meters in girth? **48,000 m³**

EXERCISES

Practice

Find the constant of variation for each relation and use it to write an equation for each statement. Then solve the equation. **13. 0.2; $y = 0.2x$; 1.2**

 13. If y varies directly as x and $y = 0.3$ when $x = 1.5$, find y when $x = 6$.

14. −50;
$xy = -50$; 1.25

14. If y varies inversely as x and $y = -2$ when $x = 25$, find x when $y = -40$.

15. Suppose y varies jointly as x and z and $y = 36$ when $x = 1.2$ and $z = 2$. Find y when $x = 0.4$ and $z = 3$. **15, $y = 15xz$; 18**

16. If y varies inversely as the square of x and $y = 9$ when $x = 2$, find y when $x = 3$. **36; $x^2y = 36$; 4**

                                           17. 16; $r = 16t^2$; 1

17. If r varies directly as the square of t and $r = 4$ when $t = \frac{1}{2}$, find r when $t = \frac{1}{4}$.

 18. Suppose y varies inversely as the square root of x and $x = 1.21$ when $y = 0.44$. Find y when $x = 0.16$. **0.484; $y = 0.484\frac{1}{\sqrt{x}}$; 1.21**

19. If y varies jointly as the cube of x and the square of z and $y = -9$ when $x = -3$ and $z = 2$, find y when $x = -4$ and $z = -3$. **$\frac{1}{12}$; $y = \frac{1}{12}x^3z^2$; −48**

20. If y varies directly as x and inversely as the square of z and $y = \frac{1}{6}$ when $x = 20$ and $z = 6$, find y when $x = 14$ and $z = 5$. **0.3; $y = \frac{0.3x}{z^2}$; 0.168**

21. Suppose y varies jointly as x and z and inversely as w and $y = -3$ when $x = 2$, $z = -3$, and $w = 4$. Find y when $x = 4$, $z = -7$ and $w = -4$. **2; $y = \frac{2xz}{w}$; 14**

22. If y varies inversely as the cube of x and directly as the square of z and $y = -6$ when $x = 3$ and $z = 9$, find y when $x = 6$ and $z = -4$. **−2; $y = \frac{-2z^2}{x^3}$; $-\frac{4}{27}$**

23. If a varies directly as the square of b and inversely as c and $a = 45$ when $b = 6$ and $c = 12$, find b when $a = 96$ and $c = 10$. **15; $a = \frac{15b^2}{c}$; ±8**

24. If y varies inversely as the square of x and $y = 2$ when $x = 4$, find x when $y = 8$. **32; $yx^2 = 32$; ±2**

Write a statement of variation relating the variables of each equation. Then name the constant of variation. **25–33. See margin.**

25. $C = \pi d$ **26.** $\frac{x}{y} = 4$ **27.** $xz^2 = \frac{3}{4}y$

 28. $V = \frac{4}{3}\pi r^3$ **29.** $4x^2 = \frac{5}{y}$ **30.** $y = \frac{2}{\sqrt{x}}$

31. $A = 0.5h(b_1 + b_2)$ **32.** $y = \frac{x}{3z^2}$ **33.** $\frac{1}{7}y = \frac{x^2}{z^3}$

 www.amc.glencoe.com/self_check_quiz

34. Write a statement of variation for the equation $y = \dfrac{kx^3 \cdot z}{w^2}$ if k is the constant of variation. **y varies jointly as the product of the cube of x and z and inversely as the square of w.**

35. **Physics** If you have observed people on a seesaw, you may have noticed that the heavier person must sit closer to the fulcrum for the seesaw to balance. In doing so, the heavier participant creates a rotational force, called *torque*. The torque on the end of a seesaw depends on the mass of the person and his or her distance from the seesaw's fulcrum. In order to reduce torque, one must either reduce the distance between the person and the fulcrum or replace the person with someone having a smaller mass.

$\vdash\!\!-d_1\!-\!\!\vdash\!\!-d_2\!-\!\dashv$

$m_1 \qquad\qquad m_2$

fulcrum

   a. What type of variation describes the relationship between torque, mass, and distance? Explain. **Joint variation; see margin for explanation.**

   b. Let m_1, d_1, and T_1 be the mass, distance, and torque on one side of a seesaw, and let m_2, d_2, and T_2 be the mass, distance, and torque on the other side. Derive an equation that represents this seesaw in balance. **See margin.**

   c. A 75-pound child and a 125-pound babysitter sit at either end of a seesaw. If the child sits 3.3 meters from the fulcrum, use the equation found in part **b** to determine how far the babysitter should sit from the pivot in order to balance the seesaw. **1.98 meters**

36. **Pool Maintenance** Kai decides to empty her pool for the winter. She knows that the time t required to empty a pool varies inversely as the rate r of pumping.

   a. Write an equation that represents this situation. Let k be the constant of variation. **$tr = k$**

   b. In the past, Kai was able to empty her pool in 45 minutes at a rate of 800 liters per minute. She now owns a new pump that can empty the pool at a rate of 1 kiloliter per minute. How long will it take Kai to empty the pool using this new pump? **36 minutes**

37. **Critical Thinking** Show that if y varies directly as x, then x varies directly as y. **See margin.**

38. **Movies** The intensity of light I, measured in *lux*, is inversely proportional to the square of the distance d between the light source and the object illuminated.

   a. Write an equation that represents this situation. **$I = \dfrac{k}{d^2}$**

   b. Using a light meter, a lighting director measures the intensity of the light from a bulb hanging 6 feet overhead a circular table at 16 lux. If the table has a 5-foot diameter, what illumination reading will the director find at the edge of the table where the actors will sit? Round to the nearest tenth. **13.6 lux**

39. **Critical Thinking** If a varies directly as the square of b and inversely as the cube of c, how is the value of a changed when the values of b and c are halved? Explain. **a is doubled; see margin for explanation.**

Lesson 3-8   Direct, Inverse, and Joint Variation   **195**

4 ASSESS

Open-Ended Assessment

Writing Have students find equations or formulas in physics books that represent direct, inverse, and joint variation. Have students write what each variable in the formula represents and how the variables are related.

 Chapter 3, Quiz D (Lessons 3-7 and 3-8) is available in the *Assessment and Evaluation Masters*, p. 70.

Additional Answers

42.

46.

---

40. **Space Science** Newton's Law of Universal Gravitation states that two objects attract one another with a force F that varies directly as the product of their masses, m_1 and m_2, and inversely as the square of the distance d between their centers. It is this force, called gravity, which pulls objects to Earth and keeps Earth in its orbit about the sun.

    a. Write an equation that represents this situation. Let G be the constant of variation. $F = G\dfrac{m_1 \cdot m_2}{d^2}$

    b. Use the chart at the right to determine the value of G if the force of attraction between Earth and the moon is 1.99×10^{20} newtons (N) and the distance between them is 3.84×10^8 meters. Be sure to include appropriate units with your answer. $6.67 \times 10^{-11} \dfrac{N \cdot m^2}{kg^2}$

	Mass
Earth	5.98×10^{24} kg
Sun	1.99×10^{30} kg
Moon	7.36×10^{22} kg

    40c. 3.53×10^{22} N

    c. Using your answers to parts **a** and **b**, find the force of attraction between Earth and the sun if the distance between them is 1.50×10^{11} meters.

    d. How many times greater is the force of attraction between the sun and Earth than between the moon and Earth? **about 178 times greater**

41. **Electricity** Wires used to connect electric devices have very small resistances, allowing them to conduct currently more readily. The resistance R of a piece of wire varies directly as its length L and inversely as its cross-sectional area A. A piece of copper wire 2 meters long and 2 millimeters in diameter has a resistance of 1.07×10^{-2} ohms (Ω). Find the resistance of a second piece of copper wire 3 meters long and 6 millimeters in diameter. (*Hint*: The cross-sectional area will be πr^2.) **$1.78 \times 10^{-3} \Omega$**

Mixed Review

42. Graph $y = \dfrac{x-1}{x^2-9}$. (*Lesson 3-7*) **See margin.**

43. Find the inverse of $f(x) = (x-3)^3 + 6$. Then state whether the inverse is a function. (*Lesson 3-4*) **$f^{-1}(x) = \sqrt[3]{x-6} + 3$; $f^{-1}(x)$ is a function.**

44. Square $ABCD$ has vertices $A(1, 2)$, $B(3, -2)$, $C(-1, -4)$, and $D(-3, 0)$. Find the image of the square after a reflection over the y-axis. (*Lesson 2-4*) **$A'(-1, 2)$, $B'(-3, -2)$, $C'(1, -4)$, $D'(3, 0)$**

45. State whether the system $4x - 2y = 7$ and $-12x + 6y = -21$ is *consistent and independent, consistent and dependent,* or *inconsistent.* (*Lesson 2-1*) **consistent and dependent**

46. Graph $g(x) = \begin{cases} 2 \text{ if } x < 1 \\ -x + 5 \text{ if } x \geq 1 \end{cases}$. (*Lesson 1-7*) **See margin.**

47. **Education** In 1995, 23.2% of the student body at Kennedy High School were seniors. By 2000, seniors made up only 18.6% of the student body. Assuming the level of decline continues at the same rate, write a linear equation in slope-intercept form to describe the percent of seniors y in the student body in year x. (*Lesson 1-4*) **$y = -0.92x + 1858.60$**

48. **SAT/ACT Practice** If $a^2b = 12^2$, and b is an odd integer, then a could be divisible by all of the following EXCEPT **D**

    A 3   B 4   C 6   D 9   E 12

Extra Credit

Complete the table with the type of variation that is shown by each formula.

	Volume	Formula	Type of Variation
Rectangular prism		$V = \ell wh$	joint
Cylinder		$V = \pi r^2 h$	joint
Cube		$V = s^3$	direct
Cone		$V = \frac{1}{3}\pi r^2 h$	joint
Pyramid		$V = \frac{1}{3}Bh$	joint
Sphere		$V = \frac{4}{3}\pi r^3$	joint

VOCABULARY

absolute maximum (p.171)	extremum (p. 171)	minimum (p. 171)
absolute minimum (p. 171)	horizontal asymptote (p. 180)	monotonicity (p. 163)
asymptotes (p. 180)	horizontal line test (p. 153)	odd function (p. 133)
constant function (pp. 137, 164)	image point (p. 127)	parent graph (p. 137)
	increasing function (p. 164)	point discontinuity (p. 159)
constant of variation (p. 189)	infinite discontinuity (p. 159)	point of inflection (p. 171)
continuous (p. 160)	inverse function (pp. 152, 155)	point symmetry (p. 127)
critical point (p. 171)	inverse process (p. 154)	rational function (p. 180)
decreasing function (p. 164)	inversely proportional (p. 191)	relative extremum (p. 172)
direct variation (p. 189)		relative maximum (p. 172)
discontinuous (p. 159)	inverse relations (p. 152)	relative minimum (p. 172)
end behavior (p. 162)	joint variation (p. 192)	slant asymptote (p. 183)
even function (p. 133)	jump discontinuity (p. 159)	symmetry with respect to
everywhere discontinuous (p. 159)	line symmetry (p. 129)	the origin (p. 128)
	maximum (p. 171)	vertical asymptote (p. 180)

UNDERSTANDING AND USING THE VOCABULARY

Choose the correct term to best complete each sentence.

1. An (odd, <u>even</u>) function is symmetric with respect to the y-axis.

2. If you can trace the graph of a function without lifting your pencil, then the graph is (<u>continuous</u>, discontinuous).

3. When there is a value in the domain for which a function is undefined, but the pieces of the graph match up, then the function has (infinite, <u>point</u>) discontinuity.

4. A function f is (<u>decreasing</u>, increasing) on an interval I if and only if for every a and b contained in I, $f(a) > f(b)$ whenever $a < b$.

5. When the graph of a function is increasing to the left of $x = c$ and decreasing to the right of $x = c$, then there is a (<u>maximum</u>, minimum) at c.

6. A (greatest integer, <u>rational</u>) function is a quotient of two polynomial functions.

7. Two relations are (direct, <u>inverse</u>) relations if and only if one relation contains the element (b, a) whenever the other relation contains the element (a, b).

8. A function is said to be (<u>monotonic</u>, symmetric) on an interval I if and only if the function is increasing on I or decreasing on I.

9. A (horizontal, <u>slant</u>) asymptote occurs when the degree of the numerator of a rational expression is exactly one greater than that of the denominator.

10. (Inverse, <u>Joint</u>) variation occurs when one quantity varies directly as the product of two or more other quantities.

interNET CONNECTION For additional review and practice for each lesson, visit: www.amc.glencoe.com

---

STUDY GUIDE AND ASSESSMENT

VOCABULARY

This section provides a listing of the new terms, properties, and phrases that were introduced in this chapter. The exercises check students' understanding of the terms by using a variety of verbal formats including matching, completion, and true/false.

Glossary A complete glossary of terms begins on page A71.

Assessment and Evaluation Masters, pp. 57–58

TEST OPTIONS

Six forms of Chapter 3 Test are available in the *Assessment and Evaluation Masters* as shown in the chart.

Chapter 3 Test, Form 1B, is shown at the right. Chapter 3 Test, Form 2B, is shown on the next page.

Form	Pages	Type	
1A	55–56	Honors	Form 1 tests are multiple choice.
1B	57–58	Average	
1C	59–60	Basic	Form 2 tests are free response.
2A	61–62	Honors	
2B	63–64	Average	
2C	65–66	Basic	

SKILLS AND CONCEPTS

Objectives and Examples This section reviews the skills and concepts of the chapter and shows completely worked examples.

Review Exercises These exercises provide practice for the corresponding objectives.

Assessment and Evaluation Masters, pp. 63–64

198 Chapter 3

SKILLS AND CONCEPTS

OBJECTIVES AND EXAMPLES

REVIEW EXERCISES

Lesson 3-1 Use algebraic tests to determine if the graph of a relation is symmetrical.

- Determine whether the graph of $f(x) = 4x - 1$ is symmetric with respect to the origin.
$$f(-x) = 4(-x) - 1$$
$$= -4x - 1$$
$$-f(x) = -(4x - 1)$$
$$= -4x + 1$$
The graph of $f(x) = 4x - 1$ is not symmetric with respect to the origin because $f(-x) \neq -f(x)$.

Determine whether the graph of each function is symmetric with respect to the origin.

11. $f(x) = -2x$ **yes**   12. $f(x) = x^2 + 2$ **no**
13. $f(x) = x^2 - x + 3$ 14. $f(x) = x^3 - 6x + 1$
    no   **no**

Determine whether the graph of each function is symmetric with respect to the x-axis, y-axis, the line $y = x$, the line $y = -x$, or none of these.

15. $xy = 4$ 16. $x + y^2 = 4$ **x-axis**
17. $x = -2y$ **none**   18. $x^2 = \dfrac{1}{y}$ **y-axis**
15. $y = x$ and $y = -x$

Lesson 3-2 Identify transformations of simple graphs.

- Describe how the graphs of $f(x) = x^2$ and $g(x) = x^2 - 1$ are related.

Since 1 is subtracted from $f(x)$, the parent function, $g(x)$ is the graph of $f(x)$ translated 1 unit down.

Describe how the graphs of **f(x)** and **g(x)** are related.   **19–22. See margin.**

19. $f(x) = x^4$ and $g(x) = x^4 + 5$
20. $f(x) = |x|$ and $g(x) = |x + 2|$
21. $f(x) = x^2$ and $g(x) = 6x^2$
22. $f(x) = [\![x]\!]$ and $g(x) = \left[\!\!\left[\dfrac{3}{4}x\right]\!\!\right] - 4$

Lesson 3-3 Graph polynomial, absolute value, and radical inequalities in two variables.

- Graph $y < x^2 + 1$.

The boundary of the inequality is the graph of $y = x^2 + 1$. Since the boundary is not included, the parabola is dashed.

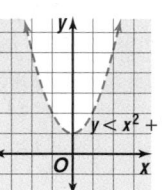

23–26. See Answer Appendix.
Graph each inequality.

23. $y > |x + 2|$ 24. $y \leq -2x^3 + 4$
25. $y < (x + 1)^2 + 2$ 26. $y \geq \sqrt{2x - 3}$

Solve each inequality.

27. $|4x + 5| > 7$ $\{x \mid x < -3 \text{ or } x > 0.5\}$
28. $|x - 3| + 2 \leq 11$ $\{x \mid -6 \leq x \leq 12\}$

Lesson 3-4 Determine inverses of relations and functions.

- Find the inverse of $f(x) = 4(x - 3)^2$.
$$y = 4(x - 3)^2 \quad \textit{Let } y = f(x).$$
$$x = 4(y - 3)^2 \quad \textit{Interchange } x \textit{ and } y.$$
$$\dfrac{x}{4} = (y - 3)^2 \quad \textit{Solve for y.}$$
$$\pm\sqrt{\dfrac{x}{4}} = y - 3$$
$$3 \pm \sqrt{\dfrac{x}{4}} = y \quad \text{So, } f^{-1}(x) = 3 \pm \sqrt{\dfrac{x}{4}}.$$

Graph each function and its inverse.

29. $f(x) = 3x - 1$ 30. $f(x) = -\dfrac{1}{4}x + 5$
31. $f(x) = \dfrac{2}{x} + 3$ 32. $f(x) = (x + 1)^2 - 4$
29–32. See Answer Appendix.
Find $f^{-1}(x)$. Then state whether $f^{-1}(x)$ is a function. $f^{-1}(x) = \sqrt[3]{x + 8} + 2$; yes
33. $f(x) = (x - 2)^3 - 8$
34. $f(x) = 3(x + 7)^4$ $f^{-1}(x) = -7 \pm \sqrt[4]{\dfrac{x}{3}}$; no

---

198   Chapter 3   The Nature of Graphs

TestCheck and Worksheet Builder

This **networkable** software has three modules.
- **Worksheet Builder** to make worksheets and tests
- **Student Module** to take tests on-screen
- **Management System** to keep student records

Three special banks are included for SAT, ACT, and TIMSS.

OBJECTIVES AND EXAMPLES

Lesson 3-5 Determine whether a function is continuous or discontinuous.

• Determine whether the function $y = \dfrac{x}{x+4}$ is continuous at $x = -4$.

Start with the first condition of the continuity test. The function is not defined at $x = -4$ because substituting -4 for x results in a denominator of zero. So the function is discontinuous at $x = -4$.

Lesson 3-5 Identify the end behavior of functions.

• Describe the end behavior of $f(x) = 3x^4$.

Make a chart investigating the value of $f(x)$ for very large and very small values of x.

x	$f(x)$
$-10{,}000$	3×10^{16}
-1000	3×10^{12}
-100	3×10^{8}
0	0
100	3×10^{8}
1000	3×10^{12}
$10{,}000$	3×10^{16}

$y \to \infty$ as $x \to \infty$,
$y \to \infty$ as $x \to -\infty$

Lesson 3-6 Find the extrema of a function.

• Locate the extrema for the graph of $y = f(x)$. Name and classify the extrema of the function.

The function has a relative minimum at $(0, -2)$ and a relative maximum at $(3, 2)$.

REVIEW EXERCISES

Determine whether each function is continuous at the given x-value. Justify your response using the continuity test.   35–37. See margin.

35. $y = x^2 + 2$; $x = 2$

36. $y = \dfrac{x - 3}{x + 1}$; $x = -1$

37. $f(x) = \begin{cases} x + 1 & \text{if } x \le 1 \\ 2x & \text{if } x > 1 \end{cases}$; $x = 1$

Describe the end behavior of each function.

38. $y = 1 - x^3$ **39.** $f(x) = x^9 + x^7 + 4$

40. $y = \dfrac{1}{x^2} + 1$ **41.** $y = 12x^5 + x^3 - 3x^2 + 4$

38–41. See margin.

Determine the interval(s) for which the function is increasing and the interval(s) for which the function is decreasing.

42. $y = -2x^3 - 3x^2 + 12x$

43. $f(x) = |x^2 - 9| + 1$

42. decreasing for $x < -2$ and $x > 1$; increasing for $-2 < x < 1$

43. decreasing for $x < -3$ and $0 < x < 3$; increasing for $-3 < x < 0$ and $x > 3$

Locate the extrema for the graph of $y = f(x)$. Name and classify the extrema of the function.

44.   **45.**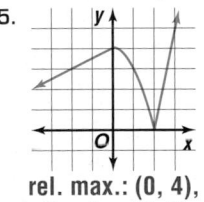

abs. max.: $(-2, 1)$ rel. max.: $(0, 4)$, rel. min.: $(2, 0)$

Determine whether the given critical point is the location of a *maximum*, a *minimum*, or a *point of inflection*.

46. $x^3 - 6x^2 + 9x, x = 3$ **47.** $4x^3 + 7, x = 0$
min.   pt. of inflection

Additional Answers
(pp. 200–201)

50. The parent graph is translated 2 units left and expanded vertically by a factor of 3. The vertical asymptote is now $x = -3$. The horizontal asymptote, $f(x) = 0$, is unchanged.

51. The parent graph is translated 3 units right and then translated 2 units up. The vertical asymptote is now $x = 3$ and the horizontal asymptote is $f(x) = 2$.

57. $140;\ y = \dfrac{140}{\sqrt{x}};\ 196$

58. $320;\ y = \dfrac{320x^2}{z};\ 8$

60a.

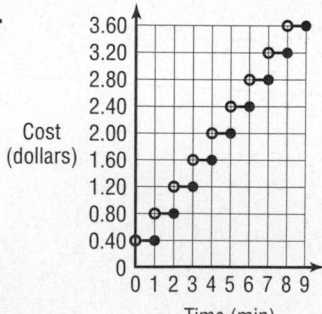

Cost (dollars) vs. Time (min)

60b. positive real numbers; positive multiples of $0.40

60c.

Time (min) vs. Cost (dollars)

Lesson 3-7 Graph rational functions.

The graph at the right shows a transformation of $f(x) = \dfrac{1}{x}$. Write an equation of the function.

The graph of $f(x) = \dfrac{1}{x}$ has been translated 2 units to the right. So, the equation of the function is $f(x) = \dfrac{1}{x - 2}$.

48. $f(x) = \dfrac{1}{x} + 1$

Lesson 3-7 Determine vertical, horizontal, and slant asymptotes.

Determine the equation of the horizontal asymptote of $g(x) = \dfrac{5x - 9}{x - 3}$.

To find the horizontal asymptote, use division to rewrite the rational expression as a quotient.

$$\begin{array}{r} 5 \\ x - 3\overline{)5x - 9} \\ \underline{5x - 15} \\ 4 \end{array}$$

Therefore, $g(x) = 5 + \dfrac{4}{x - 3}$. As $x \to \pm\infty$, the value of $\dfrac{4}{x - 3}$ approaches 0. The value of $5 + \dfrac{4}{x - 3}$ approaches 5. The line $y = 5$ is the horizontal asymptote.

Lesson 3-8 Solve problems involving direct, inverse, and joint variation.

If y varies inversely as the square of x and $y = 8$ when $x = 3$, find x when $y = 6$.

$$\dfrac{x_1{}^n}{y_2} = \dfrac{x_2{}^n}{y_1}$$

$$\dfrac{3^2}{6} = \dfrac{x_2}{8} \qquad n = 2,\ x_1 = 3,\ y_1 = 8,\ y_2 = 6$$

$6x_2{}^2 = 72$ *Cross multiply.*

$x_2{}^2 = 12$ *Divide each side by 6.*

$x_2 = \sqrt{12}$ or $2\sqrt{3}$

Each graph below shows a transformation of $f(x) = \dfrac{1}{x}$. Write an equation of each function.

48.

49.

$$f(x) = -\dfrac{2}{x}$$

Use the parent graph $f(x) = \dfrac{1}{x}$ to graph each equation. Describe the transformation(s) that have taken place. Identify the new locations of the asymptotes.   50–51. See margin.

50. $\dfrac{3}{x + 2}$

51. $\dfrac{2x - 5}{x - 3}$

Determine the equations of the vertical and horizontal asymptotes, if any, of each function.

52. $f(x) = \dfrac{x}{x - 1}$ $x = 1,\ y = 1$

53. $g(x) = \dfrac{x^2 + 1}{x + 2}$ $x = -2$

54. $h(x) = \dfrac{(x - 3)^2}{x^2 - 9}$ $x = -3,\ y = 1$

55. Does the function $f(x) = \dfrac{x^2 + 2x + 1}{x}$ have a slant asymptote? If so, find an equation of the slant asymptote. If not, explain.
yes; $y = x + 2$

Find the constant of variation and use it to write an equation for each statement. Then solve the equation.   57–58. See margin.

56. If y varies jointly as x and z and $y = 5$ when $x = -4$ and $z = -2$, find y when $x = -6$ and $z = -3$.   **0.625; $y = 0.625xz$; 11.25**

57. If y varies inversely as the square root of x and $y = 20$ when $x = 49$, find x when $y = 10$.

58. If y varies directly as the square of x and inversely as z and $y = 7.2$ when $x = 0.3$ and $z = 4$, find y when $x = 1$ and $z = 40$.

Additional Answers (p. 201)

60d. positive multiples of $0.40; positive real numbers

60e. $C^{-1}(x)$ gives the possible number of minutes spent using the scanner that costs x dollars.

61a.

APPLICATIONS AND PROBLEM SOLVING

59. Manufacturing The length of a part for a bicycle must be 6.5 ± 0.2 centimeters. If the actual length of the part is x, write an absolute value inequality to describe this situation. Then find the range of possible lengths for the part. *(Lesson 3-3)*

60. Consumer Costs A certain copy center charges users $0.40 for every minute or part of a minute to use their computer scanner. Suppose that you use their scanner for x minutes, where x is any real number greater than 0. *(Lesson 3-4)*

　　a. Sketch the graph of the function, $C(x)$, that gives the cost of using the scanner for x minutes. **a–e. See margin.**

　　b. What are the domain and range of $C(x)$?

　　c. Sketch the graph of $C^{-1}(x)$.

　　d. What are the domain and range of $C^{-1}(x)$?

　　e. What real-world situation is modeled by $C^{-1}(x)$?

59. $|x - 6.5| \le 0.2$; $6.3 \le x \le 6.7$

61. Sports One of the most spectacular long jumps ever performed was by Bob Beamon of the United States at the 1968 Olympics.

His jump of 8.9027 meters surpassed the world record at that time by over half a meter! The function $h(t) = 4.6t - 4.9t^2$ describes the height of Beamon's jump (in meters) with respect to time (in seconds). *(Lesson 3-6)*

　　a. Draw a graph of this function. **See margin.**

　　b. What was the maximum height of his jump? **1.08 m**

ALTERNATIVE ASSESSMENT

OPEN-ENDED ASSESSMENT

1. Write and then graph an equation that exhibits symmetry with respect to the
   a. x-axis.　　　　b. y-axis.
   c. line $y = x$.　　d. line $y = -x$.
   e. origin.　**See Answer Appendix.**

2. Write the equation of a parent function, other than the identity or constant function, after it has been translated right 4 units, reflected over the x-axis, expanded vertically by a factor of 2, and translated 1 unit up. **See margin.**

3. A graph has one absolute minimum, one relative minimum, and one relative maximum.
   a. Draw the graph of a function for which this is true. **See margin.**
   b. Name and classify the extrema of the function you graphed. **See margin.**

> **Additional Assessment** See page A58 for Chapter 3 practice test.

Unit 1 *interNET* Project

TELECOMMUNICATION

Sorry, You are Out of Range for Your Telephone Service . . .

- Research several cellular phone services to determine their initial start-up fee, equipment fee, the monthly service charge, the charge per minute for calls, and any other charges.

- Compare the costs of the cellular phone services you researched by writing and graphing equations.

Determine which cellular phone service would best suit your needs. Write a paragraph to explain your choice. Use graphs and area maps to support your choice.

PORTFOLIO
　Choose one of the functions you studied in this chapter. Describe the graph of the function and how it can be used to model a real-life situation.

This section provides additional practice in solving real-world problems that involve the concepts of this chapter.

ALTERNATIVE ASSESSMENT

Additional Answers
Open-Ended Assessment

2. Sample answer:
　$-2(x - 4)^2 + 1$

3a. Sample answer:

3b. abs. min.: (2, –3); rel. max.: (0, 0); rel. min.: (–2, –1)

Unit 1
***interNET* Project**

● **Sorry, You Are Out of Range for Your Telephone Service ...**
Remind students to narrow their searches to Web sites that describe cellular phone service in their state or area to find maps of the service available. Make sure that students are aware of the following peculiarities about cellular phone service plans:
- Some companies charge by a local rate for all calls in the area.
- Some companies charge roaming fees for calls in certain areas.
- Some services may not even cover all of the state.

Internet Address Book

Record useful Internet addresses in the space at right for quick reference.

www.amc.glencoe.com

Chapter 3 Skill: More Algebra

Pages 202–203 are part of a complete SAT/ACT preparation course that is described in detail on pages T8–T9 of the *Teacher's Handbook*. The test items were written in the same style as those in the SAT and ACT.

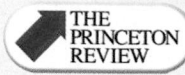 These questions were aligned and verified by The Princeton Review, the nation's leader in test preparation.

Diagnosis & Prescription

Each of the 10 SAT and ACT practice problems on page 203 is cross-referenced to the chapter where that SAT or ACT skill is covered. If students miss a particular type of problem, you can have them study that skill.

Problem Number and Skill		Page
1	More Algebra	202
2	Algebra	124
3	Arithmetic	64
4	More Algebra	202
5	Arithmetic	64
6	More Algebra	202
7	Arithmetic	64
8	More Algebra	202
9	More Algebra	202
10	More Algebra	202

More Algebra Problems

SAT and ACT tests include quadratic expressions and equations. You should be familiar with common factoring formulas, like the difference of two squares or perfect square trinomials.

$$a^2 - b^2 = (a + b)(a - b)$$
$$a^2 + 2ab + b^2 = (a + b)^2$$
$$a^2 - 2ab + b^2 = (a - b)^2$$

Some problems involve systems of equations. Simplify the equations if possible, and then add or subtract them to eliminate one of the variables.

THE PRINCETON REVIEW

TEST-TAKING TIP

If a problem seems to require lengthy calculations, look for a shortcut. There is probably a quicker way to solve it. Try to eliminate fractions and decimals. Try factoring.

ACT EXAMPLE

1. If $\dfrac{x^2 - 9}{x + 3} = 12$, then $x = ?$

A 10   B 15   C 17   D 19   E 20

HINT Look for factorable quadratics.

Solution Factor the numerator and simplify.
$$\frac{(x + 3)(x - 3)}{x + 3} = 12$$
$$x - 3 = 12 \quad \frac{x + 3}{x + 3} = 1$$
$$x = 15 \quad \textit{Add 3 to each side.}$$

The answer is choice **B**.

Alternate Solution You can also solve this type of problem, with a variable in the question and numbers in the answer choices, with a strategy called "backsolving."

Substitute each answer choice for the variable into the given expression or equation and find which one makes the statement true.

Try choice A. For $x = 10$, $\dfrac{10^2 - 9}{10 + 3} \neq 12$.

Try choice B. For $x = 15$, $\dfrac{15^2 - 9}{15 + 3} = 12$.

Therefore, choice **B** is correct.

If the number choices were large, then calculations, even using a calculator, would probably take longer than solving the problem using algebra. In this case, it is *not* a good idea to use the backsolving strategy.

SAT EXAMPLE

2. If $x = y + 1$ and $y \geq 1$, then which of the following must be equal to $x^2 - y^2$?

A $(x - y)^2$ B $x^2 - y - 1$ C $x + y$

D $x^2 - 1$ E $y^2 + 1$

HINT This difficult problem has variables in the answers. It can be solved by using algebra or the "Plug-in" strategy.

Solution Notice the word *must*. This means the relationship is true for all possible values of x and y.

To use the Plug-In strategy, choose a number greater than 1 for y, say 4. Then x must be 5. Since $x^2 - y^2 = 25 - 16$ or 9, check each expression choice to see if it is equal to 9 when $x = 5$ and $y = 4$.

Choice A: $(x - y)^2 = 1$
Choice B: $x^2 - y - 1 = 20$
Choice C: $x + y = 9$

Choice **C** is correct.

Alternate Solution You can also use algebraic substitution to find the answer. Recall that $x^2 - y^2 = (x + y)(x - y)$. The given equation, $x = y + 1$, includes y. Substitute $y + 1$ for x in the second term.

$$\begin{aligned} x^2 - y^2 &= (x + y)(x - y) \\ &= (x + y)[(y + 1) - y] \\ &= (x + y)(1) \\ &= x + y \end{aligned}$$

This is choice **C**.

 TestCheck and Worksheet Builder

Special banks of SAT, ACT, and TIMSS items are included in this CD-ROM.

After you work each problem, record your answer on the answer sheet provided or on a piece of paper.

Multiple Choice

1. For all $y \neq 3$, $\dfrac{y^2 - 9}{3y - 9} = ?$ **E**

   A y

   B $\dfrac{y + 1}{8}$

   C $y + 1$

   D $\dfrac{y}{3}$

   E $\dfrac{y + 3}{3}$

2. If $x + y = z$ and $x = y$, then all of the following are true EXCEPT   **E**

   A $2x + 2y = 2z$

   B $x - y = 0$

   C $x - z = y - z$

   D $x = \dfrac{z}{2}$

   E $z - y = 2x$

3. The Kims drove 450 miles in each direction to Grandmother's house and back again. If their car gets 25 miles per gallon and their cost for gasoline was \$1.25 per gallon for the trip to Grandmother's house, but \$1.50 per gallon for the return trip, how much *more* money did they spend for gasoline returning from Grandmother's house than they spent going to Grandmother's?   **B**

   A \$2.25

   B \$4.50

   C \$6.25

   D \$9.00

   E \$27.00

4. If $x + 2y = 8$ and $\dfrac{x}{2} - y = 10$, then $x = ?$ **D**

   A -7

   B 0

   C 10

   D 14

   E 28

5. $\dfrac{900}{10} + \dfrac{90}{100} + \dfrac{9}{1000} = $ **C**

   A 90.09

   B 90.099

   C 90.909

   D 99.09

   E 999

6. For all x, $(10x^4 - x^2 + 2x - 8) - (3x^4 + 3x^3 + 2x + 9) = ?$ **A**

   A $7x^4 - 3x^3 - x^2 - 17$

   B $7x^4 - 4x^2 - 17$

   C $7x^4 + 3x^3 - x^2 + 4x$

   D $7x^4 + 2x^2 + 4x$

   E $13x^4 - 3x^3 + x^2 + 4x$

7. If $\dfrac{n}{8}$ has a remainder of 5, then which of the following has a remainder of 7?   **B**

   A $\dfrac{n + 1}{8}$

   B $\dfrac{n + 2}{8}$

   C $\dfrac{n + 3}{8}$

   D $\dfrac{n + 5}{8}$

   E $\dfrac{n + 7}{8}$

8. If $x > 0$, then $\dfrac{\sqrt{100x^2 + 600x + 900}}{x + 3} = ?$ **B**

   A 9 B 10 C 30 D 40

   E It cannot be determined from the information given.

9. What is the value of c?   **B**

   Given: $a + b = c$
   $\quad\quad\;\, a - c = 5$
   $\quad\quad\;\, b - c = 3$

   A -10
   B -8
   C -5
   D -3
   E 3

10. **Grid-In** If $4x + 2y = 24$ and $\dfrac{7y}{2x} = 7$, then $x = ?$ **3**

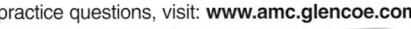 **inter NET**
CONNECTION **SAT/ACT Practice** For additional test practice questions, visit: **www.amc.glencoe.com**

THE **PRINCETON REVIEW** *SAT & ACT Preparation*   **203**

Extra Credit

TIMSS If $xy = 1$ and x is greater than 0, which of the following statements is true?   **E**
   A   When x is greater than 1, y is negative.
   B   When x is greater than 1, y is greater than 1.
   C   When x is less than 1, y is less than 1.
   D   As x increases, y increases.
   E   As x increases, y decreases.

Instructional Objectives

Lesson (pages)	Lesson Objectives	NCTM Standards 2000	State/Local Objectives
4-1 (205–212)	• Determine roots of polynomial equations. • Apply the Fundamental Theorem of Algebra.	2, 6	
4-2 (213–221)	• Solve quadratic equations. • Use the discriminant to describe the roots of quadratic equations.	2, 6	
4-3 (222–228)	• Find the factors of polynomials using the Remainder and Factor Theorems.	2, 6	
4-4 (229–235)	• Identify all possible rational roots of a polynomial equation by using the Rational Root Theorem. • Determine the number of positive and negative real roots a polynomial function has.	2, 3, 6, 10	
4-5 (236–242)	• Approximate the real zeros of a polynomial function.	2, 3, 6, 10	
4-6 (243–250)	• Solve rational equations and inequalities. • Decompose a fraction into partial fractions.	2, 6	
4-7 (251–257)	• Solve radical equations and inequalities.	2, 6	
4-8 (258–264)	• Write polynomial functions to model real-world data. • Use polynomial functions to interpret real-world data.	2, 3, 4, 6, 10	
4-8B (265–266)	• Use matrices to find a polynomial function whose graph passes through a given set of points in the coordinate plane.	1, 2, 6, 10	

Suggested Pacing *See pages T2–T4 for additional options in course-planning.*

COURSE	Lesson 4-1	Lesson 4-2	Lesson 4-3	Lesson 4-4	Lesson 4-5
Precalculus Emphasis Chapters 4–15 (45–50 minutes)	1.5 days	1.5 days	1.5 days	1.5 days	1.5 days
Precalculus Emphasis (block)	1.5 days		1.5 days		(with Lesson 4-6) 1.5 days
Trigonometric Emphasis Chapters 1–9, 11–12 (45–50 minutes)	1.5 days	1.5 days	1.5 days	1.5 days	1.5 days
Trigonometric Emphasis (block)	0.5 day	1.5 days		1.5 days	

Instructional Resources

Student Edition		Blackline Masters (page numbers)								
Lesson	Graphing Calculator Applications	Study Guide and Practice	Enrichment	Assessment and Evaluation	Graphing Calculator Keystrokes	Problem Solving & Applications	Spreadsheet Applications	TI-92 and CBL Interdisciplinary Activities	Transparencies A and B
4-1	206, 208, 211	59–60	37			7		9–12	4-1
4-2	213, 214, 216, 218	61–62	38	95					4-2
4-3	225, 226	63–64	39						4-3
4-4	230, 232, 233	65–66	40	94, 95	13				4-4
4-5	237–240	67–68	41				4		4-5
4-6		69–70	42	96					4-6
4-7	253	71–72	43			8			4-7
4-8	259, 262	73–74	44	96					4-8
4-8B	265–266				14–16				
Study Guide/ Assessment				81–93, 99					
SAT/ACT Practice				97–98					

Lesson 4-6	Lesson 4-7	Lesson 4-8	Study Guide & Assessment	Chapter Test	TOTAL
1.5 days	1.5 days	(with Lesson 4-8B) 2.5 days	1 day	1 day	15 days
(See Lesson 4-5)	1.5 days		(with Lesson 4-8B) 1 day	0.5 day	7.5 days
1.5 days	1.5 days	(with Lesson 4-8B) 2.5 days	1 day	1 day	15 days
1.5 days		(with Lesson 4-8B) 1.5 days		0.5 day	7 days

TeacherWorks™

Resource Manager

Applications

Problem Solving and Applications Masters, pp. 7, 8

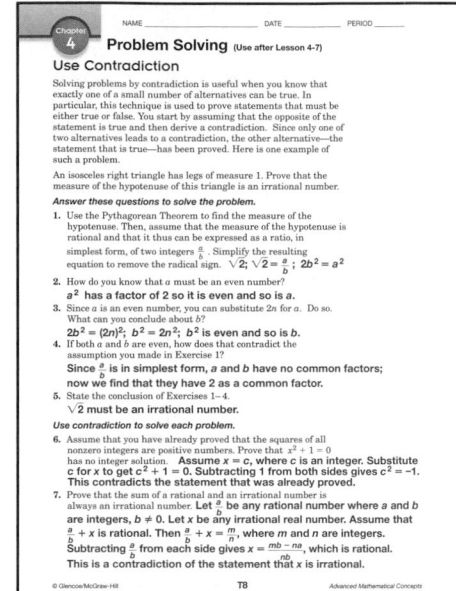

The worksheets shown on this page are a small sample of the materials available on *TeacherWorks: All-in-One Lesson Planner and Resource Center.*

This CD-ROM contains all of the blackline masters and transparencies available for this program.

It also includes a lesson planner and interactive Teacher Edition, so you can customize lesson plans and reproduce classroom resources quickly and easily, from just about anywhere.

Technology/Multimedia

Spreadsheet Applications Masters, p. 4

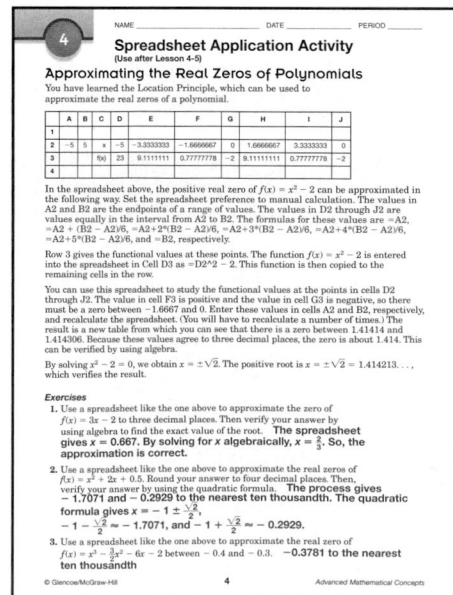

TI-92 & CBL Interdisciplinary Activities Masters, pp. 9–12

Assessment Resources

Type	Student Edition	Teacher's Wraparound Edition	Assessment and Evaluation Masters
Ongoing Assessment	Mid-Chapter Quiz, p. 235	5-Minute Check, pp. 205, 213, 222, 229, 236, 243, 251, 258	Mid-Chapter Test, p. 94 Quizzes A–D, pp. 95–96
Mixed Review	Mixed Review, pp. 212, 221, 228, 235, 242, 250, 257, 264		Cumulative Review, p. 99
Error Analysis	You Decide, pp. 226, 240	Error Analysis, pp. 209, 218, 226, 233, 239, 247, 254, 261	
Standardized Test Practice	SAT/ACT Practice, pp. 212, 221, 228, 235, 242, 250, 257, 264 SAT & ACT Preparation, pp. 272–273	TIMSS, p. 273	SAT/ACT Practice, pp. 97–98
Open-Ended Assessment	Math Journal, pp. 219, 233 Open-Ended Assessment, p. 271 Portfolio, p. 271	Modeling: pp. 242, 250, 264, 266 Speaking: pp. 221, 228, 257 Writing: pp. 212, 235	Open-Ended Assessment, p. 93
Chapter Assessment	Study Guide and Assessment, pp. 267–271 Chapter Test, p. A59		Multiple-Choice Tests (Forms 1A, 1B, 1C), pp. 81–86 Free-Response Tests (Forms 2A, 2B, 2C), pp. 87–92

TestCheck and Worksheet Builder is a networkable CD-ROM with three integrated modules. The **Worksheet Builder** creates customized worksheets, tests, and quizzes of free-response, multiple-choice, short-answer, and open-ended items. The **Student Module** gives you the option of having students take tests on-screen and get immediate feedback on their performance. Use the optional **Management System** to keep detailed student records.

Additional Chapter Resources

Student Edition
Career Choices, p. 221
Internet Project, p. 271
Extra Practice, pp. A32–A33

Teacher's Classroom Resources

Technology
Graphing Calculator
Keystrokes Masters

Teaching Aids
Answer Key Transparencies
Block Schedule Planning Guide
Lesson Planning Guide
Solutions Manual

Glencoe Instructional Technology

Instructional

Advanced MathPASS CD-ROM

Multimedia Applications CD-ROM

Assessment

TestCheck and Worksheet Builder

GLENCOE Online

Visit **www.amc.glencoe.com** for data updates, career information, graphing calculator programs, and other interactive activities.

Chapter Overview

In this chapter, students use various methods, including technology, to locate and approximate the zeros of polynomial functions. They examine the relationship of locating zeros of a function to solving for the roots of an equation. The strand of modeling real-world data is extended to include polynomial functions.

Using a Graphing Calculator

There are numerous opportunities, all optional, to use a graphing calculator in this chapter.

- Finding a value of a polynomial for a given value of x (p. 206)
- Solving quadratic equations by graphing the related function (pp. 213, 218)
- Locating the zeros of a function by using ⎸ GRAPH ⎹ (pp. 208, 211, 216, 232, 233, 238)
- Using **TABLE** to locate integral zeros (p. 225)
- Using the zero function in the **CALC** menu or **TABLE** and **TBLSET** to approximate the zeros of a function (pp. 237–239)
- Solving radical equations (p. 253)
- Determining a polynomial function that models a set of data (pp. 259–263, 265–266)

The **Graphing Calculator Appendix**, which begins on page A2, contains keystroke instructions on many of these topics.

POLYNOMIAL AND RATIONAL FUNCTIONS

> ### CHAPTER OBJECTIVES
> - **Determine roots of polynomial equations.** *(Lessons 4-1, 4-4)*
> - **Solve quadratic, rational, and radical equations and rational and radical inequalities.** *(Lessons 4-2, 4-6, 4-7)*
> - **Find the factors of polynomials.** *(Lesson 4-3)*
> - **Approximate real zeros of polynomial functions.** *(Lesson 4-5)*
> - **Write and interpret polynomial functions that model real-world data.** *(Lesson 4-8)*

SAT & ACT Preparation

This textbook contains a complete SAT/ACT preparation course described on pages T8–T9 of the *Teacher's Handbook*. This chapter contains:

- an SAT or ACT practice question at the end of each lesson
- examples of SAT and ACT problems involving coordinate geometry problems on page 272
- ten SAT and ACT practice problems on page 273

- additional SAT and ACT practice in the *Assessment and Evaluation Masters,* pp. 97–98
- SAT and ACT test practice banks in the *TestCheck and Worksheet Builder*

Polynomial Functions

OBJECTIVES
• Determine roots of polynomial equations.
• Apply the Fundamental Theorem of Algebra.

INVESTMENTS Many grandparents invest in the stock market for their grandchildren's college fund. Eighteen years ago, Della Brooks purchased $1000 worth of merchandising stocks at the birth of her first grandchild Owen. Ten years ago, she purchased $500 worth of transportation stocks, and five years ago, she purchased $250 worth of technology stocks. The stocks will be used to help pay for Owen's college education. If the stocks appreciate at an average annual rate of 12.25%, determine the current value of the college fund. *This problem will be solved in Example 1.*

Appreciation is the increase in value of an item over a period of time. The formula for compound interest can be used to find the value of Owen's college fund after appreciation. The formula is $A = P(1 + r)^t$, where P is the original amount of money invested, r is the interest rate or rate of return (written as a decimal), and t is the time invested (in years).

Example **① INVESTMENTS** The value of Owen's college fund is the sum of the current values of his grandmother's investments.

   a. Write a function in one variable that models the value of the college fund for any rate of return.

   b. Use the function to determine the current value of the college fund for an average annual rate of 12.25%.

   a. Let x represent $1 + r$ and $T(x)$ represent the total current value of the three stocks. The times invested, which are the exponents of x, are 18, 10, and 5, respectively.

Total	=	merchandising	+	transportation	+	technology
$T(x)$	=	$1000x^{18}$	+	$500x^{10}$	+	$250x^5$

   b. Since $r = 0.1225$, $x = 1 + 0.1225$ or 1.1225. Now evaluate $T(x)$ for $x = 1.1225$.

$$T(x) = 1000x^{18} + 500x^{10} + 250x^5$$
$$T(1.1225) = 1000(1.1225)^{18} + 500(1.1225)^{10} + 250(1.1225)^5$$
$$T(1.1225) \approx 10{,}038.33$$

The present value of Owen's college fund is about $10,038.33.

The expression $1000x^{18} + 500x^{10} + 250x^5$ is a **polynomial in one variable.**

Polynomial in One Variable

A polynomial in one variable, x, is an expression of the form $a_0x^n + a_1x^{n-1} + \ldots + a_{n-2}x^2 + a_{n-1}x + a_n$. The coefficients $a_0, a_1, a_2, \ldots, a_n$ represent complex numbers (real or imaginary), a_0 is not zero, and n represents a nonnegative integer.

Resource Manager

 Reproducible Masters
• *Study Guide and Practice,* pp. 59–60
• *Enrichment,* p. 37
• *Problem Solving and Applications,* p. 7
• *TI-92 & CBL Interdisciplinary Activities,* pp. 9–12

 Transparencies
• 5-Minute Check, 4-1
• Teaching, 4-1

---

1 FOCUS

 5-Minute Check
Chapter 3

1. Determine whether the graph of $f(x) = 2x^3 + 5x$ is symmetric with respect to the origin. **yes**

2. Determine whether the graph of $\dfrac{x^2}{4} + \dfrac{y^2}{9} = 1$ is symmetric with respect to the x-axis, the y-axis, the line $y = x$, the line $y = -x$, or none of these. **symmetric with respect to the x-axis and the y-axis**

3. Describe how the graphs of $f(x) = |x|$ and $g(x) = 2|x - 3| + 4$ are related. **The graph of $g(x)$ is the graph of $f(x)$ translated right 3, translated up 4, and expanded vertically by a factor of 2.**

4. Graph $y \geq (x - 6)^3 + 4$.

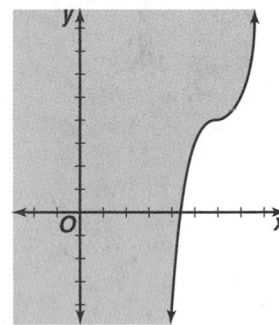

5. Determine the asymptotes for the graph of $f(x) = \dfrac{2}{x - 3}$. **vertical, $x = 3$; horizontal, $y = 0$**

Motivating the Lesson
Real-World Application Ask students to prepare a mock-up of a year in the life of a savings account that is opened on January 1 with $500 and receives 3% interest annually with interest awarded at the end of each month. If no additional deposits are made to the account, how much is in the account on December 31? **about $515.21**

Teaching Tip In Example 1, make sure that students understand that while T is a function of x, x is a function of r. Therefore, x is replaced by $1 + r$ when finding the current value.

In-Class Examples

Example 1
Hobbies A collector of Cute Cuddlies bean bags anticipated that they would appreciate in value. With that in mind, she bought 12 for $5.00 each in 1998, 20 for $6.00 each in 1999, and 22 for $6.50 each in 2000.
a. Write a function in one variable that models the value of the collection in 2001.
$f(x) = 60x^3 + 120x^2 + 143x$
b. Use the function to determine the value of the collection in 2001 if the Cuddlies appreciate at an average rate of 12% per year. **about $394.98**

Example 2
Consider the polynomial function $f(x) = 3x^4 - x^3 + x^2 + x - 1$.
a. State the degree and leading coefficient of the polynomial. **degree of 4 and leading coefficient of 3**
b. Determine whether -2 is a zero of $f(x)$. **no**

Teaching Tip In Example 2, point out that students may verify the zero graphically in order to check their work.

Teaching Tip Write the following pattern on the chalkboard.
$i = \sqrt{-1} \qquad i^2 = -1$
$i^3 = -\sqrt{1} \qquad i^4 = 1$
$i^5 = \sqrt{-1} \qquad i^6 = -1$
$i^7 = -\sqrt{-1}$
Discuss the pattern and ask students to give the next six values in the pattern. $i^8 = 1, i^9 = \sqrt{-1}, i^{10} = -1, i^{11} = -\sqrt{-1}, i^{12} = 1, i^{13} = \sqrt{-1}$

The **degree** of a polynomial in one variable is the greatest exponent of its variable. The coefficient of the variable with the greatest exponent is called the **leading coefficient.** For the expression $1000x^{18} + 500x^{10} + 250x^5$, 18 is the degree, and 1000 is the leading coefficient.

If a function is defined by a polynomial in one variable with real coefficients, like $T(x) = 1000x^{18} + 500x^{10} + 250x^5$, then it is a **polynomial function.** If $f(x)$ is a polynomial function, the values of x for which $f(x) = 0$ are called the **zeros** of the function. *If the function is graphed, these zeros are also the x-intercepts of the graph.*

Example **2** Consider the polynomial function $f(x) = x^3 - 6x^2 + 10x - 8$.
   a. State the degree and leading coefficient of the polynomial.
   b. Determine whether 4 is a zero of $f(x)$.

   a. $x^3 - 6x^2 + 10x - 8$ has a degree of 3 and a leading coefficient of 1.

   b. Evaluate $f(x) = x^3 - 6x^2 + 10x - 8$ for $x = 4$. That is, find $f(4)$.
   $f(4) = 4^3 - 6(4^2) + 10(4) - 8 \quad x = 4$
   $f(4) = 64 - 96 + 40 - 8$
   $f(4) = 0$

   Since $f(4) = 0$, 4 is a zero of $f(x) = x^3 - 6x^2 + 10x - 8$.

> **Graphing Calculator Tip**
> To find a value of a polynomial for a given value of x, enter the polynomial in the **Y=** list. Then use the **1:value** option in the **CALC** menu.

Since 4 is a zero of $f(x) = x^3 - 6x^2 + 10x - 8$, it is also a solution for the **polynomial equation** $x^3 - 6x^2 + 10x - 8 = 0$. The solution for a polynomial equation is called a **root.** The words *zero* and *root* are often used interchangeably, but technically, you find the *zero of a function* and the *root of an equation*.

A root or zero may also be an **imaginary number** such as $3i$. By definition, the imaginary unit i equals $\sqrt{-1}$. Since $i = \sqrt{-1}$, $i^2 = -1$. It also follows that $i^3 = i^2 \times i$ or $-i$ and $i^4 = i^2 \times i^2$ or 1.

The imaginary numbers combined with the real numbers compose the set of **complex numbers.** A complex number is any number of the form $a + bi$ where a and b are real numbers. If $b = 0$, then the complex number is a real number. If $a = 0$ and $b \neq 0$, then the complex number is called a **pure imaginary number.**

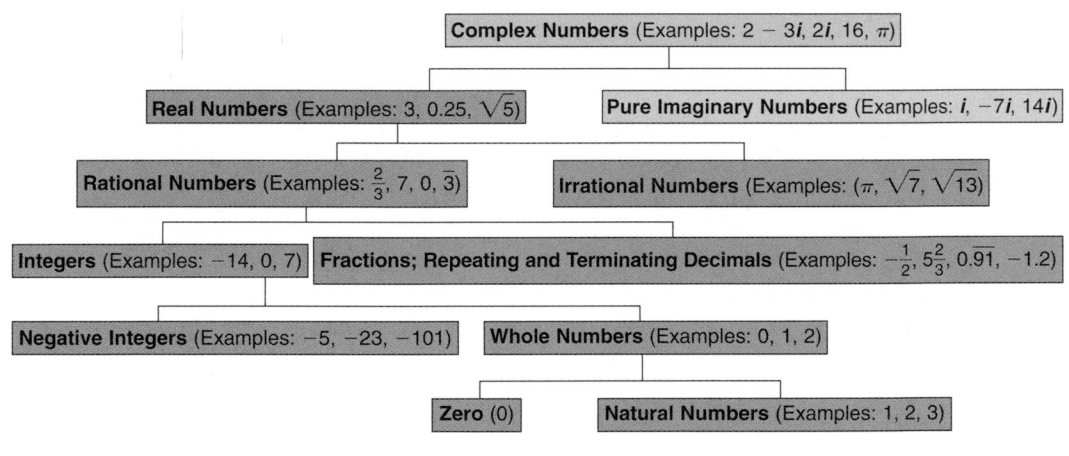

Prerequisite Skills

A review of the prerequisite skills for this lesson is provided in the *Prerequisite Skills Masters.*

Skill Number and Content	Pages
1 Properties of Real Numbers	3–4
6 Polynomials	13–14
7 Operations with Polynomials	15–16
9 Factoring by Grouping and Greatest Common Factor	19–20
10 Factoring Special Products of Polynomials	21–22
11 Factoring by Completing the Square	23–24

One of the most important theorems in mathematics is the **Fundamental Theorem of Algebra.**

Fundamental Theorem of Algebra	Every polynomial equation with degree greater than zero has at least one root in the set of complex numbers.

A corollary to the Fundamental Theorem of Algebra states that the degree of a polynomial indicates the number of possible roots of a polynomial equation.

Corollary to the Fundamental Theorem of Algebra	Every polynomial $P(x)$ of degree n ($n > 0$) can be written as the product of a constant k ($k \neq 0$) and n linear factors. $$P(x) = k(x - r_1)(x - r_2)(x - r_3) \dots (x - r_n)$$ Thus, a polynomial equation of degree n has exactly n complex roots, namely $r_1, r_2, r_3, \dots, r_n$.

The general shapes of the graphs of polynomial functions with positive leading coefficients and degree greater than 0 are shown below. These graphs also show the *maximum* number of times the graph of each type of polynomial may cross the x-axis.

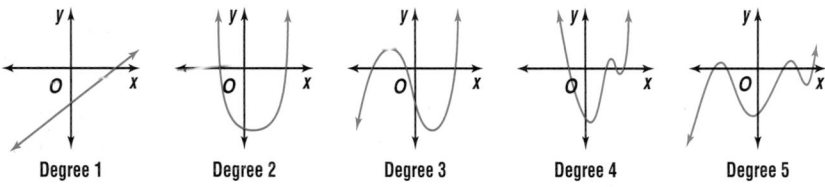

| Degree 1 | Degree 2 | Degree 3 | Degree 4 | Degree 5 |

Since the x-axis only represents real numbers, imaginary roots cannot be determined by using a graph. The graphs below have the general shape of a third-degree function and a fourth-degree function. In these graphs, the third-degree function only crosses the x-axis once, and the fourth-degree function crosses the x-axis twice or not at all.

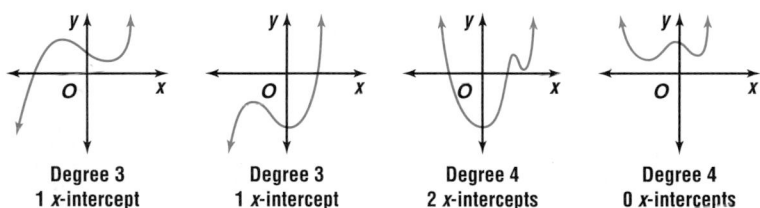

| Degree 3 1 *x*-intercept | Degree 3 1 *x*-intercept | Degree 4 2 *x*-intercepts | Degree 4 0 *x*-intercepts |

The graph of a polynomial function with odd degree must cross the x-axis at least once. The graph of a function with even degree may or may not cross the x-axis. If it does, it will cross an even number of times. Each x-intercept represents a real root of the corresponding polynomial equation.

If you know the roots of a polynomial equation, you can use the corollary to the Fundamental Theorem of Algebra to find the polynomial equation. That is, if a and b are roots of the equation, the equation must be $(x - a)(x - b) = 0$.

In-Class Examples

Example 3

a. Write a polynomial equation of least degree with roots 2, $3i$, and $-3i$.
$x^3 - 2x^2 + 9x - 18 = 0$

b. Does the equation have an odd or even degree? How many times does the graph of the related function cross the x-axis? **odd; once**

Example 4

State the number of complex roots of the equation $32x^3 - 32x^2 + 4x - 4 = 0$. Then find the roots and graph the related function.

$3; 1, -\dfrac{i\sqrt{2}}{4}, \dfrac{i\sqrt{2}}{4}$

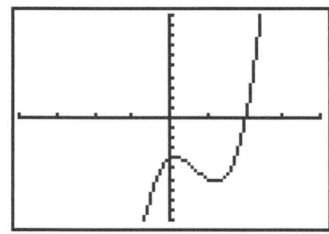

$[-2, 2]$ scl:0.5 by $[-10, 10]$ scl:1

Example 5

Sports When a golf ball is hit from a tee with a velocity of 160 ft/s at an angle of 45° with respect to the ground (horizontal), the height (in feet) of the ball above the ground is given by $h(x) = x - \dfrac{x^2}{800}$, where x is the horizontal distance from the tee.

a. How far from the tee does the ball strike the ground? **800 ft**

b. Verify your answer using a graph.

$[-100, 1000]$ scl:100 by $[-100, 400]$ scl:100

Examples

③ a. Write a polynomial equation of least degree with roots 2, $4i$, and $-4i$.

b. Does the equation have an odd or even degree? How many times does the graph of the related function cross the x-axis?

a. If $x = 2$, then $x - 2$ is a factor of the polynomial. Likewise, if $x = 4i$ and $x = -4i$, then $x - 4i$ and $x - (-4i)$ are factors of the polynomial. Therefore, the linear factors for the polynomial are $x - 2$, $x - 4i$, and $x + 4i$. Now find the products of these factors.

$$(x - 2)(x - 4i)(x + 4i) = 0$$
$$(x - 2)(x^2 - 16i^2) = 0$$
$$(x - 2)(x^2 + 16) = 0 \quad -16i^2 = -16(-1) \text{ or } 16$$
$$x^3 - 2x^2 + 16x - 32 = 0$$

A polynomial equation with roots 2, $4i$, and $-4i$ is $x^3 - 2x^2 + 16x - 32 = 0$.

b. The degree of this equation is 3. Thus, the equation has an odd degree since 3 is an odd number. Since two of the roots are imaginary, the graph will only cross the x-axis once. The graphing calculator image at the right verifies these conclusions.

$[-10, 10]$ scl:1 by $[-50, 50]$ scl:5

④ State the number of complex roots of the equation $9x^4 - 35x^2 - 4 = 0$. Then find the roots and graph the related function.

The polynomial has a degree of 4, so there are 4 complex roots.

Factor the equation to find the roots.

$$9x^4 - 35x^2 - 4 = 0$$
$$(9x^2 + 1)(x^2 - 4) = 0$$
$$(9x^2 + 1)(x + 2)(x - 2) = 0$$

To find each root, set each factor equal to zero.

$9x^2 + 1 = 0$
$x^2 = -\dfrac{1}{9}$ *Solve for x^2.*

$x = \pm\sqrt{\dfrac{1}{9} \cdot (-1)}$ *Take the square root of each side.*

$x = \pm\dfrac{1}{3}\sqrt{-1}$ or $\pm\dfrac{1}{3}i$

$x + 2 = 0$ $x - 2 = 0$
$x = -2$ $x = 2$

The roots are $\pm\dfrac{1}{3}i$, -2, and 2.

Use a table of values or a graphing calculator to graph the function. The x-intercepts are -2 and 2.

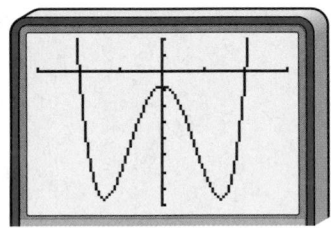

$[-3, 3]$ scl:1 by $[-40, 10]$ scl:5

The function has an even degree and has 2 real zeros.

Addressing Individual Needs

Interpersonal Have students work together in groups to sketch polynomial functions having a particular degree and number of real roots; for example, degree 3 and 3 real roots or degree 3 and only 1 real root. Then, have students experiment with coefficients in the general form of a polynomial in order to find functions whose graphs resemble student sketches.

Example **5** **METEOROLOGY** A meteorologist sends a temperature probe on a small weather rocket through a cloud layer. The launch pad for the rocket is 2 feet off the ground. The height of the rocket after launching is modeled by the equation $h = -16t^2 + 232t + 2$, where h is the height of the rocket in feet and t is the elapsed time in seconds.

Real World Application

a. When will the rocket be 114 feet above the ground?

b. Verify your answer using a graph.

a. $h = -16t^2 + 232t + 2$

$114 = -16t^2 + 232t + 2$ *Replace h with 114.*

$0 = -16t^2 + 232t - 112$ *Subtract 114 from each side.*

$0 = -8(2t^2 - 29t + 14)$ *Factor.*

$0 = -8(2t - 1)(t - 14)$ *Factor.*

$2t - 1 = 0$ or $t - 14 = 0$

$t = \frac{1}{2}$ $14 = t$

The weather rocket will be 114 feet above the ground after $\frac{1}{2}$ second and again after 14 seconds.

b. To verify the answer, graph $h(t) = -16t^2 + 232t - 112$. The graph appears to verify this solution.

[−1, 15] scl:1 by [−50, 900] scl:50

CHECK FOR UNDERSTANDING

Communicating Mathematics

Read and study the lesson to answer each question.   1–4. See margin.

1. **Write** several sentences about the relationship between zeros and roots.

2. **Explain** why zeros of a function are also the *x*-intercepts of its graph.

3. **Define** a complex number and tell under what conditions it will be a pure imaginary number. Write two examples and two **nonexamples** of a pure imaginary numbers.

4. **Sketch** the general graph of a sixth-degree function.

Guided Practice

State the degree and leading coefficient of each polynomial.

5. $a^3 + 6a + 14$ **3; 1**

6. $5m^2 + 8m^5 - 2$ **5; 8**

Determine whether each number is a root of $x^3 - 5x^2 - 3x - 18 = 0$. Explain.

7. 5   **no;** $f(5) = -33$

8. 6   **yes;** $f(6) = 0$

Lesson 4-1   Polynomial Functions   **209**

Teaching Tip In Example 5, remind students that it is sometimes helpful to look at the graph first when finding factors. Emphasize the importance of utilizing both algebraic and graphical methods together when solving the problem.

3 PRACTICE/APPLY

Error Analysis

Watch for students who focus solely on the real roots of a polynomial while ignoring any imaginary roots.

Prevent by emphasizing the importance of completing all algebraic procedures until the number of roots found matches the degree of the polynomial.

Additional Answers

3. A complex number is any number in the form *a + bi*, where *a* and *b* are real numbers and *i* is the imaginary unit. In a pure imaginary number, *a* = 0 and *b* ≠ 0. Examples: −2*i*, 3*i*; nonexamples: 5, 1 + *i*

4.

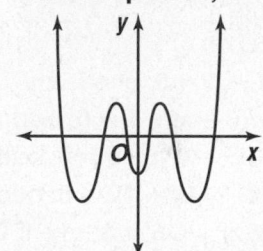

Study Guide and Practice Masters, pp. 59–60

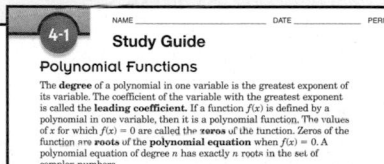

NAME _____ DATE _____ PERIOD _____

4-1   **Study Guide**

Polynomial Functions

The **degree** of a polynomial in one variable is the greatest exponent of its variable. The coefficient of the variable with the greatest exponent is called the **leading coefficient**. If a function $f(x)$ is defined by a polynomial in one variable, then it is a polynomial function. The values of x for which $f(x) = 0$ are called the **zeros** of the function. Zeros of the function are **roots** of the **polynomial equation** when $f(x) = 0$. A polynomial equation of degree n has exactly n roots in the set of complex numbers.

Example 1   State the degree and leading coefficient of the polynomial function $f(x) = 6x^5 + 8x^3 - 8x$. Then determine whether $\sqrt{\frac{2}{3}}$ is a zero of $f(x)$.

$6x^5 + 8x^3 - 8x$ has a degree of 5 and a leading coefficient of 6.
Evaluate the function for $x = \sqrt{\frac{2}{3}}$. That is, find $f\left(\sqrt{\frac{2}{3}}\right)$.

$f\left(\sqrt{\frac{2}{3}}\right) = 6\left(\sqrt{\frac{2}{3}}\right)^5 + 8\left(\sqrt{\frac{2}{3}}\right)^3 - 8\sqrt{\frac{2}{3}}$ $x = \sqrt{\frac{2}{3}}$

$= \frac{24}{9}\sqrt{\frac{2}{3}} + \frac{16}{3}\sqrt{\frac{2}{3}} - 8\sqrt{\frac{2}{3}}$

$= 0$

Since $f\left(\sqrt{\frac{2}{3}}\right) = 0$, $\sqrt{\frac{2}{3}}$ is a zero of $f(x) = 6x^5 + 8x^3 - 8x$.

Example 2   Write a polynomial equation of least degree with roots 0, $\sqrt{2}i$, and $-\sqrt{2}i$.

The linear factors for the polynomial are $x - 0$, $x - \sqrt{2}i$, and $x + \sqrt{2}i$.
Find the products of these factors.

$(x - 0)(x - \sqrt{2}i)(x + \sqrt{2}i) = 0$

$x(x^2 - 2i^2) = 0$

$x(x^2 + 2) = 0$ $-2i^2 = -2(-1)$, or 2

$x^3 + 2x = 0$

Example 3   State the number of complex roots of the equation $3x^2 + 11x - 4 = 0$. Then find the roots.

The polynomial has a degree of 2, so there are two complex roots. Factor the equation to find the roots.

$3x^2 + 11x - 4 = 0$

$(3x - 1)(x + 4) = 0$

To find each root, set each factor equal to zero.

$3x - 1 = 0$ $x + 4 = 0$

$3x = 1$ $x = -4$

$x = \frac{1}{3}$

The roots are -4 and $\frac{1}{3}$.

© Glencoe/McGraw-Hill   **59**   Advanced Mathematical Concepts

Additional Answers

1. A zero is the value of the variable for which a polynomial function in one variable equals zero. A root is a solution of a polynomial equation in one variable. When a polynomial function is the related function to the polynomial equation, the zeros of the function are the same as the roots of the equation.

2. The ordered pair (*x*, 0) represents the points on the *x*-axis. Therefore, the *x*-intercept of a graph of a function represents the point where $f(x) = 0$.

Additional Answers

11.

12.

13.

32. $x^2 - x - 6 = 0$; even; 2

33. $x^3 - 5x^2 - x + 5 = 0$; odd; 3

34. $2x^3 - 3x^2 - 18x - 8 = 0$; odd; 3

35. $x^3 + 3x^2 + 4x + 12 = 0$; odd; 1

36. $x^4 + 26x^2 + 25 = 0$; even; 0

37. $x^5 - 5x^4 - 17x^3 + 85x^2 + 16x - 80 = 0$; odd; 5

Write a polynomial equation of least degree for each set of roots. Does the equation have an odd or even degree? How many times does the graph of the related function cross the *x*-axis?

9. $-5, 7$
$x^2 - 2x - 35 = 0$; even; 2

10. $6, 2i, -2i, i, -i$
$x^5 - 6x^4 + 5x^3 - 30x^2 + 4x - 24 = 0$; odd; 1

State the number of complex roots of each equation. Then find the roots and graph the related functions. **11–13. See margin for graphs.**

11. $x^2 - 14x + 49 = 0$
2; 7, 7

12. $a^3 + 2a^2 - 8a = 0$
3; $-4, 0, 2$

13. $t^4 - 1 = 0$
4; $-1, 1, i, -i$

14. Geometry A cylinder is inscribed in a sphere with a radius of 6 units as shown.

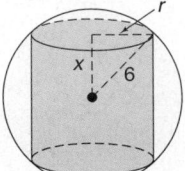

14b. $V(x) = 72\pi x - 2\pi x^3$

a. Write a function that models the volume of the cylinder in terms of *x*. (*Hint:* The volume of a cylinder equals $\pi r^2 h$.) $V(x) = \pi(36 - x^2)(2x)$

b. Write this function as a polynomial function.

c. Find the volume of the cylinder if $x = 4$.
about 502.65 units³

EXERCISES

Practice

State the degree and leading coefficient of each polynomial.

15. $5t^4 + t^3 - 7$ **4; 5**   **16.** $3x^7 - 4x^5 + x^3$ **7; 3**   **17.** $9a^2 + 5a^3 - 10$ **3; 5**

18. $14b - 25b^5$ **5; −25**   **19.** $p^5 + 7p^3 - p^6$ **6; −1**   **20.** $14y + 30 + y^2$ **2; 1**

21. Yes; the coefficients are complex numbers and the exponents of the variable are nonnegative integers.

21. Determine if $x^3 + 3x + \sqrt{5}$ is a polynomial in one variable. Explain.

22. Is $\frac{1}{a} + a^2$ a polynomial in one variable? Explain. **No; $\frac{1}{a} = a^{-1}$, which is a negative exponent.**

Determine whether each number is a root of $a^4 - 13a^2 + 12a = 0$. Explain.

23. 0   **24.** −1   **25.** 1   **26.** −4   **27.** −3   **28.** 3

23. yes; $f(0) = 0$ **25. yes; $f(1) = 0$**   **26. yes; $f(-4) = 0$**   **27. no; $f(-3) = -72$**   **28. yes; $f(3) = 0$**

24. no; $f(-1) = -24$

29. Is −2 a root of $b^4 - 3b^2 - 2b + 4 = 0$?   **no**

30. Is −1 a root of $x^4 - 4x^3 - x^2 + 4x = 0$?   **yes**

31. Each graph represents a polynomial function. State the number of complex zeros and the number of real zeros of each function.

a.   **3; 1**

b.   **2; 2**

c. 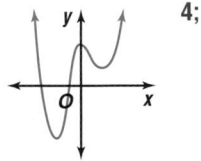   **4; 2**

Write a polynomial equation of least degree for each set of roots. Does the equation have an odd or even degree? How many times does the graph of the related function cross the *x*-axis?   **32–37. See margin.**

32. $-2, 3$ **33.** $-1, 1, 5$ **34.** $-2, -0.5, 4$

35. $-3, -2i, 2i$ **36.** $-5i, -i, i, 5i$ **37.** $-1, 1, 4, -4, 5$

38. Write a polynomial equation of least degree whose roots are -1, 1, 3, and -3.
$x^4 - 10x^2 + 9 = 0$

www.amc.glencoe.com/self_check_quiz

Additional Answers (p. 211)

39.

40.

41.

42.

42. 3; −2, −2, 2

43. 3; −3, 0, 3

44. 3; −2.5, 0, 3

48. −1, −1, 1;
See margin for
graph.

49f. not possible

**Graphing
Calculator**

State the number of complex roots of each equation. Then find the roots and graph the related function. **39–47. See margin for graphs.**

39. $x + 8 = 0$ **1; −8**
40. $a^2 − 81 = 0$ **2; −9, 9**
41. $b^2 + 36 = 0$ **2; ±6i**
42. $t^3 + 2t^2 − 4t − 8 = 0$
43. $n^3 − 9n = 0$
44. $6c^3 − 3c^2 − 45c = 0$
45. $a^4 + a^2 − 2 = 0$
    4; −1, 1, −$\sqrt{2}\,i$, $\sqrt{2}\,i$
46. $x^4 − 10x^2 + 9 = 0$
    4; −3, −1, 1, 3
47. $4m^4 + 17m^2 + 4 = 0$
    4; −0.5i, 0.5i, −2i, 2i

48. Solve $(u + 1)(u^2 − 1) = 0$ and graph the related polynomial function.

49. Sketch a fourth-degree equation for each situation.   **a–e. See margin.**
    a. no x-intercept
    b. one x-intercept
    c. two x-intercepts
    d. three x-intercepts
    e. four x-intercepts
    f. five x-intercepts

50. Use a graphing calculator to graph $f(x) = x^4 − 2x^2 + 1$.   **See margin.**
    a. What is the maximum number of x-intercepts possible for this function?   **4**
    b. How many x-intercepts are there? Name the intercept(s).   **2; −1, 1**
    c. Why are there fewer x-intercepts than the maximum number?
       (*Hint:* The factored form of the polynomial is $(x^2 − 1)^2$.)   **There are 4 real roots. However, there is a double root at −1 and a double root at 1.**

**Applications
and Problem
Solving**

51a. $V(x) =$
99,000x^3 +
55,000x^2 +
65,000x

51. **Classic Cars** Sonia Orta invests in vintage automobiles. Three years ago, she purchased a 1953 Corvette roadster for $99,000. Two years ago, she purchased a 1929 Pierce-Arrow Model 125 for $55,000. A year ago she purchased a 1909 Cadillac Model Thirty for $65,000.

    a. Let x represent 1 plus the average rate of appreciation. Write a function in terms of x that models the value of the automobiles.
    b. If the automobiles appreciate at an average annual rate of 15%, find the current value of the three automobiles.   **about $298,054.13**

52. **Critical Thinking** One of the zeros of a polynomial function is 1. After translating the graph of the function left 2 units, 1 is a zero of the new function. What do you know about the original function?   **1 and 3 are two of its zeros.**

53. **Aeronautics** At liftoff, the space shuttle *Discovery* has a constant acceleration, a, of 16.4 feet per second squared. The function $d(t) = \frac{1}{2}at^2$ can be used to determine the distance from Earth for each time interval, t, after takeoff.

53a. 7380 ft;
29,520 ft;
118,080 ft

    a. Find its distance from Earth after 30 seconds, 1 minute, and 2 minutes.

53b. It
quadruples;
$(2t)^2 = 4t^2$.

    b. Study the pattern of answers to part a. If the time the space shuttle is in flight doubles, how does the distance from Earth change? Explain.

54. **Construction** The Santa Fe Recreation Department has a 50-foot by 70-foot area for construction of a new public swimming pool. The pool will be surrounded by a concrete sidewalk of constant width. Because of water restrictions, the pool can have a maximum area of 2400 square feet. What should be the width of the sidewalk that surrounds the pool?   **5 ft**

Lesson 4-1 Polynomial Functions **211**

Additional Answers

46.

47.

48.

49a.

49b.

49c.

(continued on page 212)

Open-Ended Assessment
Writing Have students write an example of a polynomial function in their journals and explain what the corollary to the Fundamental Theorem of Algebra means for that particular function.

Additional Answers *(p. 211)*

49d.

49e.

50.

$[-5, 5]$ scl:1 by $[-2, 8]$ scl:1

Enrichment Masters, p. 37

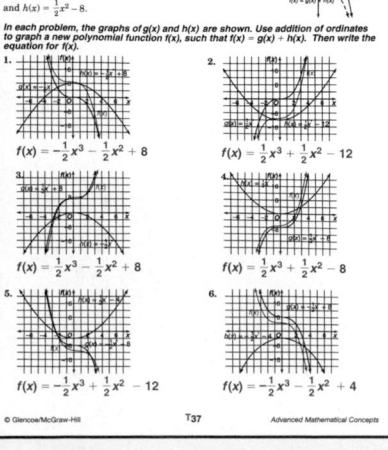

212 *Chapter 4*

---

55. **Marketing** Each week, Marino's Pizzeria sells an average of 160 large supreme pizzas for $16 each. Next week, the pizzeria plans to run a sale on these large supreme pizzas. The owner estimates that for each 40¢ decrease in the price, the store will sell approximately 16 more large pizzas. If the owner wants to sell $4,000 worth of the large supreme pizzas next week, determine the sale price.   **$10**

56. **Critical Thinking** If B and C are the real roots of $x^2 + Bx + C = 0$, where $B \neq 0$ and $C \neq 0$, find the values of B and C.   **Sample answer: 1; −2**

Mixed Review

57. Create a function in the form $y = f(x)$ that has a vertical asymptote at $x = -2$ and $x = 0$, and a hole at $x = 2$. *(Lesson 3-7)* $y = \dfrac{x-2}{x(x + 2)(x - 2)}$

58. **Construction** Selena wishes to build a pen for her animals. He has 52 yards of fencing and wants to build a rectangular pen. *(Lesson 3-6)*
   a. Find a model for the area of the pen as a function of the length and width of the rectangle. $A(x) = x(26 - x)$ b. **13 yd by 13 yd**
   b. What are the dimensions that would produce the maximum area?

59. The graph of $y = 2x^3 + 1$ is the graph of $y = 2x^3$ shifted 1 unit up.

59. Describe how the graphs of $y = 2x^3$ and $y = 2x^3 + 1$ are related. *(Lesson 3-2)*

60. Find the coordinates of P' if $P(4, 9)$ and P' are symmetric with respect to $M(-1, 9)$. *(Lesson 3-1)*   **(−6, 9)**

61. Find the determinant for $\begin{bmatrix} -15 & 5 \\ -9 & 3 \end{bmatrix}$. Tell whether an inverse exists for the matrix. *(Lesson 2-5)*   **0; no**

62. If $A = \begin{bmatrix} 2 & -1 \\ 3 & 4 \end{bmatrix}$ and $B = \begin{bmatrix} 3 & -9 & 2 \\ 5 & 7 & -6 \end{bmatrix}$, find AB. *(Lesson 2-3)* $\begin{bmatrix} 1 & -25 & 10 \\ 29 & 1 & -18 \end{bmatrix}$

63. Graph $x + 4y < 9$. *(Lesson 1-8)*   **See margin.**

64. Parallel; the lines have the same slope.

64. The slope of $\overleftrightarrow{AB}$ is 0.6. The slope of $\overleftrightarrow{CD}$ is $\dfrac{3}{5}$. State whether the lines are *parallel, perpendicular,* or *neither.* Explain. *(Lesson 1-5)*

65. Find $[f \circ g](x)$ and $[g \circ f](x)$ for the functions $f(x) = x^2 - 4$ and $g(x) = \dfrac{1}{2}x + 6$. *(Lesson 1-2)* $\dfrac{1}{4}x^2 + 6x + 32, \dfrac{1}{2}x^2 + 4$

66. **SAT Practice** In 2003, Bob's Quality Cars sold 270 more cars than in 2004. How many cars does each represent?   **A**

Year	Cars Sold by Bob's Quality Cars
2003	
2004	

A 135
B 125
C 100
D 90
E 85

Extra Credit

Three of the roots of a polynomial equation are 3, $2i$, and $-5i$.
a. What is the minimum degree of the function?   **5**
b. Write an equation with these roots.
   Sample answer: $x^5 - 3x^4 + 29x^3 - 87x^2 + 100x - 300 = 0$

Additional Answer

63.

$x + 4y < 9$

Quadratic Equations

OBJECTIVES
- Solve quadratic equations.
- Use the discriminant to describe the roots of quadratic equations.

BASEBALL On September 8, 1998, Mark McGwire of the St. Louis Cardinals broke the home-run record with his 62nd home run of the year. He went on to hit 70 home runs for the season. Besides hitting home runs, McGwire also occasionally popped out. Suppose the ball was 3.5 feet above the ground when he hit it straight up with an initial velocity of 80 feet per second. The function $d(t) = 80t - 16t^2 + 3.5$ gives the ball's height above the ground in feet as a function of time in seconds. How long did the catcher have to get into position to catch the ball after it was hit? *This problem will be solved in Example 3.*

A quadratic equation is a polynomial equation with a degree of two. Solving quadratic equations by graphing usually does not yield exact answers. Also, some quadratic expressions are not factorable over the integers. Therefore, alternative strategies for solving these equations are needed. One such alternative is solving quadratic equations by **completing the square.**

Completing the square is a useful method when the quadratic is not easily factorable. It can be used to solve any quadratic equation. Remember that, for any number b, the square of the binomial $x + b$ has the form $x^2 + 2bx + b^2$. When completing the square, you know the first term and middle term and need to supply the last term. This term equals the square of half the coefficient of the middle term. For example, to complete the square of $x^2 + 8x$, find $\frac{1}{2}(8)$ and square the result. So, the third term is 16, and the expression becomes $x^2 + 8x + 16$. Note that this technique works only if the coefficient of x^2 is 1.

Example

 Solve $x^2 - 6x - 16 = 0$.

This equation can be solved by graphing, factoring, or completing the square.

Method 1
Solve the equation by graphing the related function $f(x) = x^2 - 6x - 16$. The zeros of the function appear to be -2 and 8.

Method 2
Solve the equation by factoring.
$x^2 - 6x - 16 = 0$
$(x + 2)(x - 8) = 0$ *Factor.*

$x + 2 = 0$ or $x - 8 = 0$
$x = -2$ $\qquad$ $x = 8$

The roots of the equation are -2 and 8.

[−10, 10] scl:1 by [−30, 10] scl:5

Lesson 4-2 Quadratic Equations **213**

Resource Manager

 Reproducible Masters
- *Study Guide and Practice,* pp. 61–62
- *Enrichment,* p. 38
- *Assessment and Evaluation,* p. 95

 Transparencies
- 5-Minute Check, 4-2
- Teaching, 4-2

---

1 FOCUS

5-Minute Check
Lesson 4-1

1. State the degree and leading coefficient of $f(x) = x^2 + 9x^4 - 5x + x^3$. **4; 9**

2. Determine whether 2 is a root of $x^3 + x^2 - 4x - 4 = 0$. **yes**

3. Determine whether 3 is a zero of $f(x) = x^4 - 3x^2 - 16x - 9$. **no**

4. Write a polynomial equation of least degree with roots $-3i$, $3i$, and 4. Does the equation have an odd or an even degree? How many times does the graph of the related function cross the x-axis? $x^3 - 4x^2 + 9x - 36 = 0$; **odd; once**

5. State the number of complex roots of $4x^3 - 2x^2 + 16x - 8 = 0$. Then find the roots and graph the related function. **3; $\frac{1}{2}$, $\pm 2i$**

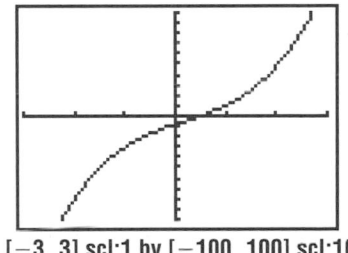

[−3, 3] scl:1 by [−100, 100] scl:10

Motivating the Lesson

Discussion Ask students what would be the quickest way to discover whether the roots of a quadratic function are real. Then list a set of quadratic equations on the chalkboard and ask students how many real roots each one has. After Example 3, point out how the discriminant can be used to determine whether the roots are real.

Teaching Tip In Example 2, remind students that when dividing the equation by the leading coefficient, both sides must be divided, not just the side containing the leading term.

Method 3
Solve the equation by completing the square.

$$x^2 - 6x - 16 = 0$$

$$x^2 - 6x = 16 \qquad \textit{Add 16 to each side.}$$

$$x^2 - 6x + 9 = 16 + 9 \qquad \textit{Complete the square by adding } \left(\frac{-6}{2}\right)^2 \textit{ or 9 to each side.}$$

$$(x - 3)^2 = 25 \qquad \textit{Factor the perfect square trinomial.}$$

$$x - 3 = \pm 5 \qquad \textit{Take the square root of each side.}$$

$$x - 3 = 5 \qquad \text{or} \qquad x - 3 = -5$$
$$x = 8 \qquad\qquad\qquad x = -2$$

The roots of the equation are 8 and -2.

Although factoring may be an easier method to solve this particular equation, completing the square can always be used to solve any quadratic equation.

When solving a quadratic equation by completing the square, the leading coefficient must be 1. When the leading coefficient of a quadratic equation is not 1, you must first divide each side of the equation by that coefficient before completing the square.

Example ❷ Solve $3n^2 + 7n + 7 = 0$ by completing the square.

Notice that the graph of the related function, $y = 3x^2 + 7x + 7$, does not cross the x-axis. Therefore, the roots of the equation are imaginary numbers. Completing the square can be used to find the roots of any equation, including one with no real roots.

[−10, 10] scl:1 by [−10, 10] scl:1

$$3n^2 + 7n + 7 = 0$$

$$n^2 + \frac{7}{3}n + \frac{7}{3} = 0 \qquad \textit{Divide each side by 3.}$$

$$n^2 + \frac{7}{3}n = -\frac{7}{3} \qquad \textit{Subtract } \frac{7}{3} \textit{ from each side.}$$

$$n^2 + \frac{7}{3}n + \frac{49}{36} = -\frac{7}{3} + \frac{49}{36} \qquad \textit{Complete the square by adding } \left(\frac{7}{6}\right)^2 \textit{ or } \frac{49}{36} \textit{ to each side.}$$

$$\left(n + \frac{7}{6}\right)^2 = -\frac{35}{36} \qquad \textit{Factor the perfect square trinomial.}$$

$$n + \frac{7}{6} = \pm i\frac{\sqrt{35}}{6} \qquad \textit{Take the square root of each side.}$$

$$n = -\frac{7}{6} \pm i\frac{\sqrt{35}}{6} \qquad \textit{Subtract } \frac{7}{6} \textit{ from each side.}$$

The roots of the equation are $-\frac{7}{6} \pm i\frac{\sqrt{35}}{6}$ or $\frac{-7 \pm i\sqrt{35}}{6}$.

From the Classroom of...

Nancy Dymott
Carthage High School
Carthage, Missouri

"Try this. Given $6x^2 + 17x + 12 = 0$, create a new quadratic equation with the leading coefficient as 1 by multiplying a times c, i.e., $6 \times 12 \to x^2 + 17x + 72 = 0$. This new equation factors to $(x + 8)(x + 9) = 0$, the roots of which are $x = -8$ and $x = -9$. Divide each of these roots by the original leading coefficient to get the original roots, i.e. $x = -\frac{4}{3}$ and $x = -\frac{3}{2}$."

Completing the square can be used to develop a general formula for solving any quadratic equation of the form $ax^2 + bx + c = 0$. This formula is called the **Quadratic Formula.**

Quadratic Formula	The roots of a quadratic equation of the form $ax^2 + bx + c = 0$ with $a \neq 0$ are given by the following formula. $$x = \frac{-b \pm \sqrt{b^2 - 4ac}}{2a}$$

The quadratic formula can be used to solve any quadratic equation. It is usually easier than completing the square.

Example **3** **BASEBALL** Refer to the application at the beginning of the lesson. How long did the catcher have to get into position to catch the ball after if was hit?

Real World Application

The catcher must get into position to catch the ball before $80t - 16t^2 + 3.5 = 0$. This equation can be written as $-16t^2 + 80t + 3.5 = 0$. Use the Quadratic Formula to solve this equation.

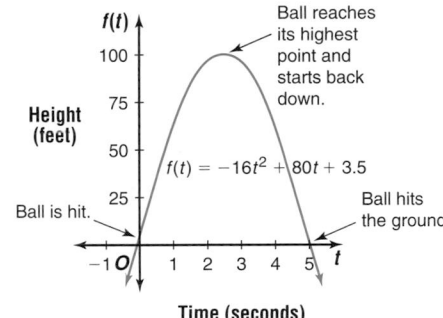

$f(t)$

Ball reaches its highest point and starts back down.

Height (feet)

$f(t) = -16t^2 + 80t + 3.5$

Ball is hit.

Ball hits the ground.

Time (seconds)

$$t = \frac{-b \pm \sqrt{b^2 - 4ac}}{2a}$$

$$t = \frac{-80 \pm \sqrt{80^2 - 4(-16)(3.5)}}{2(-16)} \qquad a = -16,\ b = 80,\ and\ c = 3.5$$

$$t = \frac{-80 \pm \sqrt{6624}}{-32}$$

$$t = \frac{-80 + \sqrt{6624}}{-32} \qquad or \qquad t = \frac{-80 - \sqrt{6624}}{-32}$$

$$t \approx -0.04 \qquad\qquad\qquad t \approx 5.04$$

The roots of the equation are about -0.04 and 5.04. Since the catcher has a positive amount of time to catch the ball, he will have about 5 seconds to get into position to catch the ball.

In the quadratic formula, the radicand $b^2 - 4ac$ is called the **discriminant** of the equation. The discriminant tells the nature of the roots of a quadratic equation or the zeros of the related quadratic function.

Lesson 4-2 Quadratic Equations **215**

216 Chapter 4

Teaching Tip When working Example 4, you may wish to remind students of the relationship between the discriminant and the graph of the related quadratic function. For example, if the discriminant is equal to 0, the graph is tangent to the x-axis.

In-Class Example

Example 4

Find the discriminant of $x^2 + 2x - 2 = 0$ and describe the nature of the roots of the equation. Then solve the equation by using the Quadratic Formula.   **12; two distinct real roots; $-1 \pm \sqrt{3}$**

Discriminant	Nature of Roots/Zeros	Graph
$b^2 - 4ac > 0$	two distinct real roots/zeros	
$b^2 - 4ac = 0$	exactly one real root/zero (The one real root is actually a double root.)	
$b^2 - 4ac < 0$	no real roots/zero (two distinct imaginary roots/zeros)	

Example **4** **Find the discriminant of $x^2 - 4x + 15 = 0$ and describe the nature of the roots of the equation. Then solve the equation by using the Quadratic Formula.**

The value of the discriminant, $b^2 - 4ac$, is $(-4)^2 - 4(1)(15)$ or -44. Since the value of the discriminant is less than zero, there are no real roots.

$$x = \frac{-b \pm \sqrt{b^2 - 4ac}}{2a}$$

$$x = \frac{-(-4) \pm \sqrt{-44}}{2(1)}$$

$$x = \frac{4 \pm 2i\sqrt{11}}{2}$$

$$x = 2 \pm i\sqrt{11}$$

The roots are $2 + i\sqrt{11}$ and $2 - i\sqrt{11}$.

The graph of $y = x^2 - 4x + 15$ verifies that there are no real roots.

[-10, 10] scl:1 by [-10, 50] scl:5

The roots of the equation in Example 4 are the complex numbers $2 + i\sqrt{11}$ and $2 - i\sqrt{11}$. A pair of complex numbers in the form $a + bi$ and $a - bi$ are called **conjugates.** Imaginary roots of polynomial equations with real coefficients always occur in conjugate pairs. Some other examples of complex conjugates are listed below.

i and $-i$ $-1 + i$ and $-1 - i$ $-i\sqrt{2}$ and $i\sqrt{2}$

Complex Conjugates Theorem

Suppose a and b are real numbers with $b \neq 0$. If $a + bi$ is a root of a polynomial equation with real coefficients, then $a - bi$ is also a root of the equation. $a + bi$ and $a - bi$ are conjugate pairs.

There are four methods used to solve quadratic equations. Two methods work for any quadratic equation. One method approximates any real roots, and one method only works for equations that can be factored over the integers.

Solution Method	Situation	Examples
Graphing	Usually, only approximate solutions are shown. If roots are imaginary (discriminant is less than zero), the graph has no *x*-intercepts, and the solutions must be found by another method. Graphing is a good method to verify solutions.	$6x^2 + x - 2 = 0$ $x = -\frac{2}{3}$ or $x = \frac{1}{2}$ $x^2 - 2x + 5 = 0$ discriminant: $(-2)^2 - 4(1)(5) = -16$ The equation has no real roots.
Factoring	When a, b, and c are integers and the discriminant is a perfect square or zero, this method is useful. It cannot be used if the discriminant is less than zero.	$g^2 + 2g - 8 = 0$ discriminant: $2^2 - 4(1)(-8) = 36$ $g^2 + 2g - 8 = 0$ $(g + 4)(g - 2) = 0$ $g + 4 = 0$ or $g - 2 = 0$ $g = -4$ $g = 2$
Completing the Square	This method works for any quadratic equation. There is more room for an arithmetic error than when using the Quadratic Formula.	$r^2 + 4r - 6 = 0$ $r^2 + 4r = 6$ $r^2 + 4r + 4 = 6 + 4$ $(r + 2)^2 = 10$ $r + 2 = \pm\sqrt{10}$ $r = -2 \pm \sqrt{10}$
Quadratic Formula	This method works for any quadratic equation.	$2s^2 + 5s + 4 = 0$ $s = \dfrac{-5 \pm \sqrt{5^2 - 4(2)(4)}}{2(2)}$ $s = \dfrac{-5 \pm \sqrt{-7}}{4}$ $s = \dfrac{-5 \pm i\sqrt{7}}{4}$

Addressing Individual Needs

Verbal/Linguistic Have students research the root words that formed the words quadratic and discriminant. **quadratic:** *quad* (Latin)—four, *quadrus* (Latin)—square; **discriminant:** *dis-* (Latin)—apart, *crimen* (Latin)—accusation; *discrimen* (Latin)—distinction Ask how the meanings of the root words relate to the mathematical meanings that they know. **A quadratic equation has a degree of two, which means that its variable of greatest power is squared. The discriminant of a quadratic equation points to (accuses) the separate (apart) roots of the equation; that is, it distinguishes the roots from the rest of the equation.**

In-Class Example

Example 5

Solve $2x^2 + 3x + 5 = 0$.

$$\frac{-3 \pm i\sqrt{31}}{4}$$

3 PRACTICE/APPLY

Error Analysis

Watch for students who incorrectly complete the square.

Prevent by having them expand several binomial squares in order to observe the pattern and then reverse the process to complete the square.

Study Guide and Practice Masters, pp. 61–62

NAME _____ DATE _____ PERIOD _____

4-2   **Study Guide**

Quadratic Equations

A quadratic equation is a polynomial equation with a degree of 2. Solving quadratic equations by graphing usually does not yield exact answers. Also, some quadratic expressions are not factorable. However, solutions can always be obtained by **completing the square.**

Example 1 Solve $x^2 - 12x + 7 = 0$ by completing the square.
$$x^2 - 12x + 7 = 0$$
$$x^2 - 12x = -7 \qquad \text{Subtract 7 from each side.}$$
$$x^2 - 12x + 36 = -7 + 36 \qquad \text{Complete the square by adding } \left[\frac{1}{2}(-12)\right]^2,$$
$$\text{or 36, to each side.}$$
$$(x - 6)^2 = 29 \qquad \text{Factor the perfect square trinomial.}$$
$$x - 6 = \pm\sqrt{29} \qquad \text{Take the square root of each side.}$$
$$x = 6 \pm\sqrt{29} \qquad \text{Add 6 to each side.}$$

The roots of the equation are $6 \pm \sqrt{29}$.

Completing the square can be used to develop a general formula for solving any quadratic equation of the form $ax^2 + bx + c = 0$. This formula is called the **Quadratic Formula** and can be used to find the roots of any quadratic equation.

| Quadratic Formula | If $ax^2 + bx + c = 0$ with $a \neq 0$, $x = \frac{-b \pm \sqrt{b^2 - 4ac}}{2a}$. |

In the Quadratic Formula, the radicand $b^2 - 4ac$ is called the **discriminant** of the equation. The discriminant tells the nature of the roots of a quadratic equation or the zeros of the related quadratic function.

Example 2 Find the discriminant of $2x^2 - 3x = 7$ and describe the nature of the roots of the equation. Then solve the equation by using the Quadratic Formula.
Rewrite the equation using the standard form $ax^2 + bx + c = 0$.
$$2x^2 - 3x - 7 = 0 \quad a = 2, b = -3, \text{ and } c = -7$$
The value of the discriminant $b^2 - 4ac$ is $(-3)^2 - 4(2)(-7)$, or 65.
Since the value of the discriminant is greater than zero, there are two distinct real roots.

Now substitute the coefficients into the quadratic formula and solve.
$$x = \frac{-(-3) \pm \sqrt{(-3)^2 - 4(2)(-7)}}{2(2)} \qquad x = \frac{-b \pm \sqrt{b^2 - 4ac}}{2a}$$
$$x = \frac{3 \pm \sqrt{65}}{4}$$
The roots are $\frac{3 + \sqrt{65}}{4}$ and $\frac{3 - \sqrt{65}}{4}$.

© Glencoe/McGraw-Hill   61   *Advanced Mathematical Concepts*

Example **⑤** Solve $6x^2 + x + 2 = 0$.

Method 1: Graphing
Graph $y = 6x^2 + x + 2$.

[−10, 10] scl:1 by [−50, 50] scl:1

The graph does not touch the x-axis, so there are no real roots for this equation. You cannot determine the roots from the graph.

Method 2: Factoring
Find the discriminant.

$$b^2 - 4ac = 1^2 - 4(6)(2) \text{ or } -47$$

The discrimimant is less than zero, so factoring cannot be used to solve the equation.

Method 3: Completing the Square
$$6x^2 + x + 2 = 0$$
$$x^2 + \frac{1}{6}x + \frac{1}{3} = 0$$
$$x^2 + \frac{1}{6}x = -\frac{1}{3}$$
$$x^2 + \frac{1}{6}x + \frac{1}{144} = -\frac{1}{3} + \frac{1}{144}$$
$$\left(x + \frac{1}{12}\right)^2 = -\frac{47}{144}$$
$$x + \frac{1}{12} = \pm i\frac{\sqrt{47}}{12}$$
$$x = -\frac{1}{12} \pm i\frac{\sqrt{47}}{12}$$
$$x = \frac{-1 \pm i\sqrt{47}}{12}$$

Completing the square works, but this method requires a lot of steps.

The roots of the equation are $\frac{-1 \pm i\sqrt{47}}{12}$.

Method 4: Quadratic Formula
For this equation, $a = 6$, $b = 1, c = 2$.

$$x = \frac{-b \pm \sqrt{b^2 - 4ac}}{2a}$$
$$x = \frac{-1 \pm \sqrt{1^2 - 4(6)(2)}}{2(6)}$$
$$x = \frac{-1 \pm \sqrt{-47}}{12}$$
$$x = \frac{-1 \pm i\sqrt{47}}{12}$$

The Quadratic Formula works and requires fewer steps than completing the square.

CHECK FOR UNDERSTANDING

Communicating Mathematics

Read and study the lesson to answer each question.   1–2. See margin.

1. **Write** a short paragraph explaining how to solve $t^2 - 6t - 4 = 0$ by completing the square.

2. **Discuss** which method of solving $5p^2 - 13p + 7 = 0$ would be most appropriate. Explain. Then solve.

218   *Chapter 4   Polynomial and Rational Functions*

Additional Answers

1. Add 4 to each side of the equation to get $t^2 - 6t = 4$. Determine the value needed to make $t^2 - 6t$ a perfect square trinomial. Add this value (9) to each side. Take the square root of each side of the equation and solve the two resulting equations. $t = 3 \pm \sqrt{13}$

2. Quadratic Formula; since the leading coefficient does not equal 1 and the discriminant equals 185 which is not a perfect square, the Quadratic Formula would be the best way to get an exact answer. Completing the square can also be used, but errors in arithmetic are more likely. A graph will give only approximate solutions. $\frac{13 \pm \sqrt{29}}{10}$

3. Describe the discriminant of the equation represented by each graph.

a.

equals 0

b.

negative number

c.

positive number

4. *Math Journal* **Solve** $x^2 + 4x - 5 = 0$ using each of the four methods discussed in this lesson. Which method do you prefer? Explain. **See margin.**

Guided Practice

Solve each equation by completing the square.

5. $x^2 + 8x - 20 = 0$ **−10, 2**

6. $2a^2 + 11a - 21 = 0$ **−7, $\frac{3}{2}$**

Find the discriminant of each equation and describe the nature of the roots of the equation. Then solve the equation by using the Quadratic Formula.

7. $m^2 + 12m + 36 = 0$
0; 1 real; −6

8. $t^2 - 6t + 13 = 0$
−16; 2 imaginary; $3 \pm 2i$

Solve each equation.

9. $p^2 - 6p + 5 = 0$ **1, 5**

10. $r^2 - 4r + 10 = 0$ **$2 \pm i\sqrt{6}$**

11. **Electricity** On a cold day, a 12-volt car battery has a resistance of 0.02 ohms. The power available to start the motor is modeled by the equation $P = 12\,I - 0.02\,I^2$, where I is the current in amperes. What current is needed to produce 1600 watts of power to start the motor?   **200 or 400 amps**

EXERCISES

Practice

Solve each equation by completing the square.   **13. −8, 11**

 A

12. $z^2 - 2z - 24 = 0$ **−4, 6**
13. $p^2 - 3p - 88 = 0$
14. $x^2 - 10x + 21 = 0$ **3, 7**

15. $d^2 - \frac{3}{4}d + \frac{1}{8} = 0$
$\frac{1}{2}, \frac{1}{4}$
16. $3g^2 - 12g = -4$
$2 \pm \frac{2\sqrt{6}}{3}$
17. $t^2 - 3t - 7 = 0$
$\frac{3}{2} \pm \frac{\sqrt{37}}{2}$

20. 121;
2 real; $-\frac{3}{2}, \frac{1}{3}$

18. What value of c makes $x^2 - x + c$ a perfect square? $\frac{1}{4}$

21. −11;
2 imaginary;
$\frac{5 \pm i\sqrt{11}}{2}$

19. Describe the nature of the roots of the equation $4n^2 + 6n + 25$. Explain.
2 imaginary ; the discriminant is negative.

Find the discriminant of each equation and describe the nature of the roots of the equation. Then solve the equation by using the Quadratic Formula.

23. −140;
2 imaginary;
$\frac{1 \pm i\sqrt{35}}{4}$

20. $6m^2 + 7m - 3 = 0$
21. $s^2 - 5s + 9 = 0$

22. $36d^2 - 84d + 49 = 0$ **0; 1 real; $\frac{7}{6}$**
23. $4x^2 - 2x + 9 = 0$

24. 112; 2 real;
$\frac{-2 \pm 2\sqrt{7}}{3}$

24. $3p^2 + 4p = 8$
25. $2k^2 + 5k = 9$ **97; 2 real; $\frac{-5 \pm \sqrt{97}}{4}$**

26. What is the conjugate of $-7 - i\sqrt{5}$?   **$-7 + i\sqrt{5}$**

27. Name the conjugate of $5 - 2i$.   **$5 + 2i$**

 www.amc.glencoe.com/self_check_quiz

Lesson 4-2 Quadratic Equations **219**

Assignment Guide

Core: 13–37 odd, 38–45
Enriched: 12–32 even, 34–45

Additional Answer
4. Graphing

Factoring
$x^2 + 4x - 5 = 0$
$(x + 5)(x - 1) = 0$
$x + 5 = 0$ or $x - 1 = 0$
    $x = -5$ $x = 1$

Completing the Square
$x^2 + 4x - 5 = 0$
$x^2 + 4x = 5$
$x^2 + 4x + 4 = 5 + 4$
$(x + 2)^2 = 9$
$x + 2 = \pm 3$
$x + 2 = 3$ or $x + 2 = -3$
  $x = 1$ $x = -5$

Quadratic Formula
$$x = \frac{-4 \pm \sqrt{4^2 - 4(1)(-5)}}{2(1)}$$
$$x = \frac{-4 \pm \sqrt{36}}{2}$$
$$x = \frac{-4 \pm 6}{2}$$
$x = -2 \pm 3$
$x = -2 + 3$ or $x = -2 - 3$
$x = 1$ $x = -5$

See students' work.

34c.

36b.

37a.

38. $ax^2 + bx + c = 0$

$x^2 + \dfrac{b}{a}x + \dfrac{c}{a} = 0$

$x^2 + \dfrac{b}{a}x = -\dfrac{c}{a}$

$x^2 + \dfrac{b}{a}x + \left(\dfrac{b}{2a}\right)^2 = -\dfrac{c}{a} + \left(\dfrac{b}{2a}\right)^2$

$\left(x + \dfrac{b}{2a}\right)^2 = -\dfrac{c}{a} + \dfrac{b^2}{4a^2}$

$\left(x + \dfrac{b}{2a}\right)^2 = \dfrac{-4ac + b^2}{4a^2}$

$x + \dfrac{b}{2a} = \pm\dfrac{\sqrt{b^2 - 4ac}}{2a}$

$x = -\dfrac{b}{2a} \pm \dfrac{\sqrt{b^2 - 4ac}}{2a}$

$x = \dfrac{-b \pm \sqrt{b^2 - 4ac}}{2a}$

Solve each equation.

28. $\dfrac{5 \pm i\sqrt{83}}{6}$ 29. $-4, 7$ 30. $-5, \dfrac{1}{4}$

28. $3s^2 - 5s + 9 = 0$ **29.** $x^2 - 3x - 28 = 0$ **30.** $4w^2 + 19w - 5 = 0$

31. $4r^2 - r = 5$ **32.** $p^2 + 2p + 8 = 0$ **33.** $x^2 - 2x\sqrt{6} - 2 = 0$
$-1, \dfrac{5}{4}$ $-1 \pm i\sqrt{7}$ $\sqrt{6} \pm 2\sqrt{2}$

Applications and Problem Solving

34. Health   Normal systolic blood pressure is a function of age. For a woman, the normal systolic pressure P in millimeters of mercury (mm Hg) is modeled by $P = 0.01A^2 + 0.05A + 107$, where A is age in years.

   a. Use this model to determine the normal systolic pressure of a 25-year-old woman.   **114.5 mm Hg**

   b. Use this model to determine the age of a woman whose normal systolic pressure is 125 mm Hg.   **40 years old**

34c. See margin for graph. As a woman gets older, the normal systolic pressure increases.

   c. Sketch the graph of the function. Describe what happens to the normal systolic pressure as a woman gets older.

35. Critical Thinking   Consider the equation $x^2 + 8x + c = 0$. What can you say about the value of c if the equation has two imaginary roots? $c > 16$

36. Interior Design   Abey Numkena is an interior designer. She has been asked to locate an oriental rug for a new corporate office. As a rule, the rug should cover $\dfrac{1}{2}$ of the total floor area with a uniform width surrounding the rug.

36a. $(12 - 2x)$
$(16 - 2x) = 96$

   a. If the dimensions of the room are 12 feet by 16 feet, write an equation to model the situation.

   b. Graph the related function.   **See margin.**

   c. What are the dimensions of the rug?   **8 ft by 12 ft**

37. Entertainment   In an action movie, a stuntwoman jumps off a building that is 50 feet tall with an upward initial velocity of 5 feet per second. The distance $d(t)$ traveled by a free falling object can be modeled by the formula $d(t) = v_0 t - \dfrac{1}{2}gt^2$, where v_0 is the initial velocity and g represents the acceleration due to gravity. The acceleration due to gravity is 32 feet per second squared.

   a. Draw a graph that relates the woman's distance traveled with the time since the jump.   **See margin.**

   b. Name the x-intercepts of the graph.   **0 and about 0.3**

37c. The x-intercepts indicate when the woman is at the same height as the beginning of the jump.

   c. What is the meaning of the x-intercepts of the graph?

   d. Write an equation that could be used to determine when the stuntwoman will reach the safety pad on the ground. (*Hint:* The ground is -50 feet from the starting point.) $-50 = 5t - 16t^2$

   e. How long will it take the stuntwoman to reach the safety pad on the ground?   **about 1.93 s**

38. Critical Thinking   Derive the quadratic formula by completing the square if $ax^2 + bx + c = 0, a \neq 0$.   **See margin.**

Extra Practice   See p. A32.

Extra Credit

Show that if a and c have different signs, then $ax^2 + bx + c = 0, a \neq 0$, has two real roots.   **If a and c have different signs, then $ac < 0$. Thus, $-4ac > 0$ and $b^2 - 4ac$ would have a positive value, indicating two real roots.**

Mixed Review

39. $2; -\dfrac{1}{3}, \dfrac{1}{6};$
See margin.

42. $(45, 60)$

39. State the number of complex roots of the equation $18a^2 + 3a - 1 = 0$. Then find the roots and graph the related function. *(Lesson 4-1)*

40. Graph $y < |x| - 2$. *(Lesson 3-5)* **See margin.**

41. Find the inverse of $f(x) = (x - 9)^2$. *(Lesson 3-4)* $f^{-1}(x) = \pm\sqrt{x + 9}$

42. Solve the system of equations, $3x + 4y = 375$ and $5x + 2y = 345$. *(Lesson 2-1)*

43. **Sales** The Computer Factory is selling a 300 MHz computer system for $595 and a 350 MHz computer system for $619. At this rate, what would be the cost of a 400 MHz computer system? *(Lesson 1-4)* **$643**

44. Find the slope of the line whose equation is $3y + 8x = 12$. *(Lesson 1-3)* $-\dfrac{8}{3}$

45. **SAT/ACT Practice** The trinomial $x^2 + x - 20$ is exactly divisible by which binomial? **A**
   A $x - 4$ **B** $x + 4$ **C** $x + 6$ **D** $x - 10$ **E** $x - 5$

CAREER CHOICES

● Environmental Engineering ●

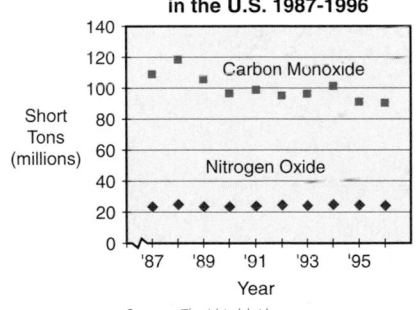

Would you like a career where you will constantly be learning and have the opportunity to work both outdoors and indoors? Environmental engineering has become an important profession in the past twenty-five years.

As an environmental engineer, you might design, build, or maintain systems for controlling wastes produced by cities or industry. These wastes can include solid waste, waste water, hazardous waste, or air pollutants. You could work for a private company, a consulting firm, or the Environmental Protection Agency. Opportunities for advancement in this field include becoming a supervisor or consultant. You might even have the opportunity to identify a new specialty area in the field of environmental engineering!

CAREER OVERVIEW

Degree Required:
Bachelor's degree in environmental engineering

Related Courses:
biology, chemistry, mathematics

Outlook:
number of jobs expected to increase though the year 2006

Emissions of Two Air Pollutants in the U.S. 1987–1996

Source: *The World Almanac*

interNET
CONNECTION   For more information about environmental engineering, visit: **www.amc.glencoe.com**

Lesson 4-2 Quadratic Equations **221**

CAREER CHOICES   ENVIRONMENTAL ENGINEERING

Math Connection
- Explain the trends in the data for both carbon monoxide and nitrogen oxide emissions. **The graph shows that carbon monoxide emissions have been primarily decreasing while nitrogen oxide emissions have remained about the same.**
- The carbon monoxide emission in 1988 was about 117,000 thousand short tons. The carbon monoxide emission in 1989 was about 104,000 thousand short tons. What was the percent decrease in emission from 1988 to 1989? **about 11%**

Open-Ended Assessment
Speaking Have students explain what the discriminant of a quadratic equation is and what its value reveals.

 Chapter 4, Quiz A (Lessons 4-1 and 4-2) is available in the *Assessment and Evaluation Masters,* p. 95.

Teaching Tip When exploring Career Choices, point out that a short ton equals 0.893 gross ton, and a gross ton equals 2240 pounds. Therefore, 1 short ton equals 2000 pounds.

Additional Answers

39.

40.

Enrichment Masters, p. 38

Lesson 4-2 **221**

1 FOCUS

5-Minute Check
Lesson 4-2

1. Solve $x^2 - 6x - 7 = 0$ by completing the square. **−1, 7**

Find the discriminant of each equation and describe the nature of the roots of the equation. Then solve the equation by using the Quadratic Formula.

2. $3x^2 - 4x + 7 = 0$ **−68; no real roots; $\dfrac{2 \pm i\sqrt{17}}{3}$**

3. $x^2 - 6x - 12 = 0$ **84; 2 real roots; $3 \pm \sqrt{21}$**

Solve each equation.

4. $x^2 - 36 = 0$ **±6**
5. $x^2 + 6x + 5 = 0$ **−5, −1**

Motivating the Lesson

Discussion Have students consider the following situation. Suppose you are told that the equation $x^4 - 23x^2 - 18x + 40 = 0$ has four real roots, and that two of them are −2 and −4. How could you find the other roots? **Sample answer: graph the function to find the other two places that it crosses the *x*-axis—at 5 and 1. Point out that students will be learning an easier method of discovering these roots in this lesson.**

4-3

The Remainder and Factor Theorems

OBJECTIVE
- Find the factors of polynomials using the Remainder and Factor Theorems.

SKIING On December 13, 1998, Olympic champion Hermann (The Herminator) Maier won the super-G at Val d'Isere, France. His average speed was 73 meters per second. The average recreational skier skis at a speed of about 5 meters per second. Suppose you were skiing at a speed of 5 meters per second and heading downhill, accelerating at a rate of 0.8 meter per second squared. How far will you travel in 30 seconds? *This problem will be solved in Example 1.*

Hermann Maier

Consider the polynomial function $f(a) = 2a^2 + 3a - 8$. Since 2 is a factor of 8, it may be possible that $a - 2$ is a factor of $2a^2 + 3a - 8$. Suppose you use long division to divide the polynomial by $a - 2$.

$$
\begin{array}{r}
2a + 7 \quad \longleftarrow quotient \\
a - 2\overline{)2a^2 + 3a - 8} \quad \longleftarrow dividend \\
\underline{2a^2 - 4a} \\
7a - 8 \\
\underline{7a - 14} \\
6 \quad \longleftarrow remainder
\end{array}
$$

divisor ⟶

From arithmetic, you may remember that the dividend equals the product of the divisor and the quotient plus the remainder. For example, $44 \div 7 = 6$ R2, so $44 = 7(6) + 2$. This relationship can be applied to polynomials.

You may want to verify that $(a - 2)(2a + 7) + 6 = 2a^2 + 3a - 8$.

Let $a = 2$.
$$
\begin{aligned}
f(a) &= (a - 2)(2a + 7) + 6 \\
f(2) &= (2 - 2)[2(2) + 7] + 6 \\
&= 0 + 6 \text{ or } 6
\end{aligned}
$$
⟷
$$
\begin{aligned}
f(a) &= 2a^2 + 3a - 8 \\
f(2) &= 2(2^2) + 3(2) - 8 \\
&= 8 + 6 - 8 \text{ or } 6
\end{aligned}
$$

Notice that the value of $f(2)$ is the same as the remainder when the polynomial is divided by $a - 2$. This example illustrates the **Remainder Theorem.**

Remainder Theorem	If a polynomial $P(x)$ is divided by $x - r$, the remainder is a constant $P(r)$, and $$P(x) = (x - r) \cdot Q(x) + P(r),$$ where $Q(x)$ is a polynomial with degree one less than the degree of $P(x)$.

The Remainder Theorem provides another way to find the value of the polynomial function $P(x)$ for a given value of r. The value will be the remainder when $P(x)$ is divided by $x - r$.

Resource Manager

 Reproducible Masters
- *Study Guide and Practice,* pp. 63–64
- *Enrichment,* p. 39
- *Spreadsheet Applications,* p. 4

 Transparencies
- 5-Minute Check, 4-3
- Teaching, 4-3

Example ❶ **SKIING** Refer to the application at the beginning of the lesson. The formula for distance traveled is $d(t) = v_0 t + \frac{1}{2}at^2$, where $d(t)$ is the distance traveled, v_0 is the initial velocity, t is the time, and a is the acceleration. Find the distance traveled after 30 seconds.

The distance formula becomes $d(t) = 5t + \frac{1}{2}(0.8)t^2$ or $d(t) = 0.4t^2 + 5t$. You can use one of two methods to find the distance after 30 seconds.

Method 1
Divide $0.4t^2 + 5t$ by $t - 30$.

$$
\begin{array}{r}
0.4t + 17 \\
t - 30 \overline{)0.4t^2 + 5t} \\
\underline{0.4t^2 - 12t} \\
17t + 0 \\
\underline{17t - 510} \\
510 \to D(30) = 510
\end{array}
$$

Method 2
Evaluate $d(t)$ for $t = 30$.

$$d(t) = 0.4t^2 + 5t$$
$$d(30) = 0.4(30^2) + 5(30)$$
$$= 0.4(900) + 5(30)$$
$$= 510$$

By either method, the result is the same. You will travel 510 meters in 30 seconds.

Long division can be very time consuming. **Synthetic division** is a shortcut for dividing a polynomial by a binomial of the form $x - r$. The steps for dividing $x^3 + 4x^2 - 3x - 5$ by $x + 3$ using synthetic division are shown below.

For x + 3, the value of r is −3.

Step 1 Arrange the terms of the polynomial in descending powers of x. Insert zeros for any missing powers of x. Then, write the coefficients as shown.

$$x^3 + 4x^2 - 3x - 5$$
$$\downarrow \quad \downarrow \quad \downarrow \quad \downarrow$$
$$1 \quad 4 \quad -3 \quad -5$$

Step 2 Write the constant r of the divisor $x - r$. In this case, write -3.

$$\underline{-3|} \quad 1 \quad 4 \quad -3 \quad -5$$

Step 3 Bring down the first coefficient.

$$\underline{-3|} \quad \underline{1 \quad 4 \quad -3 \quad -5}$$
$$1$$

Step 4 Multiply the first coefficient by r. Then write the product under the next coefficient. Add.

$$\underline{-3|} \quad 1 \quad 4 \quad -3 \quad -5$$
$$\times \quad \searrow \quad {-3}$$
$$1 \quad 1$$

Step 5 Multiply the sum by r. Then write the product under the next coefficient. Add.

$$\underline{-3|} \quad 1 \quad 4 \quad -3 \quad -5$$
$$-3 \quad -3$$
$$\times \quad 1 \quad 1 \quad -6$$

Notice that a vertical bar separates the quotient from the remainder.

Step 6 Repeat Step 5 for all coefficients in the dividend.

$$\underline{-3|} \quad 1 \quad 4 \quad -3 \quad -5$$
$$-3 \quad -3 \quad 18$$
$$\times \quad 1 \quad 1 \quad -6 | 13$$

Step 7 The final sum represents the remainder, which in this case is 13. The other numbers are the coefficients of the quotient polynomial, which has a degree one less than the dividend. Write the quotient $x^2 + x - 6$ with remainder 13. *Check the results using long division.*

Lesson 4-3 The Remainder and Factor Theorems **223**

Teaching Tip Point out that in the Remainder Theorem, $Q(x)$ represents the quotient polynomial.

Teaching Tip Students having difficulty understanding the Remainder Theorem might benefit from seeing some examples worked out in arithmetic, such as $235 \div 22 = 10$, R15 and $235 = 22 \times 10 + 15$. Working through a few examples, first with numbers and then with polynomials, should make the concepts clearer.

In-Class Example

Example 1
Real Estate Currently, 800 units of beachfront property have tenants paying $600 per week. Research indicates that for each $10 decrease in rent, 15 more units would be rented. The weekly revenue from the rentals is given by $R(x) = -150x^2 + 1000x + 480,000$, where x is the number of $10 decreases the property manager is willing to take. Find the revenue from the properties if the property manager decreases the rent by $50. **$481,250**

Teaching Tip In Example 2, emphasize the importance of leaving space in a long division problem or inserting a zero in a synthetic division problem if any powers of x in the dividend have coefficients of zero. Have each student do a few synthetic division problems independently when the concept is first presented so that the necessary steps will be learned.

In-Class Examples

Example 2
Divide $-y^6 + 4y^4 + 3y^2 + 2y$ by $y + 2$ using synthetic division.
$-y^5 + 2y^4 + 3y - 4$ with a remainder of 8

Example 3
Use the Remainder Theorem to find the remainder when $x^3 - x^2 - 5x - 3$ is divided by $x - 3$. State whether the binomial is a factor of the polynomial. Explain. **Since $f(3) = 0$, the remainder is 0. Therefore, the binomial $(x - 3)$ is a factor of the polynomial by the Remainder Theorem.**

Teaching Tip For Example 3, point out the connection between the graph of a polynomial divided by one of its binomial factors and the graph of the depressed polynomial after division by that factor. These graphs can be used to check the accuracy of students' division of polynomials by factors.

Example **2** **Divide $x^3 - x^2 + 2$ by $x + 1$ using synthetic division.**

$$
\begin{array}{r|rrrr}
-1 & 1 & -1 & 0 & 2 \\
   &   & -1 & 2 & -2 \\
\hline
   & 1 & -2 & 2 & \,|\,\, 0
\end{array}
$$

Notice there is no x term. A zero is placed in this position as a placeholder.

The quotient is $x^2 - 2x + 2$ with a remainder of 0.

In Example 2, the remainder is 0. Therefore, $x + 1$ is a factor of $x^3 - x^2 + 2$. If $f(x) = x^3 - x^2 + 2$, then $f(-1) = (-1)^3 - (-1)^2 + 2$ or 0. This illustrates the **Factor Theorem**, which is a special case of the Remainder Theorem.

Factor Theorem	The binomial $x - r$ is a factor of the polynomial $P(x)$ if and only if $P(r) = 0$.

Example **3** **Use the Remainder Theorem to find the remainder when $2x^3 - 3x^2 + x$ is divided by $x - 1$. State whether the binomial is a factor of the polynomial. Explain.**

Find $f(1)$ to see if $x - 1$ is a factor.

$f(x) = 2x^3 - 3x^2 + x$
$f(1) = 2(1^3) - 3(1^2) + 1$ *Replace x with 1.*
$\quad\;\; = 2(1) - 3(1) + 1$ or 0

Since $f(1) = 0$, the remainder is 0. So the binomial $(x - 1)$ is a factor of the polynomial by the Factor Theorem.

When a polynomial is divided by one of its binomial factors $x - r$, the quotient is called a **depressed polynomial**. A depressed polynomial has a degree less than the original polynomial. In Example 3, $x - 1$ is a factor of $2x^3 - 3x^2 + x$. Use synthetic division to find the depressed polynomial.

$$
\begin{array}{r|rrrr}
1 & 2 & -3 & 1 & 0 \\
  &   & 2 & -1 & 0 \\
\hline
  & 2 & -1 & 0 & \,|\,\, 0
\end{array}
$$

$$2x^2 - 1x + 0$$

Thus, $(2x^3 - 3x^2 + x) \div (x - 1) = 2x^2 - x$.

The depressed polynomial is $2x^2 - x$.

A depressed polynomial may also be the product of two polynomial factors, which would give you other zeros of the polynomial function. In this case, $2x^2 - x$ equals $x(2x - 1)$. So, the zeros of the polynomial function $f(x) = 2x^3 - 3x^2 + x$ are 0, $\frac{1}{2}$, and 1.

r	1	2	−16	−32
−4	1	−2	−8	0
−3	1	−1	−13	7
−2	1	0	−16	0
−1	1	1	−17	−15
0	1	2	−16	−32
1	1	3	−13	−45
2	1	4	−8	−48
3	1	5	−1	−35
4	1	6	8	0

Note that the values of r where no remainder occurs are also factors of the constant term of the polynomial.

You can also find factors of a polynomial such as $x^3 + 2x^2 - 16x - 32$ by using a shortened form of synthetic division to test several values of r. In the table, the first column contains various values of r. The next three columns show the coefficients of the depressed polynomial. The fifth column shows the remainder. Any value of r that results in a remainder of zero indicates that $x - r$ is a factor of the polynomial. The factors of the original polynomial are $x + 4$, $x + 2$, and $x - 4$.

Look at the pattern of values in the last column. Notice that when r = 1, 2, and 3, the values of f(x) decrease and then increase. This indicates that there is an x-coordinate of a relative minimum between 1 and 3.

Example **④** Determine the binomial factors of $x^3 - 7x + 6$.

Graphing Calculator Tip

The **TABLE** feature can help locate integral zeros. Enter the polynomial function as **Y₁** in the **Y=** menu and press **TABLE**. Search the **Y₁** column to find 0 and then look at the corresponding *x*-value.

Method 1 Use synthetic division.

r	1	0	−7	6
−4	1	−4	9	−30
−3	1	−3	2	0	*x + 3 is a factor.*
−2	1	−2	−3	12
−1	1	−1	−6	12
0	1	0	−7	6
1	1	1	−6	0	*x − 1 is a factor.*
2	1	2	−3	0	*x − 2 is a factor.*

Method 2 Test some values using the Factor Theorem.

$f(x) = x^3 - 7x + 6$
$f(-1) = (-1)^3 - 7(-1) + 6$ or 12
$f(1) = 1^3 - 7(1) + 6$ or 0

Because $f(1) = 0$, $x - 1$ is a factor. Find the depressed polynomial.

$$\underline{1|} \quad 1 \quad 0 \quad -7 \quad 6$$
$$\qquad\quad 1 \quad 1 \quad -6$$
$$\overline{\qquad 1 \quad 1 \quad -6 \; | \; 0}$$

The depressed polynomial is $x^2 + x - 6$. Factor the depressed polynomial.

$x^2 + x - 6 = (x + 3)(x - 2)$

The factors of $x^3 + x - 6$ are $x + 3$, $x - 1$, and $x - 2$. *Verify the results.*

The Remainder Theorem can be used to determine missing coefficients.

Example **⑤** Find the value of *k* so that the remainder of $(x^3 + 3x^2 - kx - 24) ÷ (x + 3)$ is 0.

If the remainder is to be 0, $x + 3$ must be a factor of $x^3 + 3x^2 - kx - 24$. So, $f(-3)$ must equal 0.

$f(x) = x^3 + 3x^2 - kx - 24$
$f(-3) = (-3)^3 + 3(-3)^2 - k(-3) - 24$
$\qquad 0 = -27 + 27 + 3k - 24$ *Replace f(−3) with 0.*
$\qquad 0 = 3k - 24$
$\qquad 8 = k$

The value of *k* is 8. Check using synthetic division.

$$\underline{-3|} \quad 1 \quad\;\; 3 \quad -8 \quad -24$$
$$\qquad\qquad\;\; -3 \quad\;\; 0 \quad\;\; 24$$
$$\overline{\qquad\quad 1 \quad\;\; 0 \quad -8 \; | \; 0 \;✓}$$

Error Analysis

Watch for students who make errors using the shortened form of synthetic division as shown in Example 4.

Prevent by encouraging these students to use the longer synthetic technique to eliminate the need to keep track of computations mentally.

Assignment Guide

Core: 15–39 odd, 41–43, 45–53

Enriched: 14–40 even, 41–53

Additional Answer

1. The Remainder Theorem states that if a polynomial $P(x)$ is divided by $x - r$, the remainder is $P(r)$. If a division problem has a remainder of 0, then the divisor is a factor of the dividend. This leads to the Factor Theorem, which states that the binomial $x - r$ is a factor if and only if $P(r) = 0$.

Study Guide and Practice Masters, pp. 63–64

CHECK FOR UNDERSTANDING

Communicating Mathematics

Read and study the lesson to answer each question.   **1. See margin.**

1. **Explain** how the Remainder Theorem and the Factor Theorem are related.

2. **Write** the division problem illustrated by the synthetic division. What is the quotient? What is the remainder?
   $(x^3 - 4x^2 - 7x + 8) \div (x - 5); x^2 + x - 2; -2$

3. The degree of a polynomial is one more than the degree of its depressed polynomial.

3. **Compare** the degree of a polynomial and its depressed polynomial.

4. **You Decide**   Brittany tells Isabel that if $x + 3$ is a factor of the polynomial function $f(x)$, then $f(3) = 0$. Isabel argues that if $x + 3$ is a factor of $f(x)$, then $f(-3) = 0$. Who is correct? Explain.
   Isabel; if $f(-3) = 0$, then $(x - (-3))$ or $(x + 3)$ is a factor.

Guided Practice

Divide using synthetic division.

5. $(x^2 - x + 4) \div (x - 2)$
   $x + 1$, R6

6. $(x^3 + x^2 - 17x + 15) \div (x + 5)$
   $x^2 - 4x + 3$

Use the Remainder Theorem to find the remainder for each division. State whether the binomial is a factor of the polynomial.

7. $(x^2 + 2x - 15) \div (x - 3)$ **0; yes**

8. $(x^4 + x^2 + 2) \div (x - 3)$ **92; no**

Determine the binomial factors of each polynomial.

9. $x^3 - 5x^2 - x + 5$
   $(x - 5), (x + 1), (x - 1)$

10. $x^3 - 6x^2 + 11x - 6$
    $(x - 1), (x - 2), (x - 3)$

11. Find the value of k so that the remainder of $(x^3 - 7x + k) \div (x + 1)$ is 2.   **-4**

12. Let $f(x) = x^7 + x^9 + x^{12} - 2x^2$.
    a. State the degree of $f(x)$.   **12**
    b. State the number of complex zeros that $f(x)$ has.   **12**
    c. State the degree of the depressed polynomial that would result from dividing $f(x)$ by $x - a$.   **11**
    d. Find one factor of $f(x)$.   **$x, x^2, x^{11} + x^8 + x^6 - 2x$, or $x^{10} + x^7 + x^5 - 2$**

13. **Geometry**   A cylinder has a height 4 inches greater than the radius of its base. Find the radius and the height to the nearest inch if the volume of the cylinder is 5π cubic inches.   **$r = 1$ in., $h = 5$ in.**

interNET CONNECTION

Graphing Calculator Program
For a graphing calculator program that computes the value of a function visit
www.amc. glencoe.com

EXERCISES

Practice

A

16. $x^3 + 3x^2 + 6x + 12$, R23

17. $x^3 - 2x^2 - 4x + 8$

Divide using synthetic division.   **15. $x^2 - 6x + 9$, R-1**

14. $(x^2 + 20x + 91) \div (x + 7)$ **$x + 13$**

15. $(x^3 - 9x^2 + 27x - 28) \div (x - 3)$

16. $(x^4 + x^3 - 1) \div (x - 2)$

17. $(x^4 - 8x^2 + 16) \div (x + 2)$

18. $(3x^4 - 2x^3 + 5x^2 - 4x - 2) \div (x + 1)$
    $3x^3 - 5x^2 + 10 - 14$, R12

19. $(2x^3 - 2x - 3) \div (x - 1)$
    $2x^2 + 2x$, R-3

Use the Remainder Theorem to find the remainder for each division. State whether the binomial is a factor of the polynomial.

B

20. $(x^2 - 2) \div (x - 1)$ **-1; no**

21. $(x^5 + 32) \div (x + 2)$ **0; yes**

22. $(x^4 - 6x^2 + 8) \div (x - \sqrt{2})$ **0; yes**

23. $(x^3 - x + 6) \div (x - 2)$ **12; no**

24. $(4x^3 + 4x^2 + 2x + 3) \div (x - 1)$ **13; no**

25. $(2x^3 - 3x^2 + x) \div (x - 1)$ **0; yes**

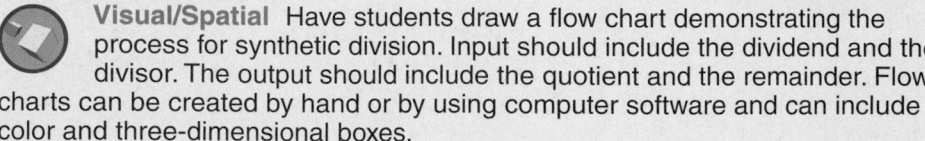

www.amc.glencoe.com/self_check_quiz

Addressing Individual Needs

Visual/Spatial   Have students draw a flow chart demonstrating the process for synthetic division. Input should include the dividend and the divisor. The output should include the quotient and the remainder. Flow charts can be created by hand or by using computer software and can include color and three-dimensional boxes.

26. Which binomial is a factor of the polynomial $x^3 + 3x^2 - 2x - 8$? **d**
  a. $x - 1$ **b.** $x + 1$ **c.** $x - 2$ **d.** $x + 2$

27. Verify that $x - \sqrt{6}$ is a factor of $x^4 - 36$. $\left(\sqrt{6}\right)^4 - 36 = 36 - 36$ or 0

28. Use synthetic division to find all the factors of $x^3 + 7x^2 - x - 7$ if one of the factors is $x + 1$. $(x - 1)(x + 1)(x + 7)$

29. $(x - 2)$, $(x + 1)$, $(x + 2)$

30. $(x - 1)$, $(x - 7)$, $(x + 7)$

31. $(x - 4)$, $(x - 2)$, $(x + 1)$

Determine the binomial factors of each polynomial.

29. $x^3 + x^2 - 4x - 4$ **30.** $x^3 - x^2 - 49x + 49$ **31.** $x^3 - 5x^2 + 2x + 8$

32. $x^3 - 2x^2 - 4x + 8$ **33.** $x^3 + 4x^2 - x - 4$ **34.** $x^3 + 3x^2 + 3x + 1$
  $(x - 2)$, $(x - 2)$, $(x + 2)$ $(x - 1)$, $(x + 1)$, $(x + 4)$ $(x + 1)$, $(x + 1)$, $(x + 1)$

35. How many times is 2 a root of $x^6 - 9x^4 + 24x^2 - 16 = 0$? **2 times**

36. Determine how many times -1 is a root of $x^3 + 2x^2 - x - 2 = 0$. Then find the other roots. **1 time; -2, 1**

Find the value of k so that each remainder is zero.

37. $(2x^3 - x^2 + x + k) \div (x - 1)$ **-2** **38.** $(x^3 - kx^2 + 2x - 4) \div (x - 2)$ **2**

39. $(x^3 + 18x^2 + kx + 4) \div (x + 2)$ **34** **40.** $(x^3 + 4x^2 - kx + 1) \div (x + 1)$ **-4**

Applications and Problem Solving

41. Bicycling Matthew is cycling at a speed of 4 meters per second. When he starts down a hill, the bike accelerates at a rate of 0.4 meter per second squared. The vertical distance from the top of the hill to the bottom of the hill is 25 meters. Use the equation $d(t) = v_0t + \frac{1}{2}at^2$ to find how long it will take Matthew to ride down the hill. **5 s**

42. $a = -5$, $b = 10$

42. Critical Thinking Determine a and b so that when $x^4 + x^3 - 7x^2 + ax + b$ is divided by $(x - 1)(x + 2)$, the remainder is 0.

43. Sculpting Esteban is preparing to start an ice sculpture. He has a block of ice that is 3 feet by 4 feet by 5 feet. Before he starts, he wants to reduce the volume of the ice by shaving off the same amount from the length, the width, and the height. **a.** $V(x) = -x^3 + 12x^2 - 47x + 60$
  a. Write a polynomial function to model the situation.
  b. Graph the function. **See margin.**
  c. He wants to reduce the volume of the ice to $\frac{3}{5}$ of the original volume. Write an equation to model the situation. **$36 = -x^3 + 12x^2 - 47x + 60$**
  d. How much should he take from each dimension? **about 0.60 ft**

44. Manufacturing An 18-inch by 20-inch sheet of cardboard is cut and folded to make a box for the Great Pecan Company.
  a. Write an polynomial function to model the volume of the box. $V(x) = 2x^3 - 38x^2 + 180x$
  b. Graph the function. **See margin.**
  c. The company wants the box to have a volume of 224 cubic inches. Write an equation to model this situation. **$224 = 2x^3 - 38x^2 + 180x$**
  d. Find a positive integer for x. **2 in.**

Additional Answers

43b.

44b.

Open-Ended Assessment

Speaking Have students take turns explaining the process and uses of synthetic division.

Additional Answer

52. $M\left(\dfrac{-7-2}{2},\dfrac{2+6}{2}\right)$ or $(-4.5, 4)$

$N\left(\dfrac{-2-2}{2},\dfrac{6-3}{2}\right)$ or $(-2, 1.5)$

slope of $\overline{MN} = \dfrac{4-1.5}{-4.5-(-2)}$ or -1

slope of $\overline{RI} = \dfrac{2-(-3)}{-7-(-2)}$ or -1

Since the slopes are the same, $\overline{MN} \parallel \overline{RI}$.

Enrichment Masters, p. 39

---

45. **Critical Thinking** Find a, b, and c for $P(x) = ax^2 + bx + c$ if $P(3 + 4i) = 0$ and $P(3 - 4i) = 0$. $a = 1, b = -6, c = 25$

Mixed Review

48. $(0, -32)$; point of inflection

49. wider than parent graph and moved 1 unit left

46. Solve $r^2 + 5r - 8 = 0$ by completing the square. *(Lesson 4-2)* $-\dfrac{5}{2} \pm \dfrac{\sqrt{57}}{2}$

47. Determine whether each number is a root of $x^4 - 4x^3 - x^2 + 4x = 0$. *(Lesson 4-1)*
   a. 2 **no**   b. 0 **yes**   c. −2 **no**   d. 4 **yes**

48. Find the critical points of the graph of $f(x) = x^5 - 32$. Determine whether each represents a *maximum*, a *minimum*, or a *point of inflection*. *(Lesson 3-6)*

49. Describe the transformation(s) that have taken place between the parent graph of $y = x^2$ and the graph of $y = 0.5(x + 1)^2$. *(Lesson 3-2)*

50. **Business** Pristine Pipes Inc. produces plastic pipe for use in newly-built homes. Two of the basic types of pipe have different diameters, wall thickness, and strengths. The strength of a pipe is increased by mixing a special additive into the plastic before it is molded. The table below shows the resources needed to produce 100 feet of each type of pipe and the amount of the resource available each week.

Resource	Pipe A	Pipe B	Resource Availability
Extrusion Dept.	4 hours	6 hours	48 hours
Packaging Dept.	2 hours	2 hours	18 hours
Strengthening Additive	2 pounds	1 pound	16 pounds

If the profit on 100 feet of type A pipe is \$34 and of type B pipe is \$40, how much of each should be produced to maximize the profit? *(Lesson 2-7)*   **300 ft of A, 600 ft of B**

51. Solve the system of equations. *(Lesson 2-2)*

$4x + 2y + 3z = 6$
$2x + 7y = 3z$
$-3x - 9y + 13 = -2z$ $\left(-\dfrac{91}{11}, \dfrac{68}{11}, \dfrac{98}{11}\right)$

52. **Geometry** Show that the line segment connecting the midpoints of sides $\overline{TR}$ and $\overline{TI}$ is parallel to $\overline{RI}$. *(Lesson 1-5)*   **See margin.**

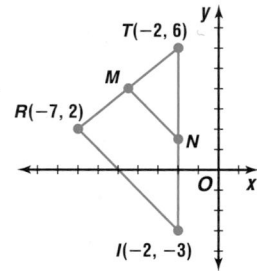

53. **SAT/ACT Practice** If $a > b$ and $c < 0$, which of the following are true?   **D**
   I. $ac < bc$
   II. $a + c > b + c$
   III. $a - c < b - c$

   A I only   **B** II only   **C** III only
   D I and II only   **E** I, II, and III

Extra Practice   See p. A32.

Extra Credit

Find the values of a, b, and c so that when $x^6 - 2x^4 + ax^2 + bx + c$ is divided by $(x - 1)$, $(x - 2)$, and $(x + 4)$, the remainder is 0.   **−125, 342, −216**

4-4

The Rational Root Theorem

OBJECTIVES
- Identify all possible rational roots of a polynomial equation by using the Rational Root Theorem.
- Determine the number of positive and negative real roots a polynomial function has.

CONSTRUCTION The longest and largest canal tunnel in the world was the Rove Tunnel on the Canal de Marseille au Rhone in the south of France. In 1963, the tunnel collapsed and excavation engineers were trying to duplicate the original tunnel. The height of the tunnel was 1 foot more than half its width. The length was 32 feet more than 324 times its width. The volume of the tunnel was 62,231,040 cubic feet. If the tunnel was a rectangular prism, find its original dimensions. *This problem will be solved in Example 4.*

The formula for the volume of a rectangular prism is $V = \ell wh$, where ℓ is the length, w is the width, and h is the height. From the information above, the height of the tunnel is $\frac{1}{2}w + 1$, and its length is $324w + 32$.

$$V = \ell wh$$
$$62{,}231{,}040 = (324w + 32)w\left(\frac{1}{2}w + 1\right) \qquad \ell = 324w + 32,\ h = \frac{1}{2}w + 1$$
$$62{,}231{,}040 = 162w^3 + 340w^2 + 32w$$
$$0 = 162w^3 + 340w^2 + 32w - 62{,}231{,}040$$
$$0 = 81w^3 + 170w^2 + 16w - 31{,}115{,}520 \qquad \textit{Divide each side by 2.}$$

We could use synthetic substitution to test possible zeros, but with such large numbers, this is not practical. In situations like this, the **Rational Root Theorem** can give direction in testing possible zeros.

Rational Root Theorem	Let $a_0x^n + a_1x^{n-1} + \ldots + a_{n-1}x + a_n = 0$ represent a polynomial equation of degree n with integral coefficients. If a rational number $\frac{p}{q}$, where p and q have no common factors, is a root of the equation, then p is a factor of a_n and q is a factor of a_0.

Example **1** **List the possible rational roots of $6x^3 + 11x^2 - 3x - 2 = 0$. Then determine the rational roots.**

According to the Rational Root Theorem, if $\frac{p}{q}$ is a root of the equation, then p is a factor of 2 and q is a factor of 6.

possible values of p: $\pm 1, \pm 2$

possible values of q: $\pm 1, \pm 2, \pm 3, \pm 6$

possible rational roots, $\frac{p}{q}$: $\pm 1, \pm 2, \pm \frac{1}{2}, \pm \frac{1}{3}, \pm \frac{1}{6}, \pm \frac{2}{3}$

(continued on the next page)

---

1 FOCUS

 5-Minute Check
Lesson 4-3

Divide by using synthetic division.
1. $(x^4 - 3x^2 + 6) \div (x + 1)$
   $x^3 - x^2 - 2x + 2;\ R\ 4$
2. $(x^3 + 2x^2 - 5x - 6) \div (x - 2)$
   $x^2 + 4x + 3;\ R\ 0$
3. $(x^4 - x^3 + 2x + 5) \div (x + 2)$
   $x^3 - 3x^2 + 6x - 10;\ R\ 25$

Use the Remainder Theorem to find the remainder for each division. State whether the binomial is a factor of the polynomial.
4. $(x^3 - 2x^2 - 33x + 10) \div (x + 5)$
   0; yes
5. $(x^3 - 7x + 5) \div (x + 3)$
   −1; no

Motivating the Lesson

Discussion Ask students what the factors of $3x^2 + 14x - 24$ might be. After some possibilities have been listed on the chalkboard, point out that they involve x or $3x$ as linear terms and factors of 24 as constant terms. Tell students that extending this fact to the case of higher-degree polynomials is one of the key ideas of this lesson.

Teaching Tip Students may find it easier to apply the theorems in this lesson if they work with a sketch of a typical polynomial of the same degree and sign of the leading coefficient as a guide. Such a sketch, showing where the function is positive or negative and where it crosses the x-axis, can help direct the search for roots.

Teaching Tip Point out that, in Example 1, students should be careful when applying the Rational Root Theorem not to confuse the roles of p and q. It is more important that they learn what p and q represent rather than memorize symbols.

In-Class Examples

Example 1
List the possible rational roots of $3x^3 - 13x^2 + 2x + 8 = 0$. Then determine the rational roots.

$\pm 1, \pm 2, \pm 4, \pm 8, \pm\frac{1}{3}, \pm\frac{2}{3}, \pm\frac{4}{3},$

$\pm\frac{8}{3}; 4, 1, -\frac{2}{3}$

Example 2
Find the roots of $x^3 + 6x^2 - 13x - 6 = 0$. $2, -4 \pm \sqrt{13}$

Teaching Tip In Example 1, remind students that inspecting the graph is the quickest way to narrow down the possible rational roots. When a root is found, the depressed polynomial can then be used to complete the problem. Remember that not all roots will necessarily be rational.

You can use a graphing utility to narrow down the possibilities. You know that all possible rational roots fall in the domain $-2 \le x \le 2$. So set your x-axis viewing window at $[-3, 3]$. Graph the related function $f(x) = 6x^3 + 11x^2 - 3x - 2$. A zero appears to occur at -2. Use synthetic division to check that -2 is a zero.

[−3, 3] scl:1 by [−10, 10] scl:1

$$\underline{-2}\,\begin{array}{r|rrrr} & 6 & 11 & -3 & -2 \\ & & -12 & 2 & 2 \\ \hline & 6 & -1 & -1 & 0 \end{array}$$

Thus, $6x^3 + 11x^2 - 3x - 2 = (x + 2)(6x^2 - x - 1)$. Factoring $6x^2 - x - 1$ yields $(3x + 1)(2x - 1)$. The roots of $6x^3 + 11x^2 - 3x - 2 = 0$ are $-2, -\frac{1}{3},$ and $\frac{1}{2}$.

A corollary to the Rational Root Theorem, called the **Integral Root Theorem,** states that if the leading coefficient a_0 has a value of 1, then any rational roots must be factors of a_n, $a_n \ne 0$.

Integral Root Theorem	Let $x^n + a_1 x^{n-1} + \ldots + a_{n-1}x + a_n = 0$ represent a polynomial equation that has a leading coefficient of 1, integral coefficients, and $a_n \ne 0$. Any rational roots of this equation must be integral factors of a_n.

Example **Find the roots of $x^3 + 8x^2 + 16x + 5 = 0$.**

There are three complex roots. According to the Integral Root Theorem, the possible rational roots of the equation are factors of 5. The possibilities are ± 5 and ± 1.

r	1	8	16	5	
5	1	13	81	410	
−5	1	3	1	0	← *There is a root at $x = -5$.*

The depressed polynomial is $x^2 + 3x + 1$. Use the quadratic formula to find the other two roots.

$$x = \frac{-b \pm \sqrt{b^2 - 4ac}}{2a}$$

$$x = \frac{-3 \pm \sqrt{3^2 - 4(1)(1)}}{2(1)} \qquad a = 1, b = 3, c = 1$$

$$x = \frac{-3 \pm \sqrt{5}}{2}$$

The three roots of the equation are $-5, \dfrac{-3 - \sqrt{5}}{2},$ and $\dfrac{-3 + \sqrt{5}}{2}$.

Descartes' Rule of Signs can be used to determine the possible number of positive real zeros a polynomial has. It is named after the French mathematician René Descartes, who first proved the theorem in 1637. *In Descartes' Rule of Signs, when we speak of a zero of the polynomial, we mean a zero of the corresponding polynomial function.*

Descartes' Rule of Signs	Suppose $P(x)$ is a polynomial whose terms are arranged in descending powers of the variable. Then the number of positive real zeros of $P(x)$ is the same as the number of changes in sign of the coefficients of the terms or is less than this by an even number. The number of negative real zeros of $P(x)$ is the same as the number of changes in sign of the coefficients of the terms of $P(-x)$, or less than this number by an even number.

Ignore zero coefficients when using this rule.

Example ❸ **Find the number of possible positive real zeros and the number of possible negative real zeros for $f(x) = 2x^5 + 3x^4 - 6x^3 + 6x^2 - 8x + 3$. Then determine the rational zeros.**

To determine the number of possible positive real zeros, count the sign changes for the coefficients.

$$f(x) = \quad 2x^5 \quad + \quad 3x^4 \quad - \quad 6x^3 \quad + \quad 6x^2 \quad - \quad 8x \quad + \quad 3$$

$$2 \qquad 3 \qquad -6 \qquad 6 \qquad -8 \qquad 3$$

no   yes   yes   yes   yes

There are four changes. So, there are four, two, or zero positive real zeros.

To determine the number of possible negative real zeros, find $f(-x)$ and count the number of sign changes.

$$f(-x) = 2(-x)^5 + 3(-x)^4 - 6(-x)^3 + 6(-x)^2 - 8(-x) + 3$$
$$f(-x) = \quad -2x^5 \quad + \quad 3x^4 \quad + \quad 6x^3 \quad + \quad 6x^2 \quad + \quad 8x \quad + \quad 3$$

$$-2 \qquad 3 \qquad 6 \qquad 6 \qquad 8 \qquad 3$$

yes   no   no   no   no

There is one change. So, there is one negative real zero.

Determine the possible zeros.
possible values of p: $\pm 1, \pm 3$
possible values of q: $\pm 1, \pm 2$
possible rational zeros, $\frac{p}{q}$: $\pm 1, \pm 3, \pm \frac{1}{2}, \pm \frac{3}{2}$

Test the possible zeros using the synthetic division and the Remainder Theorem.

r	2	3	-6	6	-8	3	
1	2	5	-1	5	-3	0	*1 is a zero.*
-1	2	1	-7	13	-21	24	
3	2	9	21	69	199	600	
-3	2	-3	3	-3	1	0	*-3 is a zero.*

(continued on the next page)

Graphing Calculator Exploration

When using **TABLE** on a graphing calculator to locate zeros, it is often helpful to first examine the graph of the function in order to get an approximation of possible values. Some graphing calculators have the split-screen capability, allowing the user to view both the graph and the table simultaneously.

Additional Answers
Graphing Calculator Exploration

4. **In the first function, there are 2 negative zeros, but according to Descartes' Rule of Signs, there should be 3 or 1 negative zeros. This is because the −2 is a double zero. In the second function, there is one negative zero, but according to Descartes' Rule of Signs, there should be 2 or 0 zeros. This is because −1 is a double root.**

5. **One number represents two zeros of the function.**

In-Class Example

Example 4
Industry A manufacturer produces boxes for a calculator company. The boxes have a volume of 240 cm³. Their height is 6 cm less than their width, while their length is 1 cm less than twice their width. Find the dimensions of such a box.
2 cm × 8 cm × 15 cm

Since there is only one negative real zero and −3 is a zero, you do not need to test any other negative possibilities.

$\frac{1}{2}$	2	4	−4	4	−6	0	$\frac{1}{2}$ is a zero.
$\frac{3}{2}$	2	6	3	$10\frac{1}{2}$	$7\frac{3}{4}$	$14\frac{5}{8}$	

All possible rational roots have been considered. There are two positive zeros and one negative zero. The rational zeros for $f(x) = 2x^5 + 3x^4 - 6x^3 + 6x^2 - 8x + 3$ are $-3, \frac{1}{2}$, and 1. Use a graphing utility to check these zeros. Note that there appears to be three x-intercepts. You can use the zero function on the **CALC** menu to verify that the zeros you found are correct.

[−4, 4] scl:1 by [−10, 85] scl:5

You can use a graphing calculator to study Descartes' Rule of Signs.

GRAPHING CALCULATOR EXPLORATION

Remember that you can determine the location of the zeros of a function by analyzing its graph.

TRY THESE Graph each function to determine how many zeros appear to exist. Use the zero function in the CALC menu to approximate each zero.

1. $f(x) = x^4 + 4x^3 + 3x^2 - 4x - 4$ **3; 1, −1, −2**
2. $f(x) = x^3 - 3x - 2$ **2; −1, 2**

3. (1) 1; 3 or 1 (2) 1; 2 or 0
WHAT DO YOU THINK?

3. Use Descartes' Rule of Signs to determine the possible positive and negative real zeros of each function.

4. How do your results from Exercise 3 compare with your results using the **TABLE** feature? Explain. **See margin.**

5. What do you think the term "double zero" means? **See margin.**

Because the zeros of a polynomial function are the roots of a polynomial equation, Descartes' Rule of Signs can be used to determine the types of roots of the equation.

Example ④ **CONSTRUCTION** Refer to the application at the beginning of the lesson. Find the original dimensions of the Rove Tunnel.

To determine the dimensions of the tunnel, we must solve the equation $0 = 81w^3 + 170w^2 + 16w - 31,115,520$. According to Descartes' Rule of Signs, there is one positive real root and two or zero negative real roots. Since dimensions are always positive, we are only concerned with the one positive real root.

Addressing Individual Needs

Logical Have students compare the products of $(x - 3)(x - 2)(x - 1)$ and $(x + 3)(x + 2)(x + 1)$. Note that one of these polynomials has only positive roots while the other has only negative roots. What features of the polynomials themselves give us a clue to this fact? **The second product has all positive coefficients, while the first has alternating signs.**

Use a graphing utility to graph the related function $V(w) = 81w^3 + 170w^2 + 16w - 31{,}115{,}520$. Since the graph has 10 unit increments, the zero is between 70 and 80.

To help determine the possible rational roots, find the prime factorization of 31,115,520.

$$31{,}115{,}520 = 2^8 \times 3^2 \times 5 \times 37 \times 73$$

[−10, 100] scl:10 by [−10, 100] scl:10

Possible rational roots between 70 and 80 are 72 ($2^3 \times 3^2$), 73, and 74 (2×37). Use the Factor Theorem until the one zero is found.

$V(w) = 81w^3 + 170w^2 + 16w - 31{,}115{,}520$
$V(72) = 81(72^3) + 170(72^2) + 16(72) - 31{,}115{,}520$
$V(72) = 30{,}233{,}088 + 881{,}280 + 1152 - 31{,}115{,}520$
$V(72) = 0$

The original width was 72 feet, the original height was $\frac{1}{2}(72) + 1$ or 37 feet, and the original length was $324(72) + 32$ or 23,360 feet, which is over 4 miles.

CHECK FOR UNDERSTANDING

Communicating Mathematics

Read and study the lesson to answer each question.

1. ±1, ±2, ±3, ±6

1. **Identify** the possible rational roots for the equation $x^4 - 3x^2 + 6 = 0$.

2. **Explain** why the Integral Root Theorem is a corollary to the Rational Root Theorem.   2–4. See margin.

3. **Write** a polynomial function $f(x)$ whose coefficients have three sign changes. Find the number of sign changes that $f(-x)$ has. Describe the nature of the zeros.

4. *Math Journal* **Describe** several methods you could use to determine the rational zeros of a polynomial function. Which would you choose to use first? Explain.

Guided Practice

6. ±1, ±3, ±$\frac{1}{2}$, ±$\frac{3}{2}$; −3, $\frac{1}{2}$, 1

List the possible rational roots of each equation. Then determine the rational roots.

5. $x^3 - 4x^2 + x + 2 = 0$ ±1, ±2; 1

6. $2x^3 + 3x^2 - 8x + 3 = 0$

Find the number of possible positive real zeros and the number of possible negative real zeros for each function. Then determine the rational zeros.

7. $f(x) = 8x^3 - 6x^2 - 23x + 6$
   2 or 0; 1; $-1\frac{1}{2}$, $\frac{1}{4}$, 2

8. $f(x) = x^3 + 7x^2 + 7x - 15$
   1; 2 or 0; −5, −3, 1

9. **Geometry** A cone is inscribed in a sphere with a radius of 15 centimeters. If the volume of the cone is 1152π cubic centimeters, find the length represented by x.   **9 cm**

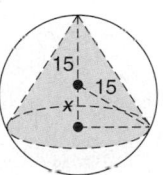

Lesson 4-4   The Rational Root Theorem   **233**

3 PRACTICE/APPLY

Error Analysis
Watch for students who incorrectly apply the Rational Root Theorem by interchanging the roles of p and q, thus giving the reciprocals of the possible roots.

Prevent by stressing the importance of constructing the rational values correctly.

Additional Answer
4. **Sample answer: you can factor the polynomial, graph the function, complete the square, or use the Quadratic Formula if it is a second-degree function, or use the Factor Theorem and the Rational Root Theorem. I would factor the polynomial if it can be factored easily. If not and it is a second-degree function, I would use the Quadratic Formula. Otherwise, I would graph the function on a graphing utility and use the Rational Root Theorem to find the exact zeros.**

Study Guide and Practice Masters, pp. 65–66

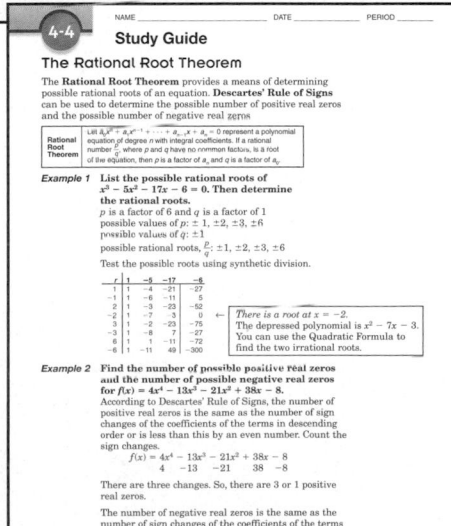

Additional Answers

2. If the leading coefficient is 1, then q must equal 1. Therefore, $\frac{p}{q}$ becomes $\frac{p}{1}$ or p, and p is defined as a factor of a_n.

3. Sample answer: $f(x) = x^3 - x^2 + x - 3$; 0; 3 or 1 possible positive zeros and no possible negative zeros

Additional Answer

23d. There are 2 negative zeros, but according to Descartes' Rule of Signs, there should be 3 or 1. This is because -1 is actually a zero twice.

EXERCISES

Practice

List the possible rational roots of each equation. Then determine the rational roots. **10.** $\pm 1, \pm 2, \pm 3, \pm 6; -3, -1, 2$ **11.** $\pm 1, \pm 2, \pm 3, \pm 6, \pm 9, \pm 18; -2$

10. $x^3 + 2x^2 - 5x - 6 = 0$ **11.** $x^3 - 2x^2 + x + 18 = 0$

12. $\pm 1, \pm 2;$
1, 2

12. $x^4 - 5x^3 + 9x^2 - 7x + 2 = 0$ **13.** $x^3 - 5x^2 - 4x + 20 = 0$

14. $2x^4 - x^3 - 6x + 3 = 0$ **15.** $6x^4 + 35x^3 - x^2 - 7x - 1 = 0$

13. $\pm 1, \pm 2, \pm 4,$
$\pm 5, \pm 10, \pm 20;$
$-2, 2, 5$

16. State the number of complex roots, the number of positive real roots, and the number of negative real roots of $x^4 - 2x^3 + 7x^2 + 4x - 15 = 0$.   **4; 3 or 1; 1**

14. $\pm 1, \pm 3, \pm \frac{1}{2},$
$\pm \frac{3}{2}, \frac{1}{2}$

Find the number of possible positive real zeros and the number of possible negative real zeros for each function. Then determine the rational zeros.

15. $\pm 1, \pm \frac{1}{2}, \pm \frac{1}{3},$
$\pm \frac{1}{6}, -\frac{1}{3}, \frac{1}{2}$

17. $f(x) = x^3 - 7x - 6$ **18.** $f(x) = x^3 - 2x^2 - 8x$ **1; 1; $-2, 0, 4$**

19. $f(x) = x^3 + 3x^2 - 10x - 24$ **20.** $f(x) = 10x^3 - 17x^2 - 7x + 2$

21. $f(x) = x^4 + 2x^3 - 9x^2 - 2x + 8$ **22.** $f(x) = x^4 - 5x^2 + 4$
    2 or 0; 2 or 0; $-4, -1, 1, 2$ **2 or 0; 2 or 0; $-2, -1, 1, 2$**

17. 1; 2 or
0; $-2, -1, 3$

23. Suppose $f(x) = (x - 2)(x + 2)(x + 1)^2$.

19. 1; 2 or 0;
$-4, -2, 3$

   a. Determine the zeros of the function.   **2, -2, -1**

   b. Write $f(x)$ as a polynomial function. $f(x) = x^4 + 2x^3 - 3x^2 - 8x - 4$

20. 2 or 0; 1;
$-\frac{1}{2}, \frac{1}{5}, 2$

   c. Use Descartes' Rule of Signs to find the number of possible positive real roots and the number of possible negative real roots.   **1; 3 or 1**

   d. Compare your answers to part **a** and part **c**. Explain.   **See margin.**

Applications and Problem Solving

24. Manufacturing   The specifications for a new cardboard container require that the width for the container be 4 inches less than the length and the height be 1 inch less than twice the length.   **b. $2208 = 2\ell^3 - 9\ell^2 + 4\ell$**

   a. Write a polynomial function that models the volume of the container in terms of its length. $V(\ell) = 2\ell^3 - 9\ell^2 + 4\ell$

   b. Write an equation if the volume must be 2208 cubic inches.

25a. Sample answer: $x^4 + x^3 + x^2 + x + 3 = 0$

   c. Find the dimensions of the new container.   **12 in. × 8 in. × 23 in.**

25. Critical Thinking   Write a polynomial equation for each restriction.

   a. fourth degree with no positive real roots

25c. Sample answer: $x^3 - x = 0$

   b. third degree with no negative real roots   **Sample answer: $x^3 - x^2 - 2 = 0$**

   c. third degree with exactly one positive root and exactly one negative real root

26. Architecture   A hotel in Las Vegas, Nevada, is the largest pyramid in the United States. Prior to the construction of the building, the architects designed a scale model.

   a. If the height of the scale model was 9 inches less than its length and its base is a square, write a polynomial function that describes the volume of the model in terms of its length. $V(\ell) = \frac{1}{3}\ell^3 - 3\ell^2$

   b. If the volume of the model is 6300 cubic inches, write an equation describing the situation. $6300 = \frac{1}{3}\ell^3 - 3\ell^2$

   c. What were the dimensions of the scale model?   **base: 30 in. by 30 in., height: 21 in.**

27. **Construction** A steel beam is supported by two pilings 200 feet apart. If a weight is placed x feet from the piling on the left, a vertical deflection d equals $0.0000008x^2(200 - x)$. How far is the weight if the vertical deflection is 0.8 feet? **100 ft**

28. **Critical Thinking** Compare and contrast the graphs and zeros of $f(x) = x^3 + 3x^2 - 6x - 8$ and $g(x) = -x^3 - 3x^2 + 6x + 8$. **The graphs are reflections of each other over the x-axis. The zeros are the same.**

Mixed Review

29. Divide $x^2 - x - 56$ by $x + 7$ using synthetic division. *(Lesson 4-3)* **$x - 8$**

30. Find the discriminant of $4x^2 + 6x + 25 = 0$ and describe the nature of the roots. *(Lesson 4-2)* **-364; 2 imaginary**

31. Write the polynomial equation of least degree whose roots are 1, -1, 2, and -2. *(Lesson 4-1)* **$x^4 - 5x^2 + 4 = 0$**

32. **Business** The prediction equation for a set of data relating the year in which a car was rented as the independent variable to the weekly car rental fee as the dependent variable is $y = 4.3x - 8424.3$. Predict the average cost of renting a car in 2008. *(Lesson 1-6)* **$210.10**

33. **SAT/ACT Practice** If $\dfrac{2x - 3}{x} = \dfrac{3 - x}{2}$, which of the following could be a value for x? **A**

　A -3　　　B -1　　　C 37　　　D 5　　　E 15

4. 441; 2 real; $\dfrac{3}{2}$, 5

MID-CHAPTER QUIZ

1. Write the polynomial equation of least degree with roots 1, -1, $2i$, and $-2i$. (Lesson 4-1) **$x^4 + 3x^2 - 4 = 0$**

2. State the number of complex roots of $x^3 - 11x^2 + 30x = 0$. Then find the roots. (Lesson 4-1) **3; 0, 5, 6**

3. Solve $x^2 + 5x = 150$ by completing the square. (Lesson 4-2) **-15, 10**

4. Find the discriminant of $6b^2 - 39b + 45 = 0$ and describe the nature of the roots of the equation. Then solve the equation by using the quadratic formula. (Lesson 4-2)

5. Divide $x^3 + 3x^2 - 2x - 8$ by $x + 2$ using synthetic division. (Lesson 4-3) **$x^2 + x - 4$**

6. Use the Remainder Theorem to find the remainder for $(x^3 - 4x^2 + 2x - 6) \div (x - 4)$. State whether the binomial is a factor of the polynomial. (Lesson 4-3) **2; no**

7. Determine the binomial factors of $x^3 - 2x^2 - 5x + 6$. (Lesson 4-3) **$(x - 3)(x - 1)(x + 2)$**

8. List the possible rational roots of $x^3 + 6x^2 + 10x + 3 = 0$. Then determine the rational roots. (Lesson 4-4) **± 1, ± 3; -3**

9. Find the number of possible positive zeros and the number of possible negative zeros for $F(x) = x^4 + 4x^3 + 3x^2 - 4x - 4$. Then determine the rational zeros. (Lesson 4-4) **1; 3 or 1; -2, -1, 1**

10. **Manufacturing** The Universal Paper Product Company makes cone-shaped drinking cups. The height of each cup is 6 centimeters more than the radius. If the volume of each cup is 27π cubic centimeters, find the dimensions of the cup. (Lesson 4-4) **$r = 3$ cm, $h = 9$ cm**

Extra Practice See p. A32.

---

4 ASSESS

Open-Ended Assessment

Writing Have students generate a polynomial function with one or more real roots. They should solve it using the techniques covered in the lesson and write a list of all of the steps used in the solution.

Mid-Chapter Quiz

The Mid-Chapter Quiz provides students with a brief review of the concepts and skills in Lessons 4-1 through 4-4. Lesson numbers are given to the right of exercises or instruction lines so students can review concepts not yet mastered.

📁 **Chapter 4, Quiz B (Lessons 4-3 and 4-4)** is available in the *Assessment and Evaluation Masters,* p. 95.

📁 **Chapter 4 Mid-Chapter Test (Lessons 4-1 through 4-4)** is available in the *Assessment and Evaluation Masters,* p. 94.

Enrichment Masters, p. 40

Extra Credit

Explain why the theorems in this lesson deal exclusively with real, rather than imaginary, roots. **Imaginary numbers are not rational. Thus, they cannot satisfy the hypotheses of the theorems.**

5-Minute Check
Lesson 4-4

List the possible roots of each equation. Then determine the rational roots.

1. $2x^4 - x^3 - 3x^2 - 31x - 15 = 0$
$\pm1, \pm3, \pm5, \pm15, \pm\frac{1}{2}, \pm\frac{3}{2},$
$\pm\frac{5}{2}, \pm\frac{15}{2}; 3, -\frac{1}{2}$

2. $x^4 - x^3 + 2x^2 - 4x - 8 = 0$
$\pm1, \pm2, \pm4, \pm8; -1, 2$

3. Find the roots of $x^3 + 7x^2 + 14x + 8 = 0$. $-1, -2, -4$

Find the number of possible positive real zeros and the number of possible negative real zeros for each function.

4. $f(x) = x^5 - x^4 - 8x^3 + 8x^2 - 9x + 9$ **0, 2, or 4; 1**

5. $g(x) = 3x^4 + 16x^3 - 7x^2 - 64x - 20$ **1; 1 or 3**

Motivating the Lesson

Hands-On Activity Have students consider the polynomial function $f(x) = x^4 + 2x^3 + 3x^2 + 2x - 2$. Ask them: *How many possible positive real zeros does it have?* **1** Then ask them to graph the function. Ask them: *Between which two integers does this positive real zero lie?* **between 0 and 1**

Locating Zeros of a Polynomial Function

OBJECTIVE
* Approximate the real zeros of a polynomial function.

Real World Application

ECONOMY Layoffs at large corporations can cause the unemployment rate to increase, while low interest rates can bolster employment. From October 1997 to November 1998, the Texas economy was strong. The Texas jobless rate during that period can be modeled by the function $f(x) = -0.0003x^4 + 0.0066x^3 - 0.0257x^2 - 0.1345x + 5.35$, where x represents the number of months since October 1997 and $f(x)$ represents the unemployment rate as a percent. Use this model to predict when the unemployment will be 2.5%.
This problem will be solved in Example 4.

The function $f(x) = -0.0003x^4 + 0.0066x^3 - 0.0257x^2 - 0.1345x + 5.35$ has four complex zeros. According to the Descartes' Rule of Signs, there are three or one positive real zeros and one negative zero. If you used a spreadsheet to evaluate the possible rational zeros, you will discover that none of the possible values is a zero of the function. This means that the zeros are not rational numbers. Another method, called the **Location Principle,** can be used to help determine the zeros of a function.

The Location Principle	Suppose $y = f(x)$ represents a polynomial function with real coefficients. If a and b are two numbers with $f(a)$ negative and $f(b)$ positive, the function has at least one real zero between a and b.

If $f(a) > 0$ and $f(b) < 0$, then the function also has at least one real zero between a and b.

This principle is illustrated by the graph. The graph of $y = f(x)$ is a continuous curve. At $x = a$, $f(a)$ is negative. At $x = b$, $f(b)$ is positive. Therefore, between the x-values of a and b, the graph must cross the x-axis. Thus, a zero exists somewhere between a and b.

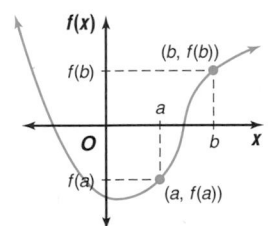

236 Chapter 4 *Polynomial and Rational Functions*

Resource Manager

Reproducible Masters
* *Study Guide and Practice,* pp. 67–68
* *Enrichment,* p. 41

Transparencies
* 5-Minute Check, 4-5
* Teaching, 4-5

Technology/Multimedia
* Advanced MathPASS, Lesson 9

Example **1** Determine between which consecutive integers the real zeros of $f(x) = x^3 - 4x^2 - 2x + 8$ are located.

There are three complex zeros for this function. According to Descartes' Rule of Signs, there are two or zero positive real roots and one negative real root. You can use substitution, synthetic division, or the **TABLE** feature on a graphing calculator to evaluate the function for consecutive integral values of x.

Method 1: Synthetic Division

r	1	−4	−2	8	
−3	1	−7	19	−49	
−2	1	−6	10	−12	*change in signs*
−1	1	−5	3	5	
0	1	−4	−2	8	
1	1	−3	−5	3	*change in signs*
2	1	−2	−6	−4	
3	1	−1	−5	−7	
4	1	0	−2	0	*zero*
5	1	1	3	23	

Method 2: Graphing Calculator

Use the **TABLE** feature.

There is a zero at 4. The changes in sign indicate that there are also zeros between −2 and −1 and between 1 and 2. This result is consistent with the Descartes' Rule of Signs.

Once you know two integers between which a zero will fall, you can use substitution or a graphing calculator to approximate the zeros.

Example **2** Approximate the real zeros of $f(x) = 12x^3 - 19x^2 - x + 6$ to the nearest tenth.

There are three complex zeros for this function. According to Descartes' Rule of Signs, there are two or zero positive real roots and one negative real root.

Use the **TABLE** feature of a graphing calculator. There are zeros between −1 and 0, between 0 and 1, and between 1 and 2. To find the zeros to the nearest tenth, use the **TBLSET** feature changing ΔTbl to 0.1.

(continued on the next page)

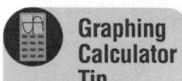

Graphing Calculator Tip

You can check these zeros by graphing the function and selecting **zero** under the **CALC** menu.

Since 0.25 is closer to zero than -2.832, the zero is about -0.5.

Since 0.106 is closer to zero than -0.816, the zero is about 0.7.

Since 0.288 is closer to zero than -1.046, the zero is about 1.4.

The zeros are about -0.5, 0.7, and 1.4.
If you need a closer approximation, change ΔTbl to 0.01.

The **Upper Bound Theorem** will help you confirm whether you have determined all of the real zeros. An **upper bound** is an integer greater than or equal to the greatest real zero.

Upper Bound Theorem	Suppose c is a positive real number and $P(x)$ is divided by $x - c$. If the resulting quotient and remainder have no change in sign, then $P(x)$ has no real zero greater than c. Thus, c is an upper bound of the zeros of $P(x)$.

Zero coefficients are ignored when counting sign changes.

The synthetic division in Example 1 indicates that 5 is an upper bound of the zeros of $f(x) = x^3 - 4x^2 - 2x + 8$ because there are no change of signs in the quotient and remainder.

A **lower bound** is an integer less than or equal to the least real zero. A lower bound of the zeros of $P(x)$ can be found by determining an upper bound for the zeros of $P(-x)$.

Lower Bound Theorem	If c is an upper bound of the zeros of $P(-x)$, then $-c$ is a lower bound of the zeros of $P(x)$.

238 Chapter 4 *Polynomial and Rational Functions*

Addressing Individual Needs

Auditory/Musical In music theory, a musical chord is named according to its root, or its first note. For example, the first note in the chord E-major is E and the whole chord is E-G-B. By knowing the root of a chord, a musician can find the rest of the notes in a chord. Play several arrangements of chords for the class, either live or tape-recorded. Have students try to identify the root of each chord. Most chords seem to "come to rest" at their roots. Then, discuss how the roots of chords are similar to roots of polynomial functions. **Sample answer: by knowing the roots of a polynomial function, one can determine other information such as the upper and lower bounds, or the solution to a real-world problem.**

Examples **③** Use the Upper Bound Theorem to find an integral upper bound and the Lower Bound Theorem to find an integral lower bound of the zeros of $f(x) = x^3 + 3x^2 - 5x - 10$.

The Rational Root Theorem tells us that ± 1, ± 2, ± 5, and ± 10 might be roots of the polynomial equation $x^3 + 3x^2 - 5x - 10 = 0$. These possible zeros of the function are good starting places for finding an upper bound.

$f(x) = x^3 + 3x^2 - 5x - 10$				
r	1	3	-5	-10
1	1	4	-1	-11
2	1	5	5	0

$f(-x) = -x^3 + 3x^2 + 5x - 10$				
r	-1	3	5	-10
1	-1	2	7	-3
2	-1	1	7	4
3	-1	0	5	5
4	-1	-1	1	-6
5	-1	-2	-5	-35

An upper bound is 2. Since 5 is an upper bound of $f(-x)$, -5 is a lower bound of $f(x)$. This means that all real zeros of $f(x)$ can be found in the interval $-5 \le x \le 2$.

④ **ECONOMICS** Refer to the application at the beginning of the lesson. Use the model to determine when the unemployment will be 2.5%.

You need to know when $f(x)$ has a value of 2.5.
$$2.5 = -0.0003x^4 + 0.0066x^3 - 0.0257x^2 - 0.1345x + 5.35$$
$$0 = -0.0003x^4 + 0.0066x^3 - 0.0257x^2 - 0.1345x + 2.85$$

Now search for the zero of the related function,
$g(x) = -0.0003x^4 + 0.0066x^3 - 0.0257x^2 - 0.1345x + 2.85$

r	-0.0003	0.0066	-0.0257	-0.1345	2.85
17	-0.0003	0.0015	-0.0002	-0.1379	0.5057
18	-0.0003	0.0012	-0.0041	-0.2083	-0.8994

There is a zero between 17 and 18 months.

Confirm this zero using a graphing calculator. The zero is about 17.4 months. So, about 17 months after October, 1997 or March, 1999, the unemployment rate would be 2.5%. *Equations dealing with unemployment change as the economic conditions change. Therefore, long range predictions may not be accurate.*

Zero
X=17.398998   Y=1.1E-12

[-10, 20] scl:1 by [-3, 5] scl:1

CHECK FOR UNDERSTANDING

Communicating Mathematics

Read and study the lesson to answer each question.

1. **Write a convincing argument** that the Location Principle works. Include a labeled graph. **See margin.**

Lesson 4-5 Locating Zeros of a Polynomial Function **239**

Additional Answer
1. If the function is negative for one value and positive for another value, the function must cross the x-axis in at least one point between the two values.

Teaching Tip In Example 3, some students may need to see a few more steps in order to find $f(-x)$. Then, have them find $f(-x)$ for another function.

In-Class Examples
Example 3
Use the Upper Bound Theorem to find an integral upper bound and the Lower Bound Theorem to find an integral lower bound of the zeros of $f(x) = 6x^3 - 7x^2 - 14x + 15$. **Sample answer: 3, -2**

Example 4
Packaging The Toaster Treats company uses a box with a square bottom to package its product. The height of the box is 3 inches more than the length of the bottom. Find the dimensions of the box if the volume is 42 in^3. **2.7 in. × 2.7 in. × 5.7 in.**

3 PRACTICE/APPLY

Error Analysis
Watch for students who treat approximate roots as exact roots.

Prevent by stressing the definition of *approximate* and showing that an approximate root will not likely evaluate as 0 when substituted into the original function.

Study Guide and Practice Masters, pp. 67–68

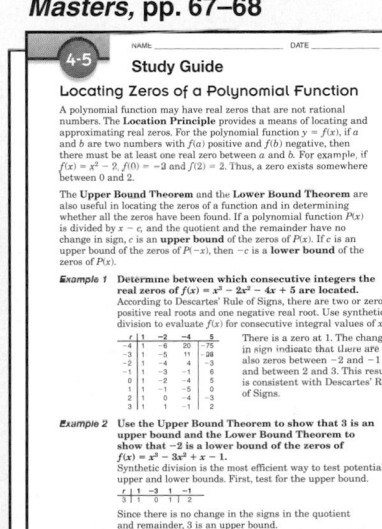

Assignment Guide

Core: 13–31 odd, 33–37,
    39–45
Enriched: 12–32 even, 33–45

Additional Answers

2. Use synthetic division to find the values of the polynomial function for consecutive integers. When the values of the function change from positive to negative or from negative to positive, there is a zero between the integers.

3. Use synthetic division to find the values of the polynomial function for consecutive integers. An integer that produces no sign change in the quotient and the remainder is an upper bound. To find a lower bound of a function, find an upper bound for the function of $-x$. The lower bound is the negative of the upper bound for the function of $-x$.

---

2. Explain how to use synthetic division to determine which consecutive integers the real zeros of a polynomial function are located.   **See margin.**

3. Describe how you find an upper bound and a lower bound for the zeros of a polynomial function.   **See margin.**

4. You Decide   After looking at the table on a graphing calculator, Tiffany tells Nikki that there is a zero between -1 and 0. Nikki argues that the zero is between -2 and -1. Who is correct? Explain.   **Nikki; the sign changes between -2 and -1.**

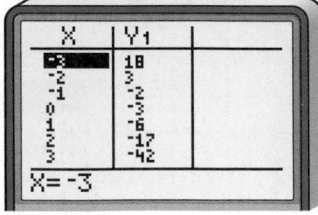

Guided Practice   Determine between which consecutive integers the real zeros of each function are located.

5. $f(x) = x^2 - 4x - 2$
4 and 5, -1 and 0

6. $f(x) = x^3 - 3x^2 - 2x + 4$
-2 and -1, at 1, 3 and 4

Approximate the real zeros of each function to the nearest tenth.

7. $f(x) = 2x^3 - 4x^2 - 3$ **2.3**

8. $f(x) = x^2 + 3x + 2$ **$-2, -1$**

Use the Upper Bound Theorem to find an integral upper bound and the Lower Bound Theorem to find an integral lower bound of the zeros of each function.   **9–10. Sample answers given.**

9. $f(x) = x^4 - 8x + 2$ **2; 0**

10. $f(x) = x^4 + x^2 - 3$ **2; -2**

11. Manufacturing   The It's A Snap Puzzle Company is designing new boxes for their 2000 piece 3-D puzzles. The old box measured 25 centimeters by 30 centimeters by 5 centimeters. For the new box, the designer wants to increase each dimension by a uniform amount.

11a. $V(x) = x^3 + 60x^2 + 1025x + 3750$

   a. Write a polynomial function that models the volume of the new box.

   b. The volume of the new box must be 1.5 times the volume of the old box to hold the increase in puzzle pieces. Write an equation that models this situation.   **$5625 = x^3 + 60x^2 + 1025x + 3750$**

   c. Find the dimensions of the new box.   **about 26.7 cm by 31.7 cm by 6.7 cm**

EXERCISES

Practice   Determine between which consecutive integers the real zeros of each function are located.

12. $f(x) = x^3 - 2$ **1 and 2**

13. $f(x) = 2x^2 - 5x + 1$

13. 0 and 1, 2 and 3

14. $f(x) = x^4 - 2x^3 + x - 2$

14. at -1, at 2

15. -3 and -2, -2 and -1, 1 and 2, 2 and 3

15. $f(x) = x^4 - 8x^2 + 10$

16. $f(x) = x^3 - 3x + 1$
-2 and -1, 0 and 1, 1 and 2

17. $f(x) = 2x^4 + x^2 - 3x + 3$
no real zeros

18. Is there a zero of $f(x) = 6x^3 + 24x^2 - 54x - 3$ between -6 and -5? Explain.
yes; $f(-6) = -111$, $f(-5) = 117$

Approximate the real zeros of each function to the nearest tenth.

19. $f(x) = 3x^4 + x^2 - 1$ **$-0.7, 0.7$**

20. $f(x) = x^2 + 3x + 1$ **$-2.6, -0.4$**

21. $f(x) = x^3 - 4x + 6$ **-2.5**

22. $f(x) = x^4 - 5x^3 + 6x^2 - x - 2$ **$-0.4, 3.4$**

23. $f(x) = 2x^4 - x^3 + x - 2$ **$-1, 1$**

24. $f(x) = x^5 - 7x^4 - 3x^3 + 2x^2 - 4x + 9$
$-1.3, 0.9, 7.4$

25. Approximate the real zero of $f(x) = x^3 - 2x^2 + 5$ to the nearest hundredth.
-1.24

 www.amc.glencoe.com/self_check_quiz

26–31. Sample answers given.

28. 6; −2

Use the Upper Bound Theorem to find an integral upper bound and the Lower Bound Theorem to find an integral lower bound of the zeros of each function.

26. $f(x) = 3x^3 - 2x^2 + 5x - 1$ **1; 0**   **27.** $f(x) = x^2 - x - 1$ **2; −1**

28. $f(x) = x^4 - 6x^3 + 2x^2 + 6x - 13$ **29.** $f(x) = x^3 + 5x^2 - 3x - 20$ **2; −6**

30. $f(x) = x^4 - 3x^3 - 2x^2 + 3x - 5$ **31.** $f(x) = x^5 + 5x^4 - 3x^3 + 20x^2 - 15$
4; −2   **1; −7**

32. Analyze the zeros of $f(x) = x^4 - 3x^3 - 2x^2 + 3x - 5$.

   a. Determine the number of complex zeros.   **4**

   b. List the possible rational zeros.   **±1, ±5**

   c. Determine the number of possible positive real zeros and the number of possible negative real zeros.   **3 or 1; 1**   d. −2 and −1, 3 and 4

   d. Determine the integral intervals where the zeros are located.

   e. Determine an integral upper bound of the zeros and an integral lower bound of the zeros.   **Sample answers: 4; −2**

   f. Determine the zeros to the nearest tenth.   **−1.4, 3.4**

Applications and Problem Solving

33a. The model is fairly close, although it is less accurate for 1950 and 1970.

33. Population   The population of Manhattan Island in New York City between 1890 to 1970 can be modeled by $P(x) = -0.78x^4 + 133x^3 - 7500x^2 + 147,500x + 1,440,000$, where $P(x)$ represents the population and x represents the number of years since 1890.

   a. According to the data at the right, how valid is the model?

   b. Use the model to predict the population in 1980.   **−253,800**

   c. According to this model, what happens between 1970 and 1980?   **The population becomes 0.**

   d. Do you think this model is valid for any time? Explain.
   No; there are still many people living in Manhattan.

Population of Manhattan Island

Year	Population
1890	1,441,216
1910	2,331,542
1930	1,867,312
1950	1,960,101
1970	1,539,233

Source: U.S. Bureau of the Census

34. Critical Thinking   Write a third-degree integral polynomial function with one zero at $\sqrt{2}$. State the zeros of the function. Draw a graph to support your answer.   **Sample answer:** $f(x) = x^3 + x^2 - 2x - 2$; $\pm \sqrt{2}$, −1;
See margin for graph.

35. Medicine   A doctor tells Masa to take 60 milligrams of medication each morning for three days. The amount of medication remaining in his body on the fourth day is modeled by $M(x) = 60x^3 + 60x^2 + 60x$, where x represents the absorption rate per day. Suppose Masa has 37.44 milligrams of medication in his body on the fourth day.

   a. Write an equation to model this situation. $37.44 = 60x^3 + 60x^2 + 60x$

   b. Write the related function for the equation. $f(x) = 60x^3 + 60x^2 + 60x - 37.44$

35c. See margin for graph; about $\frac{1}{2}$.

   c. Graph this function and estimate the absorption rate.

   d. Find the absorption rate of the medication.   **0.4**

36. Critical Thinking   Write a polynomial function with an upper bound of 1 and a lower bound of −1.   **Sample answer:** $f(x) = x^2 - 1$

Open-Ended Assessment

Modeling Give each student a cubic polynomial and have the students use **TABLE** on their graphing calculators in order to approximate the real zeros of the function. Then have them check the zeros graphically. Finally, they should use the Upper and Lower Bound Theorems to verify that all real zeros have been located.

Additional Answers

37a.

41.

Enrichment Masters, p. 41

 4-5

NAME _____ DATE _____ PERIOD _____

Enrichment

The Bisection Method for Approximating Real Zeros

The bisection method can be used to approximate zeros of polynomial functions like $f(x) = x^3 + x^2 - 3x - 3$. Since $f(1) = -4$ and $f(2) = 3$, there is at least one real zero between 1 and 2. The midpoint of this interval is $\frac{1+2}{2} = 1.5$. Since $f(1.5) = -1.875$, the zero is between 1.5 and 2. The midpoint of this interval is $\frac{1.5+2}{2} = 1.75$. Since $f(1.75) = 0.172$, the zero is between 1.5 and 1.75. $\frac{1.5+1.75}{2} = 1.625$ and $f(1.625) = -0.94$. The zero is between 1.625 and 1.75. The midpoint of this interval is $\frac{1.625+1.75}{2} = 1.6875$. Since $f(1.6875) = -0.41$, the zero is between 1.6875 and 1.75. Therefore, the zero is 1.7 to the nearest tenth. The diagram below summarizes the bisection method.

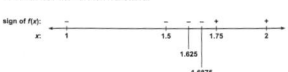

Using the bisection method, approximate to the nearest tenth the zero between the two integral values of each function.

1. $f(x) = x^3 - 4x^2 - 11x + 2, f(0) = 2, f(1) = -12$
   0.2

2. $f(x) = 2x^4 + x^2 - 15, f(1) = -12, f(2) = 21$
   1.6

3. $f(x) = x^5 - 2x^3 - 12, f(1) = -13, f(2) = 4$
   1.9

4. $f(x) = 4x^3 - 2x + 7, f(-2) = -21, f(-1) = 5$
   -1.3

5. $f(x) = 3x^3 - 14x^2 - 27x + 126, f(4) = -14, f(5) = 16$
   4.7

© Glencoe/McGraw-Hill   T41   *Advanced Mathematical Concepts*

---

37. Ecology In the early 1900s, the deer population of the Kaibab Plateau in Arizona experienced a rapid increase because hunters had reduced the number of natural predators. The food supply was not great enough to support the increased population, eventually causing the population to decline. The deer population for the period 1905 to 1930 can be modeled by $f(x) = -0.125x^5 + 3.125x^4 + 4000$, where x is the number of years from 1905.
a. Graph the function. **See margin.**
b. Use the model to determine the population in 1905. **4000 deer**
c. Use the model to determine the population in 1920. **about 67,281 deer**
d. According to this model, when did the deer population become zero? **in 1930**

38. Investments Instead of investing in the stock market, many people invest in collectibles, like baseball cards. Each year, Anna uses some of the money she receives for her birthday to buy one special baseball card. For the last four birthdays, she purchased cards for $6, $18, $24, and $18. The current value of these cards is modeled by $T(x) = 6x^4 + 18x^3 + 24x^2 + 18x$, where x represents the average rate of return plus one.
a. If the cards are worth $81.58, write an equation to model this situation. **$81.58 = 6x^4 + 18x^3 + 24x^2 + 18x$**
b. Find the value of x. **about 1.1**
c. What is the average rate of return on Anna's investments? **about 10%**

Mixed Review

39. Find the number of possible positive real zeros and the number of possible negative real zeros for $f(x) = 2x^3 - 5x^2 - 28x + 15$. Then determine the rational zeros. *(Lesson 4-4)* **2 or 0; 1; -3, 0.5, 5**

40. Physics The distance $d(t)$ fallen by a free-falling body can be modeled by the formula $d(t) = v_0 t - \frac{1}{2}gt^2$, where v_0 is the initial velocity and g represents the acceleration due to gravity. The acceleration due to gravity is 9.8 meters per second squared. If a rock is thrown upward with an initial velocity of +4 meters per second from the edge of the North Rim of the Grand Canyon, which is 1750 meters deep, determine how long it will take the rock to reach the bottom of the Grand Canyon. (*Hint:* The distance to the bottom of the canyon is -1750 meters from the rim.) *(Lesson 4-2)* **about 19.3 s**

41. Graph $y = \frac{4x}{x-1}$. *(Lesson 3-7)* **See margin.**

42. Find the value of $\begin{vmatrix} 7 & 9 \\ 3 & 6 \end{vmatrix}$. *(Lesson 2-5)* **15**

43. Find the coordinates of the midpoint of $\overline{FG}$ given its endpoints $F(-3, -2)$ and $G(8, 4)$. *(Lesson 1-5)* **(2.5, 1)**

44. $\frac{1}{2}$; -2 **44.** Name the slope and y-intercept of the graph of $x - 2y - 4 = 0$. *(Lesson 1-4)*

45. SAT/ACT Practice $\triangle ABC$ and $\triangle ABD$ are right triangles that share side $\overline{AB}$. $\triangle ABC$ has area x, and $\triangle ABD$ has area y. If $\overline{AD}$ is longer than $\overline{AC}$ and $\overline{BD}$ is longer than $\overline{BC}$, which of the following cannot be true? **B**
A $y > x$ B $y < x$ C $y \geq x$ D $y \neq x$ E $\frac{y}{x} \geq 1$

   Extra Practice See p. A33.

Extra Credit

Is it possible for a polynomial function to have a real zero greater than its greatest possible rational root? If so, give an example. **Yes; sample answer: the function $f(x) = 2x^2 - 5x + 1$ has 1 as its greatest possible rational root, while its actual root is $\frac{5 + \sqrt{17}}{4}$ or about 2.28.**

4-6

Rational Equations and Partial Fractions

SCUBA DIVING If a scuba diver goes to depths greater than 33 feet, the function $T(d) = \frac{1700}{d - 33}$ gives the maximum time a diver can remain down and still surface at a steady rate with no decompression stops. In this function, $T(d)$ represents the dive time in minutes, and d represents the depth in feet. If a diver is planning a 45-minute dive, what is the maximum depth the diver can go without decompression stops on the way back up? *This problem will be solved in Example 1.*

To solve this problem, you need to solve the equation $45 = \frac{1700}{d - 33}$. This type of equation is called a **rational equation.** A rational equation has one or more rational expressions. One way to solve a rational equation is to multiply each side of the equation by the least common denominator (LCD).

Example **1** **SCUBA DIVING** What is the maximum depth the diver can go without decompression stops on the way back up?

$$T(d) = \frac{1700}{d - 33}$$

$$45 = \frac{1700}{d - 33} \qquad \text{\textit{Replace T(d) with 45, the dive time in minutes.}}$$

$$45(d - 33) = \frac{1700}{d - 33}(d - 33) \qquad \text{\textit{Multiply each side by the LCD, d − 33.}}$$

$$45d - 1485 = 1700$$

$$45d = 3185$$

$$d \approx 70.78$$

The diver can go to a depth of about 70 feet and surface without decompression stops.

Any possible solution that results in a zero in the denominator must be excluded from your list of solutions. So, in Example 1, the solution could not be 33. Always check your solutions by substituting them into the original equation.

Lesson Notes
4-6

 5-Minute Check
Lesson 4–5

Determine between which consecutive integers the real zeros of each function are located.

1. $f(x) = x^3 - 4x^2 + 5x - 3$
   2 and 3

2. $f(x) = x^4 - 4x^3 + 8x^2 + 36x - 9$
   0 and 1, −3 and −2

3. Approximate the real zeros of the function in Exercise 2 above to the nearest tenth.
   −2.1, 0.2

4. Use the Upper Bound Theorem to find an integral upper bound and the Lower Bound Theorem to find an integral lower bound of the zeros of $f(x) = 2x^4 - 3x - 6$.
   Sample answer: 2; −2

Motivating the Lesson
Hands-On Activity Have students graph the following on the chalkboard or a graphing calculator: $y = -\frac{7}{x} + \frac{7}{x - 1} - \frac{4}{(x - 1)^2}$.

Then, have them add the polynomial terms together, forming another version of the original equation. $y = \frac{3x - 7}{x^3 - 2x^2 + x}$

Explain that in this lesson they will learn how to work with equations in the opposite direction, deriving simpler original terms from more complex sums, such as in this example. This skill will be necessary in integral calculus.

Teaching Tip For Example 2, remind students that factoring the polynomials in the denominator first can help them solve the equation more easily.

In-Class Examples

Example 1
Electricity The resistance of a wire depends on its temperature. The resistances R_1 and R_2 of a copper wire at temperatures t_1 and t_2, respectively, are related by the equation
$\dfrac{R_1}{234.5+ t_1} = \dfrac{R_2}{234.5 + t_2}$. R_1 and R_2 are measured in ohms and t_1 and t_2 are measured in °C. If the resistance of a wire is 50 ohms at 20°C, at what temperature will it have a resistance of 60 ohms?
70.9°C

Example 2
Solve $\dfrac{x}{x+4} - \dfrac{2}{x-1} = \dfrac{20}{x^2 + 3x - 4}$. **7**

Example 3
Decompose $\dfrac{6x - 2}{x^2 - 3x - 10}$ into partial fractions. $\dfrac{2}{x + 2} + \dfrac{4}{x - 5}$

Teaching Tip In Example 3 in the Student Edition, point out that any two values of x could be used to find the values of A and B, but that the given pair of values yields the quickest solution.

Example **2** Solve $a + \dfrac{a^2 - 5}{a^2 - 1} = \dfrac{a^2 + a + 2}{a + 1}$.

$$a + \frac{a^2 - 5}{a^2 - 1} = \frac{a^2 + a + 2}{a + 1}$$

$$\left(a + \frac{a^2 - 5}{a^2 - 1}\right)(a^2 - 1) = \left(\frac{a^2 + a + 2}{a + 1}\right)(a^2 - 1)$$

Multiply each side by the LCD, $(a - 1)(a + 1)$ or $a^2 - 1$.

$$a(a^2 - 1) + (a^2 - 5) = (a^2 + a + 2)(a - 1)$$
$$a^3 - a + a^2 - 5 = a^3 + a - 2$$
$$a^2 - 2a - 3 = 0$$
$$(a - 3)(a + 1) = 0 \qquad \textit{Factor.}$$

$$a - 3 = 0 \qquad\qquad a + 1 = 0$$
$$a = 3 \qquad\qquad\quad a = -1$$

When you check your solutions, you find that a cannot equal -1 because a zero denominator results. Since -1 cannot be a solution, the only solution is 3.

In order to add or subtract fractions with unlike denominators, you must first find a common denominator. Suppose you have a rational expression and you want to know what fractions were added or subtracted to obtain that expression. Finding these fractions is called decomposing the fraction into **partial fractions.** Decomposing fractions is a skill used in calculus and other advanced mathematics courses.

Example **3** Decompose $\dfrac{8y + 7}{y^2 + y - 2}$ into partial fractions.

First factor the denominator. $\qquad\qquad y^2 + y - 2 = (y - 1)(y + 2)$

Express the factored form as the sum of two fractions using A and B as numerators and the factors as denominators.
$$\frac{8y + 7}{y^2 + y - 2} = \frac{A}{y - 1} + \frac{B}{y + 2}$$

Eliminate the denominators by multiplying each side by the LCD, $(y - 1)(y - 2)$.
$$8y + 7 = A(y + 2) + B(y - 1)$$

Eliminate B by letting $y = 1$ so that $y - 1$ becomes 0.
$$8y + 7 = A(y + 2) + B(y - 1)$$
$$8(1) + 7 = A(1 + 2) + B(1 - 1)$$
$$15 = 3A$$
$$5 = A$$

Eliminate A by letting $y = -2$ so that $y + 2$ becomes 0.
$$8y + 7 = A(y + 2) + B(y - 1)$$
$$8(-2) + 7 = A(-2 + 2) + B(-2 - 1)$$
$$9 = -3B$$
$$3 = B$$

Now substitute the values for A and B to determine the partial fractions.
$$\frac{A}{y - 1} + \frac{B}{y + 2} = \frac{5}{y - 1} + \frac{3}{y + 2}$$

So, $\dfrac{8y + 7}{y^2 + y - 2} = \dfrac{5}{y - 1} + \dfrac{3}{y + 2}$.

Check to see if the sum of the two partial fractions equals the original fraction.

244 *Chapter 4 Polynomial and Rational Functions*

Prerequisite Skills

A review of the prerequisite skills for this lesson is provided in the *Prerequisite Skills Masters.*

Skill Number and Content	Pages
2 Fractions	5–6

The process used to solve rational equations can be used to solve **rational inequalities.**

Example **④** Solve $\dfrac{(x - 2)(x - 1)}{(x - 3)(x - 4)^2} < 0.$

Let $f(x) = \dfrac{(x - 2)(x - 1)}{(x - 3)(x - 4)^2}$. On a number line, mark the zeros of $f(x)$ and the excluded values for $f(x)$ with vertical dashed lines. The zeros of $f(x)$ are the same values that make $(x - 2)(x - 1) = 0$. These zeros are 2 and 1. Excluded values for $f(x)$ are the values that make $(x - 3)(x - 4)^2 = 0$. These excluded values are 3 and 4.

The vertical dashed lines separate the number line into intervals. Test a convenient value within each interval in the original rational inequality to see if the test value is a solution. If the value in the interval is a solution, all values are solutions. In this problem, it is not necessary to find the exact value of the expression.

For $x < 1$, test $x = 0$: $\quad \dfrac{(0 - 2)(0 - 1)}{(0 - 3)(0 - 4)^2} \rightarrow \dfrac{(-)(-)}{(-)(-)(-)} \rightarrow -$

So in the interval $x < 1$, $f(x) < 0$. Thus, $x < 1$ is a solution.

For $1 < x < 2$, test $x = 1.5$: $\quad \dfrac{(1.5 - 2)(1.5 - 1)}{(1.5 - 3)(1.5 - 4)^2} \rightarrow \dfrac{(-)(+)}{(-)(-)(-)} \rightarrow +$

So in the interval $1 < x < 2$, $f(x) > 0$. Thus, $1 < x < 2$ is not a solution.

For $2 < x < 3$, test $x = 2.5$: $\quad \dfrac{(2.5 - 2)(2.5 - 1)}{(2.5 - 3)(2.5 - 4)^2} \rightarrow \dfrac{(+)(+)}{(-)(-)(-)} \rightarrow -$

So in the interval $2 < x < 3$, $f(x) < 0$. Thus, $2 < x < 3$ is a solution.

For $3 < x < 4$, test $x = 3.5$: $\quad \dfrac{(3.5 - 2)(3.5 - 1)}{(3.5 - 3)(3.5 - 4)^2} \rightarrow \dfrac{(+)(+)}{(+)(-)(-)} \rightarrow +$

So in the interval $3 < x < 4$, $f(x) > 0$. Thus, $3 < x < 4$ is not a solution.

For $4 < x$, test $x = 5$: $\quad \dfrac{(5 - 2)(5 - 1)}{(5 - 3)(5 - 4)^2} \rightarrow \dfrac{(+)(+)}{(+)(+)(+)} \rightarrow +$

So in the interval $x > 4$, $f(x) > 0$. Thus, $4 < x$ is not a solution.

The solution is $x < 1$ or $2 < x < 3$. This solution can be graphed on a number line.

Lesson 4-6   Rational Equations and Partial Fractions **245**

In-Class Example

Example 4

Solve $\dfrac{(x + 2)(x - 3)}{(x - 1)(x + 1)^2} > 0$.

$-2 < x < -1, -1 < x < 1,$ or $x > 3$

Example ⑤ Solve $\frac{2}{3a} + \frac{5}{6a} > \frac{3}{4}$.

The inequality can be written as $\frac{2}{3a} + \frac{5}{6a} - \frac{3}{4} > 0$. The related function is $f(a) = \frac{2}{3a} + \frac{5}{6a} - \frac{3}{4}$. Find the zeros of this function.

$$\frac{2}{3a} + \frac{5}{6a} - \frac{3}{4} = 0$$

$$\frac{2}{3a}(12a) + \frac{5}{6a}(12a) - \frac{3}{4}(12a) = 0(12a) \quad \textit{The LCD is 12a.}$$

$$8 + 10 - 9a = 0$$

$$2 = a$$

The zero is 2. The excluded value is 0. On a number line, mark these values with vertical dashed lines. The vertical dashed lines separate the number line into intervals.

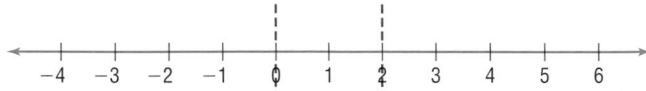

Now test a sample value in each interval to determine if the values in the interval satisfy the inequality.

For $a < 0$, test $x = -1$:

$$\frac{2}{3(-1)} + \frac{5}{6(-1)} \overset{?}{>} \frac{3}{4}$$

$$-\frac{2}{3} - \frac{5}{6} \overset{?}{>} \frac{3}{4}$$

$$-\frac{3}{2} \not> \frac{3}{4} \quad \textit{a < 0 is not a solution.}$$

For $0 < a < 2$, test $x = 1$:

$$\frac{2}{3(1)} + \frac{5}{6(1)} \overset{?}{>} \frac{3}{4}$$

$$\frac{2}{3} + \frac{5}{6} \overset{?}{>} \frac{3}{4}$$

$$\frac{3}{2} > \frac{3}{4} \quad \textit{0 < a < 2 is a solution.}$$

For $2 < a$, test $x = 3$:

$$\frac{2}{3(3)} + \frac{5}{6(3)} \overset{?}{>} \frac{3}{4}$$

$$\frac{2}{9} + \frac{5}{18} \overset{?}{>} \frac{3}{4}$$

$$\frac{1}{2} \not> \frac{3}{4} \quad \textit{2 < x is not a solution.}$$

The solution is $0 < a < 2$. This solution can be graphed on a number line.

Addressing Individual Needs

Verbal/Linguistic Write a rational equation on the chalkboard and have students suggest steps that could be taken to obtain its solution. Write the steps down as they are given and ask students to give a good, clear presentation of the techniques used.

Communicating Mathematics

Read and study the lesson to answer each question. **1–4. See margin.**

1. **Describe** the process used to solve $\frac{b+1}{3(b-2)} = \frac{5b}{6} + \frac{1}{b-2}$.

2. **Write** a sentence explaining why all solutions of a rational equation must be checked.

3. **Explain** what is meant by decomposing a fraction into partial fractions.

4. **Explain** why $x + \frac{2}{x-2} = 2 + \frac{2}{x-2}$ has no solution.

Guided Practice

Solve each equation.

5. $a - \frac{5}{a} = 4$ **−1, 5**

6. $\frac{9}{b+5} = \frac{3}{b-3}$ **7**

7. $\frac{t+4}{t} + \frac{3}{t-4} = \frac{-16}{t^2-4t}$ **−3**

8. Decompose $\frac{3p-1}{p^2-1}$ into partial fractions. $\frac{2}{p+1} + \frac{1}{p-1}$

Solve each inequality.

9. $5 + \frac{1}{x} > \frac{16}{x}$ **$x < 0, x > 3$**

10. $1 + \frac{5}{a-1} \le \frac{7}{6}$ **$a < 1, a \ge 31$**

11. **Interstate Commerce** When truckers are on long-haul drives, their driving logs must reflect their average speed. Average speed is the total distance driven divided by the total time spent driving. A trucker drove 3 hours on a freeway at 60 miles per hour and then drove 20 miles in the city. The trucker's average speed was 57.14 miles per hour.

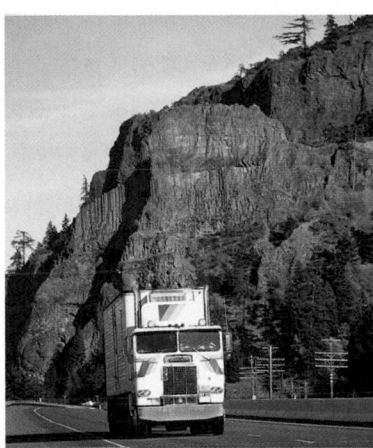

a. Write an equation that models the situation.

b. How long was the trucker driving in the city to the nearest hundredth of an hour? **0.50 h**

a. $\frac{3 \times 60 + 20}{3 + x} = 57.14$

Practice

Solve each equation.

12. $\frac{12}{t} + t - 8 = 0$ **2, 6**

13. $\frac{1}{m} = \frac{m-34}{2m^2}$ **−34**

14. $\frac{2}{y+2} + \frac{3}{y} = \frac{-y}{y+2}$ **−3**

15. $\frac{10}{n^2-1} + \frac{2n-5}{n-1} = \frac{2n+5}{n+1}$ **$\frac{5}{3}$**

16. $\frac{1}{b+2} + \frac{1}{b+2} = \frac{3}{b+1}$ **−4**

17. $\frac{7a}{3a+3} - \frac{5}{4a-4} = \frac{3a}{2a+2}$ **$-\frac{1}{2}, 3$**

18. $1 = \frac{1}{1-a} + \frac{a}{a-1}$

19. $\frac{2q}{2q+3} - \frac{2q}{2q-3} = 1$ **$\frac{-3 \pm 3\sqrt{2}}{2}$**

18. all reals except 1

20. $\frac{1}{3m} + \frac{6m-9}{3m} = \frac{3m-3}{4m}$ **$\frac{23}{15}$**

21. $\frac{-4}{x-1} = \frac{7}{2-x} + \frac{3}{x+1}$ **$\frac{5}{13}$**

www.amc.glencoe.com/self_check_quiz

Lesson 4-6 Rational Equations and Partial Fractions **247**

Error Analysis

Watch for students who sometimes forget to check that solutions are not excluded values when working with rational equations.

Prevent by stressing that the denominator can never be zero.

Assignment Guide

Core: 13–41 odd, 42, 43, 45–57
Enriched: 12–34 even, 35–57

Additional Answers

1. Multiply by the LCD, 6(b − 2). Then, solve the resulting equation.

2. If a possible solution causes a denominator to equal 0, it is not a solution of the equation.

3. Decomposing a fraction means to find two fractions whose sum or difference equals the original fraction.

4. If you solve the equation, you will get $x = 2$. However, if $x = 2$, the denominators will equal 0.

Study Guide and Practice Masters, pp. 69–70

Teaching Tip For Exercises 27–33, it may help students to understand the answers to inequalities by making a graph of each inequality.

Teaching Tip Remind students when working Exercise 35 that to make a percent into a fraction, they should divide the percent by 100, then reduce the fraction.

25. $\dfrac{2}{3y-1} + \dfrac{-2}{y-1}$

30. $x \le -4$, $-1 < x \le 4$, $x > 5$

32.
$-1 < b < -\dfrac{11}{16}$, $-\dfrac{1}{2} < b < 2$

22. Consider the equation $1 + \dfrac{n+6}{n+1} = \dfrac{4}{n-2}$.
   a. What is the LCD of the rational expressions? $(n+1)(n-2)$
   b. What values must be excluded from the list of possible solutions. $-1, 2$
   c. What is the solution of the equation? $\dfrac{1 \pm \sqrt{145}}{4}$

Decompose each expression into partial fractions.

23. $\dfrac{x-6}{x^2-2x}$ $\dfrac{3}{x} + \dfrac{-2}{x-2}$

24. $\dfrac{5m-4}{m^2-4}$ $\dfrac{3.5}{m+2} + \dfrac{1.5}{m-2}$

25. $\dfrac{-4y}{3y^2-4y+1}$

26. Find two rational expressions that have a sum of $\dfrac{9-9x}{x^2-9}$. $\dfrac{-6}{x+3}, \dfrac{-3}{x-3}$

27. Consider the inequality $\dfrac{a-2}{a} < \dfrac{a-4}{a-6}$.

   a. What is the LCD of the rational expressions? $a(a-6)$

   b. Find the zero(s) of the related function. 3

   c. Find the excluded value(s) of the related function. $0, 6$

   d. Solve the inequality. $0 < a < 3, 6 < a$

 Solve each inequality.

28. $\dfrac{2}{w} + 3 > \dfrac{29}{w}$ $w < 0, w > 9$

29. $\dfrac{(x-3)(x-4)}{(x-5)(x-6)^2} \le 0$ $x \le 3, 4 \le x < 5$

30. $\dfrac{x^2-16}{x^2-4x-5} \ge 0$

31. $\dfrac{1}{4a} + \dfrac{5}{8a} > \dfrac{1}{2}$ $0 < a < \dfrac{7}{4}$

32. $\dfrac{1}{2b+1} + \dfrac{1}{b+1} > \dfrac{8}{15}$

33. $\dfrac{7}{y+1} > 7$ $-1 < y < 0$

34. Four times the multiplicative inverse of a number is added to the number. The result is $10\dfrac{2}{5}$. What is the number? $\dfrac{2}{5}$ **or 10**

35. The ratio of $x+2$ to $x-5$ is greater than 30%. Solve for x. $x < -5$ **or** $x > 5$

Applications and Problem Solving

36. **Optics**   The lens equation is $\dfrac{1}{f} = \dfrac{1}{d_i} + \dfrac{1}{d_o}$, where f is the focal length, d_i is the distance from the lens to the image, and d_o is the distance from the lens to the object. Suppose the object is 32 centimeters from the lens and the focal length is 8 centimeters.

   a. Write a rational equation to model the situation. $\dfrac{1}{8} = \dfrac{1}{d_i} + \dfrac{1}{32}$

   b. Find the distance from the lens to the image. $10\dfrac{2}{3}$ **cm**

37. **Critical Thinking**   Write a rational equation that cannot have 3 or -2 as a solution. **Sample answer:** $\dfrac{x}{x-3} = \dfrac{1}{x+2}$

38. **Trucking**   Two trucks can carry loads of coal in a ratio of 5 to 2. The smaller truck has a capacity 3 tons less than that of the larger truck. What is the capacity of the larger truck? **5 tons**

39. Electricity   The diagram of an electric circuit shows three parallel resistors. If R represents the equivalent resistance of the three resistors, then $\frac{1}{R} = \frac{1}{R_1} + \frac{1}{R_2} + \frac{1}{R_3}$. In this circuit, R_1 represents twice the resistance of R_2, and R_3 equals 20 ohms. Suppose the equivalent resistance equals 10 ohms.

a. Write a rational equation to model the situation. $\frac{1}{10} = \frac{1}{2r} + \frac{1}{r} + \frac{1}{20}$

b. Find R_1 and R_2.   **60 ohms, 30 ohms**

40. Education   Todd has answered 11 of his last 20 daily quiz questions correctly. His baseball coach told him that he must bring his average up to at least 70% if he wants to play in the season opener. Todd vows to study diligently and answer all of the daily quiz questions correctly in the future. How many consecutive daily quiz questions must he answer correctly to bring his average up to 70%?   **10 questions**

41. Aviation   An aircraft flies 1062 miles with the wind at its tail. In the same amount of time, a similar aircraft flies against the wind 738 miles. If the air speed of each plane is 200 miles per hour, what is the speed of the wind? (*Hint:* Time equals the distance divided by the speed.)   **36 mph**

42. Critical Thinking   Solve for a if $\frac{1}{a} + \frac{1}{b} = \frac{1}{c} \cdot \frac{bc}{b-c}$

43. Statistics   A number x is said to be the *harmonic mean* of y and z if $\frac{1}{x}$ is the average of $\frac{1}{y}$ and $\frac{1}{z}$.   **a.** $\frac{1}{x} = \frac{1}{2}\left(\frac{1}{30} + \frac{1}{45}\right)$

a. Write an equation whose solution is the harmonic mean of 30 and 45.

b. Find the harmonic mean of 30 and 45.   **36**

44. Economics   Darnell drives about 15,000 miles each year. He is planning to buy a new car. The car he wants to buy averages 20 miles on one gallon of gasoline. He has decided he would buy another car if he could save at least $200 per year in gasoline expenses. Assume gasoline costs $1.20 per gallon. What is the minimum number of miles per gallon that would fulfill Darnell's criteria?   **about 25.7 mpg**

45. Navigation   The speed of the current in Puget Sound is 5 miles per hour. A barge travels with the current 26 miles and returns in $10\frac{2}{3}$ hours. What is its speed in still water?   **8 mph**

46. Critical Thinking   If $\frac{3x}{5y} = 11$, find the value of $\frac{3x - 5y}{5y}$.   **10**

Puget Sound

Open-Ended Assessment

Modeling Have students solve the inequality $\dfrac{(n+4)(n-3)}{(n-1)^2(n+2)} \leq 0$ as shown in Example 4. First, have students draw a line graph on a white sheet of paper, using a ruler if necessary. Then, as each "piece" of the graph is solved, have them use a colored pen, pencil, or colored pieces of paper to indicate the intervals that are a solution. Students will need to choose colors that represent "solution" and colors that represent "not a solution." **3, −4; $n \neq 1, -2$**

Chapter 4, Quiz C
(Lessons 4-5 and 4-6)
is available in the *Assessment and Evaluation Masters*, p. 96.

Additional Answer
56c.

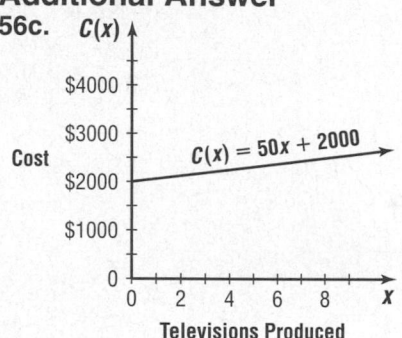

Cost vs. Televisions Produced; $C(x) = 50x + 2000$

Enrichment Masters, p. 42

Mixed Review

47. Determine between which consecutive integers the real zeros of $f(x) = x^3 + 2x^2 - 3x - 5$ are located. *(Lesson 4-5)*
−3 and −2, −2 and −1, 1 and 2

48. Use the Remainder Theorem to find the remainder for $(x^3 - 30x) \div (x + 5)$. State whether the binomial is a factor of the polynomial. *(Lesson 4-3)* **25; no**

49. State the number of complex roots of $12x^2 + 8x - 15 = 0$. Then find the roots. *(Lesson 4-1)* **$2; \dfrac{5}{6}, -\dfrac{3}{2}$**

50. Name all the values of x that are not in the domain of the function $f(x) = \dfrac{x}{|3x| - 12}$. *(Lesson 3-7)* **±4**

51. Determine if $(6, 3)$ is a solution for $y \leq \dfrac{2x + 3}{x}$. *(Lesson 3-3)* **no**

52. Determine if the graph of $y^2 = 121x^2$ is symmetric with respect to each line. *(Lesson 3-1)*
a. the x-axis **yes**   b. the y-axis **yes**
c. the line $y = x$ **no**   d. the line $y = -x$ **no**

53. **Education** The semester test in your English class consists of short answer and essay questions. Each short answer question is worth 5 points, and each essay question is worth 15 points. You may choose up to 20 questions of any type to answer. It takes 2 minutes to answer each short answer question and 12 minutes to answer each essay question. *(Lesson 2-7)*

a. You have one hour to complete the test. Assuming that you answer all of the questions that you attempt correctly, how many of each type should you answer to earn the highest score?

b. You have two hours to complete the test. Assuming that you answer all of the questions that you attempt correctly, how many of each type should you answer to earn the highest score?

53a. 18 short answer and 2 essay for a score of 120 points

53b. 12 short answer and 8 essay for a score of 180 points

54. Find matrix X in the equation $\begin{bmatrix} 1 & 1 \\ 1 & 1 \end{bmatrix}\begin{bmatrix} 3 & 5 \\ -3 & -5 \end{bmatrix} = X$. *(Lesson 2-3)* $\begin{bmatrix} 0 & 0 \\ 0 & 0 \end{bmatrix}$

55. Write the standard form of the equation of the line that is parallel to the graph of $y = 2x - 10$ and passes through the point at $(-3, 1)$. *(Lesson 1-5)*
$2x - y + 7 = 0$

56. **Manufacturing** It costs ABC Corporation $3000 to produce 20 color televisions and $5000 to produce 60 of the same color televisions. *(Lesson 1-4)*
a. Find the cost function. **$C(x) = 50x + 2000$**
b. Determine the fixed cost and the variable cost per unit. **$2000; $50**
c. Sketch the graph of the cost function. **See margin.**

57. **SAT Practice Grid-In** Find the area of the shaded region in square inches. **24**

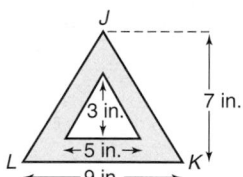

Extra Practice See p. A33.

Extra Credit

Solve for x if $\dfrac{1}{x - 1} + \dfrac{3}{y + 2} = \dfrac{1}{z}$.

$x = \dfrac{zy + 2z}{y - 3z + 2} + 1$ or $x = \dfrac{zy + y + 2 - z}{y + 2 - 3z}$

4-7

Radical Equations and Inequalities

OBJECTIVE
- Solve radical equations and inequalities.

RECREATION A pogo stick stores energy in a spring. When a jumper compresses the spring from its natural length x_0 to a length x, the maximum height h reached by the bottom of the pogo stick is related to x_0 and x by the equation $x_0 = x + \sqrt{\frac{2mgh}{k}}$, where m is the combined mass of the jumper and the pogo stick, g is the acceleration due to gravity (9.80 meters per second squared), and k is a constant that depends on the spring. If the combined mass of the jumper and the stick is 50 kilograms, the spring compressed from its natural length of 1 meter to the length of 0.9 meter. If $k = 1.2 \times 10^4$, find the maximum height reached by the bottom of the pogo stick. *This problem will be solved in Example 1.*

Equations in which radical expressions include variables, such as the equation above, are known as **radical equations.** To solve radical equations, the first step is to isolate the radical on one side of the equation. Then raise each side of the equation to the proper power to eliminate the radical expression.

The process of raising each side of an equation to a power sometimes produces **extraneous solutions.** These are solutions that do not satisfy the original equation. Therefore, it is important to check all possible solutions in the original equation to determine if any of them should be eliminated from the solution set.

Examples **1** **RECREATION** Find the maximum height reached by the bottom of the pogo stick.

$$x_0 = x + \sqrt{\frac{2mgh}{k}}$$

$$1 = 0.9 + \sqrt{\frac{2 \times 50 \times 9.80 \times h}{1.2 \times 10^4}} \qquad x_0 = 1,\ x = 0.9,\ m = 50,\\ g = 9.80,\ k = 1.2 \times 10^4$$

$$0.1 = \sqrt{\frac{980h}{1.2 \times 10^4}} \qquad \text{Isolate the radical.}$$

$$0.01 = \frac{980h}{1.2 \times 10^4} \qquad \text{Square each side.}$$

$$0.1224489796 = h \qquad \text{Solve for } h.$$

A possible solution is about 0.12. Check this solution.

(continued on the next page)

4-7

Lesson Notes

5-Minute Check
Lesson 4-6

Solve each equation.

1. $\frac{1}{2t-1} + \frac{1}{t+1} = \frac{3}{2}$ $-\frac{1}{2}, 1$

2. $\frac{2}{y} - 1 = \frac{6}{y-1}$ $-1, -2$

3. Solve $\frac{x^2 - x + 1}{x + 2} < 3$.

   $x < -2 \text{ or } -1 < x < 5$

Decompose each expression into partial fractions.

4. $\frac{5x + 7}{x^2 + x - 20}$ $\frac{2}{x+5} + \frac{3}{x-4}$

5. $\frac{5c - 17}{c^2 - 2c - 15}$ $\frac{1}{c-5} + \frac{4}{c+3}$

Motivating the Lesson

Hands-On Activity Have students work in small groups. Each group needs a stopwatch, a weight, and a piece of string. Have students use the string and weight as a pendulum and use the stopwatch to time its period for different lengths of string. Make a chart listing the period for each length, and observe how the period and the square root of the length are related. **The smaller the square root, the smaller the period.**

Resource Manager

Reproducible Masters
- *Study Guide and Practice,* pp. 71–72
- *Enrichment,* p. 43
- *Problem Solving and Applications,* p. 8

Transparencies
- 5-Minute Check, 4-7
- Teaching, 4-7

Teaching Tip In Example 1, remind students of the possibility of extraneous solutions that might occur as a result of squaring. Therefore, any possible solution must be checked.

Teaching Tip In Example 2, remind students that odd roots (cube, fifth, and so on) can be negative, while even roots cannot be negative.

In-Class Examples

Example 1

Physics The time T that it takes a pendulum to complete one swing is called its period and is given by the formula $T = 2\pi \sqrt{\dfrac{L}{384}}$, where L is the length in inches. If the period is 2.5, find the length of the pendulum. **about 60.8 in.**

Example 2
Solve $x = \sqrt{x - 3} + 3$. **3, 4**

Example 3
Solve $7 = 3 - \sqrt[3]{x + 11}$. **−75**

Check: $\quad x_0 = x + \sqrt{\dfrac{2mgh}{k}}$

$$1 \overset{?}{=} 0.9 + \sqrt{\dfrac{2 \times 50 \times 9.80 \times 0.12}{1.2 \times 10^4}}$$

$$1 \approx 0.9989949494 \quad \checkmark$$

The solution checks. The maximum height is about 0.12 meter.

❷ Solve $x = \sqrt{x + 7} + 5$.

$$x = \sqrt{x + 7} + 5$$
$$x - 5 = \sqrt{x + 7} \qquad \textit{Isolate the radical.}$$
$$x^2 - 10x + 25 = x + 7 \qquad \textit{Square each side.}$$
$$x^2 - 11x + 18 = 0$$
$$(x - 9)(x - 2) = 0 \qquad \textit{Factor.}$$
$$x - 9 = 0 \qquad x - 2 = 0$$
$$x = 9 \qquad\quad x = 2$$

Check both solutions to make sure they are not extraneous.

Check $x = 9$: $\quad x = \sqrt{x + 7} + 5$
$$9 \overset{?}{=} \sqrt{9 + 7} + 5$$
$$9 \overset{?}{=} \sqrt{16} + 5$$
$$9 \overset{?}{=} 4 + 5$$
$$9 = 9 \quad \checkmark$$

Check $x = 2$: $\quad x = \sqrt{x + 7} + 5$
$$2 \overset{?}{=} \sqrt{2 + 7} + 5$$
$$2 \overset{?}{=} \sqrt{9} + 5$$
$$2 \overset{?}{=} 3 + 5$$
$$2 \neq 8$$

One solution checks and the other solution does not check. The solution is 9.

The same method of solution works for nth root equations.

Example **❸ Solve $4 = \sqrt[3]{x + 2} + 8$.**

$$4 = \sqrt[3]{x + 2} + 8$$
$$-4 = \sqrt[3]{x + 2} \qquad \textit{Isolate the cube root.}$$
$$-64 = x + 2 \qquad \textit{Cube each side. } \left(\sqrt[3]{x + 2}\right)^3 = x + 2.$$
$$-66 = x$$

Check the solution.

Check: $\quad 4 = \sqrt[3]{x + 2} + 8$
$$4 \overset{?}{=} \sqrt[3]{-66 + 2} + 8$$
$$4 \overset{?}{=} \sqrt[3]{-64} + 8$$
$$4 \overset{?}{=} -4 + 8$$
$$4 = 4 \quad \checkmark$$

The solution checks. The solution is −66.

252 *Chapter 4 Polynomial and Rational Functions*

📁 Prerequisite Skills

A review of the prerequisite skills for this lesson is provided in the *Prerequisite Skills Masters*.

Skill Number and Content	Pages
8 Rationalizing a Denominator	17–18

If there is more than one radical in an equation, you may need to repeat the process for solving radical equations more than once until all radicals have been eliminated.

Example **4** Solve $\sqrt{x + 10} = 5 - \sqrt{3 - x}$.

Leave the radical signs on opposite sides of the equation. If you square the sum or difference of two radical expressions, the value under the radical sign in the resulting product will be more complicated.

$$\sqrt{x + 10} = 5 - \sqrt{3 - x}$$
$$x + 10 = 25 - 10\sqrt{3 - x} + 3 - x \quad \textit{Square each side.}$$
$$2x - 18 = -10\sqrt{3 - x} \quad\quad\quad \textit{Isolate the radical.}$$
$$4x^2 - 72x + 324 = 100(3 - x) \quad\quad \textit{Square each side.}$$
$$4x^2 - 72x + 324 = 300 - 100x$$
$$4x^2 + 28x + 24 = 0$$
$$x^2 + 7x + 6 = 0 \quad\quad\quad\quad \textit{Divide each side by 4.}$$
$$(x + 6)(x + 1) = 0$$
$$x + 6 = 0 \quad\quad x + 1 = 0/$$
$$x = -6 \quad\quad\quad x = -1$$

Check both solutions to make sure they are not extraneous.

Check $x = -6$:

$$\sqrt{x + 10} = 5 - \sqrt{3 - x}$$
$$\sqrt{-6 + 10} \stackrel{?}{=} 5 - \sqrt{3 - (-6)}$$
$$\sqrt{4} \stackrel{?}{=} 5 - \sqrt{9}$$
$$2 \stackrel{?}{=} 5 - 3$$
$$2 = 2 \quad ✔$$

Check $x = -1$:

$$\sqrt{x + 10} = 5 - \sqrt{3 - x}$$
$$\sqrt{-1 + 10} \stackrel{?}{=} 5 - \sqrt{3 - (-1)}$$
$$\sqrt{9} \stackrel{?}{=} 5 - \sqrt{4}$$
$$3 \stackrel{?}{=} 5 - 2$$
$$3 = 3 \quad ✔$$

Both solutions check. The solutions are -6 and -1.

Use the same procedures to solve **radical inequalities.**

Example **5** Solve $\sqrt{4x + 5} \le 10$.

$$\sqrt{4x + 5} \le 10$$
$$4x + 5 \le 100 \quad \textit{Square each side.}$$
$$4x \le 95$$
$$x \le 23.75$$

In order for $\sqrt{4x + 5}$ to be a real number, $4x + 5$ must be greater than or equal to zero.

$$4x + 5 \ge 0$$
$$4x \ge -5$$
$$x \ge -1.25$$

So the solution is $-1.25 \le x \le 23.75$. Check this solution by testing values in the intervals defined by the solution. *(continued on the next page)*

Lesson 4-7 Radical Equations and Inequalities **253**

Error Analysis

Watch for students who forget to check the constraints of the problem when working with radical inequalities.

Prevent by reminding students that they are looking for real solutions and that therefore the quantity inside the radical must be greater than or equal to zero.

Additional Answers

1. To solve the equation, you need to get rid of the radical by squaring both sides of the equation. If the radical is not isolated first, a radical will remain in the equation.

2. The process of raising to a power sometimes creates a new equation with more solutions than the original equation. These extra or extraneous solutions do not solve the original equation.

Study Guide and Practice Masters, pp. 71–72

On a number line, mark -1.25 and 23.75 with vertical dashed lines. The dashed lines separate the number line into intervals. Check the solution by testing values for x from each interval into the original inequality.

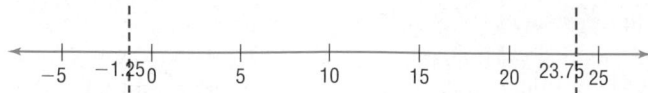

For $x \leq -1.25$, test $x = -2$: $\quad \sqrt{4(-2) + 5} \overset{?}{\leq} 10$

$\sqrt{-3} \overset{?}{\leq} 10$ *This statement is meaningless.* $x \leq -1.25$ *is not a solution.*

For $-1.25 \leq x \leq 23.75$, test $x = 0$: $\quad \sqrt{4(0) + 5} \overset{?}{\leq} 10$

$\sqrt{5} \overset{?}{\leq} 10$

$2.236 \leq 10$ $-1.25 \leq x \leq 23.75$ *is a solution.*

For $23.75 \leq x$, test $x = 25$: $\quad \sqrt{4(25) + 5} \overset{?}{\leq} 10$

$\sqrt{105} \overset{?}{\leq} 10$

$10.247 \nleq 10$ $23.75 \leq x$ *is not a solution.*

The solution checks and is graphed on the number line below.

CHECK FOR UNDERSTANDING

Communicating Mathematics

Read and study the lesson to answer each question.   **1–3. See margin.**

1. **Explain** why the first step in solving $5 + \sqrt{x + 1} = x$ should be to isolate the radical.

2. **Write** several sentences explaining why it is necessary to check for extraneous solutions in radical equations.

3. **Explain** the difference between solving an equation with one radical and solving an equation with more than one radical.

Guided Practice

Solve each equation.   **5. −733   6. no real solution**

4. $\sqrt{1 - 4t} = 2 \quad -\dfrac{3}{4}$

5. $\sqrt[3]{x + 4} + 12 = 3$

6. $5 + \sqrt{x - 4} = 2$

7. $\sqrt{6x - 4} = \sqrt{2x + 10} \quad$ **3.5**

8. $\sqrt{a + 4} + \sqrt{a - 3} = 7 \quad$ **12**

Solve each inequality.

9. $\sqrt{5x + 4} \leq 8 \quad -0.8 \leq x \leq 12$

10. $3 + \sqrt{4a - 5} \leq 10 \quad 1.25 \leq a \leq 13.5$

254   Chapter 4   *Polynomial and Rational Functions*

Additional Answer

3. When solving an equation with one radical, you isolate the radical on one side and then square each side. When there is more than one radical expression in an equation, you isolate one of the radicals and then square each side. Then, you isolate the other radical and square each side. In both cases, once you have eliminated all radical signs, you solve for the variable.

11. Amusement Parks The velocity of a roller coaster as it moves down a hill is $v = \sqrt{v_0^2 + 64h}$, where v_0 is the initial velocity and h is the vertical drop in feet. The designer of a coaster wants the coaster to have a velocity of 90 feet per second when it reaches the bottom of the hill.

a. If the initial velocity of the coaster at the top of the hill is 10 feet per second, write an equation that models the situation. $90 = \sqrt{100 + 64h}$

b. How high should the designer make the hill? **125 ft**

Practice

 Solve each equation.

A   12. $\sqrt{x + 8} = 5$ **17**   13. $\sqrt[3]{y - 7} = 4$ **71**   14. $\sqrt{8n - 5} - 1 = 2$ $\frac{7}{4}$

16. $\pm\frac{4}{3}\sqrt{3}$

15. $\sqrt{x + 16} = \sqrt{x} + 4$ **0**   16. $4\sqrt{3m^2 - 15} = 4$ 17. $\sqrt{9u - 4} = \sqrt{7u - 20}$

17. no real solution

B   18. $\sqrt[3]{6u - 5} + 2 = -3$ **−20**   19. $\sqrt{4m^2 - 3m + 2} - 2m - 5 = 0$ **−1**

20. $\sqrt{k + 9} - \sqrt{k} = \sqrt{3}$ **3**   21. $\sqrt{a + 21} - 1 = \sqrt{a + 12}$ **4**

22. $\sqrt{3x + 4} - \sqrt{2x - 7} = 3$ **4, 64**   23. $2\sqrt[3]{7b - 1} - 4 = 0$ $\frac{9}{7}$

24. $\sqrt[4]{3t} - 2 = 0$ $\frac{16}{3}$ 25. $\sqrt{x + 2} - 7 = \sqrt{x + 9}$ **no real solutions**

26. $\sqrt{2x + 1} + \sqrt{2x + 6} = 5$ $\frac{3}{2}$ 27. $\sqrt{3x + 10} = \sqrt{x + 11} - 1$ **−2**

C   28. Consider the equation $\sqrt{3t - 14} + t = 6$.

     a. Name any extraneous solutions of the equation.   **10**

     b. What is the solution of the equation?   **5**

Solve each inequality.   **30.** $-4 \le b \le 32$ **31.** $5 \le a \le 21$

29. $\sqrt{2x - 7} \ge 5$ $x \ge 16$ 30. $\sqrt{b + 4} \le 6$ 31. $\sqrt{a - 5} \le 4$

32. $\sqrt{2x - 5} \le 6$ 33. $\sqrt[4]{5y - 9} \le 2$ 34. $\sqrt{m + 2} \le \sqrt{3m + 4}$
    2.5 ≤ x ≤ 20.5   **1.8 ≤ y ≤ 5**   **m ≥ −1**

35. What values of c make $\sqrt{2c - 5}$ greater than 7?   **c > 27**

Applications and Problem Solving

36. Physics The time t in seconds that it takes an object at rest to fall a distance of s meters is given by the formula $t = \sqrt{\frac{2s}{g}}$. In this formula, g is the acceleration due to gravity in meters per second squared. On the moon, a rock falls 7.2 meters in 3 seconds.

     a. Write an equation that models the situation. $3 = \sqrt{\frac{14.4}{g}}$

     b. What is the acceleration due of gravity on the moon?   **1.6 m/s²**

Assignment Guide

Core: 13–41 odd, 43–51
Enriched: 12–34 even, 36–51

37. Critical Thinking Solve $\sqrt{x-5} = \sqrt[3]{x-3}$. **about 7.88**

38. Driving After an accident, police can determine how fast a car was traveling before the driver put on his breaks by using the equation $s = \sqrt{30fd}$. In this equation, s represents the speed in miles per hour, f represents the coefficient of friction, and d represents the length of the skid in feet. The coefficient of friction varies with road conditions. Suppose the coefficient of friction is 0.6.

a. Find the speed of a car that skids 25 feet.   **about 21.2 mph**

b. If you were driving 35 miles per hour, how many feet would it take you to stop?   **about 68 ft**   **c. No; it is not a linear function.**

c. If the speed is doubled, will the skid be twice as long? Explain.

39. Physics The period of a pendulum (the time required for one back and forth swing) can be determined by the formula $T = 2\pi\sqrt{\dfrac{\ell}{g}}$. In this formula, T represents the period, ℓ represents the length of the pendulum, and g represents acceleration due to gravity.

a. Determine the period of a 1-meter pendulum on Earth if the acceleration due to gravity at Earth's surface is 9.8 meters per second squared.   **about 2.01 s**

b. Suppose the acceleration due to gravity on the surface of Venus is 8.9 meters per second squared. Calculate the period of the pendulum on Venus.   **about 2.11 s**   **c. It must be multiplied by 4.**

c. How must the length of the pendulum be changed to double the period?

Venus

40. Astronomy Johann Kepler (1571–1630) determined the relationship of the time of revolution of two planets and their average distance from the sun. This relationship can be expressed as $\dfrac{T_a}{T_b} = \sqrt{\left(\dfrac{r_a}{r_b}\right)^3}$. In this equation, T_a represents the time it takes planet a to orbit the sun, and r_a represents the average distance between planet a and the sun. Likewise, T_b represents the time it takes planet b to orbit the sun, and r_b represents the average distance between planet b and the sun. The average distance between Venus and the sun is 67,200,000 miles, and it takes Venus about 225 days to orbit the sun. If it takes Mars 687 days to orbit the sun, what is its average distance from the sun?   **about 141,433,434 mi**

41. Critical Thinking For what values of a and b will the equation $\sqrt{2x-9} - a = b$ have no real solution?   **$a + b < 0$**

42. Engineering Engineers are often required to determine the stress on building materials. Tensile stress can be found by using the formula $T = \dfrac{t+c}{2} + \sqrt{\left(\dfrac{t-c}{2}\right)^2 + p^2}$. In this formula, T represents the tensile stress, t represents the tension, c represents the compression, and p represents the pounds of pressure per square inch. If the tensile stress is 108 pounds per square inch, the pressure is 50 pounds per square inch, and the compression is -200 pounds per square inch, what is the tension?   **about 99.88 psi**

43. Solve $\frac{a+2}{2a+1} = \frac{a}{3} + \frac{3}{4a+2}$. *(Lesson 4-6)* $\frac{3}{2}$

44. List the possible rational roots of $x^4 + 5x^3 + 5x^2 - 5x - 6 = 0$. Then determine the rational roots. *(Lesson 4-4)* $\pm 6, \pm 3, \pm 2, \pm 1; -3, -2, -1, 1$

45. Determine whether each graph has *infinite discontinuity*, *jump discontinuity*, or *point discontinuity*, or is *continuous*. *(Lesson 3-7)*

a.

point discontinuity

b.

jump discontinuity

c.

infinite discontinuity

46. Music The frequency of a sound wave is called its pitch. The pitch p of a musical tone and its wavelength w are related by the equation $p = \frac{v}{w}$, where v is the velocity of sound through air. Suppose a sound wave has a velocity of 1056 feet per second. *(Lesson 3-6)*

a. Graph the equation $p = \frac{v}{w}$. **See margin.**

b. What lines are close to the maximum values for the pitch and the wavelength? **x- and y-axes**

c. What happens to the pitch of the tone as the wavelength decreases? **c. It increases.**

d. If the wavelength is doubled, what happens to the pitch of the tone? **It is halved.**

47. Find the product of the matrices $\begin{bmatrix} 4 & -1 & 6 \\ 4 & 0 & 2 \end{bmatrix}$ and $\begin{bmatrix} 0 & 3 \\ 2 & -2 \\ 5 & 1 \end{bmatrix}$. *(Lesson 2-3)*

47. $\begin{bmatrix} 28 & 20 \\ 10 & 14 \end{bmatrix}$

48. Solve the system of equations. *(Lesson 2-2)* **(7, 1, −2)**

$a + b + c = 6$
$2a - 3b + 4c = 3$
$4a - 8b + 4c = 12$

49. Education The regression equation for a set of data is $y = -3.54x + 7107.7$, where x represents the year and y represents the average number of students assigned to each advisor in a certain business school. Use the equation to predict the number of students assigned to each adviser in the year 2005. *(Lesson 1-6)* **10 students**

50. Write the slope-intercept form of the equation that is perpendicular to $7y + 4x - 3 = 0$ and passes through the point with coordinates $(2, 5)$. *(Lesson 1-5)* $y = \frac{7}{4}x + \frac{3}{2}$

51. SAT/ACT Practice In the figure at the right, four semicircles are drawn on the four sides of a rectangle. What is the total area of the shaded regions? **C**

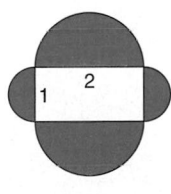

A 5π

B $\frac{5\pi}{2}$

C $\frac{5\pi}{4}$

D $\frac{5\pi}{8}$

E $\frac{5\pi}{16}$

Extra Practice See p. A33.

Lesson 4-7 Radical Equations and Inequalities **257**

Extra Credit

Solve $\sqrt{x-3} + 1 = \sqrt{1-3x} + \sqrt{4x-12}$. $x = \frac{7 \pm i\sqrt{35}}{8}$

---

4 ASSESS

Open-Ended Assessment

Speaking Have students explain the major steps involved in solving a radical equation that contains only one radical. Then, have them compare this to the steps necessary if two radicals are present.

Additional Answer
46a.

Enrichment Masters, p. 43

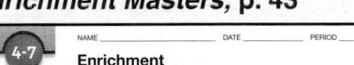

4-7
Enrichment

Discriminants and Tangents

The diagram at the right shows that through a point P outside of a circle C, there are lines that do not intersect the circle, lines that intersect the circle in one point (tangents), and lines that intersect the circle in two points (secants).

Given the coordinates for P and an equation for the circle C, how can we find the equation of a line tangent to C that passes through P?

Suppose P has coordinates $P(0, 0)$ and $\odot C$ has equation $(x-4)^2 + y^2 = 4$. Then a line tangent through P has equation $y = mx$ for some real number m.

Thus, if $T(r, s)$ is a point of tangency, then
$s = mr$ and $(r-4)^2 + s^2 = 4$.
Therefore, $(r-4)^2 + (mr)^2 = 4$.
$r^2 - 8r + 16 + m^2r^2 = 4$
$(1 + m^2)r^2 - 8r + 16 = 4$
$(1 + m^2)r^2 - 8r + 12 = 0$

The equation above has exactly one real solution for r if the discriminant is 0, that is, when $(-8)^2 - 4(1 + m^2)(12) = 0$. Solve this equation for m and you will find the slopes of the lines through P that are tangent to circle C.

1. a. Refer to the discussion above. Solve $(-8)^2 - 4(1 + m^2)(12) = 0$ to find the slopes of the two lines tangent to circle C through point P.
$\pm \frac{\sqrt{3}}{3}$

b. Use the values of m from part **a** to find the coordinates of the two points of tangency.
$(3, \pm \sqrt{3})$

2. Suppose P has coordinates $(0, 0)$ and circle C has equation $(x+9)^2 + y^2 = 9$. Let m be the slope of the tangent line to C through P.

a. Find the equations for the lines tangent to circle C through point P.
$y = \pm \frac{\sqrt{2}}{4}x$

b. Find the coordinates of the points of tangency.
$(-8, \pm 2\sqrt{2})$

© Glencoe/McGraw-Hill   T43   Advanced Mathematical Concepts

Lesson 4-7 **257**

1 FOCUS

5-Minute Check
Lesson 4-7

Solve each equation.

1. $x = \sqrt{2x - 5} + 4$ **7**
2. $\sqrt[3]{x + 3} - 6 = -5$ **−2**
3. $\sqrt{x + 3} - 2 = x - 5$ **6**

4. Solve $\sqrt{3x + 4} \le 5$.
   $-\dfrac{4}{3} \le x \le 7$

5. The ideal weight of humans is related to their height by the formula $H = \sqrt{\dfrac{W \cdot 703}{I}}$, where H is height in inches, W is weight in pounds, and I is the Body Mass Index. In order to have an index of 20, about how much should a person 5' 5" tall weigh? **about 120 lb**

Motivating the Lesson

Real-World Application Ask students to imagine that they are in charge of ordering this year's yearbooks for the school. It is important to order only the amount they can sell, because they pay the company in advance, and books are not returnable. They will be given a list of the numbers of books sold for each of the last 15 years. How can this information be displayed in order to help predict sales for this year?
Sample answer: using spreadsheet software, one can chart the sales in a 2 × 15 table with the years ordered from least to most recent. One can then use the software's chart function to make a column-style chart and add a linear trendline.

4-8

Modeling Real-World Data with Polynomial Functions

OBJECTIVES
- Write polynomial functions to model real-world data.
- Use polynomial functions to interpret real-world data.

WASTE MANAGEMENT The average daily amount of waste generated by each person in the United States is given below. This includes all wastes such as industrial wastes, demolition wastes, and sewage.

What a Waste!

Year	1980	1985	1990	1991	1992	1993	1994	1995	1996
Pounds of Waste per Person per Day	3.7	3.8	4.5	4.4	4.5	4.5	4.5	4.4	4.3

Source: Franklin Associates, Ltd.

What polynomial function could be used to model these data? *This problem will be solved in Example 3.*

In order to model real world data using polynomial functions, you must be able to identify the general shape of the graph of each type of polynomial function.

Function	Linear $y = ax + b$	Quadratic $y = ax^2 + bx + c$	Cubic $y = ax^3 + bx^2 + cx + d$	Quartic $y = ax^4 + bx^3 + cx^2 + dx + e$
Typical Graph				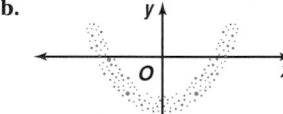
Direction Changes	0	1	2	3

Example ① Determine the type of polynomial function that could be used to represent the data in each scatter plot.

Look Back
You can refer to Lesson 1-6 to review scatter plots.

a.

The scatter plot seems to change direction two times, so a cubic function would best fit the plot.

b.

The scatter plot seems to change direction one time, so a quadratic function would best fit the plot.

Resource Manager

Reproducible Masters
- *Study Guide and Practice*, pp. 73–74
- *Enrichment*, p. 44
- *Assessment and Evaluation*, p. 96

Transparencies
- 5-Minute Check, 4-8
- Teaching, 4-8

You can use a graphing calculator to determine a polynomial function that models a set of data.

Example **2** Use a graphing calculator to write a polynomial function to model the set of data.

x	−1	−0.5	0	0.5	1	1.5	2	2.5	3	3.5	4
f(x)	−10	−6.4	−5	−5.1	−6	−6.9	−7	−5.6	−2	4.6	15

Clear the statistical memory and input the data.

Graphing Calculator Tip

You can change the appearance of the data points in the **STAT PLOT** menu. The data can appear as squares, dots, or + signs.

Using the standard viewing window, graph the data. The scatter plot seems to change direction two times, so a cubic function would best fit the scatter plot.

[−10, 10] scl:1 by [−10, 10] scl:1

Press ⎡STAT⎤ and highlight **CALC**. Since the scatter plot seems to be a cubic function, choose **6:CubicReg**.

Press ⎡2nd⎤ **[L1]** ⎡'⎤ ⎡2nd⎤ **[L2]** ⎡ENTER⎤ .

```
CubicReg
 y=ax³+bx²+cx+d
 a=1
 b=-3.001631702
 c=1.004895105
 d=-5.001864802
 R²=.9999941318
```

Rounding the coefficients to the nearest whole number, $f(x) = x^3 - 3x^2 + x - 5$ models the data. Since the value of the coefficient of determination r^2 is very close to 1, the function is an excellent fit. However it may not be the best model for the situation.

To check the polynomial function, enter the function in the **Y=** list. Then use the **TABLE** feature for x values from −1 with an interval of 0.5. Compare the values of y with those given in the data. The values are similar, and the polynomial function seems to be a good fit.

```
 X   | Y1
-1   | -10
-.5  | -6.375
 0   | -5
 .5  | -5.125
 1   | -6
 1.5 | -6.875
 2   | -7
X=-1
```

Lesson 4-8 Modeling Real-World Data with Polynomial Functions **259**

Teaching Tip In Example 2, point out that students can also graph **Y₁** in the same screen as the scatter plot in order to check the fit of the graph to the data.

In-Class Examples

Example 1
Determine the type of polynomial function that could be used to represent the data in each scatter plot.

a.

The scatter plot does not change direction. The points are decreasing to the right, so a linear function would be the best fit.

b.
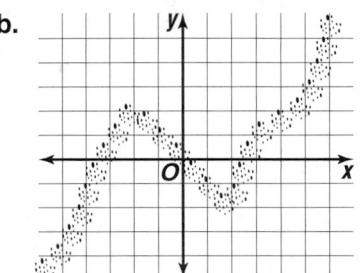

The scatter plot changes direction two times, so a cubic function would be the best fit.

Example 2
Use a graphing calculator to write a polynomial function to model the set of data.

x	f(x)
−2.0	−22.0
−1.5	−7.9
1.0	0.0
−0.5	3.5
0.0	4.1
0.5	2.9
1.0	2.1
1.5	2.5
2.0	5.8
2.5	14.1
3.0	28.0

Sample answer: $f(x) = 2x^3 - 3x^2 - x + 4$

Teaching Tip In Example 3, you may want to encourage students to generate a table using the related function and scroll the table to verify solutions to parts **b** and **c**.

Teaching Tip Remind students that recognizing the shape of the graph in Example 3 will help them select the function type. If they are not sure of the shape, they can experiment quickly with the STAT CALC function on a graphing calculator in order to find the line of best fit.

In-Class Example

Example 3
Nutrition Listed below are the number of fat grams and the corresponding Calories for single servings of several convenience items.

Item	Fat (g)	Calories
2% milk	5.0	120
Tortilla chips	8.0	160
Peanut butter	16.0	225
Popcorn	9.0	150
Coffee cake	11.0	210
Danish	11.0	260
Chicken nuggets	16.0	280
Yogurt	2.0	80
Pizza (frozen)	16.0	390
Sugar cookies	4.5	150

a. What polynomial function could be used to model these data? **Sample answer: $f(x) = 15.4x + 50.6$**

b. Use the model to predict the number of Calories for a similar food item having 10 grams of fat per serving. **Sample answer: 205 Cal**

c. Use the model to predict the number of fat grams for a similar food item having 450 Calories per serving. **about 26 g of fat**

Teaching Tip Point out that making predictions for the future using mathematical models is often inaccurate. Many outside influences can cause changes in the models as the years go by.

Example **3 WASTE MANAGEMENT** Refer to the application at the beginning of the lesson.

a. What polynomial function could be used to model these data?

b. Use the model to predict the amount of waste produced per day in 2010.

c. Use the model to predict when the amount of waste will drop to 3 pounds per day.

a. Let **L1** be the number of years since 1980 and **L2** be the pounds of solid waste per person per day. Enter this data into the calculator.

Adjust the window to an appropriate setting and graph the statistical data. The data seems to change direction one time, so a quadratic function will fit the scatter plot.

[−1, 18] scl:1 by [3, 5] scl:1

Press STAT , highlight **CALC**, and choose **5:QuadReg**.

Press 2nd [L1] ' 2nd [L2] ENTER .

Rounding the coefficients to the nearest thousandth, $f(x) = -0.004x^2 + 0.119x + 3.593$ models the data. Since the value of the coefficient of determination r^2 is close to 1, the model is a good fit.

b. Since 2010 is 30 years after 1980, find $f(30)$.

$$f(x) = -0.004x^2 + 0.119x + 3.593$$
$$f(30) = -0.004(30^2) + 0.119(30) + 3.593$$
$$= 3.563$$

According to the model, each person will produce about 3.6 pounds of waste per day in 2010.

Addressing Individual Needs

Intrapersonal Help students consider data from their science and social studies classes that could be modeled with a polynomial function. Ask students to individually make charts of their selected data. They should then write the modeling functions and draw the corresponding graphs.

c. To find when the amount of waste will be reduced to 3 pounds per person per day, solve the equation $3 = -0.004x^2 + 0.119x + 3.593$.

$$3 = -0.004x^2 + 0.119x + 3.593$$

$$0 = -0.004x^2 + 0.119x + 0.593 \qquad \textit{Add } -3 \textit{ to each side.}$$

Use the Quadratic Formula to solve for x.

$$x = \frac{-0.119 \pm \sqrt{0.119^2 - 4(-0.004)(0.593)}}{2(-0.004)} \qquad \textit{a = -0.004, b = 0.119, c = 0.593}$$

$$x = \frac{-0.119 \pm \sqrt{0.023649}}{-0.008}$$

$$x \approx -4 \text{ or } 34$$

According to the model, 3 pounds per day per person occurs 4 years before 1980 (in 1976) or 34 years after 1980 (in 2014). Since you want to know when the amount of waste will drop to 3 pounds per day, the answer is in 2014.

To check the answer, graph the related function $f(x) = -0.004x^2 + 0.119x + 0.593$. Use the **CALC** menu to determine the zero at the right. The answer of 34 years checks.

The waste should reduce to 3 pounds per person per day about 34 years after 1980 or in 2014.

[−10, 40] scl:5 by [−2, 4] scl:1

CHECK FOR UNDERSTANDING

Communicating Mathematics

Read and study the lesson to answer each question. **1–3. See margin.**

1. **Draw** an example of a scatter plot that can be represented by each type of function.

    a. quartic   **b.** quadratic   **c.** cubic

2. **Explain** why it is important to recognize the shape of the graph of each type of polynomial function.

3. **List** reasons why the amount of waste per person may vary from the model in Example 3.

Guided Practice

4. Determine the type of polynomial function that could be used to represent the data in the scatter plot. **quartic**

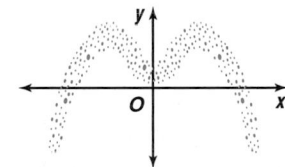

Lesson 4-8 Modeling Real-World Data with Polynomial Functions **261**

Additional Answers

1a. Sample answer:

1b. Sample answer:

1c. Sample answer:

3 PRACTICE/APPLY

Error Analysis

Watch for students who inappropriately scale the graph for the scatter plots, thus masking important graphical features, such as the linear nature of curves in cubic graphs.

Prevent by having students press ZOOM and select **9:ZoomStat** from the menu. This function will automatically size the window to fit the data.

Additional Answers

2. **You need to recognize the general shape so that you can tell the graphing calculator which type of polynomial function to use as a model.**

3. **Sample answer: if companies use less packaging materials, consumers keep items longer, and old buildings are restored instead of demolished, the amount of waste will decrease more rapidly. If consumers buy more products, companies package items in larger containers, and many old buildings are destroyed, the amount of waste will increase instead of decrease.**

Study Guide and Practice Masters, pp. 73–74

Lesson 4-8 **261**

Graphing Calculator

Use a graphing calculator to write a polynomial function to model each set of data.

5.

x	−3.5	−3	−2.5	−2	−1.5	−1	−0.5	0	0.5	1	1.5	2
f(x)	103	32	−1	−11	−9	−2	3	5	4	4	12	37

Sample answer: $f(x) = 1.98x^4 + 2.95x^3 - 5.91x^2 + 0.22x + 4.89$

6.

x	−3	−2.5	−1.5	−0.5	0	1	2	2.5	3.5
f(x)	19	11	−1	−7	−8	−5	4	11	29

Sample answer: $f(x) = 3.007x^2 + 0.001x - 7.896$

7. Population The percent of the United States population living in metropolitan areas has increased since 1950.

Year	1950	1960	1970	1980	1990	1996
Population living in metropolitan areas	56.1%	63%	68.6%	74.8%	74.8%	79.9%

Source: American Demographics

a. Write a model that relates the percent as a function of the number of years since 1950. **Sample answer: $f(x) = 0.49x + 57.7$**

b. Use the model to predict the percent of the population that will be living in metropolitan areas in 2010. **Sample answer: 87.1%**

c. Use the model to predict what year will have 85% of the population living in metropolitan areas. **Sample answer: 2006**

EXERCISES

Practice

Determine the type of polynomial function that could be used to represent the data in scatter plot.

8.

cubic

9.

quadratic

10.

linear

11. What type of polynomial function would be the best model for the set of data? **quadratic**

x	1	2	3	4	5	7	8
f(x)	15	7	2	−1	3	10	15

Graphing Calculator

Use a graphing calculator to write a polynomial function to model each set of data.

12.

x	−3	−2	−1	0	1	2	3
f(x)	8.75	7.5	6.25	5	3.75	2.5	1.25

$f(x) = -1.25x + 5$

13.

x	−2	−1	0	1	2	3
f(x)	29	2	−9	−4	17	54

$f(x) = 8x^2 - 3x - 9$

14. Sample answer: $f(x) = 1.03x^4 - 5.16x^3 + 6.08x^2 + 0.23x + 0.94$

14.

x	−1	−0.5	0	0.5	1	1.5	2	2.5	3	3.5
f(x)	13	3	1	2	3	3	1	−1	1	10

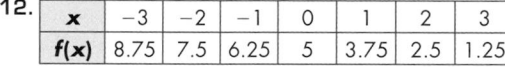 www.amc.glencoe.com/self_check_quiz

20. Sample answer:

x	f(x)
1	1
2	3
3	6
4	3
5	1
6	3
7	6
8	9
9	12

15.

x	5	7	8	10	11	12	15	16
f(x)	2	5	6	4	−1	−3	5	9

Sample answer: $f(x) = 0.09x^3 - 2.70x^2 + 24.63x - 65.21$

16.

x	30	35	40	45	50	55	60	65	70	75
f(x)	52	41	32	44	61	88	72	59	66	93

Sample answer: $f(x) = 4.05x^4 - 0.09x^3 + 6.69x^2 - 222.03x + 2697.74$

17.

x	−17	−6	−1	2	8	12	15
f(x)	51	29	−6	41	57	37	19

Sample answer: $f(x) = -0.02x^3 + 8.79x^2 + 3.35x + 27.43$

18b. Sample answer: $f(x) = -0.96x^3 + 0.56x^2 + 0.36x + 4.05$

18c. Sample answer: Cubic; the value of r^2 for the cubic function is closer to 1.

18. Consider the set of data.   **a.** Sample answer: $f(x) = 1.99x^2 - 1.74x + 2.76$

x	−2.5	−2	−1.5	−1	−0.5	0	0.5	1	1.5
f(x)	23	11	7	6	6	5	3	2	4

a. What quadratic polynomial function models the data?

b. What cubic polynomial function models the data?

c. Which model do you think is more appropriate? Explain.

Applications and Problem Solving

Real World Application

interNET CONNECTION

Data Update To compare current census data with predictions using your models, visit our website at www.amc. glencoe.com

19. Marketing   The United States Census Bureau has projected the median age of the U.S. population to the year 2080. A fast-food chain wants to target its marketing towards customers that are about the median age.

Year	1900	1930	1960	1990	2020	2050	2080
Median age	22.9	26.5	29.5	33.0	40.2	42.7	43.9

a. Write a model that relates the median age as a function of the number of years since 1900.   **Sample answer: $f(x) = 0.126x + 22.732$**

b. Use the model to predict what age the fast-food chain should target in the year 2005.   **Sample answer: 36**

c. Use the model to predict what age the fast-food chain should target in the year 2025.   **Sample answer: 38**

20. Critical Thinking   Write a set of data that could be best represented by a cubic polynomial function.   **See margin.**

21. Consumer Credit   The amount of consumer credit as a percent of disposable personal income is given below.

Year	1988	1989	1990	1991	1992	1995	1996	1997
Consumer credit	23%	24%	23%	22%	19%	21%	24%	22%

Source: *The World Almanac and Book of Facts*

21a. Sample answer: $f(x) = -0.03x^4 + 0.50x^3 - 2.79x^2 + 4.01x + 22.78$

a. Write a model that relates the percent of consumer credit as a function of the number of years since 1988.

b. Use the model to estimate the percent of consumer credit in 1994.   **Sample answer: about 16%**

22. Critical Thinking   What type of polynomial function should be used to model the scatter plot? Explain.   **A function with x^6; there are 5 changes in direction.**

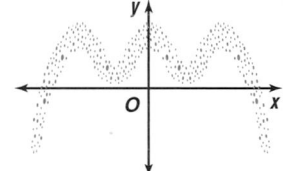

Open-Ended Assessment

Modeling Provide students with a set of data—for example, sports records—for which there appears to be a polynomial model. Ask them to find the polynomial that best fits the data and explain their response. They may also make predictions based on the model.

 Chapter 4, Quiz D (Lessons 4-7 and 4-8) is available in the *Assessment and Evaluation Masters*, p. 96.

Additional Answer

23c. Sample answer: no; according to the model, there should have been an attendance of only about 65 million. Since the actual attendance was much higher than the projected number, it is likely that the race to break the homerun record increased the attendance.

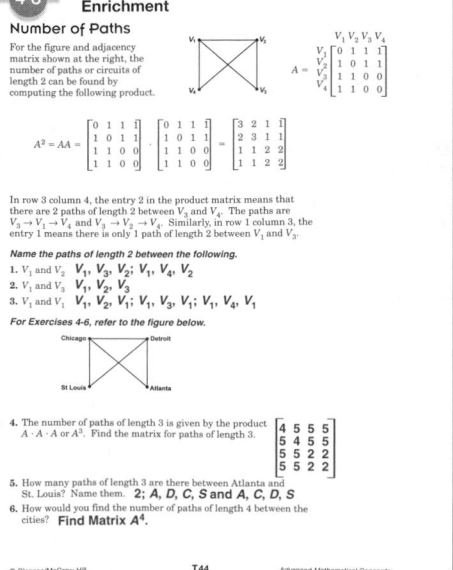
---

23. Baseball The attendance at major league baseball games for various years is listed below.

Year	1985	1987	1990	1992	1993	1994	1995	1996
Attendance (in millions)	48	53	56	57	71	50	51	62

Source: *Statistical Abstract of the United States*

23a. Sample answer:
$f(x) = 0.02x^3 - 0.46x^2 + 3.94x + 47.49$

23b. Sample answer: 2000

a. Write a model that relates the attendance in millions as a function of the number of years since 1985.

b. Use the model to predict when the attendance will reach 71 million again.

c. In 1998, Mark McGwire and Sammy Sosa raced to break the homerun record. That year the attendance reached 70.6 million. Did the attendance for 1998 follow your model? Do you think the race to break the homerun record affected attendance? Why or why not? **See margin.**

24. Communication Cellular phones are becoming more popular. The numbers of cellular subscribers are listed in below.

Year	1983	1984	1986	1989	1990	1992	1994	1996	1997
Subscribers (in millions)	0.25	3.0	6.5	3.8	5.3	11.0	24.1	44.0	55.3

Source: *Statistical Abstract of the United States*

24a. Sample answer:
$f(x) = 0.05x^3 - 0.60x^2 + 2.47x + 0.99$

24b. Sample answer: about 30.63 million

Mixed Review

a. Write a model that relates the millions of subscribers as a function of the number of years since 1983.

b. According to the model, how many subscribers were there in 1995?

c. Use the model to predict when there will be 150 million subscribers. **Sample answer: about 2001**

25. Solve $5 - \sqrt{b + 2} = 0$. *(Lesson 4-7)* **23**

26. Solve $\dfrac{6}{p + 3} + \dfrac{p}{p - 3} = 1$. *(Lesson 4-6)* **1**

27. Approximate the real zeros of $f(x) = 2x^4 - x^3 + x - 2$ to the nearest tenth. *(Lesson 4-5)* **−1, 1**

28. Agriculture If Wesley Jackson harvests his apple crop now, the yield will average 120 pounds per tree. Also, he will be able to sell the apples for $0.48 per pound. However, he knows that if he waits, his yield will increase by about 10 pounds per week, while the selling price will decrease by $0.03 per pound per week. *(Lesson 3-6)*

a. How many weeks should Mr. Jackson wait in order to maximize the profit? **2 weeks**

b. What is the maximum profit? **$58.80 per tree**

29. SAT/ACT Practice A college graduate goes to work for $x per week. After several months, the company falls on hard times and gives all the employees a 10% pay cut. A few months later, business picks up and the company gives all employees a 10% raise. What is the college graduate's new salary? **B**

A $0.90x   B $0.99x   C $x

D $1.01x   E $1.11x

264 *Chapter 4* *Polynomial and Rational Functions*

Extra Practice See p. A33.

Extra Credit

Is it always appropriate to use a polynomial model for a scatter plot in order to predict beyond the range of the data given? Give an example to support your response. **No, because realistic conditions may not exist beyond the data given; sample example: a scatter plot of sports records may show that the record time for completing a 100-meter race has decreased over the years. The model for the scatter plot will decrease indefinitely, becoming zero, then negative—however, the race will never be run in zero or negative time.**

4-8B Fitting a Polynomial Function to a Set of Points

An Extension of Lesson 4-8

Lesson Notes

1 FOCUS

Motivating the Lesson
Real-World Application When stargazers and astronomers find a comet in our solar system, they map its movement using astronomical units as it moves toward or away from the sun. A graphing calculator or computer program can be used to fit a polynomial equation to these points.

In this case, astronomers can fit points representing the comet's changing location to the polynomial equation $E = \frac{1}{2}v^2 - GMr^{-1}$, in which E is the total energy of the comet, v is the comet's velocity, r is the comet's distance from the sun, M is the mass of the sun, and G is the gravitational constant.

OBJECTIVE
• Use matrices to find a polynomial function whose graph passes through a given set of points in the coordinate plane.

The **STAT CALC** menu of a graphing calculator has built-in programs that allow you to find a polynomial function whose graph passes through as many as five arbitrarily selected points in the coordinate plane if no two points are on the same vertical line. For example, for a set of four points, you can enter the *x*-coordinates in list **L1** and the *y*-coordinates in list **L2**. Then, select **6: CubicReg** from the **STAT CALC** menu to display the coefficients for a cubic function whose graph passes through the four points.

For any finite set of points where no two points are on the same vertical line, you can perform calculations to find a polynomial function whose graph passes exactly through those points. For a set of *n* points (where $n \geq 1$), the polynomial function will have a degree of $n - 1$.

Example

Find a polynomial function whose graph passes through points $A(-2, 69)$, $B(-1.5, 19.75)$, $C(-1, 7)$, $D(2, 25)$, $E(2.5, 42)$, and $F(3, 49)$.

The equation of a polynomial function through the six points is of the form $y = ax^5 + bx^4 + cx^3 + dx^2 + ex + f$. The coordinates of each point must satisfy the equation for the function. If you substitute the *x*- and *y*-coordinates of the points into the equation, you obtain a system of six linear equations in six variables.

A: $(-2)^5a + (-2)^4b + (-2)^3c + (-2)^2d + (-2)e + 1f = 69$

B: $(-1.5)^5a + (-1.5)^4b + (-1.5)^3c + (-1.5)^2d + (-1.5)e + 1f = 19.75$

C: $(-1)^5a + (-1)^4b + (-1)^3c + (-1)^2d + (-1)e + 1f = 7$

D: $(2)^5a + (2)^4b + (2)^3c + (2)^2d + (2)e + 1f = 25$

E: $(2.5)^5a + (2.5)^4b + (2.5)^3c + (2.5)^2d + (2.5)e + 1f = 42$

F: $(3)^5a + (3)^4b + (3)^3c + (3)^2d + (3)e + 1f = 49$

Perform the following steps on a graphing calculator.

Step 1 Press $\boxed{\text{STAT}}$, select **Edit**, and enter three lists. In list **L1**, enter the *x*-coordinates of the six points in the order they appear as coefficients of *e*. In list **L2**, enter the *y*-coordinates in the order they appear after the equal signs. In list **L3**, enter six 1s.

Step 2 Notice that the columns of coefficients of *a, b, c,* and *d* are powers of the column of numbers that you entered for **L1**. To solve the linear system, you need to enter the columns of coefficients as columns of a 6×6 matrix. So, press the $\boxed{\text{MATRX}}$ key, and set the dimensions of matrix **[A]** to 6×6. Next use **9: List▶matr** from the **MATRX MATH** menu to put the columns of coefficients into matrix **[A]**. On the home screen, enter the following expression.

List▶matr (L1^5, L1^4, L1^3, L1^2, L1, L3, [A])

2 TEACH

📁 A reproduction of this lesson with graphing calculator keystrokes provided can be found in the *Graphing Calculator Keystrokes Masters,* pp. 14–16.

Teaching Tip If students have not cleared the $\boxed{\text{STAT}}$ lists of prior data, have students go to the **1:Edit** mode under $\boxed{\text{STAT}}$ and press $\boxed{▲}$ to highlight the list name, then press $\boxed{\text{CLEAR}}$ $\boxed{\text{ENTER}}$.

(continued on the next page)

Additional Answers

4. Infinitely many; suppose that you are given a set of *n* points in a coordinate plane, no two of which are on the same vertical line. You can pick an infinite number of other points with different *x*-coordinates. You could find polynomial functions that went through the original *n* points and any number of the other points.

5. There is no problem with using L1$^\wedge$0 with list L1 for the example. However, if you are using a different list which happens to have 0 as one of its elements, using L1$^\wedge$0 will result in an error message, since 0$^\wedge$0 is undefined.

4 ASSESS

Open-Ended Assessment

Modeling Have students graph models of the common polynomial function shapes (such as linear, quadratic, cubic, and quartic) that they should recognize.

Step 3 Put the column of *y*-coordinates (the numbers after the equals signs) into matrix **[B]**. Enter **List▶matr (L2, [B])** and press ENTER .

Step 4 Enter **[A]$^{-1}$[B]** and press ENTER . The calculator will display the solution of the linear system as a column matrix. The elements of this matrix are the coefficients for the polynomial being sought. The top element is the coefficient of x^5, the second element is the coefficient of x^4, and so on. The function is $f(x) = -x^5 + 3.5x^4 + 1.5x^3 - 4x^2 - x + 7$.

Step 5 Enter the polynomial function on the **Y=** list and display its graph. If you use **STAT PLOT** and display the data for the coordinates in **L1** and **L2** as a scatter plot, you will have visual confirmation that the graph of the polynomial function passes through the given points.

[5, 5] scl:1 by [−10, 80] scl:10

For problems of this kind, the coefficients in the solution matrix for the linear system are often decimals. You can enter the equation for the function on the **Y=** list efficiently and accurately by first changing the column matrix for the solution into a list. On the home screen, enter **Matr▶list ([A]$^{-1}$[B], 1, L4)**. Then go to the **Y=** list and enter the equation **Y1 = L4(1)X^5 + L4(2)X^4 + L4(3)X^3 + L4(4)X^2 + L4(5)X + L4(6)**.

TRY THESE

1. $y = 7x^3 - 4x^2 - 17x + 15$

2. Use **6: CubicReg** on the **STAT CALC** menu to fit a cubic model for the points in Exercise 1. Is the resulting cubic function the same as the one you found in Exercise 1? $y = 7x^3 - 4x^2 - 17x + 15$; **yes**

1. Use the steps given above to find a cubic polynomial function whose graph passes through points with coordinates $(-2, -23)$, $(-1, 21)$, $(0, 15)$, and $(1, 1)$.

3. Find a polynomial function whose graph passes through points with coordinates $(-3, 116)$, $(-2, 384)$, $(-0.4, -5.888)$, $(0, -4)$, $(1, 36)$, $(2, 256)$, and $(2.6, 34.1056)$.
   Sample answer: $y = -5x^6 - 2x^5 + 40x^4 - 2x^3 + x^2 + 8x - 4$

WHAT DO YOU THINK?

4. See margin.

4. How many polynomial functions have graphs that pass through a given set of points, no two of which are on the same vertical line? Explain your thinking.

5. In Step 2, why was it necessary to define the list **L3**? Would it have been possible to use **L1^0** instead of **L3** in the **List▶matr** expression used to define matrix **[A]**? Explain your thinking.   **See margin.**

VOCABULARY

completing the square (p. 213)
complex number (p. 206)
conjugate (p. 216)
degree (p. 206)
depressed polynomial (p. 224)
Descartes' Rule of Signs (p. 231)
discriminant (p. 215)
extraneous solution (p. 251)
Factor Theorem (p. 224)
Fundamental Theorem of Algebra (p. 207)
imaginary number (p. 206)

Integral Root Theorem (p. 230)
leading coefficient (p. 206)
Location Principle (p. 236)
lower bound (p. 238)
Lower Bound Theorem (p. 238)
partial fractions (p. 244)
polynomial equation (p. 206)
polynomial function (p. 206)
polynomial in one variable (p. 205)
pure imaginary number (p. 206)
Quadratic Formula (p. 215)

radical equation (p. 251)
radical inequality (p. 253)
rational equation (p. 243)
rational inequality (p. 245)
Rational Root Theorem (p. 229)
Remainder Theorem (p. 222)
root (p. 206)
synthetic division (p. 223)
upper bound (p. 238)
Upper Bound Theorem (p. 238)
zero (p. 206)

STUDY GUIDE AND ASSESSMENT

VOCABULARY

This section provides a listing of the new terms, properties, and phrases that were introduced in this chapter. The exercises check students' understanding of the terms by using a variety of verbal formats including matching, completion, and true/false.

Glossary A complete glossary of terms begins on page A71.

Assessment and Evaluation Masters, pp. 83–84

UNDERSTANDING AND USING THE VOCABULARY

Choose the correct term from the list to complete each sentence. **1. Quadratic Formula**

1. The ___?___ can be used to solve any quadratic equation.

2. The ___?___ states that if the leading coefficient of a polynomial equation a_0 has a value of 1, then any rational root must be factors of a_n. **Integral Root Theorem**

3. For a rational equation, any possible solution that results with a ___?___ in the denominator must be excluded from the list of solutions. **zero**

4. The ___?___ states that the binomial $x - r$ is a factor of the polynomial $P(x)$ if and only if $P(r) = 0$. **Factor Theorem**

5. Descartes' Rule of Signs can be used to determine the possible number of positive real zeros a ___?___ has. **polynomial function**

6. A(n) ___?___ of the zeros of $P(x)$ can be found by determining an upper bound for the zeros of $P(-x)$. **lower bound**

7. ___?___ solutions do not satisfy the original equation. **Extraneous**

8. Since the x-axis only represents real numbers, ___?___ of a polynomial function cannot be determined by using a graph. **complex roots**

9. The Fundamental Theorem of Algebra states that every polynomial equation with degree greater than zero has at least one root in the set of ___?___ **complex numbers**

10. A ___?___ is a special polynomial equation with a degree of two. **quadratic equation**

complex numbers
complex roots
discriminant
extraneous
Factor Theorem
Integral Root Theorem
lower bound
polynomial function
quadratic equation
Quadratic Formula
radical equation
synthetic division
zero

 For additional review and practice for each lesson, visit: **www.amc.glencoe.com**

Assessment Resources

Six forms of Chapter 4 Test are available in the Assessment and Evaluation Masters as shown in the chart.

Chapter 4 Test, Form 1B, is shown at the right. Chapter 4 Test, Form 2B, is shown on the next page.

TEST OPTIONS

Form	Pages	Type	
1A	81–82	Honors	Form 1 tests are multiple choice.
1B	83–84	Average	
1C	85–86	Basic	
2A	87–88	Honors	Form 2 tests are free response.
2B	89–90	Average	
2C	91–92	Basic	

SKILLS AND CONCEPTS

Objectives and Examples This section reviews the skills and concepts of the chapter and shows completely worked examples.

Review Exercises These exercises provide practice for the corresponding objectives.

Assessment and Evaluation Masters, pp. 89–90

SKILLS AND CONCEPTS

OBJECTIVES AND EXAMPLES	REVIEW EXERCISES

Lesson 4-1 Determine roots of polynomial equations.

- Determine whether 2 is a root of $x^4 - 3x^3 - x^2 - x = 0$. Explain.

$$f(2) = 2^4 - 3(2^3) - 2^2 - 2$$
$$f(2) = 16 - 24 - 4 - 2 \text{ or } -14$$

Since $f(2) \neq 0$, 2 is not a root of $x^4 - 3x^3 - x^2 - x = 0$.

Determine whether each number is a root of $a^3 - 3a^2 - 3a - 4 = 0$. Explain.

11. 0   no; $f(0) = -4$

12. 4   yes; $f(4) = 0$

13. −2   no; $f(-2) = -18$

14. Is −3 a root of $t^4 - 2t^2 - 3t + 1 = 0$?   no

15. State the number of complex roots of the equation $x^3 + 2x^2 - 3x = 0$. Then find the roots and graph the related function. **3; −3, 0, 1; See margin for graph.**

Lesson 4-2 Solve quadratic equations.

- Find the discriminant of $3x^2 - 2x - 5 = 0$ and describe the nature of the roots of the equation. Then solve the equation by using the Quadratic Formula.

The value of the discriminant, $b^2 - 4ac$, is $(-2)^2 - 4(3)(-5)$ or 64. Since the value of the discriminant is greater than zero, there are two distinct real roots.

$$x = \frac{-b \pm \sqrt{b^2 - 4ac}}{2a}$$

$$x = \frac{-(-2) \pm \sqrt{64}}{2(3)}$$

$$x = \frac{2 \pm 8}{6}$$

$$x = -1 \text{ or } \frac{5}{3}$$

Find the discriminant of each equation and describe the nature of the roots of the equation. Then solve the equation by using the Quadratic Formula.

16. $2x^2 - 7x - 4 = 0$ **81; 2 real;** $-\frac{1}{2}, 4$

17. $3m^2 - 10m + 5 = 0$ **40; 2 real;** $\frac{5 \pm \sqrt{10}}{3}$

18. $x^2 - x + 6 = 0$

19. $-2y^2 + 3y + 8 = 0$ **73; 2 real;** $\frac{3 \pm \sqrt{73}}{4}$

20. $a^2 + 4a + 4 = 0$ **0; 1 real; −2**

21. $5r^2 - r + 10 = 0$ **−199; 2 imaginary;** $\frac{1 \pm i\sqrt{199}}{10}$

18. **−23; 2 imaginary;** $\frac{1 \pm i\sqrt{23}}{2}$

Lesson 4-3 Find the factors of polynomials using the Remainder and Factor Theorems.

- Use the Remainder Theorem to find the remainder when $(x^3 + 2x^2 - 5x - 9)$ is divided by $(x + 3)$. State whether the binomial is a factor of the polynomial.

$$f(x) = x^3 + 2x^2 - 5x - 9$$
$$f(-3) = (-3)^3 + 2(-3)^2 - 5(-3) - 9$$
$$= -27 + 18 + 15 - 9 \text{ or } -3$$

Since $f(-3) = -3$, the remainder is −3. So the binomial $x + 3$ is not a factor of the polynomial by the Remainder Theorem.

Use the Remainder Theorem to find the remainder for each division. State whether the binomial is a factor of the polynomial.

22. $(x^3 - x^2 - 10x - 8) \div (x + 2)$ **0; yes**

23. $(2x^3 - 5x^2 + 7x + 1) \div (x - 5)$ **161; no**

24. $(4x^3 - 7x + 1) \div \left(x + \frac{1}{2}\right)$ **4; no**

25. $(x^4 - 10x^2 + 9) \div (x - 3)$ **0; yes**

TestCheck and Worksheet Builder

This **networkable** software has three modules.
- **Worksheet Builder** to make worksheets and tests
- **Student Module** to take tests on-screen
- **Management System** to keep student records

Three special banks are included for SAT, ACT, and TIMSS.

OBJECTIVES AND EXAMPLES

Lesson 4-4 Identify all possible rational roots of a polynomial equation by using the Rational Root Theorem.

- List the possible rational roots of $4x^3 - x^2 - x - 5 = 0$. Then determine the rational roots.

  If $\frac{p}{q}$ is a root of the equation, then p is a factor of 5 and q is a factor of 4.
  possible values of p: $\pm 1, \pm 5$
  possible values of q: $\pm 1, \pm 2, \pm 4$

  possible rational roots: $\pm 1, \pm 5, \pm \frac{1}{2}$, $\pm \frac{1}{4}, \pm \frac{5}{2}, \pm \frac{5}{4}$

  Graphing and substitution show a zero at $\frac{5}{4}$.

Lesson 4-4 Determine the number and type of real roots a polynomial function has.

- For $f(x) = 3x^4 - 9x^3 + 4x - 6$, there are three sign changes. So there are three or one positive real zeros.

  For $f(-x) = 3x^4 + 9x^3 - 4x - 6$, there is one sign change. So there is one negative real zero.

Lesson 4-5 Approximate the real zeros of a polynomial function.

- Determine between which consecutive integers the real zeros of $f(x) = x^3 + 4x^2 + x - 2$ are located.
  Use synthetic division.

r	1	4	1	-2	
4	1	0	1	-6	← change in signs
-3	1	1	2	4	
-2	1	2	-3	4	
-1	1	3	-2	0	← zero
0	1	4	1	-2	← change in signs
1	1	5	6	4	

One zero is -1. Another is located between -4 and -3. The other is between 0 and 1.

REVIEW EXERCISES

List the possible rational roots of each equation. Then determine the rational roots. **26–33. See margin.**

26. $x^3 - 2x^2 - x + 2 = 0$
27. $x^4 - x^2 + x - 1 = 0$
28. $2x^3 - 2x^2 - 2x - 4 = 0$
29. $2x^4 + 3x^3 - 6x^2 - 11x - 3 = 0$
30. $x^5 - 7x^3 + x^2 + 12x - 4 = 0$
31. $3x^3 + 7x^2 - 2x - 8 = 0$
32. $4x^3 + x^2 + 8x + 2 = 0$
33. $x^4 + 4x^2 - 5 = 0$

Find the number of possible positive real zeros and the number of possible negative real zeros for each function. Then determine the rational zeros. **34–36. See margin.**

34. $f(x) = x^3 - x^2 - 34x - 56$
35. $f(x) = 2x^3 - 11x^2 + 12x + 9$
36. $f(x) = x^4 - 13x^2 + 36$

Determine between which consecutive integers the real zeros of each function are located.

37. $g(x) = 3x^3 + 1$ **−1 and 0**
38. $f(x) = x^2 - 4x + 2$ **0 and 1, 3 and 4**
39. $g(x) = x^2 - 3x - 3$ **−1 and 0, 3 and 4**
40. $f(x) = x^3 - x^2 + 1$ **−1 and 0**
41. $g(x) = 4x^3 + x^2 - 11x + 3$ **−2 and −1, 0 and 1, 1 and 2**
42. $f(x) = -9x^3 + 25x^2 - 24x + 6$ **0 and 1**
43. Approximate the real zeros of $f(x) = 2x^3 + 9x^2 - 12x - 40$ to the nearest tenth. **−4.9, −1.8, 2.2**

Additional Answers
(pp. 268–269)

15.

$f(x) = x^3 + 2x^2 - 3x$

26. $\pm 1, \pm 2; \pm 1, 2$
27. $\pm 1; 1$
28. $\pm 1, \pm 2, \pm 4, \pm \frac{1}{2}; 2$
29. $\pm 1, \pm 3, \pm \frac{1}{2}, \pm \frac{3}{2}; -\frac{3}{2}$
30. $\pm 1, \pm 2, \pm 4; \pm 2$
31. $\pm 1, \pm 2, \pm 4, \pm 8, \pm \frac{1}{3}, \pm \frac{2}{3},$ $\pm \frac{4}{3}, \pm \frac{8}{3}; -2, -\frac{4}{3}, 1$
32. $\pm 1, \pm 2, \pm \frac{1}{4}, \pm \frac{1}{2}; -\frac{1}{4}$
33. $\pm 5, \pm 1; \pm 1$
34. 1; 2 or 0; $-4, -2, 7$
35. 2 or 0; 1; $-\frac{1}{2}, 3$
36. 2 or 0; 2 or 0; $-3, -2, 2, 3$

OBJECTIVES AND EXAMPLES	REVIEW EXERCISES

Lesson 4-6 Solve rational equations and inequalities.

- Solve $\frac{1}{9} + \frac{1}{2a} = \frac{1}{a^2}$.

$$\frac{1}{9} + \frac{1}{2a} = \frac{1}{a^2}$$

$$\left(\frac{1}{9} + \frac{1}{2a}\right)(18a^2) = \left(\frac{1}{a^2}\right)(18a^2)$$

$$2a^2 + 9a = 18$$

$$2a^2 + 9a - 18 = 0$$

$$(2a - 3)(a + 6) = 0$$

$$a = \frac{3}{2} \text{ or } -6$$

Solve each equation or inequality.

44. $n - \frac{6}{n} + 5 = 0$ **−6, 1**

45. $\frac{1}{x} = \frac{x + 3}{2x^2}$ **3**

46. $\frac{5}{6} = \frac{2m}{2m + 2} - \frac{1}{3m - 3}$ **$4 \pm \sqrt{13}$**

47. $\frac{3}{y} - 2 < \frac{5}{y}$ **$y < -1, y > 0$**

48. $\frac{2}{x + 1} < 1 - \frac{1}{x - 1}$
   $x < -1, 0 < x < 1, x > 3$

Lesson 4-7 Solve radical equations and inequalities.

- Solve $9 + \sqrt{x - 1} = 1$.

$$9 + \sqrt{x - 1} = 1$$

$$\sqrt{x - 1} = -8$$

$$x - 1 = 64$$

$$x = 65$$

Solve each equation or inequality.

49. $5 - \sqrt{x + 2} = 0$ **23**

50. $\sqrt[3]{4a - 1} + 8 = 5$ **−6.5**

51. $3 + \sqrt{x + 8} = \sqrt{x + 35}$ **1**

52. $\sqrt{x - 5} < 7$ **$5 < x < 54$**

53. $4 + \sqrt{2a + 7} \geq 6$ **$a \geq -1.5$**

Lesson 4-8 Write polynomial functions to model real-world data.

- Determine the type of polynomial function that would best fit the data in the scatter plot.

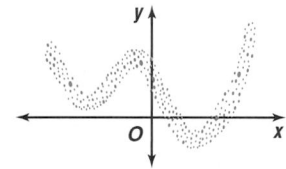

The scatter plot seems to change direction three times. So a quartic function would best fit the scatter plot.

54. Determine the type of polynomial function that would best fit the scatter plot. **cubic**

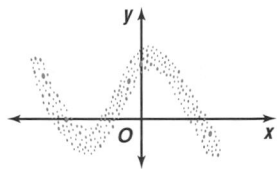

55. Write a polynomial function to model the data.

x	−3	−1	0	2	4	7
f(x)	24	6	3	9	31	94

$$f(x) = 2x^2 - x + 3$$

Additional Answers
Open-Ended Assessment (p. 271)

1. Sample answer:
   $\frac{x}{x + 3} = \frac{2}{2x + 1}$; Solutions
   are $-\frac{3}{2}$ and 2.

2a. Sample answer: $x - 4 = \sqrt{x - 2}$

2b. Sample answer: the solution is 6. Since $-1 \neq 1$, 3 is an extraneous root.

3a. Sample answer:

x	−3	−2	−1	−0.5	0	0.5	1	2
f(x)	−12	0	2	1.125	0	−0.625	0	8

3b. Sample answer: $f(x) = x^3 + x^2 - 2x$
3c. Sample answer: −2, 0, 1

APPLICATIONS AND PROBLEM SOLVING

56. Entertainment The scenery for a new children's show has a playhouse with a painted window. A special gloss paint covers the area of the window to make them look like glass. If the gloss only covers 315 square inches and the window must be 6 inches taller than it is wide, how large should the scenery painters make the window? *(Lesson 4-1)* **15 in. by 21 in.**

57. Gardening The length of a rectangular flower garden is 6 feet more than its width. A walkway 3 feet wide surrounds the outside of the garden. The total area of the walkway itself is 288 square feet. Find the dimensions of the garden. *(Lesson 4-2)* **18 ft by 24 ft**

58. Medicine Doctors can measure cardiac output in potential heart attack patients by monitoring the concentration of dye after a known amount in injected in a vein near the heart. In a normal heart, the concentration of the dye is given by $g(x) = -0.006x^4 + 0.140x^3 - 0.053x^2 + 1.79x$, where x is the time in seconds. *(Lesson 4-4)*

  a. Graph $g(x)$ **See margin. b. 0, about 23.5**

  b. Find all the zeros of this function.

59. Physics The formula $T = 2\pi\sqrt{\dfrac{\ell}{g}}$ is used to find the period T of a oscillating pendulum. In this formula, ℓ is the length of the pendulum, and g is acceleration due to gravity. Acceleration due to gravity is 9.8 meters per second squared. If a pendulum has an oscillation period of 1.6 seconds, determine the length of the pendulum. *(Lesson 4-7)* **about 0.64 m**

ALTERNATIVE ASSESSMENT

OPEN-ENDED ASSESSMENT

1. Write a rational equation that has at least two solutions, one which is 2. Solve your equation. **1–3. See margin.**

2. a. Write a radical equation that has solutions of 3 and 6, one of which is extraneous.

  b. Solve your equation. Identify the extraneous solution and explain why it is extraneous.

3. a. Write a set of data that is best represented by a cubic equation.

  b. Write a polynomial function to model the set of data.

  c. Approximate the real zeros of the polynomial function to the nearest tenth.

PORTFOLIO

Explain how you can use the leading coefficient and the degree of a polynomial equation to determine the number of possible roots of the equation.

Unit 1 *inter*NET Project

TELECOMMUNICATION

The Pen is Mightier than the Sword!

- Gather all materials obtained from your research for the mini-projects in Chapters 1, 2, and 3. Decide what types of software would help you to prepare a presentation.

- Research websites that offer downloads of software including work processing, graphics, spreadsheet, and presentation software. Determine whether the software is a demonstration version or free shareware. Select at least two different programs for each of the four categories listed above.

- Prepare a presentation of your Unit 1 project using the software that you found. Be sure that you include graphs and maps in the presentation.

> **Additional Assessment** See p. A59 for Chapter 4 practice test.

APPLICATIONS & PROBLEM SOLVING

This section provides additional practice in solving real-world problems that involve the concepts of this chapter.

Additional Answer

58a.

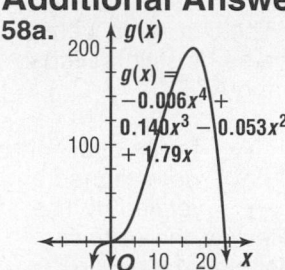

ALTERNATIVE ASSESSMENT

Unit 1 *inter*NET Project

- **The Pen Is Mightier Than the Sword!**

Students should be sure to include graphs, tables, and/or maps in their presentations. Students could present their projects in class or for a parents' night or a school open-house. If they use presentation software, they may be able to set up booths displaying their presentations in a loop on monitors or television screens, depending upon the technology available. If you wish, you may involve a technology or media arts teacher at your school in this project.

Internet Address Book

Record useful Internet addresses in the space at right for quick reference.

www.amc.glencoe.com

Chapter 4 Skill: Coordinate Geometry

Pages 272–273 are part of a complete SAT/ACT preparation course that is described in detail on pages T8–T9 of the *Teacher's Handbook*. The test items were written in the same style as those in the SAT and ACT.

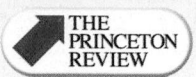 These questions were aligned and verified by The Princeton Review, the nation's leader in test preparation.

Teaching Tip To help students diagram the problems more quickly, have them review the signs of the *x*- and *y*-coordinates in the various quadrants of the coordinate plane.

Diagnosis & Prescription

Each of the 10 SAT and ACT practice problems on page 273 is cross-referenced to the chapter where that SAT or ACT skill is covered. If students miss a particular type of problem, you can have them study that skill.

Problem Number and Skill	Page
1 Coordinate Geometry	272
2 Arithmetic	64
3 Coordinate Geometry	272
4 Algebra	124
5 Coordinate Geometry	272
6 Arithmetic	64
7 More Algebra	202
8 Coordinate Geometry	272
9 More Algebra	202
10 Coordinate Geometry	272

Coordinate Geometry Problems

The ACT test usually includes several coordinate geometry problems. You'll need to know and apply these formulas for points (x_1, y_1) and (x_2, y_2):

Midpoint
$$\left(\frac{x_1 + x_2}{2}, \frac{y_1 + y_2}{2} \right)$$

Distance
$$\sqrt{(x_2 - x_1)^2 + (y_2 - y_1)^2}$$

Slope
$$\frac{y_2 - y_1}{x_2 - x_1}$$

The SAT test includes problems that involve coordinate points. But they aren't easy!

TEST-TAKING TIP
- Draw diagrams.
- Locate points on the grid or number line.
- Eliminate any choices that are clearly incorrect.

ACT EXAMPLE

1. Point $B(4,3)$ is the midpoint of line segment AC. If point A has coordinates $(0,1)$, then what are the coordinates of point C?

   A $(-4, -1)$ **B** $(4, 1)$ **C** $(4, 4)$

   D $(8, 5)$ **E** $(8, 9)$

HINT Draw a diagram. You may be able to solve the problem without calculations.

Solution Draw a diagram showing the known quantities and the unknown point *C*.

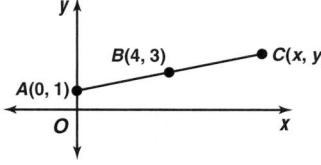

Since *C* lies to the right of *B* and the *x*-coordinate of *A* is not 4, any points with an *x*-coordinate of 4 or less can be eliminated. So eliminate choices A, B, and C.

Use the Midpoint Formula. Consider the *x*-coordinates. Write an equation in *x*.

$$\frac{0 + x}{2} = 4$$
$$x = 8$$

Do the same for *y*.
$$\frac{1 + y}{2} = 3$$
$$y = 5$$

The coordinates of C are (8, 5) The answer is choice **D**.

SAT EXAMPLE

2. What is the area of square *ABCD* in square units?

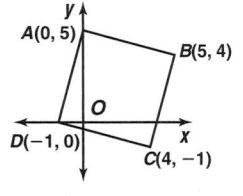

   A 25   **B** $18\sqrt{2}$ **C** 26

   D $25 + \sqrt{2}$ **E** 36

HINT Estimate the answer to eliminate impossible choices and to check your calculations.

Solution First estimate the area. Since the square's side is a little more than 5, the area is a little more than 25. Eliminate choices A and E.

To find the area, find the measure of a side and square it. Choose side *AD*, because points *A* and *D* have simple coordinates. Use the Distance Formula.

$$(\overline{AD})^2 = (-1 - 0)^2 + (0 - 5)^2$$
$$= (-1)^2 + (-5)^2$$
$$= 1 + 25 \text{ or } 26$$

The answer is choice **C**.

Alternate Solution You could also use the Pythagorean Theorem. Draw right triangle *DOA*, with the right angle at *O*, the origin. Then *DO* is 1, *OA* is 5, and *DA* is $\sqrt{26}$. So the area is 26 square units.

TestCheck and Worksheet Builder

Special banks of SAT, ACT, and TIMSS items are included in this CD-ROM.

After you work each problem, record your answer on the answer sheet provided or on a piece of paper.

Multiple Choice

1. What is the length of the line segment whose endpoints are represented on the coordinate axis by points at $(-2, 1)$ and $(1, -3)$? **C**

   A 3   B 4   C 5

   D 6   E 7

2. $\left(\frac{4}{5} \times 3\right)\left(\frac{3}{4} \times 5\right)\left(\frac{5}{3} \times 4\right) =$ **E**

   A 1   B 3   C 6   D 20   E 60

3. In the figure below, $ABCD$ is a parallelogram. What are the coordinates of point C? **E**

   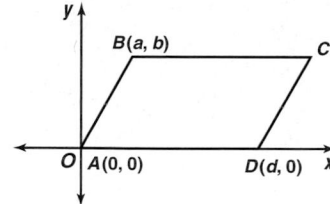

   A (x, y) B $(d + a, y)$

   C $(d - a, b)$ D $(d + x, b)$

   E $(d + a, b)$

4. A rectangular garden is surrounded by a 60-foot long fence. One side of the garden is 6 feet longer than the other. Which equation could be used to find s, the shorter side, of the garden? **E**

   A $8s + s = 60$

   B $4s = 60 + 12$

   C $s(s + 6) = 60$

   D $2(s - 6) + 2s = 60$

   E $2(s + 6) + 2s = 60$

5. What is the slope of a line perpendicular to the line represented by the equation $3x - 6y = 12$? **A**

   A -2 B $\frac{-1}{2}$ C $\frac{1}{3}$

   D $\frac{1}{2}$ E 2

6. If x is an integer, which of the following could be x^3? **C**

   A 2.7×10^{11}

   B 2.7×10^{12}

   C 2.7×10^{13}

   D 2.7×10^{14}

   E 2.7×10^{15}

7. If $0x + 5y = 14$ and $4x - y = 2$, then what is the value of $6x + 6y$? **E**

   A 2   B 7   C 12   D 16   E 24

8. What is the midpoint of the line segment whose endpoints are represented on the coordinate grid by points at $(3, 5)$ and $(-4, 3)$? **B**

   A $(-2, -5)$ B $\left(-\frac{1}{2}, 4\right)$ C $(1, 8)$

   D $\left(4, \frac{1}{2}\right)$ E $(3, 3)$

9. If $ax \neq 0$, which quantity must be non-negative? **A**

   A $x^2 - 2ax + a^2$

   B $-2ax$

   C $2ax$

   D $x^2 - a^2$

   E $a^2 - x^2$

10. **Grid-In** Points E, F, G, and H lie on a line in that order. If $EG = \frac{5}{3} EF$ and $HF = 5FG$, then what is $\frac{EF}{HG}$? **.375 or 3/8**

interNET

CONNECTION **SAT/ACT Practice** For additional test practice questions, visit: **www.amc.glencoe.com**

THE PRINCETON REVIEW *SAT & ACT Preparation* **273**

A **bubble-in answer sheet** for these practice problems is available on p. 1 of the *Assessment and Evaluation Masters.*

Additional Practice

Additional SAT and ACT practice questions and a bubble-in answer sheet are available in the *Assessment and Evaluation Masters,* pp. 2, 97, and 98.

Assessment and Evaluation Masters, pp. 97–98

Extra Credit

TIMSS Find all real values of x which satisfy the following equation.

$$\sqrt{x} - \frac{2}{\sqrt{x}} = 1$$

Show all of your work. **$x = 4$**

UNIT 2

Unit Overview

Unit 2 focuses on the relationship among various trigonometric functions and their graphs. Some of the trigonometric concepts may have been introduced in previous courses. These concepts are reviewed and extended in this unit's exploration of the characteristics of trigometric functions and graphs.

Assessment

A review and a test on the concepts in Unit 2 are provided in the *Assessment and Evaluation Masters* (pp. 401–406). The Unit 2 Review is shown below. The Unit 2 Test is shown on the next page.

Assessment and Evaluation Masters, pp. 401–402

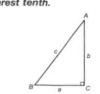

The simplest definition of trigonometry goes something like this:

Trigonometry is the study of angles and triangles. However, the study of trigonometry is really much more complex and comprehensive. In this unit alone, you will find trigonometric functions of angles, verify trigonometric identities, solve trigonometric equations, and graph trigonometric functions. Plus, you'll use vectors to solve parametric equations and to model motion. Trigonometry has applications in construction, geography, physics, acoustics, medicine, meteorology, and navigation, among other fields.

274   Unit 2   Trigonometry

Resource Manager

Reproducible Masters
- *Math and Science Activities*, pp. 6–11
- Unit 2 Review, *Assessment and Evaluation*, pp. 401–402
- Unit 2 Test, *Assessment and Evaluation*, pp. 403–406

Unit 2 *inter*NET Projects

THE CYBER CLASSROOM

From the Internet, a new form of classroom has emerged—one that exists only on the Web. Classes are now being taught exclusively over the Internet to students living across the globe. In addition, traditional classroom teachers are posting notes, lessons, and other information on web sites that can be accessed by their students. At the end of each chapter in Unit 2, you will search the Web for trigonometry learning resources.

CHAPTER 5 (page 339)

Does anybody out there know anything about trigonometry? Across the United States and the world, students are attending classes right in their own homes. What is it like to learn in this new environment? Use the Internet to find trigonometry lessons.
Math Connection: Find and compare trigonometry lessons from this textbook and the Internet. Then, select one topic from Chapter 5 and write a summary of this topic using the information you have gathered from both the textbook and the Internet.

CHAPTER 6 (page 417)

What is your sine? Mathematicians, scientists, and others share their work by means of the Internet. What are some applications of the sine or cosine function?
Math Connection: Use the Internet to find more applications of the sine or cosine function. Find data on the Internet that can be modeled by using a sine or cosine curve. Graph the data and the sine or cosine function that approximates it on the same axes.

CHAPTER 7 (page 481)

That's as clear as mud! Teachers using the Internet to deliver their courses need to provide the same clear instructions and examples to their students that they would in a traditional classroom.
Math Connection: Research the types of trigonometry sample problems given in Internet lessons. Design your own web page that includes two sample trigonometry problems.

CHAPTER 8 (page 547)

Vivid Vectors Suppose you are taking a physics class that requires you to use vectors to represent real world situations. Can you find out more about these representations of direction and magnitude?
Math Connection: Research learning sites on the Internet to find more information about vector applications. Describe three real-world situations that can be modeled by vectors. Include vector diagrams of each situation.

interNET
CONNECTION

For more information on the Unit Project, visit:
www.amc.glencoe.com

Using the *inter*NET Projects

When you or your students visit **www.amc.glencoe.com** you will find links to outside Web sites relating to these Internet projects. These links are checked frequently to ensure that they are still functioning and that the information is current.

Glencoe makes every effort to ensure that the Web site links for *Advanced Mathematical Concepts* are current and contain appropriate content. However, these Web sites are not under Glencoe's control.

In the Alternative Assessment section of each chapter's Study Guide and Assessment, you can find follow-up activities for the Internet Projects.

UNIT 2 INTERNET PROJECTS	
Chapter Number/Title	**Page**
5 *Does Anybody Out There Know Anything About Trigonometry?*	339
6 *What Is Your Sine?*	417
7 *That's as Clear as Mud!*	481
8 *Vivid Vectors!*	547

Assessment and Evaluation Masters, pp. 403–406

Unit 2   Unit 2 Test, Chapters 5–8

NAME ____ DATE ____ PERIOD ____

1. True or false: $\sin(-85°) = -\sin 85°$.   1. ____ true

2. Find the area of $\triangle ABC$ if $a = 12$, $b = 15$, and $c = 23$. Round to the nearest square unit.   2. ____ 81 units²

3. Write the equation $5x + y - 2 = 0$ 3. $\frac{5x}{\sqrt{26}} + \frac{y}{\sqrt{26}} - \frac{2}{\sqrt{26}} = 0$

4. Graph the function $y = 2\cos(\theta + 60°)$.   4. [graph]

5. Given an central angle of $60°$, find the length of its intercepted arc in a circle of radius 15 inches. Round to the nearest tenth.   5. ____ 15.7 in

6. A vector has magnitude of 18.3 centimeters and direction of $90°$. Find the magnitude of its vertical and horizontal components to the nearest tenth.   6. ____ v: 11.3 cm; h: 14.4 cm

7. Write parametric equations of $y = 5x - 2$.   7. $x = t; y = 5t - 2$

8. Find the value of $\sin^{-1}\left(\sin\frac{5\pi}{6}\right)$.   8. $\frac{\pi}{6}$

9. Use the Law of Sines to solve $\triangle ABC$ when $a = 1.43$, $b = 4.21$, and $A = 30.4°$. If no solution exists, write *none*.   9. ____ no solution

10. Use the sum or difference identity to find the exact value of $\tan 105°$.   10. ____ $-2 - \sqrt{3}$

11. Find the distance between $P(7, -4)$ and the line with equation $x - 3y + 5 = 0$. Round to the nearest tenth.   11. ____ 7.6 units

12. Find the inner product of vectors $(2, 5)$ and $(4, -2)$. Then state whether the vectors are perpendicular. Write *yes* or *no*.   12. ____ -2; no

© Glencoe/McGraw-Hill   403   *Advanced Mathematical Concepts*

Internet Address Book

Record useful Internet addresses in the space at right for quick reference.

www.amc.glencoe.com
_____
_____
_____
_____
_____

Instructional Objectives

Lesson (pages)	Lesson Objectives	NCTM Standards 2000	State/Local Objectives
5-1 (277–283)	• Convert decimal degree measures to degrees, minutes, and seconds and vice versa. • Find the number of degrees in a given number of rotations. • Identify angles that are coterminal with a given angle.	3, 6	
5-2 (284–290)	• Find the values of trigonometric ratios for acute angles of right triangles.	3, 6	
5-3 (291–298)	• Find the values of six trigonometric functions using the unit circle. • Find the values of six trigonometric functions of an angle in standard position given a point on its terminal side.	3, 6	
5-4 (299–304)	• Use trigonometry to find the measures of the sides of right triangles.	2, 3, 6	
5-5 (305–312)	• Evaluate inverse trigonometric functions. • Find missing angle measurements. • Solve right triangles.	2, 3, 6, 9	
5-6 (313–318)	• Solve triangles by using the Law of Sines if the measures of two angles and a side are given. • Find the area of a triangle if the measures of two sides and the included angle or the measures of two angles and a side are given.	3, 6, 9, 10	
5-7 (320–326)	• Determine whether a triangle has zero, one, or two solutions. • Solve triangles using the Law of Sines.	3,6,7,9,	
5-8 (327–332)	• Solve triangles by using the Law of Cosines. • Find the area of triangles if the measure of the three sides are given.	3, 6, 9	
5-8B (333–334)	• Use a program to solve triangles.	3, 6, 8	

Suggested Pacing   *See pages T2–T4 for additional options in course-planning.*

COURSE	Lesson 5-1	Lesson 5-2	Lesson 5-3	Lesson 5-4	Lesson 5-5
Precalculus Emphasis Chapters 4–15 (45–50 minutes)	1.5 days	1.5 days	1.5 days	1.5 days	1.5 days
Precalculus Emphasis (block)	0.5 day	1.5 days		1.5 days	
Trigonometric Emphasis Chapters 1–9, 11–12 (45–50 minutes)	1.5 days	1.5 days	1.5 days	1.5 days	1.5 days
Trigonometric Emphasis (block)	0.5 day	1.5 days		1.5 days	

Instructional Resources

	Student Edition	Blackline Masters (page numbers)							Transparencies A and B
Lesson	Graphing Calculator Applications	Study Guide and Practice	Enrichment	Assessment and Evaluation	Graphing Calculator Keystrokes	Problem Solving & Applications	Spreadsheet Applications	TI-92 and CBL Interdisciplinary Activities	
5-1	278	79–80	49			10			5-1
5-2	284, 286, 289	81–82	50	121	17		5		5-2
5-3		83–84	51						5-3
5-4		85–86	52	120,121				13–16	5-4
5-5		87–88	53						5-5
5-6		89–90	54	122		9			5-6
5-7	323	91–92	55		18				5-7
5-8	328, 331	93–94	56	122					5-8
5-8B	333–334				19–21				
Study Guide/ Assessment				107–119, 125					
SAT/ACT Practice				123–124					

Lesson 5-6	Lesson 5-7	Lesson 5-8	Study Guide & Assessment	Chapter Test	TOTAL
1.5 days	1.5 days	(with Lesson 5-8B) 1.5 days	1 day	1 day	14 days
1.5 days		(with Lesson 5-8B) 1.5 days		0.5 day	7 days
1.5 days	1.5 days	(with Lesson 5-8B) 1.5 days	1 day	1 day	14 days
1.5 days		(with Lesson 5-8B) 1.5 days		0.5 day	7 days

Resource Manager

The worksheets shown on this page are a small sample of the materials available on *TeacherWorks: All-in-One Lesson Planner and Resource Center.*

This CD-ROM contains all of the blackline masters and transparencies available for this program.

It also includes a lesson planner and interactive Teacher Edition, so you can customize lesson plans and reproduce classroom resources quickly and easily, from just about anywhere.

Applications

Problem Solving and Applications Masters, pp. 9, 10

Technology/Multimedia

Spreadsheet Applications Masters, p. 5

TI-92 & CBL Interdisciplinary Activities Masters, pp. 13–16

Assessment Resources

Type	Student Edition	Teacher's Wraparound Edition	Assessment and Evaluation Masters
Ongoing Assessment	Mid-Chapter Quiz, p. 304	5-Minute Check, pp. 277, 284 291, 299, 305, 313, 320, 327	Mid-Chapter Test, p. 120 Quizzes A–D, pp.121–122
Mixed Review	Mixed Review, pp. 283, 290, 298, 304, 311–312, 318, 326, 332		Cumulative Review, p. 125
Error Analysis	You Decide, pp. 308, 316	Error Analysis, pp. 280, 287, 296, 301, 308, 316, 324, 330	
Standardized Test Practice	SAT/ACT Practice, pp. 283, 290, 298, 304, 312, 318, 326, 332 SAT & ACT Preparation, pp. 340–341	TIMSS, p. 341	SAT/ACT Practice, pp. 123–124
Open-Ended Assessment	Math Journal, pp. 296, 330 Open-Ended Assessment, p. 339 Portfolio, p. 339	Modeling: pp. 298, 318, 326 Speaking: pp. 290, 312, 334 Writing: pp. 283, 304, 332	Open-Ended Assessment, p. 119
Chapter Assessment	Study Guide and Assessment, pp. 335–339 Chapter Test, p. A60		Multiple-Choice Tests (Forms 1A, 1B, 1C), pp. 107–112 Free-Response Tests (Forms 2A, 2B, 2C), pp. 113–118

TestCheck and Worksheet Builder is a networkable CD-ROM with three integrated modules. The **Worksheet Builder** creates customized worksheets, tests, and quizzes of free-response, multiple-choice, short-answer, and open-ended items. The **Student Module** gives you the option of having students take tests on-screen and get immediate feedback on their performance. Use the optional **Management System** to keep detailed student records.

Additional Chapter Resources

Student Edition
Career Choices, p. 312
History of Mathematics, p. 319
Internet Project, p. 339
Extra Practice, pp. A34–A35

Teacher's Classroom Resources

Technology
 Graphing Calculator Keystrokes Masters

Teaching Aids
 Answer Key Transparencies
Block Schedule Planning Guide
Lesson Planning Guide
Solutions Manual

Glencoe Instructional Technology

Instructional

Advanced MathPASS CD-ROM

Multimedia Applications CD-ROM

Assessment

TestCheck and Worksheet Builder

Visit **www.amc.glencoe.com** *for data updates, career information, graphing calculator programs, and other interactive activities.*

Chapter Overview

This chapter reviews angles and angle measurement in degrees and builds on that knowledge to explore trigonometric functions. To begin their study of trigonometry, students examine trigonometric ratios defined using right triangles and investigate the trigonometric functions for an angle in standard position on the unit circle. Students then solve right triangles. Students also use the Law of Sines and the Law of Cosines to solve other triangles and to find the areas of triangles.

Using a Graphing Calculator

There are numerous opportunities, all optional, to use a graphing calculator in this chapter.
- Converting decimal degree values to degrees, minutes, and seconds (p. 278)
- Finding the values of the trigonometric ratios and inverse trigonometric relations (pp. 284, 286, 289, 307)
- Solving triangles using the Law of Cosines and the Law of Sines (pp. 323, 328)
- Determining the area of a triangle using a program (p. 331)
- Using a program to solve triangles (pp. 333–334)

The **Graphing Calculator Appendix**, which begins on page A2, contains keystroke instructions on many of these topics.

THE TRIGONOMETRIC FUNCTIONS

CHAPTER OBJECTIVES

- **Convert decimal degree measures to degrees, minutes, and seconds and vice versa.** *(Lesson 5-1)*
- **Identify angles that are coterminal with a given angle.** *(Lesson 5-1)*
- **Solve triangles.** *(Lessons 5-2, 5-4, 5-5, 5-6, 5-7, 5-8)*
- **Find the values of trigonometric functions.** *(Lessons 5-2, 5-3)*
- **Find the areas of triangles.** *(Lessons 5-6, 5-8)*

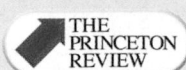 **SAT & ACT Preparation**

This textbook contains a complete SAT/ACT preparation course described on pages T8–T9 of the *Teacher's Handbook*. This chapter contains:

- an SAT or ACT practice question at the end of each lesson
- examples of SAT and ACT problems involving the Pythagorean Theorem on page 340
- ten SAT and ACT practice problems on page 341

- additional SAT and ACT practice in the *Assessment and Evaluation Masters,* pp. 123–124
- SAT and ACT test practice banks in *TestCheck and Worksheet Builder*

Angles and Degree Measure

OBJECTIVES
- Convert decimal degree measures to degrees, minutes, and seconds and vice versa.
- Find the number of degrees in a given number of rotations.
- Identify angles that are coterminal with a given angle.

NAVIGATION The sextant is an optical instrument invented around 1730. It is used to measure the angular elevation of stars, so that a navigator can determine the ship's current latitude. Suppose a navigator determines a ship in the Pacific Ocean to be located at north latitude 15.735°. How can this be written as degrees, minutes, and seconds? *This problem will be solved in Example 1.*

An angle may be generated by rotating one of two rays that share a fixed endpoint known as the **vertex.** One of the rays is fixed to form the **initial side** of the angle, and the second ray rotates to form the **terminal side.**

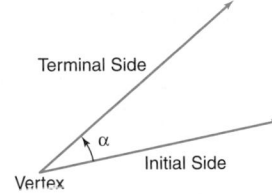

The measure of an angle provides us with information concerning the direction of the rotation and the amount of the rotation necessary to move from the initial side of the angle to the terminal side.
- If the rotation is in a counterclockwise direction, the angle formed is a *positive angle.*
- If the rotation is clockwise, it is a *negative angle.*

An angle with its vertex at the origin and its initial side along the positive *x*-axis is said to be in **standard position.** In the figures below, all of the angles are in standard position.

  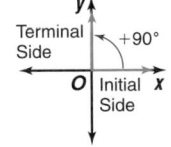

The most common unit used to measure angles is the **degree.** The concept of degree measurement is rooted in the ancient Babylonian culture. The Babylonians based their numeration system on 60 rather than 10 as we do today. In an equilateral triangle, they assigned the measure of each angle to be 60. Therefore, one sixtieth $\left(\frac{1}{60}\right)$ of the measure of the angle of an equilateral triangle was equivalent to one unit or degree (1°). The degree is subdivided into 60 equal parts known as **minutes** (1′), and the minute is subdivided into 60 equal parts known as **seconds** (1″).

Lesson 5-1 Angles and Degree Measure **277**

---

1 FOCUS

5-Minute Check
Chapter 4

1. Solve $\sqrt{x + 4} = \sqrt{3 + 2x}$. **1**

2. Solve $x^2 + 9x - 28 = 0$ by using the Quadratic Formula. $\frac{-9 \pm \sqrt{193}}{2}$

3. Find the remainder when $x^3 - x^2 + 4x - 2$ is divided by $x - 1$ by using the Remainder Theorem. State whether the binomial is a factor of the polynomial. **2; no**

4. State the number of complex roots of $x^3 - 2x^2 - 4x + 8 = 0$. Then, find the roots and graph the related function. **3; 2, –2**

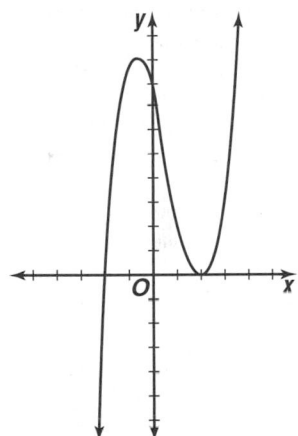

5. Determine between which consecutive integers the real zeros of the function $f(x) = 2x^3 - 4x^2 + x + 2$ are located. **–1 and 0**

Motivating the Lesson
Real-World Application An air-traffic controller tells the pilot of an airplane flying due east to turn the plane so that it has a heading of 120° east of north. What angle does the airplane turn? **30° south of due east**

Teaching Tip After discussing latitude and longitude, ask students to determine the latitude and longitude of their hometown.

In-Class Examples

Example 1
Navigation At a certain time during a flight, an airplane is on an east-west line between two signal towers.

a. If the bearing from the plane to tower A is 329.125°, change it to degrees, minutes, and seconds. **329° 7' 30"**

b. If the bearing from the plane to tower B is 35° 12' 7", write the bearing as a decimal rounded to the nearest thousandth. **35.202°**

Example 2
Give the angle measure represented by each rotation.

a. 9.5 rotations clockwise **−3420°**

b. 6.75 rotations counterclockwise **2430°**

Angles are used in a variety of real-world situations. For example, in order to locate every point on Earth, cartographers use a grid that contains circles through the poles, called *longitude lines,* and circles parallel to the equator, called *latitude lines.* Point *P* is located by traveling north from the equator through a central angle of *a*° to a circle of latitude and then west along that circle through an angle of *b*°. Latitude and longitude can be expressed in degrees as a decimal value or in degrees, minutes, and seconds.

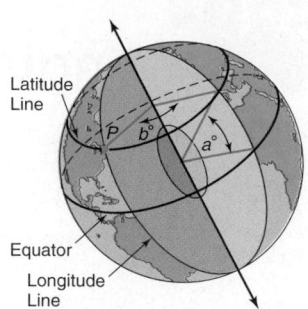

Example ① **NAVIGATION** Refer to the application at the beginning of the lesson.

a. **Change north latitude 15.735° to degrees, minutes, and seconds.**

$$15.735° = 15° + (0.735 \cdot 60)'$$ *Multiply the decimal portion of the degree*
$$= 15° + 44.1'$$ *measure by 60 to find the number of minutes.*
$$= 15° + 44' + (0.1 \cdot 60)''$$ *Multiply the decimal portion of the minute*
$$= 15° + 44' + 6''$$ *measure by 60 to find the number of seconds.*

15.735° can be written as 15° 44' 6".

Graphing Calculator Tip

▶DMS on the [ANGLE] menu allows you to convert decimal degree values to degrees, minutes, and seconds.

b. **Write north latitude 39° 5' 34" as a decimal rounded to the nearest thousandth.**

$$39° \; 5' \; 34'' = 39° + 5'\left(\frac{1°}{60'}\right) + 34''\left(\frac{1°}{3600''}\right) \text{ or about } 39.093°$$

39° 5' 34" can be written as 39.093°.

If the terminal side of an angle that is in standard position coincides with one of the axes, the angle is called a **quadrantal angle.** In the figures below, all of the angles are quadrantal.

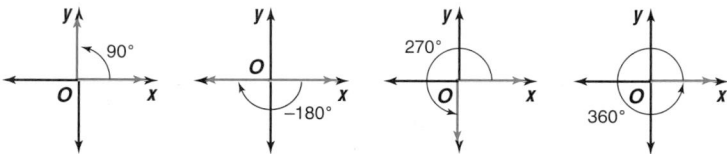

A full rotation around a circle is 360°. Measures of more than 360° represent multiple rotations.

Example ② **Give the angle measure represented by each rotation.**

a. **5.5 rotations clockwise**

5.5 × −360 = −1980   *Clockwise rotations have negative measures.*
The angle measure of 5.5 clockwise rotations is −1980°.

b. **3.3 rotations counterclockwise**

3.3 × 360 = 1188   *Counterclockwise rotations have positive measures.*
The angle measure of 3.3 counterclockwise rotations is 1188°.

278 *Chapter 5 The Trigonometric Functions*

📁 **Prerequisite Skills**

A review of the prerequisite skills for this lesson is provided in the *Prerequisite Skills Masters.*

Skill Number and Content	Pages
2 Fractions	5–6

Two angles in standard position are called **coterminal angles** if they have the same terminal side. Since angles differing in degree measure by multiples of 360° are equivalent, every angle has infinitely many coterminal angles.

Coterminal Angles	If α is the degree measure of an angle, then all angles measuring $\alpha + 360k°$, where k is an integer, are coterminal with α.

The symbol α is the lowercase Greek letter alpha.

Any angle coterminal with an angle of 75° can be written as $75° + 360k°$, where k is the number of rotations around the circle. The value of k is a positive integer if the rotations are counterclockwise and a negative integer if the rotations are clockwise.

Examples   ③ **Identify all angles that are coterminal with each angle. Then find one positive angle and one negative angle that are coterminal with the angle.**

  a. 45°

   All angles having a measure of $45° + 360k°$, where k is an integer, are coterminal with 45°. A positive angle is $45° + 360°(1)$ or 405°. A negative angle is $45° + 360°(-2)$ or −675°.

  b. 225°

   All angles having a measure of $225° + 360k°$, where k is an integer, are coterminal with 225°. A positive angle is $225° + 360(2)°$ or 945°. A negative angle is $225° + 360(-1)°$ or −135°.

  ④ **If each angle is in standard position, determine a coterminal angle that is between 0° and 360°. State the quadrant in which the terminal side lies.**

  a. 775°

   In $\alpha + 360k°$, you need to find the value of α. First, determine the number of complete rotations (k) by dividing 775 by 360.

   $$\frac{775}{360} \approx 2.152777778$$

   Then, determine the number of remaining degrees (α).

   Method 1
   $\alpha \approx 0.152777778$ rotations $\cdot 360°$
   $\quad \approx 55°$

   Method 2
   $\alpha + 360(2)° = 775°$
   $\alpha + 720° = 775°$
   $\quad\quad\quad \alpha = 55°$

   The coterminal angle (α) is 55°. Its terminal side lies in the first quadrant.

  b. −1297°

   Use a calculator.

   The angle is −217°, but the coterminal angle needs to be positive.
   $360° - 217° = 143°$

   ```
   -1297/360
              -3.602777778
   Ans+3
              -.6027777778
   Ans*360
                      -217
   ▮
   ```

   The coterminal angle (α) is 143°. Its terminal side lies in the second quadrant.

Lesson 5-1   Angles and Degree Measure **279**

3 PRACTICE/APPLY

Error Analysis

Watch for students who confuse the concepts of reference angles and coterminal angles.

Prevent by emphasizing that coterminal angles are in standard position with the same terminal side, whereas the reference angle is always an acute angle formed with the *x*-axis.

Additional Answers

1. **If an angle has a positive measure, the rotation is in a counterclockwise direction. If an angle has a negative measure, the rotation is in a clockwise direction.**

4.

Study Guide and Practice Masters, pp. 79–80

5-1 NAME _____ DATE _____ PERIOD _____

Study Guide

Angles and Degree Measure

Decimal degree measures can be expressed in **degrees**(°), **minutes**('), and **seconds**(").

Example 1 a. Change 12.520° to degrees, minutes, and seconds.
12.520° = 12° + (0.520 · 60)′   *Multiply the decimal portion of*
= 12° + 31.2′   *the degrees by 60 to find minutes.*
= 12° + 31′ + (0.2 · 60)″   *Multiply the decimal portion of*
= 12° + 31′ + 12″   *the minutes by 60 to find seconds.*
12.520° can be written as 12° 31′ 12″.

b. Write 24° 15′ 33″ as a decimal rounded to the nearest thousandth.
24° 15′ 33″ = 24° + 15′($\frac{1°}{60′}$) + 33″($\frac{1°}{3600″}$)
= 24.259°
24° 15′ 33″ can be written as 24.259°.

An angle may be generated by the rotation of one ray multiple times about the origin.

Example 2 Give the angle measure represented by each rotation.
a. 2.3 rotations clockwise
2.3 × −360 = −828   *Clockwise rotations have negative measures.*
The angle measure of 2.3 clockwise rotations is −828°.

b. 4.2 rotations counterclockwise
4.2 × 360 = 1512   *Counterclockwise rotations have positive measures.*
The angle measure of 4.2 counterclockwise rotations is 1512°.

If α is a nonquadrantal angle in standard position, its reference **angle** is defined as the acute angle formed by the terminal side of the given angle and the *x*-axis.

Reference Angle Rule | For any angle α, 0° < α < 360°, its reference angle α′ is defined by
a. α, when the terminal side is in Quadrant I,
b. 180° − α, when the terminal side is in Quadrant II,
c. α − 180°, when the terminal side is in Quadrant III, and
d. 360° − α, when the terminal side is in Quadrant IV.

Example 3 Find the measure of the reference angle for 220°.
Because 220° is between 180° and 270°, the terminal side of the angle is in Quadrant III.
220° − 180° = 40°
The reference angle is 40°.

© Glencoe/McGraw-Hill   79   Advanced Mathematical Concepts

280 *Chapter 5*

If α is a nonquadrantal angle in standard position, its **reference angle** is defined as the acute angle formed by the terminal side of the given angle and the *x*-axis. You can use the figures and the rule below to find the reference angle for any angle α where $0° < \alpha < 360°$. If the measure of α is greater than 360° or less than 0°, it can be associated with a coterminal angle of positive measure between 0° and 360°.

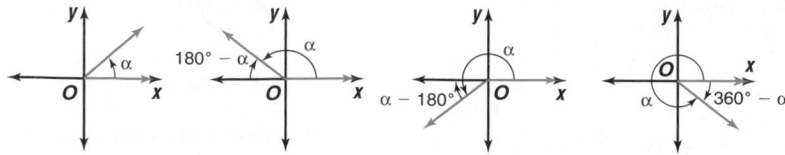

Reference Angle Rule

For any angle α, $0° < \alpha < 360°$, its reference angle α' is defined by
a. α, when the terminal side is in Quadrant I,
b. $180° - \alpha$, when the terminal side is in Quadrant II,
c. $\alpha - 180°$, when the terminal side is in Quadrant III, and
d. $360° - \alpha$, when the terminal side is in Quadrant IV.

Example ⑤ **Find the measure of the reference angle for each angle.**

a. 120°
Since 120° is between 90° and 180°, the terminal side of the angle is in the second quadrant.
$180° - 120° = 60°$
The reference angle is 60°.

b. −135°
A coterminal angle of −135 is 360 − 135 or 225. Since 225 is between 180° and 270°, the terminal side of the angle is in the third quadrant.
$225° - 180° = 45°$
The reference angle is 45°.

CHECK FOR UNDERSTANDING

Communicating Mathematics

Read and study the lesson to answer each question.

1. **Describe** the difference between an angle with a positive measure and an angle with a negative measure. **See margin.**

2. Add 29, $\frac{45}{60}$, and $\frac{26}{3600}$.

3. 270° + 360*k*° where *k* is an integer

2. **Explain** how to write 29° 45′ 26″ as a decimal degree measure.

3. **Write** an expression for the measures of all angles that are coterminal with the angle shown.

4. **Sketch** an angle represented by 3.5 counterclockwise rotations. Give the angle measure represented by this rotation. **See margin; 1260°.**

Guided Practice

Change each measure to degrees, minutes, and seconds.

5. 34.95° **34°57′**

6. −72.775° **−72° 46′ 30″**

Write each measure as a decimal to the nearest thousandth.

7. −128° 30′ 45″ **−128.513°**

8. 29° 6′ 6″ **29.102°**

280 Chapter 5 The Trigonometric Functions

From the Classroom of...

Stephanie Czarnecki
Laville Jr.–Sr. High School
Lakeville, Indiana

"After learning about navigation and surveying techniques, students get into groups and make up outdoor scavenger hunts for each other using navigation and surveying compass bearings. Here's an example: (From a starting point of the group's choice) 'Go 15 feet at a bearing of S 30° E. Next, at 40° take 20 baby steps ....' At the end, each group finds a surprise!"

Give the angle measure represented by each rotation. **10. 1620°**

9. 2 rotations clockwise **−720°**   10. 4.5 rotations counterclockwise

Identify all angles that are coterminal with each angle. Then find one positive angle and one negative angle that are coterminal with each angle.

11–12. Sample answers given.

11. 22° **22° + 360k°; 382°; −338°**   12. −170° **−170° + 360k°; 190°; −530°**

If each angle is in standard position, determine a coterminal angle that is between 0° and 360°. State the quadrant in which the terminal side lies.

13. 453° **93°; II**   14. −798° **282°; IV**

Find the measure of the reference angle for each angle.

15. 227° **47°**   16. −210° **30°**

17. **Geography** Earth rotates once on its axis approximately every 24 hours. About how many degrees does a point on the equator travel through in one hour? in one minute? in one second? **15°; 0.25° or 15′; about 0.0042° or 15″**

EXERCISES

Practice

 A

Change each measure to degrees, minutes, and seconds.

18. −16.75° **−16° 45′**   19. 168.35° **168° 21′**   20. −183.47° **−183° 28′ 12″**

21. 286.88° **286° 52′ 48″**   22. 27.465° **27° 27′ 54″**   23. 246.876° **246° 52′ 33.6″**

Write each measure as a decimal to the nearest thousandth.

24. 23° 14′ 30″ **23.242°**   25. −14° 5′ 20″ **−14.089°** 26. 233° 25′ 15″ **233.421°**

27. 173° 24′ 35″ **173.410°**   28. −405° 16′ 18″ **−405.272°**   29. 1002° 30′ 30″ **1002.508°**

Give the angle measure represented by each rotation.

 B

30. 3 rotations clockwise **−1080°**   31. 2 rotations counterclockwise **720°**

37–42. Sample answers given.

32. 1.5 rotations counterclockwise **540°** 33. 7.5 rotations clockwise **−2700°**

37. 30° + 360k°; 390°; −330°

34. 2.25 rotations counterclockwise **810°** 35. 5.75 rotations clockwise **−2070°**

38. −45° + 360k°; 315°; −405°

36. How many degrees are represented by 4 counterclockwise revolutions? **1440°**

39. 113° + 360k°; 473°; −247°

Identify all angles that are coterminal with each angle. Then find one positive angle and one negative angle that are coterminal with each angle.

37. 30°   38. −45°   39. 113°   40. 217°   41. −199°   42. −305°

40. 217° + 360k°; 577°; −143°

43. Determine the angle between 0° and 360° that is coterminal with all angles represented by 310° + 360k°, where k is any integer. **310°**

41. −199° + 360k°; 161°; −559°

44. Find the angle that is two counterclockwise rotations from 60°. Then find the angle that is three clockwise rotations from 60°. **780°; −1020°**

42. −305° + 360k°; 55°; −665°

If each angle is in standard position, determine a coterminal angle that is between 0° and 360°. State the quadrant in which the terminal side lies.

45. 400°   46. −280°   47. 940°   48. 1059°   49. −624°   50. −989°

40°; I   **80°; I**   **220°; III**   **339°; IV**   **96°; II**   **91°; II**

51. In what quadrant is the terminal side of a 1275° angle located? **III**

Find the measure of the reference angle for each angle.

52. 327°   53. 148°   54. 563°   55. −420°   56. −197°   57. 1045°
33°   **32°**   **23°**   **60°**   **17°**   **35°**
58. Name four angles between 0° and 360° with a reference angle of 20°.
20°, 160°, 200°, 340°   **59. 4500°; 270,000°**

Applications and Problem Solving

60. 90k, where k is an integer

63. 22,320°; 1,339,200°; 80,352,000°; 1,928,448,000°

64. 25° + 120k°, where k is an integer

65a. 44° 26′ 59.64″; 68° 15′ 41.76″

59. **Technology**   A computer's hard disk is spinning at 12.5 revolutions per second. Through how many degrees does it travel in a second? in a minute?

60. **Critical Thinking**   Write an expression that represents all quadrantal angles.

61. **Biking**   During the winter, a competitive bike rider trains on a stationary bike. Her trainer wants her to warm up for 5 to 10 minutes by pedaling slowly. Then she is to increase the pace to 95 revolutions per minute for 30 seconds. Through how many degrees will a point on the outside of the tire travel during the 30 seconds of the faster pace?   **17,100°**

62. **Flywheels**   A high-performance composite flywheel rotor can spin anywhere between 30,000 and 100,000 revolutions per minute. What is the range of degrees through which the composite flywheel can travel in a minute? Write your answer in scientific notation. $\mathbf{1.08 \times 10^7}$ **to** $\mathbf{3.6 \times 10^7}$ **degrees**

63. **Astronomy**   On January 28, 1998, an x-ray satellite spotted a neutron star that spins at a rate of 62 times per second. Through how many degrees does this neutron star rotate in a second? in a minute? in an hour? in a day?

64. **Critical Thinking**   Write an expression that represents any angle that is coterminal with a 25° angle, a 145° angle, and a 265° angle.

65. **Aviation**   The locations of two airports are indicated on the map.

   a. Write the latitude and longitude of the Hancock County–Bar Harbor airport in Bar Harbor, Maine, as degrees, minutes, and seconds.

   b. Write the latitude and longitude of the Key West International Airport in Key West, Florida, as a decimal to the nearest thousandth.   **24.559°; 81.760°**

Hancock County-Bar Harbor Airport
north latitude 44.4499°
west longitude 68.2616°

Key West International Airport
north latitude 24°33′32″
west longitude 81°45′34.4″

66. **Entertainment**   A tower restaurant in Sydney, Australia, is 300 meters above sea level and provides a 360° panoramic view of the city as it rotates every 70 minutes. A tower restaurant in San Antonio, Texas, is 750 feet tall. It revolves at a rate of one revolution per hour.

   a. In a day, how many more revolutions does the restaurant in San Antonio make than the one in Sydney?   **about 3.4 revolutions**

   b. In a week, how many more degrees does a speck of dirt on the window of the restaurant in San Antonio revolve than a speck of dirt on the window of the restaurant in Sydney?   **8640°**

67. Manufacturing The percent of the motor vehicles produced in the United States since 1950 is depicted in the table at the right. *(Lesson 4-8)*

Motor Vehicle Production in the United States

Year	Percent
1950	75.7
1960	47.9
1970	28.2
1980	20.8
1990	20.1
1992	20.2
1993	23.3
1994	24.8
1997	22.7

Source: American Automobile Manufacturers Association

a. Write an equation to model the percent of the motor vehicles produced in the United States as a function of the number of years since 1950.

b. According to the equation, what percent of motor vehicles will be produced in the United States in the year 2010? **Sample answer: about 32%**

67a. Sample answer: $f(x) = -0.0003x^3 + 0.0647x^2 - 3.5319x - 76.0203$

68. Solve $\sqrt[3]{6n + 5} - 15 = -10$. *(Lesson 4-7)*   **20**

69. Solve $\dfrac{x + 3}{x+2} = 2 - \dfrac{3}{x^2 + 5x + 6}$. *(Lesson 4-6)*   **0, −4**

70. Use the Remainder Theorem to find the remainder if $x^3 + 8x + 1$ is divided by $x - 2$. *(Lesson 4-3)*   **25**

71. Write a polynomial equation of least degree with roots -5, -6, and 10. *(Lesson 4-1)* $x^3 + x^2 - 80x - 300 = 0$

72. If r varies inversely as t and $r = 18$ when $t = -3$, find r when $t = -11$. *(Lesson 3-8)*   **about 4.91**

73. Determine whether the graph of $y = \dfrac{x^2-1}{x + 1}$ has infinite discontinuity, jump discontinuity, or point discontinuity, or is continuous. Then graph the function. *(Lesson 3-7)*   **Point discontinuity; see margin for graph.**

74. See margin for graph; decreasing for $x < -1$, increasing for $x > -1$.

74. Graph $f(x) = \left| (x + 1)^2 + 2 \right|$. Determine the interval(s) for which the function is increasing and the interval(s) for which the function is decreasing. *(Lesson 3-5)*

75. expanded vertically by a factor of 3, translated down 2 units

75. Use the graph of the parent function $f(x) = \dfrac{1}{x}$ to describe the graph of the related function $g(x) = \dfrac{3}{x} - 2$. *(Lesson 3-2)*

76. Solve the system of inequalities $y \le 5$, $3y \ge 2x + 9$, and $-3y \le 6x - 9$ by graphing. Name the coordinates of the vertices of the convex set. *(Lesson 2-6)*   **See margin for graph; (−1, 5), (3, 5), (0, 3)**

77. Find $[f \cdot g](x)$ if $f(x) = x - 0.2x$ and $g(x) = x - 0.3x$. *(Lesson 1-2)*   **0.56x**

78. SAT/ACT Practice $\overline{AB}$ is a diameter of circle O, and $m\angle BOD - 15°$. If $m\angle EOA = 85°$, find $m\angle ECA$.   **D**

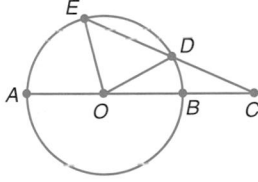

A 85°   **B** 50°   **C** 70°   **D** 35°   **E** 45°

Extra Practice See p. A34.

Extra Credit

Suppose the point (3, 3) is on the terminal side of an angle in standard position. How far is the point from the origin?   **$3\sqrt{2}$ units**

---

4 ASSESS

Open-Ended Assessment

Writing Have students write a paragraph discussing how to find the reference angle of any angle.

Additional Answers

73.

74.

76.
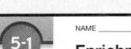

Enrichment Masters, p. 49

NAME _____ DATE _____ PERIOD _____

5-1   **Enrichment**

Reading Mathematics: If and Only If Statements

If p and q are interchanged in a conditional statement so that p becomes the conclusion and q becomes the hypothesis, the new statement, $q \rightarrow p$, is called the **converse** of $p \rightarrow q$.

If $p \rightarrow q$ is true, $q \rightarrow p$ may be either true or false.

Example   Find the converse of each conditional.
  a. $p \rightarrow q$: All squares are rectangles. (true)
     $q \rightarrow p$: All rectangles are squares. (false)
  b. $p \rightarrow q$: If a function $f(x)$ is increasing on an interval I, then for every a and b contained in I, $f(a) < f(b)$ whenever $a < b$. (true)
     $q \rightarrow p$: If for every a and b contained in an interval I, $f(a) < f(b)$ whenever $a < b$ then function $f(x)$ is increasing on I. (true)

In Lesson 3-5, you saw that the two statements in Example 2 can be combined in a single statement using the words "if and only if."

A function $f(x)$ is increasing on an interval I *if and only if* for every a and b contained in I, $f(a) < f(b)$ whenever $a < b$.

The statement "p *if and only if* q" means that p implies q and q implies p.

State the converse of each conditional. Then tell if the converse is true or false. If it is true, combine the statement and its converse into a single statement using the words "if and only if."

1. All integers are rational numbers.
   All rational numbers are integers; false

2. If for all x in the domain of a function $f(x)$, $f(-x) = -f(x)$, then the graph of $f(x)$ is symmetric with respect to the origin.
   If a function has a graph that is symmetric with respect to the origin, then $f(-x) = -f(x)$ for all x in the domain of $f(x)$; true. A function has a graph that is symmetric with respect to the origin if and only if $f(-x) = -f(x)$ for all x in the domain of $f(x)$.

3. If $f(x)$ and $f^{-1}(x)$ are inverse functions, then $[f \circ f^{-1}](x) = [f^{-1} \circ f](x) = x$.
   If $[f \circ f^{-1}](x) = [f^{-1} \circ f](x) = x$, then $f(x)$ and $f^{-1}(x)$ are inverse functions; true. Two functions, $f(x)$ and $f^{-1}(x)$, are inverse functions if and only if $[f \circ f^{-1}](x) = [f^{-1} \circ f](x) = x$.

© Glencoe/McGraw-Hill   T49   Advanced Mathematical Concepts

Lesson 5-1 **283**

5-2

Trigonometric Ratios in Right Triangles

5-Minute Check
Lesson 5-1

1. Change 34.820° to degrees, minutes, and seconds. **34° 49'12"**

2. Write 75° 54' 12" as a decimal to the nearest thousandth. **75.903°**

Give the angle measure represented by each rotation.

3. 3.75 rotations clockwise **−1350°**

4. 9 rotations counterclockwise **3240°**

5. Identify all angles that are coterminal with an angle of 172°. Then find one positive angle and one negative angle that are coterminal to that angle. **172° + 360k°; sample answers: 532°, −188°**

6. Find the measure of the reference angle for an angle of 484°. **56°**

Motivating the Lesson

Hands-On Activity Bring a glass container with a rock or other object placed on the bottom to class. Fill it with water and have a student volunteer point to the object by putting his or her finger into the water. Use a stick to demonstrate that the student may not be pointing directly at the object. Have several other students also try the experiment. Point out that finding the displacement of the light is a real-world application of trigonometry as shown in the lesson opener.

OBJECTIVE
- Find the values of trigonometric ratios for acute angles of right triangles.

PHYSICS As light passes from one substance such as air to another substance such as glass, the light is bent. The relationship between the angle of incidence θ_i and the angle of refraction θ_r is given by Snell's Law, $\frac{\sin \theta_i}{\sin \theta_r} = n$, where $\sin \theta$ represents a trigonometric ratio and n is a constant called the *index of refraction*. Suppose a ray of light passes from air with an angle of incidence of 50° to glass with an angle of refraction of 32° 16′. Find the index of refraction of the glass. *This problem will be solved in Example 2.*

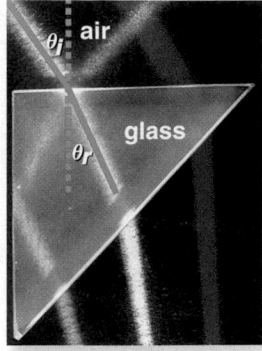

In a right triangle, one of the angles measures 90°, and the remaining two angles are *acute* and *complementary*. The longest side of a right triangle is known as the **hypotenuse** and is opposite the right angle. The other two sides are called **legs**. The leg that is a side of an acute angle is called the **side adjacent** to the angle. The other leg is the **side opposite** the angle.

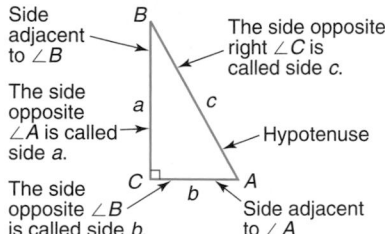

Side adjacent to ∠B → The side opposite right ∠C is called side c.

The side opposite ∠A is called side a. → Hypotenuse

The side opposite ∠B is called side b. → Side adjacent to ∠A

GRAPHING CALCULATOR EXPLORATION

Use a graphing calculator to find each ratio for the 22.6° angle in each triangle. Record each ratio as a decimal. Make sure your calculator is in degree mode.

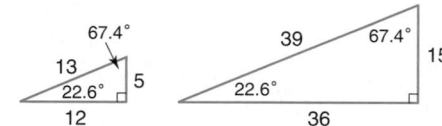

$$R_1 = \frac{\text{side opposite}}{\text{hypotenuse}} \qquad R_2 = \frac{\text{side adjacent}}{\text{hypotenuse}}$$

$$R_3 = \frac{\text{side opposite}}{\text{side adjacent}}$$

Find the same ratios for the 67.4° angle in each triangle.

TRY THESE   1–4, 6. See margin.

1. Draw two other triangles that are similar to the given triangles.

2. Find each ratio for the 22.6° angle in each triangle.

3. Find each ratio for the 67.4° angle in each triangle.

WHAT DO YOU THINK?

4. Make a conjecture about R_1, R_2, and R_3 for any right triangle with a 22.6° angle.

5. Is your conjecture true for any 67.4° angle in a right triangle? **yes**

6. Do you think your conjecture is true for any acute angle of a right triangle? Why?

Resource Manager

 Reproducible Masters
- Study Guide and Practice, pp. 81–82
- Enrichment, p. 50
- Assessment and Evaluation, p. 121
- Graphing Calculator Keystrokes, p. 17
- Spreadsheet Applications, p. 5

 Transparencies
- 5-Minute Check, 5–2
- Teaching, 5–2

If two angles of a triangle are congruent to two angles of another triangle, the triangles are similar. If an acute angle of one right triangle is congruent to an acute angle of another right triangle, the triangles are similar, and the ratios of the corresponding sides are equal. Therefore, any two congruent angles of different right triangles will have equal ratios associated with them.

In right triangles, the Greek letter θ (theta) is often used to denote a particular angle.

The ratios of the sides of the right triangles can be used to define the **trigonometric ratios.** The ratio of the side opposite θ and the hypotenuse is known as the **sine.** The ratio of the side adjacent θ and the hypotenuse is known as the **cosine.** The ratio of the side opposite θ and the side adjacent θ is known as the **tangent.**

	Words	Symbol	Definition	
Trigonometric Ratios	sine θ	sin θ	$\sin \theta = \dfrac{\text{side opposite}}{\text{hypotenuse}}$	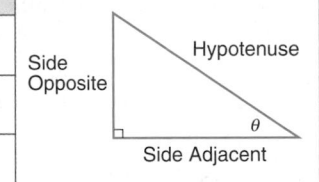
	cosine θ	cos θ	$\cos \theta = \dfrac{\text{side adjacent}}{\text{hypotenuse}}$	
	tangent θ	tan θ	$\tan \theta = \dfrac{\text{side opposite}}{\text{side adjacent}}$	

SOH-CAH-TOA is a mnemonic device commonly used for remembering these ratios.

$$\sin \theta = \frac{\text{opposite}}{\text{hypotenuse}} \qquad \cos \theta = \frac{\text{adjacent}}{\text{hypotenuse}} \qquad \tan \theta = \frac{\text{opposite}}{\text{adjacent}}$$

Example ❶ **Find the values of the sine, cosine, and tangent for ∠B.**

First, find the length of $\overline{BC}$

$(AC)^2 + (BC)^2 = (AB)^2$ *Pythagorean Theorem*
$18^2 + (BC)^2 = 33^2$ *Substitute 18 for AC and 33 for AB.*
$(BC)^2 = 765$
$BC = \sqrt{765}$ or $3\sqrt{85}$ *Take the square root of each side.*
   Disregard the negative root.

Then write each trigonometric ratio.

$\sin B = \dfrac{\text{side opposite}}{\text{hypotenuse}} \qquad \cos B = \dfrac{\text{side adjacent}}{\text{hypotenuse}} \qquad \tan B = \dfrac{\text{side opposite}}{\text{side adjacent}}$

$\sin B = \dfrac{18}{33}$ or $\dfrac{6}{11} \qquad \cos B = \dfrac{3\sqrt{85}}{33}$ or $\dfrac{\sqrt{85}}{11} \qquad \tan B = \dfrac{18}{3\sqrt{85}}$ or $\dfrac{6\sqrt{85}}{85}$

Trigonometric ratios are often simplified, but never written as mixed numbers.

In Example 1, you found the exact values of the sine, cosine, and tangent ratios. You can use a calculator to find the approximate decimal value of any of the trigonometric ratios for a given angle.

Graphing Calculator Exploration
You may want to have students use grid paper to help create similar triangles quickly. Extend this exploration by comparing the values of the ratios for the complementary angles. They can make conjectures about how these values are related and why. Emphasize that the graphing calculator *is* a scientific calculator and that students do not need to use a separate calculator for non-graphing tasks.

Teaching Tip In Example 1, remind students that they are examining right triangles. Therefore, the Pythagorean Theorem can be used to calculate the length of the third side of a triangle.

In-Class Example
Example 1
Find the values of the sine, cosine, and tangent for ∠A.

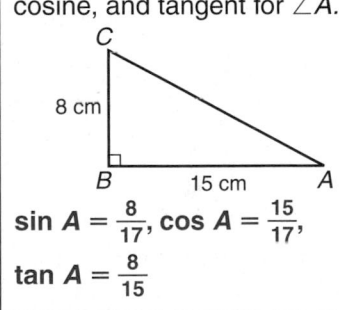

$\sin A = \dfrac{8}{17}$, $\cos A = \dfrac{15}{17}$,

$\tan A = \dfrac{8}{15}$

Teaching Tip Students may need to review how to rationalize denominators.

Additional Answers
Graphing Calculator Exploration
1. **Sample answers:**

2. $R_1 \approx 0.3846$, $R_2 \approx 0.9231$, $R_3 \approx 0.4167$
3. $R_1 \approx 0.9231$, $R_2 \approx 0.3846$, $R_3 \approx 2.4$
4. **Each ratio has the same value for all 22.6° angles.**
6. **Yes; the triangles are similar.**

Example **2** **PHYSICS** Refer to the application at the beginning of the lesson. Find the index of refraction of the glass.

$$\dfrac{\sin \theta_i}{\sin \theta_r} = n \quad \text{Snell's Law}$$

$$\dfrac{\sin 50°}{\sin 32°\ 16'} = n \quad \text{Substitute } 50° \text{ for } \theta_i \text{ and } 32°\ 16' \text{ for } \theta_r$$

$$\dfrac{0.7660444431}{0.5338605056} \approx n \quad \text{Use a calculator to find each sine ratio.}$$

$$1.434914992 \approx n \quad \text{Use a calculator to find the quotient.}$$

The index of refraction of the glass is about 1.4349.

Graphing Calculator Tip

If using your graphing calculator to do the calculation, make sure you are in degree mode.

In addition to the trigonometric ratios sine, cosine, and tangent, there are three other trigonometric ratios called **cosecant**, **secant**, and **cotangent**. These ratios are the reciprocals of sine, cosine, and tangent, respectively.

	Words	Symbol	Definition	
Reciprocal Trigonometric Ratios	cosecant θ	$\csc \theta$	$\csc \theta = \dfrac{1}{\sin \theta}$ or $\dfrac{\text{hypotenuse}}{\text{side opposite}}$	
	secant θ	$\sec \theta$	$\sec \theta = \dfrac{1}{\cos \theta}$ or $\dfrac{\text{hypotenuse}}{\text{side adjacent}}$	
	cotangent θ	$\cot \theta$	$\cot \theta = \dfrac{1}{\tan \theta}$ or $\dfrac{\text{side adjacent}}{\text{side opposite}}$	

These definitions are called the reciprocal identities.

Examples **3** a. If $\cos \theta = \dfrac{3}{4}$, find $\sec \theta$.

$$\sec \theta = \dfrac{1}{\cos \theta}$$

$$\sec \theta = \dfrac{1}{\dfrac{3}{4}} \text{ or } \dfrac{4}{3}$$

b. If $\csc \theta = 1.345$, find $\sin \theta$.

$$\sin \theta = \dfrac{1}{\csc \theta}$$

$$\sin \theta = \dfrac{1}{1.345} \text{ or about } 0.7435$$

4 Find the values of the six trigonometric ratios for $\angle P$.
First determine the length of the hypotenuse.

$$(MP)^2 + (MN)^2 = (NP)^2 \quad \text{Pythagorean Theorem}$$
$$10^2 + 7^2 = (NP)^2 \quad \text{Substitute 10 for MP and 7 for MN.}$$
$$149 = (NP)^2$$
$$\sqrt{149} = NP \quad \text{Disregard the negative root.}$$

$$\sin P = \dfrac{\text{side opposite}}{\text{hypotenuse}} \qquad \cos P = \dfrac{\text{side adjacent}}{\text{hypotenuse}} \qquad \tan P = \dfrac{\text{side opposite}}{\text{side adjacent}}$$

$$\sin P = \dfrac{7}{\sqrt{149}} \text{ or } \dfrac{7\sqrt{149}}{149} \qquad \cos P = \dfrac{10}{\sqrt{149}} \text{ or } \dfrac{10\sqrt{149}}{149} \qquad \tan P = \dfrac{7}{10}$$

$$\csc P = \dfrac{\text{hypotenuse}}{\text{side opposite}} \qquad \sec P = \dfrac{\text{hypotenuse}}{\text{side adjacent}} \qquad \cot P = \dfrac{\text{side adjacent}}{\text{side opposite}}$$

$$\csc P = \dfrac{\sqrt{149}}{7} \qquad \sec P = \dfrac{\sqrt{149}}{10} \qquad \cot P = \dfrac{10}{7}$$

Consider the special relationships among the sides of $30°-60°-90°$ and $45°-45°-90°$ triangles.

These special relationships can be used to determine the trigonometric ratios for $30°$, $45°$, and $60°$. You should memorize the sine, cosine, and tangent values for these angles.

θ	$\sin \theta$	$\cos \theta$	$\tan \theta$	$\csc \theta$	$\sec \theta$	$\cot \theta$
30°	$\dfrac{1}{2}$	$\dfrac{\sqrt{3}}{2}$	$\dfrac{\sqrt{3}}{3}$	2	$\dfrac{2\sqrt{3}}{3}$	$\sqrt{3}$
45°	$\dfrac{\sqrt{2}}{2}$	$\dfrac{\sqrt{2}}{2}$	1	$\sqrt{2}$	$\sqrt{2}$	1
60°	$\dfrac{\sqrt{3}}{2}$	$\dfrac{1}{2}$	$\sqrt{3}$	$\dfrac{2\sqrt{3}}{3}$	2	$\dfrac{\sqrt{3}}{3}$

Note that $\sin 30° = \cos 60°$ and $\cos 30° = \sin 60°$. This is an example showing that the sine and cosine are **cofunctions**. That is, if θ is an acute angle, $\sin \theta = \cos (90° - \theta)$. Similar relationships hold true for the other trigonometric ratios.

Cofunctions	$\sin \theta = \cos (90° - \theta)$	$\cos \theta = \sin (90° - \theta)$
	$\tan \theta = \cot (90° - \theta)$	$\cot \theta = \tan (90° - \theta)$
	$\sec \theta = \csc (90° - \theta)$	$\csc \theta = \sec (90° - \theta)$

CHECK FOR UNDERSTANDING

Communicating Mathematics

Read and study the lesson to answer each question. **1. See margin.**

1. **Explain** in your own words how to decide which side is opposite the given acute angle of a right triangle and which side is adjacent to the given angle.

2. **State** the reciprocal ratios of sine, cosine, and tangent.

2. cosecant; secant; cotangent

3. **Write** each trigonometric ratio for $\angle A$ in triangle ABC. **See margin.**

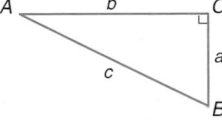

4. **Compare** $\sin A$ and $\cos B$, $\csc A$ and $\sec B$, and $\tan A$ and $\cot B$. $\sin A = \cos B$, $\csc A = \sec B$, $\tan A = \cot B$

Lesson 5-2 Trigonometric Ratios in Right Triangles **287**

Prerequisite Skills

A review of the prerequisite skills for this lesson is provided in the *Prerequisite Skills Masters*.

Skill Number and Content	Pages
8 Rationalizing a Denominator	17–18

Additional Answers

10. $\frac{3}{5}$; $\frac{4}{5}$; $\frac{3}{4}$

11. $\frac{5\sqrt{89}}{89}$; $\frac{8\sqrt{89}}{89}$; $\frac{5}{8}$

12. $\frac{3}{10}$; $\frac{\sqrt{91}}{10}$; $\frac{3\sqrt{91}}{91}$

20. $\sin R = \frac{\sqrt{527}}{24}$, $\cos R = \frac{7}{24}$,
$\tan R = \frac{\sqrt{527}}{7}$, $\csc R = \frac{24\sqrt{527}}{527}$,
$\sec R = \frac{24}{7}$, $\cot R = \frac{7\sqrt{527}}{527}$

21. $\sin R = \frac{19}{20}$, $\cos R = \frac{\sqrt{39}}{20}$,
$\tan R = \frac{19\sqrt{39}}{39}$, $\csc R = \frac{20}{19}$,
$\sec R = \frac{20\sqrt{39}}{39}$, $\cot R = \frac{\sqrt{39}}{19}$

22. $\sin R = \frac{\sqrt{154}}{44}$, $\cos R = \frac{9\sqrt{22}}{44}$,
$\tan R = \frac{\sqrt{7}}{9}$, $\csc R = \frac{2\sqrt{154}}{7}$,
$\sec R = \frac{2\sqrt{22}}{9}$, $\cot R = \frac{9\sqrt{7}}{7}$

Guided Practice

5. Find the values of the sine, cosine, and tangent for $\angle T$.
$\frac{15\sqrt{514}}{514}$; $\frac{17\sqrt{514}}{514}$; $\frac{15}{17}$

15 in.
17 in.

6. If $\sin \theta = \frac{2}{5}$, find $\csc \theta$. $\frac{5}{2}$

7. If $\cot \theta = 1.5$, find $\tan \theta$. $\frac{1}{1.5} \approx 0.6667$

8. $\sin P = \frac{3}{10}$,
$\cos P = \frac{\sqrt{91}}{10}$,
$\tan P = \frac{3\sqrt{91}}{91}$,
$\csc P = \frac{10}{3}$,
$\sec P = \frac{10\sqrt{91}}{91}$,
$\cot P = \frac{\sqrt{91}}{3}$

8. Find the values of the six trigonometric ratios for $\angle P$.

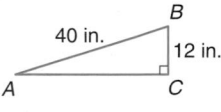
6 ft
20 ft

9. Physics You may have polarized sunglasses that eliminate glare by polarizing the light. When light is polarized, all of the waves are traveling in parallel planes. Suppose vertically polarized light with intensity I_o strikes a polarized filter with its axis at an angle of θ with the vertical. The intensity of the transmitted light I_t and θ are related by the equation $\cos \theta = \sqrt{\frac{I_t}{I_o}}$. If θ is $45°$, write I_t as a function of I_o. $I_t = 0.5 I_o$

EXERCISES

Practice

A ▶ Find the values of the sine, cosine, and tangent for each $\angle A$. **10–12. See margin.**

10.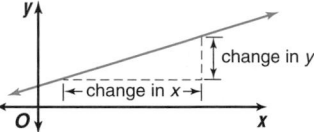
80 m
60 m
A ... B ... C

11. A
8 ft
C 5 ft B

12. B
40 in.
12 in.
A ... C

13. The slope of a line is the ratio of the change of y to the change of x. Name the trigonometric ratio of θ that equals the slope of line m.
tangent

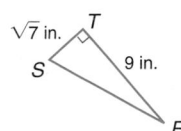
change in y
change in x

B ▶ **14.** If $\tan \theta = \frac{1}{3}$, find $\cot \theta$. **3**

15. If $\sin \theta = \frac{3}{7}$, find $\csc \theta$. $\frac{7}{3}$

16. If $\sec \theta = \frac{5}{9}$, find $\cos \theta$. $\frac{9}{5}$

17. If $\csc \theta = 2.5$, find $\sin \theta$. $\frac{1}{2.5} = 0.4$

18. $\frac{1}{0.75} \approx 1.3333$ **18.** If $\cot \theta = 0.75$, find $\tan \theta$.

19. $\frac{1}{0.125} = 8$ **19.** If $\cos \theta = 0.125$, find $\sec \theta$.

C ▶ Find the values of the six trigonometric ratios for each $\angle R$. **20–22. See margin.**

20. R
14 cm
48 cm
T ... S

21. R
40 mm
S 38 mm T

22. T
$\sqrt{7}$ in.
S
9 in.
R

23. If $\tan \theta = 1.3$, what is the value of $\cot (90° - \theta)$? **1.3**

www.amc.glencoe.com/self_check_quiz

24. Use a calculator to determine the value of each trigonometric ratio.

    a. sin 52° 47′   **b.** cos 79° 15′   **c.** tan 88° 22′ 45″

    d. cot 36° (*Hint:* Tangent and cotangent have a reciprocal relationship.)

Graphing Calculator

24a.
0.7963540136
24b. 0.186524036
24c. 35.34015106
24d. 1.37638192

25. Use the table function on a graphing calculator to complete the table. Round values to three decimal places.

θ	72°	74°	76°	78°	80°	82°	84°	86°	88°
sin	0.951	0.961	0.970	0.978	0.985	0.990	0.995	0.998	0.999
cos	0.309	0.276	0.242	0.208	0.174	0.139	0.105	0.070	0.035

    a. What value does sin θ approach as θ approaches 90°? **1**

    b. What value does cos θ approach as θ approaches 90°? **0**

26. Use the table function on a graphing calculator to complete the table. Round values to three decimal places.

θ	18°	16°	14°	12°	10°	8°	6°	4°	2°
sin	0.309	0.276	0.242	0.208	0.174	0.139	0.105	0.070	0.035
cos	0.951	0.961	0.970	0.978	0.985	0.990	0.995	0.998	0.999
tan	0.325	0.287	0.249	0.213	0.176	0.141	0.105	0.070	0.035

    a. What value does sin θ approach as θ approaches 0°? **0**

    b. What value does cos θ approach as θ approaches 0°? **1**

    c. What value does tan θ approach as θ approaches 0°? **0**

Applications and Problem Solving

27. Physics Suppose a ray of light passes from air to Lucite. The measure of the angle of incidence is 45°, and the measure of an angle of refraction is 27° 55′. Use Snell's Law, which is stated in the application at the beginning of the lesson, to find the index of refraction for Lucite. **about 1.5103**

28. Critical Thinking The sine of an acute ∠R of a right triangle is $\frac{3}{7}$. Find the values of the other trigonometric ratios for this angle.

28.
$\cos R = \frac{2\sqrt{10}}{7}$,

$\tan R = \frac{3\sqrt{10}}{20}$,

$\csc R = \frac{7}{3}$,

$\sec R = \frac{7\sqrt{10}}{20}$,

$\cot R = \frac{2\sqrt{10}}{3}$

29a. about
5.4 m/s

29b. about
5.9 m/s

29c. about
6.4 m/s

29. Track When rounding a curve, the acute angle θ that a runner's body makes with the vertical is called the angle of incline. It is described by the equation $\tan \theta = \frac{v^2}{gr}$, where v is the velocity of the runner, g is the acceleration due to gravity, and r is the radius of the track. The acceleration due to gravity is a constant 9.8 meters per second squared. Suppose the radius of the track is 15.5 meters.

    a. What is the runner's velocity if the angle of incline is 11°?

    b. Find the runner's velocity if the angle of incline is 13°.

    c. What is the runner's velocity if the angle of incline is 15°?

    d. Should a runner increase or decrease her velocity to increase his or her angle of incline? **increase**

30. Critical Thinking Use the fact that $\sin \theta = \frac{\text{side opposite}}{\text{hypotenuse}}$ and $\cos \theta = \frac{\text{side adjacent}}{\text{hypotenuse}}$ to write an expression for tan θ in terms of sin θ and cos θ. $\tan \theta = \frac{\sin \theta}{\cos \theta}$

Lesson 5-2 Trigonometric Ratios in Right Triangles **289**

Teaching Tip For Exercises 25–26, remind students to set their calculators to degree mode.

**Chapter 5, Quiz A
(Lessons 5-1 and 5-2)**
is available in the *Assessment and Evaluation Masters*, p. 121.

---

31. **Architecture** The angle of inclination of the sun affects the heating and cooling of buildings. The angle is greater in the summer than the winter. The sun's angle of inclination also varies according to the latitude. The sun's angle of inclination at noon equals $90° - L - 23.5° × \cos\left[\dfrac{(N + 10)360}{365}\right]$. In this expression, L is the latitude of the building site, and N is the number of days elapsed in the year.   **b. about 49.5°; about 2.5°**

    a. The latitude of Brownsville, Texas, is 26°. Find the angle of inclination for Brownsville on the first day of summer (day 172) and on the first day of winter (day 355).   **about 87.5°; about 40.5°**

    b. The latitude of Nome, Alaska, is 64°. Find the angle of inclination for Nome on the first day of summer and on the first day of winter.

    c. Which city has the greater change in the angle of inclination?   **neither**

32. **Biology** An object under water is not exactly where it appears to be. The displacement x depends on the angle A at which the light strikes the surface of the water from below, the depth t of the object, and the angle B at which the light leaves the surface of the water. The measure of displacement is modeled by the equation $x = t\left(\dfrac{\sin (B - A)}{\cos A}\right)$. Suppose a biologist is trying to net a fish under water. Find the measure of displacement if t measures 10 centimeters, the measure of angle A is 41°, and the measure of angle B is 60°.   **about 4.31 cm**

Mixed Review

33. Change 88.37° to degrees, minutes, and seconds. *(Lesson 5-1)*   **88° 22′ 12″**

34. Find the number of possible positive real zeros and the number of possible negative real zeros for $f(x) = x^4 + 2x^3 - 6x - 1$. *(Lesson 4-4)*   **1; 3 or 1**

35. **Business** Luisa Diaz is planning to build a new factory for her business. She hires an analyst to gather data and develop a mathematical model. In the model $P(x) = 18 + 92x - 2x^2$, P is Ms. Diaz's monthly profit, and x is the number of employees needed to staff the new facility. *(Lesson 3-6)*

    a. How many employees should she hire to maximize profits?   **23 employees**

    b. What is her maximum profit?   **$1076**

36. Find the value of $\begin{vmatrix} 7 & -3 & 5 \\ 4 & 0 & -1 \\ 8 & 2 & 0 \end{vmatrix}$. *(Lesson 2-5)*   **78**

37. Write the slope-intercept form of the equation of the line that passes through points at $(2, 5)$ and $(6, 3)$. *(Lesson 1-4)* $y = -\dfrac{1}{2} x + 6$

38. **SAT/ACT Practice** The area of a right triangle is 12 square inches. The ratio of the lengths of its legs is 2:3. Find the length of the hypotenuse.   **C**

    A $\sqrt{13}$ in.   **B** 26 in.   **C** $2\sqrt{13}$ in.   **D** 52 in.   **E** $4\sqrt{13}$ in.

Extra Practice   See p. A34.

Extra Credit

Consider the following right triangle.

a. Find θ. Explain.

$\theta = 30°$ since $\sin \theta = \dfrac{2}{4} = \dfrac{1}{2}$

and $\sin 30° = \dfrac{1}{2}$.

b. Find b. $b = 2\sqrt{3}$

Trigonometric Functions on the Unit Circle

OBJECTIVES
* Find the values of the six trigonometric functions using the unit circle.
* Find the values of the six trigonometric functions of an angle in standard position given a point on its terminal side.

FOOTBALL The longest punt in NFL history was 98 yards. The punt was made by Steve O'Neal of the New York Jets in 1969. When a football is punted, the angle made by the initial path of the ball and the ground affects both the height and the distance the ball will travel. If a football is punted from ground level, the maximum height it will reach is given by the formula $h = \frac{v_0{}^2 \sin^2 \theta}{2g}$, where v_0 is the initial velocity, θ is the measure of the angle between the ground and the initial path of the ball, and g is the acceleration due to gravity. The value of g is 9.8 meters per second squared. Suppose the initial velocity of the ball is 28 meters per second. Describe the possible maximum height of the ball if the angle is between 0° and 90°.

This problem will be solved in Example 2.

A **unit circle** is a circle of radius 1. Consider a unit circle whose center is at the origin of a rectangular coordinate system. The unit circle is symmetric with respect to the x-axis, the y-axis, and the origin.

Consider an angle θ between 0° and 90° in standard position. Let $P(x, y)$ be the point of intersection of the angle's terminal side with the unit circle. If a perpendicular segment is drawn from point P to the x-axis, a right triangle is created. In the triangle, the side adjacent to angle θ is along the x-axis and has length x. The side opposite angle θ is the perpendicular segment and has length y. According to the Pythagorean Theorem, $x^2 + y^2 = 1$. We can find values for $\sin \theta$ and $\cos \theta$ using the definitions used in Lesson 5-2.

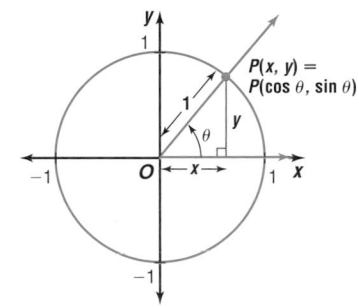

$P(x, y) = P(\cos \theta, \sin \theta)$

$$\sin \theta = \frac{\text{side opposite}}{\text{hypotenuse}} \qquad \cos \theta = \frac{\text{side adjacent}}{\text{hypotenuse}}$$

$$\sin \theta = \frac{y}{1} \text{ or } y \qquad\qquad \cos \theta = \frac{x}{1} \text{ or } x$$

Right triangles can also be formed for angles greater than 90°. In these cases, the reference angle is one of the acute angles. Similar results will occur. Thus, **sine** θ can be redefined as the y-coordinate and **cosine** θ can be redefined as the x-coordinate.

Sine and Cosine	If the terminal side of an angle θ in standard position intersects the unit circle at $P(x, y)$, then $\cos \theta = x$ and $\sin \theta = y$.

Resource Manager

Reproducible Masters
* *Study Guide and Practice,* pp. 83–84
* *Enrichment,* p. 51

 Transparencies
* 5-Minute Check, 5-3
* Teaching, 5-3

 Technology/Multimedia
* Advanced MathPASS, Lessons 10 and 11

5-Minute Check
Lesson 5-2

1. Find the values of the sine, cosine, and tangent for $\angle L$.

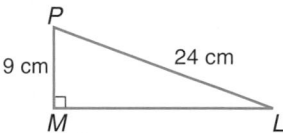

P, 9 cm, M, 24 cm, L

$\frac{3}{8}, \frac{\sqrt{55}}{8}, \frac{3\sqrt{55}}{55}$

2. If $\cos \theta = \frac{2}{9}$, find $\sec \theta$. $\frac{9}{2}$

3. If $\sin \theta = 0.525$, find $\csc \theta$. **1.905**

4. Find the values of the six trigonometric functions for $\angle R$.

R, 22 cm, S, 14 cm, T

$\sin R = \frac{7\sqrt{170}}{170}$; $\cos R = \frac{11\sqrt{170}}{170}$; $\tan R = \frac{7}{11}$;

$\csc R = \frac{\sqrt{170}}{7}$; $\sec R = \frac{\sqrt{170}}{11}$; $\cot R = \frac{11}{7}$

Motivating the Lesson

Real-World Application Ask students to describe the path of a football that has been punted. Give examples of height and horizontal distance that are important to a football team. For example, relate these ideas to the opening kickoff or the making of a field goal. At what angle should the ball be kicked to get maximum height?

Teaching Tip When working with the unit circle in Example 1, remind students that sine and cosine always range between −1 and 1, inclusively, whereas the other four trigonometric functions do not.

Teaching Tip The use of similar triangles to define sine and cosine implies that the size of the triangle (and therefore the size of the circle) does not affect the value of the sine or cosine of an angle.

Teaching Tip Note that when trigonometric functions are related to right triangles, θ must be between 0° and 90°. Therefore, the domain and range of the sine and cosine functions as related to right triangles are subsets of the domain and range of the circular functions.

In-Class Examples

Example 1
Use the unit circle to find each value.
a. $\sin(-90°)$ **−1**
b. $\cot 270°$ **0**

Example 2
Springs The displacement (in centimeters) of an oscillating spring from its point of equilibrium is $d = 12 \cos(180°t)$, where t = time in seconds. The spring is farthest from equilibrium at $t = 0$ and $t = 1$ seconds. What is the spring's largest displacement?
12 cm and −12 cm

Since there is exactly one point $P(x, y)$ for any angle θ, the relations $\cos \theta = x$ and $\sin \theta = y$ are functions of θ. Because they are both defined using the unit circle, they are often called **circular functions**.

The domain of the sine and cosine functions is the set of real numbers, since $\sin \theta$ and $\cos \theta$ are defined for any angle θ. The range of the sine and the cosine functions is the set of real numbers between −1 and 1 inclusive, since $(\cos \theta, \sin \theta)$ are the coordinates of points on the unit circle.

In addition to the sine and cosine functions, the four other **trigonometric functions** can also be defined using the unit circle.

$$\tan \theta = \frac{\text{side opposite}}{\text{side adjacent}} = \frac{y}{x} \qquad \csc \theta = \frac{\text{hypotenuse}}{\text{side opposite}} = \frac{1}{y}$$

$$\sec \theta = \frac{\text{hypotenuse}}{\text{side adjacent}} = \frac{1}{x} \qquad \cot \theta = \frac{\text{side adjacent}}{\text{side opposite}} = \frac{x}{y}$$

Since division by zero is undefined, there are several angle measures that are excluded from the domain of the tangent, cotangent, secant, and cosecant functions.

Examples ❶ **Use the unit circle to find each value.**

a. $\cos(-180°)$

The terminal side of a −180° angle in standard position is the negative x-axis, which intersects the unit circle at $(-1, 0)$. The x-coordinate of this ordered pair is $\cos(-180°)$. Therefore, $\cos(-180°) = -1$.

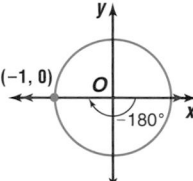

b. $\sec 90°$

The terminal side of a 90° angle in standard position is the positive y-axis, which intersects the unit circle at $(0, 1)$. According to the definition of secant, $\sec 90° = \frac{1}{x}$ or $\frac{1}{0}$, which is undefined. Therefore, $\sec 90°$ is undefined.

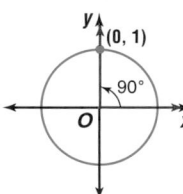

❷ **FOOTBALL** Refer to the application at the beginning of the lesson. Describe the possible maximum height of the ball if the angle is between 0° and 90°.

Real World Application

The expression $\sin^2 \theta$ means the square of the sine of θ or $(\sin \theta)^2$.

Find the value of h when $\theta = 0°$.

$$h = \frac{{v_0}^2 \sin^2 \theta}{2g}$$

$$h = \frac{28^2 \sin^2 0°}{2(9.8)} \qquad v_0 = 28, \theta = 0°, g = 9.8$$

$$h = \frac{28^2(0^2)}{2(9.8)} \qquad \sin 0° = 0$$

$$h = 0$$

Find the value of h when $\theta = 90°$.

$$h = \frac{v_0^2 \sin^2 \theta}{2g}$$

$$h = \frac{28^2 \sin^2 90°}{2(9.8)} \qquad v_0 = 28, \, \theta = 90°, \, g = 9.8$$

$$h = \frac{28^2(1^2)}{2(9.8)} \qquad \sin 90° = 1$$

$$h = 40$$

The maximum height of the ball is between 0 meters and 40 meters.

The radius of a circle is defined as a positive value. Therefore, the signs of the six trigonometric functions are determined by the signs of the coordinates of x and y in each quadrant.

Example ❸ **Use the unit circle to find the values of the six trigonometric functions for a 135° angle.**

Since 135° is between 90° and 180°, the terminal side is in the second quadrant. Therefore, the reference angle is $180° - 135°$ or 45°. The terminal side of a 45° angle intersects the unit circle at a point with coordinates $\left(\frac{\sqrt{2}}{2}, \frac{\sqrt{2}}{2} \right)$. Because the terminal side of a 135° angle is in the second quadrant, the x-coordinate is negative, and the y-coordinate is positive. The point of intersection has coordinates $\left(-\frac{\sqrt{2}}{2}, \frac{\sqrt{2}}{2} \right)$.

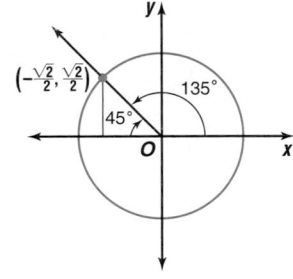

$$\sin 135° = y \qquad\qquad \cos 135° = x \qquad\qquad \tan 135° = \frac{y}{x}$$

$$\sin 135° = \frac{\sqrt{2}}{2} \qquad\qquad \cos 135° = -\frac{\sqrt{2}}{2}$$

$$\tan 135° = \frac{\frac{\sqrt{2}}{2}}{-\frac{\sqrt{2}}{2}}$$

$$\tan 135° = -1$$

$$\csc 135° = \frac{1}{y} \qquad\qquad \sec 135° = \frac{1}{x} \qquad\qquad \cot 135° = \frac{x}{y}$$

$$\csc 135° = \frac{1}{\frac{\sqrt{2}}{2}} \qquad\qquad \sec 135° = \frac{1}{-\frac{\sqrt{2}}{2}} \qquad\qquad \cot 135° = \frac{-\frac{\sqrt{2}}{2}}{\frac{\sqrt{2}}{2}}$$

$$\csc 135° = \frac{2}{\sqrt{2}} \qquad\qquad \sec 135° = -\frac{2}{\sqrt{2}} \qquad\qquad \cot 135° = -1$$

$$\csc 135° = \sqrt{2} \qquad\qquad \sec 135° = -\sqrt{2}$$

Teaching Tip The following diagram may help students remember in which quadrants the basic functions (and their reciprocals) are positive.

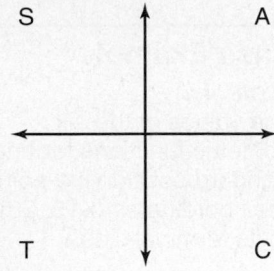

A=all
S=sin (csc)
T=tan (cot)
C=cos (sec)

Teaching Tip For Example 3, remind students to refer to the quadrant to determine whether the values of x and y are positive or negative.

In-Class Example

Example 3
Use the unit circle to find the values of the six trigonometric functions for a 210° angle.

$\sin 210° = -\frac{1}{2}$; $\cos 210° = -\frac{\sqrt{3}}{2}$;

$\tan 210° = \frac{\sqrt{3}}{3}$; $\csc 210° = -2$;

$\sec 210° = -\frac{2\sqrt{3}}{3}$; $\cot 210° = \sqrt{3}$

The sine and cosine functions of an angle in standard position may also be determined using the ordered pair of any point on its terminal side and the distance between that point and the origin.

Suppose $P(x, y)$ and $P'(x', y')$ are two points on the terminal side of an angle with measure θ, where P' is on the unit circle. Let $OP = r$. By the Pythagorean Theorem, $r = \sqrt{x^2 + y^2}$. Since P' is on the unit circle, $OP' = 1$. Triangles $OP'Q'$ and OPQ are similar. Thus, the lengths of corresponding sides are proportional.

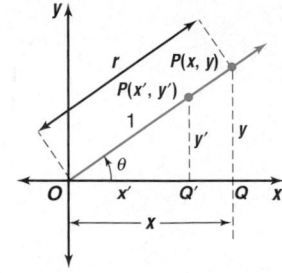

$$\frac{x'}{1} = \frac{x}{r} \qquad\qquad \frac{y'}{1} = \frac{y}{r}$$

Therefore, $\cos \theta = x'$ or $\frac{x}{r}$ and $\sin \theta = y'$ or $\frac{y}{r}$.

All six trigonometric functions can be determined using x, y, and r. The ratios do not depend on the choice of P. They depend only on the measure of θ.

Trigonometric Functions of an Angle in Standard Position	For any angle in standard position with measure θ, a point $P(x, y)$ on its terminal side, and $r = \sqrt{x^2 + y^2}$, the trigonometric functions of θ are as follows.

$$\sin \theta = \frac{y}{r} \qquad \cos \theta = \frac{x}{r} \qquad \tan \theta = \frac{y}{x}$$

$$\csc \theta = \frac{r}{y} \qquad \sec \theta = \frac{r}{x} \qquad \cot \theta = \frac{x}{y}$$

Example **4** **Find the values of the six trigonometric functions for angle θ in standard position if a point with coordinates $(5, -12)$ lies on its terminal side.**

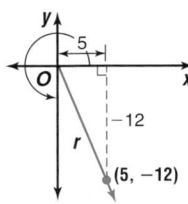

You know that $x = 5$ and $y = -12$. You need to find r.

$r = \sqrt{x^2 + y^2}$ *Pythagorean Theorem*

$r = \sqrt{5^2 + (-12)^2}$ *Substitute 5 for x and −12 for y.*

$r = \sqrt{169}$ or 13   *Disregard the negative root.*

Now write the ratios.

$\sin \theta = \frac{y}{r}$ $\cos \theta = \frac{x}{r}$ $\tan \theta = \frac{y}{x}$

$\sin \theta = \frac{-12}{13}$ or $-\frac{12}{13}$ $\cos \theta = \frac{5}{13}$ $\tan \theta = \frac{-12}{5}$ or $-\frac{12}{5}$

$\csc \theta = \frac{r}{y}$ $\sec \theta = \frac{r}{x}$ $\cot \theta = \frac{x}{y}$

$\csc \theta = \frac{13}{-12}$ or $-\frac{13}{12}$ $\sec \theta = \frac{13}{5}$ $\cot \theta = \frac{5}{-12}$ or $-\frac{5}{12}$

If you know the value of one of the trigonometric functions and the quadrant in which the terminal side of θ lies, you can find the values of the remaining five functions.

Example **5** Suppose θ is an angle in standard position whose terminal side lies in Quadrant III. If $\sin \theta = -\dfrac{4}{5}$, find the values of the remaining five trigonometric functions of θ.

To find the other function values, you must find the coordinates of a point on the terminal side of θ. Since $\sin \theta = -\dfrac{4}{5}$ and r is always positive, $r = 5$ and $y = -4$.

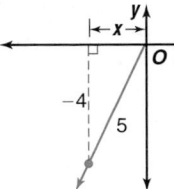

Find x.

$r^2 = x^2 + y^2$ *Pythagorean Theorem*

$5^2 = x^2 + (-4)^2$ *Substitute 5 for r and −4 for y.*

$9 = x^2$

$\pm 3 = x$ *Take the square root of each side.*

Since the terminal side of θ lies in Quadrant III, x must be negative.

Thus, $x = -3$.

Now use the values of x, y, and r to find the remaining five trigonometric functions of θ.

$\cos \theta = \dfrac{x}{r}$ $\tan \theta = \dfrac{y}{x}$

$\cos \theta = \dfrac{-3}{5}$ or $-\dfrac{3}{5}$ $\tan \theta = \dfrac{-4}{-3}$ or $\dfrac{4}{3}$

$\csc \theta = \dfrac{r}{y}$ $\sec \theta = \dfrac{r}{x}$

$\csc \theta = \dfrac{5}{-4}$ or $-\dfrac{5}{4}$ $\sec \theta = \dfrac{5}{-3}$ or $-\dfrac{5}{3}$

$\cot \theta = \dfrac{x}{y}$

$\cot \theta = \dfrac{-3}{-4}$ or $\dfrac{3}{4}$

Notice that the cosine and secant have the same sign. This will always be true since $\sec \theta = \dfrac{1}{\cos \theta}$. Similar relationships exist for the other reciprocal identities. *You will complete a chart for this in Exercise 4.*

Teaching Tip For Examples 4 and 5, remind students that a point on the terminal side of an angle defines a right triangle when a perpendicular is drawn to an axis. The Pythagorean Theorem is then used to find the hypotenuse.

In-Class Example

Example 5

Suppose θ is an angle in standard position whose terminal side lies in Quadrant IV. If $\sec \theta = \dfrac{\sqrt{29}}{5}$, find the values of the remaining five trigonometric functions of θ.

$\sin \theta = -\dfrac{2\sqrt{29}}{29}$; $\cos \theta = \dfrac{5\sqrt{29}}{29}$;

$\tan \theta = -\dfrac{2}{5}$; $\csc \theta = -\dfrac{\sqrt{29}}{2}$;

$\cot \theta = -\dfrac{5}{2}$

Addressing Individual Needs

Auditory/Musical Separate students into six teams. Using the diagram in the margin on p. 293, have each team create a rhyme for a different trigonometric value which describes the signs of that value in each of the quadrants. Then have each team perform its rhyme and teach it to the class. **Sample rhyme: Signs for secant's values make it easy—they're negative in the Quadrants numbered II and III!**

3 PRACTICE/APPLY

Error Analysis

Watch for students who ignore the negative values of the trigonometric functions in the second, third, and fourth quadrants.

Prevent by emphasizing that the first quadrant is the only quadrant where the values of all of the trigonometric functions are positive.

Assignment Guide
Core: 15–43 odd, 45–60
Enriched: 14–44 even, 45–60

Additional Answers

1. Terminal side of a 180° angle in standard position is the negative *x*-axis, which intersects the unit circle at (−1, 0). Since csc $\theta = \frac{1}{y}$, csc 180° = $\frac{1}{0}$, which is undefined.

2.
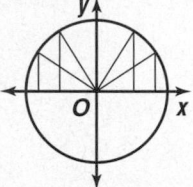

As θ goes from 0° to 90°, the *y*-coordinate increases. As θ goes from 90° to 180°, the *y*-coordinate decreases.

Study Guide and Practice Masters, pp. 83–84

296 Chapter 5

CHECK FOR UNDERSTANDING

Communicating Mathematics

Read and study the lesson to answer each question.   1–2. See margin.

1. **Explain** why csc 180° is undefined.

2. **Show** that the value of sin θ increases as θ goes from 0° to 90° and then decreases as θ goes from 90° to 180°.

3. **Confirm** that cot $\theta = \frac{\cos \theta}{\sin \theta}$.   cot $\theta = \frac{x}{y} = \frac{\cos \theta}{\sin \theta}$

4. *Math Journal* **Draw** a unit circle. Use the drawing to complete the chart below that indicates the sign of the trigonometric functions in each quadrant.

Function	Quadrant			
	I	II	III	IV
sin α or csc α	+	+	−	−
cos α or sec α	+	−	−	+
tan α or cot α	+	−	+	−

See margin for unit circle.

Guided Practice

Use the unit circle to find each value.

5. tan 180° **0**

6. sec (−90°) **undefined**

Use the unit circle to find the values of the six trigonometric functions for each angle.   7–8. See margin.

7. 30°

8. 225°

Find the values of the six trigonometric functions for angle θ in standard position if a point with the given coordinates lies on its terminal side.   9–10. See Answer Appendix.

9. (3, 4)

10. (−6, 6)

Suppose θ is an angle in standard position whose terminal side lies in the given quadrant. For each function, find the values of the remaining five trigonometric functions for θ.   11–12. See Answer Appendix.

11. tan $\theta = -1$; Quadrant IV

12. cos $\theta = -\frac{1}{2}$; Quadrant II

13. **Map Skills** The distance around Earth along a given latitude can be found using the formula $C = 2\pi r \cos L$, where r is the radius of Earth and L is the latitude. The radius of Earth is approximately 3960 miles. Describe the distances along the latitudes as you go from 0° at the equator to 90° at the poles.
The distances range from about 24,881 miles to 0 miles.

EXERCISES

Practice

Use the unit circle to find each value.

A
14. sin 90° **1**

15. tan 360° **0**

16. cot (−180°) **undefined**

17. csc 270° **−1**

18. cos (−270°) **0**

19. sec 180° **−1**

20. Find two values of θ for which sin $\theta = 0$. **Sample answers: 0°, 180°**

21. If cos $\theta = 0$, what is sec θ? **undefined**

www.amc.glencoe.com/self_check_quiz

Additional Answers

4.

7. sin 30° = $\frac{1}{2}$, cos 30° = $\frac{\sqrt{3}}{2}$, tan 30° = $\frac{\sqrt{3}}{3}$, csc 30° = 2, sec 30° = $\frac{2\sqrt{3}}{3}$, cot 30° = $\sqrt{3}$

8. sin 225° = $-\frac{\sqrt{2}}{2}$, cos 225° = $-\frac{\sqrt{2}}{2}$, tan 225° = 1, csc 225° = $-\sqrt{2}$, sec 225° = $-\sqrt{2}$, cot 225° = 1

Use the unit circle to find the values of the six trigonometric functions for each angle. **22–27. See margin.**

B

22. 45°   **23.** 150°   **24.** 315°   **25.** 210°   **26.** 330°   **27.** 420°

28. Find cot $(-45°)$.   **−1**   **29.** Find csc 390°.   **2**

Find the values of the six trigonometric functions for angle θ in standard position if a point with the given coordinates lies on its terminal side. **30–35. See margin.**

30. $(-4, -3)$ **31.** $(-6, 6)$ **32.** $(2, 0)$ **33.** $(1, -8)$ **34.** $(5, -3)$ **35.** $(-8, 15)$

36. The sine of one angle is the negative of the sine of the other angle.

36. The terminal side of one angle in standard position contains the point with coordinates $(5, -6)$. The terminal side of another angle in standard position contains the point with coordinates $(-5, 6)$. Compare the sines of these angles.

37. If $\sin \theta < 0$, where would the terminal side of the angle be located?
in Quadrant III or IV

Suppose θ is an angle in standard position whose terminal side lies in the given quadrant. For each function, find the values of the remaining five trigonometric functions for θ.   **38–43. See margin.**

C

38. $\cos \theta = -\dfrac{12}{13}$; Quadrant III   **39.** $\csc \theta = 2$; Quadrant II

40. $\sin \theta = -\dfrac{1}{5}$; Quadrant IV   **41.** $\tan \theta = 2$; Quadrant I

42. $\sec \theta = \sqrt{3}$; Quadrant IV   **43.** $\cot \theta = 1$; Quadrant III

44. If $\csc \theta = -2$ and θ lies in Quadrant III, find $\tan \theta$. $\dfrac{\sqrt{3}}{3}$

Applications and Problem Solving

45. Physics If you ignore friction, the amount of time required for a box to slide down an inclined plane is
$$\sqrt{\dfrac{2a}{g \sin \theta \cos \theta}},$$ where a is the horizontal distance defined by the inclined plane, g is the acceleration due to gravity, and θ is the angle of the inclined plane. For what values of θ is the expression undefined?   **0° or 90°**

48. $\sin \theta = \dfrac{3\sqrt{10}}{10}$, $\cos \theta = -\dfrac{\sqrt{10}}{10}$, $\tan \theta = -3$, $\csc \theta = \dfrac{\sqrt{10}}{3}$, $\sec \theta = -\sqrt{10}$, $\cot \theta = -\dfrac{1}{3}$

46. Critical Thinking For each statement, describe k.
a. $\tan (k \cdot 90°) = 0$ **k is an even integer.**
b. $\sec (k \cdot 90°)$ is undefined.   **k is an odd integer.**

47. Physics For polarized light, $\cos \theta = \sqrt{\dfrac{I_t}{I_o}}$, where θ is the angle of the axis of the polarized filter with the vertical, I_t is the intensity of the transmitted light, and I_O is the intensity of the vertically-polarized light striking the filter. Under what conditions would $I_t = I_O$?   **$\theta = 0°$**

48. Critical Thinking The terminal side of an angle θ in standard position coincides with the line $y = -3x$ and lies in Quadrant II. Find the six trigonometric functions of θ.

Additional Answers

22. $\sin 45° = \dfrac{\sqrt{2}}{2}$, $\cos 45° = \dfrac{\sqrt{2}}{2}$,
$\tan 45° = 1$, $\csc 45° = \sqrt{2}$,
$\sec 45° = \sqrt{2}$, $\cot 45° = 1$

23. $\sin 150° = \dfrac{1}{2}$, $\cos 150° = -\dfrac{\sqrt{3}}{2}$,
$\tan 150° = -\dfrac{\sqrt{3}}{3}$, $\csc 150° = 2$,
$\sec 150° = -\dfrac{2\sqrt{3}}{3}$,
$\cot 150° = -\sqrt{3}$

24. $\sin 315° = -\dfrac{\sqrt{2}}{2}$, $\cos 315° = \dfrac{\sqrt{2}}{2}$,
$\tan 315° = -1$, $\csc 315° = -\sqrt{2}$,
$\sec 315° = \sqrt{2}$, $\cot 315° = -1$

25. $\sin 210° = -\dfrac{1}{2}$, $\cos 210° = -\dfrac{\sqrt{3}}{2}$,
$\tan 210° = \dfrac{\sqrt{3}}{3}$, $\csc 210° = -2$,
$\sec 210° = -\dfrac{2\sqrt{3}}{3}$, $\cot 210° = \sqrt{3}$

Additional Answers

26. $\sin 330° = -\dfrac{1}{2}$,
$\cos 330° = \dfrac{\sqrt{3}}{2}$,
$\tan 330° = -\dfrac{\sqrt{3}}{3}$,
$\csc 330° = -2$,
$\sec 330° = \dfrac{2\sqrt{3}}{3}$,
$\cot 330° = -\sqrt{3}$

27. $\sin 420° = \dfrac{\sqrt{3}}{2}$, $\cos 420° = \dfrac{1}{2}$,
$\tan 420° = \sqrt{3}$, $\csc 420° = \dfrac{2\sqrt{3}}{3}$,
$\sec 420° = 2$, $\cot 420° = \dfrac{\sqrt{3}}{3}$

30. $\sin \theta = -\dfrac{3}{5}$, $\cos \theta = -\dfrac{4}{5}$,
$\tan \theta = \dfrac{3}{4}$, $\csc \theta = -\dfrac{5}{3}$,
$\sec \theta = -\dfrac{5}{4}$, $\cot \theta = \dfrac{4}{3}$

31. $\sin \theta = \dfrac{\sqrt{2}}{2}$, $\cos \theta = -\dfrac{\sqrt{2}}{2}$,
$\tan \theta = -1$, $\csc \theta = \sqrt{2}$,
$\sec \theta = -\sqrt{2}$, $\cot \theta = -1$

32. $\sin \theta = 0$, $\cos \theta = 1$,
$\tan \theta = 0$, $\csc \theta$ is undefined,
$\sec \theta = 1$, $\cot \theta$ is undefined.

33. $\sin \theta = -\dfrac{8\sqrt{65}}{65}$, $\cos \theta = \dfrac{\sqrt{65}}{65}$,
$\tan \theta = -8$, $\csc \theta = -\dfrac{\sqrt{65}}{8}$,
$\sec \theta = \sqrt{65}$, $\cot \theta = -\dfrac{1}{8}$

34. $\sin \theta = -\dfrac{3\sqrt{34}}{34}$, $\cos \theta = \dfrac{5\sqrt{34}}{34}$,
$\tan \theta = -\dfrac{3}{5}$, $\csc \theta = -\dfrac{\sqrt{34}}{3}$,
$\sec \theta = \dfrac{\sqrt{34}}{5}$, $\cot \theta = -\dfrac{5}{3}$

35. $\sin \theta = \dfrac{15}{17}$, $\cos \theta = -\dfrac{8}{17}$,
$\tan \theta = -\dfrac{15}{8}$, $\csc \theta = \dfrac{17}{15}$,
$\sec \theta = -\dfrac{17}{8}$, $\cot \theta = -\dfrac{8}{15}$

38. $\sin \theta = -\dfrac{5}{13}$, $\tan \theta = \dfrac{5}{12}$,
$\csc \theta = -\dfrac{13}{5}$, $\sec \theta = -\dfrac{13}{12}$,
$\cot \theta = \dfrac{12}{5}$

39. $\sin \theta = \dfrac{1}{2}$, $\cos \theta = -\dfrac{\sqrt{3}}{2}$,
$\tan \theta = -\dfrac{\sqrt{3}}{3}$, $\sec \theta = -\dfrac{2\sqrt{3}}{3}$,
$\cot \theta = -\sqrt{3}$

40. $\cos \theta = \dfrac{2\sqrt{6}}{5}$, $\tan \theta = -\dfrac{\sqrt{6}}{12}$,
$\csc \theta = -5$, $\sec \theta = \dfrac{5\sqrt{6}}{12}$,
$\cot \theta = -2\sqrt{6}$

(continued on page 298)

Additional Answers
(pp. 297–298)

41. $\sin \theta = \dfrac{2\sqrt{5}}{5}, \cos \theta = \dfrac{\sqrt{5}}{5},$
$\csc \theta = \dfrac{\sqrt{5}}{2}, \sec \theta = \sqrt{5},$
$\cot \theta = \dfrac{1}{2}$

42. $\sin \theta = -\dfrac{\sqrt{6}}{3}, \cos \theta = \dfrac{\sqrt{3}}{3},$
$\tan \theta = -\sqrt{2}, \csc \theta = -\dfrac{\sqrt{6}}{2},$
$\cot \theta = -\dfrac{\sqrt{2}}{2}$

43. $\sin \theta = -\dfrac{\sqrt{2}}{2}, \cos \theta = -\dfrac{\sqrt{2}}{2},$
$\tan \theta = 1, \csc \theta = -\sqrt{2},$
$\sec \theta = -\sqrt{2}$

55.

$f(x) = x^2 - 16$

Enrichment Masters, p. 51

5-3

NAME _____ DATE _____ PERIOD _____

Enrichment

Areas of Polygons and Circles

A regular polygon has sides of equal length and angles of equal measure. A regular polygon can be inscribed in or circumscribed about a circle. For n-sided regular polygons, the following area formulas can be used.

Area of circle $\quad A_C = \pi r^2$

Area of inscribed polygon $\quad A_I = \dfrac{nr^2}{2} \times \sin \dfrac{360°}{n}$

Area of circumscribed polygon $\quad A_C = nr^2 \times \tan \dfrac{180°}{n}$

Use a calculator to complete the chart below for a unit circle (a circle of radius 1).

	Number of Sides	Area of Inscribed Polygon	Area of Circle less Area of Polygon	Area of Circumscribed Polygon	Area of Polygon less Area of Circle
	3	1.2990381	1.8425545	5.1961524	2.0545598
1.	4	2	1.1415927	4	0.8584073
2.	8	2.8284271	0.3131655	3.3137085	0.1721158
3.	12	3	0.1415927	3.2153903	0.0737977
4.	20	3.0901699	0.0514227	3.1676888	0.0260962
5.	24	3.1058285	0.0357641	3.1596599	0.0180673
6.	28	3.1152931	0.0262996	3.1548423	0.0132497
7.	32	3.1214452	0.0201475	3.1517249	0.0101323
8.	1000	3.1415720	0.0000207	3.1416030	0.0000103

9. What number do the areas of the circumscribed and inscribed polygons seem to be approaching as the number of sides of the polygon increases?
π

© Glencoe/McGraw-Hill   T51   *Advanced Mathematical Concepts*

---

49. **Entertainment** Domingo decides to ride the Ferris wheel at the local carnival. When he gets into the seat that is at the bottom of the Ferris wheel, he is 4 feet above the ground.

 a. If the radius of the Ferris wheel is 36 feet, how far above the ground will Domingo be when his seat reaches the top? **76 ft**

 b. The Ferris wheel rotates 300° counterclockwise and stops to let another passenger on the ride. How far above the ground is Domingo when the Ferris wheel stops? **22 ft**

 c. Suppose the radius of the Ferris wheel is only 30 feet. How far above the ground is Domingo after the Ferris wheel rotates 300°? **19 ft**

 d. Suppose the radius of the Ferris wheel is r. Write an expression for the distance from the ground to Domingo after the Ferris wheel rotates 300°. $\dfrac{1}{2}r + 4$

Mixed Review

50. If $\csc \theta = \dfrac{7}{5}$, find $\sin \theta$. *(Lesson 5-2)* $\dfrac{5}{7}$

51. If a $-840°$ angle is in standard position, determine a coterminal angle that is between 0° and 360° and state the quadrant in which the terminal side lies. *(Lesson 5-1)* **240°; III**

52. Solve $5 - \sqrt{b + 2} = 0$. *(Lesson 4-7)* **23**

53. Solve $4x^2 - 9x + 5 = 0$ by using the quadratic formula. *(Lesson 4-2)* **1.25, 1**

54. If y varies directly as x and $y = 9$ when x is -15, find y when $x = 21$. *(Lesson 3-8)* **−12.6**

55. Graph the inverse of $f(x) = x^2 - 16$. *(Lesson 3-4)* **See margin.**

56. Find the multiplicative inverse of $\begin{bmatrix} 2 & 1 \\ -3 & 2 \end{bmatrix}$. *(Lesson 2-5)* $\dfrac{1}{7}\begin{bmatrix} 2 & -1 \\ 3 & 2 \end{bmatrix}$

57. Solve the system of equations. *(Lesson 2-2)*
$8m - 3n - 4p = 6$
$4m + 9n - 2p = -4$
$6m + 12n + 5p = -1$
$\left(\dfrac{3}{4}, -\dfrac{2}{3}, \dfrac{1}{2}\right)$

58. State whether each of the points at $(9, 3)$, $(-1, 2)$, and $(2, -2)$ satisfy the inequality $2x - 4y \le 7$. *(Lesson 1-8)* **yes; yes; no**

59. **Manufacturing** The length of a nail is $2\dfrac{1}{2}$ inches. The manufacturer randomly measures the nails to test if their equipment is working properly. If the discrepancy is more than $\dfrac{1}{8}$ inch, adjustments must be made. Identify the type of function that models this situation. Then write a function for the situation. *(Lesson 1-7)* **absolute value;** $f(x) = \left| 2\dfrac{1}{2} - x \right|$

60. **SAT/ACT Practice** In the figure at the right, the largest possible circle is cut out of a square piece of tin. What is the approximate total area of the remaining pieces of tin? **C**

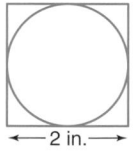

 A 0.13 in²   **B** 0.75 in²   **C** 0.86 in²
 D 1.0 in²   **E** 3.14 in²

← 2 in. →

Extra Practice See p. A34.

Extra Credit

At what points does the line $x = \dfrac{1}{2}$ intersect the unit circle? Explain how you can use these points to find $\sin \theta$ if $\cos \theta = \dfrac{1}{2}$. **The line** $x = \dfrac{1}{2}$ **intersects the circle at** $\left(\dfrac{1}{2}, \dfrac{\sqrt{3}}{2}\right)$ **and** $\left(\dfrac{1}{2}, -\dfrac{\sqrt{3}}{2}\right)$**. Because points on the unit circle have coordinates** $(\cos \theta, \sin \theta)$**,** $\sin \theta = \dfrac{\sqrt{3}}{2}$ **or** $-\dfrac{\sqrt{3}}{2}$**.**

5-4

Applying Trigonometric Functions

OBJECTIVE
- Use trigonometry to find the measures of the sides of right triangles.

Real World Application

ENTERTAINMENT The circus has arrived and the roustabouts must put up the main tent in a field near town. A tab is located on the side of the tent 40 feet above the ground. A rope is tied to the tent at this point and then the rope is placed around a stake on the ground. If the angle that the rope makes with the level ground is 50° 15′, how long is the rope? What is the distance between the bottom of the tent and the stake? *This problem will be solved in Example 2.*

Trigonometric functions can be used to solve problems involving right triangles. The most common functions used are the sine, cosine, and tangent.

Examples **1** If $P = 35°$ and $r = 14$, find q.

From the art at the right, you know the measures of an angle and the hypotenuse. You want to know the measure of the side adjacent to the given angle. The cosine function relates the side adjacent to the angle and the hypotenuse.

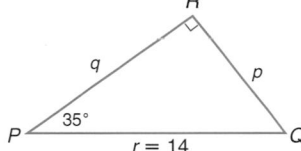

$\cos P = \dfrac{q}{r}$ $\cos = \dfrac{side\ adjacent}{hypotenuse}$

$\cos 35° = \dfrac{q}{14}$ *Substitute 35° for P and 14 for r.*

$14 \cos 35° = q$ *Multiply each side by 14.*

$11.46812862 \approx q$ *Use a calculator.*

Therefore, q is about 11.5.

Real World Application

2 **ENTERTAINMENT** Refer to the application above.

a. If the angle that the rope makes with the level ground is 52° 15′, how long is the rope?

b. What is the distance between the bottom of the tent and the stake?

a. You know the measures of an angle and the side opposite the angle. To find the length of the rope, you need to know the measure of the hypotenuse. In this case, use the sine function.

(continued on the next page)

Resource Manager

Reproducible Masters
- *Study Guide and Practice,* pp. 85–86
- *Enrichment,* p. 52
- *Assessment and Evaluation,* pp. 120, 121
- *TI–92 & CBL Interdisciplinary Activities,* pp. 13–16

Transparencies
- 5-Minute Check, 5-4
- Teaching, 5-4

Teaching Tip In Example 1, remind students that the corresponding lowercase letter is used to represent the measure of the side opposite the angle. That is, *p* is the side opposite ∠P.

In-Class Examples

Example 1
If $J = 50°$ and $j = 12$, find *r*.

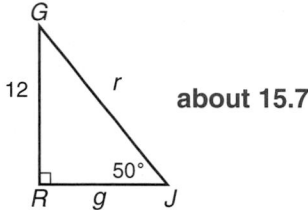

about 15.7

Example 2
Sports The chair lift at a ski resort rises at an angle of 20.75° and attains a vertical height of 1200 feet.

a. How far does the chair lift travel up the side of the mountain? **about 3387.0 ft**
b. A film crew in a helicopter records an overhead view of a skier's downhill run from where she gets off the chair lift at the top to where she gets back on the chair lift for her next run. If the helicopter follows a level flight path, what is the length of that path? **about 3167.3 ft**

Example 3
Geometry A regular hexagon is inscribed in a circle with diameter 26.6 centimeters. Find the apothem of the hexagon.

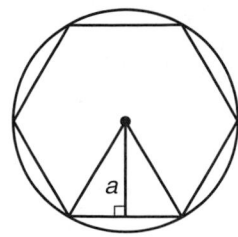

about 11.5 cm

$$\sin 52° 15' = \frac{40}{r} \qquad sin = \frac{side\ opposite}{hypotenuse}$$

$$r \sin 52° 15' = 40 \qquad Multiply\ each\ side\ by\ r.$$

$$r = \frac{40}{\sin 52° 15'} \qquad Divide\ each\ side\ by\ \sin 52° 15'.$$

$$r \approx 50.58875357 \qquad Use\ a\ calculator.$$

The rope is about 50.6 feet long.

b. To find the distance between the bottom of the tent and the stake, you need to know the length of the side adjacent to the known angle. Use the tangent function.

$$\tan 52° 15' = \frac{40}{d} \qquad tan = \frac{side\ opposite}{side\ adjacent}$$

$$d \tan 52° 15' = 40 \qquad Multiply\ each\ side\ by\ d.$$

$$d = \frac{40}{\tan 52° 15'} \qquad Divide\ each\ side\ by\ \tan 52° 15'.$$

$$d \approx 30.97130911 \qquad Use\ a\ calculator.$$

The distance between the bottom of the tent and the stake is about 31.0 feet.

You can use right triangle trigonometry to solve problems involving other geometric figures.

Example ③ **GEOMETRY** A regular pentagon is inscribed in a circle with diameter 8.34 centimeters. The *apothem* of a regular polygon is the measure of a line segment from the center of the polygon to the midpoint of one of its sides. Find the apothem of the pentagon.

First, draw a diagram. If the diameter of the circle is 8.34 centimeters, the radius is 8.34 ÷ 2 or 4.17 centimeters. The measure of α is 360° ÷ 10 or 36°.

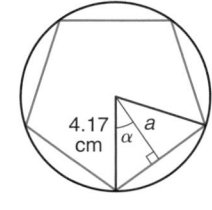

$$\cos 36° = \frac{a}{4.17} \qquad cos = \frac{side\ adjacent}{hypotenuse}$$

$$4.17 \cos 36° = a \qquad Multiply\ each\ side\ by\ 4.17.$$

$$3.373600867 \approx a \qquad Use\ a\ calculator.$$

The apothem is about 3.37 centimeters.

There are many other applications that require trigonometric solutions. For example, surveyors use special instruments to find the measures of **angles of elevation** and **angles of depression**. An angle of elevation is the angle between a horizontal line and the line of sight from an observer to an object at a higher level. An angle of depression is the angle between a horizontal line and the line of sight from the observer to an object at a lower level. The angle of elevation and the angle of depression are equal in measure because they are alternate interior angles.

300 *Chapter 5 The Trigonometric Functions*

Teaching Tip In Example 3, remind students that the inscribed figure is a regular pentagon. Therefore, the central angle formed creates an isosceles triangle and has a measure of 360° ÷ 5 or 72°. The apothem bisects the angle and the side of the pentagon, creating the right triangle to be solved.

Example

4 **SURVEYING** On May 18, 1980, Mount Saint Helens, a volcano in Washington, erupted with such force that the top of the mountain was blown off. To determine the new height at the summit of Mount Saint Helens, a surveyor measured the angle of elevation to the top of the volcano to be $37°\ 46'$. The surveyor then moved 1000 feet closer to the volcano and measured the angle of elevation to be $40°\ 30'$. Determine the new height of Mount Saint Helens.

Draw a diagram to model the situation. Let h represent the height of the volcano and x represent the distance from the surveyor's second position to the center of the base of the volcano. Write two equations involving the tangent function.

$$\tan 37°\ 46' = \frac{h}{1000 + x}$$
$$(1000 + x)\tan 37°\ 46' = h$$
$$\tan 40°\ 30' = \frac{h}{x}$$
$$x \tan 40°\ 30' = h$$

Therefore, $(1000 + x)\tan 37°\ 46' = x \tan 40°\ 30'$. Solve this equation for x.

$$(1000 + x)\tan 37°\ 46' = x \tan 40°\ 30'$$
$$1000 \tan 37°\ 46' + x \tan 37°\ 46' = x \tan 40°\ 30'$$
$$1000 \tan 37°\ 46' = x \tan 40°\ 30' - x \tan 37°\ 46'$$
$$1000 \tan 37°\ 46' = x(\tan 40°\ 30' - \tan 37°\ 46')$$
$$\frac{1000 \tan 37°\ 46'}{\tan 40°\ 30' - \tan 37°\ 46'} = x$$
$$9765.826092 \approx x \quad \textit{Use a calculator.}$$

Use this value for x and the equation $x \tan 40°\ 30' = h$ to find the height of the volcano.

$$x \tan 40°\ 30' = h$$
$$9765.826092 \tan 40°\ 30' \approx h$$
$$8340.803443 \approx h \quad \textit{Use a calculator.}$$

The new height of Mount Saint Helens is about 8341 feet.

CHECK FOR UNDERSTANDING

Communicating Mathematics

Read and study the lesson to answer each question.

1. **State** which trigonometric function you would use to solve each problem.
   a. If $S = 42°$ and $ST = 8$, find RS. **cos or sec**
   b. If $T = 55°$ and $RT = 5$, find RS. **tan or cot**
   c. If $S = 27°$ and $TR = 7$, find TS. **sin or csc**

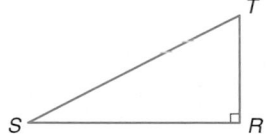

2. **Write** a problem that could be solved using the tangent function. **See margin.**

Lesson 5-4 Applying Trigonometric Functions **301**

In-Class Example

Example 4
Shore Defense An observer in the top of a lighthouse determines that the angles of depression to two sailboats directly in line with the lighthouse are 3.5° and 5.75°. If the observer is 125 feet above sea level, find the distance between the boats. **about 802 ft**

3 PRACTICE/APPLY

Error Analysis

Watch for students who use the inappropriate trigonometric ratios when solving problems.

Prevent by repeatedly showing students how to draw diagrams from the information given in applications and what ratios the given information represents.

Additional Answer

2. Sample answer: find *a*.

***Study Guide and Practice Masters*, pp. 85–86**

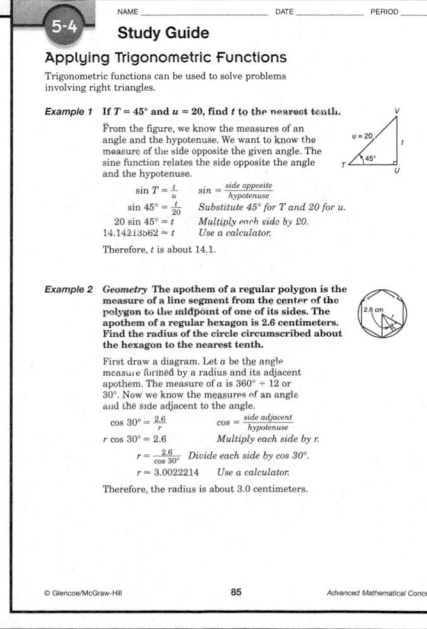

Additional Answers

3. $\angle DCB$; $\angle ABC$; the measures are equal; if parallel lines are cut by a transversal, the alternate interior angles are congruent.

4. Sample answer: if you know the angle of elevation of the sun at noon on a particular day, you can measure the length of the shadow of the building at noon on that day. The height of the building equals the length of the shadow times the tangent of the angle of elevation of the sun.

3–4. See margin.

3. **Name** the angle of elevation and the angle of depression in the figure at the right. Compare the measures of these angles. Explain.

4. **Describe** a way to use trigonometry to determine the height of the building where you live.

Guided Practice

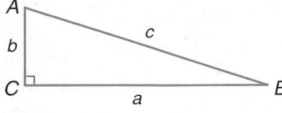

Solve each problem. Round to the nearest tenth.

5. If $b = 13$ and $A = 76°$, find a.   **52.1**
6. If $B = 26°$ and $b = 18$, find c.   **41.1**
7. If $B = 16°\ 45'$ and $c = 13$, find a.   **12.4**

8. **Geometry**   Each base angle of an isosceles triangle measures $55°\ 30'$. Each of the congruent sides is 10 centimeters long.
   a. Find the altitude of the triangle.   **about 8.2 cm**
   b. What is the length of the base?   **about 11.3 cm**
   c. Find the area of the triangle.   **about 46.7 cm²**

9. **Boating**   The Ponce de Leon lighthouse in St. Augustine, Florida, is the second tallest brick tower in the United States. It was built in 1887 and rises 175 feet above sea level. How far from the shore is a motorboat if the angle of depression from the top of the lighthouse is $13°\ 15'$?   **about 743.2 ft**

EXERCISES

Practice

Solve each problem. Round to the nearest tenth.

A
10. If $A = 37°$ and $b = 6$, find a.   **4.5**
11. If $c = 16$ and $B = 67°$, find a.   **6.3**
12. If $B = 62°$ and $c = 24$, find b.   **21.2**

B
13. If $A = 29°$ and $a = 4.6$, find c.   **9.5**
14. If $a = 17.3$ and $B = 77°$, find c.   **76.9**
15. If $b = 33.2$ and $B = 61°$, find a.   **18.4**
16. If $B = 49°\ 13'$ and $b = 10$, find a.   **8.6**
17. If $A = 16°\ 55'$ and $c = 13.7$, find a.   **4.0**
18. If $a = 22.3$ and $B = 47°\ 18'$, find c.   **32.9**

Exercises 10–18

19. 6;
10.4; 6; 8.5

C
19. Find h, n, m, and p. Round to the nearest tenth.

Exercise 19

20. **Geometry**   The apothem of a regular pentagon is 10.8 centimeters.   **a. about 13.3 cm**
   a. Find the radius of the circumscribed circle.
   b. What is the length of a side of the pentagon?   **about 15.7 cm**
   c. Find the perimeter of the pentagon.   **about 78.5 cm**

21. **Geometry**   Each base angle of an isosceles triangle measures $42°\ 30'$. The base is 14.6 meters long.
   a. Find the length of a leg of the triangle.   **about 9.9 m**
   b. Find the altitude of the triangle.   **about 6.7 m**
   c. What is the area of the triangle?   **about 48.8 m²**

 www.amc.glencoe.com/self_check_quiz

22. **Geometry**   A regular hexagon is inscribed in a circle with diameter 6.4 centimeters.

    a. What is the apothem of the hexagon?   **about 2.8 cm**

    b. Find the length of a side of the hexagon.   **3.2 cm**

    c. Find the perimeter of the hexagon.   **19.2 cm**

    d. The area of a regular polygon equals one half times the perimeter of the polygon times the apothem. Find the area of the polygon.   **about 26.6 cm^2**

Applications and Problem Solving

23. **Engineering**   The escalator at St. Petersburg Metro in Russia has a vertical rise of 195.8 feet. If the angle of elevation of the escalator is $10° 21' 36''$, find the length of the escalator.   **about 1088.8 ft**

24. **Critical Thinking**   Write a formula for the volume of the regular pyramid at the right in terms of α and s the length of each side of the base. $V = \frac{1}{6}s^3 \tan \alpha$

25. **Fire Fighting**   The longest truck-mounted ladder used by the Dallas Fire Department is 108 feet long and consists of four hydraulic sections. Gerald Travis, aerial expert for the department, indicates that the optimum operating angle of this ladder is $60°$. The fire fighters find they need to reach the roof of an 84-foot burning building. Assume the ladder is mounted 8 feet above the ground.

    a. Draw a labeled diagram of the situation.   **See margin.**

    b. How far from the building should the base of the ladder be placed to achieve the optimum operating angle?   **about 43.9 ft**

    c. How far should the ladder be extended to reach the roof?   **about 87.8 ft**

26. **Aviation**   When a 757 passenger jet begins its descent to the Ronald Reagan International Airport in Washington, D.C., it is 3900 feet from the ground. Its angle of descent is $6°$.

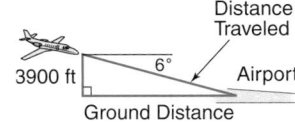

    a. What is the plane's ground distance to the airport?   **about 37,106.0 ft**

    b. How far must the plane fly to reach the runway?   **about 37,310.4 ft**

28. Let M represent the point of intersection of the altitude and $\overline{EF}$. Since $\triangle GEF$ is isosceles, the altitude bisects $\overline{EF}$. $\triangle EMG$ is a right triangle. Therefore, $\sin \theta = \frac{a}{s}$ or $s \sin \theta = a$ and $\tan \theta = \frac{a}{0.5b}$ or $0.5b \tan \theta = a$.

27. **Boat Safety**   The Cape Hatteras lighthouse on the North Carolina coast was built in 1870 and rises 208 feet above sea level. From the top of the lighthouse, the lighthouse keeper observes a yacht and a barge along the same line of sight. The angle of depression for the yacht is $20°$, and the angle of depression for the barge is $12° 30'$. For safety purposes, the keeper thinks that the two sea vessels should be at least 300 feet apart. If they are less than 300 feet, she plans to sound the horn. How far apart are these vessels? Does the keeper have to sound the horn?   **about 366.8 ft; no**

28. **Critical Thinking**   Derive two formulas for the length of the altitude a of the triangle shown at the right, given that b, s, and θ are known. Justify each of the steps you take in your reasoning.

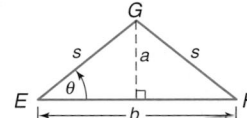

Extra Credit

If the height of a rectangular prism is 10 inches and the sides of the square base each measure 5 inches, find a trigonometric equation to represent the angle formed by the diagonal of the prism with a diagonal of a base of the prism.

Sample answers: $\sin \theta = \frac{\sqrt{6}}{3}$ or $\cos \theta = \frac{\sqrt{3}}{3}$ or $\tan \theta = \sqrt{2}$

Open-Ended Assessment
Writing Have students write an application problem that requires trigonometric ratios to solve. Students should write out complete solutions for their own problems, including explanations for each step. These problems can be used later in class for practice.

Mid-Chapter Quiz
The Mid-Chapter Quiz provides students with a brief review of the concepts and skills in Lessons 5-1 through 5-4. Lesson numbers are given to the right of exercises or instruction lines so students can review concepts not yet mastered.

 Chapter 5, Quiz B (Lessons 5-3 and 5-4) is available in the *Assessment and Evaluation Masters*, p. 121.

 Chapter 5 Mid-Chapter Test (Lessons 5-1 through 5-4) is available in the *Assessment and Evaluation Masters*, p. 120.

Additional Answer
34.

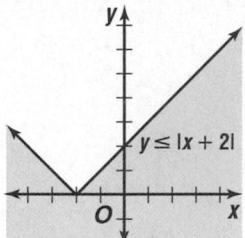

$y \le |x + 2|$

Enrichment Masters, p. 52

 5-4

Enrichment

Making and Using a Hypsometer

A **hypsometer** is a device that can be used to measure the height of an object. To construct your own hypsometer, you will need a rectangular piece of heavy cardboard that is at least 7 cm by 10 cm, a straw, transparent tape, a string about 20 cm long, and a small weight that can be attached to the string.

Mark off 1-cm increments along one short side and one long side of the cardboard. Tape the straw to the other short side. Then attach the weight to one end of the string, and attach the other end of the string to one corner of the cardboard, as shown in the figure below. The diagram below shows how your hypsometer should look.

To use the hypsometer, you will need to know the distance from the base of the object whose height you are finding to where you stand when you use the hypsometer.

Sight the top of the object through the straw. Note where the free-hanging string crosses the bottom scale. Then use similar triangles to find the height of the object.

1. Draw a diagram to illustrate how you can use similar triangles and the hypsometer to find the height of a tall object.
   See students' diagrams.

Use your hypsometer to find the height of each of the following. **See students' work.**

2. your school's flagpole
3. a tree on your school's property
4. the highest point on the front wall of your school building
5. the goal posts on a football field
6. the hoop on a basketball court
7. the top of the highest window of your school building
8. the top of a school bus
9. the top of a set of bleachers at your school
10. the top of a utility pole near your school

© Glencoe/McGraw-Hill   T52   *Advanced Mathematical Concepts*

---

29. **Recreation** Latasha and Markisha are flying kites on a windy spring day. Latasha has released 250 feet of string, and Markisha has released 225 feet of string. The angle that Latasha's kite string makes with the horizontal is 35°. The angle that Markisha's kite string makes with the horizontal is 42°. Which kite is higher and by how much? **Markisha's; about 7.2 ft**

30. **Architecture** A flagpole 40 feet high stands on top of the Wentworth Building. From a point in front of Bailey's Drugstore, the angle of elevation for the top of the pole is 54° 54′, and the angle of elevation for the bottom of the pole is 47° 30′. How high is the building? **about 131.7 ft**

47°30′   54°54′

Mixed Review

31.
$\sin 120° = \dfrac{\sqrt{3}}{2}$,

$\cos 120° = -\dfrac{1}{2}$,

$\tan 120° = -\sqrt{3}$,

$\csc 120° = \dfrac{2\sqrt{3}}{3}$,

$\sec 120° = -2$,

$\cot 120° = -\dfrac{\sqrt{3}}{3}$

31. Find the values of the six trigonometric functions for a 120° angle using the unit circle. *(Lesson 5-3)*

32. Find the sine, cosine, and tangent ratios for $\angle P$. *(Lesson 5-2)*
    $\dfrac{2\sqrt{53}}{53}, \dfrac{7\sqrt{53}}{53}, \dfrac{2}{7}$

P ———— 7 in. ———— *R*, *Q*, 2 in.

33. Write 43° 15′ 35″ as a decimal to the nearest thousandth. *(Lesson 5-1)* **43.260**

34. Graph $y \le |x + 2|$. *(Lesson 3-3)* **See margin.**

35. **Consumerism** Kareem and Erin went shopping for school supplies. Kareem bought 3 notebooks and 2 packages of pencils for $5.80. Erin bought 4 notebooks and 1 package of pencils for $6.20. What is the cost of one notebook? What is the cost of one package of pencils? *(Lesson 2-1)* **$1.32; $0.92**

36. **SAT/ACT Practice** An automobile travels *m* miles in *h* hours. At this rate, how far will it travel in *x* hours? **E**

   A $\dfrac{m}{x}$ **B** $\dfrac{m}{xh}$ **C** $\dfrac{m}{h}$ **D** $\dfrac{mh}{x}$ **E** $\dfrac{mx}{h}$

MID-CHAPTER QUIZ

1. Change 34.605° to degrees, minutes, and seconds. (Lesson 5-1) **34° 36′ 18″**

2. If a −400° angle is in standard position, determine a coterminal angle that is between 0° and 360°. State the quadrant in which the terminal side lies. (Lesson 5-1) **320°; IV**

3. Find the six trigonometric functions for $\angle G$. (Lesson 5-2) **See margin.**

   G, 12 m, *I*, 10 m, *H*

4. Find the values of the six trigonometric functions for angle θ in standard position if a point with coordinates (2, −5) lies on its terminal side. (Lesson 5-3) **See margin.**

5. **National Landmarks** Suppose the angle of elevation of the sun is 27.8°. Find the length of the shadow made by the Washington Monument, which is 550 feet tall. (Lesson 5-4) **about 1043.2 ft**

304 Chapter 5   *The Trigonometric Functions*

Extra Practice See p. A34.

Additional Answers
Mid-Chapter Quiz

3. $\sin G = \dfrac{5}{6}$, $\cos G = \dfrac{\sqrt{11}}{6}$, $\tan G = \dfrac{5\sqrt{11}}{11}$, $\csc G = \dfrac{6}{5}$, $\sec G = \dfrac{6\sqrt{11}}{11}$, $\cot G = \dfrac{\sqrt{11}}{5}$

4. $\sin \theta = -\dfrac{5\sqrt{29}}{29}$, $\cos \theta = \dfrac{2\sqrt{29}}{29}$, $\tan \theta = -\dfrac{5}{2}$, $\csc \theta = -\dfrac{\sqrt{29}}{5}$, $\sec \theta = \dfrac{\sqrt{29}}{2}$, $\cot \theta = -\dfrac{2}{5}$

5-5

Solving Right Triangles

OBJECTIVES
- Evaluate inverse trigonometric functions.
- Find missing angle measurements.
- Solve right triangles.

SECURITY A security light is being installed outside a loading dock. The light is mounted 20 feet above the ground. The light must be placed at an angle so that it will illuminate the end of the parking lot. If the end of the parking lot is 100 feet from the loading dock, what should be the angle of depression of the light? *This problem will be solved in Example 4.*

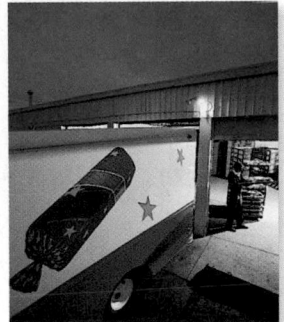

In Lesson 5-3, you learned to use the unit circle to determine the value of trigonometric functions. Some of the frequently-used values are listed below.

θ	0°	30°	45°	60°	90°	120°	135°	150°	180°
sin θ	0	$\frac{1}{2}$	$\frac{\sqrt{2}}{2}$	$\frac{\sqrt{3}}{2}$	1	$\frac{\sqrt{3}}{2}$	$\frac{\sqrt{2}}{2}$	$\frac{1}{2}$	0
cos θ	1	$\frac{\sqrt{3}}{2}$	$\frac{\sqrt{2}}{2}$	$\frac{1}{2}$	0	$-\frac{1}{2}$	$-\frac{\sqrt{2}}{2}$	$-\frac{\sqrt{3}}{2}$	-1
tan θ	0	$\frac{\sqrt{3}}{3}$	1	$\sqrt{3}$	undefined	$-\sqrt{3}$	-1	$-\frac{\sqrt{3}}{3}$	0

θ	210°	225°	240°	270°	300°	315°	330°	360°
sin θ	$-\frac{1}{2}$	$-\frac{\sqrt{2}}{2}$	$-\frac{\sqrt{3}}{2}$	-1	$-\frac{\sqrt{3}}{2}$	$-\frac{\sqrt{2}}{2}$	$-\frac{1}{2}$	0
cos θ	$-\frac{\sqrt{3}}{2}$	$-\frac{\sqrt{2}}{2}$	$-\frac{1}{2}$	0	$\frac{1}{2}$	$\frac{\sqrt{2}}{2}$	$\frac{\sqrt{3}}{2}$	1
tan θ	$\frac{\sqrt{3}}{3}$	1	$\sqrt{3}$	undefined	$-\sqrt{3}$	-1	$-\frac{\sqrt{3}}{3}$	0

Sometimes you know a trigonometric value of an angle, but not the angle. In this case, you need to use an **inverse** of the trigonometric function. The inverse of the sine function is the **arcsine relation.**

An equation such as $\sin x = \frac{\sqrt{3}}{2}$ can be written as $x = \arcsin \frac{\sqrt{3}}{2}$, which is read "$x$ is an angle whose sine is $\frac{\sqrt{3}}{2}$," or "x equals the arcsine of $\frac{\sqrt{3}}{2}$."

The solution, x, consists of all angles that have $\frac{\sqrt{3}}{2}$ as the value of sine x.

Similarly, the inverse of the cosine function is the **arccosine relation,** and the inverse of the tangent function is the **arctangent relation.**

Lesson 5-5 Solving Right Triangles **305**

Lesson Notes

5-5

1 FOCUS

5-Minute Check
Lesson 5-4

Solve each problem. Round to the nearest tenth.

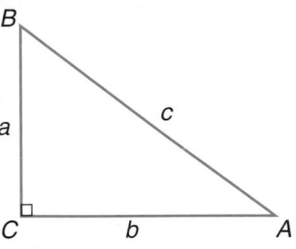

1. If $A = 38°$ and $a = 14$, find c. **22.7**
2. If $B = 58°$ and $c = 32$, find a. **17.0**
3. If $b = 45$ and $B = 72°$, find a. **14.6**
4. If $A = 50°45'$ and $c = 27$, find b. **17.1**
5. A regular hexagon has side length 8 centimeters. Find the length of the apothem to the nearest tenth. **6.9 cm**

Motivating the Lesson

Discussion Discuss with students the fact that they have learned to use trigonometric functions to solve for the sides of a right triangle. Together with the Pythagorean Theorem, the six trigonometric ratios allow students to find the sides of the triangles. Explain that they will now learn techniques to find the angles of a right triangle when information about the sides only is given.

Teaching Tip Emphasize that $\sin^{-1}$ is a relation, not a function.

Resource Manager

Reproducible Masters
- *Study Guide and Practice,* pp. 87–88
- *Enrichment,* p. 53

Transparencies
- 5-Minute Check, 5-5
- Teaching, 5-5

Teaching Tip Recall that the inverse function may be found by interchanging the coordinates of the ordered pairs of the function. In other words, the domain of the function becomes the range of its inverse and the range of the function becomes the domain of its inverse. The inverse of sine is the set of ordered pairs (sin x, x). The domain of the inverse sine is $-1 \leq x \leq 1$.

In-Class Examples

Example 1
Solve each equation.
a. tan $x = 1$ **45°, 225°, or any angles coterminal with these angles**

b. sin $x = -\dfrac{1}{2}$ **210°, 330°, or any angles coterminal with these angles**

Example 2
Evaluate each expression. Assume that all angles are in Quadrant I.

a. $\tan\left(\cos^{-1}\dfrac{4}{5}\right)$ $\dfrac{3}{4}$

b. $\cos\left(\arccos\dfrac{2}{5}\right)$ $\dfrac{2}{5}$

The equations in each row of the table below are equivalent. You can use these equations to rewrite trigonometric expressions.

Inverses of the Trigonometric Functions	**Trigonometric Function**	**Inverse Trigonometric Relation**
	$y = \sin x$	$x = \sin^{-1} y$ or $x = \arcsin y$
	$y = \cos x$	$x = \cos^{-1} y$ or $x = \arccos y$
	$y = \tan x$	$x = \tan^{-1} y$ or $x = \arctan y$

Examples **1** Solve each equation.

a. $\sin x = \dfrac{\sqrt{3}}{2}$

If $\sin x = \dfrac{\sqrt{3}}{2}$, then x is an angle whose sine is $\dfrac{\sqrt{3}}{2}$.

$x = \arcsin\dfrac{\sqrt{3}}{2}$

From the table on page 305, you can determine that x equals 60°, 120°, or any angle coterminal with these angles.

b. $\cos x = -\dfrac{\sqrt{2}}{2}$

If $\cos x = -\dfrac{\sqrt{2}}{2}$, then x is an angle whose cosine is $-\dfrac{\sqrt{2}}{2}$.

$x = \arccos -\dfrac{\sqrt{2}}{2}$

From the table, you can determine that x equals 135°, 225°, or any angle coterminal with these angles.

2 Evaluate each expression. Assume that all angles are in Quadrant I.

a. $\tan\left(\tan^{-1}\dfrac{6}{11}\right)$

Let $A = \tan^{-1}\dfrac{6}{11}$. Then $\tan A = \dfrac{6}{11}$ by the definition of inverse. Therefore, by substitution, $\tan\left(\tan^{-1}\dfrac{6}{11}\right) = \dfrac{6}{11}$.

b. $\cos\left(\arcsin\dfrac{2}{3}\right)$

Let $B = \arcsin\dfrac{2}{3}$. Then $\sin B = \dfrac{2}{3}$ by the definition of inverse. Draw a diagram of the $\angle B$ in Quadrant I.

$r^2 = x^2 + y^2$ *Pythagorean Theorem*

$3^2 = x^2 + 2^2$ *Substitute 3 for r and 2 for y.*

$\sqrt{5} = x$ *Take the square root of each side. Disregard the negative root.*

Since $\cos = \dfrac{\text{side adjacent}}{\text{hypotenuse}}$, $\cos B = \dfrac{\sqrt{5}}{3}$ and $\cos\left(\arcsin\dfrac{2}{3}\right) = \dfrac{\sqrt{5}}{3}$.

Inverse trigonometric relations can be used to find the measure of angles of right triangles. Calculators can be used to find values of the inverse trigonometric relations.

Example **③** **If $f = 17$ and $d = 32$, find E.**

In this problem, you want to know the measure of an acute angle in a right triangle. You know the side adjacent to the angle and the hypotenuse. The cosine function relates the side adjacent to the angle and the hypotenuse.

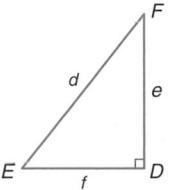

Remember that in trigonometry the measure of an angle is symbolized by the angle vertex letter.

$\cos E = \dfrac{f}{d}$	$\cos = \dfrac{\text{side adjacent}}{\text{hypotenuse}}$
$\cos E = \dfrac{17}{32}$	*Substitute 17 for f and 32 for d.*
$E = \cos^{-1} \dfrac{17}{32}$	*Definition of inverse*
$E \approx 57.91004874$	*Use a calculator.*

Therefore, E measures about $57.9°$.

Trigonometry can be used to find the angle of elevation or the angle of depression.

Example **④** ** SECURITY** Refer to the application at the beginning of the lesson. What should be the angle of depression of the light?

The angle of depression from the light and the angle of elevation to the light are equal in measure. To find the angle of elevation, use the tangent function.

$\tan \theta = \dfrac{20}{100}$	$\tan = \dfrac{\text{side opposite}}{\text{side adjacent}}$
$\theta = \tan^{-1} \dfrac{20}{100}$	*Definition of inverse*
$\theta \approx 11.30993247$	*Use a calculator.*

The angle of depression should be about $11.3°$.

You can use trigonometric functions and inverse relations to solve right triangles. To **solve a triangle** means to find all of the measures of its sides and angles. Usually, two measures are given. Then you can find the remaining measures.

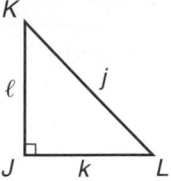
3 PRACTICE/APPLY

Error Analysis

Watch for students who confuse the roles of the trigonometric functions and the inverse trigonometric functions.

Prevent by stressing that the inverse functions are used to find the angles whereas the trigonometric functions are used to find the ratios of sides.

Additional Answer
3. Sample answer:

Study Guide and Practice Masters, pp. 87–88

Example **5** Solve each triangle described, given the triangle at the right.

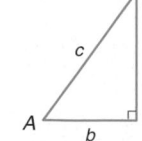

a. $A = 33°$, $b = 5.8$

Find B.

$33° + B = 90°$ *Angles A and B are complementary.*

$B = 57°$

Whenever possible, use measures given in the problem to find the unknown measures.

Find a.

$\tan A = \frac{a}{b}$

$\tan 33° = \frac{a}{5.8}$

$5.8 \tan 33° = a$

$3.766564041 \approx a$

Find c.

$\cos A = \frac{b}{c}$

$\cos 33° = \frac{5.8}{c}$

$c \cos 33° = 5.8$

$c = \frac{5.8}{\cos 33°}$

$c \approx 6.915707098$

Therefore, $B = 57°$, $a \approx 3.8$, and $c \approx 6.9$.

b. $a = 23$, $c = 45$

Find b.

$a^2 + b^2 = c^2$

$23^2 + b^2 = 45^2$

$b = \sqrt{1496}$

$b \approx 38.67815921$

Find A.

$\sin A = \frac{a}{c}$

$\sin A = \frac{23}{45}$

$A = \sin^{-1} \frac{23}{45}$

$A \approx 30.73786867$

Find B.

$30.73786867 + B \approx 90$

$B \approx 59.26213133$

Therefore, $b \approx 38.7$, $A \approx 30.7°$, and $B \approx 59.3°$.

CHECK FOR UNDERSTANDING

Communicating Mathematics

Read and study the lesson to answer each question.

1. **Tell** whether the solution to each equation is an angle measure or a linear measurement.

   a. $\tan 34° \, 15' = \frac{x}{12}$ **linear**

   b. $\tan x = 3.284$ **angle**

2. They are complementary.

2. **Describe** the relationship of the two acute angles of a right triangle.

3. **Counterexample**   You can usually solve a right triangle if you know two measures besides the right angle. Draw a right triangle and label two measures other than the right angle such that you cannot solve the triangle. **See margin.**

4. Marta; they need to find the inverse of the cosine, not $\frac{1}{\cos}$.

4. **You Decide**   Marta and Rebecca want to determine the degree measure of angle ϑ if $\cos \vartheta = 0.9876$. Marta tells Rebecca to press [2nd] [COS⁻¹] .9876 on the calculator. Rebecca disagrees. She says to press [COS] .9876 [)] [x⁻¹]. Who is correct? Explain.

Addressing Individual Needs

Logical   Logical learners will enjoy solving triangles as this requires an organized approach to piecing together given information to find missing parts. These students should be encouraged to solve a triangle in more than one way. For example, if two sides are given, the third can be found using the Pythagorean Theorem. As an alternative approach, you could also use the given sides and an inverse trigonometric function to find an angle first.

Guided Practice

Solve each equation if $0° \leq x \leq 360°$.

5. $\cos x = \dfrac{1}{2}$ **60°, 300°**

6. $\tan x = \dfrac{-\sqrt{3}}{3}$ **150°, 330°**

Evaluate each expression. Assume that all angles are in Quadrant I.

7. $\sin\left(\sin^{-1}\dfrac{\sqrt{3}}{2}\right)$ $\dfrac{\sqrt{3}}{2}$

8. $\tan\left(\cos^{-1}\dfrac{3}{5}\right)$ $\dfrac{4}{3}$

Solve each problem. Round to the nearest tenth.

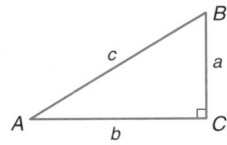

9. If $r = 7$ and $s = 10$, find R. **35.0°**

10. If $r = 12$ and $t = 20$, find S. **53.1°**

Solve each triangle described, given the triangle at the right. Round to the nearest tenth if necessary.

11. $B = 78°$, $a = 41$ **$A = 12°$, $b = 192.9$, $c = 197.2$**

12. $a = 11$, $b = 21$ **$c = 23.7$, $A = 27.6°$, $B = 62.4°$**

13. $A = 32°$, $c = 13$ **$B = 58°$, $a = 6.9$, $b = 11.0$**

14. **National Monuments** In 1906, Teddy Roosevelt designated Devils Tower National Monument in northeast Wyoming as the first national monument in the United States. The tower rises 1280 feet above the valley of the Bell Fourche River.

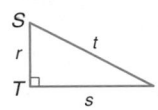

Devils Tower National Monument

a. If the shadow of the tower is 2100 feet long at a certain time, find the angle of elevation of the sun. **about 31.4°**

b. How long is the shadow when the angle of elevation of the sun is 38°? **about 1638.3 ft**

c. If a person at the top of Devils Tower sees a hiker at an angle of depression of 65°, how far is the hiker from the base of Devils Tower? **about 596.9 ft**

EXERCISES

Practice

Solve each equation if $0° \leq x \leq 360°$.

15. $\sin x = 1$ **90°**

16. $\tan x = -\sqrt{3}$ **120°, 300°**

17. $\cos x = \dfrac{\sqrt{3}}{2}$ **30°, 330°**

18. $\cos x = 0$ **90°, 270°**

19. $\sin x = -\dfrac{\sqrt{2}}{2}$ **225°, 315°**

20. $\tan x = -1$ **135°, 315°**

21. Name four angles whose sine equals $\dfrac{1}{2}$. **Sample answers: 30°, 150°, 390°, 510°**

Evaluate each expression. Assume that all angles are in Quadrant I.

22. $\cos\left(\arccos\dfrac{4}{5}\right)$ $\dfrac{4}{5}$

23. $\tan\left(\tan^{-1}\dfrac{2}{3}\right)$ $\dfrac{2}{3}$

24. $\sec\left(\cos^{-1}\dfrac{2}{5}\right)$ $\dfrac{5}{2}$

25. $\csc(\arcsin 1)$ **1**

26. $\tan\left(\cos^{-1}\dfrac{5}{13}\right)$ $\dfrac{12}{5}$

27. $\cos\left(\sin^{-1}\dfrac{2}{5}\right)$ $\dfrac{\sqrt{21}}{5}$

 www.amc.glencoe.com/self_check_quiz

Teaching Tip Triangles have been solved using a calculator. Rounding did not occur until the final answer. If a calculated value is used to find another value, the stored value in the calculator was used, *not* the rounded value. Using rounded values in calculations will result in different answers from those given.

35. about 48.8°, 48.8°, and 82.4°

37. $B = 55°$, $a = 5.6$, $c = 9.8$

38. $A = 43°$, $a = 11.7$, $c = 17.1$

39. $c = 5.7$, $A = 42.1°$, $B = 47.9°$

40. $a = 8.7$, $A = 67.1°$, $B = 22.9°$

41. $B = 38.5°$, $b = 10.6$, $c = 17.0$

Applications and Problem Solving

Real World Application

45a–c. See margin.

Solve each problem. Round to the nearest tenth.

28. If $n = 15$ and $m = 9$, find N. **59.0°**

29. If $m = 8$ and $p = 14$, find M. **34.8°**

30. If $n = 22$ and $p = 30$, find M. **42.8°**

31. If $m = 14.3$ and $n = 18.8$, find N. **52.7°**

32. If $p = 17.1$ and $m = 7.2$, find N. **65.1°**

33. If $m = 32.5$ and $p = 54.7$, find M. **36.5°**

34. Geometry If the legs of a right triangle are 24 centimeters and 18 centimeters long, find the measures of the acute angles. **about 36.9° and 53.1°**

35. Geometry The base of an isosceles triangle is 14 inches long. Its height is 8 inches. Find the measure of each angle of the triangle.

Solve each triangle described, given the triangle at the right. Round to the nearest tenth, if necessary. **36. $b = 21.4$, $A = 44.4°$, $B = 45.6°$**

C ▶ **36.** $a = 21$, $c = 30$

37. $A = 35°$, $b = 8$

38. $B = 47°$, $b = 12.5$

39. $a = 3.8$, $b = 4.2$

40. $c = 9.5$, $b = 3.7$

41. $a = 13.3$, $A = 51.5°$

42. $B = 33°$, $c = 15.2$ $A = 57°$, $a = 12.7$, $b = 8.3$

43. $c = 9.8$, $A = 14°$ $B = 76°$, $a = 2.4$, $b = 9.5$

44. Railways The steepest railway in the world is the Katoomba Scenic Railway in Australia. The passenger car is pulled up the mountain by twin steel cables. It travels along the track 1020 feet to obtain a change in altitude of 647 feet.

a. Find the angle of elevation of the railway. **about 39.4°**

b. How far does the car travel in a horizontal direction? **about 788.5 ft**

45. Critical Thinking Explain why each expression is impossible.

a. $\sin^{-1} 2.4567$

b. $\sec^{-1} 0.5239$

c. $\cos^{-1}(-3.4728)$

46. Basketball The rim of a basketball hoop is 10 feet above the ground. The free-throw line is 15 feet from the basket rim. If the eyes of a basketball player are 6 feet above the ground, what is the angle of elevation of the player's line of sight when shooting a free throw to the rim of the basket? **about 14.9°**

47. Road Safety Several years ago, a section on I-75 near Cincinnati, Ohio, had a rise of 8 meters per 100 meters of horizontal distance. However, there were numerous accidents involving large trucks on this section of highway. Civil engineers decided to reconstruct the highway so that there is only a rise of 5 meters per 100 meters of horizontal distance.

a. Find the original angle of elevation. **about 4.6°**

b. Find the new angle of elevation. **about 2.9°**

48. **Air Travel**   At a local airport, a light that produces a powerful white-green beam is placed on the top of a 45-foot tower. If the tower is at one end of the runway, find the angle of depression needed so that the light extends to the end of the 2200-foot runway.   **about 1.2°**

49. **Civil Engineering**   Highway curves are usually banked or tilted inward so that cars can negotiate the curve more safely. The proper banking angle θ for a car making a turn of radius r feet at a velocity of v feet per second is given by the equation is $\tan \theta = \dfrac{v^2}{gr}$. In this equation, g is the acceleration due to gravity or 32 feet per second squared. An engineer is designing a curve with a radius of 1200 feet. If the speed limit on the curve will be 65 miles per hour, at what angle should the curve be banked? (*Hint*: Change 65 miles per hour to feet per second.)   **about 13.3°**

50. **Physics**   According to Snell's Law, $\dfrac{\sin \theta_i}{\sin \theta_r} = n$, where θ_i is the angle of incidence, θ_r is the angle of refraction, and n is the index of refraction. The index of refraction for a diamond is 2.42. If a beam of light strikes a diamond at an angle of incidence of 60°, find the angle of refraction.   **about 21.0°**

51. **Critical Thinking**   Solve the triangle.
(*Hint*: Draw the altitude from Y.)
$y \approx 36.5$, $Z \approx 19.5°$, $Y \approx 130.5°$

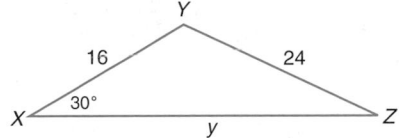

Mixed Review

52. **Aviation**   A traffic helicopter is flying 1000 feet above the downtown area. To the right, the pilot sees the baseball stadium at an angle of depression of 63°. To the left, the pilot sees the football stadium at an angle of depression of 18°. Find the distance between the two stadiums. (*Lesson 5-4*)   **about 3587.2 ft**

53. Find the six trigonometric ratios for $\angle F$.
(*Lesson 5-2*)

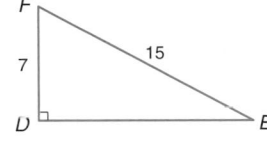

$\sin F = \dfrac{4\sqrt{11}}{15}$, $\cos F = \dfrac{7}{15}$, $\tan F = \dfrac{4\sqrt{11}}{7}$,

$\csc F = \dfrac{15\sqrt{11}}{44}$, $\sec F = \dfrac{15}{7}$, $\cot F = \dfrac{7\sqrt{11}}{44}$

54. Approximate the real zeros of the function $f(x) = 3x^3 - 16x^2 + 12x + 6$ to the nearest tenth. (*Lesson 4-5*)   **−0.3, 1.4, 4.3**

55. Determine whether the graph of $y^3 - x^2 = 2$ is symmetric with respect to the x-axis, y-axis, the graph of $y = x$, the graph of $y = -x$, or none of these. (*Lesson 3-1*)   **y-axis**

56. Use a reflection matrix to find the coordinates of the vertices of a pentagon reflected over the y-axis if the coordinates of the vertices of the pentagon are $(-5, -3)$, $(-5, 4)$, $(-3, 6)$, $(-1, 3)$, and $(-2, -2)$. (*Lesson 2-4*)
(5, −3), (5, 4), (3, 6), (1, 3), (2, −2)

57. $\begin{bmatrix} 2 & -1 & 0 \\ 3 & -1 & 1 \\ 2 & 8 & -5 \end{bmatrix}$ 57. Find the sum of the matrices $\begin{bmatrix} 4 & -3 & 2 \\ 8 & -2 & 0 \\ 9 & 6 & -3 \end{bmatrix}$ and $\begin{bmatrix} -2 & 2 & -2 \\ -5 & 1 & 1 \\ -7 & 2 & -2 \end{bmatrix}$. (*Lesson 2-3*)

Extra Practice   See p. A35.

Extra Credit

Suppose that the perimeter of a right triangle is 36 units. The legs of the triangle are 6 and 3 units less than the hypotenuse, respectively. Solve the triangle.

$a = 9$; $b = 12$; $c = 15$;
$A \approx 36.9°$; $B \approx 53.1°$

58. Write a linear equation of best fit for a set of data using the ordered pairs (1880, 42.5) and (1950, 22.2). *(Lesson 1-6)* $y = -0.29x + 587.7$

59. $y = -\frac{2}{5}x + 2$; $-\frac{2}{5}, 2$

59. Write the equation $2x + 5y - 10 = 0$ in slope-intercept form. Then name the slope and y-intercept. *(Lesson 1-5)*

60. **SAT/ACT Practice** The Natural Snack Company mixes a pounds of peanuts that cost b cents per pound with c pounds of rice crackers that cost d cents per pound to make Oriental Peanut Mix. What should the price in cents for a pound of Oriental Peanut Mix be if the company wants to make a profit of 10¢ per pound? **A**

A $\frac{ab + cd}{a + c} + 10$ **B** $\frac{b + d}{a + c} + 10$ **C** $\frac{ab + cd}{a + c} + 0.10$

D $\frac{b + d}{a + c} + 0.10$ **E** $\frac{b + d + 10}{a + c}$

CAREER CHOICES

 Architecture

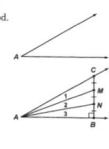

Are you creative and concerned with accuracy and detail? Do you like to draw and design new things? You may want to consider a career in architecture.

An architect plans, designs, and oversees the construction of all types of buildings, a job that is very complex. An architect needs to stay current with new construction methods and design. An architect must also be knowledgeable about engineering principles.

As an architect, you would work closely with others such as consulting engineers and building contractors. Your projects could include large structures such as shopping malls or small structures such as single-family houses. There are also specialty fields in architecture such as interior design, landscape architecture, and products and material design.

CAREER OVERVIEW

Degree Preferred:
bachelor's degree in architecture

Related Courses:
mathematics, physics, art, computer science

Outlook:
number of jobs expected to increase as fast as the average through the year 2006

Average Sale Prices of New Single-Family Houses by Region

Source: *The New York Times Almanac*

inter NET CONNECTION For more information on careers in architecture, visit: **www.amc.glencoe.com**

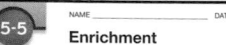

5-5

NAME _____ DATE _____ PERIOD _____

Enrichment

Disproving Angle Trisection

Most geometry texts state that it is impossible to trisect an arbitrary angle using only a compass and straightedge. This fact has been known since ancient times, but since it is usually stated without proof, some geometry students do not believe it. If the students set out to find a method for trisecting angles, they will probably try the following method. It is based on the familiar construction which allows a segment to be divided into any desired number of congruent segments. You can use inverse trigonometric functions to show that application of the method to the trisection of angles is not valid.

Given: $\angle A$
Claim: $\angle A$ can be trisected using the following method.
Method: Choose point C on one ray of $\angle A$. Through C construct a perpendicular to the other ray, intersecting it at B. Construct M and N, the points that divide $\overline{CB}$ into three congruent segments. Draw $\overline{AM}$ and $\overline{AN}$, which trisect $\angle CAB$ into the congruent angles $\angle 1$, $\angle 2$, and $\angle 3$.

The proposed method has been used to construct the figure below. $CM = MN = NB = 1$. $AB = 5$. Follow the instructions to show that the three angles $\angle 1$, $\angle 2$, and $\angle 3$, are not congruent. Find angle measures to the nearest tenth of a degree.

1. Express $m\angle 3$ as an inverse function.
   $m\angle 3 = \arctan \frac{1}{5}$
2. Find the measure of $\angle 3$.
   $m\angle 3 \approx 11.3°$
3. Write $m\angle MAB$ as an inverse function. $m\angle MAB = \arctan \frac{2}{5}$
4. Find the measure of $\angle MAB$. $m\angle MAB \approx 21.8°$
5. Find the measure of $\angle 2$. $m\angle 2 \approx 10.5°$
6. Find $m\angle CAB$ and use it to find $m \angle 1$. $m\angle 1 \approx 9.2°$
7. Explain why the proposed method for trisecting an angle fails.
   The tangent function is not linear. The ratio of the measures of two angles is not equal to the ratio of the tangents of the angles.

© Glencoe/McGraw-Hill   T53   *Advanced Mathematical Concepts*

CAREER CHOICES ARCHITECTURE

Math Connection
- Explain the trends in the data for the price of houses in the Northeast and Midwest. **The graph shows that house prices have risen in both regions but that the increase in the Northeast has been more dramatic.**
- Use a calculator to find a good model for each data set. Use 1, 2, 3, and so on for the years 1965, 1970, 1975, and so forth. **Answers may vary. Sample:**
  Northeast: $y = 4400x^2 - 507x + 15{,}742$; **Midwest:** $y = 23{,}757x - 15{,}857$;
  South: $y = 22{,}057x - 15{,}943$; **West:** $y = 1656x^2 + 15{,}849x - 4229$

5-6

The Law of Sines

OBJECTIVES
* Solve triangles by using the Law of Sines if the measures of two angles and a side are given.
* Find the area of a triangle if the measures of two sides and the included angle or the measures of two angles and a side are given.

BASEBALL A baseball fan is sitting directly behind home plate in the last row of the upper deck of Comiskey Park in Chicago. The angle of depression to home plate is 29° 54', and the angle of depression to the pitcher's mound is 24° 12'. In major league baseball, the distance between home plate and the pitcher's mound is 60.5 feet. How far is the fan from home plate? *This problem will be solved in Example 2.*

The **Law of Sines** can be used to solve triangles that are not right triangles. Consider △*ABC* inscribed in circle *O* with diameter $\overline{DB}$. Let 2*r* be the measure of the diameter. Draw $\overline{AD}$. Then ∠*D* ≅ ∠*C* since they intercept the same arc. So, sin *D* = sin *C*. ∠*DAB* is inscribed in a semicircle, so it is a right angle. sin $D = \frac{c}{2r}$. Thus, since sin *D* = sin *C*, it follows that sin $C = \frac{c}{2r}$ or $\frac{c}{\sin C} = 2r$.

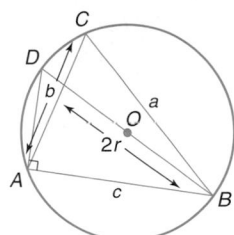

Similarly, by drawing diameters through *A* and *C*, $\frac{b}{\sin B} = 2r$ and $\frac{a}{\sin A} = 2r$.

Since each rational expression equals 2r, the following is true.

$$\frac{a}{\sin A} = \frac{b}{\sin B} = \frac{c}{\sin C}$$

These equations state that the ratio of the length of any side of a triangle to the sine of the angle opposite that side is a constant for a given triangle. These equations are collectively called the Law of Sines.

Law of Sines	Let △*ABC* be any triangle with *a*, *b*, and *c* representing the measures of the sides opposite the angles with measures *A*, *B*, and *C*, respectively. Then, the following is true. $$\frac{a}{\sin A} = \frac{b}{\sin B} = \frac{c}{\sin C}$$

From geometry, you know that a unique triangle can be formed if you know the measures of two angles and the included side (ASA) or the measures of two angles and the non-included side (AAS). Therefore, there is one unique solution when you use the Law of Sines to solve a triangle given the measures of two angles and one side. *In Lesson 5-7, you will learn to use the Law of Sines when the measures of two sides and a nonincluded angle are given.*

Resource Manager

Reproducible Masters
* *Study Guide and Practice,* pp. 89–90
* *Enrichment,* p. 54
* *Assessment and Evaluation,* p.122
* *Problem Solving and Applications,* p. 9

Transparencies
* 5-Minute Check, 5-6
* Teaching, 5-6

---

1 FOCUS

5-Minute Check
Lesson 5-5

Solve each equation if 0° ≤ x ≤ 360°.

1. tan $x = \frac{\sqrt{3}}{3}$ **30°, 210°**

2. cos $x = -\frac{1}{2}$ **120°, 240°**

Evaluate each expression. Assume that all angles are in Quadrant I.

3. $\cos\left(\arccos\frac{2}{5}\right)$ $\frac{2}{5}$

4. $\cos\left(\arcsin\frac{5}{6}\right)$ $\frac{\sqrt{11}}{6}$

5. If *q* = 38 and *r* = 22, find *S*.

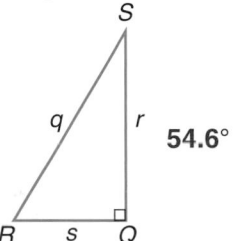

Solve each triangle described by using the triangle shown. Round to the nearest tenth, if necessary.

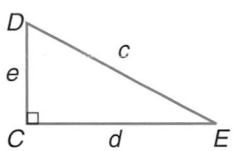

6. *D* = 48°, *e* = 25 **E = 42°, d = 27.8, c = 37.4**

7. *c* = 65, *d* = 47 **e = 44.9, D = 46.3°, E = 43.7°**

Motivating the Lesson
Real-World Application Ask students to imagine themselves in the following scenario. *You are a fan in the stands at a baseball game, sitting behind home plate. Describe your perception of where each base is located. Is it possible for you to catch a foul ball? Why or why not?*

Teaching Tip Before Example 1, verify the Law of Sines for a 45°–45°–90° triangle with sides measuring 1, 1, and $\sqrt{2}$.

In-Class Examples

Example 1
Solve $\triangle LMN$ if $L = 29°$, $M = 112°$, and $\ell = 22$.

$N = 39°$, $m \approx 42.1$, $n \approx 28.6$

Example 2
Ballooning A person in a hot-air balloon observes that the angle of depression to a building on the ground is 65.8°. After ascending vertically 500 feet, the person now observes that the angle of depression is 70.2°. How far is the balloonist now from the building? **about 2671.6 ft**

Examples

① Solve $\triangle ABC$ if $A = 33°$, $B = 105°$, and $b = 37.9$.

First, find the measure of $\angle C$.
$C = 180° - (33° + 105°)$ or $42°$

Use the Law of Sines to find a and c.

$$\frac{a}{\sin A} = \frac{b}{\sin B} \qquad\qquad \frac{c}{\sin C} = \frac{b}{\sin B}$$

$$\frac{a}{\sin 33°} = \frac{37.9}{\sin 105°} \qquad\qquad \frac{c}{\sin 42°} = \frac{37.9}{\sin 105°}$$

$$a = \frac{37.9 \sin 33°}{\sin 105°} \qquad\qquad c = \frac{37.9 \sin 42°}{\sin 105°}$$

$a \approx 21.36998397$ *Use a calculator.* $c \approx 26.25465568$ *Use a calculator.*

Therefore, $C = 42°$, $a \approx 21.4$, and $c \approx 26.3$.

② BASEBALL Refer to the application at the beginning of the lesson. How far is the fan from home plate?

Make a diagram for the problem. Remember that the angle of elevation is congruent to the angle of depression, because they are alternate interior angles.

First, find θ.

$\theta = 29° \, 54' - 24° \, 12'$ or $5° \, 42'$

Use the Law of Sines to find d.

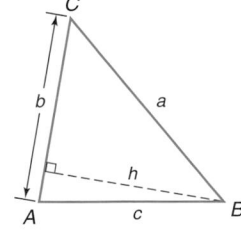

$$\frac{d}{\sin 24° \, 12'} = \frac{60.5}{\sin 5° \, 42'}$$

$$d = \frac{60.5 \sin 24° \, 12'}{\sin 5° \, 42'}$$

$d \approx 249.7020342$ *Use a calculator.*

The fan is about 249.7 feet from home plate.

The area of any triangle can be expressed in terms of two sides of a triangle and the measure of the included angle. Suppose you know the measures of $\overline{AC}$ and $\overline{AB}$ and the measure of the included $\angle A$ in $\triangle ABC$. Let K represent the measure of the area of $\triangle ABC$, and let h represent the measure of the altitude from B. Then $K = \frac{1}{2} bh$. But, $\sin A = \frac{h}{c}$ or $h = c \sin A$. If you substitute $c \sin A$ for h, the result is the following formula.

Use K for area instead of A to avoid confusion with angle A.

$$K = \frac{1}{2} bc \sin A$$

If you drew altitudes from A and C, you could also develop two similar formulas.

Area of Triangles	Let $\triangle ABC$ be any triangle with a, b, and c representing the measures of the sides opposite the angles with measurements A, B, and C, respectively. Then the area K can be determined using one of the following formulas. $\qquad K = \dfrac{1}{2}bc \sin A \qquad\qquad\qquad K = \dfrac{1}{2}ac \sin B$ $\qquad\qquad\qquad K = \dfrac{1}{2}ab \sin C$

Example ❸ Find the area of $\triangle ABC$ if $a = 4.7$, $c = 12.4$, and $B = 47°\,20'$.

$K = \dfrac{1}{2}ac \sin B$

$K = \dfrac{1}{2}(4.7)(12.4) \sin 47°\,20'$

$K \approx 21.42690449$ *Use a calculator.*

The area of $\triangle ABC$ is about 21.4 square units.

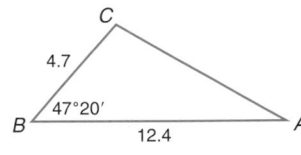

You can also find the area of a triangle if you know the measures of one side and two angles of the triangle. By the Law of Sines, $\dfrac{b}{\sin B} = \dfrac{c}{\sin C}$ or $b = \dfrac{c \sin B}{\sin C}$. If you substitute $\dfrac{c \sin B}{\sin C}$ for b in $K = \dfrac{1}{2}bc \sin A$, the result is $K = \dfrac{1}{2}c^2 \dfrac{\sin A \sin B}{\sin C}$. Two similar formulas can be developed.

Area of Triangles	Let $\triangle ABC$ be any triangle with a, b, and c representing the measures of the sides opposite the angles with measurements A, B, and C respectively. Then the area K can be determined using one of the following formulas. $\qquad K = \dfrac{1}{2}a^2 \dfrac{\sin B \sin C}{\sin A} \qquad\qquad K = \dfrac{1}{2}b^2 \dfrac{\sin A \sin C}{\sin B}$ $\qquad\qquad\qquad K = \dfrac{1}{2}c^2 \dfrac{\sin A \sin B}{\sin C}$

Example ❹ Find the area of $\triangle DEF$ if $d = 13.9$, $D = 34.4°$, and $E = 14.8°$.

First find the measure of $\angle F$.
$F = 180° - (34.4° + 14.8°)$ or $130.8°$

Then, find the area of the triangle.

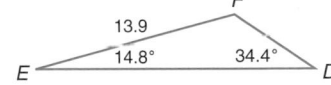

$K = \dfrac{1}{2}d^2 \dfrac{\sin E \sin F}{\sin D}$

$K = \dfrac{1}{2}(13.9)^2 \dfrac{\sin 14.8° \sin 130.8°}{\sin 34.4°}$

$K \approx 33.06497958$ *Use a calculator.*

The area of $\triangle DEF$ is about 33.1 square units.

Addressing Individual Needs

Kinesthetic Place three objects on the floor so that they form a large triangle. Use a ruler and a protractor to determine a side and two angles of the triangle. Solve the triangle using the Law of Sines. Measure the other sides and angle to verify the solution. Repeat the activity by forming a different triangle and measuring an angle and two sides.

Error Analysis

Watch for students who make errors due to mismatching angles and sides.

Prevent by stressing the importance of carefully labeling the diagrams.

Teaching Tip Triangles have been solved using a calculator. Rounding did not occur until the final answer. If a calculated value is used to find another value, the stored value in the calculator was used, *not* the rounded value. Using rounded values in calculations will result in different answers from those given.

Assignment Guide
Core: 11–35 odd, 36–42
Enriched: 12–26 even, 28–42

Additional Answer

1.

$$\frac{x}{\sin 30°} \overset{?}{=} \frac{x\sqrt{3}}{\sin 60°} \overset{?}{=} \frac{2x}{\sin 90°}$$

$$\frac{x}{\frac{1}{2}} \overset{?}{=} \frac{x\sqrt{3}}{\frac{\sqrt{3}}{2}} \overset{?}{=} \frac{2x}{1}$$

$$2x = 2x = 2x$$

Study Guide and Practice Masters, pp. 89–90

Communicating Mathematics

Read and study the lesson to answer each question. **1-2. See margin.**

1. **Show** that the Law of Sines is true for a 30°-60° right triangle.

2. **Draw and label** a triangle that has a unique solution and can be solved using the Law of Sines.

3. **Write** a formula for the area of parallelogram *WXYZ* in terms of *a*, *b*, and *X*.
   $K = ab \sin X$

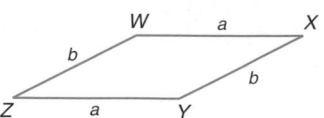

4. **You Decide** Roderick says that triangle *MNP* has a unique solution if *M*, *N*, and *m* are known. Jane disagrees. She says that a triangle has a unique solution if *M*, *N*, and *p* are known. Who is correct? Explain. **See margin.**

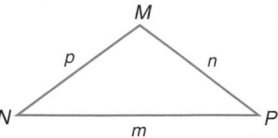

Guided Practice

5. $C = 81°$, $a = 9.1$, $b = 12.1$

6. $C = 96.8°$, $b = 15.5$, $c = 18.6$

Solve each triangle. Round to the nearest tenth.

5. $A = 40°$, $B = 59°$, $c = 14$
6. $a = 8.6$, $A = 27.3°$, $B = 55.9°$
7. If $B = 17° 55'$, $C = 98° 15'$, and $a = 17$, find c. **about 18.7**

Find the area of each triangle. Round to the nearest tenth.

8. $A = 78°$, $b = 14$, $c = 12$ **82.2 units²** 9. $A = 22°$, $B = 105°$, $b = 14$ **30.4 units²**

10. **Baseball** Refer to the application at the beginning of the lesson. How far is the baseball fan from the pitcher's mound? **about 303.7 ft**

Practice

12. $A = 30°$, $a = 19.6$, $b = 38.6$

13. $C = 120°$, $a = 8.8$, $c = 18.1$

14. $C = 65°$, $a = 12$, $b = 10.1$

15. $A = 93.9°$, $b = 3.4$, $c = 7.2$

16. $B = 76°$, $a = 13.5$, $b = 20.7$

22. 13,533.9 units²

25. about 234.8 cm²

Solve each triangle. Round to the nearest tenth. **11. $B = 70°$, $b = 29.2$, $c = 29.2$**

11. $A = 40°$, $C = 70°$, $a = 20$ **A**
12. $B = 100°$, $C = 50°$, $c = 30$
13. $b = 12$, $A = 25°$, $B = 35°$
14. $A = 65°$, $B = 50°$, $c = 12$
15. $a = 8.2$, $B = 24.8°$, $C = 61.3°$ **B**
16. $c = 19.3$, $A = 39° 15'$, $C = 64° 45'$
17. If $A = 37° 20'$, $B = 51° 30'$, and $c = 125$, find b. **about 97.8**
18. What is a if $b = 11$, $B = 29° 34'$, and $C = 23° 48'$? **about 17.9**

Find the area of each triangle. Round to the nearest tenth.

19. $A = 28°$, $b = 14$, $c = 9$ **29.6 units²** 20. $a = 5$, $B = 37°$, $C = 84°$ **8.7 units²**
21. $A = 15°$, $B = 113°$, $b = 7$ **5.4 units²** 22. $b = 146.2$, $c = 209.3$, $A = 62.2°$
23. $B = 42.8°$, $a = 12.7$, $c = 5.8$ **25.0 units²** 24. $a = 19.2$, $A = 53.8°$, $C = 65.4°$ **181.3 units²**

25. **Geometry** The adjacent sides of a parallelogram measure 14 centimeters and 20 centimeters, and one angle measures 57°. Find the area of the parallelogram. **C**

26. **Geometry** A regular pentagon is inscribed in a circle whose radius measures 9 inches. Find the area of the pentagon. **about 192.6 in²**

27. **Geometry** A regular octagon is inscribed in a circle with radius of 5 feet. Find the area of the octagon. **about 70.7 ft²**

 www.amc.glencoe.com/self_check_quiz

Additional Answers

2. Sample answer:

4. Both; if the measures of two angles and a non-included side are known or if the measures of two angles and the included side are known, the triangle is unique.

28. Landscaping A landscaper wants to plant begonias along the edges of a triangular plot of land in Winton Woods Park. Two of the angles of the triangle measure 95° and 40°. The side between these two angles is 80 feet long.

    a. Find the measure of the third angle.   **45°**   **b. about 112.7 ft and 72.7 ft**

    b. Find the length of the other two sides of the triangle.

    c. What is the perimeter of this triangular plot of land?   **about 265.4 ft**

29. Critical Thinking For △MNP and △RST, ∠M ≅ ∠R, ∠N ≅ ∠S, and ∠P ≅ ∠T. Use the Law of Sines to show △MNP ~ △RST.   **See margin.**

30. Architecture The center of the Pentagon in Arlington, Virginia, is a courtyard in the shape of a regular pentagon. The pentagon could be inscribed in a circle with radius of 300 feet. Find the area of the courtyard. **about 213,987.7 ft²**

300 ft

31. Ballooning A hot air balloon is flying above Groveburg. To the left side of the balloon, the balloonist measures the angle of depression to the Groveburg soccer fields to be 20° 15′. To the right side of the balloon, the balloonist measures the angle of depression to the high school football field to be 62° 30′. The distance between the two athletic complexes is 4 miles.

    a. Find the distance from the balloon to the soccer fields.   **about 3.6 mi**

    b. What is the distance from the balloon to the football field?   **about 1.4 mi**

32. Cable Cars The Duquesne Incline is a cable car in Pittsburgh, Pennsylvania, which transports passengers up and down a mountain. The track used by the cable car has an angle of elevation of 30°. The angle of elevation to the top of the track from a point that is horizontally 100 feet from the base of the track is about 26.8°. Find the length of the track.   **about 807.7 ft**

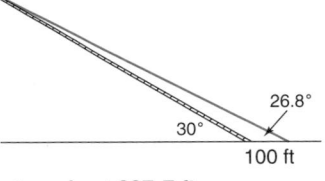

26.8°
30°
100 ft

33. Air Travel In order to avoid a storm, a pilot starts the flight 13° off course. After flying 80 miles in this direction, the pilot turns the plane to head toward the destination. The angle formed by the course of the plane during the first part of the flight and the course during the second part of the flight is 160°.

    a. What is the distance of the flight?   **about 227.7 mi**

    b. Find the distance of a direct flight to the destination.   **about 224.5 mi**

34. Architecture An architect is designing an overhang above a sliding glass door. During the heat of the summer, the architect wants the overhang to prevent the rays of the sun from striking the glass at noon. The overhang has an angle of depression of 55° and starts 13 feet above the ground. If the angle of elevation of the sun during this time is 63°, how long should the architect make the overhang?   **about 6.7 ft**

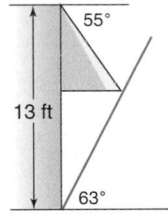

55°
13 ft
63°

Lesson 5-6 The Law of Sines   **317**

Extra Credit

Have students derive the Law of Sines from the following figure.

B
c
a
h
A
b
C

Since $\sin A = \dfrac{h}{c}$ and $\sin C = \dfrac{h}{a}$, solving each equation for h and setting the answers equal to each other gives $c \sin A = a \sin C$. Divide each side by $\sin A \sin C$ to get $\dfrac{c}{\sin C} = \dfrac{a}{\sin A}$. Since vertices B and C can be switched, $\dfrac{a}{\sin A} = \dfrac{b}{\sin B}$.

Therefore, $\dfrac{a}{\sin A} = \dfrac{b}{\sin B} = \dfrac{c}{\sin C}$.

Additional Answer

29. Applying the Law of Sines,

$$\frac{m}{\sin M} = \frac{n}{\sin N} \text{ and } \frac{r}{\sin R} = \frac{s}{\sin S}.$$

Thus, $\sin M = \dfrac{m \sin N}{n}$ and

$\sin R = \dfrac{r \sin S}{s}$. Since

$\angle M \cong \angle R$, $\sin M = \sin R$ and

$\dfrac{m \sin N}{n} = \dfrac{r \sin S}{s}$. However,

$\angle N \cong \angle S$ and $\sin N = \sin S$,

so $\dfrac{m}{n} = \dfrac{r}{s}$ and $\dfrac{m}{r} = \dfrac{n}{s}$. Similar

proportions can be derived for p and t. Therefore, $\triangle MNP \cong \triangle RST$.

Additional Answers (p. 318)

35a. $\dfrac{a}{\sin A} = \dfrac{b}{\sin B}$

$\dfrac{a}{b} = \dfrac{\sin A}{\sin B}$

35b. $\dfrac{a}{\sin A} = \dfrac{c}{\sin C}$

$\dfrac{a}{c} = \dfrac{\sin A}{\sin C}$

$\dfrac{a}{c} - 1 = \dfrac{\sin A}{\sin C} - 1$

$\dfrac{a}{c} - \dfrac{c}{c} = \dfrac{\sin A}{\sin C} - \dfrac{\sin C}{\sin C}$

$\dfrac{a - c}{c} = \dfrac{\sin A - \sin C}{\sin C}$

35c. From Exercise 34b,

$\dfrac{a - c}{c} = \dfrac{\sin A - \sin C}{\sin C}$

or $\dfrac{\sin A - \sin C}{a - c} = \dfrac{\sin C}{c}$.

$\dfrac{a}{\sin A} = \dfrac{c}{\sin C}$

$\dfrac{a}{c} = \dfrac{\sin A}{\sin C}$

$\dfrac{a}{c} + 1 = \dfrac{\sin A}{\sin C} + 1$

$\dfrac{a}{c} + \dfrac{c}{c} = \dfrac{\sin A}{\sin C} + \dfrac{\sin C}{\sin C}$

$\dfrac{a + c}{c} = \dfrac{\sin A + \sin C}{\sin C}$

$\dfrac{\sin C}{c} = \dfrac{\sin A + \sin C}{a + c}$

Therefore,

$\dfrac{\sin A - \sin C}{a - c} = \dfrac{\sin A + \sin C}{a + c}$

or $\dfrac{a + c}{a - c} = \dfrac{\sin A + \sin C}{\sin A - \sin C}.$

Lesson 5-6 **317**

4 ASSESS

Open-Ended Assessment

Modeling Give students a triangle drawn on grid paper with 2 angles and a side labeled. Have them approximate the remaining parts of the triangle as well as the area by using the grid. Then have them solve the triangle using the Law of Sines. They can also calculate the area.

Chapter 5, Quiz C
(Lessons 5-5 and 5-6)
is available in the *Assessment and Evaluation Masters*, p. 122.

Additional Answers
For Exercises 35a–35c, see page 317.

35d.
$$\frac{a}{\sin A} = \frac{b}{\sin B}$$
$$\frac{a}{b} = \frac{\sin A}{\sin B}$$
$$\frac{a}{b} + 1 = \frac{\sin A}{\sin B} + 1$$
$$\frac{a}{b} + \frac{b}{b} = \frac{\sin A}{\sin B} + \frac{\sin B}{\sin B}$$
$$\frac{a + b}{b} = \frac{\sin A + \sin B}{\sin B}$$
$$\frac{b}{a + b} = \frac{\sin B}{\sin A + \sin B}$$

41.

$-6 \le 3x - y \le 12$

Enrichment Masters, p. 54

© Glencoe/McGraw-Hill   T54   *Advanced Mathematical Concepts*

---

35. Critical Thinking Use the Law of Sines to show that each statement is true for any $\triangle ABC$. **a–d. See margin.**

a. $\dfrac{a}{b} = \dfrac{\sin A}{\sin B}$

b. $\dfrac{a - c}{c} = \dfrac{\sin A - \sin C}{\sin C}$

c. $\dfrac{a + c}{a - c} = \dfrac{\sin A + \sin C}{\sin A - \sin C}$

d. $\dfrac{b}{a + b} = \dfrac{\sin B}{\sin A + \sin B}$

Mixed Review

37. $\cos \theta = \dfrac{\sqrt{35}}{6}$,

$\tan \theta = \dfrac{-\sqrt{35}}{35}$,

$\csc \theta = -6$,

$\sec \theta = \dfrac{6\sqrt{35}}{35}$,

$\cot \theta = -\sqrt{35}$

36. Meteorology If raindrops are falling toward Earth at a speed of 45 miles per hour and a horizontal wind is blowing at a speed of 20 miles per hour, at what angle do the drops hit the ground? *(Lesson 5-5)* **about 66.0°**

37. Suppose θ is an angle in standard position whose terminal side lies in Quadrant IV. If $\sin \theta = -\dfrac{1}{6}$, find the values of the remaining five trigonometric functions for θ. *(Lesson 5-3)*

38. Identify all angles that are coterminal with an 83° angle. *(Lesson 5-1)* **$83° + 360k°$**

39. Business A company is planning to buy new carts to store merchandise. The owner believes they need at least 2 standard carts and at least 4 deluxe carts. The company can afford to purchase a maximum of 15 carts at this time; however, the supplier has only 8 standard carts and 11 deluxe carts in stock. Standard carts can hold up to 100 pounds of merchandise, and deluxe carts can hold up to 250 pounds of merchandise. How many standard carts and deluxe carts should be purchased to maximize the amount of merchandise that can be stored? *(Lesson 2-7)* **4 standard carts, 11 deluxe carts**

40. Solve the system of equations algebraically. *(Lesson 2-2)*
$4x + y + 2z = 0$
$3x + 4y - 2z = 20$
$-2x + 5y + 3z = 14$ **$(0, 4, -2)$**

41. Graph $-6 \le 3x - y \le 12$. *(Lesson 1-8)* **See margin.**

42. SAT Practice Eight cubes, each with an edge of length one inch, are positioned together to create a large cube. What is the difference in the surface area of the large cube and the sum of the surface areas of the small cubes? **A**

A 24 in^2

B 16 in^2

C 12 in^2

D 8 in^2

E 0 in^2

ANGLES

When someone uses the word "angle", what images does that conjure up in your mind? An angle seems like a simple figure, but historically mathematicians, and even philosophers, have engaged in trying to describe or define an angle. This textbook says, "an angle may be generated by the rotation of two rays that share a fixed endpoint known as the vertex." Let's look at various ideas about angles throughout history.

Autumn Borts

Early Evidence Babylonians (4000–3000 B.C.) were some of the first peoples to leave samples of their use of geometry in the form of inscribed clay tablets.

The first written mathematical work containing definitions for angles was **Euclid's** *The Elements*. Little is known about the life of **Euclid** (about 300 B.C.), but his thirteen-volume work, *The Elements,* has strongly influenced the teaching of geometry for over 2000 years. The first copy of *The Elements* was printed by modern methods in 1482 and has since been edited and translated into over 1000 editions. In *Book I* of *The Elements,* Euclid presents the definitions for various types of angles.

Euclid's definition of a plane angle differed from an earlier Greek idea that an angle was a deflection or a breaking of lines.

Greek mathematicians were not the only scholars interested in angles. **Aristotle** (384–322 B.C.) had devised three categories in which to place mathematical concepts—a quantity, a quality, or a relation. Greek philosophers argued as to which category an angle belonged. **Proclus** (410–485) felt that an angle was a combination of the three, saying "it needs the quantity involved in magnitude, thereby becoming susceptible of equality, inequality, and the like; it needs the quality given it by its form; and lastly, the relation subsisting between the lines or the planes bounding it."

The Renaissance In 1634, Pierre Herigone first used "<" as a symbol for an angle in his book *Cursus Mathematicus.* This symbol was already being used for "less than," so, in 1657, **William Oughtred** used the symbol "∠" in his book *Trigonometria.*

Modern Era Various symbols for angle, including $\wedge$, $\gtrdot$, $\widehat{ab}$, and $\widehat{ABC}$, were used during the 1700s and 1800s. In 1923, the National Committee on Mathematical Requirements recommended that "∠" be used as a standard symbol in the U.S.

Today, artists like **Autumn Borts** use angles in their creation of Native American pottery. Ms. Borts adorns water jars with carved motifs of both traditional and contemporary designs. She is carrying on the Santa Clara style of pottery and has been influenced by her mother, grandmother, and great grandmother.

1. See students' work; the sum is greater than 180°.

ACTIVITIES

1. In a previous course, you have probably drawn triangles in a plane and measured the interior angles to find the angle sum of the triangles. Triangles can also be constructed on a sphere. Get a globe. Use tape and string to form at least three different triangles. Measure the interior angles of the triangles. What appears to be true about the sum of the angles?

2. Research Euclid's famous work, *The Elements.* Find and list any postulates he wrote about angles. **See students' work.**

3. **interNET CONNECTION** Find out more about the personalities referenced in this article and others who contributed to the history of angles. Visit **www.amc.glencoe.com**

This feature provides students with an opportunity to see how the mathematics that they are studying was developed. It also gives students a glimpse of how mathematical discoveries are made.

More About the Mathematicians

The Babylonians inscribed hundreds of thousands of tablets in Mesopotamia. Scholars in the 1800s decoded the Babylonian language on the tablets and learned that the Babylonians knew that an angle inscribed in a semicircle is a right angle, knew the Pythagorean Theorem, and, had an estimate for the value of π.

Euclid was a teacher in Alexandria. He probably did not discover most of the information contained in the *Elements.* Rather, he gathered and organized all of the geometry known in the West up until his time.

Euclid's Definitions of various types of angles follow.
- *Definition 8.* "A *plane angle* is the inclination to one another of two lines in a plane which meet one another and do not lie in a straight line."
- *Definition 9.* "And when the lines containing the angle are straight, the angle is called *rectilinear.*"
- *Definition 10.* "When a straight line standing on a straight line makes the adjacent angles equal to one another, each of the equal angles is *right,* and the straight line standing on the other is called a *perpendicular* to that on which it stands."
- *Definition 11.* "An *obtuse angle* is an angle greater than a right angle."
- *Definition 12.* "An *acute angle* is an angle less than a right angle."

Additional Answers
1. **In spherical geometry, the sum of the angles of a triangle can exceed 180°.**
2. **Sample answer: Postulate 4 states that all right angles are equal to one another.**
3. **See students' work.**

Additional Activities
1. Have students compare and contrast the definitions of angles in a geometry textbook with Euclid's definitions. **See students' work.**
2. Ask students to write a paragraph explaining why they agree or disagree with Proclus's view on angles. **See students' work.**

The Ambiguous Case for the Law of Sines

OBJECTIVES
* Determine whether a triangle has zero, one, or two solutions.
* Solve triangles using the Law of Sines.

TOURISM Visitors near a certain national park can tune to a local radio station to find out about the activities that are happening in the park. The transmission tower for the radio station is along Park Road about 30 miles from the intersection of this road and the interstate. The interstate and the road form a 47° angle. If the transmitter has a range of 25 miles, how far along the interstate can the passengers in a car hear the broadcast? *This problem will be solved in Example 3.*

From geometry, you know that the measures of two sides and a nonincluded angle do not necessarily define a unique triangle. However, one of the following will be true.

1. No triangle exists.
2. Exactly one triangle exists.
3. Two triangles exist.

In other words, there may be no solution, one solution, or two solutions. A situation with two solutions is called the **ambiguous case.** Suppose you know the measures of a, b, and A. Consider the following cases.

---

1 FOCUS

5-Minute Check
Lesson 5-6

Solve each triangle. Round to the nearest tenth.
1. $A = 41°$, $B = 49°$, $a = 6.5$
   $C = 90°$, $b = 7.5$, $c = 9.9$
2. $B = 41°$, $C = 52°$, $c = 27$
   $A = 87°$, $a \approx 34.2$, $b \approx 22.5$

Find the area of each triangle. Round to the nearest tenth.
3. $a = 8.4$, $b = 10$, $C = 108°$
   about 39.9 units²
4. $a = 14.2$, $A = 18° 50'$,
   $B = 69° 18'$ **about 292.0 units²**

Motivating the Lesson
Hands-On Activity Have students use a ruler and protractor to draw $\triangle ABC$ with $A = 30°$, $a = 6$, and $b = 10$. Discuss the results. Then, have them repeat the activity with $A = 30°$, $b = 10$, and $a = 2$. Students should discover that if $A = 30°$ and $b = 10$ in both triangles, a cannot equal both 6 and 2.

Resource Manager

📁 **Reproducible Masters**
* *Study Guide and Practice,* pp. 91–92
* *Enrichment,* p. 55
* *Graphing Calculator Keystrokes,* p. 18

📽 **Transparencies**
* 5-Minute Check, 5-7
* Teaching, 5-7

Case 2: A ≥ 90°

$a \le b$	$a > b$
no solution	one solution

Example **①** Determine the number of possible solutions for each triangle.

a. $A = 30°$, $a = 8$, $b = 10$

Since $30° < 90°$, consider Case I.

$b \sin A = 10 \sin 30°$
$b \sin A = 10(0.5)$
$b \sin A = 5$

Since $5 < 8 < 10$, there are two solutions for the triangle.

b. $b = 8$, $c = 10$, $B = 118°$

Since $118° \ge 90°$, consider Case II.

In this triangle, $8 \le 10$, so there are no solutions.

Once you have determined that there are one or two solutions for a triangle given the measures of two sides and a nonincluded angle, you can use the Law of Sines to solve the triangle.

Example **②** Find all solutions for each triangle. If no solutions exist, write *none*.

a. $a = 4$, $b = 3$, $A = 112°$

Since $112° \ge 90°$, consider Case II. In this triangle, $4 > 3$, so there is one solution. First, use the Law of Sines to find B.

$$\frac{a}{\sin A} = \frac{b}{\sin B}$$

$$\frac{4}{\sin 112°} = \frac{3}{\sin B}$$

$$\sin B = \frac{3 \sin 112°}{4}$$

$$B = \sin^{-1}\left(\frac{3 \sin 112°}{4}\right)$$

$$B \approx 44.05813517 \quad \textit{Use a calculator.}$$

So, $B \approx 44.1°$.

Use the value of B to find C and c.

$C \approx 180° - (112° + 44.1°)$ or about $23.9°$

$$\frac{a}{\sin A} = \frac{c}{\sin C}$$

$$\frac{4}{\sin 112°} \approx \frac{c}{\sin 23.9°}$$

$$c \approx \frac{4 \sin 23.9°}{\sin 112°}$$

$$c \approx 1.747837108 \quad \textit{Use a calculator.}$$

The solution of this triangle is $B \approx 44.1°$, $C \approx 23.9°$, and $c \approx 1.7$.

Teaching Tip In Examples 1 and 2, you may want to have students sketch the triangles to scale in order to visualize the number of possible solutions.

Teaching Tip As an alternative approach to Example 1, part **b**, have students use the values and the Law of Sines to find C. The result is $\sin C > 1$, which is not possible in a triangle.

In-Class Examples

Example 1
Determine the number of possible solutions for each triangle.
a. $A = 63°$, $a = 18$, $b = 25$ **0**
b. $A = 105°$, $a = 73$, $b = 55$ **1**

Example 2
Find all solutions for each triangle. If no solutions exist, write *none*.
a. $A = 98°$, $a = 39$, $b = 22$
    $B \approx 34.0°$, $C \approx 48.0°$,
    $c \approx 29.3$
b. $A = 72.2°$, $a = 21$, $b = 22$
    $B \approx 85.9°$, $C \approx 21.9°$,
    $c \approx 8.2$; *or* $B \approx 94.1°$,
    $C \approx 13.7°$, $c \approx 5.2$

Teaching Tip When looking for more than one solution, stress that sine is positive in the first and second quadrants, so reference angles in both quadrants apply.

Teaching Tip Another way to determine whether there are two solutions is to find the sum of the supplement of $\angle C$ and the measure of $\angle A$; if the sum is less than 180°, there are two solutions.

b. $A = 51°, a = 40, c = 50$

Since $51° < 90°$, consider Case I.

$c \sin A = 50 \sin 51°$

≈ 38.85729807 *Use a calculator.*

Since $38.9 < 40 < 50$, there are two solutions for the triangle.

Use the Law of Sines to find C.

$$\frac{a}{\sin A} = \frac{c}{\sin C}$$

$$\frac{40}{\sin 51°} = \frac{50}{\sin C}$$

$$\sin C = \frac{50 \sin 51°}{40}$$

$$C = \sin^{-1}\left(\frac{50 \sin 51°}{40}\right)$$

$$C \approx 76.27180414 \qquad \textit{Use a calculator.}$$

Notice that the sum of the two measures for C is 180°.

So, $C \approx 76.3°$. Since we know there are two solutions, there must be another possible measurement for C. In the second case, C must be less than $180°$ and have the same sine value. Since we know that if $\alpha < 90$, $\sin \alpha = \sin (180 - \alpha)$, $180° - 76.3°$ or $103.7°$ is another possible measure for C. Now solve the triangle for each possible measure of C.

Solution I	**Solution II**

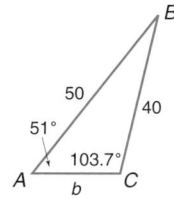

Solution I	**Solution II**
$B \approx 180° - (51° + 76.3°)$	$B \approx 180° - (51° + 103.7°)$
$B \approx 52.7°$	$B \approx 25.3°$

$$\frac{a}{\sin A} = \frac{b}{\sin B} \qquad\qquad \frac{a}{\sin A} = \frac{b}{\sin B}$$

$$\frac{40}{\sin 51°} \approx \frac{b}{\sin 52.7°} \qquad\qquad \frac{40}{\sin 51°} \approx \frac{b}{\sin 25.3°}$$

$$b \approx \frac{40 \sin 52.7°}{\sin 51°} \qquad\qquad b \approx \frac{40 \sin 25.3°}{\sin 51°}$$

$$b \approx 40.94332444 \qquad\qquad b \approx 21.99627275$$

One solution is $B \approx 52.7°$, $C \approx 76.3°$, and $b \approx 40.9$.

Another solution is $B \approx 25.3°$, $C \approx 103.7°$, and $b \approx 22.0$.

GRAPHING CALCULATOR EXPLORATION

You can store values in your calculator and use these values in other computations. In solving triangles, you can store a value for a missing part of the triangle and then use this value when solving for the other missing parts.

4. Answers will vary if rounded numbers are used to find some values.

TRY THESE

1. Rework Example 2a. Use a calculator to solve for B and store this value. Use the stored value to solve for C and c. Round the answers to the nearest tenth after you have completed all computations.
$B = 44.1°$, $C = 23.9°$, $c = 1.8$

2. Rework Example 2b. Use a calculator to solve for each possible value for C. Store these values. Use the stored values to solve for B and b in each possible triangle. Round the answers to the nearest tenth after you have completed all computations.
$B = 52.7°$, $C = 76.3°$, $b = 41.0$;
$B = 25.3°$, $C = 103.7°$, $b = 22.0$

WHAT DO YOU THINK?

3. Compare your answers with those in the examples. **Answers are slightly different.**

4. Why do you think your answers may vary slightly from those of classmates or the textbook?

Graphing Calculator Exploration

In order to store a value in the calculator, use STO→. This allows you to recall values with many decimal places easily. Values stored will remain in the memory until changed or cleared.

In-Class Example

Example 3

Construction A group of contractors is constructing a 24-foot slide on a playground. The slide inclines 45° from the horizontal. The access ladder measures 18 feet long. At what angle to the horizontal should the contractors build the ladder? **Since the ladder must make an acute angle with the horizontal, the angle cannot be 109.5°. Therefore, the solution is 70.5°.**

Example **3** **TOURISM** **Refer to the application at the beginning of the lesson. How far along the interstate can the passengers in a car hear the broadcast?**

Consider Case 1 because $47° < 90°$. Since $30 \sin 47° \approx 21.9$ and $21.9 < 25 < 30$, there are two triangles with sides 25 miles and 30 miles long and a nonincluded angle of $47°$. θ and α represent the two possible angle measures.

$$\frac{25}{\sin 47°} = \frac{30}{\sin \theta}$$

$$\sin \theta = \frac{30 \sin 47°}{25}$$

$$\theta = \sin^{-1}\left(\frac{30 \sin 47°}{25}\right)$$

$$\theta \approx 61.3571157 \quad \textit{Use a calculator.}$$

So, $\theta \approx 61.4°$ and $\alpha \approx 180° - 61.4°$ or $118.6°$.

$$\frac{25}{\sin 47°} \approx \frac{y}{\sin(180° - (47° + 61.4°))}$$

$$\frac{25}{\sin 47°} \approx \frac{y}{\sin 71.6°}$$

$$y \approx \frac{25 \sin 71.6°}{\sin 47°}$$

$$y \approx 32.4356057$$

$$\frac{25}{\sin 47°} \approx \frac{x}{\sin(180° - (47° + 118.6°))}$$

$$\frac{25}{\sin 47°} \approx \frac{x}{\sin 14.4°}$$

$$x \approx \frac{25 \sin 14.4°}{\sin 47°}$$

$$x \approx 8.5010128$$

So, $y \approx 32.4$ and $x \approx 8.5$. The passengers can hear the broadcast when the distance to the transmitter is 25 miles or less. So they could hear for about $32.4 - 8.5$ or 23.9 miles along the interstate.

Lesson 5-7 The Ambiguous Case for the Law of Sines **323**

Addressing Individual Needs

Intrapersonal Have each student write a journal entry that describes his or her approach to solving a given triangle. Ask students to include specific details about what they look for and how they start the problem. What are the important pieces of information that affect how they solve the triangle?

Error Analysis

Watch for students who ignore possible solutions in the case that two solutions exist.

Prevent by reminding students that if the given non-included angle is acute, they should check for multiple solutions.

Teaching Tip Triangles have been solved using a calculator. Rounding did not occur until the final answer. If a calculated value is used to find another value, the stored value in the calculator was used, *not* the rounded value. Using rounded values in calculations will result in different answers from those given.

Assignment Guide

Core: 11–31 odd, 33–35, 37, 39–46
Enriched: 12–32 even, 33–46

Additional Answer

1. A triangle cannot exist if $A < 90°$ and $a < b \sin A$ or if $A \geq 90°$ and $a \leq b$.

Study Guide and Practice Masters, pp. 91–92

CHECK FOR UNDERSTANDING

Communicating Mathematics

Read and study the lesson to answer each question. 1–3. See margin.

1. **Describe** the conditions where the Law of Sines indicates that a triangle cannot exist.

2. **Draw** two triangles where $A = 30°$, $a = 6$, and $b = 10$. Calculate and label the degree measure of each angle rounded to the nearest tenth.

3. **Write** the steps needed to solve a triangle if $A = 120°$, $a = 28$, and $b = 17$.

Guided Practice

Determine the number of possible solutions for each triangle.

4. $A = 113°$, $a = 15$, $b = 8$ **1**   5. $B = 44°$, $a = 23$, $b = 12$ **0**

Find all solutions for each triangle. If no solutions exist, write *none*. Round to the nearest tenth. 6. $A = 15.4°$, $B = 147.6°$, $b = 20.2$

6. $C = 17°$, $a = 10$, $c = 11$ 7. $A = 140°$, $b = 10$, $a = 3$ **none**

8. $A = 38°$, $b = 10$, $a = 8$ 9. $C = 130°$, $c = 17$, $b = 5$
$A = 37.0°$, $B = 13.0°$, $a = 13.4$

8. $B = 50.3°$, $C = 91.7°$, $c = 13.0$; $B = 129.7°$, $C = 12.3°$, $c = 2.8$

10. **Communications** A vertical radio tower is located on the top of a hill that has an angle of elevation of 10°. A 70-foot guy wire is attached to the tower 45 feet above the hill.

a. Make a drawing to illustrate the situation. **See margin.**

b. What angle does the guy wire make with the side of the hill? **about 39.3°**

c. How far from the base of the tower is the guy wire anchored to the hill? **about 46.4 ft**

EXERCISES

Practice

Determine the number of possible solutions for each triangle.

A

11. $A = 57°$, $a = 11$, $b = 19$ **0**   12. $A = 30°$, $a = 13$, $c = 26$ **1**

13. $B = 61°$, $a = 12$, $b = 8$ **0**   14. $A = 58°$, $C = 94°$, $b = 17$ **1**

15. $C = 100°$, $a = 18$, $c = 15$ **0**   16. $B = 37°$, $a = 32$, $b = 27$ **2**

17. If $A = 65°$, $a = 55$, and $b = 57$, how many possible values are there for B? **2**

Find all solutions for each triangle. If no solutions exist, write *none*. Round to the nearest tenth. 18–29. See margin.

B

18. $a = 6$, $b = 8$, $A = 150°$ 19. $a = 26$, $b = 29$, $A = 58°$

20. $A = 30°$, $a = 4$, $b = 8$ 21. $C = 70°$, $c = 24$, $a = 25$

22. $A = 40°$, $B = 60°$, $c = 20$ 23. $a = 14$, $b = 12$, $B = 90°$

24. $B = 36°$, $b = 19$, $c = 30$ 25. $A = 107.2°$, $a = 17.2$, $c = 12.2$

26. $A = 76°$, $a = 5$, $b = 20$ 27. $C = 47°$, $a = 10$, $c = 16$

28. $B = 40°$, $b = 42$, $c = 60$ 29. $b = 40$, $a = 32$, $A = 125.3°$

C

30. Copy the triangle at the right and label all measurements of the triangle.

19.3 cm **74.1°** 21.7 cm
57.4° **48.5°**
24.8 cm

www.amc.glencoe.com/self_check_quiz

Additional Answers

2.

3. **Step 1:** Determine that there is one solution for the triangle.

Step 2: Use the Law of Sines to solve for B.

Step 3: Subtract the sum of 120 and B from 180 to find C.

Step 4: Use the Law of Sines to solve for c.

31. Find the perimeter of each of the two noncongruent triangles where $a = 15$, $b = 20$ and $A = 29°$.   **about 63.9 units and 41.0 units**

32. There are two noncongruent triangles where $B = 55°$, $a = 15$, and $b = 13$. Find the measures of the angles of the triangle with the greater perimeter. $A \approx 70.9°$, $B = 55°$, $C \approx 54.1°$

33. Gears   An engineer designed three gears as shown at the right. What is the measure of θ?   **about 100.6°**

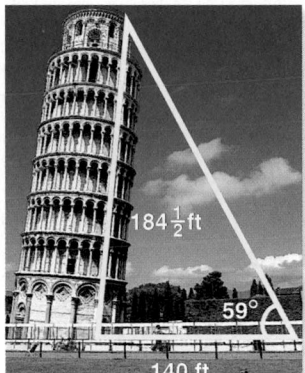

$r = 15$ cm

$r = 22$ cm

$37°$

$r = 18$ cm

34. Critical Thinking   If $b = 14$ and $A = 30°$, determine the possible values of a for each situation.

  a. The triangle has no solutions. $a < 7$

  b. The triangle has one solution. $a = 7$ or $a \geq 14$

  c. The triangle has two solutions. $7 < a < 14$

35. Architecture   The original height of the Leaning Tower of Pisa was $184\frac{1}{2}$ feet. At a distance of 140 feet from the base of the tower, the angle of elevation from the ground to the top of the tower is $59°$. How far is the tower leaning from the original vertical position?   **about 9.6°**

$184\frac{1}{2}$ ft

$59°$

140 ft

36. Navigation   The captain of the Coast Guard Cutter Pendant plans to sail to a port that is 450 miles away and $12°$ east of north. The captain first sails the ship due north to check a buoy. He then turns the ship and sails 316 miles to the port.

  a. In what direction should the captain turn the ship to arrive at the port?   **about 17.2° east of north**

36b. about 6h

  b. How many hours will it take to arrive at the turning point if the captain chooses a speed of 23 miles per hour?

  c. Instead of the plan above, the captain decides to sail 200 miles north, turn through an angle of $20°$ east of north, and then sail along a straight course. Will the ship reach the port by following this plan?   **no**

Port

316 mi

N

450 mi

$12°$

37. Communications   A satellite is orbiting Earth every 2 hours. The satellite is directly over a tracking station which has its antenna aimed $45°$ above the horizon. The satellite is orbiting 1240 miles above Earth, and the radius of Earth is about 3960 miles. How long ago did the satellite pass through the beam of the antenna? (*Hint*: First calculate θ.)   **about 4.1 min**

Beam from Antenna

Satellite

$45°$

3960 mi

θ

Open-Ended Assessment

Modeling Have each student draw an arbitrary triangle, measure one angle and two sides, determine whether the triangle can be solved, and state how many solutions exist.

Additional Answer

43.

$$\frac{\frac{3x}{x-1}+1}{3\left(\frac{3x}{x-1}\right)} = \frac{\frac{3x}{x-1}+\frac{x-1}{x-1}}{\frac{9x}{x-1}}$$

$$= \frac{\frac{4x-1}{x-1}}{\frac{9x}{x-1}}$$

$$= \frac{4x-1}{9x} \neq x$$

Enrichment Masters, p. 55

5-7

Enrichment

Spherical Triangles

Spherical trigonometry is an extension of plane trigonometry. Figures are drawn on the surface of a sphere. Arcs of great circles correspond to line segments in the plane. The arcs of three great circles intersecting on a sphere form a spherical triangle. Angles have the same measure as the tangent lines drawn to each great circle at the vertex. Since the sides are arcs, they too can be measured in degrees.

The sum of the sides of a spherical triangle is less than 360°.
The sum of the angles is greater than 180° and less than 540°.
The Law of Sines for spherical triangles is as follows.

$$\frac{\sin a}{\sin A} = \frac{\sin b}{\sin B} = \frac{\sin c}{\sin C}$$

There is also a Law of Cosines for spherical triangles.
cos a = cos b cos c + sin b sin c cos A
cos b = cos a cos c + sin a sin c cos B
cos c = cos a cos b + sin a sin b cos C

Example Solve the spherical triangle given a = 72°, b = 105°, and c = 61°.

Use the Law of Cosines.

0.3090 = (−0.2588)(0.4848) + (0.9659)(0.8746) cos A
cos A = 0.5143
A = 59°

−0.2588 = (0.3090)(0.4848) + (0.9511)(0.8746) cos B
cos B = −0.4912
B = 119°

0.4848 = (0.3090)(−0.2588) + (0.9511)(0.9659) cos C
cos C = 0.6148
C = 52°

Check by using the Law of Sines.

$$\frac{\sin 72°}{\sin 59°} = \frac{\sin 105°}{\sin 119°} = \frac{\sin 61°}{\sin 52°} = 1.1$$

Solve each spherical triangle.

1. a = 56°, b = 53°, c = 94°
A = 41°, B = 39°, C = 128°
2. a = 110°, b = 33°, c = 97°
A = 116°, B = 31°, C = 71°
3. a = 76°, b = 110°, C = 49°
A = 59°, B = 124°, c = 59°
4. b = 94°, c = 55°, A = 48°
a = 60°, B = 121°, C = 45°

© Glencoe/McGraw-Hill   T55   Advanced Mathematical Concepts

326 *Chapter 5*

---

38. **Mechanics** The blades of a power lawn mower are rotated by a two-stroke engine with a piston sliding back and forth in the engine cylinder. As the piston moves back and forth, the connecting rod rotates the circular crankshaft. Suppose the crankshaft is 5 centimeters long and the connecting rod is 15 centimeters. If the crankshaft rotates 20 revolutions per second and *P* is at the horizontal position when it begins to rotate, how far is the piston from the rim of the crankshaft after 0.01 second? **about 10.8 cm**

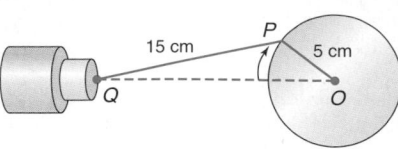

39. **Critical Thinking** If b = 12 and c = 17, find the values of B for each situation.
    a. The triangle has no solutions. **B > 44.9°**
    b. The triangle has one solution. **B ≈ 44.9°**
    c. The triangle has two solutions. **B < 44.9°**

Mixed Review

40. **Geometry** Determine the area of a rhombus if the length of a side is 24 inches and one of its angles is 32°. *(Lesson 5-6)* **about 305.2 in²**

41. **Fire Fighting** A fire is sighted from a fire tower in Wayne National Forest in Ohio. The ranger found that the angle of depression to the fire is 22°. If the tower is 75 meters tall, how far is the fire from the base of the tower? *(Lesson 5-4)* **about 185.6 m**

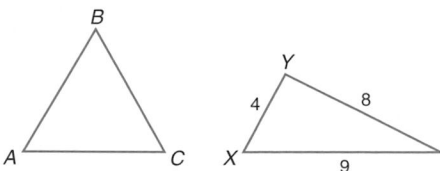

42. State the number of roots of the equation $4x^3 - 4x^2 + 13x - 6 = 0$. Then solve the equation. *(Lesson 4-2)* **3; $\frac{1}{2}$, $\frac{1 \pm i\sqrt{47}}{4}$**

43. Determine whether the functions $y = \frac{3x}{x-1}$ and $y = \frac{x+1}{3x}$ are inverses of one another. Explain. *(Lesson 3-4)* **No; see margin for explanation.**

44. Solve the system of equations algebraically. *(Lesson 2-1)*
    5x − 2y = 9
    y = 3x − 1   **(−7, −22)**

45. 5x + 2y = −22

45. Write the standard form of the equation whose graph is perpendicular to the graph of −2x + 5y = 7 and passes through the point at (−6, 4). *(Lesson 1-5)*

46. **SAT Practice Grid-In** For the triangles shown below, the perimeter of △ABC equals the perimeter of △XYZ. If △ABC is equilateral, what is the length of $\overline{AB}$? **7**

Extra Credit

If $x = \dfrac{10 \sin 120°}{\sin 41°}$, find Q. **41°**

The Law of Cosines

OBJECTIVES
- Solve triangles by using the Law of Cosines.
- Find the area of triangles if the measures of the three sides are given.

GOLF For a right-handed golfer, a *slice* is a shot that curves to the right of its intended path, and a *hook* curves off to the left. Suppose Se Ri Pak hits the ball from the seventh tee at the U.S. Women's Open and the shot is a 160-yard slice 4° from the path straight to the cup. If the tee is 177 yards from the cup, how far does the ball lie from the cup? *This problem will be solved in Example 1.*

From geometry, you know that a unique triangle can be formed if the measures of three sides of a triangle are known and the sum of any two measures is greater than the remaining measure. You also know that a unique triangle can be formed if the measures of two sides and an included angle are known. However, the Law of Sines cannot be used to solve these triangles. Another formula is needed. Consider $\triangle ABC$ with a height of h units and sides measuring a units, b units, and c units. Suppose $\overline{DC}$ is x units long. Then $\overline{BD}$ is $(a - x)$ units long.

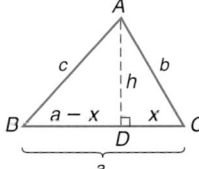

The Pythagorean Theorem and the definition of the cosine ratio can be used to show how $\angle C$, a, b, and c are related.

$c^2 = (a - x)^2 + h^2$ *Apply the Pythagorean Theorem for* $\triangle ADB$.
$c^2 = a^2 - 2ax + x^2 + h^2$ *Expand* $(a - x)^2$.
$c^2 = a^2 - 2ax + b^2$ $b^2 = x^2 + h^2$ *in* $\triangle ADC$.
$c^2 = a^2 - 2a(b \cos C) + b^2$ $\cos C = \frac{x}{b}$, *so* $b \cos C = x$.
$c^2 = a^2 + b^2 - 2ab \cos C$ *Simplify.*

By drawing altitudes from B and C, you can derive similar formulas for a^2 and b^2. All three formulas, which make up the **Law of Cosines,** can be summarized as follows.

Law of Cosines

Let $\triangle ABC$ be any triangle with a, b, and c representing the measures of sides opposite angles with measurements A, B, and C, respectively. Then, the following are true.

$$a^2 = b^2 + c^2 - 2bc \cos A$$
$$b^2 = a^2 + c^2 - 2ac \cos B$$
$$c^2 = a^2 + b^2 - 2ab \cos C$$

You can use the Law of Cosines to solve the application at the beginning of the lesson.

Lesson 5-8 The Law of Cosines **327**

Teaching Tip After stating the Law of Cosines, have students discuss why the three forms are equivalent.

In-Class Examples

Example 1
Landscaping Suppose you want to fence a triangular lot as shown below. If two sides measure 84 feet and 78 feet and the angle between the two sides is 102°, what is the length of the fence to the nearest foot?
288 ft

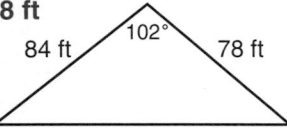

Example 2
Solve each triangle.
a. $A = 39.4°$, $b = 12$, $c = 14$
 $B \approx 58.2°$, $C \approx 82.4°$,
 $a \approx 9.0$

b. $a = 19$, $b = 24.3$, $c = 21.8$
 $A \approx 48.3°$, $B \approx 72.7°$,
 $C \approx 59.0°$

Example ❶ **GOLF** Refer to the application at the beginning of the lesson. How far does the ball lie from the cup?

In this problem, you know the measurements of two sides of a triangle and the included angle. Use the Law of Cosines to find the measure of the third side of the triangle.

$x^2 = 177^2 + 160^2 - 2(177)(160) \cos 4°$
$x^2 \approx 426.9721933$ *Use a calculator.*
$x \approx 20.66330548$

The ball is about 20.7 yards from the cup.

Many times, you will have to use both the Law of Cosines and the Law of Sines to solve triangles.

Example ❷ Solve each triangle.

a. $A = 120°$, $b = 9$, $c = 5$

$a^2 = b^2 + c^2 - 2bc \cos A$ *Law of Cosines*
$a^2 = 9^2 + 5^2 - 2(9)(5) \cos 120°$ *Substitute 9 for b, 5 for c, and 120° for A.*
$a^2 = 151$ *Use a calculator.*
$a \approx 12.28820573$

So, $a \approx 12.3$.

$\dfrac{a}{\sin A} = \dfrac{b}{\sin B}$ *Law of Sines*

$\dfrac{12.3}{\sin 120°} \approx \dfrac{9}{\sin B}$ *Substitute 12.3 for a, 9 for b, and 120° for A.*

$\sin B \approx \dfrac{9 \sin 120°}{12.3}$

$B \approx \sin^{-1}\left(\dfrac{9 \sin 120°}{12.3}\right)$

$B \approx 39.32193819$ *Use a calculator.*

So, $B \approx 39.3°$.

$C \approx 180° - (120° + 39.3°)$
$C \approx 20.7°$

The solution of this triangle is $a \approx 12.3$, $B \approx 39.3°$, and $C \approx 20.7°$.

b. $a = 24$, $b = 40$, $c = 18$

Recall that θ and $180° - \theta$ have the same sine function value, but different cosine function values. Therefore, a good strategy to use when all three sides are given is to use the Law of Cosines to determine the measure of the possible obtuse angle first. Since b is the longest side, B is the angle with the greatest measure, and therefore a possible obtuse angle.

From the Classroom of...

James R. Lee
Georgia Christian School
Valdosta, Georgia

"Using a surveyor's transit for more accurate angle measurements, I like to take my students outside to compute the distances between predetermined locations that cannot be measured directly. We apply the Law of Cosines. This gives them a working knowledge of the law and a better understanding of its usefulness."

$$b^2 = a^2 + c^2 - 2ac \cos B \qquad \text{\textit{Law of Cosines}}$$
$$40^2 = 24^2 + 18^2 - 2(24)(18) \cos B$$

$$\frac{40^2 - 24^2 - 18^2}{-2(24)(18)} = \cos B$$

$$\cos^{-1}\left(\frac{40^2 - 24^2 - 18^2}{-2(24)(18)}\right) = B$$

$$144.1140285 \approx B \qquad \text{\textit{Use a calculator.}}$$

So, $B \approx 144.1°$.

$$\frac{a}{\sin A} = \frac{b}{\sin B} \qquad \text{\textit{Law of Sines}}$$

$$\frac{24}{\sin A} \approx \frac{40}{\sin 144.1°}$$

$$\sin A \approx \frac{24 \sin 144.1°}{40}$$

$$A \approx \sin^{-1}\left(\frac{24 \sin 144.1°}{40}\right)$$

$$A \approx 20.59888389 \qquad \text{\textit{Use a calculator.}}$$

So, $A \approx 20.6°$.

$$C \approx 180 - (20.6 + 144.1)$$
$$C \approx 15.3$$

The solution of this triangle is $A \approx 20.6°$, $B \approx 144.1°$, and $C \approx 15.3°$.

If you know the measures of three sides of a triangle, you can find the area of the triangle by using the Law of Cosines and the formula $K = \frac{1}{2}bc \sin A$.

Example **3** Find the area of $\triangle ABC$ if $a = 4$, $b = 7$, and $c = 9$.

First, solve for A by using the Law of Cosines.

$$a^2 = b^2 + c^2 - 2bc \cos A$$
$$4^2 = 7^2 + 9^2 - 2(7)(9) \cos A$$

$$\frac{4^2 - 7^2 - 9^2}{-2(7)(9)} = \cos A$$

$$\cos^{-1}\left(\frac{4^2 - 7^2 - 9^2}{-2(7)(9)}\right) = A \qquad \text{\textit{Definition of} } \cos^{-1}$$

$$25.2087653 \approx A \qquad \text{\textit{Use a calculator.}}$$

So, $A \approx 25.2°$.

Then, find the area.

$$K = \frac{1}{2}bc \sin A$$

$$K \approx \frac{1}{2}(7)(9) \sin 25.2°$$

$$K \approx 13.41204768$$

The area of the triangle is about 13.4 square units.

Teaching Tip In order to help students understand the formula for area used in Example 3, remind them that the height of a triangle is expressed using the sine ratio.

Since $\sin A = \frac{h}{b}$, $h = b \sin A$.

In-Class Example

Example 3
Find the area of $\triangle ABC$ if $a = 24$, $b = 52$, and $c = 39$.
about 442.7 units²

Addressing Individual Needs

Verbal/Linguistic Have students discuss how to solve triangles knowing various combinations of sides and angles. They should tell whether there is no solution or whether the triangle can be solved using right triangle trigonometry, the Law of Sines, or the Law of Cosines.

In-Class Example

Example 4
Find the area of △*ABC*. Round to the nearest tenth.

about 978.6 units²

3 PRACTICE/APPLY

Error Analysis

Watch for students who mistakenly try to use the side of a triangle for its height when finding the area.

Prevent by stressing that the height is perpendicular to the base.

Teaching Tip Triangles have been solved using a calculator. Rounding did not occur until the final answer. If a calculated value is used to find another value, the stored value in the calculator was used, *not* the rounded value. Using rounded values in calculations will result in different answers from those given.

Study Guide and Practice Masters, pp. 93–94

If you know the measures of three sides of any triangle, you can also use **Hero's Formula** to find the area of the triangle.

Hero's Formula	If the measures of the sides of a triangle are *a*, *b*, and *c*, then the area, *K*, of the triangle is found as follows.
	$$K = \sqrt{s(s-a)(s-b)(s-c)} \text{ where } s = \frac{1}{2}(a + b + c)$$

s is called the <u>semiperimeter</u>.

Example ④ **Find the area of △*ABC*. Round to the nearest tenth.**

First, find the semiperimeter of △*ABC*.

$$s = \frac{1}{2}(a + b + c)$$

$$s = \frac{1}{2}(72 + 83 + 95)$$

$$s = 125$$

Now, apply Hero's Formula.

$$K = \sqrt{s(s - a)(s - b)(s - c)}$$

$$K = \sqrt{125(125 - 72)(125 - 83)(125 - 95)}$$

$$K = \sqrt{8,347,500}$$

$$K \approx 2889.204043 \quad \textit{Use a calculator.}$$

The area of the triangle is about 2889.2 square centimeters.

CHECK FOR UNDERSTANDING

Communicating Mathematics

Read and study the lesson to answer each question.

1. **Describe** two situations where the Law of Cosines is needed to solve a triangle.

1. See margin.

2. **Give an example** of three measurements of sides that do not form a triangle.

2. Sample answer: 1 in., 2 in., 4 in.

3. **Explain** how the Pythagorean Theorem is a special case of the Law of Cosines.

3. See margin.

4. *Math Journal* **Draw and label** a right triangle that can be solved using the trigonometric functions. Draw and label a triangle that can be solved using the Law of Sines, but not the Law of Cosines. Draw and label a triangle that can be solved using the Law of Cosines. Solve each triangle. **See margin.**

Guided Practice

Solve each triangle. Round to the nearest tenth.

5. *a* = 32, *b* = 38, *c* = 46
   A = 43.5°, *B* = 54.8°, *C* = 81.7°

6. *a* = 25, *b* = 30, *C* = 160°
   A = 9.1°, *B* = 10.9°, *c* = 54.2

7. The sides of a triangle are 18 inches, 21 inches, and 14 inches. Find the measure of the angle with the greatest measure. **about 81.0°**

Find the area of each triangle. Round to the nearest tenth.

8. *a* = 2, *b* = 7, *c* = 8 **6.4 units²**

9. *a* = 25, *b* = 13, *c* = 17 **102.3 units²**

Additional Answers

1. The Law of Cosines is needed to solve a triangle if the measures of all three sides or the measures of two sides and the included angle are given.

3. If the included angle measures 90°, the equation becomes
   $c^2 = a^2 + b^2 - 2ab \cos C$.
   Since cos 90° = 0,
   $c^2 = a^2 + b^2 - 2ab(0)$
   or $c^2 = a^2 + b^2$.

10. Softball In slow-pitch softball, the diamond is a square that is 65 feet on each side. The distance between the pitcher's mound and home plate is 50 feet. How far does the pitcher have to throw the softball from the pitcher's mound to third base to stop a player who is trying to steal third base? **about 46.1 ft**

EXERCISES

Assignment Guide
Core: 11–25 odd, 27–31, 33–38
Enriched: 12–26 even, 27–38

Additional Answers
11. $B = 44.2°$, $C = 84.8°$, $a = 7.8$
12. $A = 44.4°$, $B = 57.1°$, $C = 78.5°$
13. $A = 34.1°$, $B = 44.4°$, $C = 101.5°$
14. $A = 71.6°$, $C = 45.4°$, $b = 15.0$

Practice

 Solve each triangle. Round to the nearest tenth. **11–14. See margin.**

11. $b = 7$, $c = 10$, $A = 51°$ 12. $a = 5$, $b = 6$, $c = 7$

13. $a = 4$, $b = 5$, $c = 7$ 14. $a = 16$, $c = 12$, $B = 63°$

 15. $a = 11.4$, $b = 13.7$, $c = 12.2$ 16. $C = 79.3°$, $a = 21.5$, $b = 13$
    $A = 51.8°$, $B = 70.9°$, $C = 57.3°$ **$A = 66.9°$, $B = 33.8°$, $c = 23.0$**

17. about 13.8° 17. The sides of a triangle measure 14.9 centimeters, 23.8 centimeters, and 36.9 centimeters. Find the measure of the angle with the least measure.

Graphing Calculator Programs
For a graphing calculator program that determines the area of a triangle, given the lengths of all sides of the triangle, visit www.amc.glencoe.com

18. **Geometry** Two sides of a parallelogram measure 60 centimeters and 40 centimeters. If one angle of the parallelogram measures 132°, find the length of each diagonal. **about 91.7 cm and 44.6 cm**

Find the area of each triangle. Round to the nearest tenth.

19. $a = 4$, $b = 6$, $c = 8$ **11.6 units²**   20. $a = 17$, $b = 13$, $c = 19$ **107.8 units²**

21. $a = 20$, $b = 30$, $c = 40$ **290.5 units²**   22. $a = 33$, $b = 51$, $c = 42$ **690.1 units²**

 23. $a = 174$, $b = 138$, $c = 188$ 24. $a = 11.5$, $b = 13.7$, $c = 12.2$
    11,486.3 units²   **66.1 units²**

25. **Geometry** The lengths of two sides of a parallelogram are 48 inches and 30 inches. One angle measures 120°.

   a. Find the length of the longer diagonal. **about 68.1 in.**

   b. Find the area of the parallelogram. **about 1247.1 in²**

26. **Geometry** The side of a rhombus is 15 centimeters long, and the length of its longer diagonal is 24.6 centimeters. **b. about 110.2°, 69.8°, 110.2°, and 69.8°**

   a. Find the area of the rhombus. **about 211.2 cm²**

   b. Find the measures of the angles of the rhombus.

Applications and Problem Solving

27. **Baseball** In baseball, dead center field is the farthest point in the outfield on the straight line through home plate and second base. The distance between consecutive bases is 90 feet. In Wrigley Field in Chicago, dead center field is 400 feet from home plate. How far is dead center field from first base? **about 342.3 ft**

28. **Critical Thinking** The lengths of the sides of a triangle are 74 feet, 38 feet, and 88 feet. What is the length of the altitude drawn to the longest side? **about 31.6 ft**

www.amc.glencoe.com/self_check_quiz

Lesson 5-8 The Law of Cosines **331**

Additional Answer (p. 330)
4. Sample answers:

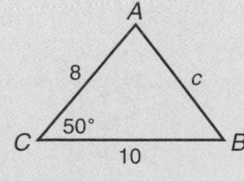

$A = 37°$, $a \approx 7.5$, $c \approx 12.5$ $C = 80°$, $b \approx 8.6$, $c \approx 12.0$ $A \approx 78.4°$, $B \approx 51.6°$, $c \approx 7.8$

Open-Ended Assessment

Writing Have each student create a real-world application problem similar to Example 1 and write a paragraph explaining how to solve the triangle involved.

 **Chapter 5, Quiz D
(Lessons 5-7 and 5-8)**
is available in the *Assessment and Evaluation Masters*, p. 122.

29. Air Travel The distance between Miami and Orlando is about 220 miles. A pilot flying from Miami to Orlando starts the flight 10° off course to avoid a storm.

29a. about 122.8 mi

  a. After flying in this direction for 100 miles, how far is the plane from Orlando?

  b. If the pilot adjusts his course after 100 miles, how much farther is the flight than a direct route? **about 2.8 mi**

30. Critical Thinking Find the area of the pentagon at the right. **about 46,468.5 ft²**

31. the player 30 ft and 20 ft from the posts

31. Soccer A soccer player is standing 35 feet from one post of the goal and 40 feet from the other post. Another soccer player is standing 30 feet from one post of the same goal and 20 feet from the other post. If the goal is 24 feet wide, which player has a greater angle to make a shot on goal?

32. Air Traffic Control A 757 passenger jet and a 737 passenger jet are on their final approaches to San Diego International Airport.

  a. The 757 is 20,000 feet from the ground, and the angle of depression to the tower is 6°. Find the distance between the 757 and the tower. **about 191,335.4 ft**

  b. The 737 is 15,000 feet from the ground, and the angle of depression to the tower is 3°. What is the distance between the 737 and the tower? **about 286,609.8 ft**

  c. How far apart are the jets? **about 96,060.0 ft**

Mixed Review

33. Determine the number of possible solutions for $\triangle ABC$ if $A = 63.2°$, $b = 18$ and $a = 17$. *(Lesson 5-7)* **2**

34. Landmarks The San Jacinto Column in Texas is 570 feet tall and, at a particular time, casts a shadow 700 feet long. Find the angle of elevation to the sun at that time. *(Lesson 5-5)* **about 39.2°**

35. Find the reference angle for $-775°$. *(Lesson 5-1)* **55°**

36. Find the value of k so that the remainder of $(x^3 - 7x^2 - kx + 6)$ divided by $(x - 3)$ is 0. *(Lesson 4-3)* **−10**

37. Find the slope of the line through points at $(2t, t)$ and $(5t, 5t)$. *(Lesson 1-3)* $\dfrac{4}{3}$

38. SAT/ACT Practice Find an expression equivalent to $\left(\dfrac{2x^2}{y}\right)^3$. **A**

  A $\dfrac{8x^6}{y^3}$ B $\dfrac{64x^6}{y^3}$ C $\dfrac{6x^5}{y^3}$ D $\dfrac{8x^5}{y^3}$ E $\dfrac{2x^5}{y^4}$

Extra Practice See p. A35.

Enrichment Masters, p. 56

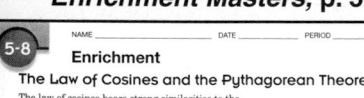

5-8

NAME _____ DATE _____ PERIOD _____

Enrichment

The Law of Cosines and the Pythagorean Theorem

The law of cosines bears strong similarities to the Pythagorean Theorem. According to the Law of Cosines, if two sides of a triangle have lengths a and b and if the angle between them has a measure of x, then the length, y, of the third side of the triangle can be found by using the equation

$$y^2 = a^2 + b^2 - 2ab \cos (x°).$$

Answer the following questions to clarify the relationship between the Law of Cosines and the Pythagorean Theorem.

1. If the value of x becomes less and less, what number is $\cos (x°)$ close to?
1

2. If the value of x is very close to zero but then increases, what happens to $\cos (x°)$ as x approaches 90°?
decreases, approaches 0

3. If x equals 90, what is the value of $\cos (x°)$? What does the equation of $y^2 = a^2 + b^2 - 2ab \cos (x°)$ simplify to if x equals 90?
0, $y^2 = a^2 + b^2$

4. What happens to the value of $\cos (x°)$ as x increases beyond 90 and approaches 180?
decreases to −1

5. Consider some particular values of a and b, say 7 for a and 19 for b. Use a graphing calculator to graph the equation you get by solving $y^2 = 7^2 + 19^2 - 2(7)(19) \cos (x°)$ for y.
See students' graphs.
  a. In view of the geometry of the situation, what range of values should you use for X on a graphing calculator?
Xmin = 0; Xmax = 360
  b. Display the graph and use the TRACE function. What do the maximum and minimum values appear to be for the function?
See students' graphs; Ymin = 12, Ymax = 26
  c. How do the answers for part **b** relate to the lengths 7 and 19? Are the maximum and minimum values from part **b** ever actually attained in the geometric situation?
min = (19 − 7); max = (19 + 7); no

© Glencoe/McGraw-Hill   T56   *Advanced Mathematical Concepts*

332 *Chapter 5*

Extra Credit

Suppose that the points at $(0, 0)$, $(6, 0)$, and $\left(3, 3\sqrt{3}\right)$ are the vertices of a triangle in the coordinate plane. Find the area of the triangle. **$9\sqrt{3}$ units²**

5-8B Solving Triangles

An Extension of Lesson 5-8

OBJECTIVE
- Use a program to solve triangles.

The following program allows you to enter the coordinates of the vertices of a triangle in the coordinate plane and display as output the side lengths and the angle measures in degrees.

- To solve a triangle using the program, you first need to place the triangle in a coordinate plane, determine the coordinates of the vertices, and then input the coordinates when prompted by the program.

- When you place the triangle in the coordinate plane, it is a good idea to choose a side whose length is given and place that side on the positive x-axis so that its left endpoint is at (0, 0). You can use the length of the side to determine the coordinates of the other endpoint.

- To locate the third vertex, you can use the given information about the triangle to write equations whose graphs intersect at the third vertex. Graph the equations and use intersect on the **CALC** menu to find the coordinates of the third vertex.

- You are now ready to run the program.

```
PROGRAM: SOLVTRI
: Disp "INPUT VERTEX A"
: Input A: Input P
: Disp "INPUT VERTEX B"
: Input B: Input Q
: Disp "INPUT VERTEX C"
: Input C: Input S
: √((A − B)² + (P − Q)²) → E
: √((B − C)² + (Q − S)²) → F
: √((A − C)² + (P − S)²) → G
: cos⁻¹ ((F² − E² − G²)/(−2EG)) → M
: cos⁻¹ ((G² − E² − F²)/(−2EF)) → N
: cos⁻¹ ((E² − F² − G²)/(−2FG)) → O
: Disp "SIDE AB" : Disp E
: Disp "SIDE BC" : Disp F
: Disp "SIDE AC" : Disp G
: Pause
: Disp "ANGLE A" : Disp M
: Disp "ANGLE B" : Disp N
: Disp "ANGLE C" : Disp O
: Stop
```

Example

Use the program to solve △ABC if A = 33°, B = 105°, and b = 37.9.

Before you use the calculator, do some advance planning. Make a sketch of how you can place the triangle in the coordinate plane to have the third vertex above the x-axis. One possibility is shown at the right.

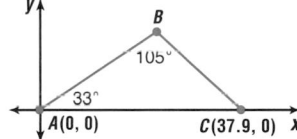

(continued on the next page)

---

1 FOCUS

Motivating the Lesson

Discussion Ask students what steps they can take to solve a triangle if they know
- the lengths of three sides
- the lengths of two sides and the measure of the included angle
- the measures of two angles and the length of an included side
- the measures of two angles and the length of a non-included side

Ask students what role the calculator could play in each case.

2 TEACH

A reproduction of this lesson with graphing calculator keystrokes provided can be found in the *Graphing Calculator Keystrokes Masters,* pp. 19–21.

Teaching Tip Make sure that the calculator is set in the degree mode, or add a line to the program to set degrees automatically.

Teaching Tip You may want students to sketch their triangles on grid paper, locating one vertex at the origin and one side along the x-axis. They can then approximate the measurements of a triangle before executing the program.

Teaching Tip In the Example, stress that the equation for the line containing side $\overline{BC}$ is obtained by shifting the line $y = \tan 138°x$ to the right by 37.9 units.

In-Class Example

Use the program to solve △DEF if $D = 62°$, $E = 70°$, and $e = 52$.
$F = 48°$, $d = 48.9$, $f = 41.1$

4 ASSESS

Open-Ended Assessment

Speaking Give students the measures of two angles and the lengths of a non-included side of a triangle. Ask them to describe the steps needed to solve the triangle by using the **SOLVTRI** program.

The slope of $\overline{AB}$ is $\frac{y - 0}{x - 0}$ or $\frac{y}{x}$. This is tan 33°. So, $\frac{y}{x} = \tan 33°$ or $y = (\tan 33°)x$.

The slope of $\overline{BC}$ is $\frac{y - 0}{x - 37.9}$. This is tangent of $(180° - 42°)$.

So, $\tan 138° = \frac{y}{x - 37.9}$ or $y = (\tan 138°)(x - 37.9)$.

Enter the equations (without the degree signs) on the **Y=** list, and select appropriate window settings. Graph the equations and use intersect in the **CALC** menu to find the coordinates of vertex B. When the calculator displays the coordinates at the bottom of the screen, go immediately to the program. Be sure you do nothing to alter the values of x and y that were displayed.

[-10, 40] scl: 5 by [-10, 40] scl: 5

When the program prompts you to input vertex A, enter 0 and 0. For vertex B, press $\boxed{X,T,\theta,n}$ $\boxed{\text{ENTER}}$ $\boxed{\text{ALPHA}}$ **[Y]** $\boxed{\text{ENTER}}$. For vertex C, enter the numbers 37.9 and 0. The calculator will display the side lengths of the triangle and pause. To display the angle measures, press $\boxed{\text{ENTER}}$.

The side lengths and angle measures agree with those in the text. Therefore, $a \approx 21.4$, $c \approx 26.3$, and $C = 42°$. *Compare the results with those in Example 1 of Lesson 5-6, on page 314.*

TRY THESE

1. Solve $\triangle ABC$ given that $AC = 6$, $BC = 8$, and $A = 35°$. (*Hint*: Place $\overline{AC}$ so that the endpoints have coordinates $(0, 0)$ and $(6, 0)$. Use the graphs of $y = (\tan 35°)x$ and $y = \sqrt{5^2 - (x - 6)^2}$ to determine the coordinates of vertex B.) $AB \approx 12.1$, $B \approx 25.5°$, $C \approx 119.5°$

2. Use **SOLVTRI** to solve the triangles in Example 2b of Lesson 5-7. $B \approx 52.7°$, $C \approx 76.3°$, $b \approx 40.9$; $B \approx 25.3°$, $C \approx 103.7°$, $b \approx 22.0$

WHAT DO YOU THINK?

3. What law about triangles is the basis of the program **SOLVTRI**? Law of Cosines

4. Suppose you want to use **SOLVTRI** to solve the triangle in Example 2a of Lesson 5-7. How would you place the triangle in the coordinate plane?
   Sample answer: Put vertex A at the origin and vertex C at (3, 0).

VOCABULARY

ambiguous case (p. 320)
angle of depression (p. 300)
angle of elevation (p. 300)
apothem (p. 300)
arccosine relation (p. 305)
arcsine relation (p. 305)
arctangent relation (p. 305)
circular function (p. 292)
cofunctions (p. 287)
cosecant (pp. 286, 292)
cosine (pp. 285, 291)
cotangent (pp. 286, 292)
coterminal angles (p. 279)

degree (p. 277)
Hero's Formula (p. 330)
hypotenuse (p. 284)
initial side (p. 277)
inverse (p. 305)
Law of Cosines (p. 327)
Law of Sines (p. 313)
leg (p. 284)
minute (p. 277)
quadrantal angle (p. 278)
reference angle (p. 280)
secant (pp. 286, 292)
second (p. 277)

side adjacent (p. 284)
side opposite (p. 284)
sine (pp. 285, 291)
solve a triangle (p. 307)
standard position (p. 277)
tangent (pp. 285, 292)
terminal side (p. 277)
trigonometric function
    (p. 292)
trigonometric ratio (p. 285)
unit circle (p. 291)
vertex (p. 277)

UNDERSTANDING AND USING THE VOCABULARY

State whether each sentence is true or false. If false, replace the underlined word(s) to make a true statement.   4. false; adjacent to

1. An angle of <u>elevation</u> is the angle between a horizontal line and the line of sight from the observer to an object at a lower level. **false; depression**

2. The inverse of the cosine function is the <u>arcsine</u> relation. **false; arccosine**

3. A degree is subdivided into 60 equivalent parts known as <u>minutes</u>. **true**

4. The leg that is a side of an acute angle of a right triangle is called the side <u>opposite</u> the angle.

5. If the terminal side of an angle θ in standard position intersects the unit circle at $P(x, y)$, the relations $\cos \theta = x$ and $\sin \theta = y$ are called <u>circular functions</u>. **true**

6. Two angles in standard position are called <u>reference</u> angles if they have the same terminal side.
false; coterminal

7. <u>Trigonometric ratios</u> are defined by the ratios of right triangles. **true**

8. The <u>Law of Sines</u> is derived from the Pythagorean Theorem. **false; Law of Cosines**

9. The ray that rotates to form an angle is called the <u>initial side</u>. **false; terminal side**

10. A circle of radius 1 is called a <u>unit circle</u>. **true**

 interNET CONNECTION   For additional review and practice for each lesson, visit: **www.amc.glencoe.com**

Six forms of Chapter 5 Test are available in the *Assessment and Evaluation Masters* as shown in the chart.

Chapter 5 Test, Form 1B, is shown at the right. Chapter 5 Test, Form 2B, is shown on the next page.

TEST OPTIONS

Form	Pages	Type	
1A	107–108	Honors	*Form 1 tests are multiple choice.*
1B	109–110	Average	
1C	111–112	Basic	
2A	113–114	Honors	*Form 2 tests are free response.*
2B	115–116	Average	
2C	117–118	Basic	

STUDY GUIDE AND ASSESSMENT

VOCABULARY

This section provides a listing of the new terms, properties, and phrases that were introduced in this chapter. The exercises check students' understanding of the terms by using a variety of verbal formats including matching, completion, and true/false.

Glossary A complete glossary of terms begins on page A71.

Assessment and Evaluation Masters, pp. 109–110

SKILLS AND CONCEPTS

Objectives and Examples This section reviews the skills and concepts of the chapter and shows completely worked examples.

Review Exercises These exercises provide practice for the corresponding objectives.

Assessment and Evaluation Masters, pp. 115–116

NAME _____ DATE _____ PERIOD _____

Chapter 5

Chapter 5 Test, Form 2B

1. Change 124.63° to degrees, minutes, and seconds. **1.** __124° 37′ 48″__

2. Write 48° 32′ 15″ as a decimal to the nearest thousandth of a degree. **2.** __48.538°__

3. State the angle measure represented by 1.25 rotations clockwise. **3.** __-450°__

4. Identify all coterminal angles between -360° and 360° for the angle 630°. **4.** __-90° and 270°__

5. Find the measure of the reference angle for 310°. **5.** __50°__

6. Find the value of the cosine for ∠A. **6.** __$\frac{\sqrt{33}}{7}$__

7. Find the value of the cosecant for ∠A. **7.** __$\frac{7}{4}$__

8. Find the value of the cotangent for ∠A. **8.** __$\frac{\sqrt{33}}{4}$__

Exercises 6–8

9. If sec θ = -4, find cos θ. **9.** __$-\frac{1}{4}$__

10. Find tan (-180)°. **10.** __0__

11. Find the exact value of sec 240°. **11.** __-2__

12. Find the exact value of sec θ for angle θ in standard position if the point at (-4, 5) lies on its terminal side. **12.** __$-\frac{\sqrt{41}}{4}$__

13. Suppose θ is an angle in standard position whose terminal side lies in Quadrant IV. If cos θ = $\frac{12}{13}$, find the value of cot θ. **13.** __$-\frac{12}{5}$__

© Glencoe/McGraw-Hill 115 Advanced Mathematical Concepts

NAME _____ DATE _____ PERIOD _____

Chapter 5

Chapter 5 Test, Form 2B (continued)

For Exercises 14 and 15, refer to the figure. The angle of elevation from the far side of the pool to the top of the waterfall is 54°, and the distance is 230 feet.

14. Find the height of the waterfall to the nearest foot. **14.** __186 feet__

15. Find the width across the pool to the nearest foot. **15.** __135 feet__

16. If 0° ≤ x ≤ 360°, solve the equation csc x = -2. **16.** __210° and 330°__

17. Assuming an angle in Quadrant I, evaluate cos (cot⁻¹ $\frac{12}{5}$). **17.** __$\frac{12}{13}$__

18. Given the triangle at the right, find m∠B to the nearest tenth of a degree if a = 12 and c = 22. **18.** __56.9°__

For Exercises 19 and 20, round answers to the nearest tenth.

19. In △ABC, A = 42°, B = 68°, and c = 15. Find a. **19.** __10.7__

20. If A = 27.2°, B = 67.4°, and a = 12.8, find the area of △ABC. **20.** __164.9 units²__

21. Determine the number of possible solutions if A = 110°, a = 5, and b = 4. **21.** __one__

For Exercises 22–25, round answers to the nearest tenth.

22. Determine the greatest possible value for side c if A = 30°, a = 5, and b = 8. **22.** __9.9__

23. In △ABC, A = 59°, b = 12, and c = 4. Find a. **23.** __10.5__

24. In △ABC, a = 4, b = 11, and c = 8. Find m∠B. **24.** __129.8°__

25. If a = 21, b = 15, and c = 28, find the area of △ABC. **25.** __154.7 units²__

Bonus The terminal side of an angle θ in standard position coincides with the line 3x − y = 0 in Quadrant III. Find cos θ to the nearest ten-thousandth. **Bonus:** __-0.3162__

© Glencoe/McGraw-Hill 116 Advanced Mathematical Concepts

SKILLS AND CONCEPTS

OBJECTIVES AND EXAMPLES

Lesson 5-1 Identify angles that are coterminal with a given angle.

- If a 585° angle is in standard position, determine a coterminal angle that is between 0° and 360°. State the quadrant in which the terminal side lies.

  First, determine the number of complete rotations (k) by dividing 585 by 360.

  $$\frac{585}{360} = 1.625$$

  Use $\alpha + 360k°$ to find the value of α.
  $$\alpha + 360(1)° = 585°$$
  $$\alpha = 225°$$

  The coterminal angle (α) is 225°. Its terminal side lies in the third quadrant.

REVIEW EXERCISES

Change each measure to degrees, minutes, and seconds.
11. 57.15° **57° 9′**   **12.** −17.125° **−17° 7′ 30″**

If each angle is in standard position, determine a coterminal angle that is between 0° and 360°. State the quadrant in which the terminal side lies.

13. 860° **140°; II**   **14.** 1146° **66°; I**

15. −156° **204°; III**   **16.** 998° **278°; IV**

17. −300° **60°; I**   **18.** 1072° **352°; IV**

19. 654° **294°; IV**   **20.** −832° **248°; III**

Find the measure of the reference angle for each angle.

21. −284° **76°**   **22.** 592° **52°**

Lesson 5-2 Find the values of trigonometric ratios for acute angles of right triangles.

- Find the values of the six trigonometric ratios for ∠M.

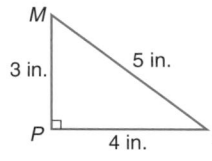

$$\sin M = \frac{4}{5} \qquad \sin M = \frac{side\ opposite}{hypotenuse}$$

$$\cos M = \frac{3}{5} \qquad \cos M = \frac{side\ adjacent}{hypotenuse}$$

$$\tan M = \frac{4}{3} \qquad \tan M = \frac{side\ opposite}{side\ adjacent}$$

$$\csc M = \frac{5}{4} \qquad \csc M = \frac{hypotenuse}{side\ opposite}$$

$$\sec M = \frac{5}{3} \qquad \sec M = \frac{hypotenuse}{side\ adjacent}$$

$$\cot M = \frac{3}{4} \qquad \cot M = \frac{side\ adjacent}{side\ opposite}$$

23. Find the values of the sine, cosine, and tangent for ∠A. **See margin.**

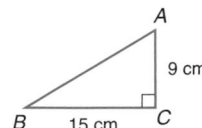

Find the values of the six trigonometric functions for each ∠M. **24–25. See margin.**

24.

25.

26. If sec θ = $\frac{7}{5}$, find cos θ. **$\frac{5}{7}$**

TestCheck and Worksheet Builder

This **networkable software** has three modules.
- **Worksheet Builder** to make worksheets and tests.
- **Student Module** to take tests on-screen.
- **Management System** to keep student records.

Three special banks are included for SAT, ACT, and TIMSS.

OBJECTIVES AND EXAMPLES

Lesson 5-3 Find the values of the six trigonometric functions of an angle in standard position given a point on its terminal side.

• Find the values of the six trigonometric functions for angle θ in standard position if a point with coordinates (3, 4) lies on its terminal side.

$$r = \sqrt{x^2 + y^2} = \sqrt{3^2 + 4^2} = \sqrt{25} \text{ or } 5$$

$$\sin\theta = \frac{y}{r} = \frac{4}{5} \qquad \cos\theta = \frac{x}{r} = \frac{3}{5}$$

$$\tan\theta = \frac{y}{x} = \frac{4}{3} \qquad \csc\theta = \frac{r}{y} = \frac{5}{4}$$

$$\sec\theta = \frac{r}{x} = \frac{5}{3} \qquad \sin\theta = \frac{x}{y} = \frac{3}{4}$$

Lesson 5-4 Use trigonometry to find the measures of the sides of right triangles.

• Refer to $\triangle ABC$ at the right. If $A = 25°$ and $b = 12$, find c.

$$\cos A = \frac{b}{c}$$

$$\cos 25° = \frac{12}{c}$$

$$c = \frac{12}{\cos 25°}$$

$$c \approx 13.2$$

Lesson 5-5 Solve right triangles.

• If $c = 10$ and $a = 9$, find A.

$$\sin A = \frac{a}{c}$$

$$\sin A = \frac{9}{10}$$

$$A = \sin^{-1}\frac{9}{10}$$

$$A \approx 64.2°$$

REVIEW EXERCISES

Find the values of the six trigonometric functions for each angle θ in standard position if a point with the given coordinates lies on its terminal side.   **27–34. See margin.**

27. (3, 3)   28. (−5, 12)
29. (8, −2)   30. (−2, 0)
31. (4, 5)   32. (−5, −9)
33. (−4, 4)   34. (5, 0)

Suppose θ is an angle in standard position whose terminal side lies in the given quadrant. For each function, find the values of the remaining five trigonometric functions for θ.
35–36. See margin.

35. $\cos\theta = -\frac{3}{8}$; Quadrant II

36. $\tan\theta = 3$; Quadrant III

Solve each problem. Round to the nearest tenth.

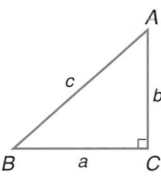

37. If $B = 42°$ and $c = 15$, find b.   **10.0**
38. If $A = 38°$ and $a = 24$, find c.   **39.0**
39. If $B = 67°$ and $b = 24$, find a.   **10.2**

Solve each equation if $0° \le x \le 360°$.

40. $\tan\theta = \frac{\sqrt{3}}{3}$ 41. $\cos\theta = -1$

   30°, 210°   **180°**

Refer to $\triangle ABC$ at the left. Solve each triangle described. Round to the nearest tenth if necessary.   **42–44. See margin.**

42. $B = 49°$, $a = 16$
43. $b = 15$, $c = 20$
44. $A = 64°$, $c = 28$

Chapter 5   Study Guide and Assessment   **337**

Additional Answers

28. $\sin\theta = \frac{12}{13}$, $\cos\theta = -\frac{5}{13}$,
    $\tan\theta = -\frac{12}{5}$, $\csc\theta = \frac{13}{12}$,
    $\sec\theta = -\frac{13}{5}$, $\cot\theta = -\frac{5}{12}$

29. $\sin\theta = -\frac{\sqrt{17}}{17}$, $\cos\theta = \frac{4\sqrt{17}}{17}$,
    $\tan\theta = -\frac{1}{4}$, $\csc\theta = -\sqrt{17}$,
    $\sec\theta = \frac{\sqrt{17}}{4}$, $\cot\theta = -4$

30. $\sin\theta = 0$, $\cos\theta = -1$,
    $\tan\theta = 0$, $\csc\theta$ is undefined,
    $\sec\theta = -1$, $\cot\theta$ is undefined.

31. $\sin\theta = \frac{5\sqrt{41}}{41}$, $\cos\theta = \frac{4\sqrt{41}}{41}$,
    $\tan\theta = \frac{5}{4}$, $\csc\theta = \frac{\sqrt{41}}{5}$,
    $\sec\theta = \frac{\sqrt{41}}{4}$, $\cot\theta = \frac{4}{5}$

32. $\sin\theta = -\frac{9\sqrt{106}}{106}$,
    $\cos\theta = -\frac{5\sqrt{106}}{106}$, $\tan\theta = \frac{9}{5}$,
    $\csc\theta = -\frac{\sqrt{106}}{9}$,
    $\sec\theta = -\frac{\sqrt{106}}{5}$, $\cot\theta = \frac{5}{9}$

33. $\sin\theta = \frac{\sqrt{2}}{2}$, $\cos\theta = -\frac{\sqrt{2}}{2}$,
    $\tan\theta = -1$, $\csc\theta = \sqrt{2}$,
    $\sec\theta = -\sqrt{2}$, $\cot\theta = -1$

34. $\sin\theta = 0$, $\cos\theta = 1$,
    $\tan\theta = 0$, $\csc\theta$ is undefined,
    $\sec\theta = 1$, $\cot\theta$ is undefined.

35. $\sin\theta = \frac{\sqrt{55}}{8}$, $\tan\theta = -\frac{\sqrt{55}}{3}$,
    $\csc\theta = \frac{8\sqrt{55}}{55}$, $\sec\theta = -\frac{8}{3}$,
    $\cot\theta = -\frac{3\sqrt{55}}{55}$

36. $\sin\theta = -\frac{3\sqrt{10}}{10}$,
    $\cos\theta = -\frac{\sqrt{10}}{10}$,
    $\csc\theta = -\frac{\sqrt{10}}{3}$,
    $\sec\theta = -\sqrt{10}$, $\cot\theta = \frac{1}{3}$

42. $A = 41°$, $b = 18.4$, $c = 24.4$

43. $a = 13.2$, $A = 41.4°$, $B = 48.6°$

44. $B = 26°$, $a = 25.2$, $b = 12.3$

Additional Answers (pp. 336–337)

23. $\sin A = \frac{5\sqrt{34}}{34}$, $\cos A = \frac{3\sqrt{34}}{34}$,
    $\tan A = \frac{5}{3}$

24. $\sin M = \frac{3\sqrt{13}}{13}$, $\cos M = \frac{2\sqrt{13}}{13}$,
    $\tan M = \frac{3}{2}$, $\csc M = \frac{\sqrt{13}}{3}$,
    $\sec M = \frac{\sqrt{13}}{2}$, $\cot M = \frac{2}{3}$

25. $\sin M = \frac{5}{6}$, $\cos M = \frac{\sqrt{11}}{6}$,
    $\tan M = \frac{5\sqrt{11}}{11}$, $\csc M = \frac{6}{5}$,
    $\sec M = \frac{6\sqrt{11}}{11}$, $\cot M = \frac{\sqrt{11}}{5}$

27. $\sin\theta = \frac{\sqrt{2}}{2}$, $\cos\theta = \frac{\sqrt{2}}{2}$,
    $\tan\theta = 1$, $\csc\theta = \sqrt{2}$,
    $\sec\theta = \sqrt{2}$, $\cot\theta = 1$

OBJECTIVES AND EXAMPLES

Lesson 5-6 Find the area of a triangle if the measures of two sides and the included angle or the measures of two angles and a side are given.

- Find the area of $\triangle ABC$ if $a = 6$, $b = 4$, and $C = 54°$.

Draw a diagram.

$K = \frac{1}{2}ab \sin C$

$K = \frac{1}{2}(6)(4) \sin 54°$

$K \approx 9.708203932$

The area of $\triangle ABC$ is about 9.7 square units.

REVIEW EXERCISES

Solve each triangle. Round to the nearest tenth. **45. $A = 52°$, $b = 100.2$, $c = 90.4$**

45. $B = 70°$, $C = 58°$, $a = 84$

46. $c = 8$, $C = 49°$, $B = 57°$
$A = 74°$, $a = 10.2$, $b = 8.9$
Find the area of each triangle. Round to the nearest tenth.

47. $A = 20°$, $a = 19$, $C = 64°$ **471.7 units^2**

48. $b = 24$, $A = 56°$, $B = 78°$ **175.6 units^2**

49. $b = 65.5$, $c = 89.4$, $A = 58.2°$ **2488.4 units^2**

50. $B = 22.6°$, $a = 18.4$, $c = 6.7$ **23.7 units^2**

Lesson 5-7 Solve triangles by using the Law of Sines.

- In $\triangle ABC$, if $A = 51°$, $C = 32°$, and $c = 18$, find a.

Draw a diagram.

$\dfrac{a}{\sin A} = \dfrac{c}{\sin C}$

$\dfrac{a}{\sin 51°} = \dfrac{18}{\sin 32°}$

$a = \dfrac{(\sin 51°)18}{\sin 32°}$

$a \approx 26.4$

Find all solutions for each triangle. If no solutions exist, write *none*. Round to the nearest tenth.

51. $A = 38.7°$, $a = 172$, $c = 203$

52. $a = 12$, $b = 19$, $A = 57°$ **none**

53. $A = 29°$, $a = 12$, $c = 15$

54. $A = 45°$, $a = 83$, $b = 79$

51. **$B = 93.7°$, $C = 47.6°$, $b = 274.5$;**
$B = 8.9°$, $C = 132.4°$, $b = 42.3$

53. **$B = 113.7°$, $C = 37.3°$, $b = 22.7$;**
$B = 8.3°$, $C = 142.7°$, $b = 3.6$

54. **$B = 42.3°$, $C = 92.7°$, $c = 117.2$**

Lesson 5-8 Solve triangles by using the Law of Cosines.

- In $\triangle ABC$, if $A = 63°$, $b = 20$, and $c = 14$, find a.

Draw a diagram.

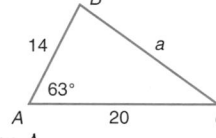

$a^2 = b^2 + c^2 - 2bc \cos A$

$a^2 = 20^2 + 14^2 - 2(20)(14) \cos 63°$

$a^2 \approx 341.77$

$a \approx 18.5$

Solve each triangle. Round to the nearest tenth. **55. $a = 36.9$, $B = 57.4°$, $C = 71.6°$**

55. $A = 51°$, $b = 40$, $c = 45$

56. $B = 19°$, $a = 51$, $c = 61$

57. $a = 11$, $b = 13$, $c = 20$

58. $B = 24°$, $a = 42$, $c = 6.5$

56. **$b = 21.0$, $A = 52.4°$, $C = 108.6°$**

57. **$A = 30.5°$, $B = 36.9°$, $C = 112.6°$**

58. **$b = 36.2$, $A = 151.8°$, $C = 4.2°$**

APPLICATIONS AND PROBLEM SOLVING

59. Camping Haloke and his friends are camping in a tent. Each side of the tent forms a right angle with the ground. The tops of two ropes are attached to each side of the tent 8 feet above the ground. The other ends of the two ropes are attached to stakes on the ground. *(Lesson 5-4)*

a. If the rope is 12 feet long, what angle does it make with the level ground?

b. What is the distance between the bottom of the tent and each stake? **about 8.9 ft**
a. **about 41.8°**

60. Navigation Hugo is taking a boat tour of a lake. The route he takes is shown on the map below. *(Lesson 5-8)*

a. How far is it from the lighthouse to the marina? **about 5.0 mi**

b. What is the angle between the route from the dock to the lighthouse and the route from the lighthouse to the marina? **about 119.5°**

1. Sample answer: about 40 cm and 10.9 cm

2a. Sample answer: $a = 10$, $b = 24$, $A = 30°$; $10 < 24$, $10 < 24 \sin 30°$

2b. Sample answer: $b = 18$; $10 < 18$, $10 > 18 \sin 30°$

ALTERNATIVE ASSESSMENT

OPEN-ENDED ASSESSMENT

1. A triangle has an area of 125 square centimeters and an angle that measures 35°. What are possible lengths of two sides of the triangle?

2. a. Give the lengths of two sides and a nonincluded angle so that no triangle exists. Explain why no triangle exists for the measures you give.

   b. Can you change the length of one of the sides you gave in part a so that two triangles exist? Explain.

PORTFOLIO

Explain how you can find the area of a triangle when you know the length of all three sides of the triangle.

Additional Assessment See p. A60 for Chapter 5 practice test.

Unit 2 *interNET* Project

THE CYBERCLASSROOM

Does anybody out there know anything about trigonometry?

- Search the Internet to find at least three web sites that offer lessons on trigonometry. Some possible sites are actual mathematics courses offered on the Internet or webpages designed by teachers.

- Compare the Internet lessons with the lessons from this chapter. Note any similarities or differences.

- Select one topic from Chapter 5. Combine the information from your textbook and the lessons you found on the Internet. Write a summary of this topic using all the information you have gathered.

APPLICATIONS & PROBLEM SOLVING

This section provides additional practice in solving real-world problems that involve the concepts of this chapter.

ALTERNATIVE ASSESSMENT

Unit 2 *interNET* Project

- **Does Anybody Out There Know Anything About Trigonometry?**

Many Web sites for trigonometry lessons are for college courses. However, some of these are introductory-level classes that have material appropriate for high school students taking an advanced mathematics class. Have students look for sites that have topics similar to what is found in Chapter 5 of their textbook.

Internet Address Book

Record useful Internet addresses in the space at right for quick reference.

www.amc.glencoe.com

Chapter 5 Skill: Pythagorean Theorem

Pages 340–341 are part of a complete SAT/ACT preparation course that is described in detail on pages T8–T9 of the *Teacher's Handbook*. The test items were written in the same style as those in the SAT and ACT.

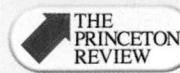 These questions were aligned and verified by The Princeton Review, the nation's leader in test preparation.

Teaching Tip Knowledge of some common Pythagorean triples may help students reduce some calculation time. Remind students that multiples of common Pythagorean triples are often used on standardized tests.
3-4-5
5-12-13
7-24-25

Diagnosis & Prescription

Each of the 10 SAT and ACT practice problems on page 341 is cross-referenced to the chapter where that SAT or ACT skill is covered. If students miss a particular type of problem, you can have them study that skill.

Problem Number and Skill	Page
1 Pythagorean Theorem	340
2 Coordinate Geometry	272
3 Arithmetic	64
4 Pythagorean Theorem	340
5 Algebra	124
6 Arithmetic	64
7 Pythagorean Theorem	340
8 More Algebra	202
9 Pythagorean Theorem	340
10 Pythagorean Theorem	340

Pythagorean Theorem

All SAT and ACT tests contain several problems that you can solve using the Pythagorean Theorem. The **Pythagorean Theorem** states that in a right triangle, the sum of the squares of the measures of the legs equals the square of the measure of the hypotenuse.

$$a^2 + b^2 = c^2$$

SAT EXAMPLE

1. A 25-foot ladder is placed against a vertical wall of a building with the bottom of the ladder standing on concrete 7 feet from the base of the building. If the top of the ladder slips down 4 feet, then the bottom of the ladder will slide out how many feet?

   A 4 ft

   B 5 ft

   C 6 ft

   D 7 ft

   E 8 ft

HINT This problem does not have a diagram. So, start by drawing diagrams.

Solution The ladder placed against the wall forms a 7-24-25 right triangle. After the ladder slips down 4 feet, the new right triangle has sides that are multiples of a 3-4-5 right triangle, 15-20-25.

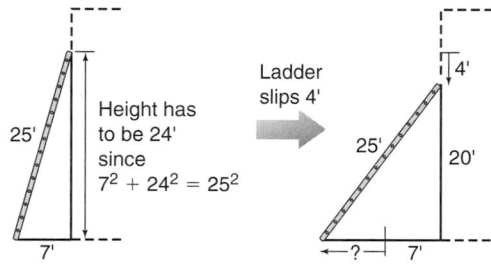

The ladder is now 15 feet from the wall. This means the ladder slipped $15 - 7$ or 8 feet. The correct answer is choice **E**.

ACT EXAMPLE

2. In the figure below, right triangles *ABC* and *ACD* are drawn as shown. If $AB = 20$, $BC = 15$, and $AD = 7$, then $CD = ?$

   A 21

   B 22

   C 23

   D 24

   E 25

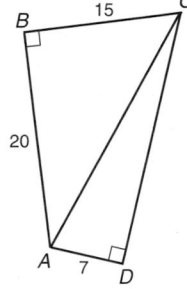

HINT Be on the lookout for problems like this one in which the application of the Pythagorean Theorem is not obvious.

Solution Notice that quadrilateral *ABCD* is separated into two right triangles, $\triangle ABC$ and $\triangle ADC$.

$\triangle ABC$ is a 15-20-25 right triangle (a multiple of the 3-4-5 right triangle). So, side $\overline{AC}$ (the hypotenuse) is 25 units long.

$\overline{AC}$ is also the hypotenuse of $\triangle ADC$. So, $\triangle ADC$ is a 7-24-25 right triangle.

Therefore, $\overline{CD}$ is 24 units long. The correct answer is choice **D**.

TestCheck and Worksheet Builder

Special banks of SAT, ACT, and TIMSS items are included in this CD-ROM.

After working each problem, record the correct answer on the answer sheet provided or use your own paper.

Multiple Choice

1. In the figure below, $y =$ **A**

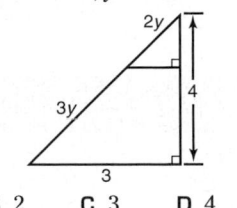

A 1   **B** 2   **C** 3   **D** 4   **E** 5

2. What graph would be created if the equation $x^2 + y^2 = 12$ were graphed in the standard (x, y) coordinate plane? **A**

A circle   **B** ellipse

C parabola   **D** straight line

E 2 rays forming a "V"

3. If $999 \times 111 = 3 \times 3 \times n^2$, then which of the following could equal n? **C**

A 9   **B** 37   **C** 111

D 222   **E** 333

4. In the figure below, $\triangle ABC$ is an equilateral triangle with $\overline{BC}$ 7 units long. If $\angle DCA$ is a right angle and $\angle D$ measures $45°$, what is the length of $\overline{AD}$ in units? **B**

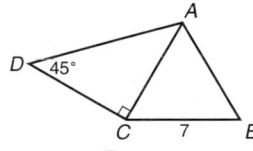

A 7   **B** $7\sqrt{2}$ **C** 14   **D** $14\sqrt{2}$

E It cannot be determined from the information given.

5. If $4 < a < 7 < b < 9$, then which of the following best defines $\frac{a}{b}$? **A**

A $\frac{4}{9} < \frac{a}{b} < 1$ **B** $\frac{4}{9} < \frac{a}{b} < \frac{7}{9}$

C $\frac{4}{7} < \frac{a}{b} < \frac{7}{9}$ **D** $\frac{4}{7} < \frac{a}{b} < 1$

E $\frac{4}{7} < \frac{a}{b} < \frac{9}{7}$

6. A swimming pool with a capacity of 36,000 gallons originally contained 9,000 gallons of water. At 10:00 A.M. water begins to flow into the pool at a constant rate. If the pool is exactly three-fourths full at 1:00 P.M. on the same day and the water continues to flow at the same rate, what is the earliest time when the pool will be completely full? **C**

A 1:40 P.M.   **B** 2:00 P.M.   **C** 2:30 P.M.

D 3:00 P.M.   **E** 3:30 P.M.

7. In the figure below, what is the length of $\overline{BC}$? **B**

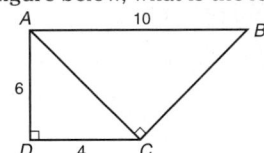

A 6   **B** $4\sqrt{3}$ **C** $2\sqrt{13}$

D 8   **E** $2\sqrt{38}$

8. If $\dfrac{x^2 + 7x + 12}{x + 4} = 5$, then $x =$ **B**

A 1   **B** 2   **C** 3   **D** 5   **E** 6

9. What is the value of $\dfrac{AC}{AD}$ if $ABCD$ is a square? **B**

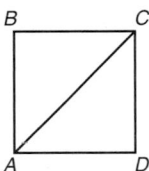

A 1

B $\sqrt{2}$

C $\sqrt{3}$

D 2

E $2\sqrt{2}$

10. Grid-In Segment AB is perpendicular to segment BD. Segment AB and segment CD bisect each other at point X. If $AB = 8$ and $CD = 10$, what is the length of BD? **3**

interNET CONNECTION **SAT/ACT Practice** For additional test practice questions, visit: **www.amc.glencoe.com**

THE PRINCETON REVIEW   SAT & ACT Preparation   **341**

A **bubble-in answer sheet** for these practice problems is available on p. 1 of the *Assessment and Evaluation Masters.*

Additional Practice

Additional SAT and ACT practice questions and a bubble-in answer sheet are available in the *Assessment and Evaluation Masters,* pp. 2, 123, and 124.

Assessment and Evaluation Masters, pp. 123–124

Extra Credit

TIMSS The ride with the cable car from Station A to Station B at the top of Mt. Glacier Takes 16 minutes. The average speed of the cable car is 2 meters per second and it moves in a straight line forming a 25° angle with the horizontal. Find the height of Mt. Glacier (measured from the level of Station A) to the nearest meter. Show all your work.   **811 m or 0.811 km**

Instructional Objectives

Lesson (pages)	Lesson Objectives	NCTM Standards 2000	State/Local Objectives
6-1 (343–351)	• Change from radian measure to degree measure, and vice versa. • Find the length of an arc given the measure of the central angle. • Find the area of a sector.	3, 4, 6	
6-2 (352–358)	• Find linear and angular velocity.	3, 4, 6	
6-3 (359–366)	• Use the graphs of the sine and cosine functions.	3, 4, 6, 10	
6-4 (368–377)	• Find the amplitude and period for sine and cosine functions. • Write equations of sine and cosine functions given the amplitude and period.	3, 4, 6, 10	
6-5 (378–386)	• Find the phase shift and the vertical translation for sine and cosine functions. • Write the equations of sine and cosine functions given the amplitude, period, phase shift, and vertical translation. • Graph compound functions.	3, 4, 6, 10	
6-6 (387–394)	• Model real-world data using sine and cosine functions. • Use sinusoidal functions to solve problems.	3, 4, 6, 10	
6-7 (395–403)	• Graph tangent, cotangent, secant, and cosecant functions. • Write equations of trigonometric functions.	3, 4, 6, 10	
6-7B (404)	• Use a graphing calculator to model beat effects produced by waves of almost equal frequencies.	3, 4, 6, 9, 10	
6-8 (405–412)	• Graph inverse trigonometric functions. • Find principal values of inverse trigonometric functions.	3, 4, 6, 10	

Suggested Pacing   *See pages T2–T4 for additional options in course-planning.*

COURSE	Lesson 6-1	Lesson 6-2	Lesson 6-3	Lesson 6-4	Lesson 6-5
Precalculus Emphasis Chapters 4–15 (45–50 minutes)	1.5 days	1.5 days	1.5 days	1.5 days	1.5 days
Precalculus Emphasis (block)	0.5 day	1.5 days		1.5 days	
Trigonometric Emphasis Chapters 1–9, 11, 12 (45–50 minutes)	1.5 days	1.5 days	1.5 days	1.5 days	1.5 days
Trigonometric Emphasis (block)	0.5 day	1.5 days		1.5 days	

Instructional Resources

Student Edition		Blackline Masters (page numbers)								Transparencies A and B
Lesson	Graphing Calculator Applications	Study Guide and Practice	Enrichment	Assessment and Evaluation	Graphing Calculator Keystrokes	Problem Solving & Applications	Spreadsheet Applications	TI-92 and CBL Interdisciplinary Activities		Transparencies A and B
6-1		99–100	61				6			6-1
6-2		101–102	62	147						6-2
6-3	364	103–104	63							6-3
6-4	369, 372, 374	105–106	64	146, 147	22					6-4
6-5	378	107–108	65		23	11		17–20		6-5
6-6	389, 390	109–110	66	148						6-6
6-7		111–112	67							6-7
6-7B	404				24–25					
6-8		113–114	68	148		12				6-8
Study Guide/ Assessment				133–145, 151						
SAT/ACT Practice				149–150						

Lesson 6-6	Lesson 6-7	Lesson 6-8	Study Guide & Assessment	Chapter Test	TOTAL
1.5 days	(with Lesson 6-7B) 1.5 days	1.5 days	1 day	1 day	14 days
(with Lesson 6-7B) 1.5 days		1.5 days		0.5 day	7 days
1.5 days	(with Lesson 6-7B) 1.5 days	1.5 days	1 day	1 day	14 days
(with Lesson 6-7B) 1.5 days		1.5 days		0.5 day	7 days

TeacherWorks™

The worksheets shown on this page are a small sample of the materials available on *TeacherWorks: All-in-One Lesson Planner and Resource Center.*

This CD-ROM contains all of the blackline masters and transparencies available for this program.

It also includes a lesson planner and interactive Teacher Edition, so you can customize lesson plans and reproduce classroom resources quickly and easily, from just about anywhere.

Applications

Problem Solving and Applications Masters, pp. 11, 12

Technology/Multimedia

Spreadsheet Applications Masters, p. 6

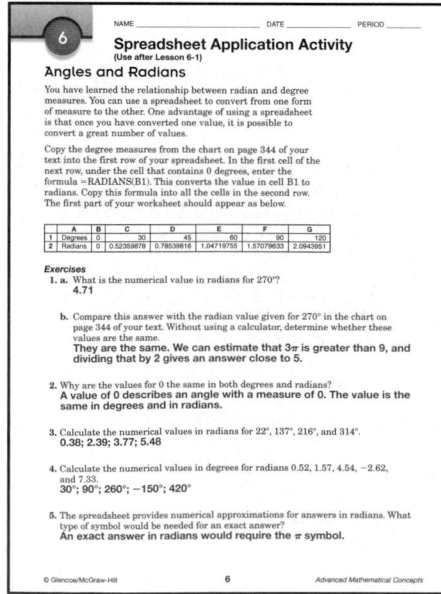

TI-92 & CBL Interdisciplinary Activities Masters, pp. 17–20

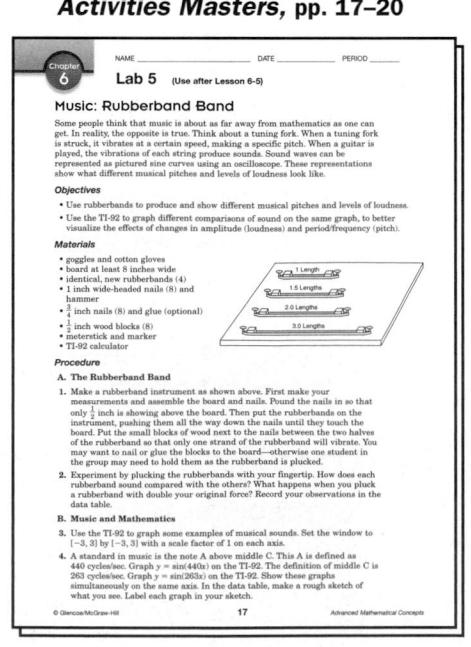

Type	Student Edition	Teacher's Wraparound Edition	Assessment and Evaluation Masters
Ongoing Assessment	Mid-Chapter Quiz, p. 377	5-Minute Check, pp. 343, 352, 359, 368, 378, 387, 395, 405	Mid-Chapter Test, p. 146 Quizzes A–D, pp. 147–148
Mixed Review	Mixed Review, pp. 350–351, 357–358, 366, 376–377, 386, 394,402–403, 412		Cumulative Review, p. 151
Error Analysis	You Decide, pp. 383, 410	Error Analysis, pp. 347, 355, 363, 372, 383, 390, 400, 410	
Standardized Test Practice	SAT/ACT Practice, pp. 351, 358, 366, 377, 386, 394, 403, 412 SAT & ACT Preparation, pp. 418–419	TIMSS, p. 419	SAT/ACT Practice, pp. 149–150
Open-Ended Assessment	Math Journal, pp. 363, 373 Open-Ended Assessment, p. 417 Portfolio, p. 417	Modeling: pp. 351, 366, 386 Speaking: pp. 377, 394, 404 Writing: pp. 358, 403, 412	Open-Ended Assessment, p. 145
Chapter Assessment	Study Guide and Assessment, pp. 413–417 Chapter Test, p. A61		Multiple-Choice Tests (Forms 1A, 1B, 1C), pp. 133–138 Free-Response Tests (Forms 2A, 2B, 2C), pp. 139–144

TestCheck and Worksheet Builder is a networkable CD-ROM with three integrated modules. The **Worksheet Builder** creates customized worksheets, tests, and quizzes of free-response, multiple-choice, short-answer, and open-ended items. The **Student Module** gives you the option of having students take tests on-screen and get immediate feedback on their performance. Use the optional **Management System** to keep detailed student records.

Additional Chapter Resources

Student Edition
Career Choices, p. 358
History of Mathematics, p. 367
Internet Project, p. 417
Extra Practice, pp. A36–A37

Teacher's Classroom Resources

Technology
Graphing Calculator Keystrokes Masters

Teaching Aids
Answer Key Transparencies
Block Schedule Planning Guide
Lesson Planning Guide
Solutions Manual

Glencoe Instructional Technology

Instructional

Advanced MathPASS CD-ROM

Multimedia Applications CD-ROM

Assessment

TestCheck and Worksheet Builder

Visit **www.amc.glencoe.com** for data updates, career information, graphing calculator programs, and other interactive activities.

Chapter Overview

This chapter reviews graphs of trigonometric functions and their inverses using radian measure. Students also learn about amplitude, period, phase shift, and vertical shift. Applications include finding linear and angular velocity and writing trigonometric equations to model real-life situations.

 Using a Graphing Calculator

There are numerous opportunities, all optional, to use a graphing calculator in this chapter.
- Graphing various forms of sine and cosine functions (pp. 364, 374, 390)
- Making conjectures about the behavior of trigonometric functions (pp. 369, 378)
- Matching equations to graphing calculator screen shots (p. 386)
- Using **SinReg** to model trigonometric functions (p. 389)
- Modeling beat effects produced by waves of almost equal frequencies (p. 404)

The **Graphing Calculator Appendix**, which begins on page A2, contains keystroke instructions on many of these topics.

GRAPHS OF TRIGONOMETRIC FUNCTIONS

CHAPTER OBJECTIVES
- Change from radian measure to degree measure, and vice versa. *(Lesson 6-1)*
- Find linear and angular velocity. *(Lesson 6-2)*
- Use and draw graphs of trigonometric functions and their inverses. *(Lessons 6-3, 6-4, 6-5, 6-6, 6-7, 6-8)*
- Find the amplitude, the period, the phase shift, and the vertical shift for trigonometric functions. *(Lessons 6-4, 6-5, 6-6, 6-7)*
- Write trigonometric equations to model a given situation. *(Lessons 6-4, 6-5, 6-6, 6-7)*

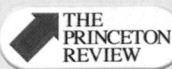 **SAT & ACT Preparation**

This textbook contains a complete SAT/ACT preparation course described on pages T8–T9 of the *Teacher's Handbook*. This chapter contains:
- an SAT or ACT practice question at the end of each lesson
- examples of ACT problems involving trigonometry on page 418
- ten ACT practice problems on page 419
- additional ACT practice in the *Assessment and Evaluation Masters,* pp. 149–150
- SAT and ACT test practice banks in the *TestCheck and Worksheet Builder*

Angles and Radian Measure

OBJECTIVES
* Change from radian measure to degree measure, and vice versa.
* Find the length of an arc given the measure of the central angle.
* Find the area of a sector.

BUSINESS Junjira Putiwuthigool owns a business in Changmai, Thailand, that makes ornate umbrellas and fans. Ms. Putiwuthigool has an order for three dozen umbrellas having a diameter of 2 meters. Bamboo slats that support each circular umbrella divide the umbrella into 8 sections or sectors. Each section will be covered with a different color fabric. How much fabric of each color will Ms. Putiwuthigool need to complete the order? *This problem will be solved in Example 6.*

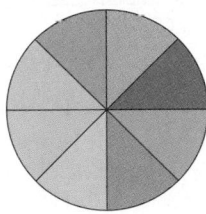

There are many real-world applications, such as the one described above, which can be solved more easily using an angle measure other than the degree. This other unit is called the **radian.**

The definition of radian is based on the concept of the unit circle. Recall that the unit circle is a circle of radius 1 whose center is at the origin of a rectangular coordinate system.

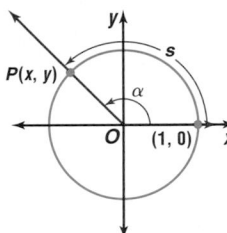

A point $P(x, y)$ is on the unit circle if and only if its distance from the origin is 1. Thus, for each point $P(x, y)$ on the unit circle, the distance from the origin is represented by the following equation.

$$\sqrt{(x - 0)^2 + (y - 0)^2} = 1$$

If each side of this equation is squared, the result is an equation of the unit circle.

$$x^2 + y^2 = 1$$

Consider an angle α in standard position, shown above. Let $P(x, y)$ be the point of intersection of its terminal side with the unit circle. The radian measure of an angle in standard position is defined as the length of the corresponding arc on the unit circle. Thus, the measure of angle α is s radians. Since $C = 2\pi r$, a full revolution correponds to an angle of $2\pi(1)$ or 2π radians.

There is an important relationship between radian and degree measure. Since an angle of one complete revolution can be represented either by $360°$ or by 2π radians, $360° = 2\pi$ radians. Thus, $180° = \pi$ radians, and $90° = \frac{\pi}{2}$ radians.

Lesson 6-1 Angles and Radian Measure **343**

The following formulas relate degree and radian measures.

Degree/ Radian Conversion Formulas	1 radian = $\frac{180}{\pi}$ degrees or about 57.3°
	1 degree = $\frac{\pi}{180}$ radians or about 0.017 radian

Angles expressed in radians are often written in terms of π. The term *radians* is also usually omitted when writing angle measures. However, the degree symbol is always used in this book to express the measure of angles in degrees.

Example **a. Change 330° to radian measure in terms of π.**

$330° = 330° \times \frac{\pi}{180°}$ *1 degree = $\frac{\pi}{180°}$*

$\quad = \frac{11\pi}{6}$

b. Change $\frac{2\pi}{3}$ radians to degree measure.

$\frac{2\pi}{3} = \frac{2\pi}{3} \times \frac{180°}{\pi}$ *1 radian = $\frac{180°}{\pi}$*

$\quad = 120°$

Angles whose measures are multiples of 30° and 45° are commonly used in trigonometry. These angle measures correspond to radian measures of $\frac{\pi}{6}$ and $\frac{\pi}{4}$, respectively. The diagrams below can help you make these conversions mentally.

You may want to memorize these radian measures and their degree equivalents to simplify your work in trigonometry.

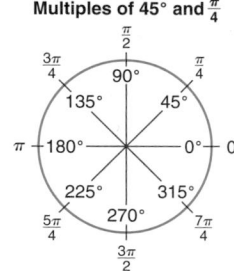

These equivalent values are summarized in the chart below.

Degrees	0	30	45	60	90	120	135	150	180	210	225	240	270	300	315	330
Radians	0	$\frac{\pi}{6}$	$\frac{\pi}{4}$	$\frac{\pi}{3}$	$\frac{\pi}{2}$	$\frac{2\pi}{3}$	$\frac{3\pi}{4}$	$\frac{5\pi}{6}$	π	$\frac{7\pi}{6}$	$\frac{5\pi}{4}$	$\frac{4\pi}{3}$	$\frac{3\pi}{2}$	$\frac{5\pi}{3}$	$\frac{7\pi}{4}$	$\frac{11\pi}{6}$

You can use reference angles and the unit circle to determine trigonometric values for angle measures expressed as radians.

344 *Chapter 6 Graphs of Trigonometric Functions*

Prerequisite Skills

A review of the prerequisite skills for this lesson is provided in the *Prerequisite Skills Masters*.

Skill Number and Content	Pages
2 Fractions	5–6
8 Rationalizing a Denominator	17–18

Example ② Evaluate $\cos \frac{4\pi}{3}$.

The reference angle for $\frac{4\pi}{3}$ is $\frac{4\pi}{3} - \pi$ or $\frac{\pi}{3}$.

Since $\frac{\pi}{3} = 60°$, the terminal side of the angle intersects the unit circle at a point with coordinates of $\left(\frac{1}{2}, \frac{\sqrt{3}}{2}\right)$.

Because the terminal side of this angle is in the third quadrant, both coordinates are negative. The point of intersection has coordinates $\left(-\frac{1}{2}, -\frac{\sqrt{3}}{2}\right)$.

Therefore, $\cos \frac{4\pi}{3} = -\frac{1}{2}$.

Radian measure can be used to find the length of a **circular arc**. A circular arc is a part of a circle. The arc is often defined by the **central angle** that intercepts it. A central angle of a circle is an angle whose vertex lies at the center of the circle.

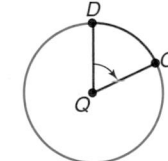

If two central angles in different circles are congruent, the ratio of the lengths of their intercepted arcs is equal to the ratio of the measures of their radii.

For example, given circles O and Q, if $\angle O \cong \angle Q$, then $\frac{m\widehat{AB}}{m\widehat{CD}} = \frac{OA}{QC}$.

Let O be the center of two concentric circles, let r be the measure of the radius of the larger circle, and let the smaller circle be a unit circle. A central angle of θ radians is drawn in the two circles that intercept $\widehat{RT}$ on the unit circle and $\widehat{SW}$ on the other circle. Suppose $\widehat{SW}$ is s units long. $\widehat{RT}$ is θ units long since it is an arc of a unit circle intercepted by a central angle of θ radians. Thus, we can write the following proportion.

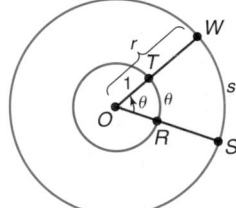

We say that an arc subtends its central angle.

$$\frac{s}{\theta} = \frac{r}{1} \text{ or } s = r\theta$$

Length of an Arc	The length of any circular arc s is equal to the product of the measure of the radius of the circle r and the radian measure of the central angle θ that it subtends. $$s = r\theta$$

From the Classroom of...

Mary Dengerud-Au
Pasco High School
Pasco, Washington

"I have students draw a circle and a diameter. Using string cut to the length of the circle's radius, students measure out an arc beginning at one end of the diameter. When the end of the arc is connected to the center of the circle, this sector forms a radian. Students continue to make radians. They discover that there are about 3 radians in a semi-circle and that the measure of a radian is about 57°."

Teaching Tip Remind students that once they locate the reference angle, they should watch the signs of each of the six trigonometric functions. These signs depend on the quadrant in which the reference angle lies.

In-Class Example
Example 2
Evaluate $\tan \frac{25\pi}{6}$. $\frac{\sqrt{3}}{3}$

Teaching Tip Remind students that the length of an arc is a part of the circumference. Thus, $\frac{\theta}{2\pi} = \frac{\text{arc length}}{2\pi r}$. On a unit circle $r = 1$, so $\frac{\theta}{2\pi} = \frac{\text{arc length}}{2\pi}$ or $\theta = \text{arc length}$.

Example ③ Given a central angle of 128°, find the length of its intercepted arc in a circle of radius 5 centimeters. Round to the nearest tenth.

First, convert the measure of the central angle from degrees to radians.

$$128° = 128° \times \frac{\pi}{180°} \quad \text{1 degree} = \frac{\pi}{180}$$
$$= \frac{32}{45}\pi \text{ or } \frac{32\pi}{45}$$

Then, find the length of the arc.

$$s = r\theta$$
$$s = 5\left(\frac{32\pi}{45}\right) \quad r = 5, \theta = \frac{32\pi}{45}$$
$$s \approx 11.17010721 \quad \text{Use a calculator.}$$

The length of the arc is about 11.2 centimeters.

You can use radians to compute distances between two cities that lie on the same longitude line.

Example ④ **GEOGRAPHY** Winnipeg, Manitoba, Canada, and Dallas, Texas, lie along the 97° W longitude line. The latitude of Winnipeg is 50° N, and the latitude of Dallas is 33° N. The radius of Earth is about 3960 miles. Find the approximate distance between the two cities.

The length of the arc between Dallas and Winnipeg is the distance between the two cities. The measure of the central angle subtended by this arc is 50° − 33° or 17°.

$$17° = 17° \times \frac{\pi}{180°} \quad \text{1 degree} = \frac{\pi}{180}$$
$$= \frac{17\pi}{180}$$

$$s = r\theta$$
$$s = 3960\left(\frac{17\pi}{180}\right) \quad r = 3960, \theta = \frac{17\pi}{180}$$
$$s \approx 1174.955652 \quad \text{Use a calculator.}$$

The distance between the two cities is about 1175 miles.

A **sector** of a circle is a region bounded by a central angle and the intercepted arc. For example, the shaded portion in the figure is a sector of circle O. The ratio of the area of a sector to the area of a circle is equal to the ratio of its arc length to the circumference.

Addressing Individual Needs

Verbal/Linguistic Have each student write a problem that gives a central angle in degree measure and the length of the radius. Then have each student show how to find the area of the sector and the length of the arc. Have the students describe each step to the class.

Let A represent the area of the sector.

$$\frac{A}{\pi r^2} = \frac{\text{length of } \overset{\frown}{RTS}}{2\pi r}$$

$$\frac{A}{\pi r^2} = \frac{r\theta}{2\pi r} \qquad \textit{The length of } \overset{\frown}{RTS} \textit{ is } r\theta.$$

$$A = \frac{1}{2}r^2\theta \qquad \textit{Solve for A.}$$

Area of a Circular Sector	If θ is the measure of the central angle expressed in radians and r is the measure of the radius of the circle, then the area of the sector, A, is as follows. $$A = \frac{1}{2}r^2\theta$$

Examples

5 Find the area of a sector if the central angle measures $\frac{5\pi}{6}$ radians and the radius of the circle is 16 centimeters. Round to the nearest tenth.

$A = \frac{1}{2}r^2\theta$ *Formula for the area of a circular sector*

$A = \frac{1}{2}(16^2)\left(\frac{5\pi}{6}\right)$ *$r = 16$, $\theta = \frac{5\pi}{6}$*

$A \approx 335.1032164$ *Use a calculator.*

The area of the sector is about 335.1 square centimeters.

6 **BUSINESS** Refer to the application at the beginning of the lesson. How much fabric of each color will Ms. Putiwuthigool need to complete the order?

There are 2π radians in a complete circle and 8 equal sections or sectors in the umbrella. Therefore, the measure of each central angle is $\frac{2\pi}{8}$ or $\frac{\pi}{4}$ radians. If the diameter of the circle is 2 meters, the radius is 1 meter. Use these values to find the area of each sector.

$A = \frac{1}{2}r^2\theta$

$A = \frac{1}{2}(1^2)\left(\frac{\pi}{4}\right)$ *$r = 1$, $\theta = \frac{\pi}{4}$*

$A \approx 0.3926990817$ *Use a calculator.*

Since there are 3 dozen or 36 umbrellas, multiply the area of each sector by 36. Ms. Putiwuthigool needs about 14.1 square meters of each color of fabric. *This assumes that the pieces can be cut with no waste and that no extra material is needed for overlapping.*

CHECK FOR UNDERSTANDING

Communicating Mathematics

Read and study the lesson to answer each question.

1. **Draw** a unit circle and a central angle with a measure of $\frac{3\pi}{4}$ radians.

2. **Describe** the angle formed by the hands of a clock at 3:00 in terms of degrees and radians.
$90°; \frac{\pi}{4}$

See margin.

Lesson 6-1 Angles and Radian Measure **347**

Additional Answer
1.

In-Class Examples
Example 5
Find the area of a sector if the central angle measures $\frac{3\pi}{7}$ radians and the radius of the circle is 11 centimeters. Round to the nearest tenth. **81.5 cm²**

Example 6
Chemistry In the universe, the most common elements are hydrogen (73.9%), helium (24.00%), oxygen (1.07%), and all others combined (1.03%). On a circle graph with a radius of 1 inch, what area of the circle should represent hydrogen? Round to the nearest tenth. **2.3 in²**

3 PRACTICE/APPLY

Error Analysis

Watch for students who forget to put square units on area-of-sector problems and linear units on length-of-arc problems.

Prevent by reminding students that area is always expressed in square units.

Study Guide and Practice Masters, pp. 99–100

Additional Answer

4. Let $R = 2r$. For the circle with radius R, $s' = R\theta$ or $2r\theta$ which is $2(r\theta)$. Thus, $s' = 2s$. For the circle with radius R, $A' = \frac{1}{2}R^2\theta$ or $\frac{1}{2}(2r)^2\theta$ which is $\frac{1}{2}(4r^2)\theta$ or $4\left(\frac{1}{2}r^2\theta\right)$. Thus, $A' = 4A$.

---

3. Divide 10 by 8.

3. **Explain** how you could find the radian measure of a central angle subtended by an arc that is 10 inches long in a circle with a radius of 8 inches.

4. **Demonstrate** that if the radius of a circle is doubled and the measure of a central angle remains the same, the length of the arc is doubled and the area of the sector is quadrupled. **See margin.**

Guided Practice

Change each degree measure to radian measure in terms of π.

5. $240°$ $\quad \frac{4\pi}{3}$

6. $570°$ $\quad \frac{19\pi}{6}$

Change each radian measure to degree measure. Round to the nearest tenth, if necessary.

7. $\frac{3\pi}{2}$ $\quad 270°$

8. -1.75 $\quad -100.3°$

Evaluate each expression.

9. $\sin\frac{3\pi}{4}$ $\quad \frac{\sqrt{2}}{2}$

10. $\tan\frac{11\pi}{6}$ $\quad -\frac{\sqrt{3}}{3}$

Given the measurement of a central angle, find the length of its intercepted arc in a circle of radius 15 inches. Round to the nearest tenth.

11. $\frac{5\pi}{6}$ $\quad 39.3$ in.

12. $77°$ $\quad 20.2$ in.

Find the area of each sector given its central angle θ and the radius of the circle. Round to the nearest tenth.

13. $\theta = \frac{2\pi}{3}, r = 1.4$ $\quad 2.1$ units^2

14. $\theta = 54°, r = 6$ $\quad 17.0$ units^2

15. **Physics** A pendulum with length of 1.4 meters swings through an angle of $30°$. How far does the bob at the end of the pendulum travel as it goes from left to right? **about 0.7 m**

EXERCISES

Practice

Change each degree measure to radian measure in terms of π.

 A

16. $135°$ $\quad \frac{3\pi}{4}$

17. $210°$ $\quad \frac{7\pi}{6}$

18. $300°$ $\quad \frac{5\pi}{3}$

19. $-450°$ $\quad -\frac{5\pi}{2}$

20. $-75°$ $\quad -\frac{5\pi}{12}$

21. $1250°$ $\quad \frac{125\pi}{18}$

Change each radian measure to degree measure. Round to the nearest tenth, if necessary.

22. $\frac{7\pi}{12}$ $\quad 105°$

23. $\frac{11\pi}{3}$ $\quad 660°$

24. 17 $\quad 974.0°$

25. -3.5 $\quad -200.5°$

26. $-\frac{\pi}{6.2}$ $\quad -29.0°$

27. 17.5 $\quad 1002.7°$

Evaluate each expression.

 B

28. $\sin\frac{5\pi}{3}$ $\quad -\frac{\sqrt{3}}{2}$

29. $\tan\frac{7\pi}{6}$ $\quad \frac{\sqrt{3}}{3}$

30. $\cos\frac{5\pi}{4}$ $\quad -\frac{\sqrt{2}}{2}$

31. $\sin\frac{7\pi}{6}$ $\quad -\frac{1}{2}$

32. $\tan\frac{14\pi}{3}$ $\quad -\sqrt{3}$

33. $\cos\left(-\frac{19\pi}{6}\right)$ $\quad -\frac{\sqrt{3}}{2}$

Given the measurement of a central angle, find the length of its intercepted arc in a circle of radius 14 centimeters. Round to the nearest tenth.

34. $\frac{2\pi}{3}$ **29.3 cm**

35. $\frac{5\pi}{12}$ **18.3 cm**

36. 150° **36.7 cm**

37. 282° **68.9 cm**

38. $\frac{3\pi}{11}$ **12.0 cm**

39. 320° **78.2 cm**

40. The diameter of a circle is 22 inches. If a central angle measures 78°, find the length of the intercepted arc. **about 15.0 in.**

41. An arc is 70.7 meters long and is intercepted by a central angle of $\frac{5\pi}{4}$ radians. Find the diameter of the circle. **about 36.0 m**

42. An arc is 14.2 centimeters long and is intercepted by a central angle of 60°. What is the radius of the circle? **about 13.6 cm**

Find the area of each sector given its central angle θ and the radius of the circle. Round to the nearest tenth.

43. 65.4 units²

44. 380.1 units²

46. 140.2 units²

47. 70.7 units²

48. 38.1 units²

43. $\theta = \frac{5\pi}{12}, r = 10$

44. $\theta = 90°, r = 22$

45. $\theta = \frac{\pi}{8}, r = 7$ **9.6 units²**

46. $\theta = \frac{4\pi}{7}, r = 12.5$

47. $\theta = 225°, r = 6$

48. $\theta = 82°, r = 7.3$

49. A sector has arc length of 6 feet and central angle of 1.2 radians.
    a. Find the radius of the circle. **5 ft**
    b. Find the area of the sector. **15 ft²**

50. A sector has a central angle of 135° and arc length of 114 millimeters.
    a. Find the radius of the circle. **about 48.4 mm**
    b. Find the area of the sector. **about 2757.8 mm²**

51. A sector has area of 15 square inches and central angle of 0.2 radians.
    a. Find the radius of the circle. **about 12.2 in.**
    b. Find the arc length of the sector. **about 2.4 in.**

52. A sector has area of 15.3 square meters. The radius of the circle is 3 meters.
    a. Find the radian measure of the central angle. **3.4 radians**
    b. Find the degree measure of the central angle. **about 194.8°**
    c. Find the arc length of the sector. **10.2 m**

Applications and Problem Solving

Real World Application

53a. about 7.9 ft

54. about 1.4 radians

53. **Mechanics** A wheel has a radius of 2 feet. As it turns, a cable connected to a box winds onto the wheel.

   a. How far does the box move if the wheel turns 225° in a counterclockwise direction?
   b. Find the number of degrees the wheel must be rotated to move the box 5 feet. **about 143.2°**

54. **Critical Thinking** Two gears are interconnected. The smaller gear has a radius of 2 inches, and the larger gear has a radius of 8 inches. The smaller gear rotates 330°. Through how many radians does the larger gear rotate?

55. **Physics** A pendulum is 22.9 centimeters long, and the bob at the end of the pendulum travels 10.5 centimeters. Find the degree measure of the angle through which the pendulum swings. **about 26.3°**

Lesson 6-1 Angles and Radian Measure **349**

56. Geography Minneapolis, Minnesota; Arkadelphia, Arkansas; and Alexandria, Louisiana lie on the same longitude line. The latitude of Minneapolis is 45° N, the latitude of Arkadelphia is 34° N, and the latitude of Alexandria is 31° N. The radius of Earth is about 3960 miles.

a. Find the approximate distance between Minneapolis and Arkadelphia.

b. What is the approximate distance between Minneapolis and Alexandria?

c. Find the approximate distance between Arkadelphia and Alexandria.

57. Civil Engineering The figure below shows a stretch of roadway where the curves are arcs of circles.

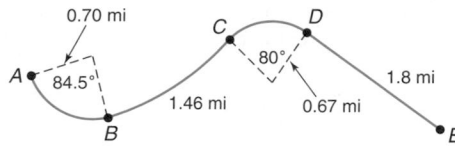

Find the length of the road from point *A* to point *E*. **about 5.23 mi**

58. Mechanics A single pulley is being used to pull up a weight. Suppose the diameter of the pulley is $2\frac{1}{2}$ feet.

a. How far will the weight rise if the pulley turns 1.5 rotations? **about 11.8 ft**

b. Find the number of degrees the pulley must be rotated to raise the weight $4\frac{1}{2}$ feet. **about 206.3°**

59. Pet Care A rectangular house is 33 feet by 47 feet. A dog is placed on a leash that is connected to a pole at the corner of the house.

a. If the leash is 15 feet long, find the area the dog has to play. **about 530.1 ft²**

b. If the owner wants the dog to have 750 square feet to play, how long should the owner make the leash? **about 17.8 ft**

60. Biking Rafael rides his bike 3.5 kilometers. If the radius of the tire on his bike is 32 centimeters, determine the number of radians that a spot on the tire will travel during the trip. **10,937.5 radians**

61. Critical Thinking A *segment* of a circle is the region bounded by an arc and its chord. Consider any minor arc. If α is the radian measure of the central angle and r is the radius of the circle, write a formula for the area of the segment. $A = \frac{1}{2} r^2(\alpha - \sin \alpha)$

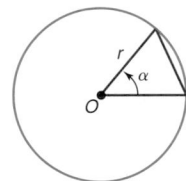

Mixed Review

62. The lengths of the sides of a triangle are 6 inches, 8 inches, and 12 inches. Find the area of the triangle. *(Lesson 5-8)* **about 21.3 in²**

63. Determine the number of possible solutions of $\triangle ABC$ if $A = 152°$, $b = 12$, and $a = 10.2$. If solutions exist, solve the triangle. *(Lesson 5-7)* **no solution**

64. Surveying Two surveyors are determining measurements to be used to build a bridge across a canyon. The two surveyors stand 560 yards apart on one side of the canyon and sight a marker C on the other side of the canyon at angles of 27° and 38°. Find the length of the bridge if it is built through point C as shown. *(Lesson 5-6)* **about 172.7 yd**

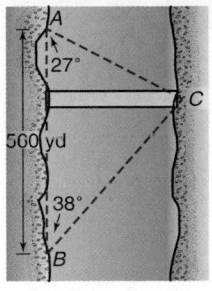

65. Suppose θ is an angle in standard position and $\tan \theta > 0$. State the quadrants in which the terminal side of θ can lie. *(Lesson 5-3)* **I, III**

66. Population The population for Forsythe County, Georgia, has experienced significant growth in recent years. *(Lesson 4-8)*

Year	1970	1980	1990	1998
Population	17,000	28,000	44,000	86,000

Source: U.S. Census Bureau

66a. Sample answer: $y = 102x^2 - 505x + 18,430$

66b. Sample answer: about 248,180

a. Write a model that relates the population of Forsythe County as a function of the number of years since 1970.
b. Use the model to predict the population in the year 2020.

67. Use the Upper Bound Theorem to find an integral upper bound and the Lower Bound Theorem to find a lower bound of the zeros of $f(x) = x^4 - 3x^3 - 2x^2 + 6x + 10$. *(Lesson 4-5)* **Sample answers: 4; −2**

68. Use synthetic division to determine if $x + 2$ is a factor of $x^3 + 6x^2 + 12x + 12$. Explain. *(Lesson 4-3)* **No; there is a remainder of 4.**

69. Determine whether the graph of $x^2 + y^2 = 16$ is symmetric with respect to the x-axis, the y-axis, the line $y = x$, or the line $y = -x$. *(Lesson 3-1)* **all**

70. Solve the system of equations algebraically. *(Lesson 2-2)*
$4x - 2y + 3z = -6$
$3x + 3y - 2z = 2$
$5x - 4y - 3z = -75$ **(−3, 9, 8)**

71. Which scatter plot shows data that has a strongly positive correlation? *(Lesson 1-6)* **b**

a.   b.   c. 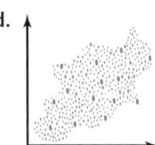   d.

72. SAT Practice If $p > 0$ and $q < 0$, which quantity must be positive? **B**
A $p + q$
B $p - q$
C $q - p$
D $p \times q$
E $p \div q$

Extra Practice See p. A36.

Extra Credit

Earth rotates on its axis once every 24 hours. How long does it take Earth to rotate through an angle of 300°? **20 h** How long does it take Earth to rotate through an angle of $\frac{2\pi}{3}$ radian? **8 h**

---

4 ASSESS

Open-Ended Assessment

Modeling Draw a picture showing a 34-inch car windshield wiper blade that sweeps at an angle of 80°. Find the surface area that the blade covers and the length of the arc that it sweeps.

80°
34 inches

807.0 in², 47.5 in.

Enrichment Masters, p. 61

5-Minute Check
Lesson 6-1

1. Change 1700° to radian measure in terms of π. $\frac{85\pi}{9}$

2. Change -5.4 radian measure to degree measure. Round to the nearest tenth. **$-309.4°$**

3. Evaluate $\sin\left(\frac{-19\pi}{4}\right)$. $-\frac{\sqrt{2}}{2}$

4. Given a central angle of 171°, find the length of its intercepted arc in a circle of radius 19 centimeters. Round to the nearest tenth. **56.7 cm**

5. Find the area of a sector if the central angle measures $\frac{3\pi}{7}$ radians and the radius of the circle is 9.5 meters. Round to the nearest tenth. **60.8 m²**

Motivating the Lesson

Real-World Application If a toy electric train is traveling on a circular track with a radius of 15 feet, determine the angular velocity if 100 revolutions are completed in 45 minutes. Express the answer in radians per second and round to the nearest hundredth. **0.23 radian/s**

Linear and Angular Velocity

ENTERTAINMENT The Children's Museum in Indianapolis, Indiana, houses an antique carousel. The carousel contains three concentric circles of animals. The inner circle of animals is approximately 11 feet from the center, and the outer circle of animals is approximately 20 feet from the center. The carousel makes $2\frac{5}{8}$ rotations per minute. Determine the angular and linear velocities of someone riding an animal in the inner circle and of someone riding an animal in the same row in the outer circle. *This problem will be solved in Examples 3 and 5.*

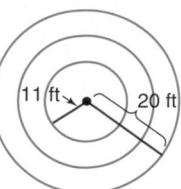

The carousel is a circular object that turns about an axis through its center. Other examples of objects that rotate about a central axis include Ferris wheels, gears, tires, and compact discs. As the carousel or any other circular object rotates counterclockwise about its center, an object at the edge moves through an angle relative to its starting position known as the **angular displacement,** or angle of rotation.

Consider a circle with its center at the origin of a rectangular coordinate system and point B on the circle rotating counterclockwise. Let the positive x-axis, or $\overrightarrow{OA}$, be the initial side of the central angle. The terminal side of the central angle is $\overrightarrow{OB}$. The angular displacement is θ. The measure of θ changes as B moves around the circle. All points on $\overrightarrow{OB}$ move through the same angle per unit of time.

Example **Determine the angular displacement in radians of 4.5 revolutions. Round to the nearest tenth.**

Each revolution equals 2π radians. For 4.5 revolutions, the number of radians is $4.5 \times 2\pi$ or 9π. 9π radians equals about 28.3 radians.

The ratio of the change in the central angle to the time required for the change is known as **angular velocity.** Angular velocity is usually represented by the lowercase Greek letter ω (omega).

Angular Velocity

If an object moves along a circle during a time of t units, then the angular velocity, ω, is given by

$$\omega = \frac{\theta}{t},$$

where θ is the angular displacement in radians.

Resource Manager

 Reproducible Masters
• *Study Guide and Practice,* pp. 101–102
• *Enrichment,* p. 62
• *Assessment and Evaluation,* p. 147

 Transparencies
• 5-Minute Check, 6-2
• Teaching, 6-2

Notice that the angular velocity of a point on a rotating object is not dependent upon the distance from the center of the rotating object.

Example **2** Determine the angular velocity if 7.3 revolutions are completed in 5 seconds. Round to the nearest tenth.

The angular displacement is $7.3 \times 2\pi$ or 14.6π radians.

$\omega = \dfrac{\theta}{t}$

$\omega = \dfrac{14.6\pi}{5}$ $\theta = 14.6\pi, t = 5$

$\omega \approx 9.173450548$ *Use a calculator.*

The angular velocity is about 9.2 radians per second.

To avoid mistakes when computing with units of measure, you can use a procedure called **dimensional analysis**. In dimensional analyses, unit labels are treated as mathematical factors and can be divided out.

Example **3** **ENTERTAINMENT**   Refer to the application at the beginning of the lesson. Determine the angular velocity for each rider in radians per second.

The carousel makes $2\frac{5}{8}$ or 2.625 revolutions per minute. Convert revolutions per minute to radians per second.

$$\dfrac{2.625 \text{ revolutions}}{1 \text{ minute}} \times \dfrac{1 \text{ minute}}{60 \text{ seconds}} \times \dfrac{2\pi \text{ radians}}{1 \text{ revolution}} \approx 0.275 \text{ radian per second}$$

Each rider has an angular velocity of about 0.275 radian per second.

The carousel riders have the same angular velocity. However, the rider in the outer circle must travel a greater distance than the one in the inner circle. The arc length formula can be used to find the relationship between the linear and angular velocities of an object moving in a circular path. If the object moves with constant **linear velocity** (v) for a period of time (t), the distance (s) it travels is given by the formula $s = vt$. Thus, the linear velocity is $v = \dfrac{s}{t}$.

As the object moves along the circular path, the radius r forms a central angle of measure θ. Since the length of the arc is $s = r\theta$, the following is true.

$s = r\theta$

$\dfrac{s}{t} = \dfrac{r\theta}{t}$ *Divide each side by t.*

$v = r\dfrac{\theta}{t}$ *Replace $\dfrac{s}{t}$ with v.*

Linear Velocity	If an object moves along a circle of radius of r units, then its linear velocity, v is given by $$v = r\dfrac{\theta}{t},$$ where $\dfrac{\theta}{t}$ represents the angular velocity in radians per unit of time.

 www.amc.glencoe.com/self_check_quiz   *Lesson 6-2   Linear and Angular Velocity*   **353**

Teaching Tip For Example 1, point out that one revolution measures 360°, and 360° equals 2π radians.

Teaching Tip For Example 2, point out that angular velocity depends on the number of revolutions completed in a unit of time and is independent of the distance from the center of the rotating object.

Teaching Tip For Example 3 below, point out that the bowl of peaches would have the same angular velocity wherever it was placed on the revolving table. Therefore, the 3-foot radius has no effect on the solution to the problem.

In-Class Examples

Example 1
Determine the angular displacement in radians of 8.7 revolutions. Round to the nearest tenth.   **54.7 radians**

Example 2
Determine the angular velocity if 5.8 revolutions are completed in 9 seconds. Round to the nearest tenth.   **4.0 radians/s**

Example 3
Furniture A circular serving table in a buffet has a radius of 3 feet. It makes 2.5 revolutions per minute. Determine the angular velocity in radians per second of a bowl of peaches sitting on the serving table.   **0.262 radian/s**

Teaching Tip For Example 4, point out that linear velocity involves the component of distance on the circle, and it is therefore dependent on the length of the radius.

Teaching Tip For Example 5, point out that angular velocity and linear velocity are two different concepts and that linear velocity is directly proportional to the radius.

In-Class Examples

Example 4
Determine the linear velocity of a point rotating at an angular velocity of 31π radians per second at a distance of 15 centimeters from the center of the rotating object. Round to the nearest tenth. **1460.8 cm/s**

Example 5
Furniture Refer to In-Class Example 3 and use its data. If a spoon is 1 foot from the center of the revolving table, and a fork is 2.5 feet from the center, determine the linear velocity for each. **0.262 ft/s, 0.654 ft/s**

Example 6
Bicycling The fastest human on a bicycle was John Howard, who achieved an incredible speed of 152.3 miles per hour in 1985. If the tires on John's bicycle have a diameter of 30 inches and during a warm-up are turning at a rate of 141 revolutions per minute, determine the bicycle's speed in miles per hour. **12.6 mph**

Since $\omega = \dfrac{\theta}{t}$, the formula for linear velocity can also be written as $v = r\omega$.

Examples

4 Determine the linear velocity of a point rotating at an angular velocity of 17π radians per second at a distance of 5 centimeters from the center of the rotating object. Round to the nearest tenth.

$v = r\omega$
$v = 5(17\pi)$ ⠀⠀⠀ $r = 5, \omega = 17\pi$
$v \approx 267.0353756$ ⠀ *Use a calculator.*

The linear velocity is about 267.0 centimeters per second.

5 **ENTERTAINMENT** Refer to the application at the beginning of the lesson. Determine the linear velocity for each rider.

From Example 3, you know that the angular velocity is about 0.275 radian per second. Use this number to find the linear velocity for each rider.

Rider on the Inner Circle
$v = r\omega$
$v \approx 11(0.275)$ ⠀⠀ $r = 11, \omega = 0.275$
$v \approx 3.025$

Rider on the Outer Circle
$v = r\omega$
$v \approx 20(0.275)$ ⠀⠀ $r = 20, \omega = 0.275$
$v \approx 5.5$

The linear velocity of the rider on the inner circle is about 3.025 feet per second, and the linear velocity of the rider on the outer circle is about 5.5 feet per second.

6 **CAR RACING** The tires on a race car have a diameter of 30 inches. If the tires are turning at a rate of 2000 revolutions per minute, determine the race car's speed in miles per hour (mph).

If the diameter is 30 inches, the radius is $\dfrac{1}{2} \times 30$ or 15 inches. This measure needs to be written in miles. The rate needs to be written in hours.

$$v = \underbrace{\qquad r \qquad}\quad \times \quad \underbrace{\qquad \omega \qquad}$$

$$v = 15 \text{ in.} \times \frac{1 \text{ ft}}{12 \text{ in.}} \times \frac{1 \text{ mi}}{5280 \text{ ft}} \times \frac{2000 \text{ rev}}{1 \text{ min}} \times \frac{2\pi}{1 \text{ rev}} \times \frac{60 \text{ min}}{1 \text{ h}}$$

$v \approx 178.4995826$ mph ⠀ *Use a calculator.*

The speed of the race car is about 178.5 miles per hour.

Addressing Individual Needs

Logical Joe is running one lap on the outside of a circular track with a diameter of 1000 feet. His constant rate of speed is 12 mph. Sunil is running one lap on the outside of a nearby circular track with a diameter of 1125 feet. He too is running at 12 mph. Find the time in minutes for each of the two runners, and compare their times with the radius of each circle. **Joe's time is 2.97 min. Sunil's time is 3.35 min. The ratio of their times is the same as the ratio of the radius of each track.**

CHECK FOR UNDERSTANDING

Communicating Mathematics

Read and study the lesson to answer each question.   1–5. See margin.

1. **Draw** a circle and represent an angular displacement of 3π radians.

2. **Write** an expression that could be used to change 5 revolutions per minute to radians per second.

3. **Compare and contrast** linear and angular velocity.

4. **Explain** how two people on a rotating carousel can have the same angular velocity but different linear velocity.

5. **Show** that when the radius of a circle is doubled, the angular velocity remains the same and the linear velocity of a point on the circle is doubled.

Guided Practice

Determine each angular displacement in radians. Round to the nearest tenth.

6. 5.8 revolutions   **36.4 radians**

7. 710 revolutions   **4461.1 radians**

Determine each angular velocity. Round to the nearest tenth.

8. 3.2 revolutions in 7 seconds
2.9 radians/s

9. 700 revolutions in 15 minutes
293.2 radians/min

Determine the linear velocity of a point rotating at the given angular velocity at a distance r from the center of the rotating object. Round to the nearest tenth.

10. $\omega = 36$ radians per second, $r = 12$ inches   **432 in./s**

11. $\omega = 5\pi$ radians per minute, $r = 7$ meters   **110.0 m/min**

12. **Space**   A geosynchronous equatorial orbiting (GEO) satellite orbits 22,300 miles above the equator of Earth. It completes one full revolution each 24 hours. Assume Earth's radius is 3960 miles.

a. How far will the GEO satellite travel in one day?   **about 164,996.4 mi**

b. What is the satellite's linear velocity in miles per hour?   **about 6874.9 mph**

EXERCISES

Practice

 A

Determine each angular displacement in radians. Round to the nearest tenth.

13. 3 revolutions   **18.8 radians**

14. 2.7 revolutions   **17.0 radians**

15. 13.2 revolutions   **82.9 radians**

16. 15.4 revolutions   **96.8 radians**

17. 60.7 revolutions   **381.4 radians**

18. 3900 revolutions   **24,504.4 radians**

 B

Determine each angular velocity. Round to the nearest tenth.

7.3 radians/min

19. 1.8 revolutions in 9 seconds

20. 3.5 revolutions in 3 minutes

19. 1.3 radians/s   21. 17.2 revolutions in 12 seconds

22. 28.4 revolutions in 19 seconds

21. 9.0 radians/s   23. 100 revolutions in 16 minutes
39.3 radians/min

24. 122.6 revolutions in 27 minutes
28.5 radians/min

22. 9.4 radians/s   25. A Ferris wheel rotates one revolution every 50 seconds. What is its angular velocity in radians per second?   **about 0.1 radian/s**

26. A clothes dryer is rotating at 500 revolutions per minute. Determine its angular velocity in radians per second.   **about 52.4 radians/s**

 www.amc.glencoe.com/self_check_quiz

Lesson 6-2   Linear and Angular Velocity   **355**

3 PRACTICE/APPLY

Error Analysis

Watch for students who have difficulty applying the procedure of dimensional analysis.

Prevent by stressing that each fraction that they set up is a factor of 1, as in $\dfrac{1 \text{ minute}}{60 \text{ seconds}}$ or $\dfrac{1 \text{ hour}}{60 \text{ minutes}}$.

Assignment Guide

Core: 13–35 odd, 37–39, 41, 43–51
Enriched: 14–34 even, 36–51

Additional Answers

4. **Both individuals would have the same change in angle during the same amount of time. However, an individual on the outside of the carousel would travel farther than an individual on the inside during the same amount of time.**

5. **Since angular velocity is $\dfrac{\theta}{t}$, the radius has no effect on the angular velocity. Let $R = 2r$. For a circle with radius R, $v' = R\dfrac{\theta}{t}$ or $(2r)\dfrac{\theta}{t}$ which is $2\left(r\dfrac{\theta}{t}\right)$. Thus, $v' = 2v$.**

***Study Guide and Practice Masters,* pp. 101–102**

NAME _____ DATE _____ PERIOD _____

6-2   Study Guide

Linear and Angular Velocity

As a circular object rotates about its center, an object at the edge moves through an angle relative to the object's starting position. That is known as the **angular displacement**, or angle of rotation. **Angular velocity**, ω, is given by $\omega = \frac{\theta}{t}$, where θ is the angular displacement in radians and t is time. **Linear velocity**, v, is given by $v = r\frac{\theta}{t}$, where $\frac{\theta}{t}$ represents the angular velocity in radians per unit of time. Since $\omega = \frac{\theta}{t}$, this formula can also be written as $v = r\omega$.

Example 1   Determine the angular displacement in radians of 3.5 revolutions. Round to the nearest tenth.

Each revolution equals 2π radians. For 3.5 revolutions, the number of radians is $3.5 \times 2\pi$, or 7π. 7π radians equals about 22.0 radians.

Example 2   Determine the angular velocity if 8.2 revolutions are completed in 3 seconds. Round to the nearest tenth.

The angular displacement is $8.2 \times 2\pi$, or 16.4π radians.
$\omega = \frac{\theta}{t}$
$\omega = \frac{16.4\pi}{3}$ $\theta = 16.4\pi, t = 3$
$\omega \approx 17.17403984$ *Use a calculator.*
The angular velocity is about 17.2 radians per second.

Example 3   Determine the linear velocity of a point rotating at an angular velocity of 13π radians per second at a distance of 7 centimeters from the center of the rotating object. Round to the nearest tenth.

$v = r\omega$
$v = 7(13\pi)$ $r = 7, \omega = 13\pi$
$v \approx 285.8849315$ *Use a calculator.*
The linear velocity is about 285.9 centimeters per second.

© Glencoe/McGraw-Hill   101   *Advanced Mathematical Concepts*

Additional Answers

1.

2. $\dfrac{5 \text{ rev}}{1 \text{ min}} \times \dfrac{2\pi \text{ radians}}{1 \text{ rev}} \times \dfrac{1 \text{ min}}{60 \text{ s}}$

3. **Linear velocity is the movement along the arc with respect to time while angular velocity is the change in the angle with respect to time.**

27. Change 85 radians per second to revolutions per minute (rpm).
    about 811.7 rpm

Determine the linear velocity of a point rotating at the given angular velocity at a distance *r* from the center of the rotating object. Round to the nearest tenth.

28. $\omega = 16.6$ radians per second, $r = 8$ centimeters   **132.8 cm/s**

29. $\omega = 27.4$ radians per second, $r = 4$ feet   **109.6 ft/s**

30. $\omega = 6.1\pi$ radians per minute, $r = 1.8$ meters   **34.5 m/min**

31. $\omega = 75.3\pi$ radians per second, $r = 17$ inches   **4021.6 in./s**

32. $\omega = 805.6$ radians per minute, $r = 39$ inches   **31,418.4 in./min**

33. $\omega = 64.5\pi$ radians per minute, $r = 88.9$ millimeters   **18,014.0 mm/min**

 34. A pulley is turned 120° per second.
   a. Find the number of revolutions per minute (rpm).   **20 rpm**
   b. If the radius of the pulley is 5 inches, find the linear velocity in inches per second.   **about 10.5 in./s**

35. Consider the tip of each hand of a clock. Find the linear velocity in millimeters per second for each hand.
   a. second hand which is 30 millimeters   **about 3.1 mm/s**
   b. minute hand which is 27 millimeters long   **about 0.05 mm/s**
   c. hour hand which is 18 millimeters long   **about 0.003 mm/s**

Applications and Problem Solving

36. Entertainment   The diameter of a Ferris wheel is 80 feet.
   a. If the Ferris wheel makes one revolution every 45 seconds, find the linear velocity of a person riding in the Ferris wheel.   **about 5.6 ft/s**
   b. Suppose the linear velocity of a person riding in the Ferris wheel is 8 feet per second. What is the time for one revolution of the Ferris wheel?   **about 31 s**

37. Entertainment   The Kit Carson County Carousel makes 3 revolutions per minute.
   a. Find the linear velocity in feet per second of someone riding a horse that is $22\frac{1}{2}$ feet from the center.   **about 7.1 ft/s**
   b. The linear velocity of the person on the inside of the carousel is 3.1 feet per second. How far is the person from the center of the carousel?   **about 9.9 ft**
   c. How much faster is the rider on the outside going than the rider on the inside?   **about 4 ft/s**

38. Critical Thinking   Two children are playing on the seesaw. The lighter child is 9 feet from the fulcrum, and the heavier child is 6 feet from the fulcrum. As the lighter child goes from the ground to the highest point, she travels through an angle of 35° in $\frac{1}{2}$ second.
   a. Find the angular velocity of each child.   **about 1.2 radians/s each**
   b. What is the linear velocity of each child?   **about 11.0 ft/s, about 7.3 ft/s**

39. Bicycling   A bicycle wheel is 30 inches in diameter.
   a. To the nearest revolution, how many times will the wheel turn if the bicycle is ridden for 3 miles?   **2017 revolutions**
   b. Suppose the wheel turns at a constant rate of 2.75 revolutions per second. What is the linear speed in miles per hour of a point on the tire?   **about 14.7 mph**

40. Space The radii and times needed to complete one rotation for the four planets closest to the sun are given at the right.

Planets

	Radius (kilometers)	Time for One Rotation (hours)
Mercury	2440	1407.6
Venus	6052	5832.5
Earth	6356	23.935
Mars	3375	24.623

Source: NASA

a. Find the linear velocity of a point on each planet's equator.

b. Compare the linear velocity of a point on the equator of Mars with a point on the equator of Earth.   **The linear velocity of Earth is about twice that of Mars.**

40a. Mercury, about 10.9 km/h; Venus, about 6.5 km/h; Earth, about 1668.5 km/h; Mars, about 861.2 km/h

41. Physics A torsion pendulum is an object suspended by a wire or rod so that its plane of rotation is horizontal and it rotates back and forth around the wire without losing energy. Suppose that the pendulum is rotated θ_m radians and released. Then the angular displacement θ at time t is $\theta = \theta_m \cos \omega t$, where ω is the angular frequency in radians per second. Suppose the angular frequency of a certain torsion pendulum is π radians per second and its initial rotation is $\frac{\pi}{4}$ radians.

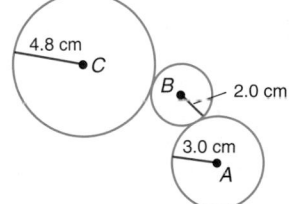

41a. $\theta = \frac{\pi}{4} \cos \pi t$

a. Write the equation for the angular displacement of the pendulum.

b. What are the first two values of t for which the angular displacement of the pendulum is 0?   **0.5 s, 1.5 s**

42. Space Low Earth orbiting (LEO) satellites orbit between 200 and 500 miles above Earth. In order to keep the satellites at a constant distance from Earth, they must maintain a speed of 17,000 miles per hour. Assume Earth's radius is 3960 miles.

a. Find the angular velocity needed to maintain a LEO satellite at 200 miles above Earth.   **about 4.1 radians/h**

b. How far above Earth is a LEO with an angular velocity of 4 radians per hour?   **290 mi**

42c. Its angular velocity is between 3.8 radians/h and 4.1 radians/h.

c. Describe the angular velocity of any LEO satellite.

43. Critical Thinking The figure at the right is a side view of three rollers that are tangent to one another.

4.8 cm C

B 2.0 cm

3.0 cm A

43a. B clockwise; C counterclockwise

a. If roller A turns counterclockwise, in which directions do rollers B and C turn?

b. If roller A turns at 120 revolutions per minute, how many revolutions per minute do rollers B and C turn?   **180 rpm; 75 rpm**

Mixed Review

44. Find the area of a sector if the central angle measures $105°$ and the radius of the circle is 7.2 centimeters. *(Lesson 6-1)*   **about 47.5 cm^2**

45. Geometry Find the area of a regular pentagon that is inscribed in a circle with a diameter of 7.3 centimeters. *(Lesson 5-4)*   **about 31.68 cm^2**

Extra Practice See p. A36.

Extra Credit

CD players use lasers to read digital codes which are translated to sound waves. If the disc moves about 200 rpm when the laser is at the outer edge of the disc, how many radians does a point on the edge of the disc turn in one second?

about $\frac{20\pi}{3}$ radians

4 ASSESS

Open-Ended Assessment

Writing Have students write a paragraph explaining the similarities and differences between angular velocity and linear velocity. Suggest that they use one or more examples to illustrate these ideas.

 Chapter 6, Quiz A (Lessons 6-1 and 6-2) is available in the *Assessment and Evaluation Masters*, p.147.

Additional Answer

49.

$y > x^3 + 1$

Enrichment Masters, p. 62

6-2

NAME _____ DATE _____ PERIOD _____

Enrichment

Angular Acceleration

An object traveling in a circular path experiences linear velocity and angular velocity. It may also experience **angular acceleration**. Angular acceleration is the rate of change in angular velocity with respect to time.

At time $t = 0$, there is an **initial angular velocity**. At the end of time t, there is a **final angular velocity**. Then the angular acceleration α of the object can be defined as

$$\alpha = \frac{\text{final angular velocity} - \text{initial angular velocity}}{\text{time}}.$$

The units for angular acceleration are usually rad/s² or rev/min².

Example A record has a small chip on its edge. If the record begins at rest and then goes to 45 revolutions per minute in 30 seconds, what is the angular acceleration of the chip?

The record starts at rest, so the initial angular velocity is 0. The final angular velocity is 45 revolutions/minute. Thus, the angular acceleration is

$$\alpha = \frac{45 - 0}{\frac{1}{2}}$$

$$= 90 \text{ rev/min}^2.$$

Solve.

1. The record in the example was playing at 45 rev/min. A power surge lasting 2 seconds caused the record to speed up to 80 rev/min. What was the angular acceleration of the chip then?
   1050 rev/min²

2. When a car enters a curve in the road, the tires are turning at an angular velocity of 50 ft/s. At the end of the curve, the angular velocity of the tires is 60 ft/s. If the curve is an arc of a circle with radius 2000 feet and central angle $\theta = \frac{\pi}{4}$, and the car travels at a constant linear velocity of 40 mph, what is the angular acceleration?
   0.37 ft/s²

© Glencoe/McGraw-Hill   T62   *Advanced Mathematical Concepts*

---

46. Write $35°\ 20'\ 55''$ as a decimal to the nearest thousandth. *(Lesson 5-1)* **35.349°**

47. Solve $10 + \sqrt{k - 5} = 8$. *(Lesson 4-7)* **no real solution**

48. Write a polynomial equation of least degree with roots -4, $3i$, and $-3i$. *(Lesson 4-1)* $x^3 + 4x^2 + 9x + 36 = 0$

49. Graph $y > x^3 + 1$. *(Lesson 3-3)* **See margin.**

50. Write the slope-intercept form of the equation of the line through points at $(8, 5)$ and $(-6, 0)$. *(Lesson 1-4)* $y = \frac{5}{14}x + \frac{15}{7}$

51. **SAT/ACT Practice** The perimeter of rectangle $QRST$ is p, and $a = \frac{3}{4}b$. Find the value of b in terms of p. **D**

    A $\frac{p}{7}$ B $\frac{4p}{7}$ C $\frac{7}{p}$ D $\frac{2p}{7}$ E $\frac{7p}{4}$

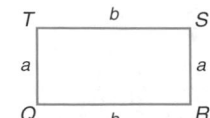

CAREER CHOICES

● Audio Recording Engineer ●

Is music your forte? Do you enjoy being creative and solving problems? If you answered yes to these questions, you may want to consider a career as an audio recording engineer. This type of engineer is in charge of all the technical aspects of recording music, speech, sound effects, and dialogue.

Some aspects of the career include controlling the recording equipment, tackling technical problems that arise during recording, and communicating with musicians and music producers. You would need to keep up-to-date on the latest recording equipment and technology. The music producer may direct the sounds you produce through use of the equipment, or you may have the opportunity to design and perfect your own sounds for use in production.

CAREER OVERVIEW

Degree Preferred:
two- or four-year degree in audio engineering

Related Courses:
mathematics, music, computer science, electronics

Outlook:
number of jobs expected to increase at a slower pace than the average through the year 2006

Sound	Decibels
Threshold of Hearing	0
Average Whisper (4 feet)	20
Broadcast Studio (no program in progress)	30
Soft Recorded Music	36
Normal Conversation (4 feet)	60
Moderate Discotheque	90
Personal Stereo	up to 120
Percussion Instruments at a Symphony Concert	up to 130
Rock Concert	up to 140

 interNET CONNECTION For more information about audio recording engineering visit: **www.amc.glencoe.com**

CAREER CHOICES   AUDIO RECORDING ENGINEER

Math Connection

- Each 10 decibels is perceived to be twice as loud as the last. How much louder does a moderate discotheque seem than a broadcast studio with no program in progress? **64 times**
- People who are regularly exposed to more than one minute of sounds registering 110 decibels or more can suffer permanent hearing loss. Compare the perceived loudness of a rock concert with a sound of 110 decibels. **The loud rock concert appears to be up to 8 times louder.**

6-3

Graphing Sine and Cosine Functions

OBJECTIVE
• Use the graphs of the sine and cosine functions.

METEOROLOGY The average monthly temperatures for a city demonstrate a repetitious behavior. For cities in the Northern Hemisphere, the average monthly temperatures are usually lowest in January and highest in July. The graph below shows the average monthly temperatures (°F) for Baltimore, Maryland, and Asheville, North Carolina, with January represented by 1.

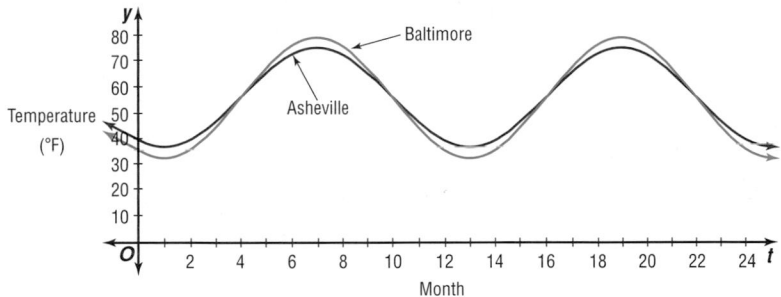

Model for Baltimore's temperature: $y = 54.4 + 22.5 \sin\left[\frac{\pi}{6}(t - 4)\right]$

Model for Asheville's temperature: $y = 54.5 + 18.5 \sin\left[\frac{\pi}{6}(t - 4)\right]$

In these equations, t denotes the month with January represented by $t = 1$. What is the average temperature for each city for month 13? Which city has the greater fluctuation in temperature?

These problems will be solved in Example 5.

Each year, the graph for Baltimore will be about the same. This is also true for Asheville. If the values of a function are the same for each given interval of the domain (in this case, 12 months or 1 year), the function is said to be **periodic.** The interval is the **period** of the function.

Periodic Function and Period	A function is *periodic* if, for some real number α, $f(x + \alpha) = f(x)$ for each x in the domain of f. The least positive value of α for which $f(x) = f(x + \alpha)$ is the *period* of the function.

 5-Minute Check
Lesson 6-2

1. Determine the angular displacement in radians of 471 revolutions. Round to the nearest tenth.
   2959.4 radians

2. Determine the angular displacement in radians of 9.3 revolutions. Round to the nearest tenth.
   58.4 radians

3. Determine the angular velocity if 450 revolutions are completed in 12 minutes. Round to the nearest tenth.
   235.6 radians/min

4. Determine the angular velocity if 7.6 revolutions are completed in 11 seconds. Round to the nearest tenth.
   4.3 radians/s

5. Determine the linear velocity of a point rotating at an angular velocity of 75 radians per second at a distance of 40 inches from the center of the rotating object. Round to the nearest tenth.   **3000 in./s**

Motivating the Lesson
Hands-On Activity Have students use calculators to find values of the sine and cosine functions in the interval 0 to π. Ask them to describe the behavior of the two functions in the interval.
Sine: increases from 0 to 1 at $\frac{\pi}{2}$, then decreases to 0 at π;
cosine: decreases from 1 to 0 at $\frac{\pi}{2}$ and continues to decrease to −1 at π.

Resource Manager

 Reproducible Masters
• *Study Guide and Practice,* pp. 103–104
• *Enrichment,* p. 63

 Transparencies
• 5-Minute Check, 6-3
• Teaching, 6-3

 Technology/Multimedia
• Advanced MathPASS, Lesson 13

Teaching Tip For Example 1, remind students that periodic functions repeat over a given interval.

In-Class Example

Example 1

Determine if each function is periodic. If so, state the period.

a.

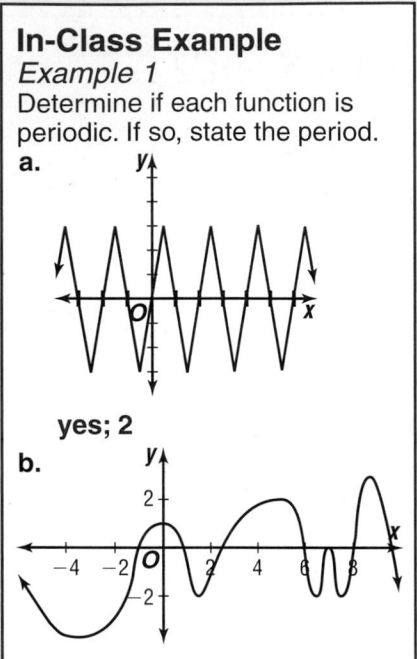

yes; 2

b.

no

Example ❶ Determine if each function is periodic. If so, state the period.

a.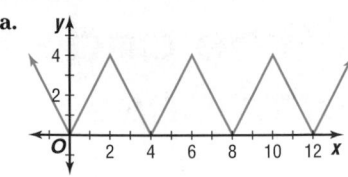

The values of the function repeat for each interval of 4 units. The function is periodic, and the period is 4.

b.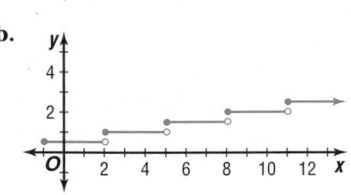

The values of the function do not repeat. The function is not periodic.

Consider the sine function. First evaluate $y = \sin x$ for domain values between -2π and 2π in multiples of $\frac{\pi}{4}$.

x	-2π	$-\frac{7\pi}{4}$	$-\frac{3\pi}{2}$	$-\frac{5\pi}{4}$	$-\pi$	$-\frac{3\pi}{4}$	$-\frac{\pi}{2}$	$-\frac{\pi}{4}$	0	$\frac{\pi}{4}$	$\frac{\pi}{2}$	$\frac{3\pi}{4}$	π	$\frac{5\pi}{4}$	$\frac{3\pi}{2}$	$\frac{7\pi}{4}$	2π
sin x	0	$\frac{\sqrt{2}}{2}$	1	$\frac{\sqrt{2}}{2}$	0	$-\frac{\sqrt{2}}{2}$	-1	$-\frac{\sqrt{2}}{2}$	0	$\frac{\sqrt{2}}{2}$	1	$\frac{\sqrt{2}}{2}$	0	$-\frac{\sqrt{2}}{2}$	-1	$-\frac{\sqrt{2}}{2}$	0

To graph $y = \sin x$, plot the coordinate pairs from the table and connect them to form a smooth curve. Notice that the range values for the domain interval $-2\pi < x < 0$ (shown in red) repeat for the domain interval between $0 < x < 2\pi$ (shown in blue). The sine function is a periodic function.

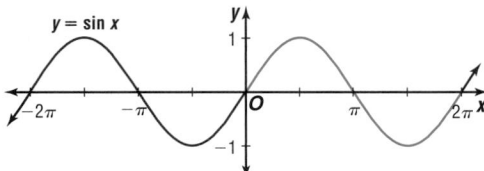

By studying the graph and its repeating pattern, you can determine the following properties of the graph of the sine function.

Properties of the Graph of $y = \sin x$

1. The period is 2π.
2. The domain is the set of real numbers.
3. The range is the set of real numbers between -1 and 1, inclusive.
4. The x-intercepts are located at πn, where n is an integer.
5. The y-intercept is 0.
6. The maximum values are $y = 1$ and occur when $x = \frac{\pi}{2} + 2\pi n$, where n is an integer.
7. The minimum values are $y = -1$ and occur when $x = \frac{3\pi}{2} + 2\pi n$, where n is an integer.

Examples

② Find $\sin \frac{9\pi}{2}$ by referring to the graph of the sine function.

Because the period of the sine function is 2π and $\frac{9\pi}{2} > 2\pi$, rewrite $\frac{9\pi}{2}$ as a sum involving 2π.

$$\frac{9\pi}{2} = 4\pi + \frac{\pi}{2}$$

$$= 2\pi(2) + \frac{\pi}{2} \quad \textit{This is a form of } \frac{\pi}{2} + 2\pi n.$$

So, $\sin \frac{9\pi}{2} = \sin \frac{\pi}{2}$ or 1.

③ Find the values of θ for which $\sin \theta = 0$ is true.

Since $\sin \theta = 0$ indicates the x-intercepts of the function, $\sin \theta = 0$ if $\theta = n\pi$, where n is any integer.

④ Graph $y = \sin x$ for $3\pi \le x \le 5\pi$.

The graph crosses the x-axis at 3π, 4π, and 5π. It has its maximum value of 1 at $x = \frac{9\pi}{2}$, and its minimum value of -1 at $x = \frac{7\pi}{2}$.
Use this information to sketch the graph.

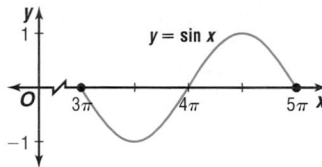

⑤ METEOROLOGY Refer to the application at the beginning of the lesson.

a. What is the average temperature for each city for month 13?
Month 13 is January of the second year. To find the average temperature of this month, substitute this value into each equation.

Baltimore

$$y = 54.4 + 22.5 \sin\left[\frac{\pi}{6}(t - 4)\right]$$

$$y = 54.4 + 22.5 \sin\left[\frac{\pi}{6}(13 - 4)\right]$$

$$y = 54.4 + 22.5 \sin \frac{3\pi}{2}$$

$$y = 54.4 + 22.5(-1)$$

$$y = 31.9$$

Asheville

$$y = 54.5 + 18.5 \sin\left[\frac{\pi}{6}(t - 4)\right]$$

$$y = 54.5 + 18.5 \sin\left[\frac{\pi}{6}(13 - 4)\right]$$

$$y = 54.5 + 18.5 \sin \frac{3\pi}{2}$$

$$y = 54.5 + 18.5(-1)$$

$$y = 36.0$$

In January, the average temperature for Baltimore is 31.9°, and the average temperature for Asheville is 36.0°.

b. Which city has the greater fluctuation in temperature? Explain.
The average temperature for January is lower in Baltimore than in Asheville. The average temperature for July is higher in Baltimore than in Asheville. Therefore, there is a greater fluctuation in temperature in Baltimore than in Asheville.

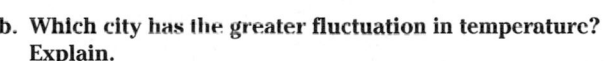

Teaching Tip For Example 2, remind students that the period for the sine function is 2π.

Teaching Tip For Example 3 below, point out that there is an infinite number of values of θ for which $\sin \theta = -1$.

In-Class Examples

Example 2
Find $\sin 3\pi$ by referring to the graph of the sine function. **0**

Example 3
Find the value of θ for which $\sin \theta = -1$ is true. $\frac{3}{2}\pi + 2\pi n$, *n* is any integer.

Example 4
Graph $y = \sin x$ for $2\pi \le x \le 4\pi$.

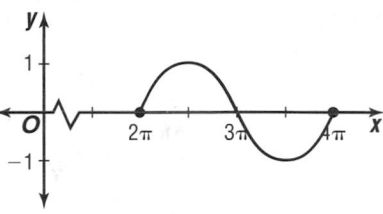

Example 5
The graph below shows the average monthly precipitation (in inches) for Seattle, Washington, and San Francisco, California, with January represented as 1.

Model for Seattle's precipitation:
$$y = 3.45 + 2.55 \cos\left(\frac{\pi}{6}t + 0.18\right)$$

Model for San Francisco's precipitation:
$$y = 2.2 + 2.2 \cos\left(\frac{\pi}{6}t - 0.41\right)$$

a. What is the average precipitation for each city for month 13? **Seattle: 5.39 in.; San Francisco: 4.39 in.**

b. Which city has the greater fluctuation in precipitation? Explain. **The average precipitation is higher in Seattle throughout the year, with the difference remaining about the same throughout the graph. Both cities have similar fluctuations in precipitation.**

Teaching Tip For Example 4 in the Student Edition, suggest that students graph two periods of $y = \cos x$ from 2π to 4π and 4π to 6π, then isolate that portion of the graph that lies between 3π and 5π.

Teaching Tip For Example 6, point out that the period of $y = \cos x$ is 2π and $-1 \leq \cos x \leq 1$.

In-Class Example

Example 6
Determine whether the graph represents $y = \sin x$, $y = \cos x$, or neither.

$y = \cos x$

Additional Answers (p. 363)

1. **Sample answer:**

   period: 6

4. **See graph below. Both functions are periodic functions with the period of 2π. The domain of both functions is the set of real numbers, and the range of both functions is the set of real numbers between −1 and 1, inclusive. The x-intercepts of the sine function are located at πn, but the x-intercepts of the cosine function are located at $\frac{\pi}{2} + \pi n$, where n is an integer. The y-intercept of the sine function is 0, but the y-intercept of the cosine function is 1. The maximum value of the sine function occurs when $x = \frac{\pi}{2} + 2\pi n$ and its minimum value occurs when $x = \frac{3\pi}{2} + 2\pi n$, where n is an integer. The maximum value of the cosine function occurs when $x = \pi n$, where n is an even integer, and its minimum value occurs when $x = \pi n$, where n is an odd integer.**

Now, consider the graph of $y = \cos x$.

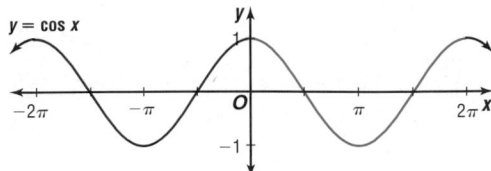

x	-2π	$-\frac{7\pi}{4}$	$-\frac{3\pi}{2}$	$-\frac{5\pi}{4}$	$-\pi$	$-\frac{3\pi}{4}$	$-\frac{\pi}{2}$	$-\frac{\pi}{4}$	0	$\frac{\pi}{4}$	$\frac{\pi}{2}$	$\frac{3\pi}{4}$	π	$\frac{5\pi}{4}$	$\frac{3\pi}{2}$	$\frac{7\pi}{4}$	2π
$\cos x$	1	$\frac{\sqrt{2}}{2}$	0	$-\frac{\sqrt{2}}{2}$	-1	$-\frac{\sqrt{2}}{2}$	0	$\frac{\sqrt{2}}{2}$	1	$\frac{\sqrt{2}}{2}$	0	$-\frac{\sqrt{2}}{2}$	-1	$-\frac{\sqrt{2}}{2}$	0	$\frac{\sqrt{2}}{2}$	1

By studying the graph and its repeating pattern, you can determine the following properties of the graph of the cosine function.

Properties of the Graph of $y = \cos x$

1. The period is 2π.
2. The domain is the set of real numbers.
3. The range is the set of real numbers between -1 and 1, inclusive.
4. The x-intercepts are located at $\frac{\pi}{2} + \pi n$, where n is an integer.
5. The y-intercept is 1.
6. The maximum values are $y = 1$ and occur when $x = \pi n$, where n is an even integer.
7. The minimum values are $y = -1$ and occur when $x = \pi n$, where n is an odd integer.

Example 6 Determine whether the graph represents $y = \sin x$, $y = \cos x$, or neither.

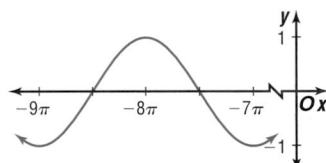

The maximum value of 1 occurs when $x = -8\pi$.   *maximum of 1 when $x = \pi n \rightarrow \cos x$*

The minimum value of -1 occurs at -9π and -7π.   *minimum of -1 when $x = \pi n \rightarrow \cos x$*

The x-intercepts are $-\frac{17\pi}{2}$ and $-\frac{15\pi}{2}$.

These are characteristics of the cosine function. The graph is $y = \cos x$.

362 Chapter 6 *Graphs of Trigonometric Functions*

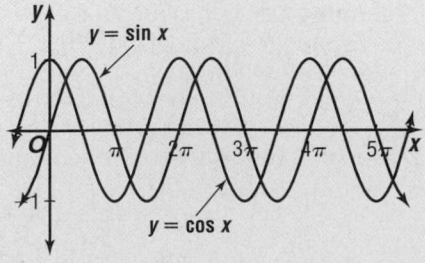

CHECK FOR UNDERSTANDING

Communicating Mathematics

Read and study the lesson to answer each question.

1. **Counterexample** Sketch the graph of a periodic function that is neither the sine nor cosine function. State the period of the function. **See margin.**

2. Sample answers:
$-\dfrac{3\pi}{2}, \dfrac{\pi}{2}, \dfrac{5\pi}{2}$

2. **Name** three values of x that would result in the maximum value for $y = \sin x$.

3. **Explain** why the cosine function is a periodic function. $\cos x = \cos(x + 2\pi)$

4. *Math Journal* **Draw** the graphs for the sine function and the cosine function. Compare and contrast the two graphs. **See margin.**

Guided Practice

5. **Determine** if the function is periodic. If so, state the period. **yes; 4**

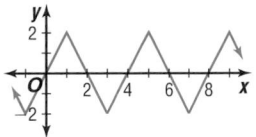

Find each value by referring to the graph of the sine or the cosine function.

6. $\cos\left(-\dfrac{\pi}{2}\right)$ **0**

7. $\sin\dfrac{5\pi}{2}$ **1**

8. Find the values of θ for which $\sin\theta = -1$ is true. $\dfrac{3\pi}{2} + 2\pi n$, **where n is an integer**

Graph each function for the given interval. **9–10. See margin.**

9. $y = \cos x, 5\pi \le x \le 7\pi$

10. $y = \sin x, -4\pi \le x \le -2\pi$

11. Determine whether the graph represents $y = \sin x$, $y = \cos x$, or neither. Explain. **Neither; the period is not 2π.**

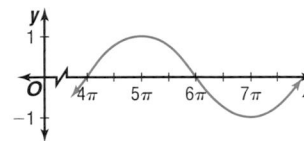

12. **Meteorology** The equation $y = 49 + 28\sin\left[\dfrac{\pi}{6}(t-4)\right]$ models the average monthly temperature for Omaha, Nebraska. In this equation, t denotes the number of months with January represented by 1. Compare the average monthly temperature for April and October. **The average temperatures are the same.**

EXERCISES

Practice

Determine if each function is periodic. If so state the period.

 13.

14.

15.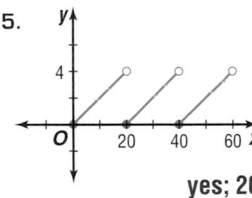

yes; 6   **no**   **yes; 20**

16. $y = |x + 5|$ **no**   17. $y = x^2$ **no**   18. $y = \dfrac{1}{x}$ **no**

 www.amc.glencoe.com/self_check_quiz   *Lesson 6-3 Graphing Sine and Cosine Functions* **363**

Addressing Individual Needs

 Auditory/Musical Have students research and possibly observe what the graphs of the sound patterns of a person's live voice, a song recorded by Frank Sinatra, a rock concert, and a healthy and an unhealthy heart would look like. Are any of these patterns periodic functions? **yes, all of them**

Additional Answers

31.

32.

33.

34.

35.
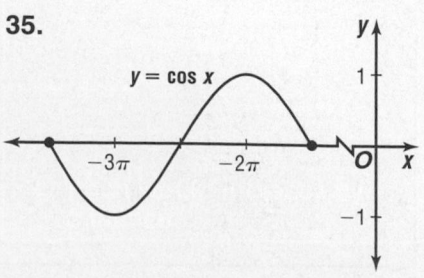

Find each value by referring to the graph of the sine or the cosine function.

19. $\cos 8\pi$ **1**   20. $\sin 11\pi$ **0**   21. $\cos \dfrac{\pi}{2}$ **0**

22. $\sin\left(-\dfrac{3\pi}{2}\right)$ **1**   23. $\sin \dfrac{7\pi}{2}$ **−1**   24. $\cos(-3\pi)$ **−1**

25. What is the value of $\sin \pi + \cos \pi$? **−1**

26. Find the value of $\sin 2\pi - \cos 2\pi$. **−1**

B Find the values of θ for which each equation is true.   **27–29.** n is an integer

27. $\cos \theta = -1$ $\pi + 2\pi n$ 28. $\sin \theta = 1$ $\dfrac{\pi}{2} + 2\pi n$ 29. $\cos \theta = 0$ $\dfrac{\pi}{2} + \pi n$

30. Under what conditions does $\cos \theta = 1$? $\theta + 2\pi n$, **where n is an integer**

Graph each function for the given interval.   **31–36. See margin.**

31. $y = \sin x, \; -5\pi \le x \le -3\pi$ 32. $y = \cos x, \; 8\pi \le x \le 10\pi$

33. $y = \cos x, \; -5\pi \le x \le -3\pi$ 34. $y = \sin x, \; \dfrac{9\pi}{2} \le x \le \dfrac{13\pi}{2}$

35. $y = \cos x, \; -\dfrac{7\pi}{2} \le x \le -\dfrac{3\pi}{2}$ 36. $y = \sin x, \; \dfrac{7\pi}{2} \le x \le \dfrac{11\pi}{2}$

37–39. See margin for explanations.
Determine whether each graph is $y = \sin x$, $y = \cos x$, or neither. Explain.

37.

$y = \cos x$

38.

neither

39.

$y = \sin x$

C 40. Describe a transformation that would change the graph of the sine function to the graph of the cosine function.   **Sample answer: a shift of $\dfrac{\pi}{2}$ to the left**

41–44. n **is an integer.**

41. Name any lines of symmetry for the graph of $y = \sin x$. $x = \dfrac{\pi}{2} + \pi n$

42. Name any lines of symmetry for the graph of $y = \cos x$. $x = \pi n$

43. Use the graph of the sine function to find the values of θ for which each statement is true.

a. $\csc \theta = 1$ $\dfrac{\pi}{2} + 2\pi n$ b. $\csc \theta = -1$ $\dfrac{3\pi}{2} + 2\pi n$ c. $\csc \theta$ is undefined. πn

44. Use the graph of the cosine function to find the values of θ for which each statement is true.
   $\dfrac{\pi}{2} + \pi n$

a. $\sec \theta = 1$ $2\pi n$ b. $\sec \theta = -1$ $\pi + 2\pi n$ c. $\sec \theta$ is undefined.

Graphing Calculator

Use a graphing calculator to graph the sine and cosine functions on the same set of axes for $0 \le x \le 2\pi$. Use the graphs to find the values of x, if any, for which each of the following is true.   **48.** $x = 0, \dfrac{\pi}{2} \le x \le \pi, \dfrac{3\pi}{2} \le x \le 2\pi$

45. $\sin x = -\cos x$ $x = \dfrac{3\pi}{4}, \dfrac{7\pi}{4}$ 46. $\sin x \le \cos x$ $0 \le x \le \dfrac{\pi}{4}, \dfrac{5\pi}{4} \le x \le 2\pi$

47. $\sin x \cos x > 1$ **none**   48. $\sin x \cos x \le 0$

49. $\sin x + \cos x = 1$ $x = 0, \dfrac{\pi}{2}, 2\pi$ 50. $\sin x - \cos x = 0$ $x = \dfrac{\pi}{4}, \dfrac{5\pi}{4}$

Additional Answers

36.

37. The maximum value of 1 occurs when $x = 4\pi$, the minimum value of -1 occurs when $x = 5\pi$, and the x-intercepts are $\dfrac{7\pi}{2}$, $\dfrac{9\pi}{2}$, and $\dfrac{11\pi}{2}$.

38. The graph does not cross the x-axis.

Applications and Problem Solving

51. **Meteorology** The equation $y = 43 + 31 \sin\left[\frac{\pi}{6}(t - 4)\right]$ models the average monthly temperatures for Minneapolis, Minnesota. In this equation, t denotes the number of months with January represented by 1.

a. What is the difference between the average monthly temperatures for July and January? What is the relationship between this difference and the coefficient of the sine term? **62; it is twice the coefficient.**

51b. 86; it is twice the constant term.

b. What is the sum of the average monthly temperatures for July and January? What is the relationship between this sum and value of constant term?

52. **Critical Thinking** Consider the graph of $y = 2 \sin x$.

a. What are the x-intercepts of the graph? **πn, where n is an integer**

b. What is the maximum value of y? **2**

c. What is the minimum value of y? **-2**

d. What is the period of the function? **2π**

e. Graph the function. **See margin.**

f. How does the 2 in the equation affect the graph? **It expands the graph vertically.**

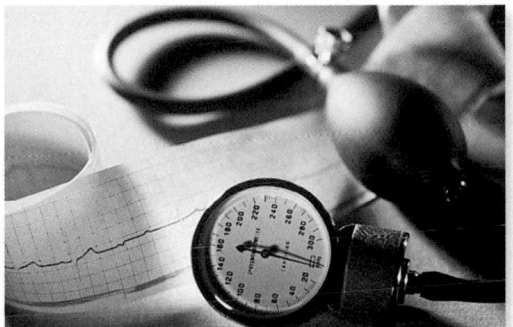

53. **Medicine** The equation $P = 100 + 20 \sin 2\pi t$ models a person's blood pressure P in millimeters of mercury. In this equation, t is time in seconds. The blood pressure oscillates 20 millimeters above and below 100 millimeters, which means that the person's blood pressure is 120 over 80. This function has a period of 1 second, which means that the person's heart beats 60 times a minute. **100; 120; 100; 80; 100**

a. Find the blood pressure at $t = 0$, $t = 0.25$, $t = 0.5$, $t = 0.75$, and $t = 1$.

b. During the first second, when was the blood pressure at a maximum? **0.25 s**

c. During the first second, when was the blood pressure at a minimum? **0.75 s**

54. **Physics** The motion of a weight on a spring can be described by a modified cosine function. The weight suspended from a spring is at its equilibrium point when it is at rest. When pushed a certain distance above the equilibrium point, the weight oscillates above and below the equilibrium point. The time that it takes for the weight to oscillate from the highest point to the lowest point and back to

the highest point is its period. The equation $v = 3.5 \cos\left(t\sqrt{\frac{k}{m}}\right)$ models the vertical displacement v of the weight in relationship to the equilibrium point at any time t if it is initially pushed up 3.5 centimeters. In this equation, k is the elasticity of the spring and m is the mass of the weight. **b. about 0.5 s**

a. Suppose $k = 19.6$ and $m = 1.99$. Find the vertical displacement after 0.9 second and after 1.7 seconds. **about -3.3 cm; about 2.0 cm**

b. When will the weight be at the equilibrium point for the first time?

c. How long will it take the weight to complete one period? **about 2.0 s**

Additional Answers
(pp. 364–365)

39. The maximum value of 1 occurs when $x = -\frac{11\pi}{2}$, the minimum value of -1 occurs when $x = -\frac{13\pi}{2}$, and the x-intercepts are -7π, -6π, and -5π.

52e.

Extra Credit

Graph $y = -\cos x$ for the interval $0 \le x \le 4\pi$. Use the graph to find the value of y when $x = 3\pi$.

$y = 1$

Open-Ended Assessment

Modeling Prepare wire models of a sine curve (0 to 2π), a tangent curve $\left(\frac{\pi}{2} \text{ to } \frac{3\pi}{2}\right)$, and a cosecant curve (0 to π). Sketch the plane for -2π to 2π on the chalkboard and have students use the wires to model the graphs of the six trigonometric functions.

 Teaching Tip As an extension to the modeling activity, have students use graphing calculators to enter $y = \sin x$. They should set the viewing window to [–7, 7] scl:1 by [–1, 1] scl:1. (This will guarantee a range of $-2\pi \le x \le 2\pi$, and the upper and lower limits of sin x need only be +1 and –1.) Select **2: zero** in the **CALC** menu and locate all of the zeros of $y = \sin x$ from -2π to 2π. –6.283185, –3.141593, 0, 3.1415927, 6.2831853

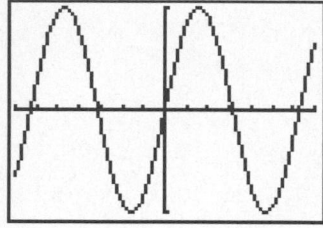

[–7, 7] scl:1 by [–1, 1] scl:1

Enrichment Masters, p. 63

366 Chapter 6

---

55. **Critical Thinking** Consider the graph of $y = \cos 2x$.
   a. What are the *x*-intercepts of the graph? $\frac{\pi}{4} + \frac{\pi n}{2}$, where *n* is an integer
   b. What is the maximum value of *y*? **1**
   c. What is the minimum value of *y*? **–1**
   d. What is the period of the function? π
   e. Sketch the graph. **See margin.**

56a. about 357 pumas, 1500 deer

56b. about 488 pumas, about 1197 deer

56c. about 545 pumas, about 1282 deer

56. **Ecology** In predator-prey relationships, the number of animals in each category tends to vary periodically. A certain region has pumas as predators and deer as prey. The equation $P = 500 + 200 \sin [0.4(t - 2)]$ models the number of pumas after *t* years. The equation $D = 1500 + 400 \sin (0.4t)$ models the number of deer after *t* years. How many pumas and deer will there be in the region for each value of *t*?

   a. $t = 0$ b. $t = 10$ c. $t = 25$

Mixed Review

57. about 52.4 radians per second

57. **Technology** A computer CD-ROM is rotating at 500 revolutions per minute. Write the angular velocity in radians per second. *(Lesson 6-2)*

58. Change -1.5 radians to degree measure. *(Lesson 6-1)* **about –85.9°**

59. Find the values of *x* in the interval $0° \le x \le 360°$ for which $\sin x = \frac{\sqrt{2}}{2}$. *(Lesson 5-5)* **45°, 135°**

60. Solve $\frac{2}{x + 2} = \frac{x}{2 - x} + \frac{x^2 + 4}{x^2 - 4}$. *(Lesson 4-6)* **no solution**

61. Find the number of possible positive real zeros and the number of negative real zeros of $f(x) = 2x^3 + 3x^2 - 11x - 6$. Then determine the rational roots. *(Lesson 4-4)* **1; 2 or 0; –3, $-\frac{1}{2}$, 2**

62. Use the Remainder Theorem to find the remainder when $x^3 + 2x^2 - 9x + 18$ is divided by $x - 1$. State whether the binomial is a factor of the polynomial. *(Lesson 4-3)* **12; no**

63. Determine the equations of the vertical and horizontal asymptotes, if any, of $g(x) = \frac{x^2}{x^2 + x}$. *(Lesson 3-7)* **$x = 0$, $x = -1$, $y = 1$**

64. reflected over the *x*-axis, expanded vertically by a factor of 3

64. Use the graph of the parent function $f(x) = x^3$ to describe the graph of the related function $g(x) = -3x^3$. *(Lesson 3-2)*

65. Find the value of $\begin{vmatrix} -2 & 4 & -1 \\ 1 & -1 & 0 \\ -3 & 4 & 5 \end{vmatrix}$. *(Lesson 2-5)* **–11**

66. $A'(-3, 2)$, $B'(-2, -4)$, $C'(-1, 6)$

66. Use a reflection matrix to find the coordinates of the vertices of $\triangle ABC$ reflected over the *y*-axis for vertices A (3, 2), B (2, –4), and C (1, 6). *(Lesson 2-4)*

67. Graph $x = \frac{3}{2} y$. *(Lesson 1-3)* **See margin.**

68. **SAT/ACT Practice** How much less is the perimeter of square *RSVW* than the perimeter of rectangle *RTUW*? **B**

   A 2 units   B 4 units
   C 9 units   D 12 units
   E 20 units

366 *Chapter 6   Graphs of Trigonometric Functions*

Extra Practice See p. A36.

Additional Answers

55e.

67.

FUNCTIONS

Mathematicians and statisticians use functions to express relationships among sets of numbers. When you use a spreadsheet or a graphing calculator, writing an expression as a function is crucial for calculating values in the spreadsheet or for graphing the function.

Early Evidence In about 2000 B.C., the Babylonians used the idea of function in making tables of values for n and $n^3 + n^2$, for $n = 1, 2, \ldots, 30$. Their work indicated that they believed they could show a correspondence between these two sets of values. The following is an example of a Babylonian table.

Johann Bernoulli

n	$n^3 + n^2$
1	2
2	12
$\vdots$	$\vdots$
30	?

The Renaissance In about 1637, **René Descartes** may have been the first person to use the term "function." He defined a function as a power of x, such as x^2 or x^3, where the power was a positive integer. About 55 years later, **Gottfried von Leibniz** defined a function as anything that related to a curve, such as a point on a curve or the slope of a curve. In 1718, **Johann Bernoulli** thought of a function as a relationship between a variable and some constants. Later in that same century, **Leonhard Euler's** notion of a function was an equation or formula with variables and constants. Euler also expanded the notion of function to include not only the written expression, but the graphical representation of the relationship as well. He is credited with the modern standard notation for function, $f(x)$.

Modern Era The 1800s brought **Joseph Lagrange's** idea of function. He limited the meaning of a function to a power series. An example of a power series is $x + x^2 + x^3 + \cdots$, where the three dots indicate that the pattern continues forever. In 1822, **Jean Fourier** determined that any function can be represented by a trigonometric series. **Peter Gustav Dirichlet** used the terminology *y is a function of x* to mean that each first element in the set of ordered pairs is different. Variations of his definition can be found in mathematics textbooks today, including this one.

Georg Cantor and others working in the late 1800s and early 1900s are credited with extending the concept of function from ordered pairs of numbers to ordered pairs of elements.

Today engineers like Julia Chang use functions to calculate the efficiency of equipment used in manufacturing. She also uses functions to determine the amount of hazardous chemicals generated during the manufacturing process. She uses spreadsheets to find many values of these functions.

ACTIVITIES

1. Make a table of values for the Babylonian function, $f(n) = n^3 + n^2$. Use values of n from 1 to 30, inclusive. Then, graph this function using paper and pencil, graphing software, or a graphing calculator. Describe the graph. **See margin.**

2. Research other functions used by notable mathematicians mentioned in this article. You may choose to explore trigonometric series. **See students' work.**

3. **inter**NET **CONNECTION** Find out more about personalities referenced in this article and others who contributed to the history of functions. Visit **www.amc.glencoe.com**

History of Mathematics **367**

This feature provides students with an opportunity to see how the mathematics that they are studying was developed. It also gives students a glimpse of how mathematical discoveries are made.

More About the Mathematicians

René Descartes was born in La Haye (now Descartes), France, in 1596. He helped create Cartesian geometry by using algebra to solve geometric problems.

Johann Bernoulli was born in Basel, Switzerland, in 1667. Bernoulli analyzed the function $y = x^x$ and discovered addition theorems for trigonometric and hyperbolic functions using differential equations.

Leonhard Euler was born in Basel, Switzerland, in 1707. Euler originated beta and gamma functions and explored integrals that can be expressed as elementary functions.

Joseph Lagrange was born in 1736 in Turin, Sardinia (now Italy), but lived and worked in France. Lagrange published the first theory of functions of a real variable.

Peter Gustav Dirichlet was born in 1805 in Duren, France (now Germany). He wrote about the convergence of trigonometric series and arbitrary functions.

Additional Answer
1. See bottom left for table.

[0, 30] scl: 5 by
[0, 30,000] scl: 5000
The graph is not a straight line. It curves upward, increasing more rapidly as the value of n increases.

(Table for Exercise 1)

n	1	2	3	4	5	6	7	8	9	10
$n^2 + n^3$	2	12	36	80	150	252	392	576	810	1100
n	11	12	13	14	15	16	17	18	19	20
$n^2 + n^3$	1452	1872	2366	2940	3600	4352	5202	6156	7220	8400
n	21	22	23	24	25	26	27	28	29	30
$n^2 + n^3$	9702	11,132	12,696	14,400	16,250	18,252	20,412	22,736	25,230	27,900

1 FOCUS

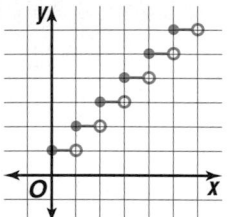

5-Minute Check
Lesson 6-3

1. Determine whether the function is periodic. If so, state the period. **no**

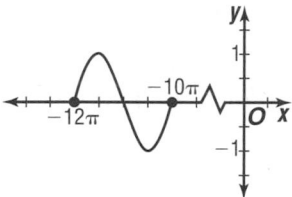

2. Find $\sin(-5\pi)$ by referring to the graph of the sine function. **0**

3. Find values of θ for which $\cos\theta = 1$ is true. **$2\pi n$, n is an integer.**

4. Graph the function $y = \sin x$ for the interval $-12\pi \leq x \leq -10\pi$.

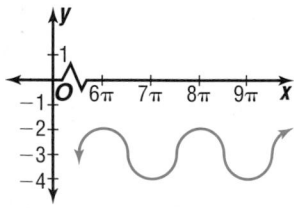

5. Determine whether the graph is $y = \sin x$, $y = \cos x$, or neither. **neither**

Motivating the Lesson
Discussion Give students a graph of $y = 3\cos x$ and ask how it differs from $y = \cos x$. Have them experiment with values of a and b in $y = a\cos bx$ in order to find an equation to match the graph.

6-4

Amplitude and Period of Sine and Cosine Functions

OBJECTIVES
- Find the amplitude and period for sine and cosine functions.
- Write equations of sine and cosine functions given the amplitude and period.

 BOATING A signal buoy between the coast of Hilton Head Island, South Carolina, and Savannah, Georgia, bobs up and down in a minor squall. From the highest point to the lowest point, the buoy moves a distance of $3\frac{1}{2}$ feet. It moves from its highest point down to its lowest point and back to its highest point every 14 seconds. Find an equation of the motion for the buoy assuming that it is at its equilibrium point at $t = 0$ and the buoy is on its way down at that time. What is the height of the buoy at 8 seconds and at 17 seconds? *This problem will be solved in Example 5.*

Recall from Chapter 3 that changes to the equation of the parent graph can affect the appearance of the graph by dilating, reflecting, and/or translating the original graph. In this lesson, we will observe the vertical and horizontal expanding and compressing of the parent graphs of the sine and cosine functions.

Let's consider an equation of the form $y = A \sin\theta$. We know that the maximum absolute value of $\sin\theta$ is 1. Therefore, for every value of the product of $\sin\theta$ and A, the maximum value of $A\sin\theta$ is $|A|$. Similarly, the maximum value of $A\cos\theta$ is $|A|$. The absolute value of A is called the **amplitude** of the functions $y = A\sin\theta$ and $y = A\cos\theta$.

Amplitude of Sine and Cosine Functions	The amplitude of the functions $y = A\sin\theta$ and $y = A\cos\theta$ is the absolute value of A, or $|A|$.

The amplitude can also be described as the absolute value of one-half the difference of the maximum and minimum function values.

$$|A| = \left| \frac{A - (-A)}{2} \right|$$

amplitude = |A|

Example **a.** State the amplitude for the function $y = 4\cos\theta$.

b. Graph $y = 4\cos\theta$ and $y = \cos\theta$ on the same set of axes.

c. Compare the graphs.

a. According to the definition of amplitude, the amplitude of $y = A\cos\theta$ is $|A|$. So the amplitude of $y = 4\cos\theta$ is $|4|$ or 4.

Resource Manager

 Reproducible Masters
- *Study Guide and Practice,* pp. 105–106
- *Enrichment,* p. 64
- *Assessment and Evaluation,* pp. 146, 147
- *Graphing Calculator Keystrokes,* p. 22

 Transparencies
- 5-Minute Check, 6-4
- Teaching, 6-4

b. Make a table of values. Then graph the points and draw a smooth curve.

θ	0	$\frac{\pi}{4}$	$\frac{\pi}{2}$	$\frac{3\pi}{4}$	π	$\frac{5\pi}{4}$	$\frac{3\pi}{2}$	$\frac{7\pi}{4}$	2π
$\cos\theta$	1	$\frac{\sqrt{2}}{2}$	0	$-\frac{\sqrt{2}}{2}$	-1	$-\frac{\sqrt{2}}{2}$	0	$\frac{\sqrt{2}}{2}$	1
$4\cos\theta$	4	$2\sqrt{2}$	0	$-2\sqrt{2}$	-4	$-2\sqrt{2}$	0	$2\sqrt{2}$	4

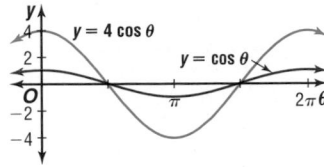

c. The graphs cross the θ-axis at $\theta = \frac{\pi}{2}$ and $\theta = \frac{3\pi}{2}$. Also, both functions reach their maximum value at $\theta = 0$ and $\theta = 2\pi$ and their minimum value at $\theta = \pi$. But the maximum and minimum values of the function $y = \cos\theta$ are 1 and -1, and the maximum and minimum values of the function $y = 4\cos\theta$ are 4 and -4. The graph of $y = 4\cos\theta$ is vertically expanded.

GRAPHING CALCULATOR EXPLORATION

▶ Select the radian mode.

▶ Use the domain and range values below to set the viewing window.

$-4.7 \le x \le 4.8$, **Xscl: 1** $-3 \le y \le 3$, **Yscl: 1**

TRY THESE
1. Graph each function on the same screen.

   a. $y = \sin x$ **b.** $y = \sin 2x$ **c.** $y = \sin 3x$
   a–c. See margin.

WHAT DO YOU THINK?

2. Describe the behavior of the graph of $f(x) = \sin kx$, where $k > 0$, as k increases.
   The graph is shrunk horizontally.

3. Make a conjecture about the behavior of the graph of $f(x) = \sin kx$, if $k < 0$. Test your conjecture.   **See margin.**

Consider an equation of the form $y = \sin k\theta$, where k is any positive integer. Since the period of the sine function is 2π, the following identity can be developed.

$$y = \sin k\theta$$
$$y = \sin(k\theta + 2\pi) \quad \textit{Definition of periodic function}$$
$$y = \sin k\left(\theta + \frac{2\pi}{k}\right) \quad k\theta + 2\pi = k\left(\theta + \frac{2\pi}{k}\right)$$

Therefore, the period of $y = \sin k\theta$ is $\frac{2\pi}{k}$. Similarly, the period of $y = \cos k\theta$ is $\frac{2\pi}{k}$.

Period of Sine and Cosine Functions

The period of the functions $y = \sin k\theta$ and $y = \cos k\theta$ is $\frac{2\pi}{k}$, where $k > 0$.

From the Classroom of...

Alan Clemens
Upper Dublin High School
Ft. Washington, Pennsylvania

"I have students go to an almanac or Internet source and find sunrise and sunset data, calculate the length of day (every 15 days), and then plot the results for a period of one year. We discuss the equation of the function that describes this plot. For students near tidal water, another activity might be to collect data on the height of the water versus time using a meterstick stuck into the sand."

---

2 TEACH

Teaching Tip When discussing the definition of amplitude, suggest that students picture the graph as an ocean wave and the x-axis as sea level. The amplitude is the height of the wave above or below sea level.

In-Class Example

Example 1

a. State the amplitude for the function $y = -2\sin\theta$. **2**

b. Graph $y = -2\sin\theta$ and $y = \sin\theta$ on the same set of axes.

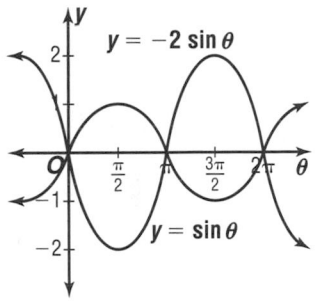

c. Compare the graphs.
The graphs cross the θ-axis at 0, π, and 2π. The graphs of $y = -2\sin\theta$ and $y = \sin\theta$ reach critical values at $\frac{\pi}{2}$ (-2 and 1, respectively) and $\frac{3\pi}{2}$ (2 and -1, respectively).

 Teaching Tip After students graph the functions in the Exploration, ask them how the behavior of $f(x) = A\sin x$ compares with that of the parent graph, $f(x) = \sin x$. Then ask them what similarities they see between $f(x) = A\sin x$ and the family of graphs described by $f(x) = Ax$.

Additional Answers
Graphing Calculator Exploration
1.

3. The graph of $f(x) = \sin kx$ for $k < 0$ is the graph of $f(x) = \sin |k| x$ reflected over the y-axis.

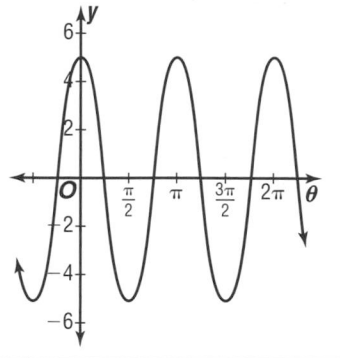
Example **2** **a.** State the period for the function $y = \cos \frac{\theta}{2}$.

b. Graph $y = \cos \frac{\theta}{2}$ and $y = \cos \theta$.

a. The definition of the period of $y = \cos k\theta$ is $\frac{2\pi}{k}$. Since $\cos \frac{\theta}{2}$ equals $\cos \left(\frac{1}{2} \theta \right)$, the period is $\frac{2\pi}{\frac{1}{2}}$ or 4π.

b.

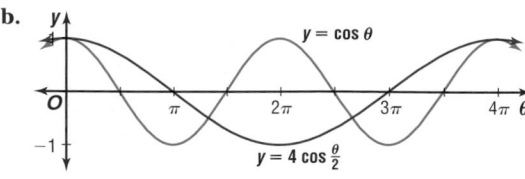

Notice that the graph of $y = \cos \frac{\theta}{2}$ is horizontally expanded.

The graphs of $y = A \sin k\theta$ and $y = A \cos k\theta$ are shown below.

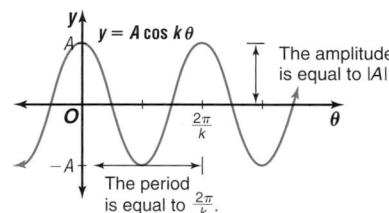

You can use the parent graph of the sine and cosine functions and the amplitude and period to sketch graphs of $y = A \sin k\theta$ and $y = A \cos k\theta$.

Example **3** State the amplitude and period for the function $y = \frac{1}{2} \sin 4\theta$. Then graph the function.

Since $A = \frac{1}{2}$, the amplitude is $\left| \frac{1}{2} \right|$ or $\frac{1}{2}$. Since $k = 4$, the period is $\frac{2\pi}{4}$ or $\frac{\pi}{2}$.

Use the basic shape of the sine function and the amplitude and period to graph the equation.

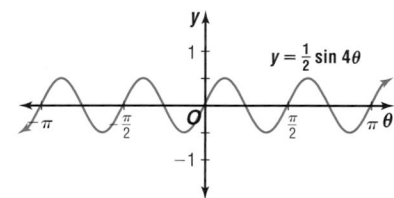

We can write equations for the sine and cosine functions if we are given the amplitude and period.

Example **4** **Write an equation of the cosine function with amplitude 9.8 and period 6π.**

The form of the equation will be $y = A \cos k\theta$. First find the possible values of A for an amplitude of 9.8.

$|A| = 9.8$
$A = 9.8$ or -9.8

Since there are two values of A, two possible equations exist.

Now find the value of k when the period is 6π.

$\dfrac{2\pi}{k} = 6\pi$ *The period of a cosine function is $\dfrac{2\pi}{k}$.*

$k = \dfrac{2\pi}{6\pi}$ or $\dfrac{1}{3}$

The possible equations are $y = 9.8 \cos \left(\dfrac{1}{3}\theta\right)$ or $y = -9.8 \cos \left(\dfrac{1}{3}\theta\right)$.

Many real-world situations have periodic characteristics that can be described with the sine and cosine functions. When you are writing an equation to describe a situation, remember the characteristics of the sine and cosine graphs. If you know the function value when $x = 0$ and whether the function is increasing or decreasing, you can choose the appropriate function to write an equation for the situation.

| If A is positive, the graph passes through the origin and heads up. | $y = A \sin \theta$ | If A is positive, the graph crosses the y-axis at its maximum. | 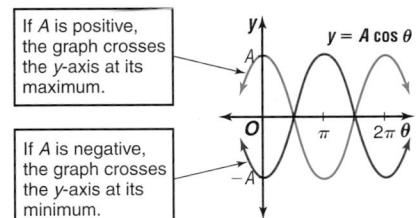 $y = A \cos \theta$ |
| If A is negative, the graph passes through the origin and heads down. | | If A is negative, the graph crosses the y-axis at its minimum. | |

Example **5** **BOATING** Refer to the application at the beginning of the lesson.

a. Find an equation for the motion of the buoy.

b. Determine the height of the buoy at 8 seconds and at 17 seconds.

a. At $t = 0$, the buoy is at equilibrium and is on its way down. This indicates a reflection of the sine function and a negative value of A. The general form of the equation will be $y = A \sin kt$, where A is negative and t is the time in seconds.

$A = -\left(\dfrac{1}{2} \times 3\dfrac{1}{2}\right)$ $\dfrac{2\pi}{k} = 14$

$A = -\dfrac{7}{4}$ or -1.75 $k = \dfrac{2\pi}{14}$ or $\dfrac{\pi}{7}$

An equation for the motion of the buoy is $y = -1.75 \sin \dfrac{\pi}{7}t$.

Lesson 6-4   Amplitude and Period of Sine and Cosine Functions **371**

Addressing Individual Needs

Auditory/Musical In their relentless search for life in outer space, scientists use sophisticated microwave listening devices tuned in to sounds originating outside of Earth. Ask students to contact people associated with a college or university whose job or hobby it is to analyze these sounds and the graphs associated with them. Have them use the search keywords SETI (Search for Extraterrestrial Intelligence) and Phoenix. Do these graphs exhibit different amplitudes and periods? **yes**

In-Class Example

Example 6

Architecture The Sears Building in Chicago sways back and forth at a vibration frequency of about 0.1 Hz. On average, it sways 6 inches from true center. Write an equation of the sine function that represents this behavior. $y = 6 \sin (0.2\pi t)$

---

3 PRACTICE/APPLY

Error Analysis

Watch for students who have difficulty drawing the basic graphs $y = \sin x$ and $y = \cos x$.

Prevent by stressing that the graph of $y = \sin x$ passes through the origin while the graph of $y = \cos x$ passes through the point at $(0, 1)$.

***Study Guide and Practice
Masters,*** pp. 105–106

---

b. Use this equation to find the location of the buoy at the given times.

At 8 seconds

$y = -1.75 \sin \left(\frac{\pi}{7} \times 8\right)$

$y \approx 0.7592965435$

At 8 seconds, the buoy is about 0.8 feet above the equilibrium point.

At 17 seconds

$y = -1.75 \sin \left(\frac{\pi}{7} \times 17\right)$

$y \approx -1.706123846$

At 17 seconds, the buoy is about 1.7 feet below the equilibrium point.

The period represents the amount of time that it takes to complete one cycle. The number of cycles per unit of time is known as the **frequency**. The period (time per cycle) and frequency (cycles per unit of time) are reciprocals of each other.

$$\text{period} = \frac{1}{\text{frequency}} \qquad \text{frequency} = \frac{1}{\text{period}}$$

The *hertz* is a unit of frequency. One hertz equals one cycle per second.

Example **6 MUSIC** Write an equation of the sine function that represents the initial behavior of the vibrations of the note G above middle C having amplitude 0.015 and a frequency of 392 hertz.

The general form of the equation will be $y = A \sin kt$, where t is the time in seconds. Since the amplitude is 0.015, $A = \pm 0.015$.

The period is the reciprocal of the frequency or $\frac{1}{392}$. Use this value to find k.

$\frac{2\pi}{k} = \frac{1}{392}$ *The period $\frac{2\pi}{k}$ equals $\frac{1}{392}$.*

$k = 2\pi(392)$ or 784π

One sine function that represents the vibration is $y = 0.015 \sin (784\pi \times t)$.

2. The graphs are a reflection of each other over the θ-axis.

CHECK FOR UNDERSTANDING

Communicating Mathematics

Read and study the lesson to answer each question.

1. **Write** a sine function that has a greater maximum value than the function $y = 4 \sin 2\theta$.   **Sample answer:** $y = 5 \sin 2\theta$

2. **Describe** the relationship between the graphs of $y = 3 \sin \theta$ and $y = -3 \sin \theta$.

Additional Answer (p. 373)

5.

All three graphs are periodic and curve above and below the **x**-axis. The amplitude of $y = 3 \cos \theta$ is 3, while the amplitude of $y = \cos \theta$ and $y = \cos 3\theta$ is 1. The period of $y = \cos 3\theta$ is $\frac{2\pi}{3}$, while the period of $y = \cos \theta$ and $y = 3 \cos \theta$ is 2π.

3. **Determine** which function has the greatest period. **C**

   A. $y = 5 \cos 2\theta$ B. $y = 3 \cos 5\theta$ C. $y = 7 \cos \dfrac{\theta}{2}$ D. $y = \cos \theta$

4. **Explain** the relationship between period and frequency.

5. *Math Journal* **Draw** the graphs for $y = \cos \theta$, $y = 3 \cos \theta$, and $y = \cos 3\theta$. Compare and contrast the three graphs. **See margin.**

Guided Practice

6–11. See margin for graphs.

6. State the amplitude for $y = -2.5 \cos \theta$. Then graph the function. **2.5**

7. State the period for $y = \sin 4\theta$. Then graph the function. $\dfrac{\pi}{2}$

State the amplitude and period for each function. Then graph each function.

8. $y = 10 \sin 2\theta$ **10; π**   9. $y = 3 \cos 2\theta$ **3; π**

10. $y = 0.5 \sin \dfrac{\theta}{6}$ **0.5; 12π**   11. $y = -\dfrac{1}{5} \cos \dfrac{\theta}{4}$ $\dfrac{1}{5}$**; 8π**

Write an equation of the sine function with each amplitude and period.

12. amplitude $= 0.8$, period $= \pi$
    $y = \pm 0.8 \sin 2\theta$

13. amplitude $= 7$, period $= \dfrac{\pi}{3}$
    $y = \pm 7 \sin 6\theta$

Write an equation of the cosine function with each amplitude and period.

14. amplitude $= 1.5$, period $= 5\pi$
    $y = \pm 1.5 \cos \dfrac{2}{5} \theta$

15. amplitude $= \dfrac{3}{4}$, period $= 6$
    $y = \pm \dfrac{3}{4} \cos \dfrac{\pi}{3} \theta$

16. **Music** Write a sine equation that represents the initial behavior of the vibrations of the note D above middle C having an amplitude of 0.25 and a frequency of 294 hertz. $y = \pm 0.25 \sin (588\pi \times t)$

EXERCISES

Practice

17–22. See margin for graphs.

State the amplitude for each function. Then graph each function.

17. $y = 2 \sin \theta$ **2**   18. $y = -\dfrac{3}{4} \cos \theta$ $\dfrac{3}{4}$ 19. $y = 1.5 \sin \theta$ **1.5**

State the period for each function. Then graph each function.

20. $y = \cos 2\theta$ **π**   21. $y = \cos \dfrac{\theta}{4}$ **8π**   22. $y = \sin 6\theta$ $\dfrac{\pi}{3}$

State the amplitude and period for each function. Then graph each function.

23–34. See Answer Appendix for graphs.

23. $y = 5 \cos \theta$ **5; 2π**   24. $y = -2 \cos 0.5\theta$ **2; 4π**

25. $y = -\dfrac{2}{5} \sin 9\theta$ $\dfrac{2}{5}$**; $\dfrac{2\pi}{9}$**   26. $y = 8 \sin 0.5\theta$ **8; 4π**

27. $y = -3 \sin \dfrac{\pi}{2}\theta$ **3; 4**   28. $y = \dfrac{2}{3} \cos \dfrac{3\pi}{7}\theta$ $\dfrac{2}{3}$**, $\dfrac{14}{3}$**

29. $y = 3 \sin 2\theta$ **3; π**   30. $y = 3 \cos 0.5\theta$ **3; 4π**

31. $y = -\dfrac{1}{3} \cos 3\theta$ $\dfrac{1}{3}$**; $\dfrac{2\pi}{3}$**   32. $y = \dfrac{1}{3} \sin \dfrac{\theta}{3}$ $\dfrac{1}{3}$**; 6π**

33. $y = -4 \sin \dfrac{\theta}{2}$ **4; 4π**   34. $y = -2.5 \cos \dfrac{\theta}{5}$ **2.5; 10π**

35. **0.5; $\dfrac{1}{349}$**

35. The equation of the vibrations of the note F above middle C is represented by $y = 0.5 \sin 698\pi t$. Determine the amplitude and period for the function.

 www.amc.glencoe.com/self_check_quiz   *Lesson 6-4*   *Amplitude and Period of Sine and Cosine Functions*   **373**

Assignment Guide

Core: 17–57 odd, 59–73
Enriched: 18–54 even, 56–73
All: Mid-Chapter Quiz, 1–10

Additional Answers

8.

9.

10.

11.

17.

18.

Additional Answers

6.

7.

(continued on page 374)

Additional Answers (p. 373)

19.

$y = 1.5 \sin \theta$

20.

$y = \cos 2\theta$

21.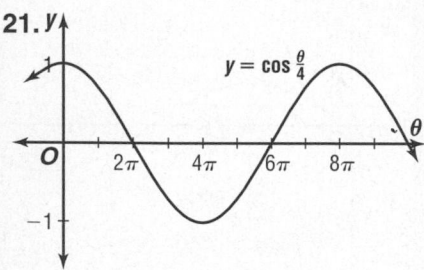

$y = \cos \frac{\theta}{4}$

22.

$y = \sin 6\theta$

Write an equation of the sine function with each amplitude and period.

36. amplitude = 0.4, period = 10π $y = \pm 0.4 \sin \frac{\theta}{5}$

37. amplitude = 35.7, period = $\frac{\pi}{4}$ $y = \pm 35.7 \sin 8\theta$

38. amplitude = $\frac{1}{4}$, period = $\frac{\pi}{3}$ $y = \pm \frac{1}{4} \sin 6\theta$

39. amplitude = 0.34, period = 0.75π $y = \pm 0.34 \sin \frac{8}{3}\theta$

40. amplitude = 4.5, period = $\frac{5\pi}{4}$ $y = \pm 4.5 \sin \frac{8}{5}\theta$

41. amplitude = 16, period = 30 $y = \pm 16 \sin \frac{\pi}{15}\theta$

Write an equation of the cosine function with each amplitude and period.

42. amplitude = 5, period = 2π $y = \pm 5 \cos \theta$

43. amplitude = $\frac{5}{8}$, period = $\frac{\pi}{7}$ $y = \pm \frac{5}{8} \cos 14\theta$

44. amplitude = 7.5, period = 6π $y = \pm 7.5 \cos \frac{\theta}{3}$

45. amplitude = 0.5, period = 0.3π $y = \pm 0.5 \cos \frac{20}{3}\theta$

46. amplitude = $\frac{2}{5}$, period = $\frac{3}{5}\pi$ $y = \pm \frac{2}{5} \cos \frac{10}{3}\theta$

47. amplitude = 17.9, period = 16 $y = \pm 17.9 \cos \frac{\pi}{8}\theta$

48. Write the possible equations of the sine and cosine functions with amplitude 1.5 and period $\frac{\pi}{2}$. $y = \pm 1.5 \sin 4\theta$, $y = \pm 1.5 \cos 4\theta$

Write an equation for each graph.

 49.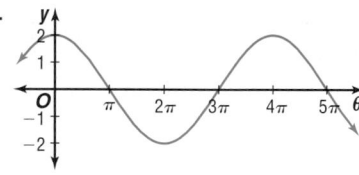

$y = 2 \cos \frac{\theta}{2}$

50.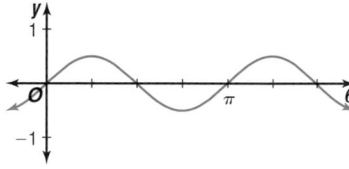

$y = 0.5 \sin 2\theta$

51.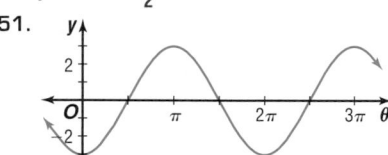

$y = -3 \cos \theta$

52.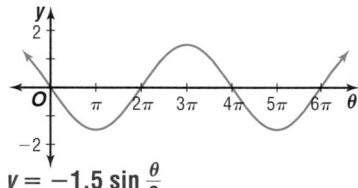

$y = -1.5 \sin \frac{\theta}{2}$

53. Write an equation for a sine function with amplitude 3.8 and frequency 120 hertz. $y = \pm 3.8 \sin (240\pi \times t)$

54. Write an equation for a cosine function with amplitude 15 and frequency 36 hertz. $y = \pm 15 \cos (72\pi \times t)$

Graphing Calculator

55. Graph these functions on the same screen of a graphing calculator. Compare the graphs.

   a. $y = \sin x$ b. $y = \sin x + 1$ c. $y = \sin x + 2$

   See margin for graphs; the graphs have the same shape, but have been translated vertically.

Additional Answer

55.

56. Boating A buoy in the harbor of San Juan, Puerto Rico, bobs up and down. The distance between the highest and lowest point is 3 feet. It moves from its highest point down to its lowest point and back to its highest point every 8 seconds. **a.** $y = -1.5 \sin \frac{\pi}{4} t$

  a. Find the equation of the motion for the buoy assuming that it is at its equilibrium point at $t = 0$ and the buoy is on its way down at that time.

  b. Determine the height of the buoy at 3 seconds. **about −1.1 ft**

  c. Determine the height of the buoy at 12 seconds. **0 ft**

57. Critical Thinking Consider the graph of $y = 2 + \sin \theta$.

  a. What is the maximum value of y? **3**

  b. What is the minimum value of y? **1**

  c. What is the period of the function? **2π**

  d. Sketch the graph. **See margin.**

58. Music Musical notes are classified by frequency. The note middle C has a frequency of 262 hertz. The note C above middle C has a frequency of 524 hertz. The note C below middle C has a frequency of 131 hertz.

  a. Write an equation of the sine function that represents middle C if its amplitude is 0.2. **$y = \pm 0.2 \sin (524\pi \times t)$**

  b. Write an equation of the sine function that represents C above middle C if its amplitude is one half that of middle C. **$y = \pm 0.1 \sin (1048\pi \times t)$**

  c. Write an equation of the sine function that represents C below middle C if its amplitude is twice that of middle C. **$y = \pm 0.4 \sin (262\pi \times t)$**

59. Physics For a pendulum, the equation representing the horizontal displacement of the bob is $y = A \cos \left(t \sqrt{\frac{g}{\ell}} \right)$. In this equation, A is the maximum horizontal distance that the bob moves from the equilibrium point, t is the time, g is the acceleration due to gravity, and ℓ is the length of the pendulum. The acceleration due to gravity is 9.8 meters per second squared.

maximum horizontal displacement (A)

ℓ

initial point

path of bob

equilibrium point

59a. $y = 1.5 \cos \left(t \sqrt{\frac{9.8}{6}} \right)$

  a. A pendulum has a length of 6 meters and its bob has a maximum horizontal displacement to the right of 1.5 meters. Write an equation that models the horizontal displacement of the bob if it is at its maximum distance to the right when $t = 0$.

  b. Find the location of the bob at 4 seconds. **about 0.6 m to the right**

  c. Find the location of the bob at 7.9 seconds. **about 1.2 m to the left**

60. Critical Thinking Consider the graph of $y = \cos (\theta + \pi)$.

  a. Write an expression for the x-intercepts of the graph. **$\frac{\pi}{2} + \pi n$, where n is an integer**

  b. What is the y-intercept of the graph? **−1**

  c. What is the period of the function? **2π**

  d. Sketch the graph. **See margin.**

57d.

$y = 2 + \sin \theta$

60d.

$y = \cos(\theta + \pi)$

61. **Physics** Three different weights are suspended from three different springs. Each spring has an elasticity coefficient of 18.5. The equation for the vertical displacement is $y = 1.5 \cos\left(t\sqrt{\dfrac{k}{m}}\right)$, where t is time, k is the elasticity coefficient, and m is the mass of the weight.

a. The first weight has a mass of 0.4 kilogram. Find the period and frequency of this spring.   **about 0.9 s per cycle; about 1.1 hertz**

b. The second weight has a mass of 0.6 kilogram. Find the period and frequency of this spring.   **about 1.1 s per cycle; about 0.9 hertz**

c. The third weight has a mass of 0.8 kilogram. Find the period and frequency of this spring.   **about 1.3 s per cycle; about 0.8 hertz**

d. As the mass increases, what happens to the period?   **It increases.**

e. As the mass increases, what happens to the frequency?   **It decreases.**

Mixed Review

62. Find $\cos\left(-\dfrac{5\pi}{2}\right)$ by referring to the graph of the cosine function. *(Lesson 6-3)*   **0**

63. Determine the angular velocity if 84 revolutions are completed in 6 seconds. *(Lesson 6-2)*   **about 88.0 radians/s**

64. Given a central angle of 73°, find the length of its intercepted arc in a circle of radius 9 inches. *(Lesson 6-1)*   **about 11.5 in.**

65. Solve the triangle if $a = 15.1$ and $b = 19.5$. Round to the nearest tenth. *(Lesson 5-5)*
$c = 24.7$, $A = 37.8°$, $B = 52.2°$

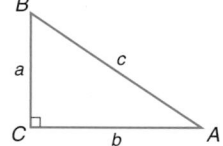

66. **Physics** The period of a pendulum can be determined by the formula $T = 2\pi\sqrt{\dfrac{\ell}{g}}$, where T represents the period, ℓ represents the length of the pendulum, and g represents the acceleration due to gravity. Determine the length of the pendulum if the pendulum has a period on Earth of 4.1 seconds and the acceleration due to gravity at Earth's surface is 9.8 meters per second squared. *(Lesson 4-7)*   **about 4.17 m**

67. Find the discriminant of $3m^2 + 5m + 10 = 0$. Describe the nature of the roots. *(Lesson 4-2)*   **−95; 2 imaginary roots**

68. **Manufacturing** Icon, Inc. manufactures two types of computer graphics cards, Model 28 and Model 74. There are three stations, A, B, and C, on the assembly line. The assembly of a Model 28 graphics card requires 30 minutes at station A, 20 minutes at station B, and 12 minutes at station C. Model 74 requires 15 minutes at station A, 30 minutes at station B, and 10 minutes at station C. Station A can be operated for no more than 4 hours a day, station B can be operated for no more than 6 hours a day, and station C can be operated for no more than 8 hours. *(Lesson 2-7)*

68a. 3 of Model 28, 10 of Model 74

a. If the profit on Model 28 is $100 and on Model 74 is $60, how many of each model should be assembled each day to provide maximum profit?

b. What is the maximum daily profit?   **$900**

Extra Credit

Graph $y = \dfrac{1}{2}\cos\dfrac{1}{8}x$ between $-8\pi \le x \le 8\pi$, and state the amplitude and the period. Then, complete the table.   **The amplitude is $\dfrac{1}{2}$ and the period is 16π.**

x	-8π	-4π	0	2π	4π	6π
$\dfrac{1}{2}\cos\dfrac{1}{8}x$	$-\dfrac{1}{2}$	0	$\dfrac{1}{2}$	$\dfrac{\sqrt{2}}{4}$	0	$-\dfrac{1}{2}$

69. (−2, 1),
(1, 1), (3, 4),
(−3, 2)

69. Use a reflection matrix to find the coordinates of the vertices of a quadrilateral reflected over the *x*-axis if the coordinates of the vertices of the quadrilateral are located at (−2, −1), (1, −1), (3, −4), and (−3, −2). *(Lesson 2-4)*

70. Graph $g(x) = \begin{cases} -3x & \text{if } x < -2 \\ 2 & \text{if } -2 \le x < 3. \\ x + 1 & \text{if } x \ge 3 \end{cases}$ *(Lesson 1-7)* **See margin.**

71. **Fund-Raising** The regression equation of a set of data is $y = 14.7x + 140.1$, where *y* represents the money collected for a fund-raiser and *x* represents the number of members of the organization. Use the equation to predict the amount of money collected by 20 members. *(Lesson 1-6)* **$434.10**

72. Given that *x* is an integer, state the relation representing $y = x^2$ and $-4 \le x \le -2$ by listing a set of ordered pairs. Then state whether this relation is a function. *(Lesson 1-1)* **{(−4, 16), (−3, 9), (−2, 4)}; yes**

73. **SAT/ACT Practice** Points *RSTU* are the centers of four congruent circles. If the area of square *RSTU* is 100, what is the sum of the areas of the four circles? **C**

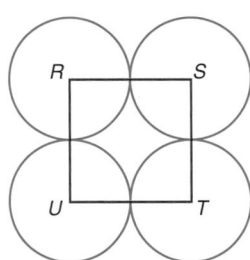

A 25π

B 50π

C 100π

D 200π

E 400π

MID-CHAPTER QUIZ

1. Change $\dfrac{5\pi}{6}$ radians to degree measure.
   (Lesson 6-1) **150°**

2. **Mechanics** A pulley with diameter 0.5 meter is being used to lift a box. How far will the box weight rise if the pulley is rotated through an angle of $\dfrac{5\pi}{3}$ radians? (Lesson 6-1) **about 1.3 m**

3. Find the area of a sector if the central angle measures $\dfrac{2\pi}{5}$ radians and the radius of the circle is 8 feet. (Lesson 6-1) **about 40.2 ft²**

4. Determine the angular displacement in radians of 7.8 revolutions. (Lesson 6-2)

5. Determine the angular velocity if 8.6 revolutions are completed in 7 seconds. (Lesson 6-2) **about 7.7 radians/s**

4. **about 49.0 radians**

6. Determine the linear velocity of a point rotating at an angular velocity of 8π radians per second at a distance of 3 meters from the center of the rotating object. (Lesson 6-2) **about 75.4 m/s**

7. Find $\sin\left(-\dfrac{7\pi}{2}\right)$ by referring to the graph of the sine function. (Lesson 6-3) **1**

8. Graph $y = \cos x$ for $7\pi \le x \le 9\pi$. (Lesson 6-3) **See margin.**

9. State the amplitude and period for the function $y = -7\cos\dfrac{\theta}{3}$. Then graph the function. (Lesson 6-4) **7; 6π; See margin.**

10. Find the possible equations of the sine function with amplitude 5 and period $\dfrac{\pi}{3}$. (Lesson 6-4) **$y = \pm 5\sin 6\theta$**

Extra Practice See p. A36.

Lesson 6-4 Amplitude and Period of Sine and Cosine Functions **377**

Additional Answers
Mid-Chapter Quiz

8.

9.

---

1 FOCUS

5-Minute Check
Lesson 6-4

State the amplitude and period for each function. Then graph each function.

1. $y = -3 \cos 2\theta$ **3, π**

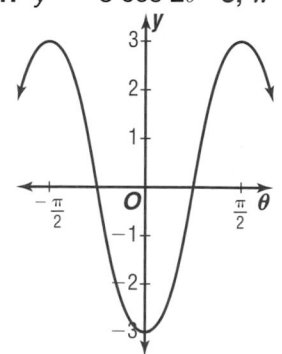

2. $y = \frac{2}{3} \cos \frac{\theta}{4}$ **$\frac{2}{3}, 8\pi$**

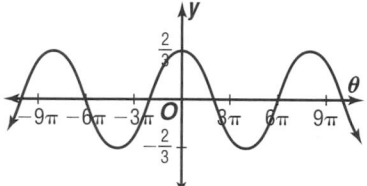

3. $y = \sin 4\theta$ **1, $\frac{\pi}{2}$**

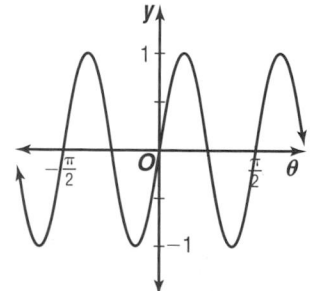

4. Write an equation of the sine function with amplitude 0.27 and period $\frac{\pi}{2}$. $y = \pm \, 0.27 \sin 4\theta$

5. Write an equation of the cosine function with amplitude $\frac{3}{5}$ and period 4.
$y = \pm \frac{3}{5} \cos \frac{\pi}{2}\theta$

6-5

Translations of Sine and Cosine Functions

OBJECTIVES
- Find the phase shift and the vertical translation for sine and cosine functions.
- Write the equations of sine and cosine functions given the amplitude, period, phase shift, and vertical translation.
- Graph compound functions.

Real World Application

TIDES One day in March in San Diego, California, the first low tide occurred at 1:45 A.M., and the first high tide occurred at 7:44 A.M. Approximately 12 hours and 24 minutes or 12.4 hours after the first low tide occurred, the second low tide occurred. The equation that models these tides is

$$h = 2.9 + 2.2 \sin \left(\frac{\pi}{6.2}t - \frac{4.85\pi}{6.2} \right),$$

where t represents the number of hours since midnight and h represents the height of the water. Draw a graph that models the cyclic nature of the tide. *This problem will be solved in Example 4.*

In Chapter 3, you learned that the graph of $y = (x - 2)^2$ is a horizontal translation of the parent graph of $y = x^2$. Similarly, graphs of the sine and cosine functions can be translated horizontally.

GRAPHING CALCULATOR EXPLORATION

- Select the radian mode.
- Use the domain and range values below to set the viewing window.
  $-4.7 \leq x \leq 4.8$, **Xscl: 1** $-3 \leq y \leq 3$, **Yscl: 1**

TRY THESE
1. Graph each function on the same screen.

 a. $y = \sin x$ **b.** $y = \sin \left(x + \frac{\pi}{4} \right)$
 c. $y = \sin \left(x + \frac{\pi}{2} \right)$ **a–c. See margin.**

WHAT DO YOU THINK?
2. Describe the behavior of the graph of $f(x) = \sin (x + c)$, where $c > 0$, as c increases. **The graph shifts farther to the left.**

3. Make a conjecture about what happens to the graph of $f(x) = \sin (x + c)$ if $c < 0$ and continues to decrease. Test your conjecture. **The graph shifts farther to the right.**

A horizontal translation or shift of a trigonometric function is called a **phase shift.** Consider the equation of the form $y = A \sin (k\theta + c)$, where $A, k, c \neq 0$. To find a zero of the function, find the value of θ for which $A \sin (k\theta + c) = 0$. Since $\sin 0 = 0$, solving $k\theta + c = 0$ will yield a zero of the function.

Resource Manager

Reproducible Masters
- *Study Guide and Practice,* pp. 107–108
- *Enrichment,* p. 65
- *Graphing Calculator Keystrokes,* p. 23
- *Problem Solving and Applications,* p. 11
- *TI-92 & CBL Interdisciplinary Activities,* pp. 17–20

Transparencies
- 5-Minute Check, 6-5
- Teaching, 6-5

$$k\theta + c = 0$$
$$\theta = -\frac{c}{k} \quad \textit{Solve for } \theta.$$

Therefore, $y = 0$ when $\theta = -\frac{c}{k}$. The value of $-\frac{c}{k}$ is the phase shift.

When $c > 0$: The graph of $y = A \sin(k\theta + c)$ is the graph of $y = A \sin k\theta$, shifted $\left|\frac{c}{k}\right|$ to the left.

When $c < 0$: The graph of $y = A \sin(k\theta + c)$ is the graph of $y = A \sin k\theta$, shifted $\left|\frac{c}{k}\right|$ to the right.

Phase Shift of Sine and Cosine Functions

The phase shift of the functions $y = A \sin(k\theta + c)$ and $y = A \cos(k\theta + c)$ is $-\frac{c}{k}$, where $k > 0$.
If $c > 0$, the shift is to the left.
If $c < 0$, the shift is to the right.

Example **1** State the phase shift for each function. Then graph the function.

a. $y = \sin(\theta + \pi)$

The phase shift of the function is $-\frac{c}{k}$ or $-\frac{\pi}{1}$, which equals $-\pi$.

To graph $y = \sin(\theta + \pi)$, consider the graph of $y = \sin \theta$. Graph this function and then shift the graph $-\pi$.

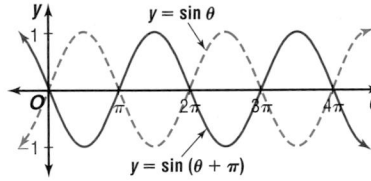

b. $y = \cos\left(2\theta - \frac{\pi}{2}\right)$

The phase shift of the function is $-\frac{c}{k}$ or $-\left(\dfrac{-\frac{\pi}{2}}{2}\right)$, which equals $\frac{\pi}{4}$.

To graph $y = \cos\left(2\theta - \frac{\pi}{2}\right)$, consider the graph of $y = \cos 2\theta$. The graph of $y = \cos 2\theta$ has amplitude of 1 and a period of $\frac{2\pi}{2}$ or π. Graph this function and then shift the graph $\frac{\pi}{4}$.

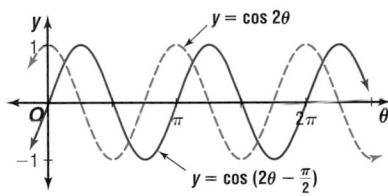

Additional Answer
Graphing Calculator Exploration
1.

Motivating the Lesson
Discussion Have students work in groups to graph the equations $y = x$, $y = x + 2$, and $y = x - 2$. Ask them to describe how the graph of $y = x$ is affected by adding 2 to x and by subtracting 2 from x. **Adding 2 raises the graph 2 units while subtracting 2 lowers it 2 units.**

2 TEACH

 Graphing Calculator Exploration

Before assigning the Exploration, you may wish to review the difference between $y = x^2$ and $y = (x + 3)^2$.

In-Class Example
Example 1
State the phase shift for each function. Then graph the function.
a. $y = \cos(\theta - \pi)$ π

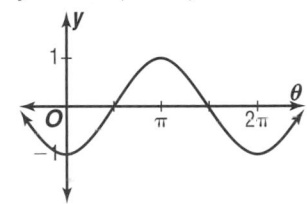

b. $y = \sin(4\theta + \pi)$ $-\frac{\pi}{4}$

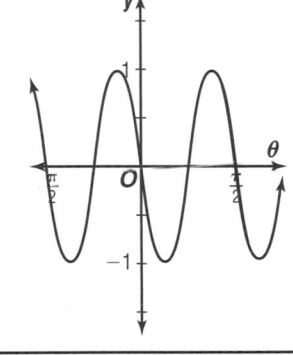

Teaching Tip For In-Class Example 1b, point out that once students graph $y = \sin(4\theta + \pi)$, they should substitute for θ some test values such as 0 or $\frac{\pi}{8}$. If $\theta = 0$, $y = \sin(0 + \pi) = \sin \pi = 0$. Show students that the graph crosses at the origin. If $\theta = \frac{\pi}{8}$, $y = \sin\left(4 \cdot \frac{\pi}{8} + \pi\right) = \sin\left(\frac{\pi}{2} + \pi\right) = \sin \frac{3\pi}{2} = -1$.

Teaching Tip For Example 2, remind students to make the vertical shift first when graphing.

In-Class Example

Example 2

State the vertical shift and the equation of the midline for the function $y = 3 \sin \theta + 2$. Then graph the function. **2; $y = 2$**

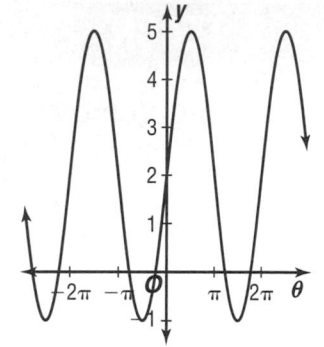

In Chapter 3, you also learned that the graph of $y = x^2 - 2$ is a vertical translation of the parent graph of $y = x^2$. Similarly, graphs of the sine and cosine functions can be translated vertically.

When a constant is added to a sine or cosine function, the graph is shifted upward or downward. If (x, y) are the coordinates of $y = \sin x$, then $(x, y + d)$ are the coordinates of $y = \sin x + d$.

A new horizontal axis known as the **midline** becomes the reference line or equilibrium point about which the graph oscillates. For the graph of $y = A \sin \theta + h$, the midline is the graph of $y = h$.

Vertical Shift of Sine and Cosine Functions	The vertical shift of the functions $y = A \sin (k\theta + c) + h$ and $y = A \cos (k\theta + c) + h$ is h. If $h > 0$, the shift is upward. If $h < 0$, the shift is downward. The midline is $y = h$.

Example 2 State the vertical shift and the equation of the midline for the function $y = 2 \cos \theta - 5$. Then graph the function.

The vertical shift is 5 units downward. The midline is the graph of $y = -5$.

To graph the function, draw the midline, the graph of $y = -5$. Since the amplitude of the function is $|2|$ or 2, draw dashed lines parallel to the midline which are 2 units above and below the midline. That is, $y = -3$ and $y = -7$. Then draw the cosine curve.

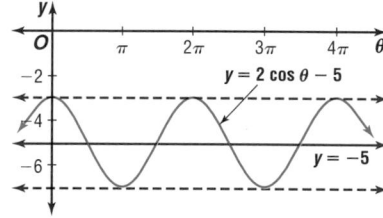

In general, use the following steps to graph any sine or cosine function.

Graphing Sine and Cosine Functions	1. Determine the vertical shift and graph the midline.
	2. Determine the amplitude. Use dashed lines to indicate the maximum and minimum values of the function.
	3. Determine the period of the function and graph the appropriate sine or cosine curve.
	4. Determine the phase shift and translate the graph accordingly.

Example ③ State the amplitude, period, phase shift, and vertical shift for

$y = 4 \cos\left(\dfrac{\theta}{2} + \pi\right) - 6$. Then graph the function.

The amplitude is $|4|$ or 4. The period is $\dfrac{2\pi}{\frac{1}{2}}$ or 4π. The phase shift is $-\dfrac{\pi}{\frac{1}{2}}$ or

-2π. The vertical shift is -6. Using this information, follow the steps for graphing a cosine function.

Step 1 Draw the midline which is the graph of $y = -6$.

Step 2 Draw dashed lines parallel to the midline, which are 4 units above and below the midline.

Step 3 Draw the cosine curve with period of 4π.

Step 4 Shift the graph 2π units to the left.

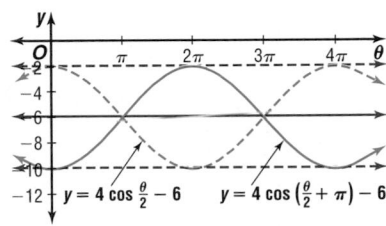

You can use information about amplitude, period, and translations of sine and cosine functions to model real-world applications.

Example ④ **TIDES** Refer to the application at the beginning of the lesson. Draw a graph that models the San Diego tide.

The vertical shift is 2.9. Draw the midline $y = 2.9$.

The amplitude is $|2.2|$ or 2.2. Draw dashed lines parallel to and 2.2 units above and below the midline.

The period is $\dfrac{2\pi}{\frac{\pi}{6.2}}$ or 12.4. Draw the sine curve with a period of 12.4.

Shift the graph $-\dfrac{\frac{-4.85\pi}{6.2}}{\frac{\pi}{6.2}}$ or 4.85 units.

You can write an equation for a trigonometric function if you are given the amplitude, period, phase shift, and vertical shift.

Lesson 6-5 Translations of Sine and Cosine Functions **381**

Teaching Tip For Example 3, remind students that the following order should be followed when graphing sine and cosine functions: vertical shift, amplitude, period, and phase shift.

In-Class Examples

Example 3
State the amplitude, period, phase shift, and vertical shift for

$y = 2 \cos\left(\dfrac{\theta}{4} + \pi\right) - 1$. Then graph the function. **2; 8π; -4π; -1**

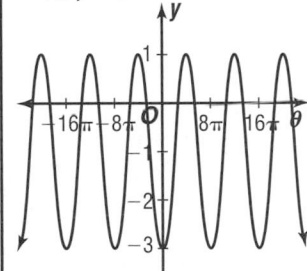

Example 4
Circuits In a particular electric circuit, at t seconds, the current is
$I(t) = 25 \sin 120\, \pi(t - \dfrac{1}{360})$.
Draw a graph that models the current in the circuit.

Teaching Tip For Example 4 in the Student Edition, remind students that tides are a function of the gravitational pull of the moon. So the time the tide takes to rise and fall is related to the orbit of the moon around Earth, which will be reflected in the graph.

Addressing Individual Needs

Visual/Spatial Complete a table for $y = 4\left(\sin x - \dfrac{\pi}{3}\right)$ for x-values in

increments of $\dfrac{\pi}{3}$. Plot the points on graph paper in one color. Then,

complete a table and plot the graph of $y = \sin x$ in another color on the

same grid. How are the two graphs related? **The graph of $y = 4\left(\sin x - \dfrac{\pi}{3}\right)$ is the**

graph of $y = \sin x$ expanded vertically to an amplitude of 4 and shifted right $\dfrac{\pi}{3}$.

Example ⑤ Write an equation of a sine function with amplitude 4, period π, phase shift $-\frac{\pi}{8}$, and vertical shift 6.

The form of the equation will be $y = A \sin (k\theta + c) + h$. Find the values of A, k, c, and h.

A: $|A| = 4$
  $A = 4$ or -4

k: $\frac{2\pi}{k} = \pi$ *The period is π.*
  $k = 2$

c: $-\frac{c}{k} = -\frac{\pi}{8}$ *The phase shift is $-\frac{\pi}{8}$.*
  $-\frac{c}{2} = -\frac{\pi}{8}$ $k = 2$
  $c = \frac{\pi}{4}$

h: $h = 6$

Substitute these values into the general equation. The possible equations are
$y = 4 \sin\left(2\theta + \frac{\pi}{4}\right) + 6$ and $y = -4 \sin\left(2\theta + \frac{\pi}{4}\right) + 6$.

Compound functions may consist of sums or products of trigonometric functions. Compound functions may also include sums and products of trigonometric functions and other functions.

Here are some examples of compound functions.

$y = \sin x \cdot \cos x$ *Product of trigonometric functions*

$y = \cos x + x$ *Sum of a trigonometric function and a linear function*

You can graph compound functions involving addition by graphing each function separately on the same coordinate axes and then adding the ordinates. After you find a few of the critical points in this way, you can sketch the rest of the curve of the function of the compound function.

Example ⑥ Graph $y = x + \cos x$.

First graph $y = \cos x$ and $y = x$ on the same axis. Then add the corresponding ordinates of the function. Finally, sketch the graph.

x	cos x	x + cos x
0	1	1
$\frac{\pi}{2}$	0	$\frac{\pi}{2} + 0 \approx 1.57$
π	-1	$\pi - 1 \approx 2.14$
$\frac{3\pi}{2}$	0	$\frac{3\pi}{2} \approx 4.71$
2π	1	$2\pi + 1 \approx 7.28$
$\frac{5\pi}{2}$	0	$\frac{5\pi}{2} \approx 7.85$
3π	-1	$3\pi - 1 \approx 8.42$

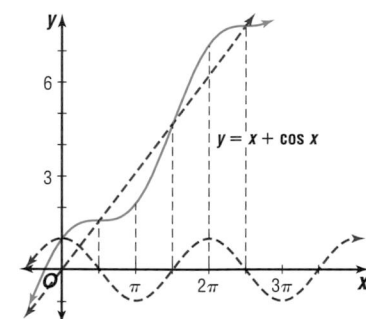

$y = x + \cos x$

Additional Answers *(p. 383)*

6.
$y = 3\cos\left(\theta - \frac{\pi}{2}\right)$

7.
$y = \sin 2\theta + 3$

Communicating Mathematics

Read and study the lesson to answer each question.   **1. See margin.**

1. **Compare and contrast** the graphs $y = \sin x + 1$ and $y = \sin (x + 1)$.

2. **Name** the function whose graph is the same as the graph of $y = \cos x$ with a phase shift of $\frac{\pi}{2}$.   **sine function**

3. **Analyze** the function $y = A \sin (k\theta + c) + h$. Which variable could you increase or decrease to have each of the following effects on the graph?
   a. stretch the graph vertically   **increase $|A|$**
   b. translate the graph downward vertically   **decrease h**
   c. shrink the graph horizontally   **increase $|k|$**
   d. translate the graph to the left.   **increase c**

4. Graph $y = \sin x$ and $y = \cos x$, and find the sum of their ordinates.   4. **Explain** how to graph $y = \sin x + \cos x$.

5. Jamal;
$-\dfrac{c}{k} = -\dfrac{-\frac{\pi}{2}}{\frac{\pi}{6}}$
or 3

5. **You Decide**   Marsha and Jamal are graphing $y = \cos \left(\frac{\pi}{6} \theta - \frac{\pi}{2} \right)$. Marsha says that the phase shift of the graph is $\frac{\pi}{2}$. Jamal says that the phase shift is 3. Who is correct? Explain.

Guided Practice

6. $\frac{\pi}{2}$; See margin.

6. State the phase shift for $y = 3 \cos \left(\theta - \frac{\pi}{2} \right)$. Then graph the function.

7. State the vertical shift and the equation of the midline for $y = \sin 2\theta + 3$. Then graph the function.   **3; $y = 3$; See margin.**

State the amplitude, period, phase shift, and vertical shift for each function. Then graph the function.   **8–9. See margin for graphs.**

8. 2; π; $-\frac{\pi}{2}$; -5 8. $y = 2 \sin (2\theta + \pi) - 5$

9. $\frac{1}{2}$; 4π; $\frac{\pi}{2}$; 3   9. $y = 3 - \frac{1}{2} \cos \left(\frac{\theta}{2} - \frac{\pi}{4} \right)$

10. Write an equation of a sine function with amplitude 20, period 1, phase shift 0, and vertical shift 100. $y = \pm 20 \sin 2\pi\theta + 100$

11. Write an equation of a cosine function with amplitude 0.6, period 12.4, phase shift -2.13, and vertical shift 7.
$$y = \pm 0.6 \cos \left(\frac{\pi}{6.2} \theta + \frac{2.13\pi}{6.2} \right) + 7$$

12. Graph $y = \sin x - \cos x$.   **See margin.**

13. **Health**   If a person has a blood pressure of 130 over 70, then the person's blood pressure oscillates between the maximum of 130 and a minimum of 70.   **b. $P = \pm 30 \sin 2\pi t + 100$**
   a. Write the equation for the midline about which this person's blood pressure oscillates. $P = 100$
   b. If the person's pulse rate is 60 beats a minute, write a sine equation that models his or her blood pressure using t as time in seconds.
   c. Graph the equation.   **See margin.**

Practice

State the phase shift for each function. Then graph each function.

14. $y = \sin (\theta - 2\pi)$ **2π**   15. $y = \sin (2\theta + \pi)$ $-\frac{\pi}{2}$ 16. $y = 2 \cos \left(\frac{\theta}{4} + \frac{\pi}{2} \right) - 2\pi$

14–16. See margin for graphs.

Error Analysis

Watch for students who have difficulty understanding the concept of phase shift.

Prevent by reminding students that phase shifts affect horizontal movement only.

Assignment Guide
Core: 15–39 odd, 41–43, 45–54
Enriched: 14–40 even, 41–54

Additional Answers

12.

13c.

14.

15.

(continued on page 384)

Additional Answers

8.

9.

16.
$y = 2 \cos\left(\frac{\theta}{4} + \frac{\pi}{2}\right)$

17.
$y = \sin\frac{\theta}{2} + \frac{1}{2}$

18.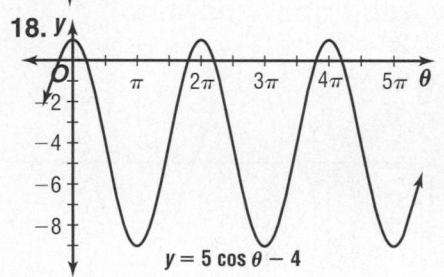
$y = 5 \cos\theta - 4$

19.
$y = 7 + \cos 2\theta$

21.
$y = 3 \cos\left(\theta - \frac{\pi}{2}\right)$

22.
$y = 6 \sin\left(\theta + \frac{\pi}{3}\right) + 2$

23.
$y = -2 + \sin\left(\frac{\theta}{3} - \frac{\pi}{12}\right)$

State the vertical shift and the equation of the midline for each function. Then graph each function.   **17–19. See margin for graphs.**

17. $\frac{1}{2}$; $y = \frac{1}{2}$

17. $y = \sin\frac{\theta}{2} + \frac{1}{2}$ **18.** $y = 5 \cos\theta - 4$ **19.** $y = 7 + \cos 2\theta$
$\quad\quad\quad\quad\quad\quad\quad\quad\quad\quad\quad -4; y = -4 \quad\quad\quad\quad\quad 7; y = 7$

20. State the horizontal and vertical shift for $y = -8 \sin(2\theta - 4\pi) - 3$.   **$2\pi$; -3**

State the amplitude, period, phase shift, and vertical shift for each function. Then graph the function.   **21–26. See margin for graphs.**

 21. $y = 3 \cos\left(\theta - \frac{\pi}{2}\right)$ $3; 2\pi; \frac{\pi}{2}; 0$ **22.** $y = 6 \sin\left(\theta + \frac{\pi}{3}\right) + 2$ $6; 2\pi; -\frac{\pi}{3}; 2$

23. $y = -2 + \sin\left(\frac{\theta}{3} - \frac{\pi}{12}\right)$ $1; 6\pi; \frac{\pi}{4}; -2$ **24.** $y = 20 + 5 \cos(3\theta + \pi)$ $5; \frac{2\pi}{3}; -\frac{\pi}{3}; 20$

25. $y = \frac{1}{4}\cos\frac{\theta}{2} - 3$ $\frac{1}{4}; 4\pi; 0; -3$ **26.** $y = 10 \sin\left(\frac{\theta}{4} - 4\pi\right) - 5$ $10; 8\pi; 16\pi; -5$

27. State the amplitude, period, phase shift, and vertical shift of the sine curve shown at the right.
$4; 4\pi; -\frac{\pi}{2}; -2$

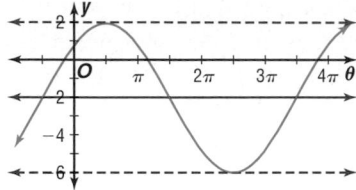

Write an equation of the sine function with each amplitude, period, phase shift, and vertical shift.

28. $y = \pm 7 \sin\left(\frac{2}{3}\theta - \frac{2\pi}{3}\right) - 7$ **28.** amplitude $= 7$, period $= 3\pi$, phase shift $= \pi$, vertical shift $= -7$

29. amplitude $= 50$, period $= \frac{3\pi}{4}$, phase shift $= \frac{\pi}{2}$, vertical shift $= -25$

29. $y = \pm 50 \sin\left(\frac{8}{3}\theta - \frac{4\pi}{3}\right) - 25$ **30.** amplitude $= \frac{3}{4}$, period $= \frac{\pi}{5}$, phase shift $= \pi$, vertical shift $= \frac{1}{4}$

Write an equation of the cosine function with each amplitude, period, phase shift, and vertical shift.

30. $y = \pm\frac{3}{4}\sin(10\theta - 10\pi) + \frac{1}{4}$ **31.** amplitude $= 3.5$, period $= \frac{\pi}{2}$, phase shift $= \frac{\pi}{4}$, vertical shift $= 7$

31. $y = \pm 3.5 \cos(4\theta - \pi) + 7$ **32.** amplitude $= \frac{4}{5}$, period $= \frac{\pi}{6}$, phase shift $= \frac{\pi}{3}$, vertical shift $= \frac{7}{5}$

33. amplitude $= 100$, period $= 45$, phase shift $= 0$, vertical shift $= -110$

32. $y = \pm\frac{4}{5}\cos(12\theta - 4\pi) + \frac{7}{5}$

 34. Write a cosine equation for the graph at the right. $y = -2 \cos\left(\frac{\theta}{2}\right) - 1$

33. $y = \pm 100 \cos\left(\frac{2\pi}{45}\theta\right) - 110$

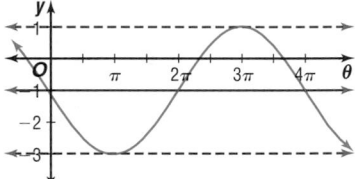

35. Write a sine equation for the graph at the right. $y = 0.5 \sin 2\theta + 3$

24.
$y = 20 + 5 \cos(3\theta + \pi)$

25.
$y = \frac{1}{4}\cos\frac{\theta}{4} - 3$

Graph each function. **36–38. See margin.**

36. $y = \sin x + x$ **37.** $y = \cos x - \sin x$ **38.** $y = \sin x + \sin 2x$

39. On the same coordinate plane, graph each function. **See margin.**

    a. $y = 2 \sin x$ **b.** $y = 3 \cos x$ **c.** $y = 2 \sin x + 3 \cos x$

40. See margin.

40. Use the graphs of $y = \cos 2x$ and $y = \cos 3x$ to graph $y = \cos 2x - \cos 3x$.

Applications and Problem Solving

41a. 3000; 1000

41b. 15,000; 5000

41d. months number 3 and 15

41e. months number 0, 12, and 24

41f. When the sheep population is at a maximum, the wolf population is on the increase because of the maximum availability of food. The upswing in wolf population leads to a maximum later.

41. Biology In the wild, predators such as wolves need prey such as sheep to survive. The population of the wolves and the sheep are cyclic in nature. Suppose the population of the wolves W is modeled by $W = 2000 + 1000 \sin\left(\dfrac{\pi t}{6}\right)$ and population of the sheep S is modeled by $S = 10{,}000 + 5000 \cos\left(\dfrac{\pi t}{6}\right)$ where t is the time in months.

  a. What are the maximum number and the minimum number of wolves?

  b. What are the maximum number and the minimum number of sheep?

  c. Use a graphing calculator to graph both equations for values of t from 0 to 24. **See margin.**

  d. During which months does the wolf population reach a maximum?

  e. During which months does the sheep population reach a maximum?

  f. What is the relationship of the maximum population of the wolves and the maximum population of the sheep? Explain.

42. See margin.

42. Critical Thinking Use the graphs of $y = x$ and $y = \cos x$ to graph $y = x \cos x$.

43. Entertainment As you ride a Ferris wheel, the height that you are above the ground varies periodically. Consider the height of the center of the wheel to be the equilibrium point. Suppose the diameter of a Ferris Wheel is 42 feet and travels at a rate of 3 revolutions per minute. At the highest point, a seat on the Ferris wheel is 46 feet above the ground.

  a. What is the lowest height of a seat?   **4 ft**

  b. What is the equation of the midline?   **$t = 25$**

  c. What is the period of the function?   **20 s**   **d.** $h = 25 + 21 \sin \dfrac{\pi t}{10}$

  d. Write a sine equation to model the height of a seat that was at the equilibrium point heading upward when the ride began.

  e. According to the model, when will the seat reach the highest point for the first time?   **5 s**

  f. According to the model, what is the height of the seat after 10 seconds?   **25 ft**

44. Electronics In electrical circuits, the voltage and current can be described by sine or cosine functions. If the graphs of these functions have the same period, but do not pass through their zero points at the same time, they are said to have a *phase difference*. For example, if the voltage is 0 at 90° and the current is 0 at 180°, they are 90° out of phase. Suppose the voltage across an inductor of a circuit is represented by $y = 2 \cos 2x$ and the current across the component is represented by $y = \cos\left(2x - \dfrac{\pi}{2}\right)$. What is the phase relationship between the signals? **There is a $\dfrac{\pi}{4}$ phase difference.**

Lesson 6-5 Translations of Sine and Cosine Functions **385**

Additional Answers

37.

38.

39.

40.

41c.

[0, 24] scl:1 by [0, 16,000] scl:1000

42.

Additional Answers (pp. 384–385)

26.

36.

4 ASSESS

Open-Ended Assessment
Modeling A Japanese company invented the first integrated radio circuit in 1966. Suppose that researchers were observing a sine curve that had an amplitude of 3 centimeters, a period of 9 centimeters, an upper shift of 2 centimeters, and a phase shift $\frac{1}{2}$ centimeter to the right. State the function that models the data.

$$y = 3 \sin\left(\frac{2\pi}{9}\theta - \frac{\pi}{9}\right) + 2$$

Additional Answers
48.

$$y = \frac{x-3}{x-2}$$

52.

$$y \le |x+4|$$

Enrichment Masters, p. 65

386 *Chapter 6*

---

45. Critical Thinking The windows for the following calculator screens are set at $[-2\pi, 2\pi]$ scl: 0.5π by $[-2, 2]$ scl: 0.5. Without using a graphing calculator, use the equations below to identify the graph on each calculator screen.

$$y = \cos x^2 \qquad y = \sqrt{\sin x} \qquad y = \frac{\cos x}{x} \qquad y = \sin \sqrt{x}$$

45a. $y = \sqrt{\sin x}$

45b. $y = \frac{\cos x}{x}$

a.

b.

45c. $y = \cos x^2$

45d. $y = \sin \sqrt{x}$

c.

d.

Mixed Review

46. Music Write an equation of the sine function that represents the initial behavior of the vibrations of the note D above middle C having amplitude 0.25 and a frequency of 294 hertz. *(Lesson 6-4)* $y = 0.25 \sin 588\pi t$

47. Determine the linear velocity of a point rotating at an angular velocity of 19.2 radians per second at a distance of 7 centimeters from the center of the rotating object. *(Lesson 6-2)* **134.4 cm/s**

48. Graph $y = \frac{x-3}{x-2}$. *(Lesson 3-7)* **See margin.**

49. Find the inverse of $f(x) = \frac{3}{x-1}$. *(Lesson 3-4)* $y = \frac{3}{x} + 1$

50. Find matrix X in the equation $\begin{bmatrix} 1 & 1 \\ 1 & 1 \end{bmatrix}\begin{bmatrix} 3 & 5 \\ -3 & -5 \end{bmatrix} = X$. *(Lesson 2-3)* $\begin{bmatrix} 0 & 0 \\ 0 & 0 \end{bmatrix}$

51. Solve the system of equations. *(Lesson 2-1)* **(1.4, −0.04)**
$$3x + 5y = 4$$
$$14x - 35y = 21$$

52. Graph $y \le |x+4|$. *(Lesson 1-8)* **See margin.**

53. Write the standard form of the equation of the line through the point at $(3, -2)$ that is parallel to the graph of $3x - y + 7 = 0$. *(Lesson 1-5)* $3x - y - 11 = 0$

54. SAT Practice Grid-In A swimming pool is 75 feet long and 42 feet wide. If 7.48 gallons equals 1 cubic foot, how many gallons of water are needed to raise the level of the water 4 inches? **7854**

386 *Chapter 6 Graphs of Trigonometric Functions*

Extra Practice See p. A37.

Extra Credit

Graph $y = 4 \cos\left(x + \frac{\pi}{8}\right) - \left(3 \sin x - 3\right) 2x$.

Modeling Real-World Data with Sinusoidal Functions

OBJECTIVES
- Model real-world data using sine and cosine functions.
- Use sinusoidal functions to solve problems.

 Real World Application

METEOROLOGY The table contains the times that the sun rises and sets on the fifteenth of every month in Brownsville, Texas.

Let $t = 1$ represent January 15.
Let $t = 2$ represent February 15.
Let $t = 3$ represent March 15.
⋮

Write a function that models the hours of daylight for Brownsville. Use your model to estimate the number of hours of daylight on September 30. *This problem will be solved in Example 1.*

Month	Sunrise A.M.	Sunset P.M.
January	7:19	6:00
February	7:05	6:23
March	6:40	6:39
April	6:07	6:53
May	5:44	7:09
June	5:38	7:23
July	5:48	7:24
August	6:03	7:06
September	6:16	6:34
October	6:29	6:03
November	6:48	5:41
December	7:09	5:41

Before you can determine the function for the daylight, you must first compute the amount of daylight for each day as a decimal value. Consider January 15. First, write each time in 24-hour time.

7:19 A.M. = 7:19

6:00 P.M. = 6:00 + 12 or 18:00

Then change each time to a decimal rounded to the nearest hundredth.

$7:19 = 7 + \dfrac{19}{60}$ or 7.32

$18:00 = 18 + \dfrac{0}{60}$ or 18.00

On January 15, there will be 18.00 − 7.32 or 10.68 hours of daylight.

Similarly, the number of daylight hours can be determined for the fifteenth of each month.

Month	Jan.	Feb.	March	April	May	June
t	1	2	3	4	5	6
Hours of Daylight	10.68	11.30	11.98	12.77	13.42	13.75

Month	July	Aug.	Sept.	Oct.	Nov.	Dec.
t	7	8	9	10	11	12
Hours of Daylight	13.60	13.05	12.30	11.57	10.88	10.53

Lesson 6-6   Modeling Real-World Data with Sinusoidal Functions **387**

Resource Manager

 Reproducible Masters
- *Study Guide and Practice,* pp. 109–110
- *Enrichment,* p. 66
- *Assessment and Evaluation,* p. 148

 Transparencies
- 5-Minute Check, 6-6
- Teaching, 6-6

---

1 FOCUS

5-Minute Check
Lesson 6-5

1. State the phase shift for $y = 2 \sin (2\theta - \pi)$. Then graph the function. $\dfrac{\pi}{2}$

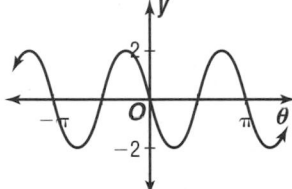

2. State the vertical shift and the equation of the midline for $y = 3 + \cos \dfrac{\theta}{2}$. Then graph the function. **3, $y = 3$**

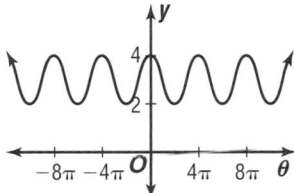

3. State the amplitude, period, phase shift, and vertical shift for $y = 2 \cos \left(\theta + \dfrac{\pi}{2}\right) + 1$. Then graph the function. **2, 2π, $-\dfrac{\pi}{2}$, 1**

4. Write an equation of the sine function with an amplitude $\dfrac{1}{2}$, period 4π, phase shift π, and vertical shift -2.
$y = \dfrac{1}{2}\left(\dfrac{\theta}{2} - \dfrac{\pi}{2}\right) - 2$;
$y = -\dfrac{1}{2}\left(\dfrac{\theta}{2} - \dfrac{\pi}{2}\right) - 2$

5. Write an equation of the cosine function with an amplitude 3, period $\dfrac{1}{2}\pi$, phase shift $-\dfrac{\pi}{4}$, and vertical shift -6.
$y = 3 \cos (4\theta + \pi) - 6$;
$y = -3 \cos (4\theta + \pi) - 6$

Since there are 12 months in a year, month 13 is the same as month 1, month 14 is the same as month 2, and so on. The function is periodic. Enter the data into a graphing calculator and graph the points. The graph resembles a type of sine curve. You can write a **sinusoidal function** to represent the data. A sinusoidal function can be any function of the form
$y = A \sin(k\theta + c) + h$ or
$y = A \cos(k\theta + c) + h$.

[−1, 13] scl:1 by [−1, 14] scl:1

Example **1** **METEOROLOGY** Refer to the application at the beginning of the lesson.

a. Write a function that models the amount of daylight for Brownsville.

b. Use your model to estimate the number of hours of daylight on September 30.

a. The data can be modeled by a function of the form $y = A \sin(kt + c) + h$, where t is the time in months. First, find A, h, and k.

A: $A = \frac{13.75 - 10.53}{2}$ or 1.61   *A is half the difference between the most daylight (13.75 h) and the least daylight (10.53 h).*

h: $h = \frac{13.75 + 10.53}{2}$ or 12.14   *h is half the sum of the greatest value and least value.*

k: $\frac{2\pi}{k} = 12$ *The period is 12.*

$k = \frac{\pi}{6}$

Substitute these values into the general form of the sinusoidal function.

$y = A \sin(kt + c) + h$

$y = 1.61 \sin\left(\frac{\pi}{6}t + c\right) + 12.14$ $A = 1.61, k = \frac{\pi}{6}, h = 12.14$

To compute c, substitute one of the coordinate pairs into the function.

$$y = 1.61 \sin\left(\frac{\pi}{6}t + c\right) + 12.14$$

$$10.68 = 1.61 \sin\left(\frac{\pi}{6}(1) + c\right) + 12.14 \quad \textit{Use } (t, y) = (1, 10.68).$$

$$-1.46 = 1.61 \sin\left(\frac{\pi}{6} + c\right) \quad \textit{Add } -12.14 \textit{ to each side.}$$

$$-\frac{1.46}{1.61} = \sin\left(\frac{\pi}{6} + c\right) \quad \textit{Divide each side by 1.61.}$$

$$\sin^{-1}\left(-\frac{1.46}{1.61}\right) = \frac{\pi}{6} + c \quad \textit{Definition of inverse}$$

$$\sin^{-1}\left(-\frac{1.46}{1.61}\right) - \frac{\pi}{6} = c \quad \textit{Add } -\frac{\pi}{6} \textit{ to each side.}$$

$$-1.659305545 \approx c \quad \textit{Use a calculator.}$$

The function $y = 1.61 \sin\left(\frac{\pi}{6}t - 1.66\right) + 12.14$ is one model for the daylight in Brownsville.

Graphing Calculator Tip

For keystroke instruction on how to find sine regression statistics, see page A25.

To check this answer, enter the data into a graphing calculator and calculate the **SinReg** statistics. Rounding to the nearest hundredth, $y = 1.60 \sin(0.51t - 1.60) + 12.12$. The models are similar. Either model could be used.

```
SinReg
 y=a*sin(bx+c)+d
 a=1.598335491
 b=.5119649711
 c=-1.603016757
 d=12.11624188
```

b. September 30 is half a month past September 15, so $t = 9.5$. Select a model and use a calculator to evaluate it for $t = 9.5$.

Model 1: Paper and Pencil

$y = 1.61 \sin\left(\frac{\pi}{6}t - 1.66\right) + 12.14$

$y = 1.61 \sin\left[\frac{\pi}{6}(9.5) - 1.66\right] + 12.14$

$y \approx 11.86349848$

Model 2: Graphing Calculator

$y = 1.60 \sin(0.51t - 1.60) + 12.12$

$y = 1.60 \sin[0.51(9.5) - 1.60] + 12.12$

$y \approx 11.95484295$

On September 30, Brownsville will have about 11.9 hours of daylight.

In general, any sinusoidal function can be written as a sine function or as a cosine function. The amplitude, the period, and the midline will remain the same. However, the phase shift will be different. To avoid a greater phase shift than necessary, you may wish to use a sine function if the function is about zero at $x = 0$ and a cosine function if the function is about the maximum or minimum at $x = 0$.

Example **2** **HEALTH** An average seated adult breathes in and out every 4 seconds. The average minimum amount of air in the lungs is 0.08 liter, and the average maximum amount of air in the lungs is 0.82 liter. Suppose the lungs have a minimum amount of air at $t = 0$, where t is the time in seconds.

a. Write a function that models the amount of air in the lungs.

b. Graph the function.

c. Determine the amount of air in the lungs at 5.5 seconds.

(continued on the next page)

 Teaching Tip The **SinReg** statistics are carried out to more decimal places than the values determined during Example 1. Therefore, the differences in rounding will cause the values to differ slightly. These equations can vary slightly because they are only models and we are not relying on them to give a precise and accurate answer.

Teaching Tip Once the students get the function, encourage them to substitute the abcissa values into the equation to see whether they get the correct answers.

In-Class Example
Example 2
Oceanography The tide in a coastal city peaks every 11.6 hours. The tide ranges from 3.9 meters to 3.3 meters. Suppose that the low tide is at $t = 0$, where t is the time in hours.
a. Write a function that models the height of the tide.
   Sample answer: $y = 0.3 \sin\left(\frac{2\pi}{11.6}t\right) + 3.6$
b. Graph the function.
   Sample answer:

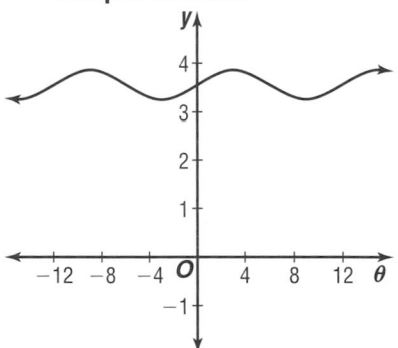

c. Determine the height of the tide at 6.2 hours. **Sample answer: about 3.54 m**

From the Classroom of...

Patty Meiergerd
Central Catholic High School
West Point, Nebraska

"I have students use a motion detector to graph the swing of a pendulum. The students then generate the equation by guessing amplitude, period, and so on. Then, students use trigonometric regression on the graphing calculator to get the actual equation."

Error Analysis

Watch for students who have difficulty translating a period of 5 units to obtain k in $y = \sin k\theta$.

Prevent by showing them that $\frac{2\pi}{k} = 5$, and that in solving for k, $k = \frac{2\pi}{5}$. Therefore, $y = \sin \frac{2\pi}{5}\theta$ has a period of 5.

Additional Answer

2. Both data that can be modeled with a polynomial function and data that can be modeled with a sinusoidal function have fluctuations. However, data that can be modeled with a sinusoidal function repeat themselves periodically, and data that can be modeled with a polynomial function do not.

***Study Guide and Practice Masters*, pp. 109–110**

---

a. Since the function has its minimum value at $t = 0$, use the cosine function. A cosine function with its minimum value at $t = 0$ has no phase shift and a negative value for A. Therefore, the general form of the model is $y = -A \cos kt + h$, where t is the time in seconds. Find A, k, and h.

    A: $A = \dfrac{0.82 - 0.08}{2}$ or 0.37 *A is half the difference between the greatest value and the least value.*

    h: $h = \dfrac{0.82 + 0.08}{2}$ or 0.45 *h is half the sum of the greatest value and the least value.*

    k: $\dfrac{2\pi}{k} = 4$ *The period is 4.*

        $k = \dfrac{\pi}{2}$

Therefore, $y = -0.37 \cos \frac{\pi}{2} t + 0.45$ models the amount of air in the lungs of an average seated adult.

b. Use a graphing calculator to graph the function.

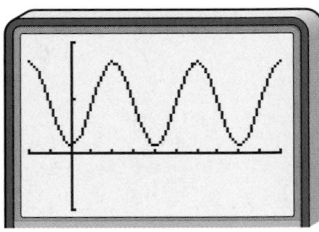

$[-2, 10]$ scl:1 by $[-0.5, 1]$ scl:0.5

c. Use this function to find the amount of air in the lungs at 5.5 seconds.

$$y = -0.37 \cos \frac{\pi}{2}t + 0.45$$

$$y = -0.37 \cos \left[\frac{\pi}{2}(5.5)\right] + 0.45$$

$$y \approx 0.711629509$$

The lungs have about 0.71 liter of air at 5.5 seconds.

CHECK FOR UNDERSTANDING

Communicating Mathematics

1. any function that can be written as a sine function or a cosine function

Read and study the lesson to answer each question.

1. **Define** sinusoidal function in your own words.

2. **Compare and contrast** real-world data that can be modeled with a polynomial function and real-world data that can be modeled with a sinusoidal function. **See margin.**

3. **Give** three real-world examples that can be modeled with a sinusoidal function. **Sample answers: the amount of daylight, the average monthly temperatures, the height of a seat on a Ferris wheel**

390   *Chapter 6   Graphs of Trigonometric Functions*

Addressing Individual Needs

Interpersonal Suppose that the temperatures were recorded in your geographic location every day all year. Have a group of students work together to determine how these data could approximate a sinusoidal curve. Discuss how period, phase shift, amplitude, and vertical shift figure into their analyses. It is not necessary that the students possess actual data, but they should be able to draw reasonable conclusions from reasonably constructed hypothetical data.

Guided Practice

4. Boating If the equilibrium point is $y = 0$, then $y = -5 \cos\left(\frac{\pi}{6}t\right)$ models a buoy bobbing up and down in the water.

    a. Describe the location of the buoy when $t = 0$. **5 units below equilibrium**

    b. What is the maximum height of the buoy? **5 units above equilibrium**

    c. Find the location of the buoy at $t = 7$. **about 4.33 units above equilibrium**

5. Health A certain person's blood pressure oscillates between 140 and 80. If the heart beats once every second, write a sine function that models the person's blood pressure. $P = 30 \sin 2\pi t + 110$

6. Meteorology The average monthly temperatures for the city of Seattle, Washington, are given below.

Jan.	Feb.	March	April	May	June	July	Aug.	Sept.	Oct.	Nov.	Dec.
41°	44°	47°	50°	56°	61°	65°	66°	61°	54°	46°	42°

    a. Find the amplitude of a sinusoidal function that models the monthly temperatures. **12.5°**

    b. Find the vertical shift of a sinusoidal function that models the monthly temperatures. **53.5°**

    c. What is the period of a sinusoidal function that models the monthly temperatures? **12 months**

    d. Write a sinusoidal function that models the monthly temperatures, using $t = 1$ to represent January.

    e. According to your model, what is the average monthly temperature in February? How does this compare to the actual average?

    f. According to your model, what is the average monthly temperature in October? How does this compare to the actual average?
    Sample answer: About 53.2°; it is close to the actual average.

6d. Sample answer:
$y = -12.5 \cos\left(\frac{\pi}{6}t - 0.5\right) + 53.5$

6e. Sample answer: About 42.8°; it is somewhat close to the actual average.

EXERCISES

Applications and Problem Solving

A **7. Music** The initial behavior of the vibrations of the note E above middle C can be modeled by $y = 0.5 \sin 660\pi t$.

    a. What is the amplitude of this model? **0.5**

    b. What is the period of this model? $\frac{1}{330}$

    c. Find the frequency (cycles per second) for this note. **330 hertz**

8. Entertainment A rodeo performer spins a lasso in a circle perpendicular to the ground. The height of the knot from the ground is modeled by $h = -3 \cos\left(\frac{5\pi}{3}t\right) + 3.5$, where t is the time measured in seconds.

    a. What is the highest point reached by the knot? **6.5 units**

    b. What is the lowest point reached by the knot? **0.5 units**

    c. What is the period of the model? $\frac{6}{5}$

    d. According to the model, find the height of the knot after 25 seconds. **2 units**

Teaching Tip For Exercise 9, tell
students that in addition to
sinusoidal functions, animal
population dynamics are also
studied in linear algebra and
matrices.

9. **Biology** In a certain region with hawks as predators and rodents as prey, the rodent population R varies according to the model $R = 1200 + 300 \sin\left(\frac{\pi}{2}t\right)$, and the hawk population H varies according to the model $H = 250 + 25 \sin\left(\frac{\pi}{2}t - \frac{\pi}{4}\right)$, with t measured in years since January 1, 1970.

   a. What was the population of rodents on January 1, 1970? **1200**

   b. What was the population of hawks on January 1, 1970? **about 232**

   c. What are the maximum populations of rodents and hawks? Do these maxima ever occur at the same time? **1500; 275; no**

9d. January 1, 1971

   d. On what date was the first maximum population of rodents achieved?

   e. What is the minimum population of hawks? On what date was the minimum population of hawks first achieved? **225; July 1, 1973**

   f. According to the models, what was the population of rodents and hawks on January 1 of the present year? **See students' work.**

B ▶ 10. **Waves** A leaf floats on the water bobbing up and down. The distance between its highest and lowest point is 4 centimeters. It moves from its highest point down to its lowest point and back to its highest point every 10 seconds. Write a cosine function that models the movement of the leaf in relationship to the equilibrium point. $y = 2 \cos\left(\frac{\pi}{5}t\right)$

11. $y = 3.55 \sin$
$\left(\frac{\pi}{6.2}t + \frac{2.34\pi}{3.1}\right)$
$+ 4.24$

11. **Tides** Write a sine function which models the oscillation of tides in Savannah, Georgia, if the equilibrium point is 4.24 feet, the amplitude is 3.55 feet, the phase shift is -4.68 hours, and the period is 12.40 hours.

12. **Meteorology** The mean average temperature in Buffalo, New York, is 47.5°. The temperature fluctuates 23.5° above and below the mean temperature. If $t = 1$ represents January, the phase shift of the sine function is 4.

12a. $y = 23.5 \sin$
$\left(\frac{\pi}{6}t - \frac{2\pi}{3}\right) + 47.5$

   a. Write a model for the average monthly temperature in Buffalo.

   b. According to your model, what is the average temperature in March?

12b. about 35.8°

   c. According to your model, what is the average temperature in August?

   about 67.9°

13. Meteorology The average monthly temperatures for the city of Honolulu, Hawaii, are given below.

Jan.	Feb.	March	April	May	June	July	Aug.	Sept.	Oct.	Nov.	Dec.
73°	73°	74°	76°	78°	79°	81°	81°	81°	80°	77°	74°

a. Find the amplitude of a sinusoidal function that models the monthly temperatures. **4°**

b. Find the vertical shift of a sinusoidal function that models the monthly temperatures. **77°**

c. What is the period of a sinusoidal function that models the monthly temperatures? **12 months**

d. Write a sinusoidal function that models the monthly temperatures, using $t = 1$ to represent January.

e. According to your model, what is the average temperature in August? How does this compare to the actual average?

f. According to your model, what is the average temperature in May? How does this compare to the actual average?

14. Critical Thinking Write a cosine function that is equivalent to $y = 3 \sin (x - \pi) + 5$.

15. Tides Burntcoat Head in Nova Scotia, Canada, is known for its extreme fluctuations in tides. One day in April, the first high tide rose to 13.25 feet at 4:30 A.M. The first low tide at 1.88 feet occurred at 10:51 A.M. The second high tide was recorded at 4:53 P.M. **a. 5.685 ft   b. 7.565 ft   c. about 12.4 h**

a. Find the amplitude of a sinusoidal function that models the tides.

b. Find the vertical shift of a sinusoidal function that models the tides.

c. What is the period of a sinusoidal function that models the tides?

d. Write a sinusoidal function to model the tides, using t to represent the number of hours in decimals since midnight.

e. According to your model, determine the height of the water at 7:30 P.M.

16. Meteorology The table at the right contains the times that the sun rises and sets in the middle of each month in New York City, New York. Suppose the number 1 represents the middle of January, the number 2 represents the middle of February, and so on.

a. Find the amount of daylight hours for the middle of each month.

b. What is the amplitude of a sinusoidal function that models the daylight hours? **2.85 h**

c. What is the vertical shift of a sinusoidal function that models the daylight hours? **12.18 h**

d. What is the period of a sinusoidal function that models the daylight hours? **12 months**

e. Write a sinusoidal function that models the daylight hours.

Month	Sunrise A.M.	Sunset P.M.	
January	7:19	4:47	**9.47 h**
February	6:56	5:24	**10.47 h**
March	6:16	5:57	**11.68 h**
April	5:25	6:29	**13.07 h**
May	4:44	7:01	**14.28 h**
June	4:24	7:26	**15.03 h**
July	4:33	7:28	**14.92 h**
August	5:01	7:01	**14 h**
September	5:31	6:14	**12.72 h**
October	6:01	5:24	**11.38 h**
November	6:36	4:43	**10.12 h**
December	7:08	4:28	**9.33 h**

Margin notes (left column):

13d. Sample answer: $y = -4 \cos \left(\frac{\pi}{6} t - 0.5 \right) + 77$

13e. Sample answer: About 80.4°; it is very close to the actual average.

13f. Sample answer: About 79.1°; it is close to the actual average.

14. Sample answer: $y = 3 \cos \left(x - \frac{3\pi}{2} \right) + 5$

15d. Sample answer: $h = 5.685 \sin \left(\frac{\pi}{6.2} t - 0.71 \right) + 7.565$

15e. Sample answer: about 8.99 ft

16e. Sample answer: $y = -2.85 \cos \left(\frac{\pi}{6} t - 0.21 \right) + 12.18$

Open-Ended Assessment

Speaking Have a student explain to the class how to write a cosine function whose period is 7 units, amplitude 3 units, and phase shift 2 units to the left.

$$y = 3 \cos\left(\frac{2\pi}{7}\theta + \frac{4\pi}{7}\right)$$

 Chapter 6, Quiz C (Lessons 6-5 and 6-6) is available in the *Assessment and Evaluation Masters*, p. 148.

Additional Answer

21.

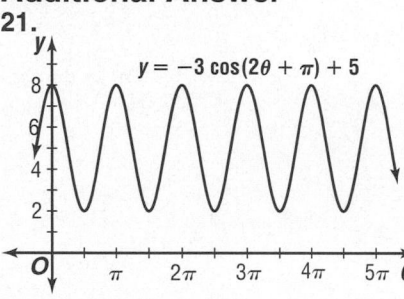

$$y = -3\cos(2\theta + \pi) + 5$$

---

 17. Critical Thinking The average monthly temperature for Phoenix, Arizona can be modeled by $y = 70.5 + 19.5 \sin\left(\frac{\pi}{6}t + c\right)$. If the coldest temperature occurs in January ($t = 1$), find the value of c. **Sample answer: about −2.09**

18. Entertainment Several years ago, an amusement park in Sandusky, Ohio, had a ride called the Rotor in which riders stood against the walls of a spinning cylinder. As the cylinder spun, the floor of the ride dropped out, and the riders were held against the wall by the force of friction. The cylinder of the Rotor had a radius of 3.5 meters and rotated counterclockwise at a rate of 14 revolutions per minute. Suppose the center of rotation of the Rotor was at the origin of a rectangular coordinate system.

a. If the initial coordinates of the hinges on the door of the cylinder are (0, −3.5), write a function that models the position of the door at t seconds. $y = -3.5 \cos\left(\frac{7\pi}{15}t\right)$

b. Find the coordinates of the hinges on the door at 4 seconds. **about (4, −3.20)**

19. Electricity For an alternating current, the instantaneous voltage V_R is graphed at the right. Write an equation for the instantaneous voltage. $V_R = 120 \sin\left(\frac{\pi}{30}t\right)$

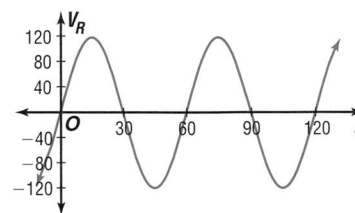

20. Meteorology Find the number of daylight hours for the middle of each month or the average monthly temperature for your community. Write a sinusoidal function to model this data. **See students' work.**

Mixed Review

21. 3; π; $-\frac{\pi}{2}$; 5; See margin.

22. $2\pi n$ where n is an integer

24. about 97.9°, 82.1°, 97.9°, 82.1°

27. $x > -1$; $x < -1$

21. State the amplitude, period, phase shift, and vertical shift for $y = -3 \cos(2\theta + \pi) + 5$. Then graph the function. *(Lesson 6-5)*

22. Find the values of θ for which $\cos \theta = 1$ is true. *(Lesson 6-3)*

23. Change 800° to radians. *(Lesson 6-1)* $\frac{40\pi}{9}$

24. Geometry The sides of a parallelogram are 20 centimeters and 32 centimeters long. If the longer diagonal measures 40 centimeters, find the measures of the angles of the parallelogram. *(Lesson 5-8)*

25. Decompose $\frac{2m + 16}{m^2 - 16}$ into partial fractions. *(Lesson 4-6)* $\frac{3}{m - 4} + \frac{-1}{m + 4}$

26. Find the value of k so that the remainder of $(2x^3 + kx^2 - x - 6) \div (x + 2)$ is zero. *(Lesson 4-3)* **5**

27. Determine the interval(s) for which the graph of $f(x) = 2|x + 1| - 5$ is increasing and the intervals for which the graph is decreasing. *(Lesson 3-5)*

28. SAT/ACT Practice If one half of the female students in a certain school eat in the cafeteria and one third of the male students eat there, what fractional part of the student body eats in the cafeteria? **E**

A $\frac{5}{12}$ B $\frac{2}{5}$ C $\frac{3}{4}$ D $\frac{5}{6}$

E not enough information given

Extra Practice See p. A37.

Extra Credit

Write a sine function that is equivalent to $y = 2 \cos\left(x - \frac{\pi}{2}\right) + 1$. **Sample answer: $y = 2 \sin x + 1$**

6-7 Graphing Other Trigonometric Functions

OBJECTIVES
- Graph tangent, cotangent, secant, and cosecant functions.
- Write equations of trigonometric functions.

SECURITY A security camera scans a long, straight driveway that serves as an entrance to an historic mansion. Suppose a line is drawn down the center of the driveway. The camera is located 6 feet to the right of the midpoint of the line. Let d represent the distance along the line from its midpoint. If t is time in seconds and the camera points at the midpoint at $t = 0$, then $d = 6 \tan\left(\frac{\pi}{30}t\right)$ models the point being scanned. In this model, the distance below the midpoint is a negative. Graph the equation for $-15 \le t \le 15$. Find the location the camera is scanning at 5 seconds. What happens when $t = 15$?
This problem will be solved in Example 4.

driveway

d

6 ft

midpoint

camera

You have learned to graph variations of the sine and cosine functions. In this lesson, we will study the graphs of the tangent, cotangent, secant, and cosecant functions. Consider the tangent function. First evaluate $y = \tan x$ for multiples of $\frac{\pi}{4}$ in the interval $-\frac{3\pi}{2} \le x \le \frac{3\pi}{2}$.

x	$-\frac{3\pi}{2}$	$-\frac{5\pi}{4}$	$-\pi$	$-\frac{3\pi}{4}$	$-\frac{\pi}{2}$	$-\frac{\pi}{4}$	0	$\frac{\pi}{4}$	$\frac{\pi}{2}$	$\frac{3\pi}{4}$	π	$\frac{5\pi}{4}$	$\frac{3\pi}{2}$
tan x	undefined	-1	0	1	undefined	-1	0	1	undefined	-1	0	1	undefined

Look Back
You can refer to Lesson 3-7 to review asymptotes.

To graph $y = \tan x$, draw the asymptotes and plot the coordinate pairs from the table. Then draw the curves.

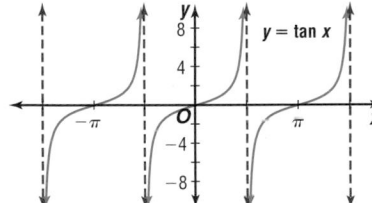
$y = \tan x$

Notice that the range values for the interval $-\frac{3\pi}{2} \le x \le -\frac{\pi}{2}$ repeat for the intervals $-\frac{\pi}{2} \le x \le \frac{\pi}{2}$ and $\frac{\pi}{2} \le x \le \frac{3\pi}{2}$. So, the tangent function is a periodic function. Its period is π.

Lesson 6-7 Graphing Other Trigonometric Functions **395**

1 FOCUS

5-Minute Check
Lesson 6-6

1. A person's blood pressure oscillates between 160 and 60. If the heart beats once every second, write a sine function that models this person's blood pressure. **$P = 50 \sin 2\pi t + 110$**

2. If the equilibrium point is $y = 0$, then $y = -4 \cos\left(\frac{\pi}{6}t\right)$ models a buoy bobbing up and down in water. Find the period of the function and the location of the buoy at $t = 10$. **12; -2**

3. A leaf floats on the water bobbing up and down. The distance between its highest and lowest points is 5 centimeters. It moves from its highest point down to its lowest point and back to its highest point every 14 seconds. Write a cosine function that models the movement of the leaf in relationship to the equilibrium point.
$y = 2.5 \cos\left(\frac{\pi}{7} \cdot t\right)$

Real-World Application If a point on an oscilloscope rides the sine curve $y = \sin x$, what path would the point on its reciprocal function $y = \csc x$ travel? **Sample answer: a point can ride the sine curve forever; however a point on the cosecant graph could ride a branch of the cosecant graph forever but could not skip from branch to branch.**

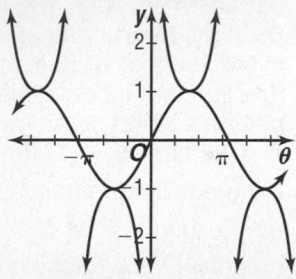

2 TEACH

Teaching Tip Remind students that the reciprocal functions tangent and cotangent have a period of π, whereas sine and its reciprocal function cosecant and cosine and its reciprocal function secant both have periods of 2π.

In-Class Example

Example 1

Find each value by referring to the graphs of the trigonometric functions.

a. $\tan \dfrac{7\pi}{2}$ **undefined**

b. $\cot \dfrac{3\pi}{2}$ **0**

By studying the graph and its repeating pattern, you can determine the following properties of the graph of the tangent function.

Properties of the Graph $y = \tan x$

1. The period is π.
2. The domain is the set of real numbers except $\dfrac{\pi}{2}n$, where n is an odd integer.
3. The range is the set of real numbers.
4. The x-intercepts are located at πn, where n is an integer.
5. The y-intercept is 0.
6. The asymptotes are $x = \dfrac{\pi}{2}n$, where n is an odd integer.

Now consider the graph of $y = \cot x$ in the interval $-\pi \le x \le 3\pi$.

x	$-\pi$	$-\dfrac{3\pi}{4}$	$-\dfrac{\pi}{2}$	$-\dfrac{\pi}{4}$	0	$\dfrac{\pi}{4}$	$\dfrac{\pi}{2}$	$\dfrac{3\pi}{4}$	π	$\dfrac{5\pi}{4}$	$\dfrac{3\pi}{2}$	$\dfrac{7\pi}{4}$	2π
cot x	undefined	1	0	-1	undefined	1	0	-1	undefined	1	0	-1	undefined

By studying the graph and its repeating pattern, you can determine the following properties of the graph of the cotangent function.

Properties of the Graph of $y = \cot x$

1. The period is π.
2. The domain is the set of real numbers except πn, where n is an integer.
3. The range is the set of real numbers.
4. The x-intercepts are located at $\dfrac{\pi}{2}n$, where n is an odd integer.
5. There is no y-intercept.
6. The asymptotes are $x = \pi n$, where n is an integer.

Example ➊ Find each value by referring to the graphs of the trigonometric functions.

a. $\tan \dfrac{9\pi}{2}$

Since $\dfrac{9\pi}{2} = \dfrac{\pi}{2}(9)$, $\tan \dfrac{9\pi}{2}$ is undefined.

b. cot $\frac{7\pi}{2}$

Since $\frac{7\pi}{2} = \frac{\pi}{2}(7)$ and 7 is an odd integer, cot $\frac{7\pi}{2} = 0$.

The sine and cosecant functions have a reciprocal relationship. To graph the cosecant, first graph the sine function and the asymptotes of the cosecant function. By studying the graph of the cosecant and its repeating pattern, you can determine the following properties of the graph of the cosecant function.

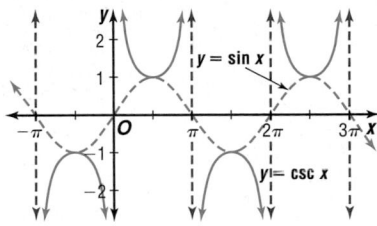

Properties of the Graph of $y = \csc x$

1. The period is 2π.
2. The domain is the set of real numbers except πn, where n is an integer.
3. The range is the set of real numbers greater than or equal to 1 or less than or equal to -1.
4. There are no x-intercepts.
5. There are no y-intercepts.
6. The asymptotes are $x = \pi n$, where n is an integer.
7. $y = 1$ when $x = \frac{\pi}{2} + 2\pi n$, where n is an integer.
8. $y = -1$ when $x = \frac{3\pi}{2} + 2\pi n$, where n is an integer.

The cosine and secant functions have a reciprocal relationship. To graph the secant, first graph the cosine function and the asymptotes of the secant function. By studying the graph and its repeating pattern, you can determine the following properties of the graph of the secant function.

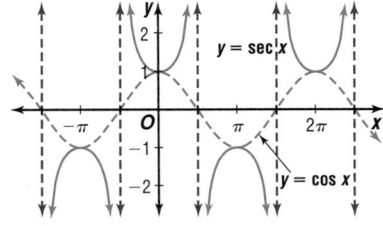

Properties of the Graph of $y = \sec x$

1. The period is 2π.
2. The domain is the set of real numbers except $\frac{\pi}{2}n$, where n is an odd integer.
3. The range is the set of real numbers greater than or equal to 1 or less than or equal to -1.
4. There are no x-intercepts.
5. The y-intercept is 1.
6. The asymptotes are $x = \frac{\pi}{2}n$, where n is an odd integer.
7. $y = 1$ when $x = \pi n$, where n is an even integer.
8. $y = -1$ when $x = \pi n$, where n is an odd integer.

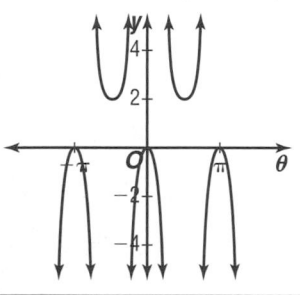
Example **2** Find the values of θ for which each equation is true.

a. $\csc \theta = 1$

From the pattern of the cosecant function, $\csc \theta = 1$ if $\theta = \dfrac{\pi}{2} + 2\pi n$, where n is an integer.

b. $\sec \theta = -1$

From the pattern of the secant function, $\sec \theta = -1$ if $\theta = \pi n$, where n is an odd integer.

The period of $y = \sin k\theta$ or $y = \cos k\theta$ is $\dfrac{2\pi}{k}$. Likewise, the period of $y = \csc k\theta$ or $y = \sec k\theta$ is $\dfrac{2\pi}{k}$. However, since the period of the tangent or cotangent function is π, the period of $y = \tan k\theta$ or $y = \cot k\theta$ is $\dfrac{\pi}{k}$. In each case, $k > 0$.

Period of Trigonometric Functions	The period of functions $y = \sin k\theta$, $y = \cos k\theta$, $y = \csc k\theta$, and $y = \sec k\theta$ is $\dfrac{2\pi}{k}$, where $k > 0$.
	The period of functions $y = \tan k\theta$ and $y = \cot k\theta$ is $\dfrac{\pi}{k}$, where $k > 0$.

The phase shift and vertical shift work the same way for all trigonometric functions. For example, the phase shift of the function $y = \tan (k\theta + c) + h$ is $-\dfrac{c}{k}$, and its vertical shift is h.

Examples **3** Graph $y = \csc\left(\dfrac{\theta}{2} - \dfrac{\pi}{4}\right) + 2$.

The period is $\dfrac{2\pi}{\frac{1}{2}}$ or 4π. The phase shift is $-\dfrac{-\frac{\pi}{4}}{\frac{1}{2}}$ or $\dfrac{\pi}{2}$. The vertical shift is 2.

Use this information to graph the function.

Step 1 Draw the midline which is the graph of $y = 2$.

Step 2 Draw dashed lines parallel to the midline, which are 1 unit above and below the midline.

Step 3 Draw the cosecant curve with period of 4π.

Step 4 Shift the graph $\dfrac{\pi}{2}$ units to the right.

4 **SECURITY**   Refer to the application at the beginning of the lesson.

  a. Graph the equation $y = 6 \tan\left(\dfrac{\pi}{30}t\right)$.

  b. Find the location the camera is scanning after 5 seconds.

  c. What happens when $t = 15$?

  a. The period is $\dfrac{\pi}{\frac{\pi}{30}}$ or 30. There are no horizontal

   or vertical shifts. Draw the asymptotes at
   $t = -15$ and $t = 15$. Graph the equation.

  b. Evaluate the equation at $t = 5$.

   $d = 6 \tan\left(\dfrac{\pi}{30}t\right)$

   $d = 6 \tan\left[\dfrac{\pi}{30}(5)\right]$ $t = 5$

   $d \approx 3.464101615$ *Use a calculator.*

   The camera is scanning a point that is about 3.5 feet above the center of the driveway.

  c. At $\tan\left[\dfrac{\pi}{30}(15)\right]$ or $\tan\dfrac{\pi}{2}$, the function is undefined. Therefore, the camera will not scan any part of the driveway when $t = 15$. It will be pointed in a direction that is parallel with the driveway.

   You can write an equation of a trigonometric function if you are given the period, phase shift, and vertical translation.

Example **5** Write an equation for a secant function with period π, phase shift $\dfrac{\pi}{3}$, and vertical shift -3.

  The form of the equation will be $y = \sec(k\theta + c) + h$. Find the values of k, c, and h.

  k: $\dfrac{2\pi}{k} = \pi$ *The period is π.*

   $k = 2$

  c: $-\dfrac{c}{k} = \dfrac{\pi}{3}$ *The phase shift is $\dfrac{\pi}{3}$.*

   $-\dfrac{c}{2} = \dfrac{\pi}{3}$ $k = 2$

   $c = -\dfrac{2\pi}{3}$

  h: $h = -3$

   Substitute these values into the general equation. The equation is

   $y = \sec\left(2\theta - \dfrac{2\pi}{3}\right) - 3$.

Addressing Individual Needs

Kinesthetic Have students demonstrate some dancing techniques from square dancing or the polka and determine whether there is a periodicity associated with these dance movements. Consider distance moved as a function of time. Ask students to show how their dance steps can model one of the six trigonometric functions.

Teaching Tip For Example 4, point out that $\tan\dfrac{\pi}{2}$ is undefined. Graphically, undefined means that the ordinate value increases rapidly as the abscissa approaches $\dfrac{\pi}{2}$. The graph approaches asymptotically the line $x = \dfrac{\pi}{2}$.

In-Class Examples

Example 4
Security A security camera is located 8 feet to the left of the midpoint of a line drawn down the center of a driveway. Let d represent the distance along the line from its midpoint. If t is time in seconds and the camera points at the midpoint at $t = 0$, then $d = 8 \tan\left(\dfrac{\pi}{50}t\right)$ models the point being scanned.

a. Graph the equation $d = 8 \tan\left(\dfrac{\pi}{50}t\right)$ for $-25 \le t \le 25$.

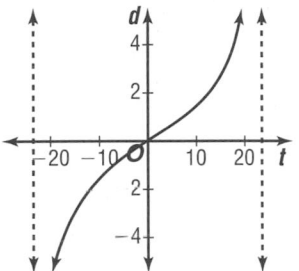

b. Find the location the camera is scanning after 15 seconds. **about 11 ft**

c. What happens when $t = 25$? **The camera is pointing parallel to the driveway.**

Example 5
Write an equation for a secant function with period 3π, phase shift $\dfrac{\pi}{2}$, and vertical shift 3.

$y = \sec\left(\dfrac{2}{3}\theta - \dfrac{\pi}{3}\right) + 3$

Teaching Tip For Example 5, tell students to use the period formula to find $k\left(\text{period} = \dfrac{2\pi}{k}\right)$, then with k, use the phase shift formula, phase shift $= -\dfrac{c}{k}$, to find c.

Error Analysis

Watch for students who have difficulty finding the period for the tangent and cotangent functions.

Prevent by stressing that the period of $y = \tan k \cdot \theta$ and $y = \cot k \cdot \theta$ is $\frac{\pi}{k}$.

Assignment Guide

Core: 13–45 odd, 47–59
Enriched: 14–42 even, 44–59

Additional Answers

2. The asymptotes of $y = \tan \theta$ and $y = \sec \theta$ are the same. The period of $y = \tan \theta$ is π and the period of $y = \sec \theta$ is 2π.

8.

11. $y = \cot\left(\frac{1}{2}\theta + \frac{\pi}{8}\right)$

Study Guide and Practice Masters, pp. 111–112

CHECK FOR UNDERSTANDING

Communicating Mathematics

Read and study the lesson to answer each question.

1. **Name** three values of θ that would result in $y = \cot \theta$ being undefined.

1. Sample answers: $-\pi, \pi, 2\pi$

2. **Compare** the asymptotes and periods of $y = \tan \theta$ and $y = \sec \theta$.   **See margin.**

3. **Describe** two different phase shifts of the secant function that would make it appear to be the cosecant function.   **Sample answers:** $\frac{\pi}{2}, -\frac{3\pi}{2}$

Guided Practice

Find each value by referring to the graphs of the trigonometric functions.

4. $\tan 4\pi$ **0**

5. $\csc\left(-\frac{7\pi}{2}\right)$ **1**

Find the values of θ for which each equation is true.

6. $\sec \theta = -1$ **πn, where n is an odd integer**

7. $\cot \theta = 1$ **$\frac{\pi}{4} + \pi n$, where n is an integer**

Graph each function.   **8–9. See margin.**

8. $y = \tan\left(\theta + \frac{\pi}{4}\right)$

9. $y = \sec(2\theta + \pi) - 1$

Write an equation for the given function given the period, phase shift, and vertical shift.

10. $y = \csc\left(\frac{2}{3}\theta - \frac{2\pi}{9}\right) - 4$

10. cosecant function, period $= 3\pi$, phase shift $= \frac{\pi}{3}$, vertical shift $= -4$

11. cotangent function, period $= 2\pi$, phase shift $= -\frac{\pi}{4}$, vertical shift $= 0$

12. **Physics**   A child is swinging on a tire swing. The tension on the rope is equal to the downward force on the end of the rope times $\sec \theta$, where θ is the angle formed by a vertical line and the rope.

a. The downward force in newtons equals the mass of the child and the swing in kilograms times the acceleration due to gravity (9.8 meters per second squared). If the mass of the child and the tire is 73 kilograms, find the downward force.   **715.4 N**

b. Write an equation that represents the tension on the rope as the child swings back and forth.   **$F = 715.4 \sec \theta$**

c. Graph the equation for $-\frac{\pi}{2} \le x \le \frac{\pi}{2}$.   **See margin.**

d. What is the least amount of tension on the rope?   **715.4 N**

e. What happens to the tension on the rope as the child swings higher and higher?   **The tension becomes greater.**

EXERCISES

Practice

Find each value by referring to the graphs of the trigonometric functions.

A

13. $\cot\left(\frac{5\pi}{2}\right)$ **0**

14. $\tan(-8\pi)$ **0**

15. $\sec\left(\frac{9\pi}{2}\right)$ **undefined**

16. $\csc\left(-\frac{5\pi}{2}\right)$ **−1**

17. $\sec 7\pi$ **−1**

18. $\cot(-5\pi)$ **undefined**

www.amc.glencoe.com/self_check_quiz

Additional Answers

9.

$y = \sec(2\theta + \pi) - 1$

12c.

$F = 715.4 \sec \theta$

19. What is the value of csc (-6π)? **undefined**

20. Find the value of tan (10π). **0**

Find the values of θ for which each equation is true.

21. $\tan \theta = 0$ **πn, where n is an integer**

22. $\sec \theta = 1$ **πn, where n is an even integer**

23. $\dfrac{3\pi}{2} + 2\pi n$, where n is an integer

23. $\csc \theta = -1$

24. $\dfrac{\pi}{4} + \pi n$, where n is an integer

24. $\tan \theta = 1$

25. $-\dfrac{\pi}{4} + \pi n$, where n is an integer

25. $\tan \theta = -1$

26. $\dfrac{3\pi}{4} + \pi n$, where n is an integer

26. $\cot \theta = -1$

27. What are the values of θ for which sec θ is undefined?

28. Find the values of θ for which cot θ is undefined.

Graph each function.   **29–34. See margin.**

29. $y = \cot\left(\theta - \dfrac{\pi}{2}\right)$

30. $y = \sec \dfrac{\theta}{3}$

31. $y = \csc \theta + 5$

32. $y = \tan\left(\dfrac{\theta}{2} - \dfrac{\pi}{4}\right) + 1$

33. $y = \csc(2\theta + \pi) - 3$

34. $y = \sec\left(\dfrac{\theta}{3} + \dfrac{\pi}{6}\right) - 2$

27. $\dfrac{\pi}{2}n$, where n is an odd integer

28. πn, where n is an integer

35. Graph $y = \cos \theta$ and $y = \sec \theta$. In the interval of -2π and 2π, what are the values of θ where the two graphs are tangent to each other?
  See margin for graph. -2π, $-\pi$, **0, π, 2π**

Write an equation for the given function given the period, phase shift, and vertical shift.   **36–41. See margin.**

36. tangent function, period $= 2\pi$, phase shift $= 0$, vertical shift $= -6$

37. cotangent function, period $= \dfrac{\pi}{2}$, phase shift $= \dfrac{\pi}{8}$, vertical shift $= 7$

38. secant function, period $= \pi$, phase shift $= -\dfrac{\pi}{4}$, vertical shift $= -10$

39. cosecant function, period $= 3\pi$, phase shift $= \pi$, vertical shift $= -1$

40. cotangent function, period $= 5\pi$, phase shift $= -\pi$, vertical shift $= 12$

41. cosecant function, period $= \dfrac{\pi}{3}$, phase shift $= -\dfrac{\pi}{2}$, vertical shift $= -5$

42. $y = \sec\left(\dfrac{2}{3}\theta + \dfrac{2\pi}{3}\right) - 8$

43. $y = \tan\left(2\theta - \dfrac{\pi}{2}\right) + 7$

42. Write a secant function with a period of 3π, a phase shift of π units to the left, and a vertical shift of 8 units downward.

43. Write a tangent function with a period of $\dfrac{\pi}{2}$, a phase shift of $\dfrac{\pi}{4}$ to the right, and a vertical shift of 7 units upward.

Applications and Problem Solving

44. **Security** A security camera is scanning a long straight fence along one side of a military base. The camera is located 10 feet from the center of the fence. If d represents the distance along the fence from the center and t is time in seconds, then $d = 10 \tan \dfrac{\pi}{40} t$ models the point being scanned.
  a. Graph the equation for $-20 \le t \le 20$.   **See margin.**
  b. about 2.4 ft from the center
  b. Find the location the camera is scanning at 3 seconds.
  c. Find the location the camera is scanning at 15 seconds.
  c. about 24.1 ft from the center

45. **Critical Thinking** Graph $y = \csc \theta$, $y = 3 \csc \theta$, and $y = -3 \csc \theta$. Compare and contrast the graphs.   **See margin.**

Lesson 6-7 Graphing Other Trigonometric Functions **401**

 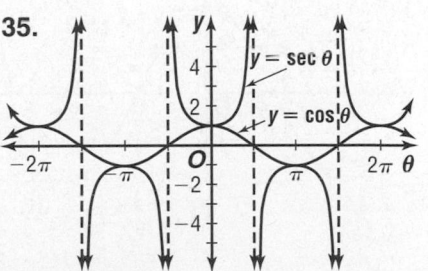
(continued on page 402)

44a.

$d = 10 \tan \frac{\pi}{40} t$

45.

The graph of $y = \csc \theta$ has no range values between -1 and 1, while the graphs of $y = 3 \csc \theta$ and $y = -3 \csc \theta$ have no range values between -3 and 3. The graphs of $y = 3 \csc \theta$ and $y = -3 \csc \theta$ are reflections of each other.

46c.

$F = 34.3 \sec \frac{\theta}{2}$

50.

$y = 2 \cos \frac{\theta}{2}$

46. Physics A wire is used to hang a painting from a nail on a wall as shown at the right. The tension on each half of the wire is equal to half the downward force times $\sec \frac{\theta}{2}$.

a. The downward force in newtons equals the mass of the painting in kilograms times 9.8. If the mass of the painting is 7 kilograms, find the downward force. **68.6 N**

b. Write an equation that represents the tension on each half of the wire. $F = 34.3 \sec \frac{\theta}{2}$

46c. See margin.   c. Graph the equation for $0 \le \theta \le \pi$.

d. What is the least amount of tension on each side of the wire? **34.3 N**

e. As the measure of θ becomes greater, what happens to the tension on each side of the wire? **The tension becomes greater.**

47. Electronics The current I measured in amperes that is flowing through an alternating current at any time t in seconds is modeled by $I = 220 \sin \left(60 \pi t - \frac{\pi}{6}\right)$.

a. What is the amplitude of the current? **220 A**

b. What is the period of the current? $\frac{1}{30}$ **s**

c. What is the phase shift of this sine function? $\frac{1}{360}$

d. Find the current when $t = 60$. **about -110 A**

48. Critical Thinking Write a tangent function that has the same graph as $y = \cot \theta$. $y = -1 \tan \left(\theta + \frac{\pi}{2}\right)$

Mixed Review

49. Tides In Daytona Beach, Florida, the first high tide was 3.99 feet at 12:03 A.M. The first low tide of 0.55 foot occurred at 6:24 A.M. The second high tide occurred at 12:19 P.M. *(Lesson 6-6)* **a. 1.72 ft   b. 2.27 ft   c. about 12.3 hr**

a. Find the amplitude of a sinusoidal function that models the tides.

b. Find the vertical shift of the sinusoidal function that models the tides.

c. What is the period of the sinusoidal function that models the tides?

49d. Sample answer:
$h = 1.72 \sin \left(\frac{2\pi}{12.3} t + 1.55\right) + 2.27$

d. Write a sinusoidal function to model the tides, using t to represent the number of hours in decimals since midnight.

e. According to your model, determine the height of the water at noon.

e. Sample answer: 3.96 ft

50. Graph $y = 2 \cos \frac{\theta}{2}$. *(Lesson 6-4)* **See margin.**

51. If a central angle of a circle with radius 18 centimeters measures $\frac{\pi}{3}$, find the length (in terms of π) of its intercepted arc. *(Lesson 6-1)* **6π cm**

52. Solve $\triangle ABC$ if $A = 62°31'$, $B = 75°18'$, and $a = 57.3$. Round angle measures to the nearest minute and side measures to the nearest tenth. *(Lesson 5-6)*
$C = 42°11'$, $b = 62.5$, $c = 43.4$

53. **Entertainment** A utility pole is braced by a cable attached to the top of the pole and anchored in a concrete block at the ground level 4 meters from the base of the pole. The angle between the cable and the ground is 73°. *(Lesson 5-4)*

   a. Draw a diagram of the problem. **See margin.**

   b. If the pole is perpendicular with the ground, what is the height of the pole? **about 13.1 m**

   c. Find the length of the cable. **about 13.7 m**

54. Find the values of the sine, cosine, and tangent for $\angle A$. *(Lesson 5-2)*

$$\sin A = \frac{7\sqrt{65}}{65},$$
$$\cos A = \frac{4\sqrt{65}}{65}, \tan A = \frac{7}{4}$$

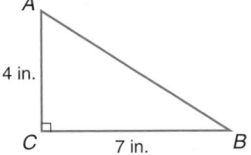

55. Solve $\dfrac{x^2 - 4}{x^2 - 3x - 10} \le 0$. *(Lesson 4-6)* **$-2 < x < 5$**

56. If r varies directly as t and $t = 6$ when $r = 0.5$, find r when $t = 10$. *(Lesson 3-8)* **about 0.83**

57. Solve the system of inequalities by graphing. *(Lesson 2-6)* **See margin.**
$$3x + 2y < 8$$
$$y < 2x + 1$$
$$-2y < -x + 4$$

58.
$y = 17.98x + 35.47; 0.88$

58. **Nutrition** The fat grams and Calories in various frozen pizzas are listed below. Use a graphing calculator to find the equation of the regression line and the Pearson product-moment correlation value. *(Lesson 1-6)*

Pizza	Fat (grams)	Calories
Cheese Pizza	14	270
Party Pizza	17	340
Pepperoni French Bread Pizza	22	430
Hamburger French Bread Pizza	19	410
Deluxe French Bread Pizza	20	420
Pepperoni Pizza	19	360
Sausage Pizza	18	360
Sausage and Pepperoni Pizza	18	340
Spicy Chicken Pizza	16	360
Supreme Pizza	18	308
Vegetable Pizza	13	300
Pizza Roll-Ups	13	250

59. **SAT/ACT Practice** The distance from City A to City B is 150 miles. From City A to City C is 90 miles. Which of the following is necessarily true? **B**

  A The distance from B to C is 60 miles.

  B Six times the distance from A to B equals 10 times the distance from A to C.

  C The distance from B to C is 240 miles.

  D The distance from A to B exceeds by 30 miles twice the distance from A to C.

  E Three times the distance from A to C exceeds by 30 miles twice the distance from A to B.

Extra Practice See p. A37.

Extra Credit

Particle physicists study the characteristics of subatomic particles by accelerating these particles in an accelerator at exceedingly high rates of speed. Suppose that a physicist observes and wishes to describe the unusual behavior of a particle as it moves. At a time that she arbitrarily assigns as 0, the particle is at a distance of $y = 1$ millimeter, then covers a great distance and disappears as it moves from $0 \le t < \frac{\pi}{2}$. Is there a trigonometric function that can model this behavior? If so, graph it.

Yes; sample answer: $y = \sec t, 0 \le t < \frac{\pi}{2}$

6-7B Sound Beats

An Extension of Lesson 6-7

1 FOCUS

Motivating the Lesson
Real-World Application Musical instruments produce a wide range of sounds and pitches. Two trumpet players could play the same note, but at slightly different intonations, and the result would produce a beat. This variation of pitch can be detected and the players can adjust their sounds to match one another. This process is called *tuning*.

2 TEACH

📁 A reproduction of this lesson with graphing calculator keystrokes provided can be found in the *Graphing Calculator Keystrokes Masters*, pp. 24–25.

Teaching Tip For Exercise 3, remind students that using the graphing calculator to find the value for a specific *x* in $f(x)$ eliminates the need to substitute $x = 187.158$ into the function $f(x) = \sin(5\pi x) + \sin(4.79\pi x)$, saving time and effort.

3 PRACTICE/APPLY

Assignment Guide

All: 1–8

4 ASSESS

Open-Ended Assessment
Speaking Ask students to work Exercise 7 at the chalkboard. Have students describe their pairs of sine functions and the resulting graphs of their sums. Students can then present to the class the interesting features that they find.

OBJECTIVE
- Use a graphing calculator to model beat effects produced by waves of almost equal frequencies.

The frequency of a wave is defined as the reciprocal of the period of the wave. If you listen to two steady sounds that have almost the same frequencies, you can detect an effect known as *beat*. Used in this sense, the word refers to a regular variation in sound intensity. This meaning is very different from another common meaning of the word, which you use when you are speaking about the rhythm of music for dancing.

A beat effect can be modeled mathematically by combination of two sine waves. The loudness of an actual combination of two steady sound waves of almost equal frequency depends on the amplitudes of the component sound waves. The first two graphs below picture two sine waves of almost equal frequencies. The amplitudes are equal, and the graphs, on first inspection, look almost the same. However, when the functions shown by the graphs are added, the resulting third graph is not what you would get by stretching either of the original graphs by a factor of 2, but is instead something quite different.

TRY THESE

1. See margin; the third graph.

1. Graph $f(x) = \sin(5\pi x) + \sin(4.79\pi x)$ using a window $[0, 10\pi]$ scl:π by $[-2.5, 2.5]$ scl:1. Which of the graphs shown above does the graph resemble?

2. Change the window settings for the independent variable to have **Xmax** $= 200\pi$. How does the appearance of the graph change? **See margin.**

3. For the graph in Exercise 2, use value on the **CALC** menu to find the value of $f(x)$ when $x = 187.158$. **0.38623583**

4. Does your graph of Exercise 2 show negative values of *y* when *x* is close to 187.158? **no**

5. −1.78043; See Answer Appendix.

5. Use value on the **CALC** menu to find $f(191.5)$. Does your result have any bearing on your answer for Exercise 4? Explain.

WHAT DO YOU THINK?

6. See Answer Appendix.

7. See students' work.

6. What aspect of the calculator explains your observations in Exercises 3-5?

7. Write two sine functions with almost equal frequencies. Graph the sum of the two functions. Discuss any interesting features of the graph.

8. Do functions that model beat effects appear to be periodic functions? Do your graphs prove that your answer is correct? **Yes; no, they only provide plausible visual evidence.**

404 Chapter 6   Graphs of Trigonometric Functions

Additional Answers

1.

2. Sample answer: the graph seems to stay above the *x*-axis for an interval of *x* values, and then stay below the *x*-axis for another interval of *x* values.

6-8

Trigonometric Inverses and Their Graphs

OBJECTIVES
- Graph inverse trigonometric functions.
- Find principal values of inverse trigonometric functions.

Look Back
You can refer to Lesson 5-5 to review the inverses of trigonometric functions.

ENTERTAINMENT Since the giant Ferris wheel in Vienna, Austria, was completed in 1897, it has been a major attraction for local residents and tourists. The giant Ferris wheel has a height of 64.75 meters and a diameter of 60.96 meters. It makes a revolution every 4.25 minutes. On her summer vacation in Vienna, Carla starts timing her ride at the midline point at exactly 11:35 A.M. as she is on her way up. When Carla reaches an altitude of 60 meters, she will have a view of the Vienna Opera House. When will she have this view for the first time? *This problem will be solved in Example 4.*

Recall that the inverse of a function may be found by interchanging the coordinates of the ordered pairs of the function. In other words, the domain of the function becomes the range of its inverse, and the range of the function becomes the domain of its inverse. For example, the inverse of $y = 2x + 5$ is $x = 2y + 5$ or $y = \frac{x-5}{2}$. Also remember that the inverse of a function may *not* be a function.

Consider the sine function and its inverse.

Relation	Ordered Pairs	Graph	Domain	Range
$y = \sin x$	$(x, \sin x)$		all real numbers	$-1 \leq y \leq 1$
$y = \arcsin x$	$(\sin x, x)$		$-1 \leq x \leq 1$	all real numbers

Notice the similarity of the graph of the inverse of the sine function to the graph of $y = \sin x$ with the axes interchanged. This is also true for the other trigonometric functions and their inverses.

Lesson 6-8 Trigonometric Inverses and Their Graphs **405**

Lesson Notes

6-8

1 FOCUS

5-Minute Check
Lesson 6-7

1. Find the value of sec (-5π) by referring to the graph of the secant function. **−1**

2. Find the values of θ for which cot $\theta = 1$ is true. $\frac{\pi}{4} + \pi \cdot n$, n **is an integer.**

3. Graph $y = 2 + \sec \theta$.

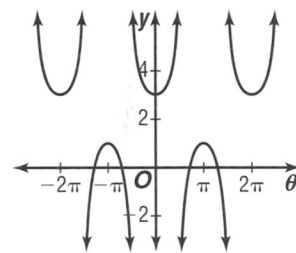

4. Write an equation for a tangent function whose period is 4π, phase shift 0, and vertical shift $+3$. $y = \tan \frac{\theta}{4} + 3$

5. Write an equation for a secant function with a period of 5π, a phase shift of 2π to the left, and a vertical shift of 3 units downward. $y = \sec \left(\frac{2}{5}\theta + \frac{4\pi}{5} \right) - 3$

Motivating the Lesson
Discussion Write the function $y = \sin x$ on the board, and have students brainstorm how they would rewrite this function in order to put the emphasis on x, the angle, itself. In other words, what would x equal?

Resource Manager

 Reproducible Masters
- *Study Guide and Practice,* pp. 113–114
- *Enrichment,* p. 68
- *Assessment and Evaluation,* p. 148
- *Problem Solving and Applications,* p. 12

 Transparencies
- 5-Minute Check, 6-8
- Teaching, 6-8

 Technology/Multimedia
- Advanced MathPASS, Lesson 14

Lesson 6-8 **405**

Relation	Ordered Pairs	Graph	Domain	Range
$y = \cos x$	$(x, \cos x)$		all real numbers	$-1 \leq y \leq 1$
$y = \arccos x$	$(\cos x, x)$		$-1 \leq x \leq 1$	all real numbers
$y = \tan x$	$(x, \tan x)$		all real numbers except $\frac{\pi}{2}n$, where n is an odd integer	all real numbers
$y = \arctan x$	$(\tan x, x)$		all real numbers	all real numbers except $\frac{\pi}{2}n$, where n is an odd integer

Notice that none of the inverses of the trigonometric functions are functions.

Consider only a part of the domain of the sine function, namely $-\frac{\pi}{2} \leq x \leq \frac{\pi}{2}$. The range then contains all of the possible values from -1 to 1. It is possible to define a new function, called Sine, whose inverse is a function.

Capital letters are used to distinguish the function with restricted domains from the usual trigonometric functions.

$$y = \text{Sin } x \text{ if and only if } y = \sin x \text{ and } -\frac{\pi}{2} \leq x \leq \frac{\pi}{2}.$$

The values in the domain of Sine are called **principal values.** Other new functions can be defined as follows.

$$y = \text{Cos } x \text{ if and only if } y = \cos x \text{ and } 0 \leq x \leq \pi.$$

$$y = \text{Tan } x \text{ if and only if } y = \tan x \text{ and } -\frac{\pi}{2} < x < \frac{\pi}{2}.$$

The graphs of $y = \text{Sin } x$, $y = \text{Cos } x$, and $y = \text{Tan } x$ are the blue portions of the graphs of $y = \sin x$, $y = \cos x$, and $y = \tan x$, respectively, shown on pages 405–406.

Note the capital "A" in the name of each inverse function.

The inverses of the Sine, Cosine, and Tangent functions are called Arcsine, Arccosine, and Arctangent, respectively. The graphs of Arcsine, Arccosine, and Arctangent are also designated in blue on pages 405–406. They are defined as follows.

Arcsine Function	Given $y = \text{Sin } x$, the inverse Sine function is defined by the equation $y = \text{Sin}^{-1} x$ or $y = \text{Arcsin } x$.
Arccosine Function	Given $y = \text{Cos } x$, the inverse Cosine function is defined by the equation $y = \text{Cos}^{-1} x$ or $y = \text{Arccos } x$.
Arctangent Function	Given $y = \text{Tan } x$, the inverse Tangent function is defined by the equation $y = \text{Tan}^{-1} x$ or $y = \text{Arctan } x$.

The domain and range of these functions are summarized below.

Function	Domain	Range
$y = \text{Sin } x$	$-\dfrac{\pi}{2} \le x \le \dfrac{\pi}{2}$	$-1 \le y \le 1$
$y = \text{Arcsin } x$	$-1 \le x \le 1$	$-\dfrac{\pi}{2} \le y \le \dfrac{\pi}{2}$
$y = \text{Cos } x$	$0 \le x \le \pi$	$-1 \le y \le 1$
$y = \text{Arccos } x$	$-1 \le x \le 1$	$0 \le y \le \pi$
$y = \text{Tan } x$	$-\dfrac{\pi}{2} < x < \dfrac{\pi}{2}$	all real numbers
$y = \text{Arctan}$	all real numbers	$-\dfrac{\pi}{2} < y < \dfrac{\pi}{2}$

Example ❶ Write the equation for the inverse of $y = \text{Arctan } 2x$. Then graph the function and its inverse.

$$y = \text{Arctan } 2x$$
$$x = \text{Arctan } 2y \qquad \textit{Exchange x and y.}$$
$$\text{Tan } x = 2y \qquad \textit{Definition of Arctan function}$$
$$\frac{1}{2}\text{Tan } x = y \qquad \textit{Divide each side by 2.}$$

Now graph the functions.

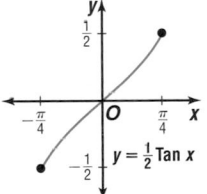

Note that the graphs are reflections of each other over the graph of $y = x$.

Lesson 6-8 *Trigonometric Inverses and Their Graphs* **407**

2 TEACH

Teaching Tip Remind students that the graph of a function and its inverse are symmetric or reflect over the line $y = x$.

Teaching Tip Strongly emphasize inverse notation with students. For example, $\sin^{-1} x \ne \dfrac{1}{\sin x}$.

In-Class Example
Example 1
Write the equation for the inverse of $y = \text{Arctan } \dfrac{1}{2}x$. Then graph the function and its inverse. **$y = 2 \text{ Tan } x$**

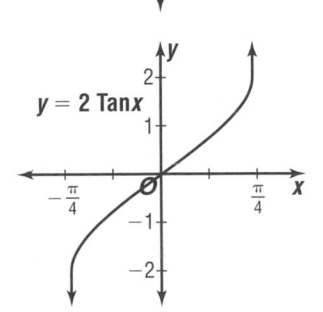

You can use what you know about trigonometric functions and their inverses to evaluate expressions.

Examples **2** **Find each value.**

a. **Arcsin** $\left(-\dfrac{\sqrt{2}}{2}\right)$

Let $\theta = \text{Arcsin}\left(-\dfrac{\sqrt{2}}{2}\right)$. *Think: Arcsin $\left(-\dfrac{\sqrt{2}}{2}\right)$ means that angle whose sin is $-\dfrac{\sqrt{2}}{2}$.*

Sin $\theta = -\dfrac{\sqrt{2}}{2}$ *Definition of Arcsin function*

$\theta = -\dfrac{\pi}{4}$ *Why is θ not $-\dfrac{3\pi}{4}$?*

b. **Sin**$^{-1}$ $\left(\cos \dfrac{\pi}{2}\right)$

If $y = \cos \dfrac{\pi}{2}$, then $y = 0$.

$\text{Sin}^{-1}\left(\cos \dfrac{\pi}{2}\right) = \text{Sin}^{-1} 0$ *Replace cos $\dfrac{\pi}{2}$ with 0.*

$= 0$

c. **sin (Tan^{-1} 1 − Sin^{-1} 1)**

Let $\alpha = \text{Tan}^{-1} 1$ and $\beta = \text{Sin}^{-1} 1$.

Tan $\alpha = 1$ Sin $\beta = 1$

$\alpha = \dfrac{\pi}{4}$ $\beta = \dfrac{\pi}{2}$

$\sin (\text{Tan}^{-1} 1 - \text{Sin}^{-1} 1) = \sin (\alpha - \beta)$

$= \sin\left(\dfrac{\pi}{4} - \dfrac{\pi}{2}\right)$ $\alpha = \dfrac{\pi}{4}, \beta = \dfrac{\pi}{2}$

$= \sin\left(-\dfrac{\pi}{4}\right)$

$= -\dfrac{\sqrt{2}}{2}$

d. **cos** $\left[\text{Cos}^{-1}\left(-\dfrac{\sqrt{2}}{2}\right) - \dfrac{\pi}{2}\right]$

Let $\theta = \text{Cos}^{-1}\left(-\dfrac{\sqrt{2}}{2}\right)$.

Cos $\theta = -\dfrac{\sqrt{2}}{2}$ *Definition of Arccosine function*

$\theta = \dfrac{3\pi}{4}$

$\cos\left[\text{Cos}^{-1}\left(-\dfrac{\sqrt{2}}{2}\right) - \dfrac{\pi}{2}\right] = \cos\left(\theta - \dfrac{\pi}{2}\right)$

$= \cos\left(\dfrac{3\pi}{4} - \dfrac{\pi}{2}\right)$ $\theta = \dfrac{3\pi}{4}$

$= \cos \dfrac{\pi}{4}$

$= \dfrac{\sqrt{2}}{2}$

3 Determine if $\text{Tan}^{-1}(\tan x) = x$ is *true* or *false* for all values of x. If false, give a counterexample.

Try several values of x to see if we can find a counterexample.

When $x = \pi$, $\text{Tan}^{-1}(\tan x) \neq x$. So $\text{Tan}^{-1}(\tan x) = x$ is not true for all values of x.

x	$\tan x$	$\text{Tan}^{-1}(\tan x)$
0	0	0
$\dfrac{\pi}{4}$	1	$\dfrac{\pi}{4}$
π	0	0

You can use a calculator to find inverse trigonometric functions. The calculator will always give the least, or principal, value of the inverse trigonometric function.

Example

Real World Application

4 **ENTERTAINMENT**   Refer to the application at the beginning of the lesson. When will Carla reach an altitude of 60 meters for the first time?

First write an equation to model the height of a seat at any time t. Since the seat is at the midline point at $t = 0$, use the sine function $y = A \sin(kt + c) + h$. Find the values of A, k, c, and h.

A: The value of A is the radius of the Ferris wheel.

$A = \dfrac{1}{2}(60.96)$ or 30.48 *The diameter is 60.96 meters.*

k: $\dfrac{2\pi}{k} = 4.25$ *The period is 4.25 minutes.*

$k = \dfrac{2\pi}{4.25}$

midline
60.96 m
64.75 m

c: Since the seat is at the equilibrium point at $t = 0$, there is no phase shift and $c = 0$.

h: The bottom of the Ferris wheel is $64.75 - 60.96$ or 3.79 meters above the ground. So, the value of h is $30.48 + 3.79$ or 34.27.

Substitute these values into the general equation. The equation is

$y = 30.48 \sin\left(\dfrac{2\pi}{4.25}t\right) + 34.27$. Now, solve the equation for $y = 60$.

$60 = 30.48 \sin\left(\dfrac{2\pi}{4.25}t\right) + 34.27$ *Replace y with 60.*

$25.73 = 30.48 \sin\left(\dfrac{2\pi}{4.25}t\right)$ *Subtract 34.27 from each side.*

$\dfrac{25.73}{30.48} = \sin\left(\dfrac{2\pi}{4.25}t\right)$ *Divide each side by 30.48.*

$\sin^{-1}\left(\dfrac{25.73}{30.48}\right) = \dfrac{2\pi}{4.25}t$ *Definition of $\sin^{-1}$*

$\dfrac{4.25}{2\pi}\sin^{-1}\left(\dfrac{25.73}{30.48}\right) = t$ *Multiply each side by $\dfrac{4.25}{2\pi}$.*

$0.6797882017 = t$ *Use a calculator.*

Carla will reach an altitude of 60 meters about 0.68 minutes after 11:35 or 11:35:41.

Addressing Individual Needs

Error Analysis

Watch for students who confuse the −1 in Sin⁻¹x with an exponent.

Prevent by emphasizing that $\text{Sin}^{-1}x$ means the same as Arcsin x.

Assignment Guide

Core: 15–41 odd, 43–54
Enriched: 14–38 even, 39–54

Additional Answers

1. $y = \sin^{-1} x$ is the inverse relation of $y = \sin x$, $y = (\sin x)^{-1}$ is the function $y = \dfrac{1}{\sin x}$, and $y = \sin(x^{-1})$ is the function $y = \sin \dfrac{1}{x}$.

2. For every y value, there are more than one x value. The graph of $y = \cos^{-1} x$ fails the vertical line test.

3. The domain of $y = \text{Sin } x$ is the set of real numbers between $-\dfrac{\pi}{2}$ and $\dfrac{\pi}{2}$, inclusive, while the domain of $y = \sin x$ is the set of all real numbers. The range of both functions is the set of all real numbers between -1 and 1, inclusive.

Study Guide and Practice Masters, pp. 113–114

CHECK FOR UNDERSTANDING

Communicating Mathematics

Read and study the lesson to answer each question. **1–3. See margin.**

1. **Compare** $y = \sin^{-1} x$, $y = (\sin x)^{-1}$, and $y = \sin(x^{-1})$.

2. **Explain** why $y = \cos^{-1} x$ is not a function.

3. **Compare and contrast** the domain and range of $y = \text{Sin } x$ and $y = \sin x$.

4. Restricted domains are denoted with a capital letter.

4. **Write** a sentence explaining how to tell if the domain of a trigonometric function is restricted.

5. **You Decide** Jake says that the period of the cosine function is 2π. Therefore, he concludes that the principal values of the domain are between 0 and 2π, inclusive. Akikta disagrees. Who is correct? Explain.
Akikta; see margin for explanation.

Guided Practice

Write the equation for the inverse of each function. Then graph the function and its inverse. **6–7. See margin for graphs.**

6. $y = \text{Arcsin } x$ $y = \text{Sin } x$

7. $y = \text{Cos}\left(x + \dfrac{\pi}{2}\right)$ $y = \text{Cos}^{-1} x - \dfrac{\pi}{2}$

Find each value.

10. $\dfrac{\sqrt{2}}{2}$

8. Arctan 1 $\dfrac{\pi}{4}$

9. $\cos(\text{Tan}^{-1} 1)$ $\dfrac{\sqrt{2}}{2}$

10. $\cos\left[\text{Cos}^{-1}\left(\dfrac{\sqrt{2}}{2}\right) - \dfrac{\pi}{2}\right]$

Determine if each of the following is *true* or *false*. If false, give a counterexample.

11. $\sin(\text{Sin}^{-1} x) = x$ for $-1 \le x \le 1$
true

12. $\text{Cos}^{-1}(-x) = -\text{Cos}^{-1} x$ for $-1 \le x \le 1$
false; sample answer: $x = 1$

13. **Geography** Earth has been charted with vertical and horizontal lines so that points can be named with coordinates. The horizontal lines are called latitude lines. The equator is latitude line 0. Parallel lines are numbered up to $\dfrac{\pi}{2}$ to the north and to the south. If we assume Earth is spherical, the length of any parallel of latitude is equal to the circumference of a great circle of Earth times the cosine of the latitude angle.

a. The radius of Earth is about 6400 kilometers. Find the circumference of a great circle. **about 40,212 km**

b. Write an equation for the circumference of any latitude circle with angle θ. **$C = 40,212 \cos \theta$**

13c. about 1.48 radians

c. Which latitude circle has a circumference of about 3593 kilometers?

d. What is the circumference of the equator? **about 40,212 km**

EXERCISES

Practice

Write the equation for the inverse of each function. Then graph the function and its inverse. **14–19. See Answer Appendix for graphs.**

A

14. $y = \arccos x$ $y = \cos x$

15. $y = \text{Sin } x$ $y = \text{Sin}^{-1} x$

16. $y = \arctan x$ $y = \tan x$

17. $y = \text{Arccos } 2x$ $y = \dfrac{1}{2} \text{Cos } x$

18. $y = \dfrac{\pi}{2} + \text{Arcsin } x$ $y = \text{Sin}\left(x - \dfrac{\pi}{2}\right)$

19. $y = \tan \dfrac{x}{2}$ $y = 2 \tan^{-1} x$

20. Is $y = \text{Tan}^{-1}\left(x + \dfrac{\pi}{2}\right)$ the inverse of $y = \text{Tan}\left(x - \dfrac{\pi}{2}\right)$? Explain. **No; the inverse is $y = \text{Tan}^{-1} x + \dfrac{\pi}{2}$.**

www.amc.glencoe.com/self_check_quiz

Additional Answers

5. There are 2 range values for each domain value between 0 and 2π. The principal values are between 0 and π, inclusive.

6.

21. The principal values of the domain of the cotangent function are $0 \le x \le \pi$. Graph $y = \text{Cot } x$ and its inverse.   **See margin.**

Find each value.

22. $\text{Sin}^{-1} 0$ **0**

23. $\text{Arccos } 0$ $\frac{\pi}{2}$

24. $\text{Tan}^{-1} \frac{\sqrt{3}}{3}$ $\frac{\pi}{6}$

25. $\text{Sin}^{-1} \left(\tan \frac{\pi}{4} \right)$ $\frac{\pi}{2}$

26. $\sin \left(2 \text{ Cos}^{-1} \frac{\sqrt{2}}{2} \right)$ **1**

27. $\cos \left(\text{Tan}^{-1} \sqrt{3} \right)$ $\frac{1}{2}$

28. $\cos \left(\text{Tan}^{-1} 1 - \text{Sin}^{-1} 1 \right)$ $\frac{\sqrt{2}}{2}$

29. $\cos \left(\text{Cos}^{-1} 0 + \text{Sin}^{-1} \frac{1}{2} \right)$ $-\frac{1}{2}$

30. $\sin \left(\text{Sin}^{-1} 1 - \text{Cos}^{-1} \frac{1}{2} \right)$ $\frac{1}{2}$

No; there is no angle with the sine of 2.

31. Is it possible to evaluate $\cos \left[\text{Cos}^{-1} \left(-\frac{1}{2} \right) - \text{Sin}^{-1} 2 \right]$? Explain.

Determine if each of the following is *true* or *false*. If false, give a counterexample.

 32. $\text{Cos}^{-1} (\cos x) = x$ for all values of x **false; $x = 2\pi$**

32–37. Sample counterexamples are given.

33. $\tan (\text{Tan}^{-1} x) = x$ for all values of x **true**

34. $\text{Arccos } x = \text{Arccos } (-x)$ for $-1 \le x \le 1$ **false; $x = -1$**

35. $\text{Sin}^{-1} x = -\text{Sin}^{-1} (-x)$ for $-1 \le x \le 1$ **true**

36. $\text{Sin}^{-1} x + \text{Cos}^{-1} x = \frac{\pi}{2}$ for $-1 \le x \le 1$ **true**

37. $\text{Cos}^{-1} x = \frac{1}{\text{Cos } x}$ for all values of x **false; $x = \frac{\pi}{2}$**

38. Sketch the graph of $y = \tan (\text{Tan}^{-1} x)$.   **See margin.**

Applications and Problem Solving

39. Meteorology   The equation $y = 54.5 + 23.5 \sin \left(\frac{\pi}{6} t - \frac{2\pi}{3} \right)$ models the average monthly temperatures of Springfield, Missouri. In this equation, t denotes the number of months with January represented by 1. During which two months is the average temperature 54.5°?   **April and October**

40. Physics   The average power P of an electrical circuit with alternating current is determined by the equation $P = VI \text{ Cos } \theta$, where V is the voltage, I is the current, and θ is the measure of the phase angle. A circuit has a voltage of 122 volts and a current of 0.62 amperes. If the circuit produces an average of 7.3 watts of power, find the measure of the phase angle.   **about 1.47 radians**

41. $\frac{\pi}{4} + \pi n$, where n is an integer

41. Critical Thinking   Consider the graphs $y = \arcsin x$ and $y = \arccos x$. Name the y coordinates of the points of intersection of the two graphs.

42. Optics   Malus' Law describes the amount of light transmitted through two polarizing filters. If the axes of the two filters are at an angle of θ radians, the intensity I of the light transmitted through the filters is determined by the equation $I = I_0 \cos^2 \theta$, where I_0 is the intensity of the light that shines on the filters. At what angle should the axes be held so that one-eighth of the transmitted light passes through the filters?   **about 1.21 radians**

7.

$y = \text{Cos} \left(x + \frac{\pi}{2} \right)$

$y = \text{Cos}^{-1} x - \frac{\pi}{2}$

21.

$y = \text{Cot } x$

$y = \text{Arccot } x$

38.

$y = \tan(\text{Tan}^{-1} x)$

Extra Credit

Given that $\tan (A + B) = \frac{\tan A + \tan B}{1 - \tan A \cdot \tan B}$, prove that: $\text{Arctan } \frac{1}{3} + \text{Arctan } \frac{1}{2} = \frac{\pi}{4}$.

Let $A = \text{Arctan } \frac{1}{3}$ and $B = \text{Arctan } \frac{1}{2}$. Therefore, Tan $A = \frac{1}{3}$ and Tan $B = \frac{1}{2}$.

$\text{Tan } (A + B) = \frac{\tan A + \tan B}{1 - \tan A \cdot \tan B} = \frac{\frac{1}{3} + \frac{1}{2}}{1 - \frac{1}{3} \cdot \frac{1}{2}} = \frac{\frac{5}{6}}{\frac{5}{6}} = 1$ So, $\tan (A + B) = 1$.

Thus, $A + B = \text{Arctan } 1$ or $\frac{\pi}{4}$. Therefore, $\text{Arctan } \frac{1}{3} + \text{Arctan } \frac{1}{2} = \frac{\pi}{4}$.

Open-Ended Assessment

Writing Have students write a paragraph explaining a function and its inverse and the need for inverse functions, particularly as they apply to the six trigonometric functions.

Chapter 6, Quiz D (Lessons 6-7 and 6-8) is available in the *Assessment and Evaluation Masters*, p. 148.

Additional Answers

44.

48.

Enrichment Masters, p. 68

---

43. Tides One day in March in Hilton Head, South Carolina, the first high tide occurred at 6:18 A.M. The high tide was 7.05 feet, and the low tide was −0.30 feet. The period for the oscillation of the tides is 12 hours and 24 minutes.

 a. Determine what time the next high tide will occur. **6:42 P.M.**

 b. Write the period of the oscillation as a decimal. **12.4 h**

 c. What is the amplitude of the sinusoidal function that models the tide? **3.675 ft**

 d. If $t = 0$ represents midnight, write a sinusoidal function that models the tide.

 e. At what time will the tides be at 6 feet for the first time that day?

43d. Sample answer: $y = 3.375 + 3.675 \sin\left(\frac{\pi}{6.2}t - 1.62\right)$

43e. Sample answer: about 4:46 A.M.

44. Critical Thinking Sketch the graph of $y = \sin(\text{Tan}^{-1} x)$. **See margin.**

45. Engineering The length L of the belt around two pulleys can be determined by the equation $L = \pi D + (d - D)\,\theta + 2C \sin\theta$, where D is the diameter of the larger pulley, d is the diameter of the smaller pulley, and C is the distance between the centers of the two pulleys. In this equation, θ is measured in radians and equals $\cos^{-1}\frac{D-d}{2C}$.

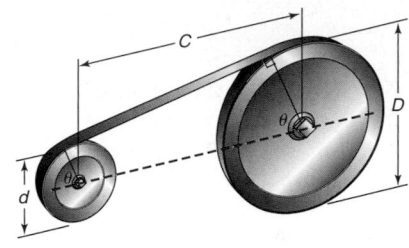

 a. If $D = 6$ inches, $d = 4$ inches, and $C = 10$ inches, find θ. **about 1.47 radians**

 b. What is the length of the belt needed to go around the two pulleys? **about 35.81 in.**

Mixed Review

46. What are the values of θ for which $\csc\theta$ is undefined? *(Lesson 6-7)* **πn, where n is an integer**

47. $y = \pm 5 \sin\left(\frac{2}{3}\theta + \frac{2\pi}{3}\right) - 8$

47. Write an equation of a sine function with amplitude 5, period 3π, phase shift $-\pi$, and vertical shift -8. *(Lesson 6-5)*

48. Graph $y = \cos x$ for $-11\pi \le x \le -9\pi$. *(Lesson 6-3)* **See margin.**

49. Geometry Each side of a rhombus is 30 units long. One diagonal makes a 25° angle with a side. What is the length of each diagonal to the nearest tenth of a unit? *(Lesson 5-6)* **25.4 units, 54.4 units**

50. Find the measure of the reference angle for an angle of 210°. *(Lesson 5-1)* **30°**

51. ± 1, ± 2, ± 3, ± 6, $\pm\frac{1}{2}$, $\pm\frac{3}{2}$

51. List the possible rational zeros of $f(x) = 2x^3 - 9x^2 - 18x + 6$. *(Lesson 4-4)*

52. Graph $y = \frac{1}{x-2} + 3$. Determine the interval(s) for which the function is increasing and the interval(s) for which the function is decreasing. *(Lesson 3-5)* **See margin; decreasing for $x < 2$ and $x > 2$.**

53. $27x^3 - 1$; $3x^3 - 3$

53. Find $[f \circ g](x)$ and $[g \circ f](x)$ if $f(x) = x^3 - 1$ and $g(x) = 3x$. *(Lesson 1-2)*

54. SAT/ACT Practice Suppose every letter in the alphabet has a number value that is equal to its place in the alphabet: the letter A has a value of 1, B a value of 2, and so on. The number value of a word is obtained by adding the values of the letters in the word and then multiplying the sum by the number of letters of the word. Find the number value of the "word" *DFGH*. **D**

 A 22 B 44 C 66 D 100 E 108

Additional Answer

52.

VOCABULARY

amplitude (p. 368)
angular displacement (p. 352)
angular velocity (p. 352)
central angle (p. 345)
circular arc (p. 345)
compound function (p. 382)
dimensional analysis (p. 353)
frequency (p. 372)
linear velocity (p. 353)

midline (p. 380)
period (p. 359)
periodic (p. 359)
phase shift (p. 378)
principal values (p. 406)
radian (p. 343)
sector (p. 346)
sinusoidal function (p. 388)

UNDERSTANDING AND USING THE VOCABULARY

Choose the correct term to best complete each sentence.

1. The (degree, <u>radian</u>) measure of an angle is defined as the length of the corresponding arc on the unit circle.

2. The ratio of the change in the central angle to the time required for the change is known as (<u>angular</u>, linear) velocity.

3. If the values of a function are (different, <u>the same</u>) for each given interval of the domain, the function is said to be periodic.

4. The (<u>amplitude</u>, period) of a function is one-half the difference of the maximum and minimum function values.

5. A central (<u>angle</u>, arc) has a vertex that lies at the center of a circle.

6. A horizontal translation of a trigonometric function is called a (<u>phase</u>, period) shift.

7. The length of a circular arc equals the measure of the radius of the circle times the (degree, <u>radian</u>) measure of the central angle.

8. The period and the (amplitude, <u>frequency</u>) are reciprocals of each other.

9. A function of the form $y = A \sin (k\theta + c) + h$ is a (<u>sinusoidal</u>, compound) function.

10. The values in the (<u>domain</u>, range) of Sine are called principal values.

 For additional review and practice for each lesson, visit: **www.amc.glencoe.com**

TEST OPTIONS

Six forms of Chapter 6 Test are available in the *Assessment and Evaluation Masters* as shown in the chart.

Chapter 6 Test, Form 1B, is shown at the right. Chapter 6 Test, Form 2B, is shown on the next page.

Form	Pages	Type
1A	133–134	Honors
1B	135–136	Average
1C	137–138	Basic
2A	139–140	Honors
2B	141–142	Average
2C	143–144	Basic

Form 1 tests are multiple choice.
Form 2 tests are free response.

STUDY GUIDE AND ASSESSMENT

VOCABULARY

This section provides a listing of the new terms, properties, and phrases that were introduced in this chapter. The exercises check students' understanding of the terms by using a variety of verbal formats including matching, completion, and true/false.

Glossary A complete glossary of terms begins on page A71.

Assessment and Evaluation Masters, pp. 135–136

SKILLS AND CONCEPTS

Objectives and Examples This section reviews the skills and concepts of the chapter and shows completely worked examples.

Review Exercises These exercises provide practice for the corresponding objectives.

Assessment and Evaluation Masters, pp. 141–142

OBJECTIVES AND EXAMPLES

Lesson 6-1 Change from radian measure to degree measure, and vice versa.

Change $-\frac{5\pi}{3}$ radians to degree measure.

$$-\frac{5\pi}{3} = \frac{5\pi}{3} \times \frac{180°}{\pi}$$
$$= -300°$$

Lesson 6-1 Find the length of an arc given the measure of the central angle.

Given a central angle of $\frac{2\pi}{3}$, find the length of its intercepted arc in a circle of radius 10 inches. Round to the nearest tenth.

$$s = r\theta$$
$$s = 10\left(\frac{2\pi}{3}\right)$$
$$s \approx 20.94395102$$

The length of the arc is about 20.9 inches.

Lesson 6-2 Find linear and angular velocity.

Determine the angular velocity if 5.2 revolutions are completed in 8 seconds. Round to the nearest tenth.

The angular displacement is $5.2 \times 2\pi$ or 10.4π radians.

$$\omega = \frac{\theta}{t}$$
$$\omega = \frac{10.4\pi}{8}$$
$$\omega \approx 4.08407045$$

The angular velocity is about 4.1 radians per second.

REVIEW EXERCISES

Change each degree measure to radian measure in terms of π.

11. 60° $\frac{\pi}{3}$ 12. −75° $-\frac{5\pi}{12}$ 13. 240° $\frac{4\pi}{3}$

Change each radian measure to degree measure. Round to the nearest tenth, if necessary.

14. $\frac{5\pi}{6}$ **150°** 15. $-\frac{7\pi}{4}$ **−315°** 16. 2.4 **137.5°**

Given the measurement of a central angle, find the length of its intercepted arc in a circle of radius 15 centimeters. Round to the nearest tenth.

17. $\frac{3\pi}{4}$ **35.3 cm** 18. 75° **19.6 cm**

19. 150° **39.3 cm** 20. $\frac{\pi}{5}$ **9.4 cm**

Determine each angular displacement in radians. Round to the nearest tenth.

21. 5 revolutions **31.4 radians**

22. 3.8 revolutions **23.9 radians**

23. 50.4 revolutions **316.7 radians**

24. 350 revolutions **2199.1 radians**

Determine each angular velocity. Round to the nearest tenth.

25. 1.8 revolutions in 5 seconds **2.3 radians/s**

26. 3.6 revolutions in 2 minutes **11.3 radians/min**

27. 15.4 revolutions in 15 seconds **6.5 radians/s**

28. 50 revolutions in 12 minutes **26.2 radians/min**

TestCheck and Worksheet Builder

This **networkable** software has three modules.
• **Worksheet Builder** to make worksheets and tests
• **Student Module** to take tests on-screen
• **Management System** to keep student records
Three special banks are included for SAT, ACT, and TIMSS.

OBJECTIVES AND EXAMPLES

Lesson 6-3 Use the graphs of the sine and cosine functions.

• Find the value of $\cos \frac{5\pi}{2}$ by referring to the graph of the cosine function.

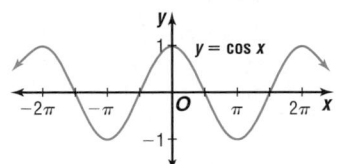

$\frac{5\pi}{2} = 2\pi + \frac{\pi}{2}$, so $\cos \frac{5\pi}{2} = \cos \frac{\pi}{2}$ or 0.

REVIEW EXERCISES

Find each value by referring to the graph of the cosine function shown at the left or sine function shown below.

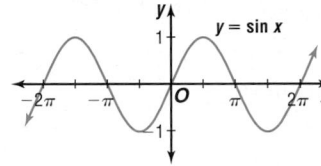

29. $\cos 5\pi$ **−1**   **30.** $\sin 13\pi$ **0**

31. $\sin \frac{9\pi}{2}$ **1**   **32.** $\cos\left(-\frac{7\pi}{2}\right)$ **0**

Lesson 6-4 Find the amplitude and period for sine and cosine functions.

• State the amplitude and period for $y = -\frac{3}{4} \cos 2\theta$.

The amplitude of $y = A \cos k\theta$ is $|A|$. Since $A = -\frac{3}{4}$, the amplitude is $\left|-\frac{3}{4}\right|$ or $\frac{3}{4}$.

Since $k = 2$, the period is $\frac{2\pi}{2}$ or π.

State the amplitude and period for each function. Then graph each function. **33–35. See margin for graphs.**

33. $y = 4 \cos 2\theta$ **4; π**

34. $y = 0.5 \sin 4\theta$ **0.5; $\frac{\pi}{2}$**

35. $y = -\frac{1}{3} \cos \frac{\theta}{2}$ **$\frac{1}{3}$; 4π**

Lesson 6-5 Write equations of sine and cosine functions, given the amplitude, period, phase shift, and vertical translation.

• Write an equation of a cosine function with an amplitude 2, period 2π, phase shift $-\pi$, and vertical shift 2.

A: $|A| = 2$, so $A = 2$ or -2.

k: $\frac{2\pi}{k} = 2\pi$, so $k = 1$.

c: $-\frac{c}{k} = -\pi$, so $-c = -\pi$ or $c = \pi$.

h: $h = 2$

Substituting into $y = A \sin(k\theta + c) + h$, the possible equations are $y = \pm 2 \cos(\theta + \pi) + 2$.

36. Write an equation of a sine function with an amplitude 4, period $\frac{\pi}{2}$, phase shift -2π, and vertical shift -1.

37. Write an equation of a sine function with an amplitude 0.5, period π, phase shift $\frac{\pi}{3}$, and vertical shift 3.

38. Write an equation of a cosine function with an amplitude $\frac{3}{4}$, period $\frac{\pi}{4}$, phase shift 0, and vertical shift 5. **36–38. See margin.**

Additional Answers

33.

34.

35.

36. $y = \pm 4 \sin(4\theta + 8\pi) - 1$

37. $y = \pm 0.5 \sin\left(2\theta - \frac{2\pi}{3}\right) + 3$

38. $y = \pm \frac{3}{4} \cos 8\theta + 5$

Additional Answers

41.

42.

43.

44.

| OBJECTIVES AND EXAMPLES | REVIEW EXERCISES |

Lesson 6-6 Use sinusoidal functions to solve problems.

• A sinusoidal function can be any function of the form

$y = A \sin (k\theta + c) + h$ or

$y = A \cos (k\theta + c) + h$.

Suppose a person's blood pressure oscillates between the two numbers given. If the heart beats once every second, write a sine function that models this person's blood pressure.

39. 120 and 80 $y = 20 \sin 2\pi t + 100$

40. 130 and 100 $y = 15 \sin 2\pi t + 115$

Lesson 6-7 Graph tangent, cotangent, secant, and cosecant functions.

• Graph $y = \tan 0.5\theta$.

The period of this function is 2π. The phase shift is 0, and the vertical shift is 0.

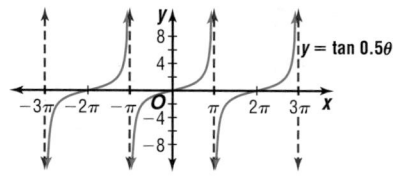

Graph each function.   **41–44. See margin.**

41. $y = \dfrac{1}{3} \csc \theta$

42. $y = 2 \tan \left(3\theta + \dfrac{\pi}{2}\right)$

43. $y = \sec \theta + 4$

44. $y = \tan \theta - 2$

Lesson 6-8 Find the principal values of inverse trigonometric functions.

• Find $\cos (\text{Tan}^{-1} 1)$.

Let $\alpha = \text{Tan}^{-1} 1$.

$\text{Tan } \alpha = 1$

$\alpha = \dfrac{\pi}{4}$

$\cos \dfrac{\pi}{4} = \dfrac{\sqrt{2}}{2}$

Find each value.

45. $\text{Arctan} \left(-1\right)$ $-\dfrac{\pi}{4}$

46. $\text{Sin}^{-1} 1$ $\dfrac{\pi}{2}$

47. $\text{Cos}^{-1} \left(\tan \dfrac{\pi}{4}\right)$ 0

48. $\sin \left(\text{Sin}^{-1} \dfrac{\sqrt{3}}{2}\right)$ $\dfrac{\sqrt{3}}{2}$

49. $\cos \left(\text{Arctan} \sqrt{3} + \text{Arcsin} \dfrac{1}{2}\right)$ 0

416   *Chapter 6   Graphs of Trigonometric Functions*

APPLICATIONS AND PROBLEM SOLVING

50. Meteorology The mean average temperature in a certain town is 64°F. The temperature fluctuates 11.5° above and below the mean temperature. If $t = 1$ represents January, the phase shift of the sine function is 3. *(Lesson 6-6)*

a. Write a model for the average monthly temperature in the town.

b. According to your model, what is the average temperature in April? **about 69.8°**

c. According to your model, what is the average temperature in July? **about 74.0°**

50a. $y = 11.5 \sin\left(\frac{\pi}{6}t - \frac{\pi}{2}\right) + 64$

51. Physics The strength of a magnetic field is called magnetic induction. An equation for *magnetic induction* is $B = \frac{F}{IL \sin \theta}$, where F is a force on a current I which is moving through a wire of length L at an angle θ to the magnetic field. A wire within a magnetic field is 1 meter long and carries a current of 5.0 amperes. The force on the wire is 0.2 newton, and the magnetic induction is 0.04 newton per ampere-meter. What is the angle of the wire to the magnetic field? *(Lesson 6-8)* $\frac{\pi}{2}$

ALTERNATIVE ASSESSMENT

OPEN-ENDED ASSESSMENT

1. The area of a circular sector is about 26.2 square inches. What are possible measures for the radius and the central angle of the sector?
Sample answer: $r = 5$ in., $\theta = \frac{2\pi}{3}$

2. a. You are given the graph of a cosine function. Explain how you can tell if the graph has been translated. Sketch two graphs as part of your explanation.

   b. You are given the equation of a cosine function. Explain how you can tell if the graph has been translated. Provide two equations as part of your explanation.
   See margin.

Additional Assessment See p. A61 for Chapter 6 practice test.

Unit 2 *inter*NET Project
THE CYBERCLASSROOM
What Is Your Sine?

- Search the Internet to find web sites that have applications of the sine or cosine function. Find at least three different sources of information.

- Select one of the applications of the sine or cosine function. Use the Internet to find actual data that can be modeled by a graph that resembles the sine or cosine function.

- Draw a sine or cosine model of the data. Write an equation for a sinusoidal function that fits your data.

PORTFOLIO
Choose a trigonometric function you studied in this chapter. Graph your function. Write three expressions whose values can be found using your graph. Find the values of these expressions.

Chapter 6   Study Guide and Assessment **417**

Internet Address Book

Record useful Internet addresses in the space at right for quick reference.

www.amc.glencoe.com

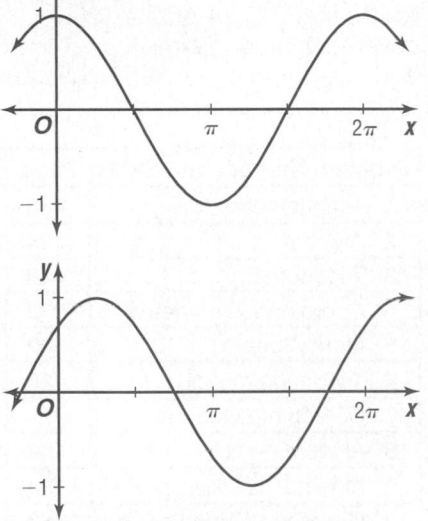

Chapter 6 Skill: Trigonometry

Pages 418–419 are part of a complete SAT/ACT preparation course that is described in detail on pages T8–T9 of the *Teacher's Handbook*. The test items were written in the same style as those in the ACT.

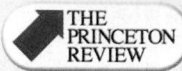 These questions were aligned and verified by The Princeton Review, the nation's leader in test preparation.

Diagnosis & Prescription

Each of the 10 ACT and SAT practice problems on page 419 is cross-referenced to the chapter where that SAT or ACT skill is covered. If students miss a particular type of problem, you can have them study that skill.

Problem Number and Skill	Page
1 Trigonometry	418
2 Algebra	124
3 Trigonometry	418
4 Coordinate Geometry	272
5 More Algebra	202
6 Trigonometry	418
7 Coordinate Geometry	272
8 Algebra	124
9 Algebra	124
10 Geometry (Triangles and Quadrilaterals)	482

Trigonometry Problems

Each ACT exam contains exactly four trigonometry problems. The SAT has none! You'll need to know the trigonometric functions in a right triangle.

$$\sin \theta = \frac{\text{opposite}}{\text{hypotenuse}} \qquad \cos \theta = \frac{\text{adjacent}}{\text{hypotenuse}} \qquad \tan \theta = \frac{\text{opposite}}{\text{adjacent}}$$

Review the reciprocal functions.

$$\csc \theta = \frac{1}{\sin \theta} \qquad \sec \theta = \frac{1}{\cos \theta} \qquad \cot \theta = \frac{1}{\tan \theta}$$

Review the graphs of trigonometric functions.

TEST-TAKING TIP

Use the memory aid SOH-CAH-TOA. Pronounce it as *so-ca-to-a*.

SOH represents **S**ine (is) **O**pposite (over) **H**ypotenuse

CAH represents **C**osine (is) **A**djacent (over) **H**ypotenuse

TOA represents **T**angent (is) **O**pposite (over) **A**djacent

---

ACT EXAMPLE

1. If $\sin \theta = \frac{1}{2}$ and $90° < \theta < 180°$, then $\theta = $?

A 100°

B 120°

C 130°

D 150°

E 160°

HINT Memorize the sine, cosine, and tangent of special angles 0°, 30°, 45°, 60°, and 90°.

Solution Draw a diagram. Use the quadrant indicated by the size of angle θ.

Recall that the $\sin 30° = \frac{1}{2}$. The angle inside the triangle is 30°. Then $\theta + 30° = 180°$.

If $\theta + 30° = 180°$, then $\theta = 150°$.

The answer is choice **D**.

ACT EXAMPLE

2. What is the least positive value for x where $y = \sin 4x$ reaches its maximum?

A $\frac{\pi}{8}$

B $\frac{\pi}{4}$

C $\frac{\pi}{2}$

D π

E 2π

HINT Review the graphs of the sine and cosine functions.

Solution The least value for x where $y = \sin x$ reaches its maximum is $\frac{\pi}{2}$. If $4x = \frac{\pi}{2}$, then $x = \frac{\pi}{8}$. The answer is choice **A**.

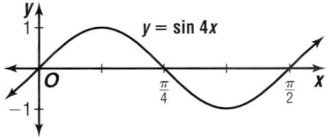

⊙ TestCheck and Worksheet Builder

Special banks of SAT, ACT, and TIMSS items are included in this CD-ROM.

After you work each problem, record your answer on the answer sheet provided or on a piece of paper.

Multiple Choice

1. What is $\sin \theta$, if $\tan \theta = \frac{4}{3}$? **B**

   A $\frac{3}{4}$　　　　　B $\frac{4}{5}$

   C $\frac{5}{4}$　　　　　D $\frac{5}{3}$

   E $\frac{7}{3}$

2. If the sum of two consecutive odd integers is 56, then the greater integer equals: **C**

   A 25　　　B 27　　　C 29

   D 31　　　E 33

3. For all θ where $\sin \theta - \cos \theta \neq 0$,

   $\dfrac{\sin^2 \theta - \cos^2 \theta}{\sin \theta - \cos \theta}$ is equivalent to **B**

   A $\sin \theta - \cos \theta$　　　B $\sin \theta + \cos \theta$

   C $\tan \theta$　　　　　D -1

   E 1

4. In the figure below, side AB of triangle ABC contains which point? **E**

   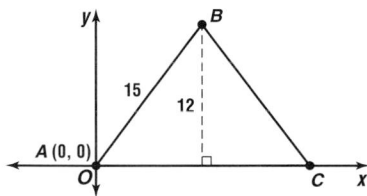

   A $(3, 2)$　　　　　B $(3, 5)$

   C $(4, 6)$　　　　　D $(4, 10)$

   E $(6, 8)$

5. Which of the following is the sum of both solutions of the equation $x^2 - 2x - 8 = 0$? **D**

   A -6　　　B -4　　　C -2

   D 2　　　E 6

6. In the figure below, $\angle A$ is a right angle, AB is 3 units long, and BC is 5 units long. If $\angle C = \theta$, what is the value of $\cos \theta$? **C**

   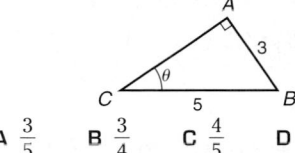

   A $\frac{3}{5}$　　B $\frac{3}{4}$　　C $\frac{4}{5}$　　D $\frac{5}{4}$　　E $\frac{5}{3}$

7. The equation $x - 7 = x^2 + y$ represents which conic? **A**

   A parabola　　　B circle　　　C ellipse

   D hyperbola　　　E line

8. If n is an integer, then which of the following must also be integers? **A**

   I. $\dfrac{16n + 16}{n + 1}$

   II. $\dfrac{16n + 16}{16n}$

   III. $\dfrac{16n^2 + n}{16n}$

   A I only　　　B II only　　　C III only

   D I and II　　　E II and III

9. For $x > 1$, which expression has a value that is less than 1? **D**

   A $x^{x - 1}$

   B $x^{x + 2}$

   C $(x + 2)^x$

   D $x^{1 - x}$

   E x^x

10. **Grid-In** In the figure, segment AD bisects $\angle BAC$, and segment DC bisects $\angle BCA$. If the measure of $\angle ADC = 100°$, then what is the measure of $\angle B$? **20**

interNET CONNECTION **SAT/ACT Practice** For additional test practice questions, visit: **www.amc.glencoe.com**

A **bubble-in answer sheet** for these practice problems is available on p. 1 of the *Assessment and Evaluation Masters.*

Additional Practice

Additional SAT and ACT practice questions and a bubble-in answer sheet are available in the *Assessment and Evaluation Masters,* pp. 2, 149, and 150.

Assessment and Evaluation Masters, pp. 149–150

Extra Credit

TIMSS A string is wound symmetrically around a cylindrical rod. The string goes exactly 4 times around the rod. The circumference of the rod is 4 centimeters and its length is 12 centimeters. Find the length of the string. **20 cm**

Trigonometric Identities and Equations

Instructional Objectives

Lesson (pages)	Lesson Objectives	NCTM Standards 2000	State/Local Objectives
7-1 (421–430)	• Identify and use reciprocal identities, quotient identities, Pythagorean identities, symmetry identities, and opposite-angle identities.	6, 7	
7-2 (431–436)	• Use the basic trigonometric identities to verify other identities. • Find numerical values of trigonometric functions.	6, 7	
7-3 (437–445)	• Use the sum and difference identities for the sine, cosine, and tangent functions.	6, 7	
7-3B (446–447)	• Identify equivalent values for trigonometric functions involving quadrantal angles.	6, 7	
7-4 (448–455)	• Use the double- and half-angle identities for the sine, cosine, and tangent functions.	6, 7	
7-5 (456–461)	• Solve trigonometric equations and inequalities.	6, 7	
7-6 (463–469)	• Write the standard form of a linear equation given the length of the normal and the angle it makes with the x-axis. • Write linear equations in normal form.	6, 7	
7-7 (470–476)	• Find the distance from a point to a line. • Find the distance between two parallel lines. • Write equations of lines that bisect angles formed by intersecting lines.	6, 7	

Suggested Pacing *See pages T2–T4 for additional options in course-planning.*

COURSE	Lesson 7-1	Lesson 7-2	Lesson 7-3	Lesson 7-4	Lesson 7-5
Precalculus Emphasis Chapters 4–15 (45–50 minutes)	1.5 days	1.5 days	(with Lesson 7-3B) 2 days	1.5 days	1.5 days
Precalculus Emphasis (block)	0.5 day	1.5 days		(with Lesson 7-3B) 1.5 days	(with Lesson 7-6) 1.5 days
Trigonometric Emphasis Chapters 1–9, 11, 12 (45–50 minutes)	1.5 days	1.5 days	(with Lesson 7-3B) 2 days	1.5 days	1.5 days
Trigonometric Emphasis (block)	0.5 day	1.5 days		(with Lesson 7-3B) 1.5 days	(with Lesson 7-6) 1.5 days

Instructional Resources

Student Edition		Blackline Masters (page numbers)							
Lesson	Graphing Calculator Applications	Study Guide and Practice	Enrichment	Assessment and Evaluation	Graphing Calculator Keystrokes	Problem Solving & Applications	Spreadsheet Applications	TI-92 and CBL Interdisciplinary Activities	Transparencies A and B
7-1		119–120	73			14			7-1
7-2	433, 435	121–122	74	173	26				7-2
7-3		123–124	75				7		7-3
7-3B	446–447				27–29				
7-4		125–126	76	172, 173					7-4
7-5	458, 460	127–128	77		30				7-5
7-6		129–130	78	174		13			7-6
7-7	473	131–132	79	174				21–24	7-7
Study Guide/ Assessment				159–171, 177					
SAT/ACT Practice				175–176					

Lesson 7-6	Lesson 7-7	Study Guide & Assessment	Chapter Test	TOTAL
1.5 days	1.5 days	1 day	1day	13 days
(See Lesson 7-5)	1.5 days		0.5 day	7 days
1.5 days	1.5 days	1 day	1 day	13 days
(See Lesson 7-5)	1.5 days		0.5 day	7 days

Resource Manager

TeacherWorks™

The worksheets shown on this page are a small sample of the materials available on *TeacherWorks: All-in-One Lesson Planner and Resource Center.*

This CD-ROM contains all of the blackline masters and transparencies available for this program.

It also includes a lesson planner and interactive Teacher Edition, so you can customize lesson plans and reproduce classroom resources quickly and easily, from just about anywhere.

Applications

Problem Solving and Applications Masters, pp. 13, 14

Technology/Multimedia

Spreadsheet Applications Masters, p. 7

TI-92 & CBL Interdisciplinary Activities Masters, pp. 21–24

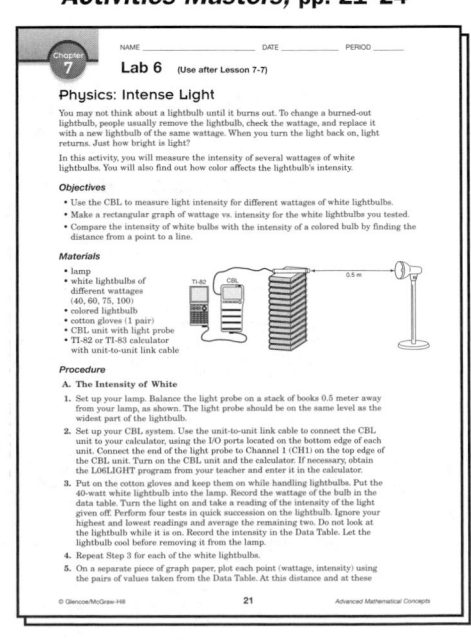

Type	Student Edition	Teacher's Wraparound Edition	Assessment and Evaluation Masters
Ongoing Assessment	Mid-Chapter Quiz, p. 445	5-Minute Check, pp. 421, 431, 437, 448, 456, 463, 470	Mid-Chapter Test, p. 172 Quizzes A–D, pp 173–174
Mixed Review	Mixed Review, pp. 429–430, 436, 444–445, 455, 461, 469, 476		Cumulative Review, p. 177
Error Analysis	You Decide, pp. 427, 453	Error Analysis, pp. 427, 433, 442, 453, 459, 467, 474	
Standardized Test Practice	SAT/ACT Practice, pp. 430, 436, 445, 455, 461, 469, 476 SAT & ACT Preparation, pp. 482–483	TIMSS, p. 483	SAT/ACT Practice, pp. 175–176
Open-Ended Assessment	Math Journal, pp. 434, 459 Open-Ended Assessment, p. 481 Portfolio, p. 481	Modeling: pp. 445, 476 Speaking: pp. 436, 455, 461 Writing: pp. 430, 447, 469	Open-Ended Assessment, p. 171
Chapter Assessment	Study Guide and Assessment, pp. 477–481 Chapter Test, p. A62		Multiple-Choice Tests (Forms 1A, 1B, 1C), pp. 159–164 Free-Response Tests (Forms 2A, 2B, 2C), pp. 165–170

TestCheck and Worksheet Builder is a networkable CD-ROM with three integrated modules. The **Worksheet Builder** creates customized worksheets, tests, and quizzes of free-response, multiple-choice, short-answer, and open-ended items. The **Student Module** gives you the option of having students take tests on-screen and get immediate feedback on their performance. Use the optional **Management System** to keep detailed student records.

Additional Chapter Resources

Student Edition
Career Choices, p. 430
History of Mathematics, p. 462
Internet Project, p. 481
Extra Practice, pp. A38–A39

Teacher's Classroom Resources

Technology
Graphing Calculator
Keystrokes Masters

Teaching Aids
Answer Key Transparencies
Block Schedule Planning Guide
Lesson Planning Guide
Solutions Manual

Glencoe Instructional Technology

Instructional

Advanced MathPASS CD-ROM

Multimedia Applications CD-ROM

Assessment

TestCheck and Worksheet Builder

GLENCOE
Online

Visit **www.amc.glencoe.com** for data updates, career information, graphing calculator programs, and other interactive activities.

Chapter Notes 7

Chapter Overview

This chapter reviews the basic trigonometric identities. Students verify trigonometric identities, use sum and difference identities, and use double- and half-angle identities. In the second half of the chapter, students solve trigonometric equations, find the normal form of a linear equation, and then determine distance from a point to a line.

 Using a Graphing Calculator

There are numerous opportunities, all optional, to use a graphing calculator in this chapter.
- Determining whether each equation is an identity (pp. 433, 435)
- Investigating families of identities called reduction identities (pp. 446–447)
- Solving difficult-to-solve trigonometric equations and inequalities by graphing (pp. 458, 460)
- Using a program to compute the distance from a point to a line (p. 470)
- Finding equations of lines that bisect the angles formed by two lines (p. 473)

The **Graphing Calculator Appendix**, which begins on page A2, contains keystroke instructions on many of these topics.

Chapter 7

TRIGONOMETRIC IDENTITIES AND EQUATIONS

CHAPTER OBJECTIVES ....●
- Use reciprocal, quotient, Pythagorean, symmetry, and opposite-angle identities. *(Lesson 7-1)*
- Verify trigonometric identities. *(Lessons 7-2, 7-3, 7-4)*
- Use sum, difference, double-angle, and half-angle identities. *(Lessons 7-3, 7-4)*
- Solve trigonometric equations and inequalities. *(Lesson 7-5)*
- Write a linear equation in normal form. *(Lesson 7-6)*
- Find the distance from a point to a line. *(Lesson 7-7)*

 Preparing for the SAT and ACT

This textbook contains a complete SAT/ACT preparation course described on pages T8–T9 of the *Teacher's Handbook*. This chapter contains:
- an SAT or ACT practice question at the end of each lesson
- examples of SAT and ACT problems involving geometry on page 482
- ten SAT and ACT practice problems on page 483
- additional SAT and ACT practice in the *Assessment and Evaluation Masters*, pp. 175–176
- SAT and ACT test practice banks in the *TestCheck and Worksheet Builder*

Basic Trigonometric Identities

OBJECTIVE
- Identify and use reciprocal identities, quotient identities, Pythagorean identities, symmetry identities, and opposite-angle identities.

OPTICS Many sunglasses have polarized lenses that reduce the intensity of light. When unpolarized light passes through a polarized lens, the intensity of the light is cut in half. If the light then passes through another polarized lens with its axis at an angle of θ to the first, the intensity of the light is again diminished.

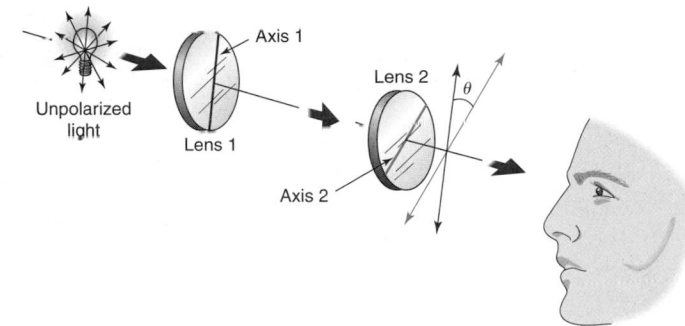

The intensity of the emerging light can be found by using the formula

$I = I_0 - \dfrac{I_0}{\csc^2 \theta}$, where I_0 is the intensity of the light incoming to the second polarized lens, I is the intensity of the emerging light, and θ is the angle between the axes of polarization. Simplify this expression and determine the intensity of light emerging from a polarized lens with its axis at a 30° angle to the original. *This problem will be solved in Example 5.*

In algebra, variables and constants usually represent real numbers. The values of trigonometric functions are also real numbers. Therefore, the language and operations of algebra also apply to trigonometry. Algebraic expressions involve the operations of addition, subtraction, multiplication, division, and exponentiation. These operations are used to form trigonometric expressions. Each expression below is a trigonometric expression.

$$\cos x - x \qquad \sin^2 a + \cos^2 a \qquad \dfrac{1 - \sec A}{\tan A}$$

A statement of equality between two expressions that is true for *all* values of the variable(s) for which the expressions are defined is called an **identity.** For example, $x^2 - y^2 = (x - y)(x + y)$ is an algebraic identity. An identity involving trigonometric expressions is called a **trigonometric identity.**

If you can show that a specific value of the variable in an equation makes the equation false, then you have produced a *counterexample*. It only takes one counterexample to prove that an equation is not an identity.

Lesson 7-1 Basic Trigonometric Identities **421**

1 FOCUS

 5-Minute Check
Chapter 6

1. Find the values of x for which $\sin x = 0$ is true. **$180k°$ or πk, for any integer k**

2. Give the amplitude and period of the function $y = 3 \sin 4\theta$. **$3; \dfrac{\pi}{2}$**

3. Find the value of x in Quadrant I that satisfies $\tan x = \dfrac{\sqrt{3}}{3}$. **$30°$ or $\dfrac{\pi}{6}$**

4. Give the range of Arccos x. **$0° \le$ Arccos $x \le 180°$ or $0 \le$ Arccos $x \le \pi$**

5. Find $\sin(\sin^{-1} 0.6)$. **0.6**

Motivating the Lesson
Discussion Ask students to compare and contrast the equations $x + 3 = 4$ and $x + 3x = 4x$. Lead them to see that although both statements are equations, the first is true for only one value of the variable, while the second is true for all values of x. A statement like the second equation is called an *identity*. Have students give some examples of identities.

Resource Manager

 Reproducible Masters
- *Study Guide and Practice,* pp. 119–120
- *Enrichment,* p. 73
- *Problem Solving and Applications,* p. 14

 Transparencies
- 5-Minute Check, 7-1
- Teaching, 7-1

 Technology/Multimedia
- Advanced MathPASS, Lesson 15

In-Class Example

Example 1
Prove that $\sin x \tan x = \cos x$ is *not* a trigonometric identity by producing a counterexample.

Suppose $x = \frac{\pi}{6}$.

$$\sin x \cdot \tan x \overset{?}{=} \cos x$$

$$\sin \frac{\pi}{6} \cdot \tan \frac{\pi}{6} \overset{?}{=} \cos \frac{\pi}{6}$$

$$\frac{1}{2} \cdot \frac{\sqrt{3}}{3} \overset{?}{=} \frac{\sqrt{3}}{2}$$

$$\frac{\sqrt{3}}{6} \neq \frac{\sqrt{3}}{2}$$

Since an inequality results, the equation is not an identity.

Teaching Tip When you discuss the quotient identities, encourage students to keep a careful list of the trigonometric identities they learn in this chapter.

Example ① Prove that $\sin x \cos x = \tan x$ is *not* a trigonometric identity by producing a counterexample.

Suppose $x = \frac{\pi}{4}$.

$$\sin x \cos x \overset{?}{=} \tan x$$

$$\sin \frac{\pi}{4} \cos \frac{\pi}{4} \overset{?}{=} \tan \frac{\pi}{4} \qquad \textit{Replace x with } \frac{\pi}{4}.$$

$$\left(\frac{\sqrt{2}}{2}\right)\left(\frac{\sqrt{2}}{2}\right) \overset{?}{=} 1$$

$$\frac{1}{2} \neq 1$$

Since evaluating each side of the equation for the same value of x produces an inequality, the equation is not an identity.

Although producing a counterexample can show that an equation is not an identity, proving that an equation is an identity generally takes more work. Proving that an equation is an identity requires showing that the equality holds for *all* values of the variable where each expression is defined. Several fundamental trigonometric identities can be verified using geometry.

Recall from Lesson 5-3 that the trigonometric functions can be defined using the unit circle. From the unit circle, $\sin \theta = \frac{y}{1}$, or y and $\csc \theta = \frac{1}{y}$. That is, $\sin \theta = \frac{1}{\csc \theta}$. Identities derived in this manner are called **reciprocal identities**.

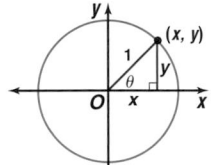

The following trigonometric identities hold for all values of θ where each expression is defined.

Reciprocal Identities

$$\sin \theta = \frac{1}{\csc \theta} \qquad \cos \theta = \frac{1}{\sec \theta}$$

$$\csc \theta = \frac{1}{\sin \theta} \qquad \sec \theta = \frac{1}{\cos \theta}$$

$$\tan \theta = \frac{1}{\cot \theta} \qquad \cot \theta = \frac{1}{\tan \theta}$$

Returning to the unit circle, we can say that $\frac{\sin \theta}{\cos \theta} = \frac{y}{x} = \tan \theta$. This is an example of a **quotient identity**.

Quotient Identities

The following trigonometric identities hold for all values of θ where each expression is defined.

$$\frac{\sin \theta}{\cos \theta} = \tan \theta \qquad \frac{\cos \theta}{\sin \theta} = \cot \theta$$

422 Chapter 7 *Trigonometric Identities and Equations*

📂 **Prerequisite Skills**

A review of the prerequisite skills for this lesson is provided in the *Prerequisite Skills Masters*.

Skill Number and Content	Pages
1 Properties of Real Numbers	3–4
2 Fractions	5–6
3 Order of Operations	7–8
9 Factoring by Grouping and Greatest Common Factor	19–20
10 Factoring Special Products of Polynomials	21–22
11 Factoring by Completing the Square	23–24

Since the triangle in the unit circle on the previous page is a right triangle, we may apply the Pythagorean Theorem: $y^2 + x^2 = 1^2$, or $\sin^2\theta + \cos^2\theta = 1$. Other identities can be derived from this one.

$$\sin^2\theta + \cos^2\theta = 1$$

$$\frac{\sin^2\theta}{\cos^2\theta} + \frac{\cos^2\theta}{\cos^2\theta} = \frac{1}{\cos^2\theta} \quad \text{\textit{Divide each side by} } \cos^2\theta.$$

$$\tan^2\theta + 1 = \sec^2\theta \quad \text{\textit{Quotient and reciprocal identities}}$$

Likewise, the identity $1 + \cot^2\theta = \csc^2\theta$ can be derived by dividing each side of the equation $\sin^2\theta + \cos^2\theta = 1$ by $\sin^2\theta$. These are the **Pythagorean identities.**

Pythagorean Identities	The following trigonometric identities hold for all values of θ where each expression is defined. $\sin^2\theta + \cos^2\theta = 1 \qquad \tan^2\theta + 1 = \sec^2\theta \qquad 1 + \cot^2\theta = \csc^2\theta$

You can use the identities to help find the values of trigonometric functions.

Example **2** Use the given information to find the trigonometric value.

a. If $\sec\theta = \dfrac{3}{2}$, find $\cos\theta$.

$$\cos\theta = \frac{1}{\sec\theta} \quad \text{\textit{Choose an identity that involves} } \cos\theta \text{ \textit{and} } \sec\theta.$$

$$= \frac{1}{\frac{3}{2}} \text{ or } \frac{2}{3} \quad \text{\textit{Substitute} } \frac{3}{2} \text{ \textit{for} } \sec\theta \text{ \textit{and evaluate.}}$$

b. If $\csc\theta = \dfrac{4}{3}$, find $\tan\theta$.

Since there are no identities relating $\csc\theta$ and $\tan\theta$, we must use two identities, one relating $\csc\theta$ and $\cot\theta$ and another relating $\cot\theta$ and $\tan\theta$.

$$\csc^2\theta = 1 + \cot^2\theta \quad \text{\textit{Pythagorean identity}}$$

$$\left(\frac{4}{3}\right)^2 = 1 + \cot^2\theta \quad \text{\textit{Substitute} } \frac{4}{3} \text{ \textit{for} } \csc\theta.$$

$$\frac{16}{9} = 1 + \cot^2\theta$$

$$\frac{7}{9} = \cot^2\theta$$

$$\pm\frac{\sqrt{7}}{3} = \cot\theta \quad \text{\textit{Take the square root of each side.}}$$

Now find $\tan\theta$.

$$\tan\theta = \frac{1}{\cot\theta} \quad \text{\textit{Reciprocal identity}}$$

$$= \pm\frac{3\sqrt{7}}{7}, \text{ or about } \pm 1.134$$

Lesson 7-1 Basic Trigonometric Identities **423**

Teaching Tip Point out that the reciprocal of the sine function is cosecant, the reciprocal of cosine is secant, and the reciprocal of tangent is cotangent.

Teaching Tip When you discuss Example 2, remind students that they also can use right triangles. Draw right triangle ABC with hypotenuse $AB = 5$ and $BC = 3$. Show that the Pythagorean Theorem and the definition of the cosecant function can be used to find $\csc\theta$.

In-Class Example

Example 2
Use the given information to find the trigonometric value.
a. If $\cos\theta = \dfrac{3}{4}$, find $\sec\theta$. $\dfrac{4}{3}$
b. If $\cos\theta = \dfrac{3}{5}$, find $\csc\theta$. $\pm\dfrac{5}{4}$

To determine the sign of a function value, you need to know the quadrant in which the angle terminates. The signs of function values in different quadrants are related according to the symmetries of the unit circle. Since we can determine the values of $\tan A$, $\cot A$, $\sec A$, and $\csc A$ in terms of $\sin A$ and/or $\cos A$ with the reciprocal and quotient identities, we only need to investigate $\sin A$ and $\cos A$.

Case	Relationship between angles *A* and *B*	Diagram	Conclusion
1	The angles differ by a multiple of 360°. $B - A = 360k°$ or $B = A + 360\,k°$		Since A and $A + 360k°$ are coterminal, they share the same value of sine and cosine.
2	The angles differ by an odd multiple of 180°. $B - A = 180°(2k - 1)$ or $B = A + 180°(2k - 1)$		Since A and $A + 180°(2k - 1)$ have terminal sides in diagonally opposite quadrants, the values of both sine and cosine change sign.
3	The sum of the angles is a multiple of 360°. $A + B = 360k°$ or $B = 360k° - A$		Since A and $360k° - A$ lie in vertically adjacent quadrants, the sine values are opposite but the cosine values are the same.
4	The sum of the angles is an odd multiple of 180°. $A + B = 180°(2k - 1)$ or $B = 180°(2k - 1) - A$		Since A and $180°(2k - 1) - A$ lie in horizontally adjacent quadrants, the sine values are the same but the cosine values are opposite.

These general rules for sine and cosine are called **symmetry identities.**

Symmetry Identities	The following trigonometric identities hold for any integer k and all values of A.

Case 1: $\quad \sin(A + 360k°) = \sin A \qquad\qquad \cos(A + 360k°) = \cos A$

Case 2: $\quad \sin(A + 180°(2k - 1)) = -\sin A \qquad \cos(A + 180°(2k - 1)) = -\cos A$

Case 3: $\quad \sin(360k° - A) = -\sin A \qquad\qquad \cos(360k° - A) = \cos A$

Case 4: $\quad \sin(180°(2k - 1) - A) = \sin A \qquad \cos(180°(2k - 1) - A) = -\cos A$

To use the symmetry identities with radian measure, replace 180° with π and 360° with 2π.

Example **3** Express each value as a trigonometric function of an angle in Quadrant I.

a. sin 600°

Relate 600° to an angle in Quadrant I.

$600° = 60° + 3(180°)$ *600° and 60° differ by an odd multiple of 180°.*

$\sin 600° = \sin (60° + 3(180°))$ *Case 2, with A = 60° and k = 2*

$ = -\sin 60°$

b. $\sin \dfrac{19\pi}{4}$

The sum of $\dfrac{19\pi}{4}$ and $\dfrac{\pi}{4}$, which is $\dfrac{20\pi}{4}$ or 5π, is an odd multiple of π.

$\dfrac{19\pi}{4} = 5\pi - \dfrac{\pi}{4}$

$\sin \dfrac{19\pi}{4} = \sin \left(5\pi - \dfrac{\pi}{4}\right)$ *Case 4, with A = $\frac{\pi}{4}$ and k = 3*

$\phantom{\sin \dfrac{19\pi}{4}} = \sin \dfrac{\pi}{4}$

c. cos (−410°)

The sum of −410° and 50° is a multiple of 360°.

$-410° = -360° - 50°$

$\cos (-410°) = \cos (-360° - 50°)$ *Case 3, with A = 50° and k = −1*

$ = \cos 50°$

d. $\tan \dfrac{37\pi}{6}$

$\dfrac{37\pi}{6}$ and $\dfrac{\pi}{6}$ differ by a multiple of 2π.

$\dfrac{37\pi}{6} = 3(2\pi) + \dfrac{\pi}{6}$ *Case 1, with A = $\frac{\pi}{6}$ and k = 3*

$\tan \dfrac{37\pi}{6} = \dfrac{\sin \dfrac{37\pi}{6}}{\cos \dfrac{37\pi}{6}}$ *Rewrite using a quotient identity since the symmetry identities are in terms of sine and cosine.*

$\phantom{\tan \dfrac{37\pi}{6}} = \dfrac{\sin \left(3(2\pi) + \dfrac{\pi}{6}\right)}{\cos \left(3(2\pi) + \dfrac{\pi}{6}\right)}$

$\phantom{\tan \dfrac{37\pi}{6}} = \dfrac{\sin \dfrac{\pi}{6}}{\cos \dfrac{\pi}{6}}$ or $\tan \dfrac{\pi}{6}$ *Quotient identity*

Teaching Tip Point out that in symmetry identities, angle *A* usually satisfies $0° \leq A \leq 90°$ or, in radian measure, $0 \leq A \leq \dfrac{\pi}{2}$. For most cases, *A* will then be a positive acute angle.

In-Class Example

Example 3
Express each value as a trigonometric function of an angle in Quadrant I.
a. cos 750°   **cos 30°**
b. $\sin \dfrac{29\pi}{6}$ $\sin \dfrac{\pi}{6}$
c. sin 930°   **−sin 30°**
d. $\tan \dfrac{25\pi}{4}$ $\tan \dfrac{\pi}{4}$

Case 3 of the Symmetry Identities can be written as the **opposite-angle identities** when $k = 0$.

Opposite-Angle Identities	The following trigonometric identities hold for all values of A. $$\sin(-A) = -\sin A$$ $$\cos(-A) = \cos A$$

The basic trigonometric identities can be used to simplify trigonometric expressions. Simplifying a trigonometric expression means that the expression is written using the fewest trigonometric functions possible and as algebraically simplified as possible. This may mean writing the expression as a numerical value.

Examples **④ Simplify $\sin x + \sin x \cot^2 x$.**

$$\begin{aligned}\sin x + \sin x \cot^2 x &= \sin x\,(1 + \cot^2 x) & \textit{Factor.}\\ &= \sin x \csc^2 x & \textit{Pythagorean identity: } 1 + \cot^2 x = \csc^2 x\\ &= \sin x \cdot \frac{1}{\sin^2 x} & \textit{Reciprocal identity}\\ &= \frac{1}{\sin x} \\ &= \csc x & \textit{Reciprocal identity}\end{aligned}$$

 ⑤ OPTICS Refer to the application at the beginning of the lesson.

a. **Simplify the formula $I = I_0 - \dfrac{I_0}{\csc^2 \theta}$.**

b. **Use the simplified formula to determine the intensity of light that passes through a second polarizing lens with axis at 30° to the original.**

a. $I = I_0 - \dfrac{I_0}{\csc^2 \theta}$

$I = I_0 - I_0 \sin^2 \theta$ *Reciprocal identity*

$I = I_0(1 - \sin^2 \theta)$ *Factor.*

$I = I_0 \cos^2 \theta$ $1 - \sin^2 \theta = \cos^2 \theta$

b. $I = I_0 \cos^2 30°$

$I = I_0 \left(\dfrac{\sqrt{3}}{2}\right)^2$

$I = \dfrac{3}{4} I_0$

The light has three-fourths the intensity it had before passing through the second polarizing lens.

Addressing Individual Needs

 Logical Have students write an expression that contains all six trigonometric functions and is equal to 3.
$\sin^2 x + \cos^2 x + \sec^2 x - \tan^2 x + \csc^2 x - \cot^2 x$

Communicating Mathematics

Read and study the lesson to answer each question. **2–5. See margin.**

1. **Find a counterexample** to show that the equation $1 - \sin x = \cos x$ is not an identity. **Sample answer: $x = 45°$**

2. **Explain** why the Pythagorean and opposite-angle identities are so named.

3. **Write** two reciprocal identities, one quotient identity, and one Pythagorean identity, each of which involves $\cot \theta$.

4. **Prove** that $\tan (-A) = -\tan A$ using the quotient and opposite-angle identities.

5. **You Decide** Claude and Rosalinda are debating whether an equation from their homework assignment is an identity. Claude says that since he has tried ten specific values for the variable and all of them worked, it must be an identity. Rosalinda explained that specific values could only be used as counterexamples to prove that an equation is not an identity. Who is correct? Explain your answer.

Guided Practice

Prove that each equation is *not* a trigonometric identity by producing a counterexample.

6. $\sin \theta + \cos \theta = \tan \theta$
   Sample answer: $\theta = 0°$

7. $\sec^2 x + \csc^2 x = 1$
   Sample answer: $x = 45°$

Use the given information to determine the exact trigonometric value.

8. $\cos \theta = \dfrac{2}{3}, 0° < \theta < 90°; \sec \theta$ $\dfrac{3}{2}$

9. $\cot \theta = -\dfrac{\sqrt{5}}{2}, \dfrac{\pi}{2} < \theta < \pi; \tan \theta$ $-\dfrac{2\sqrt{5}}{5}$

10. $-\dfrac{2\sqrt{6}}{5}$

10. $\sin \theta = -\dfrac{1}{5}, \pi < \theta < \dfrac{3\pi}{2}; \cos \theta$

11. $\dfrac{\sqrt{65}}{7}$

11. $\tan \theta = -\dfrac{4}{7}, 270° < \theta < 360°; \sec \theta$

Express each value as a trigonometric function of an angle in Quadrant I.

12. $\cos \dfrac{7\pi}{3}$ $\cos \dfrac{\pi}{3}$

13. $\csc (-330°)$ $\csc 30°$

Simplify each expression.

14. $\dfrac{\csc \theta}{\cot \theta}$ $\sec \theta$

15. $\cos x \csc x \tan x$ 1

16. $\cos x \cot x + \sin x$ $\csc x$

17. **Physics** When there is a current in a wire in a magnetic field, a force acts on the wire. The strength of the magnetic field can be determined using the formula $B = \dfrac{F \csc \theta}{I\ell}$, where F is the force on the wire, I is the current in the wire, ℓ is the length of the wire, and θ is the angle the wire makes with the magnetic field. Physics texts often write the formula as $F = I\ell B \sin \theta$. Show that the two formulas are equivalent. **See margin.**

Practice

Prove that each equation is not a trigonometric identity by producing a counterexample. **18–23 Sample answers are given.**

 A

18. $\sin \theta \cos \theta = \cot \theta$ **45°**

19. $\dfrac{\sec \theta}{\tan \theta} = \sin \theta$ **45°**

20. $\sec^2 x - 1 = \dfrac{\cos x}{\csc x}$ **30°**

21. $\sin x + \cos x = 1$ **30°**

22. $\sin y \tan y = \cos y$ **0°**

23. $\tan^2 A + \cot^2 A = 1$ **45°**

 www.amc.glencoe.com/self_check_quiz

Lesson 7-1 Basic Trigonometric Identities **427**

Error Analysis

Watch for students who do not apply the opposite-angle identities correctly.

Prevent by having students utilize the graph of the function or the unit circle in order to determine equivalent expressions.

Assignment Guide
Core: 19–59 odd, 60–69
Enriched: 18–52 even, 54–69

Additional Answers

2. Pythagorean identities are derived by applying the Pythagorean Theorem to a right triangle. The opposite angle identities are so named because $-A$ is the opposite of A.

3. $\tan \theta = \dfrac{1}{\cot \theta}, \cot \theta = \dfrac{1}{\tan \theta},$
   $\dfrac{\cos \theta}{\sin \theta} = \cot \theta,$
   $1 + \cot^2 \theta = \csc^2 \theta$

4. $\tan (-A) = \dfrac{\sin (-A)}{\cos (-A)}$
   $= \dfrac{-\sin A}{\cos A}$
   $= -\dfrac{\sin A}{\cos A}$
   $= -\tan A$

5. Rosalinda is correct; there may be other values for which the equation is not true.

Study Guide and Practice Masters, pp. 119–120

Additional Answer

17. $B = \dfrac{F \csc \theta}{I\ell}$

    $BI\ell = F \csc \theta$

    $F = \dfrac{BI\ell}{\csc \theta}$

    $F = BI\ell \left(\dfrac{1}{\csc \theta}\right)$

    $F = BI\ell \sin \theta$

Additional Answer

55. Let (x, y) be the point where the terminal side of A intersects the unit circle when A is in standard position. When A is reflected about the x-axis to obtain $-A$, the y-coordinate is multiplied by -1, but the x-coordinate is unchanged. So, $\sin(-A) = -y = -\sin A$ and $\cos(-A) = x = \cos A$.

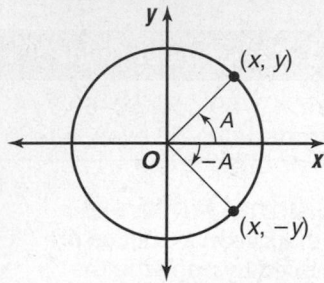

26. $\frac{4\sqrt{3}}{3}$

28. $\frac{\sqrt{5}}{3}$

29. $-\frac{\sqrt{2}}{3}$

30. $-\frac{3}{4}$

31. $\frac{\sqrt{2}}{4}$

32. $-\frac{3\sqrt{13}}{13}$

33. $-\frac{2\sqrt{6}}{7}$

34. $-3\sqrt{7}$

35. $-\frac{3}{5}$

36. $-\sqrt{65}$

37. $\frac{1}{2}$

24. Find a value of θ for which $\cos\left(\theta + \frac{\pi}{2}\right) \neq \cos\theta + \cos\frac{\pi}{2}$. **Sample answer: 0**

Use the given information to determine the exact trigonometric value.

B

25. $\sin\theta = \frac{2}{5}$, $0° < \theta < 90°$; $\csc\theta$ **$\frac{5}{2}$**

26. $\tan\theta = \frac{\sqrt{3}}{4}$, $0 < \theta < \frac{\pi}{2}$; $\cot\theta$

27. $\sin\theta = \frac{1}{4}$, $0 < \theta < \frac{\pi}{2}$; $\cos\theta$ **$\frac{\sqrt{15}}{4}$**

28. $\cos\theta = -\frac{2}{3}$, $90° < \theta < 180°$; $\sin\theta$

29. $\csc\theta = \frac{\sqrt{11}}{3}$, $\frac{\pi}{2} < \theta < \pi$; $\cot\theta$

30. $\sec\theta = -\frac{5}{4}$, $90° < \theta < 180°$; $\tan\theta$

31. $\sin\theta = -\frac{1}{3}$, $180° < \theta < 270°$; $\tan\theta$

32. $\tan\theta = \frac{2}{3}$, $\pi < \theta < \frac{3\pi}{2}$; $\cos\theta$

33. $\sec\theta = -\frac{7}{5}$, $180° < \theta < 270°$; $\sin\theta$

34. $\cos\theta = \frac{1}{8}$, $\frac{3\pi}{2} < \theta < 2\pi$; $\tan\theta$

35. $\cot\theta = -\frac{4}{3}$, $270° < \theta < 360°$; $\sin\theta$

36. $\cot\theta = -8$, $\frac{3\pi}{2} < \theta < 2\pi$; $\csc\theta$

37. If A is a second quadrant angle, and $\cos A = -\frac{\sqrt{3}}{4}$, find $\frac{\sec^2 A - \tan^2 A}{2\sin^2 A + 2\cos^2 A}$.

Express each value as a trigonometric function of an angle in Quadrant I.

38. $\sin 390°$ **$\sin 30°$**

39. $\cos\frac{27\pi}{8}$ **$-\cos\frac{3\pi}{8}$**

40. $\tan\frac{19\pi}{5}$ **$-\tan\frac{\pi}{5}$**

41. $\csc\frac{10\pi}{3}$ **$-\csc\frac{\pi}{3}$**

42. $\sec(-1290°)$ **$-\sec 30°$**

43. $\cot(-660°)$ **$\cot 60°$**

Simplify each expression.

44. $\frac{\sec x}{\tan x}$ **$\csc x$**

45. $\frac{\cot\theta}{\cos\theta}$ **$\csc\theta$**

46. $\frac{\sin(\theta + \pi)}{\cos(\theta - \pi)}$ **$\tan\theta$**

47. $(\sin x + \cos x)^2 + (\sin x - \cos x)^2$ **2**

48. $\sin x \cos x \sec x \cot x$ **$\cos x$**

49. $\cos x \tan x + \sin x \cot x$ **$\sin x + \cos x$**

C

50. $(1 + \cos\theta)(\csc\theta - \cot\theta)$ **$\sin\theta$**

51. $1 + \cot^2\theta - \cos^2\theta - \cos^2\theta\cot^2\theta$ **1**

52. $\frac{\sin x}{1 + \cos x} + \frac{\sin x}{1 - \cos x}$ **$2\csc x$**

53. $\cos^4\alpha + 2\cos^2\alpha\sin^2\alpha + \sin^4\alpha$ **1**

Applications and Problem Solving

54. **Optics** Refer to the equation derived in Example 5. What angle should the axes of two polarizing lenses make in order to block all light from passing through? **90°**

55. **Critical Thinking** Use the unit circle definitions of sine and cosine to provide a geometric interpretation of the opposite-angle identities. **See margin.**

56. Dermatology It has been shown that skin cancer is related to sun exposure. The rate W at which a person's skin absorbs energy from the sun depends on the energy S, in watts per square meter, provided by the sun, the surface area A exposed to the sun, the ability of the body to absorb energy, and the angle θ between the sun's rays and a line perpendicular to the body. The ability of an object to absorb energy is related to a factor called the *emissivity, e,* of the object. The emissivity can be calculated using the formula $e = \dfrac{W \sec \theta}{AS}$. **a. $W = eAS \cos \theta$**

a. Solve this equation for W. Write your answer using only $\sin \theta$ or $\cos \theta$.

b. Find W if $e = 0.80$, $\theta = 40°$, $A = 0.75$ m^2, and $S = 1000$ W/m^2. **459.63 W**

57. Physics A skier of mass m descends a θ-degree hill at a constant speed. When Newton's Laws are applied to the situation, the following system of equations is produced.

$$F_N - mg \cos \theta = 0$$
$$mg \sin \theta - \mu_k F_N = 0$$

where g is the acceleration due to gravity, F_N is the normal force exerted on the skier, and μ_k is the coefficient of friction. Use the system to define μ_k as a function of θ. **$\mu_k = \tan \theta$**

58. Geometry Show that the area of a regular polygon of n sides, each of length a, is given by $A = \dfrac{1}{4} na^2 \cot\left(\dfrac{180°}{n}\right)$. **See margin.**

59. Critical Thinking The circle at the right is a unit circle with its center at the origin. $\overline{AB}$ and $\overline{CD}$ are tangent to the circle. State the segments whose measures represent the ratios $\sin \theta$, $\cos \theta$, $\tan \theta$, $\sec \theta$, $\cot \theta$, and $\csc \theta$. Justify your answers. **See margin.**

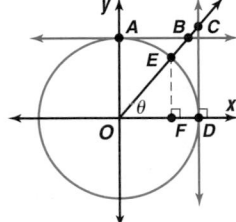

Mixed Review

60. Find $\text{Cos}^{-1}\left(-\dfrac{\sqrt{2}}{2}\right)$. *(Lesson 6-8)* **135°**

61. Graph $y = \cos\left(x - \dfrac{\pi}{6}\right)$. *(Lesson 6-5)* **See margin.**

62. Physics A pendulum 20 centimeters long swings 3°30′ on each side of its vertical position. Find the length of the arc formed by the tip of the pendulum as it swings. *(Lesson 6-1)* **2.44 cm**

63. Angle C of $\triangle ABC$ is a right angle. Solve the triangle if $A = 20°$ and $c = 35$. *(Lesson 5-4)* **$a = 12.0$, $B = 70°$, $b = 32.9$**

64. Find all the rational roots of the equation $2x^3 + x^2 - 8x - 4 = 0$. *(Lesson 4-4)* **$-2, -\dfrac{1}{2}, 2$**

65. Solve $2x^2 + 7x - 4 = 0$ by completing the square. *(Lesson 4-2)* **$-4, 0.5$**

66. Determine whether $f(x) = 3x^3 + 2x - 5$ is continuous or discontinuous at $x = 5$. *(Lesson 3-5)* **continuous**

Extra Practice See p. A38.   *Lesson 7-1 Basic Trigonometric Identities* **429**

Extra Credit

Express $\cos x$ in terms of each of the other trigonometric functions. Each expression should contain exactly one function other than the cosine. Assume that x is in Quadrant I. **Sample answer:** $\sqrt{1 - \sin^2 x}$, $\dfrac{\sqrt{\csc^2 x - 1}}{\csc x}$, $\dfrac{1}{\sec x}$, $\sqrt{\dfrac{1}{\tan^2 x + 1}}$, $\dfrac{\cot x}{\sqrt{1 + \cot^2 x}}$

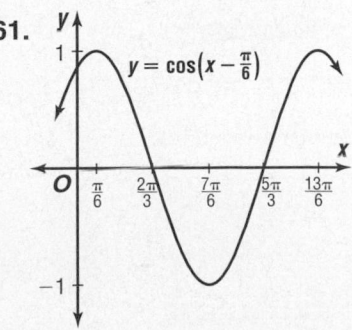

Open-Ended Assessment

Writing Have students write examples of trigonometric identities and describe how each can be used to find the value of one trigonometric function from the value of another.

---

67. Solve the system of equations algebraically. *(Lesson 2-2)* **(2, −5, −3)**
$$x + y - 2z = 3$$
$$-4x - y - z = 0$$
$$-x - 5y + 4z = 11$$

68. Write the slope-intercept form of the equation of the line that passes through points at (5, 2) and (−4, 4). *(Lesson 1-4)* $y = -\frac{2}{9}x + \frac{28}{9}$

69. **SAT/ACT Practice** Triangle ABC is inscribed in circle O, and $\overleftrightarrow{CD}$ is tangent to circle O at point C. If $m\angle BCD = 40°$, find $m\angle A$. **C**

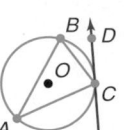

A 60°   **B** 50°   **C** 40°   **D** 30°   **E** 20°

CAREER CHOICES

● Cartographer ●

Do maps fascinate you? Do you like drawing, working with computers, and geography?

You may want to consider a career in cartography. As a cartographer, you would make maps, charts, and drawings. Cartography has changed a great deal with modern technology. Computers and satellites have become powerful new tools in making maps. As a cartographer, you may work with manual drafting tools as well as computer software designed for making maps.

The image at the right shows how a cartographer uses a three-dimensional landscape to create a two-dimensional topographic map.

There are several areas of specialization in the field of cartography. Some of these include making maps from political boundaries and natural features, making maps from aerial photographs, and correcting original maps.

CAREER OVERVIEW

Degree Preferred:
bachelor's degree in engineering or a physical science

Related Courses:
mathematics, geography, computer science, mechanical drawing

Outlook:
slower than average through 2006

inter NET CONNECTION For more information on careers in cartography, visit: **www.amc.glencoe.com**

Enrichment Masters, p. 73

7-1

NAME _____ DATE _____ PERIOD _____

Enrichment

The Physics of Soccer

Recall from Lesson 7-1 that the formula for the maximum height h of a projectile is $h = \frac{v_0^2 \sin^2 \theta}{2g}$, where θ is the measure of the angle of elevation in degrees, v_0 is the initial velocity in feet per second, and g is the acceleration due to gravity in feet per second squared.

Solve. Give answers to the nearest tenth.

1. A soccer player kicks a ball at an initial velocity of 60 ft/s and an angle of elevation of 40°. The acceleration due to gravity is 32 ft/s². Find the maximum height reached by the ball.
23.2 ft

2. With what initial velocity must you kick a ball at an angle of 35° in order for it to reach a maximum height of 20 ft?
62.4 ft/s

The distance d that a projected object travels is given by the formula $d = \frac{2v_0^2 \sin \theta \cos \theta}{g}$.

3. Find the distance traveled by the ball described in Exercise 1.
110.8 ft

In order to kick a ball the greatest possible distance at a given initial velocity, a soccer player must maximize $d = \frac{2v_0^2 \sin \theta \cos \theta}{g}$. Since 2, v_0, and g are constants, this means the player must maximize $\sin \theta \cos \theta$.

sin 0°cos 0° =	− sin 90°cos 90° = 0
sin 10°cos 10° =	− sin 80°cos 80° = 0.1710
sin 20°cos 20° =	− sin 70°cos 70° = 0.3214

4. Use the patterns in the table to hypothesize a value of θ for which $\sin \theta \cos \theta$ will be maximal. Use a calculator to check your hypothesis. At what angle should the player kick the ball to achieve the greatest distance?
45°

© Glencoe/McGraw-Hill   T73   *Advanced Mathematical Concepts*

CAREER CHOICES   CARTOGRAPHER

Math Connection
- What do you think is the first step a cartographer takes in mapping a three-dimensional landscape to a two-dimensional diagram such as a topographical map? **Overlay a grid onto the surface of the landscape.**
- How do the colors in the three-dimensional landscape correspond to the topographical map above it? What do you think the colors mean? **Each color on the three-dimensional landscape corresponds to a color "ring" in the topographical map. Each color corresponds to a specific height and all areas along the "ring" have that approximate height.**

Verifying Trigonometric Identities

OBJECTIVES
- Use the basic trigonometric identities to verify other identities.
- Find numerical values of trigonometric functions.

PROBLEM SOLVING While working on a mathematics assignment, a group of students derived an expression for the length of a ladder that, when held horizontally, would turn from a 5-foot wide corridor into a 7-foot wide corridor. They determined that the maximum length ℓ of a ladder that would fit was given by $\ell(\theta) = \dfrac{7 \sin \theta + 5 \cos \theta}{\sin \theta \cos \theta}$, where θ is the angle that the ladder makes with the outer wall of the 5-foot wide corridor. When their teacher worked the problem, she concluded that $\ell(\theta) = 7 \sec \theta + 5 \csc \theta$. Are the two expressions for $\ell(\theta)$ equivalent? *This problem will be solved in Example 2.*

Verifying trigonometric identities algebraically involves transforming one side of the equation into the same form as the other side by using the basic trigonometric identities and the properties of algebra. Either side may be transformed into the other side, or both sides may be transformed separately into forms that are the same.

Suggestions for Verifying Trigonometric Identities
- Transform the more complicated side of the equation into the simpler side.
- Substitute one or more basic trigonometric identities to simplify expressions.
- Factor or multiply to simplify expressions.
- Multiply expressions by an expression equal to 1.
- Express all trigonometric functions in terms of sine and cosine.

You cannot add or subtract quantities from each side of an unverified identity, nor can you perform any other operation on each side, as you often do with equations. An unverified identity is not an equation, so the properties of equality do not apply.

Example **1** Verify that $\sec^2 x - \tan x \cot x = \tan^2 x$ is an identity.

Since the left side is more complicated, transform it into the expression on the right.

$$\sec^2 x - \tan x \cot x \overset{?}{=} \tan^2 x$$
$$\sec^2 x - \tan x \cdot \frac{1}{\tan x} \overset{?}{=} \tan^2 x \quad \cot x = \frac{1}{\tan x}$$
$$\sec^2 x - 1 \overset{?}{=} \tan^2 x \quad Multiply.$$
$$\tan^2 x + 1 - 1 \overset{?}{=} \tan^2 x \quad \sec^2 x = \tan^2 x + 1$$
$$\tan^2 x = \tan^2 x \quad Simplify.$$

We have transformed the left side into the right side. The identity is verified.

1 FOCUS

 5-Minute Check
Lesson 7-1

1. Prove that $\sin x + \cos x = 1$ is *not* a trigonometric identity by producing a counterexample.
   Suppose $x = \frac{\pi}{4}$.
   $$\sin \frac{\pi}{4} + \cos \frac{\pi}{4} \overset{?}{=} 1$$
   $$\frac{\sqrt{2}}{2} + \frac{\sqrt{2}}{2} \overset{?}{=} 1$$
   $$\frac{2\sqrt{2}}{2} \overset{?}{=} 1$$
   $$\sqrt{2} \neq 1$$
   Therefore, $\sin x + \cos x = 1$ is not a trigonometric identity.

Use the given information to find the trigonometric value.

2. If $\csc \theta = \frac{8}{5}$, find $\sin \theta$. $\frac{5}{8}$

3. If $\cot \theta = \frac{12}{5}$, find $\sin \theta$. $\pm \frac{5}{13}$

4. Express $\cos 490°$ as a trigonometric function of an angle in Quadrant I. $-\cos 50°$

5. Simplify $\cos^2 x \tan^2 x + \cos^2 x$. **1**

Motivating the Lesson

Discussion Have students read the first paragraph of the lesson. Then, ask them to explain how trigonometric identities can be used to solve real-world problems involving triangles.

Resource Manager

Reproducible Masters
- *Study Guide and Practice,* pp. 121–122
- *Enrichment,* p. 74
- *Assessment and Evaluation,* p. 173
- *Graphing Calculator Keystrokes,* p. 26

Transparencies
- 5-Minute Check, 7-2
- Teaching, 7-2

Technology/Multimedia
- Advanced MathPASS, Lesson 15

In-Class Examples

Example 1

Verify that $\dfrac{1 + \tan^2 x}{\csc x \sec x} = \tan x$

is an identity.

$$\dfrac{1 + \tan^2 x}{\csc x \sec x} \overset{?}{=} \tan x$$

$$\dfrac{\sec^2 x}{\csc x \sec x} \overset{?}{=} \tan x$$

$$\dfrac{\sec x}{\csc x} \overset{?}{=} \tan x$$

$$\dfrac{\frac{1}{\cos x}}{\frac{1}{\sin x}} \overset{?}{=} \tan x$$

$$\dfrac{\sin x}{\cos x} \overset{?}{=} \tan x$$

$$\tan x = \tan x$$

Example 2

Problem Solving Show another verification of the identity in Example 2 in the Student Edition.

$$\dfrac{7 \sin \theta + 5 \cos \theta}{\sin \theta \cos \theta}$$

$$= \dfrac{7 \sin \theta}{\sin \theta \cos \theta} + \dfrac{5 \cos \theta}{\sin \theta \cos \theta}$$

$$= \dfrac{7}{\cos \theta} + \dfrac{5}{\sin \theta}$$

$$= 7 \sec \theta + 5 \csc \theta$$

Teaching Tip For Example 3, point out that $\sin^2 \theta + \cos^2 \theta = 1$ also can be stated as $\sin^2 \theta = 1 - \cos^2 \theta$ and $\cos^2 \theta = 1 - \sin^2 \theta$. These additional forms are useful when simplifying trigonometric identities.

Teaching Tip For Example 4, remind students that the division of a simple fraction expressed in trigonometric form by another simple fraction in trigonometric form is solved the same way as division of single fractions in arithmetic. Invert the fraction following the division sign and factor or multiply.

Graphing Calculator Exploration
Point out that by setting each side of an equation equal to y, students are creating a system of equations. When the graphs of the system match, the original equation may be an identity.

Examples

2 **PROBLEM SOLVING** Verify that the two expressions for $\ell(\theta)$ in the application at the beginning of the lesson are equivalent. That is, verify that $\dfrac{7 \sin \theta + 5 \cos \theta}{\sin \theta \cos \theta} = 7 \sec \theta + 5 \csc \theta$ is an identity.

Begin by writing the right side in terms of sine and cosine.

$$\dfrac{7 \sin \theta + 5 \cos \theta}{\sin \theta \cos \theta} \overset{?}{=} 7 \sec \theta + 5 \csc \theta$$

$$\dfrac{7 \sin \theta + 5 \cos \theta}{\sin \theta \cos \theta} \overset{?}{=} \dfrac{7}{\cos \theta} + \dfrac{5}{\sin \theta} \qquad \sec \theta = \dfrac{1}{\cos \theta}, \; \csc \theta = \dfrac{1}{\sin \theta}$$

$$\dfrac{7 \sin \theta + 5 \cos \theta}{\sin \theta \cos \theta} \overset{?}{=} \dfrac{7 \sin \theta}{\sin \theta \cos \theta} + \dfrac{5 \cos \theta}{\sin \theta \cos \theta} \qquad \textit{Find a common denominator.}$$

$$\dfrac{7 \sin \theta + 5 \cos \theta}{\sin \theta \cos \theta} = \dfrac{7 \sin \theta + 5 \cos \theta}{\sin \theta \cos \theta} \qquad \textit{Simplify.}$$

The students and the teacher derived equivalent expressions for $\ell(\theta)$, the length of the ladder.

3 Verify that $\dfrac{\sin A}{\csc A} + \dfrac{\cos A}{\sec A} = \csc^2 A - \cot^2 A$ is an identity.

Since the two sides are equally complicated, we will transform each side independently into the same form.

$$\dfrac{\sin A}{\csc A} + \dfrac{\cos A}{\sec A} \overset{?}{=} \csc^2 A - \cot^2 A$$

$$\dfrac{\sin A}{\frac{1}{\sin A}} + \dfrac{\cos A}{\frac{1}{\cos A}} \overset{?}{=} (1 + \cot^2 A) - \cot^2 A \qquad \textit{Quotient identities; Pythagorean identity}$$

$$\sin^2 A + \cos^2 A \overset{?}{=} 1 \qquad \textit{Simplify.}$$

$$1 = 1 \qquad \sin^2 A + \cos^2 A = 1$$

The techniques that you use to verify trigonometric identities can also be used to simplify trigonometric equations. Sometimes you can change an equation into an equivalent equation involving a single trigonometric function.

Example **4** Find a numerical value of one trigonometric function of x if $\dfrac{\cot x}{\cos x} = 2$.

You can simplify the trigonometric expression on the left side by writing it in terms of sine and cosine.

$$\dfrac{\cot x}{\cos x} = 2$$

$$\dfrac{\frac{\cos x}{\sin x}}{\cos x} = 2 \qquad \cot x = \dfrac{\cos x}{\sin x}$$

$$\dfrac{\cos x}{\sin x} \cdot \dfrac{1}{\cos x} = 2 \qquad \textit{Definition of division}$$

Addressing Individual Needs

Kinesthetic Have students work in groups to model signs of trigonometric functions. Each group should use masking tape or rope on the ground to model the x- and y-axes. Have one student place a meterstick on the coordinate plane with one end on the origin. Other students should state the signs of the six trigonometric functions for the angle formed.

$$\frac{1}{\sin x} = 2 \quad \textit{Simplify.}$$

$$\csc x = 2 \quad \frac{1}{\sin x} = \csc x$$

Therefore, if $\frac{\cot x}{\cos x} = 2$, then $\csc x = 2$.

You can use a graphing calculator to investigate whether an equation may be an identity.

GRAPHING CALCULATOR EXPLORATION

♦ Graph both sides of the equation as two separate functions. For example, to test $\sin^2 x = (1 - \cos x)(1 + \cos x)$, graph $y_1 = \sin^2 x$ and $y_2 = (1 - \cos x)(1 + \cos x)$ on the same screen.

♦ If the two graphs do not match, then the equation is not an identity.

♦ If the two sides appear to match in every window you try, then the equation may be an identity.

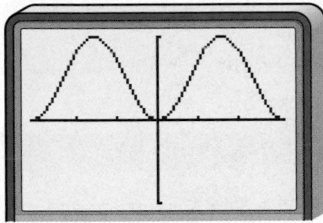

$[-\pi, \pi]$ scl:1 by $[-1, 1]$ scl:1

TRY THESE Determine whether each equation could be an identity. Write *yes* or *no*.

1. $\sin x \csc x - \sin^2 x = \cos^2 x$ **yes**

2. $\sec x + \csc x = 1$ **no**

3. $\sin x - \cos x = \frac{1}{\csc x - \sec x}$ **no**

WHAT DO YOU THINK?

4. If the two sides appear to match in every window you try, does that prove that the equation is an identity? Justify your answer.

5. Graph the function $f(x) = \frac{\sec x - \cos x}{\tan x}$. What simpler function could you set equal to $f(x)$ in order to obtain an identity? **sin x**

4. No; it is impossible to look at every window since there are an infinite number. The only way an identity can be proven is by showing algebraically that the general case is true.

CHECK FOR UNDERSTANDING

Communicating Mathematics

Read and study the lesson to answer each question. **2–3. See margin.**

1. **Write** a trigonometric equation that is not an identity. Explain how you know it is not an identity. **Answers will vary.**

2. **Explain** why you cannot square each side of the equation when verifying a trigonometric identity.

3. **Discuss** why both sides of a trigonometric identity are often rewritten in terms of sine and cosine.

Lesson 7-2 Verifying Trigonometric Identities **433**

Additional Answers

2. Sample answer: squaring each side can turn two unequal quantities into equal quantities. For example, $-1 \neq 1$, but $(-1)^2 = 1^2$.

3. Sample answer: they are the trigonometric functions with which most people are most familiar.

Additional Answer

12. $\dfrac{I\cos\theta}{R^2} \stackrel{?}{=} \dfrac{I\cot\theta}{R^2\csc\theta}$

$\dfrac{I\cos\theta}{R^2} \stackrel{?}{=} \dfrac{I\frac{\cos\theta}{\sin\theta}}{R^2\frac{1}{\sin\theta}}$

$\dfrac{I\cos\theta}{R^2} \stackrel{?}{=} \dfrac{I\frac{\cos\theta}{\sin\theta}}{R^2\frac{1}{\sin\theta}} \cdot \dfrac{\sin\theta}{\sin\theta}$

$\dfrac{I\cos\theta}{R^2} = \dfrac{I\cos\theta}{R^2}$

4. *Math Journal* **Create** your own trigonometric identity that contains at least three different trigonometric functions. Explain how you created it. Give it to one of your classmates to verify. Compare and contrast your classmate's approach with your approach. **Answers will vary.**

Guided Practice Verify that each equation is an identity. **5–9. See Answer Appendix.**

5. $\cos x = \dfrac{\cot x}{\csc x}$

6. $\dfrac{1}{\tan x + \sec x} = \dfrac{\cos x}{\sin x + 1}$

7. $\csc\theta - \cot\theta = \dfrac{1}{\csc\theta + \cot\theta}$

8. $\sin\theta\tan\theta = \sec\theta - \cos\theta$

9. $(\sin A - \cos A)^2 = 1 - 2\sin^2 A \cot A$

Find a numerical value of one trigonometric function of *x*.
Sample answer:

10. $\tan x = \dfrac{1}{4}\sec x$ $\sin x = \dfrac{1}{4}$

11. $\cot x + \sin x = -\cos x \cot x$
Sample answer: $\cos x = -1$

12. **Optics** The amount of light that a source provides to a surface is called the *illuminance*. The illuminance E in foot candles on a surface that is R feet from a source of light with intensity I candelas is $E = \dfrac{I\cos\theta}{R^2}$, where θ is the measure of the angle between the direction of the light and a line perpendicular to the surface being illuminated. Verify that $E = \dfrac{I\cot\theta}{R^2\csc\theta}$ is an equivalent formula. **See margin.**

EXERCISES

Practice Verify that each equation is an identity. **13–28. See Answer Appendix.**

A 13. $\tan A = \dfrac{\sec A}{\csc A}$

14. $\cos\theta = \sin\theta\cot\theta$

15. $\sec x - \tan x = \dfrac{1 - \sin x}{\cos x}$

16. $\dfrac{1 + \tan x}{\sin x + \cos x} = \sec x$

B 17. $\sec x \csc x = \tan x + \cot x$

18. $\sin\theta + \cos\theta = \dfrac{2\sin^2\theta - 1}{\sin\theta - \cos\theta}$

19. $(\sin A + \cos A)^2 = \dfrac{2 + \sec A \csc A}{\sec A \csc A}$

20. $(\sin\theta - 1)(\tan\theta + \sec\theta) = -\cos\theta$

21. $\dfrac{\cos y}{1 - \sin y} = \dfrac{1 + \sin y}{\cos y}$

22. $\cos\theta\cos(-\theta) - \sin\theta\sin(-\theta) = 1$

23. $\csc x - 1 = \dfrac{\cot^2 x}{\csc x + 1}$

24. $\cos B \cot B = \csc B - \sin B$

25. $\sin\theta\cos\theta\tan\theta + \cos^2\theta = 1$

26. $(\csc x - \cot x)^2 = \dfrac{1 - \cos x}{1 + \cos x}$

27. $\sin x + \cos x = \dfrac{\cos x}{1 - \tan x} + \dfrac{\sin x}{1 - \cot x}$

28. Show that $\sin\theta + \cos\theta + \tan\theta\sin\theta = \sec\theta + \cos\theta\tan\theta$.

 www.amc.glencoe.com/self_check_quiz

From the Classroom of...

Dr. Stevan R. Peters
R. McKee Voc/Tech High School
Staten Island, New York

"To strengthen students' technology skills, have students create a PowerPoint® presentation of a problem with a multi-step solution. Ask students to use transitions, bullets, and sound in their presentations."

Find a numerical value of one trigonometric function of x.

29–34. Sample answers given.

C

29. $\dfrac{\csc x}{\cot x} = \sqrt{2}$ **sec x = $\sqrt{2}$**

30. $\dfrac{1 + \tan x}{1 + \cot x} = 2$ **tan x = 2**

31. $\dfrac{1}{\cot x} - \dfrac{\sec x}{\csc x} = \cos x$ **cos x = 0**

32. $\dfrac{1 + \cos x}{\sin x} + \dfrac{\sin x}{1 + \cos x} = 4$ **sin x = $\frac{1}{2}$**

33. $\cos^2 x + 2\sin x - 2 = 0$ **sin x = 1**

34. $\csc x - \sin x \tan x + \cos x$ **cot x = 1**

35. If $\dfrac{\tan^3 \theta - 1}{\tan \theta - 1} - \sec^2 \theta - 1 = 0$, find $\cot \theta$. **1**

Graphing Calculator

Use a graphing calculator to determine whether each equation could be an identity.

36. $\dfrac{1}{\sin^2 x} + \dfrac{1}{\cos^2 x} = 1$ **no**

37. $\cos \theta(\cos \theta - \sec \theta) = -\sin^2 \theta$ **yes**

38. $2\sin A + (1 - \sin A)^2 = 2 - \cos^2 A$ **yes**

39. $\dfrac{\sin^3 x - \cos^3 x}{\sin x - \cos x} = \sin^2 x + \cos^2 x$ **no**

Applications and Problem Solving

Real World Application

40. **Electronics** When an alternating current of frequency f and peak current I_0 passes through a resistance R, then the power delivered to the resistance at time t seconds is $P = I_0^2 R \sin^2 2\pi ft$. **a. $P = I_0^2 R(1 - \cos^2 2\pi ft)$**

a. Write an expression for the power in terms of $\cos^2 2\pi ft$.

b. Write an expression for the power in terms of $\csc^2 2\pi ft$. **$P = \dfrac{I_0^2 R}{\csc^2 2\pi ft}$**

41. **Critical Thinking** Let $x = \dfrac{1}{2}\tan \theta$ where $-\dfrac{\pi}{2} < \theta < \dfrac{\pi}{2}$. Write $f(x) = \dfrac{x}{\sqrt{1 + 4x^2}}$ in terms of a single trigonometric function of θ. **$\frac{1}{2}\sin \theta$**

42. **Spherical Geometry** Spherical geometry is the geometry that takes place on the surface of a sphere. A line segment on the surface of the sphere is measured by the angle it subtends at the center of the sphere. Let a, b, and c be the sides of a right triangle on the surface of the sphere. Let the angles opposite those sides be α, β, and $\gamma = 90°$, respectively. The following equations are true:

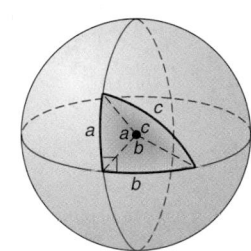

β is the Greek letter beta and γ is the Greek letter gamma.

$\sin a = \sin \alpha \sin c$

$\cos b = \dfrac{\cos \beta}{\sin \alpha}$

$\cos c = \cos a \cos b.$

Show that $\cos \beta = \tan a \cot c$. **See margin.**

43. **Physics** When a projectile is fired from the ground, its height y and horizontal displacement x are related by the equation $y = \dfrac{-gx^2}{2v_0^2 \cos^2 \theta} + \dfrac{x \sin \theta}{\cos \theta}$, where v_0 is the initial velocity of the projectile, θ is the angle at which it was fired, and g is the acceleration due to gravity. Rewrite this equation so that $\tan \theta$ is the only trigonometric function that appears in the equation.

$y = -\dfrac{gx^2}{2v_0^2}(1 + \tan^2 \theta) + x\tan \theta$

Teaching Tip Another way to verify trigonometric identities is to use boolean logic with the graphing calculator. For example, enter an equation from Exercises 36–39 in the **Y =** list. Examine the **TABLE**. Everywhere Y = 1, the equation is true. If Y = 0, the equation is false.

Additional Answer

42. $\sin a = \sin \alpha \sin c \Rightarrow \sin \alpha = \dfrac{\sin a}{\sin c}$

$\cos b = \dfrac{\cos \beta}{\sin \alpha} \Rightarrow \cos \beta = \sin \alpha \cos b$

$\cos c = \cos a \cos b \Rightarrow \cos b = \dfrac{\cos c}{\cos a}$

Then $\cos \beta = \sin \alpha \cos b$

$= \dfrac{\sin a}{\sin c} \cdot \dfrac{\cos c}{\cos a}$

$= \dfrac{\sin a}{\cos a} \cdot \dfrac{\cos c}{\sin c}$

$= \tan a \cot c$

Open-Ended Assessment
Speaking Have students explain why $\sin x = \cos x \tan x$ is an identity, but $\sin x = \frac{1}{2}$ is not.

$\sin x = \cos x \tan x$ **is true for all values of x for which the expression is defined.** $\sin x = \frac{1}{2}$ **is only true for $x = 30° + 360k°$ and $x = 150° + 360k°$.**

 Chapter 7, Quiz A (Lessons 7-1 and 7-2) is available in the *Assessment and Evaluation Masters*, p. 173.

Additional Answer
45. By the Law of Sines,
$$\frac{b}{\sin \beta} = \frac{a}{\sin \alpha}, \text{ so } b = \frac{a \sin \beta}{\sin \alpha}.$$
Then
$$A = \frac{1}{2}ab \sin \gamma$$
$$A = \frac{1}{2}a\left(\frac{a \sin \beta}{\sin \alpha}\right) \sin \gamma$$
$$A = \frac{a^2 \sin \beta \sin \gamma}{2 \sin \alpha}$$
$$A = \frac{a^2 \sin \beta \sin \gamma}{2 \sin (180° - (\beta + \gamma))}$$
$$A = \frac{a^2 \sin \beta \sin \gamma}{2 \sin (\beta + \gamma)}$$

---

44. Critical Thinking Consider a circle O with radius 1. $\overline{PA}$ and $\overline{TB}$ are each perpendicular to $\overline{OB}$. Determine the area of $ABTP$ as a product of trigonometric functions of θ.
Sample answer: $\frac{1}{2} \tan \theta \sin^2 \theta$

45. Geometry Let a, b, and c be the sides of a triangle. Let α, β, and γ be the respective opposite angles. Show that the area A of the triangle is given by
$$A = \frac{a^2 \sin \beta \sin \gamma}{2 \sin (\beta + \gamma)}.$$ **See margin.**

Mixed Review

46. Simplify $\frac{\tan x + \cos x + \sin x \tan x}{\sec x + \tan x}$. *(Lesson 7-1)* **1**

47. Write an equation of a sine function with amplitude 2, period 180°, and phase shift 45°. *(Lesson 6-5)* $y = \pm 2\sin (2x - 90°)$

48. 168°45′ **48.** Change $\frac{15\pi}{16}$ radians to degree measure to the nearest minute. *(Lesson 6-1)*

49. Solve $\sqrt[3]{3y - 1} - 2 = 0$. *(Lesson 4-7)* **3**

50. Determine the equations of the vertical and horizontal asymptotes, if any, of $f(x) = \frac{3x}{x + 1}$. *(Lesson 3-7)* $x = -1, y = 3$

51. Manufacturing The Simply Sweats Corporation makes high quality sweatpants and sweatshirts. Each garment passes through the cutting and sewing departments of the factory. The cutting and sewing departments have 100 and 180 worker-hours available each week, respectively. The fabric supplier can provide 195 yards of fabric each week. The hours of work and yards of fabric required for each garment are shown in the table below. If the profit from a sweatshirt is $5.00 and the profit from a pair of sweatpants is $4.50, how many of each should the company make for maximum profit? *(Lesson 2-7)* **40 shirts, 40 pants**

Simply Sweats Corporation
"Quality Sweatpants and Sweatshirts"

Clothing	Cutting	Sewing	Fabric
Shirt	1 h	2.5 h	1.5 yd
Pants	1.5 h	2 h	3 yd

52. {16}, {−4,4}; no; 16 is paired with two elements of the range

52. State the domain and range of the relation $\{(16,-4),(16, 4)\}$. Is this relation a function? Explain. *(Lesson 1-1)*

53. SAT/ACT Practice Divide $\frac{a - b}{a + b}$ by $\frac{b - a}{b + a}$. **D**

A 1

B $\frac{(a - b)^2}{(a + b)^2}$

C $\frac{1}{a^2 - b^2}$

D −1

E 0

436 *Chapter 7 Trigonometric Identities and Equations*

Extra Practice See p. A38.

Extra Credit

Verify that $\dfrac{\sin x}{1 -(-\cos x)} = \dfrac{1-\cos x}{-(-\sin x)}$ is an identity.

$$\frac{\sin x}{1 -(-\cos x)} \stackrel{?}{=} \frac{1 - \cos x}{-(- \sin x)}$$

$$\frac{1 - \cos x}{1 - \cos x} \cdot \frac{\sin x}{1 + \cos x} \stackrel{?}{=} \frac{1 - \cos x}{-(- \sin x)}$$

$$\frac{(1 - \cos x) \sin x}{\sin^2 x} \stackrel{?}{=} \frac{1 - \cos x}{-(-\sin x)}$$

$$\frac{1 - \cos x}{\sin x} \stackrel{?}{=} \frac{1 - \cos x}{-(-\sin x)}$$

$$\frac{1 - \cos x}{-(-\sin x)} = \frac{1 - \cos x}{-(-\sin x)}$$

Sum and Difference Identities

OBJECTIVE
- Use the sum and difference identities for the sine, cosine, and tangent functions.

 BROADCASTING Have you ever had trouble tuning in your favorite radio station? Does the picture on your TV sometimes appear blurry? Sometimes these problems are caused by *interference*. Interference can result when two waves pass through the same space at the same time. The two kinds of interference are:

- *constructive interference,* which occurs if the amplitude of the sum of the waves is greater than the amplitudes of the two component waves, and
- *destructive interference,* which occurs if the amplitude of the sum is less than the amplitudes of the component waves.

What type of interference results when a signal modeled by the equation $y = 20 \sin(3t + 45°)$ is combined with a signal modeled by the equation $y = 20 \sin(3t + 225°)$? *This problem will be solved in Example 4.*

Consider two angles α and β in standard position. Let the terminal side of α intersect the unit circle at point $A(\cos \alpha, \sin \alpha)$. Let the terminal side of β intersect the unit circle at $B(\cos \beta, \sin \beta)$. We will calculate $(AB)^2$ in two different ways.

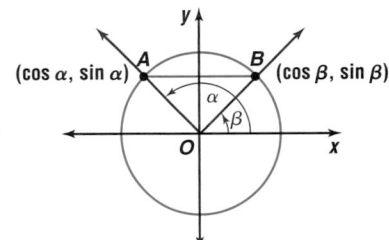

Look Back
You can refer to Lesson 5-8 to review the Law of Cosines.

First use the Law of Cosines.
$(AB)^2 = (OA)^2 + (OB)^2 - 2(OA)(OB) \cos (\alpha - \beta)$ $m\angle BOA = \alpha - \beta$
$(AB)^2 = 1^2 + 1^2 - 2(1)(1) \cos (\alpha - \beta)$ $OA = OB = 1$
$(AB)^2 = 2 - 2 \cos (\alpha - \beta)$ *Simplify.*

Now use the distance formula.
$(AB)^2 = (\cos \alpha - \cos \beta)^2 + (\sin \alpha - \sin \beta)^2$
$(AB)^2 = \cos^2 \alpha - 2 \cos \alpha \cos \beta + \cos^2 \beta + \sin^2 \alpha - 2 \sin \alpha \sin \beta + \sin^2 \beta$
$(AB)^2 = (\cos^2 \alpha + \sin^2 \alpha) + (\cos^2 \beta + \sin^2 \beta) - 2(\cos \alpha \cos \beta + \sin \alpha \sin \beta)$
$(AB)^2 = 1 + 1 - 2(\cos \alpha \cos \beta + \sin \alpha \sin \beta)$ $\cos^2 a + \sin^2 a = 1$
$(AB)^2 = 2 - 2(\cos \alpha \cos \beta + \sin \alpha \sin \beta)$ *Simplify.*

Set the two expressions for $(AB)^2$ equal to each other.
$2 - 2 \cos (\alpha - \beta) = 2 - 2(\cos \alpha \cos \beta + \sin \alpha \sin \beta)$
  $-2 \cos (\alpha - \beta) = -2(\cos \alpha \cos \beta + \sin \alpha \sin \beta)$ *Subtract 2 from each side.*
    $\cos (\alpha - \beta) = \cos \alpha \cos \beta + \sin \alpha \sin \beta$ *Divide each side by −2.*

This equation is known as the **difference identity for cosine.**

---

1 FOCUS

 5-Minute Check
Lesson 7-2

Verify that each equation is an identity.

1. $\sin x \tan x = \sec x - \cos x$

$\sin x \tan x \stackrel{?}{=} \sec x - \cos x$

$\sin x \cdot \dfrac{\sin x}{\cos x} \stackrel{?}{=} \sec x - \cos x$

$\dfrac{\sin^2 x}{\cos x} \stackrel{?}{=} \sec x - \cos x$

$\dfrac{1 - \cos^2 x}{\cos x} \stackrel{?}{=} \sec x - \cos x$

$\dfrac{1}{\cos x} - \dfrac{\cos^2 x}{\cos x} \stackrel{?}{=} \sec x - \cos x$

$\sec x - \cos x = \sec x - \cos x$

2. $1 = \tan x \cos x \csc x$

$1 \stackrel{?}{=} \tan x \cos x \csc x$

$1 \stackrel{?}{=} \dfrac{\sin x}{\cos x} \cdot \cos x \cdot \dfrac{1}{\sin x}$

$1 \stackrel{?}{=} \dfrac{\sin x}{\sin x} \cdot \dfrac{\cos x}{\cos x}$

$1 = 1$

3. $\dfrac{1}{1 - \cos^2 A} = 1 + \cot^2 A$

$\dfrac{1}{1 - \cos^2 A} \stackrel{?}{=} 1 + \cot^2 A$

$\dfrac{1}{1 - \cos^2 A} \stackrel{?}{=} 1 + \dfrac{\cos^2 A}{\sin^2 A}$

$\dfrac{1}{1 - \cos^2 A} \stackrel{?}{=} \dfrac{\sin^2 A + \cos^2 A}{\sin^2 A}$

$\dfrac{1}{1 - \cos^2 A} \stackrel{?}{=} \dfrac{1}{\sin^2 A}$

$\dfrac{1}{1 - \cos^2 A} = \dfrac{1}{1 - \cos^2 A}$

4. Find a numerical value of one trigonometric function of x if $3 \cos x \csc x = 1$.

$\cot x = \dfrac{1}{3}$ or $\tan x = 3$

Motivating the Lesson
Hands-On Activity Ask students to use calculators to check whether $\sin x + \sin y = \sin (x + y)$ and $\sin x - \sin y = \sin (x - y)$ are true with these examples.
  $\sin 50° = \sin 40° + \sin 10°$
  $\sin 30° = \sin 40° - \sin 10°$ **no**
Then, have students determine which of these expressions is true.
$\sin (a + b) = \sin a \cos b - \cos a \sin b$ **no**
$\sin (a + b) = \sin a \cos b + \cos a \sin b$ **yes**

Teaching Tip Stress that while the equation in part **b** of Example 1 is true for the given values, this is *not* a proof for all values of *x* and *y*.

In-Class Example

Example 1

a. Show by producing a counterexample that
$\cos(x - y) \neq \cos x - \cos y$.

Sample answer: let $x = \dfrac{\pi}{4}$ and $y = \dfrac{\pi}{4}$.

$\cos\left(\dfrac{\pi}{4} - \dfrac{\pi}{4}\right) \stackrel{?}{=} \cos\dfrac{\pi}{4} - \cos\dfrac{\pi}{4}$

$\cos 0 \stackrel{?}{=} \dfrac{\sqrt{2}}{2} - \dfrac{\sqrt{2}}{2}$

$1 \neq 0$

b. Show that the difference identity for cosine is true for the values used in part **a**.
From part a, we have
$\cos(x - y) = 1$.

$\cos x \cos y + \sin x \sin y$

$= \cos\dfrac{\pi}{4}\cos\dfrac{\pi}{4} + \sin\dfrac{\pi}{4}\sin\dfrac{\pi}{4}$

$= \left(\dfrac{\sqrt{2}}{2}\right)\left(\dfrac{\sqrt{2}}{2}\right) + \left(\dfrac{\sqrt{2}}{2}\right)\left(\dfrac{\sqrt{2}}{2}\right) = 1$

The **sum identity for cosine** can be derived by substituting $-\beta$ for β in the difference identity.

$\cos(\alpha + \beta) = \cos(\alpha - (-\beta))$

$\qquad\qquad = \cos\alpha\cos(-\beta) + \sin\alpha\sin(-\beta)$

$\qquad\qquad = \cos\alpha\cos\beta + \sin\alpha(-\sin\beta) \quad \cos(-\beta) = \cos\beta; \sin(-\beta) = -\sin\beta$

$\qquad\qquad = \cos\alpha\cos\beta - \sin\alpha\sin\beta$

Sum and Difference Identities for the Cosine Function	If α and β represent the measures of two angles, then the following identities hold for all values of α and β.
	$\cos(\alpha \pm \beta) = \cos\alpha\cos\beta \mp \sin\alpha\sin\beta$

Notice how the addition and subtraction symbols are related in the sum and difference identities.

You can use the sum and difference identities and the values of the trigonometric functions of common angles to find the values of trigonometric functions of other angles. Note that α and β can be expressed in either degrees or radians.

Example **1** **a.** Show by producing a counterexample that $\cos(x + y) \neq \cos x + \cos y$.

b. Show that the sum identity for cosine is true for the values used in part a.

a. Let $x = \dfrac{\pi}{3}$ and $y = \dfrac{\pi}{6}$. First find $\cos(x + y)$ for $x = \dfrac{\pi}{3}$ and $y = \dfrac{\pi}{6}$.

$\cos(x + y) = \cos\left(\dfrac{\pi}{3} + \dfrac{\pi}{6}\right)$ *Replace x with $\dfrac{\pi}{3}$ and y with $\dfrac{\pi}{6}$.*

$\qquad\qquad = \cos\dfrac{\pi}{2} \qquad \dfrac{\pi}{3} + \dfrac{\pi}{6} = \dfrac{\pi}{2}$

$\qquad\qquad = 0$

Now find $\cos x + \cos y$.

$\cos x + \cos y = \cos\dfrac{\pi}{3} + \cos\dfrac{\pi}{6}$ *Replace x with $\dfrac{\pi}{3}$ and y with $\dfrac{\pi}{6}$.*

$\qquad\qquad = \dfrac{1}{2} + \dfrac{\sqrt{3}}{2}$ or $\dfrac{1 + \sqrt{3}}{2}$

So, $\cos(x + y) \neq \cos x + \cos y$.

b. Show that $\cos(x + y) = \cos x \cos y - \sin x \sin y$ for $x = \dfrac{\pi}{3}$ and $y = \dfrac{\pi}{6}$.

First find $\cos(x + y)$. From part a, we know that $\cos\left(\dfrac{\pi}{3} + \dfrac{\pi}{6}\right) = 0$.

Now find $\cos x \cos y - \sin x \sin y$.

$\cos x \cos y - \sin x \sin y = \cos\dfrac{\pi}{3}\cos\dfrac{\pi}{6} - \sin\dfrac{\pi}{3}\sin\dfrac{\pi}{6}$ *Substitute for x and y.*

$\qquad\qquad = \left(\dfrac{1}{2}\right)\left(\dfrac{\sqrt{3}}{2}\right) - \left(\dfrac{\sqrt{3}}{2}\right)\left(\dfrac{1}{2}\right)$

$\qquad\qquad = 0$

Thus, the sum identity for cosine is true for $x = \dfrac{\pi}{3}$ and $y = \dfrac{\pi}{6}$.

Example ② **Use the sum or difference identity for cosine to find the exact value of cos 735°.**

$735° = 2(360°) + 15°$ *Symmetry identity, Case 1*
$\cos 735° = \cos 15°$

$\cos 15° = \cos (45° - 30°)$ *45° and 30° are two common angles that differ by 15°.*
$= \cos 45° \cos 30° + \sin 45° \sin 30°$ *Difference identity for cosine*
$= \dfrac{\sqrt{2}}{2} \cdot \dfrac{\sqrt{3}}{2} + \dfrac{\sqrt{2}}{2} \cdot \dfrac{1}{2}$
$= \dfrac{\sqrt{6} + \sqrt{2}}{4}$

Therefore, $\cos 735° = \dfrac{\sqrt{6} + \sqrt{2}}{4}$.

We can derive sum and difference identities for the sine function from those for the cosine function. Replace α with $\frac{\pi}{2}$ and β with s in the identities for $\cos (\alpha \pm \beta)$. The following equations result.

These equations are examples of cofunction identities.

$$\cos\left(\frac{\pi}{2} + s\right) = -\sin s \qquad \cos\left(\frac{\pi}{2} - s\right) = \sin s$$

Replace s with $\frac{\pi}{2} + s$ in the equation for $\cos\left(\frac{\pi}{2} + s\right)$ and with $\frac{\pi}{2} - s$ in the equation for $\cos\left(\frac{\pi}{2} - s\right)$ to obtain the following equations.

These equations are other cofunction identities.

$$\cos s = \sin\left(\frac{\pi}{2} + s\right) \qquad \cos s = \sin\left(\frac{\pi}{2} - s\right)$$

Replace s with $(\alpha + \beta)$ in the equation for $\cos\left(\frac{\pi}{2} - s\right)$ to derive an identity for the sine of the sum of two real numbers.

$$\cos\left[\frac{\pi}{2} - (\alpha + \beta)\right] = \sin (\alpha + \beta)$$

$$\cos\left[\left(\frac{\pi}{2} - \alpha\right) - \beta\right] = \sin (\alpha + \beta)$$

$$\cos\left(\frac{\pi}{2} - \alpha\right)\cos \beta + \sin\left(\frac{\pi}{2} - \alpha\right)\sin \beta = \sin\left(\alpha + \beta\right) \quad \textit{Identity for } \cos (\alpha - \beta)$$

$$\sin \alpha \cos \beta + \cos \alpha \sin \beta = \sin (\alpha + \beta) \quad \textit{Substitute.}$$

This equation is known as the **sum identity for sine.**

The **difference identity for sine** can be derived by replacing β with $(-\beta)$ in the sum identity for sine.

$$\sin [\alpha + (-\beta)] = \sin \alpha \cos (-\beta) + \cos \alpha \sin (-\beta)$$

$$\sin (\alpha - \beta) = \sin \alpha \cos \beta - \cos \alpha \sin \beta$$

Sum and Difference Identities for the Sine Function

If α and β represent the measures of two angles, then the following identities hold for all values of α and β.

$$\sin [\alpha \pm \beta] = \sin \alpha \cos \beta \pm \cos \alpha \sin \beta$$

Teaching Tip Remind students that sin $(x + y)$ is not sin x + sin y but rather sin x cos y + cos x sin y.

In-Class Examples

Example 3
Find the value of sin $(x + y)$ if
$0 < x < \frac{\pi}{2}$, $0 < y < \frac{\pi}{2}$,
sin $x = \frac{4}{5}$, and sin $y = \frac{5}{13}$.
$\frac{63}{65}$ or about 0.9692

Example 4
Physics In 1887 Heinrich Hertz, a German physicist, discovered electromagnetic waves. The period of a wave is the shortest interval of repetition. Show that
$\sin\left(\frac{\pi}{6} + 2\pi\right) = \sin\frac{\pi}{6}$.

See below for answer.

Examples

3 Find the value of sin $(x - y)$ if $0 < x < \frac{\pi}{2}$, $0 < y < \frac{\pi}{2}$, sin $x = \frac{9}{41}$, and sin $y = \frac{7}{25}$.

In order to use the difference identity for sine, we need to know cos x and cos y. We can use a Pythagorean identity to determine the necessary values.

$$\sin^2 \alpha + \cos^2 \alpha = 1 \Rightarrow \cos^2 \alpha = 1 - \sin^2 \alpha \quad \textit{Pythagorean identity}$$

Since we are given that the angles are in Quadrant I, the values of sine and cosine are positive. Therefore, $\cos \alpha = \sqrt{1 - \sin^2 \alpha}$.

$$\cos x = \sqrt{1 - \left(\frac{9}{41}\right)^2}$$

$$= \sqrt{\frac{1600}{1681}} \text{ or } \frac{40}{41}$$

$$\cos y = \sqrt{1 - \left(\frac{7}{25}\right)^2}$$

$$= \sqrt{\frac{576}{625}} \text{ or } \frac{24}{25}$$

Now substitute these values into the difference identity for sine.

$$\sin (x - y) = \sin x \cos y - \cos x \sin y$$

$$= \left(\frac{9}{41}\right)\left(\frac{24}{25}\right) - \left(\frac{40}{41}\right)\left(\frac{7}{25}\right)$$

$$= -\frac{64}{1025} \text{ or about } -0.0624$$

4 **BROADCASTING** Refer to the application at the beginning of the lesson. What type of interference results when signals modeled by the equations $y = 20 \sin (3t + 45°)$ and $y = 20 \sin (3t + 225°)$ are combined?

Add the two sine functions together and simplify.

$20 \sin (3t + 45°) + 20 \sin (3t + 225°)$

$= 20(\sin 3t \cos 45° + \cos 3t \sin 45°) + 20(\sin 3t \cos 225° + \cos 3t \sin 225°)$

$= 20\left[(\sin 3t)\left(\frac{\sqrt{2}}{2}\right) + (\cos 3t)\left(\frac{\sqrt{2}}{2}\right)\right] + 20\left[(\sin 3t)\left(-\frac{\sqrt{2}}{2}\right) + (\cos 3t)\left(-\frac{\sqrt{2}}{2}\right)\right]$

$= 10\sqrt{2} \sin 3t + 10\sqrt{2} \cos 3t - 10\sqrt{2} \sin 3t - 10\sqrt{2} \cos 3t$

$= 0$

The interference is destructive. The signals cancel each other completely.

You can use the sum and difference identities for the cosine and sine functions to find sum and difference identities for the tangent function.

Answer for In-Class Example 4

$$\sin\left(\frac{\pi}{6} + 2\pi\right) \stackrel{?}{=} \sin\frac{\pi}{6}$$

$$\sin\frac{\pi}{6}\cos 2\pi + \cos\frac{\pi}{6}\sin 2\pi \stackrel{?}{=} \sin\frac{\pi}{6}$$

$$\frac{1}{2} \cdot 1 + \frac{\sqrt{3}}{2} \cdot 0 \stackrel{?}{=} \frac{1}{2}$$

$$\frac{1}{2} + 0 \stackrel{?}{=} \frac{1}{2}$$

$$\frac{1}{2} = \frac{1}{2}$$

<table>
<tr><td>Sum and Difference Identities for the Tangent Function</td><td>If α and β represent the measures of two angles, then the following identities hold for all values of α and β.

$$\tan (\alpha \pm \beta) = \frac{\tan \alpha \pm \tan \beta}{1 \mp \tan \alpha \tan \beta}$$</td></tr>
</table>

You will be asked to derive these identities in Exercise 47.

Example **5** Use the sum or difference identity for tangent to find the exact value of $\tan 285°$.

$\tan 285° = \tan (240° + 45°)$ *$240°$ and $45°$ are common angles whose sum is $285°$.*

$= \dfrac{\tan 240° + \tan 45°}{1 - \tan 240° \tan 45°}$ *Sum identity for tangent*

$= \dfrac{\sqrt{3} + 1}{1 - (\sqrt{3})(1)}$ *Multiply by $\dfrac{1 + \sqrt{3}}{1 + \sqrt{3}}$ to simplify.*

$= -2 - \sqrt{3}$

You can use sum and difference identities to verify other identities.

Example **6** Verify that $\csc \left(\dfrac{3\pi}{2} + A\right) = -\sec A$ is an identity.

Transform the left side since it is more complicated.

$\csc \left(\dfrac{3\pi}{2} + A\right) \stackrel{?}{=} -\sec A$

$\dfrac{1}{\sin \left(\dfrac{3\pi}{2} + A\right)} \stackrel{?}{=} -\sec A$ *Reciprocal identity: $\csc x = \dfrac{1}{\sin x}$*

$\dfrac{1}{\sin \dfrac{3\pi}{2} \cos A + \cos \dfrac{3\pi}{2} \sin A} \stackrel{?}{=} -\sec A$ *Sum identity for sine*

$\dfrac{1}{(-1) \cos A + (0) \sin A} \stackrel{?}{=} -\sec A$ *$\sin \dfrac{3\pi}{2} = -1; \cos \dfrac{3\pi}{2} = 0$*

$-\dfrac{1}{\cos A} \stackrel{?}{=} -\sec A$ *Simplify.*

$-\sec A = -\sec A$ *Reciprocal identity*

<hr>

CHECK FOR UNDERSTANDING

Communicating Mathematics

Read and study the lesson to answer each question.   **1–2. See margin.**

1. **Describe** how you would convince a friend that $\sin (x + y) \neq \sin x + \sin y$.

2. **Explain** how to use the sum and difference identities to find values for the secant, cosecant, and cotangent functions of a sum or difference.

Lesson 7-3   Sum and Difference Identities   **441**

In-Class Examples

Example 5
Use the sum or difference identity for tangent to find the exact value of $\tan 255°$.
$2 + \sqrt{3}$

Example 6
Verify that $\sec (\pi + A) = -\sec A$ is an identity.

$\sec (\pi + A) \stackrel{?}{=} -\sec A$

$\dfrac{1}{\cos (\pi + A)} \stackrel{?}{=} -\sec A$

$\dfrac{1}{\cos \pi \cos A - \sin \pi \sin A}$

$\stackrel{?}{=} -\sec A$

$\dfrac{1}{(-1) \cos A - (0) \sin A}$

$\stackrel{?}{=} -\sec A$

$\dfrac{1}{-\cos A} \stackrel{?}{=} -\sec A$

$-\sec A = -\sec A$

Additional Answers
1. Find a counterexample, such as $x = 30°$ and $y = 60°$.
2. Find the cosine, sine, or tangent, respectively, of the sum or difference, then take the reciprocal.

<hr>

Addressing Individual Needs

Verbal/Linguistic One of the difficulties students have in mathematics is interpreting the symbolism. Have students write out each identity in this lesson using words instead of mathematical notation.

Error Analysis

Watch for students who mistakenly think that $\cos(\alpha + \beta) = \cos\alpha + \cos\beta$.

Prevent by using counterexamples to show that this is not true for all values of α and β.

Additional Answers

3. The opposite side for $90° - A$ is the adjacent side for A, so the right-triangle ratio for $\sin(90° - A)$ is the same as that for $\cos A$.

4. $\cot(\alpha + \beta)$

$= \dfrac{1}{\tan(\alpha + \beta)}$

$= \dfrac{1}{\dfrac{\tan\alpha + \tan\beta}{1 - \tan\alpha\tan\beta}}$

$= \dfrac{1 - \tan\alpha\tan\beta}{\tan\alpha + \tan\beta}$

$= \dfrac{1 - \dfrac{1}{\cot\alpha}\cdot\dfrac{1}{\cot\beta}}{\dfrac{1}{\cot\alpha} + \dfrac{1}{\cot\beta}}\cdot\dfrac{\cot\alpha\cot\beta}{\cot\alpha\cot\beta}$

$= \dfrac{\cot\alpha\cot\beta - 1}{\cot\alpha + \cot\beta}$

Study Guide and Practice Masters, pp. 123–124

3–4. See margin.

3. **Write** an interpretation of the identity $\sin(90° - A) = \cos A$ in terms of a right triangle.

4. **Derive** a formula for $\cot(\alpha + \beta)$ in terms of $\cot\alpha$ and $\cot\beta$.

Guided Practice   Use sum or difference identities to find the exact value of each trigonometric function.

5. $\cos 165°$ $-\dfrac{\sqrt{2}+\sqrt{6}}{4}$ 6. $\tan\dfrac{\pi}{12}$ $2 - \sqrt{3}$ 7. $\sec 795°$ $\sqrt{6} + \sqrt{2}$

Find each exact value if $0 < x < \dfrac{\pi}{2}$ and $0 < y < \dfrac{\pi}{2}$.

8. $\sin(x - y)$ if $\sin x = \dfrac{4}{9}$ and $\sin y = \dfrac{1}{4}$ $\dfrac{4\sqrt{15}-\sqrt{65}}{36}$

9. $\tan(x + y)$ if $\csc x = \dfrac{5}{3}$ and $\cos y = \dfrac{5}{13}$ $-\dfrac{63}{16}$

Verify that each equation is an identity.   **10–12. See Answer Appendix.**

10. $\sin(90° + A) = \cos A$ 11. $\tan\left(\theta + \dfrac{\pi}{2}\right) = -\cot\theta$

12. $\sin(x - y) = \dfrac{1 - \cot x\tan y}{\csc x\sec y}$

13. **Electrical Engineering**
Analysis of the voltage in certain types of circuits involves terms of the form $\sin(n\omega_0 t - 90°)$, where n is a positive integer, ω_0 is the frequency of the voltage, and t is time. Use an identity to simplify this expression.
$-\cos n\omega_0 t$

ω is the Greek letter omega.

Practice   Use sum or difference identities to find the exact value of each trigonometric function.

 A

14. $\cos 105°$ $\dfrac{\sqrt{2}-\sqrt{6}}{4}$ 15. $\sin 165°$ $\dfrac{\sqrt{6}-\sqrt{2}}{4}$ 16. $\cos\dfrac{7\pi}{12}$ $\dfrac{\sqrt{2}-\sqrt{6}}{4}$

17. $\sin\dfrac{\pi}{12}$ $\dfrac{\sqrt{6}-\sqrt{2}}{4}$ 18. $\tan 195°$ $2 - \sqrt{3}$ 19. $\cos\left(-\dfrac{\pi}{12}\right)$ $\dfrac{\sqrt{2}+\sqrt{6}}{4}$

 B

20. $\tan 165°$ $-2 + \sqrt{3}$ 21. $\tan\dfrac{23\pi}{12}$ $-2 + \sqrt{3}$ 22. $\sin 735°$ $\dfrac{\sqrt{6}-\sqrt{2}}{4}$

23. $\sec 1275°$ $\sqrt{2} - \sqrt{6}$ 24. $\csc\dfrac{5\pi}{12}$ $\sqrt{6} - \sqrt{2}$ 25. $\cot\dfrac{113\pi}{12}$ $2 - \sqrt{3}$

www.amc.glencoe.com/self_check_quiz

Find each exact value if $0 < x < \frac{\pi}{2}$ and $0 < y < \frac{\pi}{2}$.

26. $\sin(x+y)$ if $\cos x = \frac{8}{17}$ and $\sin y = \frac{12}{37}$ $\frac{621}{629}$

27. $\cos(x-y)$ if $\cos x = \frac{3}{5}$ and $\cos y = \frac{4}{5}$ $\frac{24}{25}$

28. $\tan(x-y)$ if $\sin x = \frac{8}{17}$ and $\cos y = \frac{3}{5}$ $-\frac{36}{77}$

29. $\cos(x+y)$ if $\tan x = \frac{5}{3}$ and $\sin y = \frac{1}{3}$ $\frac{12\sqrt{17}-5\sqrt{34}}{102}$

30. $\tan(x+y)$ if $\cot x = \frac{6}{5}$ and $\sec y = \frac{3}{2}$ $\frac{270+122\sqrt{5}}{19}$

31. $\sec(x-y)$ if $\csc x = \frac{5}{3}$ and $\tan y = \frac{12}{5}$ $\frac{65}{56}$

32. If α and β are two angles in Quadrant I such that $\sin\alpha = \frac{1}{5}$ and $\cos\beta = \frac{2}{7}$, find $\sin(\alpha-\beta)$. $\frac{2-6\sqrt{30}}{35}$

33. If x and y are acute angles such that $\cos x = \frac{1}{3}$ and $\cos y = \frac{3}{4}$, what is the value of $\cos(x+y)$? $\frac{3-2\sqrt{14}}{12}$

Verify that each equation is an identity.   **34–42. See Answer Appendix.**

34. $\cos\left(\frac{\pi}{2}+x\right) = -\sin x$

35. $\cos(60° + A) = \sin(30° - A)$

36. $\sin(A+\pi) = -\sin A$

37. $\cos(180° + x) = -\cos x$

38. $\tan(x+45°) = \frac{1+\tan x}{1-\tan x}$

39. $\sin(A+B) = \frac{\tan A + \tan B}{\sec A \sec B}$

40. $\cos(A+B) = \frac{1-\tan A \tan B}{\sec A \sec B}$

41. $\sec(A-B) = \frac{\sec A \sec B}{1 + \tan A \tan B}$

42. $\sin(x+y)\sin(x-y) = \sin^2 x - \sin^2 y$

Applications and Problem Solving

Real World Application

43. **Electronics** In an electric circuit containing a capacitor, inductor, and resistor the voltage drop across the inductor is given by $V_L = I_0\omega L \cos\left(\omega t + \frac{\pi}{2}\right)$, where I_0 is the peak current, ω is the frequency, L is the inductance, and t is time. Use the sum identity for cosine to express V_L as a function of $\sin\omega t$.
$$V_L = -I_0\omega L \sin\omega t$$

44. **Optics** The index of refraction for a medium through which light is passing is the ratio of the velocity of light in free space to the velocity of light in the medium. For light passing symmetrically through a glass prism, the index of refraction n is given by the equation $n = \dfrac{\sin\left[\frac{1}{2}(\alpha+\beta)\right]}{\sin\frac{\beta}{2}}$, where α is the

deviation angle and β is the angle of the apex of the prism as shown in the diagram. If $\beta = 60°$, show that $n = \sqrt{3}\sin\frac{\alpha}{2} + \cos\frac{\alpha}{2}$. **See margin.**

Lesson 7-3 Sum and Difference Identities **443**

Additional Answer

44.
$$n = \frac{\sin\left[\frac{1}{2}(\alpha+\beta)\right]}{\sin\frac{\beta}{2}}$$

$$n = \frac{\sin\left[\frac{1}{2}(\alpha+60°)\right]}{\sin\frac{60°}{2}}$$

$$n = \frac{\sin\left(\frac{\alpha}{2}+30°\right)}{\sin 30°}$$

$$n = \frac{\sin\frac{\alpha}{2}\cos 30° + \cos\frac{\alpha}{2}\sin 30°}{\frac{1}{2}}$$

$$n = 2\left[\left(\sin\frac{\alpha}{2}\right)\cdot\frac{\sqrt{3}}{2} + \left(\cos\frac{\alpha}{2}\right)\cdot\frac{1}{2}\right]$$

$$n = \sqrt{3}\sin\frac{\alpha}{2} + \cos\frac{\alpha}{2}$$

Additional Answers

46a. $\dfrac{\sin x \cos h + \cos x \sin h - \sin x}{h}$

46b.

$$y = \frac{\sin x \cos 0.1 + \cos x \sin 0.1 - \sin x}{0.1}$$

47. $\tan(\alpha + \beta)$

$= \dfrac{\sin(\alpha + \beta)}{\cos(\alpha + \beta)}$

$= \dfrac{\sin \alpha \cos \beta + \cos \alpha \sin \beta}{\cos \alpha \cos \beta - \sin \alpha \sin \beta}$

$= \dfrac{\dfrac{\sin \alpha \cos \beta}{\cos \alpha \cos \beta} + \dfrac{\cos \alpha \sin \beta}{\cos \alpha \cos \beta}}{\dfrac{\cos \alpha \cos \beta}{\cos \alpha \cos \beta} - \dfrac{\sin \alpha \sin \beta}{\cos \alpha \cos \beta}}$

$= \dfrac{\tan \alpha + \tan \beta}{1 - \tan \alpha \tan \beta}$

Replace β with $-\beta$ to find $\tan(\alpha - \beta)$.

$\tan(\alpha + (-\beta)) = \dfrac{\tan \alpha + \tan(-\beta)}{1 - \tan \alpha \tan(-\beta)}$

$\tan(\alpha - \beta) = \dfrac{\tan \alpha - \tan \beta}{1 + \tan \alpha \tan \beta}$

49. $\sec^2 x$

$\overset{?}{=} \dfrac{1 - \cos^2 x}{1 - \sin^2 x} + \csc^2 x - \cot^2 x$

$\overset{?}{=} \dfrac{1 - \cos^2 x}{\cos^2 x} + 1 + \cot^2 x - \cot^2 x$

$\overset{?}{=} \dfrac{1}{\cos^2 x} - \dfrac{\cos^2 x}{\cos^2 x} + 1$

$\overset{?}{=} \sec^2 x - 1 + 1$

$= \sec^2 x$

---

45. Critical Thinking Simplify the following expression without expanding any of the sums or differences. **$-\sin 2A$**

$$\sin\left(\frac{\pi}{3} - A\right)\cos\left(\frac{\pi}{3} + A\right) - \cos\left(\frac{\pi}{3} - A\right)\sin\left(\frac{\pi}{3} + A\right)$$

46a–b. See margin.

46. Calculus In calculus, you will explore the *difference quotient* $\dfrac{f(x+h) - f(x)}{h}$.

  a. Let $f(x) = \sin x$. Write and expand an expression for the difference quotient.

  b. Set your answer from part a equal to y. Let $h = 0.1$ and graph.

  c. What function has a graph similar to the graph in part b? **cos x**

47. Critical Thinking Derive the sum and difference identities for the tangent function. **See margin.**

48. Critical Thinking Consider the following theorem.

  If A, B, and C are the angles of a nonright triangle, then $\tan A + \tan B + \tan C = \tan A \tan B \tan C$.

  a. Choose values for A, B, and C. Verify that the conclusion is true for your specific values. **Answers will vary.**

  b. Prove the theorem. **See Answer Appendix.**

Mixed Review

49. See margin.

49. Verify the identity $\sec^2 x = \dfrac{1 - \cos^2 x}{1 - \sin^2 x} + \csc^2 x - \cot^2 x$. *(Lesson 7-2)*

50. If $\sin \theta = -\dfrac{1}{8}$ and $\pi < \theta < \dfrac{3\pi}{2}$, find $\tan \theta$. *(Lesson 7-1)* $\dfrac{\sqrt{7}}{21}$

51. Find $\sin\left(\text{Arctan } \sqrt{3}\right)$. *(Lesson 6-8)* $\dfrac{\sqrt{3}}{2}$

52. Find the values of θ for which $\csc \theta$ is undefined. *(Lesson 6-7)*

52. πk, where k is an integer

53. Weather The average seasonal high temperatures for Greensboro, North Carolina, are given in the table. Write a sinusoidal function that models the temperatures, using $t = 1$ to represent winter. *(Lesson 6-6)*

Winter	Spring	Summer	Fall
50°	70°	86°	71°

Source: Rand McNally & Company

$$y = 18 \sin\left(\frac{\pi}{2}t - \pi\right) + 68$$

54. State the amplitude, period, and phase shift for the function $y = 8 \cos(\theta - 30°)$. *(Lesson 6-5)* **8, 360°, 30°**

55. Find the value of $\sin(-540°)$. *(Lesson 6-3)* **0**

56. Geometry A sector has arc length of 18 feet and a central angle measuring 2.9 radians. Find the radius and the area of the sector. *(Lesson 6-1)* **6.2 ft, 55.9 ft²**

57. Navigation A ship at sea is 70 miles from one radio transmitter and 130 miles from another. The angle formed by the rays from the ship to the transmitters measures 130°. How far apart are the transmitters? *(Lesson 5-8)* **about 183 miles**

58. Determine the number of possible solutions for a triangle if $A = 120°$, $b = 12$, and $a = 4$. *(Lesson 5-7)* **0**

444 Chapter 7 *Trigonometric Identities and Equations*

59. Photography A photographer observes a 35-foot totem pole that stands vertically on a uniformly-sloped hillside and the shadow cast by it at different times of day. At a time when the angle of elevation of the sun is 37°12′, the shadow of the pole extends directly down the slope. This is the effect that the photographer is seeking. If the hillside has an angle of inclination of 6°40′, find the length of the shadow. *(Lesson 5-6)* **54.87 ft**

60. Find the roots of the equation $4x^3 + 3x^2 - x = 0$. *(Lesson 4-1)* $0, \frac{1}{4}, -1$

61. Solve $|x + 1| > 4$. *(Lesson 3-3)* 61. $\{x \mid x < -5 \text{ or } x > 3\}$

62. Find the value of the determinant $\begin{vmatrix} -1 & -2 \\ 3 & -6 \end{vmatrix}$. *(Lesson 2-5)* **12**

63. If $f(x) = 3x^2 - 4$ and $g(x) = 5x + 1$, find $f \circ g(4)$ and $g \circ f(4)$. *(Lesson 1-2)* **1319, 221**

64. SAT Practice What is the value of $(-8)^{62} \div 8^{62}$? **A**
A 1
B 0
C −1
D −8
E −62

MID-CHAPTER QUIZ

Use the given information to determine the exact trigonometric value. *(Lesson 7-1)*

1. $\sin \theta = \frac{2}{7}, 0 < \theta < \frac{\pi}{2}$; $\cot \theta$ $\frac{3\sqrt{5}}{2}$

2. $\tan \theta = -\frac{4}{3}, 90° < \theta < 180°$; $\cos \theta$ $-\frac{3}{5}$

3. Express $\cos \frac{19\pi}{4}$ as a trigonometric function of an angle in Quadrant I. *(Lesson 7-1)* $-\cos \frac{\pi}{4}$

Verify that each equation is an identity. *(Lesson 7-2)* **4–5. See margin.**

4. $\frac{1}{1 + \tan^2 x} + \frac{1}{1 + \cot^2 x} = 1$

5. $\frac{\csc^2 \theta + \sec^2 \theta}{\sec^2 \theta} = \csc^2 \theta$

Verify that each equation is an identity. *(Lessons 7-2 and 7-3)* **6–7. See Answer Appendix.**

6. $\cot x \sec x \sin x = 2 - \tan x \cos x \csc x$

7. $\tan (\alpha - \beta) = \frac{1 - \cot \alpha \tan \beta}{\cot \alpha + \tan \beta}$

8. Use a sum or difference identity to find the exact value of $\cos 75°$. *(Lesson 7-3)* $\frac{\sqrt{6} - \sqrt{2}}{4}$

Find each exact value if $0 < x < \frac{\pi}{2}$ and $0 < y < \frac{\pi}{2}$. *(Lesson 7-3)*

9. $\cos (x + y)$ if $\sin x = \frac{2}{3}$ and $\sin y = \frac{3}{4}$ 9. $\frac{\sqrt{35} - 6}{12}$

10. $\tan (x - y)$ if $\tan x = \frac{5}{4}$ and $\sec y = 2$

10. $\frac{-80 + 41\sqrt{3}}{59}$

Extra Practice See p. A38.

Lesson 7-3 Sum and Difference Identities **445**

Extra Credit

Use the identity for $\sin (x + y)$ to derive and simplify an identity for $\sin (x + y + z)$.
$\sin x \cos y \cos z + \cos x \sin y \cos z + \cos x \cos y \sin z - \sin x \sin y \sin z$

---

4 ASSESS

Open-Ended Assessment
Modeling Label six index cards as: sin 30°, sin 45°, cos 30°, cos 45°, +, and −. Have students arrange the cards in order to form the identity for sin 15°, sin 75°, cos 15°, and cos 75°.

Mid-Chapter Quiz
The Mid-Chapter Quiz provides students with a brief review of the concepts and skills in Lessons 7-1 through 7-3. Lesson numbers are given to the right of exercises or instruction lines so students can review concepts not yet mastered.

Additional Answers
Mid-Chapter Quiz

4.
$$\frac{1}{1 + \tan^2 x} + \frac{1}{1 + \cot^2 x} \overset{?}{=} 1$$
$$\frac{1}{\sec^2 x} + \frac{1}{\csc^2 x} \overset{?}{=} 1$$
$$\cos^2 x + \sin^2 x \overset{?}{=} 1$$
$$1 = 1$$

5.
$$\frac{\csc^2 \theta + \sec^2 \theta}{\sec^2 \theta} \overset{?}{=} \csc^2 \theta$$
$$\frac{\csc^2 \theta}{\sec^2 \theta} + \frac{\sec^2 \theta}{\sec^2 \theta} \overset{?}{=} \csc^2 \theta$$
$$\frac{\frac{1}{\sin^2 \theta}}{\frac{1}{\cos^2 \theta}} + 1 \overset{?}{=} \csc^2 \theta$$
$$\frac{\cos^2 \theta}{\sin^2 \theta} + 1 \overset{?}{=} \csc^2 \theta$$
$$\cot^2 \theta + 1 \overset{?}{=} \csc^2 \theta$$
$$\csc^2 \theta = \csc^2 \theta$$

Enrichment Masters, p. 75

7-3 Enrichment

Identities for the Products of Sines and Cosines

By adding the identities for the sines of the sum and difference of the measures of two angles, a new identity is obtained.

$$\sin (\alpha + \beta) = \sin \alpha \cos \beta + \cos \alpha \sin \beta$$
$$\sin (\alpha - \beta) = \sin \alpha \cos \beta - \cos \alpha \sin \beta$$
$$(i) \ \sin (\alpha + \beta) + \sin (\alpha - \beta) = 2 \sin \alpha \cos \beta$$

This new identity is useful for expressing certain products as sums.

Example Write sin 3θ cos θ as a sum.

In the right side of identity (i) let $\alpha = 3\theta$ and $\beta = \theta$ so that $2 \sin 3\theta \cos \theta = \sin (3\theta + \theta) + \sin (3\theta - \theta)$.
Thus, $\sin 3\theta \cos \theta = \frac{1}{2} \sin 4\theta + \frac{1}{2} \sin 2\theta$.

By subtracting the identities for $\sin (\alpha + \beta)$ and $\sin (\alpha - \beta)$, you obtain a similar identity for expressing a product as a difference.

$$(ii) \ \sin (\alpha + \beta) - \sin (\alpha - \beta) = 2 \cos \alpha \sin \beta$$

Example Verify the identity $\frac{\cos 2x \sin x}{\sin 2x \cos x} = \frac{(\sin 3x - \sin x)^2}{\sin^2 3x - \sin^2 x}$.

In the right sides of identities (i) and (ii) let $\alpha = 2x$ and $\beta = x$. Then write the following quotient.
$$\frac{2 \cos 2x \sin x}{2 \sin 2x \cos x} = \frac{\sin (2x + x) - \sin (2x - x)}{\sin (2x + x) + \sin (2x - x)}$$

By simplifying and multiplying by the conjugate, the identity is verified.
$$\frac{\cos 2x \sin x}{\sin 2x \cos x} = \frac{\sin 3x - \sin x}{\sin 3x + \sin x} \cdot \frac{\sin 3x - \sin x}{\sin 3x - \sin x}$$
$$= \frac{(\sin 3x - \sin x)^2}{\sin^2 3x - \sin^2 x}$$

Complete.

1. Use the identities for $\cos (\alpha + \beta)$ and $\cos (\alpha - \beta)$ to find identities for expressing the products $2 \cos \alpha \cos \beta$ and $2 \sin \alpha \sin \beta$ as a sum or difference.
$$\cos (\alpha + \beta) = \cos \alpha \cos \beta - \sin \alpha \sin \beta$$
$$2 \cos \alpha \cos \beta = \cos (\alpha + \beta) + \cos (\alpha - \beta)$$
$$\cos (\alpha - \beta) = \cos \alpha \cos \beta + \sin \alpha \sin \beta$$
$$2 \sin \alpha \sin \beta = \cos (\alpha - \beta) - \cos (\alpha + \beta)$$

2. Find the value of cos 105° cos 75° by using the identity above.
$$\cos 105° \cos 75° = \frac{1}{2}(\sin 180° + \sin 30°)$$
$$= \frac{1}{2}\left(0 + \frac{1}{2}\right) = \frac{1}{4} \text{ or } 0.25$$

© Glencoe/McGraw-Hill   T75   Advanced Mathematical Concepts

Lesson 7-3 **445**

7-3B Reduction Identities

An Extension of Lesson 7-3

1 FOCUS

Motivating the Lesson

Discussion Have students graph the following trigonometric functions on the same screen of a graphing calculator.

$y = \cos(x + 270°)$
$y = \cos(270° + x)$
$y = \cos(x - 90°)$
$y = \cos(90° - x)$
$y = \sin(180° - x)$

Then, have students discuss what these functions have in common. **They all have the same graph as $y = \sin x$.**

2 TEACH

A reproduction of this lesson with graphing calculator keystrokes provided can be found in the *Graphing Calculator Keystrokes Masters*, pp. 27–29.

Teaching Tip Remind students that one period for $y = \sin x$ or $y = \cos x$ is 360° or 2π and the amplitude is 1. This information will help define the viewing window.

OBJECTIVE
- Identify equivalent values for trigonometric functions involving quadrantal angles.

In Chapter 5, you learned that any trigonometric function of an acute angle is equal to the *cofunction* of the complement of the angle. For example, $\sin \alpha = \cos(90° - \alpha)$. This is a part of a large family of identities called the **reduction identities.** These identities involve adding and subtracting the quandrantal angles, 90°, 180°, and 270°, from the angle measure to find equivalent values of the trigonometric function. You can use your knowledge of phase shifts and reflections to find the components of these identities.

Example Find the values of the sine and cosine functions for $\alpha - 90°$, $\alpha - 180°$, and $\alpha - 270°$ that are equivalent to $\sin \alpha$.

$\alpha - 90°$
Graph $y = \sin \alpha$, $y = \sin(\alpha - 90°)$, and $y = \cos(\alpha - 90°)$, letting X in degree mode represent α. *You can select different display formats to help you distinguish the three graphs.*

You may recall from Chapter 6 that a phase shift of 90° right for the cosine function results in the sine function.

Note that the graph of $y = \cos(X - 90)$ is the same as the graph of $y = \sin X$. This suggests that $\sin \alpha = \cos(\alpha - 90°)$. *Remember that an identity must be proved algebraically. A graph does not prove an identity.*

[0, 360] scl:90 by [−2, 2] scl:1

$\alpha - 180°$
Graph $y = \sin \alpha$, $y = \sin(\alpha - 180°)$, and $y = \cos(\alpha - 180°)$ using X to represent α.

Discount $y = \cos(\alpha - 180°)$ as a possible equivalence because it would involve a phase shift, which would change the actual value of the angle being considered.

[0, 360] scl:90 by [−2, 2] scl:1

Note that the graph of $\sin(\alpha - 180°)$ is a mirror reflection of $\sin \alpha$. Remember that a reflection over the *x*-axis results in the mapping $(x, y) \rightarrow (x, -y)$. So to obtain a graph that is identical to $y = \sin \alpha$, we need the reflection of $y = \sin(\alpha - 180°)$ over the *x*-axis, or $y = -\sin(\alpha - 180°)$. Thus, $\sin \alpha = -\sin(\alpha - 180°)$. Graph the two equations to investigate this equality.

446 *Chapter 7 Trigonometric Identities and Equations*

α − 270°

In this case, sin (α − 270°) is a phase shift, so ignore it. The graph of cos (α − 270°) is a reflection of sin α over the *x*-axis. So, sin α = − cos (α − 270°).

Y1 = sin (X)
Y3 = cos (X−270)
Y2 = sin (X−270)

[0, 360] scl:90 by [−2, 2] scl:1

The family of reduction identities also contains the relationships among the other cofunctions of tangent and cotangent and secant and cosecant. In addition to α − 90°, α − 180°, and α − 270° angle measures, the reduction identities address other measures such as 90° ± α, 180° ± α, 270° ± α, and 360° + α.

TRY THESE

1. −sin, −cos, sin

2. −cot, tan, −cot

3. −tan, cot, −tan

4. −csc, −sec, csc

5. sec, −csc, −sec

7a–b. See margin.

Copy and complete each statement with the proper trigonometric functions.

1. cos α = ___?___ (α − 90°) = ___?___ (α − 180°) = ___?___ (α − 270°)

2. tan α = ___?___ (α − 90°) = ___?___ (α − 180°) = ___?___ (α − 270°)

3. cot α = ___?___ (α − 90°) = ___?___ (α − 180°) = ___?___ (α − 270°)

4. sec α = ___?___ (α − 90°) = ___?___ (α − 180°) = ___?___ (α − 270°)

5. csc α = ___?___ (α − 90°) = ___?___ (α − 180°) = ___?___ (α − 270°)

WHAT DO YOU THINK?

6. Suppose the expressions involving subtraction in Exercises 1-5 were changed to sums. **a–b. See margin.**
   a. Copy and complete each statement with the proper trigonometric functions.
   (1) sin α = ___?___ (α + 90°) = ___?___ (α + 180°) = ___?___ (α + 270°)
   (2) cos α = ___?___ (α + 90°) = ___?___ (α + 180°) = ___?___ (α + 270°)
   (3) tan α = ___?___ (α + 90°) = ___?___ (α + 180°) = ___?___ (α + 270°)
   (4) cot α = ___?___ (α + 90°) = ___?___ (α + 180°) = ___?___ (α + 270°)
   (5) sec α = ___?___ (α + 90°) = ___?___ (α + 180°) − ___?___ (α + 270°)
   (6) csc α = ___?___ (α + 90°) = ___?___ (α + 180°) = ___?___ (α + 270°)
   b. How do the identities in part a compare to those in Exercises 1-5?

7. a. Copy and complete each statement with the proper trigonometric functions.
   (1) sin α = ___?___ (90° − α) = ___?___ (180° − α) = ___?___ (270° − α)
   (2) cos α = ___?___ (90° − α) = ___?___ (180° − α) = ___?___ (270° − α)
   (3) tan α = ___?___ (90° − α) = ___?___ (180° − α) = ___?___ (270° − α)
   (4) cot α = ___?___ (90° − α) = ___?___ (180° − α) = ___?___ (270° − α)
   (5) sec α = ___?___ (90° − α) = ___?___ (180° − α) = ___?___ (270° − α)
   (6) csc α = ___?___ (90° − α) = ___?___ (180° − α) = ___?___ (270° − α)
   b. How do the identities in part a compare to those in Exercise 6a?

8. a. How did reduction identities get their name? **See margin.**
   b. If you needed one of these identities, but could not remember it, what other type(s) of identities could you use to derive it? **Sample answer: sum or difference identities**

Lesson 7-3B   Reduction Identities   **447**

Assignment Guide

All: 1–8

Additional Answers

6a. (1) −cos, −sin, cos
   (2) sin, −cos, −sin
   (3) −cot, tan, −cot
   (4) −tan, cot, −tan
   (5) csc, −sec, −csc
   (6) −sec, −csc, sec

6b. Sample answer: if a row for sin α were placed above Exercises 1–5, the entries for Exercise 6a could be obtained by interchanging the first and third columns and leaving the middle column alone.

7a. (1) cos, sin, −cos
   (2) sin, −cos, −sin
   (3) cot, −tan, cot
   (4) tan, −cot, tan
   (5) csc, −sec, −csc
   (6) sec, csc, −sec

7b. Sample answer: the entries in the rows for cos α and sec α are unchanged. All other entries are multiplied by −1.

8a. Sample answer: they can be used to reduce trigonometric functions of large positive or negative angles to those of angles in the first quadrant.

4 ASSESS

Open-Ended Assessment

Writing Have students write a list of the steps necessary to graph $y = \cos (180° − x)$ on a graphing calculator.

Double-Angle and Half-Angle Identities

ARCHITECTURE Mike MacDonald is an architect who designs water fountains. One part of his job is determining the placement of the water jets that shoot the water into the air to create arcs. These arcs are modeled by parabolic functions. When a stream of water is shot into the air with velocity v at an angle of θ with the horizontal, the model predicts that the water will travel a horizontal distance of $D = \frac{v^2}{g} \sin 2\theta$ and reach a maximum height of $H = \frac{v^2}{2g} \sin^2 \theta$, where g is the acceleration due to gravity. The ratio of H to D helps determine the total height and width of the fountain. Express $\frac{H}{D}$ as a function of θ. *This problem will be solved in Example 3.*

1 FOCUS

5-Minute Check
Lesson 7-3

1. Verify that $\cos(90° + x) = -\sin x$ is an identity.

$$\cos(90° + x) \overset{?}{=} -\sin x$$
$$\cos 90°\cos x - \sin 90°\sin x \overset{?}{=} -\sin x$$
$$0(\cos x) - 1(\sin x) \overset{?}{=} -\sin x$$
$$-\sin x = -\sin x$$

2. Use the sum or difference identity for sine to find the exact value of $\sin 375°$.
$$\frac{\sqrt{6} - \sqrt{2}}{4}$$

3. Find the value of $\cos(x - y)$ if $0 < x < \frac{\pi}{2}$, $0 < y < \frac{\pi}{2}$, $\cos x = \frac{2}{3}$, and $\cos y = \frac{4}{7}$.
$$\frac{8 + \sqrt{165}}{21} \text{ or about } 0.9926$$

4. Use the sum or difference identity for tangent to find the exact value of $\tan 465°$.
$$-2 - \sqrt{3}$$

Motivating the Lesson
Real-World Application
Using the formula given in the application, determine how far horizontally a stream of water that is shot into the air with a velocity v of 15 m/s at an angle θ of 30° with the horizontal will travel. Use $g = 9.8$ m/s². **19.88 m**

It is sometimes useful to have identities to find the value of a function of twice an angle or half an angle. We can substitute θ for both α and β in $\sin(\alpha + \beta)$ to find an identity for $\sin 2\theta$.

$$\sin 2\theta = \sin(\theta + \theta)$$
$$= \sin \theta \cos \theta + \cos \theta \sin \theta \quad \textit{Sum identity for sine}$$
$$= 2 \sin \theta \cos \theta$$

The same method can be used to find an identity for $\cos 2\theta$.

$$\cos 2\theta = \cos(\theta + \theta)$$
$$= \cos \theta \cos \theta - \sin \theta \sin \theta \quad \textit{Sum identity for cosine}$$
$$= \cos^2 \theta - \sin^2 \theta$$

If we substitute $1 - \cos^2 \theta$ for $\sin^2 \theta$ or $1 - \sin^2 \theta$ for $\cos^2 \theta$, we will have two alternate identities for $\cos 2\theta$.

$$\cos 2\theta = 2 \cos^2 \theta - 1$$
$$\cos 2\theta = 1 - 2 \sin^2 \theta$$

These identities may be used if θ is measured in degrees or radians. So, θ may represent either a degree measure or a real number.

Resource Manager

 Reproducible Masters
* *Study Guide and Practice,* pp. 125–126
* *Enrichment,* p. 76
* *Assessment and Evaluation,* pp. 172, 173

 Transparencies
* 5-Minute Check, 7-4
* Teaching, 7-4

 Technology/Multimedia
* Advanced MathPASS, Lesson 17

The tangent of a double angle can be found by substituting θ for both α and β in $\tan(\alpha + \beta)$.

$$\tan 2\theta = \tan(\theta + \theta)$$

$$= \frac{\tan\theta + \tan\theta}{1 - \tan\theta\tan\theta} \quad \textit{Sum identity for tangent}$$

$$= \frac{2\tan\theta}{1 - \tan^2\theta}$$

Double-Angle Identities

If θ represents the measure of an angle, then the following identities hold for all values of θ.

$$\sin 2\theta = 2\sin\theta\cos\theta$$

$$\cos 2\theta = \cos^2\theta - \sin^2\theta$$

$$\cos 2\theta = 2\cos^2\theta - 1$$

$$\cos 2\theta = 1 - 2\sin^2\theta$$

$$\tan 2\theta = \frac{2\tan\theta}{1 - \tan^2\theta}$$

Example ① If $\sin\theta = \frac{2}{3}$ and θ has its terminal side in the first quadrant, find the exact value of each function.

a. $\sin 2\theta$

To use the double-angle identity for $\sin 2\theta$, we must first find $\cos\theta$.

$$\sin^2\theta + \cos^2\theta = 1$$

$$\left(\frac{2}{3}\right)^2 + \cos^2\theta = 1 \quad \sin\theta = \frac{2}{3}$$

$$\cos^2\theta = \frac{5}{9}$$

$$\cos\theta = \frac{\sqrt{5}}{3}$$

Then find $\sin 2\theta$.

$$\sin 2\theta = 2\sin\theta\cos\theta$$

$$= 2\left(\frac{2}{3}\right)\left(\frac{\sqrt{5}}{3}\right) \quad \sin\theta = \frac{2}{3};\ \cos\theta = \frac{\sqrt{5}}{3}$$

$$= \frac{4\sqrt{5}}{9}$$

b. $\cos 2\theta$

Since we know the values of $\cos\theta$ and $\sin\theta$, we can use any of the double-angle identities for cosine.

$$\cos 2\theta = \cos^2\theta - \sin^2\theta$$

$$= \left(\frac{\sqrt{5}}{3}\right)^2 - \left(\frac{2}{3}\right)^2 \quad \cos\theta = \frac{\sqrt{5}}{3};\ \sin\theta = \frac{2}{3}$$

$$= \frac{1}{9}$$

Lesson 7-4 Double-Angle and Half-Angle Identities **449**

2 TEACH

Teaching Tip For Example 1, when discussing the double-angle identities, ask students to explain how they could find similar identities for the cosecant, secant, and cotangent functions.
Find and simplify reciprocals of the sine, cosine, and tangent identities, respectively.

In-Class Example

Example 1
If $\sin\theta = \frac{3}{4}$ and θ has its terminal side in the first quadrant, find the exact value of each function.

a. $\sin 2\theta$ $\frac{3\sqrt{7}}{8}$

b. $\cos 2\theta$ $-\frac{1}{8}$

c. $\tan 2\theta$ $-3\sqrt{7}$

d. $\sin 4\theta$ $-\frac{3\sqrt{7}}{32}$

Teaching Tip After working Example 1, ask students to derive another identity for $\tan 2\theta$ using the identities for $\sin 2\theta$ and $\cos 2\theta$.
Sample answer:

$$\tan 2\theta = \frac{2\sin\theta\cos\theta}{\cos^2\theta - \sin^2\theta}$$

📂 Prerequisite Skills

A review of the prerequisite skills for this lesson is provided in the *Prerequisite Skills Masters.*

Skill Number and Content	Pages
4 Algebraic Expressions	17–18

c. tan 2θ

We must find $\tan\theta$ to use the double-angle identity for $\tan 2\theta$.

$$\tan\theta = \frac{\sin\theta}{\cos\theta}$$

$$= \frac{\frac{2}{3}}{\frac{\sqrt{5}}{3}} \quad \sin\theta = \frac{2}{3}, \cos\theta = \frac{\sqrt{5}}{3}$$

$$= \frac{2}{\sqrt{5}} \text{ or } \frac{2\sqrt{5}}{5}$$

Then find $\tan 2\theta$.

$$\tan 2\theta = \frac{2\tan\theta}{1 - \tan^2\theta}$$

$$= \frac{2\left(\frac{2\sqrt{5}}{5}\right)}{1 - \left(\frac{2\sqrt{5}}{5}\right)^2} \quad \tan\theta = \frac{2\sqrt{5}}{5}$$

$$= \frac{\frac{4\sqrt{5}}{5}}{\frac{1}{5}} \text{ or } 4\sqrt{5}$$

d. cos 4θ

Since $4\theta = 2(2\theta)$, use a double-angle identity for cosine again.

$$\cos 4\theta = \cos 2(2\theta)$$

$$= \cos^2(2\theta) - \sin^2(2\theta) \quad \textit{Double-angle identity}$$

$$= \left(\frac{1}{9}\right)^2 - \left(\frac{4\sqrt{5}}{9}\right)^2 \quad \cos 2\theta = \frac{1}{9}, \sin 2\theta = \frac{4\sqrt{5}}{9} \textit{ (parts a and b)}$$

$$= -\frac{79}{81}$$

We can solve two of the forms of the identity for $\cos 2\theta$ for $\cos\theta$ and $\sin\theta$, respectively, and the following equations result.

$$\cos 2\theta = 2\cos^2\theta - 1 \quad \blacktriangleright \text{ Solve for } \cos\theta. \quad \cos\theta = \pm\sqrt{\frac{1 + \cos 2\theta}{2}}$$

$$\cos 2\theta = 1 - 2\sin^2\theta \quad \blacktriangleright \text{ Solve for } \sin\theta. \quad \sin\theta = \pm\sqrt{\frac{1 - \cos 2\theta}{2}}$$

Addressing Individual Needs

Interpersonal This lesson gives the last of the basic trigonometric identities students will need to know for this chapter. Have small groups prepare index card pairs for each identity, writing the left side of the identity on one card, the right side on the other. Groups should then shuffle the cards, deal them face down, and play a memory card-matching game.

We can replace 2θ with α and θ with $\frac{\alpha}{2}$ to derive the half-angle identities.

$$\tan\frac{\alpha}{2} = \frac{\sin\frac{\alpha}{2}}{\cos\frac{\alpha}{2}}$$

$$= \frac{\pm\sqrt{\frac{1-\cos\alpha}{2}}}{\pm\sqrt{\frac{1+\cos\alpha}{2}}} \text{ or } \pm\sqrt{\frac{1-\cos\alpha}{1+\cos\alpha}}$$

If α represents the measure of an angle, then the following identities hold for all values of α.

Half-Angle Identities

$$\sin\frac{\alpha}{2} = \pm\sqrt{\frac{1-\cos\alpha}{2}}$$

$$\cos\frac{\alpha}{2} = \pm\sqrt{\frac{1+\cos\alpha}{2}}$$

$$\tan\frac{\alpha}{2} = \pm\sqrt{\frac{1-\cos\alpha}{1+\cos\alpha}}, \cos\alpha \neq -1$$

Unlike with the double-angles identities, you must determine the sign.

Example ❷ Use a half-angle identity to find the exact value of each function.

a. $\sin\frac{7\pi}{12}$

$$\sin\frac{7\pi}{12} = \sin\frac{\frac{7\pi}{6}}{2}$$

$$= \sqrt{\frac{1-\cos\frac{7\pi}{6}}{2}} \qquad \text{Use } \sin\frac{\alpha}{2} = \pm\sqrt{\frac{1-\cos\alpha}{2}}. \text{ Since } \frac{7\pi}{12} \text{ is in}$$
$$\text{Quadrant II, choose the positive sine value.}$$

$$= \sqrt{\frac{1-\left(-\frac{\sqrt{3}}{2}\right)}{2}}$$

$$= \frac{\sqrt{2+\sqrt{3}}}{2}$$

b. $\cos 67.5°$

$$\cos 67.5° = \cos\frac{135°}{2}$$

$$= \sqrt{\frac{1+\cos 135°}{2}} \qquad \text{Use } \cos\frac{\alpha}{2} = \pm\sqrt{\frac{1+\cos\alpha}{2}}. \text{ Since } 67.5° \text{ is in}$$
$$\text{Quadrant I, choose the positive cosine value.}$$

$$= \sqrt{\frac{1-\frac{\sqrt{2}}{2}}{2}}$$

$$= \frac{\sqrt{2-\sqrt{2}}}{2}$$

Teaching Tip Point out that as students apply a half-angle identity in Example 2, the quadrant that the terminal side of the angle lies in must first be determined so that the correct sign before the radical can be chosen.

In-Class Example

Example 2
Use a half-angle identity to find the exact value of each function.

a. $\cos\frac{\pi}{8}$ $\frac{\sqrt{2+\sqrt{2}}}{2}$

b. $\tan 15°$ $2-\sqrt{3}$

Double- and half-angle identities can be used to simplify trigonometric expressions.

Example **③ ARCHITECTURE** Refer to the application at the beginning of the lesson.

a. Find and simplify $\dfrac{H}{D}$.

b. What is the ratio of the maximum height of the water to the horizontal distance it travels for an angle of $27°$?

a. $\dfrac{H}{D} = \dfrac{\dfrac{v^2}{2g} \sin^2 \theta}{\dfrac{v^2}{g} \sin 2\theta}$

$= \dfrac{\sin^2 \theta}{2 \sin 2\theta}$ *Simplify.*

$= \dfrac{\sin^2 \theta}{4 \sin \theta \cos \theta}$ $\sin 2\theta = 2 \sin \theta \cos \theta$

$= \dfrac{1}{4} \cdot \dfrac{\sin \theta}{\cos \theta}$ *Simplify.*

$= \dfrac{1}{4} \tan \theta$ *Quotient identity:* $\dfrac{\sin \theta}{\cos \theta} = \tan \theta$

Therefore, the ratio of the maximum height of the water to the horizontal distance it travels is $\dfrac{1}{4} \tan \theta$.

b. When $\theta = 27°$, $\dfrac{H}{D} = \dfrac{1}{4} \tan 27°$, or about 0.13.

For an angle of $27°$, the ratio of the maximum height of the water to the horizontal distance it travels is about 0.13.

The double- and half-angle identities can also be used to verify other identities.

Example **④** Verify that $\dfrac{\cos 2\theta}{1 + \sin 2\theta} = \dfrac{\cot \theta - 1}{\cot \theta + 1}$ is an identity.

$$\dfrac{\cos 2\theta}{1 + \sin 2\theta} \overset{?}{=} \dfrac{\cot \theta - 1}{\cot \theta + 1}$$

$$\dfrac{\cos 2\theta}{1 + \sin 2\theta} \overset{?}{=} \dfrac{\dfrac{\cos \theta}{\sin \theta} - 1}{\dfrac{\cos \theta}{\sin \theta} + 1}$$ *Reciprocal identity:* $\cot \theta = \dfrac{\cos \theta}{\sin \theta}$

$$\dfrac{\cos 2\theta}{1 + \sin 2\theta} \overset{?}{=} \dfrac{\cos \theta - \sin \theta}{\cos \theta + \sin \theta}$$ *Multiply numerator and denominator by* $\sin \theta$.

$$\dfrac{\cos 2\theta}{1 + \sin 2\theta} \overset{?}{=} \dfrac{\cos \theta - \sin \theta}{\cos \theta + \sin \theta} \cdot \dfrac{\cos \theta + \sin \theta}{\cos \theta + \sin \theta}$$ *Multiply each side by 1.*

$$\dfrac{\cos 2\theta}{1 + \sin 2\theta} \overset{?}{=} \dfrac{\cos^2 \theta - \sin^2 \theta}{\cos^2 \theta + 2\cos \theta \sin \theta + \sin^2 \theta}$$ *Multiply.*

452   *Chapter 7   Trigonometric Identities and Equations*

$$\frac{\cos 2\theta}{1 + \sin 2\theta} \stackrel{?}{=} \frac{\cos^2 \theta - \sin^2 \theta}{1 + 2\cos \theta \sin \theta} \quad \textit{Simplify.}$$

$$\frac{\cos 2\theta}{1 + \sin 2\theta} = \frac{\cos 2\theta}{1 + \sin 2\theta} \quad \textit{Double-angle identities: } \cos^2 \theta - \sin^2 \theta = \cos 2\theta,$$
$$\textit{2 cos } \theta \sin \theta = \sin 2\theta$$

CHECK FOR UNDERSTANDING

Communicating Mathematics

Read and study the lesson to answer each question.

1. **Write** a paragraph about the conditions under which you would use each of the three identities for $\cos 2\theta$. **See margin.**

2. **Derive** the identity $\sin \frac{\alpha}{2} = \pm \sqrt{\frac{1 - \cos \alpha}{2}}$ from $\cos 2\theta = 1 - 2\sin^2 \theta$. **See margin.**

3. **Name** the quadrant in which the terminal side lies.

   a. x is a second quadrant angle. In which quadrant does $2x$ lie? **III or IV**

   b. $\frac{x}{2}$ is a first quadrant angle. In which quadrant does x lie? **I or II**

   c. $2x$ is a second quadrant angle. In which quadrant does $\frac{x}{2}$ lie? **I, II, III, or IV**

4. Sample answer: $\theta = \frac{\pi}{2}$

4. **Provide a counterexample** to show that $\sin 2\theta = 2\sin \theta$ is not an identity.

5. **You Decide** Tamika calculated the exact value of $\sin 15°$ in two different ways. Using the difference identity for sine, $\sin 15°$ was $\frac{\sqrt{6} - \sqrt{2}}{4}$. When she used the half-angle identity, $\sin 15°$ equaled $\frac{\sqrt{2 - \sqrt{3}}}{2}$. Which answer is correct? Explain. **See margin.**

Guided Practice

Use a half-angle identity to find the exact value of each function.

6. $\sin \frac{\pi}{8}$ $\frac{\sqrt{2 - \sqrt{2}}}{2}$

7. $\tan 165°$ $\sqrt{3} - 2$

Use the given information to find $\sin 2\theta$, $\cos 2\theta$, and $\tan 2\theta$.

8. $\sin \theta = \frac{2}{5}$, $0° < \theta < 90°$ $\frac{4\sqrt{21}}{25}$, $\frac{17}{25}$, $\frac{4\sqrt{21}}{17}$

9. $\tan \theta = \frac{4}{3}$, $\pi < \theta < \frac{3\pi}{2}$ $\frac{24}{25}$, $-\frac{7}{25}$, $-\frac{24}{7}$

Verify that each equation is an identity. **10–12. See margin.**

10. $\tan 2\theta = \frac{2}{\cot \theta - \tan \theta}$

11. $1 + \frac{1}{2}\sin 2A = \frac{\sec A + \sin A}{\sec A}$

12. $\sin \frac{x}{2} \cos \frac{x}{2} = \frac{\sin x}{2}$

13. **Electronics** Consider an AC circuit consisting of a power supply and a resistor. If the current in the circuit at time t is $I_0 \sin \omega t$, then the power delivered to the resistor is $P = I_0^2 R \sin^2 \omega t$, where R is the resistance. Express the power in terms of $\cos 2\omega t$. $P = \frac{1}{2}I_0^2 R - \frac{1}{2}I_0^2 R \cos 2\omega t$

 www.amc.glencoe.com/self_check_quiz   *Lesson 7-4 Double-Angle and Half-Angle Identities*   **453**

Additional Answers

5. **Both answers are correct. She obtained two different representations of the same number. One way to verify this is to evaluate each expression with a calculator. To verify it algebraically, square each answer and then simplify. The same result is obtained in each case. Since each of the original answers is positive, and they have the same square, the original answers are the same number.**

10. $\tan 2\theta \stackrel{?}{=} \dfrac{2}{\cot \theta - \tan \theta}$

   $\tan 2\theta \stackrel{?}{=} \dfrac{2}{\cot \theta - \tan \theta} \cdot \dfrac{\tan \theta}{\tan \theta}$

   $\tan 2\theta \stackrel{?}{=} \dfrac{2\tan \theta}{\cot \theta \tan \theta - \tan^2 \theta}$

   $\tan 2\theta \stackrel{?}{=} \dfrac{2\tan \theta}{1 - \tan^2 \theta}$

   $\tan 2\theta = \tan 2\theta$

---

3 PRACTICE/APPLY

Error Analysis

Watch for students who confuse double- and half-angle identities.

Prevent by suggesting that they can always test an identity using known angle measures such as 30° and 60° in order to confirm that they are using the correct expression.

Additional Answers

11. $1 + \frac{1}{2}\sin 2A$

   $\stackrel{?}{=} \dfrac{\sec A + \sin A}{\sec A}$

   $\stackrel{?}{=} \dfrac{\frac{1}{\cos A} + \sin A}{\frac{1}{\cos A}}$

   $\stackrel{?}{=} \dfrac{\frac{1}{\cos A} + \sin A}{\frac{1}{\cos A}} \cdot \dfrac{\cos A}{\cos A}$

   $\stackrel{?}{=} 1 + \sin A \cos A$

   $\stackrel{?}{=} 1 + \frac{1}{2} \cdot 2 \sin A \cos A$

   $= 1 + \frac{1}{2}\sin 2A$

12. $\sin \frac{x}{2} \cos \frac{x}{2} \stackrel{?}{=} \dfrac{\sin x}{2}$

   $\dfrac{2\sin \frac{x}{2} \cos \frac{x}{2}}{2} \stackrel{?}{=} \dfrac{\sin x}{2}$

   $\dfrac{\sin 2\left(\frac{x}{2}\right)}{2} \stackrel{?}{=} \dfrac{\sin x}{2}$

   $\dfrac{\sin x}{2} = \dfrac{\sin x}{2}$

Study Guide and Practice Masters, pp. 125–126

Additional Answer

37. ∠*PBD* is an inscribed angle that subtends the same arc as the central angle ∠*POD*, so $m\angle PBD = \frac{1}{2}\theta$. By right triangle trigonometry,
$$\tan \frac{1}{2}\theta = \frac{PA}{BA} = \frac{PA}{1 + OA} = \frac{\sin \theta}{1 + \cos \theta}.$$

24. $-\frac{3\sqrt{7}}{8}, \frac{1}{8},$
$-3\sqrt{7}$

26. $-\frac{4\sqrt{21}}{25}, \frac{17}{25},$
$-\frac{4\sqrt{21}}{17}$

EXERCISES

Practice

Use a half-angle identity to find the exact value of each function.

 A

14. $\cos 15°$ $\dfrac{\sqrt{2 + \sqrt{3}}}{2}$ **15.** $\sin 75°$ $\dfrac{\sqrt{2 + \sqrt{3}}}{2}$ **16.** $\tan \dfrac{5\pi}{12}$ $2 + \sqrt{3}$

17. $\sin \dfrac{3\pi}{8}$ $\dfrac{\sqrt{2 + \sqrt{2}}}{2}$ **18.** $\cos \dfrac{7\pi}{12}$ $-\dfrac{\sqrt{2 - \sqrt{3}}}{2}$ **19.** $\tan 22.5°$ $\sqrt{2} - 1$

20. If θ is an angle in the first quadrant and $\cos \theta = \dfrac{1}{4}$, find $\tan \dfrac{\theta}{2}$. $\dfrac{\sqrt{15}}{5}$

Use the given information to find $\sin 2\theta$, $\cos 2\theta$, and $\tan 2\theta$.

 B

21. $\cos \theta = \dfrac{4}{5}, 0° < \theta < 90°$ $\dfrac{24}{25}, \dfrac{7}{25}, \dfrac{24}{7}$ **22.** $\sin \theta = \dfrac{1}{3}, 0 < \theta < \dfrac{\pi}{2}$ $\dfrac{4\sqrt{2}}{9}, \dfrac{7}{9}, \dfrac{4\sqrt{2}}{7}$

23. $\tan \theta = -2, \dfrac{\pi}{2} < \theta < \pi$ $-\dfrac{4}{5}, -\dfrac{3}{5}, \dfrac{4}{3}$ **24.** $\sec \theta = -\dfrac{4}{3}, 90° < \theta < 180°$

25. $\cot \theta = \dfrac{3}{2}, 180° < \theta < 270°$ **26.** $\csc \theta = -\dfrac{5}{2}, \dfrac{3\pi}{2} < \theta < 2\pi$
        $\dfrac{12}{13}, \dfrac{5}{13}, \dfrac{12}{5}$

27. If α is an angle in the second quadrant and $\cos \alpha = -\dfrac{\sqrt{2}}{3}$, find $\tan 2\alpha$. $\dfrac{2\sqrt{14}}{5}$

Verify that each equation is an identity.   **28–35. See Answer Appendix.**

28. $\csc 2\theta = \dfrac{1}{2} \sec \theta \csc \theta$ **29.** $\cos A - \sin A = \dfrac{\cos 2A}{\cos A + \sin A}$

30. $(\sin \theta + \cos \theta)^2 - 1 = \sin 2\theta$ **31.** $\cos x - 1 = \dfrac{\cos 2x - 1}{2(\cos x + 1)}$

32. $\sec 2\theta = \dfrac{\cos^2 \theta + \sin^2 \theta}{\cos^2 \theta - \sin^2 \theta}$ **33.** $\tan \dfrac{A}{2} = \dfrac{\sin A}{1 + \cos A}$

C

34. $\sin 3x = 3 \sin x - 4 \sin^3 x$ **35.** $\cos 3x = 4 \cos^3 x - 3 \cos x$

Applications and Problem Solving

36. Architecture   Refer to the application at the beginning of the lesson. If the angle of the water is doubled, what is the ratio of the new maximum height to the original maximum height? $4 \cos^2 \theta$

37. Critical Thinking   Circle O is a unit circle. Use the figure to prove that $\tan \dfrac{1}{2}\theta = \dfrac{\sin \theta}{1 + \cos \theta}$.   **See margin.**

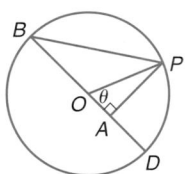

38. Physics   Suppose a projectile is launched with velocity v at an angle θ to the horizontal from the base of a hill that makes an angle α with the horizontal ($\theta > \alpha$). Then the range of the projectile, measured along the slope of the hill, is given by $R = \dfrac{2v^2 \cos \theta \sin (\theta - \alpha)}{g \cos^2 \alpha}$. Show that if $\alpha = 45°$, then

$$R = \dfrac{v^2 \sqrt{2}}{g} (\sin 2\theta - \cos 2\theta - 1). \quad \textbf{See margin.}$$

Additional Answer

38. $R = \dfrac{2v^2 \cos \theta \sin (\theta - \alpha)}{g \cos^2 \alpha}$

$R = \dfrac{2v^2 \cos \theta \, \sin (\theta - 45°)}{g \cos^2 45°}$

$R = \dfrac{2v^2 \cos \theta \, (\sin \theta \cos 45° - \cos \theta \sin 45°)}{g \cos^2 45°}$

$R = \dfrac{2v^2 \cos \theta \left((\sin\theta)\left(\frac{\sqrt{2}}{2}\right) - (\cos\theta)\left(\frac{\sqrt{2}}{2}\right)\right)}{g\left(\frac{\sqrt{2}}{2}\right)^2}$

$R = \dfrac{2 \frac{\sqrt{2}}{2} v^2 \cos \theta \, (\sin \theta - \cos \theta)}{g \cdot \frac{1}{2}}$

$R = \dfrac{\sqrt{2} v^2 \, (2 \cos \theta \sin \theta - 2 \cos^2 \theta)}{g}$

$R = \dfrac{v^2 \sqrt{2}}{g} (2 \cos \theta \sin \theta - (2 \cos^2 \theta - 1) - 1)$

$R = \dfrac{v^2 \sqrt{2}}{g} (\sin 2\theta - \cos 2\theta - 1)$

39a.
$$1 \pm \sqrt{\dfrac{1 - \cos L}{1 + \cos L}}$$
$$\overline{1 \mp \sqrt{\dfrac{1 - \cos L}{1 + \cos L}}}$$

39. Geography The Mercator projection of the globe is a projection on which the distance between the lines of latitude increases with their distance from the equator. The calculation of the location of a point on this projection involves the expression $\tan\left(45° + \dfrac{L}{2}\right)$, where L is the latitude of the point.

THE WORLD
ON MERCATOR'S PROJECTION.

  a. Write this expression in terms of a trigonometric function of L.

  b. Find the value of this expression if $L = 60°$. **$2 + \sqrt{3}$**

40. Critical Thinking Determine the tangent of angle α in the figure. **See Answer Appendix.**

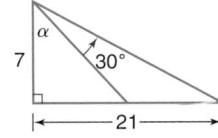

Mixed Review

41. Find the exact value of $\sec \dfrac{\pi}{12}$. *(Lesson 7-3)* **$\sqrt{6} - \sqrt{2}$**

42. Show that $\sin x^2 + \cos x^2 = 1$ is not an identity. *(Lesson 7-1)* **See Answer Appendix.**

43. Find the degree measure to the nearest tenth of the central angle of a circle of radius 10 centimeters if the measure of the subtended arc is 17 centimeters. *(Lesson 6-1)* **97.4°**

44. Surveying To find the height of a mountain peak, points A and B were located on a plain in line with the peak, and the angle of elevation was measured from each point. The angle at A was $36°40'$, and the angle at B was $21°10'$. The distance from A to B was 570 feet. How high is the peak above the level of the plain? *(Lesson 5-4)* **about 460 ft**

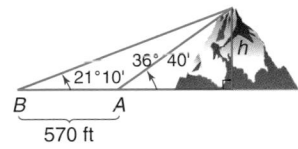

570 ft

45. Write a polynomial equation of least degree with roots $-3, 0.5, 6,$ and 2. *(Lesson 4-1)* **$2x^4 - 11x^3 - 19x^2 + 84x - 36 = 0$**

46. Graph $y = 2x + 5$ and its inverse. *(Lesson 3-4)* **See margin.**

47. Solve the system of equations. *(Lesson 2-1)* **(7, 2)**
$$x + 2y = 11$$
$$3x - 5y = 11$$

48. SAT Practice Grid-In If $(a - b)^2 = 64$, and $ab = 3$, find $a^2 + b^2$. **70**

Lesson 7-4   Double-Angle and Half-Angle Identities **455**

Extra Credit

Verify that $\sin 2\theta \cos \theta - \cos 2\theta \sin \theta = \sin \theta$ is an identity.
$$\sin 2\theta \cos \theta - \cos 2\theta \sin \theta \overset{?}{=} \sin \theta$$
$$2 \sin \theta \cos \theta \cos \theta - (1 - 2 \sin^2 \theta) \sin \theta \overset{?}{=} \sin \theta$$
$$2 \sin \theta \cos^2 \theta - \sin \theta (1 - 2 \sin^2 \theta) \overset{?}{=} \sin \theta$$
$$2 \sin \theta (1 - \sin^2 \theta) - \sin \theta + 2 \sin^3 \theta \overset{?}{=} \sin \theta$$
$$2 \sin \theta - 2 \sin^3 \theta - \sin \theta + 2 \sin^3 \theta \overset{?}{=} \sin \theta$$
$$\sin \theta = \sin \theta$$

Lesson 7-4 **455**

5-Minute Check
Lesson 7-4

If $\sin \theta = \frac{2}{5}$ and $0 < \theta < \frac{\pi}{2}$, find the exact value of each function.

1. $\cos 2\theta$ $\frac{17}{25}$

2. $\sin 2\theta$ $\frac{4\sqrt{21}}{25}$

Use a half-angle identity to find the exact value of each function.

3. $\sin 112.5°$ $\frac{\sqrt{2 + \sqrt{2}}}{2}$

4. $\cos \frac{5\pi}{12}$ $\frac{\sqrt{2 - \sqrt{3}}}{2}$

5. Verify that $2 \tan x = \frac{\sin 2x}{1 - \sin^2 x}$ is an identity.

$$2 \tan x \overset{?}{=} \frac{\sin 2x}{1 - \sin^2 x}$$

$$2 \tan x \overset{?}{=} \frac{2 \sin x \cos x}{\cos^2 x}$$

$$2 \tan x \overset{?}{=} \frac{2 \sin x}{\cos x}$$

$$2 \tan x = 2 \tan x$$

Motivating the Lesson

Hands-On Activity Certain rides at the fair exhibit periodic motion such as the Ferris wheel described in the Application. Ask students to list some other machines that they have observed that show repetitive motion.

Solving Trigonometric Equations

OBJECTIVE
- Solve trigonometric equations and inequalities.

ENTERTAINMENT

When you ride a Ferris wheel that has a diameter of 40 meters and turns at a rate of 1.5 revolutions per minute, the height above the ground, in meters, of your seat after t minutes can be modeled by the equation $h = 21 - 20 \cos 3\pi t$. How long after the ride starts will your seat first be 31 meters above the ground? *This problem will be solved in Example 4.*

So far in this chapter, we have studied a special type of trigonometric equation called an identity. Trigonometric identities are equations that are true for all values of the variable for which both sides are defined. In this lesson, we will examine another type of trigonometric equation. These equations are true for only certain values of the variable. Solving these equations resembles solving algebraic equations.

Most trigonometric equations have more than one solution. If the variable is not restricted, the periodic nature of trigonometric functions will result in an infinite number of solutions. Also, many trigonometric expressions will take on a given value twice every period.

If the variable is restricted to two adjacent quadrants, a trigonometric equation will have fewer solutions. These solutions are called **principal values.** For $\sin x$ and $\tan x$, the principal values are in Quadrants I and IV. So x is in the interval $-90° \le x \le 90°$. For $\cos x$, the principal values are in Quadrants I and II, so x is in the interval $0° \le x \le 180°$.

Example **1** Solve $\sin x \cos x - \frac{1}{2} \cos x = 0$ for principal values of x. Express solutions in degrees.

$$\sin x \cos x - \frac{1}{2} \cos x = 0$$

$$\cos x \left(\sin x - \frac{1}{2} \right) = 0 \quad \text{Factor.}$$

$$\cos x = 0 \quad \text{or} \quad \sin x - \frac{1}{2} = 0 \quad \text{Set each factor equal to 0.}$$

$$x = 90° \qquad\qquad \sin x = \frac{1}{2}$$

$$x = 30°$$

The principal values are $30°$ and $90°$.

If an equation cannot be solved easily by factoring, try writing the expressions in terms of only one trigonometric function. Remember to use your knowledge of identities.

Example **2** Solve $\cos^2 x - \cos x + 1 = \sin^2 x$ for $0 \le x < 2\pi$.

This equation can be written in terms of $\cos x$ only.

$\cos^2 x - \cos x + 1 = \sin^2 x$

$\cos^2 x - \cos x + 1 = 1 - \cos^2 x$ *Pythagorean identity: $\sin^2 x = 1 - \cos^2 x$*

$2 \cos^2 x - \cos x = 0$

$\cos x (2 \cos x - 1) = 0$ *Factor.*

$\cos x = 0$ or $2 \cos x - 1 = 0$

    $x = \dfrac{\pi}{2}$ or $x = \dfrac{3\pi}{2}$ $\cos x = \dfrac{1}{2}$

                                          $x = \dfrac{\pi}{3}$ or $x = \dfrac{5\pi}{3}$

The solutions are $\dfrac{\pi}{3}, \dfrac{\pi}{2}, \dfrac{3\pi}{2}$, and $\dfrac{5\pi}{3}$.

As indicated earlier, most trigonometric equations have infinitely many solutions. When all of the values of x are required, the solution should be represented as $x + 360k°$ or $x + 2\pi k$ for $\sin x$ and $\cos x$ and $x + 180k°$ or $x + \pi k$ for $\tan x$, where k is any integer.

Example **3** Solve $2 \sec^2 x - \tan^4 x = -1$ for all real values of x.

A Pythagorean identity can be used to write this equation in terms of $\tan x$ only.

$2 \sec^2 x - \tan^4 x = -1$

$2(1 + \tan^2 x) - \tan^4 x = -1$ *Pythagorean identity: $\sec^2 x = 1 + \tan^2 x$*

$2 + 2 \tan^2 x - \tan^4 x = -1$ *Simplify.*

$\tan^4 x - 2 \tan^2 x - 3 = 0$

$(\tan^2 x - 3)(\tan^2 x + 1) = 0$ *Factor.*

$\tan^2 x - 3 = 0$ or $\tan^2 x + 1 = 0$

     $\tan^2 x = 3$ $\tan^2 x = -1$ *This part gives no*

     $\tan x = \pm\sqrt{3}$ *solutions since*
                                                          $\tan^2 x \ge 0$.

    $x = \dfrac{\pi}{3} + \pi k$ or $x = -\dfrac{\pi}{3} + \pi k$, where k is any integer.

When a problem asks for real values of x, use radians.

The solutions are $\dfrac{\pi}{3} + \pi k$ and $-\dfrac{\pi}{3} + \pi k$.

There are times when a general expression for all of the solutions is helpful for determining a specific solution.

Teaching Tip For Example 2, point out that useful forms of the Pythagorean identity $\sin^2 \theta + \cos^2 \theta = 1$ are $\sin^2 \theta = 1 - \cos^2 \theta$ and $\cos^2 \theta = 1 - \sin^2 \theta$, and that these are especially valuable for changing some trigonometric equations in two functions into an equation in one function.

In-Class Examples

Example 1
Solve $2 \sin^2 x + \sin x - 1 = 0$ for principal values of x. Express solutions in degrees.
$x = -90°, 30°$

Example 2
Solve $\sin^2 x - \sin x + 1 = \cos^2 x$ for $0 \le x < 2\pi$.
$0, \pi, \dfrac{\pi}{6}, \dfrac{5\pi}{6}$

Example 3
Solve $4 \cos x \sin x - 2 \sin x - 2\sqrt{3} \cos x + \sqrt{3} = 0$ for all real values of x. $\dfrac{\pi}{3} + 2\pi k, \dfrac{2\pi}{3} + 2\pi k, \dfrac{5\pi}{3} + 2\pi k$

Teaching Tip For Example 3, point out that the method of factoring by grouping is at times used to solve trigonometric equations just as it is occasionally used to solve algebraic equations.

Addressing Individual Needs

Intrapersonal Ask students to create a set of trigonometric equations and inequalities and provide solutions for each. Suggest that they add a new problem to the set each day for the next several days.

Example ④ **ENTERTAINMENT** Refer to the application at the beginning of the lesson. How long after the Ferris wheel starts will your seat first be 31 meters above the ground?

$h = 21 - 20 \cos 3\pi t$

$31 = 21 - 20 \cos 3\pi t$ *Replace h with 31.*

$-\dfrac{1}{2} = \cos 3\pi t$

$\dfrac{2\pi}{3} + 2\pi k = 3\pi t$ or $\dfrac{4\pi}{3} + 2\pi k = 3\pi t$ *where k is any integer*

$\dfrac{2}{9} + \dfrac{2}{3}k = t$ or $\dfrac{4}{9} + \dfrac{2}{3}k = t$

The least positive value for *t* is obtained by letting $k = 0$ in the first expression. Therefore, $t = \dfrac{2}{9}$ of a minute or about 13 seconds.

You can solve some trigonometric inequalities using the same techniques as for algebraic inequalities. The unit circle can be useful when deciding which angles to include in the answer.

Example ⑤ **Solve $2 \sin \theta + 1 > 0$ for $0 \le \theta < 2\pi$.**

$2 \sin \theta + 1 > 0$

$\sin \theta > -\dfrac{1}{2}$ *Solve for sin θ.*

In terms of the unit circle, we need to find points with *y*-coordinates greater than $-\dfrac{1}{2}$.

The values of θ for which $\sin \theta = -\dfrac{1}{2}$ are $\dfrac{7\pi}{6}$ and $\dfrac{11\pi}{6}$. The figure shows that the solution of the inequality is $0 \le \theta < \dfrac{7\pi}{6}$ or $\dfrac{11\pi}{6} < \theta < 2\pi$.

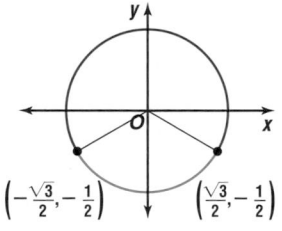

$\left(-\dfrac{\sqrt{3}}{2}, -\dfrac{1}{2}\right)$ $\left(\dfrac{\sqrt{3}}{2}, -\dfrac{1}{2}\right)$

GRAPHING CALCULATOR EXPLORATION

Some trigonometric equations and inequalities are difficult or impossible to solve with only algebraic methods. A graphing calculator is helpful in such cases.

TRY THESE   Graph each side of the equation as a separate function. 1–5. See margin.

1. $\sin x = 2 \cos x$ for $0 \le x \le 2\pi$

2. $\tan 0.5x = \cos x$ for $-2\pi \le x \le 2\pi$

3. Use the **CALC** menu to find the intersection point(s) of the graphs in Exercises 1 and 2.

WHAT DO YOU THINK?

4. What do the values in Exercise 3 represent? How could you verify this conjecture?

5. Graph $y = 2 \cos x - \sin x$ for $0 \le x \le 2\pi$.

  a. How could you use the graph to solve the equation $\sin x = 2 \cos x$? How does this solution compare with those found in Exercise 3?

  b. What equation would you use to apply this method to $\tan 0.5x = \cos x$?

4. The *x*-coordinates are the solutions of the equations. Substitute the *x*-coordinates and see that the two sides of the equation are equal.

5.

5a. The *x*-intercepts of the graph are the solutions of the equation $\sin x = 2 \cos x$. They are the same.

5b. $y = \tan 0.5x - \cos x$ or $y = \cos x - \tan 0.5x$

Communicating Mathematics

Read and study the lesson to answer each question. **1–4. See margin.**

1. **Explain** the difference between a trigonometric identity and a trigonometric equation that is not an identity.

2. **Explain** why many trigonometric equations have infinitely many solutions.

3. **Write** all the solutions to a trigonometric equation in terms of $\sin x$, given that the solutions between 0° and 360° are 45° and 135°.

4. *Math Journal* **Compare and contrast** solving trigonometric equations with solving linear and quadratic equations. What techniques are the same? What techniques are different? How many solutions do you expect? Do you use a graphing calculator in a similar manner?

Guided Practice

Solve each equation for principal values of x. Express solutions in degrees.

5. $2 \sin x + 1 = 0$ **−30°**

6. $2 \cos x - \sqrt{3} = 0$ **30°**

Solve each equation for $0° \le x < 360°$.

7. $\sin x \cot x = \dfrac{\sqrt{3}}{2}$ **30°, 330°**

8. $\cos 2x = \sin^2 x - 2$ **90°, 270°**

Solve each equation for $0 \le x < 2\pi$.

9. $3 \tan^2 x - 1 = 0$ $\dfrac{\pi}{6}, \dfrac{5\pi}{6}, \dfrac{7\pi}{6}, \dfrac{11\pi}{6}$

10. $2 \sin^2 x = 5 \sin x + 3$ $\dfrac{7\pi}{6}, \dfrac{11\pi}{6}$

Solve each equation for all real values of x.

12. $\dfrac{3\pi}{4} + \pi k$

11. $\sin^2 2x + \cos^2 x = 0$ $\dfrac{\pi}{2} + \pi k$

12. $\tan^2 x + 2 \tan x + 1 = 0$

13. $\cos^2 x + 3 \cos x = -2$ $(2k+1)\pi$

14. $\sin 2x - \cos x = 0$

14. $\dfrac{\pi}{2} + \pi k,$
$\dfrac{\pi}{6} + 2\pi k,$
$\dfrac{5\pi}{6} + 2\pi k$

15. Solve $2 \cos \theta + 1 < 0$ for $0 \le \theta < 2\pi$. $\dfrac{2\pi}{3} < \theta < \dfrac{4\pi}{3}$

16. **Physics** The work done in moving an object through a displacement of d meters is given by $W = Fd \cos \theta$, where θ is the angle between the displacement and the force F exerted. If Lisa does 1500 joules of work while exerting a 100-newton force over 20 meters, at what angle was she exerting the force? **41.41°**

Practice

A

Solve each equation for principal values of x. Express solutions in degrees.

17. $\sqrt{2} \sin x - 1 = 0$ **45°**

18. $2 \cos x + 1 = 0$ **120°**

19. $\sin 2x - 1 = 0$ **45°**

20. $\tan 2x - \sqrt{3} = 0$ **−60°, 30°**

21. $\cos^2 x = \cos x$ **0°, 90°**

22. $\sin x = 1 + \cos^2 x$ **90°**

Solve each equation for $0° \le x < 360°$.

23. $\sqrt{2} \cos x + 1 = 0$ **135°, 225°**

24. $\cos x \tan x = \dfrac{1}{2}$ **30°, 150°**

25. **0°, 45°, 180°,**
225°

25. $\sin x \tan x - \sin x = 0$

26. $2 \cos^2 x + 3 \cos x - 2 = 0$ **60°, 300°**

B

27. $\sin 2x = -\sin x$
0°, 120°, 180° 240°

28. $\cos (x + 45°) + \cos (x - 45°) = \sqrt{2}$ **0°**

 www.amc.glencoe.com/self_check_quiz

Lesson 7-5 Solving Trigonometric Equations **459**

Additional Answers

1. **A trigonometric identity is an equation that is true for all values of the variable for which each side of the equation is defined. A trigonometric equation that is not an identity is only true for certain values of the variable.**

2. **All trigonometric functions are periodic. Adding the least common multiple of the periods of the functions that appear to any solution to the equation will always produce another solution.**

3. **45° + 360x° and 135° + 360x°, where x is any integer**

(continued on page 460)

Error Analysis

Watch for students who have difficulty remembering principal values.

Prevent by having students use a visual aid for quadrants involving sine, cosine, and tangent.

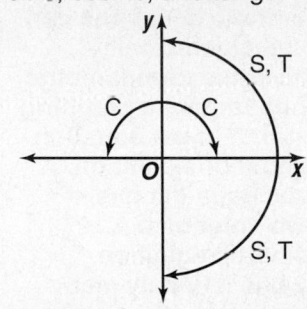

Assignment Guide

Core: 17–53 odd, 55–57, 59–72

Enriched: 18–54 even, 55–72

Study Guide and Practice Masters, pp. 127–128

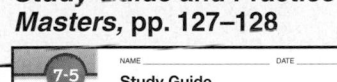

7-5 **Study Guide**

Solving Trigonometric Equations

When you solve trigonometric equations for **principal values** of x, x is in the interval $-90° \le x \le 90°$ for $\sin x$ and $\tan x$. For $\cos x$, x is in the interval $0° \le x \le 180°$. If an equation cannot be solved easily by factoring, try writing the expressions in terms of only one trigonometric function.

Example 1 Solve $\tan x \cos x - \cos x = 0$ for principal values of x. Express solutions in degrees.

$$\tan x \cos x - \cos x = 0$$
$$\cos x (\tan x - 1) = 0 \qquad Factor.$$
$$\cos x = 0 \quad or \quad \tan x - 1 = 0 \qquad Set\ each\ factor\ equal\ to\ 0.$$
$$x = 90° \qquad \tan x = 1$$
$$x = 45°$$

When $x = 90°$, $\tan x$ is undefined, so the only principal value is 45°.

Example 2 Solve $2 \tan^2 x - \sec^2 x + 3 = 1 - 2 \tan x$ for $0 \le x < 2\pi$.

This equation can be written in terms of $\tan x$ only.

$$2 \tan^2 x - \sec^2 x + 3 = 1 - 2 \tan x$$
$$2 \tan^2 x - (\tan^2 x + 1) + 3 = 1 - 2 \tan x \quad \sec^2 x = \tan^2 x + 1$$
$$\tan^2 x + 2 = 1 - 2 \tan x \qquad Simplify.$$
$$\tan^2 x + 2 \tan x + 1 = 0$$
$$(\tan x + 1)^2 = 0 \qquad Factor.$$
$$\tan x + 1 = 0 \qquad Take\ the\ square\ root\ of\ each\ side.$$
$$\tan x = -1$$
$$x = \tfrac{3\pi}{4} \ or \ x = \tfrac{7\pi}{4}$$

When you solve for all values of x, the solution should be represented as $x + 360°k$ or $x + 2\pi k$ for $\sin$ and $\cos x$ and $x + 180°k$ or $x + \pi k$ for $\tan x$, where k is any integer.

Example 3 Solve $\sin x + \sqrt{3} = -\sin x$ for all real values of x.

$$\sin x + \sqrt{3} = -\sin x$$
$$2 \sin x + \sqrt{3} = 0$$
$$2 \sin x = -\sqrt{3}$$
$$\sin x = -\dfrac{\sqrt{3}}{2}$$
$$x = \tfrac{4\pi}{3} + 2\pi k \ or \ x = \tfrac{5\pi}{3} + 2\pi k, \text{ where } k \text{ is any integer}$$

The solutions are $\tfrac{4\pi}{3} + 2\pi k$ and $\tfrac{5\pi}{3} + 2\pi k$.

© Glencoe/McGraw-Hill   127   *Advanced Mathematical Concepts*

4. Each type of equation may require adding, subtracting, multiplying, or dividing each side by the same number. Quadratic and trigonometric equations can often be solved by factoring. Linear and quadratic equations do not require identities. All linear and quadratic equations can be solved algebraically, whereas some trigonometric equations require a graphing calculator. A linear equation has at most one solution. A quadratic equation has at most two solutions. A trigonometric equation usually has infinitely many solutions unless the values of the variable are restricted.

37. $\frac{7\pi}{6} + 2\pi k, \frac{11\pi}{6} + 2\pi k$

38. $\frac{\pi}{6} + 2\pi k, \frac{5\pi}{6} + 2\pi k$

39. $\pi k, \frac{\pi}{6} + \pi k$

40. $\frac{\pi}{6} + 2\pi k, \frac{5\pi}{6} + 2\pi k$

41. πk

42. $\frac{\pi}{3} + 2\pi k, \frac{5\pi}{3} + 2\pi k$

43. $\frac{\pi}{4} + \pi k$

44. $\frac{\pi}{12} + \pi k, \frac{11\pi}{12} + \pi k$

45. $\frac{\pi}{2} + \pi k$

46. $\frac{2\pi}{3} + 2\pi k, \frac{4\pi}{3} + 2\pi k$

47. $2\pi k, \frac{\pi}{2} + 2\pi k$

48. $\frac{\pi}{6} + 2\pi k, \frac{\pi}{2} + 2\pi k, \frac{5\pi}{6} + 2\pi k$

49. $\frac{5\pi}{6} \le \theta \le \frac{7\pi}{6}$

50. $0 \le \theta < \frac{\pi}{3}$ or $\frac{5\pi}{3} < \theta < 2\pi$

51. $0 \le \theta < \frac{\pi}{4}$ or $\frac{3\pi}{4} < \theta < 2\pi$

Teaching Tip For Exercises 52–54, remind students that the window setting for the *x*-axis is given by the interval $0 \le x < 2\pi$. The *y*-axis window setting will vary.

---

29. Find all solutions to $2 \sin \theta \cos \theta + \sqrt{3} \sin \theta = 0$ in the interval $0° \le \theta < 360°$.
0°, 150°, 180°, 210°

Solve each equation for $0 \le x < 2\pi$.

30. $\frac{\pi}{6}, \frac{\pi}{4}, \frac{3\pi}{4}, \frac{5\pi}{6}, \frac{5\pi}{4}, \frac{7\pi}{4}$

30. $(2 \sin x - 1)(2 \cos^2 x - 1) = 0$

31. $4 \sin^2 x + 1 = -4 \sin x$ $\frac{7\pi}{6}, \frac{11\pi}{6}$

32. $\sqrt{2} \tan x = 2 \sin x$ $0, \frac{\pi}{4}, \pi, \frac{7\pi}{4}$

33. $\sin x = \cos 2x - 1$ $0, \pi, \frac{7\pi}{6}, \frac{11\pi}{6}$

34. $\cot^2 x - \csc x = 1$ $\frac{\pi}{6}, \frac{5\pi}{6}, \frac{3\pi}{2}$

35. $\sin x + \cos x = 0$ $\frac{3\pi}{4}, \frac{7\pi}{4}$

36. Find all values of θ between 0 and 2π that satisfy $-1 - 3 \sin \theta = \cos 2\theta$. $\frac{7\pi}{6}, \frac{11\pi}{6}$

Solve each equation for all real values of *x*.   37–48. See margin.

37. $\sin x = -\frac{1}{2}$

38. $\cos x \tan x - 2 \cos^2 x = -1$

39. $3 \tan^2 x = \sqrt{3} \tan x$

40. $2 \cos^2 x = 3 \sin x$

 41. $\frac{1}{\cos x - \sin x} = \cos x + \sin x$

42. $2 \tan^2 x - 3 \sec x = 0$

43. $\sin x \cos x = \frac{1}{2}$

44. $\cos^2 x - \sin^2 x = \frac{\sqrt{3}}{2}$

45. $\sin^4 x - 1 = 0$

46. $\sec^2 x + 2 \sec x = 0$

47. $\sin x + \cos x = 1$

48. $2 \sin x + \csc x = 3$

Solve each inequality for $0 \le \theta < 2\pi$.   49–51. See margin.

49. $\cos \theta \le -\frac{\sqrt{3}}{2}$

50. $\cos \theta - \frac{1}{2} > 0$

51. $\sqrt{2} \sin \theta - 1 < 0$

Graphing Calculator

Solve each equation graphically on the interval $0 \le x < 2\pi$.

52. $\tan x = 0.5$
0.4636, 3.6052

53. $\sin x - \frac{x}{2} = 0$
0, 1.8955

54. $\cos x = 3 \sin x$
0.3218, 3.4633

Applications and Problem Solving

Real World Application

55. Optics When light passes through a small slit, it is diffracted. The angle θ subtended by the first diffraction minimum is related to the wavelength λ of the light and the width D of the slit by the equation $\sin \theta = \frac{\lambda}{D}$. Consider light of wavelength 550 nanometers (5.5×10^{-7} m). What is the angle subtended by the first diffraction minimum when the light passes through a slit of width 3 millimeters?   **0.01°**

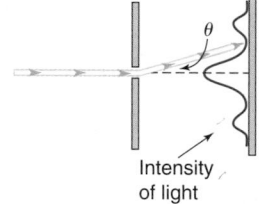

Intensity of light

56. $\frac{\pi}{3} < x < \pi$ or $\frac{5\pi}{3} < x < 2\pi$

56. Critical Thinking Solve the inequality $\sin 2x < \sin x$ for $0 \le x < 2\pi$ without a calculator.

57. Physics The range of a projectile that is launched with an initial velocity *v* at an angle of θ with the horizontal is given by $R = \frac{v^2}{g} \sin 2\theta$, where *g* is the acceleration due to gravity or 9.8 meters per second squared. If a projectile is launched with an initial velocity of 15 meters per second, what angle is required to achieve a range of 20 meters?   **30.29° or 59.71°**

58. Gemology The sparkle of a diamond is created by *refracted* light. Light travels at different speeds in different mediums. When light rays pass from one medium to another in which they travel at a different velocity, the light is bent, or refracted. According to Snell's Law, $n_1 \sin i = n_2 \sin r$, where n_1 is the index of refraction of the medium the light is exiting, n_2 is the index of refraction of the medium the light is entering, i is the angle of incidence, and r is the angle of refraction.

a. The index of refraction of a diamond is 2.42, and the index of refraction of air is 1.00. If a beam of light strikes a diamond at an angle of 35°, what is the angle of refraction?   **13.71°**

b. Explain how a gemologist might use Snell's Law to determine if a diamond is genuine.   **See margin.**

59. Music   A wave traveling in a guitar string can be modeled by the equation $D = 0.5 \sin(6.5x) \sin(2500t)$, where D is the displacement in millimeters at the position x meters from the left end of the string at time t seconds. Find the first positive time when the point 0.5 meter from the left end has a displacement of 0.01 millimeter.   **0.0013 s**

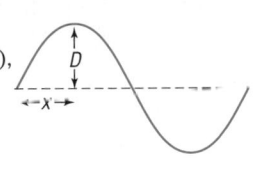

60. Critical Thinking   How many solutions in the interval $0° \le x < 360°$ should you expect for the equation $a \sin(bx + c) + d = d + \dfrac{a}{2}$, if $a \ne 0$ and b is a positive integer?   **2b**

61. Geometry   The point $P(x, y)$ can be rotated θ degrees counterclockwise about the origin by multiplying the matrix $\begin{bmatrix} x \\ y \end{bmatrix}$ on the left by the rotation matrix $R_\theta = \begin{bmatrix} \cos\theta & -\sin\theta \\ \sin\theta & \cos\theta \end{bmatrix}$. Determine the angle required to rotate the point $P(3, 4)$ to the point $P'(\sqrt{17}, 2\sqrt{2})$.   **341.32°**

Mixed Review

62. Find the exact value of cot 67.5°. *(Lesson 7-4)* $\sqrt{2} - 1$

63. Find a numerical value of one trigonometric function of x if $\dfrac{\tan x}{\sec x} = \dfrac{\sqrt{2}}{5}$. *(Lesson 7-2)*   **Sample answer:** $\sin x = \dfrac{\sqrt{2}}{5}$

64. Graph $y = \dfrac{2}{3} \cos \theta$. *(Lesson 6-4)*   **See margin.**

65. Transportation   A boat trailer has wheels with a diameter of 14 inches. If the trailer is being pulled by a car going 45 miles per hour, what is the angular velocity of the wheels in revolutions per second? *(Lesson 6-2)*   **about 18 rps**

66. Use the unit circle to find the value of csc 180°. *(Lesson 5-3)*   **undefined**

67. Determine the binomial factors of $x^3 - 3x - 2$. *(Lesson 4-3)*

67. $(x - 2)$ $(x + 1)(x + 1)$

68. Graph $y = x^3 - 3x + 5$. Find and classify its extrema. *(Lesson 3-6)*   **max: (−1, 7), min: (1, 3)**

69. Find the values of x and y for which $\begin{bmatrix} 3x + 4 \\ 6 \end{bmatrix} = \begin{bmatrix} 16 \\ 2y \end{bmatrix}$ is true. *(Lesson 2-3)*   **(4, 3)**

70. Solve the system $x - y + z = 1$, $2x + y + 3z = 5$, and $x + y - z = 11$. *(Lesson 2-2)*   **(6, 2, −3)**

71. Graph $g(x) = |x + 3|$. *(Lesson 1-7)*   **See margin.**

72. SAT/ACT Practice   If $AC = 6$, what is the area of triangle ABC?   **C**
A 1   B $\sqrt{6}$ C 3
D 6   E 12

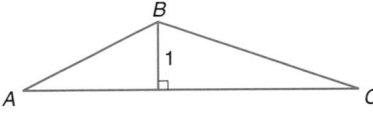

Extra Practice   See p. A39.

Lesson 7-5   Solving Trigonometric Equations   **461**

Extra Credit

Solve $\cos 2x - \cos x + 1 = 0$ for $0° \le x < 360°$.   **60°, 90°, 270°, 300°**

Lesson 7-5   **461**

This feature provides students with an opportunity to see how the mathematics they are studying was developed. It also gives students a glimpse of how mathematical discoveries are made.

More About the Mathematicians

Egyptian Mathematicians measured shadows with a vertical stick called a *gnomon*. Some historians feel that the tables relating shadow length to time of day show that early mathematicians understood the idea of functions.

Hipparchus did not use degrees when he produced his table of chords. Instead, his table used "steps," which were $\frac{1}{24}$ of a circle, equivalent to 15°.

Ptolemy (c. 85–c. 165) also used steps in his book of chords. His steps were smaller, amounting to $\frac{1}{2}$ of our degree. Ptolemy also found chord lengths by inscribing regular polygons in a circle. He could then find chord lengths for angles measuring 36°, 72°, 60°, 90°, and 120°.

Aryabhata was born in Kusumapura (now Patna), India, in 476. His table of sines was actually a table of half chords. Aryabhata was also the first person to use the versed sine in trigonometry.

Leonhard Euler was born in Basel, Switzerland, in 1707. He was the first person to view the sine and cosine as functions instead of as chords.

TRIGONOMETRY

Trigonometry was developed in response to the needs of astronomers. In fact, it was not until the thirteenth century that trigonometry and astronomy were treated as separate disciplines.

Early Evidence The earliest use of trigonometry was to calculate the heights of objects using the lengths of shadows. Egyptian mathematicians produced tables relating the lengths of shadows at particular times of day as early as the thirteenth century B.C.

The Greek mathematician **Hipparchus** (190–120 B.C.), is generally credited with laying the foundations of trigonometry. Hipparchus is believed to have produced a twelve-book treatise on the construction of a table of chords. This table related the lengths of chords of a circle to the angles subtended by those chords. In the diagram, ∠AOB would be compared to the length of chord $\overline{AB}$.

In about 100 A.D., the Greek mathematician **Menelaus**, credited with the first work on spherical trigonometry, also produced a treatise on chords in a circle. **Ptolemy**, a Babylonian mathematician, produced yet another book of chords, believed to have been adapted from Hipparchus' treatise. He used an identity similar to $\sin^2 x + \cos^2 x = 1$, except that it was relative to chords. He also used the formulas $\sin (x + y) = \sin x \cos y + \cos x \sin y$ and $\frac{a}{\sin A} = \frac{b}{\sin B} = \frac{c}{\sin C}$ as they related to chords.

In about 500 A.D., **Aryabhata**, a Hindu mathematician, was the first person to use the sine function as we know it today. He produced a table of sines and called the sine *jya*. In 628 A.D., another Hindu mathematician, **Brahmagupta**, also produced a table of sines.

The Renaissance Many mathematicians developed theories and applications of trigonometry during this time period. **Nicolas Copernicus** (1473–1543) published a book highlighting all the trigonometry necessary for astronomy at that time. During this period, the sine and versed sine were the most important trigonometric functions. Today, the versed sine, which is defined as versin $x = 1 - \cos x$, is rarely used.

Copernicus

Modern Era Mathematicians of the 1700s, 1800s, and 1900s worked on more sophisticated trigonometric ideas such as those relating to complex variables and hyperbolic functions. Renowned mathematicians who made contributions to trigonometry during this era were **Bernoulli**, **Cotes**, **DeMoivre**, **Euler**, and **Lambert**.

Today architects, such as Dennis Holloway of Santa Fe, New Mexico, use trigonometry in their daily work. Mr. Holloway is particularly interested in Native American designs. He uses trigonometry to determine the best angles for the walls of his buildings and for finding the correct slopes for landscaping.

ACTIVITIES

1. Draw a circle of radius 5 centimeters. Make a table of chords for angles of measure 10° through 90° (use 10° intervals). The table headings should be "angle measure" and "length of chord." (In the diagram of circle O, you are using ∠AOB and chord $\overline{AB}$.) **See margin.**

2. Find out more about personalities referenced in this article and others who contributed to the history of trigonometry. Visit **www.amc.glencoe.com**

Additional Answer

1.

Angle Measure	Length of Chord (cm)
10°	0.87
20°	1.74
30°	2.59
40°	3.42
50°	4.23
60°	5.00
70°	5.74
80°	6.43
90°	7.07

Normal Form of a Linear Equation

OBJECTIVES
* Write the standard form of a linear equation given the length of the normal and the angle it makes with the x-axis.
* Write linear equations in normal form.

TRACK AND FIELD When a discus thrower releases the discus, it travels in a path that is tangent to the circle traced by the discus while the thrower was spinning around. Suppose the center of motion is the origin in the coordinate system. If the thrower spins so that the discus traces the unit circle and the discus is released at (0.96, 0.28), find an equation of the line followed by the discus. *This problem will be solved in Example 2.*

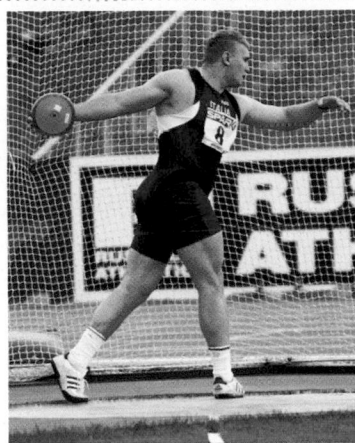

You are familiar with several forms for the equation of a line. These include slope-intercept form, point-slope form, and standard form. The usefulness of each form in a particular situation depends on how much relevant information each form provides upon inspection. In this lesson, you will learn about the **normal form** of a linear equation. The normal form uses trigonometry to provide information about the line.

In general, a **normal line** is a line that is perpendicular to another line, curve, or surface. Given a line in the *xy*-plane, there is a normal line that intersects the given line and passes through the origin. The angle between this normal line and the *x*-axis is denoted by ϕ. The normal form of the equation of the given line is written in terms of ϕ and the length p of the segment of the normal line from the given line to the origin.

ϕ is the Greek letter phi.

Suppose ℓ is a line that does not pass through the origin and p is the length of the normal from the origin. Let C be the point of intersection of ℓ with the normal, and let ϕ be the positive angle formed by the *x*-axis and $\overline{OC}$. Draw $\overline{MC}$ perpendicular to the *x*-axis. Since ϕ is in standard position, $\cos \phi = \dfrac{OM}{p}$ or $OM = p \cos \phi$ and $\sin \phi = \dfrac{MC}{p}$ or $MC = p \sin \phi$. So $\dfrac{p \sin \phi}{p \cos \phi}$ or $\dfrac{\sin \phi}{\cos \phi}$ is the slope of $\overline{OC}$. Since ℓ is perpendicular to $\overline{OC}$, the slope of ℓ is the negative reciprocal of the slope of $\overline{OC}$, or $-\dfrac{\cos \phi}{\sin \phi}$.

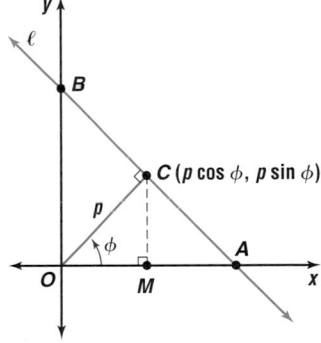

1 FOCUS

5-Minute Check
Lesson 7–5

Solve each equation for principal values of x. Express solutions in degrees.

1. $\tan x - \sqrt{3} = 0$ **60°**
2. $\sin 2x - \sin x = 0$ **0°, 60°**
3. $2 \sin^2 x - 5 \sin x + 2 = 0$ **30°**
4. Solve $\sec^2 x + 2 = 3 \sec x$ for all real values of x.
   $2\pi k, \frac{\pi}{3} + 2\pi k, \frac{5\pi}{3} + 2\pi k$, **where k is any integer**
5. Solve $\sqrt{2} \cos \theta - 1 < 0$ for $0 \le \theta < 2\pi$. $\frac{\pi}{4} < \theta < \frac{7\pi}{4}$

Motivating the Lesson
Real-World Application Have students read the first three paragraphs of the lesson. Then ask them to give everyday examples of normal lines to the following.
1. another line **adjacent sides of a rectangular wall**
2. a circle **a nail stuck in a tire**
3. a plane **a telephone line plugged into a wall**

 Resource Manager

Reproducible Masters
* *Study Guide and Practice,* pp. 129–130
* *Enrichment,* p. 78
* *Assessment and Evaluation,* p. 174
* *Problem Solving and Applications Masters,* p. 13

Transparencies
* 5-Minute Check, 7-6
* Teaching, 7-6

Teaching Tip In Example 1, point out that the line itself is perpendicular to the normal drawn from the line to the origin.

In-Class Examples

Example 1
Write the standard form of the equation of a line for which the length of the normal segment to the origin is 8 and the normal makes an angle of 45° with the positive *x*-axis.

$$\sqrt{2}x + \sqrt{2}y - 16 = 0$$

Example 2
Manufacturing When a factory worker uses a grinding wheel to sharpen a dull cutting blade, the sparks fly off the blade in a line that is tangent to the wheel. Suppose that the worker holds a blade at a 45° angle to a vertical line and the grinding wheel has a diameter of 8 inches.

a. Determine an equation of the path of the sparks.

$$x + y - 4\sqrt{2} = 0$$

b. How far above the grinding wheel must a lamp be placed so that it will not be showered with sparks?

at least $4\sqrt{2} - 4$ in., or about 1.65 in., above the wheel

---

Teaching Tip In Example 2, point out that, like the slope-intercept form and the standard form, the normal form of an equation for a straight line is another useful equation, particularly as it applies to Example 2.

Look Back
You can refer to Lesson 1-4 to review the point-slope form.

Since ℓ contains C, we can use the point-slope form to write an equation of line ℓ.

$$y - y_1 = m(x - x_1)$$

$$y - p\sin\phi = -\frac{\cos\phi}{\sin\phi}(x - p\cos\phi) \quad \textit{Substitute for m, } x_1 \textit{, and } y_1.$$

$$y\sin\phi - p\sin^2\phi = -x\cos\phi + p\cos^2\phi \quad \textit{Multiply each side by sin } \phi.$$

$$x\cos\phi + y\sin\phi = p(\sin^2\phi + \cos^2\phi)$$

$$x\cos\phi + y\sin\phi - p = 0 \quad \textit{sin}^2 \phi + \cos^2 \phi = 1$$

Normal Form

The normal form of a linear equation is

$$x\cos\phi + y\sin\phi - p = 0,$$

where p is the length of the normal from the line to the origin and ϕ is the positive angle formed by the positive *x*-axis and the normal.

You can write the standard form of a linear equation if you are given the values of ϕ and p.

Examples **1** Write the standard form of the equation of a line for which the length of the normal segment to the origin is 6 and the normal makes an angle of 150° with the positive *x*-axis.

$$x\cos\phi + y\sin\phi - p = 0 \quad \textit{Normal form}$$

$$x\cos 150° + y\sin 150° - 6 = 0 \quad \textit{ϕ = 150° and p = 6}$$

$$-\frac{\sqrt{3}}{2}x + \frac{1}{2}y - 6 = 0$$

$$\sqrt{3}x - y + 12 = 0 \quad \textit{Multiply each side by -2.}$$

The equation is $\sqrt{3}x - y + 12 = 0$.

2 **TRACK AND FIELD** Refer to the application at the beginning of the lesson.

a. Determine an equation of the path of the discus if it is released at (0.96, 0.28).

b. Will the discus strike an official at (-30, 40)? Explain your answer.

a. From the figure, we see that $p = 1$, $\sin\phi = 0.28$ or $\frac{7}{25}$, and $\cos\phi = 0.96$ or $\frac{24}{25}$.

An equation of the line is $\frac{24}{25}x + \frac{7}{25}y - 1 = 0$, or $24x + 7y - 25 = 0$.

b. The *y*-coordinate of the point on the line with an *x*-coordinate of -30 is $\frac{745}{7}$ or about 106. So the discus will not strike the official at $(-30, 40)$.

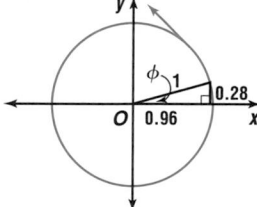

We can transform the standard form of a linear equation, $Ax + By + C = 0$, into normal form if the relationship between the coefficients in the two forms is known. The equations will represent the same line if and only if their corresponding coefficients are proportional. If $\dfrac{A}{\cos \phi} = \dfrac{B}{\sin \phi} = \dfrac{C}{-p}$, then you can solve to find expressions for $\cos \phi$ and $\sin \phi$ in terms of p and the coefficients.

$$\dfrac{A}{\cos \phi} = \dfrac{C}{-p} \quad \Rightarrow \quad -\dfrac{Ap}{C} = \cos \phi \quad \text{or} \quad \cos \phi = -\dfrac{Ap}{C}$$

$$\dfrac{B}{\sin \phi} = \dfrac{C}{-p} \quad \Rightarrow \quad -\dfrac{Bp}{C} = \sin \phi \quad \text{or} \quad \sin \phi = -\dfrac{Bp}{C}$$

We can divide $\sin \phi = -\dfrac{Bp}{C}$ by $\cos \phi = -\dfrac{Ap}{C}$, where $\cos \phi \neq 0$.

$$\dfrac{\sin \phi}{\cos \phi} = \dfrac{-\dfrac{Bp}{C}}{-\dfrac{Ap}{C}}$$

$$\tan \phi = \dfrac{B}{A} \qquad \dfrac{\sin \phi}{\cos \phi} = \tan \phi$$

Refer to the diagram at the right. Consider an angle ϕ in standard position such that $\tan \phi = \dfrac{B}{A}$.

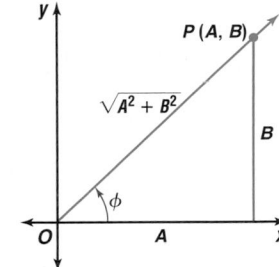

The length of $\overline{OP}$ is $\sqrt{A^2 + B^2}$.

So, $\sin \phi = \dfrac{B}{\pm\sqrt{A^2 + B^2}}$ and $\cos \phi = \dfrac{A}{\pm\sqrt{A^2 + B^2}}$.

Since we know that $\dfrac{B}{\sin \phi} = \dfrac{C}{-p}$, we can substitute

to get the result $\dfrac{B}{\dfrac{B}{\pm\sqrt{A^2 + B^2}}} = \dfrac{C}{-p}$.

Therefore, $p = \dfrac{C}{\pm\sqrt{A^2 + B^2}}$.

The $\pm$ sign is used since p is a measure and must be positive in the equation $x \cos \phi + y \sin \phi - p = 0$. Therefore, the sign must be chosen as opposite of the sign of C. That is, if C is positive, use $-\sqrt{A^2 + B^2}$, and, if C is negative, use $\sqrt{A^2 + B^2}$.

If $C = 0$, the sign is chosen so that $\sin \phi$ is positive; that is, the same sign as that of B.

Substitute the values for $\sin \phi$, $\cos \phi$, and p into the normal form.

$$\dfrac{Ax}{\pm\sqrt{A^2 + B^2}} + \dfrac{By}{\pm\sqrt{A^2 + B^2}} - \dfrac{C}{\pm\sqrt{A^2 + B^2}} = 0$$

Notice that the standard form is closely related to the normal form.

Changing the Standard Form to Normal Form

The standard form of a linear equation, $Ax + By + C = 0$, can be changed to normal form by dividing each term by $\pm\sqrt{A^2 + B^2}$. The sign is chosen opposite the sign of C.

Lesson 7-6   Normal Form of a Linear Equation   **465**

Teaching Tip For Example 3, point out that in the normal form of an equation of a straight line, p is always positive because it represents a length, and p is always preceded by a subtraction sign.

In-Class Example

Example 3
Write each equation in normal form. Then find the length of the normal and the angle it makes with the positive x-axis.

a. $2x + y - 5 = 0$
$\dfrac{2\sqrt{5}}{5}x + \dfrac{\sqrt{5}}{5}y - \sqrt{5} = 0$;
$\sqrt{5}$; $27°$

b. $3x - 7y + 9 = 0$
$-\dfrac{3\sqrt{58}}{58}x + \dfrac{7\sqrt{58}}{58}y -$
$\dfrac{9\sqrt{58}}{58} = 0$; $\dfrac{9\sqrt{58}}{58}$; $113°$

If the equation of a line is in normal form, you can find the length of the normal, p units, directly from the equation. You can find the angle ϕ by using the relation $\tan\phi = \dfrac{\sin\phi}{\cos\phi}$. However, you must find the quadrant in which the normal lies to find the correct angle for ϕ. When the equation of a line is in normal form, the coefficient of x is equal to $\cos\phi$, and the coefficient of y is equal to $\sin\phi$. Thus, the correct quadrant can be determined by studying the signs of $\cos\phi$ and $\sin\phi$. For example, if $\sin\phi$ is negative and $\cos\phi$ is positive, the normal lies in the fourth quadrant.

Example ③ **Write each equation in normal form. Then find the length of the normal and the angle it makes with the positive x-axis.**

a. $6x + 8y + 3 = 0$

Since C is positive, use $-\sqrt{A^2 + B^2}$ to determine the normal form.
$-\sqrt{A^2 + B^2} = -\sqrt{6^2 + 8^2}$ or -10

The normal form is $\dfrac{6}{-10}x + \dfrac{8}{-10}y + \dfrac{3}{-10} = 0$, or $-\dfrac{3}{5}x - \dfrac{4}{5}y - \dfrac{3}{10} = 0$.

Therefore, $\sin\phi = -\dfrac{4}{5}$, $\cos\phi = -\dfrac{3}{5}$, and $p = \dfrac{3}{10}$. Since $\sin\phi$ and $\cos\phi$ are both negative, ϕ must lie in the third quadrant.

$\tan\phi = \dfrac{-\dfrac{4}{5}}{-\dfrac{3}{5}}$ or $\dfrac{4}{3}$ *$\tan\phi = \dfrac{\sin\phi}{\cos\phi}$*

$\phi \approx 233°$ *Add 180° to the arctangent to get the angle in Quadrant III.*

The normal segment to the origin has length 0.3 unit and makes an angle of $233°$ with the positive x-axis.

b. $-x + 4y - 6 = 0$

Since C is negative, use $\sqrt{A^2 + B^2}$ to determine the normal form.
$\sqrt{A^2 + B^2} = \sqrt{(-1)^2 + 4^2}$ or $\sqrt{17}$

The normal form is $-\dfrac{1}{\sqrt{17}}x + \dfrac{4}{\sqrt{17}}y - \dfrac{6}{\sqrt{17}} = 0$ or

$-\dfrac{\sqrt{17}}{17}x + \dfrac{4\sqrt{17}}{17}y - \dfrac{6\sqrt{17}}{17} = 0$. Therefore, $\sin\phi = \dfrac{4\sqrt{17}}{17}$,

$\cos\phi = -\dfrac{\sqrt{17}}{17}$, and $p = \dfrac{6\sqrt{17}}{17}$. Since $\sin\phi > 0$ and $\cos\phi < 0$, ϕ must lie in the second quadrant.

$\tan\phi = \dfrac{\dfrac{4\sqrt{17}}{17}}{-\dfrac{\sqrt{17}}{17}}$ or -4 *$\tan\phi = \dfrac{\sin\phi}{\cos\phi}$*

$\phi \approx 104°$ *Add 180° to the arctangent to get the angle in Quadrant II.*

The normal segment to the origin has length $\dfrac{6\sqrt{17}}{17} \approx 1.46$ units and makes an angle of $104°$ with the positive x-axis.

Addressing Individual Needs

Visual/Spatial Have students use graph paper and a straightedge to draw a quadrilateral using their own coordinates for the vertices A, B, C, and D. Tell them to choose vertices such that the origin lies inside the quadrilateral. Then, have students draw normals from the origin to each side. They should find the equations of each side of the quadrilateral in standard and normal forms, and find the lengths of each normal and the angle each normal makes with the positive x-axis.

Communicating Mathematics

Read and study the lesson to answer each question. 2–4. See margin.

1. **Normal** means perpendicular.

1. **Define** the geometric meaning of the word *normal*.

2. **Describe** how to write the normal form of the equation of a line when $p = 10$ and $\phi = 30°$.

3. **Refute or defend** the following statement. *Determining the normal form of the equation of a line is like finding the equation of a tangent line to a circle of radius p.*

4. **Write** each form of the equation of a line that you have learned. Compare and contrast the information that each provides upon inspection. Create a sample problem that would require you to use each form.

Guided Practice

Write the standard form of the equation of each line given p, the length of the normal segment, and ϕ, the angle the normal segment makes with the positive x-axis.

5. $p = 10, \phi = 30°$
$\sqrt{3}x + y - 20 = 0$

6. $p = \sqrt{3}, \phi = 150°$
$\sqrt{3}x - y + 2\sqrt{3} = 0$

7. $p = 5\sqrt{2}, \phi = \frac{7\pi}{4}$
$x - y - 10 = 0$

Write each equation in normal form. Then find the length of the normal and the angle it makes with the positive x-axis. 8–10. See margin.

8. $4x + 3y = -10$

9. $y = -3x + 2$

10. $\sqrt{2}x - \sqrt{2}y = 6$

11. **Transportation** An airport control tower is located at the origin of a coordinate system where the coordinates are measured in miles. An airplane radios in to report its direction and location. The controller determines that the equation of the plane's path is $3x - 4y = 8$. **See margin.**

a. Make a drawing to illustrate the problem.

b. What is the closest the plane will come to the tower? **1.6 miles**

EXERCISES

Practice

12. $x + \sqrt{3}y - 30 = 0$

15. $\sqrt{3}x - y + 4\sqrt{3} = 0$

17. $\sqrt{3}x + y + 10 = 0$

Write the standard form of the equation of each line given p, the length of the normal segment, and ϕ, the angle the normal segment makes with the positive x-axis.

A

12. $p = 15, \phi = 60°$

13. $p = 12, \phi = \frac{\pi}{4}$
$\sqrt{2}x + \sqrt{2}y - 24 = 0$

14. $p = 3\sqrt{2}, \phi = 135°$
$x - y + 6 = 0$

15. $p = 2\sqrt{3}, \phi = \frac{5\pi}{6}$

16. $p = 2, \phi = \frac{\pi}{2}$ $y - 2 = 0$

17. $p = 5, \phi = 210°$

18. $p = 5, \phi = \frac{4\pi}{3}$
$x + \sqrt{3}y + 10 = 0$

19. $p = \frac{3}{2}, \phi = 300°$
$x - \sqrt{3}y - 3 = 0$

20. $p = 4\sqrt{3}, \phi = \frac{11\pi}{6}$
$\sqrt{3}x - y - 8\sqrt{3} = 0$

Write each equation in normal form. Then find the length of the normal and the angle it makes with the positive x-axis. 21–29. See margin.

B

21. $5x + 12y + 65 = 0$

22. $x + y = 1$

23. $3x - 4y = 15$

24. $y = 2x - 4$

25. $x = 3$

26. $-\sqrt{3}x - y = 2$

27. $y - 2 = \frac{1}{4}(x + 20)$

28. $\frac{x}{3} = y - 4$

29. $\frac{x}{20} + \frac{y}{24} = 1$

 www.amc.glencoe.com/self_check_quiz

Lesson 7-6 Normal Form of a Linear Equation **467**

Error Analysis

Watch for students who do not understand that the angle ϕ consists of the normal and the positive direction of the x-axis.

Prevent by having them draw lines on the x- and y-axes where ϕ is an acute angle, obtuse, more than 180° but less then 270°, and greater than 270° but less than 360°

Assignment Guide

Core: 13–31 odd, 33–46
Enriched: 12–30 even, 32–46

Additional Answers

2. Compute cos 30° and sin 30°. Use these as the coefficients of x and y, respectively, in the normal form. The normal form is $\frac{\sqrt{3}}{2}x + \frac{1}{2}y - 10 = 0$.

3. The statement is true. The given line is tangent to the circle centered at the origin with radius p.

8. $-\frac{4}{5}x - \frac{3}{5}y - 2 = 0$; 2; 217°

9. $\frac{3\sqrt{10}}{10}x + \frac{\sqrt{10}}{10}y - \frac{\sqrt{10}}{5} = 0$; $\frac{\sqrt{10}}{5}$; 18°

10. $\frac{\sqrt{2}}{2}x - \frac{\sqrt{2}}{2}y - 3 = 0$; 3; 315°

(continued on page 468)

Study Guide and Practice Masters, pp. 129–130

7-6 Study Guide

Normal Form of a Linear Equation

| Normal Form | The normal form of a linear equation is $x \cos \phi + y \sin \phi - p = 0$, where p is the length of the normal from the line to the origin and ϕ is the positive angle formed by the positive x-axis and the normal. |

You can write the standard form of a linear equation if you are given the values of ϕ and p.

Example 1 Write the standard form of the equation of a line for which the length of the normal segment to the origin is 5 and the normal makes an angle of 135° with the positive x-axis.

$x \cos \phi + y \sin \phi - p = 0$ *Normal form*
$x \cos 135° + y \sin 135° - 5 = 0$ $\phi = 135°$ and $p = 5$
$-\frac{\sqrt{2}}{2}x + \frac{\sqrt{2}}{2}y - 5 = 0$
$\sqrt{2}x - \sqrt{2}y + 10 = 0$ *Multiply each side by −2.*
The standard form of the equation is $\sqrt{2}x - \sqrt{2}y + 10 = 0$.

The standard form of a linear equation, $Ax + By + C = 0$, can be changed to the normal form by dividing each term of the equation by $\pm\sqrt{A^2 + B^2}$. The sign is chosen opposite the sign of C. You can then find the length of the normal, p units, and the angle ϕ.

Example 2 Write $3x + 4y - 10 = 0$ in normal form. Then find the length of the normal and the angle it makes with the positive x-axis.

Since C is negative, use $\sqrt{A^2 + B^2}$ to determine the normal form.
$\sqrt{A^2 + B^2} = \sqrt{3^2 + 4^2}$ or 5
The normal form is $\frac{3}{5}x + \frac{4}{5}y - \frac{10}{5} = 0$ or $\frac{3}{5}x + \frac{4}{5}y - 2 = 0$.
Therefore, $\cos \phi = \frac{3}{5}$, $\sin \phi = \frac{4}{5}$, and $p = 2$.
Since $\cos \phi$ and $\sin \phi$ are both positive, ϕ must lie in Quadrant I.
$\tan \phi = \frac{\frac{4}{5}}{\frac{3}{5}}$ or $\frac{4}{3}$ $\tan \phi = \frac{\sin \phi}{\cos \phi}$
$\phi \approx 53°$
The normal segment has length 2 units and makes an angle of 53° with the positive x-axis.

Additional Answer

4.

Form	Equation	Information Displayed
Slope-Intercept	$y = mx + b$	slope and y-intercept
Point-Slope	$y - y_1 = m(x - x_1)$	slope and a point on the line
Standard	$Ax + By + C = 0$	none
Normal	$x \cos \phi + y \sin \phi - p = 0$	length of the normal and the angle the normal makes with the x-axis

See students' work for sample problems.

11a.

$3x - 4y = 8$

21. $-\frac{5}{13}x - \frac{12}{13}y - 5 = 0$; 5; 247°

22. $\frac{\sqrt{2}}{2}x + \frac{\sqrt{2}}{2}y - \frac{\sqrt{2}}{2} = 0$; $\frac{\sqrt{2}}{2}$; 45°

23. $\frac{3}{5}x - \frac{4}{5}y - 3 = 0$; 3; 307°

24. $\frac{2\sqrt{5}}{5}x - \frac{\sqrt{5}}{5}y - \frac{4\sqrt{5}}{5} = 0$; $\frac{4\sqrt{5}}{5}$; 333°

25. $x - 3 = 0$; 3; 0°

26. $-\frac{\sqrt{3}}{2}x - \frac{1}{2}y - 1 = 0$; 1; 210°

27. $-\frac{\sqrt{17}}{17}x + \frac{4\sqrt{17}}{17}y - \frac{28\sqrt{17}}{17} = 0$; $\frac{28\sqrt{17}}{17}$; 104°

28. $-\frac{\sqrt{10}}{10}x + \frac{3\sqrt{10}}{10}y - \frac{6\sqrt{10}}{5} = 0$; $\frac{6\sqrt{10}}{5}$; 108°

29. $\frac{6\sqrt{61}}{61}x + \frac{5\sqrt{61}}{61}y - \frac{120\sqrt{61}}{61} = 0$; $\frac{120\sqrt{61}}{61}$; 40°

33a.

45°

1.25 ft

35b. $\phi \approx 67°$

$-\frac{12}{13}x + \frac{5}{13}y - 3 = 0$

 30. Write the standard form of the equation of a line if the point on the line nearest to the origin is at (6, 8). **3x + 4y − 50 = 0**

31. The point nearest to the origin on a line is at (4, −4). Find the standard form of the equation of the line. **x − y − 8 = 0**

Applications and Problem Solving

32. Geometry The three sides of a triangle are tangent to a unique circle called the *incircle*. On the coordinate plane, the incircle of $\triangle ABC$ has its center at the origin. The lines whose equations are $x + 4y = 6\sqrt{17}$, $2x + \sqrt{5}y = -18$, and $2\sqrt{2}x = y + 18$ contain the sides of $\triangle ABC$. What is the length of the radius of the incircle? **6 units**

33. History Ancient slingshots were made from straps of leather that cradled a rock until it was released. One would spin the slingshot in a circle, and the initial path of the released rock would be a straight line tangent to the circle at the point of release.

The rock will travel the greatest distance if it is released when the angle between the normal to the path and the horizontal is −45°. The center of the circular path is the origin and the radius of the circle measures 1.25 feet.

a. Draw a labeled diagram of the situation. **See margin.**
b. Write an equation of the initial path of the rock in standard form.
$$\sqrt{2}x - \sqrt{2}y - 2.5 = 0$$

34. Critical Thinking Consider a line ℓ with positive x- and y-intercepts. Suppose ℓ makes an angle of θ with the positive x-axis.

a. What is the relationship between θ and ϕ, the angle the normal line makes with the positive x-axis? $\theta = \phi + 90°$
b. What is the slope of ℓ in terms of θ? **tan θ**
c. What is the slope of the normal line in terms of θ? **−cot θ**
d. What is the slope of ℓ in terms of ϕ? **−cot ϕ**

35. Analytic Geometry Armando was trying to determine how to rotate the graph of a line $\delta°$ about the origin. He hypothesized that the following would be an equation of the new line.
$$x\cos(\phi + \delta) + y\sin(\phi + \delta) - p = 0$$
a. Write the normal form of the line $5x + 12y - 39 = 0$. $\frac{5}{13}x + \frac{12}{13}y - 3 = 0$
b. Determine ϕ. Replace ϕ by $\phi + 90°$ and graph the result. **See margin.**
c. Choose another value for δ, not divisible by 90°, and test Armando's conjecture. **See students' work.**
d. Write an argument as to whether Armando is correct. Include a graph in your argument. **See students' graphs; See Answer Appendix for argument.**

Extra Credit

Given the vertices of triangle ABC, $A(2, 0)$, $B(4, −2)$, and $C(0, −8)$, find the equations of $\overleftrightarrow{AB}$, $\overleftrightarrow{AC}$, and $\overleftrightarrow{BC}$ in normal form and for each find the length of the normal and the angle it makes with the positive x-axis.

$\frac{\sqrt{2}}{2}x + \frac{\sqrt{2}}{2}y - \sqrt{2} = 0$, $\frac{4\sqrt{17}}{17}x - \frac{\sqrt{17}}{17}y - \frac{8\sqrt{17}}{17} = 0$, $\frac{3\sqrt{13}}{13}x - \frac{2\sqrt{13}}{13}y - \frac{16\sqrt{13}}{13} = 0$; $\sqrt{2}$, 45°; $\frac{8\sqrt{17}}{17}$, 346°; $\frac{16\sqrt{13}}{13}$, 326°

36. Critical Thinking Suppose two lines intersect to form an acute angle α. Suppose that each line has a positive y-intercept and that the x-intercepts of the lines are on opposite sides of the origin. **a–b. See margin.**

    a. How are the angles ϕ_1 and ϕ_2 that the respective normals make with the positive x-axis related?

    b. Write an equation involving $\tan \alpha$, $\tan \phi_1$, and $\tan \phi_2$.

37. Engineering The village of Plentywood must build a new water tower to meet the needs of its residents. This means that each of the water mains must be connected to the new tower. On a map of the village, with units in hundreds of feet, the water tower was placed at the origin. Each of the existing water mains can be described by an equation. These equations are $5x - y = 15$, $3x + 4y = 36$, and $5x - 2y = -20$. The cost of laying new underground pipe is $500 per 100 feet. Find the lowest possible cost of connecting the existing water mains to the new tower. **$6927.82**

Mixed Review

38. Solve $2\cos^2 x + 7\cos x - 4 = 0$ for $0 \le x < 2\pi$. *(Lesson 7-5)* $\dfrac{\pi}{3}, \dfrac{5\pi}{3}$

39. If x and y are acute angles such that $\cos x = \dfrac{1}{6}$ and $\cos y = \dfrac{2}{3}$, find $\sin(x + y)$. *(Lesson 7-3)* $\dfrac{2\sqrt{35} + \sqrt{5}}{18}$

40. Graph $y = \sin 4\theta$. *(Lesson 6-4)* **See margin.**

41. Engineering A metallic ring used in a sprinkler system has a diameter of 13.4 centimeters. Find the length of the metallic cross brace if it subtends a central angle of $26°20'$. *(Lesson 5-8)* **3.05 cm**

cross brace

42. Solve $\dfrac{x}{x - 5} + \dfrac{17}{25 - x^2} = \dfrac{1}{x + 5}$. *(Lesson 4-6)* **−6, 2**

43. Manufacturing A porcelain company produces collectible thimble sets that contain 8 thimbles in a box that is 4 inches by 6 inches by 2 inches. To celebrate the company's 100th anniversary, they wish to market a deluxe set of 8 large thimbles. They would like to increase each of the dimensions of the box by the same amount to create a new box with a volume that is 1.5 times the volume of the original box. What should be the dimensions of the new box for the large thimbles? *(Lesson 4-5)* **4.5 in. by 6.5 in. by 2.5 in.**

44. Find the maximum and minimum values of the function $f(x, y) = 3x - y + 4$ for the polygonal convex set determined by the system of inequalities. *(Lesson 2-6)* **16, 4**
$x + y \le 8$
$y \ge 3$
$x \ge 2$

45. Solve $\begin{bmatrix} -1 & 2 \\ -4 & 3 \end{bmatrix} \cdot \begin{bmatrix} x \\ y \end{bmatrix} = \begin{bmatrix} 0 \\ 15 \end{bmatrix}$. *(Lesson 2-5)* **(−6, −3)**

46. SAT/ACT Practice If $\dfrac{a}{b} = \dfrac{4}{5}$, what is the value of $2a + b$? **E**

  A 3   **B** 13   **C** 14   **D** 26

  E cannot be determined from the given information

Additional Answer

36a.

The angles of the quadrilateral are $180° - \alpha$, $90°$, $\phi_2 - \phi_1$, and $90°$. Then $180° - \alpha + 90° + \phi_2 - \phi_1 + 90° = 360°$, which simplifies to $\phi_2 = \phi_1 + \alpha$. If the lines intersect so that α is an interior angle of the quadrilateral, the equation works out to be $\phi_2 = 180° + \phi_1 - \alpha$.

Distance From a Point to a Line

1 FOCUS

5-Minute Check
Lesson 7-6

Write the standard form of the equation for each of the following lines.

1. The length of the normal segment to the origin is 4 and the normal makes an angle of 30° with the positive *x*-axis.

$$\sqrt{3}x + y - 8 = 0$$

2. The length of the normal segment to the origin is 8 and the normal makes an angle of 135° with the positive *x*-axis.

$$x - y + 8\sqrt{2} = 0$$

Write each equation in normal form. Then find the length of the normal and the angle it makes with the positive *x*-axis.

3. $4x - 3y - 2 = 0$

$\frac{4}{5}x - \frac{3}{5}y - \frac{2}{5} = 0; \frac{2}{5}; 323°$

4. $3x + 4y - 13 = 0$

$\frac{3}{5}x + \frac{4}{5}y - \frac{13}{5} = 0; \frac{13}{5}; 53°$

5. $-8x + 9y + 7 = 0$

$\frac{8\sqrt{145}}{145}x - \frac{9\sqrt{145}}{145}y - \frac{7\sqrt{145}}{145} = 0;$

$\frac{7\sqrt{145}}{145}; 312°$

Teaching Tip For the case where the origin and *P* are on the same side of $\overleftrightarrow{RS}$, the normal to $\overleftrightarrow{TV}$ has length $-d - p$ and the angle the normal makes with the positive *x*-axis is $180° + \phi$. Then, the normal form of $\overleftrightarrow{TV}$ is $x \cos (180° + \phi) + y \sin (180° + \phi) - (-d - p) = 0$. The left side of the equation is equivalent to $-x \cos \phi - y \sin \phi + p + d$. Multiplying by -1 on both sides of the equation results in $x \cos \phi + y \sin \phi - (p + d) = 0$, which is the same as the form given in the text.

OBJECTIVES
- Find the distance from a point to a line.
- Find the distance between two parallel lines.
- Write equations of lines that bisect angles formed by intersecting lines.

OPTICS When light waves strike a surface, they are reflected in such a way that the angle of incidence equals the angle of reflection. Consider a flat mirror situated in a coordinate system such that light emanating from the point at (5, 4) strikes the mirror at (3, 0) and then passes through the point at (7, −6). Determine an equation of the line on which the mirror lies. Use the equation to determine the angle the mirror makes with the *x*-axis. *This problem will be solved in Example 4.*

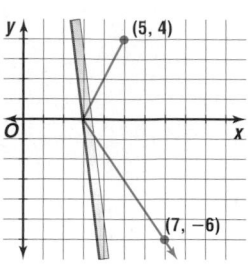

The normal form of a linear equation can be used to find the distance from a point to a line. Let $\overleftrightarrow{RS}$ be a line in the coordinate plane, and let $P(x_1, y_1)$ be a point not on $\overleftrightarrow{RS}$. *P* may lie on the same side of $\overleftrightarrow{RS}$ as the origin does or it may lie on the opposite side. If a line segment joining *P* to the origin does not intersect $\overleftrightarrow{RS}$, point *P* is on the same side of the line as the origin. Construct $\overleftrightarrow{TV}$ parallel to $\overleftrightarrow{RS}$ and passing through *P*. The distance *d* between the parallel lines is the distance from *P* to $\overleftrightarrow{RS}$. So that the derivation is valid in both cases, we will use a negative value for *d* if point *P* and the origin are on the same side of $\overleftrightarrow{RS}$.

interNET
CONNECTION

Graphing Calculator Programs
To download a graphing calculator program that computes the distance from a point to a line, visit: **www.amc.glencoe.com**

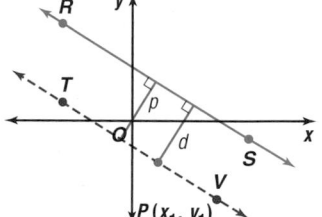

Let $x \cos \phi + y \sin \phi - p = 0$ be the equation of $\overleftrightarrow{RS}$ in normal form. Since $\overleftrightarrow{TV}$ is parallel to $\overleftrightarrow{RS}$, they have the same slope. The equation for $\overleftrightarrow{TV}$ can be written as $x \cos \phi + y \sin \phi - (p + d) = 0$. Solve this equation for *d*.

$$d = x \cos \phi + y \sin \phi - p$$

Since $P(x_1, y_1)$ is on $\overleftrightarrow{TV}$, its coordinates satisfy this equation.

$$d = x_1 \cos \phi + y_1 \sin \phi - p$$

We can use an equivalent form of this expression to find *d* when the equation of a line is in standard form.

Resource Manager

 Reproducible Masters
- *Study Guide and Practice,* pp. 131–132
- *Enrichment,* p. 79
- *Assessment and Evaluation,* p. 174
- *TI-92 & CBL Interdisciplinary Activities,* pp. 21–24

 Transparencies
- 5-Minute Check, 7-7
- Teaching, 7-7

Distance from a Point to a Line

The following formula can be used to find the distance from a point at (x_1, y_1) to a line with equation $Ax + By + C = 0$.

$$d = \frac{Ax_1 + By_1 + C}{\pm\sqrt{A^2 + B^2}}$$

The sign of the radical is chosen opposite the sign of C.

The distance will be positive if the point and the origin are on opposite sides of the line. The distance will be treated as negative if the origin is on the same side of the line as the point. If you are solving an application problem, the absolute value of d will probably be required.

Example **1** Find the distance between $P(4, 5)$ and the line with equation $8x + 5y = 20$.

First rewrite the equation of the line in standard form.

$$8x + 5y = 20 \Rightarrow 8x + 5y - 20 = 0$$

Then use the formula for the distance from a point to a line.

$$d = \frac{Ax_1 + By_1 + C}{\pm\sqrt{A^2 + B^2}}$$

$$d = \frac{8(4) + 5(5) - 20}{\pm\sqrt{8^2 + 5^2}} \qquad A = 8, B = 5, C = -20, x_1 = 4, y_1 = 5$$

$$d = \frac{37}{\sqrt{89}} \text{ or about } 3.92 \quad \textit{Since C is negative, use } +\sqrt{A^2 + B^2}.$$

Therefore, P is approximately 3.92 units from the line $8x + 5y = 20$. Since d is positive, P is on the opposite side of the line from the origin.

You can use the formula for the distance from a point to a line to find the distance between two parallel lines. To do this, choose any point on one of the lines and use the formula to find the distance from that point to the other line.

Example **2** Find the distance between the lines with equations $6x - 2y = 7$ and $y = 3x + 4$.

Since $y = 3x + 4$ is in slope-intercept form, we know that it passes through the point at $(0, 4)$. Use this point to find the distance to the other line.

The standard form of the other equation is $6x - 2y - 7 = 0$.

$$d = \frac{Ax_1 + By_1 + C}{\pm\sqrt{A^2 + B^2}}$$

$$d = \frac{6(0) - 2(4) - 7}{\pm\sqrt{6^2 + (-2)^2}} \qquad A = 6, B = -2, C = -7, x_1 = 0, y_1 = 4$$

$$d = -\frac{15}{\sqrt{40}} \text{ or about } -2.37 \quad \textit{Since C is negative, use } +\sqrt{A^2 + B^2}.$$

The distance between the lines is about 2.37 units.

An equation of the bisector of an angle formed by two lines in the coordinate plane can also be found using the formula for the distance between a point and a line. The bisector of an angle is the set of all points in the plane equidistant from the sides of the angle. Using this definition, equations of the bisectors of the angles formed by two lines can be found.

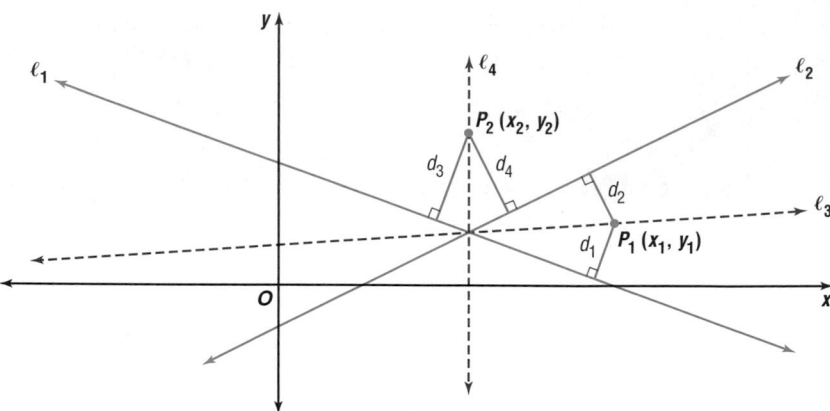

In the figure, ℓ_3 and ℓ_4 are the bisectors of the angles formed by ℓ_1 and ℓ_2. $P_1(x_1, y_1)$ is a point on ℓ_3, and $P_2(x_2, y_2)$ is a point on ℓ_4. Let d_1 be the distance from ℓ_1 to P_1, and let d_2 be the distance from ℓ_2 to P_1.

Notice that P_1 and the origin lie on opposite sides of ℓ_1, so d_1 is positive. Since the origin and P_1 are on opposite sides of ℓ_2, d_2 is also positive. Therefore, for any point $P_1(x_1, y_1)$ on ℓ_3, $d_1 = d_2$. However, d_3 is positive and d_4 is negative. *Why?* Therefore, for any point $P_2(x_2, y_2)$ on ℓ_4, $d_3 = -d_4$.

The origin is in the interior of the angle that is bisected by ℓ_3, but in the exterior of the angle bisected by ℓ_4. So, a good way for you to determine whether to equate distances or to let one distance equal the opposite of the other is to observe the position of the origin.

Relative Position of the Origin	If the origin lies within the angle being bisected or the angle vertical to it, the distances from each line to a point on the bisector have the same sign.
	If the origin does not lie within the angle being bisected or the angle vertical to it, the distances have opposite signs.

To find the equation of a specific angle bisector, first graph the lines. Then, determine whether to equate the distances or to let one distance equal the opposite of the other.

472   *Chapter 7   Trigonometric Identities and Equations*

Examples

3 Find equations of the lines that bisect the angles formed by the lines $8x - 6y = 9$ and $20x - 21y = -50$.

[−5, 15] scl:1 by [−5, 15] scl:1

Graph each equation. Note the location of the origin. The origin is in the interior of the acute angle.

Bisector of the acute angle

$d_1 = d_2$

$d_1 = \dfrac{8x_1 - 6y_1 - 9}{\sqrt{8^2 + (-6)^2}}$

$d_2 = \dfrac{20x_1 - 21y_1 + 50}{-\sqrt{20^2 + (-21)^2}}$

$\dfrac{8x_1 - 6y_1 - 9}{10} = \dfrac{20x_1 - 21y_1 + 50}{-29}$

Bisector of the obtuse angle

$d_1 = -d_2$

$d_1 - \dfrac{8x_1 - 6y_1 - 9}{\sqrt{8^2 + (-6)^2}}$

$d_2 = \dfrac{20x_1 - 21y_1 + 50}{-\sqrt{20^2 + (-21)^2}}$

$\dfrac{8x_1 - 6y_1 - 9}{10} = -\dfrac{20x_1 - 21y_1 + 50}{-29}$

Simplifying and dropping the subscripts yields $432x - 384y = -239$ and $32x + 36y = 761$, respectively.

4 **OPTICS** Refer to the application at the beginning of the lesson.

a. Determine an equation of the line on which the mirror lies.

b. Use the equation to determine the angle the mirror makes with the *x*-axis.

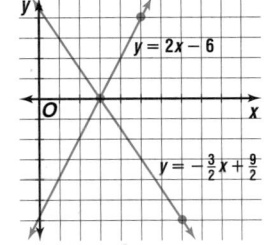

$y = 2x - 6$

$y = -\frac{3}{2}x + \frac{9}{2}$

a. Using the points at $(5, 4)$, $(7, -6)$, and $(3, 0)$ and the slope-intercept form, you can find that the light travels along lines with equations $y = 2x - 6$ and $y = -\dfrac{3}{2}x + \dfrac{9}{2}$. In standard form, these are $2x - y - 6 = 0$ and $3x + 2y - 9 = 0$.

The mirror lies on the angle bisector of the acute angles formed by these lines.

The origin is not contained in the angle we wish to bisect, so we use $d_1 = -d_2$.

$$\dfrac{2x - y - 6}{\sqrt{5}} - -\dfrac{3x + 2y - 9}{\sqrt{13}}$$

$$\sqrt{13}(2x - y - 6) = -\sqrt{5}(3x + 2y - 9)$$

$$\left(2\sqrt{13} + 3\sqrt{5}\right)x + \left(2\sqrt{5} - \sqrt{13}\right)y - 6\sqrt{13} - 9\sqrt{5} = 0$$

The mirror lies on the line with equation

$$\left(2\sqrt{13} + 3\sqrt{5}\right)x + \left(2\sqrt{5} - \sqrt{13}\right)y - 6\sqrt{13} - 9\sqrt{5} = 0.$$

Teaching Tip For Example 3, point out that $\sqrt{116} = \sqrt{4 \cdot 29} = 2\sqrt{29}$. This helps simplify the resulting equation.

In-Class Examples

Example 3
Find an equation of the line that bisects the acute angles formed by the lines $10x - 4y + 9 = 0$ and $5x + 2y - 10 = 0$.
$20x - 11 = 0$ or $x = \dfrac{11}{20}$

Example 4
Optics Refer to the application at the beginning of the lesson. Suppose that light emanates from the point at $(6, 3)$, then strikes a mirror at $(3, 0)$ and passes through the point at $(6, -7)$.

a. Determine an equation of the line on which the mirror lies.
$$\left(\sqrt{58} + 7\sqrt{2}\right)x + \left(3\sqrt{2} - \sqrt{58}\right)y - 3\sqrt{58} - 21\sqrt{2} = 0$$

b. Use the equation to determine the angle the mirror makes with the *x*-axis.
79.10°

Addressing Individual Needs

Logical Have students use the method taught in this lesson for finding the distance from a point to a line and develop a strategy for finding the coordinates of a point *D* inside triangle *ABC* where *D* is equally distant from each of the three sides of the triangle. The triangle has vertices $A(1, 5)$, $B(3, 7)$, and $C(5, 0)$.

Error Analysis

Watch for students who have difficulty finding the equations of lines that bisect angles formed by two given lines.

Prevent by suggesting that students use a straightedge and graph paper and carefully graph the conditions of the problem.

Additional Answers

1. The distance from a point to a line is the distance from that point to the closest point on the line.

2. The sign should be chosen opposite the sign of *C* where *Ax + By + C = 0* is the standard form of the equation of the line.

3. In the figure, *P* and *Q* are any points on the lines. The right triangles are congruent by AAS. The corresponding congruent sides of the triangles show that the same distance is always obtained between the two lines.

Study Guide and Practice Masters, pp. 131–132

Multiply by
$$\frac{2\sqrt{5} + \sqrt{13}}{2\sqrt{5} + \sqrt{13}}$$
to rationalize the denominator.

b. The slope of the line on which the mirror lies is $-\dfrac{2\sqrt{13} + 3\sqrt{5}}{2\sqrt{5} - \sqrt{13}}$ or $-8 - \sqrt{65}$. Recall that the slope of a line is the tangent of the angle the line makes with the *x*-axis. Since $\tan\theta = -8 - \sqrt{65}$, the positive value for θ is approximately 93.56°.

The mirror makes an angle of 93.56° with the positive *x*-axis.

CHECK FOR UNDERSTANDING

Communicating Mathematics

Read and study the lesson to answer each question. **1–4. See margin.**

1. **State** what is meant by the distance from a point to a line.

2. **Tell** how to choose the sign of the radical in the denominator of the formula for the distance from a point to a line.

3. **Explain** why you can choose any point on either line when calculating the distance between two parallel lines.

4. **Investigate** whether the formula for the distance from a point to a line is valid if the line is vertical or horizontal. Test your conjecture with some specific examples. Explain what is happening when you apply the formula in these situations.

Guided Practice

Find the distance between the point with the given coordinates and the line with the given equation.

5. $(1, 2)$, $2x - 3y = -2$ $\dfrac{2\sqrt{13}}{13}$

6. $(-2, 3)$, $6x - y = -3$ $\dfrac{12\sqrt{37}}{37}$

Find the distance between the parallel lines with the given equations.

7. $3x - 5y = 1$ $\dfrac{2\sqrt{34}}{17}$
   $3x - 5y = -3$

8. $y = -\dfrac{1}{3}x + 3$ $3\sqrt{10}$
   $y = -\dfrac{1}{3}x - 7$

9. Find equations of the lines that bisect the acute and obtuse angles formed by the graphs of $6x + 8y = -5$ and $2x - 3y = 4$. **See margin.**

10. **Navigation** Juwan drives an ATV due east from the edge of a road into the woods. The ATV breaks down and stops after he has gone 2000 feet. In a coordinate system where the positive *y*-axis points north and the units are hundreds of feet, the equation of the road is $5x - 3y = 0$. How far is Juwan from the road? **1715 ft**

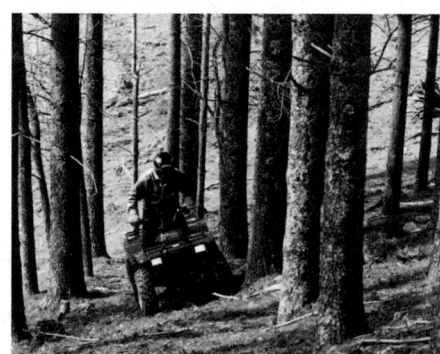

474   Chapter 7   Trigonometric Identities and Equations

www.amc.glencoe.com/self_check_quiz

Additional Answers

4. The formula is valid in either case. Examples will vary. For a vertical line, $x = a$, the formula subtracts *a* from the *x*-coordinate of the point. For a horizontal line, $y = b$, the formula subtracts *b* from the *y*-coordinate of the point.

9. $\left(20 - 6\sqrt{13}\right)x - \left(30 + 8\sqrt{13}\right)y - 40 - 5\sqrt{13} = 0$;
   $\left(20 + 6\sqrt{13}\right)x + \left(8\sqrt{13} - 30\right)y - 40 + 5\sqrt{13} = 0$

EXERCISES

Practice

Find the distance between the point with the given coordinates and the line with the given equation.

A   11. $(2, 0)$, $3x - 4y + 15 = 0$ $\dfrac{21}{5}$ 12. $(3, 5)$, $5x - 3y + 10 = 0$ $\dfrac{5\sqrt{34}}{17}$

B   13. $(0, 0)$, $-2x - y = -3$ $\dfrac{3\sqrt{5}}{5}$ 14. $(-2, -3)$, $y = 4 - \dfrac{2}{3}x$ $\dfrac{25\sqrt{13}}{13}$

15. $(3, 1)$, $y = 2x - 5$ $\mathbf{0}$ 16. $(-1, 2)$, $y = -\dfrac{4}{3}x + 6$ $\dfrac{16}{5}$

17. What is the distance from the origin to the graph of $3x - y + 1 = 0$? $\dfrac{\sqrt{10}}{10}$

Find the distance between the parallel lines with the given equations.

18. $6x - 8y = 3$ $\dfrac{4}{5}$
   $6x - 8y = -5$

19. $4x - 5y = 12$ $\dfrac{6\sqrt{41}}{41}$
   $4x - 5y = 6$

20. $y = 2x + 1$ $\dfrac{3\sqrt{5}}{5}$
   $2x - y = 2$

21. $y = -3x + 6$ $\dfrac{\sqrt{10}}{5}$
   $3x + y = 4$

22. $y = \dfrac{8}{5}x - 1$ $\dfrac{20\sqrt{89}}{89}$
   $8x + 15 = 5y$

23. $y = -\dfrac{3}{2}x$ $\dfrac{8\sqrt{13}}{13}$
   $y = -\dfrac{3}{2}x - 4$

24. What is the distance between the lines with equations $x + y - 1 = 0$ and $y = -x + 6$? $\dfrac{5\sqrt{2}}{2}$

Find equations of the lines that bisect the acute and obtuse angles formed by the lines with the given equations.   **26–27. See margin.**

C   25. $3x + 4y = 10$
      $5x - 12y = 26$
      $x + 8y = 0$; $16x - 2y - 65 = 0$

26. $4x + y = 6$
     $-15x + 8y = 68$

27. $y = \dfrac{2}{3}x + 1$
     $y = -3x - 2$

Applications and Problem Solving

Real World Application

28. **Statistics** *Prediction equations* are often used to show the relationship between two quantities. A prediction equation relating a person's systolic blood pressure y to their age x is $4x - 3y + 228 = 0$. If an actual data point is close to the graph of the prediction equation, the equation gives a good approximation for the coordinates of that point.

   a. Linda is 19 years old, and her systolic blood pressure is 112. Her father, who is 45 years old, has a systolic blood pressure of 120. For whom is the given prediction equation a better predictor?   **Linda**

   b. Refer to the graph at the right. At what age does the equation begin to predict mild hypertension?   **48**

Classification of Adult Systolic Blood Pressure

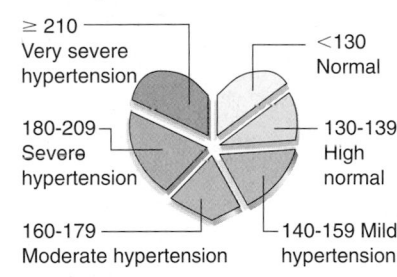

- ≥ 210 Very severe hypertension
- <130 Normal
- 180-209 Severe hypertension
- 130-139 High normal
- 160-179 Moderate hypertension
- 140-159 Mild hypertension

Source: *Archives of Internal Medicine*

Lesson 7-7   Distance From a Point to a Line   **475**

Assignment Guide

Core: 11–27 odd, 29–40
Enriched: 12–26 even, 28–40

Additional Answers
(pp. 475–476)

26. $\left(68 + 15\sqrt{17}\right)x + \left(17 - 8\sqrt{17}\right)y - 102 + 68\sqrt{17} = 0$;
$\left(68 - 15\sqrt{17}\right)x + \left(17 + 8\sqrt{17}\right)y - 102 - 68\sqrt{17} = 0$

27. $\left(2\sqrt{10} + 3\sqrt{13}\right)x + \left(\sqrt{13} - 3\sqrt{10}\right)y + 3\sqrt{10} + 2\sqrt{13} = 0$;
$\left(-2\sqrt{10} + 3\sqrt{13}\right)x + \left(\sqrt{13} + 3\sqrt{10}\right)y - 3\sqrt{10} + 2\sqrt{13} = 0$

30. The radius of the circle is $\sqrt{[(-5) - (-2)]^2 + (6 - 2)^2}$ or 5. Now find the distance from the center of the circle to the line.

$$d = \frac{Ax_1 + By_1 + C}{\pm\sqrt{A^2 + B^2}}$$

$$d = \frac{5(-5) + (-12)(6) + 32}{-\sqrt{5^2 + (-12^2)}}$$

$$d = \frac{-65}{-13}$$

$$d = 5$$

Since the distance from the center of the circle to the line is the same as the radius of the circle, the line can only intersect the circle in one point. That is, the line is tangent to the circle.

35.

$y = \csc(\theta + 60°)$

4 ASSESS

Open-Ended Assessment

Modeling The distance from a point to a line is defined as the perpendicular distance from the point to the line. Use the following figure and the distance formula to find each of the lengths AP, BP, CP and DP.

Draw other segments from P to ℓ. Do any appear shorter than $\overline{BP}$? Why do you suppose that the "shortest" length BP is considered as the distance from P to line ℓ?

 Chapter 7, Quiz D (Lesson 7-7) is available in the *Assessment and Evaluation Masters*, p. 174.

Additional Answers
Answers for Exercises 30 and 35 can be found on p. 475.

Enrichment Masters, p. 79

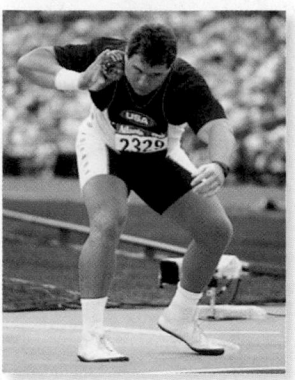

Randy Barnes

29. Track and Field In the shot put, the shot must land within a 40° sector. Consider a coordinate system where the vertex of the sector is at the origin, one side lies along the positive x-axis, and the units are meters. If a throw lands at the point with coordinates (16, 12), how far is it from being out-of-bounds? **1.09 m**

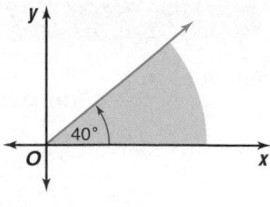

30. Critical Thinking Circle P has its center at $(-5, 6)$ and passes through the point at $(-2, 2)$. Show that the line with equation $5x - 12y + 32 = 0$ is tangent to circle P. **See margin.**

31. Geometry Find the lengths of the altitudes in the triangle with vertices $A(-3, 4)$, $B(1, 7)$, and $C(-1, -3)$. $\dfrac{34}{5}, \dfrac{34\sqrt{53}}{53}, \dfrac{17\sqrt{26}}{26}$

32. Critical Thinking In a triangle, the intersection of the angle bisectors is the center of the *inscribed circle* of the triangle. The inscribed circle is tangent to each side of the triangle. Determine the radius of the inscribed circle (the *inradius*) for the triangle with vertices at $(0, 0)$, $(10, 0)$, and $(4, 12)$. **3.33**

Mixed Review

33. Find the normal form of the equation $-2x + 7y = 5$. *(Lesson 7-6)*

34. Find $\cos 2A$ if $\sin A = \dfrac{\sqrt{3}}{6}$. *(Lesson 7-4)* $\dfrac{5}{6}$

33. $-\dfrac{2\sqrt{53}}{53}x + \dfrac{7\sqrt{53}}{53}y -$

35. Graph $y = \csc(\theta + 60°)$. *(Lesson 6-7)* **See margin.** $\dfrac{5\sqrt{53}}{53} = 0$

36. Aviation Two airplanes leave an airport at the same time. Each flies at a speed of 110 mph. One flies in the direction 60° east of north. The other flies in the direction 40° east of south. How far apart are the planes after 3 hours? *(Lesson 5-8)* **424.24 miles**

37. Physics The period of a pendulum can be determined by the formula $T = 2\pi\sqrt{\dfrac{\ell}{g}}$, where T represents the period, ℓ represents the length of the pendulum, and g represents acceleration due to gravity. Determine the period of a 2-meter pendulum on Earth if the acceleration due to gravity at Earth's surface is 9.8 meters per second squared. *(Lesson 4-7)* **about 2.8 s**

38. Find the value of k in $(x^2 + 8x + k) \div (x - 2)$ so that the remainder is zero. *(Lesson 4-3)* **−20**

39. Solve the system of equations. *(Lesson 2-2)*
$x - 2y + z = -7$
$2x + y - z = -9$
$-x + 3y - 2z = 10$ **(−6, −2, −5)**

40. SAT/ACT Practice If the area of square $BCED = 16$ and the area of $\triangle ABC = 6$, what is the length of $\overline{EF}$? **C**

A 5   B 6
C 7   D 8
E 12

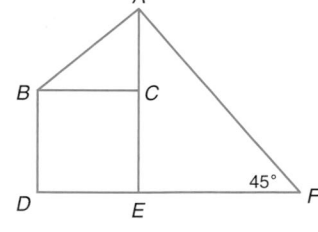

Extra Practice See p. A39.

Extra Credit

Draw a parallelogram $ABCD$ given vertices $A(5, -7)$, $B(3, 5)$, $C(-5, 9)$, and $D(-3, -3)$. Draw the diagonals $\overline{AC}$ and $\overline{BD}$ of the parallelogram and locate E, their point of intersection. What is the distance from E to side $\overline{DC}$?
$\dfrac{22\sqrt{37}}{37}$ **or about 3.62 units**

VOCABULARY

counterexample (p. 421)	principal value (p. 456)
difference identity (p. 437)	Pythagorean identity (p. 423)
double-angle identity (p. 449)	quotient identity (p. 422)
half-angle identity (p. 451)	reciprocal identity (p. 422)
identity (p. 421)	reduction identity (p. 446)
normal form (p. 463)	sum identity (p. 437)
normal line (p. 463)	symmetry identity (p. 424)
opposite-angle identity	trigonometric identity
(p. 426)	(p. 421)

UNDERSTANDING AND USING THE VOCABULARY

Choose the letter of the term that best matches each equation or phrase.

1. $\sin \frac{\alpha}{2} = \pm \sqrt{\dfrac{1 - \cos \alpha}{2}}$ **b**

2. perpendicular to a line, curve, or surface **g**

3. located in Quadrants I and IV for $\sin x$ and $\tan x$ **d**

4. $\cos (\alpha + \beta) = \cos \alpha \cos \beta - \sin \alpha \sin \beta$ **a**

5. $\cot \theta = \dfrac{1}{\tan \theta}$ **i**

6. $\dfrac{\sin \theta}{\cos \theta} = \tan \theta$ **j**

7. $\tan 2\theta = \dfrac{2 \tan \theta}{1 - \tan^2 \theta}$ **h**

8. $\sin (360k^\circ - A) = -\sin A$ **f**

9. $1 + \cot^2 \theta = \csc^2 \theta$ **e**

10. uses trigonometry to provide information about a line **c**

a. sum identity
b. half-angle identity
c. normal form
d. principal value
e. Pythagorean identity
f. symmetry identity
g. normal line
h. double-angle identity
i. reciprocal identity
j. quotient identity
k. opposite-angle identity

 For additional review and practice for each lesson, visit: **www.amc.glencoe.com**

Six forms of Chapter 7 Test are available in the *Assessment and Evaluation Masters* as shown in the chart.

Chapter 7 Test, Form 1B, is shown at the right. Chapter 7 Test, Form 2B, is shown on the next page.

TEST OPTIONS

Form	Pages	Type	
1A	159–160	Honors	Form 1 tests are multiple choice.
1B	161–162	Average	
1C	163–164	Basic	
2A	165–166	Honors	Form 2 tests are free response.
2B	167–168	Average	
2C	169–170	Basic	

VOCABULARY

This section provides a listing of the new terms, properties, and phrases that were introduced in this chapter. The exercises check students' understanding of the terms by using a variety of verbal formats including matching, completion, and true/false.

Glossary A complete glossary of terms begins on page A71.

Assessment and Evaluation Masters, pp. 161–162

NAME _____ DATE _____ PERIOD _____

Chapter 7 Test, Form 1B

Write the letter for the correct answer in the blank at the right of each problem.

1. Find an expression equivalent to $\sec \theta \sin \theta \cot \theta \csc \theta$.
A. $\tan \theta$ B. $\csc \theta$ C. $\sec \theta$ D. $\sin \theta$ 1. **B**

2. If $\sec \theta = -\frac{5}{4}$ and $180° < \theta < 270°$, find $\tan \theta$.
A. $-\frac{3}{5}$ B. $-\frac{4}{5}$ C. $\frac{3}{4}$ D. $\frac{3}{5}$ 2. **C**

3. Simplify $\frac{\tan^2(-\theta) + 1}{\tan^2 \theta}$.
A. $\csc^2 \theta$ B. -1 C. $\tan^2 \theta$ D. 1 3. **A**

4. Simplify $\frac{\tan x}{\sin x} + \frac{1}{\cos x}$.
A. $2 \tan^2 x$ B. $2 \cos x$ C. $2 \cos x - 1$ D. $2 \sec x$ 4. **D**

5. Find a numerical value of one trigonometric function of x if $\sec x \cot x = 4$.
A. $\csc x = \frac{1}{4}$ B. $\sec x = 4$ C. $\sec x = \frac{1}{4}$ D. $\csc x = 4$ 5. **D**

6. Use a sum or difference identity to find the exact value of $\sin 105°$.
A. $\frac{-\sqrt{2} - \sqrt{6}}{4}$ B. $\frac{\sqrt{6} - \sqrt{2}}{4}$ C. $\frac{\sqrt{6} + \sqrt{2}}{4}$ D. $\frac{\sqrt{2} - \sqrt{6}}{4}$ 6. **C**

7. Find the value of $\tan(\alpha - \beta)$ if $\cos \alpha = \frac{4}{5}$, $\sin \beta = -\frac{5}{13}$, $270° < \alpha < 360°$, and $270° < \beta < 360°$.
A. $\frac{16}{63}$ B. $-\frac{56}{63}$ C. $-\frac{56}{33}$ D. $\frac{56}{33}$ 7. **B**

8. Which expression is equivalent to $\cos(\pi + \theta)$?
A. $-\cos \theta$ B. $\cos \theta$ C. $-\sin \theta$ D. $\sin \theta$ 8. **A**

9. Which expression is *not* equivalent to $\cos 2\theta$ for all values of θ?
A. $2 \cos^2 \theta - 1$ B. $1 - 2 \sin^2 \theta$ C. $\cos^2 \theta + \sin^2 \theta$ D. $\cos^2 \theta - \sin^2 \theta$ 9. **C**

10. If $\sin \theta = 0.6$ and $90° < \theta < 180°$, find the exact value of $\sin 2\theta$.
A. -0.6 B. -0.96 C. 0.96 D. 0.28 10. **B**

11. If $\cos \theta = -\frac{3}{5}$ and θ has its terminal side in Quadrant II, find the exact value of $\tan 2\theta$.
A. $\frac{24}{25}$ B. $\frac{7}{25}$ C. $\frac{24}{7}$ D. $-\frac{7}{25}$ 11. **C**

12. Use a half-angle identity to find the exact value of $\cos 75°$.
A. $\frac{1}{2}\sqrt{2 + \sqrt{3}}$ B. $\frac{1}{2}\sqrt{2 - \sqrt{3}}$ C. $\frac{1}{2}\sqrt{2 + \sqrt{2}}$ D. $-\frac{1}{2}\sqrt{1 + \sqrt{3}}$ 12. **B**

© Glencoe/McGraw-Hill   161   Advanced Mathematical Concepts

NAME _____ DATE _____ PERIOD _____

Chapter 7 Test, Form 1B *(continued)*

13. Solve $\csc x + 2 = 0$ for $0 \le x < 2\pi$.
A. $\frac{\pi}{6}$ and $\frac{5\pi}{6}$ B. $\frac{\pi}{6}$ and $\frac{7\pi}{6}$ C. $\frac{4\pi}{3}$ and $\frac{5\pi}{3}$ D. $\frac{7\pi}{6}$ and $\frac{11\pi}{6}$ 13. **D**

14. Solve $2 \cos^2 x - \cos x - 1 = 0$ for principal values of x.
A. $0°$ and $120°$ B. $30°$ C. $60°$ D. $0°$ and $60°$ 14. **A**

15. Solve $4 \sin^2 x - 4 \sin x + 1 = 0$ for all real values of x.
A. $\frac{\pi}{6} + 2\pi k, \frac{7\pi}{6} + 2\pi k$ B. $\frac{7\pi}{6} + 2\pi k, \frac{11\pi}{6} + 2\pi k$ C. $\frac{5\pi}{6} + 2\pi k, \frac{11\pi}{6} + 2\pi k$ D. $\frac{\pi}{6} + 2\pi k, \frac{5\pi}{6} + 2\pi k$ 15. **D**

16. Write the equation $3x - 2y + 7 = 0$ in normal form.
A. $\frac{3\sqrt{13}}{13}x + \frac{10}{13}y + \frac{7\sqrt{13}}{13} = 0$ B. $-\frac{3\sqrt{13}}{13}x + \frac{2\sqrt{13}}{13}y - \frac{7\sqrt{13}}{13} = 0$ C. $-\frac{3\sqrt{13}}{13}x - \frac{2\sqrt{13}}{13}y - \frac{7\sqrt{13}}{13} = 0$ D. $\frac{3\sqrt{13}}{13}x + \frac{2\sqrt{13}}{13}y - \frac{7\sqrt{13}}{13} = 0$ 16. **B**

17. Write the standard form of the equation of a line for which the length of the normal is 3 and the normal makes an angle of 135° with the positive x-axis.
A. $\sqrt{2}x - \sqrt{2}y + 6 = 0$ B. $\sqrt{2}x + \sqrt{2}y - 6 = 0$ C. $\sqrt{2}x - \sqrt{2}y - 6 = 0$ D. $\sqrt{2}x + \sqrt{2}y + 6 = 0$ 17. **A**

18. Find the distance between $P(-2, 5)$ and the line with equation $x - 3y + 4 = 0$.
A. $\frac{17\sqrt{10}}{10}$ B. $\frac{9}{10}$ C. $-\frac{17\sqrt{10}}{10}$ D. $\frac{13\sqrt{10}}{10}$ 18. **D**

19. Find the distance between the lines with equations $5x + 12y = 12$ and $y = -\frac{5}{12}x + 3$.
A. $\frac{24}{13}$ B. $\frac{48}{13}$ C. $-\frac{48}{13}$ D. $\frac{46}{17}$ 19. **A**

20. Find an equation of the line that bisects the acute angles formed by the lines $2x + y - 5 = 0$ and $3x - 2y + 6 = 0$.
A. $(2\sqrt{13} - 3\sqrt{5})x + (\sqrt{13} + 2\sqrt{5})y - 5\sqrt{13} - 6\sqrt{5} = 0$
B. $(-2\sqrt{13} - 3\sqrt{5})x + (-\sqrt{13} + 2\sqrt{5})y + 5\sqrt{13} - 6\sqrt{5} = 0$
C. $(-2\sqrt{13} + 3\sqrt{5})x + (-\sqrt{13} - 2\sqrt{5})y + 5\sqrt{13} - 6\sqrt{5} = 0$
D. $(2\sqrt{13} - 3\sqrt{5})x + (-\sqrt{13} + 2\sqrt{5})y + 5\sqrt{13} + 6\sqrt{5} = 0$ 20. **B**

Bonus If $90° < \theta < 180°$, express $\cos \theta$ in terms of $\tan \theta$.
A. $-\sqrt{\frac{1}{1 + \tan^2 \theta}}$ B. $\sqrt{\frac{1}{1 + \tan^2 \theta}}$ C. $\sqrt{1 + \tan^2 \theta}$ D. $-\sqrt{1 + \tan^2 \theta}$ Bonus: **A**

© Glencoe/McGraw-Hill   162   Advanced Mathematical Concepts

SKILLS AND CONCEPTS

Objectives and Examples This section reviews the skills and concepts of the chapter and shows completely worked examples.

Review Exercises These exercises provide practice for the corresponding objectives.

Assessment and Evaluation Masters, pp. 167–168

NAME _____ DATE _____ PERIOD _____

Chapter 7 **Chapter 7 Test, Form 2B**

1. Simplify $\cos \theta \tan^2 \theta + \cos \theta$.　　　　1. _____ $\sec \theta$ _____

2. If $\cot \theta = -\frac{3}{4}$ and $90° < \theta < 180°$, find $\sin \theta$.　2. _____ $\frac{4}{5}$ _____

3. Simplify $\csc \theta - \cot \theta \cos \theta$.　　　　3. _____ $\sin \theta$ _____

4. Simplify $\frac{1 - \sin^2 \theta}{1 - \cos^2 \theta}$.　　　　4. _____ $\cot^2 \theta$ _____

5. If $\sin^2 x \sec x \cot x = 3$, find the value of $\csc x$.　5. _____ $\frac{1}{3}$ _____

6. Use a sum or difference identity to find the exact value of $\cos 255°$.　6. _____ $\frac{\sqrt{2} - \sqrt{6}}{4}$ _____

7. Find the value of $\sin(\alpha - \beta)$ if $\tan \alpha = \frac{4}{3}$, $\cot \beta = \frac{5}{12}$, $0° < \alpha < 90°$, and $0° < \beta < 90°$.　7. _____ $-\frac{16}{65}$ _____

8. Simplify $\sin\left(\frac{\pi}{2} - \theta\right)$.　　　　8. _____ $\cos \theta$ _____

9. If θ is an angle in the first quadrant and $\csc \theta = 3$, find the exact value of $\cos 2\theta$.　9. _____ $\frac{7}{9}$ _____

10. If $\sin \theta = -0.6$ and $180° < \theta < 270°$, find the exact value of $\sin 2\theta$.　10. _____ $\frac{24}{25}$ or 0.96 _____

11. If $\cos \theta = \frac{4}{5}$ and θ has its terminal side in Quadrant IV, find the exact value of $\tan 2\theta$.　11. _____ $-\frac{24}{7}$ _____

© Glencoe/McGraw-Hill　　167　　Advanced Mathematical Concepts

NAME _____ DATE _____ PERIOD _____

Chapter 7 **Chapter 7 Test, Form 2B (continued)**

12. Use a half-angle identity to find the exact value of $\cos 105°$.　12. _____ $-\frac{\sqrt{2 - \sqrt{3}}}{2}$ _____

13. Solve $\tan x - \sqrt{3} = 0$ for $0 \le x < 2\pi$.　13. _____ $\frac{\pi}{3}, \frac{4\pi}{3}$ _____

14. Solve $4 \sin^2 x - 1 = 0$ for principal values of x. Express the solution(s) in degrees.　14. _____ $-30°, 30°$ _____

15. Solve $\cos^4 x - 1 = 0$ for all real values of x.　15. _____ πk _____

16. Write the equation $2x + 5y - 3 = 0$ in normal form.　16. _____ $\frac{2\sqrt{29}}{29}x + \frac{5\sqrt{29}}{29}y - \frac{3\sqrt{29}}{29} = 0$ _____

17. Write the standard form of the equation of a line for which the length of the normal is 7 and the normal makes an angle of 150° with the positive x-axis.　17. _____ $\sqrt{3}x - y + 14 = 0$ _____

18. Find the distance between $P(-1, 4)$ and the line with equation $4x - 2y + 3 = 0$.　18. _____ $\frac{9\sqrt{5}}{10}$ _____

19. Find the distance between the lines with equations $x - 2y = 3$ and $y = \frac{1}{2}x - 2$.　19. _____ $\frac{\sqrt{5}}{5}$ _____

20. Find an equation of the line that bisects the acute angles formed by the lines with equations $3x + y + 6 = 0$ and $2x - y - 1 = 0$.　20. _____ $(3\sqrt{5} + 2\sqrt{10})x + (\sqrt{5} - \sqrt{10})y + 6\sqrt{5} - \sqrt{10} = 0$ _____

Bonus Express $\sqrt{\frac{\tan^2 \theta}{\sec^2 \theta + \cot^2 \theta \sec^2 \theta}}$ in terms of $\sin \theta$.　Bonus: _____ $\sin^2 \theta$ _____

© Glencoe/McGraw-Hill　　168　　Advanced Mathematical Concepts

OBJECTIVES AND EXAMPLES	REVIEW EXERCISES

Lesson 7-1 Identify and use reciprocal identities, quotient identities, Pythagorean identities, symmetry identities, and opposite-angle identities.

• If θ is in Quadrant I and $\cos \theta = \frac{1}{3}$, find $\sin \theta$.

$$\sin^2 \theta + \cos^2 \theta = 1$$
$$\sin^2 \theta + \left(\frac{1}{3}\right)^2 = 1$$
$$\sin^2 \theta + \frac{1}{9} = 1$$
$$\sin^2 \theta = \frac{8}{9}$$
$$\sin \theta = \frac{2\sqrt{2}}{3}$$

Use the given information to determine the trigonometric value. In each case, $0° < \theta < 90°$.

11. If $\sin \theta = \frac{1}{2}$, find $\csc \theta$.　**2**

12. If $\tan \theta = 4$, find $\sec \theta$.　$\sqrt{17}$

13. If $\csc \theta = \frac{5}{3}$, find $\cos \theta$.　$\frac{4}{5}$

14. If $\cos \theta = \frac{4}{5}$, find $\tan \theta$.　$\frac{3}{4}$

15. Simplify $\csc x - \cos^2 x \csc x$.　**sin x**

Lesson 7-2 Use the basic trigonometric identities to verify other identities.

• Verify that $\csc x \sec x = \cot x + \tan x$ is an identity.

$$\csc x \sec x \overset{?}{=} \cot x + \tan x$$
$$\frac{1}{\sin x} \cdot \frac{1}{\cos x} \overset{?}{=} \frac{\cos x}{\sin x} + \frac{\sin x}{\cos x}$$
$$\frac{1}{\sin x \cos x} \overset{?}{=} \frac{\cos^2 x + \sin^2 x}{\sin x \cos x}$$
$$\frac{1}{\sin x \cos x} = \frac{1}{\sin x \cos x}$$

Verify that each equation is an identity.

16. $\cos^2 x + \tan^2 x \cos^2 x = 1$

17. $\frac{1 - \cos \theta}{1 + \cos \theta} = (\csc \theta - \cot \theta)^2$

18. $\frac{\sec \theta + 1}{\tan \theta} = \frac{\tan \theta}{\sec \theta - 1}$

19. $\frac{\sin^4 x - \cos^4 x}{\sin^2 x} = 1 - \cot^2 x$

16–19. See margin.

Lesson 7-3 Use the sum and difference identities for the sine, cosine, and tangent functions.

• Find the exact value of $\sin 105°$.

$$\sin 105° = \sin (60° + 45°)$$
$$= \sin 60° \cos 45° + \cos 60° \sin 45°$$
$$= \frac{\sqrt{3}}{2} \cdot \frac{\sqrt{2}}{2} + \frac{1}{2} \cdot \frac{\sqrt{2}}{2}$$
$$= \frac{\sqrt{6} + \sqrt{2}}{4}$$

Use sum or difference identities to find the exact value of each trigonometric function.

20. $\cos 195°$　$\frac{-\sqrt{6} - \sqrt{2}}{4}$　21. $\cos 15°$　$\frac{\sqrt{6} + \sqrt{2}}{4}$

22. $\sin \left(-\frac{17\pi}{12}\right)$　$\frac{\sqrt{6} + \sqrt{2}}{4}$　23. $\tan \frac{11\pi}{12}$　$-2 + \sqrt{3}$

Find each exact value if $0 < x < \frac{\pi}{2}$ and $0 < y < \frac{\pi}{2}$.　24. $\frac{48 + 7\sqrt{5}}{75}$

24. $\cos (x - y)$ if $\sin x = \frac{7}{25}$ and $\cos y = \frac{2}{3}$

25. $\tan (x + y)$ if $\tan x = \frac{5}{4}$ and $\sec y = \frac{3}{2}$

25. $-\frac{180 + 82\sqrt{5}}{61}$

TestCheck and Worksheet Builder

This **networkable** software has three modules.
• **Worksheet Builder** to make worksheets and tests
• **Student Module** to take tests on-screen
• **Management System** to keep student records
Three special banks are included for SAT, ACT, and TIMSS.

OBJECTIVES AND EXAMPLES

Lesson 7-4 Use the double- and half-angle identities for the sine, cosine, and tangent functions.

- If θ is an angle in the first quadrant and $\sin \theta = \frac{3}{4}$, find $\cos 2\theta$.

$$\cos 2\theta = 1 - 2 \sin^2 \theta$$
$$= 1 - 2\left(\frac{3}{4}\right)^2$$
$$= -\frac{1}{8}$$

Lesson 7-5 Solve trigonometric equations and inequalities.

- Solve $2 \cos^2 x - 1 = 0$ for $0° \leq x < 360°$.
$$2 \cos^2 x - 1 = 0$$
$$\cos^2 x = \frac{1}{2}$$
$$\cos x = \pm\frac{\sqrt{2}}{2}$$
$$x = 45°, 135°, 225°, \text{ or } 315°$$

37. $\pi k, \frac{\pi}{4} + 2\pi k, \frac{3\pi}{4} + 2\pi k$

Lesson 7-6 Write linear equations in normal form.

- Write $3x + 2y - 6 = 0$ in normal form. Since C is negative, use the positive value of $\sqrt{A^2 + B^2}$. $\sqrt{3^2 + 2^2} = \sqrt{13}$

The normal form is
$$\frac{3}{\sqrt{13}} x + \frac{2}{\sqrt{13}} y - \frac{6}{\sqrt{13}} = 0 \text{ or}$$
$$\frac{3\sqrt{13}}{13} x + \frac{2\sqrt{13}}{13} y - \frac{6\sqrt{13}}{13} = 0.$$

REVIEW EXERCISES

Use a half-angle identity to find the exact value of each function.

26. $\cos 75°$ $\frac{\sqrt{2 - \sqrt{3}}}{2}$ 27. $\sin \frac{7\pi}{8}$ $\frac{\sqrt{2 - \sqrt{2}}}{2}$

28. $\sin 22.5°$ $\frac{\sqrt{2 - \sqrt{2}}}{2}$ 29. $\tan \frac{\pi}{12}$ $2 - \sqrt{3}$

If θ is an angle in the first quadrant and $\cos \theta = \frac{3}{5}$, find the exact value of each function.

30. $\sin 2\theta$ $\frac{24}{25}$ 31. $\cos 2\theta$ $-\frac{7}{25}$

32. $\tan 2\theta$ $-\frac{24}{7}$ 33. $\sin 4\theta$ $-\frac{336}{625}$

Solve each equation for $0° \leq x < 360°$.

34. $\tan x + 1 = \sec x$ **0°**

35. $\sin^2 x + \cos 2x - \cos x = 0$ **0°, 90°, 270°**

36. $\cos 2x + \sin x = 1$ **0°, 30°, 150°, 180°**

Solve each equation for all real values of x.

37. $\sin x \tan x - \frac{\sqrt{2}}{2} \tan x = 0$

38. $\sin 2x + \sin x = 0$ $\pi k, \frac{2\pi}{3} + 2\pi k, \frac{4\pi}{3} + 2\pi k$

39. $\cos^2 x = 2 - \cos x$ $2\pi k$

40. $x + \sqrt{3}y - 4\sqrt{3} = 0$

Write the standard form of the equation of each line given p, the length of the normal segment, and ϕ, the angle the normal segment makes with the positive x-axis.   41. $y - 5 = 0$

40. $p = 2\sqrt{3}, \phi = \frac{\pi}{3}$ 41. $p = 5, \phi = 90°$

42. $p = 3, \phi = \frac{2\pi}{3}$ 43. $p = 4\sqrt{2}, \phi = 225°$
   $-x + \sqrt{3}y - 6 = 0$ $x + y + 8 = 0$

Write each equation in normal form. Then find the length of the normal and the angle it makes with the positive x-axis.   **44–47. See margin.**

44. $7x + 3y - 8 = 0$ 45. $6x = 4y - 5$

46. $9x = -5y + 3$ 47. $x - 7y = -5$

Additional Answers
(pp. 478–479)

16. $\cos^2 x + \tan^2 x \cos^2 x \stackrel{?}{=} 1$
$$\cos^2 x + \frac{\sin^2 x}{\cos^2 x} \cos^2 x \stackrel{?}{=} 1$$
$$\cos^2 x + \sin^2 x \stackrel{?}{=} 1$$
$$1 = 1$$

17. $\frac{1 - \cos \theta}{1 + \cos \theta} \stackrel{?}{=} (\csc \theta - \cot \theta)^2$
$$\frac{1 - \cos \theta}{1 + \cos \theta} \stackrel{?}{=} \left(\frac{1}{\sin \theta} - \frac{\cos \theta}{\sin \theta}\right)^2$$
$$\frac{1 - \cos \theta}{1 + \cos \theta} \stackrel{?}{=} \frac{(1 - \cos \theta)^2}{\sin^2 \theta}$$
$$\frac{1 - \cos \theta}{1 + \cos \theta} \stackrel{?}{=} \frac{(1 - \cos \theta)^2}{1 - \cos^2 \theta}$$
$$\frac{1 - \cos \theta}{1 + \cos \theta} \stackrel{?}{=} \frac{(1 - \cos \theta)^2}{(1 - \cos \theta)(1 + \cos \theta)}$$
$$\frac{1 - \cos \theta}{1 + \cos \theta} = \frac{1 - \cos \theta}{1 + \cos \theta}$$

18. $\frac{\sec \theta + 1}{\tan \theta} \stackrel{?}{=} \frac{\tan \theta}{\sec \theta - 1}$
$$\frac{\sec \theta + 1}{\tan \theta} \stackrel{?}{=} \frac{\tan \theta (\sec \theta + 1)}{\sec^2 \theta - 1}$$
$$\frac{\sec \theta + 1}{\tan \theta} \stackrel{?}{=} \frac{\tan \theta (\sec \theta + 1)}{\tan^2 \theta}$$
$$\frac{\sec \theta + 1}{\tan \theta} = \frac{\sec \theta + 1}{\tan \theta}$$

19. (See bottom margin.)

44. $\frac{7\sqrt{58}}{58}x + \frac{3\sqrt{58}}{58}y - \frac{4\sqrt{58}}{29} = 0$;
$\frac{4\sqrt{58}}{29}$; **23°**

45. $-\frac{3\sqrt{13}}{13}x + \frac{2\sqrt{13}}{13}y - \frac{5\sqrt{13}}{26} = 0$;
$\frac{5\sqrt{13}}{26}$; **146°**

46. $\frac{9\sqrt{106}}{106}x + \frac{5\sqrt{106}}{106}y - \frac{3\sqrt{106}}{106} = 0$;
$\frac{3\sqrt{106}}{106}$; **29°**

47. $-\frac{\sqrt{2}}{10}x + \frac{7\sqrt{2}}{10}y - \frac{\sqrt{2}}{2} = 0$;
$\frac{\sqrt{2}}{2}$; **98°**

Additional Answer

19.
$$\frac{\sin^4 x - \cos^4 x}{\sin^2 x} \stackrel{?}{=} 1 - \cot^2 x$$
$$\frac{(\sin^2 x - \cos^2 x)(\sin^2 x + \cos^2 x)}{\sin^2 x} \stackrel{?}{=} 1 - \cot^2 x$$
$$\frac{\sin^2 x - \cos^2 x}{\sin^2 x} \stackrel{?}{=} 1 - \cot^2 x$$
$$1 - \frac{\cos^2 x}{\sin^2 x} \stackrel{?}{=} 1 - \cot^2 x$$
$$1 - \cot^2 x = 1 - \cot^2 x$$

OBJECTIVES AND EXAMPLES	REVIEW EXERCISES

Lesson 7-7 Find the distance from a point to a line.

Find the distance between $P(-1, 3)$ and the line with equation $-3x + 4y = -5$.

$$d = \frac{Ax_1 + By_1 + C}{\pm\sqrt{A^2 + B^2}}$$

$$= \frac{-3(-1) + 4(3) + 5}{-\sqrt{(-3)^2 + 4^2}}$$

$$= \frac{20}{-5}$$

$$= -4$$

Find the distance between the point with the given coordinates and the line with the given equation.

48. $(5, 6)$, $2x - 3y + 2 = 0$ $\dfrac{6\sqrt{13}}{13}$

49. $(-3, -4)$, $2y = -3x + 6$

50. $(-2, 4)$, $4y = 3x - 1$

51. $(21, 20)$, $y = \dfrac{1}{3}x + 6$ $\dfrac{21\sqrt{10}}{10}$

49. $\dfrac{23\sqrt{13}}{13}$ **50.** $\dfrac{23}{5}$

Lesson 7-7 Write equations of lines that bisect angles formed by intersecting lines.

Find the equations of the lines that bisect the angles formed by the lines with equations $-4x + 3y = 2$ and $x + 2y = 1$.

$$d_1 = \frac{-4x_1 + 3y_1 - 2}{\sqrt{(-4)^2 + 3^2}}$$

$$= \frac{-4x_1 + 3y_1 - 2}{5}$$

$$d_2 = \frac{x_1 + 2y_1 - 1}{\sqrt{1^2 + 2^2}}$$

$$= \frac{x_1 + 2y_1 - 1}{\sqrt{5}}$$

The origin is in the interior of one of the obtuse angles formed by the given lines.

Bisector of the acute angle: $d_1 = -d_2$

$$\frac{-4x + 3y - 2}{5} = -\frac{x + 2y - 1}{\sqrt{5}}$$

$$\left(-4\sqrt{5} + 5\right)x + \left(3\sqrt{5} + 10\right)y - 2\sqrt{5} - 5 = 0$$

Bisector of the obtuse angle: $d_1 = d_2$

$$\frac{-4x + 3y - 2}{5} = \frac{x + 2y - 1}{\sqrt{5}}$$

$$\left(-4\sqrt{5} - 5\right)x + \left(3\sqrt{5} - 10\right)y - 2\sqrt{5} + 5 = 0$$

Find the distance between the parallel lines with the given equations.

52. $y = \dfrac{x}{3} - 6$ $\dfrac{12\sqrt{10}}{5}$

  $y = \dfrac{x}{3} + 2$

53. $y = \dfrac{3}{4}x + 3$ $\dfrac{14}{5}$

  $y = \dfrac{3}{4}x - \dfrac{1}{2}$

54. $x + y = 1$ $2\sqrt{2}$

  $x + y = 5$

55. $2x - 3y + 3 = 0$ $\dfrac{9\sqrt{13}}{13}$

  $y = \dfrac{2}{3}x - 2$

Find the equations of the lines that bisect the acute and obtuse angles formed by the lines with the given equations. **56–57. See margin.**

56. $y = -3x - 2$

  $y = -\dfrac{x}{2} + \dfrac{3}{2}$

57. $-x + 3y - 2 = 0$

  $y = \dfrac{3}{5}x + 3$

Additional Answers

56. $\left(3\sqrt{5} + \sqrt{10}\right)x + \left(\sqrt{5} + 2\sqrt{10}\right)y + 2\sqrt{5} - 3\sqrt{10} = 0$;

  $\left(3\sqrt{5} - \sqrt{10}\right)x + \left(\sqrt{5} - 2\sqrt{10}\right)y + 2\sqrt{5} + 3\sqrt{10} = 0$

57. $\left(-\sqrt{34} - 3\sqrt{10}\right)x + \left(3\sqrt{34} + 5\sqrt{10}\right)y - 2\sqrt{34} - 15\sqrt{10} = 0$;

  $\left(-\sqrt{34} + 3\sqrt{10}\right)x + \left(3\sqrt{34} - 5\sqrt{10}\right)y - 2\sqrt{34} + 15\sqrt{10} = 0$

APPLICATIONS AND PROBLEM SOLVING

58. Physics While studying two different physics books, Hector notices that two different formulas are used to find the maximum height of a projectile. One formula is $h = \dfrac{v_0^2 \sin^2 \theta}{2g}$, and the other is $h = \dfrac{v_0^2 \tan^2 \theta}{2g \sec^2 \theta}$. Are these two formulas equivalent or is there an error in one of the books? Show your work. *(Lesson 7-2)*
See margin.

59. Navigation Wanda hikes due east from the edge of the road in front of a lodge into the woods. She stops to eat lunch after she has hiked 1600 feet. In a coordinate system where the positive *y*-axis points north and the units are hundreds of feet, the equation of the road is $4x - 2y = 0$. How far is Wanda from the road? *(Lesson 7-7)* **1431 ft**

60. Surveying Taigi is surveying a rectangular lot for a new office building. She measures the angle between one side of the lot and the line from her position to the opposite corner of the lot as 30°. She then measures the angle between that line and the line to a telephone pole on the edge of the lot as 45°. If Taigi stands 100 yards from the opposite corner of the lot, how far is she from the telephone pole? *(Lesson 7-3)* **about 51.76 yd**

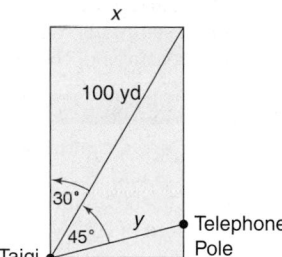

1. Sample answer: 15°; $\dfrac{\sqrt{2 - \sqrt{3}}}{2}$; $\dfrac{\sqrt{2 + \sqrt{3}}}{2}$; $2 - \sqrt{3}$

ALTERNATIVE ASSESSMENT

OPEN-ENDED ASSESSMENT

1. Give the measure θ of an angle such that trigonometric identities can be used to find exact values for $\sin \theta$, $\cos \theta$, and $\tan \theta$. Find exact values for $\sin \theta$, $\cos \theta$, and $\tan \theta$. Show your work.

2. Write an equation with $\sin x \tan x$ on one side and an expression containing one or more different trigonometric functions on the other side. Verify that your equation is an identity.
   Sample answer: $\sin x \tan x = \dfrac{1 - \cos^2 x}{\cos x}$

 PORTFOLIO
   Choose one of the identities you studied in this chapter. Explain why it is an important identity and how it is used.

 Unit 2 *inter*NET **Project**
THE CYBERCLASSROOM
That's as clear as mud!

• Search the Internet to find web sites that have lessons involving trigonometric identities. Find at least two types of identities that were not presented in Chapter 7.

• Select one of the types of identities and write at least two of your own examples or sample problems using what you learned on the Internet and in this textbook.

• Design your own web page presenting your examples. Use appropriate software to create the web page. Have another student critique your web page.

> **Additional Assessment** See p. A62 for Chapter 7 practice test.

Chapter 7 Study Guide and Assessment **481**

APPLICATIONS & PROBLEM SOLVING

This section provides additional practice in solving real-world problems that involve the concepts of this chapter.

Additional Answer
58. The formulas are equivalent.

$$\frac{v_0^2 \tan^2 \theta}{2g \sec^2 \theta} = \frac{v_0^2 \cdot \dfrac{\sin^2 \theta}{\cos^2 \theta}}{2g \cdot \dfrac{1}{\cos^2 \theta}}$$

$$= \frac{v_0^2 \cdot \dfrac{\sin^2 \theta}{\cos^2 \theta}}{2g \cdot \dfrac{1}{\cos^2 \theta}} \cdot \frac{\cos^2 \theta}{\cos^2 \theta}$$

$$= \frac{v_0^2 \sin^2 \theta}{2g}$$

ALTERNATIVE ASSESSMENT

Unit 2
*inter*NET **Project**

● **That's as Clear as Mud!**
Make sure that students know how to use bookmarks in order to track sites that they want to visit again. This is more efficient than printing out the Web page for future reference. However, if many students use the same computers or if computers could have software problems, have each student use a small notebook (3 inches by 5 inches) to record the addresses of each site. They should record the topic and the address of each site that they would like to use for reference.

Internet Address Book

Record useful Internet addresses in the space at right for quick reference.

www.amc.glencoe.com

Chapter 7 Skill: Geometry—Triangles and Quadrilaterals

Pages 482–483 are part of a complete SAT/ACT preparation course that is described in detail on pages T8–T9 of the *Teacher's Handbook*. The test items were written in the same style as those in the SAT and ACT.

 These questions were aligned and verified by The Princeton Review, the nation's leader in test preparation.

Teaching Tip Review with students the names and major characteristics of quadrilaterals such as rectangles, squares, parallelograms, and rhombi.

Diagnosis & Prescription

Each of the 10 SAT and ACT practice problems on page 483 is cross-referenced to the chapter where that SAT or ACT skill is covered. If students miss a particular type of problem, you can have them study that skill.

Problem Number and Skill	Page
1 Geometry (Triangles and Quadrilaterals)	482
2 More Algebra	202
3 Geometry (Triangles and Quadrilaterals)	482
4 Trigonometry	418
5 Coordinate Geometry	272
6 Geometry (Triangles and Quadrilaterals)	482
7 More Algebra	202
8 Geometry (Triangles and Quadrilaterals)	482
9 Geometry (Triangles and Quadrilaterals)	482
10 Geometry (Triangles and Quadrilaterals)	482

Geometry Problems — Triangles and Quadrilaterals

The ACT and SAT contain problems that deal with triangles, quadrilaterals, lengths, and angles. Be sure to review the properties of isosceles and equilateral triangles. Often several geometry concepts are combined in one problem.

Know the number of degrees in various figures.
- A straight angle measures 180°.
- A right angle measures 90°.
- The sum of the measures of the angles in a triangle is 180°.
- The sum of the measures of the angles in a quadrilateral is 360°.

THE PRINCETON REVIEW

TEST-TAKING TIP
The third side of a triangle cannot be longer than the sum of the other two sides. The third side cannot be shorter than the difference between the other two sides.

ACT EXAMPLE

1. In the figure below, O, N, and M are collinear. If the lengths of $\overline{ON}$ and $\overline{NL}$ are the same, the measure of $\angle LON$ is 30°, and $\angle LMN$ is 40°, what is the measure of $\angle NLM$?

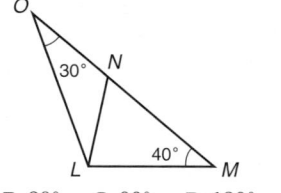

A 40° **B** 80° **C** 90° **D** 120° **E** 130°

HINT Look at *all* the triangles in a figure—large and small.

Solution Mark the angle you need to find. You may want to label the missing angles and congruent sides, as shown.

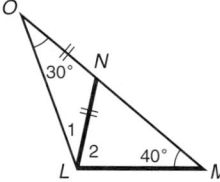

Since two sides of $\triangle ONL$ are the same length, it is isosceles. So $m\angle 1 = 30°$. Since the angles in any triangle total 180°, you can write the following equation for $\triangle OML$.
$$180° = 30° + 40° + (30° + m\angle 2)$$
$$180° = 100° + m\angle 2$$
$$80° = m\angle 2$$

The answer is choice **B**.

SAT EXAMPLE

2. If the average measure of two angles in a parallelogram is $y°$, what is the average degree measure of the other two angles?

A $180 - y$ **B** $180 - \dfrac{y}{2}$ **C** $360 - 2y$

D $360 - y$ **E** y

HINT Underline important words in the problem and the quantity you must find. Look at the answer choices.

Solution Look for key words in the problem—*average* and *parallelogram*. You need to find the average of two angles. The answer choices are expressions with the variable y. Recall that if the average of two numbers is y, then the sum of the numbers is $2y$.

So the sum of two angle measures is $2y$. Let the sum of the other two angle measures be $2x$. Find x.

The sum of the angle measures in a parallelogram is 360.

$$360 = 2y + 2x$$
$$360 = 2(y + x)$$
$$180 = y + x \qquad \textit{Divide each side by 2.}$$
$$x = 180 - y \qquad \textit{Solve for x.}$$

The answer is choice **A**.

 TestCheck and Worksheet Builder

Special banks of SAT, ACT, and TIMSS items are included in this CD-ROM.

After you work each problem, record your answer on the answer sheet provided or on a piece of paper.

Multiple Choice

1. In the figure, the measure of $\angle A$ is 80°. If the measure of $\angle B$ is half the measure of $\angle A$, what is the measure of $\angle C$? **B**

   A 40° **B** 60° **C** 80°

   D 100° **E** 120°

2. At what point (x, y) do the two lines with equations $y = 2x - 2$ and $7x - 3y = 11$ intersect? **A**

   A $(5, 8)$ **B** $(8, 5)$ **C** $\left(\dfrac{5}{8}, -1\right)$

   D $\left(\dfrac{5}{8}, 1\right)$ **E** $\left(\dfrac{25}{16}, \dfrac{9}{8}\right)$

3. In the figure below, $x + y + z =$ **D**

   A 90 **B** 180 **C** 270 **D** 360 **E** 450

4. If $x + y = 90°$ and x and y are positive, then $\dfrac{\sin x}{\cos y} =$ **D**

   A -1 **B** 0 **C** $\dfrac{1}{2}$ **D** 1

   E It cannot be determined from the information given.

5. In the figure below, what is the sum of the slopes of $\overline{AB}$ and $\overline{AD}$? **B**

   A -1 **B** 0 **C** 1 **D** 6

   E It cannot be determined from the information given.

6. In the rectangle $PQRS$ below, what is the sum $a + b$ in terms of x? **A**

   A $90 + x$

   B $190 - x$

   C $180 + x$

   D $270 - x$

   E $360 - x$

   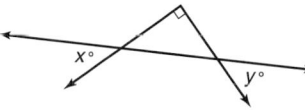

7. $\dfrac{y - \dfrac{1}{y}}{1 - \dfrac{2}{y} + \dfrac{1}{y^2}} =$ **A**

   A $\dfrac{y^2 + y}{y - 1}$ **B** $\dfrac{y^2 - y}{(y + 1)^2}$ **C** $\dfrac{y^2 - y}{y + 1}$

   D $\dfrac{y^2 + y}{(y - 1)^2}$ **E** $\dfrac{y}{y - 1}$

8. In the figure below, $\triangle ACE$ is similar to $\triangle BCD$. What is the length of $\overline{AE}$? **E**

   A 5

   B 6

   C 6.5

   D 7

   E 10

9. The number of degrees in the sum $x° + y°$ would be the same as the number of degrees in which type of angle? **C**

   A straight angle
   B obtuse angle
   C right angle
   D acute angle
   E It cannot be determined from the information given.

10. **Grid-In** A triangle has a base of 13 units, and the other two sides are congruent. If the side lengths are integers, what is the length of the shortest possible side? **7**

interNET CONNECTION SAT/ACT Practice For additional test practice questions, visit: **www.amc.glencoe.com**

THE PRINCETON REVIEW   SAT & ACT Preparation   **483**

Assessment and Evaluation Masters, pp. 175–176

Extra Credit

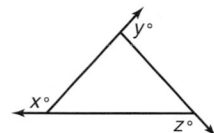

TIMSS In $\triangle ABC$, the altitudes $\overline{BN}$ and $\overline{CM}$ intersect at S, $\angle MSB = 40°$, and $\angle SBC = 20°$. Write a proof that "$\triangle ABC$ is isosceles." Give geometric reasons for statements in your proof. $\angle NSC = 40°$ (Vert. $\angle$s are =.); $\angle BSC = 140°$ (sum of suppl. $\angle$s = 180°). Since sum of angles of a triangle = 180°, $\angle MBS$ and $\angle NCS$ both equal 50° and $\angle SCB = 20°$. $\angle ABC = \angle MBS + \angle SBC = 70°$ and $\angle ACB = \angle NCS + \angle SCB = 70°$. Since $\angle ABC = \angle ACB$, $\triangle ABC$ is isosceles.

Instructional Objectives

Lesson (pages)	Lesson Objectives	NCTM Standards 2000	State/Local Objectives
8-1 (485–492)	• Find equal, opposite, and parallel vectors. • Add and subtract vectors geometrically.	2, 3	
8-2 (493–499)	• Find ordered pairs that represent vectors. • Add, subtract, multiply, and find the magnitude of vectors algebraically.	2, 3, 4	
8-3 (500–504)	• Add and subtract vectors in three-dimensional space. • Find the magnitude of vectors in three-dimensional space.	2, 3, 4, 6	
8-4 (505–511)	• Find the inner and cross products of two vectors. • Determine whether two vectors are perpendicular.	2, 3, 6	
8-4B (512)	• Use a graphing calculator program to obtain the components of the cross product of two vectors in space.	2, 3, 6	
8-5 (513–519)	• Solve problems using vectors and right triangle trigonometry.	2, 3, 6	
8-6 (520–525)	• Write vector and parametric equations of lines. • Graph parametric equations.	3, 6, 10	
8-6B (526)	• Investigate the use of parametric equations.	3, 6, 10	
8-7 (527–533)	• Model the motion of a projectile using parametric equations. • Solve problems related to the motion of a projectile, its trajectory, and range.	3, 6, 10	
8-8 (535–542)	• Transform three-dimensional figures using matrix operations to describe the transformation.	3, 6	

Suggested Pacing   See pages T2–T4 for additional options in course-planning.

COURSE	Lesson 8-1	Lesson 8-2	Lesson 8-3	Lesson 8-4	Lesson 8-5
Precalculus Emphasis Chapters 4–15 (45–50 minutes)	1.5 days	1.5 days	1.5 days	(with Lesson 8-4B) 2.5 days	1.5 days
Precalculus Emphasis (block)	0.5 day	1.5 days		(with Lesson 8-4B) 1 day	(with Lesson 8-6) 1.5 days
Trigonometric Emphasis Chapters 1–9, 11, 12 (45–50 minutes)	1.5 days	1.5 days	1.5 days	(with Lesson 8-4B) 2.5 days	1.5 days
Trigonometric Emphasis (block)	0.5 day	1.5 days		(with Lesson 8-4B) 1 day	(with Lesson 8-6) 1.5 days

Instructional Resources

Student Edition		Blackline Masters (page numbers)								
Lesson	Graphing Calculator Applications	Study Guide and Practice	Enrichment	Assessment and Evaluation	Graphing Calculator Keystrokes	Problem Solving & Applications	Spreadsheet Applications	TI-92 and CBL Interdisciplinary Activities	Transparencies A and B
8-1		137–138	85					25–28	8-1
8-2		139–140	86	199					8-2
8-3		141–142	87			16	8		8-3
8-4		143–144	88	198, 199					8-4
8-4B	512				31–32				
8-5		145–146	89			15			8-5
8-6	524	147–148	90	200					8-6
8-6B	526				33–34				
8-7	532	149–150	91						8-7
8-8		151–152	92	200					8-8
Study Guide/ Assessment				185–197, 203					
SAT/ACT Practice				201–202					

Lesson 8-6	Lesson 8-7	Lesson 8-8	Study Guide & Assessment	Chapter Test	TOTAL
(with Lesson 8-6B) 2.5 days	1.5 days	1.5 days	1 day	1 day	16 days
(See Lesson 8-5.)	(with Lesson 8-6B) 1.5 days	1.5 days		0.5 day	8 days
(with Lesson 8-6B) 2.5 days	1.5 days	1.5 days	1 day	1 day	16 days
(See Lesson 8-5.)	(with Lesson 8-6B) 1.5 days	1.5 days		0.5 day	8 days

TeacherWorks™

The worksheets shown on this page are a small sample of the materials available on *TeacherWorks: All-in-One Lesson Planner and Resource Center.*

This CD-ROM contains all of the blackline masters and transparencies available for this program.

It also includes a lesson planner and interactive Teacher Edition, so you can customize lesson plans and reproduce classroom resources quickly and easily, from just about anywhere.

Applications

Problem Solving and Applications Masters, pp. 15, 16

Technology/Multimedia

Spreadsheet Applications Masters, p. 8

TI-92 & CBL Interdisciplinary Activities Masters, pp. 25–28

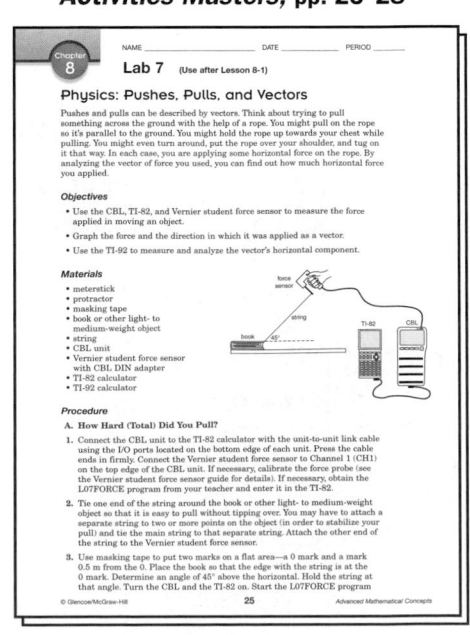

Assessment Resources

Type	Student Edition	Teacher's Wraparound Edition	Assessment and Evaluation Masters
Ongoing Assessment	Mid-Chapter Quiz, p. 511	5-Minute Check, pp. 485, 493, 500, 505, 513, 520, 527, 535	Mid-Chapter Test, p. 198 Quizzes A–D, pp.199–200
Mixed Review	Mixed Review, pp. 492, 498–499, 504, 511, 519, 525, 533, 542		Cumulative Review, p. 203
Error Analysis	You Decide, pp. 496, 502	Error Analysis, pp. 490, 496, 502, 508, 516, 523, 531, 539	
Standardized Test Practice	SAT/ACT Practice, pp. 492, 499, 504, 511, 519, 525, 533, 542 SAT & ACT Preparation, pp. 548–549	TIMSS, p. 549	SAT/ACT Practice, pp. 201–202
Open-Ended Assessment	Math Journal, pp. 508, 540 Open-Ended Assessment, p. 547 Portfolio, p. 547	Modeling: pp. 519, 542 Speaking: pp. 499, 512, 533 Writing: pp. 492, 504, 511, 525, 526	Open-Ended Assessment, p. 197
Chapter Assessment	Study Guide and Assessment, pp. 543–547 Chapter Test, p. A63		Multiple-Choice Tests (Forms 1A, 1B, 1C), pp. 185–190 Free-Response Tests (Forms 2A, 2B, 2C), pp. 191–196

TestCheck and Worksheet Builder is a networkable CD-ROM with three integrated modules. The **Worksheet Builder** creates customized worksheets, tests, and quizzes of free-response, multiple-choice, short-answer, and open-ended items. The **Student Module** gives you the option of having students take tests on-screen and get immediate feedback on their performance. Use the optional **Management System** to keep detailed student records.

Additional Chapter Resources

Student Edition
Career Choices, p. 499
History of Mathematics, p. 534
Internet Project, p. 547
Extra Practice, pp. A40–A41

Teacher's Classroom Resources

Technology
Graphing Calculator Keystrokes Masters

Teaching Aids
Answer Key Transparencies
Block Schedule Planning Guide
Lesson Planning Guide
Solutions Manual

Glencoe Instructional Technology

Instructional
Advanced MathPASS CD-ROM
Multimedia Applications CD-ROM

Assessment
TestCheck and Worksheet Builder

Visit **www.amc.glencoe.com** for data updates, career information, graphing calculator programs, and other interactive activities.

Chapter Overview

This chapter introduces vector notation; vector addition, subtraction, and multiplication; cross products; and parametric equations. Students will use matrices to model transformations in three-dimensional space.

Using a Graphing Calculator

There are numerous opportunities, all optional, to use a graphing calculator in this chapter.
- Using a graphing calculator program to obtain the components of the cross product of two vectors in space (p. 512)
- Setting up a table of values for lines and then graphing each line from its parametric form (p. 524)
- Modeling real-world situations with parametric equations (p. 526)
- Graphing a cycloid (p. 532)

The **Graphing Calculator Appendix**, which begins on page A2, contains keystroke instructions on many of these topics.

VECTORS AND PARAMETRIC EQUATIONS

CHAPTER OBJECTIVES ●

- **Add, subtract, and multiply vectors.** *(Lessons 8-1, 8-2, 8-3, 8-4)*
- **Represent vectors as ordered pairs or ordered triples and determine their magnitudes.** *(Lessons 8-2, 8-3)*
- **Write and graph vector and parametric equations.** *(Lesson 8-6)*
- **Solve problems using vectors and parametric equations.** *(Lessons 8-5, 8-6, 8-7)*
- **Use matrices to model transformations in three-dimensional space.** *(Lesson 8-8)*

484   *Chapter 8   Vectors and Parametric Equations*

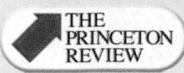

SAT & ACT Preparation

This textbook contains a complete SAT/ACT preparation course described on pages T8–T9 of the *Teacher's Handbook*. This chapter contains:

- an SAT or ACT practice question at the end of each lesson
- examples of SAT and ACT problems involving geometry topics, such as perimeter, area, and volume, on page 548
- ten SAT and ACT practice problems on page 549

- additional SAT and ACT practice in the *Assessment and Evaluation Masters,* pp. 201–202
- SAT and ACT test practice banks in the *TestCheck and Worksheet Builder*

8-1

Geometric Vectors

OBJECTIVES
- Find equal, opposite, and parallel vectors.
- Add and subtract vectors geometrically.

Real World Application

AERONAUTICS An advanced glider known as a sailplane, has high maneuverability and glide capabilities. The Ventus 2B sailplane placed first at the World Soaring Contest in New Zealand. At one competition, a sailplane traveled forward at a rate of 8 m/s, and it descended at a rate of 4 m/s. *A problem involving this situation will be solved in Example 3.*

The velocity of a sailplane can be represented mathematically by a **vector**. A vector is a quantity that has both *magnitude* and *direction*. A vector is represented geometrically by a directed line segment.

A directed line segment with an initial point at *P* and a terminal point at *Q* is shown at the right. The length of the line segment represents the **magnitude** of the vector. The direction of the arrowhead indicates the direction of the vector. The vectors can be denoted by $\vec{a}$ or $\overrightarrow{PQ}$. The magnitude of $\vec{a}$ is denoted by $|\vec{a}|$.

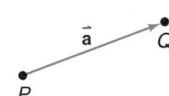

If a vector has its initial point at the origin, it is in **standard position**. The **direction** of the vector is the directed angle between the positive *x*-axis and the vector. The direction of $\vec{b}$ is 45°.

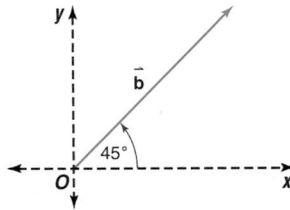

If both the initial point and the terminal point are at the origin, the vector is the **zero vector** and is denoted by $\vec{0}$. The magnitude of the zero vector is 0, and it can be in any direction.

Example ❶ Use a ruler and protractor to determine the magnitude (in centimeters) and the direction of $\vec{n}$.

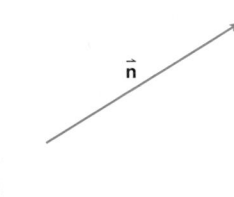

Sketch the vector in standard position and measure the magnitude and direction. The magnitude is 3.6 centimeters, and the direction is 30°.

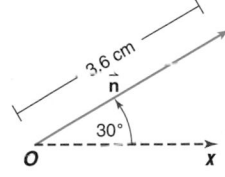

Two vectors are **equal** if and only if they have the same direction and the same magnitude.

Resource Manager

Reproducible Masters
- *Study Guide and Practice*, pp. 137–138
- *Enrichment*, p. 85
- *TI-92 & CBL Interdisciplinary Activities*, pp. 25–28

Transparencies
- 5-Minute Check, 8-1
- Teaching, 8-1

Technology/Multimedia
- Advanced MathPASS, Lesson 19

---

1 FOCUS

5-Minute Check
Chapter 7

1. Verify that $\frac{\tan^2 x}{1 + \tan^2 x} = \sin^2 x$ is an identity.
$$\frac{\tan^2 x}{1 + \tan^2 x} = \frac{\tan^2 x}{\sec^2 x}$$
$$= \left(\frac{\sin^2 x}{\cos^2 x}\right) \div \frac{1}{\cos^2 x}$$
$$= \sin^2 x$$

2. Find the exact value of $\sin(x + y)$ if $\sin x = \frac{1}{2}$, $\cos y = \frac{3}{5}$, $0 < x < \frac{\pi}{2}$, and $0 < y < \frac{\pi}{2}$. $\frac{3 + 4\sqrt{3}}{10}$

3. Use the sum or difference identity for sine to find the exact value of $\sin 15°$. $\frac{\sqrt{2 - \sqrt{3}}}{2}$ or $\frac{\sqrt{6} - \sqrt{2}}{4}$

4. Solve $\sin \theta = \sin \theta \cos \theta$ for $0 \le \theta < 2\pi$. $0, \pi$

5. Find the distance between $P(2, -3)$ and the line with equation $2x - y = -4$. $\frac{11\sqrt{5}}{5}$

Motivating the Lesson
Real-World Application Ask students: *If a pilot wanted to fly a small plane from one city to another, or a boater needed to cross a river, what would happen if each person headed directly to his or her destination?* **The wind or current would move the plane or boat off course.** Tell students that for accurate navigation, a person must take into account the direction and magnitude of all external forces acting on the craft. The mathematics involved include trigonometry and vectors and will be explored in this chapter.

In-Class Examples

Example 1
Use a ruler and protractor to determine the magnitude (in centimeters) and the direction of $\vec{v}$.

1.4 cm, 110°

Example 2
Find the sum of $\vec{u}$ and $\vec{v}$ using

a. the parallelogram method.

b. the triangle method.

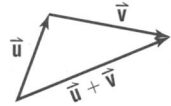

c. Compare the resultants found in both methods. **Both resultants have the same magnitude and direction.**

Teaching Tip For Example 2, remind students that two vectors originating from the same point yield a sum or resultant vector.

Six vectors are shown at the right.

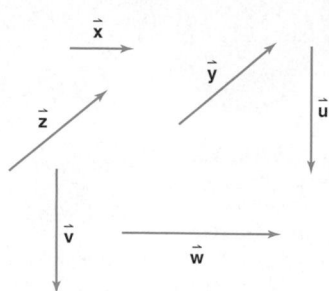

- $\vec{z}$ and $\vec{y}$ are equal since they have the same direction and $|\vec{z}| = |\vec{y}|$.

- $\vec{v}$ and $\vec{u}$ are equal.

- $\vec{x}$ and $\vec{w}$ have the same direction but $|\vec{x}| \neq |\vec{w}|$, so $\vec{x} \neq \vec{w}$.

- $|\vec{v}| = |\vec{y}|$, but they have different directions, so they are not equal.

The sum of two or more vectors is called the **resultant** of the vectors. The resultant can be found using either the *parallelogram method* or the *triangle method*.

Parallelogram Method	Triangle Method
Draw the vectors so that their initial points coincide. Then draw lines to form a complete parallelogram. The diagonal from the initial point to the opposite vertex of the parallelogram is the resultant.	Draw the vectors one after another, placing the initial point of each successive vector at the terminal point of the previous vector. Then draw the resultant from the initial point of the first vector to the terminal point of the last vector.
	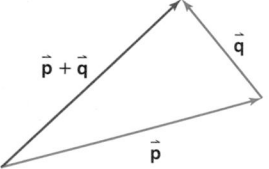
The parallelogram method cannot be used to find the sum of a vector and itself.	*This method is also called the tip-to-tail method.*

Example **2** **Find the sum of $\vec{v}$ and $\vec{w}$ using:**

a. the parallelogram method.

b. the triangle method.

c. Compare the resultants found in both methods.

a. Copy $\vec{v}$ then copy $\vec{w}$ placing the initial points together.

Form a parallelogram that has $\vec{v}$ and $\vec{w}$ as two of its sides. Draw dashed lines to represent the other two sides.

The resultant is the vector from the vertex of $\vec{v}$ and $\vec{w}$ to the opposite vertex of the parallelogram.

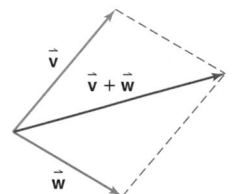

From the Classroom of...

George Rhys
William S. Hart High School
Newhall, California

"On a photocopy scale map of the campus, each group of four or five students draws vectors leading from the door of our building to a location of their choice, listing magnitude in centimeters and amplitude on a separate sheet of paper. On a subsequent day, I give a vector list to different groups who, using a ribbon cut to represent one centimeter, follow the list to find the location on the actual campus."

b. Copy $\vec{v}$, then copy $\vec{w}$ so that the initial point of $\vec{w}$ is on the terminal point of $\vec{v}$. (The tail of $\vec{v}$ connects to the tip of $\vec{w}$.)

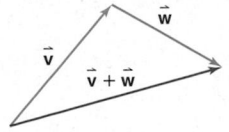

The resultant is the vector from the initial point of $\vec{v}$ to the terminal point of $\vec{w}$.

c. Use a ruler and protractor to measure the magnitude and direction of each resultant. The resultants of both methods have magnitudes of 3.5 centimeters and directions of 20°. So, the resultants found in both methods are equal.

Vectors can be used to solve real-world applications.

Example ③ **AERONAUTICS** Refer to the application at the beginning of the lesson. At the European Championships, a sailplane traveled forward at 8 m/s and descended at 4 m/s. Determine the magnitude of the resultant velocity of the sailplane.

Let 1 centimeter represent 2 m/s. Draw two vectors, $\vec{f}$ and $\vec{d}$, to represent the forward velocity and the descending velocity of the sailplane, respectively.

Use the parallelogram method. Copy $\vec{f}$. Then copy $\vec{d}$, placing the initial point at the initial point of $\vec{f}$. Draw dashed lines to represent the other two sides of the parallelogram.

The resultant velocity of the sailplane is the vector $\vec{p}$ from the vertex of $\vec{f}$ and $\vec{d}$ to the opposite vertex of the parallelogram.

Measure the resultant, 4.5 centimeters. Multiply the resultant's magnitude by 2 to determine the magnitude of the resultant velocity of the sailplane. *Why?*

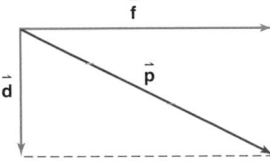

$4.5 \text{ cm} \times 2 \text{ m/s} = 9 \text{ m/s}$

The sailplane is moving at 9 m/s.

Two vectors are **opposites** if they have the same magnitude and opposite directions. In the diagram, $\vec{g}$ and $\vec{h}$ are opposites, as are $\vec{i}$ and $\vec{j}$. The opposite of $\vec{g}$ is denoted by $-\vec{g}$. You can use opposite vectors to subtract vectors. To find $\vec{g} - \vec{h}$, find $\vec{g} + (-\vec{h})$.

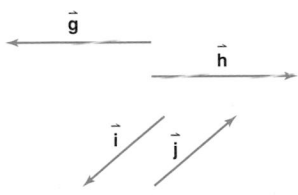

Teaching Tip In Example 3, the scale is given as 1 cm = 2 m/s. Remind students that they will determine the scales used for solving their homework problems and that whole numbers are easier to convert in the final step of a problem.

In-Class Example

Example 3

Sports In the World Rowing Championship, James Koven (USA) was the world champion for the year 1997. Suppose that while training, Koven was warming up by leisurely rowing directly across a river at the rate of 4 mph. The current was flowing at a rate of 3 mph. Use a ruler to draw each vector to scale, and draw a vector to represent the path of the boat. Determine the magnitude of the resultant velocity of the boat by measuring the vector. **5 mph**

A quantity with only magnitude is called a **scalar quantity.** Examples of scalars include mass, length, time, and temperature. The numbers used to measure scalar quantities are called **scalars.**

The product of a scalar k and a vector $\vec{a}$ is a vector with the same direction as $\vec{a}$ and a magnitude of $k|\vec{a}|$, if $k > 0$. If $k < 0$, the vector has the opposite direction of $\vec{a}$ and a magnitude of $k|\vec{a}|$. In the figure at the right, $\vec{b} = 3\vec{a}$ and $\vec{c} = -2\vec{a}$.

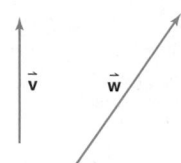

Example **4** **Use the triangle method to find $2\vec{v} - \frac{1}{2}\vec{w}$.**

Rewrite the expression as a sum.

$$2\vec{v} - \frac{1}{2}\vec{w} = 2\vec{v} + \left(-\frac{1}{2}\vec{w}\right)$$

Draw a vector twice the magnitude of $\vec{v}$ to represent $2\vec{v}$. Draw a vector with the opposite direction to $\vec{w}$ and half its magnitude to represent $-\frac{1}{2}\vec{w}$.

Place the initial point of $-\frac{1}{2}\vec{w}$ on the terminal point of $2\vec{v}$. *(Tip-to-tail method)*

Then $2\vec{v} - \frac{1}{2}\vec{w}$ has the initial point of $2\vec{v}$ and the terminal point of $-\frac{1}{2}\vec{w}$.

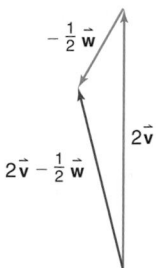

Two or more vectors are **parallel** if and only if they have the same or opposite directions. Vectors $\vec{i}$ and $\vec{j}$ have opposite directions and they are parallel. Vectors $\vec{\ell}$ and $\vec{k}$ are parallel and have the same direction.

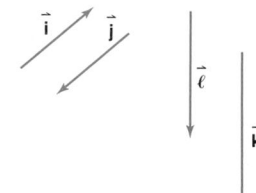

Two or more vectors whose sum is a given vector are called **components** of the given vector. Components can have any direction. Often it is useful to express a vector as the sum of two perpendicular components. In the figure at the right, $\vec{p}$ is the vertical component of $\vec{x}$, and $\vec{q}$ is the horizontal component of $\vec{x}$.

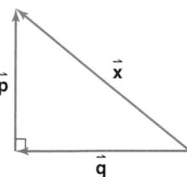

b. Let a be the measure of the angle $\vec{r}$ makes with $\vec{x}$.

The direction of the resultant can be found by using the tangent ratio.

$$\tan a = \frac{|\vec{y}|}{|\vec{x}|}$$

$$\tan a = \frac{3050}{1550}$$

$$a = 63° \qquad \textit{Take } \tan^{-1} \textit{ of each side.}$$

The resultant makes an angle of 63° with $\vec{x}$. The direction of the resultant force upon the piling is $90 - 63$ or 27° west of north. *Direction of vectors is often expressed in terms of position relative to due north, due south, due east, or due west, or as a navigational angle measures clockwise from due north.*

CHECK FOR UNDERSTANDING

Communicating
Mathematics

Read and study the lesson to answer each question.

1. **Draw** a diagram showing the resultant of two vectors, and describe how you obtained it.   **See margin.**

2. **Compare** a line segment and a vector.   **See margin**.

3. **Describe** a real-world situation involving vectors that could be represented by the diagram at the right.   **See margin.**

4. **Tell** whether $\overrightarrow{RS}$ is the same as $\overrightarrow{SR}$. Explain.   **No, they are opposites.**

Guided Practice

Use a ruler and a protractor to determine the magnitude (in centimeters) and direction of each vector.

5.

$\vec{x}$

1.2 cm, 120°

6.

$\vec{y}$

2.9 cm, 55°

7. $\vec{z}$

1.4 cm, 20°

Use $\vec{x}$, $\vec{y}$, and $\vec{z}$ above to find the magnitude and direction of each resultant.

8. $\vec{x} + \vec{y}$ **3.5 cm, 70°**   9. $\vec{x} - \vec{y}$ **2.6 cm, 210°**   10. $4\vec{y} + \vec{z}$ **12.9 cm, 51°**

11. the difference of a vector twice as long as $\vec{z}$ and a vector one third the magnitude of $\vec{x}$ **2.9 cm, 12°**

12. Find the magnitude of the horizontal and vertical components of $\vec{y}$. ≈ 1.66 cm, ≈ 2.38 cm

13. **Aviation**   An airplane is flying due west at a velocity of 100 m/s. The wind is blowing out of the south at 5 m/s.
    a. Draw a labeled diagram of the situation.   **See margin.**
    b. What is the magnitude of airplane's resultant velocity? ≈ 100.12 m/s

www.amc.glencoe.com/self_check_quiz

Addressing Individual Needs

Kinesthetic Vectors are commonly used to describe forces and how resultants are determined in real-world situations. Physicists use vectors to visually display many kinds of forces such as velocity, gravity, and torque. Have students use their bodies or objects in the classroom to model the common forces of physics. **For example, students may show the resultant vector for the gravity of a moving object by pointing one arm in the vector position for the moving car. A leg could be used to represent the gravity vector. The other arm could then display the resultant vector of the two forces.**

3 PRACTICE/APPLY

Error Analysis

Watch for students who fail to use the sine, cosine, and tangent ratios properly.

Prevent by reviewing the definition of each ratio as it applies to a right triangle.

Additional Answers

1. Sample answer:

Draw $\vec{a}$. Then draw $\vec{b}$ so that its initial point (tip) is on the terminal point (tail) of $\vec{a}$. Draw a dashed line from the initial point of $\vec{a}$ to the terminal point of $\vec{b}$. The dashed line is the resultant.

2. Sample answer: a vector has magnitude and direction. A line segment has only length. A vector can be represented by a directed line segment.

3. Sample answer: the velocities of an airplane and a wind current.

13a.

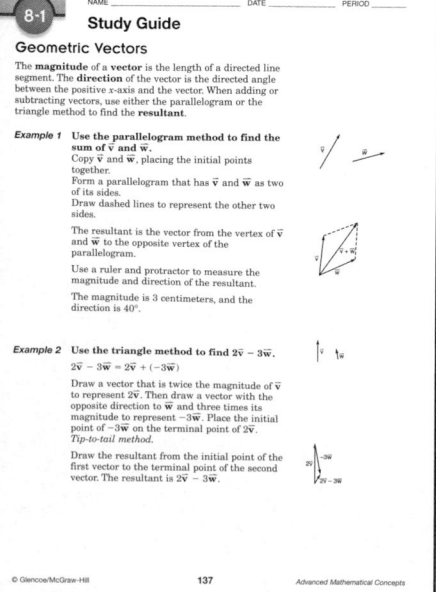

Study Guide and Practice Masters, pp. 137–138

NAME _____ DATE _____ PERIOD _____

8-1   **Study Guide**

Geometric Vectors

The **magnitude** of a **vector** is the length of a directed line segment. The **direction** of the vector is the directed angle between the positive x-axis and the vector. When adding or subtracting vectors, use either the parallelogram or the triangle method to find the **resultant**.

Example 1 Use the parallelogram method to find the sum of $\vec{v}$ and $\vec{w}$.

Copy $\vec{v}$ and $\vec{w}$, placing the initial points together.
Form a parallelogram that has $\vec{v}$ and $\vec{w}$ as two of its sides.
Draw dashed lines to represent the other two sides.
The resultant is the vector from the vertex of $\vec{v}$ and $\vec{w}$ to the opposite vertex of the parallelogram.
Use a ruler and protractor to measure the magnitude and direction of the resultant.
The magnitude is 3 centimeters, and the direction is 40°.

Example 2 Use the triangle method to find $2\vec{v} - 3\vec{w}$.

$2\vec{v} - 3\vec{w} = 2\vec{v} + (-3\vec{w})$

Draw a vector that is twice the magnitude of $\vec{v}$ to represent $2\vec{v}$. Then draw a vector with the opposite direction to $\vec{w}$ and three times its magnitude to represent $-3\vec{w}$. Place the initial point of $-3\vec{w}$ on the terminal point of $2\vec{v}$. *Tip-to-tail method.*

Draw the resultant from the initial point of the first vector to the terminal point of the second vector. The resultant is $2\vec{v} - 3\vec{w}$.

© Glencoe/McGraw-Hill   137   *Advanced Mathematical Concepts*

490 Chapter 8

---

b. Let a be the measure of the angle $\vec{r}$ makes with $\vec{x}$.

The direction of the resultant can be found by using the tangent ratio.

$$\tan a = \frac{|\vec{y}|}{|\vec{x}|}$$

$$\tan a = \frac{3050}{1550}$$

$$a = 63° \quad \textit{Take } \tan^{-1} \textit{ of each side.}$$

The resultant makes an angle of 63° with $\vec{x}$. The direction of the resultant force upon the piling is $90 - 63$ or 27° west of north. *Direction of vectors is often expressed in terms of position relative to due north, due south, due east, or due west, or as a navigational angle measures clockwise from due north.*

CHECK FOR UNDERSTANDING

Communicating Mathematics

Read and study the lesson to answer each question.

1. **Draw** a diagram showing the resultant of two vectors, and describe how you obtained it. **See margin.**

2. **Compare** a line segment and a vector. **See margin.**

3. **Describe** a real-world situation involving vectors that could be represented by the diagram at the right. **See margin.**

4. **Tell** whether $\overrightarrow{RS}$ is the same as $\overrightarrow{SR}$. Explain. **No, they are opposites.**

Guided Practice

Use a ruler and a protractor to determine the magnitude (in centimeters) and direction of each vector.

5.

1.2 cm, 120°

6.

2.9 cm, 55°

7.

1.4 cm, 20°

Use $\vec{x}$, $\vec{y}$, and $\vec{z}$ above to find the magnitude and direction of each resultant.

8. $\vec{x} + \vec{y}$ **3.5 cm, 70°**   9. $\vec{x} - \vec{y}$ **2.6 cm, 210°**   10. $4\vec{y} + \vec{z}$ **12.9 cm, 51°**

11. the difference of a vector twice as long as $\vec{z}$ and a vector one third the magnitude of $\vec{x}$ **2.9 cm, 12°**

12. Find the magnitude of the horizontal and vertical components of $\vec{y}$. ≈ 1.66 cm, ≈ 2.38 cm

13. **Aviation** An airplane is flying due west at a velocity of 100 m/s. The wind is blowing out of the south at 5 m/s.

a. Draw a labeled diagram of the situation. **See margin.**

b. What is the magnitude of airplane's resultant velocity? ≈ 100.12 m/s

www.amc.glencoe.com/self_check_quiz

EXERCISES

Practice **A** Use a ruler and a protractor to determine the magnitude (in centimeters) and direction of each vector.

14.

15. **1.4 cm, 45°**

16. **2.1 cm, 14°**

17. **3.0 cm, 340°**

2.6 cm, 128°

Use $\vec{r}$, $\vec{s}$, $\vec{t}$, and $\vec{u}$ above to find the magnitude and direction of each resultant.

18. $\vec{r} + \vec{s}$ **3 cm, 101°**
19. $\vec{s} + \vec{t}$ **3.4 cm, 25°**
20. $\vec{s} + \vec{u}$ **3.8 cm, 359°**

B
21. $\vec{u} - \vec{r}$ **5.5 cm, 324°**
22. $\vec{r} - \vec{t}$ **3.9 cm, 155°**
23. $2\vec{r}$ **5.2 cm, 128°**

24. $3\vec{s}$ **4.2 cm, 45°**
25. $3\vec{u} - 2\vec{s}$ **8.2 cm, 322°**
26. $\vec{r} + \vec{t} + \vec{u}$ **3.5 cm, 22°**

27. 5.4 cm, 133°   27. $\vec{r} + \vec{s} - \vec{u}$
28. $2\vec{s} + \vec{u} - \frac{1}{2}\vec{r}$
29. $\vec{r} - 2\vec{t} - \vec{s} + 3\vec{u}$

28. 5.5 cm, 358°   30. three times $\vec{t}$ and twice $\vec{u}$ **11.7 cm, 357°**   **3.4 cm, 301°**

Find the magnitude of the horizontal and vertical components of each vector shown for Exercises 14–17.

C
31. $\vec{r}$
32. $\vec{s}$
33. $\vec{t}$
34. $\vec{u}$

31. −1.60 cm, 2.05 cm

35. The magnitude of $\vec{m}$ is 29.2 meters, and the magnitude of $\vec{n}$ is 35.2 meters. If $\vec{m}$ and $\vec{n}$ are perpendicular, what is the magnitude of their sum?   **45.73 m**

32. 0.99 cm, 0.99 cm

36. In the parallelogram drawn for the parallelogram method, what does the diagonal between the two terminal points of the vectors represent? Explain your answer.   **The difference of the vectors; see margin.**

33. 2.04 cm, 0.51 cm

37. Is addition of vectors commutative? Justify your answer. (*Hint:* Find the sum of two vectors, $\vec{r} + \vec{s}$ and $\vec{s} + \vec{r}$, using the triangle method.)   **Yes; see margin.**

34. 2.82 cm, −1.03 cm

Additional Answers

36. Sample answer: the other diagonal would be the sum of one of the vectors and the opposite of the other vector, so it would be the difference.

37. Sample answer:

39.

$|\vec{a} + \vec{b}| = 5$
$|\vec{a}| + |\vec{b}| = 2.3 + 2.7$
$= 5$

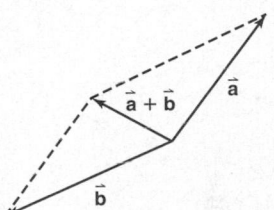

$|\vec{a} + \vec{b}| = 1.5$
$|\vec{a}| + |\vec{b}| = 2.3 + 2.7$
$= 5$

Applications and Problem Solving

  Real World Application

39. Sometimes; see margin for diagram.

38. **Physics**   Three workers are pulling on ropes attached to a tree stump as shown in the diagram. Find the magnitude and direction of the resultant force on the tree. *A newton (N) is a unit of force used in physics. A force of one newton will accelerate a one-kilogram mass at a rate of one meter per second squared.*   **61 N, 23° north of east**

39. **Critical Thinking**   Does $|\vec{a} + \vec{b}|$ always, sometimes, or never equal $|\vec{a}| + |\vec{b}|$? Draw a diagram to justify your answer.

40. **Toys**   Belkis is pulling a toy by exerting a force of 1.5 newtons on a string attached to the toy.
   a. The string makes an angle of 52° with the floor. Find the vertical and horizontal components of the force.   **≈ 1.18 N, ≈ 0.92 N**
   b. If Belkis raises the string so that it makes a 78° angle with the floor, what are the magnitudes of the horizontal and vertical components of the force?   **≈ 0.31 N, ≈ 1.47 N**

60 newtons

35 newtons

60°

40 newtons

Lesson 8-1   Geometric Vectors   **491**

Open-Ended Assessment

Writing Have students explain in a paragraph how to add and subtract two vectors. They may also include diagrams. Encourage students to write their explanations as if they were writing to an audience unfamiliar with vectors.

Additional Answers

46. $\csc\theta \ \cos\theta \ \tan\theta$

$$= \frac{1}{\sin\theta} \cdot \cos\theta \cdot \frac{\sin\theta}{\cos\theta}$$

$$= \frac{\sin\theta}{\sin\theta} \cdot \frac{\cos\theta}{\cos\theta}$$

$$= 1$$

47. $\frac{\pi}{4} + \pi n$, where n is an integer

---

41. Police Investigation Police officer Patricia Malloy is investigating an automobile accident. A car slid off an icy road at a 40° angle with the center line of the road with an initial velocity of 47 miles per hour. Use the drawing to determine the initial horizontal and vertical components of the car's velocity. **36 mph, 30 mph**

42. Critical Thinking If $\overline{a}$ is a vector and k is a scalar, is it possible that $\overline{a} = k\overline{a}$? Under what conditions is this true? **It is true when $k = 1$ or when $\overline{a}$ is the zero vector.**

43. Ranching Matsuko has attached two wires to a corner fence post to complete a horse paddock. The wires make an angle of 90° with each other. If each wire pulls on the post with a force of 50 pounds, what is the resultant force acting on the pole? **71 lb**

44. Physics Mrs. Keaton is the director of a local art gallery. She needs to hang a painting weighing 24 pounds with a wire whose parts form a 120° angle with each other. Find the pull on each part of the wire. **24 lb**

Mixed Review

45. Find the equation of the line that bisects the acute angle formed by the lines $x - y + 2 = 0$ and $y - 5 = 0$. *(Lesson 7-7)* $x - \left(1 + \sqrt{2}\right)y + 2 + 5\sqrt{2} = 0$

46. Verify that $\csc\theta \cos\theta \tan\theta = 1$ is an identity. *(Lesson 7-2)* **See margin.**

47. Find the values of θ for which $\tan\theta = 1$ is true. *(Lesson 6-7)* **See margin.**

48. Use the graphs of the sine and cosine functions to find the values of x for which $\sin x + \cos x = -1$. *(Lesson 6-5)* $x = \pi, \frac{3\pi}{2}$ for $0 \le x \le 2\pi$

49. Geometry The base angles of an isosceles triangle measure 18°29′, and the altitude to the base is 5 centimeters long. Find the length of the base and the lengths of the congruent sides. *(Lesson 5-2)* **29.9 cm; 15.8 cm**

50. Manufacturing A manufacturer produces packaging boxes for other companies. The largest box they currently produce has a height two times its width and a length one more that its width. They wish to produce a larger packaging box where the height will be increased by 2 feet and the width and length will be increased by 1 foot. The volume of the new box is 160 cubic feet. Find the dimensions of the original box. *(Lesson 4-4)* **3 ft × 4 ft × 6 ft**

51. Determine the equations of the vertical and horizontal asymptotes, if any, of $g(x) = \dfrac{x + 2}{(x - 1)(x + 3)}$. *(Lesson 3-7)* $x = -3, x = 1, y = 0$

52. SAT/ACT Practice Grid-In Three times the least of three consecutive odd integers is three greater than two times the greatest. Find the greatest of the three integers. **15**

Extra Credit

Suppose that you are given the three force vectors $\overrightarrow{a}$, $\overrightarrow{b}$, and $\overrightarrow{c}$ acting on a point. Develop a strategy for finding their resultant vector $\overrightarrow{q}$.

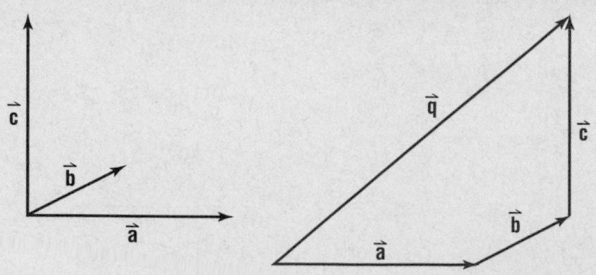

Algebraic Vectors

OBJECTIVES
- Find ordered pairs that represent vectors.
- Add, subtract, multiply, and find the magnitude of vectors algebraically.

 Real World Application

EMERGENCY MEDICINE
Paramedics Paquita Gonzalez and Trevor Howard are moving a person on a stretcher. Ms. Gonzalez is pushing the stretcher with a force of 135 newtons at 58° with the horizontal, while Mr. Howard is pulling the stretcher with a force of 214 newtons at 43° with the horizontal. What is the magnitude of the force exerted on the stretcher? *This problem will be solved in Example 3.*

We can find the magnitude and direction of the resultant by drawing vectors to represent the forces on the stretcher. However, drawings can be inaccurate and quite confusing for more complex systems of vectors. In cases where more accuracy is necessary or where the system is more complicated, we can perform operations on vectors algebraically.

Vectors can be represented algebraically using ordered pairs of real numbers. For example, the ordered pair ⟨3, 5⟩ can represent a vector in standard position. That is, its initial point is at the origin and its terminal point is at (3, 5). You can think of this vector as the resultant of a horizontal vector with a magnitude of 3 units and a vertical vector with magnitude of 5 units.

Since vectors with the same magnitude and direction are equal, many vectors can be represented by the same ordered pair. Each vector on the graph can be represented by the ordered pair ⟨3, 5⟩. The initial point of a vector can be any point in the plane. In other words, a vector does not have to be in standard position to be expressed algebraically.

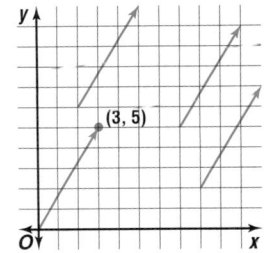

Assume that P_1 and P_2 are any two distinct points in the coordinate plane. Drawing the horizontal and vertical components of $\overrightarrow{P_1P_2}$ yields a right triangle. So, the magnitude of $\overrightarrow{P_1P_2}$ can be found by using the Pythagorean Theorem.

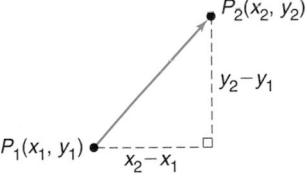

Lesson 8-2 Algebraic Vectors **493**

Resource Manager

Reproducible Masters
- *Study Guide and Practice,* pp. 139–140
- *Enrichment,* p. 86
- *Assessment and Evaluation,* p. 199

Transparencies
- 5-Minute Check, 8-2
- Teaching, 8-2

---

1 FOCUS

5-Minute Check
Lesson 8-1

1. Use a ruler and protractor to draw a vector $\vec{w}$ with magnitude 2.8 centimeters and direction 47°.

Find the sum of $\vec{a}$ and $\vec{b}$ using the following.

2. the parallelogram method

3. the triangle method

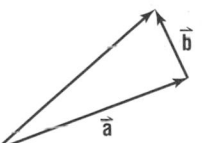

4. Compare the resultants found in Exercises 2 and 3. **same resultant**

5. Use the triangle method to find $3\vec{a} - \vec{b}$.

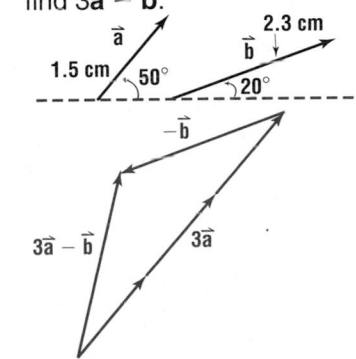

Motivating the Lesson

Discussion Ask students to decide whether the techniques learned in Lesson 8-1 are adequate to solve any problem involving vectors. Explain that while a geometric approach gives them a good intuitive feel for the subject, the more abstract algebraic method allows them to be complete and more precise in their answers, especially when they are solving problems in more than two dimensions.

2 TEACH

Teaching Tip When working Example 1, stress that the order of the coordinates representing a vector is important.

In-Class Examples

Example 1
Write the ordered pair that represents the vector from $C(7, -3)$ to $D(-2, -1)$. Then find the magnitude of $\overrightarrow{CD}$.
$\langle -9, 2 \rangle$; $\sqrt{85}$

Example 2
Let $\overline{u} = \langle 1, -4 \rangle$ and $\overline{v} = \langle 0, 8 \rangle$. Find each of the following.
a. $\overline{u} + \overline{v}$ $\langle 1, 4 \rangle$
b. $\overline{u} - \overline{v}$ $\langle 1, -12 \rangle$
c. $\frac{1}{2}\overline{v}$ $\langle 0, 4 \rangle$
d. $2\overline{u} + 3\overline{v}$ $\langle 2, 16 \rangle$

Teaching Tip Remind students that a scalar is simply a real number.

Representation of a Vector as an Ordered Pair

Let $P_1(x_1, y_1)$ be the initial point of a vector and $P_2(x_2, y_2)$ be the terminal point. The ordered pair that represents $\overrightarrow{P_1P_2}$ is $\langle x_2 - x_1, y_2 - y_1 \rangle$. Its magnitude is given by $|\overrightarrow{P_1P_2}| = \sqrt{(x_2 - x_1)^2 + (y_2 - y_1)^2}$.

Example ❶ Write the ordered pair that represents the vector from $X(-3, 5)$ to $Y(4, -2)$. Then find the magnitude of $\overrightarrow{XY}$.

First, represent $\overrightarrow{XY}$ as an ordered pair.
$\overrightarrow{XY} = \langle 4 - (-3), -2 - 5 \rangle$ or $\langle 7, -7 \rangle$

Then, determine the magnitude of $\overrightarrow{XY}$.
$|\overrightarrow{XY}| = \sqrt{[4 - (-3)]^2 + (-2 - 5)^2}$
$= \sqrt{7^2 + 7^2}$
$= \sqrt{98}$ or $7\sqrt{2}$

$\overrightarrow{XY}$ is represented by the ordered pair $\langle 7, -7 \rangle$ and has a magnitude of $7\sqrt{2}$ units.

When vectors are represented by ordered pairs, they can be easily added, subtracted, or multiplied by a scalar. The rules for these operations on vectors are similar to those for matrices. In fact, vectors in a plane can be represented as row matrices of dimension 1×2.

Vector Operations

The following operations are defined for $\overline{a}\langle a_1, a_2 \rangle$, $\overline{b}\langle b_1, b_2 \rangle$, and any real number k.
Addition: $\overline{a} + \overline{b} = \langle a_1, a_2 \rangle + \langle b_1, b_2 \rangle = \langle a_1 + b_1, a_2 + b_2 \rangle$
Subtraction: $\overline{a} - \overline{b} = \langle a_1, a_2 \rangle - \langle b_1, b_2 \rangle = \langle a_1 - b_1, a_2 - b_2 \rangle$
Scalar multiplication: $k\,\overline{a} = k\langle a_1, a_2 \rangle = \langle ka_1, ka_2 \rangle$

Example ❷ Let $\overline{m} = \langle 5, -7 \rangle$, $\overline{n} = \langle 0, 4 \rangle$, and $\overline{p} = \langle -1, 3 \rangle$. Find each of the following.

a. $\overline{m} + \overline{p}$
$\overline{m} + \overline{p} = \langle 5, -7 \rangle + \langle -1, 3 \rangle$
$= \langle 5 + (-1), -7 + 3 \rangle$
$= \langle 4, -4 \rangle$

b. $\overline{m} - \overline{n}$
$\overline{m} - \overline{n} = \langle 5, -7 \rangle - \langle 0, 4 \rangle$
$= \langle 5 - 0, -7 - 4 \rangle$
$= \langle 5, -11 \rangle$

c. $7\overline{p}$
$7\overline{p} = 7\langle -1, 3 \rangle$
$= \langle 7 \cdot (-1), 7 \cdot 3 \rangle$
$= \langle -7, 21 \rangle$

d. $2\overline{m} + 3\overline{n} - \overline{p}$
$2\overline{m} + 3\overline{n} - \overline{p} = 2\langle 5, -7 \rangle + 3\langle 0, 4 \rangle - \langle -1, 3 \rangle$
$= \langle 10, -14 \rangle + \langle 0, 12 \rangle - \langle -1, 3 \rangle$
$= \langle 11, -5 \rangle$

If we write the vectors described in the application at the beginning of the lesson as ordered pairs, we can find the resultant vector easily using vector addition. Using ordered pairs to represent vectors allows for a more accurate solution than using geometric representations.

Example

③ EMERGENCY MEDICINE Refer to the application at the beginning of the lesson. **What is the magnitude of the force exerted on the stretcher?**

Draw a diagram of the situation. Let $\vec{G}_1$ represent the force Ms. Gonzalez exerts, and let $\vec{G}_2$ represent the force Mr. Howard exerts.

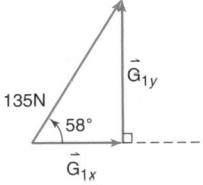

Write each vector as an ordered pair by finding its horizontal and vertical components. Let $\vec{G}_{1x}$ and $\vec{G}_{1y}$ represent the x- and y-components of $\vec{G}_1$. Let $\vec{G}_{2x}$ and $\vec{G}_{2y}$ represent the x- and y-components of $\vec{G}_2$.

$$\cos 58° = \frac{|\vec{G}_{1x}|}{135}$$

$$|\vec{G}_{1x}| = 135 \cos 58°$$

$$|\vec{G}_{1x}| \approx 71.5$$

$$\sin 58° = \frac{|\vec{G}_{1y}|}{135}$$

$$|\vec{G}_{1y}| = 135 \sin 58°$$

$$|\vec{G}_{1y}| \approx 114.5$$

$$\vec{G}_1 \approx \langle 71.5, 114.5 \rangle$$

$$\cos 43° = \frac{|\vec{G}_{2x}|}{214}$$

$$|\vec{G}_{2x}| = 214 \cos 43°$$

$$|\vec{G}_{2x}| \approx 156.5$$

$$\sin 43° = \frac{|\vec{G}_{2y}|}{214}$$

$$|\vec{G}_{2y}| = 214 \sin 43°$$

$$|\vec{G}_{2y}| \approx 145.9$$

$$\vec{G}_2 \approx \langle 156.5, 145.9 \rangle$$

Find the sum of the vectors.

$$\vec{G}_1 + \vec{G}_2 \approx \langle 71.5, 114.5 \rangle + \langle 156.5, 145.9 \rangle$$
$$\approx \langle 228.0, 260.4 \rangle$$

The net force on the stretcher is the magnitude of the sum.

$$|\vec{G}_1 + \vec{G}_2| \approx \sqrt{(228.0)^2 + (260.4)^2} \text{ or about } 346$$

The net force on the stretcher is about 346 newtons.

To convert newtons to pounds, divide the number of newtons by 4.45. So, the net force on the stretcher can also be expressed as 346 ÷ 4.45 or 77.8 pounds.

A vector that has a magnitude of one unit is called a **unit vector.** A unit vector in the direction of the positive x-axis is represented by $\vec{\mathbf{i}}$, and a unit vector in the direction of the positive y-axis is represented by $\vec{\mathbf{j}}$. So, $\vec{\mathbf{i}} = \langle 1, 0 \rangle$ and $\vec{\mathbf{j}} = \langle 0, 1 \rangle$.

Addressing Individual Needs

Interpersonal After introducing the concept of vectors represented by ordered pairs, have students work together in small groups, first calculating the sum, difference, or scalar multiplication of two vectors, then using graph paper to verify their results visually.

In-Class Example

Example 4

Write $\overrightarrow{AB}$ as the sum of unit vectors for $A(2, -7)$ and $B(-1, 5)$. $-3\vec{i} + 12\vec{j}$

3 PRACTICE/APPLY

Error Analysis

Watch for students who make errors solving problems that do not include a drawing.

Prevent by encouraging students to use a straightedge to draw neat figures that represent what they are solving algebraically.

Any vector $\vec{a} = \langle a_1, a_2 \rangle$ can be expressed as $a_1\vec{i} + a_2\vec{j}$.

$$
\begin{aligned}
a_1\vec{i} + a_2\vec{j} &= a_1\langle 1, 0 \rangle + a_2\langle 0, 1 \rangle && \vec{i} = \langle 1, 0 \rangle \text{ and } \vec{j} = \langle 0, 1 \rangle \\
&= \langle a_1, 0 \rangle + \langle 0, a_2 \rangle && \text{Scalar product} \\
&= \langle a_1 + 0, 0 + a_2 \rangle && \text{Addition of vectors} \\
&= \langle a_1, a_2 \rangle
\end{aligned}
$$

Since $\langle a_1, a_2 \rangle = \vec{a}$, $a_1\vec{i} + a_2\vec{j} = \vec{a}$. Therefore, any vector that is represented by an ordered pair can also be written as the sum of unit vectors. The zero vector can be represented as $\vec{0} = \langle 0, 0 \rangle = 0\vec{i} + 0\vec{j}$.

Example **4** Write $\overrightarrow{AB}$ as the sum of unit vectors for $A(4, -1)$ and $B(6, 2)$.

First, write $\overrightarrow{AB}$ as an ordered pair.

$$
\begin{aligned}
\overrightarrow{AB} &= \langle 6 - 4, 2 - (-1) \rangle \\
&= \langle 2, 3 \rangle
\end{aligned}
$$

Then, write $\overrightarrow{AB}$ as the sum of unit vectors.

$$\overrightarrow{AB} = 2\vec{i} + 3\vec{j}$$

CHECK FOR UNDERSTANDING

Communicating Mathematics

Read and study the lesson to answer each question. 1–3. See margin.

1. **Find a counterexample** If $|\vec{a}| = 10$ and $|\vec{b}| = 10$, then $\vec{a}$ and $\vec{b}$ are equal vectors.

2. **Describe** how to find $|\overrightarrow{XY}|$ using the graph at the right.

3. **You Decide** Lina showed Jacqui the representation of $\langle 2, -5 \rangle$ as a unit vector as follows.

$$
\begin{aligned}
\langle 2, -5 \rangle &= \langle 2, 0 \rangle + \langle 0, -5 \rangle \\
&= (5)\langle 1, 0 \rangle + (-2)\langle 0, 1 \rangle \\
&= 5\vec{i} + (-2)\vec{j} \\
&= 5\vec{i} - 2\vec{j}
\end{aligned}
$$

Jacqui said that her work was incorrect. Who is right? Explain.

Guided Practice

Write the ordered pair that represents $\overrightarrow{MP}$. Then find the magnitude of $\overrightarrow{MP}$.

4. $M(2, -1), P(-3, 4)$ $\langle -5, 5 \rangle, 5\sqrt{2}$ units

5. $M(5, 6), P(0, 5)$ $\langle -5, -1 \rangle, \sqrt{26}$ units

6. $M(-19, 4), P(4, 0)$ $\langle 23, -4 \rangle, \sqrt{545}$ units

Find an ordered pair to represent $\vec{t}$ in each equation if $\vec{u} = \langle -1, 4 \rangle$ and $\vec{v} = \langle 3, -2 \rangle$.

7. $\vec{t} = \vec{u} + \vec{v}$ $\langle 2, 2 \rangle$

8. $\vec{t} = \frac{1}{2}\vec{u} - \vec{v}$ $\langle -3\frac{1}{2}, 4 \rangle$

9. $\vec{t} = 4\vec{u} + 6\vec{v}$ $\langle 14, 4 \rangle$

10. $\vec{t} = -8\vec{u}$ $\langle 8, -32 \rangle$

496 Chapter 8 *Vectors and Parametric Equations*

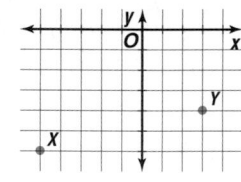
Additional Answers

1. Sample answer: $\vec{a} = \langle 8, 6 \rangle$, $\vec{b} = \langle 6, 8 \rangle$; equal vectors have the same magnitude and direction.

2. Use $|\overrightarrow{XY}| = \sqrt{(x_2 - x_1)^2 + (y_2 - y_1)^2}$ and replace the values for x and y. Answer is $\sqrt{68}$.

3. Jacqui is correct. The representation is incorrect. $\langle 2, 0 \rangle + \langle 0, -5 \rangle$ is not equal to $5\langle 1, 0 \rangle + (-2)\langle 0, 1 \rangle$. The correct expression is $2\vec{i} - 5\vec{j}$.

Find the magnitude of each vector. Then write each vector as the sum of unit vectors.

11. $\langle 8, -6 \rangle$ $10, 8\vec{i} - 6\vec{j}$

12. $\langle -7, -5 \rangle$ $\sqrt{74}, -7\vec{i} - 5\vec{j}$

13. Construction The Walker family is building a cabin for vacationing. Mr. Walker and his son Terrell have erected a scaffold to stand on while they build the walls of the cabin. As they stand on the scaffold Terrell pulls on a rope attached to a support beam with a force of 400 newtons (N) at an angle of 65° with the horizontal. Mr. Walker pulls with a force of 600 newtons at an angle of 110° with the horizontal. What is the magnitude of the combined force they exert on the log? ≈ 927 N

EXERCISES

Write the ordered pair that represents $\overline{YZ}$. Then find the magnitude of $\overline{YZ}$.

14. $Y(4, 2), Z(2, 8)$ $\langle -2, 6 \rangle, 2\sqrt{10}$ **15.** $Y(-5, 7), Z(-1, 2)$ $\langle 4, -5 \rangle, \sqrt{41}$

16. $Y(-2, 5), Z(1, 3)$ $\langle 3, -2 \rangle, \sqrt{13}$ **17.** $Y(5, 4), Z(0, -3)$ $\langle -5, -7 \rangle, \sqrt{74}$

18. $Y(3, 1), Z(0, 4)$ $\langle -3, 3 \rangle, 3\sqrt{2}$ **19.** $Y(-4, 12), Z(1, 19)$ $\langle 5, 7 \rangle, \sqrt{74}$

20. $Y(5, 0), Z(7, 6)$ $\langle 2, 6 \rangle, 2\sqrt{10}$ **21.** $Y(14, -23), Z(23, -14)$ $\langle 9, 9 \rangle, 9\sqrt{2}$

22. Find an ordered pair that represents the vector from $A(31, -33)$ to $B(36, -45)$. Then find the magnitude of $\overline{AB}$. $\langle 5, -12 \rangle, 13$

Find an ordered pair to represent $\vec{a}$ in each equation if $\vec{b} = \langle 6, 3 \rangle$ and $\vec{c} = \langle -4, 8 \rangle$.

23. $\vec{a} = \vec{b} + \vec{c}$ $\langle 2, 11 \rangle$ **24.** $\vec{a} = 2\vec{b} + \vec{c}$ $\langle 8, 14 \rangle$ **25.** $\vec{a} = \vec{b} + 2\vec{c}$ $\langle -2, 19 \rangle$

26. $\vec{a} = 2\vec{b} + 3\vec{c}$ $\langle 0, 30 \rangle$ **27.** $\vec{a} = -\vec{b} + 4\vec{c}$ $\langle -22, 29 \rangle$ **28.** $\vec{a} = \vec{b} - 2\vec{c}$ $\langle 14, -13 \rangle$

29. $\vec{a} = 3\vec{b}$ $\langle 18, 9 \rangle$ **30.** $\vec{a} = -\frac{1}{2}\vec{c}$ $\langle 2, -4 \rangle$ **31.** $\vec{a} = 6\vec{b} + 4\vec{c}$ $\langle 20, 50 \rangle$

32. $\langle 7.2, -8.4 \rangle$ **32.** $\vec{a} = 0.4\vec{b} - 1.2\vec{c}$ **33.** $\vec{a} = \frac{1}{3}\left(2\vec{b} - 5\vec{c}\right)$ **34.** $\vec{a} = \left(3\vec{b} + \vec{c}\right) + 5\vec{b}$

33. $\left\langle \frac{32}{3}, \frac{-34}{3} \right\rangle$ **35.** For $\vec{m} = \langle -5, -6 \rangle$ and $\vec{n} = \langle 6, -9 \rangle$ find the sum of the vector three times the magnitude of $\vec{m}$ and the vector two and one half times the magnitude of the opposite of $\vec{n}$. $\langle -30, 4.5 \rangle$

34. $\langle 44, 32 \rangle$

Find the magnitude of each vector. Then write each vector as the sum of unit vectors.

$\sqrt{157}, -6\vec{i} - 11\vec{j}$

36. $\langle 3, 4 \rangle$ $5, 3\vec{i} + 4\vec{j}$ **37.** $\langle 2, -3 \rangle$ $\sqrt{13}, 2\vec{i} - 3\vec{j}$ **38.** $\langle -6, -11 \rangle$

39. $\langle 3.5, 12 \rangle$ **40.** $\langle -4, 1 \rangle$ **41.** $\langle -16, -34 \rangle$
$12.5, 3.5\vec{i} + 12\vec{j}$ $\sqrt{17}, -4\vec{i} + \vec{j}$ $2\sqrt{353}, -16\vec{i} - 34\vec{j}$

Additional Answers

43. Student needs to show that
$$(\vec{v_1} + \vec{v_2}) + \vec{v_3} = \vec{v_1} + (\vec{v_2} + \vec{v_3})$$

$(\vec{v_1} + \vec{v_2}) + \vec{v_3}$

$= [\langle a, b \rangle + \langle c, d \rangle] + \langle e, f \rangle$

$= \langle a + c, b + d \rangle + \langle e, f \rangle$

$= \langle \langle a + c \rangle + e, \langle b + d \rangle + f \rangle$

$= \langle \langle a + \langle c + e \rangle, b + \langle d + f \rangle \rangle$

$= \langle a, b \rangle + \langle c + e, d + f \rangle$

$= \langle a, b \rangle + [\langle c, d \rangle + \langle e, f \rangle]$

$= \vec{v_1} + (\vec{v_2} + \vec{v_3})$

44a.

45a.

46a. They are opposites.

46b. It is a parallelogram.

48. $\cos \theta = \dfrac{(x_2 - x_1)}{|\vec{v}|} \rightarrow (x_2 - x_1) = $

$\vec{v} \cos \theta$

$\sin \theta = \dfrac{(y_2 - y_1)}{|\vec{v}|} \rightarrow (y_2 - y_1) = $

$\vec{v} \sin \theta$

---

42. Write $\overrightarrow{ST}$ as the sum of unit vectors for points $S(-9, 2)$ and $T(-4, -3)$. $5\vec{i} - 5\vec{j}$

43. Prove that addition of vectors is associative. **See margin.**

Applications and Problem Solving

Real World Application

44. Recreation In the 12th Bristol International Kite Festival in September 1997 in England, Peter Lynn set a record for flying the world's biggest kite, which had a lifting surface area of 630 square meters. Suppose the wind is blowing against the kite with a force of 100 newtons at an angle 20° above the horizontal.

a. Draw a diagram representing the situation. **See margin.**

b. How much force is lifting the kite? **34 N**

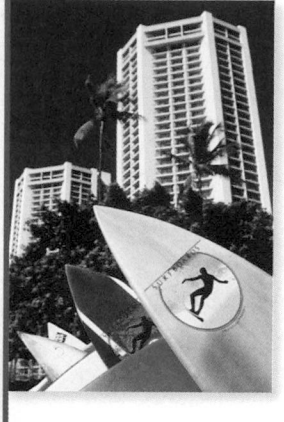

45. Surfing During a weekend surfboard competition, Kiyoshi moves at a 30° angle toward the shore. The velocity component toward the shore is 15 mph.

a. Make a labeled diagram to show Kiyoshi's velocity and the velocity components. **See margin.**

b. What is Kiyoshi's velocity? **30 mph**

46. Critical Thinking Suppose the points Q, R, S, and T are noncollinear, and $\overrightarrow{QR} + \overrightarrow{ST} = \vec{0}$. **a–b. See margin.**

a. What is the relationship between $\overrightarrow{QR}$ and $\overrightarrow{ST}$?

b. What is true of the quadrilateral with vertices Q, R, S, and T?

47. River Rafting The Soto family is rafting on the Colorado River. Suppose that they are on a stretch of the river that is 150 meters wide, flowing south at a rate of 1.0 m/s. In still water their raft travels 5.0 m/s.

a. How long does it take them to travel from one bank to the other if they head for a point directly across the river? **30 s**

b. How far downriver will the raft land? **30 m**

c. What is the velocity of the raft relative to the shore? **5.1 m/s**

48. Critical Thinking Show that any vector $\vec{v}$ can be written as $\langle |\vec{v}| \cos \theta, |\vec{v}| \sin \theta \rangle$. **See margin.**

Mixed Review

49. State whether $\overrightarrow{PQ}$ and $\overrightarrow{RS}$ are *equal*, *opposite*, *parallel*, or *none of these* for points $P(8, -7)$, $Q(-2, 5)$, $R(8, -7)$, and $S(7, 0)$. *(Lesson 8-1)* **None**

50. Find the distance from the graph of $3x - 7y - 1 = 0$ to the point at $(-1, 4)$. *(Lesson 7-7)* **4.2**

51.

$\dfrac{-\sqrt{6} - \sqrt{2}}{4}$

51. Use the sum or difference identities to find the exact value of $\sin 255°$. *(Lesson 7-3)*

52. Write an equation of the sine function that has an amplitude of 17, a period of 45°, and a phase shift of $-60°$. *(Lesson 6-5)* $y = \pm 17\sin(8x + 480°)$

53. Geometry Two sides of a triangle are 400 feet and 600 feet long, and the included angle measures 46°20′. Find the perimeter and area of the triangle. *(Lesson 5-8)* ≈**1434 ft; ≈86,751 sq ft**

54. Use the Upper Bound Theorem to find an integral upper bound and the Lower Bound Theorem to find the lower bound of the zeros of the function $f(x) = 3x^2 - 2x + 1$. *(Lesson 4-5)* **Sample answer: 2; −1**

Extra Practice See p. A40.

Extra Credit

Using vector addition, prove that $\overrightarrow{UV} = \frac{1}{2} \overrightarrow{XZ}$ given U and V are midpoints.

$\overrightarrow{UV} \parallel \overrightarrow{XZ}$

$\overrightarrow{UV} = \overrightarrow{XV} - \overrightarrow{XU}$

$= (\overrightarrow{XY} + \overrightarrow{YV}) - \overrightarrow{XU}$

$= (\overrightarrow{XY} + \tfrac{1}{2}\overrightarrow{YZ}) - \overrightarrow{XU}$

$= (\overrightarrow{XY} + \tfrac{1}{2}\overrightarrow{YZ}) - \tfrac{1}{2}\overrightarrow{XY}$

$= \tfrac{1}{2}(\overrightarrow{XY} + \overrightarrow{YZ})$

$= \tfrac{1}{2}\overrightarrow{XZ}$

55. Using a graphing calculator to graph $y = x^3 - x^2 + 3$. Determine and classify its extrema. *(Lesson 3-6)* **max: (0, 3), min: (0.67, 2.85)**

56. $y \to \infty$ as $x \to \infty$, $y \to \infty$ as $x \to -\infty$

56. Describe the end behavior of $f(x) = x^2 + 3x + 1$. *(Lesson 3-5)*

57. SAT Practice   For which values of x is $7x + 1$ greater than $7x - 1$?   **A**

   A all real numbers
   B only positive real numbers
   C only $x = 0$
   D only negative real numbers
   E no real numbers

4 ASSESS

Open-Ended Assessment
Speaking Have students take turns comparing and contrasting the geometric and algebraic methods for performing the various vector operations.

Chapter 8, Quiz A (Lessons 8-1 and 8-2) is available in the *Assessment and Evaluation Masters*, p. 199.

CAREER CHOICES

 Aerospace Engineering

Would you like to design aircraft or even spacecraft? An aerospace engineer works with other specialists to design, build, and test any vehicles that fly. As an aerospace engineer, you might concentrate on one type of air or spacecraft, or you might work on specific components of these crafts.

There are also a number of specialties within the field of aerospace engineering including analytical engineering, stress engineering, materials aerospace engineering, and marketing and sales aerospace engineering. One of the aspects of aerospace engineering is the analysis of airline accidents to determine if structural defects in design were the cause of the accident.

CAREER OVERVIEW

Degree Preferred:
bachelor's degree in aeronautical or aerospace engineering

Related Courses:
mathematics, science, computer science, mechanical drawing

Outlook:
slower than average through the year 2006

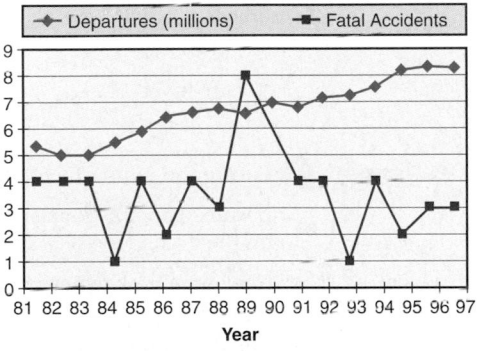

Airline Departures and Fatal Accidents

interNET CONNECTION For more information on careers in aerospace engineering, visit: **www.amc.glencoe.com**

Lesson 8-2 Algebraic Vectors **499**

CAREER CHOICES   AEROSPACE ENGINEERING

Math Connection
* Explain the trends seen in the data. **The number of departures has steadily increased. The number of fatal accidents has fluctuated year-to-year.**
* Why might this graph be misleading? **The departures are given in millions while the number of fatal accidents is not.**
* Find the number of fatal accidents per 100,000 departures for each year.
  Sample answers: 1981: 0.077; 1982: 0.060; 1983: 0.079; 1984: 0.018; 1985: 0.069; 1986: 0.016; 1987: 0.046; 1988: 0.030; 1989: 0.121; 1990: 0.087; 1991: 0.059; 1992: 0.057; 1993: 0.014; 1994: 0.053; 1995: 0.025; 1996: 0.036; 1997: 0.037

Enrichment Masters, p. 86

Lesson 8-2 **499**

Vectors in Three-Dimensional Space

5-Minute Check
Lesson 8-2

1. Write the ordered pair that represents the vector from $A(3, -1)$ to $B(-4, 0)$. Then find the magnitude of $\overrightarrow{AB}$. $\langle -7, 1 \rangle$; $5\sqrt{2}$

Find an ordered pair to represent $\vec{z}$ in each equation if $\vec{v} = \langle -1, -5 \rangle$ and $\vec{w} = \langle 6, 4 \rangle$.

2. $\vec{z} = \vec{v} + \vec{w}$ $\langle 5, -1 \rangle$

3. $\vec{z} = 2\vec{v} - \frac{1}{2}\vec{w}$ $\langle -5, -12 \rangle$

4. $\vec{z} = \vec{w} - \vec{v}$ $\langle 7, 9 \rangle$

5. Marshall's Movers specialize in moving pool tables. Mr. Marshall pushes a pool table with a force of 205 newtons at a 52° angle with the horizontal. His son pulls the pool table with a force of 185 newtons at a 48° angle with the horizontal. What is the magnitude of the force exerted on the pool table? **390 N**

6. Write $\overrightarrow{AB}$ as the sum of unit vectors of $A(0, 4)$ and $B(-3, 3)$. $-3\vec{i} - \vec{j}$

Motivating the Lesson

Real-World Application Bob is pushing a power mower up a hill with the mower tilting slightly to the right. Suppose that we wish to use vectors to determine the total force acting on the mower. Ask students what forces should be considered. **the force applied by Bob, gravity, friction, and the force of the tilt to the right** Ask students whether two-dimensional vectors can be used to demonstrate these forces all at the same time. **No, three-dimensional space must be used.**

OBJECTIVES
- Add and subtract vectors in three-dimensional space.
- Find the magnitude of vectors in three-dimensional space.

ENTOMOLOGY Entomology (en tuh MAHL uh jee) is the study of insects. Entomologists often use time-lapse photography to study the flying behavior of bees. They have discovered that worker honeybees let other workers know about new sources of food by rapidly vibrating their wings and performing a dance. Researchers have been able to create three-dimensional models using vectors to represent the flying behavior of bees. *You will use vectors in Example 3 to model a bee's path.*

Vectors in three-dimensional space can be described by coordinates in a way similar to the way we describe vectors in a plane. Imagine three real number lines intersecting at the zero point of each, so that each line is perpendicular to the plane determined by the other two. To show this arrangement on paper, a figure with the x-axis appearing to come out of the page is used to convey the feeling of depth. The axes are named the x-axis, y-axis, and z-axis.

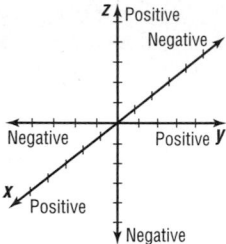

Each point in space corresponds to an ordered triple of real numbers. To locate a point P with coordinates (x_1, y_1, z_1), first find x_1 on the x-axis, y_1 on the y-axis, and z_1 on the z-axis. Then imagine a plane perpendicular to the x-axis at x_1 and planes perpendicular to the y- and z-axes at y_1 and z_1, respectively. The three planes will intersect at point P.

Example **Locate the point at $(-4, 3, 2)$.**

Locate -4 on the x-axis, 3 on the y-axis, and 2 on the z-axis.

Now draw broken lines for parallelograms to represent the three planes.

The planes intersect at $(-4, 3, 2)$.

Ordered triples, like ordered pairs, can be used to represent vectors. The geometric interpretation is the same for a vector in space as it is for a vector in a plane. A directed line segment from the origin O to $P(x, y, z)$ is called vector $\overrightarrow{OP}$, corresponding to vector $\langle x, y, z \rangle$.

An extension of the formula for the distance between two points in a plane allows us to find the distance between two points in space. The distance from the origin to a point (x, y, z) is $\sqrt{x^2 + y^2 + z^2}$. So the magnitude of vector $\langle x, y, z \rangle$ is $\sqrt{x^2 + y^2 + z^2}$. This can be adapted to represent any vector in space.

Representation of a Vector as an Ordered Triple	Suppose $P_1(x_1, y_1, z_1)$ is the initial point of a vector in space and $P_2(x_2, y_2, z_2)$ is the terminal point. The ordered triple that represents $\overrightarrow{P_1P_2}$ is $\langle x_2 - x_1, y_2 - y_1, z_2 - z_1 \rangle$. Its magnitude is given by $\lvert \overrightarrow{P_1P_2} \rvert = \sqrt{(x_2 - x_1)^2 + (y_2 - y_1)^2 + (z_2 - z_1)^2}$.

Examples

2 Write the ordered triple that represents the vector from $X(5, -3, 2)$ to $Y(4, -5, 6)$.

$$\overrightarrow{XY} = (4, -5, 6) - (5, -3, 2)$$
$$= \langle 4 - 5, -5 - (-3), 6 - 2 \rangle$$
$$= \langle -1, -2, 4 \rangle$$

3 **ENTOMOLOGY** Refer to the application at the beginning of the lesson. Suppose the flight of a honeybee passed through points at $(0, 3, 3)$ and $(5, 0, 4)$, in which each unit represents a meter. What is the magnitude of the displacement the bee experienced in traveling between these two points?

$$\text{magnitude} = \sqrt{(x_2 - x_1)^2 + (y_2 - y_1)^2 + (z_2 - z_1)^2}$$
$$= \sqrt{(5 - 0)^2 + (0 - 3)^2 + (4 - 3)^2} \qquad \langle x_1, y_1, z_1 \rangle = \langle 0, 3, 3 \rangle,$$
$$= \sqrt{25 + 9 + 1} \qquad\qquad\qquad \langle x_2, y_2, z_2 \rangle = \langle 5, 0, 4 \rangle$$
$$= 5.9$$

The magnitude of the displacement is about 5.9 meters.

Operations on vectors represented by ordered triples are similar to those on vectors represented by ordered pairs.

Example

4 Find an ordered triple that represents $3\vec{p} - 2\vec{q}$ if $\vec{p} = \langle 3, 0, 4 \rangle$ and $\vec{q} = \langle 2, 1, -1 \rangle$.

$$3\vec{p} - 2\vec{q} = 3\langle 3, 0, 4 \rangle - 2\langle 2, 1, -1 \rangle \quad \vec{p} = \langle 3, 0, 4 \rangle, \vec{q} = \langle 2, 1, -1 \rangle$$
$$= \langle 9, 0, 12 \rangle - \langle 4, 2, -2 \rangle$$
$$= \langle 5, -2, 14 \rangle$$

Lesson 8-3   Vectors in Three-Dimensional Space   **501**

2 TEACH

Teaching Tip In Example 1, remind students that the placement of the axes is different in three dimensions than in two dimensions. The *x*-coordinate, −4, appears to go back into the page to show depth. You may wish to have students rotate their books until the *x*- and *y*-axes are in a familiar orientation.

Teaching Tip In Example 2, remind students that reversing the order of the coordinates changes the vector from $\overrightarrow{AB}$ to $\overrightarrow{BA}$, which has the same magnitude but the opposite direction.

In-Class Examples

Example 1
Locate the point at $(2, -1, 6)$.

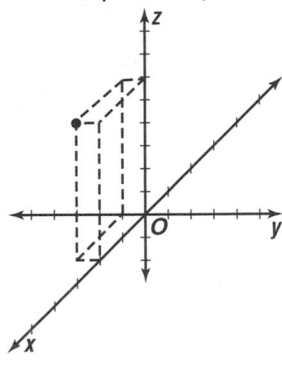

Example 2
Write the ordered triple that represents the vector from $A(-2, -5, 0)$ to $B(3, 1, 8)$. $\langle 5, 6, 8 \rangle$

Example 3
Nature A water snake swims around small rocks and brush along a river bank. Suppose that the snake's head swims through points at $(0, 2, 5)$ and $(4, 0, 3)$, in which each unit represents a meter. What is the magnitude of the displacement that the snake experienced in traveling between these two points?   **4.9 m**

Example 4
Find an ordered triple that represents $4\vec{v} + \vec{u}$ if $\vec{v} = \langle 1, 7, 1 \rangle$ and $\vec{u} = \langle -4, -2, -6 \rangle$. $\langle 0, 26, -2 \rangle$

Teaching Tip For Example 3, remind students that the magnitude of displacement is a three-dimensional vector representation of distance.

Three unit vectors are used as components of vectors in space. The unit vectors on the *x*-, *y*-, and *z*-axes are $\vec{i}, \vec{j}$, and $\vec{k}$ respectively, where $\vec{i} = \langle 1, 0, 0 \rangle$, $\vec{j} = \langle 0, 1, 0 \rangle$, and $\vec{k} = \langle 0, 0, 1 \rangle$. The vector $\vec{a} = \langle a_1, a_2, a_3 \rangle$ is shown on the graph at the right. The component vectors of $\vec{a}$ along the three axes are $a_1\vec{i}, a_2\vec{j}$, and $a_3\vec{k}$. Vector $\vec{a}$ can be written as the sum of unit vectors; that is, $\vec{a} = a_1\vec{i} + a_2\vec{j} + a_3\vec{k}$.

Example ⑤ Write $\overrightarrow{AB}$ as the sum of unit vectors for $A(5, 10, -3)$ and $B(-1, 4, -2)$.

First, express $\overrightarrow{AB}$ as an ordered triple. Then, write the sum of the unit vectors $\vec{i}, \vec{j}$, and $\vec{k}$.

$$\overrightarrow{AB} = (-1, 4, -2) - (5, 10, -3)$$
$$= \langle -1 - 5, 4 - 10, -2 - (-3) \rangle$$
$$= \langle -6, -6, 1 \rangle$$
$$= -6\vec{i} - 6\vec{j} + \vec{k}$$

CHECK FOR UNDERSTANDING

Communicating Mathematics

Read and study the lesson to answer each question.   1–3. See margin.

1. **Explain** the process you would use to locate $\overrightarrow{AB} = 2\vec{i} + 3\vec{j} + 4\vec{k}$ in space. Then sketch the vector on a coordinate system.

2. **Describe** the information you need to find the components of a three-dimensional vector from its given magnitude.

3. **You Decide**   Marshall wrote the vector $\vec{v} = \langle 1, -4, 0 \rangle$ as the sum of unit vectors $\vec{i} + 4\vec{j} + \vec{k}$. Denise said that $\vec{v}$ should be written as $\vec{i} - 4\vec{j} + \vec{k}$. Who is correct? Explain.

Guided Practice

4. Locate point $G(4, -1, 7)$ in space. Then find the magnitude of a vector from the origin to G.   **See margin for graph.** $\sqrt{66}$

Write the ordered triple that represents $\overrightarrow{RS}$. Then find the magnitude of $\overrightarrow{RS}$.

5. $R(-2, 5, 8)$, $S(3, 9, -3)$
   $\langle 5, 4, -11 \rangle, 9\sqrt{2}$

6. $R(3, 7, -1)$, $S(10, -4, 0)$
   $\langle 7, -11, 1 \rangle, 3\sqrt{19}$

Find an ordered triple to represent $\vec{a}$ in each equation if $\vec{f} = \langle 1, -3, -8 \rangle$ and $\vec{g} = \langle 3, 9, -1 \rangle$.

7. $\vec{a} = 3\vec{f} + \vec{g}$ $\langle 6, 0, -25 \rangle$

8. $\vec{a} = 2\vec{g} - 5\vec{f}$ $\langle 1, 33, 38 \rangle$

Write $\overrightarrow{EF}$ as the sum of unit vectors.

9. $E(-5, -2, 4)$, $F(6, -6, 6)$
   $11\vec{i} - 4\vec{j} + 2\vec{k}$

10. $E(-12, 15, -9)$, $F(-12, 17, -22)$
    $2\vec{j} - 13\vec{k}$

502   Chapter 8   *Vectors and Parametric Equations*

11. Physics Suppose that during a storm the force of the wind blowing against a skyscraper can be expressed by the vector $\langle 132, 3454, 0 \rangle$, where each measure in the ordered triple represents the force in newtons. What is the magnitude of this force? **≈ 3457 N**

 EXERCISES

Practice

 Locate point B in space. Then find the magnitude of a vector from the origin to B.

A ▶
12. $B(4, 1, -3)$ $\sqrt{26}$ **13.** $B(7, 2, 4)$ $\sqrt{69}$ **14.** $B(10, -3, 15)$ $\sqrt{334}$
12–14 See margin for graphs.

Write the ordered triple that represents $\overline{TM}$. Then find the magnitude of $\overline{TM}$.

15.
$\langle 1, -4, -8 \rangle, 9$

16. $\langle -1, 1, -5 \rangle,$
$3\sqrt{3}$

17. $\langle 1, -4, -4 \rangle,$
$\sqrt{33}$

18. $\langle -4, 6, -4 \rangle,$
$2\sqrt{17}$

15. $T(2, 5, 4), M(3, 1, -4)$ **16.** $T(-2, 4, 7), M(-3, 5, 2)$

17. $T(2, 5, 4), M(3, 1, 0)$ **18.** $T(3, -5, 6), M(-1, 1, 2)$

19. $T(-5, 8, 3), M(-2, -1, -6)$ **20.** $T(0, 6, 3), M(1, 4, -3)$
$\langle 3, -9, -9 \rangle, 3\sqrt{19}$ $\langle 1, -2, -6 \rangle, \sqrt{41}$

21. Write the ordered triple to represent $\overline{CJ}$. Then find the magnitude of $\overline{CJ}$.
$\langle 4, -8, -14 \rangle, 2\sqrt{69}$

Find an ordered triple to represent $\overline{u}$ in each equation if $\overline{v} = \langle 4, -3, 5 \rangle$, $\overline{w} = \langle 2, 6, -1 \rangle$, and $\overline{z} = \langle 3, 0, 4 \rangle$.

22. $\overline{u} = 6\overline{w} + 2\overline{z}$ $\langle 18, 36, 2 \rangle$ **23.** $\overline{u} = \frac{1}{2}\overline{v} - \overline{w} + 2\overline{z}$ $\langle 6, -7\frac{1}{2}, 11\frac{1}{2} \rangle$

24. $\overline{u} = \frac{3}{4}\overline{v} - \overline{w}$ $\langle 1, -8\frac{1}{4}, 4\frac{3}{4} \rangle$ **25.** $\overline{u} = 3\overline{v} - \frac{2}{3}\overline{w} + 2\overline{z}$ $\langle 16\frac{2}{3}, -13, 23\frac{2}{3} \rangle$

B ▶
26. $\overline{u} = 0.75\overline{v} + 0.25\overline{w}$
$\langle 3.5, -0.75, 3.5 \rangle$ **27.** $\overline{u} = -4\overline{w} + \overline{z}$ $\langle -5, -24, 8 \rangle$

28. Find an ordered triple to represent the sum $\frac{2}{3}\overline{f} + 3\overline{g} - \frac{2}{5}\overline{h}$, if $\overline{f} = \langle -3, 4.5, -1 \rangle$, $\overline{g} = \langle -2, 1, 6 \rangle$, and $\overline{h} = \langle 6, -3, -3 \rangle$. $\langle -\frac{52}{5}, \frac{36}{5}, \frac{278}{15} \rangle$

Write $\overline{LB}$ as the sum of unit vectors.

C ▶
29. $L(2, 2, 7), B(5, -6, 2)$ $3\overline{i} - 8\overline{j} - 5\overline{k}$ **30.** $L(-6, 1, 0), B(-4, 5, -1)$ $2\overline{i} + 4\overline{j} - \overline{k}$

31. $-2\overline{i} - 4\overline{j}$
$+ 9\overline{k}$

32. $-20\overline{i} + 5\overline{j}$
$- 11\overline{k}$

31. $L(9, 7, -11), B(7, 3, -2)$ **32.** $L(12, 2, 6), B(-8, 7, -5)$

33. $L(-1, 2, -4), B(-8, 5, -10)$ **34.** $L(-9, 12, -5), B(6, 5, -5)$
$-7\overline{i} + 3\overline{j} - 6\overline{k}$ $15\overline{i} - 7\overline{j}$

35. Show that $|\overline{G_1G_2}| = |\overline{G_2G_1}|$. **See Answer Appendix.**

36. If $\overline{m} = \langle m_1, m_2, m_3 \rangle$, then $-\overline{m}$ is defined as $\langle -m_1, -m_2, -m_3 \rangle$. Show that $|-\overline{m}| = |\overline{m}|$ **See margin.**

 www.amc.glencoe.com/self_check_quiz *Lesson 8-3 Vectors in Three-Dimensional Space* **503**

Additional Answers
(pp. 502–503)
3. Sample answer: neither is correct. The sign for the $\overline{j}$-term must be the same (−), and the coefficient for the $\overline{k}$-term is 0, so the correct way to express the vector as a sum of unit vectors is $\overline{i} - 4\overline{j}$.

4.

$G(4, -1, 7)$

12.

$B(4, 1, -3)$

13.

$B(7, 2, 4)$

14.

$B(10, -3, 15)$

36. If $\overline{m} = \langle m_1, m_2, m_3 \rangle$, then $|\overline{m}| = \sqrt{(m_1)^2 + (m_2)^2 + (m_3)^2}$. If $-\overline{m} = \langle -m_1, -m_2, -m_3 \rangle$, then $|-\overline{m}| = \sqrt{(-m_1)^2 + (-m_2)^2 + (-m_3)^2}$. Since $m_1^2 = (-m_1)^2$, $m_2^2 = (-m_2)^2$, and $m_3^2 = (-m_3)^2$, $|-\overline{m}| = |\overline{m}|$.

4 ASSESS

Open-Ended Assessment

Writing Have each student write two ordered triples on a piece of paper. Have students write a paragraph about how to add the two vectors, find the magnitude of the resultant, and graph each vector and the sum.

Additional Answers

41a.

41b. about 26 ft

41c. $\theta = 35.25°$

42. No, A and B are 2 units apart, B and C are 1.69 units apart, and A and C are 1.41 units apart.

45.

$$\frac{\sin 2X}{1 - \cos 2X} = \cot X$$

$$\frac{2 \sin X \cos X}{1 - \cos^2 X + \sin^2 X} = \cot X$$

$$\frac{2 \sin X \cos X}{2 \sin^2 X} = \cot X$$

$$\frac{\cos X}{\sin X} = \cot X$$

$$\cot X = \cot X$$

Enrichment Masters, p. 87

8-3

Enrichment

Basis Vectors in Three-Dimensional Space

The expression $\vec{v} = r\vec{u} + s\vec{w} + t\vec{z}$, the sum of two vectors each multiplied by scalars, is called a **linear combination** of the vectors $\vec{u}$, $\vec{w}$, and $\vec{z}$.

Every vector $\vec{v} \in v_3$ can be written as a linear combination of any three nonparallel vectors. The three nonparallel vectors are said to form a **basis** for v_3, which contains all vectors having 1 column and 3 rows.

Example Write the vector $\vec{v} = \begin{pmatrix} -1 \\ -4 \\ 3 \end{pmatrix}$ as a linear combination of

the vectors $\vec{u} = \begin{pmatrix} 1 \\ 3 \\ 1 \end{pmatrix}$, $\vec{w} = \begin{pmatrix} 1 \\ -2 \\ 1 \end{pmatrix}$, and $\vec{z} = \begin{pmatrix} -1 \\ -1 \\ 1 \end{pmatrix}$.

$\begin{pmatrix} -1 \\ -4 \\ 3 \end{pmatrix} = r\begin{pmatrix} 1 \\ 3 \\ 1 \end{pmatrix} + s\begin{pmatrix} 1 \\ -2 \\ 1 \end{pmatrix} + t\begin{pmatrix} -1 \\ -1 \\ 1 \end{pmatrix} = \begin{pmatrix} r + s - t \\ 3r - 2s - t \\ r + s + t \end{pmatrix}$

$-1 = r + s - t$
$-4 = 3r - 2s - t$
$3 = r + s + t$

Solving the system of equations yields the solution $r = 0$, $s = 1$, and $t = 2$. So, $\vec{v} = \vec{w} + 2\vec{z}$.

Write each vector as a linear combination of the vectors $\vec{u}$, $\vec{w}$, and $\vec{z}$.

1. $\vec{v} = \begin{pmatrix} -4 \\ -2 \\ 2 \end{pmatrix}$, $\vec{u} = \begin{pmatrix} 1 \\ 3 \\ 0 \end{pmatrix}$, $\vec{w} = \begin{pmatrix} 1 \\ 1 \\ 0 \end{pmatrix}$, and $\vec{z} = \begin{pmatrix} 0 \\ 0 \\ 1 \end{pmatrix}$

$\vec{v} = -5\vec{u} - \vec{w} + 3\vec{z}$

2. $\vec{v} = \begin{pmatrix} 5 \\ -2 \\ 0 \end{pmatrix}$, $\vec{u} = \begin{pmatrix} 1 \\ -2 \\ 3 \end{pmatrix}$, $\vec{w} = \begin{pmatrix} -1 \\ 0 \\ 1 \end{pmatrix}$, and $\vec{z} = \begin{pmatrix} 4 \\ 2 \\ -1 \end{pmatrix}$

$\vec{v} = \frac{8}{7}\vec{u} - \frac{23}{7}\vec{w} + \frac{1}{7}\vec{z}$

3. $\vec{v} = \begin{pmatrix} -1 \\ -1 \\ 2 \end{pmatrix}$, $\vec{u} = \begin{pmatrix} 1 \\ 0 \\ -1 \end{pmatrix}$, $\vec{w} = \begin{pmatrix} 2 \\ 2 \\ 0 \end{pmatrix}$, and $\vec{z} = \begin{pmatrix} 1 \\ 0 \\ 1 \end{pmatrix}$

$\vec{v} = -\frac{1}{2}\vec{u} + \frac{3}{2}\vec{z}$

© Glencoe/McGraw-Hill   T87   Advanced Mathematical Concepts

Applications and Problem Solving

37. Physics An object is in equilibrium if the magnitude of the resultant force on it is zero. Two forces on an object are represented by $\langle 3, -2, 4 \rangle$ and $\langle 6, 2, 5 \rangle$. Find a third vector that will place the object in equilibrium. **$\langle -9, 0, -9 \rangle$**

38. Critical Thinking Find the midpoint of $\vec{v}$ that extends from $V(2, 3, 6)$ to $W(4, 5, 2)$. **(3, 4, 4)**

39. Computer Games Nate Rollins is designing a computer game. In the game, a knight is standing at point $(1, 4, 0)$ watching a wizard sitting at the top of a tree. In the computer screen, the tree is one unit high, and its base is at $(2, 4, 0)$. Find the displacement vector for each situation.

   a. from the origin to the knight $\vec{i} + 4\vec{j}$
   b. from the bottom of the tree to the knight $-\vec{i}$

40. Critical Thinking Find the vector $\vec{c}$ that must be added to $\vec{a} = \langle 1, 3, 1 \rangle$ to obtain $\vec{b} = \langle 3, 1, 5 \rangle$. **$\langle 2, -2, 4 \rangle$**

Dr. Chiaki Mukai

41. Aeronautics Dr. Chiaki Mukai is Japan's first female astronaut. Suppose she is working inside a compartment shaped like a cube with sides 15 feet long. She realizes that the tool she needs is diagonally in the opposite corner of the compartment. **a–c. See margin.**

   a. Draw a diagram of the situation described above.
   b. What is the minimum distance she has to glide to secure the tool?
   c. At what angle to the floor must she launch herself?

42. Chemistry Dr. Alicia Sanchez is a researcher for a pharmaceutical firm. She has graphed the structure of a molecule with atoms having positions $A = (2, 0, 0)$, $B = \left(1, \sqrt{3}, 0\right)$, and $C = \left(1, \frac{1}{3}, \frac{2\sqrt{2}}{3}\right)$. She needs to have every atom in this molecule equidistant from each other. Has she achieved this goal? Explain why or why not. **See margin.**

Mixed Review

43. Find the sum of the vectors $\langle 3, 5 \rangle$ and $\langle -1, 2 \rangle$ algebraically. *(Lesson 8-2)* **$\langle 2, 7 \rangle$**

44. Find the coordinates of point D such that $\overline{AB}$ and $\overline{CD}$ are equal vectors for points $A(5, 2)$, $B(-3, 3)$, and $C(0, 0)$. *(Lesson 8-1)* **$D(-8, 1)$**

45. Verify that $\cot X = (\sin 2X) \div (1 - \cos 2X)$ is an identity. *(Lesson 7-4)* **See margin.**

46. If $\cos \theta = \frac{2}{3}$ and $0° \le \theta \le 90°$, find $\sin \theta$. *(Lesson 7-1)* **$\frac{\sqrt{5}}{3}$**

47. State the amplitude and period for the function $y = 6 \sin \frac{\theta}{2}$. *(Lesson 6-4)* **$6, 4\pi$**

48. Physics If a pulley is rotating at 16 revolutions per minute, what is its rate in radians per second? *(Lesson 6-2)* **$\frac{8\pi}{15}$**

49. Determine if $(7, -2)$ is a solution for $y < 4x^2 - 3x + 5$. Explain. *(Lesson 3-3)*

49. Yes, because substituting 7 for x and -2 for y results in the inequality $-2 < 180$, which is true.

50. SAT/ACT Practice You have added the same positive quantity to the numerator and denominator of a fraction. The result is **E**

   A greater than the original fraction.
   B less than the original fraction.
   C equal to the original fraction.
   D one-half the original fraction.
   E cannot be determined from the information given.

Extra Credit

An object is in equilibrium if the magnitude of the resultant force on it is zero. Three forces on an object are represented by $\langle 4, -1, 3 \rangle$, $\langle 5, 2, 3 \rangle$, and $\langle -1, 2, -6 \rangle$. Find a fourth vector that will place the object in equilibrium. **$\langle -8, -3, 0 \rangle$**

8-4

Perpendicular Vectors

Transmission   Gearshift

3rd-4th
1st-2nd
Reverse

AUTO RACING In addition to being an actor, Emilio Estevez is an avid stock car driver. Drivers in stock car races must constantly shift gears to maneuver their cars into competitive positions. A gearshift rotates about the connection to the transmission. The rate of change of the rotation of the gearshift depends on the magnitude of the force exerted by the driver and on the perpendicular distance of its line of action from the center of rotation. *You will use vectors in Example 4 to determine the force on a gearshift.*

In physics, *torque* is the measure of the effectiveness of a force in turning an object about a pivot point. We can use perpendicular vectors to find the torque of a force.

Let $\vec{a}$ and $\vec{b}$ be perpendicular vectors, and let $\overrightarrow{BA}$ be a vector between their terminal points as shown. Then the magnitudes of $\vec{a}$, $\vec{b}$, and $\overrightarrow{BA}$ must satisfy the Pythagorean Theorem.

$$|\overrightarrow{BA}|^2 = |\vec{a}|^2 + |\vec{b}|^2$$

Now use the definition of magnitude of a vector to evaluate $|\overrightarrow{BA}|^2$.

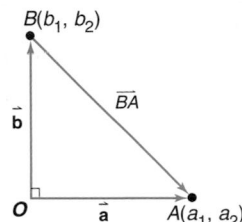

$B(b_1, b_2)$

$\overrightarrow{BA}$

$\vec{b}$

O $\vec{a}$ $A(a_1, a_2)$

$$|\overrightarrow{BA}| = \sqrt{(a_1 - b_1)^2 + (a_2 - b_2)^2} \qquad \text{\textit{Definition of magnitude}}$$
$$|\overrightarrow{BA}|^2 = (a_1 - b_1)^2 + (a_2 - b_2)^2 \qquad \text{\textit{Square each side.}}$$
$$|\overrightarrow{BA}|^2 = a_1^2 - 2a_1b_1 + b_1^2 + a_2^2 - 2a_2b_2 + b_2^2 \qquad \text{\textit{Simplify.}}$$
$$|\overrightarrow{BA}|^2 = (a_1^2 + a_2^2) + (b_1^2 + b_2^2) - 2(a_1b_1 + a_2b_2) \qquad \text{\textit{Group the squared terms.}}$$
$$|\overrightarrow{BA}|^2 = |\vec{a}|^2 + |\vec{b}|^2 - 2(a_1b_1 + a_2b_2) \qquad \begin{array}{l} |\vec{a}|^2 = a_1^2 + a_2^2 \\ |\vec{b}|^2 = b_1^2 + b_2^2 \end{array}$$

Compare the resulting equation with the original one. $|\overrightarrow{BA}|^2 = |\vec{a}|^2 + |\vec{b}|^2$ if and only if $a_1b_1 + a_2b_2 = 0$.

The expression $a_1b_1 + a_2b_2$ is frequently used in the study of vectors. It is called the **inner product** of $\vec{a}$ and $\vec{b}$.

Lesson 8-4 Perpendicular Vectors **505**

Motivating the Lesson

Discussion Show students various vectors and ask them to identify whether any of them are perpendicular. Then, ask them how they know that the vectors are perpendicular.

Lesson 8-4 **505**

In-Class Examples

Example 1
Find each inner product if $\bar{\mathbf{x}} = \langle 2, -5 \rangle$, $\bar{\mathbf{y}} = \langle 4, 1 \rangle$, and $\bar{\mathbf{z}} = \langle 10, 4 \rangle$. Are any pairs of vectors perpendicular?
a. $\bar{\mathbf{x}} \cdot \bar{\mathbf{y}}$ 3; not perpendicular
b. $\bar{\mathbf{x}} \cdot \bar{\mathbf{z}}$ 0; perpendicular
c. $\bar{\mathbf{y}} \cdot \bar{\mathbf{z}}$ 44; not perpendicular

Example 2
Find the inner product of $\bar{\mathbf{v}}$ and $\bar{\mathbf{w}}$ if $\bar{\mathbf{v}} = \langle 2, -3, -4 \rangle$ and $\bar{\mathbf{w}} = \langle 8, 3, 2 \rangle$. Are $\bar{\mathbf{v}}$ and $\bar{\mathbf{w}}$ perpendicular?
−1; no

Inner Product of Vectors in a Plane

If $\bar{\mathbf{a}}$ and $\bar{\mathbf{b}}$ are two vectors, $\langle a_1, a_2 \rangle$ and $\langle b_1, b_2 \rangle$, the inner product of $\bar{\mathbf{a}}$ and $\bar{\mathbf{b}}$ is defined as $\bar{\mathbf{a}} \cdot \bar{\mathbf{b}} = a_1 b_1 + a_2 b_2$.

$\bar{a} \cdot \bar{b}$ is read "a dot b" and is often called the <u>dot product</u>.

Two vectors are perpendicular if and only if their inner product is zero. That is, vectors $\bar{\mathbf{a}} = \langle a_1, a_2 \rangle$ and $\bar{\mathbf{b}} = \langle b_1, b_2 \rangle$ are perpendicular if $\bar{\mathbf{a}} \cdot \bar{\mathbf{b}} = 0$.

$$\bar{\mathbf{a}} \cdot \bar{\mathbf{b}} = a_1 b_1 + a_2 b_2$$

Let $a_1 b_1 + a_2 b_2 = 0$ if

$$a_1 b_1 = -a_2 b_2$$
$$\frac{a_1}{a_2} = -\frac{b_2}{b_1} \qquad a_2 \neq 0, \, b_1 \neq 0$$

Since the ratio of the components can also be thought of as the slopes of the line on which the vectors lie, the slopes are opposite reciprocals. Thus, the lines and the vectors are perpendicular.

Example ❶ Find each inner product if $\bar{\mathbf{p}} = \langle 7, 14 \rangle$, $\bar{\mathbf{q}} = \langle 2, -1 \rangle$ and $\bar{\mathbf{m}} = \langle 3, 5 \rangle$. Are any pair of vectors perpendicular?

a. $\bar{\mathbf{p}} \cdot \bar{\mathbf{q}}$

$$\begin{aligned} \bar{\mathbf{p}} \cdot \bar{\mathbf{q}} &= 7(2) + 14(-1) \\ &= 14 - 14 \\ &= 0 \end{aligned}$$

$\bar{\mathbf{p}}$ and $\bar{\mathbf{q}}$ are perpendicular.

b. $\bar{\mathbf{p}} \cdot \bar{\mathbf{m}}$

$$\begin{aligned} \bar{\mathbf{p}} \cdot \bar{\mathbf{m}} &= 7(3) + 14(5) \\ &= 21 + 70 \\ &= 91 \end{aligned}$$

$\bar{\mathbf{p}}$ and $\bar{\mathbf{m}}$ are not perpendicular.

c. $\bar{\mathbf{q}} \cdot \bar{\mathbf{m}}$

$$\begin{aligned} \bar{\mathbf{q}} \cdot \bar{\mathbf{m}} &= 2(3) + (-1)(5) \\ &= 6 - 5 \\ &= 1 \end{aligned}$$

$\bar{\mathbf{q}}$ and $\bar{\mathbf{m}}$ are not perpendicular.

The inner product of vectors in space is similar to that of vectors in a plane.

Inner Product of Vectors in Space

If $\bar{\mathbf{a}} = \langle a_1, a_2, a_3 \rangle$ and $\bar{\mathbf{b}} = \langle b_1, b_2, b_3 \rangle$, then $\bar{\mathbf{a}} \cdot \bar{\mathbf{b}} = a_1 b_1 + a_2 b_2 + a_3 b_3$.

Just as in a plane, two vectors in space are perpendicular if and only if their inner product is zero.

Example ❷ Find the inner product of $\bar{\mathbf{a}}$ and $\bar{\mathbf{b}}$ if $\bar{\mathbf{a}} = \langle -3, 1, 1 \rangle$ and $\bar{\mathbf{b}} = \langle 2, 8, -2 \rangle$. Are $\bar{\mathbf{a}}$ and $\bar{\mathbf{b}}$ perpendicular?

$$\begin{aligned} \bar{\mathbf{a}} \cdot \bar{\mathbf{b}} &= (-3)(2) + (1)(8) + (1)(-2) \\ &= -6 + 8 + (-2) \\ &= 0 \end{aligned}$$

$\bar{\mathbf{a}}$ and $\bar{\mathbf{b}}$ are perpendicular since their inner product is zero.

Another important product involving vectors in space is the **cross product.** Unlike the inner product, the cross product of two vectors is a vector. This vector does not lie in the plane of the given vectors, but is perpendicular to the plane containing the two vectors. In other words, the cross product of two vectors is perpendicular to both vectors. The cross product of $\vec{a}$ and $\vec{b}$ is written $\vec{a} \times \vec{b}$.

$\vec{a} \times \vec{b}$ is perpendicular to $\vec{a}$.
$\vec{a} \times \vec{b}$ is perpendicular to $\vec{b}$.

Teaching Tip For Example 3, review with students how to evaluate a third order determinant by expanding by minors.

In-Class Example

Example 3
Find the cross product of $\vec{a}$ and $\vec{b}$ if $\vec{a} = \langle -4, 1, 0 \rangle$ and $\vec{b} = \langle 5, 4, -2 \rangle$. Verify that the resulting vector is perpendicular to $\vec{a}$ and $\vec{b}$.
$\langle -2, -8, -21 \rangle$ **Since the inner products are zero, the cross product $\vec{a} \times \vec{b}$ is perpendicular to both $\vec{a}$ and $\vec{b}$.**

Cross Product of Vectors in Space

If $\vec{a} = \langle a_1, a_2, a_3 \rangle$ and $\vec{b} = \langle b_1, b_2, b_3 \rangle$, then the cross product of $\vec{a}$ and $\vec{b}$ is defined as follows.

$$\vec{a} \times \vec{b} = \begin{vmatrix} a_2 & a_3 \\ b_2 & b_3 \end{vmatrix} \vec{i} - \begin{vmatrix} a_1 & a_3 \\ b_1 & b_3 \end{vmatrix} \vec{j} + \begin{vmatrix} a_1 & a_2 \\ b_1 & b_2 \end{vmatrix} \vec{k}$$

Look Back
Refer to Lesson 2-5 to review determinants and expansion by minors.

An easy way to remember the coefficients of $\vec{i}, \vec{j}, \vec{k}$ is to write a determinant as shown and expand by minors using the first row.

$$\begin{vmatrix} \vec{i} & \vec{j} & \vec{k} \\ a_1 & a_2 & a_3 \\ b_1 & b_2 & b_3 \end{vmatrix}$$

Example ③ Find the cross product of $\vec{v}$ and $\vec{w}$ if $\vec{v} = \langle 0, 3, 1 \rangle$ and $\vec{w} = \langle 0, 1, 2 \rangle$. Verify that the resulting vector is perpendicular to $\vec{v}$ and $\vec{w}$.

$$\vec{v} \times \vec{w} = \begin{vmatrix} \vec{i} & \vec{j} & \vec{k} \\ 0 & 3 & 1 \\ 0 & 1 & 2 \end{vmatrix}$$

$$= \begin{vmatrix} 3 & 1 \\ 1 & 2 \end{vmatrix} \vec{i} - \begin{vmatrix} 0 & 1 \\ 0 & 2 \end{vmatrix} \vec{j} + \begin{vmatrix} 0 & 3 \\ 0 & 1 \end{vmatrix} \vec{k} \quad \textit{Expand by minors.}$$

$$= 5\vec{i} - 0\vec{j} + 0\vec{k}$$
$$= 5\vec{i} \text{ or } \langle 5, 0, 0 \rangle$$

Find the inner products $\langle 5, 0, 0 \rangle \cdot \langle 0, 3, 1 \rangle$ and $\langle 5, 0, 0 \rangle \cdot \langle 0, 1, 2 \rangle$.

$$5(0) + 0(3) + 0(1) = 0 \qquad 5(0) + 0(1) + 0(2) = 0$$

Since the inner products are zero, the cross product $\vec{v} \times \vec{w}$ is perpendicular to both $\vec{v}$ and $\vec{w}$.

In physics, the torque $\vec{T}$ about a point A created by a force $\vec{F}$ at a point B is given by $\vec{T} = \overline{AB} \times \vec{F}$. The magnitude of $\vec{T}$ represents the torque in foot-pounds.

Lesson 8-4   Perpendicular Vectors   **507**

Addressing Individual Needs

Logical Have students find the cross product of $\vec{m}$ and $\vec{n}$ if $\vec{m} = \langle 1, 4, -5 \rangle$ and $\vec{n} = \langle -2, 5, 5 \rangle$ by filling in the boxes of the equation at the right. They should also verify that the resulting vector is perpendicular to $\vec{m}$ and $\vec{n}$ by using the inner product. **Yes, the resulting vector is perpendicular to $\vec{m}$ and $\vec{n}$.**

$$\vec{m} \times \vec{n} = \begin{vmatrix} 4 & \boxed{-5} \\ 5 & \boxed{5} \end{vmatrix} \vec{i} - \begin{vmatrix} 4 & \boxed{1} \\ \boxed{-2} & 5 \end{vmatrix} \vec{j} + \begin{vmatrix} \boxed{1} & 4 \\ -2 & \boxed{5} \end{vmatrix} \vec{k}$$

$$= \boxed{45}\vec{i} - \boxed{-5}\vec{j} + 13\vec{k}$$

$$(\boxed{45}, \boxed{5}, 13) \cdot (1, 4, -5) = 0$$

$$(\boxed{45}, \boxed{5}, 13) \cdot (-2, 5, 5) = 0$$

508 Chapter 8

In-Class Example

Example 4

Sports A monster truck driver is applying a force of 28 pounds along the positive *z*-axis to the gearshift of her truck. If the center of the gearshift connection is at the point *X*, the force is applied at the point (0.70, 0, 0.31). Find the torque. **19.6 ft-lb**

3 PRACTICE/APPLY

Error Analysis

Watch for students who do not have a clear understanding of the definitions of dot product, inner product, and cross product.

Prevent by reminding students that the dot product and inner product are the same and are both derived from the distance formula. The cross product is found by expansion by minors and with determinants.

Study Guide and Practice Masters, pp. 143–144

Example **4** **AUTO RACING** Refer to the application at the beginning of the lesson. Suppose Emilio Estevez is applying a force of 25 pounds along the positive *z*-axis to the gearshift of his car. If the center of the connection of the gearshift is at the origin, the force is applied at the point (0.75, 0, 0.27). Find the torque.

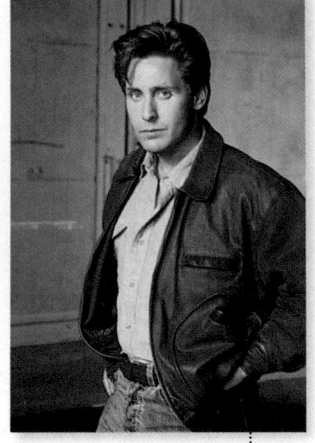

Emilio Estevez

We need to find $|\vec{\mathbf{T}}|$, the torque of the force at (0.75, 0, 0.27) where each value is the distance from the origin in feet and $\vec{\mathbf{F}}$ represents the force in pounds.

To find the magnitude of $\vec{\mathbf{T}}$, we must first find $\overrightarrow{AB}$ and $\vec{\mathbf{F}}$.

$$\overrightarrow{AB} = (0.75, 0, 0.27) - (0, 0, 0)$$
$$= \langle 0.75 - 0, 0 - 0, 0.27 - 0 \rangle \text{ or } \langle 0.75, 0, 0.27 \rangle$$

Any upward force is measured along the *z*-axis, so $\vec{\mathbf{F}} = 25\vec{\mathbf{k}}$ or $\langle 0, 0, 25 \rangle$.

Now, find $\vec{\mathbf{T}}$.

$$\vec{\mathbf{T}} = \overrightarrow{AB} \times \vec{\mathbf{F}}$$

$$= \begin{vmatrix} \vec{\mathbf{i}} & \vec{\mathbf{j}} & \vec{\mathbf{k}} \\ 0.75 & 0 & 0.27 \\ 0 & 0 & 25 \end{vmatrix}$$

$$= \begin{vmatrix} 0 & 0.27 \\ 0 & 25 \end{vmatrix} \vec{\mathbf{i}} - \begin{vmatrix} 0.75 & 0.27 \\ 0 & 25 \end{vmatrix} \vec{\mathbf{j}} + \begin{vmatrix} 0.75 & 0 \\ 0 & 0 \end{vmatrix} \vec{\mathbf{k}}$$

$$= 0\vec{\mathbf{i}} - 18.75\vec{\mathbf{j}} + 0\vec{\mathbf{k}} \text{ or } \langle 0, -18.75, 0 \rangle$$

Find the magnitude of $\vec{\mathbf{T}}$.

$$|\vec{\mathbf{T}}| = \sqrt{0^2 + (-18.75)^2 + 0^2}$$
$$= \sqrt{(-18.75)^2} \text{ or } 18.75$$

The torque is 18.75 foot-pounds.

CHECK FOR UNDERSTANDING

Communicating Mathematics

Read and study the lesson to answer each question.   1–3 See margin.

1. **Compare** $\vec{v} \times \vec{w}$ and $\vec{w} \times \vec{v}$ for $\vec{v} = \langle -1, 0, 3 \rangle$ and $\vec{w} = \langle 1, 2, 4 \rangle$.

2. **Show** that the cross product of a three-dimensional vector with itself is the zero vector.

3. *Math Journal* Could the inner product of a nonzero vector and itself ever be zero? Explain why or why not.

Guided Practice

Find each inner product and state whether the vectors are perpendicular. Write *yes* or *no*.

4. $\langle 5, 2 \rangle \cdot \langle -3, 7 \rangle$ **−1, no**   5. $\langle -8, 2 \rangle \cdot \langle 4.5, 18 \rangle$ **0, yes**   6. $\langle -4, 9, 8 \rangle \cdot \langle 3, 2, -2 \rangle$ **−10, no**

Additional Answers

1. Sample answer: vector $\vec{v} \times \vec{w}$ is the negative of vector $\vec{w} \times \vec{v}$.

3. Sample answer: no, because a vector cannot be perpendicular to itself.

2. $\vec{a} \times \vec{a} = \begin{vmatrix} \vec{\mathbf{i}} & \vec{\mathbf{j}} & \vec{\mathbf{k}} \\ a_x & a_y & a_z \\ a_x & a_y & a_z \end{vmatrix}$

$$= \begin{vmatrix} a_y & a_z \\ a_y & a_z \end{vmatrix} \vec{\mathbf{i}} - \begin{vmatrix} a_x & a_z \\ a_x & a_z \end{vmatrix} \vec{\mathbf{j}} + \begin{vmatrix} a_x & a_y \\ a_x & a_y \end{vmatrix} \vec{\mathbf{k}}$$

$$= (a_y a_z - a_y a_z)\vec{\mathbf{i}} - (a_x a_z - a_x a_z)\vec{\mathbf{j}} + (a_x a_y - a_x a_y)\vec{\mathbf{k}}$$
$$= 0\vec{\mathbf{i}} - 0\vec{\mathbf{j}} + 0\vec{\mathbf{k}} = \langle 0, 0, 0 \rangle = \vec{\mathbf{0}}$$

Find each cross product. Then verify that the resulting vector is perpendicular to the given vectors.   7–8. See margin.

7. $\langle 1, -3, 2 \rangle \times \langle -2, 1, -5 \rangle$

8. $\langle 6, 2, 10 \rangle \times \langle 4, 1, 9 \rangle$

9. Find a vector perpendicular to the plane containing the points $(0, 1, 2)$, $(-2, 2, 4)$, and $(-1, -1, -1)$.   **Sample answer: $\langle 1, -8, 5 \rangle$**

10. **Mechanics** Tikiro places a wrench on a nut and applies a downward force of 32 pounds to tighten the nut. If the center of the nut is at the origin, the force is applied at the point $(0.65, 0, 0.3)$. Find the torque.   **20.8 ft-lb**

EXERCISES

Practice

 Find each inner product and state whether the vectors are perpendicular. Write *yes* or *no*.

16. 24, no

11. $\langle 4, 8 \rangle \cdot \langle 6, -3 \rangle$ **0, yes**   12. $\langle 3, 5 \rangle \cdot \langle 4, -2 \rangle$ **2, no**   13. $\langle 5, -1 \rangle \cdot \langle -3, 6 \rangle$ **−21, no**

14. $\langle 7, 2 \rangle \cdot \langle 0, -2 \rangle$ **−4, no**   15. $\langle 8, 4 \rangle \cdot \langle 2, 4 \rangle$ **32, no**   16. $\langle 4, 9, -3 \rangle \cdot \langle -6, 7, 5 \rangle$

17. $\langle 3, 1, 4 \rangle \cdot \langle 2, 8, -2 \rangle$ 18. $\langle -2, 4, 8 \rangle \cdot \langle 16, 4, 2 \rangle$ 19. $\langle 7, -2, 4 \rangle \cdot \langle 3, 8, 1 \rangle$
   6, no   **0, yes**   **9, no**

 20. Find the inner product of $\overline{\mathbf{a}}$ and $\overline{\mathbf{b}}$, $\overline{\mathbf{b}}$ and $\overline{\mathbf{c}}$, and $\overline{\mathbf{a}}$ and $\overline{\mathbf{c}}$ if $\overline{\mathbf{a}} = \langle 3, 12 \rangle$, $\overline{\mathbf{b}} = \langle 8, -2 \rangle$, and $\overline{\mathbf{c}} = \langle 3, -2 \rangle$. Are any of the pairs perpendicular? If so, which one(s)?   **See margin.**

Find each cross product. Then verify that the resulting vector is perpendicular to the given vectors.   21–26. See margin.

21. $\langle 0, 1, 2 \rangle \times \langle 1, 1, 4 \rangle$ 22. $\langle 5, 2, 3 \rangle \times \langle -2, 5, 0 \rangle$

23. $\langle 3, 2, 0 \rangle \times \langle 1, 4, 0 \rangle$ 24. $\langle 1, -3, 2 \rangle \times \langle 5, 1, -2 \rangle$

25. $\langle -3, -1, 2 \rangle \times \langle 4, -4, 0 \rangle$ 26. $\langle 4, 0, -2 \rangle \times \langle -7, 1, 0 \rangle$

27–28. See Answer Appendix.

27. Prove that for any vector $\overline{\mathbf{a}}$, $\overline{\mathbf{a}} \times (-\overline{\mathbf{a}}) = 0$.

28. Use the definition of cross products to prove that for any vectors $\overline{\mathbf{a}}$, $\overline{\mathbf{b}}$, and $\overline{\mathbf{c}}$, $\overline{\mathbf{a}} \times (\overline{\mathbf{b}} + \overline{\mathbf{c}}) = (\overline{\mathbf{a}} \times \overline{\mathbf{b}}) + (\overline{\mathbf{a}} \times \overline{\mathbf{c}})$.

Find a vector perpendicular to the plane containing the given points.

 29. $(0, -2, 2)$, $(1, 2, -3)$, and $(4, 0, -1)$ **Sample answer: $\langle -2, -17, -14 \rangle$**

30. $(-2, 1, 0)$, $(-3, 0, 0)$, and $(5, 2, 0)$ **Sample answer: $\langle 0, 0, 6 \rangle$**

31. $(0, 0, 1)$, $(1, 0, 1)$, and $(-1, -1, -1)$ **Sample answer: $\langle 0, 2, -1 \rangle$**

32. Explain whether the equation $\overline{\mathbf{m}} \times \overline{\mathbf{n}} = \overline{\mathbf{n}} \times \overline{\mathbf{m}}$ is true. Discuss your reasoning. **See margin.**

Applications and Problem Solving

33. **Physiology** Whenever we lift an object, a torque is applied by the biceps muscle on the lower arm. The elbow acts as the axis of rotation through the joint. Suppose the muscle is attached 4 centimeters from the joint and you exert a force of 600 N lifting an object 30° to the horizontal.
   a. Make a labeled diagram showing this situation.   **See margin.**
   b. What is the torque about the elbow?   **21 N-m**

www.amc.glencoe.com/self_check_quiz   *Lesson 8-4 Perpendicular Vectors* **509**

Additional Answers

20. $\overline{\mathbf{a}} \cdot \overline{\mathbf{b}} = 0$, yes
    $\overline{\mathbf{b}} \cdot \overline{\mathbf{c}} = 28$, no
    $\overline{\mathbf{a}} \cdot \overline{\mathbf{c}} = -15$, no

21. $\langle 2, 2, -1 \rangle$, yes
    $\langle 2, 2, -1 \rangle \cdot \langle 0, 1, 2 \rangle =$
    $2(0) + 2(1) + (-1)(2) =$
    $0 + 2 - 2 = 0$
    $\langle 2, 2, -1 \rangle \cdot \langle 1, 1, 4 \rangle =$
    $2(1) + 2(1) + (-1)(4) =$
    $2 + 2 - 4 = 0$

22. $\langle -15, -6, 29 \rangle$, yes
    $\langle -15, -6, 29 \rangle \cdot \langle 5, 2, 3 \rangle =$
    $(-15)(5) + (-6)(2) + 29(3) =$
    $-75 - 12 + 87 = 0$
    $\langle -15, -6, 29 \rangle \cdot \langle -2, 5, 0 \rangle =$
    $(-15)(-2) + (-6)(5) + 29(0) =$
    $30 - 30 + 0 = 0$

23. $\langle 0, 0, 10 \rangle$, yes
    $\langle 0, 0, 10 \rangle \cdot \langle 3, 2, 0 \rangle =$
    $0(3) + 0(2) + 10(0) =$
    $0 + 0 + 0 = 0$
    $\langle 0, 0, 10 \rangle \cdot \langle 1, 4, 0 \rangle =$
    $0(1) + 0(4) + 10(0) =$
    $0 + 0 + 0 = 0$

24. $\langle 4, 12, 16 \rangle$, yes
    $\langle 4, 12, 16 \rangle \cdot \langle 1, -3, 2 \rangle =$
    $4(1) + 12(-3) + 16(2) =$
    $4 - 36 + 32 = 0$
    $\langle 4, 12, 16 \rangle \cdot \langle 5, 1, -2 \rangle =$
    $4(5) + 12(1) + 16(-2) =$
    $20 + 12 - 32 = 0$

25. $\langle 8, 8, 16 \rangle$, yes
    $\langle 8, 8, 16 \rangle \cdot \langle -3, -1, 2 \rangle =$
    $8(-3) + 8(-1) + 16(2) =$
    $-24 - 8 + 32 = 0$
    $\langle 8, 8, 16 \rangle \cdot \langle 4, -4, 0 \rangle =$
    $8(4) + 8(-4) + 16(0) =$
    $32 - 32 + 0 = 0$

26. $\langle 2, 14, 4 \rangle$, yes
    $\langle 2, 14, 4 \rangle \cdot \langle 4, 0, -2 \rangle =$
    $2(4) + 14(0) + 4(-2) =$
    $8 + 0 - 8 = 0$
    $\langle 2, 14, 4 \rangle \cdot \langle -7, 1, 0 \rangle =$
    $2(-7) + 14(1) + 4(0) =$
    $-14 + 14 + 0 = 0$

Additional Answers

7. $\langle 13, 1, -5 \rangle$, yes
   $\langle 13, 1, -5 \rangle \cdot \langle 1, -3, 2 \rangle =$
   $13(1) + 1(-3) + (-5)(2) =$
   $13 - 3 - 10 = 0$
   $\langle 13, 1, -5 \rangle \cdot \langle -2, 1, -5 \rangle =$
   $13(-2) + 1(1) + (-5)(-5) =$
   $-26 + 1 + 25 = 0$

8. $\langle 8, -14, -2 \rangle$, yes
   $\langle 8, -14, -2 \rangle \cdot \langle 6, 2, 10 \rangle =$
   $8(6) + (-14)(2) + (-2)(10) =$
   $48 - 28 - 20 = 0$
   $\langle 8, -14, -2 \rangle \cdot \langle 4, 1, 9 \rangle =$
   $8(4) + (-14)(1) + (-2)(9) =$
   $32 - 14 - 18 = 0$

34. Critical Thinking Let $\overline{x} = \langle 2, 3, 0 \rangle$ and $\overline{y} = \langle -1, 1, 4 \rangle$. Find the area of the triangle whose vertices are the origin and the endpoints of $\overline{x}$ and $\overline{y}$. $\frac{1}{2}\sqrt{233}$

35. Business Management Mr. Toshiro manages a company that supplies a variety of domestic and imported nuts to supermarkets. He received an order for 120 bags of cashews, 310 bags of walnuts, and 60 bags of Brazil nuts. The prices per bag for each type are \$29, \$18, and \$21, respectively.

 a. Represent the number of bags ordered and the cost as vectors. **See margin.**

 b. Using what you have learned in the lesson about vectors, compute the value of the order. **\$10,320**

36. Physics The work done by a force $\overline{F}$, that displaces an object through a distance d is defined as $\overline{F} \cdot \overline{d}$. This can also be expressed as $|\overline{F}||\overline{d}| \cos \theta$. Alexa is pushing a construction barrel up a ramp 4 feet long into the back of a truck. She is using a force of 120 pounds in a horizontal direction and the ramp is 45° from the horizontal.

 a. Sketch a drawing of the situation. **See margin.**

 b. How much work is Alexa doing? **339 ft-lb**

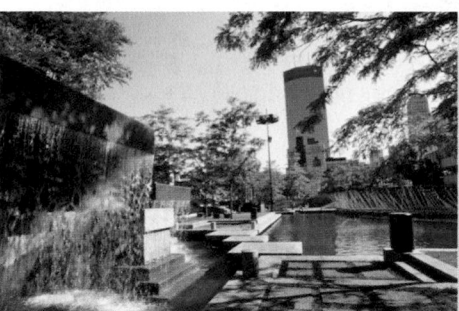

37. Architecture Steve Herr is an architect in Minneapolis, Minnesota. His latest project is designing a park. On the blueprint, the park is determined by a plane which contains the points at $(1, 0, 3)$, $(2, 5, 0)$, and $(3, 1, 4)$. One of the features of the park is a monument that must be perpendicular to the ground.

 a. Find a nonzero vector, representing the monument, perpendicular to the plane defined by the given points. $\langle 8, -7, -9 \rangle$

 b. Explain how you know that the vector is perpendicular to the plane defining the park. **See margin.**

38. Geometry A parallelepiped is a prism whose opposite faces are all parallelograms.

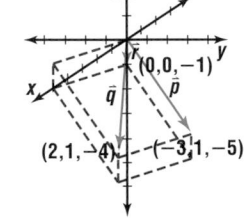

 a. Determine the volume of the parallelepiped using the expression $\overline{p} \cdot (\overline{q} \times \overline{r})$. **5 units³**

 b. Write a 3×3 matrix using the three vectors and calculate its determinant. How does this value compare to your answer in part a? **5 units³; They are the same.**

39. Critical Thinking If $\overline{v} = \langle 1, 2 \rangle$, $\overline{w} = \langle -1, 2 \rangle$, and $\overline{u} = \langle 5, 12 \rangle$, for what scalar k will the vector $k\overline{v} + \overline{w}$ be perpendicular to $\overline{u}$? $-\frac{19}{29}$

40. Proof Let $\overline{a} = \langle a_1, a_2 \rangle$ and $\overline{b} = \langle b_1, b_2 \rangle$. Use the Law of Cosines to show that if the measure of the angle between $\overline{a}$ and $\overline{b}$, θ, is any value, then the inner product $\overline{a} \cdot \overline{b} = |\overline{a}||\overline{b}| \cos \theta$. (*Hint:* Refer to the proof of inner product using the Pythagorean Theorem on page 505.) **See margin.**

Mixed Review

41. Given points $A(3, 3, -1)$ and $B(5, 3, 2)$, find an ordered triple that represents $\overrightarrow{AB}$. (Lesson 8-3) $\langle 2, 0, 3 \rangle$

42. Write the ordered pair that represents $\overrightarrow{DE}$ for the points on the graph. Then find the magnitude of $\overrightarrow{DE}$. (Lesson 8-2) $\langle -8, -5 \rangle$; $\sqrt{89}$

43. Write the equation $4x + y = 6$ in normal form. Then find p, the measure of the normal, and ϕ, the angle it makes with the positive x-axis. (Lesson 7-6)

44. Solve $\triangle ABC$ if $A = 36°$, $b = 13$, and $c = 6$. Round angle measures to the nearest minute and side measures to the nearest tenth. (Lesson 5-8)

45. Utilities A utility pole is braced by a cable attached to it at the top and anchored in a concrete block at ground level, a distance of 4 meters from the base of the pole. If the angle between the cable and the ground is 73°, find the height of the pole and the length of the cable to the nearest tenth of a meter. (Lesson 5-4)

4 m

46. Solve $3 + \sqrt{3x - 4} \geq 10$. (Lesson 4-7) $x \geq 17.67$

47. SAT/ACT Practice Let x be an integer greater than 1. What is the least value of x for which $a^2 = b^3 = x$ for some integers a and b? **B**

A 81 B 64 C 4 D 2 E 9

**Answers (left margin):**

43. $\dfrac{4\sqrt{17}}{17} x + \dfrac{\sqrt{17}}{17} y - \dfrac{6\sqrt{17}}{17} = 0$; $\dfrac{6\sqrt{17}}{17} \approx 1.46$ units; 14°

44. $a \approx 8.9$, $B = 59°25'$, $C = 84°35'$

45. 13.1 meters; 13.7 meters

MID-CHAPTER QUIZ

Use a ruler and a protractor to draw a vector with the given magnitude and direction. Then find the magnitude of the horizontal and vertical components of the vector. (Lesson 8-1) **1–2. See margin.**

1. 2.3 centimeters, 46°

2. 27 millimeters, 245°

Write the ordered pair or ordered triple that represents $\overrightarrow{CD}$. Then find the magnitude of $\overrightarrow{CD}$. (Lessons 8-2 and 8-3)

3. $C(-9, 2), D(-4, -3)$ $\langle 5, -5 \rangle, 5\sqrt{2}$

4. $C(3, 7, -1), D(5, 7, 2)$ $\langle 2, 0, 3 \rangle, \sqrt{13}$

Find an ordered pair or ordered triple to represent $\vec{r}$ in each equation if $\vec{s} = \langle 4, -3 \rangle$, $\vec{t} = \langle -6, 2 \rangle$, $\vec{u} = \langle 1, -3, -8 \rangle$, and $\vec{v} = \langle 3, 9, -1 \rangle$. (Lessons 8-2 and 8-3)

5. $\vec{r} = \vec{t} - 2\vec{s}$ $\langle -14, 8 \rangle$

6. $\vec{r} = 3\vec{u} + \vec{v}$ $\langle 6, 0, -25 \rangle$

Find each inner product and state whether the vectors are perpendicular. Write *yes* or *no*. (Lesson 8-4)

7. $\langle 3, 6 \rangle \cdot \langle -4, 2 \rangle$ **0, yes**

8. $\langle 3, -2, 4 \rangle \cdot \langle 1, -4, 0 \rangle$ **11, no**

9. Find the cross product $\langle 1, 3, 2 \rangle \times \langle 2, -1, -1 \rangle$. Then verify that the resulting vector is perpendicular to the given vectors. (Lesson 8-4) $\langle -1, 5, -7 \rangle$, **yes**

10. Entomology Suppose the flight of a housefly passed through points at $(2, 0, 4)$ and $(7, 4, 6)$, in which each unit represents a meter. What is the magnitude of the displacement the housefly experienced in traveling between these points? (Lesson 8-3) **6.7 m**

Extra Practice See p. A41.

Lesson 8-4 Perpendicular Vectors **511**

Extra Credit

When a femur is broken, sometimes traction is used to immobilize the bone. In a traction system, three forces are applied to the bone. When the femur is positioned at a 15° angle as shown, find the magnitude and direction of the resultant force, R. *Hint*: resolve each force into x- and y-components.
$R_x = 23.866$; $R_y = 8.414$; $R = $ about 25.31 lb

Open-Ended Assessment

Writing Have each student name a pair of ordered triples and then write a short explanation of how to find their inner and cross products.

Mid-Chapter Quiz

The Mid-Chapter Quiz provides students with a brief review of the concepts and skills in Lessons 8-1 through 8-4. Lesson numbers are given to the right of exercises or instruction lines so students can review concepts not yet mastered.

Additional Answers

Mid-Chapter Quiz

1.

2.3 cm 1.6 cm, 1.7 cm 46°

2.

11.4 mm, 24.5 mm 115° 2.7 cm

Chapter 8, Quiz B (Lessons 8-3 and 8-4) is available in the *Assessment and Evaluation Masters*, p. 199.

Chapter 8 Mid-Chapter Test (Lessons 8-1 through 8-4) is available in the *Assessment and Evaluation Masters*, p. 198.

Enrichment Masters, p. 88

1 FOCUS

Motivating the Lesson

Discussion Have students try to find the cross product of two vectors with very large positive or negative components, such as $\vec{u} \times \vec{v}$ if $\vec{u} = \langle 220, 34, 150 \rangle$ and $\vec{v} = \langle -25, 35, 410 \rangle$. Have students discuss the large calculations involved and why a graphing calculator program would be useful.

2 TEACH

A reproduction of this lesson with graphing calculator keystrokes provided can be found in the *Graphing Calculator Keystrokes Masters,* pp. 31–32.

Teaching Tip Remind students that this activity is not an efficient way to do cross products in two-dimensions, but it is an excellent way to reduce errors involved in calculating three-dimensional cross products.

3 PRACTICE/APPLY

Assignment Guide
All: 1–9

4 ASSESS

Open-Ended Assessment

Speaking Have students find the cross product given in Motivating the Lesson by using the graphing calculator program. Ask students working in small groups to compare and contrast the time needed to solve the problem using a calculator versus using paper and pencil.

512 *Chapter 8*

GRAPHING CALCULATOR EXPLORATION

8-4B Finding Cross Products

An Extension of Lesson 8-4

OBJECTIVE
• Use a graphing calculator program to obtain the components of the cross product of two vectors in space.

The following graphing calculator program allows you to input the components of two vectors $\vec{p} = \langle a, b, c \rangle$ and $\vec{q} = \langle x, y, z \rangle$ in space and obtain the components of their cross product.

```
PROGRAM: CROSSP
:Disp "SPECIFY (A, B, C)"
:Input A
:Input B
:Input C
:Disp "SPECIFY (X, Y, Z)"
:Input X
:Input Y
:Input Z
:Disp "CROSS PROD OF"
:Disp "(A, B, C) AND"
:Disp "(X, Y, Z) IS"
:Disp BZ - CY
:Disp CX - AZ
:Disp AY - BX
:Stop
```

interNET
CONNECTION

Graphing Calculator Programs
To download this program, visit
www.amc. glencoe.com

Enter this program on your calculator and use it as needed to complete the following exercises.

TRY THESE Use a graphing calculator to determine each cross product.

1. $\langle 7, 9, -1 \rangle \times \langle 3, -4, -5 \rangle$ $\langle -49, 32 -55 \rangle$ 2. $\langle 8, 14, 0 \rangle \times \langle -2, 6, 12 \rangle$ $\langle 168, -96, 76 \rangle$

3. $\langle 5, 5, 5 \rangle \times \langle 4, 4, 4 \rangle$ $\langle 0, 0, 0 \rangle$

4. $\langle -3, 2 -1 \rangle \times \langle 6, -3, 7 \rangle$ $\langle 11, 15, -3 \rangle$

5. $\langle 1, 6, 0 \rangle \times \langle 2, 5, 0 \rangle$ $\langle 0, 0, -7 \rangle$

6. $\langle 4, 0, -2 \rangle \times \langle 6, 0, -13 \rangle$ $\langle 0, 40, 0 \rangle$

It can be proved that if $\vec{u}$ and $\vec{v}$ are adjacent sides of a parallelogram, then the area of the parallelogram is $|\vec{u} \times \vec{v}|$. Find the area of the parallelogram having the given vectors as adjacent sides.

7. $\vec{u} = \langle -2, 4, 1 \rangle, \vec{v} = \langle 0, 6, 3 \rangle$ $\sqrt{216}$

8. $\vec{u} = \langle 1, 3, 2 \rangle, \vec{v} = \langle 9, 7, 5 \rangle$ $\sqrt{570}$

WHAT DO YOU THINK?

9. What instructions could you insert at the end of the program to have the program display the magnitude of the vectors resulting from the cross products of the vectors above? **See margin.**

512 *Chapter 8 Vectors and Parametric Equations*

Additional Answer

9. Sample answer: insert the following lines after the last line of the given program.

:Disp "LENGTH IS"
:Disp $\sqrt{((BZ - CY)^2 + (CX - AZ)^2 + (AY - BX)^2)}$

Applications with Vectors

OBJECTIVE
- Solve problems using vectors and right triangle trigonometry.

SPORTS Ty Murray was a recent World Championship Bull Rider in the National Finals Rodeo in Las Vegas, Nevada. One of the most dangerous tasks, besides riding the bulls, is the job of getting the bulls back into their pens after each event. Experienced handlers, dressed as clowns, expertly rope the animals without harming them.

Ty Murray has completed his competition ride and the two rodeo clowns are restraining the bull to return it to the paddocks. Suppose one clown is exerting a force of 270 newtons due north and the other is pulling with a force of 360 newtons due east. What is the resultant force on the bull? *This problem will be solved in Example 1.*

Vectors can represent the forces exerted by the handlers on the bull. Physicists resolve this force into component vectors. Vectors can be used to represent any quantity, such as a force, that has magnitude and direction. Velocity, weight, and gravity are some of the quantities that physicists represent with vectors.

Example

1 **SPORTS** Use the information above to describe the resultant force on the bull.

a. **Draw a labeled diagram that represents the forces.**

Let $\vec{F}_1$ and $\vec{F}_2$ represent the forces exerted by the clowns. Then $\vec{F}$ represents the resultant. Let θ represent the angle $\vec{F}$ makes with the east-west or x-axis.

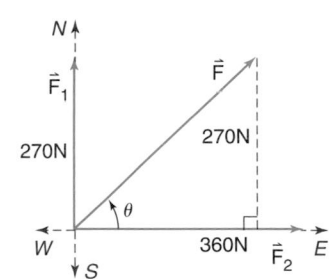

b. **Determine the resultant force exerted on the bull by the two clowns.**

$$|\vec{F}|^2 = |\vec{F}_1|^2 + |\vec{F}_2|^2$$
$$|\vec{F}|^2 = (270)^2 + (360)^2$$
$$|\vec{F}|^2 = 202{,}500$$
$$\sqrt{|\vec{F}|^2} = \sqrt{202{,}500} \text{ or } 450$$

The resultant force on the bull is 450 newtons.

1 pound ≈ 4.45 newtons so 450 N ≈ 101.12 lb

---

1 FOCUS

5-Minute Check
Lesson 8-4

Find each inner product and state whether the vectors are perpendicular.

1. $\langle 3, 0 \rangle \cdot \langle 4, 7 \rangle$ **12; no**

2. $\langle 5, 1, -2 \rangle \cdot \langle 2, 0, 5 \rangle$ **0; yes**

3. Find the cross product of $\vec{u}$ and $\vec{v}$ if $\vec{u} = \langle 0, -2, -1 \rangle$ and $\vec{v} = \langle 0, 6, 5 \rangle$. Verify that the resulting vector is perpendicular to $\vec{u}$ and $\vec{v}$.
$\langle -4, 0, 0 \rangle$; **yes, since** $\langle 0, -2, -1 \rangle \cdot \langle -4, 0, 0 \rangle$ **and** $\langle 0, 6, 5 \rangle \cdot \langle -4, 0, 0 \rangle$ **both equal 0.**

4. Suppose that the driver of an ambulance is applying a force of 30 pounds along the positive z-axis to the gearshift of his ambulance. If the center of the connection is at the point Y, the force is applied at the point (0.68, 0, 0.31). Find the torque. **20.4 ft-lb**

Motivating the Lesson
Real-World Application Have the class visualize two persons tugging at a rope in opposite directions in a straight line. Ask the class to discuss what the resultant force would be. Then, ask the class to consider the resultant force if the persons were to tug at a rope around a steel vertical pipe with the rope and the pipe forming a right angle.

Resource Manager

 Reproducible Masters
- *Study Guide and Practice,* pp. 145–146
- *Enrichment,* p. 89
- *Problem Solving and Applications,* p. 15

 Transparencies
- 5-Minute Check, 8-5
- Teaching, 8-5

Teaching Tip In all of the examples, it is important that students begin with a careful diagram. Urge students to check their answers with the diagram. This method often can prevent any errors.

In-Class Examples

Example 1
Sports Lenny Montana was a former 5-year world heavyweight wrestling champion and actor. Suppose that Lenny and a tag team wrestling partner are each pulling horizontally and at a right angle to each other on the arms of a wrestling opponent. Lenny exerts a force of 180 pounds due north while his partner exerts a force of 125 pounds due east.

a. Draw a labeled diagram that represents the forces.

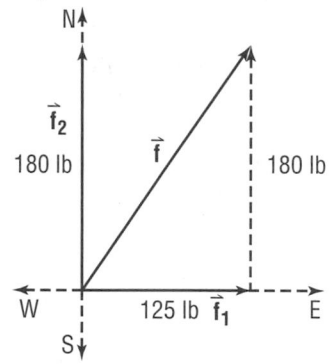

b. Determine the resultant force exerted on the opponent. **219 lb**

c. Determine the angle the resultant force makes with the east-west axis. **55.2° north of east**

Example 2
Justin works for a package delivery service. Suppose that he is pushing a cart full of packages weighing 125 pounds up a ramp 10 feet long at an incline of 20°. Find the work done by gravity as the cart moves the length of the ramp. Assume that friction is not a factor. **−427.5 ft-lb**

---

c. **Find the angle the resultant force makes with the east-west axis.**

Use the tangent ratio.

$$\tan \theta = \frac{270}{360}$$

$$\theta = \tan^{-1} \frac{270}{360}$$

$$\theta \approx 36.9° \text{ north of due east}$$

The resultant force is applied at an angle of 36.9° north of due east.

In physics, if a constant force $\vec{F}$ displaces an object an amount represented by $\vec{d}$, it does *work* on the object. The amount of work is given by $W = \vec{F} \cdot \vec{d}$.

Example **2** **Alvaro works for a package delivery service. Suppose he is pushing a cart full of packages weighing 100 pounds up a ramp 8 feet long at an incline of 25°. Find the work done by gravity as the cart moves the length of the ramp. Assume that friction is not a factor.**

First draw a labeled diagram representing the forces involved. Let $\overline{OQ}$ represent the force of gravity, or weight. The weight has a magnitude of 100 pounds and its direction is down. The corresponding unit vector is $0\vec{i} - 100\vec{j}$. So, $\vec{F} = 0\vec{i} - 100\vec{j}$. The application of the force is $\overline{OP}$, and it has a magnitude of 8 feet.

Write $\overline{OP}$ as $\vec{d} = x\vec{i} + y\vec{j}$ and use trigonometry to find x and y.

$$\cos 25° = \frac{x}{8} \qquad\qquad \sin 25° = \frac{y}{8}$$
$$x = 8 \cos 25° \qquad\qquad y = 8 \sin 25°$$
$$x \approx 7.25 \qquad\qquad y \approx 3.38$$

Then, $\vec{d} = 7.25\vec{i} + 3.38\vec{j}$

Apply the formula for determining the work done by gravity.

$$W = \vec{F} \cdot \vec{d}$$
$$W = (0\vec{i} - 100\vec{j}) \cdot (7.25\vec{i} + 3.38\vec{j})$$
$$W = 0 - 338 \text{ or } -338$$

Work done by gravity is negative when an object is lifted or raised. As the cart moves the length of the ramp, the work done by gravity is −338 ft-lb.

Sometimes there is no motion when several forces are at work on an object. This situation, when the forces balance one another, is called *equilibrium*. Recall that an object is in equilibrium if the resultant force on it is zero.

Example

❸ Ms. Davis is hanging a sign for her restaurant. The sign is supported by two lightweight support bars as shown in the diagram. If the bars make a 30° angle with each other, and the sign weighs 200 pounds, what are the magnitudes of the forces exerted by the sign on each support bar?

$\vec{F_1}$ represents the force exerted on bar 1 by the sign, $\vec{F_2}$ represents the force exerted on bar 2 by the sign, and $\vec{F_w}$ represents the weight of the sign.

Remember that equal vectors have the same magnitude and direction. So by drawing another vector from the initial point of $\vec{F_1}$ to the terminal point of $\vec{F_w}$, we can use the sine and cosine ratios to determine $|\vec{F_1}|$ and $|\vec{F_2}|$.

$$\sin 30° = \frac{200}{|\vec{F_2}|}$$

$$|\vec{F_2}| = \frac{200}{\sin 30°} \quad \textit{Divide each side by sin 30° and multiply each side by } |\vec{F_2}|.$$

$$|\vec{F_2}| = 400$$

Likewise, $|\vec{F_1}|$ can be determined by using cos 30°.

$$\cos 30° = \frac{|\vec{F_1}|}{400}$$

$$|\vec{F_1}| = 400 \cos 30°$$

$$|\vec{F_1}| \approx 346$$

The sign exerts a force of about 346 pounds on bar 1 and a force of 400 pounds on bar 2.

Teaching Tip Review tip-to-tail and point out that when you form a polygon, such as the triangle in Example 3, the last force drawn to form the polygon puts the system in equilibrium.

In-Class Example

Example 3
Dr. Masucci is hanging a sign for her medical practice. The sign is held by two support bars as shown in the figure. If the bars make a 60° angle with each other and the sign weighs 100 pounds, what are the magnitudes of the forces exerted by the sign on each support bar?

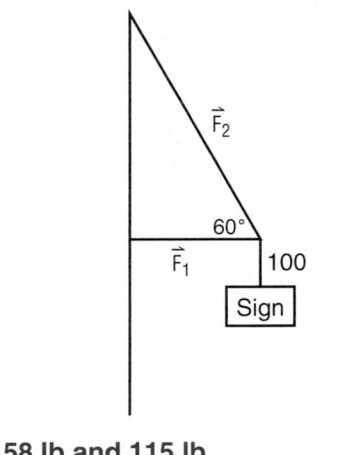

58 lb and 115 lb

Teaching Tip
For Example 4, remind students that they can review the Law of Sines in Lesson 5-6.

In-Class Example
Example 4
A large television screen inside a restaurant is supported equally on each of its sides by two cables suspended from the ceiling of the restaurant. The cables form a 100° angle with each other. If the screen weighs 800 pounds, what is the force exerted by each of the cables on the screen? **622 lb**

3 PRACTICE/APPLY

Error Analysis
Watch for students who do not make an accurate drawing of the problem in Exercise 9.

Prevent by teaching students to carefully read and label their drawings, then to reread the problem in order to check the accuracy of their drawings.

Assignment Guide

Core: 11–23 odd, 25–27, 29–31, 33–38
Enriched: 10–24 even, 25–38

Study Guide and Practice Masters, pp. 145–146

If the vectors representing forces at equilibrium are drawn tip-to-tail, they will form a polygon.

Example **4** A lighting system for a theater is supported equally by two cables suspended from the ceiling of the theater. The cables form a 140° angle with each other. If the lighting system weighs 950 pounds, what is the force exerted by each of the cables on the lighting system?

Draw a diagram of the situation. Then draw the vectors tip-to-tail.

Look Back
Refer to Lessons 5-6 through 5-8 to review Law of Sines and Law of Cosines.

Since the triangle is isosceles, the base angles are congruent. Thus, each base angle measures $\frac{180° - 40°}{2}$ or 70°. We can use the Law of Sines to find the force exerted by the cables.

$$\frac{950}{\sin 40°} = \frac{x}{\sin 70°} \quad \textit{Law of Sines}$$

$$x = \frac{950 \sin 70°}{\sin 40°}$$

$$x \approx 1388.81$$

The force exerted by each cable is about 1389 pounds.

CHECK FOR UNDERSTANDING

Communicating Mathematics

Read and study the lesson to answer each question.   **1–3. See margin.**

1. **Determine** which would require more force—pushing an object up to the top of an incline, or lifting it straight up. Assume there is no friction.

2. **Describe** how the force exerted by the cables in Example 4 is affected if the angle between the cables is increased.

3. **Explain** what it means for forces to be in equilibrium.

Guided Practice

4. Make a sketch to show the forces acting on a ship traveling at 23 knots at an angle of 17° with the current.   **See margin.**

5. Find the magnitude and direction of the resultant vector for the diagram. **421.19 N, 19.3°**

6. A 100-newton force and a 50-newton force act on the same object. The angle between the forces measures 90°. Find the magnitude of the resultant force and the angle between the resultant force and the 50-pound force.   **111.8 N, 63.43°**

Addressing Individual Needs

Kinesthetic An easy downward force vector to measure is one's weight. Have students find their weights, then think about and answer this question. *What part does gravity play in determining your weight?*
Sample answer: your weight is defined as the downward force due to gravity. Have students research the effect of gravity on the moon or other planets. Have them use the Internet to find their weights on other worlds.

7. Denzel pulls a wagon along level ground with a force of 18 newtons on the handle. If the handle makes an angle of 40° with the horizontal, find the horizontal and vertical components of the force. **13.79 N, 11.57 N**

8. A 33-newton force at 90° and a 44-newton force at 60° are exerted on an object. What is the magnitude and direction of a third force that produces equilibrium on the object? **74 N, 253°**

9. Transportation Two ferry landings are directly across a river from each other. A ferry that can travel at a speed of 12 miles per hour in still water is attempting to cross directly from one landing to the other. The current of the river is 4 miles per hour.

   a. Make a sketch of the situation. **See margin.**

   b. If a heading of 0° represents the line between the two landings, at what angle should the ferry's captain head? **≈ 19.5°**

EXERCISES

Practice

Make a sketch to show the given vectors. **10–12. See margin.**

A **10.** a force of 42 newtons acting on an object at an angle of 53° with the ground

11. an airplane traveling at 256 miles per hour at an angle of 27° from the wind

12. a force of 342 pounds acting on an object while a force of 454 pounds acts on the same object an angle of 94° with the first force

Find the magnitude and direction of the resultant vector for each diagram.

13.

390 N
425 N

576.82 N, 42.5°

14.

65 mph
300°
50 mph

99.87 mph, 334.3°

15.

115 km/h   115 km/h
120°
60°

199.19 km/h, 90°

B **16.** What would be the force required to push a 100-pound object along a ramp that is inclined 10° with the horizontal? **at least 17.36 lb**

17. What is the magnitude and direction of the resultant of a 105-newton force along the x-axis and a 110-newton force at an angle of 50° to one another?
194.87 N, 25.62°

18. To keep a 75-pound block from sliding down an incline, a 52.1-pound force is exerted on the block along the incline. Find the angle that the incline makes with the horizontal. **44°**

19. 220.5 lb, 16.7°

19. Find the magnitude and direction of the resultant of two forces of 250 pounds and 45 pounds at angles of 25° and 250° with the x-axis, respectively.

20. Three forces with magnitudes of 70 pounds, 40 pounds, and 60 pounds act on an object at angles 330°, 45°, and 135°, respectively, with the positive x-axis. Find the direction and magnitude of the resultant of these forces. **37.5°, 58.6 lb**

21. A 23-newton force acting at 60° above the horizontal and a second 23-newton force acting at 120° above the horizontal act concurrently on a point. What is the magnitude and direction of a third force that produces equilibrium? **39.8 N, 270°**

 www.amc.glencoe.com/self_check_quiz

1. Sample answer: Pushing an object up the slope requires less force because the component of the weight of the object in the direction of motion is *mg* sin θ. This is less than the weight *mg* of the object, which is the force that must be exerted to lift the object straight up.

2. The tension increases.

3. Sample answer: Forces are in equilibrium if the resultant force is $\vec{0}$.

4.

Current
17°
23 knots

9a.

4 mph
12 mph

10.

42 N
53°

11.

Wind
27°
256 mph

12.

342 lb
94°
454 lb

C

22. An object is placed on a ramp and slides to the ground. If the ramp makes an angle of 40° with the ground and the object weighs 25 pounds, find the acceleration of the object. Assume that there is no friction.   **20.6 ft/s²**

23. A force of 36 newtons pulls an object at an angle of 20° north of due east. A second force pulls on the object with a force of 48 newtons at an angle of 42° south of due west. Find the magnitude and direction of the resultant force.
19.9 N, 5.3° west of south

24. Three forces in a plane act on an object. The forces are 70 pounds, 115 pounds and 135 pounds. The 70 pound force is exerted along the positive *x*-axis. The 115 pound force is applied below the *x*-axis at a 120° angle with the 70 pound force. The angle between the 115-pound and 135-pound forces is 75°, and between the 135-pound and 70-pound forces is 165°.

   a. Make a diagram showing the forces.   **See margin.**

   b. Are the vectors in equilibrium? If not, find the magnitude and the direction of the resultant force.   **no; 134.5 lb, 28.7° south of west**

Applications and Problem Solving

Real World Application

25. **Physics**   While pulling a stalled car, a tow truck's cable makes an angle of 50° above the road. If the tension on the tow truck's cable is 1600 newtons, how much work is done by the truck on the car pulling it 1.5 kilometers down the road?   **1,542,690 N-m**

26. **Critical Thinking**   The handle of a lawnmower you are pushing makes an angle of 60° with the ground.   **a–b. See margin.**

   a. How could you increase the horizontal forward force you are applying without increasing the total force?

   b. What are some of the disadvantages of doing this?

27. **Boating**   A sailboat is headed east at 18 mph relative to the water. A current is moving the water south at 3 mph.

   a. What is the angle of the path of the sailboat?   **9.5° south of east**

   b. What is the sailboat's speed with respect to the ocean floor?   **18.2 mph**

28. **Physics**   Suzanne is pulling a wagon loaded with gardening bricks totaling 100 kilograms. She is applying a force of 100 newtons on the handle at 25° with the ground. What is the horizontal force on the wagon?   **90.63 N**

29. **Entertainment**   A unicyclist is performing on a tightrope at a circus. The total weight of the performer and her unicycle is 155 pounds. How much tension is being exerted on each part of the cable?   **left side: 760 lb; right side: 761 lb**

30. **Critical Thinking** Chaz is using a rope tied to his tractor to remove an old tree stump from a field. Which method given below—a or b—will result in the greatest force applied to the stump? Assume that the tractor will exert the same amount of force using either method. Explain your answer. **a–b. See margin.**

   a. Tie the rope to the stump and pull.

   b. Tie one end to the stump and the other end to a nearby pole. Then pull on the rope perpendicular to it at a point about halfway between the two.

31. **Travel** A cruise ship is arriving at the port of Miami from the Bahamas. Two tugboats are towing the ship to the dock. They exert a force of 6000 tons along the axis of the ship. Find the tension in the towlines if each tugboat makes a 20° angle with the axis of the ship. **3192.5 tons**

32. **Physics** A painting weighing 25 pounds is supported at the top corners by a taut wire that passes around a nail embedded in the wall. The wire forms a 120° angle at the nail. Assuming that the wire rests on the nail at its midpoint, find the pull on the wires. **25 lb**

Mixed Review

33. Find the inner product of $\vec{u}$ and $\vec{v}$ if $\vec{u} = \langle 9, 5, 3 \rangle$ and $\vec{v} = \langle -3, 2, 5 \rangle$. Are the vectors perpendicular? *(Lesson 8-4)* **−2, no**

34. If $A = (12, -5, 18)$ and $B = (0, -11, 21)$, write the ordered triple that represents $\overline{AB}$. *(Lesson 8-3)* **⟨−12, −6, 3⟩**

35. **Sports** Sybrina Floyd hit a golf ball on an approach shot with an initial velocity of 100 feet per second. The distance a golf ball travels is found by the formula $d = \dfrac{2v_0^2}{g} \sin \theta \cos \theta$, where v_0 is the initial velocity, g is the acceleration due to gravity, and θ is the measure of the angle that the initial path of the ball makes with the ground. Find the distance Ms. Floyd's ball traveled if the measure of the angle between the initial path of the ball and the ground is 65° and the acceleration due to gravity is 32 ft/s². *(Lesson 7-4)* **239.4 ft**

36. Sample answer: $y = 1.4x^2 - 2x + 3.9$

36. Write a polynomial function to model the set of data. *(Lesson 4-8)*

x	−1.5	−1	−0.5	0	0.5	1	1.5	2	2.5
f(x)	9	8	6.5	4.3	2	2.7	4.1	6.1	7.8

37. 30% beef, 20% pork; $76

37. **Food Processing** A meat packer makes a kind of sausage using beef, pork, cereal, fat, water, and spices. The minimum cereal content is 12%, the minimum fat content is 15%, the minimum water content is 6.5%, and the spices are 0.5%. The remaining ingredients are beef and pork. There must be at least 30% beef content and at least 20% pork content for texture. The beef content must equal or exceed the pork content. The cost of all of the ingredients except beef and pork is $32 per 100 pounds. Beef can be purchased for $140 per 100 pounds and pork for $90 per 100 pounds. Find the combination of beef and pork for the minimum cost. What is the minimum cost per 100 pounds? *(Lesson 2-6)*

38. **SAT/ACT Practice** Let $*x$ be defined as $*x = x^3 - x$. What is the value of $*4 - *(-3)$? **A**

   A 84 B 55 C −10

   D 22 E 4

Extra Credit

A farmer and a neighbor are removing a large boulder from a field. They are pulling on ropes attached to the boulder. The ropes make a 35° angle with each other. If the farmer is pulling with a force of 105 newtons and the neighbor is pulling with a force of 95 newtons, what is the magnitude and direction of force on the boulder? **190.8 newtons at an angle of 16.6° with the 105 newton force**

Vectors and Parametric Equations

5-Minute Check
Lesson 8-5

Two forces perpendicular to each other act on a point. One force is 40 newtons and the other is 50 newtons.

1. Draw a sketch and show the resultant force.

50 N

40 N

2. Find the magnitude of the resultant vector. **64 N**

3. Find the direction of the resultant vector. **51.3°**

4. A chandelier in a symphony hall is supported by two cables suspended from the ceiling. The cables form a 158° angle with each other. If the chandelier weighs 1200 pounds, what is the force exerted by each of the cables on the chandelier? **about 3144.5 lb**

Motivating the Lesson

Hands-On Activity Have students work in pairs to analyze the flight of a flying disk. Determine a starting point for the launch. One student throws the disk, and the other times how long it takes for the disk to hit the ground. Parametric equations can be used to show the disk's movement given its starting position and velocity.

Teaching Tip Remind students that *t* is a scalar that is also called the parameter.

OBJECTIVES
- Write vector and parametric equations of lines.
- Graph parametric equations.

AIR RACING Air racing's origins date back to the early 1900s when aviation began. One of the most important races is the National Championship Air Race held every year in Reno, Nevada. In the competition, suppose pilot Bob Hannah passes the starting gate at a speed of 410.7 mph. Fifteen seconds later, Dennis Sanders flies by at 412.9 mph. They are racing for the next 500 miles. *This information will be used in Example 4.*

The relative positions of the airplanes are dependent upon their velocities at a given moment in time and their starting positions. **Vector equations** and equations known as **parametric equations** allow us to model that movement.

In Lesson 1-4, you learned how to write the equation of a line in the coordinate plane. For objects that are moving in a straight line, there is a more useful way of writing equations for the line describing the object's path using vectors.

If a line passes through the points P_1 and P_2 and is parallel to the vector $\overline{a} = \langle a_1, a_2 \rangle$, the vector $\overline{P_1P_2}$ is also parallel to $\overline{a}$. Thus, $\overline{P_1P_2}$ must be a scalar multiple of $\overline{a}$. Using the scalar t, we can write the equation $\overline{P_1P_2} = t\overline{a}$. Notice that both sides of the equation are vectors. This is called the vector equation of the line. Since $\overline{a}$ is parallel to the line, it is called a **direction vector**. The scalar t is called a **parameter**.

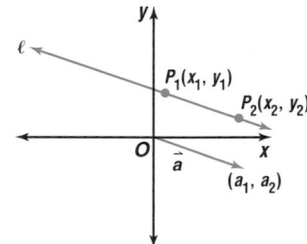

Vector Equation of a Line	A line through $P_1(x_1, y_1)$ parallel to the vector $\overline{a} = \langle a_1, a_2 \rangle$ is defined by the set of points $P_1(x_1, y_1)$ and $P_2(x_2, y_2)$ such that $\overline{P_1P_2} = t\overline{a}$ for some real number t. Therefore, $\langle x_2 - x_1, y_2 - y_1 \rangle = t\langle a_1, a_2 \rangle$.

In Lesson 8-2, you learned that the vector from (x_1, y_1) to (x, y) is $\langle x - x_1, y - y_1 \rangle$.

Example **1** Write a vector equation describing a line passing through $P_1(1, 4)$ and parallel to $\overline{a} = \langle 3, -2 \rangle$.

Let the line ℓ through $P_1(1, 4)$ be parallel to $\overline{a}$. For any point $P(x, y)$ on ℓ, $\overline{P_1P} = \langle x - 1, y - 4 \rangle$. Since P_1P is on ℓ and is parallel to $\overline{a}$, $\overline{P_1P} = t\overline{a}$, for some value t. By substitution, we have $\langle x - 1, y - 4 \rangle = t\langle 3, -2 \rangle$.

Therefore, the equation $\langle x - 1, y - 4 \rangle = t\langle 3, -2 \rangle$ is a vector equation describing all of the points (x, y) on ℓ parallel to $\overline{a}$ through $P_1(1, 4)$.

520 Chapter 8 *Vectors and Parametric Equations*

Resource Manager

 Reproducible Masters
- *Study Guide and Practice,* pp. 147–148
- *Enrichment,* p. 90
- *Assessment and Evaluation,* p. 200

 Transparencies
- 5-Minute Check, 8-6
- Teaching, 8-6

A vector equation like the one in Example 1 can be used to describe the coordinates for a point on the line for any value of the parameter t. For example, when $t = 4$ we can write the equation $\langle x - 1, y - 4\rangle = 4\langle 3, -2\rangle$ or $\langle 12, -8\rangle$. Then write the equation $x - 1 = 12$ and $y - 4 = -8$ to find the ordered pair $(13, -4)$. Likewise, when $t = 0$, the ordered pair $(1, 4)$ results. The parameter t often represents time. (In fact, that is the reason for the choice of the letter t.) An object moving along the line described by the vector equation $\langle x - 1, y - 4\rangle = t\langle 3, 2\rangle$ will be at the point $(1, 4)$ at time $t = 0$ and will be at the point $(13, -4)$ at time $t = 4$.

As you have seen, the vector equation $\langle x - x_1, y - y_1\rangle = t\langle a_1, a_2\rangle$ can be written as two equations relating the horizontal and vertical components of these two vectors separately.

$$x - x_1 = ta_1 \qquad\qquad y - y_1 = ta_2$$
$$x = x_1 + ta_1 \qquad\qquad y = y_1 + ta_2$$

The resulting equations, $x = x_1 + ta_1$ and $y = y_1 + ta_2$, are known as parametric equations of the line through $P_1(x_1, y_1)$ parallel to $\overline{a} = \langle a_1, a_2\rangle$.

Parametric Equations of a Line	A line through $P_1(x_1, y_1)$ that is parallel to the vector $\overline{a} = \langle a_1, a_2\rangle$ has the following parametric equations, where t is any real number. $x = x_1 + ta_1$ $y = y_1 + ta_2$

If we know the coordinates of a point on a line and its direction vector, we can write its parametric equations.

Example **2** Find the parametric equations for a line parallel to $\overline{q} = \langle 6, -3\rangle$ and passing through the point at $(-2, -4)$. Then make a table of values and graph the line.

Use the general form of the parametric equations of a line with $\langle a_1, a_2\rangle = \langle 6, -3\rangle$ and $\langle x_1, y_1\rangle = \langle -2, -4\rangle$.

$x = x_1 + ta_1 \qquad\qquad y = y_1 + ta_2$
$x = -2 + t(6) \qquad\qquad y = -4 + t(-3)$
$x = -2 + 6t \qquad\qquad y = -4 - 3t$

Now make a table of values for t. Evaluate each expression to find values for x and y. Then graph the line.

t	x	y
-1	-8	-1
0	-2	-4
1	4	-7
2	10	-10

Lesson 8-6 Vectors and Parametric Equations **521**

Prerequisite Skills

A review of the prerequisite skills for this lesson is provided in the *Prerequisite Skills Masters.*

Skill Number and Content	Pages
5 Solving Equations & Inequalities	11–12

2 TEACH

In-Class Examples

Example 1
Write a vector equation describing a line passing through $P_1(3, -9)$ and parallel to $\overline{a} = \langle -1, 2\rangle$.
$\langle x - 3, y + 9\rangle = t\langle -1, 2\rangle$

Example 2
Find the parametric equations of a line parallel to $\overline{b} = \langle 3, -5\rangle$ and passing through the point at $(4, 3)$. Then make a table of values and graph the line.
$x = 4 + 3t, y = 3 - 5t$

t	x	y
-1	1	8
0	4	3
1	7	-2
2	10	-7

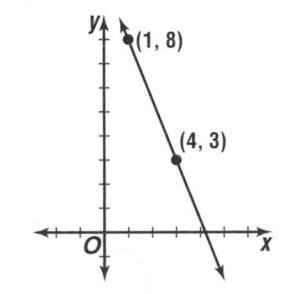

Teaching Tip For Example 2, remind students to substitute the data carefully in each of the parametric equations of a line.

Notice in Example 2 that each value of *t* establishes an ordered pair (*x, y*) whose graph is a point. As you have seen, these points can be considered the position of an object at various times *t*. Evaluating the parametric equations for a value of *t* gives us the coordinates of the position of the object after *t* units of time.

If the slope-intercept form of the equation of a line is given, we can write parametric equations for that line.

Example **3** Write parametric equations of $y = -4x + 7$.

In the equation $y = -4x + 7$, *x* is the independent variable, and *y* is the dependent variable. In parametric equations, *t* is the independent variable, and *x* and *y* are dependent variables. If we set the independent variables *x* and *t* equal, we can write two parametric equations in terms of *t*.

$x = t$
$y = -4t + 7$

Parametric equations for the line are
$x = t$ and $y = -4t + 7$.

By making a table of values for *t* and evaluating each expression to find values for *x* and *y* and graphing the line, the parametric equations $x = t$ and $y = -4t + 7$ describe the same line as $y = -4x + 7$.

We can use vector equations and parametric equations to model physical situations, such as the National Championship Air Race, where *t* represents time in hours.

Example **4** **AIR RACING**   Refer to the application at the beginning of the lesson. Use parametric equations to model the situation. Assume that both planes maintain a constant speed.

a. How long is it until the second plane overtakes the first?

b. How far have both planes traveled when the second plane overtakes the first?

a. First, write a set of parametric equations to represent each airplane's position at *t* hours.

Airplane 1: $x = 410.7t$ $x = vt$
Airplane 2: $x = 412.9(t - 0.0042)$ *15 s ≈ 0.0042 h*

Since the time at which the second plane overtakes the first is when they have traveled the same distance, set the two expressions for *x* equal to each other.

$$410.7t \approx 412.9(t - 0.0042)$$
$$410.7t \approx 412.9t - 1.73418$$
$$1.73418 \approx 2.2t$$
$$0.788 \approx t$$

In about 0.788 hour or 47 minutes, the second plane overtakes the first.

b. Use the time to find the distance traveled when the planes pass.

$x = 410.7t$

Since the speeds are given in tenths
$x = 410.7(0.788)$
of mph, we round the final answer
$x = 323.6316$ or about 323.6 miles
to the nearest tenth.

The planes have traveled about 323.6 miles when the second plane overtakes the first plane.

We can also write the equation of a line in slope-intercept or standard form if we are given the parametric equations of the line.

Example **5** Write an equation in slope-intercept form of the line whose parametric equations are $x = -5 + 4t$ and $y = 2 - 3t$.

Solve each parametric equation for t.

$x = -5 + 4t$ $\qquad\qquad$ $y = 2 - 3t$

$x + 5 = 4t$ $\qquad\qquad$ $y - 2 = -3t$

$\dfrac{x + 5}{4} = t$ $\qquad\qquad$ $\dfrac{y - 2}{3} = t$

Use substitution to write an equation for the line without the variable t.

$\dfrac{x+5}{4} = \dfrac{(y-2)}{-3}$ $\qquad$ *Substitution*

$(x + 5)(-3) = 4(y - 2)$ $\qquad$ *Cross multiply.*

$-3x - 15 = 4y - 8$ $\qquad$ *Simplify.*

$y = -\dfrac{3}{4}x - \dfrac{7}{4}$ $\qquad$ *Solve for y.*

CHECK FOR UNDERSTANDING

Communicating Mathematics

Read and study the lesson to answer each question. **1–3. See margin.**

1. **Describe** the graph of the parametric equations $x = 3 + 4t$ and $y = -1 + 2t$.

2. **Explain** how to find the parametric equations for the line through the point at $(3, 6)$, parallel to the vector $\vec{v} = \vec{i} + 2\vec{j}$.

3. **Describe** the line having parametric equations $x = 1 + t$ and $y = -t$. Include the slope of the line and the y-intercept in your description.

Guided Practice

Write a vector equation of the line that passes through point P and is parallel to $\vec{a}$. Then write parametric equations of the line. **4–5. See margin.**

4. $P(-4, 11), \vec{a} = \langle -3, 8 \rangle$ $\qquad$ 5. $P(1, 5), \vec{a} = \langle -7, 2 \rangle$

Write parametric equations of each line with the given equation.

6. $3x + 2y = 5$ $\quad x = t, y = -\dfrac{3}{2}t + \dfrac{5}{2}$ $\qquad$ 7. $4x - 6y = -12$ $\quad x = t, y = \dfrac{2}{3}t + 2$

Lesson 8-6 Vectors and Parametric Equations **523**

In-Class Example
Example 5
Write an equation in slope-intercept form of the line whose parametric equations are
$x = -2 + t$ and $y = 4 - 3t$.
$y = -3x - 2$

3 PRACTICE/APPLY

Error Analysis

Watch for students who forget to write the slope-intercept form of the equation of a line before writing parametric equations for that line.

Prevent by having students review the process of changing equations of straight lines into the slope-intercept form.

Additional Answers

1. a line through $(3, -1)$ and $(7, 1)$

2. **Sample answer: for every single unit increment of t, x increases 1 unit, y increases 2 units. Then, the parametric equations of the line are $x = 3 + t, y = 6 + 2t$.**

3. **a line with slope -1 going through $(0, 1)$ and $(1, 0)$**

4. $\langle x + 4, y - 11 \rangle = t\langle -3, 8 \rangle$
   $x = -3t - 4, y = 8t + 11$

5. $\langle x - 1, y - 5 \rangle = t\langle -7, 2 \rangle$
   $x = 1 - 7t, y = 5 + 2t$

Study Guide and Practice Masters, pp. 147–148

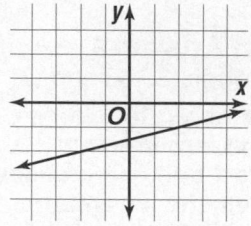
Write an equation in slope-intercept form of the line with the given parametric equations.

8. $x = −4t + 3$ $y = −\frac{5}{4}x + \frac{3}{4}$
  $y = 5t − 3$

9. $x = 9t$ $y = \frac{4}{9}x + 2$
  $y = 4t + 2$

10. Set up a table of values and then graph the line whose parametric equations are $x = 2 + 4t$ and $y = −1 + t$. **See margin.**

11. **Sports** A wide receiver catches a ball and begins to run for the endzone following a path defined by $\langle x − 5, y − 50 \rangle = t\langle 0, −10 \rangle$. A defensive player chases the receiver as soon as he starts running following a path defined by $\langle x − 10, y − 54 \rangle = t\langle −0.9, −10.72 \rangle$.

  a. Write parametric equations for the path of each player. **See margin.**

  b. If the receiver catches the ball on the 50-yard line ($y = 50$), will he reach the goal line ($y = 0$) before the defensive player catches him? **yes**

EXERCISES

Practice

Write a vector equation of the line that passes through point P and is parallel to $\bar{a}$. Then write parametric equations of the line. **12–17. See margin.**

 A

12. $P(5, 7), \bar{a} = \langle 2, 0 \rangle$ 13. $P(−1, 4), \bar{a} = \langle 6, −10 \rangle$

14. $P(−6, 10), \bar{a} = \langle 3, 2 \rangle$ 15. $P(1, 5), \bar{a} = \langle −7, 2 \rangle$

16. $P(1, 0), \bar{a} = \langle −2, −4 \rangle$ 17. $P(3, −5), \bar{a} = \langle −2, 5 \rangle$

Write parametric equations of each line with the given equation.

18. $y = 4x − 5$ 19. $−3x + 4y = 7$ 20. $2x − y = 3$ $x = t,$ $y = 2t − 3$

B 21. $9x + y = −1$ 22. $2x + 3y = 11$ 23. $−4x + y = −2$

24. Write parametric equations for the line passing through the point at $(−2, 5)$ and parallel to the line $3x − 6y = −8$. $x = t, y = \frac{1}{2}t + 6$

Write an equation in slope-intercept form of the line with the given parametric equations.

25. $x = 2t$ $y = −\frac{1}{2}x + 1$ 26. $x = −7 + \frac{1}{2}t$ 27. $y = \frac{1}{4}x + \frac{23}{4}$
  $y = 1 − t$ $y = 3t$ $y = 6x + 42$ 27. $x = 4t − 11$
    $y = t + 3$

28. $x = 4t − 8$ 29. $x = 3 + 2t$ 30. $x = 8$
  $y = 3 + t$ $y = \frac{1}{4}x + 5$ $y = −1 + 5t$ $y = \frac{5}{2}x − \frac{17}{2}$ $y = 2t + 1$ $x = 8$

C 31. A line passes through the point at $(11, −4)$ and is parallel to $\bar{a} = \langle 3, 7 \rangle$.

  a. Write a vector equation of the line. $\langle x − 11, y + 4 \rangle = t\langle 3, 7 \rangle$

  b. Write parametric equations for the line. $x = 3t + 11, y = 7t − 4$

  c. Use the parametric equations to write the equation of the line in slope-intercept form. $y = \frac{7}{3}x − \frac{89}{3}$

Graphing Calculator

Use a graphing calculator to set up a table of values and then graph each line from its parametric form. **32–34. See Answer Appendix.**

32. $x = 5t$ 33. $x = 3t + 5$ 34. $x = 1 + t$
  $y = −4 − t$ $y = 1 + t$ $y = 1 − t$

www.amc.glencoe.com/self_check_quiz

36. See margin.

**37a–b. See
margin.**

35. **Geometry** A line in a plane is defined by the parametric equations
$x = 2 + 3t$ and $y = 4 + 7t$.
   a. What part of this line is obtained by assigning only non-negative values
      to t? **right of point (2, 4)**   **b. $t < -\frac{2}{3}$**
   b. How should t be defined to give the part of the line to the left of the y-axis?

36. **Critical Thinking** Graph the parametric equations $x = \cos^2 t$ and $y = \sin^2 t$.

37. **Navigation** During a military training
   exercise, an unmanned target drone has been
   detected on a radar screen following a path
   represented by the vector equation $\langle x, y \rangle =$
   $(3, 4) + t\langle -1, 0 \rangle$. A surface to air missile is
   launched to intercept it and destroy it. This
   missile is following a trajectory represented
   by $\langle x, y \rangle = (2, 2) + t\langle 1, 2 \rangle$.
   a. Write the parametric equations for the path
      of the target drone and the missile.
   b. Will the missile intercept the target drone?

38. **Astronomy** Astronomers have traced the path of two asteroids traveling
   through space. At a particular time t, the position of asteroid Ceres can be
   represented by $(-1 + t, 4 - t, -1 + 2t)$. Asteroid Pallas' path at any time t
   can be expressed by $(-7 + 2t, -6 + 2t, -1 + t)$. **a–c. See margin.**
   a. Write the parametric equations for the path of each asteroid.
   b. Do the paths of the asteroids cross? If so, where?
   c. Following these paths, will the asteroids collide? If so, where?

**39. $x = -\frac{1}{3} + \frac{1}{3}t$,
$y = 1 + 4t$,
$z = 1 - 9t$**

39. **Critical Thinking** Find the parametric equations for the line passing through
   points at $\left(-\frac{1}{3}, 1, 1\right)$ and $(0, 5, -8)$. (*Hint:* Equations are needed for x, y, and z)

**Mixed
Review**

40. An airplane flies at 150 km/h and heads 30° south of east. A 50-km/h wind
   blows in the direction 25° west of south. Find the ground speed and direction
   of the plane. (*Lesson 8-5*) **162.2 km/h, 47° 53′ 4″SE**

41. Find the inner product $\langle 1, 3 \rangle \cdot \langle 3, -2 \rangle$ and state whether the vectors are
   perpendicular. Write *yes* or *no*. (*Lesson 8-4*) **−3, no**

42. Solve $\triangle ABC$ if $A = 40°$, $b = 16$, and $a = 9$. (*Lesson 5-7*) **No solution exists.**

43. Approximate the real zero(s) of $f(x) = x^5 + 3x^3 - 4$ to the nearest tenth.
   (*Lesson 4-5*) **1**

44. Find the inverse of the function $y = \frac{3}{2}x - 2$. (*Lesson 3-4*) **$y = \frac{2}{3}x + \frac{4}{3}$**

45. $x - y + 4 = 0$

45. Write the standard form of the equation of the line that is parallel to the
   graph of $y = x - 8$ and passes through the point at $(-3, 1)$. (*Lesson 1-5*)

46. **SAT/ACT Practice** A pulley having a 9-inch
   diameter is belted to a pulley having a
   6-inch diameter, as shown in the figure.
   If the larger pulley runs at 120 rpm
   (revolutions per minute), how fast does
   the smaller pulley run? **D**
   A 80 rpm   B 100 rpm   C 160 rpm
   D 180 rpm   E 240 rpm

Extra Credit

Ask students to consider the vector
equation $\langle x - 4, y - 3 \rangle = t\langle 5, 1 \rangle$. Have the
class select different values for t and
graph the resulting points (x, y). What sort
of function does this vector equation
represent? Have the students find an
equation in standard form in this line as
well. **a line; $x - 5y = -11$**

---

4 ASSESS

Open-Ended Assessment
Writing Have students use the
sports page in order to write word
problems that can be solved using
parametric equations. Have
students exchange problems and
take turns solving them.

**Chapter 8, Quiz C
(Lessons 8-5 and 8-6)**
is available in the *Assessment
and Evaluation Masters,* p. 200.

Additional Answers
36.

**37a. target drone: $x = 3 - t$,
$y = 4$**
   missile: $x = 2 + t$, $y = 2 + 2t$

37b. no

**38a. Pallas: $x = -7 + 2t$,
$y = -6 + 2t$, $z = -1 + t$**
   **Ceres: $x = -1 + t$, $y = 4 - t$,
$z = -1 + 2t$**

38b. yes, (1, 2, 3)

38c. no

Enrichment Masters, p. 90

526 Chapter 8

1 FOCUS

Motivating the Lesson
Real-World Application
Encourage students to search online for real data related to airplane races or a similar position, velocity, or time problem. Have students use their data with the graphing calculator lesson to model a parametric equation.

2 TEACH

📁 A reproduction of this lesson with graphing calculator keystrokes provided can be found in the *Graphing Calculator Keystrokes Masters,* pp. 33–34.

Teaching Tip Remind students that the parametric mode sets the calculator for parametric graphing, and that simultaneous mode means that the calculator will simultaneously evaluate and plot all of the selected functions for a single value of *x*, then repeat that process for each subsequent value of *x*.

3 PRACTICE/APPLY

Assignment Guide
All: 1–3

4 ASSESS

Open-Ended Assessment
Writing Have students write a short paragraph on the **TRACE** function and why it is a useful part of interpreting data from a graph.

GRAPHING CALCULATOR EXPLORATION

8-6B Modeling With Parametric Equations

An Extension of Lesson 8-6

OBJECTIVE
• Investigate the use of parametric equations.

You can use a graphing calculator and parametric equations to investigate real-world situations, such as simulating a 500-mile, two airplane race. Suppose the first plane averages a speed of 408.7 miles per hour for the entire race. The second plane starts 30 seconds after the first and averages 418.3 miles per hour.

TRY THIS

Write a set of parametric equations to represent each plane's position at *t* hours. This will simulate the race on two parallel race courses so that it is easy to make a visual comparison. Use the formula $x = vt$.

Plane 1: $x = 408.7t$ $x = vt$
            $y = 1$ *y represents the position of plane 1 on course 1.*

Plane 2: $x = 418.3(t - 0.0083)$ *Remember that plane 2 started 30 seconds or 0.0083 hour later.*
            $y = 2$ *Plane 2 is on course 2.*

Set the graphing calculator to parametric and simultaneous modes. Next, set the viewing window to the following values: **Tmin** = 0, **Tmax** = 5, **Tstep** = .005, **Xmin** = 0, **Xmax** = 500, **Xscl** = 10, **Ymin** = 0, **Ymax** = 5, and **Yscl** = 1. Then enter the parametric equations on the **Y=** menu. Then press ⬚GRAPH to "see the race."

Which plane finished first? Notice that the line on top reached the edge of the screen first. That line represented the position of the second plane, so plane 2 finished first. You can confirm the conclusion using the ⬚TRACE function. The plane with the smaller *t*-value when *x* is about 500 is the winner. Note that when x_1 is about 500, $t = 1.225$ and when x_2 is about 500, $t = 1.205$.

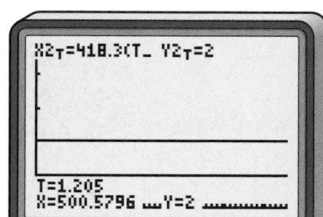

WHAT DO YOU THINK?

1. How long does it take the second plane to overtake the first? ≈ **21.7 min**

2. How far have the two planes traveled when the second plane overtakes the first? ≈ **147.8 mi**

3. How much would the pilot of the first plane have to increase the plane's speed to win the race? ≈ **6.7 mph**

526 *Chapter 8 Vectors and Parametric Equations*

Modeling Motion Using Parametric Equations

OBJECTIVES
* Model the motion of a projectile using parametric equations.
* Solve problems related to the motion of a projectile, its trajectory, and range.

SPORTS Suppose a professional football player kicks a football with an initial velocity of 29 yards per second at an angle of 68° to the horizontal. Suppose a kick returner catches the ball 5 seconds later. How far has the ball traveled horizontally and what is its vertical height at that time? *This problem will be solved in Example 2.*

Objects that are launched, like a football, are called *projectiles*. The path of a projectile is called its *trajectory*. The horizontal distance that a projectile travels is its *range*. Physicists describe the motion of a projectile in terms of its position, velocity, and acceleration. All these quantities can be represented by vectors.

The figure at the right describes the initial trajectory of a punted football as it leaves the kicker's foot. The magnitude of the initial velocity $|\vec{v}|$ and the direction θ of the ball can be described by a vector that can be expressed as the sum of its horizontal and vertical components $\vec{v}_x$ and $\vec{v}_y$.

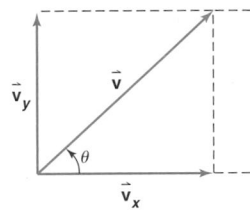

As the ball moves, gravity will act on it in the vertical direction. The horizontal component is unaffected by gravity. So, discounting air resistance, the horizontal speed is constant throughout the flight of the ball.

The vertical component of the velocity of the ball is large and positive at the beginning, decreasing to zero at the top of its trajectory, then increasing in the negative direction as it falls. When the ball returns to the ground, its vertical speed is the same as when it left the kicker's foot, but in the opposite direction.

Parametric equations can represent the position of the ball relative to the starting point in terms of the parameter of time.

Lesson 8-7 Modeling Motion Using Parametric Equations **527**

1 FOCUS

5-Minute Check
Lesson 8-6

1. Write a vector equation of the line that passes through $P_1(5, -3)$ and is parallel to $\vec{a} = \langle 1, 2 \rangle$.
   $\langle x - 5, y + 3 \rangle = t \langle 1, 2 \rangle$

2. Write parametric equations for the line parallel to $\vec{a} = \langle 2, -2 \rangle$ and passing through the point at $(1, 3)$. Then make a table of values and graph the line.
   $x = 1 + 2t, y = 3 - 2t$

t	x	y
0	1	3
1	3	1
2	5	−1

   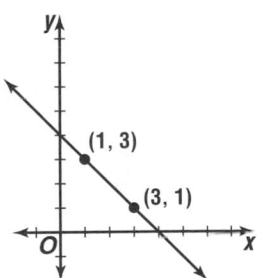

3. Write parametric equations of $y = 2x - 5$.
   $x = t$ and $y = 2t - 5$

4. Write an equation in slope-intercept form of the line whose parametric equations are $x = -4 + 2t, y = 5 - t$.
   $y = -\frac{1}{2}x + 3$

Motivating the Lesson

Hands-On Activity Have students take turns throwing a soft football at a target or into a basket. Discuss how the angle and the speed with which the ball is thrown help determine its path. How could speed and angle be measured? Are there equations for the projectile's path in terms of these parameters?

Teaching Tip Remind students that the horizontal and vertical components of a vector, $\vec{\mathbf{v}}$, are perpendicular to each other, so these results come from right triangle trigonometry.

In-Class Example

Example 1
Find the initial horizontal velocity and vertical velocity of a stone kicked with an initial velocity of 18 feet per second at an angle of 37° with the ground.
horizontal: about 14 ft/s; vertical: about 11 ft/s

In order to find parametric equations that represent the path of a projectile like a football, we must write the horizontal $\vec{\mathbf{v}}_x$ and vertical $\vec{\mathbf{v}}_y$ components of the initial velocity.

$$\cos \theta = \frac{|\vec{\mathbf{v}}_x|}{|\vec{\mathbf{v}}|} \qquad\qquad \sin \theta = \frac{|\vec{\mathbf{v}}_y|}{|\vec{\mathbf{v}}|}$$

$$|\vec{\mathbf{v}}_x| = |\vec{\mathbf{v}}| \cos \theta \qquad\qquad |\vec{\mathbf{v}}_y| = |\vec{\mathbf{v}}| \sin \theta$$

Example ❶ Find the initial horizontal velocity and vertical velocity of a stone kicked with an initial velocity of 16 feet per second at an angle of 38° with the ground.

$$|\vec{\mathbf{v}}_x| = |\vec{\mathbf{v}}| \cos \theta \qquad\qquad |\vec{\mathbf{v}}_y| = |\vec{\mathbf{v}}| \sin \theta$$

$$|\vec{\mathbf{v}}_x| = 16 \cos 38° \qquad\qquad |\vec{\mathbf{v}}_y| = 16 \sin 38°$$

$$|\vec{\mathbf{v}}_x| \approx 13 \qquad\qquad |\vec{\mathbf{v}}_y| \approx 10$$

The initial horizontal velocity is about 13 feet per second and the initial vertical velocity is about 10 feet per second.

Because horizontal velocity is unaffected by gravity, it is the magnitude of the horizontal component of the initial velocity. Therefore, the horizontal position of a projectile, after t seconds is given by the following equation.

$$\underline{horizontal\ distance} = \underline{horizontal\ velocity} \cdot \underline{time}$$

$$x \qquad = \qquad |\vec{\mathbf{v}}| \cos \theta \quad \cdot \quad t$$

$$x = t|\vec{\mathbf{v}}| \cos \theta$$

Since vertical velocity is affected by gravity, we must adjust the vertical component of initial velocity. By subtracting the vertical displacement due to gravity from the vertical displacement caused by the initial velocity, we can determine the height of the projectile after T seconds. The height, in feet or meters, of a free-falling object affected by gravity is given by the equation $h = \frac{1}{2}gt^2$, where $g \approx 9.8$ m/s^2 or 32 ft/s^2 and t is the time in seconds.

$$\underbrace{vertical\ displacement} = \underbrace{\begin{array}{c}displacement\ due\\to\ initial\ velocity\end{array}} - \underbrace{\begin{array}{c}displacement\ due\\to\ gravity\end{array}}$$

$$y \qquad = \qquad (|\vec{\mathbf{v}}| \sin \theta)t \quad - \quad \frac{1}{2}gt^2$$

$$y = t|\vec{\mathbf{v}}| \sin \theta - \frac{1}{2}gt^2$$

Therefore, the path of a projectile can be expressed in terms of parametric equations.

Parametric Equations for the Path of a Projectile

If a projectile is launched at an angle of θ with the horizontal with an initial velocity of magnitude $|\mathbf{v}|$, the path of the projectile may be described by these equations, where t is time and g is acceleration due to gravity.

$$x = t|\mathbf{v}| \cos \theta$$
$$y = t|\mathbf{v}| \sin \theta - \frac{1}{2}gt^2$$

Example **② SPORTS** Refer to the application at the beginning of the lesson.

a. **How far has the ball traveled horizontally and what is its vertical height at that time?**

b. **Suppose the kick returner lets the ball hit the ground instead of catching it. What is the hang time, the elapsed time between the moment the ball is kicked and the time it hits the ground?**

a. Write the position of the ball as a pair of parametric equations defining the path of the ball for any time t in seconds.

$$x = t\,|\mathbf{v}|\, \cos \theta \qquad\qquad y = t\,|\mathbf{v}|\, \sin \theta - \frac{1}{2}gt^2$$

The initial velocity of 29 yards per second must be expressed as 87 feet per second as gravity is expressed in terms of feet per second squared.

$x = t(87) \cos 68°$ $\qquad\qquad$ $y = t(87) \sin 68° - \frac{1}{2}(32)t^2$

$x = 87t \cos 68°$ $\qquad\qquad$ $y = 87t \sin 68° - 16t^2$

Find x and y when $t = 5$.

$x = 87(5) \cos 68°$ $\qquad\qquad$ $y = 87(5) \sin 68° - 16(5)^2$

$x \approx 163$ $\qquad\qquad$ $y \approx 3$

After 5 seconds, the football has traveled about 163 feet or $54\frac{1}{3}$ yards horizontally and is about 3 feet or 1 yard above the ground.

Height (yards) / Distance (yards)

b. Determine when the vertical height is 0 using the equation for y.

$y = 87t \sin 68° - 16t^2$

$y = 80.66t - 16t^2$

$y = t(80.66 - 16t)$ *Factor.*

(continued on the next page)

In-Class Example

Example 2

Sports Sammy Baugh of the Washington Redskins has the record for the highest average punting record for a lifetime average of 45.16 yards. Suppose that he kicked the ball with an initial velocity of 26 yards per second at an angle of 72°.

a. How far has the ball traveled horizontally and what is its vertical height at the end of 3 seconds? **about 24 yards; about 26 yards**

b. Suppose that the kick returner lets the ball hit the ground instead of catching it. What is the hang time? **about $4\frac{1}{2}$ s**

Now let $y = 0$ and solve for t.

$$0 = 80.66 - 16t$$
$$16t = 80.66$$
$$t = 5.04$$

The hang time is about 5 seconds.

The parametric equations describe the path of an object that is launched from ground level. Some objects are launched from above ground level. For example, a baseball may be hit at a height of 3 feet. So, you must add the initial vertical height to the expression for y. This accounts for the fact that at time 0, the object will be above the ground.

Example **3**

SOFTBALL Kaci Clark led the Women's Pro Softball League in strikeouts in 1998. Suppose she throws the ball at an angle of 5.2° with the horizontal at a speed of 67 mph. The distance from the pitcher's mound to home plate is 43 feet. If Kaci releases the ball 2.7 feet above the ground, how far above the ground is the ball when it crosses home plate?

First, write parametric equations that model the path of the softball. Remember to convert 67 mph to about 98.3 feet per second.

$$x = t|\vec{\mathbf{v}}| \cos \theta \qquad y = t|\vec{\mathbf{v}}| \sin \theta - \frac{1}{2}gt^2 + h$$
$$x = t(98.3)\cos 5.2° \qquad y = t(98.3)\sin 5.2° - \frac{1}{2}(32)t^2 + 2.7$$
$$x = 98.3t \cos 5.2° \qquad y = 98.3t \sin 5.2° - 16t^2 + 2.7$$

Kaci Clark

Then, find the amount of time that it will take the baseball to travel 43 feet horizontally. This will be the moment when it crosses home plate.

$$43 = 98.3t \cos 5.2°$$
$$t = \frac{43}{98.3 \cos 5.2°}$$
$$t \approx 0.439$$

The softball will cross home plate in about 0.44 second.

To find the vertical position of the ball at that time, find y when $t = 0.44$.

$$y = 98.3t \sin 5.2° - 16t^2 + 2.7$$
$$y = 98.3(0.44) \sin 5.2° - 16(0.44)^2 + 2.7$$
$$y \approx 3.522$$

The softball will be about 3.5 feet above home plate.

CHECK FOR UNDERSTANDING

Communicating Mathematics

Read and study the lesson to answer each question. **1–3. See margin.**

1. **Describe** situations in which a projectile travels vertically. At what angle with the horizontal must the projectile be launched?

2. **Describe** the vertical velocities of a projectile at its launch and its landing.

3. **Explain** the effect the angle of a golf club's head has on the angle of initial velocity of a golf ball.

Guided Practice

4. Find the initial vertical velocity of a stone thrown with an initial velocity of 50 feet per second at an angle of 40° with the horizontal. **32.14 ft/s**

5. Find the initial horizontal velocity of a discus thrown with an initial velocity of 20 meters per second at an angle of 50° with the horizontal. **12.86 m/s**

Find the initial horizontal velocity and vertical velocity for each situation.

6. A soccer ball is kicked with an initial velocity of 45 feet per second at an angle of 32° with the horizontal. **38.16 ft/s, 23.85 ft /s**

7. A stream of water is shot from a sprinkler head with an initial velocity of 7.5 meters per second at an angle of 20° with the ground. **7.05 m/s, 2.57 m/s**

8. **Meteorology** An airplane flying at an altitude of 3500 feet is dropping research probes into the eye of a hurricane. The path of the plane is parallel to the ground at the time the probes are released with an initial velocity of 300 mph. **a–d. See margin.**

   a. Write the parametric equations that represent the path of the probes.

   b. Sketch the graph describing the path of the probes.

   c. How long will it take the probes to reach the ground?

   d. How far will the probes travel horizontally before they hit the ground?

EXERCISES

Practice

9–14. See margin.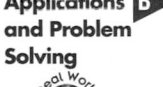

Find the initial horizontal velocity and vertical velocity for each situation.

9. a javelin thrown at 65 feet per second at an angle of 60° with the horizontal

10. an arrow released at 47 meters per second at an angle of 10.7° with the horizontal

11. a cannon shell fired at 1200 feet per second at a 42° angle with the ground

12. a can is kicked with an initial velocity of 17 feet per second at an angle of 28° with the horizontal

13. a golf ball hit with an initial velocity of 69 yards per second at 37° with the horizontal

14. a kangaroo leaves the ground at an angle of 19° at 46 kilometers per hour

Applications and Problem Solving

15. **Golf** Professional golfer Nancy Lopez hits a golf ball with a force to produce an initial velocity of 175 feet per second at an angle of 35° above the horizontal. She estimates the distance to the hole to be 225 yards. **a. See margin.**

   a. Write the position of the ball as a pair of parametric equations.

   b. Find the range of the ball. **899.32 ft or 299.77 yd**

www.amc.glencoe.com/self_check_quiz   *Lesson 8-7 Modeling Motion Using Parametric Equations*   **531**

Additional Answers

1. Sample answers: a rocket launched at 90° to the horizontal; tip-off in basketball

2. equal magnitude with opposite direction

3. The greater the angle of the head of the golf club, the greater the angle of initial velocity of the ball.

3 PRACTICE/APPLY

Error Analysis

Watch for students who forget to subtract $\frac{1}{2} gt^2$ on the parametric equation which represents the vertical height of the object.

Prevent by emphasizing that *g* represents an acceleration constant due to gravity and that gravity is a downward force on the object and thus affects only the vertical component.

Assignment Guide

Core: 9–15 odd, 17–19, 21–23, 25–29

Enriched: 10–14 even, 15–29

Additional Answers

8a. $x = 440t$, $y = -16t^2 + 3500$

8b. See students' graphs.

8c. about 14.8 s

8d. 6512 ft

9. 32.5 ft/s, 56.29 ft/s

10. 46.18 m/s, 8.73 m/s

11. 891.77 ft/s, 802.96 ft/s

12. 15.01 ft/s, 7.98 ft/s

13. 55.11 yd/s, 41.53 yd/s

14. 43.49 km/h, 14.98 km/h

15a. $x = 175t \cos 35°$, $y = 175t \sin 35° - 16t^2$

Study Guide and Practice Masters, pp. 149–150

Lesson 8-7 **531**

Additional Answers

16. The greater the angle, the greater the time the projectile stays in the air. As the angle increases from 0° to 45°, the horizontal distance increases. As the angle increases from 45° to 90°, the horizontal distance decreases.

18. No, it is a parabolic path due to the force of gravity.

19. Sample answer: no, the projectile will travel four times as far.

16. 310.87 ft

16. Projectile Motion What is the relationship between the angle at which a projectile is launched, the time it stays in the air, and the distance it covers?

17. Physics Alfredo and Kenishia are performing a physics experiment in which they will launch a model rocket. The rocket is supposed to release a parachute 300 feet in the air, 7 seconds after liftoff. They are firing the rocket at a 78° angle from the horizontal.

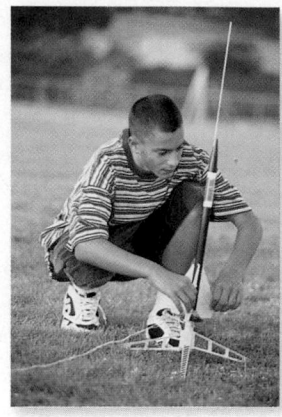

17a. 158.32 ft/s

a. Find the initial velocity of the rocket.

b. To protect other students from the falling rockets, the teacher needs to place warning signs 50 yards from where the parachute is released. How far should the signs be from the point where the rockets are launched? **127 yd**

18. Critical Thinking Is it possible for a projectile to travel in a circular arc rather than a parabolic arc? Explain your answer. **See margin.**

19. Critical Thinking Janelle fired a projectile and measured its range. She hypothesizes that if the magnitude of the initial velocity is doubled and the angle of the velocity remains the same, the projectile will travel twice as far as it did before. Do you agree with her hypothesis? Explain. **See margin.**

20. Aviation Commander Patrick Driscoll flies with the U.S. Navy's Blue Angels. Suppose he places his aircraft in a 45° dive at an initial speed of 800 km/h. **a.** $x = -222.2t \cos 45°$, $y = -222.2t \sin 45° - 4.9t^2$

a. Write parametric equations to represent the descent path of the aircraft.

b. How far has the aircraft descended after 2.5 seconds? **423.4 m**

c. What is the average rate at which the plane is losing altitude during the first 2.5 seconds? **169 m/s or 608.4 km/h**

21. Entertainment The "Human Cannonball" is shot out of a cannon with an initial velocity of 70 mph 10 feet above the ground at an angle of 35°.

a. What is the maximum range of the cannon? **323.2 ft**

b. How far from the launch point should a safety net be placed if the "Human Cannonball" is to land on it at a point 8 feet above the ground? **312.4 ft**

c. How long is the flight of the "Human Cannonball" from the time he is launched to the time he lands in the safety net? **3.71 s**

22. Critical Thinking If a circle of radius 1 unit is rolled along the *x*-axis at a rate of 1 unit per second, then the path of a point *P* on the circle is called a *cycloid*. **a. See margin.**

a. Sketch what you think a cycloid must look like. (*Hint:* Use a coin or some other circular object to simulate the situation.)

b. The parametric equations of a cycloid are $x = t - \sin t$ and $y = 1 - \cos t$ where *t* is measured in radians. Use a graphing calculator to graph the cycloid. Set your calculator in radian mode. An appropriate range is Tmin = 0, Tmax = 18.8, Tstep = 0.2, Xmin = −6.5, Xmax = 25.5, Xscl = 2, Ymin = −8.4, Ymax = 12.9, and Yscl = 1. Compare the results with your sketch. **See students' work.**

Additional Answer
22a. Sample graph:

23. **Entertainment** The Independence Day fireworks at Memorial Park are fired at an angle of 82° with the horizontal. The technician firing the shells expects them to explode about 300 feet in the air 4.8 seconds after they are fired.

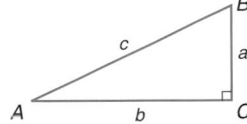

a. Find the initial velocity of a shell fired from ground level. **140.7 ft/s**

b. Safety barriers will be placed around the launch area to protect spectators. If the barriers are placed 100 yards from the point directly below the explosion of the shells, how far should the barriers be from the point where the fireworks are launched? **131.3 yd**

24. **Baseball** Derek Jeter, shortstop for the New York Yankees, comes to bat with runners on first and third bases. Greg Maddux, pitcher for the Atlanta Braves, throws a slider across the plate about waist high, 3 feet above the ground. Derek Jeter hits the ball with an initial velocity of 155 feet per second at an angle of 22° above the horizontal. The ball travels straight at the 420 foot mark on the center field wall which is 15 feet high. **a–b. See margin.**

a. Write parametric equations that describe the path of the ball.

b. Find the height of the ball after it has traveled 420 feet horizontally. Will the ball clear the fence for a home run, or will the center fielder be able to catch it?

c. If there were no outfield seats, how far would the ball travel before it hits the ground? **528.86 ft**

Mixed Review

25. Write an equation of the line with parametric equations $x = 11 - t$ and $y = 8 - 6t$ in slope-intercept form. *(Lesson 8-6)* **$y = 6x - 58$**

26. **Food Industry** Fishmongers will often place ice over freshly-caught fish that are to be shipped to preserve the freshness. Suppose a block of ice with a mass of 300 kilograms is held on an ice slide by a rope parallel to the slide. The slide is inclined at an angle of 22°. *(Lesson 8-5)*

a. What is the pull on the rope? **1101.3 N**

b. What is the force on the slide? **2725.9 N**

27. If $b = 17.4$ and $c = 21.9$, find the measure of angle A to the nearest degree. *(Lesson 5-5)* **37°**

28. Solve the system of equations. *(Lesson 2-2)* **(0.25, −0.5, 1)**

$2x - y + z = 2$

$x + 3y - 2z = -3.25$

$-4x - 5y + z = 2.5$

29. **SAT/ACT Practice** What is the area of the shaded region? **B**

A 19π B 16π

C 9π D 4π

E 2π

Extra Credit

A good golfer takes into account the wind speed and direction on the path of the ball. Suppose that when a golfer steps up to the tee, there is a 15-mph wind at his back moving directly to the pin, which is 225 yards from the tee. He sometimes makes a hole-in-one on this hole when there is no wind by hitting the ball at an initial velocity of 140 mph and an angle of about 15.4°. If he ignores the wind and hits the ball with the same initial velocity and angle, how far past the hole will the ball carry?
about 25 yards

---

4 ASSESS

Open-Ended Assessment
Speaking Have students list parametric equations for the path of a projectile given its initial velocity, $\vec{v}$, and angle, θ.

Additional Answers
24a. $x = 155t \cos 22°$
      $y = 155t \sin 22° - 16t^2 + 3$
24b. **About 36.04 ft; it will clear the fence.**

Enrichment Masters, p. 91

© Glencoe/McGraw-Hill   T91   Advanced Mathematical Concepts

Lesson 8-7 **533**

HISTORY of MATHEMATICS

This feature provides students with an opportunity to see how the mathematics that they are studying was developed. It also gives students a glimpse of how mathematical discoveries are made.

More About the Mathematicians

The Babylonians divided each hour into 60 minutes and each minute into 60 seconds. This organization of time is still followed 4000 years later.

Eratosthenes was born in Cyrene, North Africa (now Libya) in 276 B.C. He was nicknamed "Beta" because he was deemed second best in so many fields, but Eratosthenes created the first calendar with a leap year, developed a method of finding prime numbers, and used geometry and angles to estimate Earth's size.

Claudius Ptolemy was born in Egypt. An honorary citizen of Rome, Ptolemy was one of the greatest astronomers and geographers. Among his accomplishments were building a model to show the motions of the planets, determining the lengths of the seasons, and mapping the then-known world.

Gemma Frisius was born in Dokkum, Netherlands, in 1508. A professor of medicine and mathematics, Frisius made maps, invented astronomical instruments, applied trigonometric methods to astronomical problems, and described trigonometric surveying.

James Thomson was born in Annaghmore, Northern Ireland, in 1786. He taught mathematics at Queens College, Belfast, Northern Ireland. His brother, William Thomson, Lord Kelvin, was a famous physicist.

HISTORY of MATHEMATICS

ANGLE MEASURE

You know that the direction of a vector is the directed angle, usually measured in degrees, between the positive x-axis and the vector. How did that system originate? How did other angle measurement systems develop?

Early Evidence The division of a circle into 360° is based upon a unit of distance used by the **Babylonians,** which was equal to about 7 miles. They related time and miles by observing the time that it took for a person to travel one of their units of distance. An entire day was 12 "time-miles," so one complete revolution of the sky by the sun was divided into 12 units. They also divided each time-mile into 30 units. So 12 × 30 or 360 units represented a complete revolution of the sun. Other historians feel that the Babylonians used 360° in a circle because they had a base 60 number system, and thought that there were 360 days in a year.

The Greeks adopted the 360° circle. **Ptolemy** (85-165) used the degree system in his work in astronomy, introducing symbols for degrees, minutes and seconds in his mathematical work, *Almagest,* though they differed from our modern symbols. When his work was translated into Arabic, sixtieths were called "first small parts," sixtieths of sixtieths were called "second small parts," and so on. The later Latin translations were "partes minutae primae," now our "minutes," and "partes minutae secundae," now our "seconds."

The Renaissance The degree symbol (°) became more widely used after the publication of a book by Dutch mathematician, **Gemma Frisius** in 1569.

Modern Era Radian angle measure was introduced by **James Thomson** in 1873. The radian was developed to simplify formulas used in trigonometry and calculus. Radian measure can be given as an exact quantity not just an approximation.

Ptolemy

Other measurement units for angles have been developed to meet a specific need. These include mils, decimally- or centesimally-divided degrees, and millicycles.

Today mechanical engineers, like **Mark Korich,** use angle measure in designing electric and hybrid vehicles. Mr. Korich, a senior staff engineer for an auto manufacturer, calculates the draft angle that allows molded parts to pull away from the mold more easily. He also uses vectors to calculate the magnitude and direction of forces acting upon the vehicle under normal conditions.

ACTIVITIES

1. Greek mathematician, **Eratosthenes,** calculated the circumference of Earth using proportion and angle measure. By using the angle of a shadow cast by a rod in Alexandria, he determined that $\angle AOB$ was equal to 7°12′. He knew that the distance from Alexandria to Syene was 5000 stadia (singular stadium). One stadium equals 500 feet. Use the diagram to find the circumference of Earth. Compare your result to the actual circumference of Earth.

2. Research one of the famous construction problems of the Greeks—trisecting an angle using a straightedge and compass. Try to locate a solution to this problem.

3. *inter*NET **CONNECTION** Find out more about those who contributed to the history of angle measure. Visit **www.amc.glencoe.com** **See students' work.**

Additional Answers

1. **First, determine that 7° 12′ is one-fiftieth of 360°. Then, this proportion can be used:** $\frac{1}{50} = \frac{5000 \text{ stadia}}{x}$. **Then,** $x =$ **50(5000) = 250,000 stadia. 250,000(500) = 125,000,000 ft; 125,000,000 ÷ 5280 ≈ 23,674 mi. The actual circumference of Earth is 24,901.55 mi.**

2. **See students' work.**

3. **See students' work.**

Transformation Matrices in Three-Dimensional Space

OBJECTIVE
• Transform three-dimensional figures using matrix operations to describe the transformation.

COMPUTER ANIMATION Chris Wedge of Blue Sky Studios, Inc. used software to create the film that won the 1999 Academy Award for Animated Short Film. The computer software allows Mr. Wedge to draw three-dimensional objects and manipulate or transform them to create motion, color, and light direction. The mathematical processes used by the computer are very complex. *A problem related to animation will be solved in Example 2.*

Look Back
Refer to Lesson 2-3 to review ordered pairs and matrices.

Basic movements in three-dimensional space can be described using vectors and transformation matrices. Recall that a point at (x, y) in a two-dimensional coordinate system can be represented by the matrix $\begin{bmatrix} x \\ y \end{bmatrix}$. This idea can be extended to three-dimensional space. A point at (x, y, z) can be expressed as the matrix $\begin{bmatrix} x \\ y \\ z \end{bmatrix}$.

Example

You could also find the coordinates of B, G, and H first, then add 3 to the z-coordinates of A, B, and G to find the coordinates of D, C, and F.

1 Find the coordinates of the vertices of the rectangular prism and represent them as a vertex matrix.

$\overrightarrow{AE} = \langle -3 - 2, -2 - 2, 1 - (-2) \rangle$ or $\langle -5, -4, 3 \rangle$

You can use the coordinates of $\overrightarrow{AE}$ to find the coordinates of the other vertices.

$B(2, 2 + (-4), -2) = B(2, -2, -2)$
$C(2, 2 + (-4), -2 + 3) = C(2, -2, 1)$
$D(2, 2, -2 + 3) = D(2, 2, 1)$
$F(2 + (-5), 2, -2 + 3) = F(-3, 2, 1)$
$G(2 + (-5), 2, -2) = G(-3, 2, -2)$
$H(2 + (-5), 2 + (-4), -2) = H(-3, -2, -2)$

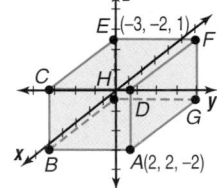

The vertex matrix for the prism is
$\begin{matrix} & A & B & C & D & E & F & G & H \\ x & 2 & 2 & 2 & 2 & -3 & -3 & -3 & -3 \\ y & 2 & -2 & -2 & 2 & -2 & 2 & 2 & -2 \\ z & -2 & -2 & 1 & 1 & 1 & 1 & -2 & -2 \end{matrix}$

In Lesson 2-4, you learned that certain matrices could transform a polygon on a coordinate plane. Likewise, transformations of three-dimensional figures, called **polyhedra,** can also be represented by certain matrices. A polyhedron (singular of *polyhedra*) is a closed three-dimensional figure made up of flat polygonal regions.

Lesson 8-8 Transformation Matrices in Three-Dimensional Space **535**

1 FOCUS

5-Minute Check
Lesson 8-7

A projectile is launched at an angle of 40° with the horizontal with an initial velocity of 250 feet per second.

1. Write equations for the horizontal distance and the vertical height at a given time of 2 seconds.

   $x = (2)(250) \cos 40°$;
   $y = (2)(250) \sin 40° - 16(2)^2$

2. Solve to find the horizontal distance and vertical distance. **about 383 ft; about 257 ft**

3. Find x and y at the end of 4 seconds. **about 766 ft; about 387 ft**

4. How long will it take the projectile to land? **about 10.0 s**

5. What is its total horizontal distance? **about 1915 ft**

Motivating the Lesson

Hands-On Activity Give students a cube that has each face painted a different color. Use a piece of string to represent a line of reflection. Set the cube on a flat surface and place the string along one edge. Ask students to identify which colors would be where if the cube were reflected over the string. Then, flip the cube to test their conjectures.

Resource Manager

Reproducible Masters
• *Study Guide and Practice,* pp. 151–152
• *Enrichment,* p. 92
• *Assessment and Evaluation,* p. 200

Transparencies
• 5-Minute Check, 8-8
• Teaching, 8-8

In-Class Examples

Example 1

Plot the coordinates of the vertices of the rectangular prism and represent them as a vertex matrix.

$A = (3, 2, -2); B = (3, -2, -2);$
$C = (3, -2, 1); D = (3, 2, 1);$
$E = (-4, -2, 1); F = (-4, 2, 1);$
$G = (-4, 2, -2); H = (-4, -2, -2)$

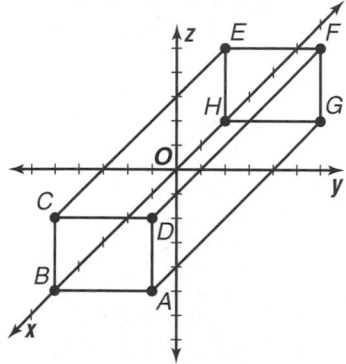

See below for answer matrix.

Example 2

Aerospace Suppose that a space engineer wishes to translate a line in space to model the path of a satellite after being struck by space debris. Suppose the line passes through points at $(1, -1, 4)$ and $(2, 1, 3)$. The engineer wants to translate the line using the vector $\vec{a} = \langle 1, 1, 1 \rangle$.

a. Write a matrix that will have such an effect on the line.

$$\begin{bmatrix} 1 & 1 \\ 1 & 1 \\ 1 & 1 \end{bmatrix}$$

b. Find the coordinates of the translated image.

$$\begin{bmatrix} 2 & 3 \\ 0 & 2 \\ 5 & 4 \end{bmatrix}$$

c. Graph the translated image.

Example **2** **COMPUTER ANIMATION** Maria is working on a computer animation project.

She needs to translate a prism using the vector $\vec{a} = \langle 3, 3, 0 \rangle$. The vertices of the prism have the following coordinates.

$A(2, 1, -4)$ $B(-1, -1, -4)$ $C(-2, 3, -4)$
$D(-2, 3, 3)$ $E(2, 1, 3)$ $F(-1, -1, 3)$

a. Write a matrix that will have such an effect on the figure.

b. Find the coordinates of the vertices of the translated image.

c. Graph the translated image.

a. To translate the prism by the vector $\vec{a} = \langle 3, 3, 0 \rangle$, we must first add 3 to each of the x- and y-coordinates. The z-coordinates remain the same. The translation matrix can be written as

$$\begin{bmatrix} 3 & 3 & 3 & 3 & 3 & 3 \\ 3 & 3 & 3 & 3 & 3 & 3 \\ 0 & 0 & 0 & 0 & 0 & 0 \end{bmatrix}.$$

b. Write the vertices of the prism in a 6×3 matrix. Then add it to the translation matrix to find the vertices of the translated image.

$$\begin{bmatrix} 3 & 3 & 3 & 3 & 3 & 3 \\ 3 & 3 & 3 & 3 & 3 & 3 \\ 0 & 0 & 0 & 0 & 0 & 0 \end{bmatrix} + \begin{matrix} A & B & C & D & E & F \\ \begin{bmatrix} 2 & -1 & -2 & -2 & 2 & -1 \\ 1 & -1 & 3 & 3 & 1 & -1 \\ -4 & -4 & -4 & 3 & 3 & 3 \end{bmatrix} \end{matrix}$$

$$= \begin{matrix} A' & B' & C' & D' & E' & F' \\ \begin{bmatrix} 5 & 2 & 1 & 1 & 5 & 2 \\ 4 & 2 & 6 & 6 & 4 & 2 \\ -4 & -4 & -4 & 3 & 3 & 3 \end{bmatrix} \end{matrix}$$

c. Draw the graph of the image.

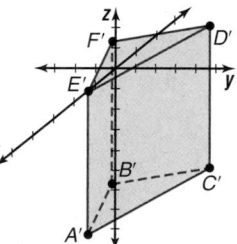

Answer matrix for Example 1

$$\begin{matrix} & A & B & C & D & E & F & G & H \\ x \\ y \\ z \end{matrix} \begin{bmatrix} 3 & 3 & 3 & 3 & -4 & -4 & -4 & -4 \\ 2 & -2 & -2 & 2 & -2 & 2 & 2 & -2 \\ -2 & -2 & 1 & 1 & 1 & 1 & -2 & -2 \end{bmatrix}$$

Teaching Tip Remind students that they can only add matrices of the same dimensions (conformal matrices) and that addition is a commutative process unlike multiplication.

Recall that certain 2×2 matrices could be used to reflect a plane figure across an axis. Likewise, certain 3×3 matrices can be used to reflect three-dimensional figures in space.

Example **③** Let *M* represent the vertex matrix of the rectangular prism in Example 1.

a. Find *TM* if $T = \begin{bmatrix} 1 & 0 & 0 \\ 0 & 1 & 0 \\ 0 & 0 & -1 \end{bmatrix}$.

b. Graph the resulting image.

c. Describe the transformation represented by matrix *T*.

a. First find *TM*.

$$TM = \begin{bmatrix} 1 & 0 & 0 \\ 0 & 1 & 0 \\ 0 & 0 & -1 \end{bmatrix} \cdot \begin{bmatrix} 2 & 2 & 2 & 2 & -3 & -3 & -3 & -3 \\ 2 & -2 & -2 & 2 & -2 & 2 & 2 & -2 \\ -2 & -2 & 1 & 1 & 1 & 1 & -2 & -2 \end{bmatrix}$$

$$TM = \begin{matrix} A' & B' & C' & D' & E' & F' & G' & H' \\ \begin{bmatrix} 2 & 2 & 2 & 2 & -3 & -3 & -3 & -3 \\ 2 & -2 & -2 & 2 & -2 & 2 & 2 & -2 \\ 2 & 2 & -1 & -1 & -1 & -1 & 2 & 2 \end{bmatrix} \end{matrix}$$

b. Then graph the points represented by the resulting matrix.

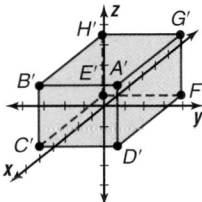

c. The transformation matrix $\begin{bmatrix} 1 & 0 & 0 \\ 0 & 1 & 0 \\ 0 & 0 & -1 \end{bmatrix}$ reflects the image of each vertex over the *xy*-plane. This results in a reflection of the prism when the new vertices are connected by segments.

The transformation matrix in Example 3 resulted in a reflection over the *xy*-plane. Similar transformations will result in reflections over the *xz*- and *yz*-planes. These transformations are summarized in the chart on the next page.

Teaching Tip In Example 3, note that the multiplication of the matrices *T* and *M* is *TM* and not *MT* since multiplication of matrices is not commutative.

In-Class Example

Example 3
Let *M* represent the vertex matrix of the rectangular prism in In-Class Example 1.

a. Find *TM* if $T = \begin{bmatrix} -1 & 0 & 0 \\ 0 & 1 & 0 \\ 0 & 0 & 1 \end{bmatrix}$.
   See below
   for answer matrix.

b. Graph the resulting image.

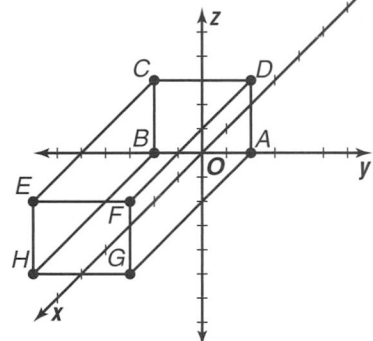

c. Describe the transformation represented by matrix *T*.
   The transformation matrix
   $\begin{bmatrix} -1 & 0 & 0 \\ 0 & 1 & 0 \\ 0 & 0 & 1 \end{bmatrix}$ **reflects the image of each vertex over the *yz*-plane.**

Answer matrix for Example 3a

$$TM = \begin{bmatrix} -2 & -2 & -2 & -2 & 3 & 3 & 3 & 3 \\ 2 & -2 & -2 & 2 & -2 & 2 & 2 & -2 \\ -2 & -2 & 1 & 1 & 1 & 1 & -2 & -2 \end{bmatrix}.$$

Teaching Tip Remind students that the result of the transformation

matrix $D = \begin{bmatrix} k & 0 & 0 \\ 0 & k & 0 \\ 0 & 0 & k \end{bmatrix}$ is a dilation

that increases the original figure by a factor of k.

In-Class Example

Example 4

A cube has vertices $A(1, 1, -1)$, $B(1, -1, -1)$, $C(1, -1, 1)$, $D(1, 1, 1)$, $E(-1, -1, 1)$, $F(-1, 1, 1)$, $G(-1, 1, -1)$, and $H(-1, -1, -1)$.

a. Find the vertex matrix for the transformation D where $k = 3$.

See below for answer matrix.

b. Draw a graph of the resulting figure.

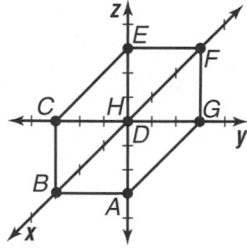

c. What effect does transformation D have on the original figure?

The transformation matrix D is a dilation. The dimensions of the cube have increased by a factor of 3.

	Reflection Matrices	
For a reflection over the:	**Multiply the vertex matrix by:**	**Resulting image**
yz-plane	$R_{yz\text{-plane}} = \begin{bmatrix} -1 & 0 & 0 \\ 0 & 1 & 0 \\ 0 & 0 & 1 \end{bmatrix}$	
xz-plane	$R_{xz\text{-plane}} = \begin{bmatrix} 1 & 0 & 0 \\ 0 & -1 & 0 \\ 0 & 0 & 1 \end{bmatrix}$	
xy-plane	$R_{xy\text{-plane}} = \begin{bmatrix} 1 & 0 & 0 \\ 0 & 1 & 0 \\ 0 & 0 & -1 \end{bmatrix}$	

One other transformation of two-dimensional figures that you have studied is the *dilation*. A dilation with scale factor k, for $k \neq 0$, can be represented by the

matrix $D = \begin{bmatrix} k & 0 & 0 \\ 0 & k & 0 \\ 0 & 0 & k \end{bmatrix}$.

Example 4 A parallelepiped is a prism whose faces are all parallelograms as shown in the graph.

a. Find the vertex matrix for the transformation D where $k = 2$.

b. Draw a graph of the resulting figure.

c. What effect does transformation D have on the original figure?

a. If $k = 2$, $D = \begin{bmatrix} 2 & 0 & 0 \\ 0 & 2 & 0 \\ 0 & 0 & 2 \end{bmatrix}$.

Find the coordinates of the vertices of the parallelepiped. Write them as vertex matrix P.

$$P = \begin{array}{cccccccc} A & B & C & D & E & F & G & H \end{array}$$
$$P = \begin{bmatrix} 1 & 0 & 0 & 1 & -2 & -2 & -3 & -3 \\ 0 & 1 & 5 & 4 & 0 & 4 & 1 & 5 \\ 0 & 5 & 5 & 0 & 0 & 0 & 5 & 5 \end{bmatrix}$$

538 Chapter 8 Vectors and Parametric Equations

Answer matrix for Example 4a

$$\begin{array}{cccccccc} A' & B' & C' & D' & E' & F' & G' & H' \end{array}$$
$$D \cdot P = \begin{bmatrix} 3 & 3 & 3 & 3 & -3 & -3 & -3 & -3 \\ 3 & -3 & -3 & 3 & -3 & 3 & 3 & -3 \\ -3 & -3 & 3 & 3 & 3 & 3 & -3 & -3 \end{bmatrix}$$

Then, find the product of D and P.

$$DP = \begin{bmatrix} 2 & 0 & 0 \\ 0 & 2 & 0 \\ 0 & 0 & 2 \end{bmatrix} \cdot \begin{bmatrix} 1 & 0 & 0 & 1 & -2 & -2 & -3 & -3 \\ 0 & 1 & 5 & 4 & 0 & 4 & 1 & 5 \\ 0 & 5 & 5 & 0 & 0 & 0 & 5 & 5 \end{bmatrix}$$

$$= \begin{matrix} & A' & B' & C' & D' & E' & F' & G' & H' \\ & \begin{bmatrix} 2 & 0 & 0 & 2 & -4 & -4 & -6 & -6 \\ 0 & 2 & 10 & 8 & 0 & 8 & 2 & 10 \\ 0 & 10 & 10 & 0 & 0 & 0 & 10 & 10 \end{bmatrix} \end{matrix}$$

b. Graph the transformation.

c. The transformation matrix D is a dilation. The dimensions of the prism have increased by a factor of 2.

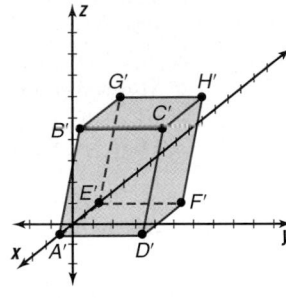

Error Analysis

Watch for students who make errors and therefore need additional practice multiplying two matrices together.

Prevent by providing students with extra examples of matrices to multiply. Also, review the process on the graphing calculator.

Additional Answers

1. It produces a reflection over the *yz*-plane, and increases the dimensions two-fold.

2. $\begin{bmatrix} 2 & 2 & 2 & 2 & 2 & 2 \\ 1 & 1 & 1 & 1 & 1 & 1 \\ -1 & -1 & -1 & -1 & -1 & -1 \end{bmatrix}$

3. The transformations are the same.

CHECK FOR UNDERSTANDING

Communicating Mathematics

Read and study the lesson to answer each question.   1–2. See margin.

1. **Describe** the transformation that matrix $T = \begin{bmatrix} -2 & 0 & 0 \\ 0 & 2 & 0 \\ 0 & 0 & 2 \end{bmatrix}$ produces on a three-dimensional figure.

2. **Write** a transformation matrix that represents the translation shown at the right.

3. **Determine** if multiplying a vertex matrix by the transformation matrix

$T = \begin{bmatrix} -1 & 0 & 0 \\ 0 & 1 & 0 \\ 0 & 0 & -1 \end{bmatrix}$ produces the same result as multiplying by

$U = \begin{bmatrix} 1 & 0 & 0 \\ 0 & 1 & 0 \\ 0 & 0 & -1 \end{bmatrix}$ and then by $V = \begin{bmatrix} -1 & 0 & 0 \\ 0 & 1 & 0 \\ 0 & 0 & 1 \end{bmatrix}$.   **See margin.**

Lesson 8-8   Transformation Matrices in Three-Dimensional Space   **539**

Addressing Individual Needs

Verbal/Linguistic Have students work in pairs. Have one student in each pair describe the graph of a three-dimensional shape by listing ordered triples while the other student determines what the three-dimensional shape looks like. Partners should then switch roles.

Study Guide and Practice Masters, pp. 151–152

Assignment Guide

Core: 7–21 odd, 23–30
Enriched: 8–18 even, 20–30

Additional Answers

5a. $\begin{bmatrix} 5 & 5 & 0 & 0 & 0 & 5 & 5 & 0 \\ 2 & 5 & 5 & 2 & 2 & 2 & 5 & 5 \\ 0 & 0 & 0 & 0 & 4 & 4 & 4 & 4 \end{bmatrix}$

5b. $\begin{bmatrix} 9 & 9 & 4 & 4 & 4 & 9 & 9 & 4 \\ 1 & 4 & 4 & 1 & 1 & 1 & 4 & 4 \\ 2 & 2 & 2 & 2 & 6 & 6 & 6 & 6 \end{bmatrix}$

5c.

Reflection over the *xz*-plane.

5d. The dimensions of the resulting figure are half the original.

6a. $D = \begin{bmatrix} 4 & 0 & 0 \\ 0 & 4 & 0 \\ 0 & 0 & 4 \end{bmatrix}$

$T = \begin{bmatrix} 2 & 2 & 2 & 2 & 2 & 2 & 2 & 2 \\ 0 & 0 & 0 & 0 & 0 & 0 & 0 & 0 \\ 0 & 0 & 0 & 0 & 0 & 0 & 0 & 0 \end{bmatrix}$

6b. Sample graph:

7. $\begin{bmatrix} 2 & 3 & 4 & 4 & 2 & 3 \\ -3 & 1 & 1 & 7 & 3 & 7 \\ 2 & 4 & -1 & -1 & 2 & 4 \end{bmatrix}$

8. $\begin{bmatrix} -3 & -3 & -3 & -3 & 4 & 4 & 4 & 4 \\ -2 & 1 & 1 & -2 & -2 & -2 & 1 & 1 \\ 2 & 2 & -2 & -2 & -2 & 2 & 2 & -2 \end{bmatrix}$

9. $\begin{bmatrix} 2 & 1 & 4 & 4 & 3 & 6 \\ -2 & 0 & -1 & -1 & 1 & 0 \\ 3 & 4 & 2 & 1 & 2 & 0 \end{bmatrix}$

---

4. *Math Journal* Using a chart, describe the effects reflection, translation, and dilation have on **a–c. See Answer Appendix.**

   a. the figure's orientation on the coordinate system.

   b. the figure's size.

   c. the figure's shape.

Guided Practice

5. Refer to the rectangular prism at the right. **a–d. See margin.**

   a. Write the matrix for the figure.

   b. Write the resulting matrix if you translate the figure using the vector $\langle 4, -1, 2 \rangle$.

   c. Transform the figure using the matrix $\begin{bmatrix} 1 & 0 & 0 \\ 0 & -1 & 0 \\ 0 & 0 & 1 \end{bmatrix}$. Graph the image and describe the result.

   d. Describe the transformation on the figure resulting from its product with the matrix $\begin{bmatrix} 0.5 & 0 & 0 \\ 0 & 0.5 & 0 \\ 0 & 0 & 0.5 \end{bmatrix}$.

6. **Architecture** Trevor is revising a design for a playground that contains a piece of equipment shaped like a rectangular prism. He needs to enlarge the prism 4 times its size and move it along the *x*-axis 2 units.

   a. What are the transformation matrices? **a–b. See margin.**

   b. Graph the rectangular prism and its image after the transformation.

EXERCISES

Practice

Write the matrix for each prism. **7–9. See margin.**

A

7.

8.

9.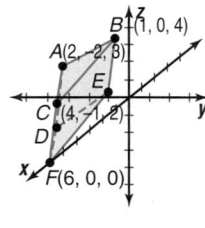

Use the prism at the right for Exercises 10-15. The vertices are $A(0, 0, 1)$, $B(0, 3, 2)$, $C(0, 5, 5)$, $D(0, 0, 4)$, $E(2, 0, 1)$, $F(2, 0, 4)$, $G(2, 3, 5)$, and $H(2, 3, 2)$. Translate the prism using the given vectors. Graph each image and describe the result.
10–12. See margin.

B

10. $\vec{a} \langle 0, -2, 4 \rangle$ 11. $\vec{b} \langle 1, -2, -2 \rangle$ 12. $\vec{c} \langle 1, 5, -3 \rangle$

 www.amc.glencoe.com/self_check_quiz

Additional Answers

10. translation −2 units along the *y*-axis, and 4 units along the *z*-axis

11. translation 1 unit along the *x*-axis, −2 units along the *y*-axis, and −2 units along the *z*-axis

Refer to the figure for Exercises 10-12. Transform the figure using each matrix. Graph each image and describe the result. **13–15. See margin.**

13. $\begin{bmatrix} 1 & 0 & 0 \\ 0 & 1 & 0 \\ 0 & 0 & 1 \end{bmatrix}$ 14. $\begin{bmatrix} 1 & 0 & 0 \\ 0 & -1 & 0 \\ 0 & 0 & -1 \end{bmatrix}$ 15. $\begin{bmatrix} -1 & 0 & 0 \\ 0 & -1 & 0 \\ 0 & 0 & -1 \end{bmatrix}$

Describe the result of the product of a vertex matrix and each matrix.

16. $\begin{bmatrix} 2 & 0 & 0 \\ 0 & 2 & 0 \\ 0 & 0 & 2 \end{bmatrix}$ 17. $\begin{bmatrix} 3 & 0 & 0 \\ 0 & 3 & 0 \\ 0 & 0 & -3 \end{bmatrix}$ 18. $\begin{bmatrix} -0.75 & 0 & 0 \\ 0 & -0.75 & 0 \\ 0 & 0 & -0.75 \end{bmatrix}$

16–18. See margin.

19. A transformation in three-dimensional space can also be represented by $T(x, y, z) \rightarrow (2x, 2y, 5z)$. **a–b. See margin.**
   a. Determine the transformation matrix.
   b. Describe the transformation that such a matrix will have on a figure.

Applications and Problem Solving

20. **Marine Biology** A researcher studying a group of dolphins uses the matrix

$$\begin{bmatrix} 20 & 136 & 247 & 302 & 351 \\ -58 & -71 & -74 & -83 & -62 \\ 27 & 53 & 59 & 37 & 52 \end{bmatrix},$$

with the ship at the origin, to track the dolphins' movement. Later, the researcher will translate the matrix to a fixed reference point using the vector $\langle 23.6, 72, 0 \rangle$. **a–c. See margin.**
   a. Write the translation matrix used for the transformation.
   b. What is the resulting matrix?
   c. Describe the result of the translation.

21. $\begin{bmatrix} -0.5 & 0 & 0 \\ 0 & 0.5 & 0 \\ 0 & 0 & 0.5 \end{bmatrix}$

21. **Critical Thinking** Write a transformational matrix that would first reflect a rectangular prism over the yz-plane and then reduce its dimensions by half.

22. **Meteorology** At the National Weather Service Center in Miami, Florida, meteorologists test models to forecast weather phenomena such as hurricanes and tornadoes. One weather disturbance, wind shear, can be modeled using a transformation on a cube. (*Hint:* You can use any size cube for the model.)

22a. Sample answer:
$\begin{bmatrix} 1 & 0 & 1 \\ 0 & 1 & 0 \\ 0 & 0 & 1 \end{bmatrix}$

   a. Write a matrix to transform a cube into a slanted parallelepiped.
   b. Graph a cube and its image after the transformation. **See Answer Appendix.**

23. **Geometry** Suppose a cube is transformed by two matrices, first by $U = \begin{bmatrix} -1 & 0 & 0 \\ 0 & -1 & 0 \\ 0 & 0 & -1 \end{bmatrix}$, and then by $T = \begin{bmatrix} 1 & 0 & 2 \\ 0 & 2 & 0 \\ 0 & 0 & 2 \end{bmatrix}$.

Describe the resulting image. **See margin.**

Additional Answers

14.
reflection across the *xy*- and *xz*-planes

15.
reflection across all three coordinate planes

16. The figure is twice the original size.

17. The figure is three times the original size and reflected over the *xy*-plane.

18. The figure is three-fourths the original size and reflected over all three coordinate planes.

19a. $\begin{bmatrix} 2 & 0 & 0 \\ 0 & 2 & 0 \\ 0 & 0 & 5 \end{bmatrix}$

19b. The transformation will magnify the *x*- and *y*-dimensions two-fold, and the *z*-dimension 5-fold.

20a. $\begin{bmatrix} 23.6 & 23.6 & 23.6 & 23.6 & 23.6 \\ 72 & 72 & 72 & 72 & 72 \\ 0 & 0 & 0 & 0 & 0 \end{bmatrix}$

20b. $\begin{bmatrix} 43.6 & 159.6 & 270.6 & 325.6 & 374.6 \\ 14 & 1 & -2 & -11 & 10 \\ 27 & 53 & 59 & 37 & 52 \end{bmatrix}$

20c. translation 23.6 units along the *x*-axis and 72 units along the *y*-axis

23. The first transformation reflects the figure over all three coordinate planes. The second transformation stretches the dimensions along *y*- and *z*-axes and skews it along the *xy*-plane.

Additional Answers (pp. 540–541)

12. translation 1 unit along the *x*-axis, 5 units along the *y*-axis, and −3 units along the *z*-axis

13.
no change

Open-Ended Assessment

Modeling Have students make up the coordinates of the vertices of a parallelepiped or a rectangular prism, and then plot these vertices to make sure that they are accurate.

 Chapter 8, Quiz D (Lessons 8-7 and 8-8) is available in the *Assessment and Evaluation Masters*, p. 200.

Additional Answers

25b.
$$\begin{bmatrix} 0 & 0 & 0 & 0 & 0 & 0 \\ 1.6 & 1.6 & 1.6 & 1.6 & 1.6 & 1.6 \\ -1.2 & -1.2 & -1.2 & -1.2 & -1.2 & -1.2 \end{bmatrix}$$

26b. Yes; both stones take 3.06 seconds to reach the ground.

Enrichment Masters, p. 92

Enrichment

Spherical Coordinates

There are many coordinate systems for locating a point in the two-dimensional plane. You have studied the two most common systems, rectangular coordinates and polar coordinates. The most commonly used three-dimensional coordinate systems are the extended rectangular system, with an added *z*-axis, and the *spherical coordinate* system, a modification of polar coordinates.

Note that the orientation of the axes shown is a different perspective than that used in your textbook.

Point $P(d, \theta, \phi)$ in three-dimensional space is located using three spherical coordinates:
  d = distance from origin
  θ = angle relative to *x*-axis
  ϕ = angle relative to *y*-axis

The figure at the right shows point Q with rectangular coordinates (2, 5, 6).

1. Find OA and AB.
   2; 6
2. Find OB by using the Pythagorean theorem.
   $2\sqrt{10}$
3. Find QB.
   5
4. Find d.
   $\sqrt{65}$
5. Use inverse trigonometric functions to find θ and ϕ to the nearest degree. Write the spherical coordinates of Q.
   $(\sqrt{65}, 72°, 52°)$

Find the spherical coordinates of the point with the given rectangular coordinates. Round distances to the nearest tenth and angles to the nearest degree.
6. (4, 12, 3)
   (13, 37°, 23°)
7. (−2, −3, −1)
   (3.7, −153°, 143°)
8. (a, b, c)
   $\left(\sqrt{a^2 + b^2 + c^2},\ \arctan\left(\frac{c}{a}\right),\ \arccos\left(\frac{b}{\sqrt{a^2 + b^2 + c^2}}\right)\right)$

© Glencoe/McGraw-Hill   T92   Advanced Mathematical Concepts

---

24. $\begin{bmatrix} 3 & 0 & 0 \\ 0 & 2 & 0 \\ 1 & 0 & -4 \end{bmatrix}$

24. Critical Thinking Matrix T maps a point $P(x, y, z)$ to the point $P'(3x, 2y, x - 4z)$. Write a 3×3 matrix for T.

25. Seismology Seismologists classify movements in the earth's crust by determining the direction and amount of movement which has taken place on a fault. Some of the classifications are shown using *block diagrams*.

dip-slip   strike-slip   oblique slip   reverse fault

a. A seismologist used the matrix below to describe a particular feature along a fault.

$$\begin{bmatrix} 123.9 & -41.3 & 201.7 & 73.8 & -129.4 & 36.4 \\ 86.4 & 144.2 & -29.9 & -84.2 & 95.5 & -125.5 \\ 206.5 & 247.8 & 262.7 & 213.2 & -165.2 & -84.1 \end{bmatrix}$$

After a series of earthquakes he determined the matrix describing the feature had changed to the following matrix.

$$\begin{bmatrix} 123.9 & -41.3 & 201.7 & 73.8 & -129.4 & 36.4 \\ 88.0 & 145.8 & -28.3 & -82.6 & 97.1 & -123.9 \\ 205.3 & 246.6 & 261.5 & 212.0 & -166.4 & -85.3 \end{bmatrix}$$

What classification of movement bests describes the transformation that has occurred? **dip-slip**

b. Determine the matrix that describes the movement that occurred. **See margin.**

Mixed Review

26. Physics Jaimie and LaShawna are standing at the edge of a cliff that is 150 feet high. At the same time, LaShawna drops a stone and Jaimie throws a stone horizontally at a velocity of 35 ft/s. *(Lesson 8-7)*

a. About how far apart will the stones be when they land? **107 ft**

b. Will the stones land at the same time? Explain. **See margin.**

27. Write an equation in slope-intercept form of the line with parametric equations $x = -5t - 1$ and $y = 2t + 10$. *(Lesson 8-6)* $y = -\frac{2}{5}x + \frac{48}{5}$

28. Evaluate $\sec\left(\cos^{-1}\frac{2}{5}\right)$ if the angle is in Quadrant I. *(Lesson 6-8)* $\frac{5}{2}$

29. Medicine Maria was told to take 80 milligrams of medication each morning for three days. The amount of medicine in her body on the fourth day is modeled by $M(x) = 80x^3 + 80x^2 + 80x$, where x represents the absorption rate per day. Suppose Maria has 24.2 milligrams of the medication present in her body on the fourth day. Find the absorption rate of the medication. *(Lesson 4-5)* ≈ 0.2

30. SAT/ACT Practice Which of the following equations are equivalent? **A**

I. $2x + 4y = 8$ II. $3x + 6y = 12$

III. $4x + 8y = 8$ IV. $6x + 12y = 16$

A I and II only   **B** I and IV only   **C** II and IV only

D III and IV only   **E** I, II, and III

Extra Practice See p. A41.

Extra Credit

What three-dimensional figure has vertices at A (0, 4, 5), B (0, 5, −1), C (4, 5, −1), D (0, 3, 3), E (0, 4, −3), and F (4, 4, −3)? Find the area of a base of the figure.

triangular prism; $2\sqrt{37}$ units²

VOCABULARY

component (p. 488)	parametric equation (p. 520)
cross product (p. 507)	polyhedron (p. 535)
direction (p. 485)	resultant (p. 486)
direction vector (p. 520)	scalar (p. 488)
dot product (p. 506)	scalar quantity (p. 488)
equal vectors (p. 485)	standard position (p. 485)
inner product (p. 505)	unit vector (p. 495)
magnitude (p. 485)	vector (p. 485)
opposite vectors (p. 487)	vector equation (p. 520)
parallel vectors (p. 488)	zero vector (p. 485)
parameter (p. 520)	

UNDERSTANDING AND USING THE VOCABULARY

Choose the correct term from the list to complete each sentence.

1. The __?__ of two or more vectors is the sum of the vectors.
   resultant

2. A(n) __?__ vector has a magnitude of one unit.   **unit**

3. Two vectors are equal if and only if they have the same direction
   and __?__.   **magnitude**

4. The __?__ product of two vectors is a vector.   **cross**

5. Two vectors in space are perpendicular if and only if their __?__
   product is zero.   **inner**

6. Velocity can be represented mathematically by a(n) __?__.   **vector**

7. Vectors with the same direction and different magnitudes are __?__.
   parallel

8. A vector with its initial point at the origin is in __?__ position.
   standard

9. A __?__ vector is used to describe the slope of a line.   **direction**

10. Two or more vectors whose sum is a given vector are called __?__ of the given vector.
    components

components
cross
direction
equal
inner
magnitude
parallel
parameter
resultant
scalar
standard
unit
vector

inter NET CONNECTION   For additional review and practice for each lesson, visit: **www.amc.glencoe.com**

Six forms of Chapter 8 Test are
available in the *Assessment and
Evaluation Masters* as shown in
the chart.

Chapter 8 Test, Form 1B, is shown at
the right. Chapter 8 Test, Form 2B, is
shown on the next page.

TEST OPTIONS

Form	Pages	Type	
1A	185–186	Honors	Form 1 tests are multiple choice.
1B	187–188	Average	
1C	189–190	Basic	
2A	191–192	Honors	Form 2 tests are free response.
2B	193–194	Average	
2C	195–196	Basic	

STUDY GUIDE AND ASSESSMENT

VOCABULARY

This section provides a listing of
the new terms, properties, and
phrases that were introduced in
this chapter. The exercises check
students' understanding of the
terms by using a variety of verbal
formats including matching,
completion, and true/false.

Glossary A complete glossary of
terms begins on page A71.

Assessment and Evaluation Masters, pp. 187–188

SKILLS AND CONCEPTS

Objectives and Examples This section reviews the skills and concepts of the chapter and shows completely worked examples.

Review Exercises These exercises provide practice for the corresponding objectives.

Assessment and Evaluation Masters, pp. 193–194

SKILLS AND CONCEPTS

OBJECTIVES AND EXAMPLES	REVIEW EXERCISES

Lesson 8-1 Add and subtract vectors geometrically.

• Find the sum of $\vec{a}$ and $\vec{b}$ using the parallelogram method.

Place the initial points of $\vec{a}$ and $\vec{b}$ together. Draw dashed lines to form a complete parallelogram. The diagonal from the initial points to the opposite vertex of the parallelogram is the resultant.

Use a ruler and a protractor to determine the magnitude (in centimeters) and direction of each vector.

11.   12.

    1.3 cm; 50°   2.9 cm; 10°

Use $\vec{p}$ and $\vec{q}$ above to find the magnitude and direction of each resultant.

13. $\vec{p} + \vec{q}$ **4.1 cm; 23°** 14. $2\vec{p} + \vec{q}$ **5.2 cm; 29°**
15. $3\vec{p} - \vec{q}$ **2.5 cm; 98°** 16. $4\vec{p} - \vec{q}$ **3.5 cm; 82°**

Find the magnitude of the horizontal and vertical components of each vector shown for Exercises 11 and 12.
17. $\vec{p}$ **1 cm, 0.8 cm**   18. $\vec{q}$ **2.9 cm, 0.5 cm**

Lesson 8-2 Find ordered pairs that represent vectors, and sums and products of vectors.

• Write the ordered pair that represents the vector from $M(3, 1)$ to $N(-7, 4)$. Then find the magnitude of $\overrightarrow{MN}$.

$\overrightarrow{MN} = \langle -7 - 3, 4 - 1 \rangle$ or $\langle -10, 3 \rangle$

$|\overrightarrow{MN}| = \sqrt{(-7 - 3)^2 + (4 - 1)^2}$ or $\sqrt{109}$

Find $\vec{a} + \vec{b}$ if $\vec{a} = \langle 1, -5 \rangle$ and $\vec{b} = \langle -2, 4 \rangle$.
$\vec{a} + \vec{b} = \langle 1, -5 \rangle + \langle -2, 4 \rangle$
$= \langle 1 + (-2), -5 + 4 \rangle$ or $\langle -1, -1 \rangle$

Write the ordered pair that represents $\overrightarrow{CD}$. Then find the magnitude of $\overrightarrow{CD}$.
19. $C(2, 3), D(7, 15)$ $\langle 5, 12 \rangle$; **13**
20. $C(-2, 8), D(4, 12)$ $\langle 6, 4 \rangle$; $2\sqrt{13}$
21. $C(2, -3), D(0, 9)$ $\langle -2, 12 \rangle$; $2\sqrt{37}$
22. $C(-6, 4), D(-5, -4)$ $\langle 1, -8 \rangle$; $\sqrt{65}$

Find an ordered pair to represent $\vec{u}$ in each equation if $\vec{v} = \langle 2, -5 \rangle$ and $\vec{w} = \langle 3, -1 \rangle$.
23. $\vec{u} = \vec{v} + \vec{w}$ $\langle 5, -6 \rangle$
24. $\vec{u} = \vec{v} - \vec{w}$ $\langle -1, -4 \rangle$
25. $\vec{u} = 3\vec{v} + 2\vec{w}$ $\langle 12, -17 \rangle$
26. $\vec{u} = 3\vec{v} - 2\vec{w}$ $\langle 0, -13 \rangle$

Lesson 8-3 Find the magnitude of vectors in three-dimensional space.

• Write the ordered triple that represents the vector from $R(-2, 0, 8)$ to $S(5, -4, -1)$. Then find the magnitude of $\overrightarrow{RS}$.

$\overrightarrow{RS} = \langle 5 - (-2), -4 - 0, -1 - 8 \rangle$
$= \langle 7, -4, -9 \rangle$

$|\overrightarrow{RS}| = \sqrt{(5 - (-2))^2 + (-4 - 0)^2 + (-1 - 8)^2}$
$= \sqrt{49 + 16 + 81}$ or about 12.1

Write the ordered triple that represents $\overrightarrow{EF}$. Then find the magnitude of $\overrightarrow{EF}$.
27. $E(2, -1, 4), F(6, -2, 1)$ $\langle 4, -1, -3 \rangle$; $\sqrt{26}$
28. $E(9, 8, 5), F(-1, 5, 11)$ $\langle -10, -3, 6 \rangle$; $\sqrt{145}$
29. $E(-4, -3, 0), F(2, -1, 7)$ $\langle 6, 2, 7 \rangle$; $\sqrt{89}$
30. $E(3, 7, -8), F(-4, 0, 5)$ $\langle -7, -7, 13 \rangle$; $\sqrt{267}$

Find an ordered triple to represent $\vec{u}$ in each equation if $\vec{v} = \langle -1, 7, -4 \rangle$ and $\vec{w} = \langle 4, -1, 5 \rangle$.
31. $\vec{u} = 2\vec{w} - 5\vec{v}$ 32. $\vec{u} = 0.25\vec{v} + 0.4\vec{w}$
    $\langle 13, -37, 30 \rangle$ $\langle 1.35, 1.35, 1 \rangle$

TestCheck and Worksheet Builder

This **networkable** software has three modules.
• **Worksheet Builder** to make worksheets and tests
• **Student Module** to take tests on-screen
• **Management System** to keep student records
Three special banks are included for SAT, ACT, and TIMSS.

OBJECTIVES AND EXAMPLES

Lesson 8-4 Find the inner and cross products of vectors.

• Find the inner product of $\vec{a}$ and $\vec{b}$ if $\vec{a} = \langle 3, -1, 7 \rangle$ and $\vec{b} = \langle 0, -2, -4 \rangle$. Are $\vec{a}$ and $\vec{b}$ perpendicular?

$\vec{a} \cdot \vec{b} = 3(0) + (-1)(-2) + 7(-4)$ or -26

$\vec{a}$ and $\vec{b}$ are not perpendicular since their inner product is not zero.

Find $\vec{c} \times \vec{d}$ if $\vec{c} = \langle -2, 1, 1 \rangle$ and $\vec{d} = \langle 1, -3, 0 \rangle$.

$$\vec{c} \times \vec{d} = \begin{vmatrix} \vec{i} & \vec{j} & \vec{k} \\ -2 & 1 & 1 \\ 1 & -3 & 0 \end{vmatrix}$$

$$= \begin{vmatrix} 1 & 1 \\ -3 & 0 \end{vmatrix} \vec{i} - \begin{vmatrix} -2 & 1 \\ 1 & 0 \end{vmatrix} \vec{j} + \begin{vmatrix} -2 & 1 \\ 1 & -3 \end{vmatrix} \vec{k}$$

$= 3\vec{i} + 1\vec{j} + 5\vec{k}$ or $\langle 3, 1, 5 \rangle$

Determine if $\vec{a}$ is perpendicular to $\vec{b}$ and $\vec{c}$ if $\vec{a} = \langle 3, 1, 5 \rangle$, $\vec{b} = \langle -2, 1, 1 \rangle$, and $\vec{c} = \langle 1, -3, 0 \rangle$.

$\vec{a} \cdot \vec{b} = 3(-2) + 1(1) + 5(1)$ or 0
$\vec{a} \cdot \vec{c} = 3(1) + 1(-3) + 5(0)$ or 0

Since the inner products are zero, $\vec{a}$ is perpendicular to both $\vec{b}$ and $\vec{c}$.

Lesson 8-5 Solve problems using vectors and right triangle trigonometry.

• Find the magnitude and direction of the resultant vector for the diagram.

$|\vec{r}|^2 = 200^2 + 280^2$
$|\vec{r}|^2 = 118{,}400$
$|\vec{r}| = \sqrt{118{,}400}$
≈ 344 N

$\tan \theta = \dfrac{200}{280}$

$\theta = \tan^{-1} \dfrac{200}{280}$

$\approx 35.6°$

REVIEW EXERCISES

Find each inner product and state whether the vectors are perpendicular. Write *yes* or *no*.

33. $\langle 5, -1 \rangle \cdot \langle -2, 6 \rangle$ **−16; no**
34. $\langle 2, 6 \rangle \cdot \langle 3, -4 \rangle$ **−18; no**
35. $\langle 4, 1, -2 \rangle \cdot \langle 3, -4, 4 \rangle$ **0; yes**
36. $\langle 2, -1, 4 \rangle \cdot \langle 6, -2, 1 \rangle$ **18; no**
37. $\langle 5, 2, -10 \rangle \cdot \langle 2, -4, -4 \rangle$ **42; no**

Find each cross product. Then verify if the resulting vector is perpendicular to the given vectors. **38–41. See margin.**

38. $\langle 5, -2, 5 \rangle \times \langle -1, 0, -3 \rangle$
39. $\langle -2, -3, 1 \rangle \times \langle 2, 3, -4 \rangle$
40. $\langle -1, 0, 4 \rangle \times \langle 5, 2, -1 \rangle$
41. $\langle 7, 2, 1 \rangle \times \langle 2, 5, 3 \rangle$

42. Find a vector perpendicular to the plane containing the points $(1, 2, 3)$, $(-4, 2, -1)$, and $(5, -3, 0)$.   **Sample answer:** $\langle -20, -31, 25 \rangle$

Find the magnitude and direction of the resultant vector for each diagram.

43.   **412.31 N, 39.09°**

260 N

320 N

44.   **41 m/s, 108.65°**

30 m/s   26°
12 m/s

Additional Answers

38. $\langle 6, 10, -2 \rangle$; **yes**
$\langle 6, 10, -2 \rangle \cdot \langle 5, -2, 5 \rangle$
$6(5) + 10(-2) + (-2)(5)$
$30 - 20 - 10 = 0$
$\langle 6, 10, -2 \rangle \cdot \langle -1, 0, -3 \rangle$
$6(-1) + 10(0) + (-2)(-3)$
$-6 + 0 + 6 = 0$

39. $\langle 9, -6, 0 \rangle$; **yes**
$\langle 9, -6, 0 \rangle \cdot \langle -2, -3, 1 \rangle$
$9(-2) + -6(-3) + (0)(1)$
$-18 + 18 + 0 = 0$
$\langle 9, -6, 0 \rangle \cdot \langle 2, 3, -4 \rangle$
$9(2) + (-6)(3) + (0)(-4)$
$18 - 18 + 0 = 0$

40. $\langle -8, 19, -2 \rangle$; **yes**
$\langle -8, 19, -2 \rangle \cdot \langle -1, 0, 4 \rangle$
$-8(-1) + 19(0) + (-2)(4)$
$8 + 0 - 8 = 0$
$\langle -8, 19, -2 \rangle \cdot \langle 5, 2, -1 \rangle$
$(-8)(5) + 19(2) + (-2)(-1)$
$-40 + 38 + 2 = 0$

41. $\langle 1, -19, 31 \rangle$; **yes**
$\langle 1, -19, 31 \rangle \cdot \langle 7, 2, 1 \rangle$
$(1)(7) + (-19)(2) + (31)(1)$
$7 - 38 + 31 = 0$
$\langle 1, -19, 31 \rangle \cdot \langle 2, 5, 3 \rangle$
$(1)(2) + (-19)(5) + (31)(3)$
$2 - 95 + 93 = 0$

Chapter 8   Study Guide and Assessment   **545**

Additional Answers

45. $\langle x - 3, y + 5 \rangle = t\langle 4, 2 \rangle$;
$x = 3 + 4t, y = -5 + 2t$

46. $\langle x + 1, y - 9 \rangle = t\langle -7, -5 \rangle$;
$x = -1 - 7t, y = 9 - 5t$

47. $\langle x - 4, y \rangle = t\langle 3, -6 \rangle$;
$x = 4 + 3t, y = -6t$

48. $x = t, y = -7 - 8t$

49. $x = t, y = -\frac{1}{2}t + \frac{5}{2}$

53.

54.

OBJECTIVES AND EXAMPLES

Lesson 8-6 Write vector and parametric equations of lines.

• Write a vector equation of the line that passes through $P(-6, 3)$ and is parallel to $\overline{\mathbf{v}} = \langle 1, 4 \rangle$. Then write parametric equations of the line.

The vector equation is $\langle x + 6, y - 3 \rangle = t\langle 1, 4 \rangle$.

Write the parametric equations of a line with $(a_1, a_2) = (-6, 3)$ and $(x_1, y_2) = (1, 4)$.
$x = x_1 + ta_1$ $\qquad$ $y = y_1 + ta_2$
$x = -6 + t(1)$ $\qquad$ $y = 3 + t(4)$
$x = -6 + t$ $\qquad$ $y = 3 + 4t$

Lesson 8-7 Model the motion of a projectile using parametric equations.

• Find the initial horizontal and vertical velocity for an arrow released at 52 m/s at an angle of 12° with the horizontal.

$|\overline{\mathbf{v}}_x| = |\overline{\mathbf{v}}| \cos \theta$ $\qquad$ $|\overline{\mathbf{v}}_y| = |\overline{\mathbf{v}}| \sin \theta$
$|\overline{\mathbf{v}}_x| = 52 \cos 12°$ $\qquad$ $|\overline{\mathbf{v}}_y| = 52 \sin 12°$
$|\overline{\mathbf{v}}_x| \approx 50.86$ $\qquad$ $|\overline{\mathbf{v}}_y| \approx 10.81$

The initial horizontal velocity is about 50.86 m/s and the initial vertical velocity is about 10.81 m/s.

Lesson 8-8 Transform three-dimensional figures using matrix operations to describe the transformation.

• A prism needs to be translated using the vector $\langle 0, 1, 2 \rangle$. Write a matrix that will have such an effect on a figure.

$$T = \begin{bmatrix} 0 & 0 & 0 & 0 & 0 & 0 & 0 & 0 \\ 1 & 1 & 1 & 1 & 1 & 1 & 1 & 1 \\ 2 & 2 & 2 & 2 & 2 & 2 & 2 & 2 \end{bmatrix}$$

53. moves 2 units along x-axis and 3 units along the z-axis. See margin for graph.

54. reflection over the xz-plane. See margin for graph.

REVIEW EXERCISES

Write a vector equation of the line that passes through point P and is parallel to $\overline{\mathbf{v}}$. Then write parametric equations of the line. **45–47. See margin.**

45. $P(3, -5), \overline{\mathbf{v}} = \langle 4, 2 \rangle$

46. $P(-1, 9), \overline{\mathbf{v}} = \langle -7, -5 \rangle$

47. $P(4, 0), \overline{\mathbf{v}} = \langle 3, -6 \rangle$

Write parametric equations of each line with the given equation. **48–49. See margin.**

48. $y = -8x - 7$

49. $y = -\frac{1}{2}x + \frac{5}{2}$

Find the initial horizontal and vertical velocity for each situation.

50. a stone thrown with an initial velocity of 15 feet per second at an angle of 55° with the horizontal **8.60 ft/s, 12.29 ft/s**

51. a baseball thrown with an initial velocity of 13.2 feet per second at an angle of 66° with the horizontal **5.37 ft/s, 12.06 ft/s**

52. a soccer ball kicked with an initial velocity of 18 meters per second at an angle of 28° with the horizontal **15.89 m/s, 8.45 m/s**

Use the prism for Exercises 53 and 54.

53. Translate the figure using the vector $\overline{\mathbf{n}} = \langle 2, 0, 3 \rangle$. Graph the image and describe the result.

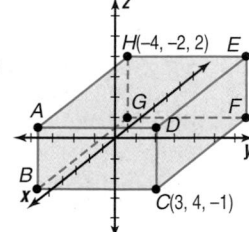

54. Transform the figure using the matrix

$$M = \begin{bmatrix} 1 & 0 & 0 \\ 0 & -1 & 0 \\ 0 & 0 & 1 \end{bmatrix}.$$ Graph the image and describe the result.

---

546 Chapter 8 Vectors and Parametric Equations

APPLICATIONS AND PROBLEM SOLVING

55. Physics Karen uses a large wrench to change the tire on her car. She applies a downward force of 50 pounds at an angle of 60° one foot from the center of the lug nut. Find the torque. *(Lesson 8-4)* **25 lb-ft**

56. Sports Bryan punts a football with an initial velocity of 38 feet per second at an angle of 40° from the horizontal. If the ball is 2 feet above the ground when it is kicked, how high is it after 0.5 second? *(Lesson 8-7)* **10.2 feet**

57. Navigation A boat that travels at 16 km/h in calm water is sailing across a current of 3 km/h on a river 250 meters wide. The boat makes an angle of 35° with the current heading into the current. *(Lesson 8-5)*
13.7 km/h
a. Find the resultant velocity of the boat.

b. How far upstream is the boat when it reaches the other shore? **275.3 m**

58. Physics Mario and Maria are moving a stove. They are applying forces of 70 N and 90 N at an angle of 30° to each other. If the 90 N force is applied along the *x*-axis find the magnitude and direction of the resultant of these forces. *(Lesson 8-5)* **154.6 N, 13.1°**

ALTERNATIVE ASSESSMENT

OPEN-ENDED ASSESSMENT

1. The ordered pair $\langle -3, 2 \rangle$ represents $\overrightarrow{XY}$.

a. Give possible coordinates for *X* and *Y*. Show that your coordinates are correct.

b. Find the magnitude of $\overrightarrow{XY}$. Did you need to know the coordinates for *X* and *Y* to find this magnitude? Explain.
1–2. See margin.

2. a. $\overrightarrow{PQ}$ and $\overrightarrow{RS}$ are parallel. Give ordered pairs for *P, Q, R*, and *S* for which this is true. Explain how you know $\overrightarrow{PQ}$ and $\overrightarrow{RS}$ are parallel.

b. $\vec{a}$ and $\vec{b}$ are perpendicular. Give ordered pairs to represent $\vec{a}$ and $\vec{b}$. Explain how you know that $\vec{a}$ and $\vec{b}$ are perpendicular.

PORTFOLIO
Devise a real-world problem that can be solved using vectors. Explain why vectors are needed to solve your problem. Solve your problem. Be sure to show and explain all of your work.

Unit 2 *inter*NET Project

THE CYBERCLASSROOM

Vivid Vectors!

• Search the Internet to find websites that have lessons about vectors and their applications. Find at least three different sources of information.

• Make diagrams of examples that use vectors to solve problems by combining the ideas you found in Internet lessons and your textbook.

• Prepare a presentation to summarize the Internet Project for Unit 2. Design the presentation as a webpage. Use appropriate software to create the webpage.

Additional Assessment See p. A63 for Chapter 8 practice test.

Chapter 8   Study Guide and Assessment **547**

Internet Address Book

Record useful Internet addresses in the space at right for quick reference.

www.amc.glencoe.com

Chapter 8 Skill: Geometry (Perimeter, Area, and Volume)

Pages 548–549 are part of a complete SAT/ACT preparation course that is described in detail on pages T8–T9 of the *Teacher's Handbook*. The test items were written in the same style as those in the SAT and ACT.

 These questions were aligned and verified by The Princeton Review, the nation's leader in test preparation.

Teaching Tip In the ACT example, notice that *ABCD* is a trapezoid. Students could use the formula for the area of a trapezoid after finding *AD*.

$$A = \frac{1}{2}\left(b_1 + b_2\right)h$$

$$A = \frac{1}{2}\left(20 + 27\right)\left(24\right)$$

$$A = 564$$

Diagnosis & Prescription

Each of the 10 SAT and ACT practice problems on page 549 is cross-referenced to the chapter where that SAT or ACT skill is covered. If students miss a particular type of problem, you can have them study that skill.

Problem Number and Skill	Page
1 Geometry (Perimeter, Area, and Volume)	548
2 Coordinate Geometry	272
3 Geometry (Perimeter, Area, and Volume)	548
4 Geometry (Triangles and Quadrilaterals)	482
5 Algebra	124
6 Geometry (Perimeter, Area, and Volume)	548
7 Arithmetic	64
8 Geometry (Perimeter, Area, and Volume)	548
9 Geometry (Perimeter, Area, and Volume)	548
10 Geometry (Perimeter, Area, and Volume)	548

Geometry Problems— Perimeter, Area, and Volume

TEST-TAKING TIP
Recall that the value of π is about 3.14. When you need to estimate a quantity that involves π, use 3 as a rough approximation.

Many SAT and ACT problems use perimeter, circumference, and area. A few problems use volume. Even though the formulas are often given on the test, it is more efficient if you memorize them.

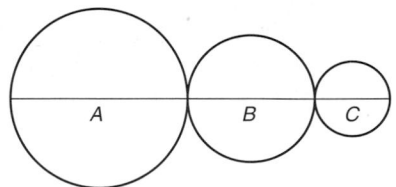

$A = \pi r^2$
$C = 2\pi r$

$A = \ell w$

$A = \frac{1}{2}bh$

SAT EXAMPLE

1. In the figure below, the radius of circle A is twice the radius of the circle B and four times the radius of C. If the sum of the areas of the circles is 84π, what is the measure of $\overline{AC}$?

HINT Use a variable when necessary. Choose it carefully.

Solution This is a grid-in problem. You must find the length of $\overline{AC}$. This segment contains the radii of the circles. Let r be the radius of circle C. Then the radius of circle B is $2r$, and the radius of circle A is $4r$.

You know the total area. Write an equation for the sum of the three areas.

$$\text{Area } A + \text{Area } B + \text{Area } C = 84\pi$$
$$\pi(4r)^2 + \pi(2r)^2 + \pi(r)^2 = 84\pi$$
$$16r^2\pi + 4r^2\pi + r^2\pi = 84\pi$$
$$21r^2\pi = 84\pi$$
$$r^2 = 4$$
$$r = 2$$

The radius of C is 2; the radius of B is 4; the radius of A is 8. Recall that the problem asked for the length of $\overline{AC}$. Use the diagram to see which lengths to add—one radius of A, two radii of B, and one radius of C.

$$AC = 8 + 4 + 4 + 2 \text{ or } 18$$

The answer is 18.

548 *Chapter 8 Vectors and Parametric Equations*

ACT EXAMPLE

2. In the figure below, if $AB = 27$, $CD = 20$, and the area of $\triangle ADC = 240$, what is the area of polygon $ABCD$ in square units?

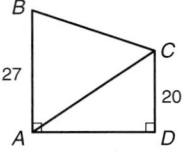

A 420   B 480   C 540   D 564   E 1128

HINT You may write in the test booklets. Mark all information you discover on the figure.

Solution The polygon is made up of two triangles, so find the area of each triangle. You know that the area of $\triangle ADC$ is 240 and its height is 20.

$$A = \frac{1}{2}bh$$
$$240 = \frac{1}{2}(b)(20)$$
$$240 = 10b$$
$$24 = b \qquad \text{So, } AD = 24.$$

Now find the area of $\triangle ABC$. The height of this triangle can be measured along $\overline{AD}$. So, $h = 24$, and the height is the measure of $\overline{AB}$, or 27. The area of $\triangle ABC = \frac{1}{2}(27)(24)$ or 324. You need to add the areas of the two triangles.

$$240 + 324 = 564$$

The answer is choice **D.**

TestCheck and Worksheet Builder

Special banks of SAT, ACT, and TIMSS items are included in this CD-ROM.

After you work each problem, record your answer on the answer sheet provided or on a piece of paper.

Multiple Choice

1. In parallelogram *ABCD* below, *BD* = 3 and *CD* = 5. What is the area of *ABCD*? **A**

   A 12   **B** 15   **C** 18   **D** 20
   E It cannot be determined from the information given.

2. In square *ABCD* below, what is the equation of circle *Q* that is circumscribed around the square? **B**

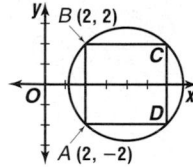

   A $(x - 4)^2 + y^2 = 4$ **B** $(x - 4)^2 + y^2 = 8$
   C $(x + 4)^2 + y^2 = 8$ **D** $(x - 4)^2 + y^2 = 32$
   E $(x + 4)^2 + y^2 = 32$

3. If the perimeter of the rectangle *ABCD* is equal to *p*, and $x = \frac{2}{3}y$, what is the value of *y* in terms of *p*? **B**

   A $\frac{p}{10}$ **B** $\frac{3p}{10}$ **C** $\frac{p}{3}$ **D** $\frac{2p}{5}$ **E** $\frac{3p}{5}$

4. If two sides of a triangle have lengths of 40 and 80, which of the following cannot be the length of the third side? **A**

   A 40   **B** 41   **C** 50   **D** 80   **E** 81

5. $\sqrt[3]{x^2} \cdot \sqrt[9]{x^3} =$ **E**

   A $x^{\frac{2}{9}}$ **B** $x^{\frac{1}{3}}$ **C** $x^{\frac{1}{2}}$ **D** $x^{\frac{2}{3}}$ **E** x

6. The figure below is made of three concentric semi-circles. What is the area of the shaded region in square units? **A**

   A 3π **B** $\frac{9}{2}\pi$ **C** 6π **D** 7π **E** 9π

7. $\frac{1}{5} + \frac{2}{25} + \frac{3}{50} =$ **D**

   A 0.170   **B** 0.240   **C** 0.320
   D 0.340   **E** 0.463

8. A 30-inch by 40-inch rectangular surface is to be completely covered with 1-inch square tiles, which cannot overlap one another and cannot overhang. If white tiles are to cover the interior and red tiles are to form a 1-inch wide border along the edge of the surface, how many red tiles will be needed? **B**

   A 70   **B** 136   **C** 140   **D** 142   **E** 144

9. Which of the following is the closest approximation of the area of a circle with radius *x*? **B**

   A $4x^2$
   B $3x^2$
   C $2x^2$
   D x^2
   E $0.75x^2$

10. **Grid-In**
    In the figure, if *AE* = 1, what is the sum of the area of △*ABC* and the area of △*CDE*? **.5 or 1/2**

SAT/ACT Practice For additional test practice questions, visit: www.amc.glencoe.com

PRINCETON REVIEW *SAT & ACT Preparation* **549**

A bubble-in answer sheet for these practice problems is available on p. 1 of the *Assessment and Evaluation Masters.*

Additional Practice
Additional SAT and ACT practice questions and a bubble-in answer sheet are available in the *Assessment and Evaluation Masters,* pp. 2, 201, and 202.

Teaching Tip For Exercise 9, remind students to draw a diagram of the triangle and a diagram of the circle before finding areas. Then the equations can be compared easily.

Assessment and Evaluation Masters, pp. 201–202

Extra Credit

TIMSS Two vectors $\vec{a}$ and $\vec{b}$ ($\vec{a}, \vec{b} \neq \vec{0}$) are related by $|\vec{a} + \vec{b}| = |\vec{a} - \vec{b}|$. What is the measure of the angle between $\vec{a}$ and $\vec{b}$? **90° or $\frac{\pi}{2}$**

UNIT
3

Unit Overview

Unit 3 focuses on the relationship of different advanced functions and their graphs. The three chapters explore three different topics united by the concept of graphing as a fundamental aspect to their study.

Assessment

A review and a test on the concepts in Unit 3 are provided in the *Assessment and Evaluation Masters* (pp. 409–414). The Unit 3 Review is shown below. The Unit 3 Test is shown on the next page.

Assessment and Evaluation Masters, pp. 409–410

UNIT 3

NAME _____ DATE _____ PERIOD _____

Unit 3 Review, Chapters 9–11

Graph the point that has the given polar coordinates. Then, name three other pairs of polar coordinates for each point.

1. $A(2, 60°)$
2. $B(-4, 45°)$
3. $C\left(1.5, \frac{\pi}{6}\right)$
4. $D\left(-2, -\frac{2\pi}{3}\right)$

Graph each polar equation. Identify the type of curve each represents.

5. $r = \sqrt{5}$
6. $\theta = 60°$
7. $r = 3\cos\theta$
8. $r = 2 + 2\sin\theta$

Find the polar coordinates of each point with the given rectangular coordinates.

9. $(-2, -2)$
10. $(2, 2)$
11. $(2, -3)$
12. $(-3, 1)$

Write each equation in rectangular form.

13. $2 = r\cos\left(\theta - \frac{\pi}{2}\right)$
14. $4 = r\cos\left(\theta + \frac{\pi}{3}\right)$

Simplify.

15. i^{35}
16. $(3 + 2i) + (3 - 2i)$
17. $i^3(3 + 3i)$
18. $(-i - 5i)(i - 5)$
19. $\frac{2 + i}{2 - 3i}$

Express each complex number in polar form.

20. $-3i$
21. $3 + 3i$
22. $-1 + 3i$
23. $4 - 5i$

Find each product. Express the product in rectangular form.

24. $2(\cos\frac{\pi}{2} + i\sin\frac{\pi}{2}) \cdot 4(\cos\frac{\pi}{2} + i\sin\frac{\pi}{2})$
25. $1.5(\cos 3.1 + i\sin 3.1) \cdot 2(\cos 0.5 + i\sin 0.5)$

Solve.

26. Find $(1 + i)^7$ using De Moivre's Theorem. Express the result in rectangular form.
27. Solve the equation $x^5 - 1 = 0$ for all roots.

Find the distance between each pair of points with the given coordinates. Then find the midpoint of the segment that has endpoints at the given coordinates.

28. $(-4, 6), (11, 2)$
29. $(5, 0), (3, -2)$
30. $(1, 9), (-4, -3)$

For the equation of each circle, identify the center and the radius. Then graph the equation.

31. $4x^2 + 4y^2 = 49$
32. $x^2 + 10x + y^2 + 8y = 20$

For the equation of each ellipse, find the coordinates of the center, foci, and vertices. Then, graph the equation.

33. $(x - 1)^2 + 2(y - 3)^2 = 25$
34. $4(x + 2)^2 + 25(y - 2)^2 = 100$

For the equation of each hyperbola, find the coordinates of the center, foci, and vertices, and the equations of the asymptotes. Then, graph the equation.

35. $4x^2 - y^2 = 27$
36. $\frac{(y + 3)^2}{4} - \frac{(x + 1)^2}{9} = 1$

© Glencoe/McGraw-Hill   409   Advanced Mathematical Concepts

You are now ready to apply what you have learned in earlier units to more complex functions. The three chapters in this unit contain very different topics; however, there are some similarities among them. The most striking similarity is that all three chapters require that you have had some experience with graphing. As you work through this unit, try to make connections between what you have already studied and what you are currently studying. This will help you use the skills you have already mastered more effectively.

Resource Manager

Reproducible Masters

- *Math and Science Activities,* pp. 12–17
- Unit 3 Review, *Assessment and Evaluation,* pp. 409–410
- Unit 3 Test, *Assessment and Evaluation,* pp. 411–414

Unit 3 *inter*NET Projects

SPACE—THE FINAL FRONTIER

People have been fascinated by space since the beginning of time. Until 1961, however, human beings have been bound to Earth, unable to feel and experience life in space. Current space programs undertaken by NASA are exploring our solar system and beyond using sophisticated unmanned satellites such as the Hubble Space Telescope in orbit around Earth and the Mars Global Surveyor in orbit around Mars. In these projects, you will look at some interesting mathematics related to space—the final frontier. At the end of each chapter in Unit 3, you will be given specific tasks to explore space using the Internet.

CHAPTER ⑨
(page 611)

From Point A to Point B In Chapter 9, you will learn about the polar coordinate system, which is quite different from the rectangular coordinate system. Do scientists use the rectangular coordinate system or the polar coordinate system as they record the position of objects in space? Or, do they use some other system?
Math Connection: Research coordinate systems by using the Internet. Write a summary that describes each coordinate system that you find. Include diagrams and any information about converting between systems.

CHAPTER ⑩
(page 691)

Out in Orbit! What types of orbits do planets, artificial satellites, or space exploration vehicles have? Can orbits be modeled by the conic sections?
Math Connection: Use the Internet to find data about the orbit of a space vehicle, satellite, or planet. Make a scale drawing of the object's orbit labeling important features and dimensions. Then, write a summary describing the orbit of the object, being sure to discuss which conic section best models the orbit.

CHAPTER ⑪
(page 753)

Kepler is Still King! Johannes Kepler (1571–1630) was an important mathematician and scientist of his time. He observed the planets and stars and developed laws for the motion of those bodies. His laws are still used today. It is truly amazing how accurate his laws are considering the primitive observation tools that he used.
Math Connection: Research Kepler's Laws by using the Internet. Kepler's Third Law relates the distance of planets from the sun and the period of each planet. Use the Internet to find the distance each planet is from the sun and to find each planet's period. Verify Kepler's Third Law.

_inter_NET **CONNECTION**
For more information on the Unit Project, visit:
www.amc.glencoe.com

Using the *inter*NET Projects

When you or your students visit **www.amc.glencoe.com** you will find links to outside Web sites relating to these Internet projects. These links are checked frequently to ensure that they are still functioning and that the information is current.

Glencoe makes every effort to ensure that the Web site links for *Advanced Mathematical Concepts* are current and contain appropriate content. However, these Web sites are not under Glencoe's control.

In the Alternative Assessment section of each chapter's Study Guide and Assessment, you can find follow-up activities for the Internet Projects.

Assessment and Evaluation Masters, pp. 411–414

UNIT 3	NAME _____ DATE _____ PERIOD _____

Unit 3 Test, Chapters 9–11

1. Evaluate $\log_6 \sqrt{6}$.   1. ___ $\frac{1}{2}$

2. Identify the conic section represented by $3x^2 - 4xy + 2y^2 - 3y = 0$.   2. ___ ellipse

3. Find the coordinates of the vertex and the equation of the axis of symmetry for the parabola with equation $2x^2 + 2x - y = -3$.   3. $\left(-\frac{1}{2}, \frac{5}{2}\right); x = -\frac{1}{2}$

4. Write the rectangular equation $x^2 + y^2 = 6$ in polar form.   4. ___ $r = \sqrt{6}$

5. Simplify $\frac{1 - i}{3 + 2i}$.   5. ___ $\frac{1}{13} - \frac{5}{13}i$

6. Graph the point with polar coordinates $\left(-2, \frac{3\pi}{2}\right)$.   6. ___

7. Solve $6 = e^{0.2t}$. Round your answer to the nearest hundredth.   7. ___ 8.96

8. Evaluate $(\sqrt{289})^{-3}$.   8. ___ $\frac{1}{4913}$

9. Find the rectangular equation of the curve whose parametric equations are $x = -2 \sin t$ and $y = \cos t$, where $0 \le t \le 2\pi$.   9. ___ $x^2 + 4y^2 = 4$

10. Find the balance after 15 years of a $2500 investment earning 5.5% interest compounded continuously.   10. ___ $5704.70

11. Find the product $2(\cos 10° + i \sin 10°) \cdot 4(\cos 20° + i \sin 20°)$. Then, express the product in rectangular form.   11. ___ $4\sqrt{3} + 4i$

12. Identify the classical curve that the graph of $r = 1 + \sin \theta$ represents.   12. ___ cardioid

13. Write the standard form of the equation of the circle that passes through the point (0, 4) and has its center at (−3, −1).   13. $(x + 3)^2 + (y + 1)^2 = 34$

© Glencoe/McGraw-Hill   411   *Advanced Mathematical Concepts*

Instructional Objectives

Lesson (pages)	Lesson Objectives	NCTM Standards 2000	State/Local Objectives
9-1 (553–560)	• Graph points in polar coordinates. • Graph simple polar equations. • Determine the distance between two points with polar coordinates.	3, 6, 10	
9-2 (561–567)	• Graph polar equations.	3, 6, 10	
9-3 (568–573)	• Convert between polar and rectangular coordinates.	3, 6, 10	
9-4 (574–579)	• Write the polar form of a linear equation. • Graph the polar form of a linear equation.	2, 3, 6, 10	
9-5 (580–585)	• Add, subtract, multiply, and divide complex numbers in rectangular form.	3, 6, 10	
9-6 (586–591)	• Graph complex numbers in the complex plane. • Convert complex numbers from rectangular to polar form and vice versa.	3, 6, 10	
9-6B (592)	• Explore geometric relationships in the complex plane.	3, 6, 10	
9-7 (593–598)	• Find the product and quotient of complex numbers in polar form.	3, 6, 10	
9-8 (599–606)	• Find powers and roots of complex numbers in polar form using De Moivre's Theorem.	3, 6, 10	

Suggested Pacing *See pages T2–T4 for additional options in course-planning.*

COURSE	Lesson 9-1	Lesson 9-2	Lesson 9-3	Lesson 9-4	Lesson 9-5
Precalculus Emphasis Chapters 4–15 (45–50 minutes)	1.5 days	1.5 days	1.5 days	1.5 days	1.5 days
Precalculus Emphasis (block)	0.5 day	1.5 days		1.5 days	
Trigonometric Emphasis Chapters 1–9, 11, 12 (45–50 minutes)	1.5 days	1.5 days	1.5 days	1.5 days	1.5 days
Trigonometric Emphasis (block)	0.5 day	1.5 days		1.5 days	

Instructional Resources

		Blackline Masters (page numbers)								
Student Edition										
Lesson	Graphing Calculator Applications	Study Guide and Practice	Enrichment	Assessment and Evaluation	Graphing Calculator Keystrokes	Problem Solving & Applications	Spreadsheet Applications	TI-92 and CBL Interdisciplinary Activities	Transparencies A and B
9-1		157–158	97						9-1
9-2	564, 566	159–160	98	225			9		9-2
9-3		161–162	99						9-3
9-4	576, 577	163–164	100	224, 225				29–32	9-4
9-5	581	165–166	101						9-5
9-6		167–168	102	226		17			9-6
9-6B	592				35–36				
9-7		169–170	103						9-7
9-8	602, 605	171–172	104	226	37	18			9-8
Study Guide/ Assessment				211–223, 229					
SAT/ACT Practice				227–228					

Lesson 9-6	Lesson 9-7	Lesson 9-8	Study Guide & Assessment	Chapter Test	TOTAL
(with Lesson 9-6B) 2.5 days	1.5 days	1.5 days	1 day	1 day	15 days
(with Lesson 9-6B) 1.5 days		1.5 days		0.5 day	7 days
(with Lesson 9-6B) 2.5 days	1.5 days	1.5 days	1 day	1 day	15 days
(with Lesson 9-6B) 1.5 days		1.5 days		0.5 day	7 days

The worksheets shown on this page are a small sample of the materials available on *TeacherWorks: All-in-One Lesson Planner and Resource Center.*

This CD-ROM contains all of the blackline masters and transparencies available for this program.

It also includes a lesson planner and interactive Teacher Edition, so you can customize lesson plans and reproduce classroom resources quickly and easily, from just about anywhere.

Applications

Problem Solving and Applications Masters, pp. 17, 18

Technology/Multimedia

Spreadsheet Applications Masters, p. 9

TI-92 & CBL Interdisciplinary Activities Masters, pp. 29–32

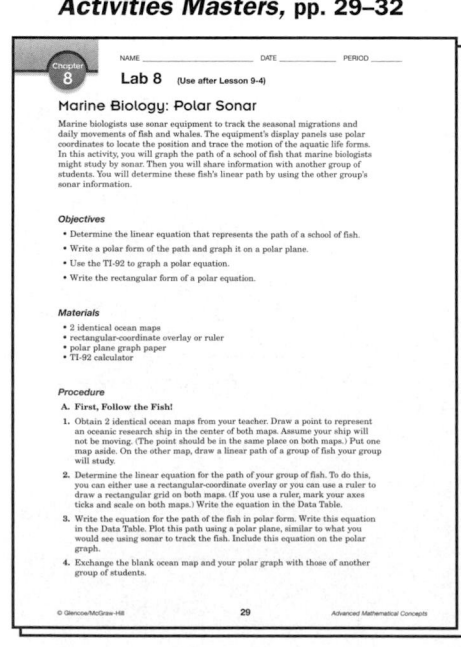

Assessment Resources

Type	Student Edition	Teacher's Wraparound Edition	Assessment and Evaluation Masters
Ongoing Assessment	Mid-Chapter Quiz, p. 579	5-Minute Check, pp. 553, 561, 568, 574, 580, 586, 593, 599	Mid-Chapter Test, p. 224 Quizzes A–D, pp. 225–226
Mixed Review	Mixed Review, pp. 560, 567, 573, 579, 584–585, 591, 597–598, 606		Cumulative Review, p. 229
Error Analysis	You Decide, pp. 565, 604	Error Analysis, pp. 557, 565, 571, 577, 583, 589, 596, 604	
Standardized Test Practice	SAT/ACT Practice, pp. 560, 567, 573, 579, 585, 591, 598, 606 SAT & ACT Preparation, pp. 612–613	TIMSS, p. 613	SAT/ACT Practice, pp. 227–228
Open-Ended Assessment	Math Journal, pp. 571, 589 Open-Ended Assessment, p. 611 Portfolio, p. 611	Modeling: pp. 567, 579, 592 Speaking: pp. 585, 591, 598 Writing: pp. 560, 573, 606	Open-Ended Assessment, p. 223
Chapter Assessment	Study Guide and Assessment, pp. 607–611 Chapter Test, p. A64		Multiple-Choice Tests (Forms 1A, 1B, 1C), pp. 211–216 Free-Response Tests (Forms 2A, 2B, 2C), pp. 217–222

TestCheck and Worksheet Builder is a networkable CD-ROM with three integrated modules. The **Worksheet Builder** creates customized worksheets, tests, and quizzes of free-response, multiple-choice, short-answer, and open-ended items. The **Student Module** gives you the option of having students take tests on-screen and get immediate feedback on their performance. Use the optional **Management System** to keep detailed student records.

Additional Chapter Resources

Student Edition
Career Choices, p. 598
Internet Project, p. 611
Extra Practice, pp. A42–A43

Teacher's Classroom Resources

Technology
Graphing Calculator Keystrokes Masters

Teaching Aids
Answer Key Transparencies
Block Schedule Planning Guide
Lesson Planning Guide
Solutions Manual

Glencoe Instructional Technology

Instructional
Advanced MathPASS CD-ROM
Multimedia Applications CD-ROM

Assessment
TestCheck and Worksheet Builder

GLENCOE
Online

Visit **www.amc.glencoe.com** for data updates, career information, graphing calculator programs, and other interactive activities.

Chapter Overview

This chapter introduces polar coordinates and continues the concepts of complex numbers from previous chapters. Students will graph polar equations, polar and rectangular coordinates, and polar forms of linear equations. In the second half of the chapter, students will investigate complex numbers in polar form. Then they will find powers and roots of complex numbers.

Using a Graphing Calculator

There are numerous opportunities, all optional, to use a graphing calculator in this chapter.
- Solving a system of polar equations (pp. 564, 566)
- Using **TABLE** in order to generate polar ordered pairs (p. 576)
- Performing complex number arithmetic (p. 581)
- Programming for complex iteration (p. 582)
- Exploring geometric relationships in the complex plane (p. 592)
- Approximating roots of complex numbers (pp. 602, 605)

The **Graphing Calculator Appendix**, which begins on page A2, contains keystroke instructions on many of these topics.

POLAR COORDINATES AND COMPLEX NUMBERS

CHAPTER OBJECTIVES

- **Graph polar equations.** *(Lessons 9-1, 9-2, 9-4)*
- **Convert between polar and rectangular coordinates.** *(Lessons 9-3, 9-4)*
- **Add, subtract, multiply, and divide complex numbers in rectangular and polar forms.** *(Lessons 9-5, 9-7)*
- **Convert between rectangular and polar forms of complex numbers.** *(Lesson 9-6)*
- **Find powers and roots of complex numbers.** *(Lesson 9-8)*

 SAT & ACT Preparation

This textbook contains a complete SAT/ACT preparation course described on pages T8–T9 of the *Teacher's Handbook*. This chapter contains:

- an SAT or ACT practice question at the end of each lesson
- examples of SAT and ACT problems involving geometry—lines, angles, and arcs—on page 612
- ten SAT and ACT practice problems on page 613

- additional SAT and ACT practice in the *Assessment and Evaluation Masters,* pp. 227–228
- SAT and ACT test practice banks in the *TestCheck and Worksheet Builder*

9-1

Polar Coordinates

Lesson Notes

9-1

OBJECTIVES
- Graph points in polar coordinates.
- Graph simple polar equations.
- Determine the distance between two points with polar coordinates.

SURVEYING Before large road construction projects, or even the construction of a new home, take place, a surveyor maps out characteristics of the land. A surveyor uses a device called a *theodolite* to measure angles. The precise locations of various land features are determined using distances and the angles measured with the theodolite. While mapping out a level site, a surveyor identifies a landmark 450 feet away and 30° to the left and another landmark 600 feet away and 50° to the right. What is the distance between the two landmarks? *This problem will be solved in Example 5.*

Recording the position of an object using the distance from a fixed point and an angle made with a fixed ray from that point uses a **polar coordinate system.** When surveyors record the locations of objects using distances and angles, they are using **polar coordinates.**

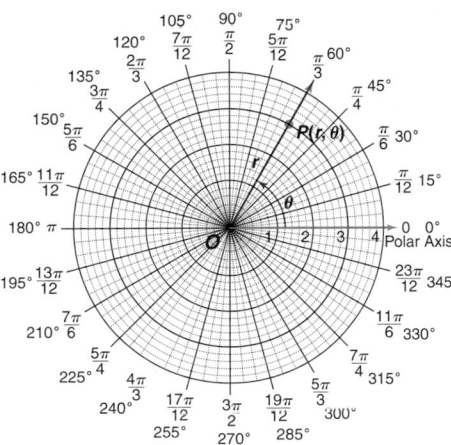

In a polar coordinate system, a fixed point *O* is called the **pole** or origin. The **polar axis** is usually a horizontal ray directed toward the right from the pole. The location of a point *P* in the polar coordinate system can be identified by polar coordinates in the form (r, θ). If a ray is drawn from the pole through point *P*, the distance from the pole to point *P* is $|r|$. The measure of the angle formed by $\overrightarrow{OP}$ and the polar axis is θ. The angle can be measured in degrees or radians. *This grid is sometimes called the polar plane.*

Consider positive and negative values of *r*.

Suppose $r > 0$. Then θ is the measure of any angle in standard position that has $\overrightarrow{OP}$ as its terminal side.

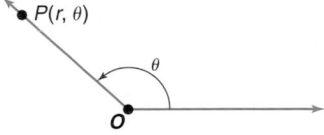

Suppose $r < 0$. Then θ is the measure of any angle that has the ray opposite $\overrightarrow{OP}$ as its terminal side.

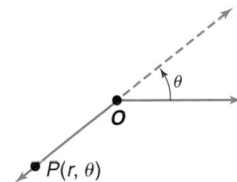

Lesson 9-1 Polar Coordinates **553**

1 FOCUS

5-Minute Check
Chapter 8

1. A vector has a magnitude of 5.6 cm and a direction of 28°. Find the magnitude of its vertical component. **2.6 cm**

2. $\vec{f} = \langle 3, 2 \rangle$ and $\vec{w} = \langle -2, 4 \rangle$. Find an ordered pair to represent $\vec{u}$ if $\vec{u} = 2\vec{f} - 3\vec{w}$. $\langle 12, -8 \rangle$

3. Find the magnitude of the vector from $A(-4, 5)$ to $B(8, 0)$. **13**

4. Find the inner product of $\vec{c}$ and $\vec{d}$ if $\vec{c} = \langle 3, -2 \rangle$ and $\vec{d} = \langle 6, 8 \rangle$. **2**

5. In still air, an airplane heads due east at 400 mph. A 40-mph wind blows from the north. Find the ground speed and direction of the plane. **402 mph; 5°43' SE**

Motivating the Lesson

Discussion Have students graph the point $P(1, \sqrt{3})$ on a coordinate plane, then draw $\overline{OP}$. Ask them to find the length of $\overline{OP}$ and the angle that $\overline{OP}$ makes with the *x*-axis. Discuss how this relates to polar coordinates. **2; 60°**

Resource Manager

Reproducible Masters
- *Study Guide and Practice,* pp. 157–158
- *Enrichment,* p. 97

Transparencies
- 5-Minute Check, 9-1
- Teaching, 9-1

Technology/Multimedia
- Advanced MathPASS, Lesson 20

Teaching Tip After discussing Example 1, ask students to relate graphing polar coordinates to the process used in determining terminal sides of angles in circular functions.

In-Class Examples

Example 1
Graph each point.
a. $S(-4, 0°)$

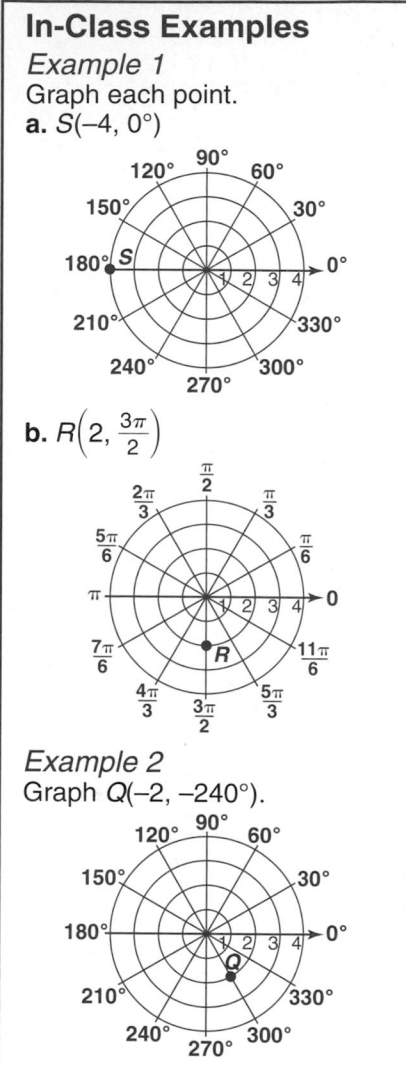

b. $R\left(2, \dfrac{3\pi}{2}\right)$

Example 2
Graph $Q(-2, -240°)$.

Example ① Graph each point.

a. $P(3, 60°)$

On a polar plane, sketch the terminal side of a 60° angle in standard position.

Since r is positive ($r = 3$), find the point on the terminal side of the angle that is 3 units from the pole.

Notice that point P is on the third circle from the pole.

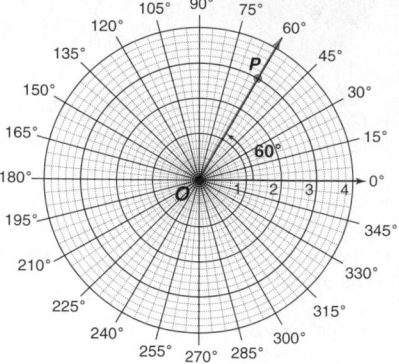

b. $Q\left(-1.5, \dfrac{7\pi}{6}\right)$

Sketch the terminal side of an angle measuring $\dfrac{7\pi}{6}$ radians in standard position.

Since r is negative, extend the terminal side of the angle in the opposite direction. Find the point Q that is 1.5 units from the pole along this extended ray.

Notice that point Q is halfway between the first and second circles from the pole.

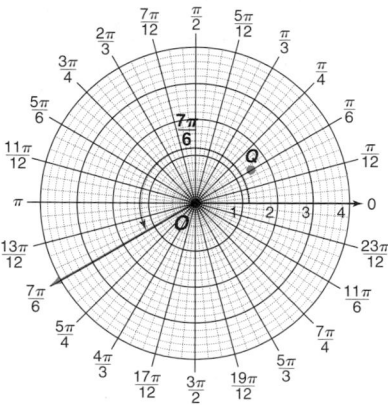

As you have seen, the r-coordinate can be any real value. The angle θ can also be negative. If $\theta > 0$, then θ is measured counterclockwise from the polar axis. If $\theta < 0$, then θ is measured clockwise from the polar axis.

Example ② Graph $R(-2, -135°)$.

Negative angles are measured clockwise. Sketch the terminal side of an angle of $-135°$ in standard position.

Since r is negative, the point $R(-2, -135°)$ is 2 units from the pole along the ray opposite the terminal side that is already drawn.

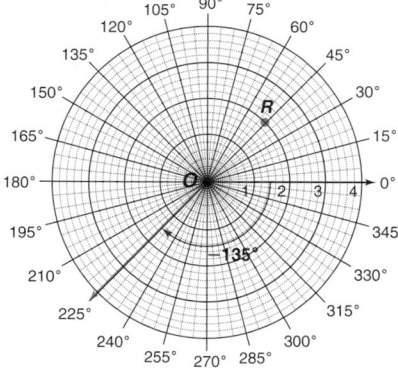

Prerequisite Skills

A review of the prerequisite skills for this lesson is provided in the *Prerequisite Skills Masters*.

Skill Number and Content	Pages
2 Fractions	5–6

In Example 2, the point $R(-2, -135°)$ lies in the polar plane 2 units from the pole on the terminal side of a 45° angle in standard position. This means that point R could also be represented by the coordinates $(2, 45°)$. In general, the polar coordinates of a point are not unique. Every point can be represented by infinitely many pairs of polar coordinates. This happens because any angle in standard position is coterminal with infinitely many other angles.

If a point has polar coordinates $(r, θ)$, then it also has polar coordinates $(r, θ + 2π)$ in radians, or $(r, θ + 360°)$ in degrees. In fact, you can add any integer multiple of $2π$ to $θ$ and find another pair of polar coordinates for the same point. If you use the opposite r-value, the angle will change by $π$, giving $(-r, θ + π)$ as another ordered pair for the same point. You can then find even more polar coordinates for the same point by adding multiples of $2π$ to $θ + π$. The following graphs illustrate six of the different ways to name the polar coordinates of the same point.

(2, 120°) or (2, −240°)

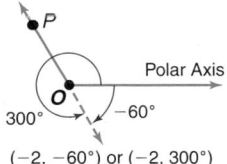

(−2, −60°) or (−2, 300°)

(2, 480°)

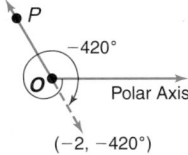

(−2, −420°)

Here is a summary of all the ways to represent a point in polar coordinates.

Multiple Representations of $(r, θ)$	If a point P has polar coordinates $(r, θ)$, then P can also be represented by polar coordinates $(r, θ + 2πk)$ or $(-r, θ + (2k + 1)π)$, where k is any integer.

In degrees, the representations are $(r, θ + 360k°)$ and $(-r, θ + (2k + 1)180°)$.
For every angle, there are infinitely many representations.

Example ③ **Name four different pairs of polar coordinates that represent point S on the graph with the restriction that $-360° ≤ θ ≤ 360°$.**

One pair of polar coordinates for point S is $(3, 150°)$.

To find another representation, use $(r, θ + 360k°)$ with $k = -1$.
$(3, 150° + 360(-1)°) = (3, -210°)$

To find additional polar coordinates, use $(-r, θ + (2k + 1)180°)$.
$(-3, 150° + 180°) = (-3, 330°)$ $k = 0$

To find a fourth pair, use $(-r, θ + (2k + 1)180°)$ with $k = -1$.
$(-3, 150° + (-1)180°) = (-3, -30°)$

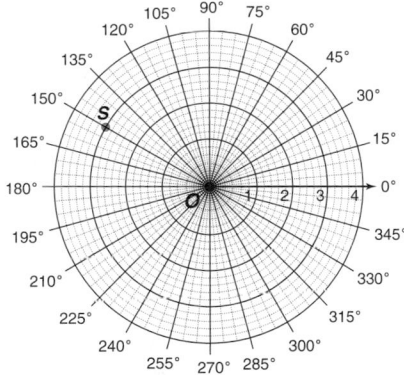

Therefore, $(3, 150°)$, $(3, -210°)$, $(-3, 330°)$, and $(-3, -30°)$ all represent the same point in the polar plane.

In-Class Example

Example 4
Graph each polar equation.
a. $r = -3$

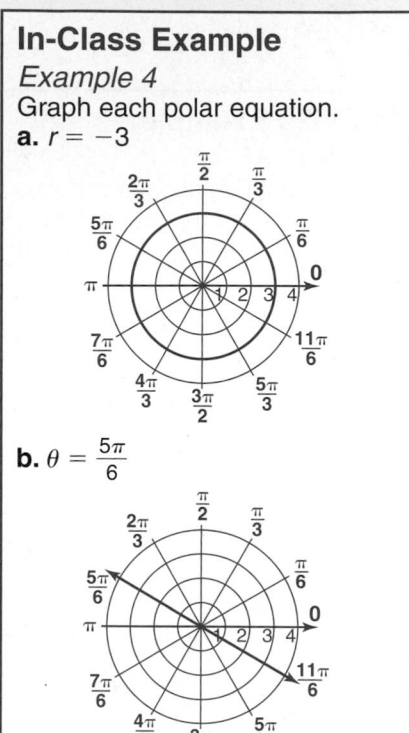

b. $\theta = \dfrac{5\pi}{6}$

An equation expressed in terms of polar coordinates is called a **polar equation.** For example, $r = 2 \sin \theta$ is a polar equation. A **polar graph** is the set of all points whose coordinates (r, θ) satisfy a given polar equation.

You already know how to graph equations in the *Cartesian,* or *rectangular, coordinate system.* Graphs of equations involving constants like $x = 2$ and $y = -3$ are considered basic in the Cartesian coordinate system. Similarly, the polar coordinate system has some basic graphs. Graphs of the polar equations $r = k$ and $\theta = k$, where k is a constant, are considered basic.

Example **4** **Graph each polar equation.**

a. $r = 3$

The solutions to $r = 3$ are the ordered pairs (r, θ) where $r = 3$ and θ is any real number. Some examples are $(3, 0)$, $\left(3, \dfrac{\pi}{4}\right)$, and $(3, \pi)$. In other words, the graph of this equation is the set of all ordered pairs of the form $(3, \theta)$. Any point that is 3 units from the pole is included in this graph. The graph is the circle centered at the origin with radius 3.

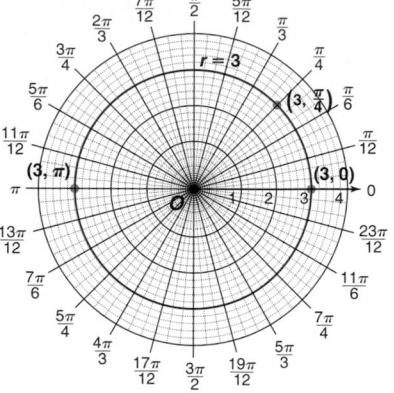

b. $\theta = \dfrac{3\pi}{4}$

The solutions to this equation are the ordered pairs (r, θ) where $\theta = \dfrac{3\pi}{4}$ and r is any real number. Some examples are $\left(1, \dfrac{3\pi}{4}\right)$, $\left(-2, \dfrac{3\pi}{4}\right)$, and $\left(3, \dfrac{3\pi}{4}\right)$. In other words, the graph of this equation is the set of all ordered pairs of the form $\left(r, \dfrac{3\pi}{4}\right)$. The graph is the line that includes the terminal side of the angle $\dfrac{3\pi}{4}$ in standard position.

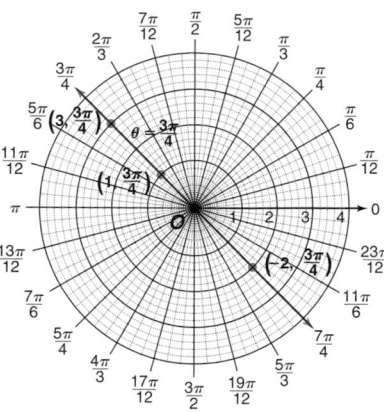

Just as you can find the distance between two points in a rectangular coordinate plane, you can derive a formula for the distance between two points in a polar plane.

556 *Chapter 9 Polar Coordinates and Complex Numbers*

Addressing Individual Needs

Kinesthetic Give each team of 4 or 5 students a rope and polar coordinate paper. One student acts as the origin and two other students stand at different points, which are marked on the graph. Measure the distance between the students using the rope and compare the result to that given by the distance formula.

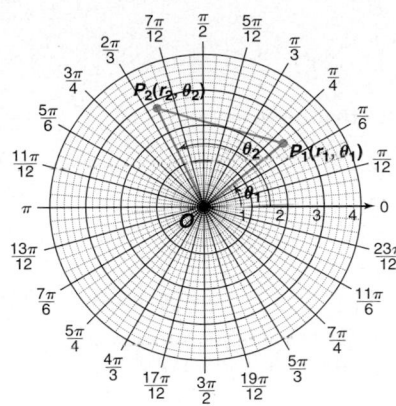

Given two points $P_1(r_1, \theta_1)$ and $P_2(r_2, \theta_2)$ in the polar plane, draw $\triangle P_1OP_2$. $\angle P_1OP_2$ has measure $\theta_2 - \theta_1$. Apply the Law of Cosines to $\triangle P_1OP_2$.

$$(P_1P_2)^2 = (OP_1)^2 + (OP_2)^2 - 2(OP_1)(OP_2) \cos (\theta_2 - \theta_1)$$

Now substitute r_1 for OP_1 and r_2 for OP_2.

$$(P_1P_2)^2 = r_1^2 + r_2^2 - 2r_1r_2 \cos (\theta_2 - \theta_1)$$
$$P_1P_2 = \sqrt{r_1^2 + r_2^2 - 2r_1r_2 \cos (\theta_2 - \theta_1)}$$

In-Class Example

Example 5
Surveying If two other landmarks are 700 feet away and 40° to the left, and 350 feet away and 35° to the right, what is the distance between the landmarks?
about 697 ft

Distance Formula in Polar Plane	If $P_1(r_1, \theta_1)$ and $P_2(r_2, \theta_2)$ are two points in the polar plane, then $$P_1P_2 = \sqrt{r_1^2 + r_2^2 - 2r_1r_2 \cos (\theta_2 - \theta_1)}.$$

3 PRACTICE/APPLY

Error Analysis
Watch for students who forget to designate θ_2 as negative in the distance formula.

Prevent by having students draw and label a picture using the information given.

Example **5** **SURVEYING** Refer to the application at the beginning of the lesson. What is the distance between the two landmarks?

Real World Application

Set up a coordinate system so that turning to the left is a positive angle and turning to the right is a negative angle. With this convention, the landmarks are at $L_1(450, 30°)$ and $L_2(600, -50°)$.

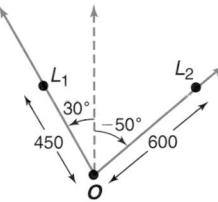

Now use the Polar Distance Formula.

$$P_1P_2 = \sqrt{r_1^2 + r_2^2 - 2r_1r_2 \cos (\theta_2 - \theta_1)} \qquad (r_1, \theta_1) = (450, 30°)$$
$$L_1L_2 = \sqrt{450^2 + 600^2 - 2(450)(600) \cos (-50° - 30°)} \qquad (r_2, \theta_2) = (600, -50°)$$
$$\approx 684.6 \text{ feet}$$

The landmarks are about 685 feet apart.

CHECK FOR UNDERSTANDING

Communicating Mathematics

Read and study the lesson to answer each question. **1–2. See margin.**

1. **Explain** why a point in the polar plane cannot be named by a unique ordered pair (r, θ).

3. Sample answer: −60° and 300°

2. **Explain** how to graph (r, θ) if $r < 0$ and $\theta > 0$.

3. **Name** two values of θ such that $(-4, \theta)$ represents the same point as $(4, 120°)$.

5. All ordered pairs of the form (r, θ) where $r = 0$.

4. **Explain** why the graph in Example 4a is also the graph of $r = -3$. **See margin.**

5. **Describe** the polar coordinates of the pole.

Lesson 9-1 Polar Coordinates **557**

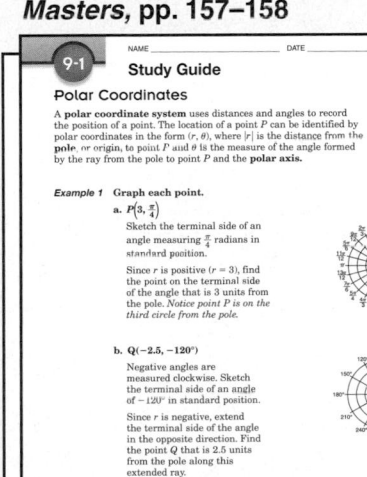

Additional Answers
1. There are infinitely many ways to represent the angle θ. Also, r can be positive or negative.

2. Draw the angle θ in standard position. Extend the terminal side of the angle in the opposite direction. Locate the point that is $|r|$ units from the pole along this extension.

4. The points 3 units from the origin in the opposite direction are on the circle where $r = 3$.

Additional Answers

6.

7.

8.

9.

11.

12.

Guided Practice Graph each point. **6–9. See margin.**

6. $A(1, 135°)$ 7. $B\left(2.5, -\frac{\pi}{6}\right)$ 8. $C(-3, -120°)$ 9. $D\left(-2, \frac{13\pi}{6}\right)$

10. Name four different pairs of polar coordinates that represent the point
at $\left(-2, \frac{\pi}{6}\right)$. **Sample answer:** $\left(-2, \frac{13\pi}{6}\right), \left(-2, \frac{25\pi}{6}\right), \left(2, \frac{7\pi}{6}\right), \left(2, \frac{19\pi}{6}\right)$

Graph each polar equation. **11–13. See margin.**

11. $r = 1$ 12. $\theta = -\frac{\pi}{3}$ 13. $r = 3.5$

14. Find the distance between $P_1\left(2.5, \frac{\pi}{6}\right)$ and $P_2\left(-3, -\frac{\pi}{4}\right)$ on the polar plane. **4.37**

15. **Gardening** A lawn sprinkler can cover the part of a circular region
determined by the polar inequalities $-30° \le \theta \le 210°$ and $0 \le r \le 20$,
where r is measured in feet.
a. Sketch a graph of the region that the sprinkler can cover. **See margin.**
b. Find the area of the region. **about 838 ft²**

EXERCISES

Practice   Graph each point. **16–27. See Answer Appendix.**

A   16. $E(2, 30°)$ 17. $F\left(1, \frac{\pi}{2}\right)$ 18. $G(5, 240°)$ 19. $H\left(\frac{1}{2}, \frac{3\pi}{4}\right)$

20. $J\left(1.5, -\frac{\pi}{4}\right)$ 21. $K\left(\frac{5}{2}, -210°\right)$ 22. $L\left(3, -\frac{\pi}{6}\right)$ 23. $M(2, -90°)$

B   24. $N(-1, 120°)$ 25. $P\left(-0.5, -\frac{11\pi}{6}\right)$ 26. $Q\left(-2, \frac{25\pi}{3}\right)$ 27. $R\left(-\frac{7}{2}, 1050°\right)$

**28–31. Sample
answers are
given.**

28. List four pairs of polar coordinates
that represent the point S in the
graph. Use both radians and
degrees.

$\left(2, \frac{\pi}{3}\right), \left(2, \frac{7\pi}{3}\right), (-2, 240°), (-2, 600°)$

29. $(1.5, 540°), (1.5, 900°), (-1.5, 0°),$
$(-1.5, 360°)$

30. $\left(-1, \frac{7\pi}{3}\right), \left(-1, \frac{13\pi}{3}\right), \left(1, \frac{4\pi}{3}\right), \left(1, \frac{10\pi}{3}\right)$

31. $(4, 675°), (4, 1035°), (-4, 135°),$
$(-4, 495°)$

Name four other pairs of polar coordinates
for each point.

29. $T(1.5, 180°)$ 30. $U\left(-1, \frac{\pi}{3}\right)$ 31. $V(4, 315°)$

Graph each polar equation. **32–40. See Answer Appendix.**

32. $r = 1.5$ 33. $\theta = \frac{5\pi}{4}$ 34. $r = 2$

35. $\theta = 30°$ 36. $\theta = -150°$ 37. $\theta = -\frac{\pi}{4}$

C   38. $\theta = 840°$ 39. $r = 0$ 40. $r = -1$

Additional Answers

13.

15a.

41. Write a polar equation for the circle centered at the origin with radius $\sqrt{2}$.
$r = \sqrt{2}$ or $r = -\sqrt{2}$

Find the distance between the points with the given polar coordinates.

42. $P_1(4, 170°)$ and $P_2(6, 105°)$ **5.63**

43. $P_1\left(1, \dfrac{\pi}{6}\right)$ and $P_2\left(5, \dfrac{3\pi}{4}\right)$ **5.35**

44. $P_1\left(-2.5, \dfrac{\pi}{8}\right)$ and $P_2\left(-1.75, -\dfrac{2\pi}{5}\right)$ **3.16**

45. $P_1(1.3, -47°)$ and $P_2(-3.6, -62°)$ **4.87**

46. Find an ordered pair of polar coordinates to represent the point whose rectangular coordinates are $(-3, 4)$.   **Sample answer: (5, 127°)**

47. Web Page Design   When designing websites with circular graphics, it is often convenient to use polar coordinates, sometimes called "pizza coordinates" in this context. If the origin is at the center of the screen, what are the polar equations of the lines that cut the region into the six congruent slices shown? $\theta = 0°, \theta = 60°, \theta = 120°$

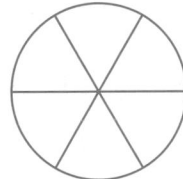

48. Critical Thinking   Prove that if $P_1(r_1, \theta)$ and $P_2(r_2, \theta)$ are two points in the polar plane, then the distance formula reduces to $P_1P_2 = |r_1 - r_2|$.   **See margin.**

49. Sailing   A graph of the maximum speed of a sailboat versus the angle of the wind is called a "performance curve" or just a "polar." The graph at the right is the polar for a particular boat when the wind speed is 20 knots. θ represents the angle of the wind in degrees, and r is the maximum speed of the boat in knots.
   a. What is the maximum speed when $\theta = 120°$?   **17 knots**
   b. What is the maximum speed when $\theta = 150°$?   **13 knots**

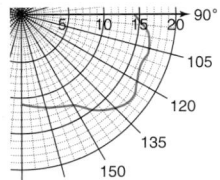

50. Acoustics   Polar coordinates can be used to model a concert amphitheater. Suppose the performer is placed at the pole and faces the direction of the polar axis. The seats have been built to occupy the region with $-\dfrac{\pi}{3} \le \theta \le \dfrac{\pi}{3}$ and $0.25 \le r \le 3$, where r is measured in hundreds of feet.
   a. Sketch this region in the polar plane.   **See margin.**
   b. How many seats are there if each person has 6 square feet of space?   **15,599**

51. Critical Thinking   Explain why the order of the points used in the distance formula is irrelevant. That is, why can you choose either point to be P_1 and the other to be P_2?   **See margin.**

---

48. $P_1P_2 = \sqrt{r_1^2 + r_2^2 - 2r_1r_2 \cos(\theta - \theta)}$
$= \sqrt{r_1^2 + r_2^2 - 2r_1r_2 \cos 0}$
$= \sqrt{r_1^2 + r_2^2 - 2r_1r_2}$
$= \sqrt{(r_1 - r_2)^2}$
$= |r_1 - r_2|$

50a.

Additional Answer

51. The distance formula is symmetric with respect to (r_1, θ_1) and (r_2, θ_2). That is,
$$\sqrt{r_2^2 + r_1^2 - 2r_2r_1 \cos(\theta_1 - \theta_2)}$$
$$= \sqrt{r_1^2 + r_2^2 - 2r_1r_2 \cos[-(\theta_2 - \theta_1)]}$$
$$= \sqrt{r_1^2 + r_2^2 - 2r_1r_2 \cos(\theta_2 - \theta_1)}.$$

Open-Ended Assessment

Writing Have students write a few sentences comparing and contrasting the polar coordinate system and the rectangular coordinate system.

Additional Answer

52a.

52. Aviation Two jets at the same altitude are being tracked on an air traffic controller's radar screen. The coordinates of the planes are $(5, 310°)$ and $(6, 345°)$, with r measured in miles.

a. Sketch a graph of this situation. **See margin.**

b. If regulations prohibit jets from passing within three miles of each other, are these planes in violation? Explain. **No; the planes are 3.44 miles apart.**

U.S. Navy Blue Angels

Mixed Review

53. Transportation Two docks are directly across a river from each other. A boat that can travel at a speed of 8 miles per hour in still water is attempting to cross directly from one dock to the other. The current of the river is 3 miles per hour. At what angle should the captain head? *(Lesson 8-5)* **about 22.0°**

54. Find the inner product $\langle 3, -2, 4 \rangle \cdot \langle 1, -4, 0 \rangle$. Then state whether the vectors are perpendicular. Write *yes* or *no*. *(Lesson 8-4)* **11; no**

55. Find the distance from the line with equation $y = 9x - 3$ to the point at $(-3, 2)$. *(Lesson 7-7)* $\dfrac{16\sqrt{82}}{41}$

56. Simplify the expression $\dfrac{1 - \sin^2 \alpha}{\sin^2 \alpha}$. *(Lesson 7-1)* **$\cot^2 \alpha$**

57. Find Arccos $\dfrac{\sqrt{3}}{2}$. *(Lesson 6-8)* **30°**

58. State the amplitude and period for the function $y = 5 \cos 4\theta$. *(Lesson 6-4)* **$5, \dfrac{\pi}{2}$**

59. one; $B = 90°$, $C = 60°$, $c = 16.1$

59. Determine the number of possible solutions and, if a solution exists, solve $\triangle ABC$ if $A = 30°$, $b = 18.6$, and $a = 9.3$. Round to the nearest tenth. *(Lesson 5-7)*

60. Find the number of possible positive real zeros and the number of possible negative real zeros of the function $f(x) = x^3 - 4x^2 + 4x - 1$. Then determine the rational zeros. *(Lesson 4-4)* **3 or 1; 0; 1**

61. Determine the slant asymptote of $f(x) = \dfrac{x^2 + 2x - 3}{x + 5}$. *(Lesson 3-7)* **$y = x - 3$**

62. Determine whether the graph of $f(x) = x^4 + 3x^2 + 2$ is symmetric with respect to the x-axis, the y-axis, both, or neither. *(Lesson 3-1)* **y-axis**

63. Find the value of the determinant $\begin{vmatrix} -2 & 4 & -1 \\ 1 & -1 & 0 \\ -3 & 4 & 5 \end{vmatrix}$. *(Lesson 2-5)* **$-11$**

64. Given that x is an integer, state the relation representing $y = 11 - x$ and $-3 \le x \le 0$ by listing a set of ordered pairs. Then state whether this relation is a function. *(Lesson 1-1)* **$\{(-3, 14), (-2, 13), (-1, 12), (0, 11)\}$, yes**

65. SAT/ACT Practice The circumference of the circle is 50π. What is the length of the diagonal $\overline{AB}$ of the square inscribed in the circle? **E**

A $12\dfrac{1}{2}$ **B** $10\sqrt{2}$ **C** 25

D $25\sqrt{2}$ **E** 50

Extra Practice See p. A42.

Enrichment Masters, p. 97

9-1 NAME _____ DATE _____ PERIOD _____

Enrichment

Distance on the Earth's Surface

As you learned in Lesson 9-1, lines of longitude on Earth's surface intersect at the North and South Poles. A line of longitude that passes completely around Earth is called a **great circle**. All great circles have the same circumference, found by calculating the circumference of a circle with Earth's radius, 3963.2 miles. (Since Earth is slightly flattened at the poles, it is not precisely spherical. The difference is so small, however, that for most purposes it can be ignored.)

1. Find the circumference of a great circle.
24,901.5 miles

On a great circle, position is measured in degrees north or south of the equator. Pittsburgh's position of 40° 26′ N means that radii from Earth's center to Pittsburgh and to the point of intersection of the equator and Pittsburgh's longitude line form an angle of 40° 26′. (See the figure above.)

2. Find the length of one degree of arc on a longitude line.
69.2 miles

3. Charleston, South Carolina (32° 46′ N), and Guayaquil, Ecuador (2° 9′ S), both lie on Pittsburgh's longitude line. Find the distance from Pittsburgh to each of the other cities.
530.3 miles; 2945.5 miles

Because circles of latitude are drawn parallel to the equator, their radii and circumferences grow steadily shorter as they approach the poles. The length of one degree of arc on a circle of latitude depends on how far north or south of the equator the circle is located. The figure at the right shows a circle of latitude of radius r located θ degrees north of the equator. Because the radii of the equator and the circle of latitude are parallel, $m\angle NEO = \theta$. Therefore, $\cos \theta = \dfrac{r}{R}$, which gives $r = R \cos \theta$, where R represents the radius of Earth.

4. Find the radius and circumference of a circle of latitude located 70° north of the equator. **1355.5 miles; 8516.8 miles**

5. Find the length of one degree of arc on the circle described in Exercise 4. **23.7 miles**

6. Bangor, Maine, and Salem, Oregon, are both located at latitude 44° 50′ N. Their respective longitudes are 68° 46′ and 123° 2′ west of Greenwich. Find the distance from Bangor to Salem.
2662.0 miles

© Glencoe/McGraw-Hill   T97   *Advanced Mathematical Concepts*

Extra Credit

Determine the rectangular coordinates for the following points:

$A\left(4, \dfrac{\pi}{2}\right)$	$A(0, 4)$
$B(2, 270°)$	$B(0, -2)$
$C(3, \pi)$	$C(-3, 0)$
$D(1, 360°)$	$D(1, 0)$

9-2

Graphs of Polar Equations

OBJECTIVE
• Graph polar equations.

AUDIO TECHNOLOGY One way to describe the ability of a microphone to pick up sounds from different directions is to examine its polar pattern. A polar coordinate system is set up with the microphone at the origin. θ is used to locate a source of sound that moves in a horizontal circle around the microphone, and r measures the amplitude of the signal that the microphone detects. The polar graph of r as a function of θ is called the *polar pattern* of the microphone. A cardioid microphone is a microphone whose polar pattern is shaped like the graph of the equation $r = 2.5 + 2.5 \cos \theta$. A graph of this polar pattern provides information about the microphone. *A problem related to this will be solved in Example 3.*

In Lesson 9-1, you learned to graph polar equations of the form $r = k$ and $\theta = k$, where k is a constant. In this lesson, you will learn how to graph more complicated types of polar equations. When you graph any type of polar equation for the first time, you can create a table of values for r and θ. Then you can use the table to plot points in the polar plane.

Example **1** Graph $r = \sin \theta$.

Make a table of values. Round the values of r to the nearest tenth. Graph the ordered pairs and connect them with a smooth curve.

θ	$\sin \theta$	(r, θ)
0°	0	(0, 0°)
30°	0.5	(0.5, 30°)
45°	0.7	(0.7, 45°)
60°	0.9	(0.9, 60°)
90°	1	(1, 90°)
120°	0.9	(0.9, 120°)
135°	0.7	(0.7, 135°)
150°	0.5	(0.5, 150°)
180°	0	(0, 180°)
210°	−0.5	(−0.5, 210°)
225°	−0.7	(−0.7, 225°)
240°	−0.9	(−0.9, 240°)
270°	−1	(−1, 270°)
300°	−0.9	(−0.9, 300°)
315°	−0.7	(−0.7, 315°)
330°	−0.5	(−0.5, 330°)
360°	0	(0, 360°)

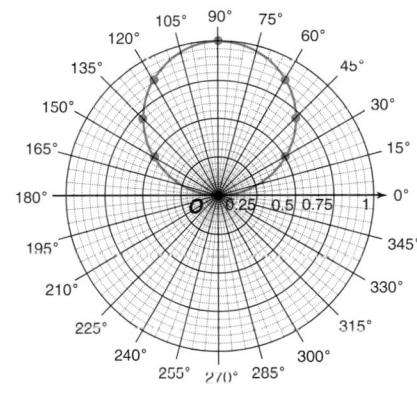

Notice that the ordered pairs obtained when $180° \leq \theta \leq 360°$ represent the same points as the ordered pairs obtained when $0° \leq \theta \leq 180°$.

Lesson 9-2 Graphs of Polar Equations **561**

Resource Manager

 Reproducible Masters
• *Study Guide and Practice,* pp. 159–160
• *Enrichment,* p. 98
• *Assessment and Evaluation,* p. 225
• *Spreadsheet Applications,* p. 9

 Transparencies
• 5-Minute Check, 9-2
• Teaching, 9-2

5-Minute Check
Lesson 9-1

Graph each of the following.

1. $A\left(3, \dfrac{\pi}{4}\right)$

2. $\theta = -210°$

3. $r = 2$

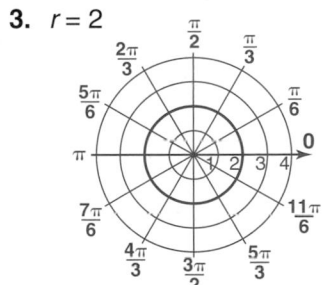

4. List four pairs of polar coordinates that represent the point P in the graph with the restriction that $-360° \leq \theta \leq 360°$.

Sample answer:
(3, −300°), (−3, 240°),
(3, 60°) (−3, −120°)

5. Find the distance between $P_1(7, 180°)$ and $P_2(9, 45°)$.
14.80

Hands-On Activity Have students work in cooperative groups, using their calculators to make a table of values for the equation $r = \sin \theta$. Have them plot the ordered pairs (r, θ) on the polar plane, and describe the graph. **The graph is a circle of diameter 1 with center at (0.5, 90°).**

2 TEACH

Teaching Tip If students have difficulty drawing polar graphs by using tables of values, suggest that they use the "wheel-and-spoke" graphing method. Have them begin at 0° and work their way around the circle counterclockwise. At each spoke (multiples of 30°), have them find the value of r and plot the point (r, θ).

In-Class Examples

Example 1
Graph $r = 2 \cos \theta$.

Example 2
Graph each polar equation.
a. $r = -2 \cos \theta$

b. $r = 1 - 2 \cos \theta$

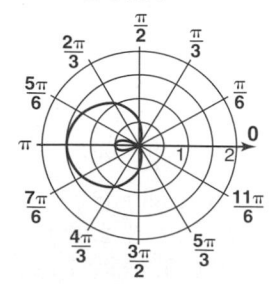

The graph of $r = \cos \theta$ is shown at the right. Notice that this graph and the graph in Example 1 are circles with a diameter of 1 unit. Both pass through the origin. As with the families of graphs you studied in Chapter 3, you can alter the position and shape of a polar graph by multiplying the function by a number or by adding to it. You can also multiply θ by a number or add a number to it in order to alter the graph. However, the changes in the graphs of polar equations can be quite different from those you studied in Chapter 3.

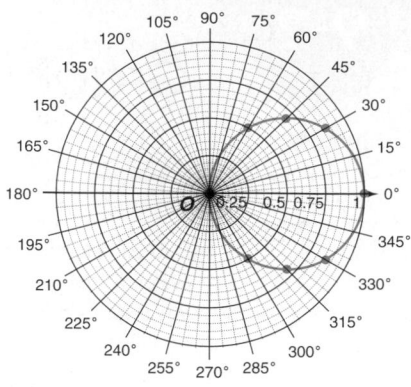

Examples ② Graph each polar equation.

    a. $r = 3 - 5 \cos \theta$

θ	$3 - 5 \cos \theta$	(r, θ)
0	-2	$(-2, 0)$
$\frac{\pi}{6}$	-1.3	$\left(-1.3, \frac{\pi}{6}\right)$
$\frac{\pi}{3}$	0.5	$\left(0.5, \frac{\pi}{3}\right)$
$\frac{\pi}{2}$	3	$\left(3, \frac{\pi}{2}\right)$
$\frac{2\pi}{3}$	5.5	$\left(5.5, \frac{2\pi}{3}\right)$
$\frac{5\pi}{6}$	7.3	$\left(7.3, \frac{5\pi}{6}\right)$
π	8	$(8, \pi)$
$\frac{7\pi}{6}$	7.3	$\left(7.3, \frac{7\pi}{6}\right)$
$\frac{4\pi}{3}$	5.5	$\left(5.5, \frac{4\pi}{3}\right)$
$\frac{3\pi}{2}$	3	$\left(3, \frac{3\pi}{2}\right)$
$\frac{5\pi}{3}$	0.5	$\left(0.5, \frac{5\pi}{3}\right)$
$\frac{11\pi}{6}$	-1.3	$\left(-1.3, \frac{11\pi}{6}\right)$
2π	-2	$(-2, 2\pi)$

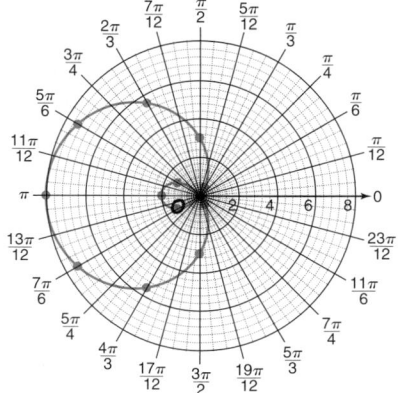

This type of curve is called a **limaçon.** Some limaçons have inner loops like this one. Other limaçons come to a point, have a dimple, or just curve outward.

Teaching Tip After working Example 2a, point out that the coefficient A in $r = A \sin \theta$ or $r = A \cos \theta$ has the effect of expanding (or contracting, if $|A| < 1$) the sine or cosine circle by a factor of $|A|$.

b. $r = 3 \sin 2\theta$

θ	$3 \sin 2\theta$	(r, θ)
0	0	$(0, 0)$
$\frac{\pi}{6}$	2.6	$\left(2.6, \frac{\pi}{6}\right)$
$\frac{\pi}{4}$	3	$\left(3, \frac{\pi}{4}\right)$
$\frac{\pi}{3}$	2.6	$\left(2.6, \frac{\pi}{3}\right)$
$\frac{\pi}{2}$	0	$\left(0, \frac{\pi}{2}\right)$
$\frac{2\pi}{3}$	-2.6	$\left(-2.6, \frac{2\pi}{3}\right)$
$\frac{3\pi}{4}$	3	$\left(-3, \frac{3\pi}{4}\right)$
$\frac{5\pi}{6}$	-2.6	$\left(-2.6, \frac{5\pi}{6}\right)$
π	0	$(0, \pi)$
$\frac{7\pi}{6}$	2.6	$\left(2.6, \frac{7\pi}{6}\right)$
$\frac{5\pi}{4}$	3	$\left(3, \frac{5\pi}{4}\right)$
$\frac{4\pi}{3}$	2.6	$\left(2.6, \frac{4\pi}{3}\right)$

θ	$3 \sin 2\theta$	(r, θ)
$\frac{3\pi}{2}$	0	$\left(0, \frac{3\pi}{2}\right)$
$\frac{5\pi}{3}$	-2.6	$\left(-2.6, \frac{5\pi}{3}\right)$
$\frac{7\pi}{4}$	-3	$\left(-3, \frac{7\pi}{4}\right)$
$\frac{11\pi}{6}$	-2.6	$\left(-2.6, \frac{11\pi}{6}\right)$

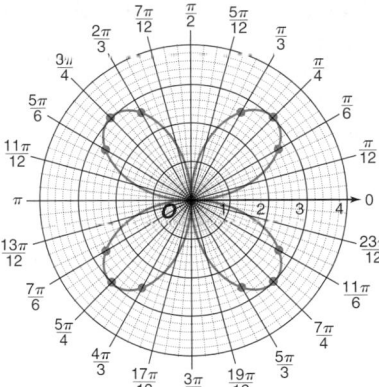

This type of curve is called a **rose.**

Real World Application

3 **AUDIO TECHNOLOGY** Refer to the application at the beginning of the lesson. The polar pattern of a microphone can be modeled by the polar equation $r = 2.5 + 2.5 \cos \theta$.

a. Sketch a graph of the polar pattern.
b. Describe what the polar pattern tells you about the microphone.

a.

θ	$2.5 + 2.5 \cos \theta$	(r, θ)
0°	5	(5, 0°)
30°	4.7	(4.7, 30°)
60°	3.8	(3.8, 60°)
90°	2.5	(2.5, 90°)
120°	1.3	(1.3, 120°)
150°	0.3	(0.3, 150°)
180°	0	(0, 180°)
210°	0.3	(0.3, 210°)
240°	1.3	(1.3, 240°)
270°	2.5	(2.5, 270°)
300°	3.8	(3.8, 300°)
330°	4.7	(4.7, 330°)

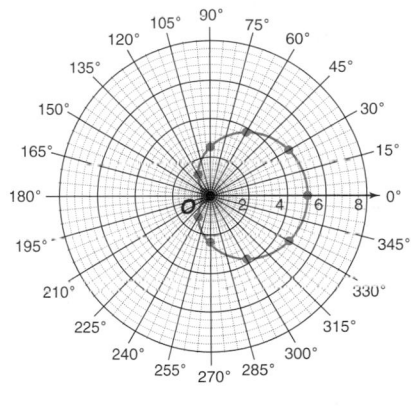

(continued on the next page)

Teaching Tip Make sure that students write ordered pairs as (r, θ) and not (θ, r).

In-Class Example

Example 3
Entertainment Suppose that a pattern created in a laser light show can be modeled by the equation $r^2 = \cos 2\theta$.
a. Sketch a graph of this equation.

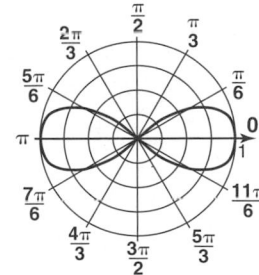

b. Discuss why you think this curve is sometimes referred to as the "figure 8" graph.
Sample answer: it resembles an 8, rotated 90°.

Addressing Individual Needs

Auditory/Musical There are many classifications of microphones, including cardioid, subcardioid, bidirectional, and omnidirectional. The polar patterns show the sensitivity of the microphones relative to the location of a source of sound. Have students research various types of microphones. If equipment is available, have students measure the sensitivity of a microphone as a function of the angle of a source of sound. Discuss with students which types of microphones would be best-suited for different situations.

Teaching Tip Remind students that it is possible to have no solutions or multiple solutions, depending on the system.

In-Class Example

Example 4
Graph the system of polar equations. Solve the system using algebra and trigonometry and compare the solutions to those on your graph.
$r = \sin \theta$
$r = 1$

The common solution is $(1, 90°)$.

Additional Answers *(p. 565)*

3. The polar coordinates of a point are not unique. A point of intersection may have one representation that satisfies one equation in a system, another representation that satisfies the other equation, but no representation that satisfies both simultaneously.

4. Barbara is correct. The interval $0 \leq \theta \leq \pi$ is not always sufficient. For example, the interval $0 \leq \theta \leq \pi$ only generates two of the four petals for the rose $r = \sin 2\theta$. $r = \sin \frac{\theta}{2}$ is an example where values of θ from 0 to 4π would have to be considered.

5.

6.

This type of curve is called a **cardioid**. A cardioid is a special type of limaçon.

b. The polar pattern indicates that the microphone will pick up only very loud sounds from behind it. It will detect much softer sounds from in front.

Limaçons, roses, and cardioids are examples of **classical curves**. The classical curves are summarized in the chart below.

			Classical Curves		
Curve	rose	lemniscate (pronounced lehm NIHS kuht)	limaçon (pronounced lee muh SOHN)	cardioid (pronounced KARD ee oyd)	spiral of Archimedes (pronounced ar kih MEED eez)
Polar Equation	$r = a \cos n\theta$ $r = a \sin n\theta$ n is a positive integer.	$r^2 = a^2 \cos 2\theta$ $r^2 = a^2 \sin 2\theta$	$r = a + b \cos \theta$ $r = a + b \sin \theta$	$r = a + a \cos \theta$ $r = a + a \sin \theta$	$r = a\theta$ (θ in radians)
General Graph					

It is possible to graph more than one polar equation at a time on a polar plane. However, the points where the graphs intersect do not always represent common solutions to the equations, since every point can be represented by infinitely many polar coordinates.

Example ④ **Graph the system of polar equations. Solve the system using algebra and trigonometry and compare the solutions to those on your graph.**

$$r = 3 - 3 \sin \theta$$
$$r = 4 - \sin \theta$$

> **Graphing Calculator Tip**
>
> A graphing calculator can graph polar equations in its polar mode. You can use the simultaneous graphing mode to help determine if the graphs pass through the same point at the same time.

To solve the system of equations, substitute $3 - 3 \sin \theta$ for r in the second equation.

$$3 - 3 \sin \theta = 4 - \sin \theta$$
$$-2 \sin \theta = 1$$
$$\sin \theta = -\frac{1}{2}$$
$$\theta = \frac{7\pi}{6} \text{ or } \theta = \frac{11\pi}{6}$$

[−6, 6] scl:1 by [−6, 6] scl:1

Substituting each angle into either of the original equations gives $r = 4.5$, so the solutions of the system are $\left(4.5, \frac{7\pi}{6}\right)$ and $\left(4.5, \frac{11\pi}{6}\right)$. Tracing the curves shows that these solutions correspond with the intersection points of the graphs.

7.

8.

9.

Communicating Mathematics

Read and study the lesson to answer each question.

1. **Write** a polar equation whose graph is a rose. **Sample answer:** $r = \sin 2\theta$.

2. **Determine** the maximum value of r in the equation $r = 3 + 5 \sin \theta$. What is the minimum value of r? **8; −2**

3. **State** the reason that algebra and trigonometry do not always find all the points of intersection of the graphs of polar equations. **See margin.**

4. **You Decide** Linh and Barbara were working on their homework together. Linh said that she thought that when graphing polar equations, you only need to generate points for which $0 \le \theta \le \pi$ because other values of θ would just generate the same points. Barbara said she thought she remembered an example from class where values of θ ranging from 0 to 4π had to be considered. Who is correct? Explain. **See margin.**

Guided Practice

5–8. See margin for graphs.

9. See margin for graph.

$\left(1, \frac{\pi}{6}\right), \left(1, \frac{5\pi}{6}\right),$

$\left(-2, \frac{3\pi}{2}\right)$

Graph each polar equation. Identify the type of curve each represents.

5. $r = 1 + \sin \theta$ **cardioid**
6. $r = 2 - 3 \sin \theta$ **limaçon**
7. $r = \cos 2\theta$ **rose**
8. $r = 1.5\theta$ **spiral of Archimedes**

9. Graph the system of polar equations $r = 2 \sin \theta$ and $r = 2 \cos 2\theta$. Solve the system using algebra and trigonometry. Assume $0 \le \theta < 2\pi$.

10. **Biology** The chambered nautilus is a mollusk whose shell can be modeled by the polar equation $r = 2\theta$.   **a. See margin.**

a. Graph this equation for $0 \le \theta \le 2\pi$.

b. Determine an approximate interval for θ that would result in a graph that models the chambered nautilus shown in the photo.
Sample answer: $0 \le \theta \le \frac{14\pi}{3}$

Practice

A

11–22. See Answer Appendix for graphs.

B

12. cardioid

18. limaçon

19. cardioid

24. Sample answer: $r = \frac{\theta}{2}$

C

Graph each polar equation. Identify the type of curve each represents.

11. $r = -3 \sin \theta$ **circle**
12. $r = 3 + 3 \cos \theta$
13. $r = 3\theta$ **spiral of Archimedes**
14. $r^2 = 4 \cos 2\theta$ **lemniscate**
15. $r = 2 \sin 3\theta$ **rose**
16. $r = -2 \sin 3\theta$ **rose**
17. $r = \frac{5}{2}\theta$ **spiral of Archimedes**
18. $r = -5 + 3 \cos \theta$
19. $r = -2 - 2 \sin \theta$
20. $r^2 = 9 \sin 2\theta$ **lemniscate**
21. $r = \sin 4\theta$ **rose**
22. $r = 2 + 2 \cos \theta$ **cardioid**

23. Write an equation for a rose with 3 petals.   **Sample answer:** $r = \sin 3\theta$

24. What is an equation for a spiral of Archimedes that passes through $A\left(\frac{\pi}{4}, \frac{\pi}{2}\right)$?

Graph each system of polar equations. Solve the system using algebra and trigonometry. Assume $0 \le \theta < 2\pi$.   **25–27. See margin.**

25. $r = 3$
    $r = 2 + \cos \theta$
26. $r = 1 + \cos \theta$
    $r = 1 - \cos \theta$
27. $r = 2 \sin \theta$
    $r = 2 \sin 2\theta$

  www.amc.glencoe.com/self_check_quiz

Lesson 9-2 Graphs of Polar Equations   **565**

Error Analysis

Watch for students who only graph part of a curve when plotting points.

Prevent by having students plot points from 0° to 360° or the period of the function, whichever is greater.

Assignment Guide

Core: 11–31 odd, 33–44
Enriched: 12–30 even, 31–44

Additional Answer

27. $(0, 0), (0, \pi), \left(\sqrt{3}, \frac{\pi}{3}\right),$
$\left(-\sqrt{3}, \frac{5\pi}{3}\right)$

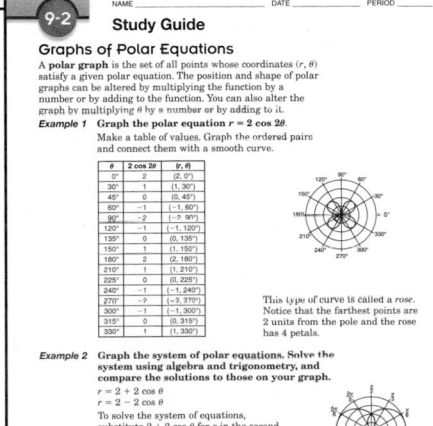

Study Guide and Practice Masters, pp. 159–160

Additional Answers

10a.

25. (3, 0)

26. $\left(1, \frac{\pi}{2}\right), \left(1, \frac{3\pi}{2}\right)$

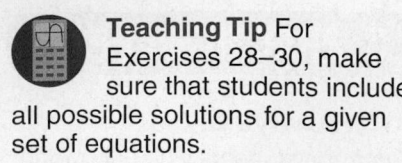
Additional Answers

28. (1, 0.5), (1, 1.0), (1, 2.1), (1, 2.6), (1, 3.7), (1, 4.2), (1, 5.2), (1, 5.8)

[−2, 2] scl:1 by [−2, 2] scl:1

29. (2, 3.5), (2, 5.9)

[−6, 6] scl:1 by [−6, 6] scl:1

30. (3.6, 0.6), (2.0, 4.7)

[−4, 4] scl:1 by [−4, 4] scl:1

32. **This microphone will pick up more sounds from behind than the cardioid microphone.**

Graphing Calculator

Applications and Problem Solving

Graph each system using a graphing calculator. Find the points of intersection. Round coordinates to the nearest tenth. Assume $0 \leq \theta < 2\pi$.

28. $r = 1$
$r = 2 \cos 2\theta$
28–30. See margin.

29. $r = 3 + 3 \sin \theta$
$r = 2$

30. $r = 2 + 2 \cos \theta$
$r = 3 + \sin \theta$

31. **Textiles** Patterns in fabric can often be created by modifying a mathematical graph. The pattern at the right can be modeled by a lemniscate.

a. Suppose the designer wanted to begin with a lemniscate that was 6 units from end to end. What polar equation could have been used? **$r^2 = 9 \cos 2\theta$ or $r^2 = 9 \sin 2\theta$**

b. What polar equation could have been used to generate a lemniscate that was 8 units from end to end? **$r^2 = 16 \cos 2\theta$ or $r^2 = 16 \sin 2\theta$**

32. **Audio Technology** Refer to the application at the beginning of the lesson and Example 3. Another microphone has a polar pattern that can be modeled by the polar equation $r = 3 + 2 \cos \theta$. Graph this polar pattern and compare it to the pattern of the cardioid microphone. **See margin.**

33. **Music** The curled part at the end of a violin is called the *scroll*. The scroll in the picture appears to curl around twice. For what interval of θ-values will the graph of $r = \theta$ model this violin scroll? **$0 \leq \theta \leq 4\pi$**

34. **Critical Thinking** **a–b. See Answer Appendix.**

a. Graph $r = \cos \dfrac{\theta}{n}$ for $n = 2, 4, 6,$ and 8. Predict the shape of the graph of $r = \cos \dfrac{\theta}{10}$.

b. Graph $r = \cos \dfrac{\theta}{n}$ for $n = 3, 5, 7,$ and 9. Predict the shape of the graph of $r = \cos \dfrac{\theta}{11}$.

35. **Communication** Suppose you want to use your computer to make a heart-shaped card for a friend. Write a polar equation that you could use to generate the shape of a heart. Make sure the graph of your equation points in the right direction. **Sample answer: $r = -1 - \sin \theta$**

36. **Critical Thinking** The general form for a limaçon is $r = a + b \cos \theta$ or $r = a + b \sin \theta$.

a. When will a limaçon have an inner loop? **$|b| > |a|$**

b. When will it have a dimple? **$|b| \leq |a| < |2b|$**

c. When will it have neither an inner loop nor a dimple? (This is called a *convex limaçon*.) **$|a| \geq |2b|$**

566 Chapter 9 Polar Coordinates and Complex Numbers

37. Critical Thinking Describe the transformation necessary to obtain the graph of each equation from the graph of the polar function $r = f(\theta)$. **See margin.**

a. $r = f(\theta - \alpha)$ (α is a constant.)   b. $r = f(-\theta)$

c. $r = -f(\theta)$ d. $r = cf(\theta)$ (c is a constant, $c > 0$.)

Mixed Review

38. Sample answer: (4, 405°), (4, 765°), (−4, −135°), (−4, 225°)

38. Find four other pairs of polar coordinates that represent the same point as $(4, 45°)$. *(Lesson 9-1)*

39. Find the cross product of $\vec{v}\langle 2, 3, 0\rangle$ and $\vec{w}\langle -1, 2, 4\rangle$. Verify that the resulting vector is perpendicular to $\vec{v}$ and $\vec{w}$. *(Lesson 8-4)* **See margin.**

40. Use a ruler and protractor to determine the magnitude (in centimeters) and direction of the vector shown at the right. *(Lesson 8-1)* **3.5 cm, 87°**

41. See margin.

42. $C = 109°5'$, $b = 23.3$, $a = 11.1$

41. Verify that $\dfrac{\sin^2 x}{\cos^4 x + \cos^2 x \sin^2 x} = \tan^2 x$ is an identity. *(Lesson 7-2)*

42. Solve $\triangle ABC$ if $A = 21°15'$, $B = 49°40'$, and $c = 28.9$. Round angle measures to the nearest minute and side measures to the nearest tenth. *(Lesson 5-6)*

43. Travel Adita is trying to decide where to go on vacation. He prefers not to fly, so he wants to take a bus or a train. The table below shows the round-trip fares for trips from his home in Kansas City, Missouri to various cities. Represent this data with a matrix. *(Lesson 2-3)* **See margin.**

Round−Trip Fares to Kansas City

	New York	Los Angeles	Miami
Bus	$240	$199	$260
Train	$254	$322	$426

44. SAT Practice Which fraction is the simplified form of $\dfrac{\frac{1}{8} + \frac{6}{4}}{\frac{3}{16}}$? **A**

A $\dfrac{26}{3}$

B $\dfrac{14}{3}$

C $\dfrac{28}{9}$

D $\dfrac{7}{3}$

E $\dfrac{4}{3}$

Extra Practice See p. A42.

Additional Answers

37a. counterclockwise rotation by an angle of α

37b. reflection about the polar axis or x-axis

37c. reflection about the origin

37d. dilation by a factor of c

Extra Credit

Graph $r = 2 \sin \theta \tan \theta$.

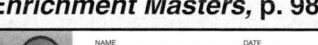

Open-Ended Assessment

Modeling Have students identify the curves represented by the following equations and explain their methods. Then, have them sketch the curves.

$r = 2 + 3 \cos \theta$ **limaçon**

$r = 4\theta$ **spiral of Archimedes**

$r = 5 + 5 \sin \theta$ **cardioid**

$r = 3 \sin 2\theta$ **rose**

$r^2 = 4 \cos 2\theta$ **lemniscate**

Chapter 9, Quiz A (Lessons 9-1 and 9-2) is available in the *Assessment and Evaluation Masters*, p. 225.

Additional Answers

39. $\langle 12, -8, 7\rangle; \langle 2, 3, 0\rangle \cdot \langle 12, -8, 7\rangle$
$= 0, \langle -1, 2, 4\rangle \cdot \langle 12, -8, 7\rangle = 0$

41. $\dfrac{\sin^2 x}{\cos^4 x + \cos^2 x \sin^2 x} \overset{?}{=} \tan^2 x$

$\dfrac{\sin^2 x}{\cos^2 x (\cos^2 x + \sin^2 x)} \overset{?}{=} \tan^2 x$

$\dfrac{\sin^2 x}{(\cos^2 x)(1)} \overset{?}{=} \tan^2 x$

$\dfrac{\sin^2 x}{\cos^2 x} \overset{?}{=} \tan^2 x$

$\tan^2 x = \tan^2 x$

43.

	NY	LA	Miami
Bus	$240	$199	$260
Train	$254	$322	$426

Enrichment Masters, p. 98

Polar and Rectangular Coordinates

5-Minute Check
Lesson 9-2

Identify the type of curve represented by each equation.

1. $r = 2\theta$ **spiral of Archimedes**

2. $r = 2 \sin 3\theta$ **rose**

3. $r = 4 + 4 \cos \theta$ **cardioid**

4. Graph the system of polar equations. Solve the system using algebra and trigonometry. Assume $0 \le \theta < 2\pi$.

$r = \cos \theta$

$r = -\sin \theta$

$\left(\dfrac{\sqrt{2}}{2}, \dfrac{7\pi}{4}\right), \left(-\dfrac{\sqrt{2}}{2}, \dfrac{3\pi}{4}\right)$

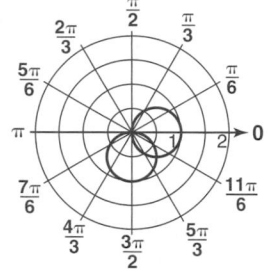

5. Graph $r = 2 \sin \theta$.

Motivating the Lesson
Real-World Application Ask students why maps often use rectangular rather than polar coordinates. **Possible answers: it is easier to cover areas with squares or rectangles than with circles. A polar map would have the center as a comparison point.**

OBJECTIVE
- Convert between polar and rectangular coordinates.

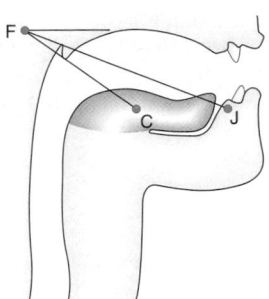

PHYSIOLOGY A laboratory has designed a voice articulation program that synthesizes speech by controlling the speech articulators: the jaw, tongue, lips, and so on. To create a mathematical model that is manageable, each of these speech articulators is identified by a point.

J = the edge of the jaw
C = the center of the tongue
F = a fixed point about which the jaw rotates

The positions of these points are given in polar coordinates with F as the pole and the horizontal as the polar axis. Changing these positions alters the sounds that are synthesized. *A problem related to this will be solved in Example 2.*

For some real-world phenomena, it is useful to be able to convert between polar coordinates and rectangular coordinates.

Suppose a rectangular coordinate system is superimposed on a polar coordinate system so that the origins coincide and the x-axis aligns with the polar axis. Let P be any point in the plane.

Polar coordinates: $P(r, \theta)$
Rectangular coordinates: $P(x, y)$

Trigonometric functions can be used to convert polar coordinates to rectangular coordinates.

Converting Polar Coordinates to Rectangular Coordinates	The rectangular coordinates (x, y) of a point named by the polar coordinates (r, θ) can be found by using the following formulas. $$x = r \cos \theta$$ $$y = r \sin \theta$$

You will derive these formulas in Exercise 42.

Resource Manager

 Reproducible Masters
- *Study Guide and Practice,* pp. 161–162
- *Enrichment,* p. 99

 Transparencies
- 5-Minute Check, 9-3
- Teaching, 9-3

Examples ❶ **Find the rectangular coordinates of each point.**

a. $P\left(5, \dfrac{\pi}{3}\right)$

For $P\left(5, \dfrac{\pi}{3}\right)$, $r = 5$ and $\theta = \dfrac{\pi}{3}$.

$x = r \cos \theta$ $y = r \sin \theta$

$\quad = 5 \cos \dfrac{\pi}{3}$ $\quad = 5 \sin \dfrac{\pi}{3}$

$\quad = 5\left(\dfrac{1}{2}\right)$ or $\dfrac{5}{2}$ $\quad = 5\left(\dfrac{\sqrt{3}}{2}\right)$ or $\dfrac{5\sqrt{3}}{2}$

The rectangular coordinates of P are $\left(\dfrac{5}{2}, \dfrac{5\sqrt{3}}{2}\right)$ or (2.5, 4.33) to the nearest hundredth.

b. $Q(-13, -70°)$

For $Q(-13, -70°)$, $r = -13$ and $\theta = -70°$.

$x = r \cos \theta$ $y = r \sin \theta$

$\quad = -13 \cos (-70°)$ $\quad = -13 \sin (-70°)$

$\quad \approx -13(0.34202)$ $\quad \approx -13(-0.93969)$

$\quad \approx -4.45$ $\quad \approx 12.22$

The rectangular coordinates of Q are approximately $(-4.45, 12.22)$.

❷ **PHYSIOLOGY** Refer to the application at the beginning of the lesson. Suppose the computer model assigns polar coordinates (7.5, 330°) to point J and (4.5, 310°) to point C in order to create a particular sound. Each unit represents a centimeter. Is the center of the tongue above or below the edge of the jaw? by how far?

Find the rectangular coordinates of each point.

For $J(7.5, 330°)$, $r = 7.5$ and $\theta = 330°$.

$x = r \cos \theta$ $y = r \sin \theta$

$\quad = 7.5 \cos 330°$ $\quad = 7.5 \sin 330°$

$\quad \approx 6.50$ $\quad = -3.75$

$J(7.5, 330°) \rightarrow J(6.50, -3.75)$

For $C(4.5, 310°)$, $r = 4.5$ and $\theta = 310°$.

$x = r \cos \theta$ $y = r \sin \theta$

$\quad = 4.5 \cos 310°$ $\quad = 4.5 \sin 310°$

$\quad \approx 2.89$ $\quad \approx -3.45$

$C(4.5, 310°) \rightarrow C(2.89, -3.45)$

Since $-3.45 > -3.75$, the center of the tongue is above the edge of the jaw when this sound is made. Subtracting the y-coordinates, we see that the center of the tongue is about 0.3 centimeter, or 3 millimeters, higher than the edge of the jaw.

2 TEACH

Teaching Tip For Example 1, remind students that cosine is always related to the *x*-direction and sine is always related to the *y*-direction.

In-Class Examples

Example 1
Find the rectangular coordinates of each point.

a. $D\left(-2, -\dfrac{5\pi}{6}\right)$ $(\sqrt{3}, 1)$

b. $C(3, 270°)$ $(0, -3)$

Example 2
Physiology Suppose that the computer model assigns polar coordinates (4.6, 320°) to point J and (3.9, 295°) to point C in order to create a particular sound. Each unit represents a centimeter. Is the center of the tongue above or below the edge of the jaw? By how far?
Center of tongue is 5.7 mm below the edge of the jaw.

Prerequisite Skills

A review of the prerequisite skills for this lesson is provided in the *Prerequisite Skills Masters*.

Skill Number and Content	Pages
9 Factoring by Grouping and Greatest Common Factor	19–20
10 Factoring Special Products of Polynomials	21–22
11 Factoring by Completing the Square	23–24

If a point is named by the rectangular coordinates (x, y), you can find the corresponding polar coordinates by using the Pythagorean Theorem and the Arctangent function. Since the Arctangent function only determines angles in the first and fourth quadrants, you must add π radians to the value of θ for points with coordinates (x, y) that lie in the second or third quadrants.

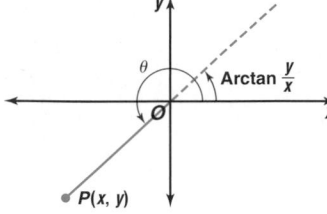

When x is zero, $\theta = \pm\frac{\pi}{2}$. Why?

When $x > 0$, $\theta = $ Arctan $\frac{y}{x}$.

When $x < 0$, $\theta = $ Arctan $\frac{y}{x} + \pi$.

Converting Rectangular Coordinates to Polar Coordinates

The polar coordinates (r, θ) of a point named by the rectangular coordinates (x, y) can be found by the following formulas.

$$r = \sqrt{x^2 + y^2}$$
$$\theta = \text{Arctan } \frac{y}{x}, \text{ when } x > 0$$
$$\theta = \text{Arctan } \frac{y}{x} + \pi, \text{ when } x < 0$$

Example ③ **Find the polar coordinates of $R(-8, -12)$.**

For $R(-8, -12)$, $x = -8$ and $y = -12$.

$r = \sqrt{x^2 + y^2}$

$= \sqrt{(-8)^2 + (-12)^2}$

$= \sqrt{208}$

≈ 14.42

$\theta = \text{Arctan } \frac{y}{x} + \pi \qquad x < 0$

$= \text{Arctan } \frac{-12}{-8} + \pi$

$= \text{Arctan } \frac{3}{2} + \pi$

≈ 4.12

The polar coordinates of R are approximately (14.42, 4.12).
Other polar coordinates can also represent this point.

The conversion equations can also be used to convert equations from one coordinate system to the other.

Examples ④ **Write the polar equation $r = 6 \cos \theta$ in rectangular form.**

$r = 6 \cos \theta$

$r^2 = 6r \cos \theta \quad$ *Multiply each side by r.*

$x^2 + y^2 = 6x \qquad r^2 = x^2 + y^2, r\cos \theta = x$

Addressing Individual Needs

Interpersonal Have each student give the polar coordinates of a point to another student. The second student converts the coordinates to rectangular form and passes them to a third student, who converts them back to polar form. Students should now compare the two polar forms. If they are not the same, ask the students whether an error necessarily occurred. Repeat, beginning with rectangular coordinates.

5 Write the rectangular equation $(x - 3)^2 + y^2 = 9$ in polar form.

$$(x - 3)^2 + y^2 = 9$$
$$(r\cos\theta - 3)^2 + (r\sin\theta)^2 = 9 \qquad x = r\cos\theta, \ y = r\sin\theta$$
$$r^2\cos^2\theta - 6r\cos\theta + 9 + r^2\sin^2\theta = 9 \qquad \textit{Multiply.}$$
$$r^2\cos^2\theta - 6r\cos\theta + r^2\sin^2\theta = 0 \qquad \textit{Subtract 9 from each side.}$$
$$r^2\cos^2\theta + r^2\sin^2\theta = 6r\cos\theta \qquad \textit{Isolate squared terms.}$$
$$r^2(\cos^2\theta + \sin^2\theta) = 6r\cos\theta \qquad \textit{Factor.}$$
$$r^2(1) = 6r\cos\theta \qquad \textit{Pythagorean Identity}$$
$$r^2 = 6r\cos\theta$$
$$r = 6\cos\theta \qquad \textit{Simplify.}$$

CHECK FOR UNDERSTANDING

Communicating Mathematics

Read and study the lesson to answer each question.

1. **Write** the polar coordinates of the point in the graph at the right. **Sample answer: $\left(2\sqrt{2}, 45°\right)$**

2. **Explain** why you have to consider what quadrant a point lies in when converting from rectangular coordinates to polar coordinates. **See margin.**

3. **Determine** the polar equation for $x = 2$. **$r = 2\sec\theta$**

4. *Math Journal* **Write** a paragraph explaining how to convert from polar coordinates to rectangular coordinates and vice versa. Include a diagram with your explanation. **See margin.**

Guided Practice

Find the polar coordinates of each point with the given rectangular coordinates. Use $0 \le \theta < 2\pi$ and $r \ge 0$.

5. $\left(-\sqrt{2}, \sqrt{2}\right)$ $\left(2, \dfrac{3\pi}{4}\right)$

6. $(-2, -5)$ **(5.39, 4.33)**

Find the rectangular coordinates of each point with the given polar coordinates.

7. $\left(-2, \dfrac{4\pi}{3}\right)$ **$(1, \sqrt{3})$**

8. $(2.5, 250°)$ **$(-0.86, -2.35)$**

Write each rectangular equation in polar form.

9. $y = 2$ **$r = 2\csc\theta$**

10. $x^2 + y^2 = 16$ **$r = 4$ or $r = -4$**

Write each polar equation in rectangular form.

11. $r = 6$ **$x^2 + y^2 = 36$**

12. $r = -\sec\theta$ **$x = -1$**

13. **Acoustics** The polar pattern of a cardioid lavalier microphone is given by $r = 2 + 2\cos\theta$.

a. Graph the polar pattern. **See margin.**

b. Will the microphone detect a sound that originates from the point with rectangular coordinates $(-2, 0)$? Explain. **See margin.**

 www.amc.glencoe.com/self_check_quiz

Lesson 9-3 Polar and Rectangular Coordinates **571**

From the Classroom of...

Kerri Darilek
Halletsville School
Halletsville, Texas

"I give my students a project whereby they create a connect-the-dot design on a polar graph. They then convert each point into rectangular coordinates and graph them on a rectangular coordinate system. Students turn in both graphs with their list of points to show that they have created the same picture."

Additional Answer

42. Drop a perpendicular from the point with polar coordinates (r, θ) to the x-axis. r is the length of the hypotenuse in the resulting right triangle. x is the length of the side adjacent to angle θ, so $\cos \theta = \frac{x}{r}$. Solving for x gives $x = r \cos \theta$. y is the length of the side opposite angle θ, so $\sin \theta = \frac{y}{r}$. Solving for y gives $y = r \sin \theta$. (The figure is drawn for a point in the first quadrant, but the signs work out correctly regardless of where in the plane the point is located.)

EXERCISES

Practice

Find the polar coordinates of each point with the given rectangular coordinates. Use $0 \le \theta < 2\pi$ and $r \ge 0$.

 A

14. $(2, -2)$ **14.** $\left(2\sqrt{2}, \frac{7\pi}{4}\right)$

15. $(0, 1)$ $\left(1, \frac{\pi}{2}\right)$

16. $\left(1, \sqrt{3}\right)$ $\left(2, \frac{\pi}{3}\right)$

B

17. $\left(-\frac{1}{4}, -\frac{\sqrt{3}}{4}\right)$ $\left(\frac{1}{2}, \frac{4\pi}{3}\right)$

18. $(3, 8)$ $(8.54, 1.21)$

19. $(4, -7)$ $(8.06, 5.23)$

Find the rectangular coordinates of each point with the given polar coordinates.

20. $\left(3, \frac{\pi}{2}\right)$ $(0, 3)$

21. $\left(\frac{1}{2}, \frac{3\pi}{4}\right)$ $\left(-\frac{\sqrt{2}}{4}, \frac{\sqrt{2}}{4}\right)$

22. $\left(-1, -\frac{\pi}{6}\right)$ $\left(-\frac{\sqrt{3}}{2}, \frac{1}{2}\right)$

24. $\left(-2\sqrt{3}, -2\right)$ **23.** $(-2, 270°)$ $(0, 2)$

24. $(4, 210°)$

25. $(14, 130°)$

$(-9.00, 10.72)$

Write each rectangular equation in polar form.

26. $x = -7$ $r = -7 \sec \theta$

27. $y = 5$ $r = 5 \csc \theta$

28. $x^2 + y^2 = 25$ $r = 5$ or $r = -5$

29. $x^2 + y^2 = 2y$ $r = 2 \sin \theta$

30. $x^2 - y^2 = 1$ $r^2 = \sec 2\theta$

31. $x^2 + (y - 2)^2 = 4$ $r = 4 \sin \theta$

Write each polar equation in rectangular form.

32. $r = 2$ $x^2 + y^2 = 4$

33. $r = -3$ $x^2 + y^2 = 9$

C   **34.** $\theta = \frac{\pi}{3}$ $y = \sqrt{3}x$

35. $r = 2 \csc \theta$ $y = 2$

36. $r = 3 \cos \theta$ $x^2 + y^2 = 3x$

37. $r^2 \sin 2\theta = 8$ $xy = 4$

38. Write the equation $y = x$ in polar form. $\theta = \frac{\pi}{4}$

39. What is the rectangular form of $r = \sin \theta$? $x^2 + y^2 = y$

Applications and Problem Solving

Real World Application

40. Surveying A surveyor identifies a landmark at the point with polar coordinates $(325, 70°)$. What are the rectangular coordinates of this point? **(111.16, 305.40)**

41. Machinery An arc of a spiral of Archimedes is used to create a disc that drives the spindle on a sewing machine. Note from the figure that the spinning disc drives the rod that moves the piston. The outline of the disc can be modeled by the graph of $r = \frac{\theta}{6}$ for $\frac{\pi}{4} \le \theta \le \frac{5\pi}{4}$ and its reflection in the line $\theta = \frac{\pi}{4}$. How far does the piston move from right to left? **0.52 unit**

42. Critical Thinking Write a convincing argument to prove that, when converting polar coordinates to rectangular coordinates, the formulas $x = r \cos \theta$ and $y = r \sin \theta$ are true. Include a labeled drawing in your answer. **See margin.**

Extra Credit

By converting to rectangular coordinates, show that $r = a \cos \theta + b \sin \theta$ is an equation for a circle. Find the center and the radius.

$r = a \cos \theta + b \sin \theta$
$r^2 = ra \cos \theta + rb \sin \theta$
$x^2 + y^2 = ax + by$
$x^2 - ax + y^2 - by = 0$
$\left(x - \frac{a}{2}\right)^2 + \left(y - \frac{b}{2}\right)^2 = \left(\frac{\sqrt{a^2 + b^2}}{2}\right)^2;$
center $\left(\frac{a}{2}, \frac{b}{2}\right)$, radius $\frac{\sqrt{a^2 + b^2}}{2}$

43. Irrigation A sod farm can use a combined Cartesian and polar coordinate system to identify points in the field. Points have coordinates of the form (x, y, r, θ). The sprinkler heads are spaced 25 meters apart. The Cartesian coordinates x and y, which are positive integers, indicate how many sprinkler heads to go across and up in the grid. The polar coordinates give the location of the point relative to the sprinkler head. If a point has coordinates $(4, 3, 2, 120°)$, then how far to the east and north of the origin is that point?   **75 m east; 118.30 m north**

44a. ⟨3.76, 1.37⟩; ⟨1.71, 4.70⟩   44b. ⟨5.47, 6.07⟩

44. Electrical Engineering When adding sinusoidal expressions like 4 sin (3.14t + 20°) and 5 sin (3.14t + 70°), electrical engineers sometimes use *phasors,* which are vectors in polar form.

  a. The phasors for these two expressions are written as 4∠20° and 5∠70°. Find the rectangular forms of these vectors.

  b. Add the two vectors you found in part a.

  c. Write the sum of the two vectors in phasor notation.   **8.17∠47.98°**

  d. Write the sinusoidal expression corresponding to the phasor in part c.   **8.17 sin(3.14t + 47.98°)**

45. Critical Thinking Identify the type of graph generated by the polar equation $r = 2a \sin \theta + 2a \cos \theta$. Write the equivalent rectangular equation.
Circle centered at (a, a) with radius $\sqrt{2} \, |a|$; $(x - a)^2 + (y - a)^2 = 2a^2$

Mixed Review

46. Graph the polar equation $r = 2 + \sin \theta$. *(Lesson 9-2)*   **See margin.**

47. Name four different pairs of polar coordinates that represent the point at $(-2, 45°)$. *(Lesson 9-1)*

48. Aviation An airplane flies at an air speed of 425 mph on a heading due south. It flies against a headwind of 50 mph from a direction 30° east of south. Find the airplane's ground speed and direction. *(Lesson 8-5)*

49. Solve $\sin^2 A = \cos A - 1$ for principal values of A. *(Lesson 7-5)*   **0°**

50. Graph $y = 2 \cos \theta$. *(Lesson 6-4)*   **See margin.**

51. Use the unit circle to find the exact value of cos 210°. *(Lesson 5-3)* $-\dfrac{\sqrt{3}}{2}$

52. Write a polynomial function to model the set of data. *(Lesson 4-8)*

x	−10	−7	−4	−1	2	5	8	11	14
f(x)	−15	−9.2	−6.9	−3	−0.1	2	1.1	−2.3	−4.5

53. Use synthetic division to divide $x^5 - 3x^2 - 20$ by $x - 2$. *(Lesson 4-3)*

54. Write the equation of best fit for a set of data using the ordered pairs $(17, 145)$ and $(25, 625)$. *(Lesson 1-6)*   **$y = 60x - 875$**

55. SAT/ACT Practice x, y, and z are different positive integers. $\dfrac{x}{y}$ and $\dfrac{y}{z}$ are also positive integers. Which of the following cannot be a positive integer?   **C**

  A $\dfrac{x}{z}$ **B** $(x)(y)$ **C** $\dfrac{z}{x}$ **D** $(x + y)z$ **E** $(x - z)y$

Extra Practice See p. A42.

Lesson 9-3   Polar and Rectangular Coordinates **573**

Mixed Review column side answers:

47.Sample answer:
(−2, 405°),
(−2, 765°),
(2, 225°),
(2, 585°)

48. 382.52 mph, 3°45′ west of south

52. Sample answer:
$y = -0.07x^2 + 0.73x - 1.36$

53. $x^4 + 2x^3 + 4x^2 + 5x + 10$

Lesson 9-3 **573**

Polar Form of a Linear Equation

5-Minute Check
Lesson 9-3

Find the rectangular coordinates of each point.

1. $A\left(3, \frac{\pi}{4}\right)$ $\left(\frac{3\sqrt{2}}{2}, \frac{3\sqrt{2}}{2}\right)$

2. $B(10, 52°)$ **(6.16, 7.88)**

3. Find the polar coordinates of $B(-2, 5)$. **(5.39, 1.95)**

4. Write the rectangular equation $x^2 + y^2 = 100$ in polar form. $r = \pm 10$

5. Write the polar equation $r \sin \theta = 5$ in rectangular form. $y = 5$

Motivating the Lesson

Discussion Ask students to give the polar form of these linear equations.

a. $y = 0$ $\theta = 0$

b. $x = 0$ $\theta = \frac{\pi}{2}$

c. $y = x$ $\theta = \frac{\pi}{4}$

Discuss the fact that in rectangular form, the above equations can be converted to polar form relatively easily. For most linear equations, however, the conversion is more difficult. This lesson will explain how to make such conversions.

OBJECTIVES
* Write the polar form of a linear equation.
* Graph the polar form of a linear equation.

BIOLOGY Scientists hope that by studying the behavior of flies, they will gain insight into the genetics and brain functions of flies. Karl Götz of the Max Planck Institute in Tübingen, Germany, designed the Buridan Paradigm, which is a device that tracks the path of a fruit fly walking between two visual landmarks. The position of the fly is recorded in both rectangular and polar coordinates. If the fly walks in a straight line from the point with polar coordinates (6, 2) to the point with polar coordinates (2, −0.5), what is the equation of the path of the fly? How close did the fly come to the origin? *This problem will be solved in Example 4.*

Look Back
You can refer to Lesson 7-6 to review normal form.

The polar form of the equation for a line ℓ is closely related to the normal form, which is $x \cos \phi + y \sin \phi - p = 0$. We have learned that $x = r \cos \theta$ and $y = r \sin \theta$. The polar form of the equation of line ℓ can be obtained by substituting these values into the normal form.

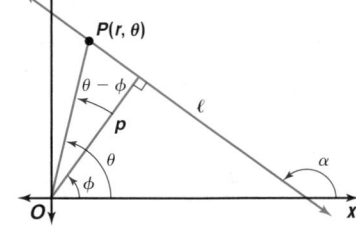

$$x \cos \phi + y \sin \phi - p = 0$$
$$(r \cos \theta) \cos \phi + (r \sin \theta) \sin \phi - p = 0$$
$$r(\cos \theta \cos \phi + \sin \theta \sin \phi) = p$$
$$r \cos (\theta - \phi) = p \quad \cos \theta \cos \phi + \sin \theta \sin \phi = \cos (\theta - \phi)$$

Polar Form of a Linear Equation	The polar form of a linear equation, where p is the length of the normal and ϕ is the positive angle between the positive x-axis and the normal, is $$p = r \cos (\theta - \phi).$$

In the polar form of a linear equation, θ and r are variables, and p and ϕ are constants. Values for p and ϕ can be obtained from the normal form of the standard equation of a line. Remember to choose the value for ϕ according to the quadrant in which the normal lies.

Resource Manager

 Reproducible Masters
* *Study Guide and Practice,* pp. 163–164
* *Enrichment,* p. 100
* *Assessment and Evaluation,* pp. 224, 225
* *TI-92 & CBL Interdisciplinary Activities,* pp. 29–32

 Transparencies
* 5-Minute Check, 9-4
* Teaching, 9-4

Example **1** Write each equation in polar form.

a. $5x + 12y = 26$

The standard form of the equation is $5x + 12y - 26 = 0$. First, write the equation in normal form to find the values of p and ϕ. To convert to normal form, we need to find the value of $\pm\sqrt{A^2 + B^2}$.

$\pm\sqrt{A^2 + B^2} = \pm\sqrt{5^2 + 12^2}$, or ± 13

Since C is negative, use $+13$. The normal form of the equation is

$\dfrac{5}{13}x + \dfrac{12}{13}y - 2 = 0$.   *The normal form is $x \cos \phi + y \sin \phi - p = 0$.*

We can see from the normal form that $p = 2$, $\cos \phi = \dfrac{5}{13}$, and $\sin \phi = \dfrac{12}{13}$.

Since $\cos \phi$ and $\sin \phi$ are both positive, the normal lies in the first quadrant.

$\tan \phi = \dfrac{\sin \phi}{\cos \phi}$

$\tan \phi = \dfrac{\frac{12}{13}}{\frac{5}{13}}$

$\tan \phi = \dfrac{12}{5}$

$\phi \approx 1.18$ *Use the Arctangent function.*

Substitute the values for p and ϕ into the polar form.

$p = r \cos (\theta - \phi) \rightarrow 2 = r \cos (\theta - 1.18)$

The polar form of $5x + 12y = 26$ is $2 = r \cos (\theta - 1.18)$.

b. $2x - 7y = -5$

The standard form of this equation is $2x - 7y + 5 = 0$.

So, $\pm\sqrt{A^2 + B^2} = \sqrt{\pm 2^2 + (-7)^2}$, or $\pm\sqrt{53}$.

Since C is positive, use $-\sqrt{53}$. Then the normal form of the equation is

$-\dfrac{2}{\sqrt{53}}x + \dfrac{7}{\sqrt{53}}y - \dfrac{5}{\sqrt{53}} = 0$.

We see that $p = \dfrac{5}{\sqrt{53}}$ or $\dfrac{5\sqrt{53}}{53}$, $\cos \phi = -\dfrac{2}{\sqrt{53}}$, and $\sin \phi = \dfrac{7}{\sqrt{53}}$.

Since $\cos \phi < 0$ but $\sin \phi > 0$, the normal lies in the second quadrant.

$\tan \phi = \dfrac{\sin \phi}{\cos \phi}$

$\tan \phi = -\dfrac{7}{2}$ $\dfrac{7}{\sqrt{53}} \div \left(-\dfrac{2}{\sqrt{53}}\right) = \dfrac{7}{\sqrt{53}} \cdot \left(-\dfrac{\sqrt{53}}{2}\right)$

$\phi \approx 106°$ *Add $180°$ to the Arctangent value.*

Substitute the values for p and ϕ into the polar form.

$p = r \cos (\theta - \phi) \rightarrow \dfrac{5\sqrt{53}}{53} = r \cos (\theta - 106°)$

The polar form of the equation is $\dfrac{5\sqrt{53}}{53} = r \cos (\theta - 106°)$.

Lesson 9-4   Polar Form of a Linear Equation **575**

📁 Prerequisite Skills

A review of the prerequisite skills for this lesson is provided in the *Prerequisite Skills Masters*.

Skill Number and Content	Pages
8 Rationalizing a Denominator	17–18

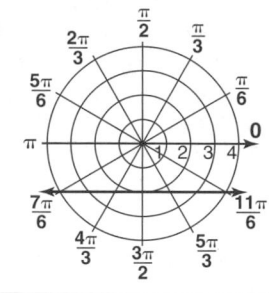
The polar form of a linear equation can be converted to rectangular form by using the angle sum and difference identities for cosine and the polar coordinate conversion equations.

Example **2** Write $2 = r \cos(\theta - 60°)$ in rectangular form.

$2 = r \cos(\theta - 60°)$	
$2 = r(\cos\theta \cos 60° + \sin\theta \sin 60°)$	*Difference identity for cosine*
$2 = r\left(\dfrac{1}{2}\cos\theta + \dfrac{\sqrt{3}}{2}\sin\theta\right)$	$\cos 60° = \dfrac{1}{2},\ \sin 60° = \dfrac{\sqrt{3}}{2}$
$2 = \dfrac{1}{2}r\cos\theta + \dfrac{\sqrt{3}}{2}r\sin\theta$	*Distributive Property*
$2 = \dfrac{1}{2}x + \dfrac{\sqrt{3}}{2}y$	*Polar to rectangular conversion equations*
$4 = x + \sqrt{3}y$	*Multiply each side by 2.*
$0 = x + \sqrt{3}y - 4$	*Subtract 4 from each side.*

The rectangular form of $2 = r \cos(\theta - 60°)$ is $x + \sqrt{3}y - 4 = 0$.

The polar form of a linear equation can be graphed by preparing a table of coordinates and then graphing the ordered pairs in the polar system.

Examples **3** Graph the equation $r = 4\sec\left(\theta + \dfrac{2\pi}{3}\right)$.

Write the equation in the form $r = \dfrac{4}{\cos\left(\theta + \frac{2\pi}{3}\right)}$.

Use your calculator to make a table of values.

θ	0	$\dfrac{\pi}{6}$	$\dfrac{\pi}{3}$	$\dfrac{\pi}{2}$	$\dfrac{2\pi}{3}$	$\dfrac{5\pi}{6}$	π
r	-8	-4.6	-4	-4.6	-8	undefined	8

Graph the ordered pairs on a polar plane.

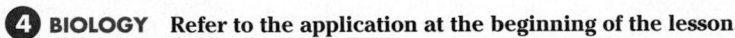

④ BIOLOGY Refer to the application at the beginning of the lesson.

a. If the fly walks in a straight line from the point with polar coordinates (6, 2) to the point with polar coordinates (2, −0.5), what is the equation of the path of the fly?

b. If *r* is measured in centimeters, how close did the fly come to the origin?

a. The two points must satisfy the equation $p = r \cos (\theta - \phi)$. Substitute both ordered pairs into this form to create a system of equations.

$p = 6 \cos (2 - \phi)$ $(r, \theta) = (6, 2)$
$p = 2 \cos (-0.5 - \phi)$ $(r, \theta) = (2, -0.5)$

Substituting either expression for *p* into the other equation results in the equation $6 \cos (2 - \phi) = 2 \cos (-0.5 - \phi)$.

A graphing calculator shows that there are two solutions to this equation between 0 and 2π: $\phi \approx 0.59$ and $\phi \approx 3.73$.

[0, 2π] scl:1 by [−6, 6] scl:1

Substituting these values into $p = 6 \cos (2 - \phi)$ yields $p \approx 0.93$ and $p \approx -0.93$, respectively. Since *p*, the length of the normal, must be positive, we use $\phi \approx 0.59$ and $p \approx 0.93$.

Therefore, the polar form of the equation of the fly's path is $0.93 = r \cos (\theta - 0.59)$.

b. The closest that the fly came to the origin is the value of *p*, 0.93 centimeter or 9.3 millimeters.

CHECK FOR UNDERSTANDING

Communicating Mathematics

Read and study the lesson to answer each question.

1. **State** the general form for the polar equation of a line. Explain the significance of each part of the equation. **See margin.**

2. **Determine** what value of θ will result in $r = p$ in the polar equation of a line. $\theta = \phi$

3. **Explain** how to find ϕ and the length of the normal for a rectangular equation of the form $x = k$. Write the polar form of the equation. **See margin.**

4. **Explain** why it is a good idea to find several ordered pairs when graphing the polar equation of a line, even though only two points are needed to determine a line. **See margin.**

Guided Practice

Write each equation in polar form. Round ϕ to the nearest degree.

5. $3x - 4y - 10 = 0$
   $2 = r \cos (\theta - 307°)$

6. $-2x + 4y = 9$ $\dfrac{9\sqrt{5}}{10} = r \cos (\theta - 117°)$

Write each equation in rectangular form.

7. $3 = r \cos(\theta - 60°)$
   $x + \sqrt{3}y - 6 = 0$

8. $r = 2 \sec\left(\theta + \dfrac{\pi}{4}\right)$
   $\sqrt{2}x - \sqrt{2}y - 4 = 0$

Lesson 9-4   Polar Form of a Linear Equation   **577**

Additional Answers

1. The polar equation of a line is $p = r \cos (\theta - \phi)$. *r* and θ are the variables. *p* is the length of the normal segment from the line to the origin and ϕ is the angle the normal makes with the positive *x*-axis.

3. The graph of the equation $x = k$ is a vertical line. Since the line is vertical, the *x*-axis is the normal line through the origin. Therefore, $\phi = 0°$ or $\phi = 180°$, depending on whether *k* is positive or negative, respectively. The origin is $|k|$ units from the given vertical line, so $p = |k|$. The polar form of the given line is $k = r \cos (\theta - 0°)$ if *k* is positive or $-k = r \cos (\theta - 180°)$ if *k* is negative. Both equations simplify to $k = r \cos \theta$.

Teaching Tip Encourage students to check their values for *p* and ϕ when solving such a system of equations.

In-Class Example
Example 4
Biology
a. If the fly walks in a straight line from the point with polar coordinates (4.5, 3.1) to the point with polar coordinates (1.2, −0.9), what is the equation of the path of the fly? $0.76 = r \cos (\theta - 4.50)$

b. If *r* is measured in centimeters, how close did the fly come to the origin? **7.6 mm**

---

3 PRACTICE/APPLY

Error Analysis

Watch for students who have their calculators set in degree mode when working problems that require radians.

Prevent by reminding students that θ is in radians.

Additional Answer

4. You can use the extra ordered pairs as a check on your work. If all the ordered pairs you plot are not collinear, then you have made a mistake.

Study Guide and Practice Masters, pp. 163–164

Additional Answers
(pp. 578–579)

9.

10.

11b.

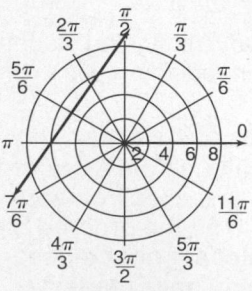

12. $4 = r \cos (\theta - 106°)$

13. $3 = r \cos (\theta - 44°)$

14. $2.1 = r \cos (\theta - 307°)$

15. $\frac{5\sqrt{13}}{13} = r \cos (\theta - 34°)$

16. $\frac{10\sqrt{41}}{41} = r \cos (\theta - 309°)$

17. $\frac{7\sqrt{10}}{10} = r \cos (\theta - 108°)$

33. Sample answer:
$2 = r \cos (\theta - 45°)$ and
$2 = r \cos (\theta - 135°)$

35a.

Graph each polar equation.   9–10. See margin.

9. $3 = r \cos \left(\theta - \frac{\pi}{3}\right)$ **10.** $r = 2 \sec (\theta + 45°)$

11. Aviation An air traffic controller is looking at a radar screen with the control tower at the origin.

   a. If the path of an airplane can be modeled by the equation $5 = r \cos \left(\theta - \frac{5\pi}{6}\right)$, then what are the polar coordinates of the plane when it comes the closest to the tower? $\left(5, \frac{5\pi}{6}\right)$

   b. Sketch a graph of the path of the plane.   **See margin.**

EXERCISES

Practice

12–17. See margin.

 A

Write each equation in polar form. Round ϕ to the nearest degree.

12. $7x - 24y + 100 = 0$ **13.** $21x + 20y = 87$

14. $6x - 8y = 21$ **15.** $3x + 2y - 5 = 0$

B

16. $4x - 5y = 10$ **17.** $-x + 3y = 7$

Write each equation in rectangular form.   **18.** $-x + \sqrt{3}y - 12 = 0$

18. $6 = r \cos (\theta - 120°)$ **19.** $4 = r \cos \left(\theta + \frac{\pi}{4}\right)$ $\sqrt{2}x - \sqrt{2}y - 8 = 0$

20. $2 = r \cos (\theta + \pi)$ $x = -2$ **21.** $1 = r \cos (\theta - 330°)$ $\sqrt{3}x - y - 2 = 0$

22. $-\sqrt{3}x + y - 22 = 0$

22. $r = 11 \sec \left(\theta + \frac{7\pi}{6}\right)$ **23.** $r = 5 \sec (\theta - 60°)$ $x + \sqrt{3}y - 10 = 0$

Graph each polar equation.   **24–29. See Answer Appendix.**

24. $6 = r \cos (\theta - 45°)$ **25.** $1 = r \cos \left(\theta - \frac{\pi}{6}\right)$

26. $2 = r \cos (\theta + 60°)$ **27.** $3 = r \cos (\theta + 90°)$

C

28. $r = 3 \sec \left(\theta + \frac{\pi}{3}\right)$ **29.** $r = 4 \sec \left(\theta - \frac{\pi}{4}\right)$

30. $\frac{5\sqrt{13}}{13} = r \cos (\theta - 56°)$

30. Write the polar form of the equation of the line that passes through points with rectangular coordinates $(4, -1)$ and $(-2, 3)$.

31. What is the polar form of the equation of the line that passes through points with polar coordinates $\left(3, \frac{\pi}{4}\right)$ and $\left(2, \frac{7\pi}{6}\right)$?   **0.31 = r cos (θ − 2.25)**

Applications and Problem Solving

Real World Application

32. Biology Refer to the application at the beginning of the lesson. Suppose the Buridan Paradigm tracks a fruit fly whose path is modeled by the polar equation $r = 6 \sec (\theta - 15°)$, where r is measured in centimeters.   **a. 6 cm**

   a. How close did the fly come to the origin?

   b. What were the polar coordinates of the fly when it was closest to the origin? **(6, 15°)**

33. Critical Thinking Write the polar forms of the equations of two lines, neither of which is vertical, such that the lines intersect at a 90° angle and have normal segments of length 2.   **See margin.**

Extra Credit

Have students find p and ϕ for the equation $Ax + By + C = 0$ and use them to write the polar form of the equation, assuming that ϕ is in Quadrant I or Quadrant IV. They can use this as a formula for writing the polar form of the equation of a line.

$$p = \frac{C}{\pm\sqrt{A^2 + B^2}}, \phi = \arctan \frac{B}{A}; \frac{C}{\pm\sqrt{A^2 + B^2}} = r \cos \left(\theta - \arctan \frac{B}{A}\right)$$

34. **Robotics** The diagram at the right shows a robot with a telescoping arm. The "hand" at the end of such an arm is called the manipulator. What polar equation should the arm be programmed to follow in order to move the manipulator in a straight line from the point with rectangular coordinates (5, 4) to the point with rectangular coordinates (15, 4)? **$4 = r\cos(\theta - 90°)$**

35. **Surveying** A surveyor records the locations of points in a plot of land by means of polar coordinates. In a circular plot of radius 500 feet, stakes are placed at (125, 130°) and (300, 70°), where r is measured in feet. The stakes are at the same elevation. **a. See margin.**   **b. $124.43 = r\cos(\theta - 135°)$**
   a. Draw a diagram of the plot of land. Include the locations of the two stakes.
   b. Find the polar equation of the line determined by the stakes.

36. See Answer Appendix.

36. **Critical Thinking** Show that $k = r\sin(\theta - \alpha)$ is also the equation of a line in polar coordinates. Identify the significance of k and α in the graph.

37. **Design** A carnival ride designer wants to create a Ferris wheel with a 40-foot radius. The designer wants the interior of the circle to have lines of lights that form a regular pentagon. Find the polar equation of the line that contains (40, 0°) and (40, 72°). **$32.36 = r\cos(\theta - 36°)$**

Mixed Review

38. Write the polar equation $r = 6$ in rectangular form. *(Lesson 9-3)* **$x^2 + y^2 = 36$**

39. rose

40. $x = t,$
   $y = \frac{1}{3}t - 2$

39. Identify the type of curve represented by the equation $r = \sin 6\theta$. *(Lesson 9-2)*

40. Write parametric equations of the line with equation $x - 3y = 6$. *(Lesson 8-6)*

41. **Gardening** A sprinkler is set to rotate 65° and spray a distance of 6 feet. What is the area of the gound being watered? *(Lesson 6-1)* **about 20.42 ft²**

42. Use the unit circle to find the value of sin 360° *(Lesson 5-3)* **0**

43. Solve $2x^3 + 5x^2 - 12x = 0$. *(Lesson 4-1)* **$0, \frac{3}{2}, -4$**

44. **SAT Practice** **Grid-In** If $c + d = 12$ and $c^2 - d^2 = 48$, then $c - d = ?$ **4**

Extra Practice See p. A42.

MID-CHAPTER QUIZ

Graph each polar equation. (Lesson 9-1)

1. $r = 4$
   1–2. See margin.

2. $\theta = \frac{2\pi}{3}$

Graph each polar equation. (Lesson 9-2)

3. $r = 3 + 3\sin\theta$
   3–4. See margin.

4. $r = \cos 2\theta$

Find the polar coordinates of each point with the given rectangular coordinates.
(Lesson 9-3) **5–6. See margin.**

5. $P(-\sqrt{2}, -\sqrt{2})$

6. $Q(0, -4)$

7. Write the polar form of the equation $x^2 + y^2 = 36$. (Lesson 9-3) **$r = 6$ or $r = -6$**

8. Find the rectangular form of $r = 2\csc\theta$. (Lesson 9-3) **$y = 2$**

Write each equation in polar form. Round ϕ to the nearest degree. (Lesson 9-4)

9. $5x - 12y = -3$ $\frac{3}{13} = r\cos(\theta - 113°)$

10. $-2x - 6y = 2$ $\frac{\sqrt{10}}{10} = r\cos(\theta - 252°)$

---

4 ASSESS

Open-Ended Assessment
Modeling Ask students to plot points to graph an equation in polar form and in rectangular form. Have them compare the two graphs.

Mid-Chapter Quiz

The Mid-Chapter Quiz provides students with a brief review of the concepts and skills in Lessons 9-1 through 9-4. Lesson numbers are given to the right of exercises or instruction lines so students can review concepts not yet mastered.

Chapter 9, Quiz B (Lessons 9-3 and 9-4) is available in the *Assessment and Evaluation Masters*, p. 225.

Chapter 9 Mid-Chapter Test (Lessons 9-1 through 9-4) is available in the *Assessment and Evaluation Masters*, p. 224.

Enrichment Masters, **p. 100**

Additional Answers
Mid-Chapter Quiz
1–4. See Answer Appendix.

5. $\left(2, \frac{5\pi}{4}\right)$

6. $\left(4, \frac{3\pi}{2}\right)$

9-5

Simplifying Complex Numbers

580 *Chapter 9*

1 FOCUS

5-Minute Check
Lesson 9-4

Write each equation in polar form. Round ϕ to the nearest degree.

1. $3x + 4y - 2 = 0$
$\frac{2}{5} = r\cos(\theta - 53°)$

2. $-2x + 5y = 10$
$\frac{10\sqrt{29}}{29} = r\cos(\theta - 112°)$

Write each equation in rectangular form.

3. $2 = r\cos(\theta - 45°)$
$\sqrt{2}x + \sqrt{2}y - 4 = 0$

4. $1 = r\cos(\theta - 270°)$ $y = -1$

5. Graph the equation
$2 = r\cos(\theta - 30°)$.

Motivating the Lesson

Hands-On Activity Have students use posterboard to create nested boxes that demonstrate how sets of numbers relate. They should include rational, irrational, real, imaginary, natural, and whole numbers as well as integers.

OBJECTIVE
• Add, subtract, multiply, and divide complex numbers in rectangular form.

Real World Application

DYNAMICAL SYSTEMS
Dynamical systems is a branch of mathematics that studies constantly changing systems like the stock market, the weather, and population. In many cases, one can catch a glimpse of the system at some point in time, but the forces that act on the system cause it to change quickly. By analyzing how a dynamical system changes over time, it may be possible to predict the behavior of the system in the future. One of the basic mathematical models of a dynamical system is iteration of a complex function. *A problem related to this will be solved in Example 4.*

A Typical Graph of the Stock Market

Recall that complex numbers are numbers of the form $a + bi$, where a and b are real numbers and i, the imaginary unit, is defined by $i^2 = -1$. The first few powers of i are shown below.

$i^1 = i$	$i^2 = -1$	$i^3 = i^2 \cdot i = -i$	$i^4 = (i^2)^2 = 1$
$i^5 = i^4 \cdot i = i$	$i^6 = i^4 \cdot i^2 = -1$	$i^7 = i^4 \cdot i^3 = -i$	$i^8 = (i^2)^4 = 1$

Notice the repeating pattern of the powers of i.

$$i, -1, -i, 1, i, -1, -i, 1$$

In general, the value of i^n, where n is a whole number, can be found by dividing n by 4 and examining the remainder as summarized in the table at the right.

You can also simplify any integral power of i by rewriting the exponent as a multiple of 4 plus a positive remainder.

To find the value of i^n, let R be the remainder when n is divided by 4.

if R = 0	$i^n = 1$
if R = 1	$i^n = i$
if R = 2	$i^n = -1$
if R = 3	$i^n = -i$

Example **1** Simplify each power of i.

a. i^{53}

Method 1
$53 \div 4 = 13 \text{ R}1$
If R = 1, $i^n = i$.
$i^{53} = i$

Method 2
$i^{53} = (i^4)^{13} \cdot i$
$= (1)^{13} \cdot i$
$= i$

b. i^{-13}

Method 1
$-13 \div 4 = -4 \text{ R}3$
If R = 3, $i^n = -i$.
$i^{-13} = -i$

Method 2
$i^{-13} = (i^4)^{-4} \cdot i^3$
$= (1)^{-4} \cdot i^3$
$= -i$

Resource Manager

 Reproducible Masters
• *Study Guide and Practice*, pp. 165–166
• *Enrichment*, p. 101

 Transparencies
• 5-Minute Check, 9-5
• Teaching, 9-5

The complex number $a + bi$, where a and b are real numbers, is said to be in **rectangular form.** a is called the **real part** and b is called the **imaginary part.** If $b = 0$, the complex number is a real number. If $b \neq 0$, the complex number is an **imaginary number.** If $a = 0$ and $b \neq 0$, as in $4i$, then the complex number is a **pure imaginary number.** Complex numbers can be added and subtracted by performing the chosen operation on both the real and imaginary parts.

Example 2 Simplify each expression.

Graphing Calculator Tip

Some calculators have a complex number mode. In this mode, they can perform complex number arithmetic.

a. $(5 - 3i) + (-2 + 4i)$

$(5 - 3i) + (-2 + 4i) = [5 + (-2)] + [-3i + 4i]$
$= 3 + i$

b. $(10 - 2i) - (14 - 6i)$

$(10 - 2i) - (14 - 6i) = 10 - 2i - 14 + 6i$
$= -4 + 4i$

The product of two or more complex numbers can be found using the same procedures you use when multiplying binomials.

Example 3 Simplify $(2 - 3i)(7 - 4i)$.

$\begin{aligned}(2 - 3i)(7 - 4i) &= 7(2 - 3i) - 4i(2 - 3i) && \textit{Distributive property} \\ &= 14 - 21i - 8i + 12i^2 && \textit{Distributive property} \\ &= 14 - 21i - 8i + 12(-1) && \textit{$i^2 = -1$} \\ &= 2 - 29i\end{aligned}$

Iteration is the process of repeatedly applying a function to the output produced by the previous input. When using complex numbers with functions, it is traditional to use z for the independent variable.

Example 4 **DYNAMICAL SYSTEMS** If $f(z) = (0.5 + 0.5i)z$, find the first five iterates of f for the initial value $z_0 = 1 + i$. Describe any pattern that you see.

$f(z) = (0.5 + 0.5i)z$

$\begin{aligned}f(1 + i) &= (0.5 + 0.5i)(1 + i) && \textit{Replace z with $1 + i$.} \\ &= 0.5 + 0.5i + 0.5i + 0.5i^2 \\ &= i && \textit{$z_1 = i$}\end{aligned}$

$\begin{aligned}f(i) &= (0.5 + 0.5i)i \\ &= 0.5i + 0.5i^2 \\ &= -0.5 + 0.5i && \textit{$z_2 = -0.5 + 0.5i$}\end{aligned}$

$\begin{aligned}f(-0.5 + 0.5i) &= (0.5 + 0.5i)(-0.5 + 0.5i) \\ &= -0.25 + 0.25i - 0.25i + 0.25i^2 \\ &= -0.5 && \textit{$z_3 = -0.5$}\end{aligned}$

(continued on the next page)

Lesson 9-5 Simplifying Complex Numbers **581**

(continued on the next page)

Teaching Tip For Example 5, make sure that students do not mistakenly try to break the fraction into real and imaginary parts and divide the *i*s.

In-Class Example

Example 5
Simplify $(-3 + i) \div (4 - 3i)$.
$-\frac{3}{5} - \frac{1}{5}i$

Graphing Calculator Programs
To download a graphing calculator program that performs complex iteration, visit:
www.amc. glencoe.com

$$f(-0.5) = (0.5 + 0.5i)(-0.5)$$
$$= -0.25 - 0.25i \qquad z_4 = -0.25 - 0.25i$$

$$f(-0.25 - 0.25i) = (0.5 + 0.5i)(-0.25 - 0.25i)$$
$$= -0.125 - 0.125i - 0.125i - 0.125i^2$$
$$= -0.25i \quad z_5 = -0.25i$$

The first five iterates of $1 + i$ are i, $-0.5 + 0.5i$, -0.5, $-0.25 - 0.25i$, and $-0.25i$. The absolute values of the nonzero real and imaginary parts (1, 0.5, 0.25) stay the same for two steps and then are halved.

Two complex numbers of the form $a + bi$ and $a - bi$ are called **complex conjugates.** Recall that if a quadratic equation with real coefficients has complex solutions, then those solutions are complex conjugates. Complex conjugates also play a useful role in the division of complex numbers. To simplify the quotient of two complex numbers, multiply the numerator and denominator by the conjugate of the denominator. The process is similar to rationalizing the denominator in an expression like $\dfrac{1}{3 + \sqrt{2}}$.

Example **5** Simplify $(5 - 3i) \div (1 - 2i)$.

$$(5 - 3i) \div (1 - 2i) = \frac{5 - 3i}{1 - 2i} \qquad \textit{Multiply by 1; } 1 + 2i \textit{ is the}$$
$$\textit{conjugate of } 1 - 2i.$$
$$= \frac{5 - 3i}{1 - 2i} \cdot \frac{1 + 2i}{1 + 2i}$$
$$= \frac{5 + 10i - 3i - 6i^2}{1 - 4i^2}$$
$$= \frac{5 + 7i - 6(-1)}{1 - (-4)} \qquad i^2 = -1$$
$$= \frac{11 + 7i}{5}$$
$$= \frac{11}{5} + \frac{7}{5}i \qquad \textit{Write the answer in the form } a + bi.$$

The list below summarizes the operations with complex numbers presented in this lesson.

For any complex numbers $a + bi$ and $c + di$, the following are true.

Operations with Complex Numbers

$$(a + bi) + (c + di) = (a + c) + (b + d)i$$
$$(a + bi) - (c + di) = (a - c) + (b - d)i$$
$$(a + bi)(c + di) = (ac - bd) + (ad + bc)i$$
$$\frac{a + bi}{c + di} = \frac{ac + bd}{c^2 + d^2} + \frac{bc - ad}{c^2 + d^2}i$$

Addressing Individual Needs

Verbal/Linguistic Have students take turns working out problems on the board, explaining the steps as they go. Make sure that everyone understands each problem before moving on. This could be done in pairs with students taking turns explaining problems to each other.

CHECK FOR UNDERSTANDING

Communicating Mathematics

Read and study the lesson to answer each question.   **1–3. See margin.**

1. **Describe** how to simplify any integral power of i.

2. **Draw** a Venn diagram to show the relationship between real, pure imaginary, and complex numbers.

3. **Explain** why it is useful to multiply by the conjugate of the denominator over itself when simplifying a fraction containing complex numbers.

4. Sample answer:
$x^2 + 1 = 0$

4. **Write** a quadratic equation that has two complex conjugate solutions.

Guided Practice

Simplify.

8. $3.5 + 10.4i$

5. i^{-6} **-1**

6. $i^{10} + i^2$ **-2**

7. $(2 + 3i) + (-6 + i)$ **$-4 + 4i$**

8. $(2.3 + 4.1i) - (-1.2 - 6.3i)$

9. $(2 + 4i) + (-1 + 5i)$ **$1 + 9i$**

10. $(-2 - i)^2$ **$3 + 4i$**

11. $\dfrac{i}{1 + 2i}$ **$\dfrac{2}{5} + \dfrac{1}{5}i$**

Look Back

You can refer to Lessons 8-1 and 8-2 to review vectors, components, and resultants.

12. **Vectors** It is sometimes convenient to use complex numbers to represent vectors. A vector with a horizontal component of magnitude a and a vertical component of magnitude b can be represented by the complex number $a + bi$. If an object experiences a force with a horizontal component of 2.5 N and a vertical component of 3.1 N as well as a second force with a horizontal component of -6.2 N and a vertical component of 4.3 N, find the resultant force on the object. Write your answer as a complex number. **$-3.7 + 7.4i$ N**

EXERCISES

Practice

 A

Simplify.

13. i^6 **-1**

14. i^{19} **$-i$**

15. i^{1776} **1**

16. $i^9 + i^{-5}$ **0**

17. $(3 + 2i) + (-4 + 6i)$ **$-1 + 8i$**

18. $(7 - 4i) + (2 - 3i)$ **$9 - 7i$**

19. $\left(\dfrac{1}{2} + i\right) - (2 - i)$ **$-\dfrac{3}{2} + 2i$**

20. $(-3 - i) - (4 - 5i)$ **$-7 + 4i$**

 B

21. $(2 + i)(4 + 3i)$ **$5 + 10i$**

22. $(1 + 4i)^2$ **$-15 + 8i$**

23. $(-2 + \sqrt{35})$ $+ (-2\sqrt{7} - \sqrt{5})i$

23. $\left(1 + \sqrt{7}i\right)\left(-2 - \sqrt{5}i\right)$

24. $\left(2 + \sqrt{-3}\right)\left(-1 + \sqrt{-12}\right)$

25. $\dfrac{2 + i}{1 + 2i}$ **$\dfrac{4}{5} - \dfrac{3}{5}i$**

26. $\dfrac{3 - 2i}{-4 - i}$ **$-\dfrac{10}{17} + \dfrac{11}{17}i$**

27. $\dfrac{5 - i}{5 + i}$ **$\dfrac{12}{13} - \dfrac{5}{13}i$**

24. $-8 +$ **C**
$3\sqrt{3}i$

28. Write a quadratic equation with solutions i and $-i$. **$x^2 + 1 = 0$**

29. Write a quadratic equation with solutions $2 + i$ and $2 - i$. **$x^2 - 4x + 5 = 0$**

32. $\left(\dfrac{1}{6} - \dfrac{\sqrt{6}}{3}\right) +$
$\left(\dfrac{\sqrt{3}}{3} + \dfrac{\sqrt{2}}{6}\right)i$

Simplify.

30. $(2 - i)(3 + 2i)(1 - 4i)$ **$12 - 31i$**

31. $(-1 - 3i)(2 + 2i)(1 - 2i)$ **$-12 - 16i$**

33. $\left(\dfrac{2}{5} - \dfrac{2\sqrt{3}}{15}\right) +$

32. $\dfrac{\frac{1}{2} + \sqrt{3}i}{1 - \sqrt{2}i}$

$\left(-\dfrac{\sqrt{2}}{5} - \dfrac{2\sqrt{6}}{15}\right)i$ 34. $\dfrac{3 + i}{(2 + i)^2}$ **$\dfrac{13}{25} - \dfrac{9}{25}i$**

33. $\dfrac{2 - \sqrt{2}i}{3 + \sqrt{6}i}$

35. $\dfrac{(1 + i)^2}{(-3 + 2i)^2}$ **$-\dfrac{24}{169} + \dfrac{10}{169}i$**

 www.amc.glencoe.com/self_check_quiz

Lesson 9-5 Simplifying Complex Numbers **583**

Additional Answers

1. Find the (positive) remainder when the exponent is divided by 4. If the remainder is 0, the answer is 1; if the remainder is 1, the answer is i; if the remainder is 2, the answer is -1; and if the remainder is 3, the answer is $-i$.

2.

Complex Numbers ($a + bi$)	
Reals ($b = 0$)	Imaginary ($b \neq 0$)
	Pure Imaginary ($a = 0$)

Applications and Problem Solving

Real World Application

36. Electricity *Impedance* is a measure of how much hindrance there is to the flow of charge in a circuit with alternating current. The impedance Z depends on the resistance R, the reactance due to capacitance X_C, and the reactance due to inductance X_L in the circuit. The impedance is written as the complex number $Z = R + (X_L - X_C)j$. (Electrical engineers use j to denote the imaginary unit.) In the first part of a particular series circuit, the resistance is 10 ohms, the reactance due to capacitance is 2 ohms, and the reactance due to inductance is 1 ohm. In the second part of the circuit, the respective values are 3 ohms, 1 ohm, and 1 ohm.

a. Write complex numbers that represent the impedances in the two parts of the circuit. $10 - j$ ohms; $3 + 0j$ ohms

b. Add your answers from part a to find the total impedance in the circuit. $13 - j$ ohms

c. The *admittance* of an AC circuit is a measure of how well the circuit allows current to flow. Admittance is the reciprocal of impedance. That is, $S = \frac{1}{Z}$. The units for admittance are siemens. Find the admittance in a circuit with an impedance of $6 + 3j$ ohms. $0.13 - 0.07j$ siemens

37. Critical Thinking

a. Solve the equation $x^2 + 8ix - 25 = 0$. $\pm 3 - 4i$

b. Are the solutions complex conjugates?   **No**

c. How does your result in part b compare with what you already know about complex solutions to quadratic equations?   **c–d. See margin.**

d. Check your solutions.

38. Critical Thinking Sometimes it is useful to separate a complex function into its real and imaginary parts. Substitute $z = x + yi$ into the function $f(z) = z^2$ to write the equation of the function in terms of x and y only. Simplify your answer.   **See margin.**

39. Dynamical Systems Find the first five iterates for the given function and initial value.

a. $f(z) = iz, z_0 = 2 - i$ $1 + 2i, -2 + i, -1 - 2i, 2 - i, 1 + 2i$

b. $f(z) = (0.5 - 0.866i)z, z_0 = 1 + 0i$ **See margin.**

40. $-\frac{11}{125} + \frac{2}{125}i$ **40. Critical Thinking** Simplify $(1 + 2i)^{-3}$.

41. Physics One way to derive the equation of motion in a spring-mass system is to solve a *differential equation*. The solutions of such a differential equation typically involve expressions of the form $\cos \beta t + i \sin \beta t$. You generally expect solutions that are real numbers in such a situation, so you must use algebra to eliminate the imaginary numbers. Find a relationship between the constants c_1 and c_2 such that $c_1(\cos 2t + i \sin 2t) + c_2(\cos 2t - i \sin 2t)$ is a real number for all values of t. $c_1 = c_2$

Mixed Review

42. $\frac{3\sqrt{10}}{20} =$

$r \cos(\theta - 162°)$

42. Write the equation $6x - 2y = -3$ in polar form. *(Lesson 9-4)*

43. Graph the polar equation $r = 4\theta$. *(Lesson 9-2)*   **See margin.**

44. Write a vector equation of the line that passes through $P(-3, 6)$ and is parallel to $\vec{v}\langle 1, -4 \rangle$. *(Lesson 8-6)* $\langle x + 3, y - 6 \rangle = t\langle 1, -4 \rangle$

45. Find an ordered triple to represent $\vec{\mathbf{u}}$ if $\vec{\mathbf{u}} = \frac{1}{4}\vec{\mathbf{v}} - 2\vec{\mathbf{w}}$, $\vec{\mathbf{v}} = \langle -8, 6, 4 \rangle$, and $\vec{\mathbf{w}} = \langle 2, -6, 3 \rangle$. *(Lesson 8-3)* $\left\langle -6, \frac{27}{2}, -5 \right\rangle$

46. If α and β are measures of two first quadrant angles, find $\cos(\alpha + \beta)$ if $\tan \alpha = \frac{4}{3}$ and $\cot \beta = \frac{5}{12}$. *(Lesson 7-3)* $-\frac{33}{65}$

47. A twig floats on the water, bobbing up and down. The distance between its highest and lowest points is 7 centimeters. It moves from its highest point down to its lowest point and back up to its highest point every 12 seconds. Write a cosine function that models the movement of the twig in relationship to the equilibrium point. *(Lesson 6-6)* $y = 3.5 \cos\left(\frac{\pi}{6}t\right)$

48. **Surveying** A surveyor finds that the angle of elevation from a certain point to the top of a cliff is 60°. From a point 45 feet farther away, the angle of elevation to the top of the cliff is 52°. How high is the cliff to the nearest foot? *(Lesson 5-4)* **221 ft**

49. What type of polynomial function would be the best model for the set of data? *(Lesson 4-8)* **quadratic**

x	−3	−1	1	3	5	7	9	11
f(x)	−4	−2	3	8	6	1	−3	−8

50. **Construction** A community wants to build a second pool at their community park. Their original pool has a width 5 times its depth and a length 10 times its depth. They wish to make the second pool larger by increasing the width of the original pool by 4 feet, increasing the length by 6 feet, and increasing the depth by 2 feet. The volume of the new pool will be 3420 cubic feet. Find the dimensions of the original pool. *(Lesson 4-4)* **15 ft by 30 ft by 3 ft**

51. If y varies jointly as x and z and $y = 80$ when $x = 5$ and $z = 8$, find y when $x = 16$ and $z = 2$. *(Lesson 3-8)* **64**

52. If $f(x) = 7 - x^2$, find $f^{-1}(x)$. *(Lesson 3-4)* $f^{-1}(x) = \pm\sqrt{7 - x}$

53. Find the maximum and minimum values of the function $f(x, y) = -2x + y$ for the polygonal convex set determined by the system of inequalities. *(Lesson 2-6)* **3, −11**
$x \leq 6$
$y \geq 1$
$y - x \leq 2$

54. Solve the system of equations. *(Lesson 2-2)* **(0, 7, 0)**
$x + 2y - 7z = 14$
$-x - 3y + 5z = -21$
$5x - y + 2z = -7$

55. **SAT/ACT Practice** If $BC = BD$ in the figure, what is the value of $x + 40$? **A**

A 100   B 80   C 60   D 40

E cannot be determined from the information given

Extra Practice See p. A13.

Lesson 9-5 *Simplifying Complex Numbers* **585**

Extra Credit

Explain the fallacy in this proof.

If $a > 0$ and $b < 0$, then $\sqrt{\dfrac{a}{b}} \neq \dfrac{\sqrt{a}}{\sqrt{b}}$.

$$i = i$$
$$\sqrt{-1} = \sqrt{-1}$$
$$\sqrt{\frac{1}{-1}} = \sqrt{\frac{-1}{1}}$$
$$\frac{\sqrt{1}}{\sqrt{-1}} = \frac{\sqrt{-1}}{\sqrt{1}}$$
$$\sqrt{1}\sqrt{1} = \sqrt{-1}\sqrt{-1}$$
$$1 = -1$$

Lesson 9-5 **585**

9-6

The Complex Plane and Polar Form of Complex Numbers

1 FOCUS

5-Minute Check
Lesson 9-5

Simplify.
1. i^{39} $-i$
2. $(3 - 6i) - (-2 + 2i)$ **$5 - 8i$**
3. $(2 + 4i)(3 - 5i)$ **$26 + 2i$**
4. $(-5 + 9i) \div (1 - i)$ **$-7 + 2i$**
5. $(5 - 4i) \div (-2 + 6i)$
$$-\frac{17}{20} - \frac{11}{20}i$$

Motivating the Lesson
Real-World Application Use a transparency of a polar plane to show the path of a plane flying from an airport (origin). Let each unit represent a hundred miles. Plot the point where the plane is 300 miles from the airport at a heading 30° north of due east. Then lay a rectangular grid with equivalent units on top of the polar grid. Ask students to redefine the position of the plane in rectangular format. Then have them represent the position of the plane as a complex number in rectanguler form.

2 TEACH

Teaching Tip In Example 1, remind students that i is not a variable, so they should not try to solve for it.

Teaching Tip In Example 2, remind students that if one of the components is missing, that means it is zero.

OBJECTIVES
- Graph complex numbers in the complex plane.
- Convert complex numbers from rectangular to polar form and vice versa.

FRACTALS One of the standard ways to generate a fractal involves iteration of a quadratic function. If the function $f(z) = z^2$ is iterated using a complex number as the initial input, there are three possible outcomes. The terms of the sequence of outputs, called the *orbit*, may

- increase in absolute value,
- decrease toward 0 in absolute value, or
- always have an absolute value of 1.

One way to analyze the behavior of the orbit is to graph the numbers in the complex plane. Plot the first five members of the orbit of $z_0 = 0.9 + 0.3i$ under iteration by $f(z) = z^2$. *This problem will be solved in Example 3.*

Recall that $a + bi$ is referred to as the rectangular form of a complex number. The rectangular form is sometimes written as an ordered pair, (a, b). Two complex numbers in rectangular form are equal if and only if their real parts are equal and their imaginary parts are equal.

Example **1** Solve the equation $2x + y + 3i = 9 + xi - yi$ for x and y, where x and y are real numbers.

$2x + y + 3i = 9 + xi - yi$

$(2x + y) + 3i = 9 + (x - y)i$ *On each side of the equation, group the real parts and the imaginary parts.*

$2x + y = 9$ and $x - y = 3$ *Set the corresponding parts equal to each other.*
$x = 4$ and $y = 1$ *Solve the system of equations.*

Complex numbers can be graphed in the **complex plane.** The complex plane has a real axis and an imaginary axis. The real axis is horizontal, and the imaginary axis is vertical. The complex number $a + bi$ is graphed as the ordered pair (a, b) in the complex plane. The complex plane is sometimes called the **Argand plane.**

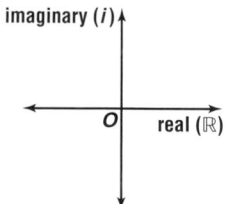

Recall that the absolute value of a real number is its distance from zero on the number line. Similarly, the **absolute value** of a complex number is its distance from zero in the complex plane. When $a + bi$ is graphed in the complex plane, the distance from zero can be calculated using the Pythagorean Theorem.

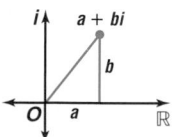

586   *Chapter 9   Polar Coordinates and Complex Numbers*

Resource Manager

 Reproducible Masters
- *Study Guide and Practice,* pp. 167–168
- *Enrichment,* p. 102
- *Assessment and Evaluation,* p. 226
- *Problem Solving and Applications,* p. 17

 Transparencies
- 5-Minute Check, 9-6
- Teaching, 9-6

Absolute Value of a Complex Number	If $z = a + bi$, then $|z| = \sqrt{a^2 + b^2}$.

Examples **2** Graph each number in the complex plane and find its absolute value.

a. $z = 3 + 2i$

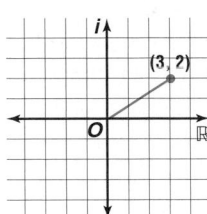

$z = 3 + 2i$
$|z| = \sqrt{3^2 + 2^2}$
$= \sqrt{13}$

b. $z = 4i$

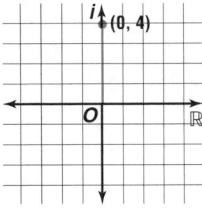

$z = 0 + 4i$
$|z| = \sqrt{0^2 + 4^2}$
$= 4$

3 **FRACTALS** Refer to the application at the beginning of the lesson. Plot the first five members of the orbit of $z_0 = 0.9 + 0.3i$ under iteration by $f(z) = z^2$.

First, calculate the first five members of the orbit. Round the real and imaginary parts to the nearest hundredth.

$z_1 = 0.72 + 0.54i \quad z_1 = f(z_0)$
$z_2 = 0.23 + 0.78i \quad z_2 = f(z_1)$
$z_3 = -0.55 + 0.35i \quad z_3 = f(z_2)$
$z_4 = 0.18 - 0.39i \quad z_4 = f(z_3)$
$z_5 = -0.12 - 0.14i \quad z_5 = f(z_4)$

Then graph the numbers in the complex plane. The iterates approach the origin, so their absolute values decrease toward 0.

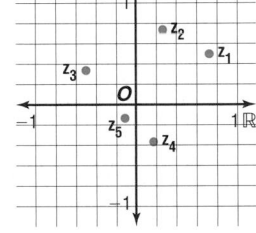

So far we have associated the complex number $a + bi$ with the rectangular coordinates (a, b). You know from Lesson 9-1 that there are also polar coordinates (r, θ) associated with the same point. In the case of a complex number, r represents the absolute value, or **modulus,** of the complex number. The angle θ is called the **amplitude** or **argument** of the complex number. Since θ is not unique, it may be replaced by $\theta + 2\pi k$, where k is any integer.

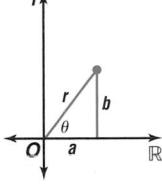

Lesson 9-6 The Complex Plane and Polar Form of Complex Numbers **587**

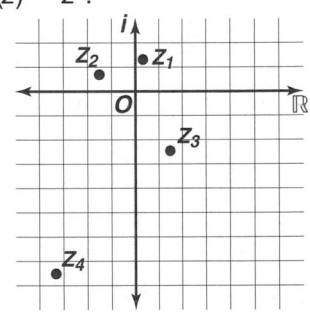

Teaching Tip For Example 4, have students expand the expression $r(\cos\theta + i\sin\theta)$ in order to confirm that it equals $a + bi$.

In-Class Example

Example 4

Express each complex number in polar form.

a. $1 - 4i$

$\sqrt{17}(\cos 4.96 + i\sin 4.96)$ or $\sqrt{17}$ cis 4.96

b. $-3 - 2i$

$\sqrt{13}(\cos 3.73 + i\sin 3.73)$ or $\sqrt{13}$ cis 3.73

Additional Answers *(p. 589)*

1. To find the absolute value of $a + bi$, square a and b, add the squares, then take the square root of the sum.

4. The conjugate of $a + bi$ is $a - bi$. $\sqrt{(a+bi)(a-bi)} = \sqrt{a^2 + b^2}$, so the friend's method gives the same answer. Sample example: the absolute value of $2 + 3i$ is $\sqrt{2^2 + 3^2} = \sqrt{13}$. Using the friend's method, the absolute value is $\sqrt{(2+3i)(2-3i)} = \sqrt{4 + 9} = \sqrt{13}$.

6.

7.

As with other rectangular coordinates, complex coordinates can be written in polar form by substituting $a = r\cos\theta$ and $b = r\sin\theta$.

$$z = a + bi$$
$$= r\cos\theta + (r\sin\theta)i$$
$$= r(\cos\theta + i\sin\theta)$$

This form of a complex number is often called the **polar** or **trigonometric form.**

Polar Form of a Complex Number	The polar form or trigonometric form of the complex number $a + bi$ is $r[\cos\theta + i\sin\theta]$.

$r(\cos\theta + i\sin\theta)$ is often abbreviated as r cis θ.

Values for r and θ can be found by using the same process you used when changing rectangular coordinates to polar coordinates. For $a + bi$, $r = \sqrt{a^2 + b^2}$ and $\theta = \text{Arctan}\frac{b}{a}$ if $a > 0$ or $\theta = \text{Arctan}\frac{b}{a} + \pi$ if $a < 0$. The amplitude θ is usually expressed in radian measure, and the angle is in standard position along the polar axis.

Example 4 Express each complex number in polar form.

a. $-3 + 4i$

First, plot the number in the complex plane.

Then find the modulus.

$r = \sqrt{(-3)^2 + 4^2}$ or 5

Now find the amplitude. Notice that θ is in Quadrant II.

$\theta = \text{Arctan}\dfrac{4}{-3} + \pi$

≈ 2.21

Therefore, $-3 + 4i \approx 5(\cos 2.21 + i\sin 2.21)$ or 5 cis 2.21.

b. $1 + \sqrt{3}i$

First, plot the number in the complex plane.

Then find the modulus.

$r = \sqrt{1^2 + \left(\sqrt{3}\right)^2}$ or 2

Now find the amplitude. Notice that θ is in Quadrant I.

$\theta = \text{Arctan}\dfrac{\sqrt{3}}{1}$ or $\dfrac{\pi}{3}$

Therefore, $1 + \sqrt{3}i = 2\left(\cos\dfrac{\pi}{3} + i\sin\dfrac{\pi}{3}\right)$ or 2 cis $\dfrac{\pi}{3}$.

Additional Answers *(p. 589)*

11. $2 + 2\sqrt{3}i$

12. $-1.98 + 0.28i$

13. $\dfrac{3}{2}$

You can also graph complex numbers in polar form.

Example **5** Graph $4\left(\cos \frac{11\pi}{6} + i \sin \frac{11\pi}{6}\right)$. **Then express it in rectangular form.**

In the polar form of this complex number, the value of r is 4, and the value of θ is $\frac{11\pi}{6}$. Plot the point with polar coordinates $\left(4, \frac{11\pi}{6}\right)$.

To express the number in rectangular form, simplify the trigonometric values:

$$4\left(\cos \frac{11\pi}{6} + i \sin \frac{11\pi}{6}\right)$$

$$= 4\left(\frac{\sqrt{3}}{2} + i\left(-\frac{1}{2}\right)\right)$$

$$= 2\sqrt{3} - 2i$$

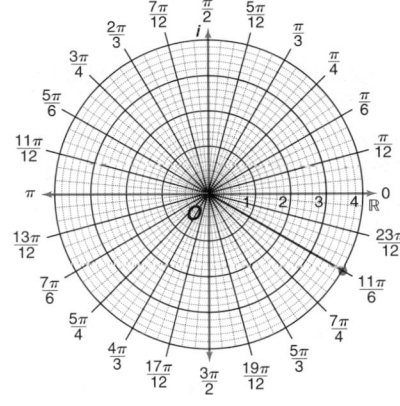

CHECK FOR UNDERSTANDING

Communicating Mathematics

Read and study the lesson to answer each question.

1. **Explain** how to find the absolute value of a complex number. **See margin.**

2. **Write** the polar form of i. $i = \cos \frac{\pi}{2} + i \sin \frac{\pi}{2}$

3. **Find a counterexample** to the statement $|z_1 + z_2| = |z_1| + |z_2|$ for all complex numbers z_1 and z_2. **Sample answer:** $z_1 = i$, $z_2 = -i$

4. *Math Journal* Your friend is studying complex numbers at another school at the same time that you are. She learned that the absolute value of a complex number is the square root of the product of the number and its conjugate. You know that this is not how you learned it. Write a letter to your friend explaining why this method gives the same answer as the method you know. Use algebra, but also include some numerical examples of both techniques. **See margin.**

Guided Practice

6–7. See margin for graphs.

8. $2\sqrt{2}$ $\left(\cos \frac{7\pi}{4} + i \sin \frac{7\pi}{4}\right)$

9. $\sqrt{41}(\cos 0.90 + i \sin 0.90)$

11–13. See margin.

5. Solve the equation $2x + y + xi + yi = 5 + 4i$ for x and y, where x and y are real numbers. $x = 1$, $y = 3$

Graph each number in the complex plane and find its absolute value.

6. $-2 - i$ $\sqrt{5}$

7. $1 + \sqrt{2}i$ $\sqrt{3}$

Express each complex number in polar form.

8. $2 - 2i$

9. $4 + 5i$

10. -2 $2(\cos \pi + i \sin \pi)$

Graph each complex number. Then express it in rectangular form.

11. $4\left(\cos \frac{\pi}{3} + i \sin \frac{\pi}{3}\right)$

12. $2(\cos 3 + i \sin 3)$

13. $\frac{3}{2}(\cos 2\pi + i \sin 2\pi)$

14. Graph the first five members of the orbit of $z_0 = -0.25 + 0.75i$ under iteration by $f(z) = z^2 + 0.5$. **See margin.**

Lesson 9-6 The Complex Plane and Polar Form of Complex Numbers **589**

Additional Answer
14.

Additional Answers

26. $3\sqrt{2}\left(\cos\frac{\pi}{4} + i\sin\frac{\pi}{4}\right)$

27. $2\left(\cos\frac{4\pi}{3} + i\sin\frac{4\pi}{3}\right)$

28. $10\left(\cos 5.36 + i\sin 5.36\right)$

29. $\sqrt{17}\left(\cos 2.90 + i\sin 2.90\right)$

30. $29\left(\cos 5.47 + i\sin 5.47\right)$

31. $2\sqrt{5}\left(\cos 2.03 + i\sin 2.03\right)$

32. $3\left(\cos 0 + i\sin 0\right)$

33. $4\sqrt{2}\left(\cos \pi + i\sin \pi\right)$

34. $2\left(\cos\frac{3\pi}{2} + i\sin\frac{3\pi}{2}\right)$

35.

36.

37.

38.

590 Chapter 9

---

15. **Vectors** The force on an object is represented by the complex number $10 + 15i$, where the components are measured in newtons.
   a. What is the magnitude of the force? **about 18.03 N**
   b. What is the direction of the force? **about 56.31°**

EXERCISES

Practice

16. $x = 6$, **A**
$y = -3$

Solve each equation for x and y, where x and y are real numbers.

16. $2x - 5yi = 12 + 15i$ 17. $1 + (x + y)i = y + 3xi$

18. $4x + yi - 5i = 2x - y + xi + 7i$ $x = -4, y = 8$ $x = \frac{1}{2}, y = 1$

Graph each number in the complex plane and find its absolute value.

**19–24. See
Answer Appendix
for graphs.**

19. $2 + 3i$ $\sqrt{13}$ 20. $3 - 4i$ **5**   21. $-1 - 5i$ $\sqrt{26}$

22. $-3i$ **3**   23. $-1 + \sqrt{5}i$ $\sqrt{6}$ 24. $4 + \sqrt{2}i$ $3\sqrt{2}$

25. Find the modulus of $z = -4 + 6i$. $2\sqrt{13}$

Express each complex number in polar form.   **26–34. See margin.**

B
26. $3 + 3i$ 27. $-1 - \sqrt{3}i$ 28. $6 - 8i$

29. $-4 + i$ 30. $20 - 21i$ 31. $-2 + 4i$

32. 3 33. $-4\sqrt{2}$ 34. $-2i$

Graph each complex number. Then express it in rectangular form.

**35–42. See
margin for
graphs.**

35. $3\left(\cos\frac{\pi}{4} + i\sin\frac{\pi}{4}\right)$ $\frac{3\sqrt{2}}{2} + \frac{3\sqrt{2}}{2}i$ 36. $\cos\left(-\frac{\pi}{6}\right) + i\sin\left(-\frac{\pi}{6}\right)$ $\frac{\sqrt{3}}{2} - \frac{1}{2}i$

37. $2\left(\cos\frac{4\pi}{3} + i\sin\frac{4\pi}{3}\right)$ $-1 - \sqrt{3}i$ 38. $10(\cos 6 + i\sin 6)$ **9.60 − 2.79i**

39. $-\sqrt{2} -$
$\sqrt{2}i$

39. $2\left(\cos\frac{5\pi}{4} + i\sin\frac{5\pi}{4}\right)$ 40. $2.5(\cos 1 + i\sin 1)$ **1.35 + 2.10i**

41. $5(\cos 0 + i\sin 0)$ **5**   42. $3(\cos \pi + i\sin \pi)$ **−3**

Graph the first five members of the orbit of each initial value under iteration by the given function.   **43–45. See Answer Appendix.**

C
43. $z_0 = -0.5 + i$, $f(z) = z^2 + 0.5$ 44. $z_0 = \frac{\sqrt{2}}{2} + \frac{\sqrt{2}}{2}i$, $f(z) = z^2$

45. Graph the first five iterates of $z_0 = 0.5 - 0.5i$ under $f(z) = z^2 - 0.5$.

**Applications
and Problem
Solving**

46. **Electrical Engineering** Refer to Exercise 44 in Lesson 9-3. Consider a circuit with alternating current that contains two voltage sources in series. Suppose these two voltages are given by $v_1(t) = 40 \sin(250t + 30°)$ and $v_2(t) = 60 \sin(250t + 60°)$, where t represents time, in seconds.
   a. The phasors for these two voltage sources are written as $40\angle 30°$ and $60\angle 60°$, respectively. Convert these phasors to complex numbers in rectangular form. (Use j as the imaginary unit, as electrical engineers do.)   **34.64 + 20j; 30 + 51.96j**
                                              64.64 + 71.96j
   b. Add these two complex numbers to find the total voltage in the circuit.
   c. Write a sinusoidal function that gives the total voltage in the circuit.
                    $v(t) = 96.73 \sin(250t + 48°)$

  www.amc.glencoe.com/self_check_quiz

Additional Answers

39.

40.

41.

47. **Critical Thinking** How are the polar forms of complex conjugates alike? How are they different? **See margin.**

48a. 7.65 + 6.44*j*; 14.04 + 7.67*j*

48b. 21.69 + 14.11*j* ohms

48. **Electricity** A series circuit contains two sources of impedance, one of $10(\cos 0.7 + j \sin 0.7)$ ohms and the other of $16(\cos 0.5 + j \sin 0.5)$ ohms.
   a. Convert these complex numbers to rectangular form.
   b. Add your answers from part a to find the total impedance in the circuit.
   c. Convert the total impedance back to polar form. **25.88(cos 0.58 + *j* sin 0.58) ohms**

49. **Transformations** Certain operations with complex numbers correspond to geometric transformations in the complex plane. Describe the transformation applied to point z to obtain point w in the complex plane for each of the following operations.
   a. $w = z + (2 - 3i)$ **Translate 2 units to the right and down 3 units.**
   b. $w = i \cdot z$ **Rotate 90° counterclockwise about the origin.**
   c. $w = 3z$ **Dilate by a factor of 3.**
   d. w is the conjugate of z **Reflect about the real axis.**

50. **Critical Thinking** Choose any two complex numbers, z_1 and z_2, in rectangular form. **a–d. See Answer Appendix.**
   a. Find the product $z_1 z_2$.
   b. Write z_1, z_2, and $z_1 z_2$ in polar form.
   c. Repeat this procedure with a different pair of complex numbers.
   d. Make a conjecture about the product of two complex numbers in polar form.

Mixed Review

51. Simplify $(6 - 2i)(-2 + 3i)$. *(Lesson 9-5)* **$-6 + 22i$**

52. Find the rectangular coordinates of the point with polar coordinates $(-3, -135°)$. *(Lesson 9-3)* $\left(\dfrac{3\sqrt{2}}{2}, \dfrac{3\sqrt{2}}{2}\right)$

53. Find the magnitude of the vector $\langle -3, 7 \rangle$, and write the vector as a sum of unit vectors. *(Lesson 8-2)* $\sqrt{58}, -3\vec{i} + 7\vec{j}$

54. Use a sum or difference identity to find $\tan 105°$. *(Lesson 7-3)* $-2 - \sqrt{3}$

55. **Mechanics** A pulley of radius 18 centimeters turns at 12 revolutions per second. What is the linear velocity of the belt driving the pulley in meters per second? *(Lesson 6-2)* **about 13.57 m/s**

56. 41.8°

56. If $a = 12$ and $c = 18$ in $\triangle ABC$, find the measure of angle A to the nearest tenth of a degree. *(Lesson 5-5)*

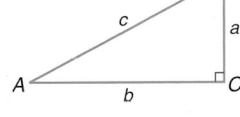

57. Solve $\sqrt{2a - 1} = \sqrt{3a - 5}$. *(Lesson 4-7)* **4**

58. Without graphing, describe the end behavior of the graph of $y = 2x^2 + 2$. *(Lesson 3-5)* **as $x \to \infty$, $y \to \infty$; as $x \to -\infty$, $y \to \infty$**

59. **SAT/ACT Practice** A person is hired for a job that pays \$500 per month and receives a 10% raise in each following month. In the fourth month, how much will that person earn? **D**
   A \$550 **B** \$600.50 **C** \$650.50 **D** \$665.50 **E** \$700

Extra Practice See p. A43.

Lesson 9-6 The Complex Plane and Polar Form of Complex Numbers **591**

9-6B Geometry in the Complex Plane

An Extension of Lesson 9-6

1 FOCUS

Motivating the Lesson
Discussion Ask students to consider the rectangular coordinate plane. Discuss how the *x*- and *y*-axes are related and how ordered pairs are determined. Then discuss the complex plane and the ℝ- and *i*-axes. Ask students to determine how a graphing calculator would simplify the analyses of these relations.

2 TEACH

📁 A reproduction of this lesson with graphing calculator keystrokes provided can be found in the *Graphing Calculator Keystrokes Masters*, pp. 35–36.

Teaching Tip Point out that setting the window in the complex plane is the same process as setting the window in the rectangular plane.

3 PRACTICE/APPLY

Assignment Guide

All: 1–8

4 ASSESS

Open-Ended Assessment
Modeling Ask students to make a conjecture for Exercise 7 if the equation were changed to $|z-(1-i)| = 9$ and then changed to $|z-(1-i)| = 12$. **The radius of the circle would change to 9 and then 12.**

Many geometric figures and relationships can be described by using complex numbers. To show points on figures, you can store the real and imaginary parts of the complex numbers that correspond to the points in lists **L1** and **L2** and use **STAT PLOT** to graph the points.

TRY THESE

1. Store $-1 + 2i$ as M and $1 + 5i$ as N. Now consider complex numbers of the form $(1 - T)M + TN$, where T is a real number. You can generate several numbers of this form and store their real and imaginary parts in **L1** and **L2**, respectively, by entering the following instructions on the home screen.

seq(is in the LIST OPS menu.
real(and imag(are in the MATH CPX menu.

Use a graphing window of $[-10, 10]$ scl:1 by $[-25, 25]$ scl:5. Turn on **Plot 1** and use a scatter plot to display the points defined in **L1** and **L2**. What do you notice about the points in the scatter plot? **They are collinear.**

2. Are the original numbers M and N shown in the scatter plot? Explain.

2. Yes. *M* is the point obtained when *T* = 0, and *N* is the point obtained when *T* = 1.

3. Repeat Exercise 1 storing $-1 + 1.5i$ as M and $-2 - i$ as N. Describe your results. **The points are again collinear, but closer together.**

4. Repeat Exercises 1 and 2 for several complex numbers M and N of your choice. (You may need to change the window settings.) Then make a conjecture about where points of the form $(1 - T)M + TN$ are located in relation to M and N. **The points are on the line through *M* and *N*.**

5. Suppose K, M, and N are three noncollinear points in the complex plane. Where will you find all the points that can be expressed in the form $aK + bM + cN$, where a, b, and c are nonnegative real numbers such that $a + b + c = 1$? Use the calculator to check your answer.
They are on or inside △*KMN*.

WHAT DO YOU THINK?

6. In Exercises 1-4, where is $(1 - T)M + TN$ in relation to M and N if the value of T is between 0 and 1? **between *M* and *N***

7. Where in the complex plane will you find the complex numbers z that satisfy the equation $|z - (1 - i)| = 5$? **on the circle of radius 5 centered at $1 - i$**

8. What equation models the points in the complex plane that lie on the circle of radius 2 that is centered at the point $2 + 3i$? $|z - (2 + 3i)| = 2$

Products and Quotients of Complex Numbers in Polar Form

OBJECTIVE
- Find the product and quotient of complex numbers in polar form.

ELECTRICITY Complex numbers can be used in the study of electricity, specifically alternating current (AC). There are three basic quantities to consider:
- the *current I*, measured in amperes,
- the *impedance Z* to the current, measured in ohms, and
- the *electromotive force E* or *voltage*, measured in volts.

These three quantities are related by the equation $E = I \cdot Z$. Current, impedance, and voltage can be expressed as complex numbers. Electrical engineers use j as the imaginary unit, so they write complex numbers in the form $a + bj$. For the total impedance $a + bj$, the real part a represents the opposition to current flow due to resistors, and the imaginary part b is related to the opposition due to inductors and capacitors. If a circuit has a total impedance of $2 - 6j$ ohms and a voltage of 120 volts, find the current in the circuit. *This problem will be solved in Example 3.*

Multiplication and division of complex numbers in polar form are closely tied to geometric transformations in the complex plane. Let $r_1(\cos \theta_1 + i \sin \theta_1)$ and $r_2(\cos \theta_2 + i \sin \theta_2)$ be two complex numbers in polar form. A formula for the product of the two numbers can be derived by multiplying the two numbers directly and simplifying the result.

$r_1(\cos \theta_1 + i \sin \theta_1) \cdot r_2(\cos \theta_2 + i \sin \theta_2)$

$= r_1 r_2(\cos \theta_1 \cos \theta_2 + i \cos \theta_1 \sin \theta_2 + i \sin \theta_1 \cos \theta_2 + i^2 \sin \theta_1 \sin \theta_2)$

$= r_1 r_2[(\cos \theta_1 \cos \theta_2 - \sin \theta_1 \sin \theta_2) + i(\sin \theta_1 \cos \theta_2 + \cos \theta_1 \sin \theta_2)]$ $i^2 = -1$

$= r_1 r_2[\cos (\theta_1 + \theta_2) + i \sin (\theta_1 + \theta_2)]$ *Sum identities for cosine and sine*

Product of Complex Numbers in Polar Form	$r_1(\cos \theta_1 + i \sin \theta_1) \cdot r_2(\cos \theta_2 + i \sin \theta_2) =$ $r_1 r_2[\cos (\theta_1 + \theta_2) + i \sin (\theta_1 + \theta_2)]$

Lesson 9-7 Products and Quotients of Complex Numbers in Polar Form **593**

1 FOCUS

5-Minute Check
Lesson 9-6

1. Solve the equation $x - 4y + 19I = -5 + 5xI + 2yI$ for x and y, where x and y are real numbers. $x = 3$ and $y = 2$

2. Graph $z = 2 + 4i$ in the complex plane and find its absolute value. $2\sqrt{5}$

3. Graph the first five members of the orbit of $z_0 = 0.4 + 0.9i$ under iteration by $f(z) = z^2$.

4. Express $-1 - i$ in polar form.
$\sqrt{2}\left(\cos \frac{5\pi}{4} + i \sin \frac{5\pi}{4}\right)$

5. Graph $6\left(\cos \frac{\pi}{6} + i \sin \frac{\pi}{6}\right)$. Then express it in rectangular form.
$3\sqrt{3} + 3i$

Discussion State that advances in mathematics often simplify previously complex or tedious tasks. Multiplication, for example, greatly simplifies the process of repeated addition. Ask students to give further examples. **Sample answer: rules of exponents simplify multiplication: $(5^4)^6 = 5^{24}$.** State that this lesson will introduce a simplified method for multiplying complex numbers in polar form.

2 TEACH

In-Class Example

Example 1
Find the product
$7\left(\cos \frac{7\pi}{12} + i \sin \frac{7\pi}{12}\right) \cdot$
$3\left(\cos \frac{\pi}{12} + i \sin \frac{\pi}{12}\right)$.
Then express the product in rectangular form.
$21\left(\cos \frac{2\pi}{3} + i \sin \frac{2\pi}{3}\right)$;
$-\frac{21}{2} + \frac{21\sqrt{3}}{2}i$

Notice that the modulus $(r_1 r_2)$ of the product of the two complex numbers is the product of their moduli. The amplitude $(\theta_1 + \theta_2)$ of the product is the sum of the amplitudes.

Example ① Find the product $3\left(\cos \frac{7\pi}{6} + i \sin \frac{7\pi}{6}\right) \cdot 2\left(\cos \frac{2\pi}{3} + i \sin \frac{2\pi}{3}\right)$. Then express the product in rectangular form.

Find the modulus and amplitude of the product.

$$r = r_1 r_2 \qquad\qquad \theta = \theta_1 + \theta_2$$
$$= 3(2) \qquad\qquad\quad = \frac{7\pi}{6} + \frac{2\pi}{3}$$
$$= 6 \qquad\qquad\qquad = \frac{11\pi}{6}$$

The product is $6\left(\cos \frac{11\pi}{6} + i \sin \frac{11\pi}{6}\right)$.

Now find the rectangular form of the product.

$$6\left(\cos \frac{11\pi}{6} + i \sin \frac{11\pi}{6}\right) = 6\left(\frac{\sqrt{3}}{2} - \frac{1}{2}i\right) \qquad \cos \frac{11\pi}{6} = \frac{\sqrt{3}}{2}, \sin \frac{11\pi}{6} = -\frac{1}{2}$$
$$= 3\sqrt{3} - 3i$$

The rectangular form of the product is $3\sqrt{3} - 3i$.

Suppose the quotient of two complex numbers is expressed as a fraction. A formula for this quotient can be derived by rationalizing the denominator. To rationalize the denominator, multiply both the numerator and denominator by the same value so that the resulting new denominator does not contain imaginary numbers.

$$\frac{r_1(\cos \theta_1 + i \sin \theta_1)}{r_2(\cos \theta_2 + i \sin \theta_2)}$$

$$= \frac{r_1(\cos \theta_1 + i \sin \theta_1)}{r_2(\cos \theta_2 + i \sin \theta_2)} \cdot \frac{(\cos \theta_2 - i \sin \theta_2)}{(\cos \theta_2 - i \sin \theta_2)} \qquad \begin{array}{l} \cos \theta_2 - i \sin \theta_2 \text{ is the} \\ \text{conjugate of } \cos \theta_2 + i \sin \theta_2 \end{array}$$

$$= \frac{r_1}{r_2} \cdot \frac{(\cos \theta_1 \cos \theta_2 + \sin \theta_1 \sin \theta_2) + i(\sin \theta_1 \cos \theta_2 - \cos \theta_1 \sin \theta_2)}{\cos^2 \theta_2 + \sin^2 \theta_2}$$

$$= \frac{r_1}{r_2}[\cos (\theta_1 - \theta_2) + i \sin (\theta_1 - \theta_2)] \qquad\qquad \textit{Trigonometric identities}$$

Quotient of Complex Numbers in Polar Form	$\dfrac{r_1(\cos \theta_1 + i \sin \theta_1)}{r_2(\cos \theta_2 + i \sin \theta_2)} = \dfrac{r_1}{r_2}[\cos (\theta_1 - \theta_2) + i \sin (\theta_1 - \theta_2)]$

Notice that the modulus $\left(\dfrac{r_1}{r_2}\right)$ of the quotient of two complex numbers is the quotient of their moduli. The amplitude $(\theta_1 - \theta_2)$ of the quotient is the difference of the amplitudes.

Example **2** Find the quotient $12\left(\cos \frac{\pi}{4} + i \sin \frac{\pi}{4}\right) \div 4\left(\cos \frac{3\pi}{2} + i \sin \frac{3\pi}{2}\right)$. Then express the quotient in rectangular form.

Find the modulus and amplitude of the quotient.

$$r = \frac{r_1}{r_2} \qquad\qquad \theta = \theta_1 - \theta_2$$

$$= \frac{12}{4} \qquad\qquad = \frac{\pi}{4} - \frac{3\pi}{2}$$

$$= 3 \qquad\qquad = -\frac{5\pi}{4}$$

The quotient is $3\left[\cos\left(-\frac{5\pi}{4}\right) + i \sin\left(-\frac{5\pi}{4}\right)\right]$.

Now find the rectangular form of the quotient.

$$3\left[\cos\left(-\frac{5\pi}{4}\right) + i \sin\left(-\frac{5\pi}{4}\right)\right] = 3\left(-\frac{\sqrt{2}}{2} + \frac{\sqrt{2}}{2}i\right) \quad \begin{array}{l} \cos\left(-\frac{5\pi}{4}\right) = \frac{\sqrt{2}}{2}, \\[6pt] \sin\left(-\frac{5\pi}{4}\right) = \frac{\sqrt{2}}{2} \end{array}$$

$$= -\frac{3\sqrt{2}}{2} + \frac{3\sqrt{2}}{2}i$$

The rectangular form of the quotient is $-\frac{3\sqrt{2}}{2} + \frac{3\sqrt{2}}{2}i$.

You can use products and quotients of complex numbers in polar form to solve the problem presented at the beginning of the lesson.

Example **3** **ELECTRICITY** If a circuit has an impedance of $2 - 6j$ ohms and a voltage of 120 volts, find the current in the circuit.

Express each complex number in polar form.

$120 = 120(\cos 0 + j \sin 0)$

$2 - 6j \approx \sqrt{40}[\cos(-1.25) + j \sin(-1.25)] \quad r = \sqrt{2^2 + (-6)^2} = \sqrt{40} \text{ or } 2\sqrt{10},$

$\qquad\quad \approx 2\sqrt{10}[\cos(-1.25) + j \sin(-1.25)] \quad \theta = \text{Arctan} \frac{-6}{2} \text{ or } -1.25$

Substitute the voltage and impedance into the equation $E = I \cdot Z$.

$$E = I \cdot Z$$

$$120(\cos 0 + j \sin 0) = I \cdot 2\sqrt{10}[\cos(-1.25) + j \sin(-1.25)]$$

$$\frac{120(\cos 0 + j \sin 0)}{2\sqrt{10}[\cos(-1.25) + j \sin(-1.25)]} = I$$

$$6\sqrt{10}(\cos 1.25 + j \sin 1.25) = I$$

Now express the current in rectangular form.

$I = 6\sqrt{10}(\cos 1.25 + j \sin 1.25)$

$\quad \approx 5.98 + 18.01j \quad$ *Use a calculator.* $6\sqrt{10} \cos 1.25 \approx 5.98,$
$\qquad\qquad\qquad\qquad\qquad 6\sqrt{10} \sin 1.25 \approx 18.01$

The current is about $6 + 18j$ amps.

Lesson 9-7   Products and Quotients of Complex Numbers in Polar Form   **595**

Addressing Individual Needs

 Logical Have teams of students write out a step-by-step guide for solving a specific problem such as in Example 2 or Example 3. Have them be very detailed as if the person reading the guide had little math experience. Then, have the teams trade guides and check the logical sequence of the steps.

Error Analysis

Watch for students who mistakenly simplify a division problem in polar form by simplifying the cosine terms and the sine terms.

Prevent by reviewing the formula for the quotient of complex numbers in polar form.

Assignment Guide

Core: 11–27 odd, 29–38
Enriched: 10–26 even, 28–38

Additional Answers

1. The modulus of the quotient is the quotient of the moduli of the two complex numbers. The amplitude of the quotient is the difference of the amplitudes of the two complex numbers.

2. Square the modulus of the given complex number and double its amplitude.

3. Addition and subtraction are easier in rectangular form. Multiplication and division are easier in polar form. See students' work for examples.

Study Guide and Practice Masters, pp. 169–170

NAME _____ DATE _____ PERIOD _____

Study Guide

Products and Quotients of Complex Numbers in Polar Form

Example 1 Find the product $2\left(\cos\frac{\pi}{2} + i\sin\frac{\pi}{2}\right) \cdot 4\left(\cos\frac{\pi}{3} + i\sin\frac{\pi}{3}\right)$. Then express the product in rectangular form.

Find the modulus and amplitude of the product.

$r = r_1 r_2$ $\theta = \theta_1 + \theta_2$
$= 2(4)$ $= \frac{\pi}{2} + \frac{\pi}{3}$
$= 8$ $= \frac{5\pi}{6}$

The product is $8\left(\cos\frac{5\pi}{6} + i\sin\frac{5\pi}{6}\right)$.

Now find the rectangular form of the product.

$8\left(\cos\frac{5\pi}{6} + i\sin\frac{5\pi}{6}\right) = 8\left(-\frac{\sqrt{3}}{2} + \frac{1}{2}i\right)$ $\cos\frac{5\pi}{6} = -\frac{\sqrt{3}}{2}, \sin\frac{5\pi}{6} = \frac{1}{2}$
$= -4\sqrt{3} + 4i$

The rectangular form of the product is $-4\sqrt{3} + 4i$.

Example 2 Find the quotient $21\left(\cos\frac{7\pi}{6} + i\sin\frac{7\pi}{6}\right) \div 7\left(\cos\frac{4\pi}{3} + i\sin\frac{4\pi}{3}\right)$. Then express the quotient in rectangular form.

Find the modulus and amplitude of the quotient.

$r = \frac{r_1}{r_2}$ $\theta = \theta_1 - \theta_2$
$= \frac{21}{7}$ $= \frac{7\pi}{6} - \frac{4\pi}{3}$
$= 3$ $= -\frac{\pi}{6}$

The quotient is $3\left[\cos\left(-\frac{\pi}{6}\right) + i\sin\left(-\frac{\pi}{6}\right)\right]$.

Now find the rectangular form of the quotient.

$3\left[\cos\left(-\frac{\pi}{6}\right) + i\sin\left(-\frac{\pi}{6}\right)\right] = 3\left(\frac{\sqrt{3}}{2} + \left(-\frac{1}{2}\right)i\right)$ $\cos\left(-\frac{\pi}{6}\right) = \frac{\sqrt{3}}{2}, \sin\left(-\frac{\pi}{6}\right) = -\frac{1}{2}$
$= \frac{3\sqrt{3}}{2} - \frac{3}{2}i$

The rectangular form of the quotient is $\frac{3\sqrt{3}}{2} - \frac{3}{2}i$.

© Glencoe/McGraw-Hill   169   Advanced Mathematical Concepts

CHECK FOR UNDERSTANDING

Communicating Mathematics

Read and study the lesson to answer each question.   1–3. See margin.

1. **Explain** how to find the quotient of two complex numbers in polar form.

2. **Describe** how to square a complex number in polar form.

3. **List** which operations with complex numbers you think are easier in rectangular form and which you think are easier in polar form. Defend your choices with examples.

Guided Practice

Find each product or quotient. Express the result in rectangular form.

4. $2\left(\cos\frac{\pi}{2} + i\sin\frac{\pi}{2}\right) \cdot 2\left(\cos\frac{3\pi}{2} + i\sin\frac{3\pi}{2}\right)$ **4**

5. $3\left(\cos\frac{\pi}{6} + i\sin\frac{\pi}{6}\right) \div 4\left(\cos\frac{2\pi}{3} + i\sin\frac{2\pi}{3}\right)$ $-\frac{3}{4}i$

6. $4\left(\cos\frac{9\pi}{4} + i\sin\frac{9\pi}{4}\right) \div 2\left[\cos\left(-\frac{\pi}{2}\right) + i\sin\left(-\frac{\pi}{2}\right)\right]$ $-\sqrt{2} + \sqrt{2}i$

7. $\frac{1}{2}\left(\cos\frac{\pi}{3} + i\sin\frac{\pi}{3}\right) \cdot 6\left(\cos\frac{5\pi}{6} + i\sin\frac{5\pi}{6}\right)$ $-\frac{3\sqrt{3}}{2} - \frac{3}{2}i$

8. Use polar form to find the product $\left(2 + 2\sqrt{3}i\right) \cdot \left(-3 + \sqrt{3}i\right)$. Express the result in rectangular form. $-12 - 4\sqrt{3}i$

9. **Electricity** Determine the voltage in a circuit when there is a current of $2\left(\cos\frac{11\pi}{6} + j\sin\frac{11\pi}{6}\right)$ amps and an impedance of $3\left(\cos\frac{\pi}{3} + j\sin\frac{\pi}{3}\right)$ ohms.

$6\left(\cos\frac{\pi}{6} + j\sin\frac{\pi}{6}\right)$ **volts**

EXERCISES

Practice

Find each product or quotient. Express the result in rectangular form.

A   10. $4\left(\cos\frac{\pi}{3} + i\sin\frac{\pi}{3}\right) \cdot 7\left(\cos\frac{2\pi}{3} + i\sin\frac{2\pi}{3}\right)$ -28

11. $6\left(\cos\frac{3\pi}{4} + i\sin\frac{3\pi}{4}\right) \div 2\left(\cos\frac{\pi}{4} + i\sin\frac{\pi}{4}\right)$ $3i$

B   12. $\frac{1}{2}\left(\cos\frac{\pi}{3} + i\sin\frac{\pi}{3}\right) \div 3\left(\cos\frac{\pi}{6} + i\sin\frac{\pi}{6}\right)$ $\frac{\sqrt{3}}{12} + \frac{1}{12}i$

13. $5(\cos\pi + i\sin\pi) \cdot 2\left(\cos\frac{3\pi}{4} + i\sin\frac{3\pi}{4}\right)$ $5\sqrt{2} - 5\sqrt{2}i$

14. $6\left[\cos\left(-\frac{\pi}{3}\right) + i\sin\left(-\frac{\pi}{3}\right)\right] \cdot 3\left(\cos\frac{5\pi}{6} + i\sin\frac{5\pi}{6}\right)$ $18i$

15. $3\left(\cos\frac{7\pi}{3} + i\sin\frac{7\pi}{3}\right) \div \left(\cos\frac{\pi}{2} + i\sin\frac{\pi}{2}\right)$ $\frac{3\sqrt{3}}{2} - \frac{3}{2}i$

16. $2(\cos 240° + i\sin 240°) \cdot 3(\cos 60° + i\sin 60°)$ $3 - 3\sqrt{3}i$

17. $\sqrt{2}\left(\cos\frac{7\pi}{4} + i\sin\frac{7\pi}{4}\right) \div \frac{\sqrt{2}}{2}\left(\cos\frac{3\pi}{4} + i\sin\frac{3\pi}{4}\right)$ -2

18. $3(\cos 4 + i\sin 4) \cdot 0.5(\cos 2.5 + i\sin 2.5)$ $1.46 + 0.32i$

19. $4[\cos(-2) + i\sin(-2)] \div (\cos 3.6 + i\sin 3.6)$ **3.10 + 2.53i**

20. $20\left(\cos\frac{7\pi}{6} + i\sin\frac{7\pi}{6}\right) \div 15\left(\cos\frac{11\pi}{3} + i\sin\frac{11\pi}{3}\right)$ $-\frac{4}{3}i$

21. $2\left(\cos\frac{3\pi}{4} + i\sin\frac{3\pi}{4}\right) \cdot \sqrt{2}\left(\cos\frac{\pi}{2} + i\sin\frac{\pi}{2}\right)$ $-2 - 2i$

22. Find the product of $2\left(\cos\frac{\pi}{3} + i\sin\frac{\pi}{3}\right)$ and $6\left[\cos\left(-\frac{\pi}{6}\right) + i\sin\left(-\frac{\pi}{6}\right)\right]$. Write the answer in rectangular form. $6\sqrt{3} + 6i$

23. If $z_1 = 4\left(\cos\frac{5\pi}{3} + i\sin\frac{5\pi}{3}\right)$ and $z_2 = \frac{1}{2}\left(\cos\frac{\pi}{3} + i\sin\frac{\pi}{3}\right)$, find $\frac{z_1}{z_2}$ and express the result in rectangular form. $-4 - 4\sqrt{3}i$

Use polar form to find each product or quotient. Express the result in rectangular form.

24. $(2 - 2i) \cdot (-3 + 3i)$ **12i**

25. $\left(\sqrt{2} - \sqrt{2}i\right) \cdot \left(-3\sqrt{2} - 3\sqrt{2}i\right)$ **−12**

26. $\left(\sqrt{3} - i\right) \div \left(2 - 2\sqrt{3}i\right)$ $\frac{\sqrt{3}}{4} + \frac{1}{4}i$ **27.** $\left(-4\sqrt{2} + 4\sqrt{2}i\right) \div (6 + 6i)$ $\frac{2\sqrt{2}}{3}i$

Applications and Problem Solving

28. Electricity   Find the current in a circuit with a voltage of 13 volts and an impedance of $3 - 2j$ ohms.   **3 + 2j amps**

29. Electricity   Find the impedance in a circuit with a voltage of 100 volts and a current of $4 - 3j$ amps.   **16 + 12j ohms**

30. Critical Thinking   Given z_1 and z_2 graphed at the right, graph z_1z_2 and $\frac{z_1}{z_2}$ without actually calculating them.
See margin.

31. Transformations

31a. The point is rotated counterclockwise about the origin by an angle of θ.

31b. The point is rotated 60° counterclockwise about the origin.

a. Describe the transformation applied to the graph of the complex number z if z is multiplied by $\cos\theta + i\sin\theta$.

b. Describe the transformation applied to the graph of the complex number z if z is multiplied by $\frac{1}{2} + \frac{\sqrt{3}}{2}i$.

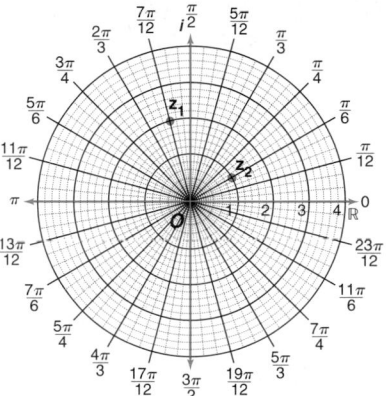

32. Critical Thinking   Find the quadratic equation $az^2 + bz + c = 0$ such that $a = 1$ and the solutions are $3\left(\cos\frac{\pi}{3} + i\sin\frac{\pi}{3}\right)$ and $2\left(\cos\frac{5\pi}{6} + i\sin\frac{5\pi}{6}\right)$.   **See margin.**

Mixed Review

33. Express $5 - 12i$ in polar form. *(Lesson 9-6)*   **13(cos 5.11 + i sin 5.11)**

34. $-\sqrt{3}x + y - 10 = 0$ **34.** Write the equation $r = 5\sec\left(\theta - \frac{5\pi}{6}\right)$ in rectangular form. *(Lesson 9-4)*

35. Physics   A prop for a play is supported equally by two wires suspended from the ceiling. The wires form a 130° angle with each other. If the prop weighs 23 pounds, what is the tension in each of the wires? *(Lesson 8-5)*   **about 27.21 lb**

Extra Practice   See p. A43.

Lesson 9-7   *Products and Quotients of Complex Numbers in Polar Form*   **597**

Additional Answers

30. Start at z_1 in the complex plane. Since the modulus of z_2 is 1, z_1z_2 and $\frac{z_1}{z_2}$ will both have the same modulus as z_1. Then z_1z_2 and $\frac{z_1}{z_2}$ can be located by rotating z_1 by $\frac{\pi}{6}$ counterclockwise and clockwise, respectively.

32. Since $a = 1$, the equation will be of form $z^2 + bz + c = 0$. The coefficient c is the product of the solutions, which is $6\left(\cos\frac{7\pi}{6} + i\sin\frac{7\pi}{6}\right)$, or $-3\sqrt{3} - 3i$ in rectangular form. The coefficient b is the opposite of the sum of the solutions, so convert the solutions to rectangular form to do the addition.

$b = -\left[3\left(\cos\frac{\pi}{3} + i\sin\frac{\pi}{3}\right) + 2\left(\cos\frac{5\pi}{6} + i\sin\frac{5\pi}{6}\right)\right]$

$= -\left[\left(\frac{3}{2} + \frac{3\sqrt{3}}{2}i\right) + \left(-\sqrt{3} + i\right)\right]$

$= \left(-\frac{3}{2} + \sqrt{3}\right) + \left(\frac{-3\sqrt{3} - 2}{2}\right)i$

Therefore, the equation is

$z^2 + \left[\left(-\frac{3}{2} + \sqrt{3}\right) + \left(\frac{-3\sqrt{3} - 2}{2}\right)i\right]z + \left(-3\sqrt{3} - 3i\right) = 0$

Extra Credit

Find the exact value of $\left(\cos\frac{\pi}{4} + i\sin\frac{\pi}{4}\right) \cdot \left(\cos\frac{\pi}{8} + i\sin\frac{\pi}{8}\right)$. Express the result in rectangular form. $\frac{\sqrt{2 - \sqrt{2}}}{2} + \frac{\sqrt{2 + \sqrt{2}}}{2}i$

Open-Ended Assessment

Speaking Have students explain in their own words how they would find the product and quotient of the complex numbers $r_1 \text{ cis } \theta_1$ and $r_2 \text{ cis } \theta_2$.

36. Solve $\cos 2x + \sin x = 1$ for principal values of x. (Lesson 7-5) **0°, 30°**

37. Write the equation for the inverse of $y = \cos x$. (Lesson 6-8) **$y = \arccos x$**

38. **SAT/ACT Practice** In the figure, the perimeter of square $BCDE$ is how much smaller than the perimeter of rectangle $ACDF$? **C**

A 2 B 3 C 4

D 6 E 16

CAREER CHOICES

● Astronomer ●

Have you ever gazed into the sky at night hoping to spot a constellation? Do you dream of having your own telescope? If you enjoy studying about the universe, then a career in astronomy may be just for you. Astronomers collect and analyze data about the universe including stars, planets, comets, asteroids, and even artificial satellites. As an astronomer, you may collect information by using a telescope or spectrometer here on earth, or you may use information collected by spacecraft and satellites.

Most astronomers specialize in one branch of astronomy such as astrophysics or celestial mechanics. Astronomers often teach in addition to conducting research. Astronomers located throughout the world are prime sources of information for NASA and other countries' space programs.

CAREER OVERVIEW

Degree Preferred:
at least a bachelor's degree in astronomy or physics

Related Courses:
mathematics, physics, chemistry, computer science

Outlook:
average through the year 2006

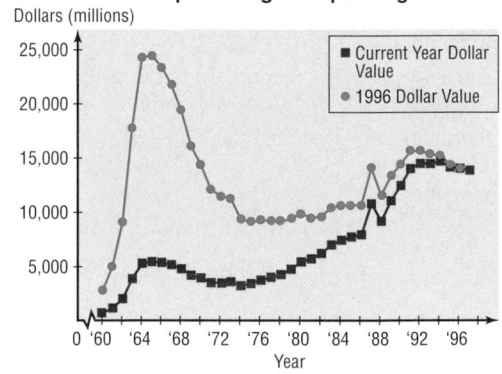

Space Program Spending

Source: National Aeronautics and Space Administration

interNET CONNECTION For more information on careers in astronomy, visit: **www.amc.glencoe.com**

598 Chapter 9 Polar Coordinates and Complex Numbers

Enrichment Masters, p. 103

9-7 NAME _____ DATE _____ PERIOD _____

Enrichment

Complex Conjugates

In Lesson 9-5, you learned that complex numbers in the form $a + bi$ and $a - bi$ are called conjugates. You can show that two numbers are conjugates by finding the appropriate values of a and b.

1. Show that the solutions of $x^2 + 2x + 3 = 0$ are conjugates. The solutions are $-1 + i\sqrt{2}$ and $-1 - i\sqrt{2}$, so $a = -1$ and $b = \sqrt{2}$.

2. Show that the solutions of $Ax^2 + Bx + C = 0$ are conjugates when $B^2 - 4AC < 0$.
By the quadratic formula, the solutions are
$-\frac{B}{2A} + i\frac{\sqrt{B^2 - 4AC}}{2A}$ and $-\frac{B}{2A} - i\frac{\sqrt{B^2 - 4AC}}{2A}$,
so $a = -\frac{B}{2A}$ and $b = \frac{\sqrt{B^2 - 4AC}}{2A}$

The conjugate of the complex number z is represented by $\bar{z}$.

3. $z = a + bi$. Use $\bar{z}$ to find the reciprocal of z.
$\frac{a}{a^2 + b^2} - i\frac{b}{a^2 + b^2} = \frac{\bar{z}}{|z|^2}$

4. $z = r(\cos \theta + i \sin \theta)$. Find $\bar{z}$. Express your answer in polar form.
$r[\cos(-\theta) + i \sin(-\theta)]$

Use your answer to Exercise 3 to solve Exercises 4 and 5.

5. Find $z \cdot \bar{z}$.
$r^2 = |z|^2$

6. Find $z \div \bar{z}$. $(z \neq 0)$
$\cos 2\theta + i \sin 2\theta = \frac{z^2}{|z|^2}$

© Glencoe/McGraw-Hill   T103   Advanced Mathematical Concepts

---

CAREER CHOICES ASTRONOMER

Math Connection
• Explain the trends seen in the data.
The amount spent on the space program increased during the 1960s but then declined sharply during the late 1960s and first half of the 1970s. Spending remained steady during the 1980s and increased somewhat during the 1990s.
• What else does this graph show in addition to the spending for the space program?
The graph shows the value of money in any particular year compared to the value in 1996.

Powers and Roots of Complex Numbers

OBJECTIVE
• Find powers and roots of complex numbers in polar form using De Moivre's Theorem.

COMPUTER GRAPHICS Many of the computer graphics that are referred to as fractals are graphs of Julia sets, which are named after the mathematician Gaston Julia. When a function like $f(z) = z^2 + c$, where c is a complex constant, is iterated, points in the complex plane can be classified according to their behavior under iteration.

• Points that escape to infinity under iteration belong to the **escape set** of the function.
• Points that do not escape belong to the **prisoner set.**

The **Julia set** is the boundary between the escape set and the prisoner set. Is the number $w = 0.6 - 0.5i$ in the escape set or the prisoner set of the function $f(z) = z^2$? *This problem will be solved in Example 6.*

You can use the formula for the product of complex numbers to find the square of a complex number.

$$[r(\cos \theta + i \sin \theta)]^2 = [r(\cos \theta + i \sin \theta)] \cdot [r(\cos \theta + i \sin \theta)]$$
$$= r^2[\cos (\theta + \theta) + i \sin (\theta + \theta)]$$
$$= r^2(\cos 2\theta + i \sin 2\theta)$$

Other powers of complex numbers can be found using De Moivre's Theorem.

De Moivre's Theorem	$[r(\cos \theta + i \sin \theta)]^n = r^n(\cos n\theta + i \sin n\theta)$

You will be asked to prove De Moivre's Theorem in Chapter 12.

Example Find $\left(2 + 2\sqrt{3}i\right)^6$.

First, write $2 + 2\sqrt{3}i$ in polar form. Note that its graph is in the first quadrant of the complex plane.

$$r = \sqrt{2^2 + \left(2\sqrt{3}\right)^2} \qquad \theta = \text{Arctan } \frac{2\sqrt{3}}{2}$$
$$= \sqrt{4 + 12} \qquad\qquad = \text{Arctan } \sqrt{3}$$
$$= 4 \qquad\qquad\qquad = \frac{\pi}{3}$$

(continued on the next page)

In-Class Examples

Example 1
Find $(1 + i)^5$. $-4 - 4i$

Example 2
Find $\left(-\sqrt{2} + \sqrt{2}i\right)^{-2}$. $\frac{1}{4}i$

Teaching Tip When students are solving problems with a negative power, make sure that they do not try to simply take the exponent and make the result negative. For example, $2^{-5} \neq -(2^5)$.

The polar form of $2 + 2\sqrt{3}i$ is $4\left(\cos\frac{\pi}{3} + i\sin\frac{\pi}{3}\right)$.

Now use De Moivre's Theorem to find the sixth power.

$$(2 + 2\sqrt{3}i)^6 = \left[4\left(\cos\frac{\pi}{3} + i\sin\frac{\pi}{3}\right)\right]^6$$

$$= 4^6\left[\cos 6\left(\frac{\pi}{3}\right) + i\sin 6\left(\frac{\pi}{3}\right)\right]$$

$$= 4096(\cos 2\pi + i\sin 2\pi)$$

$$= 4096(1 + 0i) \quad \textit{Write the result in rectangular form.}$$

$$= 4096$$

Therefore, $\left(2 + 2\sqrt{3}i\right)^6 = 4096$.

De Moivre's Theorem is valid for all rational values of n. Therefore, it is also useful for finding negative powers of complex numbers and roots of complex numbers.

Example ❷ Find $\left(\frac{\sqrt{3}}{2} - \frac{1}{2}i\right)^{-5}$.

First, write $\frac{\sqrt{3}}{2} - \frac{1}{2}i$ in polar form. Note that its graph is in the fourth quadrant of the complex plane.

$$r = \sqrt{\left(\frac{\sqrt{3}}{2}\right)^2 + \left(-\frac{1}{2}\right)^2} \qquad \theta = \text{Arctan}\ \dfrac{-\frac{1}{2}}{\frac{\sqrt{3}}{2}}$$

$$= \sqrt{\frac{3}{4} + \frac{1}{4}}\ \text{or}\ 1 \qquad\qquad = \text{Arctan}\left(-\frac{\sqrt{3}}{3}\right)\ \text{or}\ -\frac{\pi}{6}$$

The polar form of $\frac{\sqrt{3}}{2} - \frac{1}{2}i$ is $1\left[\cos\left(-\frac{\pi}{6}\right) + i\sin\left(-\frac{\pi}{6}\right)\right]$.

Use De Moivre's Theorem to find the negative 5th power.

$$\left(\frac{\sqrt{3}}{2} - \frac{1}{2}i\right)^{-5} = \left[1\left(\cos\left(-\frac{\pi}{6}\right) + i\sin\left(-\frac{\pi}{6}\right)\right)\right]^{-5}$$

$$= 1^{-5}\left[\cos(-5)\left(-\frac{\pi}{6}\right) + i\sin(-5)\left(-\frac{\pi}{6}\right)\right] \quad \textit{De Moivre's Theorem}$$

$$= 1\left(\cos\frac{5\pi}{6} + i\sin\frac{5\pi}{6}\right) \quad \textit{Simplify.}$$

$$= -\frac{\sqrt{3}}{2} + \frac{1}{2}i \quad \textit{Write the answer in rectangular form.}$$

Recall that positive real numbers have two square roots and that the positive one is called the principal square root. In general, all nonzero complex numbers have p distinct pth roots. That is, they each have two square roots, three cube roots, four fourth roots, and so on. The principal pth root of a complex number is given by:

$$(a + bi)^{\frac{1}{p}} = [r(\cos\theta + i\sin\theta)]^{\frac{1}{p}}$$

$$= r^{\frac{1}{p}}\left(\cos\frac{\theta}{p} + i\sin\frac{\theta}{p}\right).$$ *When finding a principal root, the interval $-\pi < \theta \le \pi$ is used.*

Example **3** Find $\sqrt[3]{8i}$.

$$\sqrt[3]{8i} = (0 + 8i)^{\frac{1}{3}}$$ *$a = 0$, $b = 8$*

$$= \left[8\left(\cos\frac{\pi}{2} + i\sin\frac{\pi}{2}\right)\right]^{\frac{1}{3}}$$ *Polar form; $r = \sqrt{0^2 + 8^2}$ or 8, $\theta = \frac{\pi}{2}$ since $a = 0$.*

$$= 8^{\frac{1}{3}}\left[\cos\left(\frac{1}{3}\right)\left(\frac{\pi}{2}\right) + i\sin\left(\frac{1}{3}\right)\left(\frac{\pi}{2}\right)\right]$$ *De Moivre's Theorem*

$$= 2\left(\cos\frac{\pi}{6} + i\sin\frac{\pi}{6}\right)$$

$$= 2\left(\frac{\sqrt{3}}{2} + \frac{1}{2}i\right) \text{ or } \sqrt{3} + i$$ *This is the principal cube root.*

The following formula generates all of the pth roots of a complex number. It is based on the identities $\cos\theta = \cos(\theta + 2n\pi)$ and $\sin\theta = \sin(\theta + 2n\pi)$, where n is any integer.

The p Distinct pth Roots of a Complex Number	The p distinct pth roots of $a + bi$ can be found by replacing n with $0, 1, 2, \dots, p - 1$, successively, in the following equation. $$(a + bi)^{\frac{1}{p}} = \{r[\cos(\theta + 2n\pi) + i\sin(\theta + 2n\pi)]\}^{\frac{1}{p}}$$ $$= r^{\frac{1}{p}}\left(\cos\frac{\theta + 2n\pi}{p} + i\sin\frac{\theta + 2n\pi}{p}\right)$$

Example **4** Find the three cube roots of $-2 - 2i$.

First, write $-2 - 2i$ in polar form.

$$r = \sqrt{(-2)^2 + (-2)^2} \text{ or } 2\sqrt{2} \qquad \theta = \text{Arctan}\frac{-2}{-2} + \pi \text{ or } \frac{5\pi}{4}$$

$$-2 - 2i = 2\sqrt{2}\left[\cos\left(\frac{5\pi}{4} + 2n\pi\right) + i\sin\left(\frac{5\pi}{4} + 2n\pi\right)\right]$$ *n is any integer.*

Now write an expression for the cube roots.

$$(-2 - 2i)^{\frac{1}{3}} = \left(2\sqrt{2}\left[\cos\left(\frac{5\pi}{4} + 2n\pi\right) + i\sin\left(\frac{5\pi}{4} + 2n\pi\right)\right]\right)^{\frac{1}{3}}$$

$$= \sqrt{2}\left[\cos\left(\frac{\frac{5\pi}{4} + 2n\pi}{3}\right) + i\sin\left(\frac{\frac{5\pi}{4} + 2n\pi}{3}\right)\right] \qquad \begin{aligned}(2\sqrt{2})^{\frac{1}{3}} &= \left(2^{\frac{3}{2}}\right)^{\frac{1}{3}} \\ &= 2^{\frac{1}{2}} \text{ or } \sqrt{2}\end{aligned}$$

(continued on the next page)

Lesson 9-8 *Powers and Roots of Complex Numbers* **601**

Teaching Tip For Example 3, make sure that students write the root as an expression with a rational exponent.

In-Class Examples
Example 3
Find $\sqrt[4]{2 + 6i}$. **1.51 + 0.49i**

Example 4
Find the five fifth roots of -243.
2.43 + 1.76i, -0.93 + 2.85i, -3, -0.93 − 2.85i, 2.43 − 1.76i

Teaching Tip For Example 4, emphasize that the polar form and the expression in terms of n must be found before the successive roots can be determined.

Let $n = 0$, 1, and 2 successively to find the cube roots.

$$\text{Let } n = 0. \quad \sqrt{2}\left[\cos\left(\frac{\frac{5\pi}{4} + 2(0)\pi}{3}\right) + i\sin\left(\frac{\frac{5\pi}{4} + 2(0)\pi}{3}\right)\right]$$

$$= \sqrt{2}\left(\cos\frac{5\pi}{12} + i\sin\frac{5\pi}{12}\right)$$

$$\approx 0.37 + 1.37i$$

$$\text{Let } n = 1. \quad \sqrt{2}\left[\cos\left(\frac{\frac{5\pi}{4} + 2(1)\pi}{3}\right) + i\sin\left(\frac{\frac{5\pi}{4} + 2(1)\pi}{3}\right)\right]$$

$$= \sqrt{2}\left(\cos\frac{13\pi}{12} + i\sin\frac{13\pi}{12}\right)$$

$$\approx -1.37 - 0.37i$$

$$\text{Let } n = 2. \quad \sqrt{2}\left[\cos\left(\frac{\frac{5\pi}{4} + 2(2)\pi}{3}\right) + i\sin\left(\frac{\frac{5\pi}{4} + 2(2)\pi}{3}\right)\right]$$

$$= \sqrt{2}\left(\cos\frac{21\pi}{12} + i\sin\frac{21\pi}{12}\right)$$

$$= 1 - i$$

The cube roots of $-2 - 2i$ are approximately $0.37 + 1.37i$, $-1.37 - 0.37i$, and $1 - i$. *These roots can be checked by multiplication.*

GRAPHING CALCULATOR EXPLORATION

The p distinct pth roots of a complex number can be approximated using the parametric mode on a graphing calculator. For a particular complex number $r(\cos\theta + i\sin\theta)$ and a particular value of p:

▸ Select the **Radian** and **Par** modes.

▸ Select the viewing window.
$\text{Tmin} = \frac{\theta}{p}$, $\text{Tmax} = \frac{\theta}{p} + 2\pi$, $\text{Tstep} = \frac{2\pi}{p}$,
$\text{Xmin} = -r^{\frac{1}{p}}$, $\text{Xmax} = r^{\frac{1}{p}}$, $\text{Xscl} = 1$,
$\text{Ymin} = -r^{\frac{1}{p}}$, $\text{Ymax} = r^{\frac{1}{p}}$, and $\text{Yscl} = 1$.

▸ Enter the parametric equations
$X_{1T} = r^{\frac{1}{p}}\cos T$ and $Y_{1T} = r^{\frac{1}{p}}\sin T$.

▸ Graph the equations.

▸ Use TRACE to locate the roots.

TRY THESE 1–6. See margin.
1. Approximate the cube roots of 1.
2. Approximate the fourth roots of i.
3. Approximate the fifth roots of $1 + i$.

WHAT DO YOU THINK?
4. What geometric figure is formed when you graph the three cube roots of a complex number?
5. What geometric figure is formed when you graph the fifth roots of a complex number?
6. Under what conditions will the complex number $a + bi$ have a root that lies on the positive real axis?

You can also use De Moivre's Theorem to solve some polynomial equations.

Examples **5** Solve $x^5 - 32 = 0$. Then graph the roots in the complex plane.

The solutions to this equation are the same as those of the equation $x^5 = 32$. That means we have to find the fifth roots of 32.

$32 = 32 + 0i$ $\qquad\qquad$ $a = 32, b = 0$

$\quad = 32(\cos 0 + i \sin 0)$ *Polar form;* $r = \sqrt{32^2 + 0^2}$ *or 32,* $\theta = Arctan \frac{0}{32}$ *or 0*

Now write an expression for the fifth roots.

$32^{\frac{1}{5}} = [32(\cos (0 + 2n\pi) + i \sin (0 + 2n\pi))]^{\frac{1}{5}}$

$\quad = 2\left(\cos \frac{2n\pi}{5} + i \sin \frac{2n\pi}{5}\right)$

Let $n = 0, 1, 2, 3,$ and 4 successively to find the fifth roots, $x_1, x_2, x_3, x_4,$ and x_5.

Let n = 0. $x_1 = 2(\cos 0 + i \sin 0) = 2$

Let n = 1. $x_2 = 2\left(\cos \frac{2\pi}{5} + i \sin \frac{2\pi}{5}\right) \approx 0.62 + 1.90i$

Let n = 2. $x_3 = 2\left(\cos \frac{4\pi}{5} + i \sin \frac{4\pi}{5}\right) \approx -1.62 + 1.18i$

Let n = 3. $x_4 = 2\left(\cos \frac{6\pi}{5} + i \sin \frac{6\pi}{5}\right) \approx -1.62 - 1.18i$

Let n = 4. $x_5 = 2\left(\cos \frac{8\pi}{5} + i \sin \frac{8\pi}{5}\right) \approx 0.62 - 1.90i$

The solutions of $x^5 - 32 = 0$ are 2, $0.62 \pm 1.90i$, and $-1.62 \pm 1.18i$.

The solutions are graphed at the right. Notice that the points are the vertices of a regular pentagon. The roots of a complex number are cyclical in nature. That means, when the roots are graphed on the complex plane, the roots are equally spaced around a circle.

6 **COMPUTER GRAPHICS** Refer to the application at the beginning of the lesson. Is the number $w = 0.6 - 0.5i$ in the escape set or the prisoner set of the function $f(z) = z^2$?

Iterating this function requires you to square complex numbers, so you can use De Moivre's Theorem.

Write w in polar form. $\qquad\qquad$ $r = \sqrt{0.6^2 + (-0.5)^2}$ *or about 0.78*

$w = 0.78[\cos (-0.69) + i \sin (-0.69)]$ $\quad \theta = Arctan \frac{-0.5}{0.6}$ *or about* -0.69

(continued on the next page)

Lesson 9-8 Powers and Roots of Complex Numbers **603**

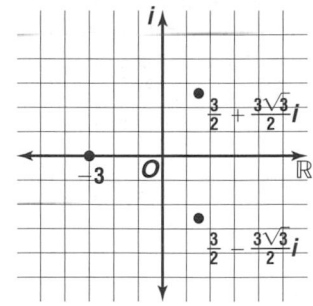

Addressing Individual Needs

Intrapersonal Have students work individually to solve some problems involving powers and roots of complex numbers. Have students write journal entries about the processes they used, carefully explaining each step.

Error Analysis

Watch for students who substitute values into the formulas incorrectly.

Prevent by having students list each variable and what its value is.

Additional Answers

2. Finding a reciprocal is the same as raising a number to the −1 power, so take the reciprocal of the modulus and multiply the amplitude by −1.

3.

4. Shembala is correct. The polar form of $a + ai$ is

$a\sqrt{2}\left(\cos \frac{\pi}{4} + i \sin \frac{\pi}{4}\right)$.

By De Moivre's Theorem, the polar form of $(a + ai)^2$ is

$2a^2\left(\cos \frac{\pi}{2} + i \sin \frac{\pi}{2}\right)$.

Since $\cos \frac{\pi}{2} = 0$, this is a pure imaginary number.

Study Guide and Practice Masters, pp. 171–172

Now iterate the function.

$w_1 = f(w)$
   $= w^2$
   $= (0.78[\cos(-0.69) + i\sin(-0.69)])^2$
   $= 0.78^2[\cos 2(-0.69) + i\sin 2(-0.69)]$ *De Moivre's Theorem*
   $= 0.12 - 0.60i$ *Use a calculator to approximate the rectangular form.*

$w_2 = f(w_1)$
   $= w_1^2$
   $= (0.78^2[\cos 2(-0.69) + i\sin 2(-0.69)])^2$ *Use the polar form of w_1.*
   $= 0.78^4[\cos 4(-0.69) + i\sin 4(-0.69)]$
   $= -0.34 - 0.14i$

$w_3 = f(w_2)$
   $= w_2^2$
   $= (0.78^4[\cos 4(-0.69) + i\sin 4(-0.69)])^2$ *Use the polar form of w_2.*
   $= 0.78^8[\cos 8(-0.69) + i\sin 8(-0.69)]$
   $= 0.10 + 0.09i$

The moduli of these iterates are 0.78^2, 0.78^4, 0.78^8, and so on. These moduli will approach 0 as the number of iterations increases. This means the graphs of the iterates approach the origin in the complex plane, so $w = 0.6 - 0.5i$ is in the prisoner set of the function.

Communicating Mathematics

Read and study the lesson to answer each question.   **2–4. See margin.**

1. **Evaluate** the product $(1 + i)(1 + i)(1 + i)(1 + i)(1 + i)$ by traditional multiplication. Compare the results with the results using De Moivre's Theorem on $(1 + i)^5$. Which method do you prefer?

1. Same results, $-4 - 4i$; answers may vary.

2. **Explain** how to use De Moivre's Theorem to find the reciprocal of a complex number in polar form.

3. **Graph** all the fourth roots of a complex number if $a + ai$ is one of the fourth roots. Assume a is positive.

4. **You Decide**   Shembala says that if $a \neq 0$, then $(a + ai)^2$ must be a pure imaginary number. Arturo disagrees. Who is correct? Use polar form to explain.

604   Chapter 9   Polar Coordinates and Complex Numbers

Guided Practice

Find each power. Express the result in rectangular form.

5. $\left(\sqrt{3} - i\right)^3$ $-8i$

6. $(3 - 5i)^4$ $-644 + 960i$

Find each principal root. Express the result in the form $a + bi$ with a and b rounded to the nearest hundredth.

7. $i^{\frac{1}{6}}$ $0.97 + 0.26i$

8. $(-2 - i)^{\frac{1}{3}}$ $0.82 - 1.02i$

Solve each equation. Then graph the roots in the complex plane.

9–10.
See margin.

9. $x^4 + i = 0$

10. $2x^3 + 4 + 2i = 0$

11. **Fractals**   Refer to the application at the beginning of the lesson. Is $w = 0.8 - 0.7i$ in the prisoner set or the escape set for the function $f(z) = z^2$? Explain.   **Escape set; the iterates escape to infinity.**

EXERCISES

Practice

 Find each power. Express the result in rectangular form.

12. $\left[3\left(\cos \frac{\pi}{6} + i \sin \frac{\pi}{6}\right)\right]^3$ $27i$

13. $\left[2\left(\cos \frac{\pi}{4} + i \sin \frac{\pi}{4}\right)\right]^5$

13. $-16\sqrt{2} - 16\sqrt{2}i$

14. $(-2 + 2i)^3$ $16 + 16i$

15. $\left(1 + \sqrt{3}i\right)^4$ $-8 - 8\sqrt{3}i$

 16. $(3 - 6i)^4$ $-567 + 1944i$

17. $(2 + 3i)^{-2}$ $-0.03 - 0.07i$

18. Raise $2 + 4i$ to the fourth power. $-112 - 384i$

Find each principal root. Express the result in the form $a + bi$ with a and b rounded to the nearest hundredth.   **19.** $1.83 + 0.81i$

19. $\left[32\left(\cos \frac{2\pi}{3} + i \sin \frac{2\pi}{3}\right)\right]^{\frac{1}{5}}$

20. $(-1)^{\frac{1}{4}}$ $0.71 + 0.71i$

21. $(-2 + i)^{\frac{1}{4}}$ $0.96 + 0.76i$

22. $(4 - i)^{\frac{1}{3}}$ $1.60 - 0.13i$

23. $(2 + 2i)^{\frac{1}{3}}$ $1.37 + 0.37i$

24. $(-1 - i)^{\frac{1}{4}}$ $0.91 - 0.61i$

25. Find the principal square root of i. $0.71 + 0.71i$

Solve each equation. Then graph the roots in the complex plane.

26. $x^3 - 1 = 0$ **26–31. See margin.**

27. $x^5 + 1 = 0$

 28. $2x^4 - 128 = 0$

29. $3x^4 + 48 = 0$

30. $x^4 - (1 + i) = 0$

31. $2x^4 + 2 + 2\sqrt{3}i = 0$

Graphing Calculator

Use a graphing calculator to find all of the indicated roots.

32. fifth roots of $10 - 9i$

33. sixth roots of $2 + 4i$

34. eighth roots of $36 + 20i$

32–34. See margin.

Applications and Problem Solving

35. **Fractals**   Is the number $\frac{1}{2} + \frac{3}{4}i$ in the escape set or the prisoner set for the function $f(z) = z^2$? Explain.   **Prisoner set; the iterates approach 0.**

36. **Critical Thinking**   Suppose $w = a + bi$ is one of the 31st roots of 1.
   a. What is the maximum value of a?   **1**
   b. What is the maximum value of b? $\sin \frac{16\pi}{31}$ or about **0.9987**

www.amc.glencoe.com/self_check_quiz   *Lesson 9-8 Powers and Roots of Complex Numbers*   **605**

Open-Ended Assessment

Writing Have students rewrite DeMoivre's Theorem in their own words. **Sample answer: the *n*th power of the complex number *r* cis θ is *r*ⁿ cis *n*θ.**

 Chapter 9, Quiz D (Lessons 9-7 and 9-8) is available in the *Assessment and Evaluation Masters*, p. 226.

Additional Answers

37. $1, \frac{1}{2} + \frac{\sqrt{3}}{2}i, -\frac{1}{2} + \frac{\sqrt{3}}{2}i, -1,$

$-\frac{1}{2} - \frac{\sqrt{3}}{2}i, \frac{1}{2} - \frac{\sqrt{3}}{2}i$

39. **The roots are the vertices of a regular polygon. Since one of the roots must be a positive real number, a vertex of the polygon lies on the positive real axis and the polygon is symmetric about the real axis. This means that the non-real complex roots occur in conjugate pairs. Since the imaginary part of the sum of two complex conjugates is 0, the imaginary part of the sum of all the roots must be 0.**

Enrichment Masters, p. 104

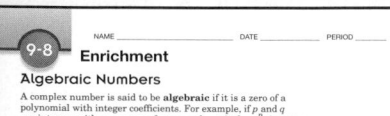

37. **Design** Gloribel works for an advertising agency. She wants to incorporate a hexagon design into the artwork for one of her proposals. She knows that she can locate the vertices of a regular hexagon by graphing the solutions to the equation $x^6 - 1 = 0$ in the complex plane. What are the solutions to this equation? **See margin.**

38. **Computer Graphics** Computer programmers can use complex numbers and the complex plane to implement geometric transformations. If a programmer starts with a square with vertices at $(2, 2)$, $(-2, 2)$, $(-2, -2)$, and $(2, -2)$, each of the vertices can be stored as a complex number in polar form. Complex number multiplication can be used to rotate the square 45° counterclockwise and dilate it so that the new vertices lie at the midpoints of the sides of the original square.

   a. What complex number should the programmer multiply by to produce this transformation? **0.5 + 0.5*i***

   38b. The square is rotated 90° counterclockwise and dilated by a factor of 0.5.

   b. What happens if the original vertices are multiplied by the square of your answer to part a?

39. **Critical Thinking** Explain why the sum of the imaginary parts of the *p* distinct *p*th roots of any positive real number must be zero. **See margin.**

Mixed Review

40. Find the product $2\left(\cos \frac{\pi}{6} + i \sin \frac{\pi}{6}\right) \cdot 3\left(\cos \frac{5\pi}{3} + i \sin \frac{5\pi}{3}\right)$. Express the result in rectangular form. *(Lesson 9-7)* **$3\sqrt{3} - 3i$**

41. Simplify $(2 - 5i) + (-3 + 6i) - (-6 + 2i)$. *(Lesson 9-5)* **$5 - i$**

42. $x = t$, $y = -2t + 7$

42. Write parametric equations of the line with equation $y = -2x + 7$. *(Lesson 8-6)*

43. Use a half-angle identity to find the exact value of $\cos 22.5°$. *(Lesson 7-4)*

43. $\dfrac{\sqrt{2 + \sqrt{2}}}{2}$

44. Solve triangle *ABC* if $A = 81°15'$ and $b = 28$. Round angle measures to the nearest minute and side measures to the nearest tenth. *(Lesson 5-4)* **$B = 8°45'$, $a = 181.9$, $c = 184.1$**

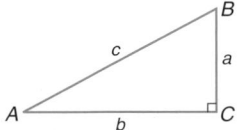

45. 800 large bears, 400 small bears

45. **Manufacturing** The Precious Animal Company must produce at least 300 large stuffed bears and 400 small stuffed bears per day. At most, the company can produce a total of 1200 bears per day. The profit for each large bear is $9.00, and the profit for each small bear is $5.00. How many of each type of bear should be produced each day to maximize profit? *(Lesson 2-7)*

46. **SAT/ACT Practice** Six quarts of a 20% solution of alcohol in water are mixed with 4 quarts of a 60% solution of alcohol in water. The alcoholic strength of the mixture is **A**

   A 36%   **B** 40%   **C** 48%   **D** 60%   **E** 80%

606 *Chapter 9 Polar Coordinates and Complex Numbers* | **Extra Practice** See p. A43. |

Extra Credit

Find the exact value of the expression $\sqrt[3]{8\left(\cos \frac{3\pi}{8} + i \sin \frac{3\pi}{8}\right)}$. Express the result in rectangular form.

$\sqrt{2 + \sqrt{2}} + \sqrt{2 - \sqrt{2}}\,i$

VOCABULARY

absolute value of a complex
  number (p. 586)
amplitude of a complex
  number (p. 587)
Argand plane (p. 586)
argument of a complex
  number (p. 587)
cardioid (p. 563)
complex conjugates (p. 582)
complex number (p. 580)
complex plane (p. 586)
escape set (p. 599)
imaginary number (p. 581)

imaginary part (p. 581)
iteration (p. 581)
Julia set (p. 599)
lemniscate (p. 564)
limaçon (p. 562)
modulus (p. 587)
polar axis (p. 553)
polar coordinates (p. 553)
polar equation (p. 556)
polar form of a complex
  number (p. 588)
polar graph (p. 556)

polar plane (p. 553)
pole (p. 553)
prisoner set (p. 599)
pure imaginary number
  (p. 581)
real part (p. 581)
rectangular form of a
  complex number (p. 581)
rose (p. 563)
spiral of Archimedes (p. 564)
trigonometric form of a
  complex number (p. 588)

UNDERSTANDING AND USING THE VOCABULARY

Choose the correct term to best complete each sentence.

1. The (<u>absolute value</u>, conjugate) of a complex number is its distance from zero in the complex plane.

2. (Complex, <u>Polar</u>) coordinates give the position of an object using distances and angles.

3. Points that do not escape under iteration belong to the (escape, <u>prisoner</u>) set.

4. The process of repeatedly applying a function to the output produced by the previous input is called (independent, <u>iteration</u>).

5. A complex number in the form bi where $b \neq 0$ is a (<u>pure imaginary</u>, real) number.

6. A (<u>cardioid</u>, rose) is a special type of limaçon.

7. The complex number $a + bi$, where a and b are real numbers, is in (polar, <u>rectangular</u>) form.

8. The (<u>spiral of Archimedes</u>, lemniscate) has a polar equation of the form $r = a\theta$.

9. The complex plane is sometimes called the (<u>Argand</u>, polar) plane.

10. The polar form of a complex number is given by $r(\cos\theta + i\sin\theta)$, where r represents the (<u>modulus</u>, amplitude) of the complex number.

 For additional review and practice for each lesson, visit: **www.amc.glencoe.com**

Six forms of Chapter 9 Test are
available in the *Assessment and
Evaluation Masters* as shown in
the chart.

Chapter 9 Test, Form 1B, is shown at
the right. Chapter 9 Test, Form 2B, is
shown on the next page.

TEST OPTIONS

Form	Pages	Type	
1A	211–212	Honors	*Form 1 tests are multiple choice.*
1B	213–214	Average	
1C	215–216	Basic	
2A	217–218	Honors	*Form 2 tests are free response.*
2B	219–220	Average	
2C	221–222	Basic	

STUDY GUIDE AND ASSESSMENT

VOCABULARY

This section provides a listing of the new terms, properties, and phrases that were introduced in this chapter. The exercises check students' understanding of the terms by using a variety of verbal formats including matching, completion, and true/false.

Glossary A complete glossary of terms begins on page A71.

***Assessment and Evaluation Masters*, pp. 213–214**

SKILLS AND CONCEPTS

Objectives and Examples This section reviews the skills and concepts of the chapter and shows completely worked examples.

Review Exercises These exercises provide practice for the corresponding objectives.

Assessment and Evaluation Masters, pp. 219–220

SKILLS AND CONCEPTS

OBJECTIVES AND EXAMPLES	REVIEW EXERCISES

Lesson 9-1 Graph points in polar coordinates.

• Graph $P\left(-2, \dfrac{5\pi}{6}\right)$.

Extend the terminal side of the angle measuring $\dfrac{5\pi}{6}$. Find the point P that is 2 units from the pole along the ray opposite the terminal side.

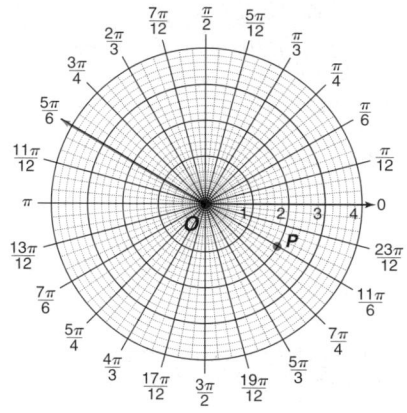

Graph each point. **11–14. See margin.**

11. $A(-3, 50°)$

12. $B(1.5, -110°)$

13. $C\left(2, \dfrac{\pi}{4}\right)$

14. $D\left(-3, \dfrac{\pi}{2}\right)$

15. Name four other pairs of polar coordinates for the point at $(4, 225°)$.
 Sample answer: (4, 585°), (4, 945°), (−4, 45°), (−4, 405°)

Graph each polar equation. **16–19. See margin.**

16. $r = \sqrt{7}$

17. $r = -2$

18. $\theta = -80°$

19. $\theta = \dfrac{3\pi}{4}$

Lesson 9-2 Graph polar equations.

• Graph $r = 3 + 3\sin\theta$.

The graph below can be made from a table of values. This graph is a cardioid.

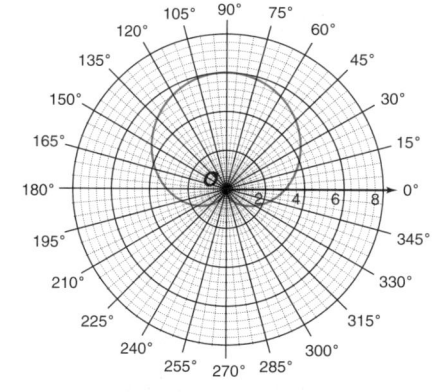

Graph each polar equation. Identify the type of curve each represents.

20. $r = 7\cos\theta$ **circle**

21. $r = 5\theta$ **spiral of Archimedes**

22. $r = 2 + 4\cos\theta$ **limaçon**

23. $r = 6\sin 2\theta$ **rose**

20–23. See margin for graphs.

608 Chapter 9 Polar Coordinates and Complex Numbers

TestCheck and Worksheet Builder

This **networkable** software has three modules.
• **Worksheet Builder** to make worksheets and tests
• **Student Module** to take tests on-screen
• **Management System** to keep student records
Three special banks are included for SAT, ACT, and TIMSS.

OBJECTIVES AND EXAMPLES

Lesson 9-3 Convert between polar and rectangular coordinates.

- Find the rectangular coordinates of $C\left(-3, \frac{\pi}{6}\right)$.

  For $C\left(-3, \frac{\pi}{6}\right)$, $r = -3$ and $\theta = \frac{\pi}{6}$.

  $x = r \cos \theta$ $\qquad$ $y = r \sin \theta$

  $= -3 \cos \frac{\pi}{6}$ $\qquad$ $= -3 \sin \frac{\pi}{6}$

  $= -\frac{3\sqrt{3}}{2}$ $\qquad$ $= -\frac{3}{2}$

  The rectangular coordinates of C are $\left(-\frac{3\sqrt{3}}{2}, -\frac{3}{2}\right)$, or about $(-2.6, -1.5)$.

REVIEW EXERCISES

Find the rectangular coordinates of each point with the given polar coordinates.

24. $(6, 45°)$ $\qquad$ 25. $(2, 330°)$ $\left(\sqrt{3}, -1\right)$

26. $\left(-2, \frac{3\pi}{4}\right)$ $\qquad$ 27. $\left(1, \frac{\pi}{2}\right)$ **(0, 1)**

$\left(\sqrt{2}, -\sqrt{2}\right)$ $\qquad$ 24. $\left(3\sqrt{2}, 3\sqrt{2}\right)$

Find the polar coordinates of each point with the given rectangular coordinates. Use $0 \leq \theta < 2\pi$ and $r \geq 0$.

28. $\left(-\sqrt{3}, -3\right)$ $\qquad$ 29. $(5, 5)$ $\left(5\sqrt{2}, \frac{\pi}{4}\right)$

30. $(-3, 1)$ $\qquad$ 31. $(4, 2)$

$\quad$ **(3.16, 2.82)** $\qquad$ **(4.47, 0.46)**

28. $\left(2\sqrt{3}, \frac{4\pi}{3}\right)$

Lesson 9-4 Write the polar form of a linear equation.

- Write $\sqrt{2}x - \sqrt{2}y = 6$ in polar form.

  The standard form of the equation is $\sqrt{2}x - \sqrt{2}y - 6 = 0$.

  Since $\sqrt{A^2 + B^2} = \sqrt{2 + 2}$ or 2, the normal form is $\frac{\sqrt{2}}{2}x - \frac{\sqrt{2}}{2}y - 3 = 0$.

  $\tan \phi = \frac{\sin \phi}{\cos \phi}$

  $\tan \phi = -1$

  $\phi = -45°$ or $315°$

  The polar form is $3 = r \cos (\theta - 315°)$.

Write each equation in polar form. Round ϕ to the nearest degree.

32. $2x + y = -3$ $\qquad$ $\frac{3\sqrt{5}}{5} = r \cos (\theta - 207°)$

33. $y = -3x - 4$ $\qquad$ $\frac{2\sqrt{10}}{5} = r \cos (\theta - 198°)$

Write each equation in rectangular form.

34. $3 = r \cos \left(\theta - \frac{\pi}{3}\right)$ $\qquad$ $x + \sqrt{3}y - 6 = 0$

35. $4 = r \cos \left(\theta + \frac{\pi}{2}\right)$ $\qquad$ $y + 4 = 0$

Lesson 9-5 Add, subtract, multiply, and divide complex numbers in rectangular form.

- Simplify $(2 - 4i)(5 + 2i)$.

  $(2 - 4i)(5 + 2i) = 2(5 + 2i) - 4i(5 + 2i)$

  $= 10 + 4i - 20i - 8i^2$

  $= 10 - 16i - 8(-1)$

  $= 18 - 16i$

Simplify.

36. $i^{10} + i^{25}$ $\quad$ $-1 + i$

37. $(2 + 3i) - (4 - 4i)$ $\quad$ $-2 + 7i$

38. $(2 + 7i) + (-3 - i)$ $\quad$ $-1 + 6i$

39. $i^3(4 - 3i)$ $\quad$ $-3 - 4i$

40. $(i - 7)(-i + 7)$ $\quad$ $-48 + 14i$

41. $\frac{4 + 2i}{5 - 2i}$ $\quad$ $\frac{16}{29} + \frac{18}{29}i$

42. $\frac{5 + i}{1 - \sqrt{2}i}$ $\quad$ $\frac{5 - \sqrt{2}}{3} + \frac{1 + 5\sqrt{2}}{3}i$

Additional Answers (p. 608)

14.

16.

17.

18.

19.

20.

(continued on page 610)

Additional Answers (p. 608)

11.

12.

13.

Additional Answers
(pp. 608, 610)

21.

22.

23.

43. $2\sqrt{2}\left(\cos\frac{\pi}{4} + i\sin\frac{\pi}{4}\right)$

44. $\sqrt{10}(\cos 5.03 + i\sin 5.03)$

45. $2\left(\cos\frac{2\pi}{3} + i\sin\frac{2\pi}{3}\right)$

46. $2\sqrt{13}(\cos 3.73 + i\sin 3.73)$

47. $\sqrt{17}(\cos 3.39 + i\sin 3.39)$

48. $4(\cos 0 + i\sin 0)$

49. $2\sqrt{2}(\cos\pi + i\sin\pi)$

50. $3\left(\cos\frac{\pi}{2} + i\sin\frac{\pi}{2}\right)$

51.

52.

OBJECTIVES AND EXAMPLES

Lesson 9-6 Convert complex numbers from rectangular to polar form and vice versa.

- Express $5 - 2i$ in polar form.

  Find the modulus.

  $r = \sqrt{5^2 + (-2)^2}$ or $\sqrt{29}$

  Find the amplitude. Since $5 - 2i$ is in Quadrant IV in the complex plane, θ is in Quadrant IV.

  $\theta = \text{Arctan}\frac{-2}{5}$ or about -0.38

  $5 - 2i = \sqrt{29}[\cos(-0.38) + i\sin(-0.38)]$

Lesson 9-7 Find the product and quotient of complex numbers in polar form.

- Find the product of $4\left(\cos\frac{\pi}{2} + i\sin\frac{\pi}{2}\right)$

  and $3\left(\cos\frac{3\pi}{4} + i\sin\frac{3\pi}{4}\right)$. Then express the product in rectangular form.

  $r = r_1 r_2 \qquad \theta = \theta_1 + \theta_2$

  $= 4(3) \qquad = \frac{\pi}{2} + \frac{3\pi}{4}$

  $= 12 \qquad = \frac{5\pi}{4}$

  The product is $12\left(\cos\frac{5\pi}{4} + i\sin\frac{5\pi}{4}\right)$.
  Now find the rectangular form of the product.

  $12\left(\cos\frac{5\pi}{4} + i\sin\frac{5\pi}{4}\right) = 12\left(-\frac{\sqrt{2}}{2} - \frac{\sqrt{2}}{2}i\right)$

  $= -6\sqrt{2} - 6\sqrt{2}i$

Lesson 9-8 Find powers and roots of complex numbers in polar form using De Moivre's Theorem.

- Find $(-3 + 3i)^4$.

  $(-3 + 3i)^4 = \left[3\sqrt{2}\left(\cos\frac{3\pi}{4} + i\sin\frac{3\pi}{4}\right)\right]^4$

  $= \left(3\sqrt{2}\right)^4(\cos 3\pi + i\sin 3\pi)$

  $= 324(-1 + 0)$ or -324

REVIEW EXERCISES

Express each complex number in polar form

43. $2 + 2i$ **44.** $1 - 3i$

45. $-1 + \sqrt{3}i$ **46.** $-6 - 4i$

47. $-4 - i$ **48.** 4

49. $-2\sqrt{2}$ **50.** $3i$

43–50. See margin.

Graph each complex number. Then express it in rectangular form. **See margin for graphs.**

51. $2\left(\cos\frac{\pi}{6} + i\sin\frac{\pi}{6}\right)$ $\sqrt{3} + i$

52. $3\left(\cos\frac{5\pi}{3} + i\sin\frac{5\pi}{3}\right)$ $\frac{3}{2} - \frac{3\sqrt{3}}{2}i$

Find each product or quotient. Express the result in rectangular form.

53. $4\left(\cos\frac{\pi}{3} + i\sin\frac{\pi}{3}\right) \cdot 3\left(\cos\frac{\pi}{3} + i\sin\frac{\pi}{3}\right)$

54. $8\left(\cos\frac{\pi}{4} + i\sin\frac{\pi}{4}\right) \cdot 4\left(\cos\frac{\pi}{2} + i\sin\frac{\pi}{2}\right)$

55. $2(\cos 2 + i\sin 2) \cdot 5(\cos 0.5 + i\sin 0.5)$

56. $8\left(\cos\frac{7\pi}{6} + i\sin\frac{7\pi}{6}\right) \div 2\left(\cos\frac{5\pi}{3} + i\sin\frac{5\pi}{3}\right)$

57. $6\left(\cos\frac{\pi}{2} + i\sin\frac{\pi}{2}\right) \div 4\left(\cos\frac{\pi}{6} + i\sin\frac{\pi}{6}\right)$

58. $2.2(\cos 1.5 + i\sin 1.5) \div$
4.4(\cos 0.6 + i\sin 0.6)$ $0.31 + 0.39i$

53. $-6 + 6\sqrt{3}i$ **54.** $-16\sqrt{2} + 16\sqrt{2}i$

55. $-8.01 + 5.98i$ **56.** $-4i$

57. $\frac{3}{4} + \frac{3\sqrt{3}}{4}i$

Find each power. Express the result in rectangular form. $-64\sqrt{3} + 64i$

59. $(2 + 2i)^8$ **4096**   **60.** $\left(\sqrt{3} - i\right)^7$

61. $(-1 + i)^4$ **62.** $(-2 - 2i)^3$
−4   **16 − 16i**

Find each principal root. Express the result in the form $a + bi$ with a and b rounded to the nearest hundredth.   **1.24 + 0.22i**

63. $i^{\frac{1}{4}}$ **0.92 + 0.38i**   **64.** $\left(\sqrt{3} + i\right)^{\frac{1}{3}}$

APPLICATIONS AND PROBLEM SOLVING

65. Chemistry An electron moves about the nucleus of an atom at such a high speed that if it were visible to the eye, it would appear as a cloud. Identify the classical curve represented by the electron cloud below. *(Lesson 9-2)* **lemniscate**

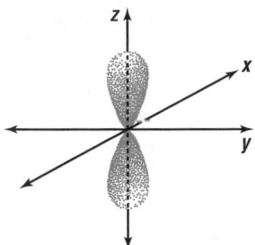

1a. Sample answer: $4 - 6i$ and $3 + 2i$

1b. No. Sample explanation: $2 - 3i$ and $5 - i$ also have this sum.

2a. Sample answer: $4 + i$

2b. No. Sample explanation: $1 + 4i$ also has this absolute value.

66. Surveying A surveyor identifies a landmark at the point with rectangular coordinates (75, 125). What are the polar coordinates of this point? *(Lesson 9-3)* **(145.77, 59.04°)**

67. Navigation A submarine sonar is tracking a ship. The path of the ship is being coded as the equation $r \cos\left(\theta - \dfrac{\pi}{2}\right) + 5 = 0$. Find the rectangular equation of the path of the ship. *(Lesson 9-4)* $y = -5$

68. Electricity Find the current in a circuit with a voltage of $50 + 180j$ volts and an impedance of $4 + 5j$ ohms. *(Lesson 9-7)* **26.83 + 11.46j amps**

ALTERNATIVE ASSESSMENT

OPEN-ENDED ASSESSMENT

1. The simplest form of the sum of two complex numbers is $7 - 4i$.

   a. Give examples of two complex numbers in which $a \neq 0$ and $b \neq 0$ with this sum.

   b. Are the two complex numbers you chose in part a the only two with this sum? Explain.

2. The absolute value of a complex number is $\sqrt{17}$.

   a. Give a complex number in which $a \neq 0$ and $b \neq 0$ with this absolute value.

   b. Is the complex number you chose in part a the only one with this absolute value? Explain.

> **Additional Assessment** See p. A64 for Chapter 9 practice test.

Unit 3 *inter*NET Project

SPACE—THE FINAL FRONTIER

Coordinates in Space—What do scientists use?

- Search the Internet to find types of coordinate systems that are used to locate objects in space. Find at least two different types.

- Write a summary that describes each coordinate system that you found. Compare them to rectangular and polar coordinates. Include diagrams illustrating how to use each coordinate system and any information you found about converting between systems.

PORTFOLIO

Choose one of the classical curves you studied in this chapter. Give the possible general forms of the polar equation for your curve and sketch the general graph. Then write and graph a specific polar equation.

Chapter 9 Study Guide and Assessment **611**

Internet Address Book

Record useful Internet addresses in the space at right for quick reference.

www.amc.glencoe.com

Chapter 9 Skill: Geometry Problems (Lines, Angles, and Arcs)

Pages 612–613 are part of a complete SAT/ACT preparation course that is described in detail on pages T8–T9 of the *Teacher's Handbook*. The test items were written in the same style as those in the SAT and ACT.

 These questions were aligned and verified by The Princeton Review, the nation's leader in test preparation.

Teaching Tip Remind students that angle relationships involving lines frequently involve parallel lines and a transversal. Review all of the types of angles that can be formed by two parallel lines and a transversal.

Diagnosis & Prescription

Each of the 10 SAT and ACT practice problems on page 613 is cross-referenced to the chapter where that SAT or ACT skill is covered. If students miss a particular type of problem, you can have them study that skill.

Problem Number and Skill	Page
1 Geometry (Lines, Angles, and Arcs)	612
2 Trigonometry	418
3 Geometry (Lines, Angles, and Arcs)	612
4 Geometry (Perimeter, Area, and Volume)	548
5 Arithmetic	64
6 Geometry (Lines, Angles, and Arcs)	612
7 More Algebra	202
8 Geometry (Lines, Angles, and Arcs)	612
9 Geometry (Lines, Angles, and Arcs)	612
10 Geometry (Lines, Angles, and Arcs)	612

Geometry Problems— Lines, Angles, and Arcs

SAT and ACT geometry problems often combine triangles and quadrilaterals with angles and parallel lines. Problems that deal with circles often include arcs and angles.

Review these concepts.
- Angles: vertical, supplementary, complementary
- Parallel Lines: transversal, alternate interior angles
- Circles: inscribed angle, central angle, arc length, tangent line

THE PRINCETON REVIEW

TEST-TAKING TIP
Use the square corner of a sheet of paper to estimate angle measure. Fold the corner in half to form a 45° angle.

SAT EXAMPLE

1. If four lines intersect as shown in the figure, $x + y =$

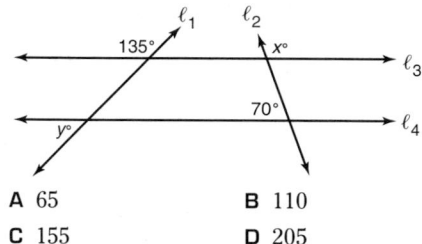

A 65 **B** 110
C 155 **D** 205
E It cannot be determined from the information given.

HINT Do not assume anything from a figure. In this figure, ℓ_3 and ℓ_4 look parallel, but that information is not given.

Solution Notice that the lines intersect to form a quadrilateral. The sum of the measures of the interior angles of a quadrilateral is 360°. Use vertical angles to determine the angles in the quadrilateral.

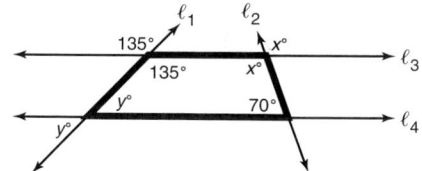

Write an equation for the sum of the angles.
$$135 + x + 70 + y = 360$$
$$205 + x + y = 360$$
$$x + y = 155$$
The answer is choice **C**.

612 *Chapter 9 Polar Coordinates and Complex Numbers*

ACT EXAMPLE

2. In the figure below, *ABCD* is a square inscribed in the circle centered at *O*. If $\overline{OB}$ is 6 units long, how many units long is minor arc $\overarc{BC}$?

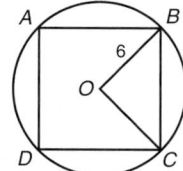

A $\dfrac{3\pi}{2}$ **B** 3π **C** 6π **D** 12π **E** 36π

HINT Consider *all* of the information given and implied. For example, a square's diagonals are perpendicular.

Solution Figure *ABCD* is a square, so its angles are each 90°, and its diagonals are perpendicular. So $\angle BOC = 90°$. This means that minor arc $\overarc{BC}$ is one-fourth of the circle.

Look at the answer choices. They all include π. Use the formula for the circumference of a circle.

$$C = 2\pi r$$
$$= 2\pi(6) \text{ or } 12\pi$$

The length of minor arc $\overarc{BC}$ is one-fourth of the circumference of the circle.

$$\tfrac{1}{4}C = \tfrac{1}{4}(12\pi) \text{ or } 3\pi$$

The answer is choice **B**.

TestCheck and Worksheet Builder

Special banks of SAT, ACT, and TIMSS items are included in this CD-ROM.

After you work each problem, record your answer on the answer sheet provided or on a piece of paper.

Multiple Choice

1. In the figure below, line L is parallel to line M. Line N intersects both L and M, with angles a, b, c, d, e, f, g, and h as shown. Which of the following lists includes all of the angles that are supplementary to $\angle a$? **A**

  A b, d, f, h **B** c, e, g **C** b, d, c

  D e, f, g, h **E** d, c, h, g

2. In the figure below, what is the area of $\triangle ABC$ in terms of x? **B**

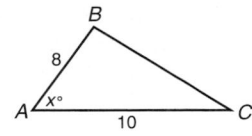

  A $10 \sin x$ **B** $40 \sin x$ **C** $80 \sin x$

  D $40 \cos x$ **E** $80 \cos x$

3. If $PQRS$ is a parallelogram and $\overline{MN}$ is a line segment, then x must equal **E**

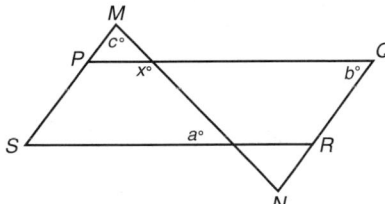

  A $180 - b$ **B** $180 - c$ **C** $a + b$

  D $a + c$ **E** $b + c$

4. If a rectangular swimming pool has a volume of 16,500 cubic feet, a depth of 10 feet, and a length of 75 feet, what is the width of the pool, in feet? **A**

  A 22   **B** 26   **C** 32   **D** 110   **E** 1650

5. $\dfrac{1}{10^{100}} - \dfrac{1}{10^{99}} =$ **A**

  A $\dfrac{-9}{10^{100}}$ **B** $\dfrac{-1}{10^{100}}$ **C** $\dfrac{1}{10^{100}}$ **D** $\dfrac{1}{10}$ **E** $\dfrac{9}{10}$

6. In the figure, what is the sum of the degree measures of the marked angles? **C**

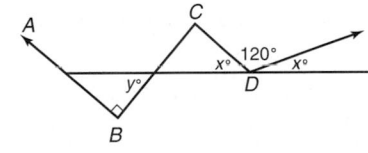

  A 180   **B** 270   **C** 360   **D** 540

  E It cannot be determined from the information given.

7. If $5x^2 + 6x = 70$ and $5x^2 - 6y = 10$, then what is the value of $10x + 10y$? **E**

  A 10   **B** 20   **C** 60   **D** 80   **E** 100

8. In the figure below, if $\overline{AB} \parallel \overline{CD}$, then what is the value of y? *Figure not drawn to scale.* **B**

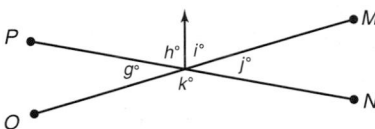

  A 30   **B** 60   **C** 90   **D** 120   **E** 150

9. Which pair must be equal? **D**

  A h and i

  B $(g + h)$ and $(i + j)$

  C $(g + i)$ and $(h + j)$

  D g and j

  E $(g + j)$ and $(h + i)$

10. **Grid-In** If ℓ_1 is parallel to ℓ_2 in the figure below, what is the value of y? **145**

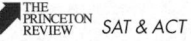

interNET CONNECTION **SAT/ACT Practice** For additional test practice questions, visit: **www.amc.glencoe.com**

THE **PRINCETON REVIEW**   *SAT & ACT Preparation*   **613**

A bubble-in answer sheet for these practice problems is available on p. 1 of the *Assessment and Evaluation Masters.*

Additional Practice

Additional SAT and ACT practice questions and a bubble-in answer sheet are available in the *Assessment and Evaluation Masters,* pp. 2, 227, and 228.

Assessment and Evaluation Masters, **pp. 227–228**

Extra Credit

TIMSS Determine all complex numbers z that satisfy the equation $z + 2\bar{z} = 3 + i$, where $\bar{z}$ denotes the conjugate of z. **$z = 1 - i$**

Instructional Objectives

Lesson (pages)	Lesson Objectives	NCTM Standards 2000	State/Local Objectives
10-1 (615–622)	• Find the distance and midpoint between two points on a coordinate plane. • Prove geometric relationships among points and lines using analytical methods.	2, 3, 10	
10-2 (623–630)	• Use and determine the standard and general forms of the equation of a circle.	2, 3, 10	
10-3 (631–641)	• Use and determine the standard and general forms of the equation of an ellipse. • Graph ellipses.	2, 3, 10	
10-4 (642–652)	• Use and determine the standard and general forms of the equation of a hyperbola. • Graph hyperbolas.	2, 3, 10	
10-5 (653–661)	• Use and determine the standard and general forms of the equation of a parabola. • Graph parabolas.	2, 3, 10	
10-6 (662–669)	• Recognize conic sections in their rectangular form by their equations. • Find a rectangular equation for a curve defined parametrically and vice versa.	2, 3, 6, 10	
10-7 (670–677)	• Find the equations of conic sections that have been translated or rotated. • Graph rotations and/or translations of conic equations. • Identify the equations of conic sections using the discriminant. • Find the angle of rotation for a given equation.	2, 3, 10	
10-8 (678–684)	• Graph and solve systems of second degree equations and inequalities.	2, 3, 10	
10-8B (685–686)	• Graph a system of second-degree inequalities using the **Shade(** command.	2, 3, 10	

Suggested Pacing   *See pages T2–T4 for additional options in course-planning.*

COURSE	Lesson 10-1	Lesson 10-2	Lesson 10-3	Lesson 10-4	Lesson 10-5
Precalculus Emphasis Chapters 4–15 (45–50 minutes)	1.5 days	1.5 days	1.5 days	1.5 days	1.5 days
Precalculus Emphasis (block)	0.5 day	1.5 days		1 day	(with Lesson 10-6) 1.5 days
Trigonometric Emphasis Chapters 1–9, 11, 12 (45–50 minutes)	OPTIONAL				
Trigonometric Emphasis (block)	OPTIONAL				

Instructional Resources

	Student Edition	Blackline Masters (page numbers)								
Lesson	Graphing Calculator Applications	Study Guide and Practice	Enrichment	Assessment and Evaluation	Graphing Calculator Keystrokes	Problem Solving & Applications	Spreadsheet Applications	TI-92 and CBL Interdisciplinary Activities	Transparencies A and B
10-1	620	177–178	109						10-1
10-2	625, 628	179–180	110	251					10-2
10-3	635, 639, 641	181–182	111		38				10-3
10-4	645	183–184	112	250, 251					10-4
10-5	655	185–186	113			20			10-5
10-6	665, 668	187–188	114	252	40		10		10-6
10-7	676	189–190	115			19		33–36	10-7
10-8	679	191–192	116	252					10-8
10-8B	685–686				41–43				
Study Guide/ Assessment				237–249, 255					
SAT/ACT Practice				253–254					

Lesson 10-6	Lesson 10-7	Lesson 10-8	Study Guide & Assessment	Chapter Test	TOTAL
1.5 days	1.5 days	(with Lesson 10-8B) 2.5 days	1 day	1 day	15 days
(See Lesson 10-5)	1.5 days		(with Lesson 10-8B) 1.5 days	0.5 day	8 days
OPTIONAL					
OPTIONAL					

The worksheets shown on this page are a small sample of the materials available on *TeacherWorks: All-in-One Lesson Planner and Resource Center.*

This CD-ROM contains all of the blackline masters and transparencies available for this program.

It also includes a lesson planner and interactive Teacher Edition, so you can customize lesson plans and reproduce classroom resources quickly and easily, from just about anywhere.

Applications

Problem Solving and Applications Masters, pp. 19, 20

Technology/Multimedia

Spreadsheet Applications Masters, p. 10

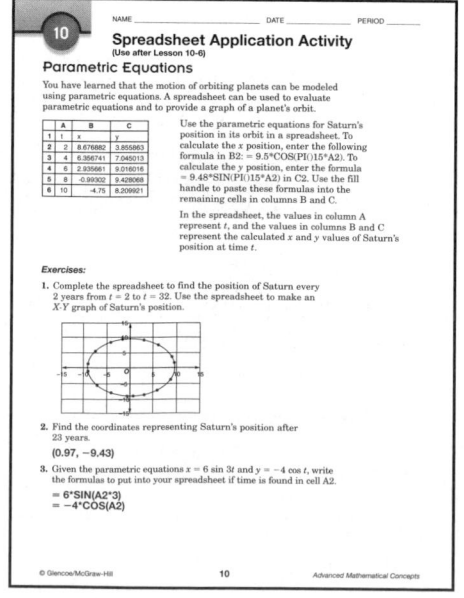

TI-92 & CBL Interdisciplinary Activities Masters, pp. 33–36

Type	Student Edition	Teacher's Wraparound Edition	Assessment and Evaluation Masters
Ongoing Assessment	Mid-Chapter Quiz, p. 661	5-Minute Check, pp. 615, 623, 631, 642, 653, 662, 670, 678	Mid-Chapter Test, p. 250 Quizzes A–D, pp. 251–252
Mixed Review	Mixed Review, pp. 621–622, 630, 640–641, 652, 661, 669, 677, 684		Cumulative Review, p. 255
Error Analysis	You Decide, pp. 627, 637, 675	Error Analysis, pp. 619, 627, 637, 649, 658, 667, 674, 682	
Standardized Test Practice	SAT/ACT Practice, pp. 622, 630, 641, 652, 661, 669, 677, 684 SAT & ACT Preparation, pp. 692–693	TIMSS, p. 693	SAT/ACT Practice, pp. 253–254
Open-Ended Assessment	Math Journal, pp. 649, 682 Open-Ended Assessment, p. 691 Portfolio, p. 691	Modeling: pp. 661, 684, 686 Speaking: pp. 622, 630, 652 Writing: pp. 641, 669, 677	Open-Ended Assessment, p. 249
Chapter Assessment	Study Guide and Assessment, pp. 687–691 Chapter Test, p. A65		Multiple-Choice Tests (Forms 1A, 1B, 1C), pp. 237–242 Free-Response Tests (Forms 2A, 2B, 2C), pp. 243–248

TestCheck and Worksheet Builder is a networkable CD-ROM with three integrated modules. The **Worksheet Builder** creates customized worksheets, tests, and quizzes of free-response, multiple-choice, short-answer, and open-ended items. The **Student Module** gives you the option of having students take tests on-screen and get immediate feedback on their performance. Use the optional **Management System** to keep detailed student records.

Additional Chapter Resources

Student Edition
Career Choices, p. 622
Internet Project, p. 691
Extra Practice, pp. A44–45

Teacher's Classroom Resources

Technology
Graphing Calculator
Keystrokes Masters

Teaching Aids
Answer Key Transparencies
Block Schedule Planning Guide
Lesson Planning Guide
Solutions Manual

Glencoe Instructional Technology

Instructional

Advanced MathPASS CD-ROM

Multimedia Applications CD-ROM

Assessment

TestCheck and Worksheet Builder

Visit **www.amc.glencoe.com** for data updates, career information, graphing calculator programs, and other interactive activities.

Chapter Overview

This chapter introduces conic sections. Students use equations to identify important characteristics of each conic section. Then they use these characteristics to graph parabolas, circles, ellipses, and hyperbolas.

Students then will find parametric forms and transformations of conic sections. Students graph systems of second-degree equations and inequalities.

 Using a Graphing Calculator

There are numerous opportunities, all optional, to use a graphing calculator in this chapter.

- Determining the distance and midpoint between two points (p. 620)
- Graphing a circle, ellipse, hyperbola, and parabola (pp. 625, 628, 635, 639, 645, 655)
- Determining the radius and the coordinates of the center of a circle from an equation written in general form (p. 628)
- Examining families of ellipses (p. 641)
- Graphing parametric equations (pp. 665, 668)
- Graphing second-degree inequalities (p. 676)
- Approximating intersections of conics (p. 679)

The **Graphing Calculator Appendix**, which begins on page A2, contains keystroke instructions on many of these topics.

CONICS

CHAPTER OBJECTIVES

- Use analytic methods to prove geometric relationships. *(Lesson 10-1)*
- Use the standard and general forms of the equations of circles, parabolas, ellipses, and hyperbolas. *(Lessons 10-2, 10-3, 10-4, and 10-5)*
- Graph circles, parabolas, ellipses, and hyperbolas. *(Lessons 10-2, 10-3, 10-4, and 10-5)*
- Find the eccentricity of conic sections. *(Lessons 10-2, 10-3, 10-4, and 10-5)*
- Recognize conic sections by their equations. *(Lesson 10-6)*
- Find parametric equations for conic sections defined by rectangular equations and vice versa. *(Lesson 10-6)*
- Find the equations of conic sections that have been translated or rotated. *(Lesson 10-7)*
- Graph and solve systems of second-degree equations and inequalities. *(Lesson 10-8)*

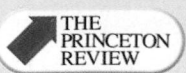 **SAT & ACT Preparation**

This textbook contains a complete SAT/ACT preparation course described on pages T8–T9 of the *Teacher's Handbook*. This chapter contains:

- an SAT or ACT practice question at the end of each lesson
- examples of SAT and ACT problems involving ratio and proportion on page 692
- ten SAT and ACT practice problems on page 693
- additional SAT and ACT practice in the *Assessment and Evaluation Masters,* pp. 253–254
- SAT and ACT test practice banks in the *TestCheck and Worksheet Builder*

Introduction to Analytic Geometry

OBJECTIVES
- Find the distance and midpoint between two points on a coordinate plane.
- Prove geometric relationships among points and lines using analytical methods.

SEARCH AND RESCUE The *Absaroka Search Dogs* has provided search teams, each of which consists of a canine and its handler, to assist in lost person searches throughout the Montana and Wyoming area since 1986. While the dogs use their highly sensitive noses to detect the lost individual, the handlers use land navigation skills to insure that they do not become lost themselves. A handler needs to be able to read a map and calculate distances. *A problem related to this will be solved in Example 2.*

The distance between two points on a number line can be found by using absolute value. Let A and B be two points with coordinates a and b, respectively.

Distance between A and B $\Rightarrow$ $|a - b|$ or $|b - a|$

The distance between two points in the coordinate plane can also be found. Consider points $J(-4, 1)$ and $K(5, 9)$. To find JK, first choose a point L such that $\overline{JL}$ is parallel to the x-axis and $\overline{KL}$ is parallel to the y-axis. In this case, L has coordinates $(5, 1)$. Since K and L lie along the line $x = 5$, KL is equal to the absolute value of the difference in the y-coordinates of K and L, $|9 - 1|$. Similarly, JL is equal to the absolute value of the difference in the x-coordinates of J and L, $|5 - (-4)|$.

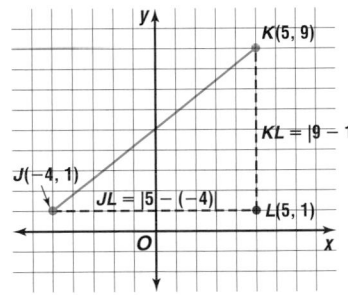

Since $\triangle JKL$ is a right triangle, JK can be found using the Pythagorean Theorem.

$$(JK)^2 = (KL)^2 + (JL)^2 \qquad \textit{Pythagorean Theorem}$$
$$JK = \sqrt{(KL)^2 + (JL)^2} \qquad \textit{Take the positive square root of each side.}$$
$$JK = \sqrt{|9 - 1|^2 + |5 - (-4)|^2} \qquad KL = |9 - 1|,\ JL = |5 - (-4)|$$
$$JK = \sqrt{8^2 + 9^2} \qquad |9 - 1|^2 = 8^2,\ |5 - (-4)|^2 = 9^2$$
$$JK = \sqrt{145} \text{ or about } 12$$

$\overline{JK}$ is about 12 units long.

---

1 FOCUS

5-Minute Check
Chapter 9

1. Find the polar coordinates of $S(\sqrt{3}, 1)$. **(2, 30°)**
2. Identify the curve represented by $r = \sin \theta$. **circle**
3. Write $x^2 + y^2 = 25$ in polar form. $r = \pm 5$
4. Simplify $(2 - i)(-3 + 4i)$. $-2 + 11i$
5. Write the polar equation $r = 3 \sin \theta$ in rectangular form. $x^2 + y^2 = 3y$

Motivating the Lesson
Real-World Application Show students a map of the area around the school. Ask students how they would determine the distance between two landmarks on the map. Have students brainstorm how to measure the distance without physically measuring it outdoors.

Resource Manager

Reproducible Masters
- *Study Guide and Practice,* pp. 177–178
- *Enrichment,* p. 109

Transparencies
- 5-Minute Check, 10-1
- Teaching, 10-1

Teaching Tip In Example 1, use the distance formula to find the distance between two points. Remind students that either ordered pair may be used for (x_1, y_1).

In-Class Examples

Example 1
Find the distance between points at (4, –2) and (8, 3).
$\sqrt{41}$

Example 2
Search and Rescue More than 3 million people visited Yellowstone National Park in Wyoming and Idaho in 1996. Sometimes visitors become lost or injured and must be rescued. Suppose that the grid in Example 2 in the Student Edition shows a section of Yellowstone. If the search team is located at (3, –1.5) and an injured visitor is located at (–2.5, 1.5), how far is the search team from the missing person? **93.97 m**

Teaching Tip In Example 2, make sure that students remember to multiply the grid distance by the kilometers-per-grid unit.

From this specific case, we can derive a formula for the distance between any two points. In the figure, assume (x_1, y_1) and (x_2, y_2) represent the coordinates of any two points in the plane.

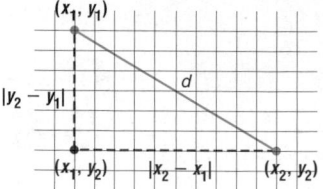

$d = \sqrt{|x_2 - x_1|^2 + |y_2 - y_1|^2}$ *Pythagorean Theorem*

$d = \sqrt{(x_2 - x_1)^2 + (y_2 - y_1)^2}$ *Why does* $|x_2 - x_1|^2 + |y_2 - y_1|^2 = (x_2 - x_1)^2 + (y_2 - y_1)^2?$

Distance Formula for Two Points	The distance, d units, between two points with coordinates (x_1, y_1) and (x_2, y_2) is given by $d = \sqrt{(x_2 - x_1)^2 + (y_2 - y_1)^2}$.

Examples   **①** **Find the distance between points at (−3, 7) and (2, −5).**

$d = \sqrt{(x_2 - x_1)^2 + (y_2 - y_1)^2}$ *Distance Formula*

$d = \sqrt{(2 - (-3))^2 + (-5 - 7)^2}$ *Let $(x_1, y_1) = (-3, 7)$ and $(x_2, y_2) = (2, -5)$.*

$d = \sqrt{5^2 + (-12)^2}$

$d = \sqrt{169}$ or 13

The distance is 13 units.

② **SEARCH AND RESCUE**   **Refer to the application at the beginning of the lesson. Suppose a backpacker lies injured in the region shown on the map at the right. Each side of a square on the grid represents 15 meters. An Absaroka team searching for the missing individual is located at (−1.5, 4.0) on the map grid while the injured person is located at (2.0, 2.8). How far is the search team from the missing person?**

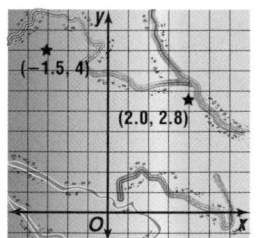

Use the distance formula to find the distance between (2.0, 2.8) and (−1.5, 4.0).

$d = \sqrt{(x_2 - x_1)^2 + (y_2 - y_1)^2}$

$d = \sqrt{(-1.5 - 2.0)^2 + (4.0 - 2.8)^2}$ *Let $(x_1, y_1) = (2.0, 2.8)$ and $(x_2, y_2) = (-1.5, 4.0)$.*

$d = 13.69$ or 3.7

The map distance is 3.7 units. Each unit equals 15 kilometers. So, the actual distance is about 3.7(15) or 55.5 kilometers.

Prerequisite Skills

A review of the prerequisite skills for this lesson is provided in the *Prerequisite Skills Masters.*

Skill Number and Content	Pages
1 Properties of Real Numbers	3–4
3 Order of Operations	7–8
9 Factoring by Grouping and Greatest Common Factor	19–20

You can use the Distance Formula and what you know about slope to investigate geometric figures on the coordinate plane.

Example

③ Determine whether quadrilateral $ABCD$ with vertices $A(3, 2)$, $B(2, -4)$, $C(-2, -3)$, and $D(-1, 3)$ is a parallelogram.

Recall that a quadrilateral is a parallelogram if one pair of opposite sides are parallel and congruent.

First, graph the figure. $\overline{DA}$ and $\overline{CB}$ are one pair of opposite sides.

To determine if $\overline{DA} \parallel \overline{CB}$, find the slopes of $\overline{DA}$ and $\overline{CB}$.

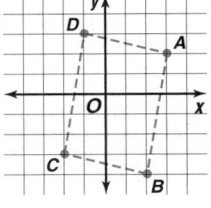

Look Back
Refer to Lesson 1-3 to review the slope formula.

slope of $\overline{DA}$

$m = \dfrac{y_2 - y_1}{x_2 - x_1}$ *Slope formula*

$= \dfrac{2 - 3}{3 - (-1)}$ *Let $(x_1, y_1) = (-1, -3)$ and $(x_2, y_2) = (3, 2)$.*

$= -\dfrac{1}{4}$

slope of $\overline{CB}$

$m = \dfrac{y_2 - y_1}{x_2 - x_1}$ *Slope formula*

$= \dfrac{-4 - (-3)}{2 - (-2)}$ *Let $(x_1, y_1) = (-2, -3)$ and $(x_2, y_2) = (2, -4)$.*

$= -\dfrac{1}{4}$

Their slopes are equal. Therefore, $\overline{DA} \parallel \overline{CB}$.

To determine if $\overline{DA} \cong \overline{CB}$, use the distance formula to find DA and CB.

$DA = \sqrt{(x_2 - x_1)^2 + (y_2 - y_1)^2}$ $CB = \sqrt{(x_2 - x_1)^2 + (y_2 - y_1)^2}$

$\quad = \sqrt{[3 - (-1)]^2 + (2 - 3)^2}$ $\quad = \sqrt{[2 - (-2)]^2 + [-4 - (-3)]^2}$

$\quad = \sqrt{17}$ $\quad = \sqrt{17}$

The measures of $\overline{DA}$ and $\overline{CB}$ are equal. Therefore, $\overline{DA} \cong \overline{CB}$.

Since $\overline{DA} \parallel \overline{CB}$ and $\overline{DA} \cong \overline{CB}$, quadrilateral $ABCD$ is a parallelogram. *You can also check your work by showing $\overline{DC} \parallel \overline{AB}$ and $\overline{DC} \cong \overline{AB}$.*

In addition to finding the distance between two points, you can use the coordinates of two points to find the midpoint of the segment between the points. In the figure at the right, the midpoint of $\overline{P_1P_2}$ is P_m. Notice that the x-coordinate of P_m is the average of the x-coordinates of P_1 and P_2. The y-coordinate of P_m is the average of the y-coordinates of P_1 and P_2.

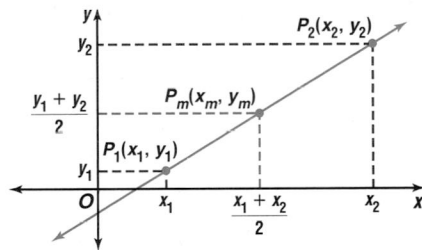

Teaching Tip For Example 3, make sure that students do not identify the figure based on appearance alone. They should verify whether one pair of opposite sides is parallel and congruent.

In-Class Example

Example 3
Determine whether quadrilateral $PQRS$ with vertices $P(-4, 2)$, $Q(-3, -2)$, $R(3, -3)$, and $S(1, 5)$ is a parallelogram. **no**

Example 4
Find the coordinates of the midpoint of the segment that has endpoints at (2, 5) and (−4, −7). **(−1, −1)**

Example 5
Prove that the diagonals of a square are perpendicular bisectors of each other. **Let A, B, C, and D be the vertices of a square as shown below.**

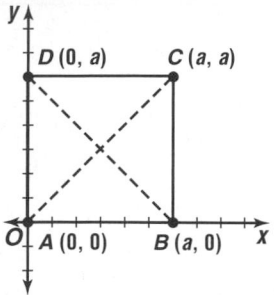

The midpoint of $\overline{AC}$ is $\left(\frac{a}{2}, \frac{a}{2}\right)$.

The midpoint of $\overline{BD}$ is $\left(\frac{a}{2}, \frac{a}{2}\right)$.

Since the midpoints coincide, the diagonals bisect each other.

slope of $\overline{AC} = \frac{a-0}{a-0} = \frac{a}{a} = 1$

slope of $\overline{BD} = \frac{a-0}{0-a} = \frac{a}{-a} = -1$

Since the slopes are negative reciprocals, $\overline{AC}$ and $\overline{BD}$ are perpendicular.

Teaching Tip Make sure that students do not draw figures to include assumptions that are not presented in the proof. For example, they should not draw an isosceles trapezoid when the proof only says *trapezoid*.

Midpoint of a Line Segment — If the coordinates of P_1 and P_2 are (x_1, y_1) and (x_2, y_2), respectively, then the midpoint of $\overline{P_1P_2}$ has coordinates $\left(\frac{x_1 + x_2}{2}, \frac{y_1 + y_2}{2}\right)$.

Example ④ Find the coordinates of the midpoint of the segment that has endpoints at (−2, 4) and (6, −5).

Let $(−2, 4)$ be (x_1, y_1) and $(6, −5)$ be (x_2, y_2). Use the Midpoint Formula.

$$\left(\frac{x_1 + x_2}{2}, \frac{y_1 + y_2}{2}\right) = \left(\frac{-2 + 6}{2}, \frac{4 + (-5)}{2}\right)$$
$$= \left(2, -\frac{1}{2}\right)$$

The midpoint of the segment is at $\left(2, -\frac{1}{2}\right)$.

Many theorems from plane geometry can be more easily proven by analytic methods. That is, they can be proven by placing the figure in a coordinate plane and using algebra to express and draw conclusions about the geometric relationships. The study of coordinate geometry from an algebraic perspective is called **analytic geometry.**

When using analytic methods to prove theorems from geometry, the position of the figure in the coordinate plane can be arbitrarily selected as long as size and shape are preserved. This means that the figure may be translated, rotated, or reflected from its original position. For polygons, one vertex is usually located at the origin, and one side coincides with the *x*-axis, as shown below.

In a right triangle, the legs are on the axes.

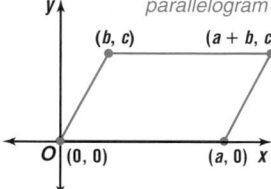

Example ⑤ Prove that the measure of the median of a trapezoid is equal to one half of the sum of the measures of the two bases.

In trapezoid *ABCD*, choose two vertices as $A(0, 0)$ and $B(a, 0)$. Since $\overline{AB} \parallel \overline{DC}$, $\overline{DC}$ lies on a horizontal grid line of the coordinate plane. Therefore, *C* and *D* must have the same *y*-coordinate. Choose arbitrary letters to represent the *y*-coordinates, and the two *x*-coordinates; in this case, $D(b, c)$ and $C(d, c)$. Let *E* be the midpoint of $\overline{AD}$, and let *F* be the midpoint of $\overline{BC}$.

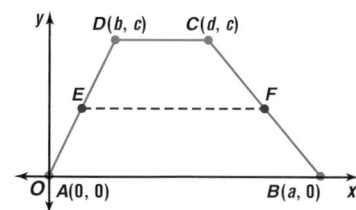

618 Chapter 10 Conics

Now, find the coordinates of E and F by using the Midpoint Formula.

The coordinates of E are $\left(\dfrac{b+0}{2}, \dfrac{c+0}{2}\right)$ or $\left(\dfrac{b}{2}, \dfrac{c}{2}\right)$.

The coordinates of F are $\left(\dfrac{d+a}{2}, \dfrac{c+0}{2}\right)$ or $\left(\dfrac{d+a}{2}, \dfrac{c}{2}\right)$.

Then find the measures of each base and the median by using the Distance Formula.

$DC = \sqrt{(d-b)^2 + (c-c)^2}$ or $d - b$

$AB = \sqrt{(a-0)^2 + (0-0)^2}$ or a

$EF = \sqrt{\left(\dfrac{d+a}{2} - \dfrac{b}{2}\right)^2 + \left(\dfrac{c}{2} - \dfrac{c}{2}\right)^2}$ or $\dfrac{1}{2}(d - b + a)$

Calculate one half of the sum of the measures of the bases.

$\dfrac{1}{2}(DC + AB) = \dfrac{1}{2}(d - b + a)$ $DC = d - b,\ AB = a$

Since both $\dfrac{1}{2}(DC + AB)$ and EF equal $\dfrac{1}{2}(d - b + a)$, it follows that $\dfrac{1}{2}(DC + AB) = EF$. Therefore, the measure of the median of a trapezoid is one half of the sum of the measures of its bases.

CHECK FOR UNDERSTANDING

Communicating Mathematics

Read and study the lesson to answer each question.

1. **Explain** why only the positive square root is considered when applying the distance formula.   **negative distances have no meaning**

2. **Describe** how can you show that a midpoint of a segment is equidistant from its endpoints given the coordinates of each point.   **See margin.**

3. **Determine** whether each diagram represents an isosceles triangle. Explain your reasoning.   **a–c. See margin for explanations.**

a.

yes

b.

yes

c.
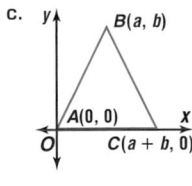
no

4. **Describe** four different ways of proving that a quadrilateral is a parallelogram if you are given the coordinates of its vertices.   **See margin.**

Guided Practice

Find the distance between each pair of points with the given coordinates. Then, find the coordinates of the midpoint of the segment that has endpoints at the given coordinates.

5. $(5, 1), (5, 11)$
   10; (5, 6)

6. $(0, 0), (-4, -3)$
   5; (−2, −1.5)

7. $(-2, 2), (0, 4)$
   $2\sqrt{2}$; (−1, 3)

Lesson 10-1   Introduction to Analytic Geometry   **619**

Additional Answers

2. Use the distance formula to show that the measure of the distance from the midpoint to either endpoint is the same.

3a. Yes; the distance from B to A is $\dfrac{a\sqrt{5}}{2}$ and the distance from B to C is also $\dfrac{a\sqrt{5}}{2}$.

3b. Yes; the distance from B to A is $\sqrt{a^2 + b^2}$ and the distance from C to A is also $\sqrt{a^2 + b^2}$.

3c. No; the distance from A to B is $\sqrt{a^2 + b^2}$, the distance from A to C is $a + b$, and the distance from B to C is $b\sqrt{2}$.

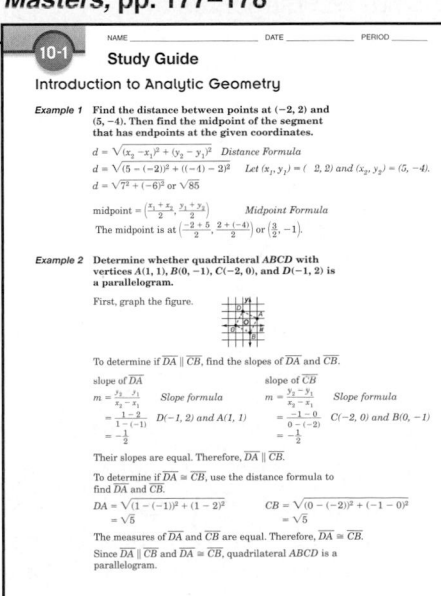

Additional Answers

8. $\overline{AB} \cong \overline{DC}$ since $AB = \sqrt{13}$ and $DC = \sqrt{13}$, and $\overline{AB} \parallel \overline{DC}$ since the slope of $\overline{AB}$ is $-\frac{2}{3}$ and the slope of $\overline{DC}$ is also $-\frac{2}{3}$.

9. $\overline{XY} \cong \overline{XZ}$, since $XY = 2\sqrt{17}$ and $DC = 2\sqrt{17}$, therefore, $\triangle XYZ$ is isosceles.

10a.

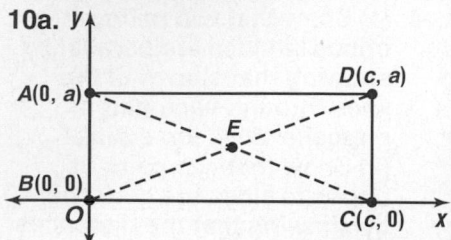

10b. $BD = \sqrt{(c-0)^2 + (a-0)^2}$
$= \sqrt{c^2 + a^2}$
$AC = \sqrt{(c-0)^2 + (0-a)^2}$
$= \sqrt{c^2 + a^2}$

Thus, $\overline{AC} \cong \overline{BD}$.

10c. The midpoint of $\overline{AC}$ is $\left(\frac{c}{2}, \frac{a}{2}\right)$.

The midpoint of $\overline{BD}$ is $\left(\frac{c}{2}, \frac{a}{2}\right)$.

Therefore, the diagonals intersect at their common midpoint, $E\left(\frac{c}{2}, \frac{a}{2}\right)$. Thus, $\overline{AE} \cong \overline{EC}$ and $\overline{BE} \cong \overline{ED}$.

10d. The diagonals of a rectangle are congruent and bisect each other.

12. 13; (1.5, 7)

13. $2\sqrt{10}$; (0, 0)

14. $8\sqrt{2}$; (4, 4)

15. $3\sqrt{5}$; (2, –4.5)

16. $4\sqrt{3}$; $(5\sqrt{2}, -3)$

17. 16; (a, –1)

18. 8; (r + 2, s)

19. $\sqrt{5}$; $\left(c + 1, d - \frac{1}{2}\right)$

20. $\sqrt{9w^2 + 4}$; $\left(w - 1, \frac{5}{2}w\right)$

26. $AB = 4$, $BC = 4$, and $CA = 4$. Thus, $\overline{AB} \cong \overline{BC} \cong \overline{CA}$. Therefore the points A, B, and C form an equilateral triangle.

8–9. See margin for justifications.

8. Determine whether the quadrilateral $ABCD$ with vertices $A(3, 4)$, $B(6, 2)$, $C(8, 7)$, and $D(5, 9)$ is a parallelogram. Justify your answer. **yes**

9. Determine whether the triangle XYZ with vertices $X(-3, 2)$, $Y(-1, -6)$, and $Z(5, 0)$ is isosceles. Justify your answer. **yes**

10. Consider rectangle $ABCD$. **a–d. See margin.**
   a. Draw and label rectangle $ABCD$ on the coordinate plane.
   b. Prove that $\overline{AC} \cong \overline{BD}$.
   c. Suppose the diagonals intersect at point E. Prove that $\overline{AE} \cong \overline{EC}$ and $\overline{BE} \cong \overline{ED}$.
   d. What can you conclude about the diagonals of a rectangle? Explain.

11. Sports The dimensions of a soccer field are 120 yards by 80 yards. A player kicks the ball from a corner to his teammate at the center of the playing field. Suppose the kicker is located at the origin.
   a. Find the ordered pair that represents the location of the kicker's teammate. **(40, 60)**
   b. Find the distance the ball travels. **$20\sqrt{13}$ or about 72 yd**

Crew Stadium, Columbus, Ohio

EXERCISES

Practice

Find the distance between each pair of points with the given coordinates. Then, find the coordinates of the midpoint of the segment that has endpoints at the given coordinates. **12–20. See margin.**

12. $(-1, 1)$, $(4, 13)$ **13.** $(1, 3)$, $(-1, -3)$ **14.** $(8, 0)$, $(0, 8)$

15. $(-1, -6)$, $(5, -3)$ **16.** $\left(3\sqrt{2}, -5\right)$, $\left(7\sqrt{2}, -1\right)$ **17.** $(a, 7)$, $(a, -9)$

18. $(6 + r, s)$, $(r - 2, s)$ **19.** (c, d), $(c + 2, d - 1)$ **20.** $(w - 2, w)$, $(w, 4w)$

21. Find all values of a so that the distance between points at $(a, -9)$ and $(-2a, 7)$ is 20 units. **$a = \pm 4$**

22. If $M\left(-3, \frac{5}{2}\right)$ is the midpoint of $\overline{CD}$ and C has coordinates $(4, -1)$, find the coordinates of D. **(–10, 6)**

Determine whether the quadrilateral having vertices with the given coordinates is a parallelogram.

23. $(-2, 3)$, $(-3, -2)$, $(2, -3)$, $(3, 2)$ **yes**   **24.** $(4, 11)$, $(8, 14)$, $(4, 19)$, $(0, 15)$ **no**

25. Collinear points lie on the same line. Find the value of k for which the points $(15, 1)$, $(-3, -8)$, and $(3, k)$ are collinear. **–5**

26. Determine whether the points $A(-3, 0)$, $B\left(-1, 2\sqrt{3}\right)$, and $C(1, 0)$ are the vertices of an equilateral triangle. Justify your answer.

26. Yes; see margin for justification.

27. Show that points $E(2, 5)$, $F(4, 4)$, $G(2, 0)$, and $H(0, 1)$ are the vertices of a rectangle. **See margin.**

www.amc.glencoe.com/self_check_quiz

Additional Answers (pp. 620–621)

27. $\overline{EF} \cong \overline{HG}$ since $EF = \sqrt{5}$ and $HG = \sqrt{5}$. $\overline{EF} \parallel \overline{HG}$ since the slope of $\overline{EF}$ is $-\frac{1}{2}$ and the slope of $\overline{HG}$ is $-\frac{1}{2}$. Thus the points form a parallelogram. $\overline{EF} \perp \overline{FG}$ since the product of the slopes of $\overline{EF}$ and $\overline{FG}$, $-\frac{1}{2} \cdot \frac{2}{1}$, is -1. Therefore, the points form a rectangle.

34a.

Prove using analytic methods. Be sure to include a coordinate diagram.

28–32. See Answer Appendix.

28. The measure of the line segment joining the midpoints of two sides of a triangle is equal to one-half the measure of the third side.

29. The diagonals of an isosceles trapezoid are congruent.

30. The medians to the congruent sides of an isosceles triangle are congruent. (*Hint:* A median of a triangle is a segment connecting a vertex to the midpoint of the side opposite the vertex.)

31. The diagonals of a parallelogram bisect each other.

32. The line segments joining the midpoints of consecutive sides of any quadrilateral form a parallelogram.

Applications and Problem Solving

Real World Application

33. **Geometry** The vertices of a rectangle are at $(-3, 1)$, $(-1, 3)$, $(3, -1)$, and $(1, -3)$. Find the area of the rectangle. **16 units²**

34. **Web Page Design** Many Internet Web pages are designed so that when the cursor is positioned over a specified area of an image, lines of text are displayed. The programming string "$x|y|$ width $|$ height" defines the location of the bottom left corner of the designated region using x and y coordinates and then defines the width and height of the region in pixels.

a. If the center of a Web page is located at the origin, graph the two regions defined by $-22|12|10|8$ and $31|-10|8|10$. **See margin.**

b. Suppose the two regions are to be no less than 40 pixels apart. Calculate the distance between the regions at their closest points to determine if this criteria is met. **about 44.64 pixels; the regions meet the criteria**

35. **Critical Thinking** Prove analytically that the segments joining midpoints of consecutive sides of an isosceles trapezoid form a rhombus. Include a coordinate diagram with your proof. **See margin.**

36. **Landscaping** The diagram shows the plans made by a landscape artist for a homeowner's 16-meter by 16-meter backyard. The homeowner has requested that the rosebushes, fountain, and garden bench be placed so that they are no more than 14 meters apart.

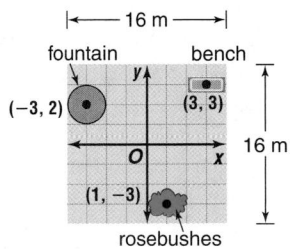

36a. Yes; see margin for explanation.

a. Has the landscape artist met the homeowner's requirements? Explain.

b. The homeowner has purchased a sundial to be placed midway between the fountain and the rosebushes. Determine the coordinates indicating where the sundial should be placed. $\left(-1, -\frac{1}{2}\right)$

37b. A line; this line is the perpendicular bisector of $\overline{AB}$.

37. **Critical Thinking** Consider point $M(t, 3t - 12)$. **a. See Answer Appendix.**

a. Prove that for all values of t, M is equidistant from $A(0, 3)$ and $B(9, 0)$.

b. Describe the figure formed by the points M for all values of t. What is the relationship between the figure and points A and B?

Mixed Review

38. Find $(-5 + 12i)^2$ *(Lesson 9-8)* $-119 - 120i$

Extra Practice See p. A44.

Additional Answers

35. Let the vertices of the isosceles trapezoid have the coordinates $A(0, 0)$, $B(2a, 0)$, $C(2a - 2c, 2b)$, $D(2c, 2b)$. The coordinates of the midpoints are: $P(a, 0)$, $Q(2a - c, b)$, $R(a, 2b)$, and $S(c, b)$.

$$PQ = \sqrt{(2a - c - a)^2 + (b - 0)^2}$$
$$= \sqrt{(a - c)^2 + b^2}$$
$$QR = \sqrt{(2a - c - a)^2 + (b - 2b)^2}$$
$$= \sqrt{(a - c)^2 + b^2}$$
$$RS = \sqrt{(a - c)^2 + (2b - b)^2}$$
$$= \sqrt{(a - c)^2 + b^2}$$
$$PS = \sqrt{(a - c)^2 + (0 - b)^2}$$
$$= \sqrt{(a - c)^2 + b^2}$$

So, all of the sides are congruent and quadrilateral $PQRS$ is a rhombus.

36a. The distance from the fountain to the rosebushes is $2\sqrt{41}$ or about 12.81 meters. The distance from the rosebushes to the bench is $4\sqrt{10}$ or about 12.65 meters. The distance from the bench to the fountain is $2\sqrt{37}$ or about 12.17 meters.

Extra Credit

Prove that the line segments joining the midpoints of the opposite sides of any quadrilateral bisect each other.

Sample answer: $M = \left(\frac{a}{2}, 0\right)$, $S = \left(\frac{a + b}{2}, \frac{c}{2}\right)$,

$N = \left(\frac{b + d}{2}, \frac{c + e}{2}\right)$, and $R = \left(\frac{d}{2}, \frac{e}{2}\right)$. The coordinates of the midpoint of $\overline{MN}$ and of $\overline{RS}$ are $\left(\frac{a + b + d}{2}, \frac{c + e}{2}\right)$.

Since $\overline{MN}$ and $\overline{RS}$ have the same midpoint, the two line segments bisect each other.

Open-Ended Assessment

Speaking Have teams of students work through the analytical proofs presented in Exercises 28–32. Then have each team explain its procedure to the class. Teams should show all necessary diagrams and equations.

39. **Physics** Suppose that during a storm, the force of the wind blowing against a skyscraper can be expressed by the vector (115, 2018, 0), where each measure in the ordered triple represents the force in Newtons. What is the magnitude of this force? *(Lesson 8-3)* **about 2021 N**

40. See Answer Appendix.

40. Verify that $2 \sec^2 x = \dfrac{1}{1 + \sin x} + \dfrac{1}{1 - \sin x}$ is an identity. *(Lesson 7-2)*

41. A circle has a radius of 12 inches. Find the degree measure of the central angle subtended by an arc 11.5 inches long. *(Lesson 6-1)* **54.9°**

42. Find sin 390°. *(Lesson 5-3)* $\dfrac{1}{2}$

43. Solve $z^2 - 8z = -14$ by completing the square. *(Lesson 4-2)* $4 \pm \sqrt{2}$

44. **SAT Practice** **Grid-In** If $x^2 = 16$ and $y^2 = 4$, what is the greatest possible value of $(x - y)^2$? **36**

CAREER CHOICES

Meteorologist

If you find the weather intriguing, then you may want to investigate a career in meteorology. Meteorologists spend their time studying weather and forecasting changes in the weather. They analyze charts, weather maps, and other data to make predictions about future weather patterns. Meteorologists also research different types of weather, such as tornadoes and hurricanes, and may even teach at universities.

As a meteorologist, you may choose to specialize in one of several areas such as climatololgy, operational meteorology, or industrial meteorology. As a meteorologist, you might even be seen on television forecasting the weather for your area!

CAREER OVERVIEW

Degree Preferred:
Bachelor's degree in meteorology

Related Courses:
mathematics, geography, physics, computer science

Outlook:
slower than average job growth through the year 2006

Heat Index for Various Temperatures and Humidities

Source: *The World Almanac* 1999

interNET CONNECTION For more information on careers in meteorology, visit: **www.amc.glencoe.com**

622 *Chapter 10 Conics*

Enrichment Masters, p. 109

NAME _____ DATE _____ PERIOD _____

10-1
Enrichment

Mathematics and History: Hypatia

Hypatia (A.D. 370–415) is the earliest woman mathematician whose life is well documented. Born in Alexandria, Egypt, she was widely known for her keen intellect and extraordinary mathematical ability. Students from Europe, Asia, and Africa flocked to the university at Alexandria to attend her lectures on mathematics, astronomy, philosophy, and mechanics.

Hypatia wrote several major treatises in mathematics. Perhaps the most significant of these was her commentary on the *Arithmetica* of Diophantus, a mathematician who lived and worked in Alexandria in the third century. In her commentary, Hypatia offered several observations about the *Arithmetica's* Diophantine problems—problems for which one was required to find only the rational solutions. It is believed that many of these observations were subsequently incorporated into the original manuscript of the *Arithmetica*.

In modern mathematics, the solutions of a **Diophantine equation** are restricted to integers. In the exercises, you will explore some questions involving simple Diophantine equations.

For each equation, find three solutions that consist of an ordered pair of integers.

1. $2x - y = 7$
   $(1, -5), (0, -7), (-1, -9)$

2. $x + 3y = 5$
   $(2, 1), (5, 0), (8, -1)$

3. $6x - 5y = -8$
   $(2, 4), (-3, -2), (-8, -8)$

4. $-11x - 4y = 6$
   $(2, -7), (-2, 4), (-6, 15)$

5. Refer to your answers to Exercises 1–4. Suppose that the integer pair (x_1, y_1) is a solution of $Ax - By = C$. Describe how to find other integer pairs that are solutions of the equation.
   Other integer pairs are of the form $(x_1 + n \cdot B, y_1 - n \cdot A)$, where n is any nonzero integer.

6. Explain why the equation $3x + 6y = 7$ has no solutions that are integer pairs.
   Rewrite $3x + 6y = 7$ as $3(x + 2y) = 7$. If x and y are integers 7 would have to be an integral multiple of 3.

7. *True or false*: Any line on the coordinate plane must pass through at least one point whose coordinates are integers. Explain.
   False; An equation like $3x + 6y = 7$ has no integer-pair solutions, so the graph of such an equation is a line that passes through no point whose coordinates are integers.

© Glencoe/McGraw-Hill   T109   *Advanced Mathematical Concepts*

CAREER CHOICES   METEOROLOGIST

Math Connection
- Explain the trends seen in the data. **At 75° F, the heat index graph looks like a straight line with a fairly steady increase for each 5°. At 85° F, the heat index increases linearly until about 90° where it begins to increase much more rapidly. At 95° and 105°, the heat index is similar to the pattern shown at 85°.**
- Explain how humidity affects the temperature our bodies feel. **At low readings of humidity, we may feel it is cooler than it actually is. Humidity greater than 40% makes us feel like it is warmer than it actually is.**

10-2

Circles

OBJECTIVES
- Use and determine the standard and general forms of the equation of a circle.
- Graph circles.

Real World Application

SEISMOLOGY Portable autonomous digital seismographs (PADSs) are used to investigate the strong ground motions produced by the aftershocks of large earthquakes. Suppose a PADS is deployed 2 miles west and 3.5 miles south of downtown Olympia, Washington, to record the aftershocks of a recent earthquake. While there, the PADS detects and records the seismic activity of another quake located 24 miles away. What are all the possible locations of this earthquake's epicenter? *This problem will be solved in Example 2.*

The pattern of the shock waves from an earthquake form **concentric** circles. A **circle** is the set of all points in the plane that are equidistant from a given point in the plane, called the **center.** The distance from the center to any point on the circle is called the **radius** of the circle. Concentric circles have the same center but not necessarily the same radius.

A circle is one type of **conic section.** Conic sections, which include circles, parabolas, ellipses and hyperbolas, were first studied in ancient Greece sometime between 600 and 300 B.C. The Greeks were largely concerned with the properties, not the applications, of conics. In the seventeenth century, applications of conics became prominent in the development of calculus.

Conic sections are used to describe all of the possible ways a plane and a double right cone can intersect. In forming the four basic conics, the plane does not pass through the vertex of the cone.

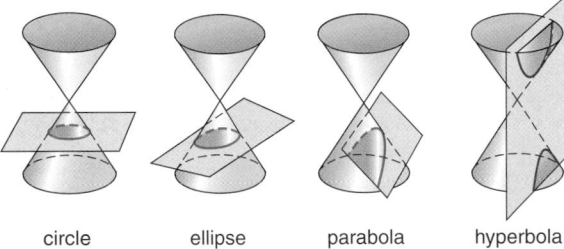

circle   ellipse   parabola   hyperbola

When the plane does pass through the vertex of a conical surface, as illustrated below, the resulting figure is called a **degenerate conic.** A degenerate conic may be a point, line, or two intersecting lines.

point
(degenerate ellipse)

line
(degenerate parabola)

intersecting lines
(degenerate hyperbola)

Resource Manager

 Reproducible Masters
- *Study Guide and Practice,* pp. 179–180
- *Enrichment,* p. 110
- *Assessment and Evaluation,* p. 251

 Transparencies
- 5-Minute Check, 10-2
- Teaching, 10-2

 5-Minute Check
Lesson 10-1

1. Find the distance between points at (10, 7) and (−4, 8). $\sqrt{197}$

2. Find the coordinates of the midpoint of the segment that has endpoints at (−5, 4) and (7, −2). **(1, 1)**

3. Determine whether quadrilateral *QRST* with the given coordinates *Q*(−4, −1), *R*(−2, −3), *S*(4, 3), and *T*(2, 5) is a parallelogram. **yes**

4. Prove that the points *A*(−7, 2), *B*(3, −4) and *C*(1, 4) are the vertices of an isosceles triangle. **The measures of two of the three sides are equal, so triangle *ABC* is isosceles.**

5. The vertices of a parallelogram are (3, 2), (0, −2), (−7, −2), and (−4, 2). Find the area of the parallelogram. **28 units²**

Motivating the Lesson

Discussion Ask students for examples of things that can be represented by circles. **Sample answers: ripples from a stone thrown in water, archery target** What information might someone want to know about a circle? **its radius, circumference, equation, graph**

Teaching Tip In Example 1, make sure that students substitute the *x*- and *y*-values of the given point into the *h* and *k* variables, not the *x* and *y* variables.

In-Class Example

Example 1
Write the standard form of the equation of the circle that is tangent to the *x*-axis and has its center at (−5, 4). Then graph the equation.
$(x + 5)^2 + (y − 4)^2 = 16$

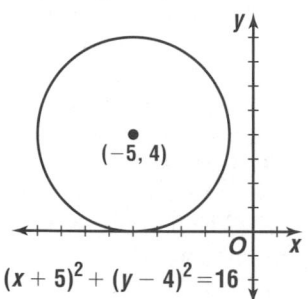

$(x + 5)^2 + (y − 4)^2 = 16$

Radius can also refer to the line segment from the center to any point on the circle.

In the figure at the right, the center of the circle is at the origin. By drawing a perpendicular from any point $P(x, y)$ on the circle but not on an axis to the *x*-axis, you form a right triangle. The Pythagorean Theorem can be used to write an equation that describes every point on a circle whose center is located at the origin.

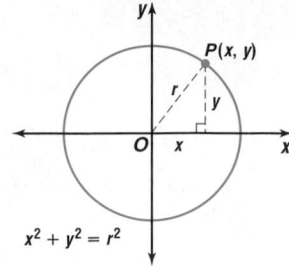

$x^2 + y^2 = r^2$

$$x^2 + y^2 = r^2 \quad \textit{Pythagorean Theorem}$$

This is the equation for the parent graph of all circles.

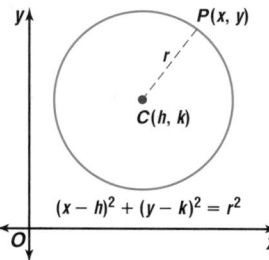

$(x − h)^2 + (y − k)^2 = r^2$

Suppose the center of this circle is translated from the origin to $C(h, k)$. You can use the distance formula to write the equation for this translated circle.

$$d = \sqrt{(x_2 − x_1)^2 + (y_2 − y_1)^2} \quad \textit{Distance formula}$$
$$r = \sqrt{(x − h)^2 + (y − k)^2} \quad \begin{array}{l}\textit{d = r, } (x_2, y_2) = (x, y),\\ \textit{and } (x_1, y_1) = (h, k)\end{array}$$
$$r^2 = (x − h)^2 + (y − k)^2 \quad \textit{Square each side.}$$

This equation is the standard form of the equation of a circle.

Standard Form of the Equation of a Circle	The standard form of the equation of a circle with radius *r* and center at (h, k) is $$(x − h)^2 + (y − k)^2 = r^2.$$

Examples Write the standard form of the equation of the circle that is tangent to the *x*-axis and has its center at (3, −2). Then graph the equation.

Since the circle is tangent to the *x*-axis, the distance from the center to the *x*-axis is the radius. The center is 2 units below the *x*-axis. Therefore, the radius is 2.

$$(x − h)^2 + (y − k)^2 = r^2 \quad \textit{Standard form}$$
$$(x − 3)^2 + [y − (−2)]^2 = 2^2 \quad \textit{h = 3, k = −2, r = 2}$$
$$(x − 3)^2 + (y + 2)^2 = 4$$

The standard form of the equation for this circle is $(x − 3)^2 + (y + 2)^2 = 4$.

$(x − 3)^2 + (y + 2)^2 = 4$

$(3, −2)$

 SEISMOLOGY Refer to the application at the beginning of the lesson.

a. **Write an equation for the set of points representing all possible locations of the earthquake's epicenter. Let downtown Olympia, Washington, be located at the origin.**

b. **Graph the equation found in part a.**

---

624 *Chapter 10 Conics*

Prerequisite Skills

A review of the prerequisite skills for this lesson is provided in the *Prerequisite Skills Masters.*

Skill Number and Content	Pages
10 Factoring Special Products of Polynomials	21–22
11 Factoring by Completing the Square	23–24

a. The location of the PADS, 2 miles west and 3.5 miles south of downtown Olympia, Washington, can be expressed as the ordered pair $(-2, -3.5)$. Since any point 24 miles from the seismograph could be the epicenter, the radius of the circle is 24.

$$(x - h)^2 + (y - k)^2 = r^2 \quad \textit{Standard form}$$
$$[x - (-2)]^2 + [y - (-3.5)]^2 = 24^2 \quad \textit{(h, k) = (−2, −3.5) and r = 24}$$
$$(x + 2)^2 + (y + 3.5)^2 = 576$$

b. The location of the epicenter lies on the circle with equation
$(x + 2)^2 + (y + 3.5)^2 = 576$.

$(x + 2)^2 + (y + 3.5)^2 = 576$

Graphing Calculator Tip

To graph a circle on a graphing calculator, first solve for *y*. Then graph the two resulting equations on the same screen by inserting $\{1, -1\}$ in front of the equation for the symbol $\pm$. Use **5:ZSquare** in the **ZOOM** menu to make the graph look like a circle

The standard form of the equation of a circle can be expanded to obtain a general form of the equation.

$$(x - h)^2 + (y - k)^2 = r^2 \quad \textit{Standard form}$$
$$(x^2 - 2hx + h^2) + (y^2 - 2ky + k^2) = r^2 \quad \textit{Expand } (x - h)^2 \textit{ and } (y - k)^2.$$
$$x^2 + y^2 + (-2h)x + (-2k)y + (h^2 + k^2) - r^2 = 0$$

Since h, k, and r are constants, let D, E, and F equal $-2h$, $-2k$ and $(h^2 + k^2) - r^2$, respectively.

$$x^2 + y^2 + Dx + Ey + F = 0$$

This equation is called the general form of the equation of a circle.

General Form of the Equation of a Circle	The general form of the equation of a circle is $$x^2 + y^2 + Dx + Ey + F = 0,$$ where D, E, and F are constants.

Notice that the coefficients of x^2 and y^2 in the general form must be 1. If those coefficients are not 1, division can be used to transform the equation so that they are 1. Also notice that there is no term containing the product of the variables, xy.

When the equation of a circle is given in general form, it can be rewritten in standard form by completing the square for the terms in x and the terms in y.

Teaching Tip Before working Example 2, point out that any situation where a distance is known, but not a direction, can be modeled by a circle.

In-Class Example

Example 2

Seismology On August 17, 1999, an earthquake of magnitude 7.4 on the Richter scale rocked Western Turkey. Had a PADS been deployed 2 miles west and 4 miles north of Izmit, Turkey, it would have indicated that the epicenter was 9 miles away.

a. Write an equation for the set of points representing all possible locations of the earthquake's epicenter. Assume that downtown Izmit, Turkey, is located at the origin.
$(x + 2)^2 + (y - 4)^2 = 81$

b. Graph the equation found in part **a.**

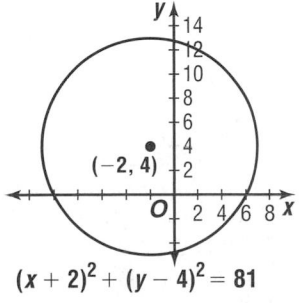

$(x + 2)^2 + (y - 4)^2 = 81$

Teaching Tip Throughout the study of conics, watch for students who have difficulty identifying the values of *h* and *k*. Until students are comfortable with these types of equations, encourage them to rewrite equations so that subtraction is used. For example, $(x + 2)^2 + (y + 5)^2 = 3^2$ becomes $[x - (-2)]^2 + [y - (-5)]^2 = 3^2$. Now *h* and *k* are more easily identified.

Teaching Tip Before working Example 3, carefully review the process of completing the square. Remind students that to complete the square of a quadratic expression in x, they should divide the coefficient of x by 2 and then square it.

In-Class Examples

Example 3
The equation of a circle is $4x^2 + 4y^2 - 24x + 16y = -51$.
a. Write the standard form of the equation.
   $(x - 3)^2 + (y + 2)^2 = 0.25$
b. Find the radius and the coordinates of the center.
   radius: 0.5 unit; center: (3, –2)
c. Graph the equation.

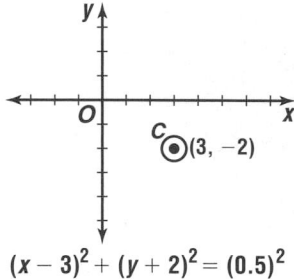

$(x - 3)^2 + (y + 2)^2 = (0.5)^2$

Example 4
Write the standard form of the equation of the circle that passes through the points at (1, 1), (1, 2), and (2, 3). Then identify the center and radius of the circle.
$$\left(x - \frac{5}{2}\right)^2 + \left(y - \frac{3}{2}\right)^2 = \frac{5}{2}$$
center = (2.5, 1.5);
radius = $\sqrt{2.5}$

Additional Answers (p. 627)
2. Sample answer:
$(x + 4)^2 + (y - 9)^2 = 1$,
$(x + 4)^2 + (y - 9)^2 = 2$,
$(x + 4)^2 + (y - 9)^2 = 3$,
$(x + 4)^2 + (y - 9)^2 = 4$,
$(x + 4)^2 + (y - 9)^2 = 5$

3. **Find the center of the circle, (h, k), by finding the midpoint of the diameter. Next find the radius of the circle, r, by finding the distance from the center to one endpoint. Then write the equation of the circle in standard form as $(x - h)^2 + (y - k)^2 = r^2$.**

4. **The equation $x^2 + y^2 + 8x + 8y + 36 = 0$ written in standard form is $(x + 4)^2 + (y + 4)^2 = -4$. Since a circle cannot have a negative radius, the graph of the equation is the empty set.**

Example **3** The equation of a circle is $2x^2 + 2y^2 - 4x + 12y - 18 = 0$.

a. **Write the standard form of the equation.**

b. **Find the radius and the coordinates of the center.**

c. **Graph the equation.**

a.
$$2x^2 + 2y^2 - 4x + 12y - 18 = 0$$
$$x^2 + y^2 - 2x + 6y - 9 = 0 \qquad \text{\textit{Divide each side by 2.}}$$
$$(x^2 - 2x + \text{?}) + (y^2 + 6y + \text{?}) = 9 \qquad \text{\textit{Group to form perfect square trinomials.}}$$
$$(x^2 - 2x + 1) + (y^2 + 6y + 9) = 9 + 1 + 9 \qquad \text{\textit{Complete the square.}}$$
$$(x - 1)^2 + (y + 3)^2 = 19 \qquad \text{\textit{Factor the trinomials.}}$$
$$(x - 1)^2 + (y + 3)^2 = \left(\sqrt{19}\right)^2 \qquad \text{\textit{Express 19 as}} \left(\sqrt{19}\right)^2 \text{\textit{to show}}$$
$$\text{\textit{that}} \ r = \sqrt{19}.$$

b. The center of the circle is located at $(1, -3)$, and the radius is $\sqrt{19}$.

c. Plot the center at $(-1, 3)$. The radius of $\sqrt{19}$ is approximately equal to 4.4.

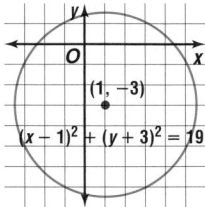

From geometry, you know that any two points in the coordinate plane determine a unique line. It is also true that any three noncollinear points in the coordinate plane determine a unique circle. The equation of this circle can be found by substituting the coordinates of the three points into the general form of the equation of a circle and solving the resulting system of three equations.

Example **4** **Write the standard form of the equation of the circle that passes through the points at (5, 3), (−2, 2), and (−1, −5). Then identify the center and radius of the circle.**

Look Back
Refer to Lesson 2-2 to review solving systems of three equations.

Substitute each ordered pair for (x, y) in $x^2 + y^2 + Dx + Ey + F = 0$, to create a system of equations.

$$(5)^2 + (3)^2 + D(5) + E(3) + F = 0 \quad (x, y) = (5, 3)$$
$$(-2)^2 + (2)^2 + D(-2) + E(2) + F = 0 \quad (x, y) = (-2, 2)$$
$$(-1)^2 + (-5)^2 + D(-1) + E(-5) + F = 0 \quad (x, y) = (-1, -5)$$

Simplify the system of equations.
$$5D + 3E + F + 34 = 0$$
$$-2D + 2E + F + 8 = 0$$
$$-D - 5E + F + 26 = 0$$

The solution to the system is $D = -4$, $E = 2$, and $F = -20$.

The general form of the equation of the circle is $x^2 + y^2 - 4x + 2y - 20 = 0$. After completing the square, the standard form is $(x - 2)^2 + (y + 1)^2 = 25$. The center of the circle is at $(2, -1)$, and its radius is 5.

Addressing Individual Needs

Verbal/Linguistic Help students see the meaning of the terms $x - h$ and $y - k$, which appear throughout this chapter. These terms indicate that the point (x, y) has been shifted horizontally by h and vertically by k. Illustrate by having students graph $(4, 6)$ and $(4 - 5, 6 - (-3))$ to see that $(4, 6)$ is shifted left 5 and up 3.

Communicating Mathematics

1. complete the square on each variable

Read and study the lesson to answer each question.   2–4. See margin.

1. **Explain** how to convert the general form of the equation of a circle to the standard form of the equation of a circle.

2. **Write** the equations of five concentric circles with different radii whose centers are at $(-4, 9)$.

3. **Describe** how you might determine the equation of a circle if you are given the endpoints of the circle's diameter.

4. **Find a counterexample** to this statement: The graph of any equation of the form $x^2 + y^2 + Dx + Ey + F = 0$ is a circle.

5. **You Decide**   Kiyo says that you can take the square root of each side of an equation. Therefore, he decides that $(x - 3)^2 + (y - 1)^2 = 49$ and $(x - 3) + (y - 1) = 7$ are equivalent equations. Ramon says that the equations are not equivalent. Who is correct? Explain.   **Ramon; The square root of a sum does not equal the sum of the square roots.**

Guided Practice

6–9. See margin for graphs.

Write the standard form of the equation of each circle described. Then graph the equation.

6. center at $(0, 0)$, radius 9
$x^2 + y^2 = 81$

7. center at $(-1, 4)$ and tangent to $x = 3$
$(x + 1)^2 + (y - 4)^2 = 16$

Write the standard form of each equation. Then graph the equation.

8. $x^2 + y^2 - 4x + 14y - 47 = 0$
$(x - 2)^2 + (y + 7)^2 = 100$

9. $2x^2 + 2y^2 - 20x + 8y + 34 = 0$
$(x - 5)^2 + (y + 2)^2 = 12$

Write the standard form of the equation of the circle that passes through points with the given coordinates. Then identify the center and radius.

10. $(0, 0), (4, 0), (0, 4)$
$(x - 2)^2 + (y - 2)^2 = 8; (2, 2); 2\sqrt{2}$

11. $(1, 3), (5, 5), (5, 3)$
$(x - 3)^2 + (y - 4)^2 = 5; (3, 4); \sqrt{5}$

Write the equation of the circle that satisfies each set of conditions.

12. $(x + 2)^2 + (y - 1)^2 = 25$

12. The circle passes through the point at $(1, 5)$ and has its center at $(-2, 1)$.

13. The endpoints of a diameter are at $(-2, 6)$ and at $(10, -10)$.
$(x - 4)^2 + (y + 2)^2 = 100$

14. **Space Science**   Apollo 8 was the first manned spacecraft to orbit the moon at an average altitude of 185 kilometers above the moon's surface. Determine an equation to model the orbit of the Apollo 8 command module if the radius of the moon is 1740 kilometers. Let the center of the moon be at the origin.
$x^2 + y^2 = 1925^2$

← Apollo 8 crew: (from left) James A. Lovell, Jr., William A. Anders, Frank Borman

EXERCISES

Practice

15–20. See margin for equations. See Answer Appendix for graphs.

Write the standard form of the equation of each circle described. Then graph the equation.

15. center at $(0, 0)$, radius 5

16. center at $(-4, 7)$, radius $\sqrt{3}$

17. center at $(-1, -3)$, radius $\dfrac{\sqrt{2}}{2}$

18. center at $(-5, 0)$, radius $\dfrac{9}{2}$

19. center at $(6, 1)$, tangent to the y-axis

20. center at $(3, -2)$, tangent to $y = 2$

 www.amc.glencoe.com/self_check_quiz

Lesson 10-2   Circles   **627**

Error Analysis

Watch for students who complete the square incorrectly.

Prevent by having them square several binomials to see the pattern used in reverse for completing the square.

Assignment Guide

Core: 15–47 odd, 49–58
Enriched: 16–42 even, 43–58

Additional Answers

9.

(5, -2 + 2√3)
(5, -2)
(5 - 2√3, -2)

15. $x^2 + y^2 = 25$

16. $(x + 4)^2 + (y - 7)^2 = 3$

17. $(x + 1)^2 + (y + 3)^2 = \dfrac{1}{2}$

18. $(x + 5)^2 + y^2 = \dfrac{81}{4}$

19. $(x - 6)^2 + (y - 1)^2 = 36$

20. $(x - 3)^2 + (y + 2)^2 = 16$

Study Guide and Practice Masters, pp. 179–180

NAME _____ DATE _____ PERIOD _____

10-2 Study Guide

Circles

The standard form of the equation of a **circle** with **radius** r and **center** at (h, k) is $(x - h)^2 + (y - k)^2 = r^2$.

Example 1   Write the standard form of the equation of the circle that is tangent to the x-axis and has its center at $(-4, 3)$. Then graph the equation.

Since the circle is tangent to the x-axis, the distance from the center to the x-axis is the radius. The center is 3 units above the x-axis. Therefore, the radius is 3.

$(x - h)^2 + (y - k)^2 = r^2$ *Standard form*
$[x - (-4)]^2 + (y - 3)^2 = 3^2$ $(h, k) = (-4, 3)$ and $r = 3$
$(x + 4)^2 + (y - 3)^2 = 9$

Example 2   Write the standard form of the equation of the circle that passes through the points at $(1, -1)$, $(5, 3)$, and $(-3, 3)$. Then identify the center and radius of the circle.

Substitute each ordered pair (x, y) in the general form $x^2 + y^2 + Dx + Ey + F = 0$ to create a system of equations.

$(1)^2 + (-1)^2 + D(1) + E(-1) + F = 0$ $(x, y) = (1, -1)$
$(5)^2 + (3)^2 + D(5) + E(3) + F = 0$ $(x, y) = (5, 3)$
$(-3)^2 + (3)^2 + D(-3) + E(3) + F = 0$ $(x, y) = (-3, 3)$

Simplify the system of equations.

$D - E + F + 2 = 0$
$5D + 3E + F + 34 = 0$
$-3D + 3E + F + 18 = 0$

The solution to the system is $D = -2, E = -6$, and $F = -6$.

The general form of the equation of the circle is $x^2 + y^2 - 2x - 6y - 6 = 0$.

$x^2 + y^2 - 2x - 6y - 6 = 0$
$(x^2 - 2x + 7) + (y^2 - 6y + 7) = 6$ *Group to form perfect square trinomials.*
$(x^2 - 2x + 1) + (y^2 - 6y + 9) = 6 + 1 + 9$ *Complete the square.*
$(x - 1)^2 + (y - 3)^2 = 16$ *Factor the trinomials.*

After completing the square, the standard form of the circle is $(x - 1)^2 + (y - 3)^2 = 16$. Its center is at $(1, 3)$, and its radius is 4.

© Glencoe/McGraw-Hill   179   *Advanced Mathematical Concepts*

Additional Answers

6.

(9, 0)
(0, -9)

7.

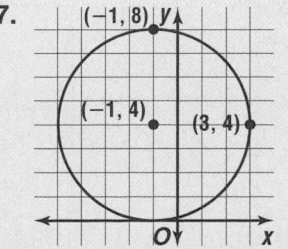

(-1, 8)
(-1, 4) (3, 4)

8.

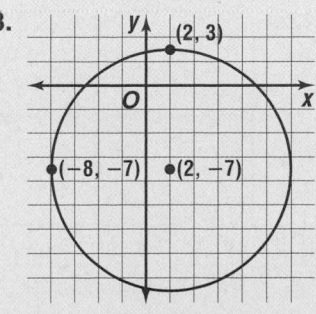

(2, 3)
(-8, -7) (2, -7)

Teaching Tip For Exercise 41, remind students of the formula for the area of a rectangle. The area equation needs to be in terms of x. Thus, students need to solve $x^2 + y^2 = 36$ for y to find a suitable measure for all sides in terms of x.

Additional Answers

21. $x^2 + y^2 = 36$

22. $x^2 + \left(y + \frac{1}{2}\right)^2 = 1$

23. $(x - 2)^2 + (y + 6)^2 = 10$

24. $\left(x + \frac{1}{2}\right)^2 + (y - 1)^2 = \frac{3}{4}$

25. $(x - 1)^2 + (y + 3)^2 = 16$

26. $\left(x - \frac{1}{4}\right)^2 + (y - 1)^2 = 9$

27.

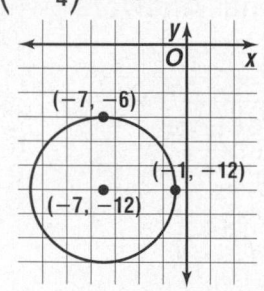

28. $(x + 3)^2 + (y - 3)^2 = 25$;
$(-3, 3)$; 5

29. $(x - 7)^2 + (y + 5)^2 = 16$;
$(7, -5)$; 4

30. $(x - 3)^2 + (y + 5)^2 = 169$;
$(3, -5)$; 13

31. $(x - 5)^2 + (y - 2)^2 = 50$;
$(5, 2)$; $5\sqrt{2}$

32. $(x - 3)^2 + (y + 2)^2 = 50$;
$(3, -2)$; $5\sqrt{2}$

33. $\left(x + \frac{1}{6}\right)^2 + \left(y - \frac{7}{6}\right)^2 = \frac{169}{18}$;
$\left(-\frac{1}{6}, \frac{7}{6}\right)$; $\frac{13\sqrt{2}}{6}$

35. $(x + 4)^2 + (y - 3)^2 = 25$

36. $(x - 2)^2 + (y - 3)^2 = 18$

37. $(x + 2)^2 + (y + 1)^2 = 32$

38. $\left(x + \frac{1}{2}\right)^2 + \left(y - \frac{5}{2}\right)^2 = \frac{17}{2}$

39. $(x - 5)^2 + (y - 1)^2 = 10$

40. $\left(x - \sqrt{2}\right)^2 + y^2 = 1$ or $x^2 + y^2 = 1$

41d.

$[0, 10]$ scl:1 by $[0, 100]$ scl:20

21–26. See margin for equations. See Answer Appendix for graphs.

*inter*NET
CONNECTION

Graphing Calculator Programs
For a graphing calculator program that determines the radius and the coordinates of the center of a circle from an equation written in general form, visit **www.amc.glencoe.com**

35–40. See margin.

Graphing Calculator

Write the standard form of each equation. Then graph the equation.

21. $36 - x^2 = y^2$

22. $x^2 + y^2 + y = \frac{3}{4}$

23. $x^2 + y^2 - 4x + 12y + 30 = 0$

24. $2x^2 + 2y^2 + 2x - 4y = -1$

25. $6x^2 - 12x + 6y^2 + 36y = 36$

26. $16x^2 + 16y^2 - 8x - 32y = 127$

27. Write $x^2 + y^2 + 14x + 24y + 157 = 0$ in standard form. Then graph the equation. $(x + 7)^2 + (y + 12)^2 = 36$; see margin for graph.

Write the standard form of the equation of the circle that passes through the points with the given coordinates. Then identify the center and radius.

28. $(0, -1), (-3, -2), (-6, -1)$

29. $(7, -1), (11, -5), (3, -5)$

30. $(-2, 7), (-9, 0), (-10, -5)$

31. $(-2, 3), (6, -5), (0, 7)$

32. $(4, 5), (-2, 3), (-4, -3)$

33. $(1, 4), (2, -1), (-3, 0)$

28–33. See margin.

34. Write the standard form of the equation of the circle that passes through the origin and points at $(2.8, 0)$ and $(5, 2)$. $(x - 1.4)^2 + (y - 3.75)^2 = 16.02$

Write the equation of the circle that satisfies each set of conditions.

35. The circle passes through the origin and has its center at $(-4, 3)$.

36. The circle passes through the point $(5, 6)$ and has its center at $(2, 3)$.

37. The endpoints of a diameter are at $(2, 3)$ and at $(-6, -5)$.

38. The points at $(-3, 4)$ and $(2, 1)$ are the endpoints of a diameter.

39. The circle is tangent to the line with equation $x + 3y = -2$ and has its center at $(5, 1)$.

40. The center of the circle is on the x-axis, its radius is 1, and it passes through the point at $\left(\frac{\sqrt{2}}{2}, \frac{\sqrt{2}}{2}\right)$.

41. A rectangle is inscribed in a circle centered at the origin with diameter 12.

 a. Write the equation of the circle that meets these conditions. $x^2 + y^2 = 36$

 b. Write the dimensions of the rectangle in terms of x. $2x$ by $2\sqrt{36 - x^2}$

 c. Write a function $A(x)$ that represents the area of the rectangle. $A(x) = 4x\sqrt{36 - x^2}$

 d. Use a graphing calculator to graph the function $y = A(x)$. **See margin.**

 e. Find the value of x, to the nearest tenth, that maximizes the area of the rectangle. What is the maximum area of the rectangle? **4.2; 72 units²**

42. Select the standard viewing window and then select **ZSquare** from the [ZOOM] menu.

 a. Select **9:Circle(** from the **DRAW** menu and then enter 2 [,] 3 [,] 4 [ )]. Then press [ENTER]. **See margin.**

 b. Describe what appears on the viewing screen. **a circle centered at (2, 3) with radius 4**

 c. Write the equation for this graph. $(x - 2)^2 + (y - 3)^2 = 16$

 d. Use what you have learned to write the command to graph the equation $(x + 4)^2 + (y - 2) = 36$. Then graph the equation. [2nd] [DRAW] 9:Circle([-] 4 [,] 2 [,] 6 [ )]; see margin for graph.

Additional Answers

42a.

$[-15.16, 15.16]$ scl:1 by $[-5, 5]$ scl:1

42d.

$[-15.16, 15.16]$ scl:1 by $[-5, 5]$ scl:1

Applications and Problem Solving

43. Sports   Cindy is taking an archery class and decides to practice her skills at home. She attaches the target shown at right to a bale of hay. The circles on the target are concentric and equally spaced apart.

24 in.

|←— 24 in. —→|

a. If the common center of the circles is located at the origin, write an equation that models the largest circle. $x^2 + y^2 = 144$

b. If the smallest circle is modeled by the equation $x^2 + y^2 = 6.25$, find the area of the region marked B.   **145.50 in.²**

44. $(x - 6.5)^2 +$ $(y - 4.5)^2 = 32.5$

44. Geometry   Write the equation of a circle that circumscribes the triangle whose sides are the graphs of $4x - 7y = 27$, $x - 5y + 3 = 0$, and $2x + 3y - 7 = 0$.

45. Critical Thinking   Consider a family of circles in which $h = k$ and the radius is 2. Let k be any real number.

Look Back
Refer to Lesson 3-2 to review families of graphs.

a. Write the equation of the family of circles. $(x - k)^2 + (y - k)^2 = 4$

b. Graph three members of this family on the same set of axes.   **See margin.**

c. Write a description of all members of this family of circles.   **See margin.**

46. Transportation   A moving truck 7 feet wide and 13 feet high is approaching a semi-circular brick archway at an apartment complex. The base of the archway is 28 feet wide. The road under the archway is divided, allowing for two-way traffic.

14 feet

28 feet

a. Write an equation, centered at the origin, of the archway that models its shape. $y = \sqrt{196 - x^2}$

46b. No, if $x = 7$, then $y = \sqrt{147} \approx$ 12.1 ft, so the truck cannot pass.

b. If the truck remains just to the right of the median, will it be able to pass under the archway without damage? Explain.

47. Critical Thinking   Find the radius and the coordinates of the center of a circle defined by the equation $x^2 + y^2 - 8x + 6y + 25 = 0$. Describe the graph of this circle.   **radius: 0; center: (4, −3); graph is a point located at (4, −3)**

48. Agriculture   One method of irrigating crops is called the center pivot system. This system rotates a sprinkler pipe from the center of the field to be irrigated. Suppose a farmer places one of these units at the center of a square plot of land 2500 feet on each side. With the center of this plot at the origin, the irrigator sends out water far enough to reach a point located at (475, 1140).

48a. $x^2 + y^2 =$ 1,525,225

48b. approximately 4,792,000 ft²

48c. about 23%

a. Find an equation representing the farthest points the water can reach.

b. Find the area of the land that receives water directly.

c. About what percent of the farmer's plot does not receive water directly?

49. Critical Thinking   Consider points $A(3, 4)$, $B(-3, -4)$, and $P(x, y)$.

a. Write an equation for all x and y for which $\overline{PA} \perp \overline{PB}$. $x^2 + y^2 = 25$

b. What is the relationship between points A, B, and P if $\overline{PA} \perp \overline{PB}$?
   If $\overline{PA} \perp \overline{PB}$, then A, P, and B are on the circle $x^2 + y^2 = 25$.

Lesson 10-2   Circles   **629**

Additional Answers

45b. Sample graph:

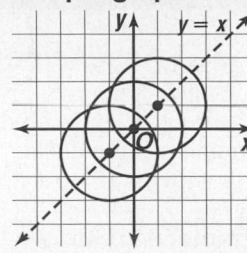

45c. All of the circles in this family have a radius of 2 and centers located on the line $y = x$.

4 ASSESS

Open-Ended Assessment

Speaking A circle has radius of n and a center at (a, b). Have students state the equation of the circle.
$(x - a)^2 + (y - b)^2 = n^2$

**Chapter 10, Quiz A
(Lessons 10-1 and 10-2)**
is available in the *Assessment and Evaluation Masters*, p. 251.

Mixed Review

50. Find the distance between points at $(4, -3)$ and $(-2, 6)$. *(Lesson 10-1)* $\sqrt{117}$

51. Simplify $(2 + i)(3 - 4i)(1 + 2i)$. *(Lesson 9-5)* $20 + 15i$

52. 15 ft horizontally, about 22 ft vertically

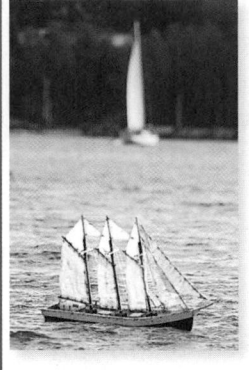

52. Sports Patrick kicked a football with an initial velocity of 60 ft/s at an angle of 60° to the horizontal. After 0.5 seconds, how far has the ball traveled horizontally and vertically. *(Lesson 8-7)*

53. Toys A toy boat floats on the water bobbing up and down. The distance between its highest and lowest point is 5 centimeters. It moves from its highest point down to its lowest point and back up to its highest point every 20 seconds. Write a cosine function that models the movement of the boat in relationship to the equilibrium point. *(Lesson 6-6)*
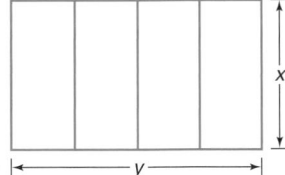 $y = 2.5 \cos\left(\dfrac{\pi}{10}\, t\right)$

54. Find the area to the nearest square unit of $\triangle ABC$ if $a = 15$, $b = 25$, and $c = 35$. *(Lesson 5-8)* **162 units²**

55. Amusement Parks The velocity of a roller coaster as it moves down a hill is $\sqrt{v_0{}^2 + 64h}$. The designer of a coaster wants the coaster to have a velocity of 95 feet per second when it reaches the bottom of the hill. If the initial velocity of the coaster at the top of the hill is 15 feet per second, how high should the designer make the hill? *(Lesson 4-7)* **137.5 ft**

56. Determine whether the graph of $y = 6x^4 - 3x^2 + 1$ is symmetric with respect to the x-axis, the y-axis, the line $y = x$, the line $y = -x$, or the origin. *(Lesson 3-1)*
y-axis

57. Business A pharmaceutical company manufactures two drugs. Each case of drug A requires 3 hours of processing time and 1 hour of curing time per week. Each case of drug B requires 5 hours of processing time and 5 hours of curing time per week. The schedule allows 55 hours of processing time and 45 hours of curing time weekly. The company must produce no more than 10 cases of drug A and no more than 9 cases of drug B. *(Lesson 2-7)*

   a. If the company makes a profit of \$320 on each case of drug A and \$500 on each case of drug B, how many cases of each drug should be produced in order to maximize profit? **10 cases of drug A, 5 cases of drug B**

   b. Find the maximum profit. **\$5700**

58. SAT/ACT Practice An owner plans to divide and enclose a rectangular property with dimensions x and y into rectangular regions as shown at the right. In terms of x and y, what is the total length of fence needed if every line segment represents a section of fence and there is no overlap between sections of fence. **A**

   A $5x + 2y$

   B $5x + 8y$

   C $4x + 2y$

   D $4xy$

   E xy

Extra Practice See p. A44.

Enrichment Masters, p. 110

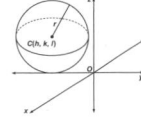

10-2 Enrichment

Spheres

The set of all points in three-dimensional space that are a fixed distance r (the **radius**), from a fixed point C (the **center**), is called a **sphere**. The equation below is an algebraic representation of the sphere shown at the right.

$(x - h)^2 + (y - k)^2 + (z - l)^2 = r^2$

A line segment containing the center of a sphere and having its endpoints on the sphere is called a **diameter** of the sphere. The endpoints of a diameter are called **poles** of the sphere. A **great circle** of a sphere is the intersection of the sphere and a plane containing the center of the sphere.

1. If $x^2 + y^2 - 4y + z^2 + 2z - 4 = 0$ is an equation of a sphere and $(1, 4, -3)$ is one pole of the sphere, find the coordinates of the opposite pole.
$(-1, 0, 1)$

2. a. On the coordinate system at the right, sketch the sphere described by the equation $x^2 + y^2 + z^2 = 9$.

   b. Is $P(2, -2, -2)$ inside, outside, or on the sphere?
   outside

   c. Describe a way to tell if a point with coordinates $P(a, b, c)$ is inside, outside, or on the sphere with equation $x^2 + y^2 + z^2 = r^2$.
   $a^2 + b^2 + c^2 < r^2$: **inside the sphere**
   $a^2 + b^2 + c^2 = r^2$: **on the sphere**
   $a^2 + b^2 + c^2 > r^2$: **outside the sphere**

3. If $x^2 + y^2 + z^2 - 4x + 6y - 2z - 2 = 0$ is an equation of a sphere, find the circumference of a great circle, and the surface area and volume of the sphere.
**8π units; 64π square units;
$\dfrac{256\pi}{3}$ cubic units**

4. The equation $x^2 + y^2 = 4$ represents a set of points in three-dimensional space. Describe that set of points in your own words. Illustrate with a sketch on the coordinate system at the right.
a cylinder

© Glencoe/McGraw-Hill   T110   *Advanced Mathematical Concepts*

Extra Credit

Write a counterexample for each statement.
1. Two circles either intersect in two places or do not intersect. **If they are tangent, they intersect at one point.**

2. If a circle and a line intersect, they intersect twice. **If the line is tangent to the circle, they intersect once.**

3. Any three points determine a circle. **If the three points are collinear, they do not determine a circle.**

Ellipses

OBJECTIVES
- Use and determine the standard and general forms of the equation of an ellipse.
- Graph ellipses.

Real World Application

MEDICAL TECHNOLOGY
To eliminate kidney stones, doctors sometimes use a medical tool called a *lithotripter*, (*LITH-uh-trip-tor*) which means "stone crusher." A lithotripter is a device that uses ultra-high-frequency shock waves moving through water to break up the stone. After x-raying a patient's kidney to precisely locate and measure the stone, the lithotripter is positioned so that the shock waves reflect off the inner surface of the elliptically-shaped tub and break up the stone. *A problem related to this will be solved in Example 1.*

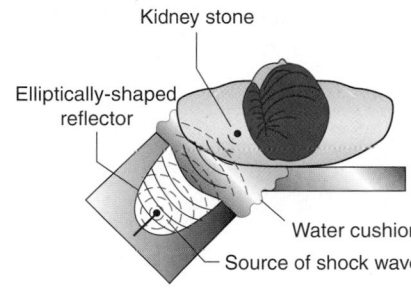

Kidney stone
Elliptically-shaped reflector
Water cushion
Source of shock wave

An **ellipse** is the set of all points in the plane, the sum of whose distances from two fixed points, called **foci**, is constant. In the figure at the right, F_1 and F_2 are the foci, and the midpoint C of the line segment joining the foci is called the **center** of the ellipse. P and Q are any two points on the ellipse. By definition, $PF_1 + PF_2 = QF_1 + QF_2$.

Foci (FOH sigh) is the plural of focus.

To help visualize this definition, imagine tacking two ends of a string at the foci and using a pencil to trace a curve as it is held tight against the string. The curve which results will be an ellipse since the sum of the distances to the foci, the total length of the string, remains constant.

The parent graph of an ellipse, shown at the right, is centered at the origin. An ellipse has two axes of symmetry, in this case, the *x*-axis and the *y*-axis. Notice that the ellipse intersects each axis of symmetry two times. The longer line segment, $\overline{AD}$, which contains the foci, is called the **major axis.** The shorter segment, $\overline{BE}$, is called the **minor axis.** The endpoints of each axis are the **vertices** of the ellipse.

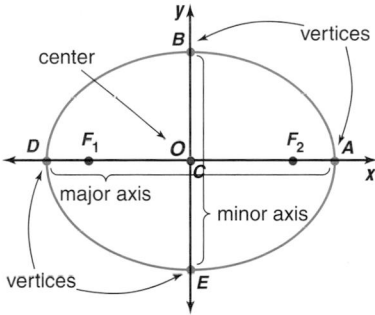

center
vertices
major axis
minor axis
vertices

Resource Manager

Reproducible Masters
- *Study Guide and Practice,* pp. 181–182
- *Enrichment,* p. 111
- *Graphing Calculator Keystrokes,* p. 38

Transparencies
- 5-Minute Check, 10-3
- Teaching, 10-3

---

1 FOCUS

5-Minute Check
Lesson 10-2

The equation of a circle is $x^2 + y^2 + 8x - 2y + 13 = 0$.

1. Write the standard form of the equation.
   $(x + 4)^2 + (y - 1)^2 = 4$

2. Find the radius and the coordinates of the center.
   2 units; (–4, 1)

3. Graph the equation.

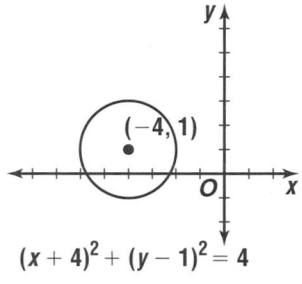

$(x + 4)^2 + (y - 1)^2 = 4$

4. Write the standard form of the equation of the circle that has its center at (1, 2) and a diameter of 6.
   $(x - 1)^2 + (y - 2)^2 = 9$

5. Graph the circle with the endpoints of a diameter at (0, –1) and (0, 13).

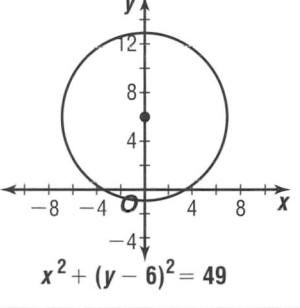

$x^2 + (y - 6)^2 = 49$

Motivating the Lesson
Hands-On Activity Have students draw an ellipse by using tack and string as shown on this page. Ask students what shape would result if both ends of string were attached to one tack. **circle**

Teaching Tip The word *lithotriptor* in the Real-World Application at the beginning of the lesson is pronounced lith'ō trip tôr.

Teaching Tip While working Example 1, caution students not to confuse the relationship that holds in an ellipse, $c^2 = a^2 - b^2$, with the relationship in the Pythagorean Theorem, $a^2 + b^2 = c^2$.

In-Class Example

Example 1
Landscape Architecture The "Ellipse" in Washington, D.C., is an elliptical grassy area between the White House and the Washington Monument. Its major axis is approximately 500 yards and its minor axis is approximately 425 yards. How far apart are the foci?
264 yards

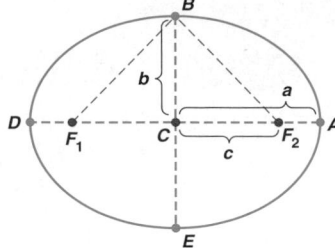

The center separates each axis into two congruent segments. Suppose we let b represent the length of the **semi-minor axis** $\overline{BC}$ and a represent the length of the **semi-major axis** $\overline{CA}$. The foci are located along the major axis, c units from the center. There is a special relationship among the values a, b, and c.

Suppose you draw $\overline{BF_1}$ and $\overline{BF_2}$. The lengths of these segments are equal because $\triangle BF_1C \cong \triangle BF_2C$. Since B and A are two points on the ellipse, you can use the definition of an ellipse to find BF_2.

$$BF_1 + BF_2 = AF_1 + AF_2 \quad \text{\textit{Definition of ellipse}}$$
$$BF_1 + BF_2 = AF_1 + DF_1 \quad \text{\textit{AF}}_2 = DF_1$$
$$BF_1 + BF_2 = AD \quad \text{\textit{Segment addition: AF}}_1 + DF_1 = AD$$
$$BF_1 + BF_2 = 2a \quad \text{\textit{Substitution: AD = 2a}}$$
$$2(BF_2) = 2a \quad \text{\textit{BF}}_1 = BF_2$$
$$BF_2 = a$$

Since $BF_2 = a$ and $\triangle BCF_2$ is a right triangle, $b^2 + c^2 = a^2$ by the Pythagorean Theorem.

Example **① MEDICAL TECHNOLOGY** Refer to the application at the beginning of this lesson. Suppose the reflector of a mobile lithotripter is 24 centimeters wide and 24 centimeters deep. How far, to the nearest hundredth of a centimeter, should the shock wave emitter be placed from a patient's kidney stone?

For the lithotripter to break up the stone, the shock wave emitter must be positioned at one focal point of the ellipse and the kidney stone at the other. To determine the distance between the emitter and the kidney stone, we must first find the lengths of the semi-major and semi-minor axes.

The semi-major axis in this ellipse is equal to the depth of the reflector, 24 centimeters. So, $a = 24$.

The semi-minor axis in this ellipse is half the width of the reflector, 12 centimeters. So, $b = 12$.

To find the focal length of the ellipse, use the formula $c^2 = a^2 - b^2$.

$$c^2 = a^2 - b^2$$
$$c^2 = (24)^2 - (12)^2 \quad \text{\textit{a = 24 and b = 12}}$$
$$c^2 = 432 \quad \text{\textit{Simplify.}}$$
$$c = \sqrt{432} \text{ or approximately } 20.8 \quad \text{\textit{Take the square root of each side.}}$$

The distance between the two foci of the ellipse is $2c$ or 41.6.

Therefore, the emitter should be placed approximately 41.6 centimeters away from the patient's kidney along the ellipse's major axis.

Addressing Individual Needs

Kinesthetic Have students use the activity in Motivating the Lesson on the previous page to make different ellipses. Have them measure the ellipses with a ruler and use the measurements to write the equations that represent the ellipses. Then, have students investigate how moving the foci and changing the length of the string change the ellipses.

The standard form of the equation of an ellipse can be derived from the definition and the distance formula. Consider the special case when the center is at the origin.

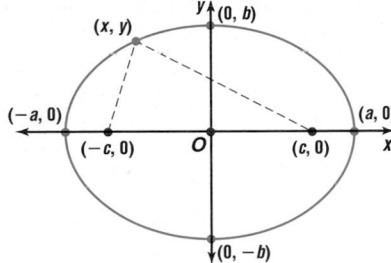

Suppose the foci are at $F_1(-c, 0)$ and $F_2(c, 0)$, and (x, y) is any point on the ellipse. By definition, the sum of the distances from a point at (x, y) to the foci is constant. To find a value for this constant, let (x, y) be one of the vertices on the x-axis, for example, $(a, 0)$. Let d_1 and d_2 be the distances from the point at $(a, 0)$ to F_1 and F_2, respectively. Use the Distance Formula.

$$d_1 = \sqrt{[a - (-c)]^2 + (0 - 0)^2} \qquad\qquad d_2 = \sqrt{(a - c)^2 + (0 - 0)^2}$$
$$= \sqrt{(a + c)^2} \qquad\qquad\qquad\qquad = \sqrt{(a - c)^2}$$
$$= a + c \qquad\qquad\qquad\qquad\qquad = a - c$$

$d_1 + d_2 = a + c + a - c$ or $2a$ Therefore, one value for the constant is $2a$.

Using the Distance Formula and this value for the constant, a general equation for any point at (x, y) is $2a = \sqrt{(x + c)^2 + y^2} + \sqrt{(x - c)^2 + y^2}$.

$$2a = \sqrt{(x + c)^2 + y^2} + \sqrt{(x - c)^2 + y^2}$$

$\sqrt{(x + c)^2 + y^2} = 2a - \sqrt{(x - c)^2 + y^2}$ *Isolate a radical.*

$(x + c)^2 + y^2 = 4a^2 - 4a\sqrt{(x - c)^2 + y^2} + (x - c)^2 + y^2$ *Square each side.*

$a^2 - xc = a\sqrt{(x - c)^2 + y^2}$ *Simplify.*

$a^4 - 2a^2xc + x^2c^2 = a^2[(x - c)^2 + y^2]$ *Square each side again.*

$x^2(a^2 - c^2) + a^2y^2 = a^2(a^2 - c^2)$ *Simplify.*

$\dfrac{x^2}{a^2} + \dfrac{y^2}{a^2 - c^2} = 1$ *Divide each side by $a^2(a^2 - c^2)$.*

$\dfrac{x^2}{a^2} + \dfrac{y^2}{b^2} = 1$ *$b^2 = a^2 - c^2$*

The resulting equation, $\dfrac{x^2}{a^2} + \dfrac{y^2}{b^2} = 1$, is the equation of an ellipse whose center is the origin and whose foci are on the x-axis. When the foci are on the y-axis, the equation is of the form $\dfrac{y^2}{a^2} + \dfrac{x^2}{b^2} = 1$.

The standard form of the equation of an ellipse with a center other than the origin is a translation of the parent graph to a center at (h, k). The table on the next page gives the standard form, graph, and general description of the equation of each type of ellipse.

Lesson 10-3 Ellipses **633**

 Prerequisite Skills

A review of the prerequisite skills for this lesson is provided in the *Prerequisite Skills Masters.*

Skill Number and Content	Pages
7 Operations with Polynomials	15–16

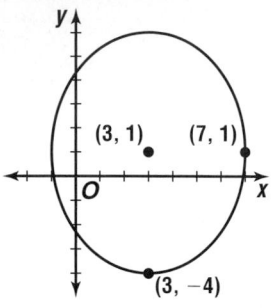
Standard Form of the Equation of an Ellipse	Orientation	Description
$\frac{(x-h)^2}{a^2} + \frac{(y-k)^2}{b^2} = 1,$ where $c^2 = a^2 - b^2$		Center: (h, k) Foci: $(h \pm c, k)$ Major axis: $y = k$ Major axis vertices: $(h \pm a, k)$ Minor axis: $x = h$ Minor axis vertices: $(h, k \pm b)$
$\frac{(y-k)^2}{a^2} + \frac{(x-h)^2}{b^2} = 1,$ where $c^2 = a^2 - b^2$		Center: (h, k) Foci: $(h, k \pm c)$ Major axis: $x = h$ Major axis vertices: $(h, k \pm a)$ Minor axis: $y = k$ Minor axis vertices: $(h \pm b, k)$

Example ❷ Consider the ellipse graphed at the right.

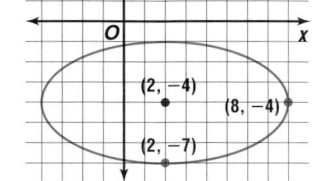

  a. **Write the equation of the ellipse in standard form.**

  b. **Find the coordinates of the foci.**

  a. The center of the graph is at $(2, -4)$. Therefore, $h = 2$ and $k = -4$.

  Since the ellipse's horizontal axis is longer than its vertical axis, a is the distance between points at $(2, -4)$ and $(8, -4)$ or 6. The value of b is the distance between points at $(2, -4)$ and $(2, -7)$ or 3.

  Therefore, the standard form of the equation of this ellipse is
  $$\frac{(x-2)^2}{6^2} + \frac{(y+4)^2}{3^2} = 1 \text{ or } \frac{(x-2)^2}{36} + \frac{(y+4)^2}{9} = 1.$$

  b. Using the equation $c = \sqrt{a^2 - b^2}$, we find that $c = 5$. The foci are located on the horizontal axis, 5 units from the center of the ellipse. Therefore, the foci have coordinates $(7, -4)$ and $(-3, -4)$.

  In all ellipses, $a^2 \geq b^2$. You can use this information to determine the orientation of the major axis from the equation. If a^2 is the denominator of the x^2 term, the major axis is parallel to the x-axis. If a^2 is the denominator of the y^2 term, the major axis is parallel to the y-axis.

Example ❸ For the equation $\dfrac{(y-3)^2}{25} + \dfrac{(x+4)^2}{9} = 1$, find the coordinates of the center, foci, and vertices of the ellipse. Then graph the equation.

Determine the values of a, b, c, h, and k.

Since $a^2 > b^2$, $a^2 = 25$ and $b^2 = 9$.

$$a^2 = 25 \qquad\qquad b^2 = 9 \qquad\qquad c = \sqrt{a^2 - b^2}$$
$$a = \sqrt{25} \text{ or } 5 \qquad b = \sqrt{9} \text{ or } 3 \qquad c = \sqrt{25 - 9} \text{ or } 4$$

$$x - h = x + 4 \qquad\qquad y - k = y - 3$$
$$h = -4 \qquad\qquad\qquad k = 3$$

Since a^2 is the denominator of the y term, the major axis is parallel to the y-axis.

Graphing Calculator Tip

You can graph an ellipse by first solving for y. Then graph the two resulting equations on the same screen.

center: $(-4, 3)$ $\qquad\qquad\qquad$ (h, k)
foci: $(-4, 7)$ and $(-4, -1)$ $\qquad$ $(h, k \pm c)$
major axis vertices: $(-4, 8)$ and $(-4, -2)$ $\quad$ $(h, k \pm a)$
minor axis vertices: $(-1, 3)$ and $(-7, 3)$ $\quad$ $(h \pm b, k)$

Graph these ordered pairs. Other points on the ellipse can be found by substituting values for x and y. Complete the ellipse.

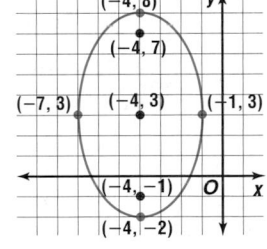

As with circles, the standard form of the equation of an ellipse can be expanded to obtain the general form. The result is a second-degree equation of the form $Ax^2 + Cy^2 + Dx + Ey + F = 0$, where $A \neq 0$ and $C \neq 0$, and A and C have the same sign. An equation in general form can be rewritten in standard form to determine the center at (h, k), the measure of the semi-major axis, a, and the measure of the semi-minor axis, b.

Example ❹ Find the coordinates of the center, the foci, and the vertices of the ellipse with the equation $4x^2 + 9y^2 - 40x + 36y + 100 = 0$. Then graph the equation.

First write the equation in standard form.

$$4x^2 + 9y^2 - 40x + 36y + 100 = 0$$
$$4(x^2 - 10x + \text{?}) + 9(y^2 + 4y + \text{?}) = -100 + \text{?} + \text{?} \quad \textit{The GCF of the x terms is 4.}$$
$$\textit{The GCF of the y terms is 9.}$$
$$4(x^2 - 10x + 25) + 9(y^2 + 4y + 4) = -100 + 4(25) + 9(4) \quad \textit{Complete the square.}$$
$$4(x - 5)^2 + 9(y + 2)^2 = 36 \qquad \textit{Factor.}$$
$$\dfrac{(x - 5)^2}{9} + \dfrac{(y + 2)^2}{4} = 1 \qquad \textit{Divide each side by 36.}$$

(continued on the next page)

Teaching Tip For Example 3, suggest that students list the values of a, b, c, h, and k as a first step.

In-Class Examples
Example 3
For the equation
$$\dfrac{(x + 3)^2}{100} + \dfrac{(y - 4)^2}{64} = 1,$$
find the coordinates of the center, foci, and vertices of the ellipse. Then graph the equation.
center: $(-3, 4)$; **foci:** $(3, 4)$ and $(-9, 4)$; **major axis vertices:** $(7, 4)$ and $(-13, 4)$; **minor axis vertices:** $(-3, 12)$ and $(-3, -4)$
Example 4

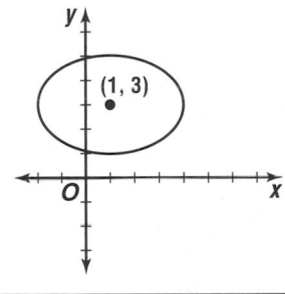

Find the coordinates of the center, the foci, and the vertices of the ellipse with the equation $4x^2 + 9y^2 - 8x - 54y + 49 = 0$. Then graph the equation.
center: $(1, 3)$; **foci:** $\left(1 + \sqrt{5}, 3\right)$, $\left(1 - \sqrt{5}, 3\right)$; **vertices:** $(-2, 3)$, $(4, 3)$, $(1, 5)$, $(1, 1)$

Teaching Tip For Example 4, make sure that students do not forget to multiply by the coefficient when balancing terms as they complete the square.

Teaching Tip Remind students that when they solve for y, they will get both a positive and a negative expression.

Additional Answers (p. 637)

2. Since the foci lie on the major axis, determine whether the major axis is horizontal or vertical. If the a^2 is the denominator of the *x* terms, the major axis is horizontal. If the a^2 is the denominator of the *y* terms, the major axis is vertical.

4. In an ellipse, $b^2 = a^2 - c^2$ and $\dfrac{c}{a} = e$.

$\dfrac{c}{a} = e \rightarrow c = ae \rightarrow c^2 = a^2 e^2$;
$b^2 = a^2 - c^2 = a^2 - a^2 e^2 =$
$a^2(1 - e^2)$

7. center: $(0, 0)$; foci: $\left(\pm 4\sqrt{2}, 0\right)$; vertices: $(\pm 6, 0)$, $(0, \pm 2)$

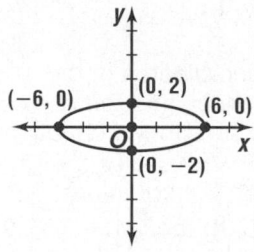

Study Guide and Practice Masters, pp. 181–182

Since $a^2 > b^2$, $a^2 = 9$ and $b^2 = 4$. Thus, $a = 3$ and $b = 2$.

Since $c^2 = a^2 - b^2$, $c = \sqrt{5}$.

Since a^2 is the denominator of the *x* term, the major axis is parallel to the *x*-axis.

center: $(5, -2)$ *(h, k)*
foci: $\left(5 \pm \sqrt{5}, -2\right)$ *(h ± c, k)*
major axis vertices: $(8, -2)$ and $(2, -2)$ *(h ± a, k)*
minor axis vertices: $(5, 0)$ and $(5, -4)$ *(h, k ± b)*

Sketch the ellipse.

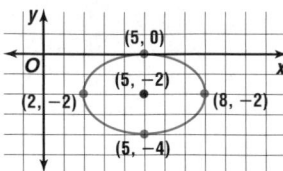

The **eccentricity** of an ellipse, denoted by *e*, is a measure that describes the shape of an ellipse. It is defined as $e = \dfrac{c}{a}$. Since $0 < c < a$, you can divide by *a* to show that $0 < e < 1$.

$0 < c < a$

$0 < \dfrac{c}{a} < 1$ *Divide by a.*

$0 < e < 1$ *Replace $\dfrac{c}{a}$ with e.*

The table shows the relationship between the value of *e*, the location of the foci, and the shape of the ellipse.

Value of *e*	Location of Foci	Graph
close to 0	near center of ellipse	$e = \frac{1}{5}$
close to 1	far from center of ellipse	$e = \frac{5}{6}$

Sometimes, you may need to find the value of *b* when you know the values of *a* and *e*. In any ellipse, $b^2 = a^2 - c^2$ and $\dfrac{c}{a} = e$. By using the two equations, it can be shown that $b^2 = a^2(1 - e^2)$. *You will derive this formula in Exercise 4.*

All of the planets in our solar system have elliptical orbits with the sun as one focus. These orbits are often described by their eccentricity.

Additional Answers (p. 637)

8. center: $(0, 4)$; foci: $\left(\pm 4\sqrt{2}, 4\right)$; vertices: $(\pm 9, 4)$, $(0, 11)$, $(0, -3)$

9. center: $(-2, 1)$; foci: $(-2, 5)$, $(-2, -3)$; vertices: $(-2, 6)$, $(-2, -4)$, $(1, 1)$, $(-5, 1)$

Example **⑤** **ASTRONOMY** Of the nine planetary orbits in our solar system, Pluto's has the greatest eccentricity, 0.248. Astronomers have determined that the orbit is about 29.646 AU (astronomical units) from the sun at its closest point to the sun (perihelion). The length of the semi-major axis is about 39.482 AU.

1 AU = the average distance between the sun and Earth, about 9.3×10^7 miles

a. Sketch the orbit of Pluto showing the sun in its position.

b. Find the length of the semi-minor axis of the orbit.

c. Find the distance of Pluto from the sun at its farthest point (aphelion).

a. The sketch at the right shows the sun as a focus for the elliptical orbit.

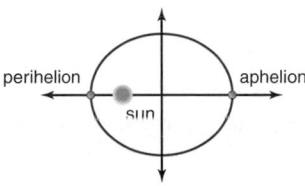

b. b is the length of the semi-minor axis.

$b^2 = a^2(1 - e^2)$

$b = \sqrt{a^2(1 - e^2)}$

$b = \sqrt{(39.482)^2(1 - 0.248^2)}$ $a = 39.482,$ $e = 0.248$

$b \approx 38.249$ AU

c. *aphelion = length of major axis − distance from sun to perihelion*

$d \quad = \quad 2(39.482) \quad - \quad 29.646$

$= 49.318$

Pluto is about 49 AU from the sun at its aphelion.

CHECK FOR UNDERSTANDING

Communicating Mathematics

1. $\dfrac{y^2}{64} + \dfrac{x^2}{25} = 1$

5. Shanice; an equation with only one squared term cannot be the equation of an ellipse.

Guided Practice

7–10. See margin.

Read and study the lesson to answer each question.

1. **Write** the equation of an ellipse centered at the origin, with $a = 8$, $b = 5$, and the major axis on the y-axis.

2. **Explain** how to determine whether the foci of an ellipse lie on the horizontal or vertical axis of an ellipse. **See margin.**

3. **Describe** the result when the foci and center of an ellipse coincide and give the eccentricity of such an ellipse. **a circle, $e = 0$**

4. **Derive** the equation $b^2 = a^2(1 - e^2)$. **See margin.**

5. **You Decide** Manuel says that the graph of $3y - 36 = 2x^2 - 18x$ is an ellipse with its major axis parallel to the y-axis. Shanice disagrees. Who is correct? Explain your answer.

6. Write the equation of the ellipse graphed at the right in standard form. Then find the coordinates of its foci.
$\dfrac{y^2}{36} + \dfrac{(x + 7)^2}{9} = 1$; **foci: $(-7, \pm 3\sqrt{3})$**

For the equation of each ellipse, find the coordinates of the center, foci, and vertices. Then graph the equation.

7. $\dfrac{x^2}{36} + \dfrac{y^2}{4} = 1$

8. $\dfrac{x^2}{81} + \dfrac{(y - 4)^2}{49} = 1$

9. $25x^2 + 9y^2 + 100x - 18y = 116$

10. $9x^2 + 4y^2 - 18x + 16y = 11$

Lesson 10-3 Ellipses **637**

Additional Answer

10. center: $(1, -2)$;
foci: $\left(1, -2 \pm \sqrt{5}\right)$;
vertices: $(1, 1)$, $(1, -5)$, $(3, -2)$, $(-1, -2)$

In-Class Example

Example 5

Astronomy The eccentricity of Uranus is 0.047. Its orbit is about 18.30 AU from the sun at its closest point to the sun. The length of the semi-major axis of the orbit is about 19.21 AU.

a. Sketch the orbit of Uranus showing the sun in its position.

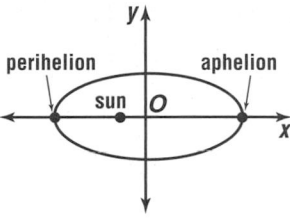

b. Find the length of the semi-minor axis of the orbit.
$b \approx 19.19$ AU

c. Find the distance of Uranus from the sun at its farthest point. **Uranus is about 20 AU from the sun at its aphelion.**

---

3 PRACTICE/APPLY

Error Analysis

Watch for students who mistakenly use $(2a)^2$ and $(2b)^2$—the lengths of the major and minor axes—instead of a and b, the lengths of the *semi*-major and *semi*-minor axes, when deriving the general form of the equation of an ellipse.

Prevent by having them sketch the graph of their resulting equation to see whether it is correct.

Additional Answers

11. $\dfrac{(y+3)^2}{16} + \dfrac{(x+2)^2}{1} = 1$

12. $\dfrac{x^2}{16} + \dfrac{y^2}{15} = 1$

13. $\dfrac{(x-1)^2}{16} + \dfrac{(y-2)^2}{4} = 1$

14. $\dfrac{(y-1)^2}{36} + \dfrac{(x-3)^2}{32} = 1$

16. $\dfrac{x^2}{49} + \dfrac{(y+5)^2}{25} = 1$; foci: $\left(\pm 2\sqrt{6}, -5\right)$

17. $\dfrac{(x+2)^2}{16} + \dfrac{y^2}{4} = 1$; foci: $\left(-2 \pm 2\sqrt{3}, 0\right)$

18. $\dfrac{(y-4)^2}{64} + \dfrac{(x+3)^2}{25} = 1$; foci: $\left(-3, 4 \pm \sqrt{39}\right)$

19. center: $(-2, 1)$; foci: $\left(-2, 1 \pm \sqrt{3}\right)$; vertices: $(-2, 3), (-2, -1), (-1, 1), (-3, 1)$

20. center: $(6, 7)$; foci: $\left(6, 7 \pm \sqrt{21}\right)$; vertices: $(-4, 7), (16, 7), (6, 18), (6, -4)$

21. center: $(4, -6)$; foci: $\left(4 \pm \sqrt{7}, -6\right)$; vertices: $(0, -6), (8, -6), (4, -3), (4, -9)$

22. center: $(0, 0)$; foci: $\left(0, \pm\sqrt{5}\right)$; vertices: $(0, \pm 3)$, $(\pm 2, 0)$

23. center: $(1, -3)$; foci: $\left(1, -3 \pm \sqrt{3}\right)$; vertices: $(2, -3), (0, -3), (1, -1), (1, -5)$

24. center: $(3, 4)$; foci: $(6, 4)$, $(0, 4)$; vertices: $(8, 4), (-2, 4)$, $(3, 0), (3, 8)$

25. center: $(-3, 1)$; foci: $(-3, 5)$, $(-3, -3)$; vertices: $\left(-3 \pm 2\sqrt{2}, 1\right)$, $\left(-3, 1 \pm 2\sqrt{6}\right)$

26. center: $(1, -3)$; foci: $(1, -3)$; vertices: $(5, -3), (-3, -3)$, $(1, 1), (1, -7)$

27. center: $(2, 4)$; foci: $\left(2 \pm \sqrt{6}, 4\right)$; vertices: $\left(2 \pm 3\sqrt{2}, 4\right), \left(2, 4 \pm 2\sqrt{3}\right)$

28. center: $(3, 1)$; foci: $\left(3, 1 \pm \sqrt{5}\right)$; vertices: $(3, 4), (3, -2), (5, 1), (1, 1)$

638 *Chapter 10*

11–14. See margin.

Write the equation of the ellipse that meets each set of conditions.

11. The center is at $(-2, -3)$, the length of the vertical major axis is 8 units, and the length of the minor axis is 2 units.

12. The foci are located at $(-1, 0)$ and $(1, 0)$ and $a = 4$.

13. The center is at $(1, 2)$, the major axis is parallel to the x-axis, and the ellipse passes through points at $(1, 4)$ and $(5, 2)$.

14. The center is at $(3, 1)$, the vertical semi-major axis is 6 units long, and $e = \dfrac{1}{3}$.

15. **Astronomy** The elliptical orbit of Mars has its foci at $(0.141732, 0)$ and $(-0.141732, 0)$, where 1 unit equals 1 AU. The length of the major axis is 3.048 AU. Determine the equation that models Mars' elliptical orbit.
$$\dfrac{x^2}{1.524^2} + \dfrac{y^2}{1.517^2} = 1$$

EXERCISES

Practice

Write the equation of each ellipse in standard form. Then find the coordinates of its foci. **16–18. See margin.**

16.

17.

18.

19–30. See margin; see Answer Appendix for graphs.

For the equation of each ellipse, find the coordinates of the center, foci, and vertices. Then graph the equation.

19. $\dfrac{(x+2)^2}{1} + \dfrac{(y-1)^2}{4} = 1$

20. $\dfrac{(x-6)^2}{100} + \dfrac{(y-7)^2}{121} = 1$

21. $\dfrac{(x-4)^2}{16} + \dfrac{(y+6)^2}{9} = 1$

22. $\dfrac{x^2}{4} + \dfrac{y^2}{9} = 1$

23. $4x^2 + y^2 - 8x + 6y + 9 = 0$

24. $16x^2 + 25y^2 - 96x - 200y = -144$

B

25. $3x^2 + y^2 + 18x - 2y + 4 = 0$

26. $6x^2 - 12x + 6y^2 + 36y = 36$

27. $18y^2 + 12x^2 - 144y - 48x = -120$

28. $4y^2 - 8y + 9x^2 - 54x + 49 = 0$

29. $49x^2 + 16y^2 + 160y - 384 = 0$

30. $9y^2 + 108y + 4x^2 - 56x = -484$

Write the equation of the ellipse that meets each set of conditions.

31–35. See margin.

31. The center is at $(-3, -1)$, the length of the horizontal semi-major axis is 7 units, and the length of the semi-minor axis is 5 units.

32. The foci are at $(-2, 0)$ and $(2, 0)$, and $a = 7$.

33. The length of the semi-minor axis is $\dfrac{3}{4}$ the length of the horizontal semi-major axis, the center is at the origin, and $b = 6$.

34. The semi-major axis has length $2\sqrt{13}$ units, and the foci are at $(-1, 1)$ and $(-1, -5)$.

35. The endpoints of the major axis are at $(-11, 5)$ and $(7, 5)$. The endpoints of the minor axis are at $(-2, 9)$ and $(-2, 1)$.

 www.amc.glencoe.com/self_check_quiz

---

29. center: $(0, -5)$; foci: $\left(0, -5 \pm \sqrt{33}\right)$; vertices: $(\pm 4, -5), (0, -12), (0, 2)$

30. center: $(7, -6)$; foci: $\left(7 \pm \sqrt{5}, -6\right)$; vertices: $(10, -6), (4, -6), (7, -4)$, $(7, -8)$

31. $\dfrac{(x+3)^2}{49} + \dfrac{(y+1)^2}{25} = 1$

32. $\dfrac{x^2}{49} + \dfrac{y^2}{45} = 1$

33. $\dfrac{x^2}{64} + \dfrac{y^2}{36} = 1$

34. $\dfrac{(y+2)^2}{52} + \dfrac{(x+1)^2}{43} = 1$

35. $\dfrac{(x+2)^2}{81} + \dfrac{(y-5)^2}{16} = 1$

36–41. See margin.

36. The foci are at $(1, -1)$ and $(1, 5)$, and the ellipse passes through the point at $(4, 2)$.

37. The center is the origin, $\frac{1}{2} = \frac{c}{a}$, and the length of the horizontal semi-major axis is 10 units.

38. The ellipse is tangent to the x- and y-axes and has its center at $(4, -7)$.

 39. The ellipse has its center at the origin, $a = 2$, and $e = \frac{3}{4}$.

40. The foci are at $(3, 5)$ and $(1, 5)$, and the ellipse has eccentricity 0.25.

41. The ellipse has a vertical major axis of 20 units, its center is at $(3, 0)$, and $e = \frac{7}{10}$.

42. The center is at $(1, -1)$, one focus is located at $\left(1, -1 + \sqrt{5}\right)$, and the ellipse has eccentricity $\frac{\sqrt{5}}{3}$. $\frac{(y + 1)^2}{9} + \frac{(x - 1)^2}{4} = 1$

Graphing Calculator

Graph each equation. Then use the TRACE function to approximate the coordinates of the vertices to the nearest integer. **43–46. See margin.**

43. $x^2 + 4y^2 - 6x + 24y = -41$

44. $4x^2 + y^2 - 8x - 2y = -1$

45. $4x^2 + 9y^2 - 16x + 18y = 11$

46. $25y^2 + 16x^2 - 150y + 32x = 159$

Applications and Problem Solving

Real World Application

48. See margin.

49a. $\frac{x^2}{2304} + \frac{y^2}{529} = 1$

49b. about 42 ft on either side of the center along the major axis

47. **Entertainment** *Elliptipool* is a billiards game that use an elliptically-shaped pool table with only one pocket in the surface. A cue ball and a target ball are used in play. The object of the game is to strike the target ball with the cue ball so that the target ball rolls into the pocket after one bounce off the side. Suppose the cue ball and target ball can be placed anywhere on the half of the table opposite the pocket. The pool table shown at the right is 4 feet wide and 6 feet long. The pocket is located $\sqrt{5}$ feet from the center of the table along the ellipse's major axis. Assuming no spin is placed on either ball and the target ball is struck squarely, where should each be placed to have the best chance of hitting the target ball into the pocket? Explain your reasoning. **See margin.**

$\sqrt{5}$ ft

6 ft

4 ft

48. **Critical Thinking** As the foci of an ellipse move farther apart with the major axis fixed, what figure does the ellipse approach? Justify your answer.

49. **History** A whispering gallery is designed using an elliptical ceiling. It operates on the principle that sound projected from one focus of an ellipse reflects off the ceiling and back to the other focus. The United States Capitol contains such an elliptical room. The room is 96 feet in length, 46 feet in width, and has a ceiling that is about 23 feet high.

   a. Write an equation modeling the shape of the room. Assume that is centered at the origin and the major axis is horizontal.

   b. John Quincy Adams is known to have overheard conversations being held at the opposing party leader's desk by standing in a certain spot in this room. Describe two possible places where Adams might have stood to overhear.

   c. About how far did Adams stand from the desk? **about 84 ft**

 Teaching Tip For Exercises 43–46, remind students to solve each equation for y, then graph the positive and negative solutions in order to see the entire ellipse.

Additional Answers

36. $\frac{(y - 2)^2}{18} + \frac{(x - 1)^2}{9} = 1$

37. $\frac{x^2}{100} - \frac{y^2}{75} = 1$

38. $\frac{(y + 7)^2}{49} + \frac{(x - 4)^2}{16} = 1$

39. $\frac{y^2}{4} + \frac{x^2}{1.75} = 1$ or $\frac{y^2}{1.75} + \frac{x^2}{4} = 1$

40. $\frac{(x - 2)^2}{16} + \frac{(y - 5)^2}{15} = 1$

41. $\frac{y^2}{100} + \frac{(x - 3)^2}{51} = 1$,

43. $(5, -3)$, $(1, -3)$, $(3, -2)$, $(3, -4)$

44. $(0, 1)$, $(2, 1)$, $(1, -1)$, $(1, 3)$

45. $(-1, -1)$, $(5, -1)$, $(2, -3)$, $(2, 1)$

46. $(4, 3)$, $(-6, 3)$, $(-1, 7)$, $(-1, -1)$

47. The target ball should be placed opposite the pocket, $\sqrt{5}$ ft from the center along the major axis of the ellipse. The cue ball can be placed anywhere on the side opposite the pocket. The ellipse has a semi-major axis of length 3 ft and a semi-minor axis of length 2 ft. Using the equation $c^2 = a^2 - b^2$, the focus of the ellipse is found to be $\sqrt{5}$ ft from the center of the ellipse. Thus the hole is located at one focus of the ellipse. The reflective properties of an ellipse should insure that a ball placed $\sqrt{5}$ ft from the center of the ellipse and hit so that it rebounds once off the wall should fall into the pocket at the other focus of the ellipse.

48. A horizontal line; see students' work.

640 Chapter 10

50. Geometry The square has an area of 1 square unit. If the square is stretched horizontally by a factor of a and compressed vertically by a factor of b, the area of the rectangle formed is ab square units. **a.** $A = \pi ab$

a. The area of a circle with equation $x^2 + y^2 = r^2$ is πr^2. Develop a formula for the area of an ellipse with equation $\frac{x^2}{a^2} + \frac{y^2}{b^2} = 1$.

b. Use the formula found in part **a** to find the area of the ellipse $\frac{x^2}{9} + \frac{y^2}{4} = 1$.   **$6\pi$ units²**

51. Critical Thinking Show that the ellipse $\frac{x^2}{a^2} + \frac{y^2}{b^2} = 1$ is symmetric with respect to the origin.   **See margin.**

52. Construction The arch of a fireplace is to be constructed in the shape of a semi-ellipse. The opening is to have a height of 3 feet at the center and a width of 8 feet along the base. To sketch the outline of the fireplace, the contractor uses an 8-foot string tied to two thumbtacks.

a. Where should the thumbtacks be placed? $\left(\pm\sqrt{7}, 0\right)$ **from the center of the arch**

b. Explain why this technique works.   **See margin.**

3 ft

8 ft

53. Astronomy The satellites orbiting Earth follow elliptical paths with the center of Earth as one focus. The table below lists data on five satellites that have orbited or currently orbit Earth.

Satellites Orbiting Earth

Name	Launch Date	Time or expected time aloft	Semi-major axis *a* (km)	Eccentricity
Sputnik I	Oct. 1957	57.6 days	6955	0.052
Vanguard I	Mar. 1958	300 years	8872	0.208
Skylab 4	Nov. 1973	84.06 days	6808	0.001
GOES 4	Sept. 1980	10^6 years	42,166	0.0003
Intelsat 5	Dec. 1980	10^6 years	41,803	0.007

53a. GOES 4; its eccentricity is closest to 0

53b. 960 km

a. Which satellite has the most circular orbit? Explain your reasoning.

b. Soviet satellite Sputnik I was the first artificial satellite to orbit Earth. If the radius of Earth is approximately 6357 kilometers, find the greatest distance Sputnik I orbited from the surface of Earth to the nearest kilometer.

Mixed Review

54. $(x + 5)^2 + (y - 3)^2 = 169$; $(-5, 3)$; 13; See margin for graph.

54. Write the standard form of the equation of the circle that passes through points at $(0, -9)$, $(7, -2)$, and $(-5, -10)$. Identify the circle's center and radius. Then graph the equation. *(Lesson 10-2)*

55. Determine whether the quadrilateral with vertices at points $(-1, -2)$, $(5, -4)$, $(4, 1)$, and $(-5, 4)$ is a parallelogram. *(Lesson 10-1)*   **no**

640 Chapter 10   Conics

56. If $\sin \theta = \frac{7}{8}$ and the terminal side of θ is in the first quadrant, find $\cos 2\theta$. *(Lesson 7-4)* $-\frac{17}{32}$

57. Write an equation of the cosine function with an amplitude of 4, a period of 180°, and a phase shift of 20°. *(Lesson 6-5)* $y = \pm 4 \cos(2x - 40°)$

58. $A = 16°23'$, $b = 9.7$, $c = 12.4$

58. Solve $\triangle ABC$ if $C = 121°32'$, $B = 42°5'$, and $a = 4.1$. Round angle measures to the nearest minute and side measures to the nearest tenth. *(Lesson 5-6)*

59. Use the Remainder Theorem to find the remainder for the quotient $(x^4 - 4x^3 - 2x^2 - 1) \div (x - 5)$. Then, state whether the binomial is a factor of the polynomial. *(Lesson 4-3)* **74, no**

60. mininum

60. Determine whether the point at $(-2, -16)$ is the location of a *minimum*, a *maximum*, or a *point of inflection* for the function $x^2 + 4x - 12$. *(Lesson 3-6)*

61. Sketch the graph of $g(x) = |x - 2|$. *(Lesson 3-2)* **See margin.**

62. **Motion** A ceiling fan has four evenly spaced blades that are each 2 feet long. Suppose the center of the fan is located at the origin and the blades of the fan lie in the x- and y-axis. Imagine a fly lands on the tip of the blade along the positive x-axis. Find the location of the fly where it landed and its location after a 90°, 180°, and 270° counterclockwise rotation of the fan. *(Lesson 2-4)* **(2, 0), (0, 2), (−2, 0), (0, −2)**

63. **SAT Practice** In parallelogram $QRST$, $a = b$ and $c = d$. What is the value of p? **C**

   A 45   B 60   C 90   D 120

   E Cannot be determined from the information given

GRAPHING CALCULATOR EXPLORATION

To graph an ellipse, such as the equation $\frac{(x-3)^2}{18} + \frac{(y+2)^2}{32} = 1$, on a graphing calculator, you must first solve for y.

$$\frac{(x-3)^2}{18} + \frac{(y+2)^2}{32} = 1$$
$$32(x-3)^2 + 18(y-2)^2 = 576$$
$$18(y-2)^2 = 576 - 32(x-3)^2$$
$$(y-2)^2 = \frac{576 - 32(x-3)^2}{18}$$

So, $y = \pm\sqrt{\dfrac{576 - 32(x-3)^2}{18}} + 2$.

The result is two equations. To graph both equations as Y₁, replace $\pm$ with $\{1, -1\}$.

Like other graphs, there are families of ellipses. Changing certain values in the equation of an ellipse creates a new member of that family.

TRY THESE For each situation, make a conjecture about the behavior of the graph. Then verify by graphing the original equation and the modified equation on the same screen using a square window.

1. x is replaced by $(x - 4)$.
2. x is replaced by $(x + 4)$.
3. y is replaced by $(y - 4)$.
4. y is replaced by $(y + 4)$.
5. 32 is switched with 18. **1–8. See margin.**

WHAT DO YOU THINK?

6. Describe the effects of replacing x in the equation of an ellipse with $(x \pm c)$ for $c > 0$.

7. Describe the effects of replacing y in the equation of an ellipse with $(y \pm c)$ for $c > 0$.

8. In the equation $\frac{x^2}{a^2} + \frac{y^2}{b^2} = 1$, describe the effect of interchanging a and b.

Extra Practice See p. A44.

Lesson 10-3 Ellipses **641**

Extra Credit

The great Indian mathematician Srinivasa Ramanujan discovered an approximation for the circumference of an ellipse, using the formula

$$C = \pi\left[3(a + b) - \sqrt{(a + 3b)(3a + b)}\right].$$

Use the formula to approximate the circumference of an ellipse (to six decimal places) if $a = 5$ and $b = 3$. **25.526986**

Hyperbolas

1 FOCUS

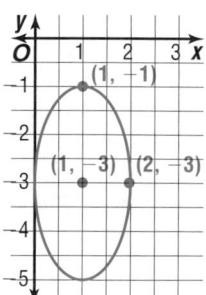

5-Minute Check
Lesson 10-3

Find each of the following for the ellipse with equation $3x^2 + 4y^2 - 6x + 16y + 7 = 0$.

1. equation in standard form
$\dfrac{(x-1)^2}{4} + \dfrac{(y+2)^2}{3} = 1$

2. coordinates of the center
$(1, -2)$

3. coordinates of the foci
$(2, -2), (0, -2)$

4. coordinates of the vertices
$(-1, -2), (3, -2),$
$\left(1, -2 + \sqrt{3}\right), \left(1, -2 - \sqrt{3}\right)$

5. Write the equation of the ellipse in standard form. Then find the coordinates of its foci.

$\dfrac{(x-1)^2}{1} + \dfrac{(y+3)^2}{4} = 1;$

$\left(1, -3 + \sqrt{3}\right),$

$\left(1, -3 - \sqrt{3}\right)$

Motivating the Lesson

Hands-On Activity Draw a set of coordinate axes on the chalkboard. Ask students to name ordered pairs satisfying the equation $xy = 16$. Remind students to think beyond just integer solutions. Then have them plot the points and show the two branches of the hyperbola $xy = 16$ in the first and third quadrants.

OBJECTIVES
* Use and determine the standard and general forms of the equation of a hyperbola.
* Graph hyperbolas.

NAVIGATION Since World War II, ships have used the LORAN (*LO*ng *RA*nge Navigation) system as a means of navigation independent of visibility conditions. Two stations, located a great distance apart, simultaneously transmit radio pulses to ships at sea. Since a ship is usually closer to one station than the other, the ship receives these pulses at slightly different times. By measuring the time differential and by knowing the speed of the radio waves, a ship can be located on a conic whose foci are the positions of the two stations.

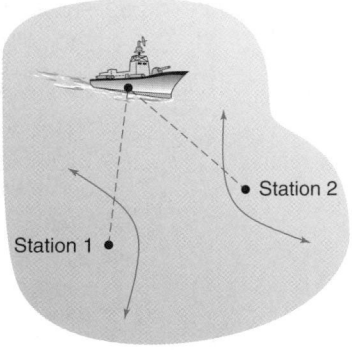

A problem related to this will be solved in Example 4.

A **hyperbola** is the set of all points in the plane in which the difference of the distances from two distinct fixed points, called **foci,** is constant. That is, if F_1 and F_2 are the foci of a hyperbola and P and Q are any two points on the hyperbola, $|PF_1 - PF_2| = |QF_1 - QF_2|$.

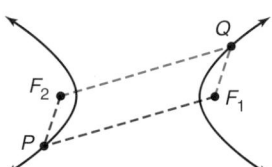

The **center** of a hyperbola is the midpoint of the line segment whose endpoints are the foci. The point on each branch of the hyperbola that is nearest the center is called a **vertex.**

Look Back
Refer to Lesson 3-7 to review asymptotes.

The **asymptotes** of a hyperbola are lines that the curve approaches as it recedes from the center. As you move farther out along the branches, the distance between points on the hyperbola and the asymptotes approaches zero.

A hyperbola has two axes of symmetry. The line segment connecting the vertices is called the **transverse axis** and has a length of $2a$ units. The segment perpendicular to the transverse axis through the center is called the **conjugate axis** and has length $2b$ units.

For a hyperbola, the relationship among a, b, and c is represented by $a^2 + b^2 = c^2$. The asymptotes contain the diagonals of the rectangle guide, which is $2a$ units by $2b$ units. The point at which the diagonals meet coincides with the center of the hyperbola.

Note that $c > a$ for the hyperbola.

642 *Chapter 10 Conics*

Resource Manager

 Reproducible Masters
* *Study Guide and Practice,* pp. 183–184
* *Enrichment,* p. 112
* *Assessment and Evaluation,* pp. 250–251

 Transparencies
* 5-Minute Check, 10-4
* Teaching, 10-4

 Technology/Multimedia
* Advanced MathPASS, Lesson 22

The standard form of the equation of a hyperbola with its origin as its center can be derived from the definition and the Distance Formula. Suppose the foci are on the *x*-axis at $(c, 0)$ and $(-c, 0)$ and the coordinates of any point on the hyperbola are (x, y).

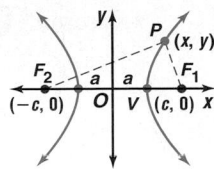

$$|PF_2 - PF_1| = |VF_2 - VF_1| \qquad \textit{Definition of hyperbola}$$

$$\left| \sqrt{(x+c)^2 + y^2} - \sqrt{(x+c)^2 + y^2} \right| = |c + a - (c - a)| \qquad \textit{Distance Formula}$$

$$\sqrt{(x+c)^2 + y^2} - \sqrt{(x+c)^2 + y^2} = 2a \qquad \textit{Simplify.}$$

$$\sqrt{(x-c)^2 + y^2} = 2a + \sqrt{(x+c)^2 + y^2} \quad \textit{Isolate a radical.}$$

$$(x-c)^2 + y^2 = 4a^2 + 4a\sqrt{(x+c)^2 + y^2} + (x+c)^2 + y^2 \quad \textit{Square each side.}$$

$$-4xc - 4a^2 - 4a\sqrt{(x+c)^2 + y^2} \qquad \textit{Simplify.}$$

$$xc + a^2 = -a\sqrt{(x+c)^2 + y^2} \qquad \textit{Divide each side by } -4.$$

$$x^2c^2 + 2a^2xc + a^4 = a^2x^2 + 2a^2xc + a^2c^2 + a^2y^2 \quad \textit{Square each side.}$$

$$(c^2 - a^2)x^2 - a^2y^2 = a^2(c^2 - a^2) \qquad \textit{Simplify.}$$

$$\frac{x^2}{a^2} - \frac{y^2}{c^2 - a^2} = 1 \qquad \textit{Divide by } a^2(c^2 - a^2).$$

$$\frac{x^2}{a^2} - \frac{y^2}{b^2} = 1 \qquad \textit{By the Pythagorean Theorem, } c^2 - a^2 = b^2.$$

If the foci are on the *y*-axis, the equation is $\dfrac{y^2}{a^2} - \dfrac{x^2}{b^2} = 1$.

As with the other graphs we have studied in this chapter, the standard form of the equation of a hyperbola with center other than the origin is a translation of the parent graph to a center at (h, k).

Standard Form of the Equation of a Hyperbola	Orientation	Description
$\dfrac{(x-h)^2}{a^2} - \dfrac{(y-k)^2}{b^2} = 1$, where $b^2 = c^2 - a^2$	$\dfrac{(x-h)^2}{a^2} - \dfrac{(y-k)^2}{b^2} = 1$	center: (h, k) foci: $(h \pm c, k)$ vertices: $(h \pm a, k)$ equation of transverse axis: $y = k$ *(parallel to x-axis)*
$\dfrac{(y-k)^2}{a^2} - \dfrac{(x-h)^2}{b^2} = 1$, where $b^2 = c^2 - a^2$	$\dfrac{(y-k)^2}{a^2} - \dfrac{(x-h)^2}{b^2} = 1$	center: (h, k) foci: $(h, k \pm c)$ vertices: $(h, k \pm a)$ equation of transverse axis: $x = h$ *(parallel to y-axis)*

Teaching Tip While discussing the standard form of the equation of a hyperbola, point out that for hyperbolas, a, b, and c are related by $a^2 + b^2 = c^2$, not $a^2 - b^2 = c^2$ as they were for ellipses.

In-Class Example

Example 1
Find the equation of the hyperbola with foci at $(1, -5)$ and $(1, 1)$ and whose transverse axis is 4 units long.

$$\frac{(y + 2)^2}{4} - \frac{(x - 1)^2}{5} = 1$$

Example **1** **Find the equation of the hyperbola with foci at (7, 1) and (−3, 1) whose transverse axis is 8 units long.**

A sketch of the graph is helpful. Let F_1 and F_2 be the foci, let V_1 and V_2 be the vertices, and let C be the center.

To locate the center, find the midpoint of $\overline{F_1F_2}$.

$$\left(\frac{7 + (-3)}{2}, \frac{1 + 1}{2}\right) \text{ or } (2, 1)$$

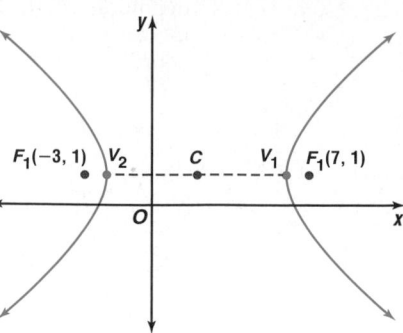

Thus, $h = 2$ and $k = 1$ since $(2, 1)$ is the center.

The transverse axis is 8 units long. Thus, $2a = 8$ or $a = 4$. So $a^2 = 16$.

Use the equation $b^2 = c^2 - a^2$ to find b^2.

$b^2 = c^2 - a^2$

$b^2 = 25 - 16 \quad c = 5, a = 4$

$b^2 = 9$

Recall that c is the distance from the center to a focus. Here $c = CF_1$ or 5.

Use the standard form when the transverse axis is parallel to the x-axis.

$$\frac{(x - h)^2}{a^2} - \frac{(y - k)^2}{b^2} = 1 \quad \Rightarrow \quad \frac{(x - 2)^2}{16} - \frac{(y - 1)^2}{9} = 1$$

Before graphing a hyperbola, it is often helpful to graph the asymptotes. As noted in the beginning of the lesson, the asymptotes contain the diagonals of the rectangle guide defined by the transverse and conjugate axes. While not part of the graph, a sketch of this $2a$ by $2b$ rectangle provides an easy way to graph the asymptotes of the hyperbola. Suppose the center of a hyperbola is the origin and the transverse axis lies along the x-axis.

From the figure at the right, we can see that the asymptotes have slopes equal to $\pm\dfrac{b}{a}$. Since both lines have a y-intercept of 0, the equations for the asymptotes are $y = \pm\dfrac{b}{a}x$.

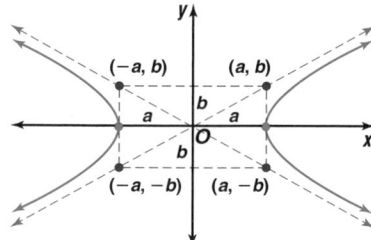

If the hyperbola were oriented so that the traverse axis was parallel to the y-axis, the slopes of the asymptotes would be $\pm\dfrac{a}{b}$. Thus, the equations of the asymptotes would be $y = \pm\dfrac{a}{b}x$.

644 *Chapter 10 Conics*

Addressing Individual Needs

Visual/Spatial Have pairs of students make models of hyperboloids. They should connect two sturdy cardboard circles with a dowel rod. Each pair of students could use circles of differing diameters. Have them punch equally spaced holes around the circumference of each circle. Then have them thread nylon string or dental floss through the holes. Have students twist the two circles in opposite directions to produce a changing hyperboloid. Discuss what aspects of a hyperboloid are affected when the circles are twisted more tightly. **The curvature of the hyperbola changes.**

The equations of the asymptotes of any hyperbola can be determined by a translation of the graph to a center at (h, k).

Equations of the Asymptotes of a Hyperbola

$y - k = \pm\dfrac{b}{a}(x - h)$, for a hyperbola with standard form $\dfrac{(x - h)^2}{a^2} - \dfrac{(y - k)^2}{b^2} = 1$	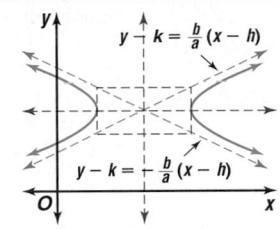
$y - k = \pm\dfrac{a}{b}(x - h)$, for a hyperbola with standard form $\dfrac{(y - k)^2}{a^2} - \dfrac{(x - h)^2}{b^2} = 1$	

Example ❷ **Find the coordinates of the center, foci, and vertices, and the equations of the asymptotes of the graph of $\dfrac{(y + 4)^2}{36} - \dfrac{(x - 2)^2}{25} = 1$. Then graph the equation.**

Since the y terms are in the first expression, the hyperbola has a vertical transverse axis.

From the equation, $h = 2$, $k = -4$, $a = 6$, and $b = 5$. The center is at $(2, -4)$.

Graphing Calculator Tip

You can graph a hyperbola on a graphing calculator by first solving for y and then graphing the two resulting equations on the same screen.

The equations of the asymptotes are $y + 4 = \pm\dfrac{6}{5}(x - 2)$.

The vertices are at $(h, k \pm a)$ or $(2, 2)$ and $(2, -10)$.

Since $c^2 = a^2 + b^2$, $c = \sqrt{61}$. Thus, the foci are at $\left(2, -4 + \sqrt{61}\right)$ and $\left(2, -4 - \sqrt{61}\right)$.

Graph the center, vertices, and the rectangle guide. Next graph the asymptotes. Then sketch the hyperbola.

By expanding the standard form for a hyperbola, you can determine the general form $Ax^2 + Cy^2 + Dx + Ey + F = 0$ where $A \neq 0$, $C \neq 0$, and A and C have different signs.

Teaching Tip For students who have difficulty graphing hyperbolas, stress the importance of the asymptotes and vertices of the hyperbola. Students can find a and b from the standard form of the equation and use them to draw the rectangle whose diagonals are the asymptotes.

In-Class Example

Example 2
Find the coordinates of the center, foci, and vertices, and the equations of the asymptotes of the graph of $\dfrac{(x - 5)^2}{25} - \dfrac{(y + 1)^2}{9} = 1$. Then graph the equation.
center: $(5, -1)$;
foci: $\left(5 \pm \sqrt{34}, -1\right)$;
vertices: $(0, -1)$, $(10, -1)$;
asymptotes: $y + 1 = \pm\dfrac{3}{5}(x - 5)$
or $\begin{cases} y = \dfrac{3}{5}x - 4 \\ y = -\dfrac{3}{5}x + 2 \end{cases}$

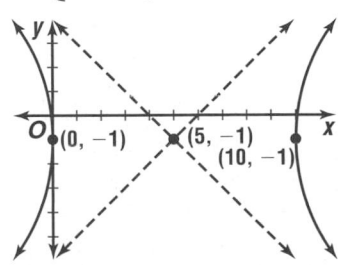

Teaching Tip Point out that in the general-form equation for an ellipse, A and C have the same sign. For a hyperbola, A and C have different signs. This is a quick way to differentiate the equations for the two shapes.

Teaching Tip Encourage students to always sketch a picture for real-world problems such as Example 4. Labeling the picture will help students keep track of data.

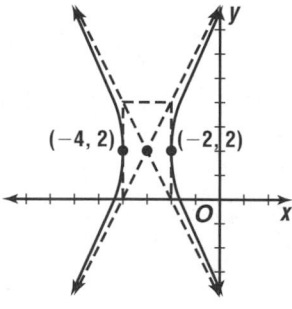
As with other general forms we have studied, the general form of a hyperbola can be rewritten in standard form. While it is important to be able to recognize the equation of a hyperbola in general form, the standard form provides important information about the hyperbola that makes it easier to graph.

Example ③ Find the coordinates of the center, foci, and vertices, and the equations of the asymptotes of the graph of $9x^2 - 4y^2 - 54x - 40y - 55 = 0$. Then graph the equation.

Write the equation in standard form. Use the same process you used with ellipses.

$$9x^2 - 4y^2 - 54x - 40y - 55 = 0$$
$$9x^2 - 54x - 4y^2 - 40y = 55 \qquad \textit{Rearrange terms.}$$
$$9(x^2 - 6x + ?) - 4(y^2 + 10y + ?) = 55 + ? + ? \qquad \textit{Factor GCF for each variable.}$$
$$9(x^2 - 6x + 9) - 4(y^2 + 10y + 25) = 55 + 9(9) - 4(25) \qquad \textit{Complete the square.}$$
$$9(x - 3)^2 - 4(y + 5)^2 = 36 \qquad \textit{Factor.}$$
$$\frac{(x - 3)^2}{4} - \frac{(y + 5)^2}{9} = 1 \qquad \textit{Divide each side by 36.}$$

The center is at $(3, -5)$. Since the x terms are in the first expression, the transverse axis is horizontal.

$a = 2, b = 3, c = \sqrt{13}$

The foci are at $\left(3 - \sqrt{13}, -5\right)$ and $\left(3 + \sqrt{13}, -5\right)$.

The vertices are at $(1, -5)$ and $(5, -5)$.

The asymptotes have equations $y + 5 = \pm\frac{3}{2}(x - 3)$.

Graph the vertices and the rectangle guide. Next graph the asymptotes. Then sketch the hyperbola.

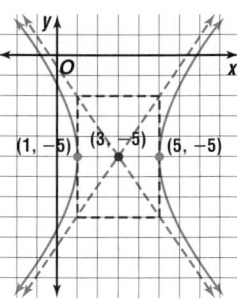

One interesting application of hyperbolas is in navigation.

Example ④ **NAVIGATION** Refer to the application at the beginning of the lesson. Suppose LORAN stations A and B are located 400 miles apart along a straight shore, with A due west of B. A ship approaching the shore receives radio pulses from the stations and is able to determine that it is 100 miles farther from station A than it is from station B.

a. Find the equation of the hyperbola on which the ship is located.

b. Find the exact coordinates of the ship if it is 60 miles from shore.

a. First set up a rectangular coordinate system with the origin located midway between station A and station B.

The stations are located at the foci of the hyperbola, so $c = 200$.

The difference of the distances from the ship to each station is 100 miles. By definition of a hyperbola, this difference equals $2a$, so $a = 50$. The vertices of the hyperbola are located on the same axis as the foci, so the vertices of the hyperbola the ship is on are at $(-50, 0)$ and $(50, 0)$.

$(-200, 0)$ O $(200, 0)$ x

Since the hyperbola's transverse axis is the x-axis, the form of the equation of this hyperbola is $\frac{x^2}{a^2} - \frac{y^2}{b^2} = 1$. Using the equation $b^2 = c^2 - a^2$, we can find the value for b^2.

$b^2 = c^2 - a^2$
$b^2 = 200^2 - 50^2 \quad$ *a = 50, c = 200*
$b^2 = 37{,}500$

Thus, the equation of the hyperbola is $\frac{x^2}{2500} - \frac{y^2}{37{,}500} = 1$

b. If the ship is 60 miles from shore, let $y = 60$ in the equation of the hyperbola and solve for x.

$$\frac{x^2}{2500} - \frac{60^2}{37{,}500} = 1$$

$$\frac{x^2}{2500} = 1 + \frac{60^2}{37{,}500}$$

$$\frac{x^2}{2500} = 1.096$$

$$x^2 = 2500(1.096)$$

$$x = \pm\sqrt{2740} \approx \pm 52.3$$

Since the ship is closer to station B than station A, we use the positive value of x to locate the ship at coordinates $(52.3, 60)$.

In the standard form of the equation of a hyperbola, if $a = b$, the graph is an **equilateral hyperbola.** Replacing a with b in the equations of the asymptotes of a hyperbola with a horizontal transverse axis reveals a property of equilateral hyperbolas.

$y - k = \frac{b}{a}(x - h)$ $y - k = -\frac{b}{a}(x - h)$

$y - k = \frac{b}{b}(x - h)$ *Let a = b* $y - k = -\frac{b}{b}(x - h)$

$y - k = (x - h)$ $y - k = -(x - h)$

The slopes of the equations of the two asymptotes are negative reciprocals, 1 and -1. Thus, the asymptotes of an equilateral hyperbola are perpendicular.

Remember that $xy = c$ is the general equation that models inverse variation.

A special case of the equilateral hyperbola is a **rectangular hyperbola,** where the coordinate axes are the asymptotes. The general equation of a rectangular hyperbola is $xy = c$, where c is a nonzero constant. The sign of the constant c determines the location of the branches of the hyperbola.

Rectangular Hyperbola: $xy = c$	
Value of c	**Location of branches of hyperbola**
Positive	Quadrants I and III
Negative	Quadrants II and IV

Example **Graph $xy = 16$.**

Since c is positive, the hyperbola lies in the first and third quadrants.

The transverse axis is along the graph of $y = x$.

The coordinates of the vertices must satisfy the equation of the hyperbola and also their graph must be points on the traverse axis. Thus, the vertices are at (4, 4) and (−4, −4).

Like an ellipse, the shape of a hyperbola is determined by its eccentricity, which is again defined as $e = \frac{c}{a}$. However, in a hyperbola, $0 < a < c$. So, $0 < 1 < e$ or $e > 1$. The table below shows the relationship between the value of the e and the shape of the hyperbola.

Value of e	Graph
close to 1	$e = \frac{9}{8}$
not close to 1	$e = \frac{13}{5}$

Since $c^2 = a^2 + b^2$ in hyperbolas, it can be shown that $b^2 = a^2(e^2 - 1)$. *You will derive this formula in Exercise 3.*

Example

6 Write the equation of the hyperbola with center at $(3, -1)$, a focus at $(3, -4)$, and eccentricity $\frac{3}{2}$.

Sketch the graph using the points given. Since the center and focus have the same x-coordinate, the transverse axis is vertical. Use the form
$$\frac{(y-k)^2}{a^2} - \frac{(x-h)^2}{b^2} = 1.$$

The focus is 3 units below the center, so $c = 3$. Now use the eccentricity to find the values of a^2 and b^2.

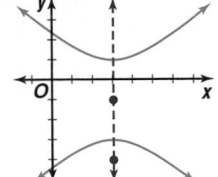

$e = \frac{c}{a}$ $b^2 = a^2(e^2 - 1)$

$\frac{3}{2} = \frac{3}{a}$ $c = 3$ and $e = \frac{3}{2}$ $b^2 = 4\left(\frac{9}{4} - \frac{4}{4}\right)$ $a^2 = 4$ and $e = \frac{3}{2}$

$2 = a$ $b^2 = 5$

$4 = a^2$

The equation is $\dfrac{(y+1)^2}{4} - \dfrac{(x-3)^2}{5} = 1$.

Teaching Tip In Example 6, make sure that students sketch the graph to help them choose the correct equation.

In-Class Example

Example 6
Write the equation of the hyperbola with center at $(-2, -4)$, a focus at $(2, -4)$, and eccentricity $\frac{4}{3}$.

$\dfrac{(x+2)^2}{9} - \dfrac{(y+4)^2}{7} = 1$

3 PRACTICE/APPLY

Error Analysis

Watch for students who use the incorrect form of the equation for the given orientation.

Prevent by having students check the table on page 645 every time they choose a form.

CHECK FOR UNDERSTANDING

Communicating Mathematics

Read and study the lesson to answer each question.   1, 3, 4. See margin.

1. **Compare and contrast** the standard forms of the equations of hyperbolas and ellipses.

2. **Determine** which of the following equations matches the graph of the hyperbola at right.   **b**

   a. $\dfrac{x^2}{4} - y^2 = 1$ **b.** $\dfrac{y^2}{4} - x^2 = 1$ **c.** $x^2 - \dfrac{y^2}{4} = 1$

3. **Derive** the equation $b^2 = a^2(e^2 - 1)$ for a hyperbola.

4. *Math Journal* **Write** an explanation of how to determine whether the transverse axis of a hyperbola is horizontal or vertical.

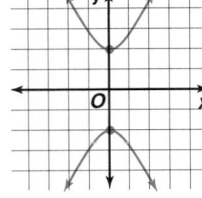

Guided Practice

For the equation of each hyperbola, find the coordinates of the center, the foci, and the vertices and the equations of the asymptotes of its graph. Then graph the equation.   5–7. See margin.

5. $\dfrac{x^2}{25} - \dfrac{y^2}{4} = 1$

6. $\dfrac{(y-3)^2}{16} - \dfrac{(x-2)^2}{4} = 1$

7. $y^2 - 5x^2 + 20x = 50$

8. Write the equation of the hyperbola graphed at the right. $\dfrac{x^2}{25} - \dfrac{(y-5)^2}{9} = 1$

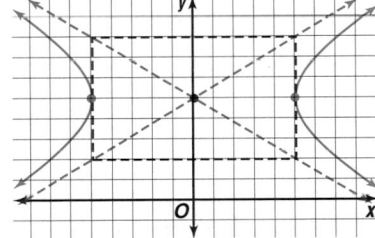

9. Graph $xy = -9$.
   See margin.

Study Guide and Practice Masters, pp. 183–184

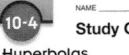

10-4   **Study Guide**

Hyperbolas

The standard form of the equation of a **hyperbola** is $\frac{(x-h)^2}{a^2} - \frac{(y-k)^2}{b^2} = 1$ when the **transverse axis** is horizontal, and $\frac{(y-k)^2}{a^2} - \frac{(x-h)^2}{b^2} = 1$ when the transverse axis is vertical. In both cases, $b^2 = c^2 - a^2$.

Example **Find the coordinates of the center, foci, and vertices, and the equations of the asymptotes of the graph of** $25x^2 - 9y^2 + 100x - 54y - 206 = 0$. **Then graph the equation.**

Write the equation in standard form.

$25x^2 - 9y^2 + 100x - 54y - 206 = 0$
$25(x^2 + 4x + ?) - 9(y^2 + 6y + ?) = 206 + ? + ?$ *GCF of x terms = 25; GCF of y terms = 9.*
$25(x^2 + 4x + 4) - 9(y^2 + 6y + 9) = 206 + 25(4) + (-9)(9)$ *Complete the square.*
$25(x+2)^2 - 9(y+3)^2 = 225$ *Factor.*
$\frac{(x+2)^2}{9} - \frac{(y+3)^2}{25} = 1$ *Divide each side by 225.*

From the equation, $h = -2$, $k = -3$.
$a = 3$, $b = 5$, and $c = \sqrt{34}$. The center is at $(-2, -3)$.

Since the x terms are in the first expression, the hyperbola has a horizontal transverse axis.

The vertices are at $(h \pm a, k)$ or $(1, -3)$ and $(-5, -3)$.

The foci are at $(h \pm c, k)$ or $(-2 \pm \sqrt{34}, -3)$.

The equations of the asymptotes are $y - k = \pm\frac{b}{a}(x-h)$ or $y + 3 = \pm\frac{5}{3}(x+2)$.

Graph the center, the vertices, and the rectangle guide, which is $2a$ units by $2b$ units. Next graph the asymptotes. Then sketch the hyperbola.

© Glencoe/McGraw-Hill   183   *Advanced Mathematical Concepts*

Additional Answers

7. center: $(2, 0)$;
   foci: $(2, \pm 6)$;
   vertices: $\left(2, \pm\sqrt{30}\right)$;
   asymptotes:
   $y = \pm\sqrt{5}\,(x - 2)$

9.

Additional Answers

10. $\dfrac{(x-1)^2}{25} - \dfrac{(y+4)^2}{4} = 1$

11. $\dfrac{(y-2)^2}{4} - \dfrac{(x-3)^2}{9} = 1$

12. $\dfrac{y^2}{18} - \dfrac{x^2}{18} = 1$

13. $\dfrac{x^2}{36} - \dfrac{y^2}{64} = 1$

15. center: $(0, 0)$; foci: $\left(\pm 2\sqrt{29}, 0\right)$; vertices: $(\pm 10, 0)$; asymptotes: $y = \pm\dfrac{2}{5}x$

16. center: $(0, 5)$; foci: $\left(\pm 3\sqrt{10}, 5\right)$; vertices: $(\pm 3, 5)$; asymptotes: $y - 5 = \pm 3x$

17. center: $(0, 0)$, foci: $\left(\pm\sqrt{53}, 0\right)$; vertices: $(\pm 2, 0)$; asymptotes: $y = \pm\dfrac{7}{2}x$

18. center: $(-1, 7)$; foci: $\left(-1, 7 \pm 2\sqrt{17}\right)$; vertices: $(-1, 15), (-1, -1)$; asymptotes: $y - 7 = \pm 4(x + 1)$

19. center: $(-3, -1)$; foci: $\left(-3 \pm 2\sqrt{5}, -1\right)$, vertices: $(1, -1), (-7, -1)$; asymptotes: $y + 1 = \pm\dfrac{1}{2}(x + 3)$

20. center: $(-3, 5)$; foci: $\left(-3, 5 \pm\sqrt{13}\right)$; vertices: $(-3, 7), (-3, 3)$; asymptotes: $y - 5 = \pm\dfrac{2}{3}(x + 3)$

21. center: $(2, 3)$; foci: $\left(2, 3 \pm\sqrt{41}\right)$; vertices: $(2, 8), (2, -2)$; asymptotes: $y - 3 = \pm\dfrac{5}{4}(x - 2)$

22. center: $(1, -3)$; foci: $\left(1 + \sqrt{85}, -3\right), \left(1 - \sqrt{85}, -3\right)$; vertices: $(8, -3), (-6, -3)$; asymptotes: $y + 3 = \pm\dfrac{6}{7}(x - 1)$

23. center: $(-4, 2)$; foci: $\left(-4, 2 \pm\sqrt{34}\right)$; vertices: $(-4, 5), (-4, -1)$; asymptotes: $y - 2 = \pm\dfrac{3}{5}(x + 4)$

10–13. See margin.

Write an equation of the hyperbola that meets each set of conditions.

10. The center is at $(1, -4)$, $a = 5$, $b = 2$, and it has a horizontal transverse axis.

11. The length of the conjugate axis is 6 units, and the vertices are at $(3, 4)$ and $(3, 0)$.

12. The hyperbola is equilateral and has foci at $(0, 6)$ and $(0, -6)$.

13. The eccentricity of the hyperbola is $\dfrac{5}{3}$, and the foci are at $(10, 0)$ and $(-10, 0)$.

14. **Aviation** Airplanes are equipped with signal devices to alert rescuers to their position. Suppose a downed plane sends out radio pulses that are detected by two receiving stations, A and B. The stations are located 130 miles apart along a stretch of I-40, with A due west of B. The two stations are able to determine that the plane is 50 miles farther from station B than from station A.

14a. $\dfrac{x^2}{625} - \dfrac{y^2}{3600} = 1$

a. Determine the equation of the hyperbola centered at the origin on which the plane is located.

b. Graph the equation, indicating on which branch of the hyperbola the plane is located. **See Answer Appendix.**

c. If the pilot estimates that the plane is 6 miles from I-40, find the exact coordinates of its position. $(-25.1, 6)$

EXERCISES

Practice

15–23. See margin; see Answer Appendix for graphs.

For the equation of each hyperbola, find the coordinates of the center, the foci, and the vertices and the equations of the asymptotes of its graph. Then graph the equation.

A 15. $\dfrac{x^2}{100} - \dfrac{y^2}{16} = 1$

16. $\dfrac{x^2}{9} - \dfrac{(y-5)^2}{81} = 1$

17. $\dfrac{x^2}{4} - \dfrac{y^2}{49} = 1$

18. $\dfrac{(y-7)^2}{64} - \dfrac{(x+1)^2}{4} = 1$

19. $x^2 - 4y^2 + 6x - 8y = 11$

20. $-4x^2 + 9y^2 - 24x - 90y + 153 = 0$

21. $16y^2 - 25x^2 - 96y + 100x - 356 = 0$

22. $36x^2 - 49y^2 - 72x - 294y = 2169$

B 23. Graph the equation $25y^2 - 9x^2 - 100y - 72x - 269 = 0$. Label the center, foci and the equations of the asymptotes.

Write the equation of each hyperbola. **24–26. See margin.**

24.

25.

26.

Graph each equation. **27–30. See Answer Appendix.**

27. $xy = 49$

28. $xy = -36$

29. $4xy = -25$

30. $9xy = 16$

www.amc.glencoe.com/self_check_quiz

Additional Answers

24. $\dfrac{(y-3)^2}{16} - \dfrac{(x-4)^2}{9} = 1$

25. $\dfrac{x^2}{9} - \dfrac{y^2}{9} = 1$

26. $\dfrac{y^2}{4} - \dfrac{(x+4)^2}{1} = 1$

Write an equation of the hyperbola that meets each set of conditions.

31–42. See margin.

31. The center is at $(4, -2)$, $a = 2$, $b = 3$, and it has a vertical transverse axis.

32. The vertices are at $(0, 3)$ and $(0, -3)$, and a focus is at $(0, -9)$.

33. The length of the transverse axis is 6 units, and the foci are at $(5, 2)$ and $(-5, 2)$.

34. The length of the conjugate axis is 8 units, and the vertices are at $(-3, 9)$ and $(-3, -5)$.

35. The hyperbola is equilateral and has foci at $(8, 0)$ and $(-8, 0)$.

36. The center is at $(-3, 1)$, one focus is at $(2, 1)$, and the eccentricity is $\frac{5}{4}$.

 37. A vertex is at $(4, 5)$, the center is at $(4, 2)$, and an equation of one asymptote is $4y + 4 = 3x$.

38. The equation of one asymptote is $3x - 11 = 2y$. The hyperbola has its center at $(3, -1)$ and a vertex at $(5, -1)$.

39. The hyperbola has foci at $(0, 8)$ and $(0, -8)$ and eccentricity $\frac{4}{3}$.

40. The hyperbola has eccentricity $\frac{6}{5}$ and foci at $(10, -3)$ and $(-2, -3)$.

41. The hyperbola is equilateral and has foci at $(9, 0)$ and $(-9, 0)$.

42. The slopes of the asymptotes are ± 2, and the foci are at $(1, 5)$ and $(1, -3)$.

Applications and Problem Solving

43. **Chemistry** According to Boyle's Law, the pressure P (in kilopascals) exerted by a gas varies inversely as the volume V (in cubic decimeters) of a gas if the temperature remains constant. That is, $PV = c$. Suppose the constant for oxygen at $25°C$ is 505.
   a. Graph the function $PV = c$ for $c = 505$. **See margin.**
   b. Determine the volume of oxygen if the pressure is 101 kilopascals. **5.0 dm³**
   c. Determine the volume of oxygen if the pressure is 50.5 kilopascals. **10.0 dm³**
   d. Study your results for parts **b** and **c**. If the pressure is halved, make a conjecture about the effect on the volume of gas. **V = 2(original V)**

44. **Critical Thinking** Prove that the eccentricity of all equilateral hyperbolas is $\sqrt{2}$. **See margin.**

45. **Nuclear Power** A nuclear cooling tower is a *hyperboloid*, that is, a hyperbola rotated around its conjugate axis. Suppose the hyperbola used to generate the hyperboloid modeling the shape of the cooling tower has an eccentricity of $\frac{5}{3}$. a. $\frac{x^2}{75^2} - \frac{y^2}{100^2} = 1$

   a. If the cooling tower is 150 feet wide at its narrowest point, determine an equation of the hyperbola used to generate the hyperboloid.
   b. If the tower is 450 feet tall, the top is 100 feet above the center of the hyperbola, and the base is 350 feet below the center, what is the radius of the top and the base of the tower?
   top: 106.07 ft; base: 273.00 ft

Lesson 10-4 Hyperbolas **651**

Additional Answers

31. $\dfrac{(y + 2)^2}{4} - \dfrac{(x - 4)^2}{9} = 1$

32. $\dfrac{y^2}{9} - \dfrac{x^2}{72} = 1$

33. $\dfrac{x^2}{9} - \dfrac{(y - 2)^2}{16} = 1$

34. $\dfrac{(y - 2)^2}{49} - \dfrac{(x + 3)^2}{16} = 1$

35. $\dfrac{x^2}{32} - \dfrac{y^2}{32} = 1$

36. $\dfrac{(x + 3)^2}{16} - \dfrac{(y - 1)^2}{9} = 1$

37. $\dfrac{(y - 2)^2}{9} - \dfrac{(x - 4)^2}{16} = 1$

38. $\dfrac{(x - 3)^2}{4} - \dfrac{(y + 1)^2}{9} = 1$

39. $\dfrac{y^2}{36} - \dfrac{x^2}{28} = 1$

40. $\dfrac{(x - 4)^2}{25} - \dfrac{(y + 3)^2}{11} = 1$

41. $\dfrac{2x^2}{81} - \dfrac{2y^2}{81} = 1$

42. $\dfrac{5(y - 1)^2}{64} - \dfrac{5(x - 1)^2}{16} = 1$

43a.

44. In an equilateral hyberbola, $a = b$ and $c^2 = a^2 + b^2$.

$c^2 = a^2 + a^2 \qquad a = b$

$c^2 = 2a^2$

$c = a\sqrt{2}$

Since $e = \frac{c}{a}$, we have

$e = \frac{c}{a}$

$e = \frac{a\sqrt{2}}{a}$

$e = \sqrt{2}$

Thus the eccentricity of any equilateral hyperbola is $\sqrt{2}$.

Extra Credit

Find the center of the conic section represented by the equation

$Ax^2 + Cy^2 + Dx + Ey + F = 0$.

$\left(-\dfrac{D}{2A}, -\dfrac{E}{2C}\right)$

Open-Ended Assessment

Speaking Ask students to compare and contrast hyperbolas and ellipses. Possible areas of discussion include standard forms of equations, graphs, axes, foci, vertices, and centers.

 Chapter 10, Quiz B (Lessons 10-3 and 10-4) is available in the *Assessment and Evaluation Masters*, p. 251.

 Chapter 10 Mid-Chapter Test (Lessons 10-1 through 10-4) is available in the *Assessment and Evaluation Masters*, p. 250.

Additional Answer
50.

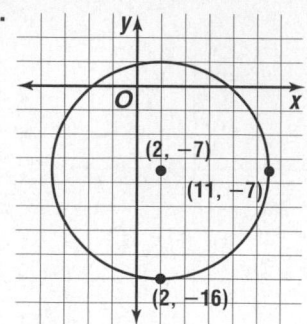

(2, −7)
(11, −7)
(2, −16)

Enrichment Masters, p. 112

---

46. approximately $\frac{x^2}{1100^2} - \frac{y^2}{10,503^2} = 1$

46. **Forestry** Two ranger stations located 4 miles apart observe a lightning strike. A ranger at station A reports hearing the sound of thunder 2 seconds prior to a ranger at station B. If sound travels at 1100 feet per second, determine the equation of the hyperbola on which the lightning strike was located. Place the two ranger stations on the x-axis with the midpoint between the two stations at the origin. The transverse axis is horizontal.

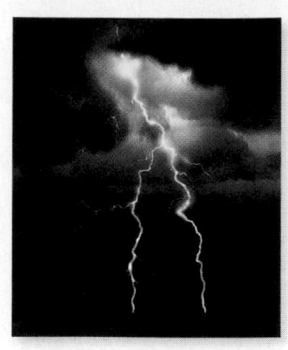

47. **Critical Thinking** A hyperbola has foci $F_1(-6, 0)$ and $F_1(6, 0)$. For any point $P(x, y)$ on the hyperbola, $|PF_1 - PF_2| = 10$. Write the equation of the hyperbola in standard form. $\frac{x^2}{25} - \frac{y^2}{11} = 1$

48a. See Answer Appendix.

48b. They are the same lines.

48c. $\frac{(y-2)^2}{25} - \frac{(x-3)^2}{16} = 1$

48. **Analytic Geometry** Two hyperbolas in which the transverse axis of one is the conjugate axis of the other are called *conjugate hyperbolas*. In equations of conjugate hyperbolas, the x^2 and y^2 terms are reversed. For example, $\frac{x^2}{16} - \frac{y^2}{9} = 1$ and $\frac{y^2}{9} - \frac{x^2}{16} = 1$ are equations of conjugate hyperbolas.

a. Graph $\frac{x^2}{16} - \frac{y^2}{9} = 1$ and $\frac{y^2}{9} - \frac{x^2}{16} = 1$ on the same coordinate plane.

b. What is true of the asymptotes of conjugate hyperbolas?

c. Write the equation of the conjugate hyperbola for $\frac{(x-3)^2}{16} - \frac{(y-2)^2}{25} = 1$.

d. Graph the conjugate hyperbolas in part **c**. **See Answer Appendix.**

Mixed Review

49. $\frac{y^2}{16} + \frac{(x-2)^2}{7} = 1$

49. Write the equation of the ellipse that has a semi-major axis length of 4 units and foci at (2, 3) and (2, −3). *(Lesson 10-3)*

50. Write $x^2 + y^2 - 4x + 14y - 28 = 0$ in standard form. Then graph the equation. *(Lesson 10-2)* $(x − 2)^2 + (y + 7)^2 = 81$; **See margin for graph.**

51. Show that the points with coordinates (−1, 3), (3, 6), (6, 2), and (2, −1) are the vertices of a square. *(Lesson 10-1)* **See margin.**

52. Name three different pairs of polar coordinates that represent point R. Assume $-360° \le \theta \le 360°$. *(Lesson 9-1)* **(90, 208°), (90, −152°), (−90, 28°)**

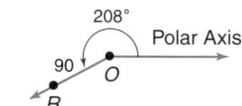

53. Find the inner product of vectors (4, −1, 8) and (−5, 2, 2). Are the vectors perpendicular? Explain. *(Lesson 8-4)* **−6; No, the inner product of the two vectors is not zero.**

54. $x + \sqrt{3}y - 6 = 0$

54. Write the standard from of the equation of the line that has a normal 3 units long and makes an angle of 60° with the positive x-axis. *(Lesson 7-6)*

55. **Aviation** An airplane flying at an altitude of 9000 meters passes directly overhead. Fifteen seconds later, the angle of elevation to the plane is 60°. How fast is the airplane flying? *(Lesson 5-4)* **about 346 m/s**

56. Approximate the real zeros of the function $f(x) = 4x^4 + 5x^3 - x^2 + 1$ to the nearest tenth. *(Lesson 4-5)* **−1.3, −0.6**

57. **SAT/ACT Practice** If r and s are integers and r + s = 0, which of the following must be true? **C**
I. $r^3 > s^3$ II. $r^3 = s^3$ III. $r^4 = s^4$

A I only   B II only   C III only
D I and II only   E I and III only

Extra Practice See p. A45.

Additional Answer

51.

D(3, 6)
A(−1, 3)
C(6, 2)
B(2, −1)

$AB = \sqrt{(2+1)^2 + (-1-3)^2} = 5$

$BC = \sqrt{(2-6)^2 + (-1-2)^2} = 5$

$CD = \sqrt{(6-3)^2 + (2-6)^2} = 5$

$AD = \sqrt{(3+1)^2 + (6-3)^2} = 5$

Thus, *ABCD* is a rhombus. The slope of $\overline{AD} = \frac{6-3}{3+1}$ or $\frac{3}{4}$

and the slope of $\overline{AB} = \frac{3+1}{-1-2}$ or $-\frac{4}{3}$. Thus, $\overline{AD}$ is

perpendicular to $\overline{AB}$ and *ABCD* is a square.

10-5

Parabolas

ENERGY The Odeillo Solar Furnace, located in southern France, uses a series of 63 flat mirrors, arranged on terraces on a hillside, to reflect the sun's rays on to a large parabolic mirror. These computer-controlled mirrors tilt to track the sun and ensure that its rays are always reflected to the central parabolic mirror.

This mirror in turn reflects the sun's rays to the focal point where a furnace is mounted on a tower. The concentrated energy generates temperatures of up to 6870°F. If the width of the Odeillo parabolic mirror is 138 feet and the furnace is located 58 feet from the center of the mirror, how deep is the mirror? *This problem will be solved in Example 2.*

In Chapter 4, you learned that the graphs of quadratic equations like $x = y^2$ or $y = x^2$ are called *parabolas*. A parabola is defined as the set of all points in a plane that are the same distance from a given point, called the **focus,** and a given line, called the **directrix.** *Remember that the distance from a point to a line is the length of the segment from the point perpendicular to the line.*

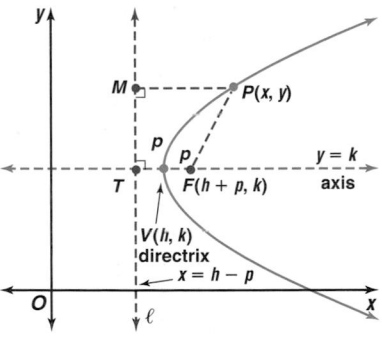

In the figure at the left, *F* is the focus of the parabola and ℓ is the directrix. This parabola is symmetric with respect to the line $y = k$, which passes through the focus. This line is called the **axis of symmetry,** or, more simply, the *axis* of the parabola. The point at which the axis intersects the parabola is called the **vertex.**

Suppose the vertex *V* has coordinates (h, k). Let p be the distance from the focus to the vertex, *FV*. By the definition of a parabola, the distance from any point on the parabola to the focus must equal the distance from that point to the directrix. So, if $FV = p$, then $VT = p$. The coordinates of *F* are $(h + p, k)$, and the equation of the directrix is $x = h - p$.

Lesson 10-5 Parabolas **653**

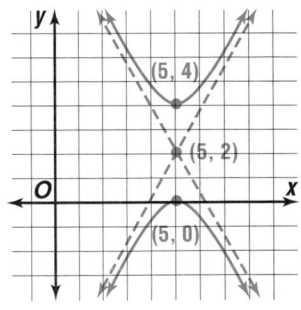

Now suppose that $P(x, y)$ is any point on the parabola other than the vertex. From the definition of a parabola, you know that $PF = PM$. Since M lies on the directrix, the coordinates of M are $(h - p, y)$.

For PF, let $F(h + p, k)$ be (x_1, y_1) and $P(x, y)$ be (x_2, y_2). Then for PM, let M be (x_1, y_1). You can use the Distance Formula to determine the equation for the parabola.

$$PF = PM$$

$$\sqrt{[x - (h + p)]^2 + (y - k)^2} = \sqrt{[x - (h - p)]^2 + (y - y)^2}$$

$$[x - (h + p)]^2 + (y - k)^2 = [x - (h - p)]^2 \quad \textit{Square each side.}$$

$$x^2 - 2x(h + p) + (h + p)^2 + (y - k)^2 = x^2 - 2x(h - p) + (h - p)^2$$

This equation can be simplified to obtain the equation

$$(y - k)^2 = 4p(x - h).$$

When p is positive, the parabola opens to the right.
When p is negative, the parabola opens to the left.

Unlike the equations of other conic sections, the equation of a parabola has only one squared term.

This is the equation of a parabola whose directrix is parallel to the y-axis. The equation of a parabola whose directrix is parallel to the x-axis can be obtained by switching the terms in the parentheses of the previous equation.

$$(x - h)^2 = 4p(y - k)$$

When p is positive, the parabola opens upward.
When p is negative, the parabola opens downward.

Standard Form of the Equation of a Parabola	Orientation when $p > 0$	Description
$(y - k)^2 = 4p(x - h)$		vertex: (h, k) focus: $(h + p, k)$ axis of symmetry: $y = k$ directrix: $x = h - p$ opening: right if $p > 0$ left if $p < 0$
$(x - h)^2 = 4p(y - k)$		vertex: (h, k) focus: $(h, k + p)$ axis of symmetry: $x = h$ directrix: $y = k - p$ opening: upward if $p > 0$ downward if $p < 0$

Example **1** Consider the equation $y^2 = 8x + 48$.

    a. Find the coordinates of the focus and the vertex and the equations of the directrix and the axis of symmetry.

    b. Graph the equation of the parabola.

    a. First, write the equation in the form $(y - k)^2 = 4p(x - h)$.

$$y^2 = 8x + 48$$
$$y^2 = 8(x + 6) \qquad Factor.$$
$$(y - 0)^2 = 4(2)(x + 6) \qquad 4p = 8, \text{ so } p = 2$$

In this form, we can see that $h = -6$, $k = 0$, and $p = 2$. We can use this to find the desired information.

Vertex: $(-6, 0)$ (h, k)
Directrix: $x = -8$ $x = h - p$
Focus: $(-4, 0)$ $(h + p, k)$
Axis of Symmetry: $y = 0$ $y = k$

The axis of symmetry is the x-axis. Since p is positive, the parabola opens to the right.

Graphing Calculator Tip

You can graph a parabola that opens to the right or to left by first solving for y and then graphing the two resulting equations on the same screen.

    b. Graph the directrix, the vertex, and the focus. To determine the shape of the parabola, graph several other ordered pairs that satisfy the equation and connect them with a smooth curve.

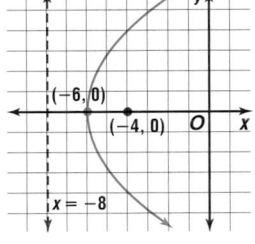

One useful property of parabolic mirrors is that all light rays traveling parallel to the mirror's axis of symmetry will be reflected by the parabola to the focus.

Example **2** **ENERGY** Refer to the application at the beginning of the lesson.

    a. Find and graph the equation of a parabola that models the shape of the Odeillo mirror.

    b. Find the depth of the parabolic mirror.

    a. The shape of the mirror can be modeled by a parabola with vertex at the origin and opening to the right. The general equation of such a parabola is $y^2 = 4px$, where p is the focal length. Given a focal length of 58 feet, we can derive the model equation.

$$y^2 = 4px$$
$$y^2 = 4(58)x \qquad p = 58$$
$$y^2 = 232x$$

Lesson 10-5 Parabolas **655**

Teaching Tip After Example 1, have students find the equation of the graph with the same shape and orientation as the graph in In-Class Example 1 but with vertex at $(-2, 5)$. $(y - 5)^2 + 2(x + 2) = 0$

In-Class Examples

Example 1
Consider the equation $y^2 + 2x = 0$.
a. Find the coordinates of the focus and the vertex and the equations of the directrix and the axis of symmetry.
    focus: $\left(-\frac{1}{2}, 0\right)$;
    vertex: $(0, 0)$;
    directrix: $x = \frac{1}{2}$;
    axis of symmetry: $y = 0$
b. Graph the equation of the parabola.

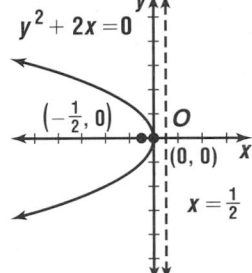

Example 2
Astronomy The parabolic mirror for the California Institute of Technology's Hale telescope at Mount Palomar in southern California measures 200 inches in diameter and has a focal length of 667 inches.
a. Find and graph the equation of a parabola that models the shape of the Palomar mirror.

b. Find the depth of the parabolic mirror. **3.75 in.**

 Teaching Tip Point out that the two equations that form the parabola on a graphing calculator are the two halves that are reflected over the axis of symmetry.

From the Classroom of...

Nancy Dymott
Carthage High School
Carthage, Missouri

"To find the vertex $V(h, k)$ of a parabola of the form $y = ax^2 + bx + c$, I teach students to solve directly for h using the formula $h = \frac{-b}{2a}$. They then find k by computing $f(h)$ or by using the formula $k = \frac{4ac - b^2}{4a}$. A similar method can be used to find the vertex $V(k, h)$ of a parabola of the form $x = ay^2 + by + c$."

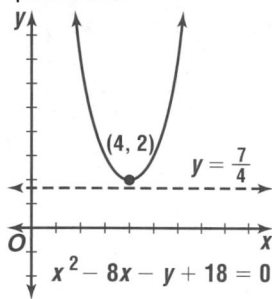
b. With the mirror's vertex at the origin, the distance from the vertex to one edge of the mirror is half the overall width of the mirror, $\frac{1}{2}(138$ feet) or 69 feet.

Use the model equation to find the depth x of the mirror when the distance from the center is 69 feet.

$$y^2 = 232x$$
$$(69)^2 = 232x \quad y = 69$$
$$4761 = 232x$$
$$x = \frac{4761}{232} \text{ or about } 20.5$$

The mirror is about 20.5 feet deep.

You can use the same process you used with circles to rewrite the standard form of the equation of a parabola in general form. *You will derive the general form in Exercise 37.*

General Form for the Equation of a Parabola	The general form of the equation of a parabola is $y^2 + Dx + Ey + F = 0$, when the directrix is parallel to the *y*-axis, or $x^2 + Dx + Ey + F = 0$, when the directrix is parallel to the *x*-axis.

It is necessary to convert an equation in general form to standard form to determine the coordinates of the vertex (h, k) and the distance from the vertex to the focus p.

Example ③ Consider the equation $2x^2 - 8x + y + 6 = 0$.

a. Write the equation in standard form.

b. Find the coordinates of the vertex and focus and the equations for the directrix and the axis of symmetry.

c. Graph the equation of the parabola.

a. Since x is squared, the directrix of this parabola is parallel to the *x*-axis.

$$2x^2 - 8x + y + 6 = 0$$
$$2x^2 - 8x = -y - 6 \qquad \textit{Isolate the x terms and the y terms.}$$
$$2(x^2 - 4x + ?) = -y - 6 + ? \qquad \textit{The GCF of the x terms is 2.}$$
$$2(x^2 - 4x + 4) = -y - 6 + 2(4) \qquad \textit{Complete the square.}$$
$$2(x - 2)^2 = -(y - 2) \qquad \textit{Simplify and factor.}$$
$$(x - 2)^2 = -\frac{1}{2}(y - 2) \qquad \textit{Divide each side by 2.}$$

The standard form of the equation is $(x - 2)^2 = -\frac{1}{2}(y - 2)$.

b. Since $4p = -\frac{1}{2}, p = -\frac{1}{8}$.

vertex: (2, 2) (h, k) focus: $\left(2, \frac{15}{8}\right)$ $(h, k + p)$

directrix: $y = \frac{17}{8}$ $y = k - p$ axis of symmetry: $x = 2$ $x = h$

c. Now sketch the graph of the parabola using the information found in part b.

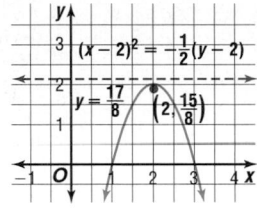

Parabolas are often used to demonstrate maximum or minimum points in real-world situations.

Example **4** **AERONAUTICS** NASA's KC-135A aircraft flies in parabolic arcs to simulate the weightlessness experienced by astronauts in space. The aircraft starts its ascent at 24,000 feet. During the ascent, all on board experience 2g's or twice the pull of Earth's gravity. As the aircraft approaches its maximum height, the engines are stopped, and the aircraft is allowed to free fall at a precisely determined angle. Zero gravity is achieved for 25 seconds as the plane reaches the top of the parabola and begins its descent. After this 25-second period, the engines are throttled to bring the aircraft out of the dive. If the height of the aircraft in feet (y) versus time in seconds (x) is modeled by the equation $x^2 - 65x + 0.11y - 2683.75 = 0$, what is the maximum height achieved by the aircraft during its parabolic flight?

First, write the equation in standard form.

$x^2 - 65x + 0.11y - 2683.75 = 0$

$x^2 - 65x = -0.11y + 2683.75$ *Isolate the x terms and y terms.*

$x^2 - 65x + 1056.25 = -0.11y + 2683.75 + 1056.25$ *Complete the square.*

$(x - 32.5)^2 = -0.11y + 3740$

$(x - 32.5)^2 = -0.11(y - 34{,}000)$

The vertex of the parabola is at (32.5, 34,000).

Remember that the vertex is the maximum or minimum point of a parabola. Since the parabola opens downward, the vertex is the maximum.

The x-coordinate of the vertex, 32.5, represents 32.5 seconds after the aircraft began the parabolic maneuver. The y-coordinate, 34,000, represents a maximum height of 34,000 feet.

Teaching Tip If students have difficulty drawing the graphs of parabolas, remind them of the value of the basic technique of plotting points. With only the vertex and two points on one side of the axis of symmetry, students can use symmetry to see the general shape of the parabola and graph it with a reasonable degree of accuracy.

In-Class Example

Example 4
Sports The path of any object moving near the surface of Earth under the influence of gravity is very nearly parabolic. Suppose a golf ball follows a path that can be modeled by the equation $x^2 - 200x + 200y = 0$. Where on the parabola is the maximum height achieved by this golf ball? **(100, 50)**

Addressing Individual Needs

Visual/Spatial Have groups of students toss a softball outside in a high arc, standing in front of a wall that is marked to indicate height or by a brick wall of the school where each brick can be measured. Have students measure the maximum height achieved by the ball and the distance between the students tossing the ball, then determine an equation to model the ball's path. Compare the results of different groups and discuss how the different-shaped parabolas resulted in different equations.

Error Analysis

Watch for students who sketch parabolas with the correct shape but in an improper direction.

Prevent by reviewing the rules for the orientation of the parabola based on p.

Additional Answers
(pp. 658–659)

1. The equation of a parabola will have only one squared term, while the equation of a hyperbola will have two squared terms.

2. $(x - 2)^2 = -16(y - 1)$

3. The vertex and focus both lie on the axis of symmetry. The directrix and axis of symmetry are perpendicular to each other. The focus and the point on the directrix collinear with the focus are equidistant from the vertex.

13. vertex: (0, 0); focus: (2, 0); directrix: $x = -2$; axis of symmetry: $y = 0$

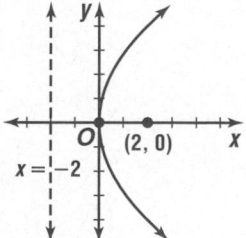

14. vertex: (0, 3); focus: (0, 2); directrix: $y = 4$; axis of symmetry: $x = 0$

All conics can be defined using the focus-directrix definition presented in this lesson. A conic section is defined to be the **locus** of points such that, for any point P in the locus, the ratio of the distance between that point and a fixed point F to the distance between that point and a fixed line ℓ, is constant. As we have seen, the point F is called the focus, and the line ℓ is the directrix. That ratio is the eccentricity of the curve, and its value can be used to determine the conic's classification. In the case of a parabola, $e = 1$. As shown previously, if $0 < e < 1$, the conic is an ellipse. If $e = 0$, the conic is a circle, and if $e > 1$, the conic is a hyperbola.

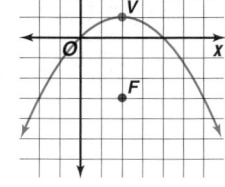

parabola	ellipse	hyperbola
$e = 1$	$e < 1, e \neq 0$	$e > 1$

For those conics having more than one focus and directrix, F' and ℓ' represent alternates that define the same conic.

CHECK FOR UNDERSTANDING

Communicating Mathematics

Read and study the lesson to answer each question. 1–3. See margin.

1. **Explain** a way in which you might distinguish the equation of a parabola from the equation of a hyperbola.

2. **Write** the equation of the graph shown at the right.

3. **Describe** the relationships among the vertex, focus, directrix and axis of symmetry of a parabola.

4. **Write** the equation in standard form of a parabola with vertex at $(-4, 5)$, opening to the left, and with a focus 5 units from its vertex. $(y - 5)^2 = -20(x + 4)$

5. **Identify** each of the following conic sections given their eccentricities.

   a. $e = \frac{1}{2}$ b. $e = 1$ c. $e = 1.25$ d. $e = 0$

      ellipse   parabola   hyperbola   circle

Guided Practice

For the equation of each parabola, find the coordinates of the vertex and focus, and the equations of the directrix and axis of symmetry. Then graph the equation. 6–8. See Answer Appendix.

6. $x^2 = 12(y - 1)$ 7. $y^2 - 4x + 2y + 5 = 0$ 8. $x^2 + 8x + 4y + 8 = 0$

Write the equation of the parabola that meets each set of conditions. Then graph the equation. 9–11. See Answer Appendix for graphs.

9. The vertex is at the origin, and the focus is at $(0, -4)$. $x^2 = -16y$

10. The parabola passes through the point at $(2, -1)$, has its vertex at $(-7, 5)$, and opens to the right. $(y - 5)^2 = 4(x + 7)$

11. The parabola passes through the point at $(5, 2)$, has a vertical axis, and has a minimum at $(4, -3)$. $(x - 4)^2 = \frac{1}{5}(y + 3)$

Additional Answers *(p.659)*

15. vertex: (0, 6); focus: (1, 6); directrix: $x = -1$; axis of symmetry: $y = 6$

16. vertex: $(-1, 1)$; focus: $(-4, 1)$; directrix: $x = 2$; axis of symmetry: $y = 1$

12. Sports In 1998, Sammy Sosa of the Chicago Cubs was in a homerun race with Mark McGwire of the St. Louis Cardinals. One day, Mr. Sosa popped a baseball straight up at an initial velocity v_0 of 56 feet per second. Its distance s above the ground after t seconds is described by $s = v_0 t - 16t^2 + 3$.

a. Graph the function $s = v_0 t - 16t^2 + 3$ for the given initial velocity. **See Answer Appendix.**

b. Find the maximum height achieved by the ball. **52 ft**

c. If the ball is allowed to fall to the ground, how many seconds, to the nearest tenth, is it in the air? **3.6 s**

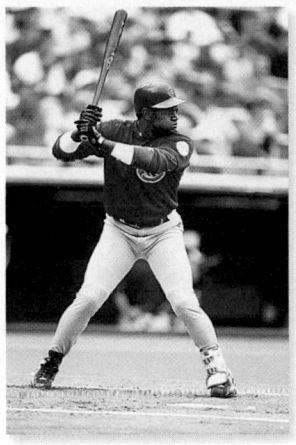

Assignment Guide

Core: 13–35 odd, 37–45
Enriched: 14–32 even, 33–45
All: Mid-Chapter Quiz, 1–10

EXERCISES

Practice

For the equation of each parabola, find the coordinates of the vertex and focus, and the equations of the directrix and axis of symmetry. Then graph the equation. **13–22. See margin.**

A

13. $y^2 = 8x$

14. $x^2 = -4(y - 3)$

15. $(y - 6)^2 = 4x$

16. $y^2 + 12x = 2y - 13$

17. $y - 2 = x^2 - 4x$

18. $x^2 + 10x + 25 = -8y + 24$

19. $y^2 - 2x + 14y = -41$

20. $y^2 - 2y - 12x + 13 = 0$

B

21. $2x^2 - 12y - 16x + 20 = 0$

22. $3x^2 - 30y - 18x + 87 = 0$

23. For graph, see Answer Appendix.

23. Consider the equation $2y^2 + 16y + 16x + 64 = 0$. Identify the coordinates of the vertex and focus and the equations of the directrix and axis of symmetry. Then graph the equation. **vertex: $(-2, -4)$; focus: $(-4, -4)$; directrix: $x = 0$; axis of symmetry: $y = -4$**

Write the equation of the parabola that meets each set of conditions. Then graph the equation. **24–32. For graphs, see Answer Appendix.**

24. The vertex is at $(-5, 1)$, and the focus is at $(2, 1)$. $(y - 1)^2 = 28(x + 5)$

25. $(y - 6)^2 = -12(x - 3)$

25. The equation of the axis is $y = 6$, the focus is at $(0, 6)$, and $p = -3$.

26. $(x - 4)^2 = 8(y + 3)$

26. The focus is at $(4, -1)$, and the equation of the directrix is $y = -5$.

27. The parabola passes through the point at $(5, 2)$, has a vertical axis, and has a maximum at $(4, 3)$. $(x - 4)^2 = -(y - 3)$

28. The parabola passes through the point at $(3, 1)$, has its vertex at $(-2, -3)$, and opens to the left. $(y + 3)^2 = -16(x + 2)$

C

29. The focus is at $(-1, 7)$, the length from the focus to the vertex is 2 units, and the function has a minimum. $(x + 1)^2 = 8(y - 5)$

30. The parabola has a vertical axis and passes through points at $(1, -7)$, $(5, -3)$, and $(7, -4)$. $(x - 5)^2 = -4(y + 3)$

31. The parabola has a horizontal axis and passes through the origin and points at $(-1, 2)$ and $(3, -2)$. $(y - 2)^2 = 4(x + 1)$

32. The parabola's directrix is parallel to the x-axis, and the parabola passes through points at $(1, 1)$, $(0, 9)$, and $(2, 1)$. $\left(x - \frac{3}{2}\right)^2 = \frac{1}{4}y$

 www.amc.glencoe.com/self_check_quiz

Additional Answers

19. vertex: $(-4, -7)$; focus: $\left(-\frac{7}{2}, -7\right)$; directrix: $x = -\frac{9}{2}$; axis of symmetry: $y = -7$

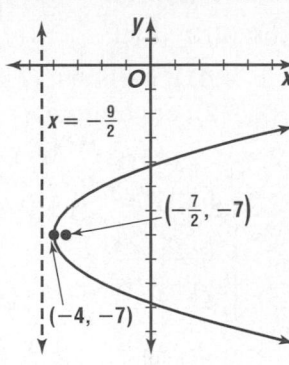

20. vertex: $(1, 1)$; focus: $(4, 1)$; directrix: $x = -2$; axis of symmetry: $y = 1$

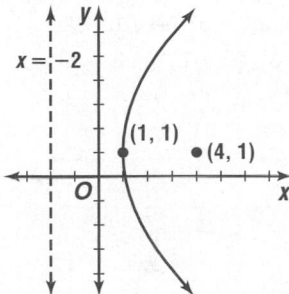

21. vertex: $(4, -1)$; focus: $\left(4, \frac{1}{2}\right)$; directrix: $y = -\frac{5}{2}$; axis of symmetry: $x = 4$

22. vertex: $(3, 2)$; focus: $\left(3, \frac{9}{2}\right)$; directrix: $y = -\frac{1}{2}$; axis of symmetry: $x = 3$

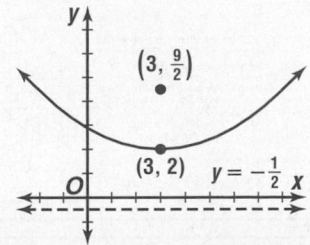

Additional Answers

17. vertex: $(2, -2)$; focus: $\left(2, -\frac{7}{4}\right)$; directrix: $y = -\frac{9}{4}$; axis of symmetry: $x = 2$

18. vertex: $(-5, 3)$; focus: $(-5, 1)$; directrix: $y = 5$; axis of symmetry: $x = -5$

37. $(y - k)^2 = 4p(x - h)$

$y^2 - 2ky + k^2 = 4px - 4ph$

$y^2 - 4px - 2ky + k^2 + 4ph = 0$

$y^2 + Dx + Ey + F = 0$

$(x - h)^2 = 4p(y - k)$

$x^2 - 2hx + h^2 = 4py - 4pk$

$x^2 - 4py - 2hx + h^2 + 4pk = 0$

$x^2 + Dx + Ey + F = 0$

39. center: (2, 3); foci:

$(2, 3 \pm \sqrt{41})$; vertices: (2, 8)

and (2, −2); asymptotes:

$y - 3 = \pm\dfrac{5}{4}(x - 2)$

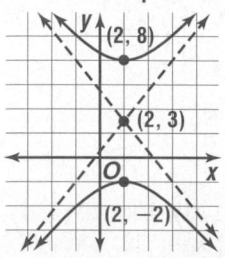

40. center: (0, −5); foci:

$\left(\pm\sqrt{21}, -5\right)$; vertices:

(0, −3), (0, −7), (−5, −5), (5, −5)

*Study Guide and Practice
Masters, pp. 185–186*

---

33. Automotive Automobile headlights contain parabolic reflectors, which work on the principle that light placed at the focus of a parabola will reflect off the mirror-like surface in lines parallel to the axis of symmetry. Suppose a bulb is placed at the focus of a headlight's reflector, which is 2 inches from the vertex.

  a. If the depth of the headlight is to be 4 inches, what should the diameter of the headlight be at its opening? **$8\sqrt{2}$ in.**

  b. Find the diameter of the headlight at its opening if the depth is increased by 25%. **$4\sqrt{10}$ in.**

34. Business An airline has been charging $140 per seat for a one-way flight. This flight has been averaging 110 passengers but can transport up to 180 passengers. The airline is considering a decrease in the price for a one-way ticket during the winter months. The airline estimates that for each $10 decrease in the ticket price, they will gain approximately 20 passengers per flight.

  a. Based on these estimates, what ticket price should the airline charge to achieve the greatest income on an average flight? **$105**

  b. New estimates reveal that the increase in passengers per flight is closer to 10 for each $10 decrease in the original ticket price. To maximize income, what should the new ticket price be? **$125**

**35a. The
opening becomes
narrower.**

**35b. The opening
becomes wider.**

35. Critical Thinking Consider the standard form of the equation of a parabola in which the vertex is known but the value of p is not known.

  a. As $|p|$ becomes greater, what happens to the shape of the parabola?

  b. As $|p|$ becomes smaller, what happens to the shape of the parabola?

36. Construction The Golden Gate Bridge in San Francisco, California, is a catenary suspension bridge, which is very similar in appearance to a parabola. The main span cables are suspended between two towers that are 4200 feet apart and 500 feet above the roadway. The cable extends 10 feet above the roadway midway between the two towers. **b. 67.6 ft**

  Sample answer:

  a. Find an equation that models the shape of the cable. **$x^2 = 9000y$**

  b. How far from the roadway is the cable 720 feet from the bridge's center?

37. Critical Thinking Using the standard form of the equation of a parabola, derive the general form of the equation of a parabola. **See margin.**

38. Critical Thinking The *latus rectum* of a parabola is the line segment through the focus that is perpendicular to the axis and has endpoints on the parabola. The length of the latus rectum is $|4p|$ units, where p is the distance from the vertex to the focus.

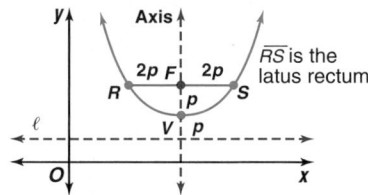

  a. Write the equation of a parabola with vertex at $(-3, 2)$, axis $y = 2$, and latus rectum 8 units long. **$(y - 2)^2 = 8(x + 3)$ or $(y - 2)^2 = -8(x + 3)$**

  b. The latus rectum of the parabola with equation $(x - 1)^2 = -16(y - 4)$ coincides with the diameter of a circle. Write the equation of the circle. **$(x - 1)^2 + y^2 = 64$**

Extra Credit

A parabola with directrix parallel to the x-axis passes through the points (1, 2), (−2, 14), and (3, 4). Find the equation of the parabola. **$y = x^2 - 3x + 4$**

Additional Answer

41.

39. Find the coordinates of the center, foci, and vertices, and the equations of the asymptotes of the graph of $\dfrac{(y-3)^2}{25} - \dfrac{(x-2)^2}{16} = 1$. Then graph the equation. *(Lesson 10-4)* **See margin.**

40. Find the coordinates of the center, foci, and vertices of the ellipse whose equation is $4x^2 + 25y^2 + 250y + 525 = 0$. Then graph the ellipse. *(Lesson 10-3)* **See margin.**

41. Graph $r = 12 \cos 2\theta$. *(Lesson 9-2)* **See margin.**

42. $2\pi n$, where n is an integer

42. Find the values of θ for which $\cos \theta = 1$ is true. *(Lesson 6-3)*

43. **Geometry** A regular hexagon is inscribed in a circle with a radius 6.4 centimeters long. Find the apothem; that is, the distance from the center of the circle to the midpoint of a side. *(Lesson 5-4)* **5.5 cm**

44. $y \to 0$ as $x \to \infty$, $y \to 0$ as $x \to -\infty$

44. Describe the end behavior of $g(x) = \dfrac{4}{x^2 + 1}$. *(Lesson 3-5)*

45. **SAT/ACT Practice** Triangle QRS has sides of lengths 14, 19, and t, where t is the length of the longest side. If t is the cube of an integer, what is the perimeter of the triangle? **C**

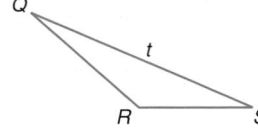

   A 41   **B** 58   **C** 60   **D** 69   **E** 76

MID-CHAPTER QUIZ

1. Given: $A(3, 3)$, $B(6, 9)$, and $C(9, 3)$ (Lesson 10-1)
   a. Show that these points form an isosceles triangle. **See margin.**
   b. Determine the perimeter of the triangle to the nearest hundredth. **19.42 units**

2. Determine the midpoint of the diagonals of the rectangle with vertices $A(-4, 9)$, $B(5, 9)$, $C(5, 5)$, and $D(-4, 5)$. (Lesson 10-1)

3. Find the coordinates of the center and radius of the circle with equation $x^2 + y^2 - 6y - 8x = -16$. Then graph the circle. (Lesson 10-2) **See margin.**

4. Write the equation of the circle with center at $(-5, 2)$ and radius $\sqrt{7}$. (Lesson 10-2)

5. **Astronomy** A satellite orbiting Earth follows an elliptical path with the center of Earth as one focus. The eccentricity of the orbit is 0.16, and the major axis is 10,440 miles long. (Lesson 10-3)
   a. If the mean diameter of Earth is 7920 miles, find the greatest and least distance of the satellite from the surface of Earth.
   b. Assuming that the center of the ellipse is the origin and the foci lie on the x-axis, write the equation of the orbit of the satellite. **See margin.**

6. Identify the center, vertices, and foci of the ellipse with equation $9x^2 + 25y^2 - 72x + 250y + 544 = 0$. Then graph the equation. (Lesson 10-3) **6–9. See margin.**

7. Identify the center, vertices, foci, and equations of the asymptotes of the graph of the hyperbola with equation $3y^2 + 24y - x^2 - 2x + 41 = 0$. Then graph the equation. (Lesson 10-4)

8. Write the equation of a hyperbola that passes through the point at $(4, 2)$ and has asymptotes with equations $y = 2x$ and $y = -2x + 4$. (Lesson 10-4)

9. Identify the vertex, focus, and equations of the axis of symmetry and directrix for the parabola with equation $y^2 - 4x + 2y + 5 = 0$. Then graph the equation. (Lesson 10-5)

10. Write the equation of the parabola that passes through the point at $(9, -2)$, has its vertex at $(5, -1)$, and opens downward. (Lesson 10-5) $(x - 5)^2 = -16(y + 1)$

2. $(0.5, 7)$
4. $(x + 5)^2 + (y - 2)^2 = 7$
5a. 2095.2 miles, 424.8 miles

Additional Answers
Mid-Chapter Quiz

1a. $AB = \sqrt{(6-3)^2 + (9-3)^2} = \sqrt{45}$
   $BC = \sqrt{(9-6)^2 + (3-9)^2} = \sqrt{45}$
   Since $AB = BC$, triangle ABC is isosceles.

3. center: $(4, 3)$; radius: 3;
   See Answer Appendix for graph.

5b. $\dfrac{x^2}{27{,}248{,}400} + \dfrac{y^2}{26{,}550{,}840.96} = 1$

6. center: $(4, -5)$; vertices: $(9, -5)$, $(-1, -5)$, $(4, -2)$, $(4, -8)$;
   foci: $(8, -5)$, $(0, -5)$; See Answer Appendix for graphs.

7. center: $(-1, -4)$; vertices: $\left(-1, -4 \pm \sqrt{2}\right)$;
   foci: $\left(-1, -4 \pm 2\sqrt{2}\right)$;
   asymptotes: $y + 4 = \pm\dfrac{\sqrt{3}}{3}(x + 1)$

Rectangular and Parametric Forms of Conic Sections

5-Minute Check
Lesson 10-5

Find each of the following for the parabola with equation
$y^2 - 6y - 4x + 17 = 0$.

1. standard form of the equation
   $(y - 3)^2 = 4(x - 2)$

2. coordinates of the focus
   (3, 3)

3. coordinates of the vertex
   (2, 3)

4. equation of the directrix
   $x = 1$

5. equation of the axis of symmetry $y = 3$

Motivating the Lesson

Discussion Write the equations of an ellipse and a hyperbola in standard form on the chalkboard. Have students rewrite each as a polynomial without fractions equal to zero. Ask them to imagine that they had been told that one polynomial represented an ellipse and one a hyperbola. Ask them to explain how they could tell which one was which. **For the ellipse, the signs of x^2 and y^2 would be the same. For the hyperbola, they would be different.**

OBJECTIVES
- Recognize conic sections in their rectangular form by their equations.
- Find a rectangular equation for a curve defined parametrically and vice versa.

TRANSPORTATION The first self-propelled boats on the western rivers of the United States were the paddlewheels. This boat used a steam engine to turn one or more circular wheels that had a paddle attached to the end of each spoke. In 1811, Robert Fulton and Nicholas Roosevelt built the first paddlewheel large enough for commercial use on the Ohio and Mississippi Rivers. By the end of the 19th century, paddlewheel boats had fought wars and carried people and cargo on nearly every river in the United States. Despite advancements in technology, paddlewheels are still in use today, though mainly for sentimental reasons. *You will solve a problem related to this in Exercise 40.*

We have determined a general equation for each conic section we have studied. All of these equations are forms of the general equation for conic sections.

General Equation for Conic Sections	The equation of a conic section can be written in the form $$Ax^2 + Bxy + Cy^2 + Dx + Ey + F = 0,$$ where A, B, and C are not all zero.

The graph of a second-degree equation in two variables always represents a conic or degenerate case, unless the equation has no graph at all in the real number plane. Most of the conic sections that we have studied have axes that are parallel to the coordinate axes. The general equations of these conics have no xy term; thus, $B = 0$. The one conic section we have discussed whose axes are not parallel to the coordinate axes is the hyperbola whose equation is $xy = k$. In its equation, $B \neq 0$.

To identify the conic section represented by a given equation, it is helpful to write the equation in standard form. However, when $B = 0$, you can also identify the conic section by how the equation compares to the general equation. The table on the next page summarizes the standard forms and differences among the general forms.

662 *Chapter 10 Conics*

Resource Manager

 Reproducible Masters
- *Study Guide and Practice,* pp. 187–188
- *Enrichment,* p.114
- *Assessment and Evaluation,* p. 252
- *Graphing Calculator Keystrokes,* p. 40
- *Spreadsheet Applications,* p. 10

 Transparencies
- 5-Minute Check, 10-6
- Teaching, 10-6

	General Form: $Ax^2 + Bxy + Cy^2 + Dx + Ey + F = 0$	
Conic Section	**Standard Form of Equation**	**Variation of General Form of Conic Equations**
circle	$(x - h)^2 + (y - k)^2 = r^2$	$A = C$
parabola	$(y - k)^2 = 4p(x - h)$ or $(x - h)^2 = 4p(y - k)$	Either A or C is zero.
ellipse	$\dfrac{(x-h)^2}{a^2} + \dfrac{(y-k)^2}{b^2} = 1$ or $\dfrac{(y-k)^2}{a^2} + \dfrac{(x-h)^2}{b^2} = 1$	A and C have the same sign and $A \neq C$.
hyperbola	$\dfrac{(x-h)^2}{a^2} - \dfrac{(y-k)^2}{b^2} = 1$ or $\dfrac{(y-k)^2}{a^2} - \dfrac{(x-h)^2}{b^2} = 1$	A and C have opposite signs.
	$xy = k$	$A = C = D = E = 0$

Remember that graphs can also be degenerate cases.

The circle is actually a special form of the ellipse, where $a^2 = b^2 = r^2$.

Example ① Identify the conic section represented by each equation.

a. $6y^2 + 3x - 4y - 12 = 0$

$A = 0$ and $C = 6$. Since $A = 0$, the conic is a parabola.

b. $3y^2 - 2x^2 + 5y - x - 15 = 0$

$A = -2$ and $C = 3$. Since A and C have different signs, the conic is a hyperbola.

c. $9x^2 + 27y^2 - 6x - 108y + 82 = 0$

$A = 9$ and $C = 27$. Since A and C have the same signs and are not equal, the conic is an ellipse.

d. $4x^2 + 4y^2 + 5x + 2y - 150 = 0$

$A = 4$ and $C = 4$. Since $A = C$, the conic is a circle.

So far we have discussed equations of conic sections in their rectangular form. Some conic sections can also be described parametrically.

The general form for a set of parametric equations is

$$x = f(t) \text{ and } y = g(t), \text{ where } t \text{ is in some interval } I.$$

As t varies over I in some order, a curve containing points (x, y) is traced out in a certain direction.

A parametric equation can be transformed into its more familiar rectangular form by eliminating the parameter t from the parametric equations.

Look Back
You can refer to Lesson 8-6 to review writing and graphing parametric equations.

Lesson 10-6 Rectangular and Parametric Forms of Conic Sections **663**

Teaching Tip After discussing the table, ask students to explain how they could transform the standard form of the equation of an ellipse to the standard form of the equation of a circle. **Substitute r^2 for a^2 and b^2, then multiply each side of the equation by r^2.**

In-Class Example

Example 1
Identify the conic section represented by each equation.
a. $3x^2 - 2y^2 - 4x + 5y = -3$
hyperbola
b. $-4x^2 - 4y^2 + 6x + y = -5$
circle
c. $8x^2 + y^2 + 3x - 7y = 0$
ellipse
d. $y^2 - x + 3y - 9 = 0$
parabola

Teaching Tip Suggest that students copy the chart on this page comparing the general forms of conic section equations into their notes so that they can refer to it when needed.

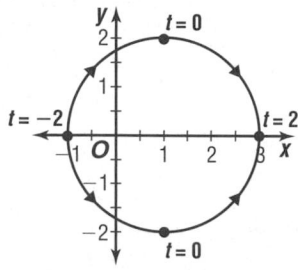
Example **2** **Graph the curve defined by the parametric equations $x = 4t^2$ and $y = 3t$, where $-2 \le t \le 2$. Then identify the curve by finding the corresponding rectangular equation.**

Make a table of values assigning values for t and evaluating each expression to find values for x and y.

t	x	y	(x, y)
-2	16	-6	$(16, -6)$
-1	4	-3	$(4, -3)$
0	0	0	$(0, 0)$
1	4	3	$(4, 3)$
2	16	6	$(16, 6)$

Then graph the curve.

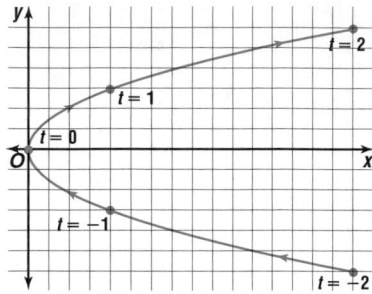

Notice the arrows indicating the direction in which the curve is traced for increasing values of t.

The graph appears to be part of a parabola. To identify the curve accurately, find the corresponding rectangular equation by eliminating t from the given parametric equations.

First, solve the equation $y = 3t$ for t.

$y = 3t$

$\dfrac{y}{3} = t$ *Solve for t.*

Then substitute $\dfrac{y}{3}$ for t in the equation $x = 4t^2$.

$x = 4t^2$

$x = 4\left(\dfrac{y}{3}\right)^2$ $t = \dfrac{y}{3}$

$x = \dfrac{4y^2}{9}$

The equation $x = \dfrac{4y^2}{9}$ is the equation of a parabola with vertex at $(0, 0)$ and its axis of symmetry along the x-axis. Notice that the domain of the rectangular equation is $x \ge 0$, which is greater than that of its parametric representation. By restricting the domain to $0 \le x \le 16$, our rectangular representation matches our parametric representation for the graph.

●

Some parametric equations require the use of trigonometric identities to eliminate the parameter t.

Example 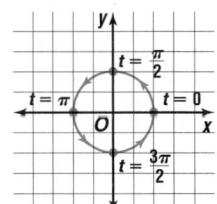 **3** **Find the rectangular equation of the curve whose parametric equations are**
$x = 2 \cos t$ and $y = 2 \sin t$, where $0 \le t \le 2\pi$. Then graph the equation using
arrows to indicate how the graph is traced.

Solve the first equation for $\cos t$ and the second equation for $\sin t$.

$$\cos t = \frac{x}{2} \text{ and } \sin t = \frac{y}{2}$$

**Graphing
Calculator
Appendix**

For keystroke
instructions on how to
graph parametric
equations, see page
A21.

Use the trigonometric identity $\cos^2 t + \sin^2 t = 1$ to rewrite the equation to
eliminate t.

$$\cos^2 t + \sin^2 t = 1$$
$$\left(\frac{x}{2}\right)^2 + \left(\frac{y}{2}\right)^2 = 1 \quad \textit{Substitution}$$
$$\frac{x^2}{4} + \frac{y^2}{4} = 1$$
$$x^2 + y^2 = 4 \quad \textit{Multiply each side by 4.}$$

This is the equation of a circle with center at $(0, 0)$ and
radius 2. As t increases from $t = 0$ to $t = 2\pi$, we see
that the curve is traced in a counterclockwise motion.

You can also use substitution to find the parametric equations for a given
conic section. If the conic section is defined as a function, $y = f(x)$, one way of
finding the parametric equations is by letting $x = t$ and $y = f(t)$, where t is in the
domain of f.

Example **4** **Find parametric equations for the equation $y = x^2 + 3$.**

Let $x = t$. Then $y = t^2 + 3$. Since the domain of the function $f(t)$ is all real
numbers, the parametric equations are $x = t$ and $y = t^2 + 3$, where $-\infty < t < \infty$.

GRAPHING CALCULATOR EXPLORATION

The graph of the parametric equations
$x = \cos t$ and $y = \sin t$, where $0 \le t \le 2\pi$
is the unit circle. Interchanging the
trigonometric functions or changing the
coefficients can alter the graph's size and
shape as well as its starting point and the
direction in which it is traced. Watch while
the graph is being drawn to see the effects.

TRY THESE

1. Graph the parametric equations
   $x = -\cos t$ and $y = \sin t$, where
   $0 \le t \le 2\pi$.
   a. Where does the graph start? **$(-1, 0)$**
   b. In which direction is the graph
   traced? **clockwise**

2. Graph the parametric equations $x = \sin t$
   and $y = \cos t$, where $0 \le t \le 2\pi$.
   a. Where does the graph start? **$(0, 1)$**
   b. In which direction is the graph
   traced? **clockwise**

3. Graph $x = 2 \cos t$, and $y = 3 \sin t$, where
   $0 \le t \le 2\pi$. What is the shape of the graph?
   an ellipse

WHAT DO YOU THINK?

4. What is the significance of the number a in
   the equations $x = a \cos t$ and $y = a \sin t$,
   where $0 \le t \le 2\pi$? **4–5. See margin.**

5. What is the result of changing the interval
   to $0 \le t \le 4\pi$ in Exercises 1-3?

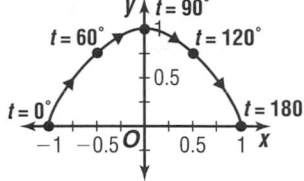

Addressing Individual Needs

Verbal/Linguistic Have groups of students choose a conic covered in the
chapter and develop a lesson to teach the class how to work with the
chosen conic section. The lessons should be presented to the rest of the
class and include visual aids and example problems. The rest of the class should
participate by asking and answering questions. The group should then submit a
detailed summary of their lessons and how they presented them.

Parametric equations are particularly useful in describing the motion of an object along a curved path.

Example **5** **ASTRONOMY** The orbit of Saturn around the sun is modeled by the equation $\frac{x^2}{(9.50)^2} + \frac{y^2}{(9.48)^2} = 1$. It takes Saturn approximately 30 Earth years to complete one revolution of its orbit.

Real World Application

a. Find parametric equations that model the motion of Saturn beginning at (9.50, 0) and moving counterclockwise around the sun.

b. Use the parametric equations to determine Saturn's position after 18 years.

a. From the given equation, you can determine that the orbital path of Saturn is an ellipse with a major axis of 9.50 AU and a minor axis of 9.48 AU.

Like a circle, the parametric representation for an ellipse involves the use of sines and cosines. The parametric representation for the given equation is an ellipse with $x = 9.5$ and $y = 0$ when $t = 0$, so the following equations are true.

$$\frac{x}{9.50} = \cos \omega t \qquad \text{and} \qquad \frac{y}{9.48} = \sin \omega t$$

You can verify that by using the equations $x = 9.50$ and $y = 0$ when $t = 0$ and $\cos^2 \omega t + \sin^2 \omega t = 1$.

To move counterclockwise, the motion will have to begin with the value of x decreasing and y increasing, so $\omega > 0$. Since Saturn completes an orbit in 30 Earth years, the sine and cosine functions have a period $\frac{2\pi}{\omega} = 30$, so $\omega = \frac{\pi}{15}$.

Thus, the parametric equations corresponding to the rectangular equation $\frac{x^2}{(9.50)^2} + \frac{y^2}{(9.48)^2} = 1$ are $x = 9.50 \cos \frac{\pi}{15} t$ and $y = 9.48 \sin \frac{\pi}{15} t$, where $0 \le t \le 30$.

You can verify the equations above using a graphing calculator to trace the ellipse.

b. The position of Saturn after 18 years is found by letting $t = 18$ in both parametric equations.

$x = 9.50 \cos\left(\frac{\pi}{15} t\right)$

$x = 9.50 \cos\left[\frac{\pi}{15}(18)\right] \quad t = 18$

$x \approx -7.69$

$y = 9.48 \sin\left(\frac{\pi}{15} t\right)$

$y = 9.48 \sin\left[\frac{\pi}{15}(18)\right] \quad t = 18$

$y \approx -5.57$

Eighteen years later, Saturn is located at (−7.69, −5.57).

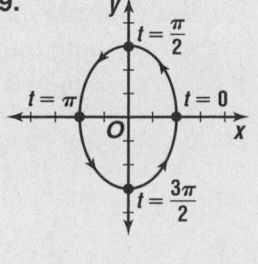

Communicating Mathematics

Read and study the lesson to answer each question.

1. **Compare and contrast** the general form of the equations of the four conic sections we have studied. **See margin.**

2. **Give** new restrictions on the parameter t in Example 2 so that the domains of the rectangular and parametric equations are the same. $-\infty < t < \infty$

3. **Write** the rectangular equation of a parabola with vertex at the origin and opening to the left. Then write the parametric equations that correspond to that parabola. **Sample answer:** $y^2 = -x$; $y = t$, $x = -t^2$, $-\infty < t < \infty$

Guided Practice

Identify the conic section represented by each equation. Then write the equation in standard form and graph the equation. **4–7. See margin.**

4. $x^2 + 9y^2 + 2x - 18y + 1 = 0$

5. $y^2 - 8x = -8$

6. $x^2 - 4x$ $y^2 - 5 - 4y = 0$

7. $x^2 - 6x + y^2 - 12y + 41 = 0$

Find the rectangular equation of the curve whose parametric equations are given. Then graph the equation using arrows to indicate orientation.

8–9. See margin for graphs.

8. $x = t$, $y = -t^2 - 6t + 2$; $-\infty < t < \infty$
8. $y = -x^2 - 6x + 2$

9. $x = 2 \cos t$, $y = 3 \sin t$; $0 \le t \le 2\pi$
$\dfrac{x^2}{4} + \dfrac{y^2}{9} = 1$

Find parametric equations for each rectangular equation.

10–11. Sample answers given.

10. $y = 2x^2 - 5x$ $x = t$,
$y = 2t^2 - 5t$, $-\infty \le t \le \infty$

11. $x^2 + y^2 = 36$ $x = 6 \cos t$,
$y = 6 \sin t$, $0 \le t \le 2\pi$

12. **Astronomy** Some comets traveling at great speeds follow parabolic paths with the sun as their focus. Suppose the motion of a certain comet is modeled by the parametric equations $x = \dfrac{t^2}{80}$, and $y = t$ for $-\infty < t < \infty$. Find the rectangular equation that models the comet's path. $y^2 = 80x$

Practice

Identify the conic section represented by each equation. Then write the equation in standard form and graph the equation. **13–24. See margin; for graphs, see Answer Appendix.**

A

13. $x^2 - 4y - 6x + 9 = 0$

14. $x^2 - 8x + y^2 + 6y + 24 = 0$

15. $x^2 - 3y^2 + 2x - 24y - 41 = 0$

16. $9x^2 + 25y^2 - 54x - 50y - 119 = 0$

17. $x^2 = y + 8x - 16$

18. $2xy = 3$

19. $5x^2 + 2y^2 - 40x - 20y + 110 = 0$

20. $x^2 - 8x + 11 = -y^2$

B

21. $8y^2 - 9x^2 - 16y + 36x - 100 = 0$

22. $4y^2 + 4y + 8x = 15$

23. Identify the conic section represented by $-4y^2 + 10x - 16y - x^2 - 5$. Write the equation in standard form and then graph the equation.

24. In the general equation of a conic, $A = C = 2$, $B = 0$, $D = -8$, $E = 12$, and $F = 6$. Write the equation in standard form. Then graph the equation.

25–30. See margin; for graphs see Answer Appendix.

Find the rectangular equation of the curve whose parametric equations are given. Then graph the equation using arrows to indicate orientation.

25. $x = t$, $y = 2t^2 - 4t + 1$; $-\infty < t < \infty$

26. $x = \cos 2t$, $y = \sin 2t$; $0 \le t \le 2\pi$

27. $x = -\cos t$, $y = \sin t$; $0 \le t \le 2\pi$

28. $x = 3 \sin t$, $y = 2 \cos t$; $0 \le t \le 2\pi$

29. $x = -\sin 2t$, $y = 2 \cos 2t$; $0 \le t \le \pi$

30. $x = 2t - 1$, $y = \sqrt{t}$; $0 \le t \le 4$

 www.amc.glencoe.com/self_check_quiz   *Lesson 10-6   Rectangular and Parametric Forms of Conic Sections*   **667**

Additional Answers

13. parabola; $(x - 3)^2 = 4y$

14. circle; $(x - 4)^2 + (y + 3)^2 = 1$

15. hyperbola; $\dfrac{(y + 4)^2}{2} - \dfrac{(x + 1)^2}{6} = 1$

16. ellipse; $\dfrac{(x - 3)^2}{25} + \dfrac{(y - 1)^2}{9} = 1$

17. parabola; $(x - 4)^2 = y$

18. hyperbola; $xy = \dfrac{3}{2}$

19. ellipse; $\dfrac{(y - 5)^2}{10} + \dfrac{(x - 4)^2}{4} = 1$

20. circle; $(x - 4)^2 + y^2 = 5$

21. hyperbola; $\dfrac{(y - 1)^2}{9} - \dfrac{(x - 2)^2}{8} = 1$

22. parabola; $\left(y + \dfrac{1}{2}\right)^2 = -2(x - 2)$

23. hyperbola; $\dfrac{(x + 5)^2}{4} - \dfrac{(y + 2)^2}{1} = 1$

24. $(x - 2)^2 + (y + 3)^2 = 10$

Error Analysis

Watch for students who make incorrect substitutions when going from parametric equations to rectangular equations.

Prevent by reminding students that parametric equations are expressed in terms of t, while rectangular equations are expressed in terms of x and y.

Assignment Guide
Core: 13–39 odd, 41–43, 45–56
Enriched: 14–38 even, 40–56

Additional Answers

25. $y = 2x^2 - 4x + 1$

26. $x^2 + y^2 = 1$

27. $x^2 + y^2 = 1$

28. $\dfrac{x^2}{9} + \dfrac{y^2}{4} = 1$

29. $x^2 + \dfrac{y^2}{4} = 1$

30. $y = \sqrt{\dfrac{x + 1}{2}}$

Study Guide and Practice Masters, pp. 187–188

Additional Answers

32. Sample answer: $x = 5 \cos t$, $y = 5 \sin t$, $0 \le t \le 2\pi$

33. Sample answer: $x = 4 \cos t$, $y = 4 \sin t$, $0 \le t \le 2\pi$

34. Sample answer: $x = 2 \cos t$, $y = 5 \sin t$, $0 \le t \le 2\pi$

35. Sample answer: $x = \cos t$, $y = 4 \sin t$, $0 \le t \le 2\pi$

36. Sample answer: $x = t$, $y = t^2 - 4t + 7$, $-\infty < t < \infty$

37. Sample answer: $x = t^2 + 2t - 1$, $y = t$, $-\infty < t < \infty$

38. Sample answer: $x = 0.25(t + 3)^2 + 2$, $y = t$, $-\infty < t < \infty$

39a. Answers will vary. Sample answers: $x = t$, $y = t^2$, $t \ge 0$; $y = \sqrt{t}$, $x = t^2$, $t \ge 0$

39b.

Tmin: [0, 5] step: 0.1
[−7.58, 7.58] scl:1 by [−5, 5] scl:1

39d. There is usually more than one parametric representation for the graph of a rectangular equation.

41a. Ellipse; point at (0, 0); the equation is that of a degenerate ellipse.

41b. Circle; point at (2, 3); the equation is that of a degenerate circle.

41c. Hyperbola; two intersecting lines $y = \pm 3x$; the equation is that of a degenerate hyperbola.

42. The substitution for x must be a function that allows x to take on all of the values stipulated by the domain of the rectangular equation. The domain of $y = x^2 - 5$ is all real numbers, but using a substitution of $x = t^2$ would only allow for values of x such that $x \ge 0$.

---

31. Find a rectangular equation for the curve whose parametric equations are $x = -3 \cos 2t$ and $y = 3 \sin 2t$, $0 \le t \le 2\pi$. **$x^2 + y^2 = 9$**

Find parametric equations for each rectangular equation.

32–38. See margin for sample answers.

32. $x^2 + y^2 = 25$

33. $x^2 + y^2 - 16 = 0$

34. $\dfrac{x^2}{4} + \dfrac{y^2}{25} = 1$

35. $\dfrac{y^2}{16} + x^2 = 1$

36. $y = x^2 - 4x + 7$

37. $x = y^2 + 2y - 1$

38. Find parametric equations for the rectangular equation $(y + 3)^2 = 4(x - 2)$.

Graphing Calculator

39. Consider the rectangular equation $x = \sqrt{y}$. **a–b, d. See margin.**
  a. By using different choices for t, find two different parametric representations of this equation.
  b. Graph the rectangular equation by hand. Then use a graphing calculator to sketch the graphs of each set of parametric equations.
  c. Are your graphs from part **b** the same? **yes**
  d. What does this suggest about parametric representations of rectangular equations?

Applications and Problem Solving

Real World Application

40b. Sample answer:
$x = 6 \sin(\pi t)$,
$y = 6 \cos(\pi t)$,
$0 \le t \le 2$

40. Transportation A riverboat's paddlewheel has a diameter of 12 feet and at full speed, makes one clockwise revolution in 2 seconds.

  a. Write a rectangular equation to model the shape of the paddlewheel. **$x^2 + y^2 = 36$**
  b. Write parametric equations describing the position of a point A on the paddlewheel for any given time t. Assume that at $t = 0$, A is at the very top of the wheel.
  c. How far will point A, which is a fixed point on the wheel, move in 1 minute? **about 1131 ft**

41. Critical Thinking Identify the graph of each equation using the method described in this lesson. Then identify the graph of each equation after first rewriting the equation in standard form and solving for y. Explain the discrepancies, if any, in your answers. **See margin.**
  a. $2x^2 + 5y^2 = 0$ b. $x^2 + y^2 - 4x - 6y + 13 = 0$ c. $y^2 - 9x^2 = 0$

42. Critical Thinking Explain why a substitution of $x = t^2$ is not appropriate when trying to find a parametric representation of $y = x^2 - 5$? **See margin.**

43b. $x = 6 \sin t$, $y = 6 \cos t$, $0 \le t \le 4\pi$

43. Timing The path traced by the tip of the second-hand of a clock can be modeled by the equation of a circle in parametric form.

  a. If the radius of the clock is 6 inches, find an equation in rectangular form that models the shape of the clock. **$x^2 + y^2 = 36$**
  b. Find parametric equations that describe the motion of the tip as it moves from 12 o'clock noon to 12 o'clock noon of the next day.
  c. Simulate the motion described by graphing the equations on a graphing calculator. **See Answer Appendix.**

Additional Answers (p.669)

45.

46.

(−5, 5)

(−5, 0)

(5, −5)

44. Framing Portraits are often framed so that the opening through which the picture is seen is an ellipse. These oval mats must be custom cut using an oval cutter whose design relies upon the parametric equations of an ellipse. The elliptical compass at the right consists of a stick with a pencil attached to one end and two pivot holes at the other. Through these holes, the stick is anchored to two small blocks, one of which can slide horizontally and the other vertically in its groove. Use the diagram of the elliptical compass at right to verify that $x = a \cos t$ and $y = b \sin t$. (*Hint:* Draw an extra vertical and an extra horizontal line to create right triangles and then use trigonometry.) **See Answer Appendix.**

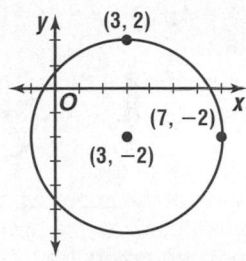

Mixed Review

45. vertex:
$(-5, 0)$; focus:
$(-5, 3)$; axis
of symmetry:
$x = -5$, directrix:
$y = -3$; see
margin for graph.

45. Find the coordinates of the vertex, focus, and the equations of the axis of symmetry and directrix of the parabola with equation $x^2 - 12y + 10x = -25$. Then graph the equation. (*Lesson 10-5*).

46. Graph $xy = -25$. (*Lesson 10-4*) **See margin.**

47. Write $3x^2 + 3y^2 - 18x + 12y = 9$ in standard form. Then graph the equation. (*Lesson 10-2*) $(x - 3)^2 + (y + 2)^2 = 16$; **see margin for graph.**

48. A 30-pound force is applied to an object at an angle of $60°$ with the horizontal. Find the magnitude of the horizontal and vertical components of the force. (*Lesson 8-1*) **15 lb, $15\sqrt{3}$ lb**

49. Statistics The prediction equation $y = -0.13x + 37.8$ gives the fuel economy y for a car with horsepower x. Is the equation a better predictor for Car 1, which has a horsepower of 135 and average 19 miles per gallon, or for Car 2, which has a horsepower of 245 and averages 16 miles per gallon? Explain. (*Lesson 7-7*) **Car 1; see margin for explanation.**

50. Find the value of $\sin\left(2 \operatorname{Sin}^{-1} \frac{1}{2}\right)$. (*Lesson 6-8*) $\dfrac{\sqrt{3}}{2}$

51. Find the area to the nearest square unit of $\triangle ABC$ if $a = 48$, $b = 32$, and $c = 44$. (*Lesson 5-8*) **685 units^2**

52. Solve $\sqrt{2y - 3} - \sqrt{2y + 3} = -1$. (*Lesson 4-7*) $\dfrac{37}{8}$

53. If y varies jointly as x and z and $y = 16$ when $x = 5$ and $z = 2$, find y when $x = 8$ and $z = 3$. (*Lesson 3-8*) **38.4**

54. Find the determinant of $\begin{bmatrix} 5 & 9 \\ 7 & -3 \end{bmatrix}$. Then state whether an inverse exists for the matrix. (*Lesson 2-5*) **−78, yes**

55. Write the point-slope form of the equation of the line through the points $(-6, 4)$ and $(3, 7)$. Then write the equation in slope-intercept form. (*Lesson 1-4*) $y - 4 = \frac{1}{3}(x + 6)$ or $y - 7 = \frac{1}{3}(x - 3)$, $y = \frac{1}{3}x + 6$

56. SAT/ACT Practice For all values where $x \neq y$, let $x \# y$ represent the lesser of the numbers x and y, and let $x @ y$ represent the greater of the number x and y. What is the value of $(1 \# 4) @ (2 \# 3)$? **B**

   A 1   **B** 2   **C** 3   **D** 4   **E** 5

Extra Practice See p. A45.

---

4 ASSESS

Open-Ended Assessment

Writing Have students write, in mixed order, equations representing a circle, a parabola, an ellipse, and a hyperbola, without specifying which equation is which. Then ask students to exchange and identify each other's equations.

📁 **Chapter 10, Quiz C (Lessons 10-5 and 10-6)** is available in the *Assessment and Evaluation Masters,* p. 252.

Additional Answers

47.

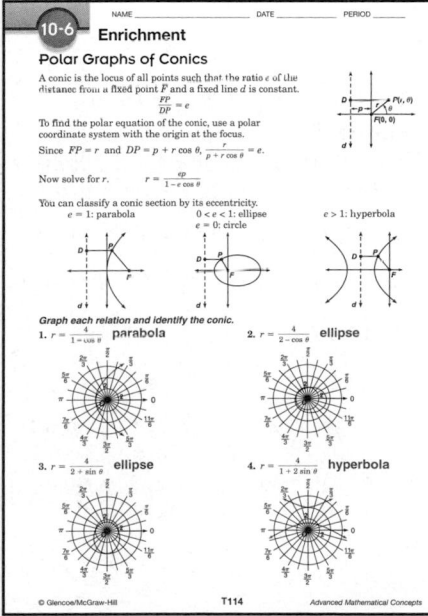

49. The point (135, 19) is about 9 units closer to the line $y = -0.13x + 37.8$ than the point (245, 16).

Enrichment Masters, p. 114

1 FOCUS

10-7

Transformations of Conics

5-Minute Check
Lesson 10-6

Identify the conic section represented by each equation.

1. $3x^2 - 2y^2 - 5x + 8y = 0$
   hyperbola

2. $2x^2 + 3x + 5y - 7 = 0$
   parabola

3. $-4x^2 - 4y^2 + 2x + y = 0$
   circle

4. $8x^2 + 5y^2 - x + 6y = 0$
   ellipse

5. Find the rectangular equation of the curve whose parametric equations are given. Then graph the equation using arrows to indicate orientation.
   $x = t$, $y = 3t^2 - 3t + 2$,
   $-\infty \leq t \leq \infty$ $y = 3x^2 - 3x + 2$

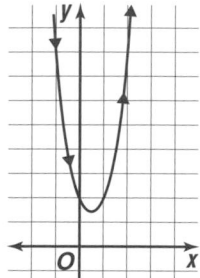

Motivating the Lesson

Hands-On Activity Ask students to find the equation of the ellipse that is the result of rotating the ellipse $\frac{x^2}{36} + \frac{y^2}{25} = 1$ by 90° about the origin. Have them graph each equation. $\frac{x^2}{25} + \frac{y^2}{36} = 1$

OBJECTIVES
- Find the equations of conic sections that have been translated or rotated.
- Graph rotations and/or translations of conic equations.
- Identify the equations of conic sections using the discriminant.
- Find the angle of rotation for a given equation.

SPORTS At the 1996 Olympics in Atlanta, Georgia, U.S. high school student Kim Rhode won the gold medal in the women's double trap shooting event, being staged at the Olympics for the first time. The double trap event consists of firing double barrel shotguns at flying clay targets that are launched two at a time out of a house located 14.6 to 24.7 meters in front of the contestants. Targets are thrown out of the house in a random arc, but always at the same height. *A problem related to this will be solved in Example 1.*

Thus far, we have used a transformation called a translation to show how the parent graph of each of the conic sections is translated to a center other than the origin. For example, the equation of the circle $x^2 + y^2 = r^2$ becomes $(x - h)^2 + (y - k)^2 = r^2$ for a center of (h, k). A translation of a set of points with respect to (h, k) is often written as follows.

$$T_{(h, k)} \quad \Longleftrightarrow \quad \text{translation with respect to } (h, k)$$

Example **1** **SPORTS** **Refer to the application above. A video game simulating the sport of double trap allows a player to shoot at two elliptically-shaped targets released from a house at the bottom of the screen. With the house located at the origin, a target at its initial location is modeled by the equation $\frac{x^2}{16} + \frac{y^2}{4} = 1$. Suppose a player misses one of the two targets released and the center of the target leaves the screen at the point (24, 30). Find an equation that models the shape and position of the target with its center translated to this point.**

To write the equation of $\frac{x^2}{16} + \frac{y^2}{4} = 1$ for $T_{(24, 30)}$, let $h = 24$ and $k = 30$. Then replace x with $x - h$ and y with $y - k$.

$x^2 \Rightarrow (x - 24)^2$

$y^2 \Rightarrow (y - 30)^2$

Thus, the translated equation is
$$\frac{(x - 24)^2}{16} + \frac{(y - 30)^2}{4} = 1$$

The graph shows the parent ellipse and its translation.

$[-35, 35]$ scl:5 by $[-8.09, 38.09]$ scl:5

670 *Chapter 10 Conics*

Resource Manager

Reproducible Masters
- *Study Guide and Practice,* pp. 189–190
- *Enrichment,* p.115
- *TI-92 & CBL Interdisciplinary Activities,* pp. 33–36

Transparencies
- 5-Minute Check, 10-7
- Teaching, 10-7

Another type of transformation you have studied is a rotation. Except for hyperbolas whose equations are of the form $xy = k$, all of the conic sections we have studied thus far have been oriented with their axes parallel to the coordinate axes. In the general form of these conics $Ax^2 + Bxy + Cy^2 + Dx + Ey + F = 0$, $B = 0$. Whenever $B \neq 0$, then the axes of the conic section are not parallel to the coordinate axes. That is, the graph is rotated.

The figures below show an ellipse whose center is the origin and its rotation. Notice that the angle of rotation has the same measure as the angles formed by the positive x-axis and the major axis and the positive y-axis and the minor axis.

 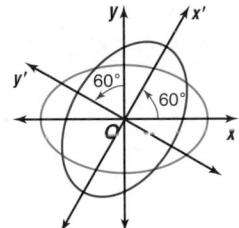

60° counterclockwise rotation

The coordinates of the points of a rotated figure can be found by using a rotation matrix.

A rotation of θ about the origin can be described by the matrix

$$\begin{bmatrix} \cos \theta & -\sin \theta \\ \sin \theta & \cos \theta \end{bmatrix}.$$

A positive value of θ indicates a counterclockwise rotation. A negative value of θ indicates a clockwise rotation.

Let $P(x, y)$ be a point on the graph of a conic section. Then let $P'(x', y')$ be the image of P after a counterclockwise rotation of θ. The values of x' and y' can be found by matrix multiplication.

$$\begin{bmatrix} x' \\ y' \end{bmatrix} = \begin{bmatrix} \cos \theta & -\sin \theta \\ \sin \theta & \cos \theta \end{bmatrix} \cdot \begin{bmatrix} x \\ y \end{bmatrix}$$

The inverse of the rotation matrix represents a rotation of $-\theta$. Multiply each side of the equation by the inverse rotation matrix to solve for x and y.

$$\begin{bmatrix} \cos \theta & \sin \theta \\ -\sin \theta & \cos \theta \end{bmatrix} \cdot \begin{bmatrix} x' \\ y' \end{bmatrix} = \begin{bmatrix} \cos \theta & \sin \theta \\ -\sin \theta & \cos \theta \end{bmatrix} \cdot \begin{bmatrix} \cos \theta & -\sin \theta \\ \sin \theta & \cos \theta \end{bmatrix} \cdot \begin{bmatrix} x \\ y \end{bmatrix}$$

$$\begin{bmatrix} x' \cos \theta + y' \sin \theta \\ -x' \sin \theta + y' \cos \theta \end{bmatrix} = \begin{bmatrix} 1 & 0 \\ 0 & 1 \end{bmatrix} \cdot \begin{bmatrix} x \\ y \end{bmatrix}$$

$$\begin{bmatrix} x' \cos \theta + y' \sin \theta \\ -x' \sin \theta + y' \cos \theta \end{bmatrix} = \begin{bmatrix} x \\ y \end{bmatrix}$$

The result is two equations that can be used to determine the equation of a conic with respect to a rotation of θ.

Rotation Equations

To find the equation of a conic section with respect to a rotation of θ, replace

x with $x' \cos \theta + y' \sin \theta$
and y with $-x' \sin \theta + y' \cos \theta$.

In-Class Example

Example 1
Astronomy Mars is about 128.4 million miles from the sun at its closest point and 154.8 million miles from the sun at its farthest point. With the sun located at the origin, the orbit of Mars is modeled by the equation $\dfrac{x^2}{20{,}051} + \dfrac{y^2}{19{,}875} = 1$.
Find an equation that models the orbit of Mars with the sun translated to the point $(-13.2, 0)$.
$\dfrac{(x - 13.2)^2}{20{,}051} + \dfrac{y^2}{19{,}875} = 1$

Teaching Tip When working Example 2, inform students that the process of rotating these equations is similar to that of finding a resultant vector. They have to take into account the cosine and sine adjustments.

In-Class Example

Example 2
Find the equation of the graph of $\frac{x^2}{25} + \frac{y^2}{4} = 1$ after it is rotated 30° about the origin. Then sketch the graph and its rotation.
$37(x')^2 - 42\sqrt{3}\,x'y' + 79(y')^2 - 400 = 0$

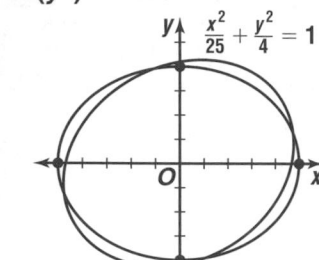

Example ❷ **Find the equation of the graph of $\frac{x^2}{16} - \frac{y^2}{9} = 1$ after it is rotated 45° about the origin. Then sketch the graph and its rotation.**

The graph of this equation is a hyperbola.

Find the expressions to replace x and y.

Replace x with $x' \cos 45° + y' \sin 45°$ or $\frac{\sqrt{2}}{2}x' + \frac{\sqrt{2}}{2}y'$.

Replace y with $-x' \sin 45° + y' \cos 45°$ or $-\frac{\sqrt{2}}{2}x' + \frac{\sqrt{2}}{2}y'$.

Computation is often easier if the equation is rewritten as an equation with denominators of 1.

$$\frac{x^2}{16} - \frac{y^2}{9} = 1$$

$$9x^2 - 16y^2 = 144 \qquad \textit{Multiply each side by 144.}$$

$$9\left(\frac{\sqrt{2}}{2}x' + \frac{\sqrt{2}}{2}y'\right)^2 - 16\left(-\frac{\sqrt{2}}{2}x' + \frac{\sqrt{2}}{2}y'\right)^2 = 144 \qquad \textit{Replace x and y.}$$

$$9\left[\frac{1}{2}(x')^2 + x'y' + \frac{1}{2}(y')^2\right] - 16\left[\frac{1}{2}(x')^2 - x'y' + \frac{1}{2}(y')^2\right] = 144 \qquad \textit{Expand the binomial.}$$

$$-\frac{7}{2}(x')^2 + 25x'y' - \frac{7}{2}(y')^2 = 144 \qquad \textit{Simplify.}$$

$$7(x')^2 - 50x'y' + 7(y')^2 = -288 \qquad \textit{Multiply each side by −2.}$$

The equation of the hyperbola after the 45° rotation is $7(x')^2 - 50x'y' + 7(y')^2 = -288$.

The graph below shows the hyperbola and its rotation.

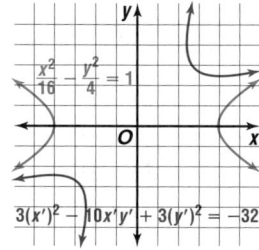

In Lesson 10-6, you learned to identify a conic from its general form $Ax^2 + Bxy + Cy^2 + Dx + Ey + F = 0$, where $B = 0$. When $B \neq 0$, the equation can be identified by examining the discriminant of the equation.

You may remember from the Quadratic Formula that the discriminant of a second-degree equation is defined as $B^2 - 4AC$ and will remain unchanged under any rotation. That is, $B^2 - 4AC = (B')^2 - 4A'C'$.

For the general equation $Ax^2 + Bxy + Cy^2 + Dx + Ey + F = 0$,
- if $B^2 - 4AC < 0$, the graph is a circle ($A = C$, $B = 0$) or an ellipse ($A \neq C$ or $B \neq 0$);
- if $B^2 - 4AC > 0$, the graph is a hyperbola;
- if $B^2 - 4AC = 0$, the graph is a parabola.

Remember that the graphs can also be degenerate cases.

Example ③ Identify the graph of the equation $x^2 - 4xy + 4y^2 + 5\sqrt{5}y + 1 = 0$.

Since the equation contains an xy-term, use the discriminant of the equation to identify the conic.

$B^2 - 4AC = (-4)^2 - 4(1)(4)$ $A = 1, B = -4, C = 4$
$ = 0$

Since $B^2 - 4AC = 0$, the graph of the equation is a parabola.

You can also use values from the general form to find the angle of rotation about the origin.

Angle of Rotation About the Origin

For the general equation $Ax^2 + Bxy + Cy^2 + Dx + Ey + F = 0$, the angle of rotation θ about the origin can be found by

$$\theta = \frac{\pi}{4}, \text{ if } A = C, \text{ or}$$
$$\tan 2\theta = \frac{B}{A - C}, \text{ if } A \neq C.$$

Example ④ Identify the graph of the equation $2x^2 + 9xy + 14y^2 - 5 = 0$. Then find θ and use a graphing calculator to draw the graph.

$B^2 - 4AC = (9)^2 - 4(2)(14)$ $A = 2, B = 9, \text{ and } C = 14$
$ = -31$

Since the discriminant is less than 0 and $A \neq C$, the graph is an ellipse.

Now find θ using $\tan 2\theta = \frac{B}{A - C}$, since $A \neq C$.

$\tan 2\theta = \frac{B}{A - C}$

$\tan 2\theta = \frac{9}{2 - 14}$

$\tan 2\theta = -0.75$

$2\theta = -36.86989765$ *Take the inverse tangent of each side.*

$\theta \approx -18°$ *Round to the nearest degree.*

To graph the equation, you must solve for y. Rewrite the equation in quadratic form, $ay^2 + by + c = 0$.

$$\begin{array}{ccc} a & b & c \\ \downarrow & \downarrow & \downarrow \end{array}$$
$$14y^2 + (9x)y + (2x^2 - 5) = 0$$

(continued on the next page)

Teaching Tip Make sure that students solve the equation in Example 4 for y, not x, when graphing.

In-Class Examples

Example 3
Identify the graph of the equation $8x^2 + 5xy - 4y^2 = -2$.
hyperbola

Example 4
Identify the graph of the equation $x^2 - 4xy - 2y^2 - 6 = 0$. Then find θ and use a graphing calculator to draw the graph.
hyperbola, $\theta \approx -27°$

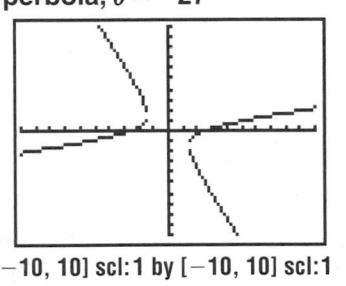

$[-10, 10]$ scl:1 by $[-10, 10]$ scl:1

Teaching Tip After working Example 4, you may wish to have students research how the formula for the angle of rotation is derived.

In-Class Example

Example 5
The graph of $3(x - 1)^2 + 4(y + 4)^2 = 0$ is a degenerate case. Identify the graph and then draw it. **a point at (1, –4)**

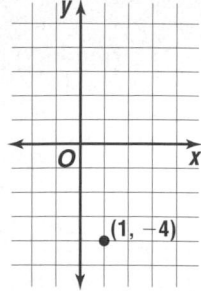

---

3 PRACTICE/APPLY

Error Analysis

Watch for students who misidentify conics when using the discriminant.

Prevent by having students copy the summary chart on page 673 into their notes and refer to it for every problem.

Study Guide and Practice Masters, pp. 189–190

10-7
NAME _____ DATE _____ PERIOD _____
Study Guide
Transformation of Conics

Translations are often written in the form $T_{(h, k)}$. To find the equation of a rotated conic, replace x with $x'\cos\theta + y'\sin\theta$ and y with $-x'\sin\theta + y'\cos\theta$.

Example Identify the graph of each equation. Write an equation of the translated or rotated graph in general form.

a. $4x^2 + y^2 = 12$ for $T_{(-2, 3)}$

The graph of this equation is an ellipse. To write the equation of $4x^2 + y^2 = 12$ for $T_{(-2, 3)}$, let $h = -2$ and $k = 3$. Then replace x with $x - h$ and y with $y - k$.

$x^2 \Rightarrow (x - (-2))^2$ or $(x + 2)^2$
$y^2 \Rightarrow (y - 3)^2$

Thus, the translated equation is $4(x + 2)^2 + (y - 3)^2 = 12$.

Write the equation in general form.

$4(x + 2)^2 + (y - 3)^2 = 12$
$4(x^2 + 4x + 4) + y^2 - 6y + 9 = 12$ *Expand the binomial.*
$4x^2 + y^2 + 16x - 6y + 25 = 12$ *Simplify.*
$4x^2 + y^2 + 16x - 6y + 13 = 0$ *Subtract 12 from both sides.*

b. $x^2 - 4y = 0, \theta = 45°$

The graph of this equation is a parabola. Find the expressions to replace x and y.

Replace x with $x'\cos 45° + y'\sin 45°$ or $\frac{\sqrt{2}}{2}x' + \frac{\sqrt{2}}{2}y'$.
Replace y with $-x'\sin 45° + y'\cos 45°$ or $-\frac{\sqrt{2}}{2}x' + \frac{\sqrt{2}}{2}y'$.

$x^2 - 4y = 0$
$\left(\frac{\sqrt{2}}{2}x' + \frac{\sqrt{2}}{2}y'\right)^2 - 4\left(-\frac{\sqrt{2}}{2}x' + \frac{\sqrt{2}}{2}y'\right) = 0$ *Replace x and y.*
$\frac{1}{2}(x')^2 + x'y' + \frac{1}{2}(y')^2 - 4\left(-\frac{\sqrt{2}}{2}x' + \frac{\sqrt{2}}{2}y'\right) = 0$ *Expand the binomial.*
$\frac{1}{2}(x')^2 + x'y' + \frac{1}{2}(y')^2 + 2\sqrt{2}x' - 2\sqrt{2}y' = 0$ *Simplify.*

The equation of the parabola after the 45° rotation is
$\frac{1}{2}x^2 + xy + \frac{1}{2}y^2 + 2\sqrt{2}x - 2\sqrt{2}y = 0$.

Now use the Quadratic Formula to solve for y.

$$y = \frac{-b \pm \sqrt{b^2 - 4ac}}{2a}$$

$$y = \frac{-(9x) \pm \sqrt{(9x)^2 - 4(14)(2x^2 - 5)}}{2(14)}$$

$$y = \frac{-9x \pm \sqrt{-31x^2 + 280}}{28}$$

Enter the equations and graph.

[−7.6, 7.6] scl:1 by [−5, 5] scl:1

This method of solving for y to graph a rotated equation is useful when trying to identify the equation of a degenerate conic.

Example ⑤ **The graph of $xy - y^2 + 2x^2 = 0$ is a degenerate case. Identify the graph and then draw it.**

In this equation, $B^2 - 4AC > 0$. At first glance, this equation may appear to be the equation of a hyperbola. A closer inspection reveals that this is a degenerate case.

Solve the equation for y by first rewriting the equation in quadratic form.

$$\begin{array}{ccc} a & b & c \\ \downarrow & \downarrow & \downarrow \end{array}$$
$$(-1)y^2 + (x)y + 2x^2 = 0$$

Now use the quadratic formula to solve for y.

$$y = \frac{-b \pm \sqrt{b^2 - 4ac}}{2a}$$

$$y = \frac{-x \pm \sqrt{x^2 - 4(-1)(2x^2)}}{2(-1)}$$

$$y = 2x \text{ or } y = -x$$

So the graph of $xy - y^2 + 2x^2$ is actually the graph of $y = 2x$ and $y = -x$, which are intersecting lines.

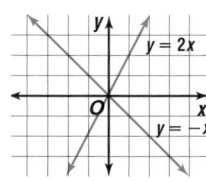

CHECK FOR UNDERSTANDING

Communicating Mathematics

Read and study the lesson to answer each question.   1–2. See margin.

1. **Sketch** a parabola with vertex at the origin and a horizontal line of symmetry. Then sketch the parabola for $T_{(-3, 3)}$. Label each graph with its equation in standard form.

2. **Write** the expressions needed to replace x and y for an equation rotated 30°.

3. **Indicate** the angle of rotation about the origin needed to transform the equation $\frac{x^2}{100} + \frac{y^2}{25} = 1$ into $\frac{x^2}{25} + \frac{y^2}{100} = 1$.   **90° or 270°**

Additional Answers

1. Sample answers:

2. Replace x with $\frac{\sqrt{3}}{2}x' + \frac{1}{2}y'$.
   Replace y with $-\frac{1}{2}x' + \frac{\sqrt{3}}{2}y'$.

4. You Decide Ebony says that $7x^2 - 6\sqrt{3}xy + 13y^2 = 0$ is the equation of an ellipse. Teisha disagrees. She says the equation is a hyperbola. Who is correct? Justify your answer. **Ebony; $B^2 - 4AC = (6\sqrt{3})^2 - 4(7)(13) < 0$ and $A \neq C$**

Guided Practice Identify the graph of each equation. Write an equation of the translated or rotated graph in general form. **5–8. See margin.**

5. $x^2 + y^2 = 7$ for $T_{(3, 2)}$
6. $y = 2x^2 - 7x + 5$ for $T_{(-4, 5)}$

7. $x^2 - y^2 = 9$, $\theta = 60°$
8. $x^2 - 5x + y^2 = 3$, $\theta = \dfrac{\pi}{4}$

Identify the graph of each equation. Then find θ to the nearest degree.

9. $9x^2 + 4xy + 4y^2 + 2 = 0$
ellipse; 19°
10. $8x^2 + 5xy - 4y^2 + 2 = 0$
hyperbola; 11°

11. The graph of $3(x - 1)^2 + 4(y + 4)^2 = 0$ is a degenerate case. Identify the graph and then draw it. **Point; see Answer Appendix for graph.**

12. Communications A satellite dish tracks a satellite directly overhead. Suppose the equation $y = \dfrac{1}{6}x^2$ models the shape of the dish when it is oriented in this position. Later in the day the dish is observed to have rotated approximately 30°.
  a. Find an equation that models the new orientation of the dish.
  b. Sketch the graphs of both equations on the same set of axes using a graphing calculator. **See margin.**

12a. $3(x')^2 + 2\sqrt{3}x'y' + (y')^2 + 12x' - 12\sqrt{3}y' = 0$

EXERCISES

Practice Identify the graph of each equation. Write an equation of the translated or rotated graph in general form. **13–24. See margin.**

A
13. $y = 3x^2 - 2x + 5$ for $T_{(2, -3)}$
14. $4x^2 + 5y^2 = 20$ for $T_{(5, -6)}$
15. $3x^2 + y^2 = 9$ for $T_{(-1, 3)}$
16. $4y^2 + 12x^2 = 24$ for $T_{(-1, 4)}$
17. $9x^2 - 25y^2 = 225$ for $T_{(0, 5)}$
18. $(x + 3)^2 = 4y$ for $T_{(-7, 2)}$

B
19. $x^2 - 8y = 0$, $\theta = 90°$
20. $2x^2 + 2y^2 = 8$, $\theta = 30°$
21. $y^2 + 8x = 0$, $\theta = \dfrac{\pi}{6}$
22. $xy = -8$, $\theta = \dfrac{\pi}{4}$
23. $x^2 - 5x + y^2 = 3$, $\theta = \dfrac{\pi}{3}$
24. $16x^2 - 4y^2 = 64$, $\theta = 60°$

25. Write the equation of the ellipse $6x^2 + 5y^2 = 30$ after a rotation of 30° about the origin. **See margin.**

26. ellipse; 23°
27. hyperbola; −6°
29. ellipse; −18°

Identify the graph of each equation. Then find θ to the nearest degree.
26. $9x^2 + 4xy + 5y^2 - 40 = 0$
27. $x^2 - xy - 4y^2 - x - y + 4 = 0$
28. $8x^2 + 8xy + 2y^2 = 0$ **parabola; 27°**
29. $2x^2 + 9xy + 14y^2 - 5 = 0$

C
30. $2x^2 + 4xy + 5y^2 + 3x - 4y - 20 = 0$ **ellipse; −27°**
31. $2x^2 + 4\sqrt{3}xy + 6y^2 + \sqrt{3}x - y = 0$ **parabola, −30°**

 www.amc.glencoe.com/self_check_quiz

Assignment Guide
Core: 13–41 odd, 43–45, 47, 49–59
Enriched: 14–42 even, 43–59

Additional Answers
5. circle; $x^2 + y^2 - 6x - 4y + 6 = 0$

6. parabola; $2x^2 + 9x - y + 14 = 0$

7. hyperbola; $(x')^2 - 2\sqrt{3}xy - (y')^2 + 18 = 0$

8. circle; $2(x')^2 + 2(y')^2 - 5\sqrt{2}x - 5\sqrt{2}y - 6 = 0$

12b.

[−10, 10] scl:1 by [−3, 10] scl:1

13. parabola; $3x^2 - 14x - y + 18 = 0$

14. ellipse; $4x^2 + 5y^2 - 40x + 60y + 260 = 0$

15. ellipse; $3x^2 + y^2 + 6x - 6y + 3 = 0$

16. ellipse; $3x^2 + y^2 + 6x - 8y + 13 = 0$

17. hyperbola; $9x^2 - 25y^2 + 250y - 850 = 0$

18. parabola; $x^2 + 20x - 4y + 108 = 0$

19. parabola; $(y')^2 + 8x' = 0$

20. circle; $(x')^2 + (y')^2 - 4 = 0$

21. parabola; $(x')^2 - 2\sqrt{3}x'y' + 3(y')^2 + 16\sqrt{3}x' + 16y' = 0$

22. hyperbola; $(x')^2 - (y')^2 - 16 = 0$

23. circle; $2(x')^2 + 2(y')^2 - 5x' - 5\sqrt{3}y' - 6 = 0$

24. hyperbola; $(x')^2 + 10\sqrt{3}x'y' + 11(y')^2 - 64 = 0$

25. $23(x')^2 + 2\sqrt{3}x'y' + 21(y')^2 - 120 = 0$

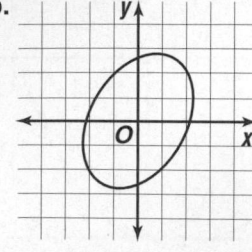

Teaching Tip For Exercises 37–42, remind students that they will need to graph positive values for *y* and negative values for *y* in order to get the total graph.

Additional Answers

44d. Circle; there is no minimum angle of rotation, since any degree of rotation will result in a graph that coincides with the original.

45. See below.

46b.

32. Identify the equation $2x^2 + 4xy + 2y^2 + 2\sqrt{2}x - 2\sqrt{2}y + 12 = 0$ as a circle, ellipse, parabola, or hyperbola. Then find θ to the nearest degree.
parabola; 45°

The graph of each equation is a degenerate case. Identify the graph and then draw it. **33–36. See Answer Appendix for graphs.**

33. $(x - 2)^2 - (x + 3)^2 = 5(y + 2)$ **line**   **34.** $2x^2 + 6y^2 + 8x - 12y + 14 = 0$ **point**

35. $y^2 - 9x^2 = 0$ **intersecting lines**   **36.** $(x - 2)^2 + (y - 2)^2 + 4(x + y) = 8$
point

Graphing Calculator

Use the Quadratic Formula to solve each equation for *y*. Then use a graphing calculator to draw the graph.   **37–42. See Answer Appendix.**

37. $x^2 - 2xy + y^2 - 5x - 5y = 0$ **38.** $2x^2 + 9xy + 14y^2 = 5$

39. $8x^2 + 5xy - 4y^2 = -2$ **40.** $2x^2 + 4\sqrt{3}xy + 6y^2 + 3x = y$

41. $2x^2 + 4xy + 2y^2 + 2\sqrt{2}x - 2\sqrt{2}y = -12$ **42.** $9x^2 + 4xy + 6y^2 = 20$

Applications and Problem Solving

Real World Application

43. Agriculture Many farmers in the Texas panhandle divide their land into smaller one square mile units. This square is then divided into four smaller units, or quadrants, of equal size. Each quadrant contains its own central pivot irrigation system. Suppose the center of the square mile of land is the origin.

  a. Determine the translation in feet needed to place an irrigation system in Quadrant I. $T_{(1320, \ 1320)}$
  b. Determine the equation for the path of the outer end of the irrigation system in Quadrant I. $(x - 1320)^2 + (y - 1320)^2 = 1{,}742{,}400$

44. Critical Thinking Identify the graph of each equation. Then determine the minimum angle of rotation needed to transform each equation so that the rotated graph coincides with its original graph.
  a. $x^2 + 6x + 9 - y = 0$ **parabola; 360°**   b. $8x^2 + 6y^2 = 24$ **ellipse; 180°**
  c. $4xy = 25$ **hyperbola; 180°**   d. $15x^2 + 15y^2 = 60$ **See margin.**

45. Critical Thinking Prove that a circle with equation $x^2 + y^2 = r^2$ remains unchanged under any rotation θ.   **See margin.**

46. Astronomy Suppose the equation $31x^2 - 10\sqrt{3}xy + 21y^2 = 144$ models the shape of a reflecting mirror in a telescope.
  a. Determine whether the reflector in the telescope is elliptical, parabolic, or hyperbolic.   **elliptical**
  b. Using a graphing calculator, sketch the graph of the equation.   **See margin.**
  c. Determine the angle through which the mirror has been rotated.   **−30°**

47. Critical Thinking Consider the equation $9x^2 - 2\sqrt{3}xy + 11y^2 - 24 = 0$.
  a. Determine the minimum angle of rotation needed to transform the graph of this equation to a graph whose axes are on the *x*- and *y*-axes.   **−30°**
  b. Use the angle of rotation in part **a** to find the new equation of the graph.

47b. $\dfrac{(x')^2}{3} + \dfrac{(y')^2}{2} = 1$

676   *Chapter 10   Conics*

Additional Answer

45. Let $x = x'\cos\theta + y'\sin\theta$ and $y = -x'\sin\theta + y'\cos\theta$.

$$x^2 + y^2 = r^2$$
$$(x'\cos\theta + y'\sin\theta)^2 + (-x'\sin\theta + y'\cos\theta)^2 = r^2$$
$$(x')^2\cos^2\theta + x'y'\cos\theta\sin\theta + (y')^2\sin^2\theta + (x')^2\sin^2\theta - x'y'\cos\theta\sin\theta + (y')^2\cos\theta = r^2$$
$$[(x')^2 + (y')^2]\cos^2\theta + [(x')^2 + (y')^2]\sin^2\theta = r^2$$
$$[(x')^2 + (y')^2](\cos^2\theta + \sin^2\theta) = r^2$$
$$[(x')^2 + (y')^2](1) = r^2$$
$$(x')^2 + (y')^2 = r^2$$

48. **Manufacturing** A cam in a punch press is shaped like an ellipse with the equation $\frac{x^2}{81} + \frac{y^2}{36} = 1$. The camshaft goes through the focus on the positive axis.

48b. $\dfrac{\left(x - 3\sqrt{5}\right)^2}{81}$

$+ \dfrac{y^2}{36} = 1$

a. Graph a model of the cam. **See margin.**

b. Find an equation that translates the model so that the camshaft is at the origin.

c. Find the equation of the model in part b when the cam is rotated to an upright position.

$$\frac{\left(x' - 3\sqrt{5}\right)^2}{36} + \frac{(y')^2}{81} = 1$$

Cam — Camshaft
Press stem
Punch

Mixed Review

49. Identify the conic section represented by the equation $5y^2 - 3x^2 + 4x - 3y - 100 = 0$. *(Lesson 10-6)* **hyperbola**

50. Write the equation of the ellipse that has its center at $(2, -3)$, $a = 1$, and $e = \frac{2\sqrt{6}}{5}$. *(Lesson 10-3)* $(x - 2)^2 + 25(y + 3)^2 = 1$ or $(y + 3)^2 + 25(x - 2)^2 = 1$

51. Graph $r = \dfrac{1}{\cos(\theta + 15°)}$. *(Lesson 9-4)* **See margin.**

52. about 9.4 m/s

52. **Boating** A boat heads due west across a lake at 8 m/s. If a current of 5 m/s moves due south, what is the boat's resultant velocity? *(Lesson 8-1)*

53. Which value is greater, $\cos 70°$ or $\cos 170°$? *(Lesson 6-3)* **cos 70°**

54. 56°15′

54. Change $\dfrac{5\pi}{16}$ radians to degree measure to the nearest minute. *(Lesson 5-1)*

55. Decompose $\dfrac{2y + 5}{y^2 + 3y + 2}$ into partial fractions. *(Lesson 4-6)* $\dfrac{-1}{y + 2} + \dfrac{3}{y + 1}$

56. If y varies inversely as x and $y = 4$ when $x = 12$, find y when $x = 5$. *(Lesson 3-8)* **9.6**

57. Solve the system of equations algebraically. *(Lesson 2-2)*
$8m - 3n - 4p = 6$
$4m + 9n - 2p = -4$ $\left(\dfrac{3}{4}, -\dfrac{2}{3}, \dfrac{1}{2}\right)$
$6m + 12n + 5p = -1$

58. Graph $h(x) = [\![x]\!] - 3$. *(Lesson 1-7)* **See margin.**

59. **SAT Practice** If $1 < b < 2$ and $2 < a < 3$, which statement is true about the expression $\dfrac{5a^8b^5}{180a^6b^2}$? **B**

A The value of the expression is never greater than 1.

B The value of the expression is always between $\dfrac{1}{9}$ and 2.

C The value of the expression is always greater than 1.

D The value of the expression is always between $\dfrac{1}{36}$ and $\dfrac{3}{2}$.

E The value of the expression is always between 0 and $\dfrac{1}{2}$.

Extra Practice See p. A45.

Lesson 10-7 *Transformations of Conics* **677**

Extra Credit

The angle of rotation about the origin of the graph of $2x^2 + 4xy + Cy^2 + 3x - 2y = 0$ is 30°. Find C. $\dfrac{6 - 4\sqrt{3}}{3}$

Lesson 10-7 **677**

1 FOCUS

5-Minute Check
Lesson 10-7

Identify the graph of each equation. Write an equation of the translated or rotated graph in general form.

1. $2x^2 + 3xy - 4y^2 - 5x = 0$ for $T_{(3, 1)}$ **hyperbola;**
   $2x^2 + 3xy - 4y^2 - 20x - y + 38 = 0$

2. $-x^2 + 2xy - y^2 + 6y = 4$ for $T_{(-2, 6)}$ **parabola;**
   $-x^2 - y^2 + 2xy - 16x + 22y - 104 = 0$

3. $x^2 + 4xy + 5y^2 - 7x - y = 0$, $\theta = 90°$ **ellipse;**
   $-5(x')^2 + (y')^2 - 4x'y' + x' - 7y' = 0$

4. Identify the graph of $x^2 + 3xy + 2y^2 - 3x + 2y + 10 = 0$. Then find θ to the nearest degree. **hyperbola; $-36°$**

5. Use the Quadratic Formula to solve $x^2 + 3xy + y^2 - 4 = 0$ for y. Then use a graphing calculator to draw the graph.
   $y = \dfrac{-3x \pm \sqrt{5x + 16}}{2}$

[−10, 10] scl:1 by [−10, 10] scl:1

Systems of Second-Degree Equations and Inequalities

OBJECTIVE
- Graph and solve systems of second degree equations and inequalities.

SEISMOLOGY The principal use of a seismograph network is to locate the epicenters of earthquakes. Seismograph stations in Chihuahua, Mazatlan, and Rosarito, Mexico, form one such network. Suppose this network detects an earthquake 622 kilometers from the Chihuahua station, 417 kilometers from the Mazatlan station, and 582 kilometers from the Rosarito station. Seismographic networks use complex software to approximate the location of the epicenter. The intersection of the three circles on the map shows the location of the epicenter to be near La Paz, Mexico. *You will solve a problem related to this in Exercise 40.*

The equation of each circle in the application above is a second-degree equation. So the three circles represent a system of second-degree equations. The coordinates of the point that satisfies all three equations is the solution to the system, which is the location of the earthquake's epicenter. You can solve this system graphically by locating the point where all three circles intersect.

The number of solutions of a system of second-degree equations equals the number of times the graphs of the equations intersect. If the system is composed of a line and a conic, there may be 0, 1, or 2 solutions. *Remember that a line is a degenerate conic.*

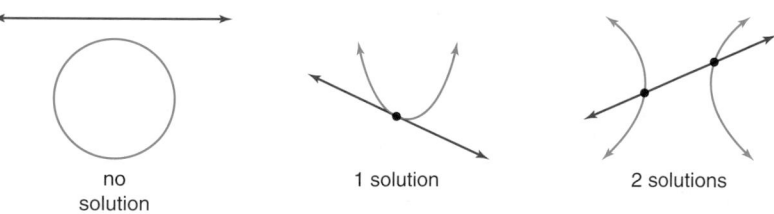

no solution   1 solution   2 solutions

Resource Manager

Reproducible Masters
- *Study Guide and Practice,* pp. 191–192
- *Enrichment,* p. 116
- *Assessment and Evaluation,* p. 252
- *Problem Solving and Applications,* p. 19

Transparencies
- 5-Minute Check, 10-8
- Teaching, 10-8

If the system is composed of two conics, there may be 0, 1, 2, 3, or 4, solutions.

| no solution | 1 solution | 2 solutions | 3 solutions | 4 solutions |

While you can determine the number of solutions by graphing the equations of a system, the exact solution is not always apparent. To find the exact solution, you must use algebra.

Example

1 a. **Graph the system of equations. Use the graph to find approximate solutions.**

b. **Solve the system algebraically.**
$$9x^2 + 25y^2 = 225$$
$$x^2 + y^2 - 2x = 15$$

Graphing Calculator Tip

You can use the CALC feature on a graphing calculator to approximate the intersection of any two functions graphed.

a. The graph of the first equation is an ellipse. The graph of the second equation is a circle. When a system is composed of an ellipse and a circle, there may be 0, 1, 2, 3, or 4 possible solutions. Graph each equation. There appear to be 3 solutions close to $(5, 0)$, $(-2, -3)$, and $(-2, 3)$.

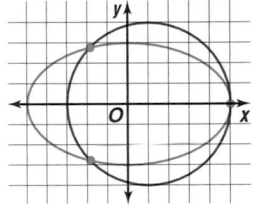

b. Since both equations contain a single term involving y, y^2, you can solve the system as follows.

First, multiply each side of the second equation by -25.
$$-25(x^2 + y^2 - 2x) = -25(15)$$
$$-25x^2 - 25y^2 + 50x = -375$$

Then, add the equations.
$$9x^2 + 25y^2 \qquad = 225$$
$$-25x^2 - 25y^2 + 50x = -375$$
$$\overline{-16x^2 \qquad + 50x = -150}$$

Using the quadratic formula, $x = 5$ or $-\dfrac{15}{8}$.

Now find y by substituting these values for x in one of the original equations.

$$x^2 + y^2 - 2x = 15$$
$$5^2 + y^2 - 2(5) = 15 \quad x = 5$$
$$25 + y^2 - 10 = 15$$
$$y^2 = 0$$
$$y = 0$$

$$x^2 + y^2 - 2x = 15$$
$$\left(-\frac{15}{8}\right)^2 + y^2 - 2\left(-\frac{15}{8}\right) = 15 \qquad x = -\frac{15}{8}$$
$$\frac{225}{64} + y^2 + \frac{15}{4} = 15$$
$$y^2 = \frac{495}{64}$$
$$y = \pm\frac{3\sqrt{55}}{8}$$

The solutions are $(5, 0)$, $\left(-\dfrac{15}{8}, \dfrac{3\sqrt{55}}{8}\right)$, and $\left(-\dfrac{15}{8}, -\dfrac{3\sqrt{55}}{8}\right)$. *How do the decimal approximations of these values compare to the approximations made in part a?*

Motivating the Lesson

Real-World Application Have students read the application at the beginning of the lesson. Then ask them whether three nonconcentric circles are sufficient to guarantee one point of intersection. Have them illustrate their answers with sketches. **No; three nonconcentric circles may not intersect at all or in 1, 2, 3, 4, 5, or 6 points.**

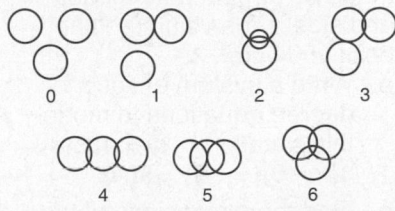

2 TEACH

Teaching Tip After discussing Example 1, stress the importance of examining a system of equations carefully before trying to solve it. Typically, a number of possible approaches will present themselves. Students can often save a considerable amount of time and effort by searching for the best method.

In-Class Example

Example 1
$$5x^2 - 4y^2 = 20$$
$$4x^2 + 16y^2 = 64$$

a. Graph the system of equations. Use the graph to find approximate solutions.

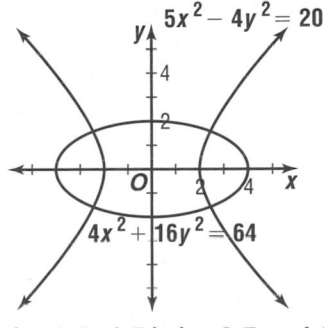

$(\pm 2.5, 1.5)$, $(\pm 2.5, -1.5)$

b. Solve the system algebraically.

$\left(\pm\sqrt{6}, \dfrac{\sqrt{10}}{2}\right), \left(\pm\sqrt{6}, -\dfrac{\sqrt{10}}{2}\right)$

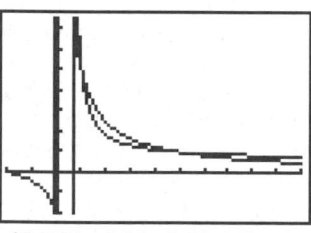
Systems of second-degree equations are useful in solving real-world problems involving more than one parameter.

Example ② **SALES** During the month of January, Photo World collected $2700 from the sale of a certain camera. After lowering the price by $15, the store sold 30 more cameras and took in $3375 from the sale of this camera the next month.

a. Write a system of second-degree equations to model this situation.

b. Find the price of the camera during each month.

c. Use a graphing calculator to check your solution.

a. From the information in the problem, we can write two equations, each of which is the equation of a conic section.

Let x = number of cameras sold
Let y = price per camera in January.

Sales in January: $xy = 2700$
Sales in February: $(x + 30)(y - 15) = 3375$

These are equations of hyperbolas.

b. To solve the system algebraically, use substitution. You can rewrite the equation of the first hyperbola as $y = \dfrac{2700}{x}$. Before substituting, expand the left-hand side of the second equation and simplify the equation.

$$(x + 30)(y - 15) - 3375 = 0$$

$$xy - 15x + 30y - 450 - 3375 = 0 \quad \textit{Expand } (x - 30)(y + 15).$$

$$xy - 15x + 30y - 3825 = 0 \quad \textit{Simplify.}$$

$$2700 - 15x + 30\left(\frac{2700}{x}\right) - 3825 = 0 \quad xy = 2700,\ y = \frac{2700}{x}$$

$$-15x - 1125 + \frac{81{,}000}{x} = 0 \quad \textit{Simplify.}$$

$$-15x^2 - 1125x + 81{,}000 = 0 \quad \textit{Multiply each side by x.}$$

$$x^2 + 75x - 5400 = 0 \quad \textit{Divide each side by } -15.$$

$$(x - 45)(x + 120) = 0 \quad \textit{Factor.}$$

$$x - 45 = 0 \quad \text{or} \quad x + 120 = 0$$
$$x = 45 \qquad\qquad\qquad x = -120$$

Since the number of cameras sold cannot be negative, the store sold 45 cameras during January.

The price of each camera sold during January was $\$\dfrac{2700}{45}$ or $60, and the price per camera in February was $60 − 15 or $45.

c. Solve each equation for y. Then, graph the equations on the same screen.

$$y = \frac{2700}{x}$$

$$y = \frac{3375}{x + 30} + 15$$

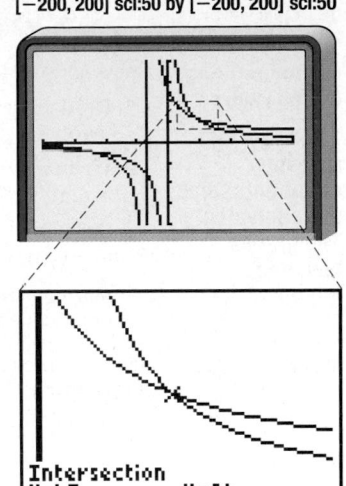

Use ⎡ZOOM⎤ to enlarge the section of the graph containing the intersection in the first quadrant. Use the **CALC: intersect** function to find the coordinates of the solution, (45, 60).

Teaching Tip For Example 3, students may want to test two points for each equation, one inside and one outside the solution set. This will help verify the shading.

In-Class Example
Example 3
Graph the solutions for the system of inequalities.
$$y - 1 > 2x^2$$
$$\frac{x^2}{9} + \frac{y^2}{4} < 1$$

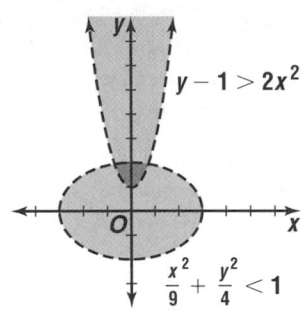

Look Back
You can refer to Lesson 2‑6 to review solving systems of linear inequalities.

Previously you learned how to graph different types of inequalities by graphing the corresponding equation and then testing points in the regions of the graph to find solutions for the inequality. The same process is used when graphing systems of inequalities involving second-degree equations.

Example ❸ **Graph the solutions for the system of inequalities.**
$$x^2 + 4y^2 \leq 4$$
$$x^2 > y^2 + 1$$

First graph $x^2 + 4y^2 \leq 4$. The ellipse should be a solid curve. Test a point either inside or outside the ellipse to see if its coordinates satisfy the inequality.

Test $(0, 0)$: $x^2 + y^2 \leq 4$
$$0^2 + 4(0)^2 \overset{?}{\leq} 4 \quad (x, y) = (0, 0)$$
$$0 \leq 4 \ \checkmark$$

Since $(0, 0)$ satisfies the inequality, shade the interior of the ellipse. Then graph $x^2 \geq y^2 + 1$. The hyperbola should be dashed. Test a point inside the branches of the hyperbola or outside its branches. *Since a hyperbola is symmetric, you need not test points within both branches.*

Test $(2, 0)$: $x^2 > y^2 + 1$
$$2^2 \overset{?}{>} 0^2 + 1 \quad (x, y) = (2, 0)$$
$$4 > 1 \ \checkmark$$

Since $(2, 0)$ satisfies the inequality, the regions inside the branches should be shaded. The intersection of the two graphs, which is shown in green, represents the solution of the system.

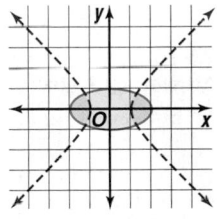

Lesson 10-8   Systems of Second-Degree Equations and Inequalities   **681**

Addressing Individual Needs

Intrapersonal Have students prepare a chart to summarize all they have learned about conic sections without using their textbooks. Then, allow them to refer to the textbooks in order to complete any information they could not recall from memory. This will point out the area(s) that they need to concentrate on before the chapter test.

Error Analysis

Watch for students who draw a solid line for a graph of a greater than or less than, but not equal to, inequality.

Prevent by having students check a point on the curve in addition to points in the plane.

Assignment Guide

Core: 13–41 odd, 43–50
Enriched: 14–36 even, 37–50

Additional Answers

1.

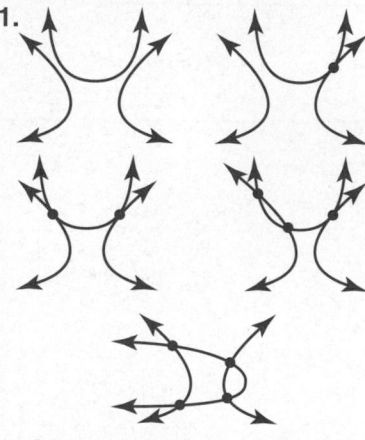

4. **Graph each second-degree inequality. The region in which the graphs overlap represents the solution to the system.**

Study Guide and Practice Masters, pp. 191–192

Communicating Mathematics

3. The system contains equation(s) that are equivalent. The graphs coincide.

Read and study the lesson to answer each question.

1. **Draw** figures illustrating each of the possible numbers of solutions to a system involving the equations of a parabola and a hyperbola. **See margin.**

2. **Write** a system of equations involving two different conic sections that has exactly one solution, the origin. **Sample answer:** $y = x^2, x^2 + (y + 3)^2 = 9$

3. **Describe** the graph of a system of second-degree equations having infinitely many solutions.

4. ***Math Journal*** Write a paragraph explaining how to solve a system of second-degree inequalities. **See margin.**

Guided Practice

Solve each system of equations algebraically. Round to the nearest tenth. Check the solutions by graphing each system. **5–8. See Answer Appendix for graphs.**

5. $\dfrac{(x - 1)^2}{20} + \dfrac{(y - 1)^2}{5} = 1$ **(3, 3), (−1, −1)**
   $x - y = 0$

6. $x^2 + y^2 = 16$ **no solution**
   $x + 2y = 10$

7. $9x^2 - 4y^2 = 36$ **(±2, 0)**
   $x^2 + y^2 = 4$

8. $x^2 = y$ **(1, 1)**
   $xy = 1$

Graph each system of inequalities. **9–11. See margin.**

9. $x^2 + y^2 \geq 16$
   $x + y \leq 2$

10. $(x - 5)^2 + 2y < 10$
    $y - 9 \geq -2x$

11. $x^2 + y^2 \leq 100$
    $x^2 + y^2 \geq 25$

12. **Gardening** A garden contains two square flowerbeds. The total area of the flowerbeds is 680 square feet, and the second bed has 288 more square feet than the first. **b. See Answer Appendix.**

12a.
$x^2 + y^2 = 680$
$x^2 - y^2 = 288$

a. Write a system of second-degree equations that models this situation.

b. Graph the system found in part **a** and estimate the solution.

c. Solve the system algebraically to find the length of each flowerbed within the garden. **22 ft and 14 ft**

Practice

Solve each system of equations algebraically. Round to the nearest tenth. Check the solutions by graphing each system.

 A

13–20. See Answer Appendix for graphs.

13. $x - 1 = 0$ **(1, ±6.9)**
    $y^2 = 49 - x^2$

14. $xy = 2$ **(2, 1), (−2, −1)**
    $x^2 = 3 + y^2$

15. $4x^2 + y^2 = 25$ **(1.5, −4), (−2, 3)**
    $-1 = 2x + y$

16. $x - y = 2$ **(8, 6), (−6, −8)**
    $x^2 = 100 - y^2$

17. $x - y = 0$ **no solution**
    $\dfrac{(x - 1)^2}{9} - y^2 = 1$

18. $3x^2 = 9 - y^2$ **(±1.3, 2.0), (±1.3, −2.0)**
    $x^2 + 2y^2 = 10$

19. $(y - 1)^2 = 4 + x$ **(0, −1), (−3, 2)**
    $x + y = -1$

20. $x^2 + y^2 = 13$ **(3, −2), (−3, 2),**
    $xy + 6 = 0$ **(2, −3), (−2, 3)**

 www.amc.glencoe.com/self_check_quiz

Additional Answers

9.

10.

11.

B 21. $x^2 + 4y^2 = 36$ **(0, 3), (±2.4, −2.8)** 22. $x^2 = 16 − y^2$ **(4.0, 0.5), (−2.8, −2.9)**
       $x^2 + y − 3 = 0$ $2y − x + 3 = 0$
       21–22. See Answer Appendix for graphs.

23. Find the coordinates of the point(s) of intersection for the graphs of
     $x^2 = 25 − 9y^2$ and $xy = −4$.   **(3, −1.3), (4, −1), (−3, 1.3), (−4, 1)**

Graph each system of inequalities.   **24–32. See margin.**

24. $x + y^2 \le 9$ 25. $x^2 + 4y^2 < 16$ 26. $x^2 + y^2 \le 36$
     $y + x^2 \le 0$ $x^2 \le y^2 + 4$ $x + y^2 > 0$

27. $y^2 < 81 − 9x^2$ 28. $y + 4 < (x − 3)^2$ 29. $16x^2 + 49y^2 \le 784$
     $16 < x^2 + y^2$ $y^2 + x \ge 5$ $49x^2 + 16y^2 \ge 784$

30. $x − (y − 1)^2 \le 0$ 31. $y − x^2 < 2$ 32. $x > \dfrac{2}{y}$
     $4y^2 \ge x^2 − 16$ $4x^2 + 9y^2 > 36$ $16x^2 − 25y^2 \ge 400$

 C 33. Graph the solution to the system $(x + 3)^2 + (y + 2)^2 \ge 36$ and $x + 3 = 0$.
       See Answer Appendix.

Write the system of equations or inequalities represented by each graph.

34. $(y + 3)^2 =$
$-2(x − 1)$,
$y = −2x − 7$

35. $x^2 + y^2 = 8$,
$xy = 4$

36. $\dfrac{y^2}{25} + \dfrac{x^2}{16} \le 1$,
$\dfrac{x^2}{9} + \dfrac{(y + 1)^2}{4} > 1$

34.   35.   36.

Applications and Problem Solving

37. **Construction**   Carrie has 150 meters of fencing material to make a pen for her bird dog. She wants to form a rectangular pen with an area of 800 square meters. What will be the dimensions of her pen?   **c. See Answer Appendix.**

    a. Let x be the width of the field and y be its length. Write a system of equations that models this situation. $2x + 2y = 150$, $xy = 800$

    b. How many solutions are possible for this type of system?   **0, 1, or 2**

    c. Graph the system to estimate the dimensions of the pen.

    d. Solve this system algebraically, rounding the dimensions to the nearest tenth of a meter.   **12.9 m by 62.1 m or 62.1 m by 12.9 m**

38. **Engineering**   The Transport and Road Research Laboratory in Great Britain proposes the use of parabolic speed bumps 4 inches in height and 1 foot in width.

    a. Write a system of second-degree inequalities that models a cross-section of this speed bump. Locate the vertex of the speed bump at (0, 4). $x^2 \le −9(y − 4)$, $y \ge 0$

    b. Graph the system found in part **a**.   **See Answer Appendix.**

    c. If the height of the speed bump is decreased to 3 inches, write a system of equations to model this new cross-section. $x^2 \le −12(y − 3)$, $y \ge 0$

39. **Critical Thinking**   Solve the system $x = −y + 1$, $xy = −12$, and $y^2 − 25 − x^2$ algebraically. Then graph the system to verify your solution(s).   **(4, −3), (−3, 4); see Answer Appendix for graph.**

26.

28.

29.

30.

31.

32.

24. 25. 27.

26. see right

Open-Ended Assessment

Modeling Two different conics can intersect in 1 point, 2 points, 3 points, or 4 points. Have students use string or rope on graph paper to illustrate each possibility.

 Chapter 10, Quiz D (Lessons 10-7 and 10-8) is available in the *Assessment and Evaluation Masters*, p. 252.

Additional Answers

40b.

42a.

Enrichment Masters, p. 116

40a. $x^2 + y^2 = 2500$, $x^2 + (y - 30)^2 = 1600$, $(x - 35)^2 + (y - 18)^2 = 169$

40. Seismology Each of three stations in a seismograph network has detected an earthquake in their region. Seismograph readings indicate that the epicenter of the earthquake is 50 kilometers from the first station, 40 kilometers from the second station, and 13 kilometers from the third station. On a map in which each grid represents one square kilometer, the first station is located at the origin, the second station at $(0, 30)$, and the third station at $(35, 18)$.

a. Write a system of second-degree equations that models this situation.

b. Graph the system and use the graph to approximate the location of the epicenter. **See margin for graph.**

c. Solve the system of equations algebraically to determine the location of the epicenter. **(40, 30)**

41. Critical Thinking Find the value of k so that the graphs of $x = 2y^2$ and $x + 3y = k$ are tangent to each other. $-\dfrac{9}{8}$

42. Entertaiment In a science fiction movie, astronomers track a large incoming asteroid and predict that it will strike Earth with disastrous results. Suppose a certain latitude of Earth's surface is modeled by $x^2 + y^2 = 40$ and the path of the asteroid is modeled by $x = 0.25y^2 + 5$.

a. Graph the two equations on the same axes. **See margin.**

b. Will the asteroid strike Earth? If so, what are the coordinates of the point of impact? **yes; (6, 2) or (6, −2)**

c. Describe this situation with parametric equations. Assume both the asteroid and Earth's surface are moving counterclockwise.

d. Graph the equations found in part **c** using a graphing calculator. Use a window that shows complete graphs of both Earth's surface and the asteroid's path. **See Answer Appendix.**

c. Earth's surface: $x_1 = 2\sqrt{10}\cos t$ and $y_1 = 2\sqrt{10}\sin t$; asteroid: $x_2 = 0.25t^2 + 5$ and $y_2 = -t$

Mixed Review

43. Find the equation of $\dfrac{x^2}{9} + y^2 = 1$ after a 30° rotation about the origin. *(Lesson 10-7)* $3(x')^2 - 4\sqrt{3}x'y' + 7(y')^2 - 9 = 0$

44. Write an equation in standard form of the line with the parametric equations $x = 4t + 1$ and $y = 5t - 7$. *(Lesson 8-6)* $5x - 4y - 33 = 0$

45. Simplify $4\csc\theta\cos\theta\tan\theta$. *(Lesson 7-1)* **4**

46. Mechanics A pulley of radius 10 centimeters turns at 5 revolutions per second. Find the linear velocity of the belt driving the pulley in meters per second. *(Lesson 6-2)* **3.14 m/s**

47. Determine between which consecutive integers the real zeros of the function $f(x) = x^3 - 4$ are located. *(Lesson 4-5)* **1 and 2**

48. Graph $y = (x + 2)^2 - 3$ and its inverse. *(Lesson 3-4)* **See margin.**

49. Is the relation $\{(4, 0), (3, 0), (5, -2), (4, -3), (0, -13)\}$ a function? Explain. *(Lesson 1-1)*

49. No; the domain value 4 is mapped to two elements in the range, 0 and −3.

50. SAT/ACT Practice In the figure at the right, two circles are tangent to each other and each is tangent to three sides of the rectangle. If the radius of each circle is 2, what is the area of the shaded region? **E**

A $32 - 12\pi$ 　　B $16 - 8\pi$ 　　C $16 - 6\pi$

D $8 - 6\pi$ 　　E $32 - 8\pi$

684 Chapter 10 Conics

Extra Practice See p. A45.

Enrichment

Intersections of Circles

Many interesting problems involving circles can be solved by using a system of equations. Consider the following problem.

Find an equation for the straight line that contains the two points of intersection of two intersecting circles whose equations are given.

You may be surprised to find that if the given circles intersect in two points, then the difference of their equations is the equation of the line containing the intersection points.

1. Circle A has equation $x^2 + y^2 = 1$ and circle B has equation $(x - 3)^2 + y^2 = 1$. Use a sketch to show that the circles do not intersect. Use an algebraic argument to show that circles A and B do not intersect.
The distance between the centers is 3. Since the sum of the radii is 2, there is 1 unit of space between the circles. Thus, the circles do not intersect.

2. Circle A has equation $(x - 2)^2 + (y + 1)^2 = 16$ and circle B has equation $(x + 3)^2 + y^2 = 9$. Use a sketch to show that the circles meet in two points. Then find an equation in standard form for the line containing the points of intersection.
$10x - 2y + 11 = 0$

3. Without graphing the equations, decide if the circles with equations $(x - 2)^2 + (y - 2)^2 = 8$ and $(x - 3)^2 + (y - 4)^2 = 4$ are tangent. Justify your answer.
The distance between the centers is $\sqrt{5} \approx 2.24$. Since the sum of the radii is $2 + 2\sqrt{2} \approx 4.83$, the circles overlap and meet in two points; they are not tangent.

© Glencoe/McGraw-Hill 　　T116 　　*Advanced Mathematical Concepts*

Extra Credit

Two points located 14 units apart on the x-axis are the centers of circles the sum of whose radii is 28. The circles intersect at (5, 12) and (5, −12). Find the equations of the circles.

$x^2 + y^2 = 169$ and $(x - 14)^2 + y^2 = 225$

Additional Answer

48.

$y = (x + 2)^2 - 3$

$y = \pm\sqrt{x + 3} - 2$

10-8B Shading Areas on a Graph

An Extension of Lesson 10-8

OBJECTIVE
• Graph a system of second-degree inequalities using the **Shade(** command.

The **Shade(** command can be used to shade areas between the graphs of two equations. To shade an area on a graph, select **7:Shade(** from the **DRAW** menu. The instruction is pasted to the home screen. The argument, or restrictive information, for this command is as follows.

Shade(*lowerfunc,upperfunc,Xleft,Xright,pattern,patres***)**

This command draws the lower function, *lowerfunc,* and the upper function, *upperfunc,* in terms of X on the current graph and shades the area that is specifically above *lowerfunc* and below *upperfunc.* This means that only the areas between the two functions defined are shaded.

Xleft and *Xright,* if included, specify left and right boundaries for the shading. *Xleft* and *Xright* must be numbers between **Xmin** and **Xmax**, which are the defaults.

The parameter *pattern* specifies one of four shading patterns.
pattern = **1**   vertical lines (default)
pattern = **2**   horizontal lines
pattern = **3**   45° lines with positive slope
pattern = **4**   45° lines with negative slope

The parameter *patres* specifies one of eight shading resolutions.

patres = **1** shades every pixel (default)	*patres* = **5** shades every fifth pixel
patres = **2** shades every second pixel	*patres* = **6** shades every sixth pixel
patres = **3** shades every third pixel	*patres* = **7** shades every seventh pixel
patres = **4** shades every fourth pixel	*patres* = **8** shades every eighth pixel

In a system of second-degree inequalities, this technique can be used to shade the interior region of a conic that is not a function.

Example

Graph the solutions for the system of inequalities below.
$$y \geq -x^2 + 2$$
$$x^2 + 9y^2 \leq 36$$

The boundary equation of the first inequality, $y = -x^2 + 2$, is a function. This inequality is graphed by first entering the equation $y = -x^2 + 2$ into the **Y=** list. Since the test point $(0, 3)$ satisfies the inequality, set the graph style to ◥ (shade above).

[−9.1, 9.1] scl:1 by [−6, 6] scl:1

(continued on the next page)

Lesson 10-8B   Shading Areas on a Graph   **685**

Lesson Notes

10-8B

1 FOCUS

Motivating the Lesson
Discussion Ask students to explain why the process of graphing inequalities contains so many steps. Discuss how graphing inequalities on a graphing calculator might reduce errors and yield more exact answers.

2 TEACH

A reproduction of this lesson with graphing calculator keystrokes provided can be found in the *Graphing Calculator Keystrokes Masters,* pp. 41–43.

Teaching Tip Have students refer to graphing systems of equations in Chapter 3. Remind students that they can use these techniques when solving systems of inequalities.

Teaching Tip Some calculators, such as the TI-83 Plus, have the ability to receive downloadable software. One of the possible new software programs will graph inequalities more easily.

Open-Ended Assessment

Modeling Have each student draw an x- and y-axis on an overhead transparency sheet. Each coordinate grid should be the same scale. Assign each student an inequality to graph on their sheet. Have students pair up their grids, laying one on top of the other to find the solution to the system of the two inequalities.

Additional Answers

1.

[−9.1, 9.1] scl:1 by [−6, 6] scl:1

2.

[−9.1, 9.1] scl:1 by [−6, 6] scl:1

3.

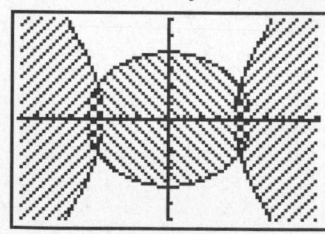

[−9.1, 9.1] scl:1 by [−6, 6] scl:1

4.

[−9.1, 9.1] scl:1 by [−6, 6] scl:1

The boundary equation for the second inequality is $x^2 + 9y^2 = 36$, which is defined using two functions, $y = -\sqrt{\dfrac{36 - x^2}{9}}$ and $y = \sqrt{\dfrac{36 - x^2}{9}}$. The lower function is $y = -\sqrt{\dfrac{36 - x^2}{9}}$, and the upper function is $y = \sqrt{\dfrac{36 - x^2}{9}}$. The two halves of the ellipse intersect at $x = -6$ and $x = 6$.

The expression to shade the area between the two halves of the ellipse is shown at the right.

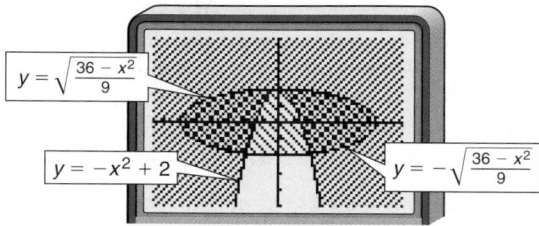

Pressing **ENTER** will compute the graph as shown below.

$y = \sqrt{\dfrac{36 - x^2}{9}}$

$y = -x^2 + 2$

$y = -\sqrt{\dfrac{36 - x^2}{9}}$

[−9.1, 9.1] scl:1 by [−6, 6] scl:1

The solution set for this system of inequalities is the darker region in which the shadings for the two inequalities over lap.

To clear the any **Shade(** commands from the viewing window, select **1:ClrDraw** from the **DRAW** menu and then press **ENTER**. Remember to also clear any functions defined in the **Y=** list.

TRY THESE

Use the shade feature to graph each system of second-degree inequalities.

1. $y \le x^2 - 5$
$9y^2 - x^2 \le 36$

2. $x^2 + y^2 \le 16$
$x \ge y^2 - 4$

3. $16x^2 + 25y^2 \le 400$
$25x^2 - 16y^2 \ge 400$

4. $8x^2 + 32y^2 \le 256$
$32x^2 + 8y^2 \le 256$

1–4. See margin.

WHAT DO YOU THINK?

5b. Find the points of intersection for the boundary equation by using the TRACE function.

5. Recall that the **Shade(** command can only shade the area between two functions.

a. To shade just the solution set for the system of inequalities in the example problem, how many regions would need a separate **Shade(** command?   **3**

b. How could you determine the approximate domain intervals for each region?

c. List and then execute three **Shade(** commands to shade the region representing the solution set for the example problem.   **See margin.**

6. Use the **Shade(** command to create a "real-world" picture. Make a list of each command needed to create the picture, as well as a sketch of what the finished picture should look like.   **See student's work.**

Additional Answer

5c. SHADE($-\sqrt{}$((36 − X²)/9), $\sqrt{}$((36 − X²)/9),−6,−2,3,4);
    SHADE($-$X² + 2, $\sqrt{}$((36 − X²)/9),−2,2,3,4);
    SHADE($-\sqrt{}$((36 − X²)/9), $\sqrt{}$((36 − X²)/9),2,6,3,4);

[−9.1, 9.1] scl:1 by [−6, 6] scl:1

VOCABULARY

analytic geometry (p. 618)	equilateral hyperbola (p. 647)
asymptotes (p. 642)	focus (p. 631, 642, 653)
axis of symmetry (p. 653)	hyperbola (p. 642)
center (p. 623, 642)	locus (p. 658)
circle (p. 623)	major axis (p. 631)
concentric (p. 623)	minor axis (p. 631)
conic section (p. 623)	radius (p. 623)
conjugate axis (p. 642)	rectangular hyperbola (p. 648)
degenerate conic (p. 623)	semi-major axis (p. 632)
directrix (p. 653)	semi-minor axis (p. 632)
eccentricity (p. 636)	transverse axis (p. 642)
ellipse (p. 631)	vertex (p. 631, 642, 653)

UNDERSTANDING AND USING THE VOCABULARY

State whether each statement is *true* or *false*. If false, replace the underlined word(s) to make a true statement.   **3. false; transverse**

1. Circles, ellipses, parabolas, and hyperbolas are all examples of <u>conic sections</u>.   **true**

2. Circles that have the same <u>radius</u> are concentric circles.   **false; center**

3. The line segment connecting the vertices of a hyperbola is called the <u>conjugate</u> axis.

4. The foci of an ellipse are located along the <u>major</u> axis of the ellipse.   **true**

5. In the general form of a <u>circle</u>, A and C have opposite signs.   **false; hyperbola**

6. A parabola is symmetric with respect to its <u>vertex</u>.   **false; axis or axis of symmetry**

7. The shape of an ellipse is described by a measure called <u>eccentricity</u>.   **true**

8. A <u>hyperbola</u> is the set of all points in a plane that are the same distance from a given point and a given line.   **false; parabola**

9. The general equation of a <u>rectangular</u> hyperbola, where the coordinate axes are the asymptotes, is $xy = c$.   **true**

10. A point is the degenerate form of a <u>parabola</u> conic.   **false; ellipse**

interNET CONNECTION   For additional review and practice for each lesson, visit: **www.amc.glencoe.com**

Six forms of Chapter 10 Test are available in the *Assessment and Evaluation Masters* as shown in the chart.

Chapter 10 Test, Form 1B, is shown at the right. Chapter 10 Test, Form 2B, is shown on the next page.

TEST OPTIONS

Form	Pages	Type	
1A	237–238	Honors	*Form 1 tests are multiple choice.*
1B	239–240	Average	
1C	241–242	Basic	
2A	243–244	Honors	*Form 2 tests are free response.*
2B	245–246	Average	
2C	247–248	Basic	

VOCABULARY

This section provides a listing of the new terms, properties, and phrases that were introduced in this chapter. The exercises check students' understanding of the terms by using a variety of verbal formats including matching, completion, and true/false.

Glossary A complete glossary of terms begins on page A71.

Assessment and Evaluation Masters, pp. 239–240

SKILLS AND CONCEPTS

Objectives and Examples This section reviews the skills and concepts of the chapter and shows completely worked examples.

Review Exercises These exercises provide practice for the corresponding objectives.

Additional Answer

13. $AB = DC = 10$ and $BC = AD = 5\sqrt{2}$. Since opposite sides of quadrilateral $ABCD$ are congruent, $ABCD$ is a parallelogram.

Assessment and Evaluation Masters, pp. 245–246

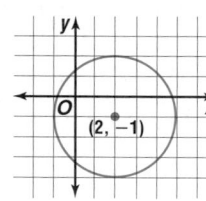

SKILLS AND CONCEPTS

OBJECTIVES AND EXAMPLES

Lesson 10-1 Find the distance and midpoint between two points on a coordinate plane.

• Find the distance between points at $(3, 8)$ and $(-5, 10)$.

$$d = \sqrt{(x_2 - x_1)^2 + (y_2 - y_1)^2}$$
$$= \sqrt{(-5 - 3)^2 + (10 - 8)^2}$$
$$= \sqrt{68} \text{ or } 2\sqrt{17}$$

Lesson 10-2 Determine the standard form of the equation of a circle and graph it.

• Write $x^2 + y^2 - 4x + 2y - 4 = 0$ in standard form. Then graph the equation.

$$x^2 + y^2 - 4x + 2y - 4 = 0$$
$$x^2 - 4x + ? + y^2 + 2y + ? = 4$$
$$x^2 - 4x + 4 + y^2 + 2y + 1 = 4 + 4 + 1$$
$$(x - 2)^2 + (y + 1)^2 = 9$$

The center of the circle is located at $(2, -1)$, and the radius is 3.

Lesson 10-3 Determine the standard form of the equation of an ellipse and graph it.

• For the equation, $\dfrac{(x + 1)^2}{9} + \dfrac{(y - 3)^2}{16} = 1$, find the coordinates of the center, foci, and vertices of the ellipse. Then graph the equation.

center: $(-1, 3)$

foci: $\left(-1, 3 + \sqrt{7}\right)$, $\left(-1, 3 - \sqrt{7}\right)$

vertices: $(2, 3)$, $(-4, 3)$, $(-1, -1)$, $(-1, 7)$

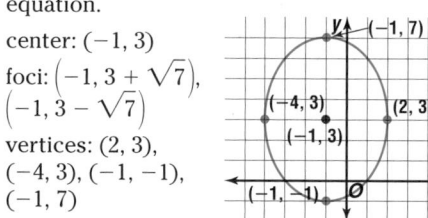

REVIEW EXERCISES

Find the distance between each pair of points with the given coordinates. Then, find the midpoint of the segment that has endpoints at the given coordinates.

11. $(1, -6), (-3, -4)$ **$2\sqrt{5}$; $(-1, -5)$**

12. $(a, b), (a + 3, b + 4)$ **5; $(a + 1.5, b + 2)$**

13. Determine whether the points $A(-5, -2)$, $B(3, 4)$, $C(10, 3)$, and $D(2, -3)$ are the vertices of a parallelogram. Justify your answer. **Yes; see margin for explanation.**

Write the standard form of the equation of each circle described. Then graph the equation. **14–15. See Answer Appendix.**

14. center at $(0, 0)$, radius $3\sqrt{3}$

15. center at $(2, 1)$, tangent to the y-axis

Write the standard form of each equation. Then graph the equation. **16–18. See margin.**

16. $x^2 + y^2 = 6y$

17. $x^2 + 14x + y^2 + 6y = 23$

18. $3x^2 + 3y^2 + 6x + 12y - 60 = 0$

19. Write the standard form of the equation of the circle that passes through points at $(1, 1)$, $(-2, 2)$, and $(-5, 1)$. Then identify the center and radius. **$(x + 2)^2 + (y + 3)^2 = 25$; $(-2, -3)$; 5**

For the equation of each ellipse, find the coordinates of the center, foci, and vertices. Then graph the equation. **20–23. See Answer Appendix.**

20. $\dfrac{(x - 5)^2}{16} + \dfrac{(y - 2)^2}{36} = 1$

21. $4x^2 + 25y^2 - 24x + 50y = 39$

22. $6x^2 + 4y^2 + 24x - 32y + 64 = 0$

23. $x^2 + 4y^2 + 124 = 8x + 48y$

24. Write the equation of an ellipse centered at $(-4, 1)$ with a vertical semi-major axis 9 units long and a semi-minor axis 6 units long.

24. $\dfrac{(y - 1)^2}{81} + \dfrac{(x + 4)^2}{36} = 1$

TestCheck and Worksheet Builder

This **networkable** software has three modules.
- **Worksheet Builder** to make worksheets and tests
- **Student Module** to take tests on-screen
- **Management System** to keep student records

Three special banks are included for SAT, ACT, and TIMSS.

OBJECTIVES AND EXAMPLES

Lesson 10-4 Determine the standard and general forms of the equation of a hyperbola and graph it.

Find the coordinates of the center, foci, and vertices, and the equations of the asymptotes of the graph of $\frac{(y-2)^2}{4} - (x-5)^2 = 1$. Then graph the equation.

center: $(5, 2)$

foci: $\left(5, 2 + \sqrt{5}\right)$, $\left(5, 2 - \sqrt{5}\right)$

vertices: $(5, 4)$, $(5, 0)$

asymptotes:
$y - 2 = \pm 2(x - 5)$

REVIEW EXERCISES

For the equation of each hyperbola, find the coordinates of the center, the foci, and the vertices and the equations of the asymptotes of its graph. Then graph the equation.

25. $\frac{x^2}{25} - \frac{y^2}{16} = 1$

26. $\frac{(y+5)^2}{36} - \frac{(x-1)^2}{9} = 1$

27. $x^2 - 4y^2 - 16y = 20$

28. $9x^2 - 16y^2 - 36x - 96y + 36 = 0$

29. Graph $xy = 9$.
25–29. See margin.
Write an equation of the hyperbola that meets each set of conditions.   **30–31. See margin.**

30. The length of the conjugate axis is 10 units, and the vertices are at $(1, -1)$ and $(1, 5)$.

31. The vertices are at $(-2, -3)$ and $(6, -3)$, and a focus is at $(-4, -3)$.

Lesson 10-5 Determine the standard and general forms of the equation of a parabola and graph it.

For the equation $(x + 1)^2 = 2(y - 3)$, find the coordinates of the vertex and focus, and the equations of the directrix and axis of symmetry. Then graph the equation.

vertex: $(-1, 3)$

focus: $\left(-1, \frac{7}{2}\right)$

directrix: $y = \frac{5}{2}$

axis of symmetry:
$x = -1$

For the equation of each parabola, find the coordinates of the vertex and focus, and the equations of the directrix and axis of symmetry. Then graph the equation.   **32–35. See Answer Appendix**

32. $(x - 5)^2 = 8(y - 3)$

33. $(y + 2)^2 = -16(x - 1)$

34. $y^2 + 6y - 4x = -25$

35. $x^2 + 4x = y - 8$

Write an equation of the parabola that meets each set of conditions.

36. The parabola passes through the point at $(-3, 7)$, has its vertex at $(-1, 3)$, and opens to the left. $(y - 3)^2 = -8(x + 1)$

37. The focus is at $(5, 2)$, and the equation of the directrix is $y = -4$. $(x - 5)^2 = 12(y + 1)$

Lesson 10-6 Recognize conic sections in their rectangular form by their equations.

Identify the conic section represented by the equation $2x^2 - 3x - y + 4 = 0$.

$A = 2$ and $C = 0$

Since $C = 0$, the conic is a parabola.

Identify the conic section represented by each equation.

38. $5x^2 - 7x + 2y^2 = 10$ **ellipse**

39. $xy = 5$ **equilateral hyperbola**

40. $2x^2 + 4x + 2y^2 - 6y + 16 = 0$ **circle**

41. $4y^2 + 6x - 5y = 20$ **parabola**

Additional Answers

27. center: $(0, -2)$; foci: $\left(\pm\sqrt{5}, -2\right)$; vertices: $(-2, -2)$, $(2, -2)$; asymptotes: $y + 2 = \pm\frac{1}{2}x$

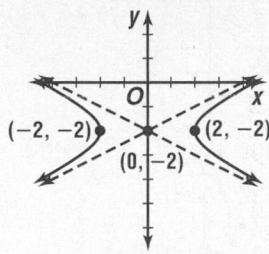

28. center: $(2, -3)$; foci: $(2, 2)$, $(2, -8)$; vertices: $(2, 0)$, $(2, -6)$; asymptotes: $y + 3 = \pm\frac{3}{4}(x - 2)$

29.

30. $\frac{(y-2)^2}{9} - \frac{(y-1)^2}{25} = 1$

31. $\frac{(x-2)^2}{16} - \frac{(y+3)^2}{20} = 1$

Additional Answers

25. center: $(0, 0)$; foci: $\left(\pm\sqrt{41}, 0\right)$; vertices: $(-5, 0)$, $(5, 0)$; asymptotes: $y = \pm\frac{4}{5}x$

26. center: $(1, -5)$; foci: $\left(1, -5 \pm 3\sqrt{5}\right)$; vertices: $(1, 1)$, $(1, -11)$; asymptotes: $y + 5 = \pm 2(x - 1)$

Additional Answers

42. $y = -x^2 + 3$

43. $x^2 + y^2 = 1$

44. $\dfrac{x^2}{4} + \dfrac{y^2}{9} = 1$

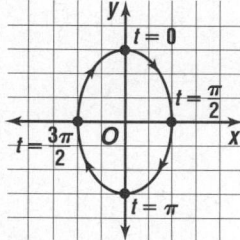

45. $y = \dfrac{x^2}{2} - 1$

46. Sample answer: $x = t$, $y = 2t^2 + 4$, $-\infty < t < \infty$

47. Sample answer: $x = 7 \sin t$, $y = 7 \cos t$, $0 \le t \le 2\pi$

48. Sample answer: $x = 6 \cos t$, $y = 9 \sin t$, $0 \le t \le 2\pi$

49. Sample answer: $x = -t^2$, $y = t$, $-\infty < t < \infty$

50. ellipse; $21(x')^2 - 10\sqrt{3}x'y' + 31(y')^2 - 144 = 0$

51. parabola; $(x')^2 - 2x'y' + (y')^2 - 4\sqrt{2}x' - 4\sqrt{2}y' = 0$

52. hyperbola; $x^2 - 4y^2 - 2x - 8y - 19 = 0$

OBJECTIVES AND EXAMPLES

Lesson 10-6 Find a rectangular equation for a curve defined parametrically and vice versa.

- Find the rectangular equation of the curve whose parametric equations are $x = 3 \sin t$ and $y = \cos t$, where $0 \le t \le 2\pi$. Then graph the equation using arrows to indicate orientation.

$$x = 3\sin t \Rightarrow \sin t = \frac{x}{3}$$
$$y = \cos t \Rightarrow \cos t = y$$
$$\sin^2 t + \cos^2 t = 1$$
$$\left(\frac{x}{3}\right)^2 + y^2 = 1$$
$$\frac{x^2}{9} + y^2 = 1$$

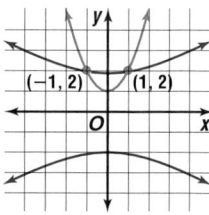

Lesson 10-7 Find the equations of conic sections that have been translated or rotated and find the angle of rotation for a given equation.

- To find the equation of a conic section with respect to a rotation of θ, replace

$$x \text{ with } x' \cos \theta + y' \sin \theta$$
$$\text{and } y \text{ with } -x' \sin \theta + y' \cos \theta.$$

Lesson 10-8 Graph and solve systems of second degree equations and inequalities.

- Solve the system of equations $x^2 - y = -1$ and $x^2 - 3y^2 = -11$ algebraically.

$$\begin{aligned} x^2 \quad\;\; - y + \;\; 1 &= 0 \\ \underline{-x^2 + 3y^2 \quad\;\; - 11} &= 0 \\ 3y^2 - y - 10 &= 0 \end{aligned}$$

$$y = -\frac{5}{3} \text{ or } y = 2$$

Substituting we find the solutions to be $(1, 2)$ and $(-1, 2)$. The graph shows these solutions to be true.

REVIEW EXERCISES

Find the rectangular equation of the curve whose parametric equations are given. Then graph the equation using arrows to indicate orientation. **42–45. See margin.**

42. $x = t$, $y = -t^2 + 3$, $-\infty \le t \le \infty$

43. $x = \cos 4t$, $y = \sin 4t$, $0 \le t \le \dfrac{\pi}{2}$

44. $x = 2 \sin t$, $y = 3 \cos t$, $0 \le t \le 2\pi$

45. $x = \sqrt{t}$, $y = \dfrac{t}{2} - 1$, $0 \le t \le 9$

Find parametric equations for each rectangular equation. **46–49. See margin.**

46. $y = 2x^2 + 4$ **47.** $x^2 + y^2 = 49$

48. $\dfrac{x^2}{36} + \dfrac{y^2}{81} = 1$ **49.** $x = -y^2$

Identify the graph of each equation. Write an equation of the translated or rotated graph in general form. **50–52. See margin.**

50. $4x^2 + 9y^2 = 36$, $\theta = \dfrac{\pi}{6}$

51. $y^2 - 4x = 0$, $\theta = 45°$

52. $4x^2 - 16(y - 1)^2 = 64$ for $T_{(1, -2)}$

Identify the graph of each equation. Then find θ to the nearest degree.

53. $6x^2 + 2\sqrt{3}xy + 8y^2 = 45$ **ellipse; $-30°$**

54. $x^2 - 6xy + 9y^2 = 7$ **parabola; $18°$**

Solve each system of equations algebraically. Round to the nearest tenth. Check the solutions by graphing each system.

55. $(x - 1)^2 + 4(y - 1)^2 = 20$ **56.** $2x - y = 0$
$\quad\; x = y$ **(3, 3), (−1, −1)** $\qquad y^2 = 49 + x^2$

57. $x^2 - 4x - 4y = 4$ **58.** $x^2 + y^2 = 12$
$\quad (x - 2)^2 + 4y = 0$ $\qquad xy = -4$

Graph each system of inequalities. **(−1.3, 3.2), (−3.3, 1.2)**

59. $x^2 + y \le 4$ **60.** $xy \ge 9$
$\quad y^2 - x \le 0$ $\qquad x^2 + y^2 < 36$

61. $x^2 \le 16 - y^2$ **62.** $x^2 - 4y \ge 8$
$\quad 36y^2 > 324 - 9x^2$ $\qquad 4y^2 - 25x^2 \ge 100$

56. (4.0, 8.1), (−4.0, −8.1)
57. (0, −1), (4, −1)
58. (−1.2, 3.2), (1.2, −3.2), (3.2, −1.2), (−3.2, 1.2)

59–62. See Answer Appendix for graphs.

Additional Answers

55.

56.

57.

APPLICATIONS AND PROBLEM SOLVING

63. Gardening Migina bought a new sprinkler that covers part or all of a circular area. With the center of the sprinkler as the origin, the sprinkler sends out water far enough to reach a point located at (12, 16). *(Lesson 10-2)*   **a. $x^2 + y^2 = 400$**

a. Find an equation representing the farthest points the sprinkler can reach.

b. Migina's backyard is 40 feet wide and 50 feet long. If Migina waters her backyard without moving the sprinkler, what percent of her backyard will not be watered directly? **about 37%**

64. Astronomy A satellite orbiting Earth follows an elliptical path with Earth at its center. The eccentricity of the orbit is 0.2, and the major axis is 12,000 miles long. Assuming that the center of the ellipse is the origin and the foci lie on the *x*-axis, write the equation of the orbit of the satellite. *(Lesson 10-6)* **See margin.**

65. Carpentry For a remodeling project, a carpenter is building a picture window that is topped with an arch in the shape of a semi-ellipse. The width of the window is to be 7 feet, and the height of the arch is to be 3 feet. To sketch the arch above the window, the carpenter uses a 7-foot string attached to two thumbtacks. Approximately where should the thumbtacks be placed? *(Lesson 10-3)* **about 1.8 feet from the center**

ALTERNATIVE ASSESSMENT

OPEN-ENDED ASSESSMENT

1. An ellipse has its center at the origin and an eccentricity of $\frac{1}{9}$. What is a possible equation for the ellipse?

2. A parabola has an axis of symmetry of $x = 2$ and a focus of (2, 5). What is a possible equation for the parabola in standard form? **Sample answer: $(x - 2)^2 = 12(y - 2)$**

1. **Sample answer: $\frac{x^2}{81} + \frac{y^2}{80} = 1$**

PORTFOLIO
Choose one of the conic sections you studied in this chapter. Explain why it is a conic section and describe how you graph it.

> **Additional Assessment** See page A65 for Chapter 10 practice test.

SPACE—THE FINAL FRONTIER

Out in Orbit!

- Search the internet for a satellite, space vehicle, or planet that travels in an orbit around a planet or star.

- Find data on the orbit of the satellite, space vehicle, or planet. This information should include the closest and farthest distance of that object from the planet it is orbiting.

- Make a scale drawing of the orbit of the satellite, space vehicle, or planet. Label important features and dimensions.

- Write a summary describing the orbit of the satellite, space vehicle, or planet. Be sure to discuss which conic section best models the orbit.

Chapter 10 Study Guide and Assessment **691**

This section provides additional practice in solving real-world problems that involve the concepts of this chapter.

Additional Answers (pp. 690–691)

58.

(graph showing a circle with points (−1.2, 3.2), (−3.2, 1.2), (3.2, −1.2), (1.2, −3.2))

64. $\dfrac{x^2}{36{,}000{,}000} + \dfrac{y^2}{34{,}560{,}000} = 1$

ALTERNATIVE ASSESSMENT

Unit 3 *inter*NET Project

● **Out in Orbit!**
Students should label the planet or star being orbited, the orbiting object, and the distance of the object at its furthest point and closest point to the planet or star. Students may also want to find the diameter of the planet or star being orbited. Students may want to attempt to put a simple coordinate system on this diagram, probably a rectangular system.

Internet Address Book

Record useful Internet addresses in the space at right for quick reference.

www.amc.glencoe.com

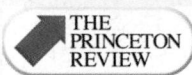
Chapter 10 Skill: Ratio and Proportion

Pages 692–693 are part of a complete SAT/ACT preparation course that is described in detail on pages T8–T9 of the *Teacher's Handbook*. The test items were written in the same style as those in the SAT and ACT.

THE PRINCETON REVIEW — These questions were aligned and verified by The Princeton Review, the nation's leader in test preparation.

Teaching Tip In addition to ratios and proportions, students should be familiar with the rules for divisibility, the slope formula, and 45°-45°-90° right triangle leg ratios.

Diagnosis & Prescription

Each of the 10 SAT and ACT practice problems on page 693 is cross-referenced to the chapter where that SAT or ACT skill is covered. If students miss a particular type of problem, you can have them study that skill.

Problem Number and Skill	Page
1 Ratio and Proportion	692
2 Algebra	124
3 Geometry (Perimeter, Area, and Volume)	548
4 Ratio and Proportion	692
5 Geometry (Lines, Angles, and Arcs)	612
6 Coordinate Geometry	272
7 Geometry (Triangles and Quadrilaterals)	482
8 Ratio and Proportion	692
9 Ratio and Proportion	692
10 Ratio and Proportion	692

Ratio and Proportion Problems

Several problems on the SAT and ACT involve ratios or proportions. The ratio of x to y can be expressed in several ways.

$$\frac{x}{y} \qquad x{:}y \qquad x \text{ to } y$$

Think of a ratio as comparing parts of a whole. If the ratio of boys to girls in a class is 2:1, then one part is 2, one part is 1, and the whole is 3. The fraction of boys in the class is $\frac{2}{3}$.

Memorize the property of proportions.

$$\text{If } \frac{a}{b} = \frac{c}{d}, \text{ then } ad = bc.$$

> **THE PRINCETON REVIEW**
>
> A ratio compares a part to a part. A fraction compares a part to a whole.
>
> Ratios only tell you the *relative* sizes of quantities, not the actual quantities.
>
> When setting up a proportion, label quantities to prevent careless errors.

ACT EXAMPLE

1. The ratio of boys to girls in a class is 4 to 5. If there are a total of 27 students in the class, how many boys are in the class?

   A 4 B 9 C 12

   D 14 E 17

   HINT Notice what question is asked. Is it a ratio, a fraction, or a number?

   Solution The ratio is 4 to 5, so the whole must be 9. The fraction of boys is $\frac{4}{9}$. The total number of students is 27, so the number of boys is $\frac{4}{9}$ of 27 or 12. The answer is choice **C**.

   Alternate Solution Another method is to use a 'ratio box' to record the numbers and guide your calculations.

Boys	Girls	Whole
4	5	9
		27

   In the Whole column, 9 must be multiplied by 3 to get the total of 27. So multiply the 4 by 3 to get the number of boys.

Boys	Girls	Whole
4	5	9
12		27

   This is answer choice **C**.

SAT EXAMPLE

2. If 2 packages contain a total of 12 doughnuts, how many doughnuts are there in 5 packages?

   A 12 B 24 C 30

   D 36 E 60

   HINT In a proportion, one ratio equals another ratio.

   Solution Write a proportion.

   packages $\longrightarrow$ $\dfrac{2}{12} = \dfrac{5}{x}$ $\longleftarrow$ packages
   doughnuts $\longrightarrow$ $\phantom{\dfrac{2}{12}}$ $\longleftarrow$ doughnuts

   Cross multiply.
   $$\frac{2}{12} = \frac{5}{x}$$
   $$2(x) = 5(12)$$
   $$x = 30$$

   The answer is choice **C**.

   Alternate Solution You can also solve this problem without using a proportion. Since 2 packages contain 12 doughnuts, each package must contain $12 \div 2$ or 6 doughnuts. Then five packages will contain 5×6 or 30 doughnuts.

   This is answer choice **C**.

 TestCheck and Worksheet Builder

Special banks of SAT, ACT, and TIMSS items are included in this CD-ROM.

After you work each problem, record your answer on the answer sheet provided or on a piece of paper.

Multiple Choice

1. In a jar of red and green jelly beans, the ratio of green jelly beans to red jelly beans is 5:3. If the jar contains a total of 160 jelly beans, how many of them are red? **C**

   A 30 B 53 C 60

   D 100 E 160

2. If $a^2b = 12^2$ and b is an odd integer, then a could be divisible by all of the following EXCEPT **D**

   A 3 B 4 C 6 D 9 E 12

3. In the figure below, $\angle A$ and $\angle ADC$ are right angles, the length of $\overline{AD}$ is 7 units, the length of $\overline{AB}$ is 10 units, and the length of $\overline{DC}$ is 6 units. What is the area, in square units, of $\triangle DCB$? **A**

   A 21

   B 24

   C $3\sqrt{149}$

   D 142

   E 210

   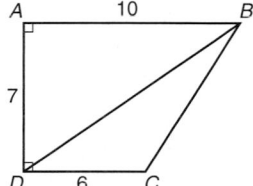

4. A science class has a ratio of girls to boys of 4 to 3. If the class has a total of 35 students, how many more girls are there than boys? **D**

   A 20 B 15 C 7 D 5 E 1

5. In the figure below, $\overline{AC} \parallel \overline{ED}$. If the length of $BD = 3$, what is the length of BE? **A**

   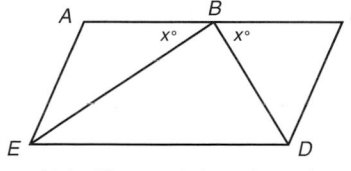

   Note: Figure not drawn to scale.

   A 3 B 4 C 5 D $3\sqrt{3}$

   E It cannot be determined from the information given.

6. What is the slope of the line that contains points at $(6, 4)$ and $(13, 5)$? **C**

   A $\frac{1}{8}$ B $-\frac{1}{9}$ C $\frac{1}{7}$ D 1 E 7

7. In $\triangle ABC$ below, if AC is equal to 8, then BC is equal to **D**

   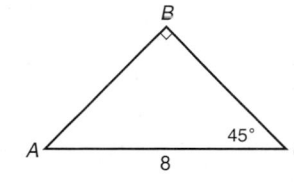

   A $8\sqrt{2}$ B 8 C 6

   D $4\sqrt{2}$ E $3\sqrt{2}$

8. The ratio of $\frac{1}{7}$ to $\frac{1}{5}$ is equal to the ratio of 100 to **E**

   A $\frac{20}{7}$ B 20 C 35 D 100 E 140

9. If there are 4 more nickels in a jar than there are dimes, which could be the ratio of dimes to nickels in the jar? **A**

   A $\frac{8}{10}$

   B 1

   C $\frac{14}{10}$

   D 4

   E None of the above

10. **Grid-In** Twenty bottles contain a total of 8 liters of apple juice. If each bottle contains the same amount of apple juice, how much juice (in liters) is in each bottle? **.4 or 2/5**

 SAT/ACT Practice For additional test practice questions, visit: **www.amc.glencoe.com**

THE PRINCETON REVIEW *SAT & ACT Preparation* **693**

A bubble-in answer sheet for these practice problems is available on p. 1 of the *Assessment and Evaluation Masters.*

Additional Practice

Additional SAT and ACT practice questions and a bubble-in answer sheet are available in the *Assessment and Evaluation Masters,* pp. 2, 253, and 254.

Assessment and Evaluation Masters, pp. 253–254

Extra Credit

TIMSS Which one of the following conics is represented by the equation $(x - 3y)(x + 3y) = 36$? **D**

A Circle

B Ellipse

C Parabola

D Hyperbola

Instructional Objectives

Lesson (pages)	Lesson Objectives	NCTM Standards 2000	State/Local Objectives
11-1 (695–703)	• Use the properties of exponents. • Evaluate and simplify expressions containing rational exponents. • Solve equations containing rational exponents.	6	
11-2 (704–711)	• Graph exponential functions and inequalities. • Solve problems involving exponential growth and decay.	2, 6, 10	
11-3 (712–717)	• Use the exponential function $y = e^x$.	2, 6, 10	
11-4 (718–725)	• Evaluate expressions involving logarithms. • Solve equations and inequalities involving logarithms. • Graph logarithmic functions and inequalities.	2, 6, 10	
11-5 (726–732)	• Find common logarithms and antilogarithms of numbers. • Solve equations and inequalities using common logarithms. • Solve real-world applications with common logarithmic functions.	2, 6, 10	
11-6 (733–737)	• Find natural logarithms of numbers. • Solve equations and inequalities using natural logarithms. • Solve real-world applications with natural logarithmic functions.	2, 6, 9, 10	
11-6B (738–739)	• Investigate the relationship between area of regions below the graph of $y = \frac{1}{x}$ and natural logarithms.	2, 3, 4, 6, 7, 8, 9, 10	
11-7 (740–748)	• Find the doubling time of an exponential quantity. • Find exponential and logarithmic functions to model real-world data. • Linearize data.	2, 6, 10	

Suggested Pacing *See pages T2–T4 for additional options in course-planning.*

COURSE	Lesson 11-1	Lesson 11-2	Lesson 11-3	Lesson 11-4	Lesson 11-5
Precalculus Emphasis Chapters 4–15 (45–50 minutes)	1.5 days	1.5 days	1.5 days	1.5 days	1.5 days
Precalculus Emphasis (block)	0.5 day	1.5 days		1.5 days	
Trigonometric Emphasis Chapters 1–9, 11, 12 (45–50 minutes)	1.5 days	1.5 days	1.5 days	1.5 days	1.5 days
Trigonometric Emphasis (block)	0.5 day	1.5 days		1.5 days	

Instructional Resources

Lesson	Graphing Calculator Applications	Study Guide and Practice	Enrichment	Assessment and Evaluation	Graphing Calculator Keystrokes	Problem Solving & Applications	Spreadsheet Applications	TI-92 and CBL Interdisciplinary Activities	Transparencies A and B
11-1	695, 699	197–198	121		44	21			11-1
11-2	705, 709, 710	199–200	122	277	45				11-2
11-3	713	201–202	123						11-3
11-4		203–204	124	276, 277		22			11-4
11-5	727, 730, 731	205–206	125					37–40	11-5
11-6	735, 736	207–208	126	278					11-6
11-6B	738–739				46–48				
11-7	742	209–210	127	278			11		11-7
Study Guide/ Assessment				263–275, 281					
SAT/ACT Practice				279–280					

Lesson 11-6	Lesson 11-7	Study Guide & Assessment	Chapter Test	TOTAL
1.5 days	(with Lesson 11-6B) 2 days	1 day	1 day	13 days
(with Lesson 11-6B) 1.5 days	1.5 days		0.5 day	7 days
1.5 days	(with Lesson 11-6B) 2 days	1 day	1 day	13 days
(with Lesson 11-6B) 1.5 days	1.5 days		0.5 day	7 days

Resource Manager

Applications

Problem Solving and Applications Masters, pp. 21, 22

Technology/Multimedia

Spreadsheet Applications Masters, p. 11

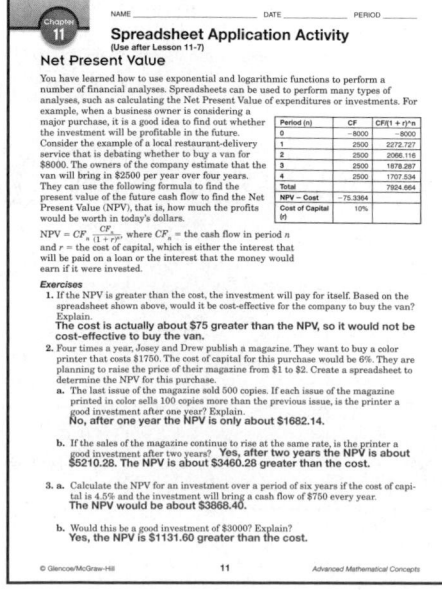

TI-92 & CBL Interdisciplinary Activities Masters, pp. 37–40

Assessment Resources

Type	Student Edition	Teacher's Wraparound Edition	Assessment and Evaluation Masters
Ongoing Assessment	Mid-Chapter Quiz, p. 717	5-Minute Check, pp. 695, 704, 712, 718, 726, 733, 740	Mid-Chapter Test, p. 276 Quizzes A–D, pp. 277–278
Mixed Review	Mixed Review, pp. 702–703, 711, 716–717, 725, 732, 737, 748		Cumulative Review, p. 281
Error Analysis	You Decide, pp. 700, 722	Error Analysis, pp. 700, 708, 714, 722, 730, 735, 744	
Standardized Test Practice	SAT/ACT Practice, pp. 703, 711, 717, 725, 732, 737, 748 SAT & ACT Preparation, pp. 754–755	TIMSS, p. 755	SAT/ACT Practice, pp. 279–280
Open-Ended Assessment	Math Journal, pp. 714, 735 Open-Ended Assessment, p. 753 Portfolio, p. 753	Modeling: pp. 711, 732 Speaking: pp. 703, 717, 737 Writing: pp. 725, 739, 748	Open-Ended Assessment, p. 275
Chapter Assessment	Study Guide and Assessment, pp. 749–753 Chapter Test, p. A66		Multiple-Choice Tests (Forms 1A, 1B, 1C), pp. 263–268 Free-Response Tests (Forms 2A, 2B, 2C), pp. 269–274

TestCheck and Worksheet Builder is a networkable CD-ROM with three integrated modules. The **Worksheet Builder** creates customized worksheets, tests, and quizzes of free-response, multiple-choice, short-answer, and open-ended items. The **Student Module** gives you the option of having students take tests on-screen and get immediate feedback on their performance. Use the optional **Management System** to keep detailed student records.

Additional Chapter Resources

Student Edition
Career Choices, p. 703
Internet Project, p. 753
Extra Practice, pp. A46–A47

Teacher's Classroom Resources

Technology
Graphing Calculator
Keystrokes Masters

Teaching Aids
Answer Key Transparencies
Block Schedule Planning Guide
Lesson Planning Guide
Solutions Manual

Glencoe Instructional Technology

Instructional
Advanced MathPASS CD-ROM
Multimedia Applications CD-ROM

Assessment
TestCheck and Worksheet Builder

GLENCOE Online

Visit **www.amc.glencoe.com** for data updates, career information, graphing calculator programs, and other interactive activities.

Chapter Overview

This chapter introduces exponential and logarithmic functions. Students simplify and evaluate exponential expressions with rational and irrational exponents and graph exponential functions and inequalities. In addition, they model real-world situations using common and natural logarithms.

Using a Graphing Calculator

There are numerous opportunities, all optional, to use a graphing calculator in this chapter.

- Investigating the properties of exponents (p. 695)
- Simplifying rational exponents (p. 699)
- Studying the graphs of exponential functions (p. 705)
- Comparing exponential functions using parent graphs and transformations (p. 709)
- Solving real-world problems involving exponents (pp. 710, 713, 715)
- Solving exponential and logarithmic equations and inequalities (pp. 727–731, 734–736, 738–739)
- Performing exponential and logarithmic regressions (pp. 741–745)

The **Graphing Calculator Appendix**, which begins on page A2, contains keystroke instructions on many of these topics.

Chapter 11

EXPONENTIAL AND LOGARITHMIC FUNCTIONS

CHAPTER OBJECTIVES
- **Simplify and evaluate expressions containing rational and irrational exponents.** *(Lessons 11-1, 11-2)*
- **Use and graph exponential functions and inequalities.** *(Lessons 11-2, 11-3)*
- **Evaluate expressions and graph and solve equations involving logarithms.** *(Lesson 11-4)*
- **Model real-world situations and solve problems using common and natural logarithms.** *(Lessons 11-5, 11-6, 11-7)*

694 Chapter 11 Exponential and Logarithmic Functions

SAT & ACT Preparation

This textbook contains a complete SAT/ACT preparation course described on pages T8–T9 of the *Teacher's Handbook.* This chapter contains:

- an SAT or ACT practice question at the end of each lesson
- examples of SAT and ACT problems involving averages on page 754
- ten SAT and ACT practice problems on page 755

- additional SAT and ACT practice in the *Assessment and Evaluation Masters,* pp. 279–280
- SAT and ACT test practice banks in the *TestCheck and Worksheet Builder*

Real Exponents

OBJECTIVES
- Use the properties of exponents.
- Evaluate and simplify expressions containing rational exponents.
- Solve equations containing rational exponents.

AEROSPACE On July 4, 1997, the Mars Pathfinder Lander touched down on Mars. It had traveled 4.013×10^8 kilometers from Earth. Two days later, the Pathfinder's Sojourner rover was released and transmitted data from Mars until September 27, 1997.

The distance Pathfinder traveled is written in **scientific notation.** A number is in scientific notation when it is in the form $a \times 10^n$ where $1 \le a < 10$ and n is an integer. Working with numbers in scientific notation requires an understanding of the definitions and properties of integral exponents. For any real number b and a positive integer n, the following definitions hold.

Definition	Example
If $n = 1$, $b^n = b$.	$17^1 = 17$
If $n > 1$, $b^n = b \cdot b \cdot b \cdot \ldots \cdot b$. n factors	$15^4 = 15 \cdot 15 \cdot 15 \cdot 15$ or $50{,}625$
If $b = 0$, $b^0 = 1$.	$400{,}785^0 = 1$
If $b \ne 0$, $b^{-n} = \dfrac{1}{b^n}$.	$7^{-3} = \dfrac{1}{7^3}$ or $\dfrac{1}{343}$

Example

1 **AEROSPACE** At their closest points, Mars and Earth are approximately 7.5×10^7 kilometers apart.

a. Write this distance in standard form.
$7.5 \times 10^7 = 7.5 (10 \times 10 \times 10 \times 10 \times 10 \times 10 \times 10)$ or $75{,}000{,}000$

b. How many times farther is the distance Mars Pathfinder traveled than the minimum distance between Earth and Mars?
Let n represent the number of times farther the distance is.
$$(7.5 \times 10^7)n = 4.013 \times 10^8$$
$$n = \frac{4.013 \times 10^8}{7.5 \times 10^7} \text{ or } 5.4$$

GRAPHING CALCULATOR EXPLORATION

Recall that if the graphs of two equations coincide, the equations are equivalent.

TRY THESE Graph each set of equations on the same screen. Use the graphs and tables to determine whether Y1 is equivalent to Y2 or Y3. **1–5. See Answer Appendix.**

1. $Y_1 = x^2 \cdot x^3$, $Y_2 = x^5$, $Y_3 = x^6$
2. $Y_1 = (x^2)^3$, $Y_2 = x^5$, $Y_3 = x^6$

WHAT DO YOU THINK?

3. Make a conjecture about the value of $a^m \cdot a^n$.
4. Make a conjecture about the value of $(a^m)^n$.
5. Use the graphing calculator to investigate the value of an expression like $\left(\dfrac{a}{b}\right)^m$. What do you observe?

Resource Manager

Reproducible Masters
- *Study Guide and Practice*, pp. 197–198
- *Enrichment*, p. 121
- *Graphing Calculator Keystrokes*, p. 44
- *Problem Solving and Applications*, p. 21

Transparencies
- 5-Minute Check, 11-1
- Teaching, 11-1

5-Minute Check
Chapter 10

1. Write the equation for the ellipse having its center at $(-2, 3)$, a major axis of length 10, and a minor axis of length 8, if its major axis is parallel to the y-axis.
$\dfrac{(x+2)^2}{16} + \dfrac{(y-3)^2}{25} = 1$

Identify the conic section represented by each equation. Write the equation in standard form and graph the equation.

2. $x^2 + y^2 - 2x - 8 = 0$
circle; $(x - 1)^2 + y^2 = 9$

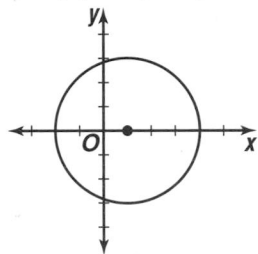

3. $9x^2 - 4y^2 + 36x + 8y - 4 = 0$
hyperbola; $\dfrac{(x+2)^2}{4} - \dfrac{(y-1)^2}{9} = 1$

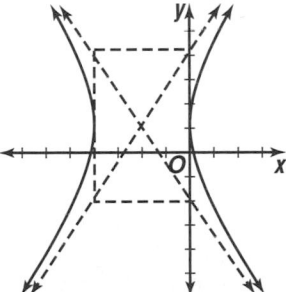

4. Find the points of intersection of the circle $x^2 + y^2 = 25$ and the parabola $y = x^2 - 5$.
$(0, -5)$, $(3, 4)$, $(-3, 4)$

Motivating the Lesson

Discussion Challenge students to use the properties of exponents to find a number x such that $2^x \cdot 2^x = 2$. Next, have them find a number y such that $3^y \cdot 3^y \cdot 3^y = 3$. After a few more such examples, they should recognize the general rule, $a^{\frac{1}{n}} = \sqrt[n]{a}$.

Teaching Tip To enter numbers in scientific notation into a graphing calculator, use the second function of LOG , [10ˣ], or the second function of , [EE]. For 2.28×10^6, enter 2.28 2nd [10ˣ] 6) or 2.28 2nd [EE] 6).

In-Class Example
Example 1
Astronomy The volume of the planet Jupiter is 1.521×10^{15} km³.
a. Write this value in standard form. **1,521,000,000,000,000**
b. If the volume of Venus is 9.868×10^{11} km³, how many times larger is Jupiter than Venus? **about 1541 times larger**

Graphing Calculator Exploration

For Exercise 5, have students graph equations with numbers substituted for *m* to see which graphs coincide. For example, let $m = 5$ and have students check $\left(\frac{a}{b}\right)^5$, $\frac{a^5}{b^5}$, $\frac{a^5}{b}$, and $\frac{a}{b^5}$.

Teaching Tip For Example 2, when introducing the properties of exponents, point out the distinction between the first, which is the product of *m* factors of *a* by *n* factors of *a*, and the second, which is the product of *n* groups each with *m* factors of *a*.

In-Class Examples
Example 2
Evaluate each expression.
a. $\frac{3^4 \cdot 3^7}{3^6}$ **3^5**
b. $\left(\frac{3}{4}\right)^{-1}$ **$\frac{4}{3}$**

Example 3
Simplify each expression.
a. $(s^4t^7)^3$ **$s^{12}t^{21}$**
b. $\frac{x^2y^5}{(x^2)^4}$ **$\frac{y^5}{x^6}$**

Teaching Tip Remind students that a product is the result of multiplication and that in Example 3, the product $(s^4t^7)^3$ equals $s^{12}t^{21}$.

The Graphing Calculator Exploration leads us to the following properties of exponents. You can use the definitions of exponents to verify the properties.

Properties of Exponents		
Suppose *m* and *n* are positive integers and *a* and *b* are real numbers. Then the following properties hold.		
Property	**Definition**	**Example**
Product	$a^m a^n = a^{m+n}$	$16^3 \cdot 16^7 = 16^{3+7}$ or 16^{10}
Power of a Power	$(a^m)^n = a^{mn}$	$(9^3)^2 = 9^{3 \cdot 2}$ or 9^6
Power of a Quotient	$\left(\frac{a}{b}\right)^m = \frac{a^m}{b^m}$, where $b \neq 0$	$\left(\frac{3}{4}\right)^5 = \frac{3^5}{4^5}$ or $\frac{243}{1024}$
Power of a Product	$(ab)^m = a^m b^m$	$(5x)^3 = 5^3 \cdot x^3$ or $125x^3$
Quotient	$\frac{a^m}{a^n} = a^{m-n}$, where $a \neq 0$	$\frac{15^6}{15^2} = 15^{6-2}$ or 15^4

Examples ② Evaluate each expression.

a. $\frac{2^4 \cdot 2^8}{2^5}$

$\frac{2^4 \cdot 2^8}{2^5} = 2^{(4+8)-5}$ *Product and Quotient Properties*

$= 2^7$

$= 128$

b. $\left(\frac{2}{5}\right)^{-1}$

$\left(\frac{2}{5}\right)^{-1} = \frac{1}{\frac{2}{5}}$ $b^{-n} = \frac{1}{b^n}$

$= \frac{5}{2}$

③ Simplify each expression.

a. $(s^2t^3)^5$

$(s^2t^3)^5 = (s^2)^5(t^3)^5$ *Power of a Product*

$= s^{(2 \cdot 5)}t^{(3 \cdot 5)}$ *Power of a Power*

$= s^{10}t^{15}$

b. $\frac{x^3y}{(x^4)^3}$

$\frac{x^3y}{(x^4)^3} = \frac{x^3y}{x^{12}}$ *Power of a Power*

$= x^{(-3-12)}y$ *Quotient Property*

$= x^{-9}y$ or $\frac{y}{x^9}$

Expressions with rational exponents can be defined so that the properties of integral exponents are still valid. Consider the expressions $3^{\frac{1}{2}}$ and $1^{\frac{1}{3}}$. Extending the properties of integral exponents gives us the following equations.

$3^{\frac{1}{2}} \cdot 3^{\frac{1}{2}} = 3^{\frac{1}{2} + \frac{1}{2}}$

$= 3^1$ or 3

By definition, $\sqrt{3} \cdot \sqrt{3} = 3$. Therefore, $3^{\frac{1}{2}}$ and $\sqrt{3}$ are equivalent.

$12^{\frac{1}{3}} \cdot 12^{\frac{1}{3}} \cdot 12^{\frac{1}{3}} = 12^{\frac{1}{3} + \frac{1}{3} + \frac{1}{3}}$

$= 12^1$ or 12

We know that $\sqrt[3]{12} \cdot \sqrt[3]{12} \cdot \sqrt[3]{12} = 12$. Therefore, $12^{\frac{1}{3}}$ and $\sqrt[3]{12}$ are equivalent.

Prerequisite Skills

A review of the prerequisite skills for this lesson is provided in the *Prerequisite Skills Masters*.

Skill Number and Content	Pages
1 Properties of Real Numbers	3–4
2 Fractions	5–6
3 Order of Operations	7–8

Exploring other expressions can reveal the following properties.

- If n is an odd number, then $b^{\frac{1}{n}}$ is the nth root of b.

- If n is an even number and $b \geq 0$, then $b^{\frac{1}{n}}$ is the non-negative nth root of b.

- If n is an even number and $b < 0$, then $b^{\frac{1}{n}}$ does not represent a real number, but a complex number.

In general, let $y = b^{\frac{1}{n}}$ for a real number b and a positive integer n. Then, $y^n = \left(b^{\frac{1}{n}}\right)^n = b^{\frac{n}{n}}$ or b. But $y^n = b$ if and only if $y = \sqrt[n]{b}$. Therefore, we can define $b^{\frac{1}{n}}$ as follows.

Definition of $b^{\frac{1}{n}}$

For any real number $b \geq 0$ and any integer $n > 1$,
$$b^{\frac{1}{n}} = \sqrt[n]{b},$$
This also holds when $b < 0$ and n is odd.

In this chapter, b will be a real number greater than or equal to 0 so that we can avoid complex numbers that occur by taking an even root of a negative number.

The properties of integral exponents given on page 696 can be extended to rational exponents.

Examples

4 **Evaluate each expression.**

a. $125^{\frac{1}{3}}$

$125^{\frac{1}{3}} = (5^3)^{\frac{1}{3}}$ *Rewrite 125 as 5^3.*

$= 5^{\frac{3}{3}}$ *Power of a Power*

$= 5$

b. $\sqrt{14} \cdot \sqrt{7}$

$\sqrt{14} \cdot \sqrt{7} = 14^{\frac{1}{2}} \cdot 7^{\frac{1}{2}}$ $\sqrt[n]{b} = b^{\frac{1}{n}}$

$= (2 \cdot 7)^{\frac{1}{2}} \cdot 7^{\frac{1}{2}}$ $14 = 2 \cdot 7$

$= 2^{\frac{1}{2}} \cdot 7^{\frac{1}{2}} \cdot 7^{\frac{1}{2}}$ *Power of a Product*

$= 2^{\frac{1}{2}} \cdot 7$ *Product Property*

$= 7\sqrt{2}$

5 **Simplify each expression.**

a. $(81c^4)^{\frac{1}{4}}$

$(81c^4)^{\frac{1}{4}} = (3^4 c^4)^{\frac{1}{4}}$ $81 = 3^4$

$= (3^4)^{\frac{1}{4}} \cdot (c^4)^{\frac{1}{4}}$ *Power of a Product*

$= 3^{\frac{4}{4}} \cdot c^{\frac{4}{4}}$ *Power of a Power*

$= 3|c|$

b. $\sqrt[6]{9x^3}$

$\sqrt[6]{9x^3} = \sqrt[6]{9} \cdot \sqrt[6]{x^3}$ *Power of a Product*

$= 9^{\frac{1}{6}} \cdot (x^3)^{\frac{1}{6}}$ $b^{\frac{1}{n}} = \sqrt[n]{b}$

$= (3^2)^{\frac{1}{6}} \cdot (x^3)^{\frac{1}{6}}$ $9 = 3^2$

$= 3^{\frac{1}{3}} \cdot x^{\frac{1}{2}}$ *Power of a Power*

$= \sqrt[3]{3} \cdot \sqrt{x}$ $b^{\frac{1}{n}} = \sqrt[n]{b}$

Teaching Tip For In-Class Example 4a, remind students that while $625^{\frac{1}{4}} = 5$, $(-625)^{\frac{1}{4}}$ is not defined in the real number system.

In-Class Examples

Example 4
Evaluate each expression.

a. $625^{\frac{1}{4}}$ **5**

b. $3^{\frac{1}{2}} \cdot 21^{\frac{1}{2}}$ $3\sqrt{7}$

Example 5
Simplify each expression.

a. $(27a^6)^{\frac{1}{3}}$ $3a^2$

b. $\sqrt[6]{8x^{12}}$ $\sqrt{2x^2}$

Teaching Tip For Example 5, remind students to watch for expressions such as $\sqrt[6]{8}$ and $\sqrt[6]{9}$; these can be further simplified to $\sqrt{2}$ and $\sqrt[3]{3}$.

Teaching Tip For In-Class Example 6b, $27^{\frac{2}{3}}$ can be written as $(3^3)^{\frac{2}{3}}$ or 3^2 or 9. It can also be written as $\sqrt[3]{27^2}$ or $\left(\sqrt[3]{27}\right)^2$. With a calculator, $\sqrt[3]{27^2}$ is $\sqrt[3]{729}$ or 9, whereas $\left(\sqrt[3]{27}\right)^2$ stated in this way is 3^2 or 9.

In-Class Examples

Example 6
Evaluate each expression.

a. $243^{\frac{4}{5}}$ **81**

b. $\dfrac{27^{\frac{5}{3}}}{27^{\frac{2}{3}}}$ **27**

Example 7

a. Express $\sqrt[4]{16s^{12}t^{16}}$ using rational exponents. $2s^3t^4$

b. Express $5x^{\frac{3}{4}}y^{\frac{1}{2}}$ using radicals. $5\sqrt[4]{x^3y^2}$

Teaching Tip For Example 7, in order to form the square root of a number, such as x^{10}, x^{40}, or x^{16}, half of each exponent yields the square root: x^5, x^{20}, or x^8.

Rational exponents with numerators other than 1 can be evaluated by using the same properties. Study the two methods of evaluating $64^{\frac{5}{6}}$ below.

Method 1	**Method 2**
$46^{\frac{5}{6}} = \left(46^{\frac{1}{6}}\right)^5$	$46^{\frac{5}{6}} = \left(46^5\right)^{\frac{1}{6}}$
$\quad = \left(\sqrt[6]{46}\right)^5$	$\quad = \sqrt[6]{46^5}$

Therefore, $\left(\sqrt[6]{46}\right)^5$ and $\sqrt[6]{46^5}$ both equal $46^{\frac{5}{6}}$.

In general, we define $b^{\frac{m}{n}}$ as $\left(b^{\frac{1}{n}}\right)^m$ or $\left(b^m\right)^{\frac{1}{n}}$. Now apply the definition of $b^{\frac{1}{n}}$ to $\left(b^{\frac{1}{n}}\right)^m$ and $\left(b^m\right)^{\frac{1}{n}}$.

$$\left(b^{\frac{1}{n}}\right)^m = \left(\sqrt[n]{b}\right)^m \qquad\qquad \left(b^m\right)^{\frac{1}{n}} = \sqrt[n]{b^m}$$

Rational Exponents

For any nonzero number b, and any integers m and n with $n > 1$, and m and n have no common factors,

$$b^{\frac{m}{n}} = \sqrt[n]{b^m} = \left(\sqrt[n]{b}\right)^m$$

except where $\sqrt[n]{b}$ is not a real number.

Examples **6** **Evaluate each expression.**

a. $625^{\frac{3}{4}}$

$$625^{\frac{3}{4}} = \left(5^4\right)^{\frac{3}{4}} \quad \textit{Write 625 as } 5^4.$$
$$\qquad\quad = 5^3 \quad\quad \textit{Power of a Product}$$
$$\qquad\quad = 125$$

b. $\dfrac{16^{\frac{3}{4}}}{16^{\frac{1}{4}}}$

$$\dfrac{16^{\frac{3}{4}}}{16^{\frac{1}{4}}} = 16^{\frac{3}{4} - \frac{1}{4}} \quad \textit{Quotient Property}$$
$$\qquad = 16^{\frac{1}{2}} \text{ or } 4$$

7 a. Express $\sqrt[3]{64s^9t^{15}}$ using rational exponents.

$$\sqrt[3]{64s^9t^{15}} = (64s^9t^{15})^{\frac{1}{3}} \quad b^{\frac{1}{n}} = \sqrt[n]{b}$$
$$\qquad\qquad = 64^{\frac{1}{3}}s^{\frac{9}{3}}t^{\frac{15}{3}} \quad \textit{Power of a Product}$$
$$\qquad\qquad = 4s^3t^5$$

b. Express $12x^{\frac{2}{3}}y^{\frac{1}{2}}$ using a radical.

$$12x^{\frac{2}{3}}y^{\frac{1}{2}} = 12(x^4y^3)^{\frac{1}{6}} \quad \textit{Power of a Product}$$
$$\qquad\quad = 12\sqrt[6]{x^4y^3}$$

Addressing Individual Needs

Verbal/Linguistic Have students explain to others in the class how to change a number written in scientific notation to standard notation.

When you simplify a radical, use the product property to factor out the *n*th roots and use the smallest index possible for the radical. Remember that you should use caution when evaluating even roots to avoid negative values that would result in an imaginary number.

In-Class Examples

Example 8
Simplify $\sqrt{r^7 s^5 t^{24}}$. $|r|^3 s^2 t^{12}\sqrt{rs}$

Example 9
Solve $616 = x^{\frac{4}{5}} - 9$. $3125 = x$

Example **8** Simplify $\sqrt{r^7 s^{25} t^3}$.

For $r^7 s^{25} t^3$ to be nonnegative, none or exactly two of the variables must be negative. Check the final answer to determine if an absolute value is needed.

$$\sqrt{r^7 s^{25} t^3} = (r^7 s^{25} t^3)^{\frac{1}{2}} \qquad b^{\frac{1}{n}} = \sqrt[n]{b}$$

$$= r^{\frac{7}{2}} s^{\frac{25}{2}} t^{\frac{3}{2}} \qquad \textit{Power of a Product}$$

$$= r^{\frac{6}{2}} r^{\frac{1}{2}} s^{\frac{24}{2}} s^{\frac{1}{2}} t^{\frac{2}{2}} t^{\frac{1}{2}} \qquad \textit{Product Property}$$

$$= |r|^3 s^{12} |t| \sqrt{rst} \qquad \textit{Use } |r| \textit{ and } |t| \textit{ since rst must be nonnegative}$$
$$\textit{and there is no indication of which variables are}$$
$$\textit{negative.}$$

You can also use the properties of exponents to solve equations containing rational exponents.

Example **9** Solve $734 = x^{\frac{3}{4}} + 5$.

$$734 = x^{\frac{3}{4}} + 5$$

$$729 = x^{\frac{3}{4}} \qquad \textit{Subtract 5 from each side.}$$

$$729^{\frac{4}{3}} = \left(x^{\frac{3}{4}}\right)^{\frac{4}{3}} \qquad \textit{Raise each side to the } \frac{4}{3} \textit{ power.}$$

$$6561 = x \qquad \textit{Use a calculator.}$$

Graphing Calculator Tip

Use the ∧ calculator key to enter rational exponents. For example, to evaluate $729^{\frac{4}{3}}$, press

729 ∧ (4
÷ 3) .

An expression can also have an irrational exponent, but what does it represent? Consider the expression $2^{\sqrt{3}}$. Since $1.7 < \sqrt{3} < 1.8$, it follows that $2^{1.7} < 2^{\sqrt{3}} < 2^{1.8}$. Closer and closer approximations for $\sqrt{3}$ allow us to find closer and closer approximations for $2^{\sqrt{3}}$. Therefore, we can define a value for a^x when x is an irrational number.

Irrational Exponents	If x is an irrational number and $b > 0$, then b^x is the real number between b^{x_1} and b^{x_2} for all possible choices of rational numbers x_1 and x_2, such that $x_1 < x < x_2$.

A calculator can be used to approximate the value of an expression with irrational exponents. For example, $2^{\sqrt{3}} \approx 3.321997085$. While operations with irrational exponents are rarely used, evaluating such expressions is useful in graphing exponential equations.

Lesson 11-1   Real Exponents   **699**

From the Classroom of...

Nancy Dymott
Carthage High School
Carthage, Missouri

"I review simplifying radicals using the following technique. When an exponent in the radicand is greater than or equal to the index number, divide the index number into the exponent. The integral part of the quotient becomes the exponent of the base in front of the radical, and the remainder becomes the exponent that stays under the radical."

Error Analysis

Watch for students who have difficulty applying the product and quotient properties.

Prevent by demonstrating real number examples until the pattern is obvious.

Assignment Guide
Core: 21–73 odd, 74–87 **Enriched:** 20–66 even, 68–87

CHECK FOR UNDERSTANDING

Communicating Mathematics

Read and study the lesson to answer each question.   1–3. See margin.

1. **State** whether -4^{-2} and $(-4)^{-2}$ represent the same quantity. Explain.

2. **Explain** why rational exponents are not defined when the denominator of the exponent in lowest terms is even and the base is negative.

3. **You Decide**   Laura says that a number written in scientific notation with an exponent of -10 is between 0 and 1. Lina says that a number written in scientific notation with an exponent of -10 is a negative number with a very large absolute value. Who is correct and why?

Guided Practice

Evaluate each expression.

4. 5^{-4} $\frac{1}{625}$ 5. $\left(\frac{9}{16}\right)^{-2}$ $\frac{256}{81}$ 6. $216^{\frac{1}{3}}$ **6**   7. $\sqrt{27}\cdot\sqrt{3}$ **9** 8. $32^{\frac{3}{5}}$ **8**

Simplify each expression.

9. $81a^{-1}$ or $\frac{81}{a}$

9. $(3a^{-2})^3\cdot 3a^5$ 10. $\sqrt{m^3n^2}\cdot\sqrt{m^4n^5}$ 11. $\sqrt{\dfrac{8^n\cdot 2^7}{4^{-n}}}$ 12. $(2x^4y^8)^{\frac{1}{2}}$ $x^2y^4\sqrt{2}$

10. $m^{\frac{7}{2}}n^{\frac{7}{2}}$ or $m^3n^3\sqrt{mn}$

Express using rational exponents.

13. $\sqrt{169x^5}$ $13x^{\frac{5}{2}}$

14. $\sqrt[4]{a^2b^3c^4d^5}$ $|a|^{\frac{1}{2}}|b|^{\frac{3}{4}}|c||d|^{\frac{5}{4}}$

11. 2^{2n+3}
$\sqrt{2^{n+1}}$

Express using a radical.

15. $6^{\frac{1}{4}}b^{\frac{3}{4}}c^{\frac{1}{4}}$ $\sqrt[4]{6b^3c}$

16. $15x^{\frac{1}{3}}y^{\frac{1}{5}}$ $15\sqrt[15]{x^5y^3}$

17. Simplify $\sqrt[3]{p^4q^6r^5}$ $pq^2r\sqrt[3]{pr^2}$

18. Solve $y^{\frac{4}{5}}=34$. **82.1**

19. **Biology**   Red blood cells are circular-shaped cells that carry oxygen through the bloodstream. The radius of a red blood cell is about 3.875×10^{-7} meters. Find the area of a red blood cell. $4.717\times 10^{-13}m^2$

EXERCISES

Practice

A

Evaluate each expression.

20. $(-6)^{-4}$ $\frac{1}{1296}$ 21. -6^{-4} $-\frac{1}{1296}$ 22. $(5\cdot 3)^2$ **225**   23. $\frac{2^4}{2^{-1}}$ **32**

24. $\left(\frac{7}{8}\right)^{-3}$ $\frac{512}{343}$ 25. $(3^{-1}+3^{-2})^{-1}$ $\frac{9}{4}$ 26. $81^{\frac{1}{2}}$ **9**   27. $729^{\frac{1}{3}}$ **9**

28. $\frac{27}{27^{\frac{2}{3}}}$ **3**   29. $2^{\frac{1}{2}}\cdot 12^{\frac{1}{2}}$ $2\sqrt{6}$ 30. $64^{\frac{1}{12}}$ $\sqrt{2}$ or $2^{\frac{1}{2}}$ 31. $16^{-\frac{1}{4}}$ $\frac{1}{2}$

B

32. $\frac{(3^7)(9^4)}{\sqrt{27^6}}$ **729**   33. $\left(\sqrt[3]{216}\right)^2$ **36**   34. $81^{\frac{1}{2}}-81^{-\frac{1}{2}}$ $\frac{8^{\frac{8}{9}}}{ }$ 35. $\frac{1}{\sqrt[7]{(-128)^4}}$ $\frac{1}{16}$

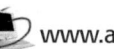 www.amc.glencoe.com/self_check_quiz

---

Study Guide and Practice Masters, pp. 197–198

Additional Answers

1. **The quantities are not the same. When the negative is enclosed inside of the parentheses and the base is raised to an even power, the answer is positive. When the negative is not enclosed inside of the parentheses and the base is raised to an even power, the answer is negative.**

2. **If the base were negative and the denominator were even, then we would be taking an even root of a negative number, which is undefined as a real number.**

3. **Laura is correct. The negative exponent of 10 represents a fraction with a numerator of 1 and a denominator of a positive power of 10. The product of this fraction and a number between 1 and 10 is between 0 and 1.**

39. $3pq^2r^{-\frac{1}{3}}$

36. $(3n^2)^3$ $27n^6$ **37.** $(y^2)^{-4} \cdot y^8$ 1 **38.** $(4y^4)^{\frac{3}{2}}$ $8y^6$ **39.** $(27p^3q^6r^{-1})^{\frac{1}{3}}$

Simplify each expression.

40. $2^{-8}x^{-8}$ or $\frac{1}{256x^8}$

40. $[(2x)^4]^{-2}$ **41.** $(36x^6)^{\frac{1}{2}}$ $6|x|^3$ **42.** $\left(\frac{b^{2n}}{b^{-2n}}\right)^{\frac{1}{2}}$ b^{2n} **43.** $\frac{2n}{4n^{\frac{1}{2}}}$ $\frac{\sqrt{n}}{2}$

45. $4f^4|g||h|^{-1}$ or $\frac{4f^4|g|}{|h|}$

44. $(3m^{\frac{1}{2}} \cdot 27n^{\frac{1}{4}})^4$ **45.** $\left(\frac{f^{-16}}{256g^4h^{-4}}\right)^{-\frac{1}{4}}$ **46.** $\sqrt[6]{x^2(x^{\frac{3}{4}} + x^{-\frac{3}{4}})}$ **47.** $(2x^{\frac{1}{4}}y^{\frac{1}{3}})(3x^{\frac{1}{4}}y^{\frac{2}{3}})$
$3^{16}m^2n$ $[x^{\frac{5}{4}}(x^{\frac{3}{2}} + 1)]^{\frac{1}{6}}$ $6x^{\frac{1}{2}}y$

48. Show that $\sqrt[m]{\sqrt[n]{a}} = \sqrt[mn]{a}$.
See margin.

Express using rational exponents.

49. $\sqrt{m^6n}$ $|m|^3n^{\frac{1}{2}}$ **50.** $\sqrt{xy^3}$ $|x|^{\frac{1}{2}}|y|^{\frac{3}{2}}$ **51.** $\sqrt[3]{8x^3y^6}$ $2xy^2$

52. $17\sqrt[7]{x^{14}y^7z^{12}}$ $17x^2yz^{\frac{12}{7}}$ **53.** $\sqrt[5]{a^{10}b^2} \cdot \sqrt[4]{c^2}$ **54.** $60\sqrt[8]{r^{80}s^{56}t^{27}}$
$a^2b^{\frac{2}{5}}|c|^{\frac{1}{7}}$ $60r^{10}|s|^7t^{\frac{27}{8}}$

Express using a radical.

55. $16^{\frac{1}{5}}$ $\sqrt[5]{16}$ **56.** $(7a)^{\frac{5}{8}}b^{\frac{3}{8}}$ $\sqrt[8]{7^5a^5b^3}$ **57.** $p^{\frac{2}{3}}q^{\frac{1}{2}}r^{\frac{1}{3}}$ $\sqrt[6]{p^4q^3r^2}$

C   **58.** $\frac{2^{\frac{2}{3}}}{2^{\frac{1}{3}}}$ $\sqrt[3]{2}$ **59.** $13a^{\frac{1}{7}}b^{\frac{1}{3}}$ $13\sqrt[21]{a^3b^7}$ **60.** $(n^3m^9)^{\frac{1}{2}}$
$|n|m^4\sqrt{mn}$

61. -0.69 **61.** What is the value of x in the equation $x = \sqrt[3]{(-245)^{-\frac{1}{5}}}$ to the nearest hundredth?

Simplify each expression.

62. $\sqrt{d^3e^2f^2}$ $d|e||f|\sqrt{d}$ **63.** $\sqrt[3]{a^5b^7c}$ $ab^2\sqrt[3]{a^2bc}$ **64.** $\sqrt{20x^3y^6}$ $2x|y|^3\sqrt{5x}$

Solve each equation.

65. $14.2 = x^{-\frac{3}{2}}$ 0.17 **66.** $724 = 15a^{\frac{5}{2}} + 12$ 4.68 **67.** $\frac{1}{8}\sqrt{x^5} = 3.5$ 3.79

Applications and Problem Solving

68. Aerospace Mars has an aproximate diameter of 6.794×10^3 kilometers. Assume that Mars is a perfect sphere. Use the formula for the volume of a sphere, $V = \frac{4}{3}\pi r^3$ to determine the volume of Mars. 1.64×10^{11} km³

69. Critical Thinking Consider the equation $y = 3^x$. Find the values of y for $x = -8, -6, -5, \frac{10}{33}, \frac{1}{2}, \frac{2}{3}, \frac{10}{9}, \frac{5}{3},$ and $\frac{7}{2}$.
a. What is the value of y if $x < 0$? $0 < y < 1$
b. What is the value of y if $0 < x < 1$? $1 < y < 3$
c. What is the value of y if $x > 1$? $y > 3$
d. Write a conjecture about the relationship between the value of the base and the value of the power if the exponent is greater than or less than 1. Justify your answer.   **See margin.**

Mars

70. Chemistry The nucleus of an atom is the center portion of the atom that contains most of its mass. A formula for the radius r of a nucleus of an atom is $r = (1.2 \times 10^{-15})A^{\frac{1}{3}}$ meters, where A is the mass number of the nucleus. Suppose the radius of a nucleus is approximately 2.75×10^{-15} meters. Is the atom boron with a mass number of 11, carbon with a mass number of 12, or nitrogen with a mass number of 14?   **carbon**

48. $\sqrt[m]{\sqrt[n]{a}} = \left(a^{\frac{1}{n}}\right)^{\frac{1}{m}}$
$= a^{\frac{1}{n} \cdot \frac{1}{m}}$
$= a^{\frac{1}{nm}} = a^{\frac{1}{mn}}$
$= \sqrt[mn]{a}$

69d. If the exponent is less than 0, the power is greater than 0 and less then 1. If the exponent is greater than 0 and less than 1, the power is greater than 1 and less than the base. If the exponent is greater than 1, the power is greater than the base.

Any number to the zero power is 1. Thus, if the exponent is less than zero, the power is less than 1. A power of a positive number is never negative, so the power is greater than 0.

Any number to the zero power is 1 and to the first power is itself. Thus, if the exponent is greater than zero and less than 1, the power is between 1 and the base.

Any number to the first power is itself. Thus, if the exponent is greater than 1, the power is greater than the base.

Additional Answers

72b. A 5-mile per hour increase in the wind speed when the wind is light has more of an effect on perceived temperature than a 5-mile per hour increase in the wind speed when the wind is heavy.

74a.
$$a^m a^n = \overbrace{a \cdot a \cdot \cdots \cdot a}^{m \text{ factors}} \cdot \overbrace{a \cdot a \cdot \cdots \cdot a}^{n \text{ factors}}$$
$$= \overbrace{a \cdot a \cdot \cdots \cdot a}^{m+n \text{ factors}}$$
$$= a^{m+n}$$

74b. See Answer Appendix.

74c.
$$(ab)^m = \overbrace{ab \cdot ab \cdot \cdots \cdot ab}^{m \text{ factors}}$$
$$= \overbrace{a \cdot a \cdot \cdots \cdot a}^{m \text{ factors}} \cdot \overbrace{b \cdot b \cdot \cdots \cdot b}^{m \text{ factors}}$$
$$= a^m b^m$$

74d.
$$\left(\frac{a}{b}\right)^m = \overbrace{\frac{a}{b} \cdot \frac{a}{b} \cdot \cdots \cdot \frac{a}{b}}^{m \text{ factors}} = \frac{a^m}{b^m}$$

74e.
$$\frac{a^m}{a^n} = \frac{\overbrace{a \cdot a \cdot \cdots \cdot a}^{m \text{ factors}}}{\underbrace{a \cdot a \cdot \cdots \cdot a}_{n \text{ factors}}} = a^{m-n}$$

75.

76.
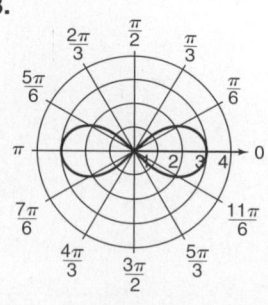

78.

---

71. Critical Thinking Find the solutions for $32^{(x^2 + 4x)} = 16^{(x^2 + 4x + 3)}$. **2, −6**

72. Meteorology Have you ever heard the meteorologist on the news tell the day's windchill factor? A model that approximates the windchill temperature for an air temperature 5°F and a wind speed of s miles per hour is $C = 69.2(0.94^s) - 50$.
  a. Copy and complete the table.
  b. How does the effect of a 5-mile per hour increase in the wind speed when the wind is light compare to the effect of a 5-mile per hour increase in the wind speed when the wind is heavy?
  See margin.

Wind Speed	Windchill
5	0.8
10	−12.7
15	−22.6
20	−29.9
25	−35.3
30	−39.2

73. Communication Geosynchronous satellites stay over a single point on Earth. To do this, they rotate with the same period of time as Earth. The distance r of an object from the center of Earth, that has a period of t seconds, is given by $r^3 = \dfrac{GM_e t^2}{4\pi^2}$, where G is 6.67×10^{-11} N m²/kg², M_e is the mass of Earth, and t is time in seconds. The mass of Earth is 5.98×10^{24} kilograms.
  a. How many meters is a geosynchronous communications satellite from the center of Earth? **42,250,474.31 meters**
  b. If the radius of Earth is approximately 6380 kilometers, how many kilometers is the satellite above the surface of Earth? **35,870 kilometers**

74. Critical Thinking Use the definition of an exponent to verify each property.
  a. $a^m a^n = a^{m+n}$ b. $(a^m)^n = a^{mn}$ c. $(ab)^m = a^m b^m$
  d. $\left(\dfrac{a}{b}\right)^m = \dfrac{a^m}{b^m}$, where $b \neq 0$ e. $\dfrac{a^m}{a^n} = a^{m-n}$, where $a \neq 0$
  74a–e. See margin.

75. See margin.

Mixed Review

75. Graph the system of inequalities $x^2 + y^2 \leq 9$ and $x^2 + y^2 \geq 4$. *(Lesson 10-8)*

76. (3, 0), $x = -3$, See margin for graph.

76. Find the coordinates of the focus and the equation of the directrix of the parabola with equation $y^2 = 12x$. Then graph the equation. *(Lesson 10-5)*

77. Find $(2\sqrt{3} + 2i)^{\frac{1}{5}}$. Express the answer in the form $a + bi$ with a and b to the nearest hundredth. *(Lesson 9-8)* **1.31 + 0.14i**

78. See margin for graph; lemniscate.

78. Graph $r^2 = 9\cos 2\theta$. Identify the classical curve it represents. *(Lesson 9-2)*

79. Baseball Lashon hits a baseball 3 feet above the ground with a force that produces an initial velocity of 105 feet per second at an angle of 42° above the horizontal. What is the elapsed time between the moment the ball is hit and the time it hits the ground? *(Lesson 8-7)* **about 4.43 s**

80. Find an ordered triple that represents $\overrightarrow{TC}$ for $T(3, -4, 6)$ and $C(2, 6, -5)$. Then find the magnitude of $\overrightarrow{TC}$. *(Lesson 8-3)* **⟨−1, 10, −11⟩, $\sqrt{222}$**

81. Find the numerical value of one trigonometric function of S if $\tan S \cos S = \dfrac{1}{2}$. *(Lesson 7-2)* **Sample answer: $\sin S = \dfrac{1}{2}$.**

82. $\dfrac{\pi}{2} + \pi n$ where n is an integer

82. Find the values of θ for which $\cot \theta = 0$ is true. *(Lesson 6-7)*

83. Agriculture A center-pivot irrigation system with a 75-meter radial arm completes one revolution every 6 hours. Find the linear velocity of a nozzle at the end of the arm. *(Lesson 6-2)* **25π m/h**

Extra Practice See p. A46.

Extra Credit

Solve the equation $\sqrt[4]{16}\left(\sqrt[4]{4} \cdot \sqrt[4]{9}\right)^3 = 12\sqrt{6}\,x$. **x = 1**

84. Find the values of x in the interval $0° \le x \le 360°$ for which $x = \arccos 0$. *(Lesson 5-5)* **90°, 270°**

85. State the number of complex roots of the equation $x^3 - 25x = 0$. Then find the roots and graph. *(Lesson 4-1)* **3; −5, 0, 5; See margin for graph.**

86. Use a graphing calculator to graph $f(x) = 2x^2 - 3x + 3$ and to determine and classify its extrema. *(Lesson 3-6)* **abs. min.: (0.75, 1.88)**

87. **SAT/ACT Practice** If all painters work at the same rate and 8 painters can paint a building in 48 hours, how many hours will it take 16 painters to paint the building? **E**

A 96   B 72   C 54   D 36   E 24

Lesson 11-1 *Real Exponents* **703**

CAREER CHOICES

Food Technologist

Everyone needs to eat properly in order to stay healthy. Food technologists help ensure that foods taste their best, provide good nutrition, and are safe. They study the composition of food and help develop methods for processing, preserving, and packaging foods. Food technologists usually specialize in some aspect of food technology such as improving taste or increasing nutritional value.

Food safety is one very important area in this field. Foods need to be handled, processed, and packaged under strict conditions to insure that they are safe for human consumption. Food technologists generally work either in private industry or for the government.

CAREER OVERVIEW

Degree Preferred:
bachelor's degree in Food Science, Food Management, or Human Nutrition

Related Courses:
mathematics, biology, chemistry, physics

Outlook:
better than average through the year 2006

Take-Out and On-Premise Meals Purchased Per Person Annually

interNET CONNECTION For more information on careers in food technology, visit: **www.amc.glencoe.com**

CAREER CHOICES   FOOD TECHNOLOGIST

Math Connection
* Explain the trends seen in the data. **The number of on-premise meals is decreasing while the number of take-out meals is increasing.**
* In what year(s) were the number of meals in each category approximately the same? **1994 and 1995**
* Estimate the number of on-premise and take-out meals at commercial restaurants that you eat per year. How do those numbers compare to the graph? **See students' answers.**

Lesson 11-1 **703**

---

1 FOCUS

5-Minute Check
Lesson 11-1

Evaluate each expression.

1. $125^{\frac{2}{3}}$ 25

2. $(3^0 - 3^2)^{\frac{1}{3}}$ −2

Simplify each expression.

3. $\sqrt[3]{81a^4b^3}$ $81^{\frac{1}{3}}a^{\frac{4}{3}}b$

4. $25^{\frac{1}{2}}x^{\frac{5}{4}}y^{\frac{1}{4}}$ $5\sqrt[4]{x^5y}$

5. $\sqrt{a^3b} \cdot \sqrt{8ab^2}$ $2a^2b\sqrt{2b}$
   or $2^{\frac{3}{2}}a^2b^{\frac{3}{2}}$

Motivating the Lesson
Real-World Application Ask students to act out a model that calculates how much money an interest-bearing account earns after so many months, or how large a bacteria colony is after so many hours of growth. Furthermore, ask them how would they predict when a bank account would double or when a bacteria colony would reach a certain critical size.

2 TEACH

Graphing Calculator Exploration

Have students graph $y = 2^x$ and discuss some of the basic features of the graph, for example, its continuity, intercepts, domain and range, and that it models a function that is increasing. Then, ask them to graph $y = \left(\frac{1}{2}\right)^x$ and $y = 3^x$ and use what they know about families of graphs to determine how the two new graphs are related to the first graph.

Exponential Functions

OBJECTIVES
- Graph exponential functions and inequalities.
- Solve problems involving exponential growth and decay.

ENTOMOLOGY In recent years, beekeepers have experienced a serious decline in the honeybee population in the United States. One of the causes for the decline is the arrival of varroa mites. Experts estimate that as much as 90% of the wild bee colonies have been wiped out.

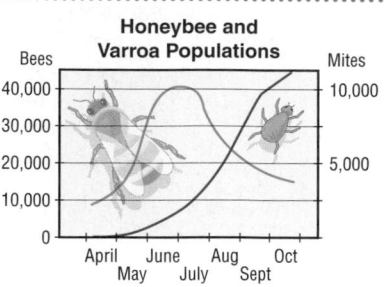

Honeybee and Varroa Populations

The graph shows typical honeybee and varroa populations over several months. A graph of the varroa population growth from April to September resembles an *exponential curve.* *A problem related to this will be solved in Example 2.*

You have evaluated functions in which the variable is the base and the exponent is any real number. For example, $y = x^5$ has x as the base and 5 as the power. Such a function is known as a **power function.** Functions of the form $y = b^x$, in which the base b is a positive real number and the exponent is a variable are known as **exponential functions.** In the previous lesson, the expression b^x was defined for integral and rational values of x. In order for us to graph $y = b^x$ with a continuous curve, we must define b^x for irrational values of x.

(5, 32)

Notice that the vertical scale is condensed.

Consider the graph of $y = 2^x$, where x is an integer. This is a function since there is a unique y-value for each x-value.

x	−4	−3	−2	−1	0	1	2	3	4	5
2^x	$\frac{1}{16}$	$\frac{1}{8}$	$\frac{1}{4}$	$\frac{1}{2}$	1	2	4	8	16	32

The graph suggests that the function is increasing. That is, for any values x_1 and x_2, if $x_1 < x_2$, then $2^{x_1} < 2^{x_2}$.

Suppose the domain of $y = 2^x$ is expanded to include all rational numbers. The additional points graphed seem to "fill in" the graph of $y = 2^x$. That is, if k is between x_1 and x_2, then 2^k is between 2^{x_1} and 2^{x_2}. The graph of $y = 2^x$, when x is a rational number, is indicated by the broken line on the graph at the right.

x	−3.5	−2.5	−1.5	−0.5	0.5	1.5	2.5	3.5	4.5
2^x	0.09	0.18	0.35	0.71	1.41	2.83	5.66	11.31	22.63

Values in the table are approximate.

704 *Chapter 11 Exponential and Logarithmic Functions*

Resource Manager

 Reproducible Masters
- *Study Guide and Practice,* pp. 199–200
- *Enrichment,* p. 122
- *Assessment and Evaluation,* p. 277
- *Graphing Calculator Keystrokes,* p. 45

 Transparencies
- 5-Minute Check, 11-2
- Teaching, 11-2

 Technology/Multimedia
- Advanced MathPASS, Lesson 23

In Lesson 11-1, you learned that exponents can also be irrational. By including irrational values of x, we can explore the graph of $y = b^x$ for the domain of all real numbers.

GRAPHING CALCULATOR EXPLORATION

TRY THESE Graph $y = b^x$ for $b = 0.5$, 0.75, 2, and 5 on the same screen.

1. What is the range of each exponential function? **positive reals**

2. What point is on the graph of each function? **(0, 1)**

3. What is the end behavior of each graph?

4. Do the graphs have any asymptotes?
3–4. See margin.

WHAT DO YOU THINK? **5–8. See margin.**

5. Is the range of every exponential function the same? Explain.

6. Why is the point at (0, 1) on the graph of every exponential function?

7. For what values of a is the graph of $y = a^x$ increasing and for what values is the graph decreasing? Explain.

8. Explain the existence or absence of the asymptotes in the graph of an exponential function.

The equations graphed in the Graphing Calculator Exploration demonstrate many properties of exponential graphs.

Characteristics of graphs of $y = b^x$		
	$b > 1$	**$0 < b < 1$**
Domain	all real numbers	all real numbers
Range	all real numbers > 0	all real numbers > 0
y-intercept	(0, 1)	(0, 1)
behavior	continuous, one-to-one, and increasing	continuous, one-to-one, and decreasing
Horizontal asymptote	negative x-axis	positive x-axis
Vertical asymptote	none	none

When $b = 1$ the graph of $y = b^x$ is the horizontal line $y = 1$.

As with other types of graphs, $y = b^x$ represents a parent graph. The same techniques used to transform the graphs of other functions you have studied can be applied to graphs of exponential functions.

Examples **1** a. **Graph the exponential functions $y = 4^x$, $y = 4^x + 2$, and $y = 4^x - 3$ on the same set of axes. Compare and contrast the graphs.**

To graph an exponential function using paper and pencil, you can use a calculator to find each range value for each domain value.

All of the graphs are continuous, increasing, and one-to-one. They have the same domain, and no vertical asymptote.

The y-intercept and the horizontal asymptotes for each graph are different from the parent graph $y = 4^x$. While $y = 4^x$ and $y = 4^x + 2$ have no x-intercept, $y = 4^x - 3$ has an x-intercept.

Lesson 11-2 Exponential Functions **705**

In-Class Example
Example 1
a. Graph the exponential functions $y = 2^x$, $y = 2^x + 3$, and $y = 2^x - 2$ on the same set of axes. Compare and contrast the graphs.

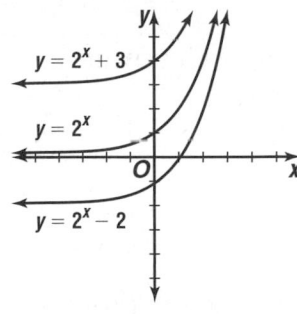

All of the graphs are continuous, increasing, and one-to-one. They have the same domain, and no vertical asymptote.

The y-intercepts and the horizontal asymptotes for each graph are different from the parent graph $y = 2^x$. While $y = 2^x$ and $y = 2^x + 3$ have no x-intercept, $y = 2^x - 2$ has an x-intercept.

b. Graph the exponential functions $y = \left(\frac{1}{3}\right)^x$, $y = 5\left(\frac{1}{3}\right)^x$, and $y = -1\left(\frac{1}{3}\right)^x$ on the same set of axes. Compare and contrast the graphs.

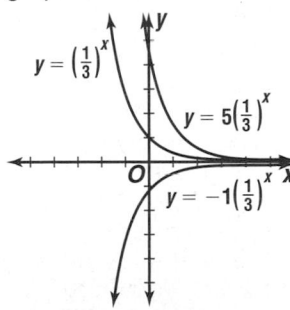

All of the graphs are decreasing, continuous, and one-to-one. They have the same domain and horizontal asymptote. They have no vertical asymptote or x-intercept. The y-intercepts for each graph are different from the parent graph $y = \left(\frac{1}{3}\right)^x$.

In-Class Example

Example 2
Business A car depreciates or loses value at the rate of 20% per year. If the car originally cost $20,000, the depreciation can be modeled by the equation $y = 20,000(0.8)^t$, where y is the depreciation and t is the time in years.

a. Find the value of the car at the end of 2 years.
   $12,800

b. Graph the depreciation function.

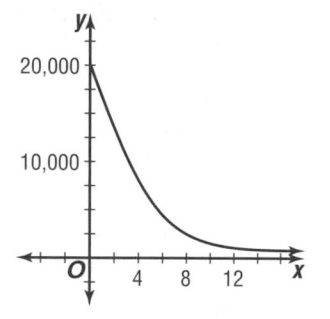

Additional Answers (p. 705)
Graphing Calculator Exploration

6. Any nonzero number raised to the zero power is 1.

7. The graph of $y = b^x$ is decreasing when $0 < b < 1$ because multiplying by a number between 0 and 1 results in a product less than the original number. The graph of $y = b^x$ is increasing when $b > 1$ because multiplying by a number greater than 1 results in a product greater than the original number.

8. There is a horizontal asymptote at $y = 0$ because a power of a positive real number is never 0 or less. As a number between 0 and 1 is raised to greater and greater powers, its value approaches 0. As a number greater than 1 is raised to powers approaching negative infinity, its value approaches 0.

---

b. Graph the exponential functions $y = \left(\frac{1}{5}\right)^x$, $y = 6\left(\frac{1}{5}\right)^x$, and $y = -2\left(\frac{1}{5}\right)^x$ on the same set of axes. Compare and contrast the graphs.

All of the graphs are decreasing, continuous, and one-to-one. They have the same domain and horizontal asymptote. They have no vertical asymptote or x-intercept.

The y-intercepts for each graph are different from the parent graph $y = \left(\frac{1}{5}\right)^x$.

2 **PHYSICS** According to Newton's Law of Cooling, the difference between the temperature of an object and its surroundings decreases in time exponentially. Suppose a certain cup of coffee is 95°C and it is in a room that is 20°C. The cooling for this particular cup can be modeled by the equation $y = 75(0.875)^t$ where y is the temperature difference and t is time in minutes.

a. Find the temperature of the coffee after 15 minutes.

b. Graph the cooling function.

a. $y = 75(0.875)^t$
   $y = 75(0.875)^{15}$ $t = 15$
   $y \approx 10.12003603$

   The difference is about 10.1° C. So the coffee is 95° C − 10.1° C or 84.9° C.

b.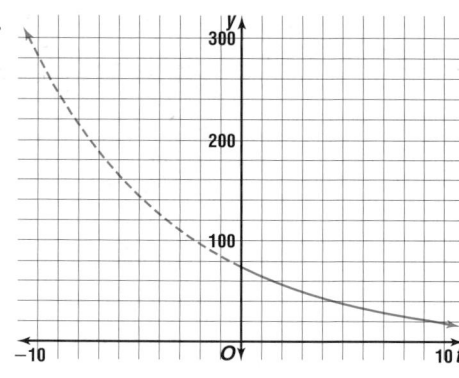

In a situation like cooling when a quantity loses value exponentially over time, the value exhibits what is called **exponential decay.** Many real-world situations involve quantities that increase exponentially over time. For example, the balance in a savings or money market account and the population of people or animals in a region often demonstrate what is called **exponential growth.** As you saw in Example 1, you can use the general form of an exponential function to describe exponential growth or decay. When you know the rate at which the growth or decay is occurring, the following equation may be used.

Exponential Growth or Decay	The equation $N = N_0(1 + r)^t$, where N is the final amount, N_0 is the initial amount, r is the rate of growth or decay per time period, and t is the number of time periods, is used for modeling exponential growth or decay.

Example **3** **ENTOMOLOGY** Refer to the application at the beginning of the lesson. Suppose that a researcher estimates that the initial population of varroa in a colony is 500. They are increasing at a rate of 14% per week. What is the expected population in 22 weeks?

$N = N_0(1 + r)^t$
$N = 500(1 + 0.14)^{22}$ *$N_0 = 500, r = 0.14, t = 22$*
$N \approx 8930.519719$ *Use a calculator.*

There will be about 8931 varroa in the colony in 22 weeks.

The general equation for exponential growth is modified for finding the balance in an account that earns compound interest.

Compound Interest	The compound interest equation is $A = P\left(1 + \dfrac{r}{n}\right)^{nt}$, where P is the principal or initial investment, A is the final amount of the investment, r is the annual interest rate, n is the number of times interest is paid, or compounded each year, and t is the number of years.

Example **4** **FINANCE** Determine the amount of money in a money market account providing an annual rate of 5% compounded daily if Marcus invested $2000 and left it in the account for 7 years.

$A = P\left(1 + \dfrac{r}{n}\right)^{nt}$

$A = 2000\left(1 + \dfrac{0.05}{365}\right)^{365 \cdot 7}$ *$P = 2000, r = 0.05, n = 365, t = 7$*

$A \approx 2838.067067$ *Use a calculator.*

After 7 years, the $2000 investment will have a value of $2838.06.
The value 2838.067067 is rounded to 2838.06 as banks generally round down when they are paying interest.

Graphing exponential inequalities is similar to graphing other inequalities.

Example **5** **Graph $y < 2^x + 1$**

First, graph $y = 2^x + 1$. Since the points on this curve are not in the solution of the inequality, the graph of $y = 2^x + 1$ is shown as a dashed curve.

(continued on the next page)

Teaching Tip For Example 3, point out that the average annual growth rate of 7.5% is applied to an ever-expanding population, therefore, its increase each succeeding year is larger than for the year before.

In-Class Examples

Example 3
Population Dynamics The average growth rate of the population of a city is 7.5% per year and is represented by the formula $y = A(1.075)^x$, where x is the number of years and y is the most recent population of the city. The city's population A is now 22,750 people. What is the expected population in 10 years? **about 46,889 people**

Example 4
Finance How much should Sabrina invest now in a money market account if she wishes to have $9000 in the account at the end of 10 years? The account provides an APR of 6% compounded quarterly. **$4961.36**

Example 5
Graph $y \geq 3^x + 1$.

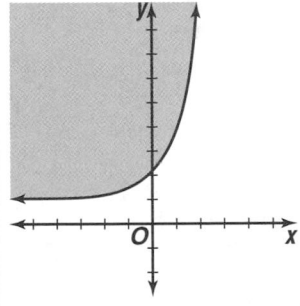

Teaching Tip For In-Class Example 5, point out that the y-intercept of the graph occurs when $x = 0$ or at $y = 3^0 + 1 = 1 + 1 = 2$. There is no x-intercept.

Addressing Individual Needs

Verbal/Linguistic Have students visit some factories or e-mail some factory representatives in order to find out whether the companies' accountants use exponential functions to calculate the depreciation of their machines and other assets used in production. Students should write and present short reports to the class on their findings.

Error Analysis

Watch for students who have difficulty distinguishing power functions from exponential functions.

Prevent by pointing out that $y = x^9$ is a power function, while $y = 9^x$, in which the exponent is a variable, is an exponential function.

Assignment Guide

Core: 11–29 odd, 30–43
Enriched: 10–22 even, 24–43

Additional Answers

1. **Power; in a power function the variable is the base; in an exponential function the variable is the exponent.**

2. **Both graphs are one-to-one, have the domain of all reals, a range of positive reals, a horizontal asymptote of $y = 0$, a y-intercept of $(0, 1)$, and no vertical asymptote. The graph of $y = b^x$ is decreasing when $0 < b < 1$ and increasing when $b > 1$.**

Study Guide and Practice Masters, pp. 199–200

Then, use $(0, 0)$ as a test point to determine which area to shade.

$y < 2^x + 1 \quad \Rightarrow \quad 0 < 2^0 + 1$
$0 < 1 + 1$
$0 < 2$

Since $(0, 0)$ satisfies the inequality, the region that contains $(0, 0)$ should be shaded.

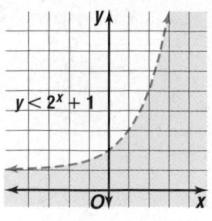

$y < 2^x + 1$

CHECK FOR UNDERSTANDING

Communicating Mathematics

Read and study the lesson to answer each question.   1–4. See margin.

1. **Tell** whether $y = x^4$ is an exponential or a power function. Justify your answer.

2. **Compare and contrast** the graphs of $y = b^x$ when $b > 1$ and when $0 < b < 1$.

3. **Write** about how you can tell whether an exponential function represents exponential growth or exponential decay.

4. **Describe** the differences between the graphs of $y = 4^x$ and $y = 4^x - 3$.

Guided Practice

Graph each exponential function or inequality.   5–7. See margin.

5. $y = 3^x$ 6. $y = 3^{-x}$ 7. $y > 2^x - 4$

8. **Business**   Business owners keep track of the value of their assets for tax purposes. Suppose the value of a computer depreciates at a rate of 25% a year. Determine the value of a laptop computer two years after it has been purchased for $3750.   **$2109.38**

9. **Demographics**   In the 1990 U.S. Census, the population of Los Angeles County was 8,863,052. By 1997, the population had increased to 9,145,219.
   a. Find the yearly growth rate by dividing the average change in population by the 1990 population.   **0.45%**
   b. Assuming the growth rate continues at a similar rate, predict the number of people who will be living in Los Angeles County in 2010.   **9,695,766**

EXERCISES

Practice

Graph each exponential function or inequality.   10–18. See Answer Appendix.

 A

10. $y = 2^x$ 11. $y = -2^x$ 12. $y = 2^{-x}$

13. $y = 2^{x + 3}$ 14. $y = -2^{x + 3}$ 15. $y > -4^x + 2$

16. $y = \left(\frac{1}{5}\right)^x$ 17. $y \leq \left(\frac{1}{2}\right)^x$ 18. $y < 2^{x - 4}$

Match each equation to its graph.

B

19. $y = 0.01^x$ **B**   A   B   C

20. $y = 5^{-x}$ **C**

21. $y = 7^{1 - x}$ **A**

Additional Answers

3. **If the base is greater than 1, the equation represents exponential growth. If the base is between 0 and 1, the equation represents exponential decay.**

4. **The graphs of $y = 4^x$ and $y = 4^x - 3$ are the same except the graph of $y = 4^x - 3$ is shifted down three units from the graph of $y = 4^x$.**

5.
$y = 3^x$

6.
$y = 3^{-x}$

Graphing Calculator

22. Graph the functions $y = 5^x$, $y = -5^x$, $y = 5^{-x}$, $y = 5^x + 2$, $y = 5^x - 2$, and $y = 10^x$ on the same screen. **a–c. See margin.**
   a. Compare $y = -5^x$, $y = 5^{-x}$ to the parent graph $y = 5^x$. Describe the transformations of the functions.
   b. What transformation occurs with the graphs of $y = 5^x + 2$ and $y = 5^x - 2$?
   c. The function $y = 10^x$ can be expressed as $y = (2 \cdot 5)^x$. Compare this function to $y = 5^x$. Is the graph of $y = 5^{2x}$ the same as $y = 10^x$? Explain.

C

23. Without graphing, describe how each pair of graphs is related. Then use a graphing calculator to check your descriptions. **a–d. See margin.**
   a. $y = 6^x$ and $y - 6^x + 4$
   b. $y = -3^x$ and $y = 3^x$
   c. $y = 7^x$ and $y = 7^{-x}$
   d. $y = 2^x$ and $y = \left(\dfrac{1}{2}\right)^x$

Applications and Problem Solving

24. **Employment** Average national teachers' salaries can be modeled using the equation $y = 9.25(1.06)^n$, where y is the salary in thousands of dollars and n is the number of years since 1970.
   a. Graph the function. **See margin.**
   b. Using this model, what can a teacher expect to have as a salary in the year 2020? **$170,400**

25. **Aviation** When kerosene is purified to make jet fuel, pollutants are removed by passing the kerosene through a special clay filter. Suppose a filter is fitted in a pipe so that 15% of the impurities are removed for every foot that the kerosene travels.
   a. Write an exponential function to model the percent of impurity left after the kerosene travels x feet. $y = (0.85)^x$
   b. Graph the function. **See margin.**
   c. About what percent of the impurity remains after the kerosene travels 12 feet? **14%**
   d. Will the impurities ever be completely removed? Explain. **See margin.**

26. **Demographics** Find the projected population of each location in 2015.
   a. In Honolulu, Hawaii, the population was 836,231 in 1990. The average yearly rate of growth is 0.7%. **995,550**

26b. 2,065,690
   b. The population in Kings County, New York has demonstrated an average decrease of 0.45% over several years. The population in 1997 was 2,240,384.
   c. Janesville, Wisconsin had a population of 139,420 in 1980 and 139,510 in 1990. **139,735**
   d. The population in Cedar Rapids, Iowa was 169,775 in the 1980 U.S. Census and 168,767 in the 1990 U.S. Census. **166,280**

27. **Biology** Scientists who study Atlantic salmon have found that the oxygen consumption of a yearling salmon O is given by the function $O - 100\left(3^{\frac{3s}{5}}\right)$, where s is the speed that the fish is traveling in feet per second.

   a. What is the oxygen consumption of a fish that is traveling at 5 feet per second? **2700 units**
   b. If a fish has traveled 4.2 miles in an hour, what is its oxygen consumption? **5800 units**

Additional Answers (pp. 708–709)

7.

22a. The graph of $y = -5^x$ is a reflection of $y = 5^x$ across the x-axis. The graph of $y = 5^{-x}$ is a reflection of $y = 5^x$ across the y-axis.

22b. The graph of $y = 5^x + 2$ is shifted up two units, while the graph of $y = 5^x - 2$ is shifted down two units.

Teaching Tip For Exercise 22, point out that in order to graph $y = 5^x$ students should enter [Y=] 5 [^] [X,T,θ,n] [GRAPH]. They should also choose an appropriate viewing window.

Additional Answers

22c. The graph of $y = 10^x$ increases more quickly than the graph of $y = 5^x$. The graphs are not the same because $5^{2x} \neq 10^x$.

23a. The graph of $y = 6^x + 4$ is shifted up four units from the graph of $y = 6^x$.

23b. The graph of $y = -3^x$ is a reflection of the graph of $y = 3^x$ across the x-axis.

23c. The graph of $y = 7^{-x}$ is a reflection of the graph of $y = 7^x$ across the y-axis.

23d. The graph of $y = \left(\dfrac{1}{2}\right)^x$ is a reflection of the graph of $y = 2^x$ across the y-axis.

24a.

25b.

25d. No; the graph has an asymptote at $y = 0$, so the percent of impurities, y, will never reach 0.

28c. 30-year: $183,578; 20-year: $108,524

28. Finance Bankers call a series of payments made at equal intervals an *annuity*. The present value of an annuity P_n is the sum of the present values of all of the periodic payments P. In other words, a lump-sum investment of P_n dollars now will provide payments of P dollars for n periods. The formula for the present value is $P_n = P\left[\dfrac{1 - (1 + i)^{-n}}{i}\right]$, where i is the interest rate for the period.

a. Use the present value formula to find the monthly payment you would pay on a home mortgage if the present value is $121,000, the annual interest rate is 7.5%, and payments will be made for 30 years. (*Hint*: First find the interest rate for the period.) **$846.05**

b. How much is the monthly mortgage payment if the borrowers choose a loan with a 20-year term and an interest rate of 7.25%? **$956.35**

c. How much will be paid in interest over the life of each mortgage?

d. Explain why a borrower might choose each mortgage. **See margin.**

29. Finance The future value of an annuity F_n is the sum of all of the periodic payments P and all of the accumulated interest. The formula for the future value is $F_n = P\left[\dfrac{(1 + i)^n - 1}{i}\right]$, where i is the interest rate for the period.

a. When Connie Hockman began her first job at the age of 22, she started saving for her retirement. Each year she places $4000 in an account that will earn an average 4.75% annual interest until she retires at 65. How much will be in the account when she retires? **$535,215.92**

b. If Ms. Hockman had invested in an account that earns an average of 5.25% annual interest, how much more would her account be worth? **$76,376.20**

30. Critical Thinking Explain when the exponential function $y = a^x$ is undefined for $a < 0$. **See margin.**

31. Investments The number of times that interest is compounded has a dramatic effect on the total interest earned on an investment.

31a. $50; $50.63; $50.94; $51.16; $51.26

a. How much interest would you earn in one year on an $1000 investment earning 5% interest if the interest is compounded once, twice, four times, twelve times, or 365 times in the year?

b. If you are making an investment that you will leave in an account for one year, which account should you choose to get the highest return? **Money Market Savings**

Account	Rate	Compounded
Statement Savings	5.1%	Yearly
Money Market Savings	5.05%	Monthly
Super Saver	5%	Daily

c. Suppose you are a bank manager determining rates on savings accounts. If the account with interest compounded annually offers 5% interest, what interest rate should be offered on an account with interest compounded daily in order for the interest earned on equal investments to be the same? **4.88%**

32. Simplify $4x^2(4x)^{-2}$. *(Lesson 11-1)* $\dfrac{1}{4}$

33. Write the rectangular equation $y = 15$ in polar form. *(Lesson 9-4)* $15 = r\sin\theta$

34. 3, no because the inner product does not equal 0.

34. Find the inner product of vectors $\langle -3, 9\rangle$ and $\langle 2, 1\rangle$. Are the vectors perpendicular? Explain. *(Lesson 8-4)*

35. See margin.

35. Find $\dfrac{1}{3}\left(\cos\dfrac{7\pi}{8} + i\sin\dfrac{7\pi}{8}\right) \cdot 3\sqrt{3}\left[\cos\left(-\dfrac{\pi}{4}\right) + i\sin\left(-\dfrac{\pi}{4}\right)\right]$. Then express the product in rectangular form with a and b to the nearest hundredth. *(Lesson 9-7)*

36. Sports Suppose a baseball player popped a baseball straight up at an initial velocity v_0 of 72 feet per second. Its distance s above the ground after t seconds is described by $s = v_0 t - 16t^2 + 4$. Find the maximum height of the ball. *(Lesson 10-5)*   **85 feet**

37. Art Leigh Ann is drawing a sketch of a village. She wants the town square to be placed midway between the library and the fire station. Suppose the ordered pair for the library is $(-7, 6)$ and the ordered pair for the fire station is $(12, 8)$. *(Lesson 10-1)*

a. Draw a graph that represents this situation.   **See margin.**

b. Determine the coordinates indicating where the town square should be placed.   **(2.5, 7)**

38. See Answer Appendix.

38. Verify that $\sin^4 A + \cos^2 A = \cos^4 A + \sin^2 A$ is an identity. *(Lesson 7-2)*

39. Mechanics A circular saw 18.4 centimeters in diameter rotates at 2400 revolutions per second. What is the linear velocity at which a saw tooth strikes the cutting surface in centimeters per second? *(Lesson 6-2)*   **139,000 cm/s**

40. Travel Martina went to Acapulco, Mexico, on a vacation with her parents. One of the sights they visited was a cliff-diving exhibition into the waters of the Gulf of Mexico. Martina stood at a lookout site on top of a 200-foot cliff. A team of medical experts was in a boat below in case of an accident. The angle of depression to the boat from the top of the cliff was $21°$. How far is the boat from the base of the cliff? *(Lesson 5-4)*   **about 521 feet**

41. Sample answer:
$y = 948.4x + 4960.6$

41. Salary Diane has had a part time job as a Home Chef demonstrator for 9 years. Her yearly income is listed in the table. Write a model that relates the income as a function of the number of years since 1990. *(Lesson 4-8)*

Year	1990	1991	1992	1993	1994	1995	1996	1997	1998
Income ($)	4012	6250	7391	8102	8993	9714	10,536	11,362	12,429

42. Use the parent graph $f(x) = \dfrac{1}{x}$ to graph $f(x) = \dfrac{1}{x+3}$. Describe the transformation(s) that have taken place. Identify the new locations of the asymptotes. *(Lesson 3-7)*   **See margin.**

43. SAT/ACT Practice In the figure, $\overline{AB}$ is the diameter of the smaller circle, and $\overline{AC}$ is the diameter of the larger circle. If the distance from B to C is 16 inches, then the circumference of the larger circle is approximately how many inches greater than the circumference of the smaller circle?   **E**

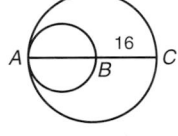

A 8   B 16   C 25   D 35   E 50

Extra Practice See p. A46.

Extra Credit

Suppose that gas leaks from a balloon at a rate such that after one hour the size of the balloon is reduced by 30%, and the leakage continues at this rate. Write a formula that models this phenomenon if the original volume of the balloon was 500 cubic inches. Also, find the balloon's volume after 3 hours.
$V = 500\,(0.7)^N$; **171.5 in³**

---

4 ASSESS

Open-Ended Assessment
Modeling Have students write an exponential function that would model how much money Sarah will have at the end of 20 years on an initial investment of $5000 at 14% compounded annually.
$y = \$5000(1.14)^{20}$ How much money will she have at the end of 20 years?   **$68,717.45** How much interest will she have earned?
$63,717.45

   Chapter 11, Quiz A (Lessons 11-1 and 11-2) is available in the *Assessment and Evaluation Masters,* p. 277.

Additional Answers
35. $\sqrt{3}\left(\cos\dfrac{5\pi}{8} + i\sin\dfrac{5\pi}{8}\right)$;
$-0.66 + 1.60\,i$

37a.

42. The parent graph is translated 3 units left. The vertical asymptote is now $x = -3$. The horizontal asymptote, $y = 0$, is unchanged.

Enrichment Masters, p. 122

1 FOCUS

5-Minute Check
Lesson 11-2

Graph each exponential function or inequality.

1. $y = 2^{x+1}$

2. $y \le 2^x - 1$

3. $y < 3^x + 1$

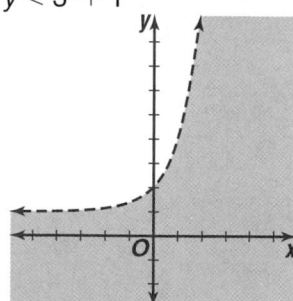

Without graphing, describe how each pair of graphs is related.

4. $y = 4^x$ and $y = 4^x - 3$
   Sample answer: $y = 4^x - 3$ is the result of a slide of $y = 4^x$ down 3 units.

5. $y = 3^x$ and $y = 3^{-x}$
   Sample answer: $y = 3^x$ and $y = 3^{-x}$ reflect or are symmetric about the *y*-axis.

The Number e

OBJECTIVES
- Use the exponential function $y = e^x$.

MEDICINE Swiss entomologist Dr. Paul Mueller was awarded the Nobel Prize in medicine in 1948 for his work with the pesticide DDT. Dr. Mueller discovered that DDT is effective against insects that destroy agricultural crops, mosquitoes that transmit malaria and yellow fever, as well as lice that carry typhus.

It was later discovered that DDT presented a risk to humans. Effective January 1, 1973, the United States Environmental Protection Agency banned all uses of DDT. More than 1.0×10^{10} kilograms of DDT had been used in the U.S. before the ban. How much will remain in the environment in 2005? *This problem will be solved in Example 1.*

DDT degrades into harmless materials over time. To find the amount of a substance that decays exponentially remaining after a certain amount of time, you can use the following formula for exponential growth or decay, which involves the number *e*.

Exponential Growth or Decay (in terms of *e*)	$N = N_0 e^{kt}$, where N is the final amount, N_0 is the initial amount, k is a constant and t is time.

The number *e* in the formula is not a variable. It is a special irrational number. This number is the sum of the infinite series shown below.

$$e = 1 + \frac{1}{1} + \frac{1}{1 \cdot 2} + \frac{1}{1 \cdot 2 \cdot 3} + \frac{1}{1 \cdot 2 \cdot 3 \cdot 4} + \cdots + \frac{1}{1 \cdot 2 \cdot 3 \cdot \cdots \cdot n} + \cdots$$

The following computation for *e* is correct to three decimal places.

$$e = 1 + \frac{1}{1} + \frac{1}{1 \cdot 2} + \frac{1}{1 \cdot 2 \cdot 3} + \frac{1}{1 \cdot 2 \cdot 3 \cdot 4} + \frac{1}{1 \cdot 2 \cdot 3 \cdot 4 \cdot 5} +$$

$$\frac{1}{1 \cdot 2 \cdot 3 \cdot 4 \cdot 5 \cdot 6} + \frac{1}{1 \cdot 2 \cdot 3 \cdot 4 \cdot 5 \cdot 6 \cdot 7}$$

$$= 1 + 1 + \frac{1}{2} + \frac{1}{6} + \frac{1}{24} + \frac{1}{120} + \frac{1}{720} + \frac{1}{5040}$$

$$= 1 + 1 + 0.5 + 0.16667 + 0.04167 + 0.00833 + 0.00139 + 0.000198$$

$$= 2.718$$

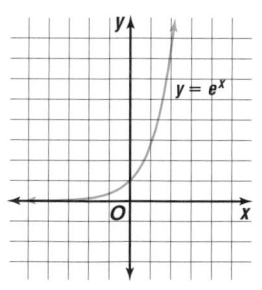

$y = e^x$

The function $y = e^x$ is one of the most important exponential functions. The graph of $y = e^x$ is shown at the right.

712 *Chapter 11 Exponential and Logarithmic Functions*

Resource Manager

 Reproducible Masters
- *Study Guide and Practice,* pp. 201–202
- *Enrichment,* p. 123

 Transparencies
- 5-Minute Check, 11-3
- Teaching, 11-3

Example

1 **MEDICINE** Refer to the application at the beginning of the lesson. Assume that there were 1.0×10^9 kilograms of DDT in the environment in 1973 and that for DDT, $k = -0.0211$.

   a. Write a function to model the amount of DDT remaining in the environment.

   b. Find the amount of DDT that will be in the environment in 2005.

   c. Graph the function and use the graph to verify your answer in part b.

 Graphing Calculator Tip

To graph an equation or evaluate an expression involving e raised to a power, use the second function of ln on a calculator.

Example 1 is an example of chemical decay.

a. $y = ne^{kt}$
$y = (1 \times 10^9)e^{-0.0211t}$
$y = (10^9)e^{-0.0211t}$

b. In 2005, it will have been 2005 − 1973 or 32 years since DDT was banned. Thus, $t = 32$.
$y = (10^9)e^{-0.0211t}$
$y = (10^9)e^{-0.0211(32)}$ $t = 32$
$y \approx (10^9)0.5090545995$

c. Use a graphing calculator to graph the function.

[−0, 200] scl:10, [0, (1 × 10⁹] scl:1 × 10⁸

If there were 1×10^9 kilograms of DDT in the environment in 1973, there will be about $0.51(1 \times 10^9)$ or 5.1×10^8 kilograms remaining in 2005.

Some banks offer accounts that compound the interest continuously. The formula for finding continuously compounded interest is different from the one used for interest that is compounded a specific number of times each year.

Continuously Compounded Interest	The equation $A = Pe^{rt}$, where P is the initial amount, A is the final amount, r is the annual interest rate, and t is time in years, is used for calculating interest that is compounded continuously.

Example

2 **FINANCE** Compare the balance after 25 years of a $10,000 investment earning 6.75% interest compounded continuously to the same investment compounded semiannually.

In both cases, $P = 10{,}000$, $r = 0.0675$, $t = 25$. When the interest is compounded semiannually, $n = 2$. Use a calculator to evaluate each expression.

Continuously

$A = Pe^{rt}$

$A = 10{,}000(e)^{(0.0675 \cdot 25)}$

$A = 54{,}059.49$

Semiannually

$A = P\left(1 + \dfrac{r}{n}\right)^{nt}$

$A = 10{,}000\left(1 + \dfrac{0.0675}{2}\right)^{2 \cdot 25}$

$A = 52{,}574.62$

The same principal invested over the same amount of time yields $54,059.49 if compounded continuously and $52,574.62 when compounded twice a year. You would earn $54059.49 − $52574.62 = $1484.87 more by choosing the account that compounds continuously.

Lesson 11-3 The Number e **713**

Teaching Tip For Example 1, remind students that to show e as a decimal on the calculator, they should press [2nd] [eˣ] 1 [ENTER].

Teaching Tip For Example 2, remind students that a useful decimal approximation of e to remember is 2.71828.

Motivating the Lesson

Hands-On Activity Ask students: *If a machine in a factory depreciates at a rate of 15% of its value each year, how much of its value remains at the end of 4 years?* **The machine retains about 52% of its value at the end of 4 years.** Have students work at the chalkboard and use calculators to come up with an exponential function to model this situation. **$y = (0.85)^x$, where x = the number of years and y represents the value for x number of years.**

Teaching Tip The number e, like π, is *transcendental*. That is, it cannot occur as the root of any polynomial function with rational coefficients.

2 TEACH

In-Class Examples

Example 1

Physics According to Newton, a beaker of liquid cools exponentially when removed from a source of heat. Assume that the initial temperature T_i is 90°F and that $k = 0.275$.

  a. Write a function to model the rate at which the liquid cools. **$T = T_i e^{-kt}$**

  b. Find the temperature T of the liquid after 4 minutes (t). **about 30°F**

  c. Graph the function and use the graph to verify your answer in part **b.**

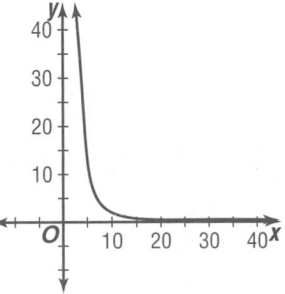

Example 2

Finance Compare the balance after 30 years of a $15,000 investment earning 12% interest compounded continuously to the same investment compounded quarterly. **5.44% more**

Error Analysis

Watch for students who have difficulty recognizing exponential functions.

Prevent by reminding them that the exponent always contains the independent variable of the exponential function.

Assignment Guide

Core: 9–13 odd, 14–28
Enriched: 8–28
All: Mid-Chapter Quiz, 1–10

Additional Answers

2. If *k* is positive, the equation models growth. If *k* is negative, the equation models decay.

3. Amount in an account with a beginning balance of $3000 and interest compounded continuously at an annual rate of 5.5%.

5. Sample answer: continuously compounded interest is a continuous function, but interest compounded monthly is a discrete function.

Study Guide and Practice Masters, pp. 201–202

NAME _____ DATE _____ PERIOD _____

Study Guide

The Number e

The number *e* is a special irrational number with an approximate value of 2.718 to three decimal places. The formula for exponential growth or decay is $N = N_0 e^{kt}$, where N is the final amount, N_0 is the initial amount, k is a constant, and t is time. The equation $A = Pe^{rt}$, where P is the initial amount, A is the final amount, r is the annual interest rate, and t is time in years, is used for calculating interest that is compounded continuously.

Example 1 *Demographics* The population of Dubuque, Iowa, declined at a rate of 0.4% between 1997 and 1998. In 1998, the population was 87,806.
a. Write a function to model the population in 1998.
b. Suppose that the rate of decline remains steady at 0.4%. Find the projected population of Dubuque in 2010.

a. $y = ne^{kt}$
$y = 87,806e^{-0.004}$ $n = 87,806; k = -0.004$

b. In 2010, it will have been 2010 − 1998, or 12 years, since the initial population figure. Thus, $t = 12$.

$y = 87,806e^{-0.004t}$
$y = 87,806e^{-0.004(12)}$ $t = 12$
$y \approx 83690.86531$ *Use a calculator.*

Given a population of 87,806 in 1998 and a steady rate of decline of 0.4%, the population of Dubuque, Iowa, will be approximately 83,691 in 2010.

Example 2 *Finance* Compare the balance after 10 years of a $5000 investment earning 8.5% interest compounded continuously to the same investment compounded quarterly.

In both cases, $P = 5000$, $r = 0.085$, and $t = 10$. When the interest is compounded quarterly, $n = 4$. Use a calculator to evaluate each expression.

Continuously	Quarterly
$A = Pe^{rt}$	$A = P\left(1 + \frac{r}{n}\right)^{nt}$
$A = 5000e^{0.085(10)}$	$A = 5000\left(1 + \frac{0.085}{4}\right)^{4 \cdot 10}$
$A \approx 11,698.23$	$A \approx 11,594.52$

You would earn $11,698.23 − $11,594.52 = $103.71 more by choosing the account that compounds continuously.

© Glencoe/McGraw-Hill   **201**   *Advanced Mathematical Concepts*

Communicating Mathematics

Read and study the lesson to answer each question.   **2, 3, 5. See margin.**

1. **Tell** which equation is represented by the graph.   **c**

  a. $y = e^x$ **b.** $y = e^{-x}$
  c. $y = -e^x$ **d.** $y = -e^{-x}$

$[-10, 10]$ scl:1 by $[-10, 10]$ scl:1

2. **Describe** the value of *k* when the equation $N = N_0 e^{kt}$ represents exponential growth and when it represents exponential decay.

3. **Describe** a situation that could be modeled by the equation $A = 3000e^{0.055t}$.

4. **State** the domain and range of the function $f(x) = e^x$.   **reals, positive reals**

5. *Math Journal* **Write** a sentence or two to explain the difference between interest compounded continuously and interest compounded monthly.

Guided Practice

6. **Demographics** Bakersfield, California was founded in 1859 when Colonel Thomas Baker planted ten acres of alfalfa for travelers going from Visalia to Los Angeles to feed their animals. The city's population can be modeled by the equation $y = 33,430e^{0.0397t}$, where *t* is the number of years since 1950.

  a. Has Bakersfield experienced growth or decline in population?   **growth**
  b. What was Bakersfield's population in 1950?   **33,430**
  c. Find the projected population of Bakersfield in 2010.   **361,931**

7. **Financial Planning** The Kwans are saving for their daughter's college education. If they deposit $12,000 in an account bearing 6.4% interest compounded continuously, how much will be in the account when Ann goes to college in 12 years?   **$25,865.41**

Applications and Problem Solving

A ▶ **8.** **Psychology** Without further study, as time passes you forget things you have learned. The Ebbinghaus model of human memory gives the percent *p* of acquired knowledge that a person retains after *t* weeks. The formula is $p = (100 - a)e^{-bt} + a$, where *a* and *b* vary from one person to another. If $a = 18$ and $b = 0.6$ for a certain student, how much information will the student retain two weeks after learning a new topic?   **43%**

B ▶ **9.** **Physics** Newton's Law of Cooling expresses the relationship between the temperature of a cooling object *y* and the time *t* elapsed since cooling began. This relationship is given by $y = ae^{-kt} + c$, where *c* is the temperature of the medium surrounding the cooling object, *a* is the difference between the initial temperature of the object and the surrounding temperature, and *k* is a constant related to the cooling object.

  a. The initial temperature of a liquid is 160°F. When it is removed from the heat, the temperature in the room is 76°F. For this object, $k = 0.23$. Find the temperature of the liquid after 15 minutes.   **78.7°F**

9b. too cold; After 5 minutes, his coffee will be about 90°F.

  b. Alex likes his coffee at a temperature of 135°. If he pours a cup of 170°F coffee in a 72°F room and waits 5 minutes before drinking, will his coffee be too hot or too cold? Explain. For Alex's cup, $k = 0.34$.

714 *Chapter 11* *Exponential and Logarithmic Functions*   **www.amc.glencoe.com/self_check_quiz**

Addressing Individual Needs

Interpersonal Present the formula for compound interest and have groups of students calculate the value of an account with a principal of $1000 after three years at 6% interest when compounded annually, quarterly, monthly, daily, and hourly. Note how the sequence of values appears to converge to a limit. This limiting value is Pe^{rt}. In this example, $Pe^{rt} = \$1197.22$.

10. **Civil Engineering** Suspension bridges can span distances far longer than any other kind of bridge. The roadway of a suspension bridge is suspended from huge cables. When a flexible cable is suspended between two points, it forms a *catenary curve*.

   a. Use a graphing calculator to graph the catenary
   $y = \dfrac{e^x + e^{-x}}{2}$. **See margin.**

   b. What kind of symmetry is displayed by the graph?
   symmetric about *y*-axis

11. **Banking** If your bank account earns interest that is compounded more than one time per year, the effective annual yield *E* is the interest rate that would give the same amount of interest earnings if the interest were compounded once per year. To find the effective annual yield, divide the interest earned by the principal.

   a. Copy and complete the table to find the effective annual yield for each account if the principal is $1000, the annual interest rate is 8%, and the term is one year.

Interest Compounded	Interest	Effective Annual Yield
Annually	$80.00	8%
Semi-annually	$81.60	8.16%
Quarterly	$82.43	8.243%
Monthly	$83.00	8.3%
Daily	$83.28	8.328%
Continuously	$83.29	8.329%

11b. continuously

b. Which type of compounding provides the greatest effective annual yield?

11c.

$E = \left(1 + \dfrac{r}{n}\right)^n - 1$

c. If *P* represents the principal and *A* is the total value of the investment, the value of an investment is $A = P(1 + E)$. Find a formula for the effective annual yield for an account with interest compounded *n* times per year.

d. Write a formula for the effective annual yield of an account with interest compounded continuously. $E = e^r - 1$

12. **Sociology** Sociologists have found that information spreads among a population at an exponential rate. Suppose that the function $y = 525(1 - e^{-0.038t})$ models the number of people in a town of 525 people who have heard news within *t* hours of its distribution.

   a. How many people will have heard about the opening of a new grocery store within 24 hours of the announcement? **314 people**

   b. Graph the function on a graphing calculator. When will 90% of the people have heard about the grocery store opening? **after about 61 h**

 13. **Customer Service** The service-time distribution describes the probability *P* that the service time of the customer will be no more than *t* hours. If *m* is the mean number of customers serviced in an hour, then $P = 1 - e^{-mt}$.

   a. Suppose a computer technical support representative can answer calls from 6 customers in an hour. What is the probability that a customer will be on hold less than 30 minutes? **95%**

   b. A credit card customer service department averages 34 calls per hour. Use a graphing calculator to determine the amount of time after which it is 50% likely that a customer has been served? **about 1.2 min**

Teaching Tip In Exercise 14, the significance of the expression $\left(\dfrac{2x+1}{2x-1}\right)^x$ is that it approximates e much faster than the better-known expression $\left(1+\dfrac{1}{x}\right)^x$.

Additional Answers

15b.

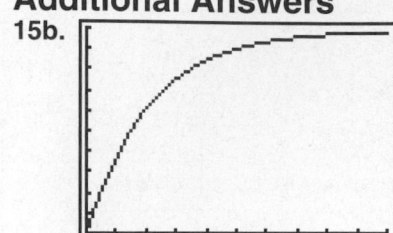

15c. Sample answer: the probability that a person who is going to respond has responded approaches 100% as t approaches infinity. New ads may be introduced after a high percentage of those who will respond have responded. The graph appears to level off after about 50 days. So, new ads can be introduced after an ad has run for about 50 days.

16c. c shifts the graph to the right or left.

18. $x\sqrt[15]{x^3 y^3 z}$

19. $6x^2 + 12xy + 6y^2 + \sqrt{2}x - \sqrt{2}y = 0$

20. $\sqrt{26}\,(\cos 3.34 + i \sin 3.34)$

14a.
2.720551414;
2.718304481;
2.718282055

14b. 2 decimal places; 4 decimal places; 6 decimal places

14. **Critical Thinking** In 1997, inventor and amateur mathematician Harlan Brothers discovered some simple algebraic expressions that approximate e.

   a. Use Brothers' expression $\left(\dfrac{2x+1}{2x-1}\right)^x$ for $x = 10$, $x = 100$, and $x = 1000$.

   b. Compare each approximation to the value of e stored in a calculator. If each digit of the calculator value is correct, how accurate is each approximation?

   c. Is the approximation always greater than e, always less than e, or sometimes greater than and sometimes less than e? **always greater**

15. **Marketing**   The probability P that a person has responded to an advertisement can be modeled by the exponential equation $P = 1 - e^{-0.047t}$ where t is the number of days since the advertisement began to appear in the media.

   a. What are the probabilities that a person has responded after 5 days, 20 days, and 90 days? **20.9%; 60.9%; 98.5%**

   b. Use a graphing calculator to graph the function. Use the graph to find when the probability that an individual has responded to the advertisement is 75%. **About 29 days; see margin for graph.**

   c. If you were planning a marketing campaign, how would you use this model to plan the introduction of new advertisements? **See margin.**

16. **Critical Thinking**   Consider $f(x) = \dfrac{e^x}{e^x + c}$ if c is a constant greater than 0.

   a. What is the domain of the function? **all reals**

   b. What is the range of the function? **$0 < f(x) < 1$**

   c. How does the value of c affect the graph? **See margin.**

Mixed Review

17. **Finance**   A *sinking fund* is a fund into which regular payments are made in order to pay off a debt when it is due. The Gallagher Construction Company foresees the need to buy a new cement truck in four years. At that time, the truck will probably cost \$120,000. The firm sets up a sinking fund in order to accumulate the money. They will pay semiannual payments into a fund with an APR of 7%.

   Use the formula for the future value of an annuity $F_n = P\left[\dfrac{(1+i)^n - 1}{i}\right]$, where F_n is the future value of the annuity, P is the payment amount, i is the interest rate for the period, and n is the number of payments, to find the payment amount. *(Lesson 11-2)* **\$13,257.20**

18. Express $x^{\frac{8}{5}} y^{\frac{3}{5}} z^{\frac{1}{5}}$ using radicals. *(Lesson 11-1)* **See margin.**

19. **Communications**   A satellite dish tracks a satellite directly overhead. Suppose the equation $y = 6x^2$ models the shape of the dish when it is oriented in this position. Later in the day, the dish is observed to have rotated approximately 45°. Find an equation that models the new orientation of the dish. *(Lesson 10-7)* **See margin.**

20. Express $-5 - i$ in polar form. *(Lesson 9-6)* **See margin.**

21. **Physics**   Tony is pushing a cart weighing 150 pounds up a ramp 10 feet long at an incline of 28°. Find the work done to push the cart the length of the ramp. Assume that friction is not a factor. *(Lesson 8-5)* **704.2 ft-lb**

22. **Safety**   A model relating the average number of crimes reported at a shopping mall as a function of the number of years since 1991 is $y = -1.5x^2 + 13.3x + 19.4$. According to this model, what is the number of crimes for the year 2001? *(Lesson 4-8)* **2.4**

Extra Practice   See p. A46.

23. Solve $\sqrt{2x + 3} = 4$. *(Lesson 4-7)* $\dfrac{13}{2}$

24. Solve $|3x + 2| \le 6$. *(Lesson 3-3)* $\left\{ x \,\middle|\, -\dfrac{8}{3} \le x \le \dfrac{4}{3} \right\}$

25. $J'(-9, -6)$, $K'(-6, 18)$, $L'(6, 15)$, $M'(9, -3)$; the dilated image has sides that are 3 times the length of the original figure.

25. Quadrilateral *JKLM* has vertices at $J(-3, -2)$, $K(-2, 6)$, $L(2, 5)$ and $M(3, -1)$. Find the coordinates of the dilated quadrilateral $J'K'L'M'$ for a scale factor of 3. Describe the dilation. *(Lesson 2-4)*

26. Find the values of *x* and *y* for which $\begin{bmatrix} 4x + y \\ x \end{bmatrix} = \begin{bmatrix} 6 \\ 2y - 12 \end{bmatrix}$ is true. *(Lesson 2-3)* **(0, 6)**

27. State the domain and range of the relation $\{(2, 7), (-4, 5), (5, 7)\}$. Is this relation a function? *(Lesson 1-1)* **{−4, 2, 5}; {5, 7}; yes**

28. **SAT/ACT Practice** An artist wants to shade exactly 18 of the 36 smallest triangles in the pattern, including the one shown. If no two shaded triangles can have a side in common, which of the triangles indicated must *not* be shaded? **D**

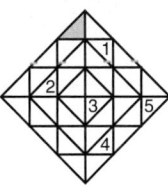

A 1 **B** 2 **C** 3 **D** 4 **E** 5

MID-CHAPTER QUIZ

Evaluate each expression. *(Lesson 11-1)*

1. $64^{\frac{1}{2}}$ **8**

2. $\left(\sqrt[3]{343}\right)^{-2}$ $\dfrac{1}{49}$

3. Simplify $\left(\dfrac{8x^3 y^{-6}}{27 w^6 z^{-9}}\right)^{\frac{1}{3}}$ $\dfrac{2}{3}w^{-2}xy^{-2}z^3$ or $\dfrac{2xz^3}{3w^2 y^2}$

4. Express $\sqrt{a^6 b^3}$ using rational exponents. *(Lesson 11-1)* $|a|^3 b^{\frac{3}{2}}$

5. Express $(125a^2 b^3)^{\frac{1}{3}}$ using radicals. *(Lesson 11-1)* $5b\sqrt[3]{a^2}$

6. **Architecture** A soap bubble will enclose the maximum space with a minimum amount of surface material. Architects have used this principle to create buildings that enclose a great amount of space with a small amount of building material. If a soap bubble has a surface area of *A*, then its volume *V* is given by the equation $V = 0.094\sqrt{A^3}$. Find the surface area of a bubble with a volume of 1.75×10^2 cubic millimeters. *(Lesson 11-1)* **1.51×10^2 mm²**

7. **Demographics** In 1990, the population of Houston, Texas was 1,637,859. In 1998, the population was 1,786,691. Predict the population of Houston in 2014. *(Lesson 11-2)* **2,126,157**

8. **Finance** Determine the amount of money in a money market account at an annual rate of 5.2% compounded quarterly if Mary invested $3500 and left it in the account for $3\frac{1}{2}$ years. *(Lesson 11-2)* **$4193.73**

9. **Forestry** The yield in millions of cubic feet *y* of trees per acre is given by $y = 6.7e^{\frac{-48.1}{t}}$ for a forest that is *t* years old.

   a. Find the yield after 15 years. **271,292 ft³**

   b. Find the yield after 50 years. *(Lesson 11-3)* **2,560,257 ft³**

10. **Biology** It has been observed that the rate of growth of a population of organisms will increase until the population is half its maximum and then the rate will decrease. If *M* is the maximum population and *b* and *c* are constants determined by the type of organism, the population *n* after *t* years is given by $n = \dfrac{M}{1 + be^{-ct}}$. A certain organism yields the values of $M = 200$, $b = 20$, and $c = 0.35$. What is the population after 2, 15, and 60 years? *(Lesson 11-3)* **18.3; 181; 200**

Extra Credit

Sharon wishes to invest enough money now in order to have $100,000 in twenty years. How much less can she invest at 10% interest in an account in which the interest rate is compounded continuously rather than in one at the same rate that is compounded annually? **$1330.83**

Logarithmic Functions

OBJECTIVES
- Evaluate expressions involving logarithms.
- Solve equations and inequalities involving logarithms.
- Graph logarithmic functions and inequalities.

 CHEMISTRY Radioactive materials decay, or change to a non-radioactive material, in a predictable way. Archeologists use the decaying property of Carbon-14 in dating artifacts like dinosaur bones. Geologists use Thorium-230 in determining the age of rock formations, and medical researchers conduct tests using Arsenic-74.

A radioactive material's half-life is the time it takes for half of a given amount of the material to decay and can range from less than a second to billions of years. The half-life for various elements is shown in the table at the right.

Element	Half-Life
Arsenic-74	17.5 days
Carbon-14	5730 years
Polonium-194	0.5 second
Thorium-230	80,000 years
Thorium-232	14 billion years
Thorium-234	25 days

How long would it take for 256,000 grams of Thorium-234 to decay to 1000 grams? Radioactive decay can be modeled by the equation $N = N_0 \left(\frac{1}{2}\right)^t$ where N is the final amount of a substance, N_0 is the initial amount, and t represents the number of half-lifes. If you want to find the number of 25-day half-lifes that will pass, you would have to use an inverse function. *This problem will be solved in Example 4.*

Since the graphs of exponential functions pass the horizontal line test, their inverses are also functions. As with inverses of other functions, you can find the inverse of an exponential function by interchanging the x- and y- values in the ordered pairs of the function.

$f(x): y = 5^x$		$f^{-1}(x): x = 5^y$	
x	**y**	**x**	**y**
−3	0.008	0.008	−3
−2	0.004	0.004	−2
−1	0.2	0.2	−1
0	1	1	0
1	0	0	1
2	25	25	2
3	125	125	3

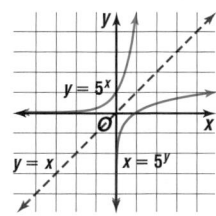

The function $f^{-1}(x)$ can be defined as $x = 5^y$. The ordered pairs can be used to sketch the graphs $y = 5^x$ and $x = 5^y$ on the same axes. *Note that they are reflections of each other over the line $y = x$.*

This example can be applied to a general statement that the inverse of $y = a^x$ is $x = a^y$. In the function $x = a^y$, y is called the **logarithm** of x. It is usually written as $y = \log_a x$ and is read "y equals the log, base a, of x." The function $y = \log_a x$ is called a **logarithmic function.**

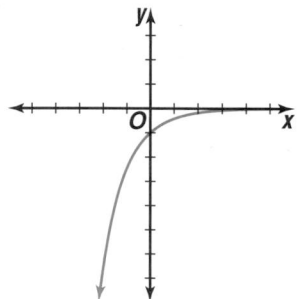

5-Minute Check
Lesson 11-3

1. Tell which equation is represented by the graph.

a. $y = e^x$ **b.** $y = e^{-x}$
c. $y = -e^x$ **d.** $y = -e^{-x}$ **d**

2. Describe a situation that could be modeled by the equation $A = 6000e^{0.06t}$. **$6000 invested for *t* years with an APR of 6% compounded continuously**

Bakersfield, California's population can be modeled by the equation $y = 33,430e^{0.039t}$, where *t* is the number of years since 1950.
3. What was Bakersfield's population in 1955? **40,628**

4. Find the projected population of Bakersfield in 2005. **285,560**

5. The Barfields are saving for their son's college education. If they deposit $31,500 in an account earning an APR of 7.6% compounded continuously, how much will be in the account when Elbert goes to college in 9 years? **$62,426.36**

Resource Manager

 Reproducible Masters
- *Study Guide and Practice,* pp. 203–204
- *Enrichment,* p. 124
- *Assessment and Evaluation,* pp. 276–277
- *Problem Solving and Applications,* p. 22

 Transparencies
- 5-Minute Check, 11-4
- Teaching, 11-4

 Technology/Multimedia
- Advanced MathPASS, Lesson 23

Logarithmic Function	The logarithmic function $y = \log_a x$, where $a > 0$ and $a \neq 1$, is the inverse of the exponential function $y = a^x$. So, $y = \log_a x$ if and only if $x = a^y$.

Each logarithmic equation corresponds to an equivalent exponential equation. That is, the logarithmic equation $y = \log_a x$ is equivalent to $a^y = x$.

exponent

$$y = \log_2 32 \longleftrightarrow 2^y = 32.$$

base

Since $2^5 = 32$, $y = 5$. Thus, $\log_2 32 = 5$.

Example ❶ **Write each equation in exponential form.**

a. $\log_{125} 25 = \dfrac{2}{3}$

The base is 125, and the exponent is $\dfrac{2}{3}$.

$25 = 125^{\frac{2}{3}}$

b. $\log_8 2 = \dfrac{1}{3}$

The base is 8, and the exponent is $\dfrac{1}{3}$.

$2 = 8^{\frac{1}{3}}$

You can also write an exponential function as a logarithmic function.

Example ❷ **Write each equation in logarithmic form.**

a. $4^3 = 64$

The base is 4, and the exponent or logarithm is 3.

$\log_4 64 = 3$

b. $3^{-3} = \dfrac{1}{27}$

The base is 3, and the exponent or logarithm is -3.

$\log_3 \dfrac{1}{27} = -3$

Using the fact that if $a^u = a^v$ then $u = v$, you can evaluate a logarithmic expression to determine its logarithm.

Example ❸ **Evaluate the expression $\log_7 \dfrac{1}{49}$.**

Let $x = \log_7 \dfrac{1}{49}$.

$$x = \log_7 \dfrac{1}{49}$$

$7^x = \dfrac{1}{49}$ *Definition of logarithm.*

$7^x = (49)^{-1}$ $a^{-m} = \dfrac{1}{a^m}$

$7^x = (7^2)^{-1}$ $7^2 = 49$

$7^x = 7^{-2}$ $(a^m)^n = a^{mn}$

$x = -2$ *If $a^u = a^v$ then $u = v$.*

Teaching Tip For Example 4, point out that the power property of logarithms enables one to solve equations made up of exponential functions.

In-Class Example

Example 4

Finance How many years will it take $700 invested at 12% interest compounded annually to amount to $47,000? **about 37.1 years**

Example **CHEMISTRY** Refer to the application at the beginning of the lesson. How long would it take for 256,000 grams of Thorium-234, with a half-life of 25 days, to decay to 1000 grams?

$$N = N_0\left(\frac{1}{2}\right)^t \qquad N = N_0(1 + r)^t \text{ for } r = \frac{1}{2}$$

$$1000 = 256{,}000\left(\frac{1}{2}\right)^t \qquad N = 1000, N_0 = 256{,}000$$

$$\frac{1}{256} = \left(\frac{1}{2}\right)^t \qquad \text{Divide each side by } 256{,}000.$$

$$\log_{\frac{1}{2}} \frac{1}{256} = t \qquad \text{Write the equation in logarithmic form.}$$

$$\log_{\frac{1}{2}}\left(\frac{1}{2}\right)^8 = t \qquad 256 = 2^8$$

$$\log_{\frac{1}{2}}\left(\frac{1}{2}\right)^8 = t \qquad \left(\frac{1}{b^n}\right) = \left(\frac{1}{b}\right)^n$$

$$\left(\frac{1}{2}\right)^8 = \left(\frac{1}{2}\right)^t \qquad \text{Definition of logarithm}$$

$$8 = t \qquad \text{It will take 8 half-lifes or 200 days.}$$

Look Back
Refer to Lesson 1-2 to review composition of functions.

Since the logarithmic function and the exponential function are inverses of each other, both of their compositions yield the identity function. Let $f(x) = \log_a x$ and $g(x) = a^x$. For $f(x)$ and $g(x)$ to be inverses, it must be true that $f(g(x)) = x$ and $g(f(x)) = x$.

$$f(g(x)) \overset{?}{=} x \qquad\qquad g(f(x)) \overset{?}{=} x$$
$$f(a^x) \overset{?}{=} x \qquad\qquad g(\log_a x) \overset{?}{=} x$$
$$\log_a a^x \overset{?}{=} x \qquad\qquad a^{\log_a x} \overset{?}{=} x$$
$$x = x \qquad\qquad x = x$$

The properties of logarithms can be derived from the properties of exponents.

Properties of Logarithms		
Suppose *m* and *n* are positive numbers, *b* is a positive number other than 1, and *p* is any real number. Then the following properties hold.		
Property	**Definition**	**Example**
Product	$\log_b mn = \log_b m + \log_b n$	$\log_3 9x = \log_3 9 + \log_3 x$
Quotient	$\log_b \frac{m}{n} = \log_b m - \log_b n$	$\log_{\frac{1}{4}} \frac{4}{5} = \log_{\frac{1}{4}} 4 - \log_{\frac{1}{4}} 5$
Power	$\log_b m^p = p \cdot \log_b m$	$\log_2 8^x = x \cdot \log_2 8$
Equality	If $\log_b m = \log_b n$, then $m = n$.	$\log_8 (3x - 4) = \log_8 (5x + 2)$ so, $3x - 4 = 5x + 2$

You will be asked to prove other properties in Exercises 2 and 61.

Each of these properties can be verified using the properties of exponents. For example, suppose we want to prove the Product Property. Let $x = \log_b m$ and $y = \log_b n$. Then by definition $b^x = m$ and $b^y = n$.

$$\log_b mn = \log_b (b^x \cdot b^y) \qquad b^x = m, b^y = n$$
$$= \log_b (b^{x + y}) \qquad \text{Product Property of Exponents}$$
$$= x + y \qquad \text{Definition of logarithm}$$
$$= \log_b m + \log_b n \qquad \text{Substitution}$$

Equations can be written involving logarithms. Use the properties of logarithms and the definition of logarithms to solve these equations.

Example **5** Solve each equation.

a. $\log_p 64^{\frac{1}{3}} = \frac{1}{2}$

$$\log_p 64^{\frac{1}{3}} = \frac{1}{2}$$
$$p^{\frac{1}{2}} = 64^{\frac{1}{3}} \quad \textit{Definition of logarithm.}$$
$$\sqrt{p} = \sqrt[3]{64} \quad b^n = \sqrt[n]{b^m}$$
$$\sqrt{p} = 4$$
$$(\sqrt{p})^2 = (4)^2 \quad \textit{Square each side.}$$
$$p = 16$$

b. $\log_4 (2x + 11) = \log_4 (5x - 4)$

$$\log_4 (2x + 11) = \log_4 (5x - 4)$$
$$2x + 11 = 5x - 4 \quad \textit{Equality Property}$$
$$-3x = -15$$
$$x = 5$$

c. $\log_{11} x + \log_{11} (x + 1) = \log_{11} 6$

$$\log_{11} x + \log_{11} (x + 1) = \log_{11} 6$$
$$\log_{11} [x(x + 1)] = \log_{11} 6 \quad \textit{Product Property}$$
$$(x^2 + x) = 6 \quad \textit{Equality Property}$$
$$x^2 + x - 6 = 0$$
$$(x - 2)(x + 3) = 0 \quad \textit{Factor.}$$
$$x - 2 = 0 \quad \text{or} \quad x + 3 = 0$$
$$x = 2 \qquad\qquad x = -3$$

By substituting $x = 2$ and $x = -3$ into the equation, we find that $x = -3$ is undefined for the equation $\log_{11} x + \log_{11} (x + 1) = \log_{11} 6$. When $x = -3$ we get an extraneous solution. So, $x = 2$ is the correct solution.

You can graph a logarithmic function by rewriting the logarithmic function as an exponential function and constructing a table of values.

Example **6** Graph $y = \log_3 (x + 1)$.

The graph of $y = \log_3 (x + 1)$ is a horizontal translation of the graph of $y = \log_3 x$.

The equation $y = \log_3 (x + 1)$ can be written as $3^y = x + 1$. Choose values for y and then find the corresponding values of x.

y	x + 1	x	(x, y)
−3	0.037	−0.963	(−0.963, −3)
−2	0.11	−0.89	(−0.89, −2)
−1	0.33	−0.67	(−0.67, −1)
0	1	0	(0, 0)
1	3	2	(2, 1)
2	9	8	(8, 2)
3	27	26	(26, 3)

3 PRACTICE/APPLY

Error Analysis

Watch for students who have difficulty correctly shading a logarithmic inequality.

Prevent by emphasizing that students use a test point to determine which region to shade.

Study Guide and Practice Masters, pp. 203–204

You can graph logarithmic inequalities using the same techniques as shown in Example 6. Choose a test point to determine which region to shade.

Example **7** Graph $y \le \log_5 x - 2$.

The boundary for the inequality $y \le \log_5 x - 2$ can be written as $y = \log_5 x - 2$. Rewrite this equation in exponential form.

$$y = \log_5 x - 2$$
$$y + 2 = \log_5 x$$
$$5^{y+2} = x$$

The graph of $y = \log_5 x - 2$ is a vertical translation of the graph of $y = \log_5 x$.

Use a table of values to graph the boundary.

y	$y + 2$	x	(x, y)
-5	-3	0.008	$(0.008, -5)$
-4	-2	0.25	$(0.25, -4)$
-3	-1	0.2	$(0.2, -3)$
-2	0	1	$(1, -2)$
-1	1	5	$(5, -1)$
0	2	25	$(25, 0)$
1	3	125	$(125, 1)$

Test a point, for example $(0, 0)$, to determine which region to shade.

$$5^{y+2} \le 0 \rightarrow 5^{0+2} \le 0 \quad \textit{False}$$

Shade the region that does not contain the point at $(0, 0)$.

Remember that you must exclude values for which the log is undefined. For example, for $y \ge \log_5 x - 2$, the negative values for x must be omitted from the domain and no shading would occur in that area.

CHECK FOR UNDERSTANDING

Communicating Mathematics

Read and study the lesson to answer each question.   **1–5. See margin.**

1. **Compare and contrast** the graphs of $y = 3^x$ and $y = \log_3 x$.

2. **Show** that the Power Property of logarithms is valid.

3. **State** the difference between the graph of $y = \log_5 x$ and $y = \log_{\frac{1}{5}} x$.

4. **You Decide**   Courtney and Sean are discussing the expansion of $\log_b mn$. Courtney states that $\log_b mn$ can be written as $\log_b m \times \log_b n$. Sean states that $\log_b mn$ can be written as $\log_b m + \log_b n$. Who is correct? Explain your answer.

5. **Explain** the relationship between the decay formula $N = N_0(1 + r)^t$ and the formula for determining half-life, $N = N_0\left(\frac{1}{2}\right)^t$.

Guided Practice

Write each equation in exponential form.

6. $\log_9 27 = \frac{3}{2}$ $9^{\frac{3}{2}} = 27$

7. $\log_{\frac{1}{25}} 5 = -\frac{1}{2}$ $\left(\frac{1}{25}\right)^{-\frac{1}{2}} = 5$

Write each equation in logarithmic form.

8. $7^{-6} = y$ $\log_7 y = -6$

9. $8^{-\frac{2}{3}} = \frac{1}{4}$ $\log_8 \frac{1}{4} = -\frac{2}{3}$

Additional Answers

1. $y = 3^x$ and $\log_3 x$ are similar in that they are both continuous, one-to-one, increasing, and inverses. $y = 3^x$ and $\log_3 x$ are not similar in that they are inverses. The domain of one is the range of another and the range of one is the domain of the other. $y = 3^x$ has a y-intercept and a horizontal asymptote, whereas $y = \log_3 x$ has an x-intercept and a vertical asymptote.

2.   Let $b^x = m$, then $\log_b m = x$.
$(b^x)^p = m^p$
$b^{xp} = m^p$
$\log_b b^{xp} = \log_b m^p$
$xp = \log_b m^p$
$p \log_b m = \log_b m^p$

Evaluate each expression.

10. $\log_2 \frac{1}{16}$ **−4**

11. $\log_{10} 0.01$ **−2**

12. $\log_7 \frac{1}{343}$ **−3**

Solve each equation.

13. $\log_2 x = 5$ **32**

14. $\log_7 n = \frac{2}{3} \log_7 8$ **4**

15. $\log_6 (4x + 4) = \log_6 64$ **15**

16. $2 \log_6 4 - \frac{1}{4} \log_6 16 = \log_6 x$ **8**

Graph each equation or inequality.   **17–18. See margin.**

17. $y = \log_{\frac{1}{2}} x$

18. $y \geq \log_6 x$

19. **Biology** The generation time for bacteria is the time that it takes for the population to double. The generation time G can be found using experimental data and the formula $G = \frac{t}{3.3 \log_b f}$, where t is the time period, b is the number of bacteria at the beginning of the experiment, and f is the number of bacteria at the end of the experiment. The generation time for mycobacterium tuberculosis is 16 hours. How long will it take four of these bacteria to multiply into 1024 bacteria?   **264 h**

EXERCISES

 Write each equation in exponential form. $7^{-4} = \frac{1}{2401}$

20. $\log_{27} 3 = \frac{1}{3}$ $27^{\frac{1}{3}} = 3$

21. $\log_{16} 4 = \frac{1}{2}$ $16^{\frac{1}{2}} = 4$

22. $\log_7 \frac{1}{2401} = -4$

23. $\log_4 32 = \frac{5}{2}$ $4^{\frac{5}{2}} = 32$

24. $\log_e 65.98 = x$ $e^x = 65.98$

25. $\log_{\sqrt{6}} 36 = 4$ $(\sqrt{6})^4 = 36$

28. $\log_{\frac{1}{8}} 512 = -3$

29. $\log_6 \frac{1}{36} = -2$

Write each equation in logarithmic form. $\log_{36} 216 = \frac{3}{2}$

26. $81^{\frac{1}{2}} = 9$ $\log_{81} 9 = \frac{1}{2}$

27. $36^{\frac{3}{2}} = 216$

28. $\left(\frac{1}{8}\right)^{-3} = 512$

29. $6^{-2} = \frac{1}{36}$

30. $16^0 = 1$ $\log_{16} 1 = 0$

31. $x^{1.238} = 14.36$ $\log_x 14.36 = 1.238$

Evaluate each expression.

32. $\log_8 64$ **2**

33. $\log_{125} 5$ $\frac{1}{3}$

34. $\log_2 32$ **5**

35. $\log_4 128$ **3.5**

36. $\log_9 9^6$ **6**

37. $\log_{49} 343$ **1.5**

38. $\log_8 16$ $\frac{4}{3}$

39. $\log_{\sqrt{8}} 4096$ **8**

40. $10^{4\log_{10} 2}$ **16**

Solve each equation.

41. $\log_x 49 = 2$ **7**

42. $\log_3 3x = \log_3 36$ **12**

43. $\log_6 x + \log_6 9 = \log_6 54$ **6**

44. $\log_8 48 - \log_8 w = \log_8 6$ **8**

45. $\log_6 216 = x$ **3**

46. $\log_5 0.04 = x$ **−2**

47. $\log_{10} \sqrt[3]{10} = x$ $\frac{1}{3}$

48. $\log_{12} x = \frac{1}{2} \log_{12} 9 + \frac{1}{3} \log_{12} 27$ **9**

49. $\log_5(x + 4) + \log_5 8 = \log_5 64$ **4**

50. $\log_4(x - 3) + \log_4(x + 3) = 2$ **5**

51. $\frac{1}{2}(\log_7 x + \log_7 8) = \log_7 16$ **32**

52. $2 \log_5(x - 2) = \log_5 36$ **8**

 www.amc.glencoe.com/self_check_quiz

Assignment Guide

Core: 21–57 odd, 59–61, 63–65, 67–78
Enriched: 20–58 even, 59–78

Additional Answers
(pp. 722–723)

3. $\log_5 x$ is an increasing function and $\log_{\frac{1}{5}} x$ is a decreasing function.

4. Sean is correct. The product property states that $\log_b mn = \log_b m + \log_b n$.

5. In half-life applications, $r = -\frac{1}{2}$. So, $(1 + r)$ becomes $\left(1 - \frac{1}{2}\right)$ or $\left(\frac{1}{2}\right)$. Thus, the formula $N = N_0 (1 + r)^t$ becomes $N = N_0 \left(\frac{1}{2}\right)^t$.

17.

18.

Graph each equation or inequality. **53–58. See margin.**

53. $y = \log_4 x$

54. $y = 3 \log_2 x$

55. $y = \log_5(x - 1)$

56. $y \le \log_2 x$

57. $y \ge 2 \log_2 x$

58. $y > \log_{10}(x + 1)$

Applications and Problem Solving

59. Public Health Inspectors for the Fulton County Health Department routinely check food samples for the presence of the E. coli bacteria. When E. coli cells are placed in a medium that provides nutrients needed for growth, the bacteria population can increase exponentially, reproducing itself every 15 minutes. If an inspector has a sample containing 1000 bacteria cells, how long will it take the population to reach 64,000 cells? **90 min**

60. Critical Thinking Using the definition of a logarithm where $y = \log_a x$, explain why the base a cannot equal 1.

60. All powers of 1 are 1, so the inverse of $y = 1^x$ is not a function.

61. Proof Prove the quotient property of logarithms, $\log_b \frac{m}{n} = \log_b m - \log_b n$, using the definition of a logarithm. **See margin.**

62. Finance Latasha plans to invest \$2500 in an account that compounds quarterly, hoping to at least double her money in 10 years.

62a.
$$2 \ge \left(1 + \frac{r}{4}\right)^{40}$$

  a. Write an inequality that can be used to find the interest rate at which Latasha should invest her money.

  b. What is the lowest interest rate that will allow her to meet her goal? **6.99%**

  c. Suppose she only wants to invest her money for 7 years. What interest rate would allow her to double her money? **≈ 10.04%**

63. Photography The aperture setting of a camera, or f-stop, controls the amount of light exposure on film. Each step up in f-stop setting allows twice as much light exposure as the previous setting. The formula $n = \log_2 \frac{1}{p}$ where p is the fraction of sunlight, represents the change in the f-stop setting n to use in less light.

  a. A nature photographer sets her camera's f-stop at $f/6.7$ while taking outdoor pictures in direct sunlight. If the amount of sunlight on a cloudy day is $\frac{1}{4}$ as bright as direct sunlight, how many f-stop settings should she move to accommodate less light? **2**

63b. less light; $\frac{1}{8}$

  b. If she moves down 3 f-stop settings from her original setting, is she allowing more or less light into the camera? What fraction of daylight is she accommodating?

64. Let $y = \log_a x$, so $x = a^y$
$x = a^y$
$\log_b x = \log_b a^y$
$\log_b x = y\log_b a$
$y = \frac{\log_b x}{\log_b a}$
$\log_a x = \frac{\log_b x}{\log_b a}$

64. Critical Thinking Show that $\log_a x = \frac{\log_b x}{\log_b a}$.

65. Meteorology Atmospheric pressure decreases as altitude above sea level increases. Atmospheric pressure P, measured in pounds per square inch, at altitude h miles can be represented by the logarithmic function $\log_{2.72} \frac{P}{14.7} = -0.02h$.

  a. Graph the function for atmospheric pressure. **See margin.**

  b. The elevation of Denver, Colorado, is about 1 mile. What is the atmospheric pressure in Denver? **14.4 psi**

  c. The lowest elevation on Earth is in the Mariana Trench in the Atlantic Ocean about 6.8 miles below sea level. If there were no water at this point, what would be the atmospheric pressure at this elevation? **16.84 psi**

65a.

$\log_{2.72}\left(\frac{P}{14.7}\right) = -0.02h$

66. Radiation Safety Radon is a naturally ocurring radioactive gas which can collect in poorly ventilated structures. Radon gas is formed by the decomposition of radium-226 which has a half-life of 1622 years. The half-life of radon is 3.82 days. Suppose a house basement contained 38 grams of radon gas when a family moved in. If the source of radium producing the radon gas is removed so that the radon gas eventually decays, how long will it take until there is only 6.8 grams of radon gas present? **about 9 days**

Mixed Review

67. Find the value of $e^{4.243}$ to the nearest ten thousandth. (Lesson 11-3) **69.6164**

68. Finance What is the monthly principal and interest payment on a home mortgage of $90,000 for 30 years at 11.5%? (Lesson 11-2) **$891.26**

69. ellipse,
$$\frac{(x-1)^2}{4} + \frac{(y-2)^2}{9} = 1;$$ **see margin for graph.**

69. Identify the conic section represented by $9x^2 - 18x + 4y^2 - 16y - 11 = 0$. Then write the equation in standard form and graph the equation. (Lesson 10-6)

70. Engineering A wheel in a motor is turning counterclockwise at 2 radians per second. There is a small hole in the wheel 3 centimeters from its center. Suppose a model of the wheel is drawn on a rectangular coordinate system with the wheel centered at the origin. If the hole has initial coordinates $(3, 0)$, what are its coordinates after t seconds? (Lesson 10-6) **$(3 \cos 2t, 3 \sin 2t)$**

71. Find the lengths of the sides of a triangle whose vertices are $A(-1, 3)$, $B(-1, -3)$, and $C(3, 0)$. (Lesson 10-1) **$AB = 6$, $BC = 5$, $AC = 5$**

72. $10\left(\cos \frac{17\pi}{12} + i \sin \frac{17\pi}{12}\right),$ $-2.59 - 9.66i$

72. Find the product $5\left(\cos \frac{3\pi}{4} + i \sin \frac{3\pi}{4}\right) \cdot 2\left(\cos \frac{2\pi}{3} + i \sin \frac{2\pi}{3}\right)$. Then express it in rectangular form. (Lesson 9-7)

73. Electricity A circuit has a current of $(3 - 4j)$ amps and an impedance of $(12 + 7j)$ ohms. Find the voltage of this circuit. (Lesson 9-5) **$64 - 27j$ volts**

74. State whether $\overrightarrow{AB}$ and $\overrightarrow{CD}$ are *opposite, parallel,* or *neither of these* for $A(-2, 5)$, $B(3, -1)$, $C(2, 6)$, and $D(7, 0)$. (Lesson 8-1) **parallel**

75. Find $\cos (A + B)$ if $\cos A = \frac{5}{13}$ and $\cos B = \frac{35}{37}$ and A and B are first quadrant angles. (Lesson 7-3) **$\frac{31}{481}$**

76. $y = 13 \sin\left(\frac{\pi}{2}t - 3.14\right) + 77$

76. Weather The maximum normal daily temperatures in each season for New Orleans, Louisiana, are given below.

Winter	Spring	Summer	Fall
64°	78°	90°	79°

Source: Rand McNally & Company

Write a sinusoidal function that models the temperatures, using $t = 1$ to represent winter. (Lesson 6-6)

77. $c = 9.5$, $A = 38°20'$, $B = 36°22'$

77. Solve $\triangle ABC$ if $C = 105°18'$, $a = 6.11$, and $b = 5.84$. (Lesson 5-8)

78. SAT/ACT Practice If $PQRS$ is a parallelogram and MN is a segment, then x must equal **E**

A $180 - b$

B $180 - c$

C $a + b$

D $a + c$

E $b + c$

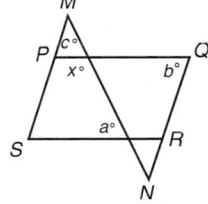

Extra Practice See p. A46.

Lesson 11-4 Logarithmic Functions **725**

Open-Ended Assessment

Writing Have students write an equation in exponential form. Then, have them exchange papers and rewrite the equation they receive in logarithmic form.

 Chapter 11, Quiz B (Lessons 11-3 and 11-4) is available in the *Assessment and Evaluation Masters*, p. 277.

 Chapter 11 Mid-Chapter Test (Lessons 11-1 through 11-4) is available in the *Assessment and Evaluation Masters*, p. 276.

Additional Answer

69.

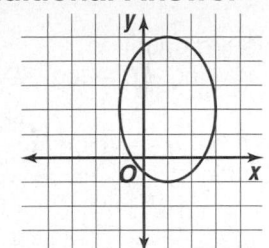

Enrichment Masters, p. 124

Extra Credit

According to Newton, a liquid cools according to the formula $T = T_i e^{-kt}$ where T is the final temperature, T_i the initial temperature, t is the time in minutes, and k is the constant which relates the variables. If $T = 40°$, $T_i = 70°$ and $k = 0.1399$, how long was the liquid cooling? **about 4 minutes** Describe how each variable changes as time passes. **Sample answer: T_i is the initial temperature and it does not change based on time. k is a constant. T will decrease as time passes.**

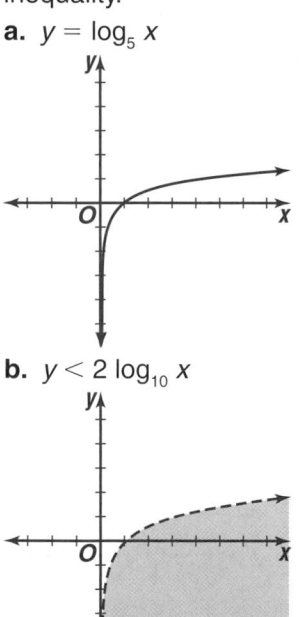

5-Minute Check
Lesson 11-4

1. Write the equation $\log_e 13.7 = x$ in exponential form. $e^x = 13.7$

2. Write the equation $\left(\frac{1}{4}\right)^{-4} = 256$ in logarithmic form. $\log_{\frac{1}{4}} 256 = -4$

3. Evaluate the expression $\log_4 4^3$. **3**

4. Solve the equation $\log_2 x + \log_2 10 = \log_2 70$. $x = 7$

5. Graph each equation or inequality.

   a. $y = \log_5 x$

   b. $y < 2 \log_{10} x$

Motivating the Lesson

Discussion Ask students which button(s) on a calculator can be used to find logarithms. **keys marked "log" or "ln"** What are the bases for these logarithmic functions? Why are they included on a calculator rather than some other logarithms? Tell students that these questions will be addressed in the remainder of Chapter 11.

11-5

Common Logarithms

OBJECTIVES
- Find common logarithms and antilogarithms of numbers.
- Solve equations and inequalities using common logarithms.
- Solve real-word applications with common logarithmic functions.

CHEMISTRY Drinking water is routinely tested for its pH level. The pH of a solution is a measure of acidity and is related to the concentration of hydrogen ions measured in moles per liter.

Testing the water's pH level may indicate the presence of a contaminant that is harmful to life forms. *This information will be used in Examples 2 and 6.*

The pH of a solution can be modeled by the equation $\text{pH} = \log \frac{1}{\text{H}^+}$, where H^+ is the number of moles of hydrogen ions per liter. Notice that no base is indicated for the logarithm in the equation for pH. When no base is indicated, the base is assumed to be 10. So, $\log \frac{1}{\text{H}^+}$ means $\log_{10} \frac{1}{\text{H}^+}$. Logarithms with base 10 are called **common logarithms.** You can easily find the common logarithms of integral powers of ten.

$$
\begin{array}{lll}
\log 1000 & = 3 & \text{since} & 1000 = 10^3 \\
\log 100 & = 2 & \text{since} & 100 = 10^2 \\
\log 10 & = 1 & \text{since} & 10 = 10^1 \\
\log 1 & = 0 & \text{since} & 1 = 10^0 \\
\log 0.1 & = -1 & \text{since} & 0.1 = 10^{-1} \\
\log 0.01 & = -2 & \text{since} & 0.01 = 10^{-2} \\
\log 0.001 & = -3 & \text{since} & 0.001 = 10^{-3}
\end{array}
$$

The common logarithms of numbers that differ by integral powers of ten are closely related. Remember that a logarithm is an exponent. For example, in the equation $y = \log x$, y is the power to which 10 is raised to obtain the value of x.

$$\log x = y \quad \text{means} \quad 10^y = x$$

$$
\begin{array}{lll}
\log 1 & = 0 & \text{since} & 10^0 = 1 \\
\log 10 & = 1 & \text{since} & 10^1 = 10 \\
\log 10^m & = m & \text{since} & 10^m = 10^m
\end{array}
$$

Example ❶ Given that $\log 7 = 0.8451$, evaluate each logarithm.

a. log 7,000,000

$$
\begin{aligned}
\log 7{,}000{,}000 &= \log(1{,}000{,}000 \times 7) \\
&= \log 10^6 + \log 7 \\
&= 6 + 0.8451 \\
&= 6.8451
\end{aligned}
$$

b. log 0.0007

$$
\begin{aligned}
\log 0.0007 &= \log(0.0001 \times 7) \\
&= \log 10^{-4} + \log 7 \\
&= -4 + 0.8451 \\
&= -3.1549
\end{aligned}
$$

Resource Manager

 Reproducible Masters
- *Study Guide and Practice,* pp. 205–206
- *Enrichment,* p. 125
- *TI-92 & CBL Interdisciplinary Activities,* pp. 37–40

 Transparencies
- 5-Minute Check, 11-5
- Teaching, 11-5

A common logarithm is made up of two parts, the **characteristic** and the **mantissa.** The mantissa is the logarithm of a number between 1 and 10. Thus, the mantissa is greater than 0 and less than 1.

In Example 1, the mantissa is log 7 or 0.8451. The characteristic is the exponent of ten that is used to write the number in scientific notation. So, in Example 1, the characteristic of 1,000,000 is 6, and the characteristic of 0.0001 is −4. Traditionally, a logarithm is expressed as the indicated sum of the mantissa and the characteristic.

You can use a calculator to solve certain equations containing common logarithms.

Example

2 **CHEMISTRY** Refer to the application at the beginning of the lesson. If the water being tested contains 7.94×10^{-9} moles of H^+ per liter, what is the pH level of the water?

$$pH = \log \frac{1}{H^+}$$

$$pH = \log \frac{1}{7.94 \times 10^{-9}} \quad H^+ = 7.94 \times 10^{-9}$$

Evaluate with a calculator.

Graphing Calculator Tip

The key marked $\boxed{\text{LOG}}$ on your calculator will display the common logarithm of a number.

$\boxed{\text{LOG}}$ 1 $\boxed{\div}$ 7.94 $\boxed{\text{2nd}}$ $\boxed{\text{[EE]}}$ $\boxed{(-)}$ 9 $\boxed{\text{ENTER}}$

$pH \approx 8.1$

The pH level of the water is about 8.1

The properties of logarithmic functions you learned in Lesson 11-4 apply to common logarithms. You can use these properties to evaluate logarithmic expressions.

Example

3 Evaluate each expression.

a. $\log 5(2)^3$

$\log 5(2)^3 = \log 5 + 3 \log 2$ *Product Property, Power Property*
$ \approx 0.6990 + 3(0.3010)$ *Use a calculator.*
$ \approx 0.6990 + 0.9031$
$ \approx 1.6021$

b. $\log \frac{19^2}{6}$

$\log \frac{19^2}{6} = 2 \log 19 - \log 6$ *Quotient Property, Power Property*
$\phantom{\log \frac{19^2}{6}} \approx 2(1.2788) - 0.7782$ *Use a calculator.*
$\phantom{\log \frac{19^2}{6}} \approx 2.5576 - 0.7782$
$\phantom{\log \frac{19^2}{6}} \approx 1.7794$

Teaching Tip For Example 1, remind students that if a base is not written in logarithmic form, it is understood to be base 10.

In-Class Examples

Example 1
Given that log 5 = 0.6990, evaluate each logarithm.
a. log 50,000   **4.6990**
b. log 0.005   **−2.3010**

Example 2
Natural Resources The annual growth rate of the moose population in Wells Gray Park in northern British Columbia, Canada, is 6.7%. This situation can be modeled by the equation

$$n = \frac{\log \frac{P_T}{P_i}}{\log 1.067}, \text{ where } P_T =$$

population today, P_i = initial population, and n = time in years. A few years ago the population was 325 moose and today it is 450 moose. Find the number of years that have elapsed between the initial population and today's population. **about 5 years**

Example 3
Evaluate each expression.
a. $y = \log 3(5)^3$ **2.5740**
b. $\log \frac{13^3}{7}$ **2.4967**

Teaching Tip For Example 3a in the Student Edition, the steps show a rounded result. The actual result from the calculator for log 5 + 3 log 2 is 1.602059991.

The graph of $y = \log x$ is shown at the right. As with other logarithmic functions, it is continuous and increasing. It has a domain of positive real numbers and a range of all real numbers. It has an x-intercept at $(1, 0)$, and there is a vertical asymptote at $x = 0$, or the y-axis, and no horizontal asymptote. The parent graph of $y = \log x$ can be transformed.

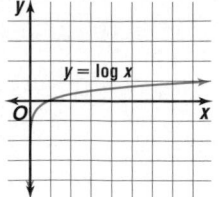

Example ❹ Graph $y > \log(x - 4)$.

The boundary of the inequality is the graph of $y = \log(x - 4)$. The graph is translated 4 units to the right of $y = \log x$ with a domain of all real numbers greater than 4, an x-intercept at $(5, 0)$ and a vertical asymptote at $x = 4$.

Choose a test point, $(6, 1)$ for example, to determine which region to shade.

$y > \log(x - 4)$
$1 \overset{?}{>} \log(6 - 1)$ *Substitution*
$1 \overset{?}{>} \log 5$
$1 \overset{?}{>} 0.6990$ *Use a calculator.*

The inequality is true, so shade the region containing the point at $(6, 1)$.

In Lesson 11-4, you evaluated logarithms in various bases. In order to evaluate these with a calculator, you must convert these logarithms in other bases to common logarithms using the following formula.

Change of Base Formula	If a, b, and n are positive numbers and neither a nor b is 1, then the following equation is true. $$\log_a n = \frac{\log_b n}{\log_b a}$$

Example ❺ Find the value of $\log_9 1043$ using the change of base formula.

$$\log_a n = \frac{\log_b n}{\log_b a}$$

$$\log_9 1043 = \frac{\log_{10} 1043}{\log_{10} 9}$$

$$\approx \frac{3.0183}{0.9542} \quad \text{Use a calculator.}$$

$$\approx 3.1632$$

The value of $\log_9 1043$ is about 3.1632.

Sometimes the logarithm of x is known to have a value of a, but x is not known. Then x is called the **antilogarithm** of a, written antilog a. If log $x = a$, then $x =$ antilog a. Remember that the inverse of a logarithmic function is an exponential function.

Example 6 **CHEMISTRY** Refer to the application at the beginning of the lesson. Technicians at a water treatment plant determine that the water supply has a pH of 6.7. What is the concentration of hydrogen ions in the tested water?

$$\text{pH} = \log \frac{1}{\text{H}^+}$$

$$6.7 = \log \frac{1}{\text{H}^+} \qquad pH = 6.7$$

$$\text{antilog } 6.7 = \frac{1}{\text{H}^+} \qquad \textit{Take the antilogarithm of each side.}$$

$$\text{H}^+ = \frac{1}{\text{antilog } 6.7}$$

$$\text{H}^+ \approx 1.9953 \times 10^{-7} \quad \textit{Use a calculator.}$$

The concentration of hydrogen ions is about 1.9953×10^{-7} moles per liter.

Logarithms can be used to solve exponential equations.

Example 7 Solve $6^{3x} = 81$.

$$6^{3x} = 81$$
$$\log 6^{3x} = \log 81 \qquad \textit{Take the logarithm of each side.}$$
$$3x \log 6 = \log 81 \qquad \log_b m^p = p \cdot \log_b m$$
$$3x = \frac{\log 81}{\log 6} \qquad \textit{Divide each side by log 6.}$$
$$x \approx 0.8175 \qquad \textit{Use a calculator.}$$

The solution is approximately 0.8175.

Graphing is an alternate way of finding approximate solutions of exponential or logarithmic equations. To do this, graph each side of the equation as a function and find the coordinates of the intersection of the graphs.

Example 8 Solve $5^{x-1} = 2^x$ by graphing.

Graph $y = 5^{x-1}$ and $y = 2^x$ on the same set of axes. The graphs appear to intersect at about (1.76, 3.38)

Therefore, $x \approx 1.76$

[−10, 10] scl:1 by [−10, 10] scl:1

Teaching Tip For In-Class Example 7, remind students that by using the power property, the values of $4x$ and ultimately x can be found readily.

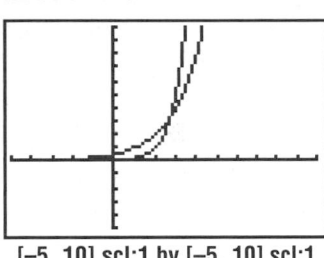

In-Class Examples

Example 6
Natural Resources Refer to In-Class Example 2. If the initial population is 325 moose, in how many years will the population triple?   **about 16.9 years**

Example 7
Solve $5^{4x} = 73$.   **0.6665**

Example 8
Solve $2^{x-1} = 5^{x-2}$ by graphing. **about 2.76**

[−5, 10] scl:1 by [−5, 10] scl:1

Teaching Tip For In-Class Example 8, remind students that the point of intersection (2.76, 3.78) of the two exponential functions is the point that solves both equations simultaneously.

Addressing Individual Needs

Visual/Spatial Have students graph $y = \log x$ and the exponential function $y = 10^x$ on the same set of axes. Then, have them observe the similarities and differences between the two graphs. **Sample answer: the two graphs are inverses of each other. Both reflect across the line $y = x$ and are functions. The equation $y = \log x$ has an x-intercept at the point (1, 0) but no y-intercept. The function $y = 10^x$ has a y-intercept at (0, 1) but no x-intercept.**

3 PRACTICE/APPLY

Error Analysis

Watch for students who mistakenly try to take the logarithm of a negative number.

Prevent by reminding students that logarithms in the real number system are undefined for negative numbers.

Assignment Guide

Core: 19–61 odd, 63–76
Enriched: 20–56 even, 58–76

 Teaching Tip For Exercise 17, point out that it requires patience to size the window properly when graphing exponential functions.

Additional Answers

1. log 1 = 0 means $\log_{10} 1 = 0$. So, $10^0 = 1$. log 10 = 1 means $\log_{10} 10 = 1$. So, $10^1 = 10$.

2. Write the number in scientific notation. The exponent of the power of 10 is the characteristic.

3. antilog 2.835 = $10^{2.835}$ = 683.9116

4. log 15 = 1.1761
log 5 = 0.6990
log 3 = 0.4771
log 5 + log 3 = 0.6990 + 0.4771 = 1.1761

Study Guide and Practice Masters, pp. 205–206

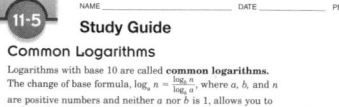

CHECK FOR UNDERSTANDING

Communicating Mathematics

Read and study the lesson to answer each question. 1–4. See margin.

1. **Demonstrate** the fact that log 1 = 0 and log 10 = 1.

2. **Write** an explanation of how you would determine the characteristic of log 38,272.65.

3. **Compare** the value of antilog 2.835 and $10^{2.835}$.

4. **Show** that log 15 = log 5 + log 3.

Guided Practice

Given that log 8 = 0.9031 and log 3 = 0.4771, evaluate each logarithm.

5. log 80,000
4.9031

6. log 0.003
−2.5229

7. log 0.0081
−2.0915

Evaluate each expression.

8. log 424 **2.6274**

9. antilog 4.8740 **74,816.95**

10. Graph $y \le \log(x + 3)$. **See margin.**

Find the value of each logarithm using the change of base formula.

11. $\log_{12} 18$ **1.1632**

12. $\log_8 15$ **1.3023**

Solve each equation or inequality.

13. $2.2^{x-5} = 9.32$ **7.83**

14. $6^{x-2} = 4^x$ **8.84**

15. $4.3^x < 76.2$ **$x < 2.97$**

16. $3^{x-3} \ge 2\sqrt[4]{4^{x-1}}$ **$x \ge 4.84$**

17. Solve $12^{x-4} = 3^{x-2}$ by graphing. **5.5850**

Graphing Calculator

18. **Seismology** The intensity of an earthquake is described by a number on the Richter scale. The Richter scale number R of an earthquake is given by the formula $R = \log\left(\dfrac{a}{T}\right) + B$, where a is the amplitude of the vertical ground motion in microns, T is the period of the seismic wave in seconds, and B is a factor that accounts for the weakening of seismic waves.

a. Find the intensity of an earthquake to the nearest tenth if a recording station measured the amplitude as 200 microns and the period as 1.6 seconds, and $B = 4.2$. **6.3**

b. How much more intense is an earthquake of magnitude 5 on the Richter scale than an earthquake that measures 4 on the Richter scale? Explain how you determined this result. **See margin.**

EXERCISES

Practice

Given that log 4 = 0.6021, log 9 = 0.9542, and log 12 = 1.0792, evaluate each logarithm.

A

19. log 400,000 **5.6021**

20. log 0.00009 **−4.0458**

21. log 1.2 **0.0792**

22. log 0.06 **−1.2218**

23. log 36 **1.5563**

24. log 108,000 **5.0334**

25. log 0.0048 **−2.3188**

26. log 4.096 **0.6124**

27. log 1800 **3.2553**

Evaluate each expression.

28. log 98.2 **1.9921**

29. log 894.3 **2.9515**

30. antilog −0.0600 **0.871**

B

31. antilog 0.3012 **2.001**

32. log 1891.91 **3.2769**

33. antilog 0.33736 **2.1745**

www.amc.glencoe.com/self_check_quiz

Additional Answers

10.

$y \le \log(x + 3)$

18b. 10 times; according to the definition of logarithms, R in the equation $R = \log\left(\dfrac{a}{T}\right) + B$ is an exponent of the base of the logarithm, 10. 10^5 is ten times greater than 10^4.

Find the value of each logarithm using the change of base formula.

34. $\log_2 8$ **3**
35. $\log_5 625$ **4**
36. $\log_6 24$ **1.7737**

37. $\log_7 4$ **0.7124**
38. $\log_{0.5} 0.0675$ **3.8890**
39. $\log_{\frac{1}{2}} 15$ **−3.9069**

Solve each equation or inequality.   **40–51. See margin.**

40. $2^x = 95$
41. $5^x = 4^{x+3}$
42. $\frac{1}{3}\log x - \log 8$

 43. $0.16^{4+3x} = 0.3^{8-x}$
44. $4\log(x+3) = 9$
45. $0.25 = \log 16^x$

46. $3^{x-1} \le 2^{x-7}$
47. $\log_x 6 > 1$
48. $4^{2x-5} \le 3^{x-3}$

49. $0.5^{2x-4} \le 0.1^{5-x}$
50. $\log_2 x = -3$
51. $x < \log_3 52.7$

Graph each equation or inequality.   **52–54. See margin.**

52. $y = 2\log(x+3)$
53. $y = \frac{1}{2}\log(x^3 - 1)$
54. $y \le 5^{x+3}$

Solve each equation by graphing.

55. $8^{2x+1} = 30.4$ **0.3210**
56. $9^{2x} - 2 = 8^x$ **−0.1975**
57. $4^{2x-1} = 2^{3x}$ **2**

Applications
and Problem
Solving

58. Aviation The altitude of an aircraft is in part affected by the outside air pressure and can be determined by the equation $h = -\frac{100}{9}\log\frac{P}{B}$, where h is the altitude in miles, P is the air pressure outside the aircraft, and B is the air pressure at sea level. Normally $B = 14.7$ pounds per square inch (psi).

   a. Suppose the air pressure outside an airplane is 10.3. What is the altitude of the plane?   **1.7 mi**

   b. If a jet's altitude is 4.3 miles above sea level, what is the air pressure outside the jet?   **6 psi**

59. Astronomy The parallax of a stellar object is the difference in direction of the object as seen from two distantly separated points. Astronomers use the parallax to determine an object's distance from Earth. The apparent magnitude of an object is its brightness as observed from Earth. Astronomers use parsecs, which stands for *par*allax of a *sec*ond, to measure distances in interstellar space. One parsec is about 3.26 light years or 19.2 trillion miles. The absolute magnitude is the magnitude an object would have if it were 10 parsecs from Earth. The greater the magnitude of an object the fainter it appears. For objects more than 30 parsecs, or 576 trillion miles from Earth, the formula relating the parallax p, the absolute magnitude M, and the apparent magnitude m, is $M = m + 5 + 5\log p$.   **a. 1.58**

   a. The object M35 in the constellation Gemini has an apparent magnitude of 5.3 and a parallax of about 0.018. Find the absolute magnitude of this object.

   b. Stars with apparent magnitudes greater than 5 can only be seen with a telescope. If a star has an apparent magnitude of 8.6 and an absolute magnitude of 5.3, find its parallax to four decimal places.   **0.0219**

60. Sales After t years, the annual sales in hundreds of thousands of units of a product q is given by $q = \left(\frac{1}{2}\right)^{0.8^t}$.

   a. Find the annual sales of a product after 9 years.   **91,116**

   b. After how many years will the annual sales be about 95,350 units?   **12 yr**

61. Critical Thinking If $x = \log 372$, the value of x is between two consecutive integers. Name these two integers and explain how you determined the values. **See margin.**

Lesson 11-5 Common Logarithms   **731**

Open-Ended Assessment
Modeling If the population of deer in a park doubled in ten years, set up a formula that would model this situation. Find the annual growth rate using your model.
$2 = 1(1 + r)^{10}$; about 7.18%

Additional Answers
66.

increasing from $-\infty$ to ∞

71.

---

62. Sound The loudness of a sound is often given on the decibel (dB) scale. The loudness of a sound L in dB is defined in terms of its intensity I by the equation $L = 10 \log \dfrac{I}{I_0}$ where I_0 is the minimum intensity audible to the average person (the "threshold of hearing"), 1.0×10^{-12} W/m².
   a. A typical rock concert can have an intensity of 1 W/m². What is the loudness in decibels of such a concert? **120 dB**
   b. A whisper has a loudness of 20 dB. What is the intensity of the whisper? 1×10^{-10} W/m²

63. Archaeology Stonehenge is an ancient megalithic site in southern England. Charcoal samples taken from a series of pits at Stonehenge have about 630 micrograms of Carbon-14 in a 1-milligram sample. Assuming the half-life of Carbon-14 is 5730 years, what is the age of the charcoal pits at Stonehenge? **3819 yr**

64. Critical Thinking If $\log_a y = \log_a p - \log_a q + \log_a r$, express y in terms of p, q, and r. $y = \dfrac{pr}{q}$

Mixed Review

65. Solve $\log_x 243 = 5$. *(Lesson 11-4)* **3**

66. Use a graphing calculator to graph $y = 4e^x$. Then describe the interval in which it is increasing or decreasing. *(Lesson 11-3)* **See margin.**

67. Express $(a^4b^2)^{\frac{1}{3}}c^{\frac{2}{3}}$ using radicals. *(Lesson 11-1)* $a\sqrt[3]{ab^2c^2}$

68. $(x - 3)^2 + (y + 1)^2 = 5$

68. Write the standard form of the equation of the circle that passes through points at $(5, 0)$, $(1, -2)$, and $(4, -3)$. *(Lesson 10-2)*

69. Find the midpoint of the segment that has endpoints at $(-2\sqrt{5}, -4)$ and $(6\sqrt{5}, -18)$. *(Lesson 10-1)* $(2\sqrt{5}, -11)$

70. Write the polar equation $r = 6$ in rectangular form. *(Lesson 9-3)* $x^2 + y^2 = 36$

71. Graph $3r = 12$ on a polar plane. *(Lesson 9-1)* **See margin.**

72. Find an ordered pair that represents $\overrightarrow{AB}$ for $A(5, -6)$, and $B(6, -5)$. Then find the magnitude of $\overrightarrow{AB}$. *(Lesson 8-2)* $\langle 1, 1 \rangle$; $\sqrt{2} \approx 1.414$

73. Geometry Find the area of a regular pentagon that is inscribed in a circle with a diameter of 7.3 centimeters. *(Lesson 5-4)* **31.68 cm²**

74. $(x - 4)(x + 3)(x - 1)$

74. Determine the binomial factors of $x^3 - 2x^2 - 11x + 12$. *(Lesson 4-3)*

75. neither; the graph of the function is not symmetric with respect to either the origin or the y-axis.

75. Is the function $y = 5x^3 - 2x + 5$ odd, even, or neither? Explain. *(Lesson 3-1)*

76. SAT/ACT Practice According to the graph, the total annual nut consumption in California projected for 2000 is approximately how many pounds? **D**

A 11,000,000
B 12,000,000
C 13,000,000
D 17,000,000
E 19,000,000

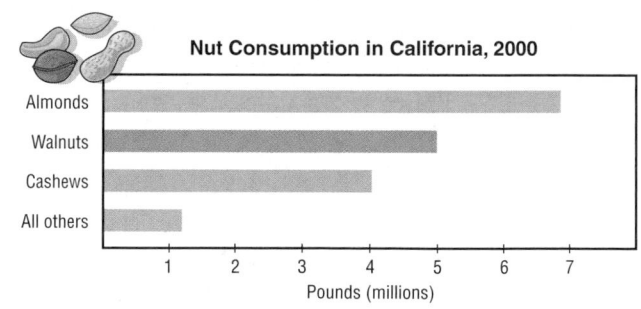

Nut Consumption in California, 2000

Almonds
Walnuts
Cashews
All others

Pounds (millions)

Extra Credit

Find the value of $7^{\log_7 10} + 3^{\log_3 10}$.   **20**

Natural Logarithms

OBJECTIVES
- Find natural logarithms of numbers.
- Solve equations and inequalities using natural logarithms.
- Solve real-world applications with natural logarithmic functions.

SOCIOLOGY Rumors often spread quickly through a group of people as one person hears something then tells other people about it, who in turn tell more people. Sociologists studying a group of people determine that the fraction of people in the group p who have heard a rumor after t days can be represented by the equation $p = \dfrac{p_0 e^{kt}}{1 - p_0(1 - e^{kt})}$, where p_0 is the fraction of people who have heard the rumor at time $t = 0$, and k is a constant. *This information will be used in Example 1.*

Often when you solve problems involving the number e, it is best to use logarithms to the base e. These are called **natural logarithms** and are usually written **ln x**. In other words, $\log_e x = \ln x$. Since e is a positive number between 2 and 3, all of the properties of logarithms also hold for natural logarithms. Note that if $\ln e = x$ and $e^x = e$, then $x = 1$. Thus, $\ln e = 1$.

Example **1** **SOCIOLOGY** **Refer to the application above. In a particular study, sociologists determine that at time $t = 0$, 5% of the people in a group have heard a certain rumor. After 2 days, 25% of the people in the group had heard the rumor. How long will it take for 80% to hear the rumor?**

First, find the value of the constant k in the formula $p = \dfrac{p_0 e^{kt}}{1 - p_0(1 - e^{kt})}$. To do this, substitute the known values.

$$p = \frac{p_0 e^{kt}}{1 - p_0(1 - e^{kt})}$$

$$0.25 = \frac{0.05 e^{2k}}{1 - 0.05(1 - e^{2k})} \qquad \text{\textit{p = 25\% or 0.25, } } p_0 = \text{\textit{5\% or 0.05, t = 2}}$$

$$0.2375 + 0.0125 e^{2k} = 0.05 e^{2k} \qquad \begin{array}{l}\textit{Multiply each side by}\\ \textit{1 - 0.05(1 - e}^{2k}\textit{).}\end{array}$$

$$0.2375 = 0.0375 e^{2k} \qquad \textit{Subtract 0.0125 } e^{2k} \textit{ from each side.}$$

$$6.3333 = e^{2k} \qquad \textit{Divide each side by 0.0375}$$

$$\ln 6.3333 = \ln e^{2k} \qquad \begin{array}{l}\textit{Take the natural logarithm}\\ \textit{of each side.}\end{array}$$

$$\ln 6.3333 = 2k \ln e \qquad \textit{ln } a^n = n \textit{ ln } a$$

$$1.8458 = 2k \qquad \textit{Use a calculator.}$$

$$0.9229 = k$$

Then, find the time when the fraction of the people in the group who have heard the rumor is 80%.

$$0.8 = \frac{0.05 e^{0.9229t}}{1 - 0.05(1 - e^{0.9229t})} \qquad \text{\textit{p = 80\% or 0.8. } } p_0 = \text{\textit{5\% or 0.05, k = 0.9229}}$$

$$0.76 + 0.04 e^{0.9229t} = 0.05 e^{0.9229t} \qquad \begin{array}{l}\textit{Multiply each side by}\\ \textit{1 - 0.05(1 - e}^{0.9229t}\textit{)}\end{array}$$

$$0.76 = 0.01 e^{0.9229t} \qquad \begin{array}{l}\textit{Subtract 0.04}e^{0.9229t}\textit{ from}\\ \textit{each side.}\end{array}$$

$$76 = e^{0.9229t} \qquad \textit{Divide each side by 0.01}$$

(continued on the next page)

 5-Minute Check
Lesson 11-5

1. Given that log 2 = 0.3010, log 3 = 0.4771, and log 7 = 0.8451, evaluate each logarithm.
   a. log 20,000 **4.3010**
   b. log 14 **1.1461**
   c. log 18 **1.2552**
2. Evaluate the expression log 752.6. **2.8766**
3. Graph the inequality $y \le 2^{x-3}$ on a graphing calculator.

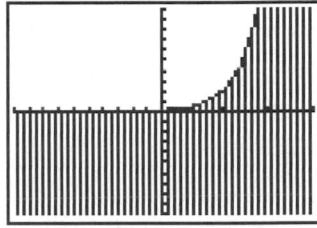

[−10, 10] scl:1 by [−10, 10] scl:1

4. Find the value of $\log_6 72$ using the change of base formula. **2.3869**
5. Solve the equation $\dfrac{1}{4} \log x = \log 15$. **$x$ = 50,629.1**

Motivating the Lesson
Real-World Application If your Aunt Emma left you an inheritance of $50,000, what average annual investment rate would be necessary for you to accumulate $2 million in a mutual fund at retirement in 45 years? **8.2%**

  Resource Manager

Reproducible Masters
- *Study Guide and Practice,* pp. 207–208
- *Enrichment,* p. 126
- *Assessment and Evaluation,* p. 278

Transparencies
- 5-Minute Check, 11-6
- Teaching, 11-6

In-Class Examples

Example 1
Physics As shown in the diagram, a rope coiled around a drum stays in place when the smaller tension T_S of the rope and the friction μ between the rope and the drum, maintain equilibrium.

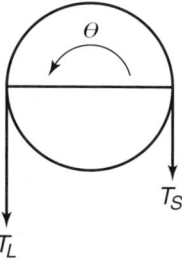

The relationship between the larger tension T_L of the drum and the rope, and the smaller tension T_S can be represented by the equation $T_L = T_S e^{\mu\theta}$, where θ is the measure of the rope in radians. When $\theta = \pi$ and $\mu = 0.3$, $T_L = 500$ pounds of force. When $\theta = 2\pi$ and T_S remains constant, $T_L =$ about 1277.7 pounds. What will the larger force be if the rope measures 5π and T_S remains constant?
about 21,595.6 pounds

Example 2
Convert $\log_5 291$ to a natural logarithm and evaluate.
about 3.5250

Example 3
Solve $7.2 = -28.8 \ln x$.
about 0.78

Example 4
Solve each equation or inequality by using natural logarithms.
a. $5^{2x} = 7^{x+1}$ **x = 1.5287**
b. $4^{x^2-1} < 12$ **$-1.67 < x < 1.67$**

$$\ln 76 = \ln e^{0.9229t} \qquad \textit{Take the natural logarithm of each side}$$
$$\ln 76 = 0.9229t \ln e \qquad \textit{ln } a^n = n \textit{ ln } a$$
$$4.3307 = 0.9229t \qquad \textit{Use a calculator.}$$
$$4.6925 = t$$

In about 4.7 days, 80% of the people in the group will hear the rumor.

Logarithms with a base other than e can be converted to natural logarithms using the change of base formula.

Example **2** **Convert $\log_6 254$ to a natural logarithm and evaluate.**

Use the change of base formula to convert the logarithm to a base e logarithm.

$$\log_6 254 = \frac{\log_e 254}{\log_e 6} \qquad \log_a n = \frac{\log_b n}{\log_b a},\ a = 6,\ b = e,\ n = 254$$

$$= \frac{\ln 254}{\ln 6} \qquad \log_e x = \ln x$$

$$\approx 3.0904 \qquad \textit{Use a calculator.}$$

So, $\log_6 254$ is about 3.0904.

Sometimes you know the natural logarithm of a number x and must find x. The antilogarithm of a natural logarithm is written **antiln x.** If $\ln x = a$, then $x = $ antiln a. Use the e^x function, which is the [2nd] function of [LN] to find the antilogarithm of a natural logarithm. *(antiln x is often written as e^x)*

Examples **3** **Solve $6.5 = -16.25 \ln x$.**

$$6.5 = -16.25 \ln x$$
$$-0.4 = \ln x \qquad \textit{Divide each side by } -16.25.$$
$$\text{antiln } (-0.4) = x \qquad \textit{Take the antilogarithm of each side.}$$
$$0.67 \approx x \qquad \textit{Use a calculator.}$$

4 **Solve each equation or inequality by using natural logarithms.**

a. $3^{2x} = 7^{x-1}$

Method 1: Algebra
$$3^{2x} = 7^{x-1}$$
$$\ln 3^{2x} = \ln 7^{x-1}$$
$$2x \ln 3 = (x-1)\ln 7$$
$$2x(1.0986) = (x-1)(1.9459)$$
$$2.1972x = 1.9459x - 1.9459$$
$$0.2513x = -1.9459$$
$$x = -7.7433$$

Method 2: Graphing
$$y_1 = 3^{2x},\ y_2 = 7^{x-1}$$

Intersection
X=-7.7429   Y=0

$[-10, 10]$ scl:1 by $[-10, 10]$ scl:1

b. $6^{x^2-2} < 48$

$$6^{x^2-2} < 48$$
$$\ln 6^{x^2-2} < \ln 48$$
$$(x^2 - 2)\ln 6 < \ln 48$$
$$x^2 - 2 < \frac{\ln 48}{\ln 6}$$
$$x^2 - 2 < 2.1606$$
$$x^2 < 4.1606$$

Intersection X=-2.0397 Y=48

[−10, 10] scl:1 by [−10, 60] scl:10

Intersection X=2.0397 Y=48

Notice that there are two intersection points, about −2.04 and 2.04. Since 6^{x^2-2} is less than 48, the solution is all points between $x = -2.04$ and $x = 2.04$. From this, we can write the inequalities $x > -2.04$ and $x < 2.04$, or $-2.04 < x < 2.04$.

CHECK FOR UNDERSTANDING

Communicating Mathematics

Read and study the lesson to answer each question. **1–4. See margin.**

1. **Explain** why $\ln e = 1$.

2. **Show** that $\log 17 \neq \ln 17$.

3. **Verify** that $\ln 64 = \ln 16 + \ln 4 = 2 \ln 8$.

4. *Math Journal* **Describe** the similarities between the equation $A = Pe^{rt}$ and the equation $N = N_0 \, e^{kt}$. How are the two equations alike?

Guided Practice

Evaluate each expression.

5. −4.7217

8. 0.4570

5. $\ln 0.0089$ 6. $\ln \frac{1}{0.32}$ **1.1394**   7. $\ln 0.21$ **−1.5606**   8. $\text{antiln}\,(-0.7831)$

Convert each logarithm to a natural logarithm and evaluate.

9. $\log_5 132$ **3.0339**

10. $\log_3 64$ **3.7856**

Use natural logarithms to solve each equation or inequality.

11. $18 = e^{3x}$ **0.9635**
12. $10 = 5e^{5k}$ **0.1386**
13. $25e^x < 100$ $x < 1.3863$
14. $4.5 \geq e^{0.031t}$ $48.5186 \geq t$

Graphing Calculator

Solve each equation or inequality by graphing. Round solutions to the nearest hundredth.

15. $6^{x+2} = 14^{x-3}$ **13.57**

16. $5^{x+3} > 10^{x-6}$ $x > 26.90$

17. **Meteorology** The atmospheric pressure varies with the altitude above the surface of Earth. Meteorologists have determined that for altitudes up to 10 kilometers, the pressure p in millimeters of mercury or torrs, is given by $p = 760e^{-0.125a}$, where a is the altitude in kilometers.

a. What is the atmospheric pressure at 3.3 kilometers?   **503.1 torrs**

b. At what altitude will the atmospheric pressure be 450 torrs?   **4.2 km**

www.amc.glencoe.com/self_check_quiz

Lesson 11-6   *Natural Logarithms*   **735**

Addressing Individual Needs

Logical Find the number of years for $3750 to appreciate to $175,000 if it is invested at 7.9% interest compounded continuously. Also, find the total interest earned over those years.   **about 49 years; $171,250**

Additional Answers

36. 2.3869
37. 4.7549
38. 4.8940
39. 1.3155
40. 3.6889
41. $40.9933 < t$
42. -0.7613
43. -0.3466
44. $y > 42.9183$
45. $x \le 1.7657$
46. $x > 9.0474$
47. $x \ge 144.9985$
48.

Intersection
X=-3.7612 Y=645

[–5, 5] scl:1, [–50, 700] scl:50

49.

Intersection
X=-7.637683 Y=.00502144

[–10, 10] scl:1, [–3, 10] scl:1

50.

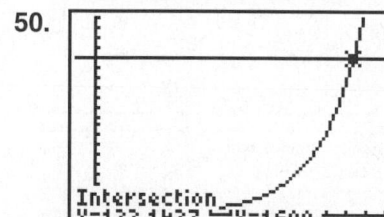

Intersection
X=133.1437 Y=1600

[–10, 15] scl:1, [–100, 2000] scl:100

51.

Intersection
X=2.1427011 Y=19.5

[–6, 6] scl:2, [–4, 24] scl:2

EXERCISES

Practice

Evaluate each expression.

A 18. $\ln 243$ **5.4931**   19. $\ln 0.763$ 20. $\ln 980$ **6.8876**   21. $\ln 125^{\frac{1}{5}}$ **0.9657**

19. **–0.2705**

22. $\ln 1$ **0**   23. $\ln 9.32$ **2.2322**   24. $\text{antiln } 2.3456$ 25. $\text{antiln } 0.1934$

24. **10.4395**

26. $\text{antiln } (-3.7612)$ 27. $\text{antiln } (-0.0034)$ 28. $\text{antiln } 4.987$ 29. $\text{antiln } (-1.42)$
    0.0233   **0.9966**   **146.4963**   **0.2417**

25. **1.2134**

Convert each logarithm to a natural logarithm and evaluate.

B 30. $\log_{12} 56$ **1.6199**   31. $\log_5 36$ **2.2266**   32. $\log_4 83$ **3.1875**

33. $\log_8 0.512$ **–0.3219**   34. $\log_6 323$ **3.2246**   35. $\log_5 \sqrt{288}$ **1.7593**

Use natural logarithms to solve each equation or inequality.   **36–47. See margin.**

C 36. $6^x = 72$ 37. $2^x = 27$ 38. $9^{x-4} = 7.13$ 39. $3^x = 3\sqrt{2}$

40. $25e^x = 1000$ 41. $60.3 < e^{0.1t}$ 42. $6.2e^{0.64t} = 3e^{t+1}$ 43. $22 = 44(1 - e^{2x})$

44. $25 < e^{0.075y}$ 45. $5^x \le 7\sqrt{6}$ 46. $12^{x-4} > 4^x$ 47. $x^{\frac{2}{3}} \ge 27.6$

Graphing Calculator

Solve each equation or inequality by graphing. Round solutions to the nearest hundredth.   **See margin for sample graphs.**
$t < 133.14$

48. $15e^{-x} = 645$ **–3.76**   49. $2^x = \sqrt{3^{x-2}}$ **–7.64**   50. $4e^{0.045t} < 1600$

51. $x = \log_4 19.5$ **2.14**   52. $10^{x-3} \ge 52$ **$x \ge 4.72$**   53. $\log_3 \sqrt[4]{5} \le x$ **$x \ge 0.37$**

Applications and Problem Solving

Real World Application

54. **Electricity** If a charged capacitor is placed in a circuit where the only other component is a resistor, then the charge will flow out of the capacitor. The charge Q on the capacitor after t seconds is given by $Q = Q_0 e^{-\frac{t}{RC}}$, where R is the resistance, C is the capacitance, and Q_0 is the initial charge on the capacitor. In a circuit with $R = 20{,}000$ ohms and $C = 4 \times 10^{-11}$ farads, how long will it take the capacitor to lose 40% of its initial charge? $\approx 4.09 \times 10^{-7}$ **s**

55. **Medicine** Nuclear medicine technologists use the iodine isotope I-131, with a half-life of 8 days, to check thyroid function of patients. After ingesting a tablet containing the iodine, the isotopes collect in a patient's thyroid, and a special camera is used to view its function. Suppose a patient ingests a tablet containing 9 microcuries of I-131. To the nearest hour, how long will it be until there are only 2.8 microcuries in the patient's thyroid? *A microcurie is a measure of radiation.* **324 hr**

56. **Temperature** According to Newton's Law of Cooling, the temperature T of an object at time t satisfies the equation $\ln |T - T_0| = -kt + C$ where T_0 is the temperature of the surrounding medium and k and C are constants. Suppose a cup of coffee with temperature of $180°$ F is placed in a $72°$ F room at time $t = 0$.

a. Find the value of C to the nearest ten thousandth. **4.6821**

b. After 2 minutes, the temperature of the coffee has dropped to $150°$ F. Find the value of k to the nearest ten thousandth. **0.1627**

c. How much longer will it take for the temperature of the coffee to drop to $100°$ F? **about 6.3 min**

57. **Critical Thinking** Solve the equation $e^{-2x} - 4e^{-x} + 3 = 0$ without graphing.
0 or –1.0986

52.

Intersection
X=4.7160033 Y=52

[–10, 10] scl:1, [–10, 75] scl:5

53.

Intersection
X=.36624338 Y=1.4953488

[–5, 5] scl:1, [–2, 5] scl:0.5

58. Finance Jenny Oiler deposited some money in an investment account that earns 6.3% interest compounded continuously.

    a. How long will it take to double the money in the account? **about 11 years**

    b. The Rule of 72 states that if you divide 72 by the interest rate of an account that compounds interest continuously, the result is the approximate number of years that it will take for the money in the account to double. Do you think the Rule of 72 is accurate? Explain your reasoning. **See students' work.**

59. Linguistics If two languages have evolved separately from a common ancestral language, the number of years since the split $n(r)$, is modeled by the formula $n(r) = -5000 \ln r$, where r is the percent of the words from the ancestral language that are common to both languages now. If two languages split off from a common ancestral language about 1800 years ago, what portion of the words from the ancestral language would you expect to find in both languages today? **≈ 70%**

60. Chemistry Radium 226, which had been used for cancer treatment, decays with a half-life of 1622 years. **a. −0.000427**

    a. Find the constant k using 1 gram as the original amount of Radium 226.

    b. Suppose there is a 2.3-gram sample of Radium 226. How long will it take to have 1.7 grams remaining? **about 708 yr**

61. Critical Thinking Using the table below, determine if y is a logarithmic function of x or if x is a logarithmic function of y. Explain your answer.

x	1	3	9	27
y	0	1	2	3

See margin.

Mixed Review

62. Use a calculator to find the common logarithm of 19.25 to the nearest ten thousandth. *(Lesson 11-5)* **1.2844**

63. Write $\log_{16} 8 = \frac{3}{4}$ in exponential form. *(Lesson 11-4)* **$16^{\frac{3}{4}} = 8$**

64. (−2.2, 0.9), (−1.7, −1.1), (2.2, 0.9), (1.7, −1.1); see margin for graph.

64. Solve the system $x^2 = y + 4$, $x^2 + 4y^2 = 8$, algebraically. Round to the nearest tenth. Check the solutions by graphing each system. *(Lesson 10-8)*

65. 0.00765 N · m

65. Chemistry Boyle's Law states that the pressure P exerted by a gas varies inversely as the volume V of the gas at constant temperature. If a specific gas is collected in a 146 cubic centimeter container and the pressure is 52.4 pascals (1 pascal $= 1$ N/m^2), use the formula $PV = c$ to find c. *(Lesson 10-4)*

66. Find the rectangular coordinates of the point with polar coordinates $(0.25, \pi)$. *(Lesson 9-3)* **(−0.25, 0)**

67. Find an ordered pair to represent $\vec{a}$ in the equation $\vec{a} = \vec{b} + 3\vec{c}$ if $\vec{b} = \langle 1, -2 \rangle$ and $\vec{c} = \langle 4, 3 \rangle$. *(Lesson 8-2)* **$\langle 13, 7 \rangle$**

68. Change $2x - 5y + 3 = 0$ to normal form. Then find p, the measure of the normal, and ϕ, the angle it makes with the positive x-axis. *(Lesson 7-6)* **See margin.**

69. Write an equation of a cosine function with amplitude 70 and period 90°. *(Lesson 6-4)* **$y = \pm 70 \cos 4\theta$**

70. SAT Practice **Grid-In** Doreen sets out on an 800-mile car trip. Her car can travel for 10 hours at 55 miles per hour on a full tank of gas. If she starts with a full tank of gas and stops only once to refuel, what is the least number of miles Doreen would have remaining after stopping to refuel? **250**

Extra Credit

It can be shown that $e^{i\theta} = \operatorname{cis} \theta$. We could use this equation to define natural logarithms for negative numbers. What would be the value of $\ln -1$, $\ln -e$, and $\ln -5$? Note: θ should be in radians. **$i\pi$; $1 + i\pi$; $\ln 5 + i\pi$**

11-6B Natural Logarithms and Area

An Extension of Lesson 11-6

1 FOCUS

Motivating the Lesson

Hands-On Activity Have students use a large posterboard to draw the function $y = \frac{1}{x}$ on a set of grid lines. Have students cut out the squares under the curve that would give an approximation of the area under the curve. Remind students that as the squares get smaller and smaller, the approximation will get closer and closer to the real values.

2 TEACH

📁 A reproduction of this lesson with graphing calculator keystrokes provided can be found in the *Graphing Calculator Keystrokes Masters*, pp. 46–48.

Teaching Tip Remind students that the answers given by their calculators are approximations, and that they will learn more about area under curves in Lesson 15-3.

OBJECTIVES
• Investigate the relationship between area of regions below the graph of $y = \frac{1}{x}$ and natural logarithms.

There is an interesting and important relationship between the function $y = \ln x$ and the areas of regions below the graph of $y = \frac{1}{x}$. (You will learn more about areas under graphs or curves in Chapter 15.) You can explore this relationship with a graphing calculator by using the ∫ **f(x) dx** feature under the **CALC** menu. Using this feature, you can have the calculator display the area of the region between the graph of a function, the x-axis, and a pair of vertical lines at $x = a$ and $x = b$, where a and b are any real numbers, called the lower and upper boundaries.

Graph the equation $y = \frac{1}{x}$ and select the ∫ **f(x) dx** feature under the **CALC** menu. Move the cursor to the left until it is at the point where $x = 0.8$. Press ENTER to set 0.8 as the lower boundary for x. Then move the cursor to the right and set 2.35 as the upper boundary for x.

[0, 4.7] scl:1, [−0.5, 2.6] scl:1

The calculator automatically shades the region between the graph, the x-axis, and the lines $x = 0.8$ and $x = 2.35$. At the bottom of the screen, the calculator displays the area of the shaded region, which is approximately 1.0775589 square units.

TRY THESE

1. Use the procedure described above to find the area of the region between the graph of $y = \frac{1}{x}$, the x-axis, and the vertical lines $x = 2$ and $x = 4$. Use $x = 2$ as the lower boundary and $x = 4$ as the upper boundary. What area does the calculator display for the shaded region? **0.69314718**

2. Press [2nd] [QUIT] to go to the home screen. Then press [2nd] [ANS] [ENTER]. What number does the calculator display? How is this number related to the area you found in Exercise 1? **0.6931471806; it is the same value as found in Exercise 1 expressed to 10 decimal places**

738 *Chapter 11 Exponential and Logarithmic Functions*

3. Press [GRAPH] to redisplay the graph. Clear the shading from the graph using the **ClrDraw** command on the **DRAW** menu to display a fresh graph.

Repeat Exercise 1 setting $x = 4$ as the *lower* limit and $x = 2$ as the *upper* limit.

a. Describe how this result is related to the answer for Exercise 1. **See margin.**

b. What kind of result do you think the calculator will display when the value of the lower limit is greater than the value of the upper limit? **See margin**

[0, 4.7] scl:1, [−0.5, 2.6] scl:1

4. Graph $y = \frac{1}{x}$. **d–e. See margin.**

a. Set $x = 1$ as the lower boundary and $x = 2$ as the upper boundary. Record the displayed area. **0.69314718**

b. Set $x = 1$ as the lower boundary and $x = 3$ as the upper boundary. Record the displayed area. **1.0986123**

c. Set $x = 1$ as the lower boundary and $x = 4$ as the upper boundary. Record the displayed area. **1.3862944**

d. Use your calculator to find the values of ln 2, ln 3, and ln 4.

e. Compare each area with each natural logarithm.

5. Find the areas of the regions under the graph of $y = \frac{1}{x}$ that result when you use $x = 1$ as the lower limit and $x = 0.6$, 0.5, and 0.4 as upper limits. How do the values compare with ln 0.6, ln 0.5, and ln 0.4? **See margin.**

WHAT DO YOU THINK?

6. Consider the region between the graph of $y = \frac{1}{x}$, the x-axis, and the vertical lines $x = 1$ and $x = k$, where $k > 0$. Make a conjecture about how the area of this region is related to ln k. **See margin.**

7. Suppose you enter the areas you found in Exercises 4 and 5 as the elements of list **L1** and the corresponding upper-boundary values for x in list **L2**. If you use the **ExpReg** on the **STAT CALC** menu to display an exponential regression equation of the form $y = ab^x$, what values would you expect the calculator to display for the constants a and b? Enter the numbers for lists **L1** and **L2** and display the regression equation to see if your prediction is correct. **See margin.**

8. One way to approach the concept of natural logarithms is to define e as the limiting value of $\left(1 + \frac{1}{n}\right)^n$ as n approaches ∞ and then define natural logarithms as logarithms to the base e. Make a conjecture of another way to define natural logarithms and e based on the ideas you have explored in the preceding exercises. **See margin.**

Lesson 11-6B Natural Logarithms and Area **739**

Assignment Guide

All: 1–8

Additional Answers

3a. The result is the opposite of the result in Exercise 1.

3b. Sample answer: a negative value

4d. 0.6931471806, 1.098612289, 1.386294361

4e. The value for each area is the same as the value of each natural logarithm.

5. −0.5108256238; −0.6931471806; −0.9162907319; these values are equal to the values of ln 0.6, ln 0.5, and ln 0.4.

6. If $k \geq 1$, then the area of the region is equal to ln k. If $0 < k < 1$, then the opposite of the area is equal to ln k.

7. The value of a should be equal to or very close to 1, and the value of b should be very close to e. This prediction is confirmed when you display the actual regression equation.

8. Sample answer: define ln k for $k > 0$ to be the area between the graph of $y = \frac{1}{x}$, the x-axis, and the vertical lines $x = 1$ and $x = k$ if $k \geq 1$ and to be the opposite of this area if $0 < k < 1$. Define e to be the value of k for which the area of the region is equal to 1.

Open-Ended Assessment

Writing Ask students to write descriptions of lower limits, upper limits, and what the shaded region of the graph represents.

Modeling Real-World Data with Exponential and Logarithmic Functions

1 FOCUS

5-Minute Check
Lesson 11-6

Evaluate each expression.
1. ln 17.9 **2.8848**

2. antiln −4.1715 **0.0154**

Convert each logarithm to a natural logarithm and evaluate.
3. $\log_5 19$ **1.8295**

4. $\log_8 0.216$ **−0.7370**

5. Use natural logarithms to solve the equation $8^{x-3} = 5.19$. **3.7919**

6. Solve $10^{x-2} \geq 20$ by graphing. Round solution to the nearest hundredth. **$x \geq 3.30$**

Motivating the Lesson

Discussion Ask students to discuss why they think it is that while $\log_{10} x$ is abbreviated by log x and $\log_e x$ is abbreviated by ln x, other logarithmic bases do not merit their own special abbreviations. **Sample answer: base e and base 10 are used more often than the other bases for logarithmic calculations.**

Teaching Tip The doubling time formula shown on this page and used in Examples 1 and 2 is often used in simplified form as the Rule of 70 by economists and demographers. Since ln 2 rounded to the nearest hundredth equals 0.69, it is often rounded further to 0.70, thus providing the basis for the Rule of 70. Furthermore, some financial experts use the Rule of 72 for doubling time. The numbers 70 and 72 are used rather than 69 in an effort to compensate for rates of growth which do not adhere strictly to a regular pattern of growth.

OBJECTIVES
- Find the doubling time of an exponential quantity.
- Find exponential and logarithmic functions to model real-world data.
- Linearize data.

INVESTMENT Latrell, a freshman at Finneytown High School, is given a gift of $4000 by his great aunt. He would like to invest the money so that he will have enough to purchase a car that costs twice that amount, or $8000, when he graduates in four years. If he invests the $4000 in an account that pays 9.5% compounded continuously, will he have enough money to buy the car? *This problem will be solved in Example 1.*

There are situations, such as investments and population, where it is of interest to know how long it takes a quantity modeled by an exponential function to reach twice its initial amount. This amount of time is called the **doubling time** of the quantity. We can derive a formula for doubling time from the exponential equation $N = N_0 e^{kt}$. If t is the doubling time, then $N = 2N_0$.

$$N = N_0 e^{kt}$$
$$2N_0 = N_0 e^{kt} \quad \text{\textit{Replace N with }} 2N_0$$
$$2 = e^{kt} \quad \text{\textit{Divide each side by }} N_0.$$
$$\ln 2 = \ln e^{kt} \quad \text{\textit{Take the natural logarithm of each side.}}$$
$$\ln 2 = kt \quad \text{\textit{ln }} e^x = x$$
$$\frac{\ln 2}{k} = t \quad \text{\textit{Divide each side by k to solve for t.}}$$

Doubling Time	The amount of time t required for a quantity modeled by the exponential equation $N = N_0 e^{kt}$ to double is given by the following equation. $$t = \frac{\ln 2}{k}$$

Example **1** **INVESTMENT** Refer to the application at the beginning of the lesson.

a. Will Latrell have enough money after 4 years to buy the car?

Find the doubling time for Latrell's investment. For continuously compounded interest, the constant k is the interest rate, written as a decimal.

$$t = \frac{\ln 2}{k}$$
$$= \frac{\ln 2}{0.095} \quad \text{\textit{The decimal for 9.5\% is 0.095.}}$$
$$\approx 7.30 \text{ years} \quad \text{\textit{Use a calculator.}}$$

Four years is not enough time for Latrell's money to double. He will not have enough money to buy the car.

740 *Chapter 11 Exponential and Logarithmic Functions*

Resource Manager

Reproducible Masters
- *Study Guide and Practice,* pp. 209–210
- *Enrichment,* p. 127
- *Assessment and Evaluation,* p. 278
- *Spreadsheet Applications,* p. 11

Transparencies
- 5-Minute Check, 11-7
- Teaching, 11-7

b. What interest rate is required for an investment with continuously compounded interest to double in 4 years?

$$t = \frac{\ln 2}{k}$$

$$4 = \frac{\ln 2}{k}$$

$$k = \frac{\ln 2}{4} \qquad \textit{Solve for k.}$$

$$k \approx 0.1733$$

An interest rate of 17.33% is required for an investment with continuously compounded interest to double in 4 years.

In Lesson 1-6, you learned how to fit a linear function to a set of data. Sometimes the scatter plot of a set of data may suggest a nonlinear model. The process of fitting an equation to nonlinear data is called **nonlinear regression.** In this lesson, we will use a graphing calculator to perform two types of nonlinear regression, exponential regression and logarithmic regression. When deciding which type of model is more appropriate, you will need to keep in mind the basic shapes of exponential and logarithmic functions.

Exponential Functions: $y = ab^x$		Logarithmic Functions: $y = a + b \ln x$	
Growth	**Decay**	**Growth**	**Decay**

Examples

Real World Application

interNET CONNECTION

Data Update
For the latest data on population and other applications studied in this lesson, visit: **www.amc. glencoe.com**

② POPULATION The table below gives the population of the world in billions for selected years during the 1900s.

Year	1900	1910	1920	1930	1940	1950	1960	1970	1980	1990	2000
Years since 1900	0	10	20	30	40	50	60	70	80	90	100
Population	1.65	1.75	1.86	2.07	2.30	2.52	3.02	3.70	4.44	5.27	6.06

Source: United Nations

a. Find a function that models the population data shown.

Enter the data on the **STAT EDIT** screen. In order to work with smaller numbers, use the number of years since 1900 as the independent variable. Then draw a scatter plot. The scatter plot suggests an exponential model.

[0, 100] scl:1 10 by [0, 6] scl:1

(continued on the next page)

Lesson 11-7 Modeling Real-World Data with Exponential and Logarithmic Functions **741**

---

2 TEACH

In-Class Examples

Example 1
Investment Hosea, a senior at Allen East High School, has saved $2000 from his summer job mowing lawns. He would like to invest the money so that he will have double the money in six years when he graduates from college. Hosea invests the $2000 in an account that pays 8% interest compounded continuously.

a. Will Hosea have doubled his money in 6 years? Explain.
No; Hosea will need 8.7 years to double his money.

b. What interest rate is required for an investment with continuously compounded interest to double in 6 years? **about 11.55%**

Example 2
Population India is expected to have the largest population in the world by 2050. The table below gives the population of India in 100 millions for selected years during the 1900s.

Year	Years Since 1950	Population
1950	0	3.58
1960	10	4.42
1970	20	5.55
1980	30	6.89
1990	40	8.51

Source: United Nations

a. Find an equation that models the population data shown.
$y = 3.5758 (1.0220)^x$

b. Use the equation to predict the population of India in 2050. **about 3.15 billion**

Lesson 11-7 **741**

In-Class Example

Example 3

Bacteriology The number of bacteria in a culture is observed for several hours with the following results recorded.

Hours (*t*)	Number of Bacteria (*N*) (thousands/cc)
2.5	10.06
5.4	14.59
6.5	20.70
9.2	27.91
9.5	31.50
11.0	40.06

Find an equation that models the data. $N = 6.5828\,(1.1769)^t$

Graphing Calculator Tip

Select **DiagnosticOn** from the **CATALOG** to display r and r^2.

To have the calculator find a regression equation of the form $y = ab^x$, use **ExpReg** from the **STAT CALC** screen.

The equation of an exponential function that models the population data is $y = 1.4457(1.0137)^x$.

```
ExpReg
 y=a*b^x
 a=1.445686606
 b=1.013691501
 r²=.970011724
 r=.9848917321
```

The equation can be written in terms of base e as follows.

$y = 1.4457(1.0137)^x$
$y = 1.4457(e^{\ln 1.0137})^x$ $e^{\ln a} = a$
$y = 1.4457 e^{(\ln 1.0137)x}$ $(a^m)^n = a^{mn}$
$y = 1.4457 e^{0.0136x}$ $\ln 1.0137 \approx 0.0136$

b. Use the equation to predict the population of the world in 2050.

To predict the population in 2050, observe that 2050 is 150 years after 1900.

$y = 1.4457(1.0137)^x$
$ = 1.4457(1.0137)^{150}$ *Replace x with 150.*
$ \approx 11.13$

The model predicts that the population of the world in 2050 will be 11.13 billion.

Real World Application

③ SKATING An ice skater begins to coast with an initial velocity of 4 meters per second. The table below gives the times required for the skater to slow down to various velocities. Find an equation that models the data.

velocity (m/s)	3.5	3	2.5	2	1.5	1	0.5
time (s)	2.40	5.18	8.46	12.48	17.66	24.95	37.43

Make a scatter plot of the data. Let the velocity be the independent variable and let time be the dependent variable.

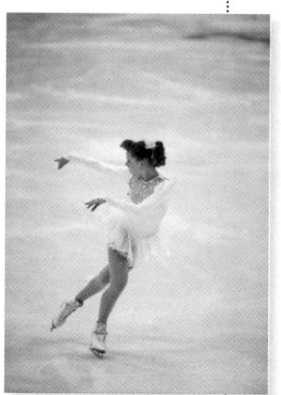

Since the scatter plot does not appear to flatten out as much as a decaying exponential function would, we will look for a logarithmic model of the form $y = a + b \ln x$ where $b < 0$ for the data.

[0, 4] scl:1 by [0, 40] scl:10

To have the calculator find a regression equation of the form $y = a + b \ln x$, use **LnReg** from the **STAT CALC** screen.

The equation of a logarithmic function that models the data is $y = 24.95 - 18 \ln x$.

```
LnReg
 y=a+blnx
 a=24.9537668
 b=-18.00000214
 r²=.999999932
 r=-.999999966
```

From the Classroom of...

Colleen Boelter
Weyauwega-Fremont High School
Weyauwega, Wisconsin

"Working in small groups, I have students choose an open-ended mathematical modeling question and answer it by writing a report or by giving an oral presentation using graphs, charts, posters, etc. One question asks students to decide when they will begin to earn money, when they will retire, how much money they will need to save prior to retirement, and how much money they need to save each month to retire with this amount."

Another approach to modeling curved data is the concept of **linearizing data**. Suppose a set of data can be modeled by the exponential function $y = ae^{bx}$. We can rewrite this equation using the properties of logarithms.

$y = ae^{bx}$

$\ln y = \ln ae^{bx}$ *Take the natural logarithm of each side.*

$\ln y = \ln a + \ln e^{bx}$ *Product property of logarithms*

$\ln y = \ln a + bx$ *$\ln e^x = x$*

The last equation shows that $\ln y$ is a linear function of x. In other words, if the ordered pairs $(x, \ln y)$ are graphed, the result will be a straight line. This means that exponential data can be analyzed by linearizing the data and then applying linear regression.

Example

Real World Application

④ ECONOMICS The Consumer Price Index (CPI) measures inflation. It is based on the average prices of goods and services in the United States, with the average for the years 1982–1984 set at an index of 100. The table below gives some CPI values from 1950 to 1996.

Year	1950	1960	1970	1980	1990	1996
CPI	24.1	29.6	38.8	82.4	130.7	156.9

Source: Bureau of Labor Statistics

a. Linearize the data. That is, make a table with x- and $\ln y$-values, where x is the number of years since 1950 and y is the CPI. Then make a scatter plot of the linearized data.

Subtract 1950 from each year and find the natural logarithm of each CPI-value.

x	0	10	20	30	40	46
ln y	3.18	3.39	3.66	4.41	4.87	5.06

The scatter plot suggests that there may be a linear relationship between x and $\ln y$.

[0, 50] scl:10 by [0, 6] scl:1

b. Find a regression equation for the linearized data.

Use **LinReg(ax+b)** on the **STAT CALC** screen to find the linear regression equation.

We can use the equation $\ln y = 0.0442x + 3.0196$ to model $\ln y$.

In-Class Example

Example 4

Economics The table below gives some CPI values from 1955 to 1997.

Year	CPI
1955	26.8
1965	31.6
1975	53.8
1985	107.8
1995	152.4
1997	160.5

Source: *Wall Street Journal Almanac 1999*

a. Linearize the data. Then make a scatter plot of the linearized data.

x	ln y
5	3.29
15	3.45
25	3.99
35	4.68
45	5.03
47	5.08

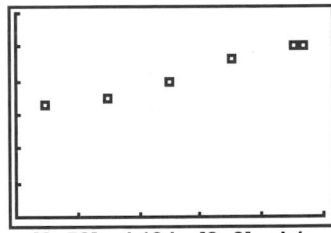

[0, 50] scl:10 by [0, 6] scl:1

b. Find a regression equation for the linearized data.
$\ln y = 0.0465x + 2.9197$

c. Use the linear regression equation to find an exponential model for the original data.
$y = 18.54e^{0.0465x}$

d. Use the exponential model to predict the CPI in 2010.
301.9

Error Analysis

Watch for students who have difficulty dealing with ln e whenever it occurs in an equation.

Prevent by stressing that ln e can be replaced by 1 since $\log_e e = 1$ or $e^1 = e$.

Additional Answers

1. Replace N by $4N_0$ in the equation $N = N_0 e^{kt}$, where N_0 is the amount invested and k is the interest rate. Then solve for t.

2. The data should be modeled with an exponential function. The points in the scatter plot approach a horizontal asymptote. Exponential functions have horizontal asymptotes, but logarithmic functions do not.

3. $y = 2e^{(\ln 4)x}$ or $y = 2e^{1.3863x}$; $\ln y = \ln 2 + (\ln 4)x$ or $\ln y = 0.6931 + 1.3863x$

Study Guide and Practice Masters, pp. 209–210

NAME _____ DATE _____ PERIOD _____

11-7

Study Guide

Modeling Real-World Data with Exponential and Logarithmic Functions

The **doubling time**, or amount of time t required for a quantity modeled by the exponential equation $N = N_0 e^{kt}$ to double, is given by $t = \frac{\ln 2}{k}$.

Example   **Finance**   Tara's parents invested $5000 in an account that earns 11.5% compounded continuously. They would like to double their investment in 5 years to help finance Tara's college education.

a. **Will the initial investment of $5000 double within 5 years?**

Find the doubling time for the investment. For continuously compounded interest, the constant k is the interest rate written as a decimal.

$t = \frac{\ln 2}{k}$

$= \frac{\ln 2}{0.115}$ *The decimal for 11.5% is 0.115.*

≈ 6.03 years   *Use a calculator.*

Five years is not enough time for the initial investment to double.

b. **What interest rate is required for an investment with continuously compounded interest to double in 5 years?**

$t = \frac{\ln 2}{k}$

$5 = \frac{\ln 2}{k}$

$\frac{1}{5} = \frac{k}{\ln 2}$ *Take the reciprocal of each side.*

$\frac{\ln 2}{5} = k$ *Multiply each side by ln 2 to solve for k.*

$0.1386 = k$

An interest rate of 13.9% is required for an investment with continuously compounded interest to double in 5 years.

© Glencoe/McGraw-Hill   209   *Advanced Mathematical Concepts*

---

c. **Use the linear regression equation to find an exponential model for the original data.**

To find a model solve the regression equation in part **b** for y.

$\ln y = 0.0442x + 3.0196$

$e^{\ln y} = e^{0.0442x + 3.0196}$ *Raise e to each side.*

$y = e^{0.0442x + 3.0196}$ $e^{\ln y} = y$

$y = e^{0.0442x} \cdot e^{3.0196}$ *Product Property of Exponents*

$y = 20.48e^{0.0442x}$ $e^{3.0196} \approx 20.48$

The Consumer Price Index between 1950 and 1996 can be modeled by the exponential function $y = 20.48e^{0.0442x}$.

d. **Use the exponential model to predict the CPI in 2020.**

The year 2020 is 70 years after 1950, so replace x with 70 in the exponential function.

$y = 20.48e^{0.0442x}$

$= 20.48e^{0.0442(70)}$ or about 451.9

The model predicts that in 2020 the CPI will be 451.9.

CHECK FOR UNDERSTANDING

Communicating Mathematics

Read and study the lesson to answer each question.   1–3. See margin.

1. **Explain** how you would determine the time needed to quadruple an investment if the interest is compounded continuously.

2. **State** whether the data in the scatter plot should be modeled with an exponential function or a logarithmic function. Explain.

3. **Write** the function $y = 2 \cdot 4^x$ as an exponential function with base e. Then write ln y as a linear function of x.

Guided Practice

Find the amount of time required for an amount to double at the given rate if the interest is compounded continuously.

4. 1.75%   **39.61 yr**

5. 8%   **8.66 yr**

6. **Radioactivity**   Exponential regression can be used in the experimental determination of the half-life of a radioactive element. A scientist starts with a 10-gram sample of uranium-239 and records the measurements shown below.

Time (min)	0	5	10	15	20
U-239 present (g)	10	8.6	7.5	6.3	5.5

a. Find a regression equation for the amount y of uranium as a function of time x. $y = 10.0170(0.9703)^x$

b. Write the regression equation in terms of base e. $y = 10.0170e^{-0.0301x}$

c. Use the equation from part b to estimate the half-life of uranium-239.

23.08 min

 www.amc.glencoe.com/self_check_quiz

Addressing Individual Needs

Auditory/Musical   Have students explore the ratio in a musical circle of fifths. A perfect fifth consists of a tone whose frequency ratio is exactly $2^{\frac{\log 3}{\log 2}}$. If your octave consists of 12 tones and fifths of 7 halfsteps, find an approximation for the exact frequency ratio. $\frac{2^{\frac{7}{12}}}{1}$ or about 1.4983

EXERCISES

Assignment Guide

Core: 7–21 odd, 22–33
Enriched: 8–12 even, 14–33

Practice

A

Find the amount of time required for an amount to double at the given rate if the interest is compounded continuously.

7. 2.25%   **30.81 yr**   8. 5%   **13.86 yr**   9. 7.125%   **9.73 yr**

Determine whether the data in each scatter plot should be modeled with an exponential function or a logarithmic function. Explain your choices.

10–13. See margin.

10.

11.

12.

13.

Additional Answers
10. Exponential; the graph has a horizontal asymptote.
11. Logarithmic; the graph has a vertical asymptote.
12. Logarithmic; the graph has a vertical asymptote.
13. Exponential; the graph has a horizontal asymptote.

Applications and Problem Solving

14. **Biology**   The data below give the number of bacteria, in millions, found in a certain culture.

Time (hours)	0	1	2	3	4
Bacteria	5	8	15	26	48

a. Find an exponential function that models the data. $y = 4.7818(1.7687)^x$

b. Write the equation from part **a** in terms of base e. $y = 4.7818e^{0.5702x}$

c. Use the model to estimate the doubling time for the culture.   **1.22 hr**

Teaching Tip Because of differences in calculators, rounding, and data points chosen, answers may vary slightly from those given.

15. **Radioactivity**   A scientist starts with a 1-gram sample of lead-211. The amount of the sample remaining after various times is shown in the table below.

Time (min)	10	20	30	40
Pb-211 present (g)	0.83	0.68	0.56	0.46

$y = 1.0091(0.9805)^x$

a. Find a regression equation for the amount y of lead as a function of time x.

b. Write the regression equation in terms of base e. $y = 1.0091e^{-0.0197x}$

c. Use the equation from part **b** to estimate the half-life of lead-211.   **35.10 min**

16. Banking Elayna found an old savings passbook in her grandparents' attic. It contained the following account balances.

Date	Balance
Jan. 1, 1955	2137.52
Jan. 1, 1956	2251.61
Jan. 1, 1957	2371.79
Jan. 1, 1958	2498.39
Jan. 1, 1959	2631.74

a. Find a function that models the amount in the account. Use the number of years after Jan. 1, 1955 as the independent variable. $y = 2137.5192(1.0534)^x$

b. Write the equation from part **a** in terms of base e. $y = 2137.5192e^{0.0520x}$

c. What was the interest rate on the account if the interest was compounded continuously and no deposits or withdrawals were made during the period in question? **5.2%**

17. Sound The human impression of the strength of a sound depends, among other things, on the frequency of the sound. The loudness level of a sound, in *phons,* is the decibel level of an equally loud 1000 Hz tone. The *sone* is another unit of comparative loudness. The table below lists some equivalent values of phons and sones. Find an equation that gives the phon level of a sound in terms of the number of sones. $y = 40 + 14.4270 \ln x$

sones	0.5	1	2	4
phons	30	40	50	60

18. Education The table below lists the number of bachelor's degrees granted in the United States for certain years.

Year	1960	1965	1970	1975	1980	1985	1990	1994
Degrees (1000s)	392	494	792	923	930	980	1052	1169

Source: U.S. National Center for Education Statistics

18a. $y = -812.9099 + 515.8976 \ln x$

18b. The year 1960 would correspond to $x = 0$ and $\ln 0$ is undefined.

a. Find a function that models the data. Let x be the number of years after 1950 and let y be the number of degrees granted, in thousands.

b. Why can x not be the number of years since 1960?

19. Critical Thinking Consider the equation $y = cx^2$, where c is a constant and $x \geq 0$. What operation could be applied to each side of the equation to obtain a linear function of x? **Take the square root of each side.**

20. Banking Tomasita deposited $1000 in an account at the beginning of 2001. The account had a 3% interest rate, compounded quarterly. The table below shows her balance after each quarter of the year.

Date	Jan. 1	Mar. 31	June 30	Sept. 30	Dec. 31
Balance	$1000	1007.50	1015.06	1022.67	1030.34

a. In an account where the interest is only compounded once per year, the balance after x years is given by $P(1 + r)^x$, where P is the amount invested and r is the interest rate as a decimal. Find the interest rate with an annual compounding that would result in the balance that Tomasita had at the end of one year. This interest rate is called the *effective rate.* **3.034%**

20b. $y = 1000.0006(1.0303)^x$

b. Find a function that models the amount in Tomasita's account. Let x be the number of years since the beginning of 2001.

c. Write the equation from part **b** in terms of base e. $y = 1000.0006e^{0.0299x}$

d. What interest rate with continuous compounding would result in the balances shown for Tomasita's account? **2.99%**

21. Population Density The table below gives the population density (persons per square mile) of the United States at various times in its history.

Year	1800	1850	1900	1950	1990
Population Density	6.1	7.9	25.6	42.6	70.3

Source: U. S. Bureau of the Census

a. Let x be the number of years since 1800 and let y be the population density in persons per square mile. Copy and complete the table below to linearize the data.

x	0	50	100	150	190
ln y	1.81	2.07	3.24	3.75	4.25

b. Find a regression equation for the linearized data. **$\ln y = 0.0137x + 1.6833$**

c. Solve the regression equation for y. **$y = e^{0.0137x + 1.6833}$**

d. Use the answer to part **c** to predict the population density of the United States in 2025. **117.4 persons per square mile**

22. Critical Thinking The graphing calculator screen shows a scatter plot of the data in the table below.

x	y	x	y
−5	2.01	−1	2.38
−4	2.02	0	3
−3	2.06	1	4.6
−2	2.15	2	8.76

a. What must you do to the data before you can use a graphing calculator to perform exponential regression? Explain. **a–b. See margin.**

b. Find a regression equation for the data.

23. Critical Thinking A *power function* is a function of the form $y = c \cdot x^a$, where c and a are constants. **ln y is a linear function of ln x.**

a. What is the relationship between $\ln y$ and x for a power function?

b. The distance that a person can see from an airplane depends on the altitude of the airplane, as shown in the table below.

Altitude (m)	500	1000	5000	10,000	15,000
Viewing Distance (km)	89	126	283	400	490

Let x be the altitude of the airplane in meters and let y be the viewing distance in kilometers. Linearize the data. **See margin.**

c. Find a linear regression equation for the linearized data.

d. Solve the linear regression equation for y to obtain a power function that gives y in terms of x. **$y = 4.0153x^{0.4994}$**

23c. ln y = 0.4994 ln x + 1.3901

Additional Answers
22a. The graph appears to have a horizontal asymptote at $y = 2$, so you must subtract 2 from each y-value before a calculator can perform exponential regression.

22b. $y = 2 + 1.0003(2.5710)^x$

23b.

ln x	ln y
6.21	4.49
6.91	4.84
8.52	5.65
9.21	5.99
9.62	6.19

4 ASSESS

Open-Ended Assessment

Writing Have students write a paragraph on how to find the doubling time of an investment when the other variables in a continuously compounded interest rate formula are given.

Chapter 11, Quiz D (Lesson 11-7) is available in the *Assessment and Evaluation Masters*, p. 278.

Enrichment Masters, p. 127

7-5 Study Guide

Solving Trigonometric Equations

When you solve trigonometric equations for **principal values** of x, x is in the interval $-90° \le x \le 90°$ for $\sin x$ and $\tan x$. For $\cos x$, x is in the interval $0° \le x \le 180°$. If an equation cannot be solved easily by factoring, try writing the epressions in terms of only one trigonometric function.

Example 1 Solve $\tan x \cos x - \cos x = 0$ for principal values of x. Express solutions in degrees.

$$\tan x \cos x - \cos x = 0$$
$$\cos x (\tan x - 1) = 0 \quad \textit{Factor.}$$
$$\cos x = 0 \quad \text{or} \quad \tan x - 1 = 0 \quad \textit{Set each factor equal to 0.}$$
$$x = 90° \qquad \tan x = 1$$
$$x = 45°$$

When $x = 90°$, $\tan x$ is undefined, so the only principal value is 45°.

Example 2 Solve $2\tan^2 x - \sec^2 x + 3 = 1 - 2\tan x$ for $0 \le x < 2\pi$.

This equation can be written in terms of $\tan x$ only.

$$2\tan^2 x - \sec^2 x + 3 = 1 - 2\tan x$$
$$2\tan^2 x - (\tan^2 x + 1) + 3 = 1 - 2\tan x \quad sec^2 x = tan^2 x + 1$$
$$\tan^2 x + 2 = 1 - 2\tan x \quad \textit{Simplify.}$$
$$\tan^2 x + 2\tan x + 1 = 0$$
$$(\tan x + 1)^2 = 0 \quad \textit{Factor.}$$
$$\tan x + 1 = 0 \quad \textit{Take the square root of each side.}$$
$$\tan x = -1$$
$$x = \frac{3\pi}{4} \text{ or } x = \frac{7\pi}{4}$$

When you solve for all values of x, the solution should be represented as $x + 360°k$ or $x + 2\pi k$ for $\sin x$ and $\cos x$ and $x + 180°k$ or $x + \pi k$ for $\tan x$, where k is any integer.

Example 3 Solve $\sin x + \sqrt{3} = -\sin x$ for all real values of x.

$$\sin x + \sqrt{3} = -\sin x$$
$$2\sin x + \sqrt{3} = 0$$
$$2\sin x = -\sqrt{3}$$
$$\sin x = -\frac{\sqrt{3}}{2}$$
$$x = \frac{4\pi}{3} + 2\pi k \text{ or } x = \frac{5\pi}{3} + 2\pi k, \text{ where } k \text{ is any integer}$$

The solutions are $\frac{4\pi}{3} + 2\pi k$ and $\frac{5\pi}{3} + 2\pi k$.

© Glencoe/McGraw-Hill   127   Advanced Mathematical Concepts

Mixed Review

24. Biology Under ideal conditions, the population of a certain bacterial colony will double in 85 minutes. How much time will it take for the population to increase 12 times? *(Lesson 11-6)*   **297.5 min or about 5 hr**

25. Use a calculator to find the antilogarithm of -2.1654 to the nearest hundredth. *(Lesson 11-5)*   **0.01**

26. Solve $\log_5(7x) = \log_5(5x + 16)$. *(Lesson 11-4)*   **8**

27. Entertainment A theater has been staging children's plays during the summer. The average attendance at each performance is 400 people, and the cost of a ticket is $3. Next summer, the theater manager would like to increase the cost of the tickets, while maximizing profits. The manager estimates that for every $1 increase in ticket price, the attendance at each performance will decrease by 20. *(Lesson 10-5)*

a. What price should the manager propose to maximize profits?   **$11.50**

b. What maximum profit might be expected?   **$2645**

28. Express $5\left(\cos \dfrac{\pi}{6} + i \sin \dfrac{\pi}{6}\right)$ in rectangular form. *(Lesson 9-6)* $\dfrac{5\sqrt{3}}{2} + \dfrac{5}{2}i$

29. Surveying A building 60 feet tall is on top of a hill. A surveyor is at a point on the hill and observes that the angle of elevation to the top of the building measures 42° and the angle of elevation to the bottom measures 18°. How far is the surveyor from the bottom of the building? *(Lesson 5-6)*   **about 109.6 ft**

30. Find the discriminant of $5x^2 - 8x + 12 = 0$. Describe the nature of its roots. Solve the equation. *(Lesson 4-2)* -176; **2 imaginary;** $\dfrac{4 \pm 2i\sqrt{11}}{5}$

31. Describe the transformation(s) of the parent graph of $f(x) = x^2$ that are required to graph $f(x) = (x + 4)^2 - 8$. *(Lesson 3-2)*   **4 units left and 8 units down**

32. Find the maximum and minimum values of the function $f(x) = 2x + 8y + 10$ for the polygonal convex set determined by the system of inequalities. *(Lesson 2-6)*   **50; 22**
$$x \ge 2$$
$$x \le 4$$
$$y \ge 1$$
$$x - 2y \le -4$$

33. SAT Practice In the figure, circles X, Y, and Z overlap to form regions a through g. How many of these regions are contained in either of the circles X or Z?   **C**

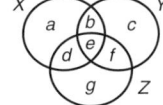

A 2

B 5

C 6

D 7

E None of the above

Extra Practice See p. A47.

Extra Credit

a. If Uranium-229 has a decay constant of -0.0119 with time in minutes, find the time needed for this element to decay from 5 grams to 2 grams.
77.0 min

b. What is the half-life of U-229?
58.2 min

VOCABULARY

antiln x (p. 734)
antilogarithm (p. 729)
characteristic (p. 727)
common logarithm (p. 726)
doubling time (p. 740)
exponential decay
  (pp. 706, 712)
exponential function (p. 704)
exponential growth
  (pp. 706, 712)
ln x (p. 733)
logarithm (p. 718)

logarithmic function (p. 718)
mantissa (p. 727)
natural logarithm (p. 733)
power function (p. 704)
scientific notation (p. 695)

Modeling

linearizing data (p. 743)
nonlinear regression (p. 741)

UNDERSTANDING AND USING THE VOCABULARY

Choose the correct term to best complete each sentence.

1. A logarithm with base 10 is called a(n) ___?___. **common logarithm**

2. A real-world situation that involves a quantity that increases exponentially over time exhibits ___?___. **exponential growth**

3. The inverse of an exponential function is a(n) ___?___. **logarithmic function**

4. A number is in ___?___ when it is in the form $a \times 10^n$ where $1 \le a < 10$ and n is an integer. **scientific notation**

5. A common logarithm is made up of two parts, the characteristic and the ___?___. **mantissa**

6. A logarithm to the base e is called a ___?___. **natural logarithm**

7. Exponential data can be analyzed by ___?___ and then applying linear regression. **linearizing data**

8. A function of the form $y = b^x$, where b is a positive real number and the exponent is a variable, is a(n) ___?___. **exponential function**

9. The process of fitting an equation to nonlinear data is called ___?___. **nonlinear regression**

10. A(n) ___?___ can be solved using logarithms. **exponential equation**

interNET
CONNECTION
For additional review and practice for each lesson, visit: **www.amc.glencoe.com**

VOCABULARY

This section provides a listing of the new terms, properties, and phrases that were introduced in this chapter. The exercises check students' understanding of the terms by using a variety of verbal formats including matching, completion, and true/false.

Glossary A complete glossary of terms begins on page A71.

Assessment and Evaluation Masters, pp. 265–266

GLENCOE'S
ASSESSMENT
ADVANTAGE

Six forms of Chapter 11 Test are available in the *Assessment and Evaluation Masters* as shown in the chart.

Chapter 11 Test, Form 1B, is shown at the right. Chapter 11 Test, Form 2B, is shown on the next page.

TEST OPTIONS

Form	Pages	Type	
1A	263–264	Honors	*Form 1 tests are multiple choice.*
1B	265–266	Average	
1C	267–268	Basic	
2A	269–270	Honors	*Form 2 tests are free response.*
2B	271–272	Average	
2C	273–274	Basic	

SKILLS AND CONCEPTS

Objectives and Examples This section reviews the skills and concepts of the chapter and shows completely worked examples.

Review Exercises These exercises provide practice for the corresponding objectives.

Assessment and Evaluation Masters, pp. 271–272

SKILLS AND CONCEPTS

OBJECTIVES AND EXAMPLES	REVIEW EXERCISES

Lesson 11-1 Evaluate and simplify expressions containing rational exponents.

• Evaluate $16^{\frac{1}{4}}$.

$$16^{\frac{1}{4}} = (2^4)^{\frac{1}{4}}$$
$$= 2^{\frac{4}{4}} \quad \textit{Power of a power}$$
$$= 2$$

Simplify $(6c^4d)^2(3c)^{-2}$.

$$(6c^4d)^2(3c)^{-2} = \frac{36c^8d^2}{9c^2}$$
$$= 4c^{8-2}d^2 \quad \textit{Quotient Property}$$
$$= 4c^6d^2$$

Evaluate each expression.

11. $\left(\frac{1}{4}\right)^{-2}$ **16**

12. $64^{\frac{1}{2}}$ **8**

13. $27^{\frac{4}{3}}$ **81**

14. $\left(\sqrt[4]{256}\right)^3$ **64**

Simplify each expression.

15. $3x^2(3x)^{-2}$ $\frac{1}{3}$

16. $\left(6a^{\frac{1}{3}}\right)^3$ **216a**

17. $\left(\frac{1}{2}x^4\right)^3$ $\frac{1}{8}x^{12}$

18. $(w^3)^4 \cdot (4w^2)^2$ **16w^{16}**

19. $\left((2a)^{\frac{1}{3}}(a^2b)^{\frac{1}{3}}\right)^3$
    $2a^3b$

20. $\left(3x^{\frac{1}{2}}y^{\frac{1}{4}}\right)(4x^2y^2)$
    $12x^{\frac{5}{2}}y^{\frac{9}{4}}$

Lesson 11-2 Graph exponential functions and inequalities.

• Graph $y = \left(\frac{1}{2}\right)^{x-1}$.

Plot the points shown in the table below. Connect the points to graph the function.

x	$\left(\dfrac{1}{2}\right)^{x-1}$
-2	8
-1	4
0	2
1	1
2	0.5
3	0.25
4	0.125

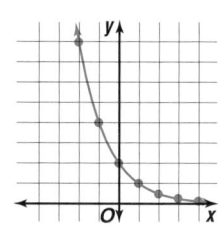

Graph each exponential function or inequality.

21. $y = 3^{-x}$ **21–26. See margin.**

22. $y = \left(\frac{1}{2}\right)^x$

23. $y = 2^{x-1}$

24. $y = 2^x + 2$

25. $y \geq -2^x + 1$

26. $y < 2^{x+2}$

Lesson 11-3 Use the exponential function $y = e^x$.

• Find the balance after 15 years of a $5000 investment earning 5.8% interest compounded continuously.

$$A = Pe^{rt}$$
$$A = 5000(e)^{(0.058 \cdot 15)}$$
$$A \approx \$11,934.55$$

Find the balance for each account after 10 years if the interest is compounded continuously.

27. $2500 invested at 6.5% **$4788.85**

28. $6000 invested at 7.25% **$12,388.39**

29. $12,000 invested at 5.9% **$21,647.86**

Left margin — Assessment Masters preview:

NAME _____ DATE _____ PERIOD _____

Chapter 11   Chapter 11 Test, Form 2B

1. Evaluate $(\sqrt[3]{81})^3$.   1. ____27____

2. Simplify $\left(\frac{25}{x^{-4}}\right)^{\frac{3}{2}}$.   2. ____125x^6____

3. Express $\sqrt[3]{32x^5y^{10}}$ using rational exponents.   3. ____2x^{\frac{3}{2}}y^2____

4. Express $4x^{\frac{1}{3}}y^{\frac{2}{3}}$ using radicals.   4. ____$4\sqrt[3]{xy^2}$____

5. Evaluate 5^π to the nearest thousandth.   5. ____156.993____

6. Sketch the graph of $y = 3^x$.   6.

7. Sketch the graph of $y \leq \left(\frac{1}{2}\right)^x$.   7.

8. A 1991 report estimated that there were 640 salmon in a certain river. If the population is decreasing exponentially at a rate of 4.3% per year, what is the expected population in 2002?   8. ____395____

9. Find the balance in an account at the end of 12 years if $4000 is invested at an interest rate of 9% that is compounded continuously.   9. ____$11,778.72____

10. Write $16^{\frac{3}{4}} = 8$ in logarithmic form.   10. ____$\log_{16} 8 = \frac{3}{4}$____

© Glencoe/McGraw-Hill   271   Advanced Mathematical Concepts

NAME _____ DATE _____ PERIOD _____

Chapter 11   Chapter 11 Test, Form 2B (continued)

11. Evaluate $\log_4 \frac{1}{64}$.   11. ____-3____

12. Solve $\log_2(x + 6) + \log_2 3 = \log_2 30$.   12. ____4____

13. Sketch the graph of $y \geq \log_4(x + 2)$.   13.

For Exercises 14-18, round your answers to the nearest thousandth.

14. Evaluate $\log \frac{8}{27}$.   14. ____0.051____

15. Find the value of $\log_3 92.4$ using the change of base formula.   15. ____4.120____

16. Solve $5^{x+2} = 7$ using common logarithms.   16. ____-0.791____

17. Convert $\log_5 156$ to a natural logarithm and evaluate.   17. ____3.138____

18. Solve $e^{4x} < 98.6$ by using natural logarithms.   18. ____$x < 1.148$____

19. Banking Find the amount of time in years required for an investment to double at a rate of 6.2% if the interest is compounded continuously.   19. ____11.180 years____

20. Biology The table below shows the population of mold spores on a given Petri dish.   20. ____$\ln y = 0.0774x + 3.8065$____

Time (days)	0	2	4	6	8
Population (thousands)	45	51	63	74	81

Let x represent the number of days and let y represent the populations in thousands. Linearize the data and find the regression equation for the linearized data.

Bonus Express $2\log_b a - \log_b c$ as a single log. Assume $b > 0$.   Bonus: ____$\log_b \frac{a^2}{c}$____

© Glencoe/McGraw-Hill   272   Advanced Mathematical Concepts

TestCheck and Worksheet Builder

This **networkable** software has three modules.
• **Worksheet Builder** to make worksheets and tests
• **Student Module** to take tests on-screen
• **Management System** to keep student records
Three special banks are included for SAT, ACT, and TIMSS.

OBJECTIVES AND EXAMPLES

Lesson 11-4 Evaluate expressions involving logarithms.

• Evaluate the expression $\log_5 25$.

Let $x = \log_5 25$.
$x = \log_5 25$
$5^x = 25$ *Definition of logarithm*
$5^x = 5^2$
$x = 2$

REVIEW EXERCISES

Write each equation in exponential form.

30. $\log_8 4 = \frac{2}{3}$ $8^{\frac{2}{3}} = 4$ 31. $\log_3 \frac{1}{81} = -4$
$3^{-4} = \frac{1}{81}$

Write each equation in logarithmic form.

32. $2^4 = 16$ $\log_2 16 = 4$ 33. $5^{-2} = \frac{1}{25}$
$\log_5 \frac{1}{25} = -2$

Evaluate each expression.

34. $\log_2 32$ **5**   35. $\log_{10} 0.001$ **−3**

36. $\log_4 \frac{1}{16}$ **−2**   37. $\log_2 0.5$ **−1**

38. $\log_6 216$ **3**   39. $\log_9 \frac{1}{9}$ **−1**

40. $\log_4 1024$ **5**   41. $\log_8 512$ **3**

Lesson 11-4 Solve equations involving logarithms.

• Solve $\log_5 4 + \log_5 x = \log_5 36$.
$\log_5 4 + \log_5 x = \log_5 36$
$\log_5 4x = \log_5 36$
$4x = 36$
$x = 9$

Solve each equation.

42. $\log_x 81 = 4$ **3**

43. $\log_{\frac{1}{2}} x = -4$ **16**

44. $\log_3 3 + \log_3 x = \log_3 45$ **15**

45. $2 \log_6 4 - \frac{1}{3} \log_6 8 = \log_6 x$ **8**

46. $\log_2 x = \frac{1}{3} \log_2 27$ **3**

47. Graph $y = \log_{10} x$ **See margin.**

Lesson 11-5 Find common logarithms and antilogarithms of numbers.

• Given that $\log 8 = 0.9031$, evaluate $\log 80,000$.
$\log 80,000 = \log(10,000 \times 8)$
$= \log 10^4 + \log 8$
$= 4 + 0.9031$
$= 4.9031$

Solve $3^{x+2} = 4^{x-1}$.
$3^{x+2} = 4^{x-1}$
$\log 3^{x+2} = \log 4^{x-1}$
$(x + 2) \log 3 = (x - 1) \log 4$
$x \log 3 + 2 \log 3 = x \log 4 - \log 4$
$x \log 3 - x \log 4 = -\log 4 - 2 \log 3$
$x (\log 3 - \log 4) = -(\log 4 + 2 \log 3)$
$x = -\frac{(\log 4 + 2 \log 3)}{(\log 3 - \log 4)}$
$x \approx 12.4565$

Given that $\log 3 = 0.4771$ and $\log 14 = 1.1461$, evaluate each logarithm.

48. $\log 300,000$ **5.4771**   49. $\log 0.0003$ **−3.5229**

50. $\log 140$ **2.1461**   51. $\log 0.014$ **−1.8539**

Solve each equation or inequality.   **−8.04**

52. $4^x = 6^{x+2}$ **−8.84**   53. $12^{0.5x} = 8^{0.1x - 4}$

54. $\left(\frac{1}{4}\right)^{3x} < 6^{x-2}$ **$x > 0.6$**   55. $0.1^{2x+8} \geq 7^{x+4}$ **$x \leq -4$**

56. $\log(2x + 3) = -\log(3 - x)$
−1.39, 2.89

Graph each equation or inequality.

57. $y = 3 \log(x - 2)$ 58. $y \geq 7^{x-2}$
57–58. See margin.

59. Solve $9^x = 4^{x-2}$ by graphing.   **−3.42**

Additional Answers
(pp. 750–751)

24.

25.

26.

47.

57.

58.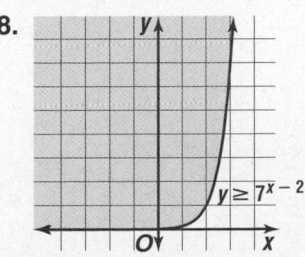

Additional Answers (pp. 750–751)

21.

23.

22.

OBJECTIVES AND EXAMPLES

Lesson 11-6 Find natural logarithms of numbers.

- Convert $\log_2 14$ to a natural logarithm and evaluate.

$$\log_2 14 = \frac{\ln 14}{\ln 2}$$
$$\approx 3.8074$$

REVIEW EXERCISES

Convert each logarithm to a natural logarithm and evaluate.

60. $\log_4 15$ **1.9534**

61. $\log_8 24$ **1.5283**

62. $\log_9 100$ **2.0959**

63. $\log_{15} 125$ **1.7829**

Lesson 11-6 Solve equations using natural logarithms.

- Use natural logarithms to solve $2^{x-1} = 5^{3x}$.

$$2^{x-1} = 5^{3x}$$
$$\ln 2^{x-1} = \ln 5^{3x}$$
$$(x-1)\ln 2 = 3x \ln 5$$
$$(x-1)(0.6931) = 3x(1.6094)$$
$$0.6931x - 0.6931 = 4.8282x$$
$$-0.6931 = 4.1351x$$
$$-0.1676 = x$$

Use natural logarithms to solve each equation or inequality.

64. $4^x = 100$ **3.3219**　　**65.** $6^{x-2} = 30$ **3.8982**

66. $3^{x+1} = 4^{2x}$ **0.6563**　**67.** $9^{4x} = 5^{x-4}$ **−0.8967**

68. $24 < e^{2x}$
$\quad x > 1.5890$

69. $15e^x \geq 200$
$\quad x \geq 2.5903$

Solve each equation or inequality by graphing. Round solutions to the nearest hundredth.

70. $2e^x = 56$ **3.33**　　**71.** $9 > e^x$ **$x < 2.20$**

Lesson 11-7 Find the doubling time of an exponential quantity.

- Find the amount of time required for an investment to double at a rate of 3.5% if the interest is compounded continuously.

Use $t = \dfrac{\ln 2}{k}$, where k is the interest rate written as a decimal.

$$t = \frac{\ln 2}{k}$$
$$= \frac{\ln 2}{0.035}$$
$$\approx 19.80 \text{ years}$$

It will take 19.8 years for the amount to double.

Find the amount of time required for an amount to double at the given rate if the interest is compounded continuously.

72. 2.8% **24.76**

73. 5.125% **13.52**

74. Finance What was the interest rate on an account that took 18 years to double if interest was compounded continuously and no deposits or withdrawals were made during the 18-year period? **3.85%**

APPLICATIONS AND PROBLEM SOLVING

75. Archaeology Carbon-14 tests are often performed to determine the age of an organism that died a long time ago. Carbon-14 has a half-life of 5730 years. If a turtle shell is found and tested to have 65% of its original carbon-14, how old is it? *(Lesson 11-4)* **3561 yr**

76. Sound The decibel, dB, is the unit of measure for intensity of sound. The equation to determine the decibel rating of a sound is $\beta = 10 \log \frac{I}{I_0}$, where $I_0 = 10^{-12}$ W/m^2, the reference intensity of the faintest audible sound. *(Lesson 11-5)*

  a. Find the decibel rating, in dB, of a whisper with an intensity of 1.15×10^{-10} W/m^2. **20.6 dB**

  b. Find the decibel rating of a teacher's voice with an intensity of 9×10^{-9} W/m^2. **39.5 dB**

  c. Find the decibel rating of a rock concert with an intensity of 8.95×10^{-3} W/m^2. **99.5 dB**

77. Population A certain city has a population of $P = 142{,}000e^{0.014t}$ where t is the time in years and $t = 0$ is the year 1990. In what year will the city have a population of 200,000? *(Lesson 11-6)* **2014**

78. Data Entry The supervisor of the data entry department found that the average number of words per minute, N, input by trainees after t weeks was $N = 65 - 30e^{-0.20t}$. *(Lesson 11-6)*

  a. What was the average number of words per minute after 2 weeks? **45**

  b. What was the average number of words per minute after 15 weeks? **64**

  c. When was the average trainee able to input 50 words per minute? **3.5 weeks**

ALTERNATIVE ASSESSMENT

OPEN-ENDED ASSESSMENT

1. The simplest form of an expression that contains at least two rational exponents is $\frac{n^2}{4m}$. If one of the exponents is negative and another is a fraction, find the original expression. **Sample answer: $(n^4)^{\frac{1}{2}}(4m)^{-1}$**

2. Melissa used the power and the product properties of logarithms to solve an equation involving logarithms. She got a solution of 1. What is the equation Melissa solved? **See margin.**

PORTFOLIO

Suggest real-world data that can be modeled with an exponential or a logarithmic function. Record the function that models the data. Explain why your function is a good model.

Unit 3 *inter*NET Project

SPACE —
THE FINAL FRONTIER

Kepler is Still King!

- Search the Internet to find out about Kepler's Laws for planetary motion.

- Kepler's Third Law relates the distance of planets from the sun and the period of each planet. Use the Internet to find the distance each planet is from the sun and to find each planet's period.

- Use the information about distance and period of the planets to verify Kepler's Third Law. You can use a spreadsheet, if you prefer.

- Write a summary describing Kepler's Third Law and how you verified his computations.

Additional Assessment See p. A66 for Chapter 11 practice test.

Chapter 11 Study Guide and Assessment **753**

This section provides additional practice in solving real-world problems that involve the concepts of this chapter.

ALTERNATIVE ASSESSMENT

Additional Answer
Open-Ended Assessment

2. Sample answer: log 2 + log $(x + 2) = \frac{1}{2}$ log 36

Unit 3 *inter*NET Project

● **Kepler Is Still King!**
Kepler's Third Law states that the square of the orbital period of any planet is the same as the cube of the planet's distance from the Sun. The distance must be in astronomical units and the period must be in Earth years. Make sure that students understand these two conversions or the computations will not work. A spreadsheet is very useful in doing the computations.

Internet Address Book

Record useful Internet addresses in the space at right for quick reference.

www.amc.glencoe.com

Chapter 11 Skill: Average Problems

Pages 754–755 are part of a complete SAT/ACT preparation course that is described in detail on pages T8–T9 of the *Teacher's Handbook*. The test items were written in the same style as those in the SAT and ACT.

 These questions were aligned and verified by The Princeton Review, the nation's leader in test preparation.

Diagnosis & Prescription

Each of the 10 SAT and ACT practice problems on page 755 is cross-referenced to the chapter where that SAT or ACT skill is covered. If students miss a particular type of problem, you can have them study that skill.

Problem Number and Skill	Page
1 Average	754
2 More Algebra	202
3 Average	754
4 Trigonometry	418
5 Ratio and Proportion	692
6 Geometry (Perimeter, Area, and Volume)	548
7 Average	754
8 Geometry (Triangles and Quadrilaterals)	482
9 Average	754
10 Algebra	124

Average Problems

Calculating the average of a set of n numbers is easy if you are given the numbers and know the definition of *average (arithmetic mean)*.

$$\text{average} = \frac{\text{sum of } n \text{ numbers}}{n} \text{ or average} = \frac{\text{sum}}{n}$$

SAT and ACT problems that use averages are more challenging. They often involve missing numbers, evenly-spaced numbers, or weighted averages.

To find *weighted average*, multiply each number in the set by the number of times it appears. Add the products and divide the sum by the total number of numbers.

SAT EXAMPLE

1. In one month, a pet shop sold 5 red parrots that weighed 2 pounds each and 4 blue parrots that weighed 3 pounds each. What was the average (arithmetic mean) weight of a parrot in pounds that this pet shop sold this month?

   A 2   **B** $2\frac{4}{9}$ **C** $2\frac{1}{2}$

   D 5   **E** 9

   HINT Eliminate choices that are impossible or unreasonable.

   Solution Estimate the average weight of a parrot. The parrots sold weigh 2 and 3 pounds. So, the average weight cannot be 5 or 9 pounds. Eliminate choices D and E.

   This is a weighted-average problem since there are different numbers of parrots with different weights. You cannot just average 2 pounds and 3 pounds. Notice that answer choice C is an obvious wrong answer as it is the average of 2 and 3.

   Find the sum of the red-parrot weights, 5×2 or 10, and the sum of the blue-parrot weights, 4×3 or 12.

   The total weight of the parrots is $10 + 12$ or 22. The total number of parrots is 9.

   So the average weight is $22 \div 9$ or $2\frac{4}{9}$ pounds.

   The correct answer is choice **B**.

ACT EXAMPLE

2. Louann computed the average of her test scores by adding the 5 scores together and dividing the sum by the number of tests. The average was 88. But she completely forgot a sixth test, which had a score of 82. What is the true average of Louann's six tests?

   A 82   **B** 85   **C** 86

   D 87   **E** 88

   HINT Use the formula sum $= n \cdot$ average to find the sum when you know the average and the number of data.

   Solution First estimate the answer. The missing score is 82, which is less than the average, 88, so the new average must be less than 88. Eliminate choice E. Also, the new average must be greater than 82, so eliminate choice A.

   Next, find the sum of the 5 scores, using the formula for average. 5×88 or 440.

   Then find the sum of all 6 scores by adding the sixth score to the sum of the first five scores. $440 + 82$ or 522.

   Finally, divide the sum of the six scores by 6. $552 \div 6$ or 87.

   The answer is choice **D**.

---

754   *Chapter 11   Exponential and Logarithmic Functions*

 TestCheck and Worksheet Builder

Special banks of SAT, ACT, and TIMSS items are included in this CD-ROM.

After you work each problem, record your answer on the answer sheet provided or on a piece of paper.

Multiple Choice

1. If the average (arithmetic mean) of four distinct positive integers is 11, what is the greatest possible value of any one of the integers? **B**

   A 35   B 38   C 40   D 41   E 44

2. Which quadratic equation has roots of $\frac{1}{2}$ and $\frac{1}{3}$? **E**

   A $5x^2 - 5x - 2 = 0$ B $5x^2 - 5x + 1 = 0$

   C $6x^2 + 5x - 1 = 0$ D $6x^2 - 6x + 1 = 0$

   E $6x^2 - 5x + 1 = 0$

3. Brad tried to calculate the average of his 5 test scores. He mistakenly divided the correct total (T) of his scores by 6. The result was 14 less than what it should have been. Which of the following equations would determine the value of T? **E**

   A $5T + 14 = 6T$ B $\frac{T}{6} = \frac{(T-14)}{5}$

   C $\frac{T}{6} - 14 = \frac{T}{5}$ D $\frac{(T-14)}{5} = \frac{T}{5}$

   E $\frac{T}{6} + 14 = \frac{T}{5}$

4. In the figure below, the ratio $\dfrac{\text{area of } \triangle ABC}{\tan A} =$ **E**

   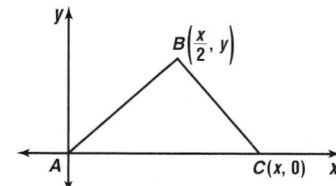

   A $\dfrac{1}{2y^2}$ B $\dfrac{1}{y^2}$ C $\dfrac{2}{x^2}$

   D $\dfrac{4}{x^2}$ E $\dfrac{x^2}{4}$

5. If the ratio of x to y is equal to the ratio of 10 to $2y$, then what is the value of x? **B**

   A $\dfrac{1}{5}$ B 5   C 8   D 12   E 20

6. If the circumference of a circle is $\frac{2\pi}{3}$, then what is half of its area? **A**

   A $\dfrac{\pi}{18}$ B $\dfrac{\pi}{9}$ C $\dfrac{4\pi}{9}$

   D $\dfrac{2\pi^2}{9}$ E $\dfrac{4\pi^2}{9}$

7. If the average (arithmetic mean) of eight numbers is 20 and the average of five of these numbers is 14, what is the average of the other three numbers? **D**

   A 14   B 17   C 20   D 30   E 34

8. In the triangle below, what is the measure of $\angle A$? **C**

   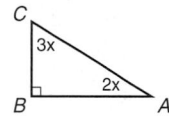

   A $9°$ B $18°$ C $36°$ D $54°$ E $108°$

9. A is the average (arithmetic mean) of three consecutive positive even integers. Which of the following could be the remainder when A is divided by 6? **C**

   A 1

   B 3

   C 4

   D 5

   E It cannot be determined from the information given.

10. **Grid-In** If b is a prime integer such that $3b > 10 > \frac{5}{6}b$, what is one possible value of b? **5, 7, or 11**

*inter*NET CONNECTION **SAT/ACT Practice** For additional test practice questions, visit: **www.amc.glencoe.com**

A bubble-in answer sheet for these practice problems is available on p. 1 of the *Assessment and Evaluation Masters.*

Additional Practice

Additional SAT and ACT practice questions and a bubble-in answer sheet are available in the *Assessment and Evaluation Masters,* pp. 2, 279, and 280.

Teaching Tip For Exercise 2, remind students that they can multiply both sides of the equation by 6 to eliminate the fractions.

Teaching Tip For Exercise 10, students should test only positive values of b, because prime numbers are not negative.

Assessment and Evaluation Masters, pp. 279–280

Extra Credit

TIMSS The number of bacteria in a colony was growing exponentially. At 1 P.M. yesterday, the number of bacteria was 1000 and at 3 P.M. yesterday, it was 4000. How many bacteria were in the colony at 6 P.M. yesterday? **32,000**

Unit Overview

Unit 4 focuses on discrete mathematics, with an emphasis on sequences, series, combinatorics, probability, statistics, and data analysis.

Assessment

A review and a test on the concepts in Unit 4 are provided in the *Assessment and Evaluation Masters* (pp. 419–424). The Unit 4 Review is shown below. The Unit 4 Test is shown on the next page.

Assessment and Evaluation Masters, pp. 419–420

Discrete Mathematics

Discrete mathematics is a branch of mathematics that deals with finite or discontinuous quantities. Usually discrete mathematics is defined in terms of its key topics. These include graphs, certain functions, logic, combinatorics, sequences, iteration, algorithms, recursion, matrices, and induction. Some of these topics have already been introduced in this book. For example, linear functions are continuous, while step functions are discrete. As you work through Unit 4, you will construct models, discover and use algorithms, and examine exciting new concepts as you solve real-world problems.

Chapter 12 Sequences and Series
Chapter 13 Combinatorics and Probability
Chapter 14 Statistics and Data Analysis

Resource Manager

Reproducible Masters
- *Math and Science Activities,* pp. 18–23
- Unit 4 Review, *Assessment and Evaluation,* pp. 419–420
- Unit 4 Test, *Assessment and Evaluation,* pp. 421–424

Unit 4 interNET Projects

THE UNITED STATES CENSUS BUREAU

Did you know that the very first United States Census was conducted in 1790? The United States Census Bureau completed this task in less than nine months. This is remarkable considering they had to do so without benefit of cars, telephones, or computers! As the U.S. population grew, it took more and more time to complete the census. The 1880 census took seven years to finish. Luckily, advancements in technology have helped the people employed by the Census Bureau analyze the data and prepare reports in a reasonable amount of time. At the end of each chapter in Unit 4, you will be given tasks to explore the data collected by the Census Bureau.

CHAPTER 12
(page 833)

That's a lot of people! In addition to determining how many people are in the United States every ten years, the Census Bureau also attempts to estimate the population at any particular time and in the future. On their web site, you can see estimates of the U. S. and world population for the current day. How does the Census Bureau estimate population?
Math Connection: Use the Internet to find population data. Model the population of the U.S. by using an arithmetic and then a geometric sequence. Predict the population for a future date using your models.

CHAPTER 13
(page 885)

Radically random! During a census year, the Census Bureau collects many type of data about people. This information includes age, ethnic background, and income, to name just a few. What types of data about the people of the United States can be found using the Internet?
Math Connection: Use data from the Internet to find the probability that a randomly-selected person in the U.S. belongs to a particular age group.

CHAPTER 14
(page 937)

More and more models! Did you know that in 1999 a person was born in the United States about every eight seconds? In that same year, about one person died every 15 seconds, and one person migrated every 20 seconds. Birth, deaths, and other factors directly affect the population. In Chapter 12, you modeled the U.S. population using sequences. What other types of population models could you use to predict population growth?
Math Connection: Use data from the Internet to write and graph several functions representing the U.S. population growth. Predict the population for a future data using your models.

interNET CONNECTION

For more information on the Unit Project, visit:
www.amc.glencoe.com

Using the interNET Projects

When you or your students visit **www.amc.glencoe.com** you will find links to outside Web sites relating to these Internet projects. These links are checked frequently to ensure that they are still functioning and that the information is current.

Glencoe makes every effort to ensure that the Web site links for *Advanced Mathematical Concepts* are current and contain appropriate content. However, these Web sites are not under Glencoe's control.

In the Alternative Assessment section of each chapter's Study Guide and Assessment, you can find follow-up activities for the Internet Projects.

<table>
<tr><th colspan="2">UNIT 4
INTERNET PROJECTS</th></tr>
<tr><th>Chapter Number/Title</th><th>Page</th></tr>
<tr><td>12   That's a Lot of People!</td><td>833</td></tr>
<tr><td>13   Radically Random!</td><td>885</td></tr>
<tr><td>14   More and More Models</td><td>937</td></tr>
</table>

Assessment and Evaluation Masters, pp. 421–424

UNIT 4	NAME _____ DATE _____ PERIOD _____	
	Unit 4 Test, Chapters 12–14	$\sum_{k=0}^{4}(23+6k)$
1. Express $23 + 29 + 35 + 41 + 47$ using sigma notation.		1. _____
2. A set of data is normally distributed with a mean of 16 and a standard deviation of 0.3. What percent of the data is between 15.2 and 16?		2. __49.6%__
3. Use mathematical induction to prove that $1^2 + 2^2 + 3^2 + \cdots + n^2 = \frac{n(n+1)(2n+1)}{6}$ for all positive integral values of n.		3. See student's work.
4. Find the twelfth term of the geometric sequence $\frac{1}{16}, \frac{1}{8}, \frac{1}{4}, \ldots$.		4. __120__
5. There are 21 wrapped packages in a grab bag at an office holiday party. Five of the packages contain $20 bills, 7 packages contain $5 bills, and 9 packages contain $1 bills. What is the probability that the first two people will choose packages with $20 bills inside?		5. __$\frac{1}{21}$__
6. Find the sum of the infinite series $1 + \frac{1}{5} + \frac{1}{25} + \cdots$.		6. __$\frac{5}{4}$__
7. Find $P(5, 2)$.		7. __20__
8. *True or false:* Choosing an entrée and choosing an appetizer from a dinner menu are independent events.		8. __true__
9. How many ways can 9 keys be arranged on a circular key ring?		9. __40,320__
10. Determine whether the series $1 + \frac{1}{2} + \frac{1}{3^2} + \frac{1}{4^2} + \cdots$ is convergent or divergent.		10. __convergent__
11. What is the probability of getting an even number on a single roll of a number cube if you roll a 3 or greater?		11. __0.5__
12. How many eight-letter patterns can be formed from the letters of the word *circular*?		12. __10,080__
© Glencoe/McGraw-Hill	421	*Advanced Mathematical Concepts*

Internet Address Book

Record useful Internet addresses in the space at right for quick reference.

www.amc.glencoe.com

_____

_____

_____

_____

Instructional Objectives

Lesson (pages)	Lesson Objectives	NCTM Standards 2000	State/Local Objectives
12-1 (759–765)	• Find the nth term and arithmetic means of an arithmetic sequence. • Find the sum of n terms of an arithmetic series.	1, 2, 6	
12-2 (766–773)	• Find the nth term and geometric means of a geometric sequence. • Find the sum of n terms of a geometric series.	1, 2, 6	
12-3 (774–783)	• Find the limit of the terms of an infinite sequence. • Find the sum of an infinite geometric series.	1, 2, 6	
12-3B (784–785)	• Explore sequences generated by continued fractions.	1, 2	
12-4 (786–793)	• Determine whether a series is convergent or divergent.	1, 2, 6	
12-5 (794–800)	• Use sigma notation.	1, 2, 6, 10	
12-6 (801–805)	• Use the Binomial Theorem to expand binomials.	1, 2, 6	
12-7 (806–814)	• Approximate e^x, trigonometric values, and logarithms of negative numbers by using series. • Use Euler's Formula to write the exponential form of a complex number.	1, 2, 6, 10	
12-8 (815–821)	• Iterate functions using real and complex numbers.	1, 2, 6, 10	
12-9 (822–828)	• Use mathematical induction to prove the validity of mathematical statements.	1, 2, 6, 7, 10	

Suggested Pacing   *See pages T2–T4 for additional options in course-planning.*

COURSE	Lesson 12-1	Lesson 12-2	Lesson 12-3	Lesson 12-4	Lesson 12-5
Precalculus Emphasis Chapters 4–15 (45–50 minutes)	1.5 days	1.5 days	(with Lesson 12-3B) 2 days	1.5 days	1.5 days
Precalculus Emphasis (block)	0.5 day	(with Lesson 12-3B) 1.5 days		1 day	(with Lesson 12-6) 1.5 days
Trigonometric Emphasis Chapters 1–9, 11, 12 (45–50 minutes)	1.5 days	1.5 days	(with Lesson 12-3B) 2 days	1.5 days	1.5 days
Trigonometric Emphasis (block)	0.5 day	(with Lesson 12-3B) 1.5 days		1 day	(with Lesson 12-6) 1.5 days

Instructional Resources

Student Edition		Blackline Masters (page numbers)								☞
Lesson	Graphing Calculator Applications	Study Guide and Practice	Enrichment	Assessment and Evaluation	Graphing Calculator Keystrokes	Problem Solving & Applications	Spreadsheet Applications	TI-92 and CBL Interdisciplinary Activities	Transparencies A and B
12-1		215–216	133				12		12-1
12-2		217–218	134			23			12-2
12-3		219–220	135	303					12-3
12-3B	784–785				48–50				
12-4		221–222	136						12-4
12-5	798	223–224	137	302, 303		24			12-5
12-6		225–226	138						12-6
12-7	809	227–228	139	304	51				12-7
12-8		229–230	140					41–44	12-8
12-9		231–232	141	304					12-9
Study Guide/ Assessment				289–301, 307					
SAT/ACT Practice				305–306					

Lesson 12-6	Lesson 12-7	Lesson 12-8	Lesson 12-9	Study Guide & Assessment	Chapter Test	TOTAL
1.5 days	1.5 days	1.5 days	1.5 days	1 day	1 day	16 days
(See Lesson 12-5.)	1.5 days		1.5 days		0.5 day	8 days
1.5 days	1.5 days	1.5 days	1.5 days	1 day	1 day	16 days
(See Lesson 12-5.)	1.5 days		1.5 days		1 day	8.5 days

Resource Manager

TeacherWorks™

The worksheets shown on this page are a small sample of the materials available on *TeacherWorks: All-in-One Lesson Planner and Resource Center.*

This CD-ROM contains all of the blackline masters and transparencies available for this program.

It also includes a lesson planner and interactive Teacher Edition, so you can customize lesson plans and reproduce classroom resources quickly and easily, from just about anywhere.

Applications

Problem Solving and Applications Masters, pp. 23, 24

Technology/Multimedia

Spreadsheet Applications Masters, p. 12

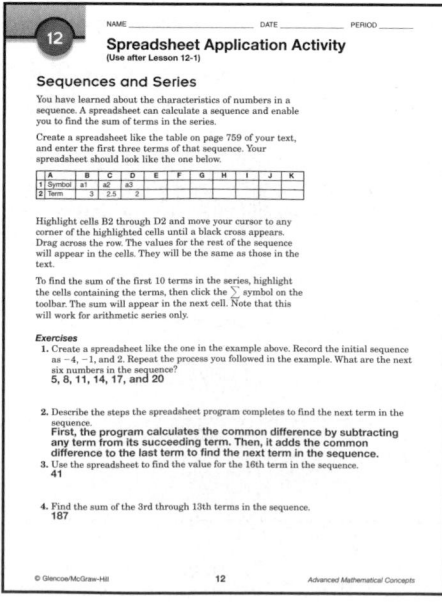

TI-92 & CBL Interdisciplinary Activities Masters, pp. 41–44

Assessment Resources

Type	Student Edition	Teacher's Wraparound Edition	Assessment and Evaluation Masters
Ongoing Assessment	Mid-Chapter Quiz, p. 793	5-Minute Check, pp. 759, 766, 774, 786, 794, 801, 806, 815, 822	Mid-Chapter Test, p. 302 Quizzes A–D, pp. 303–304
Mixed Review	Mixed Review, pp. 765, 773, 783, 793, 799–800, 805, 814, 821, 828		Cumulative Review, p. 307
Error Analysis	You Decide, pp. 763, 780	Error Analysis, pp. 762, 771, 780, 791, 797, 803, 811, 819, 826	
Standardized Test Practice	SAT/ACT Practice, pp. 765, 773, 783, 793, 800, 805, 814, 821, 828 SAT & ACT Preparation, pp. 834–835	TIMSS, p. 835	SAT/ACT Practice, pp. 305–306
Open-Ended Assessment	Math Journal, p. 791 Open-Ended Assessment, p. 833 Portfolio, p. 833	Modeling: pp. 783, 785, 805, 828 Speaking: pp. 765, 793, 821 Writing: pp. 773, 800, 814	Open-Ended Assessment, p. 301
Chapter Assessment	Study Guide and Assessment, pp. 829–833 Chapter Test, p. A67		Multiple-Choice Tests (Forms 1A, 1B, 1C), pp. 289–294 Free-Response Tests (Forms 2A, 2B, 2C), pp. 295–300

TestCheck and Worksheet Builder is a networkable CD-ROM with three integrated modules. The **Worksheet Builder** creates customized worksheets, tests, and quizzes of free-response, multiple-choice, short-answer, and open-ended items. The **Student Module** gives you the option of having students take tests on-screen and get immediate feedback on their performance. Use the optional **Management System** to keep detailed student records.

Additional Chapter Resources

Student Edition
Career Choices, p. 800
Internet Project, p. 833
Extra Practice, pp. A48–A50

Teacher's Classroom Resources

Technology
Graphing Calculator
Keystrokes Masters

Teaching Aids
Answer Key Transparencies
Block Schedule Planning Guide
Lesson Planning Guide
Solutions Manual

Glencoe Instructional Technology

Instructional
Advanced MathPASS CD-ROM
Multimedia Applications CD-ROM

Assessment
TestCheck and Worksheet Builder

Visit **www.amc.glencoe.com** for data updates, career information, graphing calculator programs, and other interactive activities.

Chapter Overview

This chapter reviews arithmetic and geometric sequences and series, and introduces infinite series and sequences. Students explore such topics as limits; sums, convergence, and divergence of series; and sigma notation. Students study Pascal's triangle, binomial expressions, and the Binomial Theorem. They also examine the work of Fibonacci and Euler, study iteration, and use mathematical induction to prove the validity of mathematical statements.

Using a Graphing Calculator

There are several opportunities, all optional, to use a graphing calculator in this chapter.

- Exploring sequences generated by continued fractions (pp. 784 –785)
- Combining **sum(** with **seq(** to obtain the sum of an infinite sequence (pp. 795, 798)
- Using ⎡TRACE⎤ to examine polynomial functions that can be used to approximate sin *x* (p. 809)

The **Graphing Calculator Appendix,** which begins on page A2, contains keystroke instructions on many of these topics.

Chapter 12

SEQUENCES AND SERIES

> ### CHAPTER OBJECTIVES
> - Identify and find *n*th terms of arithmetic, geometric, and infinite sequences. *(Lessons 12-1, 12-2)*
> - Find sums of arithmetic, geometric, and infinite series. *(Lessons 12-1, 12-2, 12-3)*
> - Determine whether a series is convergent or divergent. *(Lesson 12-4)*
> - Use sigma notation. *(Lesson 12-5)*
> - Use the Binomial Theorem to expand binomials. *(Lesson 12-6)*
> - Evaluate expressions using exponential, trigonometric, and iterative series. *(Lessons 12-7, 12-8)*
> - Use mathematical induction to prove the validity of mathematical statements. *(Lesson 12-9)*

758 *Chapter 12 Sequences and Series*

 SAT & ACT Preparation

This textbook contains a complete SAT/ACT preparation course described on pages T8–T9 of the *Teacher's Handbook.* This chapter contains:

- an SAT or ACT practice question at the end of each lesson
- examples of SAT and ACT problems involving percents on page 834
- ten SAT and ACT practice problems on page 835

- additional SAT and ACT practice in the *Assessment and Evaluation Masters,* pp. 305–306
- SAT and ACT test practice banks in the *TestCheck and Worksheet Builder*

Arithmetic Sequences and Series

OBJECTIVES
- Find the *n*th term and arithmetic means of an arithmetic sequence.
- Find the sum of *n* terms of an arithmetic series.

REAL ESTATE Ofelia Gonzales sells houses in a new development. She makes a commission of $3750 on the sale of her first house. To encourage aggressive selling, Ms. Gonzales' employer promises a $500 increase in commission for each additional house sold. Thus, on the sale of her next house, she will earn $4250 commission. How many houses will Ms. Gonzales have to sell for her total commission in one year to be at least $65,000? *This problem will be solved in Example 6.*

Look Back
Refer to Lesson 4-1 for more about natural numbers.

The set of numbers representing the amount of money earned for each house sold is an example of a **sequence.** A sequence is a function whose domain is the set of natural numbers. The **terms** of a sequence are the range elements of the function. The first term of a sequence is denoted a_1, the second term is a_2, and so on up to the *n*th term a_n.

Symbol	a_1	a_2	a_3	a_4	a_5	a_6	a_7	a_8	a_9	a_{10}
Term	3	$2\frac{1}{2}$	2	$1\frac{1}{2}$	1	$\frac{1}{2}$	0	$-\frac{1}{2}$	-1	$-1\frac{1}{2}$

The sequence given in the table above is an example of an **arithmetic sequence.** The difference between successive terms of an arithmetic sequence is a constant called the **common difference,** denoted *d*. In the example above, $d = \frac{1}{2}$.

Arithmetic Sequence

An arithmetic sequence is a sequence in which each term after the first, a_1, is equal to the sum of the preceding term and the common difference, *d*. The terms of the sequence can be represented as follows.
$$a_1, \ a_1 + d, \ a_1 + 2d, \ ...$$

To find the next term in an arithmetic sequence, first find the common difference by subtracting any term from its succeeding term. Then add the common difference to the last term to find the next term in the sequence.

Example ❶ **Find the next four terms in the arithmetic sequence $-5, -2, 1, ...$**

First, find the common difference.

$a_2 - a_1 = -2 - (-5)$ or 3   *Find the difference between pairs of consecutive*
$a_3 - a_2 = 1 - (-2)$ or 3   *terms to verify the common difference.*

The common difference is 3.
Add 3 to the third term to get the fourth term, and so on.

$a_4 = 1 + 3$ or 4 $a_5 = 4 + 3$ or 7 $a_6 = 7 + 3$ or 10 $a_7 = 10 + 3$ or 13
The next four terms are 4, 7, 10, and 13.

Lesson 12-1 Arithmetic Sequences and Series **759**

1 FOCUS

5-Minute Check
Chapter 11

1. Evaluate $256^{\frac{3}{4}}$. **64**
2. Simplify $\sqrt[3]{64x^3y^6z}$. $\mathbf{4xy^2z^{\frac{1}{3}}}$
3. Graph $y = 3^x - 2$.

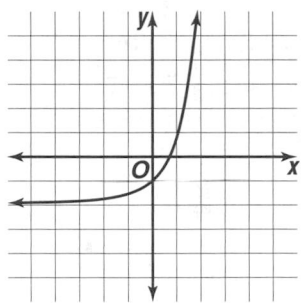

Solve.
4. $\log_x 243 = 5$ **3**
5. $3.4^x = 82.3$ **3.6039**

Motivating the Lesson
Discussion Have students try to guess how much money they would have after a month if you gave them a nickel the first day, two nickels the second day, three nickels the third day, and so on. **In 31 days, they would have 496 nickels or $24.80.**

Teaching Tip Remind students that the domain {1, 2, 3, 4, ...} represents the subscripts of *a*. The actual terms of the sequence are the members of the range.

Resource Manager

Reproducible Masters
- *Study Guide and Practice,* pp. 215–216
- *Enrichment,* p. 133
- *Spreadsheet Applications,* p. 12

Transparencies
- 5-Minute Check, 12-1
- Teaching, 12-1

Technology/Multimedia
- Advanced MathPASS, Lessons 24 and 25

In-Class Examples

Example 1
Find the next four terms in the arithmetic sequence -12, -1, 10, ... **21, 32, 43, 54**

Example 2
Find the 41st term in the arithmetic sequence 11, 4, -3, -10, ... **−269**

Example 3
Find the first term in the arithmetic sequence for which $a_{44} = 229$ and $d = 8$. **−115**

By definition, the nth term is also equal to $a_{n-1} + d$, where a_{n-1} is the $(n-1)$th term. That is, $a_n = a_{n-1} + d$. This type of formula is called a **recursive formula.** This means that each succeeding term is formulated from one or more previous terms.

The nth term of an arithmetic sequence can also be found when only the first term and the common difference are known. Consider an arithmetic sequence in which $a = -3.7$ and $d = 2.9$. Notice the pattern in the way the terms are formed.

first term	a_1	a	-3.7
second term	a_2	$a + d$	$-3.7 + 1(2.9) = -0.8$
third term	a_3	$a + 2d$	$-3.7 + 2(2.9) = 2.1$
fourth term	a_4	$a + 3d$	$-3.7 + 3(2.9) = 5.0$
fifth term	a_5	$a + 4d$	$-3.7 + 4(2.9) = 7.9$
$\vdots$	$\vdots$	$\vdots$	$\vdots$
nth term	a_n	$a + (n-1)d$	$-3.7 + (n-1)2.9$

The nth Term of an Arithmetic Sequence	The nth term of an arithmetic sequence with first term a_1 and common difference d is given by $a_n = a_1 + (n-1)d$.

Notice that the preceding formula has four variables: a_n, a_1, n, and d. If any three of these are known, the fourth can be found.

Examples **2** Find the 47th term in the arithmetic sequence -4, -1, 2, 5, ...

First, find the common difference.

$a_2 - a_1 = -1 - (-4)$ or 3　　$a_3 - a_2 = 2 - (-1)$ or 3　　$a_4 - a_3 = 5 - 2$ or 3

The common difference is 3.

Then use the formula for the nth term of an arithmetic sequence.

$a_n = a_1 + (n-1)d$
$a_{47} = -4 + (47-1)3$　　*$n = 47$, $a_1 = -4$, and $d = 3$*
$a_{47} = 134$

3 Find the first term in the arithmetic sequence for which $a_{19} = 42$ and $d = -\frac{2}{3}$.

$a_n = a_1 + (n-1)d$
$a_{19} = a_1 + (19-1)\left(-\frac{2}{3}\right)$　　*$n = 19$ and $d = -\frac{2}{3}$*
$42 = a_1 + (-12)$　　　　　　*$a_{19} = 42$*
$a_1 = 54$

Sometimes you may know two terms of an arithmetic sequence that are not in consecutive order. The terms between any two nonconsecutive terms of an arithmetic sequence are called **arithmetic means.** In the sequence below, 38 and 49 are the arithmetic means between 27 and 60.

$$5, 16, 27, 38, 49, 60$$

760 *Chapter 12 Sequences and Series*

Prerequisite Skills

A review of the prerequisite skills for this lesson is provided in the *Prerequisite Skills Masters*.

Skill Number and Content	Pages
7 Operations with Polynomials	15–16

Example **④** **Write an arithmetic sequence that has five arithmetic means between 4.9 and 2.5.**

The sequence will have the form 4.9, __?__, __?__, __?__, __?__, __?__, 2.5.
Note that 2.5 is the 7th term of the sequence or a_7.

First, find the common difference, using $n = 7$, $a_7 = 2.5$, and $a_1 = 4.9$	Then determine the arithmetic means.
$a_n = a_1 + (n - 1)d$	$a_2 = 4.9 + (-0.4)$ or 4.5
$2.5 = 4.9 + (7 - 1)d$	$a_3 = 4.5 + (-0.4)$ or 4.1
$2.5 = 4.9 + 6d$	$a_4 = 4.1 + (-0.4)$ or 3.7
$d = -0.4$	$a_5 = 3.7 + (-0.4)$ or 3.3
	$a_6 = 3.3 + (-0.4)$ or 2.9

The sequence is 4.9, 4.5, 4.1, 3.7, 3.3, 2.9, 2.5.

An indicated sum is $\overline{1 + 2 + 3 + 4}$. The sum $1 + 2 + 3 + 4$ is 10.

An **arithmetic series** is the indicated sum of the terms of an arithmetic sequence. The lists below show some examples of arithmetic sequences and their corresponding arithmetic series.

Arithmetic Sequence	**Arithmetic Series**
$-9, -3, 3, 9, 15$	$-9 + (-3) + 3 + 9 + 15$
$3, \frac{5}{2}, 2, \frac{3}{2}, 1, \frac{1}{2}$	$3 + \frac{5}{2} + 2 + \frac{3}{2} + 1 + \frac{1}{2}$
$a_1, a_2, a_3, a_4, \dots, a_n$	$a_1 + a_2 + a_3 + a_4 + \dots + a_n$

The symbol S_n, called the **nth partial sum**, is used to represent the sum of the first n terms of a series. To develop a formula for S_n for a finite arithmetic series, a series can be written in two ways and added term by term, as shown below. The second equation for S_n given below is obtained by reversing the order of the terms in the series.

$$S_n = a_1 + (a_1 + d) + (a_1 + 2d) + \dots + (a_n - 2d) + (a_n - d) + a_n$$
$$+ S_n = a_n + (a_n - d) + (a_n - 2d) + \dots + (a_1 + 2d) + (a_1 + d) + a_1$$
$$\overline{2S_n = (a_1 + a_n) + (a_1 + a_n) + (a_1 + a_n) + \dots + (a_1 + a_n) + (a_1 + a_n) + (a_1 + a_n)}$$
$$2S_n = n(a_1 + a_n) \quad \textit{There are n terms in the series, all of which are } (a_1 + a_n).$$

Therefore, $S_n = \frac{n}{2}(a_1 + a_n)$.

Sum of a Finite Arithmetic Series	The sum of the first n terms of an arithmetic series is given by $S_n = \frac{n}{2}(a_1 + a_n)$.

Example **⑤** **Find the sum of the first 60 terms in the arithmetic series $9 + 14 + 19 + \dots + 304$.**

$$S_n = \frac{n}{2}(a_1 + a_n)$$
$$S_{60} = \frac{60}{2}(9 + 304) \quad n = 60, a_1 = 9, a_{60} = 304$$
$$= 9390$$

Teaching Tip For Example 4, stress that in the formula $a_n = a_1 + (n - 1)d$, students must find d, the common difference.

In-Class Examples

Example 4
Write an arithmetic sequence that has six arithmetic means between −12 and 23. **−12, −7, −2, 3, 8, 13, 18, 23**

Example 5
Find the sum of the first 32 terms in the arithmetic series $-12 - 6 - 0 + \dots + 174$. **2592**

Teaching Tip After completing Example 4, ask students how they might determine the missing terms by working back from the last term given. **subtract *d* from each term**

Addressing Individual Needs

Kinesthetic Have students use counters, base-ten unit blocks, or other objects to model a sequence. Ask them to write the first 20 terms of the sequence being modeled. Then, have them use the model to find the sum of the first 20 terms.

In-Class Example

Example 6

Sales Anand sells clothing in a store at a local mall. His commission on the sale of new spring merchandise is $3.50 per item costing at least $25. To encourage an increase in sales, the store manager promises a $0.05 increase in commission for each additional item sold. Thus, on the sale of the next item, Anand will earn $3.55 commission. How many items worth $25 or more will Anand need to sell for his total commission on new spring merchandise to be at least $1000?   **143 items**

3 PRACTICE/APPLY

Error Analysis

Watch for students who calculate the common difference incorrectly.

Prevent by reminding them to set up the equation in the way suggested in the teaching tip for Example 4.

Study Guide and Practice Masters, pp. 215–216

Study Guide

Arithmetic Sequences and Series

A **sequence** is a function whose domain is the set of natural numbers. The **terms** of a sequence are the range elements of the function. The difference between successive terms of an **arithmetic sequence** is a constant called the **common difference**, denoted as *d*. An **arithmetic series** is the indicated sum of the terms of an arithmetic sequence.

Example 1 **a.** Find the next four terms in the arithmetic sequence $-7, -5, -3, \ldots$.
 b. Find the 38th term of this sequence.

 a. Find the common difference.
   $a_2 - a_1 = -5 - (-7)$ or 2

   The common difference is 2. Add 2 to the third term to get the fourth term, and so on.
   $a_4 = -3 + 2$ or -1 $a_5 = -1 + 2$ or 1
   $a_6 = 1 + 2$ or 3 $a_7 = 3 + 2$ or 5

   The next four terms are $-1, 1, 3,$ and 5.

 b. Use the formula for the nth term of an arithmetic sequence.
   $a_n = a_1 + (n - 1)d$
   $a_{38} = -7 + (38 - 1)2$ $n = 38, a_1 = -7, d = 2$
   $a_{38} = 67$

Example 2 Write an arithmetic sequence that has three arithmetic means between 3.2 and 4.4.

 The sequence will have the form 3.2, _?_, _?_, _?_, 4.4.

 First, find the common difference.
 $a_n = a_1 + (n - 1)d$
 $4.4 = 3.2 + (5 - 1)d$ $n = 5, a_5 = 4.4, a_1 = 3.2$
 $4.4 = 3.2 + 4d$
 $d = 0.3$

 Then, determine the arithmetic means.

a_2	a_3	a_4
$3.2 + 0.3 = 3.5$	$3.5 + 0.3 = 3.8$	$3.8 + 0.3 = 4.1$

 The sequence is 3.2, 3.5, 3.8, 4.1, 4.4.

Example 3 Find the sum of the first 50 terms in the series $11 + 14 + 17 + \cdots + 158$.

 $S_n = \frac{n}{2}(a_1 + a_n)$
 $S_{50} = \frac{50}{2}(11 + 158)$ $n = 50, a_1 = 11, a_{50} = 158$
  $= 4225$

© Glencoe/McGraw-Hill   215   *Advanced Mathematical Concepts*

When the value of the last term, a_n, is not known, you can still determine the sum of the series. Using the formula for the nth term of an arithmetic sequence, you can derive another formula for the sum of a finite arithmetic series.

$$S_n = \frac{n}{2}(a_1 + a_n)$$
$$S_n = \frac{n}{2}[a_1 + (a_1 + (n - 1)d)] \quad a_n = a_1 + (n - 1)d$$
$$S_n = \frac{n}{2}[2a_1 + (n - 1)d]$$

Example   **6 REAL ESTATE** Refer to the application at the beginning of the lesson. **How many houses will Ms. Gonzales have to sell for her total commission in one year to be at least $65,000?**

Let S_n = the amount of her desired commission, $65,000.
Let a_1 = the first commission, $3750.
In this example, $d = 500$.

We want to find n, the number of houses that Ms. Gonzales has to sell to have a total commission greater than or equal to $65,000.

$$S_n = \frac{n}{2}[2a_1 + (n - 1)d]$$
$$65{,}000 = \frac{n}{2}[2(3750) + (n - 1)(500)] \quad S_n = 65{,}000, a_1 = 3750$$
$$130{,}000 = n(7500 + 500n - 500) \quad \textit{Multiply each side by 2.}$$
$$130{,}000 = 7000n + 500n^2 \quad \textit{Simplify.}$$
$$0 = 500n^2 + 7000n - 130{,}000$$
$$0 = 5n^2 + 70n - 1300 \quad \textit{Divide each side by 100.}$$
$$n = \frac{-70 \pm \sqrt{70^2 - 4(5)(-1300)}}{2(5)} \quad \textit{Use the Quadratic Formula.}$$
$$n = \frac{-70 \pm \sqrt{30{,}900}}{10}$$
$$n \approx 10.58 \text{ and } -24.58 \quad \textit{-24.58 is not a possible answer.}$$

Ms. Gonzales must sell 11 or more houses for her total commission to be at least $65,000.

CHECK FOR UNDERSTANDING

Communicating Mathematics

Read and study the lesson to answer each question.

1. 2, −2, −6, −10, −14; yes, there is a common difference of −4.

1. **Write** the first five terms of the sequence defined by $a_n = 6 - 4n$. Is this an arithmetic sequence? Explain.

2. Consider the arithmetic sequence defined by $a_n = \frac{5 - 2n}{2}$.

   a. **Graph** the first five terms of the sequence. Let n be the x-coordinate and a_n be the y-coordinate, and connect the points. **See margin.**

   b. **Describe** the graph found in part **a**. **linear**

   c. **Find** the common difference of the sequence and determine its relationship to the graph found in part **a**. **See margin.**

Additional Answers

2a.

2c. The common difference is −1. This is the slope of the line through the points of the sequence.

3a. The number of houses sold cannot be negative.

3. Refer to Example 6.

    a. Explain why -24.58 is *not* a possible answer.

    b. Determine how much money Ms. Gonzales will make if she sells 10 houses. **$60,000**

4. Describe the common difference for an arithmetic sequence in which the terms are decreasing. **negative**

5. You Decide Ms. Brooks defined two sequences, $a_n = (-1)^n$ and $b_n = (-2)^n$, for her class. She asked the class to determine if they were arithmetic sequences. Latonya said the second was an arithmetic sequence and that the first was not. Diana thought the reverse was true. Who is correct? Explain. **See margin.**

Guided Practice

Find the next four terms in each arithmetic sequence.

6. 6, 11, 16, ...
  21, 26, 31, 36

7. $-15, -7, 1, ...$
  9, 17, 25, 33

8. $a - 6, a - 2, a + 2, ...$
  $a + 6, a + 10, a + 14, a + 18$

For Exercises 9-15, assume that each sequence or series is arithmetic.

9. Find the 17th term in the sequence for which $a_1 = 10$ and $d = -3$. **-38**

10. Find n for the sequence for which $a_n = 37$, $a_1 = -13$, and $d = 5$. **11**

11. What is the first term in the sequence for which $d = -2$ and $a_7 = 3$? **15**

12. Find d for the sequence for which $a_1 = 100$ and $a_{12} = 34$. **-6**

13. 9, 14, 19, 24   **13.** Write a sequence that has two arithmetic means between 9 and 24.

14. What is the sum of the first 35 terms in the series $7 + 9 + 11 + \cdots$? **1435**

15. Find n for a series for which $a_1 = 30$, $d = -4$, and $S_n = -210$. **21**

16. Theater Design The right side of the orchestra section of the Nederlander Theater in New York City has 19 rows, and the last row has 27 seats. The numbers of seats in each row increase by 1 as you move toward the back of the section. How many seats are in this section of the theater? **342 seats**

EXERCISES

Practice

Find the next four terms in each arithmetic sequence. **17–25. See margin.**

17. $5, -1, -7, ...$
18. $-18, -7, 4, ...$
19. $3, 4.5, 6, ...$

20. $5.6, 3.8, 2, ...$
21. $b, b + 4, b + 8, ...$
22. $-x, 0, x, ...$

23. $5n, -n, -7n, ...$
24. $5 + k, 5, 5 - k, ...$
25. $2a - 5, 2a + 2, 2a + 9, ...$

26. Determine the common difference and find the next three terms of the arithmetic sequence $3 + \sqrt{7}, 5, 7 - \sqrt{7}, ...$ $2 - \sqrt{7}; 9 - 2\sqrt{7}, 11 - 3\sqrt{7}, 13 - 4\sqrt{7}$

For Exercises 27-34, assume that each sequence or series is arithmetic.

27. Find the 25th term in the sequence for which $a_1 = 8$ and $d = 3$. **80**

28. Find the 18th term in the sequence for which $a_1 = 1.4$ and $d = 0.5$. **9.9**

29. Find n for the sequence for which $a_n = -41$, $a_1 = 19$, and $d = -5$. **13**

30. Find n for the sequence for which $a_n = 138$, $a_1 = -2$, and $d = 7$. **21**

31. What is the first term in the sequence for which $d = -3$, and $a_{15} = 38$? **80**

32. What is the first term in the sequence for which $d = \frac{1}{3}$ and $a_7 = 10\frac{2}{3}$? **$8\frac{2}{3}$**

33. Find d for the sequence in which $a_1 = 6$ and $a_{14} = 58$. **4**

34. Find d for the sequence in which $a_1 = 8$ and $a_{11} = 26$. **$1\frac{4}{5}$**

www.amc.glencoe.com/self_check_quiz

Assignment Guide

Core: 17–57 odd, 58–69
Enriched: 18–48 even, 50–69

Additional Answers

5. Neither student is correct, since neither sequence has a common difference. The difference fluctuates between -1 and 1. The second sequence has a difference that fluctuates between 2 and -2.

17. $-13, -19, -25, -31$

18. 15, 26, 37, 48

19. 7.5, 9, 10.5, 12

20. $0.2, -1.6, -3.4, -5.2$

21. $b + 12, b + 16, b + 20, b + 24$

22. $2x, 3x, 4x, 5x$

23. $-13n, -19n, -25n, -31n$

24. $5 - 2k, 5 - 3k, 5 - 4k, 5 - 5k$

25. $2a + 16, 2a + 23, 2a + 30, 2a + 37$

51. Let d be the common difference. Then, $y = x + d$, $z = x + 2d$, and $w = x + 3d$.
Substitute these values into the expression $x + w - y$ and simplify.
$x + (x + 3d) - (x + d) = x + 2d$ or z.

54b. The common difference between each consecutive term in the sequence is 180, therefore the sequence is arithmetic.

55c. Conjecture: the sum of the first n terms of the sequence of natural numbers is n^2.

Proof:

Let $a_n = 2n - 1$. The first term of the sequence of natural numbers is 1, so $a_1 = 1$.

Then, using the formula for the sum of an arithmetic series,

$S_n = \dfrac{n}{2}(a_1 + a_n)$

$S_n = \dfrac{n}{2}[1 + (2n - 1)]$

$= \dfrac{n}{2}(2n)$ or n^2

For Exercises 35–49, assume that each sequence or series is arithmetic.

35. **$17 + \sqrt{5}$** 35. What is the eighth term in the sequence $-4 + \sqrt{5}, -1 + \sqrt{5}, 2 + \sqrt{5}, \ldots$?

36. What is the twelfth term in the sequence $5 - i, 6, 7 + i, \ldots$? **$16 + 10i$**

37. Find the 33rd term in the sequence 12.2, 10.5, 8.8, ... **−42.2**

38. Find the 79th term in the sequence $-7, -4, -1, \ldots$ **227**

39. **12, 16.5, 21** 39. Write a sequence that has one arithmetic mean between 12 and 21.

40. **$-5, -2, 1, 4$** 40. Write a sequence that has two arithmetic means between -5 and 4.

41. **$\sqrt{3}$,** 41. Write a sequence that has two arithmetic means between $\sqrt{3}$ and 12.

$\dfrac{12 + 2\sqrt{3}}{3}$, 42. Write a sequence that has three arithmetic means between 2 and 5.

$\dfrac{24 + \sqrt{3}}{3}$, **12** 43. Find the sum of the first 11 terms in the series $\dfrac{3}{2} + 1 + \dfrac{1}{2} + \ldots$ **−11**

44. Find the sum of the first 100 terms in the series $-5 - 4.8 - 4.6 - \ldots$ **490**

42. **2, 2.75, 3.5,** 45. Find the sum of the first 26 terms in the series $-19 - 13 - 7 - \ldots$ **1456**

4.25, 5 46. Find n for a series for which $a_1 = -7$, $d = 1.5$, and $S_n = -14$. **8**

47. Find n for a series for which $a_1 = -3$, $d = 2.5$, and $S_n = 31.5$. **7**

48. Write an expression for the nth term of the sequence 5, 7, 9, ... **$2n + 3$**

49. Write an expression for the nth term of the sequence $6, -2, -10, \ldots$ **$-8n + 14$**

Applications and Problem Solving

Real World Application

50. **Keyboarding** Antonio has found that he can input statistical data into his computer at the rate of 2 data items faster each half hour he works. One Monday, he starts work at 9:00 A.M., inputting at a rate of 3 data items per minute. At what rate will Antonio be inputting data into the computer by lunchtime (noon)? **15 data items per min**

51. **Critical Thinking** Show that if x, y, z, and w are the first four terms of an arithmetic sequence, then $x + w - y = z$. **See margin.**

52. **Construction** The Arroyos are planning to build a brick patio that approximates the shape of a trapezoid. The shorter base of the trapezoid needs to start with a row of 5 bricks, and each row must increase by 2 bricks on each side until there are 25 rows. How many bricks do the Arroyos need to buy?
1325 bricks

53. **Critical Thinking** The measures of the angles of a convex polygon form an arithmetic sequence. The least measurement in the sequence is 85°. The greatest measurement is 215°. Find the number of sides in this polygon. **12**

54. **Geometry** The sum of the interior angles of a triangle is 180°.

54a. **360°, 540°, 720°, 900°**
   a. What are the sums of the interior angles of polygons with 4, 5, 6, and 7 sides?
   b. Show that these sums (beginning with the triangle) form an arithmetic sequence. **See margin.**
   c. Find the sum of the interior angles of a 35-sided polygon. **5940°**

Extra Credit

Give an example of an arithmetic sequence whose terms are all multiples of 3 and 5, but not multiples of 4 or 6.
Sample answer: 15, 45, 75, 105, . . .

55. Critical Thinking Consider the sequence of odd natural numbers.
   a. What is S_5? **25**
   b. What is S_{10}? **100**
   c. Make a conjecture as to the pattern that emerges concerning the sum. Write an algebraic proof verifying your conjecture. **See margin.**

56. Sports At the 1998 Winter X-Games held in Crested Butte, Colorado, Jennie Waara, from Sweden, won the women's snowboarding slope-style competition. Suppose that in one of the qualifying races, Ms. Waara traveled 5 feet in the first second, and the distance she traveled increased by 7 feet each subsequent second. If Ms. Waara reached the finish line in 15 seconds, how far did she travel? **810 feet**

57. Entertainment A radio station advertises a contest with ten cash prizes totaling $5510. There is to be a $100 difference between each successive prize. Find the amounts of the least and greatest prizes the radio station will award. **least: $101, greatest: $1001**

58. Critical Thinking Some sequences involve a pattern but are not arithmetic. Find the sum of the first 72 terms in the sequence 6, 8, 2, ..., where $a_n = a_{n-1} - a_{n-2}$. **0**

Mixed Review

59. $285.77

60. (0, −5); ($\pm\sqrt{21}$, −5); (±5, −5), (0, −3), (0, −7); see margin for graph.

59. Personal Finance If Parker Hamilton invests $100 at 7% compounded continuously, how much will he have at the end of 15 years? *(Lesson 11-3)*

60. Find the coordinates of the center, foci, and vertices of the ellipse whose equation is $4x^2 + 25y^2 + 250y + 525 = 0$. Then graph the equation. *(Lesson 10-3)*

61. Find $6\left(\cos\frac{5\pi}{8} + i\sin\frac{5\pi}{8}\right) \div 12\left(\cos\frac{\pi}{2} + i\sin\frac{\pi}{2}\right)$. Then express the quotient in rectangular form. *(Lesson 9-7)* **See margin.**

62. Find the inner product of $\overline{\mathbf{u}}$ and $\overline{\mathbf{v}}$ if $\overline{\mathbf{u}} = \langle 2, -1, 3 \rangle$ and $\overline{\mathbf{v}} = \langle 5, 3, 0 \rangle$. *(Lesson 8-4)* **7**

63. Write the standard form of the equation of a line for which the length of the normal is 5 units and the normal makes an angle of 30° with the positive x-axis. *(Lesson 7-6)* $\sqrt{3}x + y - 10 = 0$

64. Graph $y = \sec 2\theta - 3$. *(Lesson 6-7)* **See margin.**

65. $A = 70°28'$, $a = 4.2$, $b = 1.5$

65. Solve triangle ABC if $B = 19°32'$ and $c = 4.5$. Round angle measures to the nearest minute and side measures to the nearest tenth. *(Lesson 5-5)*

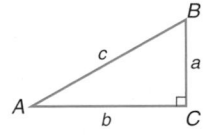

66. −23; 2 imaginary roots

66. Find the discriminant of $4p^2 - 3p + 2 = 0$. Describe the nature of its roots. *(Lesson 4-2)*

67. $y = x - 1$

67. Determine the slant asymptote of $f(x) = \dfrac{x^2 - 4x + 2}{x - 3}$. *(Lesson 3-7)*

68. Triangle ABC is represented by the matrix $\begin{bmatrix} -2 & 0 & 1 \\ 1 & 3 & -4 \end{bmatrix}$. Find the image of the triangle after a rotation of 270° counterclockwise about the origin. *(Lesson 2-4)* $A'(1, 2), B'(3, 0), C'(-4, -1)$

69. SAT/ACT Practice If $a - 4b - 15$ and $4a - b = 15$, then $a - b = ?$ **C**
   A 3   B 4   C 6   D 15   E 30

Extra Practice See p. A48.

Additional Answers

61. $0.5\left(\cos\frac{\pi}{8} + i\sin\frac{\pi}{8}\right), 0.46 + 0.19i$

64.

Enrichment Masters, p. 133

Additional Answer

60.

Geometric Sequences and Series

5-Minute Check
Lesson 12-1

1. Find the next four terms in the arithmetic sequence $-2.2, -1, 0.2, \ldots$. **1.4, 2.6, 3.8, 5**

2. Find the 33rd term in the arithmetic sequence $-8, -1, 6, 13, \ldots$. **216**

3. Find the first term in the arithmetic sequence for which $a_{24} = 58$ and $d = 5$. **−57**

4. Write an arithmetic sequence that has five arithmetic means between 21.8 and −10. **21.8, 16.5, 11.2, 5.9, 0.6, −4.7, −10**

5. Find the sum of the first 42 terms in the arithmetic series $-17 - 5 + 7 + 19 + \cdots + 475$. **9618**

Motivating the Lesson

Hands-On Activity Have each student tear in half a square piece of paper with an area of one square unit. Then, have them tear it in half again. Ask students to predict the area of one of the resulting rectangles after 5 tears. $\frac{1}{32}$ **unit²**

OBJECTIVES
- Find the *n*th term and geometric means of a geometric sequence.
- Find the sum of *n* terms of a geometric series.

ACCOUNTING Bertha Blackwell is an accountant for a small company. On January 1, 1996, the company purchased $50,000 worth of office copiers. Since this equipment is a company asset, Ms. Blackwell needs to determine how much the copiers are presently worth. She estimates that copiers depreciate at a rate of 45% per year. What value should Ms. Blackwell assign the copiers on her 2001 year-end accounting report? *This problem will be solved in Example 3.*

The following sequence is an example of a **geometric sequence**.

$$10, 2, 0.4, 0.08, 0.016, \ldots \quad \textit{Can you find the next term?}$$

The ratio of successive terms in a geometric sequence is a constant called the **common ratio,** denoted *r*.

Geometric Sequence	A geometric sequence is a sequence in which each term after the first, a_1, is the product of the preceding term and the common ratio, *r*. The terms of the sequence can be represented as follows, where a_1 is nonzero and *r* is not equal to 1 or 0. $$a_1, a_1r, a_1r^2, \ldots.$$

You can find the next term in a geometric sequence as follows.

- First divide any term by the preceding term to find the common ratio.
- Then multiply the last term by the common ratio to find the next term in the sequence.

Example Determine the common ratio and find the next three terms in each sequence.

a. $1, -\dfrac{1}{2}, \dfrac{1}{4}, \ldots$

First, find the common ratio.

$$a_2 \div a_1 = -\frac{1}{2} \div 1 \text{ or } -\frac{1}{2} \qquad\qquad a_3 \div a_2 = \frac{1}{4} \div \left(-\frac{1}{2}\right) \text{ or } -\frac{1}{2}$$

The common ratio is $-\dfrac{1}{2}$.

Multiply the third term by $-\dfrac{1}{2}$ to get the fourth term, and so on.

$$a_4 = \frac{1}{4} \cdot \left(-\frac{1}{2}\right) \text{ or } -\frac{1}{8} \qquad a_5 = -\frac{1}{8} \cdot \left(-\frac{1}{2}\right) \text{ or } \frac{1}{16} \qquad a_6 = \frac{1}{16} \cdot \left(-\frac{1}{2}\right) \text{ or } -\frac{1}{32}$$

The next three terms are $-\dfrac{1}{8}, \dfrac{1}{16}$, and $-\dfrac{1}{32}$.

Resource Manager

 Reproducible Masters
- *Study Guide and Practice,* pp. 217–218
- *Enrichment,* p. 134
- *Problem Solving and Applications,* p. 23

 Transparencies
- 5-Minute Check, 12-2
- Teaching, 12-2

 Technology/Multimedia
- Advanced MathPASS, Lessons 24 and 25

b. $r - 1, -3r + 3, 9r - 9, \ldots$

First, find the common ratio.

$a_2 \div a_1 = \dfrac{-3r + 3}{r - 1}$

$a_2 \div a_1 = \dfrac{-3(r - 1)}{r - 1}$ *Factor.*

$a_2 \div a_1 = -3$ *Simplify.*

The common ratio is -3.

$a_3 \div a_2 = \dfrac{9r - 9}{-3r + 3}$

$a_3 \div a_2 = \dfrac{9(r - 1)}{-3(r - 1)}$ *Factor.*

$a_3 \div a_2 = -3$ *Simplify.*

Multiply the third term by -3 to get the fourth term, and so on.

$a_4 = -3(9r - 9)$ or $-27r + 27$ $a_5 = -3(-27r + 27)$ or $81r - 81$

$a_6 = -3(81r - 81)$ or $-243r + 243$

The next three terms are $-27r + 27$, $81r - 81$, and $-243r + 243$.

As with arithmetic sequences, geometric sequences can also be defined recursively. By definition, the nth term is also equal to $a_{n-1}r$, where a_{n-1} is the $(n - 1)$th term. That is, $a_n = a_{n-1}r$.

Since successive terms of a geometric sequence can be expressed as the product of the common ratio and the previous term, it follows that each term can be expressed as the product of a_1 and a power of r. The terms of a geometric sequence for which $a_1 = -5$ and $r = 7$ can be represented as follows.

first term	a_1	a_1	-5
second term	a_2	$a_1 r$	$-5 \cdot 7^1 = -35$
third term	a_3	$a_1 r^2$	$-5 \cdot 7^2 = -245$
fourth term	a_4	$a_1 r^3$	$-5 \cdot 7^3 = -1715$
fifth term	a_5	$a_1 r^4$	$-5 \cdot 7^4 = -12,005$
$\vdots$	$\vdots$	$\vdots$	$\vdots$
nth term	a_n	$a r^{n-1}$	$-5 \cdot 7^{n-1}$

The nth Term of a Geometric Sequence	The nth term of a geometric sequence with first term a_1 and common ratio r is given by $a_n = a_1 r^{n-1}$.

Example **2** **Find an approximation for the 23rd term in the sequence 256, -179.2, 125.44, ...**

First, find the common ratio.

$a_2 \div a_1 = -179.2 \div 256$ or -0.7 $a_3 \div a_2 = 125.44 \div (-179.2)$ or -0.7

The common ratio is -0.7.

Then, use the formula for the nth term of a geometric sequence.

$a_n = a_1 r^{n-1}$

$a_{23} = 256(-0.7)^{23-1}$ *$n = 23$, $a_1 = 256$, $r = -0.7$*

$a_{23} \approx 0.1000914188$ *Use a calculator.*

The 23rd term is about 0.1.

Teaching Tip Remind students that the order of division when finding the common ratio is important. Divide a term by its preceding term.

Teaching Tip After giving the formula for the nth term of a geometric sequence, remind students that the common ratio is assumed not to be 1. You may also wish to present an alternative formula $a_n = a_0 r^n$. Ask students how this is equivalent to the given formula.

In-Class Examples

Example 1
Determine the common ratio and find the next three terms in each sequence.
a. 21, 4.2, 0.84 ... **0.2; 0.168, 0.0336, 0.00672**

b. $2t - 10$, $-4t + 20$, $8t - 40$ $\cdots$ **-2; $-16t + 80$, $32t - 160$, $-64t + 320$**

Example 2
Find an approximation for the 12th term in the sequence -24, 26.4, -29.04, ... **about 68.5**

Teaching Tip For Example 3 below, point out that the common ratio is $1 - 0.4 = 0.6$, which is the value of the car retained each year.

In-Class Examples

Example 3
Accounting A new car costing $26,000 depreciates at the rate of 40% each year for four years. Find the value of the car at the end of four years. **$3369.60**

Example 4
Write a sequence that has two geometric means between 128 and 54. **128, 96, 72, 54**

Teaching Tip When working In-Class Example 4, remind students that once they have found the geometric means, they should check the sequence by multiplying the first term by $\frac{3}{4}$ to get the second term, the second term by $\frac{3}{4}$ to get the third term, and the third term by $\frac{3}{4}$ to yield the last term, which should be 54.

Geometric sequences can represent growth or decay.

- For a common ratio greater than 1, a sequence may model growth. Applications include compound interest, appreciation of property, and population growth.

- For a positive common ratio less than 1, a sequence may model decay. Applications include some radioactive behavior and depreciation.

Example **3** **ACCOUNTING** Refer to the application at the beginning of the lesson. Compute the value of the copiers at the end of the year 2001.

Since the copiers were purchased at the beginning of the first year, the original purchase price of the copiers represents a_1. If the copiers depreciate at a rate of 45% per year, then they retain $100 - 45$ or 55% of their value each year.

Use the formula for the nth term of a geometric sequence to find the value of the copiers six years later or a_7.

$$a_n = a_1 r^{n-1}$$
$$a_7 = 50{,}000 \cdot (0.55)^{7-1} \quad a_1 = 50{,}000,\ r = 0.55,\ n = 7$$
$$a_7 \approx 1384.032031 \quad \text{Use a calculator.}$$

Ms. Blackwell should list the value of the copiers on her report as $1384.03.

The terms between any two nonconsecutive terms of a geometric sequence are called **geometric means**.

Example **4** Write a sequence that has two geometric means between 48 and -750.

This sequence will have the form 48, ___?___, ___?___, -750.

First, find the common ratio.

$$a_n = a_1 r^{n-1}$$
$$a_4 = a_1 r^3 \quad \text{Since there will be four terms in the sequence, } n = 4.$$
$$-750 = 48 r^3 \quad a_4 = -750 \text{ and } a_1 = 48$$
$$\frac{-125}{8} = r^3 \quad \text{Divide each side by 48 and simplify.}$$
$$\sqrt[3]{-\frac{125}{8}} = r \quad \text{Take the cube root of each side.}$$
$$-2.5 = r$$

Then, determine the geometric sequence.

$$a_2 = 48(-2.5) \text{ or } -120 \qquad a_3 = -120(-2.5) \text{ or } 300$$

The sequence is 48, -120, 300, -750.

A **geometric series** is the indicated sum of the terms of a geometric sequence. The lists below show some examples of geometric sequences and their corresponding series.

Geometric Sequence	**Geometric Series**
$3, 9, 27, 81, 243$	$3 + 9 + 27 + 81 + 243$
$16, 4, 1, \frac{1}{4}, \frac{1}{16}$	$16 + 4 + 1 + \frac{1}{4} + \frac{1}{16}$
$a_1, a_2, a_3, a_4, \ldots, a_n$	$a_1 + a_2 + a_3 + a_4 + \cdots + a_n$

To develop a formula for the sum of a finite geometric sequence, S_n, write an expression for S_n and for rS_n, as shown below. Then subtract rS_n from S_n and solve for S_n.

$$S_n = a_1 + a_1 r + a_1 r^2 + \cdots + a_1 r^{n-2} + a_1 r^{n-1}$$
$$rS_n = \phantom{a_1 + {}} a_1 r + a_1 r^2 + \cdots + a_1 r^{n-2} + a_1 r^{n-1} + a_1 r^n$$
$$\overline{S_n - rS_n = a_1 - a_1 r^n} \quad \textit{Subtract.}$$
$$S_n(1 - r) = a_1 - a_1 r^n \quad \textit{Factor.}$$
$$S_n = \frac{a_1 - a_1 r^n}{1 - r} \quad \textit{Divide each side by } 1 - r, r \neq 1.$$

Sum of a Finite Geometric Series

The sum of the first n terms of a finite geometric series is given by $S_n = \dfrac{a_1 - a_1 r^n}{1 - r}$.

Example ⑤ Find the sum of the first ten terms of the geometric series
$16 - 48 + 144 - 432 + \cdots$.

The formula for the sum of a geometric series can also be written as
$S_n = a_1 \dfrac{1 - r^n}{1 - r}.$

First, find the common ratio.

$a_2 \div a_1 = -48 \div 16$ or $-3 \qquad a_4 \div a_3 = -432 \div 144$ or -3

The common ratio is -3.

$S_n = \dfrac{a_1 - a_1 r^n}{1 - r}$

$S_{10} = \dfrac{16 - 16(-3)^{10}}{1 - (-3)} \qquad n = 10, a_1 = 16, r = -3.$

$S_{10} = -236{,}192 \qquad \textit{Use a calculator.}$

The sum of the first ten terms of the series is $-236{,}192$.

Banks and other financial institutions use compound interest to determine earnings in accounts or how much to charge for loans. The formula for compound interest is $A = P\left(1 + \dfrac{r}{n}\right)^{tn}$, where

A = the account balance,
P = the initial deposit or amount of money borrowed,
r = the annual percentage rate (APR),
n = the number of compounding periods per year, and
t = the time in years.

Suppose at the beginning of each quarter you deposit $25 in a savings account that pays an APR of 2% compounded quarterly. Most banks post the interest for each quarter on the last day of the quarter. The chart below lists the additions to the account balance as a result of each successive deposit through the rest of the year. Note that $1 + \frac{r}{n} = 1 + \frac{0.02}{4}$ or 1.005.

Date of Deposit	$A = P\left(1 + \frac{r}{n}\right)^{tn}$	1st Year Additions (to the nearest penny)
January 1	$25 (1.005)^4	$25.50
April 1	$25 (1.005)^3	$25.38
July 1	$25 (1.005)^2	$25.25
October 1	$25 (1.005)^1	$25.13
Account balance at the end of one year		$101.26

The chart shows that the first deposit will gain interest through all four compounding periods while the second will earn interest through only three compounding periods. The third and last deposits will earn interest through two and one compounding periods, respectively. The sum of these amounts, $101.26, is the balance of the account at the end of one year. This sum also represents a finite geometric series where $a_1 = 25.13$, $r = 1.005$, and $n = 4$.

$$25.13 + 25.13(1.005) + 25.13(1.005)^2 + 25.13(1.005)^3$$

Example 6 **INVESTMENTS** Hiroshi wants to begin saving money for college. He decides to deposit $500 at the beginning of each quarter (January 1, April 1, July 1, and October 1) in a savings account that pays an APR of 6% compounded quarterly. The interest for each quarter is posted on the last day of the quarter. Determine Hiroshi's account balance at the end of one year.

The interest is compounded each quarter. So $n = 4$ and the interest rate per period is 6% ÷ 4 or 1.5%. The common ratio r for the geometric series is then $1 + 0.015$, or 1.015.

The first term a_1 in this series is the account balance at the end of the first quarter. Thus, $a_1 = 500(1.015)$ or 507.5.

Apply the formula for the sum of a geometric series.

$$S_n = \frac{a_1 - a_1 r^n}{1 - r}$$

$$S_4 = \frac{507.5 - 507.5(1.015)^4}{1 - 1.015} \qquad n = 4, r = 1.015$$

$$S_4 \approx 2076.13$$

Hiroshi's account balance at the end of one year is $2076.13.

Communicating Mathematics

Read and study the lesson to answer each question.   **1–5. See margin.**

1. **Compare and contrast** arithmetic and geometric sequences.

2. **Show** that the sequence defined by $a_n = (-3)^{n+1}$ is a geometric sequence.

3. **Explain** why the first term in a geometric sequence must be nonzero.

4. **Find a counterexample** for the statement "The sum of a geometric series cannot be less than its first term."

5. **Determine** whether the given terms form a finite geometric sequence. Write *yes* or *no* and then explain your reasoning.
   a. 3, 6, 18
   b. $\sqrt{3}, 3, \sqrt{27}$
   c. $x^{-2}, x^{-1}, 1 ; x \neq 1$

6. Refer to Example 3.   **a–b. See margin.**
   a. **Make a table** to represent the situation. In the first row, put the number of years, and in the second row, put the value of the computers.
   b. **Graph** the numbers in the table. Let years be the *x*-coordinate, let value be the *y*-coordinate, and connect the points.
   c. **Describe** the graph found in part b.   **an exponential function**

Guided Practice

Determine the common ratio and find the next three terms of each geometric sequence.

8. $\dfrac{3}{2}, \dfrac{27}{4}, \dfrac{81}{8}, \dfrac{243}{16}$

7. $\dfrac{2}{3}, 4, 24, \ldots$
   6; 144, 864, 5184

8. $2, 3, \dfrac{9}{2}, \ldots$

9. $1.8, -7.2, 28.8, \ldots$
   −4; −115.2, 460.8, −1843.2

For Exercises 10–14, assume that each sequence or series is geometric.

10. Find the seventh term of the sequence 7, 2.1, 0.63, ...   **0.005103**

11. If $r = 2$ and $a_5 = 24$, find the first term of the sequence. $\dfrac{3}{2}$

12. Find the first three terms of the sequence for which $a_4 = 2.5$ and $r = 2$.   **0.3125, 0.625, 1.25**

13. Write a sequence that has two geometric means between 1 and 27.   **1, 3, 9, 27**

14. Find the sum of the first nine terms of the series $0.5 - 1 + 2 - \cdots$.   **85.5**

15. **Investment** Mika Rockwell invests in classic cars. He recently bought a 1978 convertible valued at \$20,000. The value of the car is predicted to appreciate at a rate of 3.5% per year. Find the value of the car after 10, 20, and 40 years, assuming that the rate of appreciation remains constant.   **\$28,211.98; \$39,795.78; \$79,185.19**

Practice

Determine the common ratio and find the next three terms of each geometric sequence.   **16–24. See margin.**

 A

16. 10, 2, 0.4, ...
17. 8, −20, 50, ...
18. $\dfrac{2}{9}, \dfrac{2}{3}, 2, \ldots$

19. $\dfrac{3}{4}, \dfrac{3}{10}, \dfrac{3}{25}, \ldots$
20. −7, 3.5, −1.75, ...
21. $3\sqrt{2}, 6, 6\sqrt{2}, \ldots$

 B

22. $9, 3\sqrt{3}, 3, \ldots$
23. $i, -1, -i, \ldots$
24. $t^8, t^5, t^2, \ldots$

 www.amc.glencoe.com/self_check_quiz

Lesson 12-2 Geometric Sequences and Series   **771**

Additional Answers

16. 0.2; 0.08, 0.016, 0.0032
17. −2.5; −125, 312.5, −781.25
18. 3; 6, 18, 54
19. $\dfrac{2}{5}; \dfrac{6}{125}, \dfrac{12}{625}, \dfrac{24}{3125}$
20. −0.5; 0.875, −0.4375, 0.21875
21. $\sqrt{2}$; 12, $12\sqrt{2}$, 24
22. $\dfrac{\sqrt{3}}{3}$; $\sqrt{3}, 1, \dfrac{\sqrt{3}}{3}$
23. i; 1, i, −1
24. t^{-3}; t^{-1}, t^{-4}, t^{-7}

Error Analysis

Watch for students who forget that the common ratio, like the common difference, is sometimes negative.

Prevent by reminding students that in a geometric sequence or series, the first term x_1 times r equals the second term, $x_1 r = x_2$. Solving for r sometimes yields r as a negative number.

Assignment Guide

Core: 17–41 odd, 42, 43, 45, 47–57

Enriched: 16–40 even, 41–57

Additional Answer

6b.

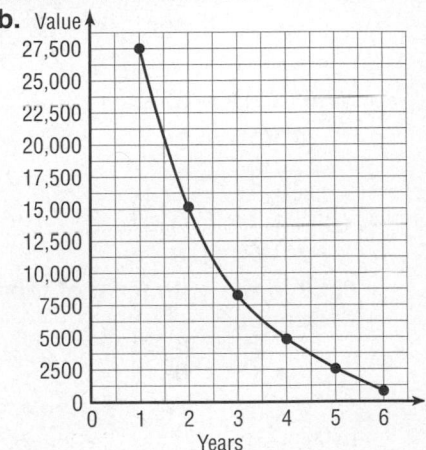

Study Guide and Practice Masters, pp. 217–218

41c. Sample answer: it is assumed that favorable conditions are maintained for the growth of the bacteria, such as an adequate food and oxygen supply, appropriate surrounding temperature, and adequate room for growth.

45. $a_2 = -3(a_1)$ by definition
$= -3(-2)$
$= 6$
$r = \dfrac{a_2}{a_1} = -3$
Then, $a_n = a_1 r^{n-1}$
So, $a_n = (-2)(-3)^{n-1}$

48. $r = \dfrac{a_2}{a_1}$

$r = \dfrac{\frac{1}{27}}{\frac{1}{81}} = 3$

$a_n = a_1 r^{n-1}$

$6561 = \dfrac{1}{81} \cdot 3^{n-1}$

$(6561)(81) = 3^{n-1}$

$(3^8)(3^4) = 3^{n-1}$

$3^{12} = 3^{n-1}$

$12 = n - 1$

$13 = n$

6561 is the 13th term of the sequence.

---

25. The first term of a geometric sequence is $\dfrac{a}{b^2}$, and the common ratio is $\dfrac{b}{a^2}$. Find the next five terms of the geometric sequence. $\dfrac{1}{ab}, \dfrac{1}{a^3}, \dfrac{b}{a^5}, \dfrac{b^2}{a^7}, \dfrac{b^3}{a^9}$

For Exercises 26–40, assume that each sequence or series is geometric.

26. Find the fifth term of a sequence whose first term is 8 and common ratio is $\dfrac{3}{2}$. $\dfrac{81}{2}$

27. Find the sixth term of the sequence $\dfrac{1}{2}, -\dfrac{3}{8}, \dfrac{9}{32}, \ldots. \quad -\dfrac{243}{2048}$

28. Find the seventh term of the sequence $40, 0.4, 0.004, \ldots. \quad 4 \times 10^{-11}$

29. Find the ninth term of the sequence $\sqrt{5}, \sqrt{10}, 2\sqrt{5}, \ldots. \quad 16\sqrt{5}$

30. If $r = 4$ and $a_6 = 192$, what is the first term of the sequence? **0.1875**

31. If $r = -\sqrt{2}$ and $a_5 = 32\sqrt{2}$, what is the first term of the sequence? $8\sqrt{2}$

32. Find the first three terms of the sequence for which $a_5 = -6$ and $r = -\dfrac{1}{3}$. **32. −486, 162, −54**

33. Find the first three terms of the sequence for which $a_5 = 0.32$ and $r = 0.2$. **200, 40, 8**

34. Write a sequence that has three geometric means between 256 and 81. **256, ±192, 144, ±108, 81**

35. Write a sequence that has two geometric means between −2 and 54. **35. −2, 6, −18, 54**

36. Write a sequence that has one geometric mean between $\dfrac{4}{7}$ and 7. **36. $\dfrac{4}{7}$, ±2, 7**

37. What is the sum of the first five terms of the series $\dfrac{5}{3} + 5 + 15 + \cdots$? **37. $\dfrac{605}{3}$**

38. What is the sum of the first six terms of the series $65 + 13 + 2.6 + \cdots$? **81.2448**

39. Find the sum of the first ten terms of the series $1 - \dfrac{3}{2} + \dfrac{9}{4} - \cdots$. $-\dfrac{11,605}{512}$

40. Find the sum of the first eight terms of the series $2 + 2\sqrt{3} + 6 + \cdots$. **40. $80(1 + \sqrt{3})$**

Applications and Problem Solving

41. **Biology** A cholera bacterium divides every half-hour to produce two complete cholera bacteria.
    a. If an initial colony contains a population of b_0 bacteria, write an equation that will determine the number of bacteria present after t hours. $b_t = b_0 \cdot 2^{2t}$
    b. Suppose a petri dish contains 30 cholera bacteria. Use the equation from part **a** to determine the number of bacteria present 5 hours later. **30,720**
    c. What assumptions are made in using the formula found in part **a**? See margin.

42. **Critical Thinking** Consider the geometric sequence with $a_4 = 4$ and $a_7 = 12$.
    a. Find the common ratio and the first term of the sequence. $\sqrt[3]{3}, \dfrac{4}{3}$
    b. Find the 28th term of the sequence. **26,244**

43. **Consumerism** High Tech Electronics advertises a weekly installment plan for the purchase of a popular brand of big screen TV. The buyer pays $5 at the end of the first week, $5.50 at the end of the second week, $6.05 at the end of the third week, and so on for one year.
    a. What will the payments be at the end of the 10th, 20th, and 40th weeks? **a. $11.79, $30.58, $205.72**
    b. Find the total cost of the TV. **$7052.15**
    c. Why is the cost found in part **b** not entirely accurate?

43c. Each payment made is rounded to the nearest penny, so the sum of the payments will actually be more than the sum found in part b.

44. **Statistics** A number x is said to be the *harmonic mean* of y and z if $\dfrac{1}{x}$ is the average of $\dfrac{1}{y}$ and $\dfrac{1}{z}$.
    a. Find the harmonic mean of 5 and 8. $\dfrac{80}{13}$
    b. 8 is the harmonic mean of 20 and another number. What is the number? **5**

772 Chapter 12 Sequences and Series

772 Chapter 12

45. Critical Thinking In a geometric sequence, $a_1 = -2$ and every subsequent term is defined by $a_n = -3a_{n-1}$, where $n > 1$. Find the nth term in the sequence in terms of n. $a_n = -2(-3)^{n-1}$; **see margin for explanation.**

46. Genealogy Wei-Ling discovers through a research of her Chinese ancestry that one of her fifteenth-generation ancestors was a famous military leader. How many descendants does this ancestor have in the fifteenth-generation, assuming each descendent had an average of 2.5 children? **372,529**

47. Personal Finance Tonisha is about to begin her junior year in high school and is looking ahead to her college career. She estimates that by the time she is ready to enter a university she will need at least $750 to purchase a computer system that will meet her needs. To avoid purchasing the computer on credit, she opens a savings account earning an APR of 2.4%, compounded monthly, and deposits $25 at the beginning of each month.
  a. Find the balance of the account at the end of the first month. **$25.05**
  b. If Tonisha continues this deposit schedule for the next two years, will she have enough money in her account to buy the computer system? Explain.
  c. Find the least amount of money Tonisha can deposit each month and still have enough money to purchase the computer. **$30.48**

47b. No. At the end of two years, she will have only $615.23 in her account.

48. Critical Thinking Use algebraic methods to determine which term 6561 is of the geometric sequence $\frac{1}{81}, \frac{1}{27}, \frac{1}{9}, \ldots$. **13th term; see margin for explanation.**

Mixed Review

49. Banking Gloria Castaneda has $650 in her checking account. She is closing out the account by writing one check for cash against it each week. The first check is for $20, the second is for $25, and so on. Each check exceeds the previous one by $5. In how many weeks will the balance in Ms. Castaneda account be $0 if there is no service charge? *(Lesson 12-1)* **13 weeks**

50. 2.3269

50. Find the value of $\log_{11} 265$ using the change of base formula. *(Lesson 11-5)*

51. Graph the system $xy \geq 2$ and $x - 3y = 2$. *(Lesson 10-8)* **See margin.**

52. Write $3x - 5y + 5 = 0$ in polar form. *(Lesson 9-4)* $\frac{5\sqrt{34}}{34} = r\cos(\theta - 121°)$

53. $x = t$, $y = -\frac{3}{4}t + \frac{5}{4}$

53. Write parametric equations of the line $3x + 4y = 5$. *(Lesson 8-6)*

54. $\frac{1}{3}$

54. If $\csc \theta = 3$ and $0° \leq \theta \leq 90°$, find $\sin \theta$. *(Lesson 7-1)* $\frac{1}{3}$

55. Weather The maximum normal daily temperatures in each season for Lincoln, Nebraska, are given below. Write a sinusoidal function that models the temperatures, using $t = 1$ to represent winter. *(Lesson 6-6)* **See margin.**

Normal Daily Temperatures for Lincoln, Nebraska

Winter	Spring	Summer	Fall
36°	61°	86°	65°

Source: Rand McNally & Company

56. Given $A = 43°$, $b = 20$, and $a = 11$, do these measurements determine one triangle, two triangles, or no triangle? *(Lesson 5-7)* **no triangle**

57. SAT Practice Grid-In If n and m are integers, and $-(n^2) \leq -\sqrt{49}$ and $m = n + 1$, what is the least possible value of mn? **6**

Extra Practice See p. A48.

Extra Credit

Find a and b if $3b$, $3ab$, 3, and $\frac{ab}{3}$ are consecutive terms of a geometric sequence. $a = \pm\frac{1}{3}$, $b = 9$

Open-Ended Assessment
Writing Have students write a paragraph explaining how to determine the value of a savings account at the end of one year if equal deposits are made quarterly and the account pays an APR of 10% compounded monthly.

Additional Answers
51.

55. $y = 25 \sin\left(\frac{\pi}{2}t - 3.14\right) + 61$

Enrichment Masters, **p. 134**

5-Minute Check
Lesson 12-2

1. Determine the common ratio and find the next three terms in each geometric sequence.

   a. $-2, -\frac{2}{3}, -\frac{2}{9}, \ldots$ $\frac{1}{3}$;
   $-\frac{2}{27}, -\frac{2}{81}, -\frac{2}{243}$

   b. $5x + 7, -10x - 14, 20x + 28, \ldots$ **−2;**
   $-40x - 56, 80x + 112, -160x - 224$

2. Find the tenth term in the sequence $\sqrt{3}, 3, 3\sqrt{3}, \ldots$.
   243

3. Write a sequence that has two geometric means between 6.4 and 12.5.
   6.4, 8, 10, 12.5

4. Find the sum of the first six terms of the geometric series $2 - 8 + 32 - 128 + \cdots$.
   −1638

5. If $r = -2$ and $a_8 = -384$, what is the first term of the geometric sequence? **3**

6. At the beginning of each year for the next five years, Marva deposits $5000 in a savings account paying an APR of 12% compounded annually. If she makes no withdrawals, how much will she have in the account at the end of five years?
   $35,575.95

12-3

Infinite Sequences and Series

OBJECTIVES
- Find the limit of the terms of an infinite sequence.
- Find the sum of an infinite geometric series.

ECONOMICS On January 28, 1999, Florida governor Jeb Bush proposed a tax cut that would allow the average family to keep an additional $96. The *marginal propensity to consume (MPC)* is defined as the percentage of a dollar by which consumption increases when income rises by a dollar. Suppose the *MPC* for households and businesses in 1999 was 75%. What would be the total amount of money spent in the economy as a result of just one family's tax savings? *This problem will be solved in Example 5.*

Governor Jeb Bush

Transaction	Expenditure	Terms of Sequence
1	$96(0.75)^1$	72
2	$96(0.75)^2$	54
3	$96(0.75)^3$	40.50
4	$96(0.75)^4$	30.76
5	$96(0.75)^5$	22.78
⋮	⋮	⋮
10	$96(0.75)^{10}$	5.41
⋮	⋮	⋮
100	$96(0.75)^{100}$	3.08×10^{-11}
⋮	⋮	⋮
500	$96(0.75)^{500}$	3.26×10^{-61}
⋮	⋮	⋮
n	$96(0.75)^n$	ar^n

Study the table at the left. Transaction 1 represents the initial expenditure of $96(0.75) or $72 by a family. The businesses receiving this money, Transaction 2, would in turn spend 75%, and so on. We can write a geometric sequence to model this situation with $a_1 = 72$ and $r = 0.75$. Thus, the geometric sequence representing this situation is

$$72, 54, 40.50, 30.76, 22.78, \ldots.$$

In theory, the sequence above can have infinitely many terms. Thus, it is called an **infinite sequence.** As n increases, the terms of the sequence decrease and get closer and closer to zero. The terms of the modeling sequence will never actually become zero; however, the terms approach zero as n increases without bound.

Consider the infinite sequence $1, \frac{1}{2}, \frac{1}{3}, \frac{1}{4}, \frac{1}{5}, \ldots$, whose nth term, a_n, is $\frac{1}{n}$. Several terms of this sequence are graphed at the right.

Notice that the terms approach a value of 0 as n increases. Zero is called the **limit** of the terms in this sequence.

Resource Manager

Reproducible Masters
- *Study Guide and Practice,* pp. 219–220
- *Enrichment,* p. 135
- *Assessment and Evaluation,* p. 303

Transparencies
- 5-Minute Check, 12-3
- Teaching, 12-3

This limit can be expressed as follows.

$$\lim_{n\to\infty} \frac{1}{n} = 0 \quad \text{∞ is the symbol for infinity.}$$

This is read "the limit of 1 over n, as n approaches infinity, equals zero."

In fact, when any positive power of n appears only in the denominator of a fraction and n approaches infinity, the limit equals zero.

$$\lim_{n\to\infty} \frac{1}{n^r} = 0, \text{ for } r > 0$$

If a general expression for the nth term of a sequence is known, the limit can usually be found by substituting large values for n. Consider the following infinite geometric sequence.

$$7, \frac{7}{4}, \frac{7}{16}, \frac{7}{64}, \frac{7}{256}, \cdots$$

This sequence can be defined by the general expression $a_n = 7\left(\frac{1}{4}\right)^{n-1}$.

$$a_{10} = 7\left(\frac{1}{4}\right)^{10-1} \approx 2.67 \times 10^{-5}$$

$$a_{50} = 7\left(\frac{1}{4}\right)^{50-1} \approx 2.21 \times 10^{-25}$$

$$a_{100} = 7\left(\frac{1}{4}\right)^{100-1} \approx 4.36 \times 10^{-60}$$

Notice that as the value of n increases, the value for a_n appears to approach 0, suggesting $\lim_{n\to\infty} 7\left(\frac{1}{4}\right)^{n-1} = 0$.

Example **1** Estimate the limit of $\frac{9}{5}, \frac{16}{4}, \frac{65}{27}, \cdots, \frac{7n^2 + 2}{2n^2 + 3n}, \cdots$.

The 50th term is $\frac{7(50)^2 + 2}{2(50)^2 + 3(50)}$, or about 3.398447.

The 100th term is $\frac{7(100)^2 + 2}{2(100)^2 + 3(100)}$, or about 3.448374.

The 500th term is $7\frac{(500)^2 + 2}{2(500)^2 + 3(500)}$, or about 3.489535.

The 1000th term is $\frac{7(1000)^2 + 2}{2(1000)^2 + 3(1000)}$, or 3.494759.

Notice that as $n \to \infty$, the values appear to approach 3.5, suggesting $\lim_{n\to\infty} \frac{7n^2 + 2}{2n^2 + 3n} = 3.5$.

Lesson 12-3   Infinite Sequences and Series   **775**

For sequences with more complicated general forms, applications of the following limit theorems, which we will present without proof, can make the limit easier to find.

If the $\lim\limits_{n\to\infty} a_n$ exists, $\lim\limits_{n\to\infty} b_n$ exists, and c is a constant, then the following theorems are true.

Theorems for Limits

Limit of a Sum	$\lim\limits_{n\to\infty} (a_n + b_n) = \lim\limits_{n\to\infty} a_n + \lim\limits_{n\to\infty} b_n$
Limit of a Difference	$\lim\limits_{n\to\infty} (a_n - b_n) = \lim\limits_{n\to\infty} a_n - \lim\limits_{n\to\infty} b_n$
Limit of a Product	$\lim\limits_{n\to\infty} a_n \cdot b_n = \lim\limits_{n\to\infty} a_n \cdot \lim\limits_{n\to\infty} b_n$
Limit of a Quotient	$\lim\limits_{n\to\infty} \dfrac{a_n}{b_n} = \dfrac{\lim\limits_{n\to\infty} a_n}{\lim\limits_{n\to\infty} b_n}$, where $\lim\limits_{n\to\infty} b_n \neq 0$
Limit of a Constant	$\lim\limits_{n\to\infty} c_n = c$, where $c_n = c$ for each n

The form of the expression for the nth term of a sequence can often be altered to make the limit easier to find.

Example ❷ **Find each limit.**

a. $\lim\limits_{n\to\infty} \dfrac{(1+3n^2)}{n^2}$

$$\lim\limits_{n\to\infty} \frac{(1+3n^2)}{n^2} = \lim\limits_{n\to\infty}\left(\frac{1}{n^2} + 3\right) \qquad \textit{Rewrite as the sum of two fractions and simplify.}$$

$$= \lim\limits_{n\to\infty} \frac{1}{n^2} + \lim\limits_{n\to\infty} 3 \qquad \textit{Limit of a Sum}$$

$$= 0 + 3 \text{ or } 3 \qquad \lim\limits_{n\to\infty} \frac{1}{n^2} = 0 \text{ and } \lim\limits_{n\to\infty} 3 = 3$$

Thus, the limit is 3.

Note that the Limit of a Sum theorem only applies here because $\lim\limits_{n\to\infty} \frac{1}{n^2}$ and $\lim\limits_{n\to\infty} 3$ each exist.

b. $\lim\limits_{n\to\infty} \dfrac{5n^2 + n - 4}{n^2 + 1}$

The highest power of n in the expression is n^2. Divide each term in the numerator and the denominator by n^2. *Why does doing this produce an equivalent expression?*

$$\lim\limits_{n\to\infty} \frac{5n^2 + n - 4}{n^2 + 1} = \lim\limits_{n\to\infty} \frac{\dfrac{5n^2}{n^2} + \dfrac{n}{n^2} - \dfrac{4}{n^2}}{\dfrac{n^2}{n^2} + \dfrac{1}{n^2}}$$

$$= \lim\limits_{n\to\infty} \frac{5 + \dfrac{1}{n} - \dfrac{4}{n^2}}{1 + \dfrac{1}{n^2}} \qquad \textit{Simplify.}$$

$$= \frac{\lim\limits_{n\to\infty} 5 + \lim\limits_{n\to\infty} \frac{1}{n} - \lim\limits_{n\to\infty} 4 \cdot \lim\limits_{n\to\infty} \frac{1}{n^2}}{\lim\limits_{n\to\infty} 1 + \lim\limits_{n\to\infty} \frac{1}{n^2}}$$ *Apply limit theorems.*

$$= \frac{5 + 0 - 4 \cdot 0}{1 + 0} \text{ or } 5$$
  $\lim\limits_{n\to\infty} 5 = 5,\ \lim\limits_{n\to\infty} \frac{1}{n} = 0,$
  $\lim\limits_{n\to\infty} 4 = 4,\ \lim\limits_{n\to\infty} 1 = 1,\ and$
  $\lim\limits_{n\to\infty} \frac{1}{n^2} = 0a$

Thus, the limit is 5.

Limits do not exist for all infinite sequences. If the absolute value of the terms of a sequence becomes arbitrarily great or if the terms do not approach a value, the sequence has no limit. Example 3 illustrates both of these cases.

Example **③** **Find each limit.**

a. $\lim\limits_{n\to\infty} \dfrac{2 + 5n + 4n^2}{2n}$

$\lim\limits_{n\to\infty} \dfrac{2 + 5n + 4n^2}{2n} = \lim\limits_{n\to\infty} \left(\dfrac{1}{n} + \dfrac{5}{2} + 2n \right)$ *Simplify.*

Note that $\lim\limits_{n\to\infty} \dfrac{1}{n} = 0$ and $\lim\limits_{n\to\infty} \dfrac{5}{2} = \dfrac{5}{2}$, but $2n$ becomes increasingly large as n approaches infinity. Therefore, the sequence has no limit.

b. $\lim\limits_{n\to\infty} \dfrac{(-1)^n n}{8n + 1}$

Begin by rewriting $\dfrac{(-1)^n n}{8n + 1}$ as $(-1)^n \cdot \dfrac{n}{8n + 1}$. Now find $\lim\limits_{n\to\infty} \dfrac{n}{8n + 1}$.

$\lim\limits_{n\to\infty} \dfrac{n}{8n + 1} = \lim\limits_{n\to\infty} \dfrac{\frac{n}{n}}{\frac{8n}{n} + \frac{1}{n}}$ *Divide the numerator and denominator by n.*

$= \lim\limits_{n\to\infty} \dfrac{1}{8 + \frac{1}{n}}$ *Simplify.*

$= \dfrac{\lim\limits_{n\to\infty} 1}{\lim\limits_{n\to\infty} 8 + \lim\limits_{n\to\infty} \frac{1}{n}}$ *Apply limit theorems.*

$= \dfrac{1}{8}$ $\lim\limits_{n\to\infty} 1 = 1,\ \lim\limits_{n\to\infty} 8 = 8,\ and\ \lim\limits_{n\to\infty} \frac{1}{n} = 0$

When n is even, $(-1)^n = 1$. When n is odd, $(-1)^n = -1$. Thus, the odd-numbered terms of the sequence described by $\dfrac{(-1)^n n}{8n + 1}$ approach $-\dfrac{1}{8}$, and the even-numbered terms approach $+\dfrac{1}{8}$. Therefore, the sequence has no limit.

Teaching Tip While working In-Class Example 3, point out that $\lim\limits_{n\to\infty} \dfrac{2}{n} = 0$, but $\lim\limits_{n\to\infty} \dfrac{4}{3}n$ becomes increasingly large as n approaches infinity. Therefore, the sequence has no limit.

In-Class Example
Example 3
Find each limit.
a. $\lim\limits_{n\to\infty} \dfrac{4n^2 - 6}{3n}$ **no limit**
b. $\lim\limits_{n\to\infty} \dfrac{(-1)^n n}{5n + 1}$ **no limit**

An **infinite series** is the indicated sum of the terms of an infinite sequence. Consider the series $\frac{1}{5} + \frac{1}{25} + \frac{1}{125} + \cdots$. Since this is a geometric series, you can find the sum of the first 100 terms by using the formula $S_n = \frac{a_1 - a_1 r^n}{1 - r}$, where $r = \frac{1}{5}$.

$$S_{100} = \frac{\frac{1}{5} - \frac{1}{5}\left(\frac{1}{5}\right)^{100}}{1 - \frac{1}{5}}$$

$$= \frac{\frac{1}{5} - \frac{1}{5}\left(\frac{1}{5}\right)^{100}}{\frac{4}{5}}$$

$$= \frac{5}{4}\left[\frac{1}{5} - \frac{1}{5}\left(\frac{1}{5}\right)^{100}\right] \text{ or } \frac{1}{4} - \frac{1}{4}\left(\frac{1}{5}\right)^{100}$$

Since $\left(\frac{1}{5}\right)^{100}$ is very close to 0, S_{100} is nearly equal to $\frac{1}{4}$. No matter how many terms are added, the sum of the infinite series will never exceed $\frac{1}{4}$, and the difference from $\frac{1}{4}$ gets smaller as $n \to \infty$. Thus, $\frac{1}{4}$ is the sum of the infinite series.

Sum of an Infinite Series	If S_n is the sum of the first n terms of a series, and S is a number such that $S - S_n$ approaches zero as n increases without bound, then the sum of the infinite series is S. $$\lim_{n \to \infty} S_n = S$$

If the sequence of partial sums S_n has a limit, then the corresponding infinite series has a sum, and the nth term a_n of the series approaches 0 as $n \to \infty$. If $\lim_{n \to \infty} a_n \neq 0$, the series has no sum. If $\lim_{n \to \infty} a_n = 0$, the series may or may not have a sum.

The formula for the sum of the first n terms of a geometric series can be written as follows.

$$S_n = a_1 \frac{(1 - r^n)}{1 - r}, r \neq 1$$

Recall that $|r| < 1$ means $-1 < r < 1$.

Suppose $n \to \infty$; that is, the number of terms increases without limit. If $|r| > 1$, r^n increases without limit as $n \to \infty$. However, when $|r| < 1$, r^n approaches 0 as $n \to \infty$. Under this condition, the above formula for S_n approaches a value of $\frac{a_1}{1 - r}$.

Sum of an Infinite Geometric Series	The sum S of an infinite geometric series for which $|r| < 1$ is given by $$S = \frac{a_1}{1 - r}.$$

Example **4** Find the sum of the series $21 - 3 + \dfrac{3}{7} - \cdots$.

In the series, $a_1 = 21$ and $r = -\dfrac{1}{7}$. Since $|r| < 1$, $S = \dfrac{a_1}{1 - r}$.

$$S = \frac{a_1}{1 - r}$$

$$= \frac{21}{1 - \left(-\dfrac{1}{7}\right)} \qquad a_1 = 21, \; r = -\frac{1}{7}$$

$$= \frac{147}{8} \text{ or } 18\frac{3}{8}$$

The sum of the series is $18\dfrac{3}{8}$.

In economics, finding the sum of an infinite series is useful in determining the overall effect of economic trends.

Example **5** **ECONOMICS** Refer to the application at the beginning of the lesson. What would be the total amount of money spent in the economy as a result of just one family's tax savings?

For the geometric series modeling this situation, $a_1 = 72$ and $r = 0.75$.

Since $|r| < 1$, the sum of the series is equal to $\dfrac{a_1}{1 - r}$.

$$S = \frac{a_1}{1 - r}$$

$$= \frac{72}{1 - 0.75} \text{ or } 288$$

Therefore, the total amount of money spent is $288.

You can use what you know about infinite series to write repeating decimals as fractions. The first step is to write the repeating decimal as an infinite geometric series.

Example **6** Write $0.\overline{762}$ as a fraction.

$$0.\overline{762} = \frac{762}{1000} + \frac{762}{1,000,000} + \frac{762}{1,000,000,000} + \cdots$$

In this series, $a_1 = \dfrac{762}{1000}$ and $r = \dfrac{1}{1000}$.

(continued on the next page)

Lesson 12-3 Infinite Sequences and Series **779**

Teaching Tip For Example 5, point out that the sum of the first n terms of a geometric series can be represented by $S_n = \dfrac{a_1 - a_1 r^n}{1 - r}$, but the sum of the terms of an infinite geometric series is given by $S = \dfrac{a_1}{1 - r}$, as long as $|r| < 1$. The inequality $|r| < 1$ is the same as $-1 < r < 1$.

In-Class Examples

Example 4
Find the sum of the series $60 + 24 + 9.6 + \cdots$. **100**

Example 5
Motion A rubber ball is dropped to the floor from a height of 27 feet. If the ball rises at each rebound to a height $\dfrac{2}{3}$ its previous height, find the total distance the ball travels before coming to rest. **81 ft**

Example 6
Write $12.\overline{981}$ as a fraction.
$12\dfrac{109}{111}$ or $\dfrac{1441}{111}$

Addressing Individual Needs

Kinesthetic Divide the class into groups. Give each group a tennis ball and a meterstick. Have the members of each group take turns dropping the tennis ball from a given height and recording the vertical distance it reaches on each bounce before coming to rest. You can use a CBL or videotape to record the movement of the ball. Discuss the series formed by the varying heights of the ball.

Error Analysis

Watch for students who use the formula $S = \frac{a_1}{1-r}$ incorrectly.

Prevent by stressing that $S = \frac{a_1}{1-r}$ is used only for an infinite geometric series whose common ratio lies between -1 and $+1$.

Additional Answers

1a.

1b. The value of a_n approaches 1 as the value of n increases.

1c. $\lim\limits_{n\to\infty} \frac{n-1}{n} = 1$

1d. $\lim\limits_{n\to\infty} \frac{n-1}{n} = 1$; they are equal.

2a. See students' work. Students should draw the following conclusions:

$\lim\limits_{n\to\infty} \left(\frac{1}{2}\right)^n = 0$

$\lim\limits_{n\to\infty} \left(\frac{1}{4}\right)^n = 0$

$\lim\limits_{n\to\infty} (1)^n = 1$

$\lim\limits_{n\to\infty} (2)^n = $ no limit

$\lim\limits_{n\to\infty} (5)^n = $ no limit

Study Guide and Practice Masters, pp. 219–220

780 *Chapter 12*

$S = \frac{a_1}{1-r}$

$= \frac{\frac{762}{1000}}{1 - \frac{1}{1000}}$

$= \frac{762}{999}$ or $\frac{254}{333}$

Thus, $0.762762\cdots = \frac{254}{333}$.   *Check this with a calculator.*

Communicating Mathematics

Read and study the lesson to answer each question.

1. Consider the sequence given by the general expression $a_n = \frac{n-1}{n}$.

   a. **Graph** the first ten terms of the sequence with the term number on the x-axis and the value of the term on the y-axis.   **a–d. See margin.**

   b. **Describe** what happens to the value of a_n as n increases.

   c. **Make a conjecture** based on your observation in part **a** as to the limit of the sequence as n approaches infinity.

   d. **Apply** the techniques presented in the lesson to evaluate $\lim\limits_{n\to\infty} \frac{n-1}{n}$. How does your answer compare to your conjecture made in part **c**?

2. Consider the infinite geometric sequence given by the general expression r^n.

   a. **Determine** the limit of the sequence for $r = \frac{1}{2}, r = \frac{1}{4}, r = 1, r = 2$, and $r = 5$.   **See margin.**

   b. **Write** a general rule for the limit of the sequence, placing restrictions on the value of r.

2b. If $|r| < 1$, then $\lim\limits_{n\to\infty} r^n = 0$.

If $|r| = 1$, then $\lim\limits_{n\to\infty} r^n = 1$.

If $|r| > 1$, then $\lim\limits_{n\to\infty} r^n$ does not exist.

3. **Give an example** of an infinite geometric series having no sum.

3. Sample answer: $2 + 4 + 8 + \cdots$

4. **You Decide** Tyree and Zonta disagree on whether the infinite sequence described by the general expression $2n - 3$ has a limit. Tyree says that after dividing by the highest-powered term, the expression simplifies to $2 - \frac{3}{n}$, which has a limit of 2 as n approaches infinity. Zonta says that the sequence has no limit. Who is correct? Explain.   **See margin.**

Guided Practice

Find each limit, or state that the limit does not exist and explain your reasoning.   **6. Does not exist; see margin for explanation.**

5. $\lim\limits_{n\to\infty} \frac{1}{5^n}$ **0**

6. $\lim\limits_{n\to\infty} \frac{5 - n^2}{2n}$

7. $\lim\limits_{n\to\infty} \frac{3n - 6}{7n}$ $\frac{3}{7}$

Write each repeating decimal as a fraction.

8. $0.\overline{7}$ $\frac{7}{9}$

9. $5.\overline{126}$ $5\frac{14}{111}$

Additional Answers

4. Zonta is correct. As n approaches infinity, the expression $2n - 3$ will continue to grow larger and larger. Tyree applied the method of dividing by the highest powered term incorrectly. Both the numerator and the denominator of the expression must be divided by the highest-powered term. It is not appropriate to apply this method here since the denominator of the expression $2n - 3$ is 1.

6. $\lim\limits_{n\to\infty} \frac{5 - n^2}{2n} = \lim\limits_{n\to\infty} \left(\frac{5}{2n} - \frac{1}{2}n\right)$

   $\lim\limits_{n\to\infty} \frac{5}{2n} = \lim\limits_{n\to\infty} \frac{5}{2} \cdot \frac{1}{n} = \frac{5}{2} \cdot 0 = 0$

   As n approaches infinity, $\frac{1}{2}n$ becomes increasingly large, so the sequence has no limit.

Find the sum of each infinite series, or state that the sum does not exist and explain your reasoning.   **12. does not exist;** $|r| = |\sqrt{3}| > 1$

10. $-6 + 3 - \frac{3}{2} + \cdots$ -4 **11.** $\frac{3}{4} + \frac{1}{4} + \frac{1}{12} + \cdots$ $1\frac{1}{8}$ **12.** $\sqrt{3} + 3 + \sqrt{27} + \cdots$

13. Entertainment   Pete's Pirate Ride operates like the bob of a pendulum. On its longest swing, the ship travels through an arc 75 meters long. Each successive swing is two-fifths the length of the preceding swing. If the ride is allowed to continue without intervention, what is the total distance the ship will travel before coming to rest?   **125 m**

75 m

EXERCISES

Practice

Find each limit, or state that the limit does not exist and explain your reasoning.   **15, 19, 22. Does not exist; see margin for explanation.**

 14. $\lim_{n\to\infty} \frac{7 - 2n}{5n}$ $-\frac{2}{5}$ **15.** $\lim_{n\to\infty} \frac{n^3 - 2}{n^2}$ **16.** $\lim_{n\to\infty} \frac{6n^2 + 5}{3n^2}$ **2**

17. $\lim_{n\to\infty} \frac{9n^3 + 5n - 2}{2n^3}$ $\frac{9}{2}$ **18.** $\lim_{n\to\infty} \frac{(3n + 4)(1 - n)}{n^2}$ -3 **19.** $\lim_{n\to\infty} \frac{8n^2 + 5n + 2}{3 + 2n}$

20. $\lim_{n\to\infty} \frac{4 - 3n + n^2}{2n^3 - 3n^2 + 5}$ **0**   **21.** $\lim_{n\to\infty} \frac{n}{3^n}$ **0**   **22.** $\lim_{n\to\infty} \frac{(-2)^n n}{4 + n}$

 23. Find the limit of the sequence described by the general expression $\frac{5n + (-1)^n}{n^2}$, or state that the limit does not exist. Explain your reasoning.   **0; see margin for explanation.**

Write each repeating decimal as a fraction.

24. $0.\overline{4}$ $\frac{4}{9}$ **25.** $0.\overline{51}$ $\frac{17}{33}$ **26.** $0.\overline{370}$ $\frac{10}{27}$

27. $6.\overline{259}$ $6\frac{7}{27}$ **28.** $0.\overline{15}$ $\frac{5}{33}$ **29.** $0.2\overline{63}$ $\frac{29}{110}$

30. Explain why the sum of the series $0.2 + 0.02 + 0.002 + \cdots$ exists. Then find the sum.   **The series is geometric, having a common ratio of 0.1. Since this ratio is less than 1, the sum of the series exists and is $\frac{2}{9}$.**

Find the sum of each series, or state that the sum does not exist and explain your reasoning.   **32, 34, 35. Does not exist, see margin for explanation.**

31. $16 + 12 + 9 + \cdots$ **64**   **32.** $5 + 7.5 + 11.25 + \cdots$ **33.** $10 + 5 + 2.5 + \cdots$ **20**

 34. $6 + 5 + 4 + \cdots$ **35.** $\frac{1}{8} + \frac{1}{4} + \frac{1}{2} + \cdots$ **36.** $-\frac{2}{3} + \frac{1}{9} - \frac{1}{54} + \cdots$ $-\frac{4}{7}$

37. $\frac{6}{5} + \frac{4}{5} + \frac{8}{15} + \cdots$ $3\frac{3}{5}$ **38.** $\sqrt{5} + 1 + \frac{\sqrt{5}}{5} + \cdots$ **39.** $8 - 4\sqrt{3} + 6 - \cdots$
$\frac{5}{4}(\sqrt{5} + 1)$ $32 - 16\sqrt{3}$

 www.amc.glencoe.com/self_check_quiz   *Lesson 12-3 Infinite Sequences and Series*   **781**

Assignment Guide

Core: 15–45 odd, 47–55
Enriched: 14–38 even, 40–55

Additional Answers

22. Dividing by the highest powered term, n, we find that
$$\lim_{n\to\infty} \frac{\frac{(-2)^n}{n}}{\frac{4}{n} + 1} = \lim_{n\to\infty} \left(\frac{(-2)^n}{n} \cdot \frac{1}{\frac{4}{n} + 1} \right)$$
which as n approaches infinity
$$\lim_{n\to\infty} \frac{1}{\frac{4}{n} + 1} = 1, \text{ but } \lim_{n\to\infty} \frac{(-2)^n}{n}$$
has no limit since $|-2| > 1$.

23. $\lim_{n\to\infty} \frac{5n + (-1)^n}{n^2} = \lim_{n\to\infty} \frac{5n}{n^2} + \lim_{n\to\infty} \frac{(-1)^n}{n^2}$
$= \lim_{n\to\infty} \frac{5}{n} + \lim_{n\to\infty} \frac{(-1)^n}{n^2}$
$= \lim_{n\to\infty} \frac{(-1)^n}{n^2}$

As n increases, the value of the numerator alternates between -1 and 1. As n approaches infinity, the value of the denominator becomes increasingly large, causing the value of the fraction to become increasingly small. Thus, the terms of the sequence alternate between smaller and smaller positive and negative values, approaching zero. So the sequence has a limit of zero.

32. This series is geometric with a common ratio of 1.5. Since this ratio is greater than 1, the sum of the series does not exist.

34. The series is arithmetic, having a general term of $7 - n$. Since $\lim_{n\to\infty} 7 - n$ does not equal zero, this series has no sum.

35. This series is geometric with a common ratio of 2. Since this ratio is greater than 1, the sum of the series does not exist.

Additional Answers

15. Simplifying the limit, we find that
$$\lim_{n\to\infty} \frac{n^3 - 2}{n^2} = \lim_{n\to\infty} \left(n - \frac{2}{n} \right)$$
$$\lim_{n\to\infty} \frac{2}{n} = \lim_{n\to\infty} 2 \cdot \frac{1}{n}$$
$$= 2 \cdot 0 \text{ or } 0,$$
but as n approaches infinity, n becomes increasingly large, so the sequence has no limit.

19. Dividing by the highest powered term, n^2, we find that
$$\lim_{n\to\infty} \frac{8 + \frac{5}{n} + \frac{2}{n^2}}{\frac{3}{n^2} + \frac{2}{n}}$$
which as n approaches infinity simplifies to $\frac{8 + 0 + 0}{0 + 0} = \frac{8}{0}$. Since this fraction is undefined, the limit does not exist.

Applications and Problem Solving

Real World Application

40. Physics A basketball is dropped from a height of 35 meters and bounces $\frac{2}{5}$ of the distance after each fall.

  a. Find the first five terms of the infinite series representing the vertical distance traveled by the ball. **35, 14, 14, 5.6, 5.6**

  b. What is the total vertical distance the ball travels before coming to rest? (*Hint:* Rewrite the series found in part **a** as the sum of two infinite geometric series.) **about 82 m**

41. Critical Thinking Consider the sequence whose *n*th term is described by $\dfrac{n^2}{2n+1} - \dfrac{n^2}{2n-1}$.

  a. Explain why $\lim\limits_{n\to\infty}\left(\dfrac{n^2}{2n+1} - \dfrac{n^2}{2n-1}\right) \neq \lim\limits_{n\to\infty}\dfrac{n^2}{2n+1} - \lim\limits_{n\to\infty}\dfrac{n^2}{2n-1}$. **See margin.**

  b. Find $\lim\limits_{n\to\infty}\left(\dfrac{n^2}{2n+1} - \dfrac{n^2}{2n-1}\right)$. $-\dfrac{1}{2}$

42. Engineering Francisco designs a toy with a rotary flywheel that rotates at a maximum speed of 170 revolutions per minute. Suppose the flywheel is operating at its maximum speed for one minute and then the power supply to the toy is turned off. Each subsequent minute thereafter, the flywheel rotates two-fifths as many times as in the preceding minute. How many complete revolutions will the flywheel make before coming to a stop? **283**

43. Critical Thinking Does $\lim\limits_{n\to\infty}\cos\dfrac{n\pi}{2}$ exist? Explain.

44. Medicine A certain drug developed to fight cancer has a half-life of about 2 hours in the bloodstream. The drug is formulated to be administered in doses of *D* milligrams every 6 hours. The amount of each dose has yet to be determined.

  a. What fraction of the first dose will be left in the bloodstream before the second dose is administered? $\dfrac{1}{8}$

  b. Write a general expression for the geometric series that models the number of milligrams of drug left in the bloodstream after the *n*th dose.

  c. About what amount of medicine is present in the bloodstream for large values of *n*? $\dfrac{8}{7}D$

  d. A level of more than 350 milligrams of this drug in the bloodstream is considered toxic. Find the largest possible dose that can be given repeatedly over a long period of time without harming the patient. **306.25 mg**

45. Geometry If the midpoints of a square are joined by straight lines, the new figure will also be a square.

  a. If the original square has a perimeter of 20 feet, find the perimeter of the new square. (*Hint:* Use the Pythagorean Theorem.) **10 $\sqrt{2}$ ft**

  b. If this process is continued to form a sequence of "nested" squares, what will be the sum of the perimeters of all the squares? **40 + 20$\sqrt{2}$ ft or about 68 ft**

46. Technology Since the mid-1980s, the number of computers in schools has steadily increased. The graph below shows the corresponding decline in the student-computer ratio.

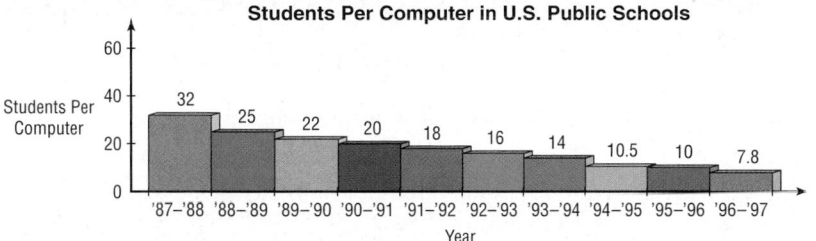

Students Per Computer in U.S. Public Schools

Students Per Computer

60 — 40 — 32, 25, 22, 20, 18, 16, 14, 10.5, 10, 7.8 — 20 — 0

'87–'88 '88–'89 '89–'90 '90–'91 '91–'92 '92–'93 '93–'94 '94–'95 '95–'96 '96–'97

Year

Source: *QED's Technology in Public Schools, 16th Edition*

Another publication states that the average number of students per computer in U.S. public schools can be estimated by the sequence model $a_n = 35.812791(0.864605)^n$, *for* $n = 1, 2, 3, \ldots$, with the 1987-1988 school year corresponding to $n = 1$.

a. Find the first ten terms of the model. Round your answers to the nearest tenth. **31.0, 26.8, 23.1, 20.0, 17.3, 15.0, 12.9, 11.2, 9.7, 8.4**

46b. 9.7; the model is 0.3 below the actual statistic.

b. Use the model to estimate the average number of students having to share a computer during the 1995-1996 school year. How does this estimate compare to the actual data given in the graph?

c. Make a prediction as to the average number of students per computer for the 2000-2001 school year. **4.7**

d. Does this sequence approach a limit? If so, what is the limit? **yes; 0**

e. Realistically, will the student computer ratio ever reach this limit? Explain. **No, the number of students per computer must be greater than zero.**

Mixed Review

47. -2, $-1\frac{1}{3}$, $-\frac{8}{9}$, $-\frac{16}{27}$

47. The first term of a geometric sequence is -3, and the common ratio is $\frac{2}{3}$. Find the next four terms of the sequence. *(Lesson 12-2)*

48. Find the 16th term of the arithmetic sequence for which $a_1 = 1.5$ and $d = 0.5$. *(Lesson 12-1)* **9**

49. $(6, -2)$; $(6 \pm \sqrt{5}, -2)$; $(8, -2)$, $(4, -2)$; $y = \frac{1}{2}x - 5$, $y = -\frac{1}{2}x + 1$

49. Name the coordinates of the center, foci, and vertices, and the equation of the asymptotes of the hyperbola that has the equation $x^2 - 4y^2 - 12x - 16y = -16$. *(Lesson 10-4)*

50. Graph $r = 6 \cos 3\theta$. *(Lesson 9-2)* **See margin.**

51. Navigation A ship leaving port sails for 125 miles in a direction 20° north of due east. Find the magnitude of the vertical and horizontal components. *(Lesson 8-1)* **42.75 miles, 117.46 miles**

52. $-\dfrac{\sqrt{2 - \sqrt{2}}}{2}$

52. Use a half-angle identity to find the exact value of cos 112.5°. *(Lesson 7-4)*

53. Graph $y = \cos x$ on the interval $-180° \le x \le 360°$. *(Lesson 6-3)* **See margin.**

54. $\pm1, \pm2, \pm\frac{1}{2}, \pm\frac{1}{4}, \pm\frac{1}{8}$

54. List all possible rational zeros of the function $f(x) = 8x^3 + 3x - 2$. *(Lesson 4-4)*

55. **SAT/ACT Practice** If $a = 4b + 26$, and b is a positive integer, then a could be divisible by all of the following EXCEPT **B**

A 2   B 4   C 5   D 6   E 7

Extra Practice See p. A49.

Lesson 12-3 Infinite Sequences and Series **783**

Extra Credit

Write $0.29\overline{25}$ as a fraction. $\dfrac{724}{2475}$

---

1 FOCUS

Motivating the Lesson
Hands-On Activity On the chalkboard, have students "discover" the pattern of the sequence

$$7,\ 7 + \frac{1}{7},\ 7 + \frac{1}{7 + \frac{1}{7}},\ 7 + \frac{1}{7 + \frac{1}{7 + \frac{1}{7}}},$$

and determine the fifth term. Then have them use a calculator to find the value of this term.

$$7 + \frac{1}{7 + \frac{1}{7 + \frac{1}{7 + \frac{1}{7}}}}\ ;\ 7.140054923$$

2 TEACH

📁 A reproduction of this lesson with graphing calculator keystrokes provided can be found in the *Graphing Calculator Keystrokes Masters*, pp. 49–51.

Teaching Tip Emphasize that the sequence of a continued fraction approaches an irrational number as a limit. For example, $1,\ 1 + \frac{1}{1}$,

$1 + \frac{1}{1 + \frac{1}{1}}$, ... approaches the

limit $\frac{1 + \sqrt{5}}{2}$, an irrational number.

12-3B Continued Fractions

An Extension of Lesson 12-3

OBJECTIVE
• Explore sequences generated by continued fractions.

An expression of the following form is called a *continued fraction*.

$$a_1 + \cfrac{b_1}{a_2 + \cfrac{b_2}{a_3 + \cfrac{b_3}{a_4 + \cdots}}}$$

By using only a finite number of "decks" and values of a_n and b_n that follow regular patterns, you can often obtain a sequence of terms that approaches a limit, which can be represented by a simple expression. For example, if all of the numbers a_n and b_n are equal to 1, then the continued fraction gives rise to the following sequence.

$$1,\ 1 + \frac{1}{1},\ 1 + \cfrac{1}{1 + \frac{1}{1}},\ 1 + \cfrac{1}{1 + \cfrac{1}{1 + \frac{1}{1}}},\ \ldots$$

The golden ratio is closely related to the Fibonnaci sequence, which you will learn about in Lesson 12-7.

It can be shown that the terms of this sequence approach the limit $\frac{1 + \sqrt{5}}{2}$. This number is often called the *golden ratio*.

Now consider the following more general sequence.

$$A,\ A + \frac{1}{A},\ A + \cfrac{1}{A + \frac{1}{A}},\ A + \cfrac{1}{A + \cfrac{1}{A + \frac{1}{A}}},\ \ldots$$

To help you visualize what this sequence represents, suppose $A = 5$. The

sequence becomes $5,\ 5 + \frac{1}{5},\ 5 + \cfrac{1}{5 + \frac{1}{5}},\ 5 + \cfrac{1}{5 + \cfrac{1}{5 + \frac{1}{5}}},\ \ldots$ or $5,\ \frac{26}{5},\ \frac{135}{26},\ \frac{701}{135},\ \ldots$.

A calculator approximation of this sequence is 5, 5.2, 5.192307692, 5,192592593, ...

interNET
CONNECTION

Graphing Calculator Programs
To download this graphing calculator program, visit our website at
www.amc.glencoe.com

Each term of the sequence is the sum of A and the reciprocal of the previous term. The program at the right calculates the value of the nth term of the above sequence for $n \geq 3$ and a specified value of A.

When you run the program it will ask you to input values for A and N.

```
PROGRAM: CFRAC
: Prompt A
: Disp "INPUT TERM"
: Disp "NUMBER N, N ≥ 3"
: Prompt N
: 1 → K
: A + 1/A → C
: Lbl 1
: A + 1/C → C
: K + 1 → K
: If K < N − 1
: Then: Goto 1
: Else: Disp C
```

Enter the program into your calculator and use it for the exercises that follow.

1. What output is given when the program is executed for $A = 1$ and $N = 10$?

2. With $A = 1$, determine the least value of N necessary to obtain an output that agrees with the calculator's nine decimal approximation of $\frac{1 + \sqrt{5}}{2}$.

3. Use algebra to show that the continued fraction $1 + \cfrac{1}{1 + \cfrac{1}{1 + \cdots}}$ has a

   value of $\frac{1 + \sqrt{5}}{2}$. (*Hint:* If $x = 1 + \cfrac{1}{1 + \cfrac{1}{1 + \cdots}}$, then $x = 1 + \frac{1}{x}$. Solve

   this last equation for x.) **See Answer Appendix.**

4. Find the exact value of $3 + \cfrac{1}{3 + \cfrac{1}{3 + \cdots}}$. $\dfrac{3 + \sqrt{13}}{2}$

5. Execute the program with $A = 3$ and $N = 40$. How does this output compare to the decimal approximation of the expression found in Exercise 4?

6. Find a radical expression for $A + \cfrac{1}{A + \cfrac{1}{A + \cdots}}$. $\dfrac{A + \sqrt{A^2 + 4}}{2}$

7. Write a modified version of the program that calculates the nth term of the following sequence for $n \geq 3$. **See margin for sample program.**

$$A, A + \frac{B}{2A}, A + \cfrac{B}{2A + \cfrac{B}{2A}}, A + \cfrac{B}{2A + \cfrac{B}{2A + \cfrac{B}{2A}}}, \ldots$$

8. For large values of N, the program output and the decimal approximation of $\sqrt{A^2 + B}$ are equal.

9. See Answer Appendix.

8. Choose several positive integer values for A and B and compare the program output with the decimal approximation of $\sqrt{A^2 + B}$ for several values of n, for $n \geq 3$. Describe your observations.

9. Use algebra to show that for $A > 0$ and $B > 0$, $A + \cfrac{B}{2A + \cfrac{B}{2A + \cdots}}$ has a

   value of $\sqrt{A^2 + B}$.

   $\left(\text{*Hint:* If } x = A + \cfrac{B}{2A + \cfrac{B}{2A + \cdots}}, \text{ then } x + A = 2A + \cfrac{B}{2A + \cfrac{B}{2A + \cdots}}.\right)$

10. If you execute the original program for $A = 1$ and $N = 20$ and then execute it for $A = -1$ and $N = 20$, how will the two outputs compare?

11. What values can you use for A and B in the program for Exercise 7 in order to approximate $\sqrt{15}$? **Sample answers: $A = 1$, $B = 14$; $A = 4$, $B = -1$.**

Additional Answer

7. Sample program:

```
Program: CCFRAC
:Prompt A :Prompt B
:Disp "INPUT TERM"
:Disp "NUMBER N, N ≥ 3"
:Prompt N
:1 → K
:B + 1/(2A) → C
:Lbl 1
:B + 1/(2A + C) → C
:K + 1 → K
:If K < N − 1
:Then: Goto 1
:Else: Disp C + A
```

Open-Ended Assessment

Modeling Have students select a number from 8 to 15 and make that number the first term of a sequence. Then have them make the second term be the first term plus $\dfrac{1}{\text{the preceding term}}$, and so forth. Finally, have them find the exact value of the sequence.

Sample answer:

$8, 8 + \dfrac{1}{8}, 8 + \cfrac{1}{8 + \cfrac{1}{8}},$

$8 + \cfrac{1}{8 + \cfrac{1}{8 + \cfrac{1}{8}}} + \cdots; 4 + \sqrt{17}$

Convergent and Divergent Series

OBJECTIVE
- Determine whether a series is convergent or divergent.

HISTORY The Greek philosopher Zeno of Elea (c. 490–430 B.C.) proposed several perplexing riddles, or paradoxes. One of Zeno's paradoxes involves a race on a 100-meter track between the mythological Achilles and a tortoise. Zeno claims that even though Achilles can run twice as fast as the tortoise, if the tortoise is given a 10-meter head start, Achilles will never catch him. Suppose Achilles runs 10 meters per second and the tortoise a remarkable 5 meters per second. By the time Achilles has reached the 10-meter mark, the tortoise will be at 15 meters. By the time Achilles reaches the 15-meter mark, the tortoise will be at 17.5 meters, and so on. Thus, Achilles is always behind the tortoise and never catches up.

1 FOCUS

5-Minute Check
Lesson 12-3

Find each limit, or state that the limit does not exist, and explain your reasoning.

1. $\lim\limits_{n \to \infty} \dfrac{2 - 2n}{6n}$ $-\dfrac{1}{3}$

2. $\lim\limits_{n \to \infty} \dfrac{6n^2 + 2n + 8}{n^2} + 3$ 6

3. $\lim\limits_{n \to \infty} \dfrac{3n^3 + 3n - 2}{9n}$ **does not exist; since** $\lim\limits_{n \to \infty} n^2$ **becomes increasingly large as it approaches infinity, the sequence has no limit.**

4. Find the limit of the sequence described by the general expression $\dfrac{4 - 5n}{3n}$. $-\dfrac{5}{3}$

5. Find the sum of the series $81 + 9 + 1 + \cdots$ or state that the sum does not exist and explain your reasoning. $91\dfrac{1}{8}$

6. Write $0.\overline{45}$ as a fraction. $\dfrac{5}{11}$

Is Zeno correct? Let us look at the distance between Achilles and the tortoise after specified amounts of time have passed. Notice that the distance between the two contestants will be zero as n approaches infinity since $\lim\limits_{n \to \infty} \dfrac{10}{2^n} = 0$.

To disprove Zeno's conclusion that Achilles will never catch up to the tortoise, we must show that there is a time value for which this 0 difference can be achieved. In other words, we need to show that the infinite series $1 + \dfrac{1}{2} + \dfrac{1}{4} + \dfrac{1}{8} + \cdots$ has a sum, or limit. *This problem will be solved in Example 5.*

Time (seconds)	Distance Apart (meters)
0	10
1	$\dfrac{10}{2} = 5$
$1 + \dfrac{1}{2} = \dfrac{3}{2}$	$\dfrac{10}{4} = 2.5$
$1 + \dfrac{1}{2} + \dfrac{1}{4} = \dfrac{7}{4}$	$\dfrac{10}{8} = 1.25$
$1 + \dfrac{1}{2} + \dfrac{1}{4} + \dfrac{1}{8} = \dfrac{15}{8}$ $\vdots$	$\dfrac{10}{16} = 0.625$ $\vdots$
$1 + \dfrac{1}{2} + \dfrac{1}{4} + \dfrac{1}{8} + \cdots$	$\dfrac{10}{2^n}$

Starting with a time of 1 second, the partial sums of the time series form the sequence $1, \dfrac{3}{2}, \dfrac{7}{4}, \dfrac{15}{18}, \ldots$. As the number of terms used for the partial sums increases, the value of the partial sums also increases. If this sequence of partial sums approaches a limit, the related infinite series is said to **converge.** If this sequence of partial sums does not have a limit, then the related infinite series is said to **diverge.**

Motivating the Lesson
Hands-On Activity Have students measure the distance between their desks and the door. Ask each student to move half the distance to the door. Then, each student should continue walking half the distance, then half the distance again, and so on. Discuss how students can get closer to the door but never actually reach the door. This type of paradox will be discussed with regards to the concept of limits.

786 *Chapter 12 Sequences and Series*

Resource Manager

Reproducible Masters
- *Study Guide and Practice*, pp. 221–222
- *Enrichment*, p. 136

Transparencies
- 5-Minute Check, 12-4
- Teaching, 12-4

Convergent and Divergent Series	If an infinite series has a sum, or limit, the series is convergent. If a series is not convergent, it is divergent.

Example ❶ **Determine whether each arithmetic or geometric series is convergent or divergent.**

There are many series that begin with the first few terms shown in this example. In this chapter, always assume that the expression for the general term is the simplest one possible.

a. $-\dfrac{1}{2} + \dfrac{1}{4} - \dfrac{1}{8} + \dfrac{1}{16} - \cdots$

This is a geometric series with $r = -\dfrac{1}{2}$. Since $|r| < 1$, the series has a limit. Therefore, the series is convergent.

b. $2 + 4 + 8 + 16 + \cdots$

This is a geometric series with $r = 2$. Since $|r| > 1$, the series has no limit. Therefore, the series is divergent.

c. $10 + 8.5 + 7 + 5.5 + \cdots$

This is an arithmetic series with $d = -1.5$. Arithmetic series do not have limits. Therefore, the series is divergent.

When a series is neither arithmetic nor geometric, it is more difficult to determine whether the series is convergent or divergent. Several different techniques can be used. One test for convergence is the **ratio test.** This test can only be used when all terms of a series are positive. The test depends upon the ratio of consecutive terms of a series, which must be expressed in general form.

Ratio Test	Let a_n and a_{n+1} represent two consecutive terms of a series of positive terms. Suppose $\lim\limits_{n\to\infty} \dfrac{a_{n+1}}{a_n}$ exists and that $r = \lim\limits_{n\to\infty}\dfrac{a_{n+1}}{a_n}$. The series is convergent if $r < 1$ and divergent if $r > 1$. If $r = 1$, the test provides no information.

The ratio test is especially useful when the general form for the terms of a series contains powers.

Example ❷ **Use the ratio test to determine whether each series is convergent or divergent.**

a. $\dfrac{1}{2} + \dfrac{2}{4} + \dfrac{3}{8} + \dfrac{4}{16} + \cdots$

First, find a_n and a_{n+1}. $\qquad a_n = \dfrac{n}{2^n} \quad$ and $\quad a_{n+1} = \dfrac{n+1}{2^{n+1}}$

Then use the ratio test. $\qquad r = \lim\limits_{n\to\infty} \dfrac{\dfrac{n+1}{2^{n+1}}}{\dfrac{n}{2^n}}$

(continued on the next page)

Lesson 12-4 Convergent and Divergent Series **787**

Teaching Tip Before discussing Example 1, stress that for $|r| > 1$, an infinite series grows without bound and is therefore divergent.

In-Class Examples

Example 1
Determine whether each arithmetic or geometric series is convergent or divergent.

a. $\dfrac{1}{8} + \dfrac{3}{20} + \dfrac{9}{50} + \dfrac{27}{125} + \cdots$. **divergent**

b. $18.75 + 17.50 + 16.25 + 15.00 + \cdots$. **divergent**

c. $65 + 13 + 2\dfrac{3}{5} + \dfrac{13}{25} + \cdots$. **convergent**

Example 2
Use the ratio test to determine whether each series is convergent or divergent.

a. $2 + \dfrac{3}{2} + \dfrac{4}{3} + \dfrac{5}{4} + \cdots$. **Test provides no information.**

b. $\dfrac{3}{4} + \dfrac{4}{16} + \dfrac{5}{64} + \dfrac{6}{256} + \cdots$. **convergent**

Prerequisite Skills

A review of the prerequisite skills for this lesson is provided in the *Prerequisite Skills Masters.*

Skill Number and Content	Pages
2 Fractions	5–6

Teaching Tip For Example 3, point out that $\frac{1}{n+2}$ appears to be a fraction less than 1 regardless of how big n gets. But it is the limiting value that counts, and $\lim\limits_{n\to\infty}\frac{1}{n+2}=1$.

In-Class Example

Example 3

Use the ratio test to determine whether the series $\frac{2}{1\cdot 2}+\frac{4}{2\cdot 3}+\frac{8}{3\cdot 4}+\frac{16}{4\cdot 5}+\cdots$ is convergent or divergent.   **divergent**

$r=\lim\limits_{n\to\infty}\dfrac{n+1}{2^{n+1}}\cdot\dfrac{2^n}{n}$ *Multiply by the reciprocal of the divisor.*

$r=\lim\limits_{n\to\infty}\dfrac{1}{2}\cdot\dfrac{n+1}{n}$ $\dfrac{2^n}{2^{n+1}}=\dfrac{1}{2}$

$r=\lim\limits_{n\to\infty}\dfrac{1}{2}\cdot\lim\limits_{n\to\infty}\dfrac{n+1}{n}$ *Limit of a Product*

$r=\dfrac{1}{2}\lim\limits_{n\to\infty}\dfrac{1+\frac{1}{n}}{1}$ *Divide by the highest power of n and then apply limit theorems.*

$r=\dfrac{1}{2}\cdot\dfrac{1+0}{1}$ or $\dfrac{1}{2}$ Since $r<1$, the series is convergent.

b. $\dfrac{1}{2}+\dfrac{2}{3}+\dfrac{3}{4}+\dfrac{4}{5}+\cdots$

$a_n=\dfrac{n}{n+1}$ and $a_{n+1}=\dfrac{n+1}{(n+1)+1}$ or $\dfrac{n+1}{n+2}$

$r=\lim\limits_{n\to\infty}\dfrac{\frac{n+1}{n+2}}{\frac{n}{n+1}}$

$r=\lim\limits_{n\to\infty}\dfrac{n^2+2n+1}{n^2+2n}$ $\dfrac{n+1}{n+2}\cdot\dfrac{n+1}{n}=\dfrac{n^2+2n+1}{n^2+2n}$

$r=\lim\limits_{n\to\infty}\dfrac{1+\frac{2}{n}+\frac{1}{n^2}}{1+\frac{2}{n}}$ *Divide by the highest power of n and apply limit theorems.*

$r=\dfrac{1+0+0}{1+0}$ or 1 Since $r=1$, the test provides no information.

The ratio test is also useful when the general form of the terms of a series contains products of consecutive integers.

Example ❸ **Use the ratio test to determine whether the series**

$$1+\frac{1}{1\cdot 2}+\frac{1}{1\cdot 2\cdot 3}+\frac{1}{1\cdot 2\cdot 3\cdot 4}+\cdots \text{ is convergent or divergent.}$$

First find the nth term and $(n+1)$th term. Then, use the ratio test.

$a_n=\dfrac{1}{1\cdot 2\cdot\,\cdots\,\cdot n}$ and $a_{n+1}=\dfrac{1}{1\cdot 2\cdot\,\cdots\,\cdot(n+1)}$

$r=\lim\limits_{n\to\infty}\dfrac{\frac{1}{1\cdot 2\cdot\,\cdots\,\cdot(n+1)}}{\frac{1}{1\cdot 2\cdot\,\cdots\,\cdot n}}$

$r=\lim\limits_{n\to\infty}\dfrac{1\cdot 2\cdot\,\cdots\,\cdot n}{1\cdot 2\cdot\,\cdots\,\cdot(n+1)}$ *Note that $1\cdot 2\cdot\cdots\cdot(n+1)=1\cdot 2\cdot\cdots\cdot n\cdot(n+1)$.*

$r=\lim\limits_{n\to\infty}\dfrac{1}{n+1}$ or 0 *Simplify and apply limit theorems.*

Since $r<1$, the series is convergent.

When the ratio test does not determine if a series is convergent or divergent, other methods must be used.

Example **4** Determine whether the series $1 + \frac{1}{2} + \frac{1}{3} + \frac{1}{4} + \frac{1}{5} + \cdots$ is convergent or divergent.

Suppose the terms are grouped as follows. Beginning after the second term, the number of terms in each successive group is doubled.

$$(1) + \left(\frac{1}{2}\right) + \left(\frac{1}{3} + \frac{1}{4}\right) + \left(\frac{1}{5} + \frac{1}{6} + \frac{1}{7} + \frac{1}{8}\right) + \left(\frac{1}{9} + \cdots + \frac{1}{16}\right) + \cdots$$

Notice that the first enclosed expression is greater than $\frac{1}{2}$, and the second is equal to $\frac{1}{2}$. Beginning with the third expression, each sum of enclosed terms is greater than $\frac{1}{2}$. Since there are an unlimited number of such expressions, the sum of the series is unlimited. Thus, the series is divergent.

A series can be compared to other series that are known to be convergent or divergent. The following list of series can be used for reference.

Summary of Series for Reference

1. **Convergent:** $a_1 + a_1 r + a_1 r^2 + \cdots + a_1 r^{n-1} + \cdots,\ |r| < 1$
2. **Divergent:** $a_1 + a_1 r + a_1 r^2 + \cdots + a_1 r^{n-1} + \cdots,\ |r| > 1$
3. **Divergent:** $a_1 + (a_1 + d) + (a_1 + 2d) + (a_1 + 3d) + \cdots$
4. **Divergent:** $1 + \frac{1}{2} + \frac{1}{3} + \frac{1}{4} + \frac{1}{5} + \cdots + \frac{1}{n} + \cdots$ *This series is known as the harmonic series.*
5. **Convergent:** $1 + \frac{1}{2^p} + \frac{1}{3^p} + \cdots + \frac{1}{n^p} + \cdots,\ p > 1$

If a series has all positive terms, the **comparison test** can be used to determine whether the series is convergent or divergent.

Comparison Test

• A series of positive terms is convergent if, for $n > 1$, each term of the series is equal to or less than the value of the corresponding term of some convergent series of positive terms.
• A series of positive terms is divergent if, for $n > 1$, each term of the series is equal to or greater than the value of the corresponding term of some divergent series of positive terms.

Example **5** Use the comparison test to determine whether the following series are convergent or divergent.

a. $\frac{4}{5} + \frac{4}{7} + \frac{4}{9} + \frac{4}{11} + \cdots$

The general term of this series is $\frac{4}{2n + 3}$. The general term of the divergent series $1 + \frac{1}{2} + \frac{1}{3} + \frac{1}{4} + \frac{1}{5} + \cdots$ is $\frac{1}{n}$. Since for all $n > 1$, $\frac{4}{2n + 3} > \frac{1}{n}$, the series $\frac{4}{5} + \frac{4}{7} + \frac{4}{9} + \frac{4}{11} + \cdots$ is also divergent.

Lesson 12-4   Convergent and Divergent Series   **789**

Teaching Tip For Example 4, point out that some infinite series are obviously divergent, such as $1 + 2 + 3 + \cdots$, which is arithmetic. In fact, all infinite arithmetic series are divergent.

In-Class Examples

Example 4
Determine whether the series
$1 + \frac{1}{\sqrt{2}} + \frac{1}{\sqrt{3}} + \frac{1}{\sqrt{4}} + \frac{1}{\sqrt{5}} + \cdots$
is convergent or divergent.
divergent

Example 5
Use the comparison test to determine whether the following series are convergent or divergent.

a. $\frac{2 \cdot 1}{2} + \frac{2 \cdot 2}{2} + \frac{2 \cdot 6}{3} + \frac{2 \cdot 24}{4} + \cdots$.   **divergent**

b. $\frac{1}{3} + \frac{1}{6} + \frac{1}{12} + \frac{1}{24} + \cdots$.
convergent

Teaching Tip After working In-Class Example 5, remind students that the given series
$\frac{1}{3} + \frac{1}{6} + \frac{1}{12} + \frac{1}{24} + \cdots$ is an infinite geometric series with a common ratio $\frac{1}{2}$. Therefore, it is convergent.

Addressing Individual Needs

Visual/Spatial Have students make a chart showing the partial sums, expressed as decimals, for terms beginning with 1 for the series $1 + \frac{1}{3} + \frac{1}{9} + \frac{1}{27} + \cdots$. They should use a calculator to find the partial sums and then find the limiting value of the series. **1.5 is the limiting value of the series.**

Term	Partial Sums
1	1
2	1.333333333
3	1.444444444
4	1.481481481
5	1.49382716
6	1.497942387

Teaching Tip After working Example 6, remind students that there is often more than one way to prove convergence or divergence for a given series. In Example 6 in the Student Edition, for example, the series $1 + 0.5 + 0.25 + \cdots$ was proven to be convergent by the ratio test. We also know it to be convergent because it is an infinite geometric series with $r = 0.5$.

Study Guide and Practice Masters, pp. 221–222

12-4

NAME _____ DATE _____ PERIOD _____

Study Guide

Convergent and Divergent Series

If an infinite series has a sum, or limit, the series is **convergent**. If a series is not convergent, it is **divergent**. When a series is neither arithmetic nor geometric and all the terms are positive, you can use the **ratio test** or the **comparison test** to determine whether the series is convergent or divergent.

Ratio Test	Let a_n and a_{n+1} represent two consecutive terms of a series of positive terms. Suppose $\lim_{n\to\infty} \frac{a_{n+1}}{a_n}$ exists and $r = \lim_{n\to\infty} \frac{a_{n+1}}{a_n}$. The series is convergent if $r < 1$ and divergent if $r > 1$. If $r = 1$, the test provides no information.
Comparison Test	• A series of positive terms is convergent if, for $n \geq 1$, each term of the series is equal to or less than the value of the corresponding term of some convergent series of positive terms. • A series of positive terms is divergent if, for $n \geq 1$, each term of the series is equal to or greater than the value of the corresponding term of some divergent series of positive terms.

Example 1 Use the ratio test to determine whether the series $\frac{1 \cdot 2}{2^1} + \frac{2 \cdot 3}{2^2} + \frac{3 \cdot 4}{2^3} + \frac{4 \cdot 5}{2^4} + \cdots$ is convergent or divergent.

First, find the nth term. Then use the ratio test.

$a_n = \frac{n(n+1)}{2^n}$ $a_{n+1} = \frac{(n+1)(n+2)}{2^{n+1}}$

$r = \lim_{n\to\infty} \frac{\frac{(n+1)(n+2)}{2^{n+1}}}{\frac{n(n+1)}{2^n}}$

$r = \lim_{n\to\infty} \frac{(n+1)(n+2)}{2^{n+1}} \cdot \frac{2^n}{n(n+1)}$ *Multiply by the reciprocal of the divisor.*

$r = \lim_{n\to\infty} \frac{n+2}{2n}$ $\frac{2^n}{2^{n+1}} = \frac{1}{2}$

$r = \lim_{n\to\infty} \frac{1 + \frac{2}{n}}{2}$ *Divide by the highest power of n and apply limit theorems.*

$r = \frac{1}{2}$ *Since $r < 1$, the series is convergent.*

Example 2 Use the comparison test to determine whether the series $\frac{1}{4^2} + \frac{1}{7^2} + \frac{1}{10^2} + \frac{1}{13^2} + \cdots$ is convergent or divergent.

The general term of the series is $\frac{1}{(3n+1)^2}$. The general term of the convergent series $1 + \frac{1}{2^2} + \frac{1}{3^2} + \frac{1}{4^2} + \cdots$ is $\frac{1}{n^2}$. Since $\frac{1}{(3n+1)^2} < \frac{1}{n^2}$ for all $n \geq 1$, the series $\frac{1}{4^2} + \frac{1}{7^2} + \frac{1}{10^2} + \frac{1}{13^2} + \cdots$ is also convergent.

© Glencoe/McGraw-Hill   221   *Advanced Mathematical Concepts*

---

b. $\frac{1}{1^2} + \frac{1}{3^2} + \frac{1}{5^2} + \frac{1}{7^2} + \cdots$

The general term of the series is $\frac{1}{(2n-1)^2}$. The general term of the convergent series $1 + \frac{1}{2^2} + \frac{1}{3^2} + \frac{1}{4^2} + \cdots$ is $\frac{1}{n^2}$. Since $\frac{1}{(2n-1)^2} \leq \frac{1}{n^2}$ for all n, the series $\frac{1}{1^2} + \frac{1}{3^2} + \frac{1}{5^2} + \frac{1}{7^2} + \cdots$ is also convergent.

With a better understanding of convergent and divergent infinite series, we are now ready to tackle Zeno's paradox.

Example **6** **HISTORY** Refer to the application at the beginning of the lesson. To disprove Zeno's conclusion that Achilles will never catch up to the tortoise, we must show that the infinite time series $1 + 0.5 + 0.25 + \cdots$ has a limit.

To show that the series $1 + 0.5 + 0.25 + \cdots$ has a limit, we need to show that the series is convergent.

The general term of this series is $\frac{1}{2^n}$. Try using the ratio test for convergence of a series.

$a_n = \frac{1}{2^n}$ and $a_{n+1} = \frac{1}{2^{n+1}}$

$r = \frac{\frac{1}{2^{n+1}}}{\frac{1}{2^n}}$

$= \frac{1}{2}$ $\frac{1}{2^{n+1}} \cdot \frac{2^n}{1} = \frac{1}{2}$

Since $r < 1$, the series converges and therefore has a sum. Thus, there is a time value for which the distance between Achilles and the tortoise will be zero. *You will determine how long it takes him to do so in Exercise 34.*

CHECK FOR UNDERSTANDING

Communicating Mathematics

1a–c. See student's work.

Read and study the lesson to answer each question.

1. a. **Write** an example, of an infinite geometric series in which $|r| > 1$.

   b. **Determine** the 25th, 50th, and 100th terms of your series.

   c. **Identify** the sum of the first 25, 50, and 100 terms of your series.

   d. **Explain** why this type of infinite geometric series does not converge. **See margin.**

2. **Estimate** the sum S_n of the series whose partial sums are graphed at the right. **16**

Additional Answers *(pp. 790–791)*

3a.

1d. In a given trigonometric series where $|r| > 1$, each succeeding term is larger than the one preceding it. Therefore, the series approaches ∞ and thus does not converge.

3. Consider the infinite series $\frac{1}{3} + \frac{2^2}{3^2} + \frac{3^2}{3^3} + \frac{4^2}{3^4} + \cdots$.

   a. **Sketch** a graph of the first eight partial sums of this series. **See margin.**

   b. **Make** a conjecture based on the graph found in part **a** as to whether the series is convergent or divergent. **convergent** $\frac{n^2}{3^n}$

   c. **Determine** a general term for this series.

   d. **Write a convincing argument** using the general term found in part **c** to support the conjecture you made in part **b**. **See margin.**

4. *Math Journal* **Make a list** of the methods presented in this lesson and in the previous lesson for determining convergence or divergence of an infinite series. Be sure to indicate any restrictions on a method's use. Then number your list as to the order in which these methods should be applied. **See margin.**

Guided Practice

6. Test provides no information.

Use the ratio test to determine whether each series is *convergent* or *divergent*.

5. $\frac{1}{2} + \frac{2}{2^2} + \frac{3}{2^3} + \cdots$ **convergent**

6. $\frac{3}{4} + \frac{7}{8} + \frac{11}{12} + \frac{15}{16} + \cdots$

7. Use the comparison test to determine whether the series $\frac{2}{1} + \frac{3}{2} + \frac{4}{3} + \cdots$ is *convergent* or *divergent*. **divergent**

Determine whether each series is *convergent* or *divergent*. **9. convergent**

8. $\frac{1}{4} + \frac{5}{16} + \frac{3}{8} + \frac{7}{16} + \cdots$ **divergent**

9. $\frac{1}{2+1^2} + \frac{1}{2+2^2} + \frac{1}{2+3^2} + \cdots$

10. $\frac{1}{1 \cdot 2} + \frac{1}{2 \cdot 2^2} + \frac{1}{3 \cdot 2^3} + \cdots$ **convergent**

11. $4 + 3 + \frac{9}{4} + \cdots$ **convergent**

12. **Ecology** An underground storage container is leaking a toxic chemical. One year after the leak began, the chemical had spread 1500 meters from its source. After two years, the chemical had spread 900 meters more, and by the end of the third year, it had reached an additional 540 meters.

   a. If this pattern continues, how far will the spill have spread from its source after 10 years? **about 3727 m**

   b. Will the spill ever reach the grounds of a school located 4000 meters away from the source? Explain. **No, the sum of the infinite series modeling this situation is 3750. Thus, the spill will spread no more than 3750 meters.**

EXERCISES

Practice

16. test provides no information

 A

 B

Use the ratio test to determine whether each series is *convergent* or *divergent*.

13. $\frac{4}{3} + \frac{4}{9} + \frac{4}{27} + \frac{4}{81} + \cdots$ **convergent**

14. $\frac{2}{5} + \frac{4}{10} + \frac{8}{15} + \cdots$ **divergent**

15. $2 + \frac{4}{2^2} + \frac{8}{3^2} + \frac{16}{4^2} + \cdots$ **divergent**

16. $\frac{2}{2 \cdot 3} + \frac{2}{3 \cdot 4} + \frac{2}{4 \cdot 5} + \cdots$

17. $1 + \frac{3}{1 \cdot 2 \cdot 3} + \frac{5}{1 \cdot 2 \cdot 3 \cdot 4 \cdot 5} + \cdots$ **convergent**

18. $5 + \frac{5^2}{1 \cdot 2} + \frac{5^3}{1 \cdot 2 \cdot 3} + \cdots$ **convergent**

19. Use the ratio test to determine whether the series $\frac{2 \cdot 4}{2} + \frac{4 \cdot 6}{4} + \frac{6 \cdot 8}{8} + \frac{8 \cdot 10}{16} + \cdots$ is convergent or divergent. **convergent**

 www.amc.glencoe.com/self_check_quiz

Lesson 12-4 Convergent and Divergent Series **791**

Error Analysis

Watch for students who have difficulty working with the comparison test.

Prevent by emphasizing that convergence can be established only by testing against a known convergent series, while divergence can be established only by testing against a known divergent series.

Assignment Guide

Core: 13–29 odd, 31–42
Enriched: 14–30 even, 31–42
All: Mid-Chapter Quiz, 1–10

Additional Answers

4. Consider the infinite series a_n.

  1. If the $\lim\limits_{n \to \infty} a_n \neq 0$, the sum of the series does not exist, and thus the series is divergent. If the $\lim\limits_{n \to \infty} a_n = 0$, the sum may or may not exist and therefore it cannot be determined from this test whether the series is convergent or divergent.

  2. If the series is arithmetic then it is divergent.

  3. If the series is geometric then the series converges for $|r| < 1$ and diverges for $|r| \geq 1$.

  4. Ratio test: the series converges for $r < 1$ and diverges for $r > 1$. If $r = 1$, the test fails. This test can only be used if all the terms of the series are positive and if the series can be expressed in general form.

  5. Comparison test: may only be used if all of the terms in the series are positive.

Additional Answer

3d. We can use the ratio test to determine whether the series is indeed convergent.

$a_n = \frac{n^2}{3^n}$ and $a_{n+1} = \frac{(n+1)^2}{3^{n+1}}$

$r = \lim\limits_{n \to \infty} \dfrac{\frac{(n+1)^2}{3^{n+1}}}{\frac{n^2}{3^n}}$

$r = \lim\limits_{n \to \infty} \frac{(n+1)^2}{3^{n+1}} \cdot \frac{3^n}{n^2}$

$r = \lim\limits_{n \to \infty} \frac{n^2 + 2n + 1}{3n^2}$

$r = \lim\limits_{n \to \infty} \frac{1 + \frac{2}{n} + \frac{1}{n}}{3}$

$r = \frac{1}{3}$

Since $r < 1$, the series is convergent.

Lesson 12-4 **791**

Use the comparison test to determine whether each series is *convergent* or *divergent*.

20. $\frac{1}{2^2} + \frac{1}{4^2} + \frac{1}{6^2} + \cdots$ **convergent**
21. $\frac{1}{2} + \frac{1}{9} + \frac{1}{28} + \frac{1}{65} + \cdots$ **convergent**

22. $\frac{1}{2} + \frac{2}{3} + \frac{3}{4} + \cdots$ **divergent**
23. $\frac{5}{3} + \frac{5}{4} + 1 + \frac{5}{6} + \cdots$ **divergent**

24. Use the comparison test to determine whether the series $\frac{1}{3} + \frac{1}{5} + \frac{1}{9} + \frac{1}{17} \cdots$ is *convergent* or *divergent*.   **convergent**

Determine whether each series is *convergent* or *divergent*.

25. $\frac{1}{2} - \frac{3}{8} + \frac{9}{32} - \cdots$ **convergent**
26. $3 + \frac{5}{3} + \frac{7}{5} + \cdots$ **divergent**

27. $\frac{1}{5 + 1^2} + \frac{1}{5 + 2^2} + \frac{1}{5 + 3^2} + \cdots$ **convergent**
28. $1 + \frac{1}{\sqrt{2}} + \frac{1}{\sqrt{3}} + \frac{1}{\sqrt{4}} + \cdots$ **divergent**

29. $\frac{4\pi}{3} + \frac{5\pi}{6} + \frac{\pi}{3} + \cdots$ **divergent**
30. $\frac{1}{4} + \frac{3}{8} + \frac{5}{16} + \frac{7}{32} + \cdots$ **convergent**

Applications and Problem Solving

31. **Economics**   The MagicSoft software company has a proposal to the city council of Alva, Florida, to relocate there. The proposal claims that the company will generate $3.3 million for the local economy by the $1 million in salaries that will be paid. The city council estimates that 70% of the salaries will be spent in the local community, and 70% of that money will again be spent in the community, and so on.

   a. According to the city council's estimates, is the claim made by MagicSoft accurate? Explain.   **No; see margin for explanation.**

   b. What is the correct estimate of the amount generated to the local economy?   **$2.3 million**

32. **Critical Thinking**   Give an example of a series $a_1 + a_2 + a_3 + \cdots + a_n + \cdots$ that diverges, but when its terms are squared, the resulting series $a_1{}^2 + a_2{}^2 + a_3{}^2 + \cdots + a_n{}^2 + \cdots$ converges.   **the harmonic series: $1 + \frac{1}{2} + \frac{1}{3} + \frac{1}{4} + \cdots$**

33. **Cellular Growth**   Leticia Cox is a biochemist. She is testing two different types of drugs that induce cell growth. She has selected two cultures of 1000 cells each. To culture A, she administers a drug that raises the number of cells by 200 each day and every day thereafter. Culture B gets a drug that increases cell growth by 8% each day and everyday thereafter.

   a. Assuming no cells die, how many cells will have grown in each culture by the end of the seventh day?   **Culture A: 1400 cells, Culture B: 713 cells**

   b. At the end of one month's time, which drug will prove to be more effective in promoting cell growth? Explain.   **Culture B, see margin for explanation.**

34. **Critical Thinking**   Refer to Example 6 of this lesson. The sequence of partial sums, $S_1, S_2, S_3, \ldots, S_n, \ldots$, for the time series is $1, \frac{3}{2}, \frac{7}{4}, \frac{15}{8}, \ldots$.

   a. Find a general expression for the nth term of this sequence. $\frac{2^n - 1}{2^{n-1}}$

   b. To determine how long it takes for Achilles to catch-up to the tortoise, find the sum of the infinite time series. (*Hint:* Recall from the definition of the sum S of an infinite series that $\lim_{n \to \infty} S_n = S$.)   **2 seconds**

35. Clocks The hour and minute hands of a clock travel around its face at different speeds, but at certain times of the day, the two hands coincide. In addition to noon and midnight, the hands also coincide at times occurring between the hours. According the figure at the right, it is 4:00.

35a. $\dfrac{1}{3}$

    a. When the minute hand points to 4, what fraction of the distance between 4 and 5 will the hour hand have traveled?

35b. $\dfrac{1}{36}$

    b. When the minute hand reaches the hour hand's new position, what additional fraction will the hour hand have traveled?

35c. $\dfrac{1}{432}, \dfrac{1}{5184}$

    c. List the next two terms of this series representing the distance traveled by the hour hand as the minute hand "chases" its position.

    d. At what time between the hours of 4 and 5 o'clock will the two hands coincide? **at $4 + \dfrac{4}{11}$ o'clock, approximately 21 min 49 s after 4:00**

Mixed Review

37. $16\sqrt{2}$

36. Evaluate $\lim\limits_{n\to\infty} \dfrac{4n^2 + 5}{3n^2 - 2n}$. *(Lesson 12-3)* $\dfrac{4}{3}$

37. Find the ninth term of the geometric sequence $\sqrt{2}, 2, 2\sqrt{2}, \ldots$. *(Lesson 12-2)*

38. Form an arithmetic sequence that has five arithmetic means between -11 and 19. *(Lesson 12-1)*   **$-11, -6, -1, 4, 9, 14, 19$**

39. Solve $45.9 = e^{0.075t}$ *(Lesson 11-6)*   **51.02**

40. Navigation A submarine sonar is tracking a ship. The path of the ship is recorded as $6 = 12r \cos(\theta - 30°)$. Find the linear equation of the path of the ship. *(Lesson 9-4)* $\sqrt{3}x + y - 1 = 0$

41. $\langle -3, 2 \rangle$

41. Find an ordered pair that represents $\overrightarrow{AB}$ for $A(8, -3)$ and $B(5, -1)$. *(Lesson 8-2)*

42. SAT/ACT Practice How many numbers from 1 to 200 inclusive are equal to the cube of an integer? **E**

    A one   **B** two   **C** three   **D** four   **E** five

---

MID-CHAPTER QUIZ

1. Find the 19th term in the sequence for which $a_1 = 11$ and $d = -2$. (Lesson 12-1) **-25**

2. Find S_{20} for the arithmetic series for which $a_1 = -14$ and $d = 6$. (Lesson 12-1) **860**

3. Form a sequence that has two geometric means between 56 and 189. (Lesson 12-2)

4. Find the sum of the first eight terms of the series $3 - 6 + 12 - \cdots$. (Lesson 12-2) **-255**

5. Find $\lim\limits_{n\to\infty} \dfrac{n^2 + 2n - 5}{n^2 - 1}$ or explain why the limit does not exist. (Lesson 12-3) **1**

6. Recreation A bungee jumper rebounds 55% of the height jumped. If a bungee jump is made using a cord that stretches 250 feet, find the total distance traveled by the jumper before coming to rest. (Lesson 12-3) **about 861 feet**

7. Find the sum of the following series. $\dfrac{1}{25} + \dfrac{1}{250} + \dfrac{1}{2500} + \cdots$. (Lesson 12-3) $\dfrac{2}{45}$

3. 56, 84, 126, 189

Determine whether each series is *convergent* or *divergent*. (Lesson 12-4)

8. $\dfrac{1}{10} + \dfrac{2}{100} + \dfrac{6}{1000} + \dfrac{24}{10,000} + \cdots$ **divergent**

9. $\dfrac{6}{5} + \dfrac{2}{5} + \dfrac{2}{15} + \cdots$ **convergent**

10. Finance Ms. Fuentes invests $500 quarterly (January 1, April 1, July 1, and October 1) in a retirement account that pays an APR of 12% compounded quarterly. Interest for each quarter is posted on the last day of the quarter. Determine the value of her investment at the end of the year. (Lesson 12-2) **$2154.57**

Extra Practice See p. A49.

Lesson 12-4 Convergent and Divergent Series   **793**

Extra Credit

Suppose that an infinite geometric series with positive terms converges to a sum, S. What is the sum of the even-numbered terms in terms of r, the common ratio? $\dfrac{r}{1+r} S$

---

4 ASSESS

Open-Ended Assessment

Speaking Ask students to describe the different tests for determining whether a series is convergent or divergent.

Mid-Chapter Quiz

The Mid-Chapter Quiz provides students with a brief review of the concepts and skills in Lessons 12-1 through 12-4. Lesson numbers are given to the right of exercises or instruction lines so students can review concepts not yet mastered.

Enrichment Masters, p. 136

Sigma Notation and the *n*th Term

OBJECTIVE
* Use sigma notation.

MANUFACTURING
Manufacturers are required by the Environmental Protection Agency to meet certain emission standards. If these standards are not met by a preassigned date, the manufacturer is fined. To encourage swift compliance, the fine increases a specified amount each day until the manufacturer is able to pass inspection. Suppose a manufacturing plant is charged $2000 for not meeting its January 1st deadline. The next day it is charged $2500, the next day $3000, and so on, until it passes inspection on January 21st. What is the total amount of fines owed by the manufacturing plant? *This problem will be solved in Example 2.*

In mathematics, the uppercase Greek letter sigma, Σ, is often used to indicate a sum or series. A series like the one indicated above may be written using **sigma notation**.

Other variables besides n may be used for the index of summation.

$$\text{maximum value of } n \longrightarrow \sum_{n=1}^{k} a_n \longleftarrow \text{expression for general term} \\ \text{starting value of } n \nearrow$$

The variable *n* used with the sigma notation is called the **index of summation**.

Sigma Notation of a Series

For any sequence $a_1, a_2, a_3, \ldots$, the sum of the first *k* terms may be written $\displaystyle\sum_{n=1}^{k} a_n$, which is read "the summation from $n = 1$ to k of a_n." Thus,

$$\sum_{n=1}^{k} a_n = a_1 + a_2 + a_3 + \cdots + a_k, \text{ where } k \text{ is an integer value.}$$

Example ❶ Write each expression in expanded form and then find the sum.

a. $\displaystyle\sum_{n=1}^{4} (n^2 - 3)$

First, write the expression in expanded form.

$$\sum_{n=1}^{4} (n^2 - 3) = \overset{n=1}{(1^2 - 3)} + \overset{n=2}{(2^2 - 3)} + \overset{n=3}{(3^2 - 3)} + \overset{n=4}{(4^2 - 3)}$$

Now find the sum.

Method 1 Simplify the expanded form.
$$(1^2 - 3) + (2^2 - 3) + (3^2 - 3) + (4^2 - 3) = -2 + 1 + 6 + 13 \text{ or } 18$$

Method 2 Use a graphing calculator.

You can combine **sum(** with **seq(** to obtain the sum of a finite sequence. In the **LIST** menu, select the **MATH** option to locate the **sum(** command. The **seq(** command can be found in the **OPS** option of the **LIST** menu.

$$\underbrace{\text{sum(seq(}}\ \overset{expression}{\underbrace{N^2 - 3}},\ \overset{\substack{starting \\ value}}{\underset{\substack{index\ of \\ summation}}{N}}\ ,\ \overset{}{\underset{}{1}}\ ,\ \overset{step\ size}{\underset{\substack{maximum \\ value}}{4}}\ ,\ \overset{}{1}\ ))$$

```
sum(seq(N²-3,N,1
,4,1))
                18
```

b. $\displaystyle\sum_{n=1}^{\infty} 5\left(-\frac{2}{7}\right)^{n-1}$

$$\sum_{n=1}^{\infty} 5\left(-\frac{2}{7}\right)^{n-1} = \overset{n=1}{5\left(-\frac{2}{7}\right)^{1-1}} + \overset{n=2}{5\left(-\frac{2}{7}\right)^{2-1}} + \overset{n=3}{5\left(-\frac{2}{7}\right)^{3-1}} + \overset{n=4}{5\left(-\frac{2}{7}\right)^{4-1}} + \cdots$$

$$= \quad 5 \quad + \quad \frac{10}{7} \quad + \quad \frac{20}{49} \quad + \quad \frac{40}{343} \quad + \cdots$$

This is an infinite geometric series. Use the formula $S = \dfrac{a_1}{1 - r}$.

$$S = \frac{5}{1 - \left(-\frac{2}{7}\right)} \qquad a_1 = 5,\ r = -\frac{2}{7}$$

$$S = \frac{35}{9}$$

Therefore, $\displaystyle\sum_{n=1}^{\infty} 5\left(-\frac{2}{7}\right)^{n-1} = \frac{35}{9}.$

A series in expanded form can be written using sigma notation if a general formula can be written for the nth term of the series.

Example ❷ **MANUFACTURING** Refer to the application at the beginning of this lesson.

    a. How much is the company fined on the 20th day?

    b. What is the total amount in fines owed by the manufacturing plant?

    c. Represent this sum using sigma notation.

    a. Since this sequence is arithmetic, we can use the formula for the nth term of an arithmetic sequence to find the amount of the fine charged on the 20th day.

$$a_n = a_1 + (n - 1)d$$
$$a_{20} = 2000 + (20 - 1)500 \quad a_1 = 2000,\ n = 20,\ and\ d = 500$$
$$= 11,500$$

The fine on the 20th day will be $11,500.

Teaching Tip Before discussing sigma notation, explain that $\sum$, the Greek capital sigma, is used like the English *S* and this may be the reason why it was chosen to represent a sum or series.

Teaching Tip Before discussing the examples, point out that the number of terms to be added is the maximum value of the index of summation k minus the starting value of the index, x, plus one, or $k - x + 1$. For example, in part **a** of In-Class Example 1, $7 - 4 + 1$, or 4, terms will be added.

In-Class Examples

Example 1
Write each expression in expanded form and then find the sum.

a. $\displaystyle\sum_{b=4}^{7} (3^b + 1)$ **3244**

b. $\displaystyle\sum_{n=1}^{\infty} 2\left(\frac{1}{3}\right)^n$ **3**

Example 2
Fundraising During a nine-hole charity golf match, one player presents the following proposition: The loser of the first hole will pay $1 to charity, and each succeeding hole will be worth twice as much as the hole immediately preceding it.

a. How much would a losing player pay on the 4th hole? **$8**

b. How much would a player pay if he or she lost all nine holes? **$511**

c. Represent this sum using sigma notation. $\displaystyle\sum_{n=1}^{9} 2^{n-1}$

In-Class Example

Example 3

Express the series $16 + 19 + 22 + 25 + \cdots + 61$ using sigma notation. $\displaystyle\sum_{n=5}^{20} (3n + 1)$

---

b. To determine the total amount owed in fines, we can use the formula for the sum of an arithmetic series. The plant will not be charged for the day it passes inspection, so it is assessed 20 days in fines.

$$S_n = \frac{n}{2}(a_1 + a_n)$$
$$= \frac{20}{2}(2000 + 11{,}500) \quad n = 20,\ a_1 = 2000,\ and\ a_{20} = 11{,}500$$
$$= 135{,}000$$

The plant must pay a total of \$135,000 in fines.

c. To determine the fine for the nth day, we can again use the formula for the nth term of an arithmetic sequence.

$$a_n = a_1 + (n - 1)d$$
$$= 2000 + (n - 1)500 \quad a_1 = 2000\ and\ d = 500$$
$$= 2000 + 500n - 500$$
$$= 500n + 1500$$

The fine on the nth day is $500n + \$1500$.

Since \$2000 is the fine on the first day and \$12,000 is the fine on the 20th day, the index of summation goes from $n = 1$ to $n = 20$.

Therefore, $2000 + 2500 + 3000 + \cdots + 11{,}500 = \displaystyle\sum_{n=1}^{20} (500n + 1500)$ or $\displaystyle\sum_{n=1}^{20} 500(n + 3)$.

When using sigma notation, it is not always necessary that the sum start with the index equal to 1.

Example ③ **Express the series $15 + 24 + 35 + 48 + \cdots + 143$ using sigma notation.**

Notice that each term is 1 less than a perfect square. Thus, the nth term of the series is $n^2 - 1$. Since $4^2 - 1 = 15$ and $12^2 - 1 = 143$, the index of summation goes from $n = 4$ to $n = 12$.

Therefore, $15 + 24 + 35 + 48 + \cdots + 143 = \displaystyle\sum_{n=4}^{12} (n^2 - 1)$.

As you have seen, not all sequences are arithmetic or geometric. Some important sequences are generated by products of consecutive integers. The product $n(n - 1)(n - 2) \cdots 3 \cdot 2 \cdot 1$ is called ***n* factorial** and symbolized $n!$.

---

n Factorial

The expression $n!$ (*n* factorial) is defined as follows for n, an integer greater than zero.

$$n! = n(n - 1)(n - 2) \cdots 1$$

By definition $0! = 1$.

Addressing Individual Needs

 Interpersonal Have students work in groups to write sample series in expanded form and using sigma notation. Have them also give examples of series that cannot be expressed in sigma notation.

Factorial notation can be used to express the general form of a series.

Example **4** Express the series $\frac{2}{2} - \frac{4}{6} + \frac{6}{24} - \frac{8}{120} + \frac{10}{720}$ using sigma notation.

The sequence representing the numerators is 2, 4, 6, 8, 10. This is an arithmetic sequence with a common difference of 2. Thus the nth term can be represented by $2n$.

Because the series has alternating signs, one factor for the general term of the series is $(-1)^{n+1}$. Thus, when n is odd, the terms are positive, and when n is even, the terms are negative.

The sequence representing the denominators is 2, 6, 24, 120, 720. This sequence is generated by factorials.

$2! = 2$
$3! = 6$
$4! = 24$
$5! - 120$
$6! = 720$

Therefore, $\frac{2}{2} - \frac{4}{6} + \frac{6}{24} - \frac{8}{120} + \frac{10}{720} = \sum_{n=1}^{5} \frac{(-1)^{n+1} 2n}{(n+1)!}$.

You can check this answer by substituting values of n into the general term.

In-Class Example
Example 4
Express the series $-\frac{4}{1} + \frac{16}{2} - \frac{64}{6} + \frac{256}{24}$ using sigma notation.
$$\sum_{n=1}^{4} \frac{(-1)^n 4^n}{n!}$$

3 PRACTICE/APPLY

Error Analysis

Watch for students who have difficulty understanding the meaning of $(-1)^n$.

Prevent by having students find the first five terms of the sequence to see the pattern.

Additional Answers
4. $(-2) + (-1) + 0 + 1 + 2 + 3$

5. $8 + 12 + 16 + 20$

6. $1 + \frac{1}{2} + \frac{1}{4} + \frac{1}{8} + \frac{1}{16}$

7. $5 + \frac{15}{4} + \frac{45}{16} + \frac{135}{64} + \cdots$

CHECK FOR UNDERSTANDING

Communicating Mathematics

Read and study the lesson to answer each question.

1. **Find a counterexample** to the following statement: "The summation notion used to represent a series is unique."

2. In Example 4 of this lesson, the alternating signs of the series were represented by a factor of $(-1)^{n+1}$.

   a. **Write** a different factor that could have been used in the general form of the nth term of the series. **Sample answer:** $(-1)^{n-1}$

   b. **Determine** a factor that could be used if the alternating signs of the series began with a negative first term. **Sample answer:** $(-1)^n$

3. Consider the series $\sum_{j=2}^{10} (-2j + 1)$.

   a. **Identify** the number of terms in this series. **9**

   b. **Write** a formula that determines the number of terms t in a finite series if the index of summation has a minimum value of a and a maximum value of b. $t = b - a + 1$

   c. **Use** the formula in part **b** to identify the number of terms in the series $\sum_{k=-2}^{3} \frac{1}{k+3}$. **6**

   d. **Verify** your answer in part **c** by writing the series $\sum_{k=-2}^{3} \frac{1}{k+3}$ in expanded form.
   $1 + \frac{1}{2} + \frac{1}{3} + \frac{1}{4} + \frac{1}{5} + \frac{1}{6}$

Guided Practice **Write each expression in expanded form and then find the sum.**

4. $\sum_{n=1}^{6} (n - 3)$ **3**   5. $\sum_{k=2}^{5} 4k$ **56**   6. $\sum_{a=0}^{4} \frac{1}{2^a}$ $1\frac{15}{16}$ 7. $\sum_{p=0}^{\infty} 5\left(\frac{3}{4}\right)^p$ **20**

4–7. **See margin for expanded form.**

Communicating Mathematics

1. Sample answer: The series $4 + 6 + 8 + 10 + 12$ can be represented by $\sum_{n=0}^{4} 2n + 4$ or by $\sum_{n=1}^{5} 2n + 2$.

Study Guide and Practice Masters, pp. 223–224

NAME _____ DATE _____ PERIOD _____

12-5 Study Guide

Sigma Notation and the nth Term

A series may be written using **sigma notation.**

maximum value of $n \to \sum_{i=1}^{k} a_n \leftarrow$ expression for general term
starting value of $n \to$
index of summation

Example 1 Write each expression in expanded form and then find the sum.

a. $\sum_{n=1}^{5} (n + 2)$

First, write the expression in expanded form.
$\sum_{n=1}^{5} (n + 2) = (1 + 2) + (2 + 2) + (3 + 2) + (4 + 2) + (5 + 2)$

Then, find the sum by simplifying the expanded form. $3 + 4 + 5 + 6 + 7 = 25$

b. $\sum_{n=1}^{\infty} 2\left(\frac{1}{4}\right)^n$

$\sum_{n=1}^{\infty} 2\left(\frac{1}{4}\right)^n = 2\left(\frac{1}{4}\right)^1 + 2\left(\frac{1}{4}\right)^2 + 2\left(\frac{1}{4}\right)^3 + \cdots$
$= \frac{1}{2} + \frac{1}{8} + \frac{1}{32} + \cdots$

This is an infinite series. Use the formula $S = \frac{a_1}{1-r}$.

$S = \frac{\frac{1}{2}}{1 - \frac{1}{4}}$ $a_1 = \frac{1}{2}, r = \frac{1}{4}$

$S = \frac{2}{3}$

Example 2 Express the series $26 + 37 + 50 + 65 + \cdots + 170$ using sigma notation.

Notice that each term is one more than a perfect square. Thus, the nth term of the series is $n^2 + 1$. Since $5^2 + 1 = 26$ and $13^2 + 1 = 170$, the index of summation goes from $n = 5$ to $n = 13$.

Therefore, $26 + 37 + 50 + 65 + \cdots + 170 = \sum_{n=5}^{13} (n^2 + 1)$.

© Glencoe/McGraw-Hill   223   Advanced Mathematical Concepts

Teaching Tip For Exercise 45, an alternate method of finding the sum of the series would be to rewrite $\dfrac{8n^3 - 2n^2 + 5}{n^4}$ as $\dfrac{8}{n} - \dfrac{2}{n^2} + \dfrac{5}{n^4}$ and find each separate summation.

Additional Answers

8. $\displaystyle\sum_{n=1}^{5} 5n$

9. $\displaystyle\sum_{n=0}^{3} (3^k + 1)$

10. $\displaystyle\sum_{n=1}^{4} (8 - 6n)$

11. $\displaystyle\sum_{n=2}^{\infty} 3\left(\dfrac{1}{2}\right)^n$

12. $\displaystyle\sum_{n=1}^{\infty} (-3)^n$

14. $(-5) + (-3) + (-1) + 1$

15. $10 + 15 + 20 + 25$

16. $(-6) + (-10) + (-14) + (-18) + (-22) + (-26)$

17. $6 + 12 + 20 + 30 + 42$

18. $5 + 3 + \dfrac{7}{3} + 2$

19. $16 + 32 + 64 + 128 + 256$

20. $\dfrac{1}{3} + 1 + 3 + 9$

21. $4\dfrac{1}{2} + 16\dfrac{1}{2} + 64\dfrac{1}{2}$

22. $8 + 16 + 32$

23. $6 + 24 + 120 + 720 + 5040$

24. $4 + 3 + 2.25 + 1.6875 + \cdots$

25. $\dfrac{8}{5} + \dfrac{16}{25} + \dfrac{32}{125} + \cdots$

27. $\displaystyle\sum_{k=1}^{4} (3k + 3)$

28. $\displaystyle\sum_{k=0}^{4} 4^k$

29. $\displaystyle\sum_{k=4}^{12} 2k$

30. $\displaystyle\sum_{k=0}^{3} (-2)^{3-k}$

31. $\displaystyle\sum_{k=1}^{4} 2 \cdot 5^k$

32. $\displaystyle\sum_{k=0}^{3} (13 - 4k)$

Express each series using sigma notation. **8–12. See margin.**

8. $5 + 10 + 15 + 20 + 25$

9. $2 + 4 + 10 + 28$

10. $2 - 4 - 10 - 16$

11. $\dfrac{3}{4} + \dfrac{3}{8} + \dfrac{3}{16} + \dfrac{3}{32} + \cdots$

12. $-3 + 9 - 27 + \cdots$

13. Aviation Each October Albuquerque, New Mexico, hosts the Balloon Fiesta. In 1998, 873 hot air balloons participated in the opening day festivities. One of these balloons rose 389 feet after 1 minute. Because the air in the balloon was not reheated, each succeeding minute the balloon rose 63% as far as it did the previous minute. **b. about 1051 ft**

a. Use sigma notation to represent the height of the balloon above the ground after one hour. Then calculate the total height of the balloon after one hour to the nearest foot.

b. What was the maximum height achieved by this balloon?

13a. $\displaystyle\sum_{n=1}^{60} 389(0.63)^{n-1}$; about 1051 ft

EXERCISES

Practice

14–25. See margin for expanded form.

Write each expression in expanded form and then find the sum.

A 14. $\displaystyle\sum_{n=1}^{4} (2n - 7)$ **−8**

15. $\displaystyle\sum_{a=2}^{5} 5a$ **70**

16. $\displaystyle\sum_{b=3}^{8} (6 - 4b)$ **−96**

17. $\displaystyle\sum_{k=2}^{6} (k + k^2)$ **110**

18. $\displaystyle\sum_{n=5}^{8} \dfrac{n}{n-4}$ **$12\dfrac{1}{3}$**

19. $\displaystyle\sum_{j=4}^{8} 2^j$ **496**

20. $\displaystyle\sum_{m=0}^{3} 3^{m-1}$ **$13\dfrac{1}{3}$**

21. $\displaystyle\sum_{r=1}^{3} \left(\dfrac{1}{2} + 4^r\right)$ **$85\dfrac{1}{2}$**

22. $\displaystyle\sum_{i=3}^{5} (0.5)^{-i}$ **56**

23. $\displaystyle\sum_{k=3}^{7} k!$ **5910**

24. $\displaystyle\sum_{p=0}^{\infty} 4(0.75)^p$ **16**

25. $\displaystyle\sum_{n=1}^{\infty} 4\left(\dfrac{2}{5}\right)^n$ **$2\dfrac{2}{3}$**

B 26. Write $\displaystyle\sum_{n=2}^{5} n + i^n$ in expanded form. Then find the sum.

$(2 + i^2) + (3 + i^3) + (4 + i^4) + (5 + i^5)$; 14

Express each series using sigma notation. **27–40. See margin.**

27. $6 + 9 + 12 + 15$

28. $1 + 4 + 16 + \cdots + 256$

29. $8 + 10 + 12 + \cdots + 24$

30. $-8 + 4 - 2 + 1$

31. $10 + 50 + 250 + 1250$

32. $13 + 9 + 5 + 1$

33. $\dfrac{1}{9} + \dfrac{1}{14} + \dfrac{1}{19} + \cdots + \dfrac{1}{49}$

34. $\dfrac{2}{3} + \dfrac{4}{5} + \dfrac{8}{7} + \dfrac{16}{9} + \cdots$

35. $4 - 9 + 16 - 25 + \cdots$

36. $5 + 5 + \dfrac{5}{2} + \dfrac{5}{6} + \dfrac{5}{24} + \cdots$

37. $-32 + 16 - 8 + 4 - \cdots$

38. $2 + \dfrac{6}{2} + \dfrac{24}{3} + \dfrac{120}{4} + \cdots$

C 39. $\dfrac{1}{5} + \dfrac{2}{7} + \dfrac{3}{11} + \dfrac{4}{19} + \dfrac{5}{35} + \cdots$

40. $\dfrac{3}{9 \cdot 2} + \dfrac{8}{27 \cdot 6} + \dfrac{15}{81 \cdot 24} + \cdots$

41. $\displaystyle\sum_{k=1}^{\infty} \dfrac{2^{\frac{k}{2}}}{3k!}$

41. Express the series $\dfrac{\sqrt{2}}{3} + \dfrac{2}{6} + \dfrac{\sqrt{8}}{18} + \dfrac{4}{72} + \dfrac{\sqrt{32}}{360} + \cdots$ using sigma notation.

43. $\dfrac{a(a+1)}{(a-1)}$

Simplify. Assume that n and m are positive integers, $a > b$, and $a > 2$.

42. $\dfrac{(a-2)!}{a!}$ — $\dfrac{1}{a(a-1)}$

43. $\dfrac{(a+1)!}{(a-2)!}$

44. $\dfrac{(a+b)!}{(a+b-1)!}$ **$a + b$**

Graphing Calculator 45. Use a graphing calculator to find the sum of $\displaystyle\sum_{n=1}^{100} \dfrac{8n^3 - 2n^2 + 5}{n^4}$. Round to the nearest hundredth. **43.64**

www.amc.glencoe.com/self_check_quiz

Additional Answers

33. $\displaystyle\sum_{k=2}^{10} \dfrac{1}{5k-1}$

37. $\displaystyle\sum_{n=0}^{\infty} \left[(-1)^{n+1} \dfrac{32}{2^n}\right]$

34. $\displaystyle\sum_{k=1}^{\infty} \dfrac{2^k}{2k+1}$

38. $\displaystyle\sum_{k=1}^{\infty} \dfrac{(k+1)!}{k}$

35. $\displaystyle\sum_{k=2}^{\infty} (-1)^k k^2$

39. $\displaystyle\sum_{k=1}^{\infty} \dfrac{k}{2^k + 3}$

36. $\displaystyle\sum_{k=0}^{\infty} \dfrac{5}{k!}$

40. $\displaystyle\sum_{k=2}^{\infty} \dfrac{k^2 - 1}{3^k k!}$

46. Advertising A popular shoe manufacturer is planning to market a new style of tennis shoe in a city of 500,000 people. Using a prominent professional athlete as their spokesperson, the company's ad agency hopes to induce 35% of the people to buy the product. The ad agency estimates that these satisfied customers will convince 35% of 35% of 500,000 to buy a pair of shoes, and those will persuade 35% of 35% of 35% of 500,000, and so on. $\sum_{n=1}^{\infty} 500,000(0.35)^n$

a. Model this situation using sigma notation.

b. Find the total number of people that will buy the product as a result of the advertising campaign. **269,230 people**

c. What percentage of the population is this? **about 53.8%**

d. What important assumption does the advertising agency make in proposing the figure found in part **b** to the shoe manufacturer?

46d. The ad agency assumes that the people who buy the tennis shoes will be satisfied with their purchase.

47. Critical Thinking Solve each equation for x.

a. $\sum_{n=1}^{6} (x - 3n) = -3$ **10**

b. $\sum_{n=0}^{5} n(n - x) = 25$ **2**

48. Critical Thinking Determine whether each equation is *true* or *false*. Explain your answer. **See margin for explanations.**

a. $\sum_{k=3}^{7} 3^k + \sum_{b=7}^{9} 3^b = \sum_{a=3}^{9} 3^a$ **false**

b. $\sum_{n=2}^{8} (2n - 3) = \sum_{m=3}^{9} (2m - 5)$ **true**

c. $2\sum_{n=3}^{7} n^2 = \sum_{n=3}^{7} 2n^2$ **true**

d. $\sum_{k=1}^{10} (5 + k) = \sum_{p=0}^{9} (4 + p)$ **false**

49. Word Play An *anagram* is a word or phrase that is made by rearranging the letters of another word or phrase. Consider the word "SILENT."

a. How many different arrangements of the letters in this word are possible? Write this number as a factorial. (*Hint:* First solve a simpler problem to see a pattern, such as how many different arrangements are there of just 2 letters? 3 letters?) **6!**

b. If a friend gives you a hint and tells you that an anagram of this word starts with "L," how many different arrangements still remain? **120**

c. Your friend gives you one more hint. The last letter in the anagram is "N." Determine how many possible arrangements remain and then determine the anagram your friend is suggesting. **24, "LISTEN"**

50b.
9-6 × 6,
16-5 × 5,
25-4 × 4,
36-3 × 3,
49-2 × 2, and
64-1 × 1

50. Chess A standard chess board contains 64 small black or white squares. These squares make up many other larger squares of various sizes.

a. How many 8 × 8 squares are there on a standard 8 × 8 chessboard? How many 7 × 7 squares? **1-8 × 8, 4-7 × 7**

b. Continue this list until you have accounted for all 8 sizes of squares.

c. Use sigma notation to represent the total number of squares found on an 8 × 8 chessboard. Then calculate this sum. $\sum_{n=1}^{8} n^2$; **204**

Mixed Review

51. Use the comparison test to determine whether the series $\frac{3}{3} + \frac{3}{4} + \frac{3}{5} + \frac{3}{6} + \cdots$ is convergent or divergent. (*Lesson 12-4*) **divergent**

52. Chemistry A vacuum pump removes 20% of the air in a sealed jar on each stroke of its piston. The jar contains 21 liters of air before the pump starts. After how many strokes will only 42% of the air remain? (*Lesson 12-3*) **4 strokes**

Extra Practice See p. A49.

Additional Answers

48a. $\sum_{k=3}^{7} 3^k = 3^3 + 3^4 + \cdots + 3^7$

$\sum_{b=7}^{9} 3^b = 3^7 + 3^8 + 3^9$

Since there are two 3^7 terms,

$\sum_{k=3}^{7} 3^k + \sum_{b=7}^{9} 3^b \neq \sum_{a=3}^{9} 3^a$.

48b. $\sum_{n=2}^{8} (2n - 3) = 1 + 3 + \cdots + 13 = 49$

$\sum_{m=3}^{9} (2m - 5) = 1 + 3 + \cdots + 13 = 49$

Since $49 = 49$, $\sum_{n=2}^{8} (2n - 3) = \sum_{m=3}^{9} (2m - 5)$.

48c. $2\sum_{n=3}^{7} n^2 = 18 + 32 + \cdots + 98 = 270$

$\sum_{n=3}^{7} 2n^2 = 18 + 32 + \cdots + 98 = 270$

Since $270 = 270$, $2\sum_{n=3}^{7} n^2 = \sum_{n=3}^{7} 2n^2$.

48d. $\sum_{k=1}^{10} (5 + n) = 6 + 7 + \cdots + 15 = 105$

$\sum_{p=0}^{9} (4 + p) = 4 + 5 + \cdots + 13 = 85$

Since $85 \neq 105$, $\sum_{k=1}^{10} (5 + n) \neq \sum_{p=0}^{9} (4 + p)$.

Extra Credit

Use sigma notation to express the total number of unit cubes needed to form a series of ever-larger cubes from 1 unit on each edge to n units on each edge.

$$\sum_{b=1}^{n} b^3$$

Open-Ended Assessment

Writing Have students write a paragraph explaining how to express a series that has a general formula for the nth term using sigma notation.

 Chapter 12, Quiz B (Lessons 12-4 and 12-5) is available in the *Assessment and Evaluation Masters,* p. 303.

 Chapter 12 Mid-Chapter Test (Lessons 12-1 through 12-5) is available in the *Assessment and Evaluation Masters,* p. 302.

Additional Answer

58. $\left(2\sqrt{17}+\sqrt{13}\right)x +$
$\left(4\sqrt{13}-3\sqrt{17}\right)y +$
$9\sqrt{17}+4\sqrt{13}=0$

Enrichment Masters, p. 137

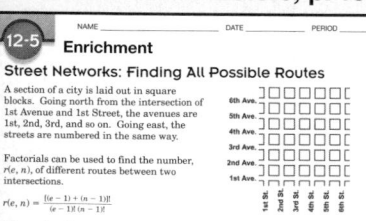

---

53. Find the first four terms of the geometric sequence for which $a_5 = 32\sqrt{2}$ and $r = -\sqrt{2}$. *(Lesson 12-2)* **$8\sqrt{2}, -16, 16\sqrt{2}, -32$**

54. Evaluate $\log_{10}0.001$. *(Lesson 11-4)* **-3**

55. Write the standard form of the equation of the circle that passes through points at $(0, 9)$, $(-7, 2)$, and $(0, -5)$. *(Lesson 10-2)* **$x^2 + (y - 2)^2 = 49$**

56. Simplify $\left(\sqrt{2} + i\right)\left(4\sqrt{2} + i\right)$. *(Lesson 9-5)* **$7 + 5i\sqrt{2}$**

57. **Sports** Find the initial vertical and horizontal velocities of a javelin thrown with an initial velocity of 59 feet per second at an angle of $63°$ with the horizontal. *(Lesson 8-7)* **52.57 ft/s, 26.79 ft/s**

58. Find the equation of the line that bisects the obtuse angle formed by the graphs of $2x - 3y + 9 = 0$ and $x + 4y + 4 = 0$. *(Lesson 7-7)* **See margin.**

59. **SAT/ACT Practice** If $\dfrac{5 + m}{9 + m} = \dfrac{2}{3}$, then $m = ?$ **D**

 A 8 B 6 C 5 D 3 E 2

CAREER CHOICES

 Operations Research Analyst

In the changing economy of today, it is difficult to start and maintain a successful business. A business operator needs to be sure that the income from the business exceeds the expenses. Sometimes, businesses need the services of an operations research analyst. If you enjoy mathematics and solving tough problems, then you may want to consider a career as an operations research analyst.

In this occupation, you would gather many types of data about a business and analyze that data using mathematics and statistics. Examples of ways you might assist a business are: help a retail store determine the best store layout, help a bank in processing deposits more efficiently, or help a business set prices. Most operations research analysts work for private industry, private consulting firms, or the government.

CAREER OVERVIEW

Degree Preferred:
bachelor's degree in applied mathematics

Related Courses:
mathematics, statistics, computer science, English

Outlook:
faster than average through the year 2006

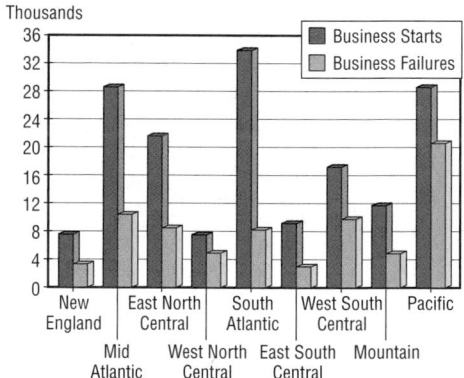

1997 Business Starts and Failures

interNET CONNECTION For more information on careers in operations research, visit: **www.amc.glencoe.com**

CAREER CHOICES | **OPERATIONS RESEARCH ANALYST**

Math Connection
- Find the approximate number of business failures in the United States in 1997. **about 83,000**
- Estimate the percent of the total failures for each of the regions shown. **Estimates will vary. Sample: New England, 4%; Middle Atlantic, 13%; East North Central, 12%; West North Central, 6%; South Atlantic, 10%; East South Central, 4%; West South Central, 12%; Mountain, 9%; Pacific, 30%**

The Binomial Theorem

OBJECTIVE
* Use the Binomial Theorem to expand binomials.

FAMILY On November 19, 1997, Bobbie McCaughey, an Iowa seamstress, gave birth by Caesarian section to seven babies. The birth of the septuplets was the first of its kind in the United States since 1985. The babies, born after just 30 weeks of pregnancy, weighed from 2 pounds 5 ounces to 3 pounds 4 ounces. *A problem related to this will be solved in Example 2.*

Recall that a binomial, such as $x + y$, is an algebraic expression involving the sum of two unlike terms, in this case x and y. Just as there are patterns in sequences and series, there are numerical patterns in the expansion of powers of binomials. Let's examine the expansion of $(x + y)^n$ for $n = 0$ to $n = 5$. You already know a few of these expansions and the others can be obtained using algebraic properties.

$$(x + y)^0 = 1x^0y^0$$
$$(x + y)^1 = 1x^1y^0 + 1x^0y^1$$
$$(x + y)^2 = 1x^2y^0 + 2x^1y^1 + 1x^0y^2$$
$$(x + y)^3 = 1x^3y^0 + 3x^2y^1 + 3x^1y^2 + 1x^0y^3$$
$$(x + y)^4 = 1x^4y^0 + 4x^3y^1 + 6x^2y^2 + 4x^1y^3 + 1x^0y^4$$
$$(x + y)^5 = 1x^5y^0 + 5x^4y^1 + 10x^3y^2 + 10x^2y^3 + 5x^1y^4 + 1x^0y^5$$

Patterns in the Binomial Expansion of $(x + y)^n$

The following patterns can be observed in the expansion of $(x + y)^n$.
1. The expansion of $(x + y)^n$ has $n + 1$ terms.
2. The first term is x^n, and the last term is y^n.
3. In successive terms, the exponent of x decreases by 1, and the exponent of y increases by 1.
4. The degree of each term (the sum of the exponents of the variables) is n.
5. In any term, if the coefficient is multiplied by the exponent of x and the product is divided by the number of that term, the result is the coefficient of the following term.
6. The coefficients are symmetric. That is, the first term and the last term have the same coefficient. The second term and the second from the last term have the same coefficient, and so on.

If just the coefficients of these expansions are extracted and arranged in a triangular array, they form a pattern called **Pascal's triangle.** As you examine the triangle shown below, note that if two consecutive numbers in any row are added, the sum is a number in the following row. These three numbers also form a triangle.

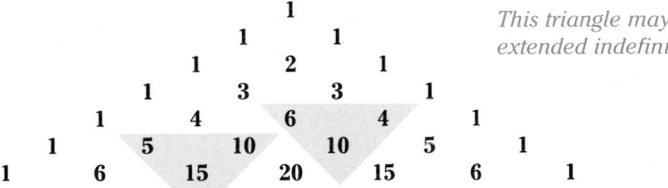

This triangle may be extended indefinitely.

 Lesson 12-6 The Binomial Theorem **801**

Resource Manager

Reproducible Masters
* *Study Guide and Practice,* pp. 225–226
* *Enrichment,* p. 138

Transparencies
* 5-Minute Check, 12-6
* Teaching, 12-6

---

1 FOCUS

5-Minute Check
Lesson 12-5

Write each expression in expanded form and then find the sum.

1. $\sum_{n=1}^{5} (6n - 5)$ $1 + 7 + 13 + 19 + 25; 65$

2. $\sum_{n=1}^{\infty} 2\left(\frac{1}{4}\right)^n$ $\frac{1}{2} + \frac{1}{8} + \frac{1}{32} + \frac{1}{128} + \cdots; \frac{2}{3}$

Express each series using sigma notation.

3. $5 + 7 + 9 + 11 + 13$ $\sum_{n=1}^{5} (2n + 3)$

4. $3 + 6 + 9 + 12 + 15 + 18$ $\sum_{n=1}^{6} 3n$

5. $3\frac{1}{3} + 4 + 4\frac{2}{3} + 5\frac{1}{3} + 6 + \cdots$ $\sum_{n=4}^{\infty} \frac{2n + 2}{3}$

Motivating the Lesson
Hands-On Activity Have students expand the product $(3x - y)(3x - y)(3x - y)(3x - y)$ by hand. Point out that a shorter method will be learned in this chapter.

Teaching Tip Introduce Pascal's triangle by writing it on the chalkboard. Have students discover the pattern before you discuss its patterns in the binomial expansion of $(x + y)^n$.

In-Class Examples

Example 1
Use Pascal's triangle to expand each binomial.

a. $(x + y)^8$ $x^8 + 8x^7y + 28x^6y^2 + 56x^5y^3 + 70x^4y^4 + 56x^3y^5 + 28x^2y^6 + 8xy^7 + y^8$

b. $(x - 5y)^7$ $x^7 - 35x^6y + 525x^5y^2 - 4375x^4y^3 + 21,875x^3y^4 - 65,625\ x^2y^5 + 109,375\ xy^6 - 78,125y^7$

Example 2
Manufacturing A quality control manager determines that out of a batch of 10 components, at most 4 are defective. How many of the possible groups of defective and non-defective components have at most 4 defective

---

Example **1** Use Pascal's triangle to expand each binomial.

a. $(x + y)^6$

First, write the series without the coefficients. Recall that the expression should have $6 + 1$ or 7 terms, with the first term being x^6 and the last term being y^6. Also note that the exponents of x should decrease from 6 to 0 while the exponents of y should increase from 0 to 6, while the degree of each term is 6.

$$x^6 + x^5y + x^4y^2 + x^3y^3 + x^2y^4 + xy^5 + y^6 \quad y^0 = 1, x^0 = 1$$

Then, use the numbers in the seventh row of Pascal's triangle as the coefficients of the terms. *Why is the seventh row used instead of the sixth row?*

$$\begin{array}{ccccccc} 1 & 6 & 15 & 20 & 15 & 6 & 1 \\ \downarrow & \downarrow & \downarrow & \downarrow & \downarrow & \downarrow & \downarrow \end{array}$$

$$(x + y)^6 = x^6 + 6x^5y + 15x^4y^2 + 20x^3y^3 + 15x^2y^4 + 6xy^5 + y^6$$

b. $(3x + 2y)^7$

Extend Pascal's triangle to the eighth row.

$$\begin{array}{ccccccc} & 1 & 6 & 15 & 20 & 15 & 6 & 1 \\ 1 & 7 & 21 & 35 & 35 & 21 & 7 & 1 \end{array}$$

Then, write the expression and simplify each term. Replace each x with $3x$ and y with $2y$.

$$(3x + 2y)^7 = (3x)^7 + 7(3x)^6(2y) + 21(3x)^5(2y)^2 + 35(3x)^4(2y)^3 + 35(3x)^3(2y)^4 + 21(3x)^2(2y)^5 + 7(3x)(2y)^6 + (2y)^7$$

$$= 2187x^7 + 10,206x^6y + 20,412x^5y^2 + 22,680x^4y^3 + 15,120x^3y^4 + 6048x^2y^5 + 1344xy^6 + 128y^7$$

You can use Pascal's triangle to solve real-world problems in which there are only two outcomes for each event. For example, you can determine the distribution of answers on true-false tests, the combinations of heads and tails when tossing a coin, or the possible sequences of boys and girls in a family.

Example **2** **FAMILY** Refer to the application at the beginning of the lesson. Of the seven children born to the McCaughey's, at least three were boys. How many of the possible groups of boys and girls have at least three boys?

Let g represent girls and b represent boys. To find the number of possible groups, expand $(g + b)^7$. Use the eighth row of Pascal's triangle for the expansion.

$$g^7 + 7g^6b + 21g^5b^2 + 35g^4b^3 + 35g^3b^4 + 21g^2b^5 + 7gb^6 + b^7$$

To have at least three boys means that there could be 3, 4, 5, 6, or 7 boys. The total number of ways to have at least three boys is the same as the sum of the coefficients of g^4b^3, g^3b^4, g^2b^5, gb^6, and b^7. This sum is $35 + 35 + 21 + 7 + 1$ or 99.

Thus, there are 99 possible groups of boys and girls in which there are at least three boys.

The general expansion of $(x + y)^n$ can also be determined by the **Binomial Theorem**.

Binomial Theorem	If n is a positive integer, then the following is true.
	$(x + y)^n = x^n + nx^{n-1}y + \dfrac{n(n-1)}{1 \cdot 2}x^{n-2}y^2 + \dfrac{n(n-1)(n-2)}{1 \cdot 2 \cdot 3}x^{n-3}y^3 + \cdots + y^n$

Example **3** Use the Binomial Theorem to expand $(2x - y)^6$.

The expansion will have seven terms. Find the first four terms using the sequence 1, $\dfrac{6}{1}$ or 6, $\dfrac{6 \cdot 5}{1 \cdot 2}$ or 15, $\dfrac{6 \cdot 5 \cdot 4}{1 \cdot 2 \cdot 3}$ or 20. Then use symmetry to find the remaining terms, 15, 6, and 1.

$(2x - y)^6 = (2x)^6 + 6(2x)^5(-y) + 15(2x)^4(-y)^2 + 20(2x)^3(-y)^3 + 15(2x)^2(-y)^4 +$
$\qquad 6(2x)(-y)^5 + (-y)^6$

$\qquad = 64x^6 - 192x^5y + 240x^4y^2 - 160x^3y^3 + 60x^2y^4 - 12xy^5 + y^6$

An equivalent form of the Binomial Theorem uses both sigma and factorial notation. It is written as follows, where n is a positive integer and r is a positive integer or zero.

$$(x + y)^n = \sum_{r=0}^{n} \frac{n!}{r!(n-r)!} x^{n-r}y^r$$

You can use this form of the Binomial Theorem to find individual terms of an expansion.

Example **4** Find the fifth term of $(4a + 3b)^7$.

$(4a + 3b)^7 = \displaystyle\sum_{r=0}^{7} \frac{7!}{r!(7-r)!} (4a)^{7-r}(3b)^r$

To find the fifth term, evaluate the general term for $r = 4$. *Since r increases from 0 to n, r is one less than the number of the term.*

$\dfrac{7!}{r!(7-r)!} (4a)^{7-r}(3b)^r = \dfrac{7!}{4!(7-4)!} (4a)^{7-4}(3b)^4$

$\qquad = \dfrac{7 \cdot 6 \cdot 5 \cdot 4!}{4! \; 3!} (4a)^3(3b)^4$

$\qquad = 181{,}440a^3b^4$

The fifth term of $(4a + 3b)^7$ is $181{,}440a^3b^4$.

CHECK FOR UNDERSTANDING

Communicating Mathematics

Read and study the lesson to answer each question.

1. a. **Calculate** the sum of the numbers in each row of Pascal's triangle for $n = 0$ to 5. **1, 2, 4, 8, 16, 32**

   b. **Write** an expression to represent the sum of the numbers in the nth row of Pascal's triangle. **2^n**

Lesson 12-6 The Binomial Theorem **803**

Addressing Individual Needs

Logical Find the value of $(0.99)^6$ by expanding $(1 - 0.01)^6$. Round to the nearest hundredth. **0.94**

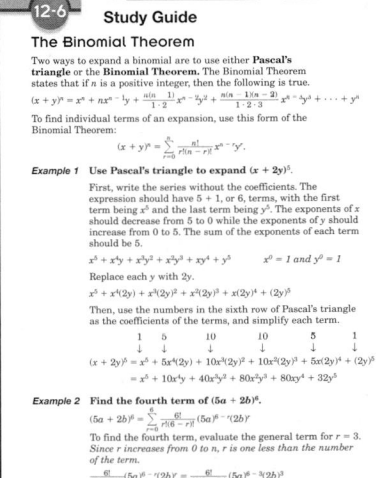

Additional Answers

4. The exponents must add to 12, so the exponent of y is $12 - 7$ or 5.

To find the coefficient of the term use the formula

$$(x + y)^n = \sum_{r=0}^{n} \frac{n!}{r!\,(n-r)!}\, x^{n-r} y^r.$$

Evaluate the general term for $n = 12$ and $r = 5$. $\frac{12!}{5!\,7!}\, x^7 y^5 = 792 x^7 y^5$.

6. $a^6 + 18a^5 + 135a^4 + 540a^3 + 1215a^2 + 1458a + 729$

7. $125 - 75y + 15y^2 - y^3$

8. $81p^4 - 216p^3q + 216p^2q^2 - 96pq^3 + 16q^4$

12. $a^8 + 8a^7b + 28a^6b^2 + 56a^5b^3 + 70a^4b^4 + 56a^3b^5 + 28a^2b^6 + 8ab^7 + b^8$

13. $n^6 - 24n^5 + 240n^4 - 1280n^3 + 3840n^2 - 6144n + 4096$

14. $81c^4 - 108c^3d + 54c^2d^2 - 12cd^3 + d^4$

16. $d^7 + 14d^6 + 84d^5 + 280d^4 + 560d^3 + 672d^2 + 448d + 128$

17. $243 - 405x + 270x^2 - 90x^3 + 15x^4 - x^5$

18. $256a^4 + 256a^3b + 96a^2b^2 + 16ab^3 + b^4$

19. $8x^3 - 36x^2y + 54xy^2 - 27y^3$

20. $81m^4 + 108\sqrt{2}m^3 + 108m^2 + 24\sqrt{2}m + 4$

21. $c^3 - 6c^2\sqrt{c} + 15c^2 - 20c\sqrt{c} + 15c - 6\sqrt{c} + 1$

22. $\frac{1}{32}n^5 + \frac{5}{8}n^4 + 5n^3 + 20n^2 + 40n + 32$

23. $81a^4 + 72a^3b + 24a^2b^2 + \frac{32}{9}ab^3 + \frac{16}{81}b^4$

24. $p^{16} + 8p^{14}q + 28p^{12}q^2 + 56p^{10}q^3 + 70p^8q^4 + 56p^6q^5 + 28p^4q^6 + 8p^2q^7 + q^8$

2c. Even-indexed terms are negative and odd-indexed terms are positive.

2. Examine the expansion of $(x - y)^n$ for $n = 3, 4,$ and 5.
  a. **Identify** the sign of the second term for each expansion.   **negative**
  b. **Identify** the sign of the third term for each expansion.   **positive**
  c. **Explain** how to determine the sign of a term without writing out the entire expansion.

3. **Restate** what is meant by the observation that each term in a binomial expansion has degree n.   **The sum of the exponents of each term is n.**

4. If ax^7y^b is a term from the expansion of $(x + y)^{12}$, **describe** how to determine its coefficient a and missing exponent b without writing the entire expansion. **See margin.**

Guided Practice

5. Use Pascal's triangle to expand $(c + d)^5$. $c^5 + 5c^4d + 10c^3d^2 + 10c^2d^3 + 5cd^4 + d^5$

Use the Binomial Theorem to expand each binomial.   **6–8. See margin.**

6. $(a + 3)^6$ **7.** $(5 - y)^3$ **8.** $(3p - 2q)^4$

Find the designated term of each binomial expansion.

9. 6th term of $(a - b)^7$ $-21a^2b^5$

10. 4th term of $(x + \sqrt{3})^9$ $252\sqrt{3}x^6$

11. Coins A coin is flipped five times. Find the number of possible sets of heads and tails that have each of the following.
  a. 0 heads   **1**
  b. 2 heads   **10**
  c. at least 4 heads   **6**
  d. at most 3 heads   **26**

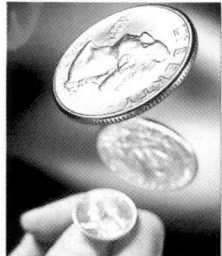

EXERCISES

Practice

Use Pascal's triangle to expand each binomial.   **12–14. See margin.**

12. $(a + b)^8$ **13.** $(n - 4)^6$ **14.** $(3c - d)^4$

15. Expand $(2 + a)^9$ using Pascal's triangle. $512 + 2304a + 4608a^2 + 5376a^3 + 4032a^4 + 2016a^5 + 672a^6 + 144a^7 + 18a^8 + a^9$

Use the Binomial Theorem to expand each binomial.   **16–24. See margin.**

16. $(d + 2)^7$ **17.** $(3 - x)^5$ **18.** $(4a + b)^4$

19. $(2x - 3y)^3$ **20.** $(3m + \sqrt{2})^4$ **21.** $(\sqrt{c} - 1)^6$

22. $\left(\frac{1}{2}n + 2\right)^5$ **23.** $\left(3a + \frac{2}{3}b\right)^4$ **24.** $(p^2 + q)^8$

25. Expand $(xy - 2z^3)^6$ using the Binomial Theorem.
  $x^6y^6 - 12x^5y^5z^3 + 60x^4y^4z^6 - 160x^3y^3z^9 + 240x^2y^2z^{12} - 192xyz^{15} + 64z^{18}$

Find the designated term of each binomial expansion.

26. 5th term of $(x + y)^9$ $126x^5y^4$ **27.** 4th term of $(a - \sqrt{2})^8$ $-112\sqrt{2}a^5$

28. 4th term of $(2a - b)^7$ $-560a^4b^3$ **29.** 7th term of $(3c + 2d)^9$ $145,152c^3d^6$

30. 8th term of $\left(\frac{1}{2}x - y\right)^{10}$ $-15x^3y^7$ **31.** 6th term of $(2p - 3q)^{11}$ $-7,185,024p^6q^5$

32. Find the middle term of the expansion of $(\sqrt{x} + \sqrt{y})^8$. $70x^2y^2$

 www.amc.glencoe.com/self_check_quiz

33. Business A company decides to form a recycling committee to find a more efficient means of recycling waste paper. The committee is to be composed of eight employees. Of these eight employees, at least four women are to be on the committee. How many of the possible groups of men and women have at least four women? **163**

34. Critical Thinking Describe a strategy that uses the Binomial Theorem to expand $(a + b + c)^{12}$. **See margin.**

35. Education Rafael is taking a test that contains a section of 12 true-false questions.

    a. How many of the possible groups of answers to these questions have exactly 8 correct answers of false? **495**

    b. How many of the possible groups of answers to these questions have at least 6 correct answers of true? **2510**

36. Critical Thinking Find a term in the expansion of $\left(3x^2 - \dfrac{1}{4x}\right)^6$ that does not contain the variable x. **See margin.**

37b. Sample answer: 1.04060401

37c. 1.04060401, the two values are equal.

37. Numerical Analysis Before the invention of modern calculators and computers, mathematicians searched for ways to shorten lengthy calculations such as $(1.01)^4$.

    a. Express 1.01 as a binomial. **Sample answer: 1 + 0.01**

    b. Use the binomial representation of 1.01 found in part **a** and the Binomial Theorem to calculate the value of $(1.01)^4$ to eight decimal places.

    c. Use a calculator to estimate $(1.01)^4$ to eight decimal places. Compare this value to the value found in part **b**.

Mixed Review

38. $1 + (-1) + (-3) + (-5) + (-7) + (-9)$; -24

38. Write $\displaystyle\sum_{k=2}^{7} 5 - 2k$ in expanded form and then find the sum. *(Lesson 12-5)*

39. Use the ratio test to determine whether the series $2 + \dfrac{2^2}{2!} + \dfrac{2^3}{3!} + \dfrac{2^4}{4!} + \cdots$ is *convergent* or *divergent*. *(Lesson 12-4)* **convergent**

40. Find the sum of $\dfrac{2}{3} + \dfrac{1}{3} + \dfrac{1}{6} + \dfrac{1}{12} + \cdots$ or explain why one does not exist. **$1\dfrac{1}{3}$** *(Lesson 12-3)*

41. Finance A bank offers a home mortgage for an annual interest rate of 8%. If a family decides to mortgage a \$150,000 home over 30 years with this bank, what will the monthly payment for the principal and interest on their mortgage be? *(Lesson 11-2)* **\$1100.65**

42. Write $\overrightarrow{MK}$ as the sum of unit vectors if $M(-2, 6, 3)$ and $K(4, 8, -2)$. *(Lesson 8-3)* **$6\vec{i} + 2\vec{j} - 5\vec{k}$**

43. Construction A highway curve, in the shape of an arc of a circle, is 0.25 mile. The direction of the highway changes 45° from one end of the curve to the other. Find the radius of the circle in feet that the curve follows. *(Lesson 6-1)* **1681 feet**

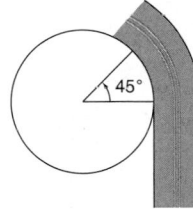

44. SAT/ACT Practice If b is a prime integer such that $3b > 10 > \dfrac{5}{6}b$, which of the following is a possible value of b? **D**

    A 2   **B** 3   **C** 4   **D** 11   **E** 13

Special Sequences and Series

1 FOCUS

5-Minute Check
Lesson 12-6

1. Use Pascal's triangle to expand $(2a + 4b)^4$. **$16a^4 + 128a^3b + 384a^2b^2 + 512ab^3 + 256b^4$**

Use the Binomial Theorem to expand each binomial.

2. $(2 - y)^6$ $64 - 192y + 240y^2 - 160y^3 + 60y^4 - 12y^5 + y^6$

3. $(5x + y)^5$ $3125x^5 + 3125x^4y + 1250x^3y^2 + 250x^2y^3 + 25xy^4 + y^5$

Find the designated term of each binomial expansion.

4. 3rd term of $(2x + y)^6$

   $240x^4y^2$

5. 4th term of $(x - 4y)^7$

   $-2240x^4y^3$

Motivating the Lesson

Real-World Application Give students pictures of items found in nature that display the Fibonacci sequence. Some examples are leaf-growth patterns, spirals in pineapples, and the population growth of some animals. Have them determine what portions of each item display the sequence.

OBJECTIVES
- Approximate e^x, trigonometric values, and logarithms of negative numbers by using series.
- Use Euler's Formula to write the exponential form of a complex number.

NATURE An important sequence found in nature can be seen in the spiral of a Nautilus shell. To see this sequence follow procedure below.

- Begin by placing two small squares with side length 1 next to each other.
- Below both of these, draw a square with side length 2.
- Now draw a square that touches both a unit square and the 2-square. This square will have sides 3 units long.
- Add another square that touches both the 2-square and the 3-square. This square will have sides of 5 units.
- Continue this pattern around the picture so that each new square has sides that are as long as the sum of the latest two square's sides.

The side lengths of these squares form what is known as the **Fibonacci sequence:** 1, 1, 2, 3, 5, 8, 13, ... A spiral that closely models the Nautilus spiral can be drawn by first rearranging the squares so that the unit squares are in the interior and then connecting quarter circles, each of whose radius is a side of a new square. This spiral is known as the *Fibonacci spiral. A problem related to the Fibonacci sequence will be solved in Example 1.*

↓ rearrange

The Fibonacci sequence describes many patterns of numbers found in nature. This sequence was presented by an Italian mathematician named Leonardo de Pisa, also known as Fibonacci (pronounced *fih buh NACH ee*), in 1201. The first two numbers in the sequence are 1, that is, $a_1 = 1$ and $a_2 = 1$. As you have seen in the example above, adding the two previous terms generates each additional term in the sequence.

The original problem that Fibonacci investigated in 1202 involved the reproductive habits of rabbits under ideal conditions.

Example ❶ **NATURE** Suppose a newly born pair of rabbits are allowed to breed in a controlled environment. How many rabbit pairs will there be after one year if the following assumptions are made?

- A male and female rabbit can mate at the age of one month.
- At the end of its second month, a female rabbit can produce another pair of rabbits (one male, one female).
- The rabbits never die.
- The female always produces one new pair every month from the second month on.

806 *Chapter 12 Sequences and Series*

Resource Manager

 Reproducible Masters
- *Study Guide and Practice*, pp. 227–228
- *Enrichment*, p. 139
- *Assessment and Evaluation*, p. 304
- *Graphing Calculator Keystrokes*, p. 52

 Transparencies
- 5-Minute Check, 12-7
- Teaching, 12-7

Based on these assumptions, at the end of the first month, there will be one pair of rabbits. And at the end of the second month, the female produces a new pair, so now there are two pairs of rabbits.

Month   Number of Pairs

Month	1	2	3	4	5	6	7	8	9	10	11	12
Number of Pairs	1	1	2	3	5	8	13	21	34	55	89	144

The following table shows the pattern.

There will be 144 pairs of rabbits during the twelfth month.

Each term in the sequence is the sum of the two previous terms.

Another important series is the series that is used to define the irrational number e. The Swiss mathematician Leonhard Euler (pronounced OY ler), published a work in which he developed this irrational number. It has been suggested that in his honor the number is called e, the Euler number. The number can be expressed as the sum of the following infinite series.

$$e = 1 + \frac{1}{1!} + \frac{1}{2!} + \frac{1}{3!} + \frac{1}{4!} + \frac{1}{5!} + \cdots + \frac{1}{n!} + \cdots$$

The Binomial Theorem can be used to derive the series for e as follows. Let k be any positive integer and apply the Binomial Theorem to $\left(1 + \frac{1}{k}\right)^k$.

$$\left(1 + \frac{1}{k}\right)^k = 1 + k\left(\frac{1}{k}\right) + \frac{k(k-1)}{2!}\left(\frac{1}{k}\right)^2 + \frac{k(k-1)(k-2)}{3!}\left(\frac{1}{k}\right)^3 + \cdots$$

$$+ \frac{k(k-1)(k-2)\cdots 1}{k!}\left(\frac{1}{k}\right)^k$$

$$= 1 + 1 + \frac{1\left(1 - \frac{1}{k}\right)}{2!} + \frac{1\left(1 - \frac{1}{k}\right)\left(1 - \frac{2}{k}\right)}{3!} + \cdots + \frac{1\left(1 - \frac{1}{k}\right)\left(1 - \frac{2}{k}\right)\cdots\frac{1}{k}}{k!}$$

Then, find the limit of $\left(1 + \frac{1}{k}\right)^k$ as k increases without bound.

Lesson 12-7 Special Sequences and Series **807**

In-Class Example

Example 1
Nature Sneezewort trees have growing points for new shoot growth that follow a pattern. Suppose that one such tree has sprouted out of the ground and is ready to begin branching. How many branches will there be in nine months if the following assumptions are made?
- When a plant produces a new shoot, that shoot needs to grow for two months before it is strong enough to support further branching on it.
- After reaching the first branching point, it branches every month.
- The branches never die or fall off.

68 branches

Teaching Tip After working Example 1, discuss the fact that Fibonacci sequences are repeated in many areas of nature. Two examples are Dudeney's cows and honeybees.

In-Class Example

Example 2
Use the first five terms of the exponential series and a calculator to approximate the value of $e^{0.65}$ to the nearest hundredth. **1.91**

Recall that $\lim\limits_{k \to \infty} \frac{1}{k} = 0$.

$$\lim_{k \to \infty}\left(1 + \frac{1}{k}\right)^k = 1 + 1 + \frac{1}{2!} + \frac{1}{3!} + \frac{1}{4!} + \frac{1}{5!} + \cdots$$

As $k \to \infty$, the number of terms in the sum becomes infinite.

Thus, e can be defined as follows.

$$e = \lim_{k \to \infty}\left(1 + \frac{1}{k}\right)^k \quad \text{or} \quad e = 1 + 1 + \frac{1}{2!} + \frac{1}{3!} + \frac{1}{4!} + \frac{1}{5!} + \cdots$$

The value of e^x can be approximated by using the following series. This series is often called the **exponential series.**

Exponential Series	$e^x = \displaystyle\sum_{n=0}^{\infty} \frac{x^n}{n!} = 1 + x + \frac{x^2}{2!} + \frac{x^3}{3!} + \frac{x^4}{4!} + \frac{x^5}{5!} + \cdots$

Example **2** **Use the first five terms of the exponential series and a calculator to approximate the value of $e^{2.03}$ to the nearest hundredth.**

$$e^x \approx 1 + x + \frac{x^2}{2!} + \frac{x^3}{3!} + \frac{x^4}{4!}$$

$$e^{2.03} \approx 1 + 2.03 + \frac{(2.03)^2}{2!} + \frac{(2.03)^3}{3!} + \frac{(2.03)^4}{4!}$$

$$\approx 1 + 2.03 + 2.06045 + 1.394237833 + 0.7075757004$$

$$\approx 7.19$$

Euler's name is associated with a number of important mathematical relationships. Among them is the relationship between the exponential series and a series called the **trigonometric series.** The trigonometric series for $\cos x$ and $\sin x$ are given below.

Trigonometric Series	$\cos x = \displaystyle\sum_{n=0}^{\infty} \frac{(-1)^n x^{2n}}{(2n)!} = 1 - \frac{x^2}{2!} + \frac{x^4}{4!} - \frac{x^6}{6!} + \frac{x^8}{8!} - \cdots$
	$\sin x = \displaystyle\sum_{n=0}^{\infty} \frac{(-1)^n x^{2n+1}}{(2n+1)!} = x - \frac{x^3}{3!} + \frac{x^5}{5!} - \frac{x^7}{7!} + \frac{x^9}{9!} - \cdots$

These two series are convergent for all values of x. By replacing x with any angle measure expressed in radians and carrying out the computations, approximate values of the trigonometric functions can be found to any desired degree of accuracy.

Example Use the first five terms of the trigonometric series to approximate the value of $\cos \frac{\pi}{3}$ to four decimal places.

$$\cos x \approx 1 - \frac{x^2}{2!} + \frac{x^4}{4!} - \frac{x^6}{6!} + \frac{x^8}{8!}$$

$$\cos \frac{\pi}{3} \approx 1 - \frac{(1.0472)^2}{2!} + \frac{(1.0472)^4}{4!} - \frac{(1.0472)^6}{6!} + \frac{(1.0472)^8}{8!} \qquad x = \frac{\pi}{3} \text{ or about } 1.0472$$

$$\cos \frac{\pi}{3} \approx 1 - 0.54831 + 0.05011 - 0.00183 + 0.00004$$

$$\cos \frac{\pi}{3} \approx 0.5004 \quad \textit{Compare this result to the actual value.}$$

GRAPHING CALCULATOR EXPLORATION

In this Exploration you will examine polynomial functions that can be used to approximate $\sin x$. The graph below shows the graphs of $f(x) = \sin x$ and $g(x) = x - \frac{x^3}{3!} + \frac{x^5}{5!}$ on the same screen.

$[-3\pi, 3\pi]$ scl:1 by $[-2, 2]$ scl:1

TRY THESE 1–6. See margin.

1. Use TRACE to help you write an inequality describing the x-values for which the graphs seem very close together.

2. In absolute value, what are the greatest and least differences between the values of $f(x)$ and $g(x)$ for the values of x described by the inequality you wrote in Exercise 1?

3. Repeat Exercises 1 and 2 using $h(x) = x - \frac{x^3}{3!} + \frac{x^5}{5!} - \frac{x^7}{7!}$ instead of $g(x)$.

4. Repeat Exercises 1 and 2 using $k(x) = x - \frac{x^3}{3!} + \frac{x^5}{5!} - \frac{x^7}{7!} + \frac{x^9}{9!}$.

WHAT DO YOU THINK?

5. Are the intervals for which you get good approximations for $\sin x$ larger or smaller for polynomials that have more terms?

6. What term should be added to $k(x)$ to obtain a polynomial with six terms that gives good approximations to $\sin x$?

Another very important formula is derived by replacing x by $i\alpha$ in the exponential series, where i is the imaginary unit and α is the measure of an angle in radians.

$$e^{i\alpha} = 1 + i\alpha + \frac{(i\alpha)^2}{2!} + \frac{(i\alpha)^3}{3!} + \frac{(i\alpha)^4}{4!} + \cdots$$

$$e^{i\alpha} = 1 + i\alpha - \frac{\alpha^2}{2!} - i\frac{\alpha^3}{3!} + \frac{\alpha^4}{4!} + \cdots \qquad i^2 = -1, \ i^3 = -i, \ i^4 = 1$$

Group the terms according to whether they contain the factor i.

$$e^{i\alpha} = \left(1 - \frac{\alpha^2}{2!} + \frac{\alpha^4}{4!} - \frac{\alpha^6}{6!} + \cdots\right) + i\left(\alpha - \frac{\alpha^3}{3!} + \frac{\alpha^5}{5!} - \frac{\alpha^7}{7!} + \cdots\right)$$

Notice that the real part is exactly $\cos \alpha$ and the imaginary part is exactly $\sin \alpha$. This relationship is called **Euler's Formula.**

Teaching Tip For Example 3, remind students that when substituting into the trigonometric series, one uses only the radian measure and not the degree measure of the angle.

In-Class Example

Example 3
Use the first five terms of the trigonometric series to find the value of $\sin \frac{\pi}{3}$ to four decimal places. **0.8660**

Graphing Calculator Exploration
After completing the Exploration, have students repeat the activity for $f(x) = \cos x$ and $g(x) = 1 - \frac{x^2}{2!} + \frac{x^4}{4!} - \frac{x^6}{6!}$.

Additional Answers
Graphing Calculator Exploration
1. Sample answer (without zooming): $-2.4 \le x \le 2.4$
2. Sample answer for greatest difference: about 0.08; the least difference is 0.
3. Sample answer: $-3.4 \le x \le 3.4$; sample answer: about 0.15; 0
4. Sample answer: $-3.8 \le x \le 3.8$; sample answer: about 0.05; 0
5. larger
6. $-\frac{x^{11}}{11!}$

Euler's Formula	$e^{i\alpha} = \cos \alpha + i \sin \alpha$

If $-i\alpha$ had been substituted for x rather than $i\alpha$, the result would have been $e^{-i\alpha} = \cos \alpha - i \sin \alpha$.

Euler's Formula can be used to write a complex number, $a + bi$, in its exponential form, $re^{i\theta}$.

$$a + bi = r(\cos \theta + i \sin \theta)$$
$$= re^{i\theta}$$

Example **4** **Write $1 + \sqrt{3}i$ in exponential form.**

Write the polar form of $1 + \sqrt{3}i$. Recall that $a + bi = r (\cos \theta + i \sin \theta)$, where $r = \sqrt{a^2 + b^2}$ and $\theta = \text{Arctan} \frac{b}{a}$ when $a > 0$.

$r = \sqrt{(1)^2 + (\sqrt{3})^2}$ or 2, and $\theta = \text{Arctan} \frac{\sqrt{3}}{1}$ or $\frac{\pi}{3}$ *$a = 1$ and $b = \sqrt{3}$*

$1 + \sqrt{3}i = 2\left(\cos \frac{\pi}{3} + i \sin \frac{\pi}{3}\right)$

$= 2e^{i\frac{\pi}{3}}$

Thus, the exponential form of $1 + \sqrt{3}i$ is $2e^{i\frac{\pi}{3}}$.

Look Back
Refer to Lesson 9-6 to review the polar form of complex numbers.

The equations for $e^{i\alpha}$ and $e^{-i\alpha}$ can be used to derive the exponential values of $\sin \alpha$ and $\cos \alpha$.

$$e^{i\alpha} - e^{-i\alpha} = (\cos \alpha + i \sin \alpha) - (\cos \alpha - i \sin \alpha)$$
$$e^{i\alpha} - e^{-i\alpha} = 2i \sin \alpha$$
$$\sin \alpha = \frac{e^{i\alpha} - e^{-i\alpha}}{2i}$$

$$e^{i\alpha} + e^{-i\alpha} = (\cos \alpha + i \sin \alpha) + (\cos \alpha - i \sin \alpha)$$
$$e^{i\alpha} + e^{-i\alpha} = 2 \cos \alpha$$
$$\cos \alpha = \frac{e^{i\alpha} + e^{-i\alpha}}{2}$$

From your study of logarithms, you know that there is no real number that is the logarithm of a negative number. However, you can use a special case of Euler's Formula to find a complex number that is the natural logarithm of a negative number.

$$e^{i\alpha} = \cos \alpha + i \sin \alpha$$
$$e^{i\pi} = \cos \pi + i \sin \pi \quad \text{Let } \alpha = \pi.$$
$$e^{i\pi} = -1 + i(0)$$
$$e^{i\pi} = -1 \qquad \text{So } e^{i\pi} + 1 = 0.$$

If you take the natural logarithm of both sides of $e^{i\pi} = -1$, you can obtain a value for $\ln(-1)$.

$$\ln e^{i\pi} = \ln(-1)$$
$$i\pi = \ln(-1)$$

Thus, the natural logarithm of a negative number $-k$, for $k > 0$, can be defined using $\ln(-k) = \ln(-1)k$ or $\ln(-1) + \ln k$, a complex number.

Example **5** **Evaluate $\ln(-270)$.**

$\ln(-270) = \ln(-1) + \ln(270)$ *Use a calculator to compute $\ln(270)$.*
$\quad\quad\quad\quad \approx i\pi + 5.5984$

Thus, $\ln(-270) \approx i\pi + 5.5984$.   *The logarithm is a complex number.*

CHECK FOR UNDERSTANDING

Communicating Mathematics

Read and study the lesson to answer each question.

1. **Explain** why in Example 2 of this lesson the exponential series gives an approximation of 7.19 for $e^{2.03}$, while a calculator gives an approximation of 7.61. **See margin.**

2. **Estimate** for what values of x the series
$1 + x + \dfrac{x^2}{2} + \dfrac{x^3}{6}$ gives a good approximation
of the function e^x using the graph at the right.
Sample answer: $-2 \le x \le 1.5$

3. **List** at least two reasons why Fibonacci's rabbit problem in Example 1 is not realistic. **See margin.**

4. **Write** a recursive formula for the terms of the Fibonacci sequence.
$a_{n+1} = a_n + a_{n-1}$ for $n \ge 2$

Guided Practice

Find each value to four decimal places.

5. $\ln(-7)$ $i\pi + 1.9459$
6. $\ln(-0.379)$ $i\pi - 0.9702$

Use the first five terms of the exponential series and a calculator to approximate each value to the nearest hundredth.

7. $e^{0.8}$ **2.22**
8. $e^{1.36}$ **3.85**

9. Use the first five terms of the trigonometric series to approximate the value of $\sin \pi$ to four decimal places. Then, compare the approximation to the actual value. **0.0069; 0**

Write each complex number in exponential form.

10. $\sqrt{2}\left(\cos \dfrac{3\pi}{4} + i \sin \dfrac{3\pi}{4}\right)$ $\sqrt{2}e^{i\frac{3\pi}{4}}$ 11. $-1 + \sqrt{3}i$ $2e^{i\frac{2\pi}{3}}$

Lesson 12-7 Special Sequences and Series **811**

Additional Answers

1. The approximation given in Example 1 only used the first five terms of the exponential series. Using more terms of the exponential series would give an approximation closer to that given by the calculator.

3. The problem seems to imply that siblings mate. Genetically, this can lead to problems. Another problem is the assumption that each birth produces only two offspring, one male and one female. Rabbits are more likely to give birth to more than two offspring and the ratio of male to female births is not guaranteed to be 1 to 1.

In-Class Example
Example 5
Evaluate $\ln(-540)$.
about $i\pi + 6.2916$

3 PRACTICE/APPLY

Error Analysis

Watch for students who have problems with the process of approximation, such as in Exercises 1, 2, and 7–9.

Prevent by reviewing graphing calculator approximations in the Graphing Calculator Exploration on p. 809.

Study Guide and Practice Masters, pp. 227–228

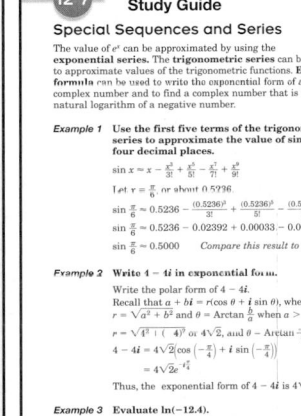

Assignment Guide

Core: 13–41 odd, 42–52
Enriched: 14–36 even, 37–52

Additional Answers

12b. No, in five years she will have increased her savings by about 34.5%, not 100%.

36. Sample answer: A transcendental number is one that cannot be the root of an algebraic equation with rational coefficients. Examples are π and e.

---

12. Investment The Cyberbank advertises an Advantage Plus savings account with a 6% interest rate compounded continuously. Morgan is considering opening up an Advantage Plus account, because she needs to double the amount she will deposit over 5 years.   **a. approximately 1.345P**

 a. If Morgan deposits P dollars into the account, approximate the return on her investment after 5 years using three terms of the exponential series. (*Hint:* The formula for continuous compounding interest is $A = Pe^{rt}$.)

12b. See margin. b. Will Morgan double her money in the desired amount of time? Explain.

 c. Check the approximation found in part **a** using a calculator. How do the two answers compare?   **The approximation is accurate to two decimal places.**

EXERCISES

Practice

Find each value to four decimal places.

$i\pi - 1.3863$

13. $\ln(-4)$ $i\pi + 1.3863$ **14.** $\ln(-3.1)$ $i\pi + 1.1314$ **15.** $\ln(-0.25)$

16. $\ln(-0.033)$ $i\pi - 3.4112$ **17.** $\ln(-238)$ $i\pi + 5.4723$ **18.** $\ln(-1207)$
                              $i\pi + 7.0959$

Use the first five terms of the exponential series and a calculator to approximate each value to the nearest hundredth.

19. $e^{1.1}$ **2.99**   **20.** $e^{-0.2}$ **0.82**   **21.** $e^{4.2}$ **39.33**

22. $e^{0.55}$ **1.73**   **23.** $e^{3.5}$ **24.02**   **24.** $e^{2.73}$ **13.16**

Use the first five terms of the trigonometric series to approximate the value of each function to four decimal places. Then, compare the approximation to the actual value.

25. $\cos \pi$ $-0.9760; -1$ **26.** $\sin \dfrac{\pi}{4}$ **0.7071; 0.7071**   **27.** $\cos \dfrac{\pi}{6}$ **0.8660; 0.8660**

28. Approximate the value of $\sin \dfrac{\pi}{2}$ to four decimal places using the first five terms of the trigonometric series.   **1.0000; 1**

Write each complex number in exponential form.

29. $5\left(\cos \dfrac{5\pi}{3} + i \sin \dfrac{5\pi}{3}\right)$ $5e^{i\frac{5\pi}{3}}$ **30.** i $e^{i\frac{\pi}{2}}$

31. $1 + i$ $\sqrt{2}e^{i\frac{\pi}{4}}$ **32.** $\sqrt{3} + i$ $2e^{i\frac{\pi}{6}}$

33. $-\sqrt{2} + \sqrt{2}i$ $2e^{i\frac{3\pi}{4}}$ **34.** $-4\sqrt{3} - 4i$ $8e^{i\frac{7\pi}{6}}$

35. Write the expression $3 + 3i$ in exponential form. $3\sqrt{2}e^{i\frac{\pi}{4}}$

See margin.

Applications and Problem Solving

36. Research Explore the meaning of the term *transcendental number*.

37. Critical Thinking Show that for all real numbers x, $\sin x = \dfrac{e^{ix} - e^{-ix}}{2i}$ and $\cos x = \dfrac{e^{ix} + e^{-ix}}{2}$.   **See margin.**

38. Research Investigate and write a one-page paper on the occurrence of Fibonacci numbers in plants.   **See students' work.**

 www.amc.glencoe.com/self_check_quiz

Additional Answer

37. $\dfrac{e^{ix} - e^{-ix}}{2i} = \dfrac{\cos x + i \sin x - (\cos x - i \sin x)}{2i} = \dfrac{2i \sin x}{2i} = \sin x$

$\dfrac{e^{ix} + e^{-ix}}{2} = \dfrac{\cos x + i \sin x + \cos x - i \sin x}{2} = \dfrac{2 \cos x}{2} = \cos x$

39. Critical Thinking Examine Pascal's triangle. What relationship can you find between Pascal's triangle and the Fibonacci sequence? **See margin.**

40. Number Theory Consider the Fibonacci sequence 1, 1, 2, 3, ..., $F_{n-2} + F_{n-1}$.

a. Find $\dfrac{F_n}{F_{n-1}}$ for the second through eleventh terms of the Fibonacci sequence. $\dfrac{1}{1}, \dfrac{2}{1}, \dfrac{3}{2}, \dfrac{5}{3}, \dfrac{8}{5}, \dfrac{13}{8}, \dfrac{21}{13}, \dfrac{34}{21}, \dfrac{55}{34}, \dfrac{89}{55}$

b. Is this sequence of ratios arithmetic, geometric, or neither? **neither**

c. Sketch a graph of the terms found in part **a**. Let n be the x-coordinate and $\dfrac{F_n}{F_{n-1}}$ be the y-coordinate, and connect the points. **See margin.**

d. Based on the graph found in part c, does this sequence appear to approach a limit? If so, use the last term found to approximate this limit to three decimal places. **yes; 1.618**

e. The *golden ratio* has a value of approximately 1.61804. How does the limit of the sequence $\dfrac{F_n}{F_{n-1}}$ compare to the golden ratio?

f. Research the term *golden ratio*. Write several paragraphs on the history of the golden ratio and describe its application to art and architecture. **See students' work.**

40e. The two ratios are equivalent to three decimal places.

41. Investment When Cleavon turned 5 years old, his grandmother decided it was time to start saving for his college education. She deposited $5000 in a special account earning 5% interest compounded continuously. By the time Cleavon begins college at the age of 18, his grandmother estimates that her grandson will need $40,000 for college tuition.

a. Approximate Cleavon's savings account balance on his 18th birthday using five terms of the exponential series. **approximately $9572.29**

b. Will the account have sufficient funds for Cleavon's college tuition by the time he is ready to start college? Explain.

c. Use a calculator to compute how long it will take the account to accumulate $40,000. How old would Cleavon be by this time?

d. To have at least $40,000 when Cleavon is 18 years old, how much should his grandmother have invested when he was 5 years old to the nearest dollar? **$20,882**

41b. No, the account will be short by more than $30,000.

41c. about 42 years; 47 years old

42. Critical Thinking Find a pattern, if one exists, for the following types of Fibonacci numbers.

a. even numbers **Every third Fibonacci number is an even number.**

b. multiples of 3 **Every fourth Fibonacci number is a multiple of 3.**

Lesson 12-7 Special Sequences and Series **813**

Open-Ended Assessment
Writing Have students write a paragraph explaining how they would approximate a trigonometric value by using the trigonometric series.

📁 **Chapter 12, Quiz C (Lessons 12-6 and 12-7)** is available in the *Assessment and Evaluation Masters,* p. 304.

Additional Answer
43. $64x^6 + 192x^5y + 240x^4y^2 + 160x^3y^3 + 60x^2y^4 + 12xy^5 + y^6$

Enrichment Masters, p. 139

Mixed Review

44. $\sum_{k=1}^{6} 2^k$

47. $y^2 + 3x + 7y = 0$

48. $4\left(\cos\dfrac{15\pi}{8} + i\sin\dfrac{15\pi}{8}\right)$

50a. 10π **radians per second**

---

43. Use the Binomial Theorem to expand $(2x + y)^6$. *(Lesson 12-6)* **See margin.**

44. Express the series $2 + 4 + 8 + \cdots + 64$ using sigma notation. *(Lesson 12-5)*

45. Number Theory If you tear a piece of paper that is 0.005 centimeter thick in half, and place the two pieces on top of each other, the height of the pile of paper is 0.01 centimeter. Let's call this the second pile. If you tear these two pieces of paper in half, the third pile will have four pieces of paper in it. *(Lesson 12-2)*
  a. How high is the third pile? the fourth pile? **0.020 cm, 0.040 cm**
  b. Write a formula to determine how high the nth pile is. **$0.005(2)^{n-1}$**
  c. Use the formula to determine in theory how high the 10th pile would be and how high the 100th pile would be. **$2.56, 3.169 \times 10^{27}$ cm**

46. Evaluate $\dfrac{8^{\frac{2}{3}}}{8^{\frac{1}{3}}}$. *(Lesson 11-1)* **2**

47. Write an equation of the parabola in general form that has a horizontal axis and passes through points at $(0, 0)$, $(2, -1)$, and $(4, -4)$. *(Lesson 10-5)*

48. Find the quotient of $16\left(\cos\dfrac{\pi}{8} + i\sin\dfrac{\pi}{8}\right)$ divided by $4\left(\cos\dfrac{\pi}{4} + i\sin\dfrac{\pi}{4}\right)$. *(Lesson 9-7)*

49. Physics Two forces, one of 30 N and the other of 50 N, act on an object. If the angle between the forces is 40°, find the magnitude and the direction of the resultant force. *(Lesson 8-5)* **75.5 N, 14°48′**

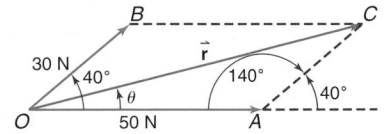

50. Carpentry Carpenters use circular sanders to smooth rough surfaces, such as wood or plaster. The disk of a sander has a radius of 6 inches and is rotating at a speed of 5 revolutions per second. *(Lesson 6-2)*
  a. Find the angular velocity of the sander disk in radians per second.
  b. What is the linear velocity of a point on the edge of the sander disk in feet per second? **15.7 ft/s**

51. Education You may answer up to 30 questions on your final exam in history class. It consists of multiple-choice and essay questions. Two 48-minute class periods have been set aside for taking the test. It will take you 1 minute to answer each multiple-choice question and 12 minutes for each essay question. Correct answers of multiple-choice questions earn 5 points and correct essay answers earn 20 points. If you are confident that you will answer all of the questions you attempt correctly, how many of each type of questions should you answer to receive the highest score? *(Lesson 2-7)* **24 multiple choice, 6 essay**

52. SAT/ACT Practice In triangle ABC, if $BD = 5$, what is the length of $\overline{BC}$? **C**
  A 3
  B 5
  C $\sqrt{39}$
  D $\sqrt{70}$
  E $\sqrt{126}$

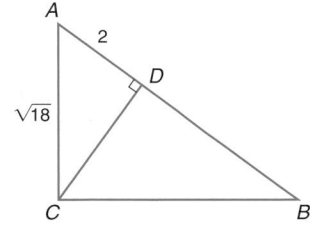

Extra Practice See p. A50.

Extra Credit

Find the total number of biological grandparents that one has beginning with one's grandparents up to and including one's great-great-great-great grandparents. **124**

12-8

Sequences and Iteration

Lesson Notes

12-8

OBJECTIVES
- Iterate functions using real and complex numbers.

ECOLOGY The population of grizzly bears on the high Rocky Mountain Front near Choteau, Montana, has a growth factor of 1.75. The maximum population of bears that can be sustained in the area is 500 bears, and the current population is 240. Write an equation to model the population. Use the equation to find the population of bears at the end of fifteen years. *This problem will be solved in Example 1.*

The population of a species in a defined area changes over time. Changes in the availability of food, good or bad weather conditions, the amount of hunting allowed, disease, and the presence or absence of predators can all affect the population of a species. We can use a mathematical equation to model the changes in a population.

One such model is the *Verhulst population model.* The model uses the recursive formula $p_{n+1} = p_n + rp_n(1 - p_n)$, where n represents the number of time periods that have passed, p_n represents the percent of the maximum sustainable population that exists at time n, and r is a growth factor.

Example **1** **ECOLOGY** Refer to the application above.

a. Write an equation to model the population.

For this problem, the Verhulst population model can be used to represent the changes in the population of a species.

$$p_{n+1} = p_n + rp_n(1 - p_n)$$
$$p_{n+1} = p_n + 1.75p_n(1 - p_n) \quad r = 1.75$$

b. Find the population of bears at the end of fifteen years.

The initial percent of the maximum sustainable population can be represented by the ratio $p_0 = \frac{240}{500}$ or 0.48. This means that the current population is 48% of the maximum sustainable population.

Now, we can find the first few iterates as follows.

$p_1 = 0.48 + 1.75(0.48)(1 - 0.48)$ or 0.9168   *n = 1, p_0 = 0.48,*
$(0.9168)(500) \approx 458$ bears   *r = 1.75*

$p_2 = 0.9168 + 1.75(0.9168)(1 - 0.9168)$ or 1.0503   *n = 2, p_1 = 0.9168,*
$(1.0503)(500) \approx 525$ bears   *r = 1.75*

$p_3 = 1.0503 + 1.75(1.0503)(1 - 1.0503)$ or 0.9578   *n = 3, p_2 = 1.0503,*
$(0.9578)(500) \approx 479$ bears   *r = 1.75*

(continued on the next page)

Lesson 12-8 Sequences and Iteration **815**

5-Minute Check
Lesson 12-7

1. Use the first five terms of the exponential series and a calculator to approximate the value of $e^{2.4}$ to the nearest hundredth. **9.97**

2. Use the first five terms of the trigonometric series to find the value of $\cos \frac{\pi}{4}$ to four decimal places. **0.7071**

3. Write $-4 - 4i$ in exponential form. $4\sqrt{2}\, e^{i\frac{5\pi}{4}}$

4. Find $\ln(-3)$ to four decimal places. $i\pi + 1.0986$

5. Write $6\left(\cos \frac{2\pi}{3} + i \sin \frac{2\pi}{3}\right)$ in exponential form. $6e^{i\frac{2\pi}{3}}$

Motivating the Lesson

Real-World Application
Christmas Clubs, offered by some banks, are a popular way to save extra money for the holidays. How much money would accumulate by the end of 52 weeks if $15 were deposited each week at an APR of 3.9% compounded weekly? **$795.70** This question can be addressed through the topic of iterated functions.

Resource Manager

 Reproducible Masters
- *Study Guide and Practice,* pp. 229–230
- *Enrichment,* p. 140
- *TI-92 & CBL Interdisciplinary Activities,* pp. 41–44

 Transparencies
- 5-Minute Check, 12-8
- Teaching, 12-8

In-Class Examples

Example 1
Population A recent study has determined the growth factor for the population of a particular bird species on an island to be 2.15. Suppose that the maximum population of that bird species which can be sustained on the island is 800 and the current population is 425.
a. Write an equation to model the population. $p_{n+1} = p_n + 2.15p_n(1 - p_n)$
b. Find the island's population of this bird species at the end of five years.   **809 birds**

Example 2
Banking Lucia has a savings account that has an annual yield of 5.8%. Find the balance in the account after the first four years if her initial balance is $1250.
$1566.22

The table shows the remainder of the values for the first fifteen years.

n	4	5	6	7	8	9	10	11	12	13	14	15
number of bears	514	489	508	494	505	497	503	498	501	499	501	499

At the end of the first 15 years, there could be 499 bears.

Notice that to determine the percent of the maximum sustainable population for a year, you must use the percent from the previous year. In Lesson 1-2, you learned that this process of composing a function with itself repeatedly is called *iteration*. Each output is called an *iterate*. To iterate a function $f(x)$, find the function value $f(x_0)$ of the initial value x_0. The second iterate is the value of the function performed on the output, that is $f(f(x_0))$.

Example ② **BANKING** Selina Anthony has a savings account that has an annual yield of 6.3%. Find the balance of the account after each of the first three years if her initial balance is $4210.

Real World Application

The balance of a savings account, using simple interest compounded at the end of a period of time, can be found by iterating the function $p_{n+1} = p_n + rp_n$, where p_n is the principal after n periods of time and r is the interest rate for a period of time.

$p_1 = p_0 + rp_0$
$p_1 = 4210 + (0.063)(4210)$ *$p_0 = 4210, r = 0.063$*
$p_1 = \$4475.23$

$p_2 = 4475.23 + (0.063)(4475.23)$ *$p_1 = 4475.23, r = 0.063$*
$p_2 \approx \$4757.17$

$p_3 = 4757.17 + (0.063)(4757.17)$ *$p_2 = 4757.17, r = 0.063$*
$p_3 \approx \$5056.87$

The balance in Ms. Anthony's account at the end of the three years is $5056.87.

Look Back
You can review graphing numbers in the complex plane in Lesson 9-6

Remember that every complex number has a real part and an imaginary part. The complex number $a + bi$ has been graphed on the complex plane at the right. The horizontal axis of the complex plane represents the real part of the number, and the vertical axis represents the imaginary part.

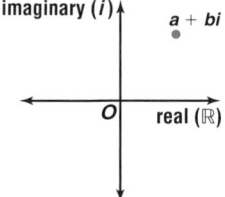

Functions can be iterated using the complex numbers as the domain.

Example **3** Find the first three iterates of the function $f(z) = 0.5z + 2$, if the initial value is $20 + 16i$.

f(z) is used to denote a function on the complex number plane.

$z_0 = 20 + 16i$
$z_1 = 0.5(20 + 16i) + 2$ or $12 + 8i$
$z_2 = 0.5(12 + 8i) + 2$ or $8 + 4i$
$z_3 = 0.5(8 + 4i) + 2$ or $6 + 2i$

We can graph this sequence of iterates on the complex plane. The graph at the right shows the **orbit,** or sequence of successive iterates, of the initial value of $z_0 = 20 + 16i$ from Example 3.

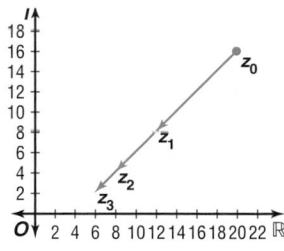

Example **4** Consider the function $f(z) = z^2 + c$, where c and z are complex numbers.

a. Find the first six iterates of the function for $z_0 = 1 + i$ and $c = -i$.

b. Plot the orbit of the initial point at $1 + i$ under iteration of the function $f(z) = z^2 - i$ for six iterations.

c. Describe the long-term behavior of the function under iteration.

a. $z_1 = (1 + i)^2 - i$
$= i$

$z_2 = (i)^2 - i$
$= -1 - i$

$z_3 = (-1 - i)^2 - i$
$= i$

$z_4 = (i)^2 - i$
$= -1 - i$

$z_5 = (-1 - i)^2 - i$
$= i$

$z_6 = (i)^2 - i$
$= -1 - i$

b.

c. The iterates repeat every two iterations.

For many centuries, people have used science and mathematics to better understand the world. However, Euclidean geometry, the familiar geometry of points, lines, and planes, is not adequate to describe the natural world. Objects like coastlines, clouds, and mountain ranges can be more easily approximated by a new type of geometry. This type of geometry is called **fractal geometry.** The function $f(z) = z^2 + c$, where c and z are complex numbers is central to the study of fractal geometry.

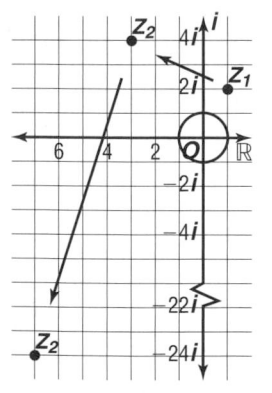
At the heart of fractal geometry are the *Julia sets*. In Chapter 9, you learned that Julia sets involve graphing the behavior of a function that is iterated in the complex plane. As a function is iterated, the iterates either escape or are held prisoner. If the sequence of iterates remains within a finite distance from $0 + 0i$, the initial point is called a **prisoner point.** If the sequence of iterates does not remain within a finite distance of $0 + 0i$, the point is called an **escaping point.**

Suppose the function $f(z) = z^2 + c$ is iterated for two different initial values. If the first one is an escaping point and the second one is a prisoner point, the graphs of the orbits may look like those shown below.

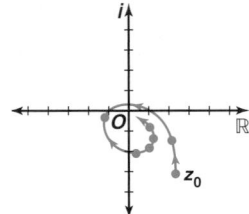

Example **5** Find the first four iterates of the function $f(z) = z^2 + c$ where $c = 0 + 0i$ for initial values whose moduli are in the regions $|z_0| < 1$, $|z_0| = 1$, and $|z_0| > 1$. Graph the orbits and describe the results.

Recall from Lesson 9-6 that the modulus of a complex number is the distance that the graph of the complex number is from the origin. So $|a + bi| = \sqrt{a^2 + b^2}$.

Choose an initial value in each of the intervals. For $|z_0| < 1$, we will use $z_0 = 0.5 + 0.5i$; for $|z_0| = 1$, $0.5 + 0.5\sqrt{3}i$; and for $|z_0| > 1$, $0.75 + 0.75i$.

For $|z_0| < 1$
$z_0 = 0.5 + 0.5i$
$z_1 = (0.5 + 0.5i)^2$ or $0.5i$
$z_2 = (0.5i)^2$ or -0.25
$z_3 = (-0.25)^2$ or 0.0625
$z_4 = 0.0625^2$ or 0.0039

For $|z_0| = 1$
$z_0 = 0.5 + 0.5\sqrt{3}i$
$z_1 = -0.5 + 0.5\sqrt{3}i$
$z_2 = -0.5 - 0.5\sqrt{3}i$
$z_3 = -0.5 + 0.5\sqrt{3}i$
$z_4 = -0.5 - 0.5\sqrt{3}i$

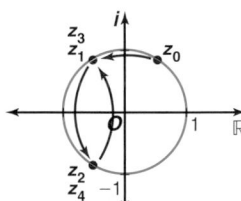

For $|z_0| > 1$
$z_0 = 0.75 + 0.75i$
$z_1 = 1.125i$
$z_2 = -1.2656$
$z_3 = 1.602$
$z_4 = 2.566$

In the first case, where $|z_0| < 1$, the iterates will approach 0, since each successive square will be smaller than the one before. Thus, the orbit in this case approaches 0, a fixed point. In the second case, where $|z_0| = 1$, the iterates are orbiting around on the unit circle. When $|z_0| > 1$, the iterates approach infinity, since each square will be greater than the one before.

The orbits in Example 5 demonstrate that the iterative behavior of various initial values in different regions behave in different ways. An initial value whose graph is inside the unit circle is a prisoner point. A point chosen on the unit circle stays on the unit circle, and a point outside of the unit circle escapes. Other functions of the form $f(z) = z^2 + c$ when iterated over the complex plane also have regions in which the points behave this way; however, they are usually not circles.

All of the initial points for a function on the complex plane are split into three sets, those that escape (called the *escape set* E), those that do not escape (called the *prisoner set* P) and the boundary between the escape set and the prisoner set (the Julia set). The escape set, the prisoner set, and the Julia set for the function in Example 5 are graphed at the right. The Julia set in Example 5 is the unit circle.

CHECK FOR UNDERSTANDING

Communicating Mathematics

Read and study the lesson to answer each question. **1–3. See margin.**

1. **Write** how iteration and composition functions are related.

2. **Describe** the orbit of a complex number under iteration.

3. **Explain** how a Julia set of a function is related to the prisoner set of the function.

Guided Practice

Find the first four iterates of each function using the given initial value. If necessary, round answers to the nearest hundredth.

4. $f(x) = x^2$; $x_0 = -1$
   1, 1, 1, 1

5. $f(x) = 2x - 5$; $x_0 = 2$
   −1, −7, −19, −43

Find the first three iterates of the function $f(z) = 0.6z + 2i$ for each initial value.

6. $z_0 = 6i$
   5.6i, 5.36i, 5.216i

7. $z_0 = 25 + 40i$
   15 + 26i, 9 + 17.6i, 5.4 + 12.56i

Find the first three iterates of the function $f(z) = z^2 + c$ for each given value of c and each initial value.

8. $c = 1 + 2i$; $z_0 = 0$
   1 + 2i, −2 + 6i, −31 − 22i

9. $c = 2 - 3i$; $z_0 = 1 + 2i$
   −1 + i, 2 − 5i, −19 − 23i

10. **Biology** The Verhulst population model describing the population of antelope in an area is $p_{n+1} = p_n + 1.75p_n(1 - p_n)$. The maximum population sustainable in the area is 40, and the current population is 24. Find the population of antelope after each of the first ten years. **41, 39, 40, 40, 40, 40, 40, 40, 40, 40**

 www.amc.glencoe.com/self_check_quiz

Lesson 12-8 Sequences and Iteration **819**

3 PRACTICE/APPLY

Error Analysis
Watch for students who make mathematical errors in the process of iteration.

Prevent by encouraging them to check carefully the accuracy of each iterate before going on to the next one.

Additional Answers
1. Iteration is the repeated composition of a function upon itself.

2. It is the sequence of iterates produced when a complex number is iterated for a function $f(z)$.

3. If the prisoner set is connected, then the Julia set is the boundary between the prisoner set and the escape set. If the prisoner set is disconnected, then the Julia set is the prisoner set.

The answers for Exercises 4 and 5 were calculated by rounding each iteration to the nearest hundreth. If rounded values are not used, students' answers may differ from those given.

Study Guide and Practice Masters, pp. 229–230

Additional Answers

The answers for Exercises 11–15 were calculated by rounding each iteration to the nearest hundredth. If rounded values are not used, students' answers may differ from those given.

11. 5, 8, 17, 44

12. 4; 16; 256; 65,536

13. 1; 16; 121; 13,456

14. 0, −1, 0, −1

15. −0.08, 0.09, −0.07, 0.08

16a. $t_1 = 2$
$t_2 = 1$
$t_3 = 2$
$t_4 = 1$
⋮
$t_{10} = 1$

16b. $t_1 = \frac{1}{2}$
$t_2 = 4$
$t_3 = \frac{1}{2}$
$t_4 = 4$
⋮
$t_{10} = 4$

16c. $t_1 = \frac{2}{7}$
$t_2 = 7$
$t_3 = \frac{2}{7}$
$t_4 = 7$
⋮
$t_{10} = 7$

16d. The values of the iterates alternate between $\frac{2}{x_0}$ and x_0.

17. $3 + 8i, 9 + 14i, 21 + 26i$

18. $11 − 2i, 25 − 6i, 53 − 14i$

19. $5 + 2i, 13 + 2i, 29 + 2i$

20. $1 − 6i, 5 − 14i, 13 − 30i$

21. $15 + 2i, 33 + 2i, 69 + 2i$

22. $3.6 − 4i, 10.2 − 10i, 23.4 − 22i$

24. $−2, 3, 8$

25. $−3i, −8 − 3i, 56 + 45i$

26. $4 + 2i, 15 + 18i, −96 + 542i$

27. $−2i, −4 − 4i, 28i$

28. $−i, −1, 1$

29. $2 + i, 5 + 7i, −22 + 73i$

EXERCISES

Practice

Find the first four iterates of each function using the given initial value. If necessary, round answers to the nearest hundredth.   **11–15. See margin.**

 11. $f(x) = 3x − 7; x_0 = 4$ 12. $f(x) = x^2; x_0 = −2$ 13. $f(x) = (x − 5)^2; x_0 = 4$

14. $f(x) = x^2 − 1; x_0 = −1$ 15. $f(x) = 2x^2 − x; x_0 = 0.1$

16. Find the first ten iterates of $f(x) = \frac{2}{x}$ for each initial value.   **a–d. See margin.**
   a. $x_0 = 1$ b. $x_0 = 4$ c. $x_0 = 7$
   d. What do you observe about the iterates of this function?

Find the first three iterates of the function $f(z) = 2z + (3 − 2i)$ for each initial value.   **17–22. See margin.**

 17. $z_0 = 5i$ 18. $z_0 = 4$ 19. $z_0 = 1 + 2i$

20. $z_0 = −1 − 2i$ 21. $z_0 = 6 + 2i$ 22. $z_0 = 0.3 − i$

23. Find the first three iterates of the function $f(z) = 3z − 2i$ for $z_0 = \frac{1}{3} + \frac{2}{3}i$.
   $1, 3 − 2i, 9 − 8i$

Find the first three iterates of the function $f(z) = z^2 + c$ for each given value of c and each initial value.   **24–29. See margin.**

 24. $c = −1; z_0 = 0 − i$ 25. $c = 1 − 3i; z_0 = i$

26. $c = 3 + 2i; z_0 = 1$ 27. $c = −4i; z_0 = 1 + i$

28. $c = 0; z_0 = \frac{\sqrt{2}}{2} − \frac{\sqrt{2}}{2}i$ 29. $c = 2 + 3i; z_0 = 1 − i$

Applications and Problem Solving

30. **Banking**   Amelia has a savings account that has an annual yield of 5.2%. Find the balance of the account after each of the first five years if her initial balance is $2000.   **$2104, $2213.41, $2328.51, $2449.59, $2576.97**

31. **Ecology**   The population of elk on the Bridger Range in the Rocky Mountains of western Montana has a growth factor of 2.5. The population in 1984 was 10% of the maximum population sustainable. What percent of the maximum sustainable population should be present in 2002?   **about 54%**

32. **Critical Thinking**   If $f(z) = z^2 + c$ is iterated with an initial value of $2 + 3i$ and $z_1 = −1 + 15i$, find c.   **$4 + 3i$**

33. **Critical Thinking**   In Exercise 16, find an initial value that produces iterates that all have the same value.   **$\pm\sqrt{2}$**

34. See students' work. Sample topics for discussion are judging soil quality and detection of heat stress in cows.

34. **Research**   Investigate the applications of fractal geometry to agriculture. How is fractal geometry being used in this field? Write several paragraphs about your findings.

35. **Critical Thinking**   **a. 1.414213562, 1.189207115, 1.090507733, 1.044273782**
   a. Use a calculator to find $\sqrt{2}, \sqrt{\sqrt{2}}, \sqrt{\sqrt{\sqrt{2}}}$, and $\sqrt{\sqrt{\sqrt{\sqrt{2}}}}$.
   b. Define an iterative function $f(z)$ that models the situation described in part a.   **$f(z) = \sqrt{z}, z_0 = 2$**
   c. Determine the limit of $f(z)$ as the number of iterations approaches infinity.   **1**
   d. Determine the limit of $f(z)$ as the number of iterations approaches infinity for integral values of $z_0 > 0$.   **1**

36. Write $2\left(\cos \dfrac{\pi}{3} + i \sin \dfrac{\pi}{3}\right)$ in exponential form. *(Lesson 12-7)*

37. Find the fifth term in the binomial expansion of $(2a - 3b)^8$. *(Lesson 12-6)*

38. State whether the series $\dfrac{1}{2} + \dfrac{1}{8} + \dfrac{1}{32} + \cdots$ is *convergent* or *divergent*. Explain your reasoning. *(Lesson 12-4)* **Convergent; see margin for explanation.**

39. **Nuclear Power** A nuclear cooling tower has an eccentricity of $\dfrac{7}{5}$. At its narrowest point the cooling tower is 130 feet wide. Determine the equation of the hyperbola used to generate the hyperboloid of the cooling tower. *(Lesson 10-4)*

40. **Optics** A beam of light strikes a diamond at an angle of $42°$. What is the angle of refraction if the index of refraction of a diamond is 2.42 and the index of refraction of air is 1.00? (Use Snell's Law, $n_1 \sin I = n_2 \sin r$, where n_1 is the index of refraction of the medium the light is exiting, n_2 is the index of refraction of the medium it is entering, I is the angle of incidence, and r is the angle of refraction.) *(Lesson 7-5)* **about 16°**

41. **Air Traffic Safety** The traffic pattern for airplanes into San Diego's airport is over the heart of downtown. Therefore, there are restrictions on the heights of new construction. The owner of an office building wishes to erect a microwave tower on top of the building. According to the architect's design, the angle of elevation from a point on the ground to the top of the 40-foot tower is $56°$. The angle of elevation from the ground point to the top of the building is $42°$. *(Lesson 5-4)*

   a. Draw a sketch of this situation. **See margin.**

   b. The maximum allowed height of any structure is 100 feet. Will the city allow the building of this tower? Explain.

42. Determine whether the graph has infinite discontinuity, jump discontinuity, or point discontinuity, or it is continuous. *(Lesson 3-5)*
infinite discontinuity

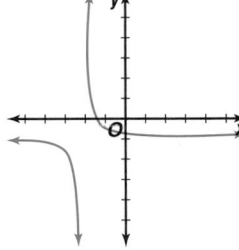

43. Find the maximum and minimum values of the function $f(x, y) = 2x + 8y + 10$ for the polygonal convex set determined by the following system of inequalities. *(Lesson 2-6)*
$$x \geq 3$$
$$x \leq 8$$
$$5 \leq y \leq 9$$
$$x + y \leq 14 \quad \textbf{92, 56}$$

44. **SAT/ACT Practice** Two students took a science test and received different scores between 10 and 100. If H equals the higher score and L equals the lower score, and the difference between the two scores equals the average of the two scores, what is the value of $\dfrac{H}{L}$? **D**

   A $\dfrac{3}{2}$ B 2   C $\dfrac{5}{2}$ D 3

   E It cannot be determined from the information given.

Extra Credit

If $x_{n+1} = \dfrac{\dfrac{N}{x_n} + x_n}{2}$ is an iterative function for $\sqrt{N}$, where $N \geq 0$, devise an iteration function for $\sqrt[3]{N}$, where $N \geq 0$.

$$x_{n+1} = \dfrac{\dfrac{N}{(x_n)^2} + 2x_n}{3}$$

Find the first four iterates of $f(x) = x^2 + 1$ using the given initial values. Round to the nearest hundredth.

1. $x_0 = 3$ 10; 101; 10,202; 104,080,805

2. $x_0 = -2$ 5; 26; 677; 458,330

Find the first three iterates of the function $f(z) = 3z + (2 - i)$ for each initial value.

3. $z_0 = 2i$ $2 + 5i$; $8 + 14i$; $26 + 41i$;

4. $z_0 = -i$ $2 - 4i$; $8 - 13i$; $26 - 40i$

Find the first three iterates of the function $f(z) = z^2 + c$ for each given value of c and each initial value.

5. $z_0 = 1 + i$, $c = 4$ $4 + 2i$; $16 + 16i$; $4 + 512i$

6. $z_0 = 1 - i$, $c = 3$ $3 - 2i$; $8 - 12i$; $-77 - 192i$

Motivating the Lesson

Real-World Application Ask students to give example of real-world situations that depend on recursive processes such as mathematical induction. **Sample answer: using an elevator to reach an upper floor of a building depends on first reaching the floor below the floor you want, and so on, down to the first floor.** Ask students to suggest other real-world examples.

12-9

Mathematical Induction

BUSINESS Felipe and Emily work a booth on the midway at the state fair. Before the fair opens, they discover that their cash supply contains only $5 and $10 bills. Rene volunteers to go and get a supply of $1s, but before leaving, Rene remarks that she could have given change for any amount greater than $4 had their cash supply contained only $2 and $5 bills. Felipe replies, "Even if we had $2 bills, you would still need $1 bills." Rene disagrees and says that she can prove that she is correct. *This problem will be solved in Example 4.*

A method of proof called **mathematical induction** can be used to prove certain conjectures and formulas. Mathematical induction depends on a recursive process that works much like an unending line of dominoes arranged so that if any one domino falls the next one will also fall.

Suppose the first domino is knocked over.	Condition 1
The first will knock down the second.	Condition 2
The second will knock down the third.	Condition 3
The third will knock down the fourth.	Condition 4
⋮	⋮

Thus, the whole line of dominos will eventually fall.

Mathematical induction operates in a similar manner. If a statement S_k implies the truth of S_{k+1} and the statement S_1 is true, then the chain reaction follows like an infinite set of tumbling dominoes.

S_1 is true.	Condition 1
S_1 implies that S_2 is true.	Condition 2
S_2 implies that S_3 is true.	Condition 3
S_3 implies that S_4 is true.	Condition 4
⋮	⋮

In general, the following steps are used to prove a conjecture by mathematical induction.

Proof by Mathematical Induction	1. First, verify that the conjecture S_n is valid for the first possible case, usually $n = 1$. *This is called the anchor step.* 2. Then, assume that S_n is valid for $n = k$, and use this assumption to prove that it is also valid for $n = k + 1$. *This is called the induction step.*

Resource Manager

Reproducible Masters
- *Study Guide and Practice,* pp. 231–232
- *Enrichment,* p. 141
- *Assessment and Evaluation,* p. 304

Transparencies
- 5-Minute Check, 12-9
- Teaching, 12-9

Technology/Multimedia
- Advanced MathPASS, Lesson 26

Thus, since S_n is valid for $n = 1$ (or any other first case), it is valid for $n = 2$. Since it is valid for $n = 2$, it is valid for $n = 3$, and so on, indefinitely.

Mathematical induction can be used to prove summation formulas.

Example ① **Prove that the sum of the first n positive integers is $\dfrac{n(n+1)}{2}$.**

Here S_n is defined as $1 + 2 + 3 + \cdots + n = \dfrac{n(n+1)}{2}$.

1. First, verify that S_n is valid for $n = 1$.

   Since the first positive integer is 1 and $\dfrac{1(1+1)}{2} = 1$, the formula is valid for $n = 1$.

2. Then assume that S_n is valid for $n = k$.

   $S_k \Rightarrow 1 + 2 + 3 + \cdots + k = \dfrac{k(k+1)}{2}$ *Replace n with k.*

   Next, prove that it is also valid for $n = k + 1$.

   $S_{k+1} \Rightarrow 1 + 2 + 3 + \cdots + k + (k+1) = \dfrac{k(k+1)}{2} + (k+1)$ *Add (k + 1) to both sides of S_k.*

   We can simplify the right side by adding $\dfrac{k(k+1)}{2} + (k+1)$.

   $S_{k+1} \Rightarrow 1 + 2 + 3 + \cdots + k + (k+1) = \dfrac{k(k+1) + 2(k+1)}{2}$ *(k + 1) is a common factor*

   $= \dfrac{(k+1)(k+2)}{2}$

   If $k + 1$ is substituted into the original formula $\left(\dfrac{n(n+1)}{2}\right)$, the same result is obtained.

   $\dfrac{(k+1)(k+1+1)}{2} = \dfrac{(k+1)(k+2)}{2}$

   Thus, if the formula is valid for $n = k$, it is also valid for $n = k + 1$. Since S_n is valid for $n = 1$, it is also valid for $n = 2$, $n = 3$, and so on. That is, the formula for the sum of the first n positive integers holds.

Mathematical induction is also used to prove properties of divisibility. Recall that an integer p is *divisible* by an integer q if $p = qr$ for some integer r.

Example ② **Prove that $4^n - 1$ is divisible by 3 for all positive integers n.**

Using the definition of divisibility, we can state the conjecture as follows:

Notice that here S_n does not represent a sum but a conjecture.

$S_n \Rightarrow 4^n - 1 = 3r$ for some integer r

1. First verify that S_n is valid for $n = 1$.

   $S_1 \Rightarrow 4^1 - 1 = 3$. Since 3 is divisible by 3, S_n is valid for $n = 1$.

2. Then assume that S_n is valid for $n = k$ and use this assumption to prove that it is also valid for $n = k + 1$.

   $S_k \Rightarrow 4^k - 1 = 3r$ for some integer r *Assume S_k is true.*
   $S_{k+1} \Rightarrow 4^{k+1} - 1 = 3t$ for some integer t *Show that S_{k+1} must follow.*

   (continued on the next page)

---

2 TEACH

In-Class Examples

Example 1
Prove that the sum of the first n even positive integers is $n(n + 1)$.
Step 1: Verify that the formula S_n is valid for $n = 1$. Since the first even positive integer is 2 and $1(1 + 1) = 2$, the formula is valid for $n = 1$.
Step 2: Assume that the formula is valid for $n = k$ and prove that it is also valid for $n = k + 1$.
$S_k \Rightarrow 2 + 4 + 6 + \cdots + 2k$
$\quad = k(k + 1)$
$S_{k+1} \Rightarrow 2 + 4 + 6 + \cdots + 2k + 2(k + 1)$
$\quad = k(k + 1) + 2(k + 1)$
$\quad = (k + 2)(k + 1)$
If $k + 1$ is substituted into the original formula, the same result is obtained. Thus, if the formula is valid for $n = k$, it is also valid for $n = k + 1$. Since the formula is valid for $n = 1$, it is also valid for $n = 2$. Since it is valid for $n = 2$, it is also valid for $n = 3$, and so on indefinitely. Thus, the formula is valid for all positive integral values of n.

Example 2
Prove that $6^n - 1$ is divisible by 5 for all positive integers n.
Step 1: Verify that $6^n - 1$ is divisible by 5, S_n, for $n = 1$. Since 5 is divisible by 5, the statement is valid for $n = 1$.
Step 2: Assume that $6^n - 1$ is divisible by 5 for $n = k$ and use this assumption to prove that it is also divisible by 5 for $n = k + 1$.
$S_k \Rightarrow 6^k - 1 = 5r$ for some integer r
$S_{k+1} \Rightarrow 6^{k+1} - 1 = 5t$ for some integer t
$\quad 6^k - 1 = 5r$
$\quad 6(6^k - 1) = 6(5r)$
$\quad 6^{k+1} - 6 = 30r$
$\quad 6^{k+1} - 1 = 30r + 5$
$\quad 6^{k+1} - 1 = 5(6r + 1)$
Let $t = 6r + 1$, an integer. Then $6^{k+1} - 1 = 5t$. We have shown that if S_k is valid, then S_{k+1} is also valid. Since S_n is valid for $n = 1$, it is also valid for $n = 2$, $n = 3$, and so on. Hence, $6^n - 1$ is divisible by 5 for all positive integers n.

Teaching Tip For Example 1, remind students that proof by mathematical induction provides for all possible numbers of terms. Without mathematical induction we can verify the sum for the first 1000 terms but can never be certain that the expression for the sum would hold true for 1001 terms.

Teaching Tip Remind students to add the exponents when applying the distributive property for $6(6^k - 1)$: $6^1(6^k - 1) = 6^{k+1} - 6$.

For this proof, rewrite the left-hand side of S_k so that it matches the left-hand side of S_{k+1}.

$4^k - 1 = 3r$	S_k
$4(4^k - 1) = 4(3r)$	*Multiply each side by 4.*
$4^{k+1} - 4 = 12r$	*Simplify.*
$4^{k+1} - 1 = 12r + 3$	*Add 3 to each side.*
$4^{k+1} - 1 = 3(4r + 3)$	*Factor.*

Let $t = 4r + 3$, an integer. Then $4^{k+1} - 1 = 3t$

We have shown that if S_k is valid, then S_{k+1} is also valid. Since S_n is valid for $n = 1$, it is also valid for $n = 2$, $n = 3$, and so on. Hence, $4^n - 1$ is divisible by 3 for all positive integers n.

There is no "fixed" way of completing Step 2 for a proof by mathematical induction. Often, each problem has its own special characteristics that require a different technique to complete the proof. You may have to multiply the numerator and denominator of an expression by the same quantity, factor or expand an expression in a special way, or see an important relationship with the distributive property.

Example **3** **Prove that $6^n - 2^n$ is divisible by 4 for all positive integers n.**

Begin by restating the conjecture using the definition of divisibility.

$S_n \Rightarrow 6^n - 2^n = 4r$ for some integer r

1. Verify that S_n is valid for $n = 1$

   $S_1 \Rightarrow 6^1 - 2^1 = 4$. Since 4 is divisible by 4, S_n is valid for $n = 1$.

2. Assume that S_n is valid for $n = k$ and use this to prove that it is also valid for $n = k + 1$.

   $S_k \Rightarrow 6^k - 2^k = 4r$ for some integer r *Assume S_k is true*

   $S_{k+1} \Rightarrow 6^{k+1} - 2^{k+1} = 4t$ for some integer t *Show that S_{k+1} must follow.*

   Begin by rewriting S_k as $6^k = 2^k + 4r$. Now rewrite the left-hand side of this expression so that it matches the left-hand side of S_{k+1}.

$6^k = 2^k + 4r$	S_k
$6^k \cdot 6 = (2^k + 4r)(2 + 4)$	*Multiply each side by a quantity equal to 6.*
$6^{k+1} = 2^{k+1} + 4(2^k) + 8r + 16r$	*Distributive Property*
$6^{k+1} - 2^{k+1} = 2^{k+1} + 4(2^k) + 8r + 16r - 2^{k+1}$	*Subtract 2^{k+1} from each side.*
$6^{k+1} - 2^{k+1} = 4 \cdot 2^k + 24r$	*Simplify.*
$6^{k+1} - 2^{k+1} = 4(2^k + 6r)$	*Factor.*

Let $t = 2^k + 6r$, an integer. Then $6^{k+1} - 2^{k+1} = 4t$.

We have shown that if S_k is valid, then S_{k+1} is also valid. Since S_n is valid for $n = 1$, it is also valid for $n = 2$, $n = 3$, and so on. Hence, $6^n - 2^n$ is divisible by 4 for all positive integers n.

Let's look at another proof that requires some creative thinking to complete.

Example ❹ **BUSINESS** **Refer to the application at the beginning of the lesson. Use mathematical induction to prove that Emily can give change in \$2s and \$5s for amounts of \$4, \$5, \$6, ..., \$n.**

Let n = the dollar amount for which a customer might need change.
Let a = the number of \$2 bills used to make the exchange.
Let b = the number of \$5 bills used to make the exchange.

$S \Rightarrow n = 2a + 5b$

Notice that the first possible case for n is not always 1.

1. First, verify that S_n is valid for $n = 4$. In other words, are there values for a and b such that $4 = 2a + 5b$?

Yes. If $a = 2$ and $b = 0$, then $4 = 2(2) + 5(0)$, so S_n is valid for the first case.

2. Then, assume that S_n is valid for $n = k$ and prove that it is also valid for $n = k + 1$.

$S_k \Rightarrow k = 2a + 5b$

$S_{k+1} \Rightarrow k + 1 = 2a + 5b + 1$ *Add 1 to each side.*

$\qquad = 2a + 5b + 6 - 5$ $1 = 6 - 5$

$\qquad = 2(a + 3) + 5(b - 1)$ $2a + 6 = 2(a + 3); 5b - 5 = 5(b - 1)$

Notice that this expression is true for all $a \geq 0$ and $b \geq 1$, since the expression $b - 1$ must be nonnegative. Therefore, we must also consider the case where $b = 0$.

If $b = 0$, then $n = 2a + 5(0) = 2a$. Assuming S_n to be true for $n = k$, we must show that it is also true for $k + 1$.

$S_k \Rightarrow k = 2a$

$S_{k+1} \Rightarrow k + 1 = 2a + 1$

$\qquad = 2(a - 2) + 4 + 1$ $2a = 2(a - 2) + 4$

$\qquad = 2(a - 2) + 1 \cdot 5$ $4 + 1 = 1 \cdot 5$

Notice that this expression is only true for all $a \geq 2$, since the expression $a - 2$ must be nonnegative. However, we have already considered the case where $a = 0$.

Thus, it can be concluded that since the conjecture is true for $k = 4$, it also valid for $n = k + 1$. Therefore, Emily can make change for amounts of \$4, \$5, \$6, ... \$n using only \$2 and \$5 bills.

Lesson 12-9 Mathematical Induction **825**

Teaching Tip When working Example 4, point out that proof by mathematical induction sometimes requires the trial-and-error method and patience in order to complete the task.

In-Class Example

Example 4
Business The summation of the first n positive integers has great utility in business. In finance, The Rule of 78 involves consecutively adding 1 to 12 months, resulting in $\frac{12 \cdot 13}{2} = 78$. Accountants add the first n consecutive years together in the sum-of-the-years depreciation method. Devise a summation formula for adding every other even integer beginning with 12, and prove it by mathematical induction.

$S_n \Rightarrow 12 + 16 + 20 + \cdots + 4(n + 2) = 2n(n + 5)$

Step 1: Verify that the formula is valid for $n = 1$.
Since the first number in the series is 12 and $2(1)(1 + 5) = 12$, the formula is valid for $n = 1$.

Step 2: Assume that the formula is valid for $n = k$ and derive a formula for $n = k + 1$.

$S_k \Rightarrow 12 + 16 + 20 + \cdots + 4(k + 2) = 2k(k + 5)$

$S_{k+1} \Rightarrow 12 + 16 + 20 + \cdots + 4(k + 2) + 4(k + 3)$

$\qquad = 2k(k + 5) + 4(k + 3)$

$\qquad = 2k^2 + 10k + 4k + 12$

$\qquad = 2k^2 + 14k + 12$

$\qquad = 2(k^2 + 7k + 6)$

$\qquad = 2(k + 1)(k + 6)$

If $(k + 1)$ is substituted into the original formula, the same result is obtained. Thus, if the formula is valid for $n = k$, it is also valid for $n = k + 1$. Since the formula is valid for $n = 1$, it is also valid for $n = 2$. Since it is valid for $n = 2$, it is also valid for $n = 3$, and so on, indefinitely. Thus, the formula is valid for all positive integral values of n.

Addressing Individual Needs

 Verbal/Linguistic Have students discuss the process of proving a statement by mathematical induction. Ask them to state the importance of verifying the first possible case.

Error Analysis

Watch for students who bypass the first step of the proof by mathematical induction.

Prevent by emphasizing that if $n = 1$ does not work, the original statement is not true and is therefore not a theorem.

Assignment Guide
Core: 11–27 odd, 29–35
Enriched: 12–22 even, 23–35

Additional Answers

1. The $n + 1$ case shows that the premise is true for an infinite number of cases.

2. Provide a counterexample.

3b. Since 3 is the first term in the sequence of partial sums and $1(1 + 2) = 3$, the formula is valid for $n = 1$.

Since 8 is the second term in the sequence of partial sums and $2(2 + 2) = 8$, the formula is valid for $n = 2$.

Since 15 is the third term in the sequence of partial sums and $3(3 + 2) = 15$, the formula is valid for $n = 3$.

Study Guide and Practice Masters, pp. 231–232

NAME _____ DATE _____ PERIOD _____

Study Guide

Mathematical Induction

A method of proof called **mathematical induction** can be used to prove certain conjectures and formulas. The following example demonstrates the steps used in proving a summation formula by mathematical induction.

Example **Prove that the sum of the first n positive even integers is $n(n + 1)$.**

Here S_n is defined as $2 + 4 + 6 + \cdots + 2n = n(n + 1)$.

1. First, verify that S_n is valid for the first possible case, $n = 1$. Since the first positive even integer is 2 and $1(1 + 1) = 2$, the formula is valid for $n = 1$.

2. Then, assume that S_n is valid for $n = k$.
$S_k \Rightarrow 2 + 4 + 6 + \cdots + 2k = k(k + 1)$ *Replace n with k.*

Next, prove that S_n is also valid for $n = k + 1$.
$S_{k+1} \Rightarrow 2 + 4 + 6 + \cdots + 2k + 2(k + 1)$
$= k(k + 1) + 2(k + 1)$ *Add $2(k + 1)$ to both sides.*

We can simplify the right side by adding $k(k + 1) + 2(k + 1)$.
$S_{k+1} \Rightarrow 2 + 4 + 6 + \cdots + 2k + 2(k + 1)$
$= (k + 1)(k + 2)$ *$(k + 1)$ is a common factor.*

If $k + 1$ is substituted into the original formula $(n(n + 1))$, the same result is obtained.
$(k + 1)[(k + 1) + 1]$ or $(k + 1)(k + 2)$

Thus, if the formula is valid for $n = k$, it is also valid for $n = k + 1$. Since S_n is valid for $n = 1$, it is also valid for $n = 2$, $n = 3$, and so on. That is, the formula for the sum of the first n positive even integers holds.

© Glencoe/McGraw-Hill   231   *Advanced Mathematical Concepts*

Communicating Mathematics

Read and study the lesson to answer each question.

1. **Explain** why it is necessary to prove the $n + 1$ case in the process of mathematical induction.   **1–2. See margin.**

2. **Describe** a method you might use to show that a conjecture is false.

3. Consider the series $3 + 5 + 7 + 9 + \cdots + 2n + 1$.
   a. **Write a conjecture** for the general formula S_n for the sum of the first n terms. $S_n \Rightarrow n(n + 2)$
   b. Verify S_n for $n = 1, 2,$ and 3.   **See margin.**
   c. Write S_k and S_{k+1}. $S_k \Rightarrow k(k + 2)$; $S_{k+1} \Rightarrow (k + 1)(k + 3)$

4. **Restate** the conjecture that $8^n - 1$ is divisible by 7 for all positive integers n using the definition of divisibility. $8^n - 1 = 7r$ **for some integer r.**

5. *Math Journal* The "domino effect" presented in this lesson is just one way to illustrate the principle behind mathematical induction. **Write** a paragraph describing another situation in real life that illustrates this principle.   **See margin.**

Guided Practice

Use mathematical induction to prove that each proposition is valid for all positive integral values of n.   **6–9. See Answer Appendix.**

6. $3 + 5 + 7 + \cdots + (2n + 1) = n(n + 2)$

7. $2 + 2^2 + 2^3 + \cdots + 2^n = 2(2^n - 1)$

8. $\dfrac{1}{2} + \dfrac{1}{2^2} + \dfrac{1}{2^3} + \cdots + \dfrac{1}{2^n} = 1 - \dfrac{1}{2^n}$

9. $3^n - 1$ is divisible by 2.

10. **Number Theory** The ancient Greeks were very interested in number patterns. Triangular numbers are numbers that can be represented by a triangular array of dots, with n dots on each side. The first three triangular numbers are 1, 3, and 6.

    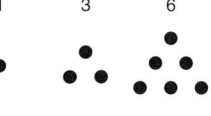

    a. Find the next five triangular numbers.   **10, 15, 21, 28, 36**
    b. Write a general formula for the nth term of this sequence. $a_n = \dfrac{n(n + 1)}{2}$
    c. Prove that the sum of first n triangular numbers can be found using the formula $\dfrac{n(n + 1)(n + 2)}{6}$.   **See margin.**

Practice

For Exercises 11-19, use mathematical induction to prove that each proposition is valid for all positive integral values of n.   **11–19. See Answer Appendix.**

 11. $1 + 5 + 9 + \cdots + (4n - 3) = n(2n - 1)$

12. $1 + 4 + 7 + \cdots + (3n - 2) = \dfrac{n(3n - 1)}{2}$

 13. $-\dfrac{1}{2} - \dfrac{1}{4} - \dfrac{1}{8} - \cdots - \dfrac{1}{2^n} = \dfrac{1}{2^n} - 1$

14. $1 + 8 + 27 + \cdots + n^3 = \dfrac{n^2(n + 1)^2}{4}$

15. $1^2 + 3^2 + 5^2 + \cdots + (2n - 1)^2 = \dfrac{n(2n - 1)(2n + 1)}{3}$

  www.amc.glencoe.com/self_check_quiz

Additional Answer

5. **Sample answer: if we wish to prove that we can climb a ladder with an indefinite number of steps, we must prove the following. First, we must show that we can climb off the ground to rung 1. Next, we must show that if we can climb to rung k, we can climb to rung $k + 1$.**

16. $1 + 2 + 4 + \cdots + 2^{n-1} = 2^n - 1$

17. $7^n + 5$ is divisible by 6

18. $8^n - 1$ is divisible by 7

19. $5^n - 2^n$ is divisible by 3

 20. Prove $S_n \Rightarrow a + (a + d) + (a + 2d) + \cdots + [a + (n-1)d] = \frac{n}{2}[2a + (n-1)d]$.

21. Prove $S_n \Rightarrow \frac{1}{1 \cdot 2} + \frac{1}{2 \cdot 3} + \frac{1}{3 \cdot 4} + \cdots + \frac{1}{n(n+1)} = \frac{n}{n+1}$.

22. Prove $S_n \Rightarrow 2^{2n+1} + 3^{2n+1}$ is divisible by 5.

Applications and Problem Solving

Real World Application

23. Historical Proof In 1730, Abraham De Moivre proposed the following theorem for finding the power of a complex number written in polar form.

$$[r(\cos \theta + i \sin \theta)]^n = r^n(\cos n\theta + i \sin n\theta)$$

See Answer Appendix.

Prove that De Moivre's Theorem is valid for any positive integer n.

24. Number Patterns Consider the pattern of dots shown below.

a. Find the next figure in this pattern. **See margin.**

b. Write a sequence to represent the number of dots added to the previous figure to create the next figure in this pattern. The first term in your sequence should be 1. **1, 3, 5, 7, ...**

c. Find the general form for the nth term of the sequence found in part **b.** **$2n - 1$**

d. Determine a formula that will calculate the total number of dots for the nth figure in this dot pattern. **n^2**

e. Prove that the formula found in part d is correct using mathematical induction. **See Answer Appendix.**

25. Critical Thinking Prove that $n^2 + 5n$ is divisible by 2 for all positive integral values of n. **See Answer Appendix.**

26. Club Activites Melissa is the activities director for her school's science club and needs to coordinate a "get to know you" activity for the group's first meeting. The activity must last at least 40 minutes but no more than 60 minutes. Melissa has chosen an activity that will require each participant to interact with every other participant in attendance only once and estimates that if her directions are followed, each interaction should take approximately 30 seconds.

26a. $\frac{n(n-1)}{2}$

a. If n people participate in Melissa's activity, develop a formula to calculate the total number of such interactions that should take place.

b. Prove that the formula found in part a is valid for all positive integral values of n. **See Answer Appendix.**

c. If 15 people participate, will Melissa's activity meet the guidelines provided to her? Explain. **See margin.**

Lesson 12-9 Mathematical Induction **827**

Additional Answers (p. 826)

10c. Step 1: Verify that the formula is valid for $n = 1$. Since 1 is the first term in the sequence and $\frac{1(1+1)(1+2)}{6} = 1$, the formula is valid for $n = 1$.

Step 2: Assume that the formula is valid for $n = k$ and derive a formula for $n = k + 1$.

$S_k \Rightarrow 1 + 3 + 6 + \cdots + \frac{k(k+1)}{2} = \frac{k(k+1)(k+2)}{6}$

$S_{k+1} \Rightarrow 1 + 3 + 6 + \cdots + \frac{k(k+1)}{2} + \frac{(k+1)(k+2)}{2} = \frac{k(k+1)(k+2)}{6} + \frac{(k+1)(k+2)}{2}$

$= \frac{k(k+1)(k+2) + 3(k+1)(k+2)}{6} = \frac{(k+1)(k+2)(k+3)}{6}$ *(continued)*

Teaching Tip Remind students that they encountered DeMoivre's Theorem in Lesson 9-8. In Exercise 23, they can now verify this theorem.

Teaching Tip For Exercise 28, you may wish to show alternate ways to verify that $0.99999 = 1$. One method is to let $n = 0.99999$ and use the process for changing repeating decimals to rational numbers in the form of $\frac{p}{q}$.

Additional Answers

24a.

26c. Yes; 15 people will require $\frac{15(14)}{2}$ or 105 interactions and last approximately 105(0.5) or 52.5 minutes.

10c. (continued) Apply the original formula for $n = k + 1$.

$\frac{(k+1)[(k+1)+1][(k+1)+2]}{6} = \frac{(k+1)(k+2)(k+3)}{6}$

The formula gives the same result as adding the $(k+1)$ term directly. Thus, if the formula is valid for $n = k$, it is also valid for $n = k + 1$. Since the formula is valid for $n = 1$, it is also valid for $n = 2$. Since it is valid for $n = 2$, it is also valid for $n = 3$, and so on indefinitely. Thus, the formula is valid for all positive integral values of n.

Lesson 12-9 **827**

4 ASSESS

Open-Ended Assessment

Modeling Describe how the **BASIC** program below is a model of mathematical induction.

```
10  INPUT N
20  PRINT N
30  LET N = N + 1
40  GOTO 20
50  END
```

Sample answer: S_1 is the input, which is printed in line 20. Since it is printed, S_1 is true. The value of the output of line 20 is incremented by one and becomes the new output. This forms an arithmetic sequence. You can assume that S_k is true, namely that the program will output the first k values of the arithmetic sequence. If S_k is printed, then S_{k+1} is printed.

 Chapter 12, Quiz D (Lessons 12-8 and 12-9) is available in the *Assessment and Evaluation Masters*, p. 304.

Enrichment Masters, p. 141

828 *Chapter 12*

---

27. **Critical Thinking** Use mathematical induction to prove that the Binomial Theorem is valid for all positive integral values of n. **See Answer Appendix.**

28. **Number Theory** Consider the following statement: $0.99999 \cdots = 1$.
   a. Write $0.\overline{9}$ as an infinite series. $0.9 + 0.09 + 0.009 + \cdots$
   b. Write an expression for the nth term of this series. $\dfrac{9}{10^n}$
   c. Write a formula for the sum of the first n terms of this series. $\dfrac{10^n - 1}{10^n}$
   d. Prove that the formula you found in part **c** is valid for all positive integral values of n using mathematical induction. **See Answer Appendix.**
   e. Use the formula found in part **c** to prove that $0.99999 \cdots = 1$. (*Hint:* Use what you know about the limits of infinite sequences.) **See Answer Appendix.**

Mixed Review

29. Find the first three iterates of the function $f(z) = 2z + i$ for $z_0 = 4 - i$. (*Lesson 12-8*) $8 - i, 16 - i, 32 - i$

30. Find d for the arithmetic sequence for which $a_1 = -6$ and $a_{29} = 64$. (*Lesson 12-1*) $\dfrac{5}{2}$

31. $\dfrac{(x-2)^2}{4} + \dfrac{(y-5)^2}{25} = 1$; ellipse

31. Write the standard form of the equation $25x^2 + 4y^2 - 100x - 40y + 100 = 0$. Then identify the conic section this equation represents. (*Lesson 10-6*)

32. **Geology** A *drumlin* is an elliptical streamlined hill whose shape can be expressed by the equation $r = \ell \cos k\theta$ for $-\dfrac{\pi}{2k} \le \theta \le \dfrac{\pi}{2k}$, where ℓ is the length of the drumlin and $k > 1$ is a parameter that is the ratio of the length to the width. Find the area in square centimeters, $A = \dfrac{\ell^2 \pi}{4k}$, of a drumlin modeled by the equation $r = 250 \cos 4\theta$. (*Lesson 9-3*) **about 12,271.85 m²**

glacier

kames   drumlins   esker

33. Write an equation of a sine function with amplitude $\dfrac{3}{4}$ and period π. (*Lesson 6-4*) $y = \pm\dfrac{3}{4} \sin 2x$

34. Find the values of x in the interval $0° \le x \le 360°$ for which $x = \cos^{-1}\left(-\dfrac{\sqrt{3}}{2}\right)$. (*Lesson 5-5*) **150°, 210°**

35. **SAT Practice** The area of *EFGH* is 25 square units. Points A, B, C, and D are on the square. *ABCD* is a rectangle, but not a square. Calculate the perimeter of *ABCD* if the distance from E to A is 1 and the distance from E to B is 1. **B**
   A 64 units
   B $10\sqrt{2}$ units
   C 10 units
   D 8 units
   E $5\sqrt{2}$ units

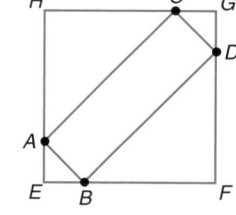

Extra Practice See p. A50.

Extra Credit

Give an example of a statement that is true for all positive integers $n \ge 2$, but is not true for $n = 1$. **Sample answer:**
$2^n > 2^{n-1} + 1$

VOCABULARY

arithmetic mean (p. 760)	Fibonacci sequence (p. 806)	nth partial sum (p. 761)
arithmetic sequence (p. 759)	fractal geometry (p. 817)	orbit (p. 817)
arithmetic series (p. 761)	geometric mean (p. 768)	Pascal's Triangle (p. 801)
Binomial Theorem (p. 803)	geometric sequence (p. 766)	prisoner point (p. 818)
common difference (p. 759)	geometric series (p. 769)	ratio test (p. 787)
common ratio (p. 766)	index of summation (p. 794)	recursive formula (p. 760)
comparison test (p. 789)	infinite sequence (p. 774)	sequence (p. 759)
convergent series (p. 786)	infinite series (p. 778)	sigma notation (p. 794)
divergent series (p. 786)	limit (p. 774)	term (p. 759)
escaping point (p. 818)	mathematical induction	trigonometric series (p. 808)
Euler's Formula (p. 809)	(p. 822)	
exponential series (p. 808)	n factorial (p. 796)	

UNDERSTANDING AND USING THE VOCABULARY

Choose the letter of the term that best matches each statement or phrase.

1. each succeeding term is formulated from one or more previous terms   **d**

2. ratio of successive terms in a geometric sequence   **i**

3. used to determine convergence   **m**

4. can have infinitely many terms   **j**

5. $e^{ix} = \cos \alpha + i \sin \alpha$ **k**

6. the terms between any two nonconsecutive terms of a geometric sequence   **f**

7. indicated sum of the terms of an arithmetic sequence   **c**

8. $n! = n(n - 1)(n - 2) \cdot \cdots \cdot 1$ **e**

9. used to demonstrate the validity of a conjecture based on the truth of a first case, the assumption of truth of a kth case, and the demonstration of truth for the $(k + 1)$th case   **b**

10. an infinite series with a sum or limit   **h**

a. term
b. mathematical induction
c. arithmetic series
d. recursive formula
e. n factorial
f. geometric mean
g. sigma notation
h. convergent
i. common ratio
j. infinite sequence
k. Euler's Formula
l. prisoner point
m. ratio test
n. limit

 interNET CONNECTION   For additional review and practice for each lesson, visit: **www.amc.glencoe.com**

Chapter 12   Study Guide and Assessment   **829**

STUDY GUIDE AND ASSESSMENT

VOCABULARY

This section provides a listing of the new terms, properties, and phrases that were introduced in this chapter. The exercises check students' understanding of the terms by using a variety of verbal formats including matching, completion, and true/false.

Glossary A complete glossary of terms begins on page A71.

GLENCOE'S ASSESSMENT ADVANTAGE

Six forms of Chapter 12 Test are available in the *Assessment and Evaluation Masters* as shown in the chart.

Chapter 12 Test, Form 1B, is shown at the right. Chapter 12 Test, Form 2B, is shown on the next page.

	TEST OPTIONS		
Form	**Pages**	**Type**	
1A	289–290	Honors	Form 1 tests are multiple choice.
1B	291–292	Average	
1C	293–294	Basic	
2A	295–296	Honors	Form 2 tests are free response.
2B	297–298	Average	
2C	299–300	Basic	

SKILLS AND CONCEPTS

Objectives and Examples This section reviews the skills and concepts of the chapter and shows completely worked examples.

Review Exercises These exercises provide practice for the corresponding objectives.

Assessment and Evaluation Masters, pp. 297–298

NAME _____ DATE _____ PERIOD _____

Chapter 12

Chapter 12 Test, Form 2B

1. Find d for the arithmetic sequence in which $a_1 = 6$ and $a_{13} = -42$. 1. ____−4____

2. Find the 40th term in the arithmetic sequence 7, 4.4, 1.8, −0.8, ⋯. 2. ____−94.4____

3. Find the sum of the first 30 terms in the arithmetic series $10 + 6 + 2 - 2 - \cdots$. 3. ____−1440____

4. Find the ninth term in the geometric sequence $\frac{1}{27}, \frac{1}{9}, \frac{1}{3}, \ldots$. 4. ____243____

5. Find the sum of the first eight terms in the geometric series $64 - 32 + 16 - 8 + \cdots$. 5. ____42.5____

6. Form a sequence that has three geometric means between −4 and −324. 6. ____−4, ±12, −36, ±108, −324____

7. Find $\lim \frac{2n-1}{n^3}$ or state that the limit does not exist. 7. ____0____

8. Find the sum of the series $12 + 8 + \frac{16}{3} + \cdots$ or state that the sum does not exist. 8. ____36____

9. Write $8.\overline{18}$ as a fraction. 9. ____$\frac{90}{11}$____

Determine whether each series is convergent or divergent.

10. $\frac{4}{1} + \frac{7}{1 \cdot 2} + \frac{10}{1 \cdot 2 \cdot 3} + \cdots$ 10. ____convergent____

11. $\frac{3}{1} + \frac{3^2}{2} + \frac{3^3}{3} + \cdots$ 11. ____divergent____

© Glencoe/McGraw-Hill   **297**   Advanced Mathematical Concepts

---

NAME _____ DATE _____ PERIOD _____

Chapter 12

Chapter 12 Test, Form 2B (continued)

12. Write $\sum_{k=0}^{3}(k+1)(k+2)$ in expanded form and then find the sum. 12. ____(1)(2) + (2)(3) + (3)(4) + (4)(5); 40____

13. Express the series $\frac{1 \cdot 0}{2} + \frac{2 \cdot 1}{3} + \frac{3 \cdot 2}{4} + \cdots + \frac{10 \cdot 9}{11}$ using sigma notation. 13. ____$\sum_{k=2}^{11} \frac{(k-1)(k-2)}{k}$____

14. Use the Binomial Theorem to expand $(2p - 3q)^4$. 14. ____$16p^4 - 96p^3q + 216p^2q^2 - 216pq^3 + 81q^4$____

15. Find the fifth term in the expansion of $(4x + 2y)^7$. 15. ____$35,840x^3y^4$____

16. Use the first five terms of the cosine series $\cos x = 1 - \frac{x^2}{2!} + \frac{x^4}{4!} - \frac{x^6}{6!} + \frac{x^8}{8!} - \cdots$ to approximate the value of $\cos \frac{\pi}{4}$ to four decimal places. 16. ____0.7071____

17. Find $\ln(-13.4)$ to four decimal places. 17. ____$i\pi + 2.5953$____

18. Find the first three iterates of the function $f(z) = 0.5z$ for $z_0 = 4 - 2i$. 18. ____$2 - i, 1 - 0.5i, 0.5 - 0.25i$____

19. Find the first three iterates of the function $f(z) = z^2 + c$ for $c = 2i$ and $z_0 = 1$. 19. ____$1 + 2i, -3 + 6i, -27 - 34i$____

20. Use mathematical induction to prove that $7 + 9 + 11 + \cdots + (2n + 5) = n(n + 6)$. Write your proof on a separate piece of paper. 20. ____See students' work.____

Bonus Find the sum of the coefficients of the expansion of $(x + y)^7$. **Bonus:** ____128____

© Glencoe/McGraw-Hill   **298**   Advanced Mathematical Concepts

---

SKILLS AND CONCEPTS

OBJECTIVES AND EXAMPLES	REVIEW EXERCISES

Lesson 12-1 Find the nth term and arithmetic means of an arithmetic sequence.

  Find the 35th term in the arithmetic sequence $-5, -1, 3, \ldots$.
Begin by finding the common difference d.
$d = -1 - (-5)$ or 4
Use the formula for the nth term.
$a_n = a_1 + (n - 1)d$
$a_{35} = -5 + (35 - 1)(4)$ or 131

11. Find the next four terms of the arithmetic sequence 3, 4.3, 5.6, …. **6.9, 8.2, 9.5, 10.8**

12. Find the 20th term of the arithmetic sequence for which $a_1 = 5$ and $d = -3$. **−52**

13. Form an arithemetic sequence that has three arithmetic means between 6 and −4. **6, 3.5, 1, −1.5, −4**

Lesson 12-1 Find the sum of n terms of an arithmetic series.

  The sum S_n of the first n terms of an arithmetic series is given by
$S_n = \frac{n}{2}(a_1 + a_n)$.

14. What is the sum of the first 14 terms in the arithmetic series $-30 - 23 - 16 - \cdots$? **217**

15. Find n for the arithmetic series for which $a_1 = 2$, $d = 1.4$, and $S_n = 250.2$. **18**

Lesson 12-2 Find the nth term and geometric means of a geometric sequence.

  Find an approximation for the 12th term of the sequence $-8, 4, -2, 1, \ldots$.
First, find the common ratio.
$a_2 \div a_1 = 4 \div (-8)$ or -0.5
Use the formula for the nth term.
$a_{12} = -8(-0.5)^{12-1} \quad a_n = a_1 r^{n-1}$
$\quad = -8(-0.5)^{11}$ or about 0.004

16. Find the next three terms of the geometric sequence 49, 7, 1, …. $\frac{1}{7}, \frac{1}{49}, \frac{1}{343}$

17. Find the 15th term of the geometric sequence for which $a_1 = 2.2$ and $r = 2$.

18. If $r = 0.2$ and $a_7 = 8$, what is the first term of the geometric sequence? **125,000**

19. Write a geometric sequence that has three geometric means between 0.2 and 125. **0.2, ±1, 5, ±25, 125**   **17. 36,044.8**

Lesson 12-2 Find the sum of n terms of a geometric series.

  Find the sum of the first 12 terms of the geometric series $4 + 10 + 25 + 62.5 + \cdots$.
First find the common ratio.
$a_2 \div a_1 = 10 \div 4$ or 2.5
Now use the formula for the sum of a finite geometric series.
$S_n = \frac{a_1 - a_1 r^n}{1 - r}$
$S_{12} = \frac{4 - 4(2.5)^{12}}{1 - 2.5} \quad n = 12, a_1 = 4, r = 2.5$
$S_{12} \approx 158{,}943.05 \quad$ *Use a calculator.*

20. What is the sum of the first nine terms of the geometric series $1.2 - 2.4 + 4.8 - \cdots$? **205.2**

21. Find the sum of the first eight terms of the geometric series $4 + 4\sqrt{2} + 8 + \cdots$. **$60(1 + \sqrt{2})$**

TestCheck and Worksheet Builder

This **networkable** software has three modules.
- **Worksheet Builder** to make worksheets and tests
- **Student Module** to take tests on-screen
- **Management System** to keep student records
Three special banks are included for SAT, ACT, and TIMSS.

OBJECTIVES AND EXAMPLES

Lesson 12-3 Find the limit of the terms and the sum of an infinite geometric series.

• Find $\lim\limits_{n\to\infty} \dfrac{2n^2 + 5}{3n^2}$.

$$\lim_{n\to\infty} \frac{2n^2 + 5}{3n^2} = \lim_{n\to\infty}\left(\frac{2}{3} + \frac{5}{3n^2}\right)$$

$$= \lim_{n\to\infty}\frac{2}{3} + \lim_{n\to\infty}\frac{5}{3}\cdot\lim_{n\to\infty}\frac{1}{n^2}$$

$$= \frac{2}{3} + \frac{5}{3}\cdot 0$$

Thus, the limit is $\dfrac{2}{3}$.

REVIEW EXERCISES

Find each limit, or state that the limit does not exist and explain your reasoning.

22. $\lim\limits_{n\to\infty} \dfrac{3n}{4n + 1}$ $\dfrac{3}{4}$ **23.** $\lim\limits_{n\to\infty} \dfrac{6n - 3}{n}$ **6**

24. $\lim\limits_{n\to\infty} \dfrac{2^n\, n^3}{3n^3}$ **See margin.**   **25.** $\lim\limits_{n\to\infty} \dfrac{4n^3 - 3n}{n^4 - 4n^3}$ **0**

26. Write $5.\overline{123}$ as a fraction. $5\dfrac{41}{333}$

27. Find the sum of the infinite series $1260 + 504 + 201.6 + 80.64 + \cdots$, or state that the sum does not exist and explain your reasoning.   **2100**

Lesson 12-4 Determine whether a series is convergent or divergent.

• Use the ratio test to determine whether the series $3 + \dfrac{3^2}{2!} + \dfrac{3^3}{3!} + \dfrac{3^4}{4!}$ is convergent or divergent.

The nth term a_n of this series has a general form of $\dfrac{3^n}{n!}$ and $a_{n+1} = \dfrac{3^{n+1}}{(n+1)!}$. Find $\lim\limits_{n\to\infty}\dfrac{a_{n+1}}{a_n}$.

$$r = \lim_{n\to\infty} \dfrac{\dfrac{3^{n+1}}{(n+1)!}}{\dfrac{3^n}{n!}} \quad\Bigg|\quad r = \lim_{n\to\infty}\frac{3}{n+1} \text{ or } 0$$

$$r = \lim_{n\to\infty}\left[\frac{3^{n+1}}{(n+1)!}\cdot\frac{n!}{3^n}\right] \quad\Big|\quad \text{Since } r < 0, \text{ the series is convergent.}$$

28. Use the ratio test to determine whether the series $\dfrac{1}{5} + \dfrac{2^2}{5^2} + \dfrac{3^2}{5^3} + \dfrac{4^2}{5^4} + \cdots$ is *convergent* or *divergent*.   **convergent**

29. Use the comparison test determine whether the series $\dfrac{6}{1} + \dfrac{7}{2} + \dfrac{8}{3} + \dfrac{9}{4} + \cdots$ is *convergent* or *divergent*.   **divergent**

30. Determine whether the series $2 + 1 + \dfrac{2}{3} + \dfrac{1}{2} + \dfrac{2}{5} + \dfrac{1}{3} + \dfrac{2}{7} + \cdots$ is *convergent* or *divergent*.   **divergent**

Lesson 12-5 Use sigma notation.

• Write $\sum\limits_{n=1}^{3}(n^2 - 1)$ in expanded form and then find the sum.

$$\sum_{n=1}^{3}(n^2 - 1) = (1^2 - 1) + (2^2 - 1) + (3^2 - 1)$$

$$= 0 + 3 + 8 \text{ or } 11$$

Write each expression in expanded form and then find the sum.

31. $\sum\limits_{a=5}^{9}(3a - 3)$ **90**   **32.** $\sum\limits_{k=1}^{\infty}(0.4)^k$ $\dfrac{2}{3}$

31–32. See margin for expanded form.

Express each series using sigma notation

33. $-1 + 1 + 3 + 5 + \cdots$ $\sum\limits_{a=0}^{\infty}(2n - 1)$

34. $2 + 5 + 10 + 17 + \cdots + 82$ $\sum\limits_{a=1}^{9}(n^2 + 1)$

Additional Answers

24. Does not exist;

$\lim\limits_{n\to\infty}\dfrac{2^n n^3}{3n^3} = \lim\limits_{n\to\infty}\dfrac{2^n}{3}$; since

$\lim\limits_{n\to\infty}\dfrac{2^n}{3}$ becomes increasingly large as n approaches infinity, the sequence has no limit.

31. $(3\cdot 5 - 3) + (3\cdot 6 - 3) + (3\cdot 7 - 3) + (3\cdot 8 - 3) + (3\cdot 9 - 3)$

32. $(0.4)^1 + (0.4)^2 + (0.4)^3 + \cdots$

Additional Answers

35. $a^6 - 24a^5 + 240a^4 - 1280a^3 + 3840a^2 - 6144a + 4096$

36. $16r^4 + 96r^3s + 216r^2s^2 + 216rs^3 + 81s^4$

47. $4, 6 - 2i, 7 - 3i$

48. $-2i, 4 - 3i, 6 - 3.5i$

49. $2 + i, 5 - 1.5i, 6.5 - 2.75i$

50. $10 - 6i, 9 - 5i, 8.5 - 4.5i$

OBJECTIVES AND EXAMPLES

REVIEW EXERCISES

Lesson 12-6 Use the Binomial Theorem to expand binomials.

Find the fourth term of $(2x - y)^6$.

$$(2x - y)^6 = \sum_{r=0}^{6} \frac{6!}{r!(6 - r)!} (2x)^{6 - r}(-y)^r$$

To find the fourth term, evaluate the general term for $r = 3$.

$$\frac{6!}{3!(6 - 3)!} (2x)^{6 - 3}(-y)^3$$

$$= \frac{6 \cdot 5 \cdot 4 \cdot 3!}{3! \cdot 3!} (2x)^3(-y^3) \text{ or } -160x^3y^3$$

Use the Binomial Theorem to expand each binomial.   35–36. See margin.

35. $(a - 4)^6$ 36. $(2r + 3s)^4$

Find the designated term of each binomial expansion.

37. 5th term of $(x - 2)^{10}$ $3360x^6$

38. 3rd term of $(4m + 1)^8$ $114{,}688m^6$

39. 8th term of $(x + 3y)^{10}$ $262{,}440x^3y^7$

40. 6th term of $(2c - d)^{12}$ $-101{,}376c^7d^5$

Lesson 12-7 Use Euler's Formula to write the exponential form of a complex number.

Write $\sqrt{3} - i$ in exponential form.

Write the polar form of $\sqrt{3} - i$.

$r = \sqrt{(\sqrt{3})^2 + (-1)^2}$ or 2, and

$\theta = \text{Arctan } \dfrac{-1}{\sqrt{3}}$ or $\dfrac{5\pi}{6}$

$\sqrt{3} - i = 2\left(\cos \dfrac{5\pi}{6} + i \sin \dfrac{5\pi}{6}\right) = 2e^{i\frac{5\pi}{6}}$

Write each expression or complex number in exponential form.

41. $2\left(\cos \dfrac{3\pi}{4} + i \sin \dfrac{3\pi}{4}\right)$ $2e^{i\frac{3\pi}{4}}$

42. $4i$ $4e^{i\frac{\pi}{2}}$

43. $2 - 2i$ $2\sqrt{2}e^{i\frac{7\pi}{4}}$

44. $3\sqrt{3} + 3i$ $6e^{i\frac{\pi}{6}}$

Lesson 12-8 Iterate functions using real and complex numbers.

Find the first three iterates of the function $f(z) = 2z$ if the initial value is $3 - i$.

$z_0 = 3 - i$
$z_1 = 2(3 - i)$ or $6 - 2i$
$z_2 = 2(6 - 2i)$ or $12 - 4i$
$z_3 = 2(12 - 4i)$ or $24 - 8i$

Find the first four iterates of each function using the given initial value. If necessary, round your answers to the nearest hundredth.

45. $f(x) = 6 - 3x, x_0 = 2$ $0, 6, -12, 42$

46. $f(x) = x^2 + 4, x_0 = -3$ $13; 173; 29{,}933;$ $895{,}984{,}493$

Find the first three iterates of the function $f(z) = 0.5z + (4 - 2i)$ for each initial value.

47. $z_0 = 4i$ 48. $z_0 = -8$

49. $z_0 = -4 + 6i$ 50. $z_0 = 12 - 8i$
47–50. See margin.

Lesson 12-9 Use mathematical induction to prove the validity of mathematical statements.

Proof by mathematical induction:
1. First, verify that the conjecture S_n is valid for the first possible case, usually $n = 1$.
2. Then, assume that S_n is valid for $n = k$ and use this assumption to prove that it is also valid for $n = k + 1$.

Use mathematical induction to prove that each proposition is valid for all positive integral values of n.   51–53. See Answer Appendix.

51. $1 + 2 + 3 + \cdots + n = \dfrac{n(n + 1)}{2}$

52. $3 + 8 + 15 + \cdots + n(n - 2) = \dfrac{n(n + 1)(2n + 7)}{6}$

53. $9^n - 4^n$ is divisible by 5.

APPLICATIONS AND PROBLEM SOLVING

54. Physics If an object starting at rest falls in a vacuum near the surface of Earth, it will fall 16 feet during the first second, 48 feet during the next second, 80 feet during the third second, and so on. *(Lesson 12-1)*

a. How far will the object fall during the twelfth second? **368 ft**

b. How far will the object have fallen after twelve seconds? **2304 ft**

55. Budgets A major corporation plans to cut the budget on one of its projects by 3 percent each year. If the current budget for the project to be cut is $160 million, what will the budget for that project be in 10 years? *(Lesson 12-2)* **$117,987,860.30**

56. Geometry If the midpoints of the sides of an equilateral triangle are joined by straight lines, the new figure will also be an equilateral triangle. *(Lesson 12-3)*

a. If the original triangle has a perimeter of 6 units, find the perimeter of the new triangle. **3 units**

b. If this process is continued to form a sequence of "nested" triangles, what will be the sum of the perimeters of all the triangles? **12 units**

ALTERNATIVE ASSESSMENT

OPEN-ENDED ASSESSMENT

1. A sequence has a common difference of 3.

a. Is this sequence arithmetic or geometric? Explain. **Arithmetic; see margin.**

b. Form a sequence that has this common difference. Write a recursive formula for your sequence. **See margin.**

2. Write a general expression for an infinite sequence that has no limit. Explain your reasoning. **See margin.**

PORTFOLIO

Explain the difference between a convergent and a divergent series. Give an example of each type of series and show why it is that type of series.

Additional Assessment See p. A67 for Chapter 12 practice test.

Unit 4 *interNET* Project

THE UNITED STATES CENSUS BUREAU

That's a lot of people!

- Use the Internet to find the population of the United States from at least 1900 through 2000. Write a sequence using the population for each ten-year interval, for example, 1900, 1910, and so on.

- Write a formula for an arithmetic sequence that provides a reasonable model for the population sequence.

- Write a formula for a geometric sequence that provides a reasonable model for the population sequence.

- Use your models to predict the U.S. population for the year 2050. Write a one-page paper comparing the arithmetic and geometric sequences you used to model the population data. Discuss which formula you think best models the data.

Chapter 12 Study Guide and Assessment **833**

APPLICATIONS & PROBLEM SOLVING

This section provides additional practice in solving real-world problems that involve the concepts of this chapter.

ALTERNATIVE ASSESSMENT

Additional Answers
Open-Ended Assessment

1a. Arithmetic sequences have common differences, while geometric sequences have common ratios.

1b. Sample answer: 1, 4, 7, 10, ...; $a_n = 1 + 3(n - 1)$

2. Sample answer: $\dfrac{6 + 5n^2}{3n}$;

$$\lim_{n \to \infty} \frac{6 + 5n^2}{3n} = \lim_{n \to \infty} \left(\frac{2}{n} + \frac{5n}{3} \right)$$

$\lim\limits_{n \to \infty} \dfrac{2}{n} = 0$, but since $\lim\limits_{n \to \infty} \dfrac{5n}{3}$ becomes increasingly large as n approaches infinity, the sequence has no limit.

Unit 4 *interNET* Project

● **That's a Lot of People!**
A possible strategy to use for writing formulas for the sequences is to average the changes in population, such as $\dfrac{\text{current population}}{\text{previous population}}$, to use as the common difference or common ratio in the sequence formulas. A spreadsheet is very useful for evaluating data. Be sure that students save the data for use in the Chapter 14 Internet project.

Internet Address Book

Record useful Internet addresses in the space at right for quick reference.

www.amc.glencoe.com

Chapter 12 Skill: Percent

Pages 834–835 are part of a complete SAT/ACT preparation course that is described in detail on pages T8–T9 of the *Teacher's Handbook*. The test items were written in the same style as those in the SAT and ACT.

 These questions were aligned and verified by The Princeton Review, the nation's leader in test preparation.

Teaching Tip Students may wish to memorize these basic equivalents:

$$0.01 = \frac{1}{100} = 1\%$$

$$0.1 = \frac{1}{10} = 10\%$$

$$0.\overline{3} = \frac{1}{3} = 33.\overline{3}\%$$

$$0.25 = \frac{1}{4} = 25\%$$

$$0.5 = \frac{1}{2} = 50\%$$

$$0.75 = \frac{3}{4} = 75\%$$

Diagnosis & Prescription

Each of the 10 SAT and ACT practice problems on page 835 is cross-referenced to the chapter where that SAT or ACT skill is covered. If students miss a particular type of problem, you can have them study that skill.

Problem Number and Skill		Page
1	Percent	64
2	Geometry (Lines, Angles, and Arcs)	612
3	Percent	834
4	Percent	834
5	Geometry (Triangles and Quadrilaterals)	482
6	Percent	834
7	Coordinate Geometry	272
8	Algebra	124
9	Percent	834
10	Percent	834

Percent Problems

A few common words and phrases used in percent problems, along with their translations into mathematical expressions, are listed below.

what ⟹ x (variable)
of ⟹ × (multiply)
is ⟹ = (equals)

What percent of A is B? ⟹ $\frac{x}{100} \cdot A = B$

What is A percent more than B? ⟹ $x = B + \frac{A}{100} \cdot B$

What is A percent less than B? ⟹ $x = B - \frac{A}{100} \cdot B$

percent of change (increase or decrease) ⟹ $\frac{\text{amount of change}}{\text{original amount}} \times 100$

THE PRINCETON REVIEW

TEST-TAKING TIP
With common percents, like 10%, 25%, or 50%, it is faster to use the fraction equivalents and mental math than a calculator.

ACT EXAMPLE

1. If c is positive, what percent of $3c$ is 9?

A $\frac{c}{100}\%$ **B** $\frac{300}{c}\%$ **C** $\frac{9}{c}\%$ **D** 3%   **E** $\frac{c}{3}\%$

HINT Use variables just as you would use numbers.

Solution Start by translating the question into an equation.

What percent of $3c$ is 9?

$$\frac{x}{100} \cdot 3c = 9$$

Now solve the equation for x, not c.

$$\frac{x}{100} = \frac{9}{3c} \quad \textit{Divide each side by 3c.}$$
$$\frac{x}{100} = \frac{3}{c} \quad \textit{Simplify.}$$
$$x = \frac{300}{c} \quad \textit{Solve for x.}$$

The answer is choice **B**.

Alternate Solution "Plug in" 3 for c. The question becomes "what percent of 9 is 9?" The answer is 100%, so check each expression choice to see if it is equal to 100% when $c = 3$.

Choice A: $\frac{c}{100}\% = \frac{3}{100}\%$

Choice B: $\frac{300}{c}\% = 100\%$

Choice **B** is correct.

834 *Chapter 12 Sequences and Series*

SAT EXAMPLE

2. An electronics store offers a 25% discount on all televisions during a sale week. How much must a customer pay for a television marked at $240?

A $60   **B** $300   **C** $230.40

D $180   **E** $215

HINT A discount is a decrease in the price of an item. So, the question asked is "What is 25% less than 240?"

Solution Start by translating this question into an equation.

What is 25% less than 240? ⟹ $x = 240 - \frac{25}{100} \cdot 240$

Now simplify the right-hand side of the equation.

$$x = 240 - \frac{1}{4} \cdot 240 \text{ or } 180$$

Choice **D** is correct.

Alternate Solution If there is a 25% discount, a customer will pay $(100 - 25)\%$ or 75% of the marked price.

What is 75% of the marked price? ⟹ $x = 0.75 \cdot 240 \text{ or } 180$

The answer is choice **D**.

TestCheck and Worksheet Builder

Special banks of SAT, ACT, and TIMSS items are included in this CD-ROM.

After you work each problem, record your answer on the answer sheet provided or on a piece of paper.

Multiple Choice

1. Shanika has a collection of 80 tapes. If 40% of her records are jazz tapes and the rest are blues tapes, how many blues tapes does she have? **D**

   A 32   **B** 40   **C** 42   **D** 48   **E** 50

2. If ℓ_1 is parallel to ℓ_2 in the figure below, what is the value of x? **C**

   A 20   **B** 50   **C** 70   **D** 80   **E** 90

3. There are k gallons of gasoline available to fill a tank. After d gallons have been pumped, then, in terms of k and d, what percent of the gasoline has been pumped? **A**

   A $\dfrac{100d}{k}\%$ **B** $\dfrac{k}{100d}\%$ **C** $\dfrac{100k}{d}\%$

   D $\dfrac{k}{100(k-d)}\%$ **E** $\dfrac{100(k-d)}{k}\%$

4. In 1985, Andrei had a collection of 48 baseball caps. Since then he has given away 13 caps, purchased 17 new caps, and traded 6 of his caps to Pierre for 8 of Pierre's caps. Since 1985, what has been the net percent increase in Andrei's collection? **B**

   A 6%   **B** $12\frac{1}{2}\%$ **C** $16\frac{2}{3}\%$

   D 25%   **E** $28\frac{1}{2}\%$

5. In the figure below, $AB = AC$ and AD is a line segment. What is the value of $x - y$? **C**

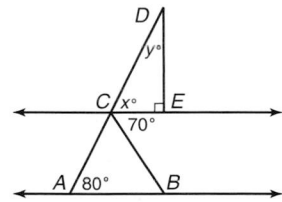

   Note: Figure is NOT drawn to scale.

   A 10   **B** 20   **C** 30   **D** 70   **E** 90

6. At the beginning of 2000, the population of Rockville was 204,000, and the population of Springfield was 216,000. If the population of each city increased by exactly 20% in 2000, how many more people lived in Springfield than in Rockville at the end of 2000? **D**

   A 9,600   **B** 10,000   **C** 12,000

   D 14,400   **E** 20,000

7. In the figure, the slope of $\overline{AC}$ is $-\frac{1}{6}$, and $m\angle C = 30°$. What is the length of $\overline{BC}$? **D**

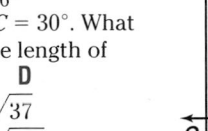

   A $\sqrt{37}$
   B $\sqrt{111}$
   C 2
   D $2\sqrt{37}$
   E It cannot be determined from the information given.

8. If $x + 6 > 0$ and $1 - 2x > -1$, then x could equal each of the following EXCEPT ? **A**

   A -6 **B** -4 **C** -2 **D** 0   **E** $\frac{1}{2}$

9. The percent increase from 99 to 100 is which of the following? **A**
   A greater than 1
   B 1
   C less than 1, but more than $\frac{1}{2}$
   D less than $\frac{1}{2}$, but more than 0
   E 0

10. **Grid-In** One fifth of the cars in a parking lot are blue, and $\frac{1}{2}$ of the blue cars are convertibles. If $\frac{1}{4}$ of the convertibles in the parking lot are blue, then what percent of the cars in the lot are neither blue nor convertibles? **50**

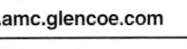

SAT/ACT Practice For additional test practice questions, visit: **www.amc.glencoe.com**

THE PRINCETON REVIEW   *SAT & ACT Preparation*   **835**

A **bubble-in answer sheet** for these practice problems is available on p. 1 of the *Assessment and Evaluation Masters.*

Additional Practice
Additional SAT and ACT practice questions and a bubble-in answer sheet are available in the *Assessment and Evaluation Masters,* pp. 2, 305, and 306.

Assessment and Evaluation Masters, pp. 305–306

Extra Credit

TIMSS The sum of the infinite geometric series $1 - \frac{1}{2} + \frac{1}{4} - \frac{1}{8} + \cdots$ is **B**

A $\frac{5}{8}$ **B** $\frac{2}{3}$ **C** $\frac{3}{5}$ **D** $\frac{3}{2}$ **E** ∞

Resource Manager

Combinatorics and Probability

Instructional Objectives

Lesson (pages)	Lesson Objectives	NCTM Standards 2000	State/Local Objectives
13-1 (837–845)	• Solve problems related to the Basic Counting Principle. • Distinguish between dependent and independent events. • Solve problems involving permutations or combinations.	1, 6, 10	
13-2 (846–851)	• Solve problems involving permutations with repetitions. • Solve problems involving circular permutations.	1, 6, 10	
13-3 (852–858)	• Find the probability of an event. • Find the odds for the success and failure of an event.	1, 5, 6	
13-4 (859–867)	• Find the probability of independent and dependent events. • Identify mutually exclusive events. • Find the probability of mutually exclusive and inclusive events.	1, 5, 6, 10	
13-5 (868–874)	• Find the probability of an event given the occurrence of another event.	1, 5, 6, 10	
13-6 (875–880)	• Find the probability of an event by using the Binomial Theorem.	1, 5, 6	

Suggested Pacing   *See pages T2–T4 for additional options in course-planning.*

COURSE	Lesson 13-1	Lesson 13-2	Lesson 13-3	Lesson 13-4	Lesson 13-5
Precalculus Emphasis Chapters 4–15 (45–50 minutes)	1.5 days	1.5 days	1.5 days	1.5 days	1.5 days
Precalculus Emphasis (block)	0.5 day	1 day		1.5 days	
Trigonometric Emphasis Chapters 1–9, 11, 12 (45–50 minutes)	OPTIONAL				
Trigonometric Emphasis (block)	OPTIONAL				

Instructional Resources

Student Edition		Blackline Masters (page numbers)								
Lesson	Graphing Calculator Applications	Study Guide and Practice	Enrichment	Assessment and Evaluation	Graphing Calculator Keystrokes	Problem Solving & Applications	Spreadsheet Applications	TI-92 and CBL Interdisciplinary Activities	Transparencies A and B
13-1		239–240	147			26	13		13-1
13-2		241–242	148	329					13-2
13-3		243–244	149	328, 329					13-3
13-4		245–246	150						13-4
13-5		247–248	151	330					13-5
13-6	877, 879	249–250	152	330	53	25			13-6
Study Guide/ Assessment				315–327, 333					
SAT/ACT Practice				331–332					

Lesson 13-6	Study Guide & Assessment	Chapter Test	TOTAL
1.5 days	1 day	1 day	11 days
1.5 days		0.5 day	5 days
OPTIONAL			
OPTIONAL			

TeacherWorks™

The worksheets shown on this page are a small sample of the materials available on *TeacherWorks: All-in-One Lesson Planner and Resource Center*.

This CD-ROM contains all of the blackline masters and transparencies available for this program.

It also includes a lesson planner and interactive Teacher Edition, so you can customize lesson plans and reproduce classroom resources quickly and easily, from just about anywhere.

Applications

Problem Solving and Applications Masters, pp. 25, 26

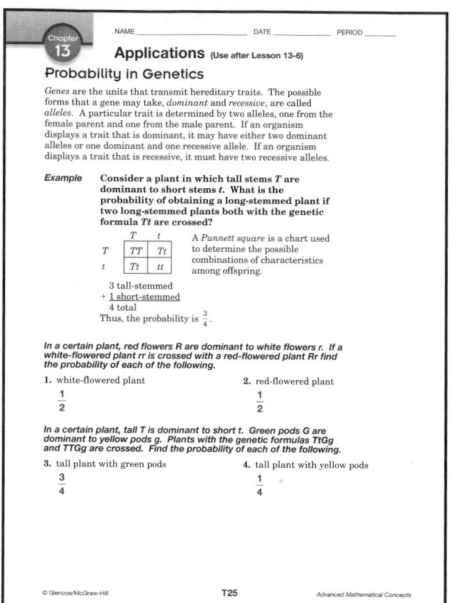

Technology/Multimedia

Spreadsheet Applications Masters, p. 13

Assessment Resources

Type	Student Edition	Teacher's Wraparound Edition	Assessment and Evaluation Masters
Ongoing Assessment	Mid-Chapter Quiz, p. 858	5-Minute Check, pp. 837, 846, 852, 859, 868, 875	Mid-Chapter Test, p. 328 Quizzes A–D, pp. 329–330
Mixed Review	Mixed Review, pp. 845, 850–851, 857–858, 867, 874, 880		Cumulative Review, p. 333
Error Analysis	You Decide, pp. 843, 855	Error Analysis, pp. 842, 848, 855, 863, 870, 878	
Standardized Test Practice	SAT/ACT Practice, pp. 845, 851, 858, 867, 874, 880 SAT & ACT Preparation, pp. 886–887	TIMSS, p. 887	SAT/ACT Practice, pp. 331–332
Open-Ended Assessment	Math Journal, pp. 863, 871 Open-Ended Assessment, p. 885 Portfolio, p. 885	Modeling: pp. 845, 880 Speaking: pp. 858, 874 Writing: pp. 851, 867	Open-Ended Assessment, p. 327
Chapter Assessment	Study Guide and Assessment, pp. 881–885 Chapter Test, p. A68		Multiple-Choice Tests (Forms 1A, 1B, 1C), pp. 315–320 Free-Response Tests (Forms 2A, 2B, 2C), pp. 321–326

TestCheck and Worksheet Builder is a networkable CD-ROM with three integrated modules. The **Worksheet Builder** creates customized worksheets, tests, and quizzes of free-response, multiple-choice, short-answer, and open-ended items. The **Student Module** gives you the option of having students take tests on-screen and get immediate feedback on their performance. Use the optional **Management System** to keep detailed student records.

Additional Chapter Resources

Student Edition
Career Choices, p. 851
Internet Project, p. 885
Extra Practice, pp. A51–A52

Teacher's Classroom Resources

Technology
Graphing Calculator
Keystrokes Masters

Teaching Aids
Answer Key Transparencies
Block Schedule Planning Guide
Lesson Planning Guide
Solutions Manual

Glencoe Instructional Technology

Instructional
Advanced MathPASS CD-ROM
Multimedia Applications CD-ROM

Assessment
TestCheck and Worksheet Builder

Visit **www.amc.glencoe.com** for data updates, career information, graphing calculator programs, and other interactive activities.

Chapter Overview

This chapter provides an introduction to some basic concepts of combinatorics and probability. Students are also exposed to permutations with repetitions and circular permutations, conditional probability, and probability involving the Binomial Theorem.

Using a Graphing Calculator

There are opportunities, all optional, to use a graphing calculator in this chapter.

- Simulating a binomial experiment (pp. 877, 879)

COMBINATORICS AND PROBABILITY

CHAPTER OBJECTIVES

- **Solve problems involving combinations and permutations.** *(Lessons 13-1, 13-2)*
- **Distinguish between independent and dependent events and between mutually exclusive and mutually inclusive events.** *(Lessons 13-1, 13-4, 13-5)*
- **Find probabilities.** *(Lessons 13-3, 13-4, 13-5, 13-6)*
- **Find odds for the success and failure of an event.** *(Lesson 13-3)*

 SAT & ACT Preparation

This textbook contains a complete SAT/ACT preparation course described on pages T8–T9 of the *Teacher's Handbook*. This chapter contains:

- an SAT or ACT practice question at the end of each lesson
- examples of SAT and ACT problems involving probability and combination on page 886
- ten SAT and ACT practice problems on page 887

- additional SAT and ACT practice in the *Assessment and Evaluation Masters,* pp. 331–332
- SAT and ACT test practice banks in the *TestCheck and Worksheet Builder*

Permutations and Combinations

OBJECTIVES
• Solve problems related to the Basic Counting Principle.
• Distinguish between dependent and independent events.
• Solve problems involving permutations or combinations.

EDUCATION Ivette is a freshman at the University of Miami. She is planning her fall schedule for next year. She has a choice of three mathematics courses, two science courses, and two humanities courses. She can only select one course from each area. How many course schedules are possible?

Let M_1, M_2, and M_3 represent the three math courses, S_1 and S_2 the science courses, and H_1 and H_2 the humanities courses. Once Ivette makes a selection from the three mathematics courses she has two choices for her science course. Then, she has two choices for humanities. A **tree diagram** is often used to show all the choices.

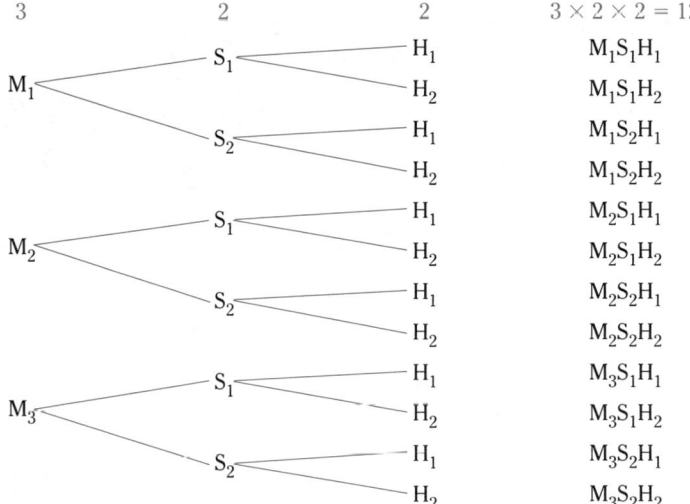

Ivette has 12 possible schedules from which to choose.

The choice of selecting a mathematics course does *not* affect the choice of ways to select a science or humanities course. Thus, these three choices are called **independent events.** Events that do affect each other are called **dependent events.** An example of dependent events would be the order in which runners finish a race. The first place runner affects the possibilities for second place, the second place runner affects the possibilities for third place, and so on.

The branch of mathematics that studies different possibilities for the arrangement of objects is called **combinatorics.** The example of choosing possible course schedules illustrates a rule of combinatorics known as the **Basic Counting Principle.**

Lesson 13-1 Permutations and Combinations **837**

1 FOCUS

 5-Minute Check
Chapter 12

1. Find the 35th term in the sequence −6, −2, 2, 6, …. **130**

2. Find the sum of the first 75 terms in the series $7 + 10 + 13 + \cdots + 229$. **8850**

3. Write a sequence that has two geometric means between −3 and −192. **−3, −12, −48, −192**

4. Find the sum of the first eight terms of the series $12 - 24 + 48 - 96 + \cdots$. **−1020**

5. Find the first three iterates of each function using the given initial value.
   a. $f(z) = (z - 2)^2$, $z_0 = 4$ **4, 4, 4**
   b. $f(z) = z(2 - z)$, $z_0 = 3$ **−3, −15, −255**

Motivating the Lesson
Real-World Application Suppose that a school has a strict dress code. If a boy in the school can select from 3 different ties and 4 pairs of pants, how many dress options does he have each day? **12** Could you think of a way to list these possibilities? **Sample answer: a tree diagram**

 Resource Manager

 Reproducible Masters
• *Study Guide and Practice,* pp. 239–240
• *Enrichment,* p. 147
• *Problem Solving and Applications,* p. 26
• *Spreadsheet Applications,* p. 13

 Transparencies
• 5-Minute Check, 13-1
• Teaching, 13-1

 Technology/Multimedia
• Advanced MathPASS, Lesson 27

Teaching Tip After defining dependent events and independent events, have students give their own examples of each kind of event.

In-Class Example

Example 1

Mr. Casey has decided to buy a new suit made either of wool or rayon. He has narrowed down the choices of available colors to gray, blue, black, or tan, and the matching tie to a floral, a stripe, or a geometric pattern.

a. Assuming that all choices make an acceptable appearance, are the choices for material, color, and tie pattern independent or dependent events?
independent

b. How many different selections of his suit and tie are possible? **24**

Basic Counting Principle	Suppose one event can be chosen in p different ways, and another independent event can be chosen in q different ways. Then the two events can be chosen successively in $p \cdot q$ ways.

This principle can be extended to any number of independent events. For example, in the previous application, the events are chosen in $p \times q \times r$ or $3 \times 2 \times 2$ different ways.

Example **1** **Vickie works for a bookstore. Her manager asked her to arrange a set of five best-sellers for a display. The display is to be set up as shown below. The display set is made up of one book from each of 5 categories. There are 4 nonfiction, 4 science fiction, 3 history, 3 romance, and 3 mystery books from which to choose.**

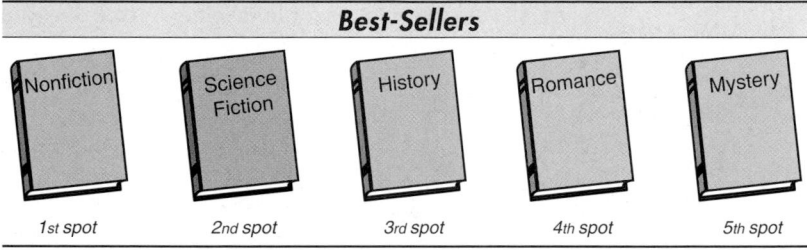

Best-Sellers

Nonfiction	Science Fiction	History	Romance	Mystery
1st spot	2nd spot	3rd spot	4th spot	5th spot

a. Are the choices for each book independent or dependent events?

Since the choice of one type of book does not affect the choice of another type of book, the events are independent.

b. How many different ways can Vickie choose the books for the display?

Vickie has four choices for the first spot in the display, four choices for the second spot, and three choices for each of the next three spots.

1ˢᵗ spot		2ⁿᵈ spot		3ʳᵈ spot		4ᵗʰ spot		5ᵗʰ spot
4	·	4	·	3	·	3	·	3

This can be represented as $4 \cdot 4 \cdot 3 \cdot 3 \cdot 3$ or 432 different arrangements.

There are 432 possible ways for Vickie to choose books for the display.

The arrangement of objects in a certain order is called a **permutation.** In a permutation, the order of the objects is very important. The symbol $P(n, n)$ denotes the number of permutations of n objects taken all at once. The symbol $P(n, r)$ denotes the number of permutations of n objects taken r at a time.

From the Classroom of...

Marilyn Knechtges
Western Reserve High School
Collins, Ohio

"I have my students make ice-cream sundaes in class to illustrate the basic counting principle. I purchase three kinds of ice cream, then only enough toppings to be equal to the number of students in the class. Students make a tree diagram of all the possible sundaes. Then, we draw names to come up and make a sundae from the list."

Permutations	The number of permutations of *n* objects, taken *n* at a time is defined as $P(n, n) = n!$.
P(n, n) and **P(n, r)**	The number of permutations of *n* objects, taken *r* at a time is defined as $P(n, r) = \dfrac{n!}{(n-r)!}$.

Recall that n! is read "n factorial" and, n! = n(n − 1)(n − 2)...(1).

Example 2

During a judging of a horse show at the Fairfield County Fair, there are three favorite horses: Rye Man, Oiler, and Sea of Gus.

a. Are the selection of first, second and third place from the three horses independent or dependent events?

b. Assuming there are no ties and the three favorites finish in the top three places, how many ways can the horses win first, second and third places?

a. The choice of a horse for first place does affect the choice for second and third places. For example, if Rye Man is first, then is impossible for it to finish second or third. Therefore, the events are dependent.

b. Since order is important, this situation is a permutation.

Method 1: Tree diagram

There are three possibilities for first place, two for second, and one for third as shown in the tree diagram below. If Rye Man finishes first, then either Oiler or Sea of Gus will finish second. If Oiler finishes second, then Sea of Gus must finish third. Likewise, if Sea of Gus finishes second, then Oiler finishes third.

There are 6 possible ways the horses can win.

Method 2: Permutation formula

This situation depicts three objects taken three at a time and can be represented as $P(3, 3)$.

$P(3, 3) = 3!$
$= 3 \cdot 2 \cdot 1$ or 6

Thus, there are 6 ways the horses can win first, second, and third place.

Example ❸ **The board of directors of B.E.L.A. Technology Consultants is composed of 10 members.**

    a. How many different ways can all the members sit at the conference table as shown?

    b. In how many ways can they elect a chairperson, vice-chairperson, treasurer, and secretary, assuming that one person cannot hold more than one office?

    a. Since order is important, this situation is a permutation. Also, the 10 members are being taken all at once so the situation can be represented as $P(10, 10)$.

$$P(10, 10) = 10!$$
$$= 10 \cdot 9 \cdot 8 \cdot 7 \cdot 6 \cdot 5 \cdot 4 \cdot 3 \cdot 2 \cdot 1 \text{ or } 3{,}628{,}800$$

There are 3,628,800 ways that the 10 board members can sit at the table.

    b. This is a permutation of 10 people being chosen 4 at a time.

$$P(10, 4) = \frac{10!}{(10 - 4)!}$$
$$= \frac{10 \cdot 9 \cdot 8 \cdot 7 \cdot 6!}{6!}$$
$$= 5040$$

There are 5040 ways in which the offices can be filled.

Suppose that in the situation presented in Example 1, Vickie needs to select three types of books from the five types available. There are $P(5, 3)$ or 60 possible arrangements. She can arrange them as shown in the table below.

Arrangement	Type		
1	nonfiction	science fiction	history
2	nonfiction	history	science fiction
3	nonfiction	romance	mystery
4	nonfiction	mystery	romance
5	science fiction	nonfiction	history
6	science fiction	history	nonfiction
7	science fiction	romance	mystery
8	science fiction	mystery	romance
9	history	nonfiction	science fiction
10	history	science fiction	nonfiction
⋮	⋮	⋮	⋮
60	romance	mystery	nonfiction

Note that arrangements 1, 2, 5, 6, 9 and 10 contain the same three types of books. In each group of three books, there are 3! or 6 ways they can be arranged. Thus, if order is disregarded, there are $\frac{60}{3!}$ or 10 different groups of three types of books that can be selected from the five types. In this situation, called a **combination,** the order in which the books are selected is *not* a consideration.

A combination of n objects taken r at a time is calculated by dividing the number of permutations by the number of arrangements containing the same elements and is denoted by $C(n, r)$.

Combination	The number of combinations of n objects taken r at a time is defined as
$C(n, r)$	$$C(n, r) = \frac{n!}{(n - r)!\, r!}.$$

The main difference between a permutation and a combination is whether order is considered (as in permutation) or not (as in combination). For example, for objects E, F, G, and H taken two at a time, the permutations and combinations are listed below.

Permutations				Combinations	
EF	FE	GE	HE	EF	FG
EG	FG	GF	HF	EG	FH
EH	FH	GH	HG	EH	GH

Note that in permutations, EF is different from FE. But in combinations, EF is the same as FE.

Example

Real World Application

4 **ART** In 1999, The National Art Gallery in Washington, D.C., opened an exhibition of the works of John Singer Sargent (1856–1925). The gallery's curator wanted to select four paintings out of twenty on display to showcase the work of the artist. How many groups of four paintings can be chosen?

Since order is not important, the selection is a combination of 20 objects taken 4 at a time. It can be represented as $C(20, 4)$.

$$C(20, 4) = \frac{20!}{(20 - 4)!\, 4!}$$

$$= \frac{20!}{16!\, 4!}$$

$$= \frac{20 \cdot 19 \cdot 18 \cdot 17 \cdot 16!}{16!\, 4!}$$ *Express 20! as* $20 \cdot 19 \cdot 18 \cdot 17 \cdot 16!$ *since* $\frac{16!}{16!} = 1$.

$$= \frac{20 \cdot 19 \cdot 18 \cdot 17}{4 \cdot 3 \cdot 2 \cdot 1}$$

$$= 4845$$

There are 4845 possible groups of paintings.

"Oyster Gatherers of Cancale," 1878

Addressing Individual Needs

Interpersonal Give groups of students different sets of manipulatives such as algebra tiles, cards, coins, and so on. Have each group write a problem involving combinations using the manipulatives, then present the problem to the class.

In-Class Example

Example 5

At Lincoln High School there are 10 names on the ballot for senior class officers. Four will be selected to form a class committee.

a. How many different committees of 4 can be formed? **210**

b. In how many ways can a committee of 4 be formed if each student has a different responsibility? **5040**

c. If there are 6 girls and 4 boys on the ballot, how many committees of 3 girls and 1 boy can be formed? **80**

3 PRACTICE/APPLY

Error Analysis

Watch for students who still have difficulty distinguishing between permutations and combinations.

Prevent by showing them some additional examples.

Teaching Tip Before assigning Exercise 2, make sure that students know the composition of a standard deck of 52 cards.

Study Guide and Practice Masters, pp. 239–240

Study Guide

Permutations and Combinations

Use the **Basic Counting Principle** to determine different possibilities for the arrangement of objects. The arrangement of objects in a certain order is called a **permutation**. A **combination** is an arrangement in which order is *not* a consideration.

Example 1 **Eight students on a student council are assigned 8 seats around a U-shaped table.**
a. **How many different ways can the students be assigned seats at the table?**
Since order is important, this situation is a permutation. The eight students are taken all at once, so the situation can be represented as P(8, 8).
P(8, 8) = 8! P(n, n) = n!
= 8 · 7 · 6 · 5 · 4 · 3 · 2 · 1 or 40,320
There are 40,320 ways the students can be seated.
b. **How many ways can a president and a vice-president be elected from the eight students?**
This is a permutation of 8 students being chosen 2 at a time.
P(8, 2) = 8!/(8 − 2)! P(n, r) = n!/(n − r)!
= 8!/6! or 56
There are 56 ways a president and vice-president can be chosen.

Example 2 **The Outdoor Environmental Club consists of 20 members, of which 9 are male and 11 are female. Seven members will be selected to form an event-planning committee. How many committees of 4 females and 3 males can be formed?**
Order is not important. There are three questions to consider.
How many ways can 3 males be chosen from 9?
How many ways can 4 females be chosen from 11?
How many ways can 3 males and 4 females be chosen together?
The answer is the product of the combinations C(9, 3) and C(11, 4).
C(9, 3) · C(11, 4) = 9!/(9 − 3)!3! · 11!/(11 − 4)!4! C(n, r) = n!/(n − r)!r!
= 9!/6!3! · 11!/7!4!
= 84 · 330 or 27,720
There are 27,720 possible committees.

© Glencoe/McGraw-Hill   239   *Advanced Mathematical Concepts*

Example ⑤ **At Grant Senior High School, there are 15 names on the ballot for junior class officers. Five will be selected to form a class committee.**

a. **How many different committees of 5 can be formed?**

b. **In how many ways can a committee of 5 be formed if each student has a different responsibility?**

c. **If there are 8 girls and 7 boys on the ballot, how many committees of 2 boys and 3 girls can be formed?**

a. Order is not important in this situation, so the selection is a combination of 15 people chosen 5 at a time.

$$C(15, 5) = \frac{15!}{(15 - 5)!\, 5!}$$

$$= \frac{15!}{10!\, 5!}$$

$$= \frac{15 \cdot 14 \cdot 13 \cdot 12 \cdot 11 \cdot 10!}{10!\, 5!} \text{ or } 3003$$

There are 3003 different ways to form the committees of 5.

b. Order has to be considered in this situation because each committee member has a different responsibility.

$$P(15, 5) = \frac{15!}{(15 - 5)!}$$

$$= \frac{15!}{10!} \text{ or } 360,360$$

There are 360,360 possible committees.

c. Order is not important. There are three questions to consider.
How many ways can 2 boys be chosen from 7?
How many ways can 3 girls be chosen from 8?
Then, how many ways can 2 boys and 3 girls be chosen together?

Since the events are independent, the answer is the product of the combinations $C(7, 2)$ and $C(8, 3)$.

$$C(7, 2) \cdot C(8, 3) = \frac{7!}{(7 - 2)!\, 2!} \cdot \frac{8!}{(8-3)!\, 3!}$$

$$= \frac{7!}{5!\, 2!} \cdot \frac{8!}{5!\, 3!}$$

$$= 21 \cdot 56 \text{ or } 1176$$

There are 1176 possible committees.

CHECK FOR UNDERSTANDING

Communicating Mathematics

Read and study the lesson to answer each question.   1–2. See margin.

1. **Compare and contrast** permutations and combinations.

2. **Write** an expression for the number of ways, out of a standard 52-card deck, that 5-card hands can have 2 jacks and 3 queens.

842   *Chapter 13   Combinatorics and Probability*

Additional Answers

1. Sample answer: both are used to determine the number of arrangements of a group of objects. However, order of the objects is important in permutations. When order of the objects is not important, combinations are computed.

2. Select 2 jacks out of 4—$C(4, 2)$.
   Select 3 queens out of 4—$C(4, 3)$.
   number of hands—$C(4, 2) \cdot C(4, 3)$

3–4. See margin.

3. You Decide Ms. Sloan asked her students how many ways 5 patients in a hospital could be assigned to 7 identical private rooms. Anita said that the problem dealt with computing $C(7, 5)$. Sam disagreed, saying that $P(7, 5)$ was the correct way to answer the question. Who is correct? Why?

4. Draw a tree diagram to illustrate all of the possible T-shirts available that come in sizes small, medium, large, and extra large and in the colors blue, green and gray.

Guided Practice

5. A restaurant offers the choice of an entrée, a vegetable, a dessert, and a drink for a lunch special. If there are 4 entrees, 3 vegetables, 5 desserts and 5 drinks available to choose from, how many different lunches are available? **300**

6. Are choosing a movie to see and choosing a snack to buy dependent or independent events? **independent**

Find each value.

7. $P(6, 6)$ **720**

8. $P(5, 3)$ **60**

9. $\dfrac{P(12, 8)}{P(6, 4)}$ **55,440**

10. $C(7, 4)$ **35**

11. $C(20, 15)$ **15,504**

12. $C(4, 3) \cdot C(5, 2)$ **40**

13. If a group of 10 students sits in the same row in an auditorium, how many possible ways can they be arranged? **3,628,800**

14. How many baseball lineups of 9 players can be formed from a team that has 15 members if all players can play any position? **5005**

15. **Postal Service** The U.S. Postal Service uses 5-digit ZIP codes to route letters and packages to their destinations. **b. 7290**
   a. How many ZIP codes are possible if the numbers 0 through 9 are used for each of the 5 digits? **100,000**
   b. Suppose that when the first digit is 0, the second, third, and fourth digits cannot be 0. How many 5-digit ZIP codes are possible if the first digit is 0?
   c. In 1983, the U.S. Postal Service introduced the ZIP +4, which added 4 more digits to the existing 5-digit ZIP codes. Using the numbers 0 through 9, how many additional ZIP codes were possible? **999,900,000**

EXERCISES

Practice 16. If you toss a coin, then roll a die, and then spin a 4-colored spinner with equal sections, how many outcomes are possible? **48**

17. How many ways can 7 classes be scheduled, if each class is offered in each of 7 periods? **5040**

18. Find the number of different 7-digit telephone numbers where:
   a. the first digit cannot be zero. **9,000,000**
   b. only even digits are used. **78,125**
   c. the complete telephone numbers are multiples of 10. **1,000,000**
   d. the first three digits are 593 in that order. **10,000**

State whether the events are *independent* or *dependent*.

19. selecting members for a team **dependent**

20. tossing a penny, rolling a die, then tossing a dime **independent**

21. deciding the order in which to complete your homework assignments **dependent**

 www.amc.glencoe.com/self_check_quiz

Assignment Guide

Core: 17–53 odd, 54–64
Enriched: 16–44 even, 46–64

Additional Answers

3. Sam is correct. The room assignments are an ordered selection of 5 rooms from the 7 rooms. A permutation should be used.

4.

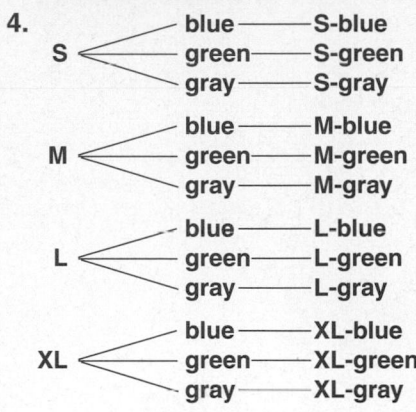

S — blue — S-blue
S — green — S-green
S — gray — S-gray

M — blue — M-blue
M — green — M-green
M — gray — M-gray

L — blue — L-blue
L — green — L-green
L — gray — L-gray

XL — blue — XL-blue
XL — green — XL-green
XL — gray — XL-gray

Additional Answer

49. $P(n, n-1) \stackrel{?}{=} P(n, n)$

$$\frac{n!}{[n-(n-1)]!} \stackrel{?}{=} \frac{n!}{(n-n)!}$$

$$\frac{n!}{1!} \stackrel{?}{=} \frac{n!}{0!} \qquad 0! = 1$$

$$n! = n!$$

Find each value.

22. $P(8, 8)$ **40,320**	23. $P(6, 4)$ **360**	24. $P(5, 3)$ **60**
25. $P(7, 4)$ **840**	26. $P(9, 5)$ **15,120**	27. $P(10, 7)$ **604,800**

28. $\dfrac{P(6, 3)}{P(4, 2)}$ **10**	29. $\dfrac{P(6, 4)}{P(5, 3)}$ **6**	30. $\dfrac{P(6, 3) \cdot P(7, 5)}{P(9, 6)}$ **5**
31. $C(5, 3)$ **10**	32. $C(10, 5)$ **252**	33. $C(4, 2)$ **6**
34. $C(12, 4)$ **495**	35. $C(9, 9)$ **1**	36. $C(14, 7)$ **3432**

39. 840

37. $C(3, 2) \cdot C(8, 3)$ **168** 38. $C(7, 3) \cdot C(8, 5)$ **1960** 39. $C(5, 1) \cdot C(4, 2) \cdot C(8, 2)$

40. A pizza shop has 14 different toppings from which to choose. How many different 4-topping pizzas can be made? **1001**

41. 2002

41. If you make a fruit salad using 5 different fruits and you have 14 different varieties from which to choose, how many different fruit salads can you make?

42. 18,564

42. How many different 12-member juries can be formed from a group of 18 people?

43. A bag contains 3 red, 5 yellow, and 8 blue marbles. How many ways can 2 red, 1 yellow, and 2 blue marbles be chosen? **420**

44. How many different ways can 11 paintings be displayed on a wall? **39,916,800**

45. From a standard 52-card deck, find how many 5-card hands are possible that have:
   a. 3 hearts and 2 clubs. **22,308**
   b. 1 ace, 2 jacks, and 2 kings. **144**
   c. all face cards. **792**

Applications and Problem Solving

Real World Application

46. **Home Security** A home security company offers a security system that uses the numbers 0 through 9, inclusive, for a 5-digit security code.
   a. How many different security codes are possible? **100,000**
   b. If no digits can be repeated, how many security codes are available? **30,240**
   c. Suppose the homeowner does not want to use 0 as one of the digits and wants only two of the digits to be odd. How many codes can be formed if the digits can be repeated? If no repetitions are allowed, how many codes are available? **1600; 480**

47. **Baseball** How many different 9-player teams can be fielded if there are 3 players who can only play catcher, 4 players who can only play first base, 6 players who can only pitch, and 14 players who can play in any of the remaining 6 positions? **216,216**

48. **Transportation** In a train yard, there are 12 flatcars, 10 tanker cars, 15 boxcars, and 5 livestock cars.
   a. If the cars must be connected according to their final destinations, how many different ways can they be arranged? $42! \approx 1.4 \times 10^{51}$
   b. How many ways can the train be made up if it is to have 30 cars? $\approx 1.1 \times 10^{10}$
   c. How many trains can be formed with 3 livestock cars, 6 flatcars, 6 tanker cars, and 5 boxcars? $\approx 5.8 \times 10^{9}$

49. **Critical Thinking** Prove $P(n, n-1) = P(n, n)$. **See margin.**

50. **Entertainment** Three couples have reserved seats for a Broadway musical. Find how many different ways they can sit if:
   a. there are no seating restrictions. **720**
   b. two members of each couple wish to sit together. **48**

51. Botany A researcher with the U.S. Department of Agriculture is conducting an experiment to determine how well certain crops can survive adverse weather conditions. She has gathered 6 corn plants, 3 wheat plants, and 2 bean plants. She needs to select four plants at random for the experiment.
  a. In how many ways can this be done? **330**
  b. If exactly 2 corn plants must be included, in how many ways can the plants be selected? **150**

52. Geometry How many lines are determined by 10 points, no 3 of which are collinear? **45**

53. Critical Thinking There are 6 permutations of the digits 1, 6, and 7.

    167   176   617   671   716   761

The average of these six numbers is $\frac{3108}{6} = 518$ which is equal to $37(1 + 6 + 7)$. If the digits are 0, 4, and 7, then the average of the six permutations is $\frac{2442}{6} = 407$ or $37(0 + 4 + 7)$. **a. 592**

  a. Use this pattern to find the average of the six permutations of 2, 5, and 9.
  b. Will this pattern hold for all sets of three digits? If so, prove it. **See margin.**

Mixed Review

54. Banking Cynthia has a savings account that has an annual yield of 5.8%. Find the balance of the account after each of the first three years if her initial balance is $2140. *(Lesson 12-8)* **$2264.12, $2395.44, $2534.38**

55. 3025

55. Find the sum of the first ten terms of the series $1^3 + 2^3 + 3^3 + \cdots$. *(Lesson 12-5)*

56. 2.26

56. Solve $7.1^x = 83.1$ using logarithms. Round to the nearest hundredth. *(Lesson 11-6)*

57. Find the value of x to the nearest tenth such that $x = e^{0.346}$. *(Lesson 11-3)* **1.4**

58. Communications A satellite dish tracks a satellite directly overhead. Suppose the graph of the equation $y = 4x^2$ models the shape of the dish when it is oriented in this position. Later in the day, the dish is observed to have rotated approximately 45°. Find an equation that models the new orientation of the dish. *(Lesson 10-7)* **$4(x')^2 + 8x'y' + 4(y')^2 + \sqrt{2}x' - \sqrt{2}y' = 0$**

59. Graph the system of polar equations $r = 2$, and $r = 2 \cos 2\theta$. Then solve the system and compare the common solutions. *(Lesson 9-2)* **(2, 180°), (2, 0°)**

60. Find the initial vertical and horizontal velocities of a rock thrown with an initial velocity of 28 feet per second at an angle of 45° with the horizontal. *(Lesson 8-7)* **19.80 ft/s, 19.80 ft/s**

61. Solve $\sin 2x + 2 \sin x = 0$ for $0° \le x \le 360°$. *(Lesson 7-5)* **0°, 180°, 360°**

62. State the amplitude, period, and phase shift for the function $y = 8 \cos(\theta - 30°)$. *(Lesson 6-5)* **8, 360°, 30°**

63. Given the triangle at the right, solve the triangle if $A = 27°$ and $b = 15.2$. Round angle measures to the nearest degree and side measures to the nearest tenth. *(Lesson 5-5)* **$B = 63°$, $a = 7.7$, $c = 17.1$**

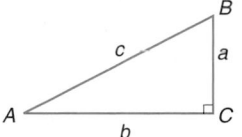

64. SAT/ACT Practice What is the number of degrees through which the hour hand of a clock moves in 2 hours 12 minutes? **A**

  A 66°   **B** 72°   **C** 126°   **D** 732°   **E** 792°

Extra Practice See p. A51.

Lesson 13-1 Permutations and Combinations **845**

Extra Credit

How many different segments can be drawn connecting any two of six points arranged in a hexagonal pattern? **15**

---

1 FOCUS

 5-Minute Check
Lesson 13-1

1. Deneen is asked to arrange a set of four best-sellers for a display in a bookstore. The display is to be set up in the following order: nonfiction, history, romance, science fiction.

   There are 3 nonfiction, 2 history, 4 romance, and 2 science fiction books from which to choose.

   a. Are the choices for each book independent or dependent events? **independent**

   b. How many different ways can Deneen choose the books for the display? **48**

2. How many ways can 6 different compact disc cases be stacked vertically? **720**

3. There are 8 children in a room with a row of 10 chairs. How many ways is it possible for the children to be seated? **1,814,400**

4. From a list of 6 books, how many groups of three books can be selected? **20**

Motivating the Lesson

Discussion Have students list all of the possible four-letter patterns with the letters *eebm*. **24** How many of the patterns have the same appearance? **12**

Permutations with Repetitions and Circular Permutations

OBJECTIVES
- Solve problems involving permutations with repetitions.
- Solve problems involving circular permutations.

 MARKETING Marketing professionals sometimes investigate the number of permutations and arrangements of letters to create company or product names. For example, the company JATACO was derived from the first initials of the owners Alan, Anthony, John, and Thomas. Suppose five high school students have developed a web site to help younger students better understand first year algebra. They decided to use the initials of their first names to create the title of their web site. The initials are: E, L, O, B, O. How many different five-letter words can be created with these letters?

Each "O" is marked a different color to differentiate it from the other. Some of the possible arrangements are listed below.

| EOLBO | EOOLB | EOOBL | EOBOL |
| LEOOB | OBELO | OBLEO | BLOOE |

The five letters can be arranged in $P(5, 5)$ or $5!$ ways. However, several of these 120 arrangements are the same unless the Os are colored. So without coloring the Os, there are repetitions in the $5!$ possible arrangements.

Permutations with Repetitions	The number of permutations of n objects of which p are alike and q are alike is $$\frac{n!}{p! \, q!}.$$

Using this formula, we find there are only $\frac{5!}{2!}$ or 60 permutations of the five letters of which 2 are Os.

Example **1** How many eight-letter patterns can be formed from the letters of the word *parabola*?

The eight letters can be arranged in $P(8, 8)$ or $8!$ ways. However, several of these 40,320 arrangements have the same appearance since *a* appears 3 times. The number of permutations of 8 letters of which 3 are the same is $\frac{8!}{3!}$ or 6720.

There are 6720 different eight-letter patterns that can be formed from the letters of the word *parabola*.

 Resource Manager

 Reproducible Masters
- *Study Guide and Practice,* pp. 241–242
- *Enrichment*, p. 148
- *Assessment and Evaluation,* p. 329

 Transparencies
- 5-Minute Check, 13-2
- Teaching, 13-2

Example **2** How many eleven-letter patterns can be formed from the letters of the word *Mississippi?*

$$\frac{11!}{4!\,4!\,2!} = \frac{11 \cdot 10 \cdot 9 \cdot 8 \cdot 7 \cdot 6 \cdot 5 \cdot 4 \cdot 3 \cdot 2 \cdot 1}{4 \cdot 3 \cdot 2 \cdot 1 \cdot 4 \cdot 3 \cdot 2 \cdot 1 \cdot 2 \cdot 1}$$ *There are 11 letters in Mississippi.*

                                                             4 i's   4 s's   2 p's

              = 34,650

There are 34,650 eleven-letter patterns.

So far, you have been studying objects that are arranged in a line. Consider the problem of making distinct arrangements of six children riding on a merry-go-round at a playground. How many riding arrangements are possible?

Let the numbers 1 through 6 represent the children. Four possible arrangements are shown below.

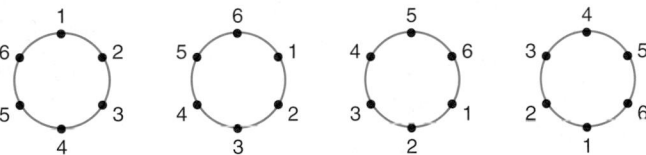

When objects are arranged in a circle, some of the arrangements are alike. In the situation above, these similar arrangements fall into groups of six, each of which can be found by rotating the circle $\frac{1}{6}$ of a revolution. Thus, the number of distinct arrangements around the circular merry-go-round is $\frac{1}{6}$ of the total number of arrangements if the children stood in a line.

$$\begin{aligned}\frac{1}{6} \cdot 6! &= \frac{6 \cdot 5 \cdot 4 \cdot 3 \cdot 2 \cdot 1}{6} \\ &= 5 \cdot 4 \cdot 3 \cdot 2 \cdot 1 \\ &= 5! \text{ or } (6 - 1)!\end{aligned}$$

Thus, there are $(6 - 1)!$ or 120 distinct arrangements of the 6 children around the merry-go-round.

Circular Permutations	If *n* objects are arranged in a circle, then there are $\frac{n!}{n}$ or $(n - 1)!$ permutations of the *n* objects around the circle.

If the circular object looks the same when it is turned over, such as a plain key ring, then the number of permutations must be divided by 2.

Example **3** At the Family Friendly Restaurant, nine bowls of food are placed on a circular, revolving tray in the center of the table. You can serve yourself from each of the bowls.

   a. Is the arrangement of the bowls on the tray a linear or circular permutation? Explain.

   The arrangement is a circular permutation since the bowls form a circle on the tray and there is no reference point.

Lesson 13-2   Permutations with Repetitions and Circular Permutations   **847**

Teaching Tip After discussing Example 2, have students formulate their own examples using words with repeated letters and give the number of letter patterns that are possible for those words.

In-Class Examples
Example 1
How many twelve-letter patterns can be formed from the letters of the word *cosmopolitan*?
79,833,600

Example 2
How many eleven-letter patterns can be formed from the letters of the word *repetitions*?
4,989,600

Prerequisite Skills

A review of the prerequisite skills for this lesson is provided in the *Prerequisite Skills Masters.*

Skill Number and Content	Pages
2 Fractions	5–6

In-Class Examples

Example 3
Martha is at a picnic that has a circular, revolving condiment server. Six condiments are placed on the table.

a. Is the arrangement of the condiments on the table a linear or circular permutation? **circular permutation**

b. How many ways can the condiments be arranged? **5! or 120**

Example 4
Thirteen children are to be seated on horses on a merry-go-round for which the attendant collects tickets, beginning with the brown horse.

a. Is the arrangement of the children on the ride a linear or circular permutation? **linear**

b. How many possible arrangements of children relative to the brown horse are there? **13! or**

Teaching Tip Stress that if *n* people are seated around a merry-go-round on which one person's position is fixed, the resulting permutations are linear permutations with *n*! linear permutations.

3 PRACTICE/APPLY

Error Analysis

Watch for students who add factors when they solve problems involving permutations with repetitions and circular permutations.

Prevent by pointing out that the two operations used almost exclusively in these problems are multiplication and division.

848 *Chapter 13*

---

b. How many ways can the bowls be arranged?

There are nine bowls so the number of arrangements can be described by (9 − 1)! or 8!
8! = 8 · 7 · 6 · 5 · 4 · 3 · 2 · 1 or 40,320

There are 40,320 ways in which the bowls can be arranged on the tray.

Suppose a CD changer holds 4 CDs on a circular platter. Let each circle below represent the platter and the labeled points represent each CD. The arrow indicates which CD will be played.

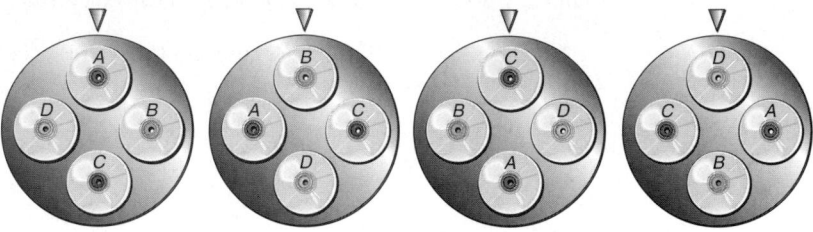

These arrangements are different. In each one, a different CD is being played. Thus, there are $P(4, 4)$ or 24 arrangements relative to the playing position.

If *n* objects are arranged relative to a fixed point, then there are *n*! permutations. Circular arrangements with fixed points of reference are treated as linear permutations.

Example **④** **Seven people are to be seated at a round table where one person is seated next to a window.**

a. Is the arrangement of the people around the table a linear or circular permutation? Explain.

b. How many possible arrangements of people relative to the window are there?

a. Since the people are seated around a table with a fixed reference point, the arrangement is a linear permutation.

b. There are seven people with a fixed reference point. So there are 7! or 5040 ways in which the people can be seated around the table.

CHECK FOR UNDERSTANDING

Communicating Mathematics

Read and study the lesson to answer each question.   1–3. See margin.

1. **Write** an explanation as to why a circular permutation is not computed the same as a linear permutation.

2. **Describe** two real-world situations involving permutations with repetitions.

3. Provide a **counterexample** for the following statement.
   The number of permutations for *n* objects in a circular arrangement is (*n* − 1)!.

Addressing Individual Needs

 Visual/Spatial Use supermarket advertisements to find examples of circular party trays made up of various food items. Have students determine the different number of arrangements of the items on the trays.

Guided Practice

How many different ways can the letters of each word be arranged?

4. *kangaroo* **10,080**

5. *classical* **22,680**

6. In how many ways can 2 red lights, 4 yellow lights, 5 blue lights, 1 green light, and 2 pink lights be arranged on a string of lights? $\dfrac{14!}{2!\,4!\,5!\,2!}$ or **7,567,560**

Determine whether each arrangement of objects is a linear or circular permutation. Then determine the number of arrangements for each situation.

7. 11 football players in a huddle **circular; 3,628,800**

8. 8 jewels on a necklace **circular; 5040**

9. 12 decorative symbols around the face of a clock **circular; 39,916,800**

10. 5 beads strung on a string arranged in a square pattern relative to a knot in the string **linear; 60**

11. **Communication** Morse code is a system of dots, dashes, and spaces that telegraphers in the United States and Canada once used to send messages by wire. How many different arrangements arc there of 5 dots, 2 dashes, and 2 spaces? **756**

EXERCISES

Practice

 A

How many different ways can the letters of each word be arranged?

12. *pizzeria* **10,080**

13. *California* **907,200**

14. *calendar* **20,160**

15. *centimeter* **302,400**

16. *trigonometry* **59,875,200**

17. *Tennessee* **3780**

18. How many different 7-digit phone numbers can have the digits 7, 3, 5, 2, 7, 3, and 2? **630**

B

19. Five country CDs and four rap CDs are to be placed in a display window. How many ways can they be arranged if they are placed by category? **126**

20. The table below shows the initial letter for each of the 50 states. How many different ways can you arrange the initial letters of the states? **≈ 1.88 × 10⁴⁹**

Initial Letter	A	C	D	F	G	H	I	K	L	M	N	O	P	R	S	T	U	V	W
Number of States	4	3	1	1	1	1	4	2	1	8	8	3	1	1	2	2	1	1	3

Determine whether each arrangement of objects is a linear or circular permutation. Then determine the number of arrangements for each situation.

21. 12 gondolas on a Ferris wheel **circular; 39,916,800**

22. a stack of 6 pennies, 3 nickels, 7 dimes, and 10 quarters **linear; ≈ 5.1 × 10¹²**

23. the placement of 9 specialty departments along the outside perimeter of a supermarket **circular; 40,320**

24. a family of 5 seated around a rectangular table **circular; 24**

25. 8 tools on a utility belt **circular; 5040**

 www.amc.glencoe.com/self_check_quiz *Lesson 13-2 Permutations with Repetitions and Circular Permutations* **849**

Assignment Guide

Core: 13–35 odd, 36–46
Enriched: 12–32 even, 33–46

Additional Answers *(p. 848)*

1. The circular permutation has no beginning or end. Therefore, the number of different arrangements is always $\dfrac{1}{n}$ of a revolution.

2. Sample answer: house or phone numbers where some of the digits repeat

3. Sample answer: the number of permutations of n charms on a bracelet with a clasp is $\dfrac{n!}{2}$.

Study Guide and Practice Masters, pp. 241–242

Determine whether each arrangement of objects is a linear or circular permutation. Then determine the number of arrangements for each situation.

 C

26. 6 houses on a cul-de-sac relative to the incoming street   **linear; 720**

27. 10 different beads on a string   **linear; 3,628,800**

28. a waiter placing 9 drinks along the edge of a circular tray   **circular; 40,320**

29. 14 keys on a key ring   **circular; 3,113,510,400**

30. 20 wooden dowels used as spokes for a wagon wheel   **circular; $\approx 1.22 \times 10^{17}$**

31. 32 horses on the outside edge of a carousel   **circular; $\approx 8.22 \times 10^{33}$**

32. 25 sections of a circular stadium relative to the main entrance
linear; $\approx 1.55 \times 10^{25}$

Applications and Problem Solving

33. **Biology**   A biologist needs to determine the number of possible arrangements of 4 kinds of molecules in a chain. If the chain contains 8 molecules with 2 of each kind, how many different arrangements are possible?   **2520**

34. **Geometry**   Suppose 7 points on the circle at the right are selected at random.
    a. Using the letters A through G, how many ways can the points be named on the circle?   **720**
    b. Relative to the point which lies on the x-axis, how many arrangements are possible?   **5040**

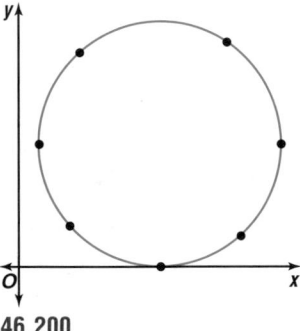

35. **Money**   Trish has a penny, 3 nickels, 4 dimes, and 3 quarters in her pocket. How many different arrangements are possible if she removes one coin at a time from her pocket?   **46,200**

36. **Critical Thinking**   An anagram is a word or phrase made from another by rearranging its letters. For example, *now* can be changed to *won*. Consider the phrase "*calculating rules.*"   **a. 4,989,600**
    a. How many different ways can the letters in *calculating* be arranged?
    b. Rearrange the letters and the space in the phrase to form the name of a branch of mathematics.   **integral calculus**

37. **Auto Racing**   Most stock car races are held on oval-shaped tracks with cars from various manufacturers. Let *F, C,* and *P* represent three auto manufacturers.
    a. Suppose for one race 20 *F* cars, 14 *C* cars, and 9 *P* cars qualified to be in a race. How many different starting line-ups based on manufacturer are possible?   **$\approx 7.85 \times 10^{17}$**
    b. If there are 43 cars racing, how many different ways could the cars be arranged on the track?   **$\approx 1.41 \times 10^{51}$**
    c. Relative to the leader of the race, how many different ways could the cars be arranged on the track?   **$\approx 6.04 \times 10^{52}$**

38. **Critical Thinking**   To break a code, Zach needs to find how many symbols there are in a particular sequence. He is told that there are 3 x's and some dashes. He is also told that there are 35 linear permutations of the symbols. What is the total number of symbols in the code sequence?   **7**

Mixed Review

39. **Food**   Classic Pizza offers pepperoni, mushrooms, sausage, onions, peppers, and olives as toppings for their 7-inch pizza. How many different 3-topping pizzas can be made? *(Lesson 13-1)*   **20**

850   *Chapter 13   Combinatorics and Probability*

Extra Practice   See p. A51.

Extra Credit

Suppose that a teacher passes out flavored lollypops in the following way: 4 lemon, 3 cherry, 8 pineapple, 2 orange, and 1 grape. How many different ways can these candies be randomly assigned to the 18 students?   **551,350,800**

40. $125x^3 - 75x^2 + 15x - 1$ **40.** Use the Binomial Theorem to expand $(5x - 1)^3$. *(Lesson 12-6)*

41. Solve $x < \log_2 413$ using logarithms. Round to the nearest hundredth. *(Lesson 11-5)* $x < 8.69$

42. Write the equation of the parabola with vertex at $(6, -1)$ and focus at $(3, -1)$. *(Lesson 10-5)* $(y + 1)^2 = -12(x - 6)$

43. Simplify $2(4 - 3i)(7 - 2i)$. *(Lesson 9-5)* $44 - 58i$

44. Find the cross product of $\vec{v}\ \langle 2, 0, 3 \rangle$ and $\vec{w}\ \langle 2, 5, 0 \rangle$. Verify that the resulting vector is perpendicular to $\vec{v}$ and $\vec{w}$. *(Lesson 8-4)* **See margin.**

45. Manufacturing A knife is held at a 45° angle to the vertical on a 16-inch diameter sharpening wheel. How far above the wheel must a lamp be placed so it will not be showered with sparks? *(Lesson 7-6)* **about 3.31 in.**

46. SAT/ACT Practice If $x^2 = 36$, then 2^{x-1} could equal **E**

A 4   **B** 6   **C** 8   **D** 16   **E** 32

CAREER CHOICES

 Actuary

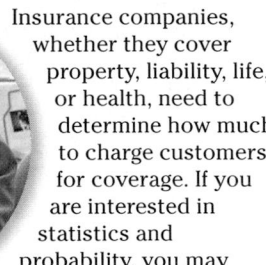

Insurance companies, whether they cover property, liability, life, or health, need to determine how much to charge customers for coverage. If you are interested in statistics and probability, you may want to consider a career as an actuary.

Actuaries use statistical methods to determine the probability of such events as death, injury, unemployment, and property damage or loss. An actuary must estimate the number and amount of claims that may be made in order for the insurance company to set its insurance coverage rates for its customers. As an actuary, you can specialize in property and liability or life and health statistics. Most actuaries work for insurance companies, consulting firms, or the government.

CAREER OVERVIEW

Degree Preferred:
bachelor's degree in actuarial science or mathematics

Related Courses:
mathematics, statistics, computer science, business courses

Outlook:
faster than average through the year 2006

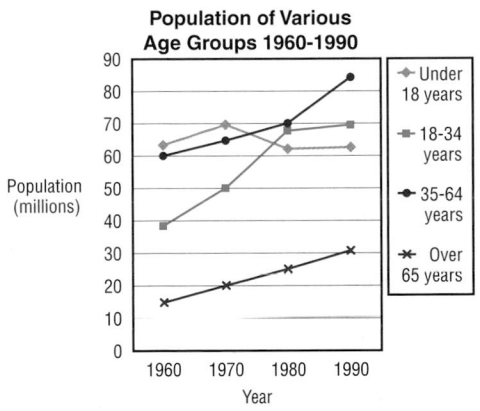

Population of Various Age Groups 1960-1990

internNET CONNECTION For more information on careers in actuarial science, visit: **www.amc.glencoe.com**

CAREER CHOICES   ACTUARY

Math Connection
- What does this line graph show? **It shows how the populations of different age groups changed from 1960 to 1990.**
- Find the approximate percent increase or decrease in population for each age group from 1960 to 1990. Which age group grew the fastest? **Estimates may vary; under 18: 0.6% decrease; 18–34: 79% increase; 35–64: 39% increase; over 65: 87% increase. The over-65 group grew the fastest.**

---

4 ASSESS

Open-Ended Assessment

Writing Have students write a short list of 5-letter words and determine the number of permutations that can be formed using all five letters.
Sample answers:
RADIO—120
TEETH—30
BOOTH—60

 Chapter 13, Quiz A (Lessons 13-1 and 13-2)
is available in the *Assessment and Evaluation Masters*, p. 329.

Additional Answer
44. $\langle -15, 6, 10 \rangle$; yes;
$\langle 2, 0, 3 \rangle \cdot \langle -15, 6, 10 \rangle = 0$,
$\langle 2, 5, 0 \rangle \cdot \langle -15, 6, 10 \rangle = 0$

Enrichment Masters, p. 148

13-2 Enrichment

Approximating Factorials

James Stirling (1692-1770) was a teacher, a friend of Sir Isaac Newton, and a mathematician who made important contributions to calculus. Today he is best remembered as the creator of a formula for approximating factorials.

Stirling's Formula $n! \approx \sqrt{2\pi n}\left(\frac{n}{e}\right)^n$, where e is the irrational number 2.7182818...

1. Stirling proved that the ratio $\frac{n!}{\sqrt{2\pi n}\left(\frac{n}{e}\right)^n}$ is less than $\frac{12n}{12n - 1}$.

As n increases, will the approximations obtained using Stirling's formula become more accurate or less accurate? Explain.

More accurate; the fraction $\frac{12n}{12n - 1}$ is always greater than 1.
As n grows larger, however, $12n$ gets closer to $12n - 1$.
The fraction $\frac{12n}{12n - 1}$ gets smaller and closer to 1.

2. Complete the chart.

n	$n!$	$\sqrt{2\pi e}\left(\frac{n}{e}\right)^n$	$\frac{n!}{\sqrt{2\pi e}\left(\frac{n}{e}\right)^n}$
10	3.6288×10^6	3.5987×10^6	1.008
20	2.4329×10^{18}	2.4228×10^{18}	1.004
30	2.6525×10^{32}	2.6452×10^{32}	1.003
40	8.1592×10^{47}	8.1422×10^{47}	1.002
50	3.0414×10^{64}	3.0363×10^{64}	1.002
60	8.3210×10^{81}	8.3094×10^{81}	1.001

© Glencoe/McGraw-Hill   T148   *Advanced Mathematical Concepts*

1 FOCUS

 5-Minute Check
Lesson 13-2

1. How many nine-letter patterns can be formed from the letters of the word *nonlinear*? **60,480**

2. How may ten-letter patterns can be formed from the letters of the word *proportion*? **151,200**

3. At a restaurant, 11 bowls of food are placed on a circular, revolving tray in the center of the table. You can serve yourself from each of the bowls. How many ways can the bowls be arranged? **10! or 3,628,800**

4. Eleven people are to be seated at a round table where one person is seated closest to the exit. How many possible arrangements of people relative to the exit are possible? **11! or 39,916,800**

Motivating the Lesson

Hands-On Activity Write the numbers 1 through 100 on slips of paper and choose one winning number. Have each student draw a number and predict the probability that it is the winning number. Repeat the experiment several times, each time decreasing the number of slips of paper.

Probability and Odds

OBJECTIVES
* Find the probability of an event.
* Find the odds for the success and failure of an event.

 MARKET RESEARCH To determine television ratings, Nielsen Media Research estimates how many people are watching any given television program. This is done by selecting a sample audience, having them record their viewing habits in a journal, and then counting the number of viewers for each program. There are about 100 million households in the U.S., and only 5000 are selected for the sample group. What is the probability of any one household being selected to participate? *This problem will be solved in Example 1.*

When we are uncertain about the occurrence of an event, we can measure the chances of its happening with **probability.** For example, there are 52 possible outcomes when selecting a card at random from a standard deck of playing cards. The set of all outcomes of an event is called the **sample space.** A desired outcome, drawing the king of hearts for example, is called a **success.** Any other outcome is called a **failure.** The probability of an event is the ratio of the number of ways an event can happen to the total number of outcomes in the sample space, which is the sum of successes and failures. There is one way to draw a king of hearts, and there are a total of 52 outcomes when selecting a card from a standard deck. So, the probability of selecting the king of hearts is $\frac{1}{52}$.

Probability of Success and of Failure	If an event can succeed in *s* ways and fail in *f* ways, then the probability of success $P(s)$ and the probability of failure $P(f)$ are as follows. $$P(s) = \frac{s}{s + f} \qquad P(f) = \frac{f}{s + f}$$

Example **1** **MARKET RESEARCH** **What is the probability of any one household being chosen to participate for the Nielsen Media Research group?**

Use the probability formula. Since 5000 households are selected to participate $s = 5000$. The denominator, $s + f$, represents the total number of households, those selected, s, and those not selected, f. So, $s + f = 100,000,000$.

$$P(5000) = \frac{5000}{100,000,000} \text{ or } \frac{1}{20,000} \qquad P(s) = \frac{s}{s + f}$$

The probability of any one household being selected is $\frac{1}{20,000}$ or 0.005%.

852 *Chapter 13 Combinatorics and Probability*

 Resource Manager

 Reproducible Masters
* *Study Guide and Practice,* pp. 243–244
* *Enrichment,* p. 149
* *Assessment and Evaluation,* pp. 328, 329

 Transparencies
* 5-Minute Check, 13-3
* Teaching, 13-3

 Technology/Multimedia
* Advanced MathPASS, Lesson 28

An event that cannot fail has a probability of 1. An event that cannot succeed has a probability of 0. Thus, the probability of success $P(s)$ is always between 0 and 1 inclusive. That is, $0 \leq P(s) \leq 1$.

Example **2** A bag contains 5 yellow, 6 blue, and 4 white marbles.
  a. **What is the probability that a marble selected at random will be yellow?**
  b. **What is the probability that a marble selected at random will *not* be white?**

  a. The probability of selecting a yellow marble is written P(yellow). There are 5 ways to select a yellow marble from the bag, and $6 + 4$ or 10 ways not to select a yellow marble. So, $s = 5$ and $f = 10$.

  $$P(\text{yellow}) = \frac{5}{5+10} \text{ or } \frac{1}{3} \quad P(s) = \frac{s}{s+f}$$

  The probability of selecting a yellow marble is $\frac{1}{3}$.

  b. There are 4 ways to select a white marble. So there are 11 ways not to select a white marble.

  $$P(\text{not white}) = \frac{11}{4+11} \text{ or } \frac{11}{15}$$

  The probability of *not* selecting a white marble is $\frac{11}{15}$.

The counting methods you used for permutations and combinations are often used in determining probability.

Example **3** **A circuit board with 20 computer chips contains 4 chips that are defective. If 3 chips are selected at random, what is the probability that all 3 are defective?**

There are $C(4, 3)$ ways to select 3 out of 4 defective chips, and $C(20, 3)$ ways to select 3 out of 20 chips.

$$P(\text{3 defective chips}) = \frac{C(4, 3)}{C(20, 3)} \quad \begin{array}{l} \leftarrow \textit{ways of selecting 3 defective chips} \\ \leftarrow \textit{ways of selecting 3 chips} \end{array}$$

$$= \frac{\frac{4!}{1!\,3!}}{\frac{20!}{17!\,3!}} \text{ or } \frac{1}{285}$$

The probability of selecting three defective computer chips is $\frac{1}{285}$.

The sum of the probability of success and the probability of failure for any event is always equal to 1.

$$P(s) + P(f) = \frac{s}{s+f} + \frac{f}{s+f}$$
$$= \frac{s+f}{s+f} \text{ or } 1$$

This property is often used in finding the probability of events. For example, the probability of drawing a king of hearts is $P(s) = \frac{1}{52}$, so the probability of not drawing the king of hearts is $P(f) = 1 - \frac{1}{52}$ or $\frac{51}{52}$. Because their sum is 1, $P(s)$ and $P(f)$ are called **complements**.

Lesson 13-3 Probability and Odds **853**

Teaching Tip For Example 2b, stress that the word *not* will involve complements, and that the sum of the complements is 1.

In-Class Examples

Example 1
History Before 2001, the United States had 42 presidents including William Jefferson Clinton. Only two of them (George Washington and John Adams) belonged to the Federalist Party. If you were to select a president at random for a history report, what is the probability that your selection would be a Federalist? $\frac{1}{21}$

Example 2
A sock drawer contains single, unmatched socks: 19 are white, 8 are black, and 5 are blue.
  a. What is the probability that a sock selected at random will be black? $\frac{1}{4}$
  b. What is the probability that a sock selected at random will *not* be white? $\frac{13}{32}$

Example 3
A box of 60 baseball cards contains 8 cards with print errors on them. If 5 cards are selected at random, what is the probability that all 5 have print errors? $\frac{1}{97,527}$

Teaching Tip Remind students that the expression *one or more* has the same meaning as *at least one.*

Teaching Tip Remind students that $P(s) + P(f)$ will always equal 1.

In-Class Examples

Example 4
A cost accountant has determined that an assembly line averages one defective whatnot for every 25 that it produces. What is the probability that an order of 100 whatnots will have one or more defective whatnots? **0.983129681 or about $\frac{49}{50}$**

Example 5
A game contains 24 cards that say "Move Ahead One Space" and 10 cards that say "Move Back One Space." A card may be chosen randomly from anywhere in the deck of 34 cards.
a. What is the probability of drawing a "Move Ahead" card? $\frac{12}{17}$
b. What are the odds that a player will move his or her token ahead one space on any given turn? $\frac{12}{5}$

Example ④ The CyberToy Company has determined that out of a production run of 50 toys, 17 are defective. If 5 toys are chosen at random, what is the probability that at least 1 is defective?

The complement of selecting at least 1 defective toy is selecting no defective toys. That is, P(at least 1 defective toy) $= 1 - P$(no defective toys).

P(at least 1 defective toy) $= 1 - P$(no defective toys).

$$= 1 - \frac{C(33, 5)}{C(50, 5)} \quad \leftarrow \text{ways of selecting 5 defective toys} \atop \leftarrow \text{ways of selecting 5 toys}$$

$$= 1 - \frac{237,336}{2,118,760}$$

$$\approx 0.8879835375 \quad \textit{Use a calculator.}$$

The probability of selecting at least 1 defective toy is about 89%.

Another way to measure the chance of an event occurring is with **odds.** The probability of success of an event and its complement are used when computing the odds of an event.

Odds	The odds of the successful outcome of an event is the ratio of the probability of its success to the probability of its failure. $$\text{Odds} = \frac{P(s)}{P(f)}$$

Example ⑤ Katrina must select at random a chip from a box to determine which question she will receive in a mathematics contest. There are 6 blue and 4 red chips in the box. If she selects a blue chip, she will have to solve a trigonometry problem. If the chip is red, she will have to write a geometry proof.

a. **What is the probability that Katrina will draw a red chip?**

b. **What are the odds that Katrina will have to write a geometry proof?**

a. The probability that Katrina will select a red chip is $\frac{4}{10}$ or $\frac{2}{5}$.

b. To find the odds that Katrina will have to write a geometry proof, you need to know the probability of a successful outcome and of a failing outcome.

Let s represent selecting a red chip and f represent not selecting a red chip.

$$P(s) = \frac{2}{5} \qquad\qquad P(f) = 1 - \frac{2}{5} \text{ or } \frac{3}{5}$$

Now find the odds.

$$\frac{P(s)}{P(f)} = \frac{\frac{2}{5}}{\frac{3}{5}} \text{ or } \frac{2}{3}$$

The odds that Katrina will choose a red chip and thus have to write a geometry proof is $\frac{2}{3}$. *The ratio $\frac{2}{3}$ is read "2 to 3."*

Addressing Individual Needs

 Logical Ask students to explain why the theoretical probability of an event may not match the experimental probability for that event. How does the number of experiments affect how close the experimental probability is to the theoretical probability? **Sample answers: experiments can include errors; the more experiments that are run, the closer the experimental probability gets to the theoretical probability.**

Sometimes when computing odds, you must find the sample space first. This can involve finding permutations and combinations.

Example ⑥ Twelve male and 16 female students have been selected as equal qualifiers for 6 college scholarships. If the awarded recipients are to be chosen at random, what are the odds that 3 will be male and 3 will be female?

First, determine the total number of possible groups.

$C(12, 3)$ *number of groups of 3 males*
$C(16, 3)$ *number of groups of 3 females*

Using the Basic Counting Principle we can find the number of possible groups of 3 males and 3 females.

$$C(12, 3) \cdot C(16, 3) = \frac{12!}{9! \, 3!} \cdot \frac{16!}{13! \, 3!} \text{ or } 123{,}200 \text{ possible groups}$$

The total number of groups of 6 recipients out of the 28 who qualified is $C(28, 6)$ or 376,740. So, the number of groups that do not have 3 males and 3 females is $376{,}740 - 123{,}200$ or $253{,}540$.

Finally, determine the odds.

$$P(s) = \frac{123{,}200}{376{,}740} \qquad\qquad P(f) = \frac{253{,}540}{376{,}740}$$

$$\text{odds} = \frac{\frac{123{,}200}{376{,}740}}{\frac{253{,}540}{376{,}740}} \quad \text{or} \quad \frac{880}{1811}$$

Thus, the odds of selecting a group of 3 males and 3 females are $\frac{880}{1811}$ or close to $\frac{1}{2}$.

CHECK FOR UNDERSTANDING

Communicating Mathematics

Read and study the lesson to answer each question.   **1–4. See margin.**

1. **Explain** how you would interpret $P(E) = \frac{1}{2}$.

2. Find two examples of the use of probability in newspapers or magazines. **Describe** how probability concepts are applied.

3. **Write** about the difference between the probability of the successful outcome of an event and the odds of the successful outcome of an event.

4. **You Decide** Mika has figured that his odds of winning the student council election are 3 to 2. Geraldo tells him that, based on those odds, the probability of his winning is 60%. Mika disagreed. Who is correct? Explain your answer.

Guided Practice

A box contains 3 tennis balls, 7 softballs, and 11 baseballs. One ball is chosen at random. Find each probability.

5. $P(\text{softball})$ $\frac{1}{3}$ 6. $P(\text{not a baseball})$ $\frac{10}{21}$ 7. $P(\text{golf ball})$ **0**

8. In an office, there are 7 women and 4 men. If one person is randomly called on the phone, find the probability the person is a woman. $\frac{7}{11}$

Lesson 13-3 Probability and Odds **855**

Additional Answers

1. The probability of the event happening is 50–50.
2. Answers will vary; see students' work.
3. Sample answer: the probability of the successful outcome of an event is the ratio of the number of successful outcomes to the total number of outcomes possible. The odds of the successful outcome of an event is the ratio of the probability of its success to the probability of its failure.
4. Geraldo is correct. $P(\text{win}) = \frac{3}{2+3} = \frac{3}{5}$ or 60%.

Teaching Tip Encourage students to read each problem carefully since there is a difference between finding probability and odds.

In-Class Example

Example 6
Four male and 2 female students have been selected as equal qualifiers for 2 college scholarships. If the awarded recipients are to be chosen at random, what are the odds that one will be male and one will be female? $\frac{8}{7}$

Teaching Tip You may want to have a discussion on how the word *odds* is misused in advertising when companies promote a game or give-away.

3 PRACTICE/APPLY

Error Analysis
Watch for students who do not have a clear understanding of odds.

Prevent by stressing that the word *odds* represents a comparison of the possible outcomes, not the probability of one outcome.

Study Guide and Practice Masters, pp. 243–244

Assignment Guide

Core: 13–39 odd, 40, 41, 43–51
Enriched: 14–36 even, 37–51
All: Mid-Chapter Quiz, 1–10

Of 7 kittens in a litter, 4 have stripes. Three kittens are picked at random. Find the odds of each event.

9. All three have stripes. $\frac{4}{31}$ 10. Only 1 has stripes. $\frac{12}{23}$ 11. One is not striped. $\frac{18}{17}$

12. **Meteorology** A local weather forecast states that the probability of rain on Saturday is 80%. What are the odds that it will not rain Saturday? (*Hint:* Rewrite the percent as a fraction.) $\frac{1}{4}$

EXERCISES

Practice

Using a standard deck of 52 cards, find each probability. *The face cards include kings, queens, and jacks.*

 A

13. P(face card) $\frac{3}{13}$ 14. P(a card of 6 or less) $\frac{6}{13}$

15. P(a black, non-face card) $\frac{5}{13}$ 16. P(not a face card) $\frac{10}{13}$

One flower is randomly taken from a vase containing 5 red flowers, 2 white flowers, and 3 pink flowers. Find each probability.

 B

17. P(red) $\frac{1}{2}$ 18. P(white) $\frac{1}{5}$

19. P(not pink) $\frac{7}{10}$ 20. P(red or pink) $\frac{4}{5}$

Jacob has 10 rap, 18 rock, 8 country, and 4 pop CDs in his music collection. Two are selected at random. Find each probability.

21. P(2 pop) $\frac{1}{130}$ 22. P(2 country) $\frac{7}{195}$

23. P(1 rap and 1 rock) $\frac{3}{13}$ 24. P(not rock) $\frac{77}{260}$

25. A number cube is thrown two times. What is the probability of rolling 2 fives? $\frac{1}{36}$

A box contains 1 green, 2 yellow, and 3 red marbles. Two marbles are drawn at random without replacement. What are the odds of each event occurring?

26. drawing 2 red marbles $\frac{1}{4}$ 27. not drawing yellow marbles $\frac{2}{3}$

28. drawing 1 green and 1 red $\frac{1}{9}$ 29. drawing two different colors $\frac{11}{4}$

Of 27 students in a class, 11 have blue eyes, 13 have brown eyes, and 3 have green eyes. If 3 students are chosen at random what are the odds of each event occurring?

 C

30. all three have blue eyes $\frac{11}{184}$ 31. 2 have brown and 1 has blue eyes $\frac{22}{53}$

32. no one has brown eyes $\frac{28}{197}$ 33. only 1 has green eyes $\frac{92}{233}$

34. The odds of winning a prize in a raffle with one raffle ticket are $\frac{1}{249}$. What is the probability of winning with one ticket? $\frac{1}{250}$

35. The probability of being accepted to attend a state university is $\frac{4}{5}$. What are the odds of being accepted to this university? $\frac{4}{1}$

 www.amc.glencoe.com/self_check_quiz

36. From a deck of 52 cards, 5 cards are drawn. What are the odds of having three cards of one suit and the other two cards be another suit? $\frac{429}{3736}$

Applications and Problem Solving

Real World Application

37. Weather During a particular hurricane, hurricane trackers determine that the odds of it hitting the South Carolina coast are 1 to 4. What is the probability of this happening? $\frac{1}{5}$

38. Baseball At one point in the 1999 season, Ken Griffey, Jr. had a batting average of 0.325. What are the odds that he would hit the ball the next time he came to bat? $\frac{13}{27}$

39. Security Kim uses a combination lock on her locker that has 3 wheels, each labeled with 10 digits from 0 to 9. The combination is a particular sequence with no digits repeating.

  a. What is the probability of someone guessing the correct combination? $\frac{1}{720}$

  b. If the digits can be repeated, what are the odds against someone guessing the combination? $\frac{999}{1}$

40. Critical Thinking Spencer is carrying out a survey of the bear population at Yellowstone National Park. He spots two bears—one has a light colored coat and the other has a dark coat.

  a. Assume that there are equal numbers of male and female bears in the park. What is the probability that both bears are male? $\frac{1}{4}$

  b. If the lighter colored bear is male, what are the odds that both are male? $\frac{1}{1}$

41. Testing Ms. Robinson gives her precalculus class 20 study problems. She will select 10 to answer on an upcoming test. Carl can solve 15 of the problems.

41a. $\frac{21}{1292}$

  a. Find the probability that Carl can solve all 10 problems on the test.

  b. Find the odds that Carl will know how to solve 8 of the problems. $\frac{225}{421}$

42. Mortality Rate During 1990, smoking was linked to 418,890 deaths in the United States. The graph shows the diseases that caused these smoking-related deaths.

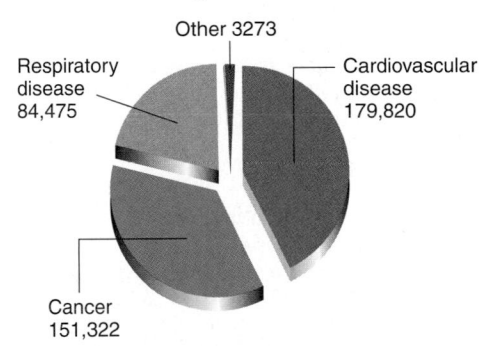

Smoking-Related Deaths

Other 3273
Respiratory disease 84,475
Cardiovascular disease 179,820
Cancer 151,322

  a. Find the probability that a smoking-related death was the result of either cardiovascular disease or cancer. $\frac{331,142}{418,890} \approx 0.791$

  b. Determine the odds against a smoking-related death being caused by cancer. $\frac{133,784}{75,661}$

43. Critical Thinking A plumber cuts a pipe in two pieces at a point selected at random. What is the probability that the length of the longer piece of pipe is at least 8 times the length of the shorter piece of pipe? $\frac{2}{9}$

Mixed Review

44. A food vending machine has 6 different items on a revolving tray. How many different ways can the items be arranged on the tray? *(Lesson 13-2)* **120**

45. The Foxtrail Condominium Association is electing board members. How many groups of 4 can be chosen from the 10 candidates who are running? *(Lesson 13-1)* **210**

46. Find S_{14} for the arithmetic series for which $a_1 = 3.2$ and $d = 1.5$. *(Lesson 12-1)* **181.3**

47. Simplify $7^{\log_7 2x}$. *(Lesson 11-4)* **2x**

48. **Landscaping** Carolina bought a new sprinkler to water her lawn. The sprinkler rotates $360°$ while spraying a stream of water. Carolina places the sprinkler in her yard so the ordered pair that represents its location is $(7, 2)$, and the sprinkler sends out water that just barely reaches the point at $(10, -8)$. Find an equation representing the farthestmost points the water can reach. *(Lesson 10-2)* $(x - 7)^2 + (y - 2)^2 = 109$

49. Find the product $3(\cos \pi + i \sin \pi) \cdot 2\left(\cos \dfrac{\pi}{4} + i \sin \dfrac{\pi}{4}\right)$. Then express it in rectangular form. *(Lesson 9-7)*
49. $6(\cos \dfrac{5\pi}{4} + i \sin \dfrac{5\pi}{4})$, $-3\sqrt{2} - 3\sqrt{2}i$

50. Find an ordered pair to represent $\vec{u}$ if $\vec{u} = \vec{v} + \vec{w}$, if $\vec{v} = \langle 3, -5 \rangle$ and $\vec{w} = \langle -4, 2 \rangle$. *(Lesson 8-2)* $\langle -1, -3 \rangle$

51. **SAT Practice** What is the area of an equilateral triangle with sides $2s$ units long? **B**

A s^2 units^2

B $\sqrt{3}s^2$ units^2

C $2s^2$ units^2

D $4s^2$ units^2

E $6s^2$ units^2

MID-CHAPTER QUIZ

Find each value. *(Lesson 13-1)*

1. $P(15, 5)$. **360,360** 2. $C(20, 9)$. **167,960**

3. Regular license plates in Ohio have three letters followed by four digits. How many different license plate arrangements are possible? *(Lesson 13-1)* **175,760,000**

4. Suppose there are 12 runners competing in the finals of a track event. Awards are given to the top five finishers. How many top-five arrangements are possible? *(Lesson 13-1)* **95,040**

5. An ice cream shop has 18 different flavors of ice cream, which can be ordered in a cup, sugar cone, or waffle cone. There is also a choice of six toppings. How many two-scoop servings with a topping are possible? *(Lesson 13-1)* **5832**

6. How many nine-letter patterns can be formed from the letters in the word *quadratic*? *(Lesson 13-2)* **181,440**

7. How many different arrangements can be made with ten pieces of silverware laid in a row if three are identical spoons, four are identical forks, and three are identical knives? *(Lesson 13-2)* **4200**

8. Eight children are riding a merry-go-round. How many ways can they be seated? *(Lesson 13-2)* **5040**

9. Two cards are drawn at random from a standard deck of 52 cards. What is the probability that both are hearts? *(Lesson 13-3)* $\dfrac{1}{17}$

10. A bowl contains four apples, three bananas, three oranges, and two pears. If two pieces of fruit are selected at random, what are the odds of selecting an orange and a banana? *(Lesson 13-3)* $\dfrac{3}{19}$

Extra Practice See p. A51.

Extra Credit

A standard number cube is rolled 100 times. About how many times should the number 4 appear? About how many times should an even number appear? About how many times should a number less than 3 appear? **17; 50; 33**

Probabilities of Compound Events

OBJECTIVES
- Find the probability of independent and dependent events.
- Identify mutually exclusive events.
- Find the probability of mutually exclusive and inclusive events.

 TRANSPORTATION According to U.S. Department of Transportation statistics, the top ten airlines in the United States arrive on time 80% of the time. During their vacation, the Hiroshi family has direct flights to Washington, D.C., Chicago, Seattle, and San Francisco on different days. What is the probability that all their flights arrived on time?

Since the flights occur on different days, the four flights represent independent events. Let A represent an on-time arrival of an airplane.

$$\begin{array}{c} \text{Flight 1} \cdot \text{Flight 2} \cdot \text{Flight 3} \cdot \text{Flight 4} \\ P(\text{all flights on time}) = P(A) \cdot P(A) \cdot P(A) \cdot P(A) \\ = (0.80)^4 \quad A = 0.80 \\ \approx 0.4096 \text{ or about } 41\% \end{array}$$

Thus, the probability of all four flights arriving on time is about 41%.

This problem demonstrates that the probability of more than one independent event is the product of the probabilities of the events.

Probability of Two Independent Events	If two events, A and B, are independent, then the probability of both events occurring is the product of each individual probability. $$P(A \text{ and } B) = P(A) \cdot P(B)$$

Example **1** **Using a standard deck of playing cards, find the probability of selecting a face card, replacing it in the deck, and then selecting an ace.**

Let A represent a face card for the first card drawn from the deck, and let B represent the ace in the second selection.

$P(A) = \dfrac{12}{52}$ or $\dfrac{3}{13}$ *12 face cards / 52 cards in a standard deck*

$P(B) = \dfrac{4}{52}$ or $\dfrac{1}{13}$ *4 aces / 52 cards in a standard deck*

The two draws are independent because when the card is returned to the deck, the outcome of the second draw is not affected by the first one.

$$P(A \text{ and } B) = P(A) \cdot P(B)$$
$$= \frac{3}{13} \cdot \frac{1}{13} \text{ or } \frac{3}{169}$$

The probability of selecting a face card first, replacing it, and then selecting an ace is $\dfrac{3}{169}$.

Lesson 13-4 Probabilities of Compound Events **859**

Resource Manager

 Reproducable Masters
- *Study Guide and Practice,* pp. 245–246
- *Enrichment,* p. 150

 Transparencies
- 5-Minute Check, 13-4
- Teaching, 13-4

1 FOCUS

Hands-On Activity Give each of 10 students 5 cards dealt randomly from a standard deck of cards. Ask students what they think the probability is of 4 of the cards having the same number. Then, see whether any students have this number in their 5 cards. $\frac{1}{4165}$

2 TEACH

Teaching Tip Before discussing the probability of two independent events, have students give examples of independent dependent events.

In-Class Examples

Example 1
Using a standard deck of playing cards, find the probability of selecting a spade, replacing it in the deck, and then selecting a face card. $\frac{3}{52}$

Example 2
History Of our first 42 American presidents, 4 were members of the Whig party and 7 were born in Ohio. Find the probability that a president selected at random will not be a Whig, but will be from Ohio. $\frac{19}{126}$ or about 15%

Example 3
Huy has a bag that contains 4 red, 6 green, and 3 brown candies. He randomly chooses 3 candies from the bag.
a. Determine whether the events are independent or dependent. **dependent**
b. What is the probability that he will choose all brown

Teaching Tip For In-Class Example 3, emphasize to students that if Huy had replaced each candy in the bag before selecting the next, the events would then be independent.

Example ② **OCCUPATIONAL HEALTH** Statistics collected in a particular coal-mining region show that the probability that a miner will develop black lung disease is $\frac{5}{11}$. Also, the probability that a miner will develop arthritis is $\frac{1}{5}$. If one health problem does not affect the other, what is the probability that a randomly-selected miner will not develop black lung disease but will develop arthritis?

The events are independent since having black lung disease does not affect the existence of arthritis.

P(not black lung disease and arthritis) = $[1 - P(\text{black lung disease})] \cdot P(\text{arthritis})$

$= \left(1 - \frac{5}{11}\right) \cdot \frac{1}{5}$ or $\frac{6}{55}$

The probability that a randomly-selected miner will not develop black lung disease but will develop arthritis is $\frac{6}{55}$.

What do you think the probability of selecting two face cards would be if the first card drawn were not placed back in the deck? Unlike the situation in Example 1, these events are dependent because the outcome of the first event affects the second event. This probability is also calculated using the product of the probabilities.

first card   *second card*

$P(\text{face card}) = \frac{12}{52}$ $P(\text{face card}) = \frac{11}{51}$

$P(\text{two face cards}) = \frac{12}{52} \cdot \frac{11}{51}$ or $\frac{11}{221}$

Notice that when a face card is removed from the deck, not only is there one less face card, but also one less card in the deck.

Thus, the probability of selecting two face cards from a deck without replacing the cards is $\frac{11}{221}$ or about $\frac{1}{20}$.

Probability of Two Dependent Events	If two events, A and B, are dependent, then the probability of both events occurring is the product of each individual probability. $P(A \text{ and } B) = P(A) \cdot P(B \text{ following } A)$

Example ③ Tasha has 3 rock, 4 country, and 2 jazz CDs in her car. One day, before she starts driving, she pulls 2 CDs from her CD carrier without looking.

a. Determine if the events are independent or dependent.

b. What is the probability that both CDs are rock?

a. The events are dependent. This event is equivalent to selecting one CD, not replacing it, then selecting another CD.

b. Determine the probability.

$P(\text{rock, rock}) = P(\text{rock}) \cdot P(\text{rock following first rock selection})$

$P(\text{rock, rock}) = \frac{3}{9} \cdot \frac{2}{8}$ or $\frac{1}{12}$

The probability that Tasha will select two rock CDs is $\frac{1}{12}$.

There are times when two events cannot happen at the same time. For example, when tossing a number cube, what is the probability of tossing a 2 *or* a 5? In this situation, both events cannot happen at the same time. That is, the events are **mutually exclusive.** The probability of tossing a 2 or a 5 is $P(2) + P(5)$, which is $\frac{1}{6} + \frac{1}{6}$ or $\frac{2}{6}$.

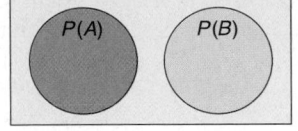

Events A and B are mutually exclusive.

Note that the two events do not overlap, as shown in the Venn diagram. So, the probability of two mutually exclusive events occurring can be represented by the sum of the areas of the circles.

Probability of Mutually Exclusive Events	If two events, A and B, are mutually exclusive, then the probability that either A or B occurs is the sum of their probabilities. $$P(A \text{ or } B) = P(A) + P(B)$$

Example ④ **Lenard is a contestant in a game where if he selects a blue ball or a red ball he gets an all-expenses paid Caribbean cruise. Lenard must select the ball at random from a box containing 2 blue, 3 red, 9 yellow, and 10 green balls. What is the probability that he will win the cruise?**

These are mutually exclusive events since Lenard cannot select a blue and a red ball at the same time. Find the sum of the individual probabilities.

$P(\text{blue or red}) = P(\text{blue}) + P(\text{red})$

$\qquad = \frac{2}{24} + \frac{3}{24} \text{ or } \frac{5}{24}$ *$P(blue) = \frac{2}{24}, P(red) = \frac{3}{24}$*

The probability that Lenard will win the cruise is $\frac{5}{24}$.

What is the probability of rolling two number cubes, in which the first number cube shows a 2 or the sum of the number cubes is 6 or 7? Since each number cube can land six different ways, and two number cubes are rolled, the sample space can be represented by making a chart. A **reduced sample space** is the subset of a sample space that contains only those outcomes that satisfy a given condition.

		Second Number Cube					
		1	2	3	4	5	6
	1	(1, 1)	(1, 2)	(1, 3)	(1, 4)	(1, 5)	(1, 6)
	2	(2, 1)	(2, 2)	(2, 3)	(2, 4)	(2, 5)	(2, 6)
First Number Cube	3	(3, 1)	(3, 2)	(3, 3)	(3, 4)	(3, 5)	(3, 6)
	4	(4, 1)	(4, 2)	(4, 3)	(4, 4)	(4, 5)	(4, 6)
	5	(5, 1)	(5, 2)	(5, 3)	(5, 4)	(5, 5)	(5, 6)
	6	(6, 1)	(6, 2)	(6, 3)	(6, 4)	(6, 5)	(6, 6)

In-Class Example
Example 4
Gina has 6 pennies, 4 nickels, and 5 dimes in her change purse. She takes one coin from the purse at random. What is the probability that it is a nickel or dime? $\frac{3}{5}$

It is possible to have the first number cube show a 2 *and* have the sum of the two number cubes be 6 or 7. Therefore, these events are not mutually exclusive. They are called **inclusive events.** In this case, you must adjust the formula for mutually exclusive events.

Note that the circles in the Venn diagram overlap. This area represents the probability of both events occurring at the same time. When the areas of the two circles are added, this overlapping area is counted twice. Therefore, it must be subtracted to find the correct probability of the two events.

Events A and B are inclusive events.

Let A represent the event "the first number cube shows a 2".
Let B represent the event "the sum of the two number cubes is 6 or 7".

$$P(A) = \frac{6}{36} \qquad\qquad P(B) = \frac{11}{36}$$

Note that (2, 4) and (2, 5) are counted twice, both as the first cube showing a 2 and as a sum of 6 or 7. To find the correct probability, you must subtract $P(2$ and sum of 6 or 7$)$.

$$\overbrace{P(2)}\ +\ \overbrace{P(\text{sum of 6 or 7})}\ -\ \overbrace{P(2 \text{ and sum of 6 or 7})}$$

$$P(2 \text{ or sum of 6 or 7}) = \frac{6}{36}\ +\ \frac{11}{36}\ -\ \frac{2}{36}\ \text{ or } \frac{15}{36}$$

The probability of the first number cube showing a 2 or the sum of the number cubes being 6 or 7 is $\frac{15}{36}$ or $\frac{5}{12}$.

Probability of Inclusive Events

If two events, A and B, are inclusive, then the probability that either A or B occurs is the sum of their probabilities decreased by the probability of both occurring.

$$P(A \text{ or } B) = P(A) + P(B) - P(A \text{ and } B)$$

Examples **5** Kerry has read that the probability for a driver's license applicant to pass the road test the first time is $\frac{5}{6}$. He has also read that the probability of passing the written examination on the first attempt is $\frac{9}{10}$. The probability of passing both the road and written examinations on the first attempt is $\frac{4}{5}$.

a. Determine if the events are mutually exclusive or mutually inclusive.

Since it is possible to pass both the road examination and the written examination, these events are mutually inclusive.

b. What is the probability that Kerry can pass either examination on his first attempt?

$P(\text{passing road exam}) = \frac{5}{6} \qquad P(\text{passing written exam}) = \frac{9}{10}$

$P(\text{passing both exams}) = \frac{4}{5}$

$P(\text{passing either examination}) = \frac{5}{6} + \frac{9}{10} - \frac{4}{5} = \frac{56}{60}$ or $\frac{14}{15}$

The probability that Kerry will pass either test on his first attempt is $\frac{14}{15}$.

Addressing Individual Needs

Visual/Spatial Suppose that a commuter must drive through three traffic lights in order to get to work. If a light is green, she is happy. If a light is red, she is angry. If more than one light is red, she is late for work. Make a chart of the eight possible combinations of green or red lights. Use the chart to find the probability that she will be late for work. $\frac{1}{2}$

6 There are 5 students and 4 teachers on the school publications committee. A group of 5 members is being selected at random to attend a workshop on school newspapers. What is the probability that the group attending the workshop will have at least 3 students?

At least 3 students means the groups may have 3, 4, or 5 students. It is not possible to select a group of 3 students, a group of 4 students, and a group of 5 students in the same 5-member group. Thus, the events are mutually exclusive.

$P(\text{at least 3 students}) = P(\text{3 students}) + P(\text{4 students}) + P(\text{5 students})$

$$= \frac{C(5, 3) \cdot C(4, 2)}{C(9, 5)} + \frac{C(5, 4) \cdot C(4, 1)}{C(9, 5)} + \frac{C(5, 5) \cdot C(4, 0)}{C(9, 5)}$$

$$= \frac{60}{126} + \frac{20}{126} + \frac{1}{126} \text{ or } \frac{9}{14}$$

The probability of at least 3 students going to the workshop is $\frac{9}{14}$.

CHECK FOR UNDERSTANDING

Communicating Mathematics

Read and study the lesson to answer each question. **1–3. See margin.**

1. **Describe** the difference between independent and dependent events.

2. a. **Draw** a Venn diagram to illustrate the event of selecting an ace or a diamond from a deck of cards.

   b. Are the events mutually exclusive? Explain why or why not.

   c. **Write** the formula you would use to determine the probability of these events.

3. *Math Journal* **Write** an example of two mutually exclusive events and two mutually inclusive events in your own life. **Explain** why the events are mutually exclusive or inclusive.

Guided Practice

Determine if each event is *independent* or *dependent*. Then determine the probability.

4. the probability of rolling a sum of 7 on the first toss of two number cubes and a sum of 4 on the second toss   **independent,** $\frac{1}{72}$

5. the probability of randomly selecting two navy socks from a drawer that contains 6 black and 4 navy socks   **dependent,** $\frac{2}{15}$

6. There are 2 bottles of fruit juice and 4 bottles of sports drink in a cooler. Without looking, Desiree chose a bottle for herself and then one for a friend. What is the probability of choosing 2 bottles of the sports drink?   **dependent,** $\frac{2}{5}$

Determine if each event is *mutually exclusive* or *mutually inclusive*. Then determine each probability.

7. the probability of choosing a penny or a dime from 4 pennies, 3 nickels, and 6 dimes   **exclusive,** $\frac{10}{13}$

8. the probability of selecting a boy or a blonde-haired person from 12 girls, 5 of whom have blonde hair, and 15 boys, 6 of whom have blonde hair   **inclusive,** $\frac{20}{27}$

9. the probability of drawing a king or queen from a standard deck of cards   **exclusive,** $\frac{2}{13}$

Lesson 13-4 Probabilities of Compound Events **863**

Lesson 13-4 **863**

Assignment Guide

Core: 17–47 odd, 48, 49, 51–65
Enriched: 16–44 even, 46–65

In a bingo game, balls numbered 1 to 75 are placed in a bin. Balls are randomly drawn and not replaced. Find each probability for the first 5 balls drawn.

10. P(selecting 5 even numbers) $\approx$ **0.025**

11. P(selecting 5 two digit numbers) $\approx$ **0.518**

12. P(5 odd numbers or 5 multiples of 4) $\approx$ **0.029**

13. P(5 even numbers or 5 numbers less than 30) $\approx$ **0.032**

14. **Business**   A furniture importer has ordered 100 grandfather clocks from an overseas manufacturer. Four clocks are damaged in shipment, but the packaging shows no signs of damage. If a dealer buys 6 of the clocks without examining them first, what is the probability that none of the 6 clocks is damaged? $\dfrac{435{,}643}{560{,}175}$

15. **Sports**   A baseball team's pitching staff has 5 left-handed and 8 right-handed pitchers. If 2 pitchers are randomly chosen to warm up, what is the probability that at least one of them is right-handed? (*Hint:* Consider the order when selecting one right-handed and one left-handed pitcher.) $\dfrac{34}{39}$

EXERCISES

Practice

Determine if each event is *independent* or *dependent*. Then determine the probability.

16. the probability of selecting a blue marble, not replacing it, then a yellow marble from a box of 5 blue marbles and 4 yellow marbles   **dependent,** $\dfrac{5}{18}$

17. the probability of randomly selecting two oranges from a bowl of 5 oranges and 4 tangerines, if the first selection is replaced   **independent,** $\dfrac{25}{81}$

18. A green number cube and a red number cube are tossed. What is the probability that a 4 is shown on the green number cube and a 5 is shown on the red number cube?   **independent,** $\dfrac{1}{36}$

19. the probability of randomly taking 2 blue notebooks from a shelf which has 4 blue and 3 black notebooks   **dependent,** $\dfrac{2}{7}$

20. A bank contains 4 nickels, 4 dimes, and 7 quarters. Three coins are removed in sequence, without replacement. What is the probability of selecting a nickel, a dime, and a quarter in that order?   **dependent,** $\dfrac{8}{195}$

21. the probability of removing 13 cards from a standard deck of cards and have all of them be red   **dependent,** $\dfrac{19}{1{,}160{,}054}$

22. dependent, $\dfrac{32}{819}$

22. the probability of randomly selecting a knife, a fork, and a spoon in that order from a kitchen drawer containing 8 spoons, 8 forks, and 12 table knives

B

23. independent, $\dfrac{35}{1024}$

23. the probability of selecting three different-colored crayons from a box containing 5 red, 4 black, and 7 blue crayons, if each crayon is replaced

24. the probability that a football team will win its next four games if the odds of winning each game are 4 to 3   **independent,** $\dfrac{256}{2401}$

25. inclusive, $\dfrac{11}{36}$

For Exercises 25-33, determine if each event is *mutually exclusive* or *mutually inclusive*. Then determine each probability.

26. inclusive, $\dfrac{7}{13}$

25. the probability of tossing two number cubes and either one shows a 4

26. the probability of selecting an ace or a red card from a standard deck of cards

27. inclusive, $\dfrac{8}{13}$

27. the probability that if a card is drawn from a standard deck it is red or a face card

 www.amc.glencoe.com/self_check_quiz

28. exclusive, $\frac{9}{14}$

29. exclusive, $\frac{1}{4}$

30. exclusive, $\frac{302}{429}$

31. exclusive, $\frac{11}{32}$

32. inclusive, $\frac{55}{221}$

33. exclusive, $\frac{19}{33}$

34. $\frac{253}{9996}$

35. $\frac{2}{221}$

36. $\left(\frac{13}{52}\right)^{10}$

37. $\frac{55}{221}$

39. $\frac{4}{7}$

28. the probability of randomly picking 5 puppies of which at least 3 are male puppies, from a group of 5 male puppies and 4 female puppies.

29. the probability of two number cubes being tossed and showing a sum of 6 or a sum of 9.

30. the probability that a group of 6 people selected at random from 7 men and 7 women will have at least 3 women

31. the probability of at least 4 tails facing up when 6 coins are dropped on the floor

32. the probability that two cards drawn from a standard deck will both be aces or both will be black

33. from a collection of 6 rock and 5 rap CDs, the probability that at least 2 are rock from 3 randomly selected

Find the probability of each event using a standard deck of cards.

34. P(all red cards) if 5 cards are drawn without replacement

35. P(both kings or both aces) if 2 cards are drawn without replacement

36. P(all diamonds) if 10 cards are selected with replacement

37. P(both red or both queens) if 2 cards are drawn without replacement

There are 5 pennies, 7 nickels, and 9 dimes in an antique coin collection. If two coins are selected at random and the coins are not replaced, find each probability.

38. P(2 pennies) $\frac{1}{21}$

39. P(2 nickels or 2 silver-colored coins)

40. P(at least 1 nickel) $\frac{17}{30}$

41. P(2 dimes or 1 penny and 1 nickel) $\frac{71}{210}$

There are 5 male and 5 female students in the executive council of the Douglas High School honor society. A committee of 4 members is to be selected at random to attend a conference. Find the probability of each group being selected.

42. P(all female) $\frac{1}{42}$

43. P(all female or all male) $\frac{1}{21}$

44. P(at least 3 females) $\frac{11}{42}$

45. P(at least 2 females and at least 1 male) $\frac{5}{7}$

Applications and Problem Solving

46. **Computers** A survey of the members of the Piper High School Computer Club shows that $\frac{2}{5}$ of the students who have home computers use them for word processing, $\frac{1}{3}$ use them for playing games, and $\frac{1}{4}$ use them for both word processing and playing games. What is the probability that a student with a home computer uses it for word processing or playing games? $\frac{29}{60}$

47. **Weather** A weather forecaster states that the probability of rain is $\frac{3}{5}$, the probability of lightning is $\frac{2}{5}$, and the probability of both is $\frac{1}{5}$. What is the probability that a baseball game will be cancelled due to rain or lightning? $\frac{4}{5}$

48. **Critical Thinking** Felicia and Martin are playing a game where the number cards from a single suit are selected. From this group, three cards are then chosen at random. What is the probability that the sum of the value of the cards will be an even number? $\frac{10}{21}$

49. **City Planning** There are six women and seven men on a committee for city services improvement. A subcommittee of five members is being selected at random to study the feasibility of modernizing the water treatment facility. What is the probability that the committee will have at least three women? $\frac{59}{143}$

52a.

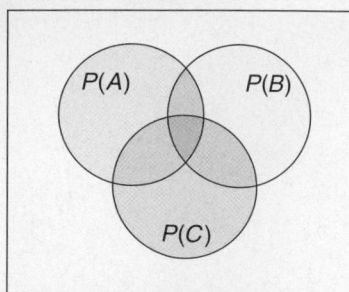

52b. $P(A \text{ or } B \text{ or } C) = P(A) + P(B) + P(C) - P(A \text{ and } B) - P(A \text{ and } C) - P(B \text{ and } C) + P(A \text{ and } B \text{ and } C)$. You must add the intersection of all three sets which have not been accounted for.

54b. $1 - \dfrac{P(366, n)}{366^n} > \dfrac{1}{2}$

---

50. Medicine   A study of two doctors finds that the probability of one doctor correctly diagnosing a medical condition is $\dfrac{93}{100}$ and the probability the second doctor will correctly diagnose a medical condition is $\dfrac{97}{100}$. What is the probability that at least one of the doctors will make a correct diagnosis? $\dfrac{9979}{10,000}$

51. $\dfrac{531}{1250}$

51. Disaster Relief   During the 1999 hurricane season, Hurricanes Dennis, Floyd, and Irene caused extensive flooding and damage in North Carolina. After a relief effort, 2500 people in one supporting community were surveyed to determine if they donated supplies or money. Of the sample, 812 people said they donated supplies and 625 said they donated money. Of these people, 375 people said they donated both. If a member of this community were selected at random, what is the probability that this person donated supplies or money?

52. Critical Thinking   If events A and B are inclusive, then $P(A \text{ or } B) = P(A) + P(B) - P(A \text{ and } B)$.   **a–b. See margin.**

   a. Draw a Venn diagram to represent $P(A \text{ or } B \text{ or } C)$.

   b. Write a formula to find $P(A \text{ or } B \text{ or } C)$.

53. Product Distribution   Ms. Kameko is the shipping manager of an Internet-based audio and video store. Over the past few months, she has determined the following probabilities for items customers might order.

Item	Probability
Action video	$\dfrac{4}{7}$
Pop/rock CD	$\dfrac{1}{2}$
Romance DVD	$\dfrac{5}{11}$
Action video and pop/rock CD	$\dfrac{2}{9}$
Pop/rock CD and romance DVD	$\dfrac{1}{7}$
Action video and romance DVD	$\dfrac{1}{4}$
Action video, pop/rock CD, and romance DVD	$\dfrac{1}{44}$

What is the probability, rounded to the nearest hundredth, that a customer will order an action video, pop/rock CD, or a romance DVD?   **0.93**

54. Critical Thinking   There are 18 students in a classroom. The students are surveyed to determine their birthday (month and day only). Assume that 366 birthdays are possible.

   a. What is the probability of any two students in the classroom having the same birthday?   **0.346**

   b. Write an inequality that can be used to determine the probability of any two students having the same birthday to be greater than $\dfrac{1}{2}$.   **See margin.**

   c. Are there enough students in the classroom to have the probability in part **a** be greater than $\dfrac{1}{2}$? If not, at least how many more students would there need to be?   **5**

55. Automotive Repairs An auto club's emergency service has determined that when club members call to report that their cars will not start, the probability that the engine is flooded is $\frac{1}{2}$, the probability that the battery is dead is $\frac{2}{5}$, and the probability that both the engine is flooded and the battery is dead is $\frac{1}{10}$.
   a. Are the events mutually exclusive or mutually inclusive? **exclusive**
   b. Draw a Venn Diagram to represent the events. **See margin.**
   c. What is the probability that the next member to report that a car will not start has a flooded engine or a dead battery? $\frac{4}{5}$

Mixed Review

56. Two number cubes are tossed and their sum is 6. Find the probability that each cube shows a 3. *(Lesson 13-3)* $\frac{1}{5}$

57. How many ways can 7 people be seated around a table? *(Lesson 13-2)* **720**

58. Sports Ryan plays basketball every weekend. He averages 12 baskets per game out of 20 attempts. He has decided to try to make 15 baskets out of 20 attempts in today's game. How many ways can Ryan make 15 out of 20 baskets? *(Lesson 12-6)* **15,504**

59. No, the spill will spread no more than 2000 meters away.

59. Ecology An underground storage container is leaking a toxic chemical. One year after the leak began, the chemical has spread 1200 meters from its source. After two years, the chemical has spread 480 meters more, and by the end of the third year it has reached an additional 192 meters. If this pattern continues, will the spill reach a well dug 2300 meters away? *(Lesson 12-4)*

60. Solve $12^{x+2} = 3^{x-4}$. *(Lesson 11-5)* **−6.7549**

61. Entertainment A theater has been staging children's plays during the summer. The average attendance at each performance is 400 people and the cost of a ticket is $3. Next summer, they would like to increase the cost of the tickets, while maximizing their profits. The director estimates that for every $1 increase in ticket price, the attendance at each performance will decrease by 20. What price should the director propose to maximize their income, and what maximum income might be expected? *(Lesson 10-5)* **$11.50, $2645**

62. Geology A drumlin is an elliptical streamlined hill whose shape can be expressed by the equation $r = \ell \cos k\theta$ for $-\frac{\pi}{2k} \leq \theta \leq \frac{\pi}{2k}$, where ℓ is the length of the drumlin and $k > 1$ is a parameter that is the ratio of the length to the width. Suppose the area of a drumlin is 8270 square yards and the formula for area is $A = \frac{\ell^2 \pi}{4k}$. Find the length of a drumlin modeled by $r = \ell \cos 7\theta$. *(Lesson 9-3)* **about 271.5 yards**

63. Write a vector equation describing a line passing through $P(1, -5)$ and parallel to $\vec{v} = \langle -2, -4 \rangle$. *(Lesson 8-6)* $\langle x - 1, y + 5 \rangle = t\langle -2, -4 \rangle$

64. Solve $2 \tan x - 4 = 0$ for principal values of x. *(Lesson 7-5)* **63°26′**

65. SAT/ACT Practice If $a = 45$, which of the following statements must be true? **B**
   I. $\overline{AD} \parallel \overline{BC}$
   II. ℓ_3 bisects $\angle ABC$.
   III. $b = 45$

   A None
   B I only
   C I and II only
   D I and III only
   E I, II, and III

Extra Practice See p. A52.

Lesson 13-4 Probabilities of Compound Events **867**

Extra Credit

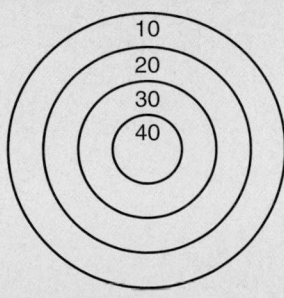

For the dart game pictured at the left, the radii of the concentric circles are 1 in., 2 in., 3 in., and 4 in. The scores you get for landing in a ring are listed on the dartboard. Suppose that you throw a dart and it lands randomly on the dartboard. What is the probability that you score over 20 points? $\frac{1}{4}$

---

4 ASSESS

Open-Ended Assessment
Writing Have students write a paragraph describing how to find the probability of inclusive events.

Additional Answer
55b.

Enrichment Masters, p. 150

Lesson 13-4 **867**

Conditional Probability

1 FOCUS

OBJECTIVE
- Find the probability of an event given the occurrence of another event.

5-Minute Check
Lesson 13-4

1. Using a standard deck of playing cards, find the probability of selecting a red card, replacing it in the deck, and then selecting a red face card. $\frac{3}{52}$

A box contains 9 tennis balls, 10 baseballs, and 5 racquet balls. Two balls are chosen from the box at random.

2. Determine whether the events are independent or dependent. **dependent**

3. What is the probability that both balls are tennis balls? $\frac{3}{23}$

4. On a game show, a contestant wins by selecting a card that shows a car or a major appliance. The cards to choose from include 2 cars, 5 major appliances, 10 trash cans, and 13 that say "Loser." What is the probability that a contestant will win? $\frac{7}{30}$

5. In a particular classroom, the probability that a class member collects baseball cards is $\frac{5}{8}$, the probability that someone collects comic books is $\frac{5}{12}$, and the probability that a class member collects both is $\frac{1}{3}$. What is the probability that a class member, chosen at random, collects either baseball cards or comic books? $\frac{17}{24}$

MEDICINE Danielle Jones works in a medical research laboratory where a drug that promotes hair growth in balding men is being tested. The results of the preliminary tests are shown in the table.

	Number of Subjects	
	Using Drug	**Using Placebo**
Hair growth	1600	1200
No hair growth	800	400

Ms. Jones needs to find the probability that a subject's hair growth was a result of using the experimental drug. *This problem will be solved in Example 1.*

The probability of an event under the condition that some preceding event has occurred is called **conditional probability.** The conditional probability that event *A* occurs given that event *B* occurs can be represented by $P(A \mid B)$. $P(A \mid B)$ *is read "the probability of A given B."*

Conditional Probability	The conditional probability of event *A*, given event *B*, is defined as $$P(A \mid B) = \frac{P(A \text{ and } B)}{P(B)} \text{ where } P(B) \neq 0.$$

Example **1** **MEDICINE** **Refer to the application above. What is the probability that a test subject's hair grew, given that he used the experimental drug?**

Let *H* represent hair growth and *D* represent experimental drug usage. We need to find $P(H \mid D)$.

$$P(H \mid D) = \frac{P(\text{used experimental drug and had hair growth})}{P(\text{used experimental drug})}$$

$$P(H \mid D) = \frac{\frac{1600}{4000}}{\frac{2400}{4000}} \quad \begin{matrix} \leftarrow P(\text{used experimental drug and had hair growth}) = \frac{1600}{4000} \\ \leftarrow P(\text{used experimental drug}) = \frac{1600 + 800}{4000} \end{matrix}$$

$$P(H \mid D) = \frac{1600}{2400} \text{ or } \frac{2}{3}$$

The probability that a subject's hair grew, given that they used the experimental drug is $\frac{2}{3}$.

Resource Manager

Reproducible Masters
- *Study Guide and Practice,* pp. 247–248
- *Enrichment,* p. 151
- *Assessment and Evaluation,* p. 330

Transparencies
- 5-Minute Check, 13-5
- Teaching, 13-5

Example **2** Denette tosses two coins. What is the probability that she has tossed 2 heads, given that she has tossed at least 1 head?

Let event A be that the two coins come up heads.

Let event B be that there is at least one head.

$P(B) = \dfrac{3}{4}$ *Three of the four outcomes have at least one head.*

$P(A \text{ and } B) = \dfrac{1}{4}$ *One of the four outcomes has two heads.*

$P(A \mid B) = \dfrac{P(A \text{ and } B)}{P(B)}$

$= \dfrac{\frac{1}{4}}{\frac{3}{4}}$

$= \dfrac{1}{4} \cdot \dfrac{4}{3}$ or $\dfrac{1}{3}$

The probability of tossing two heads, given that at least one toss was a head is $\dfrac{1}{3}$.

Sample spaces and reduced sample spaces can be used to help determine the outcomes that satisfy a given condition.

Example **3** Alfonso is conducting a survey of families with 3 children. If a family is selected at random, what is the probability that the family will have exactly 2 boys if the second child is a boy?

The sample space is $S = \{BBB, BBG, BGB, BGG, GBB, GBG, GGB, GGG\}$ and includes all of the possible outcomes for a family with three children.

Determine the reduced sample spaces that satisfy the given conditions that there are exactly 2 boys and that the second child is a boy.

The condition that there are exactly 2 boys reduces the sample space to exclude the outcomes where there are 1, 3, or no boys.

Let X represent the event that there are two boys.

$X = \{BBG, BGB, GBB\}$

$P(X) = \dfrac{3}{8}$

The condition that the second child is a boy reduces the sample space to exclude the outcomes where the second child is a girl.

Let Y represent the event that the second child is a boy.

$Y = \{BBB, BBG, GBB, GBG\}$

$P(Y) = \dfrac{4}{8}$ or $\dfrac{1}{2}$

$(X \text{ and } Y)$ is the intersection of X and Y. $(X \text{ and } Y) = \{BBG, GBB\}$.

So, $P(X \text{ and } Y) = \dfrac{2}{8}$ or $\dfrac{1}{4}$.

(continued on the next page)

Addressing Individual Needs

Interpersonal Use cooperative groups to discuss the difference between the probability of an event, the probability of inclusive events, and conditional probability. Have the groups write sample problems to fit each of the types of probabilities.

870 *Chapter 13*

In-Class Example

Example 4

A 20-sided die has the numerals 1 through 20 on its faces. The die is rolled once, and the number on the top face is recorded. What is the probability that the number is a multiple of 3 if it is known that it is odd? $\frac{3}{5}$

3 PRACTICE/APPLY

Study Guide and Practice Masters, pp. 247–248

NAME _____ DATE _____ PERIOD _____

Study Guide

Conditional Probabilities

The **conditional probability** of event *A*, given event *B*, is defined as $P(A|B) = \frac{P(A \text{ and } B)}{P(B)}$ where $P(B) \neq 0$. In some situations, event *A* is a subset of event *B*. In these situations, $P(A|B) = \frac{P(A)}{P(B)}$.

Example A box contains 3 red pencils and 4 yellow pencils. Three pencils are selected at random. What is the probability that exactly two red pencils are selected if the second pencil is red?

Sample spaces and reduced sample spaces can be used to help determine the outcomes that satisfy a given condition.

The sample space is *S* = {RRR, RRY, RYR, RYY, YRR, YRY, YYR, YYY} and includes all of the possible outcomes of selecting 3 pencils out of a box of 3 red and 4 yellow pencils.

Event *B* represents the condition that the second pencil is red.
B = {RRR, RRY, YRR, YRY}
$P(B) = \frac{4}{8}$ or $\frac{1}{2}$

Event *A* represents the condition that exactly two of the pencils are red.
A = {RRY, RYR, YRR}
$P(A) = \frac{3}{8}$

(A and B) is the intersection of A and B.
(A and B) = {RRY, YRR}.
So, P(A and B) = $\frac{2}{8}$ or $\frac{1}{4}$.
$P(A|B) = \frac{P(A \text{ and } B)}{P(B)}$
$= \frac{\frac{1}{4}}{\frac{1}{2}}$ or $\frac{1}{2}$

The probability that exactly two pencils are red given that the second pencil is red is $\frac{1}{2}$.

© Glencoe/McGraw-Hill   247   *Advanced Mathematical Concepts*

$$P(X|Y) = \frac{P(X \text{ and } Y)}{P(Y)}$$
$$= \frac{\frac{1}{4}}{\frac{1}{2}}$$
$$= \frac{1}{4} \cdot \frac{2}{1} \text{ or } \frac{1}{2}$$

The probability that a family with 3 children selected at random will have exactly 2 boys, given that the second child is a boy, is $\frac{1}{2}$.

In some situations, event *A* is a subset of event *B*. When this occurs, the probability that both event *A* and event *B*, *P*(*A* and *B*), occur is the same as the probability of event *A* occurring. Thus, in these situations $P(A|B) = \frac{P(A)}{P(B)}$.

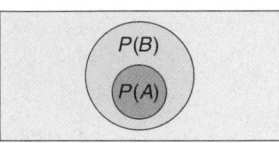

Event A is a subset of event B.

Example **4** A 12-sided dodecahedron has the numerals 1 through 12 on its faces. The die is rolled once, and the number on the top face is recorded. What is the probability that the number is a multiple of 4 if it is known that it is even?

Let *A* represent the event that the number is a multiple of 4. Thus, *A* = {4, 8, 12}.
$$P(A) = \frac{3}{12} \text{ or } \frac{1}{4}$$

Let *B* represent the event that the number is even. So, *B* = {2, 4, 6, 8, 10, 12}.
$$P(B) = \frac{6}{12} \text{ or } \frac{1}{2}$$

In this situation, *A* is a subset of *B*.

$P(A \text{ and } B) = P(A) = \frac{1}{4}$ $P(B) = \frac{1}{2}$
$P(A|B) = \frac{P(A)}{P(B)}$
$= \frac{\frac{1}{4}}{\frac{1}{2}} \text{ or } \frac{1}{2}$

The probability that a multiple of 4 is rolled, given that the number is even, is $\frac{1}{2}$.

CHECK FOR UNDERSTANDING

Communicating Mathematics

Read and study the lesson to answer each question.

1. **Explain** the relationship between conditional probability and the probability of two independent events.   **See margin.**

Additional Answer

1. Sample answer: If *A* and *B* are independent events, then $P(A | B) = P(A)$. Thus, the formula for conditional probability becomes $P(A) = \frac{P(A \text{ and } B)}{P(B)}$ or $P(A) \cdot P(B) = P(A \text{ and } B)$. This is the formula for the probability of independent events.

2–3. See margin.　**2. Describe** the sample space for *P*(face card) if the card drawn is black.

　3. *Math Journal*　Find two real-world examples that use conditional probability. Explain how you know conditional probability is used.

Guided Practice　Find each probability.

　4. Two number cubes are tossed. Find the probability that the numbers showing on the cubes match given that their sum is greater than five.　$\frac{2}{13}$

　5. One card is drawn from a standard deck of cards. What is the probability that it is a queen if it is known to be a face card?　$\frac{1}{3}$

Three coins are tossed. Find the probability that they all land heads up for each known condition.

　6. the first coin shows a head　$\frac{1}{4}$

　7. at least one coin shows a head　$\frac{1}{7}$

　8. at least two coins show heads　$\frac{1}{4}$

A pair of number cubes is thrown. Find each probability given that their sum is greater than or equal to 9.

　9. *P*(numbers match)　$\frac{1}{5}$

　10. *P*(sum is even)　$\frac{2}{5}$

　11. *P*(numbers match or sum is even)　$\frac{2}{5}$

12. Medicine　To test the effectiveness of a new vaccine, researchers gave 100 volunteers the conventional treatment and gave 100 other volunteers the new vaccine. The results are shown in the table below.

　a. What is the probability that the disease is prevented in a volunteer chosen at random?　$\frac{13}{20}$

　b. What is the probability that the disease is prevented in a volunteer who was given the new vaccine?　$\frac{17}{25}$

　c. What is the probability that the disease is prevented in a volunteer who was not given the new vaccine?　$\frac{31}{50}$

Treatment	Disease Prevented	Disease Not Prevented
New Vaccine	68	32
Conventional Treatment	62	38

13. Currency　A dollar-bill changer in a snack machine was tested with 100 $1-bills. Twenty-five of the bills were counterfeit. The results of the test are shown in the chart at the right.

Bill	Accepted	Rejected
Legal	69	6
Counterfeit	1	24

　a. What is the probability that a bill accepted by the changer is legal?　$\frac{69}{70}$

　b. What is the probability that a bill is rejected given that it is legal?　$\frac{2}{25}$

　c. What is the probability that a counterfeit bill is not rejected?　$\frac{1}{25}$

 www.amc.glencoe.com/self_check_quiz　　*Lesson 13-5　Conditional Probability*　**871**

EXERCISES

Practice

Find each probability.

A

14. Two coins are tossed. What is the probability that one coin shows heads if it is known that at least one coin is tails? $\frac{2}{3}$

15. A city council consists of six Democrats, two of whom are women, and six Republicans, four of whom are men. A member is chosen at random. If the member chosen is a man, what is the probability that he is a Democrat? $\frac{1}{2}$

16. A bag contains 4 red chips and 4 blue chips. Another bag contains 2 red chips and 6 blue chips. A chip is randomly selected from one of the bags, and found to be blue. What is the probability that the chip is from the first bag? $\frac{2}{5}$

17. Two boys and two girls are lined up at random. What is the probability that the girls are separated if a girl is at an end? $\frac{3}{5}$

18. A five-digit number is formed from the digits 1, 2, 3, 4, and 5. What is the probability that the number ends in the digits 52, given that it is even? $\frac{1}{8}$

19. Two game tiles, numbered 1 through 9, are selected at random from a box without replacement. If their sum is even, what is the probability that both numbers are odd? $\frac{5}{8}$

A card is chosen at random from a standard deck of cards. Find each probability given that the card is black.

B

20. P(ace) $\frac{1}{13}$ 21. P(4) $\frac{1}{13}$ 22. P(face card) $\frac{3}{13}$

23. P(queen of hearts) **0**   24. P(6 of clubs) $\frac{1}{26}$ 25. P(jack or ten) $\frac{2}{13}$

A container holds 3 green marbles and 5 yellow marbles. One marble is randomly drawn and discarded. Then a second marble is drawn. Find each probability.

26. the second marble is green, given that the first marble was green $\frac{2}{7}$

27. the second marble is yellow, given that the first marble was green $\frac{5}{7}$

28. the second marble is yellow, given that the first marble was yellow $\frac{4}{7}$

Three fish are randomly removed from an aquarium that contains a trout, a bass, a perch, a catfish, a walleye, and a salmon. Find each probability.

29. P(salmon, given bass) $\frac{2}{5}$

30. P(not walleye, given trout and perch) $\frac{3}{4}$

31. P(bass and perch, given not catfish) $\frac{3}{10}$

32. P(perch and trout, given neither bass nor walleye) $\frac{1}{2}$

In Mr. Hewson's homeroom, 60% of the students have brown hair, 30% have brown eyes, and 10% have both brown hair and eyes. A student is excused early to go to a doctor's appointment.

33. If the student has brown hair, what is the probability that the student also has brown eyes? $\frac{1}{6}$

34. If the student has brown eyes, what is the probability that the student does not have brown hair? $\frac{2}{3}$

35. If the student does not have brown hair, what is the probability that the student does not have brown eyes? $\frac{1}{2}$

In a game played with a standard deck of cards, each face card has a value of 10 points, each ace has a value of 1 point, and each number card has a value equal to its number. Two cards are drawn at random.

36. At least one card is an ace. What is the probability that the sum of the cards is 7 or less? $\frac{43}{99}$

37. One card is the queen of diamonds. What is the probability that the sum of the cards is greater than 18? $\frac{19}{51}$

Applications and Problem Solving

38. **Health Care** At Park Medical Center, in a sample group, there are 40 patients diagnosed with lung cancer, and 30 patients who are chronic smokers. Of these, there are 25 patients who have lung cancer and smoke.
  a. Draw a Venn diagram to represent the situation. **See margin.**
  b. If the medical center currently has 200 patients, and one of them is randomly selected for a medical study, what is the probability that the patient has lung cancer, given that the patient smokes? $\frac{5}{6}$

39. **Business** The manager of a computer software store wants to know whether people who come in and ask questions are more likely to make a purchase than the average person. A survey of 500 people exiting the store found that 250 people bought something, 120 asked questions and bought something, and 30 people asked questions but did not buy anything. Based on the survey, determine whether a person who asks questions is more likely to buy something than the average person. **See margin.**

40. **Critical Thinking** In a game using two number cubes, a sum of 10 has not turned up in the past few rolls. A player believes that a roll of 10 is "due" to come up. Analyze the player's thinking. **See margin.**

41. **Testing** Winona's chances of passing a precalculus exam are $\frac{4}{5}$ if she studies, and only $\frac{1}{5}$ if she decides to take it easy. She knows that $\frac{2}{3}$ of her class studied for and passed the exam. What is the probability that Winona studied for it? $\frac{5}{6}$

42. **Manufacturing** Three computer chip companies manufacture a product that enhances the 3-D graphic capacities of computer displays. The table below shows the number of functioning and defective chips produced by each company during one day's manufacturing cycle.

Company	Number of functioning chips	Number of defective chips
CyberChip Corp.	475	25
3-D Images, Inc.	279	21
MegaView Designs	180	20

  a. What is the probability that a randomly selected chip is defective? $\frac{33}{500}$
  b. What is the probability that a defective chip came from 3-D Images, Inc.? $\frac{7}{22}$
  c. What is the probability that a randomly selected chip is functioning? $\frac{467}{500}$
  d. If you were a computer manufacturer, which company would you select to produce the most reliable graphic chip? Why? **See margin.**

Lesson 13-5 Conditional Probability **873**

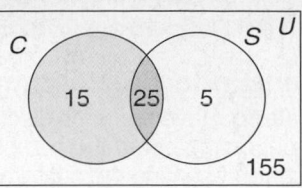

Extra Credit

Suppose that a fair coin is tossed 3 times.
  a. Find the probability that at least one tail shows if it is known that at least two heads showed. $\frac{3}{4}$
  b. Find the probability that exactly one tail shows if it is known that at least two heads showed. $\frac{3}{4}$
  c. Find the probability that exactly one head shows if it is known that exactly two tails showed. **1**
  d. Draw one card from a standard deck. Find the probability of drawing a red card given that a jack shows. $\frac{1}{2}$

Open-Ended Assessment

Speaking Have students discuss and put on the board the sample space representing the outcome of tossing two number cubes. Then, have students offer conditional probability problems they made up by using this sample space. As a class, solve each suggested problem.

11	12	13	14	15	16
21	22	23	24	25	26
31	32	33	34	35	36
41	42	43	44	45	46
51	52	53	54	55	56
61	62	63	64	65	66

Sample problem: Find the probability of having both odd numbers, given their sum is 6. $\frac{3}{5}$

 **Chapter 13, Quiz C
(Lessons 13-4 and 13-5)**
is available in the *Assessment and Evaluation Masters*, p. 330.

Enrichment Masters, p. 151

---

43. **Critical Thinking** The probability of an event A is equal to the probability of the same event, given that event B has already occurred. Prove that A and B are independent events. **See margin.**

Mixed Review

44. **City Planning** There are 6 women and 7 men on the committee for city park enhancement. A subcommittee of five members is being selected at random to study the feasibility of redoing the landscaping in one of the parks. What is the probability that the committee will have at least three women? *(Lesson 13-4)* $\frac{59}{143}$

45. Suppose there are 9 points on a circle. How many 4-sided closed figures can be formed by joining any 4 of these points? *(Lesson 13-1)* **126**

46. $1.5 + 0.75 + 0.375 + \cdots; 3$

46. Write $\sum\limits_{b=1}^{\infty} 3(0.5)^b$ in expanded form. Then find the sum. *(Lesson 12-5)*

47. Compare and contrast the graphs of $y = 3^x$ and $y = -3^x$ *(Lesson 11-2)*

47. They are reflections of each other over the x-axis.

48. Graph the system of inequalities. *(Lesson 10-8)* **See margin.**
$$x^2 + y^2 \le 81$$
$$x^2 + y^2 \ge 64$$

49. **Navigation** A submarine sonar is tracking a ship. The path of the ship is recorded as $r\cos\left(\theta - \frac{\pi}{2}\right) + 5 = 0$. Find the linear equation of the path of the ship. *(Lesson 9-4)* $y = -5$

50. Graph the line whose parametric equations are $x = 4t$, and $y = 3 + 2t$. *(Lesson 8-6)* **See margin.**

51. Find the area of the sector of a circle of radius 8 feet, given its central angle is $98°$. Round your answer to the nearest tenth. *(Lesson 6-1)* **54.7 ft²**

52. When the angle of elevation of the sun is $27°$, the shadow of a tree is 25 meters long. How tall is the tree? Round your answer to the nearest tenth. *(Lesson 5-4)* **12.7 m**

27°
25 m

53. **Photography** A photographer has a frame that is 3 feet by 4 feet. She wants to mat a group photo such that there is a uniform width of mat surrounding the photo. If the area of the photo is 6 square feet, find the width of the mat. *(Lesson 4-2)* $\frac{1}{2}$ **ft or 6 in.**

54. $x = \pm 2$; $f(x)$ is undefined when $x = \pm 2$

54. Find the value(s) of x at which $f(x) = \dfrac{5}{x^2 - 4}$ is discontinuous. Use the continuity test to justify your answer. *(Lesson 3-5)*

55. **SAT/ACT Practice** In parallelogram $ABCD$, the ratio of the shaded area to the unshaded area is **B**

A 1:2   B 1:1
C 4:3   D 2:1
E It cannot be determined from the information given.

874 Chapter 13 *Combinatorics and Probability*   **Extra Practice** See p. A52.

Additional Answers

43. $P(A|B) = \dfrac{P(A \text{ and } B)}{P(B)}$ by definition. So, if $P(A) = P(A|B)$ then by substitution $P(A) = \dfrac{P(A \text{ and } B)}{P(B)}$ or $P(A \text{ and } B) = P(A) \cdot P(B)$. Therefore, the events are independent.

48.

50.

(4, 5)
(0, 3)
(−4, 1)

The Binomial Theorem and Probability

OBJECTIVE

• Find the probability of an event by using the Binomial Theorem.

 LANDSCAPING Managers at the Eco-Landscaping Company know that a mahogany tree they plant has a survival rate of about 90% if cared for properly. If 10 trees are planted in the last phase of a landscaping project, what is the probability that 7 of the trees will survive? *This problem will be solved in Example 3.*

We can examine a simpler form of this problem. Suppose that there are only 5 trees to be planted. What is the probability that 4 will survive? The number of ways that this can happen is $C(5, 4)$ or 5.

Let S represent the probability of a tree surviving.
Let D represent the probability of a tree dying.

Look Back
Refer to Lesson 12-6 to review binomial expansions and the Binomial Theorem.

Since this situation has two outcomes, we can represent it using the binomial expansion of $(S + D)^5$. The terms of the expansion can be used to find the probabilities of each combination of the survival and death of the trees.

$$(S + D)^5 = 1S^5 + 5S^4D + 10S^3D^2 + 10S^2D^3 + 5SD^4 + 1D^5$$

coefficient	term	meaning
$C(5, 5) = 1$	$1S^5$	1 way to have all 5 trees survive
$C(5, 4) = 5$	$5S^4D$	5 ways to have 4 trees survive and 1 die
$C(5, 3) = 10$	$10S^3D^2$	10 ways to have 3 trees survive and 2 die
$C(5, 2) = 10$	$10S^2D^3$	10 ways to have 2 trees survive and 3 die
$C(5, 1) = 5$	$5SD^4$	5 ways to have 1 tree survive and 4 die
$C(5, 0) = 1$	$1D^5$	1 way to have all 5 trees die

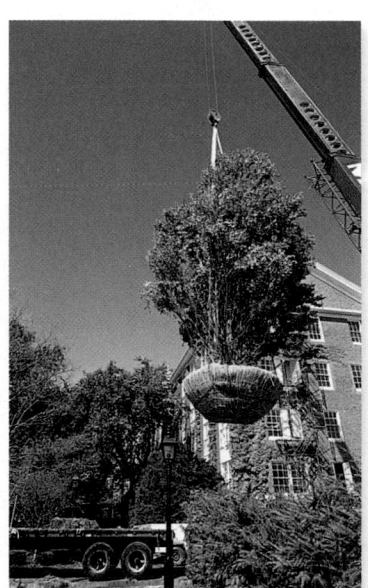

The probability of a tree surviving is 0.9. So, the probability of a tree not surviving is $1 - 0.9$ or 0.1. The probability of having 4 trees survive out of 5 can be determined as follows.

Use $5S^4D$ since this term represents 4 trees surviving and 1 tree dying.

$5S^4D = 5(0.9)^4(0.1)$ *Substitute 0.9 for S and 0.1 for D*
$5S^4D = 5(0.6561)0.1$
$5S^4D = 0.3281$ or about $\frac{1}{3}$

Thus, the probability of having 4 trees survive is about $\frac{1}{3}$.

Resource Manager

📁 **Reproducible Masters**
• *Study Guide and Practice,* pp. 249–250
• *Enrichment,* p. 152
• *Assessment and Evaluation,* p. 330
• *Graphing Calculator Keystrokes,* p. 53
• *Problem Solving and Applications,* p. 25

💻 **Transparencies**
• 5-Minute Check, 13-6
• Teaching, 13-6

---

1 FOCUS

5-Minute Check
Lesson 13-5

1. A committee of 3 girls is randomly selected to organize a school dance from among Lucia, Maria, Lori, Julie, Nancy, and Cheryl. Draw a sample space to represent all of the possible outcomes. Then find each conditional probability.
   a. P(Lucia was selected, given that Lori was selected) $\frac{2}{5}$
   b. P(Nancy was not selected, given that Cheryl and Julie were selected) $\frac{3}{4}$
2. Alyce tosses 2 coins. What is the probability that she tosses no heads, given that she tosses at least 1 tail? $\frac{1}{3}$
3. Madison is conducting a survey of families with 3 children. If a family is selected at random, what is the probability that a family will have exactly 1 boy if the second child is a girl? $\frac{1}{2}$
4. A dodecahedron solid has the numerals 1 through 12 on its faces. The dodecahedron is tossed once, and the number on the top face is recorded. What is the probability that the top number will be odd given that it is a multiple of 4? **0**
5. A pair of number cubes is thrown. Find the probability that their sum is greater than 7 if both number cubes show the same number. $\frac{1}{2}$

Motivating the Lesson
Real-World Application Near the end of the 1999 baseball season, Larry Walker of the Colorado Rockies had a batting average of .380. With that average, what is the probability that Walker will not get a hit at his next 3 consecutive times at bat? **0.238328 or about $\frac{1}{4}$**

2 TEACH

Other probabilities can be determined from the expansion of $(S + D)^5$. For example, what is the probability that at least 2 trees out of the 5 trees planted will die?

Example **1** **LANDSCAPING** Refer to the application at the beginning of the lesson. Five mahogany trees are planted. What is the probability that at least 2 trees die?

The third, forth, fifth, and sixth terms represent the conditions that two or more trees die. So, the probability of this happening is the sum of the probabilities of those terms.

P(at least 2 trees die)
$$= 10S^3D^2 + 10S^2D^3 + 5SD^4 + 1D^5$$
$$= 10(0.9)^3(0.1)^2 + 10(0.9)^2(0.1)^3 + 5(0.9)(0.1)^4 + (0.1)^5$$
$$= 10(0.729)(0.01) + 10(0.81)(0.001) + 5(0.9)(0.0001) + (0.00001)$$
$$= 0.0729 + 0.0081 + 0.00045 + 0.00001$$
$$= 0.0815$$

The probability that at least 2 trees die is about 8%.

Problems that can be solved using the binomial expansion are called **binomial experiments**.

Conditions of a Binomial Experiment

A binomial experiment exists if and only if these conditions occur.
- Each trial has exactly two outcomes, or outcomes that can be reduced to two outcomes.
- There must be a fixed number of trials.
- The outcomes of each trial must be independent.
- The probabilities in each trial are the same.

Example **2** Eight out of every 10 persons who contract a certain viral infection can recover. If a group of 7 people become infected, what is the probability that exactly 3 people will recover from the infection?

There are 7 people involved, and there are only 2 possible outcomes, recovery R or not recovery N. These events are independent, so this is a binomial experiment.

When $(R + N)^7$ is expanded, the term R^3N^4 represents 3 people recovering and 4 people not recovering from the infection. The coefficient of R^3N^4 is $C(7, 3)$ or 35.

P(exactly 3 people recovering) $= 35(0.8)^3(0.2)^4$ $R = 0.8, N = 1 - 0.8$ or 0.2
$$= 35(0.512)(0.0016)$$
$$= 0.028672$$

The probability that exactly 3 of the 7 people will recover from the infection is 2.9%.

The Binomial Theorem can be used to find the probability when the number of trials makes working with the binomial expansion unrealistic.

876 Chapter 13 Combinatorics and Probability

From the Classroom of...

Janene Russell
Diamond High School
Diamond, Missouri

"After completing the chapter on probability, I have my trigonometry students create and build games. Each game must contain an element of chance and must be original. To wrap the project up, we have a tournament. The students play all of the games and also fill out evaluations on them."

LANDSCAPING Refer to the application at the beginning of the lesson. What is the probability that 7 of the 10 trees planted will survive?

Let S be the probability that a tree will survive.
Let D be the probability that a tree will die.

Since there are 10 trees, we can use the Binomial Theorem to find any term in the expression $(S + D)^{10}$.

$$(S + D)^{10} = \sum_{r=0}^{10} \frac{10!}{r!(10 - r)!} S^{10-r}D^r$$

Look Back
Refer to Lesson 12-5 to review sigma notation.

Having 7 trees survive means that 3 will die. So the probability can be found using the term where $r = 3$, the fourth term.

$$\frac{10!}{3! (10 - 3)!} S^7 D^3 = 120 S^7 D^3$$
$$= 120(0.9)^7(0.1)^3$$
$$= 120(0.4782969)(0.001) \text{ or } 0.057395628$$

The probability of exactly 7 trees surviving is about 5.7%.

So far, the probabilities we have found have been **theoretical probabilities.** These are determined using mathematical methods and provide an idea of what to expect in a given situation. **Experimental probability** is determined by performing experiments and observing and interpreting the outcomes. One method for finding experimental probability is a **simulation.** In a simulation, a device such as a graphing calculator is used to model the event.

GRAPHING CALCULATOR EXPLORATION

You can use a graphing calculator to simulate a binomial experiment. Consider the following situation.

Robby wins 2 out of every 3 chess matches he plays with Marlene. What is the probability that he wins exactly 5 of the next 6 matches?

TRY THIS

To simulate this situation, enter **int(3*rand)** and press ENTER. Note: (**int(** and **rand** can be found in the menus accessed by pressing MATH. This will randomly generate the numbers 0, 1, or 2. Robby wins if the outcome is 0 or 1. Robby loses if 2 comes up.

```
int(3*rand)
            0
            1
            2
            2
            1
            0
```

In the simulation, one repetition of the complete binomial experiment consists of six trials or six presses of the ENTER key. Try 40 repetitions.

WHAT DO YOU THINK? 1–6. See margin.

1. What is the sample space?
2. What is P(Robby wins)?
3. In the simulation, with what probability did Robby win exactly 5 times?
4. Using the formula for computing binomial probabilities, what is the probability of Robby winning exactly five games?
5. Why do you think there is a difference between the simulation (experimental probability) and the probability computed using the formula (theoretical probability)?
6. What would you do to have the experimental probability approximate the theoretical probability?

Error Analysis

Watch for students who have difficulty interpreting the language of binomial probability problems.

Prevent by stressing that they look for key words such as *exactly 2, no more than 2, less than 2,* and *2 or more* and understand the meaning of these expressions.

Assignment Guide

Core: 13–41 odd, 42–51
Enriched: 14–34 even, 36–51

Study Guide and Practice Masters, pp. 249–250

NAME _____ DATE _____ PERIOD _____

Study Guide

The Binomial Theorem and Probability

Problems that meet the conditions of a **binomial experiment** can be solved using the Binomial Theorem. Use the Binomial Theorem to find the probability when the number of trials makes working with the binomial expansion unrealistic.

Example 1 The probability that Misha will win a word game is $\frac{3}{4}$. If Misha plays the game 5 times, what is the probability that he will win exactly 3 games?

There are 5 games and each game has only two possible outcomes, win W or lose L. These events are independent and the probability is $\frac{3}{4}$ for each game. So this is a binomial experiment.

When $(W + L)^5$ is expanded, the term W^3L^2 represents 3 wins and 2 losses. The coefficient of W^3L^2 is $C(5, 3)$, or 10.

$P(\text{exactly 3 wins}) = 10\left(\frac{3}{4}\right)^3\left(\frac{1}{4}\right)^2$ $\quad W = \frac{3}{4}, L = \frac{1}{4}$
$= 10\left(\frac{27}{64}\right)\left(\frac{1}{16}\right)$
$= \frac{270}{1024}$
$= \frac{135}{512}$ or about 26.4%

Example 2 The probability that a computer salesperson will make a sale when approaching a customer is $\frac{1}{2}$. If the salesperson approaches 12 customers, what is the probability that 8 sales will be made?

Let S be the probability of a sale.
Let N be the probability of not making a sale.

$(S + N)^{12} = \sum_{r=0}^{5} \frac{12!}{r!(12-r)!} S^{12-r} N^r$

Making 8 sales means that 4 sales will not be made. So the probability can be found using the term where $r = 4$.

$\frac{12!}{4!(12-4)!}S^8N^4 = 495S^8N^4$
$= 495\left(\frac{1}{2}\right)^8\left(\frac{1}{2}\right)^4$ $\quad S = \frac{1}{2}, N = \frac{1}{2}$
$= \frac{495}{4096}$ or 0.120849609

The probability of making exactly 8 sales is about 12.1%.

© Glencoe/McGraw-Hill   249   Advanced Mathematical Concepts

Communicating Mathematics

Read and study the lesson to answer each question.   1–3. See margin.

1. **Explain** whether or not each situation represents a binomial experiment.
   a. the probability of winning in a game where a number cube is tossed, and if 1, 2, or 3 comes up you win.
   b. the probability of drawing two red marbles from a jar containing 10 red, 30 blue, and 5 yellow marbles.
   c. the probability of drawing a jack from a standard deck of cards, knowing that the card is red.

2. **Write** an explanation of experimental probability. Give a real-world example that uses experimental probability.

3. **Describe** how to find the probability of getting exactly 2 correct answers on a true/false quiz that has 5 questions.

Guided Practice

Find each probability if a number cube is tossed five times.

4. $P(\text{only one 4})$ $\quad \frac{3125}{7776}$
5. $P(\text{no more than two 4s})$ $\quad \frac{625}{648}$
6. $P(\text{at least three 4s})$ $\quad \frac{23}{648}$
7. $P(\text{exactly five 4s})$ $\quad \frac{1}{7776}$

Jasmine Myers, a weather reporter for Channel 6, is forecasting a 30% chance of rain for today and the next four days. Find each probability.

8. $P(\text{not having rain on any day})$ $\quad \frac{2401}{10,000}$
9. $P(\text{having rain on exactly one day})$ $\quad \frac{1029}{2500}$
10. $P(\text{having rain no more than three days})$ $\quad \frac{48,461}{50,000}$

11. **Cooking** In cooking class, 1 out of 5 soufflés that Sabrina makes will collapse. She is preparing 6 soufflés to serve at a party for her parents. What is the probability that exactly 4 of them do not collapse? $\quad \frac{768}{3125}$

12. **Finance** A stock broker is researching 13 independent stocks. An investment in each will either make or lose money. The probability that each stock will make money is $\frac{5}{8}$. What is the probability that exactly 10 of the stocks will make money? **0.1372**

Practice

Isabelle carries lipstick tubes in a bag in her purse. The probability of pulling out the color she wants is $\frac{2}{3}$. Suppose she uses her lipstick 4 times in a day. Find each probability.

 A

13. $P(\text{never the correct color})$ $\quad \frac{1}{81}$
14. $P(\text{correct at least 3 times})$ $\quad \frac{16}{27}$
15. $P(\text{no more than 3 times correct})$ $\quad \frac{65}{81}$
16. $P(\text{correct exactly 2 times})$ $\quad \frac{8}{27}$

 www.amc.glencoe.com/self_check_quiz

Additional Answers

1a. Yes, it meets all three conditions.

1b. No, there are more than 2 possible outcomes.

1c. No, the events are not independent.

2. Sample answer: the probabilities derived from a simulation rather than an actual event.

3. First, determine $P(\text{right})$ and $P(\text{wrong})$. Second, set up the binomial expansion $(p_r + p_w)^5$. Third, determine the term of the expansion. Fourth, substitute the probability values for p_r and p_w. Last, compute the probability of getting exactly 2 correct answers.

Maura guesses at all 10 questions on a true/false test. Find each probability.

17. $P(7 \text{ correct})$ $\frac{15}{128}$ 18. $P(\text{at least 6 correct})$ $\frac{193}{512}$

19. $P(\text{all correct})$ $\frac{1}{1024}$ 20. $P(\text{at least half correct})$ $\frac{319}{512}$

The probability of tossing a head on a bent coin is $\frac{1}{3}$. Find each probability if the coin is tossed 4 times.

21. $P(4 \text{ heads})$ $\frac{1}{81}$ 22. $P(3 \text{ heads})$ $\frac{8}{81}$ 23. $P(\text{at least 2 heads})$ $\frac{11}{27}$

Kyle guesses at all of the 10 questions on his multiple choice test. Find each probability if each question has 4 choices.

 B

24. $P(6 \text{ correct answers})$ ≈ 0.016 25. $P(\text{half answers correct})$ ≈ 0.058

26. $P(\text{from 3 to 5 correct answers})$ ≈ 0.46

If a thumbtack is dropped, the probability of its landing point up is $\frac{2}{5}$. Mrs. Davenport drops 10 tacks while putting up the weekly assignment sheet on the bulletin board. Find each probability.

 C

27. $P(\text{all point up})$ $\approx 1.049 \times 10^{-4}$ 28. $P(\text{exactly 3 point up})$ ≈ 0.215

29. $P(\text{exactly 5 point up})$ ≈ 0.201 30. $P(\text{at least 6 point up})$ ≈ 0.166

Find each probability if three coins are tossed.

31. $P(3 \text{ heads or 3 tails})$ $\frac{1}{4}$ 32. $P(\text{at least 2 heads})$ $\frac{1}{2}$ 33. $P(\text{exactly 2 tails})$ $\frac{3}{8}$

Graphing Calculator

34. Enter the expression **6 nCr X** into the **Y=** menu. The **nCr** command is found in the probability section of the **MATH** menu. Use the **TABLE** feature to observe the results.   **a–b. See margin.**

   a. How do these results compare with the expansion of $(a + b)^6$?

   b. How would you change the expression to find the expansion of $(a + b)^8$?

35. **Sports** A football team is scheduled to play 16 games in its next season. If there is a 70% probability the team will win each game, what is the probability that the team will win at least 12 of its games? (*Hint*: Use the information from Exercise 34.) **about 45%**

Applications and Problem Solving

36. **Military Science** During the Gulf War in 1990–1991, SCUD missiles hit 20% of their targets. In one incident, six missiles were fired at a fuel storage installation.

   a. Describe what success means in this case, and state the number of trials and the probability of success on each trial. **See margin.**

   b. Find the probability that between 2 and 6 missiles hit the target. $\frac{1077}{3125}$

37. **Critical Thinking** Door prizes are given at a party through a drawing. Four out of 10 tickets are given to men who will attend, and 6 out of 10 tickets are distributed to women. Each person will receive only one ticket. Ten tickets will be drawn at random with replacement. What is the probability that all winners will be the same sex? ≈ 0.0062

38. **Medicine** Ten percent of African-Americans are carriers of the genetic disease sickle-cell anemia. Find each probability for a random sample of 20 African-Americans. **a.** 1×10^{-20} **b.** 6.4×10^{-6}

   a. $P(\text{all carry the disease})$ b. $P(\text{exactly half have the disease})$

Lesson 13-6 The Binomial Theorem and Probability **879**

Open-Ended Assessment

Modeling Have students use a spinner and a number cube to create problems based on the results of trial spins and tosses. Sample problem: *Find the probability of a prime number on the number cube given a prime number on the spinner.* $\frac{1}{2}$

📁 **Chapter 13, Quiz D (Lesson 13-6)** is available in the *Assessment and Evaluation Masters,* p. 330.

Enrichment Masters, p. 152

880 *Chapter 13*

---

39. **Airlines** A commuter airline has found that 4% of the people making reservations for a flight will not show up. As a result, the airline decides to sell 75 seats on a plane that has 73 seats (overbooking). What is the probability that for every person who shows up for the flight there will be a seat available? **0.807**

40. **Sales** Luis is an insurance agent. On average, he sells 1 policy for every 2 prospective clients he meets. On a particular day, he calls on 4 clients. He knows that he will not receive a bonus if the sales are less than or equal to three policies. What is the probability that he will not get a bonus? **≈ 0.94**

41. **Critical Thinking** Trina is waiting for her friend who is late. To pass the time, she takes a walk using the following rules. She tosses a fair coin. If it falls heads, she walks 10 meters north. If it falls tails, she walks 10 meters south. She repeats this process every 10 meters and thus executes what is called a random walk. What is the probability that after 100 meters of walking she will be at one of the following points?
    a. *P*(back at her starting point) **0.246**
    b. *P*(within 10 meters of the starting point) **0.246**
    c. *P*(exactly 20 meters from the starting point) **0.41**

Mixed Review

42. A pair of number cubes is thrown. Find the probability that their sum is less than 9 if both cubes show the same number. *(Lesson 13-5)* $\frac{2}{3}$

43. A letter is picked at random from the alphabet. Find the probability that the letter is contained in the word *house* or in the word *phone*. *(Lesson 13-4)* $\frac{7}{26}$

44. **Physical Science** Dry air expands as it moves upward into the atmosphere. For each 1000 feet that it moves upward, the air cools 5° F. Suppose the temperature at ground level is 80° F. *(Lesson 12-1)* **a. 80, 75, 70, …**
    a. Write a sequence representing the temperature decrease per 1000 feet.
    b. If *n* is the height of the air in thousands of feet, write a formula for the temperature *T* in terms of *n*. **$T = 80 - 5n$**
    c. What is the ground level temperature if the air at 40,000 feet is −125°? **75° F**

45. 0.38

45. Solve $3^{x-1} = 6^{-x}$ using logarithms. Round to the nearest hundredth. *(Lesson 11-6)*

46. (0, −3); (±2$\sqrt{6}$, −3); (7, −3), (−7, −3), (0, 2), (0, −8)

46. Name the coordinates of the center, foci, and vertices of the ellipse with the equation $\frac{x^2}{49} + \frac{(y+3)^2}{25} = 1$. *(Lesson 10-3)*

47. Express $\sqrt{2}\left(\cos -\frac{\pi}{2} + i\sin -\frac{\pi}{2}\right)$ in rectangular form. *(Lesson 9-6)* **$0 - i\sqrt{2}$**

48. Find the ordered pair that represents $\overrightarrow{WX}$ if $W(8, -3)$ and $X(6, 5)$. Then find the magnitude of $\overrightarrow{WX}$. *(Lesson 8-2)* **⟨−2, 8⟩; 2$\sqrt{17}$**

49. about 101.1 cm and 76.9 cm

49. **Geometry** The sides of a parallelogram are 55 cm and 71 cm long. Find the length of each diagonal if the larger angle measures 106°. *(Lesson 5-8)*

50. Use the Remainder Theorem to find the remainder when $x^4 + 12x^3 + 21x^2 - 62x - 72$ is divided by $x + 4$. State whether the binomial is a factor of the polynomial. *(Lesson 4-3)* **0; yes**

51. **SAT Practice** **Grid-In** A word processor uses a sheet of paper that is 9 inches wide by 12 inches long. It leaves a 1-inch margin on each side and a 1.5-inch margin on the top and bottom. What fraction of the page is used for text? **7/12**

Extra Credit

A basketball player's free-throw percentage changes with every throw. Suppose that a player has a 70% free-throw average. What is the probability that out of five shots, four will go into the basket? $\frac{7203}{20,000}$ **or about 0.36**

VOCABULARY

Basic Counting Principle (p. 837)
binomial experiments (p. 876)
circular permutation (p. 847)
combination (p. 841)
combinatorics (p. 837)
complements (p. 853)
conditional probability (p. 868)
dependent event (p. 837)
experimental probability (p. 877)
failure (p. 852)
inclusive event (p. 863)
independent event (p. 837)

mutually exclusive (p. 862)
odds (p. 854)
permutation (p. 838)
permutation with repetition (p. 846)
probability (p. 852)
reduced sample space (p. 862)
sample space (p. 852)
simulation (p. 877)
success (p. 852)
theoretical probability (p. 877)
tree diagram (p. 837)

UNDERSTANDING AND USING THE VOCABULARY

Choose the correct term to best complete each sentence.

1. Events that do not affect each other are called (dependent, independent) events. **independent**

2. In probability, any outcome other than the desired outcome is called a (failure, success). **failure**

3. The sum of the probability of an event and the probability of the complement of the event is always (0, 1). **1**

4. The (odds, probability) of an event occurring is the ratio of the number of ways the event can succeed to the sum of the number of ways the event can succeed and the number of ways the event can fail. **probability**

5. The arrangement of objects in a certain order is called a (combination, permutation). **permutation**

6. A (permutation with repetitions, circular permutation) specifically deals with situations in which some objects that are alike. **permutation with repetitions**

7. Two (inclusive, mutually exclusive) events cannot happen at the same time. **mutually exclusive**

8. A (sample space, Venn diagram) is the set of all possible outcomes of an event. **sample space**

9. The probability of an event A given that event B has occurred is called a (conditional, inclusive) probability. **conditional**

10. The branch of mathematics that studies different possibilities for the arrangement of objects is called (statistics, combinatorics). **combinatorics**

 For additional review and practice for each lesson, visit: **www.amc.glencoe.com**

Six forms of Chapter 13 Test are available in the *Assessment and Evaluation Masters* as shown in the chart.

Chapter 13 Test, Form 1B, is shown at the right. Chapter 3 Test, Form 2B, is shown on the next page.

TEST OPTIONS

Form	Pages	Type	
1A	315–316	Honors	Form 1 tests are multiple choice.
1B	317–318	Average	
1C	319–320	Basic	
2A	321–322	Honors	Form 2 tests are free response.
2B	323–324	Average	
2C	325–326	Basic	

STUDY GUIDE AND ASSESSMENT

VOCABULARY

This section provides a listing of the new terms, properties, and phrases that were introduced in this chapter. The exercises check students' understanding of the terms by using a variety of verbal formats including matching, completion, and true/false.

Glossary A complete glossary of terms begins on page A71.

Assessment and Evaluation Masters, pp. 317–318

NAME _____ DATE _____ PERIOD _____

Chapter 13 Test, Form 1B

Write the letter for the correct answer in the blank at the right of each problem.

1. Given 3 choices of sandwiches, 4 choices of chips, and 2 choices of cookies, how many different sack lunches can be prepared containing one choice of each item?
A. 12   B. 24   C. 84   D. 288 — 1. B

2. How many ways can the letters in the word *capitol* be arranged if the first letter must be *p*?
A. 120   B. 720   C. 5040   D. 40,320 — 2. B

3. How many 4-letter codes can be formed from the letters in the word *capture* if letters cannot be repeated?
A. 28   B. 840   C. 2401   D. 5040 — 3. B

4. A class consisting of 10 boys and 12 girls must select 2 boys and 2 girls to serve on a committee. How many variations of the committee can there be?
A. 2970   B. 120   C. 4800   D. 7315 — 4. A

5. On a long city block, 4-digit house numbers must begin with the digit 4 and end with either 0 or 1. How many different variations of house numbers are possible?
A. 1000   B. 144   C. 180   D. 200 — 5. D

6. How many ways can 3 identical green candles and 7 identical blue candles be arranged in a row in any variation?
A. 21   B. 3,628,800   C. 30,240   D. 120 — 6. D

7. How many ways can 8 keys be arranged on a key ring with no chain?
A. 5040   B. 40,320   C. 720   D. 64 — 7. A

8. How many ways can 9 numbers be arranged on a small rotating wheel relative to a fixed point?
A. 81   B. 5040   C. 40,320   D. 362,880 — 8. D

For Exercises 9 and 10, consider a basket that contains 15 slips of paper numbered from 1 to 15. Two slips of paper are drawn at random.

9. What is the probability of drawing 2 even numbers?
A. $\frac{49}{225}$ B. $\frac{4}{15}$ C. $\frac{1}{5}$ D. $\frac{7}{15}$ — 9. C

10. What are the odds of drawing an even number less than 7 and an odd number greater than 10?
A. $\frac{2}{35}$ B. $\frac{3}{32}$ C. $\frac{2}{5}$ D. $\frac{33}{35}$ — 10. B

11. The probability of rolling a sum of 4 when two number cubes are tossed is 1 in 12. What are the odds of rolling a sum of 4 when two number cubes are tossed?
A. $\frac{1}{11}$ B. $\frac{12}{13}$ C. $\frac{1}{12}$ D. $\frac{11}{12}$ — 11. A

© Glencoe/McGraw-Hill   317   *Advanced Mathematical Concepts*

NAME _____ DATE _____ PERIOD _____

Chapter 13 Test, Form 1B (continued)

12. The odds that it will rain in the city of Houston tomorrow are $\frac{1}{4}$. What is the probability of rain in Houston tomorrow?
A. $\frac{4}{5}$ B. $\frac{1}{4}$ C. $\frac{4}{5}$ D. $\frac{1}{5}$ — 12. C

13. Two number cubes, 1 red and 1 green, are tossed. What is the probability that the red number cube shows a number less than 3 and the green number cube shows a 6?
A. $\frac{1}{2}$ B. $\frac{1}{4}$ C. $\frac{1}{3}$ D. $\frac{1}{18}$ — 13. D

14. If two cards are drawn at random from a standard deck of cards with no replacement, find the probability that both cards are hearts.
A. $\frac{1}{4}$ B. $\frac{1}{16}$ C. $\frac{1}{17}$ D. $\frac{1}{52}$ — 14. C

15. A bucket contains 4 red, 2 yellow, and 3 green balls. If one ball is taken at random, what is the probability that it is red or green?
A. $\frac{2}{3}$ B. $\frac{7}{9}$ C. $\frac{2}{27}$ D. $\frac{7}{81}$ — 15. B

16. A school survey shows that 10% of students are in band, 12% of students are in athletics, and 6% of students are in both band and athletics. What is the probability that a student is in band or athletics?
A. $\frac{1}{60}$ B. $\frac{11}{50}$ C. $\frac{4}{25}$ D. $\frac{7}{25}$ — 16. C

17. Three coins are tossed. Find the probability that exactly 2 coins show heads if the first coin shows heads.
A. $\frac{1}{8}$ B. $\frac{3}{8}$ C. $\frac{1}{4}$ D. $\frac{1}{2}$ — 17. A

18. Given the integers 1 through 33, what is the probability that one of these integers is divisible by 4 if it is a multiple of 6?
A. $\frac{1}{4}$ B. $\frac{5}{33}$ C. $\frac{5}{8}$ D. $\frac{2}{5}$ — 18. D

19. A survey shows that 20% of all cars are white. What is the probability that exactly 3 of the next 4 cars to pass will be white?
A. $\frac{1}{4}$ B. $\frac{16}{625}$ C. $\frac{4}{625}$ D. $\frac{1}{125}$ — 19. B

20. Eight out of every 10 houses have a garage. Express as a decimal to the nearest hundredth the probability that exactly 9 out of 12 houses on a given block have a garage.
A. 0.24   B. 0.42   C. 0.56   D. 0.88 — 20. A

Bonus Two number cubes are thrown twice. What is the probability of getting a sum less than 4 on both throws?
A. $\frac{1}{36}$ B. $\frac{1}{144}$ C. $\frac{1}{12}$ D. $\frac{1}{6}$ — **Bonus:** B

© Glencoe/McGraw-Hill   318   *Advanced Mathematical Concepts*

SKILLS AND CONCEPTS

Objectives and Examples This section reviews the skills and concepts of the chapter and shows completely worked examples.

Review Exercises These exercises provide practice for the corresponding objectives.

Assessment and Evaluation Masters, pp. 323–324

SKILLS AND CONCEPTS

OBJECTIVES AND EXAMPLES	REVIEW EXERCISES

Lesson 13-1 Solve problems related to the Basic Counting Principle.

• How many possible ways can a group of eight students line up to buy tickets to a play?

There are eight choices for the first spot in line, seven choices for the second spot, six for the third spot, and so on.

$8 \cdot 7 \cdot 6 \cdot 5 \cdot 4 \cdot 3 \cdot 2 \cdot 1 = 40,320$

There are 40,320 ways for the students to line up.

11. How many different ways can three books be arranged in a row on a shelf? **6**

12. How many different ways can the digits 1, 2, 3, 4, and 5 be arranged to create a password? **120**

13. How many ways can six teachers be assigned to teach six different classes, if each teacher can teach any of the classes? **720**

Lesson 13-1 Solve problems involving permutations and combinations.

• From a choice of 3 meat toppings and 4 vegetable toppings, how many 5-topping pizzas are possible?

Since order is not important, the selection is a combination of 7 objects taken 5 at a time, or $C(7, 5)$.

$$C(7, 5) = \frac{7!}{(7-5)!\,5!} = 21$$

There are 21 possible 5-topping pizzas.

Find each value.

14. $P(6, 3)$ **120**   15. $P(8, 6)$ **20,160**

16. $C(5, 3)$ **10**   17. $C(11, 8)$ **165**

18. $\dfrac{P(6, 3)}{P(5, 3)}$ **2**   19. $C(5, 5) \cdot C(3, 2)$ **3**

20. How many ways can 6 different books be placed on a shelf if the only dictionary must be on an end? **240**

21. From a group of 3 men and 7 women, how many committees of 2 men and 2 women can be formed? **63**

Lesson 13-2 Solve problems involving permutations with repetitions.

• How many ways can the letters of *Tallahassee* be arranged?

There are 3 *a*'s, 2 *l*'s, 2 *s*'s, and 2 *e*'s. So the number of possible arrangements is

$\dfrac{11!}{3!\,2!\,2!\,2!}$ or 831,600 ways.

How many different ways can the letters of each word be arranged?

22. *level* **30**

23. *Cincinnati* **50,400**

24. *graduate* **20,160**

25. *banana* **60**

26. How many different 9-digit Social Security numbers can have the digits 2, 9, 5, 5, 0, 7, 0, 5, and 8. **30,240**

882 *Chapter 13 Combinatorics and Probability*

💿 TestCheck and Worksheet Builder

This **networkable** software has three modules.
• **Worksheet Builder** to make worksheets and tests
• **Student Module** to take tests on-screen
• **Management System** to keep student records
Three special banks are included for SAT, ACT, and TIMSS.

CHAPTER 13 • STUDY GUIDE AND ASSESSMENT

OBJECTIVES AND EXAMPLES

Lesson 13-3 Find the probability of an event.

Find the probability of randomly selecting 3 red pencils from a box containing 5 red, 3 blue, and 4 green pencils.

There are $C(5, 3)$ ways to select 3 out of 5 red pencils and $C(12, 3)$ ways to select 3 out of 12 pencils.

$$P(3 \text{ red pencils}) = \frac{C(5, 3)}{C(12, 3)}$$

$$= \frac{\frac{5!}{2!\,3!}}{\frac{12!}{9!\,3!}}$$

$$= \frac{12}{220} \text{ or } \frac{1}{22}$$

Lesson 13-3 Find the odds for the success and failure of an event.

Find the odds of randomly selecting 3 red pencils from a box containing 5 red, 3 blue, and 4 green pencils.

$$P(3 \text{ red pencils}) = P(s) = \frac{1}{22}$$

$$P(\text{not 3 red pencils}) = P(f) = 1 - \frac{1}{22} \text{ or } \frac{21}{22}$$

$$\text{Odds} = \frac{P(s)}{P(f)} = \frac{\frac{1}{22}}{\frac{21}{22}}$$

$$= \frac{1}{21} \text{ or } 1:21$$

Lesson 13-4 Find the probability of independent and dependent events.

Three yellow and 5 black marbles are placed in a bag. What is the probability of drawing a black marble, replacing it, and then drawing a yellow marble?

$$P(\text{black}) = \frac{5}{8} \qquad P(\text{yellow}) = \frac{3}{8}$$

$$P(\text{black and yellow}) = P(\text{black}) \cdot P(\text{yellow})$$

$$= \frac{5}{8} \cdot \frac{3}{8} = \frac{15}{64}$$

REVIEW EXERCISES

A bag containing 7 pennies, 4 nickels, and 5 dimes. Three coins are drawn at random. Find each probability.

27. $P(3 \text{ pennies})$ $\frac{1}{16}$

28. $P(2 \text{ pennies and 1 nickel})$ $\frac{3}{20}$

29. $P(3 \text{ nickels})$ $\frac{1}{140}$

30. $P(1 \text{ nickel and 2 dimes})$ $\frac{1}{14}$

Refer to the bag of coins used for Exercises 27-30. Find the odds of each event occurring.

31. 3 pennies $\frac{1}{15}$

32. 2 pennies and 1 nickel $\frac{3}{17}$

33. 3 nickels $\frac{1}{139}$

34. 1 nickel and 2 dimes $\frac{1}{13}$

Determine if each event is *independent* or *dependent*. Then determine the probability.

35. the probability of rolling a sum of 2 on the first toss of two number cubes and a sum of 6 on the second toss **independent,** $\frac{5}{1296}$

36. the probability of randomly selecting two yellow markers from a box that contains 4 yellow and 6 pink markers **dependent,** $\frac{2}{15}$

Chapter 13 Study Guide and Assessment **883**

SKILLS AND CONCEPTS

OBJECTIVES AND EXAMPLES

Lesson 13-4 Find the probability of mutually exclusive and inclusive events.

- On a school board, 2 of the 4 female members are over 40 years of age, and 5 of the 6 male members are over 40. If one person did not attend the meeting, what is the probability that the person was a male or a member over 40?

P(male or over 40) $= P$(male) $+ P$(over 40) $-$
$$P\text{(male \& over 40)}$$
$$= \frac{6}{10} + \frac{7}{10} - \frac{5}{10} \text{ or } \frac{4}{5}$$

REVIEW EXERCISES

A box contains slips of paper numbered from 1 to 14. One slip of paper is drawn at random. Find each probability.

37. P(selecting a prime number or a multiple of 4) $\quad \frac{11}{14}$

38. P(selecting a multiple of 2 or a multiple of 3) $\quad \frac{9}{14}$

39. P(selecting a 3 or a 4) $\quad \frac{1}{7}$

40. P(selecting an 8 or a number less than 8) $\quad \frac{4}{7}$

Lesson 13-5 Find the probability of an event given the occurrence of another event.

- A coin is tossed 3 times. What is the probability that at the most 2 heads are tossed given that at least 1 head has been tossed?

Let event A be that at most 2 heads are tossed.
Let event B be that there is at least 1 head.
$$P(A \mid B) = \frac{P(A \text{ and } B)}{P(B)}$$

$$= \frac{\frac{6}{8}}{\frac{7}{8}} \text{ or } \frac{6}{7}$$

Two number cubes are tossed.

41. What is the probability that the sum of the numbers shown on the cubes is less than 5 if exactly one cube shows a 1? $\quad \frac{2}{5}$

42. What is the probability that the numbers shown on the cubes are different given that their sum is 8? $\quad \frac{4}{5}$

43. What is the probability that the numbers shown on the cubes match given that their sum is greater than or equal to 5? $\quad \frac{2}{15}$

Lesson 13-6 Find the probability of an event by using the Binomial Theorem.

- If you guess the answers on all 8 questions of a true/false quiz, what is the probability that exactly 5 of your answers will be correct?

$$(p + q)^8 = \sum_{r=0}^{8} \frac{8!}{r!(8-r)!} \, p^{8-r}q^r$$

$$= \frac{8!}{5!(8-5)!}\left(\frac{1}{2}\right)^5\left(\frac{1}{2}\right)^3$$

$$= \frac{56}{256} \text{ or } \frac{7}{32}$$

Find each probability if a coin is tossed 4 times.

44. P(exactly 1 head) $\quad \frac{1}{4}$

45. P(no heads) $\quad \frac{1}{16}$

46. P(2 heads and 2 tails) $\quad \frac{3}{8}$

47. P(at least 3 tails) $\quad \frac{5}{16}$

APPLICATIONS AND PROBLEM SOLVING

48. Travel Five people, including the driver, can be seated in Nate's car. Nate and 6 of his friends want to go to a movie. How many different groups of friends can ride in Nate's car on the first trip if the car is full? *(Lesson 13-1)* **15**

49. Sommer has 7 different keys. How many ways can she place these keys on the key ring shown below? *(Lesson 13-2)* **2520**

50. Quality Control A collection of 15 memory chips contains 3 chips that are defective. If 2 memory chips are selected at random, what is the probability that at least one of them is good? *(Lesson 13-3)* $\frac{34}{35}$

51. Gift Exchange The Burnette family is drawing names from a bag for a gift exchange. There are 7 males and 8 females in the family. If someone draws their own name, then they must draw again before replacing their name. *(Lesson 13-4)*

   a. Reba draws the first name. What is the probability that Reba will draw a female's name that is not her own? $\frac{7}{15}$

   b. What is the probability that Reba will draw her own name, not replace it, and then draw a male's name? $\frac{1}{30}$

ALTERNATIVE ASSESSMENT

OPEN-ENDED ASSESSMENT

1. The probability of two independent events occurring is $\frac{1}{12}$. If the probability of one of the events occurring is $\frac{1}{2}$, is it possible to find the probability of the other event? If so, find the probability and give an example of a situation for which this probability could apply. If not, explain why not. **See margin.**

2. Perry says "A permutation is the same as a combination." How would you explain to Perry that his statement is incorrect? **See margin.**

PORTFOLIO
Choose one of the types of probability you studied in this chapter. Describe a situation in which this type of probability would be used. Explain why no other type of probability should be used in this situation.

Unit 4 *inter*NET **Project**

THE UNITED STATES CENSUS BUREAU

Radically Random!

• Use the Internet to find the population of the United States by age groups or ethnic background for the most recent census.

• Make a table or spreadsheet of the data.

• Suppose that a person was selected at random from all the people in the United States to answer some survey questions. Find the probability that the person was from each one of the age or ethnic groups you used for your table or spreadsheet.

• Write a summary describing how you calculated the probabilities. Include a graph with your summary. Discuss why someone might be interested in your findings.

> **Additional Assessment** See p. A68 for Chapter 13 practice test.

Chapter 13   Study Guide and Assessment **885**

Internet Address Book

Record useful Internet addresses in the space at right for quick reference.

www.amc.glencoe.com

Pages 886–887 are part of a complete SAT/ACT preparation course that is described in detail on pages T8–T9 of the *Teacher's Handbook.* The test items were written in the same style as those in the SAT and ACT.

These questions were aligned and verified by The Princeton Review, the nation's leader in test preparation.

Teaching Tip In order to distinguish such items as combinations and permutations, it is helpful to memorize a simple example of each for reference during a test.

Diagnosis & Prescription

Each of the 10 SAT and ACT practice problems on page 887 is cross-referenced to the chapter where that SAT or ACT skill is covered. If students miss a particular type of problem, you can have them study that skill.

Problem Number and Skill	Page
1 Probability and Combinations	886
2 Coordinate Geometry	272
3 Probability and Combinations	886
4 Average	754
5 Arithmetic	64
6 Probability and Combinations	886
7 Geometry (Lines, Angles, and Arcs)	612
8 Probability and Combinations	886
9 Probability and Combinations	886
10 Probability and Combinations	886

Probability and Combination Problems

Both the ACT and SAT contain probability problems. You'll need to know these concepts:

Combinations   Permutations   Tree Diagram

Outcomes   Probability

Memorize the definition of the probability of an event:

$$P(\text{event}) = \frac{\text{number of favorable outcomes}}{\text{total number of possible outcomes}}$$

SAT EXAMPLE

1. A bag contains 4 red balls, 10 green balls, and 6 yellow balls. If three balls are removed at random and no ball is returned to the bag after removal, what is the probability that all three balls will be green?

   A $\frac{1}{2}$ B $\frac{1}{8}$ C $\frac{3}{20}$ D $\frac{2}{19}$ E $\frac{3}{8}$

   HINT Calculate the probability of two independent events by multiplying the probability of the first event by the probability of the second event.

   Solution Use the definition of the probability of an event. Calculate the probability of getting a green ball each time a ball is removed. The first time a ball is removed there are a total of 20 balls and 10 of them are green. So the probability of removing a green ball as the first ball is $\frac{10}{20}$ or $\frac{1}{2}$. Now there are just 19 balls and 9 of them are green. The probability of removing a green ball is $\frac{9}{19}$. When the third ball is removed, there are 18 balls and 8 of them are green, so the probability of removing a green ball is $\frac{8}{18}$ or $\frac{4}{9}$. To find the probability of removing green balls as the first *and* the second *and* the third balls chosen, multiply the three probabilities.

   $$\frac{1}{2} \times \frac{9}{19} \times \frac{4}{19} = \frac{2}{19}$$

   The answer is choice **D**.

   886 *Chapter 13 Combinatorics and Probability*

ACT EXAMPLE

2. If you toss 3 fair coins, what is the probability of getting exactly 2 heads?

   A $\frac{1}{3}$ B $\frac{3}{8}$

   C $\frac{1}{2}$ D $\frac{2}{3}$

   E $\frac{7}{8}$

   HINT Start by listing all the possible outcomes. You can do this since the numbers are small.

   Solution Make a list and then count the possible outcomes.

   HHH, HHT, HTH, HTT,

   TTH, THT, THH, TTT

   There are 8 possible outcomes for 3 coins. Since the coins are fair, these are equally likely outcomes.

   The favorable outcomes are those that include exactly 2 heads: HHT, HTH, THH. There are 3 favorable outcomes. Give the answer.

   $$P(A) = \frac{\text{number of successful outcomes}}{\text{total number of outcomes}}$$

   $$P(\text{exactly 2 heads}) = \frac{3}{8}$$

   The answer is choice **B**.

TestCheck and Worksheet Builder

Special banks of SAT, ACT, and TIMSS items are included in this CD-ROM.

After you work each problem, record your answer on the answer sheet provided or on a piece of paper.

Multiple Choice

1. A coin was flipped 20 times and came up heads 10 times and tails 10 times. If the first and the last flips were both heads, what is the greatest number of consecutive heads that could have occurred? **D**

   A 1   B 2   C 8

   D 9   E 10

2. In the figure below, *ABCD* is a parallelogram. What must be the coordinates of Point *C*? **E**

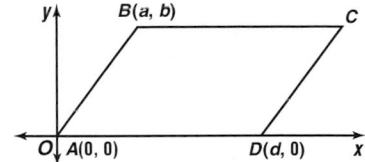

   A (x, y) B $(d + a, y)$ C $(d - a, b)$

   D $(d + x, b)$ E $(d + a, b)$

3. In a plastic jar there are 5 red marbles, 7 blue marbles, and 3 green marbles. How many green marbles need to be added to the jar in order to double the probability of selecting a green marble? **C**

   A 2   B 3   C 5   D 6   E 7

4. The average of 5 numbers is 20. If one of the numbers is 18, then what is the sum of the other four numbers? **C**

   A 2   B 20.5   C 82

   D 90   E 100

5. If the sum of *x* and *y* is an even number, and the sum of *x* and *z* is an even number, and *z* is an odd number, then which of the following must be true? **E**

   I. *y* is an even number
   II. *y* + *z* is an even number
   III. *y* is an odd number

   A I only   B II only   C III only

   D I and II   E II and III

6. A bag contains only white and blue marbles. The probability of selecting a blue marble is $\frac{1}{5}$. The bag contains 200 marbles. If 100 white marbles are added to the bag, what is the probability of selecting a white marble? **E**

   A $\frac{2}{15}$ B $\frac{7}{15}$ C $\frac{8}{15}$ D $\frac{4}{5}$ E $\frac{13}{15}$

7. In the figure below, $\ell_1 \parallel \ell_2$. Which of the labeled angles must be equal to each other? **A**

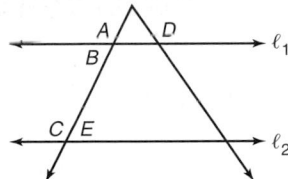

   A *A* and *C*   B *D* and *E*   C *A* and *B*

   D *D* and *B*   E *C* and *B*

8. What is the probability of drawing a diamond from a well-shuffled standard deck of playing cards? **C**

   A $\frac{1}{52}$ B $\frac{1}{13}$ C $\frac{1}{4}$

   D $\frac{4}{13}$ E 1

9. A caterer offers 7 different entrees. A customer may choose any 3 of the entrees for a dinner. How many different combinations of entrees can a customer choose? **B**

   A 6
   B 35
   C 84
   D 210
   E 840

10. **Grid-In** Six cards are numbered 0 through 5. Two are selected without replacement. What is the probability that the sum of the cards is 4? **2/15**

SAT/ACT Practice For additional test practice questions, visit: **www.amc.glencoe.com**

THE PRINCETON REVIEW   *SAT & ACT Preparation*   **887**

Extra Credit

TIMSS A set of 24 cards is numbered with the positive integers from 1 to 24. If the cards are shuffled and if only one is selected at random, what is the probability that the number of the card is divisible by 4 or 6? **D**

A $\frac{1}{6}$ B $\frac{5}{6}$ C $\frac{1}{4}$ D $\frac{1}{3}$ E $\frac{5}{12}$

A bubble-in answer sheet for these practice problems is available on p. 1 of the *Assessment and Evaluation Masters*.

Additional Practice

Additional SAT and ACT practice questions and a bubble-in answer sheet are available in the *Assessment and Evaluation Masters*, pp. 2, 331, and 332.

Assessment and Evaluation Masters, pp. 331–332

Statistics and Data Analysis

Instructional Objectives

Lesson (pages)	Lesson Objectives	NCTM Standards 2000	State/Local Objectives
14-1 (889–896)	• Draw, analyze, and use bar graphs and histograms. • Organize data into a frequency distribution table.	5, 6, 10	
14-2 (897–907)	• Find the mean, median, and mode of a set of data. • Find measures of central tendency of data organized in a stem-and-leaf plot or a frequency distribution table.	5, 6, 10	
14-3 (908–917)	• Find the interquartile range, the semi-interquartile range, mean deviation, and standard deviation of a set of data. • Organize and compare data using box-and-whisker plots.	5, 6, 10	
14-4 (918–925)	• Use the normal distribution curve.	5, 6, 10	
14-4B (926)	• Use the standard normal curve to study properties of normal distributions.	5, 6, 10	
14-5 (927–932)	• Find the standard error of the mean to predict the true mean of a population with a certain level of confidence.	5, 6, 9, 10	

Suggested Pacing
See pages T2–T4 for additional options in course-planning.

COURSE	Lesson 14-1	Lesson 14-2	Lesson 14-3	Lesson 14-4	Lesson 14-5
Precalculus Emphasis Chapters 4–15 (45–50 minutes)	1.5 days	1.5 days	1.5 days	1.5 days	(with Lesson 14-4B) 2 days
Precalculus Emphasis (block)	0.5 day	1.5 days		(with Lesson 14-4B) 1 day	(with Study Guide) 1.5 days
Trigonometric Emphasis Chapters 1–9, 11, 12 (45–50 minutes)	OPTIONAL				
Trigonometric Emphasis (block)	OPTIONAL				

Instructional Resources

Lesson	Student Edition — Graphing Calculator Applications	Study Guide and Practice	Enrichment	Assessment and Evaluation	Graphing Calculator Keystrokes	Problem Solving & Applications	Spreadsheet Applications	TI-92 and CBL Interdisciplinary Activities	Transparencies A and B
14-1	891, 892	255–256	157						14-1
14-2	900	257–258	158	355		28	14		14-2
14-3	909, 911, 912, 913, 915	259–260	159	354, 355				45–48	14-3
14-4		261–262	160	356					14-4
14-4B	926				53–54				
14-5		263–264	161	356		27			14-5
Study Guide/ Assessment				341–353, 359					
SAT/ACT Practice				357–358					

Study Guide & Assessment	Chapter Test	TOTAL
1 day	1 day	10 days
(See Lesson 14-5.)	0.5 day	5 days
OPTIONAL		
OPTIONAL		

Applications

Problem Solving and Applications Masters, pp. 27, 28

Technology/Multimedia

Spreadsheet Applications Masters, p. 14

TI-92 & CBL Interdisciplinary Activities Masters, pp. 45–48

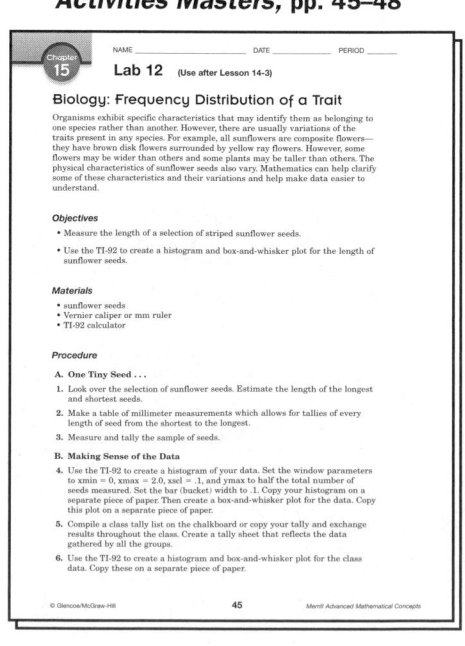

The worksheets shown on this page are a small sample of the materials available on *TeacherWorks: All-in-One Lesson Planner and Resource Center.*

This CD-ROM contains all of the blackline masters and transparencies available for this program.

It also includes a lesson planner and interactive Teacher Edition, so you can customize lesson plans and reproduce classroom resources quickly and easily, from just about anywhere.

Assessment Resources

Type	Student Edition	Teacher's Wraparound Edition	Assessment and Evaluation Masters
Ongoing Assessment	Mid-Chapter Quiz, p. 917	5-Minute Check, pp. 889, 897, 908, 918, 927	Mid-Chapter Test, p. 354 Quizzes A–D, pp. 355–356
Mixed Review	Mixed Review, pp. 896, 907, 917, 925, 932		Cumulative Review, p. 359
Error Analysis	You Decide, pp. 903, 930	Error Analysis, pp. 893, 903, 914, 922, 930	
Standardized Test Practice	SAT/ACT Practice, pp. 896, 907, 917, 925, 932 SAT & ACT Preparation, pp. 938–939	TIMSS, p. 939	SAT/ACT Practice, pp. 357–358
Open-Ended Assessment	Math Journal, pp. 893, 914 Open-Ended Assessment, p. 937 Portfolio, p. 937	Modeling: pp. 925, 926 Speaking: pp. 907, 932 Writing: pp. 896, 917	Open-Ended Assessment, p. 353
Chapter Assessment	Study Guide and Assessment, pp. 933–937 Chapter Test, p. A69		Multiple-Choice Tests (Forms 1A, 1B, 1C), pp. 341–346 Free-Response Tests (Forms 2A, 2B, 2C), pp. 347–352

TestCheck and Worksheet Builder is a networkable CD-ROM with three integrated modules. The **Worksheet Builder** creates customized worksheets, tests, and quizzes of free-response, multiple-choice, short-answer, and open-ended items. The **Student Module** gives you the option of having students take tests on-screen and get immediate feedback on their performance. Use the optional **Management System** to keep detailed student records.

Additional Chapter Resources

Student Edition
Career Choices, p. 896
Internet Project, p. 937
Extra Practice, pp. A53–A54

Teacher's Classroom Resources

Technology
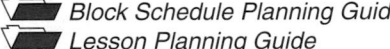 *Graphing Calculator Keystrokes Masters*

Teaching Aids
Answer Key Transparencies
 Block Schedule Planning Guide
Lesson Planning Guide
Solutions Manual

Glencoe Instructional Technology

Instructional
Advanced MathPASS CD-ROM
Multimedia Applications CD-ROM

Assessment
TestCheck and Worksheet Builder

Visit **www.amc.glencoe.com** for data updates, career information, graphing calculator programs, and other interactive activities.

Chapter Overview

This chapter introduces several methods of displaying data, including bar graphs, histograms, frequency distributions, stem-and-leaf plots, and box-and-whisker plots. Students learn to use normal distribution curves, find measures of central tendency and of variability, and predict the true mean of a population with a certain level of confidence.

 Using a Graphing Calculator

There are numerous opportunities, all optional, to use a graphing calculator in this chapter.
- Creating a histogram (p. 891)
- Creating a frequency polygon (p. 892)
- Finding the mean (p. 900)
- Using **SortA(** to reorder a list from least to greatest (p. 909)
- Using **sum(** and **abs(** to find mean deviation (p. 911)
- Finding standard deviation (pp. 912, 913)
- Drawing a box-and-whisker plot (p. 915)
- Using the standard normal curve (p. 926)

The **Graphing Calculator Appendix**, which begins on page A2, contains keystroke instructions on many of these topics.

STATISTICS AND DATA ANALYSIS

CHAPTER OBJECTIVES •

- **Make and use bar graphs, histograms, frequency distribution tables, stem-and-leaf plots, and box-and-whisker plots.** *(Lessons 14-1, 14-2, 14-3)*
- **Find the measures of central tendency and the measures of variability.** *(Lessons 14-2, 14-3)*
- **Use the normal distribution curve.** *(Lesson 14-4)*
- **Find the standard error of the mean to predict the true mean of a population with a certain level of confidence.** *(Lesson 14-5)*

 SAT & ACT Preparation

This textbook contains a complete SAT/ACT preparation course described on pages T8–T9 of the *Teacher's Handbook*. This chapter contains:

- an SAT or ACT practice question at the end of each lesson
- examples of SAT and ACT problems involving statistics and data analysis on page 938
- ten SAT and ACT practice problems on page 939

- additional SAT and ACT practice in the *Assessment and Evaluation Masters*, pp. 357–358
- SAT and ACT test practice banks in the *TestCheck and Worksheet Builder*

OBJECTIVES

- Draw, analyze, and use bar graphs and histograms.
- Organize data into a frequency distribution table.

The Frequency Distribution

FOOTBALL The AFL-NFL World Championship Game, as it was originally called, became the Super Bowl in 1969. The graph below shows the first 34 Super Bowl winners. What team has won the most Super Bowls?

Super Bowl Winners

 = 1 win

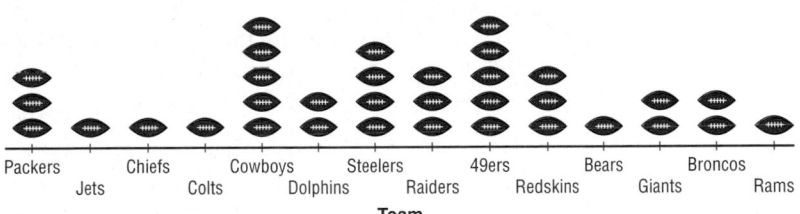

Packers		Chiefs		Cowboys		Steelers		49ers		Bears		Broncos
	Jets		Colts		Dolphins		Raiders		Redskins		Giants		Rams

Team

By looking at the graph, you can quickly determine that the Dallas Cowboys and the San Francisco 49ers have both won five Super Bowls.

A graph is often used to provide a picture of statistical data. One advantage of using a graph to show data is that a person can easily see any relationships or patterns that may exist. The number of Super Bowl victories for various teams is depicted as a **line plot**. A line plot uses symbols to show frequency. A **bar graph** can show the same information by using bars to indicate the frequency.

A **back-to-back bar graph** is a special bar graph that shows the comparisons of two sets of related data. A back-to-back graph is plotted on a coordinate system with the horizontal scale repeated in each direction from the central axis.

Example

1 ECONOMICS The following data relates the amount of education with the median weekly earnings of a full-time worker 25 years old or older for the years 1980 and 1997.

Median Weekly Earnings

	Less than 4 Years of High School	High School Diploma	1 to 3 Years of College	College Degree
1980	$222	$266	$304	$376
1997	$321	$461	$518	$779

Source: *The Wall Street Journal 1999 Almanac*

a. Make a back-to-back bar graph that represents the data.

b. Describe any trends indicated by the graph.

(continued on the next page)

5-Minute Check
Chapter 13

1. How many seven-digit phone numbers can begin with the prefix 93? **100,000**

2. How many five-letter patterns can be formed from the letters of the word *foods*? **60**

3. From a list of seven books, how many groups of four books can be selected? **35**

4. A box contains 3 tennis balls, 6 baseballs, and 5 softballs. What is the probability that a ball selected at random will be a baseball? $\frac{6}{14}$ **or** $\frac{3}{7}$

5. Find the probability of getting a sum of 7 on the first roll of two dice and a sum of 3 on the second roll. $\frac{1}{108}$

Motivating the Lesson

Real-World Application Discuss situations in which large amounts of data are given. For example, students could view a list of the winners of all of the World Series from 1903 to the current year. Ask students how they could display the data so as to quickly summarize any trends in frequency, most frequent data points, range of frequencies, and so on. Why is a graph or visual display beneficial?

Resource Manager

 Reproducible Masters
- *Study Guide and Practice,* pp. 255–256
- *Enrichment,* p. 157

Transparencies
- 5-Minute Check, 14-1
- Teaching, 14-1

Teaching Tip In Example 1, make sure that students start each side of the bar graph at 0.

In-Class Example

Example 1
Agriculture In 1997, the value of a United States farm product averaged 23% of the retail cost of the item. The other 77% consisted of processing, transporting, and retailing charges incurred after the item left the farm. The data below show the value of several farm products in 1987 and 1997.

Farm Value (%)		
Product	**Year**	
	1987	**1997**
Eggs	54	46
Fruit	26	18
Meat	47	37
Vegetables	31	21

a. Make a back-to-back bar graph that represents the data.

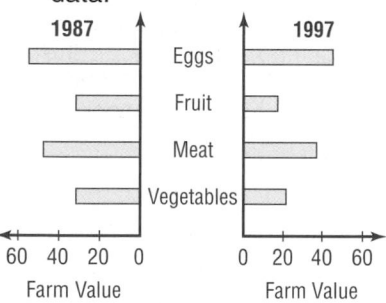

b. Describe any trends indicated by the graph.
In each case, the value decreased from 1987 to 1997, indicating that U.S. consumers are now paying more in processing, transportation and retail charges for food.

Teaching Tip After discussing the frequency distribution table in the Student Edition, emphasize that the intervals in a frequency distribution should all be equal.

a. Let the level of education be the central axis. Draw a horizontal axis that is scaled $0 to $800 in each direction. Let the left side of the graph represent the earnings from 1980 and the right side of the graph be those from 1997. Draw the bars to the appropriate length for the data.

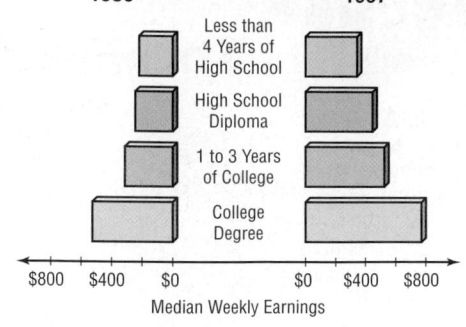

b. You can see from the graph that when you compare each level of education with the next, more education resulted in a greater increase in median weekly earnings in 1997 than in 1980.

Sometimes it is desirable to show three aspects of a set of data at the same time. To present data in this way, a **three-dimensional bar graph** is often used. The graph at the left represents the retail sales in thousands of passenger cars in the United States for three major domestic car manufacturers during the years 1994 to 1997. The grid defines the car and year. The height of each bar represents the number of cars sold each year.

Sometimes the amount of data you wish to represent in a bar graph is too great for each item of data to be considered individually. In this case, a **frequency distribution** is a convenient system for organizing the data. A number of classes are determined, and all values in a class are tallied and grouped together. To determine the number of classes, first find the **range.** The range of a set of data is the difference between the greatest and the least values in the set.

Retail Management Testing Scores	
Scores	**Frequency**
60–70	9
70–80	10
80–90	12
90–100	3

The intervals are often named by a range of values. In the table, the interval described by 60-70 means all the test scores s such that $60 \leq s < 70$. The **class interval** is the range of each class. The class intervals in a frequency distribution should all be equal. In the table, the range for each class interval is 10.

Class intervals are often multiples of 5.

The **class limits** of a set of data organized in a frequency distribution are the upper and lower values in each interval. The class limits in the testing data above are 60, 70, 80, 90, and 100. The **class marks** are the midpoints of the classes; that is, the average of the upper and lower limit for each interval. The class mark for the interval 60–70 is $\frac{60 + 70}{2}$ or 65.

The difference in consecutive class marks is the same as the class interval.

The most common way of displaying frequency distributions is by using a **histogram.** A histogram is a type of bar graph in which the width of each bar represents a class interval and the height of the bar represents the frequency in that interval. Histograms usually have fewer than ten intervals.

890 *Chapter 14   Statistics and Data Analysis*

Prerequisite Skills

A review of the prerequisite skills for this lesson is provided in the *Prerequisite Skills Masters.*

2 FOOTBALL The winning scores for the first 34 Super Bowls are 35, 33, 16, 23, 16, 24, 14, 24, 16, 21, 32, 27, 35, 31, 27, 26, 27, 38, 38, 46, 39, 42, 20, 55, 20, 37, 52, 30, 49, 27, 35, 31, 34, and 23.

a. Find the range of the data.

b. Determine an appropriate class interval.

c. Find the class marks.

d. Construct a frequency distribution of the data.

e. Draw a histogram of the data.

f. What conclusions can you determine from the graph?

Vince Lombardi Trophy

a. The range of the data is 55 − 14 or 41.

b. An appropriate class interval is 10 points, beginning with 10 points and ending with 60 points. There will be five classes.

c. The class marks are the averages of the class limits of each interval. The class marks are 15, 25, 35, 45, and 55.

d. Make a table listing class limits. Use tallies to determine the number of scores in each interval.

Winning Score	Tallies	Frequency
10–20	IIII	4
20–30	LHT LHT II	12
30–40	LHT LHT III	13
40–50	III	3
50–60	II	2

e. Label the horizontal axis with the class limits. The vertical axis should be labeled from 0 to a value that will allow for the greatest frequency. Draw the bars side by side so that the height of each bar corresponds to its interval's frequency.

Winning Scores at the Super Bowl

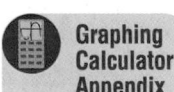

Graphing Calculator Appendix

For keystroke instruction on how to create a histogram, see pages A23-A24.

You can also use a graphing calculator to create the histogram. In statistics mode, enter the class marks in the **L1** list and the frequency in the **L2** list. Set the window using the class interval for **Xscl**, and select the histogram as the type of graph.

[0, 60] scl:10 by [0, 15] scl:1

f. The winning score at the Super Bowl tends to be between 20 and 40 points.

Lesson 14-1 The Frequency Distribution **891**

Teaching Tip In Example 2, remind students that a reasonable histogram has 5 to 8 class intervals and that other class intervals besides those used in part **b** can be used.

In-Class Example

Example 2
Basketball From 1970 to 1998, the top scorers of the NBA averaged the following points per game:
31.2, 31.7, 34.8, 34.0, 30.6, 34.5, 31.1, 31.1, 27.2, 29.6, 33.1, 30.7, 32.3, 28.4, 30.6, 32.9, 30.3, 37.1, 35.0, 32.5, 33.6, 31.5, 30.1, 32.6, 29.8, 29.3, 30.4, 29.6, and 28.7.

a. Find the range of the data. **9.9**

b–f. Sample answers are given.

b. Determine an appropriate class interval. **2 points beginning with 27.0 points and ending with 39.0 points**

c. Find the class marks. **28, 30, 32, 34, 36, 38**

d. Construct a frequency distribution of the data.

Average Points	Tallies	Frequency
27.0–29.0	III	3
29.0–31.0	LHT LHT	10
31.0–33.0	LHT IIII	9
33.0–35.0	LHT	5
35.0–37.0	I	1
37.0–39.0	I	1

e. Draw a histogram.

Average Points of NBA Leading Scorers

f. What conclusions can you determine from the graph? **Most NBA leading scorers averaged between 29 and 35 points per game.**

Civil Engineering The twenty highest dams have heights of 240, 335, 260, 300, 285, 262, 272, 221, 242, 261, 233, 250, 243, 237, 236, 243, 234, 230, 226, and 261 meters.

a–c. Sample answers are given.

a. Find an appropriate class interval. Then name the class limits and the class marks.   **20 meters; 220, 240, 260, 280, 300, 320, 340; 230, 250, 270, 290, 310, 330**

b. Construct a frequency distribution.

Height (in meters)	Frequency
220–240	7
240–260	5
260–280	5
280–300	1
300–320	1
320–340	1

c. Use a graphing calculator to draw a frequency polygon.

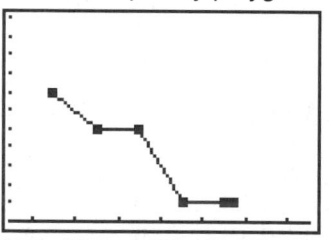

[220, 360] scl:20 by [0,11] scl:1

Study Guide and Practice Masters, pp. 255–256

 Study Guide

The Frequency Distribution
A **frequency distribution** is a convenient system for organizing large amounts of data. A number of classes are determined, and all values in a class are tallied and grouped together. The most common way of displaying frequency distributions is by using a type of bar graph called a **histogram**.

Example The number of passengers who boarded planes at 36 airports in the United States in one year are shown below.

30,526	30,372	26,623	22,722	16,287	15,246	14,807	14,117	14,054
13,547	12,916	12,616	11,906	11,622	11,489	10,828	10,653	10,008
9703	9594	9463	9348	9125	8572	7300	6772	6549
6126	5907	5712	5287	4848	4832	4820	4750	4684

Source: U.S. Department of Transportation

a. Find the range of the data.
The range of the data is 30,526 − 4684, or 25,842.

b. Determine an appropriate class interval.
An appropriate class interval is 4500 passengers, beginning with 4500 and ending with 31,500. There will be six classes.

c. Name the class limits and the class marks.
The class limits are the upper and lower values in each interval, or 4500, 9000, 13,500, 18,000, 22,500, 27,000, and 31,500. The class marks are the averages of the class limits of each interval, or 6750, 11,250, 15,750, 20,250, 24,750, and 29,250.

d. Construct a frequency distribution of the data.
Use tallies to determine the number of passengers in each interval.

Number of Passengers	Tallies	Frequency
4500-9000		13
9000-13,500		13
13,500-18,000		6
18,000-22,500		0
22,500-27,000		2
27,000-31,500		2

e. Draw a histogram of the data.
Label the horizontal axis with the class limits. The vertical axis should be labeled from 0 to a value that will allow for the greatest frequency. Draw the bars side by side so that the height of each bar corresponds to its interval's frequency.

© Glencoe/McGraw-Hill   255   Advanced Mathematical Concepts

Another type of graph can be created from a histogram. A broken line graph, often called a **frequency polygon,** can be drawn by connecting the class marks on the graph. The class marks are graphed as the midpoints of the top edge of each bar. The frequency polygon for the histogram in Example 2 is shown at the right.

Winning Scores at the Super Bowl

Example ③ **HEALTH**   A graduate student researching the effect of smoking on blood pressure collected the following readings of systolic blood pressure from 30 people within a control group.
125, 145, 110, 126, 128, 180, 177, 176, 156, 144, 182, 205, 191, 140, 138, 126, 154, 163, 172, 159, 174, 151, 142, 160, 147, 143, 158, 129, 132, 137

a. Find an appropriate class interval. Then name the class limits and the class marks.

b. Construct a frequency distribution.

c. Use a graphing calculator to draw a frequency polygon.

a. The range of the data is 205 − 110 or 95. An appropriate class interval is 15 units. The class limits are 105, 120, 135, 150, 165, 180, 195, and 210. The class marks are 112.5, 127.5, 142.5, 157.5, 172.5, 187.5, and 202.5.

b.

Systolic Blood Pressure	Tallies	Frequency
105–120			1
120–135	⊞		6
135–150	⊞ |||	8
150–165	⊞ ||	7
165–180	||||	4
180–195	|||	3
195–210			1

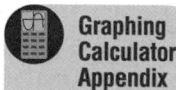 **Graphing Calculator Appendix**

For keystroke instruction on how to create a frequency polygon, see page A24.

c. In statistics mode, enter the class marks in the **L1** list and the frequency in the **L2** list. Set the window using the class interval for **Xscl**, and select the line graph as the type of graph.

[105, 210] scl:15 by [0, 10] scl:1

CHECK FOR UNDERSTANDING

Communicating Mathematics

Read and study the lesson to answer each question.   **1–2. See Answer Appendix.**

1. Compare and contrast line plot, bar graph, histogram, and frequency polygon.

2. Explain how to construct a frequency distribution.

Teaching Tip In Example 3, you may want students to sketch a histogram first, so that they can visualize the class marks before generating the frequency polygon.

 Teaching Tip If students use a graphing calculator to draw the histograms, stress the importance of clearing old data from the memory of a graphing calculator when creating a histogram. Only one set of data can be stored at a time. Keeping old data in the calculator will alter the results.

3. For explanations, see Answer Appendix.

4. See students' work.

Guided Practice

3. **Determine** which class intervals would be appropriate for the data. Explain.

55, 72, 51, 47, 73, 81, 74, 88, 83, 47, 58, 66, 64, 71, 73, 84, 61, 89, 73, 82

a. 1 **no**   b. 5 **yes**   c. 10 **yes**   d. 20 **no**   e. 30 **no**

4. *Math Journal* **Select** three graphs from newspapers or magazines. For each graph, write what conclusions might be drawn from the graph.

5. **Population** The table gives the percent of the U.S. population by age group.

Percent of U.S. Population

Age	0-9	10-19	20-29	30-39	40-49	50-59	60-69	70+
1900	14.8%	14.1%	16.3%	16.8%	12.6%	8.8%	8.6%	8.0%
1999	14.2%	14.4%	13.3%	15.6%	15.2%	10.7%	7.3%	9.3%

Source: U.S. Bureau of the Census

a. Make a back-to-back bar graph of the data. **See margin.**

b. Describe any trends indicated by the graph.

5b. In 1999, there are larger percents of older citizens than in 1900.

6. **History** The ages of the first 42 presidents when they first took office are listed.

57, 61, 57, 57, 58, 57, 61, 54, 68, 51, 49, 64, 50, 48, 65, 52, 56, 46, 54, 49, 50, 47, 55, 55, 54, 42, 51, 56, 55, 51, 54, 51, 60, 62, 43, 55, 56, 61, 52, 69, 64, 46

a. Find the range of the data. **27**

b. Determine an appropriate class interval. **Sample answer: 5**

c. What are the class limits? **Sample answer: 40, 45, 50, 55, 60, 65, 70**

d. Find the class marks. **Sample answer: 42.5, 47.5, 52.5, 57.5, 62.5, 67.5**

e. Construct a frequency distribution of the data. **See margin.**

f. Draw a histogram of the data. **See margin.**

g. Name the interval or intervals that describe the age of most presidents.
Sample answer: 50-60

EXERCISES

Applications and Problem Solving

7c. Sample answer: to determine where most of their customers live so they can target their advertising accordingly

8b. Sample answer: Men spend more hours driving than women.

7. **Sales** As customers come to the cash register at an electronics store, the sales associate asks them to give their ZIP code. During one hour, a sales associate gets the following responses.

43221, 43212, 43026, 43220, 43214, 43026, 43229, 43229, 43220, 45414, 43220, 43221, 43212, 43220, 43212, 43220, 43221, 43221, 43214, 43026

a. Make a line plot showing how many times each ZIP code was recorded. **a. See margin.**

b. Which ZIP code was recorded most frequently? **43220**

c. Why would a store want this type of information?

8. **Transportation** The average number of minutes men and women drivers spend behind the wheel daily is given below.

Age	16–19	20–34	35–49	50–64	65+
Men	58	81	86	88	73
Women	56	65	67	61	55

Source: Federal Highway Administration and the American Automobile Manufacturers Association

a. Make a back-to-back bar graph of the data. **See margin.**

b. What conclusions can you draw from the graph?

www.amc.glencoe.com/self_check_quiz

Lesson 14-1 *The Frequency Distribution* **893**

Additional Answer

8a.

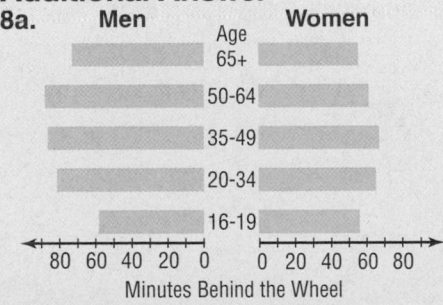

Error Analysis

Watch for students who confuse *class marks*, *limits*, and *intervals*.

Prevent by emphasizing that class marks are found by averaging the limits and that intervals are determined by the range of data.

Assignment Guide
Core: 7–11 odd, 13–22
Enriched: 7–22

Additional Answers

5a.

6e. Sample answer:

Ages	Frequency
40–45	2
45–50	6
50–55	12
55–60	12
60–65	7
65–70	3

6f. Sample answer:

7a.

Lesson 14-1 **893**

10e. Sample answer:

Grams of Fat	Frequency
0–10	7
10–20	11
20–30	10
30–40	7
40–50	2
50–60	1

10f. Sample answer:

Grams of Fat in Fast-Food Sandwiches

11e. Sample answer:

Number of Nations	Frequency
10–20	2
20–30	3
30–40	8
40–50	1
50–60	1
60–70	2
70–80	1

11f. Sample answer:

Olympic Winter Games

11g. Sample answer:

Olympic Winter Games

---

 9. Entertainment The table gives data on the rental revenue and the sale revenue of home videos as well as predictions of future revenues.

Year	Total Rental Revenue (in billions)	Total Sales Revenue (in billions)
1985	$2.55	$0.86
1990	$6.63	$3.18
1997	$7.46	$8.24
2000	$9.18	$9.76
2005	$9.26	$13.90

Source: *Video Software Dealers Association*

a. Make a back-to-back bar graph of the data.   **See margin.**

b. Which market, rental or sales, seems to have a better future? Explain.

9b. Sales; the sales revenue is growing at a faster rate than the rental revenue.

10. Nutrition The grams of fat in various sandwiches served by national fast-food restaurants are listed below.

18, 27, 15, 23, 27, 14, 15, 19, 39, 53, 31, 29, 12, 43, 38, 4, 10, 9, 21,
31, 31, 25, 28, 20, 22, 46, 15, 31, 16, 20, 30, 8, 18, 15, 7, 9, 5, 8

a. What is the range of the data?   **49**

b. Determine an appropriate class interval.   **Sample answer: 10**

c. Name the class limits.   **Sample answer: 0, 10, 20, 30, 40, 50, 60**

d. What are the class marks?   **Sample answer: 5, 15, 25, 35, 45, 55**

e. Construct a frequency distribution of the data.   **See margin.**

f. Draw a histogram of the data.   **See margin.**

g. Name the interval or intervals that describe the fat content of most sandwiches.   **Sample answer: 10-20**

11. Sports The number of nations represented at the first eighteen Olympic Winter Games are listed below.

Year	Place	Number of Nations	Year	Place	Number of Nations
1924	Chamonix, France	16	1968	Grenoble, France	37
1928	St. Moritz, Switzerland	25	1972	Sapporo, Japan	35
1932	Lake Placid, U.S.A.	17	1976	Innsbruck, Austria	37
1936	Garmisch-Partenkirchen, Germany	28	1980	Lake Placid, U.S.A.	37
1948	St. Moritz, Switzerland	28	1984	Sarajevo, Yugoslavia	49
1952	Oslo, Norway	30	1988	Calgary, Canada	57
1956	Cortina D'Ampezzo, Italy	32	1992	Albertville, France	64
1960	Squaw Valley, U.S.A.	30	1994	Lillehammer, Norway	67
1964	Innsbruck, Austria	36	1998	Nagano, Japan	72

Source: *The Complete Book of the Olympics*

a. Find the range of the data.   **56**

b. What is an appropriate class interval?   **Sample answer: 10**

c. What are the class limits?   **Sample answer: 10, 20, 30, 40, 50, 60, 70, 80**

d. What are the class marks?   **Sample answer: 15, 25, 35, 45, 55, 65, 75**

e. Construct a frequency distribution of the data.   **See margin.**

f. Draw a histogram of the data.   **See margin.**

g. Use the histogram to draw a frequency polygon.   **See margin.**

Additional Answer

9a.

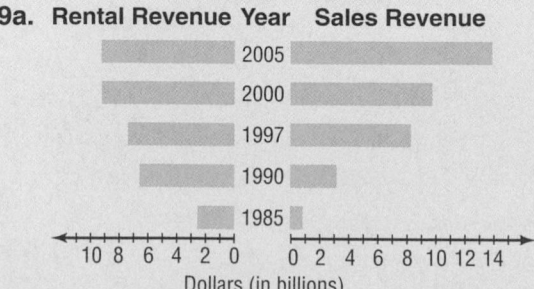

Rental Revenue   Year   Sales Revenue

Dollars (in billions)

12. Architecture The heights (in feet) of the tallest buildings in selected cities in the United States are listed below. Source: *The World Almanac, 1999*

| 1023 | 871 | 405 | 714 | 739 | 546 | 535 | 1000 | 626 | 1018 |
| 738 | 697 | 500 | 440 | 945 | 450 | 471 | 943 | 579 | 404 |

a. Find the range of the heights of these buildings.   **619**

b. What is an appropriate class interval?   **Sample answer: 100**

c. Construct a frequency distribution of the data.   **See margin.**

d. Draw a frequency polygon of the data.   **See margin.**

12e. Sample answer: 400–500

e. Which interval or intervals represent the greatest number of these buildings?

13. Baseball The greatest numbers of stolen bases for a single player are listed.

interNET CONNECTION

Data Update For the latest information about stolen bases, visit **www.amc. glencoe.com**

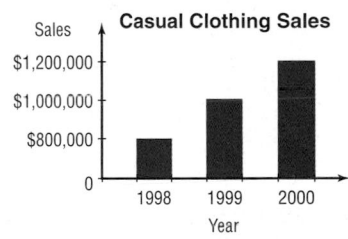

Greatest Number of Stolen Bases for a Single Player

	Year	'90	'91	'92	'93	'94	'95	'96	'97	'98	'99
Stolen Bases	American League	65	58	66	70	60	54	75	74	66	44
	National League	77	76	78	58	39	56	53	60	58	72

Source: *Information Please Almanac, 1999*

a. Make a back-to-back bar graph for the data.   **a–c. See margin.**

b. Combine data from both leagues to construct a frequency distribution.

c. Draw a histogram of the data. Then draw a frequency polygon.

13d. 7 players

d. How many players made 70 or more stolen bases to reach the record?

13e. 2 players

e. How many players made less than 50 stolen bases to reach the record?

14. Sample answer: 0.1, 0.2, 0.3, 0.4, 0.6, 0.7, 0.8, 0.9, 1.1, 1.2, 1.3, 1.4, 1.6, 1.7, 1.8, 1.9, 2.1, 2.2, 2.3, 2.4

14. Critical Thinking Create a set of data of 20 elements so that the data can be divided into five classes with class intervals of 0.5.

15. Geography The production of wheat, rice, and corn for 1997 is given in the table below. Make a three-dimensional bar graph of the data.

Country	Wheat (millions of tons)	Rice (millions of tons)	Corn (millions of tons)
China	122.6	198.5	105.4
India	68.7	123.0	9.8
United States	68.8	8.1	237.9

Source: UN Food and Agriculture Organization

See Answer Appendix.

16. Critical Thinking The graph shows a store's annual sales.

16a. See margin.

a. Why is the graph misleading?

b. Draw a graph of the data that is not misleading.   **See margin.**

16c. See students' work.

c. Find a graph in a newspaper or magazine that you believe is misleading. Explain why you believe it is misleading.

Casual Clothing Sales

Sales: $1,200,000 / $1,000,000 / $800,000 — Years: 1998, 1999, 2000

Extra Practice   See p. A53.

Extra Credit

Age Group of Internet Users	Percent of Urban Users	Percent of Rural Users
Under 25	22.0	13.3
25–34	31.6	24.2
35–44	35.4	30.2
45–54	36.5	30.8
55+	15.4	12.4

Source: *World Almanac and Book of Facts*

Have students create frequency polygons for the percents of rural and urban Internet users on the same grid. Ask students to write a statement comparing the two polygons. **See students' graphs; a higher percent of urban households than rural households used the Internet in the United States.**

12c. Sample answer:

Height (feet)	Frequency
400–500	5
500–600	4
600–700	2
700–800	3
800–900	1
900–1000	2
1000–1100	3

12d. Sample answer:

13a. Sample answer:

Greatest Number of Stolen Bases for a Single Player

13b. Sample answer:

Stolen Bases	Frequency
30–40	1
40–50	1
50–60	6
60–70	5
70–80	7

13c. Sample answer:

(continued on page 896)

Open-Ended Assessment

Writing Provide students with a hypothetical set of quiz grades and have them write a paragraph describing how to organize the data into a frequency chart and how to make a histogram of the data.

Additional Answers
(pp. 895–896)

16a. The first interval on the vertical axis represents $800,000, but the other intervals represent only $200,000. Therefore, the sales for 1999 appear to be twice the sales of 1998, but in reality they are not.

16b.

21.

Enrichment Masters, p. 157

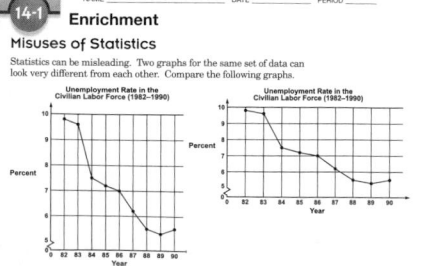

---

17. Statistics Design a survey and ask your classmates to respond to the survey. Make an appropriate graph to depict the results of the survey. **See students' work.**

Mixed Review

18. Horticulture The survival rate of a variety of mums in a certain area of the country is 80%. If 8 mums are planted, what is the probability that exactly 6 will survive? *(Lesson 13-6)* **about 29%**

19. Find the second term of $(c - 2d)^7$. *(Lesson 12-6)* $-14c^6d$

20. Solve $3.6^x = 58.9$ by using logarithms. *(Lesson 11-6)* **about 3.18**

21. Graph $9xy = -36$. *(Lesson 10-4)* **See margin.**

22. SAT Practice Grid-In If $x^2 + y^2 = 16$ and $xy = 8$, what is $(x + y)^2$? **32**

CAREER CHOICES

 Accountant

Everyone seems to like money. If you choose a career in accounting, you will be working with money, but only on paper. Accountants work with the financial records of individuals, businesses, or governments and prepare statements showing income and expenses. Accountants also prepare reports, including tax reports.

As an accountant, you may choose to specialize in an area that might include auditing or systems and procedures, or you may specialize in a particular business such as agriculture. To become a certified public accountant, you must pass an examination to be certified by the state. Accountants can work for businesses, for government, or work independently for any individual or business that desires their services.

CAREER OVERVIEW

Degree Preferred:
Bachelor's degree in accounting

Related Courses:
mathematics, communications, computer science, business courses

Outlook:
number of jobs expected to increase faster than the average through 2006

Consolidated Balance Sheets

Assets *(in thousands)*	July 4 1999	January 3 1999
Cash and cash equivalents	$ 245,159	$ 160,743
Accounts and notes receivable, net	78,865	94,689
Deferred income taxes	19,592	23,177
Inventories and other	35,181	35,085
Total current assets	378,797	313,694

interNET CONNECTION For more information on careers in accounting, visit **www.amc.glencoe.com**

CAREER CHOICES   ACCOUNTANT

Math Connection
- What is the percent of change for the cash and cash equivalents of the company from January 3, 1999 to July 4, 1999? **about 52.5%**
- What is the percent of change for the total current assets for the company from January 3, 1999 to July 4, 1999? **about 20.8%**
- Why are these two percents not near equal? **The increase in cash was proportionally greater than that for total current assets.**

Measures of Central Tendency

OBJECTIVES
- Find the mean, median, and mode of a set of data.
- Find measures of central tendency of data organized in a stem-and-leaf plot or a frequency distribution table.

ENTERTAINMENT For many years, people have attended Broadway plays in New York City. Broadway became an important theatrical district in the mid-1800s. The theatrical activity of Broadway peaked in the 1920s. Today, there are fewer new shows, but Broadway still remains a major theatrical center. The numbers of new Broadway productions in recent seasons are listed below. What is the average number of new Broadway productions for these seasons?

This problem will be solved in Example 3.

Season	New Productions	Season	New Productions	Season	New Productions
1960–1961	48	1973–1974	43	1986–1987	41
1961–1962	53	1974–1975	54	1987–1988	32
1962–1963	54	1975–1976	55	1988–1989	30
1963–1964	63	1976–1977	54	1989–1990	35
1964–1965	67	1977–1978	42	1990–1991	28
1965–1966	68	1978–1979	50	1991–1992	37
1966–1967	69	1979–1980	61	1992–1993	33
1967–1968	74	1980–1981	60	1993–1994	37
1968–1969	67	1981–1982	48	1994–1995	29
1969–1970	62	1982–1983	50	1995–1996	38
1970–1971	49	1983–1984	36	1996–1997	37
1971–1972	55	1984–1985	33	1997–1998	33
1972–1973	55	1985–1986	33	1998–1999	20

Source: The League of American Theatres and Producers, Inc.

The average number of new Broadway productions is an ambiguous term. Loosely stated, the average means the center of the distribution or the most typical case. Measures of average are also called **measures of central tendency** and include the **mean, median,** and **mode.**

The **arithmetic mean** $\overline{X}$ is often referred to as the mean. The mean is found by adding the values in a set of data and dividing the sum by the number of values in that set. Every number in a set of data affects the value of the mean. Consequently, the mean is generally a good representative measure of central tendency. However, the mean can be considerably influenced by extreme values.

Lesson 14-2 Measures of Central Tendency **897**

---

1 FOCUS

5-Minute Check
Lesson 14-1

The weights of 76 boys in a wrestling meet are tallied in the chart below.

Class Limits	Tally	f
80–90	||	2
90–100	|||	3
100–110	ЖΤΙ	6
110–120	ЖΤ ЖΤΙΙ	12
120–130	ЖΤ ЖΤ ЖΤ ЖΤ	20
130–140	ЖΤ ЖΤ ЖΤ ΙΙΙ	18
140–150	ЖΤ ЖΤ Ι	11
150–160	ΙΙΙΙ	4
160–170	Ι	1

1. Determine whether class intervals of 10 are appropriate for the data.
yes

2. Find the class marks.
85, 95, 105, 115, 125, 135, 145, 155, 165

3. Draw a histogram of the data.

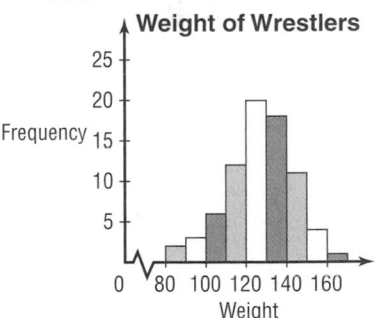

4. Draw a frequency polygon of the data.

Resource Manager

 Reproducible Masters
- *Study Guide and Practice,* pp. 257–258
- *Enrichment,* p. 158
- *Assessment and Evaluation,* p. 355
- *Problem Solving and Applications,* p. 28
- *Spreadsheet Application,* p. 14

 Transparencies
- 5-Minute Check, 14-2
- Teaching, 14-2

 Technology/Multimedia
- Advanced MathPASS, Lesson 29

Discussion Ask students to list situations in which averages are used. For example, grades and sports statistics are often of interest to students. Ask students in what ways knowing an average helps them interpret information. **Sample answer: it is easier to compare and contrast individual grades or players when you know the averages.**

2 TEACH

Teaching Tip Students should be familiar with sigma notation. Refer them to Lesson 12-5 and work the following problems for review.

1. Rewrite $\sum_{n=1}^{5} (n - 1)$ in expanded form. **$0 + 1 + 2 + 3 + 4$**

2. Evaluate $\sum_{n=1}^{5} 2n$. **30**

In-Class Example

Example 1
Find the mean of the set {36.8, 29.5, 29.1, 33.3, 30.0, 20.7, 39.5}. **about 31.3**

Example **1** Find the mean of the set {19, 21, 18, 17, 18, 22, 46}.

Notice that the mean is not necessarily a member of the set of data.

$$\overline{X} = \frac{\text{sum of the values in the set of data}}{\text{number of values in the set}}$$

$$\overline{X} = \frac{19 + 21 + 18 + 17 + 18 + 22 + 46}{7}$$

$$\overline{X} = \frac{161}{7} \text{ or } 23$$

The mean of the set of data is 23.

The general formula for the mean of any set of data can be written using sigma notation. If X is a variable used to represent any value in a set of data containing n items, then the arithmetic mean $\overline{X}$ of n values is given by the following formula.

$$\overline{X} = \frac{X_1 + X_2 + X_3 + \cdots + X_n}{n}$$

The numerator of the fraction can be abbreviated using the summation symbol Σ. *Recall that Σ is the uppercase Greek letter sigma.*

$$\sum_{i=1}^{n} X_i = X_1 + X_2 + X_3 + \cdots + X_n$$

The symbol X_i represents successive values of the set of data as i assumes successive integral values from 1 to n. Substitute the sigma notation into the formula for the mean to obtain the formula below.

$$\overline{X} = \frac{\sum_{i=1}^{n} X_i}{n} \text{ or } \overline{X} = \frac{1}{n} \sum_{i=1}^{n} X_i$$

Arithmetic Mean

If a set of data has n values given by X_i such that i is an integer and $1 \le i \le n$, then the arithmetic mean $\overline{X}$ can be found as follows.

$$\overline{X} = \frac{1}{n} \sum_{i=1}^{n} X_i$$

Another measure of central tendency is the median, symbolized by M_d.

Median

The median of a set of data is the middle value. If there are two middle values, it is the mean of the two middle values.

Notice that the median is not necessarily a member of the set of data.

Before the median can be found, the data must be arranged in an ordered sequence, usually from least to greatest. The median of the set {5, 6, 8, 11, 14} is the middle value 8. The median of the set {3, 4, 6, 7, 8, 10} is $\frac{6 + 7}{2}$ or 6.5.

The median is preferable to the mean as a measure of central tendency when there are a few extreme values or when some of the values cannot be determined. Unlike the mean, the median is influenced very little by extreme values.

| Mode | The mode of a set of data is the most frequent value. Some sets of data have multiple modes and others have no mode. |

Data with two modes are **bimodal.** Sets have no mode when each item of the set has equal frequency. The value of the mode is not affected by extreme values. Unlike the mean and median, the mode, if it exists, is always a member of the set of data.

Example

Real World Application

2 IMMIGRATION The numbers of immigrants entering the United States in a recent year are given at the right.

a. Find the mean of the data.

b. Find the median of the data.

c. Find the mode of the data.

d. State which measures of central tendency seem most representative of the set of data. Explain.

Immigrants Entering the United States

Country	Number of Immigrants
China	41,700
Cuba	26,500
Dominican Republic	39,600
India	44,900
Jamaica	19,100
Mexico	163,600
Philippines	55,900
Russia	19,700

Source: U.S. Immigration and Naturalization Service

a. Since there are 8 countries, $n = 8$.

$$\frac{1}{8}\sum_{i=1}^{8} X_i = \frac{1}{8}(41,700 + 26,500 + 39,600 + 44,900 + 19,100 + 163,600 + 55,900 + 19,700) \text{ or } 51,375$$

The mean is 51,375.

b. To find the median, order the data. Since all the numbers are multiples of 100, you can order the set by hundreds.

191   197   265   396   417   449   559   1636

Since there are an even number of data, the median is the mean of the two middle numbers, 39,600 and 41,700. The median number of immigrants is $\frac{39,600 + 41,700}{2}$ or 40,650.

c. Since all elements in the set of data have the same frequency, there is no mode.

d. Notice that the mean is affected by the extreme value 163,600 and does not accurately represent the data. The median is a more representative measure of central tendency in this case.

When you have a large number of data, it is often helpful to use a **stem-and-leaf plot** to organize your data. In a stem-and-leaf plot, each item of data is separated into two parts that are used to form a **stem** and a **leaf.** The parts are organized into two columns.

In-Class Example

Example 3

Education The following are the grades scored on a quiz with 50 possible points.

42, 49, 36, 32, 10, 19, 38, 40, 41, 50, 40, 49, 30, 20, 48, 47, 40, 41, 32, 37, 25, 41, 43, 37, 39

a. Make a stem-and-leaf plot of the quiz results.

stem	leaf
1	0 9
2	0 5
3	0 2 2 6 7 7 8 9
4	0 0 0 1 1 1 2 3
	7 8 9 9
5	0

$1|0 = 10$

b. Find the mean of the data.
about 37.04

c. Find the median of the data.
40

d. Find the mode of the data.
40, 41

e. What is a good representative number for the average of the quiz results? **40**

Stems: The column on the left shows the stems. Stems usually consist of the digits in the greatest common place value of all the data. For example, if the set of data includes the numbers 890 and 1160, the greatest common place value is hundreds. Therefore, the stem of 890 is 8, and the stem of 1160 is 11.

Leaves: The column at the right contains the leaves. The leaves are one-digit numbers, which are in the next greatest place value after the stem. The leaf of 890 is 9, and the leaf of 1160 is 6. The stems and leaves are usually arranged from least to greatest.

Example **3 ENTERTAINMENT** Refer to the application at the beginning of the lesson.

a. Make a stem-and-leaf plot of the number of new productions for the seasons listed.

b. Find the mean of the data.

c. Find the median of the data.

d. Find the mode of the data.

e. What is a good representative number for the average of the new Broadway productions for the seasons 1960-1999?

a. Since the number of productions range from 20 to 74, we will use the tens place for the stems. List the stems and draw a vertical line to the right of the stems. Then list the leaves, which in this case will be the ones digit. As shown below, it is often helpful to list the leaves as you come to them and then rewrite the plot with the leaves in order from the least to greatest.

An annotation usually accompanies a stem-and-leaf plot to give meaning to the representation.

stem	leaf
2	8 9 0
3	6 3 3 2 0 5 7 3 7 8 7 3
4	8 9 3 2 8 1
5	3 4 5 5 4 5 4 0 0
6	3 7 8 9 7 2 1 0
7	4

➡

stem	leaf
2	0 8 9
3	0 2 3 3 3 3 5 6 7 7 7 8
4	1 2 3 8 8 9
5	0 0 3 4 4 4 5 5 5
6	0 1 2 3 7 7 8 9
7	4

$2|8 = 28$

b. Enter the data in the **L1** list of a graphing calculator. Use the statistics mode of the calculator to find $\overline{X}$. The mean is 47.

```
1-Var Stats
x̄=47
Σx=1833
Σx²=93285
Sx=13.70170946
σx=13.5249058
↓n=39
■
```

c. Since the median is the middle value, it is the 20th leaf on the plot. The median is 48.

d. The stem-and-leaf plot shows the modes by repeated digits for a particular stem. There are four 3s with the stem 3. The mode is 33.

e. Although 33 is the most common number in the data, it is not a central number for the data. In this case, the mean and median seem to be more representative of the data. Therefore, a representative average number of new Broadway productions could be either 47 or 48.

In a frequency distribution containing large amounts of data, each individual value in the set of data loses its identity. The data in each class are assumed to be uniformly distributed over the class. Thus, the class mark is assumed to be the mean of the data tallied in its class. For example, the mean of the data in the class with limits 17.5-22.5 is assumed to be 20, the class mark.

In the frequency distribution, the sum of the values in a class is found by multiplying the class mark X by the frequency f of that class. The sum of all the values in a given set of data is found by adding the sums of the values of each class in the frequency distribution. The sum of all values in the set can be

Remember that the measures of central tendency are only representations of the set of data.

represented by $\sum_{i=1}^{k} (f_i \cdot X_i)$, where k is the number of classes in the frequency distribution. Thus, the arithmetic mean of n values in a frequency distribution is found by dividing the sum of the values in the set by n or an expression equivalent to n, such as $\sum_{i=1}^{k} f_i$.

Mean of the Data in a Frequency Distribution

If X_1, X_2, ..., X_k are the class marks in a frequency distribution with k classes and f_1, f_2, ..., f_k are the corresponding frequencies, then the arithmetic mean $\overline{X}$ can be approximated as follows.

$$\overline{X} = \frac{\sum_{i=1}^{k} (f_i \cdot X_i)}{\sum_{i=1}^{k} f_i}$$

Example

4 **EDUCATION** Estimate the mean of the scores of 100 students on an algebra test given the following frequency distribution.

Class Limits	Class Marks (X)	Frequency (f)	fX
97.5-102.5	100	5	500
92.5-97.5	95	9	855
87.5-92.5	90	17	1530
82.5-87.5	85	26	2210
77.5-82.5	80	22	1760
72.5-77.5	75	10	750
67.5-72.5	70	7	490
62.5-67.5	65	4	260

$$\sum_{i=1}^{8} f_i = 100 \qquad \sum_{i=1}^{8} (f_i \cdot X_i) = 8355$$

$\overline{X} = \dfrac{8355}{100}$ or 83.55   The mean is approximately 84.

Lesson 14-2   Measures of Central Tendency **901**

In-Class Example

Example 4
Basketball Estimate the mean of the scoring averages of 30 basketball players during the 1998–99 season given the following frequency distribution.

Class Limits	Class Marks (X)	Frequency (f)	fX
10–12	11	1	11
12–14	13	3	39
14–16	15	5	75
16–18	17	12	204
18–20	19	5	95
20–22	21	4	84

about 17 points per game

In-Class Example

Example 5
Basketball Estimate the median of the data in the frequency distribution in In-Class Example 4. **about 17 points per game**

The median M_d of the data in a frequency distribution is found from the **cumulative frequency distribution**. The cumulative frequency of each class is the sum of the frequency of the class and the frequencies of the previous classes. The chart shows the cumulative frequency for the data in Example 4.

It is often helpful to calculate the cumulative frequency from the last interval to the first.

Class Limits	Frequency f	Cumulative Frequency
97.5–102.5	5	100
92.5–97.5	9	95
87.5–92.5	17	86
82.5–87.5	26	69
77.5–82.5	22	43
72.5–77.5	10	21
67.5–72.5	7	11
62.5–67.5	4	4

For the class limit 77.5-82.5, the cumulative frequency equals $22 + 10 + 7 + 4$ or 43. This means that 43 algebra test scores fall below 82.5.

Since the median is the value below which 50% of the data lie, the class in which the median lies can be located. This class is called the **median class.** The median can be found by using an estimation technique called interpolation. *This method can also be used to find a score at any percent level.*

Example **⑤ EDUCATION   Estimate the median of the data in the frequency distribution in Example 4.**

Since there are 100 scores in this frequency distribution, 50 scores are below the median and 50 are above. From the chart above, find the least cumulative frequency that is greater than or equal to 50. That cumulative frequency is 69. So, the median class is 82.5-87.5.

You can use a proportion to find the value of M_d by finding the ratios of the differences in the cumulative frequencies and the upper limits of the classes.

$\dfrac{26}{5} = \dfrac{7}{x}$

$\quad x \approx 1.346153846$ *Use a calculator.*

$M_d - 82.5 = x$

$M_d - 82.5 \approx 1.3 \qquad x \approx 1.3$

$\qquad M_d \approx 83.8$

The median of the data is approximately 83.8.

Addressing Individual Needs

Kinesthetic   Use colored, interlocking building blocks to represent five different positive 2-digit numbers. Have students experiment using the blocks to calculate and model the mean, mode, median, and range of the data. For example, one color could represent 1, another 10, another 100, and another 0.5. To find the mean, they could move the blocks from larger piles to smaller piles until all of the piles were equal.

Communicating Mathematics

Read and study the lesson to answer each question.

1. **Name** three measures of central tendency. Explain how to determine each measure.   **Mean, median, mode; see margin for explanations.**

2. **Write** a set of data that is bimodal.   **Sample answer: {1, 2, 2, 3, 4, 4, 5}**

3. **Explain** how to make a stem-and-leaf plot for a set of data whose greatest value is 1430 and least value is 970.   **See margin.**

4. **You Decide**   Eight people volunteer at Central City Recreation Center. The numbers of hours the volunteers spent at the center last week are 2, 3, 15, 2, 2, 23, 19, and 2. Omar says the median is the best representative for the number of volunteer hours. Tia disagrees and claims that the mean is a better representation. Casey says the most representative number is the mode. Who is correct? Explain.   **Tia; see margin for explanation.**

Guided Practice

Find the mean, median, and mode of each set of data.

5. {10, 45, 58, 10}   **30.75; 27.5; 10**

6. {24, 28, 21, 37, 31, 29, 23, 22, 34, 31}   **28; 28.5; 31**

7. Find the mean, median, and mode of the data represented by the stem-and-leaf plot at the right.   **about 10,323; 10,500; 10,700**

stem	leaf
9	1 4 5 8 9
10	5 5 7 7 7
11	1 1 2

$9 \mid 1 = 9100$

8. The Laketown Senior Center recorded how many times its members use the center's fitness facilities during a typical month. They organized their data into the following chart.   **a. 120 members**

Visits	1–5	5–9	9–13	13–17	17–21	21–25	25–29	29–33
Members	2	8	15	6	38	31	13	7

   a. How many members used the center's fitness facilities during the month?
   b. Estimate the mean of the data.   **about 19.3**
   c. What is the median class of the data?   **17-21**
   d. Estimate the median of the data.   **about 20.1**

9. **Football**   Each December, the Liberty Bowl is played in Memphis, Tennessee. The winning scores of the first 40 Liberty Bowl games are listed below.

7, 41, 15, 6, 16, 32, 13, 14, 14, 34, 47, 17, 7, 31, 31, 7, 20, 36, 21, 20, 9, 28, 31, 21, 19, 21, 21, 21, 20, 34, 42, 23, 38, 13, 18, 30, 19, 41, 41, 23

   a. Make a stem-and-leaf plot of the winning scores.   **See margin.**
   b. What is the mean of the data?   **23.55**
   c. What is the median of the data?   **21**
   d. Find the mode of the data.   **21**
   e. What is the most representative measure of central tendency for the number of points scored by the winning team at the Liberty Bowl? Explain.

9e. Since the mean 23.55, the median 21, and the mode 21 are all representative values, any of them could be used as an average.

 www.amc.glencoe.com/self_check_quiz

Additional Answer

1. To find the mean, add the values in a set of data and divide the sum by the number of values in the set. To find the median, arrange the values in a set of data from least to greatest. If there is an odd number of values in the set, the median is the middle value. If there is an even number of values in the set, the median is the mean of the two middle values. To find the mode, find the item of data that appears more frequently than any other in the set.

Error Analysis

Watch for students who make mistakes when calculating cumulative frequencies.

Prevent by encouraging them to be consistent when setting up tables (for example, they should always add in the same direction).

Additional Answers

3. **Write the stems 9, 10, 11, 12, 13, and 14 on the left. Write the tens digits as leaves to the right of the appropriate stems. Be sure to order the leaves.**

4. **The median 2.5 and the mode 2 do not represent the greater numbers. The mean 8.5 is more representative of all 8 items in the data.**

9a.

stem	leaf
0	6 7 7 7 9
1	3 3 4 4 5 6 7 8 9 9
2	0 0 0 1 1 1 1 1 3 3 8
3	0 1 1 1 2 4 4 6 8
4	1 1 1 2 7

$1 \mid 3 = 13$

Study Guide and Practice Masters, pp. 257–258

Additional Answer

20.

stem	leaf
1	0 5 5 5 5 7 7
2	0 0 0 5 5 5 5 7 8
3	0 0 5 5 5
4	6
5	5

$1|0 = 10$

EXERCISES

Practice

Find the mean, median, and mode of each set of data.

10. {140, 150, 160, 170} **155; 155; none** 11. {3, 3, 6, 12, 3} **5.4; 3; 3**

13. 10.75; 11; 5 and 18 12. {21, 19, 17, 19} **19; 19; 19** 13. {5, 8, 18, 5, 3, 18, 14, 15}

14. {64, 87, 62, 87, 63, 98, 76, 54, 87, 58, 70, 76} **73.5; 73; 87**

15. {6, 9, 11, 11, 12, 7, 6, 11, 5, 8, 10, 6} **8.5; 8.5; 6 and 11**

16. Crates of books are being stored for later use. The weights of the crates in pounds are 142, 160, 151, 139, 145, 117, 172, 155, and 124.
   a. What is the mean of their weights? **145 lb**
   b. Find the median of their weights. **145 lb**
   c. If 5 pounds is added to each crate, how will the mean and median be affected? **Each will increase by 5 lb.**

Find the mean, median, and mode of the data represented by each stem-and-leaf plot.

17.
stem	leaf
3	5 8 8 9
4	4 5 5 5 8
5	7 7 9

$3|5 = 35$
about 45.8; 45; 45

18.
stem	leaf
5	2 4 6
6	0 1 7 8 9
7	1 6
8	0 2 6
9	1

$5|2 = 5.2$
6.95; 6.85; none

19.
stem	leaf
9	0 1 7 8 9
10	5 6 9
11	3 8 8 8
12	0 5 5

$9|0 = 900$
1088; 1090; 1180

20. Make a stem-and-leaf plot of the following ages of people attending a family picnic. **See margin.**

15, 55, 35, 46, 28, 35, 25, 17, 30, 30, 27, 35,
15, 25, 25, 20, 20, 15, 20, 17, 15, 25, 10

21. The store manager of a discount department store is studying the weekly wages of the part-time employees. The table profiles the employees.

21a. $1485, $3480, $4650, $1650, $2275, $1480, $780

   a. Find the sum of the wages in each class.
   b. What is the sum of all of the wages in the frequency distribution? **$15,800**
   c. Find the number of employees in the frequency distribution. **100 employees**
   d. What is the mean weekly wage in the frequency distribution? **about $158**
   e. Find the median class of the frequency distribution. **$150-$160**
   f. Estimate the median weekly wage in the frequency distribution. **$155**
   g. Explain why both the mean and median are good measures of central tendency in this situation. **Both values represent central values of the data.**

Weekly Wages	Frequency
$130–$140	11
$140–$150	24
$150–$160	30
$160–$170	10
$170–$180	13
$180–$190	8
$190–$200	4

22. Find the value of x so that the mean of {2, 4, 5, 8, x} is 7.5. **18.5**

23. What is the value of x so that the mean of {x, $2x - 1$, $2x$, $3x + 1$} is 6? **3**

24. Find the value of x so that the median of {11, 2, 3, 3.2, 13, 14, 8, x} is 8. **8**

25. The frequency distribution of the verbal scores on the SAT test for students at Kennedy High School is shown below.

Scores	Number of Students	Scores	Number of Students
200–250	9	500–550	18
250–300	14	550–600	12
300–350	23	600–650	7
350–400	30	650–700	3
400–450	33	700–750	1
450–500	28	750–800	1

25a. about 425.6

25c. about 420.5

  a. What is the mean of the verbal scores at Kennedy High School?

  b. What is the median class of the frequency distribution? **400–450**

  c. Estimate the median of the verbal scores at Kennedy High School.

26a. 3.92; 3.7; 3.6

26b. Only the mean would change. It would increase to 4.6.

27a–d. Sample answers are given.

26. **Weather**   The growing season in Tennessee is the period from May to September. The table at the right shows the normal rainfall for those months.

  a. Find the mean, median, and mode of this data.

  b. Suppose Tennessee received heavy rain in May totaling 8.2 inches. If this figure were used for May, how would the measures of central tendency be affected?

  c. If September were eliminated from the period, how would this affect the measures of central tendency?   **The mean increases slightly; the median increases slightly; the mode stays the same.**

Normal Rainfall for Tennessee (inches)	
May	4.8
June	3.6
July	3.9
August	3.6
September	3.7

27. **Critical Thinking**   Find a set of numbers that satisfies each list of conditions.

  a. The mean, median, and mode are all the same number.   **{1, 2, 2, 2, 3}**

  b. The mean is greater than the median.   **{4, 5, 9}**

  c. The mode is 10 and the median is greater than the mean.   **{2, 10, 10, 12}**

  d. The mean is 6, the median is $5\frac{1}{2}$, and the mode is 9.   **{3, 4, 5, 6, 9, 9}**

28. **Government**   As of 1999, the number of members in the House of Representatives for each state is given below.

AL	7	HI	2	MA	10	NM	3	SD	1
AK	1	ID	2	MI	16	NY	31	TN	9
AZ	6	IL	20	MN	8	NC	12	TX	30
AR	4	IN	10	MS	5	ND	1	UT	3
CA	52	IA	5	MO	9	OH	19	VT	1
CO	6	KS	4	MT	1	OK	6	VA	11
CT	6	KY	6	NE	3	OR	5	WA	9
DE	1	LA	7	NV	2	PA	21	WV	3
FL	23	ME	2	NH	2	RI	2	WI	9
GA	11	MD	8	NJ	13	SC	6	WY	1

  a. Make a stem-and-leaf plot of the number of representatives.   **See margin.**

  b. Find the mean of the data.   **8.7**

  c. What is the median of the data?   **6**

  d. Find the mode of the data.   **1**

  e. What is a representative average for the number of members in the House of Representatives per state? Explain.

The mean 8.7 and the median 6 are representative of the data, but the mode 1 is not representative of the data.

28a.

stem	leaf
0	1 1 1 1 1 1 1 2 2
	2 2 2 2 3 3 3 3 4
	4 5 5 5 6 6 6 6 6
	6 7 7 8 8 9 9 9 9
1	0 0 1 1 2 3 6 9
2	0 1 3
3	0 1
4	
5	2

$5|2 = 52$

interNET
CONNECTION

Data Update
For the latest information about the number of goals scored in hockey, visit **www.amc. glencoe.com**

29. Hockey A frequency distribution for the number of goals scored by teams in the National Hockey League during a recent season are given at the right.

National Hockey League Goals

Goals	Number of Teams
160–180	1
180–200	6
200–220	10
220–240	6
240–260	3
260–280	1

Source: National Hockey League

a. Use the frequency chart to estimate the mean of the number of goals scored by a team. **about 215.2**

b. What is the median class of the frequency distribution? **200-220**

c. Use the frequency chart to estimate the median of the number of goals scored by a team. **about 213**

d. The actual numbers of goals scored are listed below. Find the mean and median of the data. **about 215.9; 211**

268, 248, 245, 242, 239, 239, 237, 236, 231, 230, 217, 215, 214, 211, 210, 210, 207, 205, 202, 200, 196, 194, 192, 190, 189, 184, 179

e. How do the measures of central tendency found by using the frequency chart compare with the measures of central tendency found by using the actual data? **See margin.**

30. Critical Thinking A one-meter rod is suspended at its middle so that it balances. Suppose one-gram weights are hung on the rod at the following distances from one end.

5 cm   20 cm   37 cm   44 cm   52 cm   68 cm   71 cm   85 cm

The rod does not balance at the 50-centimeter mark.

a. Where must a one-gram weight be hung so that the rod will balance at the 50-centimeter mark? **68 cm from the end**

b. Where must a two-gram weight be hung so that the rod will balance at the 50-centimeter mark? **59 cm from the end**

31f. Median; the mean is affected by the extreme values of $162,000 and $226,000 and only two people make less than the mode.

31g. Sample answer: I have been with the company for many years, and I am still making less than the mean salary.

31. Salaries The salaries of the ten employees at the XYZ Corporation are listed below.

$54,000, $75,000, $55,000, $62,000, $226,000, $65,000, $59,000, $61,000, $162,000, $59,000

a. What is the mean of the salaries? **$87,800**

b. Find the median of the salaries. **$61,500**

c. Find the mode of the salaries. **$59,000**

d. What measure of central tendency might an employee use when asking for a raise? **mean**

e. What measure of central tendency might management use to argue against a raise for an employee? **mode**

f. What measure of central tendency do you think is most representative of the data? Why?

g. Suppose you are an employee of the company making $75,000. Write a convincing argument that you deserve a raise.

32. Education The grade point averages for a graduating class are listed in the frequency table below.

Grade Point Averages	1.75–2.25	2.25–2.75	2.75–3.25	3.25–3.75	3.75–4.25
Frequency	12	15	31	37	5

a. What is the estimated mean of the data? **about 3.04**
b. Estimate the median of the data. **about 2.96**

33. Basketball Jackson High School just announced the members of its varsity basketball team for the year. Kwan, who is 5′ 9″ tall, is the only sophomore to make the team. The other basketball team members are 5′ 11″, 6′ 0″, 5′ 7″, 6′ 3″, 6′ 1″, 6′ 6″, 5′ 8″, 5′ 9″ and 6′ 2″. How does Kwan compare with the other team members? **He is shorter than the mean (5′ 11.6″) and the median (5′ 11.5″).**

Mixed Review

34. Highway Safety The maximum speed limits in miles per hour for interstate highways for the fifty states are given below. Construct a frequency polygon of the data. *(Lesson 14-1)*
70, 65, 75, 70, 70, 75, 65, 65, 70, 70, 55, 75, 65, 65, 65, 70, 65, 70, 65, 65, 65, 70, 70, 70, 70, 65, 75, 75, 65, 65, 75, 65, 70, 70, 65, 75, 65, 65, 65, 65, 75, 65, 70, 75, 65, 65, 70, 70, 65, 75   **See margin.**

Source: National Motorists Association

35. Determine if the following event is *independent* or *dependent*. Then determine the probability. *(Lesson 13-4)*   **dependent; $\frac{3}{55}$**
the probability of randomly selecting two fitness magazines at one time from a basket containing 6 news magazines, 3 fitness magazines, and 2 sports magazines

36. Use the ratio test to determine if the series $\frac{1}{3} + \frac{2}{3^2} + \frac{3}{3^3} + \cdots + \frac{n}{3^n} + \cdots$ is *convergent* or *divergent*. *(Lesson 12-4)*   **convergent**

37. Investments An annuity pays 6%. What is the future value of the annuity if $1500 is deposited into the account every 6 months for 10 years? *(Lesson 11-2)*   **$40,305.56**

38. Graph the system of inequalities. *(Lesson 10-8)*   **See margin.**
$3x + y^2 \leq 18$
$x^2 + y^2 \geq 9$

39. SAT Practice The area of $\triangle ABC$ is between which pair of numbers?   **A**
A 16 and 17
B 15 and 16
C 12 and 13
D 10 and 11
E 9 and 10

Extra Practice See p. A53.

Extra Credit

Approximate the mean of the data shown in the histogram.
about 28.4

Open-Ended Assessment
Speaking Have students discuss the uses of measure of central tendency in business, science, and social studies.

Chapter 14, Quiz A
(Lessons 14-1 and 14-2)
is available in the *Assessment and Evaluation Masters*, p. 355.

Additional Answers
34.

38.

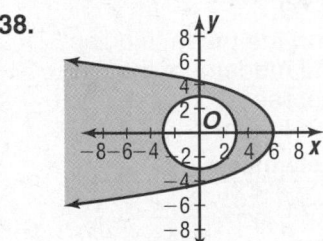

Enrichment Masters, p. 158

1 FOCUS

5-Minute Check
Lesson 14-2

1. Find the mean, median, and mode(s) for the following set of data: {10, 12, 15, 12, 16, 20, 21, 18, 16}. **mean: about 15.6; median: 16; modes: 12, 16**

2. Make a stem-and-leaf plot of the data in Exercise 1.

stem	leaf
1	0 2 2 5 6 6 8
2	0 1

$2 | 0 = 20$

3. Juan has scores of 84, 72, 91, 87, and 90 in physics. The class mean is 88. How does his mean compare? **lower**

4. Find the mean, median, and mode(s) of the data represented by the stem-and-leaf plot.

stem	leaf
7	0 3 4
8	1 1
9	0 2 8
10	1 3 5 5
11	0

$7 | 0 = 70$

mean: 91, median: 92, modes: 81, 105

5. The heights (in inches) of the students on the gymnastics team are 60, 58, 63, 62, 62.5, 61.5, 60, 60.5, and 60.

   a. What is the mean of their heights? **about 60.8 in.**

   b. What is the median of their heights? **60.5 in.**

   c. If each person grew 2 inches taller, how would the mean and median be affected? **Each would increase by 2 inches.**

14-3

Measures of Variability

OBJECTIVES
- Find the interquartile range, the semi-interquartile range, mean deviation, and standard deviation of a set of data.
- Organize and compare data using box-and-whisker plots.

***inter*NET CONNECTION**

Data Update
For the latest information about college enrollment and tuition, visit **www.amc. glencoe.com**

EDUCATION Are you planning to attend college? If so, do you know which school you are going to attend? There are several factors influencing students' decisions concerning which college to attend. Two of those factors may be the cost of tuition and the size of the school. The table lists some of the largest colleges with their total enrollment and cost for in-state tuition and fees.

College Enrollment and Tuition

College	Enrollment, 1997-1998	Tuition and Fees ($), 1997-1998
University of Texas	47,476	2866
The Ohio State University	45,462	3687
Penn State University	37,718	5832
University of Georgia	29,693	2838
Florida State University	28,285	1988
University of Southern California	27,874	20,480
Virginia Tech	24,481	4147
North Carolina State University	24,141	2232
Texas Tech University	24,075	2414
University of South Carolina	22,836	3534
University of Nebraska	22,393	2769
Colorado State University	21,970	2933
University of Illinois	21,645	4364
Auburn University (AL)	21,498	2610
University of Kentucky	20,925	2736
Kansas State University	20,325	2467
University of Oklahoma	19,886	2311
Cornell University (NY)	18,001	21,914
University of Alaska	17,090	2294

Source: College Entrance Examination Board

You will solve problems related to this in Examples 1-4.

Measures of central tendency, such as the mean, median, and mode, are statistics that describe certain important characteristics of data. However, they do not indicate anything about the variability of the data. For example, 50 is the mean of both {0, 50, 100} and {40, 50, 60}. The variability is much greater in the first set of data than in the second, since $100 - 0$ is much greater than $60 - 40$.

One **measure of variability** is the *range*. Use the information in the table above to find the range of enrollment.

$$47,476 - 17,090 = 30,386.$$

University of Texas   University of Alaska

The range of enrollment is 30,386 students.

Resource Manager

 Reproducible Masters
- *Study Guide and Practice,* pp. 259–260
- *Enrichment,* p. 159
- *Assessment and Evaluation,* pp. 354, 355
- *TI-92 & CBL Interdisciplinary Activities,* pp. 45–48

 Transparencies
- 5-Minute Check, 14-3
- Teaching, 14-3

If the median is a member of the set of data, that item of data is excluded when calculating the first and third quartile points.

If the data have been arranged in order and the median is found, the set of data is divided into two groups. Then if the median of each group is found, the data is divided into four groups. Each of these groups is called a **quartile.** There are three quartile points, Q_1, Q_2, and Q_3, that denote the breaks in the data for each quartile. The median is the second quartile point Q_2. The medians of the two groups defined by the median are the first quartile point Q_1 and the third quartile point Q_3.

One fourth of the data is less than the first quartile point Q_1, and three fourths of the data is less than the third quartile point Q_3. The difference between the first quartile point and third quartile point is called the **interquartile range.** When the interquartile range is divided by 2, the quotient is called the **semi-interquartile range.**

Semi-Interquartile Range	If a set of data has first quartile point Q_1 and third quartile point Q_3, the semi-interquartile range Q_R can be found as follows. $$Q_R = \frac{Q_3 - Q_1}{2}$$

Example ❶ **EDUCATION** Refer to the application at the beginning of the lesson.

Real World Application

a. Find the interquartile range of the college enrollments and state what it represents.

b. Find the semi-interquartile range of the college enrollments.

a. First, order the data from least to greatest, and identify Q_1, Q_2, and Q_3.

$$\overset{Q_1}{}\qquad\qquad\overset{Q_2}{}$$
17,090 18,001 19,886 20,325 20,925 21,498 21,645 21,970 22,393 22,836

$$\overset{Q_3}{}$$
24,075 24,141 24,481 27,874 28,285 29,693 37,718 45,462 47,476

The interquartile range is $28{,}285 - 20{,}925$ or 7360. This means that the middle half of the student enrollments are between 28,285 and 20,925 and are within 7360 of each other.

Graphing Calculator Tip

Enter the data into **L1** and use the **SortA(** command to reorder the list from least to greatest.

b. The semi-interquartile range is $\frac{7360}{2}$ or 3680. The halfway point between Q_1 and Q_3 can be found by adding the semi-interquartile range to Q_1. That is, $3680 + 20{,}925$ or 24,605. Since $24{,}605 > Q_2$, this indicates the data is more clustered between Q_1 and Q_2 than between Q_2 and Q_3.

Box-and-whisker plots are used to summarize data and to illustrate the variability of the data. These plots graphically display the median, quartiles, interquartile range, and extreme values in a set of data. They can be drawn vertically, as shown at the right, or horizontally. A box-and-whisker plot consists of a rectangular box with the ends, or **hinges,** located at the first and third quartiles. The segments extending from the ends of the box are called **whiskers.** The whiskers stop at the extreme values of the set, unless the set contains **outliers.** Outliers are extreme values that are more than 1.5 times the interquartile range beyond the upper or lower quartiles. Outliers are represented by single points. If an outlier exists, each whisker is extended to the last value of the data that is not an outlier.

Lesson 14-3 Measures of Variability **909**

2 TEACH

Teaching Tip After defining quartiles, encourage students to think of real-world examples of statistics given in quartiles. For example, college board test scores are often reported in quartiles.

In-Class Example

Example 1
Education Examine the table of test scores below.

Test Scores, 5th Period Class			
Juan	82	Marje	82
Lisa B.	91	Wendi	79
Ahmed	74	Jenell	99
Adam	78	José	98
Becky	94	Mark	75
Brent	68	Shiraz	61
Matt	74	Joanna	78
Chrissy	88	Alison	86
Lisa S.	64	Jason	69
Luan	42	Tai	58
Tyrone	72	Gina	82

a. Find the interquartile range of the test scores and state what it represents. **17; The middle half of the test scores are within 17 points of each other.**

b. Find the semi-interquartile range of the test scores. **8.5**

The dimensions of the box-and-whisker plot can help you characterize the data. Each whisker and each small box contains 25% of the data. If the whisker or box is short, the data are concentrated over a narrower range of values. The longer the whisker or box, the larger the range of the data in that quartile. Thus, the box-and-whisker is a pictorial representation of the variability of the data.

Example ② **EDUCATION** **Refer to the application at the beginning of the lesson. Draw a box-and-whisker plot for the enrollments.**

In Example 1, you found that Q_1 is 20,925, Q_2 is 22,836, and Q_3 is 28,285. The extreme values are the least value 17,090 and the greatest value 47,476.

Draw a number line and plot the quartiles, the median, and the extreme values. Draw a box to show the interquartile range. Draw a segment through the median to divide the box into two smaller boxes.

Before drawing the whiskers, determine if there are any outliers. From Example 1, we know that the interquartile range is 7360. An outlier is any value that lies more than 1.5(7360) or 11,040 units below Q_1 or above Q_3.

$$Q_1 - 1.5(7360) = 20,925 - 11,040 \qquad Q_3 + 1.5(7360) = 28,285 + 11,040$$
$$= 9885 \qquad\qquad\qquad\qquad = 39,325$$

The lower extreme 17,090 is within the limits. However, 47,476 and 45,462 are not within the limits. They are outliers. Graph these points on the plot. Then draw the left whisker from 17,090 to 20,925 and the right whisker from 28,285 to the greatest value that is not an outlier, 37,718.

The box-and-whisker plot shows that the two lower quartiles of data are fairly concentrated. However, the upper quartile of data is more diverse.

Another measure of variability can be found by examining deviation from the mean, symbolized by $X_i - \overline{X}$. The sum of the deviations from the mean is zero. That is, $\sum_{i=1}^{n} (X_i - \overline{X}) = 0$. For example, the mean of the data set {14, 16, 17, 20, 33} is 20. The sum of the deviations from the mean is shown in the table.

X_i	$\overline{X}$	$X_i - \overline{X}$
14	20	−6
16	20	−4
17	20	−3
20	20	0
33	20	13
$\sum_{i=1}^{5} (X_i - \overline{X})$		0

To indicate how far individual items vary from the mean, we use the absolute values of the deviation. The arithmetic mean of the absolute values of the deviations from the mean of a set of data is called the **mean deviation,** symbolized by *MD*.

Mean Deviation	If a set of data has n values given by X_i, such that $1 \le i \le n$, with arithmetic mean $\overline{X}$, then the mean deviation MD can be found as follows. $$MD = \frac{1}{n} \sum_{i=1}^{n} |X_i - \overline{X}|$$

In sigma notation for statistical data, i is always an integer and not the imaginary unit.

Example **3** **EDUCATION** Refer to the application at the beginning of the lesson. Find the mean deviation of the enrollments.

There are 19 college enrollments listed, and the mean is $\frac{1}{19} \sum_{i=1}^{19} X_i$ or about 26,093.37.

Method 1: Sigma notation

$MD \approx \frac{1}{19} \sum_{i=1}^{19} |X_i - 26{,}093.37|$

$MD \approx \frac{1}{19} \big(|47{,}476 - 26{,}093.37| + |45{,}462 -$
$\qquad 26{,}093.37| + \cdots + |17{,}090 - 26{,}093.37| \big)$

$MD \approx \frac{1}{19} \big(|21{,}382.63| + |19{,}368.63| +$
$\qquad \cdots + |-9003.37| \big)$

$MD \approx 6310.29$

The mean deviation of the enrollments is about 6310.29. This means that the enrollments are an average of about 6310.29 above or below the mean enrollment of 26,093.37.

Method 2: Graphing Calculator
Enter the data for the enrollments into **L1**. At the home screen, enter the following formula.

$\qquad$ sum(abs(L1 $-$ 26093.37))/19

The calculator determines the difference between the scores and the mean, takes the absolute value, adds the absolute values of the differences, and divides by 19. This verifies the calculation in Method 1.

Graphing Calculator Tip

The **sum(** command is located in the **MATH** section of the **LIST** menu. The **abs(** command is in the **NUM** section after pressing **MATH**.

A measure of variability that is often associated with the arithmetic mean is the **standard deviation.** Like the mean deviation, the standard deviation is a measure of the average amount by which individual items of data deviate from the arithmetic mean of all the data. Each individual deviation can be found by subtracting the arithmetic mean from each individual value, $X_i - \overline{X}$. Some of these differences will be negative, but if they are squared, the results are positive. The standard deviation is the square root of the mean of the squares of the deviation from the arithmetic mean.

Lesson 14-3 Measures of Variability **911**

Teaching Tip After defining *mean deviation*, point out that the absolute value of each deviation is taken so that the sum of the deviation is not zero.

In-Class Example
Example 3
Education Refer to the data in In-Class Example 1. Find the mean deviation of the test scores. **about 10.27**

Lesson 14-3 **911**

Teaching Tip Although traditionally the mean is represented by an uppercase X with a bar over it ($\overline{X}$), the graphing calculator uses a lowercase x with a bar over it ($\overline{x}$). This is used because of the limited number of pixels available vertically to form a capital X with a bar.

In-Class Example

Example 4

Education Refer to the data in In-Class Example 1. Find the standard deviation of the test scores. **about 13.31**

Standard Deviation

If a set of data has n values, given by X_i such that $1 \leq i \leq n$, with arithmetic mean $\overline{X}$, the standard deviation σ can be found as follows.

$$\sigma = \sqrt{\frac{1}{n} \sum_{i=1}^{n} [X_i - \overline{X}]^2}$$

σ is the lowercase Greek letter sigma.

The standard deviation is the most important and widely used measure of variability. Another statistic used to describe the spread of data about the mean is **variance.** The variance, denoted σ^2, is the mean of the squares of the deviations from $\overline{X}$. The standard deviation is the positive square root of the variance.

Example **4** **EDUCATION** Refer to the application at the beginning of the lesson. Find the standard deviation of the enrollments.

Method 1: Standard Deviation Formula

There are 19 college enrollments listed, and the mean is about 26,093.37.

$$\sigma \approx \sqrt{\frac{1}{19} \sum_{i=1}^{19} (X_i - 26{,}093.37)^2}$$

$$\sigma \approx \sqrt{\frac{1}{19} (47{,}476 - 26{,}093.37)^2 + (45{,}462 - 26{,}093.37)^2 + \cdots + (17{,}090 - 26{,}093.37)^2}$$

$$\sigma \approx \sqrt{\frac{1}{19} (21{,}382.63)^2 + (19{,}368.63)^2 + \cdots + (-9003.37)^2}$$

$$\sigma \approx 8354.59$$

The standard deviation is about 8354.59. Since the mean of the enrollments is about 26,093.37 and the standard deviation is about 8354.59, the data have a great amount of variability.

Method 2: Graphing Calculator

Enter the data in **L1**. Use the **CALC** menu after pressing STAT to find the 1-variable statistics.

The standard deviation, indicated by σx, is the fifth statistic listed.

The mean ($\overline{x}$) is 26,093.36842 and the standard deviation is 8354.5913383, which agree with the calculations using the formulas.

When studying the standard deviation of a set of data, it is important to consider the mean. For example, compare a standard deviation of 5 with a mean of 10 to a standard deviation of 5 with a mean of 1000. The latter indicates very little variation, while the former indicates a great deal of variation since 5 is 50% of 10 while 5 is only 0.5% of 1000.

The standard deviation of a frequency distribution is the square root of the mean of the squares of the deviations of the class marks from the mean of the frequency data, weighted by the frequency of each interval.

Standard Deviation of the Data in a Frequency Distribution

If X_1, X_2, ..., X_k are the class marks in a frequency distribution with k classes, and f_1, f_2, ..., f_k are the corresponding frequencies, then the standard deviation σ of the data in the frequency distribution is found as follows.

$$\sqrt{\dfrac{\sum\limits_{i=1}^{k}(X_i - \overline{X})^2 \cdot f_i}{\sum\limits_{i=1}^{k} f_i}}$$

The standard deviation of a frequency distribution is an approximate number.

Example 5 **ECONOMICS** Use the frequency distribution data below to find the arithmetic mean and the standard deviation of the price-earnings ratios of 100 manufacturing stocks.

Method 1: Using Formulas

Class Limits	Class Marks (X)	f	f · X	(X − X̄)	(X − X̄)²	(X − X̄)² · f
−0.5–4.5	2.0	5	10	−8	64	320
4.5–9.5	7.0	54	378	−3	9	486
9.5–14.5	12.0	25	300	2	4	100
14.5–19.5	17.0	13	221	7	49	637
19.5–24.5	22.0	0	0	12	144	0
24.5–29.5	27.0	1	27	17	289	289
29.5–34.5	32.0	2	64	22	484	968
		100	1000			2800

The mean $\overline{X}$ is $\dfrac{1000}{100}$ or 10.

The standard deviation σ is $\sqrt{\dfrac{2800}{100}}$ or approximately 5.29.

Since the mean number of price-earnings ratios is 10 and the standard deviation is 5.29, this indicates a great amount of variability in the data.

Method 2: Graphing Calculator

Enter the class marks in the **L1** list and the frequency in the **L2** list.

Use the **CALC** menu after pressing STAT to find the 1-variable statistics. Then type **L1, L2** and press ENTER.

The calculator confirms the standard deviation is about 5.29.

Lesson 14-3 Measures of Variability **913**

Teaching Tip Remind students that the frequency distribution table contains most of the information needed to compute mean and standard deviation, so they are simply substituting the values in order to obtain solutions.

In-Class Example
Example 5
Racing Use the frequency distribution data at the bottom of the margin to find the arithmetic mean and the standard deviation of the speeds of the winners of an annual stock car race. **about 156.43; about 11.87**

Table for In-Class Example 5

Class Limits	Class Marks	f	f · X	X − X̄	(X − X̄)²	(X − X̄)² · f
130–140	135	2	270	−21.4	457.96	915.92
140–150	145	7	1015	−11.4	129.96	909.72
150–160	155	8	1240	−1.4	1.96	15.68
160–170	165	8	1320	8.6	73.96	591.68
170–180	175	2	350	18.6	345.96	691.92
180–190	185	1	185	28.6	817.96	817.96

Error Analysis

Watch for students who have difficulty calculating standard deviation by hand.

Prevent by stressing the importance of following the order of operations when using the formula.

Additional Answers

1. The median of the data is 70, Q_1 is 60, and Q_3 is 100. The interquartile range is 40 and the semi-interquartile range is 20. The outliers are 170 and 180. The data in the first two quartiles are close together in range. The last two quartiles are more diverse.

3. Both the mean deviation and the standard deviation are measures of the average amount by which individual items of data deviate from the mean of all the data. The mean deviation uses the absolute values of the deviations. Standard deviation uses the squares of the deviations.

5.

Study Guide and Practice Masters, pp. 259–260

CHECK FOR UNDERSTANDING

Communicating Mathematics

1. See margin.

3. See margin.

4. See students' work.

Guided Practice

5. 16; 8; See margin.

Read and study the lesson to answer each question.

1. **Describe** the data shown in the box-and-whisker plot below. Include the quartile points, interquartile range, semi-interquartile range, and any outliers.

2. **Explain** how to find the variance of a set of data if you know the standard deviation.   **square the standard deviation**

3. **Compare and contrast** mean deviation and standard deviation.

4. *Math Journal*   Draw a box-and-whisker plot for data you found in a newspaper or magazine. What conclusions can you derive from the plot?

5. Find the interquartile range and the semi-interquartile range of {17, 28, 44, 37, 28, 42, 21, 41, 35, 25}. Then draw a box-and-whisker plot.

6. Find the mean deviation and the standard deviation of {$4.45, $5.50, $5.50, $6.30, $7.80, $11.00, $12.20, $17.20}   **about $3.54; about $4.11**

7. Find the arithmetic mean and the standard deviation of the frequency distribution at the right.
30,250; about 13,226.39

Class Limits	Frequency
0–10,000	15
10,000–20,000	30
20,000–30,000	50
30,000–40,000	60
40,000–50,000	30
50,000–60,000	15

8. **Meteorology**   The following table gives the normal maximum daily temperature for Los Angeles and Las Vegas.

Normal Maximum Daily Temperatures

	January	February	March	April	May	June
Los Angeles	65.7	65.9	65.5	67.4	69.0	71.9
Las Vegas	57.3	63.3	68.8	77.5	87.8	100.3

	July	August	September	October	November	December
Los Angeles	75.3	76.6	76.6	74.4	70.3	65.9
Las Vegas	105.9	103.2	94.7	82.1	67.4	57.5

Source: National Oceanic and Atmosphere Administration

a. Find the mean, median, and standard deviation for the temperatures in Los Angeles.   **70.375; 69.65; about 4.25**

b. What are the mean, median, and standard deviation for the temperatures in Las Vegas?   **about 80.48; 79.8; about 17.06**
c. See margin.

c. Draw a box-and-whisker plot for the temperatures for each city.

d. Which city has a smaller variability in temperature?   **Los Angeles**

e. What might cause one city to have a greater variability in temperature than another?   **Los Angeles is near an ocean; Las Vegas is in a desert.**

 www.amc.glencoe.com/self_check_quiz

Addressing Individual Needs

Auditory/Musical   To demonstrate standard deviation, try the following: Play a scale, for example, the key of C: C_1 D E F G A B C_2. Assign numerical values to each note as follows: $C_1 = -4$, D $= -3$, E $= -2$, F $= -1$, G $= 1$, A $= 2$, B $= 3$, $C_2 = 4$. Find the mean.   **0 or F#**   Find the standard deviation. $\sigma = 2.7$ **or about A# above the mean and D♭ below the mean**   Play the notes corresponding to the mean and the standard deviation and compare the tones.

EXERCISES

Practice

Find the interquartile range and the semi-interquartile range of each set of data. Then draw a box-and-whisker plot.

A

9. {30, 28, 24, 24, 22, 22, 21, 17, 16, 15} **7; 3.5; See margin.**

10. {7, 14, 18, 72, 13, 15, 19, 8, 17, 28, 11, 15, 24} **9.5; 4.75; See margin.**

11. {15.1, 9.0, 8.5, 5.8, 6.2, 8.5, 10.5, 11.5, 8.8, 7.6} **2.9; 1.45; See margin.**

12. Use a graphing calculator to draw a box-and-whisker plot for {7, 1, 11, 5, 4, 8, 12, 15, 9, 6, 5, 9}? **See margin.**

Find the mean deviation and the standard deviation of each set of data.

B

13. {200, 476, 721, 579, 152, 158} **211; about 223.14**

14. {5.7, 5.7, 5.6, 5.5, 5.3, 4.9, 4.4, 4.0, 4.0, 3.8} **0.672; about 0.73**

15. {369, 398, 381, 392, 406, 413, 376, 454, 420, 385, 402, 446} **20.25; about 25.31**

16. Find the variance of {34, 55, 91, 13, 22}. **774**

Find the arithmetic mean and the standard deviation of each frequency distribution.

C

17.

Class Limits	Frequency
1–5	2
5–9	8
9–13	15
13–17	6
17–21	38
21–25	31
25–29	13
29–33	7

about 19.33; about 6.48

18.

Class Limits	Frequency
53–61	3
61–69	7
69–77	11
77–85	38
85–93	19
93–101	12

81.8; about 9.69

19.

Class Limits	Frequency
70–90	2
90–110	11
110–130	39
130–150	17
150–170	9
170–190	7

about 129.65; about 23.29

Applications and Problem Solving

20. **Geography** There are seven navigable rivers that feed into the Ohio River. The lengths of these rivers are given at the right.

a. Find the median of the lengths. **259 mi**

b. Name the first quartile point and the third quartile point. **129 mi; 360 mi**

c. Find the interquartile range. **231 mi**

20g. The data in the upper quartile is more diverse than the other quartiles.

d. What is the semi-interquartile range? **115.5 mi**

e. Are there any outliers? If so, name them. **no**

f. Make a box-and-whisker plot of the lengths of the rivers. **See margin.**

g. Use the box-and-whisker plot to discuss the variability of the data.

Length of Rivers Feeding into the Ohio River

Monongahela	129 miles
Allegheny	325 miles
Kanawha	97 miles
Kentucky	259 miles
Green	360 miles
Cumberland	694 miles
Tennessee	169 miles

Source: *The Universal Almanac*

21. Sample answer: {15, 15, 15, 16, 17, 20, 24, 26, 30, 35, 45}

21. **Critical Thinking** Write a set of numerical data that could be represented by the box-and-whisker plot at the right.

Lesson 14-3 Measures of Variability **915**

Additional Answers
(pp. 914–915)

8c. Los Angeles

Las Vegas

9.

10.

11.

12.

20f.

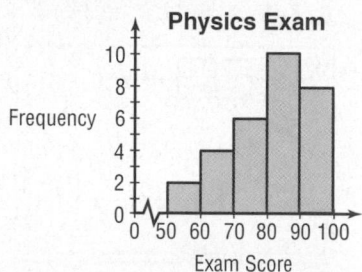
22. **Sports** During a recent season, 7684 teams played 19 NCAA women's sports. The breakdown of these teams is given below.

Sport	Teams	Sport	Teams	Sport	Teams
Basketball	966	Lacrosse	182	Swimming	432
Cross Country	838	Rowing	97	Tennis	859
Fencing	42	Skiing	40	Track, Indoor	528
Field Hockey	228	Soccer	691	Track, Outdoor	644
Golf	282	Softball	770	Volleyball	923
Gymnastics	91	Squash	26	Water Polo	23
Ice Hockey	22				

Source: The National Collegiate Athletic Association

a. What is the median of the number of women's teams playing a sport? **282**

b. Find the first quartile point and the third quartile point. **42; 770**

c. What is the interquartile range and semi-interquartile range? **728; 364**

d. Are there any outliers? If so, name them. **no**

e. Make a box-and-whisker of the number of women's teams playing a sport. **e. See margin.**

f. What is the mean of the number of women's teams playing a sport? **about 404.42**

g. Find the mean deviation of the data. **about 316.97**

h. Find the variance of the data. **about 118,712.56**

i. What is the standard deviation of the data? **about 344.55**

j. Discuss the variability of the data.

22j. There is a great variability among the number of teams in women's sports.

23. **Education** Refer to the data on the college tuition and fees in the application at the beginning of the lesson.

a. What are the quartile points of the data? **$2414, $2838, $4147**

b. Find the interquartile range. **1733**

c. Name any outliers. **$20,480, $21,914**

d. Make a box-and-whisker plot of the data. **See margin.**

e. What is the mean deviation of the data? **about 3507.18**

f. Find the standard deviation of the data. **about 5643.35**

g. Discuss the variability of the data.

23g. The data in the upper quartile is diverse.

24. **Government** The number of times the first 42 presidents vetoed bills are listed below.

2, 0, 0, 7, 1, 0, 12, 1, 0, 10, 3, 0, 0, 9, 7, 6, 29, 93, 13, 0, 12, 414, 44, 170, 42, 82, 39, 44, 6, 50, 37, 635, 250, 181, 21, 30, 43, 66, 31, 78, 44, 25

a. Make a box-and-whisker plot of the number of vetoes. **See margin.**

b. Find the mean deviation of the data. **about 67.87**

c. What is the variance of the data? **about 14,065.48**

d. What is the standard deviation of the data? **about 118.60**

e. Describe the variability of the data.

24e. The data in the upper quartile is diverse.

25. **Entertainment** The frequency distribution shows the average audience rating for the top fifty network television shows for one season.

Audience Rating	8–10	10–12	12–14	14–16	16–18	18–20	20–22
Frequency	26	12	6	2	2	0	2

Source: Nielsen Media Research

a. Find the arithmetic mean of the audience ratings. **11**

b. What is the standard deviation of the audience ratings? **about 2.94**

Extra Credit

Approximate the standard deviation of the data shown in the histogram.
about 16.28

26. yes; when the standard deviation is less than 1; when both equal 0 or 1

26. **Critical Thinking**   Is it possible for the variance to be less than the standard deviation for a set of data? If so, explain when this will occur. When would the variance be equal to the standard deviation for a set of data?

27. **Research**   Find the number of students attending each school in your county. Make a box-and-whisker plot of the data. Determine various measures of variability and discuss the variability of the data.   **See students' work**

Mixed Review

28. Consider the data represented by the stem-and-leaf plot at the right. *(Lesson 14-2)*
    a. What is the mean of the data?   **about 7.75**
    b. Find the median of the data.   **8.0**
    c. What is the mode of the data?   **8.9**

stem	leaf
4	4 4 9
5	4 5
6	2 2 4 5 9
7	1 4 5 6 7 8 9
8	0 2 4 5 6 7 8 9 9 9
9	0 2 3 3 5 6 8 9

$5 | 4 = 5.4$

29. **Fund-Raising**   Twelve students are selling programs at the Grove City High School to raise money for the athletic department. The numbers of programs sold by each student are listed below. *(Lesson 14-1)*

    51, 27, 55, 54, 68, 60, 39, 46, 46, 53, 57, 23

    a. Find the range of the number of programs sold.   **45**
    b. Determine an appropriate class interval.   **Sample answer: 10**
    c. What are the class limits?   **Sample answer: 20, 30, 40, 50, 60, 70**
    d. Construct a frequency distribution of the data.   **See margin.**
    e. Draw a histogram of the data.   **See margin.**

30. **Food Service**   Suppose nine salad toppings are placed on a circular, revolving tray. How many ways can the salad items be arranged? *(Lesson 13-2)*   **40,320 ways**

31. Find the first three iterates of the function $f(x) = 0.5x - 1$ using $x_0 = 8$. *(Lesson 12-8)*   **3, 0.5, −0.75**

32. **SAT/ACT Practice**   A carpenter divides a board that is 7 feet 9 inches long into three equal parts. What is the length of each part?

    C

    A 2 ft $6\frac{1}{3}$ in.   **B** 2 ft $8\frac{1}{3}$ in.   **C** 2 ft 7 in.

    D 2 ft 8 in.   **E** 2 ft 9 in.

MID-CHAPTER QUIZ

The scores for an exam given in physics class are given below.

82, 77, 84, 98, 93, 71, 76, 64, 89, 95, 78, 89, 65, 88, 54, 96, 87, 92, 80, 85, 93, 89, 55, 62, 79, 90, 86, 75, 99, 62

1. What is an appropriate class interval for the test scores? (Lesson 14-1)

2. Construct a frequency distribution of the test scores. (Lesson 14-1)   **See margin.**

3. Draw a histogram of the test scores. (Lesson 14-1)   **See margin.**

4. Make a stem-and-leaf plot of the test scores. (Lesson 14-2)   **See margin.**

5. What is the mean of the test scores? (Lesson 14-2)   **81.1**

6. Find the median of the test scores. (Lesson 14-2)   **84.5**

7. Find the mode of the test scores. (Lesson 14-2)   **89**

8. Make a box-and-whisker plot of the test scores. (Lesson 14-3)   **See margin.**

9. What is the mean deviation of the test scores? (Lesson 14-3)   **10.42**

10. Discuss the variability of the data. (Lesson 14-3)   **See margin.**
    1. **Sample answer: 10**

Extra Practice   See p. A54.

Additional Answers

29d. Sample answer:

Programs Sold	Frequency
20–30	2
30–40	1
40–50	2
50–60	5
60–70	2

29e. Sample answer:

---

1 FOCUS

5-Minute Check
Lesson 14-3

1. Find the interquartile range and the semi-interquartile range of the data. Then draw a box-and-whisker plot.
   52, 47, 61, 50, 55, 62, 57, 51, 50, 59 **9; 4.5**

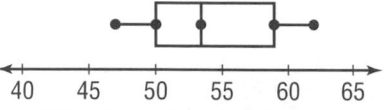

2. The following is a list of college entrance test scores for ten seniors.
   490, 560, 620, 580, 599, 740, 600, 620, 680, 500

   a. What is the interquartile range of the data? **60**

   b. What is the semi-interquartile range of the data? **30**

   c. Draw a box-and-whisker plot of the data.

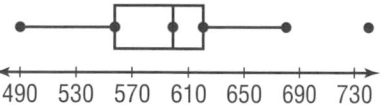

3. Find the mean and standard deviation of the frequency distribution.

Class Limits	Frequency
20–30	2
30–40	3
40–50	8
50–60	7
60–70	1
70–80	1

about 47.27; about 11.65

14-4

OBJECTIVES
- Use the normal distribution curve.

The Normal Distribution

TESTING The class of 1996 was the first class to take the adjusted Scholastic Assessment Test. The test was adjusted so that the median of the scores for the verbal section and the math section would be 500. For each section, the lowest score is 200 and the highest is 800. Suppose the verbal and math scores follow the *normal distribution*. What percent of the students taking the test would have a math score between 375 and 625? *This problem will be solved in Example 4.*

A frequency polygon displays a limited number of data and may not represent an entire population. To display the frequency of an entire population, a smooth curve is used rather than a polygon.

If the curve is symmetric, then information about the measures of central tendency can be gathered from the graph. Study the graphs below.

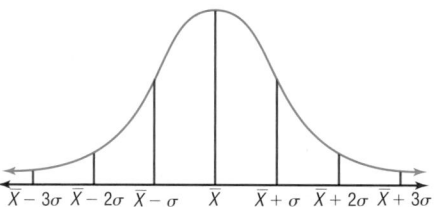

A **normal distribution** is a frequency distribution that often occurs when there is a large number of values in a set of data. The graph of this distribution is a symmetric, bell-shaped curve, shown at the left. This is known as a **normal curve.** The shape of the curve indicates that the frequencies in a normal distribution are concentrated around the center portion of the distribution. A small portion of the population occurs at the extreme values.

In a normal distribution, small deviations are much more frequent than large ones. Negative deviations and positive deviations occur with the same frequency. The points on the horizontal axis represent values that are a certain number of standard deviations from the mean $\overline{X}$. In the curve shown above, each interval represents one standard deviation. So, the section from $\overline{X}$ to $\overline{X} + \sigma$ represents those values between the mean and one standard deviation greater than the mean, the section from $\overline{X} + \sigma$ to $\overline{X} + 2\sigma$ represents the interval one standard deviation greater than the mean to two standard deviations greater than the mean, and so on. The total area under the normal curve and above the horizontal axis represents the total probability of the distribution, which is 1.

918 *Chapter 14   Statistics and Data Analysis*

Resource Manager

Reproducible Masters
- *Study Guide and Practice,* pp. 261–262
- *Enrichment,* p. 160
- *Assessment and Evaluation,* p. 356

Transparencies
- 5-Minute Check, 14-4
- Teaching, 14-4

Example

① MEDICINE The average healing time of a certain type of incision is **240 hours with a standard deviation of 20 hours.** Sketch a normal curve that represents the frequency of healing times.

First, find the values defined by the standard deviation in a normal distribution.

$$\overline{X} - 1\sigma = 240 - 1(20) \text{ or } 220$$
$$\overline{X} - 2\sigma = 240 - 2(20) \text{ or } 200$$
$$\overline{X} - 3\sigma = 240 - 3(20) \text{ or } 180$$

$$\overline{X} + 1\sigma = 240 + 1(20) \text{ or } 260$$
$$\overline{X} + 2\sigma = 240 + 2(20) \text{ or } 280$$
$$\overline{X} + 3\sigma = 240 + 3(20) \text{ or } 300$$

Sketch the general shape of a normal curve. Then, replace the horizontal scale with the values you have calculated.

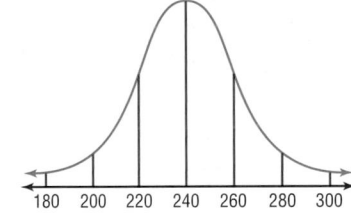

The tables below give the fractional parts of a normally distributed set of data for selected areas about the mean. The letter t represents the number of standard deviations from the mean (that is, $\overline{X} \pm t\sigma$). When $t = 1$, t represents 1 standard deviation above and below the mean.

P represents the fractional part of the data that lies in the interval $\overline{X} \pm t\sigma$. The percent of the data within these limits is $100P$.

t	P		t	P		t	P		t	P
0.0	0.000		0.9	0.632		1.7	0.911		2.5	0.988
0.1	0.080		1.0	0.683		1.8	0.929		2.58	0.990
0.2	0.159		1.1	0.729		1.9	0.943		2.6	0.991
0.3	0.236		1.2	0.770		1.96	0.950		2.7	0.993
0.4	0.311		1.3	0.807		2.0	0.955		2.8	0.995
0.5	0.383		1.4	0.838		2.1	0.964		2.9	0.996
0.6	0.451		1.5	0.866		2.2	0.972		3.0	0.997
0.7	0.516		1.6	0.891		2.3	0.979		3.5	0.9995
0.8	0.576		1.65	0.900		2.4	0.984		4.0	0.9999

The P value also corresponds to the probability that a randomly selected member of the sample lies within t standard deviation units of the mean. For example, suppose the mean of a set of data is 85 and the standard deviation is 5.

Boundaries:
$$\overline{X} - t\sigma \text{ to } \overline{X} + t\sigma$$
$$85 - t(5) \text{ to } 85 + t(5)$$
$$85 - 1(5) \text{ to } 85 + 1(5)$$
$$80 \text{ to } 90$$

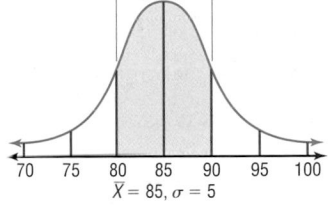

68.3% of the values in this set of data lie within one standard deviation of 85; that is, between 80 and 90.

Lesson 14-4 The Normal Distribution **919**

Motivating the Lesson

Hands-On Activity Have students perform a binomial experiment such as flipping a coin 10, 20, 30, 40 times, and so on. Have them plot the number of heads they get on each set of flips for each experiment. Note that as the number of trials increases, the plot looks more and more like the normal distribution.

2 TEACH

Teaching Tip In the Real-World Application, the median of the scores is 500, since half of the scores are above 500 and half are below. If the scores are normally distributed, the mean will equal the median.

Teaching Tip Before discussing normal distribution, emphasize that on the coordinate plane, the mean would be the maximum value of the curve.

In-Class Example

Example 1
Manufacturing The weight of a box of cereal averages 565 grams with a standard deviation of 5 grams. Sketch a normal curve that represents the frequency of box weights.

Teaching Tip Emphasize that the tables on this page are generally used for all normal distributions. The total area under the normal curve is represented by these data and is also used to represent probability distributions.

📁 Prerequisite Skills

A review of the prerequisite skills for this lesson is provided in the *Prerequisite Skills Masters.*

Skill Number and Content	Pages
12 Statistical Displays	25–26

Teaching Tip Make sure that students understand that the probability associated with the intervals of a normal distribution is fixed.

In-Class Example

Example 2
Manufacturing Refer to In-Class Example 1. Suppose that 6000 boxes are taken from the assembly line and checked. Estimate how many boxes fall within each of the following intervals.
a. 560–570 grams
   4098 boxes
b. 555–575 grams
   5730 boxes
c. 550–580 grams
   5982 boxes

If you randomly select one item from the sample, the probability that the one you pick will be between 80 and 90 is 0.683. If you repeat the process 1000 times, approximately 68.3% (about 683) of those selected will be between 80 and 90.

Thus, normal distributions have the following properties.

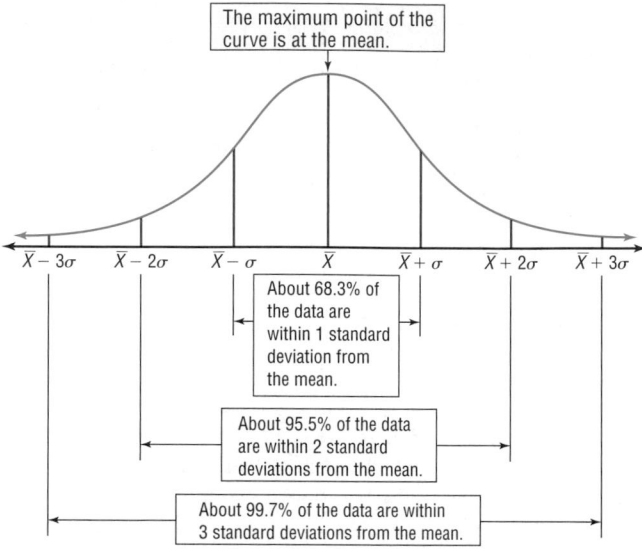

Example ② **MEDICINE** Refer to Example 1. Suppose a hospital has treated 2000 patients in the past five years having this type of incision. Estimate how many patients healed in each of the following intervals.

a. 220–260 hours

The interval 220–260 hours represents $\overline{X} \pm 1\sigma$, which represents a probability of 68.3%.

$$68.3\%(2000) = 1366$$

Approximately 1366 patients took between 220 and 260 hours to heal.

b. 200–280 hours

The interval 200–280 hours represents $\overline{X} \pm 2\sigma$, which represents a probability of 95.5%.

$$95.5\%(2000) = 1910$$

Approximately 1910 patients took between 200 and 280 hours to heal.

c. 180–300 hours

The interval 180–300 hours represents $\overline{X} \pm 3\sigma$, which represents a probability of 99.7%.

$$99.7\%(2000) = 1994$$

Approximately 1994 patients took between 180 and 300 hours to heal.

If you know the mean and the standard deviation, you can find a range of values for a given probability.

Example **3** Find the upper and lower limits of an interval about the mean within which 45% of the values of a set of normally distributed data can be found if $\overline{X} = 110$ and $\sigma = 15$.

Use the table on page 919 to find the value of t that most closely approximates $P = 0.45$. For $t = 0.6$, $P = 0.451$. Choose $t = 0.6$. Now find the limits.

$\overline{X} \pm t\sigma = 110 \pm 0.6(15)$ $\overline{X} = 110,\ t = 0.6,\ \sigma = 15$

$= 101$ and 119

The interval in which 45% of the data lies is 101–119.

If you know the mean and standard deviation, you can also find the percent of the data that lies within a given range of values.

Example **4** **TESTING**   Refer to the application at the beginning of the lesson.

   a. **Determine the standard deviation.**

   b. **What percent of the students taking the test would have a math score between 375 and 625?**

   c. **What is the probability that a senior chosen at random has a math score between 550 and 650?**

   a. For a normal distribution, both the median and the mean are 500. All of the scores are between 200 and 800. Therefore, all of the scores must be within 300 points from the mean. In a normal distribution, 0.9999 of the data is within 4 standard deviations of the mean. If the scores are to be a normal distribution, the standard deviation should be $\dfrac{300}{4}$ or 75.

   b. Write each of the limits in terms of the mean.

$$375 = 500 - 125 \text{ and } 625 = 500 + 125$$

Therefore, $\overline{X} \pm t\sigma = 500 \pm 125$ and $t\sigma = 125$. Solve for t.

$$t\sigma = 125$$
$$t(75) = 125 \qquad \sigma = 75$$
$$t \approx 1.7$$

If $t = 1.7$, then $P = 0.911$.   *Use the table on page 919.*

About 91.1% of the students taking the test would have a math score between 375 and 625.

   c. The graph shows that 550–650 does not define an interval that can be represented by $\overline{X} \pm t\sigma$. However, the interval can be defined as the difference between the intervals 500–650 and 500–550.

(continued on the next page)

Teaching Tip For Example 3, make sure that students understand that most probability ratios will be approximated using the table and not calculated exactly.

In-Class Examples

Example 3
Find the upper and lower limits of an interval about the mean within which 72% of the values of a set of normally distributed data can be found if $\overline{X} = 124$ and $\sigma = 16$.   **106.4 − 141.6**

Example 4
Physiology Margarite determined that the average woman in her math class is 63 inches tall. All of the women in her class are between 53 and 73 inches tall. Assume that the heights are normally distributed.
  a. Determine the standard deviation.   **2.5**
  b. What percent of the women in her class would be between 60 and 66 inches tall?   **77%**
  c. What is the probability that a woman chosen at random in her class is between 70.5 and 73 inches tall?   **0.145%**

Teaching Tip Remind students that they must take into account the fact that half of the probability in Example 4 falls above the mean and half below.

Addressing Individual Needs

 Logical Ask students to outline the steps needed to isolate an area under the normal curve in order to determine the probability defined by intervals of standard deviations. This multi-step process utilizes the deductive reasoning needed to analyze a diagram.

Error Analysis

Watch for students who inappropriately apply probability ratios to the interval under the normal curve.

Prevent by stressing that probability is applied to intervals around the mean.

Additional Answer

3.

Study Guide and Practice Masters, pp. 261–262

First, find the probability that the score is between the mean 500 and the upper limit 650.

$$\overline{X} + t\sigma = 650$$
$$500 + t(75) = 650$$
$$t = 2$$

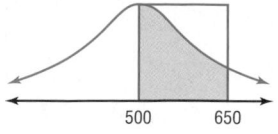

The value of P that corresponds to $t = 2$ is 0.955.

$P = 0.955$ describes the probability that a student's score falls $\pm 2(75)$ points about the mean, or between 350 and 650, but we are only considering half that interval. So, the probability that a student's score is between 500 and 650 is $\frac{1}{2}(0.955)$ or about 0.478.

Next, find the probability that a score is between the mean and the lower limit 550.

$$\overline{X} + t\sigma = 550$$
$$500 + t(75) = 550$$
$$t \approx 0.7$$

For $t = 0.7$, $P = 0.516$.

Likewise, we will only consider half of this probability or 0.258.

Now find the probability that a student's score falls in the interval 550–650.

$$P(550\text{–}650) = P(500\text{–}650) - P(500\text{–}550)$$
$$P \approx 0.478 - 0.258$$
$$P \approx 0.220 \text{ or } 22\%$$

The probability that a student's score is between 550 and 650 is about 22%.

Students who take the SAT or ACT tests will receive a score as well as a **percentile.** The percentile indicates how the student's score compares with other students taking the test.

Percentile	The *n*th percentile of a set of data is the value in the set such that *n* percent of the data is less than or equal to that value.

Therefore if a student scores in the 65th percentile, this means that 65% of the students taking the test scored the same or less than that student.

CHECK FOR UNDERSTANDING

Communicating Mathematics

1. The median, mean, and mode are the same.

Read and study the lesson to answer each question.

1. **Compare** the median, mean, and mode of a set of normally distributed data.

2. **Write** an expression for the interval that is within 1.5 standard deviations from the mean. $\overline{X} \pm 1.5\sigma$

3. **Sketch** a normal curve with a mean of 75 and a standard deviation of 10 and a normal curve with a mean of 75 and a standard deviation of 5. Which curve displays less variability? **See margin for curves; the second curve.**

922 *Chapter 14 Statistics and Data Analysis*

4. **Counterexample** Draw a curve that represents data which is not normally distributed. **See margin.**

5. **Name** the percentile that describes the median. **50th percentile**

Guided Practice

6. The mean of a set of normally distributed data is 550 and the standard deviation is 35.
   a. Sketch a curve that represents the frequency distribution. **See margin.**
   b. What percent of the data is between 515 and 585? **68.3%**
   c. Name the interval about the mean in which about 99.7% of the data are located. **445-655**
   d. If there are 200 values in the set of data, how many would be between 480 and 620? **191**

7. A set of 500 values is normally distributed with a mean of 24 and a standard deviation of 2. **d. 20.08-27.92**
   a. What percent of the data is in the interval 22-26? **68.3%**
   b. What percent of the data is in the interval 20.5-27.5? **92.9%**
   c. Find the interval about the mean that includes 50% of the data. **22.6-25.4**
   d. Find the interval about the mean that includes 95% of the data.

8c. Chemistry; the chemistry grade is 3 standard deviations above the class mean, while the speech grade is only 2 standard deviations above the class mean.

8. **Education** In her first semester of college, Salali earned a grade of 82 in chemistry and a grade of 90 in speech. **a–b. See margin.**
   a. The mean of the chemistry grades was 73, and the standard deviation was 3. Draw a normal distribution for the chemistry grades.
   b. The mean of the speech grades was 80, and the standard deviation was 5. Draw a normal distribution for the speech grades.
   c. Which of Salali's grades is relatively better based on standard deviation from the mean? Explain.

EXERCISES

Practice **A**

9. The mean of a set of normally distributed data is 12 and the standard deviation is 1.5.
   a. Sketch a curve that represents the frequency distribution. **See margin.**
   b. Name the interval about the mean in which about 68.3% of the data are located. **10.5-13.5**
   c. What percent of the data is between 7.5 and 16.5? **99.7%**
   d. What percent of the data is between 9 and 15? **95.5%**

10. Suppose 200 values in a set of data are normally distributed.
    a. How many values are within one standard deviation of the mean? **about 137**
    b. How many values are within two standard deviations of the mean? **191**
    c. How many values fall in the interval between the mean and one standard deviation above the mean? **about 68**

B

11. A set of data is normally distributed with a mean of 82 and a standard deviation of 4.
    a. Find the interval about the mean that includes 45% of the data. **79.6-84.4**
    b. Find the interval about the mean that includes 80% of the data. **76.8-87.2**
    c. What percent of the data is between 76 and 88? **86.6%**
    d. What percent of the data is between 80.5 and 83.5? **31.1%**

www.amc.glencoe.com/self_check_quiz *Lesson 14-4 The Normal Distribution* **923**

Additional Answers
4. Sample answer:

6a.

445 480 515 550 585 620 655

8a.

64 67 70 73 76 79 82

8b.

65 70 75 80 85 90 95

9a.

7.5 9 10.5 12 13.5 15 16.5

12. The mean of a set of normally distributed data is 402, and the standard deviation is 36.
   a. Find the interval about the mean that includes 25% of the data. **391.2-412.8**
   b. What percent of the data is between 387 and 417? **31.1%**
   c. What percent of the data is between 362 and 442? **72.9%**
   d. Find the interval about the mean that includes 45% of the data. **380.4-423.6**

C 13. A set of data is normally distributed with a mean of 140 and a standard deviation of 20.
   a. What percent of the data is between 100 and 150? **66.9%**
   b. What percent of the data is between 150 and 180? **28.6%**
   c. Find the value that defines the 75th percentile. **154**

14. The mean of a set of normally distributed data is 6, and the standard deviation is 0.35.
   a. What percent of the data is between 6.5 and 7? **7.8%**
   b. What percent of the data is between 5.5 and 6.2? **64.45%**
   c. Find the limit above which 90% of the data lies. **5.545**

Applications and Problem Solving

15. **Probability**   Tossing six coins is a binomial experiment.
   a. Find each probability.
      P(no tails)   P(one tail)   P(two tails)   P(three tails)
      P(four tails)   P(five tails)   P(six tails)
   b. Assume that the experiment was repeated 64 times. Make a bar graph showing how many times you would expect each outcome to occur.

15a. $\frac{1}{64}, \frac{3}{32}, \frac{15}{64},$
$\frac{5}{16}, \frac{15}{64}, \frac{3}{32}, \frac{1}{64}$

   c. Use the bar graph to determine the mean number of tails.   **3**
   d. Find the standard deviation of the number of tails.   **about 1.2**
   e. Compare the bar graph to a normal distribution.   **They are similar.**

15b. See margin.

16. **Critical Thinking**   Consider the percentile scores on a standardized test.
   a. Describe the 92nd percentile in terms of standard deviation.
   b. Name the percentile of a student whose score is 0.8 standard deviation above the mean.   **78.8 percentile**

16a. 1.4 standard deviations above the mean

17. **Health**   The lengths of babies born in City Hospital in the last year are normally distributed. The mean length is 20.4 inches, and the standard deviation is 0.8 inch. Trey was 22.3 inches long at birth.

17a. 0.8%

   a. What percent of the babies born at City Hospital were longer than Trey?

17b. 99.2%

   b. What percent of the babies born at City Hospital were shorter than Trey?

18. **Business**   The length of time a brand of CD players can be used before needing service is normally distributed. The mean length of time is 61 months, and the standard deviation is 5 months. The manufacturer plans to issue a guarantee that it will replace any CD player that breaks within a certain length of time. If the manufacturer does not want to replace any more than 2% of the CD players, how many months should they limit the guarantee?   **50.5 months**

19. **Education**   A college professor plans to grade a test on a curve. The mean score on the test is 65, and the standard deviation is 7. The professor wants 15% A's, 20% B's, 30% C's, 20% D's and 15% F's. Assume the grades are normally distributed.
   a. What is the lowest score for an A?   **72**
   b. Find the lowest passing score.   **58**
   c. What is the interval for the B's?   **68-71**

20a. a normal distribution with a small standard deviation

20b. a normal distribution with a large standard deviation

20c. a distribution where values greater than the mean are more spread out than values less than the mean

20. Critical Thinking Describe the frequency distribution represented by each graph.

 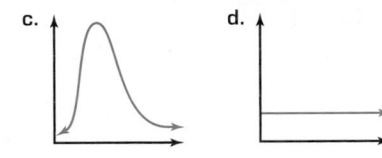

a.   b.   c.   d.

21. Industry A machine is used to fill cans of cola. The amount of cola dispensed into each can varies slightly. Suppose the amount of cola dispensed into the cans is normally distributed. **a. about 2.55 mL**

  a. If at least 95% of the cans must have between 350 and 360 milliliters of cola, find the greatest standard deviation that can be allowed.

  b. What percent of the cans will have between 353 and 357 milliliters of cola? **57.6%**

Mixed Review

20d. a distribution where all values occur with the same frequency

22. Nutrition The numbers of Calories in one serving of twenty different cereals are listed below. *(Lesson 14-3)*

    110, 110, 330, 200, 88, 110, 88, 110, 165, 390, 150, 440, 536, 200, 110, 165, 88, 147, 110, 165

  a. What is the median of the data? **148.5**

  b. Find the first quartile point and the third quartile point. **110; 200**

  c. Find the interquartile range of the data. **90**

  d. What is the semi-quartile range of the data? **45**

  e. Draw a box-and-whisker plot of the data. **See margin.**

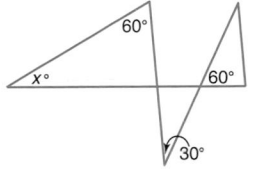

23. Find the mean, median, and mode of {33, 42, 71, 19, 42, 45, 79, 48, 55}. *(Lesson 14-2)* **about 48.2; 45; 42**

24. Write an equation for a secant function with a period of $\frac{\pi}{2}$, a phase shift of $-\pi$, and a vertical shift of 3. *(Lesson 6-7)* $y = \sec(4\theta + 4\pi) + 3$

25. Education The numbers of students attending Wilder High School during various years are listed below. *(Lesson 4-8)*

Year	1965	1970	1975	1980	1985	1990	1995	2000
Enrollment	365	458	512	468	426	401	556	629

25a. Sample answer: $y = 0.05x^3 - 2.22x^2 + 29.72x + 366.92$

25b. Sample answer: 2553 students

  a. Write an equation that models the enrollment as a function of the number of years since 1965.

  b. Use the model to predict the enrollment in 2015.

26. SAT/ACT Practice What is the value of x in the figure at the right? **E**

  A 50   B 45

  C 40   D 35

  E 30

Extra Practice See p. A54.

Lesson 14-4 The Normal Distribution **925**

---

4 ASSESS

Open-Ended Assessment

Modeling Have each student make up an example of a normal distribution, giving the mean and standard deviation. Ask them to sketch the normal curve. Then, ask them to describe how to find the probability that part of the data would fall between the second and third standard deviations below the mean.

Chapter 14, Quiz C (Lesson 14-4) is available in the *Assessment and Evaluation Masters,* p. 356.

Additional Answer

22e.

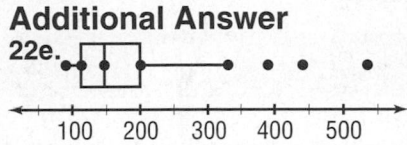

  100 200 300 400 500

Enrichment Masters, p. 160

Extra Credit

Have groups of students shake 50 pennies in a cup and record the number of heads for each of 10 throws. Then, have each group calculate the mean number of heads in each throw. Determine how many of the throws yield a number of heads within one standard deviation, two standard deviations, and three standard deviations of the mean. Compare this with the expected number of heads.
See students' work.

14-4B The Standard Normal Curve

An Extension of Lesson 14-4

1 FOCUS

Motivating the Lesson

Hands-On Activity Have students survey classmates to find the average amount of sleep each receives per night. Then, have them make a histogram of the data. Ask students to draw a curve over the data. If they expand their survey to a much larger group, it should approximate a normal curve.

2 TEACH

A reproduction of this lesson with graphing calculator keystrokes provided can be found in the *Graphing Calculator Keystrokes Masters*, pp. 54–55.

Teaching Tip Remind students that *e* is the base for the natural log and equals about 2.718.

Teaching Tip Make sure that students put the $-\frac{x^2}{2}$ term in the exponent.

3 PRACTICE/APPLY

Assignment Guide

All: 1–8

4 ASSESS

Open-Ended Assessment

Modeling Have students graph class quiz results and compare the results to the normal curve.

926 *Chapter 14*

WHAT YOU'LL LEARN
- Use the standard normal curve to study properties of normal distributions.

The graph shown at the right is known as the *standard normal curve*. The standard normal curve is the graph of

$$f(x) = \frac{1}{\sqrt{2\pi}} e^{-\frac{x^2}{2}}.$$ You can use a graphing calculator to investigate properties of this function and its graph. Enter the function for the normal curve in the **Y=** list of a graphing calculator.

[−4.7, 4.7] scl:1 by [−0.2, 0.5] scl:0.1

TRY THESE

1. See margin.

1. What can you say about the function and its graph? Be sure to include information about the domain, the range, symmetry, and end behavior.

2. The standard normal curve models a probability distribution. As a result, probabilities for intervals of *x*-values are equal to areas of regions bounded by the curve, the *x*-axis, and the vertical lines through the endpoints of the intervals. The calculator can approximate the areas of such regions. To find the area of the region bounded by the curve, the *x*-axis, and the vertical lines $x = -1$ and $x = 1$, go to the **CALC** menu and select 7: $\int$ **f(x) dx**. Move the cursor to the point where $x = -1$. Press $\boxed{\text{ENTER}}$. Then move the cursor to the point where $x = 1$ and press $\boxed{\text{ENTER}}$. The calculator will shade the region described above and display its approximate area. What number does the calculator display for the area of the shaded region? **0.68268949**

3. About 68.3%; the answer for Exercise 2 can be rounded to 0.683, which equals 68.3%.

3. Refer to the diagram on page 920. For normal distributions, about what percent of the data are within one standard deviation from the mean? How is this number related to the area you found in Exercise 2?

4. Enter $\boxed{\text{2nd}}$ [DRAW] 1. This causes the calculator to clear the shading and redisplay the graph. Find the area of the region bounded by the curve, the *x*-axis, and the vertical lines $x = -2$ and $x = 2$. **0.95449974**

5. Find the area of the region bounded by the curve, the *x*-axis, and the vertical lines $x = -3$ and $x = 3$. **0.9973002**

6. How do your answers for Exercises 4 and 5 compare to the percents in the diagram on page 920? **See margin.**

WHAT DO YOU THINK?

7. Without using a calculator, estimate the area of the region bounded by the curve, the *x*-axis, and the vertical lines $x = -4$ and $x = 4$ to four decimal places. **0.9999**

8. Change the graphing window to **Xmin** $= -47$ and **Xmax** $= 47$. Find the area of the region bounded by the curve, the *x*-axis, and the vertical lines $x = -20$ and $x = 20$. Do you think that your answer is the exact area for the region? Explain. **1; no; see margin.**

926 *Chapter 14 Statistics and Data Analysis*

Additional Answers

1. The domain of the function is the set of all real numbers. The range is the set of all real numbers *y* such that $0 < y \leq \frac{1}{\sqrt{2\pi}}$. The graph is symmetric with respect to the *y*-axis and has the *x*-axis as a horizontal asymptote. The value of *f(x)* approaches 0 as *x* approaches $\pm \infty$.

6. The answer for Exercise 3 can be rounded to 0.954, which is about 95.5%. The answer for Exercise 4 can be rounded to 0.997, which equals 99.7%.

8. Since the curve is approaching the *x*-axis asymptotically, the area is probably not exactly equal to 1.

Sample Sets of Data

OBJECTIVE

- Find the standard error of the mean to predict the true mean of a population with a certain level of confidence.

EDUCATION Rosalinda Perez is doing some research for her doctoral thesis. She wants to determine the mean amount of money spent by a school district to educate one student for a year. Since there are 14,883 school districts in the United States, she cannot contact every school district. She randomly contacts 100 of these districts to find out how much money they spend per pupil. Using these 100 values, she computes the mean expenditure to be $6130 with a standard deviation of $1410. What is the standard error of the mean? *This problem will be solved in Example 1.*

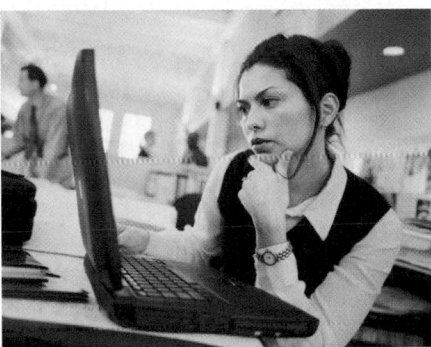

In statistics, the word **population** does not always refer to a group of people. A population is the entire set of items or individuals in the group being considered. In the application above, the population is the 14,883 school districts. Rarely will a researcher find 100% of a population accessible as a source of data. Therefore, a **random sample** of the population is selected so that it is representative of the entire population. A random sample will give each member of the population the same chance of being selected as a member of the sample.

If the sample is random, the characteristics of the population pertinent to the study should be found in the sample in about the same ratio as they exist in the total population. If Ms. Perez randomly selected the school districts in the sample, the sample will probably include some large school districts, some small school districts, some urban school districts, and some rural school districts. It would probably also include school districts from all parts of the United States.

Based upon a random sample of the population, certain inferences can be made about the population in general. The major purpose of **inferential statistics** is to use the information gathered in a sample to make predictions about a population.

Ms. Perez is not sure that the mean school district expenditure per student is a truly representative mean of all school districts. She uses another sample of 100 districts. This time she finds the mean of the per pupil expenditures to be $6080 with a standard deviation of $1390.

The discrepancies that Ms. Perez found in her two samples are common when taking random samples. Large companies and statistical organizations often take many samples to find the "average" they are seeking. For this reason, a sample mean is assumed to be near its true population mean, symbolized by μ. The standard deviation of the distribution of the sample means is known as the **standard error of the mean**.

μ is the lowercase Greek letter mu.

Resource Manager

Reproducible Masters
- *Study Guide and Practice,* pp. 263–264
- *Enrichment,* p. 161
- *Assessment and Evaluation,* p. 356
- *Problem Solving and Applications,* p. 27

 Transparencies
- 5-Minute Check, 14-5
- Teaching, 14-5

---

1 FOCUS

5-Minute Check
Lesson 14-4

1. Suppose that 300 values in a set of data have a normal distribution. How many values are within one standard deviation of the mean? **204.9**

A set of 400 values has a normal distribution with a mean of 82 and a standard deviation of 4.

2. What percent of the data is in the interval 78 to 82? **about 34.2%**

3. What percent of the data is within two standard deviations from the mean? **95.5%**

The mean of a set of normally distributed data is 25 with a standard deviation of 2.

4. Sketch the curve that represents the frequency distribution.

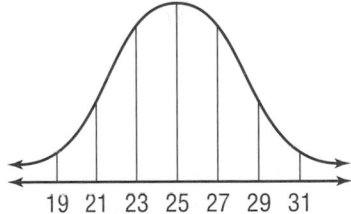

19 21 23 25 27 29 31

5. Name the interval about the mean in which about 95.5% of the data are located. **21 to 29**

6. What percent of the data are between 23 and 27? **68.3%**

Motivating the Lesson

Discussion When pollsters report their findings, they often cite a *margin of error*, for example, ±3 percentage points. Discuss the meaning of this term in relation to data samples. In this lesson, students will study a way to find the *standard error of the mean*, which tells the accuracy of a sample of data, or in other words, its margin of error.

Teaching Tip Point out that the categories Ms. Perez is including are *not* exclusive. Therefore, a school might belong to more than one category.

Teaching Tip For Example 1, after discussing the standard error of the mean, point out that the larger the set of data, the closer the sample mean is to the real mean of the data set.

In-Class Examples

Example 1
Entertainment A recent survey of 50 teens ages 13 to 17 found that the average number of hours that teens spend watching television per day is 4.5 with a standard deviation of 1 hour. What is the standard error of the mean? **about 0.14**

Example 2
Entertainment Refer to the information in In-Class Example 1. Determine the interval of television viewing times such that the probability is 90% that the mean viewing time of the entire population lies within the interval. **4.266 to 4.733 h**

Teaching Tip For Example 2, make sure that students understand that the probability applies to the true mean of the population, which may be slightly different from the sample mean.

Standard Error of the Mean	If a sample of data has N values and σ is the standard deviation, the standard error of the mean $\sigma_{\bar{X}}$ is $$\sigma_{\bar{X}} = \frac{\sigma}{\sqrt{N}}.$$

The symbol $\sigma_{\bar{X}}$ is read "sigma sub x bar."

The standard error of the mean is a measurement of how well the mean of a sample selected at random estimates the true mean.

- If the standard error of the mean is small, then the sample means are closer to (or approximately the same value as) the true mean.

- If the standard error of the mean is large, many of the sample means would be far from the true mean.

The greater the number of items or subjects in the sample, the closer the sample mean reflects the true mean and the smaller the standard error of the mean.

Example **① EDUCATION** Refer to the application at the beginning of the lesson. **What is the standard error of the mean?**

For the sample, $N = 100$. Find $\sigma_{\bar{X}}$.

$$\sigma_{\bar{X}} = \frac{1410}{\sqrt{100}} \text{ or } 141$$

The standard error of the mean of the per pupil expenditures is 141. Although 141 is a fairly large number, it is not extremely large compared to the mean 6130. The mean is a good approximation of the true mean.

The sampling error is the difference between the population mean μ and the sample mean $\bar{X}$.

The sample mean is only an estimate of the true mean of the population. Sample means of various random samples of the same population are normally distributed about the true mean with the standard error of the mean as a measure of their variability. Thus, the standard error of the mean behaves like the standard deviation. Using the standard error of the mean and the sample mean, we can state a range about the sample mean in which we think the true mean lies. Probabilities of the occurrence of sample means and true means may be determined by referring to the tables on page 919.

Example **② EDUCATION** Refer to the application at the beginning of the lesson. **Using Ms. Perez's first sample, determine the interval of per pupil expenditures such that the probability is 95% that the mean expenditure of the entire population lies within the interval.**

When $P = 95\%$ or 0.95, $t = 1.96$. *Refer to the tables on page 919.*

To find the range, use a technique similar to finding the interval for a normal distribution.

$$\bar{X} \pm t\sigma_{\bar{X}} = 6130 \pm 1.96(141)$$
$$= 5853.64 \text{ to } 6406.36$$

The probability is 95% that the true mean μ is within the interval of \$5853.64 and \$6406.36.

The probability of the true mean being within a certain range of a sample mean may be expressed as a **level of confidence**. The most commonly used levels of confidence are 1% and 5%. A 1% level of confidence means that there is less than a 1% chance that the true mean differs from the sample mean by a certain amount. That is, you are 99% confident that the true mean is within a certain range of the sample mean. A 5% level of confidence means that the probability of the true mean being within a certain range of the sample mean is 95%.

If a higher level of confidence is desired for the same number of values, accuracy must be sacrificed by providing a larger interval. However, if the number of values in the sample is larger, the interval for a given level of confidence is smaller.

Example **3** **AUTOMOTIVE ENGINEERING** The number of miles a certain sport utility vehicle can travel on open highway on one gallon of gasoline is normally distributed. You are to take a sample of vehicles, test them, and record the miles per gallon. You wish to have a 1% level of confidence that the interval containing the mean miles per gallon of the sample also contains the true mean.

Real World Application

a. **Twenty-five sports utility vehicles are selected and tested. From this sample, the average miles per gallon is 22 with a standard deviation of 4 miles per gallon. Determine the interval about the sample mean that has a 1% level of confidence.**

b. **Four hundred sports utility vehicles are randomly selected and their miles per gallon are recorded. From this sample, the average miles per gallon is 22 with a standard deviation of 4 miles per gallon. Determine the interval about the sample mean that has a 1% level of confidence.**

c. **What happens when the number of items in the sample is increased?**

a. A 1% level of confidence is given when $P = 99\%$.
When $P = 0.99$, $t = 2.58$.

Find $\sigma_{\bar{X}}$ $\sigma_{\bar{X}} = \dfrac{4}{\sqrt{25}}$ or 0.8

Find the range. $\bar{X} \pm t\sigma_{\bar{X}} = 22 \pm 2.58(0.8)$
$\bar{X} \pm t\sigma_{\bar{X}} = 19.936$ and 24.064

Thus, the interval about the sample mean is 19.936 miles per gallon to 24.064 miles per gallon.

b. Determine the range. As in part a, $t = 2.58$.

Find $\sigma_{\bar{X}}$. $\sigma_{\bar{X}} = \dfrac{4}{\sqrt{400}}$ or 0.2

Find the range. $\bar{X} \pm t\sigma_{\bar{X}} = 22 \pm 2.58(0.2)$
$\bar{X} \pm t\sigma_{\bar{X}} = 21.484$ and 22.516

Thus, the interval about the sample mean is 21.484 miles per gallon to 22.516 miles per gallon.

c. By increasing the number of items in the sample, but with the same 1% confidence level, the range decreased substantially, from about $24.064 - 19.036$ or 4.128 miles per gallon to about $22.516 - 21.484$ or 1.032 miles per gallon.

Teaching Tip In Example 3, stress the importance of sample size and its affect on level of confidence.

Teaching Tip Have students discuss how a random sample might be selected for the different situations presented in this lesson. Make a list of "Dos" and "Don'ts" for selecting a random sample.

In-Class Example

Example 3
Electrical Engineering The number of hours a certain brand of light bulb will light is normally distributed.

a. Fifty light bulbs are selected and tested. From this sample, the average life of the bulb is 960 hours with a standard deviation of 12 hours. Determine the interval about the sample mean that has a 1% level of confidence.

    955.6 h to 964.4 h

b. Three hundred bulbs are randomly selected and tested. From this sample, the average life is 960 hours with a standard deviation of 12 hours. Determine the interval about the sample mean that has a 1% level of confidence.

    958.2 h to 961.8 h

c. What happens when the number of items in the sample is increased?
    As the sample size increases from 50 to 300, the range decreases substantially from 8.8 hours to 3.6 hours.

Addressing Individual Needs

Interpersonal Have pairs of students work together to investigate the relationship between a sample size and its related standard error of the mean and level of confidence. Provide each pair with a sample mean and its standard deviation. Have one student select a desired level of confidence. The other must then determine a sample size that gives a reasonable interval about the mean giving the desired level of confidence. Repeat with the students reversing roles.

3 PRACTICE/APPLY

Error Analysis

Watch for students who mistakenly assume that a sample and a population are the same.

Prevent by emphasizing that they are trying to approximate the characteristics of the population as closely as possible.

Assignment Guide

Core: 11–27 odd, 29–31, 33–39
Enriched: 10–26 even, 28–39

Additional Answers

1. A sample is a subset of a population. However, a sample must be similar in every way to the population.

4. Every twentieth student to enter the school should produce a representative sample. The senior English class will not represent underclassmen. The track team will not represent students who prefer other types of activities.

Study Guide and Practice Masters, pp. 263–264

CHECK FOR UNDERSTANDING

Communicating Mathematics

Read and study the lesson to answer each question.   **1. See margin.**

1. **Compare and contrast** a population and a sample of the population.

2. **Describe** how to determine the standard error of the mean.

2. Divide the standard deviation of the sample by the square root of the number of values in the sample.

3. **Explain** how you can find a smaller interval for the true mean and still have the same level of confidence.   **Use a larger sample.**

4. **You Decide** As part of a class project, Karen, Tyler, and Mark need to determine the average number of hours that the students in their high school study each school night. Karen suggests they ask the students in their senior English class. Tyler suggests they ask every twentieth student as they enter the school in the morning. Mark suggests they ask the members of the track team. Whose sample should they use? Explain.   **Tyler; see margin for explanation.**

Guided Practice

Find the standard error of the mean for each sample.

5. $\sigma = 73, N = 100$ **7.3**

6. $\sigma = 3.4, N = 250$ **about 0.22**

7. If $\sigma = 5, N = 36$, and $\bar{X} = 45$, find the interval about the sample mean that has a 1% level of confidence.   **42.85–47.15**

8. If $\sigma = 5.6, N = 300$, and $\bar{X} = 55$, find the interval about the sample mean that has a 5% level of confidence.   **about 54.37–55.63**

9. **Employment** An employment agency requires all clients to take an aptitude test. They randomly selected 150 clients and recorded the amount of time each client took to complete the test. The average time needed to complete the test was 27.5 minutes with a standard deviation of 3.5 minutes.

a. What is the standard error of the mean?   **about 0.29**

b. Find the interval about the sample mean that reflects a 50% chance that the true mean lies within that interval.   **about 27.30–27.70 min**

c. Find the interval about the sample mean that has a 1% level of confidence.   **about 26.76–28.24 min**

EXERCISES

Practice

 A

Find the standard error of the mean for each sample.   **about 0.66**

10. $\sigma = 1.8, N = 81$ **0.2**

11. $\sigma = 5.8, N = 250$

12. $\sigma = 7.8, N = 140$

11. about 0.37

13. $\sigma = 14, N = 700$ **about 0.53**

14. $\sigma = 2.7, N = 130$ **about 0.24**

15. $\sigma = 13.5, N = 375$ **about 0.70**

16. If the standard deviation of a sample set of data is 5.6 and the standard error of the mean is 0.056, how many values are in the sample set?   **10,000**

For each sample, find the interval about the sample mean that has a 1% level of confidence.   **17. about 333.07–336.93**

 B

17. $\sigma = 5.3, N = 50, \bar{X} = 335$

18. $\sigma = 40, N = 64, \bar{X} = 200$ **187.1–212.9**

19. $\sigma = 12, N = 200, \bar{X} = 80$ **about 77.81–82.19**

20. $\sigma = 11.12, N = 1000, \bar{X} = 110$ **about 109.09–110.91**

www.amc.glencoe.com/self_check_quiz

21. The mean height of a sample of 100 high school seniors is 68 inches with a standard deviation of 4 inches. Determine the interval of heights such that the probability is 90% that the mean height of the entire population lies within that interval.   **67.34–68.66 in.**

For each sample, find the interval about the sample mean that has a 5% level of confidence.   **22. about 23.53–24.47   23. about 4524.21–4527.79**

22. $\sigma = 2.4, N = 100, \overline{X} = 24$

23. $\sigma = 17.1, N = 350, \overline{X} = 4526$

24. $\sigma = 28, N = 370, \overline{X} = 678$
    about 675.15–680.85

25. $\sigma = 0.67, N = 80, \overline{X} = 5.38$
    about 5.23–5.53

26. The following is a frequency distribution of the time in minutes required for a shopper to get through the checkout line at a certain discount store on a weekend. The distribution is a random sample of the weekend shoppers at the store.

Number of Minutes	3–5	5–7	7–9	9–11	11–13	13–15	15–17	17–19	19–21
Frequency	1	3	5	12	17	13	7	4	2

a. What is the mean of the data in the frequency distribution?   **12.375**

b. Find the standard deviation of the data.   **about 3.37**

c. What is the standard error of the mean?   **about 0.42**

d. Find the interval about the sample mean such that the probability is 0.95 that the true mean lies within the interval.   **about 11.55–13.20 min**

e. Find the probability that the mean of the population will be less than one minute from the mean of the sample.   **about 98.4%**

27. The standard deviation of the blood pressure of 45 women ages 40 to 50 years old is 12. What is the probability that the mean blood pressure of the random sample will differ by more than 3 points from the mean blood pressure reading for all women in that age bracket?   **about 8.9%**

Applications and Problem Solving

28. **Botany**   A botanist is studying the effects of a drought on the size of acorns produced by the oak trees. A random sample of 50 acorns reveals a mean diameter of 16.2 millimeters and a standard deviation of 1.4 millimeters.

a. Find the standard error of the mean.   **about 0.20**

b. What is the interval about the sample mean that has a 5% level of confidence?   **about 15.81–16.59 mm**

c. Find the interval about the sample mean that gives a 99% chance that the true mean lies within the interval.   **about 15.69–16.71 mm**

d. Find the interval about the sample mean such that the probability is 0.80 that the true mean lies within the interval.   **about 15.94–16.46 mm**

29. **Advertising**   The Brite Light Company wishes to include the average lifetime of its light bulbs in its advertising. One hundred light bulbs are randomly selected and illuminated. The time for each bulb to burn out is recorded. From this sample, the average life is 350 hours with a standard deviation of 45 hours.

29c. Sample answer: 338 hours; there is only 0.5% chance the mean is less than this number.

a. Find the standard error of the mean.   **4.5**

b. Determine the interval about the sample mean that has a 1% level of confidence.   **338.39–361.61 hours**

c. If you want to avoid false advertising charges, what number would you use as the average lifetime of the bulbs? Explain.

Open-Ended Assessment

Speaking Have students describe orally what the standard error of the mean is and how to find it. They can then describe its role in determining intervals about the sample mean at a given level of confidence.

📁 **Chapter 14, Quiz D (Lesson 14-5)** is available in the *Assessment and Evaluation Masters*, p. 356.

31b. With a 5% level of confidence, the average family in the town will have their televisions on from 2.98 to 5.22 hours.

31c. Sample answer: None; the sample is too small to generalize to the population of the city.

32b. about 42.38–43.02 crackers

32c. Sample answer: No; there is a 50% chance that the true mean is in the interval. However, since 43 is near one end of the interval, they may want to take another sample in the near future.

30. **Critical Thinking** There is a probability of 0.99 that the average life of a disposable hand-warming package is between 9.7936 and 10.2064 hours. The standard deviation of the sample is 0.8 hour. What is the size of the sample used to determine these values? **100 packages**

31. **Entertainment** In a certain town, a random sample of 10 families was interviewed about their television viewing habits. The mean number of hours that these families had their televisions on per day was 4.1 hours. The standard deviation was 1.8 hours.
   a. What is the standard error of the mean? **about 0.57**
   b. Make a 5% level of confidence statement about the mean number of hours that families of this town have their televisions turned on per day.
   c. What inferences about the television habits of the entire city can be drawn from this data? Explain.

32. **Quality Control** The Simply Crackers Company selects a random sample of 50 snack packages of their cheese crackers. The mean number of crackers in a package is 42.7 with a standard deviation of 3.2.
   a. Find the standard error of the mean. **about 0.45**
   b. Determine the interval of number of crackers such that the probability is 50% that the mean number of crackers lies within the interval.
   c. If the true population mean should be 43 crackers per package, should the company be concerned about this sample? Explain.

33. **Critical Thinking** The lifetimes of 1600 batteries used in radios are tested. With a 5% level of confidence, the true average life of the batteries is from 746.864 to 753.136 hours.
   a. What is the mean life of a battery in the sample? **750 h**
   b. Find the standard deviation of the life of the batteries in the sample. **64 h**

Mixed Review

34. **Tires** The lifetimes of a certain type of car tire are normally distributed. The mean lifetime is 40,000 miles with a standard deviation of 5000 miles. Consider a sample of 10,000 tires. *(Lesson 14-4)*
   a. How many tires would you expect to last between 35,000 and 45,000 miles? **6830 tires**
   b. How many tires would you expect to last between 30,000 and 40,000 miles? **4775 tires**
   c. How many tires would you expect to last less than 40,000 miles? **5000 tires**
   d. How many tires would you expect to last more than 50,000 miles? **225 tires**
   e. How many tires would you expect to last less than 25,000 miles? **15 tires**

35. Find the mean deviation and the standard deviation of $\{44, 72, 58, 61, 71, 49, 55, 68\}$. *(Lesson 14-3)* **8.25; about 9.59**

36. Find the sum of the first ten terms of the series $\frac{1}{16} + \frac{1}{4} + 1 + \cdots$. *(Lesson 12-2)* $\frac{349,525}{16}$ **or 21,845.3125**

37. Write $x = y$ in polar form. *(Lesson 9-3)* $\theta = 45°$

38. Solve $\tan x + \cot x = 2$ for principal values of x. *(Lesson 7-5)* **45°**

39. **SAT/ACT Practice** $*x$ is defined such that $*x = x^2 - 2x$. What is the value of $*2 - *1?$ **C**

   A -1 B 0 C 1 D 2 E 4

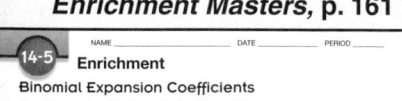

14-5 Enrichment

Binomial Expansion Coefficients

The picture at the right shows a device often used to illustrate a normal probability distribution. The device is filled with small steel marbles. The marbles roll past a series of hexagonal obstacles, collecting at the bottom in each of nine columns.

It can be shown that the number of paths from *A* to *G* is 1, *A* to *J* is 1, *A* to *H* is 3, and *A* to *I* is 3. For example, *H* can be reached by the way of *E*. Hence the number of paths to *H* is the sum of the number of paths to *D* and the number of paths to *E*. Likewise the number of paths to any point can be found by adding the number of paths to points diagonally above it. This is precisely the method by which the numbers in Pascal's triangle are obtained.

The numbers in Pascal's triangle are the coefficients in the expansion of $(x + y)^n$ where *n* is any positive integer. Therefore, the probability of a marble falling in any given column is proportional to the coefficient of the corresponding term in the binomial expansion of a power. The power is a whole number equal to the row being considered. For example, in the illustration above the columns are in the *eighth* row. Thus the probability of a marble falling in the *third* column is proportional to the coefficient of the *third* term in the binomial expansion of $(x + y)^8$.

The figure above at the right has equally-spaced vertical segments whose lengths are proportional to the numbers in the eighth row of Pascal's triangle. A smooth curve connecting the tops of these segments suggests the probable distribution of marbles in the column. Notice the similarity of this curve to the normal distribution curve.

Solve.

1. Draw a smooth curve connecting the tops of the segments whose lengths are proportional to the coefficients in the expansion of $(x + y)^{12}$.

2. A teacher decided to mark 64 tests with the grades A, B, C, D, and F (A highest) in proportion to the coefficients in the expansion of $(x + y)^4$. How many tests received each grade? **A = 4, B = 16, C = 24, D = 16, F = 4**

© Glencoe/McGraw-Hill   T161   *Advanced Mathematical Concepts*

Extra Credit

Find the sample size, *N*, if the probability is 95% that the population mean is within 12.1–13.3, and the sample standard deviation is 2. **about 43**

VOCABULARY

arithmetic mean (p. 897)	interquartile range (p. 909)	percentile (p. 922)
back-to-back bar graph (p. 889)	leaf (p. 899)	population (p. 927)
	level of confidence (p. 929)	quartile (p. 909)
bar graph (p. 889)	line plot (p. 889)	random sample (p. 927)
bimodal (p. 899)	mean (p. 897)	range (p. 890)
box-and-whisker plot (p. 909)	mean deviation (p. 910)	semi-interquartile range (p. 909)
class interval (p. 890)	measure of central tendency (p. 897)	
class limits (p. 890)		standard deviation (p. 911)
class mark (p. 890)	measure of variability (p. 908)	standard error of the mean (p. 927)
cumulative frequency distribution (p. 902)	median (p. 897)	stem (p. 899)
frequency distribution (p. 890)	median class (p. 902)	stem-and-leaf plot (p. 899)
	mode (p. 897)	three-dimensional bar graph (p. 890)
frequency polygon (p. 892)	normal curve (p. 918)	
hinge (p. 909)	normal distribution (p. 918)	variance (p. 912)
histogram (p. 890)	outlier (p. 909)	whisker (p. 909)
inferential statistics (p. 927)		

3. standard error of the mean

UNDERSTANDING AND USING THE VOCABULARY

Choose the term from the list above that best completes each statement.

1. A ____?____ is a display that visually shows the quartile points and the extreme values of a set of data. **box-and-whisker plot**

2. The ____?____ of a set of data is the middle value if there are an odd number of values.   **2. median**

3. The standard deviation of the distribution of the sample means is known as the ____?____.

4. The ____?____ of a set of data is the difference between the greatest and the least values in the set. **range**   **5. measure of central tendency**

5. A statistic that describes the center of a set of data is called an average or ____?____.

6. A ____?____ is the entire set of items or individuals in the group being considered. **population**

7. Data with two modes are ____?____. **bimodal**

8. The major purpose of ____?____ is to use the information gathered in a sample to make predictions about a population. **inferential statistics**

9. A ____?____ is the most common way of displaying a frequency distribution. **histogram**

10. A measure of variability often associated with the arithmetic mean is the ____?____. **standard deviation**

 For additional review and practice for each lesson, visit: **www.amc.glencoe.com**

STUDY GUIDE AND ASSESSMENT

VOCABULARY

This section provides a listing of the new terms, properties, and phrases that were introduced in this chapter. The exercises check students' understanding of the terms by using a variety of verbal formats including matching, completion, and true/false.

Glossary A complete glossary of terms begins on page A71.

Assessment and Evaluation Masters, pp. 343–344

Six forms of Chapter 14 Test are available in the *Assessment and Evaluation Masters* as shown in the chart.

Chapter 14 Test, Form 1B, is shown at the right. Chapter 14 Test, Form 2B, is shown on the next page.

TEST OPTIONS

Form	Pages	Type	
1A	341–342	Honors	Form 1 tests are multiple choice.
1B	343–344	Average	
1C	345–346	Basic	
2A	347–348	Honors	Form 2 tests are free response.
2B	349–350	Average	
2C	351–352	Basic	

SKILLS AND CONCEPTS

Objectives and Examples This section reviews the skills and concepts of the chapter and shows completely worked examples.

Review Exercises These exercises provide practice for the corresponding objectives.

Assessment and Evaluation Masters, pp. 349–350

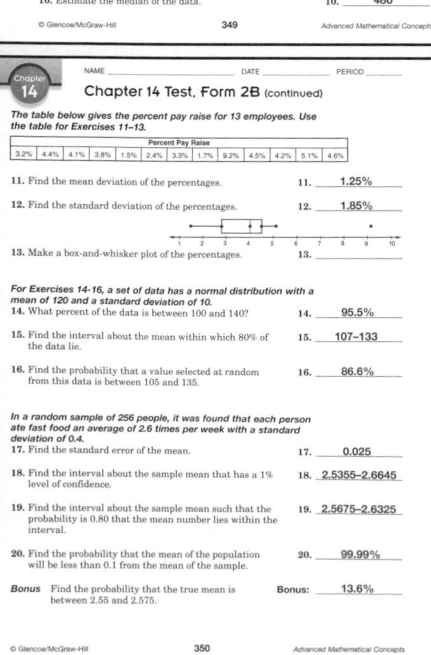

SKILLS AND CONCEPTS

OBJECTIVES AND EXAMPLES

Lesson 14-1 Draw, analyze, and use bar graphs and histograms.

- Draw a histogram of the data below.

Scores	Frequency
60–70	2
70–80	8
80–90	11
90–100	6

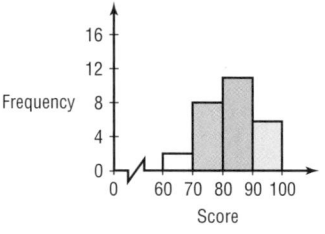

REVIEW EXERCISES

The table below gives the weight in ounces of the popular women's tennis shoes.

Weight (ounces)	Number of Shoes
9.0–10.0	2
10.0–11.0	18
11.0–12.0	5
12.0–13.0	2
13.0–14.0	3

11. What is the range of the data? **5**

12. What are the class marks?

13. Draw a histogram of the data. **See margin.**

12. **9.5, 10.5, 11.5, 12.5, 13.5**

Lesson 14-2 Find the mean, median, and mode of a set of data.

- Find the mean, median, and mode of the set {46, 47, 59, 49, 50, 48, 58, 56, 58, 54, 53}.

$$\frac{1}{11}\sum_{i=1}^{11} X_i = \frac{46 + 47 + \cdots + 54 + 53}{11} \text{ or } 53$$

The mean is 53.

To find the median, order the data.

46, 47, 48, 49, 50, 53, 54, 56, 58, 58, 64

Since there are an odd number of data, the median is the middle value. The median is 53.

The most frequent value in this set of data is 58. So, the mode is 58.

Find the mean, median, and mode of each set of data.

14. {4, 8, 2, 4, 5, 5, 6, 7, 4} **5; 5; 4**

15. {250, 200, 160, 240, 200} **210; 200; 200**

16. {19, 11, 13, 15, 16} **14.8; 15; none**

17. {6.6, 6.3, 6.8, 6.6, 6.7, 5.9, 6.4, 6.3}

18.
stem	leaf	
12	2 8	**133.25; 132; none**
13	0 1 3 5	
14	1 6	

$12 \mid 2 = 122$

17. **6.45; 6.5; 6.3 and 6.6**

TestCheck and Worksheet Builder

This **networkable** software has three modules.
- **Worksheet Builder** to make worksheets and tests
- **Student Module** to take tests on-screen
- **Management System** to keep student records

Three special banks are included for SAT, ACT, and TIMSS.

OBJECTIVES AND EXAMPLES

Lesson 14-3 Find the interquartile range, the semi-interquartile range, mean deviation, and standard deviation of data.

- interquartile range: $Q_3 - Q_1$

  semi-interquartile range: $Q_R = \dfrac{Q_3 - Q_1}{2}$

  mean deviation: $MD = \dfrac{1}{n} \sum_{i=1}^{n} |X_i - \overline{X}|$

  standard deviation: $\sigma = \sqrt{\dfrac{1}{n} \sum_{i=1}^{n} (X_i - \overline{X})^2}$

REVIEW EXERCISES

A number cube is tossed 10 times with the following results.

$$5 \quad 1 \quad 5 \quad 4 \quad 2 \quad 3 \quad 6 \quad 2 \quad 5 \quad 1$$

19. Find the interquartile range.   **3**

20. Find the semi-interquartile range.   **1.5**

21. Find the mean deviation.   **1.6**

22. Find the standard deviation.
    about 1.74

Additional Answer (p. 934)

13.

Women's Tennis Shoes

Lesson 14-4 Use the normal distribution curve.

A Normal Distribution

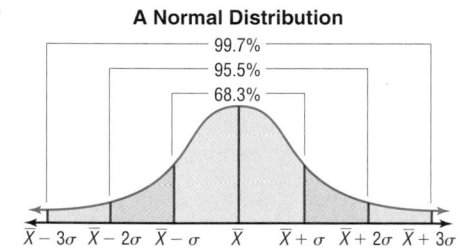

$$\overline{X} - 3\sigma \quad \overline{X} - 2\sigma \quad \overline{X} - \sigma \quad \overline{X} \quad \overline{X} + \sigma \quad \overline{X} + 2\sigma \quad \overline{X} + 3\sigma$$

A set of data is normally distributed with a mean of 75 and a standard deviation of 6. What percent of the data is between 69 and 81?

The values within one standard deviation of the mean are between $75 - 6$, or 69, and $75 + 6$, or 81. So, 68.3% of the data is between 69 and 81.

The mean of a set of normally distributed data is 88 and the standard deviation is 5.

23. What percent of the data is in the interval 78–98?   **95.5%**

24. Find the probability that a value selected at random from the data lies in the interval 86–90.   **0.311**

25. Find the interval about the mean that includes 90% of the data.   **79.75–96.25**

Suppose 150 values in a data set are normally distributed.

26. How many values are within one standard deviation of the mean?   **102.45**

27. How many values are within two standard deviations of the mean?   **143.25**

28. How many values fall in the interval between the mean and one standard deviation above the mean?   **51.225**

OBJECTIVES AND EXAMPLES	REVIEW EXERCISES

Lesson 14-5 Find the standard error of the mean to predict the true mean of a population with a certain level of confidence.

• Find the standard error of the mean for $\sigma = 12$ and $N = 100$. If $\overline{X} = 75$, find the range for a 1% level of confidence.

$$\sigma_{\overline{X}} = \frac{\sigma}{\sqrt{N}}$$

$$\sigma_{\overline{X}} = \frac{12}{\sqrt{100}} \text{ or } 1.2$$

The standard error of the mean is 1.2.

A 1% level of confidence is given when $P = 99\%$.

When $P = 0.99$, $t = 2.58$.

Use $\sigma_{\overline{X}} = 1.2$ to find the range.

$$\overline{X} \pm t\sigma_{\overline{X}} = 75 \pm (2.58)(1.2)$$

$$= 71.90 \text{ to } 78.10$$

Thus, the interval about the mean is 71.90 to 78.10.

Find the standard error of the mean for each sample.

29. $\sigma = 1.5$, $N = 90$ **about 0.16**

30. $\sigma = 4.9$, $N = 120$ **about 0.45**

31. $\sigma = 25$, $N = 400$ **1.25**

32. $\sigma = 18$, $N = 25$ **3.6**

For each sample, find the interval about the sample mean that has a 1% level of confidence.

33. $\sigma = 15$, $N = 50$, $\overline{X} = 100$ **about 94.53–105.47**

34. $\sigma = 30$, $N = 15$, $\overline{X} = 90$ **about 70.02–109.98**

35. $\sigma = 24$, $N = 200$, $\overline{X} = 40$
    about 35.62–44.38

In a random sample of 200 adults, it was found that the average number of hours per week spent cleaning their home was 1.8, with a standard deviation of 0.5.

36. Find the standard error of the mean.   **about 0.035**

37. Find the range about the mean such that the probability is 0.90 that the true mean lies within the range.   **about 1.74–1.86 h**

38. Find the range about the sample mean that has a 5% level of confidence.   **about 1.73–1.87 h**

39. Find the range about the sample mean that has a 1% level of confidence.   **about 1.71–1.89 h**

40. **Entertainment**   In a random sample of 100 families, the children watched television an average of 4.6 hours a day. The standard deviation is 1.4 hours. Find the range about the sample mean so that a probability of 0.90 exists that the true mean will lie within the range.   **about 4.37–4.83 h**

APPLICATIONS AND PROBLEM SOLVING

41. Safety The numbers of job-related injuries at a construction site for each month of 1999 are listed below. *(Lesson 14-2)*

10	13	15	39	21	24
19	16	39	17	23	25

a. Make a stem-and-leaf plot of the numbers of injuries. **See margin.**

b. What is the mean number of the data? **21.75**

c. Find the median of the data. **20**

d. Find the mode of the data. **39**

42. The height of members of the boys basketball team are normally distributed. The mean height is 75 inches, and the standard deviation is 2 inches. Randall is 80 inches tall. What percent of the boys on the basketball team are taller than Randall? *(Lesson 14-4)* **0.6%**

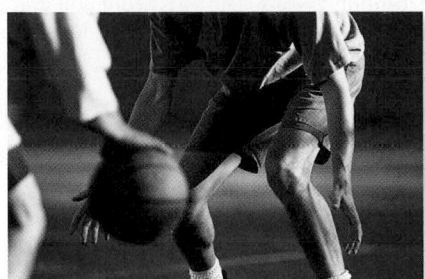

ALTERNATIVE ASSESSMENT

OPEN-ENDED ASSESSMENT

1. The mean of a set of five pieces of data is 15, and the median is 10. When one certain value is added to the set, the mean stays the same but the median changes.

   a. Find a set of data for which this is true.

   b. What value can be added to your set so that the mean stays the same but the median changes? **Sample answer: 15**

2. Find some data in a newspaper or magazine. Use what you have learned in this chapter to analyze the data. **See students' work.**
   1a. Sample answer: {2, 3, 10, 20, 40}

PORTFOLIO

Choose one of the types of data displays you studied in this chapter. Describe a situation in which this type of display would be used. Explain why the type of display you chose is the best one to use in this situation.

> **Additional Assessment** See p. A69 for a practice Chapter 14 test.

Unit 4 *interNET* Project

THE UNITED STATES CENSUS BUREAU

More and more models!

- Use the data you collected for the project in Chapter 12. Display the data in a table or use software to prepare a spreadsheet of the data.

- Use computer software or a graphing calculator to find at least three models for the population data. Draw a graph of each function.

- Compare your function models for the population data. Use your models to predict the U.S. population for the year 2050. Determine which one you think best fits the data.

- Write a one-page paper comparing the arithmetic and geometric sequences you wrote for Chapter 12 with the function models. Discuss which one model you think best fits the population data and give your estimate for the population in 2050.

Chapter 14 Study Guide and Assessment **937**

Internet Address Book

Record useful Internet addresses in the space at right for quick reference.

www.amc.glencoe.com

Chapter 14 Skill: Statistics and Data Analysis

Pages 938–939 are part of a complete SAT/ACT preparation course that is described in detail on pages T8–T9 of the *Teacher's Handbook*. The test items were written in the same style as those in the SAT and ACT.

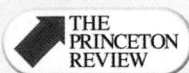 These questions were aligned and verified by The Princeton Review, the nation's leader in test preparation.

Diagnosis & Prescription

Each of the 10 SAT and ACT practice problems on page 939 is cross-referenced to the chapter where that SAT or ACT skill is covered. If students miss a particular type of problem, you can have them study that skill.

Problem Number and Skill	Page
1 Statistics	938
2 Algebra	124
3 Percent	834
4 Statistics	938
5 Geometry (Lines, Angles, and Arcs)	612
6 Statistics	938
7 Ratio and Proportion	692
8 Statistics	938
9 Statistics	938
10 Statistics	938

Statistics and Data Analysis Problems

TEST-TAKING TIP
If a problem includes a graph, look carefully at the graph including its labels and units. Then read the question.

Two or three questions may refer to the same graph.

On the SAT and ACT exams, you will calculate the mean (average), median, and mode of a data set.

The SAT and ACT exams may include a one or two questions on interpreting graphs. The most common graphs are bar graphs, circle graphs, line graphs, stem-and-leaf plots, histograms, and frequency tables.

ACT EXAMPLES

Questions 1 and 2 refer to the following graph.

Benito's Driving Speed Saturday Afternoon

1. For what percent of the time was Benito driving 40 miles per hour or faster?

   A 20   **B** 25   **C** $33\frac{1}{3}$ **D** 40   **E** 50

   HINT Watch for different units of measure.

   Solution Benito drove a total of 3 hours. He drove 40 miles per hour or faster for 1 hour and 30 minutes or $1\frac{1}{2}$ hours. The fraction of the time he drove 40 mph or more is $\dfrac{\frac{3}{2}}{3}$ or $\frac{1}{2}$, which equals 50%. The answer is choice **E**.

2. How far, in miles, did Benito drive between 1:30 and 2:00?

   A 0   **B** 15   **C** 20   **D** 30

   E It cannot be determined from the information given.

   Solution   rate × time = distance
   $$\left(30\ \text{mph}\right)\left(\frac{1}{2}\ \text{hour}\right) = 15\ \text{miles}$$
   The answer is choice **B**.

SAT EXAMPLE

3. For $x = 0$, $x = 1$, and $x = 2$, Set $A = \{x, x + 3, 3x, x^2\}$. What is the mode of Set A?

   A 0   **B** 1   **C** 2   **D** 2.5   **E** 3

   HINT Look carefully at the given information and at the form of the answer choices (numbers, variables, and so on.)

   Solution Notice that the answer choices are numbers. But Set A is defined using variable expressions. First determine the actual data of set A. Consider each value of x, one at a time. Substitute the value for x into each element of Set A.

   For $x = 0$: $x = 0$, $x + 3 = 3$, $3x = 0$, and $x^2 = 0$.

   When $x = 0$, $A = \{0, 3, 0, 0\}$.

   When $x = 1$, $A = \{1, 4, 3, 1\}$.

   When $x = 2$, $A = \{2, 5, 6, 4\}$.

   Thus, $A = \{0, 0, 0, 1, 1, 2, 3, 3, 4, 4, 5, 6\}$. The element 0 occurs three times and no other element occurs as many times. So the mode of Set A is 0. The answer is choice **A**.

   You might notice that choice D, 2.5, is the value of the median set A.

 TestCheck and Worksheet Builder

Special banks of SAT, ACT, and TIMSS items are included in this CD-ROM.

After you work each problem, record your answer on the answer sheet provided or on a piece of paper.

Multiple Choice

1. Based on the graph below, which worker had the greatest percent increase in income from week 1 to week 2? **D**

A Amy	B Brad	C Cara
D Dan	E Elsa	

2. If $a = b + bc$, then in terms of c, what does $\dfrac{b}{a}$ equal? **B**

A $\dfrac{1}{c + 2}$ B $\dfrac{1}{c + 1}$ C $\dfrac{1}{c}$

D c E $c + 1$

3. If 0.1% of m is equal to 10% of n, then m is what percent of $10n$? **D**

A 1/1000% B 10% C 100%

D 1000% E 10,000%

4. S is the set of all positive numbers n such that $n < 100$ and $\sqrt{n}$ is an integer. What is the median value of the members of set S? **C**

A 5 B 5.5 C 25

D 50 E 99

5. In the figure, D, B, and E are collinear. What is the measure of $\angle ABC$? **E**

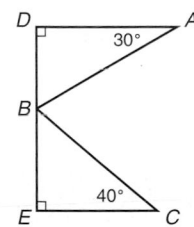

A 20° B 35° C 50° D 60° E 70°

6. How many of the scores 10, 20, 30, 35, 35, and 50 are greater than the arithmetic mean of the scores? **D**

A 0 B 1 C 2

D 3 E 4

7. In $\triangle ABC$, what is the ratio $\dfrac{\tan A}{\text{area } \triangle ABC}$? **D**

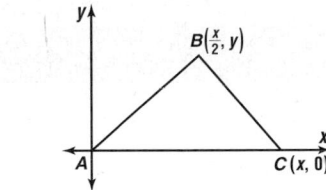

A $\dfrac{1}{2y^2}$ B $\dfrac{1}{y^2}$ C $\dfrac{2}{x^2}$ D $\dfrac{4}{x^2}$ E $\dfrac{x^2}{4}$

8. Based on the data in the table below, how many employees can this company expect to have by 2003? **D**

Year	1997	1998	1999	2000	2001
Number of Employees	1900	2200	2500	2800	3100

A 3100 B 3400 C 3550

D 3700 E 4000

9. What is the difference between the median of Set A and the mean of Set B? **C**

Set A: {2, −1, 7, −4, 11, 3}
Set B: {10, 5, −3, 4, 7, −8}

A 2
B 1.5
C 0
D −0.5
E −2

10. **Grid-In** What is the arithmetic mean of the ten numbers below? **0.6 or 6/10 or 3/5**

−820, −65, −32, 0, 1, 2, 3, 32, 65, 820

  SAT/ACT Practice For additional test practice questions, visit: **www.amc.glencoe.com**

THE PRINCETON REVIEW *SAT & ACT Preparation* **939**

A bubble-in answer sheet for these practice problems is available on p. 1 of the *Assessment and Evaluation Masters.*

Additional Practice

Additional SAT and ACT practice questions and a bubble-in answer sheet are available in the *Assessment and Evaluation Masters,* pp. 2, 357, and 358.

Teaching Tip For Exercise 9, remind students to arrange Set A from least to greatest before determining the median.

Assessment and Evaluation Masters, pp. 357–358

Extra Credit

TIMSS A warning system installation consists of two independent alarms having probabilities of operating in an emergency of 0.95 and 0.90, respectively. Find the probability that at least one alarm operates in an emergency. **A**

A 0.995 B 0.975 C 0.95 D 0.90 E 0.855

Instructional Objectives

Lesson (pages)	Lesson Objectives	NCTM Standards 2000	State/Local Objectives
15-1 (941–948)	• Calculate limits of polynomial and rational functions algebraically. • Evaluate limits of functions using a calculator.	2, 6	
15-2A (949–950)	• Approximate the slope of a curve.	2, 6	
15-2 (951–960)	• Find derivatives and antiderivatives of polynomial functions. • Use derivatives and antiderivatives in applications.	2, 6	
15-3 (961–968)	• Find values of integrals of polynomial functions. • Find areas under graphs of polynomial functions.	2, 3, 4, 6, 10	
15-4 (970–976)	• Use the Fundamental Theorem of Calculus to evaluate definite integrals of polynomial functions. • Find indefinite integrals of polynomial functions.	2, 3, 4, 6, 10	

Suggested Pacing   *See pages T2–T4 for additional options in course-planning.*

COURSE	Lesson 15-1	Lesson 15-2	Lesson 15-3	Lesson 15-4
Precalculus Emphasis Chapters 4–15 (45–50 minutes)	1.5 days	(with Lesson 15-2A) 1.5 days	1.5 days	1.5 days
Precalculus Emphasis (block)	2 days		(with Lesson 15-2A) 1.5 days	(with Study Guide) 1.5 days
Trigonometric Emphasis Chapters 1–9, 11, 12 (45–50 minutes)	OPTIONAL			
Trigonometric Emphasis (block)	OPTIONAL			

Instructional Resources

Lesson	Graphing Calculator Applications	Study Guide and Practice	Enrichment	Assessment and Evaluation	Graphing Calculator Keystrokes	Problem Solving & Applications	Spreadsheet Applications	TI-92 and CBL Interdisciplinary Activities	Transparencies A and B
	Student Edition	**Blackline Masters (page numbers)**							
15-1	944, 945, 947	269–270	165	381	56				15-1
15-2A	949–950				57–60				
15-2		271–272	166	380, 381		29	15		15-2
15-3		273–274	167	382					15-3
15-4		275–276	168	382		30			15-4
Study Guide/ Assessment				367–379, 385					
SAT/ACT Practice				383–384					

Study Guide & Assessment	Chapter Test	TOTAL
1 day	1 day	8 days
(See Lesson 15-4)	0.5 day	5.5 days
OPTIONAL		
OPTIONAL		

TeacherWorks™

The worksheets shown on this page are a small sample of the materials available on *TeacherWorks: All-in-One Lesson Planner and Resource Center*.

This CD-ROM contains all of the blackline masters and transparencies available for this program.

It also includes a lesson planner and interactive Teacher Edition, so you can customize lesson plans and reproduce classroom resources quickly and easily, from just about anywhere.

Applications

Problem Solving and Applications Masters, pp. 29, 30

Technology/Multimedia

Spreadsheet Applications Masters, p. 15

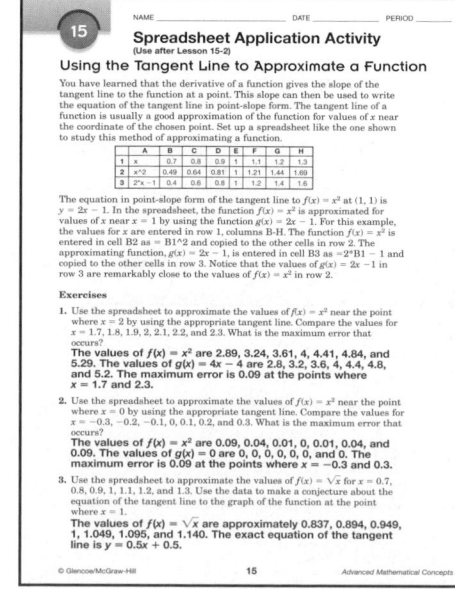

Assessment Resources

Type	Student Edition	Teacher's Wraparound Edition	Assessment and Evaluation Masters
Ongoing Assessment	Mid-Chapter Quiz, p. 960	5-Minute Check, pp. 941, 951, 961, 970	Mid-Chapter Test, p. 380 Quizzes A–D, pp. 381–382
Mixed Review	Mixed Review, pp. 948, 959–960, 968, 975–976		Cumulative Review, p. 385
Error Analysis	You Decide, pp. 966, 972	Error Analysis, pp. 946, 957, 966, 972	
Standardized Test Practice	SAT/ACT Practice, pp. 948, 960, 968, 976 SAT & ACT Preparation, pp. 982–983	TIMSS, p. 983	SAT/ACT Practice, pp. 383–384
Open-Ended Assessment	Math Journal, pp. 946, 957 Open-Ended Assessment, p. 981 Portfolio, p. 981	Modeling: p. 948 Speaking: pp. 950, 968 Writing: pp. 960, 976	Open-Ended Assessment, p. 379
Chapter Assessment	Study Guide and Assessment, pp. 977–981 Chapter Test, p. A70		Multiple-Choice Tests (Forms 1A, 1B, 1C), pp. 367–372 Free-Response Tests (Forms 2A, 2B, 2C), pp. 373–378

TestCheck and Worksheet Builder is a networkable CD-ROM with three integrated modules. The **Worksheet Builder** creates customized worksheets, tests, and quizzes of free-response, multiple-choice, short-answer, and open-ended items. The **Student Module** gives you the option of having students take tests on-screen and get immediate feedback on their performance. Use the optional **Management System** to keep detailed student records.

Additional Chapter Resources

Student Edition
Career Choices, p. 976
History of Mathematics, p. 969
Internet Project, p. 981
Extra Practice, p. A55

Teacher's Classroom Resources

Technology
Graphing Calculator Keystrokes Masters

Teaching Aids
Answer Key Transparencies
Block Schedule Planning Guide
Lesson Planning Guide
Solutions Manual

Glencoe Instructional Technology

Instructional
Advanced MathPASS CD-ROM
Multimedia Applications CD-ROM

Assessment
TestCheck and Worksheet Builder

Visit **www.amc.glencoe.com** for data updates, career information, graphing calculator programs, and other interactive activities.

Unit 5 contains only one chapter, Chapter 15. Thus, the Chapter 15 resources are also the Unit 5 resources.

Reproducible Masters
• *Math and Science Activities*, pp. 24–29

Chapter Notes   15

Chapter Overview

This chapter reviews the fundamentals of calculus. Students evaluate limits, determine derivatives and antiderivatives, and explore finding the areas under curves. Students finish the chapter by learning the Fundamental Theorem of Calculus.

Using a Graphing Calculator

There are numerous opportunities, all optional, to use a graphing calculator in this chapter.
• Using tables to generate values for functions (p. 944)
• Finding a limit (pp. 945, 947)
• Approximating the slope of a curve (pp. 949–950)
• Using a program to approximate the area under a curve (p. 961)

The **Graphing Calculator Appendix**, which begins on page A2, contains keystroke instructions on many of these topics.

UNIT 5   Calculus

Calculus is one of the most important areas of mathematics. There are two branches of calculus, *differential* calculus and *integral* calculus. Differential calculus deals mainly with variable, or changing, quantities. Integral calculus deals mainly with finding sums of infinitesimally small quantities. This generally involves finding a limit.

Chapter 15, the only chapter in Unit 5, provides an overview of some aspects and applications of calculus.

Chapter 15   Introduction to Calculus

CHAPTER OBJECTIVES
• **Evaluate limits of functions.** *(Lesson 15-1)*
• **Find derivatives and antiderivatives of polynomial functions.** *(Lessons 15-2, 15-4)*
• **Evaluate definite integrals using limits and the Fundamental Theorem of Calculus.** *(Lessons 15-3, 15-4)*

Unit 5   *inter*NET   Project

DISEASES

Did you know that many communicable diseases have been virtually eliminated as the result of vaccinations? In 1954, Jonas Salk invented a vaccine for polio. Polio was a dreaded disease from about 1942 to 1954. In 1952, there were 60,000 cases reported. As a result of Salk's miraculous discovery, there were only 5 cases of polio reported in the United States in 1996. In this project, you will look at data about dise   the United States.

CHAPTER 15 (page 981)   **Miracles of Science!** Even though many diseases that once disabled or even killed many people have been controlled, the treatment or cure for many other diseases still eludes researches. Use the Internet to find data on a particular disease.
Math Connection: Model the data with at least two functions. Predict the course of the disease in the future using your model.

   For more information on the Unit Project, visit:
www.amc.glencoe.com

THE PRINCETON REVIEW   SAT & ACT Preparation

This textbook contains a complete SAT/ACT preparation course described on pages T8–T9 of the *Teacher's Handbook*. This chapter contains:

• an SAT or ACT practice question at the end of each lesson
• examples of SAT problems involving special function and counting problems on page 982
• ten SAT practice problems on page 983

• additional SAT and ACT practice in the *Assessment and Evaluation Masters,* pp. 383–384
• SAT and ACT test practice banks in the *TestCheck and Worksheet Builder*

15-1

Limits

OBJECTIVES
* Calculate limits of polynomial and rational functions algebraically.
* Evaluate limits of functions using a calculator.

SPORTS In football, if the length of a penalty exceeds half the distance to the offending team's goal line, then the ball is moved only half the distance to the goal line. Suppose one team has the ball at the other team's 10-yard line. The other team, in an effort to prevent a touchdown, repeatedly commits penalties. After the first penalty, the ball would be moved to the 5-yard line.

The results of the subsequent penalties are shown in the table. Assuming the penalties could continue indefinitely, would the ball ever actually cross the goal line?

Penalty	1st	2nd	3rd	...
Yard Line	5	2.5	1.25	...

The ball will never reach the goal line, but it will get closer and closer after each penalty. As you saw in Chapter 12, a number that the terms of a sequence approach, without necessarily reaching it, is called a **limit.** In the application above, the limit is the goal line or 0-yard line. The idea of a limit also exists for functions.

Limit of a Function	If there is a number L such that the value of $f(x)$ gets closer and closer to L as x gets closer to a number a, then L is called the limit of $f(x)$ as x approaches a. In symbols, $L = \lim\limits_{x \to a} f(x)$.

Example **Consider the graph of the function $y = f(x)$ shown at the right. Find each pair of values.**

a. $f(2)$ and $\lim\limits_{x \to 2} f(x)$

At the point on the graph where the x-coordinate is 2, the y-coordinate is 6. So, $f(2) = 6$.

Look at points on the graph whose x-coordinates are close to, but not equal to, 2. Notice that the closer x is to 2, the closer y is to 6. So, $\lim\limits_{x \to 2} f(x) = 6$.

Lesson 15-1 Limits **941**

Resource Manager

 Reproducible Masters
* *Study Guide and Practice,* pp. 269–270
* *Enrichment,* p. 165
* *Assessment and Evaluation,* p. 381
* *Graphing Calculator Keystrokes,* p. 56

 Transparencies
* 5-Minute Check, 15-1
* Teaching, 15-1

 Technology/Multimedia
* Advanced MathPASS, Lesson 30

---

1 FOCUS

5-Minute Check
Chapter 14

The quiz scores for 15 students are 72, 69, 82, 88, 90, 69, 78, 87, 89, 69, 89, 88, 42, 72, and 54.

1. Make a histogram of the data.

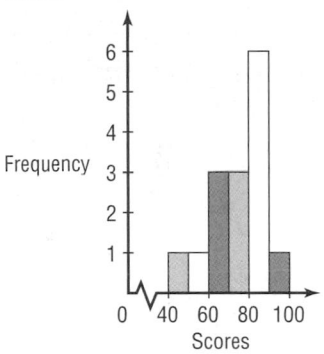

2. What are the mean and the median of the data? **75.9; 78**

3. Find the standard deviation of the data. **about 13.67**

4. Make a box-and-whisker plot of the data.

5. A set of data is normally distributed with a mean of 220 and a standard deviation of 10.

   a. What percent of the data are between 200 and 250? **97.6%**

   b. What percent of the data are between 215 and 235? **62.5%**

   c. Find the interval about the mean that includes 50% of the data. **213–227**

6. Find the interval about the sample mean that has a 5% level of confidence if $\sigma = 12.2$, $N = 500$, and $\overline{X} = 234$. **about 232.93–235.07**

Real-World Application Have students try to compress a spring and measure how long its compressed length is. It is best if the spring can be compressed partially by most students, yet can be compressed completely by only a few students. Point out that the lengths of the compressed spring approach a limiting value.

2 TEACH

Teaching Tip For Example 1 in the Student Edition, point out that the function is discontinuous only at $x = 4$.

In-Class Examples

Example 1
Consider the graph of the function $y = f(x)$. Find each pair of values.

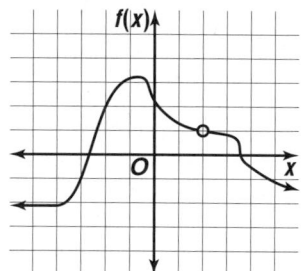

a. $f(-2)$ and $\lim\limits_{x \to -2} f(x)$
$f(-2) = 2$, $\lim\limits_{x \to -2} f(x) = 2$

b. $f(2)$ and $\lim\limits_{x \to 2} f(x)$ $f(2)$ is
undefined, $\lim\limits_{x \to 2} f(x) = 1$

Example 2
Evaluate each limit.
a. $\lim\limits_{x \to 2} (2x^3 - 6x^2 - 10x + 10)$
-18

b. $\lim\limits_{x \to \pi} \dfrac{\sin 3x}{x}$ 0

Teaching Tip For Example 2, explain that evaluating the limit of a continuous function is done by substituting for x the value that x is approaching.

---

b. $f(4)$ and $\lim\limits_{x \to 4} f(x)$

The hole in the graph indicates that the function does not have a value when $x = 4$. That is, $f(4)$ is undefined.

Look at points on the graph whose x-coordinates are close to, but not equal to, 4. The closer x is to 4, the closer y is to 3. So, $\lim\limits_{x \to 4} f(x) = 3$.

You can see from Example 1 that sometimes $f(a)$ and $\lim\limits_{x \to a} f(x)$ are the same, but at other times they are different. In Lesson 3-5, you learned about continuous functions and how to determine whether a function is continuous or discontinuous for a given value. We can use the definition of continuity to make a statement about limits.

Limit of a Continuous Function	$f(x)$ is continuous at a if and only if $$\lim\limits_{x \to a} f(x) = f(a).$$

Examples of continuous functions include polynomials as well as the functions $\sin x$, $\cos x$, and a^x. Also, $\log_a x$ is continuous if $x > 0$.

Example 2 Evaluate each limit.

a. $\lim\limits_{x \to 3} (x^3 - 5x^2 + 7x - 10)$

Since $f(x) = x^3 - 5x^2 + 7x - 10$ is a polynomial function, it is continuous at every number. So the limit as x approaches 3 is the same as the value of $f(x)$ at $x = 3$.

$\lim\limits_{x \to 3} (x^3 - 5x^2 + 7x - 10) = 3^3 - 5 \cdot 3^2 + 7 \cdot 3 - 10$ *Replace x with 3.*
$= 27 - 45 + 21 - 10$
$= -7$

The limit of $x^3 - 5x^2 + 7x - 10$ as x approaches 3 is -7.

b. $\lim\limits_{x \to \pi} \dfrac{\cos x}{x}$

Since the denominator of $\dfrac{\cos x}{x}$ is not 0 at $x = \pi$, the function is continuous at $x = \pi$.

$\lim\limits_{x \to \pi} \dfrac{\cos x}{x} = \dfrac{\cos \pi}{\pi}$ *Replace x with π.*

$= \dfrac{-1}{\pi}$ $\cos \pi = -1$

The limit of $\dfrac{\cos x}{x}$ as x approaches π is $-\dfrac{1}{\pi}$.

Limits can also be used to model real-world situations in which values approach a given value.

📂 Prerequisite Skills

A review of the prerequisite skills for this lesson is provided in the *Prerequisite Skills Masters.*

Skill Number and Content	Pages
6 Polynomials	13–14
7 Operations with Polynomials	15–16
8 Rationalizing a Denominator	17–18
9 Factoring by Grouping and Greatest Common Factor	19–20
10 Factoring Special Products of Polynomials	21–22
11 Factoring by Completing the Square	23–24

3 **PHYSICS** According to the special theory of relativity developed by Albert Einstein, the length of a moving object, as measured by an observer at rest, shrinks as its speed increases. (The difference is only noticeable if the object is moving very fast.) If L_0 is the length of the object when it is at rest, then its length L, as measured by an observer at rest, when traveling at speed v is given by the formula $L = L_0\sqrt{1 - \dfrac{v^2}{c^2}}$, where c is the speed of light. If the space shuttle were able to approach the speed of light, what would happen to its length?

We need to find $\lim\limits_{v \to c} L_0\sqrt{1 - \dfrac{v^2}{c^2}}$.

$$\lim_{v \to c} L_0\sqrt{1 - \frac{v^2}{c^2}} = L_0\sqrt{1 - \frac{c^2}{c^2}} \quad \textit{Replace } v \textit{ with } c, \textit{ the speed of light.}$$

$$= L_0\sqrt{0}$$

$$= 0$$

The closer the speed of the shuttle is to the speed of light, the closer the length of the shuttle, as seen by an observer at rest, gets to 0.

When a function is not continuous at the x-value in question, it is more difficult to evaluate the limit. Consider the function $f(x) = \dfrac{x^2 - 9}{x - 3}$. This function is not continuous at $x = 3$, because the denominator is 0 when $x = 3$. To compute $\lim\limits_{x \to 3} f(x)$, apply algebraic methods to decompose the function into a simpler one.

$$\frac{x^2 - 9}{x - 3} = \frac{(x + 3)(x - 3)}{x - 3} \quad \textit{Factor.}$$

$$= x + 3, x \neq 3 \quad \textit{Simplify.}$$

When computing the limit, we are only interested in x-values close to 3. What happens when $x = 3$ is irrelevant, so we can replace $f(x)$ with the simpler expression $x + 3$.

$$\lim_{x \to 3} \frac{x^2 - 9}{x - 3} = \lim_{x \to 3} (x + 3)$$

$$= 3 + 3 \text{ or } 6$$

The graph of $f(x)$ indicates that this answer is correct. As x gets closer to 3, the y-coordinates get closer and closer to, but never equal, 6. The limit is 6.

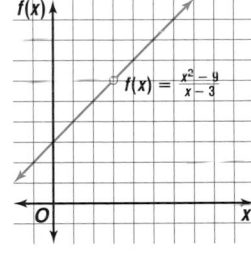

---

944 Chapter 15

Teaching Tip Remind students that the definition of a limit has x approaching a. The x-values do not actually reach a.

In-Class Example

Example 4

Evaluate each limit.

a. $\lim\limits_{x \to 2} \dfrac{x^2 - 2x}{x^2 - 4}$ $\dfrac{1}{2}$

b. $\lim\limits_{x \to 1} \dfrac{x^4 - 1}{x^2 - 1}$ 2

 **Teaching Tip** To view more decimal places than shown in the table of a graphing calculator, move the cursor to the desired y-value and look at the value shown at the bottom of the screen.

Graphing Calculator Tip

Enter the function in the **Y=** menu and set **Indpnt** to **Ask** in the **TBLSET** menu to help generate these values.

Example **4** **Evaluate each limit.**

a. $\lim\limits_{x \to 4} \dfrac{x^2 - 2x - 8}{x^2 - 4x}$

$$\lim\limits_{x \to 4} \dfrac{x^2 - 2x - 8}{x^2 - 4x} = \lim\limits_{x \to 4} \dfrac{(x + 2)(x - 4)}{x(x - 4)}$$

$$= \lim\limits_{x \to 4} \dfrac{x + 2}{x}$$

$$= \dfrac{4 + 2}{4} \text{ or } \dfrac{3}{2} \quad \textit{Replace x with 4.}$$

b. $\lim\limits_{h \to 0} \dfrac{h^3 - 4h^2 - 6h}{h}$

$$\lim\limits_{h \to 0} \dfrac{h^3 - 4h^2 - 6h}{h} = \lim\limits_{h \to 0} \dfrac{h(h^2 - 4h - 6)}{h}$$

$$= \lim\limits_{h \to 0} (h^2 - 4h - 6)$$

$$= 0^2 - 4 \cdot 0 - 6 \text{ or } -6 \quad \textit{Replace h with 0.}$$

Sometimes algebra is not sufficient to find a limit. A calculator may be useful. Consider the problem of finding $\lim\limits_{x \to 0} \dfrac{\sin x}{x}$, where x is in radians. The function is not continuous at $x = 0$, so the limit cannot be found by replacing x with 0. On the other hand, the function cannot be simplified to help make the limit easier to find. You can use a calculator to compute values of the function $\dfrac{\sin x}{x}$ for x-values that get closer and closer to 0 from either side (that is, both less than 0 and greater than 0).

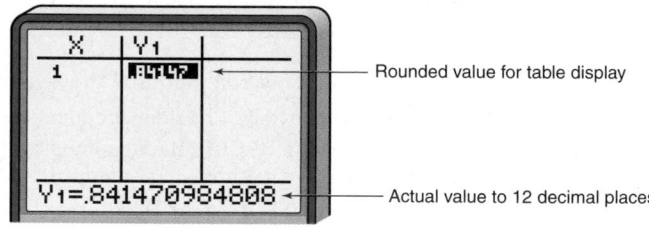

Rounded value for table display

Actual value to 12 decimal places

The tables below show the expression evaluated for values of x that approach 0.

x	$\dfrac{\sin x}{x}$
1	0.841470984808
0.1	0.998334166468
0.01	0.999983333417
0.001	0.999999833333
0.0001	0.999999998333

x	$\dfrac{\sin x}{x}$
−1	0.841470984808
−0.1	0.998334166468
−0.01	0.999983333417
−0.001	0.999999833333
−0.0001	0.999999998333

As x gets closer and closer to 0, from either side, the value of $\dfrac{\sin x}{x}$ gets closer and closer to 1. That is, $\lim\limits_{x \to 0} \dfrac{\sin x}{x} = 1$.

Example **5** Evaluate each limit.

a. $\lim\limits_{x \to 0} \dfrac{1 - \cos x}{x^2}$ (*x* is in radians.)

A graphing calculator or spreadsheet can generate more decimal places for the expression than shown here.

x	$\dfrac{1 - \cos x}{x^2}$
1	0.45970
0.1	0.49958
0.01	0.499996
0.001	0.49999996

x	$\dfrac{1 - \cos x}{x^2}$
−1	0.45970
−0.1	0.49958
−0.01	0.499996
−0.001	0.49999996

As x approaches 0, the value of $\dfrac{1 - \cos x}{x^2}$ gets closer to 0.5, so $\lim\limits_{x \to 0} \dfrac{1 - \cos x}{x^2} = 0.5$.

b. $\lim\limits_{x \to 1} \dfrac{\ln x}{x - 1}$

x	$\dfrac{\ln x}{x - 1}$
0.9	1.0536
0.99	1.0050
0.999	1.0005

x	$\dfrac{\ln x}{x - 1}$
1.1	0.95310
1.01	0.99503
1.001	0.99950

The closer x is to 1, the closer $\dfrac{\ln x}{x - 1}$ is to 1, so $\lim\limits_{x \to 1} \dfrac{\ln x}{x - 1} = 1$.

Using a calculator is not a foolproof way of evaluating $\lim\limits_{x \to a} f(x)$. You may only analyze the values of $f(x)$ for a few values of x near a. However, the function may do something unexpected as x gets even closer to a. You should use algebraic methods whenever possible to find limits.

GRAPHING CALCULATOR EXPLORATION

You can use a graphing calculator to find a limit, with less work than an ordinary scientific calculator. To find $\lim\limits_{x \to a} f(x)$, first graph the equation $y = f(x)$. Then use [ZOOM] and [TRACE] to locate a point on the graph whose x-coordinate is as close to a as you like. The y-coordinate should be close to the value of the limit.

TRY THESE Evaluate each limit.

1. $\lim\limits_{x \to 0} \dfrac{e^x - 1}{x}$ **1**

2. $\lim\limits_{x \to 2} \dfrac{x^2 - 4}{x^2 - 3x + 2}$ **4**

WHAT DO YOU THINK?

3. If you graph $y = \dfrac{\ln x}{x - 1}$ and use [TRACE], why doesn't the calculator tell you what y is when $x = 1$? **y is undefined when $x = 1$.**

4. Solve Exercise 2 algebraically. Do you get the same answer as you got from the graphing calculator? **4; yes**

5. Will the graphing calculator give you the exact answer for every limit problem? Explain. **See margin.**

Addressing Individual Needs

Visual/Spatial Using ropes, set up a grid on the floor of a large area. Have one student stand at a predetermined origin. Then have ten to fifteen students form a line on the grid that could represent a function. Have students determine the limit of the function using their locations as x-values. Ask them to determine the new x-values if they were to step out of line. Finally, have students compare their results.

Error Analysis

Watch for students who assign values to the form $\frac{0}{0}$.

Prevent by telling students that if $\frac{0}{0}$ is obtained in a rational function, the function can always be simplified.

Assignment Guide
Core: 13–41 odd, 42–58
Enriched: 12–38 even, 39–58

Additional Answers

1. **Sample answer:** The limit of $f(x)$ as x approaches a is the number that the values of $f(x)$ get closer to as the values of x get closer and closer to a.

2. **Sample answer:** $\lim_{x \to 1} f(x)$ is the number that the values of $f(x)$ approach as x approaches 1. $f(1)$ is the number that you get if you actually plug 1 into the function. They are the same if $f(x)$ is continuous at $x = 1$.

Study Guide and Practice Masters, pp. 269–270

CHECK FOR UNDERSTANDING

Communicating Mathematics

Read and study the lesson to answer each question.   1–3. See margin.

1. **Define** the expression *limit of $f(x)$ as x approaches a* in your own words.

2. **Describe** the difference between $f(1)$ and $\lim_{x \to 1} f(x)$ and explain when they would be the same number.

3. *Math Journal*   **Write** a description of the three methods in this lesson for computing $\lim_{x \to a} f(x)$. Explain when each method would be used and include examples.

Guided Practice

4. Use the graph of $y = f(x)$ to find $\lim_{x \to 0} f(x)$ and $f(0)$.   **3; 1**

Evaluate each limit.

5. $\lim_{x \to 2} (-4x^2 + 2x - 5)$ **−17**

6. $\lim_{x \to 0} (1 + x + 2^x - \cos x)$ **1**

7. $\lim_{x \to -2} \dfrac{x + 2}{x^2 - 4}$ $-\dfrac{1}{4}$

8. $\lim_{x \to 0} \dfrac{x^2 - 3x}{x^3 + 4x}$ $-\dfrac{3}{4}$

9. $\lim_{x \to 3} \dfrac{x^2 + 3x - 10}{x^2 + 5x + 6}$ $\dfrac{4}{15}$

10. $\lim_{x \to -2} \dfrac{2x^2 + 5x + 2}{x^2 + x - 2}$ **1**

11. **Hydraulics**   The velocity of a molecule of liquid flowing through a pipe depends on the distance of the molecule from the center of the pipe. The velocity, in inches per second, of a molecule is given by the function $v(r) = k(R^2 - r^2)$, where r is the distance of the molecule from the center of the pipe in inches, R is the radius of the pipe in inches, and k is a constant. Suppose for a particular liquid and a particular pipe that $k = 0.65$ and $R = 0.5$.

a. Graph $v(r)$.   **See margin.**

b. Determine the limiting velocity of molecules closer and closer to the wall of the pipe.   **0 in./s**

EXERCISES

Practice

A

Use the graph of $y = f(x)$ to find each value.

12. $\lim_{x \to -2} f(x)$ and $f(-2)$ **−1; −1**

13. $\lim_{x \to 0} f(x)$ and $f(0)$ **0; undefined**

14. $\lim_{x \to 3} f(x)$ and $f(3)$ **4; 2**

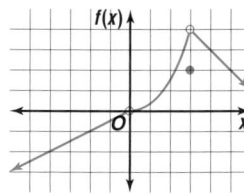

Evaluate each limit.

15. $\lim_{x \to 2} (-4x^2 - 3x + 6)$ **−16**

16. $\lim_{x \to -1} (-x^3 + 3x^2 - 4)$ **0**

B

17. $\lim_{x \to \pi} \dfrac{\sin x}{x}$ **0**

18. $\lim_{x \to 0} (x + \cos x)$ **1**

 www.amc.glencoe.com/self_check_quiz

Additional Answers

3. **Sample answer:** If $f(x)$ is continuous at $x = a$ you can plug a into the function. If the function is not continuous, you may be able to simplify it and then plug in a. If neither of these methods work, you can use a calculator. Examples will vary.

11a.

$$v(r) = 0.65(0.5^2 - r^2)$$

19. $\lim\limits_{x \to 5} \dfrac{x^2 - 25}{x - 5}$ **10**

20. $\lim\limits_{n \to 0} \dfrac{2n^2}{n}$ **0**

21. $\lim\limits_{x \to 3} \dfrac{x^2 - 3x}{x^2 + 2x - 15}$ $\dfrac{3}{8}$

22. $\lim\limits_{x \to 1} \dfrac{x^3 + 3x^2 - 4x + 8}{x + 6}$ $\dfrac{8}{7}$

23. $\lim\limits_{h \to -2} \dfrac{h^2 + 4h + 4}{h + 2}$ **0**

24. $\lim\limits_{x \to 3} \dfrac{2x^2 - 3x}{x^3 - 2x^2 + x + 6}$ $\dfrac{1}{2}$

25. $\lim\limits_{x \to 0} \dfrac{x^3 - x^2 + 2x}{x^3 + 4x^2 - 2x}$ **−1**

26. $\lim\limits_{x \to 0} \dfrac{x \cos x}{x^2 + x}$ **1**

27. $\lim\limits_{x \to 0} \dfrac{(x + 2)^2 - 4}{x}$ **4**

28. $\lim\limits_{x \to -2} \dfrac{(x + 1)^2 - 1}{x + 2}$ **−2**

29. $\lim\limits_{x \to -2} \dfrac{x^3 + 8}{x^2 - 4}$ **−3**

30. $\lim\limits_{x \to 4} \dfrac{2x - 8}{x^3 - 64}$ $\dfrac{1}{24}$

31. $\lim\limits_{x \to 1} \dfrac{\frac{1}{x} - 1}{x - 1}$ **−1**

32. $\lim\limits_{x \to 4} \dfrac{x - 4}{\sqrt{x} - 2}$ **4**

33. Find the limit as h approaches 0 of $\dfrac{2h^3 - h^2 + 5h}{h}$. **5**

34. What value does the function $g(x) = \dfrac{x + \pi}{\cos(x + \pi)}$ approach as x approaches 0? **−π**

Use a graphing calculator to find the value of each limit. (Use radians with trigonometric functions.)

35. $\lim\limits_{x \to 0} \dfrac{\tan 2x}{x}$ **2**

36. $\lim\limits_{x \to 1} \dfrac{\ln x}{\ln(2x - 1)}$ **0.5**

37. $\lim\limits_{x \to 1} \dfrac{1 - \sqrt{x}}{x - 1}$ **−0.5**

38. $\lim\limits_{x \to 0} \dfrac{3x - \sin 3x}{x^2 \sin x}$ **4.5**

39. Geometry The area of an ellipse with semi-major axis a is $\pi a \sqrt{a^2 - c^2}$, where c is the distance from the foci to the center. Find the limit of the area of the ellipse as c approaches 0. Explain why the answer makes sense. **See margin.**

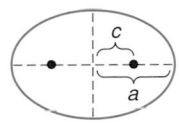

40. Biology If a population of bacteria doubles every 10 hours, then its initial hourly growth rate is $\lim\limits_{t \to 0} \dfrac{2^{\frac{1}{10}} - 1}{t}$, where t is the time in hours. Use a calculator to approximate the value of this limit to the nearest hundredth. Write your answer as a percent. **7%**

41. Critical Thinking Does $\lim\limits_{x \to 0} \sin\left(\dfrac{1}{x}\right)$ exist? That is, can you say $\lim\limits_{x \to 0} \sin\left(\dfrac{1}{x}\right) = L$ for some real number L? Explain why or why not. **See margin.**

42. Critical Thinking You saw in Example 5 that $\lim\limits_{x \to 0} \dfrac{1 - \cos x}{x^2} = 0.5$. That is, for values of x close to 0, $\dfrac{1 - \cos x}{x^2} \approx 0.5$. Solving for $\cos x$, we get $\cos x \approx 1 - \dfrac{x^2}{2}$.

a. Copy and complete the table by using a calculator. Round to six decimal places, if necessary.

x	1	0.5	0.1	0.01	0.001
$\cos x$	0.540302	0.877583	0.995004	0.999950	1.000000
$1 - \dfrac{x^2}{2}$	0.5	0.875	0.995	0.99995	1.000000

b. Is it correct to say that for values of x close to 0, the expression $1 - \dfrac{x^2}{2}$ is a good approximation for $\cos x$? Explain.

42b. yes; in the last three columns, all the decimal places of $1 - \dfrac{x^2}{2}$ agree with those of $\cos x$.

Graphing Calculator

Applications and Problem Solving

Real World Application

Teaching Tip For Exercises 35–38, make sure that students set their calculators to radian mode, not degree mode.

Additional Answers

39. πa^2; letting c approach 0 moves the foci together, so the ellipse becomes a circle. πa^2 is the area of a circle of radius a.

41. No; the graph of $f(x) = \sin\left(\dfrac{1}{x}\right)$ oscillates infinitely many times between −1 and 1 as x approaches 0, so the values of the function do not approach a unique number.

4 ASSESS

Open-Ended Assessment
Modeling Have students graph some of the functions in the exercises on the chalkboard by plotting points with *x*-coordinates closer and closer to *a* on both sides. Then, have them predict the behavior of several other limits of functions and verify their predictions by graphing.

Chapter 15, Quiz A (Lesson 15-1) is available in the *Assessment and Evaluation Masters*, p. 381.

Additional Answers
44b. He ignored the exponent. As *x* approaches 0 from the positive side, $\frac{1}{x}$ approaches infinity. A number close to 1 raised to a large power need not be close to 1. If *x* approaches 0 from the negative side, $\frac{1}{x}$ approaches negative infinity. A number close to 1 raised to a large negative power need not be close to 1, either.

50.

Enrichment Masters, p. 165

```
NAME _____ DATE _____ PERIOD _____
15-1  Enrichment

A Matter of Limits
There are many examples of limits in our world. Some of these
are absolute limits, in that they can never be exceeded. Others are
like guidelines, and still others result in a penalty if they are
exceeded. Fill in the chart below.
```

Limit	How is the limit set?	Is the limit absolute?	Penalty or consequence if the limit is exceeded
1. speed limit on a highway	The government sets the speed limits.	no	speeding ticket, fine, and so on
2. height limit on a road underpass	Height limit is set for safe clearance.	yes	Damage is done to vehicle and/or structure.
3. luggage limit on an airline flight	Airline sets limit on amount of baggage allowed.	no	Passenger must pay for any luggage beyond limit.
4. temperature of a warm object placed in a cool room	can only get as cool as its surroundings	yes	not possible
5. the speed of an accelerating space craft	physical constant, the speed of light	yes	not possible
6. credit limit on a credit card	Set by bank issuing card.	no	Financial penalty set by bank, possibly may lose card.

```
One special feature of mathematical limits is that they may be finite, infinite, or they
may not exist. Classify each limit as finite, infinite, or does not exist. If the limit is
finite, give its value.
```

7. $\lim_{x\to 0} \frac{1}{x^2+1}$ finite; 1
8. $\lim_{x\to 1} \frac{x+1}{x^2-1}$ does not exist
9. $\lim_{x\to 2} \frac{x^2-4}{x^2-x-2}$ finite; $\frac{4}{3}$
10. $\lim_{x\to 0} \ln|x|$ infinite

```
© Glencoe/McGraw-Hill          T165          Advanced Mathematical Concepts
```

948 *Chapter 15*

---

43. Physics When an object, such as a bowling ball, is dropped near Earth's surface, the distance $d(t)$ (in feet) that the object falls in *t* seconds is given by $d(t) = 16t^2$. Its velocity (in feet per second) after 2 seconds is given by $\lim_{t\to 2} \frac{d(t) - d(2)}{t - 2}$. Evaluate this limit algebraically to find the velocity of the bowling ball after 2 seconds. *You will learn more about the relationship between distance and velocity in Lesson 15-2.*   **64 ft/s**

44. Critical Thinking Yoshi decided that $\lim_{x\to 0} (1 + x)^{\frac{1}{x}}$ is 0, because as *x* approaches 0, the base of the exponential expression approaches 1, and 1 to any power is 1.
 a. Use a calculator to help deduce the exact value of $\lim_{x\to 0} (1 + x)^{\frac{1}{x}}$.   **e**
 b. Explain where Yoshi's reasoning was wrong.   **See margin.**

Mixed Review

45. Botany A random sample of fifty acorns from an oak tree in the park reveals a mean diameter of 16.2 millimeters and a standard deviation of 1.4 millimeters. Find the range about the sample mean that gives a 99% chance that the true mean lies within it. *(Lesson 14-5)*   **15.69-16.71 mm**

46. $\frac{59,049}{100,000}$

46. Tess is running a carnival game that involves spinning a wheel. The wheel has the numbers 1 to 10 on it. What is the probability of 7 never coming up in five spins of the wheel? *(Lesson 13-6)*

47. Find the third term of $(x - 3y)^5$. *(Lesson 12-6)*   **$90x^3y^2$**

48. Simplify $(16y^8)^{\frac{3}{4}}$. *(Lesson 11-1)*   **$8y^6$**

49. $\frac{(x-5)^2}{16} + \frac{(y+2)^2}{9} = 1$

49. Write the equation of the ellipse if the endpoints of the major axis are at $(1, -2)$ and $(9, -2)$ and the endpoints of the minor axis are at $(5, 1)$ and $(5, -5)$. *(Lesson 10-3)*

50. Graph the polar equation $r = -3$. *(Lesson 9-1)*   **See margin.**

51. Write the ordered pair that represents $\overrightarrow{WX}$ for $W(4, 0)$ and $X(-3, -6)$. Then find the magnitude of $\overrightarrow{WX}$. *(Lesson 8-2)* $\langle -7, -6 \rangle$; $\sqrt{85}$

52. Transportation A car is being driven at 65 miles per hour. The car's tires have a diameter of 25 inches. What is the angular velocity of the wheels in revolutions per second? *(Lesson 6-2)*   **14.6 rps**

53. Use the unit circle to find the value of csc 270°. *(Lesson 5-3)*   **−1**

54. Determine the rational roots of the equation $12x^4 - 11x^3 - 54x^2 - 18x + 8 = 0$. *(Lesson 4-4)* $\frac{1}{4}, -\frac{4}{3}$

55. Without graphing, describe the end behavior of the function $y = 4x^5 - 2x^2 + 4$. *(Lesson 3-5)*   **$y \to \infty$ as $x \to \infty$, $y \to -\infty$ as $x \to -\infty$.**

56. Find the value of the determinant $\begin{vmatrix} -1 & -2 \\ 3 & -6 \end{vmatrix}$. *(Lesson 2-5)*   **12**

57. yes; opposite sides have the same slope

57. Geometry Determine whether the figure with vertices at $(0, 3)$, $(8, 4)$, $(2, -5)$, and $(10, -4)$ is a parallelogram. Explain. *(Lesson 1-5)*

58. SAT Practice **Grid-In** If $2^n = 8$, what is the value of 3^{n+2}?   **243**

948 *Chapter 15 Introduction to Calculus*

Extra Practice See p. A55.

Extra Credit

A *removable discontinuity* occurs when a rational expression has common linear factors in both the numerator and the denominator. Because these factors can be divided, the discontinuity is "removable." At what point is the discontinuity removable for the function $f(x) = \frac{x^2 + 5x - 24}{x^2 - x - 6}$? $\left(3, \frac{11}{5}\right)$

GRAPHING CALCULATOR EXPLORATION

15-2A The Slope of a Curve

A Preview of Lesson 15-2

OBJECTIVE
• Approximate the slope of a curve.

Recall from Chapter 1 that the slope of a line is a measure of its steepness. The slope of a line is given by the formula $m = \dfrac{y_2 - y_1}{x_2 - x_1}$, where (x_1, y_1) and (x_2, y_2) are the coordinates of two distinct points on the line.

What about the slope of a *curve*? A general curve does not have the same steepness at every point, but if you look at one particular point on the graph, there will be a certain steepness at that point. How would you calculate this "slope" at a particular point?

The answer lies in an important fact about curves: the graphs of most functions are "locally linear." This means that if you look at them up close, they appear to be lines. You are familiar with this phenomenon in everyday life—the surface of Earth looks flat, even though we know it is a giant sphere.

You can use ⎡ZOOM⎤ on a graphing calculator to look very closely at the graph of a function.

Example ❶ **Find the slope of the graph of $y = x^2$ at (1, 1).**

Graph the equation $y = x^2$. Use the window $[0, 2]$ by $[0, 2]$ so that $(1, 1)$ is at the center. Zoom in on the graph four times, using $(1, 1)$ as the center each time. The graph should then look like the screen below. This graph is so straight that it has no visible curvature.

To approximate the slope of the graph, you can use ⎡TRACE⎤ to identify the approximate coordinates of two points on the curve. Then use the formula for slope. For example, use the coordinates $(1, 1)$ and $(1.0000831, 1.0001662)$.

$$m \approx \frac{1.0001662 - 1}{1.0000831 - 1}$$

$$\approx \frac{0.0001662}{0.0000831}$$

$$\approx 2$$

The slope at $(1, 1)$ is approximately 2.

X=1   .Y=1

Motivating the Lesson
Discussion Remind students how to find the slope of a line. Ask them how they might apply that method to finding the slope of a curve.

📁 A reproduction of this lesson with graphing calculator keystrokes provided can be found in the *Graphing Calculator Keystrokes Masters,* pp. 57–60.

Teaching Tip Windows are not given with the graphing calculator screens in this lesson since students are to obtain the screens by following the directions in the text.

3 PRACTICE/APPLY

Assignment Guide

All: 1–9

4 ASSESS

Open-Ended Assessment

Speaking Have students explain verbally the process used to find the slope of the graph of a function.

You can also have the calculator find its own approximation for the slope.

Example **2** Find the slope of the graph of $y = \dfrac{x^2 + 1}{x}$ at (0.5, 2.5).

Method 1: Slope Formula

Graph the equation $y = \dfrac{x^2 + 1}{x}$. Use the window [0, 1] by [2, 3] so that (0.5, 2.5) is the center. Zooming in four times results in the screen shown at the right.

The **TRACE** feature shows that the point at (0.50004156, 2.4998753) is on the graph. Use these coordinates and (0.5, 2.5) to compute an approximate slope.

$$m \approx \frac{2.4998753 - 2.5}{0.50004156 - 0.5}$$
$$\approx -3.00048123$$

Our approximation to the slope is -3.00048123, which is quite close to -3.

Method 2: Calculator Computation

To have the calculator find an approximation, apply the *dy/dx* feature from the **CALC** menu at (0.5, 2.5). The calculator display is shown at the right. This also suggests that the exact value of the slope might be -3.

When you zoom in to measure the slope, you will not always obtain the exact answer. No matter how far you zoom in on the graph of a nonlinear function, the graph is never truly straight, whether it appears to be or not. Your calculation of an approximate slope may not exactly match the calculator's value for *dy/dx*. Sometimes your algebraic approximation may be more accurate. Other times the calculator's approximation may be more accurate.

TRY THESE

1–6. Exact answers are given. Accept all reasonable approximations.

Zoom in to find the slope of the graph of each function at the given point. (Zoom in at least four times before calculating the slope.) Check your answer using the calculator's *dy/dx* feature.

1. $y = 2x^2$; (1, 2)   **4**
2. $y = \sin x$; (0, 0)   **1**
3. $y = \sqrt{x}$; (1, 1)   **0.5**
4. $y = 4x^4 - x^2$; (0.5, 0)   **1**
5. $y = \dfrac{1}{x - 3}$; (4, 1)   **−1**
6. $y = \dfrac{x + 1}{x - 2}$; (1, −2)   **−3**

WHAT DO YOU THINK?

7. For what type of function are the methods described in this lesson guaranteed to always give the exact slope?   **linear functions**

8. What is the slope of a polynomial curve at a maximum or minimum point?   **0**

9. Graph $y = e^x$. Use the *dy/dx* feature to approximate the slope of the curve at several different points. What do you notice about the values of *y* and *dy/dx*?
   They are approximately the same.

15-2

Derivatives and Antiderivatives

OBJECTIVES
* Find derivatives and antiderivatives of polynomial functions.
* Use derivatives and antiderivatives in applications.

ROCKETRY Scott and Jabbar are testing a homemade rocket in Jabbar's back yard. The boys want to keep a record of the rocket's performance so they will know if it improves when they change the design. In physics class they learned that after the rocket uses up its fuel, the rocket's height above the ground is given by the equation $H(t) = H_0 + v_0 t - 16t^2$, where H_0 is the height of the rocket (in feet) when the fuel is used up, v_0 is the rocket's velocity (in feet per second) at that time, and t is the elapsed time (in seconds) since the fuel was used up. Determine the velocity of the rocket when the fuel ran out and the maximum height the rocket reached. *This problem will be solved in Example 3.*

f'(x) is read "f-prime of x."

To solve this type of problem, we need to find the **derivative** of the function H. The derivative is related to the idea of a **tangent line** from geometry. A line tangent to a curve at a point on the curve is the line that passes through that point and has a slope equal to the slope of the curve at that point. The derivative of a function $f(x)$ is another function, $f'(x)$, that gives the slope of the tangent line to $y = f(x)$ at any point.

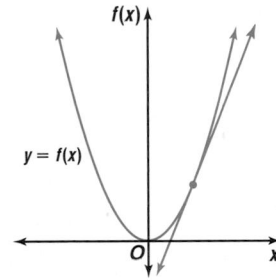

Consider the graph of $y = f(x)$ and a point $(x, f(x))$ on the graph. If the number h is close to 0, the point on the graph with x-coordinate $x + h$ will be close to $(x, f(x))$. The y-coordinate of this second point is $f(x + h)$.

Now consider the line through the points $(x, f(x))$ and $(x + h, f(x + h))$. A line that intersects a graph in two points like this is called a **secant line**. The slope of this secant line is

$$m = \frac{f(x + h) - f(x)}{(x + h) - x} \text{ or } \frac{f(x + h) - f(x)}{h}.$$

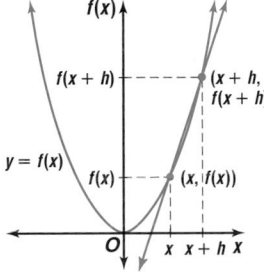

If we make h closer and closer to 0, the point $(x + h, f(x + h))$ will get closer and closer to the original point $(x, f(x))$, so the secant line will look more and more like a tangent line. This means we can compute the slope of the tangent line by finding $\lim_{h \to 0} \frac{f(x + h) - f(x)}{h}$. This limit is the derivative of the function $f(x)$.

   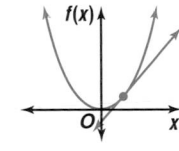

h approaches 0.

Lesson 15-2 Derivatives and Antiderivatives **951**

Resource Manager

 Reproducible Masters
* *Study Guide and Practice,* pp. 271–272
* *Enrichment,* p. 166
* *Assessment and Evaluation,* pp. 380, 381
* *Problem Solving and Applications,* p. 29
* *Spreadsheet Applications,* p. 15

 Transparencies
* 5-Minute Check, 15-2
* Teaching, 15-2

 Technology/Multimedia
* Advanced MathPASS, Lesson 31

 5-Minute Check
Lesson 15-1

1. Use the graph of $y = f(x)$ to find each value.

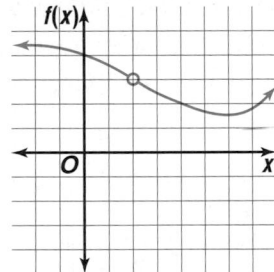

a. $\lim_{x \to 4} f(x)$ and $f(4)$ **2; 2**

b. $\lim_{x \to 2} f(x)$ and $f(2)$ **3; undefined**

Evaluate each limit.

2. $\lim_{x \to 3} \frac{x^2 - 2x - 3}{x - 3}$ **4**

3. $\lim_{x \to 0} \frac{\sin x}{x^2 + 3x}$ $\frac{1}{3}$

4. $\lim_{x \to 0} \frac{(x - 2)^2}{3}$ $\frac{4}{3}$

5. Find the limit as x approaches 2 of $\frac{x^3 - 8}{x^2 - 4}$. **3**

Motivating the Lesson
Hands-On Activity Give pairs of students stiff pieces of wire, such as straightened coat hangers, and rulers. Have each pair form the wire into a parabola and lay it on graph paper. Then, have students place the ruler so that it touches the parabola at one point and does not go inside the parabola. Ask them to determine the equation for the parabola and the slope of the ruler. Use this example to introduce the lesson.

Teaching Tip Before defining *derivative of a function,* ask those students who have studied physics whether they can recall the definition of *instantaneous velocity at time t.* Compare the definition $\lim\limits_{\Delta t \to 0} \dfrac{f(t + \Delta t) - f(t)}{\Delta t}$ to the one that will be presented.

Teaching Tip Before working Example 1, remind students that $\dfrac{dy}{dx}$ is the limit of "the change in y over the change in x," which is the limit of slope.

In-Class Example

Example 1

a. Find an expression for the slope of the tangent line to the graph of $y = 2x^2 - 3x + 4$ at any point. That is, compute $\dfrac{dy}{dx}$. **$4x - 3$**

b. Find the slopes of the tangent lines when $x = -1$ and $x = 5$. **at $x = -1$, $\dfrac{dy}{dx} = -7$; at $x = 5$, $\dfrac{dy}{dx} = 17$**

Derivative of a Function

The derivative of the function $f(x)$ is the function $f'(x)$ given by

$$f'(x) = \lim_{h \to 0} \frac{f(x + h) - f(x)}{h}.$$

$\dfrac{dy}{dx}$ is read "dy, dx." *This notation emphasizes that the derivative is a limit of slope, which is a change in y divided by a change in x.*

The process of finding the derivative is called **differentiation**. Another common notation for $f'(x)$ is $\dfrac{dy}{dx}$. The following chart summarizes the information about tangent lines and secant lines.

Type of Line	Points of Intersection with Graph	Example	Slope
Tangent	1		$\dfrac{dy}{dx} = f'(x) =$ $\lim\limits_{h \to 0} \dfrac{f(x + h) - f(x)}{h}$
Secant	2		$m = \dfrac{f(x + h) - f(x)}{h}$

Example 1

a. Find an expression for the slope of the tangent line to the graph of $y = x^2 - 4x + 2$ at any point. That is, compute $\dfrac{dy}{dx}$.

b. Find the slopes of the tangent lines when $x = 0$ and $x = 3$.

a. Find and simplify $\dfrac{f(x + h) - f(x)}{h}$, where $f(x) = x^2 - 4x + 2$.

First, find $f(x + h)$.

$f(x + h) = (x + h)^2 - 4(x + h) + 2$ *Replace x with x + h in f(x).*

$= x^2 + 2xh + h^2 - 4x - 4h + 2$

Now find $\dfrac{f(x + h) - f(x)}{h}$.

$\dfrac{f(x + h) - f(x)}{h} = \dfrac{x^2 + 2xh + h^2 - 4x - 4h + 2 - (x^2 - 4x + 2)}{h}$

$= \dfrac{2xh + h^2 - 4h}{h}$ *Simplify.*

$= \dfrac{h(2x + h - 4)}{h}$ *Factor.*

$= 2x + h - 4$ *Divide by h.*

952 *Chapter 15 Introduction to Calculus*

Prerequisite Skills

A review of the prerequisite skills for this lesson is provided in the *Prerequisite Skills Masters.*

Skill Number and Content	Pages
3 Order of Operations	7–8

Now find the limit of $2x + h - 4$ as h approaches 0 to compute $\dfrac{dy}{dx}$.

$$\frac{dy}{dx} = f'(x)$$

$$= \lim_{h \to 0} \frac{f(x + h) - f(x)}{h}$$

$$= \lim_{h \to 0} (2x + h - 4)$$

$$= 2x + 0 - 4$$

$$= 2x - 4$$

In the limit, only h approaches 0. x is fixed.

So $\dfrac{dy}{dx} = 2x - 4$.

b. At $x = 0$, $\dfrac{dy}{dx} = 2(0) - 4$ or -4. The slope of the tangent line at $x = 0$ is -4.

At $x = 3$, $\dfrac{dy}{dx} = 2(3) - 4$ or 2. The slope of the tangent line at $x = 3$ is 2.

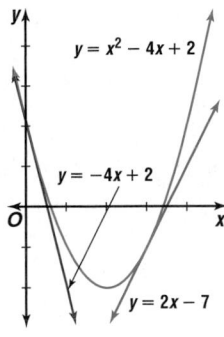

To find the derivatives of polynomials, you can use the following rules.

Derivative Rules

Constant Rule:	The derivative of a constant function is zero. If $f(x) = c$, then $f'(x) = 0$.	
Power Rule:	If $f(x) = x^n$, where n is a rational number, then $f'(x) = nx^{n-1}$.	
Constant Multiple of a Power Rule:	If $f(x) = cx^n$, where c is a constant and n is a rational number, then $f'(x) = cnx^{n-1}$.	
Sum and Difference Rule:	If $f(x) = g(x) \pm h(x)$, then $f'(x) = g'(x) \pm h'(x)$.	

Example ② **Find the derivative of each function.**

a. $f(x) = x^6$

$f'(x) = 6x^{6-1}$ *Power Rule*
$\quad\;\; = 6x^5$

b. $f(x) = x^2 - 4x + 2$

$f(x) = x^2 - 4x + 2$
$\quad\;\; = x^2 - 4x^1 + 2$ *Rewrite x as a power.*

$f'(x) = 2x^{2-1} - 4 \cdot 1x^{1-1} + 0$ *Use all four rules.*
$\quad\;\; = 2x^1 - 4x^0$
$\quad\;\; = 2x - 4$ $\qquad\qquad x^0 = 1$

c. $f(x) = 2x^4 - 7x^3 + 12x^2 - 8x - 10$

$f'(x) = 2 \cdot 4x^3 - 7 \cdot 3x^2 + 12 \cdot 2x - 8 \cdot 1 - 0$
$\quad\;\; = 8x^3 - 21x^2 + 24x - 8$

Lesson 15-2   Derivatives and Antiderivatives **953**

Teaching Tip For Example 2, make sure that students treat each term of the polynomials separately.

Teaching Tip Have students compare the results in Example 2b with those in Example 1a.

In-Class Example

Example 2
Find the derivative of each function.
a. $f(x) = x^5$ $5x^4$
b. $f(x) = x^3 + 2x$ $3x^2 + 2$
c. $f(x) = 2x^5 - x + 5$ $10x^4 - 1$
d. $f(x) = x^6 - x^5 - x^4$
$\quad 6x^5 - 5x^4 - 4x^3$
e. $f(x) = (x + 1)(x^2 - 2)$
$\quad 3x^2 + 2x - 2$

d. $f(x) = x^3 (x^2 + 5)$

$f(x) = x^3 (x^2 + 5)$

$\qquad = x^5 + 5x^3$ *Multiply to write the function as a polynomial.*

$f'(x) = 5x^4 + 5 \cdot 3x^2$

$\qquad = 5x^4 + 15x^2$

e. $f(x) = (x^2 + 4)^2$

$f(x) = (x^2 + 4)^2$

$\qquad = x^4 + 8x^2 + 16$ *Square to write the function as a polynomial.*

$f'(x) = 4x^3 + 8 \cdot 2x + 0$

$\qquad = 4x^3 + 16x$

Suppose $s(t)$ is the displacement of a moving object at time t. For example, $s(t)$ might be the object's altitude or its distance from its starting point. Then the derivative, denoted $s'(t)$ or $\frac{ds}{dt}$, is the velocity of the object at time t. Velocity is usually denoted by $v(t)$.

Example **③ ROCKETRY**   Refer to the application at the beginning of the lesson. Suppose Scott's stopwatch shows that the rocket reached its highest point 5.3 seconds after its fuel was exhausted. Jabbar's stopwatch says that the rocket hit the ground 12.7 seconds after the fuel ran out.

a. How fast was the rocket moving at the instant its fuel ran out?

b. What was the maximum height of the rocket?

a. We have to find the value of v_0. This value cannot be found directly from the height function $H(t)$ because H_0 is still unknown. Instead we use the velocity function $v(t)$ and what we can deduce about the velocity of the rocket at its highest point.

$H(t) = H_0 + v_0 t - 16t^2$

$v(t) = H'(t)$ *The velocity of the rocket is the derivative of its height.*

$\qquad = 0 + v_0 \cdot 1 - 16 \cdot 2t$ *H_0 and v_0 are constants; t is the variable.*

$\qquad = v_0 - 32t$

When the rocket was at its highest point, it was neither rising nor falling, so its velocity was 0. Substituting $v(t) = 0$ and $t = 5.3$ into the equation $v(t) = v_0 - 32t$ yields $0 = v_0 - 32(5.3)$, or $v_0 = 169.6$.

The velocity of the rocket was 169.6 ft/s when the fuel ran out.

b. We can now write the equation for the height of the rocket as $H(t) = H_0 + 169.6t - 16t^2$. When the rocket hit the ground, its height $H(t)$ was 0, so we substitute $H(t) = 0$ and $t = 12.7$ into the height equation.

$H(t) = H_0 + 169.6t - 16t^2$

$\qquad 0 = H_0 + 169.6(12.7) - 16(12.7)^2$ *$H(t) = 0, t = 12.7$*

$16(12.7)^2 - 169.6(12.7) = H_0$ *Solve for H_0.*

$\qquad\qquad H_0 = 426.72$

The height of the rocket can now be written as $H(t) = 426.72 + 169.6t - 16t^2$. To find the maximum height of the rocket, which occurred at $t = 5.3$, compute $H(5.3)$.

$$H(t) = 426.72 + 169.6t - 16t^2$$
$$H(5.3) = 426.72 + 169.6(5.3) - 16(5.3)^2 \quad \textit{Replace t with 5.3.}$$
$$= 876.16$$

The maximum height of the rocket was about 876 feet.

Finding the **antiderivative** of a function is the inverse of finding the derivative. That is, instead of finding the derivative of $f(x)$, you are trying to find a function whose derivative is $f(x)$. For a function $f(x)$, the antiderivative is often denoted by $F(x)$. The relationship between the two functions is $F'(x) = f(x)$.

Example ❹ **Find the antiderivative of the function $f(x) = 2x$.**

We are looking for a function whose derivative is $2x$. You may recall from previous examples that the function x^2 fits that description. The derivative of x^2 is $2x^{2-1}$, or $2x$.

However, x^2 is not the only function that works. The function $G(x) = x^2 + 1$ is another, since its derivative is $G'(x) = 2x + 0$ or $2x$. Another answer is $H(x) = x^2 + 17$, and still another is $J(x) = x^2 - 6$. In fact, adding any constant, positive or negative, to x^2 does not change the fact that the derivative is $2x$.

So there is an endless list of answers, all of which can be summarized by the expression $x^2 + C$, where C is any constant. So for the function $f(x) = 2x$, we say the antiderivative is $F(x) = x^2 + C$.

As with derivatives, there are rules for finding antiderivatives.

Antiderivative Rules	Power Rule:	If $f(x) = x^n$, where n is a rational number other than -1, the antiderivative is $F(x) = \dfrac{1}{n+1}x^{n+1} + C$.
	Constant Multiple of a Power Rule:	If $f(x) = kx^n$, where n is a rational number other than -1 and k is a constant, the antiderivative is $F(x) = k \cdot \dfrac{1}{n+1}x^{n+1} + C$.
	Sum and Difference Rule:	If the antiderivatives of $f(x)$ and $g(x)$ are $F(x)$ and $G(x)$, respectively, then the antiderivative of $f(x) \pm g(x)$ is $F(x) \pm G(x)$.

Example 5 Find the antiderivative of each function.

a. $f(x) = 3x^7$

$F(x) = 3 \cdot \frac{1}{7+1} x^{7+1} + C$ *Constant Multiple of a Power Rule*

$= \frac{3}{8}x^8 + C$

b. $f(x) = 4x^2 - 7x + 5$

$f(x) = 4x^2 - 7x + 5$

$= 4x^2 - 7x^1 + 5x^0$ *Rewrite the function so that each term has a power of x.*

$F(x) = 4 \cdot \frac{1}{3}x^3 + C_1 - \left(7 \cdot \frac{1}{2}x^2 + C_2\right) + 5 \cdot \frac{1}{1}x^1 + C_3$ *Constant Multiple of a Power and Sum and*

$= \frac{4}{3}x^3 - \frac{7}{2}x^2 + 5x + C$ *Let $C = C_1 - C_2 + C_3$.   Difference Rules*

c. $f(x) = x(x^2 + 2)$

$f(x) = x(x^2 + 2)$

$= x^3 + 2x$ *Multiply to write the function as a polynomial.*

$F(x) = \frac{1}{4}x^4 + C_1 + 2 \cdot \frac{1}{2}x^2 + C_2$ *Use all three antiderivative rules.*

$= \frac{1}{4}x^4 + x^2 + C$ *Let $C = C_1 + C_2$.*

In real-world situations, the derivative of a function is often called the **rate of change** of the function because it measures how fast the function changes. If you are given the derivative or rate of change of a function, you can find the antiderivative to recover the original function. If given additional information, you may also be able to find a value for the constant C.

Example 6 **CENSUS** Data on the growth of world population provided by the U. S. Census Bureau can be used to create a model of Earth's population growth. According to this model, the rate of change of the world's population since 1950 is given by $p(t) = -0.012t^2 + 48t - 47,925$, where t is the calendar year and $p(t)$ is in millions of people per year.

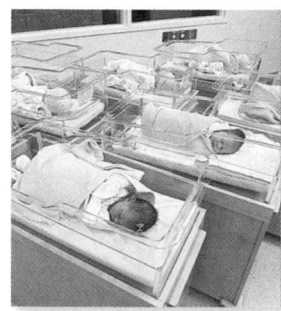

a. Given that the population in 2000 was about 6000 million people, find an equation for $P(t)$, the total population as a function of the calendar year.

b. Use the equation from part a to predict the world population in 2050.

a. $P(t)$ is the antiderivative of $p(t)$.

$p(t) = -0.012t^2 + 48t - 47,925$

$P(t) = -0.012 \cdot \frac{1}{3}t^3 + 48 \cdot \frac{1}{2}t^2 - 47,925t + C$ *Antiderivative rules*

$= -0.004t^3 + 24t^2 - 47,925t + C$

To find C, substitute 2000 for t and 6000 for $P(t)$.

$6000 = -0.004(2000)^3 + 24(2000)^2 - 47,925(2000) + C$

$6000 = -32,000,000 + 96,000,000 - 95,850,000 + C$

$C = 31,856,000$ *Solve for C.*

Substituting this value of C into our formula for $P(t)$ gives $P(t) = -0.004t^3 + 24t^2 - 47,925t + 31,856,000$. Of all the antiderivatives of $p(t)$, this is the only one that gives the proper population for the year 2000.

b. Substitute 2050 for t.

$P(t) = -0.004t^3 + 24t^2 - 47,925t + 31,856,000$

$P(2050) = -0.004(2050)^3 + 24(2050)^2 - 47,925(2050) + 31,856,000$

$= 9250$

According to the model, the world population in 2050 should be about 9250 million, or 9.25 billion.

CHECK FOR UNDERSTANDING

Communicating Mathematics

Read and study the lesson to answer each question.   1–3. See margin.

1. **Write** two different sentences that describe the relationship between the functions $4x^3$ and x^4, one using the word *derivative*, the other using the word *antiderivative*.

2. **Explain** why the Power Rule for antiderivatives is not valid when $n = -1$.

3. *Math Journal* **Write** a paragraph explaining the difference between $f(x + h)$ and $f(x) + h$. What answer would you always get if you mistakenly used $f(x) + h$ when finding a derivative using the definition?

Guided Practice

Use the definition of derivative to find the derivative of each function.

4. $f(x) = 3x + 2$ $f'(x) = 3$

5. $f(x) = x^2 + x$ $f'(x) = 2x + 1$

Use the derivative rules to find the derivative of each function.

6. $f(x) = 2x^2 - 3x + 5$ $f'(x) = 4x - 3$

7. $f(x) = -x^3 - 2x^2 + 3x + 6$ $f'(x) = -3x^2 - 4x + 3$

8. $f(x) = 3x^4 + 2x^3 - 3x - 2$ $f'(x) = 12x^3 + 6x^2 - 3$

9. Find the slope of the tangent line to the graph of $y = x^2 + 2x + 3$ at the point where $x = 1$.   **4**

Find the antiderivative of each function.

10. $f(x) = x^2$ $F(x) = \frac{1}{3}x^3 + C$

11. $f(x) = x^3 + 4x^2 - x - 3$ $F(x) = \frac{1}{4}x^4 + \frac{4}{3}x^3 - \frac{1}{2}x^2 - 3x + C$

12. $f(x) = 5x^5 + 2x^3 - x^2 + 4$ $F(x) = \frac{5}{6}x^6 + \frac{1}{2}x^4 - \frac{1}{3}x^3 + 4x + C$

Lesson 15-2   Derivatives and Antiderivatives   **957**

3 PRACTICE/APPLY

Error Analysis

Watch for students who drop the coefficient of a term when they find the antiderivative.

Prevent by having students write out every step of the problem.

Additional Answers

1. $4x^3$ is the derivative of x^4. x^4 is an antiderivative of $4x^3$.

2. Letting $n = -1$ in the expression $\frac{1}{n+1} x^{n+1}$ results in $\frac{x^0}{0}$, which is undefined.

3. $f(x + h)$ means substitute the quantity $x + h$ into the function. On the other hand, $f(x) + h$ means substitute x into the function, then add h to the result. Using $f(x) + h$ instead of $f(x + h)$ in the definition of the derivative results in:

$$\lim_{h \to 0} \frac{f(x) + h - f(x)}{h} = \lim_{h \to 0} \frac{h}{h}$$
$$= \lim_{h \to 0} 1$$
$$= 1$$

You would always get 1.

Study Guide and Practice Masters, pp. 271–272

Lesson 15-2 **957**

Additional Answers

35. $F(x) = \frac{1}{7}x^7 + C$

36. $F(x) = \frac{3}{2}x^2 + 4x + C$

37. $F(x) = \frac{4}{3}x^3 - 3x^2 + 7x + C$

38. $F(x) = 4x^3 - 3x^2 + x + C$

39. $F(x) = 2x^4 + \frac{5}{3}x^3 - \frac{9}{2}x^2 + 3x + C$

40. $F(x) = \frac{1}{20}x^5 - \frac{2}{9}x^3 + 4x + C$

41. $F(x) = 2x^3 - \frac{5}{2}x^2 - 21x + C$

42. $F(x) = \frac{1}{7}x^7 + \frac{2}{3}x^6 + \frac{4}{5}x^5 + C$

43. $F(x) = \frac{1}{3}x^3 + 2x^2 + x + C$

44. $F(x) = x^2 + x + C$

45. Any function of the form
$F(x) = \frac{1}{6}x^6 + \frac{1}{4}x^4 - \frac{1}{3}x^3 - x + C$, where C is a constant.

25. $f'(x) = 3x^2 - 4x + 5$

26. $f'(x) = 12x^3 + 21x^2 - 4x + 7$

27. $f'(x) = 6x^2 - 14x + 6$

29. $f'(x) = 81x^2 - 216x + 144$

---

13. **Business** The Better Book Company finds that the cost, in dollars, to print x copies of a book is given by the function $C(x) = 1000 + 10x - 0.001x^2$. The derivative $C'(x)$ is called the *marginal cost function*. The marginal cost is the approximate cost of printing one more book after x copies have been printed. What is the marginal cost when 1000 books have been printed?   **$8**

EXERCISES

Practice

A

Use the definition of derivative to find the derivative of each function.

14. $f(x) = 2x$ $f'(x) = 2$

15. $f(x) = 7x + 4$ $f'(x) = 7$

16. $f(x) = -3x$ $f'(x) = -3$

17. $f(x) = -4x - 9$ $f'(x) = -4$

B
18. $f(x) = 2x^2 + 5x$ $f'(x) = 4x + 5$

19. $f(x) = x^3 + 5x^2 + 6$
$f'(x) = 3x^2 + 10x$

Use the derivative rules to find the derivative of each function.

20. $f(x) = 8x$ $f'(x) = 8$

21. $f(x) = 2x + 6$ $f'(x) = 2$

22. $f(x) = \frac{1}{3}x + \frac{4}{5}$ $f'(x) = \frac{1}{3}$

23. $f(x) = -3x^2 + 2x + 9$
$f'(x) = -6x + 2$

24. $f(x) = \frac{1}{2}x^2 - x - 2$ $f'(x) = x - 1$

25. $f(x) = x^3 - 2x^2 + 5x - 6$

26. $f(x) = 3x^4 + 7x^3 - 2x^2 + 7x - 12$

27. $f(x) = (x^2 + 3)(2x - 7)$

28. $f(x) = (2x + 4)^2$ $f'(x) = 8x + 16$

29. $f(x) = (3x - 4)^3$

30. Find $f'(x)$ for the function $f(x) = \frac{2}{3}x^3 + \frac{1}{3}x^2 - x - 9$. $f'(x) = 2x^2 + \frac{2}{3}x - 1$

Find the slope of the tangent line to the graph of each equation at $x = 1$.

31. $y = x^3$ 3

32. $y = x^3 - 7x^2 + 4x + 9$ -7

33. $y = (x + 1)(x - 2)$ 1

34. $y = (5x^2 + 7)^2$ 240

Find the antiderivative of each function.   **35–44. See margin.**

35. $f(x) = x^6$

36. $f(x) = 3x + 4$

37. $f(x) = 4x^2 - 6x + 7$

38. $f(x) = 12x^2 - 6x + 1$

39. $f(x) = 8x^3 + 5x^2 - 9x + 3$

40. $f(x) = \frac{1}{4}x^4 - \frac{2}{3}x^2 + 4$

C
41. $f(x) = (2x + 3)(3x - 7)$

42. $f(x) = x^4(x + 2)^2$

43. $f(x) = \dfrac{x^3 + 4x^2 + x}{x}$

44. $f(x) = \dfrac{2x^2 - 5x - 3}{x - 3}$

45. Find a function whose derivative is $f(x) = (x^3 - 1)(x^2 + 1)$.   **See margin.**

Applications and Problem Solving

46. **Motion** Acceleration is the rate at which the velocity of a moving object changes. That is, acceleration is the derivative of velocity. If time is measured in seconds and velocity in feet per second, then acceleration is measured in feet per second squared, or ft/s^2. Suppose a car is moving with velocity $v(t) = 15 + 4t + \frac{1}{8}t^2$. *Feet per second squared is feet per second per second.*

   a. Find the car's velocity at $t = 12$.   **81 ft/s**

   b. Find the car's acceleration at $t = 12$.   **7 ft/s²**

   c. Interpret your answer to part b in words.

   c. When $t = 12$ the car's velocity is increasing at a rate of 7 ft/s per second.

   d. Suppose $s(t)$ is the car's distance, in feet, from its starting point. Find an equation for $s(t)$. $s(t) = 15t + 2t^2 + \frac{1}{24}t^3$

   e. Find the distance the car travels in the first 12 seconds.   **540 ft**

www.amc.glencoe.com/self_check_quiz

Additional Answer
53a.

170 180 190 200 210 220 230 240 250 260

47. Critical Thinking Use the definition of derivative to find the derivative of $f(x) = \dfrac{1}{x}$. $f'(x) = -\dfrac{1}{x^2}$.

**48a. About 1.6.
The total amount
spent in 2003 on
health care will
be about
$1.6 trillion.**

**48b. About 0.12.
In 2003 the
amount spent on
health care will
be increasing at
a rate of about
$120 billion per
year.**

48. Economics The graph shows
the annual spending on health
care in the U.S. for the years 1992
to 2006 (using projections for the
years after 1998.) Let $T(y)$ be the
total annual spending on health
care in year y.

a. Estimate $T(2003)$ and describe
   what it measures.

b. Estimate $T'(2003)$ and
   describe what it measures.

**Health Care Spending
(Trillions of Dollars)**

Source: Health Care Financing Administration

49. Sports Suppose a punter kicks a football so that the upward component of
its velocity is 80 feet per second. If the ball is 3 feet off the ground when it is
kicked, then the height of the ball, in feet, t seconds after it is kicked is given
by $h(t) = 3 + 80t - 16t^2$.

a. Find the upward velocity $v(t)$ of the football. $v(t) = 80 - 32t$

b. How fast is the ball travelling upward 1 second after it is kicked?   **48 ft/s**

c. Find the time when the ball reaches its maximum height. $t = 2.5$ s

d. What is the maximum height of the ball?   **103 ft**

50. Critical Thinking The derivative of the function $f(x) = e^x$ is *not* xe^{x-1}. (e^x is an
exponential function, so the Power Rule for derivatives does not apply.) Use the
definition of derivative to find the correct derivative. (*Hint*: You will need a
calculator to evaluate a limit that arises in the computation.) $f'(x) = e^x$

51. Business Joaquin and Marva are selling lemonade. The higher the price
they charge for a cup of lemonade, the fewer cups they sell. They have found
that when they charge p cents for a cup of lemonade, they sell $100 - 2p$ cups
in a day.   **a. $r(p) = p(100 - 2p)$**

a. Find a formula for the function $r(p)$ that gives their total daily revenue.

b. Find the price that Joaquin and Marva should charge to generate the highest
   possible revenue.   **25 cents**

Mixed Review

52. Evaluate $\lim\limits_{x \to 3} \dfrac{x^2 - 2x - 3}{x - 3}$. (*Lesson 15-1*)   **4**

53. Nutrition The amounts of sodium, in milligrams, present in the top brands of
peanut butter are given below. (*Lesson 14-3*)

195	210	180	225	225	225	195
225	203	225	195	195	188	191
210	233	225	248	225	210	240
180	225	240	180	225	240	240
195	189	178	255	225	225	225
194	210	225	195	188	205	

a. Make a box-and-whisker plot of the data.   **See margin.**

b. Write a paragraph describing the variability of the data.   **See students' work.**

Mid-Chapter Quiz

The Mid-Chapter Quiz provides students with a brief review of the concepts and skills in Lessons 15-1 and 15-2. Lesson numbers are given to the right of exercises or instruction lines so students can review concepts not yet mastered.

 Chapter 15, Quiz B (Lesson 15-2) is available in the *Assessment and Evaluation Masters,* p. 381.

 Chapter 15, Mid-Chapter Test (Lessons 15-1 and 15-2) is available in the *Assessment and Evaluation Masters,* p. 380.

Additional Answer
60.

$y = -3 \sin(\theta - 45°)$

Enrichment Masters, p. 166

---

54. A pair of dice is tossed. Find the probability that their sum is greater than 7 given that the numbers match. *(Lesson 13-5)* $\frac{1}{2}$

55. The first term of a geometric sequence is 9, and the common ratio is $-\frac{1}{3}$. Find the sixth term of the sequence. *(Lesson 12-2)* $-\frac{1}{27}$

56. Chemistry A beaker of water has been heated to 210°F in a room that is 74°F. Use Newton's Law of Cooling, $y = ae^{-kt} + c$, with $a = 136$°F, $k = 0.06$ min^{-1}, and $c = 74$°F to find the temperature of the water after half an hour. *(Lesson 11-3)*   **about 96°**

57. $\left(x + \frac{1}{6}\right)^2 + \left(y - \frac{7}{6}\right)^2 = \frac{169}{18}$

57. Write the standard form of the equation of the circle that passes through points at $(2, -1)$, $(-3, 0)$, and $(1, 4)$. *(Lesson 10-2)*

58. Express $5\left(\cos \frac{5\pi}{6} + i \sin \frac{5\pi}{6}\right)$ in rectangular form. *(Lesson 9-6)* $-\frac{5\sqrt{3}}{2} + \frac{5}{2}i$

59. Write parametric equations of the line passing through $P(-3, -2)$ and parallel to $\vec{v} = \langle 8, 3\rangle$. *(Lesson 8-6)* $x = 8t - 3, y = 3t - 2$

60. Graph $y = -3 \sin(\theta - 45°)$. *(Lesson 6-5)*   **See margin.**

61. about 214.9 m

61. Surveying A surveying crew is studying a housing project for possible relocation for the airport expansion. They are located on the ground, level with the houses. If the distance to one of the houses is 253 meters and the distance to the other is 319 meters, what is the distance between the houses if the angle subtended by them at the point of observation is 42°12′? *(Lesson 5-8)*

62. $\pm\frac{1}{2}, \pm 1, \pm\frac{3}{2}, \pm 3; -3, \frac{1}{2}, 1$

62. List the possible rational roots of $2x^3 + 3x^2 - 8x + 3 = 0$. Then determine the rational roots. *(Lesson 4-4)*

63. SAT/ACT Practice
In the figure, $x + y + z = ?$ **D**

A 0   B 90   C 180
D 270   E 360

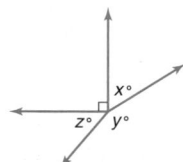

MID-CHAPTER QUIZ

Evaluate each limit. *(Lesson 15-1)*

1. $\lim\limits_{x \to -3} (2x^2 - 4x + 6)$ **36**

2. $\lim\limits_{x \to 2} \frac{x^2 - 9x + 14}{2x^2 - 7x + 6}$ **−5**

3. $\lim\limits_{x \to 0} \frac{\sin 2x}{x}$ **2**

4. f'(x) = 2x
4. Use the definition of derivative to find the derivative of $f(x) = x^2 - 3$. *(Lesson 15-2)*

Use the derivative rules to find the derivative of each function. *(Lesson 15-2)*

5. $f(x) = \pi$ **f'(x) = 0**

6. $f(x) = 3x^2 - 5x + 2$ **f'(x) = 6x − 5**

7. Medicine If $R(M)$ measures the reaction of the body to an amount M of medicine, then $R'(M)$ measures the sensitivity of the body to the medicine. Find $R'(M)$ if $R(M) = M^2\left(\frac{C}{2} - \frac{M}{3}\right)$ where C is a constant.   **See margin.**

Find the antiderivative of each function. *(Lesson 15-2)*   **8–10. See margin.**

8. $f(x) = -x^2 + 7x - 6$

9. $f(x) = 2x^3 + x^2 + 8$

10. $f(x) = -2x^4 + 6x^3 - 2x - 5$

Extra Practice See p. A55.

Extra Credit

Find the antiderivative of $\frac{(1 - 2v^2)^2}{2 - 4v^2}$.

$F(v) = \frac{1}{2}v - \frac{1}{3}v^3 + C$

Additional Answers
Mid-Chapter Quiz

7. $R'(M) = CM - M^2$

8. $F(x) = -\frac{1}{3}x^3 + \frac{7}{2}x^2 - 6x + C$

9. $F(x) = \frac{1}{2}x^4 + \frac{1}{3}x^3 + 8x + C$

10. $F(x) = -\frac{2}{5}x^5 + \frac{3}{2}x^4 - x^2 - 5x + C$

Area Under a Curve

OBJECTIVES
- Find values of integrals of polynomial functions.
- Find areas under graphs of polynomial functions.

BUSINESS The derivative of a cost function is called a *marginal cost function*. A shoe company determines that the marginal cost function for a particular type of shoe is $f(x) = 20 - 0.004x$, where x is the number of pairs of shoes manufactured and $f(x)$ is in dollars. If the company is already producing 2000 pairs of this type of shoe per day, how much more would it cost them to increase production to 3000 pairs per day? *This problem will be solved in Example 3.*

Problems like the one above can be solved using **integrals.** To understand integrals, we must first examine the area between the graph of a polynomial function and the *x*-axis for an interval from $x = a$ to $x = b$.

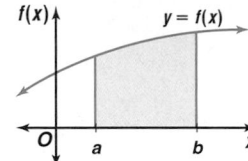

One way to estimate this area is by filling the region with rectangles, whose areas we know how to compute. If the boundary of the region is curved, the rectangles will not fit the region exactly, but you can use them for approximation. You can use rectangles of any width.

  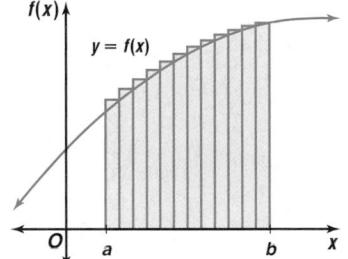

interNET
CONNECTION

Graphing Calculator Programs
To download a program that uses rectangles to approximate the area under a curve, visit:
www.amc.glencoe.com

Notice from the figures above that the thinner the rectangles are, the better they fit the region, and the better their total area approximates the area of the region. If you were to continue making the rectangles thinner and thinner, their total area would approach the exact area of the region. That is, the area of a region under the graph of a function is the limit of the total area of the rectangles as the widths of the rectangles approach 0.

Lesson 15-3 Area Under a Curve **961**

Resource Manager

Reproducible Masters
- *Study Guide and Practice,* pp. 273–274
- *Enrichment,* p. 167
- *Assessment and Evaluation,* p. 382

Transparencies
- 5-Minute Check, 15-3
- Teaching, 15-3

---

1 FOCUS

 5-Minute Check
Lesson 15-2

Find the derivative of each function.
1. $f(x) = 3x^2 + 2x + 5$
   $f'(x) = 6x + 2$
2. $f(x) = \frac{3}{5}x^5 - \frac{2}{3}x^3 + x + 2$
   $f'(x) = 3x^4 - 2x^2 + 1$
3. $f(x) = (x + 5)^2$
   $f'(x) = 2x + 10$
4. Find the slope of the tangent line to the graph of $y = 4x^2 - 6x + 1$ at $x = 3$.
   18
5. Find the antiderivative of $f(x) = 6x^3 + 12x + 11$.
   $F(x) = \frac{3}{2}x^4 + 6x^2 + 11x + C$

Motivating the Lesson
Real-World Application If a car travels 60 feet per second for 8 seconds, how far has it gone? $d = rt$, so we multiply $60 \cdot 8$ and the result is 480 feet. However, what if the car is accelerating or decelerating over those 8 seconds so that its speed is changing? How can we modify our method to find the distance traveled? Tell students that we can find the area under the graph of $v(t) = 16t - t^2$, where v is the car's velocity and t is time in seconds.

Teaching Tip
In order to illustrate the use of rectangles to approximate the area under a curve, have students graph $f(x) = 4 - x^2$ over the interval [0, 2]. Then, have them cut out this region, fold the paper accordian-style to partition it into 16 equal subintervals, and then cut it into 16 strips along the folds. Have them take the area of one rectangle and multiply by 16. Ask them how much more accurate their approximation of this area would be if they were to use 256 equal rectangles, or more.

In the figure below, the interval from a to b has been subdivided into n equal subintervals. A rectangle has been drawn on each subinterval. Each rectangle touches the graph at its upper right corner; the first touches at the x-coordinate x_1, the second touches at the x-coordinate x_2, and so on, with the last rectangle touching at the x-coordinate b, which is also denoted by x_n for consistency.

The height of the first rectangle is $f(x_1)$, the height of the second is $f(x_2)$, and so on, with the height of the last rectangle being $f(x_n)$. The length of the entire interval from a to b is $b - a$, so the width of each of the n rectangles must be $\dfrac{b - a}{n}$. This common width is traditionally denoted Δx. Δx is read "delta x."

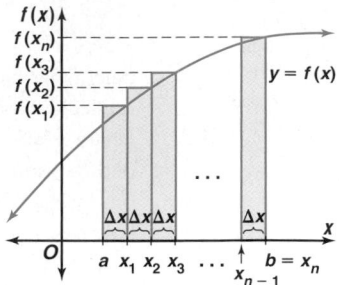

Look Back
You can refer to Lesson 12-5 to review sigma notation.

The area of the first rectangle is $f(x_1)\Delta x$, the area of the second rectangle is $f(x_2)\Delta x$, and so on. The total area A_n of the n rectangles is given by the sum of the areas.

$$A_n = f(x_1)\Delta x + f(x_2)\Delta x + \cdots + f(x_n)\Delta x$$

$$= \sum_{i=1}^{n} f(x_i)\Delta x \quad \text{\textit{i is the index of summation, not the imaginary unit.}}$$

$\displaystyle\int_a^b f(x)\,dx$ is read "the integral of $f(x)$ from a to b."

To make the width of the rectangles approach 0, we let the number of rectangles approach infinity. Therefore, the exact area of the region under the graph of the function is $\displaystyle\lim_{n\to\infty} A_n$, or $\displaystyle\lim_{n\to\infty} \sum_{i=1}^{n} f(x_i)\Delta x$. This limit is called a **definite integral** and is denoted $\displaystyle\int_a^b f(x)\,dx$.

Definite Integral	$\displaystyle\int_a^b f(x)\,dx = \lim_{n\to\infty} \sum_{i=1}^{n} f(x_i)\Delta x$ where $\Delta x = \dfrac{b-a}{n}$.

The process of finding the area under a curve is called **integration.** The following formulas will be needed in the examples and exercises.

$$1 + 2 + 3 + \cdots + n = \frac{n(n+1)}{2}$$

$$1^2 + 2^2 + 3^2 + \cdots + n^2 = \frac{n(n+1)(2n+1)}{6}$$

$$1^3 + 2^3 + 3^3 + \cdots + n^3 = \frac{n^2(n+1)^2}{4}$$

$$1^4 + 2^4 + 3^4 + \cdots + n^4 = \frac{6n^5 + 15n^4 + 10n^3 - n}{30}$$

$$1^5 + 2^5 + 3^5 + \cdots + n^5 = \frac{2n^6 + 6n^5 + 5n^4 - n^2}{12}$$

Before beginning the examples, we will derive a formula for x_i. The width Δx of each rectangle is the distance between successive x_i-values. Study the labels below the x-axis.

We see that $x_i = a + i\Delta x$. This formula will work when finding the area under the graph of any function.

Example ❶ Use limits to find the area of the region between the graph of $y = x^2$ and the x-axis from $x = 0$ to $x = 1$. That is, find $\int_0^1 x^2\, dx$.

First find Δx.

$$\Delta x = \frac{b - a}{n}$$

$$= \frac{1 - 0}{n} \text{ or } \frac{1}{n}$$

Then find x_i.

$$x_i = a + i\Delta x$$

$$= 0 + i \cdot \frac{1}{n} \text{ or } \frac{i}{n}$$

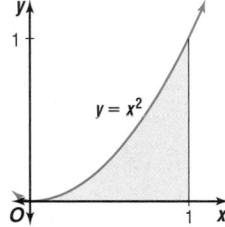

Now we can calculate the integral that gives the area.

$$\int_0^1 x^2\, dx = \lim_{n \to \infty} \sum_{i=1}^n (x_i)^2\, \Delta x \qquad f(x_i) = x_i^2$$

$$= \lim_{n \to \infty} \sum_{i=1}^n \left(\frac{i}{n}\right)^2 \left(\frac{1}{n}\right) \qquad x_i = \frac{i}{n},\ \Delta x = \frac{1}{n}$$

$$= \lim_{n \to \infty} \sum_{i=1}^n \frac{i^2}{n^3} \qquad \textit{Multiply.}$$

$$= \lim_{n \to \infty} \left(\frac{1^2}{n^3} + \frac{2^2}{n^3} + \cdots + \frac{n^2}{n^3}\right)$$

$$= \lim_{n \to \infty} \frac{1}{n^3}\, (1^2 + 2^2 + \cdots + n^2) \qquad \textit{Factor.}$$

$$= \lim_{n \to \infty} \frac{1}{n^3} \cdot \frac{n(n + 1)(2n + 1)}{6} \qquad 1^2 + 2^2 + \cdots + n^2 = \frac{n(n + 1)(2n + 1)}{6}$$

$$= \lim_{n \to \infty} \frac{2n^2 + 3n + 1}{6n^2} \qquad \textit{Multiply.}$$

$$= \lim_{n \to \infty} \frac{1}{6}\left(2 + \frac{3}{n} + \frac{1}{n^2}\right) \qquad \textit{Factor and divide by } n^2.$$

$$= \left(\lim_{n \to \infty} \frac{1}{6}\right)\left[\lim_{n \to \infty} 2 + \left(\lim_{n \to \infty} 3\right)\left(\lim_{n \to \infty} \frac{1}{n}\right) + \lim_{n \to \infty} \frac{1}{n^2}\right] \qquad \begin{array}{l}\textit{Limit theorems from}\\ \textit{Chapter 12}\end{array}$$

$$= \frac{1}{6}\, [2 + (3)(0) + 0] \text{ or } \frac{1}{3} \qquad \lim_{n \to \infty} \frac{1}{n} = 0,\ \lim_{n \to \infty} \frac{1}{n^2} = 0$$

The area of the region is $\frac{1}{3}$ square unit.

In-Class Example

Example 1
Use limits to find the area of the region between the graph of $y = x^2 + 1$ and the x-axis from $x = 0$ to $x = 2$. That is, find $\int_0^2 (x^2 + 1)\, dx$. $\dfrac{14}{3}$

Example ② Use limits to find the area of the region between the graph of $y = x^3$ and the x-axis from $x = 2$ to $x = 4$.

First, find the area under the graph from $x = 0$ to $x = 4$. Then subtract from it the area under the graph from $x = 0$ to $x = 2$. In other words,

$$\int_2^4 x^3 \, dx = \int_0^4 x^3 \, dx - \int_0^2 x^3 \, dx.$$

For $\int_0^4 x^3 \, dx$, $a = 0$ and $b = 4$, so $\Delta x = \dfrac{4}{n}$ and $x_i = \dfrac{4i}{n}$.

$$\int_0^4 x^3 \, dx = \lim_{n \to \infty} \sum_{i=1}^n (x_i)^3 \, \Delta x \qquad f(x_i) = x_i^3$$

$$= \lim_{n \to \infty} \sum_{i=1}^n \left(\frac{4i}{n}\right)^3 \cdot \frac{4}{n} \qquad x_i = \frac{4i}{n}, \, \Delta x = \frac{4}{n}$$

$$= \lim_{n \to \infty} \sum_{i=1}^n \frac{256 i^3}{n^4}$$

$$= \lim_{n \to \infty} \left(\frac{256 \cdot 1^3}{n^4} + \frac{256 \cdot 2^3}{n^4} + \cdots + \frac{256 \cdot n^3}{n^4}\right)$$

$$= \lim_{n \to \infty} \frac{256}{n^4} \cdot (1^3 + 2^3 + \cdots + n^3)$$

$$= \lim_{n \to \infty} \frac{256}{n^4} \cdot \frac{n^2(n + 1)^2}{4} \qquad 1^3 + 2^3 + \cdots + n^3 = \frac{n^2(n + 1)^2}{4}$$

$$= \lim_{n \to \infty} \frac{64n^2 + 128n + 64}{n^2}$$

$$= \lim_{n \to \infty} \left(64 + \frac{128}{n} + \frac{64}{n^2}\right) \qquad \textit{Divide by } n^2.$$

$$= 64 + 0 + 0 \text{ or } 64$$

For $\int_0^2 x^3 \, dx$, $a = 0$ and $b = 2$, so $\Delta x = \dfrac{2}{n}$ and $x_i = \dfrac{2i}{n}$.

$$\int_0^2 x^3 \, dx = \lim_{n \to \infty} \sum_{i=1}^n (x_i)^3 \, \Delta x$$

$$= \lim_{n \to \infty} \sum_{i=1}^n \left(\frac{2i}{n}\right)^3 \cdot \frac{2}{n} \qquad x_i = \frac{2i}{n}, \, \Delta x = \frac{2}{n}$$

$$= \lim_{n \to \infty} \sum_{i=1}^n \frac{16 i^3}{n^4}$$

$$= \lim_{n \to \infty} \left(\frac{16 \cdot 1^3}{n^4} + \frac{16 \cdot 2^3}{n^4} + \cdots + \frac{16 \cdot n^3}{n^4}\right)$$

$$= \lim_{n \to \infty} \frac{16}{n^4} \cdot (1^3 + 2^3 + \cdots + n^3)$$

$$= \lim_{n \to \infty} \frac{16}{n^4} \cdot \frac{n^2(n + 1)^2}{4} \qquad 1^3 + 2^3 + \cdots + n^3 = \frac{n^2(n + 1)^2}{4}$$

$$= \lim_{n \to \infty} \frac{4n^2 + 8n + 4}{n^2}$$

$$= \lim_{n\to\infty}\left(4 + \frac{8}{n} + \frac{4}{n^2}\right) \qquad \textit{Divide by } n^2.$$

$$= 4 + 0 + 0 \text{ or } 4$$

The area of the region between the graph of $y = x^3$ and the x-axis from $x = 2$ to $x = 4$ is $64 - 4$, or 60 square units.

Teaching Tip Make sure that students take note that finding the integral for a linear function does not require subtracting integrals even though the boundary does not start at zero.

In-Class Example

Example 3
Business How much would it cost the shoe company to increase production from 2000 pairs per day to 5000 pairs per day? **$18,000**

In physics, when the velocity of an object is graphed with respect to time, the area under the curve represents the displacement of the object. In business, the area under the graph of a marginal cost function from $x = a$ to $x = b$ represents the amount it would cost to increase production from a units to b units.

Example **3** **BUSINESS** Refer to the application at the beginning of the lesson. How much would it cost the shoe company to increase production from 2000 pairs per day to 3000 pairs per day?

Real World Application

The cost is given by $\int_{2000}^{3000} f(x)\,dx$ where $f(x) = 20 - 0.004x$ is the marginal cost function.

$a = 2000$ and $b = 3000$, so $\Delta x = \frac{1000}{n}$ and $x_i = 2000 + \frac{1000i}{n}$.

Since $f(x)$ is a linear function, we can calculate the value directly, without subtracting integrals as in Example 2.

$$\int_{2000}^{3000} f(x)\,dx = \lim_{n\to\infty}\sum_{i=1}^{n} f(x_i)\Delta x$$

$$= \lim_{n\to\infty}\sum_{i=1}^{n}(20 - 0.004x_i)\Delta x \quad f(x_i) = 20 - 0.004x_i$$

$$= \lim_{n\to\infty}\sum_{i=1}^{n}\left[20 - 0.004\left(2000 + \frac{1000i}{n}\right)\right]\cdot\frac{1000}{n}$$

$$= \lim_{n\to\infty}\sum_{i=1}^{n}\left(12 - \frac{4i}{n}\right)\cdot\frac{1000}{n} \quad \textit{Simplify.}$$

$$= \lim_{n\to\infty}\frac{1000}{n}\left[\left(12 - \frac{4\cdot1}{n}\right) + \left(12 - \frac{4\cdot2}{n}\right) + \cdots + \left(12 - \frac{4\cdot n}{n}\right)\right]$$

$$= \lim_{n\to\infty}\frac{1000}{n}\cdot\left[12n - \frac{4}{n}(1 + 2 + \cdots + n)\right] \quad \textit{Combine and factor.}$$

$$= \lim_{n\to\infty}\frac{1000}{n}\cdot\left[12n - \frac{4}{n}\cdot\frac{n(n+1)}{2}\right] \qquad 1 + 2 + \cdots + n = \frac{n(n+1)}{2}$$

$$= \lim_{n\to\infty}\frac{1000}{n}\cdot(10n - 2) \qquad \textit{Simplify.}$$

$$= \lim_{n\to\infty}\frac{10{,}000n - 2000}{n} \qquad \textit{Multiply.}$$

$$= \lim_{n\to\infty}\left(10{,}000 - \frac{2000}{n}\right) \qquad \textit{Divide by } n.$$

$$= 10{,}000 - 0, \text{ or } 10{,}000$$

The increase in production would cost the company $10,000.

Addressing Individual Needs

Kinesthetic Have students graph one of the exercises on large grid paper. Then, have them cut out the area under the graph and determine the number of square units involved. This will require piecing together parts of the graph. Have students compare the area found mathematically with their cut-and-paste answer.

3 PRACTICE/APPLY

Error Analysis

Watch for students who substitute the incorrect formula for $1^a + 2^a + \cdots + n^a$.

Prevent by having students copy the list of formulas into their notes for easier reference.

Assignment Guide

Core: 11–29 odd, 30, 31, 33–41
Enriched: 10–26 even, 27–41

Additional Answers

2. **Sample answer:** Subdivide the interval from *a* to *b* into *n* equal subintervals, draw a rectangle on each subinterval that touches the graph at its upper right corner, add up the areas of the rectangles, and then find the limit of the total area of the rectangles as *n* approaches infinity.

3. Lorena is correct. If the function is decreasing, then the graph will always be above the tops of the rectangles, so the total area of the rectangles will be less than the area under the graph.

Study Guide and Practice Masters, pp. 273–274

CHECK FOR UNDERSTANDING

Communicating Mathematics

1. **Sample answer:** $y = x^4$

Read and study the lesson to answer each question.

1. **Write** an equation of a function for which you would need the formula for $1^4 + 2^4 + 3^4 + \cdots + n^4$ to find the area under the graph.

2. **Describe** the steps involved in finding the area under the graph of $y = f(x)$ between $x = a$ and $x = b$. **2–3. See margin.**

3. **You Decide** Rita says that when you use rectangles that touch the graph of a function at their upper right corners, the total area of the rectangles will always be greater than the area under the curve because the rectangles stick out above the curve. Lorena disagrees. Who is correct? Explain.

Guided Practice

4. Use a limit to find the area of the shaded region in the graph at the right. $\frac{8}{3}$ units^2

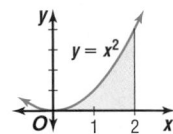

Use limits to find the area between each curve and the *x*-axis for the given interval.

5. $y = x^2$ from $x = 1$ to $x = 3$ $\frac{26}{3}$ units^2 6. $y = x^3$ from $x = 0$ to $x = 1$ $\frac{1}{4}$ unit^2

Use limits to evaluate each integral.

7. $\int_0^6 x^2 \, dx$ **72**

8. $\int_0^3 x^3 \, dx$ $\frac{81}{4}$

9. **Physics** Neglecting air resistance, an object in free fall accelerates at 32 feet per second squared. So the velocity of the object *t* seconds after being dropped is 32*t* feet per second. Suppose a ball is dropped from the top of the Sears Tower.

a. Use integration to find how far the ball would fall in the first six seconds. **576 ft**

b. Refer to the graph at the right. Would the ball hit the ground within ten seconds of being dropped? Explain your reasoning. **See margin.**

Tallest Buildings in the U.S. (ft)

1454
1250
Sears Tower
Empire State Building

EXERCISES

Practice

A Use limits to find the area of the shaded region in each graph.

10.
$y = x + 1$

4 units^2

11.
$y = x^2$

9 units^2

12.
$y = x^2$

3 units^2

 www.amc.glencoe.com/self_check_quiz

Additional Answer

9b. **Yes;** integration shows that the ball would fall 1600 ft in 10 seconds of free-fall. Since this exceeds the height of the building, the ball must hit the ground in less than 10 seconds.

Use limits to find the area between each curve and the *x*-axis for the given interval.

13. $y = x$ from $x = 1$ to $x = 3$ **4 units²**

14. $y = x^2$ from $x = 0$ to $x = 5$ $\dfrac{125}{3}$ **units²**

15. $y = 2x^3$ from $x = 1$ to $x = 5$ **312 units²**

16. $y = x^4$ from $x = 0$ to $x = 5$ **625 units²**

17. $y = x^2 + 6x$ from $x = 0$ to $x = 4$ $\dfrac{208}{3}$ **units²**

18. $y = x^2 - x + 1$ from $x = 0$ to $x = 3$ $\dfrac{15}{2}$ **units²**

19. Write a limit that gives the area under the graph of $y = \sin x$ from $x = 0$ to $x = \pi$. (Do *not* evaluate the limit.) **See margin.**

Use limits to evaluate each integral.

20. $\displaystyle\int_{0}^{2} 8x\,dx$ **16**

21. $\displaystyle\int_{1}^{4} (x + 2)\,dx$ $\dfrac{27}{2}$

22. $\displaystyle\int_{0}^{4} x^2\,dx$ $\dfrac{64}{3}$

23. $\displaystyle\int_{3}^{5} 8x^3\,dx$ **1088**

24. $\displaystyle\int_{1}^{4} (x^2 + 4x - 2)\,dx$ **45**

25. $\displaystyle\int_{0}^{2} (x^5 + x^2)\,dx$ $\dfrac{40}{3}$

26. Find the integral of x^3 from 0 to 5. $\dfrac{625}{4}$

Applications and Problem Solving

27. **Sewing** A patch in the shape of the region shown at the right is to be sewn onto a flag. If each unit in the coordinate system represents one foot, how much material is required for the patch? **10.125 ft²**

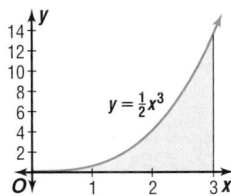

$y = \frac{1}{2}x^3$

28. **Business** Suppose the Auburn Widget Corporation finds that the marginal cost function associated with producing x widgets is $f(x) = 80 - 2x$ dollars.
   a. Refer to Exercise 13 of Lesson 15-2. Use the marginal cost function to approximate the cost for the company to produce one more widget when the production level is 20 widgets. **$40**
   b. How much would it cost the company to double its production from 20 widgets to 40 widgets? **$400**

29. **Mining** In order to distribute stress, mine tunnels are sometimes rounded. Suppose that the vertical cross sections of a tunnel can be modeled by the parabola $y = 6 - 0.06x^2$. If x and y are measured in feet, how much rock would have to be moved to make such a tunnel that is 100 feet long? **8000 ft³**

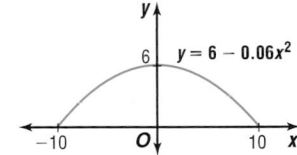

$y = 6 - 0.06x^2$

30. **Critical Thinking** Find the area of the region enclosed by the line $y = x$ and the parabola $y = x^2$. $\dfrac{1}{6}$ **unit²**

Open-Ended Assessment
Speaking Have students explain to what area $\lim\limits_{n\to\infty} \sum\limits_{i=1}^{n} \left(\dfrac{2i}{n}\right)^3 \dfrac{2}{n}$ refers.

the area under the curve $y = x^3$ from $x = 0$ to $x = 2$

📁 **Chapter 15, Quiz C (Lesson 15-3)** is available in the *Assessment and Evaluation Masters*, p. 382.

Additional Answers
31a.

32a.

Enrichment Masters, p. 167

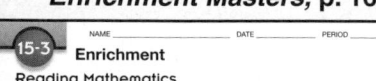

31. Budgets If the function $r(t)$ gives the rate at which a family spends money, then the total money spent between times $t = a$ and $t = b$ is $\int_a^b r(t)\, dt$. A local electric company in Alabama, where electric bills are generally low in winter and very high in summer, offers customers the option of paying a flat monthly fee for electricity throughout the year so that customers can avoid enormous summertime bills. The company has found that in past years the Johnson family's rate of electricity spending can be modeled by $r(t) = 50 + 36t - 3t^2$ dollars per month, where t is the number of months since the beginning of the year.

a. Sketch a graph of the function $r(t)$ for $0 \le t \le 12$. **See margin.**

b. Find the total amount of money the Johnsons would spend on electricity during a full year. **$1464**

c. If the Johnsons choose the option of paying a flat monthly fee, how much should the electric company charge them each month? **$122**

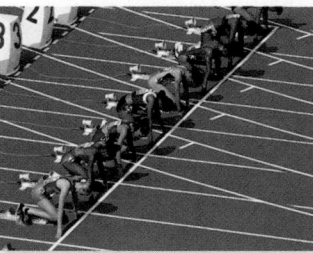

32. Sports A sprinter is trying to decide between two strategies for running a race. She can put a lot of energy into an initial burst of speed, which gives her a velocity of $v(t) = 3.5t - 0.25t^2$ meters per second after t seconds, or she can save her energy for more acceleration at the end so that her velocity is given by $v(t) = 1.2t + 0.03t^2$.

a. Graph the two velocity functions on the same set of axes for $0 \le t \le 10$. **See margin.**

b. Use integration to determine which velocity results in a greater distance covered in a 10-second race. **the first one**

33. Critical Thinking Find the value of $\int_{-r}^{r} \sqrt{r^2 - x^2}\, dx$, where r is a constant. $\dfrac{1}{2}\pi r^2$

Mixed Review

34. $f'(x) = -9x^2 + 2x - 7$

34. Find the derivative of $f(x) = -3x^3 + x^2 - 7x$. *(Lesson 15-2)*

35. Evaluate $\lim\limits_{x\to 2} \dfrac{x - 2}{x + 2}$. *(Lesson 15-1)* **0**

36. Solve the equation $\log_{\frac{1}{3}} x = -3$. *(Lesson 11-4)* **27**

37. $\overline{u} = \langle -1, -1, -10 \rangle$ $= -\overline{i} - \overline{j} - 10\overline{k}$

37. Find an ordered triple to represent $\overline{u}$ if $\overline{u} = \overline{v} + \overline{w}$, $\overline{v} = \langle 2, -5, -3 \rangle$, and $\overline{w} = \langle -3, 4, -7 \rangle$. Then write $\overline{u}$ as the sum of unit vectors. *(Lesson 8-3)*

38. If $\sin r = \dfrac{3}{5}$ and r is in the first quadrant, find $\cos 2r$. *(Lesson 7-4)* $\dfrac{7}{25}$

39. $\dfrac{1}{2}, \dfrac{\pi}{5}$

39. State the amplitude and period for the function $y = \dfrac{1}{2}\sin 10\theta$. *(Lesson 6-4)*

40. Manufacturing A cereal manufacturer wants to make a cardboard cereal box of maximum volume. The function representing the volume of the box is $v(x) = -0.7x^3 + 5x^2 + 7x$, where x is the width of the box in centimeters. Find the width of the box that will maximize the volume. *(Lesson 3-6)* **5.4 cm**

41. SAT/ACT Practice Triangle ABC has sides that are 6, 8, and 10 inches long. A rectangle that has an area equal to that of the triangle has a width of 3 inches. Find the perimeter of the rectangle in inches. **C**

  A 30   **B** 24   **C** 22   **D** 16   **E** 11

968 *Chapter 15 Introduction to Calculus*   **Extra Practice** See p. A55.

Extra Credit

Evaluate $\int_{-2}^{2} \sqrt{4 - x^2}\, dx$. 2π

CALCULUS

Calculus is fundamental to solving problems in the sciences and engineering. Two basic tools of calculus are differentiation and integration. Some of the basic ideas of calculus began to develop over 2000 years ago, but a usable form was not developed until the seventeenth century.

Early Evidence Several ideas basic to the development of calculus are the concepts of limit, infinite processes, and approximation. The Egyptians and Babylonians solved problems, such as finding the areas of circles and the volumes of pyramids, by methods resembling calculus. In about 450 B.C., **Zeno** of Elea posed problems, often called Zeno's Paradoxes, dealing with infinity. In trying to deal with these paradoxes, **Eudoxus** (about 370 B.C.), a Greek, proposed his "method of exhaustion," which is based on the idea of infinite processes. An example of this method is to show that the difference in area between a circle and an inscribed polygon can be made smaller and smaller by increasing the number of sides of the polygon.

The Renaissance Mathematicians and scientists, such as **Johann Kepler** (1571–1630), **Pierre Fermat** (1601–1665), **Gilles Roberval** (1602–1675), and **Bonaventura Cavalieri** (1598–1647), used the concept of summing an infinite number of strips to find the area under a curve. Cavalieri called this the "method of indivisibles." The use of coordinates and the development of analytic geometry by Fermat and **Renè Descartes** (1596–1650) aided in the further development of calculus.

Modern Era Most historians name **Gottfried Leibniz** (1646–1716) and **Isaac Newton** (1642–1727) as coinventors of calculus. They worked independently at approximately the same time on ideas which evolved into what is known as calculus today.

In the argument over which mathematician developed calculus first, it seems that Newton had the ideas first, but did not publish them until after Leibniz made his ideas public. However, the notation used by Leibniz was more understandable than that of Newton, and much of it is still in use.

Today aerospace engineers like **Tahani R. Amer** use calculus in many aspects of their jobs. In her job at the NASA Langley Research Center, she uses calculus for characterizing pressure measurements taken during wind tunnel tests of experimental aircraft and for working with optical measurements.

Tahani R. Amer

1–2. See margin.

ACTIVITIES

1. Demonstrate the method of exhaustion. Draw three circles of equal radii. In the first circle, inscribe a triangle, in the second a square, and in the third a pentagon. Find the difference between the area of each circle and its inscribed polygon.

2. Fermat discovered a simple method for finding the maximum and minimum points of polynomial curves. Consider the curve $y = 2x^3 - 5x^2 + 4x - 7$. If another point has abscissa $x + E$, then the ordinate is $2(x + E)^3 - 5(x + E)^2 + 4(x + E) - 7$. He set this expression equal to the original function and arrived at the equation $(6x^2 - 10x + 4)E + (6x - 5)E^2 + 2E^3 = 0$. Finish Fermat's method. Divide each term by E. Then let E be 0. What is the relationship between the roots of the resulting equation and the derivative of $2x^3 - 5x^2 + 4x - 7$?

3. **interNET CONNECTION** Find out more about persons referenced in this article and others who contributed to the history of calculus. Visit **www.amc.glencoe.com**

This feature provides students with an opportunity to see how the mathematics that they are studying was developed. It also gives students a glimpse of how mathematical discoveries are made.

More About the Mathematicians

Zeno was born around 490 B.C. in Elea, Lucania (now Italy). Zeno's dichotomy paradox says that in order to reach the end of a segment, one must first pass its midpoint, and from the beginning to the midpoint is also a segment with a midpoint, and so on, infinitely.

Eudoxus was born in 408 B.C. in Cnidus, Asia Minor (now Turkey). His method of exhaustion applies to integration.

Johann Kepler was born in Leonberg, Wurttemberg (now Germany). Calculus can be used to prove Kepler's Laws.

Pierre Fermat was born in Beaumont-de-Lomagne, France. He studied maxima and minima by examining when a curve's tangent was parallel to the x-axis.

Gilles Roberval was born in Senlis, France. He was the first to look at the area under a curve as an infinite number of rectangles.

Bonaventura Cavalieri was born in Milan, Italy. He wrote the main textbook for seventeenth-century mathematicians. Like Roberval, he studied integrals.

René Descartes was born in La Haye (now Descartes), France. He invented a method for determining normals, a study related to derivatives.

Gottfried Leibniz was born in Leipzig, Saxony (now Germany). Leibniz used the notations dx and dy to denote infinitely small changes in x and y, respectively.

Isaac Newton was born in Woolsthorpe, England. He gave the first clear statement of the Fundamental Theorem of Calculus.

Additional Answers

1. **See students' work. The difference in area should decrease as the number of sides of the polygon increases.**

2. **The roots of the resulting equation are the zeros of the derivative of the original function.**

3. **See students' work.**

The Fundamental Theorem of Calculus

1 FOCUS

5-Minute Check
Lesson 15-3

1. Use limits to find the area of the shaded region in the graph. **12**

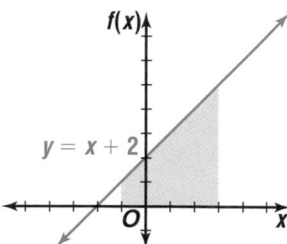

$y = x + 2$

Use limits to find the area between each curve and the x-axis for the given interval.

2. $y = x^5$ from $x = 2$ to $x = 4$
   672 units²

3. $y = x^2 + 3x + 2$ from $x = 1$ to $x = 2$ $\frac{53}{6}$ **units²**

Use limits to evaluate each integral.

4. $\int_0^3 x\, dx$ $\frac{9}{2}$

5. $\int_0^4 (3x^2 + 6x)\, dx$ **112**

Motivating the Lesson

Discussion The Fundamental Theorem of Arithmetic states that *every integer greater than one has a unique prime factorization*, and is the basis of much of the branch of mathematics known as number theory. The Fundamental Theorem of Algebra, presented in Chapter 4, is foundational to the study of modern algebra. In the same way, the Fundamental Theorem of Calculus is of major importance to the study of calculus. Ask students whether they think using limits is the most efficient way of evaluating an integral. Did they discover any sort of pattern in the previous lesson that they used as a shortcut? If possible, lead them to discover the Fundamental Theorem of Calculus.

970 *Chapter 15*

OBJECTIVES

- Use the Fundamental Theorem of Calculus to evaluate definite integrals of polynomial functions.
- Find indefinite integrals of polynomial functions.

interNET
CONNECTION

Research
For more information about the dimensions and shape of the Gateway Arch, visit:
www.amc.glencoe.com

CONSTRUCTION Two construction contractors have been hired to clean the Gateway Arch in St. Louis. The Arch is very close to a parabola in shape, 630 feet high and 630 feet across at the bottom. Using the point on the ground directly below the apex of the Arch as the origin, the equation of the Arch is approximately $y = 630 - \frac{x^2}{157.5}$. One contractor's first idea for approaching the project is to build scaffolding in the entire space under the Arch, so that the cleaning crew can easily climb up and down to any point on the Arch. The other contractor thinks there is too much space under the Arch to make the scaffolding practical. To settle the matter, the contractors want to find out how much area there is under the Arch. *This problem will be solved in Example 4.*

You have probably found the evaluation of definite integrals with limits to be a tedious process. Fortunately, there is an easier method. Consider, for example, the problem of finding the change in position of a moving object between times $t = a$ and $t = b$. In Lesson 15-3, we solved such a problem by evaluating $\int_a^b f(t)\, dt$, where $f(t)$ is the velocity of the object. Another approach would be to find the position function, which is an antiderivative of $f(t)$, for the object. Substituting a and b into the position function would give the locations of the object at those times. We could subtract those locations to find the displacement of the object. In other words, if $F(t)$ is the position function for the object, then $\int_a^b f(t)\, dt = F(b) - F(a)$.

The above relationship is actually true for *any* continuous function $f(x)$. This connection between definite integrals and antiderivatives is so important that it is called the **Fundamental Theorem of Calculus.**

Fundamental Theorem of Calculus	If $F(x)$ is the antiderivative of the continuous function $f(x)$, then $$\int_a^b f(x)\, dx = F(b) - F(a).$$

The Fundamental Theorem of Calculus provides a way to evaluate the definite integral $\int_a^b f(x)\, dx$ if an antiderivative $F(x)$ can be found. A vertical line on the right side is used to abbreviate $F(b) - F(a)$. Thus, the principal statement of the theorem may be written as follows.

$$\int_a^b f(x)\, dx = F(x)\Big|_a^b = F(b) - F(a)$$

970 *Chapter 15 Introduction to Calculus*

Resource Manager

 Reproducible Masters
- *Study Guide and Practice,* pp. 275–276
- *Enrichment,* p. 168
- *Assessment and Evaluation,* p. 382
- *Problem Solving and Applications,* p. 30

 Transparencies
- 5-Minute Check, 15-4
- Teaching, 15-4

Example **1** Evaluate $\int_2^4 x^3 \, dx$.

The antiderivative of $f(x) = x^3$ is $F(x) = \frac{1}{4}x^4 + C$.

$\int_2^4 x^3 \, dx = \frac{1}{4}x^4 + C \Big|_2^4$ *Fundamental Theorem of Calculus*

$= \left(\frac{1}{4} \cdot 4^4 + C\right) - \left(\frac{1}{4} \cdot 2^4 + C\right)$ *Let x = 4 and 2 and subtract.*

$= 64 - 4$ or 60

Notice how much easier this example was than Example 2 of Lesson 15-3. Also notice that C was eliminated during the calculation. This always happens when you use the Fundamental Theorem to evaluate a definite integral. So in this situation you can neglect the constant term when writing the antiderivative.

Due to the connection between definite integrals and antiderivatives, the antiderivative of $f(x)$ is often denoted by $\int f(x) \, dx$. $\int f(x) \, dx$ is called the **indefinite integral** of $f(x)$.

It is helpful to rewrite the antiderivative rules in terms of indefinite integrals.

Antiderivative Rules	Power Rule:	$\int x^n \, dx = \frac{1}{n+1} x^{n+1} + C$, where n is a rational number and $n \ne -1$.
	Constant Multiple of a Power Rule:	$\int kx^n \, dx = k \cdot \frac{1}{n+1} x^{n+1} + C$, where k is a constant, n is a rational number, and $n \ne -1$.
	Sum and Difference Rule:	$\int [f(x) \pm g(x)] \, dx = \int f(x) \, dx \pm \int g(x) \, dx$

Example **2** Evaluate each indefinite integral.

a. $\int 5x^2 \, dx$

$\int 5x^2 \, dx = 5 \cdot \frac{1}{3}x^3 + C$ *Constant Multiple of a Power Rule*

$= \frac{5}{3}x^3 + C$ *Simplify.*

b. $\int (4x^5 + 7x^2 - 4x) \, dx$

$\int (4x^5 + 7x^2 - 4x) \, dx = 4 \cdot \frac{1}{6}x^6 + 7 \cdot \frac{1}{3}x^3 - 4 \cdot \frac{1}{2}x^2 + C$ *Remember x = x¹.*

$= \frac{2}{3}x^6 + \frac{7}{3}x^3 - 2x^2 + C$ *Simplify.*

Lesson 15-4 The Fundamental Theorem of Calculus **971**

3 PRACTICE/APPLY

Error Analysis

Watch for students who evaluate integrals before finding the antiderivatives.

Prevent by reviewing the statement of the Fundamental Theorem of Calculus.

Study Guide and Practice Masters, pp. 275–276

972 *Chapter 15*

Examples   **3**   **Find the area of the shaded region.**

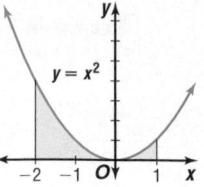

$y = x^2$

The area is given by $\int_{-2}^{1} x^2\,dx$.

The antiderivative of $f(x) = x^2$ is $F(x) = \dfrac{1}{3}x^3 + C$.

$\int_{-2}^{1} x^2\,dx = \dfrac{1}{3}x^3\Big|_{-2}^{1}$ *+ C is not needed with a definite integral.*

$= \dfrac{1}{3}(1)^3 - \dfrac{1}{3}(-2)^3$ *Let x = 1 and −2 and subtract.*

$= 3$

The area of the region is 3 square units.

Real World Application

4   **CONSTRUCTION**   **Refer to the application at the beginning of the lesson. What is the area under the Gateway Arch?**

$y = 630 - \dfrac{x^2}{157.5}$

The area is given by $\int_{-315}^{315} \left(630 - \dfrac{x^2}{157.5}\right) dx$.

315 and −315 are the x-intercepts of the parabola that models the Arch.

$\int_{-315}^{315} \left(630 - \dfrac{x^2}{157.5}\right) dx = \int_{-315}^{315} \left(630 - \dfrac{1}{157.5}x^2\right) dx$ *Rewrite the function.*

$= 630x - \dfrac{1}{157.5}\cdot\dfrac{1}{3}x^3 \Big|_{-315}^{315}$ *Antiderivative; + C not needed.*

$= \left(630\cdot 315 - \dfrac{1}{472.5}(315)^3\right)$ *Let x = 315 and −315 and subtract.*

$- \left(630\cdot(-315) - \dfrac{1}{472.5}(-315)^3\right)$

$= 132,300 - (-132,300)$ or $264,600$

The area under the Arch is 264,600 square feet.

CHECK FOR UNDERSTANDING

Communicating Mathematics

Read and study the lesson to answer each question.   **1–4. See margin.**

1. Explain the difference between $\int f(x)\,dx$ and $\int_a^b f(x)\,dx$.

2. Find a counterexample to the statement $\int_a^b f(x)g(x)\,dx = \int_a^b f(x)\,dx \cdot \int_a^b g(x)\,dx$ for all a and b and all functions $f(x)$ and $g(x)$.

3. Explain why the "$+ C$" is not needed in the antiderivative when evaluating a definite integral.

4. You Decide   Cole says that when evaluating a definite integral, the order in which you substitute a and b into the antiderivative and subtract does not matter. Rose says it does matter. Who is correct? Explain.

972   *Chapter 15   Introduction to Calculus*

Teaching Tip   For Example 4, point out that since the graph of the parabola is symmetric about the y-axis, a and b are opposites.

Guided Practice

5. $\frac{2}{3}x^3 - 2x^2 + 3x + C$

6. $\frac{1}{4}x^4 + \frac{3}{2}x^2 + x + C$

Evaluate each indefinite integral.

5. $\int (2x^2 - 4x + 3)\,dx$

6. $\int (x^3 + 3x + 1)\,dx$

7. Find the area of the shaded region in the graph at the right. $\frac{16}{3}$ units^2

$y = 4 - x^2$

Find the area between each curve and the x-axis for the given interval.

8. $\frac{32}{5}$ units^2

9. $\frac{26}{3}$ units^2

8. $y = x^4$ from $x = 0$ to $x = 2$

9. $y = x^2 + 4x + 4$ from $x = -1$ to $x = 1$

Evaluate each definite integral.

10. $\int_1^3 2x^3\,dx$ **40**

11. $\int_1^4 (x^2 - x + 6)\,dx$ $\frac{63}{2}$

12. $\int_0^2 (-2x^2 + 3x + 2)\,dx$ $\frac{14}{3}$

13. $\int_2^4 (x^3 + x - 6)\,dx$ **54**

14. **Physics** The work, in joules (J), required to stretch a certain spring a distance of ℓ meters beyond its natural length is given by $W = \int_0^\ell 500x\,dx$. How much work is required to stretch the spring 10 centimeters beyond its natural length? **2.5 J**

EXERCISES

Practice

Evaluate each indefinite integral.

A 15. $\int x^5\,dx$ $\frac{1}{6}x^6 + C$

16. $\int 6x^7\,dx$ $\frac{3}{4}x^8 + C$

B 17. $\int (x^2 - 2x + 4)\,dx$

18. $\int (-3x^2 - x + 6)\,dx$

17. $\frac{1}{3}x^3 - x^2 + 4x + C$

19. $\int (x^4 + 2x^2 - 3)\,dx$
$\frac{1}{5}x^5 + \frac{2}{3}x^3 - 3x + C$

20. $\int (4x^5 - 6x^3 + 7x^2 - 8)\,dx$
$\frac{2}{3}x^6 - \frac{3}{2}x^4 + \frac{7}{3}x^3 - 8x + C$

18. $-x^3 - \frac{1}{2}x^2 + 6x + C$

21. Find the antiderivative of $x^2 - 6x + 3$. $\frac{1}{3}x^3 - 3x^2 + 3x + C$

Find the area of the shaded region in each graph.

22.

$y = 2x^2$

23.

$y = x^2 - 2$

24.

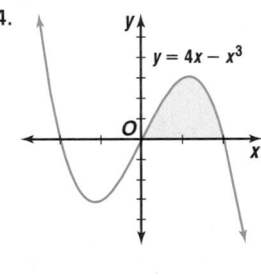

$y = 4x - x^3$

18 units^2

$\frac{13}{3}$ units^2

4 units^2

www.amc.glencoe.com/self_check_quiz

Lesson 15-4 The Fundamental Theorem of Calculus **973**

Assignment Guide

Core: 15–45 odd, 47–49, 51–60
Enriched: 16–44 even, 46–60

Additional Answers (p. 972)

1. $\int f(x)\,dx$ represents all of the functions that have $f(x)$ as their derivative. $\int_a^b f(x)\,dx$ is a number; it gives the area under the graph of $y = f(x)$ from $x = a$ to $x = b$.

2. Sample answer: let $a = 0$, $b = 1$, $f(x) = x$, and $g(x) = x$. Then
$$\int_a^b f(x)g(x)\,dx = \int_0^1 x^2\,dx = \frac{1}{3},$$
but
$$\int_a^b f(x)\,dx \cdot \int_a^b g(x)\,dx =$$
$$\int_0^1 x\,dx \cdot \int_0^1 x\,dx = \frac{1}{2} \cdot \frac{1}{2} = \frac{1}{4}.$$

3. If the "+ C" is included in the antiderivative, it will appear as a term in both $F(b)$ and $F(a)$ and will be eliminated when they are subtracted.

4. Rose is correct; the order does matter. Interchanging the order multiplies the result by –1. In symbols,
$$F(a) - F(b) = -\big(F(b) - F(a)\big).$$
So unless the answer is 0, interchanging the order will give the opposite of the right answer.

Find the area between each curve and the x-axis for the given interval.

25. $y = x^3$ from $x = 0$ to $x = 4$ **64 units²**   **26.** $y = 3x^6$ from $x = -1$ to $x = 1\frac{6}{7}$ **unit²**

27. $\frac{20}{3}$ **units²**

27. $y = x^2 - 2x$ from $x = -2$ to $x = 0$ **28.** $y = -x^2 + 2x + 3$ from $x = 1$ to $x = 3$

28. $\frac{16}{3}$ **units²**

29. $y = x^3 + x$ from $x = 0$ to $x = \frac{1}{3}$ **30.** $y = x^3 + 8x + 10$ from $x = -1$ to $x = 3$
$\frac{3}{4}$ **unit²**   **92 units²**

Evaluate each definite integral.

31. $\displaystyle\int_0^7 6x^2 \, dx$ **686**   **32.** $\displaystyle\int_2^4 3x^4 \, dx$ $\frac{2976}{5}$

33. $\displaystyle\int_{-1}^3 (x + 4) \, dx$ **20**   **34.** $\displaystyle\int_1^5 (3x^2 - 2x + 1) \, dx$ **104**

35. $\displaystyle\int_1^3 (x^3 - x^2) \, dx$ $\frac{34}{3}$ **36.** $\displaystyle\int_0^1 (x^4 + 2x^2 + 1) \, dx$ $\frac{28}{15}$

37. $\displaystyle\int_{-1}^0 (x^4 - x^3) \, dx$ $\frac{9}{20}$ **38.** $\displaystyle\int_0^2 (x^3 + x + 1) \, dx$ **8**

39. $\displaystyle\int_{-2}^5 (x^2 - 3x + 8) \, dx$ $\frac{413}{6}$ **40.** $\displaystyle\int_1^3 (x + 3)(x - 1) \, dx$ $\frac{32}{3}$

41. $\displaystyle\int_2^3 (x - 1)^3 \, dx$ $\frac{15}{4}$ **42.** $\displaystyle\int_0^1 \frac{x^2 - x - 2}{x - 2} \, dx$ $\frac{3}{2}$

43. Find the integral of $x(4x^2 + 1)$ from 0 to 2. **18**

44. What is the integral of $(x + 1)(3x + 2)$ from -1 to 1? **6**

45. The integral $\displaystyle\int_0^{n+0.5} x^k \, dx$ gives a fairly close, quick estimate of the sum of the series $\displaystyle\sum_{i=1}^n i^k$. Use the integral to estimate each sum and then find the actual sum.

a. $\displaystyle\sum_{i=1}^{20} i^3$ **44,152.52; 44,100**   **b.** $\displaystyle\sum_{i=1}^{100} i^2$ **338,358.38; 338,350**

Applications and Problem Solving

Real World Application

46. Physics   The work (in joules) required to pump all of the water out of a 10 meter by 5 meter by 2 meter swimming pool is given by $\displaystyle\int_0^2 490{,}000x \, dx$. Evaluate this integral to find the required work. **980,000 J**

47. Critical Thinking   **a. All are negative.**
   a. Suppose $f(x)$ is a function whose graph is *below* the x-axis for $a \le x \le b$. What can you say about the values of $f(x)$, $\displaystyle\sum_{i=1}^n f(x_i)\Delta x$, and $\displaystyle\int_a^b f(x) \, dx$?

   b. Evaluate $\displaystyle\int_0^2 (x^2 - 5) \, dx$. $-\frac{22}{3}$

   c. What is the area between the graph of $y = x^2 - 5$ and the x-axis from $x = 0$ to $x = 2$? $\frac{22}{3}$

48. Critical Thinking   Find the value of $\displaystyle\int_2^5 (3x - 6) \, dx$ *without* using limits or the Fundamental Theorem of Calculus. $\frac{27}{2}$

49. Stock Market The *average value* of a function $f(x)$ over the interval $a \leq x \leq b$ is defined to be $\dfrac{1}{b-a} \displaystyle\int_a^b f(x) \, dx$. A stock market analyst has determined that the price of the stock of the Acme Corporation over the year 2001 can be modeled by the function $f(x) = 75 + 8x - \dfrac{1}{2}x^2$, where x is the time, in months, since the beginning of 2001, and $f(x)$ is in dollars.

a. Sketch a graph of $f(x)$ from $x = 0$ to $x = 12$. **See margin.**

b. Find the average value of the Acme Corporation stock over the first half of 2001. **$93**

c. Find the average value of the stock over the second half of 2001. **$105**

50. Geometry The volume of a sphere of radius R can be found by slicing the sphere vertically and then integrating the areas of the resulting circular cross sections. (The cross section in the figure is a circle of radius $\sqrt{R^2 - x^2}$.) This process results in the integral $\displaystyle\int_{-R}^{R} (\pi R^2 - \pi x^2) \, dx$. Evaluate this integral to obtain the expression for the volume of a sphere of radius R. $\dfrac{4}{3}\pi R^3$

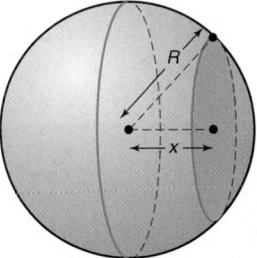

51. Space Exploration The weight of an object that is at a distance x from the center of Earth can be written as kx^{-2}, where k is a constant that depends on the mass of the object. The energy required to move the object from $x = a$ to $x = b$ is the integral of its weight, that is, $\displaystyle\int_a^b kx^{-2} \, dx$. Suppose a Lunar Surveying Module (LSM), designed to analyze the surface of the moon, weighs 1000 newtons on the surface of Earth. **a. 4.1×10^{16} Nm²**

a. Find k for the LSM. Use 6.4×10^6 meters for the radius of Earth.

b. Find the energy required to lift the LSM from Earth's surface to the moon, 3.8×10^8 meters from the center of Earth. **6.3×10^9 J**

Mixed Review

52. Use a limit to evaluate $\displaystyle\int_0^2 \dfrac{1}{2} x^2 \, dx$. *(Lesson 15-3)* $\dfrac{4}{3}$

53. Find the derivative of $f(x) = 2x^6 - 3x^2 + 2$. *(Lesson 15-2)* $f'(x) = 12x^5 - 6x$

54. Education The scores of a national achievement test are normally distributed with a mean of 500 and a standard deviation of 100. What percent of those who took the test had a score more than 100 points above or below the mean? *(Lesson 14-4)* **31.7%**

55. Fifty tickets, numbered consecutively from 1 to 50 are placed in a box. Four tickets are drawn without replacement. What is the probability that four odd numbers are drawn? *(Lesson 13-4)* $\dfrac{253}{4606}$

56. Banking Find the amount accumulated if $600 is invested at 6% for 15 years and interest is compounded continuously. *(Lesson 11-3)* **$1475.76**

57. Write an equation of the parabola with vertex at $(6, -1)$ and focus at $(3, -1)$. *(Lesson 10-5)* $(y + 1)^2 = -12(x - 6)$

58. Find $2\sqrt{2} \left(\cos \dfrac{2\pi}{3} + i \sin \dfrac{2\pi}{3} \right) \div \sqrt{2} \left(\cos \dfrac{\pi}{3} + i \sin \dfrac{\pi}{3} \right)$. Then express the result in rectangular form. *(Lesson 9-7)*

58. $2\left(\cos \dfrac{\pi}{3} + i \sin \dfrac{\pi}{3} \right)$; $1 + \sqrt{3}i$

Extra Practice See p. A55.

Additional Answer

49a.

Extra Credit

Let $f(x)$ be a continuous function and $F(x)$ be an antiderivative of f. Prove that $\displaystyle\int_b^a f(x) \, dx = -\int_a^b f(x) \, dx$ and $\displaystyle\int_a^b f(x) \, dx + \int_b^c f(x) \, dx = \int_a^c f(x) \, dx$.

$$\int_b^a f(x)\, dx = F(a) - F(b)$$
$$= -[F(b) - F(a)]$$
$$= -\int_a^b f(x)\, dx$$

$$\int_a^b f(x)\, dx + \int_b^c f(x)\, dx = [F(b) - F(a)] + [F(c) - F(b)]$$
$$= F(c) - F(a)$$
$$= \int_a^c f(x)\, dx$$

Open-Ended Assessment

Writing Have each student write a short paragraph comparing and contrasting definite and indefinite integrals.

Chapter 15, Quiz D (Lesson 15-4) is available in the *Assessment and Evaluation Masters*, p. 382.

---

59. 35.46 ft/s

59. Find the initial vertical velocity of a stone thrown with an initial velocity of 45 feet per second at an angle of 52° with the horizontal. *(Lesson 8-7)*

60. **SAT Practice** In the circle with center X, $\overset{\frown}{AE}$ is the shortest of the five unequal arcs. Which statement best describes the measure of angle AXE? **A**

A less than 72°

B equal to 72°

C greater than 72°, but less than 90°

D greater than 90°, but less than 180°

E greater than 180°

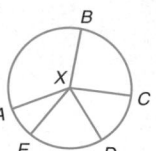

CAREER CHOICES

● Mathematician ●

Algebra, geometry, trigonometry, statistics, calculus—if you enjoy studying these subjects, then a career in mathematics may be for you. As a mathematician, you would have several options for employment.

First, a theoretical mathematician develops new principles and discovers new relationships, which may be purely abstract in nature. Applied mathematicians use new ideas generated by theoretical mathematicians to solve problems in many fields, including science, engineering and business. Mathematicians may work in related fields such as computer science, engineering, and business. As a mathematician, you can become an elementary or secondary teacher if you obtain a teaching certificate. An advanced degree is required to teach at the college level.

CAREER OVERVIEW

Degree Preferred:
bachelor's degree in mathematics

Related Courses:
mathematics, science, computer science

Outlook:
increased demand for teachers and math-related occupations through the year 2006

The Average Teacher Salary Compared to the Average Experience Level of Teachers

☐ Average Teacher Salary (1997 Dollars)
☐ Experience of Average Teacher

Source: American Federation of Teachers

inter NET CONNECTION For more information on careers in mathematics, visit: **www.amc.glencoe.com**

976 *Chapter 15 Introduction to Calculus*

Enrichment Masters, p. 168

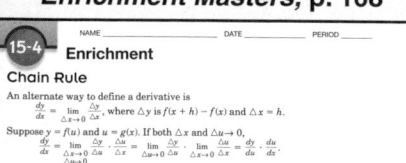

15-4 Enrichment

Chain Rule

An alternate way to define a derivative is
$\frac{dy}{dx} = \lim_{\Delta x \to 0} \frac{\Delta y}{\Delta x}$, where Δy is $f(x + h) - f(x)$ and $\Delta x = h$.

Suppose $y = f(u)$ and $u = g(x)$. If both Δx and $\Delta u \to 0$,
$\frac{dy}{dx} = \lim_{\Delta x \to 0} \frac{\Delta y}{\Delta x} \cdot \frac{\Delta u}{\Delta x} = \lim_{\Delta u \to 0} \frac{\Delta y}{\Delta u} \cdot \lim_{\Delta x \to 0} \frac{\Delta u}{\Delta x} = \frac{dy}{du} \cdot \frac{du}{dx}$.

This is known as the chain rule. It provides a short cut for finding some derivatives.

Example If $y = (x^2 - 2x)^2$, find $\frac{dy}{dx}$ by the chain rule.

Let $u = x^2 - 2x$, then $f(u) = u^2$.
$\frac{dy}{dx} = 2u \cdot (2x - 2)$ or $\frac{dy}{dx} = 2(x^2 - 2x)(2x - 2) = 4x^3 - 12x^2 + 8x$

Check: $y = (x^2 - 2x)^2 = x^4 - 4x^3 + 4x^2$
$\frac{dy}{dx} = 4x^3 - 12x^2 + 8x$ ✓

Use the chain rule to find the derivative of each function. If possible, check by expansion or multiplication.

1. $y = (x^2 - 1)^2$
$2(x^2 - 1)(2x) = 4x(x^2 - 1)$

2. $y = (\sqrt{x} + 2)^3$
$3(\sqrt{x} + 2)^2(\frac{x^{-\frac{1}{2}}}{2}) = \frac{3x^{-\frac{1}{2}}}{2}(\sqrt{x} + 2)^2$

3. $y = (x + \frac{1}{x})^2$
$2(x + \frac{1}{x})(1 - x^{-2})$

4. $y = (x^2 - 3x + 2)^2$
$2(x^2 - 3x + 2)(2x - 3) = 4x^3 - 18x^2 + 26x - 12$

5. $y = \sqrt{2x + 3}$
$\frac{1}{2}(2x + 3)^{-\frac{1}{2}}(2) = (2x + 3)^{-\frac{1}{2}}$

6. $y = (3x - 1)^{-\frac{1}{2}}$
$-\frac{1}{2}(3x - 1)^{-\frac{3}{2}}(3) = -\frac{3}{2}(3x - 1)^{-\frac{3}{2}}$

7. $y = \frac{1}{2x - 5}$
$-(2x - 5)^{-2}(2) = -2(2x - 5)^{-2}$

8. $y = \frac{2}{x^2 - 1}$
$-2(x^2 - 1)^{-2}(2x) = -4x(x^2 - 1)^{-2}$

© Glencoe/McGraw-Hill   T168   Advanced Mathematical Concepts

---

CAREER CHOICES   MATHEMATICIAN

Math Connection

● Explain any trends you see in the graph. **The average teacher's salary has fluctuated during this period with the lowest being paid in about 1962 and 1981 and the highest being paid in the 1990s, although 1972 was a relatively high year. The average experience level declined from 1962 to 1975 and then steadily rose through 1997.**

● Why is it important that teachers' salaries be given in 1997 dollars? **If inflation were not taken into account by expressing each year in terms of 1997 dollars, it would be very difficult to compare salaries fairly.**

VOCABULARY

antiderivative (p. 955)
definite integral (p. 962)
derivative (p. 951)
differentiation (p. 952)
Fundamental Theorem of Calculus (p. 970)
indefinite integral (p. 971)
integral (p. 961)

integration (p. 962)
limit (p. 941)
rate of change (p. 956)
secant line (p. 951)
slope of a curve (p. 949)
tangent line (p. 951)

UNDERSTANDING AND USING THE VOCABULARY

State whether each sentence is *true* or *false*. If false, replace the underlined word(s) to make a true statement.

1. $f(a)$ and $\lim\limits_{x \to a} f(x)$ are <u>always</u> the same. **false; sometimes**

2. The process of finding the area under a curve is called <u>integration</u>. **true**

3. The inverse of finding the derivative of a function is finding the <u>definite</u> integral. **false; indefinite**

4. The <u>Fundamental Theorem of Calculus</u> can be used to evaluate a definite integral. **true**

5. A line that intersects a graph in two points is called a <u>tangent</u> line. **false; secant**

6. A line that passes through a point on a curve and has a slope equal to the slope of the curve at that point is called a <u>secant</u> line. **false; tangent**

7. The <u>conjugate</u> of a function $f(x)$ is another function $f'(x)$ that gives the slope of the tangent line to $y = f(x)$ at any point. **false; derivative**

8. If you look at one particular point on the graph of a curve, there is a certain steepness, called the <u>slope</u>, at that point. **true**

9. The derivative of a function can also be called the <u>domain</u> of the function because it measures how fast the function changes. **false; rate of change**

10. The process of finding a <u>limit</u> is called differentiation. **false; derivative**

 CONNECTION For additional review and practice for each lesson, visit: **www.amc.glencoe.com**

Six forms of Chapter 15 Test are available in the *Assessment and Evaluation Masters* as shown in the chart.

Chapter 15 Test, Form 1B, is shown at the right. Chapter 15 Test, Form 2B, is shown on the next page.

TEST OPTIONS

Form	Pages	Type	
1A	367–368	Honors	Form 1 tests are multiple choice.
1B	369–370	Average	
1C	371–372	Basic	
2A	373–374	Honors	Form 2 tests are free response.
2B	375–376	Average	
2C	377–378	Basic	

STUDY GUIDE AND ASSESSMENT

VOCABULARY

This section provides a listing of the new terms, properties, and phrases that were introduced in this chapter. The exercises check students' understanding of the terms by using a variety of verbal formats including matching, completion, and true/false.

Glossary A complete glossary of terms begins on page A71.

Assessment and Evaluation Masters, pp. 369–370

SKILLS AND CONCEPTS

Objectives and Examples This section reviews the skills and concepts of the chapter and shows completely worked examples.

Review Exercises These exercises provide practice for the corresponding objectives.

Assessment and Evaluation Masters, pp. 375–376

SKILLS AND CONCEPTS

OBJECTIVES AND EXAMPLES	REVIEW EXERCISES

Lesson 15-1 Calculate limits of polynomial and rational functions algebraically.

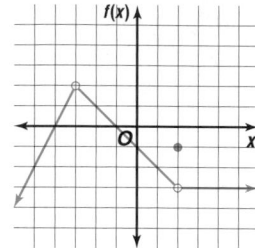

Consider the graph of the function $y = f(x)$ shown below. Find $f(-3)$ and $\lim\limits_{x \to -3} f(x)$.

There is no point on the graph with an x-coordinate of -3, so $f(-3)$ is undefined.

Look at points on the graph whose x-coordinates are close to, but not equal to, -3. The closer x is to -3, the closer y is to 2. So, $\lim\limits_{x \to -3} f(x) = 2$.

Evaluate each limit.

a. $\lim\limits_{x \to 2} \dfrac{x^2 + x - 6}{x^2 + 3x}$

$\lim\limits_{x \to 2} \dfrac{x^2 + x - 6}{x^2 + 3x} = \lim\limits_{x \to 2} \dfrac{(x + 3)(x - 2)}{x(x + 3)}$

$= \lim\limits_{x \to 2} \dfrac{x - 2}{x}$

$= \dfrac{2 - 2}{2}$

$= 0$

b. $\lim\limits_{x \to 0} \dfrac{x \cos x}{x}$

$\lim\limits_{x \to 0} \dfrac{x \cos x}{x} = \lim\limits_{x \to 0} \cos x$

$= \cos 0$

$= 1$

11. Refer to the graph of $y = f(x)$ at the left. Find $f(2)$ and $\lim\limits_{x \to 2} f(x)$. **−1, −3**

Evaluate each limit.

12. $\lim\limits_{x \to -2} (x^3 - x^2 - 5x + 6)$ **4**

13. $\lim\limits_{x \to 0} (2x - \cos x)$ **−1**

14. $\lim\limits_{x \to 1} \dfrac{x^2 - 36}{x + 6}$ **−5**

15. $\lim\limits_{x \to 0} \dfrac{5x^2}{2x}$ **0**

16. $\lim\limits_{x \to 4} \dfrac{x^2 + 2x}{x^2 - 3x - 10}$ **−4**

17. $\lim\limits_{x \to 0} (x + \sin x)$ **0**

18. $\lim\limits_{x \to 0} \dfrac{x^2 + x \cos x}{2x}$ **$\dfrac{1}{2}$**

19. $\lim\limits_{x \to 2} \dfrac{x^3 + 2x^2 - 4x - 8}{x^2 - 4}$ **4**

20. $\lim\limits_{x \to 0} \dfrac{(x - 3)^2 - 9}{2x}$ **−3**

21. $\lim\limits_{x \to 5} \dfrac{x^2 - 9x + 20}{x^2 - 5x}$ **$\dfrac{1}{5}$**

TestCheck and Worksheet Builder

This **networkable** software has three modules.
- **Worksheet Builder** to make worksheets and tests
- **Student Module** to take tests on-screen
- **Management System** to keep student records

Three special banks are included for SAT, ACT, and TIMSS.

OBJECTIVES AND EXAMPLES	REVIEW EXERCISES

Lesson 15-2 Find derivatives of polynomial functions.

• Find the derivative of each function.

a. $f(x) = 3x^4 + 2x^3 - 7x - 5$

$$f'(x) = 3 \cdot 4x^3 + 2 \cdot 3x^2 - 7 \cdot 1 - 0$$
$$= 12x^3 + 6x^2 - 7$$

b. $f(x) = 2x^3(x^2 + 1)$

First, multiply to write the function as a polynomial.

$$f(x) = 2x^3(x^2 + 1)$$
$$= 2x^5 + 2x^3$$

Then find the derivative.

$$f'(x) = 10x^4 + 6x^2$$

Use the definition of derivative to find the derivative of each function.

22. $f(x) = 2x + 1 \quad f'(x) = 2$

23. $f(x) = 4x^2 + 3x - 5 \quad f'(x) = 8x + 3$

24. $f(x) = x^3 - 3x \quad f'(x) = 3x^2 - 3$

Use the derivative rules to find the derivative of each function.

25. $f(x) = 2x^6 \quad f'(x) = 12x^5$

26. $f(x) = -3x + 7 \quad f'(x) = -3$

27. $f(x) = 3x^2 - 5x \quad f'(x) = 6x - 5$

28. $f(x) = \frac{1}{4}x^2 - x + 4 \quad f'(x) = \frac{1}{2}x - 1$

29. $f(x) = \frac{1}{2}x^4 - 2x^3 + \frac{1}{3}x - 4$

30. $f(x) = (x + 3)(x + 4) \quad f'(x) = 2x + 7$

31. $f(x) = 5x^3(x^4 - 3x^2) \quad f'(x) = 35x^6 - 75x^4$

32. $f(x) = (x - 2)^3 \quad f'(x) = 3x^2 - 12x + 12$

---

29. $f'(x) = 2x^3 - 6x^2 + \frac{1}{3}$

Lesson 15-2 Find antiderivatives of polynomial functions.

• Find the antiderivative of each function.

a. $f(x) = 5x^2$

$$F(x) = 5 \cdot \frac{1}{2+1}x^{2+1} + C$$
$$= \frac{5}{3}x^3 + C$$

b. $f(x) = -2x^3 + 6x^2 - 5x + 4$

$$F(x) = -2 \cdot \frac{1}{4}x^4 + 6 \cdot \frac{1}{3}x^3 -$$
$$5 \cdot \frac{1}{2}x^2 + 4 \cdot x + C$$
$$= -\frac{1}{2}x^4 + 2x^3 - \frac{5}{2}x^2 + 4x + C$$

Find the antiderivative of each function.

33. $f(x) = 8x \quad F(x) = 4x^2 + C$

34. $f(x) = 3x^2 + 2 \quad F(x) = x^3 + 2x + C$

35. $f(x) = -\frac{1}{2}x^3 + 2x^2 - 3x - 2$

36. $f(x) = x^4 - 5x^3 + 2x - 6$

37. $f(x) = (x - 4)(x + 2)$

38. $f(x) = \frac{x^2 - x}{x} \quad F(x) = \frac{1}{2}x^2 - x + C$

35. $F(x) = -\frac{1}{8}x^4 + \frac{2}{3}x^3 - \frac{3}{2}x^2 - 2x + C$

36. $F(x) = \frac{1}{5}x^5 - \frac{5}{4}x^4 + x^2 - 6x + C$

37. $F(x) = \frac{1}{3}x^3 - x^2 - 8x + C$

OBJECTIVES AND EXAMPLES

Lesson 15-3 Find areas under graphs of polynomial functions.

• Use limits to find the area of the region between the graph of $y = 3x^2$ and the x-axis from $x = 0$ to $x = 1$.

$$\int_0^1 3x^2 \, dx = \lim_{n \to \infty} \sum_{i=1}^n 3(x_i)^2 \Delta x$$

$$= \lim_{n \to \infty} \sum_{i=1}^n 3\left(\frac{i}{n}\right)^2\left(\frac{1}{n}\right)$$

$$= \lim_{n \to \infty} \sum_{i=1}^n \frac{3i^2}{n^3}$$

$$= \lim_{n \to \infty} \frac{3}{n^3}(1^2 + 2^2 + \cdots + n^2)$$

$$= \lim_{n \to \infty} \frac{3}{n^3} \cdot \frac{n(n+1)(2n+1)}{6}$$

$$= \lim_{n \to \infty} \frac{1}{2}\left(2 + \frac{3}{n} + \frac{1}{n^2}\right)$$

$$= 1 + 0 + 0 \text{ or } 1 \text{ unit}^2$$

REVIEW EXERCISES

Use limits to find the area between each curve and the x-axis for the given interval.

39. $y = 2x$ from $x = 0$ to $x = 2$ **4 units²**

40. $y = x^3$ from $x = 0$ to $x = 1$ $\frac{1}{4}$ **unit²**

41. $y = x^2$ from $x = 3$ to $x = 4$ $\frac{37}{3}$ **units²**

42. $y = 6x^2$ from $x = 1$ to $x = 2$ **14 units²**

Lesson 15-4 Use the Fundamental Theorem of Calculus to evaluate definite integrals of polynomial functions.

• Evaluate $\int_4^7 (x^2 - 3) \, dx$.

$$\int_4^7 (x^2 - 3) \, dx$$

$$= \frac{1}{3}x^3 - 3x \Big|_4^7$$

$$= \left(\frac{1}{3} \cdot 7^3 - 3 \cdot 7\right) - \left(\frac{1}{3} \cdot 4^3 - 3 \cdot 4\right)$$

$$= 84$$

Evaluate each definite integral.

43. $\int_2^4 6x \, dx$ **36**

44. $\int_{-3}^2 3x^2 \, dx$ **35**

45. $\int_{-2}^2 (3x^2 - x + 3) \, dx$ **28**

46. $\int_0^4 (x + 2)(2x + 3) \, dx$ $\frac{368}{3}$

Lesson 15-4 Find indefinite integrals of polynomial functions.

• Evaluate $\int (6x^2 - 4x) \, dx$.

$$\int (6x^2 - 4x) \, dx = 6 \cdot \frac{1}{3}x^3 - 4 \cdot \frac{1}{2}x^2 + C$$

$$= 2x^3 - 2x^2 + C$$

Evaluate each indefinite integral.

47. $\int 6x^4 \, dx$ $\frac{6}{5}x^5 + C$

48. $\int (-3x^2 + 2x) \, dx$ $-x^3 + x^2 + C$

49. $\int (x^2 + 5x - 2) \, dx$ $\frac{1}{3}x^3 + \frac{5}{2}x^2 - 2x + C$

50. $\int (3x^5 + 4x^4 - 7x) \, dx$ $\frac{1}{2}x^6 + \frac{4}{5}x^5 - \frac{7}{2}x^2 + C$

APPLICATIONS AND PROBLEM SOLVING

51. Physics The kinetic energy of an object with mass m is given by the formula $k(t) = \frac{1}{2} m \cdot v(t)^2$, where $v(t)$ is the velocity of the object at time t. Suppose $v(t) = \frac{50}{1 + t^2}$ for all $t \geq 0$. What does the kinetic energy of the object approach as time approaches 100? *(Lesson 15-1)* **0.0000125m**

52. Business The controller for an electronics company has used the production figures for the last few months to determine that the function $c(x) = -9x^5 + 135x^3 + 10{,}000$ approximates the cost of producing x thousands of one of their products. Find the marginal cost if they are now producing 2600 units. *(Lesson 15-2)* **$681.41**

53. Motion An advertisement for a sports car claims that the car can accelerate from 0 to 60 miles per hour in 5 seconds. *(Lesson 15-2)*

a. Find the acceleration of the sports car in feet per second squared, assuming that it is constant. **17.6 ft/s^2**

b. Write an equation for the velocity of the sports car at t seconds. **$v(t) = 17.6t$**

c. Write an equation for the distance traveled in t seconds. **$d(t) = 8.8t^2$**

ALTERNATIVE ASSESSMENT

OPEN-ENDED ASSESSMENT

1. The limit of a continuous function $f(x)$ as x approaches 1 is 5. Give an example of a function for which this is true. Show why the limit of your function as x approaches 1 is 5. **See margin.**

2. The area of the region between the graph of the function $g(x)$ and the x-axis from $x = 0$ to $x = 1$ is 4. Give an example of a function for which this is true. Show that the area of the region between the graph of your function and the x-axis from $x = 0$ to 1 is in fact 4. **See margin.**

 PORTFOLIO

Explain the difference between a definite integral and an indefinite integral. Give an example of each.

Now that you have completed your work in this book, review your portfolio entries for each chapter. Make any necessary changes or corrections. Add a table of contents to your portfolio at this time.

 Unit 5 *inter*NET **Project**

DISEASES

- Use the Internet to find the number of cases reported or the number of deaths for one particular disease for a period of at least 10 years. Some possible diseases you might choose to research are measles, tuberculosis, or AIDS. Make a table or spreadsheet of the data.

- Use computer software or a graphing calculator to find at least two polynomial functions that model the data. Find the derivative for each of your function models. What does the derivative represent?

- Use each model to predict the cases or deaths from the disease in the year 2010. Write a one-page paper comparing the models. Discuss which model you think best fits the data. Include any limitations of the model.

Additional Assessment See p. A70 for Chapter 15 Practice Test.

Chapter 15 Study Guide and Assessment **981**

Internet Address Book

Record useful Internet addresses in the space at right for quick reference.

www.amc.glencoe.com

Chapter 15 Skill: Special Functions and Counting

Pages 982–983 are part of a complete SAT/ACT preparation course that is described in detail on pages T8–T9 of the *Teacher's Handbook*. The test items were written in the same style as those in the SAT.

 These questions were aligned and verified by The Princeton Review, the nation's leader in test preparation.

Diagnosis & Prescription

Each of the ten SAT practice problems on page 983 is cross-referenced to the chapter where that SAT skill is covered. If students miss a particular type of problem, you can have them study that skill.

Problem Number and Skill		Page
1	Special Functions and Counting	982
2	Geometry (Triangles and Quadrilaterals)	482
3	Special Functions and Counting	982
4	Geometry (Perimeter, Area, and Volume)	548
5	More Algebra	202
6	Special Functions and Counting	982
7	Average	754
8	Special Functions and Counting	982
9	Special Functions and Counting	982
10	Special Functions and Counting	982

Special Function and Counting Problems

The SAT includes function problems that use special symbols like $\oplus$ or # or $\otimes$. (The ACT does not contain this type of problem.)

Here's a simple example: If $x \# y = x + 2y$, then what is $2 \# 5$? To find $2 \# 5$, replace x with 2 and y with 5. Thus, $2 \# 5 = 2 + 2(5) = 12$. The SAT may also include problems that involve counting regions, surfaces, or intersections. The questions usually ask for the maximum or minimum number.

SAT EXAMPLE

1. Let $\boxed{x}$ be defined for all positive integers x as the product of the distinct prime factors of x. What is the value of

$$\frac{\langle 6 \rangle}{\langle 81 \rangle}?$$

HINT The SAT often combines two mathematical concepts in one problem. For example, this problem combines a special function and prime factors.

Solution Carefully read the definition of $\boxed{x}$. Recall the meaning of "distinct prime factors." Write the prime factorization of each number, identify which prime factors are distinct, and then find the product.

Start with the first number, 6. $6 = 2 \times 3$. Both 2 and 3 are distinct prime factors. The product of the distinct prime factors is 6.

Do the same with 81. $81 = 3 \times 3 \times 3 \times 3$. There is just one distinct prime factor, 3. So the product of the distinct prime factors is also 3.

Finally, substitute the values for $\langle 6 \rangle$ and $\langle 81 \rangle$ into the fraction.

$\frac{\langle 6 \rangle}{\langle 81 \rangle} = \frac{6}{3} = 2$. The answer is 2.

Grid-in this answer on your answer sheet.

SAT EXAMPLE

2. The figure below is a square separated into two non-overlapping regions. What is the greatest number of non-overlapping regions that can be made by drawing any two additional straight lines?

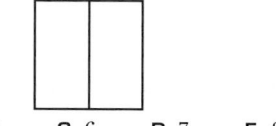

A 4   **B** 5   **C** 6   **D** 7   **E** 8

HINT Watch out for "obvious" answers on difficult problems (those numbered 18 or higher). They are usually wrong answers.

Solution Draw right on your test booklet.

The most obvious ways to draw two more lines are shown at the right.

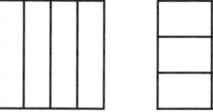

The first figure has 4 regions; the second figure has 6 regions. So you can immediately eliminate answer choices A and B.

For the maximum number of regions, it is likely that the lines will *not* be parallel, as they are in the figures above.

Draw the two lines with the fewest possible criteria: not horizontal, not vertical, not parallel, and not perpendicular.

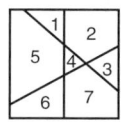

There are 7 regions. The answer is choice **D**.

 TestCheck and Worksheet Builder

Special banks of SAT, ACT, and TIMSS items are included in this CD-ROM.

After you work each problem, record your answer on the answer sheet provided or on a piece of paper.

Multiple Choice

1. If $x \otimes y = \dfrac{1}{x-y}$, what is the value of $\dfrac{1}{2} \otimes \dfrac{1}{3}$? **A**

A 6 B $\dfrac{6}{5}$ C $\dfrac{1}{6}$

D -1 E -6

2. If one side of a triangle is twice as long as a second side of length x, then the perimeter of the triangle can be: **D**

A $2x$ B $3x$ C $4x$

D $5x$ E $6x$

3. If 3 parallel lines are cut by 3 nonparallel lines, what is the maximum number of intersections possible? **D**

A 9 B 10 C 11

D 12 E 13

4. In the figure below, if segment $\overline{WZ}$ and segment $\overline{XY}$ are diameters with lengths of 12, what is the area of the shaded region? **B**

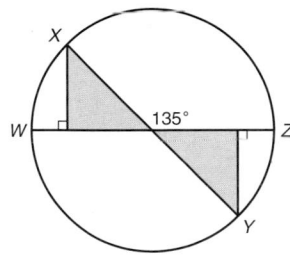

A 9 B 18 C 36

D 54 E 108

5. Which of the following represents the values of x that are solutions of the inequality $x^2 < x + 6$? **C**

A $x > -2$

B $x < 3$

C $-2 < x < 3$

D $-3 < x < 2$

E $x < -2 \;\cup\; x > 3$

6. $\boxed{x} = \dfrac{1}{2}x$ if x is composite.

$\boxed{x} = 3x$ if x is prime.

What is the value of $\boxed{5} + \boxed{16}$? **B**

A 21 B 23 C 31

D 46 E 69

7. What is the average of all the integers from 1 to 20 inclusive? **C**

A 9.5 B 10 C 10.5

D 20 E 21

8. All faces of a cube with a 4-meter edge are painted blue. If the cube is then cut into cubes with 1-meter edges, how many of the 1-meter cubes have blue paint on exactly one face? **A**

A 24 B 36 C 48

D 60 E 72

9. For all numbers n, let $\{n\}$ be defined as $n^2 - 1$. What is the value of $\{\{x\}\}$? **D**

A $x^2 - 1$

B $x^4 - 1$

C $x^4 - 2x^2 - 1$

D $x^4 - 2x^2$

E x^4

10. Grid-In Let $\boxed{x}$ be defined for all positive integers x as the product of the distinct prime factors of x. What is the value of

$$\dfrac{\boxed{20}}{\boxed{16}}\,? \quad \mathbf{5}$$

inter NET CONNECTION **SAT/ACT Practice** For additional test practice questions, visit: **www.amc.glencoe.com**

THE PRINCETON REVIEW *SAT & ACT Preparation* **983**

Extra Credit

TIMSS The value of $\displaystyle\lim_{h \to 0} \dfrac{\sqrt{2+h} - \sqrt{2}}{h}$ is ? **B**

A 0

B $\dfrac{1}{2\sqrt{2}}$

C $\dfrac{1}{2}$

D $\dfrac{1}{\sqrt{2}}$

E ∞

A bubble-in answer sheet for these practice problems is available on p. 1 of the *Assessment and Evaluation Masters.*

Additional Practice

Additional SAT and ACT practice questions and a bubble-in answer sheet are available in the *Assessment and Evaluation Masters,* pp. 2, 383, and 384.

Teaching Tip For Exercise 7, remind students that the word *inclusive* in this case means that the numbers 1 and 20 are included as well as the integers in between.

Assessment and Evaluation Masters, pp. 383–384

STUDENT HANDBOOK

Skills

Reference

1 Introduction to the Graphing Calculator

This section introduces you to some commonly-used keys and menus of the calculator.

Setting Preferences

All of the instructions in the Graphing Calculator Appendix are TI-83 Plus specific.

Another version of the appendix for the Casio CFX-9850 GaPLUS is available in the *Graphing Calculator Keystroke Masters.*

MODE   The MODE key allows you to select your preferences in many aspects of calculation and graphing. Many of these settings are rarely changed in common usage. This screen shows the default mode settings.

← type of numeric notation
← number of decimal places in results
← unit of angle measure used
← type of graph (function, parametric, polar, sequence)
← whether to connect graphed points

← real, rectangular complex, or polar complex number system
← graph occupies full screen, top of screen with HOME screen below, or left side of screen with TABLE on right

To change the preferences, use the arrow keys to highlight your choice and press ENTER .

FORMAT   The **FORMAT** menu is the second function of ZOOM and sets preferences for the appearance of your graphing screen. The default screen is shown below.

← rectangular or polar coordinate system
← whether to display the cursor coordinates on screen
← whether to show a grid pattern on screen
← whether to show the axes
← whether to label the axes
← whether to show the equation being graphed

You can change your preferences in the **FORMAT** menu in the same way you change MODE settings.

Using Menus

Many keys on the calculator access menus from which you can select a function, command, or setting. Some keys access multiple menus. You can use the right and left arrow keys to scroll through the different menu names located at the top of the screen. As each menu name is highlighted, the choices on the screen change. The screens on the next page show various menus accessed by using MATH .

Math menu

Number menu

Complex Number menu

Probability menu

To select a choice in a menu, either use the arrow keys to highlight your choice and press ENTER or simply press the number or letter of your selection. Notice that entry 7 in the first screen has a down arrow instead of a colon after the 7. This signifies there are more entries in the menu.

Alternate Function Keys

Whenever an alternate function is indicated in the keystrokes of this appendix, we will use brackets to show that the function is listed above a key.

Above most keys are one or two additional labels representing commands, menus, letters, lists, or operational symbols. These are accessed by using 2nd or ALPHA.

- 2nd accesses the commands on the left above the key. Note that these commands and 2nd are the same color.

- ALPHA accesses the commands on the right above each key. These commands and ALPHA are also the same color.

- Pressing 2nd ALPHA engages the **[A-LOCK]** or Alpha Lock command. This enables you to select consecutive ALPHA commands without pressing ALPHA before each command. This is especially useful when entering programs.

Each letter accessed by using ALPHA can be used to enter words or labels on the screen, but can also be used as a variable. A value can be stored to each variable.

Computation

A graphing calculator is also a scientific calculator. That is, it follows the order of operations when evaluating entries. Unlike some scientific calculators, the graphing calculator displays every entry in the expression.

Before pressing ENTER to evaluate the expression, you can use the arrow keys to scroll through the expression to make corrections. Corrections can be made in three ways.

- Use DEL to delete any unwanted entries.

- Use 2nd **[INS]** to insert omitted entries.

- "Type" over an incorrect entry. This overprints any entries and does not shift the entries to the right as a word processor does.

If you have an expression that you wish to evaluate repeatedly with a change in one part of the expression, you can press ⎣2nd⎦ **[ENTRY]** after you have pressed ⎣ENTER⎦ and the expression will reappear. You can edit it for your next computation. The **ENTRY** command always repeats the last entered expression. You cannot scroll back through previous expressions you have evaluated.

Example ❶ **Evaluate** $\dfrac{\sqrt{3^2 - 4(6) + [5 - (-12)]^3}}{6}$.

Press: ⎣2nd⎦ [√‾] 3 ⎣x²⎦ ⎣−⎦ 4 ⎣(⎦ 6 ⎣)⎦ ⎣+⎦ ⎣(⎦
5 ⎣−⎦ ⎣(−)⎦ 12 ⎣)⎦ ⎣∧⎦ 3 ⎣)⎦ ⎣÷⎦ 6 ⎣ENTER⎦

The minus key and the negative key are different keys.

Note that the square root function automatically includes a left parenthesis. You must enter the right parenthesis to indicate the end of the expression under the radical sign. If you have the decimal in the **Float** mode, as many as 10 digits may appear in the answer.

```
√(3²−4(6)+(5−ˉ12
)^3)/6
          11.66428547
```

❷ **Evaluate each expression if** $a = 4$, $b = -5$, $c = 2$, $d = \dfrac{2}{3}$, **and** $e = -1.5$.

a. $abc - 3de^4$ **b.** $\dfrac{e + 4a}{c^2 + 8b}$

For a series of expressions that use the same values for the variables, it is often helpful to store the value for each variable into the calculator. You can combine several commands in one line by using the colon after each command. The following commands save the values for variables a, b, c, d and e.

Press: 4 ⎣STO►⎦ ⎣ALPHA⎦ **[A]** ⎣ALPHA⎦ **[:]** ⎣(−)⎦ 5 ⎣STO►⎦ ⎣ALPHA⎦ **[B]** ⎣ALPHA⎦ **[:]** 2 ⎣STO►⎦
⎣ALPHA⎦ **[C]** ⎣ALPHA⎦ **[:]** 2 ⎣÷⎦ 3 ⎣STO►⎦ ⎣ALPHA⎦ **[D]** ⎣ALPHA⎦ **[:]** ⎣(−)⎦ 1.5 ⎣STO►⎦
⎣ALPHA⎦ **[E]** ⎣ENTER⎦

a. Method 1: Using stored values

⎣ALPHA⎦ **[A]** ⎣ALPHA⎦ **[B]** ⎣ALPHA⎦ **[C]** ⎣−⎦ 3 ⎣ALPHA⎦
[D] ⎣ALPHA⎦ **[E]** ⎣∧⎦ 4 ⎣ENTER⎦

Method 2: Entering computations

4 ⎣×⎦ ⎣(−)⎦ 5 ⎣×⎦ 2 ⎣−⎦ 3 ⎣×⎦ 2 ⎣÷⎦ 3 ⎣×⎦ ⎣(⎦ ⎣(−)⎦
1.5 ⎣)⎦ ⎣∧⎦ 4 ⎣ENTER⎦

```
ABC−3DE^4
            -50.125
4*ˉ5*2−3*2/3*(ˉ1
.5)^4
            -50.125
```

b. Method 1: Using stored values

⎣(⎦ ⎣ALPHA⎦ **[E]** ⎣+⎦ 4 ⎣ALPHA⎦ **[A]** ⎣)⎦ ⎣÷⎦ ⎣(⎦
⎣ALPHA⎦ **[C]** ⎣x²⎦ ⎣+⎦ 8 ⎣ALPHA⎦ **[B]** ⎣)⎦ ⎣ENTER⎦

Method 2: Entering computations

⎣(⎦ ⎣(−)⎦ 1.5 ⎣+⎦ 4 ⎣×⎦ 4 ⎣)⎦ ⎣÷⎦ ⎣(⎦ 2 ⎣x²⎦ ⎣+⎦
8 ⎣×⎦ ⎣(−)⎦ 5 ⎣)⎦ ⎣ENTER⎦

```
(E+4A)/(C²+8B)
        -.4027777778
(ˉ1.5+4*4)/(2²+8
*ˉ5)
        -.4027777778
```

2 Graphing Functions

Most functions can be graphed by using the [Y=] key. The viewing window most often used for non-trigonometric functions is the standard viewing window $[-10, 10]$ scl:1 by $[-10, 10]$ scl:1, which can be accessed by selecting **6:ZStandard** on the **ZOOM** menu. Then the window can be adjusted so that a **complete graph** can be viewed. A complete graph is one that shows the basic characteristics of the parent graph.

Example ❶ **Linear Functions** A complete linear graph shows the *x*- and *y*-intercepts.

a. Graph $y = 3x - 4$ in the standard viewing window.

Press: [Y=] 3 [X,T,θ,*n*] [—] 4 [ZOOM] 6

If your calculator is already set for the standard viewing window, press [GRAPH] *instead of* [ZOOM] *6.*

Both the *x*- and *y*-intercepts of the linear graph are viewable in this window, so the graph is complete.

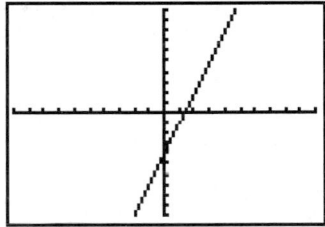

$[-10, 10]$ scl:1 by $[-10, 10]$ scl:1

b. Graph $y = -2(x + 5) - 2$.

Press: [Y=] [(-)] 2 [(] [X,T,θ,*n*] [+] 5 [)] [—] 2 [GRAPH]

When this equation is graphed in the standard viewing window (Figure 1), a complete graph is not visible. The graph indicates that the *y*-intercept is less than -10. You can experiment with the **Ymin** setting or you can rewrite the equation in $y = mx + b$ form, which would be $y = -2x - 12$. The *y*-intercept is -12, so **Ymin** should be less than -12. Remember that **Xmax** and **Ymax** can be less than 10 so that your screen is less compressed. Use the **WINDOW** menu to change the parameters, or settings, and press [GRAPH] to view the result. There are many windows that will enable you to view the complete graph (Figure 2).

Figure 1	Figure 2
$[-10, 10]$ scl:1 by $[-10, 10]$ scl:1	$[-10, 10]$ scl:1 by $[-15, 5]$ scl:1

❷ **Quadratic Functions** When graphing quadratic functions, a complete graph includes the vertex of the parabola and enough of the graph to determine if it opens upward or downward.

(continued on the next page)

Graph $y = 4(x - 3)^2 + 4$.

Press: [Y=] 4 [(] [X,T,θ,n] [−] 3 [)] [x²] [+] 4 [GRAPH]

While the standard viewing window shows a complete graph, you may want to change the viewing window to see more of the graph.

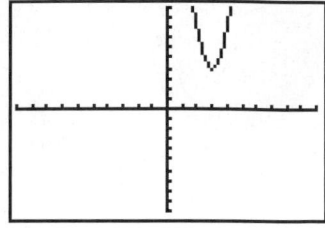

$[-10, 10]$ scl:1 by $[-10, 10]$ scl:1

Example ❸ **Polynomial Functions**   The graphs of other polynomial functions are complete when their maxima, minima, and x-intercepts are visible.

a. Graph $y = 5x^3 + 4x^2 - 2x + 4$.

Press: [Y=] 5 [X,T,θ,n] [∧] 3 [+] 4 [X,T,θ,n] [x²] [−] 2 [X,T,θ,n] [+] 4 [GRAPH]

A complete graph is shown in the standard viewing window. You may want to redefine your window to observe the intercepts, maximum, and minimum points more closely.

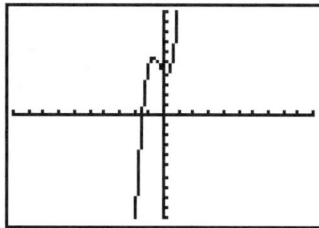

$[-10, 10]$ scl:1 by $[-10, 10]$ scl:1

b. Graph $y = x^4 - 13x^2 + 36$.

Press: [Y=] [X,T,θ,n] [∧] 4 [−] 13 [X,T,θ,n] [x²] [+] 36 [GRAPH]

The standard viewing window (Figure 1) does not show a complete graph. It seems that only the y parameters need to be adjusted. Experiment to find a window that is suitable. Figure 2 shows a sample.

Figure 1

$[-10, 10]$ scl:1 by $[-10, 10]$ scl:1

Figure 2

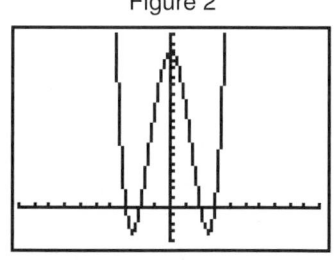

$[-10, 10]$ scl:1 by $[-8, 40]$ scl:2

❹ **Exponential Functions**   A complete graph of an exponential function shows the curvature of the graph and the y-value that it approaches.

Graph $y = 9^{2 + x}$.

Press: [Y=] 9 [∧] [(] 2 [+] [X,T,θ,n] [)] [GRAPH]

Note that you must use parentheses to group the terms that make up the exponent.

A complete graph seems to appear in the second quadrant of the standard viewing window. Vary the **WINDOW** settings to view the graph more closely.

$[-7, 1]$ scl:1 by $[-1, 9]$ scl:1

Example **5** **Logarithmic Functions** A complete graph of a logarithmic function shows the curvature of the graph and the values, or locations, of the asymptotes that the curve approaches.

a. Graph $y = \log (x + 6)$.

Press: Y= LOG X,T,θ,n + 6)
GRAPH

An entire graph appears in the standard viewing window, but is very small. Redefine the **WINDOW** parameters for y, so that the graph is more visible.

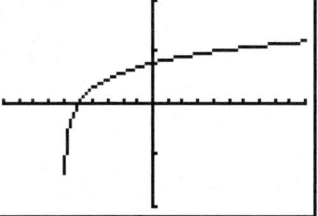

[−10, 10] scl:1 by [−2, 2] scl:1

b. Graph $y = \log_4 x$.

To graph a logarithmic function with a base other than 10, you must first change the function by using the change of base formula, $\log_a x = \dfrac{\log x}{\log a}$.

Press: Y= LOG X,T,θ,n) ÷ LOG
4) GRAPH

An entire graph appears in the standard viewing window, but is very small. Redefine the **WINDOW** parameters, so that the graph is more visible.

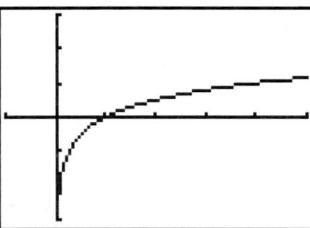

[−1, 5] scl:1 by [−3, 3] scl:1

You can graph multiple functions on a single screen. Each function is denoted by **Y1=, Y2=, Y3=**, and so on, in the **Y=** menu. To graph more than one function, press ENTER at the end of each function you are entering and the cursor will move to the next function to be entered.

Example **6** **Systems of Equations**

Graph $y = 0.5x + 4$ and $y = 2x^2 - 5x + 1$.

Press: Y= 0.5 X,T,θ,n + 4 ENTER 2 X,T,θ,n x²
− 5 X,T,θ,n + 1 GRAPH

The standard viewing window shows that the line and parabola intersect in two points.

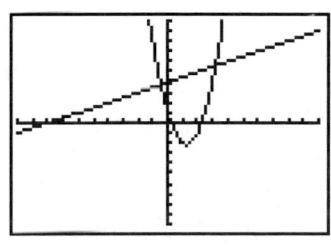

[−10, 10] scl:1 by [−10, 10] scl:1

GRAPHING CALCULATOR APPENDIX

3 Analyzing Functions

In addition to graphing a function, you can use other tools on a graphing calculator to analyze functions. One of those tools is a function table.

Example ➊ **How to Use a Table**   You may complete a table manually or automatically. To create a table for one or more functions, you must first enter each function into the **Y=** list. Then set up and create the table.

The function values are the dependent variable values.

a. Use a table to evaluate the function $y = 4x^2 - 2x + 7$ for $\{-9, -4, 0, 1, 5\}$.

In this case you only need to evaluate the function for selected values, so use the **TBLSET** menu to have the calculator ask for the values of the independent variable (domain) and find the function value (range) automatically.

Press: $\boxed{\text{Y=}}$ 4 $\boxed{\text{X,T,}\theta,n}$ $\boxed{x^2}$ $\boxed{-}$ 2 $\boxed{\text{X,T,}\theta,n}$ $\boxed{+}$

7 $\boxed{\text{2nd}}$ $\boxed{\text{[TBLSET]}}$ $\boxed{\blacktriangledown}$ $\boxed{\blacktriangledown}$ $\boxed{\blacktriangleright}$ $\boxed{\text{ENTER}}$

$\boxed{\text{2nd}}$ $\boxed{\text{[TABLE]}}$ $\boxed{(-)}$ 9 $\boxed{\text{ENTER}}$ $\boxed{(-)}$

4 $\boxed{\text{ENTER}}$ 0 $\boxed{\text{ENTER}}$ 1 $\boxed{\text{ENTER}}$ 5 $\boxed{\text{ENTER}}$

X	Y1	
-9	349	
-4	79	
0	7	
1	9	
5	97	
X=		

b. Use a table to evaluate the functions $y = 5x^2 - x + 1$ and $y = 6 - x^3$ for the integers from -3 to 3, inclusive.

When you want to evaluate a function for a range of values, have the calculator find both the values of the independent variable and the function values automatically. In **Table Setup**, enter the initial number of the domain as the **TblStart** value and the increment between the values of the independent variable as **ΔTbl**. Entering more than one function in the **Y=** list allows you to evaluate all of the functions in one table.

Press: $\boxed{\text{Y=}}$ 5 $\boxed{\text{X,T,}\theta,n}$ $\boxed{x^2}$ $\boxed{-}$ $\boxed{\text{X,T,}\theta,n}$ $\boxed{+}$

1 $\boxed{\text{ENTER}}$ 6 $\boxed{-}$ $\boxed{\text{X,T,}\theta,n}$ $\boxed{\wedge}$ 3 $\boxed{\text{2nd}}$

$\boxed{\text{[TBLSET]}}$ $\boxed{(-)}$ 3 $\boxed{\text{ENTER}}$ 1 $\boxed{\text{ENTER}}$ $\boxed{\text{ENTER}}$

$\boxed{\text{2nd}}$ $\boxed{\text{[TABLE]}}$

X	Y1	Y2
-3	49	33
-2	23	14
-1	7	7
0	1	6
1	5	5
2	19	-2
3	43	-21
X= -3		

Once you create a table, you can scroll through the values using the arrow keys.

ZOOM allows you to quickly adjust the viewing window of a graph in different ways. The effect of each choice on the **ZOOM** menu is shown on the next page.

1: ZBox   Allows you to draw a box to define the viewing window

2: Zoom In   Magnifies the graph around the cursor

3: Zoom Out   Views more of a graph around the cursor

4: ZDecimal   Sets ΔX and ΔY to 0.1

5: ZSquare   Sets equal-sized pixels on the x-and y-axes

6: ZStandard   Sets the standard viewing window, $[-10, 10]$ scl:1 by $[-10, 10]$ scl:1

7: ZTrig   Sets the built-in trig window, $\left[-\dfrac{47}{24}\pi, \dfrac{47}{24}\pi\right]$ scl: $\dfrac{\pi}{2}$ by $[-4, 4]$ scl:1 for radians or $[-352.5.\ 352.5]$ scl:90 by $[-4, 4]$ scl: 1 for degrees

8: ZInteger   Sets integer values on both the x-and y-axes

9: ZoomStat   Sets values for displaying all of the data in the current stat lists

0: ZoomFit   Fits Ymin and Ymax to show all function values for Xmin to Xmax

Example ❷ Using ZOOM **to Graph in the Standard and Square Windows**

Graph the circle with equation $x^2 + y^2 = 16$ in the standard viewing window. Then use ZSquare to view the graph in a square screen.

First solve the equation for y in order to enter it into the **Y=** list.

$$x^2 + y^2 = 16 \rightarrow y = \pm\sqrt{16 - x^2}$$

The two pieces of the graph can be entered at one time using $\{-1, 1\}$. This expression tells the calculator to graph -1 and 1 times the function.

Press: Y= | 2nd | [{] | (−) | 1 | , | 1 | 2nd | [}] | 2nd | [√] 16 | − | X,T,θ,n | x² |) | ZOOM | 6

The circle is distorted when viewed in the standard viewing window.

Press: ZOOM 5

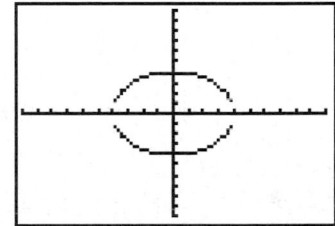

$[-10, 10]$ scl:1 by $[-10, 10]$ scl:1

Using **ZSquare** makes the circle appear as a circle.

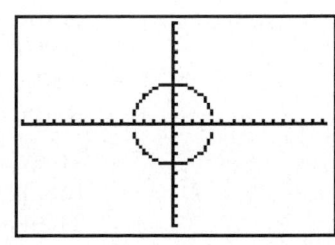

$[-15.16, 15.16]$ scl:1 by $[-10, 10]$ scl:1

Example ❸ **Using** [ZOOM] **to Zoom In and Out** **Graph** $y = 0.5x^3 - 3x^2 - 12$ **in the standard viewing. Zoom out to view a complete graph. Then zoom in to approximate the y-intercept of the graph to the nearest whole number.**

Press: [Y=] 0.5 [X,T,θ,n] [∧] 3 [−] 3 [X,T,θ,n] [x²] [−] 12 [ZOOM] 6

The complete graph is not shown in the standard viewing window. (Figure 1) When you zoom out or in, the calculator allows you to choose the point around which it will zoom. Zooming out around the origin once allows a complete graph to be shown. (Figure 2)

Press: [ZOOM] 3 [ENTER]

Figure 1

[−10, 10] scl:1 by [−10, 10] scl:1

Figure 2

[−40, 40] scl:1 by [−40, 40] scl:1

Now zoom in to approximate the y-intercept. Choose a point close to the intercept by using the arrow keys.

Press: [ZOOM] 2 [ENTER]

The y-intercept appears to be about -12. Zooming in again may allow you to make a closer approximation.

[−10, 10] scl:1 by [−24.19, −4.19] scl:1

The [TRACE] feature allows you to move the cursor along a graph and display the coordinates of the points on the graph.

Example ❹ **Using** [TRACE] **Graph** $y = 4x + 2$ **and** $y = -3x^2 - x + 5$. **Use the TRACE feature to approximate the coordinates of the intersection of the graphs in the first quadrant. Then evaluate** $y = -3x^2 - x + 5$ **for** $x = 1.7$.

Press: [Y=] 4 [X,T,θ,n] [+] 2 [ENTER] [(−)] 3 [X,T,θ,n] [x²] [−] [X,T,θ,n] [+] 5 [TRACE]

Move the cursor along the graphs using [◄] and [►].

Make sure that CoordOn is highlighted in the FORMAT menu to display the cursor coordinates as you trace.

Pressing [2nd] [◄] or [2nd] [►] moves the cursor more quickly. If your cursor moves off of the screen, the calculator will automatically update the viewing window so that the cursor is visible. Use [▼] and [▲] to move from one function to the other. The intersection is at about (0.4, 4).

To evaluate a function for a value and move to that point, place the cursor on the function graph. Then enter the value and press [ENTER]. When $x = 1.7$, $y = -5.37$ for $y = -3x^2 - x + 5$.

[−10, 10] scl:1 by [−10, 10] scl:1

Using ⌜ TRACE ⌟ to locate the intersection points of the graphs of two functions gives you an approximation of the coordinates. For more accurate coordinates, you can use the **intersect** option on the **CALC** menu.

Example ⑤ **Finding Intersection Points** Use 5:intersection on the CALC menu to find the coordinates of the intersection of the graphs of $y = 4x + 2$ and $y = -3x^2 - x + 5$.

If you do not have the functions graphed, enter the functions into the **Y=** list and press ⌜ GRAPH ⌟. Then find the coordinates of the intersection.

Press: ⌜ 2nd ⌟ [CALC] 5

The intersection of the graphs must appear on the screen to find the coordinates when using intersect.

Place the cursor on one graph and press ⌜ ENTER ⌟. Then move the cursor to the other graph and press ⌜ ENTER ⌟. To guess at the intersection, move the cursor to a point close to the intersection or enter an *x*-value and press ⌜ ENTER ⌟. If there is more than one intersection point, the calculator will find the one closest to your guess. The cursor will move to the intersection point and the coordinates will be displayed.

[−10, 10] scl:1 by [−10, 10] scl:1

The **CALC** menu also allows you to find the zeros of a function.

Example ⑥ **Finding Zeros** Find the zeros of $f(x) = -2x^4 + 3x^2 + 2x + 5$.

Press: ⌜Y=⌟ ⌜(−)⌟ 2 ⌜X,T,θ,n⌟ ⌜∧⌟ 4 ⌜+⌟ 3 ⌜X,T,θ,n⌟ ⌜x²⌟ ⌜+⌟ 2 ⌜X,T,θ,n⌟ ⌜+⌟ 5 ⌜ 2nd ⌟ [CALC] 2

The calculator can find one zero at a time. Use the arrow keys or enter a value to choose the left bound for the interval in which the calculator will search for the zero and press ⌜ ENTER ⌟. Choose the right bound and press ⌜ ENTER ⌟. Select a point near the zero using the arrow keys or by entering a value and press ⌜ ENTER ⌟. Repeat with another interval to find the other zero. The zeros of this function are about −1.42 and 1.71.

[−10, 10] scl:1 by [−10, 10] scl:1

Real-world application problems often require you to find the relative minimum or maximum of a function. You can use **3:minimum** or **4:maximum** features on the **CALC** menu of a graphing calculator to solve these problems.

Example **Finding Maxima and Minima**   Determine the relative minimum and the relative maximum for the graph of $f(x) = 4x^3 - 6x + 5$.

First graph the function.

Press: [Y=] 4 [X,T,θ,*n*] [∧] 3 [−] 6 [X,T,θ,*n*] [+] 5 [ZOOM] 6

To find the relative minimum press [2nd] [CALC] 3.

Similar to finding a zero, choose the left and right bound of the interval and guess the minimum or maximum. The point at about (0.71, 2.17) is a relative minimum.

[−10, 10] scl:1 by [−10, 10] scl:1

Use a similar method to find the relative maximum, by pressing [2nd] [CALC] 4. The point at about (−0.71, 7.83) is a relative maximum.

[−10, 10] scl:1 by [−10, 10] scl:1

4 Graphing Inequalities

Most linear and nonlinear inequalities can be graphed using the $\boxed{Y=}$ key and selecting the appropriate graph style in the **Y=** editor. To select the appropriate graph style, select the graph style icon in the first column of the **Y=** editor and press $\boxed{ENTER}$ repeatedly to rotate through the graph styles.

- To shade the area above a graph, select the Above style icon, ◥.

- To shade the area below a graph select the Below style icon, ◣.

Before graphing an inequality, clear any functions in the **Y=** list by pressing $\boxed{Y=}$ and then using the arrow keys and the $\boxed{CLEAR}$ key to select and clear all functions. If you do not wish to clear a function, you can turn that particular graph off by using the arrow keys to position the cursor over that function's = sign and then pressing $\boxed{ENTER}$ to change the selection status.

Example **❶** **Linear Inequalities**

a. Graph $y \leq 2x - 3$ in the standard viewing window.

First enter the boundary equation $y = 2x - 3$ into the **Y=** list.

Press: $\boxed{Y=}$ 2 $\boxed{X,T,\theta,n}$ $\boxed{-}$ 3

Next, press the $\boxed{\blacktriangleleft}$ key until the icon before = flashes. Press $\boxed{ENTER}$ until the icon changes to the Below style icon, ◣, for "$y \leq$".

Finally, if your calculator is not already set for the standard viewing window, press $\boxed{ZOOM}$ 6. Otherwise, press $\boxed{GRAPH}$.

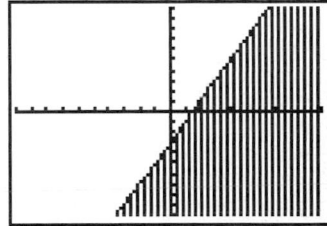

[−10, 10] scl:1 by [−10, 10] scl:1

b. Graph $y \geq -4x + 5$ in the standard viewing window.

Press: $\boxed{Y=}$ $\boxed{(-)}$ 4 $\boxed{X,T,\theta,n}$ $\boxed{+}$ 5

Next, press the $\boxed{\blacktriangleleft}$ key until the icon before = flashes. Then press $\boxed{ENTER}$ until the icon changes to the Above style icon, ◥, since the inequality asks for "$y \geq$".

Finally, press $\boxed{GRAPH}$.

[−10, 10] scl:1 by [−10, 10] scl:1

Example ❷ **Nonlinear Inequalities**   The procedure for graphing nonlinear inequalities is the same as that of graphing linear inequalities.

a. Graph $y \leq 0.25x^2 - 4$ in the standard viewing window.

Press: $\boxed{\text{Y=}}$ 0.25 $\boxed{\text{X,T,}\theta\text{,}n}$ $\boxed{x^2}$ $\boxed{-}$ 4

Next, select the Below style icon, ◣, since the inequality asks for "$y \leq$", and then press $\boxed{\text{GRAPH}}$.

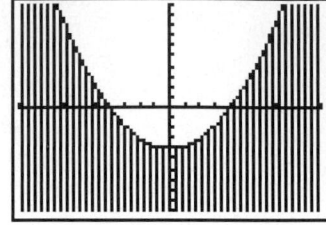

[−10, 10] scl:1 by [−10, 10] scl:1

b. Graph $y \geq 0.2x^4 - 3x^2 + 4$.

Press: $\boxed{\text{Y=}}$ 0.2 $\boxed{\text{X,T,}\theta\text{,}n}$ $\boxed{\wedge}$ 4 $\boxed{-}$
$\boxed{\text{X,T,}\theta\text{,}n}$ $\boxed{\wedge}$ 2 $\boxed{+}$ 4

Next, select the Above style icon, ◤, since the inequality asks for "$y \geq$". Then press $\boxed{\text{GRAPH}}$.

[−10, 10] scl:1 by [−10, 10] scl:1

c. Graph $y \leq \sqrt{x + 2} + 4$.

Press: $\boxed{\text{Y=}}$ $\boxed{\text{2nd}}$ $[\sqrt{\ }]$ $\boxed{\text{X,T,}\theta\text{,}n}$ $\boxed{+}$ 2 $\boxed{)}$ $\boxed{+}$ 4

Next, select the Below style icon, ◣, since the inequality asks for "$y \leq$". Then press $\boxed{\text{GRAPH}}$.

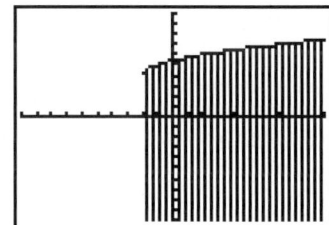

[−10, 10] scl:1 by [−10, 10] scl:1

d. Graph $y \geq 3^x - 5$.

Press: $\boxed{\text{Y=}}$ 3 $\boxed{\wedge}$ $\boxed{\text{X,T,}\theta\text{,}n}$ $\boxed{-}$ 5

Next, select the Above style icon, ◤, since the inequality asks for "$y \geq$". Then press $\boxed{\text{GRAPH}}$.

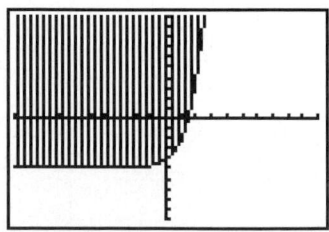

[−10, 10] scl:1 by [−10, 10] scl:1

Graphing systems of inequalities on a graphing calculator is similar to graphing systems of equations.

Example ❸ **Graph the system of inequalities.**
$$y \geq 2x - 5$$
$$y \leq x^2 - 4x + 1$$

Method 1: Shading Options in **Y=**

Press: Y= 2 X,T,θ,n — 5 ENTER X,T,θ,n x² — 4 X,T,θ,n + 1

Select the Above style icon, ◣, for $y \geq 2x - 5$ and the Below style icon, ◤, for $y \leq x^2 - 4x + 1$. Then press GRAPH .

Notice that the first inequality is indicated using vertical lines and the second inequality uses horizontal lines. The solution to the system is shown by the intersection of the shaded areas.

[−10, 10] scl:1 by [−10, 10] scl:1

The **Shade(** *command can only be used with two inequalities which can be written with "y ≤" in one inequality and "y ≥" in the other.*

Method 2: Using the Shade Command

Some systems of inequalities can be graphed by using the **Shade(** command and entering a function for a lower boundary and a function for the upper boundary of the inequality. The calculator first graphs both functions and then shades above the first function entered and below the second function entered.

Before graphing an inequality using the **Shade(** command, clear any graphics from the viewing window by pressing 2nd [DRAW] 1 ENTER . Also clear any equations in the **Y=** list. If not already there, return to the home screen by pressing 2nd [QUIT].

Press: 2nd [DRAW] 7 2 X,T,θ,n — 5 , X,T,θ,n x² — 4 X,T,θ,n + 1) ENTER

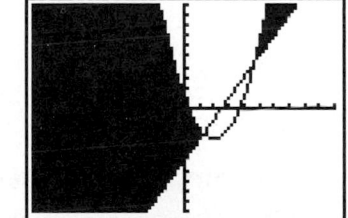

[−10, 10] scl:1 by [−10, 10] scl:1

For more shading pattern options, see page 685.

GRAPHING CALCULATOR APPENDIX

5 Matrices

A graphing calculator can perform operations with matrices. It can also find determinants and inverses of matrices. The **MATRX** menus are accessed using 2nd [MATRX].

On a TI-83, the MATRX **key accesses the matrix operations menus.**

There are three menus in the **MATRX** menu.

- The **NAMES** menu lists the matrix locations available. There are ten matrix variables, [A] through [J].

- The **MATH** menu lists the matrix functions available.

- The **EDIT** menu allows you to define matrices.

A matrix with dimension 2×3 indicates a matrix with 2 rows and 3 columns. Depending on available memory, a matrix may have up to 99 rows or columns.

Example **1** **Entering a Matrix** Enter matrix $A = \begin{bmatrix} 1 & 3 \\ 2 & -2 \end{bmatrix}$.

To enter a matrix into your calculator, choose the **EDIT** menu and select the matrix name. Then enter the dimensions and elements of the matrix.

Press: 2nd [MATRX] ◀ ENTER 2 ENTER
2 ENTER 1 ENTER 3 ENTER 2 ENTER (-) 2

Press 2nd [QUIT] to return to the **HOME** screen. Then press 2nd [MATRX] ENTER ENTER to display the matrix.

```
[A]
         [[1  3 ]
          [2 -2]]
```

You can find the determinant and inverse of a matrix very quickly with a graphing calculator.

Example **2** **Determinant and Inverse of a Matrix**

a. Find the determinant of matrix A.

Press: 2nd [MATRX] ▶ 1 2nd [MATRX]
1 ENTER

The determinant of matrix A is -8.

```
det([A]
              -8
```

b. Find the inverse of matrix A.

Press: 2nd [MATRX] 1 x^{-1} ENTER

$A^{-1} = \begin{bmatrix} 0.25 & 0.375 \\ 0.25 & -0.125 \end{bmatrix}$

```
[A]⁻¹
   [[.25  .375 ]
    [.25 -.125]]
```

Example **3** **Operations with Matrices**

Enter matrix $B = \begin{bmatrix} 2 & 3 & 6 \\ 4 & -8 & 5 \end{bmatrix}$. Then perform each operation.

a. $\frac{1}{2}B$

First, enter matrix B.

Press: [2nd] [MATRX] [◄] 2 2 [ENTER]

3 [ENTER] 2 [ENTER] 3 [ENTER] 6 [ENTER]

4 [ENTER] [(−)] 8 [ENTER] 5 [2nd] [QUIT]

Then find $\frac{1}{2}B$.

Press: .5 [2nd] [MATRX] 2 [ENTER]

$\frac{1}{2}B = \begin{bmatrix} 1 & 1.5 & 3 \\ 2 & -4 & 2.5 \end{bmatrix}$

```
.5[B]
   [[1 1.5 3 ]
    [2 -4 2.5]]
```

b. AB

Press: [2nd] [MATRX] 1 [2nd] [MATRX] 2 [ENTER]

$AB = \begin{bmatrix} 14 & -21 & 21 \\ -4 & 22 & 2 \end{bmatrix}$

```
[A][B]
  [[14 -21 21]
   [-4 22  2 ]]
```

c. A^2

Press: [2nd] [MATRX] 1 [x^2] [ENTER]

$A^2 = \begin{bmatrix} 7 & -3 \\ -2 & 10 \end{bmatrix}$

```
[A]²
   [[7  -3]
    [-2 10]]
```

d. $AB + B$

Press: [2nd] [MATRX] 1 [2nd] [MATRX] 2 [+]

[2nd] [MATRX] 2 [ENTER]

$AB + B = \begin{bmatrix} 16 & -18 & 27 \\ 0 & 14 & 7 \end{bmatrix}$

```
[A][B]+[B]
   [[16 -18 27]
    [0  14  7 ]]
```

6 Graphing Trigonometric Functions

Trigonometric functions and the inverses of trigonometric functions can be graphed using $\boxed{Y=}$. The functions and their inverses can be graphed in degrees or radians. You must set the calculator in **Radian** or **Degree** mode. The standard viewing window for trigonometric functions can be set by pressing $\boxed{\text{ZOOM}}$ **7:Trig**, which automatically adjusts the *x*- and *y*-axes scales for degrees or radians.

Example **1** **a. Using Degrees** Graph $y = \cos x$.

First, set the calculator in degree mode by pressing $\boxed{\text{MODE}}$ $\boxed{\blacktriangledown}$ $\boxed{\blacktriangledown}$ $\boxed{\blacktriangleright}$ $\boxed{\text{ENTER}}$.

Now enter and graph the function. Press $\boxed{Y=}$ $\boxed{\text{COS}}$ $\boxed{\text{X,T,}\theta,n}$ $\boxed{)}$ $\boxed{\text{ZOOM}}$ 7.

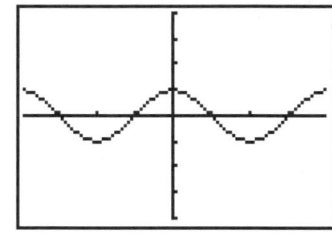

[−352.5, 352.5] scl:90 by [−4, 4] scl:1

b. Using Radians Graph $y = \sin x$.

Change to radian mode by pressing $\boxed{\text{MODE}}$ $\boxed{\blacktriangledown}$ $\boxed{\blacktriangledown}$ $\boxed{\text{ENTER}}$. Press $\boxed{Y=}$ $\boxed{\text{CLEAR}}$ to delete the function entered in part **a**. Then press $\boxed{\text{SIN}}$ $\boxed{\text{X,T,}\theta,n}$ $\boxed{)}$ to enter the new function. Next, press $\boxed{\text{ZOOM}}$ 7 to set the viewing window to accommodate radian mode and graph the function.

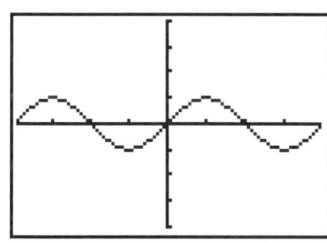

$[-2\pi, 2\pi]$ scl:$\frac{\pi}{2}$ by [−4, 4] scl:1

c. Amplitude, Period, and Phase Shift Graph $y = 2 \sin\left(\frac{1}{2}x - 60°\right)$ using the viewing window [−540, 540] scl:90 by [−3, 3] scl:1. Then state the amplitude, period, and phase shift.

Make sure the calculator is in degree mode.

Press: $\boxed{Y=}$ 2 $\boxed{\text{SIN}}$ $\boxed{\text{X,T,}\theta,n}$ $\boxed{÷}$ 2 $\boxed{-}$ 60 $\boxed{)}$ $\boxed{\text{GRAPH}}$

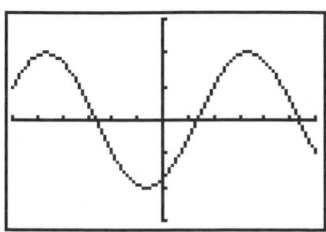

[−540, 540] scl:90 by [−3, 3] scl:1

The amplitude of the function is $\dfrac{|-2-2|}{2}$ or 2. The period is $\dfrac{360}{\frac{1}{2}}$ or 720°, and the phase shift is $\dfrac{-60}{\frac{1}{2}}$ or −120°.

d. Graph $y = 3x + \sin x$.

Use radian mode and the viewing window $[-2\pi, 2\pi]$ scl: $\frac{\pi}{2}$ by $[-6\pi, 6\pi]$ scl: $\frac{\pi}{2}$. You can enter the viewin\g window using
2nd $[\pi]$. The calculator will determine a decimal approximation.

Press: Y= 3 X,T,θ,*n* + SIN X,T,θ,*n*
GRAPH

$[-2\pi, 2\pi]$ scl: $\frac{\pi}{2}$ by $[-6\pi, 6\pi]$ scl: $\frac{\pi}{2}$

Because a graphing calculator only graphs functions, the graph of each inverse is limited to the domain for which the inverse of the function is defined.

Example ❷ **Inverses of Trigonometric Functions**

a. Graph $y = \text{Arcsin } x$.

Choose an appropriate viewing window for degree or radian mode.

Press: Y= 2nd [sin⁻¹] X,T,θ,*n*)
GRAPH

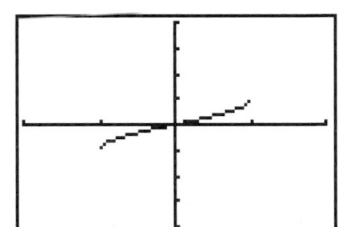

$[-2, 2]$ scl:1 by $[-360, 360]$ scl:90 or
$[-2, 2]$ scl:1 by $[-2\pi, 2\pi]$ scl: $\frac{\pi}{2}$

b. Graph $y = \text{Cos}^{-1}\left(\dfrac{\sqrt{2}}{2} + x\right)$.

Press: Y= 2nd [cos⁻¹] 2nd
[√] 2) ÷ 2 + X,T,θ,*n*)
GRAPH

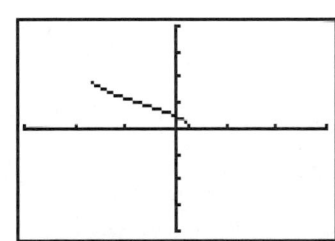

$[-3, 3]$ scl:1 by $[-360, 360]$ scl:90 or
$[-3, 3]$ scl:1 by $[-2\pi, 2\pi]$ scl: $\frac{\pi}{2}$

c. Graph $y = \cos(2 \text{ Sin}^{-1} x)$.

Use the viewing window $[-3, 3]$ scl:1 by $[-2, 2]$ scl:1.

Press: Y= COS 2 2nd [sin⁻¹]
X,T,θ,*n*)) GRAPH

The result is the same whether using degree or radian mode.

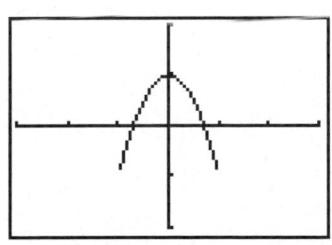

$[-3, 3]$ scl:1 by $[-2, 2]$ scl:1

7 Graphing Special Functions

Most special functions can be graphed using the [Y=] key. The absolute value function **abs(** and the greatest integer function **int(** can be found in the **MATH NUM** menu.

Example **1** **Absolute Value**   Graph $y = 2|x - 4|$.

Press [Y=] 2 [MATH] [▶] 1 [X,T,θ,n] [—] 4 [ )]

[ZOOM] 6 to graph the function in the standard viewing window.

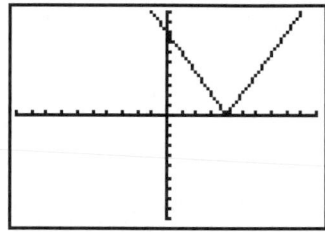

[−10, 10] scl:1 by [−10, 10] scl:1

2 **Greatest Integer Function**   Graph $y = [\![x - 1.5]\!]$.

First, make sure the calculator is set for dot plotting rather than the connected plotting used in most other functions. Press [MODE], highlight **Dot**, and press [ENTER].

Then, enter the function. Press [Y=] [MATH] [▶] 5 [X,T,θ,n] [—] 1.5 [ )] [ZOOM] 6. If your calculator is already set for the standard viewing window, press [GRAPH] instead of [ZOOM] 6.

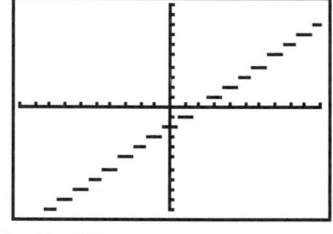

[−10, 10] scl:1 by [−10, 10] scl:1

The **TEST** menu allows you to graph other piecewise functions. Enter the pieces of the function as a sum of the products of each piece of the function and its domain. For example, $y = \begin{cases} 5 \text{ if } x < 2 \\ 4x \text{ if } x > 2 \end{cases}$ is entered as **(5)(X<2) + (4X)(X>2)** in the **Y=** menu.

Example **3** **Piecewise Function**   Graph $y = \begin{cases} 3 \text{ if } x \le -3 \\ 1 + x \text{ if } -3 < x \le 2. \\ 9 - 2x \text{ if } x > 2 \end{cases}$

Place the calculator in **Dot** mode. Then enter the function in the **Y=** list using the **TEST** menu options.

Press: [Y=] [( ] 3 [ )] [( ] [X,T,θ,n] [2nd] [TEST]

6 [(-)] 3 [ )] [+] [( ] [( ] 1 [+] [X,T,θ,n] [ )]

[( ] [(-)] 3 [2nd] [TEST] 5 [X,T,θ,n] [ )]

[( ] [X,T,θ,n] [2nd] [TEST] 6 2 [ )] [+]

[( ] 9 [—] 2 [X,T,θ,n] [ )] [( ] [X,T,θ,n] [2nd]

[TEST] 3 2 [ )] [ZOOM] 6

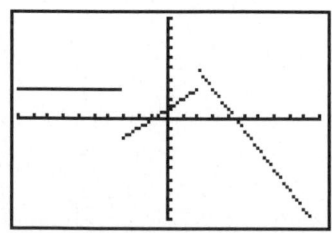

[−10, 10] scl:1 by [−10, 10] scl:1

8 Parametric and Polar Equations

Parametric and polar equations can be graphed using the [Y=] key by selecting the **Par** or **Pol** setting in the **MODE** menu.

Example

❶ Parametric Equations Graph the parametric equations $x = -3 + 2t$ and $y = 5 - 4t$.

First, set the calculator to parametric mode by pressing [MODE] [▼] [▼] [▼] [▶] [ENTER] . Then press [WINDOW] 0 [ENTER] 10 [ENTER] 0.5 [ENTER] [(−)] 10 [ENTER] 10 [ENTER] 1 [ENTER] [(−)] 10 [ENTER] 10 [ENTER] 1 [ENTER] to set the viewing window.

Enter the parametric equations.

Press: [Y=] [(−)] 3 [+] 2 [X,T,θ,n] [ENTER] 5 [−] 4 [X,T,θ,n] [GRAPH]

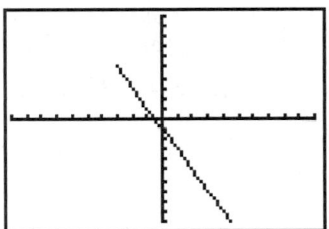

[−10, 10] tstep:0.5, [−10, 10] scl:1 by [−10, 10] scl:1

❷ Polar Equations Graph $r = 2 \cos \theta$.

Set the calculator in polar mode. Press [MODE] [▼] [▼] [▼] [▶] [▶] [ENTER] , and make sure the calculator is in radian mode. Then set the viewing window by pressing [WINDOW] 0 [ENTER] [2nd] [π] [ENTER] 0.05 [ENTER] [(−)] 5 [ENTER] 5 [ENTER] 1 [ENTER] [(−)] 2 [ENTER] 2 [ENTER] 1 [ENTER] .

Enter the equation. Press [Y=] 2 [COS] [X,T,θ,n] [ )] [GRAPH] .

[0, 2π] θstep:0.05, [−5, 5] scl:1 by [−2, 2] scl:1

The graph appears to be an ellipse, but pressing [ZOOM] 5 to set a square viewing window shows that it is actually a circle.

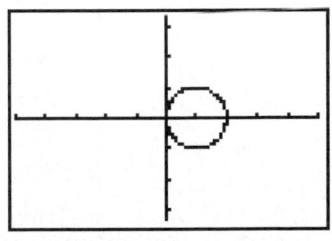

[0, π] θstep:0.05, [−5, 5] scl:1 by [−3.29787234, 3.29787234] scl:1

GRAPHING CALCULATOR APPENDIX

9 Statistics and Statistical Graphs

A graphing calculator allows you to enter a set of data and generate statistics and statistical graphs. Before you enter data values, make sure you clear the **Y=** list, L1 and L2, and the graphics screen. Clear the **Y=** list by pressing [Y=] [CLEAR] . Use the [▼] key to select additional equations and clear them also. To clear **L1** and **L2**, press [STAT] 4 [2nd] **[L1]** [,] **[L2]** [ENTER] . If you need to clear the graphics screen, press [2nd] **[DRAW]** 1 [ENTER] .

Example ❶ **Enter Data into Lists** **Enter the following data into a graphing calculator.**

49 53 54 54 56 55 57 61 51 58 41 59 54 50 60 44

Press: [STAT] 1 53 [ENTER] 54 [ENTER] 54 [ENTER]
56 [ENTER] 55 [ENTER] 57 [ENTER] 61 [ENTER]
51 [ENTER] 58 [ENTER] 41 [ENTER] 59 [ENTER]
54 [ENTER] 50 [ENTER] 60 [ENTER] 44 [ENTER]

You can use the up and down arrow keys to scroll through the list.

L1	L2	L3	1
41			
59			
54			
50			
60			
44			
L1(17) =			

❷ **Find Mean, Median, and Mode** **Find the mean, median, and mode of the data in Example 1.**

Press [STAT] [▶] 1 [ENTER] . This function displays many statistics about the data. $\overline{X}$ denotes the mean. Scroll down to find the median.

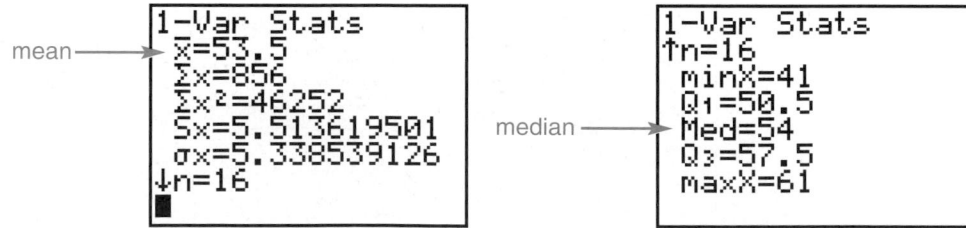

mean →
```
1-Var Stats
x̄=53.5
Σx=856
Σx²=46252
Sx=5.513619501
σx=5.338539126
↓n=16
■
```

median →
```
1-Var Stats
↑n=16
 minX=41
 Q₁=50.5
 Med=54
 Q₃=57.5
 maxX=61
```

The calculator does not have a function to determine the mode. You can find the mode by examining the data. First sort the data to write them in order from least to greatest.

Press: [STAT] 2 [2nd] **[L1]** [)] [ENTER]

Then scroll through the data by pressing [STAT] 1 and using the [▲] and [▼] keys. You will find that the mode is 54.

Examples

3 Box-and-Whisker Plots

a. Draw a box-and-whisker plot for the data.

30.2	29.0	26.2	25.8	23.8	43.0	19.8	19.4
26.0	46.6	26.8	22.8	35.4	25.2	12.2	31.4

Set the viewing window. Next, set the plot type. Press ⟨2nd⟩ [STAT PLOT] 1 ⟨ENTER⟩ ⟨▼⟩

⟨▶⟩ ⟨▶⟩ ⟨▶⟩ ⟨▶⟩ to highlight ⊟ and press

⟨ENTER⟩ . Make sure L1 is entered in **Xlist:**. If not, move the cursor to highlight and press ⟨2nd⟩ [L1] ⟨ENTER⟩ . Then, enter the data into L1. Press ⟨STAT⟩ 1 30.2 ⟨ENTER⟩ 29 ⟨ENTER⟩ … 31.4 ⟨ENTER⟩ ⟨GRAPH⟩ .

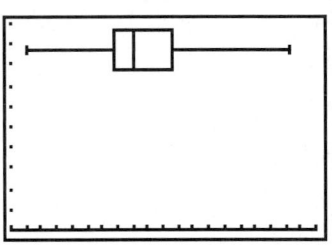

[10, 50] scl:2 by [0, 10] scl:1

b. Draw a box-and-whisker plot with outliers using the data.

Without clearing the lists or graphic screen, press ⟨2nd⟩ [STAT PLOT] 2 ⟨ENTER⟩ ⟨▼⟩ ⟨▶⟩ ⟨▶⟩

⟨▶⟩ , to highlight ⊟⋯, and press ⟨ENTER⟩ .

Make sure L1 is entered in **Xlist:**. Press ⟨GRAPH⟩ .

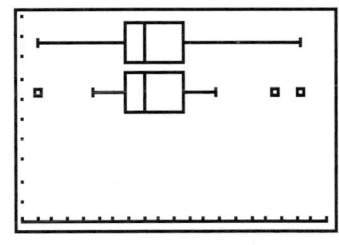

[10, 50] scl:2 by [0, 10] scl:1

c. Find the upper and lower quartiles, the median, and the outliers.

Press ⟨TRACE⟩ and use ⟨◀⟩ and ⟨▶⟩ to move the cursor along the graph. The values will be displayed. For this data, the upper quartile is 30.8, the lower quartile is 23.3, the median is 26.1, and the outliers are 43 and 46.6.

4 Histograms

a. Use the data on the number of public libraries in each state and Washington, D.C., to make a histogram.

273	102	159	196	1030	235	244	30	27	428	366	49	141
772	427	554	372	188	322	273	187	491	659	361	243	346
110	283	78	238	455	92	1067	352	86	684	192	201	640
74	180	134	284	753	96	204	308	309	174	451	74	

Enter the data in L1. Press ⟨STAT⟩ 1 273 ⟨ENTER⟩ 102 ⟨FNTER⟩ … 74 ⟨ENTER⟩ .

Set the viewing window. Choose **Xmin**, **Xmax**, and **Xscl** to determine the number of bars in the histogram. For this data, the least value is 27 and the greatest is 1067. If **Xmin** = 0, **Xmax** = 1100, and **Xscl** = 100, the histogram will have 11 bars each representing an interval of 100.

Choose the type of graph. Press ⟨2nd⟩ [STAT PLOT] 1 ⟨ENTER⟩ ⟨▼⟩ ⟨▶⟩ ⟨▶⟩ ⟨ENTER⟩ ⟨▼⟩ ⟨2nd⟩ [L1] ⟨ENTER⟩ ⟨▼⟩ 1 ⟨ENTER⟩ . Then press ⟨GRAPH⟩ to draw the histogram.

[0, 1100] scl:100 by [0, 12] scl:1

b. Use the data in the table to draw a histogram and its frequency polygram.

Class Limits	Frequency
155–170	7
170–185	15
185–200	34
200–215	38
215–230	42
230–245	35
245–260	33
260–275	21
275–290	18
290–305	6
305–320	3

Make sure you have cleared the Y= *list,* L1 *and* L2, *and the graphic screen.*

Enter the class marks as L1. Press STAT 1 162.5 ENTER 177.5 ENTER … 312.5 ENTER .

Move the cursor to L2. Enter the frequencies. Press 7 ENTER 15 ENTER … 3 ENTER 2nd [QUIT].

Set the viewing window. Use the minimum and maximum of the class limits for **Xmin** and **Xmax**. Use the size of the intervals for **Xscl**. Choose the *y*-axis values to show the complete histogram.

Set the plot type. Press 2nd [STAT PLOT] 1 ENTER ▼ ▶ ▶ ENTER ▼ 2nd [L1] ENTER ▼ 2nd [L2] ENTER . Then press GRAPH .(Figure 1)

Without clearing the lists or graphic screen, press 2nd [STAT PLOT] 1 ENTER ▼ ▶ to highlight ⌁, and press ENTER . Make sure L1 is entered in **Xlist:** and L2 is the **Ylist**. Choose □ as the mark. Press GRAPH .

Figure 1

[155, 320] scl:15 by [0, 50] scl:5

Figure 2

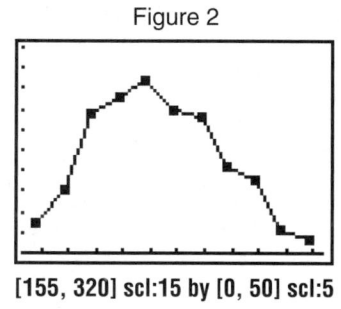

[155, 320] scl:15 by [0, 50] scl:5

Example ❺ Scatter Plot, Connected Line Scatter Plot, and Regression Line

a. Use these data to draw a scatter plot: (20.0, 5.2), (10.2, 1.9), (7.3, 1.6), (6.8, 2.6), (5.9, 1.0), (2.6, 0.7), (2.8, 0.35), (2.7, 0.15).

Clear previous data and graphs and set the viewing window. Enter the *x*-values into L1 and the *y*-values into L2. Then draw the scatter plot by pressing 2nd [STAT PLOT] 1 ENTER ▼ to highlight ⋰ and press ENTER . Make sure L1 is the **Xlist:** and L2 is the **Ylist:**. Then press GRAPH .

[0, 25] scl:5 by [0, 6] scl:1

b. Use the data to draw a line graph.

Press 2nd [STAT PLOT] 1 ▼ ▶ to highlight ⌁ and press ENTER GRAPH .

[0, 25] scl:5 by [0, 6] scl:1

c. Draw a regression line for the data in the table.

Set the plot to display a scatter plot by pressing [2nd] [STAT PLOT] 1 [▼] [ENTER] [2nd] [QUIT].

To calculate the coefficients of regression press [STAT] [▶] 4 [ENTER] .

Then, write the equation of the regression line. You can automatically enter the regression equation in the **Y=** list.

Press [Y=] [VARS] 5 [▶] [▶] 1. Finally, graph the regression line by pressing [GRAPH] .

[0, 25] scl:5 by [0, 6] scl:1

There are also regression models for analyzing data that are not linear built into the calculator.

Example 6

Nonlinear Regression Find a sine regression equation to model the data in the table. Graph the data and the regression equation.

x	1	2	3	4	5	6	7	8	9	10	11	12
y	39	42	45	48	54	59	63	64	59	52	44	40

Enter the data into lists **L1** and **L2**. Press [STAT] 1 [ENTER] 2 [ENTER] … 12 [ENTER] [▶] 39 [ENTER] 42 [ENTER] … 40 [ENTER] .

Find the regression statistics.

Press: [STAT] [▶] [ALPHA] [C] [ENTER]

Enter the regression equation into the **Y=** list.

Press: [Y=] [VARS] 5 [▶] [▶] 1

Then format the scatter plot to graph the data by pressing [2nd] [STAT PLOT] 1 [▼] [ENTER] . Make sure that **L1** is chosen as the **Xlist** and **L2** is chosen as the **Ylist**. Set the viewing window. Press [GRAPH] to see the scatter plot and the graph of the regression equation.

[0, 13] scl:1 by [30, 70] scl:5

Also see pages 389, 739, 741–744 for other examples of nonlinear regression.

Additional Answers
Lesson 1-3

1.

$f(x) = x - 2$

2.

$f(x) = -3x + 4$

3.

$f(x) = -1$

4.

$f(x) = 4x$

5.

$f(x) = 2x - 1$

6.

$f(x) = -x - 5$

EXTRA PRACTICE

Lesson 1-1 *(Pages 5–11)*

State the domain and range of each relation. Then state whether the relation is a function. Write *yes* or *no*. **1.** {−2, 1, 2}; {−4, 2, 4}; no **2.** {−3, −0.5, 0.5, 3}; {0.5, 3}; yes

1. {(1, 4), (−2, 2), (2, 2), (1, −4)}

2. {(0.5, 3), (−0.5, 3), (3, 0.5), (−3, 0.5)}

3. {(2, 2), (5, 7), (−1, 1), (0, 3), (7, 5)}
{−1, 0, 2, 5, 7}; {1, 2, 3, 5, 7}; yes

4. {(3.2, 4), (2.3, −4), (2, 3), (3.2, −1)}
{2, 2.3, 3.2}; {−4, −1, 3, 4}; no

Evaluate each function for the given value.

5. $f(4)$ if $f(x) = 4x - 2$ **14**

6. $g(-3)$ if $g(x) = 2x^2 - x + 5$ **26**

7. $h(1.5)$ if $h(x) = \dfrac{3}{2x}$ **1**

8. $k(5m)$ if $k(x) = \left|3x^2 - 3\right|$ $\left|75m^2 - 3\right|$

Lesson 1-2 *(Pages 13–19)*

1. Given $f(x) = 2x - 1$ and $g(x) = x^2 + 3x - 1$, find $f(x) + g(x)$, $f(x) - g(x)$, $f(x) \cdot g(x)$, and $\left(\dfrac{f}{g}\right)(x)$.
$x^2 + 5x - 2$; $-x^2 - x$; $2x^3 + 5x^2 - 5x + 1$; $\dfrac{2x - 1}{x^2 + 3x - 1}$

Find $[f \circ g](x)$ and $[g \circ f](x)$ for each $f(x)$ and $g(x)$.

2. $f(x) = 3 - x$
$g(x) = 4x^2$ $3 - 4x^2$;
$36 - 24x + 4x^2$

3. $f(x) = \dfrac{1}{3}x - 1$ $\dfrac{1}{3}x + 2$;
$g(x) = x + 9$ $\dfrac{1}{3}x + 8$

4. $f(x) = -2x$
$g(x) = 2x^3 - x^2 + x - 1$
$-4x^3 + 2x^2 - 2x + 2$;
$-16x^3 - 4x^2 - 2x - 1$

Lesson 1-3 *(Pages 20–25)*   **1–6. See margin for graphs.**

Find the zero of each function. If no zero exists, write *none*. Then graph the function.

1. $f(x) = x - 2$ **2**

2. $f(x) = -3x + 4$ $\dfrac{4}{3}$

3. $f(x) = -1$ **none**

4. $f(x) = 4x$ **0**

5. $f(x) = 2x - 1$ $\dfrac{1}{2}$

6. $f(x) = -x - 5$ **−5**

Lesson 1-4 *(Pages 26–31)*

Write an equation in slope-intercept form for each line described.   **5.** $y = -\dfrac{1}{2}x + 2$

1. slope = 2, *y*-intercept = 1 $y = 2x + 1$

2. slope = −1, passes through (1, 2) $y = -x + 3$

3. slope = $-\dfrac{1}{4}$, *y*-intercept = −3 $y = -\dfrac{1}{4}x - 3$

4. slope = 0, passes through (−2, −4) $y = -4$

5. passes through $A(2, 1)$ and $B(-2, 3)$

6. *x*-intercept = −1, *y*-intercept = 6 $y = 6x + 6$

7. the *x*-axis $y = 0$

8. slope = 1.5, *x*-intercept = 10 $y = 1.5x - 15$

Additional Answers
Lesson 1-7

1.

2.

$g(x) = |x + 2|$

3.

Lesson 1-5 *(Pages 32–37)*

1. Are the graphs of $y = 4x - 2$ and $y = -4x + 3$ *parallel, coinciding, perpendicular,* or *none of these*? Explain.   **None of these; the slopes are neither the same nor opposite reciprocals.**

Write the standard form of the equation of the line that is parallel to the graph of the given equation and passes through the point with the given coordinates.

2. $y = x + 1$; $(0, -2)$
$x - y - 2 = 0$

3. $y = 2x - 2$; $(1, 3)$
$2x - y + 1 = 0$

4. $y = -1$; $(-4, 12)$
$y - 12 = 0$

Write the standard form of the equation of the line that is perpendicular to the graph of the given equation and passes through the point with the given coordinates.

5. $3x - y = 5$; $(-2, 6)$
$x + 3y - 16 = 0$

6. $x = 10$; $(12, -15)$
$y + 15 = 0$

7. $5x - 2y = 1$; $(3, -7)$
$2x + 5y + 29 = 0$

Lesson 1-6 *(Pages 38–44)*

The table shows the number of students enrolled in U.S. public elementary and secondary schools for several school years.

Students Enrolled in U.S. Public Elementary and Secondary Schools								
School Year	1989–1990	1990–1991	1991–1992	1992–1993	1993–1994	1994–1995	1995–1996	2010–2011
Enrollment (thousands)	40.5	41.2	42.0	42.8	43.5	44.1	44.8	?

Source: The World Almanac, 1999

1. Graph the data on a scatter plot. Use the last year of the school year for your graph.   **See margin.**

2. Use two ordered pairs to write the equation of a best-fit line.   **Sample answer:** $y = 0.7167x - 1385.7$

3. Use a graphing calculator to find an equation of the regression line for the data. What is the correlation value?   **Sample answer:** $y = 0.7214x - 1395$; $r \approx 0.99$

4. If the equation of the regression line shows a moderate or strong relationship, predict the missing value. Explain whether you think the prediction is reliable.   **Sample answer: 55.7 thousand; yes; the correlation value shows a strong correlation.**

Lesson 1-7 *(Pages 45–51)*
Graph each function.   **1–6. See margin.**

1. $f(x) = \begin{cases} x + 2 \text{ if } x \le -1 \\ 2x \text{ if } x > -1 \end{cases}$

2. $g(x) = |x + 2|$

3. $h(x) = [\![x]\!] - 1$

4. $f(x) = \begin{cases} -2 \text{ if } -2 \le x \le -1 \\ x \text{ if } -1 < x \le 2 \\ 3x \text{ if } x > 2 \end{cases}$

5. $g(x) = |5 - 2x|$

6. $k(x) = 2[\![x]\!]$

Lesson 1-8 *(Pages 52–56)*
Graph each inequality.   **1–6. See margin.**

1. $y > 2$

2. $x + y \le 3$

3. $-y > x + 1$

4. $4x + 2y \le 6$

5. $-1 \le x + y \le 4$

6. $y \le |x|$

Additional Answers
Lesson 1-8

1.
$y > 2$

2.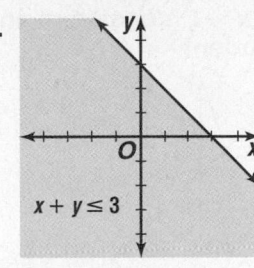
$x + y \le 3$

3.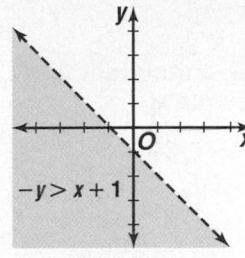
$-y > x + 1$

4.
$4x + 2y \le 6$

5.
$-1 \le x + y \le 4$

6.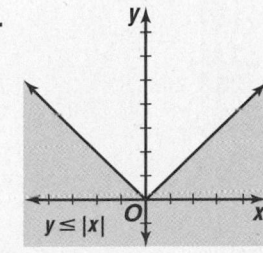
$y \le |x|$

Additional Answers
Lesson 1-7 *(continued)*

4.
$f(x)$

5.
$g(x)$
$g(x) = |5 - 2x|$

6.
$k(x)$

Additional Answers
Lesson 2-1

1.

$y = 2x - 1$; consistent and dependent

2.

$(-1, 5)$; consistent and independent

3.

no solution; inconsistent

Additional Answers
Lesson 2-3

4. $\begin{bmatrix} 11 & -6 \\ 1 & 6 \\ 10 & 10 \end{bmatrix}$

5. $\begin{bmatrix} -5 & 6 \\ 5 & 1 \end{bmatrix}$

6. $\begin{bmatrix} 8 & 0 & -12 \\ 16 & -12 & 8 \end{bmatrix}$

7. impossible

8. $\begin{bmatrix} 26 & -13 \\ 3 & 17 \\ 22 & 26 \end{bmatrix}$

9. impossible

10. $\begin{bmatrix} -14 & -14 \\ 16 & 3 \end{bmatrix}$

11. $\begin{bmatrix} -6 & -23 \\ 42 & -18 \end{bmatrix}$

Lesson 2-1 *(Pages 67–72)*

Solve each system of equations by graphing. Then state whether the system is *consistent and independent, consistent and dependent,* or *inconsistent.* 1–3. See margin.

1. $2x - y - 1 = 0$
$3y = 6x - 3$

2. $y - 3x = 8$
$x + y = 4$

3. $3x - y = -1$
$2y - 6x = -4$

Solve each system of equations algebraically.

4. $5x + 2y = 1$ $(-1, 3)$
$x + 2y = 5$

5. $2x + 4y = 8$ $(4, 0)$
$2x + 3y = 8$

6. $8x + 2y = 2$ $(-1, 5)$
$3x - 4y = -23$

Lesson 2-2 *(Pages 73–77)*

Solve each system of equations.

1. $x + y = 6$ $(1, 5, -3)$
$x + z = -2$
$y + z = 2$

2. $x + 2y - z = -7$ $(1, -3, 2)$
$2x - 2y - z = 6$
$x + y - 2z = -6$

3. $2x - 3y + z = 1$ $(-1, 0, 3)$
$x + y - z = -4$
$3x - 2y + 2z = 3$

Lesson 2-3 *(Pages 78–86)*

Find the values of x and y for which each of the following equations is true.

1. $\begin{bmatrix} 2x + y \\ x + 2y \end{bmatrix} = \begin{bmatrix} -1 \\ 1 \end{bmatrix}$ $(-1, 1)$

2. $\begin{bmatrix} x + 2y \\ 2x - 2y \end{bmatrix} = \begin{bmatrix} 5 \\ -2 \end{bmatrix}$ $(1, 2)$

3. $\begin{bmatrix} 3x \\ 4y \end{bmatrix} = \begin{bmatrix} y - 7 \\ 5x \end{bmatrix}$ $(-4, -5)$

Use matrices *A, B, C, D,* and *E* to find each of the following. If the matrix does not exist, write *impossible.* 4–11. See margin.

$$A = \begin{bmatrix} 4 & -1 \\ 1 & 5 \\ 2 & 6 \end{bmatrix} \quad B = \begin{bmatrix} 2 & 0 & -3 \\ 4 & -3 & 2 \end{bmatrix} \quad C = \begin{bmatrix} 7 & -5 \\ 0 & 1 \\ 8 & 4 \end{bmatrix} \quad D = \begin{bmatrix} -4 & 1 \\ 2 & 3 \end{bmatrix} \quad E = \begin{bmatrix} 1 & -5 \\ -3 & 2 \end{bmatrix}$$

4. $A + C$
5. $D - E$
6. $4B$
7. $D + B$
8. $2C + 3A$
9. AC
10. ED
11. $BC - D$

Lesson 2-4 *(Pages 88–96)* 1–5. See margin for graphs.

2. $J'(4, -5)$, $K'(-5, -8)$, $L'(-6, 4)$, $M'(-1, 2)$

Use matrices to perform each transformation. Then graph the pre-image and image on the same coordinate grid. 4. $W'(3, -3)$, $X'(6, -2)$, $Y'(5, 1)$, $Z'(2, 0)$

1. Triangle *ABC* has vertices $A(5, -3)$, $B(2, 4)$, and $C(-2, -6)$. Use scalar multiplication to find the coordinates of the triangle after a dilation of scale factor 0.5. $A'\left(2\frac{1}{2}, -1\frac{1}{2}\right)$, $B'(1, 2)$, $C'(-1, -3)$

2. Quadrilateral *JKLM* has vertices $J(6, -2)$, $K(-3, -5)$, $L(-4, 7)$, and $M(1, 5)$. Find the coordinates of the quadrilateral after a translation of 2 units left and 3 units down.

3. Triangle *NPQ* is represented by the matrix $\begin{bmatrix} -5 & 3 & 0 \\ 2 & 8 & -4 \end{bmatrix}$. Find the coordinates of the image of the triangle after a reflection over the *y*-axis. $N'(5, 2)$, $P'(-3, 8)$, $Q'(0, -4)$

4. Square *WXYZ* has vertices $W(-3, 3)$, $X(-6, 2)$, $Y(-5, -1)$, and $Z(-2, 0)$. Find the coordinates of the image of the square after a rotation of 180° counterclockwise about the origin.

5. Triangle *FGH* has vertices $F(-4, 2)$, $G(2, -3)$, and $H(-6, -7)$. Find the coordinates of the image of $\triangle FGH$ after $Rot_{90} \circ R_{y = x}$. $F'(4, 2)$, $G'(-2, -3)$, $H'(6, -7)$

Additional Answers
Lesson 2-4

1.

2.

3.

Lesson 2-5 (Pages 98–105)

Find the value of each determinant.

1. $\begin{vmatrix} 3 & 7 \\ -11 & 2 \end{vmatrix}$ **83**

2. $\begin{vmatrix} -3 & -5 \\ -7 & -2 \end{vmatrix}$ **−29**

3. $\begin{vmatrix} -5 & 0 \\ -\frac{1}{2} & -6 \end{vmatrix}$ **30**

4. $\begin{vmatrix} -1 & 0 & 2 \\ -3 & 1 & -2 \\ 5 & -1 & -3 \end{vmatrix}$ **1**

5. $\begin{vmatrix} -1 & 3 & 2 \\ 4 & -2 & 1 \\ 3 & -3 & -4 \end{vmatrix}$ **34**

6. $\begin{vmatrix} 4 & 0 & -1 \\ 5 & 3 & 6 \\ -2 & -5 & 2 \end{vmatrix}$ **163**

Find the multiplicative inverse of each matrix, if it exists. 7–10. See margin.

7. $\begin{bmatrix} 2 & 1 \\ 1 & 3 \end{bmatrix}$

8. $\begin{bmatrix} 10 & 0 \\ 5 & 4 \end{bmatrix}$

9. $\begin{bmatrix} 5 & -6 \\ -3 & 4 \end{bmatrix}$

10. $\begin{bmatrix} 3 & -5 \\ 6 & 1 \end{bmatrix}$

Solve each system by using a matrix equation.

11. $3x + 2y = 22$ **(4, 5)**
    $x - 2y = -6$

12. $4x - 2y = -6$ **(−2, −1)**
    $3x + y = -7$

13. $2x + y = 4$ $\left(\dfrac{5}{2}, -1\right)$
    $4x - 3y = 13$

Lesson 2-6 (Pages 107–111)   **1–3. See margin for graphs.**

Solve each system of inequalities by graphing. Name the vertices of each polygonal convex set. Then, find the maximum and minimum values of each function for the set.

1. $x \le 3$ **(3, 1), (0, 1), (3, 7);**
   $y \ge 1$ **33; 3**
   $y \le 2x + 1$
   $f(x, y) = 4x + 3y$

2. $y \ge 2$ **(0, 3), (4, 7),**
   $0 \le x \le 4$ **(4, 2), (0, 2);**
   $y \le x + 3$ **6; −3**
   $f(x, y) = 2x - y$

3. $y \le -x$
   $y \ge 2x - 4$
   $|x| \le 1$
   $f(x, y) = -x - y$
   (−1, 1), (−1, −6),
   (1, −2), (1, −1);
   7; 0

Lesson 2-7 (Pages 112–118)

Solve each problem, if possible. If not possible, state whether the problem is *infeasible*, has *alternate optimal solutions*, or is *unbounded*.

1. **Business**   The area of a parking lot is 600 square meters. A car requires 6 square meters of space and a bus requires 30 square meters. The attendant can handle only 60 vehicles. If a car is charged $3.00 and a bus is charged $8.00, find how many of each type of vehicle should be accepted to maximize income.   **50 cars, 10 buses**

2. **Manufacturing**   Based on survey results, Yummy Ice Cream concluded that it should make at least twice as many gallons of black walnut flavor as chocolate mint flavor. One distributor wants to order at least 20,000 gallons of the chocolate mint flavor. The company has all of the ingredients to produce both flavors, but it has only 45,000 gallon size containers available. If each gallon of ice cream sells for $2.95, how many gallons of each type flavor should the company produce?   **alternate optimal solutions**

3. **Business**   Cathy places greeting cards from two different companies on a display rack that can hold 90 cards. She has agreed to display at least 40 of company A's cards on the rack and at least 25 of company B's cards. Cathy makes a profit of $0.30 on each card she sells from company A and $0.32 on each card she sells from company B. How many cards should Cathy display from each company to maximize her profit?   **A: 40 cards, B: 50 cards**

4. **Personal Finance**   Kristin makes $5 an hour at a video store and $7 an hour at a landscaping company. She must work at least 4 hours per week at the video store, and the total number of hours she works at both jobs in a week cannot be greater than 15. What is the maximum amount Kristin could earn (before deductions) in a week?   **$97**

Additional Answers
Lesson 2-5

7. $\begin{bmatrix} \dfrac{3}{5} & -\dfrac{1}{5} \\ -\dfrac{1}{5} & \dfrac{2}{5} \end{bmatrix}$

8. $\begin{bmatrix} \dfrac{1}{10} & 0 \\ -\dfrac{1}{8} & \dfrac{1}{4} \end{bmatrix}$

9. $\begin{bmatrix} 2 & 3 \\ \dfrac{3}{2} & \dfrac{5}{2} \end{bmatrix}$

10. $\begin{bmatrix} \dfrac{1}{33} & \dfrac{5}{33} \\ -\dfrac{2}{11} & \dfrac{1}{11} \end{bmatrix}$

Additional Answers
Lesson 2-6

1.

2.

3.

Additional Answers
Lesson 2-4 *(continued)*

4.

5.

Additional Answers
Lesson 3-2

1. $g(x)$ is a translation of $f(x)$ up 2 units.

2. $g(x)$ is the graph of $f(x)$ expanded vertically by a factor of 3.

3. $g(x)$ is a translation of $f(x)$ left 4 units and down 3 units.

4. $g(x)$ is a translation of $f(x)$ right 1 unit and compressed vertically by a factor of 2.

Additional Answers
Lesson 3-3

1.

$y \le x^2 + 1$

2.

$y > |x + 4|$

3.

$y \le -\sqrt{x - 2}$

4.
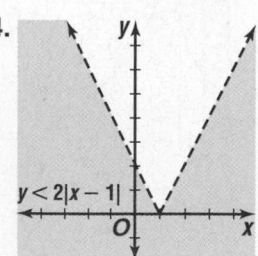
$y < 2|x - 1|$

Lesson 3-1 *(Pages 127–136)*

Determine whether the graph of each function is symmetric with respect to the origin.

1. $f(x) = -4x$ **yes**　　2. $f(x) = x^2 + 3$ **no**　　3. $f(x) = \dfrac{1}{3x^3}$ **yes**

Determine whether the graph of each equation is symmetric with respect to the *x*-axis, *y*-axis, the line $y = x$, the line $y = -x$, or none of these.

4. $xy = 2$ **$y = x$, $y = -x$**　　5. $y + x^2 = 3$ **y-axis**　　6. $y^2 = \dfrac{2x^2}{7} + 1$ **x-axis, y-axis**

7. $|x| = 4y$ **y-axis**　　8. $y = 3x$ **none of these**　　9. $y = \pm\sqrt{x^2 - 1}$ **x-axis, y-axis**

Lesson 3-2 *(Pages 137–145)*

Describe how the graphs of *f(x)* and *g(x)* are related.　**1–4. See margin.**

1. $f(x) = \dfrac{1}{x}$ and $g(x) = \dfrac{1}{x} + 2$

2. $f(x) = x^2$ and $g(x) = 3x^2$

3. $f(x) = |x|$ and $g(x) = |x + 4| - 3$

4. $f(x) = x^3$ and $g(x) = \dfrac{1}{2}(x - 1)^3$

5. Write the equation of the graph obtained when the graph of $y = [\![x]\!]$ is expanded horizontally by a factor of 2 reflected over the *x*-axis, and then translated 5 units down.　$y = -\left[\!\!\left[\dfrac{1}{2}x\right]\!\!\right] - 5$

Lesson 3-3 *(Pages 146–151)*

Graph each inequality.　**1–6. See margin.**

1. $y \le x^2 + 1$　　2. $y > |x + 4|$　　3. $y \le -\sqrt{x - 2}$

4. $y < 2|x - 1|$　　5. $y \le |x| + 2$　　6. $y > (x + 2)^2 - 1$

Solve each inequality.　**8. $\{x \mid x \le -4 \text{ or } x \ge 5\}$　9. $\{x \mid -2 < x < 7\}$**

7. $|x - 2| \le 3$ **$\{x \mid -1 \le x \le 5\}$**　　8. $|4x - 2| \ge 18$　　9. $|5 - 2x| < 9$

10. $|x + 1| - 3 > 1$　　11. $|2x + 3| < 27$　　12. $|3x + 4| - 3x \ge 0$

　　$\{x \mid x < -5 \text{ or } x > 3\}$　　$\{x \mid -15 < x < 12\}$　　**all real numbers**

Lesson 3-4 *(Pages 152–158)*

Graph each function and its inverse.　**1–3. See margin.**

1. $f(x) = |x| - 2$　　2. $f(x) = x^2 + 1$　　3. $f(x) = -1$

Find $f^{-1}(x)$. Then state whether $f^{-1}(x)$ is a function.　**4–12. See margin.**

4. $f(x) = 4x - 5$　　5. $f(x) = -2x + 2$　　6. $f(x) = x^2 + 6$

7. $f(x) = (x - 2)^2$　　8. $f(x) = -\dfrac{x}{2}$　　9. $f(x) = \dfrac{1}{x - 4}$

10. $f(x) = x^2 + 8x - 2$　　11. $f(x) = x^3 + 4$　　12. $f(x) = -\dfrac{3}{(x + 1)^2}$

A30　*Extra Practice*

5.

$y \le |x| + 2$

6.
$y > (x + 2)^2 - 1$

Lesson 3-5 *(Pages 159–168)*

Determine whether each function is continuous at the given *x*-value. Justify your response using the continuity test.

1. $y = x^2 - 3$; $x = 2$

2. $y = \dfrac{x}{x + 3}$; $x = -3$ **No; the function is undefined when $x = -3$.**

3. $f(x) = \dfrac{x - 2}{(x - 1)^2}$; $x = 1$ **No; the function is undefined when $x = 1$.**

4. $f(x) = \begin{cases} x - 2 \text{ if } x \geq 3 \\ 2x - 5 \text{ if } x < 3 \end{cases}$; $x = 3$

5. Determine whether the graph at the right has infinite discontinuity, jump discontinuity, or point discontinuity, or is continuous.
jump discontinuity
   1. Yes; the function approaches 1 as *x* approaches 2 from both sides.
   2. Yes; the function approaches 1 as *x* approaches 3 from both sides.

Describe the end behavior of each function.   **6–8. See margin.**

6. $y = x^3 + 3x^2 + x - 2$

7. $y = 5 - x^4$

8. $f(x) = \dfrac{3}{x^2}$

Lesson 3-6 *(Pages 171–179)*

Locate the extrema for the graph of $y = f(x)$. Name and classify the extrema of the function.

1.

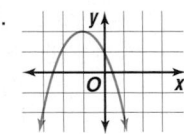

abs. max.: $(-1, 2)$

2.

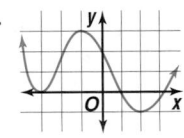

rel. min.: $(-3, 0)$
rel. max.: $(-1, 3)$
abs. min.: $(2, -1)$

Lesson 3-7 *(Pages 180–188)*

Determine the equations of the vertical and horizontal asymptotes, if any, of each function.

1. $f(x) = \dfrac{3x}{x - 2}$ **$x = 2$, $y = 3$**

2. $g(x) = \dfrac{2x^2}{x + 3}$ **$x = -3$**

3. $h(x) = \dfrac{x - 5}{x^2 + 6x + 5}$ **$x = -5$, $x = -1$, $y = 0$**

4. Does the function $f(x) = \dfrac{x^2 + 2x + 1}{x - 3}$ have a slant asymptote? If so, find an equation of the slant asymptote. If not, explain.   **yes; $y = x + 5$**

Lesson 3-8 *(Pages 189–196)*

Find the constant of variation for each relation and use it to write an equation for each statement. Then solve the equation.   **4. 3; $y = 3xz$; 150   5. 243; $yx^2 = 243$; 9.72**

1. If *y* varies directly as *x*, and $y = 8$ when $x = 2$, find *y* when $x = 9$.   **4; $y = 4x$; 36**

2. If *g* varies directly as *w*, and $g = 10$ when $w = -3$, find *w* when $g = 4$.   **$-\dfrac{10}{3}$; $g = -\dfrac{10}{3}w$; $-\dfrac{6}{5}$**

3. If *t* varies inversely as *r*, and $r = 14$ when $t = -6$, find *r* when $t = -7$.   **-84; $rt = -84$; 12**

4. If *y* varies jointly as *x* and *z*, and $y = 60$ when $x = 5$, and $z = 4$, find *y* when $x = 5$ and $z = 10$.

5. Suppose *y* varies inversely as the square of *x* and $x = 3$ when $y = 27$. Find *y* when $x = 5$.

6. Suppose *a* varies jointly as *b* and the cube of *c* and $a = -36$ when $b = 3$ and $c = 2$. Find *a* when $b = 5$ and $c = 3$.   **-1.5; $a = -1.5bc^3$; -202.5**

Additional Answers
Lesson 3-4

1.

2.

3.

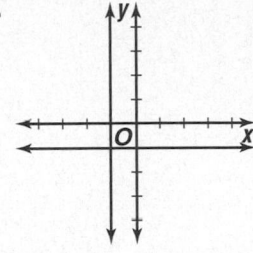

4. $f^{-1}(x) = \dfrac{x + 5}{4}$; **yes**

5. $f^{-1}(x) = \dfrac{x - 2}{-2}$; **yes**

6. $f^{-1}(x) = \pm\sqrt{x - 6}$; **no**

7. $f^{-1}(x) = 2 \pm \sqrt{x}$; **no**

8. $f^{-1}(x) = -2x$; **yes**

9. $f^{-1}(x) = \dfrac{1}{x} + 4$; **yes**

10. $f^{-1}(x) = -4 \pm \sqrt{x + 18}$; **no**

11. $f^{-1}(x) = \sqrt[3]{x - 4}$; **yes**

12. $f^{-1}(x) = -1 \pm \sqrt{\dfrac{-3}{x}}$; **no**

Additional Answers
Lesson 3-5

6. $y \to -\infty$ as $x \to -\infty$, $y \to \infty$ as $x \to \infty$

7. $y \to -\infty$ as $x \to -\infty$, $y \to -\infty$ as $x \to \infty$

8. $y \to 0$ as $x \to -\infty$, $y \to 0$ as $x \to \infty$

Lesson 4-1 *(Pages 205–212)*

Determine whether each number is a root of $x^3 - 7x^2 + 2x + 40 = 0$. Explain.

1. -2 yes; $f(-2) = 0$ 2. 1 no; $f(1) = 36$ 3. 2 no; $f(2) = 24$ 4. 5 yes; $f(5) = 0$

5. $x^2 - 7x + 12 = 0$; even; 2 6. $x^3 + x^2 - 4x - 4 = 0$; odd; 3 7. $x^3 + 1.5x^2 - x - 1.5 = 0$; odd; 3

Write the polynomial equation of least degree for each set of roots. Does the equation have an odd or even degree? How many times does the graph of the related function cross the x-axis?

5. $3, 4$

6. $-2, -1, 2$

7. $-1.5, -1, 1$

8. $-2, -i, i$

9. $-3i, -i, i, 3i$

10. $-1, 1, 2, 3$

$x^3 + 2x^2 + x + 2 = 0$; odd; 1 $x^4 + 10x^2 + 9 = 0$; even; 0 $x^4 - 5x^3 + 5x^2 + 5x - 6 = 0$; even; 4

Lesson 4-2 *(Pages 213–221)*

Solve each equation by completing the square.

1. $x^2 - 4x - 5 = 0$ $-1, 5$ 2. $x^2 + 6x + 8 = 0$ $-4, -2$ 3. $m^2 + 3m - 2 = 0$ $-\frac{3}{2} \pm \frac{\sqrt{17}}{2}$

4. $2a^2 - 8a - 6 = 0$ $2 \pm \sqrt{7}$ 5. $h^2 - 12h = 4$ $6 \pm 2\sqrt{10}$ 6. $x^2 - 9x + 1 = 0$ $\frac{9}{2} \pm \frac{\sqrt{77}}{2}$

7. 121; 2 real; $-1, \frac{7}{4}$ 8. 44; 2 real; $-1 \pm \sqrt{11}$ 9. -263; 2 imaginary; $\frac{5 \pm i\sqrt{263}}{24}$

Find the discriminant of each equation and describe the nature of the roots of the equation. Then solve the equation by using the Quadratic Formula.

7. $4x^2 - 3x - 7 = 0$

8. $w^2 + 2w - 10 = 0$

9. $12t^2 - 5t + 6 = 0$

10. $x^2 + 6x - 13 = 0$

11. $4n^2 - 4n + 1 = 0$

12. $4x^2 + 6x = 15$

88; 2 real; $-3 \pm \sqrt{22}$ 0; 1 real; $\frac{1}{2}$ 276; 2 real; $\frac{-3 \pm \sqrt{69}}{4}$

Lesson 4-3 *(Pages 222–228)*

Divide using synthetic division.

1. $(x^2 + 10x + 8) \div (x + 2)$ $x + 8$, R -8 2. $(x^3 - 3x^2 + 4x - 1) \div (x - 1)$ $x^2 - 2x + 2$, R1

3. $(x^3 - 3x - 5) \div (x + 1)$ $x^2 - x - 2$, R -3 4. $(x^4 - 2x^3 - 7x^2 - 3x - 4) \div (x - 4)$

$x^3 + 2x^2 + x + 1$

Use the Remainder Theorem to find the remainder for each division. State whether the binomial is a factor of the polynomial.

5. $(x^2 + 2x - 8) \div (x + 4)$ **0; yes**

6. $(x^3 + 12) \div (x - 1)$ **13; no**

7. $(4x^3 + 2x^2 + 6x + 1) \div (x + 1)$ -7; no

8. $(x^4 - 4x^2 + 16) \div (x - 4)$ **208; no**

Lesson 4-4 *(Pages 229–235)* 1. $\pm 1, \pm 2, \pm 3, \pm 6$; $-3, -1, 2$

List the possible rational roots of each equation. Then determine the rational roots.

1. $x^3 + 2x^2 - 5x - 6 = 0$

2. $2x^4 - x^3 + 2x^2 - 3x + 1 = 0$ $\pm 1, \pm \frac{1}{2}$; $\frac{1}{2}$

3. $x^3 + x^2 - 2 = 0$ $\pm 1, \pm 2$; 1

4. $6x^4 + x^3 + 22x^2 + 4x - 8 = 0$ $\pm 1, \pm 2, \pm 4$, $\pm 8, \pm \frac{1}{2}, \pm \frac{1}{3}, \pm \frac{2}{3}, \pm \frac{4}{3}, \pm \frac{8}{3}, \pm \frac{1}{6}$; $-\frac{2}{3}, \frac{1}{2}$

Find the number of possible positive real zeros and the number of possible negative real zeros for each function. Then determine the rational zeros.

5. $f(x) = x^3 - 4x^2 - x + 4$ **2 or 0; 1; $-1, 1, 4$** 6. $f(x) = x^4 + x^3 + 3x^2 - 5x + 10$ 2 or 0; 2 or 0; none

7. $f(x) = 4x^3 - 7x + 3$ **2 or 0; 1; $-\frac{3}{2}, \frac{1}{2}, 1$** 8. $f(x) = x^4 - x^3 + 4x - 4$

3 or 1; 1; 1

A32 *Extra Practice*

Lesson 4-5 *(Pages 236–242)*

Determine between which consecutive integers the real zeros of each function are located.

1. $f(x) = 2x^2 - 4x - 5$
−1 and 0, 2 and 3

2. $f(x) = x^3 - 5$
1 and 2

3. $f(x) = x^4 - x^2 + 4x + 2$
−2 and −1, −1 and 0

Approximate the real zeros of each function to the nearest tenth.

4. $f(x) = 4x^4 - 6x^2 - 2x + 1$
0.3, 1.3

5. $f(x) = 2x^5 + 3x^4 - 12x + 4$
−2.2, 0.3, 1.2

6. $f(x) = -2x^4 + 5$
−1.3, 1.3

Lesson 4-6 *(Pages 243–250)*

Solve each equation or inequality.

1. $\dfrac{6}{x} + x = 5$ **2, 3**

2. $\dfrac{7}{y-1} - \dfrac{4}{y} = \dfrac{y}{y-1}$ **−1, 4**

3. $\dfrac{5}{r+1} - \dfrac{4}{r-1} = \dfrac{1}{r^2-1}$ **10**

4. $2 = \dfrac{1}{2-t} + \dfrac{4}{t-2}$ **3.5**

5. $\dfrac{1}{3w} + \dfrac{4}{5w} \leq \dfrac{1}{15}$
$w < 0$ or $w \geq 17$

6. $\dfrac{x-2}{x} < \dfrac{x-4}{x-6}$
$0 < x < 3$ or $x > 6$

Lesson 4-7 *(Pages 251–257)*

Solve each equation or inequality.

1. $\sqrt{2+3t} = 4$ $\dfrac{14}{3}$

2. $4 - \sqrt{x-2} = 1$ **11**

3. $\sqrt[3]{y-7} + 10 = 2$ **−505**

4. $\sqrt{a+1} - 5 = \sqrt{a-6}$
no solution

5. $\sqrt{2x+3} \leq 2$
$-\dfrac{3}{2} \leq x \leq \dfrac{1}{2}$

6. $\sqrt[4]{6a-2} > 4$
$a > 43$

Lesson 4-8 *(Pages 258–264)*

Use a graphing calculator to write a polynomial function to model each set of data.

1.

x	−4	−2	0	2	4	6	8
f(x)	−5	−3.5	−2	−0.5	1	2.5	4

$f(x) = 0.75x - 2$

2.

x	−2	−1.5	−1	−0.5	0	0.5	1
f(x)	0	2.375	3	2.625	2	1.875	3

$f(x) = x^3 + x^2 - x + 2$

3.

x	−3	−1.5	−1	0	0.5	2	2.5
f(x)	3.81	0.44	−0.56	−0.7	−0.56	1.31	2.44

Sample answer: $f(x) = 0.51x^2 + 0.02x - 0.79$

4. Families The average numbers of households in the United States from 1991 to 1997 are listed below.

Year	1991	1992	1993	1994	1995	1996	1997
Number of Households (in thousands)	94.3	95.7	96.4	97.1	98.9	99.6	100.0

Source: *The 1999 World Almanac and Book of Facts*

a. Write a model that relates the number of households as a function of the number of years since 1991. **Sample answer:** $y = 0.979x + 94.493$

b. Use the model to predict the number of U.S. households in the year 2006.
Sample answer: about 109.178 thousand

Additional Answers
Lesson 5-2

3. $\sin A = \dfrac{8\sqrt{113}}{113}$, $\cos A =$

$\dfrac{7\sqrt{113}}{113}$, $\tan A = \dfrac{8}{7}$, $\csc A =$

$\dfrac{\sqrt{113}}{8}$, $\sec A = \dfrac{\sqrt{113}}{7}$,

$\cot A = \dfrac{7}{8}$

4. $\sin A = \dfrac{\sqrt{159}}{28}$, $\cos A = \dfrac{25}{28}$,

$\tan A = \dfrac{\sqrt{159}}{25}$, $\csc A =$

$\dfrac{28\sqrt{159}}{159}$, $\sec A = \dfrac{28}{25}$,

$\cot A = \dfrac{25\sqrt{159}}{159}$

5. $\sin A = \dfrac{2\sqrt{13}}{13}$, $\cos A =$

$\dfrac{3\sqrt{13}}{13}$, $\tan A = \dfrac{2}{3}$, $\csc A =$

$\dfrac{\sqrt{13}}{2}$, $\sec A = \dfrac{\sqrt{13}}{3}$,

$\cot A = \dfrac{3}{2}$

Additional Answers
Lesson 5-3

3. $\sin \theta = -\dfrac{2\sqrt{5}}{5}$, $\cos \theta =$

$-\dfrac{\sqrt{5}}{5}$, $\tan \theta = 2$, $\csc \theta =$

$-\dfrac{\sqrt{5}}{2}$, $\sec \theta = -\sqrt{5}$,

$\cot \theta = \dfrac{1}{2}$

4. $\sin \theta = \dfrac{\sqrt{2}}{2}$, $\cos \theta = -\dfrac{\sqrt{2}}{2}$,

$\tan \theta = -1$, $\csc \theta = \sqrt{2}$,

$\sec \theta = -\sqrt{2}$, $\cot \theta = -1$

5. $\sin \theta = \dfrac{2\sqrt{29}}{29}$, $\cos \theta =$

$\dfrac{5\sqrt{29}}{29}$, $\tan \theta = \dfrac{2}{5}$, $\csc \theta =$

$\dfrac{\sqrt{29}}{2}$, $\sec \theta = \dfrac{\sqrt{29}}{5}$,

$\cot \theta = \dfrac{5}{2}$

Lesson 5-1 *(Pages 277–283)*

Change each measure to degrees, minutes, and seconds.

1. $13.75°$ **13° 45′**
2. $75.72°$ **75° 43′ 12″**
3. $-29.44°$ **−29° 26′ 24″**
4. $87.81°$ **87° 48′ 36″**

Write each measure as a decimal to the nearest thousandth.

5. $144° \; 12′ \; 30″$
   144.208°
6. $-38° \; 15′ \; 10″$
   −38.253°
7. $-107° \; 12′ \; 45″$
   −107.213°
8. $51° \; 14′ \; 32″$
   51.242°

If each angle is in standard position, determine a coterminal angle that is between 0° and 360°. State the quadrant in which the terminal side lies.

9. $850°$ **130°; II**
10. $-65°$ **295°; IV**
11. $1012°$ **292°; IV**
12. $578°$ **218°; III**

Find the measure of the reference angle for each angle.

13. $126°$ **54°**
14. $-480°$ **60°**
15. $642°$ **78°**
16. $1154°$ **74°**

Lesson 5-2 *(Pages 284–290)*

1. If $\tan \theta = \dfrac{5}{6}$, find $\cot \theta$. $\dfrac{6}{5}$
2. If $\csc \theta = 2.5$, find $\sin \theta$. **0.4**

Find the values of the six trigonometric functions for each $\angle A$. **3–5. See margin.**

3.

4.

5.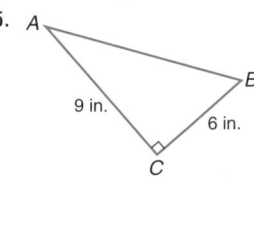

Lesson 5-3 *(Pages 291–298)*

1. If $\tan \theta = 0$, what is $\cot \theta$? **undefined**
2. Find two values of θ for which $\cos \theta = 0$.
   Sample answers: 90°, 270°

Find the values of the six trigonometric functions for angle θ **in standard position if a point with the given coordinates lies on its terminal side.** **3–6. See margin.**

3. $(-1, -2)$
4. $(-2, 2)$
5. $(5, 2)$
6. $(-4, 3)$

Lesson 5-4 *(Pages 299–304)*

Solve each problem. Round to the nearest tenth.

1. If $A = 38°$ and $b = 15$, find a. **11.7**
2. If $c = 19$ and $B = 87°$, find a. **1.0**
3. If $a = 16.5$ and $B = 65.4°$, find c. **39.6**
4. If $B = 42° \; 30′$ and $b = 12$, find a. **13.1**
5. If $B = 75°$ and $c = 5.8$, find b. **5.6**

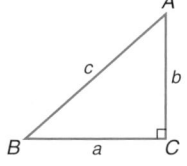

6. A statue 20 feet high stands on top of a base. From a point in front of the statue, the angle of elevation to the top of the statue is 48°, and the angle of elevation to the bottom of the statue is 42°. How tall is the base?
   about 85.7 ft

6. $\sin \theta = \dfrac{3}{5}$, $\cos \theta = -\dfrac{4}{5}$,

$\tan \theta = -\dfrac{3}{4}$, $\csc \theta = \dfrac{5}{3}$,

$\sec \theta = -\dfrac{5}{4}$, $\cot \theta = -\dfrac{4}{3}$

Lesson 5-5 *(Pages 305–312)*
Evaluate each expression. Assume that all angles are in Quadrant I.

1. $\sin\left(\arcsin\frac{3}{4}\right)$ **$\frac{3}{4}$**
2. $\sec\left(\cos^{-1}\frac{1}{2}\right)$ **2**
3. $\cot(\tan^{-1} 1)$ **1**

Solve each problem. Round to the nearest tenth.

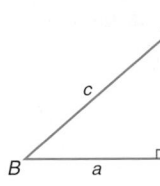

4. If $a = 38$ and $b = 25$, find A. **56.7°**
5. If $c = 19$ and $b = 17$, find B. **63.5°**
6. If $a = 24$ and $c = 30$, find B. **36.9°**
7. If $c = 12.6$ and $a = 9.2$, find B. **43.1°**
8. If $b = 36.5$ and $a = 28.4$, find A. **37.9°**

Lesson 5-6 *(Pages 313–318)*
Solve each triangle. Round to the nearest tenth. **2. $B = 41°$, $a = 17.8$, $b = 11.8$**

1. $A = 75°$, $B = 50°$, $a = 7$ **$C = 55°$, $b = 5.6$, $c = 5.9$**
2. $A = 97°$, $C = 42°$, $c = 12$
3. $B = 49°$, $C = 32°$, $a = 10$ **$A = 99°$, $b = 7.6$, $c = 5.4$**
4. $A = 22°$, $C = 41°$, $b = 25$ **$B = 117°$, $a = 10.5$, $c = 18.4$**

Find the area of each triangle. Round to the nearest tenth.

5. $A = 34°$, $b = 12$, $c = 6$ **20.1 units²**
6. $B = 56.8°$, $A = 87°$, $c = 6.8$ **32.7 units²**
7. $a = 8$, $B = 60°$, $C = 75°$ **37.9 units²**
8. $A = 43°$, $b = 16$, $c = 12$ **65.5 units²**

Lesson 5-7 *(Pages 320–326)* **2. $B = 44.8°$, $C = 110.2°$, $c = 13.3$; $B = 135.2°$, $C = 19.8°$, $c = 4.8$**
Find all solutions for each triangle. If no solution exists, write *none*. Round to the nearest tenth.

1. $a = 5$, $b = 10$, $A = 145°$ **none**
2. $A = 25°$, $a = 6$, $b = 10$
3. $B = 56°$, $b = 34$, $c = 50$ **none**
4. $A = 45°$, $B = 85°$, $c = 15$ **$C = 50°$, $a = 13.8$, $b = 19.5$**

Lesson 5-8 *(Pages 327–332)*
Solve each triangle. Round to the nearest tenth. **2. $A = 48.2°$, $B = 35.4°$, $C = 96.4°$**

1. $b = 6$, $c = 8$, $A = 62°$ **$a = 7.4$, $B = 45.6°$, $C = 72.4°$**
2. $a = 9$, $b = 7$, $c = 12$
3. $B = 48°$, $c = 18$, $a = 14$ **$b = 13.5$, $A = 50.3°$, $C = 81.7°$**
4. $a = 14.2$, $b = 24.5$, $C = 85.3°$ **$c = 27.3$, $A = 31.2°$, $B = 63.5°$**

Find the area of each triangle. Round to the nearest tenth.

5. $a = 4$, $b = 7$, $c = 10$ **10.9 units²**
6. $a = 4$, $b = 6$, $c = 5$ **9.9 units²**
7. $a = 12.4$, $b = 8.6$, $c = 14.2$ **52.9 units²**
8. $a = 150$, $b = 124$, $c = 190$ **9289.4 units²**

9. **Geometry** The lengths of two sides of a parallelogram are 35 inches and 28 inches. One angle measures 110°.
   a. Find the length of the longest diagonal. **51.8 in.**
   b. Find the area of the parallelogram. **920.9 in²**

Additional Answers
Lesson 6-3

4.

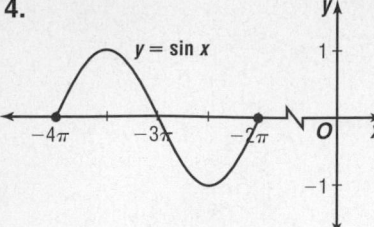

$y = \sin x$

5.

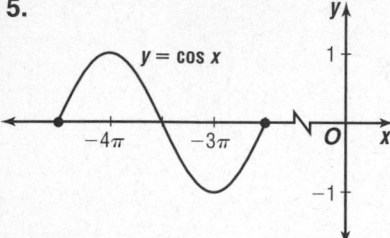

$y = \cos x$

Additional Answers
Lesson 6-4

1.

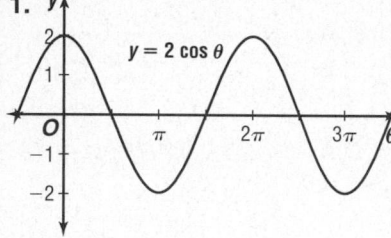

$y = 2 \cos \theta$

2.

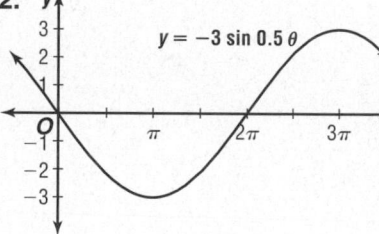

$y = -3 \sin 0.5\theta$

3.

$y = \frac{1}{2} \cos \frac{\theta}{4}$

Lesson 6-1 *(Pages 343–351)*

Change each degree measure to radian measure in terms of π.

1. $120°$ $\frac{2\pi}{3}$
2. $280°$ $\frac{14\pi}{9}$
3. $-440°$ $-\frac{22\pi}{9}$
4. $-150°$ $-\frac{5\pi}{6}$

Change each radian measure to degree measure. Round to the nearest tenth.

5. $\frac{8\pi}{3}$ $480°$
6. $\frac{5\pi}{12}$ $75°$
7. -2 $-114.6°$
8. 10.5 $601.6°$

Evaluate each expression.

9. $\sin \frac{5\pi}{6}$ $\frac{1}{2}$
10. $\sin \frac{4\pi}{3}$ $-\frac{\sqrt{3}}{2}$
11. $\cos \frac{9\pi}{4}$ $\frac{\sqrt{2}}{2}$
12. $\cos\left(-\frac{3\pi}{2}\right)$ 0

13. The diameter of a circle is 10 inches. If a central angle measures 80°, find the length of the intercepted arc.   **about 7.0 in.**

Lesson 6-2 *(Pages 352–358)*

Determine each angular displacement in radians. Round to the nearest tenth.

1. 5 revolutions
 31.4 radians
2. 3.8 revolutions
 23.9 radians
3. 14.2 revolutions
 89.2 radians

Determine each angular velocity. Round to the nearest tenth.   **5. 4.7 radians/min**

4. 2.1 revolutions in 5 seconds   **2.6 radians/s**
5. 1.5 revolutions in 2 minutes
6. 15.8 revolutions in 18 seconds   **5.5 radians/s**
7. 140 revolutions in 20 minutes
 44.0 radians/min

8. A children's Ferris wheel rotates one revolution every 30 seconds. What is its angular velocity in radians per second?   **about 0.2 radian/s**

Lesson 6-3 *(Pages 359–366)*

Find each value by referring to the graph of the sine or cosine function.

1. $\cos 4\pi$ **1**
2. $\sin 8\pi$ **0**
3. $\sin \frac{3\pi}{2}$ **−1**

Graph each function for the given interval.   **4–5. See margin.**

4. $y = \sin x, \; -4\pi \le x \le -2\pi$
5. $y = \cos x, \; -\frac{9\pi}{2} \le x \le -\frac{5\pi}{2}$

Lesson 6-4 *(Pages 368–377)*   **1–3. See margin for graphs.**

State the amplitude and period for each function. Then graph each function.

1. $y = 2 \cos \theta$ **2; 2π**
2. $y = -3 \sin 0.5\theta$ **3; 4π**
3. $y = \frac{1}{2} \cos \frac{\theta}{4}$ **$\frac{1}{2}$; 8π**

Write an equation of the sine function with each amplitude and period.

4. amplitude = 0.5, period = 6π
 $y = \pm 0.5 \sin \frac{\theta}{3}$
5. amplitude = 2, period = $\frac{\pi}{3}$
 $y = \pm 2 \sin 6\theta$

Write an equation of the cosine function with each amplitude and period.

6. amplitude = $\frac{3}{5}$, period = 4π
 $y = \pm \frac{3}{5} \cos \frac{\theta}{2}$
7. amplitude = 0.25, period = 8
 $y = \pm 0.25 \cos \frac{\pi}{4} \theta$

Lesson 6-5   *(Pages 378–386)*   **1–3. See margin for graphs.**
State the phase shift for each function. Then graph each function.

1. $y = \sin(2\theta - \pi)$ $\dfrac{\pi}{2}$ 2. $y = 2\cos(\theta + 2\pi)$ -2π 3. $y = \sin\left(\dfrac{\theta}{2} + \dfrac{\pi}{2}\right)$ $-\pi$

Write an equation of the sine function with each amplitude, period, phase shift, and vertical shift.

4. amplitude = 2, period = 2π, phase shift = π, vertical shift = -1 $y = \pm 2\sin(\theta - \pi) - 1$

5. amplitude = 0.5, period = $\dfrac{\pi}{4}$, phase shift = 0, vertical shift = 3 $y = \pm 0.5\sin 8\theta + 3$

Write an equation of the cosine function with each amplitude, period, phase shift, and vertical shift.

6. amplitude = 20, period = $\dfrac{\pi}{2}$, phase shift = 2π, vertical shift = 4 $y = \pm 20\cos(4\theta - 8\pi) + 4$

7. amplitude = $\dfrac{3}{4}$, period = 10, phase shift = 0, vertical shift = $\dfrac{1}{2}$ $y = \pm\dfrac{3}{4}\cos\dfrac{\pi}{5}\theta + \dfrac{1}{2}$

Lesson 6-6   *(Pages 387–394)*

1. **Meteorology**   The equation $d = 2.7\sin(0.5m - 1.4) + 12.1$ models the amount of daylight in Cincinnati, Ohio, for any given day. In this equation $m = 1$ represents the middle of January, $m = 2$ represents the middle of February, etc.

   a. What is the least amount of daylight in Cincinnati?   **9.4 h**

   b. What is the greatest amount of daylight in Cincinnati?   **14.8 h**

   c. Find the number of hours of daylight in the middle of October.   **about 10.9 h**

2. **Waves**   A buoy floats on the water bobbing up and down. The distance between its highest and lowest point is 6 centimeters. It moves from its highest point down to its lowest point and back to its highest point every 14 seconds. Write a cosine function that models the movement to the equilibrium point. $y = 3\cos\left(\dfrac{\pi}{7}t\right)$

Lesson 6-7   *(Pages 395–403)*
Graph each function.   **1–3. See margin.**

1. $y = \cot\left(\theta - \dfrac{\pi}{4}\right)$ 2. $y = \sec\theta + 2$ 3. $y = \csc(2\theta + 2\pi)$

Lesson 6-8   *(Pages 405–412)*
Find each value.

1. $\text{Cos}^{-1}\,0$ $\dfrac{\pi}{2}$ 2. $\text{Arcsin}\,0$ **0**   3. $\cos(\text{Tan}^{-1}\,1)$ $\dfrac{\sqrt{2}}{2}$

4. $\text{Cos}^{-1}\left(\tan\dfrac{3\pi}{4}\right)$ π 5. $\sin\left(\text{Cos}^{-1}\dfrac{1}{2} + \text{Sin}^{-1}\,0\right)$ $\dfrac{\sqrt{3}}{2}$ 6. $\cos\left(2\,\text{Sin}^{-1}\dfrac{\sqrt{3}}{2}\right)$ $-\dfrac{1}{2}$

Additional Answers
Lesson 6-5

1.

2.

3.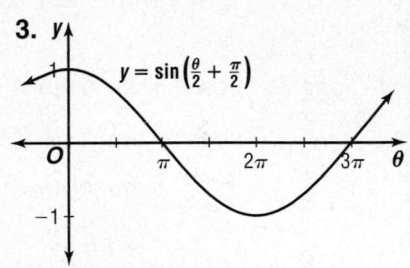

Additional Answers
Lesson 6-7

1.

2.

3.

Additional Answers
Lesson 7-2

1. $\csc^2 \theta \overset{?}{=} \cot^2 \theta + \sin \theta \csc \theta$

$\csc^2 \theta \overset{?}{=} \cot^2 \theta + \sin \theta \cdot \dfrac{1}{\sin \theta}$

$\csc^2 \theta \overset{?}{=} \cot^2 \theta + 1$

$\csc^2 \theta = \csc^2 \theta$

2. $\dfrac{\sec \theta - \csc \theta}{\csc \theta \sec \theta} \overset{?}{=} \sin \theta - \cos \theta$

$\dfrac{\sec \theta}{\csc \theta \sec \theta} - \dfrac{\csc \theta}{\csc \theta \sec \theta} \overset{?}{=} \sin \theta - \cos \theta$

$\dfrac{1}{\csc \theta} - \dfrac{1}{\sec \theta} \overset{?}{=} \sin \theta - \cos \theta$

$\sin \theta - \cos \theta = \sin \theta - \cos \theta$

3. $\sin^2 x + \cos^2 x$

$\overset{?}{=} \sec^2 x - \tan^2 x$

$1 \overset{?}{=} \tan^2 x + 1 - \tan^2 x$

$1 = 1$

4. $\sec A - \cos A \overset{?}{=} \tan A \sin A$

$\dfrac{1}{\cos A} - \cos A \overset{?}{=} \tan A \sin A$

$\dfrac{1}{\cos A} - \dfrac{\cos^2 A}{\cos A} \overset{?}{=} \tan A \sin A$

$\dfrac{1 - \cos^2 A}{\cos A} \overset{?}{=} \tan A \sin A$

$\dfrac{\sin^2 A}{\cos A} \overset{?}{=} \tan A \sin A$

$\dfrac{\sin A}{\cos A} \sin A \overset{?}{=} \tan A \sin A$

$\tan A \sin A = \tan A \sin A$

Lesson 7-1 *(Pages 421–430)*

Use the given information to determine the exact trigonometric value.

1. $\cos \theta = \dfrac{1}{4}, 0° < \theta < 90°; \csc \theta$ $\quad \dfrac{4\sqrt{15}}{15}$

2. $\cot \theta = -\dfrac{\sqrt{6}}{3}, \dfrac{\pi}{2} < \theta < \pi; \tan \theta$ $\quad -\dfrac{\sqrt{6}}{2}$

Express each value as a trigonometric function of an angle in Quadrant I.

3. $\cos \dfrac{13\pi}{6}$ $\quad \cos \dfrac{\pi}{6}$

4. $\tan (-315°)$ $\quad \tan 45°$

5. $\csc (-930°)$ $\quad \csc 30°$

Simplify each expression.

6. $\dfrac{\tan \theta}{\sin \theta}$ $\quad \sec \theta$

7. $\cot \theta \tan \theta + \sin \theta \sec \theta$ $\quad 1 + \tan \theta$

8. $(1 + \sin x)(1 - \sin x)$ $\quad \cos^2 x$

9. $\dfrac{\cot x \sin x}{\csc x \cos x}$ $\quad \sin x$

Lesson 7-2 *(Pages 431–436)*

Verify that each equation is an identity. **1–4. See margin.**

1. $\csc^2 \theta = \cot^2 \theta + \sin \theta \csc \theta$

2. $\dfrac{\sec \theta - \csc \theta}{\csc \theta \sec \theta} = \sin \theta - \cos \theta$

3. $\sin^2 x + \cos^2 x = \sec^2 x - \tan^2 x$

4. $\sec A - \cos A = \tan A \sin A$

Find a numerical value of one trigonometric function of x.

5. $\dfrac{\cot x}{\csc x} = 1$ **Sample answer:** $\cos x = 1$

6. $2 \tan x \sin x + 2 \cos x = \csc x$ **Sample answer:** $\cot x = 2$

Lesson 7-3 *(Pages 437–445)*

Use sum or difference identities to find the exact value of each trigonometric function.

1. $\cos 75°$ $\quad \dfrac{\sqrt{6} - \sqrt{2}}{4}$

2. $\sin 105°$ $\quad \dfrac{\sqrt{6} + \sqrt{2}}{4}$

3. $\tan \dfrac{\pi}{12}$ $\quad 2 - \sqrt{3}$

4. $\tan \dfrac{7\pi}{12}$ $\quad -2 - \sqrt{3}$

5. $\sec \dfrac{29\pi}{12}$ $\quad \sqrt{6} + \sqrt{2}$

6. $\cot 375°$ $\quad 2 + \sqrt{3}$

Find each exact value if $0 < x < \dfrac{\pi}{2}$ and $0 < y < \dfrac{\pi}{2}$.

7. $\sin (x + y)$ if $\cos x = \dfrac{2}{5}$ and $\sin y = \dfrac{3}{4}$ $\quad \dfrac{6 + 7\sqrt{3}}{20}$

8. $\cos (x - y)$ if $\cos x = \dfrac{5}{12}$ and $\cos y = \dfrac{11}{12}$ $\quad \dfrac{55 + \sqrt{2737}}{144}$

9. $\tan (x + y)$ if $\cot x = \dfrac{4}{3}$ and $\sec y = \dfrac{5}{4}$ $\quad \dfrac{24}{7}$

10. $\sec (x - y)$ if $\tan x = \dfrac{7}{6}$ and $\csc y = \dfrac{8}{5}$ $\quad \dfrac{48\sqrt{3315} - 280\sqrt{85}}{179}$

Lesson 7-4   (Pages 448–455)

Use a half-angle identity to find the exact value of each function.

1. $\sin 15°$ $\dfrac{\sqrt{2-\sqrt{3}}}{2}$

2. $\cos 75°$ $\dfrac{\sqrt{2-\sqrt{3}}}{2}$

3. $\tan \dfrac{\pi}{12}$ $2-\sqrt{3}$

4. $\cos 22.5°$ $\dfrac{\sqrt{2+\sqrt{2}}}{2}$

5. $\sin \dfrac{5\pi}{12}$ $\dfrac{\sqrt{2+\sqrt{3}}}{2}$

6. $\tan 112.5°$ $-1-\sqrt{2}$

Use the given information to find $\sin 2\theta$, $\cos 2\theta$, and $\tan 2\theta$.

7. $\cos \theta = \dfrac{2}{7},\, 0° < \theta < 90°$ $\dfrac{12\sqrt{5}}{49};\, -\dfrac{41}{49};\, -\dfrac{12\sqrt{5}}{41}$

8. $\sin \theta = \dfrac{2}{3},\, 0 < \theta < \dfrac{\pi}{2}$ $\dfrac{4\sqrt{5}}{9};\, \dfrac{1}{9};\, 4\sqrt{5}$

9. $\tan \theta = -3,\, 90° < \theta < 180°$ $-\dfrac{3}{5};\, -\dfrac{4}{5};\, \dfrac{3}{4}$

10. $\csc \theta = -\dfrac{3}{2},\, \dfrac{3\pi}{2} < \theta < 2\pi$ $-\dfrac{4\sqrt{5}}{9};\, \dfrac{1}{9};\, -4\sqrt{5}$

Lesson 7-5   (Pages 456–461)

Solve each equation for $0° \le x < 360°$.

1. $4\cos^2 x - 2 = 0$ **45°, 135°, 225°, 315°**

2. $\sin^2 x \csc x - 1 = 0$ **90°**

3. $\sqrt{3}\cot x = 2\cos x$ **60°, 90°, 120°, 270°**

4. $3\cos^2 x = 6\cos x - 3$ **0°**

Lesson 7-6   (Pages 463–469)

Write the standard form of the equation of each line given p, the length of the normal segment, and ϕ, the angle the normal segment makes with the positive x-axis.

1. $p = 12,\, \phi = 30°$
$\sqrt{3}x + y - 24 = 0$

2. $p = 2,\, \phi = \dfrac{\pi}{3}$
$x + \sqrt{3}y - 4 = 0$

3. $p = \dfrac{1}{2},\, \phi = 150°$
$-\sqrt{3}x + y - 1 = 0$

Write each equation in normal form. Then find the length of the normal and the angle it makes with the positive x-axis.   **4–6. See margin.**

4. $4x + 10y + 10 = 0$

5. $x - y = 2$

6. $2x + 3y = 12$

Lesson 7-7   (Pages 470–476)

Find the distance between the point with the given coordinates and the line with the given equation.

1. $(2, 1),\, 3x - 2y = 2$ $\dfrac{2\sqrt{13}}{13}$

2. $(3, 0),\, 2x + 4y - 2 = 0$ $\dfrac{2\sqrt{5}}{5}$

3. $(-1, -4),\, -3x + y = 1$ $\dfrac{\sqrt{10}}{5}$

4. $(4, -2),\, y = \dfrac{2}{3}x - 2$ $\dfrac{8\sqrt{13}}{13}$

Find the distance between the parallel lines with the given equations.

5. $x + 2y = -2$ $\dfrac{6\sqrt{5}}{5}$
$x + 2y = 4$

6. $y = \dfrac{2}{3}x + 3$ $\dfrac{2\sqrt{13}}{13}$
$3y - 2x = 7$

7. Geometry   Find the height to the nearest tenth of a unit of a trapezoid with parallel bases that lie on lines with equations $2x + 5y = -1$ and $2x + 5y = 4$.   **0.9 unit**

Additional Answers
Lesson 7-6

4. $-\dfrac{2\sqrt{29}}{29}x - \dfrac{5\sqrt{29}}{29}y -$

$\dfrac{5\sqrt{29}}{29} = 0;\ \dfrac{5\sqrt{29}}{29};$

about 248°

5. $\dfrac{\sqrt{2}}{2}x - \dfrac{\sqrt{2}}{2}y - \sqrt{2} = 0;$

$\sqrt{2};\ \mathbf{315°}$

6. $\dfrac{2\sqrt{13}}{13}x + \dfrac{3\sqrt{13}}{13}y -$

$\dfrac{12\sqrt{13}}{13} = 0;\ \dfrac{12\sqrt{13}}{13};$

about 56°

Lesson 8-1 *(Pages 485–492)*

Use a ruler and a protractor to determine the magnitude (in centimeters) and direction of each vector.

1.

2.

3.

     1.7 cm; 70°   **2.1 cm; 104°**   **3.6 cm; 330°**

Use $\vec{a}$, $\vec{b}$, and $\vec{c}$ above to find the magnitude and direction of each resultant.

4. $\vec{a} + \vec{b}$ **3.6 cm; 89°**   5. $\vec{b} + \vec{c}$ **2.6 cm; 23°**   6. $\vec{a} + \vec{c}$ **3.7 cm; 357°**

7. $\vec{a} - \vec{b}$ **1.2 cm; 342°**   8. $2\vec{c}$ **7.2 cm; 330°**   9. $2\vec{c} - \vec{b}$ **8.8 cm; 340°**

Find the magnitude of the horizontal and vertical components of each vector shown for Exercises 1–3.

10. $\vec{a}$ **0.58; 1.60**   11. $\vec{b}$ **0.51; 2.04**   12. $\vec{c}$ **3.12; 1.8**

Lesson 8-2 *(Pages 493–499)*

Find the ordered pair that represents $\overrightarrow{AB}$. Then find the magnitude of $\overrightarrow{AB}$.

1. $A(3, 6)$, $B(4, 1)$ $\langle 1, -5 \rangle$; $\sqrt{26}$ 2. $A(-1, 3)$, $B(-2, 2)$ $\langle -1, -1 \rangle$; $\sqrt{2}$

3. $A(0, -4)$, $B(-1, -8)$ $\langle -1, -4 \rangle$; $\sqrt{17}$ 4. $A(1, 10)$, $B(3, -9)$ $\langle 2, -19 \rangle$; $\sqrt{365}$

5. $A(-6, 0)$, $B(-3, -6)$ $\langle 3, -6 \rangle$; $3\sqrt{5}$ 6. $A(4, -5)$, $B(0, 7)$ $\langle -4, 12 \rangle$; $4\sqrt{10}$

Find the magnitude of each vector and write each vector as the sum of unit vectors.

7. $\langle 5, 6 \rangle$ $\sqrt{61}$; $5\vec{i} + 6\vec{j}$ 8. $\langle -2, 4 \rangle$ $2\sqrt{5}$; $-2\vec{i} + 4\vec{j}$ 9. $\langle -10, -5 \rangle$ $5\sqrt{5}$; $-10\vec{i} - 5\vec{j}$

10. $\langle 2.5, 6 \rangle$ **6.5**; $2.5\vec{i} + 6\vec{j}$ 11. $\langle 2, -6 \rangle$ $2\sqrt{10}$; $2\vec{i} - 6\vec{j}$ 12. $\langle -15, -12 \rangle$

                                                                 $3\sqrt{41}$; $-15\vec{i} - 12\vec{j}$

Lesson 8-3 *(Pages 500–504)*

Find an ordered triple to represent $\vec{p}$ in each equation if $\vec{q} = \langle 1, 2, -1 \rangle$, $\vec{r} = \langle -2, 2, 4 \rangle$, and $\vec{s} = \langle -4, -3, 0 \rangle$.

1. $\vec{p} = 2\vec{q} + 3\vec{s}$ $\langle -10, -5, -2 \rangle$ 2. $\vec{p} = \vec{q} - \frac{1}{2}\vec{r} + \vec{s}$ $\langle -2, -2, -3 \rangle$

3. $\vec{p} = -2\vec{r} + \vec{s}$ $\langle 0, -7, -8 \rangle$ 4. $\vec{p} = \frac{3}{4}\vec{s} + 2\vec{q}$ $\left\langle -1, 1\frac{3}{4}, -2 \right\rangle$

5. **Physics** If vectors working on an object are in equilibrium, then their resultant is zero. Two forces on an object are represented by $\langle 2, -4, 1 \rangle$ and $\langle 5, 4, 3 \rangle$. Find a third vector that will place the object in equilibrium. $\langle -7, 0, -4 \rangle$

Lesson 8-4 *(Pages 505–511)*

Find each inner product and state whether the vectors are perpendicular. Write *yes* or *no*.

1. $\langle 3, 4 \rangle \cdot \langle 2, 5 \rangle$ **26; no**
2. $\langle -3, 2 \rangle \cdot \langle 4, 6 \rangle$ **0; yes**
3. $\langle -5, 3 \rangle \cdot \langle 2, -3 \rangle$ **−19; no**
4. $\langle 8, 6 \rangle \cdot \langle -2, -3 \rangle$ **−34; no**
5. $\langle 3, 4, 0 \rangle \cdot \langle 4, -3, 6 \rangle$ **0; yes**
6. $\langle 4, 5, 1 \rangle \cdot \langle -1, -2, 3 \rangle$ **−11; no**

Find each cross product. Then verify that the resulting vector is perpendicular to the given vectors.   **7–10. See margin for verification.**

7. $\langle 1, 0, 3 \rangle \times \langle 1, 1, 2 \rangle$ $\langle \mathbf{-3, 1, 1} \rangle$
8. $\langle 3, 0, 4 \rangle \times \langle -1, 5, 2 \rangle$ $\langle \mathbf{-20, -10, 15} \rangle$
9. $\langle -1, 1, 0 \rangle \times \langle 2, 1, 3 \rangle$ $\langle \mathbf{3, 3, -3} \rangle$
10. $\langle -1, -3, 2 \rangle \times \langle 6, -1, -2 \rangle$ $\langle \mathbf{8, 10, 19} \rangle$

Lesson 8-5 *(Pages 513–519)*

Find the magnitude and direction of the resultant vector for each diagram.

1. **297.32 N; 42.3°**

2. **86.60 mph; 30°**

3. **582.66 N; 19.9°**

4. A 90 Newton force and a 110 Newton force act on the same object. The angle between the forces measures 90°. Find the magnitude of the resulting force.   **about 142.13 N**

Lesson 8-6 *(Pages 520–525)*   **1.** $\langle x - 2, y - 3 \rangle = t\langle 1, 0 \rangle;\ x = 2 + t,\ y = 3$

Write a vector equation of the line that passes through point *P* and is parallel to $\overline{a}$. Then write parametric equations of the line.   **2.** $\langle x + 1, y + 4 \rangle = t\langle 5, 2 \rangle;\ x = -1 + 5t,\ y = -4 + 2t$

1. $P(2, 3), \overline{a} = \langle 1, 0 \rangle$
2. $P(-1, -4), \overline{a} = \langle 5, 2 \rangle$
3. $P(-3, 6), \overline{a} = \langle -2, 4 \rangle$
4. $P(3, 0), \overline{a} = \langle 0, -1 \rangle$

$\langle x + 3, y - 6 \rangle = t\langle -2, 4 \rangle;\ x = -3 - 2t,\ y = 6 + 4t$ $\langle x - 3, y \rangle = t\langle 0, -1 \rangle;\ x = 3,\ y = -t$

Write an equation in slope-intercept form of the line with the given parametric equations.

5. $x = 3t$ $y = \frac{1}{3}x + 2$
   $y = 2 + t$
6. $x = -1 + 2t$ $y = 2x + 2$
   $y = 4t$
7. $x = 3t - 10$ $y = \frac{1}{3}x + \frac{7}{3}$
   $y = t - 1$

Lesson 8-7 *(Pages 527–533)*   **1. about 39.14 yd/s; about 58.03 yd/s**

1. **Sports**   A golf ball is hit with an initial velocity of 70 yards per second at 34° with the horizontal. Find the initial vertical and horizontal velocity for the ball.

2. **Sports**   An outfielder catches a fly ball and then throws it to third base to tag the runner. The outfielder releases the ball at an initial velocity of 75 feet per second at an angle of 25° with the horizontal. Assume the ball is released 5 feet above the ground.

   a. Write two parametric equations that represent the path of the ball.

   b. How far will the ball travel horizontally before hitting the ground?   **about 145 ft**

   c. What is the maximum height of the trajectory?   **about 20.6 ft**
      2a. $x = 75t \cos 25°,\ y = 5 + 75t \sin 25° - 16t^2$

Lesson 8-8 *(Pages 535–542)*

Describe the result of the product of a vertex matrix and each matrix.   **1–3. See margin.**

1. $\begin{bmatrix} 4 & 0 & 0 \\ 0 & 4 & 0 \\ 0 & 0 & -4 \end{bmatrix}$
2. $\begin{bmatrix} 0.5 & 0 & 0 \\ 0 & 0.5 & 0 \\ 0 & 0 & 0.5 \end{bmatrix}$
3. $\begin{bmatrix} -1.5 & 0 & 0 \\ 0 & 1.5 & 0 \\ 0 & 0 & 1.5 \end{bmatrix}$

Additional Answers

Lesson 8-4

7. $\langle -3, 1, 1 \rangle; \langle -3, 1, 1 \rangle \cdot \langle 1, 0, 3 \rangle = 0,$
   $\langle -3, 1, 1 \rangle \cdot \langle 1, 1, 2 \rangle = 0$

8. $\langle -20, -10, 15 \rangle;$
   $\langle -20, -10, 15 \rangle \cdot \langle 3, 0, 4 \rangle = 0,$
   $\langle -20, -10, 15 \rangle \cdot \langle -1, 5, 2 \rangle = 0$

9. $\langle 3, 3, -3 \rangle;$
   $\langle 3, 3, -3 \rangle \cdot \langle -1, 1, 0 \rangle = 0,$
   $\langle 3, 3, -3 \rangle \cdot \langle 2, 1, 3 \rangle = 0$

10. $\langle 8, 10, 19 \rangle;$
    $\langle 8, 10, 19 \rangle \cdot \langle -1, -3, 2 \rangle = 0,$
    $\langle 8, 10, 19 \rangle \cdot \langle 6, -1, -2 \rangle = 0$

Additional Answers

Lesson 8-8

1. The figure is 4 times the original size and reflected over the *xy*-plane.

2. The figure is half the original size.

3. The figure is 1.5 times the original size and reflected over the *yz*-plane.

Additional Answers
Lesson 9-1

1.

2.

3.

4.

5.

6.

Lesson 9-1 *(Pages 553–560)*
Graph each point.   1–4. See margin.

1. $K(4, 45°)$

2. $M\left(2, \frac{\pi}{6}\right)$

3. $N\left(\frac{3}{2}, -240°\right)$

4. $P\left(-1.5, \frac{5\pi}{6}\right)$

Graph each polar equation.   5–8. See margin.

5. $r = 2$

6. $\theta = 60°$

7. $r = -2.5$

8. $\theta = \frac{7\pi}{6}$

9. Write a polar equation for the circle centered at the origin with radius $\sqrt{5}$.

$r = \sqrt{5}$ or $r = -\sqrt{5}$

Lesson 9-2 *(Pages 561–567)*   1–3. See margin for graphs.
Graph each polar equation. Identify the type of curve each represents.

1. $r = -2 \sin \theta$ **circle**

2. $r = 4\theta$ **spiral of Archimedes**

3. $r = 2 + 2 \cos \theta$ **cardioid**

4. Write an equation for a rose with 5 petals.   **Sample answer: $r = \sin 5\theta$**

Lesson 9-3 *(Pages 568–573)*
Find the polar coordinates of each point with the given rectangular coordinates. Use $0 \le \theta < 2\pi$ and $r \ge 0$.

1. $(1, -1)$ $\left(\sqrt{2}, \frac{7\pi}{4}\right)$

2. $(3, 0)$ **(3, 0)**

3. $\left(2, \sqrt{2}\right)$ **(2.45, 0.62)**

Find the rectangular coordinates of each point with the given polar coordinates.

4. $\left(2, \frac{\pi}{4}\right)$ $\left(\sqrt{2}, \sqrt{2}\right)$

5. $\left(\frac{1}{4}, \frac{\pi}{2}\right)$ $\left(0, \frac{1}{4}\right)$

6. $(5, 240°)$ **(−2.5, −4.33)**

Write each rectangular equation in polar form.

7. $x = -2$ **$r = -2 \sec \theta$**

8. $y = 6$ **$r = 6 \csc \theta$**

9. $x^2 + y^2 = 36$ **$r = 6$ or $r = -6$**

10. $x^2 + y^2 = 3y$ **$r = 3 \sin \theta$**

Write each polar equation in rectangular form.

11. $r = 4$ **$x^2 + y^2 = 16$**

12. $r = 4 \cos \theta$ **$x^2 + y^2 = 4x$**

Lesson 9-4 *(Pages 574–579)*
Write each equation in polar form. Round ϕ to the nearest degree.

1. $6x - 5y + 6 = 0$
$\frac{6\sqrt{61}}{61} = r \cos (\theta - 140°)$

2. $3x + 9y = 90$
$3\sqrt{10} = r \cos (\theta - 72°)$

Write each equation in rectangular form.

3. $8 = r \cos (\theta - 30°)$ $\sqrt{3}x + y - 16 = 0$

4. $1 = r \cos (\theta + \pi)$ **$x = -1$**

5. Graph the polar equation $3 = r \cos (\theta - 30°)$.   **See margin.**

7.

8.

Lesson 9-5 *(Pages 580–585)*

Simplify.

1. i^{-10} **−1**
2. i^{17} ***i***
3. i^{1000} **1**
4. $i^{12} + i^{-4}$ **2**
5. $(4 - i) + (-3 + 5i)$ **1 + 4*i***
6. $(6 + 6i) - (2 + 4i)$ **4 + 2*i***
7. $(3 + i)(5 - 3i)$ **18 − 4*i***
8. $(2 + 5i)^2$ **−21 + 20*i***
9. $(1 - \sqrt{2}i)(-3 - \sqrt{8}i)$ **−7 + $\sqrt{2}i$**
10. $\dfrac{4 + i}{1 - i}$ $\mathbf{\dfrac{3}{2} + \dfrac{5}{2}i}$
11. $\dfrac{6 + 2i}{-2 + i}$ **−2 − 2*i***
12. $\dfrac{(i - 2)^2}{4 + 2i}$ $\mathbf{\dfrac{1}{5} - \dfrac{11}{10}i}$

Lesson 9-6 *(Pages 586–591)*

1. Solve $4x - 6yi = 14 + 12i$ for x and y, where x and y are real numbers.   ***x* = 3.5, *y* = −2**

Graph each number in the complex plane and find its absolute value.

2. $4 + i$ $\sqrt{17}$
3. $-5i$ **5**
4. $2 - \sqrt{3}i$ $\sqrt{7}$

2–4. See margin for graphs.

Express each complex number in polar form.

5. $4 + 4i$
6. $-2 + i$
7. $4 - \sqrt{2}i$

$4\sqrt{2}\left(\cos\dfrac{\pi}{4} + i\sin\dfrac{\pi}{4}\right)$ $\sqrt{5}(\cos 2.68 + i\sin 2.68)$ $3\sqrt{2}(\cos 5.94 + i\sin 5.94)$

8. **Electricity** The impedance in one part of a series circuit is $5(\cos 0.9 + j\sin 0.9)$ ohms and in the second part of the circuit it is $8(\cos 0.4 + j\sin 0.4)$ ohms.

  a. Convert these complex numbers to rectangular form.   **3.11 + 3.92*j*, 7.37 + 3.12*j***

  b. Add your answers from part **a** to find the total impedance in the circuit.   **10.48 + 7.04j ohms**

  c. Convert the total impedance back to polar form.   **12.63(cos 0.59 + *j* sin 0.59) ohms**

Lesson 9-7 *(Pages 593–598)*

Find each product or quotient. Express the result in rectangular form.

1. $6\left(\cos\dfrac{\pi}{2} + i\sin\dfrac{\pi}{2}\right) \cdot 4\left(\cos\dfrac{\pi}{4} + i\sin\dfrac{\pi}{4}\right)$ **$-12\sqrt{2} + 12\sqrt{2}i$**

2. $3\left(\cos\dfrac{3\pi}{4} + i\sin\dfrac{3\pi}{4}\right) \div \dfrac{1}{2}(\cos\pi + i\sin\pi)$ **$3\sqrt{2} - 3\sqrt{2}i$**

3. $5(\cos 135° + i\sin 135°) \cdot 2(\cos 45° + i\sin 45°)$ **−10**

Lesson 9-8 *(Pages 599–606)*

Find each power. Express the result in rectangular form.

1. $\left[4\left(\cos\dfrac{\pi}{2} + i\sin\dfrac{\pi}{2}\right)\right]^4$ **256**
2. $(12i - 5)^3$ **2035 − 828*i***

Find each principal root. Express the result in the form $a + bi$ with a and b rounded to the nearest hundredth.

3. $(1 + i)^{\frac{1}{3}}$ **1.08 + 0.29*i***
4. $(-1)^{\frac{1}{5}}$ **0.81 + 0.59*i***

Additional Answer *(p. A42)*
Lesson 9-4

5.

Additional Answers
Lesson 9-6

2.

(4, 1)

3.

(0, −5)

4.

$(2, -\sqrt{3})$

Additional Answers *(p. A42)*
Lesson 9-2

1.

2.

3.

Additional Answers
Lesson 10-2

1.

$(x + 2)^2 + (y - 2)^2 = 2$

2.

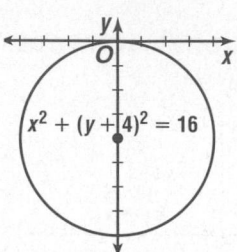

$x^2 + (y + 4)^2 = 16$

3.

$x^2 + y^2 = 49$

4.

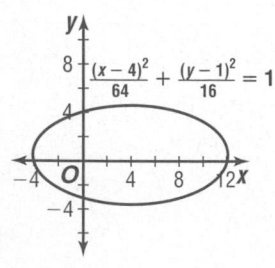

$(x + 3)^2 + (y - 4)^2 = 7$

Additional Answer
Lesson 10-3

3.

$\dfrac{(x - 4)^2}{64} + \dfrac{(y - 1)^2}{16} = 1$

Lesson 10-1 *(Pages 615–622)*   **4.** $(-16, 14)$

Find the distance between each pair of points with the given coordinates. Then, find the coordinates of the midpoint of the segment that has endpoints at the given coordinates.

1. $(-2, 2), (4, 5)$
$3\sqrt{5}; (1, 3.5)$

2. $(-3, 6), (8, -1)$
$\sqrt{170}; (2.5, 2.5)$

3. $(r, 6), (r, -2)$
$8; (r, 2)$

4. If $M(-5, 8)$ is the midpoint of $\overline{AB}$ and B has coordinates $(6, 2)$, find the coordinates of A.

5. Determine whether the quadrilateral having vertices at $(5, 10), (5, 2), (2, 8),$ and $(2, 5)$ is a parallelogram.   **no**

6. Andrea's garden is 50 feet long and 40 feet wide. Andrea makes a path from one corner of the garden to a water fountain at the center. Suppose the corner of the garden is the origin and the garden is in the first quadrant.

   a. Find the ordered pair that represents the location of the water fountain.   **(25, 20)**

   b. Find the length of the path.   **$5\sqrt{41}$ or about 32 ft**

Lesson 10-2 *(Pages 623–630)*   **1–4. See margin for graphs.**

Write the standard form of the equation of each circle described. Then graph the equation.

1. center at $(-2, 2)$, radius $\sqrt{2}$
$(x + 2)^2 + (y - 2)^2 = 2$

2. center at $(0, -4)$, tangent to the x-axis
$x^2 + (y + 4)^2 = 16$

Write the standard form of each equation. Then graph the equation.

3. $x^2 = 49 - y^2$
$x^2 + y^2 = 49$

4. $x^2 + y^2 + 6x - 8y + 18 = 0$
$(x + 3)^2 + (y - 4)^2 = 7$

Write the standard form of the equation of the circle that passes through the points with the given coordinates. Then identify the center and the radius.

5. $(2, -2), (0, -4), (-2, -2)$
$x^2 + (y + 2)^2 = 4; (0, -2); 2$

6. $(-1, 3), (-4, 6), (-7, 3)$
$(x + 4)^2 + (y - 3)^2 = 9; (-4, 3); 3$

7. Write the equation of the circle that passes through the point $(2, -5)$ and has its center at $(4, 0)$.
$(x - 4)^2 + y^2 = 29$

Lesson 10-3 *(Pages 631–641)*

Write the equation of each ellipse in standard form. Then find the coordinates of its foci.

1.

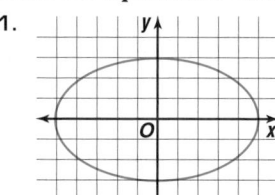

$\dfrac{x^2}{25} + \dfrac{y^2}{9} = 1;$
$(\pm 4, 0)$

2.

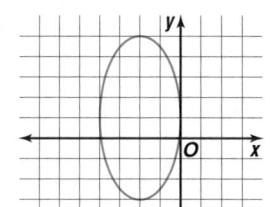

$\dfrac{(y - 1)^2}{16} + \dfrac{(x + 2)^2}{4} = 1;$
$(-2, 1 \pm 2\sqrt{3})$

3. For the equation of the ellipse $\dfrac{(x - 4)^2}{64} + \dfrac{(y - 1)^2}{16} = 1$, find the coordinates of the center, foci, and vertices. Then graph the equation. $(4, 1); (4 \pm 4\sqrt{5}, 1); (12, 1), (-4, 1), (4, 5), (4, -3);$
See margin for graph.

Lesson 10-4 *(Pages 642–652)* **1–2. See margin.**

1. Graph the equation $\dfrac{(y-2)^2}{49} - \dfrac{(x-1)^2}{9} = 1$. Label the center, foci, and the equations of the asymptotes.

2. Graph the equation $xy = 16$.

Write an equation of the hyperbola that meets each set of conditions.

3. The center is at $(-4, 3)$, $a = 3$, $b = 2$, and it has a horizontal transverse axis.

4. The foci are at $(2, -3)$, and $(2, 7)$ and the vertices are at $(2, -1)$ and $(2, 5)$.

3. $\dfrac{(x+4)^2}{9} - \dfrac{(y-3)^2}{4} = 1$ 4. $\dfrac{(y-2)^2}{9} - \dfrac{(x-2)^2}{16} = 1$

Lesson 10-5 *(Pages 653–661)* **1–2. See margin for graph.**
For the equation of each parabola, find the coordinates of the vertex and focus, and the equations of the directrix and axis of symmetry. Then graph the equation.

1. $y^2 = 4x$
   $(0, 0)$; $(1, 0)$; $x = -1$; $y = 0$

2. $x^2 - 4x + 4 = 12y - 12$
   $(2, 1)$; $(2, 4)$; $y = -2$; $x = 2$

Write the equation of the parabola that meets each set of conditions.

3. The vertex is at $(-2, 3)$ and the focus is at $(0, 3)$. $(y - 3)^2 = 8(x + 2)$

4. The focus is at $(0, -2)$ and the equation of the directrix is $y = -3$. $x^2 = 2(y + 2.5)$

Lesson 10-6 *(Pages 662–669)*
Identify the conic section represented by each equation. Then write the equation in standard form. **1. circle;** $(x - 4)^2 + (y + 1)^2 = 4$ **2. hyperbola;** $\dfrac{(x+5)^2}{4} - \dfrac{(y+2)^2}{1} = 1$

1. $x^2 + y^2 - 8x + 2y + 13 = 0$

2. $x^2 - 4y^2 + 10x - 16y = -5$

3. $y^2 - 5x - 6y + 9 = 0$
   parabola; $(y - 3)^2 = 5x$

4. $x^2 + 2y^2 + 2x + 8y = 15$
   ellipse; $\dfrac{(x+1)^2}{24} + \dfrac{(y+2)^2}{12} = 1$

Lesson 10-7 *(Pages 670–677)*
Identify the graph of each equation. Write an equation of the translated or rotated graph in general form. **1. circle;** $x^2 + y^2 - 2x + 2y - 7 = 0$ **2. ellipse;** $4x^2 + y^2 + 24x + 4y + 24 = 0$

1. $x^2 + y^2 = 9$ for $T_{(1, -1)}$

2. $4x^2 + y^2 = 16$ for $T_{(-3, -2)}$

3. $49x^2 - 16y^2 = 784$; $\theta = \dfrac{\pi}{4}$
   hyperbola; $33(x')^2 + 130x'y' + 33(y')^2 - 1568 = 0$

4. $4x^2 - 25y^2 = 64$; $\theta = 90°$
   hyperbola; $4(y')^2 - 25(x')^2 - 64 = 0$

Identify the graph of each equation. Then find θ to the nearest degree.

5. $2y^2 + 3y - 2\sqrt{2}xy + x^2 - 1 = 0$ **parabola; $-35°$** 6. $15x^2 + 5xy + 5y^2 + 9 = 0$ **ellipse; $13°$**

Lesson 10-8 *(Pages 678–684)*
Solve each system of equations algebraically. Round to the nearest tenth. Check the solutions by graphing each system. **1–2. See margin for graphs.**

1. $xy = 3$
   $x^2 - y^2 = 8$ **$(3, 1)$, $(-3, -1)$**

2. $x - y = 4$
   $x^2 = 10y^2 + 10$ **$(5.4, 1.4)$, $(3.5, -0.5)$**

Additional Answers
Lesson 10-5

1.

2.
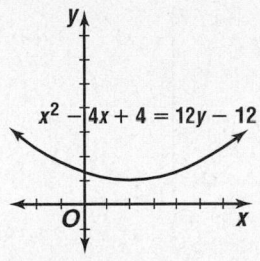

Additional Answers
Lesson 10-8

1.

2.
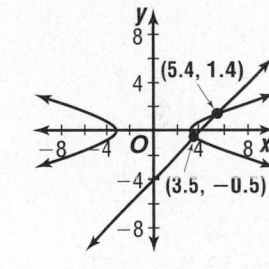

Additional Answers
Lesson 10-4

1.

2.

Additional Answers
Lesson 11-2

1.

$y = 3^x$

2.

$y = 3^{-x}$

3.

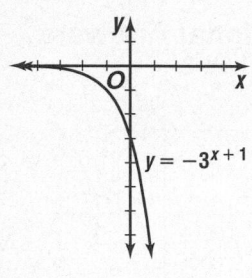

$y = -3^{x+1}$

Lesson 11-1 *(Pages 695–703)*

Evaluate each expression.

1. $(-12)^{-2}$ $\dfrac{1}{144}$

2. -12^{-2} $-\dfrac{1}{144}$

3. $(4 \cdot 6)^3$ **13,824**

4. $\left(\dfrac{2}{3}\right)^4$ $\dfrac{16}{81}$

5. $\dfrac{16}{16^{\frac{1}{2}}}$ **4**

6. $27^{\frac{1}{2}} \cdot 20^{\frac{1}{2}}$ $6\sqrt{15}$

7. $\left(\sqrt[4]{625}\right)^2$ **25**

8. $\dfrac{1}{\sqrt[3]{(15)^6}}$ $\dfrac{1}{225}$

Simplify each expression.

9. $(2a^4)^2$ $4a^8$

10. $(x^4)^3 \cdot x^5$ x^{17}

11. $((3f)^{-2})^3$ $\dfrac{1}{729\,f^6}$

12. $\left(\dfrac{c^{-3a}}{c^{4a}}\right)^2$ $\dfrac{1}{c^{14a}}$

13. $\left(2n^{\frac{1}{3}} \cdot 3n^{\frac{1}{3}}\right)^6$ $46{,}656n^5$

14. $\left(\dfrac{h^6}{216h^{-3}}\right)^{-\frac{1}{3}}$ $\dfrac{6}{h^3}$

15. $\sqrt[3]{z^4(z^4)^{\frac{1}{2}}}$ z^2

16. $(4r^2t^5)(16r^4t^8)^{\frac{1}{4}}$ $8\,|r|^3\,|t|^7$

Express using rational exponents.

17. $\sqrt{a^3b^5}$ $|a|^{\frac{3}{2}}|b|^{\frac{5}{2}}$

18. $\sqrt[3]{64m^9n^6}$ $4m^3n^2$

19. $15\sqrt[3]{r^{12}t^2}$ $15r^4t^{\frac{2}{3}}$

20. $\sqrt[8]{256x^2y^{16}}$ $2\,|x|^{\frac{1}{4}}y^2$

Lesson 11-2 *(Pages 704–711)*

Graph each exponential function.   **1–3. See margin.**

1. $y = 3^x$

2. $y = 3^{-x}$

3. $y = -3^{x+1}$

Lesson 11-3 *(Pages 712–717)*

1. **Psychology**   The Ebbinghaus Model for human memory gives the percent p of acquired knowledge that a person retains after an amount of time. The formula is $p = (100 - a)e^{-bt} + a$, where t is the time in weeks, and a and b vary from one person to another. If $a = 18$ and $b = 0.6$ for a certain student, how much information will the student retain two weeks after learning a new topic?   **about 42.7%**

2. **Physics**   Newton's Law of Cooling expresses the relationship between the temperature in degrees Fahrenheit of a cooling object y and the time elapsed since cooling began t in minutes. This relationship is given by $y = ae^{-kt} + c$, where c is the temperature surrounding the medium. Suppose vegetable soup is heated to 210°F in the microwave. If the room temperature is 70°F, what will the temperature of the soup be after 10 minutes? Assume that $a = 140$ and $k = 0.01$.   **about 197°F**

3. **Forestry**   The yield y in millions of cubic feet of trees per acre for a forest stand that is t years old is given by $y = 6.7\,e^{\frac{-48.1}{t}}$.

   a. Find the yield after 15 years.   **about 271,292 ft³**

   b. Find the yield after 50 years.   **about 2,560,257 ft³**

4. **Banking**   Compare the balance after 20 years of a $5000 investment earning 5.8% compounded continuously to the same investment compounded semiannually.   **The account that compounds continuously would earn $261.04 more than the account compounded semiannually.**

Lesson 11-4 *(Pages 718–725)*

Write each equation in exponential form.

1. $\log_{16} 2 = \frac{1}{4}$ $16^{\frac{1}{4}} = 2$
2. $\log_{\frac{1}{2}} 8 = -3$ $\left(\frac{1}{2}\right)^{-3} = 8$
3. $\log_4 \frac{1}{4} = -1$ $4^{-1} = \frac{1}{4}$

Write each equation in logarithmic form.

4. $8^{-2} = x$ $\log_8 x = -2$
5. $x^5 = 32$ $\log_x 32 = 5$
6. $\left(\frac{1}{4}\right)^{-2} = 16$ $\log_{\frac{1}{4}} 16 = -2$

Evaluate each expression.

7. $\log_5 \frac{1}{5}$ -1
8. $\log_3 27$ 3
9. $\log_{36} 6$ $\frac{1}{2}$

Solve each equation.

10. $\log_3 y = 4$ 81
11. $\log_5 r = \log_5 8$ 8
12. $\log_5 35 - \log_5 d = \log_5 5$ 7
13. $\log_4 \sqrt{4} = x$ $\frac{1}{2}$
14. $\log_4 (2x + 3) = \log_4 15$ 6
15. $4 \log_8 2 + \frac{1}{3}\log_8 27 = \log_8 a$ 48

Lesson 11-5 *(Pages 726–732)*

Given that log 5 = 0.6990, log 8 = 0.9031, and log 14 = 1.1461, evaluate each logarithm.

1. log 5000   **3.6990**
2. log 0.0008   **−3.0969**
3. log 0.14   **−0.8539**

Find the value of each logarithm using the change of base formula.

4. $\log_3 81$ **4**
5. $\log_6 12$ **1.3869**
6. $\log_5 29$ **2.0922**

Solve each equation.

7. $3^x = 45$ **3.4650**
8. $6^x = 2^{x-1}$ **−0.6309**
9. $5 \log y = \log 32$ **2**

Lesson 11-6 *(Pages 733–737)*

Evaluate each expression.

1. ln 35   **3.5553**
2. ln 0.562   **−0.5763**
3. antiln 1.2354   **3.4398**

Convert each logarithm to a natural logarithm and evaluate.

4. $\log_{15} 10$ **0.8503**
5. $\log_3 14$ **2.4022**
6. $\log_8 350$ **2.8171**

Use natural logarithms to solve each equation or inequality.

7. $5^x = 90$ **2.7959**
8. $7^{x+2} = 5.25$ **−1.1478**
9. $4^x = 4\sqrt{3}$ **1.3962**
10. $6e^x = 48$ **2.0794**
11. $50.2 < e^{0.2x}$ $x >$ **19.5801**
12. $16 = 10(1 + e^x)$ **−0.5108**

Lesson 11-7 *(Pages 740–748)*

Find the amount of time required for an amount to double at the given rate if the interest is compounded continuously.

1. 4.5%   **15.40 yr**
2. 6%   **11.55 yr**
3. 8.125%   **8.53 yr**

4. **Biology**   The data below give the number of bacteria found in a certain culture.

Time (hours)	0	1	2	3	4
Bacteria	6	7	12	20	32

a. Find an exponential model for the data. $y = 5.2449(1.5524)^x$

b. Write the equation from part **a** in terms of base *e*. $y = 5.2449e^{0.4398x}$

c. Use the model to estimate the doubling time for the culture.   **about 1.58 hr**

Lesson 12-1 *(Pages 759–765)*

Find the next four terms in each arithmetic sequence. **2.** $-4, -5.5, -7, -8.5$

1. $7, 3, -1, ...$ **$-5, -9, -13, -17$** 2. $0.5, -1, -2.5, ...$ 3. $-14, -8, -2, ...$ **4, 10, 16, 22**

4. $3, 2.8, 2.6, ...$ 5. $4x, -x, -6x, ...$ 6. $2y - 4, 2y - 2, 2y, ...$
   2.4, 2.2, 2, 1.8   **$-11x, -16x, -21x, -26x$**   **$2y + 2, 2y + 4, 2y + 6, 2y + 8$**

For Exercises 7–13, assume that each sequence or series is arithmetic.

7. Find the 16th term in the sequence for which $a_1 = 2$ and $d = 5$.   **77**

8. Find n for the sequence for which $a_n = -20$, $a_1 = 6$ and $d = -2$.   **14**

9. What is the first term in the sequence for which $d = 4$ and $a_{12} = 42$?   **-2**

10. Find d for the sequence in which $a_1 = 7$ and $a_{13} = 30$.   **$1\frac{11}{12}$**

11. What is the 24th term in the sequence $10.5, 10, 9.5, ...$?   **-1**

12. Find the sum of the first 12 terms in the series $2 + 2.8 + 3.6 + \cdots$.   **76.8**

13. Find n for the series for which $a_1 = -4$, $d = 4$, and $S_n = 80$.   **8**

Lesson 12-2 *(Pages 766–773)*   **1. 0.5; 1.75, 0.875, 0.4375**

Determine the common ratio and find the next three terms of each geometric sequence.

1. $14, 7, 3.5, ...$ 2. $-2, 4, -8, ...$ **$-2; 16, -32, 64$** 3. $\frac{2}{3}, \frac{1}{2}, \frac{3}{8}, ...$ **3. $\frac{3}{4}; \frac{9}{32}, \frac{27}{128}, \frac{81}{512}$**

4. $10, -5, 2.5, ...$ 5. $8, 8\sqrt{2}, 16, ...$ 6. $a^{10}, a^8, a^6, ...$
   $-0.5; -1.25, 0.625, -0.3125$ **$\sqrt{2}; 16\sqrt{2}, 32, 32\sqrt{2}$**   **$\frac{1}{a^2}; a^4, a^2, 1$**

For Exercises 7–11, assume that each sequence or series is geometric.

7. Find the sixth term of a sequence whose first term is 9 and common ratio is 2.   **288**

8. If $r = 4$ and $a_8 = 100$, what is the first term of the sequence?   **$\frac{25}{4096}$**

9. Find the first three terms of the sequence for which $a_5 = 10$ and $r = -\frac{1}{2}$.   **160, -80, 40**

10. Write a sequence that has two geometric means between 4 and 256.   **4, 16, 64, 256**

11. What is the sum of the first six terms of the series $3 + 9 + 27 + \cdots$?   **1092**

12. **Biology** A certain bacteria divides every 15 minutes to produce two complete bacteria.

   a. If an initial colony contains a population of b_0 bacteria, write an equation that will determine the number of bacteria b_t present after t hours.   **$b_t = b_0 \cdot 16^t$**

   b. Suppose a petri dish contains 12 bacteria. Use the equation found in part **a** to determine the number of bacteria present 4 hours later.   **786,432**

Lesson 12-3 *(Pages 774–783)* 2. Limit does not exist; see margin for explanation.
Find each limit, or state that the limit does not exist and explain your reasoning.

1. $\lim\limits_{n\to\infty} \dfrac{4+2n}{3n}$ $\dfrac{2}{3}$

2. $\lim\limits_{n\to\infty} \dfrac{n^4-3n}{n^3}$

3. $\lim\limits_{n\to\infty} \dfrac{8n^2+6n-2}{4n^2}$ 2

4. $\lim\limits_{n\to\infty} \dfrac{4n^2-2n+1}{n^2+2}$ 4

5. $\lim\limits_{n\to\infty} \dfrac{n^3-n^2+4}{5+2n^3}$ $\dfrac{1}{2}$

6. $\lim\limits_{n\to\infty} \dfrac{2^n n}{2+n}$

6. Limit does not exist; see margin for explanation.

Write each repeating decimal as a fraction.

7. $0.\overline{09}$ $\dfrac{1}{11}$

8. $0.1\overline{3}$ $\dfrac{2}{15}$

9. $7.\overline{407}$ $7\dfrac{11}{27}$

Find the sum of the series, or state that the sum does not exist and explain your reasoning.

10. $\dfrac{1}{20}+\dfrac{1}{40}+\dfrac{1}{80}+\cdots$ $\dfrac{1}{10}$

11. $\dfrac{2}{7}+\dfrac{4}{7}+\dfrac{8}{7}+\cdots$ S_n does not exist; see margin for explanation.

Lesson 12-4 *(Pages 786–793)*
Use the ratio test to determine whether each series is *convergent* or *divergent*.

1. $1^2+2^2+4^2+8^2+\cdots$ divergent

2. $\dfrac{1}{3}+\dfrac{2}{3}+1+\dfrac{4}{3}\cdots$ test provides no information

3. $1+\dfrac{1}{3\cdot8}+\dfrac{1}{9\cdot27}+\dfrac{1}{27\cdot64}+\cdots$ convergent

4. $4+2+1+\dfrac{1}{2}+\cdots$ convergent

Use the comparison test to determine whether each series is *convergent* or *divergent*.

5. $\dfrac{7}{7}+\dfrac{7}{13}+\dfrac{7}{19}+\dfrac{7}{25}+\cdots$ divergent

6. $\dfrac{1}{2^2}+\dfrac{1}{4^2}+\dfrac{1}{6^2}+\dfrac{1}{8^2}+\cdots$ convergent

Determine whether each series is *convergent* or *divergent*.

7. $1+\dfrac{1}{2^0+1}+\dfrac{1}{2^1+1}+\dfrac{1}{2^2+1}+\dfrac{1}{2^3+1}+\cdots$ convergent

8. $\dfrac{2}{3}+\dfrac{4}{4}+\dfrac{8}{5}+\dfrac{16}{6}+\cdots$ divergent

Lesson 12-5 *(Pages 794–800)* 1. $2+5+8+11+14$; 40
Write each expression in expanded form and then find the sum. 3. $7+14+23+34+47$; 125

1. $\sum\limits_{n=1}^{5}(3n-1)$

2. $\sum\limits_{a=3}^{6}4a$ 12 + 16 + 20 + 24; 72

3. $\sum\limits_{k=3}^{7}(k^2-2)$

4. $\sum\limits_{j=4}^{8}\dfrac{j}{j+3}$

5. $\sum\limits_{p=0}^{4}3^p$

6. $\sum\limits_{n=1}^{\infty}2\cdot\left(\dfrac{3}{4}\right)^n$

$\dfrac{4}{7}+\dfrac{5}{8}+\dfrac{6}{9}+\dfrac{7}{10}+\dfrac{8}{11}$; $3\dfrac{2683}{9240}$ $1+3+9+27+81$; 121 $1\dfrac{1}{2}+1\dfrac{1}{8}+\dfrac{27}{32}+\cdots+2\cdot\left(\dfrac{3}{4}\right)^{\infty}$; 6

Express each series using sigma notation.

7. $5+8+11+14$ $\sum\limits_{n=1}^{4}(3n+2)$

8. $-8-12-16-\cdots-40$ $\sum\limits_{r=2}^{10}-4r$

9. $\dfrac{1}{4}+0+4+\cdots+65{,}536$ $\sum\limits_{m=0}^{9}4^{m-1}$

10. $1+2+6+24+\cdots$ $\sum\limits_{w=1}^{\infty}w!$

Additional Answers
Lesson 12-3

2. $\lim\limits_{n\to\infty}\dfrac{n^4-3n}{n^3}=\lim\limits_{n\to\infty}\dfrac{1-\frac{3}{n^3}}{\frac{1}{n}}$,

which as n approaches infinity simplifies to $\dfrac{1-0}{0}$. Since this fraction is undefined, the limit does not exist.

6. $\lim\limits_{n\to\infty}\dfrac{2^n n}{2+n}=$

$\lim\limits_{n\to\infty}2^n\cdot\dfrac{\lim\limits_{n\to\infty}1}{\lim\limits_{n\to\infty}\frac{2}{n}+\lim\limits_{n\to\infty}1}=\lim\limits_{n\to\infty}2^n$

As n approaches infinity, 2^n becomes increasingly large, so there is no limit.

11. a_n is getting increasingly larger.

Additional Answers
Lesson 12-9

1. **Step 1:** Verify that the formula is valid for $n = 1$.

   Since $S_1 = 2$ and $1(1 + 1) = 2$, the formula is valid for $n = 1$.

   Step 2: Assume that the formula is valid for $n = k$ and show that it is also valid for $n = k + 1$. $2 + 4 + 6 + \cdots + 2k = k(k - 1)$

   Derive the formula for $n = k + 1$ by adding $2(k + 1)$ to each side.

   $2 + 4 + 6 + \cdots + 2k + 2(k + 1) = k(k + 1) + 2(k + 1)$

   $2 + 4 + 6 + \cdots + 2k + 2(k + 1) = (k + 1)(k + 2)$

   Apply the original formula for $n = k + 1$.

   $S_{k+1} = (k + 1)[(k + 1) + 1]$ or $(k + 1)(k + 2)$

   The formula gives the same result as adding the $(k + 1)$ term directly. Thus, if the formula is valid for $n = k$, it is also valid for $n = k + 1$. Since S_n is valid for $n = 2$, it is also valid for $n = 2$, $n = 3$, and so on, indefinitely. Hence, the formula is valid for all positive integral values of n.

2. **Step 1:** Verify that the formula is valid for $n = 1$.

   Since $S_1 = 1$ and $\dfrac{1(1 + 1)(1 + 2)}{6} = 1$, the formula is valid for $n = 1$.

   Step 2: Assume that the formula is valid for $n = k$ and show that it is also valid for $n = k + 1$.

   $1 + 3 + 6 + \cdots + \dfrac{k(k + 1)}{2} = \dfrac{k(k + 1)(k + 2)}{6}$

   Derive the formula for $n = k + 1$ by adding $\dfrac{(k + 1)(k + 2)}{2}$ to each side.

   $1 + 3 + 6 + \cdots + \dfrac{k(k + 1)}{2} + \dfrac{(k + 1)(k + 2)}{2} = \dfrac{k(k + 1)(k + 2)}{6} + \dfrac{(k + 1)(k + 2)}{2}$

Lesson 12-6 *(Pages 801–805)*

Use Pascal's triangle to expand each binomial.   2. $n^5 + 5n^4m + 10n^3m^2 + 10n^2m^4 + 5nm^4 + m^5$

1. $(2 + x)^4$
   $16 + 32x + 24x^2 + 8x^3 + x^4$

2. $(n + m)^5$

3. $(4a - b)^3$
   $64a^3 - 48a^2b + 12ab^2 - b^3$

Use the Binomial Theorem to expand each binomial.   5. $16r^4 + 32r^3s + 24r^2s^2 + 8rs^3 + s^4$

4. $(m - 3)^6$
   $m^6 - 18m^5 + 135m^4 - 540m^3 + 1215m^2 - 1458m + 729$

5. $(2r + s)^4$

6. $(5x - 4y)^3$
   $125x^3 - 300x^2y + 240xy^2 - 64y^3$

Find the designated term of each binomial expression.

7. 6th term of $(x + y)^8$ $56x^3y^5$

8. 5th term of $(b + \sqrt{3})^7$ $315b^3$

9. 3rd term of $(4z - w)^{10}$ $2{,}949{,}120z^8w^2$

10. 8th term of $(2h - k)^{12}$ $-25{,}344h^5k^7$

Lesson 12-7 *(Pages 806–814)*

Find each value to four decimal places.

1. $\ln(-3)$ $i\pi + 1.0986$

2. $\ln(-4.6)$ $i\pi + 1.5261$

3. $\ln(-0.75)$ $i\pi - 0.2877$

Use the first five terms of the exponential series and a calculator to approximate each value to the nearest hundredth.

4. $e^{1.2}$ 3.29

5. $e^{-0.7}$ 0.50

6. $e^{3.65}$ 26.81

Use the first five terms of the trigonometric series to approximate the value of each function to four decimal places. Then, compare the approximation to the actual value.

7. $\cos \dfrac{\pi}{4}$ $0.7071; 0.7071$

8. $\sin \dfrac{\pi}{6}$ $0.5000; 0.5$

9. $\cos \dfrac{\pi}{3}$ $0.5000; 0.5$

Lesson 12-8 *(Pages 815–821)*

Find the first four iterates of each function using the given initial value. If necessary, round your answers to the nearest hundredth.

1. $f(x) = 2x; x_0 = -2$ $-4, -8, -16, -32$

2. $f(x) = x^2; x_0 = 4$ $16, 256, 65{,}536, 4{,}294{,}967{,}296$

Find the first three iterates of the function $f(z) = 0.5z + i$ for each initial value.

3. $z_0 = 2i$ $2i, 2i, 2i$

4. $z_0 = 4 + 4i$ $2 + 3i, 1 + 2.5i, 0.5 + 2.25i$

5. **Banking**   Mendella has a savings account that has an annual yield of 5.4%. Find the balance of the account after each of the first five years if her initial balance is $4000.
   $4216, $4443.66, $4683.62, $4936.54, $5203.11

Lesson 12-9 *(Pages 822–828)*

Use mathematical induction to prove that each proposition is valid for all positive integral values of n.   1–2. See margin.

1. $2 + 4 + 6 + \cdots + 2n = n(n + 1)$

2. $1 + 3 + 6 + \cdots + \dfrac{n(n + 1)}{2} = \dfrac{n(n + 1)(n + 2)}{6}$

3. Prove that $5^n - 1$ is even for all positive integers n.   **See margin.**

$1 + 3 + 6 + \cdots + \dfrac{k(k + 1)}{2} + \dfrac{(k + 1)(k + 2)}{2} = \dfrac{k(k + 1)(k + 2) + 3(k + 1)(k + 2)}{6}$

$1 + 3 + 6 + \cdots + \dfrac{k(k + 1)}{2} + \dfrac{(k + 1)(k + 2)}{2} = \dfrac{(k + 1)(k + 2)(k + 3)}{6}$

Apply the original formula for $n = k + 1$.

$S_{k+1} = \dfrac{(k + 1)[(k + 1) + 1][(k + 1) + 2]}{6}$ or $\dfrac{(k + 1)(k + 2)(k + 3)}{6}$

The formula gives the same result as adding the $k + 1$ term directly. Thus, if the formula is valid for $n = k$, it is also valid for $n = k + 1$. Since the formula is valid for $n = 1$, it is also valid for $n = 2$, $n = 3$, and so on, indefinitely. Hence, the formula is valid for all positive integral values of n.

Lesson 13-1 *(Pages 837–845)*

1. If you roll two 6-sided number cubes and then spin a 6-colored spinner with equal sections, how many outcomes are possible? **216**
2. How many ways can 8 books be arranged on a shelf? **40,320**

State whether the events are *independent* or *dependent*.

3. tossing three coins, then rolling a die   **independent**
4. selecting members for a committee   **dependent**
5. deciding the order in which to answer your e-mail messages   **dependent**

Find each value.

6. $P(5, 5)$ **120**
7. $P(8, 3)$ **336**
8. $P(4, 1)$ **4**
9. $P(10, 9)$ **3,628,800**
10. $P(9, 6)$ **60,480**
11. $P(7, 3)$ **210**
12. $\dfrac{P(5, 2)}{P(2, 1)}$ **10**
13. $\dfrac{P(8, 6)}{P(7, 4)}$ **24**
14. $\dfrac{P(5, 2) \cdot P(8, 4)}{P(10, 1)}$ **3360**
15. $C(4, 2)$ **6**
16. $C(10, 7)$ **120**
17. $C(6, 5)$ **6**
18. $C(4, 3) \cdot C(7, 3)$ **140**
19. $C(3, 1) \cdot C(8, 7)$ **24**
20. $C(9, 5) \cdot C(4, 3)$ **504**

Lesson 13-2 *(Pages 846–851)*

How many different ways can the letters of each word be arranged?

1. *mailbox* **5040**
2. *textbook* **10,080**
3. *almanac* **840**
4. *dictionary* **1,814,400**

5. How many different 4-digit access codes can have the digits 5, 7, 2, and 7? **12**

Determine whether each arrangement of objects is a *linear* or *circular* permutation. Then determine the number of arrangements for each situation.

6. 4 friends seated around a square table   **circular; 6**
7. 9 charms on a charm bracelet with no clasp   **circular; 20,160**
8. a stack of 5 books on a table   **linear; 120**

Lesson 13-3 *(Pages 852–859)*

Using a standard deck of 52 cards, find each probability.

1. $P(\text{ace})$ $\dfrac{1}{13}$
2. $P(\text{a card of 5 or less})$ $\dfrac{5}{13}$
3. $P(\text{a red face card})$ $\dfrac{3}{26}$
4. $P(\text{not a queen})$ $\dfrac{12}{13}$

One pencil is randomly taken from a box containing 8 red pencils, 4 green pencils, and 2 blue pencils. Find each probability.

5. $P(\text{blue})$ $\dfrac{1}{7}$
6. $P(\text{green})$ $\dfrac{2}{7}$
7. $P(\text{not red})$ $\dfrac{3}{7}$
8. $P(\text{red or green})$ $\dfrac{6}{7}$

A bag contains 2 white, 4 yellow, and 10 red markers. Two markers are drawn at random without replacement. What are the odds of each event occurring?

9. drawing 2 yellow markers $\dfrac{1}{19}$
10. not drawing red markers $\dfrac{1}{7}$
11. drawing 1 white and 1 yellow $\dfrac{1}{14}$
12. drawing two different colors $\dfrac{17}{13}$

Extra Practice **A51**

Lesson 13-4 *(Pages 860–867)*

Determine if each event is *independent* or *dependent*. Then determine the probability.

1. the probability of selecting a red marble, not replacing it, then a green marble from a box of 6 red marbles and 2 green marbles **dependent,** $\frac{3}{14}$

2. the probability of randomly selecting two dimes from a bag containing 10 dimes and 8 pennies, if the first selection is replaced **independent,** $\frac{25}{81}$

3. There are two traffic lights along the route that Laura drives from home to work. One traffic light is red 50% of the time. The next traffic light is red 60% of the time. The lights operate on separate timers. Find the probability that these lights will both be red on Laura's way from home to work. **independent;** $\frac{3}{10}$

Determine if each event is *mutually exclusive* or *mutually inclusive*. Then determine each probability.

4. the probability of tossing two number cubes and either one shows a 5 **inclusive,** $\frac{11}{36}$

5. the probability of selecting a card from a standard deck of cards and the card is a 10 or an ace **exclusive,** $\frac{2}{13}$

Lesson 13-5 *(Pages 868–874)*

A jar contains 4 blue paper clips and 8 red paper clips. One paper clip is randomly drawn and discarded. Then a second paper clip is drawn. Find each probability.

1. the second paper clip is blue, given that the first paper clip was red $\frac{4}{11}$

2. the second paper clip is blue, given that the first paper clip was blue $\frac{3}{11}$

3. A pair of number cubes is thrown. Find the probability that the numbers of the dice match given that their sum is greater than 7. $\frac{1}{5}$

4. A pair of number cubes is thrown. Find the probability that their sum is greater than 7 given that the numbers match. $\frac{1}{2}$

5. One box contains 3 red balls and 4 white balls. A second box contains 5 red balls and 3 white balls. A box is selected at random and one ball is withdrawn. If the ball is white, what is the probability that it was taken from the second box? $\frac{21}{53}$

Lesson 13-6 *(Pages 875–882)*

Find each probability if a coin is tossed three times.

1. P(all heads) $\frac{1}{8}$ 2. P(exactly 2 tails) $\frac{3}{8}$ 3. P(at least 2 heads) $\frac{1}{2}$

Jojo MacMahon plays for the Worthington Wolves softball team. She is now batting 0.200 (meaning 200 hits in 1000 times at bat). Find the probability for the next five times at bat.

4. P(exactly 1 hit) $\frac{256}{625}$ or 0.4096   5. P(exactly 3 hits)$\frac{32}{625}$ or 0.0512   6. P(at least 4 hits)$\frac{21}{3125}$ or 0.00672

Lesson 14-1 *(Pages 889–896)*

The daily grams of fat consumed by 30 adults who participated in a random survey are listed below.

$$45, 22, 36, 30, 59, 29, 28, 45, 55, 38, 36, 40, 35, 62, 69,$$
$$28, 45, 38, 39, 45, 40, 42, 62, 51, 42, 60, 29, 26, 60, 70$$

1. What is the range of the data? **48**
2. Determine an appropriate class interval. **Sample answer: 10**
3. Name the class limits. **Sample answer: 20, 30, 40, 50, 60, 70, 80**
4. What are the class marks? **Sample answer: 25, 35, 45, 55, 65, 75**
5. Construct a frequency distribution of the data. **See margin.**
6. Draw a histogram of the data. **See margin.**
7. Name the interval or intervals that describe the grams of fat consumed by most adults who participated in the survey. **Sample answer: 40–50**

Lesson 14-2 *(Pages 897–907)*

Find the mean, median, and mode of each set of data.

1. {130, 190, 180, 150} **162.5; 165; none**
2. {18, 19, 18, 16, 17, 15} **about 17.2; 17.5; 18**
3. {25, 38, 36, 42, 30, 28} **about 33.2; 33; none**
4. {2, 5, 9, 10, 3, 4, 6, 9, 5, 1} **5.4; 5; 5 and 9**
5. {2.5, 5.6, 6, 7, 2.3, 6.4, 6.5, 7, 8, 10, 4, 5.6} **about 5.9; 6.2; 5.6 and 7**

Find the mean, median, and mode of the data represented by each stem-and-leaf plot.

6.
stem	leaf
1	4 5 5 6
2	0 1 4 7 8
3	6 9 9

$1|4 = 14$

24.5; 22.5; 15 and 39

7.
stem	leaf
3	0 4 6
5	2 4 4 6 6 7
6	2 3 6 8
7	0 1 6 7
8	2

$3|0 = 3.0$

about 5.9; 5.95; 5.4 and 5.6

8.
stem	leaf
8	0 2 3 9
9	6 7 8
10	4 5 5 8
11	1 7 8

$8|0 = 800$

995; 1010; 1050

9. Make a stem-and-leaf plot of the following number of hours worked by employees at a restaurant. **See margin.**

$$36, 17, 24, 39, 44, 37, 28, 29, 40, 55,$$
$$35, 34, 42, 29, 26, 24, 12, 19, 34, 23$$

10. Find the value of x so that the mean of {4, 5, 6, 9, 10, x} is 8. **14**
11. Find the value of x so that the median of {4, 3, 19, 16, 4, 7, 12, x} is 7.5. **8**

Additional Answers
Lesson 14-1

5.
Grams of Fat	Frequency
20–30	6
30–40	7
40–50	8
50–60	3
60–70	5
70–80	1

6.

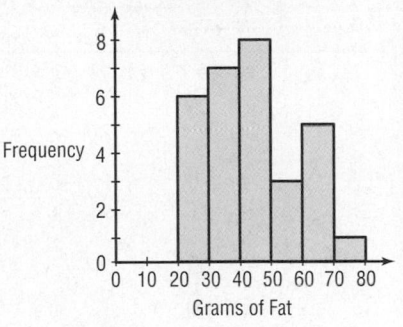

Grams of Fat Consumed by Adults

Additional Answer
Lesson 14-2

9.
stem	leaf
1	2 7 9
2	3 4 4 6 8 9 9
3	4 4 5 6 7 9
4	0 2 4
5	5

$1|2 = 12$

Additional Answers

Lesson 14-3

1.

2.

5f.

Lesson 14-3 *(Pages 908–917)*

Find the interquartile range and semi-interquartile range of each set of data. Then draw a box-and-whisker plot.   1–2. See margin for box-and-whisker plots.

1. {45, 39, 44, 39, 51, 38, 59, 35, 58, 79, 40}   **19; 9.5**
2. {2, 6, 4.5, 4, 3, 8, 3, 8, 10, 4, 2.5, 7.3, 4, 8, 1, 2.2}   **4.9; 2.45**

Find the mean deviation and the standard deviation of each set of data.

3. {150, 220, 180, 200, 175, 180, 250, 212, 195}   **about 21.98; about 27.56**
4. {3.5, 4.2, 3.7, 5.5, 2.9, 1.4, 2.4, 2, 3, 5.3, 4.6}   **about 1.05; about 1.26**

5. **Sports**   The numbers of hours per week members of the North High School basketball team spent practicing, either as a team or individually, are listed below.

    15, 18, 16, 20, 22, 18, 19, 20, 24, 18, 16, 18

   a. Find the median number of practice hours.   **18**
   b. Name the first quartile point and the third quartile point.   **17; 20**
   c. Find the interquartile range.   **3**
   d. What is the semi-interquartile range?   **1.5**
   e. Are there any outliers? If so, name them.   **no**
   f. Make a box-and-whisker plot of the data.   **See margin.**

Lesson 14-4 *(Pages 918–925)*

1. The mean of a set of normally distributed data is 10 and the standard deviation is 2.
   a. Find the interval about the mean that includes 25% of the data.   **9.4–10.6**
   b. What percent of the data is between 8 and 14?   **81.9%**
   c. What percent of the data is between 7 and 10?   **43.3%**
   d. Find the interval about the mean that includes 80% of the data.   **7.4–12.6**
2. Suppose 400 values in a set of data are normally distributed.
   a. How many values are within one standard deviation of the mean?   **273.2**
   b. How many values are within two standard deviations of the mean?   **382**
   c. How many values fall in the interval between the mean and one standard deviation above the mean?   **136.6**

Lesson 14-5 *(Pages 927–932)*

Find the standard error of the mean for each sample.

1. $\sigma = 1.2, N = 90$ **about 0.13**   2. $\sigma = 3.4, N = 100$ **0.34**   3. $\sigma = 12.4, N = 240$ **about 0.80**

For each sample, find the interval about the sample mean that has a 1% level of confidence.

4. $\sigma = 4.2, N = 40, \overline{X} = 150$
   about 148.29–151.71

5. $\sigma = 10, N = 78, \overline{X} = 320$
   about 317.08–322.92

Lesson 15-1 *(Pages 941–948)*
Evaluate each limit.

1. $\lim\limits_{x\to 4} (x^2 + 2x - 2)$ **22**

2. $\lim\limits_{x\to -1} (-x^4 + x^3 - 2x + 1)$ **1**

3. $\lim\limits_{x\to 0} (x + \sin x)$ **0**

4. $\lim\limits_{x\to -4} \dfrac{x^2 - 16}{x + 4}$ **−8**

5. $\lim\limits_{x\to -2} \dfrac{x^2 + 5x + 6}{x^2 + x - 2}$ $-\dfrac{1}{3}$

6. $\lim\limits_{x\to 2} \dfrac{3x + 9}{x^2 - 5x - 24}$ $-\dfrac{1}{2}$

Lesson 15-2 *(Pages 951–960)*
Use the definition of derivative to find the derivative of each function.

1. $f(x) = 5x$ $\quad f'(x) = 5$

2. $f(x) = 9x - 2$ $\quad f'(x) = 9$

Use the derivative rules to find the derivative of each function.

3. $f(x) = \dfrac{1}{2}x + \dfrac{2}{3}$ $\quad f'(x) = \dfrac{1}{2}$

4. $f(x) = x^2 + 4x + 8$ $\quad f'(x) = 2x + 4$

Find the antiderivative of each function.

5. $f(x) = x^5$ $\quad F(x) = \dfrac{1}{6}x^6 + C$

6. $f(x) = 2x^2 - 8x + 2$ $\quad F(x) = \dfrac{2}{3}x^3 - 4x^2 + 2x + C$

7. $f(x) = \dfrac{1}{5}x^3 - \dfrac{3}{4}x - 1$

$\quad F(x) = \dfrac{1}{20}x^4 - \dfrac{3}{8}x^2 - x + C$

8. $f(x) = \dfrac{x^3 - 2x^2 + x}{x}$

$\quad F(x) = \dfrac{1}{3}x^3 - x^2 + x + C$

Lesson 15-3 *(Pages 961–968)*
Use limits to evaluate each integral.

1. $\displaystyle\int_0^3 5x\,dx$ $\quad \dfrac{45}{2}$

2. $\displaystyle\int_1^5 (x + 1)\,dx$ **16**

3. $\displaystyle\int_0^2 (x^2 + 4x + 4)\,dx$ $\quad \dfrac{56}{3}$

4. **Business** A T-shirt company determines that the marginal cost function for one of their T-shirts is $f(x) = 6 - 0.002x$, where x is the number of T-shirts manufactured and $f(x)$ is in dollars. If the company is already producing 1500 of this type of T-shirt per day, how much more would it cost them to increase production to 2000 T-shirts per day? **$1250**

Lesson 15-4 *(Pages 970–976)*
Evaluate each indefinite integral.

1. $\displaystyle\int x^6\,dx$ $\quad \dfrac{1}{7}x^7 + C$

2. $\displaystyle\int 5x^4\,dx$ $\quad x^5 + C$

3. $\displaystyle\int (x^2 - x + 5)\,dx$ $\quad \dfrac{1}{3}x^3 - \dfrac{1}{2}x^2 + 5x + C$

4. $\displaystyle\int (-4x^4 + x^2 - 6)\,dx$ $\quad -\dfrac{4}{5}x^5 + \dfrac{1}{3}x^3 - 6x + C$

Evaluate each definite integral.

5. $\displaystyle\int_{-2}^2 14x^6\,dx$ **512**

6. $\displaystyle\int_0^6 (x + 2)\,dx$ **30**

7. $\displaystyle\int_2^4 (x^2 - 4)\,dx$ $\quad \dfrac{32}{3}$

8. $\displaystyle\int_4^5 (x - 4)(x + 2)\,dx$ $\quad \dfrac{10}{3}$

10.

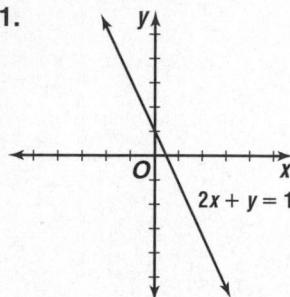

$y = 3x - 6$

11.

$2x + y = 1$

12.

$y + 4x \leq 12$

13.

$y > |x| + 2$

18.

$g(x) = [\![x + 1]\!]$

CHAPTER 1 TEST

State the domain and range of each relation. Then state whether the relation is a function. Write *yes* or *no*.

1. $\{(-1, 2), (0, 5), (3, 4), (2, 4)\}$
{−1, 0, 2, 3}; {2, 4, 5}; yes

2. $\{(6, 7), (7, 2), (4, -2), (6, -3), (-5, 0)\}$
{−5, 4, 6, 7}; {−3, −2, 0, 2, 7}; no

Find each value if $f(x) = x - 3x^2$.

3. $f(4)$ **−44**

4. $f(-7)$ **−154**

5. $f(a + 2)$ **$-3a^2 - 11a - 10$**

6. Physics The equation of the illuminance, E, of a small light source is $E = \dfrac{P}{4\pi d^2}$, where P is the luminous flux (in lumens, lm) of the source and d is its distance from the surface. Suppose the luminous flux of a desk lamp is 1200 lm. **a. 117.89 lm/m²**

a. What is the illuminance (in lm/m²) on a desktop if the lamp is 0.9 m above it?

b. Name the values of d, if any, that are not in the domain of the given function. **d cannot be negative or zero.**

7. If $f(x) = x^2 - 7$ and $g(x) = x + 3$, find $f(x) + g(x)$, $f(x) - g(x)$, $f(x) \cdot g(x)$, and $\left(\dfrac{f}{g}\right)(x)$.

$x^2 + x - 4$; $x^2 - x - 10$; $x^3 + 3x^2 - 7x - 21$; $\dfrac{x^2 - 7}{x + 3}$

Find $[f \circ g](x)$ and $[g \circ f](x)$ for each $f(x)$ and $g(x)$.

8. $f(x) = x + 1$ **$4x - 4$; $4x - 1$**
$g(x) = 4x - 5$

9. $f(x) = 5x$ **$10x^2 + 30$; $50x^2 + 6$**
$g(x) = 2x^2 + 6$

Graph each equation or inequality.   **10–13. See margin.**

10. $y = 3x - 6$

11. $2x + y = 1$

12. $y + 4x \leq 12$

13. $y > |x| + 2$

Write an equation in slope-intercept form for each line described.

14. slope $= \dfrac{5}{3}$, passes through $(-1, 3)$
$y = \dfrac{5}{3}x + \dfrac{14}{3}$

15. passes through $(0, 4)$ and $(8, -2)$
$y = -\dfrac{3}{4}x + 4$

Write the standard forms of the equations of the lines that are parallel and perpendicular to the graph of the given equation and pass through the point with the given coordinates.

16. $y = 4x - 1$; $(0, 2)$
$4x - y + 2 = 0$; $x + 4y - 8 = 0$

17. $x - 5y = 3$; $(-1, 2)$
$x - 5y + 11 = 0$; $5x + y + 3 = 0$

18. Graph $g(x) = [\![x + 1]\!]$ **See margin.**

19. Graph $f(x) = \begin{cases} 2 \text{ if } x \leq -3 \\ x + 1 \text{ if } -3 < x \leq 1 \\ \dfrac{1}{2}x \text{ if } x > 1 \end{cases}$ **See margin.**

20. Grades The table below shows the statistics grades and the economics grades for a group of college students at the end of a semester.

Statistics Grade	95	51	79	47	82	52	67	78	66
Economics Grade	88	70	70	67	71	80	68	85	90

a. Graph the data on a scatter plot.   **See margin.**

b. Use two ordered pairs to write the equation of a best-fit line.   **Sample answer: Using (47, 67) and (95, 88), $y = 0.44x + 46.44$**

www.amc.glencoe.com/chapter_test

19.

$f(x)$

20a.

$f(x)$

$f(x) = -x - 5$

AAP01-06C-834176

CHAPTER 2 TEST

1.

1. Solve the system of equations by graphing. Then state whether the system is *consistent and independent, consistent and dependent,* or *inconsistent.*
$2x - y = -1$ **(1, 3); consistent and independent; see margin for graph.**
$x + y = 4$

Solve each system of equations algebraically.   **(−1, 2, −4)**

2. $3x + y = 7$
$5x + 2y = 12$
(2, 1)

3. $4x - 5y = -2$
$3x + 2y = -13$
(−3, −2)

4. $4x + 6y - 3z = 20$
$x - 5y + z = -15$
$-7x + y + 2z = 1$

5. Find the values of x and y for which $\begin{bmatrix} y \\ 3x - 4 \end{bmatrix} = \begin{bmatrix} 12 - 2x \\ 2y \end{bmatrix}$ is true.   **$x = 4, y = 4$**

Use matrices A, B, and C to find each of the following. If the matrix does not exist, write *impossible.*

$$A = \begin{bmatrix} -5 & 1 \\ 4 & 2 \end{bmatrix} \qquad B = \begin{bmatrix} -2 & 3 \\ -3 & 0 \end{bmatrix} \qquad C = \begin{bmatrix} -1 & 0 & 3 \\ 3 & 2 & -4 \end{bmatrix}$$

6. $A + B$ $\begin{bmatrix} -7 & 4 \\ 1 & 2 \end{bmatrix}$ **7.** $-3C$ $\begin{bmatrix} 3 & 0 & -9 \\ -9 & -6 & 12 \end{bmatrix}$ **8.** $2B - A$ $\begin{bmatrix} 1 & 5 \\ -10 & -2 \end{bmatrix}$ **9.** BC $\begin{bmatrix} 11 & 6 & -18 \\ 3 & 0 & -9 \end{bmatrix}$

10.

Use matrices to perform each transformation. Then graph the pre-image and image on the same coordinate grid.   **10–11. See margin for graphs.**

10. Triangle ABC has vertices $A(1, -3)$, $B(-4, 3)$, and $C(6, 2)$. Use scalar multiplication to find the coordinates of the triangle after a dilation of scale factor 1.5. **$A'(1.5, -4.5)$, $B'(-6, 4.5)$, $C'(9, 3)$**

11.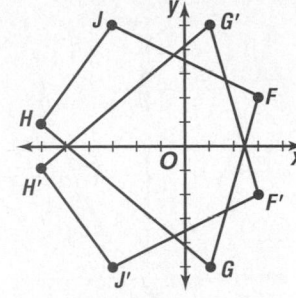

11. Quadrilateral $FGHJ$ has vertices $F(3, 2)$, $G(1, -5)$, $H(-6, 1)$, and $J(-3, 5)$. Find the coordinates of the quadrilateral after a reflection over the x-axis.   **$F'(3, -2)$, $G'(1, 5)$, $H'(-6, -1)$, $J'(-3, -5)$**

Find the value of each determinant.

12. $\begin{vmatrix} 5 & 7 \\ 3 & 6 \end{vmatrix}$ **9**

13. $\begin{vmatrix} 2 & 4 \\ -10 & 9 \end{vmatrix}$ **58**

14. $\begin{vmatrix} 1 & 2 & -1 \\ 3 & 0 & 1 \\ 4 & 1 & -2 \end{vmatrix}$ **16**

Find the inverse of each matrix, if it exists.

15. $\begin{bmatrix} 2 & 1 \\ 1 & 3 \end{bmatrix}$ **See margin.**

16. $\begin{bmatrix} 3 & -4 \\ 4 & -2 \end{bmatrix}$ $\begin{bmatrix} -\frac{1}{5} & \frac{2}{5} \\ -\frac{2}{5} & \frac{3}{10} \end{bmatrix}$

17. $\begin{bmatrix} -5 & 4 \\ -15 & 12 \end{bmatrix}$ **does not exist**

15. $\begin{bmatrix} \frac{3}{5} & -\frac{1}{5} \\ -\frac{1}{5} & \frac{2}{5} \end{bmatrix}$

18. Solve the system by using a matrix equation. $2x - y = 7$
(2, −3) $3x + y = 3$

19. Find the maximum and minimum values of $f(x, y) = 2x - y$ for the polygonal convex set determined by the system of inequalities.
$x \leq 2, y \geq 0, y \leq \frac{1}{2}x + 3, y \geq 3x - 4$ $\frac{8}{3}$, **−12**

20. **Advertising** Ruff, a dog food manufacturer, wants to advertise both in a magazine and on radio. The magazine charges $100 per ad and requires the purchase of at least three ads. The radio station charges $200 per commercial minute and requires the purchase of at least four minutes. Each magazine ad reaches 12,000 people while each commercial minute reaches 16,000 people. Ruff can spend at most $1300 on advertising. How many ads and commercial minutes should the manufacturing company purchase to reach the most people?
5 ads; 4 commercial minutes

www.amc.glencoe.com/chapter_test

Additional Answers

5. The graph of $g(x)$ is the graph of $f(x)$ translated left 3 units and reflected over the x-axis.

6. The graph of $g(x)$ is the graph of $f(x)$ stretched vertically by a factor of 4 and then translated down 2 units.

7.

$y > |x - 4|$

8.

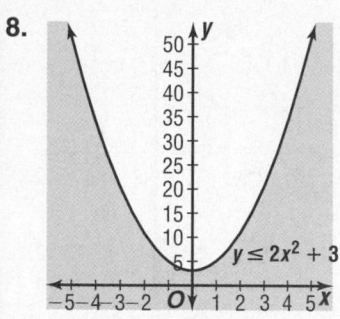

$y \le 2x^2 + 3$

9.

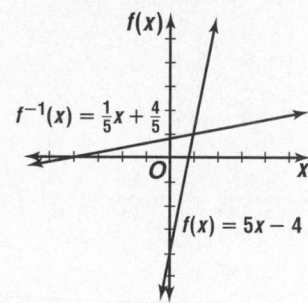

$f^{-1}(x) = \frac{1}{5}x + \frac{4}{5}$

$f(x) = 5x - 4$

10.

$f(x) = \frac{3}{x-2}$

$f^{-1}(x) = \frac{3}{x} + 2$

12. The function is defined when $x = 0$, the function approaches 1 as x approaches 0 from both sides, and $f(0) = 1$.

CHAPTER 3 TEST

Determine whether the graph of each equation is symmetric with respect to the x-axis, y-axis, the line $y = x$, the line $y = -x$, or none of these.

1. $y = 2x + 1$
  none of these

2. $y = -\frac{2}{x^2}$
  y-axis

3. $x = y^2 + 3$
  x-axis

4. $xy = 5$
  $y = x, y = -x$

Describe how the graphs of $f(x)$ and $g(x)$ are related.   **5–6. See margin.**

5. $f(x) = x^2$ and $g(x) = -(x + 3)^2$

6. $f(x) = |x|$ and $g(x) = 4|x| - 2$

Graph each inequality.   **7–8. See margin.**

7. $y > |x - 4|$

8. $y \le 2x^2 + 3$

Find $f^{-1}(x)$ and state whether $f^{-1}(x)$ is a function. Then graph the function and its inverse.

9. $f(x) = 5x - 4$ $f^{-1}(x) = \frac{1}{5}x + \frac{4}{5}$; yes
9–10. See margin for graphs.

10. $f(x) = \frac{3}{x-2}$ $f^{-1}(x) = \frac{3}{x} + 2$; yes

Determine whether each function is continuous at the given x-value. Justify your answer using the continuity test.

11. $\frac{x^2}{x-2}; x = 2$
  No; the function is undefined when $x = 2$.

12. $f(x) = \begin{cases} x + 1 \text{ if } x > 0 \\ 1 \text{ if } x \le 0 \end{cases}; x = 0$
  Yes; see margin for justification.

13. Describe the end behavior of $y = x^4 - 2x^2 - 1$. $y \to \infty$ as $x \to \infty$, $y \to \infty$ as $x \to -\infty$

Determine whether the given critical point is the location of a *maximum*, a *minimum*, or a *point of inflection*.

14. $y = x^2 - 8x + 4, x = 4$
  minimum

15. $y = x^3 - 3x^2 + x - 1, x = 1$
  point of inflection

Determine the equations of the vertical and horizontal asymptotes, if any, of each function.

16. $f(x) = \frac{4x}{x-1}$ $x = 1, y = 4$

17. $g(x) = \frac{x}{x^2 - 4}$ $x = \pm 2, y = 0$

18. **Chemistry**   A chemist pours 4 molar acid solution into 250 mL of a 1 molar acid solution. After x mL of the 4 molar solution have been added, the concentration of the mixture is given by $C(x) = \frac{0.25 + 0.004x}{0.25 + 0.001x}$.
  a. Find the horizontal asymptote of the graph of $C(x)$. $y = 4$
  b. What is the chemical interpretation of the horizontal asymptote?   **As the amount of 4 molar solution added increases, the molarity of the mixture approaches 4.**

Find the constant of variation for each relation and use it to write an equation for each statement. Then solve the equation.

19. If y varies directly as x and $y = 0.5$ when $x = 2$, find y when $x = 10$.   **0.25; $y = 0.25x$; 2.5**

20. If y varies inversely as the square of x and $y = 8$ when $x = 3$, find x when $y = 18$.
  72; $y = \frac{72}{x^2}$; ± 2

www.amc.glencoe.com/chapter_test

CHAPTER 4 TEST

1. Write a polynomial equation of least degree with roots 4, i, and $-i$. $x^3 - 4x^2 + x - 4 = 0$

Find the discriminant of each equation and describe the nature of the roots of the equation. Then solve the equation by using the Quadratic Formula.

2. $n^2 - 5n + 4 = 0$
   9; 2 real; 1, 4

3. $z^2 - 7z - 3 = 0$
   61; 2 real; $\dfrac{7 \pm \sqrt{61}}{2}$

4. $2a^2 - 5a + 4 = 0$
   -7; 2 imaginary; $\dfrac{5 \pm \sqrt{7}i}{4}$

Divide using synthetic division.

5. $(2x^3 - 3x^2 + 3x - 4) \div (x - 2)$
   $2x^2 + x + 5$, R6

6. $(x^4 - 5x^3 - 13x^2 + 53x + 60) \div (x + 1)$
   $x^3 - 6x^2 - 7x + 60$

Use the Remainder Theorem to find the remainder for each division. State whether the binomial is a factor of the polynomial.

7. $(x^3 + 8x^2 + 2x - 11) \div (x + 2)$ 9; no

8. $(4x^4 - 2x^2 + x - 3) \div (x - 1)$ 0; yes

Find the number of possible positive real zeros and the number of possible negative real zeros for each function. Then determine the rational zeros.

9. $f(x) = 6x^3 + 11x^2 - 3x - 2$
   1; 2 or 0; $-2, -\dfrac{1}{3}, \dfrac{1}{2}$

10. $f(x) = x^4 + x^3 - 9x^2 - 17x - 8$
    1; 3 or 1; -1

Approximate the real zeros of each function to the nearest tenth.

11. $f(x) = x^2 - 3x - 3$ $-0.8, 3.8$

12. $f(x) = x^3 - x + 1$ -1.3

Use the Upper Bound Theorem to find an integral upper bound and the Lower Bound Theorem to find an integral lower bound of the zeros of each function.

13. $f(x) = x^3 + 3x^2 - 5x - 9$
    Sample answer: 2; -5

14. $f(x) = 2x^4 + 3x^3 - x^2 + x + 1$
    Sample answer: 1; -2

Solve each equation or inequality. 16. $-8 \pm 4\sqrt{3}$ 17. $x < -17, -2 < x < 0$

15. $\dfrac{1}{80} + \dfrac{1}{a} = \dfrac{1}{10}$ $\dfrac{80}{7}$

16. $\dfrac{4}{x - 2} = \dfrac{3}{x^2 - 4} - \dfrac{1}{4}$

17. $\dfrac{5}{x + 2} > \dfrac{5}{x} + \dfrac{2}{3x}$

18. $\sqrt{y - 2} - 3 = 0$ 11

19. $\sqrt{2x + 2} = \sqrt{3x - 5}$ 7

20. $\sqrt{11 - 10m} > 9$ $m < -7$

Decompose each expression into partial fractions.

21. $\dfrac{5z - 11}{2z^2 + z - 6}$ $\dfrac{3}{z + 2} - \dfrac{1}{2z - 3}$

22. $\dfrac{7x^2 + 18x - 1}{(x^2 - 1)(x + 2)}$ $\dfrac{6}{x + 1} - \dfrac{3}{x + 2} + \dfrac{4}{x - 1}$

23. What type of polynomial function would be the best model for the set of data?   **quadratic**

x	−3	−2	−1	0	1	2	3
f(x)	26	16	6	4	3	8	14

24. **Manufacturing** The volume of a fudge tin must be 120 cubic centimeters. The tin is 7 centimeters longer than it is wide and six times longer than it is tall. Find the dimensions of the tin.   **12 cm by 5 cm by 2 cm**

25. **Travel** A car travels 500 km in the same time that a freight train travels 350 km. The speed of the car is 30 km/h more than the speed of the train. Find the speed of the freight train.   **70 km/h**

 www.amc.glencoe.com/chapter_test

CHAPTER TESTS

CHAPTER 5 TEST

If each angle is in standard position, determine a coterminal angle that is between 0° and 360°. State the quadrant in which the terminal side lies.

1. 995°   **275°; IV**
2. −234°   **126°; II**
3. 410°   **50°; I**
4. −1245°   **195°; III**

5. Find the values of the six trigonometric ratios for ∠R.

   $\sin R = \frac{4}{5}$, $\cos R = \frac{3}{5}$, $\tan R = \frac{4}{3}$,

   $\csc R = \frac{5}{4}$, $\sec R = \frac{5}{3}$, $\cot R = \frac{3}{4}$

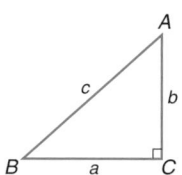

Use the unit circle to find each value.

6. tan 60° $\sqrt{3}$
7. sec 270°   **undefined**
8. sin (−405°) $-\frac{\sqrt{2}}{2}$

Find the values of the six trigonometric functions for angle θ in standard position if a point with the given coordinates lies on its terminal side.   9–11. See margin.

9. (3, 5)
10. (−4, 2)
11. (0, −3)

Solve each problem. Round to the nearest tenth.

12. If $b = 42$ and $A = 77°$, find c.   **186.7**
13. If $a = 13$ and $B = 27°$, find b.   **6.6**
14. If $c = 14$ and $A = 32° \, 17'$, find a.   **7.5**
15. If $a = 23$ and $c = 37$, find B.   **51.6°**
16. If $a = 3$ and $b = 11$, find A.   **15.3°**

17. **Recreation**   A kite is fastened to the ground by a string that is 65 meters long. If the angle of elevation of the kite is 70°, how far is the kite above the ground?   **61.1 m**

Find the area of each triangle. Round to the nearest tenth.

18. $A = 36°$, $a = 24$, $C = 87°$ **410.4 units²**
19. $b = 56.4$, $c = 92.5$, $A = 58.4°$ **2221.7 units²**

20. **Geometry**   An isosceles triangle has a base of 22 centimeters and a vertex angle measuring 36°. Find its perimeter.   **93.2 cm**

Find all solutions for each triangle. If no solutions exist, write *none*. Round to the nearest tenth.

21. $a = 64$, $c = 90$, $C = 98°$
    $A = 44.8°$, $B = 37.2°$, $b = 55.0$

22. $a = 9$, $b = 20$, $A = 31°$ **none**

Solve each triangle. Round to the nearest tenth.

23. $a = 13$, $b = 7$, $c = 15$
    $A = 60°$, $B = 27.8°$, $C = 92.2°$

24. $a = 20$, $c = 24$, $B = 47°$
    $b = 17.9$, $C = 78.3°$, $A = 54.7°$

25. **Navigation**   A ship at sea is 70 miles from one radio transmitter and 130 miles from another. The measurement of the angle between signals is 130°. How far apart are the transmitters?   **183.0 miles**

 www.amc.glencoe.com/chapter_test

CHAPTER 6 TEST

Change each degree measure to radian measure in terms of π.

1. $225°$ $\dfrac{5\pi}{4}$

2. $480°$ $\dfrac{8\pi}{3}$

Evaluate each expression.

3. $\sin\dfrac{5\pi}{6}$ $\dfrac{1}{2}$

4. $\tan\dfrac{5\pi}{4}$ 1

5. Determine the linear velocity of a point rotating at an angular velocity of 7.1 radians per second at a distance of 12 centimeters from the center of the rotating object.   **85.2 cm/s**

Graph each function for the given interval.   **6–7. See margin.**

6. $y = \sin x,\ -\pi \le x \le \pi$

7. $y = \cos x,\ -\pi \le x \le \pi$

State the amplitude and period for each function.

8. $y = 3\cos 4\theta$ $3;\ \dfrac{\pi}{2}$

9. $y = -2\sin\dfrac{3\pi}{2}\theta$ $2;\ \dfrac{4}{3}$

10. Write an equation of the sine function with amplitude 4, period 4π, phase shift 0, and vertical shift -2. $y = \pm 4\sin 0.5\theta - 2$

11. Write an equation of the cosine function with amplitude 0.5, period $\dfrac{\pi}{2}$, phase shift $\dfrac{\pi}{4}$, and vertical shift 1. $y = \pm 0.5\cos(4\theta - \pi) + 1$

Graph each function.   **12–13. See margin.**

12. $y = 3\cos\dfrac{\theta}{2}$

13. $y = \tan\left(2\theta - \dfrac{\pi}{4}\right)$

Write the equation for the inverse of each function. Then graph the function and its inverse.

14. $y = \text{Arccsc } x$ $y = \text{Csc } x$

15. $y = \tan x$ $y = \arctan x$

14–15. See margin for graphs.

Find each value.

16. $\sin\left(\text{Arccos }\dfrac{1}{2}\right)$ $\dfrac{\sqrt{3}}{2}$

17. $\tan\left(\pi + \text{Sin}^{-1}\dfrac{1}{2}\right)$ $\dfrac{\sqrt{3}}{3}$

18. **Astronomy** The linear velocity of Earth's moon is about 2300 mph. If the average distance from the center of Earth to the center of the Moon is 240,000 miles, how long does it take the Moon to make one revolution about Earth? Assume the orbit is circular.   **about 656 hours or 27.3 days**

19. **Meteorology** The average monthly temperatures for the city of Chicago, Illinois, are given below.

Jan.	Feb.	March	April	May	June	July	Aug.	Sept.	Oct.	Nov.	Dec.
21°	25°	37°	49°	59°	69°	73°	72°	64°	53°	40°	27°

Write a sinusoidal function that models the monthly temperatures, using $t = 1$ to represent January.   **Sample answer:** $y = 26\sin\left(\dfrac{\pi}{6}t - \dfrac{2\pi}{3}\right) + 47$

20. **Civil Engineering** Jake is designing a curve for a new highway. He uses the equation $\tan\theta = \dfrac{v^2}{rg}$, where the radius r will be 1200 feet. The curve will be designed for a maximum velocity v of 65 mph, and the acceleration g due to gravity is 32 ft/s². At what angle θ should the curve be banked? (Express your answer in radians.)   **about 0.23 radians**

 www.amc.glencoe.com/chapter_test

6.

7.

12.

13.

14.

15.

Additional Answers

6. $\tan \theta (\cot \theta + \tan \theta) \overset{?}{=} \sec^2 \theta$

$\tan \theta \cot \theta + \tan^2 \theta \overset{?}{=} \sec^2 \theta$

$1 + \tan^2 \theta \overset{?}{=} \sec^2 \theta$

$\sec^2 \theta = \sec^2 \theta$

7. $\sin^2 A \cot^2 A$

$\overset{?}{=} (1 - \sin A)(1 + \sin A)$

$\sin^2 A \cdot \dfrac{\cos^2 A}{\sin^2 A}$

$\overset{?}{=} (1 - \sin A)(1 + \sin A)$

$\cos^2 A$

$\overset{?}{=} (1 - \sin A)(1 + \sin A)$

$1 - \sin^2 A$

$\overset{?}{=} (1 - \sin A)(1 + \sin A)$

$(1 - \sin A)(1 + \sin A)$

$= (1 - \sin A)(1 + \sin A)$

8. $\dfrac{\sec x}{\sin x} - \dfrac{\sin x}{\cos x} \overset{?}{=} \cot x$

$\dfrac{\sec x \cos x - \sin^2 x}{\sin x \cos x} \overset{?}{=} \cot x$

$\dfrac{1 - \sin^2 x}{\sin x \cos x} \overset{?}{=} \cot x$

$\dfrac{\cos x}{\sin x} \overset{?}{=} \cot x$

$\cot x = \cot x$

9. $\dfrac{\cos x}{1 + \sin x} + \dfrac{\cos x}{1 - \sin x} \overset{?}{=} 2 \sec x$

$\dfrac{\cos x - \cos x \sin x + \cos x + \sin x \cos x}{1 - \sin^2 x}$

$\overset{?}{=} 2 \sec x$

$\dfrac{2 \cos x}{1 - \sin^2 x} \overset{?}{=} 2 \sec x$

$\dfrac{2 \cos x}{\cos^2 x} \overset{?}{=} 2 \sec x$

$\dfrac{2}{\cos x} \overset{?}{=} 2 \sec x$

$2 \sec x = 2 \sec x$

10. $\csc (A - B)$

$\overset{?}{=} \dfrac{\sec B}{\sin A - \cos A \tan B}$

$\csc (A - B)$

$\overset{?}{=} \dfrac{\dfrac{1}{\cos B}}{\sin A - \cos A \dfrac{\sin B}{\cos B}}$

$\csc (A - B)$

$\overset{?}{=} \dfrac{1}{\sin A \cos B - \cos A \sin B}$

$\csc (A - B)$

$\overset{?}{=} \dfrac{1}{\sin (A - B)}$

$\csc (A - B)$

$= \csc (A - B)$

11. $\cot 2\theta \overset{?}{=} \dfrac{1}{2} \cot \theta - \dfrac{1}{2} \tan \theta$

$\cot 2\theta \overset{?}{=} \dfrac{\cos \theta}{2 \sin \theta} - \dfrac{\sin \theta}{2 \cos \theta}$

$\cot 2\theta \overset{?}{=} \dfrac{\cos^2 \theta - \sin^2 \theta}{2 \sin \theta \cos \theta}$

$\cot 2\theta \overset{?}{=} \dfrac{\cos 2\theta}{\sin 2\theta}$

$\cot 2\theta = \cot 2\theta$

CHAPTER 7 TEST

Use the given information to determine the exact trigonometric value.

1. $\sin \theta = \dfrac{1}{3}$, $0° < \theta < 90°$; $\cos \theta$ $\dfrac{2\sqrt{2}}{3}$

2. $\sec \theta = -2$, $\dfrac{\pi}{2} < \theta < \pi$; $\tan \theta$ $-\sqrt{3}$

3. $\sin \theta = -\dfrac{4}{5}$, $\pi < \theta < \dfrac{3\pi}{2}$; $\sec \theta$ $-\dfrac{5}{3}$

4. $\csc \theta = -\dfrac{5}{3}$, $270° < \theta < 360°$; $\cos \theta$ $\dfrac{4}{5}$

5. Express $\tan (-420°)$ as a trigonometric function of an angle in Quadrant I. $-\tan 60°$

Verify that each equation is an identity.   6–11. See margin.

6. $\tan \theta (\cot \theta + \tan \theta) = \sec^2 \theta$

7. $\sin^2 A \cot^2 A = (1 - \sin A)(1 + \sin A)$

8. $\dfrac{\sec x}{\sin x} - \dfrac{\sin x}{\cos x} = \cot x$

9. $\dfrac{\cos x}{1 + \sin x} + \dfrac{\cos x}{1 - \sin x} = 2 \sec x$

10. $\csc (A - B) = \dfrac{\sec B}{\sin A - \cos A \tan B}$

11. $\cot 2\theta = \dfrac{1}{2} \cot \theta - \dfrac{1}{2} \tan \theta$

Use sum or difference identities to find the exact value of each trigonometric function.

12. $\sin 255°$ $-\dfrac{\sqrt{2} + \sqrt{6}}{4}$

13. $\tan \dfrac{5\pi}{12}$ $2 + \sqrt{3}$

14. If $\cos \theta = \dfrac{3}{4}$ and $270° < \theta < 360°$, find $\sin 2\theta$, $\cos 2\theta$, and $\tan 2\theta$. $-\dfrac{3\sqrt{7}}{8}$, $\dfrac{1}{8}$, $-3\sqrt{7}$

15. Use a half-angle identity to find the exact value of $\cos 22.5°$. $\dfrac{\sqrt{2 + \sqrt{2}}}{2}$

Solve each equation for principal values of x. Express solutions in degrees.

16. $\tan^2 x = \sqrt{3} \tan x$ $0°, 60°$

17. $\cos 2x - \cos x = 0$ $0°, 120°$

Solve each equation for $0° \le x < 360°$.

18. $\sin x - \cos x = 0$ $45°, 225°$

19. $2 \cos^2 x + 3 \sin x = 3$ $30°, 90°, 150°$

Write each equation in normal form. Then find the length of the normal and the angle it makes with the positive x-axis.

20. $y = x + 3$

$-\dfrac{\sqrt{2}}{2} x + \dfrac{\sqrt{2}}{2} y - \dfrac{3\sqrt{2}}{2} = 0$; $\dfrac{3\sqrt{2}}{2}$; $135°$

21. $5 + 5y = 10x$

$\dfrac{2\sqrt{5}}{5} x - \dfrac{\sqrt{5}}{5} y - \dfrac{\sqrt{5}}{5} = 0$; $\dfrac{\sqrt{5}}{5}$; $333°$

Find the distance between the point with the given coordinates and the line with the given equation.

22. $(-5, 8)$, $2x + y = 6$ $\dfrac{8\sqrt{5}}{5}$

23. $(-6, 8)$, $3x + 4y + 2 = 0$ $\dfrac{16}{5}$

24. Find an equation of the line that bisects the acute angles formed by the graphs of $5x + 2y = 7$ and $y = -\dfrac{3}{4} x + 1$. $\left(25 + 3\sqrt{29}\right)x + \left(10 + 4\sqrt{29}\right)y - 35 - 4\sqrt{29} = 0$

25. Physics The range of a projected object is the distance that it travels from the point where it is released. In the absence of air resistance, a projectile released at an angle of inclination θ with an initial velocity of v_0 has a range of $R = \dfrac{v_0{}^2}{g} \sin 2\theta$, where g is the acceleration due to gravity. Find the range of a projectile with an initial velocity of 88 feet per second if $\sin \theta = \dfrac{3}{5}$ and $\cos \theta = \dfrac{4}{5}$. The acceleration due to gravity is 32 feet per second squared.   **232.32 ft**

 www.amc.glencoe.com/chapter_test

CHAPTER 8 TEST

Use a ruler and a protractor to determine the magnitude (in centimeters) and direction of each vector.

1. **2.5 cm, 60°**

2. $\vec{b}$ **1.6 cm, 25°**

Use $\vec{a}$ and $\vec{b}$ above to find the magnitude and direction of each resultant.

3. $\vec{a} + \vec{b}$ **3.9 cm, 46°**

4. $2\vec{b} - \vec{a}$ **1.8 cm, 334°**

Write the order pair or ordered triple that represents $\overline{AB}$. Then find the magnitude of $\overline{AB}$.

5. $A(3, 6), B(-1, 9)$ **⟨−4, 3⟩; 5**

6. $A(-2, 7), B(3, 10)$ **⟨5, 3⟩; $\sqrt{34}$**

7. $A(2, -4, 5), B(9, -3, 7)$ **⟨7, 1, 2⟩; $3\sqrt{6}$**

8. $A(-4, -8, -2), B(-8, -10, 2)$ **⟨−4, −2, 4⟩; 6**

Let $\vec{r} = \langle -1, 3, 4\rangle$ and $\vec{s} = \langle 4, 3, -6\rangle$.

9. Find $\vec{r} - \vec{s}$.   **⟨−5, 0, 10⟩**

10. Find $3\vec{s} - 2\vec{r}$.   **⟨14, 3, −26⟩**

11. Find $\vec{r} + 3\vec{s}$.   **⟨11, 12, −14⟩**

12. Find $|\vec{r}|$.   **$\sqrt{26}$**

13. Find $|\vec{s}|$.   **$\sqrt{61}$**

     15. $4\vec{i} + 3\vec{j} - 6\vec{k}$

     14. $-\vec{i} + 3\vec{j} + 4\vec{k}$

14. Write $\vec{r}$ as the sum of unit vectors.

15. Write $\vec{s}$ as the sum of unit vectors.

16. Find $\vec{r} \cdot \vec{s}$.   **−19**

17. Find $\vec{r} \times \vec{s}$.   **⟨−30, 10, −15⟩**

18. Is $\vec{r}$ perpendicular to $\vec{s}$?   **no**

19. Find the magnitude and direction of the resultant of two forces of 125 pounds and 60 pounds at angles of 40° and 165° with the x-axis, respectively.   **103.06 lb, 68.48°**

20. Write parametric equations for the line that passes through the point $P(3, 11)$ and is parallel to $\vec{a} = \langle 2, -5\rangle$.   **$x = 2t + 3, y = -5t + 11$**

21. Write an equation in slope-intercept form of the line with the parametric equations.

$$x = 2t + 3 \qquad \mathbf{y = \tfrac{1}{2}x - \tfrac{1}{2}}$$
$$y = t + 1$$

22. Find the initial horizontal velocity and vertical velocity of a hockey puck shot at 100 mph at an angle of 2° with the ice.   **99.94 mph, 3.49 mph**

23. Describe the result of the product of a vertex matrix and the matrix $\begin{bmatrix} -4 & 0 & 0 \\ 0 & 4 & 0 \\ 0 & 0 & 4 \end{bmatrix}$.
The figure is four times the original size and reflected over the yz-plane.

24. **Physics** A downward force of 110 pounds is applied to the end of a 1.5-foot lever. What is the resulting torque about the axis of rotation if the angle of the lever to the horizontal is 60°?   **82.5 lb-ft**

25. **Gardening** Tei uses a sprinkler to water his garden. The sprinkler discharges water with a velocity of 28 feet per second. If the angle of the water with the ground is 35°, how far will the water travel in the horizontal direction? The acceleration due to gravity is 32 feet per second squared.   **23.02 feet**

1.

2.

3.

4.

5.

6.

CHAPTER 9 TEST

Graph each point.   1–3. See margin.

1. $A(2.5, 140°)$

2. $B\left(-2, \dfrac{5\pi}{4}\right)$

3. $C\left(3, -\dfrac{\pi}{6}\right)$

Graph each polar equation.   4–7. See margin.

4. $\theta = \dfrac{3\pi}{2}$

5. $r = -4$

6. $r = 6 \cos 3\theta$

7. $r = 2 + 2 \sin \theta$

Find the polar coordinates of each point with the given rectangular coordinates. Use $0 \le \theta < 2\pi$ and $r \ge 0$.

8. $(2, 2)$ $\left(2\sqrt{2}, \dfrac{\pi}{4}\right)$

9. $(-6, 0)$ **$(6, \pi)$**

10. $(-2, -3)$ **$(3.61, 4.12)$**

Find the rectangular coordinates of each point with the given polar coordinates.

11. $\left(3, -\dfrac{5\pi}{4}\right)$ $\left(-\dfrac{3\sqrt{2}}{2}, \dfrac{3\sqrt{2}}{2}\right)$

12. $\left(2, \dfrac{7\pi}{6}\right)$ **$(-\sqrt{3}, -1)$**

13. $(-4, 1.4)$ **$(-0.68, -3.94)$**

Write each rectangular equation in polar form.

14. $y = -3$ **$r = -3 \csc \theta$**

15. $x^2 + y^2 = 3x$ **$r = 3 \cos \theta$**

Write each polar equation in rectangular form.

16. $r = 7$ **$x^2 + y^2 = 49$**

17. $5 = r \cos(\theta - 45°)$ **$\sqrt{2}x + \sqrt{2}y - 10 = 0$**

Write each equation in polar form. Round ϕ to the nearest degree.

18. $5x + 3y = -3$ $\dfrac{3\sqrt{34}}{34} = r \cos(\theta - 211°)$

19. $2x - 4y = 1$ $\dfrac{\sqrt{5}}{10} = r \cos(\theta - 297°)$

Simplify.

20. i^{93} **i**

21. $(2 - 5i) + (-2 + 4i)$ **$-i$**

22. $-6i - (-3 + 2i)$ **$3 - 8i$**

23. $(3 + 5i)(3 - 2i)$ **$19 + 9i$**

24. $(1 - 3i)(2 - i)(1 + 2i)$ **$13 - 9i$**

25. $\dfrac{6 - 2i}{2 + i}$ **$2 - 2i$**

Express each complex number in polar form.

26. $-4 + 4i$ $4\sqrt{2}\left(\cos \dfrac{3\pi}{4} + i \sin \dfrac{3\pi}{4}\right)$

27. -5 **$5(\cos \pi + i \sin \pi)$**

Find each product or quotient. Express the result in rectangular form.

28. $4\left(\cos \dfrac{3\pi}{2} + i \sin \dfrac{3\pi}{2}\right) \cdot 3\left(\cos \dfrac{\pi}{4} + i \sin \dfrac{\pi}{4}\right)$ $6\sqrt{2} - 6\sqrt{2}i$

29. $2\sqrt{3}\left(\cos \dfrac{2\pi}{3} + i \sin \dfrac{2\pi}{3}\right) \div \sqrt{3}\left(\cos \dfrac{\pi}{6} + i \sin \dfrac{\pi}{6}\right)$ **$2i$**

30. Find $(1 - i)^8$. **16**

31. Find $\sqrt[3]{-27i}$. $\dfrac{3\sqrt{3}}{2} - \dfrac{3}{2}i$

32. $\dfrac{\sqrt{3}}{2} + \dfrac{1}{2}i, -\dfrac{\sqrt{3}}{2} + \dfrac{1}{2}i, -i$; see margin for graph.

32. Solve the equation $x^3 - i = 0$. Then graph the roots in the complex plane.

33. **Electricity** The current in a circuit is $8(\cos 307° + j \sin 307°)$ amps and the impedance is $20(\cos 115° + j \sin 115°)$ ohms. Find the polar form of the voltage in the circuit. (*Hint:* $E = I \cdot Z$)
$160(\cos 62° + j \sin 62°)$ volts

 www.amc.glencoe.com/chapter_test

7.

32.

CHAPTER 10 TEST

1. $\sqrt{17}; \left(1, \frac{3}{2}\right)$

Find the distance between each pair of points with the given coordinates. Then, find the midpoint of the segment that has endpoints at the given coordinates.

$\sqrt{k^2 + 4}; \left(\frac{5}{2}k, k\right)$

1. $(-1, 2), (3, 1)$

2. $(3k, k + 1), (2k, k - 1)$

3. Write the standard form of the equation of the circle that passes through $(-6, -4)$ and has its center at $(-8, 3)$. $(x + 8)^2 + (y - 3)^2 = 53$ 4. $(0, 0); (0, \pm 2); (0, \pm \sqrt{10}), (\pm\sqrt{6}, 0)$

4. Find the coordinates of the center, foci, and vertices of the ellipse with equation $\frac{x^2}{6} + \frac{y^2}{10} = 1$.

5. Write an equation of the ellipse centered at the origin that has a horizontal major axis, $e = \frac{1}{2}$, and $c = \frac{1}{2}$. $x^2 + \frac{4y^2}{3} = 1$ 6. $(-4, 2); \left(-4, 2 \pm \sqrt{23}\right); (-4, -6), (-4, -2); y - 2 = \pm\frac{4\sqrt{7}}{7}(x + 4)$

6. Find the coordinates of the center, foci, and the vertices and the equations of the asymptotes of the graph of the hyperbola with equation $\frac{(y - 2)^2}{16} - \frac{(x + 4)^2}{7} - 1$.   7. $\frac{(y - 1)^2}{4} - \frac{(x + 5)^2}{5} = 1$

7. Write an equation of the hyperbola that has eccentricity $\frac{3}{2}$ and foci at $(-5, -2)$ and $(-5, 4)$.

8. Find the coordinates of the vertex and focus, and the equations of the directrix and axis of symmetry of the parabola with equation $(y + 3)^2 = 8x$. $(0, -3); (2, -3); x = -2; y = -3$

9. Write an equation of the parabola that has a focus at $(3, -5)$ and a directrix with equation $y = -2$. $(x - 3)^2 = -6\left(y + \frac{7}{2}\right)$

Identify the conic section represented by each equation. Then write the equation in standard form and graph the equation.   10–17. See margin for equations and graphs.

10. $x^2 + 6x - 8y - 7 = 0$ **parabola**

11. $4x^2 - y^2 = 1$ **hyperbola**

12. $x^2 + y^2 + 4x - 12y + 36 = 0$ **circle**

13. $3x^2 - 16y^2 - 18x + 128y - 37 = 0$ **hyperbola**

14. $9x^2 - y^2 - 90x + 8y + 200 = 0$ **hyperbola**

15. $2x^2 - 13y^2 + 5 = 0$ **hyperbola**

16. $y^2 - 2x + 10y + 27 = 0$ **parabola**

17. $x^2 + 2y^2 + 2x - 12y + 11 = 0$ **ellipse**

Find the rectangular equation of the curve whose parametric equations are given.

18. $x = t, y = 2t^2 + t, -\infty < t < \infty$ $y = 2x^2 + x$ 19. $x = 2 \cos t, y = 2 \sin t, 0 \le t \le 2\pi$ $x^2 + y^2 = 4$

Identify the graph of each equation. Write an equation of the translated or rotated graph in general form.   20. ellipse; $4x^2 + y^2 - 16x + 4y - 16 = 0$

20. $4(x + 1)^2 + (y - 3)^2 = 36$ for $T_{(3, -5)}$ 21. $2x^2 - y^2 = 8, \theta = 60°$ **hyperbola; $(x')^2 - 6\sqrt{3}\,x'y' - 5(y')^2 + 32 = 0$**

22. Solve the system of equations algebraically. Round to the nearest tenth.
$x^2 + 4y^2 = 4$
$(x - 1)^2 + y^2 = 1$ **(2, 0), (0.7, ±0.9)**

23. Graph the system of inequalities.   **See margin.**
$x^2 + y^2 - 2x - 4y + 1 \ge 0$
$x^2 - 4y - 2x + 5 \ge 0$

24. **Geometry**   Write the standard form of the equation of the circle that passes through $(0, 1)$, $(-2, 3)$, and $(4, 5)$. Then identify the center and radius. $(x - 1)^2 + (y - 4)^2 = 10; (1, 4); \sqrt{10}$

25. **Technology**   Mirna bought a motion detector light and installed it in the center of her backyard. It can detect motion within a circle defined by the equation $x^2 + y^2 = 90$. If a person walks northeast through the backyard along a line defined by the equation $y = 2x - 3$, at what point will the person set off the motion detector?   **(-3, -9)**

 www.amc.glencoe.com/chapter_test

Additional Answers

10. $(x + 3)^2 = 8(y + 2)$

11. $\frac{x^2}{\frac{1}{4}} - \frac{y^2}{1} = 1$

Additional Answers

12. $(x + 2)^2 + (y - 6)^2 = 4$
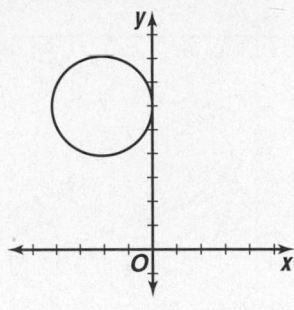

13. $\frac{(y - 4)^2}{12} - \frac{(x - 3)^2}{64} = 1$

14. $\frac{(x - 5)^2}{1} - \frac{(y - 4)^2}{9} = 1$
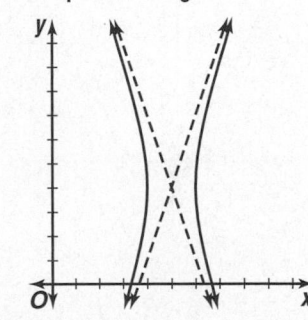

15. $\frac{y^2}{\frac{5}{13}} - \frac{x^2}{\frac{5}{2}} = 1$
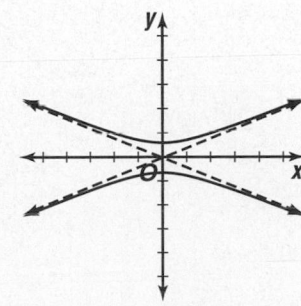

(continued on page A66)

CHAPTER TESTS

16. $(y + 5)^2 = 2(x - 1)$

17. $\dfrac{(x + 1)^2}{8} + \dfrac{(y - 3)^2}{4} = 1$

23.

CHAPTER 11 TEST

Evaluate each expression.

1. $343^{\frac{2}{3}}$ **49**

2. $64^{-\frac{1}{3}}$ $\dfrac{1}{4}$

Simplify each expression.

3. $\left((2a)^3\right)^{-2}$ $\dfrac{1}{64a^6}$

4. $\left(x^{\frac{3}{2}}y^2z^{\frac{5}{4}}\right)^4$ $x^6y^8z^5$

5. Express $\sqrt[3]{27a^6b^{12}}$ using rational exponents. $3a^2b^4$

6. Express $m^{\frac{1}{2}}n^{\frac{2}{3}}$ using a radical. $\sqrt[6]{m^3n^4}$

Graph each exponential function.   **7–8. See margin.**

7. $y = \left(\dfrac{1}{3}\right)^{x - 2}$

8. $y = 5^{x + 1}$

Write each equation in exponential form.

9. $\log_4 2 = \dfrac{1}{2}$ $4^{\frac{1}{2}} = 2$

10. $\log_{\frac{1}{6}} 216 = -3$ $\left(\dfrac{1}{6}\right)^{-3} = 216$

Write each equation in logarithmic form.

11. $5^4 = 625$ $\log_5 625 = 4$

12. $8^5 = m$ $\log_8 m = 5$

Solve each equation.

13. $\log_x 32 = -5$ $\dfrac{1}{2}$

14. $\log_5 (2x) = \log_5 (3x - 4)$ **4**

15. $3.6^x = 72.4$ **3.3430**

16. $6^{x - 1} = 8^{2 - x}$ **1.5372**

Find the value of each logarithm using the change of base formula.

17. $\log_4 15$ **1.9534**

18. $\log_3 0.9375$ **−0.0587**

Evaluate each expression.

19. $\log_{81} 3$ $\dfrac{1}{4}$

20. $\log 542$ **2.7340**

21. $\ln 0.248$ **−1.3943**

22. $\operatorname{antiln} (-1.9101)$ **0.1481**

23. Find the amount of time required for an amount to double at a rate of 5.4% if the interest is compounded continuously.   **12.84 yr**

24. **Biology**   A certain bacteria will triple in 6 hours. If the final count is 8 times the original count, how much time has passed?   **11 hours, 21 minutes**

25. **Electricity**   The charge on a discharging capacitor is given by $q(t) = Qe^{-\frac{t}{RC}}$, where Q is the initial charge on the capacitor with capacitance C, R is the resistance in the circuit, and t is time, in seconds. If $Q = 5 \times 10^{-6}$ coulomb, $R = 3$ ohms, and $C = 2 \times 10^{-6}$ farad, find the time required for the charge on the capacitor to decrease to 1×10^{-6} coulomb.   **9.66 × 10⁻⁶ s**

www.amc.glencoe.com/chapter_test

Additional Answers

7.

8.

CHAPTER 12 TEST

For Exercises 1–4, assume that each sequence or series is arithmetic.

1. Find the next four terms of the sequence 2, 4.5, 7, … . **9.5, 12, 14.5, 17**

2. Find the 24th term of the sequence $-6, -1, 4, …$. **109**

3. Write a sequence that has three arithmetic means between -4 and 8. **$-4, -1, 2, 5, 8$**

4. Find n for the series in which $a_1 = 12$, $d = 5$, and $S_n = 345$. **10**

For Exercises 5–7, assume that each sequence or series is geometric. 5. See margin.

5. Determine the common ratio and find the next three terms of the sequence $\frac{1}{4}, \frac{1}{10}, \frac{1}{25}, …$.

6. Write a sequence that has three geometric means between 16 and 1. **16, ±8, 4, ±2, 1**

7. Find the sum of the first 10 terms of the series $\frac{5}{2} + 5 + 10 + \cdots$. **$\frac{5115}{2}$**

Find each limit, or state that the limit does not exist and explain your reasoning.

8. $\lim\limits_{n\to\infty} \dfrac{n^3 + 3}{3n^2 + 1}$ **does not exist; see margin for explanation.**

9. $\lim\limits_{n\to\infty} \dfrac{n^3 + 4}{2n^3 + 3n}$ **$\frac{1}{2}$**

Determine whether each series is *convergent* or *divergent*.

10. $\dfrac{1}{3 \cdot 1^2} + \dfrac{1}{3 \cdot 2^2} + \dfrac{1}{3 \cdot 3^2} + \cdots$ **convergent**

11. $\dfrac{1}{6} + \dfrac{1}{3} + \dfrac{1}{2} + \dfrac{2}{3} + \cdots$ **divergent**

Express each series using sigma notation. 12-13. Sample answers are given.

12. $5 + 10 + 15 + \cdots + 95$ **$\sum\limits_{k=1}^{19} 5k$**

13. $6 + 9 + \dfrac{27}{2} + \dfrac{81}{4} + \cdots$ **$\sum\limits_{k=1}^{\infty} 6\left(\dfrac{3}{2}\right)^{k-1}$**

14. Use the Binomial Theorem to expand $(2a - 3b)^5$.
    $32a^5 - 240a^4b + 720a^3b^2 - 1080a^2b^3 + 810ab^4 - 243b^5$

Find the designated term of each binomial expansion.

15. sixth term of $(a + 2)^{10}$ **$8064a^5$**

16. fifth term of $(3x - y)^8$ **$5670x^4y^4$**

17. Write $-2 + 2i$ in exponential form. **$2\sqrt{2}e^{i\frac{3\pi}{4}}$**

18. Find the first three iterates of the function $f(z) = 3z + (2 - i)$ for the initial value $z_0 = 2i$.
    $2 + 5i, 8 + 14i, 26 + 41i$

19. Use mathematical induction to prove that the following proposition is valid for all positive integral values of n. **See margin.**
    $$2 \cdot 3 + 4 \cdot 5 + 6 \cdot 7 + \cdots + 2n(2n + 1) = \frac{n(n + 1)(4n + 5)}{3}$$

20. **Finance** Melissa deposits $200 into an investment every 3 months. The investment pays an APR of 8% and interest is compounded quarterly. Melissa makes each payment at the beginning of the quarter and the interest is posted at the end of the quarter. What will be the total value of the investment at the end of 10 years? **$12,322.00**

Chapter 12 Test **A67**

Additional Answer

19. **Step 1: Verify that the formula is valid for $n = 1$.** Since $2 \cdot 1(2 \cdot 1 + 1) = 6$ and $\dfrac{1(1 + 1)(4 \cdot 1 + 5)}{3} = 6$, the formula is valid for $n = 1$.

Step 2: Assume that the formula is valid for $n = k$ and prove that it is valid for $n = k + 1$.

$2 \cdot 3 + 4 \cdot 5 + 6 \cdot 7 + \cdots + 2k(2k + 1) = \dfrac{k(k + 1)(4k + 5)}{3}$

$2 \cdot 3 + 4 \cdot 5 + 6 \cdot 7 + \cdots + 2k(2k + 1) + 2(k + 1)(2k + 3)$

$= \dfrac{k(k + 1)(4k + 5)}{3} + 2(k + 1)(2k + 3)$

$= \dfrac{k(k + 1)(4k + 5)}{3} + \dfrac{6(k + 1)(2k + 3)}{3}$

$= \dfrac{k(k + 1)(4k + 5) + 6(k + 1)(2k + 3)}{3}$

$= \dfrac{(k + 1)[k(4k + 5) + 6(2k + 3)]}{3}$

$= \dfrac{(k + 1)(4k^2 + 17k + 18)}{3}$

$= \dfrac{(k + 1)(k + 2)(4k + 9)}{3}$

Evaluate the original formula for $n = k + 1$.

$\dfrac{(k + 1)[(k + 1) + 1][4(k + 1) + 5]}{3} = \dfrac{(k + 1)(k + 2)(4k + 9)}{3}$

The formula gives the same result as adding the $(k + 1)$ term directly. Thus, if the formula is valid for $n = k$, it is also valid for $n = k + 1$. Since the formula is valid for $n = 1$, it is also valid for $n = 2$. Since it is valid for $n = 2$, it is also valid for $n = 3$, and so on, indefinitely. Thus, the formula is valid for all positive integral values of n.

Chapter 12 Test **A67**

Additional Answers

5. $\dfrac{2}{5}; \dfrac{2}{125}, \dfrac{4}{625}, \dfrac{8}{3125}$

8. $\lim\limits_{n\to\infty} \dfrac{n^3 + 3}{3n^2 + 1} = \lim\limits_{n\to\infty} \dfrac{n + \dfrac{3}{n^2}}{3 + \dfrac{1}{n^2}}$,

$\lim\limits_{n\to\infty} 3 = 3$, and $\lim\limits_{n\to\infty} \dfrac{1}{n^2} = 0$, so the denominator approaches 3. As n approaches infinity, the n term in the numerator makes the whole numerator approach infinity, so the entire fraction has no limit.

Find each value.

1. $P(6, 2)$ **30**
2. $P(7, 5)$ **2520**
3. $C(8, 3)$ **56**
4. $C(5, 4)$ **5**

5. The letters r, s, t, u, and v are to be used to form five-letter patterns. How many patterns can be formed if repetitions are not allowed? **120**

6. **Employment** Five people have applied for three different positions in a store. If each person is qualified for each position, in how many ways can the positions be filled? **60**

7. How many ways can 7 people be seated at a round table relative to each other? **720**

8. **Sports** How many baseball teams can be formed from 15 players if 3 only pitch and the others play any of the remaining 8 positions? **1485**

A bag contains 4 red and 6 white marbles.

9. How many ways can 5 marbles be selected if exactly 2 must be red? **120**

10. If two marbles are chosen at random, find $P(2 \text{ white})$. $\frac{1}{3}$

11. **Sports** The probability that the Pirates will win a game against the Hornets is $\frac{1}{4}$. What are the odds that the Pirates will beat the Hornets? $\frac{1}{3}$

12. Five cards are dealt from a standard deck of cards. What is the probability that they are all from the same suit? $\frac{33}{16,660}$

13. Find the probability of getting a sum of 8 on the first throw of two number cubes and a sum of 4 on the second throw. $\frac{5}{432}$

14. **Communication** Callie is having a new phone line installed. What is the probability that the final 3 digits in the telephone number will all be odd? $\frac{1}{8}$

15. A bag contains 3 red, 4 white, and 5 blue marbles. If 3 marbles are selected at random, what is the probability that all are red or all are blue? $\frac{1}{20}$

16. A card is drawn from a standard deck of cards. What is the probability of selecting an ace or a black card? $\frac{7}{13}$

17. A four-digit number is formed from the digits 7, 3, 3, 2, and 2. If the number formed is odd, what is the probability that the 3s are together? $\frac{5}{18}$

18. Two different numbers are selected at random from the numbers 1 through 9. If their product is even, what is the probability that both numbers are even? $\frac{3}{13}$

19. Five bent coins are tossed. The probability of heads is $\frac{2}{3}$ for each of them. What is the probability that no more than 2 coins will show heads? $\frac{17}{81}$

20. **Archery** While shooting arrows, Akira can hit the center of the target 4 out of 5 times. What is the probability that he will hit it exactly 4 out of the next 7 times? $\frac{1792}{15,625}$ **or 0.114688**

 www.amc.glencoe.com/chapter_test

CHAPTER 14 TEST

The number of absences for a random sample of 80 high school students at Dover High School last school year are given below.

6	16	12	7	7	9	13	12	7	7
19	4	9	6	4	11	13	10	16	20
10	17	11	12	6	9	10	14	3	8
8	12	13	8	8	11	12	1	11	5
11	16	13	5	10	1	10	8	15	10
13	9	20	5	9	15	11	18	12	14
10	16	8	10	2	11	19	10	12	17
14	6	9	12	10	14	8	9	7	9

1. What is the range of the data? **19**
2. What is an appropriate class interval? **Sample answer: 3**
3. What are the class limits? **Sample answer: 0, 3, 6, 9, 12, 15, 18, 21**
4. What are the class marks? **Sample answer: 1.5, 4.5, 7.5, 10.5, 13.5, 16.5, 19.5**
5. Construct a frequency distribution of the data. **See margin.**
6. Draw a histogram of the data. **See margin.**
7. Find the mean of the data. **about 10.44**
8. Find the median of the data. **10**

A small metal object is weighed on a laboratory balance by each of 15 students in a physics class. The weight of the object in grams is reported as 2.341, 2.347, 2.338, 2.350, 2.344, 2.342, 2.345, 2.348, 2.340, 2.345, 2.343, 2.344, 2.347, 2.341, and 2.344.

9. Find the median of the data. **2.344**
10. Find the first quartile point and the third quartile point. **2.341, 2.347**
11. What is the semi-interquartile range? **0.003**
12. Make a box-and-whisker plot of the data. **See margin.**
13. What is the mean deviation of the data? **about 0.0025**
14. What is the standard deviation of the data? **about 0.0031**

The mean of a set of normally distributed data is 24 and the standard deviation is 2.8.

15. Find the interval about the mean that includes 68.3% of the data. **21.2–26.8**
16. Find the interval about the mean that includes 90% of the data. **19.38–28.62**
17. What percent of the data is between 18.4 and 32.4? **97.6%**
18. What percent of the data is between 24 and 29.6? **47.75%**

Suppose a random sample of data has $\sigma = 3.6$, $N = 400$, and $\overline{X} = 57$.

19. Find the standard error of the mean. **0.18**
20. Find the interval about the sample mean that has a 5% level of confidence. **56.65–57.35**

 www.amc.glencoe.com/chapter_test

Additional Answers

6.

12.

Additional Answer

5.

Number of Absences	Tallies	Frequency
0–3	III	3
3–6	HHH I	6
6–9	HHH HHH HHH I	16
9–12	HHH HHH HHH HHH HHH	25
12–15	HHH HHH HHH II	17
15–18	HHH III	8
18–21	HHH	5

CHAPTER 15 TEST

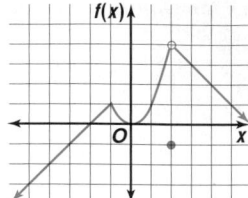

Use the graph of $y = f(x)$ to find each value.

1. $\lim\limits_{x \to -1} f(x)$ and $f(-1)$ **1; 1**

2. $\lim\limits_{x \to 2} f(x)$ and $f(2)$ **4; −1**

Evaluate each limit.

3. $\lim\limits_{x \to 0} \dfrac{x^2 + 3x}{x}$ **3**

4. $\lim\limits_{x \to 1} \dfrac{x^2 - 3x + 2}{x - 1}$ **−1**

5. $\lim\limits_{x \to 3} \dfrac{x^2 - 9}{x^3 - 27}$ **$\frac{2}{9}$**

6. $\lim\limits_{x \to 1} \dfrac{x^2 - 2x + 3}{3x^2 - 5}$ **−1**

7. Use the definition of derivative to find the derivative of $f(x) = x^2 - 2x$.   **$f'(x) = 2x - 2$**

Use the derivative rules to find the derivative of each function.

8. $f(x) = \frac{1}{2}x^2 - 7x + 1$ **$f'(x) = x - 7$**

9. $f(x) = 4x^3 - 4$ **$f'(x) = 12x^2$**

10. $f(x) = 6x^4 - 2x^2 - 30$ **$f'(x) = 24x^3 - 4x$**

11. $f(x) = 2x^5 - 4x^3 + \frac{2}{5}x^2 - 6$ **See margin.**

12. $f(x) = 2x^4(x^3 + 3x^2)$ **$f'(x) = 14x^6 + 36x^5$**

13. $f(x) = (x + 3)^2$ **$f'(x) = 2x + 6$**

Find the antiderivative of each function.

14. $f(x) = -2x + 6$ **$F(x) = -x^2 + 6x + C$**

15. $f(x) = -x^3 + 4x^2 - x + 4$ **See margin.**

16. $f(x) = \frac{1}{2}x^3 - \frac{2}{7}x + 5$ **See margin.**

17. $f(x) = \dfrac{x^3 - 4x^2 + x}{x}$ **$F(x) = \frac{1}{3}x^3 - 2x^2 + x + C$**

18. Use a limit to find the area between the graph of $y = x^3$ and the x-axis from $x = 0$ to $x = 2$.   **4 units^2**

19. Find the area of the shaded region in the graph at the right.   **26 units^2**

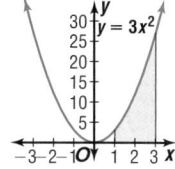

Evaluate each definite integral.

20. $\displaystyle\int_0^1 (2x + 3)\, dx$ **4**

21. $\displaystyle\int_1^3 (-x^2 + x + 6)\, dx$ **$\frac{22}{3}$**

Evaluate each indefinite integral.

22. $\displaystyle\int (1 - 2x)\, dx$ **$x - x^2 + C$**

23. $\displaystyle\int (3x^2 + 4x + 7)\, dx$ **$x^3 + 2x^2 + 7x + C$**

24. **Baseball**   A baseball is hit so that its height, in feet, above the ground after t seconds is given by $h(t) = 3 + 95t - 16t^2$. Find the vertical velocity of the ball after 2 seconds.   **31 ft/s**

25. **Geometry**   The volume of a cone of radius r and height h can be calculated using the formula $V = 2\pi \displaystyle\int_0^r \left(-\dfrac{h}{r}x^2 + hx\right) dx$. Use this formula to find the volume of a cone with radius 3 units and height 2 units.   **6π units^3**

 www.amc.glencoe.com/chapter_test

GLOSSARY

For commonly-used formulas, see inside back cover.

A

abscissa (5)   the first element of an ordered pair

absolute maximum (171)   a point that represents the maximum value a function assumes over its domain

absolute minimum (171)   a point that represents the minimum value a function assumes over its domain

absolute value function (47)   a piecewise function, written as $f(x) = |x|$, where $f(x) \geq 0$ for all values of x

absolute value of a complex number (586)   The absolute value of the complex number $a + bi$ is $|a + bi| = \sqrt{a^2 + b^2}$.

additive identity matrix (80)   If an $m \times n$ matrix I is a zero matrix, then I is an additive identity matrix, such that for any $m \times n$ matrix A, $A + I = A$.

alternate optimal solutions (114)   When there are two or more optimal solutions for a linear programming problem, the problem is said to have alternate optimal solutions.

ambiguous case (320)   Given two sides and a nonincluded angle, there exist two possible triangles with the given measures.

amplitude (368) $|A|$ for functions in the form $y = A \sin k\alpha$ or $y = A \cos k\alpha$

amplitude of a complex number (587)   the angle θ when a complex number is written in the form $r(\cos \theta + i \sin \theta)$

analytic geometry (618)   the study of coordinate geometry from an algebraic perspective

angle of depression (300)   the angle between a horizontal line and the line of sight from the observer to an object at a lower level

angle of elevation (300)   the angle between a horizontal line and the line of sight from the observer to an object at a higher level

angular displacement (352)   As any circular object rotates counterclockwise about its center, an object at the edge moves through an angle relative to its starting position known as the angular displacement, or angle of rotation.

angular velocity (352)   the change in the central angle with respect to time as an object moves along a circular path

antiderivative (955) $F(x)$ is an antiderivative of $f(x)$ if and only if $F'(x) = f(x)$.

antilogarithm (729)   If $\log x = a$, then x is called the antilogarithm of a, abbreviated antilog a.

apothem (300)   a segment from the center of a regular polygon perpendicular to a side of the polygon

arccosine (305)   the inverse of $y = \cos x$, written as $x = \arccos y$

arcsine (305)   the inverse of $y = \sin x$, written as $x = \arcsin y$

arctangent (305)   the inverse of $y = \tan x$ written as $x = \arctan y$

argument of a complex number (587)   The argument of the complex number $r(\cos \theta + i \sin \theta)$ is the angle θ.

arithmetic mean (758; 897)   1. the terms between any two nonconsecutive terms of an arithmetic sequence   2. a measure of central tendency found by dividing the sum of all values by the number of values

arithmetic sequence (757)   a sequence in which the difference between successive terms is a constant

arithmetic series (759)   the indicated sum of the terms of an arithmetic sequence

asymptote (180)   a line that a graph approaches but never intersects

axis of symmetry (653)   a line about which a figure is symmetric

B

back-to-back bar graph (889)   a graph plotted on a two-quadrant coordinate system with the horizontal scale repeated in each direction from the central axis used to show comparisons

bar graph (889)   a graphic form using bars to make comparisons of statistics

Basic Counting Principle (837)   If event M can occur in m ways and is followed by an event N that can occur in n ways, then the event M followed by the event N can occur in $m \cdot n$ ways.

best-fit line (38)   the graph of a prediction equation

bimodal data (899)   data with two modes

binomial experiment (878)   a problem that can be solved using binomial expansion

boundary (52)   a line or curve that separates the coordinate plane into two regions

box-and-whisker plot (909)   a diagram that graphically displays the median, quartiles, extreme values, and outliers in a set of data

C

cardioid (563)   the graph of a polar equation of the form $r = a \pm a \sin \theta$ or $r = a \pm a \cos \theta$

center of a circle   See circle.

center of a hyperbola (642)   the midpoint of the segment whose endpoints are the foci

central angle (345)   an angle whose vertex lies at the center of a circle

characteristic (727)   the part of the logarithm of a number which is the exponent of 10 used to write the number in scientific notation

circle (623)   the locus of all points in a plane at a given distance, called the radius, from a fixed point on the plane, called the center

circular arc (345)   a part of a circle that is intercepted by a central angle of the circle

circular functions (292)   functions defined using a unit circle

circular permutation (847)   a circular arrangement of objects in a certain order

classic curves (564)   special curves formed by graphing polar equations

class interval (890)   the range of each class in a frequency distribution

class limits (890)   the upper and lower values in each class in a frequency distribution

class marks (890)   the means of the class limits in a frequency distribution

coinciding lines (32)   The graphs of two equations that represent the same line are coinciding lines.

column matrix (78)   a matrix that has only one column

combination (841)   an arrangement of objects where the order is not a consideration

combinatorics (837)   the investigation of the different possibilities for the arrangement of objects

common difference (757)   the difference between the successive terms of an arithmetic sequence

common logarithms (726)   logarithms that use 10 as the base

common ratio (764)   the ratio of successive terms of a geometric sequence

comparison test (787)   a method to test convergence in an infinite series

complements (853)   Two events are complements if and only if the sum of their probabilities is 1.

completing the square (213)   a process used to create a perfect square trinomial

complex conjugates (216, 582)   The conjugate of the complex number $a + bi$ is $a - bi$.

complex number (206, 580)   any number that can be written in the form $a + bi$, where a and b are real numbers and i is the imaginary unit

complex plane (586)   The complex number $a + bi$ is graphed as the ordered pair (a, b) in the complex plane. The real axis is horizontal, and the imaginary axis is vertical.

components of a vector (488)   two or more vectors whose sum is the given vector

composite (15)   Given functions f and g, the composite function $f \circ g$ can be described by $[f \circ g](x) = f(g(x))$.

composition of functions   See composite.

compound function (382)   a function consisting of the sums or products of trigonometric functions

concentric circles (623)   circles with the same center

conditional probability (868)   the probability of an event under the condition that some preceding event has occurred

conic section (623)   a curve determined by the intersection of a plane with a double right cone

conjugate   See complex conjugates.

conjugate axis (642)   the segment perpendicular to the transverse axis of a hyperbola through its center

consistent (67)   A system of equations that has at least one solution is called consistent.

constant function (22, 137, 164)   a function of the form $f(x) = b$

constant of variation (189)   the constant k used with direct or inverse variation

constraints (112)   conditions given to variables, often expressed as linear inequalities

continuous (160)   A function is said to be continuous at point (x_1, y_1) if it is defined at that point and passes through that point without a break.

converge (784)   If a sequence has a limit, then the related infinite series is said to converge.

convergent series (784)   an infinite series that has a sum or limit

correlation coefficient (40)   a value that describes the nature of a set of data. The more closely the data fit a line, the closer the correlation coefficient, r, approaches 1 or -1.

cosecant (286, 292)   For any angle, with measure α, a point $P(x, y)$ on its terminal side, $r = \sqrt{x^2 + y^2}$, $\csc \alpha = \dfrac{r}{y}$.

cosine (285, 291)   For any angle, with measure α, a point $P(x, y)$ on its terminal side, $r = \sqrt{x^2 + y^2}$, $\cos \alpha = \dfrac{x}{r}$.

cotangent (286, 292)   For any angle, with measure α, a point $P(x, y)$ on its terminal side, $r = \sqrt{x^2 + y^2}$, $\cot \alpha = \dfrac{x}{y}$.

coterminal angles (279)   two angles in standard position that have the same terminal side

counterexample (421)   an example used to show that a given statement is not always true

critical points (171)   points at which the nature of a graph changes

cross product (507)   The cross product of $\vec{a}$ and $\vec{b}$ if $\vec{a} = (a_1, a_2, a_3)$ and $\vec{b} = (b_1, b_2, b_3)$ is defined as follows.

$$\vec{a} \times \vec{b} = \begin{vmatrix} a_2 & a_3 \\ b_2 & b_3 \end{vmatrix}\vec{i} - \begin{vmatrix} a_1 & a_3 \\ b_1 & b_3 \end{vmatrix}\vec{j} + \begin{vmatrix} a_1 & a_2 \\ b_1 & b_2 \end{vmatrix}\vec{k}$$

cumulative frequency distribution (902)   the sum of the frequency of a class and the frequencies of previous classes

D

decreasing (164)   A function f is decreasing on an interval I if and only if for every a and b contained in I, $f(a) > f(b)$ whenever $a < b$.

definite integral (962)   an integral that has lower and upper bounds

degenerate conic (623)   the intersection of a plane with a double right cone resulting in a point, a line, or two intersecting lines

degree (277)   the measure of an angle that is $\frac{1}{360}$ of a complete rotation in the positive direction

degree of a polynomial in one variable (206)   the greatest exponent of the variable of the polynomial

dependent events (837)   events that affect each other

dependent system (67)   A system of equations that has infinitely many solutions is called dependent.

depressed polynomial (224)   the quotient when a polynomial is divided by one of its factors

derivative of $f(x)$ (953)   the function $f'(x)$, which is defined as $f'(x) = \lim\limits_{h \to 0} \dfrac{f(x + h) - f(x)}{h}$

determinant (98)   a square array of numbers having a numerical value; the numerical value of the square array of numbers

differentiation (952)   the process of finding derivatives

dilation (88)   a transformation in which a figure is enlarged or reduced

dimensional analysis (353)   a procedure in which unit labels are treated as mathematical factors and can be divided out

dimensions of a matrix (78)   the number of rows, m, and the number of columns, n, of the matrix written as $m \times n$

direction of a vector (485)   the directed angle between the positive x-axis and the vector

direction vector (520)   a vector used to describe the slope of a line

directrix   See parabola.

direct variation (189) y varies directly as x^n if there is some nonzero constant k such that $y = kx^n$, $n > 0$. The variable k is called the constant of variation.

discontinuous (159)   A function is said to be discontinuous at point (x_1, y_1) if there is a break in the graph of the function at that point.

discriminant (215)   in the quadratic formula, the expression under the radical sign, $b^2 - 4ac$

diverge (784)   If a sequence does not have a limit, then the related infinite series is said to diverge.

divergent series (784)   an infinite series that does not have a sum or limit

domain (5)   the set of all abscissas of the ordered pairs of a relation

dot product   See inner product.

doubling time (740)   the amount of time it takes a quantity to reach twice its initial amount

E

eccentricity (636)   the ratio of the distance between any point of a conic section and a fixed point to the distance between the same point of the conic section to a fixed line

element of a matrix (78)   any value in the array of values

elimination method (68)   a technique used to solve a system of equations

ellipse (631)   the locus of all points in a plane such that the sum of the distances from two given points in the plane, called foci, is constant

end behavior (162)   the behavior of $f(x)$ as $|x|$ becomes very large

equilateral hyperbola (647)   a hyperbola with perpendicular asymptotes

equal matrices (79)   Two matrices are equal if and only if they have the same dimensions and their corresponding elements are identical.

equal vectors (485)   Two vectors are equal if and only if they have the same direction and the same magnitude.

escape set (599)   the set of initial values for which the iterates of a function approach infinity

escaping point (816)   If the iterates of a function approach infinity for some initial value, the initial point is called an escaping point.

even function (133)   a function hose graph is symmetric with respect to the y-axis

everywhere discontinuous (159)   A function that is impossible to graph in the real number system is said to be everywhere discontinuous.

experimental probability (877)   a probability determined by performing tests or experiments and observing the outcomes

exponential decay (706, 712)   Exponential decay occurs when a quantity decreases exponentially over time.

exponential function (704)   a function in the form $y = a^x$, where a is a positive real number

exponential growth (706, 712)   Exponential growth occurs when a quantity increases exponentially over time.

exponential series (806)   the series by which e^x may be approximated;

$$e^x = 1 + x + \frac{x^2}{2!} + \frac{x^3}{3!} + \frac{x^4}{4!} + \frac{x^5}{5!} \cdots$$

extraneous solution (251)   a solution that does not satisfy the original equation

extremum (171)   a maximum or minimum value of a function

failure (852)   any outcome other than the desired outcome of an event

family of graphs (26)   a group of graphs that displays one or more similar characteristics

Fibonacci sequence (804)   a sequence in which the first two terms are 1 and each of the additional terms is generated by adding the two previous terms

focus (pl. **foci**)   See parabola, ellipse, hyperbola.

fractal geometry (815)   the study of the properties of fractals and their application to natural phenomena

frequency (372)   the number of cycles per unit of time

frequency distribution (890)   a system of organizing data by determining classes and the frequency of values in each class

frequency polygon (892)   a broken line graph drawn by connecting the class marks on a histogram

function (6)   a relation in which each element of the domain is paired with exactly one element in the range

function notation (7)   In function notation, an equation of y in terms of x can be rewritten so that $y = f(x)$.

geometric mean (766)   the terms between any two nonconsecutive terms of a geometric sequence

geometric sequence (764)   a sequence in which the ratio between successive terms is a constant

geometric series (767)   the indicated sum of the terms of a geometric sequence

goodness of fit (40)   the degree to which data fits a regression line

greatest integer function (46)   a step function, written as $f(x) = [x]$, where $f(x)$ is the greatest integer not greater than x

half plane (52)   The graph of $y = mx + b$ separates the coordinate plane into two regions, called half planes.

hinges (909)   in a box-and-whisker plot, the ends of the rectangular box which are located at the first and third quartiles

histogram (890)   a type of bar graph in which the width of each bar represents a class interval and the height of the bar represents the frequency in that interval

horizontal asymptote (180)   The line $y = b$ is a horizontal asymptote for a function $f(x)$ if $f(x) \to b$ as $x \to \infty$ or as $x \to -\infty$.

horizontal line test (153)   a test used to determine if the inverse of a relation is a function

hyperbola (642)   the locus of all points in the plane such that the absolute value of the difference of the distances from two given points in the plane, called foci, is constant.

hypotenuse (284)   the side opposite the right angle in a right triangle

identity (421)   a statement of equality between two expressions that is true for all values of the variable for which the expression is defined

identity matrix (99)   The identity matrix of nth order, I_n, is the square matrix whose elements in the main diagonal, from upper left to lower right, are 1s, while all other elements are 0s.

image (88)   the graph of an object after a transformation

image point (127)   When applying point symmetry to a set of points, each point P in the set must have an image point P' which is also in the set.

imaginary number (206, 581)   a complex number of the form $a = bi$ where $b \neq 0$ and i is the imaginary unit

imaginary part of a complex number (581) b in the complex number $a + bi$

inclusive events (863)   two events whose outcomes may be the same

inconsistent system (67)   a system of equations that has no solutions

increasing (164)   A function f is increasing on an interval I if and only if for every a and b contained in I, $f(a) < f(b)$ whenever $a < b$.

indefinite integral (973)   The antiderivative of $f(x)$, denoted by $\int f(x)\,dx$ is called the indefinite integral of $f(x)$.

independent events (837)   events that do not affect each other

independent system (67)   a system of equations that has exactly one solution

index of summation (792)   the variable used with the summation symbol

infeasible (113)   When the constraints of a linear programming problem cannot be satisfied simultaneously, then the system is infeasible.

inferential statistics (927)   statistics based on information gathered in a sample to make predictions about a population

infinite discontinuity (159)   as the graph of $f(x)$ approaches a given value of x, $|f(x)|$ becomes increasingly greater

infinite sequence (772)   a sequence which has infinitely many terms

infinite series (776)   the indicated sum of the terms of an infinite sequence

initial side of an angle (277)   the fixed ray of an angle

inner product (505)   If $\vec{a}$ and $\vec{b}$ are two vectors, $\langle a_1, a_2, a_3 \rangle$ and $\langle b_1, b_2, b_3 \rangle$, then the inner product of $\vec{a}$ and $\vec{b}$ is defined as follows.
$$\vec{a} \cdot \vec{b} = a_1b_1 + a_2b_2 + a_3b_3$$
This is also called a dot product.

integral (961)   the area between the curve of a function, the x-axis, and the lines $x = a$ and $x = b$

integration (962)   the process of finding the area under a curve

interquartile range (909)   the difference between the first quartile point and the third quartile point

inverse function (152, 155)   Two functions are inverse functions if and only if $[f \circ g](x) = [g \circ f](x)$ for all values of x.

inversely proportional (191) y varies inversely as x^n if there is some nonzero constant k such that $x^n y = k$ or $y = \dfrac{k}{x^n}$, $n > 0$.

inverse matrix (99)   If $A = \begin{bmatrix} a_1 & b_1 \\ a_2 & b_2 \end{bmatrix}$ and
$\begin{vmatrix} a_1 & b_1 \\ a_2 & b_2 \end{vmatrix} \neq 0$, then $A^{-1} = \dfrac{1}{\begin{vmatrix} a_1 & b_1 \\ a_2 & b_2 \end{vmatrix}} \begin{bmatrix} b_2 & -b_1 \\ -a_2 & a_1 \end{bmatrix}$.

inverse of a trigonometric function (305)   the arccosine, arcsine, and arctangent relations

inverse process (154)   To find the inverse of a function, an inverse process is used to solve for y after switching variables.

inverse relations (152)   Two relations are inverse relations if and only if one relation contains the element (b, a), whenever the other relation contains the element (a, b).

iterate (16)   each output of an iterated function

iteration (16, 581)   the composition of a function and itself

joint variation (192) y varies jointly as x^n and z^n if there is some nonzero constant k such that $y = kx^n z^n$, where $x \neq 0$, $z \neq 0$, and $n > 0$.

Julia set (599)   the boundary between the escape set and the prisoner set for the iteration of complex numbers through the quadratic equation $f(z) = z^2 + c$ where c is a complex number

jump discontinuity (159)   The graph of $f(x)$ stops and then begins again with an open circle at a different range value for a given value of the domain.

leading coefficient (206)   the coefficient of the term with the highest degree

leaf   See stem-and-leaf plot.

legs (284)   the two sides of a right triangle forming the right angle

lemniscate (564)   The graph of a polar equation of the form $r^2 = a^2 \sin 2\theta$ or $r^2 = a^2 \cos 2\theta$.

level of confidence (929)   The probability of the true mean being within a certain range of a sample mean may be expressed as a level of confidence.

limaçon (562)   The graph of a polar equation of the form $r = a \pm b \sin \theta$ or $r = a \pm b \cos \theta$.

limit (772, 943)   If there is a number L such that the value of $f(x)$ gets closer and closer to L as x gets closer to a number a, then L is called the limit of $f(x)$ as x approaches a. In symbols, $L = \lim\limits_{x \to a} f(x)$.

linear equation (20)   an equation of the form $Ax + By = 0$, where A and B are not both 0

linear function (22)   a function defined by $f(x) = mx + b$, where m and b are real numbers

linear inequality (52)   a relation whose boundary is a straight line

linearizing data (743)   a method for modeling curve or exponential data that can be analyzed by linearizing the data and then applying a linear regression

linear programming (112)   a procedure for finding the maximum or minimum value of a function in two variables subject to given constraints on the variables

GLOSSARY

linear velocity (353)   distance traveled per unit of time

line plot (889)   a display of statistical data on a number line so that patterns and variability in data can be determined

line symmetry (129)   Two distinct points P and P' are symmetric with respect to a line ℓ if and only if ℓ is the perpendicular bisector of $\overline{PP'}$. A point P is symmetric to itself with respect to a line ℓ if and only if P is on ℓ.

locus (658)   a set of points and only those points that satisfy a given set of conditions

logarithm (718)   In the function $x = a^y$, y is called the logarithm, base a, of x.

logarithmic function (718) $y = \log_a x$, $a > 0$ and $a \neq 1$, which is the inverse of the exponential function $y = a^x$.

lower bound (238)   the integer less than or equal to the least real zero of the polynomial $P(x)$

M

$m \times n$ matrix (78)   a matrix with m rows and n columns

magnitude of a vector (485)   the length of the directed line segment

major axis (631)   the axis of symmetry of an ellipse which contains the foci

mantissa (727)   the common logarithm of a number between 1 and 10

mathematical induction (820)   a method of proof that depends on a recursive process

matrix (78)   any rectangular array of terms called elements

maximum (121)   a critical point of a graph where the curve changes from an increasing curve to a decreasing curve

mean (897)   a measure of central tendency found by dividing the sum of all values by the number of values

mean deviation (910)   the arithmetic mean of the absolute value of the deviations from the mean of a set of data

measure of central tendency (897)   a number that represents the center or middle of a set of data

measure of variability (908)   a number that represents the variability or diversity of a set of data

median (897)   the middle value of a set of data that has been arranged into an ordered sequence

median class (902)   in a frequency distribution, the class in which the median of the data is located

midline (380)   a horizontal axis used as the reference line for vertical shifts of the graphs of sine and cosine functions

minimum (171)   a critical point of a graph where the curve changes from a decreasing curve to an increasing curve

minor (98)   The minor of an element of an nth-order matrix is the determinant of $(n - 1)$th order found by deleting the row and column containing the element.

minor axis (631)   the axis of symmetry of an ellipse which does not contain the foci

minute (277)   a unit of angle measure that is $\frac{1}{60}$ of a degree

mode (897)   the item of data that appears more frequently than any other in the set

model (27)   an equation used to approximate a real-world set of data

modulus (587)   the number r when a complex number is written in the form $r(\cos \theta + i \sin \theta)$

monotonicity (163)   A function is said to be monotonic on an interval I if and only if the function is increasing on I or decreasing on I.

mutually exclusive events (862)   two events whose outcomes can never be the same

N

natural logarithm (733)   logarithms that use e as the base, written $\ln x$

n factorial (794)   written $n!$, for n, an integer greater than zero, the product $n(n - 1)(n - 2) \cdots 1$

nonlinear regression (741)   the process of fitting an equation to nonlinear data

nonsingular matrix (98)   a matrix with a nonzero determinant

normal curve (918)   a symmetric bell-shaped graph of a normal distribution

normal distribution (918)   A frequency distribution that often occurs when there is a large number of values in a set of data: about 68% of the values are within one standard deviation of the mean, 95% of the values are within two standard deviations from the mean, and 99% of the values are within three standard deviations.

normal form (463)   the equation of a line that is written in terms of the length of the normal from the line to the origin

normal line (463)   a line that is perpendicular to another line, curve, or surface

nth order matrix (78)   a square matrix with n rows and n columns

nth partial sum (759)   the sum of the first n terms of a series

odd function (133)   a function whose graph is symmetric with respect to the origin

odds (854)   the ratio of the probability of the success of an event to the probability of its compliment

opposite vectors (487)   two vectors that have the same magnitude and opposite directions

orbit (815)   the graph of the sequence of successive iterates

ordered triple (74; 500)   1. the solution of a system of equations in three variables   2. coordinates of the location of a point in space

ordinate (5)   the second element of an ordered pair

outlier (909)   a value of a set of data that is more than 1.5 interquartile ranges beyond the upper or lower quartiles

P

parabola (653)   the locus of all points in a given plane that are the same distance from a given point, called the focus, and a given line, called the directrix

parallel lines (32)   nonvertical coplanar lines that have equal slopes; any two coplanar vertical lines

parallel vectors (488)   two vectors that have the same or opposite directions

parameter (520)   the independent variable t in the vector equation of a line

parametric equation of a line (520)   the vector equation $\langle x - x_1, y - y_1 \rangle = t\langle a_1, a_2 \rangle$ written as the two equations $x = x_1 + ta_1$ and $y = y_1 + ta_2$

parent graph (137)   an anchor graph from which other graphs in the family are derived

partial fraction (244)   one of the fractions that was added or subtracted to result in a given rational expression

Pascal's triangle (799)   a triangular array of numbers such that the $(n + 1)^{th}$ row is the coefficient of the terms of the expansion $(x + y)^n$ for $n = 0, 1, 2 \ldots$

percentile (922)   The nth percentile of a set of data is the value of the data that is equal to or greater than n percent of the data.

periodic function (359)   A function is periodic if, for some real number α, $f(x + \alpha) = f(x)$ for each x in the domain of f. The least positive value of α for which $f(x) = f(x + \alpha)$ is the *period* of the function.

permutation (838)   the arrangement of objects in a certain order

permutation with repetition (846)   the arrangement of objects in a certain order in which some of the objects are alike

perpendicular lines (34)   any two nonvertical lines, the product of whose slopes is -1; any vertical line and any horizontal line

phase shift (378)   the least value of $|k\theta + c|$, for which the trigonometric function $f(k\theta + c) = 0$

piecewise function (45)   a function in which different equations are used for different intervals of the domain

point discontinuity (159)   When there is a value in the domain for which $f(x)$ is undefined, but the pieces of the graph match up, then $f(x)$ has point discontinuity.

point of inflection (171)   a critical point of a graph where the graph changes its curvature from concave down to concave up or vise versa

point-slope form (28)   the equation of the line that contains the point with coordinates (x_1, y_1) and having slope m written in the form $y - y_1 = m(x - x_1)$

point symmetry (127)   Two distinct points P and P' are symmetric with respect to point M, if and only if M is the midpoint of $\overline{PP'}$. Point M is symmetric with respect to itself.

polar axis (553)   a ray whose initial point is the pole

polar coordinate system (553)   a grid of concentric circles and their center, which is called the pole, whose radii are integral multiples of 1

polar equation (556)   an equation that uses polar coordinates

polar form (588)   the complex number $x + y\boldsymbol{i}$ written as $r(\cos \theta + \boldsymbol{i} \sin \theta)$ where $r = \sqrt{x^2 + y^2}$ and $\theta = \text{Arctan} \frac{y}{x}$ where $x > 0$ and $\theta = \text{Arctan} \frac{y}{x} + \pi$ where $x < 0$

polar graph (556)   the representation of the solution set which is the set of points whose coordinates (r, θ) satisfy a given polar equation

polar plane   See polar coordinate system.

pole (180)   1. See polar coordinate system. 2. vertical asymptote of a rational function

polygonal convex set (108)   the solution of a system of linear inequalities

polyhedron (535)   a closed three-dimensional figure made up of flat polygonal regions

polynomial equation (206)   a polynomial that is set equal to zero

polynomial function (206)   a function $y = P(x)$ where $P(x)$ is a polynomial in one variable

polynomial in one variable (205)   an expression of the form $a_0 x^n + a_1 x^{n-1} + \cdots + a_{n-1} x + a_n$ where the coefficients $a_0, a_1, \ldots, a_n$ represent complex numbers, a_0 is not zero, and n represents a nonnegative integer

population (927)   the entire set of items or individuals in the group being considered

power function (704) a function in the form $y = x^b$, where b is a real number

prediction equation (38) an equation suggested by the points of a scatter plot used to predict other points

pre-image (88) the graph of an object before a transformation

principal values (406, 456) the unique solutions of a trigonometric equation if the values of the function are restricted to two adjacent quadrants

prisoner point (816) If the iterates do not approach infinity for some initial value, that point is called a prisoner point.

prisoner set (599) the set of initial values for which the iterates of a function do not approach infinity

probability (852) the measure of the chance of a desired outcome happening

pure imaginary number (206, 581) the complex number $a + bi$ when $a = 0$ and $b \neq 0$

Q

quadrantal angle (278) an angle in standard position whose terminal side coincides with one of the axes

quartile (909) one of four groupings of a set of data determined by the median of the set and the medians of the sets determined by the median

R

radian (343) the measure of a central angle whose sides intercept an arc that is the same length as the radius of the circle

radical equation (251) an equation that contains a radical expression with the variable in the radicand

radical inequality (253) an inequality that contains a radical expression with the variable in the radicand

radius See circle.

random sample (927) a sample in which every member of the population has an equal chance to be selected

range (890) the difference of the greatest and least values in a set of data

range of a relation (5) the set of all ordinates of the ordered pairs of a relation

rate of change (956) the derivative of a function when applied to real-world applications

rational equation (243) an equation that consists of one or more rational expressions

rational function (180) the quotient of two polynomials in the form $f(x) = \frac{g(x)}{h(x)}$, where $h(x) \neq 0$

rational inequality (245) an inequality that consists of one or more rational expressions

ratio test (785) a method to test convergence in an infinite series

real part of a complex number (581) a in the complex number $a + bi$

rectangular form (581) a complex number written as $x + yi$, where x is the real part and yi is the imaginary part

rectangular hyperbola (648) A special case of the equilateral hyperbola, where the coordinate axes are the asymptotes. The general equation of a rectangular hyperbola is $xy = c$, where c is a nonzero constant.

recursive formula (758) a formula used for determining the next term of a sequence using one or more of the previous terms

reduced sample space (862) the subset of a sample space that contains only those outcomes that satisfy a given condition

reduction identity (446) identity that involves adding and subtracting the quandrantal angles, $90°$, $180°$, and $270°$, from the angle measure to find equivalent values of a trigonometric function

reference angle (280) the acute angle formed by the terminal side of an angle in standard position and the x-axis

reflection (88) a linear transformation that flips a figure over a line called the line of symmetry

reflection matrix (89) a matrix used to reflect an object over a line or plane

regression line (40) a best-fit line

relation (5) a set of ordered pairs

relative extremum (172) a point that represents the maximum or minimum for a certain interval

relative maximum (172) a point that represents the maximum for a certain interval

relative minimum (172) a point that represents the minimum for a certain interval

resultant of vectors (486) the sum of two or more vectors

root (206) a solution of the equation $P(x) = 0$

rose (563) The graph of a polar equation of the form $r = a \cos n\theta$ or $r = a \sin n\theta$.

rotation (88) a transformation in which an object is moved around a center point

rotation matrix (91) a matrix used to rotate an object

row matrix (78) a matrix that has only one row

S

sample space (852) the set of all possible outcomes of an event

scalar (80, 488)   a real number

scalar quantity (488)   A quantity with only magnitude is called a scalar quantity.

scatter plot (38)   a visual representation of data

scientific notation (695)   the expression of a number in the form $a \times 10n$, where $1 \le a < 10$ and n is an integer

secant (286, 292)   For any angle, with measure α, a point $P(x, y)$ on its terminal side, $r = \sqrt{x^2 + y^2}$, $\sec \alpha = \dfrac{r}{x}$.

secant line (953)   a line that intersects a curve at two or more points

second (277)   a unit of angle measure that is $\dfrac{1}{60}$ of a minute

sector of a circle (346)   a region bounded by a central angle and the intercepted arc

semi-interquartile range (909)   one-half the interquartile range of a set of data

semi-major axis (632)   one of the two segments into which the center of an ellipse divides the major axis

semi-minor axis (632)   one of the two segments into which the center of an ellipse divides the minor axis

sequence (757)   a set of numbers in a specific order

side adjacent (284)   the side of a right triangle that is a side of an acute angle in the triangle

side opposite (284)   the side of a right triangle that is opposite an acute angle in the triangle

sigma notation (792)   For any sequence $a_1, a_2, a_3,$ ..., the sum of the first k terms may be written $\displaystyle\sum_{n=1}^{k} a_n$, which is read "the summation from $n = 1$ to k of a_n." Thus, $\displaystyle\sum_{n=1}^{k} a_n = a_1 + a_2 + a_3 + \cdots + a_k$, where k is an integer value.

simulation (877)   a technique used to model probability experiments for real-world applications

sine (285, 291)   For any angle, with measure α, a point $P(x, y)$ on its terminal side, $r = \sqrt{x^2 + y^2}$, $\cos \alpha = \dfrac{y}{r}$.

sinusoidal function (388)   a function of the form $y = A \sin (k\theta + c) + h$ or $y = A \cos (k\theta + c) + h$

slant asymptote (183)   The oblique line ℓ is a slant asymptote for a function $f(x)$ if the graph of $f(x)$ approaches ℓ as $x \to \infty$ or as $x \to -\infty$.

slope-intercept form (21)   the equation of a line with slope, m, and y-intercept, b, written in the form $y = mx + b$

slope of a curve (949)   the slope of a line tangent to a particular point on the graph of a curve

slope of a line (20-21)   the value $m = \dfrac{y_2 - y_1}{x_2 - x_1}$ where (x_1, y_1) and (x_2, y_2), $x_2 \ne x_1$, are two points of the line

solution (67)   an ordered pair representing the solution common to both equations in a system of equations

solve a triangle (307)   to find all of the measures of a triangle's sides and angles

spiral of Archimedes (564)   a function of the form $r = n\theta$

square matrix (78)   a matrix with the same number of rows as columns

standard deviation (911)   a measure of the average amount by which individual items of data deviate from the arithmetic mean of all the data

standard error of the mean (927)   the standard deviation of the distribution of a sample mean

standard form (21)   a linear equation written in the form $Ax + By + C = 0$, where A, B, and C are real numbers and A and B are not both zero

standard position (277)   an angle with its vertex at the origin and its initial side along the positive x-axis

standard position of a vector (485)   If a vector has its initial point at the origin, it is in standard position.

stem-and-leaf plot (899)   a display of numerical data for which each value is separated into two numbers, the stem which consists of the digits in the greatest common place value, and the leaf which contains the other digits of each item of data

step function (46)   a function whose graph is a series of disjoint lines or steps

substitution method (68)   a method for solving a system of equations

success (850)   the desired outcome of an event

symmetry identity (424)   trigonometric identities related to the symmetries of the unit circle

symmetry with respect to the origin (128)   The graph of a relation S is symmetric with respect to the origin if and only if $(a, b) \in S$ implies that $(-a, -b) \in S$. A function has a graph that is symmetric with respect to the origin if and only if $f(-x) = -f(x)$ for all x in the domain of f.

synthetic division (223)   a method used to divide a polynomial by a binomial

system of equations (67)   a set of equations with the same variables

system of linear inequalities (107)   a set of inequalities with the same variables

T

tangent (953; 285, 292)   1. a line that intersects a curve at exactly one point   2. For any angle, with measure α, a point $P(x, y)$ on its terminal side, $r = \sqrt{x^2 + y^2}$, $\tan \alpha = \dfrac{y}{x}$.

term (757)   a number in a sequence or series

terminal side of an angle (277)   a ray of an angle that rotates about the vertex

theoretical probability (877)   probability determined using mathematical methods to model outcomes of a given situation

three-dimensional bar graph (890)   a graphic form using bars to make comparisons among multiple aspects of statistical data

transformation (88)   functions that map points of a graph onto its image

translation (88)   a linear transformation that slides the graph vertically and/or horizontally on the coordinate plane, but does not change its shape

translation matrix (88)   the matrix used to represent the translation of a set of points with respect to (h, k) which is equal to $\begin{bmatrix} h & h & h & h \\ k & k & k & k \end{bmatrix}$

transverse axis (642)   the line segment that has as its endpoints the vertices of a hyperbola

tree diagram (837)   a diagram used to show the total number of possible outcomes of an event

trigonometric form   See polar form.

trigonometric functions (292)   For any angle, with measure α, a point $P(x, y)$ on its terminal side, $r = \sqrt{x^2 + y^2}$, the trigonometric functions of α are as follows.

$$\sin \alpha = \frac{y}{r} \qquad \cos \alpha = \frac{x}{r} \qquad \tan \alpha = \frac{y}{x}$$
$$\csc \alpha = \frac{r}{y} \qquad \sec \alpha = \frac{x}{r} \qquad \cot \alpha = \frac{x}{y}$$

trigonometric identity (421)   an equation involving a trigonometric function that is true for all values of the variable. *See inside back cover for complete list of identities.*

trigonometric ratio (285)   a ratio of the sides of a right triangle used to define the sine, cosine, and tangent ratios of the triangle

trigonometric series (806)   infinite series that define the trigonometric functions sine and cosine

U

unbounded (113)   The solution of a linear programming problem is unbounded if the region defined by the constraints is infinitely large.

unit circle (291)   a circle of radius 1 unit whose center is at the origin of a rectangular coordinate system

unit vector (496)   a vector of length 1 that is parallel to the x-, y-, or z-axis

upper bound (238)   the integer greater than or equal to the greatest zero of the polynomial function

variance (912)   the mean of the squares of the deviations from the arithmetic mean

vector (485)   a quantity, or directed distance, that has both magnitude and direction

vector equation (520)   A line through $P(x, y)$ parallel to the vector $\widehat{\mathbf{a}} = \langle a_1, a_2 \rangle$ is defined by the set of points $P_1(x_1, y_1)$ and $P(x, y)$ such that $\overrightarrow{P_1 P} = t\vec{\mathbf{a}}$ for some real number t. Therefore, $\langle x - x_1, y - y_1 \rangle = t\langle a_1, a_2 \rangle$.

vertex (277; 535)   1. the common endpoint of two rays forming an angle   2. the common endpoints of the sides of a polygon or where three or more edges of a polyhedron intersect

vertex matrix (88)   a matrix used to represent the coordinates of the vertices of an object

vertex of a conic section (631, 642, 653)   a point at which a conic section intersects its axis of symmetry

vertical asymptote (180)   The line $x = a$ is a vertical asymptote for a function $f(x)$ if $f(x) \to \infty$ or $f(x) \to -\infty$ as $x \to a$ from either the left or the right.

vertical line test (7)   a test used to determine if a relation is a function

whisker (909)   the segments extending from the ends of the box in a box-and-whisker plot

x-intercept (20)   the x-coordinate of the point at which the graph of an equation crosses the x-axis

y-intercept (20)   the y-coordinate of the point at which the graph of an equation crosses the y-axis

Z

zero (22, 206)   a value of x for which $f(x) = 0$

zero matrix (80)   a matrix whose elements are all zero

zero vector (485)   a vector with initial and terminal points at the origin

SELECTED ANSWERS

CHAPTER 1   LINEAR RELATIONS AND FUNCTIONS

Pages 9–12   Lesson 1-1

5. $y = x - 4$

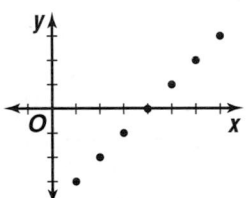

x	y
1	−3
2	−2
3	−1
4	0
5	1
6	2
7	3

7. $\{(-6, 1), (-4, 0), (-2, -4), (1, 3), (4, 3)\}$; $D = \{-6, -4, -2, 1, 4\}$; $R = \{-4, 0, 1, 3\}$

9.

x	y
1	−5
2	−5
3	−5
4	−5
5	−5
6	−5
7	−5
8	−5

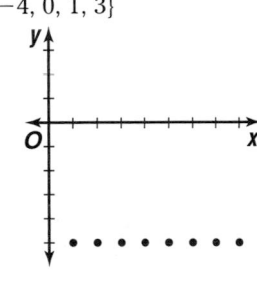

11. $\{-3, 3, 6\}$; $\{-6, -2, 0, 4\}$; no; 6 is matched with two members of the range.   **13.** −84

15. $x \geq -1$

17. $y = 3x$

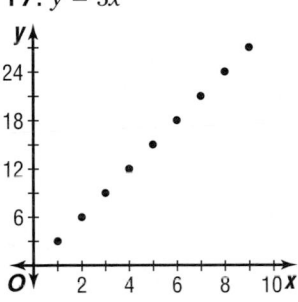

x	y
1	3
2	6
3	9
4	12
5	15
6	18
7	21
8	24
9	27

19. $y = 8 + x$

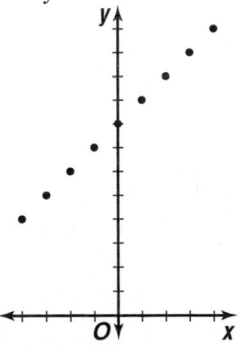

x	y
−4	4
−3	5
−2	6
−1	7
0	8
1	9
2	10
3	11
4	12

21. $\{(-10, 0), (-5, 0), (0, 0), (5, 0)\}$; $D = \{-10, -5, 0, 5\}$; $R = \{0\}$ **23.** $\{(-3, -2), (-1, 1), (0, 0), (1, 1)\}$; $D = \{-3, -1, 0, 1\}$; $R = \{-2, 0, 1\}$

25. $\{(3, -4), (3, -2), (3, 0), (3, 1), (3, 3)\}$; $D = \{3\}$; $R = \{-4, -2, 0, 1, 3\}$

27.

x	y
1	−1
2	−2
3	−3
4	−4
5	−5
6	−6

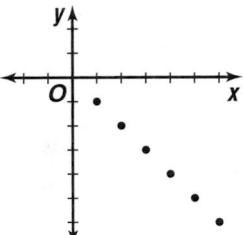

29.

x	y
1	0
2	3
3	6
4	9
5	12

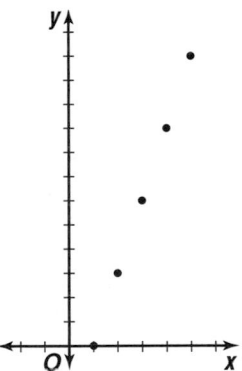

31.

x	y
4	2
4	−2

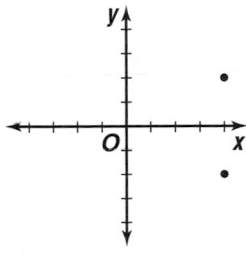

33. $\{1\}$; $\{-6, -2, 0, 4\}$; no; The x-value 1 is paired with more than one y-value.   **35.** $\{0, 2, 5\}$; $\{-8, -2, 0, 2, 8\}$; no; The x-values 2 and 5 are paired with more than one y-value.   **37.** $\{-9, 2, 8, 9\}$; $\{-3, 0, 8\}$; yes; Each x-value is paired with exactly one y-value.

39. domain: $\{-3, -2, -1, 1, 2, 3\}$; range: $\{-1, 1, 2, 3\}$; A function because each x-value is paired with exactly one y-value.   **41.** 9   **43.** 2

45. $2n^2 - 5n + 12$ **47.** $|25m^2 - 13|$ **49.** $x \leq -3$ or $x \geq 3$ **51a.** $x \neq 1$ **51b.** $x \neq -5$

51c. $x \neq -2, 2$ **53.** $3x^3 + 4x - 7$

55a. 14,989,622.9 m; 59,958,491.6 m; 419,709,441.2 m; 1,768,775,502 m   **55b.** 23,983,396.64 m   **57.** B

SELECTED ANSWERS

Pages 17–19 Lesson 1-2

5. $3x^2 + 6x + 4$; $3x^2 + 2x - 14$; $6x^3 + 35x^2 + 26x - 45$; $\dfrac{3x^2 + 4x - 5}{2x + 9}$, $x \neq -\dfrac{9}{2}$ **7.** $2x^2 - 4x - 3$; $4x^2 - 16x + 15$ **9.** 5, 11, 23 **11.** $x^2 - x + 9$; $x^2 - 3x - 9$; $x^3 + 7x^2 - 18x$; $\dfrac{x^2 - 2x}{x + 9}$, $x \neq -9$

13. $\dfrac{x^3 - 2x^2 - 35x + 3}{x - 7}$, $x \neq 7$ $-\dfrac{x^3 - 2x^2 - 35x - 3}{x - 7}$, $x \neq 7$; $\dfrac{3x^2 + 15x}{x - 7}$, $x \neq 7$; $\dfrac{3}{x^3 - 2x^2 - 35x}$, $x \neq -5$, 0, or 7

15. $x^2 + 8x + 7$; $x^2 - 5$ **17.** $3x^2 - 4$; $3x^2 - 24x + 48$ **19.** $2x^3 + 2x^2 + 2$; $8x^3 + 4x^2 + 1$

21. $\dfrac{x}{x - 1}$, $x \neq 1$; $\dfrac{1}{x}$, $x \neq 0$ **23.** $x \neq 7$ **25.** 7, 2, 7

27. 2, 2, 2 **29.** Yes; If $f(x)$ and $g(x)$ are both lines, they can be represented as $f(x) = m_1x + b_1$ and $g(x) = m_2x + b_2$. Then $[f \circ g](x) = m_1(m_2x + b_2) + b_1 = m_1m_2x + m_1b_2 + b_1$. Since m_1 and m_2 are constants, m_1m_2 is a constant. Similarly, m_1, b_2, and b_1 are constants, so $m_1b_2 + b_1$ is a constant. Thus, $[f \circ g](x)$ is a linear function if $f(x)$ and $g(x)$ are both linear. **31a.** $h[f(x)]$, because you must subtract before figuring the bonus.

31b. \$3750 **33a.** $v(p) = \dfrac{7p}{47}$ **33b.** $r(v) = 0.84v$

33c. $r(p) = \dfrac{147p}{1175}$ **33d.** \$52.94, \$28.23, \$99.72

35. $\{(-1, 8), (0, 4), (2, -6), (5, -9)\}$; $D = \{-1, 0, 2, 5\}$; $R = \{-9, -6, 4, 8\}$ **37.** $3\dfrac{11}{16}$ **39.** C

13.

15.

17.

19.

21.

23.
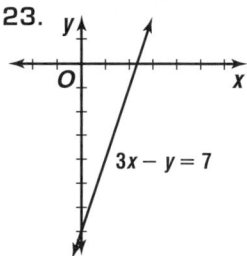

Pages 23–25 Lesson 1-3

5.

7.

9. -12

25. 3

27. 0

29. $\dfrac{8}{5}$

31. 2
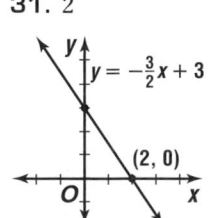

11a. (38.500, 173), (44.125, 188) **11b.** 2.667
11c. For each 1-centimeter increase in the length of a man's tibia, there is a 2.667-centimeter increase in the man's height.

33a. 0.4 ohm **33b.** 2.4 volts **35a.** $\dfrac{1}{4}$ **35b.** For each 1-degree increase in the temperature, there is a $\dfrac{1}{4}$-pascal increase in the pressure.

35c.

37a. 36; The software has no monetary value after 36 months. **37b.** −290; For every 1-month change in the number of months, there is a $290 decrease in the value of the software.

37c.

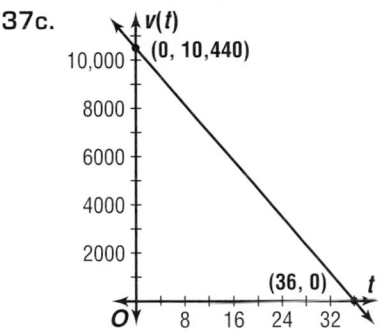

39a. 0.86 **39b.** $1552.30 **39c.** 0.14
39d. $252.70 **41a.** $d(p) = 0.88p$ **41b.** $r(d) = d − 100$ **41c.** $r(p) = 0.88p − 100$ **41d.** $603.99, $779.99, $1219.99 **43.** −671 **45.** {(−3, 14), (−2, 13), (−1, 12), (0, 11)}, yes

Pages 29–31   Lesson 1-4
7. $y = 4x − 10$ **9.** $y = 2$ **11.** $y = 5x − 2$
13. $y = -\frac{3}{4}x$ **15.** $y = 6x − 19$ **17.** $y = -\frac{4}{9}x + \frac{49}{9}$
19. $y = 1$ **21.** $x = 0$ **23.** $x + 2y + 10 = 0$
25a. $t = 2 + \frac{x − 7000}{2000}$ **25b.** about 5.7 weeks
27a. Sample answer: Using (20, 28) and (27, 37), $y = \frac{9}{7}x + \frac{16}{7}$ **27b.** Using sample answer from part a, 26.7 mpg **27c.** Sample answer: The estimate is close but not exact since only two points were used to write the equation. **29.** Yes; the slope of the line through (5, 9) and (−3, 3) is $\frac{3 − 9}{−3 − 5}$ or $\frac{3}{4}$. The slope of the line through (−3, 3), and (1, 6) is $\frac{6 − 3}{1 − (−3)}$ or $\frac{3}{4}$. Since these two lines would have the same slope and would share a point, their equations would be the same. Thus, they are the same line and all three points are collinear. **31a.** $6111 billion **31b.** The rate is the slope. **33.** $x^5 − 3x^4 + 7x^3$, $\frac{x^3}{x^2 − 3x + 7}$ **35.** A

Pages 35–37   Lesson 1-5
5. none of these **7.** parallel **9.** $5x − y − 16 = 0$
11. parallelogram **13.** parallel **15.** perpendicular
17. perpendicular **19.** coinciding **21.** None of these; the slopes are neither the same nor opposite reciprocals. **23.** $4x − 9y − 183 = 0$ **25.** $x + 5y + 15 = 0$ **27.** $y + 13 = 0$ **29a.** 4 **29b.** $-\frac{49}{4}$
31. $x − 5y − 29 = 0$; $x = 7$; $x + 5y + 15 = 0$
33a. No; the lines that represent the situation do not coincide. **33b.** Yes; the lines that represent the situation coincide. **35.** $y = −2x + 7$

37.

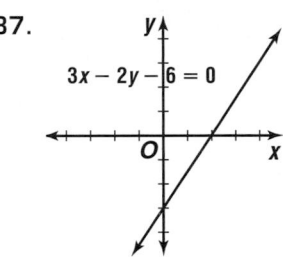

39. Sample answer: {(2, 4), (2, −4), (1, 2), (1, −2), (0, 0)}; because the x-values 1 and 2 are paired with more than one y-value

Pages 41–44   Lesson 1-6
5a.

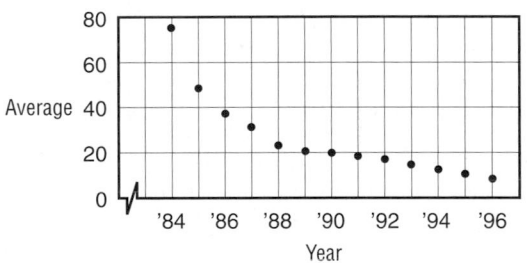

5b. Sample answer: Using (1987, 32) and (1996, 7.8), $y = −2.69x + 5377.03$ **5c.** $y = −6.28x + 12,530.14$; $r ≈ −0.82$ **5d.** 1995; No; In 1995 there were 10 students per computer.

7a.

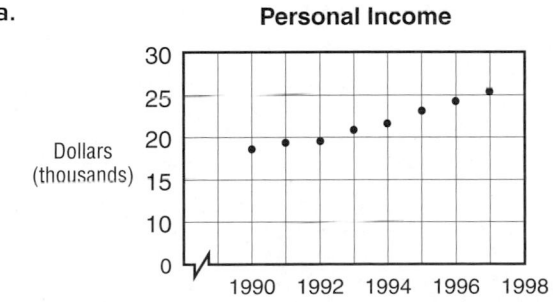

7b. Sample answer: Using (1991, 19,100) and (1995, 23,233), $y = 1058.25x − 2,087,875.75$
7c. $y = 1052.32x − 2,076,129.64$; $r ≈ 0.99$
7d. $33,771.96; Yes, r shows a strong relationship.

9a.

Acorn Size and Range

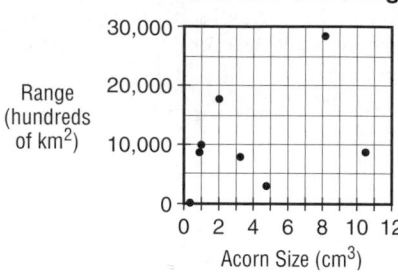

Range (hundreds of km²)

Acorn Size (cm³)

9b. Sample answer: Using (0.3, 233) and (3.4, 7900), $y = 2473.23x - 508.97$ **9c.** $y = 885.82 + 6973.14$; $r \approx 0.38$ **9d.** The correlation value does not show a strong or moderate relationship.

11a.

World Population

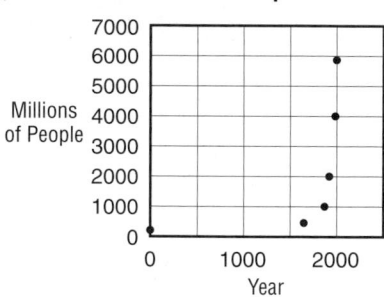

Millions of People

Year

11b. Using (1, 200) and (1998, 5900), $y = 2.85x + 197.14$ **11c.** $y = 1.62x - 277.53$; $r \approx 0.56$
11d. 2979 million; No, the correlation value is not showing a very strong relationship. **13.** The rate of growth, which is the slope of the graphs of the regression equations, for the women is less than that of the men's rate of growth. If that trend continues, the men's median salary will always be more than the women's. **15.** $6x + y + 22 = 0$ **17.** $x^3 + 3x^2 + 3x + 1$; $x^3 + 1$ **19.** C

Pages 48–51 Lesson 1-7

5.

7.

9. greatest integer function; h is hours, $c(h)$ is the cost, $c(h) = \begin{cases} 50h \text{ if } \llbracket h \rrbracket = h \\ 50\llbracket h + 1 \rrbracket \text{ if } \llbracket h \rrbracket < h \end{cases}$

11.

13.

15.

17.

19.

21.

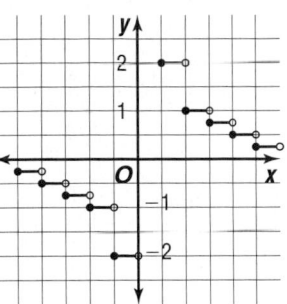

23. step; t is the time in hours, $c(t)$ is the cost in dollars, $c(t) = \begin{cases} 6 \text{ if } t \leq \frac{1}{2} \\ 10 \text{ if } \frac{1}{2} < t \leq 1 \\ 16 \text{ if } 1 < t \leq 2 \\ 24 \text{ if } 2 < t \leq 24 \end{cases}$

25. w is the weight in pounds, $d(w)$ is the discrepancy, $d(w) = |1 - w|$

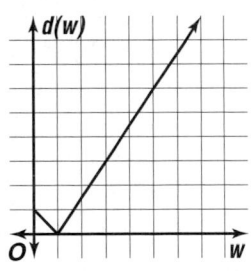

27. If n is any integer, then all ordered pairs (x, y) where x and y are both in the interval $[n, n + 1)$ are solutions. **29a.** step

29b. $t(x) = \begin{cases} 6\% \text{ if } x \leq \$10{,}000 \\ 8\% \text{ if } \$10{,}000 < x \leq \$20{,}000 \\ 9.5\% \text{ if } x > \$20{,}000 \end{cases}$

29c.

29d. 9.5%

31a.

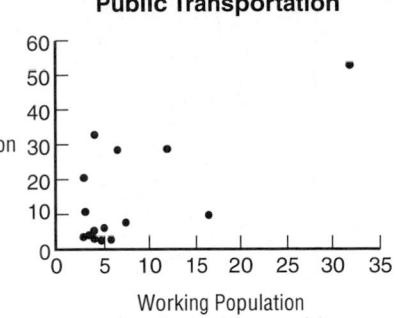

Public Transportation

31b. Sample answer: Using (3,183,088, 53.4) and (362,777, 3.3), $y = 0.0000178x - 3.26$
31c. $y = 0.0000136x + 4.55$, $r \approx 0.68$ **31d.** 8.73%; No, the actual value is 22%. **33a.** (39, 29), (32, 15)
33b. 2 **33c.** The average number of points scored each minute. **35.** \$47.92 **37.** A

Pages 55–56 Lesson 1-8

5.

7.

9.

11.

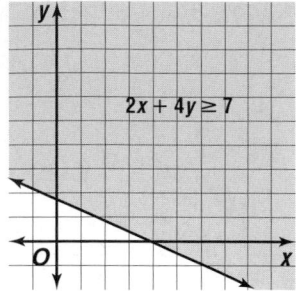

$2x + 4y \geq 7$

13.

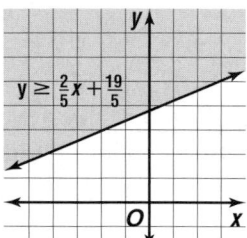

$y \geq \frac{2}{5}x + \frac{19}{5}$

15.

$y \geq |x|$

17.

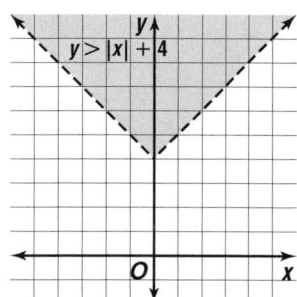

$y > |x| + 4$

19.

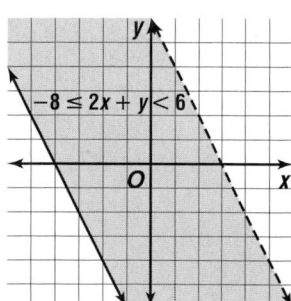

$-8 \leq 2x + y < 6$

21.

$x \geq 0$
$y \geq 0$

23a. $8x + 10y \leq 480$

23b.

23c. Sample answer: (0, 48), (60, 0), (45, 6)
23d. Sample answer: Using complex computer programs and systems of inequalities. **25a.** points in the first and third quadrants **25b.** If x and y satisfy the inequality, then either $x \geq 0$ and $y \geq 0$ or $x \leq 0$ and $y \leq 0$. If $x \geq 0$ and $y \geq 0$, then $|x| = x$ and $|y| = y$. Thus, $|x| + |y| = x + y$. Since $x + y$ is positive, $|x + y| = x + y$. If $x \leq 0$ and $y \leq 0$, then $|x| = -x$ and $|y| = -y$. Then $|x| + |y| = -x + (-y)$ or $-(x + y)$. Since both x and y are negative, $(x + y)$ is negative, and $|x + y| = -(x + y)$.
27a. $0.6(220 - a) \leq r \leq 0.9(220 - a)$

27b.

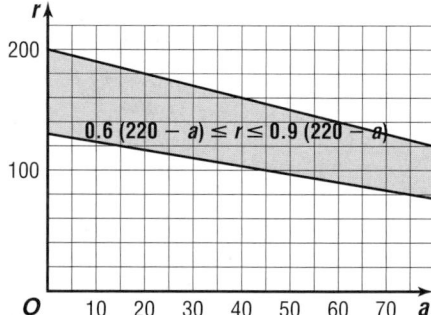

$0.6(220 - a) \leq r \leq 0.9(220 - a)$

29a. $3x - y - 2 = 0$ **29b.** $x + 3y + 6 = 0$
31a. (0, 23), (16, 48); 1.5625 **31b.** the average change in the temperature per hour

Pages 57–61 Chapter 1 Study Guide and Assessment

1. c **3.** d **5.** i **7.** h **9.** e **11.** 10 **13.** 57
15. $\frac{6}{5}$ **17.** $|m^2 + 5m + 4|$ **19.** $x^2 + 5x - 2$;
$x^2 + 3x + 2$; $x^3 + 2x^2 - 8x$; $\frac{x^2 + 4x}{x - 2}$, $x \neq 2$
21. $x^2 + 8x + 16$; $x^2 + 6x + 8$; $x^3 + 11x^2 + 40x + 48$; $x + 3$, $x \neq -4$ **23.** $\frac{x^3 - 8x^2 + 16x + 4}{x - 4}$,

$x \neq 4$; $\frac{x^3 - 8x^2 + 16x - 4}{x - 4}$, $x \neq 4$; $4x$, $x \neq 4$;

$\frac{x^3 - 8x^2 + 16x}{4}$, $x \neq 4$ **25.** $1.5x^2 + 5$;
$0.75x^2 + 15x + 75$ **27.** $x^2 - x + 7$; $x^2 + 11x + 31$
29. $-2x^2 - 7$; $2x^2 - 12x + 28$

31.

33.

35.

37.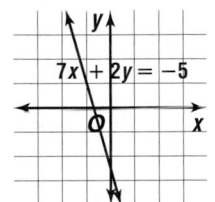

39. $y = 2x - 3$ **41.** $y = \frac{1}{2}x + \frac{9}{2}$ **43.** $y = 4x - 4$
45. $y = 0$ **47.** $x - y = 0$ **49.** $2x + y + 4 = 0$
51. $x + 2y - 9 = 0$

53a.

Overseas Visitors

53b. Sample answer: Using (1987, 10,434) and (1996, 12,909), $y = 275x - 535{,}991$
53c. $y = 147.8x - 282{,}157.4$; $r \approx 0.61$
53d. 14,181,600 visitors; Sample answer: This is not a good prediction, because the r-value does not indicate a strong relationship.

55.

57.

59.

61.

63.

65.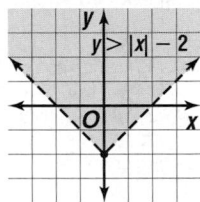

67a. 10 m, 40 m, 90 m, 160 m, 250 m   **67b.** Yes, each element of the domain is paired with exactly one element of the range.   **69.** $y = -0.284x + 12.964$; The correlation is moderately negative, so the regression line is somewhat representative of the data.

Page 65   Chapter 1 SAT and ACT Practice
1. D   **3.** B   **5.** A   **7.** A   **9.** C

Chapter 2   Systems of Equations and Inequalities

Pages 70–72   Lesson 2-1
5. (1, 3)   **7.** (2, −5)   **9.** (6, 4)
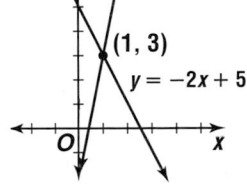
11. consistent and independent   **13.** consistent and dependent

15. (4, −3)
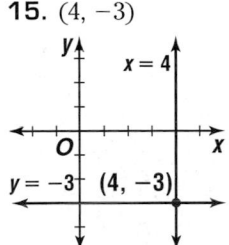

17. no solution   **19.** (0, 3)
   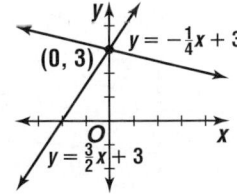

21. (3, 21)   **23.** (5.25, 0.75)   **25.** (5, 2)
27. $\left(\frac{1}{3}, \frac{2}{3}\right)$ **29.** $\left(-\frac{6}{43}, -\frac{64}{43}\right)$

31. Sample answer: Elimination could be considered easiest since the first equation multiplied by 2 added to the second equation eliminates b; Substitution could also be considered easiest since the first equation can be written as $a = b$, making substitution very easy; $(-3, -3)$. **33a.** 6, 6, 8; 6, 6, 8 **33b.** isosceles **35a.** $(7, 5.95)$ **35b.** If you drink 7 servings of soft drink, the price for each option is the same. If you drink fewer than 7 servings of soft drink during that week, the disposable cup price is better. If you drink more than 7 servings of soft drink, the refillable mug price is better. **35c.** Over a year's time, the refillable mug would be more economical. **37.** $1500

39.

41. $y = 2x + 6$
43. $3x + 1$
45. A

Pages 76–77   Lesson 2-2

5. $(7, -1, 1)$ **7.** acceleration: -32 ft/s^2, initial velocity: 56 ft/s, initial height: 35 ft **9.** $(-2, 2, 4)$ **11.** $(-6, -4, 7)$ **13.** no solution **15.** $(11, -17, 14)$ **17.** $(-4, 10, 7)$ **19.** International Fund = $1200; Fixed Assets Fund = $200; company stock = $600 **21.** $(-32, 138, 2)$ **23.** $(1, 1, 1), (-2, -2, -2)$ **25.**

27a. $C(x) = 50x + 2000$
27b. $2000, $50

27c.

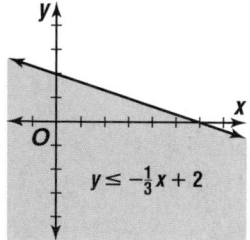

Pages 82–86   Lesson 2-3

5. $(7, 2)$ **7.** $(4, 0)$ **9.** impossible **11.** $\begin{bmatrix} 16 & 4 \\ -8 & 24 \end{bmatrix}$

13. $[6 \;\; -18]$ **15.** $(6, 11)$ **17.** $(5, 2.5)$ **19.** $(7, 9)$
21. $(-5, -15)$ **23.** $(-1, 1)$ **25.** $(5, 3, 2)$

27. $\begin{bmatrix} 8 & 12 \\ -7 & 9 \end{bmatrix}$ **29.** impossible **31.** $\begin{bmatrix} -2 & -2 \\ 5 & 7 \end{bmatrix}$

33. $\begin{bmatrix} 0 & 4 & 8 \\ -8 & 12 & 0 \\ 16 & 16 & -8 \end{bmatrix}$ **35.** $\begin{bmatrix} -14 & 3 & -2 \\ -2 & 3 & 5 \end{bmatrix}$

37. $\begin{bmatrix} 25 & 35 \\ -30 & 5 \end{bmatrix}$ **39.** impossible

41. $\begin{bmatrix} 16 & 4 & 12 \\ -22 & -14 & 16 \end{bmatrix}$ **43.** $\begin{bmatrix} 10 & -13 & -10 \\ -5 & 14 & -3 \end{bmatrix}$

45. $\begin{bmatrix} -42 & 86 & -160 \\ -421 & 213 & -111 \end{bmatrix}$ **47.** $\begin{bmatrix} 78 & 30 & 12 \\ 12 & -120 & 168 \\ 72 & 90 & -72 \end{bmatrix}$

49. Sample answer:

	1996	2000	2006
18 to 24	8485	8526	8695
25 to 34	10,102	9316	9078
35 to 44	8766	9039	8433
45 to 54	6045	6921	7900
55 to 64	2444	2741	3521
65 and older	2381	2440	2572

51a. $a = 1, b = 0, c = 0, d = 1$ **51b.** a matrix equal to the original one **53.** The numbers in the first row are the triangular numbers. If you look at the diagonals in the matrix, the triangular numbers are the end numbers. To find the diagonal that contains 2001, find the smallest triangular number that is greater than or equal to 2001. The formula for the nth triangular number is $\dfrac{n(n + 1)}{2}$. Solve $\dfrac{n(n + 1)}{2} \geq 2001$. The solution is 63. So the 63rd entry in the first row is $\dfrac{63(63 + 1)}{2} = 2016$. Since $2016 - 2001 = 15$, we must count 15 places backward along the diagonal to locate 2001 in the matrix. This movement takes us from the position (row, column) = (1, 63) to $(1 + 15, 63 - 15) =$ (16, 48). **55.** $\left(\dfrac{1}{2}, \dfrac{1}{3}, \dfrac{1}{4} \right)$

57.

59. Sample answer: $y = 0.36x + 61.4$ **61.** $\dfrac{3}{5}$
63. -2656

Pages 93–96   Lesson 2-4

5. $J'(-3, 7.5)$,
$K'(1.5, 4.5)$, $L'(0, -3)$

7. $A'(1, 2)$, $B'(4, 1)$,
$C'(3, -2)$, $D'(0, -1)$

9. $L'(-6, -4)$,
$M'(-3, -2)$, $N'(-1, 2)$

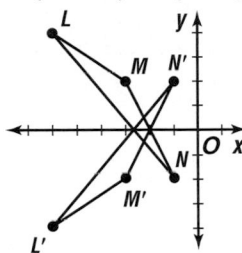

11. $A'(3, 3)$, $B'(3, 12)$,
$C'(15, 3)$

13. $P'(-6, 0)$,
$Q'(-4, 4)$,
$R'(2, 6)$,
$S'(8, 4)$

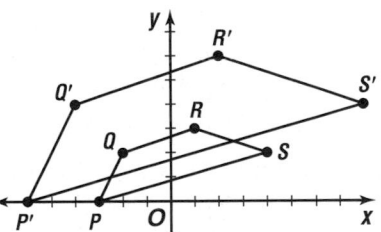

15. $W'(1, -2)$, $X'(4, 3)$,
$Y'(6, -3)$

17. $C'(0, 5)$, $D'(4, 9)$,
$E'(8, 5)$, $F'(4, 1)$

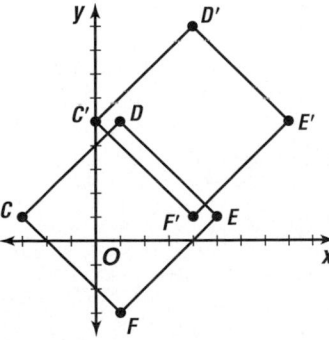

19. $A'(-1, 2)$,
$B'(0, 4)$, $C'(2, 3)$

21. $H(-2, -1)$, $I'(1, -3)$, **23.** $O'(0, 0)$, $P'(-4, 0)$,
$J'(5, -1)$, $K'(4, 2)$ $\qquad$ $Q'(-4, -4)$, $R'(0, -4)$

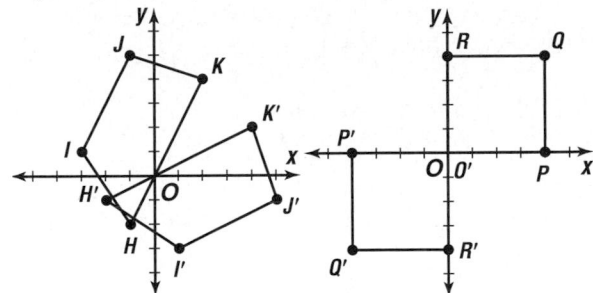

25a. Let $\begin{bmatrix} a & b \\ c & d \end{bmatrix} = R_{x\text{-axis}}$.

$\begin{bmatrix} a & b \\ c & d \end{bmatrix} \cdot \begin{bmatrix} 1 & -2 & -1 \\ 3 & -1 & -3 \end{bmatrix} = \begin{bmatrix} 1 & -2 & -1 \\ -3 & 1 & 3 \end{bmatrix}$

$\begin{bmatrix} a + 3b & -2a - b & -a - 3b \\ c + 3d & -2c - d & -c - 3d \end{bmatrix} = \begin{bmatrix} 1 & -2 & -1 \\ -3 & 1 & 3 \end{bmatrix}$

$a + 3b = 1 \quad -2a - b = -2 \quad -a - 3b = -1$
$c + 3d = -3 \quad -2c - d = 1 \quad -c - 3d = 3$

Thus, $a = 1$, $b = 0$, $c = 0$, and $d = -1$. By
substitution, $R_{x\text{-axis}} = \begin{bmatrix} 1 & 0 \\ 0 & -1 \end{bmatrix}$.

25b. Let $\begin{bmatrix} a & b \\ c & d \end{bmatrix} = R_{y\text{-axis}}$.

$\begin{bmatrix} a & b \\ c & d \end{bmatrix} \cdot \begin{bmatrix} 1 & -2 & -1 \\ 3 & -1 & -3 \end{bmatrix} = \begin{bmatrix} -1 & -2 & 1 \\ 3 & -1 & -3 \end{bmatrix}$

$\begin{bmatrix} a + 3b & -2a - b & -a - 3b \\ c + 3d & -2c - d & -c - 3d \end{bmatrix} = \begin{bmatrix} -1 & 2 & 1 \\ 3 & -1 & -3 \end{bmatrix}$

$a + 3b = -1 \quad -2a - b = 2 \quad -a - 3b = 1$
$c + 3d = 3 \quad -2c - d = -1 \quad -c - 3d = -3$

Thus, $a = -1$, $b = 0$, $c = 0$, and $d = 1$. By
substitution, $R_{y\text{-axis}} = \begin{bmatrix} -1 & 0 \\ 0 & 1 \end{bmatrix}$.

25c. Let $\begin{bmatrix} a & b \\ c & d \end{bmatrix} = R_{y = x}$.

$\begin{bmatrix} a & b \\ c & d \end{bmatrix} \cdot \begin{bmatrix} 1 & -2 & -1 \\ 3 & -1 & -3 \end{bmatrix} = \begin{bmatrix} 3 & -1 & -3 \\ 1 & -2 & -1 \end{bmatrix}$

$\begin{bmatrix} a + 3b & -2a - b & -a - 3b \\ c + 3d & -2c - d & -c - 3d \end{bmatrix} = \begin{bmatrix} 3 & -1 & -3 \\ 1 & -2 & -1 \end{bmatrix}$

$a + 3b = 3 \quad -2a - b = -1 \quad -a - 3b = -3$
$c + 3d = 1 \quad -2c - d = -2 \quad -c - 3d = -1$

Thus, $a = 0$, $b = 1$, $c = 1$, and $d = 0$. By substitution,
$R_{y = x} = \begin{bmatrix} 0 & 1 \\ 1 & 0 \end{bmatrix}$.

25d. Let $\begin{bmatrix} a & b \\ c & d \end{bmatrix} = Rot_{90}$.

$$\begin{bmatrix} a & b \\ c & d \end{bmatrix} \cdot \begin{bmatrix} 1 & -2 & -1 \\ 3 & -1 & -3 \end{bmatrix} = \begin{bmatrix} -3 & 1 & 3 \\ 1 & -2 & -1 \end{bmatrix}$$

$$\begin{bmatrix} a + 3b & -2a - b & -a - 3b \\ c + 3d & -2c - d & -c - 3d \end{bmatrix} = \begin{bmatrix} -3 & 1 & 3 \\ 1 & -2 & -1 \end{bmatrix}$$

$a + 3b = -3 \quad -2a - b = -1 \quad -a - 3b = 3$
$c + 3d = 1 \quad -2c - d = -2 \quad -c - 3d = -1$

Thus, $a = 0$, $b = -1$, $c = 1$, and $d = 0$. By

substitution, $Rot_{90} = \begin{bmatrix} 0 & -1 \\ 1 & 0 \end{bmatrix}$.

25e. Let $\begin{bmatrix} a & b \\ c & d \end{bmatrix} = Rot_{180}$.

$$\begin{bmatrix} a & b \\ c & d \end{bmatrix} \cdot \begin{bmatrix} 1 & -2 & -1 \\ 3 & -1 & -3 \end{bmatrix} = \begin{bmatrix} -1 & 2 & 1 \\ -3 & 1 & 3 \end{bmatrix}$$

$$\begin{bmatrix} a + 3b & -2a - b & -a - 3b \\ c + 3d & -2c - d & -c - 3d \end{bmatrix} = \begin{bmatrix} -1 & 2 & 1 \\ -3 & 1 & 3 \end{bmatrix}$$

$a + 3b = -1 \quad -2a - b = 2 \quad -a - 3b = 1$
$c + 3d = -3 \quad -2c - d = 1 \quad -c - 3d = 3$

Thus, $a = -1$, $b = 0$, $c = 0$, and $d = -1$. By

substitution, $Rot_{180} = \begin{bmatrix} -1 & 0 \\ 0 & -1 \end{bmatrix}$.

25f. Let $\begin{bmatrix} a & b \\ c & d \end{bmatrix} = Rot_{270}$.

$$\begin{bmatrix} a & b \\ c & d \end{bmatrix} \cdot \begin{bmatrix} 1 & -2 & -1 \\ 3 & -1 & -3 \end{bmatrix} = \begin{bmatrix} 3 & -1 & -3 \\ -1 & 2 & 1 \end{bmatrix}$$

$$\begin{bmatrix} a + 3b & -2a - b & -a - 3b \\ c + 3d & -2c - d & -c - 3d \end{bmatrix} = \begin{bmatrix} 3 & -1 & -3 \\ -1 & 2 & 1 \end{bmatrix}$$

$a + 3b = 3 \quad -2a - b = -1 \quad -a - 3b = -3$
$c + 3d = -1 \quad -2c - d = 2 \quad -c - 3d = 1$

Thus, $a = 0$, $b = 1$, $c = -1$, and $d = 0$. By

substitution, $Rot_{270} = \begin{bmatrix} 0 & 1 \\ -1 & 0 \end{bmatrix}$.

27. $J''(6, -4)$, $K''(3, -2)$, $L''(1, 2)$

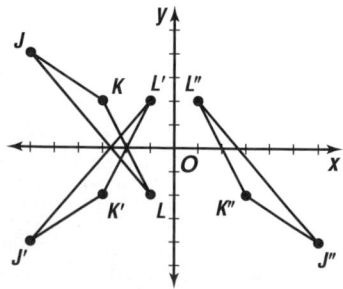

29a. The bishop moves along a diagonal until it encounters the edge of the board or another piece. The line along which it moves changes vertically and horizontally by 1 unit with each square moved, so the translation matrices are scalars. Sample matrices are $c\begin{bmatrix} 1 & 1 \\ 1 & 1 \end{bmatrix}$, $c\begin{bmatrix} 1 & 1 \\ -1 & -1 \end{bmatrix}$, $c\begin{bmatrix} -1 & -1 \\ 1 & 1 \end{bmatrix}$, and $c\begin{bmatrix} -1 & -1 \\ -1 & -1 \end{bmatrix}$, where c is the number of squares moved.

29b. The knight moves in combinations of 2 vertical-1 horizontal or 1 vertical-2 horizontal squares. These can be either up or down, left or right. Sample matrices are $\begin{bmatrix} 1 & 1 \\ 2 & 2 \end{bmatrix}$, $\begin{bmatrix} 1 & 1 \\ -2 & -2 \end{bmatrix}$, $\begin{bmatrix} -1 & -1 \\ 2 & 2 \end{bmatrix}$, $\begin{bmatrix} -1 & -1 \\ -2 & -2 \end{bmatrix}$, $\begin{bmatrix} 2 & 2 \\ 1 & 1 \end{bmatrix}$, $\begin{bmatrix} 2 & 2 \\ -1 & -1 \end{bmatrix}$, $\begin{bmatrix} -2 & -2 \\ 1 & 1 \end{bmatrix}$, and $\begin{bmatrix} -2 & -2 \\ -1 & -1 \end{bmatrix}$.

29c. The king can move 1 unit in any direction. The matrices describing this are $\begin{bmatrix} 1 & 1 \\ 0 & 0 \end{bmatrix}$, $\begin{bmatrix} -1 & -1 \\ 0 & 0 \end{bmatrix}$, $\begin{bmatrix} 0 & 0 \\ 1 & 1 \end{bmatrix}$, $\begin{bmatrix} 0 & 0 \\ -1 & -1 \end{bmatrix}$, $\begin{bmatrix} 1 & 1 \\ 1 & 1 \end{bmatrix}$, $\begin{bmatrix} 1 & 1 \\ -1 & -1 \end{bmatrix}$, $\begin{bmatrix} -1 & -1 \\ 1 & 1 \end{bmatrix}$, and $\begin{bmatrix} -1 & -1 \\ -1 & -1 \end{bmatrix}$.

31. $(0, -125)$; $(125, 0)$, $(0, 125)$, $(-125, 0)$ **33.** The repeated dilations animate the growth of something from small to larger similar to a lens zooming into the origin. **35.** $\begin{bmatrix} 4 & 13 \\ -4 & 12 \end{bmatrix}$

37. hardbacks \$1, paperbacks \$0.25
39. $4x - y + 9 = 0$ **41.** $x^5 - 3x^4 + 7x^3$, $\dfrac{x^3}{x^2 - 3x + 7}$

Pages 102–105 Lesson 2-5

5. 10 **7.** -413 **9.** $-\dfrac{1}{29}\begin{bmatrix} 7 & -3 \\ -5 & -2 \end{bmatrix}$

11. $\left(-\dfrac{111}{13}, \dfrac{129}{13}\right)$ **13.** 8 kg of the metal with 55% aluminum and 12 kg of the metal with 80% aluminum
15. 4 **17.** 4 **19.** 48 **21.** -37 **23.** 1

25. 175.668 **27.** $-\dfrac{1}{10}\begin{bmatrix} -2 & 3 \\ 2 & 2 \end{bmatrix}$ **29.** $\dfrac{1}{6}\begin{bmatrix} 2 & -2 \\ -1 & 4 \end{bmatrix}$

31. does not exist **33.** $\begin{bmatrix} \frac{1}{2} & \frac{1}{8} \\ -5 & \frac{3}{4} \end{bmatrix}$ **35.** $(0, -2)$

37. $\left(\dfrac{7}{12}, \dfrac{7}{12}\right)$ **39.** $\left(\dfrac{1}{3}, -\dfrac{2}{3}\right)$ **41.** $\left(\dfrac{2}{9}, -\dfrac{4}{3}, -\dfrac{1}{3}\right)$

43. 30,143 **45.** $(2, -1, 3)$

47. Let $A = \begin{bmatrix} a_1 & b_1 \\ a_2 & b_2 \end{bmatrix}$ and $I = \begin{bmatrix} 1 & 0 \\ 0 & 1 \end{bmatrix}$.

$$A^{-1} = \begin{bmatrix} \dfrac{b_2}{a_1b_2 - a_2b_1} & \dfrac{-b_1}{a_1b_2 - a_2b_1} \\[2ex] \dfrac{-a_2}{a_1b_2 - a_2b_1} & \dfrac{a_1}{a_1b_2 - a_2b_1} \end{bmatrix}$$

$$AA^{-1} = \begin{bmatrix} \dfrac{a_1b_2 - a_2b_1}{a_1b_2 - a_2b_1} & \dfrac{-a_1b_1 + b_1a_1}{a_1b_2 - a_2b_1} \\[2ex] \dfrac{a_2b_2 - a_2b_2}{a_1b_2 - a_2b_1} & \dfrac{a_1b_2 - a_2b_1}{a_1b_2 - a_2b_1} \end{bmatrix}$$

$$= \begin{bmatrix} 1 & 0 \\ 0 & 1 \end{bmatrix} = I$$

Thus, $AA^{-1} = I$.

49. Yes

$A = \begin{bmatrix} a & b \\ c & d \end{bmatrix}$. Does $(A^2)^{-1} = (A^{-1})^2$?

$$A^2 = \begin{bmatrix} a^2 + bc & ab + bd \\ ac + cd & bc + d^2 \end{bmatrix}$$

$$(A^2)^{-1} = \frac{1}{a^2d^2 - 2abcd + b^2d^2} \begin{bmatrix} bc + d^2 & -ab - bd \\ -ac - cd & a^2 + bc \end{bmatrix}$$

$$A^{-1} = \frac{1}{ad - bc}\begin{bmatrix} d & -b \\ -c & a \end{bmatrix} = \begin{bmatrix} \dfrac{d}{ad - bc} & \dfrac{-b}{ad - bc} \\[2ex] \dfrac{-c}{ad - bc} & \dfrac{a}{ad - bc} \end{bmatrix}$$

$$(A^{-1})^2 = \frac{1}{a^2d^2 - 2abcd + b^2d^2} \begin{bmatrix} bc + d^2 & -ab - bd \\ -ac - cd & a^2 + bc \end{bmatrix}$$

Thus, $(A^2)^{-1} = (A^{-1})^2$.

51. computer system: $959, printer: $239
53. $H'(5, 9)$, $I'(1, 5)$, $J'(-3, 9)$, $K'(1, 13)$
55. infinitely many solutions   **57.** $x - 2y + 8 = 0$
59a. $\frac{1}{12}$ or approximately 0.0833   **59b.** 1.5 ft
61. No, more than one member of the range is paired with the same member of the domain.

Pages 109–111   Lesson 2-6
5. $(-1, 0)$, $(-1, 3)$, $(0, 4)$, $(7, 0.5)$, $(7, 0)$
7. 3, -11

9.

11.

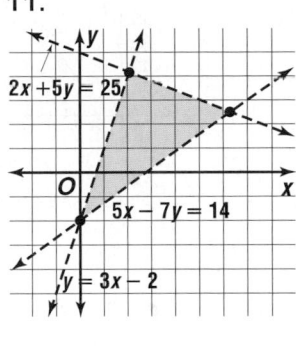

13. $\left(\frac{3}{5}, 3\frac{1}{5}\right)$, $\left(-\frac{2}{5}, 1\frac{1}{5}\right)$, $\left(1\frac{3}{5}, \frac{1}{5}\right)$

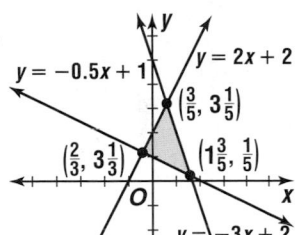

15. $(2, 5)$, $(7, 0)$, $(4, 0)$, $(1, 5)$

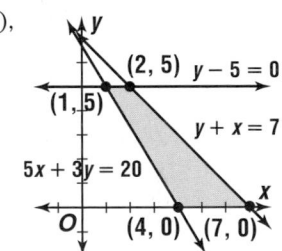

17. 19, 2   **19.** 16, 2   **21.** 9, -4 **23.** $x \leq 4$, $x \geq -4$, $y \leq 4$, $y \geq -4$ **25a.** vertices: $\left(5\frac{1}{2}, 0\right)$, $\left(6\frac{1}{2}, 0\right)$, $\left(9\frac{2}{3}, 6\frac{1}{3}\right)$, $\left(7\frac{1}{2}, 8\frac{1}{2}\right)$, $\left(2\frac{1}{2}, 8\frac{1}{2}\right)$, $\left(1\frac{1}{5}, 4\frac{3}{5}\right)$, $\left(2\frac{1}{2}, 2\right)$ **25b.** max at $\left(7\frac{1}{2}, 8\frac{1}{2}\right) = 88\frac{1}{2}$; min at $\left(2\frac{1}{2}, 2\right) = 24\frac{1}{2}$

27a.

27b. $f(x, y) = 30x + 40y$ **27c.** 80 ft^2 at the Main St. site and 20 ft^2 at the High St. site   **27d.** The maximum number of customers can be reached by renting 120 ft^2 at Main St.   **29.** $\frac{1}{7}\begin{bmatrix} 2 & -1 \\ 3 & 2 \end{bmatrix}$

31.

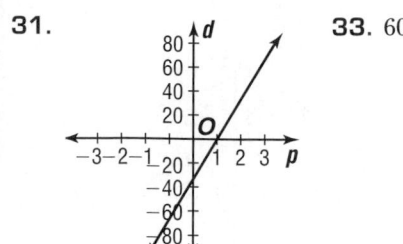

33. 60

Pages 115–118 Lesson 2-7

5a. $25x + 50y \leq 4200$ **5b.** $3x + 5y \leq 480$
5c.

5d. $P(x, y) = 5x + 8y$ **5e.** 160 small packages, 0 large packages **5f.** $800 **5g.** No. If revenue is maximized, the company will not deliver any large packages, and customers with large packages to ship will probably choose another carrier for all of their business. **7.** 225 Explorers, 0 Grande Expeditions
9. infeasible **11.** alternate optimal solutions

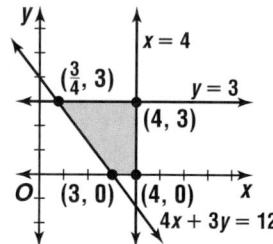

13a. Let d = the number of day-shift workers and n = the number of night-shift workers. $d \geq 5$; $n \geq 6$; $d + n \geq 14$
13b.

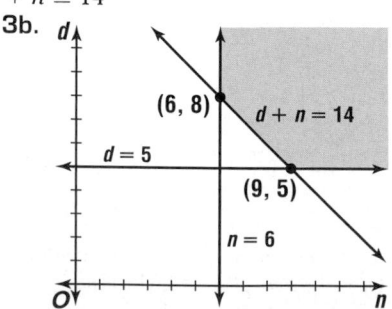

13c. $C(n, d) = 52d + 60n$ **13d.** 8 day-shift and 6 night-shift workers **13e.** $776 **15.** 10 section-I questions, 2 section-II questions **17.** $4000 in First Bank, $7000 in City Bank **19.** 600 units of snack size, 1800 units of family-size

21. alternate optimal solutions **23a.** $720
23b. Sample answer: Spend more than 30 hours per week on these services. **25.** (0, 6) **27.** Sample answer: $C = \$13.65 + \$0.15(n - 30)$; $15.45

Pages 119–123 Chapter 2 Study Guide and Assessment

1. translation **3.** determinant **5.** scalar multiplication **7.** polygonal convex set
9. element **11.** $(2, -4)$ **13.** $\left(-\dfrac{5}{11}, -\dfrac{2}{11}\right)$
15. $(1, 2)$ **17.** $(-10, -6, 0)$ **19.** $(2, -1, 3)$
21. $\begin{bmatrix} -10 & -13 \\ 2 & 2 \end{bmatrix}$ **23.** $\begin{bmatrix} -8 \\ 20 \end{bmatrix}$ **25.** impossible
27. impossible

29. $W'(-2, 3)$, $X'(-1, -2)$, $Y'(0, -4)$, $Z'(1, 2)$
31. $P'(6, -8)$, $Q'(2, 4)$, $R'(-2, 2)$

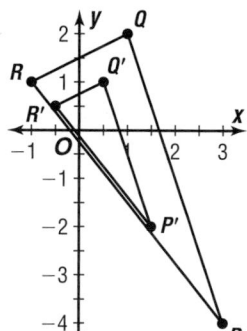

33. $\begin{bmatrix} 1 & 1 & 1 \\ 3 & 3 & 3 \end{bmatrix}$ **35.** 0 **37.** 160 **39.** $\dfrac{1}{23}\begin{bmatrix} 5 & -8 \\ 1 & 3 \end{bmatrix}$
41. $\dfrac{1}{7}\begin{bmatrix} -4 & -5 \\ -1 & -3 \end{bmatrix}$ **43.** $\dfrac{1}{32}\begin{bmatrix} 1 & 5 \\ -6 & 2 \end{bmatrix}$ **45.** $(13, -5)$
47. $(-7, -4)$ **49.** $17, -4$ **51.** 22 gallons in the truck and 6 gallons in the motorcycle **53.** 39 in., 31 in., 13 in.

Page 125 Chapter 2 SAT and ACT Practice
1. D **3.** D **5.** C **7.** D **9.** C

Chapter 3 The Nature of Graphs

Pages 134–136 Lesson 3-1
7. yes **9.** $y = x$ **11.** y-axis
13. x-intercept: (5, 0); other points: $\left(-6, \dfrac{3\sqrt{11}}{5}\right)$, $\left(6, -\dfrac{3\sqrt{11}}{5}\right)$, $\left(-6, -\dfrac{3\sqrt{11}}{5}\right)$

15. no
17. yes **19.** no
21. $y = x$ and $y = -x$

23. none of these **25.** all **27.** x-axis and y-axis, $y = x$, and $y = -x$

29.

31. both

33. x-axis

35. both

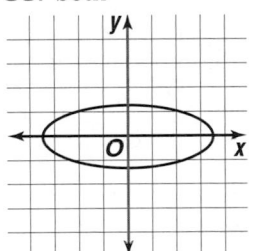

37. The equation $|y| = x^3 - x$ is symmetric about the x-axis.

39. Sample answer: $y = 0$
41. $\left(4\sqrt{2}, 6\right)$ or $\left(-4\sqrt{2}, 6\right)$
43. 50 bicycles, 75 tricycles
45. $(-2, -3, 7)$

47.

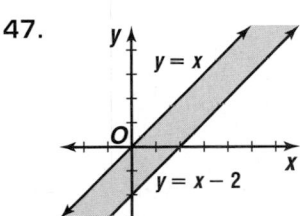

49. $-2x + 23, -2x + 5$

Pages 142–145 Lesson 3-2

7. $g(x)$ is the graph of $f(x)$ compressed horizontally by a factor of $\frac{1}{3}$, reflected over the x-axis.

9a. translated up 3 units, portion of graph below x-axis reflected over the x-axis **9b.** reflected over the x-axis, compressed horizontally by a factor of $\frac{1}{2}$

9c. translated left 1 unit, compressed vertically by a factor of 0.75

11.

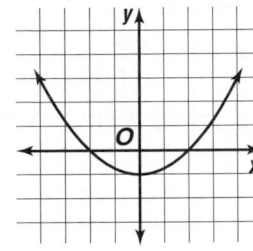

13. The graph of $g(x)$ is a translation of the graph of $f(x)$ up 6 units.
15. The graph of $g(x)$ is the graph of $f(x)$ compressed horizontally by a factor of $\frac{1}{5}$.
17. The graph of $g(x)$ is the graph of $f(x)$ expanded vertically by a factor of 3. **19.** $g(x)$ is the graph of $f(x)$ reflected over the x-axis, expanded horizontally by a factor of 2.5, translated up 3 units. **21a.** expanded horizontally by a factor of 5 **21b.** expanded vertically by a factor of 7, translated down 0.4 units **21c.** reflected across the x-axis, expanded vertically by a factor of 9, translated left 1 unit **23a.** compressed vertically by a factor of $\frac{1}{3}$, translated left 2 units **23b.** reflected over the y-axis, translated down 7 units **23c.** translated right 3 units and up 4 units, expanded vertically by a factor of 2 **25a.** compressed horizontally by a factor of $\frac{2}{5}$, translated down 3 units

25b. reflected over the y-axis, compressed vertically by a factor of 0.75 **25c.** The portion of the parent graph on left of the y-axis is replaced by a reflection of the portion on the right of the y-axis. The new image is then translated 4 units right.

27. $y = \dfrac{0.25}{x - 4} + 3$ **29.**

31.

33.

35a. 0

$[-7.6, 7.6]$ scl:1 by $[-5, 5]$ scl:1

35b. 0.5

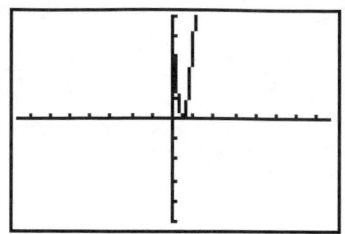

[−7.6, 7.6] scl:1 by [−5, 5] scl:1

35c. −1.5

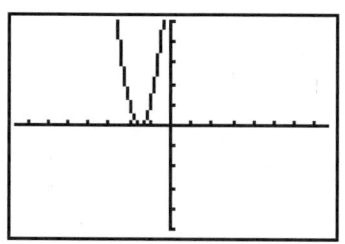

[−7.6, 7.6] scl:1 by [−5, 5] scl:1

37a. 0

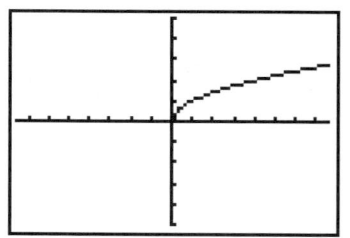

[−7.6, 7.6] scl:1 by [−5, 5] scl:1

37b. −2.5

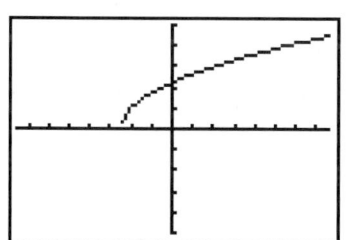

[−7.6, 7.6] scl:1 by [−5, 5] scl:1

37c. 0.6

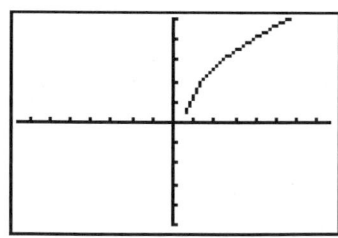

[−7.6, 7.6] scl:1 by [−5, 5] scl:1

39. The x-intercept will be $-\dfrac{b}{a}$.

41a. 25 units^2

41b.

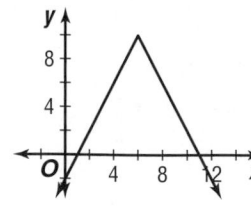

The area of the triangle is $\frac{1}{2}(10)(10)$ or 50 units^2. Its area is twice as large as that of the original triangle. The area of the triangle formed by $y = c \cdot f(x)$ would be 25c units^2.

41c.

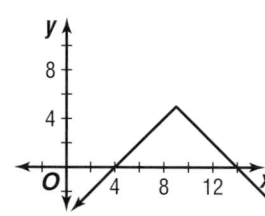

The area of the triangle is $\frac{1}{2}(10)(5)$ or 25 units^2. Its area is the same as that of the original triangle. The area of the triangle formed by $y = f(x + c)$ would be 25 units^2.

43a. reflection over the x-axis, reflection over the y-axis, vertical translation, horizontal compression or expansion, and vertical expansion or compression **43b.** horizontal translation **45.** 30 preschoolers and 20 school-age **47.** $x = \pm 5, y = 9, z = 6$ **49.** The graph implies a negative linear relationship. **51.** −250 **53.** A

Pages 149–151 Lesson 3-3
5. yes **7.**

9.

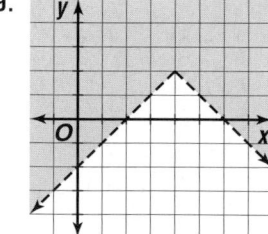

11. $\{x \mid 1 \le x \le 2\}$
13. no **15.** yes
17. yes **19.** (0, 0), (1, 1), and (1, −1); if these points are in the shaded region and the other points are not, then the graph is correct.

21.

23.

25.

27.

29.

31.
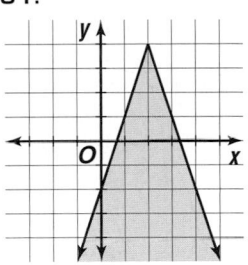

33. $\{x \mid x < -9 \text{ or } x > 1\}$ **35.** $\{x \mid -2 < x < 9\}$
37. no solution   **39.** $\{x \mid -17 \le x \le 7\}$
41. $\{x \mid 5.5 < x < 10\}$ **43.** $x \ge 83\frac{2}{3}$

45a.
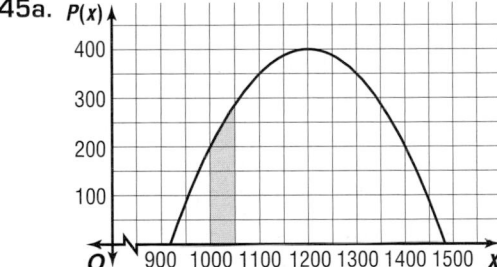

45b. The shaded region shows all points (x, y) where x represents the number of cookies sold and y represents the possible profit made for a given week.

47. y-axis
49. $\begin{bmatrix} 6 & \dfrac{-21}{4} \\ -3 & 0 \end{bmatrix}$
53. 10

51.
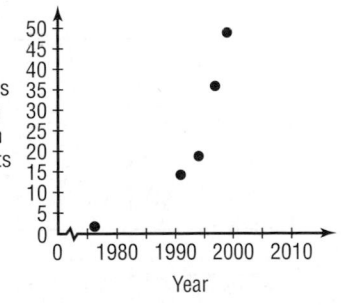
States with Teen Courts
Year

Pages 156–158   Lesson 3-4

7.

9. $f^{-1}(x) = -\dfrac{1}{3}x + \dfrac{2}{3}$; $f^{-1}(x)$ is a function.
11. $f^{-1}(x) = -2 \pm \sqrt{x - 6}$; $f^{-1}(x)$ is not a function.

13. $f^{-1}(x) = 2x + 10$; $[f \circ f^{-1}](x) = f(2x + 10) = \dfrac{1}{2}(2x + 10) - 5 = x$

$[f^{-1} \circ f](x) = f^{-1}\left(\dfrac{1}{2}x - 5\right) = 2\left(\dfrac{1}{2}x - 5\right) + 10 = x$

Since $[f \circ f^{-1}](x) = [f^{-1} \circ f](x) = x$, f and f^{-1} are inverse functions.

15.

17.

19.

21.

23.

25. $f^{-1}(x) = \dfrac{x - 7}{2}$; $f^{-1}(x)$ is a function.
27. $f^{-1}(x) = \dfrac{1}{x}$; $f^{-1}(x)$ is a function.
29. $f^{-1}(x) = 3 \pm \sqrt{x - 7}$; $f^{-1}(x)$ is not a function.

31. $f^{-1}(x) = \dfrac{1}{x} - 2$; $f^{-1}(x)$ is a function.
33. $f^{-1}(x) = 2 - \sqrt[3]{\dfrac{2}{x}}$; $f^{-1}(x)$ is a function.

35.

37.
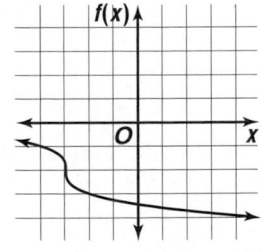

39. $f^{-1}(x) = -\dfrac{3}{2}x + \dfrac{1}{4}$

$[f \circ f^{-1}](x) = f\left(-\dfrac{3}{2}x + \dfrac{1}{4}\right)$

$\qquad = -\dfrac{2}{3}\left(-\dfrac{3}{2}x + \dfrac{1}{4}\right) + \dfrac{1}{6}$

$\qquad = x - \dfrac{1}{6} + \dfrac{1}{6}$

$\qquad = x$

$[f^{-1} \circ f](x) = f^{-1}\left(-\dfrac{2}{3}x + \dfrac{1}{6}\right)$

$\qquad = -\dfrac{3}{2}\left(-\dfrac{2}{3}x + \dfrac{1}{6}\right) + \dfrac{1}{4}$

$\qquad = x - \dfrac{1}{4} + \dfrac{1}{4}$

$\qquad = x$

Since $[f \circ f^{-1}](x) = [f^{-1} \circ f](x) = x$, f and f^{-1} are inverse functions.

41a.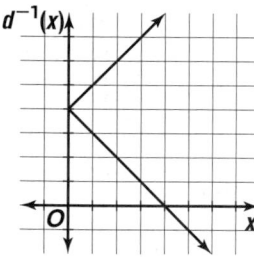

41b. No; the graph of $d(x)$ fails the horizontal line test. **41c.** $d^{-1}(x)$ gives the numbers that are 4 units from x on the number line. There are always two such numbers, so d^{-1} associates two values with each x-value. Hence, $d^{-1}(x)$ is not a function.   **43a.** Sample answer: $y = -x$.   **43b.** The graph of the function must be symmetric about the line $y = x$.   **43c.** Yes, because the line $y = x$ is the axis of symmetry *and* the reflection line.   **45.** It must be translated up 6 units <u>and</u> 5 units to the left; $y = (x - 6)^2 - 5$, $y = 6 \pm \sqrt{x + 5}$.   **47a.** Yes. If the encoded message is not unique, it may not be decoded properly.   **47b.** The inverse of the encoding function must be a function so that the encoded message may be decoded.
47c. $(x + 2)^2 - 3$
47d. FUNCTIONS ARE FUN   **49.** both
51. $(-1, 7)$
55. $y = -x + 7$

53.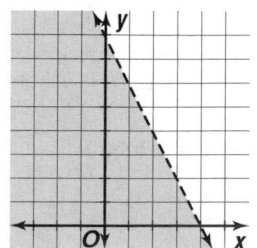

Pages 165–168   Lesson 3-5

Pages 165–168   Lesson 3-5
5. No. y is undefined when $x = -3$.
7. $y \to \infty$ as $x \to \infty$, $y \to -\infty$ as $x \to -\infty$

9. 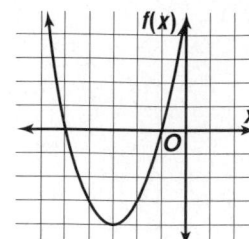   Decreasing for $x < -3$. Increasing for $x > -3$.

11a. $t = 4$ **11b.** when $t = 4$ **11c.** 10 amps
13. No. The function is undefined when $x = 2$.
15. Yes. The function is defined when $x = 3$; the function approaches 1 (in fact is equal to 1) as x approaches 3 from both sides; and $y = 1$ when $x = 3$.   **17.** Yes. The function is defined when $x = 1$; $f(x)$ approaches 3 as x approaches 1 from both sides; and $f(1) = 3$.   **19.** Sample answer: $x = 0$. $g(x)$ is undefined when $x = 0$.   **21.** $y \to -\infty$ as $x \to \infty$, $y \to -\infty$ as $x \to -\infty$ **23.** $y \to \infty$ as $x \to \infty$, $y \to \infty$ as $x \to -\infty$ **25.** $f(x) \to 2$ as $x \to \infty$, $f(x) \to 2$ as $x \to -\infty$

27. 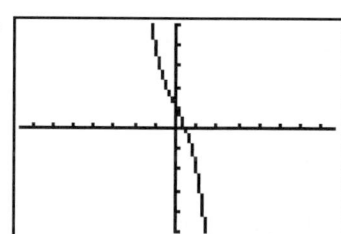   decreasing for all x

$[-7.6, 7.6]$ scl:1 by $[-5, 5]$ scl:1

29. 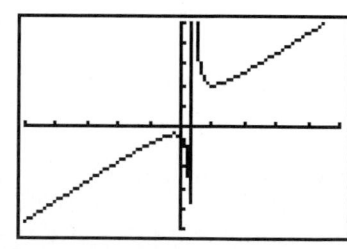   increasing for $x < -1$ and $x > 5$; decreasing for $-1 < x < 2$ and $2 < x < 5$

$[-25, 25]$ scl:5 by $[-25, 25]$ scl:5

31.   decreasing for $x < -\dfrac{3}{2}$ and $0 < x < \dfrac{3}{2}$; increasing for $-\dfrac{3}{2} < x < 0$ and $x > \dfrac{3}{2}$

$[-7.6, 7.6]$ scl:1 by $[-1, 9]$ scl:1

33a. f is decreasing for $-2 < x < 0$ and increasing for $x < -2$. f has jump discontinuity when $x = -3$ and $f(x) \to -\infty$ as $x \to -\infty$.

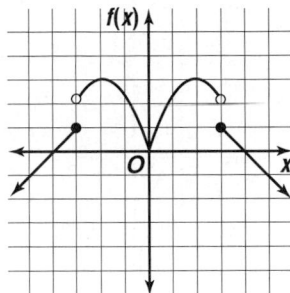

33b. f is increasing for $-2 < x < 0$ and decreasing for $x < -2$. f has a jump discontinuity when $x = -3$ and $f(x) \to \infty$ as $x \to -\infty$.

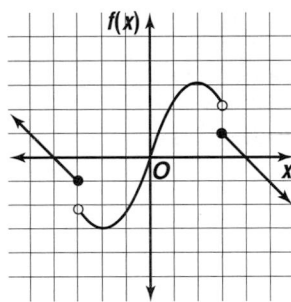

35a. 1954–1956, 1958–1959, 1960–1961, 1962–1963, 1966–1968, 1973–1974, 1975–1976, 1977–1978, 1989–1991 **35b.** 1956–1958, 1959–1960, 1961–1962, 1963–1966, 1968–1973, 1974–1975, 1976–1977, 1978–1989, 1991–1996 **37a.** The function must be monotonic. **37b.** The inverse must be monotonic.
39. $a = 4$, $b = 2$ **41.** The graph of $g(x)$ is the graph of $f(x)$ translated left 2 units and down 4 units.
43. 42 **45.** 20

Pages 176–179 Lesson 3-6
5. rel. min.: $(-1, -3)$; rel. max.: $(3, 3)$
7. rel. min.: $(-2.25, -10.54)$

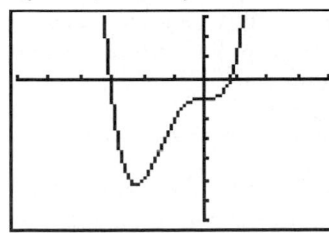

[−6, 4] scl:1 by [−14, 6] scl:2

9. min. **11.** min. **13.** abs. max.: $(-4, 1)$
15. rel. max.: $(-2, 7)$; abs. min.: $(3, -3)$
17. abs. min.: $(3, -8)$; rel. max.: $(5, -2)$; rel. min.: $(8, -5)$

19. abs. max.: $(1.5, -1.75)$

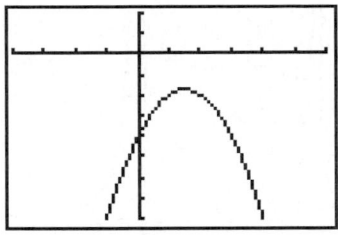

[−5, 5] scl:1 by [−8, 2] scl:1

21. rel. max.: $(-0.59, 0.07)$, rel. min.: $(0.47, -3.51)$

[−5, 5] scl:1 by [−5, 5] scl:1

23. rel. max.: $(-1, 1)$; rel. min.: $(0.25, -3.25)$

[−5, 5] scl:1 by [−5, 5] scl:1

25. abs. min.: $(-3.18, -15.47)$; rel. min.: $(0.34, -0.80)$; rel. max.: $(-0.91, 3.04)$ **27.** max. **29.** max.
31. pt. of inflection **33.** min. **35a.** $V(x) = 2x(12.5 - 2x)(17 - 2x)$ **35b.** 2.37 cm by 2.37 cm
37a. $f(x) = 5000\sqrt{x^2 + 4} + 3500(10 - x)$
37b. 1.96 km from point B **39.** The particle is at rest when $t \approx 0.14$ and when $t \approx 3.52$. Its position at these times are $s(0.14) \approx -8.79$ and $s(3.52) \approx -47.51$. **41.** No; the function is undefined when $x = 5$. **43.** 120 units of notebook and 80 units of newsprint **45.** -1, yes **47.** 5 free throws, 9 2-point field goals, 3 3-point field goals
49. perpendicular **51.** D

Pages 185–188 Lesson 3-7
5. $x = 5$, $y = 1$ **7.** $f(x) = \dfrac{1}{x + 1} - 2$
9. The parent graph is translated 2 units to the left and down 1 unit. The vertical asymptote is now at $x = -2$ and the horizontal asymptote is now $y = -1$.

11.

13a.

37.

39.

13b. $P = 0$, $V = 0$ **13c.** The pressure approaches 0. **15.** $x = -6$ **17.** $x = -1$, $x = -3$, $y = 0$ **19.** $x = 1$, $y = 1$ **21.** $f(x) = \dfrac{1}{x + 3} + 1$ **23.** $f(x) = -\dfrac{1}{x} + 1$

25. The parent graph is translated 4 units right and expanded vertically by a factor of 2. The vertical asymptote is now $x = 4$. The horizontal asymptote, $y = 0$, is unchanged.

27. The parent graph is expanded vertically by a factor of 3, reflected about the x-axis, and translated 2 units up. The vertical asymptote, $x = 0$, is unchanged. The horizontal asymptote is now $y = 2$.

29. The parent graph is translated 5 units left. The translated graph is expanded vertically by a factor of 22 and then translated 4 units down. The vertical asymptote is $x = -5$ and the horizontal asymptote is $y = -4$.

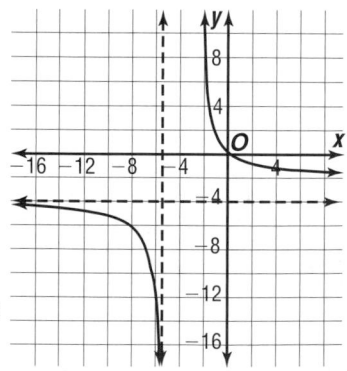

31. $y = x + 3$
33. $y = \dfrac{1}{2} x - \dfrac{5}{4}$

35.

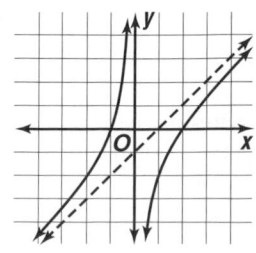

41a. $C(t) = \dfrac{480 + 3t}{40 + t}$ **41b.** 11.43 L **43.** Sample answer: $f(x) = \dfrac{(x - 2)(x + 3)(x + 5)^2}{(x - 4)(x + 5)}$ **45.** Sample answer: $f(x) = \dfrac{x}{x^2 + 1}$ **47a.** $\dfrac{a^2 - 9}{a - 3}$ **47b.** The slope approaches 6. **49.** $y = \pm\sqrt{x + 9}$ **51.** $\begin{bmatrix} 24 & -20 \\ -32 & 16 \end{bmatrix}$ **53.** $(3, 2)$ **55.** $16 - 8x^2$, $2 - 64x^2$

Pages 193–196 Lesson 3-8
5. 12, $xy = 12$; $\dfrac{4}{5}$ **7.** 0.5; $y = 0.5xz^3$; 108
9. y varies directly as x^4, $\dfrac{1}{7}$ **11.** y varies inversely as x; -3 **13.** 0.2; $y = 0.2x$; 1.2 **15.** 15, $y = 15xz$; 18 **17.** 16; $r = 16t^2$; 1 **19.** $\dfrac{1}{12}$; $y = \dfrac{1}{12} x^3z^2$; -48
21. 2; $y = \dfrac{2xz}{w}$; 14 **23.** 15; $a = \dfrac{15b^2}{c}$; ± 8 **25.** C varies directly as d; π **27.** y varies jointly as x and the square of z; $\dfrac{4}{3}$ **29.** y varies inversely as the square of x, $\dfrac{5}{4}$ **31.** A varies jointly as h and the quantity $b_1 + b_2$; 0.5 **33.** y varies directly as x^2 and inversely as the cube of z; 7 **35a.** Joint variation; to reduce torque one must either reduce the distance or reduce the mass on the end of the fulcrum. Thus, torque varies directly as the mass and the distance from the fulcrum. Since there is more than one quantity in direct variation with the torque on the seesaw, the variation is joint.
35b. $T_1 = km_1d_1$ and $T_2 = km_2d_2$
$\quad T_1 = T_2$
$\quad km_1d_1 = km_2d_2$ *Substitution property of equality*
$\quad m_1d_1 = m_2d_2$
35c. 1.98 meters **37.** If y varies directly as x then there is a nonzero constant k such that $y = kx$. Solving for x, we find $x = \dfrac{1}{k} y$. $\dfrac{1}{k}$ is a nonzero constant, so x varies directly as y. **39.** a is doubled
$a = \dfrac{kb^2}{c^3}$; $a = \dfrac{k\left(\frac{1}{2} b\right)^2}{\left(\frac{1}{2} c\right)^3}$; $\dfrac{a = \frac{1}{4} kb^2}{\frac{1}{8} c^3}$; $a = 2\,\dfrac{kb^2}{c^3}$

41. $1.78 \times 10^{-3}\ \Omega$ **43.** $f^{-1}(x) = \sqrt[3]{x - 6} + 3$; $f^{-1}(x)$ is a function. **45.** consistent and dependent **47.** $y = -0.92x + 1858.60$

Pages 197–201 Chapter 3 Study Guide and Assessment

1. even **3.** point **5.** maximum **7.** inverse **9.** slant **11.** yes **13.** no **15.** $y = x$ and $y = -x$ **17.** none **19.** $g(x)$ is a translation of the graph of $f(x)$ up 5 units. **21.** $g(x)$ is the graph of $f(x)$ expanded vertically by a factor of 6.

23.   **25.**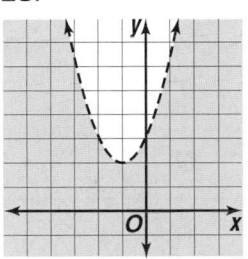

27. $\{x \mid x < -3 \text{ or } x > 0.5\}$

29.   **31.**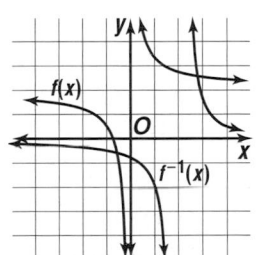

33. $f^{-1}(x) = \sqrt[3]{x + 8} + 2$; yes **35.** Yes. The function is defined when $x = 2$; the function approaches 6 as x approaches 2 from both sides; and $y = 6$ when $x = 2$. **37.** Yes. The function is defined when $x = 1$; the function approaches 2 as x approaches 1 from both sides; and $y = 2$ when $x = 1$. **39.** $y \to \infty$ as $x \to \infty$, $y \to -\infty$ as $x \to -\infty$ **41.** $y \to \infty$ as $x \to \infty$, $y \to -\infty$ as $x \to -\infty$ **43.** decreasing for $x < -3$ and $0 < x < 3$; increasing for $-3 < x < 0$ and $x > 3$ **45.** rel. max.: $(0, 4)$, rel. min.: $(2, 0)$ **47.** pt. of inflection **49.** $f(x) = -\dfrac{2}{x}$

51. The parent graph is translated 3 units right and then translated 2 units up. The vertical asymptote is now $x = 3$ and the horizontal asymptote is $y = 2$.

53. $x = -2$ **55.** yes; $y = x + 2$ **57.** 140; $y = \dfrac{140}{\sqrt{x}}$; 196 **59.** $|x - 6.5| \le 0.2$; $6.3 \le x \le 6.7$ **61b.** 1.08 m

61a.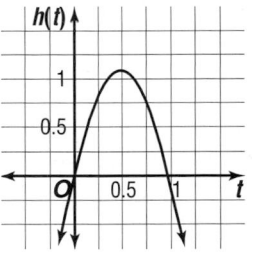

Page 203 Chapter 3 SAT and ACT Practice

1. E **3.** B **5.** C **7.** B **9.** B

Chapter 4 Polynomial and Rational Functions

Pages 209–212 Lesson 4-1

5. $3; 1$ **7.** no; $f(5) = -33$ **9.** $x^2 - 2x - 35 = 0$; even; 2 **11.** $2; 7, 7$ **13.** $4; -1, 1, i, -i$

  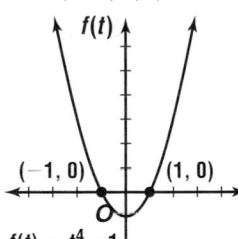

15. $4; 5$ **17.** $3; 5$ **19.** $6; -1$ **21.** Yes; the coefficients are complex numbers and the exponents of the variable are nonnegative integers. **23.** yes; $f(0) = 0$ **25.** yes; $f(1) = 0$ **27.** no; $f(-3) = -72$ **29.** no **31a.** $3; 1$ **31b.** $2; 2$ **31c.** $4; 2$ **33.** $x^3 - 5x^2 - x + 5 = 0$; odd; 3 **35.** $x^3 + 3x^2 + 4x + 12 = 0$; odd; 1 **37.** $x^5 - 5x^4 - 17x^3 + 85x^2 + 16x - 80 = 0$; odd; 5 **39.** $1; -8$ **41.** $2; \pm 6i$

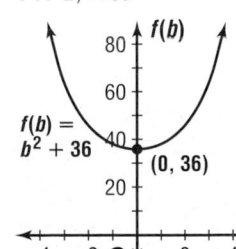

43. 3; $-3, 0, 3$

$(-3, 0)$ $(0, 0)$ $(3, 0)$

$f(n) = n^3 - 9n$

45. 4; $-1, 1, -\sqrt{2}i, \sqrt{2}i$

$(-1, 0)$ $(1, 0)$

$f(a) = a^4 + a^2 - 2$ $(0, -2)$

63.

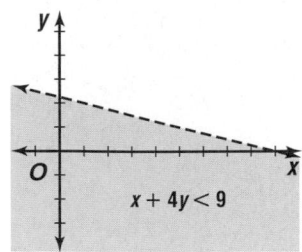

$x + 4y < 9$

65. $\frac{1}{4}x^2 + 6x + 32$; $\frac{1}{2}x^2 + 4$

47. 4; $-0.5i, 0.5i, -2i, 2i$ **49a.**

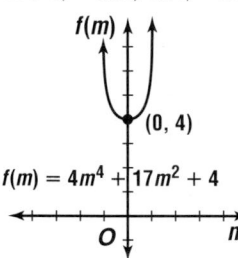

$(0, 4)$

$f(m) = 4m^4 + 17m^2 + 4$

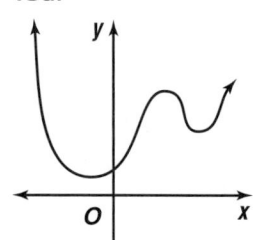

Pages 219–221 Lesson 4-2
5. $-10, 2$ **7.** 0; 1 real; -6 **9.** 1, 5 **11.** 200 or 400 amps **13.** $-8, 11$ **15.** $\frac{1}{2}, \frac{1}{4}$ **17.** $\frac{3}{2} \pm \frac{\sqrt{37}}{2}$
19. 2 imaginary; the discriminant is negative.
21. -11; 2 imaginary; $\frac{5 \pm i\sqrt{11}}{2}$ **23.** -140;
2 imaginary; $\frac{1 \pm i\sqrt{35}}{4}$ **25.** 97; 2 real; $\frac{-5 \pm \sqrt{97}}{4}$
27. $5 + 2i$ **29.** $-4, 7$ **31.** $-1, \frac{5}{4}$
33. $\sqrt{6} \pm 2\sqrt{2}$ **35.** $c > 16$
37a.

$d(t) = 5t - 16t^2$

37b. 0 and about 0.3
37c. The x-intercepts indicate when the woman is at the same height as the beginning of the jump. **37d.** $-50 = 5t - 16t^2$
37e. about 1.93 s

49b.

49c.

39. 2; $-\frac{1}{3}, \frac{1}{6}$

$f(a) = 18a^2 + 3a - 1$

41. $f^{-1}(x) = \pm\sqrt{x} + 9$
43. $643
45. A

49d.

49e.

49f. not possible **51a.** $V(x) = 99,000x^3 + 55,000x^2 + 65,000x$ **51b.** about $298,054.13 **53a.** 7380 ft; 29,520 ft; 118,080 ft **53b.** It quadruples; $(2t)^2 = 4t^2$.
55. $10 **57.** $y = \dfrac{x - 2}{x(x + 2)(x - 2)}$ **59.** The graph of $y = 2x^3 + 1$ is the graph of $y = 2x^3$ shifted 1 unit up.
61. 0; no

Pages 226–228 Lesson 4-3
5. $x + 1$, R6 **7.** 0; yes **9.** $(x - 5), (x + 1), (x - 1)$
11. -4 **13.** $r = 1$ in., $h = 5$ in. **15.** $x^2 - 6x + 9$, R-1 **17.** $x^3 - 2x^2 - 4x + 8$ **19.** $2x^2 + 2x$, R -3
21. 0; yes **23.** 12; no **25.** 0; yes **27.** $(\sqrt{6})^4 - 36 = 36 - 36$ or 0 **29.** $(x - 2), (x + 1), (x + 2)$
31. $(x - 4), (x - 2), (x + 1)$ **33.** $(x - 1), (x + 1), (x + 4)$ **35.** 2 times **37.** -2 **39.** 34 **41.** 5 s
43a. $V(x) = -x^3 + 12x^2 - 47x + 60$

43b.

43c. $36 = -x^3 + 12x^2 - 47x + 60$ **43d.** about 0.60 ft

45. $a = 1, b = -6, c = 25$ **47a.** no **47b.** yes
47c. no **47d.** yes **49.** wider than parent graph
and moved 1 unit left **51.** $\left(-\dfrac{91}{11}, \dfrac{68}{11}, \dfrac{98}{11}\right)$ **53.** D

Page 233–235 Lesson 4-4
5. $\pm 1, \pm 2; 1$ **7.** 2 or 0; 1; $-1\frac{1}{2}, \frac{1}{4}, 2$ **9.** 9 cm
11. $\pm 1, \pm 2, \pm 3, \pm 6, \pm 9, \pm 18; -2$ **13.** $\pm 1, \pm 2, \pm 4,$
$\pm 5, \pm 10, \pm 20; -2, 2, 5$ **15.** $\pm 1, \pm\frac{1}{2}, \pm\frac{1}{3}, \pm\frac{1}{6}; -\frac{1}{3}, \frac{1}{2}$
17. 1; 2 or 0; $-2, -1, 3$ **19.** 1; 2 or 0; $-4, -2, 3$
21. 2 or 0; 2 or 0; $-4, -1, 1, 2$ **23a.** $2, -2, -1$
23b. $f(x) = x^4 + 2x^3 - 3x^2 - 8x - 4$ **23c.** 1; 3 or 1
23d. There are 2 negative zeros, but according to
Descartes' rule of signs, there should be 3 or 1. This
is because, -1 is actually a zero twice.
25a. Sample answer: $x^4 + x^3 + x^2 + x + 3 = 0$
25b. Sample answer: $x^3 - x^2 - 2 = 0$ **25c.** Sample
answer: $x^3 - x = 0$ **27.** 100 ft **29.** $x - 8$
31. $x^4 - 5x^2 + 4 = 0$ **33.** A

Pages 240–242 Lesson 4-5
5. 4 and 5, -1 and 0 **7.** 2.3 **9.** Sample answers:
2; 0 **11a.** $V(x) = x^3 + 60x^2 + 1025x + 3750$
11b. $5625 = x^3 + 60x^2 + 1025x + 3750$ **11c.**
about 26.7 cm by 31.7 cm by 6.7 cm **13.** 0 and 1, 2
and 3 **15.** -3 and $-2, -2$ and $-1, 1$ and 2, 2 and 3
17. no real zeros **19.** $-0.7, 0.7$ **21.** -2.5
23. $-1, 1$ **25.** -1.24 **27.** Sample answers: 2; -1
29. Sample answers: 2; -6 **31.** Sample answers:
1; -7 **33a.** The model is fairly close, although it is
less accurate for 1950 and 1970. **33b.** $-253,800$
33c. The population becomes 0. **33d.** No;
there are still many people living in Manhattan.
35a. $37.44 = 60x^3 + 60x^2 + 60x$ **35b.** $f(x) = 60x^3 + 60x^2 + 60x - 37.44$

35c.   about $\frac{1}{2}$ **35d.** 0.4

37a.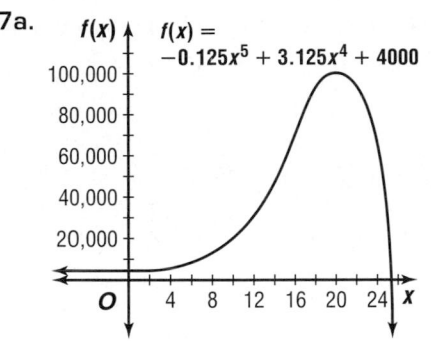

37b. 4000 deer **37c.** about 67,281 deer **37d.** in
1930 **39.** 2 or 0; 1; $-3, 0.5, 5$
41.   **43.** (2.5, 1) **45.** B

Pages 247–250 Lesson 4-6
5. $-1, 5$ **7.** -3 **9.** $x < 0, x > 3$
11a. $\dfrac{3 \times 60 + 20}{3 + x} = 57.14$ **11b.** 0.50 h **13.** -34
15. $\dfrac{5}{3}$ **17.** $-\dfrac{1}{2}, 3$ **19.** $\dfrac{-3 \pm 3\sqrt{2}}{2}$ **21.** $\dfrac{5}{13}$
23. $\dfrac{3}{x} + \dfrac{-2}{x - 2}$ **25.** $\dfrac{2}{3y - 1} + \dfrac{-2}{y - 1}$ **27a.** $a(a - 6)$
27b. 3 **27c.** 0, 6 **27d.** $0 < a < 3, 6 < a$
29. $x \le 3, 4 \le x < 5$ **31.** $0 < a < \dfrac{7}{4}$
33. $-1 < y < 0$ **35.** $x < -5$, or $x > 5$ **37.** Sample
answer: $\dfrac{x}{x - 3} = \dfrac{1}{x + 2}$ **39a.** $\dfrac{1}{10} = \dfrac{1}{2r} + \dfrac{1}{r} + \dfrac{1}{20}$
39b. 60 ohms, 30 ohms **41.** 36 mph
43a. $\dfrac{1}{x} = \dfrac{1}{2}\left(\dfrac{1}{30} + \dfrac{1}{45}\right)$ **43b.** 36 **45.** 8 mph
47. -3 and $-2, -2$ and $-1, 1$ and 2 **49.** 2; $\dfrac{5}{6}, -\dfrac{3}{2}$
51. no **53a.** 18 short answer and 2 essay for a
score of 120 points **53b.** 12 short answer and
8 essay for a score of 180 points **55.** $2x - y + 7 = 0$ **57.** 24

Pages 254–257 Lesson 4-7
5. -733 **7.** 3.5 **9.** $-0.8 \le x \le 12$
11a. $90 = \sqrt{100 + 64h}$ **11b.** 125 ft **13.** 71
15. 0 **17.** no real solution **19.** -1 **21.** 4
23. $\dfrac{9}{7}$ **25.** no real solution **27.** -2 **29.** $x \ge 16$
31. $5 \le a \le 21$ **33.** $1.8 \le y \le 5$ **35.** $c > 27$
37. about 7.88 **39a.** about 2.01 s **39b.** about
2.11 s **39c.** It must be multiplied by 4.
41. $a + b < 0$ **43.** $\dfrac{3}{2}$ **45a.** point discontinuity

45b. jump discontinuity **45c.** infinite discontinuity
47. $\begin{bmatrix} 28 & 20 \\ 10 & 14 \end{bmatrix}$ **49.** 10 students **51.** C

Pages 262–264 Lesson 4-8
5. Sample answer: $f(x) = 1.98x^4 + 2.95x^3 - 5.91x^2 + 0.22x + 4.89$ **7a.** Sample answer: $f(x) = 0.49x + 57.7$ **7b.** Sample answer: 87.1% **7c.** Sample answer: 2006 **9.** quadratic **11.** quadratic
13. $f(x) = 8x^2 - 3x - 9$ **15.** Sample answer: $f(x) = 0.09x^3 - 2.70x^2 + 24.63x - 65.21$
17. Sample answer: $f(x) = -0.02x^3 + 8.79x^2 + 3.35x + 27.43$ **19a.** Sample answer: $f(x) = 0.126x + 22.732$ **19b.** Sample answer: 36 **19c.** Sample answer: 38 **21a.** Sample answer: $f(x) = -0.03x^4 + 0.50x^3 - 2.79x^2 + 4.01x + 22.78$ **21b.** Sample answer: about 16% **23a.** Sample answer: $f(x) = 0.02x^3 - 0.46x^2 + 3.94x + 47.49$ **23b.** Sample answer: 2000 **23c.** Sample answer: No; according to the model, there should have been an attendance of only about 65 million. Since the actual attendance was much higher than the projected number, it is likely that the race to break the homerun record increased the attendance. **25.** 23
27. $-1, 1$ **29.** B

Pages 267–271 Chapter 4 Study Guide and Assessment
1. Quadratic Formula **3.** zero **5.** polynomial function **7.** Extraneous **9.** complex numbers
11. no; $f(0) = -4$ **13.** no; $f(-2) = -18$
15. $3; -3, 0, 1$

17. 40; 2 real; $\dfrac{5 \pm \sqrt{10}}{3}$

19. 73; 2 real; $\dfrac{3 \pm \sqrt{73}}{4}$

$f(x) = x^3 + 2x^2 - 3x$

21. -199; 2 imaginary; $\dfrac{1 \pm i\sqrt{199}}{10}$

23. 161; no **25.** 0; yes **27.** $\pm 1; 1$ **29.** $\pm 1, \pm 3, \pm\dfrac{1}{2}, \pm\dfrac{3}{2}; -\dfrac{3}{2}$ **31.** $\pm 1, \pm 2, \pm 4, \pm 8, \pm\dfrac{1}{3}, \pm\dfrac{2}{3}, \pm\dfrac{4}{3}, \pm\dfrac{8}{3}; -2, -\dfrac{4}{3}, 1$ **33.** $\pm 5, \pm 1; \pm 1$ **35.** 2 or 0; 1; $-\dfrac{1}{2}, 3$ **37.** -1 and 0 **39.** -1 and 0, 3 and 4
41. -2 and -1, 0 and 1, 1 and 2 **43.** $-4.9, -1.8, 2.2$ **45.** 3 **47.** $y < -1, y > 0$ **49.** 23 **51.** 1
53. $a \geq -1.5$ **55.** $f(x) = 2x^2 - x + 3$ **57.** 18 ft by 24 ft **59.** about 0.64 m

Page 273 Chapter 4 SAT and ACT Practice
1. C **3.** E **5.** A **7.** E **9.** A

Chapter 5 The Trigonometric Functions

Pages 280–283 Lesson 5-1
5. $34°\ 57'$ **7.** $-128.513°$ **9.** $-720°$
11. $22° + 360k°$; Sample answers: $382°$; $-338°$
13. $93°$; II **15.** $47°$ **17.** $15°$; $0.25°$ or $15'$; about $0.0042°$ or $15''$ **19.** $168°\ 21'$ **21.** $286°\ 52'\ 48''$
23. $246°\ 52'\ 33.6''$ **25.** $-14.089°$ **27.** $173.410°$
29. $1002.508°$ **31.** $720°$ **33.** $-2700°$
35. $-2070°$ **37.** $30° + 360k°$; Sample answers: $390°$; $-330°$ **39.** $113° + 360k°$; Sample answers: $473°$; $-247°$ **41.** $-199° + 360k°$; Sample answers: $161°$; $-559°$ **43.** $310°$ **45.** $40°$; I **47.** $220°$; III
49. $96°$; II **51.** III **53.** $32°$ **55.** $60°$
57. $35°$ **59.** $4500°$; $270,000°$ **61.** $17,100°$
63. $22,320°$; $1,339,200°$; $80,352,000°$; $1,928,448,000°$
65a. $44°\ 26'\ 59.64''$; $68°\ 15'\ 41.76''$ **65b.** $24.559°$; $81.760°$ **67a.** Sample answer: $f(x) = -0.0003x^3 + 0.0647x^2 - 3.5319x + 76.0203$ **67b.** Sample answer: about 32% **69.** $0, -4$ **71.** $x^3 + x^2 - 80x - 300 = 0$
73. point discontinuity

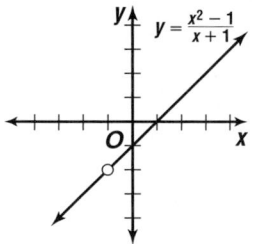

$y = \dfrac{x^2 - 1}{x + 1}$

75. expanded vertically by a factor of 3, translated down 2 units **77.** $0.56x$

Pages 288–290 Lesson 5-2
5. $\dfrac{15\sqrt{514}}{514}$; $\dfrac{17\sqrt{514}}{514}$; $\dfrac{15}{17}$ **7.** $\dfrac{1}{1.5} \approx 0.6667$
9. $I_t = 0.5I_o$ **11.** $\dfrac{5\sqrt{89}}{89}$; $\dfrac{8\sqrt{89}}{89}$; $\dfrac{5}{8}$ **13.** tangent
15. $\dfrac{7}{3}$ **17.** $\dfrac{1}{2.5} = 0.4$ **19.** $\dfrac{1}{0.125} = 8$
21. $\sin R = \dfrac{19}{20}$, $\cos R = \dfrac{\sqrt{39}}{20}$, $\tan R = \dfrac{19\sqrt{39}}{39}$, $\csc R = \dfrac{20}{19}$, $\sec R = \dfrac{20\sqrt{39}}{39}$, $\cot R = \dfrac{\sqrt{39}}{19}$
23. 1.3
25.

θ	72°	74°	76°	78°
sin	0.951	0.961	0.970	0.978
cos	0.309	0.276	0.242	0.208
80°	**82°**	**84°**	**86°**	**88°**
0.985	0.990	0.995	0.998	0.999
0.174	0.139	0.105	0.070	0.035

25a. 1 **25b.** 0 **27.** about 1.5103 **29a.** about 5.4 m/s **29b.** about 5.9 m/s **29c.** about 6.4 m/s
29d. increase **31a.** about 87.5°; about 40.5°
31b. about 49.5°; about 2.5° **31c.** neither
33. 88° 22′ 12″ **35a.** 23 employees **35b.** $1076
37. $y = -\frac{1}{2}x + 6$

Pages 296–298 Lesson 5-3

5. 0 **7.** $\sin 30° = \frac{1}{2}$, $\cos 30° = \frac{\sqrt{3}}{2}$,

$\tan 30° = \frac{\sqrt{3}}{3}$, $\csc 30° = 2$, $\sec 30° = \frac{2\sqrt{3}}{3}$,

$\cot 30° = \sqrt{3}$ **9.** $\sin \theta = \frac{4}{5}$, $\cos \theta = \frac{3}{5}$, $\tan \theta = \frac{4}{3}$,

$\csc \theta = \frac{5}{4}$, $\sec \theta = \frac{5}{3}$, $\cot \theta = \frac{3}{4}$ **11.** $\sin \theta = -\frac{\sqrt{2}}{2}$,

$\cos \theta = \frac{\sqrt{2}}{2}$, $\csc \theta = -\sqrt{2}$, $\sec \theta = \sqrt{2}$, $\cot \theta = -1$

13. The distances range from about 24,881 miles to 0 miles. **15.** 0 **17.** −1 **19.** −1 **21.** undefined

23. $\sin 150° = \frac{1}{2}$, $\cos 150° = -\frac{\sqrt{3}}{2}$,

$\tan 150° = -\frac{\sqrt{3}}{3}$, $\csc 150° = 2$, $\sec 150° = -\frac{2\sqrt{3}}{3}$,

$\cot 150° = -\sqrt{3}$ **25.** $\sin 210° = -\frac{1}{2}$,

$\cos 210° = -\frac{\sqrt{3}}{2}$, $\tan 210° = \frac{\sqrt{3}}{3}$, $\csc 210° = -2$,

$\sec 210° = -\frac{2\sqrt{3}}{3}$, $\cot 210° = \sqrt{3}$

27. $\sin 420° = \frac{\sqrt{3}}{2}$, $\cos 420° = \frac{1}{2}$, $\tan 420° = \sqrt{3}$,

$\csc 420° = \frac{2\sqrt{3}}{3}$, $\sec 420° = 2$, $\cot 420° = \frac{\sqrt{3}}{3}$

29. 2 **31.** $\sin \theta = \frac{\sqrt{2}}{2}$, $\cos \theta = -\frac{\sqrt{2}}{2}$, $\tan \theta = -1$,

$\csc \theta = \sqrt{2}$, $\sec \theta = -\sqrt{2}$, $\cot \theta = -1$

33. $\sin \theta = -\frac{8\sqrt{65}}{65}$, $\cos \theta = \frac{\sqrt{65}}{65}$, $\tan \theta = -8$,

$\csc \theta = -\frac{\sqrt{65}}{8}$, $\sec \theta = \sqrt{65}$, $\cot \theta = -\frac{1}{8}$

35. $\sin \theta = \frac{15}{17}$, $\cos \theta = -\frac{8}{17}$, $\tan \theta = -\frac{15}{8}$,

$\csc \theta = \frac{17}{15}$, $\sec \theta = -\frac{17}{8}$, $\cot \theta = -\frac{8}{15}$ **37.** in

Quadrant III or IV **39.** $\sin \theta = \frac{1}{2}$, $\cos \theta = -\frac{\sqrt{3}}{2}$,

$\tan \theta = -\frac{\sqrt{3}}{3}$, $\sec \theta = -\frac{2\sqrt{3}}{3}$, $\cot \theta = -\sqrt{3}$

41. $\sin \theta = \frac{2\sqrt{5}}{5}$, $\cos \theta = \frac{\sqrt{5}}{5}$, $\csc \theta = \frac{\sqrt{5}}{2}$,

$\sec \theta = \sqrt{5}$, $\cot \theta = \frac{1}{2}$ **43.** $\sin \theta = -\frac{\sqrt{2}}{2}$,

$\cos \theta = -\frac{\sqrt{2}}{2}$, $\tan \theta = 1$, $\csc \theta = -\sqrt{2}$,

$\sec \theta = -\sqrt{2}$ **45.** 0° or 90° **47.** $\theta = 0°$ **49a.** 76 ft
49b. 22 ft **49c.** 19 ft **49d.** $\frac{1}{2}r + 4$ **51.** 240°; III
53. 1.25, 1
55.

57. $\left(\frac{3}{4}, -\frac{2}{3}, \frac{1}{2}\right)$

59. absolute value; $f(x) = \left|2\frac{1}{2} - x\right|$

Pages 302–304 Lesson 5-4

5. 52.1 **7.** 12.4 **9.** about 743.2 ft **11.** 6.3
13. 9.5 **15.** 18.4 **17.** 4.0 **19.** 6; 10.4; 6; 8.5
21a. about 9.9 m **21b.** about 6.7 m **21c.** about 48.8 m² **23.** about 1088.8 ft
25a.

25b. about 43.9 ft
25c. about 87.8 ft

27. about 366.8 ft; no **29.** Markisha's; about 7.2 ft
31. $\sin 120° = \frac{\sqrt{3}}{2}$, $\cos 120° = -\frac{1}{2}$,

$\tan 120° = -\sqrt{3}$, $\csc 120° = \frac{2\sqrt{3}}{3}$, $\sec 120° = -2$,

$\cot 120° = -\frac{\sqrt{3}}{3}$ **33.** 43.260 **35.** $1.32; $0.92

Pages 309–312 Lesson 5-5

5. 60°, 300° **7.** $\frac{\sqrt{3}}{2}$ **9.** 35.0° **11.** $A = 12°$,

$b = 192.9$, $c = 197.2$ **13.** $B = 58°$, $a = 6.9$, $b = 11.0$
15. 90° **17.** 30°, 330° **19.** 225°, 315° **21.** Sample
answers: 30°, 150°, 390°, 510° **23.** $\frac{2}{3}$ **25.** 1

27. $\frac{\sqrt{21}}{5}$ **29.** 34.8° **31.** 52.7° **33.** 36.5°

35. about 48.8°, 48.8°, and 82.4° **37.** $B = 55°$,
$a = 5.6$, $c = 9.8$ **39.** $c = 5.7$, $A = 42.1°$, $B = 47.9°$
41. $B = 38.5°$, $b = 10.6$, $c = 17.0$ **43.** $B = 76°$,
$a = 2.4$, $b = 9.5$ **45a.** Since the sine function is the
side opposite divided by the hypotenuse, the sine
cannot be greater than 1. **45b.** Since the secant
function is the hypotenuse divided by the side
opposite, the secant cannot be between 1 and −1.
45c. Since cosine function is the side adjacent
divided by the hypotenuse, the cosine cannot be less
than −1. **47a.** about 4.6° **47b.** about 2.9°

49. about 13.3° **51.** $y = 36.5, Z \approx 19.5°, Y \approx 130.5°$
53. $\sin F = \dfrac{4\sqrt{11}}{15}$, $\cos F = \dfrac{7}{15}$, $\tan F = \dfrac{4\sqrt{11}}{7}$,

$\csc F = \dfrac{15\sqrt{11}}{44}$, $\sec F = \dfrac{15}{7}$, $\cot F = \dfrac{7\sqrt{11}}{44}$

55. y-axis **57.** $\begin{bmatrix} 2 & -1 & 0 \\ 3 & -1 & 1 \\ 2 & 8 & -5 \end{bmatrix}$ **59.** $y = -\dfrac{2}{5}x + 2$;

$-\dfrac{2}{5}$; 2

Pages 316–318   Lesson 5-6
5. $C = 81°, a = 9.1, b = 12.1$ **7.** about 18.7
9. 30.4 units^2 **11.** $B = 70°, b = 29.2, c = 29.2$
13. $C = 120°, a = 8.8, c = 18.1$ **15.** $A = 93.9°$,
$b = 3.4, c = 7.2$ **17.** about 97.8 **19.** 29.6 units^2
21. 5.4 units^2 **23.** 25.0 units^2 **25.** about
234.8 cm^2 **27.** about 70.7 ft^2

29. Applying the Law of Sines, $\dfrac{m}{\sin M} = \dfrac{n}{\sin N}$ and

$\dfrac{r}{\sin R} = \dfrac{s}{\sin S}$. Thus, $\sin M = \dfrac{m \sin N}{n}$ and

$\sin R = \dfrac{r \sin S}{s}$. Since $\angle M \cong \angle R$, $\sin M = \sin$

R and $\dfrac{m \sin N}{n} = \dfrac{r \sin S}{s}$. However, $\angle N \cong \angle S$ and

$\sin N = \sin S$, so $\dfrac{m}{n} = \dfrac{r}{s}$ and $\dfrac{m}{r} = \dfrac{n}{s}$.
Similar proportions can be derived for p and t.
Therefore, $\triangle MNP \cong \triangle RST$.
31a. about 3.6 mi **31b.** about 1.4 mi **33a.** about
227.7 mi **33b.** about 224.5 mi

35a. $\dfrac{a}{\sin A} = \dfrac{b}{\sin B}$

$\dfrac{a}{b} = \dfrac{\sin A}{\sin B}$

35b. $\dfrac{a}{\sin A} = \dfrac{c}{\sin C}$

$\dfrac{a}{c} = \dfrac{\sin A}{\sin C}$

$\dfrac{a}{c} - 1 = \dfrac{\sin A}{\sin C} - 1$

$\dfrac{a}{c} - \dfrac{c}{c} = \dfrac{\sin A}{\sin C} - \dfrac{\sin C}{\sin C}$

$\dfrac{a - c}{c} = \dfrac{\sin A - \sin C}{\sin C}$

35c. From Exercise 34b, $\dfrac{a - c}{c} = \dfrac{\sin A - \sin C}{\sin C}$ or

$\dfrac{\sin A - \sin C}{a - c} = \dfrac{\sin C}{c}$.

$\dfrac{a}{\sin A} = \dfrac{c}{\sin C}$

$\dfrac{a}{c} = \dfrac{\sin A}{\sin C}$

$\dfrac{a}{c} + 1 = \dfrac{\sin A}{\sin C} + 1$

$\dfrac{a}{c} + \dfrac{c}{c} = \dfrac{\sin A}{\sin C} + \dfrac{\sin C}{\sin C}$

$\dfrac{a + c}{c} = \dfrac{\sin A + \sin C}{\sin C}$

$\dfrac{\sin C}{c} = \dfrac{\sin A + \sin C}{a + c}$

Therefore, $\dfrac{\sin A - \sin C}{a - c} = \dfrac{\sin A + \sin C}{a + c}$ or $\dfrac{a + c}{a - c} =$

$\dfrac{\sin A + \sin C}{\sin A - \sin C}$.

35d. $\dfrac{a}{\sin A} = \dfrac{b}{\sin B}$

$\dfrac{a}{b} = \dfrac{\sin A}{\sin B}$

$\dfrac{a}{b} + 1 = \dfrac{\sin A}{\sin B} + 1$

$\dfrac{a}{b} + \dfrac{b}{b} = \dfrac{\sin A}{\sin B} + \dfrac{\sin B}{\sin B}$

$\dfrac{a + b}{b} = \dfrac{\sin A + \sin B}{\sin B}$

$\dfrac{b}{a + b} = \dfrac{\sin B}{\sin A + \sin B}$

37. $\cos \theta = \dfrac{\sqrt{35}}{6}$, $\tan \theta = -\dfrac{\sqrt{35}}{35}$, $\csc \theta = -6$,

$\sec \theta = \dfrac{6\sqrt{35}}{35}$, $\cot \theta = -\sqrt{35}$ **39.** 4 standard

carts, 11 deluxe carts **41.**

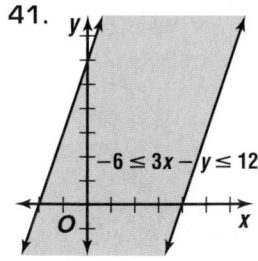

Pages 324–326   Lesson 5-7
5. 0 **7.** none **9.** $A = 37.0°, B = 13.0°, a = 13.4$
11. 0 **13.** 0 **15.** 0 **17.** 2 **19.** $B = 71.1°$,
$C = 50.9°, c = 23.8; B = 108.9°, C = 13.1°, c = 6.9$
21. $A = 78.2°, B = 31.8°, b = 13.5; A = 101.8°$,
$B = 8.2°, b = 3.6$ **23.** none **25.** $B = 30.1°$,
$C = 42.7°, b = 9.0$ **27.** $A = 27.2°, B = 105.8°$,
$b = 21.1$ **29.** none **31.** about 63.9 units and
41.0 units **33.** about 100.6° **35.** about 9.6°

37. about 4.1 min   **39a.** $B > 44.9°$ **39b.** $B \approx 44.9°$
39c. $B < 44.9°$ **41.** about 185.6 m   **43.** no;

$$\frac{\frac{3x}{x-1}+1}{3\left(\frac{3x}{x-1}\right)} = \frac{\frac{3x}{x-1}+\frac{x-1}{x-1}}{\frac{9x}{x-1}} =$$

$$\frac{\frac{4x-1}{x-1}}{\frac{9x}{x-1}} = \frac{4x-1}{9x} \neq x \qquad \textbf{45. } 5x + 2y = 22$$

Pages 330–332   Lesson 5-8
5. $A = 43.5°$, $B = 54.8°$, $C = 81.7°$ **7.** about 81.0°
9. 102.3 units^2 **11.** $B = 44.2°$, $C = 84.8°$, $a = 7.8$
13. $A = 34.1°$, $B = 44.4°$, $C = 101.5°$ **15.** $A = 51.8°$,
$B = 70.9°$, $C = 57.3°$ **17.** about 13.8°
19. 11.6 units^2 **21.** 290.5 units^2 **23.** 11,486.3 units^2
25a. about 68.1 in.   **25b.** about 1247.1 in^2
27. about 342.3 ft   **29a.** about 122.8 mi
29b. about 2.8 mi   **31.** the player 30 ft and 20 ft
from the posts   **33.** 2   **35.** 55°   **37.** $\frac{4}{3}$

**Pages 335–339   Chapter 5 Study Guide and
Assessment**
1. false; depression   **3.** true   **5.** true   **7.** true
9. false; terminal side   **11.** $57° \, 9'$ **13.** 140°; II
15. 204°; III   **17.** 60°; I   **19.** 294°; IV   **21.** 76°
23. $\sin A = \frac{5\sqrt{34}}{34}$, $\cos A = \frac{3\sqrt{34}}{34}$, $\tan A = \frac{5}{3}$

25. $\sin M = \frac{5}{6}$, $\cos M = \frac{\sqrt{11}}{6}$, $\tan M = \frac{5\sqrt{11}}{11}$,
$\csc M = \frac{6}{5}$, $\sec M = \frac{6\sqrt{11}}{11}$, $\cot M = \frac{\sqrt{11}}{5}$

27. $\sin \theta = \frac{\sqrt{2}}{2}$, $\cos \theta = \frac{\sqrt{2}}{2}$, $\tan \theta = 1$,
$\csc \theta = \sqrt{2}$, $\sec \theta = \sqrt{2}$, $\cot \theta = 1$

29. $\sin \theta = -\frac{\sqrt{17}}{17}$, $\cos \theta = \frac{4\sqrt{17}}{17}$, $\tan \theta = -\frac{1}{4}$,
$\csc \theta = -\sqrt{17}$, $\sec \theta = \frac{\sqrt{17}}{4}$, $\cot \theta = -4$

31. $\sin \theta = \frac{5\sqrt{41}}{41}$, $\cos \theta = \frac{4\sqrt{41}}{41}$, $\tan \theta = \frac{5}{4}$,
$\csc \theta = \frac{\sqrt{41}}{5}$, $\sec \theta = \frac{\sqrt{41}}{4}$, $\cot \theta = \frac{4}{5}$

33. $\sin \theta = \frac{\sqrt{2}}{2}$, $\cos \theta = -\frac{\sqrt{2}}{2}$, $\tan \theta = -1$,
$\csc \theta = \sqrt{2}$, $\sec \theta = -\sqrt{2}$, $\cot \theta = -1$

35. $\sin \theta = \frac{\sqrt{55}}{8}$, $\tan \theta = -\frac{\sqrt{55}}{3}$, $\csc \theta = \frac{8\sqrt{55}}{55}$,
$\sec \theta = -\frac{8}{3}$, $\cot \theta = -\frac{3\sqrt{55}}{55}$ **37.** 10.0   **39.** 10.2

41. 180°   **43.** $a = 13.2$, $A = 41.4°$, $B = 48.6°$
45. $A = 52°$, $b = 100.2$, $c = 90.4$ **47.** 471.7 units^2
49. 2488.4 units^2 **51.** $B = 93.7°$, $C = 47.6°$,
$b = 274.5$; $B = 8.9°$, $C = 132.4°$, $b = 42.3$
53. $B = 113.7°$, $C = 37.3°$, $b = 22.7$; $B = 8.3°$,
$C = 142.7°$, $b = 3.6$ **55.** $a = 36.9$, $B = 57.4°$,
$C = 71.6°$ **57.** $A = 30.5°$, $B = 36.9°$, $C = 112.6°$
59a. about 41.8°   **59b.** about 8.9 ft

Page 341   Chapter 5 SAT and ACT Practice
1. C   **3.** A   **5.** C   **7.** C   **9.** B

Chapter 6   Graphs of Trigonometric Functions

Pages 348–351   Lesson 6-1
5. $\frac{4\pi}{3}$ **7.** 270°   **9.** $\frac{\sqrt{2}}{2}$ **11.** 39.3 in.
13. 2.1 units^2 **15.** about 0.7 m   **17.** $\frac{7\pi}{6}$
19. $-\frac{5\pi}{2}$ **21.** $\frac{125\pi}{18}$ **23.** 660°   **25.** $-200.5°$
27. 1002.7°   **29.** $\frac{\sqrt{3}}{3}$ **31.** $-\frac{1}{2}$ **33.** $-\frac{\sqrt{3}}{2}$
35. 18.3 cm   **37.** 68.9 cm   **39.** 78.2 cm
41. about 36.0 m   **43.** 65.4 units^2 **45.** 9.6 units^2
47. 70.7 units^2 **49a.** 5 ft   **49b.** 15 ft^2
51a. about 12.2 in.   **51b.** about 2.4 in.
53a. about 7.9 ft   **53b.** about 143.2°   **55.** about
26.3°   **57.** about 5.23 mi   **59a.** about 530.1 ft^2
59b. about 17.8 ft   **61.** $A = \frac{1}{2} r^2(\alpha - \sin \alpha)$
63. no solution   **65.** I, III   **67.** Sample answers:
4; -2 **69.** all   **71.** b

Pages 355–358   Lesson 6-2
7. 4461.1 radians   **9.** 293.2 radians/min
11. 110.0 m/min   **13.** 18.8 radians
15. 82.9 radians   **17.** 381.4 radians
19. 1.3 radians/s   **21.** 9.0 radians/s
23. 39.3 radians/min   **25.** about 0.1 radian/s
27. about 811.7 rpm   **29.** 109.6 ft/s
31. 4021.6 in./s   **33.** 18,014.0 mm/min
35a. about 3.1 mm/s   **35b.** about 0.05 mm/s
35c. about 0.003 mm/s   **37a.** about 7.1 ft/s
37b. about 9.9 ft   **37c.** about 4 ft/s   **39a.** 2017
revolutions   **39b.** about 14.7 mph
41a. $\theta = \frac{\pi}{4} \cos \pi t$ **41b.** 0.5 s, 1.5 s
43a. B clockwise; C counterclockwise
43b. 180 rpm; 75 rpm   **45.** about 31.68 cm^2
47. no real solution

49.

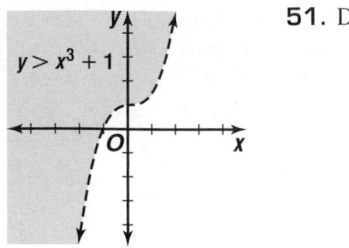

$y > x^3 + 1$

51. D

Pages 363–366   Lesson 6-3
5. yes; 4   **7.** 1
9.

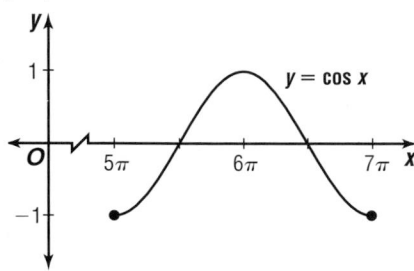

$y = \cos x$

11. Neither; the period is not 2π **13.** yes; 6
15. yes; 20   **17.** no   **19.** 1   **21.** 0   **23.** −1
25. −1   **27.** $\pi + 2\pi n$, where n is an integer
29. $\frac{\pi}{2} + \pi n$, where n is an integer
31.

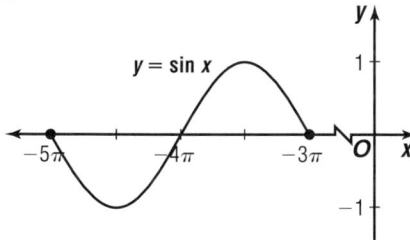

$y = \sin x$

33.

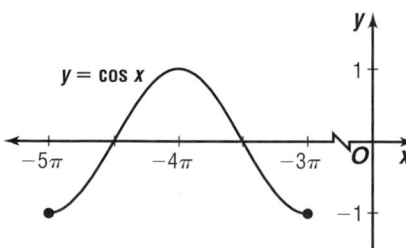

$y = \cos x$

35.

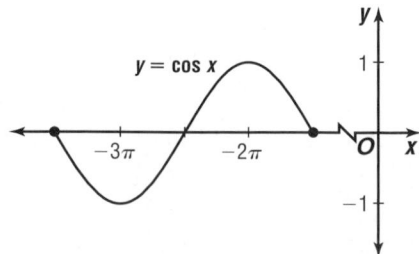

$y = \cos x$

---

37. $y = \cos x$; the maximum value of 1 occurs when $x = 4\pi$, the minimum value of −1 occurs when $x = 5\pi$, and the x-intercepts are $\frac{7\pi}{2}$, $\frac{9\pi}{2}$, and $\frac{11\pi}{2}$.
39. $y = \sin x$; the maximum value of 1 occurs when $x = -\frac{11\pi}{2}$, the minimum value of −1 occurs when $x = -\frac{13\pi}{2}$, and the x-intercepts are -7π, -6π, and -5π.   **41.** $x = \frac{\pi}{2} + \pi n$, where n is an integer
43a. $\frac{\pi}{2} + 2\pi n$, where n is an integer
43b. $\frac{3\pi}{2} + 2\pi n$, where n is an integer   **43c.** πn, where n is an integer   **45.** $x = \frac{3\pi}{4}, \frac{7\pi}{4}$ **47.** none
49. $x = 0, \frac{\pi}{2}, 2\pi$ **51a.** 62; it is twice the coefficient.   **51b.** 86; it is twice the constant term.
53a. 100; 120; 100; 80; 100   **53b.** 0.25 s
53c. 0.75 s   **55a.** $\frac{\pi}{4} + \frac{\pi n}{2}$, where n is an integer
55b. 1   **55c.** −1   **55d.** π

55e.

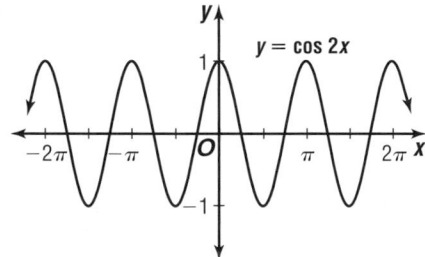

$y = \cos 2x$

57. about 52.4 radians per second
59. 45°, 135°
61. 1; 2 or 0; $-3, -\frac{1}{2}, 2$
63. $x = 0, x = -1, y = 1$
65. −11

67.

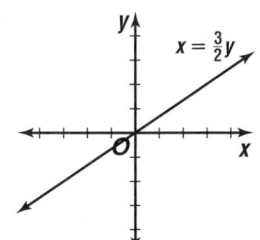

$x = \frac{3}{2}y$

Pages 373–377   Lesson 6-4
7. $\frac{\pi}{2}$

$y = \sin 4\theta$

---

9. $3; \pi$

11. $\frac{1}{5}; 8\pi$

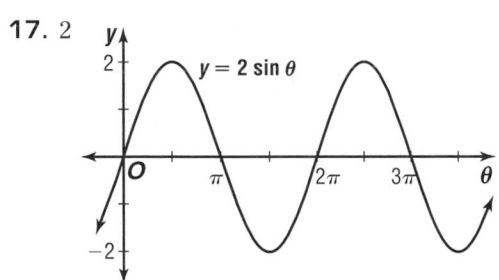

13. $y = \pm 7 \sin 6\theta$ **15.** $y = \pm \frac{3}{4} \cos \frac{\pi}{3} \theta$

17. 2

19. 1.5

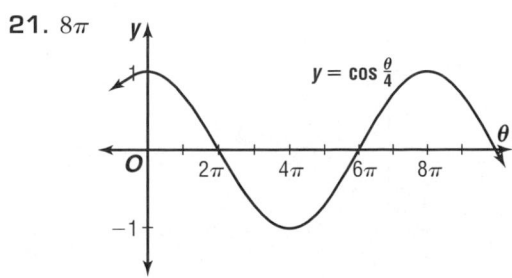

21. 8π

23. $5; 2\pi$

25. $\frac{2}{5}; \frac{2\pi}{9}$

27. $3; 4$

29. $3; \pi$

31. $\frac{1}{3}; \frac{2\pi}{3}$

33. $4; 4\pi$

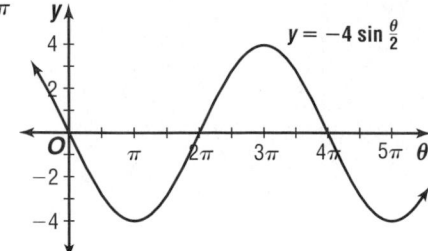

35. $0.5; \dfrac{1}{349}$ **37.** $y = \pm 35.7 \sin 8\theta$
39. $y = \pm 0.34 \sin \dfrac{8}{3}\theta$ **41.** $y = \pm 16 \sin \dfrac{\pi}{15}\theta$
43. $y = \pm \dfrac{5}{8} \cos 14\theta$ **45.** $y = \pm 0.5 \cos \dfrac{20}{3}\theta$
47. $y = \pm 17.9 \cos \dfrac{\pi}{8}\theta$ **49.** $y = 2 \cos \dfrac{\theta}{2}$
51. $y = -3 \cos \theta$ **53.** $y = \pm 3.8 \sin (240\pi \times t)$

55.

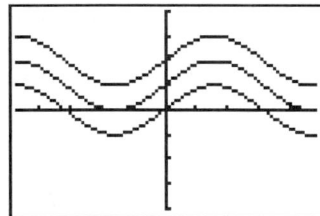

The graphs have the same shape, but have been translated vertically.

57a. 3   **57b.** 1   **57c.** 2π

57d.

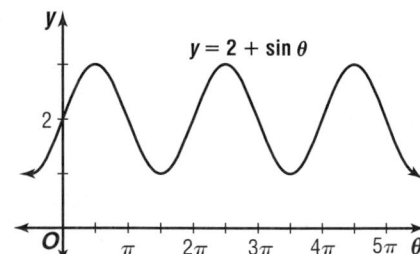

59a. $y = 1.5 \cos \left(t \sqrt{\dfrac{9.8}{6}} \right)$ **59b.** about 0.6 m to the
right   **59c.** about 1.2 m to the left   **61a.** about
0.9 s/cycle; about 1.1 hertz   **61b.** about 1.1 s/cycle;
about 0.9 hertz   **61c.** about 1.3 s/cycle; about
0.8 hertz   **61d.** It increases.   **61e.** It decreases.
63. about 88.0 radians/s   **65.** $c = 24.7, A = 37.8°$,
$B = 52.2°$ **67.** -95; 2 imaginary roots   **69.** $(-2, 1)$,
$(1, 1), (3, 4), (-3, 2)$ **71.** \$434.10   **73.** C

Pages 383–386   Lesson 6-5
7. $3; y = 3$

9. $\dfrac{1}{2}; 4\pi; \dfrac{\pi}{2}; 3$

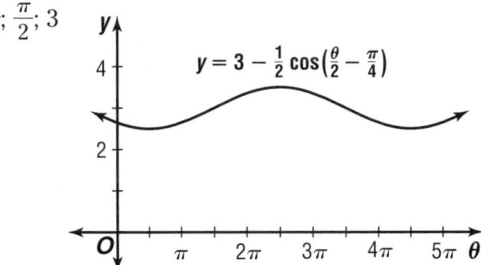

11. $y = \pm 0.6 \cos \left(\dfrac{\pi}{6.2}\theta + \dfrac{2.13\pi}{6.2} \right) + 7$
13a. $P = 100$ **13b.** $P = \pm 30 \sin 2\pi t + 100$

13c.

15. $-\dfrac{\pi}{2}$

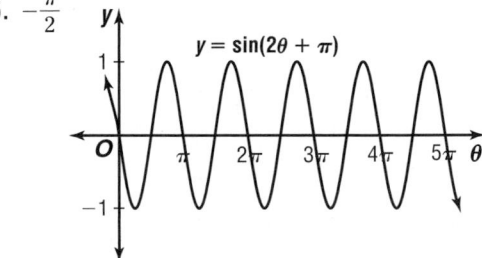

17. $\dfrac{1}{2}; y = \dfrac{1}{2}$

19. $7; y = 7$

21. $3; 2\pi; \frac{\pi}{2}; 0$

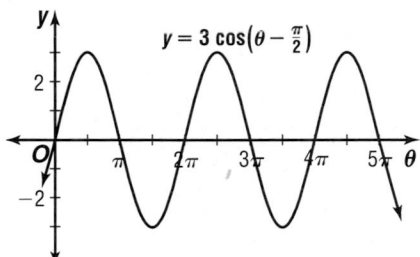

23. $1; 6\pi; \frac{\pi}{4}; -2$

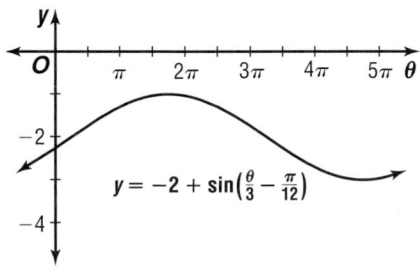

25. $\frac{1}{4}; 4\pi; 0; -3$

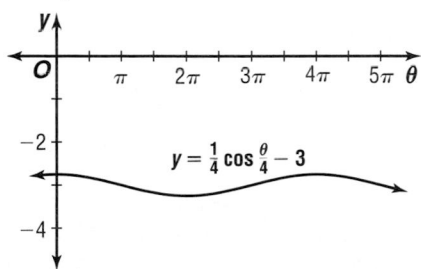

27. $4; 4\pi; -\frac{\pi}{2}; -2$ **29.** $y = \pm 50 \sin$
$\left(\frac{8}{3}\theta - \frac{4\pi}{3}\right) - 25$ **31.** $y = \pm 3.5 \cos(4\theta - \pi) + 7$
33. $y = \pm 100 \cos\left(\frac{2\pi}{45}\theta\right) - 110$
35. $y = 0.5 \sin 2\theta + 3$
37.

39.

41a. 3000; 1000 **41b.** 15,000; 5000

41c.

[0, 24] scl:1 by [0, 16,000] scl:1000

41d. months number 3 and 15 **41e.** months number 0, 12, 24 **41f.** When the sheep population is at a maximum, the wolf population is on the increase because of the maximum availability of food. The upswing in wolf population leads to a maximum later. **43a.** 4 ft **43b.** $t = 25$ **43c.** 20 s
43d. $h = 25 + 21 \sin\left(\frac{\pi t}{10}\right)$ **43e.** 5 s **43f.** 25 ft

45a. $y = \sqrt{\sin x}$ **45b.** $y = \frac{\cos x}{x}$ **45c.** $y = \cos x^2$
45d. $y = \sin \sqrt{x}$ **47.** 134.4 cm/s **49.** $y = \frac{3}{x} + 1$
51. $(1.4, -0.04)$ **53.** $3x - y - 11 = 0$

Pages 391–394 Lesson 6-6
5. $P = 30 \sin 2\pi t + 110$ **7a.** 0.5 **7b.** $\frac{1}{330}$
7c. 330 hertz **9a.** 1200 **9b.** about 232 **9c.** 1500; 275; no **9d.** January 1, 1971 **9e.** 225; July 1, 1973
11. $y = 3.55 \sin\left(\frac{\pi}{6.2} t + \frac{2.34\pi}{3.1}\right) + 4.24$ **13a.** 4°
13b. 77° **13c.** 12 months **13d.** Sample answer:
$y = -4 \cos\left(\frac{\pi}{6}t - 0.5\right) + 77$ **13e.** Sample answer: About 80.4°; it is very close to the actual average.
13f. Sample answer: About 79.1°; it is close to the actual average. **15a.** 5.685 ft **15b.** 7.565 ft
15c. about 12.4 h **15d.** Sample answer:
$h = 5.685 \sin\left(\frac{\pi}{6.2}t - 0.71\right) + 7.565$ **15e.** Sample answer: about 8.99 ft **17.** Sample answer: about -2.09 **19.** $V_R = 120 \sin\left(\frac{\pi}{30}t\right)$

21. $3; \pi; -\frac{\pi}{2}; 5$

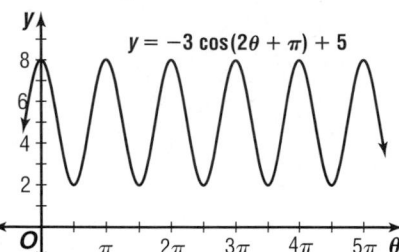

$y = -3\cos(2\theta + \pi) + 5$

23. $\frac{40\pi}{9}$ **25.** $\frac{3}{m-4} + \frac{-1}{m+4}$ **27.** $x > -1; x < -1$

Pages 400–403   Lesson 6-7

5. 1   **7.** $\frac{\pi}{4} + \pi n$, where n is an integer

9.

$y = \sec(2\theta + \pi)$

11. $y = \cot\left(\frac{1}{2}\theta + \frac{\pi}{8}\right)$ **13.** 0   **15.** undefined

17. -1 **19.** undefined   **21.** πn, where n is an integer   **23.** $\frac{3\pi}{2} + 2\pi n$, where n is an integer

25. $-\frac{\pi}{4} + \pi n$, where n is an integer

27. $\frac{\pi}{2}n$, where n is an odd integer

29.

$y = \cot\left(\theta - \frac{\pi}{2}\right)$

31.

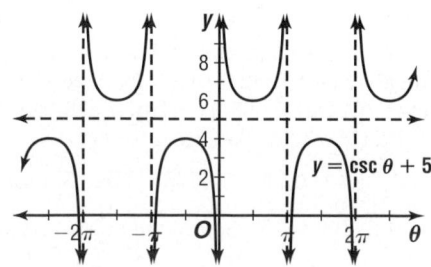

$y = \csc\theta + 5$

33.

$y = \csc(2\theta + \pi) - 3$

35.

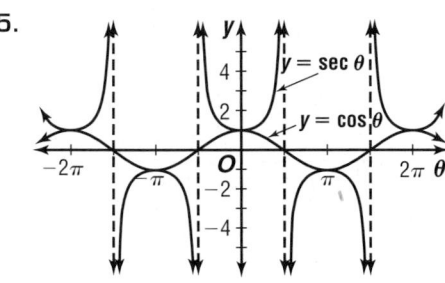

$y = \sec\theta$
$y = \cos\theta$

$-2\pi, -\pi, 0, \pi, 2\pi$

37. $y = \cot\left(2\theta - \frac{\pi}{4}\right) + 7$

39. $y = \csc\left(\frac{2}{3}\theta - \frac{2\pi}{3}\right) - 1$

41. $y = \csc(6\theta + 3\pi) - 5$

43. $y = \tan\left(2\theta - \frac{\pi}{2}\right) + 7$

45.

The graph of $y = \csc\theta$ has no range values between -1 and 1, while the graphs of $y = 3\csc\theta$ and $y = -3\csc\theta$ have no range values between -3 and 3. The graphs of $y = 3\csc\theta$ and $y = -3\csc\theta$ are reflections of each other.

47a. 220 A   **47b.** $\frac{1}{30}$ s   **47c.** $\frac{1}{360}$

47d. about -110 A   **49a.** 1.72 ft   **49b.** 2.27 ft

49c. about 12.3 hr   **49d.** Sample answer:

$h = 1.72\sin\left(\frac{2\pi}{12.3}t + 1.55\right) + 2.27$ **49e.** Sample

answer: 3.96 ft   **51.** 6π cm

53a.

53b. about 13.1 m

53c. about 13.7 m

55. $-2 < x < 5$

73°

4 m

57.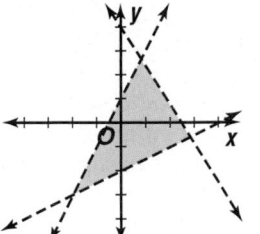

59. B

19. $y = 2 \tan^{-1} x$

21.

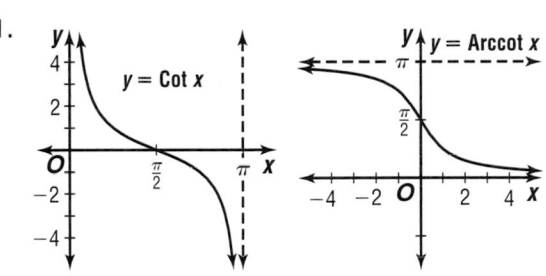

Pages 410–412   Lesson 6-8

7. $y = \text{Cos}^{-1} x - \dfrac{\pi}{2}$

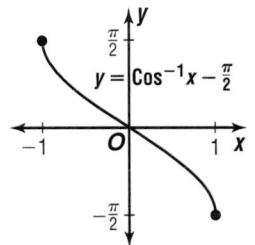

9. $\dfrac{\sqrt{2}}{2}$ **11.** true   **13a.** about 40,212 km

13b. $C = 40{,}212 \cos \theta$ **13c.** about 1.48 radians
13d. about 40,212 km

15. $y = \text{Sin}^{-1} x$

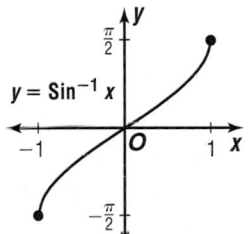

17. $y = \dfrac{1}{2} \text{Cos} \, x$

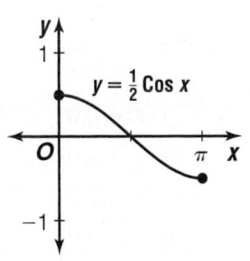

23. $\dfrac{\pi}{2}$ **25.** $\dfrac{\pi}{2}$ **27.** $\dfrac{1}{2}$ **29.** $-\dfrac{1}{2}$ **31.** No; there is
no angle with the sine of 2.   **33.** true   **35.** true
37. false; sample answer: $x = \dfrac{\pi}{2}$ **39.** April and
October   **41.** $\dfrac{\pi}{4} + \pi n$, where n is an integer
43a. 6:42 P.M.   **43b.** 12.4 h   **43c.** 3.675 ft
43d. Sample answer: $y = 3.375 + 3.675 \sin$
$\left(\dfrac{\pi}{6.2} t - 1.62 \right)$ **43e.** Sample answer: about 4:46 A.M.
45a. about 1.47 radians   **45b.** about 35.81 in.
47. $y = \pm 5 \sin \left(\dfrac{2}{3} \theta + \dfrac{2\pi}{3} \right) - 8$ **49.** 25.4 units,
54.4 units   **51.** $\pm 1, \pm 2, \pm 3, \pm 6, \pm \dfrac{1}{2}, \pm \dfrac{3}{2}$
53. $27x^3 - 1; 3x^3 - 3$

**Pages 413–417   Chapter 6 Study Guide and
Assessment**

1. radian   **3.** the same   **5.** angle   **7.** radian
9. sunusoidal   **11.** $\dfrac{\pi}{3}$ **13.** $\dfrac{4\pi}{3}$ **15.** $-315°$
17. 35.3 cm   **19.** 39.3 cm   **21.** 31.4 radians
23. 316.7 radians   **25.** 2.3 radians/s
27. 6.5 radians/s   **29.** -1 **31.** 1

33. $4; \pi$

35. $\frac{1}{3}$; 4π

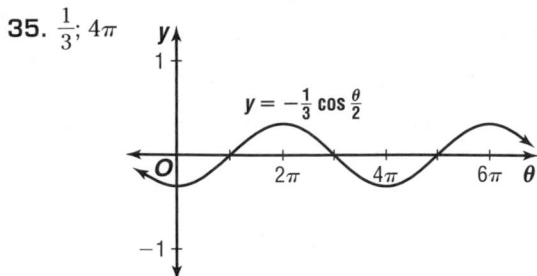

$y = -\frac{1}{3} \cos \frac{\theta}{2}$

37. $y = \pm 0.5 \sin\left(2\theta - \frac{2\pi}{3}\right) + 3$

39. $y = 20 \sin 2\pi t + 100$

41.

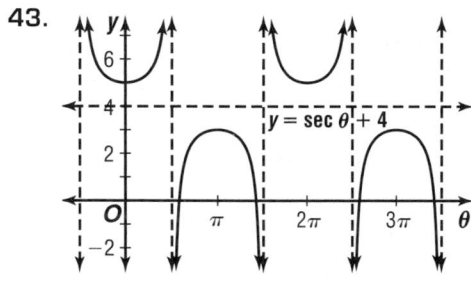

$y = \frac{1}{3} \csc \theta$

43.

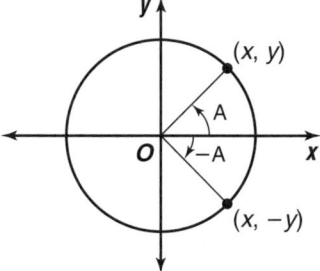

$y = \sec \theta + 4$

45. $-\frac{\pi}{4}$ **47.** 0 **49.** 0 **51.** $\frac{\pi}{2}$

Page 419 Chapter 6 SAT and ACT Practice
1. B **3.** B **5.** D **7.** A **9.** D

Chapter 7 Trigonometric Identities and Equations

Pages 427–430 Lesson 7-1
7. Sample answer: $x = 45°$ **9.** $-\frac{2\sqrt{5}}{5}$ **11.** $\frac{\sqrt{65}}{7}$
13. $\csc 30°$ **15.** 1

17. $B = \frac{F \csc \theta}{I\ell}$
$BI\ell = F \csc \theta$
$F = \frac{BI\ell}{\csc \theta}$
$F = BI\ell\left(\frac{1}{\csc \theta}\right)$
$F = BI\ell \sin \theta$

19. Sample answer: 45° **21.** Sample answer: 30°

23. Sample answer: 45° **25.** $\frac{5}{2}$ **27.** $\frac{\sqrt{15}}{4}$

29. $-\frac{\sqrt{2}}{3}$ **31.** $\frac{\sqrt{2}}{4}$ **33.** $-\frac{2\sqrt{6}}{7}$ **35.** $-\frac{3}{5}$

37. $\frac{1}{2}$ **39.** $-\cos \frac{3\pi}{8}$ **41.** $-\csc \frac{\pi}{3}$ **43.** $\cot 60°$
45. $\csc \theta$ **47.** 2 **49.** $\sin x + \cos x$ **51.** 1
53. 1 **55.** Let (x, y) be the point where the terminal side of A intersects the unit circle when A is in standard position. When A is reflected about the x-axis to obtain $-A$, the y-coordinate is multiplied by -1, but the x-coordinate is unchanged. So, $\sin(-A) = -y = -\sin A$ and $\cos(-A) = x = \cos A$.

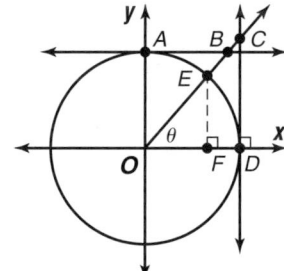

57. $\mu_k = \tan \theta$ **59.** $\sin \theta = EF$ and $\cos \theta = OF$ since the circle is a unit circle. $\tan \theta = \frac{CD}{OD} = \frac{CD}{1} = CD$.
$\sec \theta = \frac{CO}{OD} = \frac{CO}{1} = CO$.
$\Delta EOF \sim \Delta OBA$, so
$\frac{OF}{EF} = \frac{BA}{OA} = \frac{BA}{1} = BA$. Then $\cot \theta = \frac{\cos \theta}{\sin \theta} = \frac{OF}{EF} = BA$. Also by similar triangles, $\frac{EO}{EF} = \frac{OB}{OA}$, or $\frac{1}{EF} = \frac{OB}{1}$.
Then $\csc \theta = \frac{1}{\sin \theta} = \frac{1}{EF} = \frac{OB}{1} = OB$.

61.

$y = \cos\left(x - \frac{\pi}{6}\right)$

63. $a = 12.0$,
$B = 70°$,
$b = 32.9$
65. -4, 0.5
67. $(2, -5, -3)$
69. C

5. $\cos x \overset{?}{=} \dfrac{\cot x}{\csc x}$

$\cos x \overset{?}{=} \dfrac{\dfrac{\cos x}{\sin x}}{\dfrac{1}{\sin x}}$

$\cos x \overset{?}{=} \dfrac{\cos x}{1}$

$\cos x = \cos x$

7. $\csc \theta - \cot \theta \overset{?}{=} \dfrac{1}{\csc \theta + \cot \theta}$

$\csc \theta - \cot \theta \overset{?}{=} \dfrac{1}{\csc \theta + \cot \theta} \cdot \dfrac{\csc\theta - \cot \theta}{\csc\theta - \cot \theta}$

$\csc \theta - \cot \theta \overset{?}{=} \dfrac{\csc \theta - \cot \theta}{\csc^2 \theta - \cot^2 \theta}$

$\csc \theta - \cot \theta \overset{?}{=} \dfrac{\csc \theta - \cot \theta}{(1 + \cot^2 \theta) - \cot^2 \theta}$

$\csc \theta - \cot \theta \overset{?}{=} \dfrac{\csc \theta - \cot \theta}{1}$

$\csc \theta - \cot \theta = \csc \theta - \cot \theta$

9. $(\sin A - \cos A)^2 \overset{?}{=} 1 - 2 \sin^2 A \cot A$

$\sin^2 A - 2 \sin A \cos A + \cos^2 A$
$\overset{?}{=} 1 - 2 \sin^2 A \cot A$

$1 - 2 \sin A \cos A \overset{?}{=} 1 - 2 \sin^2 A \cot A$

$1 - 2 \sin A \cos A \dfrac{\sin A}{\sin A} \overset{?}{=} 1 - 2 \sin^2 A \cot A$

$1 - 2 \sin^2 A \dfrac{\cos A}{\sin A} \overset{?}{=} 1 - 2 \sin^2 A \cot A$

$1 - 2 \sin^2 A \cot A = 1 - 2 \sin^2 A \cot A$

11. Sample answer: $\cos x = -1$

13. $\tan A \overset{?}{=} \dfrac{\sec A}{\csc A}$

$\tan A \overset{?}{=} \dfrac{\dfrac{1}{\cos A}}{\dfrac{1}{\sin A}}$

$\tan A \overset{?}{=} \dfrac{\sin A}{\cos A}$

$\tan A = \tan A$

15. $\sec x - \tan x \overset{?}{=} \dfrac{1 - \sin x}{\cos x}$

$\sec x - \tan x \overset{?}{=} \dfrac{1}{\cos x} - \dfrac{\sin x}{\cos x}$

$\sec x - \tan x = \sec x - \tan x$

17. $\sec x \csc x \overset{?}{=} \tan x + \cot x$

$\sec x \csc x \overset{?}{=} \dfrac{\sin x}{\cos x} + \dfrac{\cos x}{\sin x}$

$\sec x \csc x \overset{?}{=} \dfrac{\sin x}{\cos x} \cdot \dfrac{\sin x}{\sin x} + \dfrac{\cos x}{\sin x} \cdot \dfrac{\cos x}{\cos x}$

$\sec x \csc x \overset{?}{=} \dfrac{\sin^2 x}{\cos x \sin x} + \dfrac{\cos^2 x}{\sin x \cos x}$

$\sec x \csc x \overset{?}{=} \dfrac{\sin^2 x + \cos^2 x}{\cos x \sin x}$

$\sec x \csc x \overset{?}{=} \dfrac{1}{\cos x \sin x}$

$\sec x \csc x \overset{?}{=} \dfrac{1}{\cos x} \cdot \dfrac{1}{\sin x}$

$\sec x \csc x = \sec x \csc x$

19. $(\sin A + \cos A)^2 \overset{?}{=} \dfrac{2 + \sec A \csc A}{\sec A \csc A}$

$(\sin A + \cos A)^2 \overset{?}{=} \dfrac{2}{\sec A \csc A} + \dfrac{\sec A \csc A}{\sec A \csc A}$

$(\sin A + \cos A)^2 \overset{?}{=} 2 \dfrac{1}{\sec A} \cdot \dfrac{1}{\csc A} + 1$

$(\sin A + \cos A)^2 \overset{?}{=} 2 \cos A \sin A + 1$

$(\sin A + \cos A)^2 \overset{?}{=} 2 \cos A \sin A +$
$\sin^2 A + \cos^2 A$

$(\sin A + \cos A)^2 = (\sin A + \cos A)^2$

21. $\dfrac{\cos y}{1 - \sin y} \overset{?}{=} \dfrac{1 + \sin y}{\cos y}$

$\dfrac{\cos y}{1 - \sin y} \cdot \dfrac{1 + \sin y}{1 + \sin y} \overset{?}{=} \dfrac{1 + \sin y}{\cos y}$

$\dfrac{\cos y (1 + \sin y)}{1 - \sin^2 y} \overset{?}{=} \dfrac{1 + \sin y}{\cos y}$

$\dfrac{\cos y (1 + \sin y)}{\cos^2 y} \overset{?}{=} \dfrac{1 + \sin y}{\cos y}$

$\dfrac{1 + \sin y}{\cos y} = \dfrac{1 + \sin y}{\cos y}$

23. $\csc x - 1 \overset{?}{=} \dfrac{\cot^2 x}{\csc x + 1}$

$\csc x - 1 \overset{?}{=} \dfrac{\csc^2 x - 1}{\csc x + 1}$

$\csc x - 1 \overset{?}{=} \dfrac{(\csc x + 1)(\csc x - 1)}{\csc x + 1}$

$\csc x - 1 = \csc x - 1$

25. $\sin \theta \cos \theta \tan \theta + \cos^2 \theta \overset{?}{=} 1$

$\sin \theta \cos \theta \dfrac{\sin \theta}{\cos \theta} + \cos^2 \theta \overset{?}{=} 1$

$\sin^2 \theta + \cos^2 \theta \overset{?}{=} 1$

$1 = 1$

27. $\sin x + \cos x \overset{?}{=} \dfrac{\cos x}{1 - \tan x} + \dfrac{\sin x}{1 - \cot x}$

$\sin x + \cos x \overset{?}{=} \dfrac{\cos x}{1 - \dfrac{\sin x}{\cos x}} + \dfrac{\sin x}{1 - \dfrac{\cos x}{\sin x}}$

$\sin x + \cos x \overset{?}{=} \dfrac{\cos x}{1 - \dfrac{\sin x}{\cos x}} \cdot \dfrac{\cos x}{\cos x} +$

$\dfrac{\sin x}{1 - \dfrac{\cos x}{\sin x}} \cdot \dfrac{\sin x}{\sin x}$

$\sin x + \cos x \overset{?}{=} \dfrac{\cos^2 x}{\cos x - \sin x} + \dfrac{\sin^2 x}{\sin x - \cos x}$

$\sin x + \cos x \overset{?}{=} -\dfrac{\cos^2 x}{\sin x - \cos x} + \dfrac{\sin^2 x}{\sin x - \cos x}$

$\sin x + \cos x \overset{?}{=} \dfrac{\sin^2 x - \cos^2 x}{\sin x - \cos x}$

$\sin x + \cos x \overset{?}{=} \dfrac{(\sin x + \cos x)(\sin x - \cos x)}{\sin x - \cos x}$

$\sin x + \cos x = \sin x + \cos x$

29. Sample answer: $\sec x = \sqrt{2}$ **31.** Sample answer: $\cos x = 0$ **33.** Sample answer: $\sin x = 1$

35. 1   **37.** yes   **39.** no   **41.** $\dfrac{1}{2} \sin \theta$

43. $y = -\dfrac{gx^2}{2v_0^2}(1 + \tan^2\theta) + x\tan\theta$ **45.** By the

Law of Sines, $\dfrac{b}{\sin\beta} = \dfrac{a}{\sin\alpha}$, so $b = \dfrac{a\sin\beta}{\sin\alpha}$. Then

$A = \dfrac{1}{2}ab\sin\gamma$

$A = \dfrac{1}{2}a\left(\dfrac{a\sin\beta}{\sin\alpha}\right)\sin\gamma$

$A = \dfrac{a^2\sin\beta\sin\gamma}{2\sin\alpha}$

$A = \dfrac{a^2\sin\beta\sin\gamma}{2\sin\left(180° - (\beta + \gamma)\right)}$

$A = \dfrac{a^2\sin\beta\sin\gamma}{2\sin(\beta + \gamma)}$

47. $y = \pm 2\sin(2x - 90°)$ **49.** 3   **51.** 40 shirts, 40 pants   **53.** D

Pages 442–445   Lesson 7-3

5. $-\dfrac{\sqrt{2} + \sqrt{6}}{4}$ **7.** $\sqrt{6} + \sqrt{2}$ **9.** $-\dfrac{63}{16}$

11.

$\tan\left(\theta + \dfrac{\pi}{2}\right) \overset{?}{=} -\cot\theta$

$\dfrac{\sin\left(\theta + \dfrac{\pi}{2}\right)}{\cos\left(\theta + \dfrac{\pi}{2}\right)} \overset{?}{=} -\cot\theta$

$\dfrac{\sin\theta\cos\dfrac{\pi}{2} + \cos\theta\sin\dfrac{\pi}{2}}{\cos\theta\cos\dfrac{\pi}{2} - \sin\theta\sin\dfrac{\pi}{2}} \overset{?}{=} -\cot\theta$

$\dfrac{(\sin\theta)\cdot 0 + (\cos\theta)\cdot 1}{(\cos\theta)\cdot 0 - (\sin\theta)\cdot 1} \overset{?}{=} -\cot\theta$

$-\dfrac{\cos\theta}{\sin\theta} \overset{?}{=} -\cot\theta$

$-\cot\theta = -\cot\theta$

13. $-\cos n\omega_0 t$ **15.** $\dfrac{\sqrt{6} - \sqrt{2}}{4}$ **17.** $\dfrac{\sqrt{6} - \sqrt{2}}{4}$

19. $\dfrac{\sqrt{2} + \sqrt{6}}{4}$ **21.** $-2 + \sqrt{3}$ **23.** $\sqrt{2} - \sqrt{6}$

25. $2 - \sqrt{3}$ **27.** $\dfrac{24}{25}$ **29.** $\dfrac{12\sqrt{17} - 5\sqrt{34}}{102}$

31. $\dfrac{65}{56}$ **33.** $\dfrac{3 - 2\sqrt{14}}{12}$

35.

$\cos(60° + A) \overset{?}{=} \sin(30° - A)$

$\cos 60°\cos A - \sin 60°\sin A \overset{?}{=}$

$\qquad\qquad \sin 30°\cos A - \cos 30°\sin A$

$\dfrac{1}{2}\cos A - \dfrac{\sqrt{3}}{2}\sin A = \dfrac{1}{2}\cos A - \dfrac{\sqrt{3}}{2}\sin A$

37.

$\cos(180° + x) \overset{?}{=} -\cos x$

$\cos 180°\cos x - \sin 180°\sin x \overset{?}{=} -\cos x$

$-1\cdot\cos x - 0\cdot\sin x \overset{?}{=} -\cos x$

$-\cos x = -\cos x$

39. $\sin(A + B) \overset{?}{=} \dfrac{\tan A + \tan B}{\sec A \sec B}$

$\sin(A + B) \overset{?}{=} \dfrac{\dfrac{\sin A}{\cos A} + \dfrac{\sin B}{\cos B}}{\dfrac{1}{\cos A}\cdot\dfrac{1}{\cos B}}$

$\sin(A + B) \overset{?}{=} \dfrac{\dfrac{\sin A}{\cos A} + \dfrac{\sin B}{\cos B}}{\dfrac{1}{\cos A}\cdot\dfrac{1}{\cos B}}\cdot\dfrac{\cos A\cos B}{\cos A\cos B}$

$\sin(A + B) \overset{?}{=} \dfrac{\sin A\cos B + \cos A\sin B}{1}$

$\sin(A + B) = \sin(A + B)$

41. $\sec(A - B) \overset{?}{=} \dfrac{\sec A\sec B}{1 + \tan A\tan B}$

$\sec(A - B) \overset{?}{=} \dfrac{\dfrac{1}{\cos A}\cdot\dfrac{1}{\cos B}}{1 + \dfrac{\sin A}{\cos A}\cdot\dfrac{\sin B}{\cos B}}$

$\sec(A - B) \overset{?}{=} \dfrac{\dfrac{1}{\cos A}\cdot\dfrac{1}{\cos B}}{1 + \dfrac{\sin A}{\cos A}\cdot\dfrac{\sin B}{\cos B}}\cdot\dfrac{\cos A\cos B}{\cos A\cos B}$

$\sec(A - B) \overset{?}{=} \dfrac{1}{\cos A\cos B + \sin A\sin B}$

$\sec(A - B) \overset{?}{=} \dfrac{1}{\cos(A - B)}$

$\sec(A - B) = \sec(A - B)$

43. $V_L = -I_0\omega L\sin\omega t$ **45.** $-\sin 2A$

47. $\tan(\alpha + \beta) = \dfrac{\sin(\alpha + \beta)}{\cos(\alpha + \beta)}$

$\tan(\alpha + \beta) = \dfrac{\sin\alpha\cos\beta + \cos\alpha\sin\beta}{\cos\alpha\cos\beta - \sin\alpha\sin\beta}$

$\tan(\alpha + \beta) = \dfrac{\dfrac{\sin\alpha\cos\beta}{\cos\alpha\cos\beta} + \dfrac{\cos\alpha\sin\beta}{\cos\alpha\cos\beta}}{\dfrac{\cos\alpha\cos\beta}{\cos\alpha\cos\beta} - \dfrac{\sin\alpha\sin\beta}{\cos\alpha\cos\beta}}$

$\tan(\alpha + \beta) = \dfrac{\tan\alpha + \tan\beta}{1 - \tan\alpha\tan\beta}$

Replace β with $-\beta$ to find $\tan(\alpha - \beta)$.

$\tan(\alpha + (-\beta)) = \dfrac{\tan\alpha + \tan(-\beta)}{1 - \tan\alpha\tan(-\beta)}$

$\tan(\alpha - \beta) = \dfrac{\tan\alpha - \tan\beta}{1 + \tan\alpha\tan\beta}$

49. $\sec^2 x \overset{?}{=} \dfrac{1 - \cos^2 x}{1 - \sin^2 x} + \csc^2 x - \cot^2 x$

$\sec^2 x \overset{?}{=} \dfrac{1 - \cos^2 x}{\cos^2 x} + 1 + \cot^2 x - \cot^2 x$

$\sec^2 x \overset{?}{=} \dfrac{1}{\cos^2 x} - \dfrac{\cos^2 x}{\cos^2 x} + 1$

$\sec^2 x \overset{?}{=} \sec^2 x - 1 + 1$

$\sec^2 x = \sec^2 x$

51. $\dfrac{\sqrt{3}}{2}$ **53.** $y = 18\sin\left(\dfrac{\pi}{2}t - \pi\right) + 68$ **55.** 0

57. about 183 miles   **59.** 54.87 ft
61. $\{x \mid x < -5 \text{ or } x > 3\}$ **63.** 1319, 221

Pages 453-455   Lesson 7-4

7. $\sqrt{3} - 2$ **9.** $\dfrac{24}{25}, -\dfrac{7}{25}, -\dfrac{24}{7}$

11. $1 + \dfrac{1}{2}\sin 2A \stackrel{?}{=} \dfrac{\sec A + \sin A}{\sec A}$

$$1 + \frac{1}{2}\sin 2A \stackrel{?}{=} \frac{\dfrac{1}{\cos A} + \sin A}{\dfrac{1}{\cos A}}$$

$$1 + \frac{1}{2}\sin 2A \stackrel{?}{=} \frac{\dfrac{1}{\cos A} + \sin A}{\dfrac{1}{\cos A}} \cdot \frac{\cos A}{\cos A}$$

$1 + \dfrac{1}{2}\sin 2A \stackrel{?}{=} 1 + \sin A \cos A$

$1 + \dfrac{1}{2}\sin 2A \stackrel{?}{=} 1 + \dfrac{1}{2} \cdot 2\sin A \cos A$

$1 + \dfrac{1}{2}\sin 2A = 1 + \dfrac{1}{2}\sin 2A$

13. $P = \dfrac{1}{2}I_0{}^2 R - \dfrac{1}{2}I_0{}^2 R \cos 2\omega t$ **15.** $\dfrac{\sqrt{2 + \sqrt{3}}}{2}$

17. $\dfrac{\sqrt{2 + \sqrt{2}}}{2}$ **19.** $\sqrt{2} - 1$ **21.** $\dfrac{24}{25}, \dfrac{7}{25}, \dfrac{24}{7}$

23. $-\dfrac{4}{5}, -\dfrac{3}{5}, \dfrac{4}{3}$ **25.** $\dfrac{12}{13}, \dfrac{5}{13}, \dfrac{12}{5}$ **27.** $\dfrac{2\sqrt{14}}{5}$

29. $\cos A - \sin A \stackrel{?}{=} \dfrac{\cos 2A}{\cos A + \sin A}$

$\cos A - \sin A \stackrel{?}{=} \dfrac{\cos^2 A - \sin^2 A}{\cos A + \sin A}$

$\cos A - \sin A \stackrel{?}{=} \dfrac{(\cos A - \sin A)(\cos A + \sin A)}{\cos A + \sin A}$

$\cos A - \sin A = \cos A - \sin A$

31. $\cos x - 1 \stackrel{?}{=} \dfrac{\cos 2x - 1}{2(\cos x + 1)}$

$\cos x - 1 \stackrel{?}{=} \dfrac{2\cos^2 x - 1 - 1}{2(\cos x + 1)}$

$\cos x - 1 \stackrel{?}{=} \dfrac{2\cos^2 x - 2}{2(\cos x + 1)}$

$\cos x - 1 \stackrel{?}{=} \dfrac{2(\cos^2 x - 1)}{2(\cos x + 1)}$

$\cos x - 1 \stackrel{?}{=} \dfrac{2(\cos x - 1)(\cos x + 1)}{2(\cos x + 1)}$

$\cos x - 1 = \cos x - 1$

33. $\tan \dfrac{A}{2} \stackrel{?}{=} \dfrac{\sin A}{1 + \cos A}$

$$\tan \frac{A}{2} \stackrel{?}{=} \frac{\sin 2\left(\dfrac{A}{2}\right)}{1 + \cos 2\left(\dfrac{A}{2}\right)}$$

$$\tan \frac{A}{2} \stackrel{?}{=} \frac{2\sin \dfrac{A}{2}\cos \dfrac{A}{2}}{1 + 2\cos^2 \dfrac{A}{2} - 1}$$

$$\tan \frac{A}{2} \stackrel{?}{=} \frac{2\sin \dfrac{A}{2}\cos \dfrac{A}{2}}{2\cos^2 \dfrac{A}{2}}$$

$$\tan \frac{A}{2} \stackrel{?}{=} \frac{\sin \dfrac{A}{2}}{\cos \dfrac{A}{2}}$$

$\tan \dfrac{A}{2} = \tan \dfrac{A}{2}$

35.

$\cos 3x \stackrel{?}{=} 4\cos^3 x - 3\cos x$

$\cos(2x + x) \stackrel{?}{=} 4\cos^3 x - 3\cos x$

$\cos 2x \cos x - \sin 2x \sin x \stackrel{?}{=} 4\cos^3 x - 3\cos x$

$(2\cos^2 x - 1)\cos x -$
$\qquad 2\sin^2 x \cos x \stackrel{?}{=} 4\cos^3 x - 3\cos x$

$(2\cos^2 x - 1)\cos x -$
$\qquad 2(1 - \cos^2 x)\cos x \stackrel{?}{=} 4\cos^3 x - 3\cos x$

$2\cos^3 x - \cos x -$
$\qquad 2\cos x + 2\cos^3 x \stackrel{?}{=} 4\cos^3 x - 3\cos x$

$4\cos^3 x - 3\cos x = 4\cos^3 x - 3\cos x$

37. $\angle PBD$ is an inscribed angle that subtends the same arc as the central angle $\angle POD$, so $m\angle PBD = \dfrac{1}{2}\theta$. By right triangle trigonometry,
$\tan \dfrac{1}{2}\theta = \dfrac{PA}{BA} = \dfrac{PA}{1 + OA} = \dfrac{\sin \theta}{1 + \cos \theta}$.

39a. $\dfrac{1 \pm \sqrt{\dfrac{1 - \cos L}{1 + \cos L}}}{1 \mp \sqrt{\dfrac{1 - \cos L}{1 + \cos L}}}$ **39b.** $2 + \sqrt{3}$

41. $\sqrt{6} - \sqrt{2}$ **43.** $97.4°$ **45.** $2x^4 - 11x^3 - 19x^2 + 84x - 36 = 0$ **47.** $(7, 2)$

Pages 459–461   Lesson 7-5

5. $-30°$ **7.** $30°, 330°$ **9.** $\dfrac{\pi}{6}, \dfrac{5\pi}{6}, \dfrac{7\pi}{6}, \dfrac{11\pi}{6}$

11. $\dfrac{\pi}{2} + \pi k$ **13.** $(2k + 1)\pi$ **15.** $\dfrac{2\pi}{3} < \theta < \dfrac{4\pi}{3}$

17. $45°$ **19.** $45°$ **21.** $0°, 90°$ **23.** $135°, 225°$

25. $0°, 45°, 180°, 225°$ **27.** $0°, 120°, 180°, 240°$

29. $0°, 150°, 180°, 210°$ **31.** $\dfrac{7\pi}{6}, \dfrac{11\pi}{6}$ **33.** $0, \pi, \dfrac{7\pi}{6},$
$\dfrac{11\pi}{6}$ **35.** $\dfrac{3\pi}{4}, \dfrac{7\pi}{4}$ **37.** $\dfrac{7\pi}{6} + 2\pi k, \dfrac{11\pi}{6} + 2\pi k$

39. $\pi k, \dfrac{\pi}{6} + \pi k$ **41.** πk **43.** $\dfrac{\pi}{4} + \pi k$ **45.** $\dfrac{\pi}{2} + \pi k$

47. $2\pi k, \dfrac{\pi}{2} + 2\pi k$ **49.** $\dfrac{5\pi}{6} \le \theta \le \dfrac{7\pi}{6}$

51. $0 \le \theta < \dfrac{\pi}{4}$ or $\dfrac{3\pi}{4} < \theta < 2\pi$ **53.** 0, 1.8955

55. $0.01°$ **57.** $30.29°$ or $59.71°$ **59.** 0.0013 s

61. $341.32°$ **63.** Sample answer: $\sin x = \dfrac{\sqrt{2}}{5}$

65. about 18 rps **67.** $(x-2)(x+1)(x+1)$

69. $(4, 3)$ **71.**

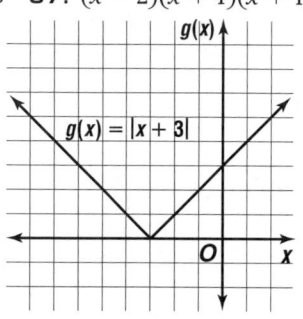

Pages 467–469 Lesson 7-6

5. $\sqrt{3}x + y - 20 = 0$ **7.** $x - y - 10 = 0$

9. $\dfrac{3\sqrt{10}}{10}x + \dfrac{\sqrt{10}}{10}y - \dfrac{\sqrt{10}}{5} = 0$; $\dfrac{\sqrt{10}}{5}$; $18°$

11a.

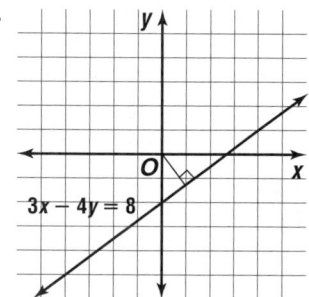

11b. 1.6 miles **13.** $\sqrt{2}x + \sqrt{2}y - 24 = 0$

15. $\sqrt{3}x - y + 4\sqrt{3} = 0$ **17.** $\sqrt{3}x + y + 10 = 0$

19. $x - \sqrt{3}y - 3 = 0$ **21.** $-\dfrac{5}{13}x - \dfrac{12}{13}y - 5 = 0$;

5; $247°$ **23.** $\dfrac{3}{5}x - \dfrac{4}{5}y - 3 = 0$; 3; $307°$

25. $x - 3 = 0$; 3; $0°$ **27.** $-\dfrac{\sqrt{17}}{17}x + \dfrac{4\sqrt{17}}{17}y -$

$\dfrac{28\sqrt{17}}{17} = 0$; $\dfrac{28\sqrt{17}}{17}$; $104°$ **29.** $\dfrac{6\sqrt{61}}{61}x +$

$\dfrac{5\sqrt{61}}{61}y - \dfrac{120\sqrt{61}}{61} = 0$; $\dfrac{120\sqrt{61}}{61}$; $40°$

31. $x - y - 8 = 0$

33a.

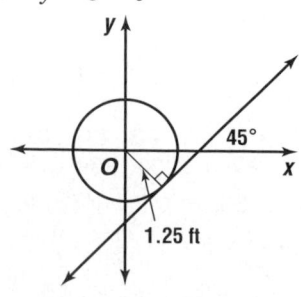

33b. $\sqrt{2}x - \sqrt{2}y - 2.5 = 0$ **35a.** $\dfrac{5}{13}x + \dfrac{12}{13}y -$

$3 = 0$ **35b.** $\theta = 67°$

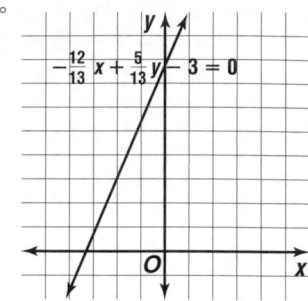

35d. The line with normal form $x \cos \phi + y \sin \phi - p = 0$ makes an angle of ϕ with the positive x-axis and has a normal of length p. The graph of Armando's equation is a line whose normal makes an angle of $\phi + \delta$ with the x-axis and also has length p. Therefore, the graph of Armando's equation is the graph of the original line rotated $\delta°$ counterclockwise about the origin. Armando is correct. **37.** \$6927.82 **39.** $\dfrac{2\sqrt{35} + \sqrt{5}}{18}$

41. 3.05 cm **43.** 4.5 in. by 6.5 in. by 2.5 in.

45. $(-6, -3)$

Pages 474–476 Lesson 7-7

5. $\dfrac{2\sqrt{13}}{13}$ **7.** $\dfrac{2\sqrt{34}}{17}$ **9.** $\left(20 - 6\sqrt{13}\right)x -$

$\left(30 + 8\sqrt{13}\right)y - 40 - 5\sqrt{13} = 0$; $\left(20 + 6\sqrt{13}\right)x +$

$\left(8\sqrt{13} - 30\right)y - 40 + 5\sqrt{13} = 0$ **11.** $\dfrac{21}{5}$

13. $\dfrac{3\sqrt{5}}{5}$ **15.** 0 **17.** $\dfrac{\sqrt{10}}{10}$ **19.** $\dfrac{6\sqrt{41}}{41}$

21. $\dfrac{\sqrt{10}}{5}$ **23.** $\dfrac{8\sqrt{13}}{13}$ **25.** $x + 8y = 0$; $16x -$

$2y - 65 = 0$ **27.** $\left(2\sqrt{10} + 3\sqrt{13}\right)x +$

$\left(\sqrt{13} - 3\sqrt{10}\right)y + 3\sqrt{10} + 2\sqrt{13} = 0$;

$\left(-2\sqrt{10} + 3\sqrt{13}\right)x + \left(\sqrt{13} + 3\sqrt{10}\right)y - 3\sqrt{10} +$

$2\sqrt{13} = 0$ **29.** 1.09 m **31.** $\dfrac{34}{5}$, $\dfrac{34\sqrt{53}}{53}$, $\dfrac{17\sqrt{26}}{26}$

33. $-\dfrac{2\sqrt{53}}{53}x + \dfrac{7\sqrt{53}}{53}y - \dfrac{5\sqrt{53}}{53} = 0$

35.

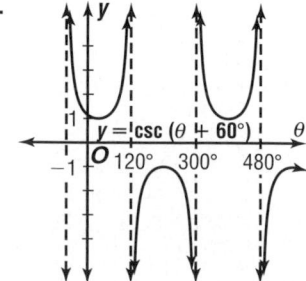

37. about 2.8 s **39.** $(-6, -2, -5)$

Pages 477–481   Chapter 7 Study Guide and Assessment

1. b   **3.** d   **5.** i   **7.** h   **9.** e   **11.** 2   **13.** $\frac{4}{5}$
15. $\sin x$

17. $\frac{1 - \cos\theta}{1 + \cos\theta} \stackrel{?}{=} (\csc\theta - \cot\theta)^2$

$\frac{1 - \cos\theta}{1 + \cos\theta} \stackrel{?}{=} \left(\frac{1}{\sin\theta} - \frac{\cos\theta}{\sin\theta}\right)^2$

$\frac{1 - \cos\theta}{1 + \cos\theta} \stackrel{?}{=} \frac{(1 - \cos\theta)^2}{\sin^2\theta}$

$\frac{1 - \cos\theta}{1 + \cos\theta} \stackrel{?}{=} \frac{(1 - \cos\theta)^2}{1 - \cos^2\theta}$

$\frac{1 - \cos\theta}{1 + \cos\theta} \stackrel{?}{=} \frac{(1 - \cos\theta)^2}{(1 - \cos\theta)(1 + \cos\theta)}$

$\frac{1 - \cos\theta}{1 + \cos\theta} = \frac{1 - \cos\theta}{1 + \cos\theta}$

19.

$\frac{\sin^4 x - \cos^4 x}{\sin^2 x} \stackrel{?}{=} 1 - \cot^2 x$

$\frac{(\sin^2 x - \cos^2 x)(\sin^2 x + \cos^2 x)}{\sin^2 x} \stackrel{?}{=} 1 - \cot^2 x$

$\frac{\sin^2 x - \cos^2 x}{\sin^2 x} \stackrel{?}{=} 1 - \cot^2 x$

$1 - \frac{\cos^2 x}{\sin^2 x} \stackrel{?}{=} 1 - \cot^2 x$

$1 - \cot^2 x = 1 - \cot^2 x$

21. $\frac{\sqrt{6} + \sqrt{2}}{4}$ **23.** $-2 + \sqrt{3}$ **25.** $-\frac{180 + 82\sqrt{5}}{61}$

27. $\frac{\sqrt{2 - \sqrt{2}}}{2}$ **29.** $2 - \sqrt{3}$ **31.** $-\frac{7}{25}$

33. $-\frac{336}{625}$ **35.** $0°, 90°, 270°$ **37.** $\pi k, \frac{\pi}{4} + 2\pi k,$
$\frac{3\pi}{4} + 2\pi k$ **39.** $2\pi k$ **41.** $y - 5 = 0$ **43.** $x + y +$
$8 = 0$ **45.** $-\frac{3\sqrt{13}}{13} x + \frac{2\sqrt{13}}{13} y - \frac{5\sqrt{13}}{26} = 0;$

$\frac{5\sqrt{13}}{26}; 146°$ **47.** $-\frac{\sqrt{2}}{10} x + \frac{7\sqrt{2}}{10} y - \frac{\sqrt{2}}{2} = 0;$

$\frac{\sqrt{2}}{2}; 98°$ **49.** $\frac{23\sqrt{13}}{13}$ **51.** $\frac{21\sqrt{10}}{10}$ **53.** $\frac{14}{5}$

55. $\frac{9\sqrt{13}}{13}$ **57.** $\left(-\sqrt{34} - 3\sqrt{10}\right)x +$
$\left(3\sqrt{34} + 5\sqrt{10}\right)y - 2\sqrt{34} - 15\sqrt{10} = 0;$
$\left(-\sqrt{34} + 3\sqrt{10}\right)x + \left(3\sqrt{34} - 5\sqrt{10}\right)y -$
$2\sqrt{34} + 15\sqrt{10} = 0$ **59.** 1431 ft

Page 483   Chapter 7 SAT and ACT Practice
1. B   **3.** D   **5.** B   **7.** A   **9.** C

Chapter 8   Vectors and Parametric Equations

Pages 490–492   Lesson 8-1
5. 1.2 cm, 120°   **7.** 1.4 cm, 20°   **9.** 2.6 cm, 210°
11. 2.9 cm, 12°
13a.

5 ⌐_____ 100

13b. ≈100.12 m/s   **15.** 1.4 cm, 45°   **17.** 3.0 cm,
340°   **19.** 3.4 cm, 25°   **21.** 5.5 cm, 324°
23. 5.2 cm, 128°   **25.** 8.2 cm, 322°   **27.** 5.4 cm,
133°   **29.** 3.4 cm, 301°   **31.** −1.60 cm, 2.05 cm
33. 2.04 cm, 0.51 cm   **35.** 45.73 m
37. Yes; Sample answer:

39. Sometimes;

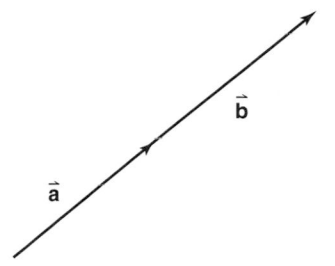

$|\vec{a} + \vec{b}| = 5$
$|\vec{a}| + |\vec{b}| = 2.3 + 2.7$
$= 5$

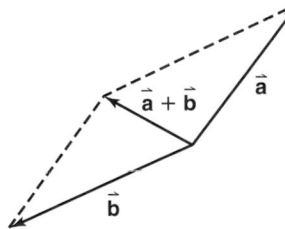

$|\vec{a} + \vec{b}| = 1.5$
$|\vec{a}| + |\vec{b}| = 2.3 + 2.7$
$= 5$

41. 36 mph, 30 mph   **43.** 71 lb
45. $x - \left(1 + \sqrt{2}\right)y + 2 + 5\sqrt{2} = 0$ **47.** $\frac{\pi}{4} + \pi n$
where n is an integer   **49.** 15.8 cm; 29.9 cm
51. $x = -3, x = 1, y = 0$

Pages 496–499   Lesson 8-2
5. $\langle -5, -1 \rangle, \sqrt{26}$ **7.** $\langle 2, 2 \rangle$ **9.** $\langle 14, 4 \rangle$ **11.** 10,
$8\vec{i} - 6\vec{j}$ **13.** ≈927 N   **15.** $\langle 4, -5 \rangle, \sqrt{41}$
17. $\langle -5, -7 \rangle, \sqrt{74}$ **19.** $\langle 5, 7 \rangle, \sqrt{74}$ **21.** $\langle 9, 9 \rangle,$
$9\sqrt{2}$ **23.** $\langle 2, 11 \rangle$ **25.** $\langle -2, 19 \rangle$ **27.** $\langle -22, 29 \rangle$

29. $\langle 18, 9 \rangle$ **31.** $\langle 20, 50 \rangle$ **33.** $\left\langle \frac{32}{3}, -\frac{34}{3} \right\rangle$

35. $\langle -30, 4.5 \rangle$ **37.** $\sqrt{13}, 2\vec{i} - 3\vec{j}$

39. 12.5, $3.5\vec{i} + 12\vec{j}$ **41.** $2\sqrt{353}, -16\vec{i} - 34\vec{j}$

43. $(\vec{v}_1 + \vec{v}_2) + \vec{v}_3 = [\langle a, b \rangle + \langle c, d \rangle] + \langle e, f \rangle$
$= \langle a + c, b + d \rangle + \langle e, f \rangle$
$= \langle\langle a + c \rangle + e, \langle b + d \rangle + f \rangle\rangle$
$= \langle a + \langle c + e \rangle, b + \langle d + f \rangle\rangle$
$= \langle a, b \rangle + \langle c + e, d + f \rangle$
$= \langle a, b \rangle + [\langle c, d \rangle + \langle e, f \rangle]$
$= \vec{v}_1 + (\vec{v}_2 + \vec{v}_3)$

45a.

45b. 30 mph **47a.** 30 s **47b.** 30 m **47c.** 5.1 m/s
49. None **51.** $\dfrac{-\sqrt{6}-\sqrt{2}}{4}$ **53.** ≈ 1434 ft;
$\approx 86{,}751$ sq ft **55.** max: $(0, 3)$, min: $(0.67, 2.85)$
57. A

Pages 502–504 Lesson 8-3
5. $\langle 5, 4, -11\rangle$, $9\sqrt{2}$ **7.** $\langle 6, 0, -25\rangle$ **9.** $11\vec{\mathbf{i}} - 4\vec{\mathbf{j}} + 2\vec{\mathbf{k}}$ **11.** ≈ 3457 N

13.

$\sqrt{69}$

15. $\langle 1, -4, -8\rangle$, 9 **17.** $\langle 1, -4, -4\rangle$, $\sqrt{33}$
19. $\langle 3, -9, -9\rangle$, $3\sqrt{19}$ **21.** $\langle 4, -8, -14\rangle$, $2\sqrt{69}$
23. $\left\langle 6, -7\frac{1}{2}, 11\frac{1}{2}\right\rangle$ **25.** $\left\langle 16\frac{2}{3}, -13, 23\frac{2}{3}\right\rangle$
27. $\langle -5, -24, 8\rangle$ **29.** $3\vec{\mathbf{i}} - 8\vec{\mathbf{j}} - 5\vec{\mathbf{k}}$ **31.** $-2\vec{\mathbf{i}} - 4\vec{\mathbf{j}} + 9\vec{\mathbf{k}}$ **33.** $-7\vec{\mathbf{i}} + 3\vec{\mathbf{j}} - 6\vec{\mathbf{k}}$ **35.** $\left|\overrightarrow{G_1G_2}\right| =$
$\sqrt{(x_2 - x_1)^2 + (y_2 - y_1)^2 + (z_2 - z_1)^2} =$
$\sqrt{(x_1 - x_2)^2 + (y_1 - y_2)^2 + (z_1 - z_2)^2} = \left|\overrightarrow{G_2G_1}\right|$
because $(x - y)^2 = (y - x)^2$ for all real numbers x
and y. **37.** $\langle -9, 0, -9\rangle$
39a. $\vec{\mathbf{i}} + 4\vec{\mathbf{j}}$ **39b.** $-\vec{\mathbf{i}}$

41a.

41b. About 26 ft **41c.** $\theta = 35.25°$ **43.** $\langle 2, 7\rangle$

45. $\dfrac{\sin 2X}{1 - \cos 2X} = \cot X$

$\dfrac{2\sin X \cos X}{1 - \cos^2 X + \sin^2 X} = \cot X$

$\dfrac{2\sin X \cos X}{2\sin^2 X} = \cot X$

$\dfrac{\cos X}{\sin X} = \cot X$

$\cot X = \cot X$

47. $6, 4\pi$ **49.** yes, because substituting 7 for x and -2 for y results in the inequality $-2 < 180$ which is true.

Pages 508–511 Lesson 8-4
5. 0, yes **7.** $\langle 13, 1, -5\rangle$, yes; $\langle 13, 1, -5\rangle \cdot \langle 1, -3, 2\rangle = 13(1) + 1(-3) + (-5)(2) = 13 - 3 - 10 = 0$;
$\langle 13, 1, -5\rangle \cdot \langle -2, 1, -5\rangle = 13(-2) + 1(1) + (-5)(-5) = -26 + 1 + 25 = 0$ **9.** Sample answer: $\langle 1, -8, 5\rangle$ **11.** 0, yes **13.** -21, no **15.** 32, no
17. 6, no **19.** 9, no **21.** $\langle 2, 2, -1\rangle$, yes; $\langle 2, 2, -1\rangle \cdot \langle 0, 1, 2\rangle = 2(0) + 2(1) + (-1)(2) = 2 + 2 - 2 = 0$; $\langle 2, 2, -1\rangle \cdot \langle 1, 1, 4\rangle = 2(1) + 2(1) + (-1)(4) = 2 + 2 - 4 = 0$ **23.** $\langle 0, 0, 10\rangle$, yes; $\langle 0, 0, 10\rangle \cdot \langle 3, 2, 0\rangle = 0(3) + 0(2) + 10(0) = 0 + 0 + 0 = 0$; $\langle 0, 0, 10\rangle \cdot \langle 1, 4, 0\rangle = 0(1) + 0(4) + 10(0) = 0 + 0 + 0 = 0$ **25.** $\langle 8, 8, 16\rangle$, yes; $\langle 8, 8, 16\rangle \cdot \langle -3, -1, 2\rangle = 8(-3) + 8(-1) + 16(2) = -24 - 8 + 32 = 0$; $\langle 8, 8, 16\rangle \cdot \langle 4, -4, 0\rangle = 8(4) + 8(-4) + 16(0) = 32 - 32 + 0 = 0$ **27.** Sample answer: Let $\vec{\mathbf{v}} = \langle v_1, v_2, v_3\rangle$ and $-\vec{\mathbf{v}} = \langle -v_1, -v_2, -v_3\rangle$

$\vec{\mathbf{v}} \times (-\vec{\mathbf{v}}) = \begin{bmatrix} \vec{\mathbf{i}} & \vec{\mathbf{j}} & \vec{\mathbf{k}} \\ v_1 & v_2 & v_3 \\ -v_1 & -v_2 & -v_3 \end{bmatrix}$

$= \begin{bmatrix} v_2 & v_3 \\ -v_2 & -v_3 \end{bmatrix}\vec{\mathbf{i}} - \begin{bmatrix} v_1 & v_3 \\ -v_1 & -v_3 \end{bmatrix}\vec{\mathbf{j}} + \begin{vmatrix} v_1 & v_2 \\ -v_1 & -v_2 \end{vmatrix}\vec{\mathbf{k}}$

$= 0\vec{\mathbf{i}} - 0\vec{\mathbf{j}} + 0\vec{\mathbf{k}} = \vec{0}$

29. Sample answer: $\langle -2, -17, -14\rangle$ **31.** Sample answer: $\langle 0, 2, -1\rangle$
33a.

33b. 21 N · m

35a. $\vec{\mathbf{o}} = \langle 120, 310, 60\rangle$, $\vec{\mathbf{c}} = \langle 29, 18, 21\rangle$
35b. $10,320 **37a.** Sample answer: $\langle 8, -7, -9\rangle$
37b. The cross product of two vectors is always a vector perpendicular to the two vectors and the plane in which they lie. **39.** $-\dfrac{19}{29}$ **41.** $\langle 2, 0, 3\rangle$

43. $\frac{4}{\sqrt{17}}x + \frac{1}{\sqrt{17}}y - \frac{6}{\sqrt{17}} = 0$;

$\frac{6}{\sqrt{17}} \approx 1.46$ units; $76°$ **45.** 13.1 meters;

13.7 meters **47.** B

Pages 516–519   Lesson 8-5
5. 421.19 N, 19.3° **7.** 13.79 N, 11.57 N

9a.   **9b.** $\approx 19.5°$

11.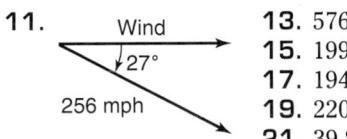
13. 576.82 N, 42.5°
15. 199.19 km/h, 90°
17. 194.87 N, 25.62°
19. 220.5 lb, 16.7°
21. 39.8 N, 270°
23. 19.9 N, 5.3° west of south **25.** 1,542,690 N · m
27a. 9.5° south of east **27b.** 18.2 mph
29. Left side: 760 lb, Right side: 761 lb
31. 3192.5 tons **33.** −2, no **35.** 239.4 ft
37. 30% beef, 20% pork; $76

Pages 523–525   Lesson 8-6
5. $\langle x - 1, y - 5 \rangle = t\langle -7, 2 \rangle$; $x = 1 - 7t$, $y = 5 + 2t$
7. $x = t$, $y = \frac{2}{3}t + 2$ **9.** $y = \frac{4}{9}x + 2$ **11a.** Receiver:
$x = 5$, $y = 50 - 10t$; Defensive player:
$x = 10 - 0.9t$, $y = 54 - 10.72t$ **11b.** yes
13. $\langle x + 1, y - 4 \rangle = t\langle 6, -10 \rangle$; $x = -1 + 6t$,
$y = 4 - 10t$ **15.** $\langle x - 1, y - 5 \rangle = t\langle -7, 2 \rangle$; $x = 1 - 7t$,
$y = 5 + 2t$ **17.** $\langle x - 3, y + 5 \rangle = t\langle -2, 5 \rangle$; $x = 3 - 2t$,
$y = -5 + 5t$ **19.** $x = t$, $y = \frac{3}{4}t + \frac{7}{4}$ **21.** $x = t$,
$y = 9t - 1$ **23.** $x = t$, $y = 4t - 2$
25. $y = -\frac{1}{2}x + 1$ **27.** $y = \frac{1}{4}x + \frac{23}{4}$ **29.** $y = \frac{5}{2}x - \frac{17}{2}$ **31a.** $\langle x - 11, y + 4 \rangle = t\langle 3, 7 \rangle$ **31b.** $x = 3t + 11$,
$y = 7t - 4$ **31c.** $y = \frac{7}{3}x - \frac{89}{3}$

33.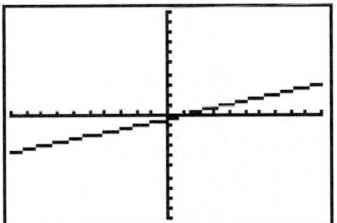

$[-10, 10]$ tstep:1 $[-20, 20]$ Xscl:2 $[-20, 20]$ Yscl:2
35a. Right of point (2, 4) **35b.** $t < -\frac{2}{3}$
37a. Target drone: $x = 3 - t$, $y = 4$; Missile: $x = 2 + t$,
$y = 2 + 2t$ **37b.** No **39.** $x = -\frac{1}{3} + \frac{1}{3}t$, $y = 1 + 4t$,
$z = 1 - 9t$ **41.** −3, no **43.** 1 **45.** $x - y + 4 = 0$

Pages 531–533   Lesson 8-7
5. 12.86 m/s **7.** 7.05 m/s, 2.57 m/s **9.** 32.5 ft/s,
56.29 ft/s **11.** 891.77 ft/s, 802.96 ft/s
13. 55.11 yd/s, 41.53 yd/s **15a.** $x = 175t \cos 35°$,
$y = 175t \sin 35° - 16t^2$ **15b.** 899.32 ft or 299.77 yd
17a. 158.32 ft/s **17b.** 127 yd. **19.** Sample
answer: No, the projectile will travel four times as
far. **21a.** 323.2 ft **21b.** 312.4 ft **21c.** 3.71 s
23a. 140.7 ft/s **23b.** 131.3 yd **25.** $y = 6x - 58$
27. 37° **29.** B

Pages 540–542   Lesson 8-8
5a. $\begin{bmatrix} 5 & 5 & 0 & 0 & 0 & 5 & 5 & 0 \\ 2 & 5 & 5 & 2 & 2 & 2 & 5 & 5 \\ 0 & 0 & 0 & 0 & 4 & 4 & 4 & 4 \end{bmatrix}$

5b. $\begin{bmatrix} 9 & 9 & 4 & 4 & 4 & 9 & 9 & 4 \\ 1 & 4 & 4 & 1 & 1 & 1 & 4 & 4 \\ 2 & 2 & 2 & 2 & 6 & 6 & 6 & 6 \end{bmatrix}$

5c.   Reflection over the xz-plane.

5d. The dimensions of the resulting figure are half the original.

7. $\begin{bmatrix} 2 & 3 & 4 & 4 & 2 & 3 \\ -3 & 1 & 1 & 7 & 3 & 7 \\ 2 & 4 & -1 & -1 & 2 & 4 \end{bmatrix}$

9. $\begin{bmatrix} 2 & 1 & 4 & 4 & 3 & 6 \\ -2 & 0 & -1 & -1 & 1 & 0 \\ 3 & 4 & 2 & 1 & 2 & 0 \end{bmatrix}$

11.

Translation 1 unit along the x-axis, -2 units along the y-axis, and -2 units along the z-axis.

13.

No change

15.

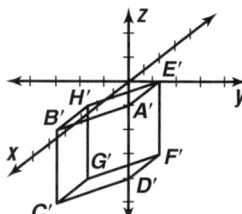

Reflection across all three coordinate planes

17. The figure is three times the original size and reflected over the xy-plane.

19a. $\begin{bmatrix} 2 & 0 & 0 \\ 0 & 2 & 0 \\ 0 & 0 & 5 \end{bmatrix}$ **19b.** The transformation will magnify the x- and y-dimensions two-fold, and the z-dimension 5-fold.

21. $\begin{bmatrix} -0.5 & 0 & 0 \\ 0 & 0.5 & 0 \\ 0 & 0 & 0.5 \end{bmatrix}$ **23.** The first transformation reflects the figure over all three coordinate planes. The second transformation stretches the dimensions along y- and z-axes and skews it along the xy-plane.

25a. dip-slip

25b. $\begin{bmatrix} 0 & 0 & 0 & 0 & 0 & 0 \\ 1.6 & 1.6 & 1.6 & 1.6 & 1.6 & 1.6 \\ -1.2 & -1.2 & -1.2 & -1.2 & -1.2 & -1.2 \end{bmatrix}$

27. $y = -\frac{2}{5}x + \frac{48}{5}$ **29.** ≈ 0.2

Pages 543–547 **Chapter 8 Study Guide and Assessment**

1. resultant **3.** magnitude **5.** inner **7.** parallel
9. direction **11.** 1.5 cm; 50° **13.** 4.1 cm; 23°
15. 2.5 cm; 98° **17.** 0.8 cm; 1 cm **19.** $\langle 5, 12 \rangle$; 13
21. $\langle -2, 12 \rangle$; $2\sqrt{37}$ **23.** $\langle 5, -6 \rangle$ **25.** $\langle 12, -17 \rangle$
27. $\langle 4, -1, -3 \rangle$; $\sqrt{26}$ **29.** $\langle 6, 2, 7 \rangle$; $\sqrt{89}$
31. $\langle 13, -37, 30 \rangle$ **33.** -16; no **35.** 0; yes
37. 42; no **39.** $\langle 9, -6, 0 \rangle$; yes; $\langle 9, -6, 0 \rangle \cdot \langle -2, -3, 1 \rangle$
$= 9(-2) + (-6)(-3) + 0(1) = -18 + 18 + 0 = 0$
$\langle 9, -6, 0 \rangle \cdot \langle 2, 3, -4 \rangle = 9(2) + (-6)(3) + 0(-4) =$
$18 - 18 + 0 = 0$ **41.** $\langle 1, -19, 31 \rangle$; yes; $\langle 1, -19, 31 \rangle \cdot$
$\langle 7, 2, 1 \rangle = 1(7) + (-19)(2) + 31(1) = 7 - 38 + 31 =$
0; $\langle 1, -19, 31 \rangle \cdot \langle 2, 5, 3 \rangle = 1(2) + (-19)(5) + 31(3) =$
$2 - 95 + 95 = 0$ **43.** 412.31 N; 39.09°
45. $\langle x - 3, y + 5 \rangle = t\langle 4, 2 \rangle$; $x = 3 + 4t, y = -5 + 2t$
47. $\langle x - 4, y \rangle = t\langle 3, -6 \rangle$; $x = 4 + 3t, y = -6t$
49. $x = t, y = -\frac{1}{2}t + \frac{5}{2}$ **51.** 5.37 ft/s, 12.06 ft/s
53.

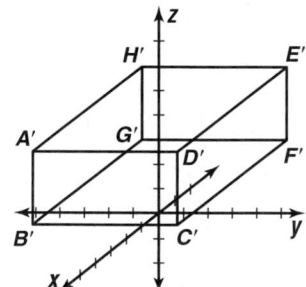

moves 2 units along x-axis and 3 units along the z-axis. **55.** 25 lb-ft **57a.** 13.7 km/h **57b.** 275.3 m

Page 549 **Chapter 8 SAT and ACT Practice**
1. A **3.** B **5.** E **7.** D **9.** B

Chapter 9 Polar Coordinates and Complex Numbers

Pages 558–560 **Lesson 9-1**
7.

9.

11.

13.

15a. **15b.** about 838 ft²

17. **19.**

21. **23.**

25. **27.**

29. Sample answer: $(1.5, 540°)$, $(1.5, 900°)$, $(-1.5, 0°)$, $(-1.5, 360°)$ **31.** Sample answer: $(4, 675°)$, $(4, 1035°)$, $(-4, 135°)$, $(-4, 495°)$

33. **35.**

37. **39.**

41. $r = \sqrt{2}$ or $r = -\sqrt{2}$ **43.** 5.35 **45.** 4.87
47. $\theta = 0°$, $\theta = 60°$, $\theta = 120°$ **49a.** 17 knots
49b. 13 knots **51.** The distance formula is
symmetric with respect to (r_1, θ_1) and (r_2, θ_2).
That is,

$$\sqrt{r_2{}^2 + r_1{}^2 - 2r_2r_1 \cos(\theta_1 - \theta_2)} =$$
$$\sqrt{r_1{}^2 + r_2{}^2 - 2r_1r_2 \cos[-(\theta_2 - \theta_1)]} =$$
$$\sqrt{r_1{}^2 + r_2{}^2 - 2r_1r_2 \cos(\theta_2 - \theta_1)}$$

53. about 22.0° **55.** $\dfrac{16\sqrt{82}}{41}$ **57.** 30° **59.** one;

$B = 90°$, $C = 60°$, $c = 16.1$ **61.** $y = x - 3$
63. -11 **65.** E

Pages 565–567 Lesson 9-2
5. cardioid **7.** rose

9. $\left(1, \frac{\pi}{6}\right), \left(1, \frac{5\pi}{6}\right), \left(-2, \frac{3\pi}{2}\right)$

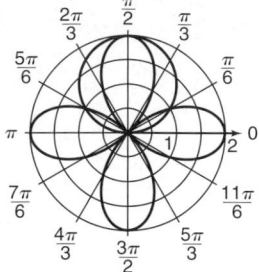

11. circle   **13.** spiral of Archimedes

15. rose   **17.** spiral of Archimedes

19. cardioid   **21.** rose

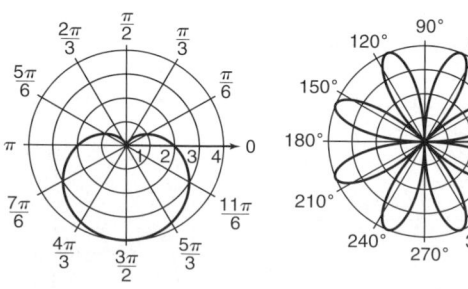

23. Sample answer: $r = \sin 3\theta$

25. $(3, 0)$ **27.** $(0, 0), (0, \pi),$ $\left(\sqrt{3}, \frac{\pi}{3}\right), \left(-\sqrt{3}, \frac{5\pi}{3}\right)$

29. $(2, 3.5), (2, 5.9)$

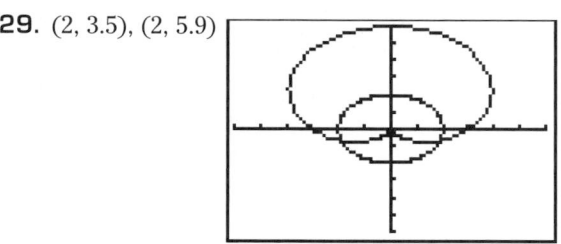

[−6, 6] scl:1 by [−6, 6] scl:1

31a. $r^2 = 9\cos 2\theta$ or $r^2 = 9\sin 2\theta$
31b. $r^2 = 16\cos 2\theta$ or $r^2 = 16\sin 2\theta$
33. $0 \le \theta \le 4\pi$ **35.** Sample answer: $r = -1 - \sin\theta$
37a. counterclockwise rotation by an angle of α
37b. reflection about the polar axis or x-axis
37c. reflection about the origin   **37d.** dilation by a factor of c
39. $\langle 12, -8, 7\rangle$; $\langle 2, 3, 0\rangle \cdot \langle 12, -8, 7\rangle = 0$, $\langle -1, 2, 4\rangle \cdot \langle 12, -8, 7\rangle = 0$

41.
$$\frac{\sin^2 x}{\cos^4 x + \cos^2 x \sin^2 x} \overset{?}{=} \tan^2 x$$

$$\frac{\sin^2 x}{\cos^2 x\,(\cos^2 x + \sin^2 x)} \overset{?}{=} \tan^2 x$$

$$\frac{\sin^2 x}{(\cos^2 x)(1)} \overset{?}{=} \tan^2 x$$

$$\frac{\sin^2 x}{\cos^2 x} \overset{?}{=} \tan^2 x$$

$$\tan^2 x = \tan^2 x$$

43.

	NY	LA	Miami
Bus	$240	$199	$260
Train	$254	$322	$426

Pages 571–573   Lesson 9-3
5. $\left(2, \frac{3\pi}{4}\right)$ **7.** $\left(1, \sqrt{3}\right)$ **9.** $r = 2\csc\theta$
11. $x^2 + y^2 = 36$

13a.

13b. No. The given point is on the negative x-axis, directly behind the microphone. The polar pattern indicates that the microphone does not pick up any sound from this direction.

15. $\left(1, \frac{\pi}{2}\right)$ **17.** $\left(\frac{1}{2}, \frac{4\pi}{3}\right)$ **19.** $(8.06, 5.23)$

21. $\left(-\frac{\sqrt{2}}{4}, \frac{\sqrt{2}}{4}\right)$ **23.** $(0, 2)$ **25.** $(-9.00, 10.72)$

27. $r = 5 \csc \theta$ **29.** $r = 2 \sin \theta$ **31.** $r = 4 \sin \theta$

33. $x^2 + y^2 = 9$ **35.** $y = 2$ **37.** $xy = 4$

39. $x^2 + y^2 = y$ **41.** 0.52 unit **43.** 75 m east; 118.30 m north **45.** circle centered at (a, a) with radius $\sqrt{2}\,|a|$; $(x - a)^2 + (y - a)^2 = 2a^2$.

47. Sample answer: $(-2, 405°)$, $(-2, 765°)$, $(2, 225°)$, $(2, 585°)$ **49.** $0°$ **51.** $-\frac{\sqrt{3}}{2}$ **53.** $x^4 + 2x^3 + 4x^2 + 5x + 10$ **55.** C

Pages 577–579 Lesson 9-4

5. $2 = r \cos (\theta - 307°)$ **7.** $x + \sqrt{3}y - 6 = 0$

9.

11a. $\left(5, \frac{5\pi}{6}\right)$

11b.

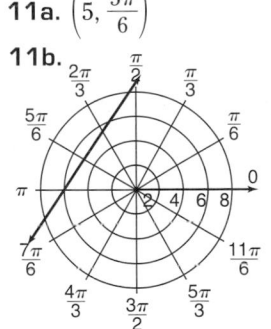

13. $3 = r \cos (\theta - 44°)$ **15.** $\frac{5\sqrt{13}}{13} = r \cos (\theta - 34°)$

17. $\frac{7\sqrt{10}}{10} = r \cos (\theta - 108°)$ **19.** $\sqrt{2}x - \sqrt{2}y - 8 = 0$ **21.** $\sqrt{3}x - y - 2 = 0$

23. $x + \sqrt{3}y - 10 = 0$

25.   **27.**

29.

31. $0.31 = r \cos (\theta - 2.25)$

33. Sample answer: $2 = r \cos (\theta - 45°)$ and $2 = r \cos (\theta - 135°)$

35a.

35b. $124.43 = r \cos (\theta - 135°)$

37. $32.36 = r \cos (\theta - 36°)$

39. rose

41. about 20.42 ft^2

43. $0, \frac{3}{2}, -4$

Pages 583–585 Lesson 9-5

5. -1 **7.** $-4 + 4i$ **9.** $1 + 9i$ **11.** $\frac{2}{5} + \frac{1}{5}i$

13. -1 **15.** 1 **17.** $-1 + 8i$ **19.** $-\frac{3}{2} + 2i$

21. $5 + 10i$ **23.** $\left(-2 + \sqrt{35}\right) + \left(-2\sqrt{7} - \sqrt{5}\right)i$

25. $\frac{4}{5} - \frac{3}{5}i$ **27.** $\frac{12}{13} - \frac{5}{13}i$ **29.** $x^2 - 4x + 5 = 0$

31. $-12 - 16i$ **33.** $\left(\frac{2}{5} - \frac{2\sqrt{3}}{15}\right) + \left(-\frac{\sqrt{2}}{5} - \frac{2\sqrt{6}}{15}\right)i$ **35.** $-\frac{24}{169} + \frac{10}{169}i$

37a. $\pm 3 - 4i$ **37b.** No. **37c.** The solutions need not be complex conjugates because the coefficients in the equation are not all real.

37d. $(3 - 4i)^2 + 8i(3 - 4i) - 25 \overset{?}{=} 0$
$-7 - 24i + 24i + 32 - 25 \overset{?}{=} 0$
$0 = 0$

$(-3 - 4i)^2 + 8i(-3 - 4i) - 25 \overset{?}{=} 0$
$-7 + 24i - 24i + 32 - 25 \overset{?}{=} 0$
$0 = 0$

39a. $1 + 2i, -2 + i, -1 - 2i, 2 - i, 1 + 2i$

39b. $0.5 - 0.866i, -0.500 - 0.866i, -1.000 - 0.000i, -0.500 + 0.866i, 0.500 + 0.866i$ **41.** $c_1 = c_2$

43.

45. $\left\langle -6, \frac{27}{2}, -5 \right\rangle$

47. $y = 3.5 \cos \left(\frac{\pi}{6}t\right)$

49. quadratic

51. 64 **53.** 3, -11

55. A

Pages 589–591   Lesson 9-6

5. $x = 1, y = 3$ **7.** $\sqrt{3}$
9. $\sqrt{41}(\cos 0.90 +$
$i \sin 0.90)$

11. $2 + 2\sqrt{3}i$

13. $\dfrac{3}{2}$

15a. about 18.03 N   **15b.** about 56.31°
17. $x = \dfrac{1}{2}, y = 1$

19. $\sqrt{13}$

21. $\sqrt{26}$

23. $\sqrt{6}$

25. $2\sqrt{13}$ **27.** $2\left(\cos \dfrac{4\pi}{3} + i \sin \dfrac{4\pi}{3}\right)$
29. $\sqrt{17}(\cos 2.90 + i \sin 2.90)$
31. $2\sqrt{5}(\cos 2.03 + i \sin 2.03)$
33. $4\sqrt{2}(\cos \pi + i \sin \pi)$

35. $\dfrac{3\sqrt{2}}{2} + \dfrac{3\sqrt{2}}{2}i$ **37.** $-1 - \sqrt{3}i$

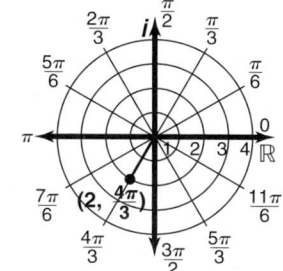

39. $-\sqrt{2} - \sqrt{2}i$ **41.** 5

43.

45.

47. The moduli are the same, but the amplitudes are opposites. **49a.** Translate 2 units to the right and down 3 units. **49b.** Rotate 90° counterclockwise about the origin. **49c.** Dilate by a factor of 3. **49d.** Reflect about the real axis. **51.** $-6 + 22i$ **53.** $\sqrt{58}$, $-3\vec{i} + 7\vec{j}$ **55.** about 13.57 m/s **57.** 4 **59.** D

Pages 596–598 Lesson 9-7

5. $-\dfrac{3}{4}i$ **7.** $-\dfrac{3\sqrt{3}}{2} - \dfrac{3}{2}i$

9. $6\left(\cos\dfrac{\pi}{6} + j\sin\dfrac{\pi}{6}\right)$ volts **11.** $3i$

13. $5\sqrt{2} - 5\sqrt{2}i$ **15.** $\dfrac{3\sqrt{3}}{2} - \dfrac{3}{2}i$

17. -2 **19.** $3.10 + 2.53i$ **21.** $-2 - 2i$

23. $-4 - 4\sqrt{3}i$ **25.** -12 **27.** $\dfrac{2\sqrt{2}}{3}i$

29. $16 + 12j$ ohms **31a.** The point is rotated counterclockwise about the origin by an angle of θ. **31b.** The point is rotated 60° counterclockwise about the origin. **33.** $13(\cos 5.11 + i \sin 5.11)$ **35.** about 27.21 lb **37.** $y = \arccos x$

Pages 605–606 Lesson 9-8

5. $-8i$ **7.** $0.97 + 0.26i$
9. $0.38 + 0.92i$,
$-0.92 + 0.38i$,
$-0.38 - 0.92i$,
$0.92 - 0.38i$

11. Escape set; the iterates escape to infinity. **13.** $-16\sqrt{2} - 16\sqrt{2}i$ **15.** $-8 - 8\sqrt{3}i$ **17.** $-0.03 - 0.07i$ **19.** $1.83 + 0.81i$ **21.** $0.96 + 0.76i$ **23.** $1.37 + 0.37i$ **25.** $0.71 + 0.71i$

27. $0.81 + 0.59i$,
$-0.31 + 0.95i$,
-1,
$-0.31 - 0.95i$,
$0.81 - 0.59i$

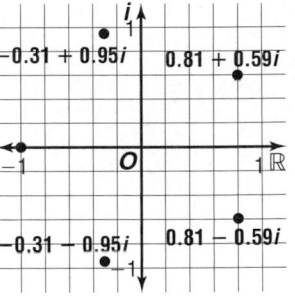

29. $\sqrt{2} + \sqrt{2}i$,
$-\sqrt{2} + \sqrt{2}i$,
$-\sqrt{2} - \sqrt{2}i$,
$\sqrt{2} - \sqrt{2}i$

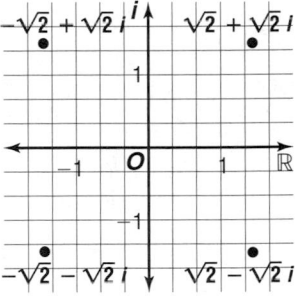

31. $0.59 + 1.03i$,
$-1.03 + 0.59i$,
$-0.59 - 1.03i$,
$1.03 - 0.59i$

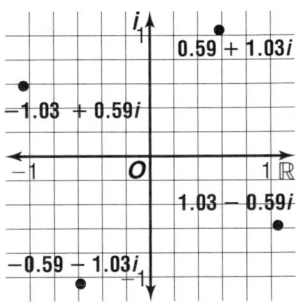

33. $1.26 + 0.24i$, $0.43 + 1.21i$, $-0.83 + 0.97i$, $-1.26 - 0.24i$, $-0.43 - 1.21i$, $0.83 - 0.97i$ **35.** Prisoner set; the iterates approach 0. **37.** 1, $\dfrac{1}{2} + \dfrac{\sqrt{3}}{2}i$, $-\dfrac{1}{2} + \dfrac{\sqrt{3}}{2}i$, -1, $-\dfrac{1}{2} - \dfrac{\sqrt{3}}{2}i$, $\dfrac{1}{2} - \dfrac{\sqrt{3}}{2}i$

39. The roots are the vertices of a regular polygon. Since one of the roots must be a positive real number, a vertex of the polygon lies on the positive real axis and the polygon is symmetric about the real axis. This means the non-real complex roots occur in conjugate pairs. Since the imaginary part of the sum of two complex conjugates is 0, the imaginary part of the sum of all the roots must be 0. **41.** $5 - i$

43. $\dfrac{\sqrt{2 + \sqrt{2}}}{2}$ **45.** 800 large bears, 400 small bears

Pages 607–611 Chapter 9 Study Guide and Assessment

1. absolute value **3.** prisoner **5.** pure imaginary **7.** rectangular **9.** Argand

11.

13.

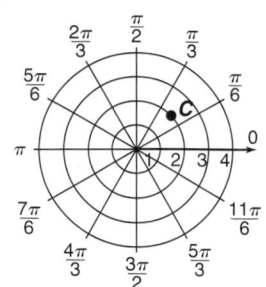

15. Sample answer: $(4, 585°), (4, 945°), (-4, 45°),$ $(-4, 405°)$

17.

19.

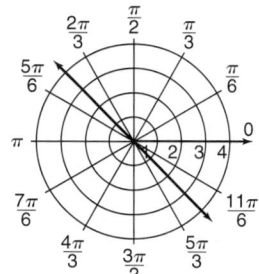

21. spiral of Archimedes **23.** rose

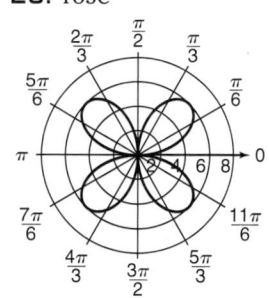

25. $(\sqrt{3}, -1)$ **27.** $(0, 1)$ **29.** $\left(5\sqrt{2}, \dfrac{\pi}{4}\right)$

31. $(4.47, 0.46)$ **33.** $\dfrac{2\sqrt{10}}{5} = r\cos(\theta - 198°)$

35. $y + 4 = 0$ **37.** $-2 + 7i$ **39.** $-3 - 4i$

41. $\dfrac{16}{29} + \dfrac{18}{29}i$ **43.** $2\sqrt{2}\left(\cos\dfrac{\pi}{4} + i\sin\dfrac{\pi}{4}\right)$

45. $2\left(\cos\dfrac{2\pi}{3} + i\sin\dfrac{2\pi}{3}\right)$ **47.** $\sqrt{17}(\cos 3.39 +$ $i\sin 3.39)$ **49.** $2\sqrt{2}(\cos\pi + i\sin\pi)$

51. $\sqrt{3} + i$

53. $-6 + 6\sqrt{3}i$
55. $-8.01 + 5.98i$
57. $\dfrac{3}{4} + \dfrac{3\sqrt{3}}{4}i$
59. 4096 **61.** -4
63. $0.92 + 0.38i$
65. lemniscate
67. $y = -5$

1. A **3.** E **5.** A **7.** E **9.** D

Chapter 10 Conics

Pages 620–622 Lesson 10-1
5. 10, $(5, 6)$ **7.** $2\sqrt{2}, (-1, 3)$ **9.** yes; $\overline{XY} \cong \overline{XZ}$, since $XY = 2\sqrt{17}$ and $DC = 2\sqrt{17}$, therefore $\triangle XYZ$ is isosceles. **11a.** $(40, 60)$ **11b.** $20\sqrt{13}$ yd or about 72 yards **13.** $2\sqrt{10}; (0, 0)$ **15.** $3\sqrt{5};$ $(2, -4.5)$ **17.** $16; (a, -1)$ **19.** $\sqrt{5}; \left(c + 1, d - \dfrac{1}{2}\right)$
21. $a = \pm 4$ **23.** yes **25.** -5 **27.** $\overline{EF} \cong \overline{HG}$ since $EF = \sqrt{5}$ and $HG = \sqrt{5}$. $\overline{EF} \parallel \overline{HG}$ since the slope of $\overline{EF}$ is $-\dfrac{1}{2}$ and the slope of $\overline{HG}$ is $-\dfrac{1}{2}$. Thus the points form a parallelogram. $\overline{EF} \perp \overline{FG}$ since the product of the slopes of $\overline{EF}$ and $\overline{FG}$, $-\dfrac{1}{2} \cdot \dfrac{2}{1}$, is -1. Therefore, the points form a rectangle. **29.** In trapezoid $ABCD$, let A and B have coordinates $(0, 0)$ and $(b, 0)$, respectively. To make the trapezoid isosceles, let C have coordinates $(b - a, c)$ and let D have coordinates (a, c).

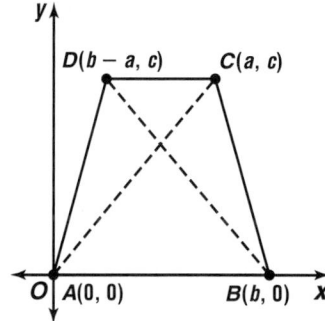

$AC = \sqrt{(a - 0)^2 + (c - 0)^2} = \sqrt{a^2 + c^2}$
$BD = \sqrt{(b - a - b)^2 + (c - 0)^2} = \sqrt{a^2 + c^2}$
$AC = \sqrt{a^2 + c^2} = \sqrt{a^2 + c^2} = BD$, so the diagonals of an isosceles trapezoid are congruent.
31. Let A and B have coordinates $(0, 0)$ and $(b, 0)$ respectively. To make a parallelogram, let C have coordinates $(a + b, c)$ and let D have coordinates (a, c). The midpoint of $\overline{BD}$ is $\left(\dfrac{a + b}{2}, \dfrac{0 + c}{2}\right)$ or $\left(\dfrac{a + b}{2}, \dfrac{c}{2}\right)$.

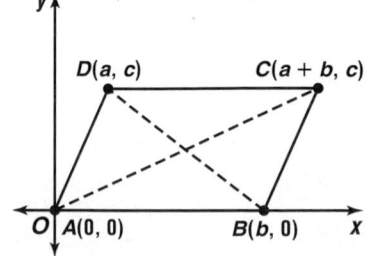

The midpoint of $\overline{AC}$ is $\left(\dfrac{a + b + 0}{2}, \dfrac{c + 0}{2}\right)$ or $\left(\dfrac{a + b}{2}, \dfrac{c}{2}\right)$. Since the diagonals have the same midpoint, the diagonals bisect each other.

33. 16 square units **35.** Let the vertices of the isosceles trapezoid have the coordinates $A(0, 0)$, $B(2a, 0)$, $C(2a - 2c, 2b)$, $D(2c, 2b)$. The coordinates of the midpoints are: $P(a, 0)$, $Q(2a - c, b)$, $R(a, 2b)$, and $S(c, b)$.

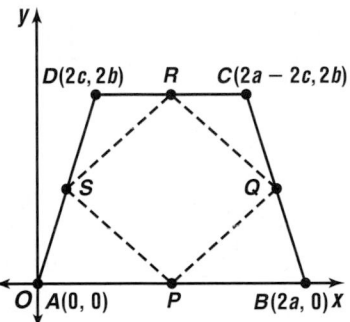

$$PQ = \sqrt{(2a - c - a)^2 + (b - 0)^2}$$
$$= \sqrt{(a - c)^2 + b^2}$$
$$QR = \sqrt{(2a - c - a)^2 + (b - 2b)^2}$$
$$= \sqrt{(a - c)^2 + b^2}$$
$$RS = \sqrt{(a - c)^2 + (2b - b)^2}$$
$$= \sqrt{(a - c)^2 + b^2}$$
$$PS = \sqrt{(a - c)^2 + (0 - b)^2}$$
$$= \sqrt{(a - c)^2 + b^2}$$

So, all of the sides are congruent and quadrilateral *PQRS* is a rhombus.

37a.
$$MA = \sqrt{t^2 + (3t - 15)^2}$$
$$= \sqrt{t^2 + 9t^2 - 90t + 225}$$
$$= \sqrt{10t^2 - 90t + 225}$$
$$MB = \sqrt{(t - 9)^2 + (3t - 12)^2}$$
$$= \sqrt{t^2 - 18t + 81 + 9t^2 - 72t + 144}$$
$$= \sqrt{10t^2 - 90t + 225}$$
$$MA = MB$$
$$\sqrt{10t^2 - 90t + 225} = \sqrt{10t^2 - 90t + 225}$$

Since the above equation is a true statement, *t* can take on any real values. **37b.** A line; this line is the perpendicular bisector of $\overline{AB}$. **39.** about 2021 N
41. $54.9°$ **43.** $4 \pm \sqrt{2}$

Pages 627–630   Lesson 10-2
7. $(x + 1)^2 + (y - 4)^2 = 16$

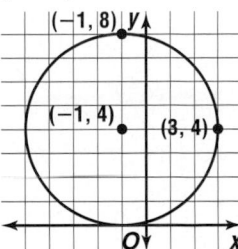

9. $(x - 5)^2 + (y + 2)^2 = 12$

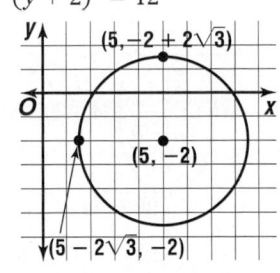

11. $(x - 3)^2 + (y - 4)^2 = 5$; $(3, 4)$; $\sqrt{5}$
13. $(x - 4)^2 + (y + 2)^2 = 100$
15. $x^2 + y^2 = 25$ **17.** $(x + 1)^2 + (y + 3)^2 = \frac{1}{2}$

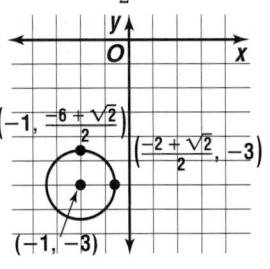

19. $(x - 6)^2 + (y - 1)^2 = 36$

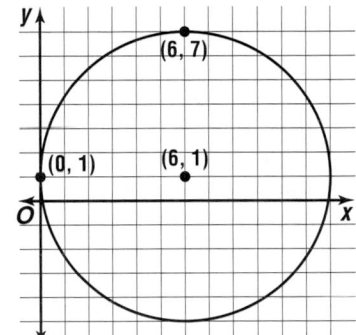

21. $x^2 + y^2 = 36$ **23.** $(x - 2)^2 + (y + 6)^2 = 10$

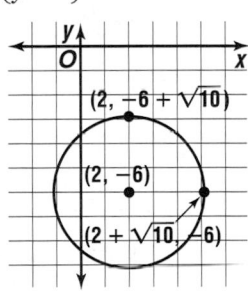

25. $(x - 1)^2 + (y + 3)^2 = 16$ **27.** $(x + 7)^2 + (y + 12)^2 = 36$

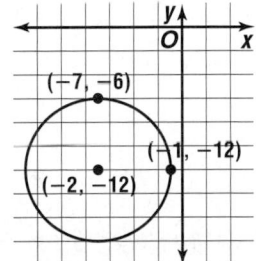

29. $(x - 7)^2 + (y + 5)^2 = 16$; $(7, -5)$; 4
31. $(x - 5)^2 + (y - 2)^2 = 50$; $(5, 2)$, $5\sqrt{2}$
33. $\left(x + \frac{1}{6}\right)^2 + \left(y - \frac{7}{6}\right)^2 = \frac{169}{18}$; $\left(-\frac{1}{6}, \frac{7}{6}\right)$; $\frac{13\sqrt{2}}{6}$

35. $(x + 4)^2 + (y - 3)^2 = 25$ **37.** $(x + 2)^2 + (y + 1)^2 = 32$ **39.** $(x - 5)^2 + (y - 1)^2 = 10$
41a. $x^2 + y^2 = 36$ **41b.** $2x$ by $2\sqrt{36 - x^2}$
41c. $A(x) = 4x\sqrt{36 - x^2}$

41d.

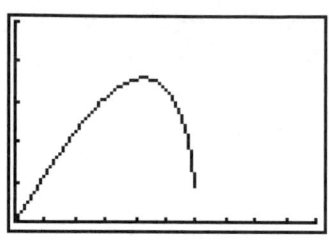

[0, 10] scl:1 by [0, 100] scl:20

41e. 4.2; 72 units^2 **43a.** $x^2 + y^2 = 144$
43b. about 145.50 in.^{2} **45a.** $(x - k)^2 + (y - k)^2 = 4$
45b. Sample graph:

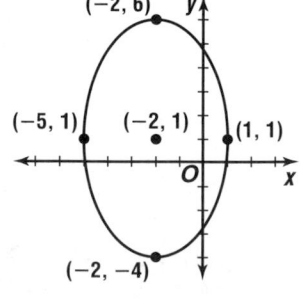

45c. All the circles in this family have a radius of 2 and centers located on the line $y = x$. **47.** radius: 0; center: $(4, -3)$; graph is a point located at $(4, -3)$
49a. $x^2 + y^2 = 25$

49b. If $\overline{PA} \perp \overline{PB}$, then A, P, and B are on the circle $x^2 + y^2 = 25$. **51.** $20 + 15i$ **53.** $y = 2.5 \cos\left(\frac{\pi}{10}t\right)$
55. 137.5 ft **57a.** 10 cases of drug A, 5 cases of drug B **57b.** $5700

Pages 637–641 Lesson 10-3
7. center: $(0, 0)$; foci: $(\pm 4\sqrt{2}, 0)$; vertices: $(\pm 6, 0)$, $(0, \pm 2)$
9. center: $(-2, 1)$; foci: $(-2, 5)$, $(-2, -3)$; vertices: $(-2, 6)$, $(-2, -4)$, $(1, 1)$, $(-5, 1)$

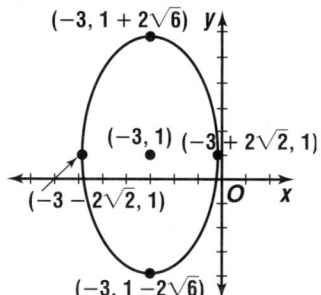

11. $\frac{(y + 3)^2}{16} + \frac{(x + 2)^2}{1} = 1$ **13.** $\frac{(x - 1)^2}{16} + \frac{(y - 2)^2}{4} = 1$ **15.** $\frac{x^2}{1.524^2} + \frac{y^2}{1.517^2} = 1$

17. $\frac{(x + 2)^2}{16} + \frac{y^2}{4} = 1$; foci: $\left(-2 \pm 2\sqrt{3}, 0\right)$

19. center: $(-2, 1)$; foci: $\left(-2, 1 \pm \sqrt{3}\right)$; vertices: $(-2, 3)$, $(-2, -1)$, $(-1, 1)$, $(-3, 1)$

21. center: $(4, -6)$, foci: $(4 \pm \sqrt{7}, -6)$; vertices: $(0, -6)$, $(8, -6)$, $(4, -3)$, $(4, -9)$

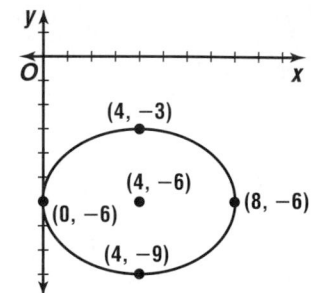

23. center: $(1, -3)$; foci: $\left(1, -3 \pm \sqrt{3}\right)$; vertices: $(2, -3)$, $(0, -3)$, $(1, -1)$, $(1, -5)$

25. center: $(-3, 1)$; foci: $(-3, 5)$, $(-3, -3)$; vertices: $\left(-3 \pm 2\sqrt{2}, 1\right)$, $\left(-3, 1 \pm 2\sqrt{6}\right)$

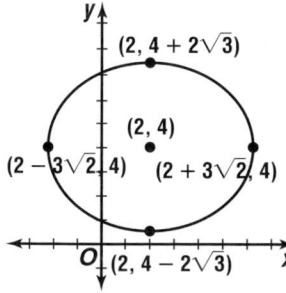

27. center: $(2, 4)$; foci: $\left(2 \pm \sqrt{6}, 4\right)$; vertices: $\left(2 \pm 3\sqrt{2}, 4\right)$, $\left(2, 4 \pm 2\sqrt{3}\right)$

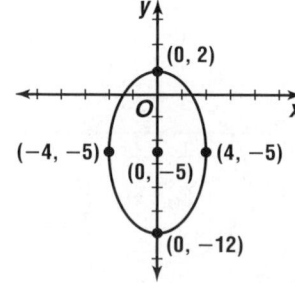

29. center: $(0, -5)$; foci: $\left(0, -5 \pm \sqrt{33}\right)$; vertices: $(\pm 4, -5)$, $(0, -12)$, $(0, 2)$

31. $\dfrac{(x + 3)^2}{49} + \dfrac{(y + 1)^2}{25} = 1$ **33.** $\dfrac{x^2}{64} + \dfrac{y^2}{36} = 1$

35. $\dfrac{(x + 2)^2}{81} + \dfrac{(y - 5)^2}{16} = 1$ **37.** $\dfrac{x^2}{100} + \dfrac{y^2}{75} = 1$

39. $\dfrac{y^2}{4} + \dfrac{x^2}{1.75} = 1$ or $\dfrac{y^2}{1.75} + \dfrac{x^2}{4} = 1$

41. $\dfrac{y^2}{100} + \dfrac{(x - 3)^2}{51} = 1$ **43.** $(5, -3), (1, -3),$ $(3, -2), (3, -4)$ **45.** $(-1, -1), (5, -1), (2, -3),$ $(2, 1)$ **47.** The target ball should be placed opposite the pocket, $\sqrt{5}$ feet from the center along the major axis of the ellipse. The cue ball can be placed anywhere on the side opposite the pocket. The ellipse has semi-major axis of length 3 ft and a semi-minor axis of length 2 ft. Using the equation $c^2 = a^2 - b^2$, the focus of the ellipse is found to be $\sqrt{5}$ ft from the center of the ellipse. Thus the hole is located at one focus of the ellipse. The reflective properties of an ellipse should insure that a ball placed $\sqrt{5}$ ft from the center of the ellipse and hit so that it rebounds once off the wall, should fall into the pocket at the other focus of the ellipse.

49a. $\dfrac{x^2}{2304} + \dfrac{y^2}{529} = 1$ **49b.** about 42 ft on either side of the center along the major axis   **49c.** about 84 ft   **51.** If (x, y) is a point on the ellipse, then show that $(-x, -y)$ is also on the ellipse.

$$\dfrac{x^2}{a^2} + \dfrac{y^2}{b^2} = 1$$

$$\dfrac{(-x)^2}{a^2} + \dfrac{(-y)^2}{b^2} = 1 \quad \textit{Replace x with } -x \textit{ and y with } -y.$$

$$\dfrac{x^2}{a^2} + \dfrac{y^2}{b^2} = 1 \quad (-x)^2 = x^2 \textit{ and } (-y)^2 = y^2$$

Thus $(-x, -y)$ is also a point on the ellipse and the ellipse is therefore symmetric with respect to the origin.

53a. GOES 4; its eccentricity is closest to 0   **53b.** 960 km
55. no
57. $y = \pm 4 \cos (2x - 40°)$
59. 74, no

61.

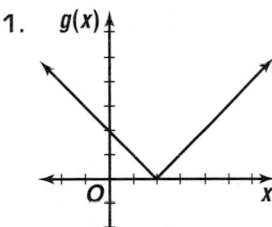

63. C

Pages 649–652   Lesson 10-4
5. center: $(0, 0)$; foci: $\left(\pm\sqrt{29}, 0\right)$; vertices: $(\pm 5, 0)$; asymptotes: $y = \pm\dfrac{2}{5}x$

7. center: $(2, 0)$; foci: $(2, \pm 6)$; vertices: $\left(2, \pm\sqrt{30}\right)$; asymptotes: $y = \pm\sqrt{5}(x - 2)$

9.

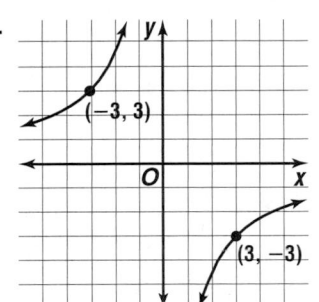

11. $\dfrac{(y - 2)^2}{4} - \dfrac{(x - 3)^2}{9} = 1$ **13.** $\dfrac{x^2}{36} - \dfrac{y^2}{64} = 1$

15. center: $(0, 0)$; foci: $\left(\pm 2\sqrt{29}, 0\right)$; vertices: $(\pm 10, 0)$; asymptotes: $y = \pm\dfrac{2}{5}x$

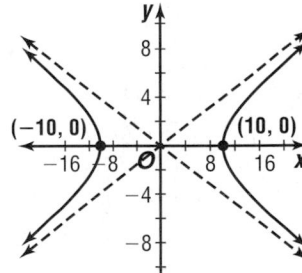

17. center: $(0, 0)$, foci: $\left(\pm\sqrt{53}, 0\right)$; vertices: $(\pm 2, 0)$; asymptotes: $y = \pm\dfrac{7}{2}x$

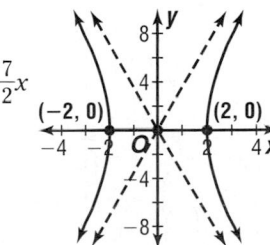

19. center: $(-3, -1)$; foci: $\left(-3 \pm 2\sqrt{5}, -1\right)$, vertices: $(1, -1),$ $(-7, -1)$; asymptotes: $y + 1 = \pm\dfrac{1}{2}(x + 3)$

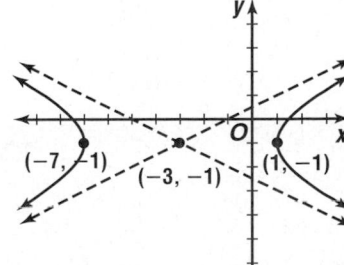

21. center: $(2, 3)$; foci: $\left(2, 3 \pm \sqrt{41}\right)$, vertices: $(2, 8)$, $(2, -2)$; asymptotes: $y - 3 = \pm\frac{5}{4}(x - 2)$

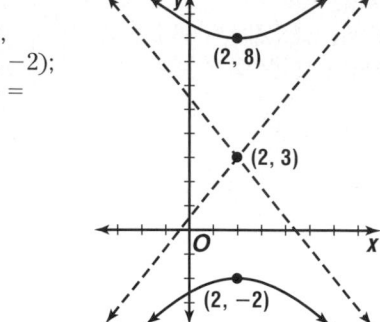

23. center: $(-4, 2)$; foci: $(-4, 2 \pm \sqrt{34})$; vertices: $(-4, 5)$, $(-4, -1)$; asymptotes: $y - 2 = \pm\frac{3}{5}(x - 4)$

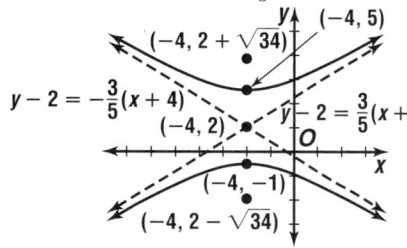

25. $\frac{x^2}{9} - \frac{y^2}{9} = 1$

27.

29.

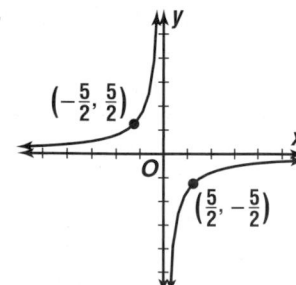

31. $\frac{(y + 2)^2}{4} - \frac{(x - 4)^2}{9} = 1$ **33.** $\frac{x^2}{9} - \frac{(y - 2)^2}{16} = 1$

35. $\frac{x^2}{32} - \frac{y^2}{32} = 1$ **37.** $\frac{(y - 2)^2}{9} - \frac{(x - 4)^2}{16} = 1$

39. $\frac{y^2}{36} - \frac{x^2}{28} = 1$ **41.** $\frac{2x^2}{81} - \frac{2y^2}{81} = 1$

43a.

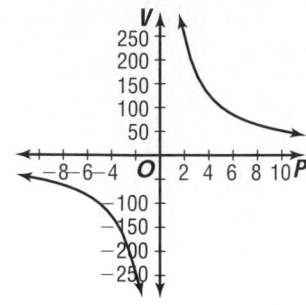

43b. 5.0 dm^3 **43c.** 10.0 dm^3 **43d.** $V = 2(\text{original } V)$ **45a.** $\frac{x^2}{75^2} - \frac{y^2}{100^2} = 1$ **45b.** top: 106.07 ft; base: 273.00 ft   **47.** $\frac{x^2}{25} - \frac{y^2}{11} = 1$

49. $\frac{y^2}{16} + \frac{(x - 2)^2}{7} = 1$

51.

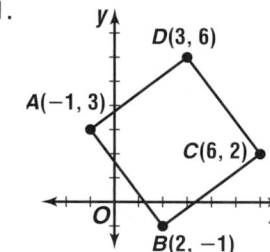

$AB = \sqrt{(2 + 1)^2 + (-1 - 3)^2} = 5$
$BC = \sqrt{(2 - 6)^2 + (-1 - 2)^2} = 5$
$CD = \sqrt{(6 - 3)^2 + (2 - 6)^2} = 5$
$AD = \sqrt{(3 + 1)^2 + (6 - 3)^2} = 5$
Thus, $ABCD$ is a rhombus. The slope of $\overline{AD} = \frac{6 - 3}{3 + 1}$ or $\frac{3}{4}$ and the slope of $\overline{AB} = \frac{3 + 1}{-1 - 2}$ or $-\frac{4}{3}$. Thus, $\overline{AD}$ is perpendicular to $\overline{AB}$ and $ABCD$ is a square.   **53.** -6; No, the inner product of the two vectors is not zero.   **55.** about 346 m/s   **57.** C

Pages 658–661   Lesson 10-5
7. vertex: $(1, -1)$; focus: $(2, -1)$; directrix: $x = 0$; axis of symmetry: $y = -1$

9. $x^2 = -16y$

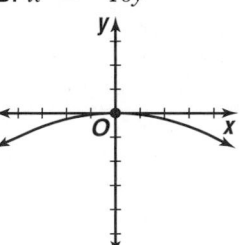

11. $(x - 4)^2 = \frac{1}{5}(y + 3)$

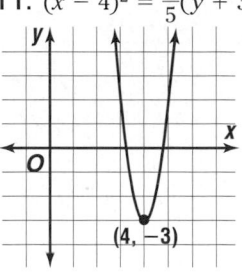

13. vertex: $(0, 0)$; focus: $(2, 0)$; directrix: $x = -2$; axis of symmetry: $y = 0$

15. vertex: $(0, 6)$; focus: $(1, 6)$; directrix: $x = -1$; axis of symmetry: $y = 6$

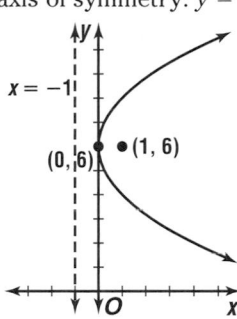

17. vertex: $(2, -2)$; focus: $\left(2, -\frac{7}{4}\right)$; directrix: $y = -\frac{9}{4}$; axis of symmetry: $x = 2$

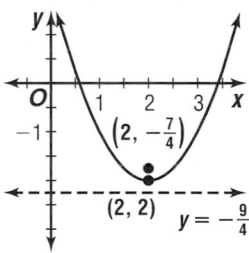

19. vertex: $(-4, -7)$; focus: $\left(-\frac{7}{2}, -7\right)$; directrix: $x = -\frac{9}{2}$; axis of symmetry: $y = -7$

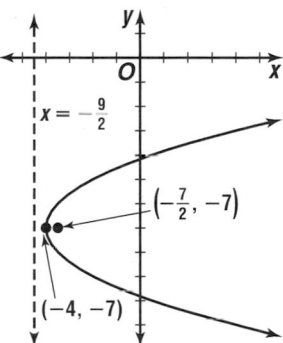

21. vertex: $(4, -1)$; focus: $\left(4, \frac{1}{2}\right)$, directrix: $y = -\frac{5}{2}$; axis of symmetry: $x = 4$

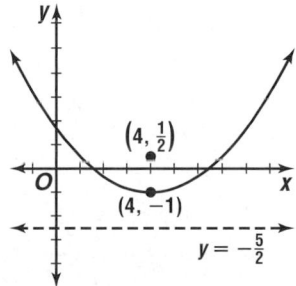

23. vertex: $(-2, -4)$; focus: $(-4, -4)$; directrix: $x = 0$; axis of symmetry: $y = -4$

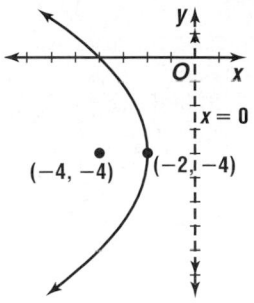

25. $(y - 6)^2 = -12(x - 3)$

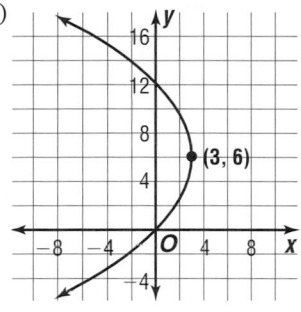

27. $(x - 4)^2 = -(y - 3)$

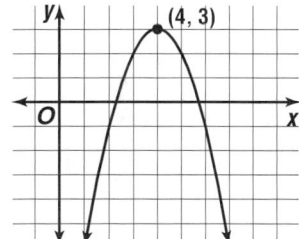

29. $(x + 1)^2 = 8(y - 5)$

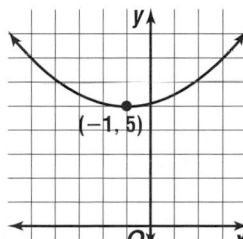

31. $(y - 2)^2 = 4(x + 1)$

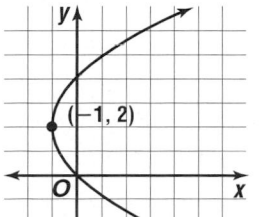

33a. $8\sqrt{2}$ in. **33b.** $4\sqrt{10}$ in. **35a.** The opening becomes narrower. **35b.** The opening becomes wider.

SELECTED ANSWERS

39. center: $(2, 3)$,
foci: $\left(2, 3 \pm \sqrt{41}\right)$;
vertices:
$(2, 8)$ and $(2, -2)$;
asymptotes:
$y - 3 = \pm\dfrac{5}{4}(x - 2)$

41.

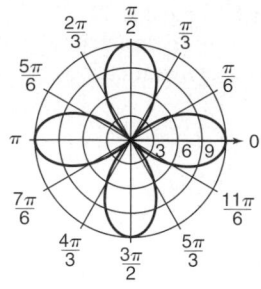

43. 5.5 cm
45. C

15. hyperbola;
$\dfrac{(y + 4)^2}{2} - \dfrac{(x + 1)^2}{6} = 1$

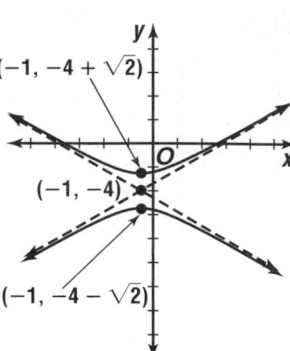

17. parabola;
$(x - 4)^2 = y$

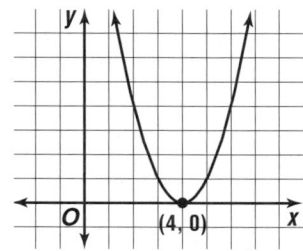

Pages 667–669 Lesson 10-6

5. parabola;
$y^2 = 8(x - 1)$

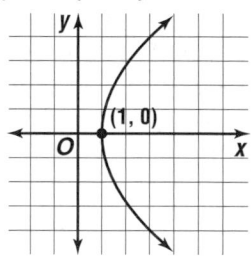

7. circle; $(x - 3)^2 +$
$(y - 6)^2 = 4$

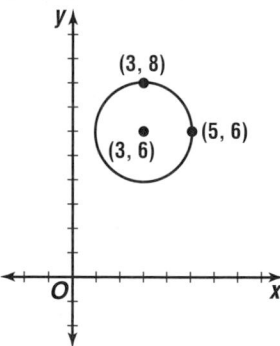

19. ellipse;
$\dfrac{(y - 5)^2}{10} + \dfrac{(x - 4)^2}{4} = 1$

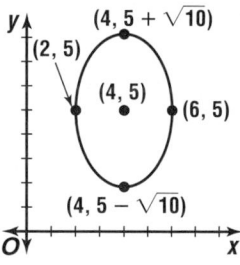

21. hyperbola;
$\dfrac{(y - 1)^2}{9} - \dfrac{(x - 2)^2}{8} = 1$

23. hyperbola;
$\dfrac{(x + 5)^2}{4} - \dfrac{(y + 2)^2}{1} = 1$

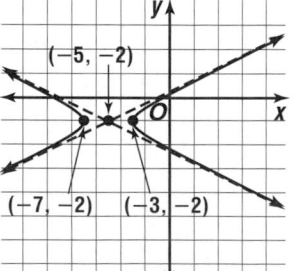

9. $\dfrac{x^2}{4} + \dfrac{y^2}{9} = 1$

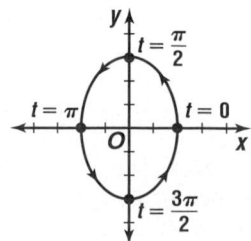

11. Sample answer:
$x = 6 \cos t$,
$y = 6 \sin t, 0 \le t \le 2\pi$
13. parabola;
$(x - 3)^2 = 4y$

25. $y = 2x^2 - 4x + 1$

27. $x^2 + y^2 = 1$

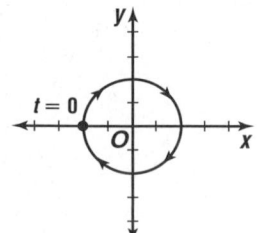

A132 *Selected Answers*

29. $x^2 + \dfrac{y^2}{4} = 1$ **31.** $x^2 + y^2 = 9$

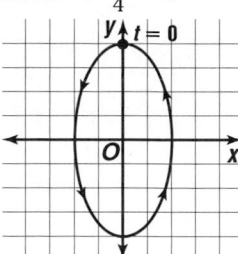

33. Sample answer: $x = 4 \cos t$, $y = 4 \sin t$, $0 \le t \le 2\pi$ **35.** Sample answer: $x = \cos t$, $y = 4 \sin t$, $0 \le t \le 2\pi$ **37.** Sample answer: $x = t^2 + 2t - 1$, $y = t$, $-\infty < t < \infty$ **39a.** Answers will vary. Sample answers: $x = t$, $y = t^2$, $t \ge 0$; $x = \sqrt{t}$, $y = t$, $t \ge 0$.

39b.

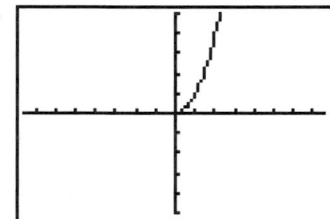

Tmin: [0, 5] step: 0.1
[−7.58, 7.58] scl:1 by [−5, 5] scl:1

39c. yes   **39d.** There is usually more than one parametric representation for the graph of a rectangular equation.   **41a.** Ellipse; point at $(0, 0)$; the equation is that of a degenerate ellipse. **41b.** Circle; point at $(2, 3)$; the equation is that of a degenerate circle.   **41c.** Hyperbola; two intersecting lines $y = \pm 3x$; the equation is that of a degenerate hyperbola.   **43a.** $x^2 + y^2 = 36$ **43b.** $x = 6 \sin t$, $y = 6 \cos t$, $0 \le t \le 4\pi$

43c.

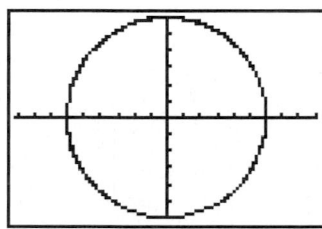

Tmin: [0, 4π] step: 0.1
[29.10, 9.10] scl:1 by [26, 6] scl:1

45. vertex: $(-5, 0)$;
focus: $(-5, 3)$; axis of
symmetry: $x = -5$,
directrix: $y = -3$

47.

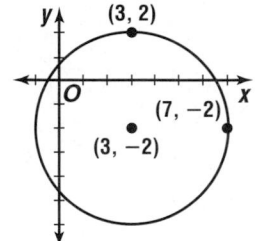

49. Car 1; the point $(135, 19)$ is about 9 units closer to the line $y = -0.13x + 37.8$ than the point $(245, 16)$.   **51.** 685 units^2 **53.** 38.4

55. $y - 4 = \dfrac{1}{3}(x + 6)$ or $y - 7 = \dfrac{1}{3}(x - 3)$, $y = \dfrac{1}{3}x + 6$

Pages 675–677   Lesson 10-7
5. circle; $x^2 + y^2 - 6x - 4y + 6 = 0$ **7.** hyperbola; $(x')^2 - 2\sqrt{3}x'y' - (y')^2 + 18 = 0$ **9.** ellipse; 19°
11. point

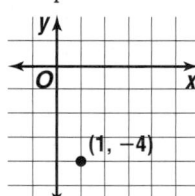

13. parabola; $3x^2 - 14x - y + 18 = 0$ **15.** ellipse; $3x^2 + y^2 + 6x - 6y + 3 = 0$
17. hyperbola; $9x^2 - 25y^2 + 250y - 850 = 0$ **19.** parabola; $(y')^2 + 8x' = 0$

21. parabola; $(x')^2 - 2\sqrt{3}x'y' + 3(y')^2 + 16\sqrt{3}x' + 16y' = 0$ **23.** circle; $2(x')^2 + 2(y')^2 - 5x' - 5\sqrt{3}y' - 6 = 0$ **25.** $23(x')^2 + 2\sqrt{3}x'y' + 21(y')^2 - 120 = 0$ **27.** hyperbola; −6°
29. ellipse; −18°   **31.** parabola; −30°

33. line

35. intersecting lines

37.

[−6.61, 14.6] scl:1 by [−2, 12] scl:1

39.

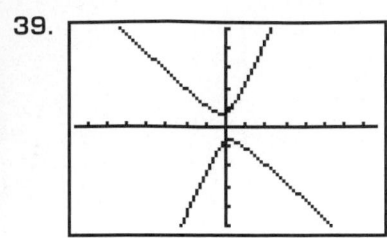

[−7.58, 7.58] scl:1 by [−5, 5] scl:1

41.

[−10.58, 4.58] scl:1 by [−2, 8] scl:1

43a. $T_{(1320, 1320)}$ **43b.** $(x − 1320)^2 + (y − 1320)^2 = 1,742,400$ **45.** Let $x = x' \cos \theta + y' \sin \theta$ and $y = −x' \sin \theta + y' \cos \theta$.

$$x^2 + y^2 = r^2$$
$$(x' \cos \theta + y' \sin \theta)^2 + (−x' \sin \theta + y' \cos \theta)^2 = r^2$$
$$(x')^2 \cos^2 \theta + x'y' \cos \theta \sin \theta + (y')^2 \sin^2 \theta$$
$$+ (x')^2 \sin^2 \theta − x'y' \cos \theta \sin \theta + (y')^2 \cos \theta = r^2$$
$$[(x')^2 + (y')^2] \cos^2 \theta + [(x')^2 + (y')^2] \sin^2 \theta = r^2$$
$$[(x')^2 + (y')^2](\cos^2 \theta + \sin^2 \theta) = r^2$$
$$[(x')^2 + (y')^2](1) = r^2$$
$$(x')^2 + (y')^2 = r^2$$

47a. −30° **47b.** $\dfrac{(x')^2}{3} + \dfrac{(y')^2}{2} = 1$

49. hyperbola **51.**

53. $\cos 70°$

55. $\dfrac{−1}{y + 2} + \dfrac{3}{y + 1}$

57. $\left(\dfrac{3}{4}, −\dfrac{2}{3}, \dfrac{1}{2}\right)$

59. B

Pages 682–684   Lesson 10-8
5. $(3, 3)$, $(−1, −1)$ **7.** $(\pm 2, 0)$

9.

11.

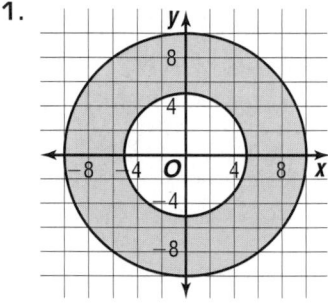

13. $(1, \pm 6.9)$ **15.** $(1.5, −4)$, $(−2, 3)$

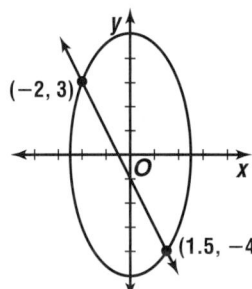

17. no solution **19.** $(0, −1)$, $(−3, 2)$

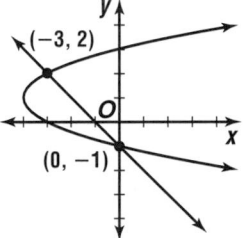

21. $(0, 3)$, $(\pm 2.4, −2.8)$ **23.** $(3, −1.3)$, $(4, −1)$, $(−3, 1.3)$, $(−4, 1)$

25.

27.

29.

31.

33.

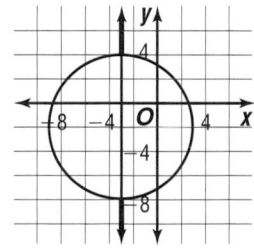

35. $x^2 + y^2 = 8$, $xy = 4$ **37a.** $2x + 2y = 150$; $xy = 800$
37b. 0, 1, 2

37c.

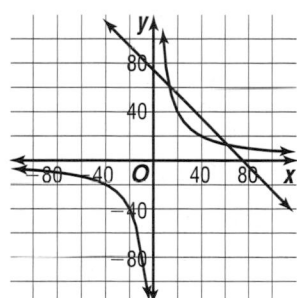

37d. 12.9 m by 62.1 m or 62.1 m by 12.9 m

39. $(4, -3)$, $(-3, 4)$

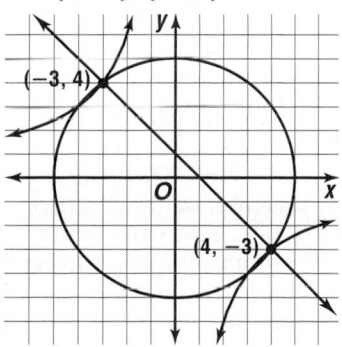

41. $-\frac{9}{8}$ **43.** $3(x')^2 - 4\sqrt{3}x'y' + 7(y')^2 - 9 = 0$
45. 4 **47.** 1 and 2 **49.** No; the domain value 4 is mapped to two elements in the range, 0 and -3.

Pages 687–691   Chapter 10 Study Guide and Assessment

1. true **3.** false; transverse **5.** false, hyperbola
7. true **9.** true **11.** $2\sqrt{5}$; $(-1, -5)$ **13.** yes;
$AB = DC = 10$ and $BC = AD = 5\sqrt{2}$. Since opposite sides of quadrilateral $ABCD$ are congruent, $ABCD$ is a parallelogram.

15. $(x - 2)^2 + (y - 1)^2 = 4$

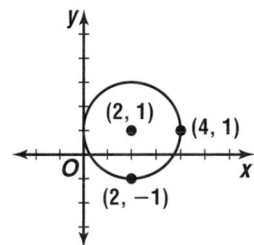

17. $(x + 7)^2 + (y + 3)^2 = 81$

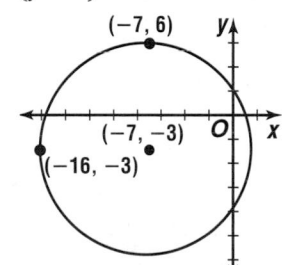

19. $(x + 2)^2 + (y + 3)^2 = 25$; $(-2, -3)$; 5
21. center: $(3, -1)$, foci: $(3 \pm \sqrt{21}, -1)$, vertices: $(3, 1)$, $(8, -1)$, $(3, -3)$, $(-2, -1)$

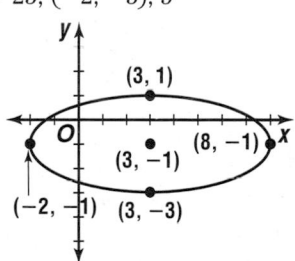

23. center: $(4, 6)$; foci: $\left(4 \pm 3\sqrt{3}\right)$; vertices: $(-2, 6)$, $(10, 6)$, $(4, 3)$, $(4, 9)$

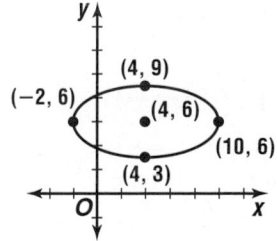

25. center: $(0, 0)$; foci: $\left(\pm\sqrt{41}, 0\right)$; vertices: $(-5, 0)$, $(5, 0)$; asymptotes: $y = \pm\frac{4}{5}x$

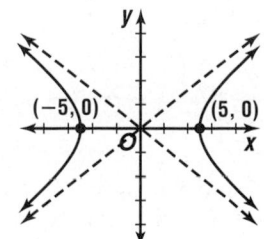

27. center: $(0, -2)$;
foci: $\left(\pm\sqrt{5}, -2\right)$;
vertices: $(-2, -2)$, $(2, -2)$
asymptotes: $y + 2 = \pm\frac{1}{2}x$

29.

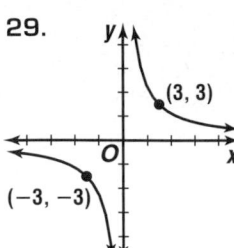

31. $\dfrac{(x - 2)^2}{16} - \dfrac{(y + 3)^2}{20} = 1$

33. vertex: $(-3, -2)$,
focus: $(-3, -2)$,
directrix: $x = 5$; axis
of symmetry: $y = -2$

35. vertex: $(-2, 4)$,
focus: $(-2, 4.25)$,
directrix: $y = 3.75$; axis
of symmetry: $x = -2$

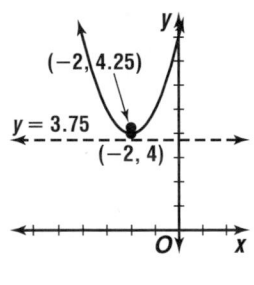

37. $(x - 5)^2 = 12(y + 1)$ **39.** equilateral
hyperbola   **41.** parabola

43. $x^2 + y^2 = 1$

45. $y = \dfrac{x^2}{2} - 1$

47. Sample answer: $x = 7 \sin t$, $y = 7 \cos t$,
$0 \le t \le 2\pi$ **49.** Sample answer: $x = -t^2$, $y = t$,
$-\infty \le t < \infty$ **51.** parabola; $(x')^2 - 2x'y' + (y')^2 -$
$4\sqrt{2}x' - 4\sqrt{2}y' = 0$ **53.** ellipse; $-30°$

55. $(3, 3)$, $(-1, -1)$

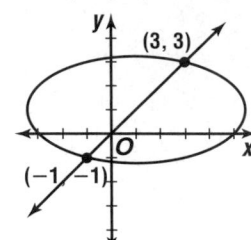

57. $(0, -1)$ $(4, -1)$

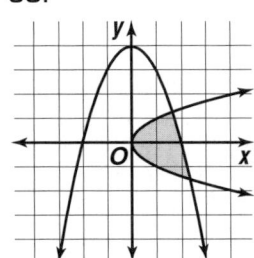

59.

61.

63a. $x^2 + y^2 = 400$ **63b.** about 37%   **65.** about
1.8 feet from the center

Page 693   Chapter 10 SAT and ACT Practice
1. C   **3.** A   **5.** A   **7.** D   **9.** A

Chapter 11   Exponential and Logarithmic Functions

Pages 700–703   Lesson 11-1
5. $\dfrac{256}{81}$ **7.** 9   **9.** $81a^{-1}$ or $\dfrac{81}{a}$ **11.** $2^{2n + 3}\sqrt{2^{n + 1}}$
13. $13x^{\frac{5}{2}}$ **15.** $\sqrt[4]{6b^3c}$ **17.** $pq^2r\sqrt[3]{pr^2}$
19. 4.717×10^{-13} m^2 **21.** $-\dfrac{1}{1296}$ **23.** 32   **25.** $\dfrac{9}{4}$
27. 9   **29.** $2\sqrt{6}$ **31.** $\dfrac{1}{2}$ **33.** 36   **35.** $\dfrac{1}{16}$
37. 1   **39.** $3pq^2r^{-\frac{1}{3}}$ **41.** $6|x|^3$ **43.** $\dfrac{\sqrt{n}}{2}$
45. $4f^4|g||h|^{-1}$ or $\dfrac{4f^4|g|}{|h|}$ **47.** $6x^{\frac{1}{2}}y$ **49.** $|m|^3n^{\frac{1}{2}}$
51. $2xy^2$ **53.** $a^2b^{\frac{2}{5}}|c|^{\frac{1}{2}}$ **55.** $\sqrt[5]{16}$ **57.** $\sqrt[6]{p^4q^3r^2}$
59. $13\sqrt[21]{a^3b^7}$ **61.** -0.69 **63.** $ab^2\sqrt[3]{a^2bc}$
65. 0.17   **67.** 3.79   **69a.** $0 < y < 1$
69b. $1 < y < 3$ **69c.** $y > 3$ **69d.** If the exponent
is less than 0, the power is greater than 0 and less
than 1. If the exponent is greater than 0 and less than
1, the power is greater than 1 and less than the base.
If the exponent is greater than 1, the power is greater
than the base. Any number to the zero power is 1.
Thus, if the exponent is less than zero, the power is
less than 1. A power of a positive number is never

negative, so the power is greater than 0. Any number to the zero power is 1 and to the first power is itself. Thus, if the exponent is greater than zero and less than 1, the power is between 1 and the base. Any number to the first power is itself. Thus, if the exponent is greater than 1, the power is greater than the base. **71.** 2, −6 **73a.** 42,250,474.31 m **73b.** 35,870 km

75.

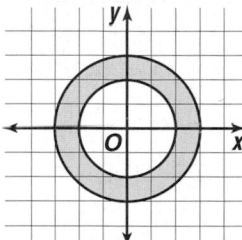

77. 1.31 + 0.14i **79.** about 4.43 s

81. Sample answer: $\sin S = \dfrac{1}{2}$ **83.** 25π m/h

85. 3; −5, 0, 5 **87.** E

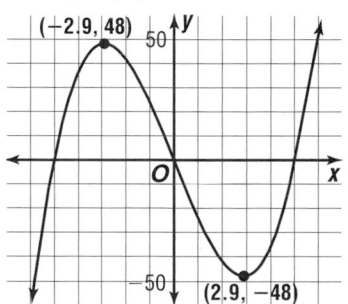

Pages 708–711 Lesson 11-2

5. **7.**

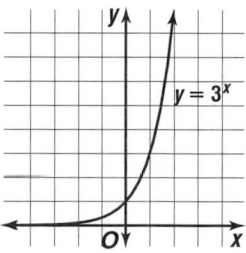

9a. 0.45% **9b.** 9,695,766

11. **13.**

15. **17.**

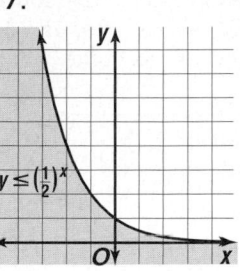

19. B **21.** A **23a.** The graph of $y = 6^x + 4$ is shifted up four units from the graph of $y = 6^x$. **23b.** The graph of $y = -3^x$ is a reflection of the graph of $y = 3^x$ across the x-axis. **23c.** The graph of $y = 7^{-x}$ is a reflection of the graph of $y = 7^x$ across the y-axis. **23d.** The graph of $y = \left(\dfrac{1}{2}\right)^x$ is a reflection of the graph of $y = 2^x$ across the y-axis. **25a.** $y = (0.85)^x$

25b.

25c. 14% **25d.** No; the graph has an asymptote at $y = 0$, so the percent of impurities y will never reach 0. **27a.** 2700 units **27b.** 5800 units **29a.** $535,215.92 **29b.** $76,376.20 **31a.** $50; $50.63; $50.94; $51.16; $51.26 **31b.** Money Market Savings **31c.** 4.88% **33.** 15 = $r \sin \theta$

35. $\sqrt{3}\left(\cos \dfrac{5\pi}{8} + i \sin \dfrac{5\pi}{8}\right)$; −0.66 + 1.60$i$

37a.

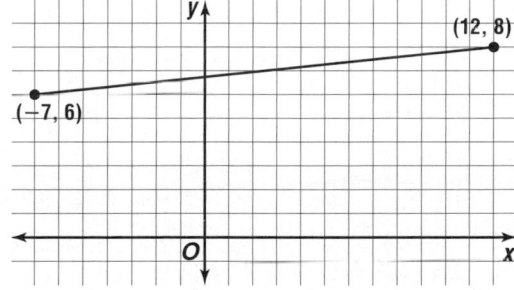

37b. (2.5, 7) **39.** 139,000 cm/s **41.** Sample answer: $y = 948.4x + 4960.6$ **43.** E

Pages 714–717   Lesson 11-3
7. \$25,865.41   **9a.** 78.7°F   **9b.** Too cold; after 5 minutes, his coffee will be about 90°F.
11a.

Interest Compounded	Interest	Effective Annual Yield
Annually	\$80.00	8%
Semi-annually	\$81.60	8.16%
Quarterly	\$82.43	8.243%
Monthly	\$83.00	8.3%
Daily	\$83.28	8.328%
Continually	\$83.29	8.329%

11b. continuously   **11c.** $E = \left(1 + \dfrac{r}{n}\right)^n - 1$

11d. $E = e^r - 1$ **13a.** 95%   **13b.** about 1.2 min
15a. 20.9%; 60.9%; 98.5%

15b. about 29 days

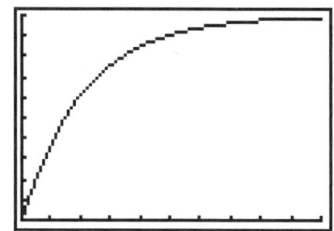

15c. Sample answer: The probability that a person who is going to respond has responded approaches 100% as t approaches infinity. New ads may be introduced after a high percentage of those who will respond have responded. The graph appears to level off after about 50 days. So, new ads can be introduced after an ad has run about 50 days.
17. \$13,257.20   **19.** $6x^2 + 12xy + 6y^2 + \sqrt{2}x - \sqrt{2}y = 0$ **21.** 704.2 ft · lb   **23.** $\dfrac{13}{2}$ **25.** $J'(-9, -6)$, $K'(-6, 18)$, $L'(6, 15)$, $M'(9, -3)$; the dilated image has sides that are 3 times the length of the original figure.   **27.** {−4, 2, 5}; {5, 7}; yes

Pages 722–724   Lesson 11-4
7. $\left(\dfrac{1}{25}\right)^{-\frac{1}{2}} = 5$ **9.** $\log_8 \dfrac{1}{4} = -\dfrac{2}{3}$ **11.** −2   **13.** 32
15. 15
17.

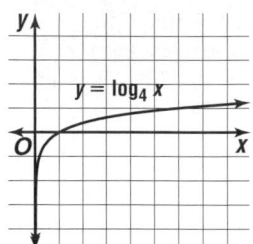

19. 264 h   **21.** $16^{\frac{1}{2}} = 4$ **23.** $4^{\frac{5}{2}} = 32$
25. $\left(\sqrt{6}\right)^4 = 36$ **27.** $\log_{36} 216 = \dfrac{3}{2}$
29. $\log_6 \dfrac{1}{36} = -2$ **31.** $\log_x 14.36 = 1.238$ **33.** $\dfrac{1}{3}$

35. 3.5   **37.** 1.5   **39.** 8   **41.** 7   **43.** 6   **45.** 3
47. $\dfrac{1}{3}$ **49.** 4   **51.** 32
53.

55.

57.

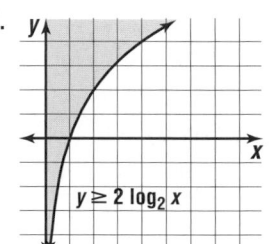

59. 90 min

61. Let $\log_b m = x$ and $\log_b n = y$.
So, $b^x = m$ and $b^y = n$.
$$\dfrac{m}{n} = \dfrac{b^x}{b^y} = b^{x-y}$$
$$\dfrac{m}{n} = b^{x-y}$$
$$\log_b \dfrac{m}{n} = x - y$$
$$\log_b \dfrac{m}{n} = \log_b m - \log_b n$$
63a. 2   **63b.** less light; $\dfrac{1}{8}$

65a.

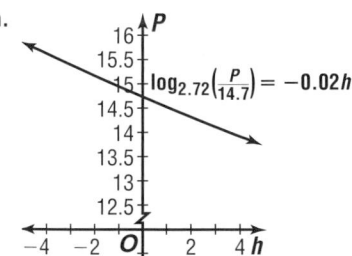

65b. 14.4 psi   **65c.** 16.84 psi   **67.** 69.6164
69. ellipse, $\dfrac{(x-1)^2}{4} + \dfrac{(y-2)^2}{9} = 1$

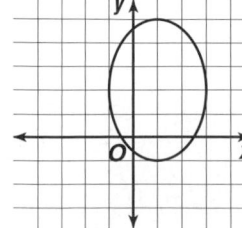

71. $AB = 6$, $BC = 5$, $AC = 5$ **73.** $64 - 27j$ volts
75. $\dfrac{31}{481}$ **77.** $c = 9.5$, $A = 38°20'$, $B = 36°22'$

Pages 730–732 Lesson 11-5
5. 4.9031 **7.** −2.0915 **9.** 74,816.95 **11.** 1.1632
13. 7.83 **15.** $x < 2.97$ **17.** 5.5850 **19.** 5.6021
21. 0.0792 **23.** 1.5563 **25.** −2.3188 **27.** 3.2553
29. 2.9515 **31.** 2.001 **33.** 2.1745 **35.** 4 **37.**
0.7124 **39.** −3.9069 **41.** 18.6377 **43.** 0.3434
45. 0.2076 **47.** $1 < x < 6$ **49.** $x \geq 3.8725$
51. $x < 3.6087$
53.

[−1, 10] scl:1, [−1, 3] scl:1

55. 0.3210 **57.** 2 **59a.** 1.58 **59b.** 0.0219 **61.**
Sample answer: x is between 2 and 3 because 372
is between 100 and 1000, and log 100 = 2 and
log 1000 = 3. **63.** 3819 yr **65.** 3 **67.** $a\sqrt[3]{ab^2c^2}$
69. $\left(2\sqrt{5}, -11\right)$
71.

73. 31.68 cm²
75. Neither; the graph
of the function is not
symmetric with respect
to either the origin or
the y-axis.

Pages 735–737 Lesson 11-6
5. −4.7217 **7.** −1.5606 **9.** 3.0339 **11.** 0.9635
13. $x < 1.3863$
15. 13.57

[−20, 20] scl:2, [−4, 20] scl:2

17a. 503.1 torrs
17b. 4.2 km
19. −0.2705
21. 0.9657
23. 2.2322
25. 1.2134
27. 0.9966
29. 0.2417
31. 2.2266

33. −0.3219 **35.** 1.7593 **37.** 4.7549 **39.** 1.3155
41. $40.9933 < t$ **43.** −0.3466 **45.** $x \leq 1.7657$
47. $x \geq 144.9985$ **49.** −7.64

[−10, 10] scl:1, [−3, 10] scl:1

51. 2.14

[−6, 6] scl:2, [−4, 24] scl:2

53. $x \geq 0.37$

[−5, 5] scl:1, [−2, 5] scl:0.5

55. 324 hr
57. 0 or −1.0986
59. ≈ 70%

61. y is a logarithmic function of x. The pattern in
the table can be determined by $3^y = x$ which
can be expressed as $\log_3 x = y$. **63.** $16^{\frac{3}{4}} = 8$
65. 0.00765 N · m **67.** ⟨13, 7⟩ **69.** $y = \pm 70 \cos 4\theta$

Pages 744–748 Lesson 11-7
5. 8.66 yr **7.** 30.81 yr **9.** 9.73 yr
11. logarithmic; the graph has a vertical asymptote
13. exponential; the graph has a horizontal
asymptote **15a.** $y − 1.0091(0.9805)^x$
15b. $y = 1.0091e^{-0.0197x}$ **15c.** 35.10 min
17. $y = 40 + 14.4270 \ln x$ **19.** Take the square
root of each side.
21a.

x	0	50	100	150	190
$\ln y$	1.81	2.07	3.24	3.75	4.25

21b. $\ln y = 0.0137x + 1.6833$
21c. $y = e^{0.0137x + 1.6833}$ **21d.** 117.4 persons per
square mile **23a.** $\ln y$ is a linear function of $\ln x$.
23b. The result of part **a** indicates that we should
take the natural logarithms of both the x- and
y-values.

$\ln x$	6.21	6.91	8.52	9.21	9.62
$\ln y$	4.49	4.84	5.65	5.99	6.19

23c. $\ln y = 0.4994 \ln x + 1.3901$
23d. $y = 4.0153x^{0.4994}$ **25.** 0.01 **27a.** $11.50
27b. $2645 **29.** about 109.6 ft **31.** 4 units left
and 8 units down **33.** C

Pages 749–753   Chapter 11 Study Guide and Assessment

1. common logarithm   **3.** logarithmic function
5. mantissa   **7.** linearizing data   **9.** nonlinear regression   **11.** 16   **13.** 81   **15.** $\frac{1}{3}$ **17.** $\frac{1}{8}x^{12}$
19. $2a^3b$

21.

23.

25.
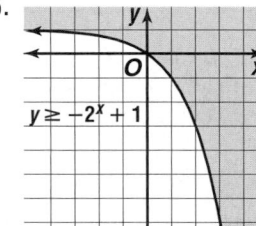

27. $4788.85
29. $21,647.86
31. $3^{-4} = \dfrac{1}{81}$
33. $\log_5 \dfrac{1}{25} = -2$
35. -3

37. -1 **39.** -1 **41.** 3   **43.** 16   **45.** 8

47.

49. -3.5229
51. -1.8539
53. -8.04
55. $x \le -4$

57.

59. -3.42
61. 1.5283
63. 1.7829

65. 3.8982   **67.** -0.8967 **69.** $x \ge 2.5903$
71. $x < 2.20$ **73.** 13.52   **75.** 3561 yr   **77.** 2014

Page 755   Chapter 11 SAT and ACT Practice
1. B   **3.** E   **5.** B   **7.** D   **9.** C

Chapter 12   Sequences and Series

Page 763–765   Lesson 12-1
7. 9, 17, 25, 33   **9.** -38 **11.** 15   **13.** 9, 14, 19, 24
15. 21   **17.** $-13, -19, -25, -31$ **19.** 7.5, 9, 10.5,
12   **21.** $b + 12, b + 16, b + 20, b + 24$ **23.** $-13n,$
$-19n, -25n, -31n$ **25.** $2a + 16, 2a + 23, 2a + 30,$
$2a + 37$ **27.** 80   **29.** 13   **31.** 80   **33.** 4
35. $17 + \sqrt{5}$ **37.** -42.2 **39.** 12, 16.5, 21
41. $\sqrt{3}, \dfrac{12 + 2\sqrt{3}}{3}, \dfrac{24 + \sqrt{3}}{3}, 12$ **43.** -11

45. 1456   **47.** 7   **49.** $-8n + 14$ **51.** Let d be the
common difference. Then, $y = x + d$, $z = x + 2d$, and
$w = x + 3d$. Substitute these values into the
expression $x + w - y$ and simplify. $x + (x + 3d) -$
$(x + d) = x + 2d$ or z.   **53.** 12   **55a.** 25
55b. 100   **55c.** Conjecture: The sum of the first n
term of the sequence of natural numbers is n^2. Proof:
Let $a_n = 2n - 1$. The first term of the sequence of
natural numbers is 1, so $a_1 = 1$. Then, using the
formula for the sum of an arithmetic series,

$$S_n = \frac{n}{2}(a_1 + a_n)$$
$$S_n = \frac{n}{2}[1 + (2n - 1)]$$
$$= \frac{n}{2}(2n) \text{ or } n^2$$

57. least: $101, greatest: $1001   **59.** $285.77
61. $0.5\left(\cos \dfrac{\pi}{8} + i \sin \dfrac{\pi}{8}\right)$, $0.46 + 0.19i$

63. $\sqrt{3}x + y - 10 = 0$ **65.** $A = 70°28'$, $a = 4.2$,
$b = 1.5$ **67.** $y = x - 1$ **69.** C

Pages 771–773   Lesson 12-2
7. 6; 144, 864, 5184   **9.** -4; -115.2, 460.8, -1843.2
11. $\dfrac{3}{2}$ **13.** 1, 3, 9, 27   **15.** $28,211.98; $39,795.78;
$79,185.19   **17.** -2.5; -125, 312.5, -781.25
19. $\dfrac{2}{5}$; $\dfrac{6}{125}, \dfrac{12}{625}, \dfrac{24}{3125}$ **21.** $\sqrt{2}$; 12, $12\sqrt{2}$, 24
23. i; 1, i, -1 **25.** $\dfrac{1}{ab}, \dfrac{1}{a^3}, \dfrac{b}{a^5}, \dfrac{b^2}{a^7}, \dfrac{b^3}{a^9}$ **27.** $-\dfrac{243}{2048}$
29. $16\sqrt{5}$ **31.** $8\sqrt{2}$ **33.** 200, 40, 8   **35.** -2, 6,
-18, 54   **37.** $\dfrac{605}{3}$ **39.** $-\dfrac{11,605}{512}$ **41a.** $b_t = b_0 \cdot 2^{2t}$
41b. 30,720   **41c.** Sample answer: It is assumed
that favorable conditions are maintained for the
growth of the bacteria, such as an adequate food
and oxygen supply, appropriate surrounding
temperature, and adequate room for growth.
43a. $11.79, $30.58, $205.72   **43b.** $7052.15
43c. Each payment made is rounded to the
nearest penny, so the sum of the payments will
actually be more than the sum found in part **b**.
45. $a_n = (-2)(-3)^{n-1}$ **47a.** $25.05   **47b.** No. At
the end of two years, she will have only $615.23 in
her account.   **47c.** $30.54   **49.** 13 weeks

51.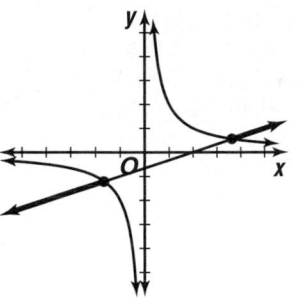

53. $x = t$, $y = -\frac{3}{4}t + \frac{5}{4}$

55. $y = 25 \sin\left(\frac{\pi}{2}t - 3.14\right) + 61$ **57.** 6

Pages 780–783   Lesson 12-3

5. 0; as $n \to \infty$, 5^n becomes increasingly large and thus the value $\frac{1}{5^n}$ becomes smaller and smaller, approaching zero. So the sequence has a limit of zero.

7. $\frac{3}{7}$; $\lim\limits_{n\to\infty} \frac{3n - 6}{7n} = \lim\limits_{n\to\infty}\left(\frac{3}{7} - \frac{6}{7} \cdot \frac{1}{n}\right)$
$= \lim\limits_{n\to\infty} \frac{3}{7} - \lim\limits_{n\to\infty} \frac{6}{7} \cdot \lim\limits_{n\to\infty} \frac{1}{n}$
$= \frac{3}{7} - \frac{6}{7} \cdot 0$ or $\frac{3}{7}$

9. $5\frac{14}{111}$ **11.** $1\frac{1}{8}$ **13.** 125 m   **15.** does not exist; simplifying the limit, we find that
$\lim\limits_{n\to\infty} \frac{n^3 - 2}{n^2} = \lim\limits_{n\to\infty}\left(n - \frac{2}{n}\right)$, $\lim\limits_{n\to\infty} \frac{2}{n} = \lim\limits_{n\to\infty} 2 \cdot \frac{1}{n} = 2 \cdot 0$
or 0, but as n approaches infinity, n becomes increasingly large, so the sequence has no limit.

17. $\frac{9}{2}$; $\lim\limits_{n\to\infty} \frac{9n^3 + 5n - 2}{2n^3} = \lim\limits_{n\to\infty}\left(\frac{9}{2} + \frac{5}{2n^2} - \frac{1}{n^3}\right)$
$= \lim\limits_{n\to\infty} \frac{9}{2} + \lim\limits_{n\to\infty} \frac{5}{2} \cdot \lim\limits_{n\to\infty} \frac{1}{n^2} -$
$\lim\limits_{n\to\infty} \frac{1}{n^3}$
$= \frac{9}{2} + \frac{5}{2} \cdot 0 - 0$ or $\frac{9}{2}$

19. Does not exist; dividing by the highest powered term, n^2, we find $\lim\limits_{n\to\infty} \dfrac{8 + \frac{5}{n} + \frac{2}{n^2}}{\frac{3}{n^2} + \frac{2}{n}}$ which as n approaches infinity simplifies to $\frac{8 + 0 + 0}{0 + 0} = \frac{8}{0}$. Since this fraction is undefined, the limit does not exist.

21. 0; as $n \to \infty$, 3^n becomes increasingly large and thus the value $\frac{1}{3^n}$ becomes smaller and smaller, approaching zero. So the sequence has a limit of zero.

23. 0,
$\lim\limits_{n\to\infty} \frac{5n + (-1)^n}{n^2} = \lim\limits_{n\to\infty} \frac{5n}{n^2} + \lim\limits_{n\to\infty} \frac{(-1)^n}{n^2}$
$= \lim\limits_{n\to\infty} \frac{5}{n} + \lim\limits_{n\to\infty} \frac{(-1)^n}{n^2}$
$= \lim\limits_{n\to\infty} \frac{(-1)^n}{n^2}$

As n increases, the value of the numerator alternates between -1 and 1. As n approaches infinity, the value of the denominator becomes increasingly large, causing the value of the fraction to become increasingly small. Thus the terms of the sequence alternate between smaller and smaller positive and negative values, approaching zero. So the sequence has a limit of zero.

25. $\frac{17}{33}$ **27.** $6\frac{7}{27}$ **29.** $\frac{29}{110}$ **31.** 64   **33.** 20

35. Does not exist; this series is geometric with a common ratio of 2. Since this ratio is greater than 1, the sum of the series does not exist.   **37.** $3\frac{3}{5}$

39. $32 - 16\sqrt{3}$ **41a.** The limit of a difference equals the difference of the limits only if the two limits exist. Since neither $\lim\limits_{n\to\infty} \frac{n^2}{2n + 1}$ nor $\lim\limits_{n\to\infty} \frac{n^2}{2n - 1}$ exists, this property of limits does not apply.

41b. $-\frac{1}{2}$ **43.** No; if n is even, $\lim\limits_{n\to\infty} \cos \frac{n\pi}{2} = \frac{1}{2}$, but if n is odd, $\lim\limits_{n\to\infty} \cos \frac{n\pi}{2} = -\frac{1}{2}$.   **45a.** $10\sqrt{2}$ ft

45b. $40 + 20\sqrt{2}$ ft or about 68 ft   **47.** $-2, -1\frac{1}{3}$, $-\frac{8}{9}, -\frac{16}{27}$ **49.** $(6, -2)$; $\left(6 \pm \sqrt{5}, -2\right)$; $(8, -2)$, $(4, -2)$; $y = -\frac{1}{2}x + 5$, $y = \frac{1}{2}x - 1$ **51.** 42.75 miles, 117.46 miles

53.   **55.** B

Pages 791–793   Lesson 12-4

5. convergent   **7.** divergent   **9.** convergent
11. convergent   **13.** convergent   **15.** divergent
17. convergent   **19.** convergent   **21.** convergent
23. divergent   **25.** convergent   **27.** convergent
29. divergent   **31a.** No, MagicSoft let $a_1 = 1{,}000{,}000$ to arrive at their figure. The first term of this series is $1{,}000{,}000 \cdot 0.70$ or 700,000.
31b. \$2.3 million   **33a.** Culture A: 1400 cells, Culture B: 713 cells   **33b.** Culture B; at the end of one month, culture A will have produced 6000 cells while culture B will have produced 9062 cells.

35a. $\frac{1}{3}$ **35b.** $\frac{1}{36}$ **35c.** $\frac{1}{432}, \frac{1}{5184}$ **35d.** at $4 + \frac{4}{11}$ o'clock, approximately 21 min 49 s after 4:00
37. $16\sqrt{2}$ **39.** 51.02 **41.** $\langle -3, 2 \rangle$

Pages 798–800 Lesson 12-5
5. $8 + 12 + 16 + 20$ **7.** $5 + \frac{15}{4} + \frac{45}{16} + \frac{135}{64} + \cdots$

9. $\displaystyle\sum_{k=0}^{3} (3^k + 1)$ **11.** $\displaystyle\sum_{n=2}^{\infty} 3\left(\frac{1}{2}\right)^n$

13a. $\displaystyle\sum_{n=1}^{60} 389(0.63)^{n-1}$; about 1051 ft **13b.** about

1051 ft **15.** $10 + 15 + 20 + 25$ **17.** $6 + 12 + 20 + 30 + 42$ **19.** $16 + 32 + 64 + 128 + 256$
21. $4\frac{1}{2} + 16\frac{1}{2} + 64\frac{1}{2} +$ **23.** $6 + 24 + 120 +$
$720 + 5040$ **25.** $\frac{8}{5} + \frac{16}{25} + \frac{32}{125} + \cdots$

27. $\displaystyle\sum_{k=1}^{4} (3k + 3)$ **29.** $\displaystyle\sum_{k=4}^{12} 2k$ **31.** $\displaystyle\sum_{k=1}^{4} 2 \cdot 5^k$

33. $\displaystyle\sum_{k=2}^{10} \frac{1}{5k - 1}$ **35.** $\displaystyle\sum_{k=2}^{\infty} (-1)^k k^2$

37. $\displaystyle\sum_{n=0}^{\infty} \left[(-1)^{n+1} \frac{32}{2^n} \right]$ **39.** $\displaystyle\sum_{k=1}^{\infty} \frac{k}{2^k + 3}$

41. $\displaystyle\sum_{k=1}^{\infty} \frac{2^{\frac{k}{2}}}{3k!}$ **43.** $a(a+1)(a-1)$ **45.** 43.64

47a. $(x - 3) + (x - 6) + (x - 9) + (x - 12) +$
$(x - 15) + (x - 18) = -3$
$6x - 63 = -3$
$6x = 60$
$x = 10$

47b. $0 + 1(1 - x) + 2(2 - x) + 3(3 - x) +$
$4(4 - x) + 5(5 - x) = 25$
$1 - x + 4 - 2x + 9 - 3x + 16 - 4x + 25 - 5x = 25$
$55 - 15x = 25$
$-15x = -30$
$x = 2$

49a. 6! **49b.** 120 **49c.** 24, "LISTEN"
51. divergent **53.** $8\sqrt{2}, -16, 16\sqrt{2}, -32$
55. $x^2 + (y - 2)^2 = 49$ **57.** 52.57 ft/s, 26.79 ft/s
59. D

Pages 804–805 Lesson 12-6
5. $c^5 + 5c^4d + 10c^3d^2 + 10c^2d^3 + 5cd^4 + d^5$
7. $125 - 75y + 15y^2 - y^3$ **9.** $-21a^2b^5$ **11a.** 1
11b. 10 **11c.** 6 **11d.** 26 **13.** $n^6 - 24n^5 + 240n^4 - 1280n^3 + 3840n^2 - 6144n + 4096$
15. $512 + 2304a + 4608a^2 + 5376a^3 + 4032a^4 + 2016a^5 + 672a^6 + 144a^7 + 18a^8 + a^9$ **17.** $243 - 405x + 270x^2 - 90x^3 + 15x^4 - x^5$ **19.** $8x^3 - 36x^2y + 54xy^2 - 27y^3$ **21.** $c^3 - 6c^2\sqrt{c} + 15c^2 - 20c\sqrt{c} + 15c - 6\sqrt{c} + 1$ **23.** $81a^4 + 72a^3b + 24a^2b^2 + \frac{32}{9}ab^3 + \frac{16}{81}b^4$ **25.** $x^6y^6 - 12x^5y^5z^3 + 60x^4y^4z^6 - 160x^3y^3z^9 + 240x^2y^2z^{12} - 192xyz^{15} + 64z^{18}$

27. $-112\sqrt{2}a^5$ **29.** $145,152c^3d^6$
31. $-7,185,024p^6q^5$ **33.** 163 **35a.** 495
35b. 2510 **37a.** Sample answer: $1 + 0.01$
37b. Sample answer: 1.04060401 **37c.** 1.04060401; the two values are equal. **39.** convergent
41. $1100.65 **43.** 1681 feet

Pages 811–814 Lesson 12-7
5. $i\pi + 1.9459$ **7.** 2.22 **9.** 0.0069; 0 **11.** $2e^{i\frac{2\pi}{3}}$
13. $i\pi + 1.3863$ **15.** $i\pi - 1.3863$ **17.** $i\pi + 5.4723$
19. 2.99 **21.** 39.33 **23.** 24.02 **25.** $-0.9760; -1$
27. 0.8660; 0.8660 **29.** $5e^{i\frac{5\pi}{3}}$ **31.** $\sqrt{2}e^{i\frac{\pi}{4}}$
33. $2e^{i\frac{3\pi}{4}}$ **35.** $3\sqrt{2}e^{i\frac{\pi}{4}}$

37.
$$\frac{e^{ix} - e^{-ix}}{2i} = \frac{\cos x + i \sin x - (\cos x - i \sin x)}{2i}$$
$$= \frac{2i \sin x}{2i}$$
$$= \sin x$$
$$\frac{e^{ix} + e^{-ix}}{2} = \frac{\cos x + i \sin x + \cos x - i \sin x}{2}$$
$$= \frac{2 \cos x}{2}$$
$$= \cos x$$

39. If you add the numbers on the diagonal lines as shown, the sums are the terms of the Fibonacci sequence.

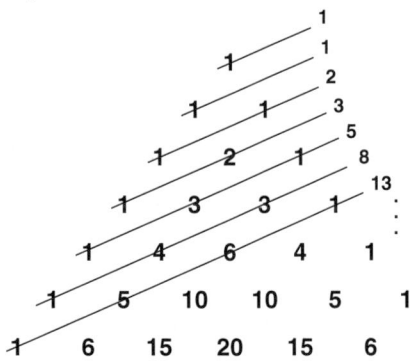

41a. approximately $9572.29 **41b.** No, she will be short by more than $30,000! **41c.** about 42 years; 47 years old **41d.** $20,882 **43.** $64x^6 + 192x^5y + 240x^4y^2 + 160x^3y^3 + 60x^2y^4 + 12xy^5 + y^6$
45a. 0.020 cm, 0.040 cm **45b.** $0.005(2)^{n-1}$
45c. $2.56, 3.169 \times 10^{27}$ cm **47.** $y^2 + 3x + 7y = 0$
49. 75.5 N, 14°48′ **51.** 24 multiple choice, 6 essay

Pages 819–821 Lesson 12-8
5. $-1, -7, -19, -43$ **7.** $15 + 26i, 9 + 17.6i, 5.4 + 12.56i$ **9.** $-1 + i, 2 - 5i, -19 - 23i$ **11.** 5, 8, 17, 44 **13.** 1; 16; 121; 13,456 **15.** $-0.08, 0.09, -0.07, 0.08$ **17.** $3 + 8i, 9 + 14i, 21 + 26i$
19. $5 + 2i, 13 + 2i, 29 + 2i$ **21.** $15 + 2i, 33 + 2i, 69 + 2i$ **23.** $1, 3 - 2i, 9 - 8i$ **25.** $-3i, -8 - 3i, 56 + 45i$ **27.** $-2i, -4 - 4i, 28i$ **29.** $2 + i, 5 + 7i, -22 + 73i$ **31.** about 54% **33.** $\pm\sqrt{2}$

35a. 1.414213562, 1.189207115, 1.090507733, 1.044273782 **35b.** $f(z) = \sqrt{z}$, $z_0 = 2$ **35c.** 1

36d. 1 **37.** $90{,}720a^4b^4$ **39.** $\dfrac{x^2}{4225} - \dfrac{y^2}{4056} = 1$

41a.

41b. No, the height of the building is about 62 feet for a total of about 102 feet with the tower.
43. 92, 56

Pages 826–828   Lesson 12-9
7. Step 1: Verify that the formula is valid for $n = 1$. Since 2 is the first term in the sequence and $2(2^1 - 1) = 2$, the formula is valid for $n = 1$. Step 2: Assume that the formula is valid for $n = k$ and derive a formula for $n = k + 1$.
$$S_k \Rightarrow 2 + 2^2 + 2^3 + \cdots + 2^k = 2(2^k - 1)$$
$$S_{k+1} \Rightarrow 2 + 2^2 + 2^3 + \cdots +$$
$$2^k + 2^{k+1} = 2(2^k - 1) + 2^k + 1$$
$$= 2 \cdot 2^{k+1} - 2$$
$$= 2(2^{k+1} - 1)$$
When the original formula is applied for $n = k + 1$, the same result is obtained. Thus if the formula is valid for $n = k$, it is also valid for $n = k + 1$. Since the formula is valid for $n = 1$, it is also valid for $n = 2$, $n = 3$, and so on indefinitely. Thus, the formula is valid for all positive integral values of n.
9. S_n: $3^n - 1 = 2r$ for some integer r. Step 1: Verify that S_n is valid for $n = 1$. $S_1 \Rightarrow 3^1 - 1$ or 2. Since $2 = 2 \cdot 1$, S_n is valid for $n = 1$. Step 2: Assume that S_n is valid for $n = k$ and show that it is also valid for $n = k + 1$.
$$S_k \Rightarrow 3^k - 1 = 2r \text{ for some integer } r$$
$$S_{k+1} \Rightarrow 3^{k+1} - 1 = 2t \text{ for some integer } t$$
$$3^k - 1 = 2r$$
$$3(3^k - 1) = 3 \cdot 2r$$
$$3^{k+1} - 3 = 6r$$
$$3^{k+1} - 1 = 6r + 2$$
$$3^{k+1} - 1 = 2(3r + 1)$$

Thus, $3^{k+1} - 1 = 2t$, where $t = 3r + 1$ is an integer, and we have shown that if S_n is valid, then S_{k+1} is also valid. Since S_n is valid for $n = 1$, it is also valid for $n = 2$, $n = 3$, and so on indefinitely. Hence, $3n - 1$ is divisible by 2 for all integral values of n.

11. Step 1: Verify that the formula is valid for $n = 1$. Since 1 is the first term in the sequence and $(1)[2(1) - 1] = 1$, the formula is valid for $n = 1$. Step 2: Assume that the formula is valid for $n = k$ and derive a formula for $n = k + 1$.
$$S_k \Rightarrow 1 + 5 + 9 + \cdots + (4k - 3) = k(2k - 1)$$
$$S_{k+1} \Rightarrow 1 + 5 + 9 + \cdots + (4k - 3) + (4k + 1)$$
$$= k(2k - 1) + (4k + 1)$$
$$= 2k^2 + 3k + 1$$
$$= (k + 1)(2k + 1)$$
Apply the original formula for $n = k + 1$.
$$(k + 1)[2(k + 1) - 1] = (k + 1)(2k + 1)$$
The formula gives the same result as adding the $(k + 1)$ term directly. Thus if the formula is valid for $n = k$, it is also valid for $n = k + 1$. Since the formula is valid for $n = 1$, it is also valid for $n = 2$, $n = 3$, and so on indefinitely. Thus, the formula is valid for all positive integral values of n.

13. Step 1: Verify that the formula is valid for $n = 1$. Since $-\dfrac{1}{2}$ is the first term in the sequence and $\dfrac{1}{2^1} - 1 = -\dfrac{1}{2}$, the formula is valid for $n = 1$. Step 2: Assume that the formula is valid for $n = k$ and derive a formula for $n = k + 1$.
$$S_k \Rightarrow -\frac{1}{2} - \frac{1}{4} - \frac{1}{8} - \cdots - \frac{1}{2^k} = \frac{1}{2^k} - 1$$
$$S_{k+1} \Rightarrow \frac{1}{2} - \frac{1}{4} - \frac{1}{8} - \cdots - \frac{1}{2^k} - \frac{1}{2^{k+1}}$$
$$= \frac{1}{2^k} - 1 - \frac{1}{2^{k+1}}$$
$$= \frac{2}{2 \cdot 2^k} - 1 - \frac{1}{2^{k+1}}$$
$$= \frac{2}{2^{k+1}} - 1 - \frac{1}{2^{k+1}}$$
$$= \frac{1}{2^{k+1}} - 1$$
When the original formula is applied for $n = k + 1$, the same result is obtained. Thus if the formula is valid for $n = k$, it is also valid for $n = k + 1$. Since the formula is valid for $n = 1$, it is also valid for $n = 2$, $n = 3$, and so on indefinitely. Thus, the formula is valid for all positive integral values of n.

15. Step 1: Verify that the formula is valid for $n = 1$. Since 1 is the first term in the sequence and $\dfrac{1[2(1) - 1][2(1) + 1]}{3} = 1$, the formula is valid for $n = 1$. Step 2: Assume that the formula is valid for $n = k$ and derive a formula for $n = k + 1$.

$S_k \Rightarrow 1^2 + 3^2 + 5^2 + \cdots +$

$(2k-1)^2 = \dfrac{k(2k-1)(2k+1)}{3}$

$S_{k+1} \Rightarrow 1^2 + 3^2 + 5^2 + \cdots + (2k-1)^2 +$

$(2k+1)^2 = \dfrac{k(2k-1)(2k+1)}{3} + (2k+1)^2$

$= \dfrac{k(2k-1)(2k+1) + 3(2k+1)^2}{3}$

$= \dfrac{[k(2k-1) + 3(2k+1)](2k+1)}{3}$

$= \dfrac{(2k^2 + 5k + 3)(2k+1)}{3}$

$= \dfrac{(2k+3)(k+1)(2k+1)}{3}$

Apply the original formula for $n = k + 1$.

$\dfrac{(k+1)[2(k+1)-1][2(k+1)+1]}{3}$

$= \dfrac{(k+1)(2k+1)(2k+3)}{3}$

The formula gives the same result as adding the $(k+1)$ term directly. Thus if the formula is valid for $n = k$, it is also valid for $n = k + 1$. Since the formula is valid for $n = 1$, it is also valid for $n = 2$, $n = 3$, and so on indefinitely. Thus, the formula is valid for all positive integral values of n.

17. $S_n \Rightarrow 7^n + 5 = 6r$ for some integer r. Step 1: Verify that S_n is valid for $n = 1$. $S_1 \Rightarrow 7^1 + 5$ or 12. Since $12 = 6 \cdot 2$, S_n is valid for $n = 1$. Step 2: Assume that S_n is valid for $n = k$ and show that it is also valid for $n = k + 1$.

$S_k \Rightarrow 7^k + 5 = 6r$ for some integer r

$S_{k+1} \Rightarrow 7^{k+1} + 5 = 6t$ for some integer t

$7^k + 5 = 6r$

$7(7^k + 5) = 7 \cdot 6r$

$7^{k+1} + 35 = 42r$

$7^{k+1} + 5 = 42r - 30$

$7^{k+1} + 5 = 6(7r - 5)$

Thus, $7^{k+1} + 5 = 6t$, where $t = 7r - 5$ is an integer, and we have shown that if S_n is valid, then S_{k+1} is also valid. Since S_n is valid for $n = 1$, it is also valid for $n = 2$, $n = 3$, and so on indefinitely. Hence, $7^n + 5$ is divisible by 6 for all integral values of n.

19. $S_n \Rightarrow 5^n - 2^n = 3r$ for some integer r. Step 1: Verify that S_n is valid for $n = 1$. $S_1 \Rightarrow 5^1 - 2^1$ or 3. Since $3 = 3 \cdot 1$, S_n is valid for $n = 1$. Step 2: Assume that S_n is valid for $n = k$ and show that it is also valid for $n = k + 1$.

$S_k \Rightarrow 5^k - 2^k = 3r$ for some integer r

$S_{k+1} \Rightarrow 5^{k+1} - 2^{k+1} = 3t$ for some integer t

$5^k - 2^k = 3r$

$5^k = 2^k + 3r$

$5^k \cdot 5 = (2^k + 3r)(2 + 3)$

$5^{k+1} = 2^{k+1} + 3(2^k) + 6r + 9r$

$5^{k+1} - 2^{k+1} = 2^{k+1} + 3(2^k) + 6r + 9r - 2^{k+1}$

$= 3(2^k) + 15r$

$= 3(2^k + 5r)$

Thus, $5^{k+1} - 2^{k+1} = 3t$, where $t = 2^k + 5r$ is an integer, and we have shown that if S_n is valid, then S_{k+1} is also valid. Since S_n is valid for $n = 1$, it is also valid for $n = 2$, $n = 3$, and so on indefinitely. Hence, $5^n - 2^n$ is divisible by 3 for all integral values of n.

21. Step 1: Verify that the formula is valid for $n = 1$. Since $\dfrac{1}{2}$ is the first term in the sequence and $\dfrac{1}{1+1} = \dfrac{1}{2}$, the formula is valid for $n = 1$. Step 2: Assume that the formula is valid for $n = k$ and derive a formula for $n = k + 1$.

$S_k \Rightarrow \dfrac{1}{1 \cdot 2} + \dfrac{1}{2 \cdot 3} + \dfrac{1}{3 \cdot 4} + \cdots + \dfrac{1}{k(k+1)} = \dfrac{k}{k+1}$

$S_{k+1} \Rightarrow \dfrac{1}{1 \cdot 2} = \dfrac{1}{2 \cdot 3} + \dfrac{1}{3 \cdot 4} + \cdots + \dfrac{1}{k(k+1)} +$

$\dfrac{1}{(k+1)(k+2)} = \dfrac{k}{k+1} + \dfrac{1}{(k+1)(k+2)}$

$= \dfrac{k(k+2) + 1}{(k+1)(k+2)}$

$= \dfrac{k^2 + 2k + 1}{(k+1)(k+2)}$

$= \dfrac{(k+1)^2}{(k+1)(k+2)}$

$= \dfrac{k+1}{k+2}$

Apply the original formula for $n = k + 1$.

$\dfrac{(k+1)}{(k+1)+1} = \dfrac{k+1}{k+2}$

The formula gives the same result as adding the $(k+1)$ term directly. Thus if the formula is valid for $n = k$, it is also valid for $n = k + 1$. Since the formula is valid for $n = 1$, it is also valid for $n = 2$, $n = 3$, and so on indefinitely. Thus, the formula is valid for all positive integral values of n.

23. Step 1: Verify that the formula is valid for $n = 1$. Since $S_1 \Rightarrow [r(\cos \theta + i \sin \theta)]^1$ or $r(\cos \theta + i \sin \theta)$ and $r^1[\cos(1)\theta + i \sin(1)\theta] = r(\cos \theta + i \sin \theta)$, the formula is valid for $n = 1$. Step 2: Assume that the formula is valid for $n = k$ and derive a formula for $n = k + 1$. That is, assume that $[r(\cos \theta + i \sin \theta)]^k = r^k(\cos k\theta + i \sin k\theta)$. Multiply each side of the equation by $r(\cos \theta + i \sin \theta)$.

$[r(\cos \theta + i \sin \theta)]^{k+1}$

$= [r^k(\cos k\theta + i \sin k\theta)] \cdot [r(\cos \theta + i \sin \theta)]$

$= r^{k+1}[\cos k\theta \cos \theta + (\cos k\theta)(i \sin \theta) + i \sin k\theta \cos \theta + i^2 \sin k\theta \sin \theta]$

$= r^{k+1}[(\cos k\theta \cos \theta - \sin k\theta \sin \theta) + i(\sin k\theta \cos \theta + \cos k\theta \sin \theta)]$

$= r^{k+1}[\cos(k+1)\theta + i \sin(k+1)\theta]$

When the original formula is applied for $n = k + 1$, the same result is obtained. Thus if the formula is valid for $n = k$, it is also valid for $n = k + 1$. Since the formula is valid for $n = 1$, it is also valid for

$n = 2, n = 3$ and so on indefinitely. Thus, the formula is valid for all positive integral values of n.

25. $S_1 \Rightarrow n^2 + 5n = 2r$ for some positive integer r. Step 1: Verify that S_1 is valid for $n = 1$. $S_1 \Rightarrow 1^2 + 5 \cdot 1$ or 6. Since $6 = 2 \cdot 3$, S_1 is valid for $n = 1$. Step 2: Assume that S_n is valid for $n = k$ and show that it is valid for $n = k + 1$.

$S_k \Rightarrow k^2 + 5k = 2r$ for some positive integer r
$S_{k+1} \Rightarrow (k + 1)^2 + 5(k + 1) = 2t$ for some positive integer t

$$
\begin{aligned}
(k + 1)^2 + 5(k + 1) &= k^2 + 2k + 1 + 5k + 5 \\
&= (k^2 + 5k) + (2k + 6) \\
&= 2r + 2(k + 3) \\
&= 2(r + k + 3)
\end{aligned}
$$

Thus, if $k^2 + 5k = 2t$, where $t = r + k + 3$ is an integer, and we have shown that if S_n is valid, then S_{k+1} is also valid. Since S_n is valid for $n + 1$, it is also valid for $n = 2$, $n = 3$, and so on indefinitely. Hence, $n^2 + 5n$ is divisible by 2 for all positive integral values of n.

27. Step 1: Verify that $S_n \Rightarrow (x + y)^n = x^n + nx^{n-1}y + \dfrac{n(n-1)}{2!}x^{n-2}y^2 + \dfrac{n(n-1)(n-2)}{3!}x^{n-3}y^3 + \cdots + y^n$ is valid for n = 1. Since $S_1 \Rightarrow (x - y)^1 = x^1 = 1x^0y^1$ or $x + y$, S_n is valid for $n = 1$. Step 2: Assume that the formula is valid for $n = k$ and derive a formula for $n = k + 1$.

$$S_k \Rightarrow (x + y)^k = x^k + kx^{k-1}y + \dfrac{k(k-1)}{2!}x^{k-2}y^2 + \dfrac{k(k-1)(k-2)}{3!}x^{k-3}y^3 + \cdots + y^k$$

$$S_{k+1} \Rightarrow (x + y)^k(x + y) = (x + y)\Big(x^k + kx^{k-1}y + \dfrac{k(k-1)}{2!}x^{k-2}y^2 + \dfrac{k(k-1)(k-2)}{3!}x^{k-3}y^3 + \cdots + y^k\Big)$$

$$(x + y)^{k+1} = x\Big(x^k + kx^{k-1}y + \dfrac{k(k-1)}{2!}x^{k-2}y^2 + \dfrac{k(k-1)(k-2)}{3!}x^{k-3}y^3 + \cdots + y^k + y\big(x^k + kx^{k-1}y + \dfrac{k(k-1)}{2!}x^{k-2}y^2 + \dfrac{k(k-1)(k-2)}{3!}x^{k-3}y^3 + \cdots + y^k\big)\Big)$$

$$= x^{k+1} + kx^ky + \dfrac{k(k-1)}{2!}x^{k-1}y^2 + \cdots + xy^k + x^ky + kx^{k-1}y^2 + \dfrac{k(k-1)}{2!}x^{k-2}y^3 + \cdots + y^{k+1}$$

$$= x^{k+1} + (k + 1)x^ky + kx^{k-1}y^2 + \dfrac{k(k-1)}{2!}x^{k-1}y^2 + \cdots + y^{k+1}$$

$$= x^{k+1} + (k + 1)x^ky + \dfrac{k(k+1)}{2!}x^{k-1}y^2 + \cdots + y^{k+1}$$

When the original formula is applied for $n = k + 1$, the same result is obtained. Thus if the formula is valid for $n = k$, it is also valid for $n = k + 1$. Since the formula is valid for $n = 1$, it is also valid for $n = 2$, $n = 3$, and so on indefinitely. Thus, the formula is valid for all positive integral values of n.

29. $8 - i$, $16 - i$, $32 - i$ **31.** $\dfrac{(x-2)^2}{4} + \dfrac{(y-5)^2}{25} = 1$; ellipse **33.** $y = \pm\dfrac{3}{4}\sin 2x$ **35.** B

Pages 829–833 Chapter 12 Study Guide and Assessment

1. d **3.** m **5.** k **7.** c **9.** b **11.** 6.9, 8.2, 9.5, 10.8 **13.** 6, 3.5, 1, -1.5, -4 **15.** 18 **17.** 36,044.8 **19.** 0.2, 1, 5, 25, 125 **21.** $62\big(1 + \sqrt{2}\big)$ **23.** 6 **25.** 0 **27.** 2100 **29.** divergent **31.** $(3 \cdot 5 - 3) + (3 \cdot 6 - 3) + (3 \cdot 7 - 3) + (3 \cdot 8 - 3) + (3 \cdot 9 - 3)$

33. $\displaystyle\sum_{a=0}^{\infty}(2n - 1)$ **35.** $a^6 - 24a^5 + 240a^4 - 1280a^3 + 3840a^2 - 6144a + 4096$

37. $3360x^6$ **39.** $102{,}400m^6$ **41.** $2e^{i\frac{3\pi}{4}}$ **43.** $2\sqrt{2}e^{i\frac{7\pi}{4}}$ **45.** 0, 6, -12, 42 **47.** 4, $6 - 2i$, $7 - 3i$ **49.** $2 + i$, $5 - 1.5i$, $6.5 - 2.75i$

51. Step 1: Verify that the formula is valid for $n = 1$. Since the first term in the sequence is 1 and $\dfrac{1(1 + 1)}{2} = 1$, the formula is valid for $n = 1$. Step 2: Assume that the formula is valid for $n = k$ and derive a formula for $n = k + 1$.

$$S_k \Rightarrow 1 + 2 + 3 + \cdots + k = \dfrac{k(k + 1)}{2}$$

$$S_{k+1} \Rightarrow 1 + 2 + 3 + \cdots + k + (k + 1)$$

$$
\begin{aligned}
&= \dfrac{k(k + 1)}{2} + \dfrac{2(k + 1)}{2} \\
&= \dfrac{k^2 + k}{2} + \dfrac{2k + 2}{2} \\
&= \dfrac{k^2 + k + 2k + 2}{2} \\
&= \dfrac{k^2 + 3k + 2}{2} \\
&= \dfrac{(k + 1)(k + 2)}{2}
\end{aligned}
$$

Apply the original formula for $n = k + 1$.

$$\dfrac{(k + 1)[(k + 1) + 1]}{2} = \dfrac{(k + 1)(k + 2)}{2}$$

The formula gives the same result as adding the $(k + 1)$ term directly. Thus, if the formula is valid for $n = k$, it is also valid for $n = k + 1$. Since the formula is valid for $n = 1$, it is also valid for $n = 2$, $n = 3$, and so on indefinitely. Thus, the formula is valid for all positive integral values of n.

53. $S_n \Rightarrow 9^n - 4^n = 5r$ for some integer r. Step 1: Verify that S_n is valid for $n = 1$. $S_1 \Rightarrow 9^1 - 4^1$ or 5. Since $5 = 5 \cdot 1$, S_n is valid for $n = 1$. Step 2: Assume that S_n is valid for $n = k$ and show that it is also valid for $n = k + 1$.

$S_k \Rightarrow 9^k - 4^k = 5r$ for some integer r
$S_{k+1} \Rightarrow 9^{k+1} - 4^{k+1} = 5t$ for some integer t

$$9^k - 4^k = 5r$$
$$9^k = 4^k + 5r$$
$$9(9^k) = (4^k + 5r)(4 + 5)$$
$$9^{k+1} = 4^{k+1} + 5(4^k) + 20r + 25r$$
$$9^{k+1} - 4^{k+1} = 4^{k+1} + 5(4^k) + 20r + 25r - 4^{k+1}$$
$$= 5(4^k) + 45r$$
$$= 5(4^k + 9r)$$

Thus, $9^{k+1} - 4^{k+1} = 5t$, where $t = 4^k + 9r$ is an integer, and we have shown that if S_n is valid, then S_{k+1} is also valid. Since S_n is valid for $n = 1$, it is also valid for $n = 2$, $n = 3$, and so on indefinitely. Hence, $9^n - 4^n$ is divisible by 5 for all integral values of n.
55. $117,987,860.30

Page 835   Chapter 12 SAT and ACT Practice
1. D   **3.** A   **5.** C   **7.** D   **9.** A

Chapter 13 Combinatorics and Probability

Pages 843–845   Lesson 13-1
5. 300   **7.** 720   **9.** 55,440   **11.** 15,504
13. 3,628,800   **15a.** 100,000   **15b.** 7290
15c. 999,900,000   **17.** 5040   **19.** dependent
21. dependent   **23.** 360   **25.** 840   **27.** 604,800
29. 6   **31.** 10   **33.** 6   **35.** 1   **37.** 168   **39.** 840
41. 2002   **43.** 420   **45a.** 22,308   **45b.** 144
45c. 792   **47.** 216,216   **49.** $P(n, n-1) \overset{?}{=} P(n, n)$

$$\frac{n!}{[n-(n-1)]!} \overset{?}{=} \frac{n!}{(n-n)!}$$
$$\frac{n!}{1!} \overset{?}{=} \frac{n!}{0!}$$
$$n! = n!$$

51a. 330   **52b.** 150   **53a.** 592
53b. Yes. Let h, t, and u be the digits.
$$100h + 10t + u$$
$$100h + 10u + t$$
$$100t + 10h + u$$
$$100t + 10u + h$$
$$100u + 10t + h$$
$$+\ 100u + 10h + t$$
$$\overline{200(h+t+u) + 20(h+t+u) + 2(h+t+u) =}$$
$$222(h+t+u)$$
$$\frac{222(h+t+u)}{6} = 37(h+t+u)$$
55. 3025   **57.** 1.4   **59.** $(2, 180°), (2, 0°)$ **61.** $0°$,
$180°, 360°$ **63.** $B = 63°, a = 7.7, c = 17.1$

Pages 849–851   Lesson 13-2
5. 22,680   **7.** circular; 3,628,800   **9.** circular;
39,916,800   **11.** 756   **13.** 907,200   **15.** 302,400
17. 3780   **19.** 126   **21.** circular; 39,916,800
23. circular; 40,320   **25.** circular; 5040
27. linear; 3,628,800   **29.** circular; 3,113,510,400
31. circular; $\approx 8.22 \times 10^{33}$ **33.** 2520   **35.** 46,200
37a. $\approx 7.85 \times 10^{17}$ **37b.** $\approx 1.41 \times 10^{51}$
37c. $\approx 6.04 \times 10^{52}$ **39.** 20   **41.** $x < 8.69$
43. $44 - 58i$ **45.** about 3.31 inches

Pages 855–858   Lesson 13-3
5. $\frac{1}{3}$ **7.** 0   **9.** $\frac{4}{31}$ **11.** $\frac{18}{17}$ **13.** $\frac{3}{13}$ **15.** $\frac{5}{13}$
17. $\frac{1}{2}$ **19.** $\frac{7}{10}$ **21.** $\frac{1}{130}$ **23.** $\frac{3}{13}$ **25.** $\frac{1}{36}$
27. $\frac{2}{3}$ **29.** $\frac{11}{4}$ **31.** $\frac{22}{53}$ **33.** $\frac{92}{233}$ **35.** $\frac{4}{1}$
37. $\frac{1}{5}$ **39a.** $\frac{1}{720}$ **39b.** $\frac{999}{1}$ **41a.** $\frac{21}{1292}$
41b. $\frac{225}{421}$ **43.** $\frac{2}{9}$ **45.** 210   **47.** $2x$
49. $6\left(\cos\frac{5\pi}{4} + i\sin\frac{5\pi}{4}\right), -3\sqrt{2} - 3\sqrt{2}i$ **51.** B

Pages 863–867   Lesson 13-4
5. dependent, $\frac{2}{15}$ **7.** exclusive, $\frac{10}{13}$ **9.** exclusive,
$\frac{2}{13}$ **11.** ≈ 0.518 **13.** ≈ 0.032 **15.** $\frac{34}{39}$
17. independent, $\frac{25}{81}$ **19.** dependent, $\frac{2}{7}$
21. dependent, $\frac{19}{1,160,054}$ **23.** independent,
$\frac{35}{1024}$ **25.** inclusive, $\frac{11}{36}$ **27.** inclusive, $\frac{8}{13}$
29. exclusive, $\frac{1}{4}$ **31.** exclusive, $\frac{11}{32}$ **33.** exclusive,
$\frac{19}{33}$ **35.** $\frac{2}{221}$ **37.** $\frac{55}{221}$ **39.** $\frac{4}{7}$ **41.** $\frac{71}{210}$
43. $\frac{1}{21}$ **45.** $\frac{5}{7}$ **47.** $\frac{4}{5}$ **49.** $\frac{59}{143}$ **51.** $\frac{531}{1250}$
53. 0.93   **55a.** exclusive
55b.

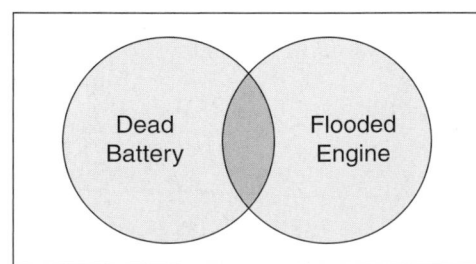

55c. $\frac{4}{5}$ **57.** 720   **59.** No, the spill will spread no more than 2000 meters away.   **61.** $11.50, $2645
63. $\langle x-1, y+5 \rangle = t\langle -2, -4 \rangle$ **65.** B

Pages 871–874 Lesson 13-5

5. $\frac{1}{3}$ **7.** $\frac{1}{7}$ **9.** $\frac{1}{5}$ **11.** $\frac{2}{5}$ **13a.** $\frac{69}{70}$ **13b.** $\frac{2}{25}$

13c. $\frac{1}{25}$ **15.** $\frac{1}{2}$ **17.** $\frac{3}{5}$ **19.** $\frac{5}{8}$ **21.** $\frac{1}{13}$ **23.** 0

25. $\frac{2}{13}$ **27.** $\frac{5}{7}$ **29.** $\frac{2}{5}$ **31.** $\frac{3}{10}$ **33.** $\frac{1}{6}$ **35.** $\frac{1}{2}$

37. $\frac{19}{51}$

39. A = person buys something
B = person asks questions

$$P(A\,|\,B) = \frac{\frac{120}{500}}{\frac{150}{500}} \text{ or } \frac{4}{5}$$

Four out of five people who ask questions will make a purchase. Therefore, they are more likely to buy something if they ask questions. **41.** $\frac{5}{6}$

43. $P(A\,|\,B) = \dfrac{P(A \text{ and } B)}{P(B)}$ by definition. So, if $P(A) = P(A\,|\,B)$ then by substitution $P(A) = \dfrac{P(A \text{ and } B)}{P(B)}$ or $P(A \text{ and } B) = P(A) \cdot P(B)$. Therefore, the events are independent. **45.** 126 **47.** They are reflections of each other over the x-axis.

49. $y = -5$ **51.** 54.7 ft^2 **53.** $\frac{1}{2}$ ft or 6 in. **55.** B

Pages 878–880 Lesson 13-6

5. $\frac{625}{648}$ **7.** $\frac{1}{7776}$ **9.** $\frac{1029}{2500}$ **11.** $\frac{768}{3125}$ **13.** $\frac{1}{81}$

15. $\frac{65}{81}$ **17.** $\frac{15}{128}$ **19.** $\frac{1}{1024}$ **21.** $\frac{1}{81}$ **23.** $\frac{11}{27}$

25. ≈ 0.058 **27.** $\approx 1.049 \times 10^{-4}$ **29.** ≈ 0.201

31. $\frac{1}{4}$ **33.** $\frac{3}{8}$ **35.** about 45% **37.** ≈ 0.0062

39. 0.807 **41a.** 0.246 **41b.** 0.246 **41c.** 0.41

43. $\frac{7}{26}$ **45.** 0.38 **47.** $0 - i\sqrt{2}$ **49.** about 101.1 cm and 76.9 cm **51.** 7/12

Pages 881–885 Chapter 13 Study Guide and Assessment

1. independent **3.** 1 **5.** permutation
7. mutually exclusive **9.** conditional **11.** 6
13. 720 **15.** 20,160 **17.** 165 **19.** 3 **21.** 63
23. 50,400 **25.** 60 **27.** $\frac{1}{16}$ **29.** $\frac{1}{140}$ **31.** $\frac{1}{15}$

33. $\frac{1}{139}$ **35.** independent, $\frac{5}{1296}$ **37.** $\frac{9}{14}$ **39.** $\frac{1}{7}$

41. $\frac{2}{5}$ **43.** $\frac{2}{15}$ **45.** $\frac{1}{16}$ **47.** $\frac{5}{16}$ **49.** 2520

51a. $\frac{7}{15}$ **51b.** $\frac{1}{30}$

Page 887 Chapter 13 SAT and ACT Practice
1. D **3.** C **5.** E **7.** A **9.** B

Chapter 14 Statistics and Data Analysis

Pages 893–896 Lesson 14-1

5a.

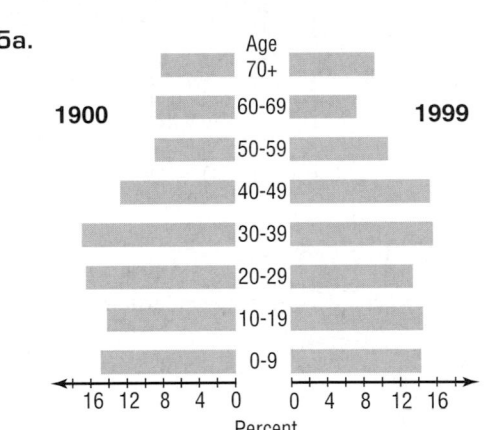

5b. In 1999, there are larger percents of older citizens than in 1990.

7a.

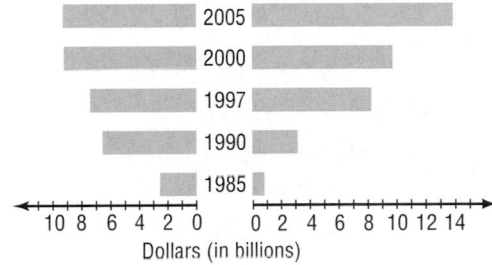

7b. 43220 **7c.** Sample answer: to determine where most of their customers live so they can target their advertising accordingly

9a.

9b. Sales; the sales revenue is growing at a faster rate than the rental revenue. **11a.** 56
11b. Sample answer: 10 **11c.** Sample answer: 10, 20, 30, 40, 50, 60, 70, 80 **11d.** Sample answer: 15, 25, 35, 45, 55, 65, 75

11e. Sample answer:

Number of Nations	Frequency
10–20	2
20–30	3
30–40	8
40–50	1
50–60	1
60–70	2
70–80	1

11f. Sample answer:

11g. Sample answer:

13a.

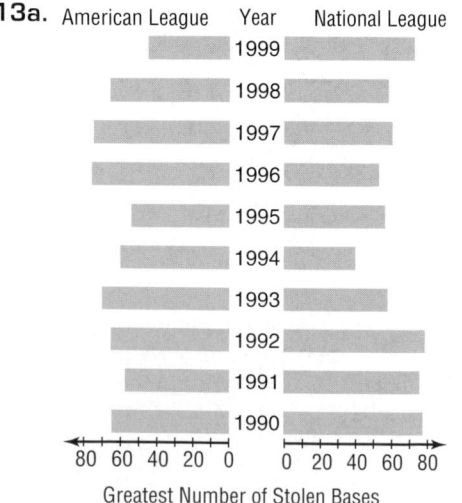

13b. Sample answer:

Stolen Bases	Frequency
30–40	1
40–50	1
50–60	6
60–70	5
70–80	7

13c. Sample answer:

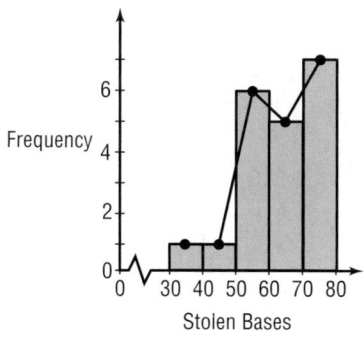

13d. 7 players   **13e.** 2 players

15.

19. $-14c^6d$ **21.**

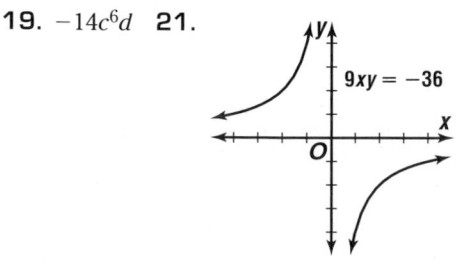

Pages 903–907   Lesson 14-2
5. 30.75; 27.5; 10   **7.** about 10,323; 10,500; 10,700

9a.

stem	leaf
0	6 7 7 7 9
1	3 3 4 4 5 6 7 8 9 9
2	0 0 0 1 1 1 1 1 3 3 8
3	0 1 1 1 2 4 4 6 8
4	1 1 1 2 7

$1|3 = 13$

9b. 23.55 **9c.** 21 **9d.** 21

9e. Since the mean 23.55, the median 21, and the mode 21 are all representative values, any of them could be used as an average. **11.** 5.4; 3; 3 **13.** 10.75; 11; 5 and 18 **15.** 8.5; 8.5; 6 and 11 **17.** about 45.8; 45; 45 **19.** 1088; 1090; 1180 **21a.** $1485, $3480, $4650, $1650, $2275, $1480, $780 **21b.** $15,800 **21c.** 100 employees **21d.** about $158 **21e.** $150–$160 **21f.** $155 **21g.** Both values represent central values of the data. **23.** 3 **25a.** about 425.6 **25b.** 400–450 **25c.** about 420.5 **27a.** Sample answer: {1, 2, 2, 2, 3} **27b.** Sample answer: {4, 5, 9} **27c.** Sample answer: {2, 10, 10, 12} **27d.** Sample answer: {3, 4, 5, 6, 9, 9} **29a.** about 215.2 **29b.** 200–220 **29c.** about 213 **29d.** about 215.9; 211 **29e.** The mean calculated using the frequency distribution is very close to the one calculated with the actual data. The median calculated with the actual data is less than the one calculated with the frequency distribution. **31a.** $87,800 **31b.** $61,500 **31c.** $59,000 **31d.** mean **31e.** mode **31f.** Median; the mean is affected by the extreme values of $162,000 and $226,000, and only two people make less than the mode. **31g.** Sample answer: I have been with the company for many years, and I am still making less than the mean salary. **33.** He is shorter than the mean (5'11.6") and the median (5'11.5"). **35.** dependent; $\frac{3}{55}$ **37.** $40,305.56 **39.** A

Pages 914–917   Lesson 14-3
5. 16; 8

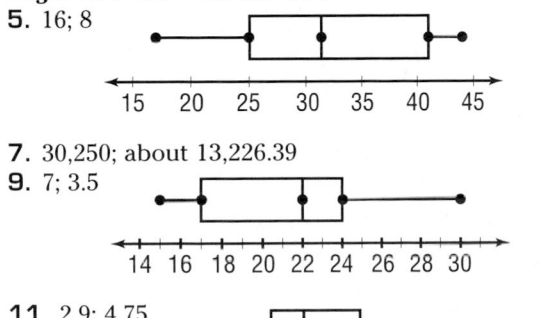

7. 30,250; about 13,226.39
9. 7; 3.5

11. 2.9; 4.75

13. 211; about 223.14 **15.** 20.25; about 25.31
17. about 19.33; about 6.48 **19.** about 129.65;

about 23.29 **21.** Sample answer: {15, 15, 15, 16, 17, 20, 24, 26, 30, 35, 45} **23a.** $2414, $2838, $4147 **23b.** 1733 **23c.** $20,480, $21,914
23d.

23e. about 3507.18 **23f.** about 5643.35 **23g.** The data in the upper quartile is diverse. **25a.** 11 **25b.** about 2.94 **29a.** 45 **29b.** Sample answer: 10 **29c.** Sample answer: 20, 30, 40, 50, 60, 70
29d. Sample answer:

Programs Sold	Frequency
20–30	2
30–40	1
40–50	2
50–60	5
60–70	2

29e. Sample answer:

31. 3, 0.5, −0.75

Pages 923–925   Lesson 14-4
7a. 68.3% **7b.** 92.9% **7c.** 22.6–25.4
7d. 20.08–27.92
9a.

9b. 10.5–13.5
9c. 99.7%
9d. 95.5%
11a. 79.6–84.4
11b. 76.8–87.2
11c. 86.6%
11d. 31.1%

13a. 66.9% **13b.** 28.6% **13c.** 154
15a. $\frac{1}{64}, \frac{3}{32}, \frac{15}{64}, \frac{5}{16}, \frac{15}{64}, \frac{3}{32}, \frac{1}{64}$
15b.

15c. 3
15d. about 1.2
15e. They are similar.

17a. 0.8% **17b.** 99.2% **19a.** 72 **19b.** 58
19c. 68–71 **21a.** about 2.55 mL **21b.** 57.6%

23. about 48.2; 45; 42   **25a.** Sample answer:
$y = 0.05x^3 - 2.22x^2 + 29.72x + 366.92$
25b. Sample answer: 2553 students

Pages 930–932   Lesson 14-5
5. 7.3   **7.** 42.85–47.15   **9a.** about 0.29   **9b.** about
27.30–27.70 min   **9c.** about 26.76–28.24 min
11. about 0.37   **13.** about 0.53   **15.** about 0.70
17. about 333.07–336.93   **19.** about 77.81–82.19
21. 67.34–68.66 in.   **23.** about 4524.21–4527.79
25. about 5.23–5.53   **27.** about 8.9%   **29a.** 4.5
29b. 338.39–361.61 hours   **29c.** Sample answer:
338 hours; there is only 0.5% chance the mean is less
than this number.   **31a.** about 0.57   **31b.** With a
5% level of confidence, the average family in the
town will have their televisions on from 2.98 to
5.22 hours.   **31c.** Sample answer: None; the sample
is too small to generalize to the population of the
city.   **33a.** 750 h   **33b.** 64 h   **35.** 8.25; about 9.59
37. $\theta = 45°$ **39.** C

**Pages 933–937   Chapter 14 Study Guide and
Assessment**
1. box-and-whisker plot   **3.** standard error of the
mean   **5.** measure of central tendency   **7.** bimodal
9. histogram   **11.** 5
13.

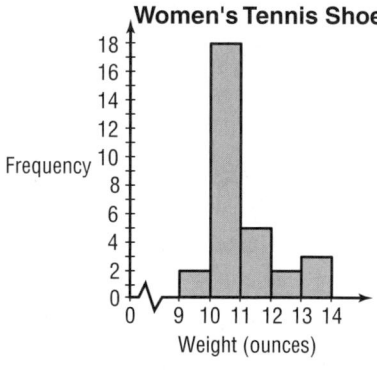

Women's Tennis Shoes

15. 210; 200; 200   **17.** 6.45; 6.5; 6.3 and 6.6   **19.** 3
21. 1.6   **23.** 95.5%   **25.** 79.75–96.25   **27.** 143.25
29. about 0.16   **31.** 1.25   **33.** about 94.53–105.47
35. about 35.62–44.38   **37.** about 1.74–1.86 h
39. about 1.71–1.89 h

41a.

stem	leaf
1	0 3 5 6 7 9
2	1 3 4 5
3	9 9

$1 | 0 = 10$

41b. 21.75
41c. 20
41d. 39

Page 939   Chapter 14 SAT and ACT Practice
1. D   **3.** D   **5.** E   **7.** D   **9.** C

Chapter 15   Introduction to Calculus

Pages 946–948   Lesson 15-1
5. -17 **7.** $-\dfrac{1}{4}$ **9.** $\dfrac{4}{15}$

11a.

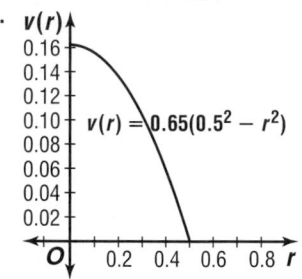

$v(r) = 0.65(0.5^2 - r^2)$

11b. 0 in./s   **13.** 0; undefined   **15.** -16 **17.** 0
19. 10   **21.** $\dfrac{3}{8}$ **23.** 0   **25.** -1 **27.** 4   **29.** -3
31. -1 **33.** 5   **35.** 2   **37.** -0.5 **39.** πa^2; letting
c approach 0 moves the foci together, so the ellipse
becomes a circle. πa^2 is the area of a circle of radius a.
41. No; the graph of $f(x) = \sin\left(\dfrac{1}{x}\right)$ oscillates
infinitely many times between -1 and 1 as x
approaches 0, so the values of the function do
not approach a unique number.   **43.** 64 ft/s
45. 15.684–16.716 mm   **47.** $90x^3y^2$
49. $\dfrac{(x-5)^2}{16} + \dfrac{(y+2)^2}{9} = 1$ **51.** $\langle -7, -6 \rangle$; $\sqrt{85}$
53. -1 **55.** $y \to \infty$ as $x \to \infty$, $y \to -\infty$ as $x \to -\infty$
57. Yes; opposite sides have the same slope.

Pages 957–960   Lesson 15-2
5. $f'(x) = 2x + 1$ **7.** $f'(x) = -3x^2 - 4x + 3$ **9.** 4
11. $F(x) = \dfrac{1}{4}x^4 + \dfrac{4}{3}x^3 - \dfrac{1}{2}x^2 - 3x + C$ **13.** \$8
15. $f'(x) = 7$ **17.** $f'(x) = -4$ **19.** $f'(x) = 3x^2 + 10x$ **21.** $f'(x) = 2$ **23.** $f'(x) = -6x + 2$
25. $f'(x) = 3x^2 - 4x + 5$ **27.** $f'(x) = 6x^2 - 14x + 6$ **29.** $f'(x) = 81x^2 - 216x + 144$ **31.** 3
33. 1   **35.** $F(x) = \dfrac{1}{7}x^7 + C$ **37.** $F(x) = \dfrac{4}{3}x^3 - 3x^2 + 7x + C$ **39.** $F(x) = 2x^4 + \dfrac{5}{3}x^3 - \dfrac{9}{2}x^2 + 3x + C$ **41.** $F(x) = 2x^3 - \dfrac{5}{2}x^2 - 21x + C$
43. $F(x) = \dfrac{1}{3}x^3 + 2x^2 + x + C$ **45.** Any function
of the form $F(x) = \dfrac{1}{6}x^6 + \dfrac{1}{4}x^4 - \dfrac{1}{3}x^3 - x + C$,
where C is a constant.   **47.** $f'(x) = -\dfrac{1}{x^2}$
49a. $v(t) = 80 - 32t$ **49b.** 48 ft/s   **49c.** $t = 2.5$ s
49d. 103 ft   **51a.** $r(p) = p(100 - 2p)$
51b. 25 cents
53a.

170 180 190 200 210 220 230 240 250 260

55. $-\frac{1}{27}$ **57.** $\left(x + \frac{1}{6}\right)^2 + \left(y - \frac{7}{6}\right)^2 = \frac{169}{18}$
59. $x = 8t - 3, y = 3t - 2$ **61.** about 214.9 m
63. D

Pages 966–968   Lesson 15-3

5. $\frac{26}{3}$ units^2 **7.** 72 **9a.** 576 ft **9b.** Yes;
integration shows that the ball would fall 1600 ft in
10 seconds of free-fall. Since this exceeds the height
of the building, the ball must hit the ground in less
than 10 seconds. **11.** 9 units^2 **13.** 4 units^2

15. 312 units^2 **17.** $\frac{208}{3}$ units^2

19. $\lim\limits_{n\to\infty} \sum\limits_{i=1}^{n} \left(\sin i \, \frac{\pi}{n}\right) \cdot \frac{\pi}{n}$ **21.** $\frac{27}{2}$ **23.** 1088

25. $\frac{40}{3}$ **27.** 10.125 ft^2 **29.** 8000 ft^3
31a.

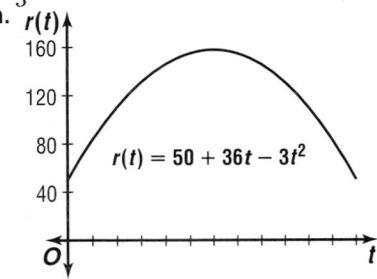

$r(t) = 50 + 36t - 3t^2$

31b. \$1464 **31c.** \$122 **33.** $\frac{1}{2}\pi r^2$ **35.** 0
37. $\vec{u} = \langle -1, -1, -10 \rangle = -\vec{i} - \vec{j} - 10\vec{k}$
39. $\frac{1}{2}, \frac{\pi}{5}$ **41.** C

Pages 973–976   Lesson 15-4

5. $\frac{2}{3}x^3 - 2x^2 + 3x + C$ **7.** $\frac{16}{3}$ units^2 **9.** $\frac{26}{3}$ units^2

11. $\frac{63}{2}$ **13.** 54 **15.** $\frac{1}{6}x^6 + C$ **17.** $\frac{1}{3}x^3 - x^2 +$

$4x + C$ **19.** $\frac{1}{5}x^5 + \frac{2}{3}x^3 - 3x + C$ **21.** $\frac{1}{3}x^3 - 3x^2 +$

$3x + C$ **23.** $\frac{13}{3}$ units^2 **25.** 64 units^2

27. $\frac{20}{3}$ units^2 **29.** $\frac{3}{4}$ unit^2 **31.** 686 **33.** 20

35. $\frac{34}{3}$ **37.** $\frac{9}{20}$ **39.** $\frac{413}{6}$ **41.** $\frac{15}{4}$ **43.** 18
45a. 44,152.52; 44,100 **45b.** 338,358.38; 338,350
47a. All are negative. **47b.** $-\frac{22}{3}$ **47c.** $\frac{22}{3}$

49a.

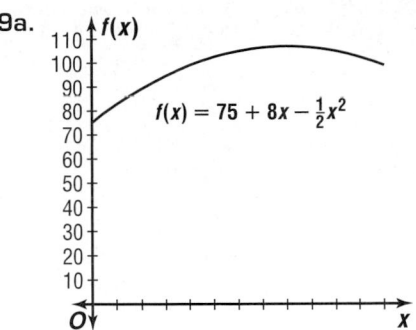

$f(x) = 75 + 8x - \frac{1}{2}x^2$

49b. \$93 **49c.** \$105 **51a.** 4.1×10^{16} Nm^2
51b. 6.3×10^9 J **53.** $f'(x) = 12x^5 - 6x$
55. $\frac{253}{4606}$ **57.** $(y + 1)^2 = -12(x - 6)$
59. 35.46 ft/s

**Pages 977–981   Chapter 15 Study Guide and
Assessment**

1. false; sometimes **3.** false; indefinite **5.** false;
secant **7.** false; derivative **9.** false; rate of change
11. $-1, -3$ **13.** -1 **15.** 0 **17.** 0 **19.** 4
21. $\frac{1}{5}$ **23.** $f'(x) = 8x + 3$ **25.** $f'(x) = 12x^5$
27. $f'(x) = 6x - 5$ **29.** $f'(x) = 2x^3 - 6x^2 + \frac{1}{3}$
31. $f'(x) = 35x^6 - 75x^4$ **33.** $F(x) = 4x^2 + C$
35. $F(x) = -\frac{1}{8}x^4 + \frac{2}{3}x^3 - \frac{3}{2}x^2 - 2x + C$
37. $F(x) = \frac{1}{3}x^3 - x^2 - 8x + C$ **39.** 4 units^2
41. $\frac{37}{3}$ units^2 **43.** 36 **45.** 28 **47.** $\frac{6}{5}x^5 + C$
49. $\frac{1}{3}x^3 + \frac{5}{2}x^2 - 2x + C$ **51.** $0.0000125m$
53a. 17.6 ft/s^2 **53b.** $v(t) = 17.6t$
53c. $d(t) = 8.8t^2$

Page 983   Chapter 15 SAT and ACT Practice
1. A **3.** D **5.** C **7.** C **9.** D

PHOTO CREDITS

Turner/The Image Bank; **630** The Image Bank; **639** Catherine Karnow/CORBIS; **647** CORBIS; **651** James Westwater; **652** Rob Matheson/The Stock Market; **657** NASA; **659** Bryan Yablonsky/DUOMO; **662** John Elk/Stock Boston; **666** World Perspectives/Stone; **668** Stephen Frisch/Stock Boston; **670** Rick Stewart/Allsport; **675** Telegraph Colour Library/FPG; **676** Louis Bencze/Stone; **680** Jon Riley/Stone; **683** David Young-Wolff/PhotoEdit; **684** Photofest; **694** Mike Brown/Liaison Agency, (inset)NASA; **700** SuperStock; **701** Photodisc; **703** Stephen Frisch/Stock Boston; **707** Klaus Hackenberg/Masterfile; **709** Bill Brooks/Masterfile; **712** CORBIS/ Bettmann; **715** Glen Allison/Stone; **720** A. Ramey/Stock Boston; **724** Tom Bean/PNI; **731** VCG/FPG; **732** Hugh Sitton/Stone; **736** Icon Images; **742** Karl Weatherly/CORBIS; **748** Joanna B. Pinneo/PNI; **756–757** Bruce Rowell/Masterfile; **759** Jeff Zaruba/Stone; **765** Nathan Bilow/Allsport; **768** Llewellyn/Pictor; **773** Jon Riley/Stone; **774** Michael Burchfield/AP/Wide World Photo; **779** Mark Burnett; **781** David Batterbury/CORBIS; **794** Telegraph Colour Library/FPG; **798** SuperStock; **800** Llewellyn/Pictor; **804** Laszlo Stern/Photex; **806** Kevin Schafer/CORBIS; **813** George Shelley/The Stock Market; **815** Daniel J. Cox/Stone; **822** Jonathan Meyers/FPG; **827** Tom Stewart/The Stock Market;

836 Kevin Horan/Stone; **839** Arthur Grace/Stock Boston; **841** Francis G. Mayer/CORBIS; **846** Icon Images; **851** Frank Siteman/Stone; **852** Jose L. Pelaez/The Stock Market; **854** Gamma Presse Images/Liaison Agency; **856** Mark Burnett; **859** file photo; **866** Buck Campbell/FPG; **869** Photodisc; **870** Icon Images; **872** Alvin E. Staffan; **875** Bill Horsman/Stock Boston; **878** Jeff Bates; **885** Icon Images; **888** Doug Wilson/CORBIS, (inset)John Biever/Sports Illustrated/TIME; **891** Joe Traver/Liaison Agency; **896** Elena Rooraid/PhotoEdit; **897** Archive Photos/The Image Bank; **900** Hiroyuki Matsumoto/Stone; **905** AFP/CORBIS; **907** Icon Images; **911** Bob Daemmrich/Stock Boston; **916** Rick Rickman/DUOMO; **917** Aaron Haupt; **920** Roy Ooms/Masterfile; **925** Rod Joslin photo; **927** R.W. Jones/CORBIS; **929** Imtek Imagineering/Masterfile; **937** David Madison/Stone; **940** Oliver Meckes/EOS/Gelderblom/Photo Researchers; **943** NASA; **946** G.W. Willis/BPS/Stone; **948** Lonny Kalfus/Stone; **954** Photodisc; **956** Bruce Forster/Stone; **961** Photodisc; **968** Andy Lyons/Allsport; **969** NASA; **970** Vladmir Pcholkin/FPG; **975** Jean Miele/The Stock Market; **976** Tom Stewart/The Stock Market; **981** Imtek Imagineering/Masterfile; **A1** (t)SuperStock, (b) The Lowe Art Museum, The U. of Miami/SuperStock.

INDEX

Red type denotes items only in the Teacher's Wraparound Edition.

 A

abscissa, 5, 21, 57

absolute maximum, 170, 172, 176

absolute minimum, 170, 172, 201

absolute value
functions, 7, 10–11, 47–48, 50, 56, 60, 135, 137, 140, 142–144, 151–152, 156, 164, 166, 168, 170, 198, 283, 298, 304, 394, 445, 641
graphs, 72, 86, 118, 221, 386, 461
inequalities, 53, 147–148, 150–151, 158, 198, 201, 221
of complex numbers, 587, 589

acceleration, 76–77, 157, 220, 223, 242, 255, 289, 291, 297, 311

accounting, 24, 45, 47, 50, 766, 768, 896

acoustics, 559, 571

acute angle, 280, 284, 308, 310, 336, 469, 492

addition, 421
of complex numbers in rectangular form, 609
of functions, 14, 17, 18, 25, 58
of matrices, 79, 120, 212
of terms, 64
of vectors, 496

additive
identity matrix, 80
inverse, 80

Addressing Individual Needs, *see Prerequisite Skills and Multiple Learning Styles*

advertising, 83, 799, 931

aeronautics/aerospace, 211, 485, 487, 499, 504, 657, 695, 701

agriculture, 20, 21, 105, 117, 264, 629, 676, 702, 896

algebraic vectors, 493–499

alternate interior angles, 300, 314, 612–613

alternate optimal solutions, 114–117

altitudes of triangles, 36, 302, 731

ambiguous case, 320

Amer, Tahani R., 969

amplitudes, 368, 370–374, 378, 381–382, 389, 391, 393–394, 402, 412–413, 415, 498, 504, 560, 587, 594–595, 607, 641, 845

amusement parks, 255, 630

anagram, 797, 850

analytic geometry, 157, 167, 188, 468, 615, 618, 652

angles
bisector of, 419, 473
coterminal, 279–281, 298, 304, 318, 336, 555
measures of, 419, 461

of convex polygons, 331, 394, 762
of depression, 300, 302–303, 306, 309, 311, 314, 317, 326
of elevation, 300–302, 306, 309–310, 314, 317, 332, 445, 455, 821, 874
of incidence, 284, 289, 311, 821
of incline, 289, 290, 297
of refraction, 284, 289, 311, 461, 821
of rotation, 670, 673–674, 676
sum of triangles, 319
vertex, 307

angular displacement, 352–353, 355, 357, 377, 414

angular velocity, 352–357, 376–377, 386, 413–414, 461

animation, 88, 90, 91, 95, 535–536

antiderivatives, 951–960, 970–971, 972

antilogarithms, 726, 729, 734, 748, 751

antilogarithms of natural logorithms (antiln x), 734

apex of the prism, 444

apothem, 300, 302–303, 661

Applications, 4c, 66c, 126c, 204c, 276c, 342c, 420c, 484c, 552c, 614c, 694c, 758c, 836c, 888c, 940c

approximating
cube roots, 602
e, 716
fifth roots, 602
fourth roots, 602
intersections of graphs, 679
relative maximum and minimum, 173
zeros of functions, 232, 236–237, 240, 264, 271, 311, 525, 652

arc length, 349, 444, 612

arccosine, 305–306, 335, 406–412, 416, 703

archaeology, 23, 732, 753

architecture, 234, 290, 304, 312, 317, 325, 448, 452, 454, 510, 540, 717, 895

arcsine, 305–306, 335, 406–412, 416

arctangent, 305, 406–412, 416

area, 317, 548, 963
of circles, 77, 549, 579, 969
of circular cross sections, 975
of circular sectors, 347, 358, 377, 417, 444, 874
of ellipses, 640, 947
of octagons, 316
of parallelograms, 316, 331, 549
of pentagons, 316, 322, 357, 732
of polygons, 303, 548
of rectangles, 179, 341, 621, 961–962
of regions bounded by curves, 926
of regular polygons, 303, 429
of rhombi, 326, 331
of shaded regions, 257, 533, 549, 684, 966, 972–973, 983
of squares, 77, 272, 341, 476, 828, 858

of triangles, 104, 144, 302, 314–316, 329–330, 338, 350, 461, 548–549, 613, 630, 669, 858, 907
under a curve, 961–968, 977, 980

Argand plane, 586, 607

argument of complex numbers, 587

Aristotle, 319

arithmetic means, 31, 754–755, 757–759, 761–762, 791, 897, 901, 911, 914–915, 939

arithmetic sequences, 757–762, 765, 781, 791, 828, 830

arithmetic series, 759, 791, 829–830, 858

arrangements, *see permutations*

art, 711, 841

Aryabhata, 462

Assessment Resources, 2–3, 4d, 62–65, 66d, 124–125, 126d, 198, 202–203, 204d, 268, 272–273, 274–275, 276d, 336, 340–341, 342d, 414, 418–419, 420d, 478, 482–483, 484d, 548–549, 550–551, 552d, 612–613, 614d, 688, 692–693, 694d, 750, 754–757, 758d, 830, 834–835, 836d, 882, 886–887, 888d, 934, 938–939, 940d, 978, 982–983

Associative Property, 86

astronomy, 256, 282, 525, 637–638, 640, 661, 666–667, 676, 691, 731

asymptotes
horizontal, 180–184, 186–187, 200, 436, 492, 705–706, 728
of conjugate hyperbolas, 652
of hyperbolas, 642–652
slant, 183–184, 186–187, 197, 200, 560, 765
vertical, 180–187, 200, 492, 705, 728

audio technology, 358, 561, 563, 566

Auditory/Musical Learning Style, *see Multiple Learning Styles*

augmented matrices, 97, 106

automobiles/travel, 189–190, 211, 256, 310, 354, 505, 508, 660, 850, 867, 907, 929

averages, 482, 617, 753–755, 887, 983
class marks, 890–891, 893, 895, 901, 913, 934
mean, 50, 111, 150, 661, 715, 897–900, 903–907, 914, 916–917, 920–921, 923, 925, 934, 937, 939, 943
measures of central tendency, 897–908, 933
median, 44, 619, 897–900, 903–907, 910, 914–917, 925, 933–934, 937, 939
mode, 897, 899–900, 903–906, 917, 925, 934, 937

aviation, 11, 61, 249, 282, 303, 311, 317, 332, 476, 520, 532, 560, 573, 578, 650, 652, 709, 731, 798, 821, 880

axis of symmetry, 137, 691
  for ellipses, 631
  for hyperbolas, 137, 653–661, 691
  for parabolas, 653–661

Babylonians, 97, 367, 534

back-to-back bar graph, 889

backsolving, 202

banking, 715, 746, 773, 816, 820, 845, 975

ballooning, 317

bar graphs, 889, 924, 934, 937
  back-to-back bar graph, 889
  histogram, 890, 892, 933–934, 938

base, 71, 192
  angles, 302, 492, 516
  of exponential expression, 704
  of isosceles triangles, 310
  of a logarithm, 718–719, 726, 728, 733

basic counting principle, 837, 855, 882

Bernouilli, Jakob, 462

Bernoulli, Johann, 367

best-fit lines, 38, 40–41, 51, 60, 258–259, 270, 573

bimodal, 899

binomial
  expansion, 802, 804, 821, 832, 875–876, 948
  experiments, 876, 878, 924
  factors, 225, 235, 461, 732
  representation, 805
  Theorem, 803–805, 807, 814, 828, 832, 851, 875–880, 884

biology, 167, 290, 385, 392, 574, 577–578, 709, 717, 723, 745, 748, 772, 819, 850, 947

biomedical engineering, 136

bisectors
  of acute angles, 470, 472
  of obtuse angles, 473, 480
  of segments, 419

block diagrams, 542

blueprint, 56, 510

boating, 302, 303, 368, 371, 375, 391, 518, 677

Boolean operators and statements, 169

Borts, Autumn, 319

botany, 29, 845, 931, 948

boundary, 52–53, 147, 685–686, 722, 728

bounded region, 108, 110

box-and-whisker plots, 908–910, 914–915, 933

Boyle's Law, 651, 737

Brahmagupta, 462

break-even point, 67, 72

broadcasting, 437, 440

Brothers' expression, 716

budgets, 9, 118, 833

business, 13, 14, 19, 25, 28, 31, 37, 44, 48, 49, 51, 55, 56, 72, 110, 111, 114, 115, 116, 144, 171, 175, 178, 228, 235, 247, 290, 318, 343, 347, 510, 630, 660, 708, 805, 822, 825, 864, 873, 924, 958–959, 961, 965, 967, 981

Calculators, *see Technology*

calculus, 172, 444, 534

camping, 339

Canter, Georg, 367

capacity, 248

cardioid classical curves, 564–566, 571, 607–608

Career Choices, 12, 105, 136, 221, 312, 358, 420, 499, 598, 622, 703, 800, 851, 896, 976

carpentry, 691, 814

Cartesian coordinate system, 556, 573

cartographer, 430

catenary curve, 715

cause and effect relationship, 41

Cavalieri, Bonaventura, 969

Cayley, Arthur, 97

CD-ROM, *see Technology*

cellular growth, 792

Celsius temperature, 17, 152

census, 956

center, 623, 626, 763, 880
  of circles, 628, 688
  of ellipses, 631–641
  of hyperbolas, 642–652, 781

central angles, 345–348, 357, 376–377, 413–414, 417, 444, 445, 612, 622, 874

change
  from degree measure to radian measure, 344, 348, 394, 414
  from radian measure to degree measure, 414, 436, 671
  from standard form to normal form, 465
  in signs of polynomial function, 237–238, 424
  of base formula, 728, 730–731, 734, 771

Chapter Resources, *see Resource Manager*

chemistry, 24, 103, 150, 180, 187, 504, 611, 651, 701, 718, 720, 726–727, 729, 737, 799, 960

chess, 799

childcare, 145

circle graphs, 857, 938

circles, 105, 376, 419, 532, 608, 622–630, 658, 663, 684, 687
  central angle of, 345–348, 357, 376–377, 413–414, 417, 444, 445, 612, 874
  circumference of, 410, 560

  circumscribed around polygons, 302, 549, 629
  concentric, 345, 352, 622, 623
  diameter of, 95, 283, 300, 303, 313, 349, 412, 623–630, 660
  inscribed in polygons, 298
  radii of, 105, 192, 210, 226, 289, 293, 298, 300, 302, 311, 316–317, 325, 335, 349, 357, 377, 413, 417, 476, 532, 548, 622–630, 684, 805
  sector of, 346, 349, 444
  segment of, 350
  standard form of the equation of, 624–625, 627
  tangent to lines, 628
  unit, 291–292, 294, 296, 304–305, 335, 343, 345, 422–423, 428, 454, 458, 461, 579, 819, 948

circuits, 249

circular arcs, 345

circular functions, 292, 335

circular permutations, 847–851, 867

circumference, 296, 410, 534, 548, 560, 711, 755

circumscribed circle, 302, 549, 629

cis, 588

civil engineering, 311, 350, 715, 865, 874

classical curves, 564
  cardioid, 564–566, 571, 607–608
  lemniscate, 564–566
  limacon, 562, 565–566, 607–608, 611
  Spiral of Archimedes, 564–565, 572, 607–608

class interval, 890–895, 917

class limits, 890, 893–894, 901, 917

class marks, 890–891, 893, 895, 901, 913, 934

Classroom Vignettes, 39, 46, 108, 117, 129, 139, 155, 214, 237, 280, 328, 345, 369, 389, 423, 434, 486, 501, 571, 655, 699, 742, 775, 838, 876

clocks, 793

clockwise rotations, 92, 671

club activities, 827

coefficient, 40, 68, 73, 106, 205–206, 213, 231, 236, 260, 265–266, 365
  of friction, 256, 429
  matrix, 101, 103
  symmetric, 801

cofunctions, 287, 439, 446

coinciding, 32–33, 35–36, 68, 179

coin flips, 124, 804

collinear points, 31, 273, 845, 939

columns of a matrix, 78–79, 81, 85, 119

column matrix, 78

combinations, 837–845, 881–882, 885–886

common denominators, 432

common difference, 757–761, 833

common factor, 184

common logarithms, 726–728, 737, 749, 751

common ratios of geometric sequences, 764–770, 783, 795, 829–830, 960

logarithmic, 718–719, 727–728, 737, 741, 744–745, 749
notation, 7
odd, 133, 135, 163, 166, 732
operations with, 13, 58
periodic functions, 360
phase shift, 378–379, 381–383, 389, 394, 398–402, 412–413, 417, 446, 498, 641, 845, 925
point discontinuity, 159–160, 165–166, 169, 184–185, 187, 197, 257, 283
pole, 180
polynomial, 137, 151, 156, 162, 164–167, 170, 172–173, 176, 179–180, 182, 197–199, 206, 210, 222, 231, 233, 236, 258–262, 265–267, 269–271, 585, 980
quadratic, 137, 140–143, 160, 164, 168, 170, 176, 198, 215, 258, 260, 261–262, 298
radical, 8, 11, 143, 156, 198
range, 5–6, 8–9, 15, 19, 31, 47–48, 51, 57, 72, 111, 149, 157, 201, 395–397, 405, 407, 436, 705–706, 714, 716–717, 890–891, 893–895, 908, 926, 929, 933–934, 948
rational, 11, 137, 180, 182, 184, 186–187, 196–199
secant, 167, 188, 286, 292, 295, 301, 311, 336–337, 364, 395–403, 416, 418, 421–481
sine, 285–289, 291, 294–295, 298–299, 301, 303, 311, 336–337, 359, 361, 364, 368, 370, 377, 389, 385, 389, 391, 406, 412, 415, 418, 421–481, 515, 519
step, 46, 50, 60, 104, 143, 156, 162, 166, 168, 198
table, 138, 141, 152, 160–161, 175, 184, 199, 295, 362, 704
tangent, 171, 188, 285–289, 292, 295, 299, 301, 304, 311, 334, 336–337, 395–403, 406–412, 416, 418, 421–481, 492, 684
trigonometric, 293, 295–298, 301, 304, 318, 337, 382, 425, 428, 431, 434–436, 702
undefined, 399
vertical displacement, 365, 528
vertical line test, 7–8, 25
zeros of, 22–23, 24–25, 57, 86, 206, 232, 271, 378, 498

Fundamental Theorem of Algebra, 205, 207
corollary to, 207

Fundamental Theorem of Calculus, 970–977, 980

fund-raising, 111, 145, 377, 917

games, 95, 795

gap discontinuity, 169–170

gardening, 271, 579, 682, 691

Gauss, Carl Friedrich, 97

gears, 325

gemology, 461

general form, 409
of circles, 625–627, 687

of conic sections, 662, 667
of hyperbolas, 646, 687
of parabolas, 655–656
of parametric equations, 521
of sequences, 775
of series, 797
of sinusoidal functions, 388

geography, 11, 281, 346, 350, 410, 455, 895, 915

geology, 828, 867

geometry, 35, 36, 71, 104, 110, 123, 135, 144, 145, 179, 187, 192, 210, 226, 228, 233, 300, 302–303, 310, 316, 326, 331, 357, 394, 412, 429, 436, 461, 468, 476, 492, 498, 510, 525, 541, 621, 629, 640, 661, 732, 764, 782, 833, 845, 850, 880, 947–948, 975

geometric means, 766, 768, 772, 793, 829

geometric sequences, 766–769, 771–773, 783, 793, 800, 830, 960
common ratios, 764–770, 783, 795, 829–830, 960
geometric means, 766, 768, 772, 793, 829
nth term, 760

geometric series, 769, 778, 830

girth, 109

golden ratio, 784, 813

goodness of fit, 38, 40

government, 905, 916

graphical iteration, 586–587, 589, 599, 604, 607

graphing
absolute value inequalities, 221
absolute value, 72, 86, 118, 386
asymptotes, 645–646
boundaries, 54, 146
catenary curves, 715
circles, 623, 669
complex numbers, 586–589
compound functions, 378, 382
conic sections, 623–691
constant functions, 46
dilations, 93
ellipses, 641, 661
even functions, 134
exponential functions, 708, 718
functions, 365, 373, 377, 379–381, 384–385, 390, 394, 400–401, 416–417, 659, 709, 713
geometric transformations, 539
greatest integer functions, 677,
inequalities, 52–53, 55–56, 60, 72, 77, 111, 146, 149–151, 179, 198, 358, 708, 722
inverses, 22, 24, 48–50, 56, 133, 156, 165, 177, 183, 199, 221, 298, 405, 455, 684
lines, 20, 31, 37, 59, 455, 521, 524, 874
logarithmic functions, 718
orbits, 818
ordered pairs, 10, 561, 576, 635
parabolas, 656
parametric equations, 525, 665
piecewise functions, 45, 170, 377
polar equations, 552–557, 561–565, 573–574, 578–579, 584, 608, 677
polygonal convex regions, 111
rational functions, 200, 242, 386
relations, 131
statistical

back-to-back bar graph, 889
bar, 889, 924, 934, 937
circle graphs, 857, 938
cumulative frequency, 902
cumulative frequency distribution, 902
frequency polygons, 892, 895, 918
histograms, 890, 892, 933–934, 938
scatter plots, 38, 41–42, 51, 60, 145, 151, 258–263, 270, 351, 592, 741, 744–745, 747
stem-and-leaf plot, 899–900, 903–904, 917, 938
step functions, 104
systems of equations and inequalities, 107, 112–116, 122, 681–685, 690, 702, 773, 874, 907
translations, 88, 92–94, 104–119, 137–140, 142, 145, 540, 618
trigonometric functions, 429, 469, 573

Graphing Calculator
Appendix, A2–A25
Exercises, 11, 103, 143, 166, 177, 211, 262, 289, 364, 374, 435, 460, 524, 566, 605, 628, 639, 676, 709, 730, 731, 735, 798, 879, 915, 947
Exploration, 13, 69, 86, 106, 133, 232, 265, 284, 323, 333, 369, 378, 404, 433, 458, 512, 526, 592, 602, 641, 665, 685, 695, 705, 738, 784, 809, 877, 926, 945, 949
Tips, 46, 67, 99, 138, 153, 164, 173, 181, 206, 225, 238, 253, 259, 278, 286, 328, 372, 389, 564, 576, 581, 625, 635, 645, 655, 679, 699, 713, 727, 742, 909, 911, 944

Graphing Calculators, *see Technology*

gravitational potential energy, 166

gravity, 289, 291

greatest common factors, 646, 656

greatest integer function, 46, 48, 50, 60, 137, 139, 156, 166, 170

growth formulas, 706, 712

half angle-identities, 451–454, 477, 479, 606, 783

half-life, 718, 722, 737

half planes, 52, 53

Hands-On Activity, *see Motivating the Lesson*

harmonic means, 249, 772

harmonic series, 792

health, 212, 383, 389, 391, 724, 813, 860, 892, 924

height, 76, 226

Herigone, Pierre, 319

Hero's Formula, 330

hertz, 372

Hipparachus, 462

histograms, 890, 892, 933–934, 938

history, 468, 639, 786, 790, 827, 893

INDEX

228, 283, 368, 380, 624, 631, 643, 670, 705–706, 709, 711, 728

partial fractions, 248, 677

partial sums, 786, 790

Pascal's triangle, 801–802, 804, 813

Pearson Product-Moment Correlation, 40–41, 403

pendulum, 348–349, 375

pentagon, 94, 300, 311, 316

percent, 20, 179, 283, 341, 606, 799, 835, 937, 939, 947

perfect square, 219

perfect square trinomial, 202, 214,

perimeter, 71, 303, 316, 341, 366, 548–549, 833
  of hexagons, 303
  of pentagons, 302
  of rectangles, 188, 358, 366, 598, 828, 968
  of squares, 105, 598, 782
  of triangles, 105, 123, 326, 498, 661, 907, 983

periodic functions, 359–360, 363, 365, 395, 456

periods, 361–362, 365, 368–376, 378, 381–382, 385, 388–389, 393–394, 396–402, 412–413, 415–416, 498, 504, 560, 641, 968
  of pendulums, 256, 376, 476
  of sinusoidal functions, 391

Permutation Formula, 839

permutations, 837–851, 881–882, 885–886
  circular, 847–851, 867
  with repetitions, 846–851, 881

perpendicular, 34–36, 59, 61, 158, 179, 212, 326, 341, 463, 491, 510, 519
  bisector, 35, 621
  components, 488
  lines, 24, 34, 59, 483
  segments, 291
  vectors, 505, 519, 525, 547, 560, 566, 711

pharmacology, 127, 146, 148

phase angle, 411

phase difference, 385

phase shift, 378–379, 381–383, 389, 394, 398–402, 412–413, 417, 446, 498, 641, 845, 925

phasor notation, 573

phons, 746

photography, 445, 724, 874

physics, 18, 76, 77, 93, 135, 157, 166, 179, 188, 195, 242, 255–256, 271, 284, 286, 288–289, 297, 311, 349, 357, 365, 375–376, 402, 411, 417, 429, 435, 454, 459–460, 476, 481, 492, 503–504, 510, 518–519, 532, 542, 547, 584, 622, 706, 714, 716, 782, 814, 833, 880, 941, 948, 974, 981

physiology, 509, 568, 569

pi (π), 548

piecewise function, 45–48, 60–61, 132, 161, 170, 196, 199

plane geometry, 618

planes, 509
  intersecting, 73–74, 500, 623
  perpendicular, 500

plot
  box-and-whisker, 908–910, 914–915, 933
  line, 167, 889
  stem-and-leaf, , 899–900, 903–904, 917, 938

plug-in strategy, 124, 202

point discontinuity of functions, 159–160, 165–166, 169, 184–185, 187, 197, 257, 283, 821

points, 86, 158
  critical, 164, 171–179
  escape, 818
  of inflection, 171–178, 199, 228, 641
  maximum, 171–177, 188, 197, 199, 207, 211, 228, 257, 392, 413, 469, 641
  minimum, 171–177, 188, 197, 199, 228, 249, 413, 469, 641
  prisoner, 818–819, 827
  relative maximum, 172–174, 176–177, 179, 199, 201
  relative minimum, 172–174, 176–177, 199, 201, 225

point-slope form of linear equations, 28–29, 34–35, 39, 57, 59, 463–454

point symmetry, 127

polar coordinate system, 553, 556, 568
  graphing polar functions, 556, 562, 579, 661, 948
  polar axis, 553, 554, 568, 588
  polar coordinates, 553, 555–556, 559, 564, 567–568, 570–574, 576–579, 607, 611, 652, 737
  polar inequalities, 558

polar distance formula, 557

polar equations, 556, 562, 567, 579, 948

polar form, 571–572, 574, 577, 588, 591, 593, 595–597, 600–601, 603–604, 606–607, 609–610, 716, 810, 932
  of complex numbers, 588
  of rectangular coordinates, 570

phasors, 573

polarized filter, 297

polarized light, 297

pole of rational function, 180

poles, 553, 555, 568

polygonal convex set, 107–110, 114, 118–119, 122–123, 168, 188, 469, 585, 748, 821

polygonal regions, 535

polygons, 88, 109, 112, 122, 516, 618

polyhedra, 535

polynomial, 224, 227, 250, 268, 366, 641, 809, 880
  curves, 969
  division, 183–184
  equation, 206–209, 213, 229, 239, 603
  functions, 137, 151, 156, 162, 164–167, 170, 172–173, 176, 179–180, 182, 197–199, 206, 210, 222, 231, 233, 236, 258–262, 265–267, 269–271, 585, 980

inequalities, 146–147, 149–150, 179

population, 241, 707, 709, 741, 742, 747, 753, 893, 927, 933, 947

possible outcomes, 886

postal service, 49, 159, 162, 843

power
  function, 704, 708, 747
  of complex numbers, 599
  of complex numbers written in polar forms, 827
  of power rules, 136, 697
  of products, 696–699
  property, 722
  property of logarithms, 722
  rule, 953, 955, 971
  rule for antiderivatives, 957
  rule for derivatives, 959

prediction equation models, 38–39, 475, 717, 742, 956

pre-image, 88–89, 91–95, 142

Prerequisite Skills, 6, 14, 21, 68, 128, 138, 147, 153, 181, 206, 244, 252, 278, 287, 344, 422, 449, 521, 554, 569, 575, 616, 624, 633, 696, 760, 787, 847, 890, 919, 942, 952

Previewing the Chapter, 4a, 66a, 126a, 204a, 276a, 342a, 420a, 484a, 552a, 614a, 694a, 758a, 836a, 888a, 940a

prime, 983
  factors, 65, 233, 983
  integers, 755, 805
  numbers, 64

principle roots, 601, 605

prism, 538

prisoner points, 818–819, 827

prisoner set, 599, 603, 605, 607, 819

probability, 716, 852–858, 881, 884, 886, 924, 936, 948, 957, 975
  binomial experiments, 876, 878, 924
  complements, 853
  conditional, 868–874
  of compound events, 859–867
  of dependent events, 67–68, 70–72, 86, 136, 196, 837–838, 843, 883, 907
  of failure, 852, 881, 883
  of inclusive events, 862
  of independent events, 837–838, 870, 874, 881
  of mutually exclusive events, 861–863, 867, 881, 884
  of mutually inclusive events, 862
  of two events, 859–860
  odds, 125, 341, 854–857, 883

Problem Solving, 4c, 66c, 126c, 204c, 276c, 342c, 420c, 484c, 552c, 614c, 694c, 758c, 836c, 888c, 940c

product
  of functions, 13–14
  of complex numbers, 606
  of complex numbers in polar form, 593–598, 610
  of consecutive integers, 796
  of matrices, 81, 257
  property, 699, 720–721, 727
  property of exponents, 744

projectiles, 527–529, 531–532

S

INDEX

ANSWER APPENDIX

CHAPTER 1 LINEAR RELATIONS AND FUNCTIONS

Pages 8–12 Lesson 1-1

16b.

26.

x	y
−4	−9
−3	−8
−2	−7
−1	−6
0	−5
1	−4

27.

x	y
1	−1
2	−2
3	−3
4	−4
5	−5
6	−6

28.

x	y
−5	5
−4	4
−3	3
−2	2
−1	1
0	0
1	1

29.

x	y
1	0
2	3
3	6
4	9
5	12

30.

x	y
11	3
11	−3

31.

x	y
4	2
4	−2

Page 13 Lesson 1-2
Graphing Calculator Exploration

1.

X	Y₁	Y₂
0	-1	2
1	1	5
2	3	8
3	5	11
4	7	14
5	9	17
6	11	20

Y₁=-1

X	Y₂	Y₃
0	2	-3
1	5	-4
2	8	-5
3	11	-6
4	14	-7
5	17	-8
6	20	-9

Y₃=-3

2.

X	Y₁	Y₂
0	-1	2
1	1	5
2	3	8
3	5	11
4	7	14
5	9	17
6	11	20

X=0

X	Y₂	Y₃
0	2	-2
1	5	5
2	8	24
3	11	55
4	14	98
5	17	153
6	20	220

Y₃=-2

3.

X	Y₁	Y₂
0	-1	2
1	1	5
2	3	8
3	5	11
4	7	14
5	9	17
6	11	20

X=0

X	Y₂	Y₃
0	2	-.5
1	5	.2
2	8	.375
3	11	.45455
4	14	.5
5	17	.52941
6	20	.55

Y₃=-.5

Pages 23–25   Lesson 1-3

12.

$y = 4x - 9$

13.

$y = 3$

14.
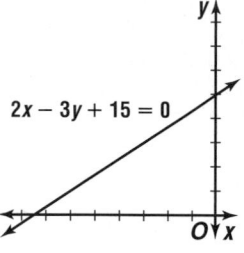
$2x - 3y + 15 = 0$

15.

$x - 4 = 0$

16.

$y = 6x - 1$

17.

$y = 5 - 2x$

18.

$y + 8 = 0$

19.

$2x + y = 0$

20.
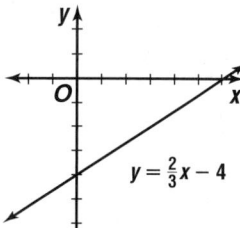
$y = \frac{2}{3}x - 4$

21.
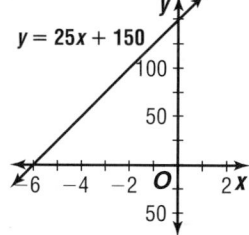
$y = 25x + 150$

22.

$2x + 5y = 8$

23.
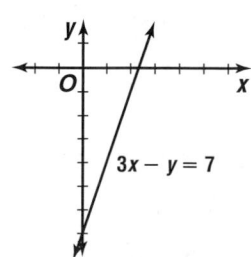
$3x - y = 7$

Pages 29–31   Lesson 1-4

2. Sample answer:
Use point-slope form.
$$y - y_1 = m(x - x_1)$$
$$y - (-4) = \frac{1}{4}(x - 3)$$
$$y + 4 = \frac{1}{4}x - \frac{3}{4}$$
$$x - 4y - 19 = 0$$

Use slope-intercept form.
$$y = mx + b$$
$$-4 = \frac{1}{4}(3) + b$$
$$-\frac{19}{4} = b$$

Substitute the slope and intercept into the general form.
$$y = \frac{1}{4}x - \frac{19}{4}$$
Write in standard form.
$$x - 4y - 19 = 0$$

Pages 41–44   Lesson 1-6

4a.
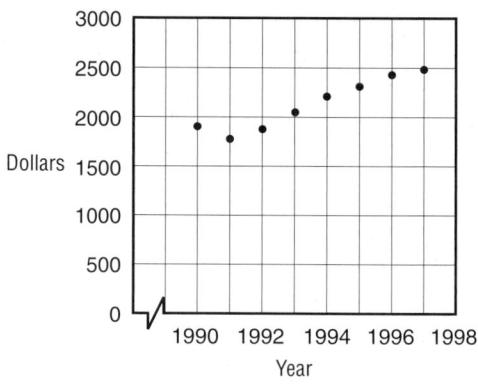
Personal Consumption on Durable Goods

8a.

Car Weight and Mileage

9a.
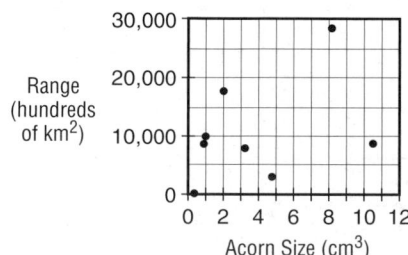
Acorn Size and Range

10a.

Working Women

11a.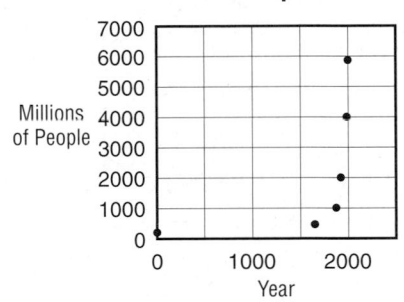

World Population

Pages 48–51 Lesson 1-7

5.

6.

7.

8.

11.

12.

13.

14.

15.

16.

17.

18.

19.

20.

21.

22.

31a.

Public Transportation

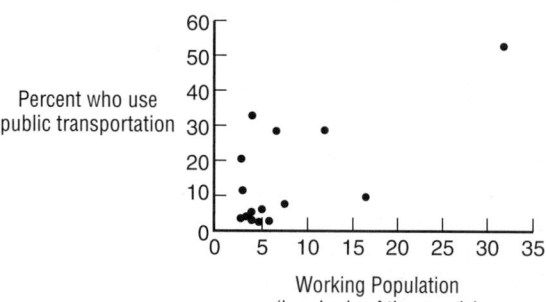

Percent who use public transportation (y-axis)

Working Population (hundreds of thousands)

Pages 54–56   Lesson 1-8

4.

$x + y < 4$

5.
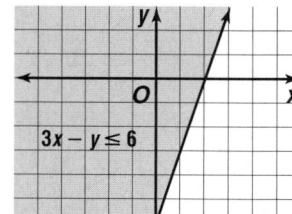

$3x - y \le 6$

6.

$7 < x + y \le 9$

7.
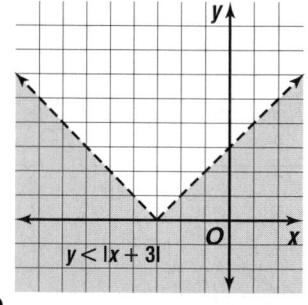

$y < |x + 3|$

9.
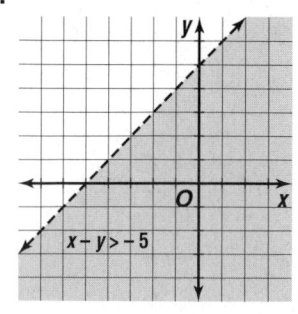

$y < 3$

10.

$x - y > -5$

11.

$2x + 4y \ge 7$

12.
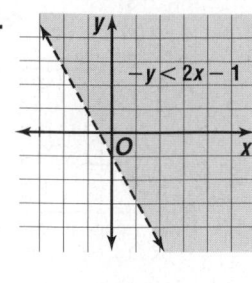

$-y < 2x - 1$

13.
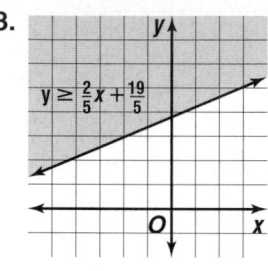

$y \ge \frac{2}{5}x + \frac{19}{5}$

14.
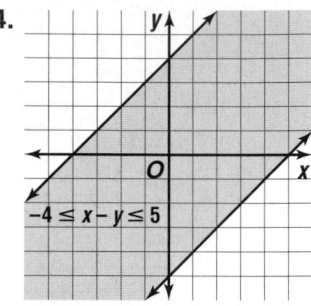

$-4 \le x - y \le 5$

15.

$y \ge |x|$

16.
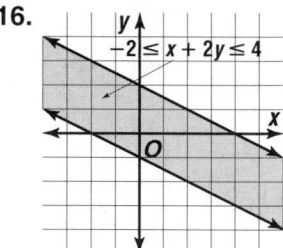

$-2 \le x + 2y \le 4$

17.
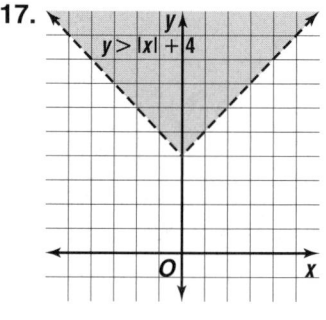

$y > |x| + 4$

18.
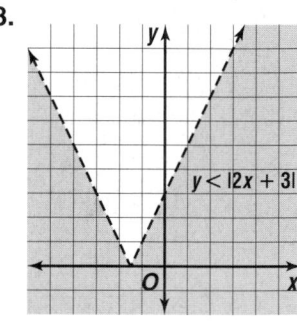

$y < |2x + 3|$

19.
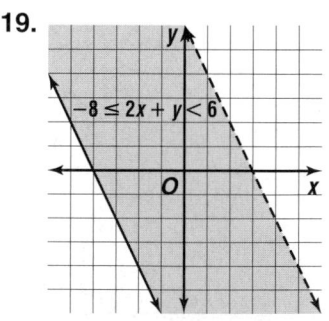

$-8 \le 2x + y < 6$

20.
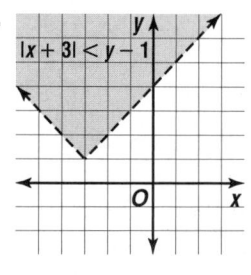

$|x + 3| < y - 1$

21.
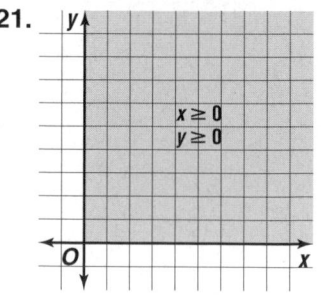

$x \ge 0$
$y \ge 0$

22.

26b.

27b.

28c.

Pages 57–61 Chapter 1 Study Guide and Assessment

31.

32.

33.

34.

35.

36.

37.

38.

53a.

59.

60.

61.

62.

63.

64.

65.

66.

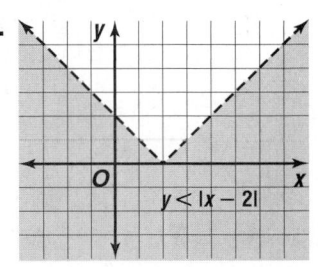

CHAPTER 2   SYSTEM OF LINEAR EQUATIONS AND INEQUALITIES

Pages 70–72   Lesson 2-1

39.

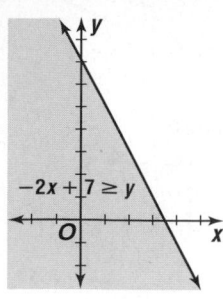

$-2x + 7 \geq y$

40.

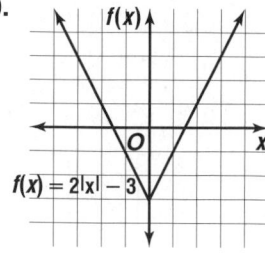

$f(x) = 2|x| - 3$

Page 77   Lesson 2-2

27c.

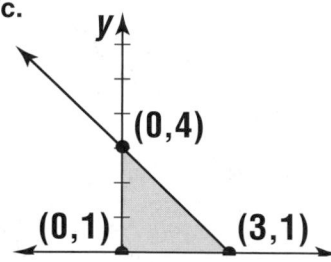

(0,4)   (0,1)   (3,1)

Page 86   Lesson 2-3
Graphing Calculator Exploration

2a. Let $A = \begin{bmatrix} a_{11} & a_{12} \\ a_{21} & a_{22} \end{bmatrix}$ and $B = \begin{bmatrix} b_{11} & b_{12} \\ b_{21} & b_{22} \end{bmatrix}$.

$$A + B = \begin{bmatrix} a_{11} & a_{12} \\ a_{21} & a_{22} \end{bmatrix} + \begin{bmatrix} b_{11} & b_{12} \\ b_{21} & b_{22} \end{bmatrix}$$

$$= \begin{bmatrix} a_{11} + b_{11} & a_{12} + b_{12} \\ a_{21} + b_{21} & a_{22} + b_{22} \end{bmatrix}$$

$$= \begin{bmatrix} b_{11} + a_{11} & b_{12} + a_{12} \\ b_{21} + a_{21} & b_{22} + a_{22} \end{bmatrix}$$

$$= \begin{bmatrix} b_{11} & b_{12} \\ b_{21} & b_{22} \end{bmatrix} + \begin{bmatrix} a_{11} & a_{12} \\ a_{21} & a_{22} \end{bmatrix}$$

$$= B + A$$

2b. Let $A = \begin{bmatrix} a_{11} & a_{12} \\ a_{21} & a_{22} \end{bmatrix}$, $B = \begin{bmatrix} b_{11} & b_{12} \\ b_{21} & b_{22} \end{bmatrix}$, $C = \begin{bmatrix} c_{11} & c_{12} \\ c_{21} & c_{22} \end{bmatrix}$.

$$(A + B) + C = \begin{bmatrix} a_{11} + b_{11} & a_{12} + b_{12} \\ a_{21} + b_{21} & a_{22} + b_{22} \end{bmatrix} + \begin{bmatrix} c_{11} & c_{12} \\ c_{21} & c_{22} \end{bmatrix}$$

$$= \begin{bmatrix} (a_{11} + b_{11}) + c_{11} & (a_{12} + b_{12}) + c_{12} \\ (a_{21} + b_{21}) + c_{21} & (a_{22} + b_{22}) + c_{22} \end{bmatrix}$$

$$= \begin{bmatrix} a_{11} + (b_{11} + c_{11}) & a_{12} + (b_{12} + c_{12}) \\ a_{21} + (b_{21} + c_{21}) & a_{22} + (b_{22} + c_{22}) \end{bmatrix}$$

$$= \begin{bmatrix} a_{11} & a_{12} \\ a_{21} & a_{22} \end{bmatrix} + \begin{bmatrix} b_{11} + c_{11} & b_{12} + c_{12} \\ b_{21} + c_{21} & b_{22} + c_{22} \end{bmatrix}$$

$$= A + (B + C)$$

2c. Let $A = \begin{bmatrix} a_{11} & a_{12} \\ a_{21} & a_{22} \end{bmatrix}$ and $B = \begin{bmatrix} b_{11} & b_{12} \\ b_{21} & b_{22} \end{bmatrix}$.

$$AB = \begin{bmatrix} a_{11}b_{11} + a_{12}b_{21} & a_{11}b_{12} + a_{12}b_{22} \\ a_{21}b_{11} + a_{22}b_{21} & a_{21}b_{12} + a_{22}b_{22} \end{bmatrix}$$

$$BA = \begin{bmatrix} b_{11}a_{11} + b_{12}a_{21} & b_{11}a_{12} + b_{12}a_{22} \\ b_{21}a_{11} + b_{22}a_{21} & b_{21}a_{12} + b_{22}a_{22} \end{bmatrix}$$

Thus, $AB \neq BA$, since $a_{12}b_{21} \neq b_{12}a_{21}$.

2d. Let $A = \begin{bmatrix} a_{11} & a_{12} \\ a_{21} & a_{22} \end{bmatrix}$, $B = \begin{bmatrix} b_{11} & b_{12} \\ b_{21} & b_{22} \end{bmatrix}$, and $C = \begin{bmatrix} c_{11} & c_{12} \\ c_{21} & c_{22} \end{bmatrix}$.

$$(AB)C = \begin{bmatrix} a_{11}b_{11} + a_{12}b_{21} & a_{11}b_{12} + a_{12}b_{22} \\ a_{21}b_{11} + a_{22}b_{21} & a_{21}b_{12} + a_{22}b_{22} \end{bmatrix} \begin{bmatrix} c_{11} & c_{12} \\ c_{21} & c_{22} \end{bmatrix}$$

$$= \begin{bmatrix} a_{11}b_{11}c_{11} + a_{11}b_{12}c_{21} + a_{12}b_{21}c_{11} + a_{12}b_{22}c_{21} \\ a_{21}b_{11}c_{11} + a_{21}b_{12}c_{21} + a_{22}b_{21}c_{11} + a_{22}b_{22}c_{21} \end{bmatrix}$$

$$\begin{bmatrix} a_{11}b_{11}c_{12} + a_{11}b_{12}c_{22} + a_{12}b_{21}c_{12} + a_{12}b_{22}c_{22} \\ a_{21}b_{11}c_{12} + a_{21}b_{12}c_{22} + a_{22}b_{21}c_{12} + a_{22}b_{22}c_{22} \end{bmatrix}$$

$$A(BC) = \begin{bmatrix} a_{11} & a_{12} \\ a_{21} & a_{22} \end{bmatrix} \begin{bmatrix} b_{11}c_{11} + b_{12}c_{21} & b_{11}c_{12} + b_{12}c_{22} \\ b_{21}c_{11} + b_{22}c_{21} & b_{21}c_{12} + b_{22}c_{22} \end{bmatrix}$$

$$= \begin{bmatrix} a_{11}b_{11}c_{11} + a_{11}b_{12}c_{21} + a_{12}b_{21}c_{11} + a_{12}b_{22}c_{21} \\ a_{21}b_{11}c_{11} + a_{21}b_{12}c_{21} + a_{22}b_{21}c_{11} + a_{22}b_{22}c_{21} \end{bmatrix}$$

$$\begin{bmatrix} a_{11}b_{11}c_{12} + a_{11}b_{12}c_{22} + a_{12}b_{21}c_{12} + a_{12}b_{22}c_{22} \\ a_{21}b_{11}c_{12} + a_{21}b_{12}c_{22} + a_{22}b_{21}c_{12} + a_{22}b_{22}c_{22} \end{bmatrix}$$

Therefore, $(AB)C = A(BC)$.

Page 87   Lesson 2-4A
1. The new figure is a 90° counterclockwise rotation of $\triangle LMN$.
2. The new figure is a 180° counterclockwise rotation of $\triangle LMN$.
3. The new figure is a 270° counterclockwise rotation of $\triangle LMN$.

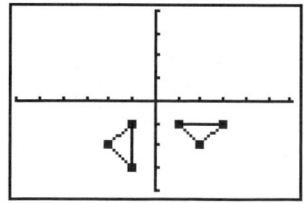

Pages 92–96   Lesson 2-4

5.

6.

7.

8.

9.

25a. Let $\begin{bmatrix} a & b \\ c & d \end{bmatrix} = R_{x\text{-axis}}$

$$\begin{bmatrix} a & b \\ c & d \end{bmatrix} \cdot \begin{bmatrix} 1 & -2 & -1 \\ 3 & -1 & -3 \end{bmatrix} = \begin{bmatrix} 1 & -2 & -1 \\ -3 & 1 & 3 \end{bmatrix}$$

$$\begin{bmatrix} a+3b & -2a-b & -a-3b \\ c+3d & -2c-d & -c-3d \end{bmatrix} = \begin{bmatrix} 1 & -2 & -1 \\ -3 & 1 & 3 \end{bmatrix}$$

$a+3b=1 \quad -2a-b=-2 \quad -a-3b=-1$
$c+3d=-3 \quad -2c-d=1 \quad -c-3d=3$
Thus, $a=1$, $b=0$, $c=0$, and $d=-1$.

By substitution, $R_{x\text{-axis}} = \begin{bmatrix} -1 & 0 \\ 0 & 1 \end{bmatrix}$.

25b. Let $\begin{bmatrix} a & b \\ c & d \end{bmatrix} = R_{y\text{-axis}}$

$$\begin{bmatrix} a & b \\ c & d \end{bmatrix} \cdot \begin{bmatrix} 1 & -2 & -1 \\ 3 & -1 & -3 \end{bmatrix} = \begin{bmatrix} -1 & 2 & 1 \\ 3 & -1 & -3 \end{bmatrix}$$

$$\begin{bmatrix} a+3b & -2a-b & -a-3b \\ c+3d & -2c-d & -c-3d \end{bmatrix} = \begin{bmatrix} -1 & 2 & 1 \\ 3 & -1 & -3 \end{bmatrix}$$

$a+3b=-1 \quad -2a-b=2 \quad -a-3b=1$
$c+3d=3 \quad -2c-d=-1 \quad -c-3d=-3$
Thus, $a=-1$, $b=0$, $c=0$, and $d=1$.

By substitution, $R_{y\text{-axis}} = \begin{bmatrix} -1 & 0 \\ 0 & 1 \end{bmatrix}$.

25c. Let $\begin{bmatrix} a & b \\ c & d \end{bmatrix} = R_{y=x}$.

$$\begin{bmatrix} a & b \\ c & d \end{bmatrix} \cdot \begin{bmatrix} 1 & -2 & -1 \\ 3 & -1 & -3 \end{bmatrix} = \begin{bmatrix} 3 & -1 & -3 \\ 1 & -2 & -1 \end{bmatrix}$$

$$\begin{bmatrix} a+3b & -2a-b & -a-3b \\ c+3d & -2c-d & -c-3d \end{bmatrix} = \begin{bmatrix} 3 & -1 & -3 \\ 1 & -2 & -1 \end{bmatrix}$$

$a+3b=3 \quad -2a-b=-1 \quad -a-3b=-3$
$c+3d=1 \quad -2c-d=-2 \quad -c-3d=-1$
Thus, $a=0$, $b=1$, $c=1$, and $d=0$.

By substitution, $R_{y=x} = \begin{bmatrix} 0 & 1 \\ 1 & 0 \end{bmatrix}$.

25d. Let $\begin{bmatrix} a & b \\ c & d \end{bmatrix} = Rot_{90}$.

$$\begin{bmatrix} a & b \\ c & d \end{bmatrix} \cdot \begin{bmatrix} 1 & -2 & -1 \\ 3 & -1 & -3 \end{bmatrix} = \begin{bmatrix} -3 & 1 & 3 \\ 1 & -2 & -1 \end{bmatrix}$$

$$\begin{bmatrix} a+3b & -2a-b & -a-3b \\ c+3d & -2c-d & -c-3d \end{bmatrix} = \begin{bmatrix} -3 & 1 & 3 \\ 1 & -2 & -1 \end{bmatrix}$$

$a+3b=-3 \quad -2a-b=-1 \quad -a-3b=3$
$c+3d=1 \quad -2c-d=-2 \quad -c-3d=-1$
Thus, $a=0$, $b=-1$, $c=1$, and $d=0$.

By substitution, $R_{Rot\,90} = \begin{bmatrix} 0 & -1 \\ 1 & 0 \end{bmatrix}$.

25e. Let $\begin{bmatrix} a & b \\ c & d \end{bmatrix} = Rot_{180}$.

$$\begin{bmatrix} a & b \\ c & d \end{bmatrix} \cdot \begin{bmatrix} 1 & -2 & -1 \\ 3 & -1 & -3 \end{bmatrix} = \begin{bmatrix} -1 & 2 & 1 \\ -3 & 1 & 3 \end{bmatrix}$$

$$\begin{bmatrix} a+3b & -2a-b & -a-3b \\ c+3d & -2c-d & -c-3d \end{bmatrix} = \begin{bmatrix} -1 & 2 & 1 \\ -3 & 1 & 3 \end{bmatrix}$$

$a+3b=-1 \quad -2a-b=2 \quad -a-3b=1$
$c+3d=-3 \quad -2c-d=1 \quad -c-3d=3$
Thus, $a=-1$, $b=0$, $c=0$, and $d=-1$.

By substitution, $Rot_{180} = \begin{bmatrix} -1 & 0 \\ 0 & -1 \end{bmatrix}$.

25f. Let $\begin{bmatrix} a & b \\ c & d \end{bmatrix} = Rot_{270}$.

$$\begin{bmatrix} a & b \\ c & d \end{bmatrix} \cdot \begin{bmatrix} 1 & -2 & -1 \\ 3 & -1 & -3 \end{bmatrix} = \begin{bmatrix} 3 & -1 & -3 \\ -1 & 2 & 1 \end{bmatrix}$$

$$\begin{bmatrix} a+3b & -2a-b & -a-3b \\ c+3d & -2c-d & -c-3d \end{bmatrix} = \begin{bmatrix} 3 & -1 & -3 \\ -1 & 2 & 1 \end{bmatrix}$$

$a+3b=3 \quad -2a-b=-1 \quad -a-3b=-3$
$c+3d=-1 \quad -2c-d=2 \quad -c-3d=1$
Thus, $a=0$, $b=1$, $c=-1$, and $d=0$.

By substitution, $Rot_{270} = \begin{bmatrix} 0 & 1 \\ -1 & 0 \end{bmatrix}$.

29a. The bishop moves along a diagonal until it encounters the edge of the board or another piece. The line along which it moves changes vertically and horizontally by 1 unit with each square moved, so the translation matrices are scalars. Sample matrices are

$$c\begin{bmatrix} 1 & 1 \\ 1 & 1 \end{bmatrix},\ c\begin{bmatrix} 1 & 1 \\ -1 & -1 \end{bmatrix},\ c\begin{bmatrix} -1 & -1 \\ 1 & 1 \end{bmatrix},\ \text{and } c\begin{bmatrix} -1 & -1 \\ -1 & -1 \end{bmatrix},$$

where c is the number of squares moved.

29b. The knight moves in combinations of 2 vertical–1 horizontal or 1 vertical–2 horizontal squares. These can be either up or down, left or right. Sample

matrices are $\begin{bmatrix} 1 & 1 \\ 2 & 2 \end{bmatrix}, \begin{bmatrix} 1 & 1 \\ -2 & -2 \end{bmatrix}, \begin{bmatrix} -1 & -1 \\ 2 & 2 \end{bmatrix}, \begin{bmatrix} -1 & -1 \\ -2 & -2 \end{bmatrix}, \begin{bmatrix} 2 & 2 \\ 1 & 1 \end{bmatrix},$

$\begin{bmatrix} 2 & 2 \\ -1 & -1 \end{bmatrix}, \begin{bmatrix} -2 & -2 \\ 1 & 1 \end{bmatrix}$ and $\begin{bmatrix} -2 & -2 \\ -1 & -1 \end{bmatrix}.$

29c. The king can move 1 unit in any direction. The matrices describing this are $\begin{bmatrix} 1 & 1 \\ 0 & 0 \end{bmatrix}$, $\begin{bmatrix} -1 & -1 \\ 0 & 0 \end{bmatrix}$, $\begin{bmatrix} 0 & 0 \\ 1 & 1 \end{bmatrix}$,

$\begin{bmatrix} 0 & 0 \\ -1 & -1 \end{bmatrix}$, $\begin{bmatrix} 1 & 1 \\ 1 & 1 \end{bmatrix}$, $\begin{bmatrix} 1 & 1 \\ -1 & -1 \end{bmatrix}$, $\begin{bmatrix} -1 & -1 \\ 1 & 1 \end{bmatrix}$, and $\begin{bmatrix} -1 & -1 \\ -1 & -1 \end{bmatrix}$.

Page 96   Lesson 2-4   *Mid-Chapter Quiz*

10. The result is the original figure. The original figure is represented by $\begin{bmatrix} a & b & c \\ d & e & f \end{bmatrix}$. The reflection over the x-axis is found by $\begin{bmatrix} 1 & 0 \\ 0 & -1 \end{bmatrix} \cdot \begin{bmatrix} a & b & c \\ d & e & f \end{bmatrix} = \begin{bmatrix} a & b & c \\ -d & -e & -f \end{bmatrix}$.

The reflection of the image over the x-axis is found by $\begin{bmatrix} 1 & 0 \\ 0 & -1 \end{bmatrix} \cdot \begin{bmatrix} a & b & c \\ d & e & f \end{bmatrix} = \begin{bmatrix} a & b & c \\ d & e & f \end{bmatrix}$. The matrix for the final image is the same as that of the original figure.

Pages 102–105   Lesson 2-5
49. yes

$A = \begin{bmatrix} a & b \\ c & d \end{bmatrix}$. Does $(A^2)^{-1} = (A^{-1})^2$?

$A^2 = \begin{bmatrix} a^2 + bc & ab + bd \\ ac + cd & bc + d^2 \end{bmatrix}$

$(A^2)^{-1} = \dfrac{1}{a^2d^2 - 2abcd + b^2d^2} \begin{bmatrix} bc + d^2 & -ab - bd \\ -ac - cd & a^2 + bc \end{bmatrix}$

$A^{-1} = \dfrac{1}{ad - bc} \begin{bmatrix} d & -b \\ -c & a \end{bmatrix} = \begin{bmatrix} \dfrac{d}{ad - bc} & \dfrac{-b}{ad - bc} \\ \dfrac{-c}{ad - bc} & \dfrac{a}{ad - bc} \end{bmatrix}$

$(A^{-1})^2 = \dfrac{1}{a^2d^2 - 2abcd + b^2d^2} \begin{bmatrix} bc + d^2 & -ab - bd \\ -ac - cd & a^2 + bc \end{bmatrix}$

Thus, $(A^2)^{-1} = (A^{-1})^2$.

Pages 109–111   Lesson 2-6
13.

14.

15.

17.

18.

19.

20.

21.

22.

27a.

30.
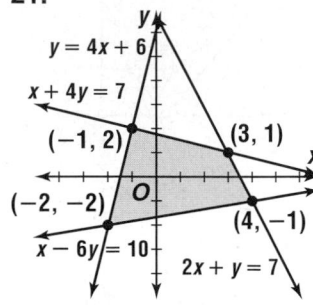

31.

Pages 119–123   Chapter 2   Study Guide and Assessment

39. $\dfrac{1}{23} \begin{bmatrix} 5 & -8 \\ 1 & 3 \end{bmatrix}$

40. no inverse

41. $\dfrac{1}{7} \begin{bmatrix} -4 & -5 \\ -1 & -3 \end{bmatrix}$

42. $\dfrac{1}{11} \begin{bmatrix} 7 & -2 \\ -5 & 3 \end{bmatrix}$

43. $\dfrac{1}{32} \begin{bmatrix} 1 & 5 \\ -6 & 2 \end{bmatrix}$

44. no inverse

CHAPTER 3   THE NATURE OF GRAPHS

Pages 133–136   Lesson 3-1
31.

32.

33.

34.

35.

36.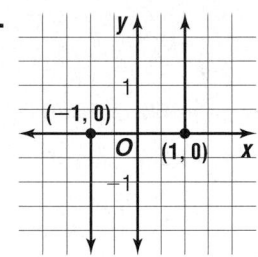

Pages 142–145   Lesson 3-2

12a.

12b.

26.

28.

29.

30.

31.

32.

33.

34.

35a.

$[-7.6, 7.6]$ scl:1 by $[-5, 5]$ scl:1

35b.

$[-7.6, 7.6]$ scl:1 by $[-5, 5]$ scl:1

35c.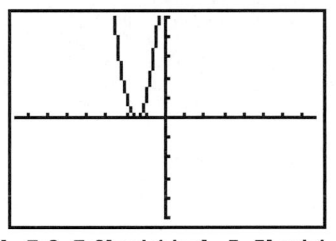

$[-7.6, 7.6]$ scl:1 by $[-5, 5]$ scl:1

36a.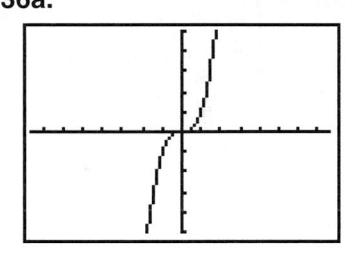

$[-7.6, 7.6]$ scl:1 by $[-5, 5]$ scl:1

36b.

$[-7.6, 7.6]$ scl:1 by $[-5, 5]$ scl:1

36c.

$[-7.6, 7.6]$ scl:1 by $[-5, 5]$ scl:1

37a.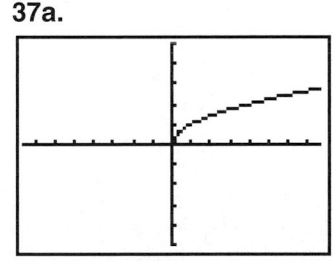

$[-7.6, 7.6]$ scl:1 by $[-5, 5]$ scl:1

37b.

$[-7.6, 7.6]$ scl:1 by $[-5, 5]$ scl:1

37c.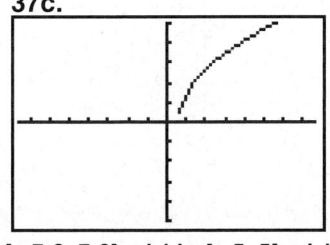

$[-7.6, 7.6]$ scl:1 by $[-5, 5]$ scl:1

Pages 149–151   Lesson 3-3

20.

21.

22.

23.

24.

25.

26.

27.

28.

29.

30.

31.
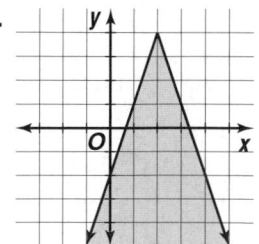

Page 151   Lesson 3-3
Mid-Chapter Quiz

7.

8.
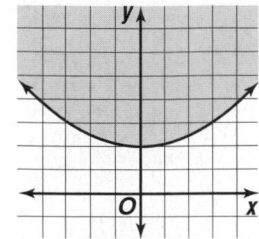

Page 155–158   Lesson 3-4

6.

7.

8.

15.

16.

17.

18.

19.

20.

21.

22.

23.

35.

36.

37.

38.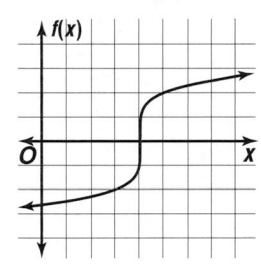

Pages 165–168, Lesson 3–5

9.

10.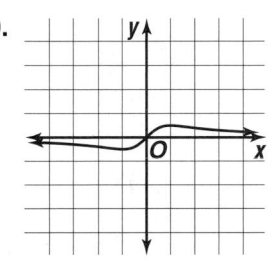

decreasing for $x < -3$;
increasing for $x > -3$

decreasing for $x < -1$
and $x > 1$; increasing
for $-1 < x < 1$

Page 170   Lesson 3–5B

1.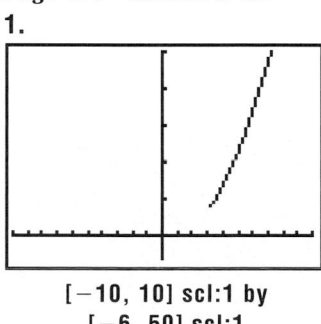

$[-10, 10]$ scl:1 by
$[-6, 50]$ scl:1

2.

$[-18.8, 18.8]$ scl:1 by
$[-12.4, 12.4]$ scl:1

3.

$[-9.4, 9.4]$ scl:1 by
$[-6.2, 6.2]$ scl:1

4.

$[-4.7, 4.7]$ scl:1 by
$[-25, 25]$ scl:10

5.

$[-4.7, 4.7]$ scl:1 by
$[-3.1, 3.1]$ scl:1

6.

$[-9,4, 9.4]$ scl:1 by
$[-6.2, 6.2]$ scl:1

7.

$[-9.4, 9.4]$ scl:1 by
$[-3, 9.4]$ scl:1

8.

$[-4.7, 4.7]$ scl:1 by
$[-2, 8]$ scl:1

Pages 176–179   Lesson 3-6

6.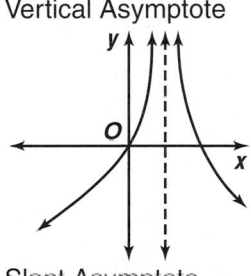

$[-4, 6]$ scl:1 by
$[-20, 20]$ scl:5

7.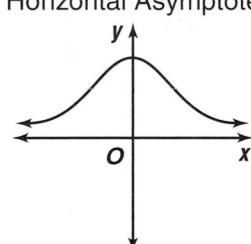

$[-6, 4]$ scl:1 by
$[-14, 6]$ scl:2

Pages 185–188   Lesson 3-7

2. Sample Graphs:
Vertical Asymptote   Horizontal Asymptote

Slant Asymptote

35.

36.

(2, 4)

37.

38.

39.

40.

44b.

Pages 197–201   Chapter 3   Study Guide and Assessment

23.

24.

25.

26.

29.

30.

31.

32.

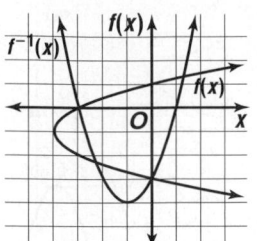

Page 201   Chapter 3   Study Guide and Assessment
Open-Ended Assessment

1a. Sample answer: $x = y^2$ **1b.** Sample answer: $y = x^2$

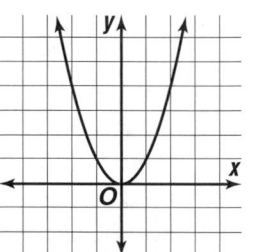

1c. Sample answer: $-xy = 1$

1d.   Sample answer: $xy = 1$

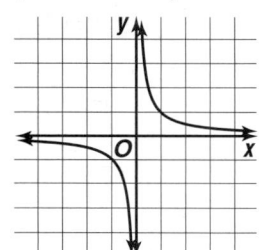

1e. Sample answer: $y = x^3$

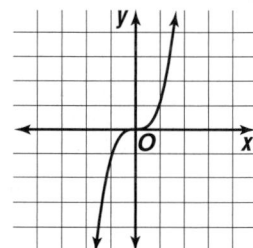

CHAPTER 5   THE TRIGONOMETRIC FUNCTIONS

Pages 296–298   Lesson 5-3

9. $\sin \theta = \frac{4}{5}$, $\cos \theta = \frac{3}{5}$, $\tan \theta = \frac{4}{3}$, $\csc \theta = \frac{5}{4}$,
   $\sec \theta = \frac{5}{3}$, $\cot \theta = \frac{3}{4}$

10. $\sin \theta = \frac{\sqrt{2}}{2}$, $\cos \theta = -\frac{\sqrt{2}}{2}$, $\tan \theta = -1$, $\csc \theta = \sqrt{2}$,
   $\sec \theta = -\sqrt{2}$, $\cot \theta = -1$

11. $\sin \theta = -\frac{\sqrt{2}}{2}$, $\cos \theta = \frac{\sqrt{2}}{2}$, $\csc \theta = -\sqrt{2}$, $\sec \theta = \sqrt{2}$,
   $\cot \theta = -1$

12. $\sin \theta = \frac{\sqrt{3}}{2}$, $\tan \theta = -\sqrt{3}$, $\csc \theta = \frac{2\sqrt{3}}{3}$, $\sec \theta = -2$,
   $\cot \theta = -\frac{\sqrt{3}}{3}$

Pages 372–377   Lesson 6-4

23.

$y = 5 \cos \theta$

24.
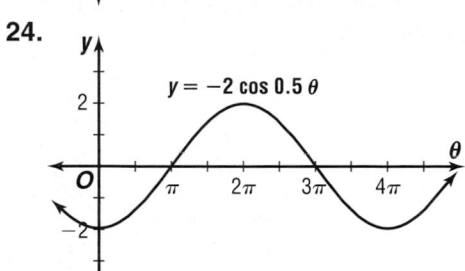
$y = -2 \cos 0.5\,\theta$

25.
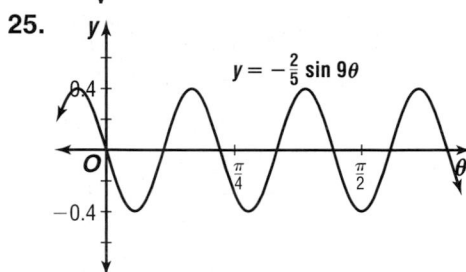
$y = -\frac{2}{5} \sin 9\theta$

26.
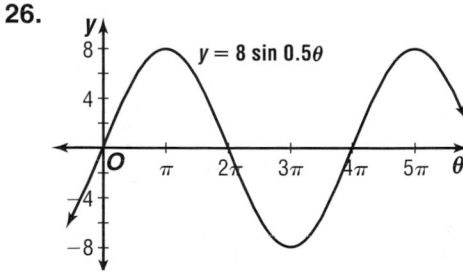
$y = 8 \sin 0.5\theta$

27.
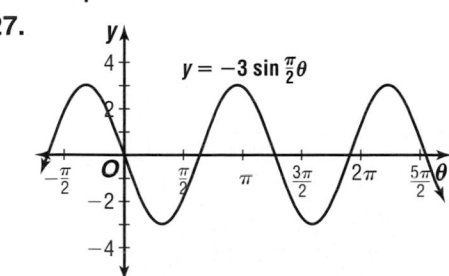
$y = -3 \sin \frac{\pi}{2}\theta$

28.
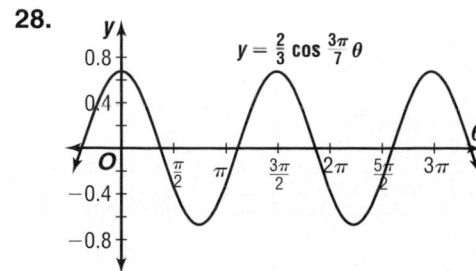
$y = \frac{2}{3} \cos \frac{3\pi}{7}\,\theta$

29.
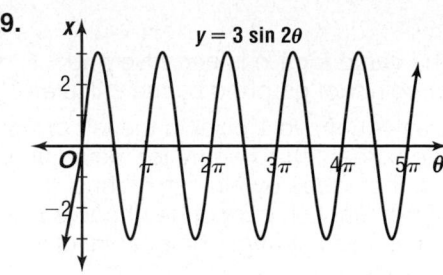
$y = 3 \sin 2\theta$

30.
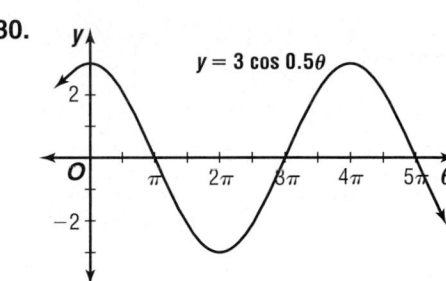
$y = 3 \cos 0.5\theta$

31.
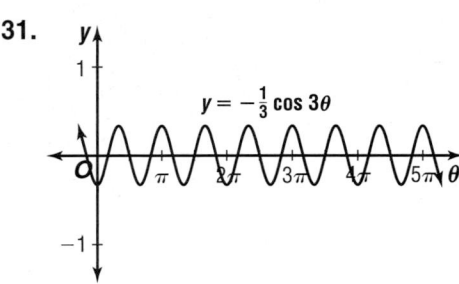
$y = -\frac{1}{3} \cos 3\theta$

32.
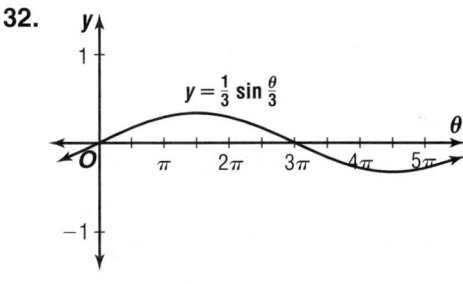
$y = \frac{1}{3} \sin \frac{\theta}{3}$

33.
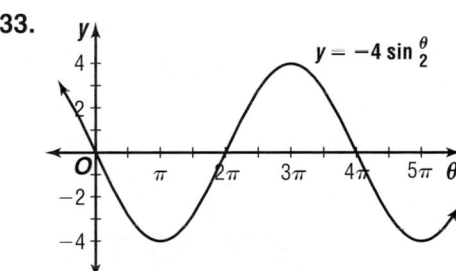
$y = -4 \sin \frac{\theta}{2}$

34.
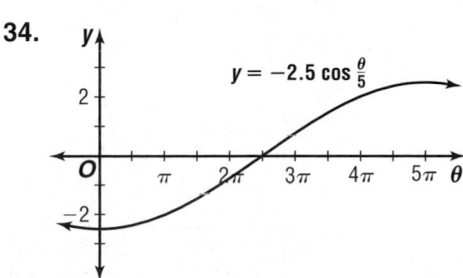
$y = -2.5 \cos \frac{\theta}{5}$

Page 404   Lesson 6–7B

5. −1.78043; yes; the value for $f(x)$ is negative and corresponds to a point not graphed by the calculator.

6. Sample answer: as you move 1 pixel to the left or right of any pixel on the screen, the x-value for the adjacent pixel decreases or increases by almost 7. Thus, the "find" behavior of the function cannot be observed from the graph unless you change the interval of numbers for the x-axis.

Pages 410–412   Lesson 6-8

14.

15.

16.

17.

18.

19.

CHAPTER 7   TRIGONOMETRIC IDENTITIES AND EQUATIONS

Pages 433–436   Lesson 7-2

5. $\cos x \overset{?}{=} \dfrac{\cot x}{\csc x}$

$\cos x \overset{?}{=} \dfrac{\frac{\cos x}{\sin x}}{\frac{1}{\sin x}}$

$\cos x \overset{?}{=} \dfrac{\cos x}{1}$

$\cos x = \cos x$

6. $\dfrac{1}{\tan x + \sec x} \overset{?}{=} \dfrac{\cos x}{\sin x + 1}$

$\dfrac{1}{\frac{\sin x}{\cos x} + \frac{1}{\cos x}} \overset{?}{=} \dfrac{\cos x}{\sin x + 1}$

$\dfrac{1}{\frac{\sin x + 1}{\cos x}} \overset{?}{=} \dfrac{\cos x}{\sin x + 1}$

$\dfrac{\cos x}{\sin x + 1} = \dfrac{\cos x}{\sin x + 1}$

7. $\csc \theta - \cot \theta \overset{?}{=} \dfrac{1}{\csc \theta + \cot \theta}$

$\csc \theta - \cot \theta \overset{?}{=} \dfrac{1}{\csc \theta + \cot \theta} \cdot \dfrac{\csc \theta - \cot \theta}{\csc \theta - \cot \theta}$

$\csc \theta - \cot \theta \overset{?}{=} \dfrac{\csc \theta - \cot \theta}{\csc^2 \theta - \cot^2 \theta}$

$\csc \theta - \cot \theta \overset{?}{=} \dfrac{\csc \theta - \cot \theta}{(1 + \cot^2 \theta) - \cot^2 \theta}$

$\csc \theta - \cot \theta \overset{?}{=} \dfrac{\csc \theta - \cot \theta}{1}$

$\csc \theta - \cot \theta = \csc \theta - \cot \theta$

8. $\sin \theta \tan \theta \overset{?}{=} \sec \theta - \cos \theta$

$\sin \theta \tan \theta \overset{?}{=} \dfrac{1}{\cos \theta} - \cos \theta$

$\sin \theta \tan \theta \overset{?}{=} \dfrac{1}{\cos \theta} - \dfrac{\cos^2 \theta}{\cos \theta}$

$\sin \theta \tan \theta \overset{?}{=} \dfrac{1 - \cos^2 \theta}{\cos \theta}$

$\sin \theta \tan \theta \overset{?}{=} \dfrac{\sin^2 \theta}{\cos \theta}$

$\sin \theta \tan \theta \overset{?}{=} \sin \theta \dfrac{\sin \theta}{\cos \theta}$

$\sin \theta \tan \theta = \sin \theta \tan \theta$

9.
$$(\sin A - \cos A)^2 \overset{?}{=} 1 - 2\sin^2 A \cot A$$
$$\sin^2 A - 2\sin A \cos A + \cos^2 A \overset{?}{=} 1 - 2\sin^2 A \cot A$$
$$1 - 2\sin A \cos A \overset{?}{=} 1 - 2\sin^2 A \cot A$$
$$1 - 2\sin A \cos A \frac{\sin A}{\sin A} \overset{?}{=} 1 - 2\sin^2 A \cot A$$
$$1 - 2\sin^2 A \frac{\cos A}{\sin A} \overset{?}{=} 1 - 2\sin^2 A \cot A$$
$$1 - 2\sin^2 A \cot A = 1 - 2\sin^2 A \cot A$$

13.
$$\tan A \overset{?}{=} \frac{\sec A}{\csc A}$$
$$\tan A \overset{?}{=} \frac{\frac{1}{\cos A}}{\frac{1}{\sin A}}$$
$$\tan A \overset{?}{=} \frac{\sin A}{\cos A}$$
$$\tan A = \tan A$$

14.
$$\cos\theta \overset{?}{=} \sin\theta \cot\theta$$
$$\cos\theta \overset{?}{=} \sin\theta \frac{\cos\theta}{\sin\theta}$$
$$\cos\theta = \cos\theta$$

15.
$$\sec x - \tan x \overset{?}{=} \frac{1 - \sin x}{\cos x}$$
$$\sec x - \tan x \overset{?}{=} \frac{1}{\cos x} - \frac{\sin x}{\cos x}$$
$$\sec x - \tan x = \sec x - \tan x$$

16.
$$\frac{1 + \tan x}{\sin x + \cos x} \overset{?}{=} \sec x$$
$$\frac{1 + \frac{\sin x}{\cos x}}{\sin x + \cos x} \overset{?}{=} \sec x$$
$$\frac{\cos x}{\cos x} \cdot \frac{1 + \frac{\sin x}{\cos x}}{\sin x + \cos x} \overset{?}{=} \sec x$$
$$\frac{\cos x + \sin x}{\cos x(\sin x + \cos x)} \overset{?}{=} \sec x$$
$$\frac{1}{\cos x} \overset{?}{=} \sec x$$
$$\sec x = \sec x$$

17.
$$\sec x \csc x \overset{?}{=} \tan x + \cot x$$
$$\sec x \csc x \overset{?}{=} \frac{\sin x}{\cos x} + \frac{\cos x}{\sin x}$$
$$\sec x \csc x \overset{?}{=} \frac{\sin x}{\cos x} \cdot \frac{\sin x}{\sin x} + \frac{\cos x}{\sin x} \cdot \frac{\cos x}{\cos x}$$
$$\sec x \csc x \overset{?}{=} \frac{\sin^2 x}{\cos x \sin x} + \frac{\cos^2 x}{\sin x \cos x}$$
$$\sec x \csc x \overset{?}{=} \frac{\sin^2 x + \cos^2 x}{\cos x \sin x}$$
$$\sec x \csc x \overset{?}{=} \frac{1}{\cos x \sin x}$$
$$\sec x \csc x \overset{?}{=} \frac{1}{\cos x} \cdot \frac{1}{\sin x}$$
$$\sec x \csc x = \sec x \csc x$$

18.
$$\sin\theta + \cos\theta \overset{?}{=} \frac{2\sin^2\theta - 1}{\sin\theta - \cos\theta}$$
$$\sin\theta + \cos\theta \overset{?}{=} \frac{2\sin^2\theta - (\sin^2\theta + \cos^2\theta)}{\sin\theta - \cos\theta}$$
$$\sin\theta + \cos\theta \overset{?}{=} \frac{\sin^2\theta - \cos^2\theta}{\sin\theta - \cos\theta}$$
$$\sin\theta + \cos\theta \overset{?}{=} \frac{(\sin\theta - \cos\theta)(\sin\theta + \cos\theta)}{\sin\theta - \cos\theta}$$
$$\sin\theta + \cos\theta = \sin\theta + \cos\theta$$

19.
$$(\sin A + \cos A)^2 \overset{?}{=} \frac{2 + \sec A \csc A}{\sec A \csc A}$$
$$(\sin A + \cos A)^2 \overset{?}{=} \frac{2}{\sec A \csc A} + \frac{\sec A \csc A}{\sec A \csc A}$$
$$(\sin A + \cos A)^2 \overset{?}{=} 2\frac{1}{\sec A} \cdot \frac{1}{\csc A} + 1$$
$$(\sin A + \cos A)^2 \overset{?}{=} 2\cos A \sin A + 1$$
$$(\sin A + \cos A)^2 \overset{?}{=} 2\cos A \sin A + \sin^2 A + \cos^2 A$$
$$(\sin A + \cos A)^2 = (\sin A + \cos A)^2$$

20.
$$(\sin\theta - 1)(\tan\theta + \sec\theta) \overset{?}{=} -\cos\theta$$
$$\sin\theta \tan\theta - \tan\theta + \sin\theta \sec\theta - \sec\theta \overset{?}{=} -\cos\theta$$
$$\sin\theta \frac{\sin\theta}{\cos\theta} - \frac{\sin\theta}{\cos\theta} + \sin\theta \frac{1}{\cos\theta} - \frac{1}{\cos\theta} \overset{?}{=} -\cos\theta$$
$$\frac{\sin^2\theta - \sin\theta + \sin\theta - 1}{\cos\theta} \overset{?}{=} -\cos\theta$$
$$\frac{\sin^2\theta - 1}{\cos\theta} \overset{?}{=} -\cos\theta$$
$$\frac{-\cos^2\theta}{\cos\theta} \overset{?}{=} -\cos\theta$$
$$-\cos\theta = -\cos\theta$$

21.
$$\frac{\cos y}{1 - \sin y} \overset{?}{=} \frac{1 + \sin y}{\cos y}$$
$$\frac{\cos y}{1 - \sin y} \cdot \frac{1 + \sin y}{1 + \sin y} \overset{?}{=} \frac{1 + \sin y}{\cos y}$$
$$\frac{\cos y(1 + \sin y)}{1 - \sin^2 y} \overset{?}{=} \frac{1 + \sin y}{\cos y}$$
$$\frac{\cos y(1 + \sin y)}{\cos^2 y} \overset{?}{=} \frac{1 + \sin y}{\cos y}$$
$$\frac{1 + \sin y}{\cos y} = \frac{1 + \sin y}{\cos y}$$

22.
$$\cos\theta \cos(-\theta) - \sin\theta \sin(-\theta) \overset{?}{=} 1$$
$$\cos\theta \cos\theta - \sin\theta(-\sin\theta) \overset{?}{=} 1$$
$$\cos^2\theta + \sin^2\theta \overset{?}{=} 1$$
$$1 = 1$$

23.
$$\csc x - 1 \overset{?}{=} \frac{\cot^2 x}{\csc x + 1}$$
$$\csc x - 1 \overset{?}{=} \frac{\csc^2 x - 1}{\csc x + 1}$$
$$\csc x - 1 \overset{?}{=} \frac{(\csc x + 1)(\csc x - 1)}{\csc x + 1}$$
$$\csc x - 1 = \csc x - 1$$

24.
$$\cos B \cot B \overset{?}{=} \csc B - \sin B$$
$$\cos B \cot B \overset{?}{=} \frac{1}{\sin B} - \sin B$$
$$\cos B \cot B \overset{?}{=} \frac{1}{\sin B} - \frac{\sin^2 B}{\sin B}$$
$$\cos B \cot B \overset{?}{=} \frac{1 - \sin^2 B}{\sin B}$$
$$\cos B \cot B \overset{?}{=} \frac{\cos^2 B}{\sin B}$$
$$\cos B \cot B \overset{?}{=} \cos B \frac{\cos B}{\sin B}$$
$$\cos B \cot B = \cos B \cot B$$

25.
$$\sin\theta \cos\theta \tan\theta + \cos^2\theta \overset{?}{=} 1$$
$$\sin\theta \cos\theta \frac{\sin\theta}{\cos\theta} + \cos^2\theta \overset{?}{=} 1$$
$$\sin^2\theta + \cos^2\theta \overset{?}{=} 1$$
$$1 = 1$$

26.

$$(\csc x - \cot x)^2 \overset{?}{=} \frac{1 - \cos x}{1 + \cos x}$$

$$\csc^2 x - 2\csc x \cot x + \cot^2 x \overset{?}{=} \frac{1 - \cos x}{1 + \cos x}$$

$$\frac{1}{\sin^2 x} - 2\frac{1}{\sin x} \cdot \frac{\cos x}{\sin x} + \frac{\cos^2 x}{\sin^2 x} \overset{?}{=} \frac{1 - \cos x}{1 + \cos x}$$

$$\frac{1 - 2\cos x + \cos^2 x}{\sin^2 x} \overset{?}{=} \frac{1 - \cos x}{1 + \cos x}$$

$$\frac{(1 - \cos x)^2}{1 - \cos^2 x} \overset{?}{=} \frac{1 - \cos x}{1 + \cos x}$$

$$\frac{(1 - \cos x)^2}{(1 - \cos x)(1 + \cos x)} \overset{?}{=} \frac{1 - \cos x}{1 + \cos x}$$

$$\frac{1 - \cos x}{1 + \cos x} = \frac{1 - \cos x}{1 + \cos x}$$

27. $\sin x + \cos x \overset{?}{=} \dfrac{\cos x}{1 - \tan x} + \dfrac{\sin x}{1 - \cot x}$

$$\sin x + \cos x \overset{?}{=} \frac{\cos x}{1 - \frac{\sin x}{\cos x}} + \frac{\sin x}{1 - \frac{\cos x}{\sin x}}$$

$$\sin x + \cos x \overset{?}{=} \frac{\cos x}{1 - \frac{\sin x}{\cos x}} \cdot \frac{\cos x}{\cos x} + \frac{\sin x}{1 - \frac{\cos x}{\sin x}} \cdot \frac{\sin x}{\sin x}$$

$$\sin x + \cos x \overset{?}{=} \frac{\cos^2 x}{\cos x - \sin x} + \frac{\sin^2 x}{\sin x - \cos x}$$

$$\sin x + \cos x \overset{?}{=} -\frac{\cos^2 x}{\sin x - \cos x} + \frac{\sin^2 x}{\sin x - \cos x}$$

$$\sin x + \cos x \overset{?}{=} \frac{\sin^2 x - \cos^2 x}{\sin x - \cos x}$$

$$\sin x + \cos x \overset{?}{=} \frac{(\sin x + \cos x)(\sin x - \cos x)}{\sin x - \cos x}$$

$$\sin x + \cos x = \sin x + \cos x$$

28.

$$\sin \theta + \cos \theta + \tan \theta \sin \theta \overset{?}{=} \sec \theta + \cos \theta \tan \theta$$

$$\sin \theta + \cos \theta + \frac{\sin \theta}{\cos \theta} \sin \theta \overset{?}{=} \sec \theta + \cos \theta \tan \theta$$

$$\sin \theta + \cos \theta + \frac{\sin^2 \theta}{\cos \theta} \overset{?}{=} \sec \theta + \cos \theta \tan \theta$$

$$\sin \theta + \frac{\cos^2 \theta}{\cos \theta} + \frac{\sin^2 \theta}{\cos \theta} \overset{?}{=} \sec \theta + \cos \theta \tan \theta$$

$$\sin \theta + \frac{\cos^2 \theta + \sin^2 \theta}{\cos \theta} \overset{?}{=} \sec \theta + \cos \theta \tan \theta$$

$$\sin \theta + \frac{1}{\cos \theta} \overset{?}{=} \sec \theta + \cos \theta \tan \theta$$

$$\sin \theta + \sec \theta \overset{?}{=} \sec \theta + \cos \theta \tan \theta$$

$$\sin \theta \frac{\cos \theta}{\cos \theta} + \sec \theta \overset{?}{=} \sec \theta + \cos \theta \tan \theta$$

$$\cos \theta \frac{\sin \theta}{\cos \theta} + \sec \theta \overset{?}{=} \sec \theta + \cos \theta \tan \theta$$

$$\cos \theta \tan \theta + \sec \theta \overset{?}{=} \sec \theta + \cos \theta \tan \theta$$

$$\sec \theta + \cos \theta \tan \theta = \sec \theta + \cos \theta \tan \theta$$

Pages 441–445 Lesson 7-3

10.

$$\sin (90° + A) \overset{?}{=} \cos A$$

$$\sin (90°) \cos A + \cos (90°) \sin A \overset{?}{=} \cos A$$

$$1 \cdot \cos A + 0 \cdot \sin A \overset{?}{=} \cos A$$

$$\cos A = \cos A$$

11.

$$\tan \left(\theta + \frac{\pi}{2}\right) \overset{?}{=} -\cot \theta$$

$$\frac{\sin \left(\theta + \frac{\pi}{2}\right)}{\cos \left(\theta + \frac{\pi}{2}\right)} \overset{?}{=} -\cot \theta$$

$$\frac{\sin \theta \cos \frac{\pi}{2} + \cos \theta \sin \frac{\pi}{2}}{\cos \theta \cos \frac{\pi}{2} - \sin \theta \sin \frac{\pi}{2}} \overset{?}{=} -\cot \theta$$

$$\frac{(\sin \theta) \cdot 0 + (\cos \theta) \cdot 1}{(\cos \theta) \cdot 0 - (\sin \theta) \cdot 1} \overset{?}{=} -\cot \theta$$

$$-\frac{\cos \theta}{\sin \theta} \overset{?}{=} -\cot \theta$$

$$-\cot \theta = -\cot \theta$$

12. $\sin (x - y) \overset{?}{=} \dfrac{1 - \cot x \tan y}{\csc x \sec y}$

$$\sin (x - y) \overset{?}{=} \frac{1 - \frac{\cos x}{\sin x} \cdot \frac{\sin y}{\cos y}}{\frac{1}{\sin x} \cdot \frac{1}{\cos y}}$$

$$\sin (x - y) \overset{?}{=} \frac{1 - \frac{\cos x}{\sin x} \cdot \frac{\sin y}{\cos y}}{\frac{1}{\sin x} \cdot \frac{1}{\cos y}} \cdot \frac{\sin x \cos y}{\sin x \cos y}$$

$$\sin (x - y) \overset{?}{=} \frac{\sin x \cos y - \cos x \sin y}{1}$$

$$\sin (x - y) = \sin (x - y)$$

34.

$$\cos \left(\frac{\pi}{2} + x\right) \overset{?}{=} -\sin x$$

$$\cos \frac{\pi}{2} \cos x - \sin \frac{\pi}{2} \sin x \overset{?}{=} -\sin x$$

$$0 \cdot \cos x - 1 \cdot \sin x \overset{?}{=} -\sin x$$

$$-\sin x = -\sin x$$

35.

$$\cos (60° + A) \overset{?}{=} \sin (30° - A)$$

$$\cos 60° \cos A - \sin 60° \sin A \overset{?}{=} \sin 30° \cos A - \cos 30° \sin A$$

$$\frac{1}{2} \cos A - \frac{\sqrt{3}}{2} \sin A = \frac{1}{2} \cos A - \frac{\sqrt{3}}{2} \sin A$$

36.

$$\sin (A + \pi) \overset{?}{=} -\sin A$$

$$\sin A \cos \pi + \cos A \sin \pi \overset{?}{=} -\sin A$$

$$(\sin A)(-1) + (\cos A)(0) \overset{?}{=} -\sin A$$

$$-\sin A = -\sin A$$

37.

$$\cos (180° + x) \overset{?}{=} -\cos x$$

$$\cos 180° \cos x - \sin 180° \sin x \overset{?}{=} -\cos x$$

$$-1 \cdot \cos x - 0 \cdot \sin x \overset{?}{=} -\cos x$$

$$-\cos x = -\cos x$$

38.

$$\tan (x + 45°) \overset{?}{=} \frac{1 + \tan x}{1 - \tan x}$$

$$\frac{\tan x + \tan 45°}{1 - \tan x \tan 45°} \overset{?}{=} \frac{1 + \tan x}{1 - \tan x}$$

$$\frac{\tan x + 1}{1 - (\tan x)(1)} \overset{?}{=} \frac{1 + \tan x}{1 - \tan x}$$

$$\frac{1 + \tan x}{1 - \tan x} = \frac{1 + \tan x}{1 - \tan x}$$

39. $\sin(A+B) \overset{?}{=} \dfrac{\tan A + \tan B}{\sec A \sec B}$

$\sin(A+B) \overset{?}{=} \dfrac{\dfrac{\sin A}{\cos A} + \dfrac{\sin B}{\cos B}}{\dfrac{1}{\cos A} \cdot \dfrac{1}{\cos B}}$

$\sin(A+B) \overset{?}{=} \dfrac{\dfrac{\sin A}{\cos A} + \dfrac{\sin B}{\cos B}}{\dfrac{1}{\cos A} \cdot \dfrac{1}{\cos B}} \cdot \dfrac{\cos A \cos B}{\cos A \cos B}$

$\sin(A+B) \overset{?}{=} \dfrac{\sin A \cos B + \cos A \sin B}{1}$

$\sin(A+B) = \sin(A+B)$

40. $\cos(A+B) \overset{?}{=} \dfrac{1 - \tan A \tan B}{\sec A \sec B}$

$\cos(A+B) \overset{?}{=} \dfrac{1 - \dfrac{\sin A}{\cos A} \cdot \dfrac{\sin B}{\cos B}}{\dfrac{1}{\cos A} \cdot \dfrac{1}{\cos B}}$

$\cos(A+B) \overset{?}{=} \dfrac{1 - \dfrac{\sin A}{\cos A} \cdot \dfrac{\sin B}{\cos B}}{\dfrac{1}{\cos A} \cdot \dfrac{1}{\cos B}} \cdot \dfrac{\cos A \cos B}{\cos A \cos B}$

$\cos(A+B) \overset{?}{=} \dfrac{\cos A \cos B - \sin A \sin B}{1}$

$\cos(A+B) = \cos(A+B)$

41. $\sec(A-B) \overset{?}{=} \dfrac{\sec A \sec B}{1 + \tan A \tan B}$

$\sec(A-B) \overset{?}{=} \dfrac{\dfrac{1}{\cos A} \cdot \dfrac{1}{\cos B}}{1 + \dfrac{\sin A}{\cos A} \cdot \dfrac{\sin B}{\cos B}}$

$\sec(A-B) \overset{?}{=} \dfrac{\dfrac{1}{\cos A} \cdot \dfrac{1}{\cos B}}{1 + \dfrac{\sin A}{\cos A} \cdot \dfrac{\sin B}{\cos B}} \cdot \dfrac{\cos A \cos B}{\cos A \cos B}$

$\sec(A-B) \overset{?}{=} \dfrac{1}{\cos A \cos B + \sin A \sin B}$

$\sec(A-B) \overset{?}{=} \dfrac{1}{\cos(A-B)}$

$\sec(A-B) = \sec(A-B)$

42.
$$\sin(x+y)\sin(x-y) \overset{?}{=} \sin^2 x - \sin^2 y$$

$(\sin x \cos y + \cos x \sin y)(\sin x \cos y - \cos x \sin y)$
$$\overset{?}{=} \sin^2 x - \sin^2 y$$

$(\sin x \cos y)^2 - (\cos x \sin y)^2 \overset{?}{=} \sin^2 x - \sin^2 y$

$\sin^2 y \cos^2 y - \cos^2 x \sin^2 y \overset{?}{=} \sin^2 x - \sin^2 y$

$\sin^2 x \cos^2 y + \sin^2 x \sin^2 y - \sin^2 x \sin^2 y - \cos^2 x \sin^2 y$
$$\overset{?}{=} \sin^2 x - \sin^2 y$$

$\sin^2 x (\cos^2 y + \sin^2 y) - \sin^2 y (\sin^2 x + \cos^2 x)$
$$\overset{?}{=} \sin^2 x - \sin^2 y$$

$(\sin^2 x)(1) - (\sin^2 y)(1) \overset{?}{=} \sin^2 x - \sin^2 y$

$\sin^2 x - \sin^2 y = \sin^2 x - \sin^2 y$

48b. $\tan A + \tan B + \tan C$
$$\overset{?}{=} \tan A \tan B \tan C$$

$\tan A + \tan B + \tan(180° - (A+B))$
$$\overset{?}{=} \tan A \tan B \tan(180° - (A+B))$$

$\tan A + \tan B + \dfrac{\tan 180° - \tan(A+B)}{1 + \tan 180° \tan(A+B)}$
$$\overset{?}{=} \tan A \tan B \, \dfrac{\tan 180° - \tan(A+B)}{1 + \tan 180° \tan(A+B)}$$

$\tan A + \tan B + \dfrac{0 - \tan(A+B)}{1 + 0 \cdot \tan(A+B)}$
$$\overset{?}{=} \tan A \tan B + \dfrac{0 - \tan(A+B)}{1 + 0 \cdot \tan(A+B)}$$

$\tan A + \tan B - \tan(A+B)$
$$\overset{?}{=} -\tan A \tan B \tan(A+B)$$

$(\tan A \tan B) \cdot \dfrac{1 - \tan A \tan B}{1 - \tan A \tan B} - \tan(A+B)$
$$\overset{?}{=} -\tan A \tan B \tan(A+B)$$

$\tan(A+B)(1 - \tan A \tan B) - \tan(A+B)$
$$\overset{?}{=} -\tan A \tan B \tan(A+B)$$

$(1 - \tan A \tan B - 1)\tan(A+B)$
$$\overset{?}{=} -\tan A \tan B \tan(A+B)$$

$-\tan A \tan B \tan(A+B)$
$$= -\tan A \tan B \tan(A+B)$$

Page 445   Lesson 7-3
Mid-Chapter Quiz

6. $\cot x \sec x \sin x \overset{?}{=} 2 - \tan x \cos x \csc x$

$\dfrac{\cos x}{\sin x} \cdot \dfrac{1}{\cos x} \cdot \sin x \overset{?}{=} 2 - \dfrac{\sin x}{\cos x} \cdot \cos x \cdot \dfrac{1}{\sin x}$

$1 \overset{?}{=} 2 - 1$

$1 = 1$

7. $\tan(\alpha - \beta) \overset{?}{=} \dfrac{1 - \cot \alpha \tan \beta}{\cot \alpha + \tan \beta}$

$\tan(\alpha - \beta) \overset{?}{=} \dfrac{1 - \dfrac{1}{\tan \alpha} \cdot \tan \beta}{\dfrac{1}{\tan \alpha} + \tan \beta}$

$\tan(\alpha - \beta) \overset{?}{=} \dfrac{1 - \dfrac{1}{\tan \alpha} \cdot \tan \beta}{\dfrac{1}{\tan \alpha} + \tan \beta} \cdot \dfrac{\tan \alpha}{\tan \alpha}$

$\tan(\alpha - \beta) \overset{?}{=} \dfrac{\tan \alpha - \tan \beta}{1 + \tan \alpha \tan \beta}$

$\tan(\alpha - \beta) = \tan(\alpha - \beta)$

Pages 453–455   Lesson 7-4

28. $\csc 2\theta \overset{?}{=} \dfrac{1}{2}\sec\theta \csc\theta$

$\dfrac{1}{\sin 2\theta} \overset{?}{=} \dfrac{1}{2}\sec\theta \csc\theta$

$\dfrac{1}{2\sin\theta \cos\theta} \overset{?}{=} \dfrac{1}{2}\sec\theta \csc\theta$

$\dfrac{1}{2} \cdot \dfrac{1}{\sin\theta} \cdot \dfrac{1}{\cos\theta} \overset{?}{=} \dfrac{1}{2}\sec\theta \csc\theta$

$\dfrac{1}{2}\csc\theta \sec\theta \overset{?}{=} \dfrac{1}{2}\sec\theta \csc\theta$

$\dfrac{1}{2}\sec\theta \csc\theta = \dfrac{1}{2}\sec\theta \csc\theta$

29. $\cos A - \sin A \overset{?}{=} \dfrac{\cos 2A}{\cos A + \sin A}$

$\cos A - \sin A \overset{?}{=} \dfrac{\cos^2 A - \sin^2 A}{\cos A + \sin A}$

$\cos A - \sin A \overset{?}{=} \dfrac{(\cos A - \sin A)(\cos A + \sin A)}{\cos A + \sin A}$

$\cos A - \sin A = \cos A - \sin A$

30. $(\sin \theta + \cos \theta)^2 - 1 \overset{?}{=} \sin 2\theta$

$\sin^2 \theta + 2 \sin \theta \cos \theta + \cos^2 \theta - 1 \overset{?}{=} \sin 2\theta$

$2 \sin \theta \cos \theta + 1 - 1 \overset{?}{=} \sin 2\theta$

$2 \sin \theta \cos \theta \overset{?}{=} \sin 2\theta$

$\sin 2\theta = \sin 2\theta$

31. $\cos x - 1 \overset{?}{=} \dfrac{\cos 2x - 1}{2(\cos x + 1)}$

$\cos x - 1 \overset{?}{=} \dfrac{2\cos^2 x - 1 - 1}{2(\cos x + 1)}$

$\cos x - 1 \overset{?}{=} \dfrac{2\cos^2 x - 2}{2(\cos x + 1)}$

$\cos x - 1 \overset{?}{=} \dfrac{2(\cos^2 x - 1)}{2(\cos x + 1)}$

$\cos x - 1 \overset{?}{=} \dfrac{2(\cos x - 1)(\cos x + 1)}{2(\cos x + 1)}$

$\cos x - 1 = \cos x - 1$

32. $\sec 2\theta \overset{?}{=} \dfrac{\cos^2 \theta + \sin^2 \theta}{\cos^2 \theta - \sin^2 \theta}$

$\sec 2\theta \overset{?}{=} \dfrac{1}{\cos 2\theta}$

$\sec 2\theta = \sec 2\theta$

33. $\tan \dfrac{A}{2} \overset{?}{=} \dfrac{\sin A}{1 + \cos A}$

$\tan \dfrac{A}{2} \overset{?}{=} \dfrac{\sin 2\left(\frac{A}{2}\right)}{1 + \cos 2\left(\frac{A}{2}\right)}$

$\tan \dfrac{A}{2} \overset{?}{=} \dfrac{2 \sin \frac{A}{2} \cos \frac{A}{2}}{1 + 2\cos^2 \frac{A}{2} - 1}$

$\tan \dfrac{A}{2} \overset{?}{=} \dfrac{2 \sin \frac{A}{2} \cos \frac{A}{2}}{2\cos^2 \frac{A}{2}}$

$\tan \dfrac{A}{2} \overset{?}{=} \dfrac{\sin \frac{A}{2}}{\cos \frac{A}{2}}$

$\tan \dfrac{A}{2} = \tan \dfrac{A}{2}$

34. $\sin 3x \overset{?}{=} 3 \sin x - 4 \sin^3 x$

$\sin (2x + x) \overset{?}{=} 3 \sin x - 4 \sin^3 x$

$\sin 2x \cos x + \cos 2x \sin x \overset{?}{=} 3 \sin x - 4 \sin^3 x$

$2 \sin x \cos^2 x + (1 - 2\sin^2 x) \sin x$
$\overset{?}{=} 3 \sin x - 4 \sin^3 x$

$2 \sin x(1 - \sin^2 x) + (1 - 2\sin^2 x) \sin x$
$\overset{?}{=} 3 \sin x - 4 \sin^3 x$

$2 \sin x - 2 \sin^3 x + \sin x - 2 \sin^3 x$
$\overset{?}{=} 3 \sin x - 4 \sin^3 x$

$3 \sin x - 4 \sin^3 x = 3 \sin x - 4 \sin^3 x$

35. $\cos 3x \overset{?}{=} 4 \cos^3 x - 3 \cos x$

$\cos (2x + x) \overset{?}{=} 4 \cos^3 x - 3 \cos x$

$\cos 2x \cos x - \sin 2x \sin x$
$\overset{?}{=} 4 \cos^3 x - 3 \cos x$

$(2 \cos^2 x - 1) \cos x - 2 \sin^2 x \cos x$
$\overset{?}{=} 4 \cos^3 x - 3 \cos x$

$(2 \cos^2 x - 1) \cos x - 2(1 - \cos^2 x) \cos x$
$\overset{?}{=} 4 \cos^3 x - 3 \cos x$

$2 \cos^3 x - \cos x - 2 \cos x + 2 \cos^3 x$
$\overset{?}{=} 4 \cos^3 x - 3 \cos x$

$4 \cos^3 x - 3 \cos x = 4 \cos^3 x - 3 \cos x$

40. $\tan (\alpha + 30°) = \dfrac{21}{7}$

$\tan (\alpha + 30°) = 3$

$\dfrac{\tan \alpha + \tan 30°}{1 - \tan \alpha \tan 30°} = 3$

$\tan \alpha + \dfrac{\sqrt{3}}{3} = 3 - \sqrt{3} \tan \alpha$

$\tan \alpha + \sqrt{3} \tan \alpha = 3 - \dfrac{\sqrt{3}}{3}$

$\left(1 + \sqrt{3}\right) \tan \alpha = 3 - \dfrac{\sqrt{3}}{3}$

$\tan \alpha = \dfrac{3 - \frac{\sqrt{3}}{3}}{1 + \sqrt{3}}$

$\tan \alpha = \dfrac{9 - \sqrt{3}}{3 + 3\sqrt{3}}$

$\tan \alpha = \dfrac{-6 + 5\sqrt{3}}{3}$

42. Sample answer:

$\sin\left(\sqrt{\pi}\right)^2 + \cos\left(\sqrt{\pi}\right)^2 = \sin \pi + \cos \pi$
$= 0 + (-1)$
$= -1$
$\neq 1$

Pages 467–469   Lesson 7-6

35d. The line with normal form $x \cos \phi + y \sin \phi - p = 0$ makes an angle of ϕ with the positive x-axis and has a normal length of p. The graph of Armando's equation is a line whose normal makes an angle of $\phi + \delta$ with the x-axis and also has length p. Therefore, the graph of Armando's equation is the graph of the original line rotated $\delta°$ counterclockwise about the origin. Armando is correct. See students' graphs.

CHAPTER 8   VECTORS AND PARAMETRIC EQUATIONS

Pages 502–504   Lesson 8-3

35. $|G_1 G_2| = \sqrt{(x_2 - x_1)^2 + (y_2 - y_1)^2 + (z_2 - z_1)^2}$
$= \sqrt{(x_1 - x_2)^2 + (y_1 - y_2)^2 + (z_2 - z_1)^2}$
$= |G_2 G_1|$

because $(x - y)^2 = (y - x)^2$ for all real numbers x and y.

27. Sample answer:

Let $\vec{a} = \langle a_1, a_2, a_3 \rangle$ and $-\vec{a} = \langle -a_1, -a_2, -a_3 \rangle$.

$$\vec{a} \times (-\vec{a}) = \begin{vmatrix} \vec{i} & \vec{j} & \vec{k} \\ a_1 & a_2 & a_3 \\ -a_1 & -a_2 & -a_3 \end{vmatrix}$$

$$= \begin{vmatrix} a_2 & a_3 \\ -a_2 & -a_3 \end{vmatrix}\vec{i} - \begin{vmatrix} a_1 & a_3 \\ -a_1 & -a_3 \end{vmatrix}\vec{j} + \begin{vmatrix} a_1 & a_2 \\ -a_1 & -a_2 \end{vmatrix}\vec{k}$$

$$= 0\,\vec{i} - 0\,\vec{j} + 0\,\vec{k} = \vec{0}$$

28. $\vec{a} \times (\vec{b} + \vec{c})$

$$= \begin{vmatrix} \vec{i} & \vec{j} & \vec{k} \\ a_1 & a_2 & a_3 \\ (b_1 + c_1) & (b_2 + c_2) & (b_3 + c_3) \end{vmatrix}$$

$$= \begin{vmatrix} a_2 & a_3 \\ (b_2 + c_2) & (b_3 + c_3) \end{vmatrix}\vec{i} - \begin{vmatrix} a_1 & a_3 \\ (b_1 + c_1) & (b_3 + c_3) \end{vmatrix}\vec{j} + \begin{vmatrix} a_1 & a_2 \\ (b_1 + c_1) & (b_2 + c_2) \end{vmatrix}\vec{k}$$

$$= [a_2(b_3 + c_3) - a_3(b_2 + c_2)]\vec{i} - [a_1(b_3 + c_3) - a_3(b_1 + c_1)]\vec{j} + [a_1(b_2 + c_2) - a_2(b_1 + c_1)]\vec{k}$$

$$= [(a_2 b_3 + a_2 c_3) - (a_3 b_2 + a_3 c_2)]\vec{i} - [(a_1 b_3 + a_1 c_3) - (a_3 b_1 + a_3 c_1)]\vec{j} + [(a_1 b_2 + a_1 c_2) - (a_2 b_1 + a_2 c_1)]\vec{k}$$

$$= [(a_2 b_3 - a_3 b_2) + (a_2 c_3 - a_3 c_2)]\vec{i} - [(a_1 b_3 - a_3 b_1) + (a_1 c_3 - a_3 c_1)]\vec{j} + [(a_1 b_2 - a_2 b_1) + (a_1 c_2 - a_2 c_1)]\vec{k}$$

$$= (a_2 b_3 - a_3 b_2)\vec{i} + (a_2 c_3 - a_3 c_2)\vec{i} - (a_1 b_3 - a_3 b_1)\vec{j} + (a_1 c_3 - a_3 c_1)\vec{j} + (a_1 b_2 - a_2 b_1)\vec{k} + (a_1 c_2 - a_2 c_1)\vec{k}$$

$$= [(a_2 b_3 - a_3 b_2)\vec{i} - (a_1 b_3 - a_3 b_1)\vec{j} + (a_1 b_2 - a_2 b_1)\vec{k}] + [(a_2 c_3 - a_3 c_2)\vec{i} - (a_1 c_3 - a_3 c_1)\vec{j} + (a_1 c_2 - a_2 c_1)\vec{k}]$$

$$= \left[\begin{vmatrix} a_2 & a_3 \\ b_2 & b_3 \end{vmatrix}\vec{i} - \begin{vmatrix} a_1 & a_3 \\ b_1 & b_3 \end{vmatrix}\vec{j} + \begin{vmatrix} a_1 & a_2 \\ b_1 & b_2 \end{vmatrix}\vec{k} \right] + \left[\begin{vmatrix} a_2 & a_3 \\ c_2 & c_3 \end{vmatrix}\vec{i} - \begin{vmatrix} a_1 & a_3 \\ c_1 & c_3 \end{vmatrix}\vec{j} + \begin{vmatrix} a_1 & a_2 \\ c_1 & c_2 \end{vmatrix}\vec{k} \right]$$

$$= (\vec{a} \times \vec{b}) + (\vec{a} \times \vec{c})$$

40. $|\overrightarrow{BA}|^2 = |\vec{a}|^2 + |\vec{b}|^2 -$

$\qquad 2|\vec{a}||\vec{b}|\cos\theta \quad \left(\sqrt{(a_1 - b_1)^2 + (a_2 + b_2)^2} \right)^2$

$= \left(\sqrt{a_1{}^2 + a_2{}^2} \right)^2 + \left(\sqrt{b_1{}^2 + b_2{}^2} \right)^2 -$

$\qquad 2\left(\sqrt{a_1{}^2 + a_2{}^2} \right)\left(\sqrt{b_1{}^2 + b_2{}^2} \right)\cos\theta \quad (a_1 - b_1)^2 +$

$\qquad (a_2 - b_2)^2$

$= a_1{}^2 + a_2{}^2 + b_1{}^2 + b_2{}^2 -$

$\qquad 2\left(\sqrt{a_1{}^2 + a_2{}^2} \right)\left(\sqrt{b_1{}^2 + b_2{}^2} \right)\cos\theta \quad a_1{}^2 -$

$\qquad 2a_1 b_1 + b_1{}^2 + a_2{}^2 - 2a_2 b_2 + b_2{}^2$

$= a_1{}^2 + a_2{}^2 + b_1{}^2 + b_2{}^2 -$

$\qquad 2\left(\sqrt{a_1{}^2 + a_2{}^2} \right)\left(\sqrt{b_1{}^2 + b_2{}^2} \right)\cos\theta - 2a_1 b_1 -$

$\qquad 2a_2 b_2$

$= -2\left(\sqrt{a_1{}^2 + a_2{}^2} \right)\left(\sqrt{b_1{}^2 + b_2{}^2} \right)\cos\theta \quad a_1 b_1 +$

$\qquad a_2 b_2$

$= \sqrt{a_1{}^2 + a_2{}^2}\ \sqrt{b_1{}^2 + b_2{}^2}\ \cos\theta \quad a_1 b_1 + a_2 b_2$

$= |\vec{a}||\vec{b}|\cos\theta \quad \vec{a} \cdot \vec{b}$

$= |\vec{a}||\vec{b}|\cos\theta$

32.

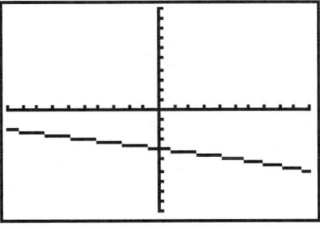

T	X₁ᴛ	Y₁ᴛ
1.00000	5.0000	-5.000
2.0000	10.000	-6.000
3.0000	15.000	-7.000
4.0000	20.000	-8.000
5.0000	25.000	-9.000
6.0000	30.000	-10.00
14.000	70.000	-18.00

T=1

[−5, 5] Tstep:1 [−10, 10]
Xscl:1 [−10, 10] Yscl:1

33.

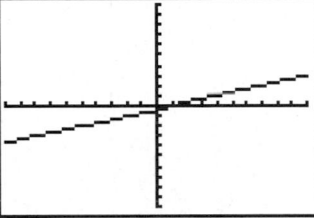

T	X₁ᴛ	Y₁ᴛ
1.0000	8.0000	**2.00000**
2.0000	11.000	3.0000
3.0000	14.000	4.0000
4.0000	17.000	5.0000
5.0000	20.000	6.0000
6.0000	23.000	7.0000
14.000	47.000	15.000

Y₁ᴛ=2

[−10, 10] Tstep:1 [−20, 20]
Xscl:2 [−20, 20] Yscl:2

34.

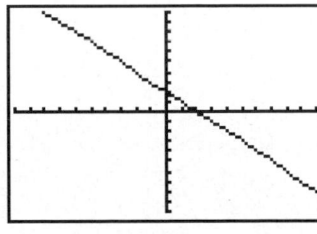

T	X₁ᴛ	Y₁ᴛ
1.00000	2.0000	0.0000
2.0000	3.0000	-1.000
3.0000	4.0000	-2.000
4.0000	5.0000	-3.000
5.0000	6.0000	-4.000
6.0000	7.0000	-5.000
14.000	15.000	-13.00

T=1

[−10, 10] Tstep:1 [−10, 10]
Xscl:1 [−10, 10] Yscl:1

Pages 539–542   Lesson 8-8

4a–4c.

Transformation	Orientation	Size	Shape
reflection	yes	no	no
translation	no	no	no
dilation	no	yes	no

22b.

CHAPTER 9   POLAR COORDINATES AND COMPLEX NUMBERS

Pages 558–560   Lesson 9-1

16.

17.

18.

19.

20.

21.

22.

23.

24.

25.

26.

27.

32.

33.

34.

35.

36.

37.

38.

39.

40.

11.

12.

13.

14.

15.

16.

17.

18.

19.

20.

21.

22.

34a.

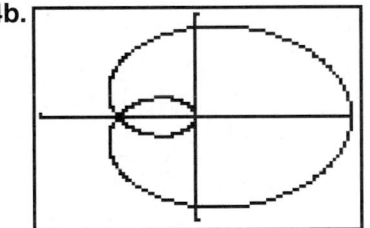

When n = 10, two more outer rings will appear.

34b.

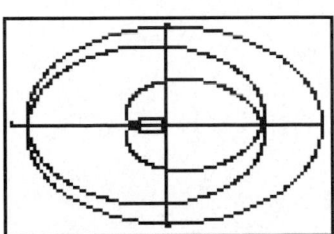

(continued on next page)

ANSWER APPENDIX

34b. *(continued)*

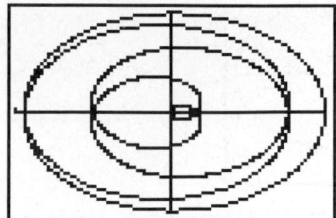

When n = 11, the innermost loop will be on the left and there will be an additional outer ring.

Pages 577–579   Lesson 9-4

24.

25.

26.

27.

28.

29.

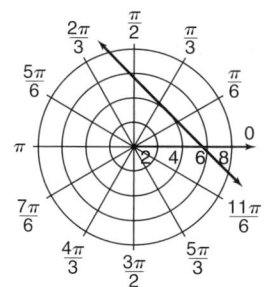

36. $k = r \sin(\theta - \alpha)$

$k = r[\sin\theta\cos\alpha - \cos\theta\sin\alpha]$

$k = r\sin\theta\cos\alpha - r\cos\theta\sin\alpha$

$k = y\cos\alpha - x\sin\alpha$

This is the equation of a line in rectangular coordinates. Solving the last equation for y yields $y = (\tan\alpha)x + \dfrac{k}{\cos\alpha}$.

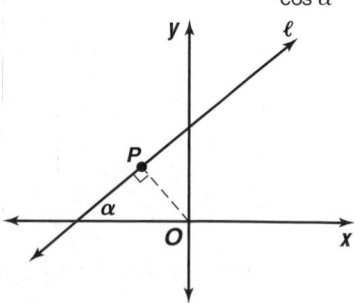

The slope of the line shows that α is the angle the line makes with the *x*-axis. To find the length of the normal segment in the figure, observe that the complementary angle to α in the right triangle is $90° - \alpha$, so the θ-coordinate of *P* in polar coordinates is $180° - (90° - \alpha) = \alpha + 90°$. Substitute into the original polar equation to find the *r*-coordinate of *P*:

$k = r\sin(\alpha + 90° - \alpha)$

$k = r\sin 90°$

$k = r$

Therefore, k is the length of the normal segment.

Page 579   Lesson 9-4
Mid-Chapter Quiz

1.

2.

3.

4.

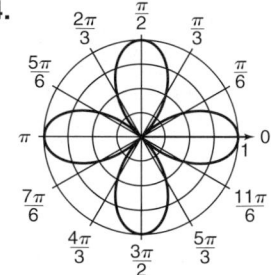

Pages 589–591   Lesson 9-6

19.

20.

21.

22.

23.

24.

43.

44.

45.

50a. Sample answer: let $z_1 = 1 + i$ and $z_2 = 3 + 4i$. Then $z_1z_2 = (1 + i)(3 + 4i) = -1 + 7i$

50b. $z_1 = \sqrt{2}\left(\cos \frac{\pi}{4} + i \sin \frac{\pi}{4}\right)$
$\approx 1.41(\cos 0.79 + i \sin 0.79)$

$z_2 = 5(\cos 0.93 + i \sin 0.93)$

$z_1z_2 = 5\sqrt{2}\left(\cos 1.71 + i \sin 1.71\right)$
$\approx 7.07\left(\cos 1.71 + i \sin 1.71\right)$

50c. Sample answer: let $z_1 = 2 - 4i$ and $z_2 = -1 + 3i$. Then $z_1 = 2\sqrt{5}(\cos 5.18 + i \sin 5.18)$
$\approx 4.47(\cos 5.18 + i \sin 5.18)$,

$z_2 = \sqrt{10}(\cos 1.89 + i \sin 1.89)$
$\approx 3.16(\cos 1.89 + i \sin 1.89)$,

and $z_1z_2 = (2 - 4i)(-1 + 3i)$
$= 10 + 10i$
$= 10\sqrt{2}\left(\cos \frac{\pi}{4} + i \sin \frac{\pi}{4}\right)$
$\approx 14.14(\cos 0.79 + i \sin 0.79)$.

50d. To multiply two complex numbers in a polar form, multiply the moduli and add the amplitudes. (In the sample answer for 50c, note that $5.18 + 1.89 = 7.07$, which is coterminal with 0.79.

Pages 604–606   Lesson 9–8

26.

27.

28.

29.

30.

31.

Pages 619–622 Lesson 10-1

28. Let $A(0, 0)$, $B(b, c)$, and $C(a, 0)$ be the vertices of a triangle. Let D be the midpoint of $\overline{AB}$ and E be the midpoint of $\overline{BC}$.

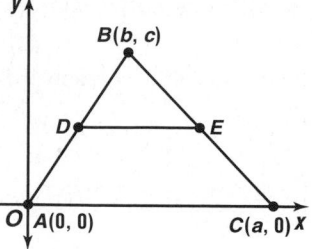

The coordinates of D are $\left(\dfrac{0 + b}{2}, \dfrac{0 + c}{2}\right)$ or $\left(\dfrac{b}{2}, \dfrac{c}{2}\right)$.

The coordinates of E are $\left(\dfrac{b + a}{2}, \dfrac{c + 0}{2}\right)$ or $\left(\dfrac{b + a}{2}, \dfrac{c}{2}\right)$.

$AC = \sqrt{(a - 0)^2 + (0 - 0)^2} = \sqrt{a^2}$ or a

$DE = \sqrt{\left(\dfrac{b + a}{2} - \dfrac{b}{2}\right)^2 + \left(\dfrac{c}{2} - \dfrac{c}{2}\right)^2} = \sqrt{\dfrac{a^2}{4}}$ or $\dfrac{a}{2}$

Since $DE = \dfrac{1}{2}AC$,

the line segment joining the midpoints of two sides of a triangle is equal in length to one-half the third side.

29. In trapezoid $ABCD$, let A and B have coordinates $(0, 0)$ and $(b, 0)$, respectively. To make the trapezoid isosceles, let C have coordinates $(b - a, c)$ and let D have coordinates (a, c).

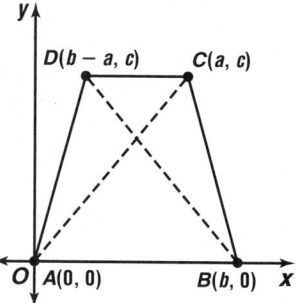

$AC = \sqrt{(a - 0)^2 + (c - 0)^2} = \sqrt{a^2 + c^2}$

$BD = \sqrt{(b - a - b)^2 + (c - 0)^2} = \sqrt{a^2 + c^2}$

$AC = \sqrt{a^2 + c^2} = \sqrt{a^2 + c^2} = BD$, so the diagonals of an isosceles trapezoid are congruent.

30. In $\triangle ABC$, let the vertices be $A(0, 0)$ and $B(a, 0)$. Since $\overline{AC}$ and $\overline{BC}$ are congruent sides, let the third vertex be $C\left(\dfrac{a}{2}, b\right)$. Let D be the midpoint of $\overline{AC}$ and let E be the midpoint of $\overline{BC}$.

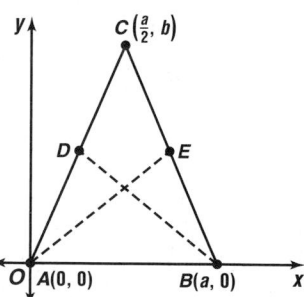

The coordinates of D are $\left(\dfrac{\frac{a}{2} + 0}{2}, \dfrac{b + 0}{2}\right)$ or $\left(\dfrac{a}{4}, \dfrac{b}{2}\right)$.

The coordinates of E are $\left(\dfrac{\frac{a}{2} + a}{2}, \dfrac{b + 0}{2}\right)$ or $\left(\dfrac{3a}{4}, \dfrac{b}{2}\right)$.

$AE = \sqrt{\left(\dfrac{3a}{4} - 0\right)^2 + \left(\dfrac{b}{2} - 0\right)^2} = \dfrac{1}{2}\sqrt{\dfrac{9a^2}{4} + b^2}$

$BD = \sqrt{\left(a - \dfrac{a}{4}\right)^2 + \left(0 - \dfrac{b}{2}\right)^2} = \dfrac{1}{2}\sqrt{\dfrac{9a^2}{4} + b^2}$

Since $AE = BD$, the medians to the congruent sides of an isosceles triangle are congruent.

31. Let A and B have coordinates $(0, 0)$ and $(b, 0)$, respectively. To make a parallelogram, let C have coordinates $(b + a, c)$ and let D have coordinates (a, c).

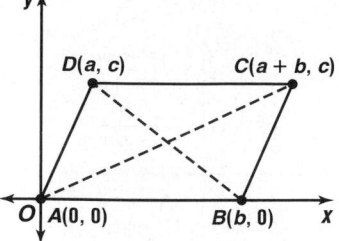

The midpoint of $\overline{BD}$ is $\left(\dfrac{a + b}{2}, \dfrac{0 + c}{2}\right)$ or $\left(\dfrac{a + b}{2}, \dfrac{c}{2}\right)$.

The midpoint of $\overline{AC}$ is $\left(\dfrac{a + b + 0}{2}, \dfrac{c + 0}{2}\right)$ or $\left(\dfrac{a + b}{2}, \dfrac{c}{2}\right)$.

Since the diagonals have the same midpoint, the diagonals bisect each other.

32. Let the vertices of quadrilateral $ABCD$ be $A(a, e)$, $B(b, f)$, $C(c, g)$, and $D(d, h)$. The midpoints of $\overline{AB}$, $\overline{BC}$, $\overline{CD}$, and $\overline{DA}$, respectively, are $L\left(\dfrac{a + b}{2}, \dfrac{e + f}{2}\right)$, $M\left(\dfrac{b + c}{2}, \dfrac{f + g}{2}\right)$, $N\left(\dfrac{c + d}{2}, \dfrac{g + h}{2}\right)$, and $P\left(\dfrac{a + d}{2}, \dfrac{e + h}{2}\right)$.

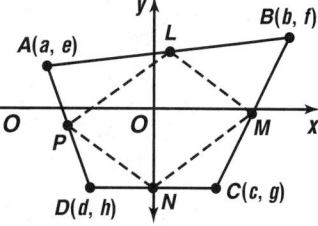

The slope of $\overline{LM}$ is $\dfrac{\frac{f + g}{2} - \frac{e + f}{2}}{\frac{b + c}{2} - \frac{a + b}{2}}$ or $\dfrac{g - e}{c - a}$.

The slope of $\overline{NP}$ is $\dfrac{\frac{e + h}{2} - \frac{g + h}{2}}{\frac{a + d}{2} - \frac{c + d}{2}}$ or $\dfrac{e - g}{a - c}$.

These slopes are equal, so $\overline{LM} \parallel \overline{NP}$.

The slope of $\overline{MN}$ is $\dfrac{\frac{f + g}{2} - \frac{g + h}{2}}{\frac{b + c}{2} - \frac{c + d}{2}}$ or $\dfrac{f - h}{b - d}$.

The slope of $\overline{PL}$ is $\dfrac{\frac{e + h}{2} - \frac{e + f}{2}}{\frac{a + d}{2} - \frac{a + b}{2}}$ or $\dfrac{h - f}{d - b}$.

These slopes are equal, so $\overline{MN} \parallel \overline{PL}$. Since $\overline{LM} \parallel \overline{NP}$ and $\overline{MN} \parallel \overline{PL}$, $PLMN$ is a parallelogram.

37a. Find a representation for MA and for MB.

$MA = \sqrt{t^2 + (3t - 15)^2}$

$\quad\;\; = \sqrt{t^2 + 9t^2 - 90t + 225}$

$\quad\;\; = \sqrt{10t^2 - 90t + 225}$

$MB = \sqrt{(t - 9)^2 + (3t - 12)^2}$

$\quad\;\; = \sqrt{t^2 - 18t + 81 + 9t^2 - 72t + 144}$

$\quad\;\; = \sqrt{10t^2 - 90t + 225}$

By setting these representations equal to each other, you find a value for t that would make the two distances equal.

$$MA = MB$$

$$\sqrt{10t^2 - 90t + 225} = \sqrt{10t^2 - 90t + 225}$$

Since the above equation is a true statement, t can take on any real values.

40. $2 \sec^2 x \overset{?}{=} \dfrac{1}{1 + \sin x} + \dfrac{1}{1 - \sin x}$

$2 \sec^2 x \overset{?}{=} \dfrac{(1 - \sin x) + (1 + \sin x)}{(1 + \sin x)(1 - \sin x)}$

$2 \sec^2 x \overset{?}{=} \dfrac{2}{1 - \sin^2 x}$

$2 \sec^2 x \overset{?}{=} \dfrac{2}{\cos^2 x}$

$2 \sec^2 x = 2 \sec^2 x$

Pages 627–630   Lesson 10-2

15.

16.

17.

18.

19.

20.

21.

22.

23.

24.

25.

26.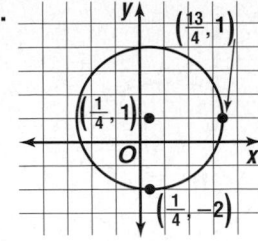

Pages 637–641   Lesson 10-3

19.

20.

21.

22.

23.

24.

25.

26.

27.

28.

29.

30.

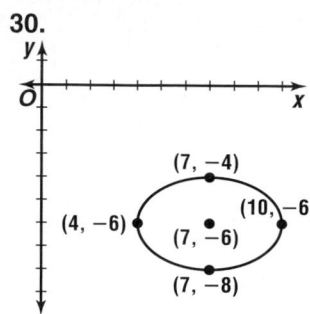

Pages 649–652   Lesson 10-4

14b.

15.

16.

17.

18.

19.

20.

21.

22.

23.

27.

28.

29.

30.

48a. 48d.

 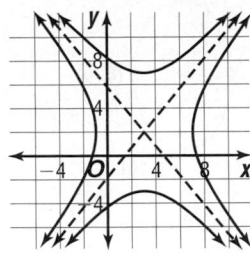

Pages 658–661   Lesson 10-5

6. vertex: (0, 1);
focus: (0, 4);
directrix: $y = -2$;
axis of symmetry: $x = 0$

7. vertex: (1, −1);
focus: (2, −1);
directrix: $x = 0$;
axis of symmetry: $y = -1$

 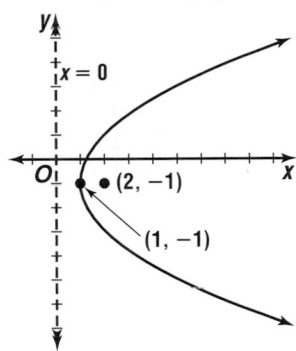

8. vertex: (−4, 2);
focus: (−4, 1);
directrix: $y = 3$;
axis of symmetry: $x = -4$

9.

10.

11.

12a. 23.

24.

25.

26.

27.

28.

29.

30.

31.

32.

Page 661   Lesson 10-5
Mid-Chapter Quiz

3.

6.

7.

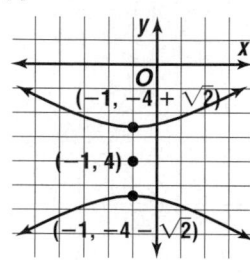

Pages 667–669   Lesson 10-6

13.

14.

15.

16.

17.

18.

19.

20.

21.

22.

23.

24.

25.

26.

27.

28.

29.

30.

43c.

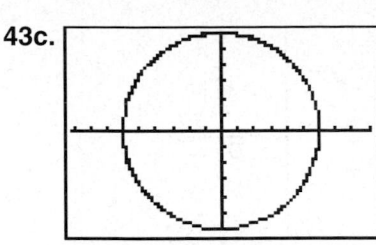

Tmin: [0, 4π] step: 0.1
[−9.10, 9.10] scl:1 by [−6, 6] scl:1

44. After drawing a vertical line through (x, y) and a horizontal line through the endpoint opposite (x, y), two right triangles are formed. Both triangles contain an angle t, since corresponding angles are congruent when two parallel lines are cut by a transversal.

Using the larger triangle, $\cos t = \frac{x}{a}$ or $x = a \cos t$.

Using the smaller triangle, $\sin t = \frac{y}{b}$ or $y = b \sin t$.

Pages 674–677   Lesson 10-7

11.

33.

34.

35.

36.

37.

[−6.61, 14.6] scl:1
by [−2, 12] scl:1

38.

[−7.58, 7.58] scl:1
by [−5, 5] scl:1

39.

[−7.58, 7.58] scl:1
by [−5, 5] scl:1

40.

[−8.31, 2.31] scl:1 by
[−2, 5] scl:1

41.

[−10.58. 4.58] scl:1 by
[−2, 8] scl:1

42.

[−7.85, 7.85] scl:1 by
[−5, 5] scl:1

Pages 682–684   Lesson 10-8

5.

6.

7.

8.

12b.

13.

14.

15.

16.

17.

18.

19.

20.

21.

22.

33.

37c.

38b.

39.

42d.

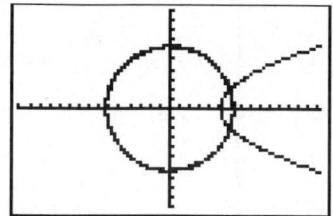

Tmin = −8, Tmax = 8 Tstep = 0.13
[−15.16, 15.16] scl:1 by [−10, 10] scl:1

Pages 687–691 Chapter 10 Study Guide and Assessment

14. $x^2 + y^2 = 27$

15. $(x − 2)^2 + (y − 1)^2 = 4$

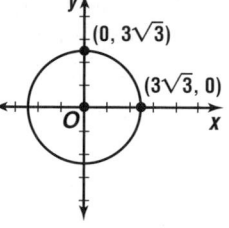

16. $x^2 + (y − 3)^2 = 9$

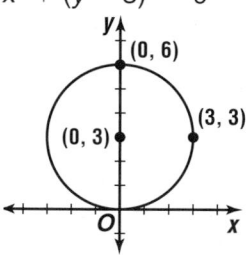

17. $(x + 7)^2 + (y + 3)^2 = 81$

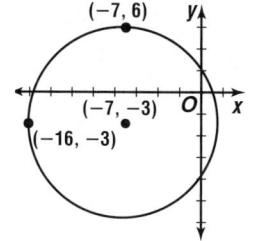

18. $(x + 1)^2 + (y + 2)^2 = 25$

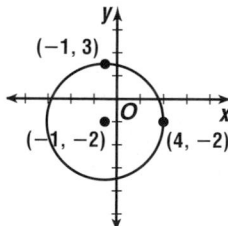

20. center: (5, 2),
foci: $\left(5, 2 ± 2\sqrt{5}\right)$;
vertices: (5, 8), (5, −4),
(9, 2), (1, 2)

21. center: (3, −1),
foci: $\left(3 ± \sqrt{21}, −1\right)$,
vertices: (3, 1), (8, −1),
(3, −3), (−2, −1)

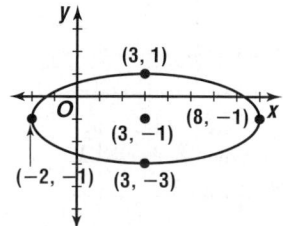

22. center: $(-2, 4)$;
foci: $\left(-2, 4 \pm \sqrt{2}\right)$;
vertices: $\left(-2, 4 \pm \sqrt{6}\right)$,
$(-4, 4), (0, 4)$

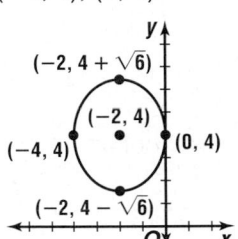

23. center: $(4, 6)$;
foci: $\left(4 \pm 3\sqrt{3}, 6\right)$;
vertices: $(-2, 6), (10, 6)$,
$(4, 3), (4, 9)$

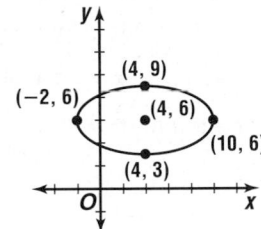

32. vertex: $(5, 3)$,
focus: $(5, 5)$,
directrix: $y = 1$;
axis of symmetry: $x = 5$

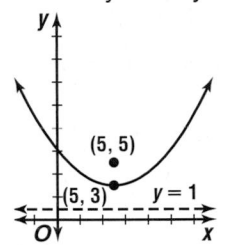

33. vertex: $(1, -2)$,
focus: $(-3, -2)$,
directrix: $x = 5$;
axis of symmetry: $y = -2$

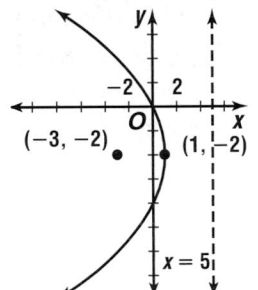

34. vertex: $(4, -3)$,
focus: $(5, -3)$,
directrix: $x = 3$;
axis of symmetry: $y = -3$

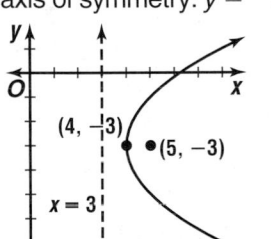

35. vertex: $(-2, 4)$,
focus: $(-2, 4.25)$,
directrix: $y = 3.75$;
axis of symmetry: $x = -2$

59.

60.

61.

62.
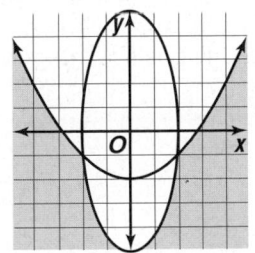

CHAPTER 11 EXPONENTIAL AND LOGARITHMIC FUNCTIONS

Page 695 Lesson 11-1
Graphing Calculator Exploration

1.

2.
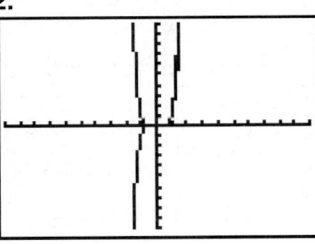

$[-10, 10]$ scl:1 by $[-10, 10]$ scl:1 $[-5, 5]$ scl:1 by $[-1, 20]$ scl:1

3. $a^m a^n = a^{m+n}$

4. $(a^m)^n = a^{mn}$

5. $\left(\dfrac{a}{b}\right)^m = \dfrac{a^m}{b^m}$, where $b \neq 0$

Pages 700–703 Lesson 11-1
74b.

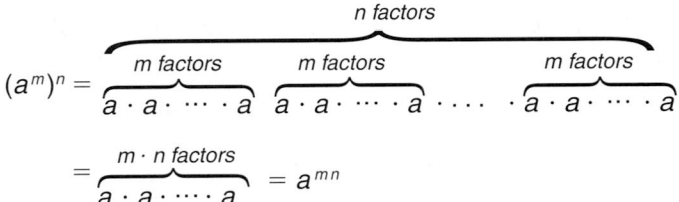

Pages 708–711 Lesson 11-2

10.

11.

12.

13.

14.

15.

16.

17.

18.

38.
$$\sin^4 A + \cos^2 A = \cos^4 A + \sin^2 A$$
$$(\sin^2 A)^2 + \cos^2 A = \cos^4 A + \sin^2 A$$
$$(1 - \cos^2 A)^2 + \cos^2 A = \cos^4 A + \sin^2 A$$
$$(1 - 2\cos^2 A + \cos^4 A) + \cos^2 A = \cos^4 A + \sin^2 A$$
$$\cos^4 A + 1 - \cos^2 A = \cos^4 A + \sin^2 A$$
$$\cos^4 A + 1 + \sin^2 A = \cos^4 A + \sin^2 A$$

CHAPTER 12   SEQUENCES AND SERIES

Page 785   Lesson 12-3B

3.
$$x = 1 + \frac{1}{x}$$
$$x^2 = x + 1$$
$$x^2 - x - 1 = 0$$
$$x = \frac{-(-1) \pm \sqrt{(-1)^2 - 4(1)(-1)}}{2(1)}$$
$$x = \frac{1 \pm \sqrt{5}}{2}$$

Since the sum $1 + \cfrac{1}{1 + \cfrac{1}{1 + \cdots}}$ cannot be negative,

the value of $1 + \cfrac{1}{1 + \cfrac{1}{1 + \cdots}}$ is $\dfrac{1 + \sqrt{5}}{2}$.

9.
$$x = A + \cfrac{B}{2A + \cfrac{B}{2A + \cdots}}$$
$$x + A = 2A + \frac{B}{x + A}$$
$$(x + A)^2 = 2A(x + A) + B$$
$$x^2 + 2Ax + A^2 = 2Ax + 2A^2 + B$$
$$x^2 = A^2 + B$$
$$x = \pm\sqrt{A^2 + B}$$

Since the sum $A + \cfrac{B}{2A + \cfrac{B}{2A + \cdots}}$ cannot be negative,

the value of $A + \cfrac{B}{2A + \cfrac{B}{2A + \cdots}}$ is $\sqrt{A^2 + B}$.

Page 826–828   Lesson 12-9

6. Step 1: Verify that the formula is valid for $n = 1$.
Since 3 is the first term in the sequence and $1(1 + 2) = 3$, the formula is valid for $n = 1$.

Step 2: Assume that the formula is valid for $n = k$ and derive a formula for $n = k + 1$.
$$S_k \Rightarrow 3 + 5 + 7 + \cdots + (2k + 1) = k(k + 2)$$
$$S_{k+1} \Rightarrow 3 + 5 + 7 + \cdots + (2k + 1) + (2k + 3)$$
$$= k(k + 2) + (2k + 3)$$
$$= k^2 + 4k + 3$$
$$= (k + 1)(k + 3)$$
Apply the original formula for $n = k + 1$.
$$(k + 1)[(k + 1) + 2] = (k + 1)(k + 3)$$
The formula gives the same result as adding the $(k + 1)$ term directly. Thus, if the formula is valid for $n = k$, it is also valid for $n = k + 1$. Since the formula is valid for $n = 1$, it is also valid for $n = 2$, $n = 3$, and so on, indefinitely. Thus, the formula is valid for all positive integral values of n.

7. Step 1: Verify that the formula is valid for $n = 1$.
Since 2 is the first term in the sequence and $2(2^1 - 1) = 2$, the formula is valid for $n = 1$.

Step 2: Assume that the formula is valid for $n = k$ and derive a formula for $n = k + 1$.
$$S_k \Rightarrow 2 + 2^2 + 2^3 + \cdots + 2^k = 2(2^k - 1)$$
$$S_{k+1} \Rightarrow 2 + 2^2 + 2^3 + \cdots + 2^k + 2^{k+1}$$
$$= 2(2^k - 1) + 2^{k+1}$$
$$= 2 \cdot 2^{k+1} - 2$$
$$= 2(2^{k+1} - 1)$$
When the original formula is applied for $n = k + 1$, the same result is obtained. Thus, if the formula is valid for $n = k$, it is also valid for $n = k + 1$. Since the formula is valid for $n = 1$, it is also valid for $n = 2$, $n = 3$, and so on, indefinitely. Thus, the formula is valid for all positive integral values of n.

8. Step 1: Verify that the formula is valid for $n = 1$.
Since $\frac{1}{2}$ is the first term in the sequence and
$1 - \frac{1}{2^1} = \frac{1}{2}$, the formula is valid for $n = 1$.
Step 2: Assume that the formula is valid for $n = k$ and derive a formula for $n = k + 1$.

$$S_k \Rightarrow \frac{1}{2} + \frac{1}{2^2} + \frac{1}{2^3} + \cdots + \frac{1}{2^k} = 1 - \frac{1}{2^k}$$

$$S_{k+1} \Rightarrow \frac{1}{2} + \frac{1}{2^2} + \frac{1}{2^3} + \cdots + \frac{1}{2^k} + \frac{1}{2^{k+1}}$$

$$= 1 - \frac{1}{2^k} + \frac{1}{2^{k+1}}$$

$$= 1 - \frac{2}{2 \cdot 2^k} + \frac{1}{2^{k+1}}$$

$$= 1 - \frac{2}{2^{k+1}} + \frac{1}{2^{k+1}}$$

$$= 1 - \frac{1}{2^{k+1}}$$

When the original formula is applied for $n = k + 1$, the same result is obtained. Thus if the formula is valid for $n = k$, it is also valid for $n = k + 1$. Since the formula is valid for $n = 1$, it is also valid for $n = 2$, $n = 3$, and so on, indefinitely. Thus, the formula is valid for all positive integral values of n.

9. $S_n : 3^n - 1 = 2r$ for some integer r

Step 1: Verify that S_n is valid for $n = 1$.

$S_1 \Rightarrow 3^1 - 1$ or 2. Since $2 = 2 \cdot 1$, S_n is valid for $n = 1$.

Step 2: Assume that S_n is valid for $n = k$ and show that it is also valid for $n = k + 1$.

$S_k \Rightarrow 3^k - 1 = 2r$ for some integer r

$S_{k+1} \Rightarrow 3^{k+1} - 1 = 2t$ for some integer t

$$3^k - 1 = 2r$$

$$3(3^k - 1) = 3 \cdot 2r$$

$$3^{k+1} - 3 = 6r$$

$$3^{k+1} - 1 = 6r + 2$$

$$3^{k+1} - 1 = 2(3r + 1)$$

Thus, $3^{k+1} - 1 = 2t$, where $t = 3r + 1$ is an integer, and we have shown that if S_n is valid, then S_{k+1} is also valid. Since S_n is valid for $n = 1$, it is also valid for $n = 2$, $n = 3$, and so on indefinitely. Hence, $3^n - 1$ is divisible by 2 for all integral values of n.

11. Step 1: Verify that the formula is valid for $n = 1$.
Since 1 is the first term in the sequence and $(1)[2(1) - 1] = 1$, the formula is valid for $n = 1$.
Step 2: Assume that the formula is valid for $n = k$ and derive a formula for $n = k + 1$.

$$S_k \Rightarrow 1 + 5 + 9 + \cdots + (4k - 3) = k(2k - 1)$$

$$S_{k+1} \Rightarrow 1 + 5 + 9 + \cdots + (4k - 3) + (4k + 1)$$

$$= k(2k - 1) + (4k + 1)$$

$$= 2k^2 + 3k + 1$$

$$= (k + 1)(2k + 1)$$

Apply the original formula for $n = k + 1$.

$$(k + 1)[2(k + 1) - 1] = (k + 1)(2k + 1)$$

The formula gives the same result as adding the $(k + 1)$ term directly. Thus if the formula is valid for $n = k$, it is also valid for $n = k + 1$. Since the formula is valid for $n = 1$, it is also valid for $n = 2$, $n = 3$, and so on, indefinitely. Thus, the formula is valid for all positive integral values of n.

12. Step 1: Verify that the formula is valid for $n = 1$.
Since 1 is the first term in the sequence and
$\frac{(1)[3(1) - 1]}{2} = 1$, the formula is valid for $n = 1$.
Step 2: Assume that the formula is valid for $n = k$ and derive a formula for $n = k + 1$.

$$S_k \Rightarrow 1 + 4 + 7 + \cdots + (3k - 2) = \frac{k(3k - 1)}{2}$$

$$S_{k+1} \Rightarrow 1 + 4 + 7 + \cdots + (3k - 2) + (3k + 1)$$

$$= \frac{k(3k - 1)}{2} + (3k + 1)$$

$$= \frac{k(3k - 1)}{2} + \frac{2(3k + 1)}{2}$$

$$= \frac{3k^2 + 5k + 2}{2} = \frac{(k + 1)(3k + 2)}{2}$$

Apply the original formula for $n = k + 1$.
$$\frac{(k + 1)[3(k + 1) - 1]}{2} = \frac{(k + 1)(3k + 2)}{2}$$

The formula gives the same result as adding the $(k + 1)$ term directly. Thus, if the formula is valid for $n = k$, it is also valid for $n = k + 1$. Since the formula is valid for $n = 1$, it is also valid for $n = 2$, $n = 3$, and so on, indefinitely. Thus, the formula is valid for all positive integral values of n.

13. Step 1: Verify that the formula is valid for $n = 1$.
Since $-\frac{1}{2}$ is the first term in the sequence and
$\frac{1}{2^1} - 1 = -\frac{1}{2}$, the formula is valid for $n = 1$.
Step 2: Assume that the formula is valid for $n = k$ and derive a formula for $n = k + 1$.

$$S_k \Rightarrow -\frac{1}{2} - \frac{1}{4} - \frac{1}{8} - \cdots - \frac{1}{2^k} = \frac{1}{2^k} - 1$$

$$S_{k+1} \Rightarrow -\frac{1}{2} - \frac{1}{4} - \frac{1}{8} - \cdots - \frac{1}{2^k} - \frac{1}{2^{k+1}}$$

$$= \frac{1}{2^k} - 1 - \frac{1}{2^{k+1}}$$

$$= \frac{2}{2 \cdot 2^k} - 1 - \frac{1}{2^{k+1}}$$

$$= \frac{2}{2^{k+1}} - 1 - \frac{1}{2^{k+1}} = \frac{1}{2^{k+1}} - 1$$

When the original formula is applied for $n = k + 1$, the same result is obtained. Thus, if the formula is valid for $n = k$, it is also valid for $n = k + 1$. Since the formula is valid for $n = 1$, it is also valid for $n = 2$, $n = 3$, and so on, indefinitely. Thus, the formula is valid for all positive integral values of n.

14. Step 1: Verify that the formula is valid for $n = 1$.
Since 1 is the first term in the sequence and
$\frac{1^2(1 + 1)^2}{4} = 1$, the formula is valid for $n = 1$.
Step 2: Assume that the formula is valid for $n = k$ and derive a formula for $n = k + 1$.

$$S_k \Rightarrow 1 + 8 + 27 + \cdots + k^3 = \frac{k^2(k + 1)^2}{4}$$

$$S_{k+1} \Rightarrow 1 + 8 + 27 + \cdots + k^3 + (k + 1)^3$$

$$= \frac{k^2(k + 1)^2}{4} + (k + 1)^3$$

$$= \frac{k^2(k + 1)^2 + 4(k + 1)^3}{4}$$

$$= \frac{(k + 1)^2[k^2 + 4(k + 1)]}{4}$$

$$= \frac{(k + 1)^2(k^2 + 4k + 4)}{4}$$

$$= \frac{(k + 1)^2(k + 2)^2}{4}$$

(continued)

Apply the original formula for $n = k + 1$.

$$\frac{(k+1)^2[(k+1)+1]^2}{4} = \frac{(k+1)^2(k+2)^2}{4}$$

The formula gives the same result as adding the $(k + 1)$ term directly. Thus, if the formula is valid for $n = k$, it is also valid for $n = k + 1$. Since the formula is valid for $n = 1$, it is also valid for $n = 2$, $n = 3$, and so on, indefinitely. Thus, the formula is valid for all positive integral values of n.

15. Step 1: Verify that the formula is valid for $n = 1$.
Since 1 is the first term in the sequence and $\frac{1[2(1)-1][2(1)+1]}{3} = 1$, the formula is valid for $n = 1$.

Step 2: Assume that the formula is valid for $n = k$ and derive a formula for $n = k + 1$.

$S_k \Rightarrow 1^2 + 3^2 + 5^2 + \cdots + (2k-1)^2 = \frac{k(2k-1)(2k+1)}{3}$

$S_{k+1} \Rightarrow 1^2 + 3^2 + 5^2 + \cdots + (2k-1)^2 + (2k+1)^2$

$= \frac{k(2k-1)(2k+1)}{3} + (2k+1)^2$

$= \frac{k(2k-1)(2k+1) + 3(2k+1)^2}{3}$

$= \frac{[k(2k-1) + 3(2k+1)](2k+1)}{3}$

$= \frac{(2k^2 + 5k + 3)(2k+1)}{3}$

$= \frac{(2k+3)(k+1)(2k+1)}{3}$

Apply the original formula for $n = k + 1$.

$\frac{(k+1)[2(k+1)-1][2(k+1)+1]}{3}$

$= \frac{(k+1)(2k+1)(2k+3)}{3}$

The formula gives the same result as adding the $(k + 1)$ term directly. Thus, if the formula is valid for $n = k$, it is also valid for $n = k + 1$. Since the formula is valid for $n = 1$, it is also valid for $n = 2$, $n = 3$, and so on, indefinitely. Thus, the formula is valid for all positive integral values of n.

16. Step 1: Verify that the formula is valid for $n = 1$.
Since $S_1 \Rightarrow 1$ and $2^1 - 1 = 1$, the formula is valid for $n = 1$.

Step 2: Assume that the formula is valid for $n = k$ and derive a formula for $n = k + 1$.

$S_k \Rightarrow 1 + 2 + 4 + \cdots + 2^{k-1} = 2^k - 1$

$S_{k+1} \Rightarrow 1 + 2 + 4 + \cdots + 2^{k-1} + 2^k = 2^k - 1 + 2^k$

$= 2(2^k) - 1$

$= 2^{k+1} - 1$

When the original formula is applied for $n = k + 1$, the same result is obtained. Thus, if the formula is valid for $n = k$, it is also valid for $n = k + 1$. Since the formula is valid for $n = 1$, it is also valid for $n = 2$, $n = 3$, and so on, indefinitely. Thus, the formula is valid for all positive integral values of n.

17. $S_n \Rightarrow 7^n + 5 = 6r$ for some integer r
Step 1: Verify that S_n is valid for $n = 1$.
$S_1 \Rightarrow 7^1 + 5$ or 12. Since $12 = 6 \cdot 2$, S_n is valid for $n = 1$.
Step 2: Assume that S_n is valid for $n = k$ and show that it is also valid for $n = k + 1$.

$S_k \Rightarrow 7^k + 5 = 6r$ for some integer r

$S_{k+1} \Rightarrow 7^{k+1} + 5 = 6t$ for some integer t

$7^k + 5 = 6r$

$7(7^k + 5) = 7 \cdot 6r$

$7^{k+1} + 35 = 42r$

$7^{k+1} + 5 = 42r - 30$

$7^{k+1} + 5 = 6(7r - 5)$

Thus, $7^{k+1} + 5 = 6t$, where $t = 7r - 5$ is an integer, and we have shown that if S_n is valid, then S_{k+1} is also valid. Since S_n is valid for $n = 1$, it is also valid for $n = 2$, $n = 3$, and so on, indefinitely. Hence, $7^n + 5$ is divisible by 6 for all integral values of n.

18. $S_n \Rightarrow 8^n - 1 = 7r$ for some integer r
Step 1: Verify that S_n is valid for $n = 1$.
$S_1 \Rightarrow 8^1 - 1$ or 7. Since $7 = 7 \cdot 1$, S_n is valid for $n = 1$.
Step 2: Assume that S_n is valid for $n = k$ and show that it is also valid for $n = k + 1$.

$S_k \Rightarrow 8^k - 1 = 7r$ for some integer r

$S_{k+1} \Rightarrow 8^{k+1} - 1 = 7t$ for some integer t

$8^k - 1 = 7r$

$8(8^k - 1) = 8 \cdot 7r$

$8^{k+1} - 8 = 56r$

$8^{k+1} - 1 = 56r + 7$

$8^{k+1} - 1 = 7(8r + 1)$

Thus, $8^{k+1} - 1 = 7t$, where $t = 8r + 1$ is an integer, and we have shown that if S_n is valid, then S_{k+1} is also valid. Since S_n is valid for $n = 1$, it is also valid for $n = 2$, $n = 3$, and so on, indefinitely. Hence, $8^n - 1$ is divisible by 7 for all integral values of n.

19. $S_n \Rightarrow 5^n - 2^n = 3r$ for some integer r
Step 1: Verify that S_n is valid for $n = 1$.
$S_1 \Rightarrow 5^1 - 2^1$ or 3. Since $3 = 3 \cdot 1$, S_n is valid for $n = 1$.
Step 2: Assume that S_n is valid for $n = k$ and show that it is also valid for $n = k + 1$.

$S_k \Rightarrow 5^k - 2^k = 3r$ for some integer r

$S_{k+1} \Rightarrow 5^{k+1} - 2^{k+1} = 3t$ for some integer t

$5^k - 2^k = 3r$

$5^k = 2^k + 3r$

$5^k \cdot 5 = (2^k + 3r)(2 + 3)$

$5^{k+1} = 2^{k+1} + 3(2^k) + 6r + 9r$

$5^{k+1} - 2^{k+1} = 2^{k+1} + 3(2^k) + 6r + 9r - 2^{k+1}$

$= 3(2^k) + 15r$

$= 3(2^k + 5r)$

Thus, $5^{k+1} - 2^{k+1} = 3t$, where $t = 2^k + 5r$ is an integer, and we have shown that if S_n is valid, then S_{k+1} is also valid. Since S_n is valid for $n = 1$, it is also valid for $n = 2$, $n = 3$, and so on, indefinitely. Hence, $5^n - 2^n$ is divisible by 3 for all integral values of n.

20. Step 1: Verify that the formula is valid for $n = 1$.
Since a is the first term in the sequence and $\frac{1}{2}[2a + (1 - 1)d] = a$, the formula is valid for $n = 1$.

Step 2: Assume that the formula is valid for $n = k$ and derive a formula for $n = k + 1$.

$S_k \Rightarrow a + (a + d) + (a + 2d) + \cdots + [a + (k - 1)d]$
$\qquad = \frac{k}{2}[2a + (k - 1)d]$

$S_{k+1} \Rightarrow a + (a + d) + (a + 2d) + \cdots + [a + (k - 1)d] +$
$\qquad (a + kd)$

$$= \frac{k}{2}[2a + (k - 1)d] + (a + kd)$$

$$= \frac{k[2a + (k - 1)d] + 2(a + kd)}{2}$$

$$= \frac{2ak + k(k - 1)d + 2a + 2kd}{2}$$

$$= \frac{(k + 1)2a + [k(k - 1) + 2k]d)}{2}$$

$$= \frac{(k + 1)2a + (k^2 + k)d}{2}$$

$$= \frac{(k + 1)2a + k(k + 1)d}{2}$$

$$= \frac{(k + 1)}{2}(2a + kd)$$

Apply the original formula for $n = k + 1$.
$$\frac{(k + 1)}{2}\left(2a + [(k + 1) - 1]d\right) = \frac{(k + 1)}{2}(2a + kd)$$

The formula gives the same result as adding the $(k + 1)$ term directly. Thus, if the formula is valid for $n = k$, it is also valid for $n = k + 1$. Since the formula is valid for $n = 1$, it is also valid for $n = 2$, $n = 3$, and so on, indefinitely. Thus, the formula is valid for all positive integral values of n.

21. Step 1: Verify that the formula is valid for $n = 1$.
Since $\frac{1}{2}$ is the first term in the sequence and $\frac{1}{1 + 1} = \frac{1}{2}$, the formula is valid for $n = 1$.

Step 2: Assume that the formula is valid for $n = k$ and derive a formula for $n = k + 1$.

$S_k \Rightarrow \frac{1}{1 \cdot 2} + \frac{1}{2 \cdot 3} + \frac{1}{3 \cdot 4} + \cdots + \frac{1}{k(k + 1)} = \frac{k}{k + 1}$

$S_{k+1} \Rightarrow \frac{1}{1 \cdot 2} + \frac{1}{2 \cdot 3} + \frac{1}{3 \cdot 4} + \cdots + \frac{1}{k(k + 1)} + \frac{1}{(k + 1)(k + 2)}$

$$= \frac{k}{k + 1} + \frac{1}{(k + 1)(k + 2)}$$

$$= \frac{k(k + 2) + 1}{(k + 1)(k + 2)}$$

$$= \frac{k^2 + 2k + 1}{(k + 1)(k + 2)}$$

$$= \frac{(k + 1)^2}{(k + 1)(k + 2)}$$

$$= \frac{k + 1}{k + 2}$$

Apply the original formula for $n = k + 1$.
$$\frac{(k + 1)}{(k + 1) + 1} = \frac{k + 1}{k + 2}$$
The formula gives the same result as adding the $(k + 1)$ term directly. Thus, if the formula is valid for $n = k$, it is also valid for $n = k + 1$. Since the formula is valid for $n = 1$, it is also valid for $n = 2$, $n = 3$, and so on, indefinitely. Thus, the formula is valid for all positive integral values of n.

22. $S_n \Rightarrow 2^{2n+1} + 3^{2n+1} = 5r$ for some integer r
Step 1: Verify that S_n is valid for $n = 1$.
$S_1 \Rightarrow 2^{2(1)+1} + 3^{2(1)+1}$ or 35. Since $35 = 5 \cdot 7$, S_n is valid for $n = 1$.

Step 2: Assume that S_n is valid for $n = k$ and show that it is also valid for $n = k + 1$.

$S_k \Rightarrow 2^{2k+1} + 3^{2k+1} = 5r$ for some integer r
$S_{k+1} \Rightarrow 2^{2k+3} + 3^{2k+3} = 5t$ for some integer t
$2^{2k+1} + 3^{2k+1} = 5r$
$\qquad 2^{2k+1} = 5r - 3^{2k+1}$
$\qquad 2^{2k+1} \cdot 2^2 = (5r - 3^{2k+1})(3^2 - 5)$
$\qquad 2^{2k+3} = 45r - 25r - 3^{2k+3} + 5(3^{2k+1})$
$2^{2k+3} + 3^{2k+3} = 45r - 25r - 3^{2k+3} + 5(3^{2k+1}) + 3^{2k+3}$
$\qquad\qquad = 20r + 5(3^{2k+1})$
$\qquad\qquad = 5(4 + 3^{2k+1})$

Thus, $2^{2k+3} + 3^{2k+3} = 5t$, where $t = 4 + 3^{2k+1}$ is an integer, and we have shown that if S_n is valid, then S_{k+1} is also valid. Since S_n is valid for $n = 1$, it is also valid for $n = 2$, $n = 3$, and so on, indefinitely. Hence, $2^{2n+1} + 3^{2k+1}$ is divisible by 5 for all integral values of n.

23. Step 1: Verify that the formula is valid for $n = 1$.
Since $S_1 \Rightarrow [r(\cos\theta + i\sin\theta)]^1$ or $r(\cos\theta + i\sin\theta)$ and $r^1[\cos(1)\theta + i\sin(1)\theta] = r(\cos\theta + i\sin\theta)$, the formula is valid for $n = 1$.

Step 2: Assume that the formula is valid for $n = k$ and derive a formula for $n = k + 1$. That is, assume that $[r(\cos\theta + i\sin\theta)]^k = r^k(\cos k\theta + i\sin k\theta)$.

Multiply each side of the equation by $r(\cos\theta + i\sin\theta)$.
$[r(\cos\theta + i\sin\theta)]^{k+1}$
$\quad = [r^k(\cos k\theta + i\sin k\theta)] \cdot [r(\cos\theta + i\sin\theta)]$
$\quad = r^{k+1}[\cos k\theta \cos\theta + (\cos k\theta)(i\sin\theta) +$
$\quad\quad i\sin k\theta \cos\theta + i^2 \sin k\theta \sin\theta]$
$\quad = r^{k+1}[(\cos k\theta \cos\theta - \sin k\theta \sin\theta) +$
$\quad\quad i(\sin k\theta \cos\theta + \cos k\theta \sin\theta)]$
$\quad = r^{k+1}[\cos(k + 1)\theta + i\sin(k + 1)\theta]$

When the original formula is applied for $n = k + 1$, the same result is obtained. Thus, if the formula is valid for $n = k$, it is also valid for $n = k + 1$. Since the formula is valid for $n = 1$, it is also valid for $n = 2$, $n = 3$, and so on, indefinitely. Thus, the formula is valid for all positive integral values of n.

24e. Step 1: Verify that the formula is valid for $n = 1$.
Since 1 is the first term in the sequence and $1^2 = 1$, the formula is valid for $n = 1$.
Step 2: Assume that the formula is valid for $n = k$ and derive a formula for $n = k + 1$.
$S_k \Rightarrow 1 + 3 + 5 + 7 + \cdots + (2k - 1) = k^2$
$S_{k+1} \Rightarrow 1 + 3 + 5 + 7 + \cdots + (2k - 1) + (2k + 1)$
$\qquad\qquad = k^2 + (2k + 1)$
$\qquad\qquad = k^2 + 2k + 1$
$\qquad\qquad = (k + 1)^2$

(continued)

When the original formula is applied for $n = k + 1$, the same result is obtained. Thus, if the formula is valid for $n = k$, it is also valid for $n = k + 1$. Since the formula is valid for $n = 1$, it is also valid for $n = 2$, $n = 3$, and so on, indefinitely. Thus, the formula is valid for all positive integral values of n.

25. $S_1 \Rightarrow n^2 + 5n = 2r$ for some integer r.

Step 1: Verify that S_1 is valid for $n = 1$.

$S_1 \Rightarrow 1^2 + 5 \cdot 1$ or 6

Since $6 = 2 \cdot 3$, S_1 is valid for $n = 1$.

Step 2: Assume that S_n is valid for $n = k$ and show that it is valid for $n = k + 1$.

$S_k \Rightarrow k^2 + 5k = 2r$ for some integer r

$S_{k+1} \Rightarrow (k + 1)^2 + 5(k + 1) = 2t$ for some integer t

$(k + 1)^2 + 5(k + 1)$

$= k^2 + 2k + 1 + 5k + 5$

$= (k^2 + 5k) + (2k + 6)$

$= 2r + 2(k + 3)$

$= 2(r + k + 3)$

Thus, $k^2 + 5k = 2t$, where $t = r + k + 3$ is an integer, and we have shown that if S_n is valid, then S_{k+1} is also valid. Since S_n is valid for $n = 1$, it is also valid for $n = 2$, $n = 3$, and so on, indefinitely. Hence, $n^2 + 5n$ is divisible by 2 for all positive integral values of n.

26b. Step 1: Verify that $S_n \Rightarrow 0 + 1 + 2 + 3 + \cdots + (n - 1) = \frac{n(n - 1)}{2}$ is valid for $n = 1$.

Since 0 is the first term in the sequence and $\frac{1(1 - 1)}{2} = 0$, the formula is valid for $n = 1$.

Step 2: Assume that the formula is valid for $n = k$ and derive a formula for $n = k + 1$.

$S_k \Rightarrow 0 + 1 + 2 + \cdots + (k - 1) = \frac{k(k - 1)}{2}$

$S_{k+1} \Rightarrow 1 + 3 + 5 + 7 + \cdots + (k - 1) + k$

$= \frac{k(k - 1)}{2} + k$

$= \frac{k(k - 1) + 2k}{2}$

$= \frac{k^2 - k + 2k}{2}$

$= \frac{k(k + 1)}{2}$

Apply the original formula for $n = k + 1$.

$\frac{(k + 1)(k + 1) - 1}{2} = \frac{k(k + 1)}{2}$

The formula gives the same result as adding the $(k + 1)$ term directly. Thus, if the formula is valid for $n = k$, it is also valid for $n = k + 1$. Since the formula is valid for $n = 1$, it is also valid for $n = 2$, $n = 3$, and so on, indefinitely. Thus, the formula is valid for all positive integral values of n.

27. Step 1: Verify that $S_n \Rightarrow (x + y)^n = x^n + nx^{n-1}y + \frac{n(n-1)}{2!} x^{n-2}y^2 + \frac{n(n-1)(n-2)}{3!} x^{n-3}y^3 + \cdots + y^n$ is valid for $n = 1$. Since $S_1 \Rightarrow (x + y)^1 = x^1 + 1x^0y^1$, or $x + y$, S_n is valid for $n = 1$.

Step 2: Assume that the formula is valid for $n = k$ and derive a formula for $n = k + 1$.

$S_k \Rightarrow (x + y)^k = x^k + kx^{k-1}y + \frac{k(k - 1)}{2!} x^{k-2}y^2 + \frac{k(k - 1)(k - 2)}{3!} x^{k-3}y^3 + \cdots + y^k$

$S_{k+1} \Rightarrow (x + y)^k(x + y)$

$= (x + y)(x^k + kx^{k-1}y + \frac{k(k - 1)}{2!} x^{k-2}y^2 + \frac{k(k - 1)(k - 2)}{3!} x^{k-3}y^3 + \cdots + y^k(x + y)^{k+1}$

$= x(x^k + kx^{k-1}y + \frac{k(k - 1)}{2!} x^{k-2}y^2 + \frac{k(k - 1)(k - 2)}{3!}x^{k-3}y^3 + \cdots + y^k) + y(x^k + kx^{k-1}y) + \frac{k(k - 1)}{2!} x^{k-2}y^2 + \frac{k(k - 1)(k - 2)}{3!} x^{k-3}y^3 + \cdots + y^k)$

$= x^{k+1} + kx^ky + \frac{k(k - 1)}{2!} x^{k-1}y^2 + \cdots + xy^k + x^ky + kx^{k-1}y^2 + \frac{k(k - 1)}{2!} x^{k-2}y^3 + \cdots + y^{k+1}$

$= x^{k+1} + (k + 1)x^ky + kx^{k-1}y^2 + \frac{k(k - 1)}{2!} x^{k-1}y^2 + \cdots + y^{k+1}$

$= x^{k+1} + (k + 1)x^ky + \frac{k(k - 1)}{2!}x^{k-1}y^2 + \cdots + y^{k+1}$

When the original formula is applied for $n = k + 1$, the same result is obtained. Thus, if the formula is valid for $n = k$, it is also valid for $n = k + 1$. Since the formula is valid for $n = 1$, it is also valid for $n = 2$, $n = 3$, and so on, indefinitely. Thus, the formula is valid for all positive integral values of n.

28d. Step 1: Verify that S_n: $0.9 + 0.09 + 0.009 + \cdots + \frac{9}{10^n} = \frac{10^1 - 1}{10^1}$ is valid for $n = 1$.

Since 0.9 is the first term in the sequence and $\frac{10^1 - 1}{10^1} = 0.9$, the formula is valid for $n = 1$.

Step 2: Assume that the formula is valid for $n = k$ and derive a formula for $n = k + 1$.

$S_k \Rightarrow 0.9 + 0.09 + 0.009 + \cdots + \frac{9}{10^k} = \frac{10^k - 1}{10^k}$

$S_{k+1} \Rightarrow 0.9 + 0.09 + 0.009 + \cdots + \frac{9}{10^k} + \frac{9}{10^{k+1}}$

$= \frac{10^k - 1}{10^k} + \frac{9}{10^{k+1}}$

$= \frac{10(10^k - 1) + 9}{10^{k+1}}$

$= \frac{10^{k+1} - 10 + 9}{10^{k+1}}$

$= \frac{10^{k+1} - 1}{10^{k+1}}$

When the original formula is applied for $n = k + 1$, the same result is obtained. Thus if the formula is valid for $n = k$, it is also valid for $n = k + 1$. Since the formula is valid for $n = 1$, it is also valid for $n = 2$, $n = 3$, and so on, indefinitely. Thus, the formula is valid for all positive integral values of n.

28e. $\lim\limits_{n \to \infty} \frac{10^n - 1}{10^n} = \lim\limits_{n \to \infty} 1 - \frac{1}{10^n}$

$= 1 - 0$ or 1

Thus, $0.999\ldots = 1$.

51. Step 1: Verify that the formula is valid for $n = 1$.

Since the first term in the sequence is 1 and $\frac{1(1 + 1)}{2} = 1$, the formula is valid for $n = 1$.

Step 2: Assume that the formula is valid for $n = k$ and derive a formula for $n = k + 1$.

$S_k \Rightarrow 1 + 2 + 3 + \cdots + k = \frac{k(k + 1)}{2}$

$S_{k+1} \Rightarrow 1 + 2 + 3 + \cdots + k + (k + 1)$

$\quad = \frac{k(k + 1)}{2} + \frac{2(k + 1)}{2}$

$\quad = \frac{k^2 + k}{2} + \frac{2k + 2}{2}$

$\quad = \frac{k^2 + k + 2k + 2}{2}$

$\quad = \frac{k^2 + 3k + 2}{2}$

$\quad = \frac{(k + 1)(k + 2)}{2}$

Apply the original formula for $n = k + 1$.

$\frac{(k + 1)[(k + 1) + 1]}{2} = \frac{(k + 1)(k + 2)}{2}$

The formula gives the same result as adding the $(k + 1)$ term directly. Thus, if the formula is valid for $n = k$, it is also valid for $n = k + 1$. Since the formula is valid for $n = 1$, it is also valid for $n = 2$, $n = 3$, and so on, indefinitely. Thus, the formula is valid for all positive integral values of n.

52. Step 1: Verify that the formula is valid for $n = 1$. Since the first term of the sequence is 3 and $1(1 + 2) = 3$, the formula is valid for $n = 1$.

Step 2: Assume that the formula is valid for $n = k$ and derive a formula for $n = k + 1$.

$S_k \Rightarrow 3 + 8 + 15 + \cdots + k(k + 2) = \frac{k(k + 1)(2k + 7)}{6}$

$S_{k+1} \Rightarrow 3 + 8 + 15 + \cdots + k(k + 2) + (k + 1)(k + 3)$

$\quad = \frac{k(k + 1)(2k + 7)}{6} + (k + 1)(k + 3)$

$\quad = \frac{k(k + 1)(2k + 7)}{6} + \frac{(k + 1)(k + 3)}{6}$

$\quad = \frac{k(k + 1)(2k + 7) + 6(k + 1)(k + 3)}{6}$

$\quad = \frac{(k + 1)[k(2k + 7) + 6(k + 3)]}{6}$

$\quad = \frac{(k + 1)(2k^2 + 7k + 6k + 18)}{6}$

$\quad = \frac{(k + 1)(2k^2 + 13k + 18)}{6}$

$\quad = \frac{(k + 1)(k + 2)(2k + 9)}{6}$

Apply the original formula for $n = k + 1$.

$\frac{(k + 1)[(k + 1) + 1][(2(k + 1) + 7]}{6} = \frac{(k + 1)(k + 2)(2k + 9)}{6}$

The formula gives the same result as adding the $(k + 1)$ term directly. Thus, if the formula is valid for $n = k$, it is also valid for $n = k + 1$. Since the formula is valid for $n = 1$, it is also valid for $n = 2$, $n = 3$, and so on, indefinitely. Thus, the formula is valid for all positive integral values of n.

53. $S_n \Rightarrow 9^n - 4^n = 5r$ for some integer r

Step 1: Verify that S_n is valid for $n = 1$.

$S_1 \Rightarrow 9^1 - 4^1$ or 5. Since $5 = 5 \cdot 1$, S_n is valid for $n = 1$.

Step 2: Assume that S_n is valid for $n = k$ and show that it is also valid for $n = k + 1$.

$S_k \Rightarrow 9^k - 4^k = 5r$ for some integer r

$S_{k+1} \Rightarrow 9^{k+1} - 4^{k+1} = 5t$ for some integer t

$\quad 9^k - 4^k = 5r$

$\quad\quad 9^k = 4^k + 5r$

$\quad 9(9^k) = (4^k + 5r)(4 + 5)$

$\quad 9^{k+1} = 4^{k+1} + 5(4^k) + 20r + 25r$

$9^{k+1} - 4^{k+1} = 4^{k+1} + 5(4^k) + 20r + 25r - 4^{k+1}$

$\quad\quad\quad = 5(4^k) + 45r$

$\quad\quad\quad = 5(4^k + 9r)$

Thus, $9^{k+1} - 4^{k+1} = 5t$, where $t = 4^k + 9r$ is an integer, and we have shown that if S_n is valid, then S_{k+1} is also valid. Since S_n is valid for $n = 1$, it is also valid for $n = 2$, $n = 3$, and so on, indefinitely. Hence, $9^n - 4^n$ is divisible by 5 for all integral values of n.

CHAPTER 14 STATISTICS AND DATA ANALYSIS

Pages 892–896 Lesson 14-1

1. A line plot, a bar graph, a histogram, and a frequency polygon all show data visually. A line plot shows the frequency of specific quantities by using symbols, and a bar graph shows the frequency of specific quantities by using bars. A histogram is a special bar graph in which the width of each bar represents a class interval. A frequency polygon shows the frequency of a class interval using a broken line graph.

2. Choose an appropriate class interval. Use tally marks to determine the number of elements in each class interval.

3a. There would be too many classes.

3b. There would be 9 classes.

3c. There would be 5 classes.

3d. There would only be 3 classes.

3de. There would only be 2 classes.

15. Sample answer:

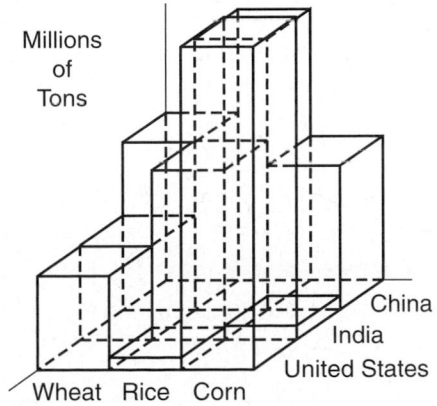

Millions of Tons

China
India
United States
Wheat   Rice   Corn

EXTRA PRACTICE

Page A27   Lesson 1-6

DATE DUE

AUG 1 3 2007			
GAYLORD			PRINTED IN U.S.A.

TRIGONOMETRIC IDENTITIES

Reciprocal Identities

$$\sin \theta = \frac{1}{\csc \theta} \qquad \csc \theta = \frac{1}{\sin \theta}$$

$$\cos \theta = \frac{1}{\sec \theta} \qquad \sec \theta = \frac{1}{\cos \theta}$$

$$\tan \theta = \frac{1}{\cot \theta} \qquad \cot \theta = \frac{1}{\tan \theta}$$

Quotient Identities

$$\frac{\sin \theta}{\cos \theta} = \tan \theta \qquad \frac{\cos \theta}{\sin \theta} = \cot \theta$$

Pythagorean Identities

$$\sin^2 \theta + \cos^2 \theta = 1$$

$$\tan^2 \theta + 1 = \sec^2 \theta$$

$$1 + \cot^2 \theta = \csc^2 \theta$$

Cofunction Identities

$$\sin \theta = \cos (90° - \theta) \qquad \cos \theta = \sin (90° - \theta)$$

$$\tan \theta = \cot (90° - \theta) \qquad \cot \theta = \tan (90° - \theta)$$

$$\sec \theta = \csc (90° - \theta) \qquad \csc \theta = \sec (90° - \theta)$$

Sum and Difference Identities

$$\sin (\alpha \pm \beta) = \sin \alpha \cos \beta \pm \cos \alpha \sin \beta$$

$$\cos (\alpha \pm \beta) = \cos \alpha \cos \beta \mp \sin \alpha \sin \beta$$

$$\tan (\alpha \pm \beta) = \frac{\tan \alpha \pm \tan \beta}{1 \mp \tan \alpha \tan \beta}$$

Opposite-Angle Identities

$$\sin (-A) = -\sin A \qquad \cos (-A) = \cos A$$

Double-Angle Identities

$$\sin 2\theta = 2 \sin \theta \cos \theta$$

$$\cos 2\theta = \cos^2 \theta - \sin^2 \theta$$

$$= 2 \cos^2 \theta - 1$$

$$= 1 - 2 \sin^2 \theta$$

$$\tan 2\theta = \frac{2 \tan \theta}{1 - \tan^2 \theta}$$

Half-Angle Identities

$$\sin \frac{\alpha}{2} = \pm \sqrt{\frac{1 - \cos \alpha}{2}}$$

$$\cos \frac{\alpha}{2} = \pm \sqrt{\frac{1 + \cos \alpha}{2}}$$

$$\tan \frac{\alpha}{2} = \pm \sqrt{\frac{1 - \cos \alpha}{1 + \cos \alpha}}, \cos a \neq -1$$

Symmetry Identities

The following trigonometric identities hold for any integer k and all values of A.

Case 1: $\sin (A + 360k°) = \sin A$
$\cos (A + 360k°) = \cos A$

Case 2: $\sin (A + 180°(2k - 1)) = -\sin A$
$\cos (A + 180°(2k - 1)) = -\cos A$

Case 3: $\sin (360k° - A) = -\sin A$
$\cos (360k° - A) = \cos A$

Case 4: $\sin (180°(2k - 1) - A) = \sin A$
$\cos (180° (2k - 1) - A) = -\cos A$

FORMULAS

CHAPTER 1 (pp. 4–65)

Slope $\quad m = \dfrac{y_1 - y_2}{x_1 - x_2}$

CHAPTER 2 (pp. 66–125)
Determinant

$$\begin{vmatrix} a_1 & b_1 \\ a_2 & b_2 \end{vmatrix} = a_1 b_2 - a_2 b_1$$

$$\begin{vmatrix} a_1 & b_1 & c_1 \\ a_2 & b_2 & c_2 \\ a_3 & b_3 & c_3 \end{vmatrix} = a_1 \begin{vmatrix} b_2 & c_2 \\ b_3 & c_3 \end{vmatrix} -$$

$$b_1 \begin{vmatrix} a_2 & c_2 \\ a_3 & c_3 \end{vmatrix} + c_1 \begin{vmatrix} a_2 & b_2 \\ a_3 & b_3 \end{vmatrix}$$

Inverse of a Second Order Matrix

$$A^{-1} = \dfrac{1}{\begin{vmatrix} a_1 & b_1 \\ a_2 & b_2 \end{vmatrix}} \begin{bmatrix} b_2 & -b_1 \\ -a_2 & a_1 \end{bmatrix}$$

CHAPTER 3 (pp. 126–203)
Direct Variation $\quad y = kx^n, n > 0$

Inverse Variation

$x^n y = k$ or $y = \dfrac{k}{x^n}, n > 0$

Joint Variation $\quad y = kx^n z^n$, where
$x \neq 0, z \neq 0$, and $n > 0$

CHAPTER 4 (pp. 204–273)
Quadratic Formula

$x = \dfrac{-b \pm \sqrt{b^2 - 4ac}}{2a}$

CHAPTER 5 (pp. 276–341)
Trigonometric Ratios

$\sin \theta = \dfrac{\text{side opposite}}{\text{hypotenuse}}$

$\cos \theta = \dfrac{\text{side adjacent}}{\text{hypotenuse}}$

$\tan \theta = \dfrac{\text{side opposite}}{\text{side adjacent}}$

Inverse Trigonometric Ratios

$\csc \theta = \dfrac{1}{\sin \theta}$ or $\dfrac{\text{hypotenuse}}{\text{side adjacent}}$

$\sec \theta = \dfrac{1}{\cos \theta}$ or $\dfrac{\text{hypotenuse}}{\text{side adjacent}}$

$\cot \theta = \dfrac{1}{\tan \theta}$ or $\dfrac{\text{side adjacent}}{\text{side opposite}}$

**Trigonometric Functions
of an Angle in Standard Position**

$\sin \theta = \dfrac{y}{r} \qquad \csc \theta = \dfrac{r}{y}$

$\cos \theta = \dfrac{x}{r} \qquad \sec \theta = \dfrac{r}{x}$

$\tan \theta = \dfrac{y}{x} \qquad \cot \theta = \dfrac{x}{y}$

Law of Sines

$\dfrac{a}{\sin A} = \dfrac{b}{\sin B} = \dfrac{c}{\sin C}$

Area of a Triangle

$K = \dfrac{1}{2} bc \sin A$

$K = \dfrac{1}{2} c^2 \dfrac{\sin A \sin B}{\sin C}$

Law of Cosines
$a^2 = b^2 + c^2 - 2bc \cos A$
$b^2 = a^2 + c^2 - 2ac \cos B$
$c^2 = a^2 + b^2 - 2ab \cos C$

Hero's Formula
$K = \sqrt{s(s - a)(s - b)(s - c)}$, where
$s = \dfrac{1}{2}(a + b + c)$

CHAPTER 6 (pp. 342–419)
Degree/Radian Conversion

$1 \text{ radian} = \dfrac{180}{\pi} \text{ degrees}$

$1 \text{ degree} = \dfrac{\pi}{180} \text{ radians}$

Length of an Arc $\quad s = r\theta$

Area of a Circular Sector

$A = \dfrac{1}{2} r^2 \theta$

Angular Velocity $\quad \omega = \dfrac{\theta}{t}$

Linear Velocity $\quad \nu = r\dfrac{\theta}{t}$

CHAPTER 7 (pp. 420–483)
Distance from a Point to a Line

$d = \dfrac{Ax_1 + By_1 + C}{\pm\sqrt{A^2 + B^2}}$

CHAPTER 8 (pp. 484–549)
**Magnitude of $\overrightarrow{P_1 P_2}$ in Two
Dimensions**

$|\overrightarrow{P_1 P_2}| = \sqrt{(x_2 - x_1)^2 + (y_2 - y_1)^2}$

**Magnitude of $\overrightarrow{P_1 P_2}$ in Three
Dimensions**

$|\overrightarrow{P_1 P_2}| =$
$\sqrt{(x_2 - x_1)^2 + (y_2 - y_1)^2 + (z_2 - z_1)^2}$

Inner Product of Vectors in a Plane
$\vec{\mathbf{a}} \cdot \vec{\mathbf{b}} = a_1 b_1 + a_2 b_2$

Inner Product of Vectors in Space
$\vec{\mathbf{a}} \cdot \vec{\mathbf{b}} = a_1 b_1 + a_2 b_2 + a_3 b_3$

Cross Product of Vectors in Space

$\vec{\mathbf{a}} \times \vec{\mathbf{b}} = \begin{vmatrix} a_2 & a_3 \\ b_2 & b_3 \end{vmatrix} \vec{\mathbf{i}} - \begin{vmatrix} a_1 & a_3 \\ b_1 & b_3 \end{vmatrix} \vec{\mathbf{j}} +$

$\begin{vmatrix} a_1 & a_2 \\ b_1 & b_2 \end{vmatrix} \vec{\mathbf{k}}$

**Parametric Equations for the
Path of a Projectile**
$x = t|\vec{\mathbf{v}}| \cos \theta$ and
$y = t|\vec{\mathbf{v}}| \sin \theta - \dfrac{1}{2} gt^2$

CHAPTER 9 (pp. 552–613)
Distance Formula in Polar Plane
$P_1 P_2 =$
$\sqrt{r_1^2 + r_2^2 - 2r_1 r_2 \cos(\theta_2 - \theta_1)}$

**Conversion from Polar Coordinates
to Rectangular Coordinates**
$x = r \cos \theta, y = r \sin \theta$

**Conversion from Rectangular
Coordinates to Polar Coordinates**
$r = \sqrt{x^2 + y^2}$
$\theta = \text{Arctan } \dfrac{y}{x}$, when $x > 0$
$\theta = \text{Arctan } \dfrac{y}{x} + \pi$, when $x < 0$

**Absolute Value of a Complex
Number**
$|a + bi| = \sqrt{a^2 + b^2}$

Polar Form of a Complex Number
$r(\cos \theta + i \sin \theta)$

**Product of Complex Numbers
in Polar Form**
$r_1(\cos \theta_1 + i \sin \theta_1) \cdot$
$r_2(\cos \theta_2 + i \sin \theta_2) =$
$r_1 r_2 [\cos(\theta_1 + \theta_2) + i \sin(\theta_1 + \theta_2)]$

(continued)

FORMULAS

Quotient of Complex Numbers in Polar Form

$$\frac{r_1(\cos\theta_1 + i\sin\theta_1)}{r_2(\cos\theta_2 + i\sin\theta_2)} =$$
$$\frac{r_1}{r_2}[\cos(\theta_1 - \theta_2) + i\sin(\theta_1 - \theta_2)]$$

De Moivre's Theorem

$$[r(\cos\theta + i\sin\theta)]^n =$$
$$r^n(\cos n\theta + i\sin n\theta)$$

CHAPTER 10 (pp. 614–693)

Distance Formula for Two Points

$$d = \sqrt{(x_2 - x_1)^2 + (y_2 - y_1)^2}$$

Midpoint of a Line Segment

$$\left(\frac{x_1 + x_2}{2}, \frac{y_1 + y_2}{2}\right)$$

Rotation Equations

$$x = x'\cos\theta + y'\sin\theta$$

$$y = -x'\sin\theta + y'\cos\theta$$

Angle of Rotation

$$\theta = \frac{\pi}{4}, \text{ if } A = C \text{ or } \tan 2\theta = \frac{B}{A - C}, \text{ if } A \neq C$$

CHAPTER 11 (pp. 694–755)

Exponential Growth or Decay Formulas

$$N = N_o(1 + r)^t \qquad N = N_o e^{kt}$$

Compound Interest Formula

$$A = P\left(1 + \frac{r}{n}\right)^{nt}$$

Continuously Compounded Interest Formula

$$A = Pe^{rt}$$

Change of Base Formula

$$\log_a n = \frac{\log_b n}{\log_b a}$$

Doubling Time $\quad t = \dfrac{\ln 2}{k}$

CHAPTER 12 (pp. 758–835)

nth Term of an Arithmetic Sequence

$$a_n = a_1 + (n - 1)d$$

Sum of an Arithmetic Sequence

$$S_n = \frac{n}{2}(a_1 + a_n)$$

nth Term of a Geometric Sequence

$$a_n = a_1 r^{n-1}$$

Sum of a Finite Geometric Sequence

$$S_n = \frac{a_1 - a_1 r^n}{1 - r}$$

Sum of an Infinite Geometric Series

$$S_n = \frac{a_1}{1 - r}$$

Euler's Formula

$$e^{i\alpha} = \cos\alpha + i\sin\alpha$$

CHAPTER 13 (pp. 836–887)

Definitions of Permutations

$$P(n, n) = n! \qquad P(n, r) = \frac{n!}{(n - r)!}$$

Definition of Combination

$$C(n, r) = \frac{n!}{(n - r)!\, r!}$$

Permutation of n objects with repetitions (p alike and q alike)

$$\frac{n!}{p!\, q!}$$

Circular Permutations of n Objects

$$\frac{n!}{n} \text{ or } (n - 1)!$$

Probability of Two Independent Events

$$P(A \text{ and } B) = P(A) \cdot P(B)$$

Probability of Two Dependent Events

$$P(A \text{ and } B) = P(A) \cdot P(B \text{ following } A)$$

Probability of Mutually Exclusive Events

$$P(A \text{ or } B) = P(A) + P(B)$$

Probability of Inclusive Events

$$P(A \text{ or } B) = P(A) + P(B) - P(A \text{ and } B)$$

Conditional Probability

$$P(A\,|\,B) = \frac{P(A \text{ and } B)}{P(B)} \text{ where } P(B) \neq 0$$

CHAPTER 14 (pp. 888–939)

Arithmetic Mean $\quad \overline{X} = \dfrac{1}{n}\sum_{i=1}^{n} X_i$

Mean of the Data in Frequency Distribution

$$\overline{X} = \frac{\sum_{i=1}^{k}(f_i \cdot X_i)}{\sum_{i=1}^{k} f_i}$$

Semi-Interquartile Range

$$Q_R = \frac{Q_3 - Q_1}{2}$$

Mean Deviation

$$MD = \frac{1}{n}\sum_{i=1}^{n} |X_i - \overline{X}|$$

Standard Deviation

$$\sigma = \sqrt{\frac{1}{n}\sum_{i=1}^{n}(X_i - \overline{X})^2}$$

Standard Deviation of the Data in a Frequency Distribution

$$\sigma = \sqrt{\frac{\sum_{i=1}^{n}(X_i - \overline{X})^2 \cdot f_i}{\sum_{i=1}^{n} f_i}}$$

Standard Error of the Mean

$$\sigma_{\overline{X}} = \frac{\sigma}{\sqrt{N}}$$

CHAPTER 15 (pp. 941–983)

Constant Multiple of a Power Rule for Derivatives

If $f(x) = cx^n$, where c is a constant and n is a rational number, then $f'(x) = cnx^{n-1}$.

Constant Multiple of a Power Rule for Antiderivatives

If $f(x) = kx^n$, where n is a rational number other than -1 and k is a constant, the antiderivative is $F(x) = k \cdot \dfrac{1}{n + 1}x^{n+1} + C$.

Definite Integral

$$\int_a^b f(x)\, dx = \lim_{n\to\infty}\sum_{i=1}^{n} f(x_i)\,\Delta x \text{ where}$$

$$\Delta x = \frac{b - a}{n}$$

Fundamental Theorem of Calculus

$$\int_a^b f(x)\, dx = F(b) - F(a), \text{ where}$$

$F(x)$ is the antiderivative of $f(x)$.